HANDBOOK OF DERIVATIZATION REACTIONS FOR HPLC

George Lunn and Louise C. Hellwig

A WILEY-INTERSCIENCE PUBLICATION

JOHN WILEY & SONS, INC.

New York / Chichester / Weinheim / Brisbane / Singapore / Toronto

This book is printed on acid-free paper. ⊗

Copyright © 1998 by John Wiley & Sons, Inc.

All rights reserved. Published simultaneously in Canada.

Library of Congress Cataloging in Publication Data:
Lunn, George.
    Handbook of derivatization reactions for HPLC / by George Lunn and
 Louise C. Hellwig
        p. cm.
    "A Wiley-Interscience publication."
    Includes bibliographical references (p. 1615) and index.
    ISBN 0-471-16458-5 (cloth/CD-ROM : alk. paper).—ISBN
 0-471-23888-0 (cloth : alk. paper.—ISBN 0-471-23889-9 (CD-ROM)
    1. High performance liquid chromatography—Handbooks, manuals,
 etc. 2. Derivatization—Handbooks, manuals, etc. I. Hellwig,
 Louise C., 1949–   II. Title.
 QD79.C454L86 1998
 543'.0894—dc21   97-22137
                                                                   CIP

Printed in the United States of America

10 9 8 7 6 5 4 3 2

CONTENTS

v

PREFACE

This book is a collection of derivatization procedures that can be used in high-performance liquid chromatography (HPLC) and capillary electrophoresis (CE). For each functional group various techniques are described in sufficient detail that the analyst can replicate the procedure without reference to the original publication. Within the section for each functional group the reagents are grouped by type, generally with a reaction scheme. Since detailed procedures for the same reagent are listed together, it is very easy for the researcher to combine features of different methods, e.g., the derivatization procedure from one paper and the chromatographic procedure from another paper, to provide methods tailored to the researcher's requirements. In addition, synthetic procedures are provided for virtually all the reagents that are not commercially available. It has been our experience that these procedures can be hard to locate and various sections of the procedures are frequently scattered amongst a number of papers. This book contains more than 2100 abstracts taken from over 1900 papers. This field continues to be very active; more than half of the papers abstracted have been published since 1991.

Although the spread of computers has led to readily available laboratory-based searches of the literature, this resource is not exploited as much as it might be. One reason for this reluctance is, of course, that a computer search merely produces a listing of possibly relevant references. Tedious and time-consuming searches in the library are necessary to find the most relevant reference that can be turned into a practical analytical procedure in the searcher's own laboratory. The reference finally chosen will depend on the individual circumstances such as availability of equipment. In addition, it is difficult to devise a satisfactory search for derivatization procedures. Searches may be incomplete or have an excessive number of false positives. This book circumvents this lengthy process by providing fully abstracted and evaluated procedures. The analyst can rapidly identify a relevant procedure and put it into practice without having to consult the original literature. The amount of literature in this field is so great that it would be impossible to abstract all of the relevant papers. For this reason we have added bibliographies throughout the book. Some sections have bibliographies that list papers that describe the use of the same reagent to derivatize the same functional group. At the end of the book we have added a bibliography of more than 3000 papers describing derivatization and post-column reaction procedures. These papers were identified using computer searches and have not, for the most part, been evaluated further. Finally we have added a bibliography of reviews of derivatization procedures.

Readers familiar with *HPLC Methods for Pharmaceutical Analysis* by George Lunn and Norman R. Schmuff, also published by John Wiley, will note many similarities. The basic layout of each abstract is essentially the same although some changes have been made, e.g., in the capillary electrophoresis abstracts. In addition this book is organized differently. In this case the organization depends on the functional group that is derivatized rather than the drug to be analyzed. As before, features such as analytical matrix, chromatographic conditions, sensitivity, mode of detection, other compounds that interfere with the analysis, and other compounds that may be determined at the same time are described in the abstracts.

This book is the successor to *Handbook of Analytical Derivatization Reactions* by Daniel R. Knapp, published by John Wiley in 1979. The explosive growth in this area has forced us to confine the present volume to derivatization procedures used in HPLC and capillary electrophoresis (CE). All the HPLC procedures from the previous volume have been incorporated in this book but the original volume should still be consulted for procedures that can be used in gas chromatography (GC) and mass spectroscopy (MS). Procedures for GC and MS constitute, by far, the majority of the procedures in the previous book.

We thank the Chemical Abstracts Service of the American Chemical Society for generously supporting this work and Dr. Norman Schmuff, FDA, for much advice and help. The use of the National Library of Medicine (NLM) and the NIH Library, National Institutes of Health, Bethesda, MD, is greatly appreciated. We would also like to thank Ken Niles of NLM for extending our library privileges and Perry King of John Wiley for help with the electronic version. Special thanks are due to our editor, Betty Sun, who brought the whole project together. Dr. Paul Sadek drew the structures. Although many people have helped with the preparation of this work, the mistakes are our own. We would appreciate hearing from anyone who has corrections, comments, or suggestions. We can be reached at lunng@cder.fda.gov or lhellwig@morgan.edu.

The content of this publication does not necessarily reflect the views or policies of the Food and Drug Administration, nor does mention of trade names, commercial products, or organizations imply endorsement by the U.S. Government.

ABOUT THIS BOOK

SCOPE

In general derivatization is defined as chemical modification to improve the chromatography or detectability of a compound. Thus ion-pair chromatography, where a new compound is not formed, is not covered. Chemiluminescence detection is also generally not covered unless a derivative is formed prior to detection. (A listing of papers describing chemiluminescence detection can be found in Appendix I.) On the other hand both pre-column and post-column derivatization are covered, as is post-column extraction.

STRUCTURE OF THE BOOK

The book is organized according to the functional group that is derivatized. Within the chapter for each functional group chemically similar reagents are grouped together in sections and within each section all the abstracts for each reagent are grouped together. In most cases reaction schemes are given. In some sections bibliographies detail other papers that refer to derivatization of the same functional group with that reagent. Occasionally the nature of the derivatization reaction is not clear and these procedures are gathered together in the Miscellaneous Reactions chapter. Within this chapter the procedures are further divided according to the nature of the reaction (e.g., post-column reaction) and then grouped according to the reagent used. Post-column photochemical reactions are grouped according to the nature of the detector employed. Papers that describe more than one derivatization procedure have abstracts in each appropriate section. A common combination is the use of a pre-column derivatization procedure together with post-column reaction detection.

Two appendices are placed at the end of the book. A bibliography details other papers that deal with derivatization or with post-column reaction. For the most part these papers were identified by computer searches and have not been examined individually. The references in this bibliography are grouped according to the compound or group of compounds that is derivatized. For reasons of space each reference is given only once and references that refer to more than one group of compounds are arbitrarily placed in one category. For more complete searches it might be better to search in the electronic version. Finally, a bibliography lists reviews of derivatization procedures. To help in locating appropriate procedures involving particular reagents, a Molecular Formula Index and a Name Index are also provided.

ABSTRACT STRUCTURE

The detailed procedures for HPLC methods normally contain the following sections. Of course, not all papers give full details, so some sections may be missing.

Matrix
Analyte
Sample Preparation
Guard Column
Column
Mobile Phase
Column Temperature (if not ambient)
Flow Rate
Injection Volume
Retention Time
Detector
Internal Standard
Limit of Detection
Limit of Quantitation
Compounds that Are Extracted under These Conditions
Compounds that Are Chromatographed Simultaneously under These Conditions
Compounds that Are Also Chromatographed under These Conditions
Compounds that Are Non-interfering
Compounds that Are Interfering
Key Words
Reference

For capillary electrophoresis the sections are very similar but a few changes are made. Instead of the column section, sections for the capillary and the capillary preparation procedure are added, running buffer replaces mobile phase, migration time replaces retention time, capillary temperature replaces column temperature, and voltage and/or current replaces flow rate. In HPLC only the injection volume (in μL) is given whereas in capillary electrophoresis more details are given for the injection procedure. Additionally, capillary electrophoresis abstracts list the type of instrument used.

SELECTION OF A METHOD

Different strategies are available for the selection of a method. Using the bound volume the chemist selects the reactive group to be derivatized and then the type of reagent to be used. Within this section there may be a choice of one reagent or many and for any given reagent a variety of abstracted procedures may be available. Using the electronic version a similar strategy may be employed but more options are available. For example, a particular analyte may be selected rather than a generic group. Thus valproic acid could be selected rather than just carboxylic acid. Additionally, equipment preferences, such as fluorescence detection, can be accommodated. Users will find that the system is very flexible.

Regardless of the technique employed the user will often be left with a number of very similar procedures. Unfortunately papers do not come with a numerical "quality" score that could be listed with the abstract. Plate counts are seldom given and can rarely be accurately estimated from reproduced chromatograms. However, methods that give unacceptably broad peaks are not abstracted (and seldom published) so the reader can be sure that the peak shape will be reasonable. Key features that may be useful in selecting a method (and which are abstracted when given in the paper) are limit of detection/quantitation, matrix in which the analyte is present, and other compounds that may be analyzed at the same time. In this last section the user can find listed compounds that may interfere with the analysis and compounds that can be determined at the same time.

ABSTRACT CONVENTIONS

In general, we use the reagent names that were used by the authors of the original paper or the reagent names used by chemical supply houses even though these may not always be strictly in accord with IUPAC or Chemical Abstracts rules. Alternative names are frequently given in parentheses. If reagents are not widely available (e.g., Aldrich, Sigma), a supplier is given. If they are not commercially available a synthetic procedure is given in virtually all cases.

If not otherwise indicated, the detailed HPLC procedures describe procedures carried out at ambient temperature using stainless steel columns, and biological fluids are derived from humans. In some cases, these parameters may be specified. For example, if **both** human blood and rat blood are analyzed, **both** human and rat will be indicated in the key words section. Note that the noun is used instead of the adjective, e.g., cow **not** bovine. For capillary electrophoresis similar conventions are used and procedures are carried out at ambient temperature, and biological fluids are derived from humans. In addition, the detector is at the cathode and the voltage is fixed while the current varies, unless otherwise specified.

Note that for HPLC the Injection Volume may be either the volume actually injected or the volume of the injection loop. If it is the volume actually injected, this value is also given in the Sample Preparation section. If the actual injection volume is not given in the Sample Preparation section, the Injection Volume given is that of the injection loop.

HPLC gradients are linear and mobile phases are v/v, unless otherwise noted. Times given when describing gradient elution, and other procedures such as column switching, are the times for each step, e.g., "MeOH:water 15:85 for 4 min, to 50:50 over 2 min, maintain at 50:50 for 4 min." If we were to include the cumulative times (t) in the example above it would read: "MeOH:water 15:85 for 4 min (t = 4), to 50:50 over 2 min (t = 6), maintain at 50:50 for 4 min (t = 10)."

For the sake of consistency, conditioning procedures for solid-phase extraction (SPE) cartridges are always described at the beginning of the sample preparation sections. Bear in mind, however, that the conditioning procedure should be carried out just prior to use. Thus, if sample preparation is a lengthy procedure, it may be necessary to delay SPE cartridge conditioning until the step requiring the cartridge.

Retention times and migration times are frequently estimated from reproduced chromatograms and so the accuracy may not be high. In particular, differences in retention times between adjacent peaks, e.g., enantiomers, may have a high margin of error.

This book uses non-English characters that are available in extended ASCII. Other non-English characters will be represented by the closest English equivalent or will be spelled out. This is particularly important with the names of authors. For example, Bronnum, where the o has a stroke through it, will be printed as shown because an o with a stroke through it is not available in extended ASCII. On the other hand, Carratù will be printed as shown because u with an accent is available in extended ASCII (# 151). A similar situation applies with Greek characters. Thus α, β, and so on, are available and will be printed as such, but a capital delta is not available and will be represented by "delta." The extended ASCII characters which are used include the following: α, β, δ, ε, Θ, μ, π, σ, Σ, τ, á, à, â, ä, Ä, å, Å, ç, Ç, é, É, è, ê, ë, í, ì, î, ï, ñ, Ñ, ó, ò, ô, ö, Ö, ú, ù, û, ü, Ü, $\leq$, $\pm$, $\geq$, and °.

LITERATURE SEARCHES

Computer searches were conducted for derivatization reactions using Medline and Toxline from 1980 to the present and *Chemical Abstracts* from 1978 to the present.

In Medline and Toxline the search strategy was:

HPLC **and** derivatization (tw) where tw = text word.

In *Chemical Abstracts*, the search strategy was somewhat more complex:

> DERIVATIZ? or POSTCOLUMN? or POST COLUMN?
> and CHROMATOG? (L) (LIQ OR LIQUID)
> and JOURNAL/DT
> and ENG/LA
> and PY > = 1978
> not GAS (W) CHROMATOG? or GC or GLC
> not MS
> not MASS (W) SPEC?

(Initial studies showed that searches tended to produce a number of irrelevant citations involving gas chromatography and mass spectrometry, so it was necessary to remove these terms. The search was also narrowed by specifying only journal articles in English. Even so, 4548 citations were identified.)

In addition to computer searches, some journals were routinely surveyed for relevant articles. These journals were:

Analyst
Analytical Chemistry
Antimicrobial Agents and Chemotherapy
Arzneimittelforschung

Biochemical Pharmacology
Biological and Pharmaceutical Bulletin
Biomedical Chromatography
Biopharmaceutics and Drug Disposition
Chemical and Pharmaceutical Bulletin
Chromatographia
Clinical Chemistry
Clinical Pharmacology and Therapeutics
Drug Metabolism and Disposition
Electrophoresis
Farmaco
Journal of Analytical Toxicology
Journal of AOAC International (formerly *Journal of the Association of Official Analytical Chemists*)
Journal of Chromatographic Science
Journal of Chromatography (Part A and Part B)
Journal of Clinical Pharmacology
Journal of Forensic Sciences
Journal of Liquid Chromatography & Related Technology (formerly *Journal of Liquid Chromatography*)
Journal of Medicinal Chemistry
Journal of Pharmaceutical and Biomedical Analysis
Journal of Pharmaceutical Sciences
Journal of Pharmacology and Experimental Therapeutics
Pharmaceutical Research
Pharmazie
Therapeutic Drug Monitoring
Xenobiotica

Many other journals were consulted when relevant articles were identified by computer searches or by references in other papers. In general, the literature was covered from 1978 to the present.

EXTRACTION FROM BIOLOGICAL MATRICES

In this book certain terms concerning extraction from biological matrices have highly specific meanings. These terms are as follows:

Also   Compounds that can be analyzed under the same conditions as the target compound. It is not specified whether they interfere or can be extracted from the biological matrix in question.

Extracted   Compounds that can be extracted from the biological matrix in
                      question, can be chromatographed under the same condi-
                      tions, and do not interfere with the determination of the tar-
                      get compound.

Interfering   Compounds that interfere with the analysis of the target com-
                      pound. Compounds that interfere with the chromatography
                      of the internal standard are not listed in this category be-
                      cause another internal standard can always be selected or
                      an external standard procedure can be used.

Non-interfering   Compounds that do not interfere with the analysis because no
                      peaks appear on the chromatogram.

Simultaneous   Compounds that can be chromatographed at the same time as
                      the target compound and do not interfere with the determi-
                      nation of the target compound. Note that the compound can-
                      not necessarily be extracted from the biological matrix in
                      question (although this may be possible).

MATRICES

In an attempt to simplify searching procedures, we have made an effort to minimize
the variety of terms used in the matrix heading. However, in a number of cases,
the matrix is associated with various key words that can be used to narrow the
search. For example, the term "formulations" has the key words tablets, creams,
ointments, and injections associated with it. Thus, to find references applicable to
tablets, search first for formulations under the matrix heading and then tablets
under the key word heading. Note that the term "bulk" is used instead of "raw
materials." Some of the more common matrix terms and their associated key words
are given below.

Matrix Term	Associated Key Word
bile	
blood	plasma, serum, whole blood
bulk	
CSF	
dialysate	
formulations	capsules, injections, tablets, creams, ointment, etc.
microsomal incubations	
milk	
perfusate	
reaction mixtures	
saliva	
solutions	
tissue	muscle, kidney, liver, heart, spleen, brain, etc.
urine	

UNITS

The units used are as follows:

HPLC column dimensions in mm (length × internal diameter)
HPLC flow rates in mL/min
HPLC injection volume in μL
HPLC retention times and capillary electrophoresis migration times in min
temperatures in °C
wavelengths in nm

Note that 1 pound/square inch (psi) = 6894.76 Pascals (Pa)

ABBREVIATIONS

α	separation factor; defined by k_2'/k_1' where k_2' is the capacity factor of the second peak and k_1' is the capacity factor of the first peak.
BHT	2,6-*di-tert*-butyl-4-methylphenol; butylated hydroxytoluene
CE	capillary electrophoresis
DMF	dimethylformamide
DMSO	dimethyl sulfoxide
E	electrochemical detector
ELSD	evaporative light-scattering detector
em	emission wavelength
EtOH	ethanol
ex	excitation wavelength
F	fluorescence detector
GPC	gel permeation chromatography
h	hour
HPLC	high-performance liquid chromatography
IS	internal standard
k'	capacity factor; defined by $(t_R - t_0)/t_0$ where t_R is the retention time and t_0 is the column dead time
L	liter
LOD	limit of detection or some other description, indicating that this is the smallest concentration or quantity that can be detected or analyzed
LOQ	lower limit of quantitation, either given as such in the paper or taken as the lower limit of the linear quantitation range
M	molar (i.e., moles/L)
MeCN	acetonitrile
MeOH	methanol
min	minutes
mL	milliliter
mM	milli-molar (i.e., milli-moles/L)
MS	mass spectrometric detector
MTBE	methyl *tert*-butyl ether

nM	nano-molar (i.e., nano-moles/L)
RI	refractive index detector
RT	retention time
s	seconds
SEC	size exclusion chromatography
SFC	supercritical fluid chromatography
SFE	supercritical fluid extraction
SIM	selected-ion monitoring
SPE	solid phase extraction
Temp	temperature
U	units
UV	ultraviolet detector

PIC REAGENTS

These reagents are offered by Waters as buffered solutions containing the following compounds:

> PIC A is tetrabutylammonium sulfate
>
> PIC B5 is pentanesulfonic acid
>
> PIC B6 is hexanesulfonic acid
>
> PIC B7 is heptanesulfonic acid
>
> PIC B8 is 1-octanesulfonic acid
>
> PIC D4 is dibutylamine phosphate

WORKING PRACTICES

In general, good working practice, e.g., filtering and degassing mobile phases, using in-line filters, HPLC or analytical grade materials, and high-quality water is assumed. Solutions containing compounds should be protected from light and silanized glassware should be used, unless you have good reason to believe that these precautions are not necessary. A number of excellent texts discuss good working practices and procedures in HPLC[1-7] and capillary electrophoresis,[8-14] and these should be consulted.

Details of solution preparation are generally not given. It should be remembered that the preparation of a dilute aqueous solution of a relatively water-insoluble compound can frequently be made by dissolving the compound in a small volume of a water-miscible organic solvent and diluting this solution with water.

It is also assumed that safe working practices are observed. In particular, capillary electrophoresis involves the use of lethal voltages. Do not disable safety interlocks! Organic solvents should only be evaporated in a properly functioning chemical fume hood, correct protective equipment should be worn when dealing with potentially hazardous chemical or biological materials, and waste solutions should be disposed of in accordance with all applicable regulations. This book contains a number of procedures for the synthesis of reagents that are not commercially available or are difficult to obtain. These procedures should only be undertaken by trained personnel who are knowledgeable about the safety precautions that are necessary.

A number of solvents used in HPLC are particularly hazardous. For example, benzene is a human carcinogen[15]; chloroform,[16] dichloromethane,[17] dioxane,[18] and carbon tetrachloride[19] are carcinogenic in experimental animals; and DMF[20] and MTBE[21,22] may be carcinogenic. Recent work has shown that n-hexane is surprisingly toxic.[23,24] Organic solvents are, in general, flammable and toxic by inhalation, ingestion, and skin absorption. Sodium azide is carcinogenic and toxic and liberates explosive, volatile, toxic hydrazoic acid with acid. Sodium azide can form explosive heavy metal azides, e.g., with plumbing fixtures, and so should not be discharged down the drain.[25] Disposal procedures have been described for a number of hazardous compounds[25] and recent papers describe a procedure for the hydrolysis of acetonitrile in waste solvent to the much less toxic acetic acid and ammonia.[26,27]

SUPPLIERS

Suppliers of critical items such as columns are given in the abstracts but the suppliers for widely available items are not listed. These suppliers are as follows:

Item	Supplier
Adsorbosphere	Alltech Associates
Asahipak	Asahi Chemical
Bakerbond	J.T. Baker
Bond Elut	Varian
μBondapak	Waters
Chiralcel	Daicel
Co:Pell	Whatman
Corasil	Waters
Cyclobond	Advanced Separation Technologies
Econosil	Alltech Associates
Econosphere	Alltech Associates
Extrelut	E. Merck
Hypersil	Shandon
Inertsil	MetaChem
LiChroprep	E. Merck
LiChrosorb	E. Merck
LiChrosphere	E. Merck
Micropak	Varian
Microsorb	Rainin
NewGuard	Applied Biosystems
Nova-Pak	Waters
Nucleosil	Macherey Nagel
Partisil	Whatman
Pecosphere	Perkin-Elmer
Porasil	Waters
Sep-Pak	Waters
Spheri-5	Applied Biosystems

Item	Supplier
Spheri-10	Applied Biosystems
Spherisorb	Phase Separations
SPICE	Analtech
Supelcosil	Supelco
Ultrasphere	Beckman
Ultremex	Phenomenex
Vydac	The Separations Group
Zorbax	Mac-Mod Analytical

This list is not intended to be definitive. Many other companies supply these pieces of equipment.

TRADEMARKS

The following trademarks are used:

Trademark	Company
Adsorbosphere	Alltech Associates, Inc.
Asahipak	Asahi Chemical Industry Co. Ltd.
Bakerbond	J.T. Baker
Bond Elut	Varian Associates, Inc.
μBondapak	Waters Associates, Inc.
Chiralcel	Daicel Chemical Industries, Ltd.
Co:Pell	Whatman Chemical Separation Co.
Corasil	Waters Associates, Inc.
Cyclobond	Advanced Separation Technologies, Inc.
Econosil	Alltech Associates, Inc.
Econosphere	Alltech Associates, Inc.
Extrelut	E. Merck
Hypersil	Shandon Scientific, Ltd.
Inertsil	GL Sciences Inc.
LiChroprep	E. Merck
LiChrosorb	E. Merck
LiChrosphere	E. Merck
Micropak	Varian Associates, Inc.
Microsorb	Rainin Instrument Co. Inc.
NewGuard	Applied Biosystems
Nova-Pak	Waters Associates, Inc.
Nucleosil	Macherey Nagel
Partisil	Whatman Chemical Separation Co.
Pecosphere	Perkin-Elmer
PIC	Waters Associates, Inc.
Porasil	Waters Associates, Inc.
Resolve	Waters Associates, Inc.
Sep-Pak	Waters Associates, Inc.
Spheri-5	Applied Biosystems
Spheri-10	Applied Biosystems

Trademark	Company
Spherisorb	Phase Separations, Ltd.
SPICE	Analtech
Supelcosil	Supelco, Inc.
Ultrasphere	Beckman Instruments, Inc.
Ultremex	Phenomenex, Inc.
Vydac	The Separations Group
Zorbax	DuPont Company

REFERENCES

1. Meyer, V.R. *Practical High-Performance Liquid Chromatography*, 2nd edition, John Wiley: New York, 1994.

2. McMaster, M.C. *HPLC, a Practical User's Guide*, VCH: New York, 1994.

3. Bidlingmeyer, B.A. *Practical HPLC Methodology and Applications*, John Wiley: New York, 1993.

4. Lough, W.J.; Wainer, I.W. (Eds.) *High Performance Liquid Chromatography*, Routledge, Chapman and Hall: New York, 1992.

5. Snyder, L.R.; Glajch, J.L.; Kirkland, J.J. *Practical HPLC Method Development*, John Wiley: New York, 1988.

6. Lawrence, J.F. *Organic Trace Analysis by Liquid Chromatography*, Academic Press: New York, 1981.

7. Snyder, L.R.; Kirkland, J.J. *Introduction to Modern Liquid Chromatography*, 2nd edition, John Wiley: New York, 1979.

8. Landers, J.P. *Handbook of Capillary Electrophoresis*, 2nd edition, CRC Press: Boca Raton, NJ, 1996.

9. Lunte, S.M.; Radzik, D.M. *Pharmaceutical and Biomedical Applications of Capillary Electrophoresis*, Elsevier Science: New York, 1996.

10. Baker, D.R. *Capillary Electrophoresis*, John Wiley: New York, 1995.

11. Altria, K.D. (Ed.) *Capillary Electrophoresis Guidebook: Principles, Operation, and Applications*, Humana Press: Totowa, NJ, 1995.

12. Righetti, P.G.; Hancock, W. (Eds.) *Capillary Electrophoresis in Analytical Biotechnology*, CRC Press: Boca Raton, FL, 1995.

13. Baker, D. *Capillary Electrophoresis*, Prentice Hall: Englewood Cliffs NJ, 1994.

14. Foret, F.; Krivankov, L.; Bocek, P. *Capillary Zone Electrophoresis*, VCH: New York, 1994.

15. Lewis, R.J., Sr. *Sax's Dangerous Properties of Industrial Materials,* 8th edition, Van Nostrand-Reinhold: New York, 1992, pp. 356–358.

16. Reference 15, pp. 815–816.

17. Reference 15, pp. 2311–2312.

18. Reference 15, pp. 1449–1450.

19. Reference 15, pp. 701–702.

20. Reference 15, p. 1378.

21. Belpoggi, F.; Soffritti, M.; Maltoni, C. Methyl-tertiary-butyl ether (MTBE)—a gasoline additive—causes testicular and lympho-haematopoietic cancers in rats. *Toxicol. Ind. Health* **1995**, *11*, 119–149.

22. Mehlman, M.A. Dangerous and cancer-causing properties of products and chemicals in the oil refining and petrochemical industry: Part XV. Health hazards and health risks

from oxygenated automobile fuels (MTBE): Lessons not heeded. *Int. J. Occup. Med. Toxicol.* **1995**, *4*, 219–236.

23. Meyer, V. A safer solvent. *Anal. Chem.* **1997**, *69*, 18A.

24. Hayes, A.. (Ed.) *Principles and Methods of Toxicology*, 3rd edition, Raven: New York, 1994, p. 383.

25. Lunn, G.; Sansone, E.B. *Destruction of Hazardous Chemicals in the Laboratory*, 2nd edition, John Wiley: New York, 1994.

26. Gilomen, K.; Stauffer, H.P.; Meyer, V.R. Detoxification of acetonitrile—water wastes from liquid chromatography. *Chromatographia* **1995**, *41*, 488–491.

27. Gilomen, K.; Stauffer, H.P.; Meyer, V.R. Management and detoxification of acetonitrile wastes from liquid chromatography. *LC.GC* **1996**, *14*, 56–58.

GUIDE TO THE PERPLEXED: SOME HINTS ON USING THE ELECTRONIC VERSION

Although the use of the software should be intuitively obvious the following hints are offered to enable the user to quickly become comfortable using the program.

Two search options are available: Simple Search and Complex Search. They may be used at any time by clicking on the appropriate icon, either from the opening screen or from within the program.

SIMPLE SEARCH

The Simple Search screen mimics the form of the abstracts. You can move through it to the appropriate section and enter a search term. Many of the sections have buttons that will deploy lists of all the possible values. For example, clicking the button next to Reagent Type will produce a list of reagent types, e.g., acyl halide, thiol. Typing the initial letter or two will bring you to the appropriate term which may then be selected. Clicking on Go will initiate the search. Note that the search can be restricted to either HPLC or capillary electrophoresis (CE) references by checking the appropriate box on the first screen. The default value is both HPLC and CE.

COMPLEX SEARCH

The Complex Search feature can be used to combine two search terms, e.g., papers published in 1994 that use dansyl chloride. However, it may be advantageous to use the Complex Search feature for only one search term since it avoids scrolling through all the Simple Search screens. Clicking on the initial button gives you a list of the fields that may be searched in a Complex Search, i.e., Reagent Name, Matrix, First Author, Year, etc. When you select a category it may give you a box into which you can enter the appropriate search term or it may deploy a list of available terms. Most categories give a list of available terms but some (First Author, Year, Also Analyzed, Extracted, Interfering, Key Words, Mobile Phase, Non-

Interfering, Reference, Running Buffer, Sample Preparation, and Simultaneously Analyzed) would give unmanageably long lists, so these categories give you a box for entering the term.

SEARCHING

The program will pick up all instances when the search term occurs in the appropriate box. Thus (in Matrix) "bile" will find all instances of bile, "blood" all instances of blood, and "bile, blood" all instances of EITHER bile OR blood. The terms do not need to be contiguous. Thus "blood, urine" will bring up all references having either blood or urine and including "blood, saliva, urine."

Particular care has been taken to limit the number of categories in the Detector field. Although the detectors can be defined in various ways, e.g., UV 254, UV 214, searching is limited to the five main types of detectors: Chemiluminescence, Electrochemical, Fluorescence, Mass Spectrometric, and UV. Similarly in the Column field the list has been simplified to the basic types, e.g., anion-exchange, C18. In many cases the type of chromatography can be selected by searching for the appropriate column type.

POST-COLUMN REACTIONS

Where the chemistry is well defined, e.g., phthalaldehyde/2-mercaptoethanol, then post-column reactions will be listed at the appropriate place under the group that is being derivatized. However, when the chemistry is not well defined (e.g., aluminum trichloride) these reactions are listed under Miscellaneous Reactions, Post-Column Reaction. Searching for "post-column reaction" under Key Words will find either type (411 hits) but searching under Reagent Type will find only those references in the Miscellaneous Reactions section (191 hits).

RESULTS

A list of hits is displayed. The list can be sorted by Reagent, Reactive Group, Year, or First Author. Highlight an appropriate reference and double click or hit enter. The appropriate reference will then be displayed and you can move through it to see all the data or print it out.

ACYL HALIDE

AMINE

1-(2-Pyridyl)piperazine

SAMPLE
Matrix: air
Analyte: phosgene and isocyanates
Sample preparation: Pack 100 mg coated Chromosorb in a 75 × 6.5 glass tube and secure
with plugs of coated glass wool, pull air through the tube at 1 L/min for at least 20 min,
add 1-2 mL MeCN to the tube, sonicate for 10 min, filter (0.45 μm) the liquid, inject a 20
μL aliquot. (Prepare coated Chromosorb by adding 10 g 80/100 mesh Chromosorb W, acid
washed (Alltech) to 10 mL 2.5% 1-(2-pyridyl)piperazine in hexane, mix, evaporate to dryness under a stream of nitrogen. Prepare coated glass wool by adding 10 g glass wool to
10 mL 2.5% 1-(2-pyridyl)piperazine in hexane, mix, evaporate to dryness under a stream
of nitrogen.)

HPLC VARIABLES
Column: 250 × 5 5 μm Supelcosil LC18-DB octadecyl
Mobile phase: MeCN:3% pH 6.0 ammonium acetate 30:70
Flow rate: 1.5
Injection volume: 20
Detector: UV 270

CHROMATOGRAM
Retention time: 16 (phosgene), 21 (2,6-toluenediisocyanate), 24 (2,4-toluenediisocyanate)
Limit of detection: 5 ppb

REFERENCE
Rando, R.J.; Poovey, N.G.; Chang, S.-N. Collection and chemical derivatization of airborne phosgene
with 1-(2-pyridyl)-piperazine and determination by high performance liquid chromatography,
J.Liq.Chromatogr., **1993**, *16*, 3291–3309.

Tryptamine

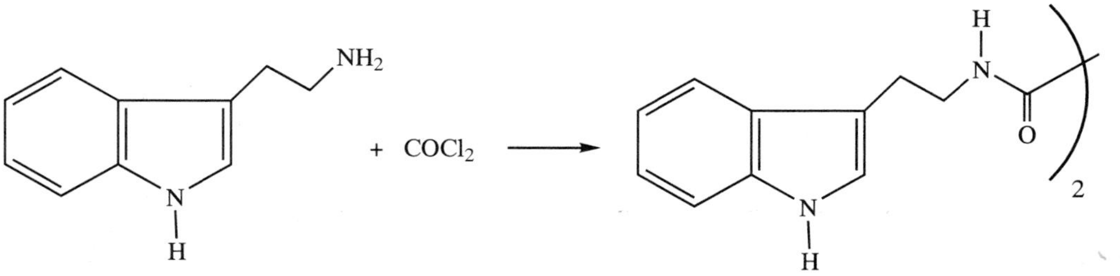

SAMPLE
Matrix: air
Analyte: phosgene
Sample preparation: Pass air through 20 mL reagent at 100 mL/min, evaporate to dry-
ness, reconstitute with 8 mL MeCN, add 200 μL MeCN:acetic acid 99.9:0.1, let stand for

30 min, add 100 µL water, let stand for 30 min, dilute 10-fold with mobile phase, inject an aliquot. (Prepare reagent by dissolving 30 mg tryptamine in 5 mL MeCN, dilute a 500 µL aliquot to 200 mL with 2,2,4-trimethylpentane, agitate until the MeCN disappears.)

HPLC VARIABLES
Column: 250 × 4.6 5 µm CSC-Hypersil-ODS
Mobile phase: MeCN:0.6% ammonium acetate 50:50
Flow rate: 0.8
Injection volume: 20
Detector: F ex 275 em 360; E, ESA Coulochem Model 5100A, ESA Model 5010 analytical cell, porous graphite working electrodes, detector 1 +0.3 V, detector 2 +0.6 V (monitored)

CHROMATOGRAM
Retention time: 7.5
Limit of detection: 8 ng/mL (liquid; F or E), 0.04 mg/cubic m (air; F or E)

REFERENCE
Wu, W.S.; Gaind, V.S. Determination of phosgene (carbonyl chloride) in air by high-performance liquid chromatography with a dual selective detection system, *Analyst*, **1993**, *118*, 1285–1287.

SOLID PHASE REAGENT

Polymer-supported 8-Amino-2-naphthoxide

SAMPLE
Matrix: solutions
Analyte: acyl halides
Sample preparation: Inject a 20 µL aliquot onto column A and start the gradient. Initially elute column A onto column B and after 4 min remove column B from the circuit.

HPLC VARIABLES
Column: A 33 × 3.9 polymer-supported 8-amino-2-naphthoxide; B 300 × 3.9 µBondapak C18 (Preparation of polymer-supported 8-amino-2-naphthoxide is as follows. Extract -400

mesh AG 1-X8 anion exchange resin (Cl⁻ form, Bio-Rad) in a Soxhlet extractor with MeCN for 24 h. Pack the resin in a 170 × 12 glass column and wash with 500 mL water, wash with 2.5 L 1 M NaOH. (Check that the effluent is chloride free by acidifying an aliquot with nitric acid and adding silver nitrate.) Wash with 1 L water. Stir 10 g of this resin in basic water with a 0.5 molar excess of 8-amino-2-naphthol for 3 h, add 1 equivalent of NaOH, filter, wash three times with 200 mL portions of water, wash with 200 mL MeCN, dry in air, store in the freezer. 8-Amino-2-napthol is available from Tokyo Kasei (TCI America, Portland OR). Synthesis is as follows (Caution! Molten NaOH is extremely corrosive, use appropriate protective equipment!). Melt NaOH with the addition of a small quantity of water, gradually add 8-amino-2-naphthalenesulfonic acid so as to avoid foaming, when addition is complete heat at 265-275° for 1 h, heat at 305° for 5 min, cool to 200°, cautiously dilute with water, neutralize with HCl, boil, filter, add NaOH until no more precipitate of 8-amino-2-naphthol is obtained (J. Am. Chem. Soc. 1929, 51, 1766).)

Mobile phase: Gradient. MeCN:water from 10:90 to 55:45 over 12 min (Waters curve no. 8 [initial concentration for about 3 min then rises steeply to final concentration]), maintain at 55:45.

Column temperature: 60
Flow rate: 2
Injection volume: 20
Detector: UV 214

CHROMATOGRAM
Retention time: 13 (acetyl chloride), 15.5 (valeryl chloride), 17 (cinnamoyl chloride)
Limit of detection: 0.84-1.5 ppb

REFERENCE
Colgan, S.T.; Krull, I.S.; Dorschel, C.; Bidlingmeyer, B.A. Derivatization of alkyl halides, acid chlorides, and other electrophiles with polymer-immobilized 8-amino-2-naphthoxide, *J.Chromatogr.Sci.*, **1988**, 26, 501−512.

ALCOHOL

ACTIVATED HALIDE

3-Bromomethyl-7-methoxy-1,4-benzoxazin-2-one

+ ROH $\longrightarrow$

SAMPLE
Matrix: blood
Analyte: methanol
Sample preparation: 270 μL Plasma + 30 μL water, mix for 10 s, add 300 μL 50 mM KOH, vortex for 30 s. Remove a 400 μL aliquot and filter (Millipore Ultrafree-MC, 30000 NWML polysulfone PTTK membrane) while centrifuging at 2900 g for 1 h. Remove a 200 μL aliquot of the ultrafiltrate and add it to 100 μL 500 mM benzyl dimethyl n-tetracdecylammonium chloride in water, add 300 μL 80 μM 1-nitronaphthalene in dichloromethane, add 200 μL 20 mM 3-bromomethyl-7-methoxy-1,4-benzoxazin-2-one in dichloromethane, shake mechanically at 30° for 2 h, add 3 mL water, shake, inject a 10 μL aliquot of the dichloromethane layer. (3-Bromomethyl-7-methoxy-1,4-benzoxazin-2-one is also available from Tokyo Kasei (TCI America, 911 North Harborgate St., Portland, OR 97203; 800-423-8616,503-283-1681, (fax) 503-283-1987; www.tciamerica.com). Synthesis of 3-bromomethyl-7-methoxy-1,4-benzoxazin-2-one is as follows. Add 36 g sodium hydrosulfite to 5.07 g 5-methoxy-2-nitrophenol in 60 mL water, reflux under an inert gas for 30 min, cool in ice, filter to obtain 2-amino-5-methoxyphenol as amber prisms. Add 3.48 g ethyl pyruvate to a solution of 4.17 g 2-amino-5-methoxyphenol in 45 mL EtOH and 11 mL acetic acid, stir at room temperature in the dark overnight, filter, recrystallize the solid from EtOH to obtain 3-methyl-7-methoxy-1,4-benzoxazin-2-one as orange-red crystals (mp 126°). Add 3.76 g phenyltrimethylammonium tribromide to a solution of 1.09 g 3-methyl-7-methoxy-1,4-benzoxazin-2-one in 20 mL THF stirred at 0°, stir gently overnight in a refrigerator, dilute with 50 mL diethyl ether, wash with 25 mL 100 mM sodium bisulfite, wash with 25 mL 100 mM sodium bicarbonate, wash with water. Dry the organic layer over anhydrous sodium sulfate, evaporate to dryness under reduced pressure, chromatograph the residue on 50 g silica gel with n-hexane:dichloromethane 50:50, recrystallize the product from n-hexane:dichloromethane 50:50 to obtain 3-bromomethyl-7-methoxy-1,4-benzoxazin-2-one as faint yellow prisms (mp 146°) (J. Chromatogr. 1992, 591, 159).)

HPLC VARIABLES
Column: 250 × 4 5 μm LiChrospher diol
Mobile phase: n-Hexane:dichloromethane 90:10
Flow rate: 1.2
Injection volume: 10
Detector: UV 350

CHROMATOGRAM
Retention time: 13
Internal standard: 1-nitronaphthalene (4)
Limit of detection: 60 μM

OTHER SUBSTANCES
Extracted: ethanol

KEY WORDS
plasma; ultrafiltrate

REFERENCE
Chen, S.-H.; Yen, C.-H.; Wu, H.-L.; Wu, S.-M.; Kou, H.-S.; Lin, S.-J. Derivatization-high performance
   liquid chromatographic determination of methanol in human plasma, *J.Liq.Chromatogr.Rel.Technol.*,
   1997, *20*, 1967–1978.

4-Chloro-7-nitro-2,1,3-benzoxadiazole

SAMPLE
Matrix: urine
Analyte: estrogens
Sample preparation: Condition a Sep-Pak C18 SPE cartridge with 10 mL water, 5 mL
   MeOH, and 10 mL water. 1 mL Urine + 2 nmoles equilin + 100 µL 1.5 M pH 3 acetate
   buffer, add to the SPE cartridge, wash with 10 mL 150 mM pH 3 acetate buffer, elute
   with 3 mL MeOH. Add HCl to the eluate so that the concentration of HCl is 500 mM,
   heat at 100° for 1.5 h, neutralize with sodium bicarbonate, extract with 2 mL chloroform.
   Evaporate the organic layer to dryness, reconstitute with 1 mL 5 µM 4-chloro-7-nitro-
   2,1,3-benzoxadiazole (4-chloro-7-nitrobenzo-2-oxa-1,3-diazole; NBD-Cl) in MeCN containing 25 nM 18-crown-6 and 15 mM potassium carbonate, heat at 80° for 30 min, filter,
   inject a 10-15 µL aliquot.

HPLC VARIABLES
Column: 5 µm Hypersil ODS
Mobile phase: MeOH:water 75:25
Flow rate: 1
Injection volume: 10-15
Detector: UV 380

CHROMATOGRAM
Retention time: 6.50 (estrone), 8.39 (estradiol), 3.23 (estriol)
Internal standard: equilin (4.95)
Limit of detection: 30-50 nM

KEY WORDS
SPE; derivatives are not fluorescent

REFERENCE
Tirendi, S.; Lancetta, T.; Bousquet, E. Estrogens determination in urine by RP-HPLC with UV detection,
   Farmaco, **1994**, *49*, 427–430.

1-(Ethylthio)-3-(dichloro-1,3,5-triazinyl)-2-n-propylisoindole

+ ROH ⟶

SAMPLE
Matrix: solutions
Analyte: estrogens
Sample preparation: Mix 20 μL of a 1.25 μM solution of estrogens in acetone with 20 μL
3 mM 1-(ethylthio)-3-(dichloro-1,3,5-triazinyl)-2-n-propylisoindole in acetone, add 10 μL
50 mM NaOH in water, mix, let stand at room temperature for 5 s, add 10 μL 174 mM
acetic acid, mix thoroughly, inject a 10 μL aliquot. (Synthesis of 1-(ethylthio)-3-(dichloro-
1,3,5-triazinyl)-2-n-propylisoindole is as follows. Dissolve 268 mg o-phthalaldehyde in 10
mL anhydrous ether at room temperature, add 150 μL ethanethiol, mix, add 165 μL n-
propylamine, mix, cool to 0°, stir, add 20 mL 18.4 mg/mL cyanuric chloride in ether dropwise, stir for 1 min, evaporate under reduced pressure to remove the solvent. Dissolve
the residue in 10 mL acetone, purify by preparative TLC on silica by elution with benzene
(Caution! Benzene is a carcinogen!). Extract the appropriate band with acetone, evaporate
the extract to dryness under reduced pressure. Purify by dissolving the crude product in
a small amount of acetone and precipitating the pure compound by the dropwise addition
of water, dry under reduced pressure to yield 1-(ethylthio)-3-(dichloro-1,3,5-triazinyl)-2-n-
propylisoindole as orange needles (mp 108-109°).)

HPLC VARIABLES
Column: 250 × 4.6 5 μm TSKgel ODS-120 A (Tosoh)
Mobile phase: MeOH:water 90:6
Flow rate: 1
Injection volume: 10
Detector: F ex 415 em 445

CHROMATOGRAM
Retention time: 36 (ethinylestradiol), 40 (17α-estradiol), 42 (estrone), 52 (17β-estradiol)
Limit of detection: 1.1 pmole (estrone)

REFERENCE
Fujino, H.; Goya, S. 1-(Ethylthio)-3-(dihalo-1,3,5-triazinyl)-2-n-propylisoindole as fluorescent labeling reagent for estrogens, *Anal.Sci.*, **1989**, 5, 105–106.

1-(Ethylthio)-3-(difluoro-1,3,5-triazinyl)-2-n-propylisoindole

+ ROH ⟶

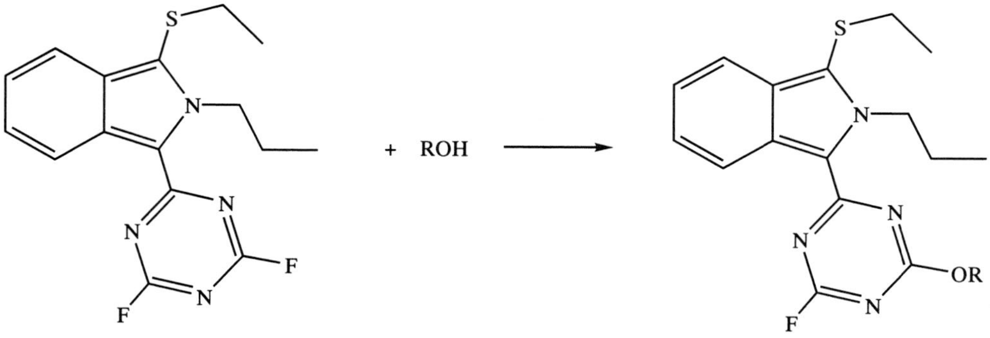

SAMPLE

Matrix: solutions

Analyte: estrogens

Sample preparation: Mix 20 μL of a 1.25 μM solution of estrogens in acetone with 20 μL 3 mM 1-(ethylthio)-3-(difluoro-1,3,5-triazinyl)-2-n-propylisoindole in acetone, add 10 μL 10 mM NaOH in water, mix, let stand at room temperature for 5 s, add 10 μL 17.4 mM acetic acid, mix thoroughly, inject a 10 μL aliquot. (Synthesis of 1-(ethylthio)-3-(difluoro-1,3,5-triazinyl)-2-n-propylisoindole is as follows. Dissolve 268 mg o-phthalaldehyde in 10 mL anhydrous ether at room temperature, add 150 μL ethanethiol, mix, add 165 μL n-propylamine, mix, cool to 0°, stir, add 20 mL 18.4 mg/mL cyanuric fluoride (Fluka) in ether dropwise, stir for 1 min, evaporate under reduced pressure to remove the solvent. Dissolve the residue in 10 mL acetone, purify by preparative TLC on silica by elution with benzene (Caution! Benzene is a carcinogen!). Extract the appropriate band with acetone, evaporate the extract to dryness under reduced pressure. Purify by dissolving the crude product in a small amount of acetone and precipitating the pure compound by the dropwise addition of water, dry under reduced pressure to yield 1-(ethylthio)-3-(difluoro-1,3,5-triazinyl)-2-n-propylisoindole as orange needles (mp 105-106°).)

HPLC VARIABLES

Column: 250 × 4.6 5 μm TSKgel ODS-120 A (Tosoh)

Mobile phase: MeOH:water 90:6

Flow rate: 1

Injection volume: 10

Detector: F ex 415 em 445

CHROMATOGRAM

Retention time: 26 (estrone), 28 (ethinyl estradiol), 30 (17α-estradiol), 37 (17β-estradiol)

Limit of detection: 270 fmole (estrone)

REFERENCE

Fujino, H.; Goya, S. 1-(Ethylthio)-3-(dihalo-1,3,5-triazinyl)-2-*n*-propylisoindole as fluorescent labeling reagent for estrogens, *Anal.Sci.*, **1989**, *5*, 105–106.

FEBPI

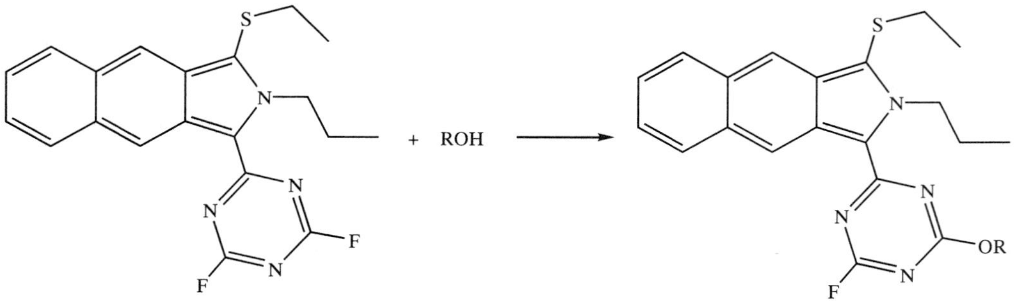

SAMPLE

Matrix: solutions

Analyte: estrogens

Sample preparation: Mix 20 μL of a 500 nM solution of estrogens in acetone with 20 μL 1.5 mM FEBPI in acetone, add 10 μL 20 mM NaOH in water, let stand at room temperature for 5 s, add 10 μL 174 mM acetic acid in water, mix thoroughly, inject a 10 μL aliquot. (FEBPI is 3-(difluoro-1,3,5-triazinyl)-1-(ethylthio)-2-n-propylbenz[f]isoindole. Preparation is as follows. Stir 200 mg 2,3-naphthalenedicarboxaldehyde in 20 mL anhydrous ether at room temperature, add 81 μL ethanethiol, add 90 μL n-propylamine, cool in ice-water, slowly add 146 μL cyanuric fluoride (Fluka), stir for 1 min, remove the solvent by evaporation under reduced pressure, dissolve the residue in 20 mL acetone, purify by preparative TLC on silica with benzene (Caution! Benzene is a carcinogen!), extract the appropriate band with acetone. Dissolve the crude product in a little acetone, precipitate by the dropwise addition of water, dry under reduced pressure to yield 3-(difluoro-1,3,5-triazinyl)-1-(ethylthio)-2-n-propylbenz[f]isoindole as dark-purple needles (mp 158-160°).)

HPLC VARIABLES

Column: 250 × 4.6 5 μm TSK gel ODS-120A (Tosoh)

Mobile phase: MeOH

Flow rate: 1

Injection volume: 10

Detector: F ex 490 em 520

CHROMATOGRAM

Retention time: 17 (ethinyl estradiol), 21 (estrone), 27 (17β-estradiol)

Limit of detection: 80 fmole

REFERENCE

Fujino, H.; Goya, S. 3-(Difluoro-1,3,5-triazinyl)-1-(ethylthio)-2-n-propylbenz[f]isoindole as a fluorescence derivatization reagent for estrogens in high-performance liquid chromatography, *Chem.Pharm.Bull.*, **1989**, *37*, 1939–1940.

1-(Ethylthio)-3-(difluoro-1,3,5-triazinyl)-2-n-propylisoindole

+ ROH ⟶

SAMPLE
Matrix: solutions
Analyte: estrogens
Sample preparation: Mix 20 μL of a 1.25 μM solution of estrogens in acetone with 20 μL 3 mM 1-(ethylthio)-3-(difluoro-1,3,5-triazinyl)-2-n-propylisoindole in acetone, add 10 μL 10 mM NaOH in water, mix, let stand at room temperature for 5 s, add 10 μL 17.4 mM acetic acid, mix thoroughly, inject a 10 μL aliquot. (Synthesis of 1-(ethylthio)-3-(difluoro-1,3,5-triazinyl)-2-n-propylisoindole is as follows. Dissolve 268 mg o-phthalaldehyde in 10 mL anhydrous ether at room temperature, add 150 μL ethanethiol, mix, add 165 μL n-propylamine, mix, cool to 0°, stir, add 20 mL 18.4 mg/mL cyanuric fluoride (Fluka) in ether dropwise, stir for 1 min, evaporate under reduced pressure to remove the solvent. Dissolve the residue in 10 mL acetone, purify by preparative TLC on silica by elution with benzene (Caution! Benzene is a carcinogen!). Extract the appropriate band with acetone, evaporate the extract to dryness under reduced pressure. Purify by dissolving the crude product in a small amount of acetone and precipitating the pure compound by the dropwise addition of water, dry under reduced pressure to yield 1-(ethylthio)-3-(difluoro-1,3,5-triazinyl)-2-n-propylisoindole as orange needles (mp 105-106°).)

HPLC VARIABLES
Column: 250 × 4.6 5 μm TSKgel ODS-120 A (Tosoh)
Mobile phase: MeOH:water 90:6
Flow rate: 1
Injection volume: 10
Detector: F ex 415 em 445

CHROMATOGRAM
Retention time: 26 (estrone), 28 (ethinyl estradiol), 30 (17α-estradiol), 37 (17β-estradiol)
Limit of detection: 270 fmole (estrone)

REFERENCE
Fujino, H.; Goya, S. 1-(Ethylthio)-3-(dihalo-1,3,5-triazinyl)-2-n-propylisoindole as fluorescent labeling reagent for estrogens, *Anal.Sci.*, **1989**, *5*, 105–106.

FEBPI

SAMPLE
Matrix: solutions
Analyte: estrogens
Sample preparation: Mix 20 µL of a 500 nM solution of estrogens in acetone with 20 µL 1.5 mM FEBPI in acetone, add 10 µL 20 mM NaOH in water, let stand at room temperature for 5 s, add 10 µL 174 mM acetic acid in water, mix thoroughly, inject a 10 µL aliquot. (FEBPI is 3-(difluoro-1,3,5-triazinyl)-1-(ethylthio)-2-n-propylbenz[f]isoindole. Preparation is as follows. Stir 200 mg 2,3-naphthalenedicarboxaldehyde in 20 mL anhydrous ether at room temperature, add 81 µL ethanethiol, add 90 µL n-propylamine, cool in ice-water, slowly add 146 µL cyanuric fluoride (Fluka), stir for 1 min, remove the solvent by evaporation under reduced pressure, dissolve the residue in 20 mL acetone, purify by preparative TLC on silica with benzene (Caution! Benzene is a carcinogen!), extract the appropriate band with acetone. Dissolve the crude product in a little acetone, precipitate by the dropwise addition of water, dry under reduced pressure to yield 3-(difluoro-1,3,5-triazinyl)-1-(ethylthio)-2-n-propylbenz[f]isoindole as dark-purple needles (mp 158-160°).)

HPLC VARIABLES
Column: 250 × 4.6 5 µm TSK gel ODS-120A (Tosoh)
Mobile phase: MeOH
Flow rate: 1
Injection volume: 10
Detector: F ex 490 em 520

CHROMATOGRAM
Retention time: 17 (ethinyl estradiol), 21 (estrone), 27 (17β-estradiol)
Limit of detection: 80 fmole

REFERENCE
Fujino, H.; Goya, S. 3-(Difluoro-1,3,5-triazinyl)-1-(ethylthio)-2-n-propylbenz[f]isoindole as a fluorescence derivatization reagent for estrogens in high-performance liquid chromatography, *Chem.Pharm.Bull.*, **1989**, *37*, 1939–1940.

ACYL AZIDE

3-(2-Phthalimidyl)benzoyl Azide

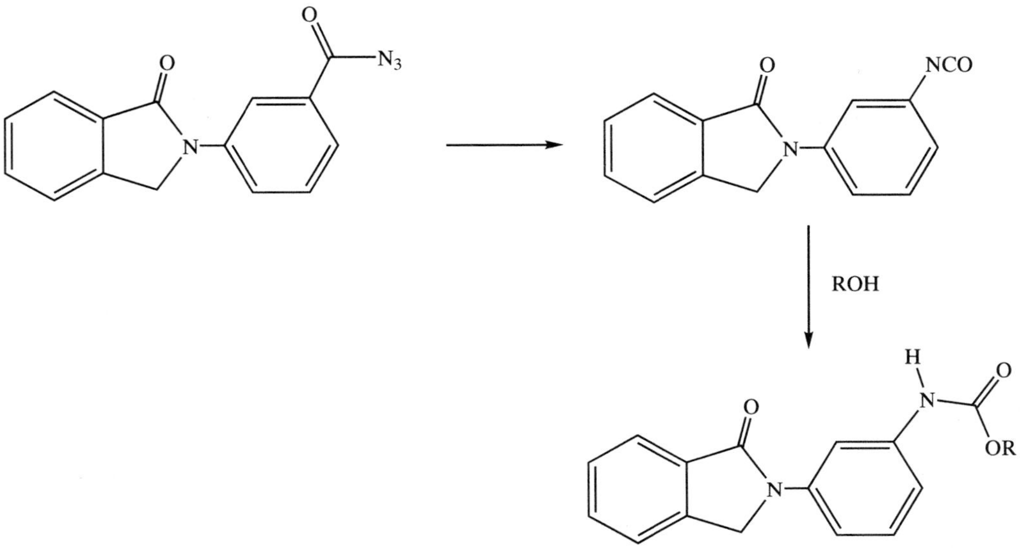

SAMPLE
Matrix: blood
Analyte: alcohols
Sample preparation: 5 μL Serum + 45 μL water + 200 mM NaOH in EtOH:water 95:
5, heat at 100° for 15 min, cool, add 1 mL water, add 3 mL hexane, vortex for 2 min,
centrifuge at 3000 rpm for 5 min. Remove the organic layer and evaporate it to dryness
under a stream of nitrogen, reconstitute the residue in 20 μL acetone, add 50 μL 5 mM
4-dimethylaminopyridine in acetone, mix well, add 100 μL 20 mM 3-(2-phthalimi-
dyl)benzoyl azide in acetone, mix well, heat at 125° in a stoppered vial for 30 min, cool
in a water bath, inject a 10 μL aliquot. (Omit 4-dimethylaminopyridine for cholesterol
determinations. Synthesis of 3-(2-phthalimidyl)benzoyl azide is as follows. Mix 680 mg
m-aminobenzoic acid in 50 mL diethyl ether with 670 mg o-phthalaldehyde in 100 mL
diethyl ether, mix, stir at room temperature for 2 days, filter to obtain 3-(2-phthalimi-
dyl)benzoic acid as a white solid. Dissolve 500 mg 3-(2-phthalimidyl)benzoic acid in 15
mL DMF, add 560 mg diphenylphosphoryl azide in 4 mL DMF, add 200 mg triethylamine,
stir at 0° for 2 h, add 60 mL 5% sodium bicarbonate in water, add 100 mL diethyl ether,
shake. Wash the organic layer twice with 100 mL portions of cold water. Collect the solid
at the interface and combine it with the residue obtained when the organic layer is evaporated under reduced pressure. Take up the crude product in acetone, add water to precipitate pure product, repeat this procedure twice to obtain 3-(2-phthalimidyl)benzoyl
azide as white needles (mp 123-126°).)

HPLC VARIABLES
Column: 100 × 6 ERC-ODS-1161 (Erma, Tokyo)
Mobile phase: MeCN:water 70:30 (A) or MeCN:EtOH 70:30 (B)
Flow rate: 1
Injection volume: 10
Detector: F ex 302 em 440

CHROMATOGRAM
Retention time: 10 (octanol (A)), 12 (l-menthol (A)), 14 (nonanol (A)), 21 (decanol (A)), 10
(cholesterol (B))

Limit of detection: 100-400 fmole

KEY WORDS
serum; serum extraction only validated for cholesterol

REFERENCE
Tsuruta, Y.; Date, Y.; Kohashi, K. (2-Phthalimidyl)benzoylazides as fluorescence labeling reagents for alcohols in high-performance liquid chromatography, *Anal.Sci.*, **1991**, *7*, 411−414.

ACYL HALIDE

Acetyl Chloride

SAMPLE
Matrix: blood
Analyte: β-tigogenin cellobioside
Sample preparation: 1 mL Serum + 50 μL 100 μg/mL IS, vortex for 5 s, add 5 mL hexane, vortex for 10 s, centrifuge at 1500 g for 3 min, discard the hexane layer, add 2 mL MeCN, vortex for 10 min, centrifuge at 2500 g for 10 min. Remove the supernatant and evaporate it to dryness in a vortex evaporator at 50°, reconstitute the residue in 100 μL MeCN, vortex for 5 s, sonicate for 1 min, vortex for 10 min, add 200 μL acetyl chloride, vortex for 1 min, let stand at room temperature overnight, evaporate to remove excess reagent and solvent in a vortex evaporator at 50°, reconstitute the residue in 500 μL MeCN, vortex for 5 s, sonicate for 1 min, vortex for 10 min, centrifuge at 2500 g for 3 min. Remove the supernatant and evaporate it to dryness in a vortex evaporator at 50°, reconstitute the residue in 60 μL mobile phase, inject a 50 μL aliquot.

HPLC VARIABLES
Column: 150 × 3.9 NovaPak C18
Mobile phase: MeOH:isopropanol:water 50:40:10
Flow rate: 1
Injection volume: 50
Detector: MS, SCIEX API III triple quadrupole, APCI, nebulizer probe 450°, corona discharge 5 μA, collision induced dissociation with argon at 5.5×10^{14} atoms/cm², collision energy 18 eV, multiple reaction monitoring, m/z 1035

CHROMATOGRAM
Retention time: 1.9
Internal standard: pentadeutero β-tigogenin cellobioside
Limit of quantitation: 10 ng/mL

KEY WORDS
serum

REFERENCE
Avery, M.J.; Fouda, H.G. Development of a high-performance liquid chromatographic-atmospheric pressure chemical ionization-tandem mass spectrometric assay for β-tigogenin cellobioside in human serum, *J.Chromatogr.B*, **1997**, *689*, 365−370.

p-Anisoyl Chloride

SAMPLE
Matrix: solutions
Analyte: hexachlorophene
Sample preparation: Dissolve 4-20 μg in 1 mL 5% NaOH, add 30 μL p-anisoyl chloride, vortex for 1 min, let stand at room temperature for 20 min, add 9 mL water, vortex for 2 min, extract three times with 10 mL portions of hexane. Combine the extracts and evaporate them to dryness under a stream of nitrogen, reconstitute the residue in 1 mL butyl chloride, inject a 10 μL aliquot.

HPLC VARIABLES
Column: 610 × 2.3 36-40 μm Sil-X silica (Nester-Faust)
Mobile phase: Hexane:n-butyl chloride 55:45
Flow rate: 0.7
Injection volume: 10
Detector: UV 254

CHROMATOGRAM
Retention time: 10.5
Limit of detection: 30 ppb

KEY WORDS
normal phase

REFERENCE
Porcaro, P.J.; Shubiak, P. Detection of nanogram quantities of hexachlorophene by ultraviolet liquid chromatography, *Anal.Chem.*, **1972**, *44*, 1865–1867.

9-Anthroyl Chloride

RELATED REFERENCES
Bayliss, M.A.J.; Homer, R.B.; Shepherd, M.J. Anthracene-9-carbonyl chloride as a fluorescence and ultraviolet derivatizing reagent for the high-performance liquid chromatographic analysis of hydroxy compounds. *J.Chromatogr.* **1988**, *445*, 393-402.
Bayliss, M.A.J.; Homer, R.B.; Shepherd, M.J. Determination of diethylene glycol in wine by high-performance liquid chromatography using anthracene-9-carbonyl chloride as a derivatizing reagent. *J.Chromatogr.* **1988**, *445*, 403-408.

SAMPLE
Matrix: endotoxins
Analyte: β-hydroxymyristic acid and β-hydroxylauric acid
Sample preparation: Condition a Lichrolut RP 18 SPE cartridge (Merck) with 1 mL MeCN:water 60:40, do not allow to run dry. Mix 100 μL of a 1 μg/mL solution of endo-

toxins in 0.05% triethylamine in water with 4.9 mL 2 M sulfuric acid in MeOH, heat at 110° for 4 h, cool, add 2 mL water, add 1.5 mL n-hexane, shake vigorously for 2 min, centrifuge at 1500 g for 10 min, add the upper hexane layer to a Lichrolut Si (Merck) SPE cartridge, wash with 4 mL MTBE:n-hexane 8:92, elute with two 700 μL portions of MTBE:n-hexane 25:75. Evaporate the eluate to dryness under a stream of nitrogen at room temperature, reconstitute the residue in 100 μL 5 mg/mL 9-anthroyl chloride in MeCN, shake vigorously let stand at room temperature for 20 min or heat at 60° for 10 min, add 40 μL water, let stand at room temperature for 5 min, add to the C18 SPE cartridge, let stand for 5 min, wash with 5 mL MeCN:water 60:40, dry under vacuum for 10 min, elute with 1 mL MeCN. Evaporate the eluate to dryness under a stream of nitrogen at room temperature, reconstitute the residue in 150 μL MeCN, inject a 10 μL aliquot. (The carboxylic acid groups are esterified to the corresponding methyl esters and the hydroxy groups react with 9-anthroyl chloride. Prepare 9-anthroyl chloride by reacting 9-anthracenecarboxylic acid with thionyl chloride in dried toluene, evaporate to dryness, recrystallize 3 times from dried hexane with hot filtration through a dried ceramic filter to give 9-anthroyl chloride as a bright-yellow solid (mp 95°).)

HPLC VARIABLES
Guard column: 20 × 3 stationary phase not specified
Column: 250 × 3 5 μm NuSil C18 (Macherey-Nagel)
Mobile phase: MeCN:water 95:5
Flow rate: 1
Injection volume: 10
Detector: F ex 250 em 462

CHROMATOGRAM
Retention time: β-hydroxylauric acid methyl ester (5.37), β-hydroxymyristic acid methyl ester (7.87)
Internal standard: β-hydroxytridecanoic acid methyl ester (6.45), β-hydroxypentadecanoic acid methyl ester (9.72)
Limit of detection: 0.5 pg

KEY WORDS
SPE; comparison with other derivatizing reagents

REFERENCE
Parlesak, A.; Bode, C. Lipopolysaccharide determination by reversed-phase high-performance liquid chromatography after fluorescence labeling, *J.Chromatogr.A*, **1995**, *711*, 277–288.

2-(9-Anthryl)ethyl Chloroformate

SAMPLE
Matrix: solutions
Analyte: phenols
Sample preparation: Filter (0.2 μm) water, remove a 100 μL aliquot and add it to 500 μL 2.59 μM 2-(9-anthryl)ethyl chloroformate in MeCN and 200 μL 25 mM pH 9.6 borate buffer, heat at 43° for 35 min, inject a 10 μL aliquot. (Prepare 2-(9-anthryl)ethyl chloroformate as follows. Stir a solution of 3 g of 9-bromoanthracene in 100 mL ether at 0°

under argon or nitrogen, add 9 mL 1.6 M n-butyllithium over 5 min, stir for 30 min, add an ice-cold solution of 3 g ethylene oxide (Caution! Ethylene oxide is a carcinogen!) in 16 mL ether, stir for 1 h, add 70 mL water, add 50 mL ether, remove the organic layer, extract the aqueous layer with 100 mL dichloromethane. Combine the organic layers and wash them with water, dry over anhydrous sodium sulfate, evaporate to dryness, chromatograph on silica gel with dichloromethane to give 2-(9-anthryl)ethanol as pale yellow crystals (mp 106-8°) (J.Org.Chem. 1986, 51, 2956). Stir a solution of 2-(9-anthryl)ethanol in ether in the presence of pyridine (as an HCl scavenger) at 0°, add a solution of phosgene in toluene. 2-(9-anthryl)ethyl chloroformate is obtained as colorless crystals (mp 86-87° from pentane). Protect stock solutions from light and store them in the refrigerator (Anal.Chem. 1991, 63, 292).)

HPLC VARIABLES
Column: 125 × 4 5 μm LiChrospher 100 RP-18
Mobile phase: Gradient. MeCN:water from 70:30 to 100:0 over 10 min, maintain at 100:0 for 10 min
Flow rate: 0.75
Injection volume: 10
Detector: F ex 256 em 418 (cut-off filter)

CHROMATOGRAM
Retention time: 8.57 (phenol), 9.64 (4-methylphenol), 10.58 (3,4-dimethylphenol), 12.32 (4-tert-butylphenol)
Limit of detection: 7-10 nM

KEY WORDS
wastewater

REFERENCE
Landzettel, W.J.; Hargis, K.J.; Caboot, J.B.; Adkins, K.L.; Strein, T.G.; Veening, H.; Becker, H.-D. High-performance liquid chromatographic separation and detection of phenols using 2-(9-anthrylethyl) chloroformate as a fluorophoric derivatizing agent, *J.Chromatogr.A*, **1995**, *718*, 45−51.

Benoxaprofen Chloride

SAMPLE
Matrix: blood, urine
Analyte: trospium chloride
Sample preparation: Plasma. 5 mL Plasma + 1 mL 1 M NaOH, heat at 140-145° for 105 min, cool to 40°, add 200 μL 25% HCl, mix, centrifuge at 5000 g for 10 min, heat at 120° for 10 min, cool, centrifuge. Remove a 5 mL aliquot and add it to 500 μL reagent 1, mix. Remove a 5 mL aliquot and add it to 4.8 mL chloroform, shake vigorously for 3 min, centrifuge at 6000 g for 30 min. Remove a 4 mL aliquot of the organic phase and add it to 2.3 mL 100 mM HCl, shake vigorously, centrifuge at 5000 g for 10 min. Remove a 2 mL aliquot of the aqueous layer and add it to 2 mL MeOH, evaporate to dryness under reduced pressure, add MeOH, evaporate to dryness under a stream of nitrogen at 80-90°, repeat process several times with 10-11 mL MeOH total, add 200 μL 10 mg/mL benoxaprofen chloride in dry MeCN, heat at 140-145° for 30 min, evaporate to dryness under

reduced pressure, add 1 mL ethyl acetate, add 1.2 mL water, shake mechanically for 5 min, centrifuge at 5000 g for 10 min, discard the organic layer, wash the aqueous phase again with 1 mL ethyl acetate. Evaporate the aqueous phase to dryness under reduced pressure, reconstitute with 100 μL MeCN:water 31:69, inject a 20 μL aliquot. Urine. 5 mL Urine + 1 mL 1 M NaOH, heat at 140-145° for 90 min, cool, add 200 μL 25% HCl, add 600 μL reagent 2, shake, centrifuge at 5000 g for 10 min. Remove a 5 mL aliquot and add it to 4.8 mL chloroform, shake vigorously for 3 min, centrifuge at 6000 g for 30 min. Remove a 4 mL aliquot of the organic phase and add it to 2.3 mL 100 mM HCl, shake vigorously, centrifuge at 5000 g for 10 min. Remove a 2 mL aliquot of the aqueous layer and add it to 2 mL MeOH, evaporate to dryness under reduced pressure, add MeOH, evaporate to dryness under a stream of nitrogen at 80-90°, repeat process several times with 10-11 mL MeOH total, add 200 μL 10 mg/mL benoxaprofen chloride in dry MeCN, heat at 140-145° for 30 min, evaporate to dryness under reduced pressure, add 1 mL ethyl acetate, add 1.2 mL water, shake mechanically for 5 min, centrifuge at 5000 g for 10 min, discard the organic layer, wash the aqueous phase again with 1 mL ethyl acetate. Evaporate the aqueous phase to dryness under reduced pressure, reconstitute with 100 μL MeCN:water 40:60, inject a 20 μL aliquot. (Trospium chloride is hydrolyzed to nortropane-8-spiro-1'-pyrrolidinium chloride (which is also a metabolite) and this compound is derivatized with benoxaprofen chloride. Reagent 1 was a mixture of 98.7 mg dipicrylamine containing 50% water, 10 mL 100 mM NaOH, and 600 mg anhydrous sodium carbonate. Reagent 2 was a mixture of 32.9 mg dipicrylamine containing 50% water and 10 mL 100 mM NaOH. Prepare dipicrylamine as follows (Caution! Dipicrylamine is potentially explosive and highly toxic, store moistened with 50% water!). Add 50 g 2,4-dinitrodiphenylamine to 420 g nitric acid (36° Bé., 52%, d = 1.33) heated to 62° over 2 h, heat at 62-90° for another 3 h, cool, filter, wash the product until it is free of acid, dry to obtain 2,2',4,4'-tetranitrodiphenylamine as a yellow solid (mp 187.4°). Add 50 g tetranitrodiphenylamine over 1 h to 500 g of a mixture of equal parts 92% sulfuric acid and 93% nitric acid at room temperature, after 4.5 h add to a large volume of ice water, filter, recrystallize the product from acetone to obtain dipicrylamine (2,2',4,4',6,6'-hexanitrodiphenylamine) as yellow crystals (mp 242.9°) (J.Am.Chem.Soc. 1919, 41, 1013). Prepare benoxaprofen chloride as follows. Dissolve 600 mg benoxaprofen in 50 mL dry toluene, slowly add 5 mL thionyl chloride (freshly distilled from linseed oil), reflux for 30 min, evaporate to dryness, recrystallize from dichloromethane (if necessary) to give benoxaprofen chloride (mp 91.5°) (J. Chromatogr.1984, 310, 167).)

HPLC VARIABLES
Column: 125 × 4.6 5 μm Nucleosil C8 (plasma) or 125 × 4.6 5 μm LiChrosorb RP-8 (urine)
Mobile phase: MeCN:water 31:69 (plasma) or 40:60 (urine) containing 80 mM NaCl, 31 mM choline chloride, and 10 mL/L 1 M HCl
Column temperature: 55 (plasma), 50 (urine)
Flow rate: 2
Injection volume: 20
Detector: F ex 313 em 370

CHROMATOGRAM
Retention time: 7.5 (plasma), 6 (urine)
Limit of quantitation: 0.5-1 ng/mL (plasma), 3 ng/mL (urine)

KEY WORDS
plasma; pharmacokinetics

REFERENCE
Schladitz-Keil, G.; Spahn, H.; Mutschler, E. Fluorimetric determination of the quaternary compound trospium and its metabolite in biological material after derivatization with benoxaprofen chloride, *J.Chromatogr.*, **1985**, *345*, 99−110.

SAMPLE
Matrix: bulk
Analyte: alcohols
Sample preparation: Heat 500 nmole alcohol with 500 μL 1 mg/mL benoxaprofen chloride in MeCN or toluene at 60-80° for 1 h, inject a 10 μL aliquot. (Prepare benoxaprofen chloride as follows. Dissolve 600 mg benoxaprofen in 50 mL dry toluene, slowly add 5 mL

thionyl chloride (freshly distilled from linseed oil), reflux for 30 min, evaporate to dryness, recrystallize from dichloromethane (if necessary) to give benoxaprofen chloride (mp 91.5°).)

HPLC VARIABLES
Column: 250 × 4.6 7 μm Zorbax-sil (A) or 120 × 4.6 5 μm LiChrosorb RP-8 (B)
Mobile phase: Cyclohexane:dichloromethane:THF 5:1:1 (A) or acetone:10 mM sodium heptanesulfonate:phosphoric acid 60:40:0.15 (B)
Column temperature: ambient (A), 55 (B)
Flow rate: 1 (A) or 2 (B)
Injection volume: 10
Detector: F ex 312 em 365

CHROMATOGRAM
Retention time: 6.7 (isoamyl alcohol (A)), 7.3 (butanol (A)), 7.5 (choline (B)), 8.3 (scopolamine N-butylbromide (B)), 8.5 (propanol (A)), 9.9 (EtOH (A)), 11.4 (MeOH (A))

KEY WORDS
normal phase; reverse phase

REFERENCE
Spahn, H.; Weber, H.; Mutschler, E.; Möhrke, W. α-Alkyl-α-arylacetic acid derivatives as fluorescence markers for thin-layer chromatographic and high-performance liquid chromatographic assay of amines and alcohols, *J.Chromatogr.*, **1984**, *310*, 167–178.

Benzoyl Chloride

RELATED REFERENCES
Bremer, E.G.; Gross, S.K.; McCluer, R.H. Quantitative analysis of monosialogangliosides by high-performance liquid chromatography of their perbenzoyl derivatives. *J.Lipid Res.* **1979**, *20*, 1028-1035.

Daniel, P.F. Separation of benzoylated oligosaccharides by reversed-phase high-pressure liquid chromatography: application to high-mannose type oligosaccharides. *Methods Enzymol.* **1987**, *138*, 94-116.

Galensa, R.; Landt, K.; Herres, W. Application of the flow-cell HPLC-FTIR technique to the analysis of benzoylated saccharides and compounds with similar polarity. *Z.Lebensm.-Unters.Forsch.* **1987**, *185*, 36-38.

Gross, S.K.; McCluer, R.H. High-performance liquid chromatographic analysis of neutral glycosphingolipids as their per-O-benzoyl derivatives. *Anal.Biochem.* **1980**, *102*, 429-433.

Ioneda, T.; Ono, S.S. Chromatographic and mass spectrometric analyses of 1-monomycoloyl glycerol fraction from Rhodococcus lentifragmentus as per-O-benzoyl derivatives. *Chem.Phys.Lipids* **1996**, *81*, 11-19.

Ullman, M.D.; McCluer, R.H. High-pressure liquid chromatography analysis of neutral glycosphingolipids: perbenzoylated mono-, di-, tri-, and tetraglycosylceramides. *Methods Enzymol.* **1987**, *138*, 117-125.

Vollmer, P.A.; Harty, D.C.; Erickson, N.B.; Balhon, A.C.; Dean, R.A. Serum ethylene glycol by high-performance liquid chromatography. *J.Chromatogr.B* **1996**, *685*, 370-374.

SAMPLE
Matrix: blood
Analyte: chloral hydrate (as the active metabolite trichloroethanol)
Sample preparation: 250 μL Plasma + 250 μL 150 μg/mL 4-chloro-1-butanol in water, vortex, add 20 μL benzoyl chloride dropwise on the surface, vortex, add 250 μL 4 M NaOH, rotate at medium sped for 10 min, add 10 mL pentane, rotate for 5-7 min, cen-

trifuge at 2000 g for 5 min. Remove the organic layer and evaporate it to dryness at 45-50°, reconstitute the residue in 100 μL MeOH, inject a 5-10 μL aliquot.

HPLC VARIABLES
Guard column: 7 μm RP-18 (Brownlee)
Column: 150 × 4.6 5 μm Ultrasphere ODS
Mobile phase: MeCN:MeOH:water 30:30:40
Flow rate: 2
Injection volume: 5-10
Detector: UV 237

CHROMATOGRAM
Retention time: 12 (for trichloroethanol, the active metabolite)
Internal standard: 4-chloro-1-butanol (7)
Limit of quantitation: 7 μg/mL

OTHER SUBSTANCES
Non-interfering: acetaminophen, barbiturates, salicylic acid

KEY WORDS
plasma

REFERENCE
Gupta, R.N. Determination of trichloroethanol, the active metabolite of chloral hydrate, in plasma by liquid chromatography, *J.Chromatogr.*, **1990**, *500*, 655–659.

SAMPLE
Matrix: blood
Analyte: tromethamine
Sample preparation: 100 μL Plasma + 50 μL 750 μg/mL 2,3-butanediol in water + 200 μL 4 M NaOH, vortex briefly, add 40 μL benzoyl chloride, vortex for 3 min, add 8 mL MTBE:MeOH 99:1, vortex for 3 min, centrifuge at 1200 g for 5 min. Remove the organic layer and evaporate it to dryness in a vortex evaporator at 55° for 30 min, cool to room temperature, reconstitute the residue in 500 μL MeOH, inject a 10 μL aliquot.

HPLC VARIABLES
Column: 100 × 4.6 5 μm Ultrasphere octyl
Mobile phase: Gradient. MeCN:25 mM pH 6.5 potassium phosphate buffer 40:60 45:55 for 10 min, to 73:27 over 7 min (concave gradient), return to initial conditions over 1 min.
Flow rate: 3
Injection volume: 10
Detector: UV 237

CHROMATOGRAM
Retention time: 15.0
Internal standard: 2,3-butanediol (7.1)
Limit of detection: 282 ng/mL

KEY WORDS
plasma

REFERENCE
Gumbhir, K.; Mason, W.D. High-performance liquid chromatographic method for the determination of tris(hydroxymethyl)aminomethane (tromethamine) in human plasma, *J.Chromatogr.*, **1992**, *583*, 99–104.

SAMPLE
Matrix: blood
Analyte: sugars
Sample preparation: 70 μL Serum + 10 μL 1 mM D-glucosamine.HCl + 20 μL 1 M K$_2$HPO$_4$ + 10 μL benzoyl chloride + 25 μL 8 M NaOH, vortex at 2500 vibrations/min for 5 min, add 10 μL 1.4 M phosphoric acid and 100 μL ethyl acetate, vortex at 2500 vibrations/min for 1 min. Remove 25 μL of the ethyl acetate phase and add it to 100 μL MeCN:water 70:30, inject an aliquot.

HPLC VARIABLES
Guard column: 5 μm Kromasil 100 C18
Column: 250 × 4 5 μm Kromasil 100 C18
Mobile phase: Gradient. MeCN:water from 70:30 to 95:5 over 30 min.
Flow rate: 1
Injection volume: 50
Detector: UV 228; MS, electrospray, Finnigan MAT, TSQ 700, flow rate 1 μL/min, 2.8 kV, drying gas 140°

CHROMATOGRAM
Retention time: 6.1 (benzyl alcohol), 7.5 (tetrabenzoyl dextrose), 7.7 (myoinositol), 7.9 (tetrabenzoyl dextrose), 8.1 (tetrabenzoyl dextrose), 10.9 (adenosine), 11.3 (pentabenzoyl mannitol), 11.5 (pentabenzoyl mannitol), 11.6 (2-desoxy-D-glucose), 12.1 (2-desoxy-D-glucose), 12.1 (3-O-methylglucose), 12.3 (cytidine), 14.0 (sucrose), 14.7 (sucrose), 15.6 (pentabenzoyl dextrose), 15.9 (pentabenzoyl dextrose), 18.7 (hexabenzoyl mannitol), 19.5 (sucrose), 20.0 (sucrose), 20.8 (sucrose), 26.2 (sucrose)
Internal standard: D-glucosamine (9.7)
Limit of detection: 1-5 pmol

KEY WORDS
serum; fetal bovine serum

REFERENCE
Oehlke, J.; Brudel, M.; Blasig, I.E. Benzoylation of sugars, polyols and amino acids in biological fluids for high-performance liquid chromatographic analysis, *J.Chromatogr.B*, **1994**, *655*, 105−111.

SAMPLE
Matrix: bulk
Analyte: hydroxy steroids
Sample preparation: Dissolve 0.5-50 mg compound in 4 mL pyridine, add a 3 molar excess of benzoyl chloride, shake for 5 min, heat at 80° for 15 min, add to 50 mL 100 mM HCl, extract with 50 mL ether. Wash the organic layer three times with dilute acid, wash with water, wash twice with 50 mL portions of saturated sodium carbonate solution, wash with water. Evaporate the organic layer to dryness, reconstitute with chloroform, inject an aliquot.

HPLC VARIABLES
Column: 1000 × 2 Permaphase ODS (DuPont)
Mobile phase: MeOH:water 66:34
Flow rate: 0.33
Detector: UV 254

CHROMATOGRAM
Retention time: 7.5 (androsterone), 10.5 (dehydroepiandrosterone), 12 (epiandrosterone), 26 (delta^5-pregnenolone), 32 (androstanolone)

REFERENCE
Fitzpatrick, F.A.; Siggia, S. High resolution liquid chromatography of derivatized non-ultraviolet absorbing hydroxy steroids, *Anal.Chem.*, **1973**, *45*, 2310−2314.

SAMPLE
Matrix: bulk
Analyte: glucosylceramide
Sample preparation: Dissolve 1 mg compound in 1 mL pyridine, add 200 μL benzoyl chloride, heat at 60° for 1 h, evaporate to dryness under reduced pressure at 50°, reconstitute with 2 mL hexane and 2 mL MeOH:water 95:5 saturated with sodium carbonate. Wash the hexane layer with 2 mL MeOH:water 95:5, inject an aliquot of the hexane layer.

HPLC VARIABLES
Column: 305 mm × 6.3 mm o.d. μBondapack C18
Mobile phase: MeOH

Flow rate: 3
Detector: UV 254

CHROMATOGRAM
Retention time: 15-25 (multiple peaks)

REFERENCE
Suzuki, A.; Handa, S.; Yamakawa, T. Separation of molecular species of glucosylceramide by high performance liquid chromatography of their benzoyl derivatives, *J.Biochem.*, **1976**, *80*, 1181–1183.

SAMPLE
Matrix: bulk
Analyte: sapogenins
Sample preparation: 25 mg Sapogenins + 2 mL pyridine + 100 µL benzoyl chloride, heat at 80° for 30 min, cool to room temperature, add 10 mL dichloromethane, add 10 mL water, add 2 mL concentrated HCl, shake for 15 s, discard the aqueous layer, wash the organic layer with 10 mL water, inject an aliquot.

HPLC VARIABLES
Column: 250 × 4 10 µm LiChrosorb RP8
Mobile phase: MeCN:water 80:20
Flow rate: 3.9
Injection volume: 10
Detector: UV 235

CHROMATOGRAM
Retention time: 3 (delta9(11)-dehydrohecogenin), 3.5 (hecogenin), 8 (delta9(11)-dehydrotigogenin), 8.5 (sarsasapogenin), 10 (tigogenin, diosgenin)

REFERENCE
Higgins, J.W. A high-performance liquid chromatographic analysis of the benzoate esters of sapogenins isolated from Agave, *J.Chromatogr.*, **1976**, *11*, 329–334.

SAMPLE
Matrix: bulk
Analyte: C_{32} hopanoid
Sample preparation: Equilibrate a Sep Pak C18 SPE cartridge with MeCN. Dissolve in 100 µL pyridine, add 50 µL redistilled benzoyl chloride, heat at 80° for 1 h, cool to room temperature, add 1 mL 100 mM HCl, extract three times with 500 µL portions of ether. Combine the ether layers and wash them three times with 1.5 mL portions of 100 mM sodium bicarbonate, wash three times with 1.5 mL portions of 100 mM HCl, evaporate to dryness, reconstitute with 200 µL chloroform, add to the SPE cartridge, wash with 5 mL MeCN, elute with 2 mL THF.

HPLC VARIABLES
Column: 5 µm Econosil C18
Mobile phase: MeCN:THF 60:40
Flow rate: 1
Detector: UV 240

CHROMATOGRAM
Retention time: 10
Internal standard: cholesterol benzoate (5.2)

KEY WORDS
SPE

REFERENCE
Barrow, K.D.; Chuck, J.-A. Determination of hopanoid levels in bacteria using high-performance liquid chromatography, *Anal.Biochem.*, **1990**, *184*, 395–399.

SAMPLE
Matrix: bulk
Analyte: heterocyst-type glycolipids

Sample preparation: Mix 60 μg heterocyst-type glycolipids with 100 μL freshly-prepared pyridine:benzoyl chloride 10:2, flush with nitrogen, heat at 60° for 1.5 h, evaporate to dryness under a stream of nitrogen, reconstitute with 2 mL n-heptane. Vortex with 2 mL portions of buffer for 1 min, centrifuge, discard the aqueous layer, repeat the wash twice more, vortex with MeOH:water 80:20 for 1 min, centrifuge, discard the aqueous layer. Remove a 1.5 mL aliquot of the organic layer and evaporate it to dryness at 30°, reconstitute the residue in 100 μL n-heptane, inject an aliquot. (Buffer was MeOH:water 80:20 saturated with sodium carbonate (4 mg/mL).)

HPLC VARIABLES
Column: 150 × 4.6 3 μm RoSil spherical silica (Alltech)
Mobile phase: Gradient. n-Heptane:isopropanol from 95.5:0.5 to 98:2 over 12 min.
Column temperature: 30
Flow rate: 1.5
Detector: UV 230

CHROMATOGRAM
Retention time: 2-7 (depending on structure)
Limit of detection: 20 pmole

KEY WORDS
normal phase

REFERENCE
Davey, M.W.; Lambein, F. Quantitative derivatization and high-performance liquid chromatographic analysis of cyanobacterial heterocyst-type glycolipids, *Anal.Biochem.*, **1992**, *206*, 323–327.

SAMPLE
Matrix: food
Analyte: hydroxybenzoic acid esters
Sample preparation: Dissolve 300 mg mustard in 10 mL MeCN with shaking, sonicate for 15 min, filter (P 4 frit), wash through with 10 mL MeCN, evaporate to dryness under reduced pressure at 35-50°, take up the residue in 2 mL pyridine, add 200 μL benzoyl chloride, heat at 75° for at least 1.5 h, let stand overnight, add 20 mL water, shake, let stand for 5 h, filter (0.2 μm), rinse the flask with three 5 mL portions of water, pass the rinses through the filter. Rinse the flask with three 10 mL portions of isooctane:diethyl ether:MeCN 500:65:1 and wash the contents of the filter through a Sep-Pak silica SPE cartridge with these rinses, inject an aliquot of the eluate.

HPLC VARIABLES
Column: 300 × 3 5 μm LiChrosorb Si 60 in a glass column
Mobile phase: Isooctane:diethyl ether:MeCN 500:35:0.3
Flow rate: 1.2
Injection volume: 10
Detector: UV 240

CHROMATOGRAM
Retention time: 5.8 (propyl p-hydroxybenzoate), 6.8 (ethyl p-hydroxybenzoate), 8 (methyl p-hydroxybenzoate)
Limit of detection: 0.2 ppm

KEY WORDS
mustard; normal phase

REFERENCE
Galensa, R.; Schäfers, F.-I. Hochleistungsflüssigskeits- und capillargaschromatographische Bestimmung von para-Hydroxibenzoesäureestern in Lebensmitteln mit komplexer Matrix [High performance liquid and capillary gas chromatographic determination of esters of p-hydroxybenzoic acid in foodstuffs with complex matrix], *Z.Lebensm.Unters.Forsch.*, **1981**, *173*, 279–284.

SAMPLE
Matrix: food
Analyte: hydroxybenzoic acid esters

Sample preparation: Dissolve 300 mg mustard in 10 mL MeCN with shaking, sonicate for 15 min, filter (P 4 frit), wash through with 10 mL MeCN, evaporate to dryness under reduced pressure at 35-50°, take up the residue in 2 mL pyridine, add 200 µL benzoyl chloride, heat at 75° for at least 1.5 h, let stand overnight, add 20 mL water, shake, let stand for 5 h, filter (0.2 µm), rinse the flask with three 5 mL portions of water, pass the rinses through the filter. Rinse the flask with three 10 mL portions of isooctane:diethyl ether:MeCN 500:65:1 and wash the contents of the filter through a Sep-Pak silica SPE cartridge with these rinses, evaporate the eluate, reconstitute, inject an aliquot.

HPLC VARIABLES
Column: 300 × 3 7 µm RP-18 (Riedel de Haen)
Mobile phase: MeOH:water 80:20
Flow rate: 0.6
Injection volume: 10
Detector: UV 240

CHROMATOGRAM
Retention time: 5 (methyl p-hydroxybenzoate), 6 (ethyl p-hydroxybenzoate), 7.5 (propyl p-hydroxybenzoate)
Limit of detection: 0.2 ppm

KEY WORDS
mustard

REFERENCE
Galensa, R.; Schäfers, F.-I. Hochleistungsflüssigkeits- und capillargaschromatographische Bestimmung von para-Hydroxibenzoesäureestern in Lebensmitteln mit komplexer Matrix [High performance liquid and capillary gas chromatographic determination of esters of p-hydroxybenzoic acid in foodstuffs with complex matrix], *Z.Lebensm.Unters.Forsch.*, **1981**, *173*, 279–284.

SAMPLE
Matrix: food
Analyte: sugar alcohols
Sample preparation: Freeze chewing gum, pulverize. Sonicate 1 g with 80 mL EtOH:water 96:4 at 60° for 20 min, cool, filter, rinse the filter, make up the filtrate to 100 mL. Remove a 5 mL aliquot and evaporate it to dryness under reduced pressure, add 4 mL pyridine, add 500 µL benzoyl chloride, sonicate at 60° for 1 h with swirling every 15 min, add 500 µL MeOH, swirl, let stand for 10 min, add 50 mL water (Z. Lebensm. Unters. Forsch. 1984, 178, 199), shake, add to a Sep-Pak RP-18 SPE cartridge (conditioning of cartridge is not necessary), rinse flask four times with 5 mL portions of water, add the rinses to the SPE cartridge, push 5 mL volumes of air through cartridge 3 times, elute with five 10 mL portions of isooctane:ether:MeCN 60:32:8, make up the volume of the eluate to 50 mL, inject an aliquot.

HPLC VARIABLES
Column: 300 × 3 5 µm LiChrosorb Si 60 (glass column)
Mobile phase: Isooctane:ether:MeCN 150:60:10
Flow rate: 0.9
Injection volume: 10
Detector: UV 230

CHROMATOGRAM
Retention time: 10 (xylitol), 13 (dextrose), 16 (sorbitol), 17.5 (mannitol), 35 (saccharose)
Limit of detection: 0.1 ppm

KEY WORDS
SPE; chewing gum; mayonnaise; normal phase

REFERENCE
Galensa, R. Hochleistungs-flüssigchromatographische Bestimmung von Zuckeralkoholen mit UV-Detektion im ppm-Bereich in Lebensmitteln. I. [High-performance liquid chromatographic determination of sugar alcohols with UV-detection in the ppm-range in food. I.], *Z.Lebensm.Unters.Forsch.*, **1983**, *176*, 417–420.

SAMPLE

Matrix: urine

Analyte: galactosylceramides

Sample preparation: Condition a 400 μL C18 SPE cartridge (Sepralyte) with 2.5 mL chloroform:MeOH 2:1, 2.5 mL chloroform:MeOH 50:50, 2.5 mL EtOH:water 65:35, and 2.5 mL water (A). Condition a 2 mL column of DEAE-52 (Whatman) with 800 mM sodium acetate in MeOH, with 8 mL MeOH, and with 8 mL chloroform:MeOH 2:1 (B). Condition a 400 μL C18 SPE cartridge with 2.5 mL 100 mM KCl in MeOH:water 50:50 and 3.5 mL water (C). Lyophilize 10-20 mL urine at -70°, suspend residue in 3 mL 100 mM KCl, add 4 mL MeOH, add 8 mL chloroform, mix vigorously for at least 2 h, centrifuge at 400 g for 10 min. Filter the lower organic layer through glass wool and evaporate it to dryness under a stream of nitrogen, reconstitute the residue in 1 mL 600 mM NaOH in MeOH, let stand at room temperature for 1 h, add 1.5 mL 400 mM HCl, add 1 mL MeOH, add 500 μL water, pass twice through the SPE cartridge (A), wash the cartridge (A) with 3.5 mL 100 mM KCl in MeOH:water 50:50, elute with 5 mL MeOH. Evaporate the eluate to dryness under a stream of nitrogen, reconstitute the residue in 1 mL chloroform:MeOH 2:1, add to the column (B), rinse the tube with two 1 mL portions of chloroform:MeOH 2:1, add the rinses to the column (B), wash the column (B) with 15 mL chloroform:MeOH 2:1, elute with 20 mL 500 mM potassium acetate in chloroform:MeOH4:1 containing 140 mM ammonium hydroxide. Evaporate the eluate to dryness and reconstitute with 20 mL MeOH:water 50:50, pass it twice through the cartridge (C), wash the cartridge (C) with 2.5 mL MeOH:water 50:50, elute with 5 mL MeOH. Evaporate the eluate to dryness and lyophilize the residue, reconstitute with 1 mL dry 50 mM HCl in MeOH (Supelco), let stand for 16 h, add 3 mL chloroform, add 500 μL MeOH, add 1.1 mL 2 g/L sodium bicarbonate, centrifuge. Remove the lower organic phase and wash it with 2 mL chloroform:MeOH:100 mM aqueous KCl 3:48:47, wash the lower phase with 2 mL chloroform:MeOH:water 3:48:47. Evaporate the lower organic phase to dryness and lyophilize the residue, reconstitute with 500 μL 100 mL/L benzoyl chloride in dry pyridine, heat at 37° for 16 h, evaporate to dryness, reconstitute with 3 mL hexane, vigorously wash the hexane layer five times with 2 mL portions of 400 mg/L sodium carbonate in MeOH:water 80:20, vigorously wash the hexane layer three times with 2 mL portions of MeOH:water 80:20. Evaporate the upper hexane layer to dryness at room temperature and reconstitute with carbon tetrachloride (Caution! Carbon tetrachloride is a carcinogen!), inject an aliquot.

HPLC VARIABLES

Column: 500 × 2.1 Zipax silica (Rockland Technologies)

Mobile phase: Gradient. Hexane:dioxane from 99:1 to 80:20 over 16 min (Caution! Dioxane is a carcinogen!).

Flow rate: 2

Detector: UV 229

CHROMATOGRAM

Retention time: 6 (NFA), 7.5 (HFA)

Limit of detection: 100 pmole

KEY WORDS

SPE; normal phase

REFERENCE

Natowicz, M.R.; Prence, E.M.; Chaturvedi, P.; Newburg, D.S. Urine sulfatides and the diagnosis of metachromatic leukodystrophy, *Clin.Chem.*, **1996**, *42*, 232–238.

(S)-(+)-2-tert-Butyl-2-methyl-1,3-benzodioxole-4-carboxyl Chloride

SAMPLE
Matrix: solutions
Analyte: diacylglycerols
Sample preparation: Mix 5 mL of a 3.4 mg/mL solution in dichloromethane with 500 μL reagent, stir at room temperature for 2 h, add 5 mL dichloromethane, wash 3 times with 10 mL portions of saturated sodium bicarbonate solution, wash with 20 mL water, dry over anhydrous magnesium sulfate, evaporate to dryness under reduced pressure at 40°, purify by TLC on a 50 × 50 mm silica gel plate using n-hexane:ethyl acetate 10:1, extract the appropriate band with mobile phase, inject an aliquot. (Reagent was 24 mg (S)-(+)-2-tert-butyl-2-methyl-1,3-benzodioxole-4-carboxyl chloride in 500 μL pyridine:4-dimethylaminopyridine 90:10. Preparation of (S)-(+)-2-tert-butyl-2-methyl-1,3-benzodioxole-4-carboxyl chloride is described below.)

HPLC VARIABLES
Column: 250 × 4.6 Develosil 60-3 silica gel (Nomura)
Mobile phase: n-Hexane:n-butanol 300:1
Flow rate: 0.6
Detector: F ex 310 em 370

CHROMATOGRAM
Retention time: 19 (sn-1,2-dipalmitolylglycerol), 20 (sn-2,3-dipalmitolylglycerol), 22 (sn-1,3-dipalmitolylglycerol (optically inactive))

KEY WORDS
normal phase; chiral

REFERENCE
Kim, J.-H.; Nishida, Y.; Ohrui, H.; Meguro, H. Simple and highly sensitive high-performance liquid chromatographic method for separating enantiomeric diacylglycerols by direct derivatization with a fluorescent chiral agent, (S)-(+)-2-*tert*.-butyl-2-methyl-1,3-benzodioxole-4-carboxylic acid, *J.Chromatogr.A*, **1995**, *693*, 241–249.

SAMPLE
Matrix: solutions
Analyte: monoacylglycerols
Sample preparation: Mix 2 mL of a 3.5 mg/mL solution in dichloromethane with 200 μL reagent, stir at room temperature for 2 h, add 10 mL dichloromethane, wash 3 times with 10 mL portions of saturated sodium bicarbonate solution, wash with 20 mL water, dry over anhydrous magnesium sulfate, evaporate to dryness under reduced pressure at 40°, purify by TLC on silica gel using n-hexane:ethyl acetate 10:1, extract the appropriate band with mobile phase, inject an aliquot. (Reagent was 20 mg (S)-(+)-2-tert-butyl-2-methyl-1,3-benzodioxole-4-carboxyl chloride in 200 μL pyridine:4-dimethylaminopyridine 90:10. Preparation of (S)-(+)-2-tert-butyl-2-methyl-1,3-benzodioxole-4-carboxyl chloride is as follows. Add 370 mg thionyl chloride to a solution of 33 mg (S)-(+)-2-tert-butyl-2-methyl-1,3-benzodioxole-4-carboxylic acid in 5 mL dry benzene (Caution! Benzene is a carcinogen!), heat at 60° for 10 min, remove the solvents by evaporation to give (S)-(+)-2-tert-butyl-2-methyl-1,3-benzodioxole-4-carboxyl chloride. Synthesize (S)-(+)-2-tert-bu-

tyl-2-methyl-1,3-benzodioxole-4-carboxylic acid as follows. Reflux 3-methylcatechol (2,3-dihydroxytoluene) and pinacolone (t-butylmethylketone) in toluene in the presence of p-toluenesulfonic acid for 64 h to obtain 2-t-butyl-2,4-dimethyl-1,3-benzodioxole (mp 27°). Oxidize 2-t-butyl-2,4-dimethyl-1,3-benzodioxole with potassium permanganate in pyridine:water 50:50 at 70-80° for 6 h to obtain 2-tert-butyl-2-methyl-1,3-benzodioxole-4-carboxylic acid (mp 179-180°), resolve with cinchonidine (Tet. Lett 1989, 30, 5277).)

HPLC VARIABLES

Column: 50 × 4.6 Develosil 60-3 silica gel (Nomura)
Mobile phase: n-Hexane:t-butanol 100:0.4
Flow rate: 0.6
Detector: F ex 310 em 370

CHROMATOGRAM

Retention time: k' 18.38 (sn-2-monopalmitoylglycerol), k' 29.75 (sn-1-monostearoylglycerol), k' 31.01 (sn-3-monostearoylglycerol), k' 31.03 (sn-1-monooleoylglycerol), k' 31.11 (sn-1-monopalmitoylglycerol), k' 32.40 (sn-3-monooleoylglycerol), k' 32.56 (sn-3-monopalmitoylglycerol), k' 32.88 (sn-1-monomyristoylglycerol), k' 33.16 (sn-1-monolinoleoylglycerol), k' 34.54 (sn-3-monomyristoylglycerol), k' 34.64 (sn-1-monolauroylglycerol), k' 34.90 (sn-3-monolinoleoylglycerol), k' 35.77 (sn-1-monolinolenoylglycerol), k' 36.33 (sn-3-monolauroylglycerol), k' 37.57 (sn-3-monolinolenoylglycerol)

KEY WORDS

normal phase; chiral

REFERENCE

Kim, J.-H.; Nishida, Y.; Ohrui, H.; Meguro, H. Highly sensitive high-performance liquid chromatographic method to discriminate enantiomeric monoacylglycerols based on fluorescent chiral derivatization with (S)-(+)-2-*tert.*-butyl-2-methyl-1,3-benzodioxole-4-carboxylic acid, *J.Chromatogr.A*, **1995**, *709*, 375–380.

Camphanic Chloride

SAMPLE

Matrix: urine
Analyte: 1-(4-methoxyphenyl)propane-1,2-diol
Sample preparation: Hydrolyse urine with β-glucuronidase (bovine liver) in pH 5.0 acetate buffer at 37° overnight, purify by chromatography on 250 × 4.6 5 μm Spherisorb ODS-2 Excel with MeOH:1% acetic acid 100:0 for 3 min, to 80:20 over 12 min, to 70:30 over 15 min, to 50:50 over 10 min, maintain at 50:50 for 5 min, to 40:60 over 5 min, to 0:100 over 5 min, maintain at 0:100 for 5 min at 1 mL/min with detection at 275 nm. Collect peaks at 28 and 35 min and evaporate to dryness. Dissolve about 20 mg in 1 mL pyridine, add 10 mg (-)-(S)-camphanic chloride, mix, let stand at room temperature for 24 h.

HPLC VARIABLES

Column: 100 × 4.6 5 μm Si60 (Merck)
Mobile phase: n-Hexane:ethyl acetate 80:20

Flow rate: 1
Detector: UV 275

CHROMATOGRAM
Retention time: 26.7, 32.5, 35.8, 45.3 (stereoisomers)

KEY WORDS
rat; chiral

REFERENCE
Ishida, T.; Bounds, S.V.; Caldwell, J. Stereochemical aspects of the hydration of trans-anethole epoxide in the rat, *Chirality*, **1995**, *7*, 278–284.

7-[(Chlorocarbonyl)methoxy]-4-methylcoumarin

SAMPLE
Matrix: bulk
Analyte: steroids
Sample preparation: Dissolve sample in 100 μL 23 mM 7-[(chlorocarbonyl)methoxy]-4-methylcoumarin in dichloromethane, add 15 μL 100 mM 4-(dimethylamino)pyridine in dichloromethane, evaporate to dryness reconstitute with 300 μL MeCN, add 700 μL water, mix, add to a C18 SPE cartridge, wash with 20 mL MeCN:water 30:70, elute with 3 mL MeCN. Evaporate the eluate to dryness, reconstitute with 30 μL MeCN:water 75:25, inject an aliquot. (Dry dichloromethane by passing 20 mL through a 50 × 4 column packed with anhydrous sodium carbonate. Synthesis of 7-[(chlorocarbonyl)methoxy]-4-methylcoumarin is as follows. Neutralize 13.09 g bromoacetic acid in 40 mL water with 8.7 g sodium bicarbonate, add a solution of 15.08 g 7-hydroxy-4-methylcoumarin in 85 mL 1.1 M NaOH, reflux for 1 h, add concentrated HCl dropwise until the intense yellow color vanishes, stir overnight, acidify with HCl to obtain a pasty solid, recrystallize from water to obtain 7-[(carboxy)methoxy]-4-methylcoumarin as white crystals (mp 208-211°). 7-[(Carboxy)methoxy]-4-methylcoumarin is also available from Aldrich. Reflux a solution of 8.9 g 7-[(carboxy)methoxy]-4-methylcoumarin in 57 g freshly-distilled thionyl chloride for 1 h, remove most of the thionyl chloride by distillation, cool, add pentane, recrystallize the precipitate from ethyl acetate:thionyl chloride 95:5, dry under vacuum to obtain 7-[(chlorocarbonyl)methoxy]-4-methylcoumarin as crystals (mp129-130°). Phillips et al. have reported that 7-[(chlorocarbonyl)methoxy]-4-methylcoumarin contains 3-chloro-7-[(chlorocarbonyl)methoxy]-4-methylcoumarin as a minor but highly fluorescent impurity. Syntheses of 3-chloro-7-[(chlorocarbonyl)methoxy]-4-methylcoumarin have now been reported (Syn. Comm. 1996, 26, 1805) and the compound is available from Aldrich (cat. no. 46,739-

1). 3-Chloro-7-[(chlorocarbonyl)methoxy]-4-methylcoumarin has the advantage of being much more fluorescent although it is less soluble. It is also significantly less polar as reflected in much longer retention times under reversed-phase conditions (Phillips, L.R., personal communication).)

HPLC VARIABLES
Column: 1500 × 0.24 3 μm Spherisorb ODS-2 in a fused-silica tube
Mobile phase: Gradient. MeCN:water from 75:25 to 100:0 over 120 min (non-linear gradient).
Flow rate: 0.001
Injection volume: 0.2
Detector: F ex 315 em 389 (cut-off filter)

CHROMATOGRAM
Retention time: 48 (11β-hydroxyandrosterone), 54 (11β-hydroxyetiocholanolone), 60 (tetrahydrocortisol), 62 (β-cortolone), 64 (β-cortol), 65 (α-cortolone), 66 (α-cortol), 73 (androsterone), 78 (dehydroepiandrosterone), 82 (pregnanetriol), 94 (androstanediol), 120 (pregnanediol)

KEY WORDS
SPE; microbore

REFERENCE
Karlsson, K.E.; Wiesler, D.; Alasandro, M.; Novotny, M. 7-[(Chlorocarbonyl)methoxy]-4-methylcoumarin: a novel fluorescent reagent for the precolumn derivatization of hydroxy compounds in liquid chromatography, *Anal.Chem.*, **1985**, *57*, 229–234.

2-(5-Chlorocarbonyl-2-oxazolyl)-5,6-methylenedioxybenzofuran

SAMPLE
Matrix: blood
Analyte: didanosine
Sample preparation: Condition a Toyopak ODS M SPE cartridge (Tosoh, Tokyo) with 6 mL MeOH, 12 mL water, and 2 mL 100 mM pH 4.5 phosphate buffer. 100 μL Plasma + 10 μL 85 μM IS + 890 μL 100 mM pH 4.5 phosphate buffer, mix, add to the SPE cartridge at 120-150 μL/min, wash with 2 mL 100 mM pH 4.5 phosphate buffer, wash with 2 mL water, elute with 1 mL MeOH. Evaporate the eluate to dryness under a stream of nitrogen at 37°, reconstitute the residue in 100 μL dry pyridine, add 900 μL 3 mM reagent in dry benzene (Caution! Benzene is a carcinogen!), heat at 100° in the dark for 50 min, cool, evaporate to dryness under a stream of nitrogen at 60°, reconstitute the residue in 1 mL mobile phase, inject a 20 μL aliquot. (The reagent was 2-(5-chlorocarbonyl-2-oxazolyl)-5,6-methylenedioxybenzofuran (Dojindo Laboratories, Kumamoto, Japan (Dojindo Molecular Technologies, Inc., 3 Bethesda Metro Center, Suite 700, Bethesda MD 20814; (301) 664-8448; www.dojindo.co.jp)). Synthesis is as follows. Add ethyl oxalyl chloride in ether

to a solution of diazomethane in ether at 0° to give ethyl diazopyruvate (Caution! Diazo compounds are explosive and toxic!) (cf. Buehler,C.A.; Pearson,D.E. Survey of Organic Syntheses, Wiley, New York, 1970, p. 179). Heat 100 mg ethyl diazopyruvate, a few mg copper(II) acetylacetonate, and 400 μL chloroacetonitrile in benzene at 60° overnight (Caution! Benzene is a carcinogen!), cool, add to sodium bicarbonate solution, extract with ether, dry the organic layer, evaporate, chromatograph on silica with petroleum ether: ethyl acetate 90:10, distil the product at 90°/12 mm Hg to give ethyl 2-chloromethyl-5-oxazolecarboxylate as an oil in 18% yield (US Patent 4 603 209 (July 29, 1986)). Add 2 mL phosphorus oxychloride dropwise to a solution of 2 g sesamol in 3 mL DMF at 0°, heat on a steam bath with frequent shaking for 1 h, cool in ice, add 50 mL saturated sodium acetate solution, heat on a steam bath for 30 min, cool, filter, recrystallize the solid from EtOH to give 2-hydroxy-4,5-methylenedioxybenzaldehyde as colorless needles (mp 125-126°) (Bull. Chem. Soc. Jpn. 1962, 35, 1321). Stir 1.4 g ethyl 2-chloromethyl-5-oxazolecarboxylate, 1.5 g 2-hydroxy-4,5-methylenedioxybenzaldehyde, 2 g potassium carbonate, and 50 mL anhydrous DMF at 120° overnight, cool, filter. Evaporate the filtrate to dryness under reduced pressure to give 2-(5-ethoxycarbonyl-2-oxazolyl)-5,6-methylenedioxybenzofuran as a colorless crystalline powder (mp 186°) (yield 39%). Reflux 260 mg 2-(5-ethoxycarbonyl-2-oxazolyl)-5,6-methylenedioxybenzofuran, 100 mg KOH, 20 mL EtOH, and 30 mL water for 2 h, concentrate under reduced pressure, dissolve the residue in 100 mL water, wash with ethyl acetate, treat the aqueous layer with activated carbon, acidify the aqueous layer to pH 2 with 2 M HCl. Filter the precipitate and recrystallize it from EtOH to give 2-(2-oxazole-5-carboxylic acid)-5,6-methylenedioxybenzofuran as a colorless crystalline powder (mp 294-295°). Reflux 150 mg 2-(2-oxazole-5-carboxylic acid)-5,6-methylenedioxybenzofuran and 5 mL thionyl chloride for 2 h, pour the reaction mixture into 300 mL petroleum ether. Filter the precipitate and dry it over KOH to give 2-(5-chlorocarbonyl-2-oxazolyl)-5,6-methylenedioxybenzofuran (mp 290°) (Anal. Sci. 1989, 5, 525). Preparation of diazomethane is as follows. Caution! Diazomethane is toxic, explosive, and carcinogenic! A face shield and a safety screen should always be used and the preparation should only be carried out in a properly functioning chemical fume hood. Only smooth glass apparatus with rubber stoppers and plastic tubing should be used. Scratched glassware, ground glass joints, and sharp edges should be avoided (Org. Syn., Coll. Vol. VI; Wiley:New York, 1988, pp. 432-435). Procedures have been reported for the synthesis of diazomethane from N-methyl-N-nitrosourea (Org. Syn., Coll. Vol. II; Wiley: New York, 1943, pp. 165-167), N-nitroso-β-methylaminoisobutyl methyl ketone (Org. Syn., Coll. Vol. III; Wiley: New York, 1955, pp. 244-248), and N,N'-dimethyl-N,N'-dinitrosoterephthalamide (Org. Syn., Coll. Vol. V; Wiley: New York, 1973, pp. 351-355). Probably the most convenient starting material is N-methyl-N-nitroso-p-toluenesulfonamide (Diazald) (Aldrichimica Acta 1983, 16, 3-10; Org. Syn., Coll. Vol. IV, Wiley: New York, 1963, pp. 250-253). Add 10 mL 95% EtOH to a solution of 5 g KOH in 8 mL water, warm to 65° using a water bath, add a solution of 5 g N-methyl-N-nitroso-p-toluenesulfonamide in 45 mL ether over 20 min at such a rate as to keep the reaction volume constant. Collect the ether and diazomethane that distil in an ice-cooled receiving flask under a dry ice/acetone condenser. When all the N-methyl-N-nitroso-p-toluenesulfonamide has been used up, slowly add 10 mL ether to the reaction flask and continue distillation until the distillate is colorless. A purpose-built distillation apparatus can be purchased from Aldrich (Aldrichimica Acta 1983, 16, 3-10). Excess quantities of diazomethane can be destroyed by adding acetic acid until the yellow color of the diazomethane is discharged. The safe disposal of the nitroso compounds used to generate diazomethane has been discussed (Lunn,G.; Sansone,E.B. Destruction of Hazardous Chemicals in the Laboratory, Second Edition. Wiley: New York, 1994, pp. 277-289).)

HPLC VARIABLES
Column: 150 × 4.6 5 μm TSKgel ODS-80TM
Mobile phase: MeCN:100 mM pH 7.0 phosphate buffer 35:65
Flow rate: 1
Injection volume: 20
Detector: F ex 360 em 475

CHROMATOGRAM
Retention time: 14.6

Internal standard: 3'-deoxythymidine (21.5)
Limit of detection: 1.3 pmole

OTHER SUBSTANCES
Extracted: 2',3'-dideoxyadenosine

KEY WORDS
rat; plasma; pharmacokinetics

REFERENCE
Nagaoka, H.; Nohta, H.; Saito, M.; Ohkura, Y. Determination of 2',3'-dideoxyinosine and 2',3'-dideoxy-adenosine in rat plasma by high-performance liquid chromatography with precolumn fluorescence derivatization, *Chem.Pharm.Bull.*, **1992**, *40*, 2202–2204.

SAMPLE
Matrix: solutions
Analyte: alcohols
Sample preparation: Mix 3.5 mL of a 40 nM-50 µM solution in benzene with 500 µL 3 mM reagent in benzene and 100 µL 3.7 mM pyridine in benzene (Caution! Benzene is a carcinogen!), heat in the dark at 100° for 40 min (primary and secondary alcohols) or at 140° for 50 min (tertiary alcohols), cool, dilute 100-fold with mobile phase, inject a 20 µL aliquot. (The reagent is 2-(5-chlorocarbonyl-2-oxazolyl)-5,6-methylenedioxybenzofuran, available from Dojindo Laboratories, Kumamoto, Japan (Dojindo Molecular Technologies, Inc., 3 Bethesda Metro Center, Suite 700, Bethesda MD 20814; (301) 664-8448; www.dojindo.co.jp). Synthesis is described above.)

HPLC VARIABLES
Column: 150 × 4.6 5 µm Cosmosil 5C18 (Nacalai Tesque)
Mobile phase: MeCN:water 70:30 (A) or 50:50 (B)
Column temperature: 40
Flow rate: 1
Injection volume: 20
Detector: F ex 360 em 460

CHROMATOGRAM
Retention time: 4.3 (2-propanol (A)), 4.3 (phenol (A)), 4.5 (1-propanol (A)), 5.0 (benzyl alcohol (A)), 5.3 (2-methyl-2-propanol (A)), 5.5 (1-butanol (A)), 5.5 (2-methyl-1-propanol (A)), 6.8 (2-methyl-2-butanol (A)), 7.3 (3-methyl-1-butanol (A)), 7.8 (cyclohexanol (A)), 10.0 (MeOH (B)), 10.2 (1-hexanol (A)), 12.0 (EtOH (B))
Limit of detection: 3-900 fmole

OTHER SUBSTANCES
Non-interfering: aldehydes, amino acids, aromatic amines, carboxylic acids, ketones, sulf-hydryl compounds

REFERENCE
Nagaoka, H.; Nohta, H.; Kaetsu, Y.; Saito, M.; Ohkura, Y. 2-(5-Chlorocarbonyl-2-oxazolyl)-5,6-methyl-enedioxybenzofuran as fluorescence derivatization reagent for alcohols in high performance liquid chromatography, *Anal.Sci.*, **1989**, *5*, 525–530.

SAMPLE
Matrix: solutions
Analyte: nucleosides
Sample preparation: Lyophilize 0.1-1 mL of an aqueous solution, reconstitute with 100 µL pyridine, add 900 µL 1.7 mM reagent in benzene (Caution! Benzene is a carcinogen!), heat at 100° in the dark for 50 min, cool, dilute 10-100-fold with MeCN:100 mM pH 4.5 phosphate buffer 50:50 (for low concentrations lyophilize reaction mixture and reconsti-tute with 1 mL MeCN/buffer), inject a 20 µL aliquot. (The reagent is 2-(5-chlorocarbonyl-2-oxazolyl)-5,6-methylenedioxybenzofuran, available from Dojindo Laboratories, Kuma-moto, Japan (Dojindo Molecular Technologies, Inc., 3 Bethesda Metro Center, Suite 700, Bethesda MD 20814; (301) 664-8448; www.dojindo.co.jp). Synthesis is described above.)

HPLC VARIABLES
Column: 150 × 4.6 5 µm TSKgel ODS-80TM (Tosoh)

Mobile phase: MeCN:100 mM pH 4.5 phosphate buffer 25:75
Flow rate: 1
Injection volume: 20
Detector: F ex 360 em 475

CHROMATOGRAM
Retention time: 8.5 (xanthosine), 9.5 (inosine), 11 (guanosine), 12.5 (cytidine), 19 (uridine), 21 (adenosine), 30 (2'-deoxyguanosine), 34.5 (2'-deoxycytidine), 41.5 (thymidine), 51 (2'-deoxyadenosine)
Limit of detection: 1.9-12 pmole

REFERENCE
Nagaoka, H.; Nohta, H.; Saito, M.; Ohkura, Y. High-performance liquid chromatographic determination of ribonucleosides and 2'-deoxyribonucleosides based on precolumn fluorescence derivatization of the sugar moieties, *Anal.Sci.*, **1992**, *8*, 345–349.

SAMPLE
Matrix: solutions
Analyte: nucleotides
Sample preparation: Mix 0.1-1 mL of an aqueous solution (containing 0.05-10 nmoles) with 300 μL 150 mM sodium azide (Caution! Sodium azide is carcinogenic and toxic!), lyophilize, add 500 μL pyridine:3 mM reagent in benzene 25:75 (Caution! Benzene is a carcinogen!), heat at 100° in the dark for 1.5 h, cool, evaporate to dryness under a stream of nitrogen at 60°, reconstitute the residue in 1 mL mobile phase, inject an aliquot. (The reagent is 2-(5-chlorocarbonyl-2-oxazolyl)-5,6-methylenedioxybenzofuran, available from Dojindo Laboratories, Kumamoto, Japan (Dojindo Molecular Technologies, 3 Bethesda Metro Center, Suite 700, Bethesda MD 20814; (301) 664 8448; www.dojindo.co.jp). Synthesis is described above.)

HPLC VARIABLES
Column: 150 × 4.6 5 μm TSKgel ODS-80TM (Tosoh)
Mobile phase: MeCN:50 mM pH 3.0 citric acid containing 50 mM Na$_2$HPO$_4$ 40:60
Flow rate: 1
Detector: F ex 340 em 420

CHROMATOGRAM
Retention time: 2.4 (uridine-3'-monophosphate), 2.6 (cytidine-3'-monophosphate, guanosine-3'-monophosphate), 2.8 (adenosine-3'-monophosphate), 3.3 (thymidine-3'-monophosphate), 3.5 (2'-deoxyguanosine-3'-monophosphate), 3.8 (2'-deoxyadenosine-3'-monophosphate, 2'-deoxycytidine-3'-monophosphate)
Limit of detection: 0.8-6 pmole

REFERENCE
Nagaoka, H.; Nohta, H.; Saito, M.; Ohkura, Y. Precolumn derivatization of nucleotides based on fluorescent carbamate formation of the sugar moieties in high-performance liquid chromatography, *Chem.Pharm.Bull.*, **1992**, *40*, 2559–2561.

4-(N-Chloroformylmethyl-N-methyl)amino-7-N,N-dimethylaminosulfonyl-2,1,3-benzoxadiazole

SAMPLE

Matrix: solutions

Analyte: estrone

Sample preparation: Mix 10 μL 0.5 mM compound in anhydrous benzene containing 100 mM quinuclidine with 10 μL 25 mM DBD-COCl in anhydrous benzene (Caution! Benzene is a carcinogen!), heat at 60° for 15 min, add 980 μL MeCN:water:acetic acid 50:50:1, inject a 2 μL aliquot. (Purify quinuclidine by sublimation. DBD-COCl is 4-(N-chloroformylmethyl-N-methyl)amino-7-N,N-dimethylaminosulfonyl-2,1,3-benzoxadiazole. DBD-COCl is available from Tokyo Kasei (TCI America, Portland OR). Synthesis is as follows. Dissolve 0.5 g magnesium sulfate heptahydrate and 6 g NaOH in 60 mL water, throughout the reaction keep the flask at about 20° with cold water cooling, add 15 mL 30% hydrogen peroxide, add 75 mL MeOH, add 12.1 g powdered benzoyl peroxide in one go, stir for 10 min, pour into 150 mL 20% sulfuric acid, extract three times with 50 mL portions of chloroform, determine peroxybenzoic acid concentration by iodometric titration (Tetrahedron 1967, 23, 3327). Slowly add 110 mL 1 M peroxybenzoic acid in chloroform to 7 g 2,6-difluoroaniline dissolved in 100 mL chloroform, stir at room temperature, when reaction is complete (iodometric titration) wash with 2% sodium thiosulfate, wash with 5% sodium carbonate, wash with water, dry over anhydrous sodium sulfate, evaporate to dryness under reduced pressure, recrystallize 2,6-difluoronitrosobenzene from EtOH (mp 108.5-109.5). Stir 8.5 g 2,6-difluoronitrosobenzene in 85 mL DMSO at room temperature and add a solution of 3.91 g sodium azide in 85 mL DMSO dropwise, let stand for about 1 h, add to a large volume of water, extract with ether, dry the extracts over anhydrous sodium sulfate, evaporate to dryness under reduced pressure and distil to give 4-fluoro-2,1,3-benzoxadiazole as a colorless oil (bp 83°/12 mm Hg) (J.Chem.Soc.(C) 1970, 1433). Add 11 mL chlorosulfonic acid dropwise to 3 g 4-fluoro-2,1,3-benzoxadiazole in 10 mL chloroform at 0-10° (use a calcium chloride drying tube), stir at room temperature for 1 h, reflux for 2 h, cool, slowly pour into ice water, remove the organic layer, extract the aqueous layer with chloroform, combine the organic layers, wash, dry over anhydrous magnesium sulfate, evaporate under reduced pressure, take up the residue in 5 mL benzene (Caution! Benzene is a carcinogen!), chromatograph on a 150 × 30 column of silica gel (100-200 mesh Kanto Chemical) with n-hexane:benzene 50:50, evaporate the appropriate fractions to give 4-(chlorosulfonyl)-7-fluoro-2,1,3-benzoxadiazole (CBD-F) as pale yellow needles (mp 64-66°) (Anal. Chem. 1984. 56, 2461). Stir 0.76 g CBD-F in 70 mL MeCN at 0-10° and add 1 g dimethylamine hydrochloride in 10 mL 100 mM pH 10 borax dropwise, adjust pH to 5 with 1 M HCl, concentrate to about 10 mL under reduced pressure, extract three times with 200 mL portions of diethyl ether, wash with water, dry over anhydrous magnesium sulfate, evaporate under reduced pressure, chromatograph on a 500 × 20 column of silica gel with chloroform, isolate the appropriate fraction and re-chromatograph on the same column with ethyl acetate:benzene 1:2 to give 4-(N,N-dimethylaminosulfonyl)-7-fluoro-2,1,3-benzoxadiazole (DBD-F) as white needles (mp 124-125°) (yield = 1% !) (Analyst 1989, 114, 413). On a Merck no. 5714 60F$_{254}$ tlc plate eluted with chloroform DBD-F has Rf 0.32 and lies between two other reaction products. DBD-F can also be purchased from Tokyo Kasei. Stir N-methylglycine and 2.3 g sodium carbonate in water at room temperature, add 880 mg DBD-F in 40 mL MeCN dropwise, stir for 1 h, evaporate to remove the MeCN, wash twice with 50 mL portions of ethyl acetate. Acidify the aqueous phase with HCl and extract it twice with 300 mL portions of ethyl acetate. Wash the organic layer twice with 100 mL portions of saturated aqueous NaCl, dry over anhydrous magnesium sulfate, evaporate to dryness under reduced pressure, recrystallize from ethyl acetate to give 4-(N-carboxymethyl-N-methyl)amino-7-dimethylaminosulfonyl-2,1,3-benzoxadiazole (DBD-COOH) as orange-yellow crystals (mp 209-210°). Add 3.5 mL oxalyl chloride and 24 µL DMF to 1 g DBD-COOH in anhydrous benzene, stir at room temperature for 30 min, reflux for 1 h, evaporate to dryness, add 20 mL dry benzene to the residue, filter, evaporate the filtrate to give 4-(N-chloroformylmethyl-N-methyl)amino-7-N,N-dimethylaminosulfonyl-2,1,3-benzoxadiazole (DBD-COCl) as yellow crystals (mp 102°).)

HPLC VARIABLES
Column: 150 × 4.6 5 µm Cosmosil 5C18
Mobile phase: Gradient. MeCN:water from 50:50 to 100:0 over 20 min, maintain at 100:0 for 1 h.
Flow rate: 1
Injection volume: 2
Detector: F ex 440 em 543

CHROMATOGRAM
Retention time: 15
Limit of detection: 40 fmole

REFERENCE
Imai, K.; Fukushima, T.; Yokosu, H. A novel electrophilic reagent, 4-(N-chloroformylmethyl-N-methyl)amino-7-N,N-dimethylaminosulphonyl-2,1,3-benzoxadiazole (DBD-COCl) for fluorometric detection of alcohols, phenols, amines and thiols, *Biomed.Chromatogr.*, **1994**, *8*, 107−113.

SAMPLE
Matrix: solutions
Analyte: hydroxy acids
Sample preparation: Mix 10 µL 0.5 mM compound in anhydrous benzene containing 100 mM quinuclidine with 10 µL 25 mM DBD-COCl in anhydrous benzene (Caution! Benzene is a carcinogen!), heat at 60° for 15 min, add 980 µL MeCN:water:acetic acid 50:50:1, inject a 2 µL aliquot. (Purify quinuclidine by sublimation. DBD-COCl is 4-(N-chloroformylmethyl-N-methyl)amino-7-N,N-dimethylaminosulfonyl-2,1,3-benzoxadiazole. DBD-COCl is available from Tokyo Kasei (TCI America, Portland OR). Synthesis is described above.)

HPLC VARIABLES
Column: 150 × 4.6 5 µm Cosmosil 5C18
Mobile phase: Gradient. MeCN:0.1% acetic acid in water from 20:80 to 60:40 over 20 min, maintain at 60:40 for 1 h.
Flow rate: 1
Injection volume: 2
Detector: F ex 442 em 551

CHROMATOGRAM
Retention time: 14 (lactic acid), 19 (mandelic acid)
Limit of detection: 125-145 fmole

REFERENCE
Imai, K.; Fukushima, T.; Yokosu, H. A novel electrophilic reagent, 4-(N-chloroformylmethyl-N-methyl)amino-7-N,N-dimethylaminosulphonyl-2,1,3-benzoxadiazole (DBD-COCl) for fluorometric detection of alcohols, phenols, amines and thiols, *Biomed.Chromatogr.*, **1994**, *8*, 107−113.

SAMPLE
Matrix: solutions
Analyte: alcohols
Sample preparation: Add 25 µL 10 mM 4-(N-chloroformylmethyl-N-methyl)amino-7-(N,N-dimethylaminosulfonyl)-2,1,3-benzoxadiazole in anhydrous benzene (Caution! Benzene is a carcinogen!) to 25 µL of a solution of the alcohol in anhydrous benzene, heat at 60° for

40 min, add 950 μL 1% acetic acid in benzene. Remove a 50 μL aliquot and evaporate it to dryness, reconstitute the residue in 50 μL MeCN, inject a 1 μL aliquot. (Dry benzene over molecular sieve before use.) 4-(N-Chloroformylmethyl-N-methyl)amino-7-(N,N-dimethylaminosulfonyl)-2,1,3-benzoxadiazole (DBD-COCl) is available from Tokyo Kasei (TCI America Portland OR). Synthesis is described above.)

HPLC VARIABLES
Column: 150 × 4.6 5 μm TSKgel ODS 80TM (Tosoh)
Mobile phase: MeCN:water 80:20
Flow rate: 1
Injection volume: 1
Detector: F ex 450 em 560; MS, Hitachi M-1200, APCI, vaporizer 180°, desolvation at 400°, drift voltage 40 V, focus voltage 120 V, multiplier voltage 1.8 kV

CHROMATOGRAM
Retention time: 7.2 (eicosapentaenoylethanolamide), 9.8 (arachidonylethanolamide (anandamide)), 14.0 (palmitylethanolamide), 15.4 (oleinylethanolamide)
Limit of detection: 20 fmole (F)

REFERENCE
Koga, D.; Santa, T.; Hagiwara, K.; Imai, K.; Takizawa, H.; Nagano, T.; Hirobe, M.; Ogawa, M.; Sato, T.; Inoue, K. High-performance liquid chromatography and fluorometric detection of arachidonylethanolamide (anandamide) and its analogues, derivatized with 4-(N-chloroformylmethyl-N-methyl)amino-7-N,N-dimethylaminosulphonyl-2,1,3-benzoxadiazole (DBD-COCl), *Biomed.Chromatogr.*, **1995**, *9*, 56−57.

4-(2-Chloroformylpyrrolidin-1-yl)-7-(N,N-dimethylaminosulfonyl)-2,1,3-benzoxadiazole

SAMPLE
Matrix: bulk
Analyte: alcohols
Sample preparation: Dissolve 1 mg alcohol in 1 mL 1 mM 4-(2-chloroformylpyrrolidin-1-yl)-7-(N,N-dimethylaminosulfonyl)-2,1,3-benzoxadiazole in anhydrous benzene:pyridine 99:1 (Caution! Benzene is a carcinogen!), let stand for 3 h, inject a 5 μL aliquot. (4-(2-Chloroformylpyrrolidin-1-yl)-7-(N,N-dimethylaminosulfonyl)-2,1,3-benzoxadiazole the DBD-Pro-COCl, is available from Tokyo Kasei (TCI America, Portland OR. Synthesis is as follows. Dissolve 0.5 g magnesium sulfate heptahydrate and 6 g NaOH in 60 mL water, throughout the reaction keep the flask at about 20° with cold water cooling, add 15 mL 30% hydrogen peroxide, add 75 mL MeOH, add 12.1 g powdered benzoyl peroxide in one go, stir for 10 min, pour into 150 mL 20% sulfuric acid, extract three times with 50 mL portions of chloroform, determine peroxybenzoic acid concentration by iodometric titration (Tetrahedron 1967, 23, 3327). Slowly add 110 mL 1 M peroxybenzoic acid in chloroform to 7 g 2,6-difluoroaniline dissolved in 100 mL chloroform, stir at room temperature, when reaction is complete (iodometric titration) wash with 2% sodium thiosulfate, wash with 5% sodium carbonate, wash with water, dry over anhydrous sodium sulfate, evaporate to

dryness under reduced pressure, recrystallize 2,6-difluoronitrosobenzene from EtOH (mp 108.5-109.5). Stir 8.5 g 2,6-difluoronitrosobenzene in 85 mL DMSO at room temperature and add a solution of 3.91 g sodium azide in 85 mL DMSO dropwise, let stand for about 1 h, add to a large volume of water, extract with ether, dry the extracts over anhydrous sodium sulfate, evaporate to dryness under reduced pressure and distil to give 4-fluoro-2,1,3-benzoxadiazole as a colorless oil (bp 83°/12 mm Hg) (J.Chem.Soc.(C) 1970, 1433). Add 11 mL chlorosulfonic acid dropwise to 3 g 4-fluoro-2,1,3-benzoxadiazole in 10 mL chloroform at 0-10° (use a calcium chloride drying tube), stir at room temperature for 1 h, reflux for 2 h, cool, slowly pour into ice water, remove the organic layer, extract the aqueous layer with chloroform, combine the organic layers, wash, dry over anhydrous magnesium sulfate, evaporate under reduced pressure, take up the residue in 5 mL benzene (Caution! Benzene is a carcinogen!), chromatograph on a 150 × 30 column of silica gel (100-200mesh Kanto Chemical) with n-hexane:benzene 50:50, evaporate the appropriate fractions to give 4-(chlorosulfonyl)-7-fluoro-2,1,3-benzoxadiazole (CBD-F) as pale yellow needles (mp 64-66°) (Anal. Chem. 1984. 56, 2461). Stir 0.76 g CBD-F in 70 mL MeCN at 0-10° and add 1 g dimethylamine hydrochloride in 10 mL 100 mM pH 10 borax dropwise, adjust pH to 5 with 1 M HCl, concentrate to about 10 mL under reduced pressure, extract three times with 200 mL portions of diethyl ether, wash with water, dry over anhydrous magnesium sulfate, evaporate under reduced pressure, chromatograph on a 500 × 20 column of silica gel with chloroform, isolate the appropriate fraction and re-chromatograph on the same column with ethyl acetate:benzene1:2 to give 4-(N,N-dimethylaminosulfonyl)-7-fluoro-2,1,3-benzoxadiazole (DBD-F) as white needles (mp 124-125°) (yield = 1% !). On a Merck no. 5714 60F$_{254}$ tlc plate eluted with chloroform DBD-F has Rf 0.32 and lies between two other reaction products (Analyst 1989, 114, 413). DBD-F can also be purchased from Tokyo Kasei. Add 100 mg DBD-F in 10 mL MeCN to 47 mg (S)-(-)proline in 20 mL 250 mM pH 11.5 sodium carbonate solution, stir at room temperature for 30 min, wash with ethyl acetate, adjust the pH of the aqueous layer to 1-2 with 2 M HCl, extract three times with 30 mL ethyl acetate. Combine the extracts and evaporate them under reduced pressure, recrystallize from benzene/ethyl acetate to give(S)-(-)-(N,N-dimethylaminosulfonyl)-7-(2-carboxypyrrolidin-1-yl)-2,1,3-benzoxadiazole (DBD-Pro) as yellow needles (mp 187-9° d) (Analyst 1989, 114, 1233). Suspend 55 mg (S)-(-)-DBD-Pro in 55 mL anhydrous diethyl ether at 0°, add 110 mg phosphorus pentachloride, stir at 5° for 1 h, filter quickly, evaporate to dryness under reduced pressure, dry under vacuum over phosphorus pentoxide for 12 h to give (S)-(-)-4-(2-chloroformylpyrrolidin-1-yl)-7-(N,N-dimethylaminosulfonyl)-2,1,3-benzoxadiazole (DBD-Pro-Cl) as yellow crystals (mp 116-17°).)

HPLC VARIABLES
Column: 150 × 4.6 5 µm Inertsil SIL
Mobile phase: Hexane:ethyl acetate 80:20
Column temperature: 40
Flow rate: 1
Injection volume: 5
Detector: F ex 450 em 560

CHROMATOGRAM
Retention time: 10.90 ((S)-hexan-2-ol), 12.83 ((R)-hexan-2-ol), 10.00 ((S)-heptan-2-ol), 12.08 ((R)-heptan-2-ol), 9.07 ((S)-nonan-2-ol), 11.33 ((R)-nonan-2-ol), 15.85 ((S)-1-phenylethanol), 19.04 ((R)-1-phenylethanol)

KEY WORDS
chiral; normal phase

REFERENCE
Toyo'oka, T.; Ishibashi, M.; Terao, T.; Imai, K. 4-(N,N-Dimethylaminosulfonyl)-7-(2-chloroformylpyrrolidin-1-yl)-2,1,3-benzoxadiazole: Novel fluorescent chiral derivatization reagents for the resolution of alcohol enantiomers by high-performance liquid chromatography, *Analyst*, **1993**, *118*, 759–763.

4-(2-Chloroformylpyrrolidin-1-yl)-7-nitro-2,1,3-benzoxadiazole

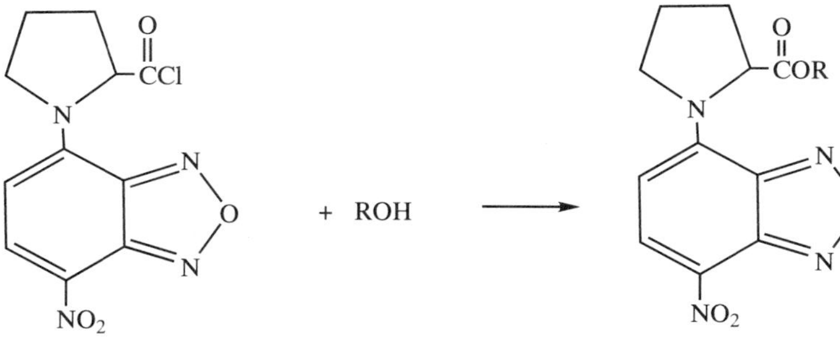

SAMPLE

Matrix: solutions

Analyte: alcohols

Sample preparation: Mix 50 μL 2 mM alcohol in anhydrous benzene:pyridine 98:2 with 50 μL 10 mM R-(+)-NBD-Pro-Cl in anhydrous benzene (Caution! Benzene is a carcinogen!), heat at 80° for 4 h, cool in ice water, add 900 μL MeCN:methylamine 99:1, inject a 5 μL aliquot. (R-(+)-NBD-Pro-Cl is available from Tokyo Kasei (TCI America, Portland OR). Synthesis is as follows. Add a solution of 1 g 4-chloro-7-nitrobenzofurazan in 50 mL EtOH to a solution of 575 mg proline in 100 mL 5% sodium acetate solution, reflux for 10 min, acidify to pH 1.5 with 4 M HCl, evaporate to dryness. Dissolve the residue in water and extract with ethyl acetate. Evaporate the organic layer to dryness and recrystallize from benzene/ethyl acetate to give NBD-Pro as orange needles (mp 156-7°) (Anal.Chim.Acta 1983, 149, 305). Dissolve 2.6 g R-(+)-NBD-Pro in 200 mL anhydrous dichloromethane, add 10 mL oxalyl chloride and 200 μL DMF, stir at room temperature for 1 h, evaporate to dryness under reduced pressure, dissolve the residue in 100 mL anhydrous benzene, filter, evaporate the filtrate to dryness, dry over phosphorus pentoxide under vacuum to give R-(+)-NBD-Pro-Cl (R-(+)-4-(2-chloroformylpyrrolidin-1-yl)-7-nitrobenzofurazan) as red-orange crystals (mp 103-4° d).)

HPLC VARIABLES

Column: 150 × 4.6 5 μm Inertsil SIL

Mobile phase: n-Hexane:ethyl acetate 70:30

Column temperature: 40

Flow rate: 1

Injection volume: 5

Detector: F ex 485 em 530

CHROMATOGRAM

Retention time: 8.64 (S-2-hexanol), 7.48 (R-2-hexanol), 8.28 (S-2-heptanol), 7.01 (R-2-heptanol), 7.74 (S-2-nonanol), 6.43 (R-2-nonanol), 12.26 (S-1-phenylethanol), 10.34 (R-1-phenylethanol)

KEY WORDS

chiral; normal phase; can also be separated by reverse-phase with MeCN:water 50:50 but separation of enantiomers is less

REFERENCE

Toyo'oka, T.; Liu, Y.-M.; Hanioka, N.; Jinno, H.; Ando, M.; Imai, K. Resolution of enantiomers of alcohols and amines by high-performance liquid chromatography after derivatization with a novel fluorescent chiral reagent, *J.Chromatogr.A*, **1994**, *675*, 79–88.

Coumarin-3-carbonyl Chloride

+ ROH ⟶

SAMPLE
Matrix: solutions
Analyte: tricothecenes
Sample preparation: Condition a BDH silica gel SPE cartridge with 5 mL chloroform: MeOH 70:30 and 5 mL benzene (Caution! Benzene is a carcinogen!). Evaporate 10 μL of a toluene solution to dryness under a stream of nitrogen, add 10 μL 6.5 μg/mL 4-di methylaminopyridine in toluene, add 10 μL 3.25 μg/mL coumarin-3-carbonyl chloride in toluene, heat at 80° for 20 min, cool in ice water, add to the SPE cartridge, rinse the tube 5 times with 1 mL portions of benzene, add the rinses to the SPE cartridge, wash 5 times with 2 mL portions of hexane:ethyl acetate 20:80, elute with two 7 mL portions of hexane:ethyl acetate 10:90. Evaporate the eluate to dryness under a stream of nitrogen, reconstitute the residue in 1 mL mobile phase, filter (4.5 μm), inject a 10 μL aliquot. (Rinse all glassware with toluene before use. Preparation of coumarin-3-carbonyl chloride is as follows. Reflux 5 g coumarin-3-carboxylic acid in 30 g dry dichloromethane under a calcium chloride drying tube, add 10 mL thionyl chloride dropwise while refluxing, reflux for 40 min (add dichloromethane to maintain the level if necessary), evaporate to dryness under reduced pressure at 50°, dissolve the residue in anhydrous chloroform, heat, add hexane until the solution turns cloudy, cool to obtain coumarin-3-carbonyl chloride as pale yellow crystals (mp 143°).)

HPLC VARIABLES
Guard column: 15 3.2 7 μm RP-18
Column: 250 × 4.6 5 μm RP-18
Mobile phase: MeCN:water:acetic acid 65:35:0.75
Flow rate: 1
Injection volume: 10
Detector: F ex 292 em 425

CHROMATOGRAM
Retention time: 12 (T-2), 13 (T-4), 15 (HT-2), 18 (T-3)
Limit of detection: 0.83-2 ng

KEY WORDS
SPE

REFERENCE
Cohen, H.; Boutin-Muma, B. Fluorescence detection of trichothecene mycotoxins as coumarin-3-carbonyl chloride derivatives by high-performance liquid chromatography, *J.Chromatogr.*, **1992**, *595*, 143–148.

3,4-Dihydro-6,7-dimethoxy-4-methyl-3-oxoquinoxaline-2-carboxyl Chloride

SAMPLE
Matrix: solutions
Analyte: alcohols
Sample preparation: Mix 500 μL of a 0.04-50 μM solution in benzene with 500 μL 3 mM 3,4-dihydro-6,7-dimethoxy-4-methyl-3-oxoquinoxaline-3-carboxyl chloride in benzene (Caution! Benzene is a carcinogen!), heat at 40° in the dark for 40 min, cool, add 20 μL of the reaction mixture to 2 mL MeOH, inject a 10 μL aliquot. (3,4-Dihydro-6,7-dimethoxy-4-methyl-3-oxoquinoxaline-3-carboxyl chloride is available from Dojindo Molecular Technologies, Inc., 3 Bethesda Metro Center, Suite 700, Bethesda MD 20814; (301) 664-8448; www.dojindo.co.jp. Synthesis is as follows. Stir 483 g veratrole in 1.45 L acetic acid at 15°, add 683 g concentrated nitric acid (s.g. 1.05) over 1 h keeping the temperature below 40° (cool if necessary), add 2.127 L fuming nitric acid (s.g.1.50) over 1 h keeping the temperature below 30°, allow to stand for 2 h, pour into a large volume of cold water, filter, wash the solid until it is free from acid, recrystallize from EtOH to give 4,5-dinitroveratrole (mp 129.5-130.5°) (J. Am. Chem. Soc. 1946, 68, 1536). Reflux 5 g 4,5-dinitroveratrole in 200 mL benzene (Caution! Benzene is a carcinogen!), add 100 g 60 mesh iron powder and 20 mL concentrated HCl in small portions over 1 h, reflux for 4 h, add 10 mL water, reflux for 2 h, cool, make alkaline with 2.5 M NaOH, extract several times with 200 mL portions of benzene. Combine the extracts and evaporate to dryness, add 10 mL concentrated HCl, recrystallize from EtOH to give 1,2-diamino-4,5-dimethoxybenzene hydrochloride as slightly pink needles (mp 240° d) (Anal. Chim. Acta 1982, 134, 39). Another synthesis of 1,2-diamino-4,5-dimethoxybenzene is as follows. Add 9.66 g veratrole in 9.5 mL glacial acetic acid dropwise to 26.6 mL fuming nitric acid over 30 min using an ice bath, stir at room temperature for 30 min, pour into 500 mL ice-cold water, filter to obtain 1,2-dimethoxy-4,5-dinitrobenzene, recrystallize repeatedly from EtOH. Dissolve 1 g 1,2-dimethoxy-4,5-dinitrobenzene in 50 mL EtOH, add 100 mg 10% palladium on charcoal, stir under hydrogen at atmospheric pressure and room temperature for 4 days, filter through Celite under an atmosphere of nitrogen, saturate the filtrate with hydrogen chloride gas, filter under nitrogen, dry the solid under vacuum to obtain 1,2-diamino-4,5-dimethoxybenzene hydrochloride (mp 240° d (Anal. Chim. Acta 1982, 134, 39)) (Anal. Chim. Acta 1992, 263, 137). 1,2-Diamino-4,5-dimethoxybenzene is available from Molecular Probes, Eugene OR or Dojindo Molecular Technologies, Inc., 3 Bethesda Metro Center, Suite 700, Bethesda MD 20814; (301) 664-8448; www.dojindo.co.jp. Dissolve 8 g 1,2-diamino-4,5-dimethoxybenzene monohydrochloride and 8 g α-ketomalonic acid in 20 mL 500 mM HCl, heat in a boiling water bath for 2 h, cool in ice, filter, wash the precipitate with water, recrystallize from dioxane:water 90:10 (Caution! Dioxane is a carcinogen!) to give 3,4-dihydro-6,7-dimethoxy-3-oxoquinoxaline-2-carboxylic acid as orange needles (mp 268°). Treat 5.5 g 3,4-dihydro-6,7-dimethoxy-3-oxoquinoxaline-2-carboxylic acid in 50 mL anhydrous MeOH with ethereal diazomethane, evaporate to dryness under reduced pressure, dissolve the residue in 30 mL chloroform, chromatograph on a 250 × 57 column of 130 g 70-230 mesh silica gel 60 (Merck) with n-hexane:ethyl acetate 50:50 to give methyl 3,4-dihydro-6,7-dimethoxy-4-methyl-3-oxoquinoxaline-2-carboxylate as yellow needles (mp 164°). Dissolve 2.5 g methyl 3,4-dihydro-6,7-dimethoxy-4-methyl-3-oxoquinoxaline-2-carboxylate in 200 mL 1 M NaOH, let stand at rom temperature for about 70 min, wash 5 times with 200 mL portions of ethyl acetate. Neutralize the aqueous layer

with dilute HCl, filter, recrystallize the precipitate from dioxane:water 80:20 to give 3,4-dihydro-6,7-dimethoxy-4-methyl-3-oxoquinoxaline-2-carboxylic acid as yellow needles (mp 222°). Dissolve 1 g 3,4-dihydro-6,7-dimethoxy-4-methyl-3-oxoquinoxaline-2-carboxylic acid in 20 mL freshly distilled thionyl chloride, reflux for 1 h, cool, add 50 mL light petroleum (bp 30-60°), filter, recrystallize the precipitate from benzene:light petroleum 90:10 (Caution! Benzene is a carcinogen!) to give 3,4-dihydro-6,7-dimethoxy-4-methyl-3-oxoquinoxaline-2-carboxyl chloride as orange needles (mp 261°). Preparation of diazomethane is as follows. Caution! Diazomethane is toxic, explosive, and carcinogenic! A face shield and a safety screen should always be used and the preparation should only be carried out in a properly functioning chemical fume hood. Only smooth glass apparatus with rubber stoppers and plastic tubing should be used. Scratched glassware, ground glass joints, and sharp edges should be avoided (Org. Syn., Coll. Vol. VI; Wiley: New York, 1988, pp. 432-435). Procedures have been reported for the synthesis of diazomethane from N-methyl-N-nitrosourea (Org. Syn., Coll. Vol. II; Wiley: New York, 1943, pp. 165-167), N-nitroso-β-methylaminoisobutyl methyl ketone (Org. Syn., Coll. Vol. III; Wiley: New York, 1955, pp. 244-248), and N,N'-dimethyl-N,N'-dinitrosoterephthalamide (Org. Syn., Coll. Vol. V; Wiley: New York, 1973, pp. 351-355). Probably the most convenient starting material is N-methyl-N-nitroso-p-toluenesulfonamide (Diazald) (Aldrichimica Acta 1983, 16, 3-10; Org. Syn., Coll. Vol. IV, Wiley: New York, 1963, pp. 250-253). Add 10 mL 95% EtOH to a solution of 5 g KOH in 8 mL water, warm to 65° using a water bath, add a solution of 5 g N-methyl-N-nitroso-p-toluenesulfonamide in 45 mL ether over 20 min at such a rate as to keep the reaction volume constant. Collect the ether and diazomethane that distil in an ice-cooled receiving flask under a dry ice/acetone condenser. When all the N-methyl-N-nitroso-p-toluenesulfonamide has been used up, slowly add 10 mL ether to the reaction flask and continue distillation until the distillate is colorless. A purpose-built distillation apparatus can be purchased from Aldrich (Aldrichimica Acta 1983, 16, 3-10). Excess quantities of diazomethane can be destroyed by adding acetic acid until the yellow color of the diazomethane is discharged. The safe disposal of the nitroso compounds used to generate diazomethane has been discussed (Lunn,G.; Sansone,E.B. Destruction of Hazardous Chemicals in the Laboratory, Second Edition. Wiley: New York, 1994, pp. 277-289).)

HPLC VARIABLES
Column: 150 × 6 10 μm YMC Pack C8 (Yamamura Chemical Laboratories, Kyoto)
Mobile phase: MeOH:water 70:30
Flow rate: 2
Injection volume: 10
Detector: F ex 400 em 500

CHROMATOGRAM
Retention time: 3 (benzyl alcohol), 4.5 (cyclohexanol), 6 (n-hexanol)
Limit of detection: 2-3 fmole

REFERENCE
Iwata, T.; Yamaguchi, M.; Hara, S.; Nakamura, M.; Ohkura, Y. 3,4-Dihydro-6,7-dimethoxy-4-methyl-3-oxo-quinoxaline-2-carbonyl chloride as a highly sensitive fluorescence derivatization reagent for alcohols in high-performance liquid chromatography, *J.Chromatogr.*, **1986**, *362*, 209−216.

3,5-Dinitrobenzoyl Chloride

When digoxin and related compounds are derivatized all the hydroxy groups except that at C14 are derivatized.

RELATED REFERENCES
Goncalves, J.C.S.; Sclavons, M.; Poupaert, J.H.; Dumont, P. High-performance liquid chromatography of 1,2-diacyl-rac-glycerols and 1,3-diacylglycerols through chromogenic derivatization with 3,5-dinitrobenzoyl. *J.Chromatogr.* **1987**, *411*, 472-475.
Valdez, D.; Reier, J.C. A simplified procedure for the derivatization of alcohols at dilute levels in aqueous solutions with 3,5-dinitrobenzoyl chloride. *J.Chromatogr.Sci.* **1986**, *24*, 356-360.

Valdez, D.; Reier, J.C. Derivatization of glycols, hydroxyamines, and polyols at trace levels with 3,5-dinitrobenzoyl chloride utilizing aqueous to nonaqueous phase transfer on a reverse phase cartridge. *J.Liq.Chromatogr.* **1987**, *10*, 863-880.

SAMPLE
Matrix: blood
Analyte: choline
Sample preparation: 250 μL Plasma + 25 μL 0.41 mM IS + 750 μL ice-cold 1 M formic acid in acetone, mix, centrifuge for 15 min. Remove the supernatant and add it to the SPE column, wash with 1 mL 100 mM pH 4.0 ammonium acetate, elute with 1 mL 2 M NaCl in MeOH:water 50:50, force out all liquid under pressure. Evaporate the eluate under a stream of nitrogen and keep under vacuum for 1 h, add 1 mL MeCN to the residue, mix, centrifuge, remove the supernatant, repeat the extraction. Combine the supernatants and evaporate them to dryness under a stream of nitrogen, reconstitute the residue in 300 μL freshly prepared 21.7 mM 3,5-dinitrobenzoyl chloride in pyridine (dry pyridine over KOH), heat at 105° for 1 h, evaporate the pyridine under a stream of nitrogen, extract the residue with 300 μL water then with 200 μL water, combine the extracts, filter (0.3 μm), inject a 100 μL aliquot. (Preparation of SPE column. Let 300 mg AG 50W-X12 cation-exchange resin sit overnight in 1 mL 100 mM pH 4.0 ammonium acetate. Add the mixture to a Pasteur pipette with a glass wool plug, wash column with 1 mL 2 M NaCl in MeOH:water 50:50, activate column with 1 mL 100 mM pH 4.0 ammonium acetate.)

HPLC VARIABLES
Column: Two 300 × 4 10 μm μBondapak C18 in series
Mobile phase: MeCN:water 50:50 containing 5 mM sodium dodecyl sulfate and 0.1% acetic acid
Flow rate: 2-2.3
Injection volume: 100
Detector: UV 254

CHROMATOGRAM
Retention time: 13
Internal standard: 3-hydroxy-N,N,N-trimethylpropanaminium iodide (15) (Prepare by adding 18.5 g iodomethane to 10 g 3-hydroxy-N,N-dimethylaminopropane in 24 mL EtOH, stir, filter, add cold diethyl ether to the filtrate, filter. Combine precipitates, recrystallize from EtOH/diethyl ether, mp 203-4°)
Limit of quantitation: 1 μM

KEY WORDS
plasma; SPE

REFERENCE
Buchanan, D.N.; Fucek, F.R.; Domino, E.F. Paired-ion high-performance liquid chromatographic assay for plasma choline, *J.Chromatogr.*, **1980**, *181*, 329–335.

SAMPLE
Matrix: bulk
Analyte: RS-93522-004 (2-[4-(2,3-dihydroxypropoxy)phenyl]ethyl methyl 1,4-dihydro-2,6-dimethyl-4-(3-nitrophenyl)-3,5-pyridinedicarboxylate)
Sample preparation: Dissolve 10 mg RS-93522-004 and 44 mg 3,5-dinitrobenzoyl chloride in 3 mL dichloromethane, slowly add 1 mL 26 μL/mL triethylamine in dichloromethane, shake gently intermittently for 10-15 min, evaporate to dryness, reconstitute with 5 mL

MeCN, slowly pass through a Sep-Pak C18 SPE cartridge, dilute the eluate to 25 mL with MeCN, dilute to 100 mL with mobile phase, inject an aliquot.

HPLC VARIABLES
Column: 250 × 4.6 5 μm Spherisorb C8
Mobile phase: MeOH:water 63:37
Column temperature: 40
Flow rate: 2
Injection volume: 25
Detector: UV 254

CHROMATOGRAM
Retention time: 41 ((R,R)(S,S)), 44 ((R,S)(S,R))
Limit of detection: 0.3% of major isomer

REFERENCE
Kern, J.R.; Lokensgard, D.M.; Yang, T.Y. Chromatographic separation of the diastereomers of a dihydropyridine-type calcium channel antagonist as the bis-3,5-dinitrobenzoates, *J.Chromatogr.*, **1988**, *457*, 309–316.

SAMPLE
Matrix: bulk
Analyte: digoxin and metabolites
Sample preparation: Dissolve a small amount in 200 μL dry pyridine, add 15 mg 3,5-dinitrobenzoyl chloride, shake for 2 h, evaporate to dryness under nitrogen under reduced pressure. Reconstitute with 1.5 mL ethyl acetate, wash 4 times with 1 mL portions of 5% sodium bicarbonate containing 2.5 mg/mL 4-dimethylaminopyridine, wash 4 times with 1 mL portions of 1% HCl, wash 4 times with 1 mL portions of water, evaporate to dryness under a stream of nitrogen, reconstitute with mobile phase, inject an aliquot (J.Chromatogr.Sci. 1983, 21, 495).

HPLC VARIABLES
Guard column: 15 × 3.2 Brownlee ODS
Column: 150 × 4.6 3 μm Spherisorb ODS II
Mobile phase: MeOH:EtOH:MeCN:isopropanol:100 mM pH 4.6 sodium acetate buffer 40:3:60:2:22
Flow rate: 1
Detector: UV 254; E, ESA Coulochem Model 5100A, Model 5020 guard cell -0.8 V (placed before the injector), Model 5010 dual-electrode analytical cell with glassy-carbon electrodes (-0.8 V first electrode, +0.8 V second electrode)

CHROMATOGRAM
Retention time: 4.5 (digoxigenin), 5.5 (digoxigenin monodigitoxoside), 8 (digoxigenin bis-digitoxoside), 12 (dihydrodigoxin), 12.5 (digoxin)
Limit of detection: 0.39 ng (E)

REFERENCE
Embree, L.; McErlane, K.M. Electrochemical detection of 3,5-dinitrobenzoyl derivatives of digoxin by high-performance liquid chromatography, *J.Chromatogr.*, **1990**, *526*, 439–446.

SAMPLE
Matrix: solutions
Analyte: glycols
Sample preparation: Add 0.5 mmole propylene glycol to a solution of 0.5 g 3,5-dinitrobenzoyl chloride in 30 mL pyridine, mix well, heat at 60° for 15 min, adjust pH to 2.5 with 2 M HCl (Caution! Exothermic reaction!), cool to room temperature, add 25 mL butyl acetate, shake for 3 min. Remove the organic layer and add it to 50 mL 1% sodium carbonate solution, shake for 2 min. Remove the organic layer and add it to 25 mL 0.25 M sulfuric acid, shake for 2 min. Remove the organic layer and add it to 10 mL water, shake for 1 min, inject a 10 μL aliquot of the organic layer.

HPLC VARIABLES
Column: 2000 × 2.1 Corasil II
Mobile phase: Heptane:ethyl acetate 75:25

Flow rate: 1.7
Injection volume: 10
Detector: UV 254

CHROMATOGRAM
Retention time: 5 (propylene glycol), 8 (ethylene glycol), 17 (diethylene glycol)

KEY WORDS
normal phase

REFERENCE
Carey, M.A.; Persinger, H.E. Liquid chromatographic determination of traces of aliphatic carbonyl compounds and glycols as derivatives that contain the dinitrophenyl group, *J.Chromatogr.Sci.*, **1972**, *10*, 537–543.

SAMPLE
Matrix: solutions
Analyte: cardiac glycosides
Sample preparation: Add 15 mg 3,5-dinitrobenzoyl chloride to 200 μL of a solution in dry pyridine, shake for 2 h, evaporate to dryness under a stream of nitrogen, reconstitute with 1.5 mL ethyl acetate. Wash this solution four times with 1 mL portions of 2.5 mg/mL 4-dimethylaminopyridine in 5% sodium bicarbonate solution, wash four times with 1 mL portions of 1% HCl, wash four times with 1 mL portions of water, evaporate to dryness under a stream of nitrogen, reconstitute with mobile phase, inject a 0.2 μL aliquot.

HPLC VARIABLES
Column: 150 × 0.5 5 μm SC-01 octadecylsilyl in a PTFE column (Japan Spectroscopic)
Mobile phase: MeCN:water 75:25
Flow rate: 0.008
Injection volume: 0.2
Detector: UV 230

CHROMATOGRAM
Retention time: 10 (digoxigenin), 15 (digoxigenin monodigitoxoside), 12 (3-epidigoxigenin), 32 (β-methyldigoxin), 40 (digoxin)
Internal standard: gitoxin (48)
Limit of quantitation: 2 ng
Limit of detection: 0.6 ng

OTHER SUBSTANCES
Also analyzed: digitoxigenin (MeCN:water 20:7), digitoxigenin monodigitoxoside (MeCN:water 20:7), 3-epidigitoxigenin (MeCN:water 20:7)

KEY WORDS
microbore

REFERENCE
Fujii, Y.; Oguri, R.; Mitsuhashi, A.; Yamazaki, M. Micro HPLC separation of 3,5-dinitrobenzoyl derivatives of cardiac glycosides and their metabolites, *J.Chromatogr.Sci.*, **1983**, *21*, 495–499.

SAMPLE
Matrix: solutions
Analyte: glycerophospholipids
Sample preparation: Hydrolyze 0.5 mg choline glycerophospholipids or ethanolamine glycerophospholipids in 1 mL 30 U/mL phospholipase C in 100 mM pH 7.6 Tris HCl in the presence of 2 mL diethyl ether for 30 (choline) or 45 (ethanolamine) min. Remove the organic layer and extract the aqueous layer twice more with diethyl ether, combine the organic layers and evaporate them to dryness under a stream of nitrogen, dry under vacuum, add 1 mL pyridine, add 30 mg dinitrobenzoyl chloride, add 10 mg dimethylaminopyridine, heat at 60° for 6 min, add 1 mL 0.1% bicarbonate solution, extract with

hexane, purify by TLC (20 cm × 20 cm × 250 µm plates developed with benzene (Caution! Benzene is a carcinogen!)), inject an aliquot.

HPLC VARIABLES
Guard column: 25 × 4 4 µm Superspher 100 RP-18
Column: 250 × 4 4 µm Superspher 100 RP-18
Mobile phase: MeCN
Column temperature: 30
Flow rate: 1.5
Detector: UV 230

CHROMATOGRAM
Retention time: 20-70 (depending on structure)

KEY WORDS
protect from light

REFERENCE
Menguy, L.; Christon, R.; Van Dorsselaer, A.; Léger, C.L. Apparent relative retention of the phosphatidylethanolamine molecular species 18:0-20:5(n-3), 16:0-22:6(n-3) and the sum 16:0-20:4(n-6) plus 16:0-20:3(n-9) in the liver microsomes of pig on an essential fatty acid deficient diet, *Biochim.Biophys.Acta*, **1992**, *1123*, 41−50.

SAMPLE
Matrix: solutions
Analyte: diradylglycerol
Sample preparation: Condition a Sep-Pak C18 SPE cartridge with 10 mL diethyl ether, 15 mL MeOH, and 15 mL MeOH:water 80:20. Add 500 µL 50 mg/mL 3,5-dinitrobenzoyl chloride in pyridine to the sample, heat at 60° for 15 min, add 2 mL MeOH:water 80:20, add 2 mL water, add to the SPE cartridge, wash with two 15 mL aliquots of MeOH:water 80:20, elute with 25 mL freshly distilled diethyl ether, evaporate to dryness under a stream of nitrogen, reconstitute with 20 µL MeCN:isopropanol 50:50, inject an aliquot.

HPLC VARIABLES
Column: 250 × 4.6 5 µm Spherisorb S5ODS2
Mobile phase: Gradient. MeCN:isopropanol from 90:10 to 50:50 over 45 min.
Flow rate: 1
Detector: UV 254

CHROMATOGRAM
Retention time: 15-40 (depending on species)

KEY WORDS
SPE

REFERENCE
Pettitt, T.R.; Wakelam, M.J. Bombesin stimulates distinct time-dependent changes in the sn-1,2-diradylglycerol molecular species profile from Swiss 3T3 fibroblasts as analysed by 3,5-dinitrobenzoyl derivatization and h.p.l.c. separation, *Biochem.J.*, **1993**, *289*, 487−495.

SAMPLE
Matrix: urine
Analyte: digoxin
Sample preparation: 10 mL Urine + 0.5 mL 20 µg/mL digitoxigenin in dichloromethane + 20 mL dichloromethane, shake for 15 min, centrifuge for 20 min. Remove the organic phase and add it to 15 mL 5% sodium bicarbonate, shake for 15 min, centrifuge for 20 min. Remove the organic phase and evaporate it to dryness at 50° under a stream of nitrogen, add 200 µL derivatizing solution to the residue, shake gently at room temperature for 10 min, evaporate to dryness under a stream of nitrogen at 50°, add 2 mL 2 mg/mL 4-dimethylaminopyridine in 5% sodium bicarbonate, shake for 5 min, add 1 mL chloroform, rock on an Aliquot Mixer. Remove the organic phase and add it to 2 mL 5% sodium bicarbonate solution, mix for 2 min. Remove the organic phase and add it to 3 mL 50 mM HCl containing 5% NaCl, mix for 2 min. Remove the organic phase and repeat the acid wash 3 more times, inject a 100 µL aliquot of the organic phase. (The derivatizing

solution was 85 mg/mL 3,5-dinitrobenzoyl chloride in pyridine, prepared with gentle warming to help the solid dissolve.)

HPLC VARIABLES
Column: 250 × 4.6 10 μm Partisil 10
Mobile phase: Hexane:dichloromethane:MeCN 60:20:20
Flow rate: 1.8
Injection volume: 100
Detector: UV 254

CHROMATOGRAM
Retention time: 54
Internal standard: digitoxigenin (17)
Limit of detection: 100 ng/mL

OTHER SUBSTANCES
Simultaneously analyzed: digoxigenin, digoxigenin bisdigitoxoside, digoxigenin monodigitoxoside

KEY WORDS
normal phase

REFERENCE
Bockbrader, H.N.; Reuning, R.H. Digoxin and metabolites in urine: A derivatization-high-performance liquid chromatographic method capable of quantitating individual epimers of dihydrodigoxin, *J.Chromatogr.*, **1984**, *310*, 85–95.

(S)-(+)-Flunoxaprofen Chloride

SAMPLE
Matrix: blood, urine
Analyte: ciclotropium bromide
Sample preparation: Plasma. 4 mL Plasma + 5.5 mL chloroform + 500 μL reagent A + 250 μL 1 M HCl, shake vigorously for 5 min, centrifuge at 10° at 2500 g for 30 min. Remove a 4 mL aliquot of the organic layer and add it to 2.4 mL 100 mM HCl, shake for 15 min, centrifuge at 20° at 2500 g for 10 min. Remove a 2 mL aliquot of the aqueous layer and add it to 1 mL 1 M NaOH, heat at 140-5° for 90 min (to hydrolyse ciclotropium), cool to room temperature, add 500 μL reagent A, adjust pH to 8.5-9.5 with 5 M HCl, add 2.6 mL chloroform, shake for 15 min, centrifuge at 20° at 2500 g for 10 min. Remove a 2 mL aliquot of the chloroform layer and add it to 1.4 mL 100 mM HCl, shake for 15 min, centrifuge at 20° at 2500 g for 10 min. Remove a 1 mL aliquot of the aqueous layer and add it to 1 mL MeOH, evaporate to dryness under reduced pressure, add three 3 mL portions of MeOH and evaporate to dryness each time, take up the residue in 200 μL 10 mg/mL flunoxaprofen chloride in MeCN (freshly prepared), heat at 110° for 15 min, evaporate to dryness under reduced pressure, add 1 mL ethyl acetate, add 1.3 mL 10 mM HCl, shake vigorously for 15 min, centrifuge at 20° at 2500 g for 10 min, discard the organic layer, wash the aqueous layer twice more with ethyl acetate, evaporate the aqueous layer to dryness under reduced pressure, reconstitute with 100 μL MeCN:MeOH:water 1:1:1, inject a 20 μL aliquot. Urine. 1 mL Urine + 100 μL 1 μg/mL IS in water + 2 mL 0.5 M NaOH, heat at 140-5° for 90 min (to hydrolyse ciclotropium), cool to room temperature, add 500 μL reagent B, adjust pH to 8.5-9.5 with 5 M HCl, add 2.6 mL chloroform, shake for 15 min, centrifuge at 20° at 2500 g for 10 min. Remove a 2 mL aliquot of the chloroform layer and add it to 1.4 mL 100 mM HCl, shake for 15 min, centrifuge at 20° at 2500 g for 10 min. Remove a 1 mL aliquot of the aqueous layer and

add it to 1 mL MeOH, evaporate to dryness under reduced pressure, add three 3 mL portions of MeOH and evaporate to dryness each time, take up the residue in 200 μL 10 mg/mL flunoxaprofen chloride in MeCN (freshly prepared), heat at 110° for 15 min, evaporate to dryness under reduced pressure, add 1 mL ethyl acetate, add 1.3 mL 10 mM HCl, shake vigorously for 15 min, centrifuge at 20° at 2500 g for 10 min, discard the organic layer, wash the aqueous layer twice more with ethyl acetate, evaporate the aqueous layer to dryness under reduced pressure, reconstitute with 500 μL MeCN:MeOH: water 1:1:1, inject a 20 μL aliquot. (Prepare reagent A by mixing 100 mg dipicrylamine, 600 mg anhydrous sodium carbonate, and 10 mL water. Prepare reagent B by mixing 35 mg dipicrylamine and 10 mL 100 mM NaOH. Prepare dipicrylamine as follows (Caution! Dipicrylamine is potentially explosive and highly toxic, store moistened with 50% water!). Add 50 g 2,4-dinitrodiphenylamine to 420 g nitric acid (36° Bé., 52%, d = 1.33) heated to 62° over 2 h, heat at 62-90° for another 3 h, cool, filter, wash the product until it is free of acid, dry to obtain 2,2',4,4'-tetranitrodiphenylamine as a yellow solid (mp 187.4°). Add 50 g tetranitrodiphenylamine over 1 h to 500 g of a mixture of equal parts 92% sulfuric acid and 93% nitric acid at room temperature, after 4.5 h add to a large volume of ice water, filter, recrystallize the product from acetone to obtain dipicrylamine (2,2',4,4',6,6'-hexanitrodiphenylamine) as yellow crystals (mp 242.9°) (J.Am.Chem.Soc. 1919, 41, 1013). Prepare (S)-flunoxaprofen chloride as follows. Dissolve 1 mmole (S)-flunoxaprofen in 25 mL toluene, add a trace of DMF (J.Chromatogr. 1990, 528, 55), add 2.5 mL thionyl chloride, reflux for 30 min, remove solvent by evaporation, dry the residue under vacuum over KOH, recrystallize from dichloromethane (mp 73°) (J.Chromatogr. 1988, 427, 131).)

HPLC VARIABLES
Column: 150 × 3.9 5 μm Suplex pkb-100 (Supelco)
Mobile phase: MeCN:water 55:45 containing 1 mL/L 50% phosphoric acid and 0.6 g/L dodecyl sulfate.
Flow rate: 1
Injection volume: 20
Detector: F ex 310 em 365

CHROMATOGRAM
Retention time: 8.0
Internal standard: N-butyltropinium (11.1)
Limit of detection: 0.5 ng/mL (plasma), 10 ng/mL (urine)

KEY WORDS
plasma; pharmacokinetics

REFERENCE
Liebmann, B.; Henke, D.; Spahn-Langguth, H.; Mutschler, E. Determination of the quaternary compound ciclotropium in human biological material after hydrolysis and derivatization with the fluorophor flunoxaprofen chloride, *J.Chromatogr.*, **1991**, *572*, 181–193.

(+)-1-(9-Fluorenyl)ethyl Chloroformate

SAMPLE
Matrix: blood
Analyte: carnitine
Sample preparation: 250 μL Plasma + 100 μL 50 mM pH 10.4 carbonate buffer, mix, add 200 μL 15 mM (+)-1-(9-fluorenylethyl)chloroformate in acetone, heat at 45° for 1 h, add 200 μL acetic acid buffer, add 250 μL water, inject an aliquot.

CAPILLARY ELECTROPHORESIS
Capillary: 67 cm × 50 μm CElect H-150 coated capillary (60 cm to detector) (Supelco)
Running buffer: 20 mM pH 4.35 Phosphate buffer containing 20 mM heptakis(2,6-di-O-methyl)-β-cyclodextrin
Voltage/Current: 17.5 kV
Injection: Pressure injection for 3 s
Detector: UV 214
Model: Beckman P/ACE 2100 or 5510
Migration time: 13 (D), 14 (L)
Internal standard: 2-amino-4,6-dimethylpyrimidine (12)
Limit of detection: 20 μM

KEY WORDS
chiral; plasma

REFERENCE
Vogt, C.; Kiessig, S. Separation of D/L-carnitine enantiomers by capillary electrophoresis, *J.Chromatogr.A*, **1996**, *745*, 53–60.

SAMPLE
Matrix: solutions
Analyte: carnitine
Sample preparation: 30 μL Carnitine in water + 30 μL 50 mM pH 10.4 carbonate buffer + 80 μL 15 mM (+)-1-(9-fluorenyl)ethyl chloroformate in acetone, heat at 45° for 1 h, add 90 μL 50 mM pH 4.2 acetate buffer, inject an aliquot.

HPLC VARIABLES
Column: 240 × 4.6 5 μm RP18
Mobile phase: MeCN:buffer 28.5:71.5 (Buffer was 6.8 mL triethylamine in 1 L water, adjust pH to 2.6 with 85% phosphoric acid.)
Flow rate: 2
Injection volume: 20
Detector: F ex 260 em 310

CHROMATOGRAM
Retention time: 15 (D), 17 (L)
Limit of detection: 0.5% of other enantiomer

KEY WORDS
chiral

REFERENCE

Vogt, C.; Georgi, A.; Werner, G. Enantiomeric separation of D/L-carnitine using HPLC and CZE after derivatization, *Chromatographia*, **1995**, *40*, 287–295.

SAMPLE
Matrix: solutions
Analyte: carnitine
Sample preparation: Prepare a solution of carnitine in 50 mM tetrabutylammonium hydroxide in water. 50 μL Carnitine solution + 200 μL 4.5 mM (+)-1-(9-fluorenyl)ethyl chloroformate in acetone, heat at 80° for 25 min, cool, dilute with 4 mL MeCN:buffer 25:75, inject an aliquot. (Buffer was 50 mM KH_2PO_4 containing 5 mM tetrabutylammonium hydroxide adjusted to pH 7.0 with 1 M KOH.)

CAPILLARY ELECTROPHORESIS
Capillary: 60 cm × 50 μm fused-silica
Running buffer: 50 mM KH_2PO_4 adjusted to pH 3.40 with concentrated phosphoric acid
Voltage/Current: 15 kV
Detector: UV 214
Migration time: 33 (D), 36 (L)

KEY WORDS
chiral

REFERENCE

De Witt, P.; Deias, R.; Muck, S.; Galletti, B.; Meloni, D.; Celletti, P.; Marzo, A. High-performance liquid chromatography and capillary electrophoresis of L- and D-carnitine by precolumn diastereomeric derivatization, *J.Chromatogr.B*, **1994**, *657*, 67–73.

4-Fluorobenzoyl Chloride

SAMPLE
Matrix: blood
Analyte: ecgonine methyl ester
Sample preparation: Condition a 300 mg Bond Elut Certify SPE cartridge with 6 mL MeOH, 3 mL water, and 5 mL 10 mM pH 2.0 NaH_2PO_4. 1 mL Plasma + 100 μL 20 μg/mL tropacocaine in water + 3 mL 10 mM pH 2.0 NaH_2PO_4 vortex briefly, add to the SPE cartridge at 0.5 mL/min, dry under vacuum for 1-2 min, wash with 3 mL water at 1 mL/min, dry under vacuum for 3 min, wash with 3 mL 100 mM HCl, dry under vacuum for 3 min, wash rapidly with 6 mL MeOH, dry under vacuum for 5 min, elute with 6 mL dichloromethane:isopropanol:ammonium hydroxide 80:20:2 without vacuum. Evaporate the eluate to dryness under a stream of nitrogen at 50°, reconstitute the residue in 2 mL dichloromethane:isopropanol 80:20. Remove a 1 mL aliquot and evaporate it to dryness under a stream of nitrogen, reconstitute with 1 mL benzene (Caution! Benzene is a carcinogen!), add 200 μL pyridine, add 100 μL 4-fluorobenzoyl chloride, heat at 85° for 1 h, cool to room temperature, add 3 mL 1 M HCl, vortex for 5 min, centrifuge at 800 g for 10 min. Remove the lower aqueous layer and add it to 2.5 mL 2 M sodium carbonate, add 1 mL chloroform, vortex for 5 min, centrifuge at 800 g for 10 min. Remove the organic layer and evaporate it to dryness under a stream of nitrogen at 50°, reconstitute the residue in 300 μL mobile phase, inject a 100 μL aliquot.

HPLC VARIABLES
Column: 250 × 4.6 5 μm Bakerbond Cyanopropyl + 150 × 4.6 5 μm Microsorb silica in series
Mobile phase: MeCN:10 mM pH 3.0 NaH_2PO_4 30:70
Flow rate: 1.1

Injection volume: 100
Detector: UV 235

CHROMATOGRAM
Retention time: 20.5
Internal standard: tropacocaine (16.0)
Limit of quantitation: 260 ng/mL
Limit of detection: 90 ng/mL

OTHER SUBSTANCES
Extracted: cocaine (unaffected by derivatization)

KEY WORDS
rat; plasma; SPE

REFERENCE
Virag, L.; Mets, B.; Jamdar, S. Determination of cocaine, norcocaine, benzoylecgonine and ecgonine methyl ester in rat plasma by high-performance liquid chromatography with ultraviolet detection, *J.Chromatogr.B*, **1996**, *681*, 263–269.

Menthoxyacetyl Chloride

SAMPLE
Matrix: solutions
Analyte: (±)-trans-7,8-dihydroxy-7,8-dihydrobenzo[a]pyrene
Sample preparation: Derivatize with (-)-menthoxyacetyl chloride. (In a related procedure 2 moles (-)-menthoxyacetyl chloride was added slowly with cooling to 1 mole 9,10-dihydroxy-9,10-dihydrophenanthrene in 20 mL pyridine. After standing overnight the mixture was poured into dilute HCl and the derivative collected, washed, and dried (J. Chem. Soc. 1949, 2808). Prepare (-)-menthoxyacetyl chloride by dissolving 90 g (-)-menthoxyacetic acid in 150 mL thionyl chloride, heat at 50° for 3 h (J. Am. Chem. Soc. 1934, 56, 2093), evaporate under reduced pressure to obtain (-)-menthoxyacetyl chloride (bp 100-101°/1.7-1.8 mm Hg).)

HPLC VARIABLES
Column: 300 × 4 μBondapak C18
Mobile phase: MeOH:water 75:25
Detector: UV 254

CHROMATOGRAM
Retention time: 7, 10 (enantiomers)

KEY WORDS
chiral

REFERENCE
Harvey, R.G.; Cho, H. Efficient resolution of the dihydrodiol derivatives of benzo[a]pyrene by high-pressure liquid chromatography of the related (-)-dimenthoxyacetates, *Anal.Biochem.*, **1977**, *80*, 540–546.

Menthyl Chloroformate

+ ROH $\longrightarrow$

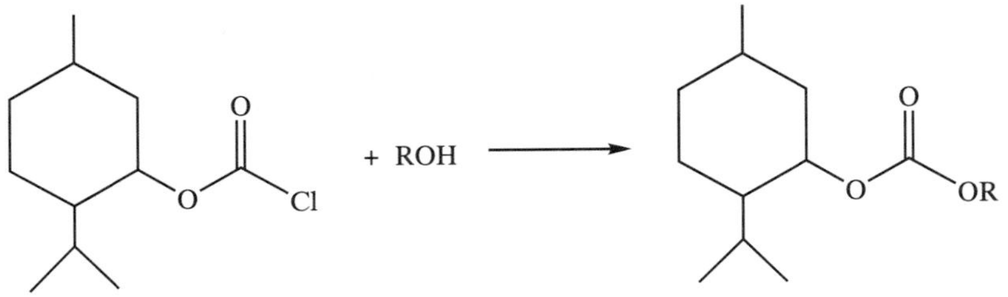

SAMPLE
Matrix: blood
Analyte: warfarin
Sample preparation: 1 mL Plasma + 100 μL 1 M HCl + 5 mL dichloromethane, stir for 10 min, centrifuge. Remove the organic layer and evaporate it to dryness under a stream of nitrogen at 40°, add 100 μL 1 M (-)-1-menthylchloroformate in dichloromethane, add 20 μL triethylamine, heat at 30° for 20 min, centrifuge, wash the supernatant with 3 mL 1 M HCl, evaporate to dryness, reconstitute the residue in 100 μL mobile phase, inject a 50 μL aliquot.

HPLC VARIABLES
Column: 150 × 4.6 5 μm Nucleosil silica
Mobile phase: Heptane:ethyl acetate 93:7 (water concentration 51 ppm)
Flow rate: 1
Injection volume: 50
Detector: UV 310

CHROMATOGRAM
Retention time: 10 (S), 11.5 (R)

KEY WORDS
plasma; chiral; normal phase

REFERENCE
Aycard, M.; Letellier, S.; Maupas, B.; Guyon, F. Determination of (R) and (S) warfarin in plasma by high performance liquid chromatography using precolumn derivatization, *J.Liq.Chromatogr.*, **1992**, *15*, 2175–2182.

7-Methoxycoumarin-3-carbonyl Chloride

+ ROH $\longrightarrow$

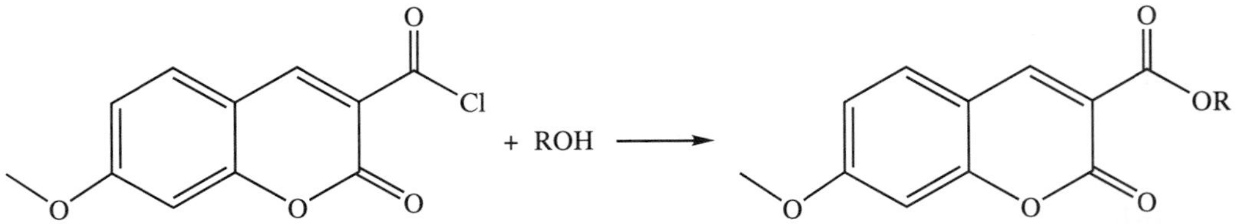

SAMPLE
Matrix: solutions
Analyte: alcohols
Sample preparation: Mix 50 μL of a solution in acetone with 50 μL 10 mM 7-methoxycoumarin-3-carbonyl chloride in benzene (Caution! Benzene is a carcinogen!), heat at 100° for 20 min, cool, add 900 μL MeOH, inject a 20 μL aliquot. (Preparation of 7-methoxycoumarin-3-carbonyl chloride (7-methoxycoumarin-3-carboxylic acid chloride) is as follows. Add a solution of 1.9 g sodium cyanoacetate in 20 mL water to 1.5 g 2-hydroxy-4-methoxybenzaldehyde in 10 mL water and 10 mL 2 M NaOH, heat at 35° for 1 h, filter,

dilute and acidify the filtrate. Collect the solid and wash it, dry to obtain trans-2-hydroxy-4-methoxybenzylidenecyanoacetic acid (mp 196°). Reflux 1.9 g trans-2-hydroxy-4-methoxybenzylidenecyanoacetic acid in 40 mL water for 5 min, filter, acidify and cool the filtrate. Recrystallize the precipitate from dilute EtOH to give 7-methoxycoumarin-3-carboxylic acid as faint greenish-yellow plates (mp 195°). 7-Methoxycoumarin-3-carboxylic acid is also available from Molecular Probes, Eugene OR. Reflux 0.5 g 7-methoxycoumarin-3-carboxylic acid and 1 mL thionyl chloride in 3 mL chloroform until the reaction ceases, add light petroleum (bp 60-80°). Collect the precipitate and recrystallize it from chloroform/light petroleum to give 7-methoxycoumarin-3-carbonyl chloride as pale yellow needles (mp 143°) (J. Chem. Soc. 1949, S12).)

HPLC VARIABLES
Column: 150 × 4 5 μm TSK gel ODS-120A (Toyo Soda)
Mobile phase: MeOH:water:acetic acid 70:30:2 (A) or MeOH:THF:acetic acid 92:8:2 (B)
Flow rate: 0.5 (A), 1 (B)
Injection volume: 20
Detector: F ex 355 em 400

CHROMATOGRAM
Retention time: 10 (benzyl alcohol, cyclohexanol (A)), 12 (2-pentanol (A)), 14 (1-pentanol (A)), 10 (cholesterol (B)), 10.5 (cholestanol (B))
Limit of detection: 0.12-0.50 pmole

REFERENCE
Hamada, C.; Iwasaki, M.; Kuroda, N.; Ohkura, Y. 3-Chloroformyl-7-methoxycoumarin as a fluorescent derivatization reagent for alcoholic compounds in liquid chromatography and its use for the assay of 17-oxosteroids in urine, *J.Chromatogr.*, **1985**, *341*, 426–431.

SAMPLE
Matrix: urine
Analyte: 17-oxosteroids
Sample preparation: 500 μL Urine + 200 μL 2 M pH 5.2 acetate buffer + 10 μL glucuronidase/arylsulfatase (95400 U/mL β-glucuronidase and 5110 U/mL arylsulfatase, Sigma), heat at 37° overnight, add 5 mL dichloromethane, vortex for 1 min. Remove a 2 mL aliquot of the organic layer and evaporate it to dryness under reduced pressure below 30°, reconstitute the residue in 500 μL acetone, vortex for 1 min. Remove a 50 μL aliquot and add it to 50 μL 10 mM 7-methoxycoumarin-3-carbonyl chloride in benzene (Caution! Benzene is a carcinogen!), heat at 100° for 20 min, cool, add 900 μL MeOH, inject a 20 μL aliquot. (Preparation of 7-methoxycoumarin-3-carbonyl chloride (7-methoxycoumarin-3-carboxylic acid chloride) is described above.)

HPLC VARIABLES
Column: 150 × 4 5 μm TSK gel ODS-120A (Toyo Soda)
Mobile phase: MeOH:water:acetic acid 80:20:2
Flow rate: 1
Injection volume: 20
Detector: F ex 355 em 400

CHROMATOGRAM
Retention time: 10 (eticholanolone), 12 (androsterone), 16 (dehydroepiandrosterone)
Limit of detection: 87-162 ng/mL

REFERENCE
Hamada, C.; Iwasaki, M.; Kuroda, N.; Ohkura, Y. 3-Chloroformyl-7-methoxycoumarin as a fluorescent derivatization reagent for alcoholic compounds in liquid chromatography and its use for the assay of 17-oxosteroids in urine, *J.Chromatogr.*, **1985**, *341*, 426–431.

6-Methoxy-2-methylsulfonylquinoline-4-carbonyl Chloride

SAMPLE
Matrix: solutions
Analyte: alcohols
Sample preparation: Mix a 200 μL aliquot of a 100 μM solution in benzene with 600 μL 4 mM 6-methoxy-2-methylsulfonylquinoline-4-carbonyl chloride in benzene and 100 μL 1 M pyridine in benzene (Caution! Benzene is a carcinogen!), heat at 100° for 1 h, cool, dilute 20-fold with MeCN, inject a 10 μL aliquot. (Preparation of 6-methoxy-2-methyl-sulfonylquinoline-4-carbonyl chloride is as follows. Dissolve 10 g p-anisidine and 15.6 g diethyl ketomalonate (ethyl mesoxalate) in 10 mL acetic acid, boil gently for 3 h, add 10 mL EtOH while warm, let stand in the refrigerator overnight, filter. (It may be necessary to use the hydrate of diethyl ketomalonate, diethyl dihydroxymalonate. Prepare this compound from equimolar amounts of diethyl ketomalonate and water, recrystallize from chloroform (mp 56-57°) (Org. Syn. 1932, I, 266).) Wash the solid with 15 mL EtOH and 15 mL ether, recrystallize several times from EtOH:water 75:25 with charcoal to yield ethyl 2,3-dihydro-3-hydroxy-5-methoxy-2-oxoindole-3-carboxylate as colorless crystals (mp 193-194°). Dissolve ethyl 2,3-dihydro-3-hydroxy-5-methoxy-2-oxoindole-3-carboxylate in 12 times as much 1 M KOH, heat on a water bath, filter. Acidify the filtrate with HCl, carbon dioxide is evolved. Reflux this weakly acid solution and add a small amount of a dilute solution of iron chloride, vigorously pull air through the mixture (?) until the solution turns deep red, cool, recrystallize the product from water to obtain 5-methoxyisatin as deep-brown-red needles (mp 201-202°). Gently heat 5-methoxyisatin with 2 equivalents of acetic anhydride for 30 min, while still warm add an equal volume of benzene (Caution! Benzene is a carcinogen!), pour into copious amounts of boiling benzene, extract the residue repeatedly with hot benzene. Treat the benzene solution with charcoal, filter, concentrate to a small volume, cool in ice, recrystallize from chloroform/benzene to obtain 1-acetyl-5-methoxyisatin as red crystals (mp 144-145°) (Ber. 1921, 54B, 3079; Chem. Abs. 1922, 16, 1771). Add 3 g 1-acetyl-5-methoxyisatin to a boiling mixture of 10 g caustic soda solution (d = 1.17) and 100 mL water, heat gently, cool, acidify with sulfuric acid, stir, filter. Wash the solid with acetone and ether, recrystallize from acetic acid to obtain 1,2-dihydro-6-methoxy-2-oxoquinoline-4-carboxylic acid (mp 326° d) (Ber. 1921, 54B, 3090; Chem. Abs. 1922, 16, 1772). Reflux 1,2-dihydro-6-methoxy-2-oxoquinoline-4-carboxylic acid in phosphorus oxychloride to obtain 2-chloro-6-methoxyquinoline-4-carboxylic acid. Boil 2-chloro-6-methoxyquinoline-4-carboxylic acid with 1.2 equivalents thiourea in isopropanol for 15-30 min, dilute with water to obtain 1,2-dihydro-6-methoxy-2-thioxoquinoline-4-carboxylic acid (mp 271-273° d) (cf. Chem. Pharm. Bull. 1980, 28, 49). Dissolve 1 g 1,2-dihydro-6-methoxy-2-thioxoquinoline-4-carboxylic acid in 10 mL 1 M NaOH, add 2.3 g iodomethane, reflux for 1.5 h, cool, acidify with 10% HCl to obtain 6-methoxy-2-methylthioquinoline-4-carboxylic acid (mp 205-206°). React 6-methoxy-2-methylthioquinoline-4-carboxylic acid with 6.6% potassium permanganate solution (cf. Yakugaku Zasshi 1969, 89, 74) to obtain 6-methoxy-2-methylsulfonylquinoline-4-carboxylic acid (mp 215-216°) (Chem. Pharm. Bull. 1992, 40, 1322). Add 400 mg 6-methoxy-2-methylsulfonylquinoline-4-carboxylic acid to 5 mL thionyl chloride, reflux at 80° with stirring for 30 min, evaporate to dryness under a stream of nitrogen under reduced pressure, dissolve the residue in anhydrous benzene, filter, add petroleum ether, collect the precipitate, recrystallize from benzene/petroleum ether to obtain 6-methoxy-2-methylsulfonylquinoline-4-carbonyl chloride as yellow needles (mp 175-176°).)

HPLC VARIABLES
Column: 150 × 4.6 5 μm Cosmosil C18-AR (Nacalai Tesque, Kyoto)
Mobile phase: MeCN:water 70:30
Flow rate: 0.5
Injection volume: 10
Detector: F ex 355 em 457

CHROMATOGRAM
Retention time: 6.6 (2-propanol), 6.8 (1-propanol), 6.9 (phenol), 7.9 (benzyl alcohol), 8.0 (2-methyl-2-propanol), 8.5 (2-methyl-1-propanol), 8.7 (1-butanol), 9.9 (2-methyl-2-butanol), 10.1 (3-pentanol), 10.6 (3-methyl-1-butanol), 11.1 (1-pentanol), 11.5 (cyclohexanol), 14.9 (1-hexanol)
Limit of detection: 0.07-50 pmole

REFERENCE
Yoshida, T.; Moriyama, Y.; Taniguchi, H. 6-Methoxy-2-methylsulfonylquinoline-4-carbonyl chloride as a fluorescence derivatization reagent for alcohols in high-performance liquid chromatography, *Anal.Sci.*, **1992**, *8*, 355–359.

1-Naphthoyl Chloride

SAMPLE
Matrix: blood
Analyte: digoxin and metabolites
Sample preparation: Condition a 1 mL Cyclobond I β-cyclodextrin SPE cartridge (Astec) with 2 mL MeOH, 2 mL MeCN, 2 mL isopropanol, and 2 mL water (SPE cartridge A). Condition a 1 mL Cyclobond I β-cyclodextrin SPE cartridge (Astec) with 2 mL MeOH, 2 mL MeCN, and 2 mL dichloromethane (SPE cartridge B). Condition a 1 mL Bond Elut C1 SPE cartridge with 2 mL MeOH and 2 mL MeCN (SPE cartridge C). 1 mL Serum + 10 ng digitoxin + 1 mL water, add to SPE cartridge A, wash with 2 mL water, wash with 1 mL MeOH:7.5 mM pH 7.0 potassium phosphate buffer 20:80, wash with 3 mL water, wash with 1 mL isopropanol:water 10:90, dry under vacuum for 5 min, wash with ten 100 μL aliquots of dichloromethane, dry under vacuum for 5 min, elute with 1 mL isopropanol. Evaporate the eluate to dryness under a stream of nitrogen at room temperature, add 50 μL 10% 4-dimethylaminopyridine in MeCN then 50 μL 4% 1-naphthoyl chloride in MeCN under nitrogen in a glove box (relative humidity <26%), mix thoroughly, heat at 50° for 1 h, centrifuge briefly, evaporate under a stream of nitrogen, add 2 mL 5% pH 10.0 sodium bicarbonate solution, shake for 1 min, add 2 mL chloroform, shake, centrifuge. Remove the organic layer and wash it with 2 mL 5% sodium bicarbonate, wash twice with 2 mL portions of 50 mM HCl, evaporate to dryness under a stream of nitrogen at room temperature, reconstitute with 200 μL dichloromethane, add to SPE cartridge B, wash with eight 100 μL aliquots of dichloromethane, elute with 1 mL MeOH. Evaporate the eluate to dryness under a stream of nitrogen, reconstitute with 250 μL MeCN, add to SPE cartridge C, rinse container with 250 μL MeCN, add rinse to the SPE cartridge, add 500 μL MeCN to the SPE cartridge. Collect all the eluates and evaporate them to dryness under a stream of nitrogen, reconstitute with mobile phase, inject an aliquot. (Purify 4-dimethylaminopyridine by passing a 30% solution in MeCN through a layer of silica gel covered with a layer of activated charcoal, evaporate the filtrate under reduced pressure, store the residue in a desiccator. Immerse glassware in sulfuric acid:nitric acid

80:20 for 24 h, wash with water, treat with 1% Surfasil (Pierce) in toluene, rinse with water,dry in an oven.)

HPLC VARIABLES
Guard column: 15 × 3.2 7 μm silica (Applied Biosystems)
Column: 150 × 4.6 3 μm Spherisorb silica
Mobile phase: Hexane:dichloromethane:MeCN:MeOH 36:6.3:5.4:0.2
Flow rate: 1.6
Injection volume: 20
Detector: F ex 217 em 340

CHROMATOGRAM
Retention time: 5.5 (digoxigenin), 6.5 (digoxigenin monodigitoxoside), 8.5 (digoxigenin bis-digitoxoside), 10 (digoxin), 12 (dihydrodigoxin)
Internal standard: digitoxin (9.5)
Limit of detection: 0.25 ng/mL

OTHER SUBSTANCES
Extracted: metabolites
Non-interfering: acetaminophen, acetazolamide, acyclovir, albuterol, allopurinol, amioda-rone, amitriptyline, amoxicillin, ampicillin, aspirin, atenolol, atropine, azathioprine, bu-metanide, calcitriol, captopril, carbamazepine, cefazolin, cefoperazone, ceftazidime, cefti-zoxime, cephalexin, chlordiazepoxide, ciprofloxacin, clavulanic acid, clindamycin, clonidine, clotrimazole, codeine, conjugated estrogens, cyclophosphamide, diazepam, di-phenhydramine, dipyridamole, dobutamine, docusate sodium, dopamine, enalapril, eryth-romycin, famotidine, fluconazole, furosemide, gemfibrozil, gentamicin, glyburide, heparin, hydralazine, hydrochlorothiazide, ibuprofen, ipratropium bromide, isosorbide dinitrate, isradipine, labetalol, lidocaine, lorazepam, lovastatin, medroxyprogesterone acetate, me-peridine, metoclopramide, metolazone, metoprolol, midazolam, minoxidil, morphine, nic-otine, nifedipine, nitroglycerin, norepinephrine, nystatin, oxybutynin, oxycodone, pentox-iphylline, phenytoin, piroxicam, prednisone, procainamide, procaine, promethazine, propoxyphene, ranitidine, sotalol, spironolactone, sulbactam, sulfamethoxazole, sulfisox-azole, temazepam, tetracycline, timolol, tobramycin, triamcinolone acetonide, triamter-ene, trimethoprim, vancomycin, verapamil, warfarin

KEY WORDS
normal phase; serum; SPE; pharmacokinetics

REFERENCE
Tzou, M.-C.; Sams, R.A.; Reuning, R.H. Specific and sensitive determination of digoxin and metabolites in human serum by high-performance liquid chromatography with cyclodextrin solid-phase extrac-tion and precolumn fluorescence derivatization, *J.Pharm.Biomed.Anal.*, **1995**, *13*, 1531–1540.

SAMPLE
Matrix: plant extract
Analyte: alkaloids
Sample preparation: Prepare a solution containing 1.2 mg plant extract, 0.08 μL thionyl chloride, 4.22 μL 1-naphthoyl chloride, and 25 μg IS in 1 mL MeCN, heat at 88° for 2 h, cool to room temperature for 15 min, inject an aliquot.

HPLC VARIABLES
Guard column: 20 × 4 5 μm HP C18
Column: 150 × 3.9 4 μm Nova-Pak C18
Mobile phase: MeOH:diethylamine 100:0.2
Flow rate: 1.2
Injection volume: 20
Detector: UV 224

CHROMATOGRAM
Retention time: 2 (verticinone), 3 (ebeiedinone), 6 (isoverticine), 8 (verticine), 10 (ebeiedine)
Internal standard: solanidine (18)

REFERENCE

Ding, K.; Lin, G.; Ho, Y.-P.; Cheng, T.Y.; Li, P. Prederivatization and high-performance liquid chromatographic analysis of alkaloids of bulbs of Fritillaria, *J.Pharm.Sci.*, **1996**, *85*, 1174–1179.

SAMPLE

Matrix: urine, feces

Analyte: digoxin

Sample preparation: Urine. Place 1 mL 100 ng/mL digitoxin in isopropanol in a tube and evaporate. Add 1 mL urine + 2 mL dichloromethane, shake by hand 4 times, centrifuge 1650 g. Remove organic layer and wash it twice with 2 mL 5% sodium bicarbonate solution, evaporate under nitrogen at 50°. Add 25 mg 4-dimethylaminopyridine and 10 μL 1-naphthoyl chloride, add 100 μL MeCN, vortex thoroughly, place in water bath at 50° for 1 h, centrifuge, evaporate at 50° under nitrogen. Add 2 mL 5% sodium bicarbonate solution, shake mechanically for 5 min, add 2 mL chloroform, shake by hand. Remove organic layer and wash it twice with 2 mL 5% sodium bicarbonate solution, wash three times with 0.05 M HCl containing 5% NaCl, evaporate chloroform, dissolve residue in mobile phase. Feces. Dilute 5:1 (v/w) with 5 μg/mL clindamycin in water to stop bacterial metabolism, homogenize with mechanical shaking for 15 min. Evaporate 1 mL 100 ng/mL digitoxin in isopropanol into a tube, weigh ca. 1 g homogenate into the tube, add 1 mL water, vortex 30 s, shake 15 min, centrifuge 1 h. Pour off supernatant and extract it with 2 mL dichloromethane. Wash the extract twice with 2 mL 5% sodium bicarbonate solution, evaporate under nitrogen at 50°. Add 25 mg 4-dimethylaminopyridine and 10 μL 1-naphthoyl chloride, add 100 μL MeCN, vortex thoroughly, place in water bath at 50° for 1 h, centrifuge, evaporate at 50° under nitrogen. Add 2 mL 5% sodium bicarbonate solution, shake mechanically for 5 min, add 2 mL chloroform, shake by hand. Remove organic layer and wash it twice with 2 mL 5% sodium bicarbonate solution, wash three times with 0.05 M HCl containing 5% NaCl, evaporate chloroform, dissolve residue in mobile phase.

HPLC VARIABLES

Column: 150 × 4.6 3 μm Adsorbosphere SI

Mobile phase: Hexane:dichloromethane:MeCN 6:1:1

Flow rate: 1.8-2

Injection volume: 20-175

Detector: F ex 217 em 340 cut-off filter (372 nm max)

CHROMATOGRAM

Retention time: 9.4

Internal standard: digitoxin (8.1)

Limit of detection: 5 ng/mL (urine), 50 ng/g (feces)

OTHER SUBSTANCES

Simultaneously analyzed: metabolites

KEY WORDS

normal phase

REFERENCE

Shepard, T.A.; Hui, J.; Chandrasekaran, A.; Sams, R.A.; Reuning, R.H.; Robertson, L.W.; Caldwell, J.H.; Donnerberg, R.L. Digoxin and metabolites in urine and feces: a fluorescence derivatization-high-performance liquid chromatographic technique, *J.Chromatogr.*, **1986**, *380*, 89–98.

d-2-(2-Naphthyl)propionyl Chloride

SAMPLE
Matrix: bulk
Analyte: diltiazem
Sample preparation: Heat 300 mg diltiazem hydrochloride in 10 mL 5 mM HCl at 100° for 50 min, cool, make up to 20 mL with water. Remove a 4 mL aliquot and add it to 1.6 g NaCl, add 1.5 mL 1 M NaOH, add 20 mL dichloromethane, shake. Remove the organic layer and filter it, evaporate a 10 mL aliquot of the filtrate to dryness under reduced pressure, dry the residue under reduced pressure for 1 h, reconstitute the residue in dry dichloromethane, add 250 μL dry pyridine, add 75 mg d-2-(2-naphthyl)propionyl chloride in dry dichloromethane, let stand at room temperature for 15 min, add 5 mL MeOH, make up to 50 mL with MeCN, inject an aliquot. (Diltiazem is hydrolyzed to deacetyl-diltiazem and this compound is then acylated on the OH with d-2-(2-naphthyl)propionyl chloride. Synthesis of d-2-(2-naphthyl)propionyl chloride is as follows. Prepare methyl 2-naphthylacetate from 2-naphthylacetic acid in a conventional fashion. Add 4.8 g 50% NaH in oil to 20.1 g methyl 2-naphthylacetate and 10 mL MeI in 200 mL dimethoxyethane over 45 min, stir overnight, add 5 mL EtOH, evaporate to about two thirds volume under reduced pressure, add a solution of 10 g NaOH in 20 mL water, heat at 50° for 5 h, add water, wash with diethyl ether, acidify the aqueous layer with dilute HCl, extract with heptane:diethyl ether 50:50, evaporate the organic layer to dryness to obtain 2-(2-naphthyl)propionic acid as white crystals. Add 32.6 g cinchonidine suspended in 100 mL MeOH to 22.2 g 2-(2-naphthyl)propionic acid in 100 mL acetone, boil down to 120 mL, cool, add 130 mL diethyl ether, filter, recrystallize the crystals twice more to obtain the cinchonidine salt (mp 146-147°). Suspend the cinchonidine salt in diethyl ether:heptane 50:50, add dilute HCl, shake. Remove the organic layer and wash it with water, evaporate to dryness, recrystallize from acetone/heptane to obtain d-2-(2-naphthyl)propionic acid (mp 140-142°). Suspend 4 g d-2-(2-naphthyl)propionic acid in benzene (Caution! Benzene is a carcinogen!), add 6 mL oxalyl chloride, stir at 35-40° for 1 h, evaporate to dryness under reduced pressure below 50°, dissolve the residue in benzene, evaporate to dryness, repeat this procedure, recrystallize from hexane to obtain d-2-(2-naphthyl)propionyl chloride (mp 39-41°, $[\alpha]_D^{20}$ +73.6° (c = 1, chloroform)).)

HPLC VARIABLES
Column: 200 × 4 Nucleosil 5C18
Mobile phase: MeCN:10 mM pH 6.6 ammonium acetate 90:10
Flow rate: 1.5
Injection volume: 20
Detector: UV 254

CHROMATOGRAM
Retention time: 12 (l), 15 (d)
Limit of detection: 0.1%

KEY WORDS
chiral

REFERENCE
Shimizu, R.; Ishii, K.; Tsumagari, N.; Tanigawa, M.; Matsumoto, M.; Harrison, I.T. Determination of optical isomers in diltiazem hydrochloride by high-performance liquid chromatography, *J.Chromatogr.*, **1982**, *253*, 101–108.

N-1-(2-Naphthylsulfonyl)-2-pyrrolidinecarbonyl Chloride

+ ROH ⟶

SAMPLE
Matrix: bulk
Analyte: diltiazem
Sample preparation: Mix 100 mg diltiazem hydrochloride with 5 mL 1 M NaOH, make
up to 50 mL with MeOH, shake, let stand at room temperature for 30 min. Remove a 10
mL aliquot and remove the MeOH by evaporation under reduced pressure, add 10 mL
chloroform, add 20 mL water, shake. Remove a 1.5 mL aliquot of the organic layer and
evaporate it to dryness under reduced pressure, dry the residue under reduced pressure
for 1 h, reconstitute the residue in 500 μL dry dichloromethane, add 20 μL pyridine, add
20 mg (S)-(-)-N-1-(2-naphthylsulfonyl)-2-pyrrolidinecarbonyl chloride, mix, let stand at
room temperature for 15 min, make up to 20 mL with chloroform, inject an aliquot. (The
diltiazem is deacetylated with NaOH and the free alcohol reacts with the reagent. Prepare
(S)-(-)-N-1-(2-naphthylsulfonyl)-2-pyrrolidinecarbonyl chloride as follows. Slowly add a solution of 24.5 g 2-naphthalenesulfonyl chloride in 186 mL diethyl ether to a solution of
10.35 g L-proline and 37.25 g potassium carbonate in 216 mL water stirred at 0°, stir for
2 days. Remove the aqueous layer and wash it with diethyl ether, acidify to pH 2 with
10% HCl, extract with 500 mL ethyl acetate. Wash the organic layer with saturated
aqueous NaCl and dry it over anhydrous sodium sulfate, evaporate to dryness, recrystallize from benzene to give (S)-(-)-N-1-(2-naphthylsulfonyl)-2-pyrrolidinecarboxylic acid
(mp 133-135°). Add 17.5 mL oxalyl chloride to a solution of 15.6 g (S)-(-)-N-1-(2-naphthyl-
sulfonyl)-2-pyrrolidinecarboxylic acid in benzene, stir at 40-50° overnight, evaporate to
dryness under reduced pressure, recrystallize from benzene/hexane to give (S)-(-)-N-1-(2-
naphthylsulfonyl)-2-pyrrolidinecarbonyl chloride (mp 107-109°; $[\alpha]_D^{20}$ -81.6° (c = 1,
chloroform)).)

HPLC VARIABLES
Column: 150 × 4.6 Zorbax Sil
Mobile phase: Chloroform:dichloromethane:MeOH:diethylamine 100:25:15:0.05
Injection volume: 20
Detector: UV 254

CHROMATOGRAM
Retention time: 11 (SS), 13 (SR), 15 (RR), 17 (RS)
Limit of detection: 0.1%

KEY WORDS
normal phase; chiral

REFERENCE
Shimizu, R.; Kakimoto, T.; Ishii, K.; Fujimoto, Y.; Nishi, H.; Tsumagari, N. New derivatization reagent
for the resolution of optical isomers in diltiazem hydrochloride by high-performance liquid chromatography, *J.Chromatogr.*, **1986**, *357*, 119–125.

Naproxen Chloride

SAMPLE

Matrix: blood

Analyte: venlafaxine

Sample preparation: 1 mL Plasma + 50 μL 22.5 μg/mL IS in water + 200 μL saturated sodium borate, vortex, add 5 mL isopropyl ether (Caution! Isopropyl ether readily forms explosive peroxides!), shake for 15 min, centrifuge at 2500 rpm for 10 min. Remove a 4.5 mL aliquot of the organic phase and add it to 400 μL 10 mM HCl, shake for 15 min, centrifuge at 2500 rpm for 10 min. Discard the organic phase and add 1 mL saturated sodium borate solution to the aqueous layer, vortex, add 2 mL isopropyl ether, shake for 15 min, centrifuge at 2500 rpm for 5 min. Remove the organic layer and evaporate it to dryness under a stream of nitrogen, add 5-10 mg anhydrous sodium carbonate powder to the residue, add 200 μL 150 μg/mL naproxen chloride in dichloromethane, vortex, let stand in the dark for 20 h, evaporate to dryness under a stream of nitrogen, reconstitute with 1 mL water, vortex, add 2 mL chloroform, shake for 10 min, centrifuge at 2500 rpm for 5 min. Remove the chloroform layer and add it to 2 mL water, shake for 10 min, centrifuge at 2500 rpm for 5 min. Remove the organic layer and evaporate it to dryness under a stream of nitrogen, reconstitute the residue in 400 μL initial mobile phase, inject an 11 μL aliquot. (Prepare naproxen chloride as follows. React 8.1 g naproxen in 40 mL dichloromethane with 25 g oxalyl chloride at room temperature for 22 h, evaporate to dryness under reduced pressure at 42°, reconstitute the residue with 35 mL dichloromethane, slowly add 1.12 L n-hexane, chill to precipitate crystals, decant solvent carefully and rapidly, dry crystals rapidly under nitrogen, dry crystals of naproxen chloride (mp 95-7°) over calcium chloride in a vacuum desiccator.)

HPLC VARIABLES

Guard column: 20 × 4.6 Supelguard LC-8-DB (Supelco)

Column: 150 × 4.6 5 μm Supelcosil LC-8-DB

Mobile phase: Gradient. MeCN:buffer from 50:50 to 54:46 over 30 min, to 55:45 over 6 min, to 57:43 over 14 min, maintain at 57:43 for 3 min, return to initial conditions over 3 min. (Buffer was 100 mM KH_2PO_4 adjusted to pH 3.0 with 85% orthophosphoric acid, add 0.07% triethylamine to achieve a final pH of 3.25.)

Flow rate: 0.8 for 30 min, to 2.0 over 6 min, maintain at 2.0 for 17 min, return to 0.8 over 3 min

Injection volume: 11

Detector: UV 229

CHROMATOGRAM

Retention time: 16.53 (R), 18.03 (S)

Internal standard: 1-[2-(dimethylamino)-1-(2-chlorophenyl)ethyl]cyclohexanol (Wy-45,818) (19.74, 22.38 (enantiomers))

Limit of quantitation: 50 ng/mL

Limit of detection: 25 ng/mL (22 μL injection)

OTHER SUBSTANCES

Extracted: metabolites

KEY WORDS
chiral; dog; rat; human; plasma; pharmacokinetics

REFERENCE
Wang, C.P.; Howell, S.R.; Scatina, J.; Sisenwine, S.F. The disposition of venlafaxine enantiomers in dogs, rats, and humans receiving venlafaxine, *Chirality*, **1992**, *4*, 84–90.

4-Nitrobenzoyl Chloride

When digoxin and related compounds are derivatized all the hydroxy groups except that at C14 are derivatized.

SAMPLE
Matrix: blood, tissue
Analyte: carbohydrates
Sample preparation: Homogenize lens tissue in 4 (human) or 1 (rat) mL 2 mg/mL sodium fluoride. Dilute 600 μL frozen and thawed erythrocytes with 400 μL water. Filter (Amicon Centrifree) 1 mL homogenate, plasma, or diluted erythrocytes while centrifuging at 2400 g for 30 min. 200 μL Filtrate + 10 μL 1 mg/mL IS, mix, lyophilize, add 200 μL 100 mg/mL p-nitrobenzoyl chloride in pyridine, heat at 60° for 1 h, add 1 drop of water, add 2 mL chloroform. Wash mixture twice with 2 mL 5% sodium bicarbonate and twice with 3 mL 1 M HCl by vortexing for 1 min and centrifuging for 30 s, inject a 50 μL aliquot of the organic layer.

HPLC VARIABLES
Column: 250 × 4.6 6 μm Zorbax SIL
Mobile phase: Hexane:chloroform:MeCN 10:3:1.9 containing 0.1% water
Column temperature: 35
Flow rate: 1.5
Injection volume: 50
Detector: UV 260

CHROMATOGRAM
Retention time: 13 (dextrose), 17 (myo-inositol), 17.5 (mannitol), 19 (D-sorbitol), 20 (fructose)
Internal standard: perseitol (α-mannoheptitol) (26)
Limit of detection: 1-2 ng

KEY WORDS
plasma; erythrocytes; lens; human; rat; normal phase

REFERENCE
Petchey, M.; Crabbe, M.J.C. Analysis of carbohydrates in lens, erythrocytes, and plasma by high-performance liquid chromatography of nitrobenzoate derivatives, *J.Chromatogr.*, **1984**, *307*, 180–184.

SAMPLE
Matrix: bulk, formulations
Analyte: digitalis glycosides
Sample preparation: Ampoules. Add the contents of 1 ampoule (2 mL) to 15 mL 2% sodium bicarbonate solution, extract 5 times with 10 mL portions of chloroform:isopropanol 60:40, wash each extract with the same 10 mL portion of water, wash with another 10 mL portion of water. Combine the organic layers and evaporate them to dryness, transfer the residue to another tube with two 1 mL portions of chloroform:pyridine 10:1, evaporate to dryness under reduced pressure at 50°, add 200 μL reagent, shake well, let stand at room temperature for 10 min, evaporate to dryness under reduced pressure at 50°,

flush the tube with a stream of air or nitrogen, add 2 mL 5% sodium carbonate solution containing 2.5 mg/mL 4-dimethylaminopyridine, shake or sonicate for 5 min, extract with 2 mL chloroform. Wash the extract with 2 mL 5% sodium bicarbonate solution, wash twice with 3 mL portions of 50 mM HCl containing 5% NaCl, inject a 20 μL aliquot. Bulk. Prepare a solution in pyridine containing ⩽10 mg/mL. Add 150 μL reagent to 50 μL solution, shake well, let stand at room temperature for 10 min, evaporate to dryness under reduced pressure at 50°, flush the tube with a stream of air or nitrogen, add 2 mL 5% sodium carbonate solution containing 2.5 mg/mL 4-dimethylaminopyridine, shake or sonicate for 5 min, extract with 2 mL chloroform. Wash the extract with 2 mL 5% sodium bicarbonate solution, wash twice with 3 mL portions of 50 mM HCl containing 5% NaCl, inject a 20 μL aliquot. (Prepare reagent fresh each day by dissolving 100 mg 4-nitrobenzoyl chloride in 1 mL pyridine with gentle warming.)

HPLC VARIABLES
Column: 200 × 3 5 μm Merckosorb SI 60
Mobile phase: n-Hexane:chloroform:MeCN 30:10:9
Flow rate: 1.5
Detector: UV 254

CHROMATOGRAM
Retention time: 1.6 (gitoxigenin), 1.8 (digitoxigenin), 2.1 (diginatigenin), 2.7 (digoxigenin), 2.8 (gitaloxigenin), 4 (gitoxin), 4.3 (digitoxin), 4.8 (diginatin), 5.9 (digoxin), 6.5 (gitaloxin), 8.6 (lanatoside B), 9.1 (lanatoside A), 10.2 (lanatoside D), 12.7 (lanatoside C), 14.1 (lanatoside E)
Limit of detection: 11 ng/mL (100 μL injection)

OTHER SUBSTANCES
Interfering: 20

KEY WORDS
ampoules; normal phase

REFERENCE
Nachtmann, F.; Spitzy, H.; Frei, R.W. Rapid and sensitive high-resolution procedure for digitalis glycoside analysis by derivatization liquid chromatography, *J.Chromatogr.*, **1976**, *122*, 293–303.

SAMPLE
Matrix: formulations
Analyte: carbohydrates
Sample preparation: Remove the water from 10 μL syrup under reduced pressure for 10 min, reconstitute with 2 mL pyridine. Remove a 25 μL aliquot and add it to 75 μL reagent, shake well, let stand at room temperature for 10 min, evaporate to dryness under reduced pressure at room temperature, flush the tube with a stream of air or nitrogen, add 2 mL 5% sodium carbonate solution containing 2.5 mg/mL 4-dimethylaminopyridine, shake or sonicate for 5 min, extract with 2 mL chloroform. Wash the extract with 2 mL 5% sodium bicarbonate solution, wash twice with 3 mL portions of 50 mM HCl containing 5% NaCl, inject an aliquot. (Prepare reagent by dissolving 100 mg 4-nitrobenzoyl chloride in pyridine with gentle warming.)

HPLC VARIABLES
Column: 150 × 3 5 μm LiChrosorb SI 60
Mobile phase: n-Hexane:chloroform:MeCN 10:3:1.9 containing 0.1% water
Flow rate: 1.4
Injection volume: 50
Detector: UV 260

CHROMATOGRAM
Retention time: 1.5 (propylene glycol), 3 (glycerin), 5 (dextrose), 6.5 (fructose), 8 (sorbitol), 9 (fructose), 15 (saccharose)

KEY WORDS
syrup; normal phase

REFERENCE

Nachtmann, F.; Budna, K.W. Sensitive determination of derivatized carbohydrates by high-performance liquid chromatography, *J.Chromatogr.*, **1977**, *136*, 279–287.

SAMPLE

Matrix: urine

Analyte: digoxin

Sample preparation: 10 mL Urine + 2 mL 1 M HCl (check pH is 1-2), heat at 37° for 3 h, add 5 mL pH 6.5 phosphate buffer, add 2 mL 1 M NaOH (check pH is 6.5-7.0). Add to a 20 cm Extrelut column, rinse flask with 3 mL water, add rinsings to column, dry for 15 min, elute with 40 mL dichloromethane, evaporate eluent to dryness, dry over concentrated sulfuric acid. Prepare a 100 mg/mL solution of 4-nitrobenzoyl chloride (4-NBP) in dry pyridine with gentle heating. Use immediately. Dissolve residue from column in 30 µL dry pyridine, add 20 µL 2 mg/mL digitoxigenin in pyridine, add 300 µL 4-NBP solution, shake well, heat at 70° for 1 h, add 2 mL 5% sodium bicarbonate, shake until precipitate has dissolved, add 2 mL chloroform, shake, centrifuge, repeat extraction twice. Combine chloroform layers, wash three times with 2 mL 1 M HCl, inject an aliquot of chloroform solution directly.

HPLC VARIABLES

Column: 200 × 4 Hibar 5 µm Lichrosorb Si 60

Mobile phase: n-Hexane:dichloromethane:methanol 82.9:14.2:2.9

Flow rate: 1.2

Injection volume: 20

Detector: UV 258

CHROMATOGRAM

Retention time: 12

Internal standard: digitoxigenin (8)

Limit of detection: 1 µg/mL

KEY WORDS

normal phase; SPE; Digoxin is hydrolysed to digoxigenin and determined as its 4-NBP derivative.

REFERENCE

Jakobsen, P.; Waldorff, S. Determination of digoxin, digoxigenin and dihydrodigoxigenin in urine by extraction, derivatization and high-performance liquid chromatography, *J.Chromatogr.*, **1986**, *382*, 349–354.

2-Quinoxaloyl Chloride

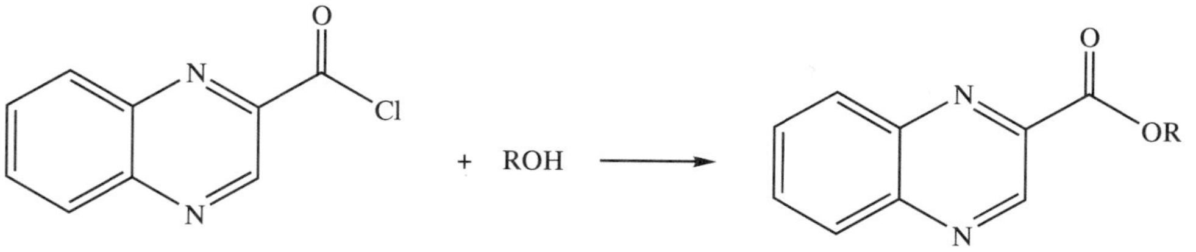

SAMPLE

Matrix: bulk

Analyte: α-hydroxycarboxylic acids

Sample preparation: Mix 500 µg hydroxyacid with 500 µg 2-quinoxaloyl chloride in 1 mL MeCN (if the solubility is poor use MeCN:water 50:50), add 1 drop triethylamine, mix, let stand at room temperature for 30 min, dilute 5-fold with MeCN, inject a 5 µL aliquot.

HPLC VARIABLES
Column: 250 × 4.6 5 μm Cyclobond I 2000 β-cyclodextrin-bonded (Technicol, Stockport UK)
Mobile phase: MeCN:MeOH:triethylamine:glacial acetic acid 100:0:0.5:0.25 (A) or 75:25:1:0.5 (B)
Flow rate: 1
Injection volume: 5
Detector: UV 315

CHROMATOGRAM
Retention time: k' 1.94 (hexahydromandelic acid, α = 1.46 (B)), k' 2.23 (α-hydroxyhexa-cosanoic acid, α = 1.12 (A)), k' 2.53 (α-hydroxybehenic acid, α = 1.11 (A)), k' 2.82 (α-hydroxyarachidic acid, α = 1.12 (A)), k' 3.01 (α-hydroxystearic acid, α = 1.13 (A)), k' 3.20 (α-hydroxypalmitic acid, α = 1.14 (A)), k' 3.33 (α-hydroxymyristic acid, α = 1.13 (A)), k' 3.40 (malic acid, α = 1.11 (B)), k' 3.44 (α-hydroxylauric acid, α = 1.14 (A)), k' 3.50 (α-hydroxycapric acid, α = 1.14 (A)), k' 3.61 (α-hydroxyvaleric acid, α = 1.17 (A)), k' 3.61 (α-hydroxycaprylic acid, α = 1.14 (A)), k' 3.72 (α-hydroxybutyric acid, α = 1.17 (A)), k' 3.77 (α-hydroxyisovaleric acid, α = 1.15 (A)), k' 3.77 (α-hydroxycaproic acid, α = 1.15 (A)), k' 3.88 (lactic acid, α = 1.16 (A)), k' 4.04 (α-hydroxyisocaproic acid, α = 1.13 (A)) [k' of first eluted (L) enantiomer]
Limit of detection: 0.1%

KEY WORDS
chiral

REFERENCE
Brightwell, M.; Pawlowska, M.; Zukowski, J. HPLC resolution of hydroxy carboxylic acid enantiomers using 2-quinoxaloyl chloride as a new precolumn derivatizing agent, *J.Liq.Chromatogr.*, **1995**, *18*, 2765–2781.

Salicyl Chloride

SAMPLE
Matrix: blood, urine
Analyte: testosterone
Sample preparation: Condition a Bond-Elut C18 SPE cartridge with two 500 μL portions of MeOH and two 500 μL portions of water. 1 mL Plasma or urine + 1 mL water, add to the SPE cartridge, let stand for 2 min, wash with two 1 mL portions of water, wash with two 1 mL portions of MeOH:water 10:90, elute with 1 mL MeOH. Evaporate the eluate to dryness under a stream of nitrogen, reconstitute with 50 μL dry benzene (Caution! Benzene is a carcinogen!), add 5 mg potassium carbonate, add 50 μL 200 mM salicyl chloride in dry benzene, add 50 μL 250 mM 18-crown-6 in dry benzene, shake, heat at 70° for 1 h, cool, centrifuge, evaporate to dryness under a stream of nitrogen, reconstitute with 100 μL mobile phase, inject a 20 μL aliquot on to column A and elute to waste with mobile phase A, after 8 min elute the contents of column A on to column B with mobile phase B, elute with mobile phase B, monitor the effluent from column B. (Prepare salicyl chloride (salicylic acid chloride) by stirring 27.5 g freshly distilled thionyl chloride in 30 mL dry benzene at 0° (Caution! Benzene is a carcinogen!), protect the reaction with a calcium chloride drying tube and a nitrogen atmosphere, add 25 g sodium salicylate, stir at 0° for 1 h, remove solvent by vacuum distillation, take up the residue in 50 mL dry petroleum ether, stir for 15 min, centrifuge. Remove the petroleum ether layer and evaporate it to give salicyl chloride.)

HPLC VARIABLES
Column: A 5 × 4 35-40 μm RP8 Perisorb (Merck); B 100 × 4.6 Spheri 5 RP8
Mobile phase: A MeOH:water 30:70; B MeOH:water 70:30 containing 2 g/L lithium perchlorate trihydrate and 2 mL/L glacial acetic acid
Flow rate: A 0.8; B 1
Injection volume: 20
Detector: E, LKB (Bromma) 2143, glassy carbon electrode +1.0 V, palladium reference electrode

CHROMATOGRAM
Retention time: 10 (testosterone), 10 (androsterone)
Limit of quantitation: 12.5 ng/mL

KEY WORDS
plasma; SPE; column-switching

REFERENCE
Wintersteiger, R.; Sepulveda, M.J. Electrochemical detection of anabolics in human plasma and urine, *Anal.Chim.Acta*, **1993**, *273*, 383–390.

ACYL IMIDAZOLE

7-[(Imidazolemethanoyl)methoxy]-4-methylcoumarin

SAMPLE
Matrix: bulk
Analyte: benzyl alcohol
Sample preparation: Dissolve 600 μg 7-[(imidazolemethanoyl)methoxy]-4-methylcoumarin in 2.5 mL toluene and 750 μL MeCN, add 20 μL benzyl alcohol, add 150 μL 93.3 μg/mL 4-dimethylaminopyridine in MeCN, mix vigorously, heat at 60° for 1.5 h, cool, evaporate to dryness under a stream of nitrogen at room temperature, reconstitute the residue in 300 μL mobile phase, inject a 5 μL aliquot. (Preparation of 7-[(imidazolemethanoyl)methoxy]-4-methylcoumarin is as follows. Stir 102.3 mg 7-(carboxymethoxy)-4-methylcoumarin in 7 mL THF, add 70.9 mg 1,1'-carbonyldiimidazole in one portion, reflux for 30 min, stir at room temperature for 5 h. Filter and dry the solid under reduced pressure to obtain 7-[(imidazolemethanoyl)methoxy]-4-methylcoumarin as a white solid (mp 161-162°). Fluorescence detection can also be used.)

HPLC VARIABLES
Column: 150 × 3.9 4 μm Nova-Pak C18
Mobile phase: MeCN:MeOH:100 mM pH 5.5 ammonium acetate buffer 50:1.5:48.5
Flow rate: 0.7
Injection volume: 5
Detector: MS, Hewlett-Packard 5989A, thermospray interface, filament-assisted ionization mode, ion source 280°, probe stem 112°, probe tip 235-245°

CHROMATOGRAM
Retention time: 7
Limit of detection: 0.8 ng

REFERENCE

Phillips, L.R.; Supko, J.G.; Wolfe, T.L.; Malspeis, L. Precolumn derivatization of hydroxy compounds with 7-[(imidazolemethanoyl)methoxy]-4-methylcoumarin (IMMC) and LC/TSP-MS of the resulting esters, *Proc.Am.Soc.Mass Spectrom.*, **1995**, *43*, 163–164.

Naphthoylimidazole

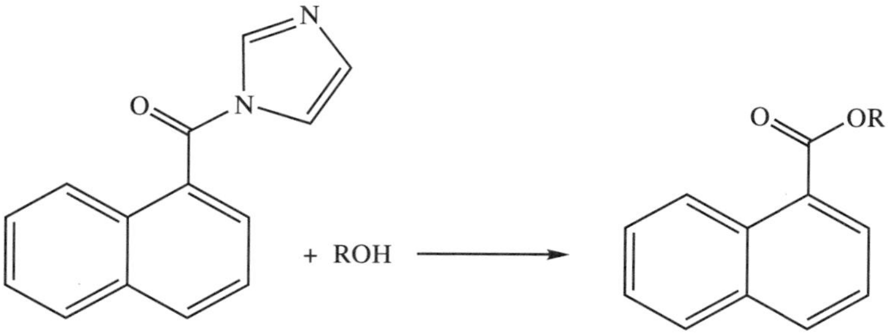

SAMPLE

Matrix: solutions

Analyte: ouabain

Sample preparation: Dry a solution containing 500 pmole material, dissolve in 250 μL anhydrous MeCN, add 1.5 mg naphthoylimidazole (Fluka), add 0.4 μL 1,8-diazabicyclo[5.4.0]undec-7-ene (DBU), stir at room temperature for 3 h, add 1 mL MeCN:water 20:80, add to a SepPak C18 SPE cartridge, wash with 2 mL MeCN:water 20:80, wash with 8 mL MeCN:water 40:60, wash with 5 mL MeCN:water 50:50, elute with 5 mL MeCN, inject an aliquot of the eluate.

HPLC VARIABLES

Column: 250 × 4.6 10 μm Phenomenex C18

Mobile phase: MeOH:water 92:8

Flow rate: 1

Detector: F ex 234 em 374

CHROMATOGRAM

Retention time: 10

KEY WORDS

silylate glassware; SPE

REFERENCE

Tymiak, A.A.; Norman, J.A.; Bolgar, M.; Didonato, G.C.; Lee, H.; Parker, W.L.; Lo, L.-C.; Berova, N.; Nakanishi, K.; Haber, E.; Haupert, G.T. Jr. Physicochemical characterization of a ouabain isomer isolated from bovine hypothalamus, *Proc.Natl.Acad.Sci.U.S.A.*, **1993**, *90*, 8189–8193.

ACYL NITRILE

1-Anthroylnitrile

SAMPLE
Matrix: blood
Analyte: bile acids
Sample preparation: Condition a BondElut SPE cartridge with 5 mL EtOH and 5 mL water. 100 μL Serum + 250 ng deoxycholic acid 12-propionate + 1 mL 500 mM pH 7.0 phosphate buffer, mix, add to the SPE cartridge, wash with 2 mL water, wash with 1 mL 1.5% EtOH, elute with 2 mL 90% EtOH. Evaporate a 400 μL aliquot of the eluate, add 100 μL 2 mg/mL 1-anthroylnitrile in MeCN, add 0.16% quinuclidine in MeCN, heat at 60° for 20 min, add 50 μL MeOH, evaporate under nitrogen. Dissolve the residue in 1 mL 90% EtOH, add to a 18 × 6 100 mg column of PHP-LH-20 Sephadex at 0.2 mL/min, wash with 1 mL 90% EtOH, elute with 5 mL 100 mM acetic acid in 90% EtOH (free bile acids), elute with 5 mL 200 mM formic acid in 90% EtOH (glycine-conjugated bile acids), elute with 5 mL 300 mM pH 6.3 acetic acid-potassium acetate in 90% EtOH (taurine-conjugated bile acids). Evaporate each fraction, dissolve the residue in 100-200 μL MeOH, inject a 5-10 μL aliquot. (Preparation of PHP-LH-20 Sephadex is as follows. Suspend 75.7 g Sephadex LH-20 in 200 mL dichloromethane using a glass stirring rod (not a magnetic stirrer) for 30 min, add 19 mL boron trifluoride ethyl etherate, after 15 min add 50 mL 35% epichlorohydrin in dichloromethane at 1-2 mL/min (Caution! Epichlorohydrin is a carcinogen!), stir for another 30 min, filter, wash with EtOH, dry chlorohydroxypropyl Sephadex LH-20 at 50° (J.Chromatogr. 1971, 59, 45). Stir 27.2 g chlorohydroxypropyl Sephadex LH-20 in 100.5 mL piperidine at room temperature for 30min, add 5.74 g KOH in 302 mL MeOH, heat at 50-60° for 3 h with occasional shaking, filter, wash with EtOH:water 50:50, wash with 200 mM acetic acid in EtOH:water 70:30, wash with EtOH:water 90:10 until washings become neutral, store in EtOH:water 90:10 (Clin. Chim. Acta 1978 87 141). 1-Anthroylnitrile is available from Wako Chemicals, Richmond VA. Synthesis is as follows. Dissolve 50 g benzanthrone in 500 mL concentrated sulfuric acid with gentle warming, pour this solution cautiously into 4 L hot water with vigorous stirring. Boil the suspension and slowly add 200 g chromium(VI) oxide (Caution! Chromium oxide is a carcinogen and highly corrosive!), after 6 h cool the mixture, filter, wash the precipitate with hot water. Dissolve the precipitate in dilute ammonia and precipitate with acid, crystallize from boiling concentrated nitric acid to give anthraquinone-1-carboxylic acid (Ber. 1924, 57, 1775). Warm, on a water bath, anthraquinone-1-carboxylic acid in dilute ammonia with twice the amount of zinc dust, when the reaction has ceased (30 min ?) filter the reaction the reaction mixture, add HCl to the filtrate to obtain anthracene-1-carboxylic acid as yellow needles, recrystallize from EtOH (mp 245°) (Ber. 1897, 30, 1118). Stir 1 g anthracene-1-carboxylic acid in 15 mL anhydrous dichloromethane, add 2 mL oxalyl chloride, reflux for 1 h, evaporate to give 1-anthroyl chloride as an oily residue. Dissolve 1-anthroyl chloride in 15 mL dichloromethane, add 3 mL trimethylsilyl cyanide, add 1 mg zinc iodide, stir at room temperature for 2 h, evaporate to dryness, recrystallize from hexane/dichloromethane to give 1-anthroylnitrile as orange-yellow needles (mp 164-5°) (Anal.Chim.Acta 1983, 147, 397).)

HPLC VARIABLES
Column: 150 × 4 5 μm Cosmosil 5C18

Mobile phase: MeOH:0.3% pH 6.0 potassium phosphate buffer 5:1
Flow rate: 1.8
Injection volume: 10
Detector: F ex 370 em 470

CHROMATOGRAM
Retention time: 6 (cholic acid), 10 (ursodiol), 13 (chenodiol), 15 (deoxycholic acid)
Internal standard: deoxycholic acid 12-propionate (20)
Limit of detection: 50 nM

KEY WORDS
serum ; SPE

REFERENCE
Goto, J. ; Saito, M. ; Chikai, T. ; Goto, N. ; Nambara, T. Studies on Steroids. CLXXXVII. Determination of serum bile acids by high-performance liquid chromatography with fluorescence labeling, *J.Chromatogr.*, **1983**, *276*, 289–300.

SAMPLE
Matrix: bulk
Analyte: alcohol ethoxylates
Sample preparation: Mix 1-250 μg alcohol ethoxylate, 5 mL 0.2% triethylamine in MeCN, and 5 mg 1-anthroylnitrile, heat at 45° for 2 h, cool to room temperature, dilute to 0.1-2.5 μg/mL with mobile phase, inject a 250 μL aliquot. (1-Anthroylnitrile is available from Wako Chemicals, Richmond VA. Synthesis is described above.)

HPLC VARIABLES
Column: 150 × 6 3 μm Hypersil ODS
Mobile phase: MeCN:water 70:30
Flow rate: 2
Injection volume: 250
Detector: F ex 395 em 450

CHROMATOGRAM
Limit of detection: 50 ppb

REFERENCE
Kudoh, M.; Ozawa, H.; Fudano, S.; Tsuji, K. Determination of trace amounts of alcohol and alkyl-phenol ethoxylates by high-performance liquid chromatography with fluorimetric detection, *J.Chromatogr.*, **1984**, *287*, 337–344.

SAMPLE
Matrix: bulk
Analyte: monohydroxy fatty acids
Sample preparation: Dissolve hydroxy acids in 50 μL 2 mg/mL 1-anthroylnitrile in dry MeCN, add 50 μL 1.6 mg/mL quinuclidine in MeCN, mix well, heat at 60° for 30 min, add 100 μL MeOH, evaporate to dryness under a stream of nitrogen, reconstitute with MeOH, purify by TLC (Merck silica gel G using toluene:ethyl acetate 9:3) with detection using primuline, remove the appropriate band by scraping, elute compounds with chloroform:MeOH 1:2. (1-Anthroylnitrile is available from Wako Chemicals, Richmond VA. Synthesis is described above.)

HPLC VARIABLES
Column: LiChrosorb Si-60
Mobile phase: Hexane:cyclohexane:diethyl ether:acetic acid 150:150:30:0.4
Flow rate: 1
Detector: F ex 380 em 460

CHROMATOGRAM
Retention time: 20 (15-hydroxyeicosatetraenoic acid), 23 (12-hydroxyeicosatetraenoic acid), 23 (13-hydroxyoctadecadienoic acid), 27 (9-hydroxyoctadecadienoic acid), 34 (5-hydroxyeicosatetraenoic acid)

KEY WORDS
normal phase

REFERENCE
Metori, A.; Ogamo, A.; Nakagawa, Y. Quantitation of monohydroxy fatty acids by high-performance liquid chromatography with fluorescence detection, *J.Chromatogr.*, **1993**, *622*, 147–151.

SAMPLE
Matrix: formulations
Analyte: atropine
Sample preparation: Grind tablets to a fine powder, weigh out amount containing 25 μg atropine, extract with 40 mL chloroform, filter, wash filter with chloroform, make up filtrate to 50 mL with chloroform, mix. Remove a 1 mL aliquot and evaporate it to dryness under reduced pressure at 40°, reconstitute with 500 μL 4 mg/mL quinuclidine in acetone, add 500 μL 2 mg/mL 1-anthroylnitrile in acetone, heat at 30° for 10 min, add 1 mL 2% phosphoric acid, cool to room temperature, make up to 10 mL with acetone, inject a 10 μL aliquot. (1-Anthroylnitrile is available from Wako Chemicals, Richmond VA. Synthesis is described above.)

HPLC VARIABLES
Column: 150 × 4.6 5 μm Cosmocil 5C-18 (Nacalai Tesque, Tokyo)
Mobile phase: MeCN:buffer 60:40 (Buffer was 20 mM sodium dodecyl sulfate adjusted to pH 3.5 with phosphoric acid.)
Column temperature: 40
Flow rate: 1
Injection volume: 10
Detector: F ex 255 em 474

CHROMATOGRAM
Retention time: 12
Limit of quantitation: 50 ng/mL
Limit of detection: 10 ng/mL

OTHER SUBSTANCES
Non-interfering: albumin, amomum seed, caffeine, chlorpheniramine, cinnamon bark, cloves, fennel, geranium herb, glycyrrhiza, lysozyme, swertia herb, vitamin B1, vitamin B2

KEY WORDS
tablets

REFERENCE
Takahashi, M.; Nagashima, M.; Shigeoka, S.; Nishijima, M.; Kamata, K. Determination of atropine in pharmaceutical preparations by liquid chromatography with fluorescence detection, *J.Chromatogr.A*, **1997**, *775*, 137–141.

SAMPLE
Matrix: tissue
Analyte: pregnenolone
Sample preparation: Homogenize 1 g tissue with 2 mL isotonic saline at 0°, add 2 mL MeOH, add 50 μL 1 μg/mL IS in EtOH, mix, centrifuge at 2800 rpm for 30 min, remove the supernatant, suspend the precipitate in 2 mL MeOH, centrifuge at 2800 rpm for 30 min. Combine the supernatants, centrifuge at 2800 rpm for 30 min. Dilute the supernatant with 45 mL water, add to a 500 mg Bond Elut C8 SPE cartridge, wash with 6 mL water, elute with 5 mL EtOH:water 90:10. Add the eluate to a 6 mm ID column of piperidinohydroxypropyl Sephadex LH-20. Concentrate the eluate to 1 mL under reduced pressure, add 4 mL water, add to a 500 mg Bond Elut C18 SPE cartridge, wash with 6 mL water, elute with 5 mL MeOH:water 80:20. Evaporate the eluate to dryness under reduced pressure, reconstitute the residue in 100 μL 5 mg/mL 1-anthroylnitrile (1-anthroyl cyanide) in MeCN and 50 μL 0.24% quinuclidine in MeCN, heat at 60° for 30 min, add 2 drops MeOH, evaporate under a stream of nitrogen, add the residue to a 30 × 6 silica gel column, wash with 3 mL hexane, wash with 6 mL hexane:ethyl acetate 20:1, elute with 20 mL hexane:ethyl acetate. Add the eluate to a 30 × 6 silica gel column, wash with 3 mL hexane, wash with 9 mL hexane:acetone 60:1, elute with 13 mL hexane:acetone 60:1. Evaporate the eluate, reconstitute with EtOH, inject an aliquot. (Preparation of piperidinohydroxypropyl Sephadex LH-20 is as follows. Suspend 75.7g Sephadex LH-20 in 200 mL dichloromethane using a glass stirring rod (not a magnetic

stirrer) for 30 min, add 19 mL boron trifluoride ethyl etherate, after 15 min add 50 mL 35% epichlorohydrin in dichloromethane at 1-2 mL/min (Caution! Epichlorohydrin is a carcinogen!), stir for another 30 min, filter, wash with EtOH, dry chlorohydroxypropyl Sephadex LH-20 at 50° (J. Chromatogr. 1971, 59, 45). Stir 27.2 g chlorohydroxypropyl Sephadex LH-20 in 100.5 mL piperidine at room temperature for 30 min, add 5.74 g KOH in 302 mL MeOH, heat at 50-60° for 3 h with occasional shaking, filter, wash with EtOH:water 50:50, wash with 200 mM acetic acid in EtOH:water 70:30, wash with EtOH:water 90:10 until washings become neutral, store in EtOH:water 90:10 (Clin. Chim. Acta 1978 87 141). 1-Anthroylnitrile is available from Wako Chemicals, Richmond VA. Synthesis is described above.)

HPLC VARIABLES
Column: 150 × 4.6 4 μm J'sphere ODS-L80 (YMC, Kyoto)
Mobile phase: MeCN:water 90:12
Flow rate: 1
Detector: F ex 370 em 470

CHROMATOGRAM
Retention time: 13
Internal standard: 3β-hydroxy-16-methylpregna-5,16-dien-20-one (15)
Limit of quantitation: 10 ng/g

KEY WORDS
rat; brain; SPE

REFERENCE
Shimada, K.; Nakagi, T. Studies on neurosteroids. IV. Quantitative determination of pregnenolone in rat brains using high-performance liquid chromatography, *J.Liq.Chromatogr.& Rel.Technol.*, **1996**, *19*, 2593–2602.

9-Anthroylnitrile

SAMPLE
Matrix: formulations
Analyte: carnitine
Sample preparation: Weigh out syrup, injections, or finely ground tablets containing 50 mg carnitine, add 40 mL water, sonicate for 20 min, cool, make up to 50 mL with water, centrifuge at 2000 rpm for 10 min. Remove a 1 mL aliquot and add it to 1 mL 100 μg/mL quinuclidine in MeCN, evaporate to dryness under reduced pressure at 50°, add 1 mL 2 mg/mL 9-anthroylnitrile in DMSO, heat at 80° for 1.5 h, cool to room temperature, make up to 5 mL with DMSO, add a 100 μL aliquot to a 1 mL Bond Elut silica gel SPE cartridge, wash with 10 mL MeCN:MeOH 90:10, elute with 20 mL water, inject a 10 μL aliquot. (Prepare 9-anthroyl nitrile as follows. Stir 1 g 9-anthracenecarboxylic acid in 15 mL anhydrous dichloromethane, add 2 mL oxalyl chloride, reflux for 1 h, evaporate to give 9-anthroyl chloride as an oily residue. Dissolve 9-anthroyl chloride in 15 mL dichloromethane, add 3 mL trimethylsilyl cyanide, add 1 mg zinc iodide, stir at room temperature for 2 h, evaporate to dryness, recrystallize from hexane/dichloromethane to give 9-anthroyl nitrile as orange-yellow needles (mp 143-4°) (Anal.Chim.Acta 1983, 147, 397). 9-Anthroylnitrile is also available from Wako Chemicals, Richmond VA or Molecular Probes, Eugene OR.)

HPLC VARIABLES

Column: 150 × 4.6 Ultron ES-OVM ovomucoid-conjugated (Shinwa, Kyoto)
Mobile phase: MeCN:buffer 17:83 (Buffer was 20 mM KH_2PO_4 adjusted to pH 4.5 with phosphoric acid.)
Column temperature: 35
Flow rate: 1
Injection volume: 10
Detector: UV 254

CHROMATOGRAM

Retention time: 7 (D), 10 (L)
Limit of detection: 50 μg/mL

KEY WORDS

chiral; SPE; tablets; syrup; injections

REFERENCE

Takahashi, M.; Terashima, K.; Nishijima, M.; Kamata, K. Separation of carnitine enantiomers as the 9-anthroylnitrile derivatives and high-performance liquid chromatographic analysis on an ovomucoid-conjugated column, *J.Pharm.Biomed.Anal.*, **1996**, *14*, 1579−1584.

SAMPLE

Matrix: solutions
Analyte: steroids
Sample preparation: Evaporate solution (eluate from preparative HPLC) to dryness under a stream of nitrogen, reconstitute with 10 μL 2 μg/mL 9-anthroylnitrile in MeCN and 10 μL triethylamine:MeCN 30:70 under nitrogen, let stand at room temperature for 20 min, add 5 μL water, after 6 min add 50 μL 600 mM acetic acid in MeCN, evaporate to dryness under a stream of nitrogen at 37°, reconstitute with 90 μL $MeOH:0.4$ N NaH_2PO_4 60:40, add to a Cyclobond I silica-bonded β-cyclodextrin SPE cartridge (Astec), wash with 1 mL water, wash with 8 mL MeOH:water 25:75 containing 7.5 mM pH 7.0 phosphate buffer, elute with 1 mL MeOH, evaporate to dryness under a stream of nitrogen, reconstitute with mobile phase, inject an aliquot on to column A and elute to waste with mobile phase, after the solvent front has passed through divert the effluent from column A on to column B, monitor the effluent from column B. (9-Anthroylnitrile is available from Wako Chemicals, Richmond VA or Molecular Probes, Eugene OR. Synthesis is as described above.)

HPLC VARIABLES

Column: A 30 × 2.1 silica (Brownlee); B 150 × 2 Hypersil
Mobile phase: Hexane:ethyl acetate 67:33 (half-saturated with water)
Flow rate: 0.5
Detector: F ex 305-395 em 430-470

CHROMATOGRAM

Retention time: 5.48 (hydrocortisone), 7.21 (prednisolone), 10.2 (cortisone)
Limit of detection: 9 pg

KEY WORDS

SPE; column-switching; normal phase

REFERENCE

Haegele, A.D.; Wade, S.E. Ultrasensitive differential measurement of cortisol and cortisone in biological samples using fluorescent ester derivatives in normal phase HPLC, *J.Liq.Chromatogr.*, **1991**, *14*, 1133−1148.

SAMPLE

Matrix: urine
Analyte: 6β-hydroxycortisol
Sample preparation: Condition a Bond Elut C18 SPE cartridge with 10 mL EtOH, 10 mL water, and 1 mL pH 7 phosphate buffer. Condition a Clin Elut CE1000M SPE cartridge with 300 μL 15% ammonium carbonate. Condition a Clin Elut CE1000M SPE cartridge with 300 μL 100 mM NaOH. 500 μL Urine + 1 mL pH 7 phosphate buffer, mix, add to

the Bond Elut SPE cartridge, wash with 3 mL water, wash with 3 EtOH:water 5:95, elute with 2 mL EtOH:water 70:30. Evaporate the eluate to dryness under reduced pressure, reconstitute the residue in 300 μL ethyl acetate, add to the Clin Elut SPE cartridge conditioned with ammonium carbonate, elute with 4 mL ethyl acetate, evaporate eluate to dryness (?), add 100 μL 1 mg/mL 9-anthroyl nitrile in MeCN:triethylamine 95:5, mix, let stand at room temperature for 1 h. Evaporate to dryness under a stream of nitrogen, reconstitute the residue in 100 μL acetone, add 2 mL hexane, mix, heat at 70-80° for a few min, add to the Clin Elut SPE cartridge conditioned with NaOH, wash with 4 mL hexane, elute with 4 mL dichloromethane. Evaporate the eluate to dryness under reduced pressure, reconstitute with 500 μL ethyl acetate, inject a 10 μL aliquot. (9-Anthroylnitrile is available from Wako Chemicals, Richmond VA or Molecular Probes, Eugene OR. Synthesis is as described above.)

HPLC VARIABLES
Column: 150 × 4 5 μm Cosmosil 5SL (Nakarai Kagaku, Kyoto)
Mobile phase: Hexane:ethyl acetate 1:2
Flow rate: 1
Injection volume: 10
Detector: F ex 360 em 460

CHROMATOGRAM
Retention time: 8.5
Internal standard: 6α-hydroxycortisol (10)
Limit of detection: 25 pg

KEY WORDS
SPE; normal phase

REFERENCE
Goto, J.; Shamsa, F.; Nambara, T. Studies on steroids. CLXXXII. Determination of 6β-hydroxycortisol in urine by high-performance liquid chromatography with fluorescence detection, *J.Liq.Chromatogr.*, **1983**, *6*, 1977–1985.

(aS)-2'-Methoxy-1,1'-binaphthalene-2-carbonyl Cyanide

SAMPLE
Matrix: blood
Analyte: penbutolol
Sample preparation: 100-500 μL Plasma + 5 ng bufarolol, mix, add 4 mL 500 mM pH 7.0 potassium phosphate buffer, add to a Sep-Pak C18 SPE cartridge, wash with 5 mL water, wash with 5 mL EtOH:water 30:70, elute with 5 mL EtOH:methylamine 99.9:0.1, evaporate to dryness, add 100 μL 2 mg/mL (aS)-2'-methoxy-1,1'-binaphthalene-2-carbonyl cyanide in MeCN containing 0.1% quinuclidine, heat at 60° for 20 min, add 50 μL MeOH, evaporate to dryness under a stream of nitrogen, reconstitute with 1 mL in EtOH:water 90:10, add to an 18 × 6 column packed with 100 mg carboxymethyl Sephadex LH-200, wash with EtOH:water 90:10 at 0.2 mL/min, elute with 5 mL 100 mM methylamine in EtOH:water 90:10. Evaporate the eluate to dryness, reconstitute with

50-100 μL mobile phase, inject an aliquot. (Derivatization occurs on the alcohol. Preparation of (aS)-2'-methoxy-1,1'-binaphthalene-2-carbonyl cyanide is as follows. Treat 1-bromo-2-naphthol with sodium hydride in DMF, add iodomethane, stir at room temperature overnight to obtain 1-bromo-2-methoxynaphthalene (mp 85-86°). Add a solution of 37.7 g 1-bromo-2-methylnaphthalene in 200 mL ether over 1 h to a sonicated mixture of 7 g magnesium turnings in 50 mL ether, the mixture should reflux rapidly (Caution! There may be in an induction period!), sonicate for 2 h after addition is complete, add 200 mL benzene (Caution! Benzene is a carcinogen!), add this mixture dropwise to a stirred mixture of 100 mmoles 1-bromo-2-methoxynaphthalene and 655 mg bis(triphenylphosphine)nickel(II) chloride (NiCl$_2$(PPh$_3$)$_2$) in 150 mL benzene at room temperature over 1 h, stir at room temperature overnight, reflux for 3 h, remove the ether by distillation through a short Vigreux column, remove the solvent by evaporation under reduced pressure, remove excess 1-bromo-2-methylnaphthalene by heating at 150°/0.1 mm Hg, cool, dissolve the residue in hexane, pass through silica gel, evaporate to dryness, recrystallize from hexane to obtain 1-methoxy-2'-methylbinaphthalene (mp 118-121°). Reflux 10 mmoles 1-methoxy-2'-methyl-binaphthalene, 1.96 g N-bromosuccinimide, and 100 mg benzoyl peroxide in 70 mL carbon tetrachloride for 3 h, filter, evaporate the filtrate to obtain crude 1-bromomethyl-2'-methoxy-binaphthalene. Dissolve the crude 1-bromomethyl-2'-methoxy-binaphthalene in 60 mL DMSO under nitrogen, slowly add a sodium ethoxide/nitropropane mixture, stir at room temperature for 3 h, stir at 60° for 3 h, pour into 300 mL ice-water, extract with dichloromethane, wash with 2 M HCl, wash with 1 M sodium carbonate, wash with water, dry over anhydrous sodium sulfate, evaporate to obtain crude 2'-methoxy-1,1'-binaphthalene-2-carboxaldehyde. (Prepare the sodium ethoxide/nitropropane mixture by dissolving 580 mg sodium in 35 mL EtOH, add 3.25 g 2-nitropropane.) Reflux the crude 2'-methoxy-1,1'-binaphthalene-2-carboxaldehyde in 60 mL acetone, add a solution of 2.36 g potassium permanganate in 60 mL hot water dropwise over 1 h, heat for an additional hour, pass sulfur dioxide through the solution until it becomes clear (sodium metabisulfite may work). Filter off the precipitate and dissolve it in 200 mL hot toluene, add a small amount of activated charcoal, filter while hot, concentrate to about a third of the volume, recrystallize from EtOH:water 1:2 to obtain 2'-methoxy-1,1'-binaphthalene-2-carboxylic acid (mp 258.5-260°) (Bull. Chem. Soc. Japan 1986, 59, 2044). Reflux 9.15 g racemic 2'-methoxy-1,1'-binaphthalene-2-carboxylic acid in 55 mL freshly distilled thionyl chloride for 5 h, evaporate under reduced pressure, add a little benzene, evaporate under reduced pressure, repeat the benzene evaporation twice more to obtain 2'-methoxy-1,1'-binaphthalene-2-carbonyl chloride as a brown solid. Dissolve the acid chloride in 70 mL benzene, add dropwise to 12.8 g (-)-menthol in 100 mL benzene containing 1 g 4-dimethylaminopyridine and 5 mL pyridine, stir overnight at room temperature, heat at 70° for 3 h, cool, dilute with benzene, wash with 2 M HCl, wash with 1 M sodium carbonate, wash with water, dry over anhydrous magnesium sulfate in the presence of activated charcoal, evaporate to dryness, remove as much menthol as possible by sublimation under vacuum, chromatograph twice on a column of silica gel with toluene to obtain the (aS,R) menthol ester (mp 145-146° from hexane) and the (aR,R) menthol ester (mp 126-129° from hexane) as well as a mixture of diastereomers. Reflux the (aS,R) menthol ester with KOH in aqueous EtOH for 8-10 h to obtain (aS)-2'-methoxy-1,1'-binaphthalene-2-carboxylic acid (Bull. Chem. Soc. Japan 1989, 62, 1528). Add 1.5 mL oxalyl chloride to a solution of (aS)-2'-methoxy-1,1'-binaphthalene-2-carboxylic acid in 10 mL anhydrous benzene, reflux for 10 h, evaporate to dryness under reduced pressure. Take up the residue in 10 mL anhydrous benzene, add 1 mL trimethylsilyl cyanide, add 1 mg zinc iodide, stir at room temperature for 5 h, evaporate to dryness, recrystallize from hexane/acetone to obtain (aS)-2'-methoxy-1,1'-binaphthalene-2-carbonyl cyanide as orange-yellow needles (mp 143-146°).)

HPLC VARIABLES
Column: 150 × 4.6 5 μm Cosmosil 5SL (Nacalai Tesque, Kyoto)
Mobile phase: Hexane:ethyl acetate:triethylamine 83.3:16.7:0.005
Flow rate: 2
Detector: F ex 290 em 405

CHROMATOGRAM
Retention time: 10.5 (R), 13.5 (S)

Internal standard: bufarolol (8.5 (R), 12 (S))
Limit of detection: 30 pg

KEY WORDS
plasma; chiral; normal phase; dog; SPE; pharmacokinetics

REFERENCE
Goto, J.; Shao, G.; Ito, M.; Kuriki, T.; Nambara, T. High-performance liquid chromatographic determination of penbutolol enantiomers in plasma with fluorescence detection, *Anal.Sci.*, **1991**, *7*, 723–726.

SAMPLE
Matrix: bulk
Analyte: β-blockers
Sample preparation: Mix the compound with (aS)-2'-methoxy-1,1'-binaphthalene-2-carbonyl cyanide in MeCN containing 0.1% quinuclidine, heat at 60° for 20 min, take up the reaction mixture in EtOH:water 90:10, add to an 18 × 6 column packed with 100 mg carboxymethyl Sephadex LH-200, elute with 100 mM methylamine in EtOH:water 90:10. Evaporate the eluate to dryness, reconstitute with ethyl acetate, inject an aliquot. (Derivatization occurs on the alcohol. Preparation of (aS)-2'-methoxy-1,1'-binaphthalene-2-carbonyl cyanide is described above.)

HPLC VARIABLES
Column: 150 × 4.6 5 μm Cosmosil 5SL (Nacalai Tesque, Kyoto)
Mobile phase: Hexane:ethyl acetate:triethylamine 66.6:33.3:0.1 (A) or 83.3:16.7:0.1 (B)
Flow rate: 2
Detector: F ex 330 em 420

CHROMATOGRAM
Retention time: 6 ((+)-propranolol (A)), 7.5 ((-)-propranolol (A)), 6.5 ((+)-bufarolol (B)), 8 ((-)-bufarolol (B)), 6.5 ((+)-penbutolol (B)), 8.5 ((-)-penbutolol (B))
Limit of detection: 100 fmole

KEY WORDS
chiral; normal phase

REFERENCE
Goto, J.; Shao, G.; Fukasawa, M.; Nambara, T.; Miyano, S. A chiral axis derivatization reagent for the resolution of β-adrenergic blockers by liquid chromatography with fluorescence detection, *Anal.Sci.*, **1991**, *7*, 645–647.

2-Methyl-1,1'-binaphthalene-2'-carbonyl Cyanide

RELATED REFERENCE
Goto, J.; Goto, N.; Nambara, T. New type of derivatization reagents for liquid chromatographic resolution of enantiomeric hydroxyl compounds. *Chem.Pharm.Bull.* **1982**, *30*, 4597-4599.

SAMPLE
Matrix: blood
Analyte: propranolol

Sample preparation: 1 mL Plasma + 10 ng (+)-bufuralol + 4 mL 500 mM pH 7.0 sodium phosphate buffer, mix, add to a Sep-Pak C18 SPE cartridge, wash with 5 mL water, wash with 5 mL EtOH:water 30:70, elute with 8 mL EtOH. Evaporate the eluate to dryness, reconstitute the residue in 100 μL 2 mg/mL (-)-2-methyl-1,1'-binaphthalene-2'-carbonyl cyanide in MeCN containing 0.01% quinuclidine, heat at 60° for 20 min, add 50 μL MeOH, evaporate to dryness, reconstitute with 1 mL EtOH:water 90:10, add to an 18 × 6 column containing 100 mg carboxymethyl Sephadex LH-20, wash with EtOH:water 90:10 at 0.2 mL/min, elute with 3 mL 100 mM methylamine in EtOH:water 90:10. Evaporate the eluate to dryness, reconstitute the residue in 50-100 μL mobile phase, inject a 10-20 μL aliquot. (Synthesis of (-)-2-methyl-1,1'-binaphthalene-2'-carbonyl cyanide is as follows. Reflux 210 g 1-bromo-2-methylnaphthalene, 160 g N-bromosuccinimide, 1 g benzoyl peroxide, and 250 mL carbon tetrachloride for 2.5 h, add 250 mL carbon tetrachloride, filter while warm, wash the residue several times with solvent. Concentrate and cool the filtrate to give 1-bromo-2-bromomethylnaphthalene (mp 230-240°) (J. Org. Chem. 1949, 14, 375). Dissolve 90 g 1-bromo-2-bromomethylnaphthalene in 400 mL chloroform, reflux, add 46.5 g powdered hexamine in portions, remove the hexaminium salt by filtration. Reflux this salt in 650 mL 50% acetic acid for 1 h, add 105 mL concentrated HCl, reflux for 5 min, cool, obtain 1-bromo-2-naphthaldehyde (mp 119-120°) by filtration. Heat 11 g 1-bromo-2-naphthaldehyde in 275 mL acetone at 60-68°, add a hot solution of 14 g potassium permanganate in 330 mL water over 30 min, heat for another 30 min, pass in sulfur dioxide (sodium metabisulfite ?) until the solution is clear, pour into water to give 1-bromo-2-naphthoic acid, purify by forming the ammonium salt and reprecipitating. Reflux 1-bromo-2-naphthoic acid in MeOH in the presence of sulfuric acid to give methyl 1-bromo-2-naphthoate. Heat methyl-1-bromo-2-naphthoate with copper bronze at 270-280° for 20 min, while still hot extract with toluene, cool to obtain dimethyl 1,1'-binaphthalene-2,2'-dicarboxylate, obtain more crystals by evaporating some of the solvent, recrystallize from EtOH to give dimethyl 1,1'-binaphthalene-2,2'-dicarboxylate (mp 158°) (J. Chem. Soc. 1955, 1242). Add 8 g lithium tri-tert-butoxyaluminohydride in portions to 2.8 g dimethyl 1,1'-binaphthalene-2,2'-dicarboxylate in 150 mL anhydrous benzene:ether 50:50 (Caution! Benzene is a carcinogen!), heat at 80° for 2 h, acidify with 5% HCl. Remove the organic layer and dry it over anhydrous sodium sulfate, evaporate to dryness, chromatograph on 50 g silica gel with hexane:ethyl acetate 80:20, recrystallize the product from hexane/acetone to give methyl 2-hydroxymethyl-1,1'-binaphthalene-2'-dicarboxylate (mp 117.5-118.5°). Add 5 mL 30% hydrogen bromide in acetic acid to 2 g methyl 2-hydroxymethyl-1,1'-binaphthalene-2'-dicarboxylate in 10 mL acetic acid, stir at 50° for 10 min, pour into ice-water, filter, chromatograph the solid on 40 g silica gel with hexane:ethyl acetate 30:1 to give methyl 2-bromomethyl-1,1'-binaphthalene-2'-dicarboxylate as pale yellow needles (mp 137-138°). Add 400 mg sodium borohydride to 1.9 g methyl 2-bromo-methyl-1,1'-binaphthalene-2'-dicarboxylate in 10 mL DMSO, stir at 60° for 15 min, pour into ice-water, acidify with concentrated HCl, chromatograph the crude product on 40 g silica gel with hexane:ethyl acetate 10:1, recrystallize from MeOH to give methyl 2-methyl-1,1'-binaphthalene-2'-carboxylate as colorless needles mp 97-98°. Add 30 mL 10% KOH to 1.2 g methyl 2-methyl-1,1'-binaphthalene-2'-carboxylate in 50 mL MeOH, reflux for 3 h, pour into ice-water, filter, recrystallize from hexane/ethyl acetate to give 2-methyl-1,1'-binaphthalene-2'-carboxylic acid as colorless needles (mp 232-233°). Add 4.1 g (-)-brucine in 20 mL EtOH to 3.3 g 2-methyl-1,1'-binaphthalene-2'-carboxylic acid dissolved in 60 mL EtOH, allow to stand overnight, filter, recrystallize the precipitate several times from EtOH. Add 5% HCl to the salt and extract with ethyl acetate, wash the organic layer with water, dry over anhydrous sodium sulfate, evaporate to dryness, recrystallize from hexane/acetone to give (-)-2-methyl-1,1'-binaphthalene-2'-carboxylic acid as colorless needles (mp 229-229.5°; $[\alpha]_D^{20}$ -41.3°(c = 0.58 in chloroform). Add 3 mL oxalyl chloride to 500 mg (-)-2-methyl-1,1'-binaphthalene-2'-carboxylic acid in 30 mL anhydrous dichloromethane, stir at room temperature for 2 h, evaporate to give an oily residue, take up in 10 mL dichloromethane, add 2 mL trimethylsilyl cyanide, add 1 mg zinc iodide, stir at room temperature for 2 h, evaporate to dryness, chromatograph on 5 g silica gel with hexane to give (-)-2-methyl-1,1'-binaphthalene-2'-carbonyl cyanide as a yellow oil ($[\alpha]_D^{20}$ -42.8° (c = 1.05 in chloroform) (Anal. Sci. 1990, 6, 261).)

HPLC VARIABLES
Column: 150 × 4.6 5 μm spherical silica (Waters)

Mobile phase: Hexane:ethyl acetate:MeOH 90:6:1.8
Injection volume: 10-20
Detector: F ex 318 em 408

CHROMATOGRAM
Retention time: 9 (-), 11 (+)
Internal standard: (+)-bufuralol (5)
Limit of detection: 100 pg

KEY WORDS
plasma; SPE; chiral; normal phase

REFERENCE
Shao, G.; Goto, J.; Nambara, T. Separation and determination of propranolol enantiomers in plasma by high-performance liquid chromatography with fluorescence detection, *J.Liq.Chromatogr.*, **1991**, *14*, 753–763.

SAMPLE
Matrix: solutions
Analyte: penbutolol
Sample preparation: Dissolve 100 ng penbutolol and 200 μg (-)-2-methyl-1,1'-binaphthalene-2'-carbonyl cyanide in 200 μL 0.01% quinuclidine in MeCN, heat at 60° for 10 min, inject an aliquot. (Synthesis of (-)-2-methyl-1,1'-binaphthalene-2'-carbonyl cyanide is described above.)

HPLC VARIABLES
Column: 150 × 4.6 5 μm spherical silica (Waters)
Mobile phase: Hexane:chloroform:MeOH 100:5:0.3
Detector: F ex 342 em 420

CHROMATOGRAM
Retention time: 11 (+), 12 (-)
Limit of detection: 200 pg

KEY WORDS
chiral; normal phase

REFERENCE
Goto, J.; Goto, N.; Shao, G.; Ito, M.; Hongo, A.; Nakamura, S.; Nambara, T. Fluorescence chiral derivatization reagents for high performance liquid chromatographic resolution of enantiomeric hydroxyl compounds, *Anal.Sci.*, **1990**, *6*, 261–264.

Pyrene-1-carbonyl Cyanide

RELATED REFERENCE
Goto, J.; Komatsu, S.; Inada, M.; Nambara, T. New sensitive fluorescence labeling reagent for high performance liquid chromatography of hydroxysteroids. *Anal.Sci.* **1986**, *2*, 585-586

SAMPLE
Matrix: formulations

Analyte: carnitine
Sample preparation: Powder tablets, weigh out an appropriate amount, add 40 mL water, sonicate for 20 min, cool, make up to 50 mL, centrifuge at 1300 g for 10 min. remove a 2 mL aliquot of the supernatant and add it to a 150 × 10 column containing 2 g 100-200 mesh Amberlite CG-120 cation-exchange resin (Na$^+$ form), wash with 25 mL water, elute with 20 mL 2% ammonia solution, adjust the volume of the eluate to 25 mL with water. Remove a 500 µL aliquot and add it to 500 µL 50 µg/mL triamterene in MeCN:DMSO 99:1, evaporate to dryness under reduced pressure at 50°, reconstitute with 1 mL 200 µg/mL pyrene-1-carbonyl cyanide (Wako Chemicals, Richmond VA) in DMSO, heat at 80° for 30 min, inject a 1 µL aliquot.

HPLC VARIABLES
Column: 250 × 4.6 5 µm TSKgel SP-2SW (Tosoh)
Mobile phase: MeCN:buffer 25:75 (Buffer was 10 mM (NH$_4$)$_2$HPO$_4$ adjusted to pH 7.5 with phosphoric acid.)
Column temperature: 40
Flow rate: 1
Injection volume: 1
Detector: F ex 355 em 420

CHROMATOGRAM
Retention time: 5
Internal standard: triamterene (8)
Limit of detection: 500 ng/mL

OTHER SUBSTANCES
Non-interfering: caffeine, cinnamon bark extract, coptis rhizome extract, EtOH, gentian extract, ginseng extract, glucuronolactone, glycyrrhizia extract, inositol, niacinamide, pantothenol, sucrose, vitamin B1, vitamin B2, vitamin B6

KEY WORDS
tablets; SPE

REFERENCE
Kamata, K.; Takahashi, M.; Terasima, K.; Nishijima, M. Liquid chromatographic determination of carnitine by precolumn derivatization with pyrene-1-carbonyl cyanide, *J.Chromatogr.A*, **1994**, *667*, 113–118.

AMINE

Veratrylamine

SAMPLE
Matrix: blood

Analyte: 5-hydroxyindoles

Sample preparation: 40 μL Human platelet-poor plasma + 20 μL 20 nM 5-hydroxyindole-3-acetamide in water + 140 μL 100 mM pH 9.0 Tris-HCl buffer, mix, filter (92 × 15 cellulose Ultracent-30 cartridge, Tosoh) while centrifuging at 1500 g for 30 min. Remove a 100 μL aliquot of the filtrate and add it to 50 μL 100 mM pH 9.0 Tris-HCl buffer, add 100 μL 20 mM veratrylamine (3,4-dimethoxybenzylamine) in DMSO:water 50:50, add 50 μL 30 mM potassium ferricyanide in DMSO:water 80:20, mix, let stand at room temperature for 2 min, inject a 100 μL aliquot. (Recrystallize veratrylamine hydrochloride from EtOH before use. Discard veratrylamine solution after 1 week. Discard potassium ferricyanide solution after 1 day.)

HPLC VARIABLES

Column: 150 × 4.6 5 μm Wakosil II 5C18 RS (Wako)

Mobile phase: MeCN:10 mM pH 6.0 phosphate buffer:50 mM sodium 1-hexanesulfonate 35:45:20

Flow rate: 0.8

Injection volume: 100

Detector: F ex 345 em 475

CHROMATOGRAM

Retention time: 4.4 (5-hydroxyindole-3-acetic acid), 4.4 (5-hydroxytryptophan), 9.2 (serotonin), 16.0 (N-acetyl-5-hydroxytryptamine), 17.5 (5-hydroxytryptophol)

Internal standard: 5-hydroxyindole-3-acetamide (13.5)

Limit of detection: 450 pM

KEY WORDS

plasma; platelet-poor plasma; ultrafiltrate

REFERENCE

Ishida, J.; Takada, M.; Yamaguchi, M. 3,4-Dimethoxybenzylamine as a sensitive pre-column fluorescence derivatization reagent for the determination of serotonin in human platelet-poor plasma, *J.Chromatogr.B*, **1997**, *692*, 31–36.

ANHYDRIDE

Acetic Anhydride

RELATED REFERENCE

Mackay, L.G.; Croft, M.Y.; Selby, D.S.; Wells, R.J. Determination of nonylphenol and octylphenol ethoxylates in effluent by liquid chromatography with fluorescence detection. *J.AOAC Int.* **1997**, *80*, 401-407.

SAMPLE

Matrix: barley

Analyte: sterigmatocystin

Sample preparation: Prepare a column of 10 g 63-200 μm silica gel (Merck Type 60) in cyclohexane in a 500 × 22 column, add 15 g anhydrous sodium sulfate to the top of the column. Shake 50 g ground barley, 180 mL MeCN, and 4% KCl in water using a wrist-action shaker at 50% amplitude for 30 min, filter (paper), wash the filtrate twice with 50 mL portions of hexane. Add 25 mL water and 50 mL chloroform to the filtrate, shake, repeat extraction again with 25 mL chloroform. Combine the organic layers and evaporate them to dryness under reduced pressure. Take up the residue in 10 mL benzene (Caution! Benzene is a carcinogen!) and add it to the column, rinse the flask with 10 mL benzene, add the rinse to the column, elute with 200 mL cyclohexane:ethyl acetate 80:20. Collect the entire eluate and evaporate it to dryness under reduced pressure, reconstitute in 5 mL chloroform. Remove a 2.5 mL aliquot and evaporate it to dryness under a stream of nitrogen, add 500 μL pyridine, add 100 μL acetic anhydride, heat at 100° for 3 h, evaporate to dryness under a stream of nitrogen, reconstitute with 1 mL MeCN, sonicate, centrifuge for 1 min, inject a 50 μL aliquot of the supernatant.

HPLC VARIABLES
Column: 250 × 4 10 μm Lichrospher Si-100 RP-18
Mobile phase: Gradient. MeOH:water 50:50 for 9 min, to 100:0 over 5 min, maintain at 100:0 for 7 min.
Column temperature: 50
Flow rate: 1.5
Injection volume: 50
Detector: F ex 256 em 418

CHROMATOGRAM
Retention time: 7.6
Limit of quantitation: 20 ng/g

REFERENCE
Abramson, D.; Thorsteinson, T. Determinatiom of sterigmatocystin in barley by acetylation and liquid chromatography, *J.Assoc.Off.Anal.Chem.*, **1989**, 72, 342–344.

SAMPLE
Matrix: beer
Analyte: 2-acetyl-4(5)-tetrahydroxybutylimidazole
Sample preparation: Soak 100 g Amberlite CG-50(H) type 1 weak cation-exchange resin in water overnight, rinse several times with water, pour into a 10 mm dia column to a bed height of 100 mm, wash with 20 mL water. Soak 100 g 100-200 mesh Dowex 50W-X8 strong cation-exchange resin in water overnight, rinse several times with water, pour into a 10 mm dia column to a bed height of 100 mm, wash with 20 mL water. Add 50 mL degassed beer to the Amberlite column, elute with 75 mL water, collect all the eluate and add it to the Dowex column, wash with 100 mL water, elute with 100 mL 300 mM HCl. Evaporate the eluate to dryness at 45° for 30 min, reconstitute with 5-10 mL water, add a 1 mL aliquot to a Sep-Pak C18 SPE cartridge, elute with 3 mL water. Collect all the eluate and make up to 4 mL. Remove a 1 mL aliquot and evaporate it to dryness under a stream of nitrogen at 50°, add 10 μL pyridine, add 150 μL acetic anhydride, mix, heat at 90° for 10 min, evaporate to dryness under a stream of nitrogen at room temperature (evaporate for an additional 5 min after dryness is achieved), reconstitute with 1 mL water, add 1 mL chloroform, shake gently, repeat the extraction twice with 500 μL aliquots of chloroform. Combine the organic layers and dry them over anhydrous sodium sulfate, evaporate to dryness under a stream of nitrogen at room temperature, reconstitute the residue in 1 mL mobile phase (allow to stand at room temperature for several h or overnight), inject a 50 μL aliquot.

HPLC VARIABLES
Column: 150 × 4.6 5 μm Supelcosil LC-18
Mobile phase: THF:20 mM pH 6.0 KH_2PO_4 12:88
Flow rate: 1
Injection volume: 50
Detector: UV 287

CHROMATOGRAM
Retention time: 6
Limit of detection: 10 ng/mL

KEY WORDS
SPE

REFERENCE

Lawrence, J.F.; Ménard, C. Determination of 2-acetyl-4(5)-tetrahydroxybutylimidazole in beers by high-performance liquid chromatography with confirmation by chemical derivatization, *J.Chromatogr.*, **1989**, *466*, 421–426.

SAMPLE

Matrix: blood
Analyte: budesonide
Sample preparation: Condition a Bond Elut C18 SPE cartridge. Add plasma + IS to SPE cartridge, wash with aqueous EtOH, wash with water, wash with heptane, elute with ethyl acetate in heptane, esterify with acetic anhydride and triethylamine in MeCN, evaporate, reconstitute in mobile phase, inject an aliquot.

HPLC VARIABLES

Guard column: 10 × 3 Chromguard
Column: 33 × 4.6 3 μm Supelcosil LC-8-DB
Mobile phase: MeOH:100 mM pH 5 ammonium acetate 64:36
Flow rate: 1.4
Injection volume: 100
Detector: MS, Finnigan 4500 quadrupole, thermospray, scan time 40 ms, source block 220°, repeller 45 V, vaporizer 105°, jet block 180°, aerosol 220°

CHROMATOGRAM

Retention time: 3.6 (as budesonide 21-acetate)
Internal standard: octadeutero budesonide
Limit of quantitation: 0.1 nM

KEY WORDS

plasma; LC-MS; SPE

REFERENCE

Lindberg, C.; Paulson, J.; Blomqvist, A. Evaluation of an automated thermospray liquid chromatography-mass spectrometry system for quantitative use in bioanalytical chemistry, *J.Chromatogr.*, **1991**, *554*, 215–226.

SAMPLE

Matrix: bulk, formulations
Analyte: iodochlorhydroxyquin
Sample preparation: Weigh out 30 mg of bulk drug or an amount of cream equivalent to 30 mg iodochlorhydroxyquin, add 70 mL THF, shake vigorously until the cream has dissolved, make up to 100 mL with THF. Remove a 5 mL aliquot and add it to 1 mL pyridine and 1 mL acetic anhydride, heat at 60° for 15 min, cool, add 15 mL 450 μg/mL testosterone acetate in butyl chloride:THF 94:6, mix thoroughly. Remove a 3 mL aliquot and evaporate it to dryness under a stream of nitrogen at 40°, reconstitute the residue in 15 mL mobile phase with gentle warming and vigorous shaking, inject an aliquot.

HPLC VARIABLES

Column: 300 × 4 10 μm μPorasil
Mobile phase: Butyl chloride:water-saturated butyl chloride:THF:glacial acetic acid 55:55:3:2
Flow rate: 2-3
Detector: UV 254

CHROMATOGRAM

Retention time: 6
Internal standard: testosterone acetate (8)

OTHER SUBSTANCES

Simultaneously analyzed: impurities

KEY WORDS

cream; normal phase

REFERENCE

Kubiak, E.J.; Munson, J.W. Analysis of iodochlorhydroxyquin in cream formulations and bulk drugs by high-performance liquid chromatography, *J.Pharm.Sci.*, **1982**, *71*, 872−875.

SAMPLE

Matrix: milk

Analyte: ivermectin

Sample preparation: Prepare a SPE cartridge by adding 2 g 40 μm Bondesil C18 18% load endcapped (Varian) to a 25 mL syringe barrel fitted with a 20 μm frit, wash with 5 mL petroleum ether, 5 mL acetone, and two 5 mL aliquots of MeOH, aspirate with full vacuum for <5 s (A). Condition a 500 mg Bond Elut LRC silica SPE cartridge with 3 mL hexane:ethyl acetate 60:40 (B). Condition a 500 mg Bond Elut LRC silica SPE cartridge with 4 mL chloroform (C). 25 mL Milk + 200 μL 500 ng/mL abamectin (avermectins) in MeOH, mix, add 5 mL to the SPE cartridge (A), mix milk with C18 material, let stand for 2 min, wash spatula with water, wash with two 5 mL portions of water, elute with 10 mL ethyl acetate, allow eluate to pass through a 5 cm layer of anhydrous sodium sulfate. Evaporate the eluate to dryness under a stream of nitrogen below 50°, add 2 mL hexane:ethyl acetate 60:40 to the oily residue, vortex, sonicate for 1 min, add mixture to SPE cartridge (B), rinse in with 1 mL hexane:ethyl acetate 60:40, wash with 5 mL hexane:ethyl acetate 60:40, elute with 5 mL MeOH:ethyl acetate 50:50. Evaporate the eluate to dryness under a stream of nitrogen below 60° (this residue should have no moisture in it), reconstitute the residue in 100 μL reagent, vortex gently for a few s, heat at 95° for 1 h, cool, add 1 mL chloroform, vortex, add to SPE cartridge (C), wash in with three 1 mL portions of chloroform,elute with 2 mL chloroform. Collect all the eluate and evaporate it to dryness under a stream of nitrogen below 60°, reconstitute in 500 μL MeOH, inject a 50 μL aliquot. (Prepare reagent by sequentially mixing 900 μL DMF, 300 μL acetic anhydride, and 200 μL N-methylimidazole just before use.)

HPLC VARIABLES

Guard column: Newguard RP-18 (Brownlee)

Column: 250 × 4.6 5 μm Econosil C18

Mobile phase: MeOH:THF:water 85:15:5

Flow rate: 1

Injection volume: 50

Detector: F ex 364 em 455

CHROMATOGRAM

Retention time: 15

Internal standard: abamectin (avermectins) (10.5)

Limit of detection: <1 ppb

KEY WORDS

cow; silylate glassware

REFERENCE

Schenck, F.J.; SPE; MSPD Isolation and quantification of ivermectin in bovine milk by matrix solid phase dispersion (MSPD) extraction and liquid chromatographic determination, *J.Liq.Chromatogr.*, **1995**, *18*, 349−362.

SAMPLE

Matrix: solutions

Analyte: monosaccharides

Sample preparation: Dissolve 50 mg sugars in 700 μL pyridine, add 700 μL 720 mM hydroxylamine hydrochloride in pyridine, heat at 60° for 10 min, add 250 μL acetic anhydride, heat at 75° for 10 min, evaporate to dryness under reduced pressure, reconstitute with 3 mL chloroform. Wash the organic layer three times with 6 mL portions of water and dry it over anhydrous sodium sulfate, evaporate to dryness under reduced pressure, take up in chloroform, pass through silica gel using chloroform, evaporate the eluate to dryness, reconstitute, inject a 5 μL aliquot.

HPLC VARIABLES

Column: 250 × 4 5 μm μBondapak C18

Mobile phase: Gradient. MeCN:water from 35:75 to 50:50 over 15 min.
Flow rate: 1
Injection volume: 5
Detector: UV 207

CHROMATOGRAM

Retention time: k' 3.1 (xylose), k' 3.3 (lyxose), k' 3.4 (ribose), k' 3.4 (arabinose), k' 4.3 (allose), k' 4.3 (altrose), k' 3.9 (dextrose), k' 4.2 (mannose), k' 4.0 (gulose), k' 3.8 (idose), k' 4.8 (talose), k' 4.2 (galactose), k' 3.0, k' 3.2 (fructose, syn and anti isomers)
Limit of detection: 3 μg

REFERENCE

Velasco, D.; Castells, J.; Lopez-Calahorra, F. High-performance liquid chromatographic separation of monosaccharides as their peracetylated ketoximes and aldononitriles, *J.Chromatogr.*, **1990**, *519*, 228–236.

SAMPLE

Matrix: tissue
Analyte: dinoprost
Sample preparation: Condition two Sep-Pak C18 SPE cartridges with water. Homogenize 1.5 g tissue with 22 mL 52.6 mM pH 7.52 Tris-HCl buffer. Mix 13 mL homogenate with 550 μL 70 μg/mL indomethacin in 1 mM sodium carbonate solution, vortex, shake at 120 rpm at 37° for 1 h, add 278 pmole dinoprost-d_4, add 50 μL 2 M HCl, add 430 μL EtOH, centrifuge at 4° at 1500 g for 10 min, add the supernatant to a SPE cartridge, wash with 5 mL MeCN:water 5:95, wash with 10 mL water, elute with 6 mL MeCN. Evaporate the eluate to dryness under reduced pressure, reconstitute with 200 μL pyridine, add 80 μL acetic anhydride, let stand at 5° under argon overnight, add 3.5 mL 10% acetic acid, add 500 μL MeCN, add the mixture to a SPE cartridge, wash with 3mL MeCN:water 5:95, wash with 10 mL water, elute with 5 mL MeCN, inject an aliquot of the eluate.

HPLC VARIABLES

Column: 150 × 4.6 5 μm Nucleosil 100-5C18
Mobile phase: MeCN:100 mM formic acid:100 mM ammonium formate 60:8:32
Flow rate: 1
Injection volume: 20
Detector: MS, Vestec Model 750B HPLC-TSP-MS interface, positive ion mode, vaporizer control 146°, vaporizer tip 280°, vapour 323°, block 346°, tip heater 348°, m/z 301

CHROMATOGRAM

Retention time: 8
Internal standard: dinoprost-d_4 (8 min, m/z 305)
Limit of detection: 0.2 pmole

KEY WORDS

rat; brain; SPE; Method can be used for many related compounds (J.Chromatogr. 1991, 568, 11).

REFERENCE

Yamane, M.; Abe, A. High-performance liquid chromatography-thermospray mass spectrometry of hydroxy-polyunsaturated fatty acid acetyl derivatives, *J.Chromatogr.*, **1992**, *575*, 7–18.

SAMPLE

Matrix: tissue
Analyte: ivermectin
Sample preparation: Condition a 6 mL 500 mg Bond Elut C18 SPE cartridge with 5 mL MeCN and 5 mL MeCN:water:triethylamine 30:70:0.1. Homogenize (Polytron) 5 g tissue and 15 mL MeCN for 20 s, rinse probe with 5 mL MeCN, shake mechanically at high speed for 5 min, centrifuge at 2000 g for 5 min. Re-extract the solid with 10 mL MeCN. Add the supernatants to the alumina column. Combine the eluates, add 70 mL water, add 100 μL triethylamine, mix, add to the C18 SPE cartridge, pull air through the SPE cartridge for 3 min, elute with 5 mL MeCN. Evaporate the eluate to dryness under a stream of nitrogen at 60°, reconstitute the residue in 100 μL freshly prepared reagent, vortex for 15 s, heat at 95-100° for 45 min, cool, add 1 mL chloroform, vortex, add to a

2.8 mL 500 mg Bond Elut silica SPE cartridge, elute with three 3 mL aliquots of chloroform. Combine the eluates and evaporate them to dryness under a stream of nitrogen at 60°, reconstitute the residue in 1 mL MeOH, filter, inject a 40 μL aliquot. (Prepare alumina column as follows. Shake 94 g Brockman Activity I neutral alumina (Fisher) and 6 mL water for 45 min, add 4.5 g alumina to an 8 mL column with a frit. Reagent was 200 μL 1-methylimidazole, 600 μL acetic anhydride, and 900 μL DMF.)

HPLC VARIABLES
Guard column: 30 × 4.6 RP-18 (Brownlee)
Column: 250 × 4.6 RP-18 OD-224 (Brownlee)
Mobile phase: MeOH:water 97:3
Flow rate: 1.8
Injection volume: 40
Detector: F ex 365 em 425

CHROMATOGRAM
Retention time: 9.3
Limit of detection: 2 ppb

KEY WORDS
SPE; cow; pig; sheep; fish; liver; muscle

REFERENCE
Salisbury, C.D.C. Modified method for the determination of ivermectin residues in animal tissues, *J.AOAC Int.*, **1993**, *76*, 1149–1151.

SAMPLE
Matrix: tissue
Analyte: ivermectin
Sample preparation: Condition a 6 mL 500 mg Bakerbond C18 SPE cartridge with three 5 mL portions of MeOH, 5 mL MeCN, and three 5 mL portions of MeCN:water:triethylamine 30:70:0.1. Condition a Waters silica SPE cartridge with 8 mL chloroform. Homogenize (Ultraturrax) 5 g minced tissue with 15 mL MeCN at high speed for 3 min, rinse blade with 2 mL MeCN, sonicate for 15 min, centrifuge at 3000 rpm for 5 min, filter (paper), extract the residue again with 10 mL MeCN, wash the filter with 3 mL MeCN. Combine the organic layers and add 70 mL water and 100 μL triethylamine, stir thoroughly, add to the C18 SPE cartridge, wash with two 5 mL portions of MeCN:water 50:50, elute with 7 mL MTBE at 2 mL/min. Store the eluate overnight at -20°, remove the organic layer and evaporate it to dryness under a stream of nitrogen at 50°, reconstitute the residue in 3 mL MeOH, add 100 μL water, add 3 mL hexane, vortex, remove the hexane layer, repeat the hexane wash. Extract the combined hexane layers with 1 mL MeOH. Combine the MeOH layers and evaporate them to dryness under a stream of nitrogen at 50°, heat in a vacuum oven at 50° for 30 min, reconstitute the residue in 150 μL 1-methylimidazole:acetic anhydride:DMF 2:3:9 (freshly prepared), vortex for 30 s, heat at 100° for 1 h, cool, add 1 mL chloroform, vortex, add to the silica SPE cartridge, elute with three 3 mL portions of chloroform. Evaporate the eluate to dryness under a stream of nitrogen at 50°, reconstitute the residue in 400 μL MeOH, vortex, inject a 20 μL aliquot.

HPLC VARIABLES
Guard column: 20 × 4.6 5 μm Supelcosil LC-18
Column: 150 × 4.6 5 μm Supelcosil LC-18
Mobile phase: MeOH:water 95:5
Flow rate: 1.8
Injection volume: 20
Detector: F ex 360 em 470

CHROMATOGRAM
Retention time: 7
Limit of detection: 2 ng/g

KEY WORDS
SPE; liver; muscle; fat; guinea pig; cow; pig; horse; sheep; pharmacokinetics

REFERENCE

Dusi, G.; Curatolo, M.; Fierro, A.; Faggionato, E. Determination of the antiparasitic drug ivermectin in liver, muscle and fat tissue samples from swine, cattle, horses and sheep using HPLC with fluorescence detection, *J.Liq.Chromatogr.Rel.Technol.*, **1996**, *19*, 1607–1616.

SAMPLE

Matrix: urine

Analyte: 4-hydroxymerbarone

Sample preparation: 100 µL Urine + 25 µL 16 µg/mL IS in DMSO + 25 µL 1.2 M potassium carbonate, mix, add 10 µL acetic anhydride, vortex immediately for 1 min, let stand for 5 min, add 400 µL buffer, vortex, centrifuge for 5 min, inject a 30 µL aliquot. (Buffer was 67 mM ammonium acetate, 33 mM acetic acid, 40 mM magnesium sulfate, and 1 mM sodium dodecyl sulfate in MeOH:water 35:65.)

HPLC VARIABLES

Column: 150 × 3.9 4 µm Nova-Pak C18

Mobile phase: MeOH:water 25:75 containing 67 mM ammonium acetate, 33 mM acetic acid, 40 mM magnesium sulfate, and 1 mM sodium dodecyl sulfate (Every 100 samples reverse flush column with 300-500 mL water, 300-500 mL MeOH, then with MeOH:water 25:75 for 1 h, re-equilibrate with mobile phase overnight.)

Flow rate: 1

Injection volume: 30

Detector: UV 293

CHROMATOGRAM

Retention time: 6.0

Internal standard: 3'-fluoromerbarone (NSC 372106) (15)

Limit of quantitation: 250 ng/mL

OTHER SUBSTANCES

Extracted: metabolites

KEY WORDS

merbarone is chromatographed in an underivatized form under these conditions

REFERENCE

Supko, J.G.; Malspeis, L. Concurrent determination of merbarone and its urinary metabolites by reversed-phase HPLC with precolumn phenolic acetylation, *J.Liq.Chromatogr.*, **1991**, *14*, 2169–2188.

SAMPLE

Matrix: urine

Analyte: glucuronides

Sample preparation: Condition a 500 mg Bakerbond C18 SPE cartridge with MeOH and 0.1% HCl. Adjust pH of 10 mL urine to 2 with 6 M HCl, add to the SPE cartridge, wash with 5 mL 0.1% HCl, elute with 500 µL MeOH. Evaporate the eluate to dryness under a stream of nitrogen, add 1 mL 0.5% diazomethane in ether, let stand at room temperature for 20 min, evaporate to dryness under a stream of nitrogen, add 200 µL acetic anhydride, add 200 µL pyridine, add 4 mg N,N-dimethylaminopyridine, let stand at room temperature for 5 h, add 5 mL water, extract three times with 3 mL portions of diethyl ether. Combine the extracts and wash them with three 3 mL portions of 100 mM HCl, dry over sodium sulfate, evaporate to dryness under a stream of nitrogen, reconstitute with 1 mL mobile phase, inject a 20 µL aliquot. (Preparation of diazomethane is as follows. Caution! Diazomethane is toxic, explosive, and carcinogenic! A face shield and a safety screen should always be used and the preparation should only be carried out in a properly functioning chemical fume hood. Only smooth glass apparatus with rubber stoppers and plastic tubing should be used. Scratched glassware, ground glass joints, and sharp edges should be avoided (Org. Syn., Coll. Vol. VI; Wiley: New York, 1988, pp. 432-435). Procedures have been reported for the synthesis of diazomethane from N-methyl-N-nitrosourea (Org. Syn., Coll. Vol. II; Wiley: New York, 1943, pp. 165-167), N-nitroso-β-methylamino-isobutyl methyl ketone (Org. Syn., Coll. Vol. III; Wiley:New York, 1955, pp. 244-248), and N,N'-dimethyl-N,N'-dinitrosoterephthalamide (Org. Syn., Coll. Vol. V; Wiley: New York, 1973, pp. 351-355). Probably the most convenient starting material is N-methyl-N-nitroso-p-toluenesulfonamide (Diazald) (Aldrichimica Acta 1983, 16, 3-10; Org. Syn., Coll. Vol. IV, Wiley: New York, 1963, pp. 250-253). Add 10 mL 95% EtOH to a solution of 5 g KOH in 8 mL water, warm to 65° using a water bath, add a solution of 5 g N-methyl-N-nitroso-p-toluenesulfonamide in 45 mL ether over 20 min at such a rate as to keep the reaction volume constant. Collect the ether and diazomethane that distil in an ice-cooled receiving flask under a dry ice/acetone condenser. When all the N-methyl-N-nitroso-p-toluenesulfonamide has been used up, slowly add 10 mL ether to the reaction flask and continue distillation until the distillate is colorless. A purpose-built distillation apparatus can be purchased from Aldrich (Aldrichimica Acta 1983, 16, 3-10). Excess quantities of diazomethane can be destroyed by adding acetic acid until the yellow color of the diazomethane is discharged. The safe disposal of the nitroso compounds used to generate diazomethane has been discussed (Lunn,G.; Sansone,E.B. Destruction of Hazardous Chemicals in the Laboratory, Second Edition. Wiley: New York, 1994, pp. 277-289).)

HPLC VARIABLES
Column: 250 × 4 Eurospher 80-5 vertex C18 (Knauer)
Mobile phase: MeCN:water 35:65 (A), 50:50 (B), 40:60 (C), or 34:55 (D)
Column temperature: 20
Flow rate: 2
Injection volume: 20
Detector: UV 280 (A,C,D); UV 220 (B)

CHROMATOGRAM
Retention time: 10.25 (4-hydroxyphenazone glucuronide (A)), 15.0 (clofibrate glucuronide (B)), 16.5 (N,N-diethyl dithiocarbamate glucuronide (D)), 11.75 (N^4-acetylsulfamethoxazole N^1-glucuronide (C))

REFERENCE
Kohl, C.; Oelschläger, H.; Rothley, D. Identification of drug glucuronides in human urine by RP-HPLC after derivatization, *J.Pharm.Biomed.Anal.*, **1994**, *12*, 249–254.

SAMPLE
Matrix: vegetables
Analyte: tomatine
Sample preparation: Condition a Sep-Pak C18 SPE cartridge with 10 mL MeOH and 10 mL 1% acetic acid (SPE cartridge 1). Condition a Sep-Pak C18 SPE cartridge with 10 mL MeOH and 10 mL MeOH:water 50:50 (SPE cartridge 2). Homogenize (Nippon Seiki universal homogenizer) 5 g homogenized tomato with 50 mL 1% acetic acid and 2.5 g diatomaceous earth (HiFlo Super-Cel (Wako)) for 3 min, filter (Toyo No. 5A paper), re-extract the residue with 40 mL 1% acetic acid. Combine the filtrates and make up to 100 mL with 1% acetic acid, add a 40 mL aliquot to SPE cartridge 1, wash with 5 mL 1% acetic acid, wash with 10 mL MeOH:water 20:80, elute with 5 mL MeOH. Evaporate the eluate to dryness and reconstitute the residue with 200 μL pyridine and 500 μL acetic anhydride, reflux for 1 h, cool to room temperature, add 30 mL MeOH:water 50:50, add to SPE cartridge 2, rinse the reaction flask with 5 mL MeOH:water 50:50, add the rinse to SPE cartridge 2, wash with 10 mL MeOH:water 70:30, elute with 5 mL MeOH. Evaporate the eluate to dryness under reduced pressure, reconstitute the residue with 500 μL MeCN, inject a 20 μL aliquot. (Also for tomato leaves, purée, ketchup, and juice.)

HPLC VARIABLES
Column: 250 × 4.6 5 μm Inertsil ODS-2
Mobile phase: MeCN:water 90:10
Flow rate: 1
Injection volume: 20
Detector: UV 205

CHROMATOGRAM
Retention time: 12
Limit of detection: 1 μg/g

KEY WORDS
tomato; SPE; fruit; leaves; purée; ketchup; juice

REFERENCE
Takagi, K.; Toyoda, M.; Shimizu, M.; Satoh, T.; Saito, Y. Determination of tomatine in foods by liquid chromatography after derivatization, *J.Chromatogr.A*, **1994**, *659*, 127–131.

SAMPLE
Matrix: water
Analyte: phenols
Sample preparation: Prepare an SPE cartridge by adding 500 mg 120-400 mesh Carbograph 4 graphitized carbon black (210 m²/g, Carbochimica Romana, Rome) to a 65 × 13 polypropylene tube using polyethylene frits. Condition with 10 mL 10 mM tetrabutylammonium chloride in dichloromethane:MeOH 80:20, 2 mL MeOH, and 14 mL water acidified to pH 2 with HCl. Filter (Whatman GF/C 1.5 μm glass fiber) river water, pass 4 L through the SPE cartridge at 100 mL/min, wash with 7 mL water at 5-7 mL/min, pull air through the SPE cartridge for 1 min, wash with 800 μL MeOH, dry under vacuum for 1 min, elute in a reverse fashion with 6 mL 10 mM tetrabutylammonium chloride in dichloromethane:MeOH 80:20 at 6 mL/min. Remove a 3 mL aliquot of the eluate and evaporate it to dryness under a stream of nitrogen at 27°, reconstitute the residue in 150 μL 100 mM sodium carbonate in MeCN:water 20:80, add 40 μL acetic anhydride, heat at 50° for 6 min, inject a 50 μL aliquot. (Phenols can also be determined without derivatization. Derivatization provides confirmation of peak identity.)

HPLC VARIABLES
Column: 250 × 4.6 5 μm Alltima LC-18 (Alltech)
Mobile phase: Gradient. A was 0.025% trifluoroacetic acid in water. B was 0.0125% trifluoroacetic acid in MeCN. A:B from 78:22 to 10:90 over 27 min.
Flow rate: 1
Injection volume: 50
Detector: UV 280 for 18 min then UV 220

CHROMATOGRAM
Retention time: 12.4 (phenol), 13.8 (2-nitrophenol), 14.4 (4-nitrophenol), 16.2 (2,4-dinitrophenol), 17.0 (2-chlorophenol), 17.9 (2,4-dimethylphenol), 18.8 (4,6-dinitro-2-methylphenol), 19.9 (4-chloro-3-methylphenol), 20.6 (2,4-dichlorophenol), 23.6 (2,4,6-trichlorophenol), 28.4 (pentachlorophenol)
Limit of detection: <50 pg/mL

KEY WORDS
SPE; river water

REFERENCE
Di Corcia, A.; Bellioni, A.; Madbouly, M.D.; Marchese, S. Trace determination of phenols in natural waters. Extraction by a new graphitized carbon black cartridge followed by liquid chromatography and re-analysis after phenol derivatization, *J.Chromatogr.A*, **1996**, *733*, 383–393.

Benzoic Anhydride

+ ROH ⟶

SAMPLE
Matrix: bulk
Analyte: saccharides
Sample preparation: Evaporate hydrolysates of glycosaminoglycans to dryness, reconstitute in 500 μL 10% benzoic anhydride in pyridine containing 5% 4-dimethylaminopyridine, heat at 37° for 1.5 h, add 4.5 mL water, shake vigorously, pass through a Sep-Pak

C18 SPE cartridge three times, wash with 10 mL pyridine:water 10:90, wash with 5 mL water, reverse the direction of flow and elute with 2.5 mL MeCN, evaporate the eluate to dryness, reconstitute with MeCN, centrifuge at 11000 g for 5 min, inject an aliquot.

HPLC VARIABLES
Guard column: 30 × 4.6 RP-18
Column: 250 × 4.6 Supelcosil LC-18
Mobile phase: MeCN:water 75:25
Flow rate: 1
Injection volume: 20
Detector: UV 230

CHROMATOGRAM
Retention time: 4.3 (N-acetylgalactosamine, N-acetylglucosamine), 5.7 (1-methylfucose), 5.8 (1,6-anhydroidose), 5.9 (galactosamine), 6.1 (glucosamine), 7.0 (1-methylxylose), 8.0 (1-methylmannose), 8.9 (fucose), 9.1 (1-methylgalactose), 9.4 (xylose, 1-methylglucose), 10.5 (mannose), 12.2 (dextrose, galactose)

KEY WORDS
SPE

REFERENCE
Karamanos, N.K.; Hjerpe, A.; Tsegenidis, T.; Engfeldt, B.; Antonopoulos, C.A. Determination of iduronic acid and glucuronic acid in glycosaminoglycans after stoichiometric reduction and depolymerization using high-performance liquid chromatography and ultraviolet detection, *Anal.Biochem.*, **1988**, *172*, 410−419.

Diacetyl-L-tartaric Anhydride

SAMPLE
Matrix: blood
Analyte: propranolol
Sample preparation: 1 mL Plasma + 1 mL 0.5 N pH 10 sodium bicarbonate/sodium carbonate buffer + 100 μL 300 ng/mL IS, vortex for 10 s, add to a 150 × 14 column packed with a 45 mm layer of Extrelut on top of a 25 mm layer of anhydrous sodium sulfate. elute with 15 mL diethyl ether. Add the eluate to 100 μL 10 mM trichloroacetic acid in dry dichloromethane, add 100 μL 250 mM (+)-diacetyl-L-tartaric anhydride ((R,R)-O,O-diacetyltartaric acid anhydride) in acetic acid:dichloromethane 20:80, mix, heat at 40° for 4 h, evaporate to dryness under a stream of nitrogen, wash the tube down with 1 mL MeOH, evaporate to dryness under a stream of nitrogen, reconstitute with 20 μL acetic acid, add 20 μL MeOH, add 60 μL water, inject a 50μL aliquot.

HPLC VARIABLES
Guard column: 15 × 4 10 μm Lichrosorb RP18
Column: 125 × 4 5 μm C18 Hypersil
Mobile phase: MeCN:2% aqueous acetic acid 70:30, adjusted to pH 4.0 with concentrated ammonia
Flow rate: 1
Injection volume: 50
Detector: F ex 290 em 335

CHROMATOGRAM
Retention time: 4 (R), 6 (S)
Internal standard: N-tert-butylpropranolol (Synthesize as follows. Reflux 2.9 g 1-naphthol, 30 mL epichlorohydrin (Caution! Epichlorohydrin is a carcinogen!), and 4.4 g (ca. 22 mequiv OH$^-$, Merck) ion-exchange resin for 4 h, filter, evaporate to dryness, take the residue up in toluene, evaporate to dryness, take the residue up in toluene, evaporate to dryness, take up the residue in hot petroleum ether, evaporate to dryness. Reflux the residue with 30 mL tert-butylamine for 16 h, evaporate to dryness, take up the residue in 30 mL diethyl ether, wash with two 15 mL portions of water, add 4.5 mL 4 M HCl. Remove the organic phase and the product crystallizes after several h, recrystallize from water to give N-tert-butylpropranolol hydrochloride (mp 180°).) (5 (R),7 (S))
Limit of detection: 0.5 ng/mL (R), 1 ng/mL (S)

KEY WORDS
plasma; chiral; pharmacokinetics

REFERENCE
Lindner, W.; Rath, M.; Stoschitzky, K.; Uray, G. Enantioselective drug monitoring of (R)- and (S)-propranolol in human plasma via derivatization with optically active (R,R)-O,O-diacetyl tartaric acid anhydride, *J.Chromatogr.*, **1989**, *487*, 375–383.

SAMPLE
Matrix: blood
Analyte: halofantrine
Sample preparation: 500 µL Plasma + 30 µL 100 µg/mL IS in MeCN:water 80:20, vortex at high speed and add 2 mL MeCN, centrifuge at 1800 g for 3 min. Remove the supernatant and add it to 500 µL ammonium hydroxide and 5 mL MTBE:hexane 50:50, vortex at high speed for 90 s, centrifuge at 1800 g. Remove the upper organic layer and evaporate it to dryness under a stream of nitrogen at 25°, reconstitute the residue in 300 µL 250 mM diacetyl-L-tartaric anhydride ((+)-di-O-acetyl-L-tartaric acid anhydride) in acetic acid:dichloromethane 20:80 (freshly prepared), heat at 45° for 30 min, add 300 µL MeOH, evaporate to dryness under a stream of nitrogen at 25°, reconstitute with 170 µL mobile phase, inject a 30-100 µL aliquot.

HPLC VARIABLES
Guard column: Guard-Pak ODS (Waters)
Column: 250 × 4.6 Ultrasphere ODS
Mobile phase: MeCN:buffer 53.5:46.5 containing 0.9 g/L sodium dodecyl sulfate (Buffer was 25 mM KH$_2$PO$_4$ containing 1.5 mL/L 2 M sulfuric acid and 0.5 mL/L triethylamine, pH 5.0.)
Flow rate: 1.2
Injection volume: 30-100
Detector: UV 254

CHROMATOGRAM
Retention time: 11.5 (+), 13.3 (-)
Internal standard: (±)-2,4-dichloro-α-[2-(dibutylamino)ethyl]-6-(trifluoromethyl)-9-phenanthrenemethanol (SK&F 99123) (17.0, 20.7 (enantiomers))
Limit of quantitation: 12.5 ng/mL

KEY WORDS
plasma; pharmacokinetics; chiral

REFERENCE
Brocks, D.R.; Dennis, M.J.; Schaefer, W.H. A liquid chromatographic assay for the stereospecific quantitative analysis of halofantrine in human plasma, *J.Pharm.Biomed.Anal.*, **1995**, *13*, 911–918.

O,O-Dibenzoyltartaric acid anhydride

SAMPLE
Matrix: bulk
Analyte: 1-methyl-3-pyrrolidinol
Sample preparation: Dissolve 10 mg 1-methyl-3-pyrrolidinol in 1 mL THF, add 35 mg, add 80 mg O,O-dibenzoyltartaric acid anhydride, heat at 50° for 4 h, dilute 10-fold with MeOH, inject a 3 μL aliquot. (Synthesize (+)-O,O'-dibenzoyl-L-tartaric anhydride as follows. Heat 1 mole L-(+)-tartaric acid and 2.7 mole benzoyl chloride at 150° for 4 h, cool, wash with ligroin, boil with xylene, remove the organic phase, repeat several times, filter, dry in a desiccator (Electrophoresis 1994, 15, 769).)

HPLC VARIABLES
Column: 150 × 4.6 5 μm Zorbax C8
Mobile phase: MeOH:0.1% triethylanine in water 50:50, adjusted to pH 4.2 with glacial acetic acid
Flow rate: 2
Injection volume: 3
Detector: UV 254

CHROMATOGRAM
Retention time: 2.3 (S), 3.3 (R)

KEY WORDS
chiral

REFERENCE
Demian, I.; Gripshover, D.F. High-performance liquid chromatographic determination of enantiomeric purity of 1-methyl-3-pyrrolidinol via derivatization with (R,R)-O,O-dibenzoyltartaric acid anhydride, *J.Chromatogr.*, **1987**, *387*, 532–535.

Di-p-toluoyltartaric Acid Anhydride

+ ROH ⟶

SAMPLE
Matrix: bulk
Analyte: atenolol
Sample preparation: Dissolve 6 g atenolol in 100 mL water, adjust pH to 4.0 with 500 mM trichloroacetic acid, freeze-dry. Dissolve in 200 mL chloroform, add 16 g (R,R)-O,O-di-p-toluoyltartaric acid anhydride, reflux for 1 h, cool, wash twice with 200 mL portions of 200 mM pH 6.9 ammonium acetate. Evaporate the chloroform layer to dryness under reduced pressure, reconstitute with mobile phase, inject an aliquot. (Synthesis of (R,R)-O,O-di-p-toluoyltartaric acid anhydride is as follows. Slowly heat 85 g finely pulverized (+)-tartaric acid and 325 p-toluoyl chloride to 120° using an oil bath, after about 30 min when the evolution of HCl is almost over the melt will turn into a thick crystal mash, heat at 140° for 2 h, cool, mash with 300 mL cold benzene (Caution! Benzene is a carcinogen!), filter. Crystallize the solid from xylene, recrystallize twice from ethyl acetate to give (R,R)-O,O-di-p-toluoyltartaric acid anhydride (mp 197-198° d) (Helv. Chim. Acta 1943, 26, 922). For preparative work, collect the fractions containing the diastereomers separately and evaporate them to dryness under reduced pressure, reconstitute with 100 mL 1 M pH 11.8 aqueous ammonia, heat at 60° for 1 h, adjust to pH 4.0 with 2 M HCl, wash twice with 200 mL portions of chloroform. Make the aqueous phase alkaline with ammonia, freeze-dry, take up the residue in chloroform, filter, evaporate the filtrate to dryness under a stream of nitrogen to obtain the enantiomers.)

HPLC VARIABLES
Guard column: 50 × 21.4 8 μm silica (Rainin Dynamax axial compression)
Column: two 250 × 21.4 8 μm silica columns in series (Rainin Dynamax axial compression)
Mobile phase: Chloroform:MTBE:MeOH:hexane:acetic acid 100:100:80:30:1
Flow rate: 10
Detector: UV 280

CHROMATOGRAM
Retention time: 7 (-), 12 (+)

KEY WORDS
chiral; preparative; normal phase

REFERENCE
Wilson, M.J.; Ballard, K.D.; Walle, T. Preparative resolution of the enantiomers of the beta-blocking drug atenolol by chiral derivatization and high performance liquid chromatography, *J.Chromatogr.*, **1988**, *431*, 222–227.

Phthalic Anhydride

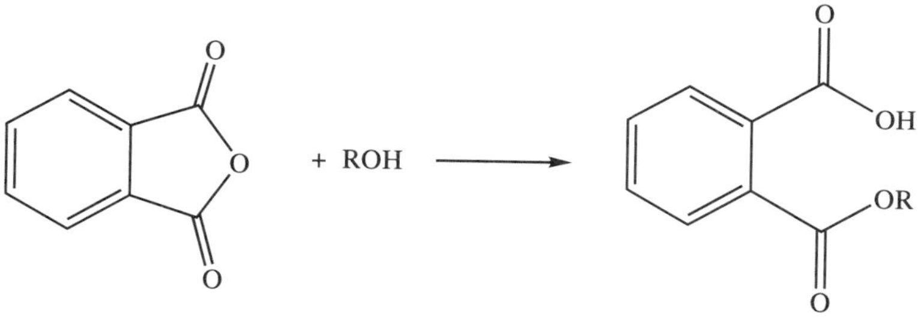

+ ROH ⟶

SAMPLE
Matrix: bulk
Analyte: polyethylene glycol 600
Sample preparation: Add 1 mL 1 M phthalic anhydride in pyridine containing 30 mM imidazole to 100 mg polymer, heat at 100° for 1 h, cool, add 2 mL water, heat at 50° for 30 min. Remove a 500 μL aliquot and add it to 2 mL water, mix, inject an aliquot.

CAPILLARY ELECTROPHORESIS
Capillary: 45 cm × 75 μm μPage-3 3%T/3%C gel-filled with NO urea (40.1 cm to detector) (J&W Scientific)
Capillary preparation: Operate at -100 v/cm for 5 min, ramp to -250 V/cm over 30 min, maintain at -250 V/cm for 5 min
Running buffer: pH 8.3 Tris-borate
Voltage/Current: -243 V/cm
Injection: Electrokinetic injection at -10 kV for 1 min
Detector: UV 275
Model: Dionex CES-1
Migration time: 12-26 (depending on degree of polymerization)

OTHER SUBSTANCES
Also analyzed: AP3S (sulfated 3 mol ethylene oxide derivative of phenol), AP7P (phosphated 7 mol ethylene oxide derivative of phenol), AP30 (30 mol ethylene oxide derivative of benzene), AP40P (phosphated 40 mol ethylene oxide derivative of phenol), surfactants

KEY WORDS
Some changes in CE conditions are needed for ionic surfactants

REFERENCE
Wallingford, R.A. Oligomeric separation of ionic and nonionic ethoxylated polymers by capillary gel electrophoresis, *Anal.Chem.*, **1996**, *68*, 2541–2548.

Trifluoroacetic Anhydride

RELATED REFERENCES
Cobin, J.A.; Johnson, N.A. Determination of total avermectin B_1 and 8,9-Z-avermectin B_1 residues in wine by liquid chromatography. *J.AOAC Int.* **1996**, *79*, 1158-1161.
Lin, C.C.; Matuszewski, B.K.; Zagrobelny, J.; Dobrinska, M.R. Picogram determination of an avermectin analog in dog plasma by high-performance liquid chromatography with fluorescence detection. *J.Liq.Chromatogr.Rel.Technol.* **1997**, *20*, 443-458.

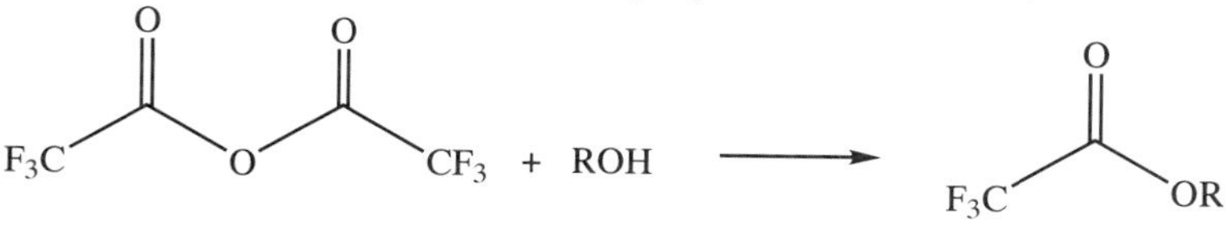

+ ROH ⟶

SAMPLE
Matrix: blood
Analyte: levomoprolol
Sample preparation: Condition a 1 mL Bond Elut CN SPE cartridge with 1 volume dichloromethane, 1 volume MeOH, and 1 volume pH 10 carbonate buffer, do not allow to go dry. 1 mL Plasma + 10 μL 100 ng/mL IS in water + 500 μL pH 10 carbonate buffer, mix, add to the SPE cartridge, wash with two volumes of pH 10 carbonate buffer, centrifuge at 4000 rpm for 15 s, elute with two volumes of dichloromethane while centrifuging at 1000 rpm for 3 min. Evaporate the eluate to dryness under a stream of nitrogen at room temperature, reconstitute the residue in 1 mL diethyl ether, add 250 μL trifluoroacetic anhydride, let stand at room temperature for 45 min, evaporate to dryness under a stream of nitrogen, reconstitute with 500 μL mobile phase, inject a 100 μL aliquot.

HPLC VARIABLES
Column: 250 × 4 5 μm LiChrosorb CN
Mobile phase: n-Pentane:diethyl ether 55:45
Flow rate: 1
Injection volume: 100
Detector: UV 223

CHROMATOGRAM
Retention time: 4.7
Internal standard: 1-(3-chloroisoxazol-5-yl)-2-(tert-butylamino)ethanol (Zambon Group, Milan) (5)

KEY WORDS
SPE; plasma; an aliquot of the effluent can be further analyzed by GC using an electron capture detector

REFERENCE
Gianesello, V.; Brenn, E.; Figini, G.; Gazzaniga, A. Determination by coupled high-performance liquid chromatography-gas chromatography of the β-blocker levomoprolol in plasma following ophthalmic administration, *J.Chromatogr.*, **1989**, *473*, 343–352.

SAMPLE
Matrix: vegetables
Analyte: 4"-deoxy-4"-(epimethylamino)avermectin B$_1$ benzoate (MK-0244)
Sample preparation: Condition a 1 g Bond Elut C8 SPE cartridge with 5 mL MeOH and 5 mL water. Condition a 500 mg Bond Elut propylsulfonyl SPE cartridge with 5 mL 1% ammonium acetate in MeOH, 5 mL 1% phosphoric acid in MeOH, 5 mL water, 5 mL MeOH, and 5 mL ethyl acetate. Homogenize (Polytron) 10 g ground vegetable with 25 mL MeOH, filter (Whatman No. 50 paper) the supernatant, rinse apparatus twice with 25 mL portions of MeOH, filter rinse, rinse apparatus with 50 mL water, filter rinse, make up filtrate to 500 mL with water, add to the C8 SPE cartridge at 10 mL/min, elute with 10 mL 1% ammonium acetate in MeOH, evaporate the eluate to about 1 mL, dilute to 5 mL with 1% ammonium acetate in water, add 5 mL ethyl acetate, shake for 1 min, centrifuge at 2000 rpm for 2 min, repeat the extraction. Combine the ethyl acetate layers and add them to the propylsulfonyl SPE cartridge, elute with 5 mL 1% ammonium acetate in MeOH. Make up the volume of the eluate to 5 mL with 1% ammonium acetate in MeOH, vortex for 10 s, centrifuge for 1 min. Remove a 2.5 mL aliquot of the supernatant and evaporate it to dryness under a stream of nitrogen for 70-85° (make sure that no moisture is left), reconstitute with MeCN, add 100 μL 1-methylimidazole, vortex, sonicate for 5-10 s, centrifuge at 2000 rpm for 2 min, cool in ice, add 300 μL trifluoroacetic anhydride:MeCN 1:2 (Caution! Corrosive!), vortex, centrifuge for 1 min, let stand at room temperature for 10 min, make up to 5 mL with MeCN, vortex, centrifuge briefly, inject an aliquot.

HPLC VARIABLES
Guard column: 15 × 3.2 7 μm RP-18 OD-GU C18 (Brownlee)
Column: 150 × 4.6 Chromegabond C18 (ES Industries)
Mobile phase: MeOH:water 93:7
Column temperature: 30

Flow rate: 1.5
Detector: F ex 365 em 418 (cutoff filter)

CHROMATOGRAM
Retention time: 10 (MK-244 B1$_b$), 12 (MK-244 B1$_a$)
Limit of quantitation: 5 ng/g
Limit of detection: 2 ng/g

KEY WORDS
celery; lettuce; silanize glassware with dichlorodimethylsilane; SPE

REFERENCE
Prabhu, S.V.; Wehner, T.A.; Egan, R.S.; Tway, P.C. Determination of 4"-deoxy-4"-(epimethylam-ino)avermectin B$_1$ benzoate (MK-0244) and its delta 8,9-isomer in celery and lettuce by HPLC with fluorescence detection, *J.Agric.Food Chem.*, **1991**, *39*, 2226–2230.

Trimethylacetic Anhydride

$(H_3C)_3C$ ... O ... O ... $C(CH_3)_3$ + ROH $\longrightarrow$ $(H_3C)_3C$ ... O ... OR

SAMPLE
Matrix: bulk
Analyte: liothyronine
Sample preparation: Dissolve 1 mg compound in 1 mL EtOH and 10 μL 1 M KOH in EtOH:water 50:50 (freshly prepared). Remove a 10 μL aliquot and evaporate it to dryness at <0.05 Torr at 45° for 30 min, add 50 μL 80 mM 4-dimethylaminopyridine in dry MeCN, add 5 μL EtOH, vortex thoroughly, add 50 μL trimethylacetic anhydride, vortex for 10 s, heat at 65-70° for 50 min, add 100 μL EtOH, heat at 65-70° for 10 min, evaporate to dryness, add 100 μL toluene, add 100 μL 100 mM pH 6 phosphate buffer, vortex, centrifuge. Remove the organic layer and evaporate it to dryness, add 100 μL MeCN, sonicate for 2 min, add 100 μL pH 2.1 phosphate buffer, mix, inject an aliquot. (Pass MeCN through an aluminum oxide column before use.)

HPLC VARIABLES
Column: 150 × 4.6 Supelcosil LC-8
Mobile phase: Gradient. MeCN:10 mM KH$_2$; PO$_4$; adjusted to pH 2.1 with phosphoric acid from 30:70 to 87:13 over 10 min.
Flow rate: 2
Detector: UV 214

CHROMATOGRAM
Retention time: 10.7

OTHER SUBSTANCES
Simultaneously analyzed: 3,5-diiodothyronine, thyroxine

REFERENCE
Joppich, M.; Joppich-Kuhn, R.; Sentissi, A.; Giese, R.W. Single-step, quantitative derivatization of amino, carboxyl, and hydroxyl groups in iodothyronine amino acids with ethanolic pivalic anhydride containing 4-dimethylaminopyridine, *Anal.Biochem.*, **1986**, *153*, 159–165.

CARBOXYLIC ACID

9-Anthracenecarboxylic Acid

+ ROH ⟶

SAMPLE
Matrix: blood
Analyte: dolichols
Sample preparation: Dilute plasma with an equal volume of water. Mix 2 mL diluted plasma with 250 μL 50% KOH, add 3 mL 1% pyrogallol in MeOH, heat at 70° for 1 h, cool to room temperature, add IS in n-hexane, extract twice with 4 mL portions of n-hexane. Combine the organic layers and evaporate them to dryness under a stream of nitrogen at 40°, reconstitute the residue in 50 μL 200 mM 9-anthracenecarboxylic acid in distilled THF containing 200 mM triphenylphosphine, add 50 μL 200 mM diethyl azodicarboxylate in THF, let stand at room temperature for 10 min, evaporate to dryness under a stream of nitrogen at 40°, reconstitute the residue in 400 μL distilled benzene (Caution! Benzene is a carcinogen!), add to a Sep-Pak Si SPE cartridge, elute with 2.5 mL benzene. Evaporate the eluate to dryness, reconstitute the residue in 500 μL MeCN:ethyl acetate 30:70, inject a 100 μL aliquot.

HPLC VARIABLES
Column: 300 × 4.6 5 μm Nucleosil C18
Mobile phase: MeCN:ethyl acetate:water 32:68:2
Flow rate: 1.4
Injection volume: 100
Detector: F ex 360 em 460

CHROMATOGRAM
Retention time: 17 (dolichol-17), 20 (dolichol-18), 22 (dolichol-19), 25 (dolichol-20)
Internal standard: 2,2-didecaprenylethanol (33)
Limit of detection: 5 ng

KEY WORDS
SPE; plasma

REFERENCE
Yamada, K.; Yokohama, H.; Abe, S.; Katayama, K.; Sato, T. High-performance liquid chromatographic method for the determination of dolichols in tissues and plasma, *Anal.Biochem.*, **1985**, *150*, 26−31.

SAMPLE
Matrix: bulk
Analyte: dihydroqinghaosu
Sample preparation: 250 mg Dihydroqinghaosu + 10 mL 40 mg/mL 9-anthracenecarboxylic acid in dichloromethane containing 30 μL/mL triethylamine, shake well, add 5 mL 80 mg/mL mesitylenesulfonyl chloride in dichloromethane, add 250 mg 4-dimethylaminopyridine, let stand at room temperature for 15 h, evaporate to dryness under a stream of nitrogen, add 20 mL pH 11 sodium carbonate buffer, heat on a boiling water bath for 30 min, filter, reconstitute the solid, inject an aliquot. (Before use warm solution of mesitylenesulfonyl chloride in dichloromethane briefly at 60° to ensure dissolution.)

HPLC VARIABLES
Column: 150 × 4.6 Ultrasphere C8

Mobile phase: MeCN:water 80:20
Flow rate: 3
Injection volume: 10
Detector: UV 254

CHROMATOGRAM
Retention time: 7

OTHER SUBSTANCES
Simultaneously analyzed: metabolites

REFERENCE
Idowu, O.R.; Grace, J.M.; Leo, K.U.; Brewer, T.G.; Peggins, J.O. Rapid derivatization of alcohols with
  carboxylic-sulphonic mixed anhydrides for HPLC-UV/fluorescence analysis: Application to the detection of dihydroqinghaosu (DQHS) and its metabolites in biological samples,
  J.Liq.Chromatogr.Rel.Technol., **1997**, *20*, 1553–1577.

(-)-2-tert-Butyl-2-methyl-1,3-benzodioxole-4-carboxylic Acid

SAMPLE
Matrix: bulk
Analyte: monosaccharides
Sample preparation: Dissolve 10 μg of a reducing sugar in 100 μL 0.01% perchloric acid
  in acetic anhydride, let stand for 5 min, add 100 μL EtOH, let stand at room temperature
  for 30 min, add 1 mg potassium carbonate, add 100 μL dichloromethane, centrifuge. Remove the supernatant and evaporate it to dryness under reduced pressure, reconstitute
  the residue in 100 μL dichloromethane, add 200 μL 33% HBr in acetic acid, let stand for
  1 h, concentrate under reduced pressure below 60°, dissolve the residue in 100 μL acetone,
  add an excess of (-)-2-tert-butyl-2-methyl-1,3-benzodioxole-4-carboxylic acid, add an equimolar amount of potassium bicarbonate, heat at 60° for 1 h, purify an aliquot by TLC (50
  × 50 silica gel on aluminum sheet (Merck DC-Alufolien) developed with toluene:ethyl

acetate 2:1), remove the fluorescent band at R_f 0.5, extract with MeCN, inject an aliquot of the extract. (Synthesize (-)-2-tert-butyl-2-methyl-1,3-benzodioxole-4-carboxylic acid by analogy with the synthesis of the enantiomer. Reflux 3-methylcatechol (2,3-dihydroxytoluene) and pinacolone (t-butylmethylketone) in toluene in the presence of p-toluenesulfonic acid for 64 h to obtain 2-t-butyl-2,4-dimethyl-1,3-benzodioxole (mp 27°). Oxidize 2-t-butyl-2,4-dimethyl-1,3-benzodioxole with potassium permanganate in pyridine:water 50:50 at 70-80° for 6 h to obtain 2-tert-butyl-2-methyl-1,3-benzodioxole-4-carboxylic acid (mp 179-180°), resolve with cinchonidine (Tet. Lett 1989, 30, 5277).)

HPLC VARIABLES
Column: 150 × 4.6 ODS
Mobile phase: MeCN:water:isopropanol 40:40:10
Flow rate: 0.8
Detector: F

CHROMATOGRAM
Retention time: 22.3 (L-1,2-cis-mannose), 22.8 (D-1,2-cis-mannose), 24.8 (D-arabinose), 25.7 (L-arabinose), 27.1 (D-1,2-trans-mannose), 27.5 (L-galactose), 28.3 (L-xylose), 28.5 (L-1,2-trans-mannose), 28.5 (L-glucose), 29.3 (D-galactose), 29.8 (D-xylose), 30.0 (D-1,2-cis-rhamnose), 30.0 (L-1,2-cis-rhamnose), 30.1 (D-glucose), 30.6 (L-fucose), 32.4 (D-fucose), 36.9 (D-1,2-trans-rhamnose), 40.3 (L-1,2-trans-rhamnose)
Limit of detection: "a few pmoles"

KEY WORDS
chiral

REFERENCE
Nishida, Y.; Bai, C.; Ohrui, H.; Meguro, H. A highly sensitive method to identify the DL-configurations of monosaccharides based on (-)-TBMB carboxylic acid and HPLC, *J.Carbohydr.Chem.*, **1994**, *13*, 1003-1008.

N-(Carbobenzyloxy)-L-phenylalanine

SAMPLE
Matrix: formulations
Analyte: betamethasone and dexamethasone
Sample preparation: Grind tablet in 5 mL water, sonicate for 20 min, extract with 9 mL dichloromethane, extract with three 5 mL portions of dichloromethane. Filter the extracts and make up the filtrate to 25 mL with dichloromethane. Mix a 500 μL aliquot with 200 μL 1.2 mM phenacetin in dichloromethane, 100 μL 0.5 mM 4-dimethylaminopyridine in dichloromethane, 100 μL 100 mM N-(carbobenzyloxy)-L-phenylalanine (N-CBZ-L-Phe) in dichloromethane, and 100 μL 100 mM N,N'-dicyclohexylcarbodiimide, shake mechanically at 30° for 1 h, inject a 10 μL aliquot.

HPLC VARIABLES
Guard column: 10 μm Resolve silica (Waters)
Column: 75 × 3.9 4 μm Nova-Pak silica
Mobile phase: n-Hexane:dichloromethane:isopropanol 50:50:2
Flow rate: 1
Injection volume: 10
Detector: UV 240

CHROMATOGRAM
Retention time: 5 (dexamethasone), 6.5 (betamethasone)
Internal standard: phenacetin (9.5)
Limit of detection: 2.1-4.2 pmole

KEY WORDS
normal phase; tablets

REFERENCE
Chen, S.-H.; Wu, S.-M.; Wu, H.-L. Stereochemical analysis of betamethasone and dexamethasone by derivatization and high-performance liquid chromatography, *J.Chromatogr.*, **1992**, *595*, 203–208.

SAMPLE
Matrix: solutions
Analyte: acebutolol
Sample preparation: 100 μL 55 mM N-(Carbobenzyloxy)-L-phenylalanine (N-benzyloxy-carbonyl-L-phenylalanine, N-CBZ-L-Phe) in dichloromethane + 100 μL 14 mM N,N-dimethylaminopyrine (dimethylaminopyridine (?)) in dichloromethane + 100 μL 9-acetylanthracene in dichloromethane, cool in an ice bath, add 500 μL of a solution of acebutolol in dichloromethane, add 100 μL 240 mM dicyclohexylcarbodiimide in dichloromethane, shake mechanically at 0° for 30 min, add 100 μL 1.06 M acetic anhydride in dichloromethane, shake mechanically at 30° for 15 min, add 1 mL MeOH, mix, inject an aliquot. (N-CBZ-L-Phe derivatizes the alcohol and acetic anhydride derivatizes the secondary amine.)

HPLC VARIABLES
Guard column: 10 μm Nova-Pak C18 precolumn
Column: 150 × 3.9 4 μm Nova-Pak C18
Mobile phase: MeOH:water 60:40
Flow rate: 1.3
Detector: UV 254

CHROMATOGRAM
Retention time: 14.13 (S), 15.74 (R)
Internal standard: 9-acetylanthracene (6.52)

KEY WORDS
chiral

REFERENCE
Wen, Y.H.; Wu, S.S.; Wu, H.L. Chiral separation of acebutolol by derivatization and high-performance liquid chromatography, *J.Liq.Chromatogr.*, **1995**, *18*, 3329–3345.

Carbobenzyloxy-L-proline

$+ ROH \longrightarrow$

SAMPLE
Matrix: blood
Analyte: warfarin
Sample preparation: 200 μL Plasma + 100 μL 8.46 μg/mL IS in water + 1 mL 100 mM potassium carbonate + 4 mL ether, shake for 3 min, centrifuge at 3000 rpm for 5 min, discard the organic layer. Acidify the aqueous layer with 1.5 mL 1 M HCl, add 6 mL ether, shake for 3 min, centrifuge at 3000 rpm for 3 min, freeze by immersion in liquid nitrogen for 40-60 s. Remove the organic layer and evaporate it to dryness under a stream of nitrogen at 45°, reconstitute the residue in 10 μL 100 mg/mL carbobenzyloxy-L-proline in MeCN, 10 μL 1 mg/mL imidazole in MeCN, and 10 μL 100 mg/mL dicyclohexylcar-bodiimide in MeCN, vortex for 10 s, let stand for 2 h, centrifuge at 3000 rpm for 5 min, inject a 3-10 μL aliquot.

HPLC VARIABLES
Column: 250 × 5 5 μm Spherisorb Si
Mobile phase: Hexane:ethyl acetate:MeOH:acetic acid 74.75:25:0.25:0.4
Flow rate: 1
Injection volume: 3-10
Detector: UV 313

CHROMATOGRAM
Retention time: 14 (SS), 17 (RS)
Internal standard: 4'-fluorowarfarin (15, 19 (enantiomers))
Limit of detection: 160 ng (S), 96 ng (R)

KEY WORDS
plasma; chiral; pharmacokinetics; normal phase

REFERENCE
Banfield, C.; Rowland, M. Stereospecific high-performance liquid chromatographic analysis of warfarin in plasma, *J.Pharm.Sci.*, **1983**, 72, 921–924.

SAMPLE
Matrix: blood
Analyte: warfarin
Sample preparation: Acidify 1 mL plasma to pH <2 with 100 μL 2 M HCl, extract with 5 mL diethyl ether:hexane 50:50. Remove the organic layer and evaporate it to dryness, reconstitute the residue in 20 μL 100 mg/mL carbobenzyloxy-L-proline in MeCN, 20 μL 1 mg/mL imidazole in MeCN, and 20 μL 100 mg/mL dicyclohexylcarbodiimide in MeCN, vortex for 10 s, let stand for 5-16 h, make up to 100 μL with MeCN, inject a 20 μL aliquot (J.Pharm.Sci. 1983, 72, 921; J.Clin.Pharmacol. 1995, 35, 1008).

HPLC VARIABLES
Column: 100 × 4.6 3 μm Microspher C18 (Chrompack)
Mobile phase: MeCN:isopropanol:pH 6.6 phosphate buffer (I = 0.017) 36:12:52
Flow rate: 1.5
Injection volume: 20
Detector: UV 313

CHROMATOGRAM
Retention time: 14 (R), 17 (S)
Internal standard: p-chlorowarfarin (26)
Limit of detection: 60 ng/mL

KEY WORDS
plasma; chiral; pharmacokinetics

REFERENCE
Sutfin, T.; Balmer, K.; Boström, H.; Eriksson, S.; Höglund, P.; Paulsen, O. Stereoselective interaction of omeprazole with warfarin in healthy men, *Ther.Drug Monit.*, **1989**, *11*, 176–184.

2-(4-Carboxyphenyl)-5,6-dimethylbenzimidazole

SAMPLE
Matrix: blood
Analyte: corticosteroids
Sample preparation: 100 µL Plasma + 10 µL IS in water, extract twice by shaking for 1 min with 1.2 mL dichloromethane, evaporate organic layer below 40° under reduced pressure, dissolve residue in 100 µL MeCN. Add 10 µL reagent 1, add 10 µL reagent 2, heat at 70° for 20 min, cool to room temperature, add 100 µL water, add 200 µL MeOH:water 1:1, add to Sep-Pak C18 cartridge, wash vial with 2 mL MeOH:water 1:1 and add washings to cartridge, wash cartridge with 40 mL MeOH:water 1:1, elute with 5 mL MeOH. Concentrate eluent to 500 µL by evaporation at 40° under reduced pressure, inject 20 µL aliquot. (Reagent 1 was 30 mg 2-(4-carboxyphenyl)-5,6-dimethylbenzimidazole in 3 mL pyridine, add 700 mg 4-piperidinopyridine, dilute to 10 mL with MeCN. Reagent 2 was 700 mg 1-isopropyl-3-(3-dimethylaminopropyl)carbodiimide perchlorate in 10 mL MeCN. Prepare 2-(4-carboxyphenyl)-5,6-dimethylbenzimidazole as described below. 4-Piperidinopyridine is not commercially available but 4-dimethylaminopyridine or 4-pyrrolidinopyridine can be used instead although interferences are greater (J. Chromatogr. 1991, 585, 219). Alternatively 4-piperidinopyridine can be synthesized as described below. Prepare 1-isopropyl-3-(3-dimethylaminopropyl)carbodiimide perchlorate as described below.)

HPLC VARIABLES
Guard column: 50 × 4.6 7 µm Zorbax ODS
Column: 250 × 4.6 7 µm Zorbax ODS
Mobile phase: MeOH:water 75:25 containing 5 mM tetramethylammonium hydrogen sulfate
Flow rate: 0.4
Injection volume: 20
Detector: F ex 334 em 418

CHROMATOGRAM
Retention time: 21.1 (aldosterone), 25.2 (cortisone), 26.5 (hydrocortisone), 33.1 (dexamethasone), 43.7 (triamcinolone), 49.4 (triamcinolone acetonide)

Internal standard: fluocinolone acetonide (40.7)
Limit of detection: 0.6-3 pg/mL

KEY WORDS
plasma

REFERENCE
Katayama, M.; Masuda, Y.; Taniguchi, H. Determination of corticosteroids in plasma by high-performance liquid chromatography after pre-column derivatization with 2-(4-carboxyphenyl)-5,6-dimethylbenzimidazole, *J.Chromatogr.*, **1993**, *612*, 33–39.

SAMPLE
Matrix: blood
Analyte: estrogens
Sample preparation: 100 µL Plasma + 10 µL IS in water, extract twice by shaking for 1 min with 1.2 mL dichloromethane, evaporate organic layer below 40° under reduced pressure, dissolve residue in 100 µL MeCN. Add 10 µL reagent 1, add 10 µL reagent 2, heat at 50° for 15 min, cool to room temperature, add 100 µL water, add 200 µL MeOH:water 1:1, add to Sep-Pak C18 cartridge, wash vial with 2 mL MeOH:water 1:1 and add washings to cartridge, wash cartridge with 40 mL MeOH:water 1:1, elute with 5 mL MeOH. Concentrate eluent to 500 µL by evaporation at 40° under reduced pressure, inject 20 µL aliquot. (Reagent 1 was 30 mg 2-(4-carboxyphenyl)-5,6-dimethylbenzimidazole in 3 mL pyridine, add 700 mg 4-piperidinopyridine, dilute to 10 mL with MeCN. Reagent 2 was 700 mg 1-isopropyl-3-(3-dimethylaminopropyl)carbodiimide perchlorate in 10 mL MeCN. Prepare 2-(4-carboxyphenyl)-5,6-dimethylbenzimidazole as follows. Add 13 g 4-carboxybenzaldehyde (terephthalaldehydic acid) in 400 mL EtOH dropwise to 4,5-dimethyl-1,2-phenylenediamine in 400 mL EtOH in an ice bath, after 1 h reflux for 8 h, cool to room temperature, collect the precipitate, recrystallize three times from MeOH:water 50:50 to give 2-(4-carboxyphenyl)-5,6-dimethylbenzimidazole as a white amorphous product (mp >300°) (J.Chromatogr. 1991, 585, 219). 4-Piperidinopyridine is not commercially available but 4-dimethylaminopyridine or 4-pyrrolidinopyridine can be used instead although interferences are greater (J. Chromatogr. 1991, 585, 219). Alternatively 4-piperidinopyridine can be synthesized as follows. Add 200 mmoles piperidine dropwise with stirring to 15 g phosphorus pentoxide and 9.51 g 4-hydroxypyridine, heat at 250° for 7 h, cautiously pour onto 200 g ice, add 400 mL 1 M NaOH, add 200 mL ether. Remove the ether layer and extract the aqueous layer three times with 100 mL portions of ether. Combine the organic layers and dry them over anhydrous potassium carbonate, evaporate, distil the residue, recrystallize from petroleum ether (bp 80-100°) to give 4-piperidinopyridine (bp 167-170°/11 mm Hg; mp 79-80°) (Synthesis 1978, 844). Alternatively, add 1.94 g 4-bromopyridine hydrochloride to 5 mL 50% NaOH, add 5 mL piperidine, add 2.72 g benzyltriethylammonium bromide, heat at 100° for 5 h, remove excess piperidine by distillation, add 25 mL water, extract four times with 25 mL portions of benzene. Combine the organic layers and dry them over anhydrous sodium sulfate, boil the residue with petroleum ether to give 4-piperidinopyridine (mp 80°) (Syn. Commun. 1979, 9, 251). Prepare 1-isopropyl-3-(3-dimethylaminopropyl)carbodiimide perchlorate as follows. Stir 1.41 moles isopropylisocyanate in 750 mL dichloromethane at 5°, add 144 g 3-dimethylaminopropylamine (N,N-dimethyl-1,3-propanediamine) in 250 mL dichloromethane at such a rate that the temperature does not exceed 10°, add 500 mL triethylamine, add 300 g p-toluenesulfonyl chloride in 300 mL dichloromethane at such a rate that the temperature does not exceed 10°, reflux for 3 h, add 400 g anhydrous sodium carbonate, add 3.5 L ice water, stir vigorously for 30 min, remove the organic phase. Extract the aqueous phase three times with 500 mL portions of dichloromethane. Combine the organic layers and dry them over anhydrous sodium sulfate, evaporate under reduced pressure, distil the residue to give 1-isopropyl-3-(3-dimethylaminopropyl)carbodiimide (bp 91-92°/10 mm Hg (Ber. 1941, 74B, 1285)) (cf. Org. Syn. 1973, Coll. Vol. V, 555). Prepare pyridine perchlorate from pyridine and 20% perchloric acid, crystallize from EtOH (Ber. 1926, 59, 446). Add 18 g pyridine perchlorate in portions to 100 mmoles1-isopropyl-3-(3-dimethylaminopropyl)carbodiimide stirred in 200 mL dichloromethane at 0°, let stand for 30 min, filter, add 200 mL anhydrous diethyl ether to the filtrate. Filter off the precipitate and recrystallize it from dichloromethane/diethyl ether to give 1-isopropyl-3-(3-dimethylaminopropyl)carbodiimide perchlorate (mp 88-90°) (Chem. Pharm. Bull. 1985, 33, 5375).)

HPLC VARIABLES
Guard column: 50 × 4 5 μm Wakosil 5C18 (Wako, Osaka)
Column: 300 × 4 5 μm Wakosil 5C18 (Wako, Osaka)
Mobile phase: MeOH:water 90:10
Flow rate: 0.7
Injection volume: 20
Detector: F ex 336 em 440

CHROMATOGRAM
Retention time: 10.5 (estriol), 15.4 (ethinyl estradiol), 16.5 (equilenin, equilin), 17.2 (estrone), 18.2 (estradiol), 19.2 (estetrol), 28.1 (4-hydroxyestradiol), 35.5 (2-hydroxyestradiol)
Internal standard: sec-butyl p-hydroxybenzoate (14.3)
Limit of detection: 1-2 pg/mL

OTHER SUBSTANCES
Non-interfering: cortisone, prednisolone, triamcinolone

KEY WORDS
plasma; SPE

REFERENCE
Katayama, M.; Taniguchi, H. Determination of estrogens in plasma by high-performance liquid chromatography after pre-column derivatization with 2-(4-carboxyphenyl)-5,6-dimethylbenzimidazole, *J.Chromatogr.*, **1993**, *616*, 317–322.

SAMPLE
Matrix: solutions
Analyte: fatty alcohols
Sample preparation: Prepare a solution of the alcohol by dissolving 10 mg alcohol in 200 μL pyridine and diluting to 10 mL in MeCN. Mix 1 mL solution with 100 μL reagent and 100 μL 2% 1-isopropyl-3-(3-dimethylaminopropyl)carbodiimide perchlorate in MeCN, heat at 80° for 20 min, cool to room temperature, add 1 mL water, add 2 mL isopropanol:water 50:50, add to a Sep-Pak ODS SPE cartridge, rinse the tube with 3 mL isopropanol:water 50:50, add the rinse to the SPE cartridge, wash with 3 mL isopropanol:water 50:50, elute with 2 mL isopropanol, inject a 20 μL aliquot of the eluate. (Reagent was 10 mg 2-(4-carboxyphenyl)-5,6-dimethylbenzimidazole in 1 mL pyridine, add 700 mg 4-piperidinopyridine, dilute to 10 mL with MeCN. Prepare 2-(4-carboxyphenyl)-5,6-dimethylbenzimidazole as described above. 4-Piperidinopyridine is not commercially available but 4-dimethylaminopyridine or 4-pyrrolidinopyridine can be used instead although interferences are greater. Alternatively 4-piperidinopyridine can be synthesized as described above. Prepare 1-isopropyl-3-(3-dimethylaminopropyl) carbodiimide perchlorate as described above.)

HPLC VARIABLES
Guard column: 50 × 4.6 7 μm Zorbax ODS
Column: 250 × 4.6 7 μm Zorbax ODS
Mobile phase: MeOH:isopropanol 85:15
Flow rate: 1
Injection volume: 20
Detector: F ex 338 em 428

CHROMATOGRAM
Retention time: 6.2 (dodecyl alcohol), 7.2 (tetradecyl alcohol), 9.4 (cetyl alcohol), 12.5 (stearyl alcohol), 16.5 (eicosyl alcohol)
Limit of detection: 10-20 pg/mL

KEY WORDS
SPE

REFERENCE
Katayama, M.; Masuda, Y.; Taniguchi, H. Determination of alcohols by high-performance liquid chromatography after pre-column derivatization with 2-(4-carboxyphenyl)-5,6-dimethylbenzimidazole, *J.Chromatogr.*, **1991**, *585*, 219–224.

2-(4-Carboxyphenyl)-6-methoxybenzofuran

SAMPLE
Matrix: solutions
Analyte: alcohols
Sample preparation: Prepare a 5 ng/mL solution in MeCN. 2 mL Solution + 2 mL 4% 4-dimethylaminopyridine in MeCN + 2 mL 2% 1-(3-dimethylaminopropyl)-3-ethylcarbodi-imide in MeCN + 2 mL reagent solution, heat at 60° for 30 min, cool to room temperature, inject a 10 μL aliquot directly. Alternatively, add the derivatized reaction mixture to 2 mL 100 mM NaOH and 5 mL n-hexane, vortex for 1 min, centrifuge at 1000 g for 2 min. Remove the organic layer and evaporate it to dryness under a stream of nitrogen, recon-stitute the residue in 2 mL MeCN:acetic acid 99.9:0.1, inject a 10 μL aliquot. (Prepare the reagent solution by dissolving 1 mg 2-(4-carboxyphenyl)-6-methoxybenzofuran in 200 μL pyridine, dilute to 10 mL with MeCN. Prepare 2-(4-carboxyphenyl)-6-methoxybenzo-furan as follows. Immediately before use prepare a solution of sodium methoxide by dis-solving 1.2 g sodium metal in 20 mL MeOH. Dissolve 3.04 g 2-hydroxy-4-methoxyben-zaldehyde and 4.9 g α-bromo-p-toluonitrile in 25 mL DMF, add the sodium methoxide solution dropwise, heat at 120° for 4 h, remove the MeOH by distillation, pour the reaction mixture into a mixture of 60 g of ice-cold water and 10 mL MeOH, stir at 0° for 1 h, filter, wash the crystals with water, dry under vacuum, recrystallize from isopropanol to give 6-methoxy-2-(4-cyanophenyl)benzofuran (mp 157°). Dissolve 2.49 g 6-methoxy-2-(4-cyano-phenyl)benzofuran and 10 g powdered KOH in 100 mL propylene glycol, reflux for 8 h, cool to room temperature, pour into a mixture of 100 g ice-water and 30 mL concentrated HCl, filter, wash the crystals with water, dissolve the solid in a minimum of DMF and recrystallize from EtOH to give 2-(4-carboxyphenyl)-6-methoxybenzofuran (mp >300°).)

HPLC VARIABLES
Column: 150 × 4.6 5 μm Ultrasphere C8
Mobile phase: Gradient. A was MeCN:water:acetic acid 35:65:0.1. B was MeCN:acetic acid 99.9:0.1. A:B from 70:30 to 0:100 over 20 min.
Flow rate: 1
Injection volume: 10
Detector: F ex 315 em 390

CHROMATOGRAM
Retention time: 7.5 (MeOH), 8.5 (EtOH), 10 (propanol), 11 (butanol), 12.5 (pentanol), 14 (hexanol), 15 (heptanol), 16 (octanol), 17.5 (nonanol), 18.5 (decanol), 19.5 (undecanol), 20.5 (dodecanol), 21 (tridecanol), 22 (tetradecanol), 23 (pentadecanol), 24 (hexadecanol), 24.5 (heptadecanol), 25.5 (octadecanol), 26 (nonadecanol), 27.5 (eicosanol)
Limit of detection: 0.1-0.5 pg

REFERENCE
Haj-Yehia, A.I.; Benet, L.Z. Determination of alcohols by high-performance liquid chromatography with fluorimetric detection after precolumn derivatization with 2-(4-carboxyphenyl)-6-methoxybenzofuran, *J.Chromatogr.A*, **1996**, *724*, 107–115.

Dihydrofluorescein Diacetate

+ ROH ⟶

SAMPLE
Matrix: solutions
Analyte: dihydroartemisinin
Sample preparation: Stir 1 μmole dihydroartemisinin, 1.5 μmoles dihydrofluorescein diacetate (diacetyldihydrofluorescein), 5 μmoles 4-dimethylaminopyridine, and 15 μmoles N,N'-dicyclohexylcarbodiimide in 750 μL dichloromethane at room temperature for 8 h (or at 40° for 4 h), inject a 10 μL aliquot.

HPLC VARIABLES
Guard column: silica
Column: 100 × 8 Radial-PAK M-Porasil
Mobile phase: Hexane:isopropanol 95:5
Flow rate: 1.5
Injection volume: 10
Detector: UV 235

CHROMATOGRAM
Retention time: 5
Limit of detection: 0.1 ng

KEY WORDS
normal phase

REFERENCE
Luo, X.-d.; Xie, M.; Zou, A.-q. Sub-nanogram detection of dihydroartemisinin after chemical derivatization with diacetyldihydrofluorescein followed by high-performance liquid chromatography and UV absorption, *Chromatographia*, **1987**, *23*, 112–114.

9-Fluoreneacetic Acid

SAMPLE
Matrix: blood
Analyte: dihydroqinghaosu
Sample preparation: 1 mL Plasma + 5 mL MTBE, vortex for 3 min, centrifuge at 2500 rpm for 15 min. Remove the organic layer and evaporate it to dryness under a stream of nitrogen, reconstitute the residue in 200 μL reagent, add 2 mL 3 mg/mL 2,4,6-triisopropylbenzenesulfonyl chloride in dichloromethane, add 2 mL 2.5 mg/mL 4-dimethylaminopyridine in dichloromethane, mix, let stand at room temperature for 2 h, evaporate to dryness under a stream of nitrogen, add 2.5 mL pH 11 sodium carbonate buffer, shake, let stand at room temperature for 15 min, add 4 mL MTBE, vortex for 2 min. Remove the organic layer and wash it with 4 mL water, wash with 2.5 mL 2 M HCl, wash with 4 mL water, dry the organic layer over anhydrous sodium sulfate, evaporate to dryness under a stream of nitrogen,reconstitute with 1 mL MeOH, inject a 30 μL aliquot. (Prepare reagent by dissolving 20 mg 9-fluoreneacetic acid and 20 μL triethylamine in 2 mL dichloromethane.)

HPLC VARIABLES
Column: 150 × 4.6 Ultrasphere C8
Mobile phase: MeCN:water 80:20
Flow rate: 3
Injection volume: 10
Detector: UV 254

CHROMATOGRAM
Retention time: 8

KEY WORDS
sheep; plasma

REFERENCE
Idowu, O.R.; Grace, J.M.; Leo, K.U.; Brewer, T.G.; Peggins, J.O. Rapid derivatization of alcohols with carboxylic-sulphonic mixed anhydrides for HPLC-UV/fluorescence analysis: Application to the detection of dihydroqinghaosu (DQHS) and its metabolites in biological samples, *J.Liq.Chromatogr.Rel.Technol.*, **1997**, *20*, 1553–1577.

SAMPLE
Matrix: blood
Analyte: dihydroqinghaosu
Sample preparation: 1 mL Plasma + 5 mL MTBE, vortex for 3 min, centrifuge at 2500 rpm for 15 min. Remove the organic layer and evaporate it to dryness under a stream of nitrogen, reconstitute the residue in 200 μL reagent, add 2 mL 3 mg/mL 2,4,6-mesitylenesulfonyl chloride in dichloromethane, add 2 mL 2.5 mg/mL 4-dimethylaminopyridine in dichloromethane, mix, let stand at room temperature for 2 h, evaporate to dryness under a stream of nitrogen, add 2.5 mL pH 11 sodium carbonate buffer, shake, let stand at room temperature for 15 min, add 4 mL MTBE, vortex for 2 min. Remove the organic layer and wash it with 4 mL water, wash with 2.5 mL 2 M HCl, wash with 4 mL water, dry the organic layer over anhydrous sodium sulfate, evaporate to dryness under a stream of

nitrogen, reconstitute with 1 mL MeOH, inject a 30 μL aliquot. (Prepare reagent by dissolving 20 mg 9-fluoreneacetic acid and 20 μL triethylamine in 2 mL dichloromethane.)

HPLC VARIABLES
Column: 150 × 4.6 Ultrasphere C8
Mobile phase: MeCN:water 80:20
Flow rate: 3
Injection volume: 10
Detector: UV 254

CHROMATOGRAM
Retention time: 8

KEY WORDS
sheep; plasma

REFERENCE
Idowu, O.R.; Grace, J.M.; Leo, K.U.; Brewer, T.G.; Peggins, J.O. Rapid derivatization of alcohols with carboxylic-sulphonic mixed anhydrides for HPLC-UV/fluorescence analysis: Application to the detection of dihydroqinghaosu (DQHS) and its metabolites in biological samples, *J.Liq.Chromatogr.Rel.Technol.*, **1997**, *20*, 1553–1577.

SAMPLE
Matrix: microsomal incubations
Analyte: dihydroqinghaosu
Sample preparation: 1 mL Microsomal incubation + 100 μL 50 mM NaOH + 5 mL MTBE, vortex for 3 min, centrifuge at 2500 rpm for 15 min. Remove the organic layer and evaporate it to dryness under a stream of nitrogen, reconstitute the residue in 200 μL reagent, add 2 mL 3 mg/mL 2,4,6-trichlorobenzoyl chloride in dichloromethane, add 2 mL 2.5 mg/mL 4-dimethylaminopyridine in dichloromethane, mix, let stand at room temperature for 2 h, evaporate to dryness under a stream of nitrogen, add 2.5 mL pH 11 sodium carbonate buffer, shake, let stand at room temperature for 15 min, add 4 mL MTBE, vortex for 2 min. Remove the organic layer and wash it with 4 mL water, wash with 2.5 mL 2 M HCl, wash with 4 mL water, dry the organic layer over anhydrous sodium sulfate, evaporate to dryness under a stream of nitrogen, reconstitute with 1 mL MeOH, inject a 30 μL aliquot. (Prepare reagent by dissolving 20 mg 9-fluoreneacetic acid and 20 μL triethylamine in 2 mL dichloromethane. This procedure works well with unreactive alcohols but with more reactive alcohols there is a tendency to form the trichlorobenzoyl ester.)

HPLC VARIABLES
Column: 150 × 4.6 Ultrasphere C8
Mobile phase: MeCN:water 80:20
Flow rate: 3
Injection volume: 10
Detector: UV 254

CHROMATOGRAM
Retention time: 8

OTHER SUBSTANCES
Extracted: metabolites

KEY WORDS
rat; liver

REFERENCE
Idowu, O.R.; Grace, J.M.; Leo, K.U.; Brewer, T.G.; Peggins, J.O. Rapid derivatization of alcohols with carboxylic-sulphonic mixed anhydrides for HPLC-UV/fluorescence analysis: Application to the detection of dihydroqinghaosu (DQHS) and its metabolites in biological samples, *J.Liq.Chromatogr.Rel.Technol.*, **1997**, *20*, 1553–1577.

Hexachlorobicyclo[2.2.1]hept-5-ene-2-carboxylic Acid

SAMPLE

Matrix: blood

Analyte: warfarin

Sample preparation: Prepare a chromatographic column of 250 mg silica gel H in a Pasteur pipette, wash with three 1 mL portions of diethyl ether and three 1 mL portions of hexane:ether 4:1. Dry 40-500 μL 1 μg/mL (+)-p-chlorowarfarin in MeOH in a tube under a stream of nitrogen, add 1 mL citrated plasma, add 100 μL 1 M HCl, add 5 mL diethyl ether, vortex for 30 s, centrifuge at 1000 g for 5 min. Remove the ether layer and evaporate it at 40° under a stream of nitrogen, reconstitute in 1 mL hexane:ether 4:1, add to chromatographic column, wash with two 1 mL portions of hexane:ether 4:1, elute with 1 mL ether. Discard first 200 μL eluate, evaporate remainder at 40°, add 15 μL 50 mg/mL HCA in MeCN:water 25:1, add 10 μL 100 mg/mL N,N′-dicyclohexylcarbodiimide in MeCN:water 25:1, let stand for 10 min, inject a 20 μL aliquot within 4 h. (Diethyl ether was freshly distilled before use. HCA was (-)-(1S,2R,4R)-endo-1,4,5,6,7,7-hexachlorobicyclo[2.2.1]hept-5-ene-2-carboxylic acid, preparation is as follows. Reflux 137.3 g hexachlorocyclopentadiene and 43.3 g methyl acrylate by heating at 90-95° with stirring for 24 h, cool, dissolve the product in 625 mL MeOH, add 530 mL 10% NaOH, stir overnight at room temperature, evaporate under reduced pressure to remove the MeOH, acidify with 2 M HCl, extract with dichloromethane. Evaporate the organic layer and recrystallize the residue from hexane to give endo-1,4,5,6,7,7-hexachlorobicyclo[2.2.1]hept-5-ene-2-carboxylic acid as colorless crystals (mp 156-170° (d)). Heat 10 g endo-1,4,5,6,7,7-hexachlorobicyclo[2.2.1]hept-5-ene-2-carboxylic acid with 10 mL thionyl chloride at 100° for 1.5 h, remove the excess thionyl chloride under reduced pressure, add a solution of 5.46 g 2,3-O-isopropylidene-D-ribonic gamma-lactone (2,3-O-isopropylidene-D(+)-ribonic acid 1,4-lactone) in 50 mL THF to the residue, slowly add 3 mL pyridine, let stand for 3 h, filter, wash the crystals with THF. Evaporate the filtrate under reduced pressure, mix the residue with ethyl acetate, wash with 1 M aqueous HCl saturated with NaCl, wash with NaCl solution, evaporate under reduced pressure, chromatograph on a short column under vacuum on Merck silica gel H (TLC grade) with hexane:ethyl acetate 80:20 to separate the diastereomeric esters. Recrystallize the (-)-ester from hexane/ethyl acetate to give colorless plates (mp 129-133°; $[\alpha]_D^{29}$ -47.3° (c, 2.8 in chloroform)). Dissolve 3.58 g of the (-)-ester in 25 mL MeOH and 7 mL 10% NaOH in water, let stand overnight, add 150 mL water, acidify with 1 M HCl, filter, recrystallize the solid from hexane to obtain (-)-(1S,2R,4R)-endo-1,4,5,6,7,7-hexachlorobicyclo[2.2.1]hept-5-ene-2-carboxylic acid (HCA) as colorless crystals (mp 150-186° (d); $[\alpha]_D^{29}$ -9.8° (c = 5.5 in chloroform)) (Aust. J. Chem. 1987, 40, 1641).)

HPLC VARIABLES

Guard column: 15 × 4 10 μm LiChrosorb C18

Column: 250 × 4 10 μm LiChrosorb C18

Mobile phase: MeCN:water 80:20

Flow rate: 1.6

Injection volume: 20

Detector: F ex 313 em 370 (cut-off filter) following postcolumn reaction. The column effluent mixed with 200 mM NaOH pumped at 0.5 mL/min and flowed through a 1 m reaction coil to the detector.

CHROMATOGRAM
Retention time: 6 (S), 7 (R)
Internal standard: (+)-p-chlorowarfarin (9)
Limit of detection: 5 ng/mL

KEY WORDS
plasma; SPE; chiral; post-column reaction

REFERENCE
Carter, S.R.; Duke, C.C.; Cutler, D.J.; Holder, G.M. Sensitive stereospecific assay of warfarin in plasma: reversed-phase high-performance liquid chromatographic separation using diastereoisomeric esters of (-)-(1S,2R,4R)-endo-1,4,5,6,7,7-hexachlorobicyclo[2.2.1]hept-5-ene-2-carboxylic acid, *J.Chromatogr.*, **1992**, *574*, 77–83.

9-Phenanthrenecarboxylic Acid

SAMPLE
Matrix: bile
Analyte: dihydroqinghaosu
Sample preparation: 1 mL Bile + 5 mL MTBE, vortex for 3 min, centrifuge at 2500 rpm for 15 min. Remove the organic layer and evaporate it to dryness under a stream of nitrogen, reconstitute the residue in 200 µL reagent, add 2 mL 3 mg/mL 2,4,6-triisopropylbenzenesulfonyl chloride in dichloromethane, add 2 mL 2.5 mg/mL 4-dimethylaminopyridine in dichloromethane, mix, let stand at room temperature for 2 h, evaporate to dryness under a stream of nitrogen, add 2.5 mL pH 11 sodium carbonate buffer, shake, let stand at room temperature for 15 min, add 4 mL MTBE, vortex for 2 min. Remove the organic layer and wash it with 4 mL water, wash with 2.5 mL 2 M HCl, wash with 4 mL water, dry the organic layer over anhydrous sodium sulfate, evaporate to dryness under a stream of nitrogen, reconstitute with 1 mL MeOH, inject a 30 µL aliquot. (Prepare reagent by dissolving 20 mg 9-phenanthrenecarboxylic acid and 20 µL triethylamine in 2 mL dichloromethane. Prepare 9-phenanthrenecarboxylic acid by refluxing 9-cyanophenanthrene in 10 M NaOH for 28 h.)

HPLC VARIABLES
Column: 150 × 4.6 Ultrasphere C8
Mobile phase: MeCN:water 80:20
Flow rate: 3
Injection volume: 10
Detector: UV 254

CHROMATOGRAM
Retention time: 5

OTHER SUBSTANCES
Extracted: metabolites

KEY WORDS
rat

REFERENCE
Idowu, O.R.; Grace, J.M.; Leo, K.U.; Brewer, T.G.; Peggins, J.O. Rapid derivatization of alcohols with carboxylic-sulphonic mixed anhydrides for HPLC-UV/fluorescence analysis: Application to the detection of dihydroqinghaosu (DQHS) and its metabolites in biological samples, *J.Liq.Chromatogr.Rel.Technol.*, **1997**, *20*, 1553–1577.

Trolox Methyl Ether

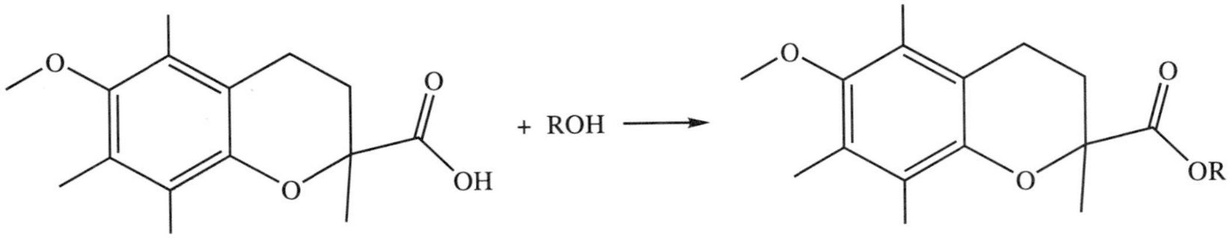

SAMPLE
Matrix: solutions
Analyte: temazepam
Sample preparation: Dissolve the compound, S-trolox methyl ether (Fluka), dicyclohexylcarbodiimide, and 4-dimethylaminopyridine in dichloromethane, stir at room temperature for 1 h (10 h for ethyl loflazepate), filter (0.45 μm), inject an aliquot.

HPLC VARIABLES
Column: 300 × 0.32 5 μm LiChrosorb Diol
Mobile phase: Carbon dioxide:MeOH 91.5:8.5 or 90:10
Column temperature: 80
Injection volume: 0.2
Detector: UV 254

CHROMATOGRAM
Retention time: 11.8 (ethyl loflazepate, second peak, 90:10), 19.1 (oxazepam, second peak, 90:10), 50.2 (lorazepam, second peak, 91.5:8.5), 64.2 (temazepan, second peak, 91.5:8.5)

KEY WORDS
subcritical fluid chromatography; chiral; resolution (R_s) 1.1-1.4

REFERENCE
Almquist, S.R.; Petersson, P.; Walther, W.; Markides, K.E. Direct and indirect approaches to enantiomeric separation of benzodiazepines using micro column techniques, *J.Chromatogr.A*, **1994**, *679*, 139–146.

DIAZO COMPOUND

4-Cyanophenyldiazonium Chloride

SAMPLE
Matrix: solutions
Analyte: phenols
Sample preparation: Mix solution with 5 mL 200 g/L sodium acetate trihydrate and 650 μL 4-cyanophenyldiazonium chloride solution, let stand for 1 min, add 5 mL 160 g/L sodium carbonate monohydrate, add 150 mg tetrabutylammonium bromide, extract with 2 mL n-butanol, centrifuge, inject an aliquot of the organic layer. (Prepare 4-cyanophenyldiazonium chloride solution by mixing 2.566 g p-aminobenzonitrile and 108 mL concentrated HCl in 1 L water. Cool a 25 mL aliquot in an ice bath, slowly add with stirring 3 mL of a 25 g/L sodium nitrite solution. Use reagent within 1 h.)

HPLC VARIABLES
Column: 250 × 2.6 HC ODS/SIL-X C18 (Perkin-Elmer)
Mobile phase: MeOH:water 64:36
Flow rate: 1
Injection volume: 10
Detector: UV 370

CHROMATOGRAM
Retention time: 1.40 (catechol), 1.40 (salicylic acid), 3.00 (3-hydroxybenzoic acid), 3.20 (phenol), 4.30 (m-cresol), 4.80 (o-cresol), 5.20 (3-nitrophenol), 5.50 (3-trifluoromethylphenol), 6.20 (3,5-dimethylphenol), 6.60 (2,5-dimethylphenol), 6.90 (2,6-dimethylphenol), 7.80 (2,3-dimethylphenol), 8.40 (3-tert-butylphenol), 9.60 (p-cresol), 9.85 (2-sec-butylphenol), 13.40 (2-tert-butylphenol), 18.90 (2,3,5,6-tetramethylphenol)
Limit of detection: 10 ppb

REFERENCE
Baiocchi, C.; Campi, E.; Gennaro, M.; Mentasti, E.; Mirti, P. Reversed phase liquid chromatographic separation of phenolic compounds with a new derivatizing reaction, *Chromatographia*, **1982**, *15*, 660–664.

Trimethylsilyldiazomethane

SAMPLE
Matrix: blood

Analyte: 7-hydroxygranisetron (1-methyl-N-(endo-9-methyl-9-azabicyclo[3.3.1]non-3-yl)-7-hydroxy-1H-indazole-3-carboxamide)

Sample preparation: Condition a 100 mg Bond Elut C2 SPE cartridge with two 1 mL portions of MeOH, 1 mL water, and two 1 mL portions of buffer. 1 mL Plasma + 100 μL 40 ng/mL IS in water + 500 μL buffer, add to the SPE cartridge, wash with 1 mL MeCN:water 40:60, remove all wash solvent, elute with 800 μL MeOH. Evaporate the eluate to dryness under a stream of nitrogen at 50°. Add 30 μL 10% trimethylsilyldiazomethane in hexane and 30 μL 2% N,N-diisopropylethylamine in MeOH to the residue, heat at 50° for 20 min, cool to room temperature, evaporate to dryness under a stream of nitrogen at 50°, reconstitute in 300 μL MeOH:water 10:90, centrifuge at 1700 g for 5 min, inject a 150 μL aliquot. (Phosphate buffer was 4.33 g Na_2HPO_4 and 3.04 g NaH_2PO_4 .$2H_2O$ in 50 mL water.)

HPLC VARIABLES

Guard column: 15 × 3.2 NewGuard RP-18
Column: 250 × 4.6 5 μm Develosil ODS-5 (Nomura Chemical)
Mobile phase: MeOH:buffer 30:70 (Buffer was 15.4 g ammonium acetate and 20 mL tetra-n-butylammonium hydroxide in 1.7 L water, adjust pH to 4.70 with glacial acetic acid, make up to 2 L with water.)
Column temperature: 45
Flow rate: 1
Injection volume: 150
Detector: F ex 310 em 420

CHROMATOGRAM

Retention time: 9.5
Internal standard: 2-methyl-M-(endo-9-methyl-9-azabicyclo[3.3.1]non-3-yl)-2H-indazole-3-carboxamide hydrochloride (not derivatized) (6)
Limit of detection: 42 pg/mL

OTHER SUBSTANCES

Extracted: granisteron (not derivatized)

KEY WORDS

plasma; SPE; only 7-hydroxygranisetron is derivatized

REFERENCE

Kudoh, S.; Sato, T.; Okada, H.; Kumakura, H.; Nakamura, H. Simultaneous determination of granisetron and 7-hydroxygranisetron in human plasma by high-performance liquid chromatography with fluorescence detection, *J.Chromatogr.B*, **1994**, *660*, 205–210.

HYDRAZONE

3-Methyl-2-benzothiazolinone Hydrazone

SAMPLE

Matrix: water
Analyte: phenols
Sample preparation: For water containing 3'-aminoacetophenone, 4-aminobenzoic acid, 4-amino-5-hydroxynaphthalene-2,7-disulfonic acid, 4-amino-2-hydroxybenzoic acid, 4-aminophenol, aniline, 4-tert-butylphenol, 4-chloroaniline, 4-chloro-2-methylphenol, 1,2-dia-

minobenzene, 2,6-di-tert-butylphenol, 2,4-di-tert-butylphenol, 1,2-dihydroxybenzene, 1,4-dihydroxybenzene, 2,4-dihydroxybenzoic acid, 2,6-dimethylphenol, 2,4-dinitrophenol, 4-hydroxybenzoic acid, 5-hydroxybenzothiazole, 4-hydroxycinnamic acid, 2-hydroxy-5-methoxybenzoic acid, 2-hydroxynaphthalene-1,4-disulfonic acid, 1-hydroxynaphthalene-4-sulfonic acid, 4-hydroxyphenylacetic acid, 8-hydroxyquinoline, 2-mercaptobenzothiazole, 2-methoxyphenol, 2-methylphenol, 3-methylphenol, 4-methylphenol, 1-naphthol, 1-naphthylamine, 2-naphthylthiol,4-nitrophenol, pentachlorophenol, phenol, thiophenol, and 1,3,5-trihydroxybenzene.

HPLC VARIABLES

Column: 250 × 3 5 µm Eurosphere 100 C8 (Knauer, Berlin)

Mobile phase: Gradient. A was 1 mM NaH_2PO_4 adjusted to pH 3.05 with phosphoric acid. B was MeCN:water 95:5. A:B 80:20 for 5 min, to 30:70 over 5 min

Flow rate: 0.7

Injection volume: 2-50

Detector: UV 500 following post-column reaction. The column effluent mixed with 0.15% 3-methyl-2-benzothiazolinone hydrazone hydrochloride in water pumped at 0.2 mL/min and with 0.4% ceric ammonium sulfate in 5% sulfuric acid pumped at 0.2 mL/min and the mixture flowed directly to the detector.

CHROMATOGRAM

Limit of detection: 2-20 ng

KEY WORDS

post-column reaction; wastewater

REFERENCE

Fiehn, O.; Jekel, M. Analysis of phenolic compounds in industrial wastewater with high-performance liquid chromatography and post-column reaction detection, *J.Chromatogr.A*, **1997**, *769*, 189–200.

ISOCYANATE

7-Diethylaminocoumarin-3-carbonyl Azide

The azide forms the corresponding isocyanate on heating.

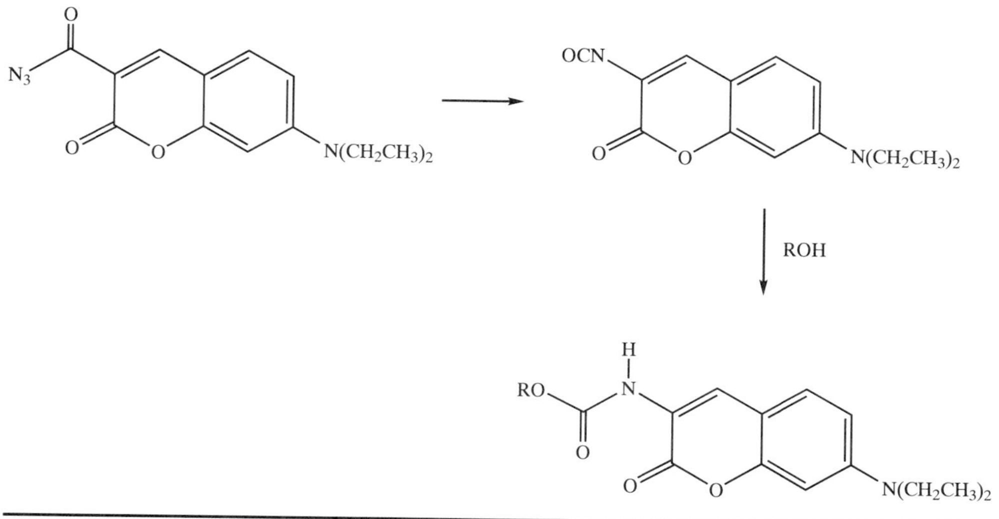

SAMPLE

Matrix: blood

Analyte: 1-O-hexadecyl-2-sn-lysoglyceryl-3-phosphorylcholine

Sample preparation: 25 µL Serum + 5 µL 1 mg/mL IS in Tris-Tyrode's buffer containing 0.25% lipid-free bovine serum albumin + 500 µL water-saturated butanone, shake for a few min, centrifuge at 800 g for 10 min. Remove a 250 µL aliquot of the organic layer and evaporate it to dryness under reduced pressure, dry under reduced pressure at room temperature until all traces of water are removed (about 30 min). Reconstitute with 100 µL 1 mg/mL 7-diethylaminocoumarin-3-carbonyl azide (Molecular Probes, Eugene OR) in anhydrous toluene, heat at 80° for about 3 h, cool, inject an aliquot.

HPLC VARIABLES
Column: 300 × 3.9 4 µm Nova-Pack C18
Mobile phase: Gradient. A was MeOH:water 80:20 containing 250 µg/mL choline chloride. B was chloroform. A:B from 100:0 to 45:55 over 22 min, re-equilibrate at initial conditions for 10 min.
Flow rate: 1
Detector: F ex 400 em 480

CHROMATOGRAM
Retention time: 15.01
Internal standard: 1-O-octadecyl-2-sn-lysoglyceryl-3-phosphorylcholine (16.31)
Limit of detection: 0.5 pmole

KEY WORDS
serum

REFERENCE
Balestrieri, C.; Camussi, G.; Giovane, A.; Iorio, E.L.; Quagliuolo, L.; Servillo, L. Measurement of platelet-activating factor acetylhydrolase activity by quantitative high-performance liquid chromatography determination of coumarin-derivatized 1-O-alkyl-2-sn-lysoglyceryl-3-phosphorylcholine, *Anal. Biochem.*, **1996**, *233*, 145–150.

3,4-Dihydro-6,7-dimethoxy-4-methyl-3-oxoquinoxaline-2-carbonyl Azide

RELATED REFERENCE
Iwata, T.; Yamaguchi, M.; Hanazono, H.; Imazato, Y.; Nakamura, M.; Ohkura, Y. Determination of vitamin D3 and 25-hydroxyvitamin D3 in sera by column-switching high performance liquid chromatography with fluorescence detection. *Anal.Sci.* **1990**, *6*, 361-366

SAMPLE
Matrix: solutions
Analyte: alcohols

Sample preparation: Mix a 100 μL aliquot of a 0.12-15 μM solution of an alcohol in benzene with 900 μL 3 mM 3,4-dihydro-6,7-dimethoxy-4-methyl-3-oxoquinoxaline-2-carbonyl azide in benzene (Caution! Benzene is a carcinogen!), heat in the dark at 130° for 1 h, cool. Remove a 20 μL aliquot of the reaction mixture and add it to 100 μL MeOH, mix, inject a 10 μL aliquot. (Toluene may be substituted for benzene but the peaks are 40% less. On heating the azide forms the corresponding isocyanate which then reacts with the alcohol. Synthesis of 3,4-dihydro-6,7-dimethoxy-4-methyl-3-oxoquinoxaline-2-carbonyl azide is as follows. Stir 483 g veratrole in 1.45 L acetic acid at 15°, add 683 g concentrated nitric acid (s.g. 1.05) over 1 h keeping the temperature below 40° (cool if necessary), add 2.127 L fuming nitric acid (s.g. 1.50) over 1 h keeping the temperature below 30°, allow to stand for 2 h, pour into a large volume of cold water, filter, wash the solid until it is free from acid, recrystallize from EtOH to give 4,5-dinitroveratrole (mp 129.5-130.5°) (J. Am. Chem. Soc. 1946, 68, 1536). Reflux 5 g 4,5-dinitroveratrole in 200 mL benzene (Caution! Benzene is a carcinogen!), add 100 g 60 mesh iron powder and 20 mL concentrated HCl in small portions over 1 h, reflux for 4 h, add 10 mL water, reflux for 2 h, cool, make alkaline with 2.5 M NaOH, extract several times with 200 mL portions of benzene. Combine the extracts and evaporate to dryness, add 10 mL concentrated HCl, recrystallize from EtOH to give 1,2-diamino-4,5-dimethoxybenzene hydrochloride as slightly pink needles (mp 240° d) (Anal. Chim. Acta 1982, 134, 39). Another synthesis of 1,2-diamino-4,5-dimethoxybenzene is as follows. Add 9.66 g veratrole in 9.5 mL glacial acetic acid dropwise to 26.6 mL fuming nitric acid over 30 min using an ice bath, stir at room temperature for 30 min, pour into 500 mL ice-cold water, filter to obtain 1,2-dimethoxy-4,5-dinitrobenzene, recrystallize repeatedly from EtOH. Dissolve 1 g 1,2-dimethoxy-4,5-dinitrobenzene in 50 mL EtOH, add 100 mg 10% palladium on charcoal, stir under hydrogen at atmospheric pressure and room temperature for 4 days, filter through Celite under an atmosphere of nitrogen, saturate the filtrate with hydrogen chloride gas, filter under nitrogen, dry the solid under vacuum to obtain 1,2-diamino-4,5-dimethoxybenzene hydrochloride (mp 240° d (Anal. Chim. Acta 1982, 134, 39)) (Anal. Chim. Acta 1992, 263, 137). 1,2-Diamino-4,5-dimethoxybenzene is also available from Molecular Probes, Eugene OR or Dojindo Molecular Technologies, Inc., 3 Bethesda Metro Center, Suite 700, Bethesda MD 20814; (301) 664-8448; www.dojindo.co.jp. Dissolve 8 g 1,2-diamino-4,5-dimethoxybenzene monohydrochloride and 8 g α-ketomalonic acid in 20 mL 500 mM HCl, heat in a boiling water bath for 2 h, cool in ice, filter, wash the precipitate with water, recrystallize from dioxane:water 90:10 (Caution! Dioxane is a carcinogen!) to give 3,4-dihydro-6,7-dimethoxy-3-oxoquinoxaline-2-carboxylic acid as orange needles (mp 268°). Treat 5.5 g 3,4-dihydro-6,7-dimethoxy-3-oxoquinoxaline-2-carboxylic acid in 50 mL anhydrous MeOH with ethereal diazomethane, evaporate to dryness under reduced pressure, dissolve the residue in 30 mL chloroform, chromatograph on a 250 × 57 column of 130 g 70-230 mesh silica gel 60 (Merck) with n-hexane:ethyl acetate 50:50 to give methyl 3,4-dihydro-6,7-dimethoxy-4-methyl-3-oxoquinoxaline-2-carboxylate as yellow needles (mp 164°). Dissolve 2.5 g methyl 3,4-dihydro-6,7-dimethoxy-4-methyl-3-oxoquinoxaline-2-carboxylate in 200 mL 1 M NaOH, let stand at room temperature for about 70 min, wash 5 times with 200 mL portions of ethyl acetate. Neutralize the aqueous layer with dilute HCl, filter, recrystallize the precipitate from dioxane:water 80:20 to give 3,4-dihydro-6,7-dimethoxy-4-methyl-3-oxoquinoxaline-2-carboxylic acid as yellow needles (mp 222°). Dissolve 1 g 3,4-dihydro-6,7-dimethoxy-4-methyl-3-oxoquinoxaline-2-carboxylic acid in 20 mL freshly distilled thionyl chloride, reflux for 1 h, cool, add 50 mL light petroleum (bp 30-60°), filter, recrystallize the precipitate from benzene:light petroleum 90:10 (Caution! Benzene is a carcinogen!) to give 3,4-dihydro-6,7-dimethoxy-4-methyl-3-oxoquinoxaline-2-carboxyl chloride as orange needles (mp 261°) (J. Chromatogr. 1986, 362, 209). 3,4-Dihydro-6,7-dimethoxy-4-methyl-3-oxoquinoxaline-2-carboxyl chloride is available from Dojindo Molecular Technologies, Inc., 3 Bethesda Metro Center, Suite 700, Bethesda MD 20814; (301) 664-8448; www.dojindo.co.jp. Dissolve 500 mg 3,4-dihydro-6,7-dimethoxy-4-methyl-3-oxoquinoxaline-2-carboxyl chloride in 200 mL dry acetone, add 160 mg activated sodium azide (Caution! Sodium azide is highly toxic!), stir at 0° for 2 h, pour into 100 mL icewater. Collect the precipitate and recrystallize it from hexane:benzene 50:50 to give 3,4-dihydro-6,7-dimethoxy-4-methyl-3-oxoquinoxaline-2-carbonyl azide (mp 272°). Preparation of diazomethane is as follows. Caution! Diazomethane is toxic, explosive, and carcinogenic! A face shield and a safety screen should always be used and the preparation should only be carried out in a properly functioning chemical fume hood. Only smooth glass apparatus with rubber stoppers and plastic tubing should be used. Scratched glassware, ground glass joints, and sharp edges should be avoided (Org. Syn., Coll. Vol. VI; Wiley: New York, 1988, pp. 432-435). Procedures have been reported for the synthesis of diazomethane from N-methyl-N-nitrosourea (Org. Syn., Coll. Vol. II; Wiley: New York, 1943, pp. 165-167), N-nitroso-β-methylaminoisobutyl methyl ketone (Org. Syn., Coll. Vol. III; Wiley: New York, 1955, pp. 244-248), and N,N'-dimethyl-N,N'-dinitrosoterephthalamide (Org. Syn., Coll. Vol. V; Wiley: New York, 1973, pp. 351-355). Probably the most convenient starting material is N-methyl-N-nitroso-p-toluenesulfonamide (Diazald) (Aldrichimica Acta 1983, 16, 3-10; Org. Syn., Coll. Vol.IV, Wiley: New York, 1963, pp. 250-253). Add 10 mL 95% EtOH to a solution of 5 g KOH in 8 mL water, warm to 65° using a water bath, add a solution of 5 g N-methyl-N-nitroso-p-toluenesulfonamide in 45 mL ether over 20 min at such a rate as to keep the reaction volume constant. Collect the ether and diazomethane that distil in an ice-cooled receiving flask under a dry ice/acetone condenser. When all the N-methyl-N-nitroso-p-toluenesulfonamide has been used up, slowly add 10 mL ether to the reaction flask and continue distillation until the distillate is colorless. A purpose-built distillation apparatus can be purchased from Aldrich (Aldrichimica Acta 1983, 16, 3-10). Excess quantities of diazomethane can be destroyed by adding acetic acid until the yellow color of the diazomethane is discharged. The safe disposal of the nitroso compounds used to generate diazomethane has been discussed (Lunn,G.; Sansone,E.B. Destruction of Hazardous Chemicals in the Laboratory, Second Edition. Wiley: New York, 1994, pp. 277-289).)

HPLC VARIABLES
Column: 150 × 6 10 μm YMC Pack C8 (Yamamura, Kyoto)
Mobile phase: MeOH:water 70:30 (A) or 100:0 (B) or 60:40 (C)
Flow rate: 2
Injection volume: 10
Detector: F ex 360 em 440

CHROMATOGRAM
Retention time: 3.5 (benzyl alcohol (A)), 3.7 (11-dehydrocorticosterone (A)), 4 (cyclohexanol (A)), 4.3 (cholesterol (B)), 4.5 (n-hexanol (A)), 4.7 (cholestanol (B)), 6.8 (2-methyl-2-propanol (C)), 7 (2-methyl-2-butanol (A)), 7.4 (deoxycorticosterone (A)), 7.9 (2-methyl-2-pentanol (A)), 11.5 (ethisterone (A)), 18.2 (dehydroisoandrosterone (A)), 40.6 (pregnenolone (A))
Limit of quantitation: 2-45 fmole

OTHER SUBSTANCES
Non-interfering: aldehydes, amines, amino acids, carboxylic acids, ketones, lactic acid, malic acid, phenols, thiols

REFERENCE
Yamaguchi, M.; Iwata, T.; Nakamura, M.; Ohkura, Y. 3,4-Dihydro-6,7-dimethoxy-4-methyl-3-oxoquinoxaline-2-carbonyl azide as a highly sensitive fluorescence derivatization reagent for primary, secondary, and tertiary alcohols in high-performance liquid chromatography, *Anal.Chim.Acta*, **1987**, *193*, 209−217.

3,5-Dinitrophenyl Isocyanate

SAMPLE
Matrix: solutions

Analyte: 1-phenylethanol

Sample preparation: Reflux a 10% excess of 3,5-dinitrobenzoyl azide in toluene for 6-10 min, add the alcohol, cool, dilute, inject an aliquot. (Dissolve 3,5-dinitrobenzoyl chloride in the minimum amount of glacial acetic acid, add one equivalent of sodium azide in portions, after 30 min dilute with water, filter to obtain 3,5-dinitrobenzoyl azide. Refluxing in toluene causes the azide to form the isocyanate which then reacts with the alcohol to give the urea.)

HPLC VARIABLES

Column: Chiral stationary phase CSP 2 (J. Chromatogr. 1984, 316, 585)
Mobile phase: Hexane:isopropanol 20:80
Flow rate: 2

CHROMATOGRAM

Retention time: k' 3.4 (α = 1.36)

OTHER SUBSTANCES

Also analyzed: other chiral alcohols in homologous series

KEY WORDS

chiral

REFERENCE

Pirkle, W.H.; Mahler, G.; Hyun, M.H. Separation of the enantiomers of 3,5-dinitrophenyl carbamates and 3,5-dinitrophenyl ureas, *J.Liq.Chromatogr.*, **1986**, *9*, 443–453.

2-[2-(Isocyanate)ethyl]-3-methyl-1,4-naphthoquinone

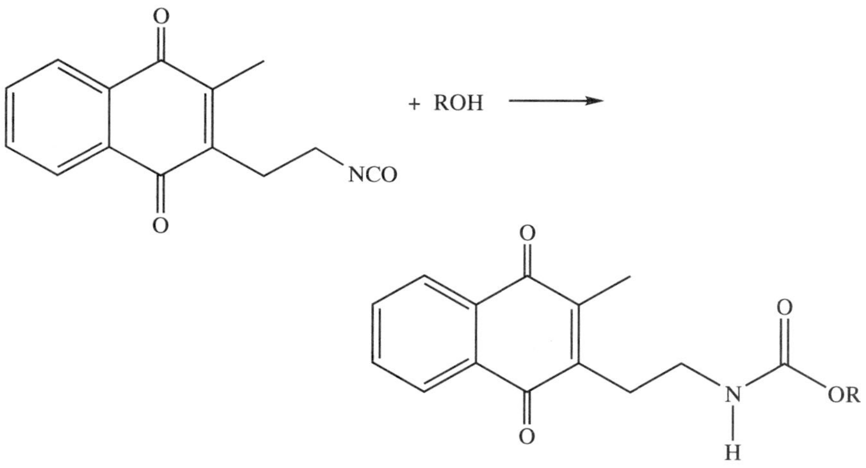

SAMPLE

Matrix: blood
Analyte: cholesterol
Sample preparation: 5 µL Serum + 5 µL 1 mM 1-eicosanol in benzene (Caution! Benzene is a carcinogen!) + 500 µL 200 mM KOH in 95% EtOH, heat at 100° for 15 min, cool, add 1 mL water, add 3 mL hexane, vortex for 10 min, centrifuge at 1000 g for 5 min. Remove the organic layer and evaporate it to dryness under a stream of nitrogen, reconstitute the residue in 200 µL 0.1% 2-[2-(isocyanate)ethyl]-3-methyl-1,4-naphthoquinone in acetone, heat at 100° for 15 min until the solvent has almost evaporated, cool, add 1 mL MeOH, inject a 10 µL aliquot. (Synthesis of 2-[2-(isocyanate)ethyl]-3-methyl-1,4-naph-thoquinone is as follows. Prepare disuccinyl peroxide by adding 10 g succinic anhydride

to 25 mL 7.5% hydrogen peroxide below 30°, filter after 35 min, wash several times with small portions of water, dry (Am. Chem. J. 1904, 32, 43). Add 3.04 g disuccinyl peroxide in small portions to 2.24 g 2-methyl-1,4-naphthoquinone in 5 mL acetic acid at a temperature just below the boiling point, dilute with water, neutralize with 10% sodium carbonate containing enough sodium dithionite to convert the quinone to the hydroquinone. Wash the aqueous layer with ether and acidify it with acetic acid. Extract the aqueous layer with ether and shake the organic layer with 3 g silver oxide and 4 g magnesium sulfate. Treat the organic layer with activated carbon (Norit), filter, evaporate to dryness, recrystallize from benzene/light petroleum to give 2-carboxyethyl-3-methyl-1,4-naphthoquinone as microprisms (mp 145.9-146.5°, yield 20%) (J. Am Chem. Soc. 1947, 69, 2338). Add 2 g oxalyl chloride to 3.35 g 2-carboxyethyl-3-methyl-1,4-naphthoquinone in 10 mL dry benzene (Caution! Benzene is a carcinogen!), stir at 60° for 15 min, evaporate to dryness under reduced pressure, take up the residue in 10 mL dry acetone, stir at 0°, add 1.3 g sodium azide (Caution! Sodium azide is highly toxic!), stir at 0° for 2 h, evaporate to dryness under reduced pressure, dissolve the residue in ethyl acetate, chromatograph on silica gel with n-hexane:ethyl acetate 10:1 to give 2-[2-(azidocarbonyl)ethyl]-3-methyl-1,4-naphthoquinone (AMQ) (mp 56.5-58°) (Anal. Sci. 1991, 7 Supplement, 173). Heat 500 mg 2-[2-(azidocarbonyl)ethyl]-3-methyl-1,4-naphthoquinone in 10 mL benzene at 100° for 1 h, cool, evaporate to dryness under reduced pressure, dissolve the residue in ethyl acetate and chromatograph on silica gel with hexane:ethyl acetate 10:1 to give 2-[2-(isocyanate)ethyl]-3-methyl-1,4-naphthoquinone (mp 70-71°) (J. Chromatogr. 1993, 641, 176).)

HPLC VARIABLES
Column: 150 × 4 5 μm Inertsil C8
Mobile phase: MeOH:EtOH:water 10:7:3 containing 50 mM sodium perchlorate
Flow rate: 1
Injection volume: 10
Detector: E, Toa ICA 3060, glassy carbon working electrode +0.7 V, Ag/AgCl reference electrode following post-column reaction. The column effluent passed through a 10 × 4.6 column packed with 10 μm 5% platinum on alumina catalyst and flowed to the detector.

CHROMATOGRAM
Retention time: 25.7 (cholesterol), 29.1 (cholestanol)
Internal standard: 1-eicosanol (21.8)
Limit of detection: 6.6 pg (cholesterol), 7.4 pg (cholestanol)

KEY WORDS
serum; post-column reaction

REFERENCE
Nakajima, M.; Yamato, S.; Wakabayashi, H.; Shimada, K. High-performance liquid chromatographic determination of cholesterol and cholestanol in human serum by precolumn derivatization with 2-[2-(isocyanate)ethyl]-3-methyl-1,4-naphthoquinone combined with platinum catalyst reduction and electrochemical detection, *Biol.Pharm.Bull.*, **1995**, *18*, 1762–1764.

SAMPLE
Matrix: solutions
Analyte: alcohols
Sample preparation: Add 200 μL 0.1% 2-[2-(azidocarbonyl)ethyl]-3-methyl-1,4-naphthoquinone in acetone to 5 μL of a 2.5 μL/mL solution of alcohol in benzene (Caution! Benzene is a carcinogen!), heat at 100° for 15 min, cool, add 1 mL MeOH, mix, inject a 10 μL aliquot. (Synthesis of 2-[2-(azidocarbonyl)ethyl]-3-methyl-1,4-naphthoquinone is described above. On heating to 100° the azide forms the corresponding isocyanate.)

HPLC VARIABLES
Column: 150 × 4 5 μm Inertsil C8
Mobile phase: MeOH:water 90:10 containing 50 mM sodium perchlorate
Flow rate: 1
Injection volume: 10
Detector: UV 280; F ex 320 em 430; E, Toa ICA 3060, glassy carbon working electrode +0.7 V, Ag/AgCl reference electrode following post-column reaction. The column effluent passed

through a catalyst column and flowed to the detector. (Prepare the catalyst column by tap filling a 10 × 4.6 column with 10 μm 5% platinum on alumina catalyst (Toa electronics, Tokyo), purge with water at 10 mL/min for 5 min.)

CHROMATOGRAM

Retention time: 6.7 (n-butanol (mobile phase 70:30)), 6.4 (i-butanol (mobile phase 70:30)), 6.1 (sec-butanol (mobile phase 70:30)), 6.0 (t-butanol (mobile phase 70:30)), 6.2 (benzyl alcohol (mobile phase 70:30)), 8.7 (cyclohexyl alcohol (mobile phase 70:30)), 4.2 (1-decanol), 4.7 (1-undecanol), 5.9 (1-dodecanol), 6.3 (1-tridecanol), 7.1 (2-tetradecanol), 7.6 (1-tetradecanol), 8.8 (1-pentadecanol), 10.4 (1-hexadecanol), 12.3 (1-heptadecanol), 14.6 (1-stearyl alcohol), 21.2 (1-eicosanol)

Limit of quantitation: 8-23 fmole

OTHER SUBSTANCES

Non-interfering: aldehydes, amino acids, carboxylic acids, ketones

KEY WORDS

post-column reaction; no reaction with aldehydes, amino acids, carboxylic acids, ketones; reagent also reacts with hydroxycarboxylic acid and amines

REFERENCE

Nakajima, M.; Wakabayashi, H.; Yamato, S.; Shimada, K. New labeling agent for high performance liquid chromatographic determination of alcohols with fluorescence and electrochemical detection, *Anal.Sci.*, **1991**, *7 Supplement*, 173−176.

(S)-(+)-1-(1-Naphthyl)ethyl Isocyanate

RELATED REFERENCES

Hirata, H.; Yamashina, T.; Higuchi, K.; Sakaki, K.; Iida, I. Determination of enantiomer content of secondary alkanol as diastereomeric N-[1-(1-naphthyl)ethyl]carbamate by normal phase HPLC (high performance liquid chromatography). *Yukagaku* **1991**, *40*, 995-1001.

Ruettimann, A.; Schiedt, K.; Vecchi, M. Separation of (3R,3'R)-, (3R,3'S; meso)-, (3S,3'S)-zeaxanthin, (3R,3'R,6'R)-, (3R,3'S,6'S)- and (3S,3'S,6'S)-lutein via the dicarbamates of (S)-(+)-α-(1-naphthyl)ethyl isocyanate. *HRC CC,J.High Resolut.Chromatogr.Chromatogr.Commun.* **1983**, *6*, 612-616.

SAMPLE

Matrix: blood, urine

Analyte: methocarbamol

Sample preparation: Dilute urine 1:100. Plasma or diluted urine + 10 μg/mL (R)-(-)-flecainide acetate in water + 1 mL 1 M NaOH + 5 mL ethyl acetate, shake mechanically for 20 min, centrifuge at 1000 g for 5 min. Remove the organic layer and evaporate it to dryness under a stream of nitrogen, reconstitute the residue in 1 mL distilled ethyl acetate, add 80 μL 0.2% (S)-(+)-1-(1-naphthyl)ethyl isocyanate in ethyl acetate, vortex for 5 s, heat at 85° for 12 h, evaporate, reconstitute in 200 μL ethyl acetate, inject a 30-150 μL aliquot.

HPLC VARIABLES

Column: 250 × 4.6 Partisil 5 silica

Mobile phase: Hexane:isopropanol 95:5

Flow rate: 1.6

Injection volume: 30-150
Detector: UV 280

CHROMATOGRAM
Retention time: 33 (S), 41 (R)
Internal standard: (R)-(-)-flecainide (12)
Limit of quantitation: 500 ng/mL
Limit of detection: 10 ng/mL

OTHER SUBSTANCES
Simultaneously analyzed: guaifenesin

KEY WORDS
rat; human; plasma; normal phase; chiral; pharmacokinetics

REFERENCE
Alessi-Severini, S.; Coutts, R.T.; Jamali, F.; Pasutto, F.M. High-performance liquid chromatographic analysis of methocarbamol enantiomers in biological fluids, *J.Chromatogr.*, **1992**, *582*, 173–179.

SAMPLE
Matrix: blood, urine
Analyte: eliprodil
Sample preparation: Plasma. 200 μL Plasma + 200 μL pH 6.5 buffer + 20 μL 200 U/mL β-glucuronidase (from E. coli, Sanofi-Pasteur), mix, heat at 37° for 24 h, make up to 1 mL with water, add 20 μL 1.25 μg/mL IS, add 1 mL pH 12 buffer, vortex, add 6 mL n-hexane, shake at 40 rpm for 10 min, centrifuge at 15° at 900 g for 5 min, centrifuge at -20° at 900 g for 8 min. Remove the organic layer and evaporate it to dryness under a stream of nitrogen at 70°, add 200 μL 0.1% (S)-(+)-1-(1-naphthyl)ethyl isocyanate in MeCN under a stream of nitrogen, vortex at 70° for 40 min, evaporate to dryness under a stream of nitrogen, add 250 μL MeCN:pH 4.5 buffer 40:60, vortex, inject a 150 μL aliquot on to column A and elute to waste with mobile phase A, after 5 min elute the contents of column A on to column B with mobile phase B, after 2.5 min remove column A from the circuit, elute column B with mobile phase B, monitor the effluent from column B. (Backflush column A with MeCN:water 70:30, MeCN, MeOH, and THF, then re-equilibrate with MeCN:water 50:50 at 2 mL/min.) Urine. 100 μL Urine + 100 μL pH 6.5 buffer + 50 μL 200 U/mL β-glucuronidase (from E. coli, Sanofi-Pasteur), mix, heat at 37° for 24 h, make up to 1 mL with water, add 20 μL 20 μg/mL IS, add 1 mL pH 12 buffer, vortex, add 6 mL n-hexane, shake at 40 rpm for 10 min, centrifuge at 15° at 900 g for 5 min, centrifuge at -20° at 900 g for 8 min. Remove the organic layer and evaporate it to dryness under a stream of nitrogen at 70°, add 200 μL 0.1% (S)-(+)-1-(1-naphthyl)ethyl isocyanate in MeCN under a stream of nitrogen, vortex for 30 s, heat at 70° for 40 min, evaporate to dryness under a stream of nitrogen, add 250 μL MeCN:pH 4.5 buffer 40:60, vortex, inject a 50 μL aliquot on to column A and elute to waste with mobile phase A, after 5 min elute the contents of column A on to column B with mobile phase B, after 2.5 min remove column A from the circuit, elute column B with mobile phase B, monitor the effluent from column B. (Backflush column A with MeCN:water 70:30, MeCN, MeOH, and THF, then re-equilibrate with MeCN:water 50:50 at 2 mL/min.) (Prepare pH 6.5 buffer by adding 20 mL 136.08 mg/mL KH_2PO_4 to 780 mL water, adjusting pH to 6.5 with 1 M KOH, and making up to 1 L with water. Prepare pH 12 buffer by dissolving 26.5 g sodium carbonate and 21 g sodium bicarbonate in 800 mL water, adjust pH to 12 with about 30 mL 30% NaOH, make up to 1 L with water. Prepare pH 4.5 buffer by dissolving 6.8 g KH_2PO_4 in 1 L water.)

HPLC VARIABLES
Column: A 20 × 4.6 5 μm Supelguard LC8; B 20 × 4.6 40 μm Pelliguard LC8 (Supelco) + 150 × 4.6 5 μm Hypersil C8 BDS
Mobile phase: A MeCN:water 50:50; B MeCN:MeOH:buffer 56:2:42 (Prepare buffer by dissolving 6.8 g KH_2PO_4 and 3.4 mL orthophosphoric acid in 4 L water, pH 2.6.)
Flow rate: A 2; B 1.2
Injection volume: 50-150
Detector: F ex 275 em 336

CHROMATOGRAM
Retention time: 16 (S-(+)), 17 (R-(-))
Internal standard: (+)-α-(3,4-dichlorophenyl)-4[(4-fluorophenyl)methyl]piperidine-1-ethanol hydrochloride (SL83.0601-10, Synthélabo Recherche, Bagneux, France) (19, 21)
Limit of quantitation: 0.75 ng/mL (plasma), 50 ng/mL (urine)

KEY WORDS
plasma; column-switching; pharmacokinetics; chiral

REFERENCE
Malavasi, B.; Ripamonti, M.; Rouchouse, A.; Ascalone, V. Stereoselective determination of unchanged and glucuroconjugated eliprodil, a new anti-ischaemic drug, in human plasma and urine by precolumn derivatization and column-switching high-performance liquid chromatography with fluorescence detection, *J.Chromatogr.A*, **1996**, *729*, 323–333.

1-Naphthyl Isocyanate

RELATED REFERENCES
Hsu, C.L.; Walters, R.R. Chiral separation of ibutilide enantiomers by derivatization with 1-naphthyl isocyanate and high-performance liquid chromatography on a Pirkle column. *J.Chromatogr.* **1991**, *550*, 621-628.
Krueger, J.; Rabe, H.; Reichmann, G.; Ruestow, B. Separation and determination of diacyl glycerols as their naphthylurethanes by high-performance liquid chromatography. *J.Chromatogr.* **1984**, *307*, 387-392.
Lemr, K.; Zanette, M.; Marcomini, A. Reversed-phase high-performance liquid chromatographic separation of 1-naphthyl isocyanate derivatives of linear alcohol polyethoxylates. *J.Chromatogr.A* **1994**, *686*, 219-224.

SAMPLE
Matrix: blood
Analyte: ibutilide
Sample preparation: Condition a 1 mL 100 mg Bond-Elut C18 SPE cartridge with 1 mL MeCN:acetone:triethylamine 50:50:0.2, with 1 mL 0.1% triethylamine in water, and 1 mL water. 0.1-1 mL Plasma + 1 mL buffer + 100 µL 100-250 ng/mL IS in MeCN:10 mM ammonium acetate 30:70, mix, add to the SPE cartridge, wash with 1 mL water, wash with 2 mL MeCN:MeOH:water 25:25:50, wash with 1 mL water, dry under vacuum for 10 min, wash with 300 µL hexane, dry under vacuum for 5 min, elute with 500 µL MeCN:acetone:triethylamine 50:50:0.2. Evaporate the eluate to dryness under a stream of nitrogen at 30°, reconstitute the residue in 10 µL 0.1% acetic acid in MeCN and 100 µL 0.1% 1-naphthyl isocyanate in MeCN, mix, heat at 30° for 10 min, add 600 MeOH: water:trifluoroacetic acid 10:90:0.1, inject a 600 µL aliquot onto column A and elute to waste with mobile phase A, after 8 min backflush the contents of column A onto column B with mobile phase C, after 3.1 min elute column B to waste with mobile phase D, after 6.1 min divert the fraction containing the drugs onto column C, after 2 min elute the contents of column C onto column D with mobile phase E, elute column E with mobile phase E, monitor the effluent from column E. Before the next injection flush column A with mobile phase B for 22.1 min and re-equilibrate with mobile phase A for 2 min. (Buffer was 50 mM pH 7.0 NaH$_2$PO$_4$ containing 0.1% triethylamine.)

HPLC VARIABLES

Column: A two 15 × 3.2 7 μm Newguard in series; B 20 × 4 4 μm Sentry Nova-Pak C18 + 75 × 4.6 3 μm Ultremex 3 C8; C 30 × 4.6 Spheri-5 RP-18; D 100 × 4.6 3 μm Pirkle covalent dinitrophenyl-D-phenylglycine (Regis)

Mobile phase: A MeOH:water 40:60; B MeCN:MeOH:water:trifluoroacetic acid:triethylamine 40:40:20:0.1:0.1; C MeCN:MeOH:water:trifluoroacetic acid:triethylamine 25:5:70:0.1:0.1; D MeCN:water:trifluoroacetic acid:triethylamine 52:48:0.1:0.1; E MeOH:trifluoroacetic acid:triethylamine 100:0.3:0.3

Flow rate: 1
Injection volume: 600
Detector: F ex 290 em 345

CHROMATOGRAM

Retention time: 11.7 (+), 12.6 (-)
Internal standard: N-[4-[4-(ethyloctylamino)-1-hydroxybutyl]phenyl]methanesulfonamide (Upjohn U-74747) (IS detected using F ex 224 em 340 (cut-off filter) detector between columns B and C) (4)
Limit of quantitation: 17 pg/mL

KEY WORDS

SPE; plasma; column-switching; heart-cut; human; rat; rabbit; dog; chiral

REFERENCE

Hsu, C.-y.L.; Walters, R.R. Assay of the enantiomers of ibutilide and artilide using solid-phase extraction, derivatization, and achiral-chiral column-switching high-performance liquid chromatography, *J.Chromatogr.B*, **1995**, *667*, 115–128.

Phenyl Isocyanate

RELATED REFERENCE

Robles, M.D.; Niell, F.X.; Matés, J.M. Separation and determination of α- and β-galactose from agar-type polysaccharides by liquid chromatography. *J.Chromatogr.Sci.* **1996**, *34*, 517-520.

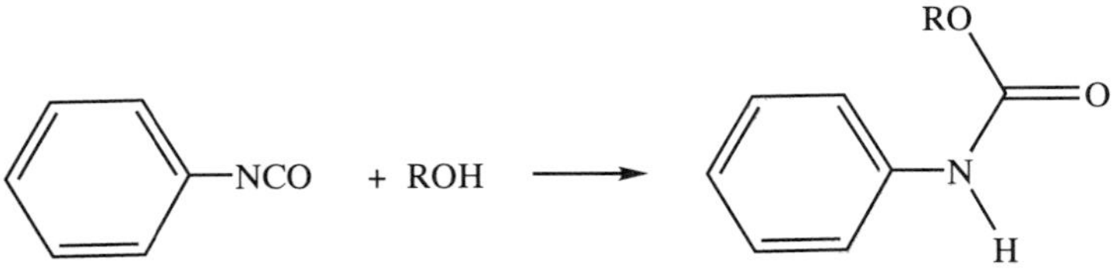

SAMPLE

Matrix: blood
Analyte: stiripentol
Sample preparation: 300 μL Blood + 15 μg piperonyl alcohol + 3 mL ethyl acetate, extract, centrifuge at 3000 rpm for 10 min. Remove the organic layer and evaporate it to dryness under a stream of nitrogen, reconstitute the residue with three 50 μL portions of anhydrous toluene, add 2 μL triethylamine, add 0.5 μL di-n-butyltin dilaurate, add 5 μL phenyl isocyanate, let stand overnight at room temperature, add EtOH, inject an aliquot.

HPLC VARIABLES

Column: 250 × 4.6 5 μm Bakerbond DNBPG chiral
Mobile phase: Hexane:isopropanol 95:5
Flow rate: 2
Detector: F ex 290 em 355

CHROMATOGRAM

Retention time: 6.4 (S), 7.3 (R)
Internal standard: piperonyl alcohol (9.0)
Limit of detection: 8 ng/mL

KEY WORDS
rat; whole blood; pharmacokinetics; chiral

REFERENCE
Zhang, K.; Tang, C.; Rashed, M.; Cui, D.; Tombret, F.; Botte, H.; Lepage, F.; Levy, R.H.; Baillie, T.A. Metabolic chiral inversion of stiripentol in the rat. I. Mechanistic studies, *Drug Metab.Dispos.*, **1994**, *22*, 544–553.

SAMPLE
Matrix: bulk
Analyte: fatty alcohols
Sample preparation: Prepare a 2 mg/mL solution in DMF, mix a 1 mL aliquot with 100 μL phenyl isocyanate, stir at 50° for 30 min, cool, make up to 10 mL with MeOH, inject a 20 μL aliquot.

HPLC VARIABLES
Column: 150 × 4.5 5 μm Hypersil C18
Mobile phase: MeCN:water 97:3
Flow rate: 1
Injection volume: 20
Detector: UV 230

CHROMATOGRAM
Retention time: 10 (hexadecanol), 16 (octadecanol), 24 (eicosanol)

REFERENCE
Andrisano, V.; Gotti, R.; Di Pietra, A.M.; Cavrini, V. Comparative evaluation of three chromatographic methods in the quality control of fatty alcohols for pharmaceutical and cosmetic use, *Farmaco*, **1994**, *49*, 387–391.

SAMPLE
Matrix: sewage
Analyte: alcohol ethoxylates
Sample preparation: Preserve sewage with 10 mM sodium azide (Caution! Sodium Azide is carcinogenic and highly toxic!). Filter (glass fiber), dilute filtrate with water if necessary, extract four times by solvent sublation (passing fine bubbles of nitrogen or helium through a column of the aqueous solution to transport the surfactant to the ethyl acetate layer) with 50-100 mL portions of ethyl acetate for 10-20 min each time. Combine the extracts and filter them through 5 g anhydrous sodium sulfate, concentrate to 5 mL under reduced pressure, evaporate to dryness under a stream of nitrogen, emulsify with 1 mL hexane: dichloromethane 50:50, add to a 200 × 10 glass column packed with 7 g alumina (150 mesh neutral alumina, Brockman I, standard grade, deactivate with 5% w/w water before use), rinse container with two 1 mL portions of hexane:dichloromethane 50:50, add the rinses to the column, wash with 90 mL hexane:dichloromethane 50:50, elute with 90 mL dichloromethane:MeOH 100:1, evaporate the eluate to dryness under reduced pressure and using nitrogen, add 3 μg 1-eicosanol, add 10 μL phenyl isocyanate, add 250 μL dichloromethane, mix, heat at 55° for 45 min (with a loose cap on the vial), dissolve the residue in 250 μL mobile phase, inject a 20 μL aliquot. Add 200 mg sodium hydroxide to the solid removed by filtration, Soxhlet extract with MeOH for 16 h. Evaporate the extract to 25 mL, dilute to 1 L with water, extract as above.

HPLC VARIABLES
Column: 125 × 4 Lichrocart 5 μm C18
Mobile phase: Gradient. MeOH:water from 80:20 to 100:0 over 25 min.
Flow rate: 2
Injection volume: 20
Detector: UV 235

CHROMATOGRAM
Retention time: 10-20 (depending on chain length)
Internal standard: 1-eicosanol (25)
Limit of detection: 3 ng/mL

KEY WORDS
Neodol; Dobanol

REFERENCE
Kiewiet, A.T.; van der Steen, J.M.D.; Parsons, J.R. Trace analysis of ethoxylated nonionic surfactants in samples of influent and effluent of sewage treatment plants by high-performance liquid chromatography, *Anal.Chem.*, **1995**, *67*, 4409–4415.

SAMPLE
Matrix: solutions
Analyte: monosaccharides
Sample preparation: Add 55 μL phenyl isocyanate to a 1 mg/mL solution in DMF, heat at 55° for 95 min, cool, add 500 μL MeOH, let stand for 5 min, make up to 6 mL with DMF, dilute an aliquot 10-fold with DMF, inject an aliquot.

HPLC VARIABLES
Column: 220 × 4.6 5 μm ODS 224 RP18 (Brownlee)
Mobile phase: MeCN:water 60:40
Flow rate: 2
Injection volume: 10
Detector: UV 240

CHROMATOGRAM
Retention time: 5.0 (deoxyribose), 7.9 (deoxyglucose), 8.8 (methylglucoside), 8.9 (ribose), 9.4 (xylose), 10.0 (arabinose), 10.7 (methylmannoside), 10.8 (methylgalactoside), 12 (lyxose), 12.2 (fucose), 13.6 (rhamnose), 15.7 (dextrose), 18.5 (allose), 19.1 (mannose), 19.7 (galactose)
Limit of detection: 0.2-1 ng

KEY WORDS
More than one derivative was observed for each analyte; retention times are for major derivatives.

REFERENCE
Rakotomanga, S.; Baillet, A.; Pellerin, F.; Baylocq-Ferrier, D. Liquid chromatographic analysis of monosaccharides with phenylisocyanate derivatization, *J.Pharm.Biomed.Anal.*, **1992**, *10*, 587–591.

SAMPLE
Matrix: tissue
Analyte: sugar alcohols
Sample preparation: Homogenize sciatic nerve or eye lenses from 1 rat with 1 mL 150-600 mM IS in water, add 2.4 mL ice-cold EtOH, let stand in ice for 5 min, centrifuge at 4° at 10000 g for 5 min. Remove a 25-100 μL aliquot of the supernatant and lyophilize it, reconstitute with 70 μL pyridine (dried over NaOH), add 20 μL phenyl isocyanate, heat at 55° for 1 h, cool in ice for 1 min, add 20 μL MeOH, heat at 55° for 5 min, add 90 μL pyridine, inject a 5 μL aliquot.

HPLC VARIABLES
Column: 150 × 4.6 5 μm TSK-gel ODS-80TM (Toyo Soda)
Mobile phase: MeCN:EtOH:water 50:20:30
Flow rate: 1
Injection volume: 5
Detector: UV 240

CHROMATOGRAM
Retention time: 11.8 (myo-inositol), 14.0 (galactitol), 16.1 (sorbitol)
Internal standard: glucose diethyl mercaptal (21.7)
Limit of quantitation: 500 pmole

KEY WORDS
lens; nerves; rat

REFERENCE
Miwa, I.; Kanbara, M.; Wakazono, H.; Okuda, J. Analysis of sorbitol, galactitol, and *myo*-inositol in lens and sciatic nerve by high-performance liquid chromatography, *Anal.Biochem.*, **1988**, *173*, 39–44.

SAMPLE
Matrix: urine
Analyte: gluconolactone, galactonolactone, and galactitol
Sample preparation: Lyophilize a 5 mL aliquot of urine at -50° for 18 h, reconstitute the residue in 5 mL DMF, centrifuge at 3000 g for 10 min. 1 mL Supernatant + 300 μL phenyl isocyanate, heat at 100° for 1 h, cool, add 500 μL MeOH, inject an aliquot.

HPLC VARIABLES
Column: 220 × 4.6 5 μm ODS 224 RP18 (Brownlee)
Mobile phase: MeCN:water 60:40
Flow rate: 2
Injection volume: 10
Detector: UV

CHROMATOGRAM
Retention time: 9 (gluconolactone), 10.5 (galactonolactone), 20.5 (galactitol)
Limit of detection: 0.4 ng

OTHER SUBSTANCES
Simultaneously analyzed: allose, dextrose, galactose, mannitol, myoinositol, sorbitol

REFERENCE
Rakotomanga, S.; Baillet, A.; Pellerin, F.; Baylocq-Ferrier, D. Simultaneous determination of gluconolactone, galactonolactone and galactitol in urine by reversed-phase liquid chromatography: application to galactosemia, *J.Chromatogr.*, **1991**, *570*, 277–284.

SAMPLE
Matrix: wastewater
Analyte: ethoxylated alcohol surfactants
Sample preparation: Fill a 12 mm ID glass column to a depth of 180 mm with a methanol slurry of Amberlite XAD-2 resin, wash with 40 mL 50 mM NaOH in MeOH, wash with 55 mL MeOH:chloroform:concentrated HCl 50:50:10, wash with 25 mL MeOH, wash with 250 mL water. Add 20 g 50-100 mesh AG1-X2 chloride form anion-exchange resin (Bio-Rad) in a water slurry to a 300 × 15 glass column, wash with 50 mL 1 M NaOH at 1-2 drops/s, wash with water until the washings are neutral, wash with 50 mL MeOH, place a glass wool plug on top of the bed, add 16 g 50-100 mesh AG50W-X8 acid-form cation-exchange resin (Bio-Rad) in a MeOH slurry, wash with 50 mL MeOH. Pass a 0.5-10 L sample of wastewater through the XAD-2 column at 1 drop/s, wash with 30 mL petroleum ether, cautiously apply pressure to force out the petroleum ether, elute with 40 mL ethyl ether at 1 drop/s, elute with 60 mL ethyl ether:MeOH 50:50, elute with 30 mL MeOH. Collect all the eluate and evaporate it to about 5 mL under a stream of nitrogen on a steam bath, add 5 mL water, evaporate to about 5 mL, add 45 mL 5 M NaCl, add 50 mL ethyl acetate, shake vigorously for 1 min, remove the aqueous layer, wash the organic layer with 50 mL 5 M NaCl. Combine the aqueous layers and extract them with 50 mL ethyl acetate. Combine the ethyl acetate layers and add 30 g anhydrous sodium sulfate, shake for 10-15 s, filter the organic layer through 30 g anhydrous sodium sulfate, wash through with two 25 mL portions of ethyl acetate, evaporate just to dryness under a stream of nitrogen using a steam bath, dissolve the residue in 10 mL MeOH, add to the ion-exchange column, rinse the flask with two 10 mL portions of MeOH, add the rinses to the column, elute with MeOH at 1-2 mL/min until 125 mL effluent is collected. Evaporate the effluent to dryness under a stream of nitrogen using a steam bath, reconstitute with 10 mL dichloromethane, add 5 mL cobalt thiocyanate solution, shake vigorously for 1 min. Remove the dichloromethane layer and repeat the extraction twice more, wash the glassware with 10 mL dichloromethane. Combine all the dichloromethane layers and filter them through 10-15 g anhydrous sodium sulfate, wash through with 20 mL dichloromethane, evaporate to dryness under a stream of nitrogen using a steam bath. Add 50 μL 2.5 mg/mL 1-octanol in ethylene dichloride containing 2.5 mg/mL 1-eicosanol and 10 μL phenyl isocyanate, rinse the walls of the vial with 50 μL ethylene dichloride. Cap the vial loosely and heat it at 55 ± 2° in a vacuum oven at 70-100 kPa below atmospheric pressure for 45 min, reconstitute with 250 μL ethylene dichloride, inject an

aliquot. (Prepare cobalt thiocyanate solution by dissolving 15 g cobalt(II) nitrate hexa-hydrate and 100 g ammonium thiocyanate in 500 mL water.)

HPLC VARIABLES
Column: 300 × 3.9 µBondapak C18
Mobile phase: Gradient. MeOH:water from 80:20 to 100:0 over 30 min.
Flow rate: 2
Detector: UV 240

CHROMATOGRAM
Retention time: 8-20 (depending on degree of ethoxylation)
Internal standard: 1-octanol (4), 1-eicosanol (22)
Limit of quantitation: 100 ppb

KEY WORDS
SPE

REFERENCE
Schmitt, T.M.; Allen, M.C.; Brain, D.K.; Guin, K.F.; Lemmel, D.E.; Osborn, Q.W. HPLC determination of ethoxylated alcohol surfactants in wastewater, *J.Am.Oil Chem.Soc.*, **1990**, *67*, 103–109.

Propyl Isocyanate

SAMPLE
Matrix: blood
Analyte: oxiracetam
Sample preparation: Condition a 100 mg phenylboronic acid SPE cartridge (Analytichem) with one column volume of 100 mM pH 8.5 potassium phosphate buffer. Condition a silica SPE cartridge with one column volume of MeCN. 200 µL Plasma + 50 µL water + 50 µL 25 µg/mL IS in water + 500 µL MeCN, vortex, centrifuge at 8800 g for 10 min. Remove the supernatant and add it to 500 µL 100 mM pH 8.5 potassium phosphate buffer, vortex, add to the phenylboronic acid SPE cartridge. Collect the eluate and evaporate it to dryness under a stream of nitrogen at 50°, reconstitute with 1 mL MeOH, sonicate, vortex, centrifuge at 1500 g for 5 min. Evaporate the supernatant to dryness under a stream of nitrogen at 50°, reconstitute the residue in 200 µL anhydrous pyridine, add 100 µL propyl isocyanate, vortex, heat at 50° for 1 h, evaporate to dryness under a stream of nitrogen. Reconstitute the residue in two 500 µL aliquots of MeCN, add to the silica SPE cartridge, wash with 1 mL MeCN, elute with 1 mL MeOH:MeCN 50:50. Evaporate the eluate to dryness under a stream of nitrogen at 50°, reconstitute the residue in 200 µL water, inject a 10-25 µL aliquot.

HPLC VARIABLES
Column: 250 × 2 5 µm Ultrasphere octadecylsilica
Mobile phase: Gradient. MeOH:50 mM pH 6.0 acetate buffer 10:90 for 5 min, to 20:80 over 4 min, maintain at 20:80 for 1 min, to 50:50 over 1 min, maintain at 50:50 for 5 min, return to initial conditions over 1 min, re-equilibrate for 14 min.
Column temperature: 50
Flow rate: 0.3
Injection volume: 10-25
Detector: F ex 340 em 455 following post-column reaction. The column effluent mixed with the reagent pumped at 0.2 mL/min and the mixture flowed through a reaction coil (1 mL volume, ABI Analytical PCRS Model 520) at 90° to the detector. (Reagent was prepared

by adding 2 mL 2 mg/mL o-phthalaldehyde in MeOH and 80 µL 3-mercaptopropionic acid to 1 L 2 g/L NaOH in water, filter (0.45 µm), use within 48 h.)

CHROMATOGRAM
Retention time: 10.9
Internal standard: 4-hydroxy-2-oxo-1-pyrrolidinepropionamide (ISF 2839) (12.9)
Limit of quantitation: 40 ng/mL
Limit of detection: 20 ng/mL

KEY WORDS
plasma; SPE; post-column reaction

REFERENCE
Simpson, R.C.; Boppana, V.K.; Hwang, B.Y.; Rhodes, G.R. Determination of oxiracetam in human plasma by reversed-phase high-performance liquid chromatography with fluorimetric detection, *J.Chromatogr.*, **1993**, *631*, 227–232.

N-(p-Toluenesulfonyl)pyrrolidinyl Isocyanate

SAMPLE
Matrix: bulk
Analyte: phenylpropanol
Sample preparation: Reflux a 10% excess of reagent in toluene for 10 min, add the drug, heat at 100° for 1 h, cool, dilute, inject an aliquot. (The reagent was N-(p-toluenesul-fonyl)prolyl azide and was prepared as follows. Mix 40-45 mmoles L-(-)-proline, 40 mL THF, and 200 mL 10% potassium carbonate, add 37-43 mmoles p-toluenesulfonyl chloride in 40 mL THF dropwise, heat at 50° and maintain at pH 8 or above for 3 h, cool, acidify to pH 2, extract with chloroform. Extract the organic layers with potassium carbonate in water. Acidify the aqueous layer and extract it with chloroform. Dry the chloroform layer and evaporate it to dryness, recrystallize the resulting 1-[(p-toluene)sulfonyl]proline from petroleum ether and benzene (Caution! Benzene is a carcinogen!) (Anal.Chem. 1984, 56, 958). Suspend 86 mmoles 1-[(p-toluene)sulfonyl]proline in 15 mL water and add sufficient acetone to give a clear solution, cool to 0°, add 10.2 g triethylamine in 175 mL acetone, slowly add 12.5 g ethyl chloroformate in 45 mL acetone while maintaining the tempera-ture at 0°, stir at 0° for 30 min, add dropwise 8.6 g sodium azide in 30 mL water, stir at 0° for 1 h, pour into ice water, extract with ether, dry over anhydrous magnesium sulfate, evaporate under reduced pressure at room temperature to give N-(p-toluenesul-fonyl)prolyl azide (cf J.Org.Chem. 1961, 26, 3511). Heating N-(p-toluenesulfonyl)prolyl azide in refluxing toluene produces N-(p-toluenesulfonyl)pyrrolidinyl isocyanate in situ.)

HPLC VARIABLES
Column: 300 × 4 7-9 μm silica gel
Mobile phase: Petroleum ether:isopropanol 99:1
Flow rate: 1.5
Detector: UV 254

CHROMATOGRAM
Retention time: 23.3, 27.8 (enantiomers)

KEY WORDS
chiral; normal phase

REFERENCE
Zhou, Y.; Sun, Z.P.; Lin, D.K. Liquid chromatographic evaluation of a new chiral derivatizing agent for enantiomeric resolution of amine and alcohol drugs, *J.Liq.Chromatogr.*, **1990**, *13*, 875−885.

MISCELLANEOUS REACTIONS

Amination

SAMPLE
Matrix: blood, tissue
Analyte: serotonin and 5-hydroxyindole-3-acetic acid
Sample preparation: Plasma. 200 μL Plasma + 40 μL 500 nM 5-hydroxyindole-3-acetamide + 80 μL 1.5 M perchloric acid, centrifuge, inject a 200 μL aliquot of the supernatant. Tissue. Homogenize rat brain with 5 volumes of 200 μg/mL ascorbic acid in 500 mM perchloric acid. Remove a 500 μL aliquot and add it to 40 μL 6.25 μM 5-hydroxyindole-3-acetamide, centrifuge, inject a 100 μL aliquot.

HPLC VARIABLES
Guard column: TSKgel ODS-80Tm (Tosoh)
Column: 150 × 4.6 5 μm TSKgel ODS-80Tm (Tosoh)
Mobile phase: MeCN:10 mM pH 4.7 acetate buffer 5:95
Flow rate: 1
Injection volume: 100-200
Detector: F ex 345 em 481 following post-column reaction. The column effluent mixed with the reagent pumped at 0.5 mL/min and the mixture flowed through a 7 m × 0.5 mm i.d. PTFE coil at 70° then a 0.5 m × 0.5 mm i.d. coil at 0° to the detector (Analyst 1993, 118, 165). (Reagent was 20 mM benzylamine and 3 mM potassium hexacyanoferrate(III) in MeCN:25 mM pH 10.0 borate buffer 50:50.)

CHROMATOGRAM
Retention time: 6.5 (serotonin), 14 (5-hydroxyindole-3-acetic acid)

Internal standard: 5-hydroxyindole-3-acetamide (10)
Limit of detection: 1.6-3.3 nM

KEY WORDS

post-column reaction; human; rat; brain; plasma

REFERENCE

Ishida, J.; Iizuka, R.; Yamaguchi, M. Serotonin and 5-hydroxyindole-3-acetic acid in human plasma and rat brain determined by liquid chromatography with post-column derivatization and fluorescence detection [letter], *Clin.Chem.*, **1993**, *39*, 2355–2356.

SAMPLE

Matrix: blood, tissue, urine
Analyte: 5-hydroxyindoles
Sample preparation: Urine. Dilute urine 20 times with water, filter (0.45 μm). Remove a 100 μL aliquot and add it to 100 μL 1.5 μM 5-hydroxytryptophol, mix, cool in ice, inject a 100 μL aliquot. Plasma. 200 μL Plasma + 40 μL 500 nM 5-hydroxyindole-3-acetamide + 80 μL 1.5 M perchloric acid, mix, centrifuge, inject a 100 μL aliquot (Clin.Chem. 1993, 39, 2355). Tissue. Homogenize rat brain tissue with five volumes 500 mM perchloric acid containing 300 μg/mL ascorbic acid. Remove a 500 μL aliquot and add it to 40 μL 6.25 μM 5-hydroxyindole-3-acetamide + 80 μL 1.5 M perchloric acid, mix, centrifuge, inject a 100 μL aliquot (Clin.Chem. 1993, 39, 2355).

HPLC VARIABLES

Guard column: TSKgel ODS-80Tm (Tosoh)
Column: 150 × 4.6 5 μm TSKgel ODS-80Tm (Tosoh)
Mobile phase: MeCN:10 mM pH 4.7 acetate buffer 5:95
Flow rate: 1
Injection volume: 100
Detector: F ex 345 em 481 following post-column reaction. The column effluent mixed with the reagent pumped at 0.5 mL/min and the mixture flowed through a 7 m × 0.5 mm i.d. PTFE coil at 70° then a 0.5 m × 0.5 mm i.d. coil at 0° to the detector. (Reagent was 20 mM benzylamine and 3 mM potassium hexacyanoferrate(III) in MeCN:25 mM pH 10.0 borate buffer 50:50).

CHROMATOGRAM

Retention time: 5.5 (5-hydroxytryptophan), 8.0 (5-hydroxytryptamine), 13.0 (5-hydroxyindol-3-ylacetic acid), 39.0 (N-acetyl-5-hydroxytryptamine)
Internal standard: 5-hydroxytryptophol (23.0), 5-hydroxyindole-3-acetamide (10)
Limit of detection: 140-470 fmole

KEY WORDS

post-column reaction; plasma

REFERENCE

Ishida, J.; Iizuka, R.; Yamaguchi, M. High-performance liquid chromatographic determination of 5-hydroxyindoles by post-column fluorescence derivatization, *Analyst*, **1993**, *118*, 165–169.

SAMPLE

Matrix: enzyme incubations
Analyte: 5-hydroxyindoles
Sample preparation: 100 μL Enzyme incubation + 100 μL 500 mM perchloric acid, mix, centrifuge at 1800 g for 10 min. Filter (0.45 μm cellulose acetate) the supernatant, inject a 100 μL aliquot of the supernatant.

HPLC VARIABLES

Column: 150 × 4.6 5 μm TSKgel ODS-80Tm
Mobile phase: MeOH:1 mM pH 3.0 glycine buffer 10:90
Flow rate: 1
Injection volume: 100
Detector: F ex 345 em 481 following post-column reaction. The column effluent mixed with reagent pumped at 0.5 mL/min and flowed through a 7 m × 0.5 mm ID PTFE tube at 70° and a length of 0.5 mm ID PTFE tube at 0° to the detector. (Reagent was 20 mM

benzylamine in MeCN:25 mM pH 10.0 borate buffer 50:50 containing 3 mM potassium ferricyanide.)

CHROMATOGRAM
Retention time: 4.0 (5-hydroxytryptamine), 7.0 (5-hydroxytryptophan), 13.0 (5-hydroxyindole-3-acetamide), 24.8 (5-hydroxytryptophol), 33.1 (5-hydroxyindole-3-acetic acid), 44.2 (N-acetyl-5-hydroxytryptamine)
Limit of detection: 100 fmole

KEY WORDS
post-column reaction

REFERENCE
Iizuka, R.; Ishida, J.; Yoshitake, T.; Nakamura, M.; Yamaguchi, M. Assay for tryptophan hydroxylase activity in rat brain by high-performance liquid chromatography with fluorescence detection, *Biol.Pharm.Bull.*, **1996**, *19*, 762–764.

Aromatization

RELATED REFERENCES
Alvinerie, M.; Sutra, J.F.; Capela, D.; Galtier, P.; Fernandez-Saurez, A.; Horne, E.; O'Keeffe, M. Matrix solid-phase dispersion technique for the determination of moxidectin in bovine tissues. *Analyst* **1996**, *121*, 1469-1472.
Nordlander, I.; Johnsson, H. Determination of ivermectin residues in swine tissues—an improved clean-up procedure using solid-phase extraction. *Food Addit.Contam.* **1990**, *7*, 79-82.
Sundaram, K.M.S.; Curry, J. Determination of abamectin in some forest matrices by liquid chromatography with fluorescence detection. *J.Liq.Chromatogr.Rel.Technol.* **1997**, *20*, 1757-1772

SAMPLE
Matrix: blood
Analyte: ivermectin
Sample preparation: Condition a 500 mg Sep-Pak C18 SPE cartridge with 4 mL MeCN, 5 mL chloroform, 4 mL MeCN, and 4 mL water. 1 mL Plasma + 500 μL MeCN + 500 μL 1 ng/mL IS in MeCN, mix for 15 s, centrifuge at 2500 g for 10 min, add the supernatant to the SPE cartridge, dry under vacuum for 15 min, elute with 5 mL chloroform. Evaporate the eluate to dryness under a stream of nitrogen at <50°, reconstitute with 100 μL N-methylimidazole:MeCN 1:1, add 150 μL trifluoroacetic anhydride:MeCN 1:2, let stand for <30 s, inject a 100 μL aliquot.

HPLC VARIABLES
Column: 250 × 4.6 5 μm Zorbax C8
Mobile phase: MeCN:THF:water 40:40:20
Column temperature: 30
Flow rate: 1
Injection volume: 100
Detector: F ex 365 em 475

CHROMATOGRAM
Retention time: 18 (ivermectin B_{1a})
Internal standard: avermectin B_{1a} (12.5)
Limit of detection: 20 pg/mL

KEY WORDS
plasma; SPE; cow

REFERENCE
de Montigny, P.; Shim, J.S.; Pivnichny, J.V. Liquid chromatographic determination of ivermectin in animal plasma with trifluoroacetic anhydride and *N*-methylimidazole as the derivatization reagent, *J.Pharm.Biomed.Anal.*, **1990**, *8*, 507–511.

SAMPLE
Matrix: blood
Analyte: ivermectin
Sample preparation: Condition a 3 mL C18 SPE cartridge (J.T. Baker) with 4 mL MeCN, 5 mL chloroform, 4 mL MeCN, and 4 mL water. 1 mL Plasma + 50 µL 86-285 ng/mL IS in MeCN, vortex for 15 s, add 1 mL MeCN, mix, centrifuge at 1130 g for 15 min, add the supernatant to the SPE cartridge. Reconstitute the residue in 3.5 mL MeCN:water 1:2, vortex for 15 s, centrifuge at 1130 g for 15 min, add the supernatant to the SPE cartridge. Wash the SPE cartridge with 4 mL MeCN:water 1:2, dry under vacuum for 1 h, elute with 5 mL MeCN:chloroform 50:50. Evaporate the eluate to dryness under a stream of nitrogen at 45°, reconstitute with two 100 µL portions of MeCN, evaporate to dryness under a stream of nitrogen at room temperature. Reconstitute with 100 µL N-methylimidazole:MeCN 1:1, add 150 µL trifluoroacetic anhydride:MeCN 1:2, let stand for 1.7 min, inject a 150 µL aliquot.

HPLC VARIABLES
Column: 100 × 2 3 µm MOS-Hypersil-2
Mobile phase: Gradient. MeCN:water from 72:28 to 92:8 over 15 min.
Flow rate: 0.3
Injection volume: 150
Detector: F ex 365 em 475

CHROMATOGRAM
Retention time: 22.5
Internal standard: ivermectin monosaccharide (20)
Limit of detection: 10 pg/mL

KEY WORDS
plasma; SPE; dog; narrow bore

REFERENCE
Rabel, S.R.; Stobaugh, J.F.; Heinig, R.; Bostick, J.M. Improvements in detection sensitivity for the determination of ivermectin in plasma using chromatographic techniques and laser-induced fluorescence detection with automated derivatization, *J.Chromatogr.*, **1993**, *617*, 79–86.

SAMPLE
Matrix: blood
Analyte: moxidectin
Sample preparation: Condition a 1 mL 100 mg Supelclean LC18 SPE cartridge with 5 mL MeOH and 5 mL water at 6 mL/min. 1 mL Plasma + 1 mL MeCN + 250 µL water, mix for 20 min, centrifuge at 2000 g for 2 min, add 2.2 mL supernatant to the SPE cartridge at 3 mL/min, wash with 2 mL water, wash with 1 mL MeOH:water 25:75 at 3 mL/min, dry under nitrogen at 6 mL/min for 10 s, elute with 1.2 mL MeOH at 3 mL/min. Evaporate the eluate to dryness under a stream of nitrogen at 50°, reconstitute with 100 µL N-methylimidazole in MeCN, add 150 µL trifluoroacetic anhydride in MeCN, let stand for <30 s, inject a 100 µL aliquot.

HPLC VARIABLES
Column: Supelcosil C18
Mobile phase: MeCN:MeOH:0.2% acetic acid 62:30:8
Flow rate: 1.5
Injection volume: 100
Detector: F ex 383 em 447

CHROMATOGRAM
Retention time: 9.2
Limit of quantitation: 0.1 ng/mL

KEY WORDS
plasma; SPE; cow; pharmacokinetics

REFERENCE
Alvinerie, M.; Sutra, J.F.; Badri, M.; Galtier, P. Determination of moxidectin in plasma by high-perfor-
mance liquid chromatography with automated solid-phase extraction and fluorescence detection,
J.Chromatogr.B, **1995**, *674*, 119–124.

SAMPLE
Matrix: tissue
Analyte: ivermectin
Sample preparation: Condition a Bond Elut C8 SPE cartridge with 5 mL MeCN and 5
mL MeCN:water:triethylamine 30:70:0.1. Homogenize (Silverson) 5 g frozen minced tissue with 15 mL MeCN at full speed for 1 min, centrifuge at 4° at 2000 g for 10 min.
Remove a 13 mL aliquot of the supernatant and add it to 35 mL water and 50 µL triethylamine, elute with 5 mL MeCN. Evaporate the eluate to dryness under a stream of
nitrogen, reconstitute with 200 µL 1-methylimidazole:MeCN 1:1, add 300 µL trifluoroacetic anhydride:MeCN 1:2, mix, store cold, inject an aliquot.

HPLC VARIABLES
Column: 250 × 4.6 Partisil 5 ODS-3
Mobile phase: MeOH:water 96:4
Flow rate: 1.8
Detector: F ex 364 em 470

CHROMATOGRAM
Retention time: 6 (ivermectin B_{1a})
Limit of detection: 1 ng/g

KEY WORDS
SPE; salmon; brain; gill; kidney; liver; muscle; skin; spleen

REFERENCE
Kennedy, D.G.; Cannavan, A.; Hewitt, S.A.; Rice, D.A.; Blanchflower, W.J. Determination of ivermectin
residues in the tissues of Atlantic salmon (*Salmo salar*) using HPLC with fluorescence detection,
Food Addit.Contam., **1993**, *10*, 579–584.

SAMPLE
Matrix: tissue
Analyte: ivermectin
Sample preparation: Condition a 3 mL Bakerbond C8 SPE cartridge with 5 mL MeCN
and 5 mL MeCN:water:triethylamine 30:70:0.1. Prepare a 55 mm column of 70-230
mesh Kiselgel 60 (Merck) in a Pasteur pipette, condition with 3 mL hexane:isopropanol
60:40. Homogenize (Polytron) 5 g blended tissue with 12 mL MeCN, rinse homogenizer
with 3 mL MeCN, centrifuge the mixture at 4000 rpm for 10 min. Remove the supernatant
and make up to 50 mL with water, add 50 µL triethylamine, shake, add to the SPE
cartridge, elute with 5 mL MeCN at 1 drop/s. Evaporate the eluate to 300 µL under a
stream of nitrogen at 40°, transfer to a smaller vial with MeCN, evaporate to dryness
under a stream of nitrogen at 60°, reconstitute the residue in 100 µL 1-methylimidazole:
MeCN 50:50, add 150 µL trifluoroacetic anhydride:MeCN 1:2, shake
at room temperature for 1 min, add to the column, rinse the vial with 500 µL hexane:
isopropanol 60:40, add the rinse to the column, elute with 1 mL hexane:isopropanol
60:40. Evaporate the eluate to dryness under a stream of nitrogen with heating, reconstitute with 250 µL MeCN, inject a 20 µL aliquot.

HPLC VARIABLES
Guard column: 5 µm LiChrospher 100 RP-18
Column: 125 × 4 5 µm LiChrospher 60 RP-select B
Mobile phase: MeOH:water 95:5

Column temperature: 40
Flow rate: 1
Injection volume: 20
Detector: F ex 365 em 465

CHROMATOGRAM
Retention time: 3.5
Limit of quantitation: 2.5 ppb
Limit of detection: 1 ppb

KEY WORDS
muscle; liver; pig; cow; SPE

REFERENCE
Guggisberg, D.; Sievi, M.; Koch, H. Methode zur quantitativen Bestimmung von Ivermectin in Fleisch und Leber mit HPLC und Vorsäulenderivatisation [Method for the quantitative determination of ivermectin in meat and liver by HPLC and pre-column derivatization], *Mitteilungen aus dem Gebiete der Lebensmitteluntersuchung und Hygiene*, **1994**, *85*, 395−405.

Dehydration

SAMPLE
Matrix: urine
Analyte: prostaglandins
Sample preparation: Urine. Make up 24 h volume of urine to 20 mL with water, add 200 ng IS, acidify to pH 4-5 with 2 M HCl, centrifuge, add 5 mL Amberlite XAD-2 (wetted form, wash with MeOH and water before use), stir at 0° for 30 min, place slurry in a sintered-glass funnel, wash with 20 mL water, force out residual water with nitrogen, elute with 10 mL MeOH. Concentrate the eluate to about 500 μL (mostly water) under a stream of nitrogen, extract twice with 1 mL ethyl acetate, evaporate to dryness, dissolve the residue in 30 μL diethyl ether:MeOH:acetic acid 90:10:0.5, add 200 μL diethyl ether:MeOH:acetic acid 100:1:0.5, add to SPE column, wash with 6 mL diethyl ether: MeOH:acetic acid 100:1:0.5, elute with 6 mL diethyl ether:MeOH:acetic acid 90:10: 0.5. Evaporate the eluate under a stream of nitrogen, dissolve the residue in 200 μL 0.5 M KOH, let stand at 20° for 1 h, adjust pH to 4 with 2 M acetic acid, extract with 2 mL diethyl ether, evaporate to dryness, add to another SPE column, wash with 6 mL hexane:diethyl ether:acetic acid 20:80:0.1, elute with 6 mL diethyl ether:MeOH:acetic acid 100:1:0.5, evaporate to dryness under a stream of nitrogen, reconstitute in 30 μL MeOH, inject a 5-10 μL aliquot. (Prepare SPE column (1 mL bed volume) by adding a slurry of 70-230 mesh Kieselgel 60 (Merck) in diethyl ether:acetic acid 100:0.5 to a Pasteur pipette, wash with 6 mL diethyl ether:acetic acid 100:0.5. The prostaglandin E's are dehydrated to the corresponding prostaglandin B's.)

HPLC VARIABLES
Column: 250 × 4.6 Lichrospher 100 CH-18/2 RP-18
Mobile phase: MeCN:water:acetic acid 62:38:0.1
Flow rate: 1
Injection volume: 5-10
Detector: UV 278

CHROMATOGRAM
Retention time: 10 (delta^{17}-tetranor-prostaglandin E$_1$ (as delta^{17}-tetranor-prostaglandin B$_1$)), 15 (tetranor-prostaglandin E$_1$ (as tetranor-prostaglandin B$_1$)), 24 (prostaglandin E$_3$ (as prostaglandin B$_3$)), 38 (dinoprostone (as prostaglandin B$_2$)), 43 (alprostadil (as prostaglandin B$_1$))
Internal standard: omega-nor-prostaglandin E$_2$ (as omega-nor-prostaglandin B$_2$) (21)

KEY WORDS
rat; SPE

REFERENCE
Kivits, G.A.A.; Nugteren, D.H. The urinary excretion of prostaglandins E and their corresponding tetranor metabolites by rats fed a diet rich in eicosapentaenoate, *Biochim.Biophys.Acta*, **1988**, *958*, 289–299.

Diazo Coupling

X = SO$_3$H, NO$_2$

SAMPLE
Matrix: air
Analyte: phenol
Sample preparation: Pull 5-150 L air through 10 mL 0.06% NaOH in water at 1-2 L/min. Add 1 mL buffer and 3 mL 0.1% 4-nitrobenzenediazonium tetrafluoroborate in water, make up to 20 mL with water, let stand for 15 min, inject a 2-40 μL aliquot. (Buffer was 0.51% sodium bicarbonate containing 0.21% sodium carbonate, pH 11.5.)

HPLC VARIABLES
Column: 200 × 4.6 Polygosil 60-5 C18 (Macherey-Nagel)
Mobile phase: MeOH:water 85:15
Flow rate: 1.1
Injection volume: 2-40
Detector: UV 365

CHROMATOGRAM
Retention time: 4
Limit of detection: 50 pg

OTHER SUBSTANCES
Also analyzed: o-cresol, m-cresol, p-cresol, 1-naphthol, 2-naphthol, 2,3-xylenol, 2,4-xylenol, 2,5-xylenol, 2,6-xylenol, 3,4-xylenol, 3,5-xylenol

REFERENCE
Kuwata, K.; Uebori, M.; Yamazaki, Y. Determination of phenol in polluted air as *p*-nitrobenzeneazophenol derivative by reversed phase high performance liquid chromatography, *Anal.Chem.*, **1980**, *52*, 857–860.

SAMPLE

Matrix: air

Analyte: phenols

Sample preparation: Pull 5-150 L air through 10 mL 0.12% NaOH in water at 1-2 L/min. Remove a 5 mL aliquot, add 1 mL buffer, add 3 mL 0.1% 4-nitrobenzenediazonium tetrafluoroborate in water, make up to 20 mL with water, let stand for 30 min, add 1 mL 1% NaOH, add 1 mL carbon tetrachloride, shake, centrifuge, inject a 2-40 μL aliquot of the aqueous (p-unsubstituted phenols) layer or a 2-10 μL aliquot of the organic (p-substituted phenols) layer. (Buffer was 0.51% sodium bicarbonate containing 0.21% sodium carbonate, pH 11.5.)

HPLC VARIABLES

Column: 200 × 4.6 5 μm LiChrosorb RP-18

Mobile phase: MeOH:water 85:15

Flow rate: 1.1

Injection volume: 2-40

Detector: UV 365

CHROMATOGRAM

Retention time: 3.46 (o-chlorophenol), 3.89 (phenol), 4.60 (m-cresol), 4.72 (m-chlorophenol), 4.94 (α-naphthol), 4.97 (o-cresol), 5.09 (m-ethylphenol), 5.58 (3,5-xylenol), 5.70 (o-ethylphenol), 5.89 (2,5-xylenol), 6.16 (2,6-xylenol), 6.18 (2,3-xylenol), 6.66 (p-chlorophenol), 7.34 (p-cresol), 8.79 (p-ethylphenol), 9.59 (β-naphthol), 10.81 (3,4-xylenol), 12.68 (2,4-xylenol)

Limit of detection: 0.05-2 ng

REFERENCE

Kuwata, K.; Uebori, M.; Yamazaki, Y. Reversed-phase liquid chromatographic determination of phenols in auto exhaust and tobacco smoke as p-nitrobenzeneazophenol derivatives, *Anal.Chem.*, **1981**, *53*, 1531–1534.

SAMPLE

Matrix: solutions

Analyte: phenols

Sample preparation: Mix 10 mL of a 50 μg/mL solution with 2 mL 100 mg/mL NaOH and 5 mL reagent, mix, let stand for 15 min, add 3.4 mL 15 mg/mL tetrabutylammonium bromide in butanol (saturated with water), extract, inject a 10 μL aliquot of the organic layer. (Prepare reagent by mixing 5 volumes 7.6 mg/mL sulfanilic acid with 1 volume 470 mg/mL sulfuric acid, cool in an ice bath, slowly add 5 volumes 3.4 mg/mL sodium nitrite. Discard the reagent after 10 min.)

HPLC VARIABLES

Column: 300 × 3.9 10 μm μBondapak RP phenyl

Mobile phase: MeOH:water 48:52 containing 3 mM tetrabutylammonium bromide

Flow rate: 2

Injection volume: 10

Detector: UV 370

CHROMATOGRAM

Retention time: 4.15 (4-nitrophenol), 5.12 (phenol), 7.12 (2-methylphenol), 7.12 (3-methylphenol), 8.24 (3,5-xylenol), 9.15 (3-chlorophenol), 9.57 (2,6-xylenol), 10.24 (2,5-xylenol), 10.42 (2,3-xylenol), 17.12 (3-t-butylphenol), 24.24 (2-sec-butylphenol), >26 (2-t-butylphenol), >26 (4-methylphenol)

Limit of quantitation: 100 ppb

REFERENCE

Baiocchi, C.; Gennaro, M.C.; Campi, E.; Mentasti, E.; Aruga, R. HPLC identification and separation of phenolic compounds derivatized with diazotized sulfanilic acid. Structural effects on retention times, *Anal.Lett.*, **1982**, *15*, 1539–1548.

Emerson Reaction

For a discussion of the Emerson reaction see *J. Org. Chem.* **1943**, *8*, 417.

SAMPLE
Matrix: solutions
Analyte: phenols
Sample preparation: Mix 100 mL of an aqueous solution with 10 mL buffer, 4 mL 20 mM (?) 4-aminoantipyrine, and 500 mL 80 mM (?) potassium ferricyanide solution, let stand for 45 min, extract with 5 mL dichloromethane, inject a 10 µL aliquot of the organic layer (cf J. Org. Chem. 1943, 8, 417). (Prepare buffer by adjusting pH of 5.30 g/L ammonium chloride solution to 8.5 with ammonia.)

HPLC VARIABLES
Guard column: 50 × 4.6 7 µm LiChrosorb RP-C18
Column: 250 × 4.6 7 µm LiChrosorb RP-C18
Mobile phase: Gradient. MeOH:water 50:50 for 24 min, to 70:30 over 1 min
Flow rate: 1
Injection volume: 20
Detector: UV 470

CHROMATOGRAM
Retention time: 19 (2,4-dichlorophenol), 23 (3,4-dichlorophenol), 30 (2,4,5-trichlorophenol)
Limit of quantitation: 5 ppb

REFERENCE
Fayyad, M.K.; Alawi, M.A.; El-Ahmed, T.J. HPLC determination of the phenolic metabolites of phenoxy alkanoic acid herbicides, *Chromatographia*, **1989**, *28*, 465−472.

SAMPLE
Matrix: water
Analyte: phenols
Sample preparation: Add disodium EDTA to sample. 100 mL Water + 10 mL pH 8-9 Britton-Robinson buffer (µ = 0.09) + 1 mL 1.5% 4-aminoantipyrine in water + 5 mL 2% potassium ferricyanide in water + 10 mL chloroform, stir for 10 min, inject a 10 µL aliquot of the organic layer (cf J. Org. Chem. 1943, 8, 417).

HPLC VARIABLES
Column: 300 × 3.9 µBondapak phenyl
Mobile phase: MeOH:water 60:40
Flow rate: 1
Injection volume: 10
Detector: UV 480

CHROMATOGRAM

Retention time: k' 1.36 (phenol), k' 1.79 (2-methylphenol), k' 1.88 (3-methylphenol), k' 2.45 (2-ethylphenol), k' 3.07 (3-ethylphenol), k' 2.43 (2,3-dimethylphenol), k' 2.85 (2,5-dimethylphenol), k' 3.03 (2,6-dimethylphenol), k' 2.00 (3,5-dimethylphenol), k' 2.60 (2,3,5-trimethylphenol), k' 3.92 (2,3,6-trimethylphenol), k' 3.72 (2,3,5,6-tetramethylphenol), k' 2.14 (2-chlorophenol), k' 3.07 (3-chlorophenol), k' 1.36 (4-chlorophenol), k' 1.88 (4-chloro-3-methylphenol), k' 3.71 (2,3-dichlorophenol), k' 2.14 (2,4-dichlorophenol), k' 4.04 (2,6-dichlorophenol), k' 4.04 (2,4,6-trichlorophenol), k' 1.63 (2-nitrophenol), k' 2.13 (3-nitrophenol)

Limit of detection: 20 ng/mL

REFERENCE

Blo, G.; Dondi, F.; Betti, A.; Bighi, C. Determination of phenols in water samples as 4-aminoantipyrine derivatives by high-performance liquid chromatography, *J.Chromatogr.*, **1983**, *257*, 69–79.

SAMPLE

Matrix: water

Analyte: phenol

Sample preparation: 100 mL Water + 2 mL 50 mM KCl in water + 1 mL 0.6% dextrin in water + 1 mL 80 mM silver nitrate in water + 2 mL pH 9 borax buffer + 500 μL 2% 4-aminoantipyrine in water + 10 mL chloroform, stir at 40° for 40 min, inject an aliquot of the organic layer (cf J. Org. Chem. 1943, 8, 417).

HPLC VARIABLES

Column: 300 × 3.9 μBondapak phenyl

Mobile phase: MeOH:water 60:40

Flow rate: 1

Injection volume: 10

Detector: UV 254

CHROMATOGRAM

Retention time: 7

Limit of quantitation: 0.5 ppm

REFERENCE

Blo, G.; Dondi, F.; Bighi, C. High-performance liquid chromatographic determination of phenols as 4-aminoantipyrine derivatives; silver chloride as oxidizing agent in the derivatization reaction, *J.Chromatogr.*, **1984**, *295*, 231–235.

Formylation

SAMPLE

Matrix: solutions

Analyte: peptides

Sample preparation: 200 μL Solution + 100 μL chloroform + 50 μL 3 M KOH, heat at 60° for 10 min, cool in ice-water for 1 min, add 50 μL 14 M acetic acid, add 300 μL freshly prepared 265 μg/mL 1,2-diamino-4,5-dimethoxybenzene monohydrochloride in water (with cooling in ice-water), heat at 60° for 18 min, cool, inject a 100 μL aliquot. (1,2-Diamino-4,5-dimethoxybenzene is available from Molecular Probes, Eugene OR or Dojindo Molecular Technologies, Inc., 3 Bethesda Metro Center, Suite 700, Bethesda MD 20814; (301) 664-8448; www.dojindo.co.jp. Synthesis is as follows. Stir 483 g veratrole in 1.45 L acetic acid at 15° for 1 h, add 683 g concentrated nitric acid (d 1.05) over 1 h (maintain the temperature below 40° by cooling and regulating the rate of addition of the nitric acid). Continue stirring and add 2.127 L fuming nitric acid (d 1.50) over 1 h while maintaining the temperature below 30°, let stand for 2 h, pour into a large volume of cold water, filter, wash the solid with water until the washings are neutral, recrystallize from EtOH to give 4,5-dinitroveratrole (mp 129.5-130.5°) (J. Am. Chem. Soc. 1946, 68, 1536). Reflux 5 g 4,5-dinitroveratrole in 200 mL benzene (Caution! Benzene is a carcinogen!), add 100 g 60 mesh iron powder and 20 mL concentrated HCl in small portions over 1 h, reflux for 4 h, add 10 mL water, reflux for 2 h, cool, make alkaline with 2.5 M NaOH, extract several times with 200 mL portions of benzene. Combine the organic layers and evaporate them to dryness, add 10 mL concentrated HCl, recrystallize from EtOH to give 1,2-diamino-4,5-dimethoxybenzene monohydrochloride as very slightly pink needles (mp 240°) (Anal. Chim. Acta 1982, 134, 39). Another synthesis of 1,2-diamino-4,5-dimethoxybenzene is as follows. Add 9.66 g veratrole in 9.5 mL glacial acetic acid dropwise to 26.6 mL fuming nitric acid over 30 min using an ice bath, stir at room temperature for 30 min, pour into 500 mL ice-cold water, filter to obtain 1,2-dimethoxy-4,5-dinitrobenzene, recrystallize repeatedly from EtOH. Dissolve 1 g 1,2-dimethoxy-4,5-dinitrobenzene in 50 mL EtOH, add 100 mg 10% palladium on charcoal, stir under hydrogen at atmospheric pressure and room temperature for 4 days, filter through Celite under an atmosphere of nitrogen, saturate the filtrate with hydrogen chloride gas, filter under nitrogen, dry the solid under vacuum to obtain 1,2-diamino-4,5-dimethoxybenzene hydrochloride (mp 240° d (Anal. Chim. Acta 1982, 134, 39)) (Anal. Chim. Acta 1992, 263, 137).)

HPLC VARIABLES
Column: 150 × 4 5 μm LiChrosorb RP-18
Mobile phase: MeCN:buffer:50 mM sodium 1-hexanesulfonate 26:64:10 (Prepare buffer by dissolving 14.9 g KCl in 950 mL water, adjusting pH to 2.2 with concentrated HCl, and making up to 1 L with water.)
Flow rate: 0.8
Injection volume: 100
Detector: F ex 350 em 425

CHROMATOGRAM
Retention time: 10.9 (methionine enkephalin), 14.5 (angiotensin II), 16.8 (angiotensin III), 18.4 (leucine enkephalin), 31.5 (angiotensin I)
Limit of detection: 6.8-26.2 pmole

KEY WORDS
specific for tyrosine-containing peptides

REFERENCE
Ishida, J.; Kai, M.; Ohkura, Y. High-performance liquid chromatography of tyrosine-containing peptides by pre-column derivatization involving formylation followed by fluorescence reaction with 1,2-di-amino-4,5-dimethoxybenzene, *J.Chromatogr.*, **1986**, *356*, 171–177.

SAMPLE
Matrix: tissue
Analyte: enkephalins
Sample preparation: Condition a Bond Elut C18 SPE cartridge with 3 mL water and 3 mL MeOH (in this order ?). Homogenize 200 mg tissue with 3 mL 100 mM HCl at 0-4°, add 50 μL 1 μM IS, add 1 mL 100 mM HCl, add 500 μL 2 M perchloric acid, mix, centrifuge at 800 g for 10 min, remove the supernatant, suspend the precipitate in 2 mL 200 mM perchloric acid, centrifuge. Combine the supernatants and adjust pH to 7-8 with about 2 mL 1 M sodium bicarbonate, add to the SPE cartridge, wash with 1 mL water, wash with 2 mL dichloromethane, wash with 1 mL water, wash with 3 mL 100 mM HCl, wash with 1 mL water, wash with 3 mL 100 mM pH 8.5 borate buffer, wash with 1 mL water, elute with 1 mL MeOH:water 90:10. Evaporate the eluate to dryness under vacuum at 30°, reconstitute with 200 μL water. Remove a 100 μL aliquot and add it to 50 μL chloroform, add 25 μL 3 M KOH, heat at 60° for 10 min, cool in ice-water for 1 min, add 25 μL 14 M acetic acid, add 25 μL 4.6 mM 1,2-diamino-4,5-dimethoxybenzene monohydrochloride in water (with cooling in ice-water), heat at 60° for 18 min, cool, inject a 100 μL aliquot. (1,2-Diamino-4,5-dimethoxybenzene is available from Molecular Probes, Eugene OR or Dojindo Molecular Technologies, Inc., 3 Bethesda Metro Center, Suite 700, Bethesda MD 20814; (301) 664-8448; www.dojindo.co.jp. Synthesis is described above.)

HPLC VARIABLES
Column: 250 × 4.6 5 μm TSK gel ODS-120T (Tosoh)
Mobile phase: MeCN:50 mM pH 2.2 phosphate buffer:50 mM sodium 1-hexanesulfonate 26:64:10 (Wash with MeCN:water 50:50 for 20 min after the last peak elutes.)
Flow rate: 1
Injection volume: 100
Detector: F ex 350 em 425

CHROMATOGRAM
Retention time: 16.8 (methionine enkephalin), 25.8 (leucine enkephalin)
Internal standard: [Ala^2,Ala^3]methionine enkephalin (24.0)
Limit of detection: 5.6 pmole/g

KEY WORDS
rat; brain; SPE

REFERENCE
Kai, M.; Ishida, J.; Ohkura, Y. High-performance liquid chromatographic determination of leucine-enkephalin-like peptide in rat brain by pre-column fluorescence derivatization involving formylation followed by reaction with 1,2-diamino-4,5-dimethoxybenzene, *J.Chromatogr.*, **1988**, *430*, 271–278.

Glucuronidation

SAMPLE
Matrix: bulk
Analyte: N-0437 (2-(N-propyl-N-2-thienylethylamino)-5-hydroxytetralin)
Sample preparation: Shake 2 mg cow liver microsomes in 1 mL 80 mM pH 7.4 Tris-HCl buffer containing 50 mM magnesium chloride at 37° for 10 min, add 300 nmoles N-0427, add 6 μmoles uridine 5'-diphosphoglucuronic acid, vortex for 5 s, shake at 37° for 2 h, cool on ice, centrifuge at 8000 g for 10 min, inject an aliquot of the supernatant. (The enzyme involved is uridine 5'-diphosphoglucuronyltransferase (UDPGT). Prepare microsomes by homogenizing (Potter-Elvehjem, PTFE pestle) cow liver at 5° with four volumes of 154 mM KCl (adjusted to pH 7.0 with 154 mM KOH)), centrifuge at 9000 g for 20 min, centrifuge the supernatant at 100000 g for 1 h, suspend the microsomal pellet in 154 mM KCl to a protein concentration of 10 mg/mL (Anal.Biochem. 1988, 171, 382).)

HPLC VARIABLES
Column: 150 × 4.6 3 μm Rosil C18 (Alltech)
Mobile phase: MeCN:10 mM pH 6.8 phosphate buffer 12:88
Flow rate: 1.5
Injection volume: 50
Detector: UV 225

CHROMATOGRAM
Retention time: 39 (-), 44 (+)

KEY WORDS
chiral

REFERENCE
Gerding, T.K.; Drenth, B.F.H.; Van de Grampel, V.J.M.; Niemeijer, N.R.; de Zeeuw, R.A.; Tepper, P.G.; Horn, A.S. Determination of enantiomeric purity of the new D-2 dopamine agonist 2-(N-propyl-N-2-thienylethylamino)-5-hydroxytetralin (N-0437) by reversed-phase high-performance liquid chromatography after pre-column derivatization with D(+)-glucuronic acid, *J.Chromatogr.*, **1989**, *487*, 125–134.

Hydroxylation

SAMPLE
Matrix: solutions
Analyte: peptides
Sample preparation: Prepare peptide solution in PBS. 46 μL Solution + 5 μL 10 mg/mL tyrosinase (mushroom, Sigma T7755) in PBS + 5 μL 18 mM (?) ascorbic acid solution, let stand at room temperature for 20 min, add 135 μL ferricyanide solution, add 100 μL MeCN, add 25 μL 20 mg/mL 1,2-diamino-1,2-diphenylethane in 100 mM HCl, let stand at room temperature for 50 min, inject a 100 μL aliquot. (Prepare ferricyanide solution by dissolving 63 mg potassium ferricyanide and 119 mg KCl in 5.5 mL PBS. Prepare PBS by dissolving 160 g NaCl, 4 g KH_2PO_4, 72.7 g $Na_2HPO_4.7H_2O$, and 4 g KCl in 20 L water. Prepare 1,2-diamino-1,2-diphenylethane (1,2-diphenylethylenediamine) by refluxing 1.9 equivalents benzaldehyde with 1 equivalent ammonium acetate for 3 h and collecting the precipitate, wash the precipitate with EtOH. Hydrolyse the precipitate with 33% sulfuric acid, steam distil to remove benzoic acid and benzaldehyde, neutralize with ammonia to precipitate the product. Recrystallize 1,2-diamino-1,2-diphenylethane from petroleum ether (mp 118-119°). Another synthesis of meso-1,2-diphenylethylenediamine is as follows. Reflux 40 g ammonium acetate and 100 mL benzaldehyde for 3 h, cool, collect the precipitate, wash the precipitate with EtOH until it is white, recrystallize from n-butanol to obtain N-benzoyl-N'-benzylidene-meso-1,2-diphenylethylenediamine (mp 259°). Cautiously add 54 mL concentrated sulfuric acid to 100 mL water (Caution! Extremely exothermic!), cautiously add 10 g unpurified N-benzoyl-N'-benzylidene-meso-1,2-diphenylethylenediamine, heat, pass steam through the reaction mixture to remove benzaldehyde and benzoic acid, when the distillate is no longer acid (about 4 h) cool, filter. Cautiously neutralize the filtrate with concentrated ammonium hydroxide with ice-cooling, extract with ether, dry over solid KOH, evaporate to dryness, recrystallize from petroleum ether (bp 60-80°) to obtain meso-1,2-diphenylethylenediamine as white crystals (mp 120°) (J. Inorg. Nucl. Chem. 1965, 27, 270).)

HPLC VARIABLES
Column: 150 × 4.6 5 μm Nucleosil C18
Mobile phase: MeCN:50 mM pH 6.8 Tris 2:5
Flow rate: 1

Injection volume: 100
Detector: F ex 345 em 417

CHROMATOGRAM
Retention time: 3.8 (Tyr-Gly), 5 (Leu-enkephalin)
Limit of detection: 200 fmole

KEY WORDS
analysis is specific for tyrosine-containing peptides

REFERENCE
Tellier, M.; Prankerd, R.J.; Hochhaus, G. A new fluorogenic assay for tyrosine-containing peptides, *J.Pharm.Biomed.Anal.*, **1991**, *9*, 557–563.

Oxidation

SAMPLE
Matrix: blood
Analyte: salinomycin
Sample preparation: Extract plasma with isooctane, add the organic layer to a silica SPE cartridge, wash with dichloromethane, wash with dichloromethane:MeOH (?) 98.5:1.5, elute with dichloromethane:MeOH 90:10, evaporate to dryness, reconstitute with dichloromethane, oxidize with pyridinium dichromate, concentrate, inject an aliquot on to column A and elute to waste with mobile phase A, after 1.85 min divert the effluent from column A on to column B (?), after another 1.8 min remove column A from the circuit, elute column B with mobile phase B, monitor the effluent from column B.

HPLC VARIABLES
Column: A 75 mm long C18 (Waters); B 250 mm long C18 (Beckman)
Mobile phase: A MeCN:0.01% HCl 90:10; B MeCN:0.01% HCl 96:4
Detector: UV 225

CHROMATOGRAM
Limit of detection: 5 ng/mL

KEY WORDS
plasma; column-switching; heart cut; SPE

REFERENCE
Wei, A.T.; Dimenna, G.P.; Karnes, H.T. HPLC analysis of sodium salinomycin in human plasma using derivatization and heart cut column switching, *Pharm.Res.*, **1992**, *9*, S21.

SAMPLE
Matrix: seminal fluid
Analyte: prostaglandins
Sample preparation: Condition a Sep-Pak C18 SPE cartridge with 20 mL EtOH and 20 mL water. Dilute seminal fluid to 1 mL, adjust to pH 3.5 with aqueous formic acid, centrifuge, add the supernatant to the SPE cartridge, wash with 20 mL EtOH:water 15:85, wash with 20 mL water, remove excess water mechanically, wash with 20 mL hexane, elute with 4 mL methyl formate. Dry the eluate under a stream of nitrogen, add 10-20 µL reagent, vortex for 1 min, let stand at room temperature for 8 min, add 90-180 µL water, extract with an equal volume of ethyl acetate, centrifuge at 2000 g for 3 min, evaporate the organic layer to dryness under a stream of nitrogen, reconstitute, inject an aliquot (J. Chromatogr. 1985, 349, 431). (Prepare the reagent by stirring 100 mg pyridinium dichromate in 50 mL MeCN at room temperature for 1 h, centrifuge, use the

supernatant (5 mM; 1.9 mg/mL), store at 5°, discard after 2 days (J. Chromatogr. 1983, 282, 435).)

HPLC VARIABLES

Column: 220 × 2.1 5 μm Spheri-5 C18
Mobile phase: Gradient. MeCN:0.5 mM formic acid 30:70 for 4.5 min, to 40:60 (step gradient).
Flow rate: 0.4
Injection volume: 1
Detector: UV 229

CHROMATOGRAM

Retention time: 2.5 (19-hydroxyprostaglandin E_2), 3 (oxoprostaglandin E_2), 3.2 (19-hydroxyprostaglandin E_1), 3.8 (oxoprostaglandin E_1), 9.2 (dinoprostone), 10 (alprostadil)

KEY WORDS

SPE

REFERENCE

Doehl, J.; Greibrokk, T. Determination of prostaglandins in human seminal fluid by solid-phase extraction, pyridinium dichromate derivatization and high-performance liquid chromatography, *J.Chromatogr.*, **1990**, *529*, 21–32.

SAMPLE

Matrix: solutions
Analyte: prostaglandins
Sample preparation: Dry solution under a stream of nitrogen, add 10 equivalents of reagent, vortex for 1 min, let stand at room temperature for 8 min, add a volume of water equivalent to one tenth the volume of the reaction mixture, inject a 5 μL aliquot. (Prepare the reagent by stirring 100 mg pyridinium dichromate in 50 mL MeCN at room temperature for 1 h, centrifuge, use the supernatant (5 mM; 1.9 mg/mL), store at 5°, discard after 2 days.)

HPLC VARIABLES

Guard column: 50 × 4.6 40 μm pellicular C18 (Supelco)
Column: 200 × 4.6 5 μm RP-18 (Brownlee)
Mobile phase: Gradient. MeCN:10 mM formic acid from 40:60 to 60:40 over 10 min
Flow rate: 1.5
Injection volume: 5
Detector: UV 228 for 10 min then UV 298

CHROMATOGRAM

Retention time: 6 (dinoprostone), 6.5 (alprostadil), 9.5 (prostaglandin A_2), 10.5 (prostaglandin A_1), 11 (prostaglandin B_2), 11.5 (prostaglandin B_1)
Limit of detection: 30-80 pmole

REFERENCE

Dohl, J.; Greibrokk, T. High-performance liquid chromatographic separation and ultraviolet detection of prostaglandins, oxidized by pyridinium dichromate, *J.Chromatogr.*, **1983**, *282*, 435–442.

SAMPLE

Matrix: solutions
Analyte: prostaglandins
Sample preparation: Dry solution under a stream of nitrogen, add 10-20 μL reagent, vortex for 1 min, let stand at room temperature for 8 min, add 90-180 μL water, inject a 1 μL aliquot. Alternatively, extract with an equal volume of ethyl acetate, centrifuge at 2000 g for 3 min, evaporate the organic layer to dryness under a stream of nitrogen, reconstitute, inject an aliquot. (Prepare the reagent by stirring 100 mg pyridinium dichromate in 50 mL MeCN at room temperature for 1 h, centrifuge, use the supernatant (5 mM; 1.9 mg/mL), store at 5°, discard after 2 days.)

HPLC VARIABLES

Column: 250 × 1.3 8 μm C18 (Chrompack)
Mobile phase: MeCN:10 mM pH 2.7 phosphoric acid 38:62

Flow rate: 0.06
Injection volume: 1
Detector: UV 229

CHROMATOGRAM
Retention time: 20 (dinoprost), 22 (prostaglandin $F_{1\alpha}$, 25 (dinoprostone), 28 (alprostadil)
Limit of detection: 0.14 pmole

KEY WORDS
microbore

REFERENCE
Doehl, J.; Greibrokk, T. High-performance liquid chromatographic separation and determination of prostaglandins, oxidized by pyridinium dichromate. Optimization and applications, *J.Chromatogr.*, **1985**, *349*, 431–438.

Phenolic Coupling

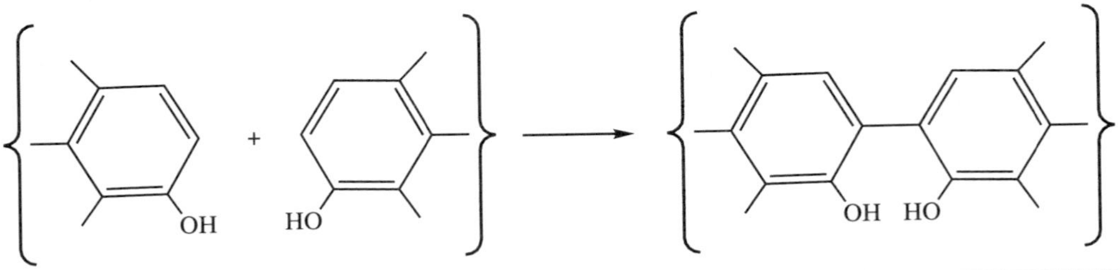

SAMPLE
Matrix: urine
Analyte: 6-acetylmorphine
Sample preparation: Centrifuge urine at 3000 rpm for 5 min. Remove an 8 mL aliquot of the supernatant and add it to 12 mL 1 M pH 9.0 ammonium chloride buffer, mix, add to an 85 × 27 Extrelut SPE cartridge, allow to drain for 10 min, elute with 40 mL dichloromethane:isopropanol 85:15. Extract the eluate with 10 mL 50 mM sulfuric acid, discard the organic layer. Add the aqueous layer to an 85 × 27 Extrelut SPE cartridge, allow to drain for 10 min, elute with 40 mL dichloromethane:isopropanol 85:15. Evaporate the eluate to dryness under a stream of nitrogen at 40°, reconstitute the residue in 200 μL 125 μg/mL morphine hydrochloride in 15 mM HCl, add 10 μL 300 mM pH 8.5 Tris buffer, add 105 μL 15 mM potassium ferricyanide in water, mix, let stand for 2 min, inject a 50 μL aliquot. Between runs make three consecutive injections of MeCN:water:triethylamine 40:60:0.1 to prevent carryover. (Oxidation of 6-acetylmorphine and morphine forms a fluorescent dimer by phenolic coupling. Product is acetylpseudomorphine; stucture given in J. Chromatogr. 1975, 109, 37.)

HPLC VARIABLES
Column: 150 × 4.6 5 μm Hypersil ODS
Mobile phase: MeCN:water:triethylamine 16:84:0.1
Column temperature: 30
Flow rate: 1.5
Injection volume: 50
Detector: F ex 320 em 436

CHROMATOGRAM
Retention time: 10
Limit of detection: 1 ng/mL

KEY WORDS
SPE

REFERENCE
Derks, H.J.G.M.; van Twillert, K.; Pereboom-de Fauw, D.P.K.H.; Zomer, G.; Loeber, J.G. Determination of the heroin metabolite 6-acetylmorphine by high-performance liquid chromatography using automated pre-column derivatization and fluorescence detection, *J.Chromatogr.*, **1986**, *370*, 173–178.

SILYLATING REAGENT

Bis(trimethylsilyl)trifluoroacetamide

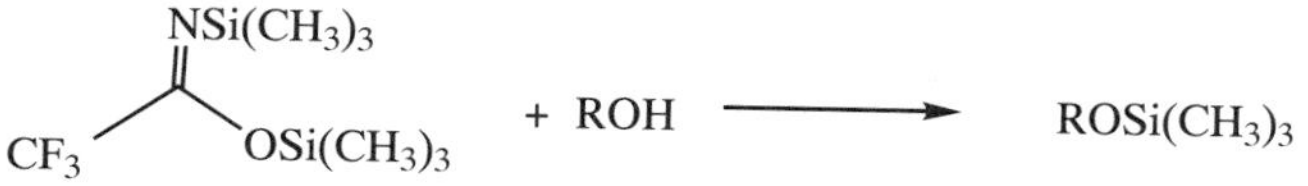

$$CF_3 \begin{matrix} NSi(CH_3)_3 \\ OSi(CH_3)_3 \end{matrix} + ROH \longrightarrow ROSi(CH_3)_3$$

SAMPLE
Matrix: bulk
Analyte: 24(R)-hydroxyvitamin D3
Sample preparation: Dissolve compound in 30 μL pyridine, add 25 μL bis(trimethylsilyl)trifluoroacetamide containing 1% trimethylsilyl chloride, heat at 55° for 45 min, evaporate to dryness under a stream of nitrogen, reconstitute with 50 μL hexane, inject an aliquot.

HPLC VARIABLES
Column: 250 × 4.5 microparticulate silica
Mobile phase: Hexane:ethyl acetate 99.85:0.15
Flow rate: 2
Detector: UV 254

CHROMATOGRAM
Retention time: 7.5

KEY WORDS
normal phase

REFERENCE
Wichmann, J.; Schnoes, H.K.; DeLuca, H.F. Isolation of identification of 24(R)-hydroxyvitamin D3 from chicks give large doses of vitamin D3, *Biochem.*, **1981**, *20*, 2350–2353.

SAMPLE
Matrix: bulk
Analyte: 24,25,26,27-tetranor-23-hydroxyvitamin D3
Sample preparation: Dissolve in 15 μL pyridine, add 10 μL N,O-bis(trimethylsilyl)trifluoroacetamide containing 1% trimethylsilyl chloride, heat at 55° for 45 min, evaporate to dryness under a stream of nitrogen.

HPLC VARIABLES
Column: 250 × 6.2 Zorbax-SIL
Mobile phase: Hexane:isopropanol 99.75:0.25
Detector: UV

CHROMATOGRAM
Retention time: 4.45

KEY WORDS
normal phase

REFERENCE
Jones, G.; Kano, K.; Yamada, S.; Furusawa, T.; Takayama, H.; Suda, T. Identification of 24,25,26,27-tetranor-23-hydroxyvitamin D3 as a product of the renal metabolism of 24,25-dihydroxyvitamin D3, *Biochem.*, **1984**, *23*, 3749–3754.

SAMPLE
Matrix: bulk
Analyte: 1,23-dihydroxy-24,25,26,27-tetranorvitamin D3
Sample preparation: Dissolve 1 μg compound in 15 μL pyridine, add 10 μL bis(trimethylsilyl)trifluoroacetamide containing 1% trimethylsilyl chloride, heat at 55° for 45 min, evaporate to dryness under a stream of nitrogen.

HPLC VARIABLES
Column: Zorbax-SIL
Mobile phase: Hexane:ethyl acetate 99:1
Flow rate: 2

CHROMATOGRAM
Retention time: 4

KEY WORDS
normal phase

REFERENCE
Reddy, G.S.; Tserng, K.Y.; Thomas, B.R.; Dayal, R.; Norman, A.W. Isolation and identification of 1,23-dihydroxy-24,25,26,27-tetranorvitamin D3, a new metabolite of 1,25-dihydroxyvitamin D3 produced in rat kidney, *Biochem.*, **1987**, *26*, 324–331.

tert-Butyldimethylsilyl Chloride

$$(H_3C)_3C-\underset{\underset{CH_3}{|}}{\overset{\overset{CH_3}{|}}{Si}}Cl \quad + \quad ROH \quad \longrightarrow \quad (H_3C)_3C-\underset{\underset{CH_3}{|}}{\overset{\overset{CH_3}{|}}{Si}}OR$$

SAMPLE
Matrix: bulk
Analyte: cannabinoids
Sample preparation: Dissolve 1 mmole compound, 4 mmole imidazole, and 2 mmole t-butyldimethylsilyl chloride in 500 μL DMF, heat on a steam bath for 1 h, add 2 mL 10% NaOH, extract 3 times with 2 mL portions of petroleum ether bp (30-60). Combine the extracts and dry them over anhydrous magnesium sulfate, evaporate to dryness.

HPLC VARIABLES
Column: 305 mm × 6.4 mm (o.d.) 10 μm μBondapak C18
Mobile phase: MeOH:water 90:10
Flow rate: 1
Detector: UV 254

CHROMATOGRAM
Retention time: 13.9 (cannabinol), 16.1 (delta^8-tetrahydrocannabinol), 16.2 (dronabinol), 35 (cannabidiol)

REFERENCE
Knaus, E.E.; Coutts, R.T.; Kazakoff, C.W. The separation, identification, and quantitation of cannabinoids and their t-butyldimethylsilyl, trimethylsilylacetate, and diethylphosphate derivatives using high-pressure liquid chromatography, gas-liquid chromatography, and mass spectrometry, *J.Chromatogr.Sci.*, **1976**, *14*, 525–530.

Chlorodimethylphenylsilane

SAMPLE
Matrix: bulk
Analyte: monosaccharides
Sample preparation: Dissolve 1-10 mg carbohydrate in 150 μL DMF, add 200 μL 330 μg/mL imidazole in DMF, heat at 100° for 1 h, cool in ice, add 70 μL chlorodimethyl-phenylsilane, let stand at room temperature for 6-18 h (or heat at 100° for 1 h), extract twice with 200 μL aliquots of hexane, inject a 3 μL aliquot of the hexane layer. (Dry DMF over molecular sieve. Procedure tends to produce multiple peaks. Retention times are those of the major peak.)

HPLC VARIABLES
Column: 250 × 4.6 Partisil 5
Mobile phase: Hexane:ethyl acetate 99.5:0.5
Flow rate: 1.5
Injection volume: 3
Detector: UV 254

CHROMATOGRAM
Retention time: k' 2.6 (fucose), k' 2.8 (tagatose), k' 2.8 (glucose), k' 2.9 (xylose), k' 3.0 (mannose), k' 3.0 (galactose), k' 3.1 (fructose), k' 3.7 (rhamnose), k' 3.7 (altrose), k' 4.0 (ribose), k' 4.1 (lyxose), k' 5.9 (arabinose)

KEY WORDS
normal phase

REFERENCE
White, C.A.; Vass, S.W.; Kennedy, J.F.; Large, D.G. High-pressure liquid chromatography of dimethyl-phenylsilyl derivatives of some monosaccharides, *Carbohydr.Res.*, **1983**, *119*, 241−247.

SAMPLE
Matrix: solutions
Analyte: glycopyranosides
Sample preparation: Mix 150 μL of a saccharide solution in DMF with 200 μL 330 mg/mL imidazole in DMF, heat at 100° for 1 h, cool in ice, add 70 μL chlorodimethylphenylsilane (phenyldimethylsilyl chloride), mix, let stand at room temperature for 6-18 h (or heat at 100° for 1 h), extract twice with 200 μL portions of hexane, inject a 3 μL aliquot of the extracts.

HPLC VARIABLES
Column: 250 × 4.6 Partisil 5
Mobile phase: Hexane.ethyl acetate 98:2
Flow rate: 1.5
Injection volume: 3
Detector: UV 260

CHROMATOGRAM
Retention time: 5 (methyl β-D-glucopyranoside), 8 (methyl β-D-xylopyranoside), 8.5 (methyl β-D-galactopyranoside), 12 (methyl α-D-glucopyranoside), 15 (methyl α-D-xylo-pyranoside), 19 (methyl α-D-galactopyranoside)
Limit of detection: 250 ng

KEY WORDS
derivatization of monosaccharides and disaccharides produces multiple peaks; normal phase

REFERENCE
White, C.A.; Vass, S.W.; Kennedy, J.F.; Large, D.G. Analysis of phenyldimethylsilyl derivatives of mon-
osaccharides and their role in high-performance liquid chromatography of carbohydrates,
J.Chromatogr., **1983**, *264*, 99–109.

SOLID PHASE REAGENT

Polymer-bound Dimethylaminopyridinium Fluorenylmethyl Carbamate

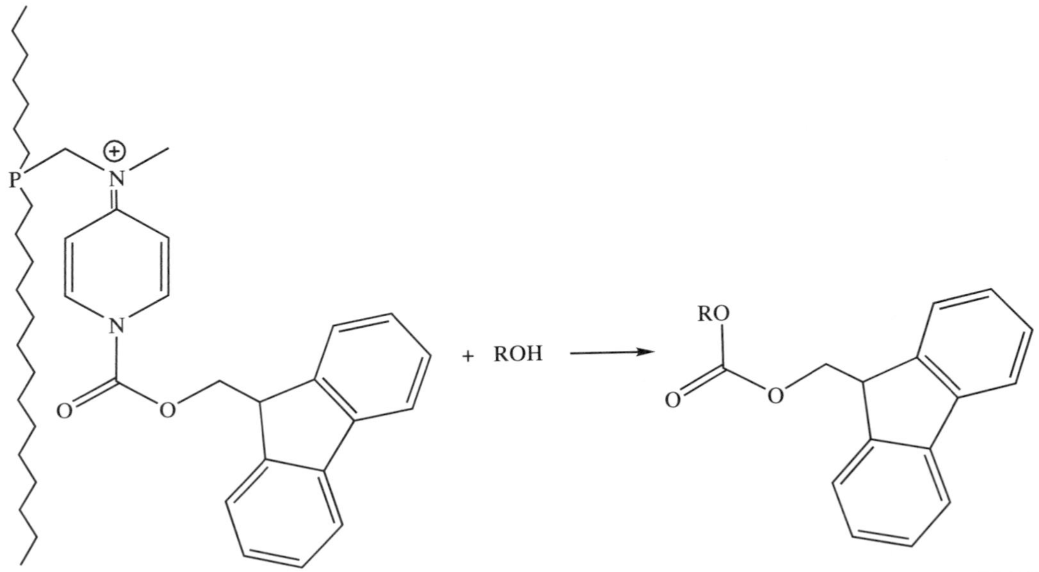

SAMPLE
Matrix: flour
Analyte: alcohols
Sample preparation: Extract 1 g wheat flour with 100 mL chloroform, filter (0.5 μm), mix
an aliquot with 30 mg polymer-bound dimethylaminopyridinium fluorenylmethyl carbamate, heat at 60° for 20 min, elute with 1 mL MeCN, inject a 20 μL aliquot. (Prepare
polymer-bound dimethylaminopyridinium fluorenylmethyl carbamate as follows. Suspend
10 g dry 200-400 mesh 4% cross-linked 4.2 mequiv Cl/g macroporous chloromethylstyrene-
divinylbenzene copolymer (Bio-Rad) in 15 mL DMF, saturate with methylamine gas at 0°,
rotate in a sealed vessel for 1 day, wash with three 30 mL portions of dioxane (Caution!
Dioxane is a carcinogen!), wash with three 30 mL portions of ethanol, wash with three
30 mL portions of isopropanol:2 M NaOH 50:50, wash with water until neutral, wash
with three 30 mL portions of EtOH, wash with two 30 mL portions of diethyl ether, dry
under vacuum to obtain white polymeric N-methylene methylamine. Add 3.5 g of this
polymer to 1 mL water, 2 mL EtOH, and 5 mL triethylamine and allow it to swell, add
4-chloropyridine hydrochloride (?), evaporate until the polymer is slightly wet, heat in a
pressure vessel at 140° for 4 days, wash as before, stir with 5 mL acetic anhydride:
dichloromethane 50:50 at room temperature for 1 h, wash as before with two 10 mL
portions, dry to constant weight at 140° to give light-brown polymer bound dimethylaminopyridine. Allow 3 g of this polymer to swell in anhydrous dichloromethane, add 3 g 9-
fluorenylmethyl chloroformate, stir at room temperature for 1 h, filter, wash with dichloromethane under anhydrous conditions until the washings contain negligible amounts

of 9-fluorenylmethyl chloroformate (by HPLC), dry under vacuum at room temperature to obtain light-brown polymer-bound dimethylaminopyridinium fluorenylmethyl carbamate, store at 5° under anhydrous conditions.)

HPLC VARIABLES

Column: 250 × 4 5 μm LiChrospher C18
Mobile phase: MeCN:water 60:40 (A) or MeCN:MeOH:water 50:10:40 (B)
Flow rate: 1.4
Injection volume: 20
Detector: F ex 265 em 320

CHROMATOGRAM

Retention time: 4 (2-ethyl-2-nitro-1-propanol (A)), 5.6 (EtOH (A)), 6 (2-methyl-2-nitro-1-propanol (A)), 6 (methoxyethanol (B)), 7 (2-chloro-1-propanol (A)), 8 (EtOH (B)), 9.5 (iso-propanol (B)), 11 (1-pentanol (B)), 16 (1,8-octanediol (B))
Limit of detection: 150 ppb

REFERENCE

Gao, C.X.; Krull, I.S. Polymeric dimethylaminopyridinium reagents for derivatization of weak nucleophiles in high-performance liquid chromatography-ultraviolet/fluorescence detection, *J.Chromatogr.*, **1990**, *515*, 337–356.

SUCCINIMIDYL ESTER

6-Aminoquinolyl-N-hydroxysuccinimidyl Carbamate

SAMPLE

Matrix: beverages
Analyte: alcohols
Sample preparation: Vacuum filter beverage. 40 μL Sample + 130 μL 500 mM pH 7.5 borate buffer + 30 μL 3 mg/mL 6-aminoquinolyl-N-hydroxysuccimidyl carbamate in far UV MeCN, agitate for 2 min, inject a 5 μL aliquot. (6-Aminoquinolyl-N-hydroxysuccinimidyl carbamate can be purchased from Waters or synthesized as follows. Reflux 3 g N,N'-succinimidyl carbonate in 100 mL dry MeCN, add 1.5 g 6-aminoquinoline in 50 mL dry MeCN dropwise over 30 min, reflux for 30 min, evaporate to half volume under reduced pressure, cool for 24 h, filter, wash the solid with cold MeCN, recrystallize from MeCN to obtain 6-aminoquinolyl-N-hydroxysuccinimidyl carbamate as off-white crystals (mp 210-215° (d)) (Anal. Biochem. 1993, 211, 279).)

HPLC VARIABLES

Guard column: 10 × 2 3 μm Microspher C18 (Chrompack)
Column: two 100 × 4.6 3 μm Microspher C18 (Chrompack) in series

Mobile phase: Gradient. A was MeCN:pH 6 sodium acetate buffer 5:95. B was MeCN:pH 6 sodium acetate buffer 95:5. A:B from 80:20 to 27:73 over 40 min, to 0:100 over 1 min, maintain at 0:100 for 2 min, return to initial conditions, re-equilibrate for 10 min.
Column temperature: 33
Flow rate: 1
Injection volume: 5
Detector: F ex 290 em 345 (bandpass filter)

CHROMATOGRAM
Retention time: 8.5 (MeOH), 12 (EtOH), 16.5 (isopropanol), 17 (n-propanol), 21 (2-butanol), 21.7 (2-methyl-1-propanol), 22 (1-butanol), 25 (3-pentanol), 25.5 (2-pentanol), 26 (1-pentanol), 26.5 (2-methyl-1-butanol), 31 (1-hexanol), 35.5 (1-heptanol)
Limit of detection: 100 pmoles (primary alcohols), 250 pmoles (secondary alcohols)

KEY WORDS
tertiary alcohols are not derivatized; beer; spirits

REFERENCE
Motte, J.C.; Windey, R.; Delafortie, A. High-sensitivity fluorescence derivatization for the determination of hydroxy compounds in aqueous solution by high-performance liquid chromatography, *J.Chromatogr.A*, **1996**, *728*, 333–341.

SULFONATE

2-(N-Phthalimido)ethyl 2-(Dimethylamino)ethanesulfonate

SAMPLE
Matrix: solutions
Analyte: chlorophenols
Sample preparation: Mix a 200 μL aliquot of a 50 μM solution of chlorophenol in toluene with 50 mg potassium carbonate, add 100 μL 100 mM 18-crown-6 in toluene, add 500 μL 60 mM 2-(N-phthalimido)ethyl 2-(dimethylamino)ethanesulfonate in toluene, shake at 95° for 4 h, cool. Remove a 400 μL aliquot of the toluene layer and add it to 1 mL 1 M sulfuric acid, vortex for 30 s. Remove a 100 μL aliquot of the toluene layer and evaporate it to near dryness under a stream of nitrogen, reconstitute with 100 μL 40 μM diphenyl in MeCN, inject a 15 μL aliquot. (Preparation of 2-(N-phthalimido)ethyl 2-(dimethylamino)ethanesulfonate is as follows. Stir 2.4 g N-hydroxethylphthalimide in 60 mL chloroform at 60°, add 2.65 mL 2-chloroethanesulfonyl chloride, add 7.79 mL 45% trimethylamine in water (Fluka), stir at 0° for 1.5 h, wash 3 times with 60 mL portions of water, wash twice with 60 mL portions of 10% sodium carbonate solution, dry over 2.5 g anhydrous sodium sulfate, evaporate to dryness under reduced pressure. Take up the residue in 4

mL dichloromethane and chromatograph on a 400 × 30 column of 120 g silica gel 60 with dichloromethane to give 2-(N-phthalimido)ethyl ethanesulfonate as a white powder (mp 87-88°). Stir 1.42 g 2-(N-phthalimido)ethyl ethanesulfonate and 890 µL 40% dimethylamine in water 100 mL dichloromethane at 0° for 1.5 h, add 2.5 g anhydrous sodium sulfate, filter, evaporate to dryness under reduced pressure, recrystallize the residue from n-hexane:chloroform 60:40 to give 2-(N-phthalimido)ethyl 2-(dimethylamino) ethanesulfonate as colorless plates (mp 106-107°).)

HPLC VARIABLES
Column: 150 × 3.9 4 µm Nova-Pak C18
Mobile phase: MeCN:water 55:45
Flow rate: 0.9
Injection volume: 15
Detector: UV 225

CHROMATOGRAM
Retention time: 6 (4-chlorophenol), 8 (2,4-dichlorophenol), 13 (2,4,6-trichlorophenol)
Internal standard: diphenyl (10)
Limit of detection: 10-20 pmole

REFERENCE
Kou, H.-S.; Wu, H.-L.; Wang, Y.-F.; Chen, S.-H.; Wu, S.-S. Chemically removable derivatization reagent for liquid chromatography. I. 2-(N-Phthalimido)ethyl 2-(dimethylamino)ethanesulfonate, *J.Chromatogr.A*, **1995**, *710*, 267–272.

SULFONYL HALIDE

Dabsyl Chloride

SAMPLE
Matrix: bulk
Analyte: alcohols
Sample preparation: Mix 1 mL alcohol with 10 mL 100 mM sodium bicarbonate and 20 mL 10 mM dabsyl chloride, reflux for 10 min, cool, extract with diethyl ether, wash with pH 3 citrate buffer, evaporate the organic layer, reconstitute, inject an aliquot.

HPLC VARIABLES
Column: 250 × 4 10 µm LiChrosorb Si 60
Mobile phase: n-Heptane:ethyl acetate 85:15
Flow rate: 1.2
Detector: UV 254

CHROMATOGRAM
Retention time: 7 (heptanol), 7.5 (hexanol), 8 (pentanol), 9 (butanol), 10 (propanol), 12.5 (EtOH), 15 (MeOH)

KEY WORDS
normal phase

REFERENCE
Wolski, T.; Golkiewicz, W.; Bartuzi, G. Chromatographic analysis of 4-dimethylaminoazobenzene-4'-sulfonyl and 4-naphthalene-1-azo-(4'-dimethylaminobenzene)sulfonyl derivatives of aliphatic alcohols, *J.Chromatogr.*, **1986**, *362*, 217−226.

Dansyl Chloride

Primary amines, secondary amines, and phenolic hydroxyls are all dansylated at about the same rate. Under these conditions tertiary amines and alkyl hydroxyls are not dansylated.

RELATED REFERENCES
Chen, S.S-H.; Kou, A.Y.; Chen, H-H.Y. Measurement of ethanolamine- and serine-containing phospholipids by high-performance liquid chromatography with fluorescence detection of their Dns derivatives. *J.Chromatogr.* **1981**, *208*, 339-346.

de Ruiter, C.; Otten, R.R.; Brinkman, U.A.T.; Frei, R.W. Rapid and simple dansylation of phenolic steroids using a two-phase system and phase-transfer catalysis. *J.Chromatogr.* **1988**, *436*, 429−436.

Tagliaro, F.; Dorizzi, R.; Plescia, M.; Pradella, M.; Ferrari, S.; Lo Cascio, V. Determination of morphine in biological fluids by HPLC with pre-column dansyl derivatization and fluorescence detection. *Fresenius' Z.Anal.Chem.* **1984**, *317*, 678−679.

Takadate, A.; Hiraga, H.; Fujino, H.; Goya, S. A convenient derivatization with anion exchange resin catalysts for high-performance liquid chromatographic analysis. I. Derivatization of estrogens with dansyl chloride. *Chem.Pharm.Bull.* **1985**, *33*, 5092−5095.

SAMPLE
Matrix: blood

Analyte: estradiol

Sample preparation: Condition a 1 mL octadecylsilica (C18) SPE cartridge (Baker) with 3 mL MeOH and 10 mL water. Wash an amino-bonded Bond Elut (NH2) SPE cartridge with dichloromethane and dry it with nitrogen. 1 mL Serum + 100 μL 21% phosphoric acid, add to the C18 SPE cartridge, wash with 3 mL water, elute with 600 μL dichloromethane, add 100 μL 1 mg/mL dansyl chloride in chloroform, add 100 μL 30 mg/mL pH 12 tetrabutylammonium bromide in water, add 500 μL water (adjusted to pH 12 with NaOH), vortex vigorously for 2 min, slowly add a 500 μL aliquot of the organic phase to the NH2 SPE cartridge, allow to stand for 10 min, elute with 3 mL dichloromethane. Evaporate the eluate to dryness, reconstitute the residue in 500 μL MeOH:water 50:50, inject a 100 μL aliquot.

HPLC VARIABLES
Column: 200 × 3.1 3 μm LiChrosorb RP-18

Mobile phase: MeOH:100 mM pH 7.0 imidazole buffer:2.5 mM pH 7.0 imidazole buffer 83.25:2.125:14.625, containing 1 mM tetrapentylammonium bromide

Flow rate: 0.5

Injection volume: 100

Detector: Chemiluminescence (470 nm cutoff filter) following post-column reaction. The column effluent mixed with 50 mM hydrogen peroxide in MeCN containing 5 mM bis(2-nitrophenyl)oxalate pumped at 0.3 mL/min and the mixture flowed into the detector. (Prepare bis(2-nitrophenyl)oxalate by dissolving 13.9 g 2-nitrophenol in 250 mL benzene (Caution! Benzene is a carcinogen!), remove 50 mL benzene by azeotropic distillation, cool to 10°, add 10.1 g freshly distilled triethylamine, add 7 g oxalyl chloride dropwise, allow to warm to room temperature, let stand overnight, evaporate to dryness under reduced pressure, recrystallize to give bis(2-nitrophenyl)oxalate (J. Chem. Educ. 1974, 51, 529).)

CHROMATOGRAM
Retention time: 20
Limit of detection: 0.06 ng/mL

KEY WORDS
serum; SPE

REFERENCE
Kwakman, P.J.M.; Kamminga, D.A.; Brinkman, U.A.T.; de Jong, G.J. Liquid chromatographic determination of oestradiol in serum by pre-column derivatization with dansyl chloride or laryl chloride and peroxyoxalate chemiluminescence detection, *J.Pharm.Biomed.Anal.*, **1991**, *9*, 753–759.

SAMPLE
Matrix: blood
Analyte: morphine and acetylmorphine
Sample preparation: Condition a Bond Elut SPE cartridge with 1 mL MeOH, 1 mL water, and 1 mL 100 mM pH 9.0 sodium bicarbonate buffer, do not allow to go dry. 100 μL Plasma + 20 μL 1 μg/mL nalorphine in MeOH:water 30:70 + 750 μL 100 mM pH 9.0 sodium bicarbonate buffer, vortex for 10 s, add to the SPE cartridge, wash with 1 mL 100 mM pH 9.0 sodium bicarbonate buffer, wash with 1 mL water, wash with 100 μL MeOH:water 50:50, dry under vacuum for 10 min, elute with three 250 μL aliquots of MeOH. Evaporate the eluate to dryness at 45° in a vacuum centrifuge for 1 h, reconstitute with 25 μL 100 mM pH 11.4 sodium carbonate or 25 μL pH 9.5 sodium bicarbonate, add 25 μL 1 mg/mL dansyl chloride in acetone, vortex, let stand in the dark at 45° for 20 min, add 250 μL toluene, vortex for 2 min, centrifuge at 12500 g for 1 min, inject a 100-200 μL aliquot of the upper organic layer.

HPLC VARIABLES
Guard column: 10 × 4.6 3 μm silica
Column: 150 × 4.6 3 μm Spherisorb 3CN
Mobile phase: n-Hexane:isopropanol:ammonia 95:5:0.25 (Place a silica column between the pump and the injector.)
Flow rate: 1.5
Injection volume: 100-200
Detector: F ex 340 em 500

CHROMATOGRAM
Retention time: 3.1 (6-acetylmorphine), 5.7 (morphine)
Internal standard: nalorphine (2.8)
Limit of quantitation: 10 ng/mL

KEY WORDS
plasma; pharmacokinetics; SPE; normal phase

REFERENCE
Barrett, D.A.; Shaw, P.N.; Davis, S.S. Determination of morphine and 6-acetylmorphine in plasma by high-performance liquid chromatography with fluorescence detection, *J.Chromatogr.*, **1991**, *566*, 135–145.

SAMPLE
Matrix: blood, urine
Analyte: steroids
Sample preparation: Hydrolyse serum, plasma, or urine with β-glucuronidase and sulfatase, extract with diethyl ether. Remove the organic layer and evaporate it to dryness under a stream of nitrogen at room temperature, reconstitute the residue in 40 μL buffer, add 100 μL 1.5 mg/mL dansyl chloride in acetone, shake vigorously for 30 s, heat at 100° for 5 min, inject a 20 μL aliquot. (Prepare buffer by adjusting the pH of 4 g/L sodium bicarbonate in water to 10.5 with 5 M NaOH.)

HPLC VARIABLES
Column: 250 × 2.6 PAH-10 C18 (Perkin-Elmer)
Mobile phase: Gradient. MeCN:water from 60:40 to 95:5 over 15 min (Perkin-Elmer curve 1), maintain at 95:5 for 10 min. (Flush column with MeCN at 0.1 mL/min overnight.)
Flow rate: 1
Injection volume: 20
Detector: F ex 335 em 540

CHROMATOGRAM
Retention time: 4 (estriol), 10 (estrone), 11.5 (zeranol), 12.5 (zanone, zenone), 16.5 (hexestrol), 18 (diethylstilbestrol)
Limit of detection: 5-80 ng

KEY WORDS
cow; sheep; plasma; serum; Endogenous peaks in urine interfere with all compounds except hexestrol and diethylstilbestrol.; LOD is too high for practical detection of compounds in serum and plasma.

REFERENCE
Rhys Williams, A.T.; Winfield, S.A.; Belloli, R.C. Dns derivatization of anabolic agents with high-performance liquid chromatographic separation and fluorescence detection, *J.Chromatogr.*, **1982**, *240*, 224–229.

SAMPLE
Matrix: blood, urine
Analyte: morphine
Sample preparation: 10 mL Serum, plasma, whole blood, or urine + 10 mL nalorphine in water + 25 mL saturated ammonium sulfate solution + 500 µL concentrated HCl, heat at 120° for 30 min, filter (Whatman No. 1 paper), adjust the pH of the filtrate to 9.0 with 25% NaOH, extract with 125 mL chloroform:isopropanol 80:20, repeat extraction with 50 mL chloroform:isopropanol 80:20. Combine the organic phases and wash them twice with 15 mL portions of 50 mM sodium borate solution, extract the organic phase twice with 10 mL portions of 1 M sulfuric acid. Combine the aqueous extracts and add 4 mL saturated ammonium sulfate, adjust pH to 9.0 with 25% NaOH, extract twice with 5 mL portions of chloroform:isopropanol 80:20. Combine the organic layers and evaporate them to dryness under a stream of nitrogen, reconstitute the residue in 50 µL water, add 100 µL 0.1% dansyl chloride in acetone, add 50 µL 200 mM sodium carbonate, let stand at room temperature in the dark for 3 h, add 1 mL toluene, vortex for 2 min. Remove the organic layer and evaporate it to dryness under a stream of nitrogen, reconstitute the residue in mobile phase, inject a 20 µL aliquot.

HPLC VARIABLES
Column: 150 × 4.5 3 µm Spherisorb S3W silica
Mobile phase: n-Hexane:isopropanol:ammonia 97:2.7:0.3
Flow rate: 1.5
Injection volume: 20
Detector: F ex 330-380 (filters) em 410-500 (filters)

CHROMATOGRAM
Retention time: 9.5
Internal standard: nalorphine (5)
Limit of detection: 0.2 pmole

OTHER SUBSTANCES
Non-interfering: acetaminophen, amitriptyline, amphetamine, atropine, benzoylecgonine, caffeine, carbamazepine, carisoprodol, chlorpromazine, chlorprothixene, cimetidine, cocaine, codeine, cyclizine, dextromethorphan, diazepam, dihydrocodeine, diphenoxilate, disopyramide, doxepin, doxylamine, emetine, erythromycin, ethylmorphine, flurazepam, glutethimide, hydrocodone, hydrocortisone, hydromorphone, hydroxyzine, imipramine, lidocaine, loxapine, meperidine, metapyrilene, methadone, methamphetamine, methocarbamol, methylphenidate, naloxone, nicotine, nordiazepam, nortriptyline, orphenadrine, oxycodone, papaverine, pentazocine, phenacetin, phencyclidine, phenmetrazine, phenol-

phthalein, phentermine, phenytoin, prazepam, procainamide, propoxyphene, propranolol, protriptyline, pyrilamine, quinine, spironolactone, strychnine, terpin hydrate, thioridazine, thiothixene, triamterene, trifluoperazine, triflupromazine, trihexyphenidyl, trimethoprim, trimetobenzamide, tripelennamine

KEY WORDS
serum; plasma; whole blood; normal phase

REFERENCE
Tagliaro, F.; Frigerio, A.; Dorizzi, R.; Lubli, G.; Marigo, M. Liquid chromatography with pre-column dansyl derivatisation and fluorimetric detection applied to the assay of morphine in biological samples, *J.Chromatogr.*, **1985**, *330*, 323–331.

SAMPLE
Matrix: bulk
Analyte: fenthion
Sample preparation: Dissolve in 1 mL 0.5 M NaOH, heat at 75° for 45 min, cool, add 2 drops methyl isobutyl ketone, add 2 drops 0.1% dansyl chloride in acetone, shake well, heat at 65° for 30 min, cool, acidify with 10% HCl, add 300 µL benzene (Caution! Benzene is a carcinogen!), extract, inject a 1-10 µL aliquot. ("Dansylation of phenyl moiety" occurs.)

HPLC VARIABLES
Column: 1000 × 2.4 Zipax coated with 0.5% β,β'-oxydipropionitrile
Mobile phase: Hexane:MeOH 95:5
Flow rate: 0.78
Injection volume: 1-10
Detector: F ex Turner filter no. 811 em Turner filter no. 817

CHROMATOGRAM
Retention time: 3

KEY WORDS
normal phase

REFERENCE
Frei, R.W.; Lawrence, J.F. Fluorigenic labelling in high-speed liquid chromatography, *J.Chromatogr.*, **1973**, *83*, 321–330.

SAMPLE
Matrix: formulations
Analyte: estrogens, conjugated
Sample preparation: Powder tablets, weigh out amount corresponding to 6.9 mg conjugated estrogens, add 6 g Celite 545, add 4 mL water, mix, add to a mixture of 2 g Celite and 1 mL water in a 150 × 25 tube, dry rinse container with 1 g Celite and add this to the tube, elute with 100 mL water-saturated ether, collect this eluate (A), elute with 5 mL 20 mg/mL dicyclohexylamine acetate in chloroform, elute with 145 mL chloroform (B). Combine the chloroform eluates and evaporate them to dryness under a stream of air on a steam bath, reconstitute with 20 mL MeOH, add 6 mL 5% HCl, reflux for 12 min, cool in an ice bath, add 5 mL 400 µg/mL ethinyl estradiol in MeOH, add 70 mL water, add 50 mL benzene (Caution! Benzene is a carcinogen!), shake for 1 min. Remove the organic layer and wash it with 10 mL water, three 15 mL portions of 2% sodium carbonate solution, and two 10 mL portions of water. Pass the organic layer through 30 g anhydrous sodium sulfate in a column to give C. Evaporate a 2 mL aliquot of the solution (or an aliquot of eluate A) to dryness under a stream of air, reconstitute with 10 mL 200 µg/mL dansyl chloride in acetone, add 15 mL buffer, let stand in the dark for 30 min, add 50 mL water, add 50 mL ether, shake for several min, extract the aqueous layer with 25 mL ether. Combine the ether layers and wash them with two 25 mL portions of water, pass the organic layer through a 150 × 25 column containing 50 g anhydrous sodium sulfate, wash the column with 25 mL ether. Combine the eluates and evaporate them to dryness under a stream of air on a steam bath, reconstitute with 10 mL chloroform, inject an aliquot. (Under these conditions estrone, equilin, and equilenin co-elute under these conditions. They can be reduced to β-estradiol, β-dihydroequilin, and β-dihydroequilenin, respectively, as follows. Evaporate a 10 mL aliquot of eluate (A) or solution (C) to dryness

under a stream of air, reconstitute with 20 mL MeOH, add 150 mg sodium borohydride (Caution! Flammable hydrogen gas is evolved!), let stand for 45 min, add 70 mL water, add 50 mL benzene, shake for 1 min. Remove the organic layer and wash it with four 20 mL portions of water, pass through 30 g of anhydrous sodium sulfate in a 150 × 25 tube, evaporate a 10 mL aliquot to dryness and proceed with the derivatization as described above. (Prepare the buffer by dissolving 366.7 mg anhydrous sodium carbonate in 300 mL water and adding 150 mL acetone. Note that the initial elution with ether (A) gives free estrogens and the elution with chloroform (B) gives 3-sulfate derivatives which are then hydrolyzed.)

HPLC VARIABLES
Column: 250 × 4.6 5 μm Zorbax-Sil
Mobile phase: n-Heptane:chloroform:EtOH 50:49.5:0.5
Flow rate: 2
Injection volume: 10
Detector: F ex 240-420 (filter) em 440 (cutoff filter)

CHROMATOGRAM
Retention time: 4 (estrone), 4 (equilin), 4 (equilenin), 12.5 (α-estradiol), 15 (α-dihydro-equilin), 16 (α-dihydroequilenin), 18 (β-estradiol), 19.5 (β-dihydroequilin), 22 (β-dihydroequilenin)
Internal standard: ethinyl estradiol (9)

KEY WORDS
normal phase; tablets

REFERENCE
Roos, R.W.; Lau-Cam, C.A. Liquid chromatographic analysis of conjugated and esterified estrogens in tablets, *J.Pharm.Sci.*, **1985**, *74*, 201–204.

SAMPLE
Matrix: hair
Analyte: morphine
Sample preparation: Wash 100-200 mg hair with 10 mL ethyl ether and 12 mL 10 mM HCl, add 3 mL 100 mM HCl, heat at 45° for 12 h, neutralize with 100 μL 3 M NaOH, add to an extraction tube containing sodium carbonate, sodium bicarbonate, and organic solvents (Toxi-Tubes A, Analytical Systems), add 2 mL water, vortex for 2 min, centrifuge at 759 g for 10 min, remove the organic layer, add dichloromethane:dichloroethane:heptane 18:18:64 to the aqueous layer, extract. Combine the organic layers and evaporate them to dryness, reconstitute with 50 μL water, add 50 μL 1 mg/mL dansyl chloride in acetone, add 50 μL 100 mM sodium carbonate, let stand in the dark at room temperature for at least 1.5 h, add 1 mL toluene, vortex for 2 min. Remove the organic layer and evaporate it to dryness under a stream of nitrogen, reconstitute the residue in mobile phase, inject a 20 μL aliquot.

HPLC VARIABLES
Column: 250 × 4.6 5 μm Spherisorb silica
Mobile phase: Hexane:isopropanol:ammonia 95:4.5:0.5
Flow rate: 2
Injection volume: 20
Detector: F ex 330-380 (bandpass filter) em 410-500 (bandpass filter)

CHROMATOGRAM
Retention time: 13
Limit of detection: 60 pg

OTHER SUBSTANCES
Non-interfering: codeine

KEY WORDS
derivatization at C3 hydroxy group; normal phase

REFERENCE

Marigo, M.; Tagliaro, F.; Poiesi, C.; Lafisca, S.; Neri, C. Determination of morphine in the hair of heroin addicts by high performance liquid chromatography with fluorimetric detection, *J.Anal.Toxicol.*, **1986**, *10*, 158–161.

SAMPLE

Matrix: tablets
Analyte: estrogens
Sample preparation: Powder tablets (60 mesh), take powder equivalent to about 3.2 mg conjugated estrogens, add 50 mL MeOH, shake 30 min, dilute to 100 mL with MeOH, mix, filter, discard first 20 mL filtrate. Take a 25 mL aliquot, add 1 mL HCl, add boiling chips, heat on a steam bath for 5 min, cool, add 70 mL water, extract with 75 mL benzene (Caution! Benzene is a carcinogen!). Wash the benzene layer with 15 mL water, four times with 15 mL 2% sodium carbonate in water, and twice with 10 mL water. Pass the benzene through a tube containing 30 g anhydrous sodium sulfate, wash the tube with 25 mL benzene, evaporate to dryness. Add 10 mL 200 µg/mL dansyl chloride in acetone, swirl to dissolve, add 15 mL base solution, mix, stopper, allow to stand in the dark for 30 min. Extract twice with 50 mL ether, wash each extract twice with 25 mL water, pass the ether through a 150 × 25 mm tube containing 50 g anhydrous sodium sulfate, wash the column with 25 mL ether, evaporate the ether layers to dryness, dissolve residue in 5 mL chloroform, inject a 10 µL aliquot. (Prepare base solution by dissolving 366.7 mg anhydrous sodium carbonate in 300 mL water and adding 150 mL acetone.)

HPLC VARIABLES

Column: 250 × 3.2 5 µm LiChrosorb Si-60
Mobile phase: n-Heptane:chloroform 50:50
Flow rate: 0.6
Injection volume: 10
Detector: F ex 240-420 em 440 (cut-off)

CHROMATOGRAM

Retention time: 8 (estrone, equilin, equilenin), 22 (α-estradiol), 25 (α-dihydroequilin), 27 (α-dihydroequilenin), 30 (β-estradiol), 32 (β-dihydroequilin), 37 (β-dihydroequilenin)

KEY WORDS

normal phase; estrone, equilin, equilenin not resolved; ethinyl estradiol can be used as IS (J.Pharm.Sci. 1985, 74, 201)

REFERENCE

Roos, R.W.; Medwick, T. Application of dansyl derivatization to the high pressure liquid chromatographic identification of equine estrogens, *J.Chromatogr.Sci.*, **1980**, *18*, 626–630.

SAMPLE

Matrix: urine
Analyte: hydroxybiphenyls
Sample preparation: 5 mL Urine + 1 mL chloroform, shake for 5 min, centrifuge for 20 min. Remove 800 µL of the organic layer and evaporate it to dryness under a stream of nitrogen, reconstitute the residue in 400 µL 0.1% dansyl chloride in acetone, add 30 µL 100 mM sodium carbonate, mix briefly, heat at 42-45° for 15-20 min, cool to room temperature, add 2 drops of 1 M NaOH, extract with 500 µL n-hexane. Remove the organic layer and evaporate it to dryness under a stream of nitrogen, reconstitute the residue in 500 µL n-hexane, inject a 5 µL aliquot.

HPLC VARIABLES

Column: 400 × 2.4 7-18 µm silica (Brinkmann)
Mobile phase: Hexane:chloroform 90:10
Injection volume: 5
Detector: UV 254; F (wavelengths not given)

CHROMATOGRAM

Retention time: 5.2 (p-hydroxybiphenyl), 7.1 (o-hydroxybiphenyl)

KEY WORDS

LOD 0.1 ng (F); normal phase

REFERENCE
Cassidy, R.M.; LeGay, D.S.; Frei, R.W. Analysis of phenols by derivatization and high-speed liquid chromatography, *J.Chromatogr.Sci.*, **1974**, *12*, 85–89.

2-Fluorenesulfonyl Chloride

SAMPLE
Matrix: solutions
Analyte: phenols
Sample preparation: 1 mL Aqueous phenol solution + 2 mL MeCN + 5 mL buffer, stir at 40°, add 2 mL 3.78 mM 2-fluorenesulfonyl chloride in MeCN, stir at 40° for 1 h, add 50 μL 500 mM NaOH, add 2 mL 3.78 mM 2-fluorenesulfonyl chloride in MeCN, stir at 40° for 1 h, cool, inject an aliquot. (Buffer was 30 mM sodium bicarbonate adjusted to pH 6.9 with 1.8 M sulfuric acid. Preparation of 2-fluorenesulfonyl chloride is as follows. Dissolve 33.3 g fluorene in 400 mL dichloromethane, stir at 5°, add a solution of 13.2 g chlorosulfonic acid in 100 mL dichloromethane over 2 h, filter, wash the solid with hexane to obtain 2-fluorenesulfonic acid. Dissolve the 2-fluorenesulfonic acid in water, add 1 M KOH until basic to litmus. Filter off the potassium salt and slurry it in hot acetone, chill in an ice-bath, repeat this process 2 or 3 times until the product appears dry. Mix the salt with 110 mmoles phosphorus trichloride and enough chloroform to produce a slurry, stir vigorously, add 130 mmoles phosphorus oxychloride in chloroform over 30 min, add 100-200 mL chloroform, reflux for 2.5 h, isolate the product, recrystallize from acetic acid:acetic anhydride 75:25, recrystallize from dichloromethane/hexane to obtain 2-fluorenesulfonyl chloride (mp 162-164°).)

HPLC VARIABLES
Column: 150 × 4.6 5 μm Microsorb C18
Mobile phase: Gradient. MeCN:water from 70:30 to 100:0 over 15 min.
Flow rate: 1
Injection volume: 10-40
Detector: UV 254; F ex 280 em 325

CHROMATOGRAM
Retention time: 6 (guaicol), 6.5 (p-nitrophenol), 8 (m-cresol, p-cresol), 8.5 (o-cresol), 9 (p-chlorophenol), 10 (2,4-dichlorophenol), 11 (2,6-dichlorophenol), 15 (2,3,5,6-tetrachlorophenol), 17.5 (pentachlorophenol)
Limit of detection: 50 pg (F)

REFERENCE
Carlson, R.M.; Swanson, T.A.; Oyler, A.R.; Lukasewycz, M.T.; Liukkonen, R.J.; Voelkner, K.S. Phenol analysis using 2-fluorenesulfonyl chloride as a UV-fluorescent derivatizing agent, *J.Chromatogr.Sci.*, **1984**, *22*, 272–275.

Lissamine

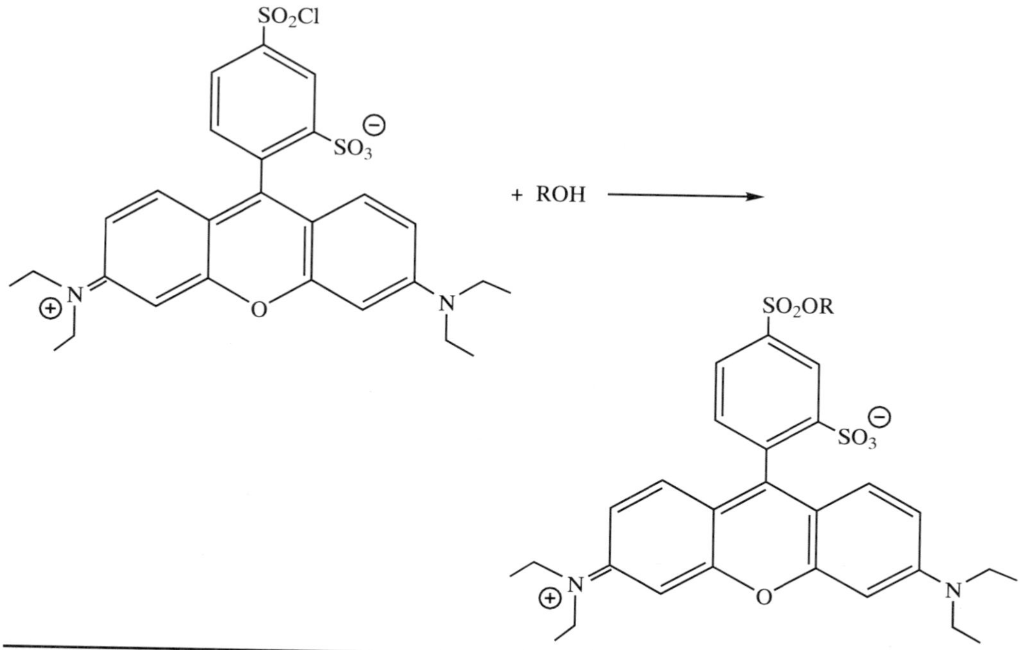

SO_2Cl

$SO_3^{\ominus}$

+ ROH ⟶

SO_2OR

$SO_3^{\ominus}$

SAMPLE
Matrix: solutions
Analyte: estradiol
Sample preparation: Wash an amino-bonded Bond Elut (NH2) SPE cartridge with di-chloromethane and dry it with nitrogen. Adjust an aqueous solution to pH 12 with 1 M NaOH. 500 μL Solution + 100 μL 30 mg/mL pH 12 tetrabutylammonium bromide in water + 600 μL 20 μg/mL laryl chloride (Lissamine; Rhodamine B sulfonyl chloride (Ko-dak)) in dichloromethane, vortex vigorously for 2 min, slowly add a 500 μL aliquot of the organic phase to the SPE cartridge, allow to stand for 10 min, elute with 3 mL dichloromethane:MeOH 96:4. Evaporate the eluate to dryness, reconstitute the residue in 500 μL MeOH:water 50:50, inject a 100 μL aliquot.

HPLC VARIABLES
Column: 200 × 3.1 3 μm LiChrosorb RP-18
Mobile phase: MeOH:100 mM pH 7.0 imidazole buffer:2.5 mM pH 7.0 imidazole buffer 78.5:2:19.5, containing 1 mM tetrapentylammonium bromide
Flow rate: 0.5
Injection volume: 100
Detector: Chemiluminescence (550 nm cutoff filter) following post-column reaction. The column effluent mixed with 50 mM hydrogen peroxide in MeCN containing 5 mM bis(2-nitrophenyl)oxalate pumped at 0.3 mL/min and the mixture flowed into the detector.

CHROMATOGRAM
Retention time: 16
Limit of detection: 0.5 nM

KEY WORDS
SPE

REFERENCE
Kwakman, P.J.M.; Kamminga, D.A.; Brinkman, U.A.T.; de Jong, G.J. Liquid chromatographic deter-mination of oestradiol in serum by pre-column derivatization with dansyl chloride or laryl chloride and peroxyoxalate chemiluminescence detection, *J.Pharm.Biomed.Anal.*, **1991**, 9, 753–759.

4-Naphthalene-1-azo-(4'-dimethylaminobenzene)sulfonyl Chloride

SAMPLE
Matrix: bulk
Analyte: alcohols
Sample preparation: Mix 1 mL alcohol with 10 mL 100 mM sodium bicarbonate and 20 mL 10 mM 4-naphthalene-1-azo-(4'-dimethylaminobenzene)sulfonyl chloride, reflux for 10 min, cool, extract with diethyl ether, wash with pH 3 citrate buffer, evaporate the organic layer, reconstitute, inject an aliquot. (Prepare 4-naphthalene-1-azo-(4'-dimethylamino-benzene)sulfonyl chloride by diazo coupling N,N-dimethylaniline with 1-naphthalenesulfonic acid to give sodium 4-naphthalene-1-azo-(4'-dimethylaminobenzene)sulfonate (Chem. Abs. 1986, 104, 168798f). Stir 20 g sodium 4-naphthalene-1-azo-(4'-dimethylami-nobenzene)sulfonate, 30 g phosphorus pentachloride, and 30 mL phosphorus oxychloride for 1 h, let stand overnight, remove excess phosphorus oxychloride by distillation under reduced pressure, rinse the residue with water, dry at 60°, extract with acetone to give 4-naphthalene-1-azo-(4'-dimethylaminobenzene)sulfonyl chloride (mp 172-174°) (Chem. Abs. 1990, 112, 55277 g).)

HPLC VARIABLES
Column: 250 × 4 10 μm LiChrosorb RP-18
Mobile phase: MeOH:water 10:90
Flow rate: 1.2
Detector: UV 254

CHROMATOGRAM
Retention time: 3.5 (MeOH), 4 (EtOH), 5 (propanol), 6 (butanol), 7 (pentanol), 8.5 (hexanol), 10 (heptanol)

REFERENCE
Wolski, T.; Golkiewicz, W.; Bartuzi, G. Chromatographic analysis of 4-dimethylaminoazobenzene-4'-sulfonyl and 4-naphthalene-1-azo-(4'-dimethylaminobenzene)sulfonyl derivatives of aliphatic alcohols, *J.Chromatogr.*, **1986**, *362*, 217–226.

1-Pyrenesulfonyl Chloride

SAMPLE
Matrix: blood
Analyte: estradiol
Sample preparation: 1 mL Serum + 1 mL 1 mM tetrapentylammonium bromide in 1 M NaOH, mix, add 5 mL 1 mM 1-pyrenesulfonyl chloride (Molecular Probes, Eugene OR) in dichloromethane, vortex for 10 min, centrifuge at 1800 rpm for 10 min. Remove the organic layer and evaporate it to dryness under reduced pressure, reconstitute the residue in mobile phase, inject a 20 μL aliquot.

HPLC VARIABLES
Column: 250 × 4.6 5 μm Ultramex C8
Mobile phase: MeCN:water 75:25
Flow rate: 1.5
Injection volume: 20
Detector: UV 348; F ex 350 em 385; F ex 325 (Ar laser)

CHROMATOGRAM
Retention time: 8

OTHER SUBSTANCES
Simultaneously analyzed: equilin, estrone

KEY WORDS
serum

REFERENCE
DeSilva, K.H.; Vest, F.B.; Karnes, H.T. Pyrene sulphonyl chloride as a reagent for quantitation of oestrogens in human serum using HPLC with conventional and laser-induced fluorescence detection, *Biomed.Chromatogr.*, **1996**, *10*, 318–324.

ALDEHYDE

ACTIVE METHYLENE

2-Cyanoacetamide

SAMPLE
Matrix: carbohydrates
Analyte: monosaccharides
Sample preparation: Mix 10 nmoles total monosaccharides with 200 μL 2 M trifluoroacetic acid, flush with nitrogen for a few min, seal, heat at 100° for 6 h, evaporate to dryness under reduced pressure in a desiccator over NaOH pellets, reconstitute with water, inject an aliquot.

HPLC VARIABLES
Column: 80 × 8 11 μm Hitachi No. 2633 resin (quaternary ammonium)
Mobile phase: Gradient. A was 250 mM pH 8.2 borate buffer. B was 400 mM pH 7.4 borate buffer. C was 600 mM pH 9.3 borate buffer. A:B:C from 100:0:0 to 0:100:0 over 15 min, maintain at 0:100:0 for 20 min, to 0:0:100 over 10 min, maintain at 0:0:100.
Column temperature: 65
Flow rate: 1
Injection volume: 20
Detector: F ex 331 em 383 following post-column reaction. The column effluent mixed with 10% 2-cyanoacetamide in water pumped at 0.25 mL/min and 600 mM pH 9.3 borate buffer pumped at 0.25 mL/min and the mixture flowed through a 10 m × 0.5 mm ID PTFE coil at 100 ± 0.5° to the detector.

CHROMATOGRAM
Retention time: 20 (rhamnose), 33 (lyxose), 36 (ribose), 40 (mannose), 45 (fucose), 48 (arabinose), 50 (galactose), 58 (xylose), 65 (dextrose)
Limit of detection: 0.1-1 nmole

KEY WORDS
post-column reaction

REFERENCE
Honda, S.; Takahashi, M.; Kakehi, K.; Ganno, S. Rapid, automated analysis of monosaccharides by high-performance anion-exchange chromatography of borate complexes with fluorimetric detection using 2-cyanoacetamide, *Anal.Biochem.*, **1981**, *113*, 130–138.

SAMPLE
Matrix: glycoconjugates
Analyte: aldoses
Sample preparation: Mix 0.1-1.5 mg glycoconjugate with 200 μL 2 M trifluoroacetic acid, flush with nitrogen for a few min, seal, heat at 100° for 6 h, evaporate to dryness under reduced pressure in a desiccator over NaOH pellets, reconstitute with 200 μL water, inject a 20 μL aliquot.

HPLC VARIABLES
Column: Hitachi No. 2633 resin

Mobile phase: Gradient. A was 250 mM pH 8.2 borate buffer. B was 400 mM pH 7.4 borate buffer. C was 600 mM pH 9.3 borate buffer. A:B:C from 100:0:0 to 0:100:0 over 15 min, maintain at 0:100:0 for 20 min, to 0:0:100 over 11 min, maintain at 0:0:100.
Column temperature: 65 ± 1
Flow rate: 1
Injection volume: 20
Detector: UV 276 following post-column reaction. The column effluent mixed with 1% 2-cyanoacetamide pumped at 0.5 mL/min and 600 mM pH 10.5 borate buffer pumped at 0.5 mL/min and the mixture flowed through a 10 m × 0.5 mm ID PTFE coil at 100 ± 0.2° and a 1 m × 0.5 mm ID PTFE cooling coil to the detector.

CHROMATOGRAM
Retention time: 20 (rhamnose), 34 (lyxose), 315 (ribose), 38 (mannose), 42 (fucose), 47 (arabinose), 50 (galactose), 57 (xylose), 64 (dextrose)
Limit of detection: 1 nmole

KEY WORDS
post-column reaction

REFERENCE
Honda, S.; Takahashi, M.; Nishimura, Y.; Kakehi, K.; Ganno, S. Sensitive ultraviolet monitoring of aldoses in automated borate complex anion-exchange chromatography with 2-cyanoacetamide, *Anal.Biochem.*, **1981**, *118*, 162−167.

SAMPLE
Matrix: solutions
Analyte: carbohydrates

HPLC VARIABLES
Column: 11 μm Hitachi No. 2633 pellicular quaternary ammonium anion-exchange resin
Mobile phase: 200 mM pH 7.2 Borate buffer for 22 min then 500 mM pH 9.6 borate buffer
Column temperature: 65
Flow rate: 0.35
Detector: F ex 331 em 385 following post-column reaction. The column effluent mixed with the reagent pumped at 0.5 mL/min and the mixture flowed through a 10 m × 0.5 mm ID PTFE coil at 100° to the detector. (Reagent was 2-cyanoacetamide in 300 mM pH 7.5 phosphate/borate buffer.)

CHROMATOGRAM
Retention time: 16 (cellobiose), 23 (maltose), 31 (lactose), 36 (rhamnose), 48 (ribose), 52 (mannose), 62 (fructose), 71 (galactose), 80 (xylose), 94 (dextrose)

KEY WORDS
post-column reaction

REFERENCE
Honda, S.; Matsuda, Y.; Takahashi, M.; Kakehi, K.; Ganno, S. Fluorimetric determination of reducing carbohydrates with 2-cyanoacetamide and application to automated analysis of carbohydrates as borate complexes, *Anal.Chem.*, **1980**, *52*, 1079−1082.

SAMPLE
Matrix: solutions
Analyte: reducing sugars
Sample preparation: Inject a 1 μL aliquot.

HPLC VARIABLES
Column: 150 × 4.6 MicroPak NH2 (Convert column to phosphate form by conditioning with 1% phosphoric acid.)
Mobile phase: MeCN:water 70:30
Flow rate: 0.8
Injection volume: 1
Detector: F ex 334 em (5-58 band filter) following post-column reaction. The column effluent mixed with 5% 2-cyanoacetamide in 100 mM pH 10.4 potassium borate buffer pumped

at 0.4 mL/min and the mixture flowed through a 10 m × 0.23 mm ID coil at 120° to the detector.

CHROMATOGRAM
Retention time: 5 (ribose), 6.5 (fructose), 8 (dextrose), 12.5 (maltose)

KEY WORDS
post-column reaction

REFERENCE
Schlabach, T.D.; Robinson, J. Improvements in sensitivity and resolution with the cyanoacetamide reaction for the detection of chromatographically separated reducing sugars, *J.Chromatogr.*, **1983**, *282*, 169–177.

SAMPLE
Matrix: solutions
Analyte: aldoses
Sample preparation: Inject a 20 µL aliquot of a solution in mobile phase.

HPLC VARIABLES
Column: 150 × 6 6 µm Shodex RSPak DC-613 sulfonated polystyrene 55% cross-linked with divinylbenzene sodium form (Showa Denko)
Mobile phase: MeCN:water 80:20
Column temperature: 4
Flow rate: 0.5
Injection volume: 20
Detector: UV 280 following post-column reaction. The column effluent mixed with 500 mM pH 8.5 borate buffer pumped at 0.5 mL/min and 1% 2-cyanoacetamide in water pumped at 0.5 mL/min and the mixture flowed through a 5 m × 0.5 mm ID PTFE coil at 100 ± 1° and a 1 m × 0.5 mm PTFE cooling coil to the detector.

CHROMATOGRAM
Retention time: k' 5.14 (α-D-allose), k' 4.84 (β-D-allose), k' 4.08 (α-D-altrose), k' 3.66 (β-D-altrose), k' 4.36 (α-D-arabinose), k' 3.48 (β-D-arabinose), k' 3.48 (α-L-arabinose), k' 4.36 (β-L-arabinose), k' 4.90 (α-D-dextrose), k' 5.00 (β-D-dextrose), k' 2.93 (α-L-fucose), k' 3.24 (β-L-fucose), k' 5.66 (α-D-galactose), k' 6.67 (β-D-galactose), k' 4.54 (α-D-gulose), k' 4.54 (β-D-gulose), k' 3.32 (α-D-idose), k' 3.44 (β-D-idose), k' 2.43 (α-D-lyxose), k' 3.26 (β-D-lyxose), k' 3.86 (α-D-mannose), k' 6.73 (β-D-mannose), k' 1.61 (α-L-rhamnose), k' 2.69 (β-L-rhamnose), k' 5.04 (α-D-ribose), k' 2.75 (β-D-ribose), k' 5.40 (α-D-talose), k' 5.40 (β-D-talose), k' 2.86 (α-D-xylose), k' 2.86 (β-D-xylose)

KEY WORDS
post-column reaction

REFERENCE
Honda, S.; Suzuki, S.; Kakehi, K. Improved analysis of aldose anomers by high-performance liquid chromatography on cation-exchange columns, *J.Chromatogr.*, **1984**, *291*, 317–325.

SAMPLE
Matrix: solutions
Analyte: reducing carbohydrates

HPLC VARIABLES
Column: 250 × 4 6 µm Shodex RSPak DC-613 sulfonated polystyrene 55% cross-linked with divinylbenzene (H$^+$) (Showa Denko)
Mobile phase: MeCN:water 90:10
Flow rate: 0.6
Injection volume: 20
Detector: E, Irika E-502, glassy carbon working electrode 0.30 V, Ag/AgCl reference electrode, following post-column reaction. The column effluent mixed with 200 mM pH 9.5 borate buffer pumped at 0.25 mL/min and 1.5% 2-cyanoacetamide in water pumped at 0.25 mL/min and the mixture flowed through a 10 m × 0.5 mm ID PTFE coil at 100° and a 1 m × 0.5 mm PTFE cooling coil to the detector.

CHROMATOGRAM
Retention time: 12 (rhamnose), 13 (xylose), 15 (fucose), 26 (galactose)
Limit of detection: 20 pmole

KEY WORDS
post-column reaction

REFERENCE
Honda, S.; Konishi, T.; Suzuki, S. Electrochemical detection of reducing carbohydrates in high-performance liquid chromatography after post-column derivatization with 2-cyanoacetamide, *J.Chromatogr.*, **1984**, *299*, 245–251.

SAMPLE
Matrix: solutions
Analyte: oligosaccharides

HPLC VARIABLES
Column: 150 × 4.1 5 μm PRP-1 polymeric C18 (Hamilton)
Mobile phase: MeCN:buffer 20:80 (Buffer was 40 mM tetrabutylammonium hydroxide adjusted to pH 9.0 with 85% phosphoric acid.)
Flow rate: 0.8
Injection volume: 20
Detector: UV 276 following post-column reaction. The column effluent mixed with the reagent pumped at 0.4 mL/min and the mixture flowed through a 10 m × 0.5 mm ID PEEK coil at 100° then a 2 m × 0.25 mm ID PEEK coil in a room temperature water bath to the detector. (Reagent was 1% 2-cyanoacetamide in 200 mM sodium borate buffer, adjusted to pH 9.0 with 85% phosphoric acid. Prepare fresh daily.)

CHROMATOGRAM
Retention time: 3-20 (depending on degree of oligomerization)
Limit of detection: 20 ng

KEY WORDS
post-column reaction

REFERENCE
Cramer, J.A.; Bailey, L.C. A reversed-phase ion-pair high-performance liquid chromatography method for bovine testicular hyaluronidase digests using postcolumn derivatization with 2-cyanoacetamide and ultraviolet detection, *Anal.Biochem.*, **1991**, *196*, 183–191.

SAMPLE
Matrix: solutions
Analyte: hyaluronic acid
Sample preparation: Inject a 10-20 μL aliquot.

HPLC VARIABLES
Column: 250 × 4.6 TSKgel NH2-60
Mobile phase: MeCN:buffer 54:46 (Buffer was 40 mM Tris-HCl borate buffer adjusted to pH 7.5 with HCl containing 5 mM sodium sulfate.)
Flow rate: 0.5
Injection volume: 10-20
Detector: F ex 346 em 410 following post-column reaction. The effluent from the column mixed with 300 mM NaOH pumped at 0.25 mL/min and with 1% 2-cyanoacetamide pumped at 0.25 mL/min. The mixture flowed through a 10 m × 0.5 mm ID PTFE coil at 105° and a 2 m × 0.25 mm ID PTFE coil at 25° to the detector.

CHROMATOGRAM
Limit of detection: 100 ng/mL

OTHER SUBSTANCES
Simultaneously analyzed: chondroitin sulfate, dermatan sulfate

KEY WORDS
post-column reaction

REFERENCE

Akiyama, H.; Saito, M.; Qiu, G.; Toida, T.; Imanari, T. Analytical studies on hyaluronic acid synthesis by normal human epidermal keratinocytes cultured in a serum-free medium, *Biol.Pharm.Bull.*, **1994**, *17*, 361–364.

SAMPLE

Matrix: urine

Analyte: disaccharides from hyaluronic acid

Sample preparation: Adjust 9 mL urine to pH 5 with 2 M HCl, add 600 μL 5% hexadecylpyridinium chloride, let stand at 0° for 4 h, centrifuge at 2300 g for 15 min. Remove the precipitate and wash it with 1.5 mL 0.1% hexadecylpyridinium chloride. Dissolve the precipitate in 1 mL 2.5 M NaCl and centrifuge at 2300 g for 15 min. Remove the supernatant and mix it with 11 mL EtOH:water 85:15, let stand at 0° overnight (Chem. Pharm. Bull. 1989, 37, 1627). Remove the precipitate and lyophilize it, dissolve in 20 μL water, add 0.01 unit hyaluronidase SD (from *Streptococcus dysgalactiae* (EC 4.2.2), Seikagaku Kogyo, Tokyo) in 20 μL pH 6.2 phosphate buffer, heat at 37° for 3 h, centrifuge at 3000 g for 5 min, inject a 20 μL aliquot.

HPLC VARIABLES

Column: 250 × 4.6 TSKgel NH2-60

Mobile phase: MeCN:buffer 64:36 (Buffer was 100 mM boric acid containing 10 mM sodium sulfate.)

Flow rate: 0.5

Injection volume: 20

Detector: F ex 346 em 410 following post-column reaction. The effluent from the column was mixed with 300 mM NaOH (pumped at 0.2 mL/min) and 1% 2-cyanoacetamide in 1 mM EDTA (pumped at 0.2 mL/min). The mixture passed through a 10 m × 0.5 mm i.d. PTFE coil at 100° and a 2 m × 0.25 mm i.d. PTFE coil at 25° to the detector.

CHROMATOGRAM

Retention time: 20 (2-acetamide-2-deoxy-3-O-(β-D-gluco-4-enepyranosyluronic acid)-D-glucose), 23 (2-acetamide-2-deoxy-3-O-(β-D-gluco-4-enepyranosyluronic acid)-D-galactose)

Limit of detection: 2 pmoles

KEY WORDS

post-column reaction

REFERENCE

Akiyama, H.; Toyoda, H.; Yamanashi, S.; Sagehashi, Y.; Toida, T.; Imanari, T. Microdetermination of hyaluronic acid in human urine by high performance liquid chromatography, *Biomed.Chromatogr.*, **1991**, *5*, 189–192.

Diphenylthiobarbituric Acid

SAMPLE
Matrix: blood
Analyte: malondialdehyde
Sample preparation: 10-100 μL Plasma or serum + 1 mL reagent, mix, make up to 1.1 mL with water (if necessary), mix well, heat at 95° for 40 min, cool in tap water for 5 min, add 500 μL MeCN:pyridine 80:20, vortex for 1 min, centrifuge at 3000 rpm for 10 min, inject a 50 μL aliquot of the supernatant. (Prepare reagent by dissolving 296 mg 1,3-diphenyl-2-thiobarbituric acid in 50 mL 200 mM Na_2HPO_4 adjust pH to 3 with 4% phosphoric acid, make up to 100 mL with water. 1,3-Diphenyl-2-thiobarbituric acid is available from Dojindo Molecular Technologies, Inc., 3 Bethesda Metro Center, Suite 700, Bethesda MD 20814; (301) 664-8448; www.dojindo.co.jp. Prepare 1,3-diphenyl-2-thiobarbituric acid (mp 245°) by refluxing 20 g thiocarbanilide(1,3-diphenyl-2-thiourea), 12 g malonic acid, and 18 mL acetyl chloride on a water bath for 30 min (Proc. Indian Acad. Sci. 1938, 8A, 145; Chem. Abstr. 1939, 33, 620[3]).)

HPLC VARIABLES
Column: 150 × 4 5 μm LiChrosorb RP-18
Mobile phase: MeCN:100 mM NaCl 50:50
Flow rate: 0.7
Injection volume: 50
Detector: UV 537

CHROMATOGRAM
Retention time: 7
Limit of detection: 100 nM

OTHER SUBSTANCES
Simultaneously analyzed: bilirubin

KEY WORDS
rat; plasma; human; serum

REFERENCE
Nakashima, K.; Ando, T.; Akiyama, S. High performance liquid chromatographic determination of lipoperoxides in rat plasma following derivatization to 1,3-diphenyl-2-thiobarbituric acid condensate, *Chem.Pharm.Bull.*, **1984**, *32*, 1654−1657.

Thiobarbituric Acid

RELATED REFERENCE
Seto, H. Determination of malonaldehyde and pyrimidopurinones by high performance liquid chromatography with 2-thiobarbituric acid-reaction detector. *Eisei Kagaku* **1987**, *33*, 436-441.

SAMPLE
Matrix: blood
Analyte: malondialdehyde
Sample preparation: 500 μL Plasma + 750 μL 440 mM phosphoric acid + 250 μL 42 mM thiobarbituric acid (4,6-dihydroxy-2-mercaptopyrimidine), heat at 100° for 1 h, cool to 0°. Remove a 500 μL aliquot and add it to 500 μL MeOH:1 M NaOH 91:9, vortex, centrifuge, inject a 30 μL aliquot of the supernatant.

HPLC VARIABLES
Column: 125 × 4.6 5 μm Nucleosil C18
Mobile phase: MeOH:50 mM pH 6.8 potassium phosphate buffer 40:60
Flow rate: 1.5
Injection volume: 30
Detector: F ex 532 em 553

CHROMATOGRAM
Retention time: 1.75
Limit of detection: 100 nM

KEY WORDS
plasma

REFERENCE
Londero, D.; Lo Greco, P. Automated high-performance liquid chromatographic separation with spectrofluorometric detection of a malondialdehyde-thiobarbituric acid adduct in plasma, *J.Chromatogr.A*, **1996**, *729*, 207–210.

SAMPLE
Matrix: blood
Analyte: malondialdehyde
Sample preparation: 100 μL Plasma + 700 μL 1% orthophosphoric acid, vortex for 10 s, add 200 μL reagent, vortex for 10 s, heat at 100° for 1 h, cool in ice until 10 min before analysis, vortex for 10 s. Remove a 200 μL aliquot and add it to 200 μL MeOH:2 M NaOH 12:1, vortex for 10 s, centrifuge at 13000 g for 3 min, inject a 50 μL aliquot of the supernatant. (Prepare reagent by warming 600 mg 2-thiobarbituric acid in 80 mL water to 35-40° with stirring, cool to room temperature, make up to 100 mL with water. Prepare glassware by machine washing with detergent, rinsing with water, soaking in 1% nitric acid, rinsing with water, flushing with 96% EtOH, and drying in an oven.)

HPLC VARIABLES
Guard column: 4 × 4 5 μm LiChrospher 100 RP-18
Column: 250 × 4 5 μm LiChrospher 100 RP-18
Mobile phase: MeOH:buffer 40:60 (Buffer was 10 mM KH_2PO_4 adjusted to pH 6.8 with 2 M KOH. At the end of each day flush column with water for 30 min and with MeOH for 30 min.)
Flow rate: 0.5

Injection volume: 50
Detector: UV 532

CHROMATOGRAM
Retention time: 9.67
Limit of detection: 20 nM

KEY WORDS
plasma

REFERENCE
Nielsen, F.; Mikkelsen, B.B.; Nielsen, J.B.; Andersen, H.R.; Grandjean, P. Plasma malondialdehyde as biomarker for oxidative stress: reference interval and effects of life-style factors, *Clin.Chem.*, **1997**, *43*, 1209–1214.

AMINE

2-Aminoacridone

RELATED REFERENCES
Guttman, A. Analysis of monosaccharide composition by capillary electrophoresis. *J.Chromatogr.A* **1997**, *763*, 271-277.

Okafo, G.; Burrow, L.; Carr, S.A.; Roberts, G.D.; Johnson, W.; Camilleri, P. A coordinated high-performance liquid chromatographic, capillary electrophoretic, and mass spectrometric approach for the analysis of oligosaccharide mixtures derivatized with 2-aminoacridone. *Anal.Chem.* **1996**, *68*, 4424-4430.

SAMPLE
Matrix: solutions
Analyte: oligosaccharides
Sample preparation: Lyophilize a solution containing 1 nmole oligosaccharide, add 5 μL 100 mM 2-aminoacridone in glacial acetic acid:DMSO 15:85, add 5 μL 1 M sodium cyanoborohydride in water, heat at 37° for 16 h or 45° for 2 h, evaporate to dryness, reconstitute in 10 μL DMSO:water 50:50, dilute with 2 mL 16 mM NaH$_2$PO$_4$, inject an aliquot. (2-Aminoacridone can be purchased from Molecular Probes, Eugene OR or Lambda Probes, Graz, Austria. Synthesis is as follows. Reflux 40 g 2-bromobenzoic acid, 30 g 4'-aminoacetanilide, 30 g potassium carbonate, and 200 mg copper powder in 200 mL amyl alcohol at 155° for 4 h, remove the amyl alcohol by steam distillation, cool, neutralize with dilute HCl, filter, wash the precipitate with water until the washings are acid free. Recrystallize from EtOH and three times from xylene to give 4'-acetamidodiphenylamine-

2-carboxylic acid as creamy-white needles (mp 240°). Heat 40.5 g 4'-acetamidodiphenyl-amine-2-carboxylic acid in 200 mL 96% sulfuric acid at 100° for 3 h, pour onto 400 g ice, heat at 100° for 2 h, cool. Partially neutralize to precipitate crystals of 2-aminoacridone sulfate. Recrystallize the free base from EtOH (mp 297°) (J. Chem. Soc. 1934, 433).)

HPLC VARIABLES
Column: 250 × 4.6 PA03 amine-bonded silica (YMC)
Mobile phase: Gradient. 16 mM NaH_2PO_4 : 798 mM NaH_2PO_4 100:0 for 10 min, to 16.75 : 83.25 over 50 min, to 0:100 over 20 min.
Flow rate: 1
Detector: F ex 405 em 525

CHROMATOGRAM
Retention time: 25-70 (depending on structure)
Limit of detection: 50 pmole

REFERENCE
Kitagawa, H.; Kinoshita, A.; Sugahara, K. Microanalysis of glycosaminoglycan-derived disaccharides labeled with the fluorophore 2-aminoacridone by capillary electrophoresis and high-performance liquid chromatography, *Anal.Biochem.*, **1995**, *232*, 114–121.

4-Aminobenzoic Acid

SAMPLE
Matrix: plants
Analyte: saccharides
Sample preparation: Stir 40 g dried plant material with 600 mL water at 500 rpm for 24 h, filter (grid mesh 0.3 mm), centrifuge the filtrate at 17700 g for 15 min. Remove an 80 mL aliquot of the supernatant to 400 mL EtOH, let stand at 4° for 48 h, centrifuge at 17700 g for 30 min, discard the supernatant, suspend the pellet in 100 mL ether, evaporate the ether to dryness at 40°. Add 2.5 mL 6 M trifluoroacetic acid to 20 mg of the precipitate, dissolve with gentle swirling, reflux at 110° for 4 h, lyophilize to dryness, add reagent so that the concentration of saccharide is 2 mg/mL, vortex gently, heat at 50° for 2 h, cool to room temperature, dilute 10-100-fold with MeOH, inject an aliquot. (Prepare reagent by dissolving 10 mg sodium cyanoborohydride in 1 mL MeOH containing 7% 4-aminobenzoic acid and 10% acetic acid.)

CAPILLARY ELECTROPHORESIS
Capillary: 72 cm × 50 μm fused-silica (50 cm to detector)
Capillary preparation: Between runs wash capillary with 1 M NaOH for 3-4 min and with 1 mM NaOH for 2 min, equilibrate with running buffer for 3-6 min. Store capillaries in 1 mM NaOH overnight. Flush new capillaries with 1 M NaOH for 1 h and with 1 mM NaOH for 5 min.
Running buffer: 150 mM pH 10.0 Borate buffer
Capillary temperature: 30
Voltage/Current: 28 kV
Injection: Vacuum injection at 16.9 kPa for 1 s
Detector: UV 285
Model: Applied Biosystems Model 270A
Migration time: 9.2 (cellobiose), 9.4 (2-deoxy-D-ribose), 9.6 (melibiose), 10 (lactose), 11.2 (sorbose), 11.5 (xylose), 11.7 (dextrose), 11.9 (fructose), 12.3 (arabinose), 13.1 (fucose), 14 (galactose), 20 (glucuronic acid), 21 (galacturonic acid)

REFERENCE

Oefner, P.J.; Vorndran, A.E.; Grill, E.; Huber, C.; Bonn, G.K. Capillary zone electrophoretic analysis of carbohydrates by direct and indirect UV detection, *Chromatographia*, **1992**, *34*, 308–316.

SAMPLE

Matrix: solutions

Analyte: carbohydrates

Sample preparation: For each 1 mg saccharides add 500 μL reagent, vortex gently, heat at 50° for 2 h, cool to room temperature, dilute 10-100 fold with MeOH, inject an aliquot. (Prepare reagent by dissolving 10 mg sodium cyanoborohydride in MeOH containing 7% 4-aminobenzoic acid and 10% acetic acid.)

CAPILLARY ELECTROPHORESIS

Capillary: 72 cm × 50 μm fused-silica (50 cm to detector)

Capillary preparation: Between runs wash capillary with 1 M NaOH for 3 min, wash with 1 mM NaOH for 2 min, equilibrate with running buffer for 3 min. Store capillary in 1 mM NaOH overnight. Flush a new capillary with 1 M NaOH for 1 h then with 1 mM NaOH for 5 min.

Running buffer: 150 mM Boric acid adjusted to pH 10.0 with 2 M NaOH

Capillary temperature: 30

Voltage/Current: 28 kV/ 79 μA

Injection: Vacuum injection at 16.9 kPa for 1 s

Detector: UV 285

Model: Applied Biosystems Model 270A

Migration time: 9 (cellobiose), 9.3 (2-deoxy-D-ribose), 9.4 (melibiose), 10 (lactose), 11.3 (sorbose), 11.5 (xylose), 11.8 (dextrose), 11.9 (fructose), 12.1 (arabinose), 12.9 (fucose), 14 (galactose), 20 (glucuronic acid), 21 (galacturonic acid)

Limit of detection: 4-80 μM

REFERENCE

Grill.E.; Huber, C.; Oefner, P.; Vorndran, A.; Bonn, G. Capillary zone electrophoresis of p-aminobenzoic acid derivatives of aldoses, ketoses and uronic acids, *Electrophoresis*, **1993**, *14*, 1004–1010.

4-Aminobenzonitrile

SAMPLE

Matrix: solutions

Analyte: carbohydrates

Sample preparation: Add reagent to the carbohydrate solution to make a total volume of 2 mL, vortex gently, heat at 90° for 15 min, cool to room temperature, dilute 20-1000-fold with running buffer, inject an aliquot. (Just prior to use prepare reagent by dissolving 10 mg sodium cyanoborohydride in 1 mL 5% acetic acid containing 6% 4-aminobenzonitrile.)

CAPILLARY ELECTROPHORESIS

Capillary: 55 cm × 50 μm fused-silica (35 cm to detector)

Running buffer: 25 mM Tris containing 100 mM sodium dodecyl sulfate, adjusted to pH 7.5 with phosphoric acid

Capillary temperature: 30

Voltage/Current: 30 kV/55 μA

Injection: Vacuum injection at 16.9 kPa for 1 s

Detector: UV 285

Model: Applied Biosystems Model 270A

Migration time: 2.7 (fructose), 2.8 (sorbose), 2.95 (lactose), 3.2 (melibiose), 3.25 (cellobiose), 3.3 (maltotriose), 3.45 (maltose), 3.65 (mannose), 3.7 (dextrose), 3.73 (galactose), 3.77 (ribose), 3.9 (lyxose), 4 (arabinose), 3.1 (xylose)

REFERENCE

Schwaiger, H.; Oefner, P.J.; Huber, C.; Grill.E.; Bonn, G.K. Capillary zone electrophoresis and micellar electrokinetic chromatography of 4-aminobenzonitrile carbohydrate derivatives, *Electrophoresis*, **1994**, *15*, 941–952.

N-(4-Aminobenzoyl)-L-glutamic Acid

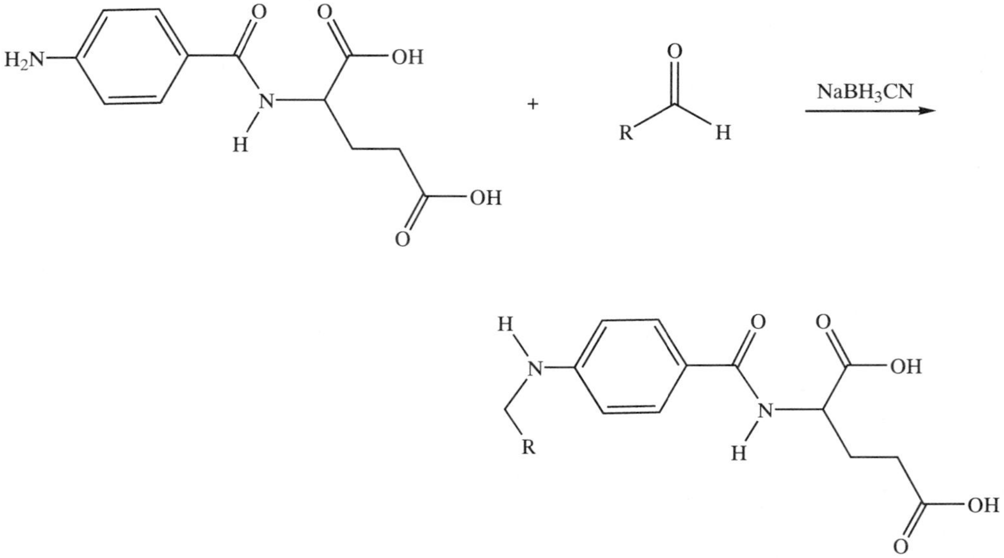

SAMPLE

Matrix: solutions

Analyte: saccharides

Sample preparation: Mix 3 parts of a sample solution (containing up to 2 mg/mL) with 1 part reagent, heat at 90° for 3 h, cool, dilute, inject an aliquot. (Prepare reagent by adding 20 mg sodium cyanoborohydride to 1 mL 4% N-(4-aminobenzoyl)-L-glutamic acid (Fluka) in 50% acetic acid.)

CAPILLARY ELECTROPHORESIS

Capillary: 70 (?) cm × 75 μm (50 cm to detector) (Polymicro Technologies)

Running buffer: 100 mM pH 10.0 Sodium borate buffer

Voltage/Current: 10 kV

Injection: Electrokinetic injection at 5 kV for 5 s

Detector: UV 291

Model: Isco Model 3850

Migration time: 19 (maltohexaose), 20 (maltopentaose)

REFERENCE

Plocek, J.; Novotny, M.V. Capillary electrophoresis of oligosaccharides derivatized with N-(4-aminobenzoyl)-L-glutamic acid for ultraviolet absorbance detection, *J.Chromatogr.A*, **1997**, *757*, 215–223.

8-Aminonaphthalene-1,3,6-trisulfonic acid

SAMPLE

Matrix: solutions

Analyte: malto-oligosaccharides

Sample preparation: 1 µmole Oligosaccharide + 200 µL 200 mM 8-aminonaphthalene-1,3,6-trisulfonic acid (Molecular Probes, Eugene OR) in water:acetic acid 85:15 + 200 µL 1 M sodium cyanoborohydride in DMSO + 20 µL 500 mM 1,3,5-benzenetrisulfonic acid in water, vortex, heat at 40° for 15 h, dilute 20-40-fold with water, filter (0.22 µm cellulose), inject an aliquot of the filtrate.

CAPILLARY ELECTROPHORESIS

Capillary: 47 cm × 20 µm (40 cm to detector) (Quadrex, New Haven CT)

Capillary preparation: Between runs flush capillary with 5 capillary volumes of running buffer.

Running buffer: 200 mM pH 2.0 Phosphate buffer

Capillary temperature: 25

Voltage/Current: 15 kV

Injection: Pressure injection at 0.5 psi for 3-7 s

Detector: UV 214; F ex 325 (2.8 mW He-Cd laser) em 520 (bandpass filter)

Model: Beckman P/ACE Model 2100

Migration time: 7-14 (depending on structure)

Internal standard: 1,3,5-benzenetrisulfonic acid

KEY WORDS

detector at anode

REFERENCE

Chiesa, C.; Horváth, C. Capillary zone electrophoresis of malto-oligosaccharides derivatized with 8-aminonapthalene-1,3,6-trisulfonic acid, *J.Chromatogr.*, **1993**, *645*, 337–352.

SAMPLE

Matrix: solutions

Analyte: oligosaccharides

Sample preparation: Freeze dry 5-10 µL aliquots of 1 mM solutions, reconstitute with 5 µL 20-50 mM 8-aminonaphthalene-1,3,6-trisulfonic acid (Molecular Probes, Eugene OR) in water:acetic acid 97:3, add 5 µL 0.2 M sodium cyanoborohydride in DMSO, vortex, centrifuge briefly at 10000 g, heat at 90° for 1 h, dry under vacuum at 45° for 4 h, reconstitute with running buffer, inject an aliquot.

CAPILLARY ELECTROPHORESIS

Capillary: 35 cm × 50 µm coated fused-silica (35 cm to detector) (Polymicro Technologies)

Capillary preparation: Coat the capillaries as follows. Treat capillary with 100 mM NaOH for 1 h, rinse with water, rinse with MeOH, fill capillary under nitrogen pressure with 10 µL (gamma-methacryloxypropyl)trimethoxysilane dissolved in dichloromethane containing 20 mM acetic acid, after 1 h rinse with MeOH, rinse with water, pass 4% acrylamide solution containing 1 µL/mL tetramethylethylenediamine and 1 mg/mL ammonium persulfate through the capillary under nitrogen pressure for 30 min (Caution! Acrylamide is a carcinogen!), rinse with water, dry under a stream of nitrogen.

Running buffer: 100 mM pH 8.65 Borate-tris buffer

Voltage/Current: -500 V/cm

Injection: Hydrodynamic

Detector: F ex 325 (He-Cd laser) em 514

Model: laboratory constructed
Migration time: 3-10 (depending on degree of polymerization)

REFERENCE

Stefansson, M.; Novotny, M. Separation of complex oligosaccharide mixtures by capillary electrophoresis in the open-tubular format, *Anal.Chem.*, **1994**, *66*, 1134–1140.

SAMPLE

Matrix: solutions
Analyte: oligosaccharides
Sample preparation: 5 µg Complex oligosaccharide + 1 µL 150 mM 8-aminonaphthalene-1,3,6-trisulfonic acid (Molecular Probes, Eugene OR) in water:acetic acid 85:15 + 1 µL 1 M sodium cyanoborohydride in DMSO, vortex, heat at 40° for 15 h, dilute 100-200-fold, inject an aliquot.

CAPILLARY ELECTROPHORESIS

Capillary: 27 cm × 50 µm (20 cm to detector)
Capillary preparation: Between runs flush capillary with 100 mM NaOH for 2 min and with running buffer for 2 min.
Running buffer: 50 mM Phosphate buffer adjusted to pH 2,5 with 2 M HCl
Capillary temperature: 25
Voltage/Current: 10 kV
Injection: Hydrodynamic injection at 35 mbar for 2-6 s (4-12 nL)
Detector: F ex 325 (2 mW He-Cd laser) em 520 (bandpass filter)
Model: Beckman P/ACE 2100
Migration time: 4.5-7.3 (depending on structure)
Limit of detection: 50 nM

KEY WORDS

detector at anode

REFERENCE

Klockow, A.; Amadò, R.; Widmer, H.M.; Paulus, A. Separation of 8-aminonaphthalene-1,3,6-trisulfonic acid-labelled neutral and sialylated N-linked complex oligosaccharides by capillary electrophoresis, *J.Chromatogr.A*, **1995**, *716*, 241–257.

SAMPLE

Matrix: solutions
Analyte: sugars
Sample preparation: 56 µmoles Dextrose + 100 µL 150 mM 8-aminonaphthalene-1,3,6-trisulfonic acid (Molecular Probes, Eugene OR) in acetic acid:water 15:85 + 100 µL 1 M sodium cyanoborohydride in DMSO, vortex, heat at 40° for 15 h, dilute 1000-fold with water, inject an aliquot.

CAPILLARY ELECTROPHORESIS

Capillary: 27 cm × 50 µm (20.5 cm to detector) (Polymicro Technologies)
Capillary preparation: Between runs flush with 100 mM NaOH for 2 min and with running buffer for 2 min.
Running buffer: 150 mM Boric acid adjusted to pH 9.5 with 2 M NaOH
Capillary temperature: 25
Voltage/Current: 10 kV
Injection: Hydrodynamic injection
Detector: F ex 325 (2 mW He-Cd laser) em 520 (bandpass filter)
Model: Beckman P/ACE 2100
Migration time: 6.46 (N-acetylgalactosamine), 7.98 (N-acetylglucosamine), 10.4 (dextrose), 11.3 (fucose), 10.8 (galactose), 9.91 (mannose), 12.7 (xylose)

REFERENCE

Klockow, A.; Amadò, R.; Widmer, H.M.; Paulus, A. The influence of buffer composition on separation efficiency and resolution on capillary electrophoresis of 8-aminonaphthalene-1,3,6-trisulfonic acid labeled monosaccharides and complex carbohydrates, *Electrophoresis*, **1996**, *17*, 110–119.

Aminopyrazine

SAMPLE
Matrix: solutions
Analyte: monosaccharides
Sample preparation: Freeze dry 10 µL of a 100 µM-10 mM solution in a glass tube, reconstitute with 50 µL 750 mM aminopyrazine in acetic acid, heat at 90° for 30 min, add 50 µL 700 mM borane-dimethylamine complex in acetic acid, heat at 90° for 5 min, cool to room temperature. Remove a 10 µL aliquot and dry under reduced pressure at 40° for 10 min to remove the acetic acid, reconstitute with 5 mL 700 mM pH 9.0 potassium borate buffer, inject a 5 µL aliquot.

HPLC VARIABLES
Column: 150 × 4.6 PALPAK Type A (Takara Shuzo, Kyoto)
Mobile phase: MeCN:700 mM pH 9.0 borate buffer 10:90
Column temperature: 65
Flow rate: 0.3
Injection volume: 5
Detector: F ex 245 em 410

CHROMATOGRAM
Retention time: 11 (N-acetylgalactosamine), 19 (rhamnose), 25 (xylose), 28 (N-acetylglucosamine), 40 (glucose), 44 (mannose), 50 (fucose), 65 (galactose)
Limit of detection: 25-1000 fmole

REFERENCE
Tachiki, K.; Yoshida, H.; Hamase, K.; Zaitsu, K. Aminopyrazine as a precolumn derivatizing reagent for the fluorescence detection of monosaccharides in high-performance liquid chromatography, *Anal.Sci.*, **1997**, *13*, 509–512.

9-Aminopyrene-1,4,6-trisulfonic acid

RELATED REFERENCES
Evangelista, R.A.; Chen, F.-T.A.; Guttman, A. Reductive amination of N-linked oligosaccharides using organic acid catalysts. *J.Chromatogr.A* **1996**, *745*, 273-280.
Guttman, A. Analysis of monosaccharide composition by capillary electrophoresis. *J.Chromatogr.A* **1997**, *763*, 271-277.
Guttman, A.; Chen, F.T.; Evangelista, R.A.; Cooke, N. High-resolution capillary gel electrophoresis of reducing oligosaccharides labeled with 1-aminopyrene-3,6,8-trisulfonate. *Anal.Biochem.* **1996**, *233*, 234-242.

SAMPLE
Matrix: bulk
Analyte: monosaccharides
Sample preparation: Add the material to 2 µL 200 mM 8-aminopyrene-1,3,6-trisulfonic acid in 15% acetic acid, add 2 µL 1 M sodium cyanoborohydride in THF, heat at 37° overnight, dilute 2500-fold with water, inject an aliquot. (9-Aminopyrene-1,4,6-trisulfonic acid is the same as 8-aminopyrene-1,3,6-trisulfonic acid. It is available from Molecular Probes, Eugene OR. Synthesis is as follows. Rapidly add 300 g anhydrous sodium sulfate to 1300 g concentrated sulfuric acid, cool to 58°, add 202 g finely powdered pyrene over 5 min without cooling, stir for 15 min, cool to 50-55°, add 800 g 65% oleum (with cooling with 12° water) over 20 min, stir for 5 h without cooling, dilute with ice-water, neutralize with calcium carbonate, filter, reduce the filtrate to 10 L, exchange the cation with soda, salt out with 20% sodium chloride to obtain sodium pyrenetetrasulfonate. Heat 61 g sodium pyrenetetrasulfonate with 610 mL 22% aqueous ammonia in a rotating autoclave at 200-210° for 18 h (the pressure rises to 45 atmospheres), remove the excess ammonia by distillation under vacuum, precipitate oxypyrenetrisulfonic acid with a little NaCl, precipitate 9-aminopyrene-1,4,6-trisulfonic acid by saturating with NaCl, recrystallize from dilute NaCl (Liebig's Annalen der Chemie 1939, 540, 189).)

CAPILLARY ELECTROPHORESIS
Capillary: 30 cm × 25 µm fused-silica (30 cm to detector)
Running buffer: 25 mM pH 10 Tetraborate buffer
Capillary temperature: 20
Voltage/Current: 750 V/cm/17 µA
Injection: Pressure injection at 3.45 kPa for 5-10 s.
Detector: F ex 488 (4 mW argon-ion laser) em 520
Model: Beckman P/ACE 5500
Migration time: 5.7 (N-acetylgalactosamine), 6.3 (N-acetylglucosamine), 7.1 (mannose), 7.4 (glucose), 8 (fucose), 8.3 (galactose)

KEY WORDS
detector at anode

REFERENCE
Guttman, A.; Brunet, S.; Cooke, N. Capillary electrophoresis fingerprinting of carbohydrates in the biopharmaceutical and food and beverage industries, *LC.GC*, **1996**, *14*, 788–792.

SAMPLE
Matrix: solutions
Analyte: monosaccharides
Sample preparation: Mix 2 µL of a 5 mM solution with 2 µL 100 mM 9-aminopyrene-1,4,6-trisulfonic acid in 4.2 M acetic acid, add 4 µL 1 M sodium cyanoborohydride in THF, centrifuge, heat at 75° for 1 h, dilute 100- to 1000-fold, inject an aliquot. (Purify reaction mixtures containing oligosaccharides on an 80 × 5 Sephadex G-10 column with 50 mM pH 8.5 bicarbonate buffer. 9-Aminopyrene-1,4,6-trisulfonic acid is the same as 8-aminopyrene-1,3,6-trisulfonic acid and is available from Molecular Probes, Eugene OR. Synthesis is described above.)

CAPILLARY ELECTROPHORESIS
Capillary: 27 cm × 20 µm fused-silica (20 cm to detector) (Polymicro Technologies)
Capillary preparation: Between runs wash capillary with 1 M NaOH at 15 psi for 12 s, wash with water at 15 psi for 12 s, condition with running buffer for 4 min.
Running buffer: 120 mM pH 7.0 MOPS
Voltage/Current: 25 kV/19 µA
Detector: F ex 488 (2.5 mW argon ion laser) em 520 ± 9 (narrow-bandfilter; notch filter at 488 nm)
Model: Beckman P/ACE 2100
Migration time: 4.6 (N-acetylgalactosamine), 4.8 (N-acetylglucosamine), 5.9 (mannose), 6 (galactose), 6.3 (dextrose), 6.5 (rhamnose), 6.6 (fucose), 8.4 (ribose), 8.6 (arabinose), 9.2 (xylose)

OTHER SUBSTANCES
Also analyzed: oligosaccharides (different CE conditions)

REFERENCE
Chen, F.-T.A.; Evangelista, R.A. Analysis of mono- and oligosaccharide isomers derivatized with 9-aminopyrene-1,4,6-trisulfonate by capillary electrophoresis with laser-induced fluorescence, *Anal.Biochem.*, **1995**, *230*, 273–280.

SAMPLE
Matrix: solutions
Analyte: sugars
Sample preparation: Evaporate 5 μL of a 1 mM solution in water to dryness under reduced pressure, add 2 μL 100 mM 9-aminopyrene-1,4,6-trisulfonic acid, add 2 μL 1.8 M citric acid in water, add 2 μL 1 M sodium cyanoborohydride in THF, heat at 75° for 1 h, dilute to 200 μL with water, dilute 25-fold, inject an aliquot. (9-Aminopyrene-1,4,6-trisulfonic acid is the same as 8-aminopyrene-1,3,6-trisulfonic acid and is available from Molecular Probes, Eugene OR or Lambda Fluoreszenztechnologie, Graz, Austria. Synthesis is as described above.)

CAPILLARY ELECTROPHORESIS
Capillary: 25 cm × 19 μm fused-silica
Capillary preparation: Between runs rinse capillary with 1 M NaOH at 15 psi for 12 s and with running buffer at 15 psi for 1.2 min.
Running buffer: 120 mM pH 10.2 Borate buffer
Voltage/Current: 30 kV/26 μA
Injection: Pressure injection at 0.5 psi for 20 s
Detector: F ex 488 (laser) em 520
Model: Beckman P/ACE 2100
Migration time: 3.1 (N-acetylgalactosamine),3.6 (N-acetylglucosamine), 4.4 (dextrose), 4.9 (fucose), 5.3 (galactose), 4.2 (mannose), 4.8 (xylose)

REFERENCE
Evangelista, R.A.; Guttman, A.; Chen, F.-T.A. Acid-catalyzed reductive amination of aldoses with 8-aminopyrene-1,3,6-trisulfonate, *Electrophoresis*, **1996**, *17*, 347–351.

2-Aminopyridine

RELATED REFERENCES
Kodama, C.; Ototani, N.; Isemura, M.; Yosizawa, Z. High-performance liquid chromatography of pyridylamino derivatives of unsaturated disaccharides produced from chondroitin sulfate isomers by chondroitinases. *J.Biochem.(Tokyo)* 1984, *96*, 1283-1287.

Maness, N.O.; Miranda, E.T.; Mort, A.J. Recovery of sugar derivatives from 2-aminopyridine labeling mixtures for high-performance liquid chromatography using UV or fluorescence detection. *J.Chromatogr.* 1991, *587*, 177-183.

SAMPLE
Matrix: bulk
Analyte: monosaccharides
Sample preparation: Heat 200 nmoles compound with 10 μL 2.76 g/mL 2-aminopyridine at 90° for 1 h, add 10 μL 20% dimethylamine-borane complex in acetic acid, heat at 80° for 50 min, add 400 μL water, add the mixture to a 30 × 9 Dowex 50X8 (NH_4^+) column equilibrated with 20 mM pH 6.8 ammonium acetate buffer, elute with 20 mM pH 8.5 buffer, inject an aliquot of the eluate. (Recrystallize 2-aminopyridine from n-hexane.)

HPLC VARIABLES
Column: 250 × 4.6 TSK-GEL Amide-80 (Tosoh)
Mobile phase: Gradient. MeCN:buffer from 65:35 to 90:10 over 30 min. (Buffer was 3% pH 7.3 triethylamine/acetic acid.)
Column temperature: 40
Flow rate: 1
Detector: F ex 320 em 390

CHROMATOGRAM
Retention time: 7 (L-fucose), 12 (N-acetyl-D-glucosamine)

KEY WORDS
SPE; Dowex column gives much faster results.

REFERENCE
Fan, J.-Q.; Huynh, L.H.; Lee, Y.C. Purification of 2-aminopyridine derivatives of oligosaccharides and related compounds by cation-exchange chromatography, *Anal.Biochem.*, **1995**, *232*, 65–68.

SAMPLE
Matrix: carbohydrates
Analyte: reducing monosaccharides
Sample preparation: Dissolve 1 mg carbohydrate in 100 μL 2 M trifluoroacetic acid, flush with nitrogen, seal tube, heat at 100° for 6 h, evaporate to dryness under reduced pressure in a desiccator containing solid NaOH. Add reagent solution so that the concentration of saccharide is 10-100 mM, vortex gently, heat at 50° for 2 h, inject an aliquot. (Reagent solution was 10 mg/mL sodium cyanoborohydride in MeOH containing 10% 2-aminopyridine and 10% acetic acid.)

CAPILLARY ELECTROPHORESIS
Capillary: 65 cm × 50 μm fused-silica (50 cm to detector) (Scientific Glass Engineĉcp
Capillary preparation: Rinse capillary with running buffer for 5 min before each run. After 20 runs wash capillary with MeOH for 30 s.
Running buffer: 200 mM pH 10.5 Borate buffer
Voltage/Current: 15 kV
Injection: Siphon injection at 5 cm for 5 s
Detector: UV 240
Migration time: 11.2 (N-acetylgalactosamine), 11.8 (lyxose), 12 (rhamnose), 12.8 (xylose), 13 (ribose), 13.4 (N-acetylglucosamine), 13.8 (dextrose), 14.4 (arabinose), 15.8 (fucose), 16.8 (cinnamic acid), 22.6 (glucuronic acid), 26.6 (galacturonic acid)
Internal standard: galactose (9)

REFERENCE
Honda, S.; Iwase, S.; Makino, A.; Fujiwara, S. Simultaneous determination of reducing monosaccharides by capillary zone electrophoresis as the borate complexes of N-2-pyridylglycamines, *Anal.Biochem.*, **1989**, *176*, 72–77.

SAMPLE
Matrix: glycoproteins
Analyte: sugars
Sample preparation: Heat 10 nmole with 200 μL anhydrous hydrazine (Caution! Hydrazine is a carcinogen!) at 100° for 10 h in an evacuated sealed tube, evaporate to dryness under reduced pressure, reconstitute with toluene, evaporate to dryness under reduced pressure, repeat toluene evaporation several times. Add 200 μL freshly-prepared 9.8% sodium bicarbonate solution and 8 μL acetic anhydride to the residue, let stand for 5 min, add 200 μL freshly-prepared 9.8% sodium bicarbonate solution, add 8 μL acetic anhydride, let stand at room temperature with occasional stirring for 30 min, adjust pH to 3 with 100-200 mesh Dowex 50WX2 (H⁺ form), add the mixture to a 20 × 5 column of Dowex 50WX2 (H⁺ form), wash with 5 bed volumes of water. Evaporate the eluate to dryness, reconstitute the residue in 40 μL reagent 1, heat in a sealed tube at 100° for 13 min, cool, open, add 2 μL freshly prepared reagent 2, seal tube, heat at 90° for 15 h. Dilute the reaction mixture with 300 μL water and add it to a 500 × 9 column of Sephadex G-15 made up in 10 mM pH 6.0 ammonium acetate buffer, elute with 10 mM pH 6.0

ammonium acetate buffer at 6 mL/min, collect fractions, inject an aliquot. (Prepare reagent 1 by dissolving 1 g 2-aminopyridine in 760 μL concentrated HCl, store at -20° before use. The pH of the reagent should be 6.2 after dilution with 2 parts water. If sialic acid is present use a reagent prepared by mixing 1 g 2-aminopyridine, 300 μL water, and 100 μL concentrated HCl. Prepare reagent 2 by mixing 10 mg sodium cyanoborohydride, 20 μL reagent 1, and 30 μL water.)

HPLC VARIABLES
Column: 300 × 4 5 μm TSK-Gel LS410 (Toyo Soda)
Mobile phase: 1-Butanol:100 mM pH 4.0 ammonium acetate 0.25:99.75
Flow rate: 1.6
Detector: F ex 320 em 400

CHROMATOGRAM
Retention time: 3-10 (depending on structure)

KEY WORDS
SPE

REFERENCE
Hase, S.; Ibuki, T.; Ikenaka, T. Reexamination of the pyridylamination used for fluorescence labeling of oligosaccharides and its application to glycoproteins, *J.Biochem.*, **1984**, *95*, 197–203.

SAMPLE
Matrix: plants
Analyte: saccharides
Sample preparation: Stir 40 g dried plant material with 600 mL water at 500 rpm for 24 h, filter (grid mesh 0.3 mm), centrifuge the filtrate at 17700 g for 15 min. Remove an 80 mL aliquot of the supernatant and add it to 400 mL EtOH, let stand at 4° for 48 h, centrifuge at 17700 g for 30 min, discard the supernatant, suspend the pellet in 100 mL ether, evaporate the ether to dryness at 40°. Add 2.5 mL 6 M trifluoroacetic acid to 20 mg of the precipitate, dissolve with gentle swirling, reflux at 110° for 4 h, lyophilize to dryness, add reagent so that the concentration of saccharide is 2 mg/mL, vortex gently, heat at 50° for 2 h, cool to room temperature, dilute 10-100-fold with MeOH, inject an aliquot. (Prepare reagent by dissolving 10 mg sodium cyanoborohydride in 1 mL MeOH containing 10% 2-aminopyridine and 10% acetic acid.)

CAPILLARY ELECTROPHORESIS
Capillary: 72 cm × 50 μm fused-silica (50 cm to detector)
Capillary preparation: Between runs wash capillary with 1 M NaOH for 3-4 min and with 1 mM NaOH for 2 min, equilibrate with running buffer for 3-6 min. Store capillaries in 1 mM NaOH overnight. Flush new capillaries with 1 M NaOH for 1 h and with 1 mM NaOH for 5 min.
Running buffer: 150 mM pH 10.5 Borate buffer
Capillary temperature: 30
Voltage/Current: 20 kV
Injection: Vacuum injection at 16.9 kPa for 1 s
Detector: UV 237
Model: Applied Biosystems Model 270A
Migration time: 10.8 (2-deoxy-D-ribose), 11.2 (maltotriose), 12.1 (maltose), 12.3 (rhamnose), 12.5 (lyxose), 13.1 (xylose), 13.3 (ribose), 14.2 (dextrose), 14.8 (arabinose), 16.5 (fucose), 17.5 (galactose), 22.9 (glucuronic acid), 27.3 (galacturonic acid)

REFERENCE
Oefner, P.J.; Vorndran, A.E.; Grill, E.; Huber, C.; Bonn, G.K. Capillary zone electrophoretic analysis of carbohydrates by direct and indirect UV detection, *Chromatographia*, **1992**, *34*, 308–316.

SAMPLE
Matrix: solutions
Analyte: neutral and amino sugars
Sample preparation: Heat 100-200 pmole sample with 20 μL 4 M trifluoroacetic acid and 20 μL 4 M HCl in a tube sealed under vacuum at 100° for 6 h, add 500 pmole L-rhamnose, evaporate to dryness under reduced pressure at 50°, add 50 μL 9.8% sodium bicarbonate

solution (freshly prepared), add 2 μL acetic anhydride, let stand at room temperature with occasional stirring for 30 min, add 200 μL 100-200 mesh Dowex 50W-X2 (H$^+$), check that pH is about 3. Add the mixture to a 100 × 5 column and wash it with 5 bed volumes of water, evaporate to dryness under reduced pressure, add 5 μL reagent, seal tube, heat at 100° for 13-15 min, add 2 μL 20 mg/mL sodium cyanoborohydride in water (freshly prepared), reseal the tube, heat at 90° for 8 h, dilute with 20 μL water, inject the whole amount on to a 600 × 7.5 10 μm TSK-GEL G2000PW column (Toyo Soda) and elute with 20 mM pH 7.5 ammonium acetate buffer at 0.5 mL/min, collect the sugar fraction at 40-55 min. Evaporate the eluate to dryness and reconstitute it with 250 μL water, inject a 5 μL aliquot. (Prepare reagent by mixing 500 mg 2-aminopyridine, 400 μL concentrated HCl, and 11 mL water.)

HPLC VARIABLES
Column: two 250 × 4.6 5 μm Ultrasphere-ODS column in series
Mobile phase: MeCN:250 mM pH 4.0 sodium citrate buffer 1:99
Flow rate: 0.5
Injection volume: 5
Detector: F ex 320 em 400

CHROMATOGRAM
Retention time: 34.5 (galactose), 36 (dextrose), 37.5 (mannose), 39 (xylose), 45 (ribose), 57 (fucose), 72 (N-acetyl-D-mannosamine), 82 (2-deoxy-D-ribose), 98 (N-acetylglucosamine), 107 (N-acetylgalactosamine)
Internal standard: L-rhamnose (65)
Limit of quantitation: 10 pmoles

KEY WORDS
SPE

REFERENCE
Takemoto, H.; Hase, S.; Ikenaka, T. Microquantitative analysis of neutral and amino sugars as fluorescent pyridylamino derivatives by high-performance liquid chromatography, *Anal.Biochem.*, **1985**, *145*, 245–250.

SAMPLE
Matrix: solutions
Analyte: maltooligosaccharides
Sample preparation: Mix 1 g saccharide, 3 g 2-aminopyridine, 500 mg sodium cyanoborohydride, 100 μL acetic acid, and 16 mL MeOH, heat at 37° overnight, inject an aliquot (J. Biochem. 1979, 85, 217).

CAPILLARY ELECTROPHORESIS
Capillary: 80 cm × 50 μm fused-silica (50 cm to detector) (Polymicro Technologies)
Capillary preparation: Flush capillary with running buffer before each separation. Flush new capillaries with 1 M NaOH, water, and running buffer.
Running buffer: 100 mM pH 4.0 Phosphate buffer
Voltage/Current: 20 kV/ca. 60 μA
Injection: Electromigration at 18 kV for 15 s
Detector: UV 240
Model: laboratory constructed
Migration time: 18-22 (depending on degree of polymerization)

REFERENCE
Nashabeh, W.; El Rassi, Z. Capillary zone electrophoresis of pyridylamino derivatives of maltooligosaccharides, *J.Chromatogr.*, **1990**, *514*, 57–64.

SAMPLE
Matrix: solutions
Analyte: oligosaccharides
Sample preparation: 1 mg Oligosaccharides + 10 μL water + 40 μL reagent, heat at 80° for 15 h, add to a 500 × 10 column of Sephadex G-15, elute with MeOH:20 mM ammonium acetate 25:75 at 0.5 mL/min, monitor by fluorescence (ex 316 em 395). Collect the 20 to 36 mL fraction and evaporate it to dryness under reduced pressure, reconstitute

with 2 mL water, add this solution to a 2 mL column of CM-52 (proton form), wash with 20 mL water, elute with 20 mL 50 mM ammonium acetate. Evaporate the eluate to dryness, reconstitute with 200 μL water, inject an aliquot. (Prepare reagent by dissolving 184 mg 2-aminopyridine, 35 mg sodium cyanoborohydride, and 80 μL acetic acid in 350 μL MeOH.)

CAPILLARY ELECTROPHORESIS
Capillary: 20 cm × 25 μm polyacrylamide coated (Bio-Rad)
Running buffer: 100 mM pH 2.5 Phosphate buffer
Voltage/Current: 8 kV
Injection: Electromigration at 8 kV for 30 s
Detector: UV 240
Model: Bio-Rad HPE 100
Migration time: 5-30 (depending on degree of polymerization)

REFERENCE
Honda, S.; Makino, A.; Suzuki, S.; Kakehi, K. Analysis of the oligosaccharides in ovalbumin by high-performance capillary electrophoresis, *Anal.Biochem.*, **1990**, *191*, 228–234.

6-Aminoquinoline

SAMPLE
Matrix: solutions
Analyte: monosaccharides
Sample preparation: Prepare a solution containing 100 μM monosaccharide, 30 mM 6-aminoquinoline, 5 mM sodium cyanoborohydride, and 150 mM acetic acid, heat at 80° for 45 min (40° for 2 h for oligosaccharides), cool to room temperature, filter, inject an aliquot.

CAPILLARY ELECTROPHORESIS
Capillary: 43 cm × 30 μm fused silica (38 cm to detector) (Skandinaviska GeneTech, Kungsbacka, Sweden)
Running buffer: 420 mM Boric acid adjusted to pH 9 with 1 M NaOH
Voltage/Current: 55 μA (const. pwr. 1200 mW)
Injection: Hydrodynamic injection at 75 m for 10 s
Detector: UV 245
Model: Dionex
Migration time: 6.5 (rhamnose), 7 (xylose), 8 (dextrose), 8.1 (mannose), 8.2 (arabinose), 9.9 (galactose), 11.5 (4-O-methylglucuronic acid), 12.6 (glucuronic acid), 15.3 (galacturonic acid)
Limit of detection: 1 μM

REFERENCE
Rydlund, A.; Dahlman, O. Efficient capillary zone electrophoretic separation of wood-derived neutral and acidic mono- and oligosaccharides, *J.Chromatogr.A*, **1996**, *738*, 129–140.

Ammonia

$$R-\underset{H}{\overset{O}{\|}} \quad + \quad NH_3 \quad \xrightarrow{\;NaBH_3CN\;} \quad RCH_2NH_2$$

SAMPLE
Matrix: solutions
Analyte: saccharides
Sample preparation: Dissolve saccharides in water, add excess of 2 M ammonium sulfate or 4 M ammonium chloride, add excess 400 mM sodium cyanoborohydride, mix well, heat at 100° for 100-120 min, cool in an ice bath. Mix an aliquot with 10-20 μL 20 mM KCN in water, add 5-10 μL 10 mM reagent in MeOH, let stand at room temperature for 1 h, inject an aliquot. (Reagent was 3-(4-carboxybenzoyl)-2-quinolinecarboxaldehyde. Synthesis is as follows. Mix 9.69 g p-toluidine, 10.89 g o-nitrobenzaldehyde, and 25 mL EtOH and allow to react for 5 min, filter, wash the solid with EtOH to give 2-nitro-N-(p-tolyl)benzaldimine (Talanta 1989, 36, 321). (More 2-nitro-N-(p-tolyl)benzaldimine can be obtained by adding water to the filtrate.) Add, in portions in a thin stream, a hot solution of 46 g sodium sulfide in 23 mL water and 23 mL EtOH to a stirred refluxing solution of 100 mmole 2-nitro-N-(p-tolyl)benzaldimine in 50 mL EtOH. After a vigorous reaction 2-amino-N-(p-tolyl)benzaldimine crystallizes from the cooling solution (mp 102-103° after recrystallization from dilute MeOH) (Ber. 1943, 76, 1099). Wash 950 mg of a commercial 50% slurry of sodium hydride with pentane, add 7 mL dry THF (distilled from lithium aluminum hydride), add 1.51 g methyl 4-cyanobenzoate in 10 mL THF, add dropwise 1.38 mL acetone (distilled from calcium chloride), reflux for 1.5 h, cool, acidify with 3 M HCl. Remove the organic layer and wash it with brine and sodium bicarbonate solution, dry over anhydrous magnesium sulfate, evaporate to give (4-cyanobenzoyl)acetone. Reflux 433 mg (4-cyanobenzoyl)acetone, 486 mg 2-amino-N-(p-tolyl)benzaldimine, 69 mL piperidine, and EtOH:water 95:5 for 18 h, remove volatiles by steam distillation, add the residue to water and dichloromethane. Remove the organic layer and dry it, evaporate to give 3-(4-cyanobenzoyl)-2-methylquinoline. Suspend 547 mg 3-(4-cyanobenzoyl)-2-methylquinoline in 13 mL EtOH:water 95:5, add 500 mg KOH, reflux for 6 h, cool, concentrate, add the residue to ether and water. Remove the aqueous layer and adjust the pH to 5 with tartaric acid, let stand for 15 min, filter, wash the precipitate with water, dry under vacuum to give 3-(4-carboxybenzoyl)-2-methylquinoline. Dissolve 266 mg 3-(4-carboxybenzoyl)-2-methylquinoline in 6 mL acetic acid, add 112 mg selenium dioxide, stir at 80° for 2 h, filter through Celite, wash the precipitate with several volumes of hot MeOH, dilute the filtrate with water, allow to stand, filter, wash the solid with water, dry under vacuum to give 3-(4-carboxybenzoyl)-2-quinolinecarboxaldehyde (Anal. Chem. 1991, 63, 408).)

CAPILLARY ELECTROPHORESIS
Capillary: 88 cm × 50 μm (58 cm to detector) (Polymicro Technologies)
Running buffer: 10 mM pH 9.40 Na_2HPO_4 containing 10 mM sodium tetraborate
Voltage/Current: 20 kV/12 μA
Injection: Hydrodynamic injection for 5 s
Detector: F ex 457 (argon laser) em 552
Model: laboratory constructed
Migration time: 10.6 (glucosamine), 11.2 (galactosamine), 11.6 (erythrose), 12.4 (ribose), 13.2 (talose), 13.4 (mannose), 14 (dextrose), 14.2 (galactose), 17.2 (galacturonic acid), 18.2 (glucuronic acid), 18.8 (glucosaminic acid), 21 (glucose 6-phosphate), 11-14 (oligosaccharides, depending on degree of polymerization)
Limit of detection: 0.5-2.3 amole

REFERENCE
Liu, J.; Shirota, O.; Wiesler, D.; Novotny, M. Ultrasensitive fluorometric detection of carbohydrates as derivatives in mixtures separated by capillary electrophoresis, *Proc.Nat.Acad.Sci.USA*, **1991**, *88*, 2302–2306.

SAMPLE

Matrix: solutions

Analyte: monosaccharides

Sample preparation: Add 25 nmoles 3-O-methylglucose, evaporate the solution to dryness, add 50 µL 300 mM sodium cyanoborohydride in 2 M pH 7.0 ammonium acetate (freshly prepared), heat at 105° for 4 h, add 100 µL water, add 40 µL 6 M formic acid, evaporate to dryness under reduced pressure, add 500 µL MeOH, evaporate to dryness, repeat MeOH evaporation twice more, add 100 µL EtOH:water:triethylamine 40:40:20, evaporate to dryness, add 100 µL EtOH:triethylamine:water:phenyl isothiocyanate 70:10:10:10, let stand at room temperature for 20 min, evaporate to dryness under reduced pressure, reconstitute with 20 µL MeCN:water 60:40, add 180 µL MeCN:5 mM pH 7.4 sodium phosphate buffer 5:95, filter, inject an aliquot.

HPLC VARIABLES

Column: 250 × 4.6 5 µm Microsorb C18

Mobile phase: Gradient. A was 50 mM pH 6.8 ammonium acetate. B was 100 mM pH 6.8 ammonium acetate in MeCN:MeOH:water 44:10:46. A:B 78:22 until the run is over, to 0:100 over 5 min, maintain at 0:100 for 6 min, return to initial conditions over 5 min

Column temperature: 30

Flow rate: 0.8

Detector: UV 254

CHROMATOGRAM

Retention time: 11 (galactose), 12 (dextrose), 13 (mannose), 13.5 (xylose), 15 (ribose), 16.5 (fucose)

Internal standard: 3-O-methylglucose (20)

Limit of detection: 50 pmole

REFERENCE

Spiro, M.J.; Spiro, R.G. Monosaccharide determination of glycoconjugates by reverse-phase high-performance liquid chromatography of their phenylthiocarbamyl derivatives, *Anal.Biochem.*, **1992**, *204*, 152–157.

Aniline

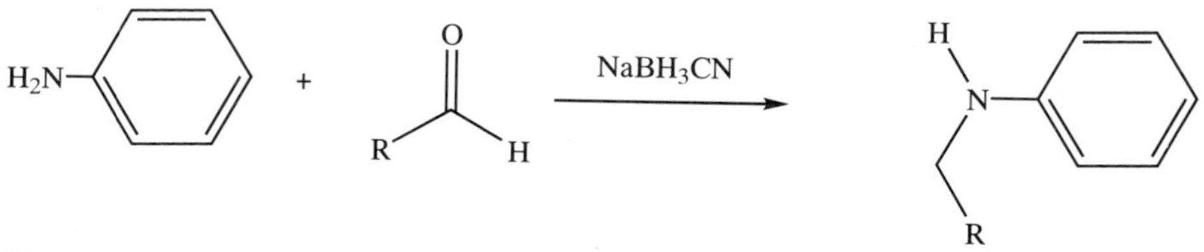

SAMPLE

Matrix: bulk

Analyte: oligosaccharides

Sample preparation: Mix 1 mmole aniline, 35 mg sodium cyanoborohydride, 41 µL glacial acetic acid, and 350 µL MeOH, add 10 µL water, add 1 mg oligosaccharide, mix, heat at 80° for 30 min, cool, add 1 mL water, add 1 mL chloroform, mix, inject an aliquot of the upper aqueous phase.

HPLC VARIABLES

Column: 250 × 4.6 LiChrosorb Si60

Mobile phase: Gradient. A was 0.05% 1,4-diaminobutane in water. B was 0.05% 1,4-diaminobutane in MeCN. A:B from 15:85 to 40:60 over 50 min.

Flow rate: 0.5

Detector: UV 229; UV 254

CHROMATOGRAM

Retention time: 10-26 (depending on degree of oligomerization)

REFERENCE
Wang, W.T.; LeDonne, N.C. Jr.; Ackerman, B.; Sweeley, C.C. Structural characterization of oligosaccharides by high-performance liquid chromatography, fast-atom bombardment-mass spectrometry,
and exoglycosidase digestion, *Anal.Biochem.*, **1984**, *141*, 366–381.

Ethyl 4-Aminobenzoate

RELATED REFERENCES
Akiyama, T. Separation of neutral mono- and oligosaccharides derivatized with ethyl p-aminobenzoate
by high-performance liquid chromatography on an amine-bonded vinyl alcohol copolymer column.
J.Chromatogr. **1991**, *588*, 53-59.
Matsuura, F.; Imaoka, A. Chromatographic separation of asparagine-linked oligosaccharides labeled
with an ultraviolet-absorbing compound, p-aminobenzoic acid ethyl ester. *Glycoconjugate J.* **1988**, *5*,
13-26.
Ohta, M.; Kobatake, M.; Matsumura, A.; Matsuura, F. Separation of Asn-linked sialyloligosaccharides
labeled with p-aminobenzoic acid ethyl ester by high-performance liquid chromatography.
Agric.Biol.Chem. **1990**, *54*, 1045-1047.

SAMPLE
Matrix: bulk
Analyte: oligosaccharides
Sample preparation: Mix 1 mmole ethyl 4-aminobenzoate, 35 mg sodium cyanoborohydride, 41 μL glacial acetic acid, and 350 μL MeOH, add 10 μL water, add 1 mg oligosaccharide, mix, heat at 80° for 30 min, cool, add 1 mL water, add 1 mL chloroform, mix,
inject an aliquot of the upper aqueous phase.

HPLC VARIABLES
Column: 250 × 4.6 LiChrosorb Si60
Mobile phase: Gradient. A was 0.05% 1,4-diaminobutane in water. B was 0.05% 1,4-dia-
minobutane in MeCN. A:B from 35:65 to 55:45 over 40 min.
Flow rate: 0.5
Detector: UV 229; UV 254

CHROMATOGRAM
Retention time: 5-23 (depending on degree of oligomerization)
Limit of detection: 50 pmole (UV 229)

REFERENCE
Wang, W.T.; LeDonne, N.C. Jr.; Ackerman, B.; Sweeley, C.C. Structural characterization of oligosaccharides by high-performance liquid chromatography, fast-atom bombardment-mass spectrometry,
and exoglycosidase digestion, *Anal.Biochem.*, **1984**, *141*, 366–381.

SAMPLE
Matrix: glycoproteins
Analyte: monosaccharides
Sample preparation: 200 μg Glycoprotein + 100 μL water + 100 μL 4 M trifluoroacetic
acid, heat at 100° for 6 h, cool to room temperature, evaporate to dryness under reduced

pressure at 35°, add 40 μL reagent, heat at 80° for 1 h, cool to room temperature, add 200 μL water, add 200 μL chloroform, vortex vigorously, centrifuge for 1 min, inject an aliquot of the upper aqueous layer. (Prepare the reagent by mixing 165 mg ethyl 4-aminobenzoate, 35 mg sodium cyanoborohydride, 41 μL glacial acetic acid, and 350 μL glacial acetic acid.)

HPLC VARIABLES
Column: 150 × 3.9 Pico.Tag (Waters)
Mobile phase: MeCN:MeOH:50 mM pH 4.5 sodium acetate 10:5:85
Column temperature: 45
Flow rate: 1.2
Detector: UV 254

CHROMATOGRAM
Retention time: 3.4 (glucosamine), 3.8 (galactosamine), 4.8 (lactose), 5.8 (maltose), 6.6 (galactose, dextrose), 6.9 (mannose), 8 (xylose), 8.6 (N-acetylglucosamine), 10 (N-acetylgalactosamine), 10.4 (fucose), 12.5 (2-deoxyglucose)

REFERENCE
Kwon, H.; Kim, J. Determination of monosaccharides in glycoproteins by reverse-phase high-performance liquid chromatography, *Anal.Biochem.*, **1993**, *215*, 243–252.

SAMPLE
Matrix: plants
Analyte: saccharides
Sample preparation: Stir 40 g dried plant material with 600 mL water at 500 rpm for 24 h, filter (grid mesh 0.3 mm), centrifuge the filtrate at 17700 g for 15 min. Remove an 80 mL aliquot of the supernatant to 400 mL EtOH, let stand at 4° for 48 h, centrifuge at 17700 g for 30 min, discard the supernatant, suspend the pellet in 100 mL ether, evaporate the ether to dryness at 40°. Add 2.5 mL 6 M trifluoroacetic acid to 20 mg of the precipitate, dissolve with gentle swirling, reflux at 110° for 4 h, lyophilize to dryness, add reagent so that the concentration of saccharide is 2 mg/mL, vortex gently, heat at 50° for 2 h, cool to room temperature, dilute 40-fold with MeOH, inject an aliquot. (Prepare reagent by dissolving 10 mg sodium cyanoborohydride in 1 mL MeOH containing 10% ethyl 4-aminobenzoate and 10% acetic acid.)

CAPILLARY ELECTROPHORESIS
Capillary: 72 cm × 50 μm fused-silica (50 cm to detector)
Capillary preparation: Between runs wash capillary with 1 M NaOH for 4 min and with 1 mM NaOH for 2 min, equilibrate with running buffer for 4 min. Flush new capillaries with 1 M NaOH for 1 h and with 1 mM NaOH for 5 min.
Running buffer: 175 mM pH 10.5 Borate buffer
Capillary temperature: 30
Voltage/Current: 25 kV
Injection: Vacuum injection at 16.9 kPa for 1 s
Detector: UV 305
Model: Applied Biosystems Model 270A
Migration time: 7.8 (2-deoxy-D-ribose), 8.2 (maltotriose), 8.7 (rhamnose), 9 (cellobiose), 9.2 (xylose), 9.5 (ribose), 9.8 (lactose), 10 (dextrose), 10.2 (arabinose), 11.2 (fucose), 12 (galactose), 14 (mannuronic acid), 14.3 (glucuronic acid), 16.5 (galacturonic acid)
Limit of detection: 2 μM

REFERENCE
Vorndran, A.E.; Grill, E.; Huber, C.; Oefner, P.J.; Bonn, G.K. Capillary zone electrophoresis of aldoses, ketoses and uronic acids derivatized with ethyl p-aminobenzoate, *Chromatographia*, **1992**, *34*, 109–114.

SAMPLE
Matrix: solutions
Analyte: oligosaccharides
Sample preparation: Mix 100 nmole oligosaccharides with 70 μL reagent, heat at 80° for 1.5 h, cool, evaporate to dryness under a stream of nitrogen at 20°, reconstitute with chloroform:MeOH 2:1, add to a 250 μL column of phenylboronate agarose (Amicon

PBA60) made up in chloroform:MeOH 2:1, wash with 2.5 mL chloroform:MeOH 2:1, elute with chloroform:MeOH:water 2:8:5. (Reagent was 2.8 mg ethyl 4-aminobenzoate, 280 μg sodium cyanoborohydride, and 1.6 μL acetic acid in 700 μL MeOH.)

HPLC VARIABLES
Column: 250 × 4.6 TSK-GEL Amide 80 (TOSOH)
Mobile phase: Gradient. MeCN:water from 75:25 to 50:50 over 20 min
Flow rate: 1
Injection volume: 20
Detector: UV 304

CHROMATOGRAM
Retention time: 10-15 (depending on structure)

KEY WORDS
SPE; other oligosaccharides are determined with different chromatographic conditions

REFERENCE
Higashi, H.; Ito, M.; Fukaya, N.; Yamagata, S.; Yamagata, T. Two-dimensional mapping by the high-performance liquid chromatography of oligosaccharides released from glycosphingolipids by endo-glycoceramidase, *Anal.Biochem.*, **1990**, *186*, 355–362.

n-Hexyl p-Aminobenzoate

SAMPLE
Matrix: solutions
Analyte: oligosaccharides
Sample preparation: Dissolve 1-20 μg oligosaccharides in 10 μL water, add 40 μL reagent, make up to 200 μL with MeOH, heat at 80° in a vial sealed with a PTFE cap for 45 min, cool, add 1 mL water, add 1 mL chloroform, vortex, centrifuge, remove the aqueous phase, re-extract the organic phase with 1 mL water. Combine the aqueous layers and lyophilize them, reconstitute, inject an aliquot. (Reagent was 100 μmoles n-hexyl p-aminobenzoate, 35 mg sodium cyanoborohydride, 41 μL glacial acetic acid, and 350 μL MeOH. Prepare n-hexyl p-aminobenzoate by refluxing 10 mmoles p-aminobenzoic acid with 100-150 moles n-hexanol in the presence of 15 mmoles 48% boron trifluoride ethyl etherate for 10-24 h until all the acid is consumed, remove the excess alcohol by distillation under reduced pressure. Take up the residue in 25-40 mL ether, wash with 80 mL 5% sodium carbonate, wash twice with 80 mL portions of water, add 300-500 mL hexane to the organic phase and allow it to crystallize at -20°, recrystallize n-hexyl p-aminobenzoate (mp 60-61°) from hexane.)

HPLC VARIABLES
Column: 250 × 4.6 C18 (Vydac)
Mobile phase: Gradient. MeCN:water 0:100 for 10 min, to 50:50 over 50 min.
Flow rate: 1
Detector: UV 304

CHROMATOGRAM
Retention time: 20-45 (depending on structure)

REFERENCE
Poulter, L.; Karrer, R.; Burlingame, A.L. *n*-Alkyl *p*-aminobenzoates as derivatizing agents in the isolation, separation, and characterization of submicrogram quantities of oligosaccharides by liquid secondary ion mass spectrometry, *Anal.Biochem.*, **1991**, *195*, 1–13.

1-Maltoheptaosyl-1,5-diaminonaphthalene

SAMPLE
Matrix: bulk
Analyte: heparin
Sample preparation: Dissolve 5 mg heparin and 10 mg 1-maltoheptaosyl-1,5-diaminonaphthalene in 25 μL 10 mM phosphoric acid, add 5 μL 2 M sodium cyanoborohydride, heat at 65° for 3 h, make up to 500 μL with water, inject an aliquot. (Synthesis of 1-maltoheptaosyl-1,5-diaminonaphthalene is as follows. Heat 500 mg maltoheptaose and 362.5 mg 1,5-diaminonaphthalene in 10 mL EtOH:20 mM phosphoric acid 50:50 at 85° for 30 min, add sodium cyanoborohydride to a final concentration of 300 mM, heat at 85° for 3 h, remove the solvent by evaporation under a stream of nitrogen, centrifuge, purify the supernatant three times by precipitation from 90% acetone with drying under nitrogen, store at 4°.)

CAPILLARY ELECTROPHORESIS
Capillary: 60 cm × 50 μm (50 cm to detector) (Polymicro Technologies)
Capillary preparation: Coat column as follows. Adjust the pH of 20 mL water to 3.5 with acetic acid, add 80 μL 3-(trimethoxysilyl)propyl methacrylate (3-methacryloxypropyltrimethoxysilane), mix, suck into capillary, let stand at room temperature for 1 h, remove the solution, wash with water. Fill the capillary with a deaerated 3-4% acrylamide solution containing 1 μL/mL N,N,N',N'-tetramethylethylenediamine and 1 mg/mL potassium

persulfate, let stand for 30 min, remove excess solution by aspiration, rinse with water, remove water by aspiration, dry at 35° (J. Chromatogr. 1985, 347, 191).
Running buffer: 10 mM pH 3.1 Sodium citrate
Voltage/Current: 21 kV
Detector: F ex 325 (He-Cd laser) em >360 (Cut-off filter)
Model: laboratory-constructed
Migration time: 15-42 (depending on structure)

KEY WORDS
coated capillary

REFERENCE
Sudor, J.; Novotny, M.V. End-label free-solution electrophoresis of the low molecular weight heparins, *Anal.Chem.*, **1997**, *69*, 3199–3204.

Octylamine

SAMPLE
Matrix: solutions
Analyte: oligosaccharides
Sample preparation: 650 μg Maltodextrins + 1 μL octylamine + 3 equivalents sodium cyanoborohydride + 200 μL 5% acetic acid in MeOH:water 75:25, heat at 70° for 1 h, evaporate to dryness under a stream of nitrogen, add 2 μL triethylamine, add 1 μL 2,4-dinitrofluorobenzene, add 200 μL MeOH:water 80:20, heat at 45° overnight, inject an aliquot.

HPLC VARIABLES
Column: 250 × 4.6 5 μm Ultrasphere ODS
Mobile phase: MeCN:water 30:70
Column temperature: 50
Flow rate: 1
Detector: UV 392

KEY WORDS
retention time depends on degree of polymerization from 1 to 16

REFERENCE
Zhang, Y.; Cedergren, R.A.; Nieuwenhuis, T.J.; Hollingsworth, R.I. *N,N*-(2,4-Dinitrophenyl)octylamine derivatives for the isolation, purification, and mass spectrometric characterization of oligosaccharides, *Anal.Biochem.*, **1993**, *208*, 363–371.

L-Phenylalanine Methyl Ester

SAMPLE
Matrix: blood
Analyte: gossypol
Sample preparation: 400 μL Serum + 500 μL saturated disodium EDTA, mix, let stand for 10 min, adjust to pH 8.0 with concentrated NaOH solution using a micro-electrode, add 50 μL 1 g/mL L-phenylalanine methyl ester in MeCN, let stand for 10 min, adjust pH to 7.00 with concentrated sulfuric acid, add 2 mL ether, vortex, centrifuge, repeat extraction. Combine the organic layers and evaporate them to dryness under reduced pressure, reconstitute the residue in 500 μL MeCN, inject a 50 μL aliquot.

HPLC VARIABLES
Column: 250 × 4.5 5 μm Hypersil ODS
Mobile phase: MeCN:THF:buffer 76:2:22 (Buffer was 10 mM KH_2PO_4 adjusted to pH 2.35 with phosphoric acid.)
Flow rate: 2.5
Injection volume: 50
Detector: UV 250

CHROMATOGRAM
Retention time: 11 (-), 17 (+)
Limit of detection: 30 ng/mL

KEY WORDS
serum; chiral

REFERENCE
Matlin, S.A.; Belenguer, A.; Vince, P.M.; Stein, R. Analysis of gossypol enantiomers in human serum, *J.Liq.Chromatogr.*, **1990**, *13*, 2261–2268.

Procaine

SAMPLE
Matrix: solutions
Analyte: oligosaccharides
Sample preparation: Add 3.5 μL 300 mM procaine hydrochloride (2-(diethylamino)ethyl 4-aminobenzoate hydrochloride) in MeOH to 350 μg sodium cyanoborohydride, add 0.4 μL glacial acetic acid. Add this mixture to 4 μL 250 μM oligosaccharide in water, make up to 20 μL with MeOH, vortex, heat at 80° for 1 h, cool, add 100 μL water, inject an aliquot.

HPLC VARIABLES
Column: 250 × 10 5 μm Cosmosil 5C18-AR (Nacalai Tesque)
Mobile phase: Gradient. MeCN:0.1% trifluoroacetic acid from 5:95 to 20:80 over 30 min.
Flow rate: 2
Detector: UV 310

CHROMATOGRAM
Retention time: 21 (maltohexaose)

KEY WORDS
Derivatives can be isolated from HPLC effluent. Derivatization enhances ionization efficiency in electrospray MS.

REFERENCE
Yoshino, K.; Takao, T.; Murata, H.; Shimonishi, Y. Use of the derivatizing agent 4-aminobenzoic acid 2-(diethylamino)ethyl ester for high-sensitivity detection of oligosaccharides by electrospray ionization mass spectrometry, *Anal.Chem.*, **1995**, *67*, 4028–4031.

Tyramine

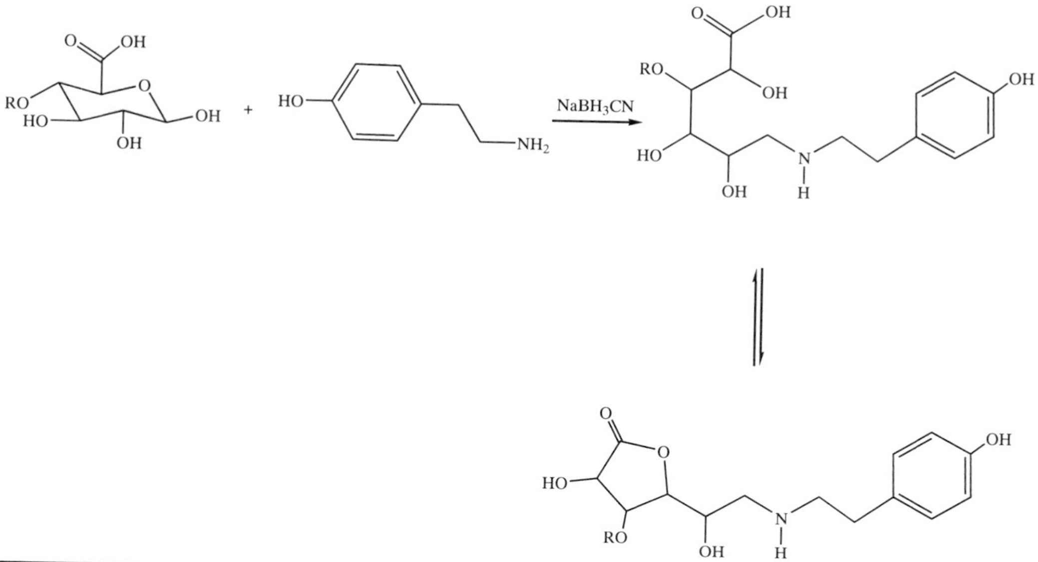

SAMPLE

Matrix: bulk

Analyte: trigalacturonic acid

Sample preparation: Dissolve 10 mg trigalacturonic acid in 100 μL water, add 350 μL reagent, heat at 80° for 1 h, cool to room temperature, add glacial acetic acid until the evolution of hydrogen ceases, evaporate to dryness under a stream of air at 25°, reconstitute with 500 μL water, add to a 500 × 10 column of Sephadex G-25, elute with water at 0.4 mL/min, collect 1.2 mL fractions, assay for derivatized product at UV 274. Concentrate the appropriate fractions to 1 mL under reduced pressure, inject a 100 μL aliquot. (Prepare reagent by dissolving 137 mg tyramine and 35 mg sodium cyanoborohydride in 400 μL acetic acid:MeOH 10:90 at 100°.)

HPLC VARIABLES

Column: CarboPac PA-1 (Dionex)

Mobile phase: Gradient. 200 mM pH 8 sodium acetate:500 mM pH 8 sodium acetate from 100:0 to 0:100 over 30 min.

Flow rate: 1

Injection volume: 100

Detector: UV 274; E, Dionex, gold working electrode, pulsed amperometric mode, E1 150 mV, E2 700 mV, E3 -300 mV, T1 480 ms, T2 120 ms, T3 360 ms, 3 μA sensitivity following post-column reaction. The column effluent mixed with 400 mM NaOH pumped at 0.5 mL/min and the mixture flowed to the detector.

CHROMATOGRAM

Retention time: 20 (lactone form), 27 (free acid form)

REFERENCE

Spiro, M.D.; Ridley, B.L.; Glushka, J.; Darvill, A.G.; Albersheim, P. Synthesis and characterization of tyramine-derivatized (1→4)-linked α-D-oligogalacturonides, *Carbohydr.Res.*, **1996**, *290*, 147–157.

AMINOPHENOL

2-Amino-4,5-ethylenedioxyphenol

SAMPLE
Matrix: solutions
Analyte: aldehydes
Sample preparation: Mix a 500 μL aliquot of a solution of an aldehyde in MeOH with 500 μL 1.5 mM 2-amino-4,5-ethylenedioxyphenol in MeOH and 500 μL 5 mM dicyclohexylcarbodiimide in MeOH, heat at 100° in a tightly closed vial for 45 min, cool, add 500 μL 6 M perchloric acid, let stand at room temperature for more than 1 min, adjust pH to 6-7.5 with about 500 μL 6 M NaOH, inject a 20 μL aliquot. (2-Amino-4,5-ethylenedioxyphenol can be synthesized by analogy to the synthesis of 3,4-dimethoxy-6-aminophenol (cf. Can. J. Chem. 1966, 44, 1875). Add 145 g 20% peracetic acid in 20-30 mL portions with occasional shaking to 54.7 g 1,4-benzodioxan-6-carboxaldehyde (allow the temperature to fall to 32° before adding the next portion), let stand at room temperature for 24 h, evaporate the acetic acid under reduced pressure, add 250 mL toluene, stir for 15 min. Remove the solution and remove the toluene by evaporation, distil under reduced pressure (about 120°/2 mm Hg) to give crude product. Shake the crude product with 150 mL 10% NaOH, wash 3 times with 50 mL portions of ether (the washes contain unreacted starting material). Acidify the aqueous layer with 20% sulfuric acid, extract with 200 mL ether. Wash the ether layer 3 times with 50 mL portions of 5% sodium bicarbonate, wash with water, dry over potassium sulfate, evaporate to obtain 6-hydroxy-1,4-benzodioxan. Dissolve 100 mmoles 6-hydroxy-1,4-benzodioxan in 60 mL 10% NaOH, cool, add 15 g acetic anhydride with stirring over 10 min, extract with carbon tetrachloride. Wash the organic layer with sodium carbonate solution, dry over anhydrous magnesium sulfate, evaporate to dryness, distil under reduced pressure (about 130°/2 mm Hg) to yield 6-acetoxy-1,4-benzodioxan. Dissolve 100 mmoles 6-acetoxy-1,4-benzodioxan in 60 mL glacial acetic acid, slowly add a solution of 11 g concentrated nitric acid in 20 mL glacial acetic acid, stir for 3 h, pour into 400 mL water, let stand for 1 h, filter to obtain 6-acetoxy-7-nitro-1,4-benzodioxan, recrystallize from EtOH. Heat 100 mmoles 6-acetoxy-7-nitro-1,4-benzodioxan, 220 mL EtOH, and 125 mL 20% sulfuric acid on a water bath for 40 min, cool, filter to obtain 6-hydroxy-7-nitro-1,4-benzodioxan. Hydrogenate 6-hydroxy-7-nitro-1,4-benzodioxan in 96% EtOH or methoxyethanol over platinum oxide or palladium on charcoal at room temperature under 3 atmospheres pressure to obtain 2-amino-4,5-ethylenedioxyphenol (6-amino-7-hydroxy-1,4-benzodioxan).)

HPLC VARIABLES
Column: 150 × 4.6 5 μm TSKgel ODS-80Tm (Tosoh)
Mobile phase: MeOH:water 70:30
Flow rate: 0.8
Injection volume: 20
Detector: F ex 330 em 390

CHROMATOGRAM
Retention time: 5.9 (3,4-dihydroxybenzaldehyde), 8.3 (isovanillin), 9.2 (vanillin), 17.3 (benzaldehyde), 20 (4-methoxybenzaldehyde), 27.9 (4-methylbenzaldehyde)
Limit of detection: 5-10 pmole

OTHER SUBSTANCES
Non-interfering: acetaldehyde, acetic acid, adenine, adenosine, amino acids, ascorbic acid, benzoic acid, butyraldehyde, cholesterol, creatine, creatinine, cytidine, cytosine, dopa-

mine, epinephrine, formaldehyde, fructose, galactose, glucosamine, glucose, guanine, guanosine, histamine, homovanillic acid, 5-hydroxyindoleacetic acid, imidazoleacetic acid, α-ketoglutaric acid, maltose, norepinephrine, oxalic acid, phenylpyruvic acid, propionaldehyde, ribose, salicylic acid, serotonin, sucrose, thymidine, thymine, uracil, uric acid, uridine, vanillylmandelic acid

REFERENCE

Nohta, H.; Sakai, F.; Kai, M.; Ohkura, Y.; Saito, M. 2-Amino-4,5-ethylenedioxyphenol as fluorescence derivatization reagent for aromatic aldehydes in liquid chromatography, *Anal.Chim.Acta*, **1994**, *287*, 223–227.

AMINOTHIOL

2,2'-Dithiobis(1-amino-4,5-dimethoxybenzene)

SAMPLE

Matrix: solutions

Analyte: aldehydes

Sample preparation: Mix a 1 mL aliquot of a solution in water with 1 mL 8 mM sodium sulfite in 2.8 M disodium hydrogen phosphite and 1 mL 1.1 mM (?) 2,2'-dithiobis(1-amino-4,5-dimethoxybenzene) solution, heat at 37° for 1 h, adjust pH to 6-7 with about 1 mL 800 mM NaOH, inject a 20 μL aliquot. (Disodium hydrogen phosphite does not appear to be commercially available, the authors may mean disodium hydrogen phosphate. On the other hand disodium hydrogen phosphate solutions of a concentration greater than about 1 M cannot be made. Disodium hydrogen phosphite can be prepared by neutralizing a solution of phosphorous acid with sodium carbonate (J.W. Mellor. A Comprehensive Treatise on Inorganic and Theoretical Chemistry, Longmans, Green: London, 1931). An alternative procedure is as follows. Add 20 mL 130 mM 2,2'-dithiobis(1-amino-4,5-dimethoxybenzene) in MeOH containing 0.8 M disodium hydrogen phosphite and 30 g/L tri-n-butylphosphine to 10 mL of a 400 mM solution of aldehyde in MeOH, stir at 37° for 1 h, stir at room temperature for 6 h, filter. Dissolve the solid in a suitable solvent, inject an aliquot. Synthesis of 2,2'-dithiobis(1-amino-4,5-dimethoxybenzene) is as follows. Stir 20 g 4-bromoveratrole in 60 mL acetic acid, add 10 mL concentrated nitric acid dropwise while keeping the temperature at 10-30° with occasional cooling, pour into ice-water. Collect the precipitate and dissolve it in 500 mL hot EtOH, add activated charcoal, filter, add 40 mL water to the filtrate to obtain 4-bromo-5-nitroveratrole as light-yellow crystals (mp 121-122°). Dissolve 5 g 4-bromo-5-nitroveratrole in 50 mL 95% EtOH, add sodium sulfide (prepare immediately before use by melting 5 g sodium sulfide nonahydrate and 700 mg sulfur together), reflux for 30 min, pour into ice-water. Collect the solid and recrystallize it from dichloromethane to obtain 2,2'-dithiobis(1-nitro-4,5-dimethoxybenzene) as yellow needles (mp 231-232°). Dissolve 2 g 2,2'-dithiobis(1-nitro-4,5-dimethoxybenzene) in 300 mL EtOH, add 8 g tin powder, add 30 mL concentrated HCl dropwise, make alkaline with 4 M NaOH, filter. Dilute the filtrate with 200 mL water, extract twice with 100 mL portions of benzene (Caution! Benzene is a carcinogen!). Combine the extracts and evaporate them to dryness under reduced pressure, mix the residue with 10 mL benzene and 2 mL 10% hydrogen peroxide, stir for 30 min, recrystallize the precipitate from EtOH to give 2,2'-dithiobis(1-amino-4,5-dimethoxybenzene) as colorless crystals (mp 155-156°).)

HPLC VARIABLES
Column: 250 × 4.6 5 μm L-column ODS (Chemical Inspection and Testing Institute, Tokyo)
Mobile phase: MeOH:water 70:30
Flow rate: 0.8
Injection volume: 20
Detector: F ex 335 em 430

CHROMATOGRAM
Retention time: 7 (3,4-dihydroxybenzaldehyde), 9.5 (isovanillin), 10 (vanillin), 22 (benzaldehyde), 24 (4-methoxybenzaldeyde), 35 (4-methylbenzaldehyde)
Limit of detection: 8-20 fmole

OTHER SUBSTANCES
Non-interfering: acetaldehyde, acrolein, butyraldehyde, cinnamaldehyde, crotonaldehyde, p-dimethylaminocinnamaldehyde, formaldehyde, isovaleraldehyde, phenylacetaldehyde, propionaldehyde

REFERENCE
Hara, S.; Nakamura, M.; Sakai, F.; Nohta, H.; Ohkura, Y.; Yamaguchi, M. 2,2'-Dithiobis(1-amino-4,5-dimethoxybenzene) as a highly sensitive, selective and stable fluorescence derivatization reagent for aromatic aldehydes in liquid chromatography, *Anal.Chim.Acta*, **1994**, *291*, 189−195.

DIAMINE

1,2-Diamino-4,5-ethylenedioxybenzene

SAMPLE
Matrix: blood, tissue
Analyte: forphenicine (L-(4-formyl-3-hydroxyphenyl)glycine)
Sample preparation: Serum. 100 μL Serum + 100 μL 1.5 M perchloric acid + 50 μL water, mix, centrifuge at 1000 g for 15 min. Remove a 100 μL aliquot of the supernatant and add it to 200 μL 15 mM 1,2-diamino-4,5-ethylenedioxybenzene, heat at 60° for 30 min, inject a 100 μL aliquot. Muscle. Homogenize 100 mg muscle with 150 μL 8.4 mM acetic acid, add 150 μL water, add 250 μL 1.5 M perchloric acid, mix, centrifuge at 1000 g for 20 min. Remove a 100 μL aliquot of the supernatant and add it to 200 μL 15 mM 1,2-diamino-4,5-ethylenedioxybenzene, heat at 60° for 30 min, inject a 100 μL aliquot. (Prepare 1,2-diamino-4,5-ethylenedioxybenzene as follows. Dissolve 2 g 1,4-benzodioxan in 20 mL acetic acid, add 800 mg nitric acid (d 1.51) in 10 mL acetic acid dropwise, let stand at room temperature for 2 h, pour into water, recrystallize from EtOH to give 1-

nitro-3,4-ethylenedioxybenzene as yellow needles (mp 121°) (Annalen 1894, 280, 167). Dissolve 13 g 1-nitro-3,4-ethylenedioxybenzene (6-nitro-1,4-benzodioxan) in 225 mL anhydrous glacial acetic acid, add 75 mL concentrated sulfuric acid, add 3.3 mL nitric acid (d = 1.50), heat at 90-95° for 2 h, pour into water, collect the precipitate, wash with water, wash with a little EtOH:water 96:4, recrystallize several times from glacial acetic acid: EtOH 80:20 to give 1,2-dinitro-4,5-ethylenedioxybenzene as light yellow needles (mp 131-132°) (Rec. Trav. Chim. Pays-Bas 1941, 60, 569). Dissolve 5.3 g 1,2-dinitro-4,5-ethylene-dioxybenzene in 200 mL benzene (Caution! Benzene is a carcinogen!), while refluxing add 100 g 80 mesh iron powder and 20 mL concentrated HCl in small portions over 1 h, reflux for 4 h, add 10 mL water, reflux for 2 h, cool, make alkaline with 2.6 M NaOH, extract three times with 200 mL portions of benzene, combine the extracts, remove the benzene by evaporation, add 10 mL concentrated HCl, recrystallize from EtOH to give 1,2-diamino-4,5-ethylenedioxybenzene dihydrochloride as colorless needles (mp 215-218° d) (Chem. Pharm. Bull. 1987, 35, 687).)

HPLC VARIABLES

Column: 250 × 4 5 μm TSK gel ODS-120T (Tosoh)
Mobile phase: MeCN:30 mM pH 6.5 phosphate buffer 1:5
Flow rate: 1
Injection volume: 100
Detector: F ex 350 em 420

CHROMATOGRAM
Retention time: 18.8
Limit of detection: 7.35 nM (serum), 5.36 pmole/g (muscle)

OTHER SUBSTANCES
Non-interfering: acetaldehyde, N-acetylneuramic acid, amino acids, epinephrine, estriol, estrone, formaldehyde, forphenicol, fructose, glucose, α-ketoglutaric acid, norepinephrine, α-phenylpyruvic acid, pyruvic acid

KEY WORDS
mouse; serum; muscle; pharmacokinetics

REFERENCE
Chao, W.-F.; Kai, M.; Ohkura, Y. High-performance liquid chromatographic determination of forpheni-cine in mouse serum and muscle by pre-column fluorescence derivatization using 1,2-diamino-4,5-ethylenedioxybenzene as fluorogenic reagent, *J.Chromatogr.*, **1988**, *430*, 361–367.

SAMPLE
Matrix: solutions
Analyte: aldehydes
Sample preparation: 1 mL Aqueous solution + 1 mL 100 mM pH 3.0 phosphate buffer + 1 mL 3 mM 1,2-diamino-4,5-ethylenedioxybenzene in water, heat at 60° for 30 min, inject a 100 μL aliquot. (Preparation of 1,2-diamino-4,5-ethylenedioxybenzene is as described above.)

HPLC VARIABLES
Column: 150 × 4 5 μm TSK gel ODS-120T (Tosoh)
Mobile phase: MeCN:50 mM pH 7.0 phosphate buffer 28:72
Flow rate: 1
Injection volume: 100
Detector: F ex 350 em 400

CHROMATOGRAM
Retention time: 16 (benzaldehyde), 25 (2-chlorobenzaldehyde), 40 (cinnamaldehyde), 29 (2,3-dimethoxybenzaldehyde), 8 (4-hydroxybenzaldehyde), 9 (4-hydroxy-3-methoxybenzal-dehyde), 22 (3-methoxybenzaldehyde), 18 (4-methoxybenzaldehyde), 15 (phenylacet-aldehyde)
Limit of detection: 0.01-10 pmole

REFERENCE

Chao, W.-f.; Kai, M.; Ishida, J.; Ohkura, Y.; Hara, S.; Yamaguchi, M. 1,2-Diamino-4,5-ethylenedioxyben-zene as a highly sensitive fluorogenic reagent for aromatic aldehydes, *Anal.Chim.Acta*, **1988**, *215*, 259–266.

4,5-Dimethyl-1,2-phenylenediamine

SAMPLE

Matrix: solutions
Analyte: aldehydes
Sample preparation: Mix 1 mL solution in EtOH:water 20:80 with 1 mL reagent, shake well, heat in a boiling water bath for 50 min, cool to room temperature, add 500 μL 500 mM NaOH, mix, inject a 20 μL aliquot. (Prepare the reagent by dissolving 30 mg 4,5-dimethyl-1,2-phenylenediamine in 300 μL 500 mM sulfuric acid, make up to 100 mL with water, store in the dark, discard after 2 days.)

HPLC VARIABLES

Guard column: 50 × 4 7 μm Zorbax ODS
Column: 250 × 4 7 μm Zorbax ODS
Mobile phase: MeOH:20 mM pH 4.0 acetate buffer 50:50 containing 5 mM n-dodecyltri-methylammonium bromide
Column temperature: 30
Flow rate: 1
Injection volume: 20
Detector: F ex 330 em 400

CHROMATOGRAM

Retention time: 6.0 (p-hydroxybenzaldehyde), 7.6 (vanillin), 9.2 (5-(hydroxymethyl)-2-furfural, 10.0 (o-methoxybenzaldehyde), 11.2 (m-hydroxybenzaldehyde), 13.0 (p-methoxyben-zaldehyde), 13.6 (3,4-dimethoxybenzaldehyde), 16.4 (benzaldehyde), 18.0 (furfural), 26.5 (m-methoxybenzaldehyde), 39.0 (o-hydroxybenzaldehyde)
Limit of detection: 0.05-2.5 ng/mL

REFERENCE

Katayama, M.; Mukai, Y.; Taniguchi, H. Determination of aromatic aldehydes by high-performance liq-uid chromatography after precolumn fluorescent derivatization with 4,5-dimethyl-*o*-phenylenedi-amine, *Anal.Sci.*, **1987**, *3*, 565–568.

DIKETONE

Acetylacetone

RELATED REFERENCES

Chiavari, G.; Facchini, M.C.; Fuzzi, S. Determination of formaldehyde as its lutidine derivative in the atmospheric liquid phase by high-performance liquid chromatography. *J.Chromatogr.* **1985**, *333*, 262-268.

Taatjes, D.J.; Gaudiano, G.; Resing, K.; Koch, T.H. Redox pathway leading to the alkylation of DNA by the anthracycline, antitumor drugs adriamycin and daunomycin. *J.Med.Chem.* **1997**, *40*, 1276-1286

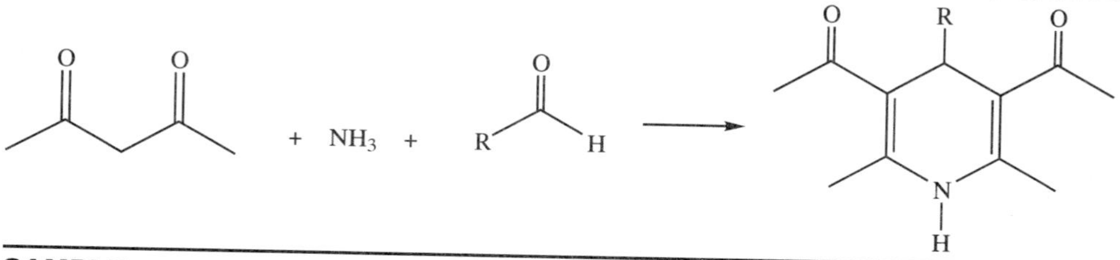

SAMPLE

Matrix: solutions

Analyte: formaldehyde and formaldehyde-releasing preservatives

HPLC VARIABLES

Column: 150 × 4.6 5 μm PRP-1 polymeric reversed-phase (Hamilton)

Mobile phase: Gradient. MeCN:water:buffer 30:70:0 for 5 min, to 30:60:10 over 0.1 min, maintain at 30:60:10 for 4.9 min, return to initial conditions over 0.1 min, re-equilibrate for 20 min. (Prepare buffer by dissolving 140 mg diaminopropionic acid monohydrochloride and 4.15 mL concentrated HCl in 1 L water.)

Flow rate: 0.5

Injection volume: 50

Detector: UV 410 following post-column reaction. The column effluent mixed with the reagent and the mixture flowed through a 1.5 mL reaction coil at 100° to the detector. (Prepare reagent by dissolving 154 g ammonium acetate, 3 mL glacial acetic acid, and 2 mL 2,4-pentanedione in 1 L water, pH 6.0. Prepare fresh each week.)

CHROMATOGRAM

Retention time: 3.8 (N,N"-methylenebis[N'-[3-(hydroxymethyl)-2,5-dioxo-4-imidazolidinyl]urea]; Germall-115), 4.6 (formaldehyde), 5.6 (1,3-(dihydroxymethyl)-5,5-dimethylhydantoin; Glydant), 13.8 (trans-1-(3-chloroallyl)-3,5,7-triaza-1-azoniaadamantane chloride; Quaternium 15; Dowicil-75)

Limit of detection: 400 ppb

KEY WORDS

post-column reaction; derivatives are also reported to fluoresce at 510 nm

REFERENCE

Summers, W.R. Characterization of formaldehyde and formaldehyde-releasing preservatives by combined reversed-phase cation-exchange high-performance liquid chromatography with postcolumn derivatization using Nash's reagent, *Anal.Chem.*, **1990**, *62*, 1397–1402.

1,3-Cyclohexanedione

RELATED REFERENCES

Rees, S.A.; Austin, P.M. Determination of U-89968E, a 5HT1a agonist in rat plasma using solid-phase extraction, precolumn derivatization and reversed-phase high-performance liquid chromatography. *J.Pharm.Biomed.Anal.* **1997**, *15*, 739-748.

Stahovec, W.L.; Mopper, K. Trace analysis of aldehydes by pre-column fluorigenic labeling with 1,3-cyclohexanedione and reversed-phase high-performance liquid chromatography. *J.Chromatogr.* **1984**, *298*, 399-406.

SAMPLE

Matrix: blood

Analyte: aldehydes

Sample preparation: 1 mL Plasma or hemolysate + 1 mL reagent, heat at 60° for 1 h, cool, add isooctane, vortex, centrifuge at 4° at 20000 g for 15 min. Remove the aqueous layer, centrifuge at 4° at 20000 g for 20 min, filter, inject an aliquot. (Prepare reagent by dissolving 10 g ammonium acetate, 400 mg 1,3-cyclohexanedione, and 3.2 mL concentrated HCl in 30 mL water, heat in a capped bottle at 60° for 1 h, cool in ice, pass through a Bond Elut C-18 SPE cartridge and a 1 g Sep-Pak SPE cartridge in sequence.)

HPLC VARIABLES

Guard column: C-18

Column: 250 mm long Ultrasphere ODS

Mobile phase: MeCN:water 30:70

Flow rate: 1

Injection volume: 20

Detector: F ex 305-395 (filter) em 450 ± 3.5 (filter)

CHROMATOGRAM

Retention time: 3.8 (formaldehyde), 4.5 (acetaldehyde)

Limit of detection: 2 pmole

KEY WORDS

plasma; hemolysate

REFERENCE

Peterson, C.M.; Polizzi, C.M. Improved method for acetaldehyde in plasma and hemoglobin-associated acetaldehyde: Results in teetotalers and alcoholics reporting for treatment, *Alcohol*, **1987**, *4*, 477–480.

SAMPLE

Matrix: blood

Analyte: acetaldehyde

Sample preparation: 200 μL Blood + 200 μL reagent, heat at 70° for 1 h, cool on ice, centrifuge at 10000 g for 5 min, dialyze (15 kD MW cutoff) against 1 mL water, pass the water through column A, elute the contents of column A on to column B with mobile phase, elute with mobile phase, monitor the effluent from column B. (Prepare reagent by dissolving 10 g ammonium acetate, 2 mL 2 M 1,3-cyclohexanedione in MeCN:water 50:50, and 6.85 g ammonium chloride in water, make up to 40 mL with water, heat in a capped bottle at 70° for 1 h, cool in ice, pass twice through a C-18 Mega Bond Elut SPE cartridge preconditioned with MeOH and water. EtOH in blood can be enzymatically converted to acetaldehyde before analysis (details in paper).)

HPLC VARIABLES
Column: A 5 × 1.6 Hypersil ODS C18; B 150 × 4.6 Zorbax RX-C18
Mobile phase: MeCN:water 20:80
Flow rate: 2
Detector: F ex 305-395 (filter) em 450 ± 3.5 (filter)

CHROMATOGRAM
Retention time: 2

KEY WORDS
dialysis; column-switching; whole blood

REFERENCE
Chen, H.-M.; Peterson, C.M. Quantifying ethanol by high performance liquid chromatography with pre-column enzymatic conversion and derivatization with fluorimetric detection, *Alcohol*, **1994**, *11*, 577–582.

SAMPLE
Matrix: blood
Analyte: aldehydes
Sample preparation: Add 50 μL plasma to an ice-cold mixture of 500 μL reagent and 450 μL water, heat at 60° for 1 h, centrifuge at 0° at 1300 g for 5 min. Remove a 100 μL aliquot of the supernatant and add it to 50 μL 25 μg/mL IS in EtOH and 250 μL water, mix, inject a 50 μL aliquot. (Prepare the reagent by dissolving 30 g ammonium acetate, 750 mg 1,3-cyclohexanedione, and 15 mL acetic acid in 150 mL water, dilute to 300 mL with water, heat at 60° for 1 h, cool in ice-water, pass through a 3 g Bond Elut C18 SPE cartridge.)

HPLC VARIABLES
Guard column: LichroCart RP-18
Column: 150 × 4.6 Cosmosil 5 C18 (Nacalai Tesque, Kyoto)
Mobile phase: Gradient. A was MeCN:water 10:90. B was MeCN:water 60:40. A:B 100:0 for 3 min, to 0:100 over 25 min, maintain at 0:100 for 5 min, re-equilibrate at initial conditions for 7 min.
Injection volume: 50
Detector: F ex 395 em 457

CHROMATOGRAM
Retention time: 12 (formaldehyde), 14 (acetaldehyde), 16 (propionaldehyde), 19 (butyraldehyde), 20 (valeraldehyde), 23 (hexaldehyde), 25 (heptaldehyde)
Internal standard: octaldehyde cyclohexanedione derivative (Prepare as follows. Condition a Bond Elut C18 SPE cartridge with 5 mL aldehyde-free EtOH and 15 mL water. Heat 1 mg octylamine with 10 mL reagent at 60° for 1 h, add to the SPE cartridge, wash with 5 mL water, wash with 5 mL MeCN:water 30:70, wash with 3 mL MeCN:water 40:60, elute with 10 mL MeCN. Evaporate the eluate to dryness under a stream of nitrogen at 40°, reconstitute the residue in EtOH to give a 25 μg/mL solution.) (28)
Limit of detection: 100 ng/mL

KEY WORDS
plasma; human; monkey; dog; rat; mouse

REFERENCE
Matsuoka, M.; Imado, N.; Maki, T.; Banno, K.; Sato, T. Determination of free aliphatic aldehydes in plasma by high-performance liquid chromatography of the 1,3-cyclohexanedione derivatives, *Chromatographia*, **1996**, *43*, 501–506.

SAMPLE
Matrix: tissue
Analyte: josamycin
Sample preparation: Blend (Virtis model 45 with U-shaped blades) 2.5 g tissue with 20 mL MeCN:10 mM pH 6.0 phosphate buffer 65:35 for 10 min, centrifuge at 5° at 8500 g for 5 min. Remove the supernatant and adjust the volume to 25 mL with MeCN:10 mM pH 6.0 phosphate buffer 65:35. Remove a 1.5 mL aliquot and add it to 5 mL isooctane,

shake for 10 min, centrifuge at 5° at 3000 g for 5 min, discard the organic layer. Add 500 μL reagent to the aqueous layer, mix, heat at 90° for 2 h, cool, inject a 100 μL aliquot. (Prepare reagent by dissolving 1 g 1,3-cyclohexanedione and 25 g ammonium acetate in 60 mL water and 8 mL concentrated HCl, make up to 100 mL with water. Store at 5°, discard after 1 month.)

HPLC VARIABLES
Guard column: 4×4 5 μm LiChrospher 100 RP-18 end capped
Column: 125×4 5 μm LiChrospher 100 RP-18 end capped
Mobile phase: MeCN:MeOH:10 mmole pH 6.0 phosphate buffer 45:5:50
Column temperature: 45
Flow rate: 1.5
Injection volume: 100
Detector: F ex 375 em 450

CHROMATOGRAM
Retention time: 10.3
Limit of detection: 25 ng/g

OTHER SUBSTANCES
Simultaneously analyzed: spiramycin, tylosin
Non-interfering: acetaldehyde, benzaldehyde, erythromycin, formaldehyde

KEY WORDS
pig; muscle; liver; kidney; fat

REFERENCE
Leroy, P.; Decolin, D.; Nicolas, A.; Archimbault, P. Determination of josamycin residues in porcine tissues using high-performance liquid chromatography with pre-column derivatization and spectrofluorometric detection, *Analyst*, **1994**, *119*, 2743–2747.

5,5-Dimethyl-1,3-cyclohexanedione

SAMPLE
Matrix: solutions
Analyte: aldehydes
Sample preparation: Mix 2 mL of an aqueous solution with 1 mL reagent and 250 μL 9 M sulfuric acid, heat in a boiling water bath for 20 min, cool in an ice bath, inject a 5-200 μL aliquot. (Make the 9 M sulfuric acid carbonyl free by refluxing for 2 h, cool before use. Prepare reagent by mixing 60 g ammonium acetate and 17.5 mL 120 mg/mL 5,5-dimethyl-1,3-cyclohexanedione (dimedone) in isopropanol, make up to 100 mL with water, store in an amber bottle, discard after 2 weeks. Let stand for 1 day before use. Before use heat at 100° for 30 min in a covered vessel, cool to room temperature, wash twice

with 10 mL portions of carbonyl-free dichloromethane. Alternatively pass through a C18 Sep-Pak SPE cartridge (conditioned with 2 mL MeOH and 5 mL water) at 3-4 mL/min.)

HPLC VARIABLES

Column: 250 × 4.6 5 μm Ultrasphere ODS

Mobile phase: Gradient. MeCN:water from 50:50 to 55:45 over 4 min, to 80:20 over 3 min, maintain at 80:20 for 7 min, to 100:0 in 1 min, maintain at 100:0 for 4 min, return to initial conditions over 2 min.

Flow rate: 1

Injection volume: 20-200

Detector: F ex 385 em 460

CHROMATOGRAM

Retention time: 5 (formaldehyde), 5.5 (acetaldehyde), 6 (propionaldehyde), 6.5 (benzaldehyde), 8 (butyraldehyde), 9.5 (pentanal), 11 (hexanal), 12.5 (heptanal), 14 (octanal), 15 (nonanal)

Limit of detection: 30 fmole

REFERENCE

Mopper, K.; Stahovec, W.L.; Johnson, L. Trace analysis of aldehydes by reversed-phase high-performance liquid chromatography and precolumn fluorigenic labeling with 5,5-dimethyl-1,3-cyclohexanedione, *J.Chromatogr.*, **1983**, *256*, 243–252.

ENONE

1-(4-Methoxy)phenyl-3-methyl-5-pyrazolone

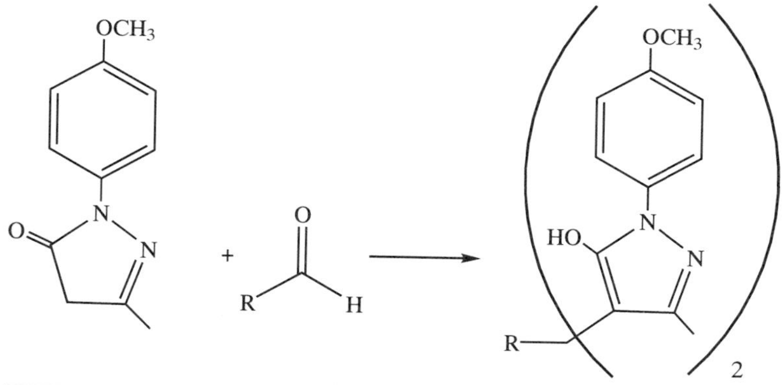

SAMPLE

Matrix: bulk

Analyte: oligosaccharides

Sample preparation: Mix 0.5 pmole-5 nmole oligosaccharide with 20 μL 300 mM NaOH and 20 μL 500 mM 1-(4-methoxy)phenyl-3-methyl-5-pyrazolone in MeOH, heat at 70° for 20 min, cool, add 20 μL 300 mM HCl, add 200 μL water, add 200 μL ethyl acetate saturated with water, shake vigorously, discard the organic phase, repeat the ethyl acetate wash four more times. Evaporate the aqueous phase to dryness and reconstitute the residue in 200 μL MeCN:water 15:85, inject a 20 μL aliquot. (Synthesis of 1-(4-methoxy)phenyl-3-methyl-5-pyrazolone is as follows. Reflux 5.6 g 4-methoxyphenylhydrazine hydrochloride, 5.45 g sodium acetate trihydrate, and 4.16 g ethyl acetoacetate in 40 mL EtOH for 2 h, cool, evaporate to dryness, dissolve the residue in 10 mL EtOH, filter, evaporate the filtrate to dryness, dissolve the residue in a small volume of benzene:ethyl acetate 80:20 (Caution! Benzene is a carcinogen!), chromatograph on a column of 150 g silica gel 60 (Merck) equilibrated with benzene:ethyl acetate 80:20, collect 5 mL fractions (monitor by TLC using Merck silica gel 60 F_{254} eluted with benzene:ethyl acetate 80:20, UV detection, R_f 0.41). Combine the appropriate fractions and evaporate them to dryness, recrystallize the residue from MeOH to give 1-(4-methoxy)phenyl-3-methyl-5-pyrazolone.)

HPLC VARIABLES
Column: 250 × 4.6 Capcell Pak C-18 (Shiseido)
Mobile phase: MeCN:100 mM pH 7.0 phosphate buffer 15:85
Flow rate: 0.6
Injection volume: 20
Detector: UV 249

CHROMATOGRAM
Retention time: 18-65 (depending on degree of polymerization)

REFERENCE
Kakehi, K.; Suzuki, S.; Honda, S.; Lee, Y.C. Precolumn labeling of reducing carbohydrates with 1-(p-methoxy)phenyl-3-methyl-5-pyrazolone: analysis of neutral and sialic acid-containing oligosaccharides found in glycoproteins, *Anal.Biochem.*, **1991**, *199*, 256−268.

SAMPLE
Matrix: bulk
Analyte: hyaluronic acid
Sample preparation: Mix 100 μL of a 1-500 μg/mL solution of hyaluronic acid in water with 40 μL 100 mM pH 5.2 citrate/phosphate buffer, add 10 500 U/mL hyaluronate 4-glycanohydrolase (sheep testis, Type V, Sigma) in water, heat at 37° for 5 h, heat in a boiling water bath for 3 min, evaporate to dryness. Reconstitute with 20 μL 300 mM NaOH, add 20 μL 500 mM 1-(4-methoxy)phenyl-3-methyl-5-pyrazolone in MeOH, heat at 70° for 20 min, add 20 μL 300 mM HCl, add 200 μL water, add 200 μL ethyl acetate saturated with water, shake vigorously, discard the organic phase, repeat the ethyl acetate wash twice more. Evaporate the aqueous phase to dryness and reconstitute the residue in 200 μL MeCN:water 15:85, inject a 20 μL aliquot. (Synthesis of 1-(4-methoxy)phenyl-3-methyl-5-pyrazolone is as described above.)

HPLC VARIABLES
Column: 150 × 6 Cosmosil 5C18-AR
Mobile phase: MeCN:100 mM pH 7.0 phosphate buffer 15:85
Flow rate: 0.8
Injection volume: 20
Detector: UV 249

CHROMATOGRAM
Retention time: 9 (hexasaccharide), 10 (tetrasaccharide), 16 (disaccharide)

REFERENCE
Kakehi, K.; Ueda, M.; Suzuki, S.; Honda, S. Determination of hyaluronic acid by high-performance liquid chromatography of the oligosaccharides derived therefrom as 1-(4-methoxy)phenyl-3-methyl-5-pyrazolone derivatives, *J.Chromatogr.*, **1993**, *630*, 141−146.

3-Methyl-1-phenylpyrazolin-5-one

RELATED REFERENCES

Chiesa, C.; Oefner, P.J.; Zieske, L.R.; O'Neill, R.A. Micellar electrokinetic chromatography of monosaccharides derivatized with 1-phenyl-3-methyl-2-pyrazolin-5-one. *J.Capillary Electrophor.* **1995**, *2*, 175-183.

Strydom, D.J. Chromatographic separation of 1-phenyl-3-methyl-5-pyrazolone-derivatized neutral, acidic and basic aldoses. *J.Chromatogr.A* **1994**, *678*, 17-23.

SAMPLE
Matrix: bulk
Analyte: oligosaccharides
Sample preparation: Add 20 μL 500 mM 3-methyl-1-phenylpyrazolone in MeOH and 20 μL 300 mM NaOH to 5-20 μg dried oligosaccharides, heat at 70° for 30 min, cool to room temperature, dilute to 250 μL with water, dialyze for 4 h with a 1000 molecular weight cut-off membrane to remove excess reagents, evaporate to dryness under reduced pressure, reconstitute with 200 μL water, inject a 50 μL aliquot.

HPLC VARIABLES
Column: 150 × 4.6 3 μm Spherisorb S3-ODS 2
Mobile phase: Gradient. MeCN:100 mM pH 7 sodium phosphate buffer 12:88 for 16 min, to 16:84 over 24 min, maintain at 16:84 for 4 min.
Flow rate: 0.7
Injection volume: 50
Detector: UV 245

CHROMATOGRAM
Retention time: 20-36 (depending on structure)

REFERENCE .
Lines, A.C. High-performance liquid chromatographic mapping of the oligosaccharides released from the humanised immunoglobulin, CAMPATH ™1H, *J.Pharm.Biomed.Anal.*, **1996**, *14*, 601–608.

SAMPLE
Matrix: solutions
Analyte: saccharides

Sample preparation: Add 50 μL 500 mM 3-methyl-1-phenyl-2-pyrazolin-5-one in MeOH and 50 μL 300 mM NaOH to 10-500 pmole saccharides, heat at 70° for 30 min, cool to room temperature, neutralize with 100 mM HCl, evaporate to dryness under reduced pressure, add 200 μL water, add 200 μL chloroform, shake vigorously. Remove the aqueous layer and evaporate it to dryness, reconstitute with 0.1-1 mL mobile phase, inject an aliquot.

HPLC VARIABLES
Column: 250 × 4.6 Capcell Pak C18 (Shiseido, Tokyo)
Injection volume: 20
Detector: UV 245; E, Irika E-502, glassy carbon working electrode 600 mV, Ag/AgCl reference electrode

CHROMATOGRAM
Retention time: 11 (mannose), 13 (lyxose), 14 (rhamnose), 20 (N-acetylglucosamine), 21 (dextrose), 23 (N-acetylgalactosamine), 25 (galactose), 28 (arabinose), 30 (fucose)
Limit of detection: 1 pmole (UV), 100 fmole (E)

REFERENCE
Honda, S.; Akao, E.; Suzuki, S.; Okuda, M.; Kakehi, K.; Nakamura, J. High-performance liquid chromatography of reducing carbohydrates as strongly ultraviolet-absorbing and electrochemically sensitive 1-phenyl-3-methyl-5-pyrazolone derivatives, *Anal.Biochem.*, **1989**, *180*, 351–357.

SAMPLE
Matrix: solutions
Analyte: saccharides
Sample preparation: Add 50 μL 500 mM 3-methyl-1-phenyl-2-pyrazolin-5-one in MeOH and 50 μL 300 mM NaOH to a dried sample, heat at 70° for 30 min, cool to room temperature, add 50 μL 300 mM HCl, evaporate to dryness under reduced pressure, add 200 μL water, add 200 μL chloroform, shake vigorously. Remove the aqueous layer and evaporate it to dryness, reconstitute with a small volume of MeOH, inject an aliquot.

CAPILLARY ELECTROPHORESIS
Capillary: 78 cm × 50 μm fused-silica (63 cm to detector) (Scientific Glass Engineering)
Capillary preparation: Before each run rinse with 100 mM NaOH and with running buffer. After 10 runs rinse with MeOH.
Running buffer: 200 mM pH 9.5 Borate buffer
Voltage/Current: 15 kV
Injection: Siphon at 5 cm for 5 s
Detector: UV 245
Migration time: 20 (xylose), 20.2 (allose), 20.5 (arabinose), 20.7 (altrose), 21 (lyxose), 21.3 (mannose), 21.3 (dextrose), 21.7 (idose), 22.3 (gulose), 23.3 (talose), 23.7 (galactose), 16-20 (oligoglucans, depending on degree of polymerization)
Internal standard: amobarbital (17.5)

REFERENCE
Honda, S.; Suzuki, S.; Nose, A.; Yamamoto, K.; Kakehi, K. Capillary zone electrophoresis of reducing mono- and oligo-saccharides as the borate complexes of their 3-methyl-1-phenyl-2-pyrazolin-5-one derivatives, *Carbohydrate Res.*, **1991**, *215*, 193–198.

HYDRAZINE

4-Aminobenzoic Hydrazide

SAMPLE

Matrix: beverages, food
Analyte: carbohydrates
Sample preparation: Dilute beverages 10-fold, filter (0.45 μm), inject a 10 μL aliquot. Cereal, grain, tobacco. Pulverize 1 g cereal, grain, or tobacco, blend with 100 mL water for 10 min, filter, inject a 10 μL aliquot of the filtrate. Extract fatty samples with hexane:water 50:50, filter the aqueous layer, inject a 10 μL aliquot of the filtrate. Dilute dairy samples 100-fold, filter, inject a 10 μL aliquot of the filtrate.

HPLC VARIABLES

Guard column: H^+ guard column (Bio-Rad)
Column: Carbohydrate column in Pb^{2+} mode (Kratos)
Mobile phase: water
Column temperature: 85
Flow rate: 0.4
Injection volume: 10
Detector: UV 410 post-column reaction. The column effluent mixed with the reagent pumped at 0.5 mL/min and the mixture flowed through a 1 mL reaction coil (ABI Analytical) at 100° to the detector. (Prepare the reagent as follows. Sonicate 2.5 g finely ground 4-aminobenzoic hydrazide and 2.46 mL concentrated HCl in 100 mL water until the mixture is homogeneous (Solution A). Sonicate 9.6 g NaOH and 1.46 g NaCl in 200 mL water until the mixture is homogeneous (Solution B). Combine Solution A and Solution B, sparge with helium, maintain at 0-5°.)

CHROMATOGRAM

Retention time: 17 (sucrose), 19 (lactose), 21 (dextrose), 24 (galactose), 27 (fructose)
Limit of detection: 20 ng

KEY WORDS

post-column reaction; cereal; grain; tobacco; dairy products

REFERENCE

Femia, R.A.; Weinberger, R. Determination of reducing and non-reducing carbohydrates in food products by liquid chromatography with post-column catalytic hydrolysis and derivatization. Comparison with refractive index detection, *J.Chromatogr.*, **1987**, *402*, 127–134.

4-(Aminosulfonyl)-7-hydrazino-2,1,3-benzoxadiazole

SAMPLE
Matrix: solutions
Analyte: propanal
Sample preparation: Mix 10 μL of a 2-3 μM solution in MeCN with 20 μL 500 μM 4-(aminosulfonyl)-7-hydrazino-2,1,3-benzoxadiazole in MeCN and 10 μL 1% trifluoroacetic acid in MeCN, let stand at room temperature for 5 h, inject an aliquot. (Synthesis of 4-(aminosulfonyl)-7-hydrazino-2,1,3-benzoxadiazole is as follows. Dissolve 0.5 g magnesium sulfate heptahydrate and 6 g NaOH in 60 mL water, throughout the reaction keep the flask at about 20° with cold water cooling, add 15 mL 30% hydrogen peroxide, add 75 mL MeOH, add 12.1 g powdered benzoyl peroxide in one go, stir for 10 min, pour into 150 mL 20% sulfuric acid, extract three times with 50 mL portions of chloroform, determine peroxybenzoic acid concentration by iodometric titration (Tetrahedron 1967, 23, 3327). Slowly add 110 mL 1 M peroxybenzoic acid in chloroform to 7 g 2,6-difluoroaniline dissolved in 100 mL chloroform, stir at room temperature, when reaction is complete (iodometric titration) wash with 2% sodium thiosulfate, wash with 5% sodium carbonate, wash with water, dry over anhydrous sodium sulfate, evaporate to dryness under reduced pressure, recrystallize 2,6-difluoronitrosobenzene from EtOH (mp 108.5-109.5). Stir 8.5 g 2,6-difluoronitrosobenzene in 85 mL DMSO at room temperature and add a solution of 3.91 g sodium azide in 85 mL DMSO dropwise, let stand for about 1 h, add to a large volume of water, extract with ether, dry the extracts over anhydrous sodium sulfate, evaporate to dryness under reduced pressure and distil to give 4-fluoro-2,1,3-benzoxadiazole as a colorless oil (bp 83°/12 mm Hg) (J.Chem.Soc.(C) 1970, 1433). Add 11 mL chlorosulfonic acid dropwise to 3 g 4-fluoro-2,1,3-benzoxadiazole in 10 mL chloroform at 0-10° (use a calcium chloride drying tube), stir at room temperature for 1 h, reflux for 2 h, cool, slowly pour into ice water, remove the organic layer, extract the aqueous layer with chloroform, combine the organic layers, wash, dry over anhydrous magnesium sulfate, evaporate under reduced pressure, take up the residue in 5 mL benzene (Caution! Benzene is a carcinogen!), chromatograph on a 150 × 30 column of silica gel (100-200 mesh Kanto Chemical) with n-hexane:benzene 50:50, evaporate the appropriate fractions to give 4-(chlorosulfonyl)-7-fluoro-2,1,3-benzoxadiazole (CBD-F) as pale yellow needles (mp 64-66°). Add 1 g CBD-F dropwise to 100 mL 6% ammonium hydroxide, neutralize with 10% HCl, evaporate under reduced pressure, add 200 mL MeCN to the residue, filter. Evaporate the filtrate and chromatograph on a 300 × 20 column of 100-200 mesh silica with chloroform, collect the appropriate fractions and evaporate them to give ABD-F (4-(aminosulfonyl)-7-fluoro-2,1,3-benzoxadiazole) as white needles (mp 145-6°) after recrystallization from n-hexane/benzene (Caution! Benzene is a carcinogen!) (Anal. Chem. 1984, 56, 2461). 4-(Aminosulfonyl)-7-fluoro-2,1,3-benzoxadiazole is also available from Wako Chemicals, Richmond VA; Molecular Probes, Eugene OR; or Dojindo Molecular Technologies, Inc., 3 Bethesda Metro Center, Suite 700, Bethesda MD 20814, (301) 664-8448, www.dojindo.co.jp. Dissolve 24 mg 4-(aminosulfonyl)-7-fluoro-2,1,3-benzoxadiazole in 3 mL MeCN, add 10 μL 98% hydrazine hydrate, heat in the dark at 50-55° for 20 min, evaporate to dryness, recrystallize from MeOH to give 4-(aminosulfonyl)-7-hydrazino-2,1,3-benzoxadiazole as yellow-orange needles (mp 184-185° d).)

HPLC VARIABLES
Column: 150 × 4.6 5 μm TSK-LS 80Tm (Tosoh)
Mobile phase: MeCN:water:trifluoroacetic acid 35:65:0.05

Flow rate: 1
Detector: F ex 450 em 550

CHROMATOGRAM
Retention time: 6
Limit of detection: 6.7 μM

OTHER SUBSTANCES
Also analyzed: acetone, butyraldehyde, 4'-ethylacetophenone, heptan-4-one, p-hydroxybenzaldehyde

REFERENCE
Uzu, S.; Kanda, S.; Imai, K.; Nakashima, K.; Akiyama, S. Fluorogenic reagents: 4-Aminosulphonyl-7-hydrazino-2,1,3-benzoxadiazole, 4-(N,N-dimethylaminosulphonyl)-7-hydrazino-2,1,3-benzoxadiazole and 4-hydrazino-7-nitro-benzoxadiazole hydrazine for aldehydes and ketones, *Analyst*, **1990**, *115*, 1477–1482.

Benzoic Hydrazide

SAMPLE
Matrix: solutions
Analyte: monosaccharides
Sample preparation: Mix a 20 μL aliquot of a 5 mM solution in EtOH with 90 μL 1% sodium cyanoborohydride in EtOH containing 1-2% acetic acid, add 10 μL benzoic hydrazide in EtOH:water 75:25, heat at 60° for 5 h, cool to room temperature, inject an aliquot. (Adjust the concentration of benzoic hydrazide so that it is present in 20-fold excess for standards and 200-fold excess for hydrolyzed glycoproteins.)

CAPILLARY ELECTROPHORESIS
Capillary: 70 cm × 50 μm fused-silica (52 cm to detector) (Yongnian, Hebei, China)
Capillary preparation: Flush with running buffer before each injection. Flush a new capillary with 1 M NaOH, water, and running buffer.

Running buffer: 200 mM pH 10.8 Borate buffer
Capillary temperature: 30
Voltage/Current: 20 kV
Injection: Vacuum injection for 8 s.
Detector: UV 220
Model: Spectra PHORESIS 1000 (Thermo Separation Products)
Migration time: 16.8 (N-acetylgalactosamine), 17.5 (rhamnose), 19.5 (lyxose), 21.5 (xylose), 22.5 (mannose), 23 (glucose), 23.5 (arabose), 25.5 (fucose), 27 (galactose)
Limit of detection: 30-40 fmole

REFERENCE
Liu, Q.; Zhang, R.; Liu, G. Use of benzoyl hydrazine reagent for monosaccharide determination by high performance capillary electrophoresis, *J.Liq.Chromatogr.Rel.Technol.*, **1997**, *20*, 1123–1137.

4-(Biotinamido)phenylacetylhydrazide

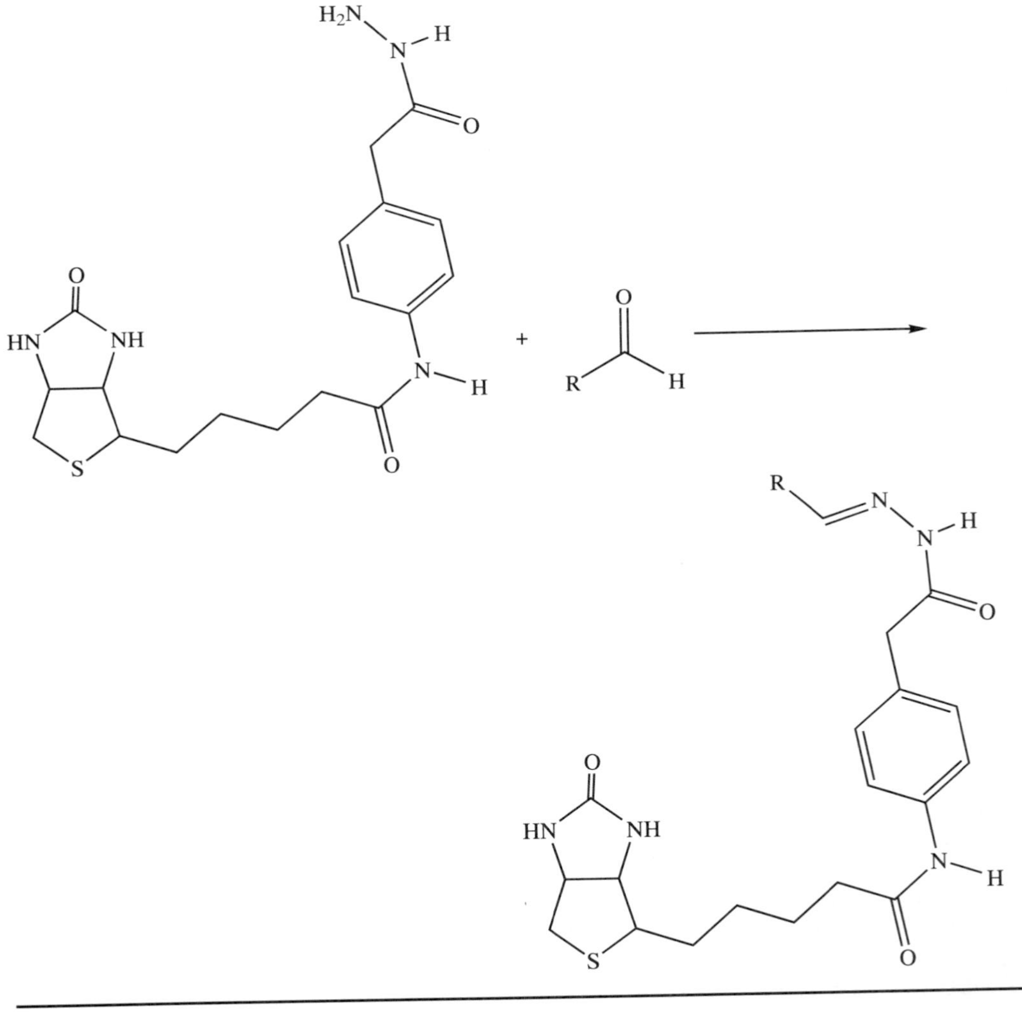

SAMPLE
Matrix: solutions
Analyte: oligosaccharides

Sample preparation: Heat 10 μL of a ≤1 mM solution in water with 10 μL 0.3-8.3 mM reagent in MeCN:water 30:70 at 90° for 1 h (use a 2-fold molar excess of reagent), cool in an ice bath, add 20 μL 50 mM pH 3.5 formate buffer, let stand at 4° for 5 h, inject an aliquot. (Prepare the reagent, 4-(biotinamido)phenylacetylhydrazide, as follows. Dissolve 6.69 mL thionyl chloride in 50 mL ice-cold MeOH, stir for 30 min, add 3 g p-aminophenylacetic acid, stir at 0° for 30 min, stir at room temperature for 20 h, evaporate to dryness, dissolve in MeOH, crystallize from ether to give 4-aminophenylacetic acid methyl ester hydrochloride. Suspend 3 g biotin in 80 mL DMF, add 3.4 mL triethylamine, add 2.1 mL isobutyl chloroformate, stir at 0° for 30 min, add 2.3 g 4-aminophenylacetic acid methyl ester hydrochloride, stir at room temperature for 16 h, evaporate to dryness, chromatograph on silica gel 60 (Merck) to obtain 4-(biotinamido)phenylacetic acid methyl ester. Suspend 2 g 4-(biotinamido)phenylacetic acid methyl ester in 64 mL hydrazine hydrate, stir at room temperature for 24 h, crystallize from ether to give 4-(biotinamido)phenylacetylhydrazide.)

HPLC VARIABLES
Column: 250 × 4.6 TSKgel ODS 80 (Tosoh)
Mobile phase: Gradient. A was MeCN:70 mM pH 6.8 phosphate buffer 10:90. B was MeCN:70 mM pH 6.8 phosphate buffer 30:70. A:B from 100:0 to 80:20 over 30 min, to 50:50 over 30 min.
Column temperature: 40
Flow rate: 0.5
Detector: UV 252

CHROMATOGRAM
Retention time: 15-50 (depending on degree of oligomerization)
Limit of detection: 330 fmole

REFERENCE
Shinohara, Y.; Sota, H.; Gotoh, M.; Hasebe, M.; Tosu, M.; Nakao, J.; Hasegawa, Y.; Shiga, M. Bifunctional labeling reagent for oligosaccharides to incorporate both chromophore and biotin groups, *Anal.Chem.*, **1996**, *68*, 2573–2579.

(R)-(+)-4-(2-Carbazolylpyrrolidin-1-yl)-7-(N,N-dimethylaminosulfonyl)-2,1,3-benzoxadiazole

SAMPLE
Matrix: solutions
Analyte: saccharides
Sample preparation: Mix a 50 μL aliquot of a 500 μM saccharide solution in MeCN:water 30:70 with 50 μL 10 mM reagent in MeCN and 100 μL 0.5% trichloroacetic acid in MeCN, heat at 65° in the dark for 3 h. Remove a 50 μL aliquot of the reaction mixture, add 200 μL water, add 200 μL ethyl acetate, mix, centrifuge at 3000 rpm for 2 min, repeat the ethyl acetate wash twice more. Dry the aqueous layer under reduced pressure, reconstitute with 200 μL MeCN, inject an aliquot. (Synthesis of reagent, R-(+)-DBD-ProCZ, is as follows. Dissolve 0.5 g magnesium sulfate heptahydrate and 6 g NaOH in 60 mL water,

throughout the reaction keep the flask at about 20° with cold water cooling, add 15 mL 30% hydrogen peroxide, add 75 mL MeOH, add 12.1 g powdered benzoyl peroxide in one go, stir for 10 min, pour into 150 mL 20% sulfuric acid, extract three times with 50 mL portions of chloroform, determine peroxybenzoic acid concentration by iodometric titration (Tetrahedron 1967, 23, 3327). Slowly add 110 mL 1 M peroxybenzoic acid in chloroform to 7 g 2,6-difluoroaniline dissolved in 100 mL chloroform, stir at room temperature, when reaction is complete (iodometric titration) wash with 2% sodium thiosulfate, wash with 5% sodium carbonate, wash with water, dry over anhydrous sodium sulfate, evaporate to dryness under reduced pressure, recrystallize 2,6-difluoronitrosobenzene from EtOH (mp 108.5-109.5). Stir 8.5 g 2,6-difluoronitrosobenzene in 85 mL DMSO at room temperature and add a solution of 3.91g sodium azide in 85 mL DMSO dropwise, let stand for about 1 h, add to a large volume of water, extract with ether, dry the extracts over anhydrous sodium sulfate, evaporate to dryness under reduced pressure and distil to give 4-fluoro-2,1,3-benzoxadiazole as a colorless oil (bp 83°/12 mm Hg) (J.Chem.Soc.(C) 1970, 1433). Add 11 mL chlorosulfonic acid dropwise to 3 g 4-fluoro-2,1,3-benzoxadiazole in 10 mL chloroform at 0-10° (use a calcium chloride drying tube), stir at room temperature for 1 h, reflux for 2 h, cool, slowly pour into ice water, remove the organic layer, extract the aqueous layer with chloroform, combine the organic layers, wash, dry over anhydrous magnesium sulfate, evaporate under reduced pressure, take up the residue in 5 mL benzene (Caution! Benzene is a carcinogen!), chromatograph on a 150 × 30 column of silica gel (100-200 mesh Kanto Chemical) with n-hexane:benzene 50:50, evaporate the appropriate fractions to give 4-(chlorosulfonyl)-7-fluoro-2,1,3-benzoxadiazole (CBD-F) as pale yellow needles (mp 64-66°) (Anal. Chem. 1984. 56, 2461). Stir 0.76 g CBD-F in 70 mL MeCN at 0-10° and add 1 g dimethylamine hydrochloride in 10 mL 100 mM pH 10 borax dropwise, adjust pH to 5 with 1 M HCl, concentrate to about 10 mL under reduced pressure, extract three times with 200 mL portions of diethyl ether, wash with water, dry over anhydrous magnesium sulfate, evaporate under reduced pressure, chromatograph on a 500 × 20 column of silica gel with chloroform, isolate the appropriate fraction and re-chromatograph on the same column with ethyl acetate:benzene 1:2 to give 4-(N,N-dimethylaminosulfonyl)-7-fluoro-2,1,3-benzoxadiazole (DBD-F) as white needles (mp 124-125°) (yield = 1% !). On a Merck no. 5714 60F$_{254}$ tlc plate eluted with chloroform DBD-F has Rf 0.32 and lies between two other reaction products (Analyst 1989, 114, 413). DBD-F can be purchased from Tokyo Kasei (TCI America, 911 North Harborgate St., Portland, OR 97203; 800-423-8616, 503-283-1681, (fax) 503-283-1987; www.tciamerica.com). Add 100 mg DBD-F in 10 mL MeCN to 47 mg R-(+)-proline in 20 mL 250 mM pH 11.5 sodium carbonate solution, stir at room temperature for 30 min, wash with ethyl acetate, adjust the pH of the aqueous layer to 1-2 with 2 M HCl, extract three times with 30 mL ethyl acetate. Combine the extracts and evaporate them under reduced pressure, recrystallize from benzene/ethyl acetate to give R-(+)-4-(N,N-dimethylaminosulfonyl)-7-(2-carboxypyrrolidin-1-yl)-2,1,3-benzoxadiazole (DBD-Pro) as yellow needles (mp 187-9° d) (Analyst 1989, 114, 1233). Suspend 55 mg (R)-(+)-DBD-Pro in 55 mL anhydrous diethyl ether at 0°, add 110 mg phosphorus pentachloride, stir at 5° for 1 h, filter quickly, evaporate to dryness under reduced pressure, dry under vacuum over phosphorus pentoxide for 12 h to give R-(+)-4-(N,N-dimethylaminosulfonyl)-7-(2-chloroformylpyrrolidin-1-yl)-2,1,3-benzoxadiazole (DBD-Pro-Cl) as yellow crystals (mp 116-17°) (Analyst 1993, 118, 759). DBD-Pro-Cl is available from Tokyo Kasei (TCI America, Portland OR). Add 130 mg DBD-Pro-Cl dissolved in 25 mL anhydrous benzene dropwise to 100 mL MeOH containing 70 mg hydrazine hydrate, stir for 30 min at room temperature, evaporate under reduced pressure, recrystallize from ethyl acetate:MeOH 90:10 to give R-(+)-4-(2-carbazolylpyrrolidin-1-yl)-7-(N,N-dimethylaminosulfonyl)-2,1,3-benzoxadiazole (R-(+)-DBD-ProCZ) as orange crystals (mp 107-109°) (Anal. Proc. 1994, 31, 265).)

HPLC VARIABLES

Column: 150 × 4.6 5 μm Inertsil ODS-80A
Mobile phase: MeCN:water 15:85
Column temperature: 40
Flow rate: 1
Detector: F ex 450 em 540

CHROMATOGRAM
Retention time: 11.07 (glucose), 10.41 (galactose), 12.36 (N-acetyl-D-glucosamine), 11.27 (mannose), 14.49 (arabinose), 15.81 (xylose)

REFERENCE
Toyo'oka, T.; Kuze, A. Determination of saccharides labelled with a fluorescent reagent, DBD-ProCZ, by liquid chromatography, *Biomed.Chromatogr.*, **1997**, *11*, 132–136.

4-(2-Carbazolylpyrrolidin-1-yl)-7-nitro-2,1,3-benzoxadiazole

SAMPLE
Matrix: food
Analyte: aldehydes
Sample preparation: 1 g Soybean oil or 250 mg sesame oil + 1 mL reagent, vortex for 1 min, heat at 60° for 10 min, cool on ice, centrifuge at 5° at 3000 g for 10 min, inject a 5 μL aliquot of the supernatant. (Reagent was 0.5 mM NBD-ProCZ in MeOH:water 80:20 containing 0.25% trichloroacetic acid and 150 μg/mL BHT. Prepare NBD-ProCZ as follows. Add a solution of 1 g 4-chloro-7-nitrobenzofurazan in 50 mL EtOH to a solution of 575 mg proline in 100 mL 5% sodium acetate solution, reflux for 10 min, acidify to pH 1.5 with 4 M HCl, evaporate to dryness. Dissolve the residue in water and extract with ethyl acetate. Evaporate the organic layer to dryness and recrystallize from benzene/ethyl acetate (Caution! Benzene is a carcinogen!) to give NBD-Pro as orange needles (mp 156-7°) (Anal.Chim.Acta 1983, 149, 305). Dissolve 2.6 g NBD-Pro in 200 mL anhydrous dichloromethane, add 10 mL oxalyl chloride and 200 μL DMF, stir at room temperature for 1 h, evaporate to dryness under reduced pressure, dissolve the residue in 100 mL anhydrous benzene, filter, evaporate the filtrate to dryness, dry over phosphorus pentoxide under vacuum to give NBD-ProCl as red-orange crystals (mp 103-4° d) (J.Chromatogr.A 1994, 675, 79). NBD-Pro-Cl is also available from Tokyo Kasei (TCI America, Portland OR). Add 50 mg NBD-Pro-Cl dissolved in 15 mL anhydrous benzene dropwise to 100 mL MeOH containing 35 mg hydrazine hydrate (Caution! Hydrazine hydrate is a carcinogen and may explode on distillation!), stir for 30 min at room temperature, evaporate under reduced pressure, suspend the residue with 20 mL water, extract three times with 15 mL portions of ethyl acetate, combine the extracts and evaporate them to dryness under reduced pressure, recrystallize from ethyl acetate:MeOH 50:50 to give NBD-ProCZ (4-(2-carbazolylpyrrolidin-1-yl)-7-nitro-2,1,3-benzoxadiazole as red crystals (mp 90-94° d) (Anal.Proc. 1994, 31, 265).)

HPLC VARIABLES
Column: 150 × 4.6 5 μm Inertsil ODS-80A
Mobile phase: Gradient. MeCN:MeOH:water 15:40:45 for 10 min, to 40:40:20 over 10 min, maintain at 40:40:20 for 10 min, return to initial conditions over 0.1 min, re-equilibrate for 9.9 min.
Column temperature: 40
Flow rate: 1

Injection volume: 5
Detector: F ex 488 (10 mW argon ion laser) em 540 ± 20 (interference filter) or F ex 490 em 540 (conventional F detector)

CHROMATOGRAM
Retention time: 9 (butanal), 15 (pentanal), 17 (4-hydroxy-2-nonenal), 20 (hexanal), 23 (heptanal), 25 (octanal), 27.5 (nonanal)
Limit of detection: 10 fmole (laser F), 2.5 pmole (conventional F)

KEY WORDS
soybean oil; sesame oil; protect from light; laser

REFERENCE
Liu, Y.-M.; Miao, J.-R.; Toyo'oka, T. Determination of 4-hydroxy-2-nonenal by precolumn derivatization and liquid chromatography with laser fluorescence detection, *J.Chromatogr.A*, **1996**, *719*, 450–456.

Dansyl Hydrazine

RELATED REFERENCES
Alpenfels, W.F. A rapid and sensitive method for the determination of monosaccharides as their dansyl hydrazones by high-performance liquid chromatography. *Anal.Biochem.* **1981**, *114*, 153-157.
Hull, S.R.; Turco, S.J. Separation of dansyl hydrazine-derivatized oligosaccharides by liquid chromatography. *Anal.Biochem.* **1985**, *146*, 143-149.
Mainka, A.; Bächmann, K. UV detection of derivatized carbonyl compounds in rain samples in capillary electrophoresis using sample stacking and a Z-shaped flow cell. *J.Chromatogr.A* **1997**, *767*, 241-247.
Shinomiya, K.; Yamanashi, S.; Imanari, T. Fluorometric analysis of urinary chondroitin sulfate isomers by HPLC using dansylhydrazine as a prelabeling reagent. *Biomed.Chromatogr.* **1987**, *2*, 169-172.

SAMPLE
Matrix: air
Analyte: aldehydes
Sample preparation: Slurry Extrelut silica gel with MeCN, let stand for 1 h, filter, dry under vacuum, slurry with dichloromethane, let stand for 1 h, filter, dry under vacuum. Purify dansylhydrazine by injecting 500 μL aliquots of a 20 mg/mL solution in MeCN onto a 500 × 10 10 μm RSI column (Alltech) and eluting with MeCN:water from 20:80 to 50:50 over 5 min, to 58:42 over 5 min, to 80:20 (step gradient), maintain at 80:20 for 15 min, return to initial conditions over 5 min. Elute at 4 mL/min for 10 min, 6 mL/min for 15 min, then 4 mL/min for 5 min. Collect the fraction between 10 and 13.5 min, remove the MeCN at 30° under vacuum, pour onto silica gel, extract with dichloromethane. Adjust the dansyl hydrazine concentration to 1 mg/mL and add a 1.5 mL aliquot to 700 mg silica gel, dry under vacuum, pack in a 4 mm ID quartz tube. Pull air through a tube at 2 L/min. Add 1.4 mL rain to a tube, let stand for at least 30 min. Extract tubes with 4 mL dichloromethane, evaporate the eluate to dryness under a stream of nitrogen at 35°, reconstitute the residue in 200 μL MeCN, inject a 10 μL aliquot.

HPLC VARIABLES
Column: 250 × 4 5 μm Superspher RP 18 (Merck)
Mobile phase: Gradient. MeCN:water from 20:80 to 50:50 over 5 min, to 65:35 over 15 min, to 80:20 over 13 min, to 75:25 over 7 min, return to initial conditions over 10 min.

Flow rate: 1
Injection volume: 10
Detector: F ex 355 em 525

CHROMATOGRAM
Retention time: 10 (acetaldehyde), 12 (acetone), 13 (propanal), 15 (crotonaldehyde), 16 (n-butanal), 18 (cyclohexanone), 20 (n-pentanal), 20.5 (isopentanal)
Limit of detection: 10-150 pg

KEY WORDS
SPE

REFERENCE
Schmied, W.; Przewosnik, M.; Bachmann, K. Determination of traces of aldehydes and ketones in the troposphere via solid phase derivatization with DNSH, *Fresenius' Z.Anal.Chem.*, **1989**, *335*, 464–468.

SAMPLE
Matrix: bulk
Analyte: hexanal and hydroxybutanone
Sample preparation: Dissolve in 1 mL MeOH containing a 2-fold molar excess of dansyl hydrazine, add 2 drops glacial acetic acid, heat at 70° for 15 min, evaporate to dryness under reduced pressure, reconstitute with 500 μL benzene (Caution! Benzene is a carcinogen!), inject a 1-10 μL aliquot.

HPLC VARIABLES
Column: 1000 × 2.4 Zipax coated with 0.5% β,β'-oxydipropionitrile
Mobile phase: Diisopropyl ether (Caution! Diisopropyl ether readily forms explosive peroxides!)
Flow rate: 0.91
Injection volume: 1-10
Detector: F ex Turner filter no. 811 em Turner filter no. 817

CHROMATOGRAM
Retention time: 2 (hexanal), 2.5 (3-hydroxy-2-butanone)

KEY WORDS
normal phase

REFERENCE
Frei, R.W.; Lawrence, J.F. Fluorigenic labelling in high-speed liquid chromatography, *J.Chromatogr.*, **1973**, *83*, 321–330.

SAMPLE
Matrix: dihydroxyacetone matrix
Analyte: formaldehyde
Sample preparation: 500 mg Dihydroxyacetone matrix + 4 mL reagent, mix, let stand for 25 min, inject an aliquot. (Prepare reagent by adjusting the pH of 6 mg/mL dansyl-hydrazine in MeOH:water 80:20 to 3.0 with 50 mM sulfuric acid.)

CAPILLARY ELECTROPHORESIS
Capillary: 57 cm × 50 μm (50 cm to detector)
Running buffer: MeOH:buffer 20:80 (Prepare buffer by adding 75 mM NaH_2PO_4 to 75 mM Na_2HPO_4 until a pH of 6.5 is achieved.)
Capillary temperature: 24
Voltage/Current: 25 kV/30 μA
Injection: Hydrodynamic injection for 15 s
Detector: UV 214
Model: Beckman P/ACE System 2100
Migration time: 8.15
Limit of detection: 200 ppb

OTHER SUBSTANCES
Non-interfering: dihydroxyacetone

REFERENCE

Feige, K.; Ried, T.; Bächmann, K. Determination of formaldehyde by capillary electrophoresis in the presence of a dihydroxyacetone matrix, *J.Chromatogr.A*, **1996**, *730*, 333–336.

SAMPLE

Matrix: solutions
Analyte: sugars
Sample preparation: Condition a Sep-Pak C18 SPE cartridge with 3 mL MeCN and 5 mL water. Mix 250 μL of an aqueous solution with 225 μL 1% dansylhydrazine in EtOH and 45 μL 10% trichloroacetic acid in water, heat at 65° for 20 min, dilute with water to an organic solvent concentration of ≤5%, add a 5 mL aliquot to the SPE cartridge, wash with 5 mL MeCN:water 5:95 at ≤2 mL/min, elute with 6 mL MeCN:water 20:80 at ≤2 mL/min (J. Chromatogr. 1983, 256, 27), lyophilize the eluate, reconstitute with MeCN: water 20:80, inject a 20 μL aliquot.

HPLC VARIABLES

Column: 250 × 4.6 10 μm 600 RPB C18 (Alltech)
Mobile phase: MeCN:water 20:80 containing 10 mM formic acid, 40 mM acetic acid, and 1 mM triethylamine. (After each run flush column with MeCN:MeOH 20:80 for 5 min.)
Flow rate: 1
Injection volume: 20
Detector: UV 254

CHROMATOGRAM

Retention time: 8 (galactose), 9 (dextrose), 10 (mannose), 12 (xylose), 13 (lyxose), 16 (fucose)
Limit of detection: 200-300 pmole

KEY WORDS

SPE

REFERENCE

Eggert, F.M.; Jones, M. Measurement of neutral sugars in glycoproteins as dansyl derivatives by automated high-performance liquid chromatography, *J.Chromatogr.*, **1985**, *333*, 123–131.

SAMPLE

Matrix: tears
Analyte: dextrose
Sample preparation: Evaporate 3-5 μL tear fluid to dryness under a stream of nitrogen, reconstitute with 18 μL 25 mM sodium tetraborate solution, add 2 μL 13.4 mM trifluoroacetic acid in water, add 50 μL 2-10 mM dansylhydrazine in EtOH, add 30 μL EtOH, heat at 68° for 15-18 min, store at -20°.

CAPILLARY ELECTROPHORESIS

Capillary: 90 cm × 50 μm fused-silica (60 cm to detector) (Polymicro Technologies)
Capillary preparation: Before each injection flush capillary with MeOH for 30 s, with 100 mM NaOH for 30 s, and with running buffer for 30 s. Store overnight in water or running buffer. Flush new capillaries with 100 mM NaOH, water, and running buffer.
Running buffer: 25 mM pH 9.2 sodium tetraborate
Voltage/Current: 22 kV/23 μA
Injection: Hydrodynamic injection at 19 cm for 10 s (2 nL)
Detector: F ex 325 (He-Cd laser) em 500 (long-pass filter)
Migration time: 13.75 (dextrose), 13.9 (fructose), 14.2 (fucose), 14.75 (galactose), 12.6 (lactose), 12.4 (maltose), 13.4 (mannose)
Limit of detection: 50 nM (S/N 3)

REFERENCE

Perez, S.A.; Colón, L.A. Determination of carbohydrates as their dansylhydrazine derivatives by capillary electrophoresis with laser-induced fluorescence detection, *Electrophoresis*, **1996**, *17*, 352–358.

SAMPLE

Matrix: urine
Analyte: sugars

Sample preparation: 100 μL Urine + 10 μL 10% trichloroacetic acid in water + 50 μL 5% dansyl hydrazine in MeCN, heat at 65° for 20 min, cool in ice, add an equal volume of water, inject a 10-300 μL aliquot.

HPLC VARIABLES
Column: 250 × 4.6 5 μm Nucleosil ODS
Mobile phase: MeCN:80 mM acetic acid 21:79
Flow rate: 1
Injection volume: 10
Detector: F ex 360 em >470

CHROMATOGRAM
Retention time: 6 (gentobiose), 7.5 (lactose), 8 (maltose), 8 (cellobiose), 8.5 (galactose), 9 (dextrose), 10 (mannose), 11.5 (xylose), 11.5 (arabinose), 11.5 (fructose), 12.5 (2-deoxyglucose), 14 (ribose), 15 (fucose), 16 (rhamnose), 17 (2-deoxyribose), 19 (fructose)
Limit of detection: 5-15 pmole

REFERENCE
Mopper, K.; Johnson, L. Reversed-phase liquid chromatographic analysis of Dns-sugars. Optimization of derivatization and chromatographic procedures and applications to natural samples, *J.Chromatogr.*, **1983**, *256*, 27–38.

2,5-Dihydroxybenzohydrazide

SAMPLE
Matrix: solutions
Analyte: aldehydes
Sample preparation: Dissolve 80 mg 2,5-dihydroxybenzohydrazide and 90 μL cyclohexanecarboxaldehyde in 5 mL MeOH and 1 mL acetic acid, stir at room temperature for 1 h, add water, collect the precipitate, recrystallize from MeOH/water. Inject a 5 μL aliquot of a solution. (This procedure is for preparing an analytical sample but it is the only procedure provided. Preparation of 2,5-dihydroxybenzohydrazide is as follows. Prepare methyl 2,5-dihydroxybenzoate from 2,5-dihydroxybenzoic acid (gentisic acid) in a conventional fashion. Heat 35 g methyl 2,5-dihydroxybenzoate and 18 mL 85% hydrazine hydrate on a steam bath (Caution! Hydrazine hydrate is a carcinogen and explodes on distillation in air!), after the mixture has liquified and solidified recrystallize the solid from water to obtain 2,5-dihydroxybenzohydrazide as colorless needles (mp 209-210°) (J. Org. Chem. 1952, 17, 1653).)

HPLC VARIABLES
Guard column: 4 × 4 5 μm Hypersyl ODS RP-18
Column: 100 × 4.6 3 μm Adsorbosphere
Mobile phase: MeCN:MeOH:50 mM phosphate buffer 39:22:39, adjusted to pH 7.0 with KOH
Flow rate: 1
Injection volume: 5
Detector: E, ESA Model 5100A Coulochem, Model 5010 analytical cell, porous graphite electrodes +0.3 V, oxidative mode

CHROMATOGRAM
Retention time: 2.54 (cyclohexanecarboxaldehyde), 4.23 (octanal), 9.01 (decanal)
Limit of quantitation: 1 ng/mL
Limit of detection: 100-200 fmole

REFERENCE
Bousquet, E.; Tirendi, S.; Prezzavento, O.; Tateo, F. 2,5-Dihydroxybenzohydrazide as electroactive labeling reagent for aliphatic aldehydes by high performance liquid chromatography with electrochemical and ultraviolet detection, *J.Liq.Chromatogr.*, **1995**, *18*, 1933–1945.

4-(N,N-Dimethylaminosulfonyl)-7-hydrazino-2,1,3-benzoxadiazole

SAMPLE
Matrix: blood
Analyte: acetaldehyde
Sample preparation: 50 μL Plasma + 10 μL 112.4 μM propionaldehyde in water + 40 μL 10 mM 4-(N,N-dimethylaminosulfonyl)-7-hydrazino-2,1,3-benzoxadiazole in MeCN + 20 μL 1% trifluoroacetic acid in MeCN, vortex for 30 s, let stand at room temperature for 30 min, centrifuge at 4° at 400 g for 10 min, filter (0.45 μm) the upper layer, inject a 20 μL aliquot of the filtrate. (Prepare IS solution by diluting a 56.2 mM solution of propionaldehyde in MeCN 500-fold with water. 4-(N,N-Dimethylaminosulfonyl)-7-hydrazino-2,1,3-benzoxadiazole can be purchased from Tokyo Kasei (TCI America, Portland OR). Synthesis is as follows. Dissolve 0.5 g magnesium sulfate heptahydrate and 6 g NaOH in 60 mL water, throughout the reaction keep the flask at about 20° with cold water cooling, add 15 mL 30% hydrogen peroxide, add 75 mL MeOH, add 12.1 g powdered benzoyl peroxide in one go, stir for 10 min, pour into 150 mL 20% sulfuric acid, extract three times with 50 mL portions of chloroform, determine peroxybenzoic acid concentration by iodometric titration (Tetrahedron 1967, 23, 3327). Slowly add 110 mL 1 M peroxybenzoic acid in chloroform to 7 g 2,6-difluoroaniline dissolved in 100 mL chloroform, stir at room temperature, when reaction is complete (iodometric titration) wash with 2% sodium thiosulfate, wash with 5% sodium carbonate, wash with water, dry over anhydrous sodium sulfate, evaporate to dryness under reduced pressure, recrystallize 2,6-difluoronitrosobenzene from EtOH (mp 108.5-109.5). Stir 8.5 g 2,6-difluoronitrosobenzene in 85 mL DMSO at room temperature and add a solution of 3.91 g sodium azide in 85 mL DMSO dropwise, let stand for about 1 h, add to a large volume of water, extract with ether, dry the extracts over anhydrous sodium sulfate, evaporate to dryness under reduced pressure and distil to give 4-fluoro-2,1,3-benzoxadiazole as a colorless oil (bp 83°/12 mm Hg) (J.Chem.Soc.(C) 1970, 1433). Add 11 mL chlorosulfonic acid dropwise to 3 g 4-fluoro-2,1,3-benzoxadiazole in 10 mL chloroform at 0-10° (use a calcium chloride drying tube), stir at room temperature for 1 h, reflux for 2 h, cool, slowly pour into ice water, remove the organic layer, extract the aqueous layer with chloroform, combine the organic layers, wash, dry over anhydrous magnesium sulfate, evaporate under reduced pressure, take

up the residue in 5 mL benzene (Caution! Benzene is a carcinogen!), chromatograph on a 150 × 30 column of silica gel (100-200 mesh Kanto Chemical) with n-hexane:benzene 50:50, evaporate the appropriate fractions to give 4-(chlorosulfonyl)-7-fluoro-2,1,3-benzoxadiazole (CBD-F) as pale yellow needles (mp 64-66°) (Anal. Chem. 1984. 56, 2461). Stir 0.76 g CBD-F in 70 mL MeCN at 0-10° and add 1 g dimethylamine hydrochloride in 10 mL 100 mM pH 10 borax dropwise, adjust pH to 5 with 1 M HCl, concentrate to about 10 mL under reduced pressure, extract three times with 200 mL portions of diethyl ether, wash with water, dry over anhydrous magnesium sulfate, evaporate under reduced pressure, chromatograph on a 500 × 20 column of silica gel with chloroform, isolate the appropriate fraction and re-chromatograph on the same column with ethyl acetate:benzene 1:2 to give 4-(N,N-dimethylaminosulfonyl)-7-fluoro-2,1,3-benzoxadiazole (DBD-F) as white needles (mp 124-125°) (yield = 1% !) (Analyst 1989, 114, 413). On a Merck no. 5714 60F$_{254}$ tlc plate eluted with chloroform DBD-F has Rf 0.32 and lies between two other reaction products. DBD-F can also be purchased from Tokyo Kasei. Dissolve 80 mg 4-(N,N-dimethylaminosulfonyl)-7-fluoro-2,1,3-benzoxadiazole in 12 mL MeCN, add 40 μL 98% hydrazine hydrate, heat in the dark at 50-55° for 20 min, evaporate to dryness, recrystallize from MeOH to give 4-(N,N-dimethylaminosulfonyl)-7-hydrazino-2,1,3-benzoxadiazole as reddish-brown crystals (mp 138-139° d) (Analyst 1990, 115, 1477).)

HPLC VARIABLES
Column: 250 × 4.6 5 μm SP-120-ODS (Daiso, Osaka)
Mobile phase: MeCN:water 46:54
Flow rate: 1
Injection volume: 20
Detector: F ex 445 em 560

CHROMATOGRAM
Retention time: 11
Internal standard: propionaldehyde (18)
Limit of detection: 300 nM

OTHER SUBSTANCES
Extracted: formaldehyde
Simultaneously analyzed: butyraldehyde, cinnamaldehyde, p-dimethylaminobenzaldehyde, furfural, glutaraldehyde, p-hydroxybenzaldehyde, malondialdehyde, terephthalaldehyde

KEY WORDS
plasma

REFERENCE
Nakashima, K.; Hidaka, Y.; Yoshida, T.; Kuroda, N.; Akiyama, S. High-performance liquid chromatographic determination of short-chain aliphatic aldehydes using 4-(N,N-dimethylaminosulphonyl)-7-hydrazino-2,1,3-benzoxadiazole as a fluorescence reagent, *J.Chromatogr.B*, **1994**, *661*, 205–210.

SAMPLE
Matrix: bulk
Analyte: oxidized choline glycerophospholipids
Sample preparation: Dissolve 5 pmole oxidized choline glycerophospholipids in 75 μL MeCN, add 75 μL 0.1% trifluoroacetic acid, add 150 μL 500 μM 4-(N,N-dimethylaminosulfonyl)-7-hydrazino-2,1,3-benzoxadiazole in MeCN, mix, let stand in the dark at room temperature for 1 h, evaporate to dryness under a stream of nitrogen, purify by TLC using chloroform:MeOH:water 10:5:1. (4-(N,N-Dimethylaminosulfonyl)-7-hydrazino-2,1,3-benzoxadiazole can be purchased from Tokyo Kasei (TCI America, Portland OR). Synthesis is as described above.)

HPLC VARIABLES
Column: LiChrosorb RP-18
Mobile phase: MeCN:MeOH:water 2.5:90.5:7 containing 20 mM choline chloride

Flow rate: 1
Detector: F ex 450 em 550

CHROMATOGRAM
Retention time: 6 (1-palmitoyl-2-(5-oxopentanoyl)-sn-glycero-3-phosphocholine), 9 (1-palmitoyl-2-(9-oxononanoyl)-sn-glycero-3-phosphocholine)

REFERENCE
Ou, Z.; Ogamo, A.; Guo, L.; Konda, Y.; Harigaya, Y.; Nakagawa, Y. Identification and quantitation of choline glycerophospholipids that contain aldehyde residues by fluorometric high-performance liquid chromatography, *Anal.Biochem.*, **1995**, *227*, 289–294.

SAMPLE
Matrix: solutions
Analyte: propanal
Sample preparation: Mix 10 μL of a 2-3 μM solution in MeCN with 20 μL 500 μM 4-(N,N-dimethylaminosulfonyl)-7-hydrazino-2,1,3-benzoxadiazole in MeCN and 10 μL 0.1% trifluoroacetic acid in MeCN, let stand at room temperature for 1 h, inject an aliquot. (4-(N,N-Dimethylaminosulfonyl)-7-hydrazino-2,1,3-benzoxadiazole is available from Tokyo Kasei (TCI America, Portland OR). Synthesis is as described above.)

HPLC VARIABLES
Column: 150 × 4.6 5 μm TSK-LS 80Tm (Tosoh)
Mobile phase: MeCN:water:trifluoroacetic acid 46:54:0.05
Flow rate: 1
Detector: F ex 450 em 545

CHROMATOGRAM
Retention time: 6.5
Limit of detection: 7.5 μM

OTHER SUBSTANCES
Also analyzed: acetone, butyraldehyde, 4'-ethylacetophenone, heptan-4-one, p-hydroxybenzaldehyde

REFERENCE
Uzu, S.; Kanda, S.; Imai, K.; Nakashima, K.; Akiyama, S. Fluorogenic reagents: 4-Aminosulphonyl-7-hydrazino-2,1,3-benzoxadiazole, 4-(*N*,*N*-dimethylaminosulphonyl)-7-hydrazino-2,1,3-benzoxadiazole and 4-hydrazino-7-nitro-benzoxadiazole hydrazine for aldehydes and ketones, *Analyst*, **1990**, *115*, 1477–1482.

SAMPLE
Matrix: solutions
Analyte: aldehydes
Sample preparation: Mix 20 μL of a solution of the aldehyde in MeCN with 40 μL 10 mM 4-(N,N-dimethylaminosulfonyl)-7-hydrazino-2,1,3-benzoxadiazole in MeCN, add 20 μL 0.5% trifluoroacetic acid in MeCN, mix, let stand at room temperature for 30 min, inject a 20 μL aliquot. (4-(N,N-Dimethylaminosulfonyl)-7-hydrazino-2,1,3-benzoxadiazole can be purchased from Tokyo Kasei, TCI America, Portland OR). Synthesis is as described above.)

HPLC VARIABLES
Column: 150 × 4.6 5 μm SP-120-5-ODS (Daiso, Osaka)
Mobile phase: MeCN:5 mM pH 7.5 imidazole buffer 80:20 (A) or MeCN containing 5 mM imidazole (B)
Flow rate: 1
Injection volume: 20
Detector: Chemiluminescence following post-column reaction. The column effluent mixed with the reagent pumped at 1 mL/min and the mixture flowed to the detector. (Reagent was 100 mM hydrogen peroxide in MeCN containing 0.6 mM bis(2,4,6-trichlorophenyl) oxalate.); F ex 450 em 559

CHROMATOGRAM
Retention time: 7 (hexanal (A)), 8.5 (heptanal (A)), 11 (octanal (A)), 14 (nonanal (A)), 25.5 (undecanal (A)), 6.5 (undecanal (B)), 8 (cis-7-tetradecenal (B)), 10 (cis-9-hexadecenal (B)), 14.5 (cis-13-octadecenal (B))
Limit of detection: 14-18 fmole

KEY WORDS
post-column reaction

REFERENCE
Nakashima, K.; Yoshida, T.; Kuroda, N.; Akiyama, S. High-performance liquid chromatography of long chain aliphatic aldehydes with peroxyoxalate chemiluminescence detection utilizing a fluorogenic reagent, 4-(N,N-dimethylaminosulphonyl)-7-hydrazino-2,1,3-benzoxadiazole, *Biomed.Chromatogr.*, **1996**, *10*, 99–101.

2,4-Dinitrophenylhydrazine

RELATED REFERENCES
Barnes, A.R. Determination of glutaraldehyde in solution as its bis-2,4-dinitrophenylhydrazone derivative; determination of geometrical isomer ratios. *Pharm.Acta Helv.* **1993**, *68*, 113-119.

Barry, J.L.; Tome, D. Formaldehyde content of milk in goats fed formaldehyde-treated soybean oil-meal. *Food Addit.Contam.* **1991**, *8*, 633-640.

Beacham, D.B. Analyses of aldehydes and ketones using the DNPH derivatization process. *Proc.,Annu.Meet.- Air Waste Manage.Assoc.* **1994**, *87th*, 17pp 94-TA2CO.11P.

Chiavari, G.; Bergamini, C. High-performance liquid chromatography of carbonyl compounds as 2,4-dinitrophenylhydrazones with electrochemical detection. *J.Chromatogr.* **1985**, *318*, 427-432.

Coutrim, M.X.; Nakamura, L.A.; Collins, C.H. Quantification of 2,4-dinitrophenylhydrazones of low molecular mass aldehydes and ketones using HPLC. *Chromatographia* **1993**, *37*, 185-190.

Dahlgran, J.R.; Jameson, M.N. Determination of formaldehyde and other aldehydes in industrial surfactants by liquid chromatographic separation of their respective 2,4-dinitrophenylhydrazine derivatives. *J.Assoc.Off.Anal.Chem.* **1988**, *71*, 560-563.

Foster, P.; Ferrari, C.; Jacob, V.; Roche, A.; Baussand, P.; Delachaume, J.C. Determination of C1-C5 carbonyls in the atmosphere by 2,4-dinitrophenylhydrazine-coated sorbent sampling and HPLC analysis. The influence of water on ketone analysis. *Analusis* **1996**, *24*, 71-73.

Jacobs, W.A.; Kissinger, P.T. Determination of carbonyl 2,4-dinitrophenylhydrazones by liquid chromatography/electrochemistry. *J.Liq.Chromatogr.* **1982**, *5*, 669-676.

Karamanos, N.K.; Tsegenidis, T.; Antonopoulos, C.A. Analysis of neutral sugars as dinitrophenyl-hydrazones by high-performance liquid chromatography. *J.Chromatogr.* **1987**, *405*, 221-228.

Lange, J.; Eckhoff, S. Determination of carbonyl compounds in exhaust gas by using a modified DNPH-method. *Fresenius' J.Anal.Chem* **1996**, *356*, 385-389.

Levin, J-O.; Lindahl, R.; Heeremans, C.E.M.; van Oosten, K. Certification of reference materials related to the monitoring of aldehydes in air by derivatization with 2,4-dinitrophenylhydrazine. *Analyst* **1996**, *121*, 1273-1278.

Liebezeit, G. HPLC gradient elution of dinitrophenylhydrazones of aldehydes and ketones from aqueous samples. *HRC CC,J.High Resolut.Chromatogr.Chromatogr.Commun.* **1982**, *5*, 215-216.

Lo Coco, F.; Valentini, C.; Novelli, V.; Ceccon, L. High-performance liquid chromatographic determination of 2-furaldehyde and 5-hydroxymethyl-2-furaldehyde in honey. *J.Chromatogr.A* **1996**, *749*, 95-102.

Menet, M.-C.; Gueylard, D.; Fiever, M.-H.; Thuillier, A. Fast specific separation and sensitive quantification of bactericidal and sporicidal aldehydes by high-performance liquid chromatography: example of glutaraldehyde determination. *J.Chromatogr.B* **1997**, *692*, 79-86.

Possanzini, M.; Di Palo, V. Short-term measurements of acrolein in ambient air. *Chromatographia* **1996**, *43*, 433-435.

Reindl, B.; Stan, H.J. Determination of volatile aldehydes in meat as 2,4-dinitrophenylhydrazones using reversed-phase high-performance liquid chromatography. *J.Agric.Food Chem.* **1982**, *30*, 849-854.

Schmidt, R.H.; Davidson, S.M.; Bates, R.P. Acetaldehyde determination in fermented food products by direct 2,4-dinitrophenylhydrazine derivatization, extraction and high-performance liquid chromatography. *J.Food Sci.* **1983**, *48*, 1556-1557.

Smith, D.F.; Kleindienst, T.E.; Hudgens, E.E. Improved high-performance liquid chromatographic method for artifact-free measurements of aldehydes in the presence of ozone using 2,4-dinitrophenylhydrazine. *J.Chromatogr.* **1989**, *483*, 431-436.

Tuss, H.; Neitzert, V.; Seiler, W.; Neeb, R. Method for determination of formaldehyde in air in the pptv-range by HPLC after extraction as 2,4-dinitrophenylhydrazone. *Fresenius' Z.Anal.Chem.* **1982**, *312*, 613-617.

Viras, L.G.; Kotzias, D.; Duane, M. Application of the 2,4-dinitrophenylhydrazine method for measuring carbonyl compounds in a semi-remote and in an urban area. *Fresenius Environ.Bull.* **1992**, *1*, S73-S78,

SAMPLE

Matrix: air

Analyte: aldehydes

Sample preparation: Pull 5-30 L of air through two 10 mL aliquots of reagent in series at 0.5-1.5 L/min. Combine the samples and extract them twice with 5 mL portions of chloroform. Combine the chloroform layers and wash them with 20 mL 2 M HCl, wash with 20 mL water, evaporate to dryness under reduced pressure with gentle heating, reconstitute with 2 mL MeCN, inject a 4 μL aliquot. (Prepare reagent by dissolving 500 mg 2,4-dinitrophenylhydrazine (recrystallized from EtOH) in 500 mL 2 M HCl and washing twice with 5 mL portions of chloroform.)

HPLC VARIABLES

Column: 200 × 4.6 5 μm LiChrosorb RP-18

Mobile phase: MeCN:water 62:38

Flow rate: 1.5

Injection volume: 4

Detector: UV 254

CHROMATOGRAM

Retention time: 4.03 (formaldehyde), 5.12 (acetaldehyde), 6.69 (acrolein), 6.85 (acetone), 7.42 (propionaldehyde), 9.70 (crotonaldehyde), 10.40 (methyl ethyl ketone), 10.60 (isobutyraldehyde, n-butyraldehyde), 12.41 (benzaldehyde), 14.83 (methyl isopropyl ketone, methyl n-propyl ketone, diethyl ketone), 14.83 (isovaleraldehyde), 15.80 (n-valeraldehyde), 17.53 (o-tolualdehyde), 18.26 (m-tolualdehyde), 18.94 (p-tolualdehyde), 20.55 (methyl isobutyl ketone, methyl sec-butyl ketone), 21.20 (methyl t-butyl ketone), 22.57 (methyl n-butyl ketone), 23.94 (n-caproaldehyde), 34.58 (methyl n-amyl ketone)

Limit of detection: 1.5-2.6 ppb

REFERENCE

Kuwata, K.; Uebori, M.; Yamasaki, Y. Determination of aliphatic and aromatic aldehydes in polluted airs as their 2,4-dinitrophenylhydrazones by high performance liquid chromatography, *J.Chromatogr.Sci.*, **1979**, *17*, 264–268.

SAMPLE

Matrix: air

Analyte: aldehydes

Sample preparation: Condition a Sep-Pak C18 SPE cartridge with 2 mL MeCN and 2 mL 2 mg/mL 2,4-dinitrophenylhydrazine in MeCN:85% orthophosphoric acid 100:1, dry under nitrogen at 100 mL/min for 30 min, pull air through the cartridge at 0.5-1.5 mL/min for 24 h, elute with 1 mL MeCN, elute with 1 mL water, filter (Zetapor) the eluate, inject a 20 μL aliquot of the filtrate. (Purify nitrogen by passing it over activated charcoal.)

HPLC VARIABLES

Column: two 100 × 4.6 3 μm Microsphere C18 columns in series (Chrompack)

Mobile phase: Gradient. MeCN:MeOH:water 7:60:33 for 6.4 min, to 65:7:28 over 13.6 min, to 100:0:0 over 10 min.
Flow rate: 1
Injection volume: 20
Detector: UV 360

CHROMATOGRAM

Retention time: 4 (formaldehyde), 5.5 (acetaldehyde), 7.5 (acrolein), 7.5 (acetone), 8.5 (propanal), 11 (crotonaldehyde), 12.5 (2-butanone, butanal), 14.5 (benzaldehyde), 15.5 (trans-2-pentenal), 16.9 (cyclohexanone), 17.2 (valeraldehyde), 20.5 (trans-2-hexenal), 21.5 (2-hexanone), 22 (hexanal)

KEY WORDS

SPE

REFERENCE

Dye, C.; Oehme, M. Comments concerning the HPLC separation of acrolein from other C_3 carbonyl compounds as 2,4-dinitrophenylhydrazones; a proposal for improvement, *J.High Res.Chromatogr.*, **1992**, *15*, 5–8.

SAMPLE

Matrix: air
Analyte: formaldehyde
Sample preparation: Wash 15 cm dia filters (Whatman No. 41) with hot (70-80°) water for 30 min, wash with 1.5 L water, dry at 40-50° for 1-1.5 h, soak in 10% sodium bisulfite solution, dry at 60-70° for 1-1.5 h, store in a desiccator. Pull air through two filters in a 47 mm dia holder at 83 mL/min for 30 min, sonicate filters with 4 mL 1 M HCl for 5 min. Remove a 1 mL aliquot and heat it at 50-60° for 5 min, adjust pH to 11.5-12.5 with 1 M NaOH, add this solution dropwise to 3 mL 2,4-dinitrophenylhydrazine solution, make up to 10 mL inject an aliquot. (Recrystallize 2,4-dinitrophenylhydrazine twice from EtOH before use. Prepare a solution by dissolving 102.8 mg 2,4-dinitrophenylhydrazine in 500 mL MeCN containing 1 mL concentrated sulfuric acid.)

HPLC VARIABLES

Column: 250 × 4.6 5 µm C18 (Alltech)
Mobile phase: MeCN:water 60:40
Flow rate: 1.5
Injection volume: 10
Detector: UV 360

CHROMATOGRAM

Limit of detection: 0.2 ppb

REFERENCE

De Andrade, J.B.; Tanner, R.L. Determination of formaldehyde by HPLC as the DNPH derivative following high-volume air sampling onto bisulfite-coated cellulose filters, *Atmos.Environ.Part A Gen.Top.*, **1992**, *26*, 819–825.

SAMPLE

Matrix: air
Analyte: aldehydes
Sample preparation: Pull air at 2.3 L/min through 6 mL reagent solution passing through a helical-coil scrubber at 0.3 mL/min (design details in paper) for about 20 min, inject a 1.23 mL aliquot. (Prepare stock reagent solution by dissolving 1.5 g 2,4-dinitrophenylhydrazine (recrystallized twice from MeCN) in 300 mL concentrated HCl:water:MeCN 10:40:50. Prepare the working reagent solution by diluting 1.6 mL of the stock solution with 450 mL water, extract overnight with three 10 mL portions of carbon tetrachloride to remove traces of hydrazones.)

HPLC VARIABLES

Column: 250 mm long 5 µm Hypersil ODS
Mobile phase: Gradient. MeCN:MeOH:water (details not given)
Flow rate: 1.6

Injection volume: 1230
Detector: UV 360 for 20 min then UV 400

CHROMATOGRAM
Retention time: 8.2 (glycolaldehyde), 12.5 (formaldehyde), 24.0 (glyoxal), 25.0 (methyl-glyoxal)
Limit of detection: 5-20 ppb (in gas phase)

REFERENCE
Lee, Y.-N.; Zhou, X. Methods for the determination of some soluble atmospheric carbonyl compounds, *Environ.Sci.Technol.*, **1993**, *27*, 749–756.

SAMPLE
Matrix: air
Analyte: aldehydes
Sample preparation: Pull air through impregnated silica at 200 mL/min for 10 min, elute with three 1 mL portions of MeOH, inject a 10 μL aliquot of the eluate. (Prepare impregnated silica by mixing 50 g Chromosorb P, 50 mL 0.5% 2,4-dinitrophenylhydrazine in MeCN, and 300 μL concentrated phosphoric acid.)

HPLC VARIABLES
Column: 100×3 5 μm SCG C18 (Tessek)
Mobile phase: MeOH:water 70:30
Column temperature: 25
Flow rate: 0.5
Injection volume: 10
Detector: UV 355

CHROMATOGRAM
Retention time: 5 (formaldehyde), 6.3 (acetaldehyde), 8 (propanal), 11 (butanal)
Limit of detection: <1 mg/cu m

REFERENCE
Lehotay, J.; Halmo, F. Determination of aliphatic aldehydes C_1-C_4 in waste gas by HPLC, *J.Liq.Chromatogr.*, **1994**, *17*, 847–854.

SAMPLE
Matrix: air
Analyte: acetaldehyde
Sample preparation: Pull air through a coated filter at 200 mL/min, shake the filter with 3-10 mL MeCN for 1 min, inject a 10 μL aliquot of the solution. (Coat the filter as follows. Mix 300 mg recrystallized 2,4-dinitrophenylhydrazine, 500 μL concentrated phosphoric acid, 1.5 mL glycerol:EtOH 20:80, and 9 mL MeCN, dip a 2 cm × 2 cm section of 0.3 μm pore size glass-fiber filter (SKC Type AE) in the solution, allow to dry on a glass surface.)

HPLC VARIABLES
Column: 150×4.6 Spherisorb ODS2
Mobile phase: MeOH:water 65:35
Flow rate: 1
Injection volume: 10
Detector: UV 365

CHROMATOGRAM
Retention time: 6.8, 7.5 (cis/trans isomers)
Limit of detection: 0.4 mg/m^3

REFERENCE
Lindahl, R.; Levin, J.-O.; Mårtensson, M. Validation of a diffusive sampler for the determination of acetaldehyde in air, *Analyst*, **1996**, *121*, 1177–1181.

SAMPLE
Matrix: automobile exhaust
Analyte: aldehydes

Sample preparation: Immediately before use add 2 drops 1 M perchloric acid to 10 mL 1.6-3.1 mM 2,4-dinitrophenylhydrazine (recrystallized from MeOH) in MeCN, pull 20 L air through the solution at 0.5-1.5 L/min, inject an aliquot.

HPLC VARIABLES
Column: 250 × 4.6 Zorbax ODS
Mobile phase: Gradient. MeCN:water 67:33 for 8 min, to 90:10 over 17 min, to 100:0 over 3 min.
Flow rate: 0.7 for 8 min then 1.0
Injection volume: 30
Detector: UV 365

CHROMATOGRAM
Retention time: 6.2 (formaldehyde), 7.5 (acetaldehyde), 8.3 (furfural), 8.6 (acrolein), 9 (acetone), 9.3 (propionaldehyde), 9.6 (salicaldehyde), 10.5 (crotonaldehyde), 11.5 (butyraldehyde), 12 (glyoxal), 12.5 (benzaldehyde), 14 (glutaraldehyde), 14.5 (valeraldehyde), 16 (p-tolualdehyde), 18 (hexanaldehyde), 21.5 (3-heptanone), 22 (heptanaldehyde), 25.5 (octanaldehyde), 28.5 (nonanaldehyde)
Limit of detection: 4-20 ppb

REFERENCE
Lipari, F.; Swarin, S.J. Determination of formaldehyde and other aldehydes in automobile exhaust with an improved 2,4-dinitrophenylhydrazine method, *J.Chromatogr.*, **1982**, *247*, 297–306.

SAMPLE
Matrix: beverages
Analyte: 2-furaldehyde
Sample preparation: Mix 2.5 mL brandy or whisky with 2.5 mL EtOH:water 40:60, add 4 mL 2.5 mM 2,4-dinitrophenylhydrazine in MeCN, adjust pH to 2 with a few drops of 70% perchloric acid, make up to 10 mL with 2.5 mM 2,4-dinitrophenylhydrazine in MeCN, stir for at least 25 min, inject a 10 μL aliquot. (Recrystallize 2,4-dinitrophenylhydrazine from MeOH.)

HPLC VARIABLES
Column: 250 × 4.6 5 μm Supelcosil LC-18
Mobile phase: MeCN:water 65:35
Flow rate: 1
Injection volume: 10
Detector: UV 388

CHROMATOGRAM
Retention time: 7
Limit of detection: 13 nM

KEY WORDS
brandy; whisky

REFERENCE
Lo Coco, F.; Ceccon, L.; Valentini, C.; Novelli, V. High-performance liquid chromatographic determination of 2-furaldehyde in spirits, *J.Chromatogr.*, **1992**, *590*, 235–240.

SAMPLE
Matrix: blood
Analyte: acetaldehyde and acetone
Sample preparation: 500 μL Rapidly separated plasma + 300 μL 3 M perchloric acid + 80 μL freshly prepared 12.5 mM 2,4-dinitrophenylhydrazine in 6 M HCl + 800 μL 3 M sodium acetate (vary amount of sodium acetate, if necessary, to maintain pH at 4.5-5.0), add 2 mL isooctane, shake vigorously, centrifuge at 10000 g for 5 min, repeat extraction. Combine the organic layers and evaporate them to dryness under a stream of nitrogen at 25°, reconstitute with MeCN:water 50:50, inject an aliquot. (If necessary, add 10 mM sodium azide to blood to minimize artifactual formation of acetaldehyde.)

HPLC VARIABLES
Column: 150 × 3.9 10 μm μBondapak C18

Mobile phase: MeCN:water 50:50
Flow rate: 3
Detector: UV 365

CHROMATOGRAM
Retention time: 2.31 (acetaldehyde), 3.2 (acetone)
Limit of detection: 20 pmole

KEY WORDS
plasma

REFERENCE
Di Padova, C.; Alderman, J.; Lieber, C.S. Improved methods for the measurement of acetaldehyde concentrations in plasma and red blood cells, *Alcohol Clin.Exp.Res.*, **1986**, *10*, 86−89.

SAMPLE
Matrix: blood
Analyte: 4-hydroxycyclophosphamide
Sample preparation: Prepare plasma rapidly and keep cold. 500 μL Plasma + 1 mL cold MeCN, shake for 30 s, centrifuge at 6000 g for 3 min. 1 mL Supernatant + 70 μL 1 M HCl + 200 μL 3.8 mg/mL 2,4-dinitrophenylhydrazine in MeCN, vortex for 30 s, heat at 50° for 5 min, inject a 40 μL aliquot. (Prepare 2,4-dinitrophenylhydrazine solution fresh each 14 days. 4-Hydroxycyclophosphamide ring opens to an aldehyde which is then derivatized.)

HPLC VARIABLES
Guard column: 10 × 3.2 3 μm Spherisorb S3ODS2
Column: 100 × 4.6 3 μm Spherisorb S3ODS2
Mobile phase: MeCN:pH 7 phosphate buffer (ionic strength, μ = 0.02) 40:60
Flow rate: 1
Injection volume: 40
Detector: UV 357

CHROMATOGRAM
Retention time: 9.6
Limit of detection: 22 ng/mL

KEY WORDS
plasma; pharmacokinetics

REFERENCE
Johansson, M.; Bielenstein, M. Determination of 4-hydroxycyclophosphamide in plasma, as 2,4-dinitrophenylhydrazone derivative of aldophosphamide, by liquid chromatography, *J.Chromatogr.B*, **1994**, *660*, 111−120.

SAMPLE
Matrix: blood
Analyte: allantoin
Sample preparation: 300 μL Plasma + 600 μL MeCN, mix vigorously, centrifuge at 2500 g for 10 min, remove the supernatant, add 600 μL MeCN:water 2:1 to the residue, mix vigorously, centrifuge. Combine the supernatants and add them to the SPE cartridge, recover the eluate, elute with 2 mL MeCN:1 mM phosphoric acid 50:50. Combine the eluates and evaporate them to dryness under a stream of nitrogen at 60°, reconstitute the residue in 400 μL 100 mM NaOH, heat on a boiling water bath for 20 min, cool, add 600 μL 1.5 mM 2,4-dinitrophenylhydrazine in 2.5 M HCl, heat at 50° for 1 h, centrifuge at 10000 g for 30 min, inject a 50 μL aliquot of the supernatant. (Prepare the SPE cartridges by making a slurry of AG1-X8 (chloride form) strong ion-exchange resin (Bio-Rad) in water, let stand for 30 min, add 0.5 mL of the slurry to a 65 × 5 polypropylene column fitted with a 20 μm polyethylene frit, add another frit to the top, wash the column with water until no chloride ions are eluted, equilibrate with 3 mL MeCN:1 mM phosphoric acid 50:50. Test for the presence of chloride ions by adding 1 mL eluate to 300 μL 300 mM silver nitrate solution, a white precipitate indicates the presence of chloride ions. Regenerate columns by washing with 5 mL 1 M HCl, wash with 20 mL water until no

chloride ions are detected. Allantoin is hydrolyzed to glyoxylic acid which is then derivatized.)

HPLC VARIABLES
Guard column: Reversed-phase guard column (Whatman)
Column: 100 × 4.6 Partisphere C18 (Whatman)
Mobile phase: Gradient. A was MeCN:100 mM pH 6.0 KH_2PO_4 5:95. B was MeCN:100 mM pH 6.0 KH_2PO_4 50:50. A:B 100:0 for 1 min, to 84:16 over 20 min, to 0:100 over 9 min, re-equilibrate at the initial conditions for 10 min.
Flow rate: 1
Injection volume: 50
Detector: UV 360

CHROMATOGRAM
Retention time: 16.6
Limit of quantitation: 4 μM

KEY WORDS
plasma; SPE

REFERENCE
Lagendijk, J.; Ubbink, J.B.; Vermaak, W.J.H. The determination of allantoin, a possible indicator of oxidant status, in human plasma, *J.Chromatogr.Sci.*, **1995**, *33*, 186–193.

SAMPLE
Matrix: blood
Analyte: aldehydes
Sample preparation: Wet 300 mg Amberlite XAD-2 resin with 300 μL MeCN, add 1 mL of plasma, add 4 mL 4 μM 2,4-dinitrophenylhydrazine in 1-1.5 M HCl, shake for 10 min, filter, wash the resin with four 5 mL portions of water (or until the washings are neutral), dry under vacuum, elute the resin with 5 mL MeCN, inject a 25 μL aliquot.

HPLC VARIABLES
Column: Symmetry octylsilica (Waters)
Mobile phase: Gradient. MeCN:water from 35:65 to 80:20 over 20 min.
Flow rate: 1
Injection volume: 25
Detector: UV (wavelength not given)

CHROMATOGRAM
Retention time: 4.5 (formaldehyde), 6 (acetaldehyde)
Limit of quantitation: 3 μg/mL

OTHER SUBSTANCES
Simultaneously analyzed: acetone

KEY WORDS
plasma

REFERENCE
Breckenridge, S.M.; Yin, X.; Rosenfeld, J.M.; Yu, Y.H. Analytical derivatizations of volatile and hydrophilic carbonyls from aqueous matrix onto a solid phase of a polystyrene-divinylbenzene macroreticular resin, *J.Chromatogr.B*, **1997**, *694*, 289–296.

SAMPLE
Matrix: blood, urine
Analyte: allantoin
Sample preparation: Urine. Adjust pH to <3 with 1 M sulfuric acid, dilute so that the concentration of allantoin was 0.03-0.32 mM. 500 μL Acidified urine + 50 μL indicator + 100 μL 600 mM NaOH, if the color did not change to blue (pH >9.2) add 600 mM NaOH in 50 μL increments, heat at 85° for 1 h, add 200 μL reagent, heat at 85° for 20 min, cool, centrifuge at 35000 g for 15 min, inject a 10 μL aliquot of the supernatant. Plasma. 2 mL Plasma + 2 mL 10% trichloroacetic acid, centrifuge at 35000 g for 20 min. 500 μL Supernatant + 50 μL indicator + 100 μL 600 mM NaOH, if the color did not

change to blue (pH >9.2) add 600 mM NaOH in 50 μL increments, heat at 85° for 1 h, add 200 μL reagent, heat at 85° for 20 min, cool, centrifuge at 35000 g for 15 min, inject a 10 μL aliquot of the supernatant. (Indicator was 600 mM NaOH and 0.04% thymol blue in water, filter (Whatman No. 1 paper). Reagent was 1 g/L 2,4-dinitrophenylhydrazine in 2 M HCl, filter (Whatman No. 1 paper). Allantoin is hydrolyzed to glyoxylic acid which is then hydrolyzed.)

HPLC VARIABLES
Guard column: 25 × 2 30-40 μm pellicular reversed-phase C18
Column: 150 × 3.9 4 μm Nova-Pak C18
Mobile phase: Gradient. A was MeCN:10 mM acetic acid, pH adjusted to 6.1 with ammonia 15:85. B was MeCN. A:B from 100:0 to 82:18 over 3 min (Waters concave no. 7) to 18:82 over 2 min (Waters concave no. 7), stay at 18:82 for 3 min, to 100:0 over 1 min (linear), re-equilibrate at 100:0 for 15 min.
Flow rate: 1
Injection volume: 10
Detector: UV 360

CHROMATOGRAM
Retention time: 3, 7.3 (syn- and anti-derivatization products)
Limit of quantitation: 5 μM

KEY WORDS
plasma

REFERENCE
Chen, X.B.; Kyle, D.J.; Orskov, E.R. Measurement of allantoin in urine and plasma by high-performance liquid chromatography with pre-column derivatization, *J.Chromatogr.*, **1993**, *617*, 241−247.

SAMPLE
Matrix: cells
Analyte: aldehydes
Sample preparation: Mix 4.5 mL of a suspension of red blood cells with 500 μL EDTA solution and 5 mL reagent in the dark for 2 h, let stand in the dark at 0° for 1 h, extract with 7 mL dichloromethane, centrifuge 3 times at 900 g, evaporate to dryness, dissolve the residue in 1 mL dichloromethane, chromatograph on a 20 cm × 20 cm × 0.2 mm silica gel F_{254} plate (Merck) with dichloromethane for 5 cm and with benzene for 15 cm (Caution! Benzene is a carcinogen!). Remove the appropriate band and extract with three 10 mL portions of MeOH, evaporate the extract to dryness, reconstitute with 1 mL MeOH, inject an aliquot. (Prepare reagent by dissolving 2,4-dinitrophenylhydrazine in 1 M HCl, extracting with 15 mL n-hexane, and adjusting concentration to 1.8 mM by measuring absorbance.)

HPLC VARIABLES
Guard column: 50 × 4 Nucleosil 5C18
Column: 250 × 4 Nucleosil 5C18
Mobile phase: MeOH:water 80:20
Flow rate: 0.9
Detector: UV 350

CHROMATOGRAM
Retention time: 6.3 (4-hydroxyhexenal), 8.8 (4-hydroxyoctenal), 11.3 (4-hydroxynonenal), 12.5 (pentanal), 16 (2,4-hexadienal), 18.4 (hexanal), 28.5 (2,4-heptadienal), 30.5 (octenal), 33.5 (2,5-nonadienal), 46.5 (2,4-decadienal)

KEY WORDS
rabbit

REFERENCE
Werner, A.; Siems, W.; Grune, T.; Schreiter, C. Interrelation between nucleotide degradation and aldehyde formation in red blood cells. Influence of xanthine oxidase on metabolism: an application of nucleotide and aldehyde analyses by high-performance liquid chromatography, *J.Chromatogr.*, **1990**, *507*, 311−319.

SAMPLE

Matrix: cosmetics

Analyte: formaldehyde

Sample preparation: Dilute 1 g cosmetic to 10 mL with THF or THF:water 90:10. Remove a 1 mL aliquot and add it to 400 μL reagent, vortex for 1 min, let stand at room temperature for 2 min, add 400 μL 100 mM pH 6.8 phosphate buffer, add 700 μL 1 M NaOH, inject a 6 μL aliquot. (Prepare reagent by dissolving 250 mg 2,4-dinitrophenylhydrazine in 100 mL concentrated HCl, make up to 250 mL with water.)

HPLC VARIABLES

Column: 250 × 4 10 μm LiChrosorb RP-8

Mobile phase: MeCN:water

Flow rate: 1

Injection volume: 6

Detector: UV 345

CHROMATOGRAM

Retention time: 6.5

Limit of detection: 200 ng/mL

REFERENCE

Benassi, C.A.; Semenzato, A.; Bettero, A. High-performance liquid chromatographic determination of free formaldehyde in cosmetics, *J.Chromatogr.*, **1989**, *464*, 387–393.

SAMPLE

Matrix: microsomal incubations

Analyte: aldehydes

Sample preparation: Centrifuge 48 mL microsomal incubation at 100000 g for 40 min. Mix an aliquot of the supernatant with an equal volume of the reagent or suspend the precipitate in 8 mL reagent, let stand in the dark at room temperature for 12 h, extract twice with 3 mL portions of chloroform, freeze at -20° for 2 h, filter, chromatograph on a 20 cm × 20 cm silica gel plate (Merck) with dichloromethane for 5 cm and with benzene for 15 cm (Caution! Benzene is a carcinogen!). Remove the appropriate band and extract with two 5 mL portions of MeOH, evaporate the extract to dryness, reconstitute with 0.5 mL MeOH, inject a 50 μL aliquot. (Prepare reagent by dissolving 50 mg 2,4-dinitrophenylhydrazine (recrystallized from n-butanol) in 100 mL 1 M HCl, heat at 50° for 1 h, cool, extract twice with 50 mL portions of n-hexane, flush with nitrogen, store in the dark.)

HPLC VARIABLES

Column: 250 × 4.6 Zorbax ODS

Mobile phase: Gradient. MeOH:water 80:20 for 20 min, to 100:0 over 8 min, maintain at 100:0 for 20 min.

Flow rate: 0.9

Injection volume: 50

Detector: UV 378

CHROMATOGRAM

Retention time: 9 (propanal), 10 (4-hydroxyoctenal), 11.5 (butanal), 12.5 (4-hydroxynonenal), 14 (pent-2-enal), 15 (pentanal), 20 (hex-2-enal), 22 (hexanal), 25 (hydroxyundecanal), 27 (hept-2-enal), 32 (oct-2-enal)

KEY WORDS

rat; liver

REFERENCE

Esterbauer, H.; Cheeseman, K.H.; Dianzani, M.U.; Poli, G.; Slater, T.F. Separation and characterization of the aldehydic products of lipid peroxidation stimulated by ADP-Fe^{2+} in rat liver microsomes, *Biochem.J.*, **1982**, *208*, 129–140.

SAMPLE

Matrix: microsomal incubations

Analyte: malonaldehyde

Sample preparation: 300 μL Microsomal incubation + 300 μL 2.5 mg/mL 2,4-dinitrophenylhydrazine in 1 M HCl containing 2-nitroresorcinol, let stand at 25° for 1 h, centrifuge at 5000 g for 10 min, inject a 20 μL aliquot of the supernatant.

HPLC VARIABLES
Column: 250 × 4.6 Cosmosil 5 C18 (Nacalai Tesque)
Mobile phase: MeCN:10 mM HCl 45:55
Flow rate: 1.5
Injection volume: 20
Detector: UV 310

CHROMATOGRAM
Retention time: 7.8
Internal standard: 2-nitroresorcinol (3.5)
Limit of detection: 25 pmole

KEY WORDS
rat; liver

REFERENCE
Tomita, M.; Okuyama, T.; Kawai, S. Determination of malonaldehyde in oxidized biological materials by high-performance liquid chromatography, *J.Chromatogr.*, **1990**, *515*, 391−397.

SAMPLE
Matrix: microsomal incubations
Analyte: phenylacetaldehyde
Sample preparation: 500 μL Microsomal incubation + 100 μL 25% zinc sulfate:800 mM semicarbazide 50:50, neutralize with saturated barium hydroxide solution, vortex, centrifuge. Remove the supernatant and add it to 1 mL water, 100 μL 0.25% 2,4-dinitrophenylhydrazine in 6 M HCl, and 2 mL hexane, mix for 30 min. Remove the hexane layer and add it to 350 μL MeCN, mix, inject an aliquot of the MeCN layer.

HPLC VARIABLES
Column: 250 × 4.6 5 μm Spherisorb ODS-2
Mobile phase: MeCN:water 60:40
Flow rate: 1.5
Detector: UV 230

KEY WORDS
human; liver; pharmacokinetics

REFERENCE
Tateishi, T.; Wood, A.J.J.; Guengerich, F.P.; Wood, M. Biotransformation of tritiated fentanyl in human liver microsomes. Monitoring metabolism using phenylacetic acid and 2-phenylethanol, *Biochem.Pharmacol.*, **1995**, *50*, 1921−1924.

SAMPLE
Matrix: microsomal incubations
Analyte: chloral hydrate, acetone, and aldehydes
Sample preparation: 1 mL Microsomal incubation + 200 μL 3 mg/mL 2,4-dinitrophenylhydrazine in 2 M HCl (freshly prepared), shake for 10 min, extract twice with 1 mL portions of ethyl acetate. Combine the extracts and evaporate them to dryness under reduced pressure, reconstitute with 500 μL MeCN, inject an aliquot.

HPLC VARIABLES
Column: 250 × 4.6 5 μm Hypersil ODS
Mobile phase: MeCN:water 55:45
Flow rate: 2
Detector: UV 330

CHROMATOGRAM
Retention time: 5.7 (acetaldehyde), 7.3 (acetone), 11.1 (chloral hydrate), 4.4 (formaldehyde), 3.8 (malondialdehyde), 8.2 (propionaldehyde)

KEY WORDS
mouse; liver; chloral hydrate is not necessarily derivatized

REFERENCE
Ni, Y.-C.; Wong.T.-Y. ; Lloyd, R.V.; Heinze, T.M.; Shelton, S.; Casciano, D.; Kadlubar, F.F.; Fu, P.P. Mouse liver microsomal metabolism of chloral hydrate, trichloroacetic acid, and trichloroethanol leading to induction of lipid peroxidation via a free radical mechanism, *Drug Metab.Dispos.*, **1996**, *24*, 81–90.

SAMPLE
Matrix: perfusate
Analyte: aldehydes
Sample preparation: 1.5 mL Perfusate + 100 µL 3.1 mg/mL 2,4-dinitrophenylhydrazine in 2 M HCl + 500 µL water, vortex for 15 min, add 10 mL pentane, shake intermittently for 30 min, remove the organic layer and extract the aqueous layer with 20 mL pentane. Combine the pentane layers and evaporate them to dryness under a stream of nitrogen at 30°, reconstitute the residue in 200 µL MeCN, filter (0.2 µm Nylon-66), inject a 25 µL aliquot of the filtrate.

HPLC VARIABLES
Guard column: Bondapak C18 Guard-Pak
Column: 75 × 4.6 3 µm Ultrasphere ODS C18
Mobile phase: MeCN:water:acetic acid 40:60:0.1 (At the start of each day wash column with MeCN:acetic acid 100:0.1.)
Flow rate: 1
Injection volume: 25
Detector: UV 356

CHROMATOGRAM
Retention time: 5.3 (malonaldehyde (UV 307)), 6.6 (formaldehyde), 10.3 (acetaldehyde), 16.5 (acetone), 20.5 (propionaldehyde)

REFERENCE
Cordis, G.A.; Bagchi, D.; Maulik, N.; Das, D.K. High-performance liquid chromatographic method for the simultaneous detection of malonaldehyde, acetaldehyde, formaldehyde, acetone and propional-dehyde to monitor the oxidative stress in heart, *J.Chromatogr.A*, **1994**, *661*, 181–191.

SAMPLE
Matrix: perfusate
Analyte: malonaldehyde
Sample preparation: 1.5 mL Perfusate + 100 µL 3.1 mg/mL 2,4-dinitrophenylhydrazine in 2 M HCl + 500 µL water, vortex for 15 min, add 10 mL pentane, shake intermittently for 30 min, remove the organic layer, add 20 mL pentane, extract. Combine the organic layers and evaporate them to dryness under a stream of nitrogen at 30°, reconstitute the residue in 200 µL MeCN, filter (0.2 µm Nylon-66), inject a 25 µL aliquot of the filtrate.

HPLC VARIABLES
Guard column: Nova-Pak C18 Guard-Pak
Column: 75 × 4.6 3 µm Ultrasphere ODS C18
Mobile phase: MeCN:water:acetic acid 40:60:0.1 (At the beginning of each day wash the column with MeCN:acetic acid 100:0.1.)
Flow rate: 1
Injection volume: 25
Detector: UV 307

CHROMATOGRAM
Retention time: 5.3
Limit of detection: 10 pmole

REFERENCE
Cordis, G.A.; Maulik, N.; Das, D.K. Detection of oxidative stress in heart by estimating the dinitro-phenylhydrazine derivative of malonaldehyde, *J.Mol.Cell Cardiol.*, **1995**, *27*, 1645–1653.

SAMPLE
Matrix: reaction mixtures

Analyte: aldehydes
Sample preparation: Add reaction mixture to 1 mL 0.25% 2,4-dinitrophenylhydrazine in 6 M HCl, let stand at room temperature for 30 min, extract with dichloromethane. Remove the organic layer and wash it with water, dry under reduced pressure, reconstitute with MeCN, inject an aliquot.

HPLC VARIABLES
Column: 150 × 4.6 Hitachi Gel ODS
Mobile phase: MeCN:water 55:45
Flow rate: 1
Detector: UV 360

CHROMATOGRAM
Retention time: 4 (formaldehyde), 5 (acetaldehyde), 16 (glyoxal)

REFERENCE
Sayato, Y.; Nakamuro, K.; Ueno, H. Mutagenicity of products formed by ozonation of naphthoresorcinol in aqueous solutions, *Mutat.Res.*, **1987**, *189*, 217–222.

SAMPLE
Matrix: solutions
Analyte: aldehydes
Sample preparation: Mix 1 mL of a 0.1-10 µg/mL solution in water with 20 µL 5 N phosphoric acid, add 200 µL 1 mg/mL 2,4-dinitrophenylhydrazine in MeCN, let stand at room temperature for 10 min, inject an aliquot.

CAPILLARY ELECTROPHORESIS
Capillary: 72 cm × 50 µm fused-silica (50 cm to detector) (GL Sciences)
Running buffer: 20 mM pH 9 NaH_2PO_4 containing 20 mM sodium tetraborate and 50 mM sodium dodecyl sulfate
Capillary temperature: 30
Voltage/Current: 20 kV
Injection: Vacuum injection at 5 inch Hg for 1 s (3 nL)
Detector: UV 360
Model: Applied Biosystems Model 270A
Migration time: 11.8 (formaldehyde), 14.2 (acetaldehyde), 17.7 (acrolein), 18.3 (propionaldehyde)

REFERENCE
Takeda, S.; Wakida, S.-i.; Yamane, M.; Higashi, K. Analysis of lower aliphatic aldehydes in water by micellar electrokinetic chromatography with derivatization to 2,4-dinitrophenylhydrazones, *Electrophoresis*, **1994**, *15*, 1332–1334.

SAMPLE
Matrix: solutions
Analyte: mannose
Sample preparation: 4 mL 300 µg/mL 2,4-dinitrophenylhydrazine + 300 µL 2 M HCl + 3 mL 15-30 mg/mL mannitol in water, mix, let stand at room temperature for 1.5 h, make up to 10 mL with water, inject an aliquot.

HPLC VARIABLES
Column: Ultrasphere C18
Mobile phase: MeCN:10 mM HCl 35:65
Flow rate: 1
Detector: UV 365

OTHER SUBSTANCES
Non-interfering: mannitol

REFERENCE
Dubost, D.C.; Kaufman, M.J.; Zimmerman, J.A.; Bogusky, M.J.; Coddington, A.B.; Pitzenberger, S.M. Characterization of a solid state reaction product from a lyophilized formulation of a cyclic heptapeptide. A novel example of an excipient-induced oxidation, *Pharm.Res.*, **1996**, *13*, 1811–1814.

SAMPLE
Matrix: tissue
Analyte: 4-hydroxynonenal
Sample preparation: Prepare a 30% homogenate in ice cold 154 mM pH 7.4 containing 3 mM EDTA, extract with chloroform, evaporate the organic layer to dryness, reconstitute with 0.048% 2,4-dinitrophenylhydrazine in chloroform:acetic acid 90:10, let stand in the dark for 12 h, evaporate to dryness under reduced pressure, reconstitute with chloroform, purify by TLC using dichloromethane and benzene (Caution! Benzene is a carcinogen!), remove the silica containing the hydrazone, elute with MeOH, purify this fraction by TLC using dichloromethane, remove the silica containing the hydrazone, elute with MeOH, evaporate, reconstitute with 500 μL MeOH, inject an aliquot.

HPLC VARIABLES
Column: 250 × 4.6 Zorbax ODS
Mobile phase: MeCN:water 70:30
Column temperature: 36
Flow rate: 1
Detector: UV 350

CHROMATOGRAM
Retention time: 10

KEY WORDS
mouse; liver

REFERENCE
Benedetti, A.; Pompella, A.; Fulceri, R.; Romani, A.; Comporti, M. 4-Hydroxynonenal and other aldehydes produced in the liver in vivo after bromobenzene intoxication, *Toxicol.Pathol.*, **1986**, *14*, 457–461.

DMEQ-hydrazide

SAMPLE
Matrix: solutions
Analyte: aldehydes
Sample preparation: Mix 100 μL of an aldehyde with 100 μL 5 mM DMEQ-hydrazide in DMF and 50 μL 10% trichloroacetic acid in water, heat at 60° for 15 min, inject a 10 μL aliquot. (DMEQ-hydrazide is 6,7-dimethoxy-1-methyl-2-oxo-1,2-dihydroquinoxalin-3-yl-propionohydrazide. It is available from Wako, Richmond VA.)

HPLC VARIABLES
Column: 250 × 4.6 5 μm L-column ODS (Chemicals Inspection and Testing Institute, Tokyo)
Mobile phase: MeOH:water 70:30 (A) or 85:15 (B) or 50:50 (C)
Injection volume: 20
Detector: F ex 362 em 442

CHROMATOGRAM
Retention time: 7.4 (cyclohexanecarbaldehyde (A)), 8.2 (2-naphthaldehyde (A)), 9.4 (heptanal (A)), 11.4 (dodecanal (B)), 14 (octanal (A)), 18.6 (benzaldehyde (C)), 20.2 (vanillin (C)), 34.8 (decanal (A)), 41.4 (isovanillin (C))
Limit of detection: 13-55 fmole

OTHER SUBSTANCES
Non-interfering: alcohols, aldosterone, amines, amino acids, carboxylic acids, fatty acids, α-ketoglutaric acid, ketones, phenols, phenylpyruvic acid, pyruvic acid, sugars, thiols

REFERENCE
Iwata, T.; Hirose, T.; Nakamura, M.; Yamaguchi, M. 6,7-Dimethoxy-1-methyl-2-oxo-1,2-dihydroquinoxalin-3-ylpropiono-hydrazide as a fluorescence derivatization reagent for aldehydes in high-performance liquid chromatography, *Analyst*, **1993**, *118*, 517–519.

9-Fluorenylmethoxycarbonyl Hydrazide

RELATED REFERENCE
Zhang, R.; Zhang, Z.; Liu, G.; Hidaka, Y.; Shimonishi, Y. High-performance liquid chromatographic determination of neutral and amino monosaccharides by ultraviolet and fluorescence detection of sugar 9-fluorenylmethoxycarbonyl hydrazones and 9-fluorenylmethoxycarbonyl amino sugars at picomole and sub-picomole levels. *J.Chromatogr.* **1993**, *646*, 45-52.

SAMPLE
Matrix: solutions
Analyte: xylose
Sample preparation: 10 μL EtOH containing sugars + 110 μL EtOH:acetic acid 99.9:0.1 + 100 μL 9-fluorenylmethoxycarbonyl hydrazide in MeCN, mix, heat at 65° for 3 h, cool to room temperature, dilute with EtOH or EtOH:acetic acid 99.9:0.1, inject an aliquot. (9-Fluorenylmethoxycarbonyl hydrazide (N-(9-fluorenylmethoxycarbonyl) hydrazine) is available from Molecular Probes, Eugene OR. Synthesis is as follows. Dissolve 100 mg 9-fluorenylmethylchloroformate in 25 mL MeCN, add this solution dropwise with stirring to 1 mL hydrazine hydrate (Caution! Hydrazine hydrate is a carcinogen!), stir for 30 min, evaporate under reduced pressure, use the crude product or recrystallize from EtOH or MeCN (mp 173-5°). Prepare a solution of 9-fluorenylmethoxycarbonyl hydrazide in MeCN so that the hydrazine:sugar ratio is 10:1.)

HPLC VARIABLES
Column: 150 × 4.6 5 μm Zorbax ODS
Mobile phase: Gradient. A was MeCN:water 27:73 containing 80 mM acetic acid. B was MeCN:water 30:70 containing 80 mM acetic acid. A:B from 100:0 to 0:100 over 30 min.
Flow rate: 1
Detector: F ex 270 em 320

CHROMATOGRAM
Retention time: 15
Limit of detection: 0.1 pmole

OTHER SUBSTANCES
Simultaneously analyzed: fructose, galactose, lactose, maltose, mannose, ribose

REFERENCE
Zhang, R.-E.; Cao, Y.-L.; Hearn, M.W. Synthesis and application of Fmoc-hydrazine for the quantitative determination of saccharides by reversed-phase high-performance liquid chromatography in the low and subpicomole range, *Anal.Biochem.*, **1991**, *195*, 160–167.

SAMPLE
Matrix: solutions
Analyte: oligosaccharides
Sample preparation: Mix 1-100 nmoles oligosaccharide in 10 μL MeOH:water 60:40 with 110 μL 0.5-5% acetic acid and 100 μL 50 mM 9-fluorenylmethoxycarbonyl hydrazide in MeCN, heat at 37° for 6 h, dilute with mobile phase, inject an aliquot. (9-Fluorenylmethoxycarbonyl hydrazide (FMOC hydrazine) is available from Molecular Probes, Eugene OR. Synthesis is as described above.)

HPLC VARIABLES
Column: 150 × 4.6 Cosmosil 5 C18
Mobile phase: MeOH:water 60:40
Injection volume: 10
Detector: UV 254; F ex 270 em 320

CHROMATOGRAM
Retention time: 7-50 (depending on degree of polymerization)

REFERENCE
Zhang, Z.; Zhang, R.; Liu, G. Analysis of the 9-fluorenylmethoxycarbonyl hydrazide labelling of neutral and sialic acid-containing oligosaccharides by reversed-phase high-performance liquid chromatography, *J.Chromatogr.A*, **1996**, *728*, 343–350.

SAMPLE
Matrix: urine
Analyte: sugars
Sample preparation: 10 μL Urine + 200 μL reagent, heat at 65° for 16 h, cool to room temperature, inject a 5 μL aliquot of the clear supernatant. (Prepare reagent by dissolving 5 mg 9-fluorenylmethoxycarbonyl hydrazide in 1 mL MeCN, add 10 μL buffer. Buffer was 1.44 M formic acid containing 600 mM NaOH. 9-Fluorenylmethoxycarbonyl hydrazide (FMOC hydrazine) is available from Molecular Probes, Eugene OR. Synthesis is as described above.)

HPLC VARIABLES
Guard column: 10 × 4.6 3 μm Spherisorb ODS II
Column: 125 × 4.6 3 μm Spherisorb ODS II
Mobile phase: Gradient. Isopropanol:isobutyl alcohol:water 6:6:88 for 13 min, to 80:0:20 (step gradient), maintain at 80:0:20 for 6 min, re-equilibrate at initial conditions.
Column temperature: 50
Injection volume: 5
Detector: F ex 270 em 315

CHROMATOGRAM
Retention time: 5 (lactulose), 9.8 (xylose), 10.5 (rhamnose), 11 (3-O-methyl-D-glucose)
Limit of detection: 20-110 nM

REFERENCE
Rooyakkers, D.R.; van Eijk, H.M.H.; Deutz, N.E.P. Simple and sensitive multi-sugar-probe gut permeability test by high-performance liquid chromatography with fluorescence labelling, *J.Chromatogr.A*, **1996**, *730*, 99–105.

7-Hydrazino-4-nitro-2,1,3-benzoxadiazole

SAMPLE
Matrix: solutions
Analyte: aldehydes
Sample preparation: Mix 10 μL of a 10 mM solution in MeOH with 10-50 μL MeOH: water 75:25 containing a 5-fold molar excess of NBD-H, heat at 50° under nitrogen in the dark for 30 min, cool, add 100 μL water, extract with 100 μL benzene (Caution! Benzene is a carcinogen!), centrifuge, inject a 20 μL aliquot of the organic layer. (7-Hydrazino-4-nitrobenzo-2-oxa-1,3-diazole (NBD-H) is available from Tokyo Kasei (TCI America, Portland OR). Synthesis is as follows. Dissolve 10 mg 7-chloro-4-nitrobenzo-2-oxa-1,3-diazole (NBD-Cl) in 5 mL chloroform, add 5 mL hydrazine solution (Caution! Hydrazine is a carcinogen!), let stand in the dark under nitrogen for 1 h, wash the precipitate with benzene, dry at room temperature, store in the refrigerator under nitrogen. Prepare hydrazine solution by mixing 200 μL hydrazine with 5 mL MeOH.)

HPLC VARIABLES
Column: 250 × 4.6 10 μm LiChrosorb RP 8
Mobile phase: MeCN:water 50:50
Flow rate: 1
Injection volume: 20
Detector: UV 254

CHROMATOGRAM
Retention time: 2 (propionaldehyde), 3 (butyraldehyde), 4.5 (valeraldehyde), 5.5 (caproaldehyde)
Limit of detection: 5-10 ng

REFERENCE
Guebitz, G.; Wintersteiger, R.; Frei, R.W. Fluorogenic labelling of carbonylcompounds with 7-hydrazino-4-nitrobenz-2-oxa-1,3-diazole (NBD-H), *J.Liq.Chromatogr.*, **1984**, *7*, 839–854.

Luminarin 3

SAMPLE
Matrix: solutions
Analyte: aldehydes
Sample preparation: Mix 100 μL of a 5-500 μM solution in 100 mM sulfuric acid with 100 μL 100 mM sulfuric acid and 10 μL 10 mM luminarin 3 in DMSO, let stand at room temperature for 1 h, add 100 μL 600 mM sodium bicarbonate solution to adjust the pH to 7.0, agitate gently until gas evolution ceases, add 1 mL dichloromethane, vortex for 2 min, centrifuge. Remove the lower organic layer and evaporate it to dryness under a stream of nitrogen at room temperature, reconstitute the residue in 100 μL MeCN, inject an aliquot. (Luminarin 3, 2,3,6,7-tetrahydro-11-oxo-1H,5H,11H-[1]benzopyrano[6,7,8-ij]quinolizine-9-acetic acid hydrazide, is available from Eurobio, Les Ulis, France. Synthesis is as follows. Reflux (with protection from moisture and with stirring) 2.12 g 8-hydroxyjulolidine, 2.22 g diethyl 1,3-acetonedicarboxylate (oxo-3-glutaric acid ethyl ester, Fluka), 1.71 g anhydrous zinc chloride, and 6 mL EtOH for 24 h, cool, add to 200 mL water, extract with 200 mL ethyl acetate, extract with 100 mL ethyl acetate. Combine the organic layers and wash them with water, dry over magnesium sulfate, evaporate to dryness, recrystallize from 5 parts ethyl acetate to give ethyl 2,3,6,7-tetrahydro-11-oxo-1H,5H,11H-[1]benzopyrano[6,7,8-ij]quinolizine-9-acetate. Stir 5 g of this compound with 8 mL hydrazine hydrate in 100 mL MeOH for 4 h, filter, wash the solid with 10 mL MeOH, wash with 10 mL dichloromethane to give luminarin 3 (World Pat. 89 12,052; Chem. Abstr. 1990, 113, 23889n).)

HPLC VARIABLES
Column: 250 × 4.6 5 μm Nucleosil ODS
Mobile phase: MeCN:10 mM pH 7.5 imidazole nitrate buffer 40:60
Flow rate: 1.5
Injection volume: 20
Detector: F ex 399 em 485

CHROMATOGRAM

Retention time: k' 3.05 (acetaldehyde), k' 10.28 (acetoacetaldehyde), k' 5.58 (acrolein), k' 2.93 (formaldehyde), k' 7.30 (furfural), k' 0.74 (glyceraldehyde), k' 2.62 (hydroxymethylfurfural), k' 3.16 (malonaldehyde), k' 4.72 (methylmalonaldehyde), k' 5.33 (succinaldehyde)

Limit of detection: 158-1950 fmole

OTHER SUBSTANCES

Also analyzed: ketones

REFERENCE

Traoré, F.; Tod, M.; Chalom, J.; Farinotti, R.; Mahuzier, G. 1H,5H,11H-[1]Benzopyrano[6,7,8-ij]quinolizine-9-acetic acid 2,3, 6,7-tetrahydro-11-oxohydrazide fluorogenic reagent for liquid chromatographic determination of aldehydes and ketones, *Anal.Chim.Acta*, **1992**, *269*, 211–222.

3-Methyl-2-benzothiazolinone Hydrazone

RELATED REFERENCE

Chiavari, G.; Laghi, M.C.; Torsi, G. High-performance liquid chromatographic analysis of aldehydes at trace level as their 3-methylbenzothiazolone azine derivatives. *J.Chromatogr.* **1989**, *475*, 343-351.

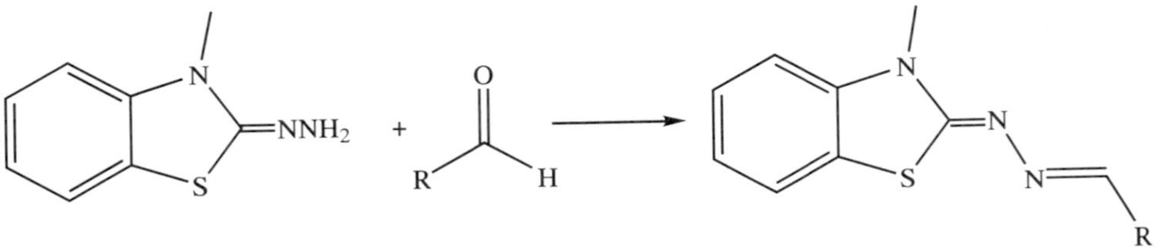

SAMPLE

Matrix: solutions

Analyte: aldehydes

Sample preparation: Derivatize aldehydes and ketones following this example. Heat 2 g 4-dimethylaminobenzaldehyde and 2 g 3-methyl-2-benzothiazolinone hydrazone (Fluka) in 50 mL EtOH and 5 mL acetic acid at 100° for 30 min (Annalen 1957, 609, 172), dissolve the product in EtOH, inject an aliquot.

HPLC VARIABLES

Column: 250 × 4.6 Erbasil C18 (Carlo Erba)

Mobile phase: MeOH:20 mM KH_2PO_4 70:30

Flow rate: 1.2

Detector: E, Metrohm 656, glassy carbon electrode +1.05 V, Ag/AgCl reference electrode

CHROMATOGRAM

Retention time: k' 2.22 (formaldehyde), k' 2.67, k' 3.00 (acetaldehyde (Z/E isomers)), k' 3.89, k' 4.89 (propionaldehyde (Z/E isomers)), k' 5.44, k' 6.89 (butyraldehyde (Z/E isomers)), k' 3.30, k' 4.10 (acrolein (Z/E isomers)), k' 4.56, k' 6.22 (crotonaldehyde (Z/E isomers)), k' 7.4, k' 17.8 (benzaldehyde (Z/E isomers)), k' 3.55 (acetone), k' 5.11, k' 6.22 (methyl ethyl ketone (Z/E isomers)), k' 8.22, k' 19.11 (acetophenone (Z/E isomers))

Limit of detection: 0.3-13.5 pmole

REFERENCE

Chiavari, G.; Facchini, M.C.; Fuzzi, S. Behaviour of 3-methyl-2-benzothiazolone azines of carbonyl compounds in high-performance liquid chromatography, *J.Chromatogr.*, **1987**, *387*, 459–466.

1-Methyl-1-(2,4-dinitrophenyl)hydrazine

SAMPLE
Matrix: air
Analyte: aldehydes
Sample preparation: Pull air through 80 mL reagent at 250 mL/min for 2 min, make up to 100 mL with MeCN, inject a 10 μL aliquot. (Prepare reagent by adding 105 mg 1-methyl-1-(2,4-dinitrophenyl)hydrazine to a solution of 500 μL concentrated sulfuric acid in 99.5 mL MeCN. Synthesis of 1-methyl-1-(2,4-dinitrophenyl)hydrazine is as follows. Add 6 mL methylhydrazine (Caution! Methylhydrazine is a carcinogen!) in 30 mL EtOH to a solution of 10.8 g potassium acetate in 50 mL water, heat to reflux, add a solution of 20.2 g 2,4-dinitrochlorobenzene in 100 mL EtOH dropwise with stirring, reflux for 8 h, cool to 0°, filter. Wash the solid with warm (60°) EtOH, wash with hot water to obtain 1-methyl-1-(2,4-dinitrophenyl)hydrazine (mp 143°) as a yellow-orange solid. The progress of the reaction may be monitored by silica TLC using toluene:MeCN 70:30.)

HPLC VARIABLES
Column: 150 × 4.6 5 μm Deltabond AK
Mobile phase: Gradient. MeCN:water 30:70 for 1 min, to 42:58 over 5.5 min, to 100:0 over 3 min, maintain at 100:0 for 0.5 min, return to initial conditions over 2 min.
Flow rate: 1.5
Injection volume: 10
Detector: UV 368

CHROMATOGRAM
Retention time: 4.7 (formaldehyde), 5.5 (acetaldehyde), 7 (acrolein), 7.2 (propanal), 8.6 (butanal), 9 (benzaldehyde), 9.3 (p-tolualdehyde)

OTHER SUBSTANCES
Simultaneously analyzed: acetone, 2-butanone

REFERENCE
Büldt, A.; Karst, U. 1-Methyl-1-(2,4-dinitrophenyl)hydrazine as a new reagent for the HPLC determination of aldehydes, *Anal.Chem.*, **1997**, *69*, 3617–3622.

4-(1-Methylphenanthro[9,10-d]imidazol-2-yl)benzohydrazide

SAMPLE

Matrix: solutions

Analyte: aldehydes

Sample preparation: Mix 100 μL of an aqueous solution with 50 μL 5 mM MPIB-hydrazide in DMF and 50 μL 2 mM sulfuric acid, heat at 40° for 15 min, inject a 10 μL aliquot. (Synthesis of MPIB-hydrazide, 4-(1-methylphenanthro[9,10-d]imidazol-2-yl)benzohydrazide, is as follows. Stir 1 g 9,10-diaminophenanthrene and 800 mg methyl 4-formylbenzoate (terephthaldehydic acid methyl ester) in 200 mL EtOH at room temperature for 1 h, add 5 mL MeOH saturated with HCl, reflux under an inert gas for 2 h, cool, concentrate to 50 mL under reduced pressure, chromatograph the precipitate on a 200 × 35 column of 70-230 mesh silica gel (ca. 100 g; Merck) with chloroform, recrystallize from MeOH to give methyl 4-(phenanthro[9,10-d]imidazol-2-yl)benzoate as colorless needles (mp 312-315°). Dissolve 500 mg methyl 4-(phenanthro[9,10-d]imidazol-2-yl)benzoate in 100 mL anhydrous MeOH, treat with a solution of diazomethane in ether, evaporate to dryness under reduced pressure, dissolve the residue in 20 mL chloroform, chromatograph on a 200 × 60 column of about 250 g 100 mesh silica gel with chloroform to give methyl 4-(1-methylphenanthro[9,10-d]imidazol-2-yl)benzoate as colorless needles (mp 199-201°). Dissolve 2 g methyl 4-(1-methylphenanthro[9,10-d]imidazol-2-yl)benzoate in 100 mL aqueous hydrazine hydrate (45%) (Caution! Hydrazine hydrate is a carcinogen and explodes on distillation in air!), heat at 100° for 1 h, recrystallize the precipitate from 95% EtOH to give MPIB-hydrazide(4-(1-methylphenanthro[9,10-d]imidazol-2-yl)benzohydrazide) (mp 291-293°). Preparation of diazomethane is as follows. Caution! Diazomethane is toxic, explosive, and carcinogenic! A face shield and a safety screen should always be used and the preparation should only be carried out in a properly functioning chemical fume hood. Only smooth glass apparatus with rubber stoppers and plastic tubing should be used. Scratched glassware, ground glass joints, and sharp edges should be avoided (Org. Syn., Coll. Vol. VI; Wiley: New York, 1988, pp. 432-435). Procedures have been reported for the synthesis of diazomethane from N-methyl-N-nitrosourea (Org. Syn., Coll. Vol. II; Wiley: New York, 1943, pp. 165-167), N-nitroso-β-methylamino-isobutyl methyl ketone (Org. Syn., Coll. Vol. III; Wiley: New York, 1955, pp. 244-248), and N,N'-dimethyl-N,N'-dinitrosoterephthalamide (Org. Syn., Coll. Vol. V; Wiley: New

York, 1973, pp. 351-355). Probably the most convenient starting material is N-methyl-N-nitroso-p-toluenesulfonamide (Diazald) (Aldrichimica Acta 1983, 16, 3-10; Org. Syn., Coll. Vol. IV, Wiley: New York, 1963, pp. 250-253). Add 10 mL 95% EtOH to a solution of 5 g KOH in 8 mL water, warm to 65° using a water bath, add a solution of 5 g N-methyl-N-nitroso-p-toluenesulfonamide in 45 mL ether over 20 min at such a rate as to keep the reaction volume constant. Collect the ether and diazomethane that distil in an ice-cooled receiving flask under a dry ice/acetone condenser. When all the N-methyl-N-nitroso-p-toluenesulfonamide has been used up, slowly add 10 mL ether to the reaction flask and continue distillation until the distillate is colorless. A purpose-built distillation apparatus can be purchased from Aldrich (Aldrichimica Acta 1983, 16, 3-10). Excess quantities of diazomethane can be destroyed by adding acetic acid until the yellow color of the diazomethane is discharged. The safe disposal of the nitroso compounds used to generate diazomethane has been discussed (Lunn,G.; Sansone,E.B. Destruction of Hazardous Chemicals in the Laboratory, Second Edition. Wiley: New York, 1994, pp. 277-289).)

HPLC VARIABLES
Column: 250 × 4.6 5 μm L-column ODS (Chemical Inspection and Testing Institute, Tokyo)
Mobile phase: MeOH:water 80:20 (A) or 87:13 (B) or 75:25 (C)
Flow rate: 1
Injection volume: 10
Detector: F ex 325 (10 mW He-Cd laser) em 460

CHROMATOGRAM
Retention time: 15 (cyclohexanecarboxaldehyde (A)), 18.2 (decanal (B)), 19.4 (2-naphthaldehyde (A)), 20 (heptanal (A)), 20 (benzaldehyde (C)), 27.6 (octanal (A))
Limit of detection: 1.4-4.4 fmole (conventional F), 0.18-0.80 fmole (laser F)

KEY WORDS
detection limits for ketones are much higher

REFERENCE
Iwata, T.; Ishimaru, T.; Yamaguchi, M. 4-(1-Methyl-2-phenanthro[9,10-*d*]imidazol-2-yl)-benzohydrazide as a derivatization reagent for aldehydes in high-performance liquid chromatography with conventional and laser-induced fluorescence detection, *Anal.Sci.*, **1997**, *13*, 501−504.

2-Nitrophenylhydrazine

SAMPLE
Matrix: urine
Analyte: formaldehyde
Sample preparation: Condition a 3 mL C18 SPE cartridge (Baker) with 3 mL MeOH and 2 mL cyclohexane. 150 μL Urine + 200 μL reagent + 1 mL cyclohexane, shake in the dark at 60 rpm overnight, centrifuge at 10000 g for 5 min. Add a 900 μL aliquot of the cyclohexane layer to the SPE cartridge, dry under vacuum, elute with two 500 μL aliquots of MeCN, inject a 10 μL aliquot of the eluate. (Prepare reagent by making a 30.9 mM solution of 2-nitrophenylhydrazine in 1 M HCl, wash 5 volumes of reagent with 1 volume chloroform, store in the dark at 4°, prepare fresh each day.)

HPLC VARIABLES
Column: 83 × 4.6 3 μm Pecosphere 3CR-C18
Mobile phase: MeCN:water 55:45
Flow rate: 1.5

Injection volume: 10
Detector: UV 254

CHROMATOGRAM
Retention time: 2.5 (formaldehyde), 3.5 (acetone), 7.3 (methyl ethyl ketone)
Internal standard: cyclohexanone (10)
Limit of detection: 100 ng/mL

KEY WORDS
SPE; pharmacokinetics

REFERENCE
Van Doorn, J.E.; De Cock, J.; Kezic, S.; Monster, A.C. Determination of methyl ethyl ketone in human urine after derivatization with o-nitrophenylhydrazine, using solid-phase extraction and reversed-phase high-performance liquid chromatography and ultraviolet detection, *J.Chromatogr.*, **1989**, *489*, 419–424.

Phenylhydrazine

SAMPLE
Matrix: urine
Analyte: glyoxylic acid
Sample preparation: Preserve urine with chlorhexidine gluconate, filter (0.22 μm cellulose). 30 μL Urine + 2 mL buffer + 100 μL 5% phenylhydrazine in water, shake, let stand at room temperature for 10 min, inject an aliquot. (Prepare buffer by dissolving 4.35 g K_2HPO_4 in 250 mL water and adjusting the pH to 8.3 with phosphoric acid.)

HPLC VARIABLES
Guard column: 30 × 4 30-40 μm Perisorb RP-18 (Merck)
Column: 250 × 4 10 μm LiChrospher RP-18
Mobile phase: MeOH:buffer 11:89 (Buffer was 150 mM acetic acid adjusted to pH 6.80 with ammonia solution.)
Flow rate: 1
Injection volume: 50
Detector: UV 324

CHROMATOGRAM
Retention time: 11
Limit of detection: 500 nM

OTHER SUBSTANCES
Simultaneously analyzed: α-ketoglutaric acid, oxaloacetic acid, pyruvic acid
Non-interfering: amino acids, ascorbate, dextrose, glutarate, glycolate, malate, maleate, malonate, mesoxalate, oxalate, succinate, tartrate, tartronate

REFERENCE
Petrarulo, M.; Pellegrino, S.; Bianco, O.; Marangella, M.; Linari, F.; Metasti, E. High-performance liquid chromatographic determination of glyoxylic acid in urine, *J.Chromatogr.*, **1988**, *432*, 37–46.

SAMPLE
Matrix: urine
Analyte: glycolic acid

Sample preparation: Preserve urine collected over 24 h with 2 mL chlorhexidine gluconate. 50 μL Urine + 50 μL 600 μM oxalacetic acid in water (prepare fresh each day) + 2 mL 100 mM pH 8.3 phosphate buffer + 100 μL 100 mM L-cysteine hydrochloride in water (prepare fresh each week) + 100 μL 494 mM phenyl hydrazine in water (prepare fresh each day), vortex, add 50 μL 25 U/mL glycolate oxygen oxidoreductase (from spinach, EC 1.1.3.15, Sigma) suspended in 100 mM pH 8.3 phosphate buffer (prepare fresh each week), mix, let stand at room temperature for 10 min, inject an aliquot. (Recrystallize phenylhydrazine hydrochloride twice from water, dry overnight at 37°, store in the dark.)

HPLC VARIABLES
Guard column: 30 × 4 30-40 μm Perisorb RP-18 (Merck)
Column: 250 × 4 10 μm LiChrospher RP-18
Mobile phase: MeOH:150 mM acetic acid 12:88, adjusted to pH 6.80 with ammonia
Flow rate: 2
Injection volume: 50
Detector: UV 324

CHROMATOGRAM
Retention time: 5.15
Internal standard: oxalacetic acid (2.95)
Limit of detection: 10 μM

OTHER SUBSTANCES
Simultaneously analyzed: ketoglutaric acid, pyruvic acid
Non-interfering: amino acids, ascorbic acid, citric acid, dextrose, glutaric acid, hydroxyphenylpyruvic acid, lactic acid, maleic acid, malic acid, malonic acid, mesoxalate, oxalic acid, β-phenylpyruvuc acid, succinic acid, tartaric acid, tartronic acid

REFERENCE
Petrarulo, M.; Pellegrino, S.; Bianco, O.; Marangella, M.; Linari, F.; Mentasti, E. Derivatization and high-performance liquid chromatographic determination of urinary glycolic acid, *J.Chromatogr.*, **1989**, *465*, 87–93.

Purpald

SAMPLE
Matrix: solutions
Analyte: monosaccharides

HPLC VARIABLES
Column: 300 × 7.8 Aminex HPX-87P cation-exchange (Bio-Rad)
Mobile phase: Water
Column temperature: 85
Flow rate: 0.8
Injection volume: 20
Detector: UV 550 following post-column reaction. The column effluent mixed with the reagent pumped at 0.4 mL/min and the mixture flowed through a 20 m × 0.3 mm ID PTFE coil at 90° and a coil at 0° to the detector. (Reagent was prepared from 4000 ppm Purpald (4-amino-3-hydrazino-5-mercapto-1,2,4-triazole) in 2 M NaOH (A) and 40 mM hydrogen peroxide (B). A:B 70:30.)

CHROMATOGRAM
Retention time: 15 (glucose), 16.5 (xylose), 17 (galactose), 19.5 (mannose), 25 (arabinose), 30 (fructose), 42 (ribose)
Limit of detection: 15-80 ng

KEY WORDS
post-column reaction

REFERENCE
Del Nozal, M.J.; Bernal, J.L.; Hernandez, V.; Toribio, L.; Mendez, R. Purpald (4-amino-3-hydrazino-5-mercapto-1,2,4-triazole) as a reagent for post-column derivatization of neutral monosaccharides in high pressure liquid chromatography, *J.Liq.Chromatogr.*, **1993**, *16*, 1105–1116.

Semicarbazide

SAMPLE
Matrix: blood
Analyte: pyridoxal
Sample preparation: 1 mL Plasma + 150 μL 3 M perchloric acid, vortex, centrifuge at >1500 g. Remove a 500 μL aliquot of the supernatant and add it to 300 μL buffer, mix, add 100 μL 10 mg/mL acid phosphatase (2 U/mg, grade II, Boehringer-Mannheim) in water, heat at 40° for 16 h, add 150 μL 3 M perchloric acid, vortex, centrifuge at >1500 g, inject a 20 μL aliquot of the supernatant. (Buffer was 1 M sodium acetate/acetic acid containing 24 g/L NaOH, pH 4.6.)

HPLC VARIABLES
Column: 125 × 4 Nucleosil 120 5 C18
Mobile phase: 50 mM Perchloric acid containing 20 mM triethylamine
Flow rate: 2
Injection volume: 20
Detector: F ex 365 em 480 following post-column reaction. The column effluent mixed with the 3.35 g/L semicarbazide hydrochloride in 1.5 M NaOH pumped at 0.5 mL/min and the mixture flowed through a 10 m × 0.3 mm ID crocheted coil of PTFE tubing at 70° to the detector.

CHROMATOGRAM
Retention time: 1.75
Limit of detection: 2 ng/mL

KEY WORDS
post-column reaction; plasma; pharmacokinetics

REFERENCE
Mascher, H. Determination of total pyridoxal in human plasma following oral administration of vitamin B6 by high-performance liquid chromatography with post-column derivatization, *J.Pharm.Sci.*, **1993**, *82*, 972–974.

HYDROXYLAMINE

3,4-Dinitrobenzyloxamine

SAMPLE
Matrix: bulk
Analyte: 4'-monophosphoryl lipid A
Sample preparation: Heat 1 mg compound with 200 μL 10 mg/mL 3,4-dinitrobenzylox-amine (Regis Technologies, Inc., Morton Grove IL) in anhydrous pyridine at 60 ± 1° for 3 h, evaporate to dryness under a stream of nitrogen at 60 ± 1°, reconstitute with 200 μL chloroform:MeOH 2:1 containing IS, inject a 10 μL aliquot.

HPLC VARIABLES
Guard column: C18 Guard-Pak
Column: 300 × 3.9 4 μm NovaPak C18
Mobile phase: Gradient. A was 5 mM tetrabutylammonium dihydrogen phosphate in MeCN:water 95:5. B was 5 mM tetrabutylammonium dihydrogen phosphate in isopropanol:water 95:5. A:B from 90:10 to 20:80 over 45 min.
Column temperature: 50
Flow rate: 1
Injection volume: 10
Detector: UV 254

CHROMATOGRAM
Retention time: 14-43 (depending on structure)
Internal standard: S-444 (derivatized as above) (41)
Limit of detection: 7 ng

REFERENCE
Hagen, S.R.; Thompson, J.D.; Snyder, D.S.; Myers, K.R. Analysis of a monophosphoryl lipid A immunostimulant preparation from *Salmonella minnesota* R595 by high-performance liquid chromatography, *J.Chromatogr.A,* **1997,** *767,* 53–61.

Hydroxylamine

SAMPLE
Matrix: solutions
Analyte: monosaccharides
Sample preparation: Dissolve 50 mg sugars in 700 μL pyridine, add 700 μL 720 mM hydroxylamine hydrochloride in pyridine, heat at 60° for 10 min, add 250 μL acetic anhydride, heat at 75° for 10 min, evaporate to dryness under reduced pressure, reconstitute with 3 mL chloroform. Wash the organic layer three times with 6 mL portions of water and dry it over anhydrous sodium sulfate, evaporate to dryness under reduced pressure,

take up in chloroform, pass through silica gel using chloroform, evaporate the eluate to dryness, reconstitute, inject a 5 μL aliquot.

HPLC VARIABLES
Column: 250 × 4 5 μm μBondapak C18
Mobile phase: Gradient. MeCN:water from 35:75 to 50:50 over 15 min.
Flow rate: 1
Injection volume: 5
Detector: UV 207

CHROMATOGRAM
Retention time: k' 3.1 (xylose), k' 3.3 (lyxose), k' 3.4 (ribose), k' 3.4 (arabinose), k' 4.3 (allose), k' 4.3 (altrose), k' 3.9 (dextrose), k' 4.2 (mannose), k' 4.0 (gulose), k' 3.8 (idose), k' 4.8 (talose), k' 4.2 (galactose), k' 3.0, k' 3.2 (fructose, syn and anti isomers)
Limit of detection: 3 μg

REFERENCE
Velasco, D.; Castells, J.; Lopez-Calahorra, F. High-performance liquid chromatographic separation of monosaccharides as their peracetylated ketoximes and aldononitriles, *J.Chromatogr.*, **1990**, *519*, 228–236.

4-Nitrobenzylhydroxylamine

SAMPLE
Matrix: bulk
Analyte: xyloglucan oligosaccharides
Sample preparation: Dissolve 5-5000 pmole oligosaccharides in 200 μL water, add 200 μL 10 mg/mL 4-nitrobenzylhydroxylamine hydrochloride in pyridine, evaporate to dryness under a stream of nitrogen, dissolve in 500 μL toluene, evaporate to dryness, repeat the toluene evaporation procedure twice more, add 100 μL 20 mg/mL sodium cyanoborohydride in MeOH:glacial acetic acid 90:10, sonicate for 5 min, evaporate to dryness under a stream of nitrogen, reconstitute with 500 μL MeOH:glacial acetic acid 90:10, evaporate to dryness under a stream of nitrogen, repeat this process twice more. Reconstitute with 100 μL 20 mg/mL sodium cyanoborohydride in MeOH:glacial acetic acid 90:10, sonicate for 5 min, evaporate to dryness under a stream of nitrogen, reconstitute with 500 μL MeOH:glacial acetic acid 90:10, evaporate to dryness under a stream of nitrogen, repeat this process twice more. Reconstitute with 500 μL MeOH, evaporate to dryness under a stream of nitrogen, repeat this process twice more, add 500 μL THF, vortex, centrifuge at 2000 g for 15 min, discard the supernatant, repeat the THF wash 3 more times. Dry the residue under a stream of nitrogen, dissolve in 100 μL water, inject an aliquot.

HPLC VARIABLES
Column: 250 × 4 CarboPac PA1 (Dionex)
Mobile phase: Gradient. A was 100 mM NaOH containing 50 mM sodium acetate. B was 100 mM NaOH containing 100 mM sodium acetate. A:B from 100:0 to 0:100 over 25 min.

Flow rate: 1
Detector: UV 275; E, Dionex sequential pulsed amperometric detection (PAD)

CHROMATOGRAM
Retention time: 15-25 (depending on structure)
Limit of detection: 10 pmole

REFERENCE
Pauly, M.; York, W.S.; Guillen, R.; Albersheim, P.; Darvill, A.G. Improved protocol for the formation of *N*-(*p*-nitrobenzyloxy)aminoalditol derivatives of oligosaccharides, *Carbohydr.Res.*, **1996**, *282*, 1–12.

MISCELLANEOUS REACTIONS

Oxidation

SAMPLE
Matrix: tissue
Analyte: pyridoxal-5-phosphate
Sample preparation: Homogenize tissue with 10 volumes 10 mM pH 7.4 potassium phosphate buffer. Remove a 50 μL aliquot and add it to 50 μL 10% trichloroacetic acid, mix, heat at 50° for 15 min, add 35 μL 3.3 M K_2HPO_4, add 1 μL 80 mM KCN, heat at 50° for 25 min, add 12.5 μL 28% phosphoric acid, centrifuge at 13000 g for 10 min, filter (0.45 μm) the supernatant, inject a 5 μL aliquot of the filtrate.

HPLC VARIABLES
Column: 150 × 4 5 μm STR ODS-H (Shimadzu)
Mobile phase: 2 M Acetic acid containing 1 mM sodium 1-heptanesulfonate, adjusted to pH 3.75 with solid KOH
Flow rate: 0.8
Injection volume: 5
Detector: F ex 318 em 418

CHROMATOGRAM
Retention time: 3
Limit of detection: 50 fmole

KEY WORDS
brain

REFERENCE
Naoi, M.; Ichinose, H.; Takahashi, T.; Nagatsu, T. Sensitive assay for determination of pyridoxal-5-phosphate in enzymes using high-performance liquid chromatography after derivatization with cyanide, *J.Chromatogr.*, **1988**, *434*, 209–214.

ALKENE

MISCELLANEOUS REACTIONS

Aromatization

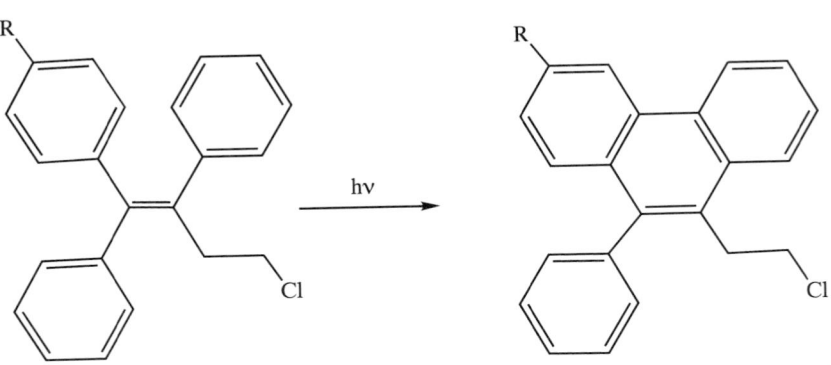

SAMPLE
Matrix: blood
Analyte: toremifene
Sample preparation: 1 mL Plasma + 10 mL hexane:n-butanol 98:2, vortex for 1 min, centrifuge at 1000 g for 10 min. Remove the organic layer and evaporate it to dryness under a stream of nitrogen at 37°, reconstitute the residue in 300 μL MeOH, filter (0.2 μm). Place the filtrate in a quartz cuvette, irradiate with a mercury vapor lamp (15 W, peak wavelength 254 nm, General Electric No. G15T8) for 2 min (Caution! Protect personnel from UV radiation with aluminum foil shielding!), inject a 100 μL aliquot of the irradiated sample.

HPLC VARIABLES
Column: Ultrasphere ODS C18
Mobile phase: MeOH:water:triethylamine 92.9:7:0.1 (At the end of each day wash column with MeOH at 2 mL/min for 30 min.)
Flow rate: 2
Injection volume: 100
Detector: F ex 266

CHROMATOGRAM
Retention time: 6.03
Limit of detection: 8 ng/mL

OTHER SUBSTANCES
Extracted: metabolites

KEY WORDS
plasma; UV irradiation

REFERENCE
Holleran, W.M.; Gharbo, S.A.; DeGregorio, M.W. Quantitation of toremifene and its major metabolites in human plasma by high-performance liquid chromatography following fluorescent activation, *Anal.Lett.*, **1987**, *20*, 871–879.

Bromination

SAMPLE
Matrix: feed
Analyte: aflatoxins
Sample preparation: 25 g Feed + 12.5 g Celite + 12.5 mL water + 125 mL chloroform, shake for 30 min, filter (paper). Evaporate the filtrate to dryness under a stream of nitrogen at 50°, reconstitute the residue in 200 μL chloroform, vortex, inject a 20 μL aliquot. (Treat Celite 545 overnight with HCl:water 1:1, filter (paper), wash with water until the washings are neutral, dry at 110°.)

HPLC VARIABLES
Column: two 100 × 3 5 μm LiChrosorb RP-18 columns in series
Mobile phase: MeCN:MeOH:water 4:7:13 containing 1 mM KBr and 1 mM nitric acid
Flow rate: 0.5
Injection volume: 20
Detector: F ex 360 (filter) em >420 (filter) following post-column reaction. The column effluent passed through an electrochemical cell (10 V, 100 kilohm resistance, construction details in Anal. Chim. Acta 1984, 162, 19) and the bromide was oxidized to bromine. The mixture flowed through a 4 s reaction coil to the detector.

CHROMATOGRAM
Retention time: 9 (G_2), 10 (G_1), 11.5 (B_2), 14 (B_1)
Limit of detection: 0.4-0.8 ppb

KEY WORDS
post-column reaction; protect from light

REFERENCE
Kok, W.T.; van Neer, T.C.H.; Traag, W.A.; Tuinstra, L.G.M.T. Determination of aflatoxins in cattle feed by liquid chromatography and post-column derivatization with electrochemically generated bromine, *J.Chromatogr.*, **1986**, *367*, 231–236.

Hydration

SAMPLE
Matrix: corn
Analyte: aflatoxins
Sample preparation: Condition a Sep-Pak silica SPE cartridge with 10 mL hexane. 50 g Ground corn + 25 mL water + 25 g Celite 545 + 250 mL chloroform, shake on a wrist-action shaker for 30 min, filter (coarse paper), make up the filtrate to 50 mL with hexane, add a 15 mL aliquot to the SPE cartridge, wash with two 15 mL portions of hexane:chloroform 75:25, elute with 15 mL hexane:acetone 50:50. Evaporate the eluate to dryness under a stream of air or nitrogen at 35°, reconstitute with 1.5 mL trifluoroacetic acid:acetic acid:water 10:10:80, shake briefly, sonicate for 2 min, heat at 55° for 15 min, sonicate briefly, allow to cool for 5 min, filter (0.45 μm), inject a 20 μL aliquot.

HPLC VARIABLES
Guard column: 30 × 4.6 5 μm RP-18 (Brownlee)
Column: 100 × 4.6 5 μm RP-18 (Brownlee)
Mobile phase: MeCN:THF:water 10:6:84
Flow rate: 0.8
Injection volume: 20
Detector: F ex 365 (bandpass filter) em >418 (longpass filter)

CHROMATOGRAM
Retention time: 5.4 (aflatoxin G_{2a} from aflatoxin G_1), 7.1 (aflatoxin B_{2a} from aflatoxin B_1), 10.7 (aflatoxin G_2), 15.7 (aflatoxin B_2)
Limit of detection: <1 ng/g

KEY WORDS
SPE

REFERENCE
Hutchins, J.E.; Lee, Y.J.; Tyczkowska, K.; Hagler, W.M. Jr. Evaluation of silica cartridge purification and hemiacetal formation for liquid chromatographic determination of aflatoxins in corn, *Arch.Environ.Contam.Toxicol.*, **1989**, *18*, 319–326.

SAMPLE
Matrix: milk
Analyte: aflatoxin M_1
Sample preparation: Condition a Sep-Pak C18 SPE cartridge with two 10 mL portions of MeOH and with two 10 mL portions of water. Mix 20 mL milk with 30 mL water, add to the SPE cartridge over 15 min, wash with 10 mL water, dry, wash with 10 mL hexane, dry, elute with 4 mL dichloromethane:acetone 80:20. Remove a 5 mL aliquot of the eluate and evaporate it to dryness under a stream of nitrogen in a silanized vial, add 200 μL hexane, add 200 μL trifluoroacetic acid, heat at 40° for 10 min, evaporate to dryness under a stream of nitrogen, reconstitute with 40-100 μL MeCN:water 25:75, inject a 20 μL aliquot.

HPLC VARIABLES
Column: 250 × 4.6 4 μm Nova-Pak C18
Mobile phase: MeCN:water 25:75
Flow rate: 0.7
Injection volume: 20
Detector: F ex 365 em 420

CHROMATOGRAM
Retention time: 5
Limit of detection: 5 pg/mL

KEY WORDS
comparison with ELISA

REFERENCE
Markaki, P.; Melissari, E. Occurrence of aflatoxin M_1 in commercial pasteurized milk determined with ELISA and HPLC, *Food Addit.Contam.*, **1997**, *14*, 451–456.

SAMPLE
Matrix: tissue
Analyte: aflatoxin B_1
Sample preparation: Homogenize 1 g chick liver with 5 mL water, adjust pH to 2 with 1 M HCl, add 10-20 g aflatoxin B_2 in benzene:MeCN 90:10 (Caution! Benzene is a carcinogen!), add 10 mL chloroform:acetone 80:20, shake at slow speed for 20 min, freeze at -20°, centrifuge. Remove the organic layer and evaporate it to dryness under reduced pressure, reconstitute the residue with 50 μL trifluoroacetic acid and 200 μL MeCN:water 10:90, let stand at room temperature for 40-50 min, inject a 50 μL aliquot.

HPLC VARIABLES
Column: 250 × 4.6 10 μm Spherisorb ODS
Mobile phase: MeOH:water 65:35

Flow rate: 0.8
Injection volume: 50
Detector: F ex 360 em 420

CHROMATOGRAM
Retention time: 4.4
Internal standard: aflatoxin B_2 (6.4)
Limit of detection: 1 ng/g

KEY WORDS
chicken; liver

REFERENCE
Espada, Y.; Guitart, R.; Arboix, M. Quantitative determination of aflatoxin B_1 in chick liver, *Food Addit.Contam.*, **1991**, *8*, 163–169.

AMIDE

ACTIVATED HALIDE

2-Bromo-2'-acetonaphthone

SAMPLE
Matrix: solutions
Analyte: barbiturates
Sample preparation: Mix 50 μL of a 20-200 μg/mL solution in acetone with 50 μL of a 0.4-1.6 mg/mL solution of 2-bromo-2'-acetonaphthone in acetone, add 5-10 mg cesium carbonate, heat at 30° for 30 min, add 50 μL glacial acetic acid, mix, inject an aliquot.

HPLC VARIABLES
Column: 300 × 4 μBondapak C18
Mobile phase: MeOH:water 80:20
Flow rate: 2
Detector: UV 249

CHROMATOGRAM
Retention time: 3.1 (mephobarbital), 3.3 (hexobarbital), 4.2 (heptobarbital), 4.9 (barbital), 4.95 (phenobarbital), 6.85 (butobarbital), 8 (amobarbital), 8.75 (pentobarbital), 9.5 (secobarbital)
Limit of detection: 1 ng

REFERENCE
Hulshoff, A.; Roseboom, H.; Renema, J. Improved detectability of barbiturates in high-performance liquid chromatography by pre-column labelling and ultraviolet detection, *J.Chromatogr.*, **1979**, *186*, 535–541.

3-Bromomethyl-6,7-dimethoxy-1-methyl-2(1H)-quinoxalinone

SAMPLE
Matrix: blood
Analyte: 5-fluorouracil and floxuridine
Sample preparation: 100 μL Serum + 5 μL 42 μM 5-chlorouracil in acetone, mix, let stand at room temperature for 3 min, add 500 μL 100 mM pH 3.5 potassium phosphate buffer, add 2 mL ethyl acetate, vortex for 2 min, centrifuge at 1000 g for 5 min. Remove a 1.4 mL aliquot of the organic layer and evaporate it to dryness under reduced pressure. Add 20 mg solid potassium bicarbonate:anhydrous sodium sulfate 1:7 to the residue, add 50 μL 1.3 mM 3-bromomethyl-6,7-dimethoxy-1-methyl-2(1H)-quinoxalinone in acetone, add 50 μL 1.5 mM 18-crown-6 in acetone, heat at 50° for 20 min, cool, inject a 10 μL aliquot. (Silanize all glassware. 3-Bromomethyl-6,7-dimethoxy-1-methyl-2(1H)-quinoxalinone is available from Dojindo Molecular Technologies, Inc., 3 Bethesda Metro Center, Suite 700, Bethesda MD 20814; (301) 664-8448; www.dojindo.co.jp. Synthesis is as follows. Stir 483 g veratrole in 1.45 L acetic acid at 15° for 1 h, add 683 g concentrated nitric acid (d 1.05) over 1 h (maintain the temperature below 40° by cooling and regulating the rate of addition of the nitric acid). Continue stirring and add 2.127 L fuming nitric acid (d 1.50) over 1 h while maintaining the temperature below 30°, let stand for 2 h, pour into a large volume of cold water, filter, wash the solid with water until the washings are neutral, recrystallize from EtOH to give 4,5-dinitroveratrole (mp 129.5-130.5°) (J. Am. Chem. Soc. 1946, 68, 1536). Reflux 5 g 4,5-dinitroveratrole in 200 mL benzene (Caution! Benzene is a carcinogen!), add 100 g 60 mesh iron powder and 20 mL concentrated HCl in small portions over 1 h, reflux for 4 h, add 10 mL water, reflux for 2 h, cool, make alkaline with 2.5 M NaOH, extract several times with 200 mL portions of benzene. Combine the organic layers and evaporate them to dryness, add 10 mL concentrated HCl, recrystallize from EtOH to give 1,2-diamino-4,5-dimethoxybenzene monohydrochloride as very slightly pink needles (mp 240°) (Anal. Chim. Acta 1982, 134, 39). Heat 2.5 mmoles 1,2-diamino-4,5-dimethoxybenzene hydrochloride and 2.4 mmoles pyruvic acid in 30 mL 500 mM HCl on a boiling water bath for 2 h, cool with ice-water, filter. Wash the precipitate with water and dry it under vacuum, recrystallize from MeOH:water 90:10 to give 6,7-dimethoxy-3-methyl-2(1H)-quinoxalinone as yellow needles (mp 255°) (Chem. Pharm. Bull. 1985, 33, 3493). Treat 1 g 6,7-dimethoxy-3-methyl-2(1H)-quinoxalinone dissolved in 50 mL anhydrous MeOH with a solution of diazomethane in ether, evaporate to dryness under reduced pressure, dissolve the residue in 5 mL ethyl acetate, chromatograph on a 250 × 35 column filled with 130 g 70-230 mesh silica gel 60 (Merck) using n-hexane: ethyl acetate 25:75 to give 6,7-dimethoxy-1,3-dimethyl-2(1H)-quinoxalinone as yellow needles (mp 170-171°). Dissolve 350 mg 6,7-dimethoxy-1,3-dimethyl-2(1H)-quinoxalinone

in 3 mL acetic acid, add 350 mg anhydrous sodium acetate, add 2 mL 1.5 M bromine in acetic acid, heat at 100° for 15 min, cool, add 10 mL ether, filter, wash the solid 2 or 3 times with small portions of ether. Combine the filtrate and washings and evaporate them to dryness, dissolve the residue in 5 mL ethyl acetate, chromatograph on a 250 × 35 column filled with 130 g 70-230 mesh silica gel 60 (Merck) using ether, evaporate the main fraction to dryness, recrystallize the residue from n-hexane:ethyl acetate 50:50 to give 3-bromomethyl-6,7-dimethoxy-1-methyl-2(1H)-quinoxalinone as yellow needles (mp 161-163°) (J. Chromatogr. 1985, 346, 227).)

HPLC VARIABLES
Column: 100 × 8 10 μm Radial Pak C18 (Waters) (Wash with MeOH at 2 mL/min for 20 min at the end of each day.)
Mobile phase: Gradient. MeOH:water 35:65 for 15 min, 50:50 for 25 min (step gradient), re-equilibrate at initial conditions for 20 min.
Flow rate: 1.5
Injection volume: 10
Detector: F ex 370 em 455

CHROMATOGRAM
Retention time: 10.2 (floxuridine), 28.1 (5-fluorouracil)
Internal standard: 5-chlorouracil (32.5)
Limit of detection: 12.5-25 ng/mL

KEY WORDS
serum; pharmacokinetics

REFERENCE
Yamaguchi, M.; Nakamura, M.; Kuroda, N.; Ohkura, Y. Determination of 5-fluorouracil and 5-fluoro-2'-deoxyuridine in human serum by high-performance liquid chromatography with fluorescence detection, *Anal.Sci.*, **1987**, *3*, 75−79.

4-(Bromomethyl)-7-methoxycoumarin

SAMPLE
Matrix: blood
Analyte: ethosuximide
Sample preparation: 400 µL Plasma + 100 µL water, mix for 10 s, add 1 mL isopropanol, vortex for 30 s, centrifuge at 1800 g for 5 min. Remove a 1 mL aliquot of the supernatant and add it to 300 µL 10 mM NaOH, evaporate to dryness under a stream of nitrogen at 50°, add 200 µL 2.5 mM 4-(bromomethyl)-7-methoxycoumarin in MeCN, add 300 µL of a solution of 2,2'-dinitrobiphenyl in MeCN, add 100 mg potassium carbonate, shake at 70° for 1.5 h, inject a 15 µL aliquot.

HPLC VARIABLES
Column: 150 × 3.9 4 µm Nova-Pak C18
Mobile phase: MeCN:MeOH:water 20:20:60
Flow rate: 1.3
Injection volume: 15
Detector: UV 320

CHROMATOGRAM
Retention time: 10
Internal standard: 2,2'-dinitrobiphenyl (14)
Limit of detection: 7 pmole

OTHER SUBSTANCES
Non-interfering: acetazolamide, carbamazepine, phenobarbital, primidone, valproic acid

KEY WORDS
plasma

REFERENCE
Chen, S.-H.; Wu, H.-L.; Wu, J.-K.; Kou, H.-S.; Wu, S.-M. Determination of ethosuximide in plasma by derivatization and high performance liquid chromatography, *J.Liq.Chromatogr.Rel.Technol.*, **1997**, 20, 1579–1589.

N-Chloromethyl-4-nitrophthalimide

SAMPLE
Matrix: solutions
Analyte: barbiturates
Sample preparation: Evaporate a solution in water, MeOH, or diethyl ether to dryness, add a 3-fold molar excess of triethylamine, add 0.5-3 mL MeCN, add a 3-fold molar excess of N-chloromethyl-4-nitrophthalimide, heat at 60° for 1 h, inject an aliquot. (N-Chloromethyl-4-nitrophthalimide is available from Tokyo Kasei (TCI America, Portland OR). Synthesis is as follows. Suspend 130 g 4-nitrophthalimide in 80 mL 40% formaldehyde solution, add 200 mL water, reflux for 4 h, filter while hot, N-hydroxymethyl-4-nitrophthalimide crystallizes on cooling (cf. J. Am. Chem. Soc. 1922, 44, 817). Mix a suspension of 2.26 g N-hydroxymethyl-4-nitrophthalimide in 10-15 mL ether with a suspension of 2.1 g phosphorus pentachloride in 10-15 mL ether, after 10 min heat on a water bath, cool in an ice-salt mixture, add ice-water dropwise with shaking, filter to obtain N-chloromethyl-4-nitrophthalimide, dry under vacuum (cf. Chem. Ber. 1959, 9, 1258).)

HPLC VARIABLES
Column: 7 μm LiChrosorb RP8
Mobile phase: MeCN:water 60:40
Flow rate: 1.5
Detector: UV 254

CHROMATOGRAM
Retention time: 3.5 (methylphenobarbital), 4.5 (phenobarbital), 6 (cyclobarbital), 7.2 (amobarbital), 7.6 (secobarbital)
Limit of detection: 4 ng

REFERENCE
Lindner, W.; Santi, W. N-Chloromethylphthalimides as derivatization reagents for high-performance liquid chromatography, *J.Chromatogr.*, **1979**, *176*, 55–64.

ACYL HALIDE

9-Fluorenylmethyl Chloroformate

SAMPLE
Matrix: solutions
Analyte: hexosamines and hexosaminitols
Sample preparation: Mix 10 μL of a 0.1-1 mM solution with 100 μL 200 mM pH 7.0 borate buffer, add 100 μL 0.5-5 mM 9-fluorenylmethyl chloroformate in MeCN, let stand at room temperature for 5-10 min, inject an aliquot.

HPLC VARIABLES
Column: 250 × 5 Sinopak-S 5 C8 (Institute of Chemistry, Academia Sinica)
Mobile phase: MeCN:water 30:70
Detector: UV 254; F ex 270 em 320

CHROMATOGRAM
Retention time: 7 (glucosamine anomer), 7.5 (glucosamine anomer), 8 (glucosaminitol), 11 (N-acetylglucosamine), 12 (N-acetylglucosaminitol)
Limit of detection: 0.4-3.8 pmole (F), 4.5-20 pmole (UV)

REFERENCE
Zhang, Z.; Renen, Z.; Liu, G. High-performance liquid chromatographic analysis of hexosamines, hexosaminitols, N-acetylhexosamines and N-acetylhexosaminitols by ultraviolet and fluorescence detection at picomole levels, *J.Chromatogr.A*, **1996**, *730*, 107–114.

4-Nitrobenzoyl Chloride

SAMPLE
Matrix: bulk
Analyte: glucosylceramide
Sample preparation: Add 2-10 mg compound to 500 μL pyridine:acetic anhydride 50:50, heat at 80° for 2 h, dry in a desiccator over phosphorus pentoxide, add 50-120 mg 4-nitrobenzoyl chloride in 300 μL pyridine, heat at 60° for 4 h, cool in an ice bath, add 2

mL chloroform, add 2 mL 3% sodium bicarbonate solution, mix, centrifuge at 600 g for 3 min. Remove the lower organic layer and wash it twice with 3% sodium bicarbonate solution and 3 times with 2 mL portions of water. Evaporate the organic layer to dryness and suspend the residue in 50 µL chloroform:MeOH 50:50. Chromatograph on a 500 × 10 column of Sephadex LH-20 equilibrated with chloroform:MeOH 50:50, inject an aliquot of the purified compound as a solution in chloroform.

HPLC VARIABLES
Column: 305 mm × 6.3 mm o.d. µBondapak C18
Mobile phase: MeCN
Flow rate: 1
Detector: UV 254

CHROMATOGRAM
Retention time: 10-60 (multiple peaks)

OTHER SUBSTANCES
Also analyzed: globoside, hematoside, lactosylceramide, trihexosylceramide

REFERENCE
Suzuki, A.; Handa, S.; Yamakawa, T. Separation of molecular species of higher glycolipids by high performance liquid chromatography of their O-acetyl-N-p-nitrobenzoyl derivatives, *J.Biochem.*, **1977**, *82*, 1185–1187.

MISCELLANEOUS REACTIONS

Chlorination

SAMPLE
Matrix: solutions
Analyte: proteins
Sample preparation: Inject an aliquot of a solution in 50 mM pH 7.5 phosphate buffer.

HPLC VARIABLES
Column: 300 × 7.5 TSKgel-G3000SW (Tosoh)
Mobile phase: 100 mM pH 7.5 Phosphate buffer containing 100 mM sodium sulfate
Flow rate: 0.8
Injection volume: 20
Detector: F ex 370 em 440 following post-column reaction. The column effluent mixed with the oxidant pumped at 0.2 mL/min and this mixture flowed through a 3 m × 0.5 mm ID PTFE coil at 70°. The effluent from this coil mixed with the reagent pumped at 0.2 mL/min and this mixture flowed through a 5 m × 0.5 mm ID PTFE coil at 70° and a 1 m × 0.5 mm ID PTFE cooling coil to the detector. (Prepare oxidant by diluting 10% sodium hypochlorite solution with 50 mM NaH_2PO_4 containing 0.1% Brij-35 and 50 mM Na_2HPO_4 containing 0.1% Brij-35 to a final chlorine concentration of 0.8% and final pH of 7.5. Prepare reagent by dissolving 8 g sodium nitrite and 40 mg thiamine hydrochloride in 100 mL 50 mM pH 7.5 phosphate buffer, adjust to pH 7.5 with 50 mM Na_2HPO_4 or 50 mM NaH_2PO_4 make up to 200 mL with 50 mM pH 7.5 phosphate buffer. The chlorinated proteins oxidize thiamine to fluorescent thiochrome.)

CHROMATOGRAM

Retention time: 8 (thyroglobulin), 12 (bovine serum albumin), 12 (ovalbumin), 14 (myoglobin), 14 (cytochrome c), 16 (lysozyme)
Limit of detection: 10 ng

KEY WORDS

post-column reaction

REFERENCE

Yokoyama, T.; Kinoshita, T. High-performance liquid chromatographic determination of proteins by post-column fluorescence derivatization with thiamine reagent, *J.Chromatogr.*, **1990**, *518*, 141–148.

Cyclization

SAMPLE

Matrix: urine
Analyte: aminolevulinic acid
Sample preparation: 50 μL Urine + 3.5 mL acetylacetone:EtOH:water 15:10:75 containing 4 g/L NaCl + 450 μL formaldehyde solution, heat in a boiling water bath for 30 min, cool, inject a 50 μL aliquot. (Formaldehyde solution was 85 mL formalin in 1 L water.)

HPLC VARIABLES

Column: TSK-gel 80 TM (Tosoh)
Mobile phase: MeOH:water:acetic acid 60:39:1
Column temperature: 40
Flow rate: 0.8
Injection volume: 50
Detector: F ex 363 em 473

CHROMATOGRAM

Retention time: 7.6
Limit of detection: 10 ng/mL

REFERENCE

Okayama, A.; Fujii, S.; Miura, R. Optimized fluorometric determination of urinary delta-aminolevulinic acid by using pre-column derivatization, and identification of the derivative, *Clin.Chem.*, **1990**, *36*, 1494–1497.

Hydrolysis

SAMPLE
Matrix: blood
Analyte: urea
Sample preparation: Inject a 500 μL aliquot of serum ultrafiltrate onto a 50 × 4.6 column packed with immobilized urease, let stand at room temperature for 10-15 min, inject another 100 μL ultrafiltrate to force ultrafiltrate into the 10 μL sample loop, inject this aliquot. (Prepare the urease column by suspending 1 g Eupergit C (epoxyacrylic resin granules, Röhm Pharma, Weiterstadt, Germany) in 2.5 mL 4 mg/mL urease (EC 3.5.1.5, Type IV, Jack Beans, Sigma) in water, filter after 1 h (Analyst 1984, 109, 147), suspend in water, slurry pack in a 50 × 4.6 column. The enzyme converts serum urea to ammonium ions which are then detected by the HPLC system. Wash column with pH 7 phosphate buffer after use, store in the refrigerator.)

HPLC VARIABLES
Column: 250 × 4.6 Wescan cation-exchange (Bio-Rad)
Mobile phase: pH 2.28 Phosphoric acid
Flow rate: 1
Injection volume: 10
Detector: Conductivity

CHROMATOGRAM
Retention time: 4

KEY WORDS
horse; serum; ultrafiltrate

REFERENCE
Shintani, H.; Ube, S. Simultaneous determination of serum cations, anions and uremic toxins by ion chromatography using an immobilized enzyme, *J.Chromatogr.*, **1985**, *344*, 145–156.

SAMPLE
Matrix: blood, urine
Analyte: urea and ammonia
Sample preparation: Dilute 100 (serum) or 1000 (urine) fold with water, filter (0.2 μm), adjust pH to 10-11 with KOH, add a 2 mL aliquot to the SPE column, discard the first 1.5 mL eluate, inject a 10 μL aliquot of the next eluate fraction. (Prepare the SPE column by adding 50-100 mesh Dowex 1-X2 strongly basic anion exchange resin to a Pasteur pipette (40 mm bed depth), add a few mL water to the column.)

HPLC VARIABLES
Guard column: 60 × 2 C18
Column: 150 × 3 5 μm Spherisorb ODS-2
Mobile phase: 50 mM pH 6.9 potassium phosphate buffer containing 5 mM sodium octylsulfonate
Flow rate: 0.5
Injection volume: 10
Detector: F ex 340 em 455 following post-column reaction. The column effluent passed through a 60 × 2 immobilized urease solid-phase reactor and then mixed with the reagent pumped at 0.5 mL/min. This mixture flowed through a coil of 0.2 mm ID PTFE tubing (volume 600 μL) to the detector. (Prepare reagent by dissolving 24.7 g boric acid in 1 L water, adjust pH to 10.2 with KOH, add 10 mL 80 mg/mL o-phthalaldehyde in EtOH, add 1 mL mercaptoethanol, store under nitrogen at 4°, stable for at least 1 week. Prepare the solid-phase reactor by treating 10 μm LiChrospher SI 500 with 3-aminopropyltri-

ethoxysilane, couple urease (urea amidohydrolase EC 3.5.1.5, U-2000 (Sigma)) to the silica using 25% glutaraldehyde solution. Slurry the urease-silica in water and pack the reactor with 10 mM pH 6.9 potassium phosphate buffer at 110 bar for 15 min (J. Chromatogr. 1985, 325, 255). To maintain reactor performance flush system for 30 min with 50 mM pH 6.9 potassium phosphate buffer containing 5 mM EDTA every 2 weeks. Store reactor at 4° when not in use.)

CHROMATOGRAM
Retention time: 3.4 (urea), 6.9 (ammonia)
Limit of detection: 30 ppb

OTHER SUBSTANCES
Non-interfering: amino acids

KEY WORDS
serum; SPE; post-column reaction

REFERENCE
Jansen, H.; van der Velde, E.G.; Brinkman, U.A.T.; Frei, R.W. Liquid chromatographic determination of urea and ammonia in body fluids using a post-column enzymatic reactor, *J.Chromatogr.*, **1986**, *378*, 215–221.

SAMPLE
Matrix: bulk
Analyte: herbicides
Sample preparation: Dissolve in 1 mL 2 M NaOH, heat at 75° for 45 min, cool, add 2 drops methyl isobutyl ketone, add 2 drops 0.1% dansyl chloride in acetone, shake well, heat at 65° for 30 min, cool, acidify with 10% HCl, add 300 μL benzene (Caution! Benzene is a carcinogen!), extract, inject a 1-10 μL aliquot.

HPLC VARIABLES
Column: 1000 × 2.4 Zipax coated with 0.5% β,β'-oxydipropionitrile
Mobile phase: Hexane:MeOH 95:5
Flow rate: 0.78
Injection volume: 1-10
Detector: F ex Turner filter no. 811 em Turner filter no. 817

CHROMATOGRAM
Retention time: 5 (linuron), 5.4(chloropropham), 6.5 (propham)

KEY WORDS
normal phase

REFERENCE
Frei, R.W.; Lawrence, J.F. Fluorigenic labelling in high-speed liquid chromatography, *J.Chromatogr.*, **1973**, *83*, 321–330.

SULFONYL HALIDE

Dansyl Chloride

SAMPLE
Matrix: blood
Analyte: barbiturates
Sample preparation: Extract 20 μL whole blood with three 20 μL portions of acetone: diethyl ether 50:50. Combine the organic layers and add 5 μL ethyl acetate, dry over 4 Å molecular sieve, evaporate to about 5 μL (mostly ethyl acetate), add 15 μg dansyl chloride, add 2 mg potassium carbonate, reflux for 2 h, dilute to 100 μL, inject a 5 μL aliquot.

HPLC VARIABLES
Column: 500 × 2.5 30 μm pellicular C18
Mobile phase: Gradient. MeOH:water 0:100 for 8 min, to 65:35 over 20 min.
Flow rate: 1
Injection volume: 5
Detector: F ex 360 em 520

CHROMATOGRAM
Retention time: 16 (aprobarbital), 21 (barbital), 32 (heptabarbital)

KEY WORDS
whole blood

REFERENCE
Dünges, W.; Naundorf, G.; Seiler, N. High pressure liquid chromatographic analysis of barbiturates in the picomole range by fluorometry of their DANS-derivatives, *J.Chromatogr.Sci.*, **1974**, *12*, 655–657.

AMINE

ACTIVATED HALIDE

2-Bromoacetophenone

SAMPLE
Matrix: blood
Analyte: cidofovir
Sample preparation: Briefly vortex 100 μL plasma with 300 μL 500 ng/mL IS in MeCN: water:acetic acid 80:19:1, centrifuge at 15000 g for 5 min. Remove the supernatant and add it to 100 μL 1.25 mM 2-bromoacetophenone (phenacyl bromide) in MeCN, heat at 80° for 45 min, evaporate to dryness under reduced pressure at room temperature, reconstitute with 60 μL water, vortex briefly, centrifuge at 15000 g for 5 min, inject a 20 μL aliquot of the supernatant.

HPLC VARIABLES
Column: 100 × 4.6 5 μm Prodigy ODS-2 (Phenomenex)
Mobile phase: MeCN:water 30:70 containing 12 mM phosphoric acid and 6 mM dodecyl-triethylammonium phosphate (Q12) (Bodman) (final pH 3.0-3.1)
Column temperature: 45
Flow rate: 3
Injection volume: 20
Detector: F ex 305 em 370

CHROMATOGRAM
Retention time: 3.1
Internal standard: cytidine-5'-monophosphate (5.3)
Limit of detection: 5 ng/mL

OTHER SUBSTANCES
Non-interfering: (-)-cis-5-(4-amino-1,2-dihydro-2-oxo-1-pyrimidinyl)-1,3-oxothiolane-2-methanol (3-TC)), zalcitabine

KEY WORDS
monkey; plasma

REFERENCE
Eisenberg, E.J.; Cundy, K.C. High-performance liquid chromatographic determination of cytosine-containing compounds by precolumn fluorescence derivatization with phenacyl bromide: application to antiviral nucleosides and nucleotides, *J.Chromatogr.B*, **1996**, *679*, 119-127.

SAMPLE

Matrix: solutions

Analyte: cytosine containing compounds

Sample preparation: Briefly vortex 100 µL of an aqueous solution with 300 µL MeCN: water:acetic acid 80:19:1, centrifuge at 15000 g for 5 min. Remove the supernatant and add it to 100 µL 1.25 mM 2-bromoacetophenone (phenacyl bromide) in MeCN, heat at 80° for 45 min, evaporate to dryness under reduced pressure at room temperature, reconstitute with 60 µL water, vortex briefly, centrifuge at 15000 g for 5 min, inject a 20 µL aliquot of the supernatant.

HPLC VARIABLES

Column: 100 × 4.6 5 µm Prodigy ODS-2 (Phenomenex)

Mobile phase: MeCN:water:trifluoroacetic acid 16:84:0.1

Column temperature: 45

Flow rate: 2.2

Injection volume: 20

Detector: F ex 305 em 370

CHROMATOGRAM

Retention time: 1.4 (cytosine monophosphate), 2 (cytosine, zalcitabine), 2.7 (cytidine), 3.3 (deoxycytidine), 9 ((-)-cis-5-(4-amino-1,2-dihydro-2-oxo-1-pyrimidinyl)-1,3-oxothiolane-2-methanol (3-TC))

OTHER SUBSTANCES

Non-interfering: guanidine, hypoxanthine, thymine, uracil, xanthine

REFERENCE

Eisenberg, E.J.; Cundy, K.C. High-performance liquid chromatographic determination of cytosine-containing compounds by precolumn fluorescence derivatization with phenacyl bromide: application to antiviral nucleosides and nucleotides, *J.Chromatogr.B*, **1996**, *679*, 119–127.

4-(Bromomethyl)-6,7-dimethoxycoumarin

RELATED REFERENCE

Uematsu, T.; Nakashima, M.; Fujii, M.; Hamano, K.; Yasutomi, M.; Kodaira, S.; Kato, T.; Kotake, K.; Oka, H.; Masuike, T. Measurement of 5-fluorouracil in scalp hair: a possible index of patient compliance with oral adjuvant chemotherapy. *Eur.J.Clin.Pharmacol.* **1996**, *50*, 109-113.

SAMPLE

Matrix: blood

Analyte: fluorinated uracils

Sample preparation: Add 250 µL serum to a 20 × 7 DEAE-Cellulofine AM anion-exchange column (Seikagaku Tokyo), elute with 3.5 mL 1 mM HCl, discard the first 0.5 mL eluate, collect the next 3 mL eluate. Evaporate the eluate to 0.5 mL under reduced pressure, add 15 mL ethyl acetate, shake, centrifuge. Remove the organic layer and evaporate it to dryness, reconstitute the residue in 800 µL anhydrous acetone, add 100 µL 750 µg/mL 4-(bromomethyl)-6,7-dimethoxycoumarin in acetone, add 100 µL 250 µg/mL 18-crown-6 in acetone, add 1.5 mg anhydrous potassium carbonate, heat at 70° for 15 min (protect from atmospheric moisture with a calcium chloride drying tube), cool, inject an aliquot

HPLC VARIABLES
Column: 200 × 4 5 μm Nucleosil 5 C18
Mobile phase: MeOH:water 60:40
Flow rate: 0.8
Detector: F ex 340 em 420

CHROMATOGRAM
Retention time: 5 (floxuridine), 7 (ftorafur), 10 (5-fluorouracil)
Limit of quantitation: 6-100 ng/mL

KEY WORDS
serum; protect from light

REFERENCE
Yoshida, S.; Adachi, T.; Hirose, S. 4-Bromomethyl-6,7-dimethoxycoumarin as a fluorescence reagent for precolumn derivatization of 5-fluorouracil compounds in high-performance liquid chromatography, *J.Chromatogr.*, **1988**, *430*, 156–162.

4-(Bromomethyl)-7-methoxycoumarin

SAMPLE
Matrix: blood
Analyte: fluorouracil and tegafur
Sample preparation: 500 μL Serum + 500 μL physiological saline + 100 μL 500 mM NaH$_2$PO$_4$ + 8 mL ethyl acetate, extract, centrifuge. Remove the organic layer and evaporate it to dryness under reduced pressure, reconstitute the residue in 1 mL 500 μg/mL 4-(bromomethyl)-7-methoxycoumarin in acetone:MeCN 1:2 containing 100 μg/mL 18-crown-6 and 1 mg/mL potassium carbonate, reflux in the dark for 45 min, cool, add valeric acid, reflux for 5 min, dilute with acetone, inject an aliquot.

HPLC VARIABLES
Column: 200 × 4 5 μm Nucleosil 5 C18
Mobile phase: MeOH:water 70:30
Flow rate: 0.8
Detector: F ex 346 em 395

CHROMATOGRAM
Retention time: 4.5 (tegafur (ftorafur)), 7 (fluorouracil)
Internal standard: valeric acid (8)
Limit of detection: 100-384 fmole

KEY WORDS
serum

REFERENCE
Iwamoto, M.; Yoshida, S.; Hirose, S. Fluorescence determination of 5-fluorouracil and 1-(tetrahydro-2-furanyl)-5-fluorouracil in blood serum by high-performance liquid chromatography, *J.Chromatogr.*, **1984**, *310*, 151–157.

SAMPLE
Matrix: blood, saliva, urine
Analyte: pilocarpine

Sample preparation: Mix 50 μL urine with 5 μL 10% sodium bicarbonate solution add 50 μL 500 ng/mL pilosine in MeOH. Mix 1.5 mL saliva with 90 μL 10% sodium bicarbonate, add 50 μL 500 ng/mL pilosine in MeOH. Mix 3 mL Plasma with 50 μL 500 ng/mL pilosine in MeOH. Extract these samples twice with 3 mL portions of chloroform. Combine the organic layers and evaporate them to dryness, reconstitute the residue in 200 μL 0.08% 4-(bromomethyl)-7-methoxycoumarin in acetone, heat at 37° for 48 h, evaporate to dryness, reconstitute with 1 mL mobile phase, centrifuge at 20000 g for 5 min, inject an aliquot of the supernatant.

HPLC VARIABLES
Guard column: 20 × 4.6 cyanopropyl silica (Brownlee)
Column: 220 × 4.6 cyanopropyl silica (Brownlee)
Mobile phase: MeCN:buffer 30:70 (Buffer was 3 mM diethylamine adjusted to pH 3.5 with 1 M phosphoric acid.)
Column temperature: 37
Flow rate: 1
Injection volume: 100
Detector: F ex 324 em 400

CHROMATOGRAM
Retention time: 13.86
Internal standard: pilosine (12.36)
Limit of detection: 1 ng/mL

OTHER SUBSTANCES
Interfering: isopilocarpine

KEY WORDS
silanize all glassware with dichlorodimethylsilane for 12 h; plasma; pharmacokinetics

REFERENCE
Aromdee, C.; Fawcett, J.P.; Ledger, R. Sensitive high-performance liquid chromatographic assay for pilocarpine in biological fluids using fluorescence derivatization, *J.Chromatogr.B*, **1996**, *677*, 313–318.

SAMPLE
Matrix: bulk
Analyte: 5-fluorouracil
Sample preparation: Dissolve 1-10 ng 5-fluorouracil in 100 μL DMSO, add 400 μL 2.5 mg/mL 4-(bromomethyl)-7-methoxycoumarin in DMSO, add 5 mg potassium carbonate, shake at room temperature for 15 min, add 500 μL water, centrifuge at 15600 g for 5 min, inject a 100 μL aliquot of the supernatant onto column A and elute to waste with mobile phase A, after 9.5 min divert the effluent containing the derivatized 5-fluorouracil onto column B, after another 2 min elute column B with mobile phase B, monitor the effluent from column B.

HPLC VARIABLES
Column: A 150 × 4.6 5 μm CPS-Hypersil cyanopropyl; B 150 × 4.6 5 μm ODS-Hypersil
Mobile phase: A MeOH:water 50:50; B MeOH:water 60:40
Column temperature: 35 (column A only)
Flow rate: 1
Injection volume: 100
Detector: F ex 325 em 395

CHROMATOGRAM
Retention time: 17
Limit of detection: 5 pg

KEY WORDS
column-switching; heart-cut

REFERENCE

Kindberg, C.G.; Slavik, M.; Riley, C.M.; Stobaugh, J.F. High-performance liquid chromatography of 5-fluorouracil after derivatization with 4-bromomethyl-7-methoxycoumarin. Characterization of the derivative and the use of column switching for the improvement of resolution and the enhancement of sensitivity, *J.Pharm.Biomed.Anal.*, **1989**, *7*, 459–469.

SAMPLE

Matrix: enzyme incubations
Analyte: bromodeoxyuridine and thymidine
Sample preparation: Condition a 100 mg Sep-Pak C18 SPE cartridge with 2 mL MeCN, 1 mL water, and 1 mL buffer. Centrifuge enzyme incubation in buffer, add a 200 μL aliquot to the SPE cartridge, wash with 200 μL water, elute with 600 μL MeCN:water 60:40. Evaporate the eluate to dryness under a stream of nitrogen, reconstitute the residue in 50 μL DMSO, add 2 mg powdered potassium carbonate, mix thoroughly, add 20 μL 2.5 mM 18-crown-6 in MeCN, mix thoroughly, add 200 μL 2.5 mM 4-(bromomethyl)-7-methoxycoumarin in MeCN, mix thoroughly, evaporate to dryness under a stream of nitrogen, reconstitute with 200 μL MeCN:water 75:25, centrifuge. Remove a 190 μL aliquot of the supernatant and add it to 10 μL 500 mM acetic acid, inject an aliquot. (Buffer was 50 mM pH 7.6 potassium phosphate buffer containing 5 mM magnesium sulfate.)

HPLC VARIABLES

Column: 100 × 5 Novapak C18
Mobile phase: Gradient. A was water. B was MeCN:water 75:25. A:B from 72:28 to 68:32 over 4 min, maintain at 68:32 for 4 min, to 20:80 over 1 min
Flow rate: 2
Detector: F ex 320 em 394

CHROMATOGRAM

Retention time: 5.5 (thymidine), 8.5 (bromodeoxyuridine)

KEY WORDS

SPE

REFERENCE

Stratford, M.R.L.; Dennis, M.F. Measurement of incorporation of bromodeoxyuridine into DNA by high performance liquid chromatography using a novel fluorescent labelling technique, *Int.J.Radiat.Oncol.Biol.Phys.*, **1992**, *22*, 485–487.

SAMPLE

Matrix: solutions
Analyte: pyrimidines
Sample preparation: Add 10 mg of 4-(bromomethyl)-7-methoxycoumarin and 10 mg potassium carbonate to 5 mL of a solution in DMSO, after 5 min at room temperature add 4-nitrobenzoic acid.

HPLC VARIABLES

Column: 200 × 4 5 μm Nucleosil 5 C18
Mobile phase: MeCN:MeOH:water 5:29:66
Flow rate: 0.6
Detector: F ex 346 em 395

CHROMATOGRAM

Retention time: 5 (inosine), 6 (uridine), 7 (uracil), 8 (thymine), 12.5 (floxuridine (fluorodeoxyuridine))
Internal standard: chlorodeoxyuridine (16.5)

KEY WORDS

some details of serum extraction in paper

REFERENCE

Yoshida, S.; Hirose, S.; Iwamoto, M. Use of 4-bromomethyl-7-methoxycoumarin for derivatization of pyrimidine compounds in serum analysed by high-performance liquid chromatography with fluorimetric detection, *J.Chromatogr.*, **1986**, *383*, 61–68.

SAMPLE

Matrix: tissue

Analyte: 5-fluorouracil

Sample preparation: Homogenize (Ultra-Turrax) 1 g tissue with 0.05-1 μg IS and 15 mL ice-cold MeCN, rinse homogenizer with 10 mL ice-cold MeCN, centrifuge at 6000 rpm for 15 min, remove the supernatant, wash the pellet with 5 mL MeCN, centrifuge for 5 min. Combine the supernatants, remove a 5 mL aliquot, evaporate to dryness under a stream of nitrogen at room temperature, reconstitute the residue in 50 μL n-hexane:ethyl acetate:water 40:80:1, chromatograph on a 125 × 4 5 μm Spherisorb Si column (A) and a 200 × 4.6 5 μm Spherisorb Si column (B) with n-hexane:ethyl acetate:water 40:80:1 at 1 mL/min using column-switching. Initially elute the columns in series, after 3 min (after 5-fluorouracil has eluted from column A to column B) elute only column B, monitor the effluent from column B at 266 nm, collect the appropriate fraction. (Backflush column A to waste for 6 min before the next injection.) Evaporate the eluate to dryness under a stream of nitrogen, reconstitute with 10 μL acetone, add 1 mg freshly-powdered potassium carbonate, add 5 μL 200 μg/mL 18-crown-6 in acetone, add 20 μL 750 μg/mL 4-(bromomethyl)-7-methoxycoumarin in acetone, heat at 70° for 25 min, inject a 1 μL aliquot.

HPLC VARIABLES

Guard column: 20 × 1.6 3 μm Nukleosil-120 (MZ Analysentechnik)

Column: 125 × 1.6 3 μm Nukleosil-120 (MZ Analysentechnik)

Mobile phase: MeCN:MeOH:water 30:15:50

Flow rate: 0.06

Injection volume: 1

Detector: F ex 305 em 407

CHROMATOGRAM

Retention time: 27

Internal standard: 5-chlorouracil (31)

Limit of quantitation: 30 ng/g

Limit of detection: 3 ng/g

KEY WORDS

microbore; human; pig; liver

REFERENCE

Jochheim, C.; Janning, P.; Marggraf, U.; Löffler, T.M.; Hasse, F.; Linscheid, M. A procedure for the determination of 5-fluorouracil in tissue using microbore HPLC and fluorescence detection, *Anal.Biochem.*, **1994**, *217*, 285−291.

9-Chloro-10-methylacridinium Triflate

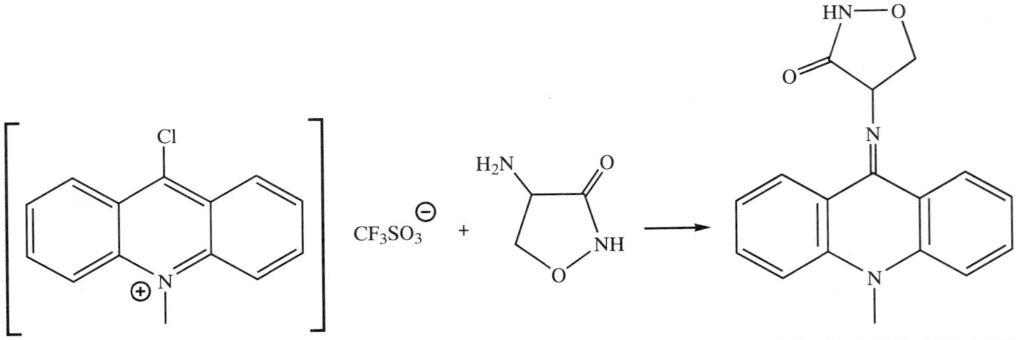

SAMPLE

Matrix: urine

Analyte: cycloserine

Sample preparation: Filter (0.45 μm) urine. Remove a 1 mL aliquot and add a 20-fold molar excess of 2 mM 9-chloro-10-methylacridinium triflate in MeCN:pH 5.0 phosphate buffer 50:50, vortex for 10 s, heat at 70° for 30 min, add 1 mL glacial acetic acid, mix, inject a 10 μL aliquot. (Synthesis of 9-chloro-10-methylacridinium triflate is as follows. Dissolve 6.07 g 9-chloroacridine (Eastman) in 55 mL dry dichloromethane, add 5 g methyl trifluoromethanesulfonate, stir for 3 h, filter, wash the solid with cold dichloromethane, dry in air overnight, recrystallize from MeCN to obtain 9-chloro-10-methylacridinium triflate as yellow crystals (mp 227-229°).)

HPLC VARIABLES
Column: 250 × 4.6 Partisil silica
Mobile phase: MeCN:EtOH:glacial acetic acid 50:30:20
Flow rate: 2
Injection volume: 10
Detector: F ex 257 em 475

CHROMATOGRAM
Retention time: 5.3
Limit of quantitation: 150 ng/mL

REFERENCE
Yoo, G.S.; Choi, K.; Stewart, J.T. Second derivative ultraviolet spectrophotometry and high performance liquid chromatography with fluorometric detection of cycloserine using 9-chloro-10-methylacridinium triflate as a new UV and fluorescent labeling agent, *Anal.Lett.*, **1990**, *23*, 1245–1263.

4-Chloro-7-nitro-2,1,3-benzoxadiazole

RELATED REFERENCE
Nishikawa, Y.; Kuwata, K. Liquid chromatographic determination of low molecular weight aliphatic amines in air via derivatization with 7-chloro-4-nitro-2,1,3-benzoxadiazole. *Anal.Chem.* **1984**, *56*, 1790-1793.

SAMPLE
Matrix: beverages
Analyte: amines
Sample preparation: Adjust pH of beer to 9.0 with 1 M NaOH, mix with equal volumes of MeOH saturated with sodium acetate and 10 mg/mL 4-chloro-7-nitro-2,1,3-benzoxadiazole in MeOH, heat at 60° in a sealed container for 2 h, dilute 1:4, inject an aliquot.

CAPILLARY ELECTROPHORESIS
Capillary: 81 cm × 75 μm (74.5 cm to detector) fused-silica
Capillary preparation: Between runs wash with 100 mM NaOH, rinse with water, equilibrate with running buffer for 5 min.
Running buffer: 10 mM Boric acid containing 50 mM sodium dodecyl sulfate, adjusted to pH 9.3 with NaOH
Voltage/Current: 24 kV/68 μA
Injection: Vacuum injection at -3.4 Pa for 2 s.
Detector: F ex 488 (5 mW argon ion laser) em 540
Model: Isco Model 3850

Migration time: 4.2 (ammonia), 6.1 (ethylamine), 9.6 (isoamylamine), 10.8 (dibutylamine)
Limit of detection: 9-130 μM

KEY WORDS
beer

REFERENCE
Preston, L.M.; Weber, M.L.; Murray, G.M. Micellar electrokinetic capillary chromatography with laser-induced fluorimetric detection of amines in beer, *J.Chromatogr.B*, **1997**, *695*, 175–180.

SAMPLE
Matrix: blood
Analyte: pseudoephedrine
Sample preparation: 1 mL Plasma + 0.1-1 μg IS + 500 μL 25% potassium carbonate + 10 mL diethyl ether, shake for 10 min, centrifuge at 1000 g for 5 min, freeze in MeOH/dry ice. Remove the organic layer and add it to 1 mL 50 mM HCl, shake for 5 min, centrifuge at 1000 g for 2 min, freeze in MeOH/dry ice, discard the organic layer, remove traces of ether with a stream of nitrogen. Thaw the aqueous layer and add 40 μL 1 M NaOH and 500 μL 200 mM pH 8.5 borate buffer, add 400 μL 0.5% 4-chloro-7-nitro-2,1,3-benzoxadiazole (NBD-Cl) in methylisobutylketone, heat at 79° for 1 h (mix at 5 min intervals), cool on ice, add 3 mL cyclohexane, vortex, centrifuge at 1000 g for 3 min, inject a 200 μL aliquot on to column A and elute to waste with mobile phase A, after 3 min elute the contents of column A on to column B with mobile phase B, after 2 min remove column A from the circuit, elute column B with mobile phase B, monitor the effluent from column B.

HPLC VARIABLES
Column: A Guard Pak CN (Waters); B 250 × 4.6 silica (Alltech)
Mobile phase: A Cyclohexane; B Cyclohexane:toluene:MeOH:butanol 50:47:1:2
Flow rate: 2
Injection volume: 200
Detector: F ex 460-500 (bandpass filter) em > 500 (cutoff filter)

CHROMATOGRAM
Retention time: 12.0
Internal standard: 2-phenylglycinol (2-amino-2-phenylethanol) (9.7)
Limit of quantitation: 10 ng/mL
Limit of detection: 2 ng/mL

OTHER SUBSTANCES
Simultaneously analyzed: dimethoxyphenethylamine, ephedrine, norpsdeudoephedrine, phenylephrine, phenylethanolamine, phenylethylamine, phenylpropanolamine, tyramine
Non-interfering: acetaminophen, aspirin, caffeine, salicylamide, theophylline

KEY WORDS
column-switching; normal phase; plasma; pharmacokinetics

REFERENCE
Veals, J.; Kim, H.; Korduba, C.; Curtis, D.; Durante, E.; Lin, C. Determination of plasma pseudoephedrine by fluorescence detection and high performance liquid chromatography, *J.Liq.Chromatogr.*, **1988**, *11*, 417–433.

SAMPLE
Matrix: blood, urine
Analyte: amphetamines
Sample preparation: Adjust pH of 10 mL plasma or 20 mL urine to 11.4 with 5 M NaOH, add to a column containing 1.5 g Amberlite XAD-2, wash with 10 mL water, elute with 20 (plasma) or 40 (urine) mL chloroform:isopropanol 75:25, add 100 μL 6 M HCl in EtOH to the eluate, evaporate to dryness under reduced pressure, reconstitute with 4 mL 100 mM sodium bicarbonate, add 2 mL 1% 4-chloro-7-nitro-2,1,3-benzoxadiazole in methyl isobutyl ketone, shake briefly, heat at 80° for 30 min, inject a 20 μL aliquot of the upper organic layer.

HPLC VARIABLES
Column: 200 × 5 Partisil 5
Mobile phase: Cyclohexane:ethyl acetate 60:40
Column temperature: 20
Flow rate: 2
Injection volume: 20
Detector: F ex 465 em 515

CHROMATOGRAM
Retention time: 1 (amphetamine), 2 (methamphetamine)
Internal standard: methylamine (5)
Limit of quantitation: 500 ng/mL

KEY WORDS
plasma; normal phase; comparison with other derivatization reagents and with ion-pair chromatography; SPE

REFERENCE
Farrell, B.M.; Jefferies, T.M. An investigation of high-performance liquid chromatographic methods for the analysis of amphetamines, *J.Chromatogr.*, **1983**, *272*, 111–128.

SAMPLE
Matrix: blood, urine
Analyte: taurine
Sample preparation: Condition a Sep-Pak C18 SPE cartridge with 5 mL MeOH and 10 mL water. 500 μL Plasma + 500 μL 10% trichloroacetic acid, mix, centrifuge at 4° at 9000 g for 10 min. 1 mL Urine + 1 mL 10% trichloroacetic acid, mix, centrifuge at 4° at 9000 g for 10 min. Add a 1 mL aliquot of the supernatant to the SPE cartridge, elute with 2 mL 5% trichloroacetic acid. Collect all the effluent from the SPE cartridge and make up to 10 mL with 5% trichloroacetic acid. Remove a 50 μL aliquot and add it to 150 μL 400 mM pH 9.0 sodium borate buffer, mix, add 50 μL MeOH, add 100 μL 30 mM 4-chloro-7-nitro-2,1,3-benzoxadiazole in MeOH, heat at 60° in the dark for 40 min, make up to 1 mL with cold mobile phase, inject a 25-100 μL aliquot.

HPLC VARIABLES
Guard column: CN Guard-PAK (Waters)
Column: 250 × 4 10 μm Partisil SAX
Mobile phase: MeCN:25 mM citric acid 10:90, pH adjusted to 2.9 with 1 M NaOH
Flow rate: 1.3
Injection volume: 25-100
Detector: F ex 470 em 530

CHROMATOGRAM
Retention time: 10
Limit of detection: 5 pmole

KEY WORDS
plasma

REFERENCE
Palmerini, C.A.; Fini, C.; Cantelmi, M.G.; Floridi, A. Assessment of taurine in plasma and urine by anion-exchange high-performance liquid chromatography with pre-column derivatization, *J.Chromatogr.*, **1987**, *423*, 292–296.

SAMPLE
Matrix: blood, urine
Analyte: baclofen
Sample preparation: 500 μL Plasma + 2 mL MeOH, centrifuge. 1 mL Supernatant or 100 μL urine + 500 μL pH 9 sodium tetraborate buffer + 250 μL 0.2% 4-chloro-7-nitro-2,1,3-benzoxadiazole in MeOH, heat at 60° for 45 min, acidify with 100 mM HCl, extract with 5 mL ethyl acetate. Evaporate 3 mL of the extract to about 500 μL, dry with anhydrous sodium sulfate, pass through SPE column, elute with ethyl acetate to give a final volume of 2 mL. Evaporate it to dryness under a stream of nitrogen at 50°, reconstitute

the residue in 1 mL mobile phase, inject a 10 μL aliquot. (Prepare the 25 × 6 SPE column with 63-200 μm silica gel (Merck), wet it with ethyl acetate.)

HPLC VARIABLES
Column: 300 × 3.9 10 μm Bondapak C18
Mobile phase: MeOH:water 45:55
Flow rate: 0.4
Injection volume: 10
Detector: F ex 463 em 524

CHROMATOGRAM
Retention time: 4.0
Limit of detection: 20 ng/mL (plasma), 100 ng/mL (urine)

KEY WORDS
plasma; pharmacokinetics; SPE

REFERENCE
Tosunoglu, S.; Ersoy, L. Determination of baclofen in human plasma and urine by high-performance liquid chromatography with fluorescence detection, *Analyst*, **1995**, *120*, 373–375.

SAMPLE
Matrix: solutions
Analyte: N-nitrosoproline
Sample preparation: Evaporate aqueous solutions to dryness under reduced pressure at 50°, add 1 mL 15% HBr in glacial acetic acid, heat at 50° for 30 min, add 1 mL water, evaporate to dryness under reduced pressure at 50°, reconstitute with 100 μL water, add sufficient sodium carbonate to saturate the solution, add 900 μL 3.33 mg/mL 4-chloro-7-nitro-2,1,3-benzoxadiazole (NBD chloride) in MeCN, heat at 50° for 30 min, cool to room temperature, dilute 1:20, inject an aliquot.

HPLC VARIABLES
Column: 250 × 2 MicroPak LiChrosorb Si 60-10 (Varian)
Mobile phase: n-Hexane:ethyl acetate:acetic acid 50:50:0.5
Flow rate: 2
Detector: F (wavelengths not given)

CHROMATOGRAM
Retention time: 2.1
Limit of detection: 0.5 ng

KEY WORDS
normal phase

REFERENCE
Wolfram, J.H.; Feinberg, J.I.; Doerr, R.C.; Fiddler, W. Determination of N-nitrosoproline at the nanogram level, *J.Chromatogr.*, **1977**, *132*, 37–43.

3-(4,6-Difluorotriazinyl)amino-7-methoxycoumarin

SAMPLE
Matrix: urine
Analyte: amantadine
Sample preparation: 100 μL Urine + 200 μL 50 nM n-decylamine in benzene + 100 μL 1 M NaOH (Caution! Benzene is a carcinogen!), shake mechanically for 15 min, centrifuge at 6400 rpm for 6 min. Remove a 20 μL aliquot of the organic layer and add it to 10 μL 250 μM 3-(4,6-difluorotriazinyl)amino-7-methoxycoumarin in benzene, heat at 140° for 15 min, cool to room temperature, inject a 10 μL aliquot. (Synthesis of 3-(4,6-difluorotriazinyl)amino-7-methoxycoumarin is as follows. Heat 2-hydroxy-4-methoxybenzaldehyde with 0.5 equivalents glycine, 1.25 equivalents fused sodium acetate, and 5 parts freshly distilled acetic anhydride at 140° for 1 h and at 160° for 1 h, cool to 100°, pour into water containing cracked ice, let stand overnight. Wash the solid with very dilute NaOH, cold water, and a few mL of hot MeOH. Crystallize from acetic acid to obtain 3-acetamido-7-methoxycoumarin (mp 230°). Suspend 1 g 3-acetamido-7-methoxycoumarin in 25 mL acetic acid and 25 mL 50% sulfuric acid, heat at 50-60° for 30-45 min, pour into an equal volume of cold water, neutralize with sodium bicarbonate, recrystallize from EtOH/water to obtain 3-amino-7-methoxycoumarin (mp 154°) (J. Ind. Chem. Soc. 1971, 48, 371). Stir 300 mg 3-amino-7-methoxycoumarin in 100 mL anhydrous ether, slowly add 1 mL cyanuric fluoride (Fluka), stir at room temperature for 1 h, evaporate to dryness under reduced pressure, recrystallize from n-hexane to give 3-(4,6-difluorotriazinyl)amino-7-methoxycoumarin as colorless prisms (mp 212-215°). 3-(4,6-Difluorotriazinyl)amino-7-methoxycoumarin reacts with primary and secondary alkyl amines but not with aromatic amines or alcohols.)

HPLC VARIABLES
Column: 250 × 4.6 5 μm TSK gel ODS-120T (Tosoh)
Mobile phase: MeCN:water 50:10
Flow rate: 1
Injection volume: 10
Detector: F ex 345 em 410

CHROMATOGRAM
Retention time: 14
Internal standard: n-decylamine (26)
Limit of detection: 2.5 nM

REFERENCE
Fujino, H.; Goya, S. A fluorogenic reagent, 3-(4,6-difluorotriazinyl)amino-7-methoxycoumarin, for the determination of amantadine by high-performance liquid chromatography, *Chem.Pharm.Bull.*, **1990**, *38*, 544–545.

4-(N,N-Dimethylaminosulfonyl)-7-fluoro-2,1,3-benzoxadiazole

RELATED REFERENCES
Nakashima, K.; Suetsugu, K.; Yoshida, K.; Akiyama, S.; Uzu, S.; Imai, K. High performance liquid chromatography with chemiluminescence detection of methamphetamine and its related compounds using 4-(N,N-dimethylaminosulphonyl)-7-fluoro-2,1,3-benzoxadiazole. *Biomed.Chromatogr.* **1992**, *6*, 149-154.

Uzu, S.; Imai, K.; Nakashima, K.; Akiyama, S. 4-(N,N-dimethylaminosulfonyl)-7-fluoro-2,1,3-benzoxadiazole as chemilumigenic reagent for high performance liquid chromatographic peroxyoxalate chemiluminescence detection. *Biomed.Chromatogr.* **1991**, *5*, 184-185.

SAMPLE
Matrix: blood

Analyte: metoprolol

Sample preparation: Condition a Bond-Elut C18 SPE cartridge with 2 mL MeCN and 1 mL MeCN:water 40:60. Add 20 μL serum to 500 μL MeCN:water 40:60, vortex, add to SPE cartridge, wash with 2 mL MeCN:water 40:60, elute with 1 mL MeOH:water 1:1 containing 0.05% trifluoroacetic acid. Evaporate eluate to dryness under reduced pressure at 40°, reconstitute with 50 μL 100 mM pH 9.0 borate buffer containing 2 mM EDTA, add 50 μL 50 mM 4-(N,N-dimethylaminosulfonyl)-7-fluoro-2,1,3-benzoxadiazole (DBD-F) in MeCN, heat at 45° for 8 h, add 100 μL 100 mM acetic acid in MeCN:water 50:50, inject a 20 μL aliquot. (4-(N,N-Dimethylaminosulfonyl)-7-fluoro-2,1,3-benzoxadiazole (DBD-F) can be purchased from Tokyo Kasei (TCI America, 911 North Harborgate St., Portland, OR 97203; 800-423-8616, 503-283-1681, (fax) 503-283-1987; www.tciamerica.com). Synthesis of DBD-F is as follows. Dissolve 0.5 g magnesium sulfate heptahydrate and 6 g NaOH in 60 mL water, throughout the reaction keep the flask at about 20° with cold water cooling, add 15 mL 30% hydrogen peroxide, add 75 mL MeOH, add 12.1 g powdered benzoyl peroxide in one go, stir for 10 min, pour into 150 mL 20% sulfuric acid, extract three times with 50 mL portions of chloroform, determine peroxybenzoic acid concentration by iodometric titration (Tetrahedron 1967, 23, 3327). Slowly add 110 mL 1 M peroxybenzoic acid in chloroform to 7 g 2,6-difluoroaniline dissolved in 100 mL chloroform, stir at room temperature, when reaction is complete (iodometric titration) wash with 2% sodium thiosulfate, wash with 5% sodium carbonate, wash with water, dry over anhydrous sodium sulfate, evaporate to dryness under reduced pressure, recrystallize 2,6-difluoronitrosobenzene from EtOH (mp 108.5-109.5). Stir 8.5 g 2,6-difluoronitrosobenzene in 85 mL DMSO at room temperature and add a solution of 3.91 g sodium azide in 85 mL DMSO dropwise, let stand for about 1 h, add to a large volume of water, extract with ether, dry the extracts over anhydrous sodium sulfate, evaporate to dryness under reduced pressure and distil to give 4-fluoro-2,1,3-benzoxadiazole as a colorless oil (bp 83°/12 mm Hg) (J.Chem.Soc.(C) 1970, 1433). Add 11 mL chlorosulfonic acid dropwise to 3 g 4-fluoro-2,1,3-benzoxadiazole in 10 mL chloroform at 0-10° (use a calcium chloride drying tube), stir at room temperature for 1 h, reflux for 2 h, cool, slowly pour into ice water, remove the organic layer, extract the aqueous layer with chloroform, combine the organic layers, wash, dry over anhydrous magnesium sulfate, evaporate under reduced pressure, take up the residue in 5 mL benzene (Caution! Benzene is a carcinogen!), chromatograph on a 150 × 30 column of silica gel (100-200 mesh Kanto Chemical) with n-hexane:benzene 50:50, evaporate the appropriate fractions to give 4-(chlorosulfonyl)-7-fluoro-2,1,3-benzoxadiazole (CBD-F) as pale yellow needles (mp 64-66°) (Anal. Chem. 1984.56, 2461). Stir 0.76 g CBD-F in 70 mL MeCN at 0-10° and add 1 g dimethylamine hydrochloride in 10 mL 100 mM pH 10 borax dropwise, adjust pH to 5 with 1 M HCl, concentrate to about 10 mL under reduced pressure, extract three times with 200 mL portions of diethyl ether, wash with water, dry over anhydrous magnesium sulfate, evaporate under reduced pressure, chromatograph on a 500 × 20 column of silica gel with chloroform, isolate the appropriate fraction and re-chromatograph on the same column with ethyl acetate:benzene 1:2 to give 4-(N,N-dimethylaminosulfonyl)-7-fluoro-2,1,3-benzoxadiazole (DBD-F) as white needles (mp 124-125°) (yield = 1% !). On a Merck no. 5714 60F_{254} tlc plate eluted with chloroform DBD-F has Rf 0.32 and lies between two other reaction products (Analyst 1989, 114, 413).)

HPLC VARIABLES

Column: 250 × 4.6 5 µm TSK gel ODS 80Tm (Tosoh)
Mobile phase: MeCN:THF:50 mM pH 6.0 imidazole nitrate buffer 28:20:52
Column temperature: 40
Flow rate: 0.8
Injection volume: 20
Detector: Chemiluminescence following post-column reaction. The column effluent mixed with the reagent pumped at 1.4 mL/min and the mixture flowed to the detector. (Reagent was 0.25 mM bis[4-nitro-2-(3,6,9-trioxadecyloxycarbonyl)phenyl] oxalate (Wako) and 37.5 mM hydrogen peroxide in MeCN:ethyl acetate 50:50.)

CHROMATOGRAM
Retention time: 9
Limit of detection: 0.8 ng/mL

KEY WORDS
serum; SPE

REFERENCE
Uzu, S.; Imai, K.; Nakashima, K.; Akiyama, S. Use of 4-(N,N-dimethylaminosulphonyl)-7-fluoro-2,1,3-benzoxadiazole as a labelling reagent for peroxyoxalate chemiluminescence detection and its application to the determination of the β-blocker metoprolol in serum by high-performance liquid chromatography, *Analyst*, **1991**, *116*, 1353–1357.

SAMPLE
Matrix: blood
Analyte: ebiratide
Sample preparation: Condition a 100 mg C18 SPE cartridge (Varian) with 2 mL MEOH and 2 mL water. 400 µL Plasma + 800 µL water, mix, add to the SPE cartridge, wash with 1 mL water, elute with 1 mL 5 mM HCl in MeOH:water 30:70. Evaporate the eluate to dryness in a freeze dryer, reconstitute with 100 µL 100 mM pH 9.0 borate buffer, add 100 µL 4-(N,N-dimethylaminosulfonyl)-7-fluoro-2,1,3-benzoxadiazole in MeCN, mix, heat at 50° for 30 min, inject a 100 µL aliquot. (4-(N,N-Dimethylaminosulfonyl)-7-fluoro-2,1,3-benzoxadiazole (DBD-F) can be purchased from Tokyo Kasei (TCI America, 911 North Harborgate St., Portland, OR 97203; 800-423-8616, 503-283-1681, (fax) 503-283-1987; www.tciamerica.com). Synthesis of DBD-F is as described above.)

HPLC VARIABLES
Column: 150 × 4.6 5C18 (Vydac)
Mobile phase: Gradient. A MeCN:MeOH:50 mM pH 6.0 imidazole nitrate 26.7:13.3:60. B MeCN:MeOH:50 mM pH 6.0 imidazole nitrate 40:20:40. A:B from 100:0 to 55:45 over 15 min, to 30:70 over 45 min, re-equilibrate at initial conditions for 15 min.
Flow rate: 1
Injection volume: 100

Detector: Chemiluminescence. The column effluent mixed with the reagent pumped at 1.2 mL/min and the mixture flowed through a coil at 30° to the detector. (Reagent was 100 mM hydrogen peroxide containing 0.5 mM bis[4-nitro-2-(3,6,9-trioxadecyloxycarbonyl)phenyl]oxalate (Wako, Richmond VA).)

CHROMATOGRAM
Retention time: 33
Limit of detection: 250 fmole

KEY WORDS
rat; plasma; SPE; fluorescence detection is more sensitive

REFERENCE
Hamachi, Y.; Nakashima, K.; Akiyama, S. High performance liquid chromatography with peroxyoxalate chemiluminescence detection of synthetic peptide, ebiratide, *J.Liq.Chromatogr.Rel.Technol.*, **1997**, *20*, 2377–2387.

SAMPLE
Matrix: corn
Analyte: fumonisins
Sample preparation: Condition a 3 mL Bond Elut SAX SPE cartridge with 5 mL MeOH and 5 mL MeOH:water 75:25. 10 g Ground corn + 50 mL MeOH:water 75:25, shake vigorously for 15 min, filter (Whatman No. 5 paper), add 10 mL of the filtrate to the SPE cartridge at 2 mL/min, wash with 8 mL MeOH:water 75:25, wash with 3 mL MeOH, elute with 10 mL MeOH:acetic acid 99:1. Evaporate the eluate to dryness at 50°, reconstitute with 200 μL 100 mM pH 9.3 borate buffer containing 1 mM EDTA. Remove a 50 μL aliquot and add it to 50 μL 100 mM pH 9.3 borate buffer containing 1 mM EDTA, add 50 μL 5 mg/mL 4-(N,N-dimethylaminosulfonyl)-7-fluoro-2,1,3-benzoxadiazole (DBD-F) in MeCN, heat in the dark at 60° for 1 h, cool in ice, inject a 50 μL aliquot. (DBD-F can be purchased from Tokyo Kasei (TCI America, 911 North Harborgate St., Portland, OR 97203; 800-423-8616, 503-283-1681, (fax) 503-283-1987; www.tciamerica.com). Synthesis of DBD-F is as described above.)

HPLC VARIABLES
Guard column: present but not specified
Column: 150 × 4.6 5 μm TSKgel ODS 80Ts (Tosoh)
Mobile phase: Gradient. A was MeOH:50 mM $NaH_2;PO_4$; 50:50. B was MeCN:water 75:25. A:B from 100:0 to 85:15 over 5 min, to 10:90 over 13 min, maintain at 10:90 for 7 min.
Flow rate: 0.8
Injection volume: 50
Detector: F ex 450 em 590

CHROMATOGRAM
Retention time: 19 (fumonisin B_1), 21 (fumonisin B_2)
Limit of detection: 10 ng/g

KEY WORDS
SPE; comparison with phthalaldehyde derivatization

REFERENCE
Akiyama, H.; Miyahara, M.; Toyoda, M.; Saito, Y. Liquid chromatographic determination of fumonisins B_1 and B_2 in corn by precolumn derivatization with 4-(N,N-dimethylaminosulfonyl)-7-fluoro-2,1,3-benzoxadiazole (DBD-F), *J.Food Hyg.Soc.Japan*, **1995**, *36*, 77–81.

SAMPLE
Matrix: solutions
Analyte: calcitonin
Sample preparation: 5 μL Calcitonin in MeCN:0.2% aqueous phosphoric acid containing 100 mM sodium perchlorate 30:70 + 20 μL 250 mM pH 8.5 borate buffer containing 2.5 mM sodium dodecyl sulfate + 2.5 μL 30 mM 4-(N,N-dimethylaminosulfonyl)-7-fluoro-2,1,3-benzoxadiazole in MeCN, add 11 μL MeCN and 11.5 μL water, heat at 50° for 3 h, cool, add 5 μL 1 M HCl, add 40 μL MeCN, add 5 μL water, inject a 50 μL aliquot. (4-(N,N-Dimethylaminosulfonyl)-7-fluoro-2,1,3-benzoxadiazole (DBD-F) can be purchased

from Tokyo Kasei (TCI America, 911 North Harborgate St., Portland, OR 97203; 800-423-8616, 503-283-1681, (fax) 503-283-1987; www.tciamerica.com). Synthesis of DBD-F is as described above.)

HPLC VARIABLES
Column: 150 × 4.6 5 μm TSKgel ODS 80Tm (Tosoh)
Mobile phase: Gradient. A was MeCN containing 0.05% trifluoroacetic acid. B was water containing 0.05% trifluoroacetic acid. A:B 33:67 for 8 min, to 67:33 over 17 min.
Column temperature: 40
Flow rate: 1
Injection volume: 50
Detector: F ex 430 em 558

CHROMATOGRAM
Retention time: 26
Limit of detection: 71 pg

KEY WORDS
salmon

REFERENCE
Fukuda, T.; Ishikawa, K.; Imai, K. Sensitive determination of salmon calcitonin, by means of pre-column derivatization, HPLC and fluorometric determination, *Biomed.Chromatogr.*, **1995**, *9*, 52–55.

SAMPLE
Matrix: solutions
Analyte: ebiratide
Sample preparation: Prepare a 100 μM solution in 100 mM pH 9.0 borate buffer. 200 μL Solution + 200 μL 30 mM DBD-F in MeCN, mix, heat at 50° for 30 min, inject an aliquot. (DBD-F is 4-(N,N-dimethylaminosulfonyl)-7-fluoro-2,1,3-benzoxadiazole. It can be purchased from Tokyo Kasei Kogyo (TCI America, 911 North Harborgate St., Portland, OR 97203; 800-423-8616, 503-283-1681, (fax) 503-283-1987; www.tciamerica.com) or synthesized as described above.)

HPLC VARIABLES
Column: 150 × 4.6 5C18 protein and peptide (Vydac)
Mobile phase: MeCN:50 mM Na_2HPO_4 40:60 adjusted to pH 7.0 with 20% phosphoric acid
Flow rate: 1
Injection volume: 10
Detector: F ex 440 em 580; UV 220

CHROMATOGRAM
Retention time: 10
Limit of detection: 0.25 pmole (F)

REFERENCE
Hamachi, Y.; Tsujiyama, T.; Nakashima, K.; Akiyama, S. High-performance liquid chromatography with fluorescence detection of ebiratide using 4-(N,N-dimethylamino-sulphonyl)-7-fluoro-2,1,3-benzoxadiazole as a fluorgenic reagent, *Biomed.Chromatogr.*, **1995**, *9*, 216–220.

2,4-Dinitrofluorobenzene

RELATED REFERENCES
Barends, D.M.; Brouwers, J.C.A.M.; Hulshoff, A. Fast pre-column derivatization of aminoglycosides with 1-fluoro-2,4-dinitrobenzene and its application to pharmaceutical analysis. *J.Pharm.Biomed.Anal.* **1987**, *5*, 613-617.
Bethune, C.; Bui, T.; Liu, M.L.; Kay, M.A.; Ho, R.J.Y. Development of a high-performance liquid chromatographic assay for G418 sulfate (Geneticin). *Antimicrob.Agents Chemother.* **1997**, *41*, 661-664.
Chen, Z.; Xu, G.; Specht, K.; Yang, R.; She, S. Determination of taurine in biological samples by reversed-phase liquid chromatography with precolumn derivatization with dinitrofluorobenzene. *Anal.Chim.Acta* **1994**, *296*, 249-253.

Elrod, L., Jr.; White, L.B.; Wong, C.F. Determination of fortimicin A sulfate by high-performance liquid chromatography after derivatization with 2,4-dinitrofluorobenzene. *J.Chromatogr.* **1981**, *208*, 357-363.

Holleschau, A.M.; Rathbun, W.B.; Nagasawa, H.T. An HPLC radiotracer method for assessing the ability of L-cysteine prodrugs to maintain glutathione levels in the cultured rat lens. *Curr.Eye Res.* **1996**, *15*, 501-510.

Wong, L.T.; Beaubien, A.R.; Pakuts, A.P. Determination of amikacin in microlitre quantities of biological fluids by high-performance liquid chromatography using 1-fluoro-2,4-dinitrobenzene derivatization. *J.Chromatogr.* **1982**, *231*, 145-154.

SAMPLE
Matrix: blood
Analyte: sulfonylurea drugs
Sample preparation: 2 mL Plasma + 300 μL 1 μg/mL tolbutamide in water + 500 μL 1 M HCl + 8 mL chloroform, shake on a reciprocal shaker for 10 min, centrifuge at 2000 g for 15 min. Remove 7 mL of the organic layer and evaporate it to dryness under nitrogen at 60°. Dissolve residue in 100 μL 3 mg/mL 2,4-dinitrofluorobenzene in n-butyl acetate, heat at 120° for 30 min, evaporate to dryness under nitrogen at 60°, dissolve residue in 100 μL mobile phase, inject 30-70 μL aliquot. (The compound is thermolyzed to the corresponding amine which is then derivatized.)

HPLC VARIABLES
Column: 125 × 4.6 5 μm C8 Perkin-Elmer
Mobile phase: MeCN:water 50:50 containing 0.15% phosphoric acid
Flow rate: 1.5
Injection volume: 30-70
Detector: UV 350

CHROMATOGRAM
Retention time: 3.4 (glyburide (glibenclamide)), 6.2 (chlorpropamide)
Internal standard: tolbutamide (4.5)
Limit of detection: 2-5 ng/mL

OTHER SUBSTANCES
Non-interfering: acetaminophen, aspirin, diazepam, chlordiazepoxide, quinidine, phenytoin, theophylline, phenobarbital

KEY WORDS
plasma

REFERENCE
Zecca, L.; Trivulzio, S.; Pinelli, A.; Colombo, R.; Tofanetti, O. Determination of glibenclamide, chlorpropamide and tolbutamide in plasma by high-performance liquid chromatography with ultraviolet detection, *J.Chromatogr.*, **1985**, *339*, 203–209.

SAMPLE
Matrix: blood
Analyte: netilmicin
Sample preparation: 50 μL Serum + 20 μL 10 μg/mL gentamicin C1a in water + 50 μL buffer, vortex for 15 s, add 200 μL MeCN, vortex for 15 s, centrifuge at 2000 g for 5 min. Filter (0.45 μm, Millex-HV4) the supernatant and add 300 μL of the filtrate to 20 μL 250 mg/mL 2,4-dinitrofluorobenzene in MeCN. Heat at 80° for 2 h, cool rapidly to room temperature, filter (0.45 μm, Millex-HV4), inject a 50 μL aliquot of the filtrate. (Buffer was prepared by dissolving 3.81 g disodium tetraborate decahydrate in water, adjusting pH to 10 with NaOH, and making up to 100 mL with water.)

HPLC VARIABLES
Guard column: 33 × 4.6 5 μm C18 (Perkin-Elmer)

Column: 300 × 3.9 10 μm μBondapak C18
Mobile phase: MeCN:water:acetic acid 70:30:0.1
Flow rate: 2.2
Injection volume: 50
Detector: UV 365

CHROMATOGRAM
Retention time: 13.5
Internal standard: gentamicin C1a (11.0)
Limit of quantitation: 500 ng/mL

KEY WORDS
serum; guinea pig; human

REFERENCE
Dionisotti, S.; Bamonte, F.; Gamba, M.; Ongini, E. High-performance liquid chromatographic determination of netilmicin in guinea-pig and human serum by fluorodinitrobenzene derivatization with spectrophotometric detection, *J.Chromatogr.*, **1988**, *434*, 169–176.

SAMPLE
Matrix: blood
Analyte: sulfonylurea drugs
Sample preparation: 2 mL Serum + 2 mL water + 200 μL 1 (?) M HCl + 200 μL 2.5 μg/mL glibornuride in MeOH + 7 mL diethyl ether, mix, centrifuge at 2000 rpm for 5 min. Remove 6.5 mL of the organic layer and evaporate it to dryness under a stream of nitrogen, reconstitute the residue in 500 μL 2 mg/mL 2,4-dinitrofluorobenzene in butyl acetate, heat at 120° for 1 h, cool, evaporate to dryness under a stream of nitrogen, reconstitute the residue in 150 μL mobile phase, inject a 120 μL aliquot. (The compound is thermolyzed to the corressponding amine which is then derivatized.)

HPLC VARIABLES
Column: 250 × 4.6 5 μm Spherisorb ODS 2
Mobile phase: MeCN:0.4% aqueous phosphoric acid 75:25
Column temperature: 40
Flow rate: 1.2
Injection volume: 120
Detector: UV 360

CHROMATOGRAM
Retention time: 4.2 (chlorpropamide), 4.9 (tolbutamide), 6.5 (glyburide (glibenclamide)), 6.5 (glipizide), 7.3 (tolazamide)
Internal standard: glibornuride (5.8)
Limit of detection: 40 ng/mL

KEY WORDS
serum

REFERENCE
Starkey, B.J.; Mould, G.P.; Teale, J.D. The determination of sulphonylurea drugs by HPLC and its clinical application, *J.Liq.Chromatogr.*, **1989**, *12*, 1889–1896.

SAMPLE
Matrix: blood
Analyte: glimepiride
Sample preparation: 1 mL Serum + 40 μL 10 μg/mL IS1 in MeOH containing 10 μg/mL IS2 + 1 mL 50 mM pH 1 HCl/KCl buffer + 5 mL diethyl ether, shake for 20 min, centrifuge at 2500 g for 5 min. Remove 4 mL of the organic layer and evaporate it to dryness under a stream of nitrogen at 30°, reconstitute the residue in 100 μL reagent, heat at 100° for 20 min, evaporate to dryness under a stream of nitrogen at 60°, reconstitute with 200 μL initial mobile phase, inject a 100 μL aliquot. (Prepare reagent by dissolving 30 μL 2,4-dinitrofluorobenzene in 10 mL n-butyl acetate.)

HPLC VARIABLES
Column: 125 × 4.6 5 μm Spherisorb ODS

Mobile phase: Gradient. MeCN:50 mM perchloric acid 40:60 for 6 min, to 58:42 (step gradient), maintain at 58:42 for 8 min, re-equilibrate at initial conditions for 2 min.
Flow rate: 2
Injection volume: 100
Detector: UV 350

CHROMATOGRAM
Retention time: 11.3
Internal standard: IS1 (1-[4-[2-(3-ethyl-4-methyl-2-oxo-3-pyrroline-1-carboxamido)ethyl]phenylsulfonyl]-3-(4-ethylcyclohexyl)urea) (13.4), IS2 (1-[4-[2-(5-chloro-2-methoxyphenyl-1-carboxamido)ethyl]phenylsulfonyl]-3-(4-hydroxycyclohexyl)urea)(3.4)
Limit of detection: 5 ng/mL

OTHER SUBSTANCES
Extracted: metabolites
Simultaneously analyzed: glibornuride, glyburide, tolbutamide

KEY WORDS
serum; silanize glassware with dichlorodimethylsilane; pharmacokinetics

REFERENCE
Lehr, K.H.; Damm, P. Simultaneous determination of the sulphonylurea glimepiride and its metabolites in human serum and urine by high-performance liquid chromatography after pre-column derivatization, *J.Chromatogr.*, **1990**, *526*, 497–505.

SAMPLE
Matrix: blood
Analyte: tolbutamide
Sample preparation: 1 mL Serum + 200 μL 1 M HCl + 10 μg glyburide + 5 mL toluene, shake gently for 15 min, centrifuge at 1500 g for 3 min. Remove the organic layer and evaporate it to dryness, reconstitute the residue in 25 μL 6 mg/mL 2,4-dinitrofluorobenzene in n-butyl acetate (prepare fresh each week, store at 4° in the dark), heat at 120° for 30 min, evaporate to dryness, reconstitute with 50 μL mobile phase, inject a 25-50 μL aliquot. Alternatively, filter (Amicon YMT membrane, 30000 MW cutoff) 200 μL 100 mM NaOH while centrifuging at 4°, rinse filter with 500 μL water, filter 1 mL serum in the same unit while centrifuging at 4° at 2500 g for 1.5 h. Remove a 700 μL aliquot of the ultrafiltrate, add 200 μL 1 M HCl, add 10 μg glyburide, add 5 mL toluene, shake gently for 15 min, centrifuge at 1500 g for 3 min. Remove the organic layer and evaporate it to dryness, reconstitute the residue in 25 μL 6 mg/mL dinitrofluorobenzene in n-butyl acetate (prepare fresh each week, store at 4° in the dark), heat at 120° for 30 min, evaporate to dryness, reconstitute with 50 μL mobile phase, inject a 25-50 μL aliquot. (Tolbutamide is thermolyzed to butylamine which is then derivatized.)

HPLC VARIABLES
Column: 250 × 4.6 7 μm LiChrosorb RP18
Mobile phase: MeCN:10 mM pH 3.5 phosphate buffer 80:20
Flow rate: 1
Injection volume: 25-50
Detector: UV 360

CHROMATOGRAM
Retention time: 5
Internal standard: glyburide (7)
Limit of detection: 2 ng/mL

KEY WORDS
serum; ultrafiltrate; pharmacokinetics

REFERENCE
Arcelloni, C.; Fermo, I.; Calderara, A.; Pacchioni, M.; Pontiroli, A.E.; Paroni, R. Glibenclamide and tolbutamide in human serum: Rapid measurement of the free fraction, *J.Liq.Chromatogr.*, **1990**, *13*, 175–189.

SAMPLE
Matrix: blood
Analyte: isepamicin
Sample preparation: 50 μL Plasma + 20 μL 10 mg/mL gentamicin C1a in water + 50 μL buffer, vortex for 15 s, add 200 μL MeCN, vortex for 20 s, centrifuge at 2000 g for 5 min. Filter (Millex-HV4) the supernatant. Heat 200 μL filtrate and 20 μL 250 mg/mL 2,4-dinitrofluorobenzene in MeCN at 80° for 1 h, cool, inject a 50 μL aliquot. (Buffer was 3.81 g disodium tetraborate decahydrate in water, adjust pH to 10 with NaOH, make up to 100 mL with water.)

HPLC VARIABLES
Guard column: 25 × 4 10 μm LiChroCART RP 18
Column: 250 × 4 5 μm LiChrosorb RP 18
Mobile phase: MeCN:water 70:30 containing 1 mL/L acetic acid
Flow rate: 2
Injection volume: 50
Detector: UV 365

CHROMATOGRAM
Retention time: 13.0
Internal standard: gentamicin C1a (10.0)
Limit of quantitation: 500 ng/mL

OTHER SUBSTANCES
Non-interfering: ampicillin, aspirin, captopril, cefazolin, cefotaxime, ceftazidime, ceftriaxone, cephalosporins, chlorpromazine, diazepam, heparin, propranolol, sulfamethoxazole, sulpiride, trimethoprim, verapamil

KEY WORDS
plasma; guinea pig; human

REFERENCE
Dionisotti, S.; Bamonte, F.; Scaglione, F.; Ongini, E. Simple measurement of isepamicin, a new aminoglycoside antibiotic, in guinea pig and human plasma, using high-performance liquid chromatography with ultraviolet detection, *Ther.Drug Monit.*, **1991**, *13*, 73–78.

SAMPLE
Matrix: blood
Analyte: amikacin
Sample preparation: Place 400 μL plasma in an Amicon Centrifree Micropartition system unit and centrifuge at 800 g for 50 min. Mix a 200 μL aliquot of plasma ultrafiltrate or urine with 200 μL 1% tris(hydroxymethyl)aminomethane (TRIS) in water, 400 μL DMSO, and 400 μL 1.5% 2,4-dinitrofluorobenzene in 95% EtOH. Heat at 55° for 30 min with mixing at 5 min intervals. Cool at room temperature for 30 min, inject a 10 μL aliquot.

HPLC VARIABLES
Guard column: 25 × 4 37-50 μm C18/Corasil
Column: 300 × 3.9 10 μm μBondapak C18
Mobile phase: MeCN:2-methoxyethanol:THF:glacial acetic acid:1% tris(hydroxymethyl)aminomethane (TRIS) in water 41:4.52:4.24:0.21:50, pH 7.00 ± 0.02
Column temperature: 58
Flow rate: 1 (0-8.5 min), 0.5 (8.5-19 min), 3 (19-25 min), 1 (25-30 min)
Injection volume: 10
Detector: UV 340

CHROMATOGRAM
Retention time: 16.5
Limit of quantitation: 500 ng/mL (plasma), 10000 ng/mL (urine)

OTHER SUBSTANCES
Non-interfering: cefepime

REFERENCE

Papp, E.A.; Knupp, C.A.; Barbhaiya, R.H. High-performance liquid chromatographic assays for the quantification of amikacin in human plasma and urine, *J.Chromatogr.*, **1992**, *574*, 93–99.

SAMPLE

Matrix: blood

Analyte: Ro 42-5892 ((S)-α-[(S)-α-[(tert-butylsulfonyl)methyl]hydrocinnamido]-N-[(1S,2R,3S)-1- (cyclohexylmethyl) - 3 - cyclopropyl - 2,3 - dihydroxylpropyl]imidazole - 4 - propionamide methanesulfonate)

Sample preparation: 1 mL Plasma + 85 μL 1 M NaOH + 800 μL 8.75 ng/mL IS in MeCN, mix, let stand for 10 min, centrifuge at 1500 g for 10 min. Remove the liquid phase and add it to 2 mL dichloromethane, rotate head-over-head at 20 rpm for 10 min, centrifuge at 1700 g for 20 min. Remove the lower organic layer and evaporate it to dryness under a stream of helium at 70°, reconstitute the residue in 70 μL MeCN:water 50:50, centrifuge at 4600 g for 10 min. Remove a 65 μL aliquot and cool it to 10°, add 15 μL 200 mM pH 8.0 sodium borate buffer at 10°, add 15 μL 133 mM 2,4-dinitrofluorobenzene in MeCN at 10°, heat at 80° for 25 min, inject a 50 μL aliquot while still hot.

HPLC VARIABLES

Guard column: 4 × 4 5 μm LiChrospher 100 RP-18

Column: 150 × 3.9 4 μm Novapak C18

Mobile phase: MeCN:buffer 85:100 (Buffer was 92.5 mg/L disodium EDTA in 100 mM acetic acid, adjusted to pH 7.0 with 1 M NaOH.)

Flow rate: 1.2

Injection volume: 25

Detector: E, ESA Coulochem II, Model 5011 measuring cell, first electrode 300 mV second electrode 550 mV, Model 5020 guard cell 800 mV (before injector) following post-column reaction. The column effluent flowed through a 1 m × 0.3 mm ID PTFE coil irradiated at 254 nm to the detector.

CHROMATOGRAM

Retention time: 10.5

Internal standard: (S)-α-[(S)-α-[(tert-butylsulfonyl)methyl]hydrocinnamido]-N-[(1S,2R,3S)-1-(cyclohexylmethyl)-3-isopropyl-2,3-dihydroxylpropyl]imidazole-4-propionamide (Ro 42-4661/000) (16.5)

Limit of quantitation: 0.3 ng/mL

KEY WORDS

post-column reaction; plasma

REFERENCE

Leube, J.; Fischer, G Determination of the renin inhibitor Ro 42-5892 in human plasma by automated pre-column derivatization, reversed-phase high-performance liquid chromatographic separation and electrochemical detection after post-column irradiation, *J.Chromatogr.B*, **1995**, *665*, 373–381.

SAMPLE

Matrix: blood, urine

Analyte: paromomycin

Sample preparation: Plasma. 300 μL Plasma + 30 μL 101.4 μg/mL kanamycin B in water + 100 μL 2 M perchloric acid, vortex for 2-3 s, centrifuge at 1000 g for 5 min. Remove the supernatant and neutralize it with 1.5 M NaOH, add 300 μL buffer, add 400 μL DMSO, add 100 μL 2% 2,4-dinitrofluorobenzene in EtOH, vortex, heat at 64° for 30 min, add 3 mL toluene, vortex, centrifuge, discard the upper toluene layer, add 3 mL MeCN: toluene 50:50, vortex for 5-10 s. Remove the upper organic layer and evaporate it to dryness under a stream of nitrogen at 30-40°, reconstitute the residue in 1 mL MeCN: water 50:50, inject a 20 μL aliquot. Urine. Dilute urine 100-fold with water. 300 μL Diluted urine + 30 μL 101.4 μg/mL kanamycin B in water + 300 μL buffer + 400 μL DMSO + 100 μL 2% 2,4-dinitrofluorobenzene in EtOH, vortex, heat at 64° for 30 min, add 3 mL toluene, vortex, centrifuge, discard the upper toluene layer, add 3 mL MeCN: toluene 50:50, vortex for 5-10 s. Remove the upper organic layer and evaporate it to dryness under a stream of nitrogen at 30-40°, reconstitute the residue in 1 mL MeCN:

water 50:50, inject a 20 μL aliquot. (Prepare buffer by mixing 80 mL 100 mM Na_2HPO_4 and 20 mL 100 mM NaH_2PO_4, adding 1 g Tris HCl, and adjusting the pH to 7.8 with 6 M HCl.)

HPLC VARIABLES

Column: 250 × 4.6 5 μm Zorbax SB-C18
Mobile phase: MeOH:water 64:36, adjusted to pH 3.0 with phosphoric acid
Column temperature: 50
Flow rate: 2
Injection volume: 10-20
Detector: UV 350

CHROMATOGRAM

Retention time: 14.0
Internal standard: kanamycin B (24.0)
Limit of quantitation: 500 ng/mL (plasma), 1 μg/mL (urine)
Limit of detection: 200 ng/mL (plasma), 500 ng/mL (urine)

KEY WORDS

plasma; pharmacokinetics

REFERENCE

Lu, J.; Cwik, M.; Kanyok, T. Determination of paromomycin in human plasma and urine by reversed-phase high-performance liquid chromatography using 2,4-dinitrofluorobenzene derivatization, *J.Chromatogr.B*, **1997**, *695*, 329–335.

SAMPLE

Matrix: bulk
Analyte: ethanolamine glycerophospholipids
Sample preparation: Dissolve 0.1-2.0 mg ethanolamine glycerophospholipids in 2 mL benzene (Caution! Benzene is a carcinogen!), add 50 μL triethylamine, add 5 μL 2,4-dinitrofluorobenzene, let stand at room temperature for 2 h, evaporate to dryness, extract with chloroform/MeOH, methylate with diazomethane at room temperature for 10 min, evaporate to dryness. Purify by TLC on Kieselgel 60 with seven ascending runs of n-hexane: chloroform 30:70, scrape off the yellow bands, elute compounds with chloroform:MeOH 1:2, dilute with an equal volume of chloroform:water 1:1, evaporate the chloroform layer, reconstitute with MeOH, inject an aliquot. (Preparation of diazomethane is as follows. Caution! Diazomethane is toxic, explosive, and carcinogenic! A face shield and a safety screen should always be used and the preparation should only be carried out in a properly functioning chemical fume hood. Only smooth glass apparatus with rubber stoppers and plastic tubing should be used. Scratched glassware, ground glass joints, and sharp edges should be avoided (Org. Syn., Coll. Vol. VI; Wiley: New York, 1988, pp. 432-435). Procedures have been reported for the synthesis of diazomethane from N-methyl-N-nitrosourea (Org. Syn., Coll. Vol. II; Wiley: New York, 1943, pp. 165-167), N-nitroso-β-methylamino-isobutyl methyl ketone (Org. Syn., Coll. Vol. III; Wiley: New York, 1955, pp. 244-248), and N,N'-dimethyl-N,N'-dinitrosoterephthalamide (Org. Syn., Coll. Vol. V; Wiley: New York, 1973, pp. 351-355). Probably the most convenient starting material is N-methyl-N-nitroso-p-toluenesulfonamide (Diazald) (Aldrichimica Acta 1983, 16, 3-10; Org. Syn., Coll. Vol. IV, Wiley: New York, 1963, pp. 250-253). Add 10 mL 95% EtOH to a solution of 5 g KOH in 8 mL water, warm to 65° using a water bath, add a solution of 5 g N-methyl-N-nitroso-p-toluenesulfonamide in 45 mL ether over 20 min at such a rate as to keep the reaction volume constant. Collect the ether and diazomethane that distil in an ice-cooled receiving flask under a dry ice/acetone condenser. When all the N-methyl-N-nitroso-p-toluenesulfonamide has been used up, slowly add 10 mL ether to the reaction flask and continue distillation until the distillate is colorless. A purpose-built distillation apparatus can be purchased from Aldrich (Aldrichimica Acta 1983, 16, 3-10). Excess quantities of diazomethane can be destroyed by adding acetic acid until the yellow color of the diazomethane is discharged. The safe disposal of the nitroso compounds used to generate diazomethane has been discussed (Lunn,G.; Sansone,E.B. Destruction of Hazardous Chemicals in the Laboratory, Second Edition. Wiley: New York, 1994, pp. 277-289).)

HPLC VARIABLES
Column: 250 × 4 LiChrosorb RP-18
Mobile phase: MeCN:MeOH:isopropanol:water 85:3:4:2
Flow rate: 1
Detector: UV 235

CHROMATOGRAM
Retention time: 15-65 (depending on structure)

REFERENCE
Nakagawa, Y.; Waku, K. Separation of the molecular species of diacyl and alkenylacyl subclasses of the methyl ester of dinitrophenylethanolamine glycerophospholipid by high-performance liquid chromatography, *J.Chromatogr.*, **1989**, *487*, 239–245.

SAMPLE
Matrix: fermentation broth
Analyte: aminoglycoside antibiotics
Sample preparation: 5 mL Fermentation broth + 5 mL saturated aqueous solution of Tris + 20 mL MeCN, centrifuge at 3000 rpm for 10 min. Remove a 1 mL aliquot of the supernatant and add it to 3 mL 150 mM 2,4-dinitrofluorobenzene in MeOH, heat at 100° under a reflux condenser for 45 min, make up to 4 mL with mobile phase, inject an aliquot.

HPLC VARIABLES
Column: 200 × 4.6 10 μm LiChrosorb RP-8
Mobile phase: MeCN:water:acetic acid 55:45:0.15
Flow rate: 1.2
Injection volume: 20
Detector: UV 350

CHROMATOGRAM
Retention time: 9.56 (apramycin), 11.28 (kanamycin B), 14.86 (tobramycin)

REFERENCE
Harangi, J.; Deák, M.; Nánási, P.; Bacsa, G. Determination of the major factors of fermentation of the nebramycin complex by high performance liquid chromatography, *J.Liq.Chromatogr.*, **1984**, *7*, 83–93.

SAMPLE
Matrix: formula, media, tissue
Analyte: taurine
Sample preparation: Boil 500 g tuna or squid meat in 1 L water for 2 h, centrifuge an aliquot at 15000 g for 3 min. Dilute 1 g infant formula with 1 mL water, centrifuge at 15000 g for 3 min. Centrifuge an aliquot of liquid media containing bacteria at 15000 g for 3 min. 100 μL Supernatant + 100 μL 4% NaOH, mix, add 20 μL 10% 2,4-dinitrofluorobenzene in acetone, shake vigorously, let stand for 15 min, add 100 μL 10% orthophosphoric acid, add 500 μL chloroform, shake, centrifuge at 15000 g for 15 s. Remove 200 μL of the upper aqueous layer, add 500 μL chloroform, shake, centrifuge at 15000 g for 15 s, inject an aliquot of the aqueous layer.

HPLC VARIABLES
Column: 50 × 4 10 μm Silosorb C18 (Elsico, Moscow)
Mobile phase: MeOH:water:glacial acetic acid:triethylamine 25:75:1:0.1
Flow rate: 5
Injection volume: 1
Detector: UV 350

CHROMATOGRAM
Retention time: 0.2 (taurine), 0.8 (cysteic acid)
Limit of detection: 10 pmole

KEY WORDS
fish; tuna; squid

REFERENCE
Polanuer, B.; Ivanov, S.; Sholin, A. Rapid assay of dinitrophenyl derivative of taurine by high-performance liquid chromatography, *J.Chromatogr.B*, **1994**, *656*, 81–85.

SAMPLE
Matrix: formulations
Analyte: neomycin
Sample preparation: Powder. Dissolve 50 mg powder in 25 mL 20 mM pH 9.0 borate buffer, remove a 5 mL aliquot and add it to 15 mL 150 mM 2,4-dinitrofluorobenzene in MeOH (prepare fresh daily), heat at 100° for 45 min, cool, make up to 250 mL with mobile phase, discard the upper aqueous phase, inject a 20 μL aliquot of the lower organic phase. Ointment. Weigh out 5 g ointment, add 25 mL chloroform, heat at 60° until the ointment dissolves, shake vigorously for 15 min, centrifuge at 4000 g for 15 min, discard the chloroform, add 25 mL chloroform to the solid, centrifuge, discard the chloroform, add 10 mL 20 mM pH 9.0 borate buffer and 15 mL n-heptane to the solid, shake vigorously for 10 min, remove a 5 mL aliquot of the aqueous phase and add it to 15 mL 150 mM 2,4-dinitrofluorobenzene in MeOH (prepare fresh daily), heat at 100° for 45 min, cool, make up to 250 mL with mobile phase, discard the upper aqueous phase, inject a 20 μL aliquot of the lower organic phase.

HPLC VARIABLES
Column: 250 × 4.6 5 μm LiChrosorb SI-100
Mobile phase: Chloroform:THF:water 60:39.2:0.8
Flow rate: 1
Injection volume: 20
Detector: UV 350

CHROMATOGRAM
Retention time: 10 (neomycin C), 14 (neomycin B)
Limit of detection: 1 ng

KEY WORDS
powders; ointment; normal phase

REFERENCE
Tsuji, K.; Goetz, J.F.; VanMeter, W.; Gusciora, K.A. Normal-phase high-performance liquid chromatographic determination of neomycin sulfate derivatized with 1-fluoro-2,4-dinitrobenzene, *J.Chromatogr.*, **1979**, *175*, 141–152.

SAMPLE
Matrix: formulations
Analyte: neomycin
Sample preparation: 5 g Ointment + 3 mL MeOH, heat at 55° for 5 min, vortex twice for 20 s, centrifuge at 2000 g for 2 min, discard the supernatant, repeat the MeOH wash twice more, add 30 mL chloroform, heat at 55°, vortex for 15 s, add 10 mL MeOH:water 20:80, shake vigorously for 20 min, centrifuge at 1000 g for 3 min, remove the upper aqueous layer, repeat the MeOH/water extraction twice more. Combine the aqueous layers and make up to 50 mL with 20 mM pH 9.0 borate buffer. Remove a 10 mL aliquot and add it to 15 mL 150 mM 2,4-dinitrofluorobenzene in MeOH, heat at 100° for 45 min, cool, make up to 250 mL with mobile phase, inject an aliquot of the yellow organic layer.

HPLC VARIABLES
Column: 250 × 4.6 5 μm LiChrosorb silica SI-100
Mobile phase: THF:chloroform:water:glacial acetic acid 39.2:59.8:0.8:0.2
Flow rate: 1
Injection volume: 40
Detector: UV 254

CHROMATOGRAM
Retention time: 4 (neomycin C), 12 (neomycin B)

OTHER SUBSTANCES
Simultaneously analyzed: hydrocortisone acetate

Non-interfering: cortisone acetate, fluorometholone, methylprednisolone, prednisolone acetate

KEY WORDS
normal phase; ointment

REFERENCE
Binns, R.B.; Tsuji, K. High-performance liquid chromatographic analysis of neomycin in petrolatum-based ointments and in veterinary formulations, *J.Pharm.Sci.*, **1984**, *73*, 69–72.

SAMPLE
Matrix: formulations, tissue
Analyte: tobramycin
Sample preparation: Homogenize (Polytron) kidney or lung tissue with 2.5 volumes cold sterile PBS for 30 s. 100 μL Tissue homogenate or liposome encapsulations + 1 mL MeOH, vortex for 1 min, heat at 65° for 30 min, add 900 μL PBS, vortex for 1 min, centrifuge at 4° at 5000 g for 20 min. Remove a 170 μL aliquot of the supernatant and add it to 90 μL 180 mg/mL 2,4-dinitrofluorobenzene in MeOH, add 60 μL 100 mM pH 9.3 borate buffer, add 670 μL MeOH, vortex, heat at 85° for 30 min, cool to room temperature, inject an aliquot.

HPLC VARIABLES
Column: 250 × 4.6 4 μm Zorbax SB-C18
Mobile phase: MeCN:MeOH:10 mM potassium phosphate buffer 65:10:25, pH 3.5
Flow rate: 1.3
Injection volume: 10
Detector: UV 350

CHROMATOGRAM
Limit of detection: 200 ng/mL (PBS), 300 ng/mg (lung), 500 ng/mg (kidney)

KEY WORDS
rat; lung; kidney; liposome encapsulations

REFERENCE
Beaulac, C.; Clément-Major, S.; Hawari, J.; Lagace, J. Eradication of mucoid *Pseudomonas aeriginosa* with fluid liposome-encapsulated tobramycin in an animal model of chronic pulmonary infection, *Antimicrob.Agents Chemother.*, **1996**, *40*, 665–669.

SAMPLE
Matrix: protein
Analyte: lysine
Sample preparation: Weigh out amount of protein containing about 2.5 mg total lysine, add 10 mL 8% sodium bicarbonate solution, stir for 10 min, add 15 mL EtOH, add 100 μL 2,4-dinitrofluorobenzene, stir in the dark at room temperature for 2 h, evaporate at 90° to remove the EtOH, add enough 6 M HCl to neutralize the sodium bicarbonate and to achieve a level of 1 mL for each 1 mg of protein, sonicate for 20 min, heat at 110° under nitrogen for 24 h, cool to room temperature, filter (Whatman no. 52 paper), wash out the flask with 125 mL water, add 100 mL MeCN to the filtrate, make up to 250 mL with water. Evaporate an aliquot to dryness under reduced pressure at 40°, reconstitute with MeCN:water 20:80, filter (0.22 μm nylon), inject an aliquot of the filtrate. (Only available lysine in the protein is determined because it is derivatized on the side chain and then the protein is hydrolyzed. Free lysine, N-terminal lysine, and cross-linked lysine give the bis-derivative and are not determined by this method.)

HPLC VARIABLES
Column: 300 × 3.9 10 μm μBondapak C18
Mobile phase: MeCN:10 mM pH 5 acetate buffer 22:78
Column temperature: 50
Flow rate: 2
Injection volume: 20
Detector: UV 360

CHROMATOGRAM
Retention time: 6.5

REFERENCE
Castillo, G.; Sanz, M.A.; Serrano, M.A.; Hernández, T.; Hernández, A. An isocratic high-performance liquid chromatographic method for determining the available lysine in foods, *J.Chromatogr.Sci.*, **1997**, *35*, 423–429.

SAMPLE
Matrix: urine
Analyte: amantadine
Sample preparation: Prepare a derivatization vial by adding 160 μL 300 mM cetyltrimethylammonium bromide in MeOH to a vial (Eppendorf no. 3813) and evaporating the MeOH at room temperature. 5 mL Urine adjusted to about pH 11 with 10 M NaOH, centrifuge at 3000 g for 10 min. Add 450 μL supernatant and 50 μL 400 mM pH 11 sodium borate buffer to a derivatization vial, vortex for 15 s, add 25 μL 800 mM 2,4-dinitrofluorobenzene in acetone, mix. Remove a 100 μL aliquot and allow it to pass back and forth through a 100 mm length of 0.5 mm ID PTFE tubing at 60° for 4 min, inject a 25 μL aliquot. Backflush PTFE tubing with 1 mL MeCN and 1 mL water.

HPLC VARIABLES
Column: 150 × 3 5 μm Hypersil ODS
Mobile phase: MeCN:10 mM pH 2.5 citrate buffer 75:25 containing 20 mM tetramethylammonium bromide
Injection volume: 25
Detector: UV 350

CHROMATOGRAM
Retention time: 5
Limit of detection: 75 ng/mL

REFERENCE
van der Horst, F.A.L.; Teeuwsen, J.; Holthuis, J.J.M.; Brinkman, U.A.T. High-performance liquid chromatographic determination of amantadine in urine after micelle-mediated pre-column derivatization with 1-fluoro-2,4-dinitrobenzene, *J.Pharm.Biomed.Anal.*, **1990**, *8*, 799–804.

SAMPLE
Matrix: urine
Analyte: amantadine
Sample preparation: Adjust 5 mL urine to pH 11 with 10 M NaOH, centrifuge at 3000 g for 10 min. Remove a 450 μL aliquot of the supernatant and add it to 50 μL 400 mM pH 11 sodium borate buffer, vortex for 15 s, add 25 μL 800 mM 2,4-dinitrofluorobenzene in acetone, add 225 μL 300 mM cetyltrimethylammonium bromide in MeOH (?), mix, heat a 100 μL aliquot at 60° for 4 min, inject a 25 μL aliquot.

HPLC VARIABLES
Column: 150 × 3 5 μm Hypersil ODS
Mobile phase: MeCN:10 mM pH 2.5 citrate buffer 75:25 containing 20 mM tetramethylammonium bromide
Injection volume: 25
Detector: UV 350

CHROMATOGRAM
Retention time: 5
Limit of detection: 75 ng/mL

REFERENCE
van der Horst, F.A.L.; Teeuwsen, J.; Holthuis, J.J.M.; Brinkman, U.A.T. High-performance liquid chromatographic determination of amantadine in urine after micelle-mediated pre-column derivatization with 1-fluoro-2,4-dinitrobenzene, *J.Pharm.Biomed.Anal.*, **1990**, *8*, 799–804.

4-Fluoro-3-nitrobenzotrifluoride

SAMPLE
Matrix: solutions
Analyte: amines
Sample preparation: Evaporate to dryness, add diaminooctane, reconstitute with 100 μL 1 M sodium carbonate, add 300 μL 10 μL/mL 4-fluoro-3-nitrobenzotrifluoride in DMSO, mix, heat at 60° for 20 min, add 40 μL 1 M histidine in 1 M sodium carbonate, heat at 60° for 5 min, cool, extract with 2 mL 2-methylbutane, centrifuge at 1500 g for 5 min, repeat the extraction. Combine the organic layers and evaporate them to dryness, reconstitute the residue in 50 μL MeOH, inject a 10 μL aliquot.

HPLC VARIABLES
Column: 250 × 4.5 10 μm Spherisorb ODS
Mobile phase: MeCN:water 80:20
Column temperature: 40
Flow rate: 3
Injection volume: 10
Detector: UV 242

CHROMATOGRAM
Retention time: 2.8 (putrescine), 4.5 (spermidine), 8 (spermine)
Internal standard: diaminooctane (6.2)
Limit of detection: 5-25 pmole

REFERENCE
Spragg, B.P.; Hutchings, A.D. High-performance liquid chromatographic determination of putrescine, spermidine and spermine after derivatization with 4-fluoro-3-nitrobenzotrifluoride, *J.Chromatogr.*, **1983**, *258*, 289-291.

4-Fluoro-7-nitro-2,1,3-benzoxadiazole

RELATED REFERENCES
Andreeva, M.; Niedmann, P.D.; Schütz, E.; Wieland, E.; Armstrong, V.W.; Oellerich, M. Determination of MEGX by HPLC with fluorescence detection. *Clin.Chem.* **1997**, *43*, 1081-1083.

Beijersten, I.; Westerlund, D. Derivatization of dipeptides with 4-fluoro-7-nitro-2,1,3-benzoxadiazole for laser-induced fluorescence and separation by micellar electrokinetic chromatography. *J.Chromatogr.A* **1995**, *716*, 389-399.

Bryan, P.D.; Emry, M.L.; el-Shourbagy, T.; Hui, Y.; Chu, S. Determination of ABT-089 in human plasma by high performance liquid chromatography using in situ precolumn derivatization with 7-fluoro-4-nitrobenzo-2-oxa-1,3-diazole (Abstract APQ 1029). *Pharm.Res.* **1996**, *13*, S10.

Hauck, M.; Köbler, H. Bestimmung von Cyclamat in komplexer Matrix mittels HPLC nach Vorsäulen-derivatisierung mit 4-fluoro-7-nitrobenzofurazan [Determination of cyclamate in complex matrix using HPLC after column derivatization with 4-fluoro-7-nitrobenzofurazan]. *Z.Lebensm.Unters.Forsch.* **1990**, *191*, 322-324.

Imai, K.; Watanabe, Y.; Toyooka, T. Fluorometric assay of amino acids and amines by use of 7-fluoro-4-nitrobenzo-2-oxa-1,3-diazole (NBD-F) in high-performance liquid chromatography. *Chromatographia* **1982**, *16*, 214-215.

Imai, K.; Ueda, E.; Toyooka, T. High-performance liquid chromatography with photochemical fluorometric detection of tryptophan based on 4-fluoro-7-nitrobenzo-2-oxa-1,3-diazole. Total protein amino acid analysis. *Anal.Chim.Acta* **1988**, *205*, 7-14.

Imanari, T.; Taguchi, K.; Tanabe, S.; Shinomiya, K.; Kunitomo, A.; Suzuki, H. High-performance liquid chromatography of hexosamines and hexosaminols derived by treatment with 7-fluoro-4-nitrobenzo-2-oxa-1,3-diazole. *Chem.Pharm.Bull.* **1985**, *33*, 3057-3058.

Jiménez, M.; Mateo, R. Determination of mycotoxins produced by *Fusarium* isolates from banana fruits by capillary gas chromatography and high-performance liquid chromatography. *J.Chromatogr.A* **1997**, *778*, 363-372.

Kamisaki, Y.; Wada, K.; Nakamoto, K.; Kishimoto, Y.; Kitano, M.; Itoh, T. Sensitive determination of nitrotyrosine in human plasma by isocratic high-performance liquid chromatography. *J.Chromatogr.B* **1996**, *685*, 343-347.

Kanazawa, H.; Nagatsuka, T.; Miyazaki, M.; Matsushima, Y. Determination of peptides by high-performance liquid chromatography with laser-induced fluorescence detection. *J.Chromatogr.A* **1997**, *763*, 23-29.

Kobayashi, R.; Nakashima, K.; Sugano, K.; Nishikawa, M.; Akiyama, S. Gas phase derivatization of ammonia with 4-fluoro-7-nitrobenzo-2-oxa-1,3-diazole and its application to urease assay. *Chem.Pharm.Bull.* **1992**, *40*, 1327-1328.

Miyano, H.; Toyooka, T.; Imai, K. Further studies on the reaction of amines and proteins with 4-fluoro-7-nitrobenzo-2-oxa-1,3-diazole. *Anal.Chim.Acta* **1985**, *170*, 81-87.

Scott, P.M.; Lawrence, G.A. Liquid chromatographic determination of fumonisins with 4-fluoro-7-nitro-benzofurazan. *J.AOAC Int.* **1992**, *75*,829-834.

Scott, P.M.; Lawrence, G.A. Determination of hydrolysed fumonisin B$_1$ in alkali-processed corn foods. *Food Addit.Contam.* **1996**, *13*, 823-832.

SAMPLE
Matrix: blood
Analyte: A-82695 (3-methyl-5-(2-pyrollidinyl)isoxazole)
Sample preparation: 1 mL Plasma + 100 μL 2.22 μg/mL IS in water + 500 μL 500 mM sodium carbonate + 5 mL ethyl acetate, vortex for 30 s, centrifuge at 1819 g for 10 min. Remove the organic layer and add it to 300 μL 10 mM HCl, vortex for 30 s, centrifuge at 1918 g for 10 min. Remove the aqueous layer and add it to 50 μL 500 mM pH 9.86 phosphate buffer, add 300 μL 500 μg/mL 4-fluoro-7-nitro-2,1,3-benzoxadiazole in EtOH, mix briefly, heat at 50° for 5 min, cool in an ice bath, add 5 mL ethyl acetate, vortex, centrifuge. Remove the organic layer and evaporate it to dryness under a stream of air at room temperature, reconstitute the residue in 200 μL MeOH:water, inject a 100 μL aliquot.

HPLC VARIABLES
Column: 50 × 4.6 3 μm Spherisorb ODS-2
Mobile phase: MeCN:water 30:70
Flow rate: 1
Injection volume: 100
Detector: F ex 480-487 em 529-530; MS, Perkin-Elmer Sciex API III+ LC-MS-MS, APCI, nebulizer 450° 70 psi, make-up gas air, 70 eV, argon collision gas, m/z 316

CHROMATOGRAM
Retention time: 8.5
Internal standard: A-83030 (18.5)
Limit of quantitation: 0.01 ng/mL

OTHER SUBSTANCES
Extracted: A-79814

KEY WORDS
dog; monkey; plasma; pharmacokinetics

REFERENCE
Hui, Y.-h.; Marsh, K.C. Sensitive detection of selected cholinergic channel activators derivatized with 7-fluoro-4-nitrobenzo-2-oxa-1,3-diazole, *J.Pharm.Biomed.Anal.*, **1996**, *14*, 131–142.

SAMPLE
Matrix: blood, tissue
Analyte: ABT-089 (2-methyl-3-(2-(S)-pyrrolidinylmethoxy)pyridine)
Sample preparation: Homogenize rat brain tissue with 2 volumes of 0.9% NaCl. Mix 1 (monkey, dog) or 0.2 (rat) mL plasma or 0.2 mL brain homogenate with 100 µL 2 µg/mL IS in water, add 500 µL 500 mM sodium carbonate, add 5 mL hexane:ethyl acetate 10:90, vortex for 30 s, centrifuge at 1819 g for 10 min. Remove the organic layer and add it to 300 µL 10 mM HCl, vortex for 30 s, centrifuge at 1819 g for 10 min. Remove the aqueous layer and add it to 100 µL 500 mM pH 9.86 phosphate buffer, add 300 µL freshly-prepared 100 µg/mL 4-fluoro-7-nitro-2,1,3-benzoxadiazole in EtOH, mix briefly, heat at 50° for 30 min, cool in ice, add 5 mL hexane:ethyl acetate 50:50, vortex, centrifuge. Remove the organic layer and evaporate it to dryness under a stream of air at room temperature, reconstitute the residue in 200 µL MeOH:water 50:50, inject an aliquot.

HPLC VARIABLES
Column: 100 × 4.6 5 µm YMC-basic (YMC)
Mobile phase: MeCN:MeOH:10 mM tetramethylammonium perchlorate:trifluoroacetic acid 18:10:72:0.1
Flow rate: 1
Detector: F ex 495 em 533

CHROMATOGRAM
Retention time: 8
Internal standard: A-93840 (11)
Limit of quantitation: 0.1 ng/mL

KEY WORDS
plasma; dog; monkey; rat; brain; pharmacokinetics

REFERENCE
Hui, Y.-h.; Carroll, S.; Marsh, K.C. Development of a sensitive method for quantitation of ABT-089 in plasma using fluorescence labeling with 7-fluoro-4-nitrobenzo-2-oxa-1,3-diazole, *J.Chromatogr.B*, **1997**, *695*, 337–347.

SAMPLE
Matrix: blood, urine
Analyte: tromethamine
Sample preparation: Plasma. 100 µL Plasma + 100 µL 400 µg/mL 2-amino-2-ethyl-1,3-propanediol + 100 µL 7% perchloric acid, vortex, centrifuge at 2000 g for 5 min. Remove a 200 µL aliquot of the supernatant and add it to 300 µL 200 mM pH 9.2 borate buffer, vortex briefly, add 40 µL 4 mg/mL 4-fluoro-7-nitro-2,1,3-benzoxadiazole in MeCN, vortex briefly, heat at 80° for 30 min, cool to room temperature in a water bath for 10 min, add 500 µL 5 M NaOH, vortex for 10 s, add 100 µL benzoyl chloride, vortex for 1 min, add 2 mL ethyl acetate:MeOH 90:10, rotate for 10 min, centrifuge at 2000 g for 5 min. Remove the supernatant and evaporate it to dryness under vacuum for 1 h, reconstitute the residue in 1 mL MeCN:10 mM phosphoric acid 80:20, inject a 50 µL aliquot. Urine. 200 µL Urine + 100 µL 200 µg/mL 2-amino-2-ethyl-1,3-propanediol + 100 µL water + 100 µL 10 mM pH 8.5 borate buffer + 200 µL 4 mg/mL 4-fluoro-7-nitro-2,1,3-benzoxadiazole in MeCN, vortex briefly, heat at 80° for 30 min, cool to room temperature in a water bath for 10 min, add 500 µL 5 M NaOH, vortex for 10 s, add 100 µL benzoyl chloride, vortex for 1 min, add 2 mL ethyl acetate:MeOH 90:10, rotate for 10 min, centrifuge at 2000 g for 5 min. Remove the supernatant and evaporate it to dryness under vacuum for 1 h, reconstitute the residue in 1 mL MeCN:10 mM phosphoric acid 80:20, inject a 50 µL aliquot.

HPLC VARIABLES
Guard column: 20 × 4.6 Supelcosil octadecylsilane silica column
Column: 250 × 4.6 Supelcosil octadecylsilane

Mobile phase: MeCN:10 mM phosphoric acid 70:30, adjusted to pH 2.5 with 10 M KOH
Flow rate: 1
Injection volume: 50
Detector: F ex 460 em 520

CHROMATOGRAM
Retention time: 12
Internal standard: 2-amino-2-ethyl-1,3-propanediol (10)
Limit of quantitation: 1 µg/mL (plasma), 5 µg/mL (urine)

KEY WORDS
plasma

REFERENCE
Morris, M.J.; Hsieh, J.Y.-K. Determination of tris(hydroxymethyl)aminomethane (tromethamine) in human plasma and urine by high-performance liquid chromatography with fluorescence detection, J.Chromatogr., **1993**, 622, 87–92.

SAMPLE
Matrix: blood, urine
Analyte: ethambutol
Sample preparation: Plasma. 200 µL Plasma + 300 µL 5 M NaOH + 5 mL diethyl ether, vortex for 1 min, centrifuge at 1200 g for 3 min, remove the organic layer, repeat the extraction. Combine the organic layers and add them to 200 µL 10 mM phosphoric acid, vortex for 1 min, centrifuge at 1200 g for 3 min. Remove the aqueous phase and add it to 300 µL 200 mM pH 7.5 borate buffer, vortex for 10 s, add 40 µL 4 mg/mL 4-fluoro-7-nitro-2,1,3-benzooxadiazole in MeCN, heat at 80° for 30 min, add 50 µL 1 M phosphoric acid, cool in dry ice/acetone for 1 min, add 2 mL ethyl acetate, vortex for 1 min, centrifuge at 1200 g for 3 min. Discard the upper organic layer and add 50 µL 5 M NaOH to the lower aqueous phase, add 2 mL ethyl acetate:MeOH 90:10, mix for 1 min, centrifuge at 1200 g for 3 min. Remove the upper organic layer and evaporate it to dryness under a stream of nitrogen at 37°, reconstitute the residue in 250 µL 10 mM pH 2.5 phosphoric acid, inject a 200 µL aliquot. Urine. Dilute 100 µL urine to 20 mL with water. 200 µL Diluted urine + 300 µL 200 mM pH 7.5 borate buffer, vortex for 10 s, add 40 µL 4 mg/mL 4-fluoro-7-nitro-2,1,3-benzooxadiazole in MeCN, heat at 80° for 30 min, add 50 µL 1 M phosphoric acid, cool in dry ice/acetone for 1 min, add 2 mL ethyl acetate, vortex for 1 min, centrifuge at 1200 g for 3 min. Discard the upper organic layer and add 50 µL 5 M NaOH to the lower aqueous phase, add 2 mL ethyl acetate:MeOH 90:10, mix for 1 min, centrifuge at 1200 g for 3 min. Remove the upper organic layer and evaporate it to dryness under a stream of nitrogen at 37°, reconstitute the residue in 1 mL 10 mM pH 2.5 phosphoric acid, inject a 200 µL aliquot.

HPLC VARIABLES
Guard column: 37-53 µm Pellicular ODS (Whatman)
Column: 250 × 4.6 5 µm Spherisorb CN
Mobile phase: MeCN:buffer 30:70 (Buffer was 10 mM phosphoric acid adjusted to pH 2.5 with 10 M KOH.)
Flow rate: 1
Injection volume: 200
Detector: F ex 490 em 540

CHROMATOGRAM
Retention time: 14
Limit of quantitation: 10 ng/mL (plasma), 10 µg/mL (urine)

KEY WORDS
plasma; 4-chloro-7-nitro-2,1,3-benzooxadiazole is a less reactive derivatizing reagent

REFERENCE
Breda, M.; Marrari, P.; Pianezzola, E.; Strolin Benedetti, M. Determination of ethambutol in human plasma and urine by high-performance liquid chromatography with fluorescence detection, J.Chromatogr.A, **1996**, 729, 301–307.

SAMPLE
Matrix: solutions
Analyte: polyamines
Sample preparation: Mix 100 μL of a solution in 66 mM pH 7.0 phosphate buffer with 100 μL 600 μM 4-fluoro-7-nitro-2,1,3-benzoxadiazole in EtOH, heat at 50° in the dark for 1 min, add 25 μL 500 mM HCl, inject a 10 μL aliquot.

HPLC VARIABLES
Column: 5 μm TSK LS-160 polyethylene glycol dimethacrylate gel (Toyo Soda)
Mobile phase: MeCN:MeOH:35% HCl 30:70:0.9
Column temperature: 50
Flow rate: 0.6
Injection volume: 10
Detector: F ex 470 em 530

CHROMATOGRAM
Retention time: 15 (putrescine), 20 (spermidine), 25 (spermidine)
Limit of detection: 0.076-1.2 pmole

REFERENCE
Toyo'oka, T.; Watanabe, Y.; Imai, K. Reaction of amines of biological importance with 4-fluoro-7-nitrobenzo-2-oxa-1,3-diazole, *Anal.Chim.Acta*, **1983**, *149*, 305–312.

SAMPLE
Matrix: solutions
Analyte: amines
Sample preparation: 100 μL 100 μM Solution + 400 μL buffer + 500 μL 80 mM 4-fluoro-7-nitro-2,1,3-benzoxadiazole in EtOH, mix, heat at 60° in the dark for 1 min, dilute with MeCN:water 50:50, inject a 20 μL aliquot. (Prepare buffer by adjusting 100 mM boric acid to pH 8.0 with NaOH.)

HPLC VARIABLES
Guard column: 30 × 4.6 Spheri-5 RP-18
Column: 250 × 4.6 Inertsil ODS-2
Mobile phase: MeCN:water 60:40 containing 1 mM imidazole, pH adjusted to 7.0 with nitric acid
Flow rate: 1
Injection volume: 20
Detector: Chemiluminescence following post-column reaction. The column effluent mixed with the reagent pumped at 1 mL/min and the mixture flowed through a 300 mm × 0.25 mm ID coil to the detector. (Prepare the reagent by dissolving 112 mg bis(2,4,6-trichlorophenyl) oxalate in 500 mL MeCN, add 8.6 mL 30% hydrogen peroxide, sonicate.); F ex 470 em 530

CHROMATOGRAM
Retention time: 7 (phenylpropanolamine), 8 (ephedrine), 9 (benzylamine), 11.5 (phenylethylamine), 15 (phenylpropylamine, N-methylphenethylamine), 17 (methamphetamine), 19 (N-isopropylbenzylamine), 20 (phenylbutylamine)
Limit of detection: 400-30000 fmole (chemiluminescence), 200-30000 fmole (F)

KEY WORDS
post-column reaction; comparison with other derivatization reagents

REFERENCE
Hayakawa, K.; Hasegawa, K.; Imaizumi, N.; Wong, O.S.; Miyazaki, M. Determination of amphetamine-related compounds by high-performance liquid chromatography with chemiluminescence and fluorescence detections, *J.Chromatogr.*, **1989**, *464*, 343–352.

SAMPLE
Matrix: solutions
Analyte: mexiletine
Sample preparation: 10 μL 100 μM Mexiletine hydrochloride in 200 mM pH 8.0 borate buffer containing 4 mM disodium EDTA + 30 μL 50 mM 4-fluoro-7-nitro-2,1,3-benzox-

adiazole in MeCN, mix, heat at 60° for 5 min, add 960 μL MeOH:acetic acid 99:1, inject a 10 μL aliquot.

HPLC VARIABLES

Column: 150 × 60 (sic) 5 μm Ultron ES-phCD phenylcarbamylated β-cyclodextrin (Sinwa Kakou)
Mobile phase: MeCN:MeOH:water 10:40:50
Flow rate: 0.8
Injection volume: 10
Detector: F ex 470 em 530

CHROMATOGRAM

Retention time: 60 (R-(-)), 65 (S-(+))

KEY WORDS

chiral

REFERENCE

Fukushima, T.; Kato, M.; Santa, T.; Imai, K. Enantiomeric separation and spectrofluorometric detection of the racemic drugs, (±)-1-(2,5-dimethylphenoxy)-2-propamine (mexiletine) and (3RS)-4-amino-3-hydroxybutanoic acid (GABOB), derivatized with 4-fluoro-7-nitro-2,1,3-benzoxadiazole on a phenyl-carbamylated cyclodextrin bonded stationary phase, *Analyst*, **1995**, *120*, 381–383.

SAMPLE

Matrix: water
Analyte: anatoxin-a
Sample preparation: Condition a 3 mL WCX weak cation-exchange SPE cartridge (Supelco) with 6 mL MeOH and 6 mL water. Adjust the pH of a 10 mL sample to 7, add to the SPE cartridge, wash with 3 mL MeOH:water 50:50, dry with air, elute with 10 mL MeOH:trifluoroacetic acid 99.8:0.2. Evaporate the eluate to dryness under a stream of nitrogen at 50°, reconstitute the residue in MeCN, evaporate to dryness, add 50 μL 1 mg/mL 4-fluoro-7-nitro-2,1,3-benzoxadiazole in MeCN, add 100 μL 100 mM sodium borate, mix for 1 min, add 50 μL 1 M HCl, inject a 20 μL aliquot.

HPLC VARIABLES

Column: 250 × 3.2 Ultremex 5 C18
Mobile phase: MeCN:water 45:55
Column temperature: 30
Flow rate: 0.5
Injection volume: 20
Detector: F ex 470 em 530

CHROMATOGRAM

Retention time: 15.8
Limit of detection: 0.1 ng

KEY WORDS

SPE

REFERENCE

James, K.J.; Sherlock, I.R. Determination of the cyanobacterial neurotoxin, anatoxin-a, by derivatization using 7-fluoro-4-nitro-2,1,3-benzoxadiazole (NBD-F) and HPLC analysis with fluorimetric detection, *Biomed.Chromatogr.*, **1996**, *10*, 46–47.

Marfey's Reagent

SAMPLE
Matrix: solutions
Analyte: baclofen
Sample preparation: 200 μL 10 mM Baclofen in 100 mM sodium bicarbonate + 200 μL 10 mM (Marfey's Reagent (Nα-(2,4-dinitro-5-fluorophenyl)-L-alaninamide) in acetone (freshly prepared), stir at 40° for 1 h, cool, add 100 μL 200 mM HCl, inject an aliquot (Anal. Biochem. 1992, 202, 210).

HPLC VARIABLES
Column: 150 × 4.6 5 μm YMC.GEL C8 (YMC)
Mobile phase: MeOH:5% KH$_2$PO$_4$ (pH 4.0) 11:8
Flow rate: 1
Detector: UV 340

CHROMATOGRAM
Retention time: 17 (R-(-)), 20 (S-(+))

KEY WORDS
chiral

REFERENCE
Shimada, K.; Mitamura, K.; Morita, M.; Hirakata, K. Separation of the diastereomers of baclofen by high performance liquid chromatography using cyclodextrin as a mobile phase additive, *J.Liq.Chromatogr.*, **1993**, *16*, 3311–3320.

4-Nitrobenzyl Bromide

SAMPLE
Matrix: aqueous humor
Analyte: pilocarpine
Sample preparation: 100 μL Aqueous humor + 500 μL 300 mM pH 8.4 potassium bicarbonate + 1 mL dichloromethane, vortex for 1 min, centrifuge at 2000 rpm for 5 min, repeat extraction. Combine the organic layers and evaporate them to dryness under a

stream of nitrogen at 40°, reconstitute the residue in 200 μL 250 μg/mL 4-nitrobenzyl bromide in MeCN, heat in a sealed tube at 40° for 24 h, cool, inject an aliquot.

HPLC VARIABLES
Column: 300 × 3.9 10 μm μBondapak C18
Mobile phase: MeOH:water 80:20 containing 1 mM sodium octanesulfonate
Flow rate: 1.6
Detector: UV 254

CHROMATOGRAM
Retention time: 11
Limit of detection: <50 ng/mL

OTHER SUBSTANCES
Simultaneously analyzed: isopilocarpine

KEY WORDS
rabbit; silanize glassware with dimethyldichlorosilane

REFERENCE
Mitra, A.K.; Baustian, C.L.; Mikkelson, T.J. High-performance liquid chromatographic determination of pilocarpine in aqueous humor: derivatization by quaternization of methylimidazole tertiary amine group, *J.Pharm.Sci.*, **1980**, *69*, 257–261.

ACYL HALIDE

2-Anthroyl Chloride

SAMPLE
Matrix: blood
Analyte: mexiletine
Sample preparation: 200 μL Serum + 30 μL 1 μg/mL IS in water + 200 μL 300 mM barium hydroxide, vortex, add 200 μL 300 mM zinc sulfate, add 200 μL 2 M NaOH, add 5 mL diethyl ether, vortex, centrifuge at 2000 g for 5 min, repeat extraction. Combine the organic layers and evaporate them to 1 mL under a stream of nitrogen at 37°, add 300 μL 100 mM HCl, vortex, discard the ether layer. Wash the aqueous layer with 2 mL diethyl ether, add 300 μL 2 M NaOH to the aqueous layer, add 10 μL 1 mg/mL 2-anthroyl chloride in dry dichloromethane, vortex for 3 min, extract with 1 mL dichloromethane. Remove the organic layer and evaporate it to dryness under a stream of nitrogen at 37°, reconstitute the residue in 200 μL mobile phase, inject an aliquot. (Prepare 2-anthroyl chloride as follows. Reflux 500 mg anthraquinone-2-carboxylic acid with 30 mL 14% aqueous ammonia, add 2 g zinc dust in small portions over 30 min, reflux with stirring for 1 h, filter, acidify the filtrate to pH 3 with dilute HCl, filter, recrystallize from hot EtOH to give 2-anthracenecarboxylic acid (mp 278-279°, yield 23%). 2-Anthracenecarboxylic acid is available from Tokyo Kasei (TCI America, Portland OR). Add 100 mg 2-anthracenecarboxylic acid in 100 mL dry benzene (Caution! Benzene is a carcinogen!) to 500 μL oxalyl chloride with stirring, heat slowly on a water bath for 30 min, add 500 μL oxalyl chloride, heat for 30 min, cool, filter. Evaporate the filtrate to obtain 40 mg pale yellow 2-anthroyl chloride. Purify by dissolving in dichloromethane and precipitating with hexane. Purify further by HPLC by injecting 100 μL aliquots of a 50 mg/mL solution in MeCN onto a 250 × 9.4 10 μm ODS-2 Magnum-9 (Whatman) column and eluting with MeCN at 0.7 mL/min. Using a UV 230 detector collect the appropriate fraction (about 11 min) and evaporate to dryness to obtain pure compound.)

HPLC VARIABLES
Guard column: 150 × 4.6 5 μm silica (Alltech)
Column: 250 × 4.6 5 μm Pirkle 1-A phenylglycine (Regis)
Mobile phase: Hexane:isopropanol:chloroform 78:7:15
Flow rate: 0.8
Injection volume: 20
Detector: F ex 270 em 420

CHROMATOGRAM
Retention time: 11.5 (R-(-)), 12.5 (S-(+))
Internal standard: 1-(2',6'-dimethylphenoxy)-2-ethanamine hydrochloride (Boehringer Ingelheim KPE-2963) (14.5)
Limit of quantitation: 2.5 ng/mL
Limit of detection: 0.5 ng/mL

KEY WORDS
serum; chiral; pharmacokinetics

REFERENCE
Kwok, D.K.W.; Igwemezie, L.; Kerr, C.R.; McErlane, K.M. High-performance liquid chromatographic analysis using a highly sensitive fluorogenic reagent, 2-anthroyl chloride, and stereoselective determination of the enantiomers of mexiletine in human serum, *J.Chromatogr.B*, **1994**, *661*, 271–280.

SAMPLE
Matrix: blood, saliva, urine
Analyte: mexiletine
Sample preparation: Serum, saliva. 1 mL Serum or saliva + 30 ng IS + 200 μL 150 mM barium hydroxide solution, vortex, add 200 μL 150 mM zinc sulfate solution, mix, add 200 μL 2 M NaOH, extract twice with 5 mL portions of diethyl ether. Combine the organic layers and evaporate them to 1 mL under a stream of nitrogen at 37°, add 300 μL 100 mM HCl, extract. Remove the aqueous layer and wash it twice with 2 mL portions of diethyl ether, add 300 μL 2 M NaOH to the aqueous layer, add 10 μL 1 mg/mL 2-anthroyl chloride, vortex for 3 min, extract with 1 mL dichloromethane. Remove the organic layer and evaporate it to dryness under a stream of nitrogen at 37°, reconstitute the residue in 200 μL mobile phase, inject an aliquot. Urine. Make urine alkaline with 2 M NaOH, extract twice with 5 mL portions of diethyl ether. Combine the organic layers and evaporate them to 1 mL under a stream of nitrogen at 37°, add 300 μL 100 mM HCl, extract. Remove the aqueous layer and wash it twice with 2 mL portions of diethyl ether, add 300 μL 2 M NaOH to the aqueous layer, add 10 μL 1 mg/mL 2-anthroyl chloride, vortex for 3 min, extract with 1 mL dichloromethane. Remove the organic layer and evaporate it to dryness under a stream of nitrogen at 37°, reconstitute the residue in 200 μL mobile phase, inject an aliquot. (Prepare 2-anthroyl chloride as described above.)

HPLC VARIABLES
Column: 250 × 4.6 5 μm Pirkle 1-A phenylglycine (Regis) + 150 × 4.6 5 μm silica (Alltech) in series
Mobile phase: Hexane:isopropanol:chloroform 39:3.5:75.5
Flow rate: 0.8
Injection volume: 20
Detector: F ex 270 em 420

CHROMATOGRAM
Internal standard: 1-(2',6'-dimethylphenoxy)-2-ethanamine (KOE 2963)

KEY WORDS
chiral; serum; pharmacokinetics

REFERENCE
Kwok, D.W.; Kerr, C.R.; McErlane, K.M. Pharmacokinetics of mexiletine enantiomers in healthy human subjects. A study of the in vivo serum protein binding, salivary excretion and red blood cell distribution of the enantiomers, *Xenobiotica*, **1995**, *25*, 1127–1142.

2-(9-Anthryl)ethyl Chloroformate

SAMPLE
Matrix: blood
Analyte: polyamines
Sample preparation: Reflux 5 mL serum with 5 mL 6 M HCl for 10 h, evaporate to dryness under reduced pressure at 50°, neutralize with dilute KOH, evaporate to dryness, reconstitute with 5 mL water, centrifuge at 3400 rpm for 10 min, filter (0.45 μm). Remove a 100 μL aliquot of the filtrate and add it to 400 μL 25 mM pH 9.0 borate buffer and 500 μL 5.8 mM 2-(9-anthryl)ethyl chloroformate in MeCN, heat at 36° for 5 min in the dark, inject an aliquot. (Prepare 2-(9-anthryl)ethyl chloroformate as follows. Stir a solution of 3 g of 9-bromoanthracene in 100 mL ether at 0° under argon or nitrogen, add 9 mL 1.6 M n-butyllithium over 5 min, stir for 30 min, add an ice-cold solution of 3 g ethylene oxide (Caution! Ethylene oxide is a carcinogen!) in 16 mL ether, stir for 1 h, add 70 mL water, add 50 mL ether, remove the organic layer, extract the aqueous layer with 100 mL dichloromethane. Combine the organic layers and wash them with water, dry over anhydrous sodium sulfate, evaporate to dryness, chromatograph on silica gel with dichloromethane to give 2-(9-anthryl)ethanol as pale yellow crystals (mp 106-8°) (J.Org.Chem. 1986, 51, 2956). Stir a solution of 2-(9-anthryl)ethanol in ether in the presence of pyridine (as an HCl scavenger) at 0°, add a solution of phosgene in toluene. 2-(9-anthryl)ethyl chloroformate is obtained as colorless crystals (mp 86-87° from pentane). Protect stock solutions from light and store them in the refrigerator.)

HPLC VARIABLES
Column: 125 × 4 5 μm LiChrospher 100 RP-18
Mobile phase: Gradient. MeCN:water from 80:20 to 100:0 over 10 min, maintain at 100:0 for 15 min.
Flow rate: 1
Injection volume: 20
Detector: F ex 258 em 418 (cut-off filter)

CHROMATOGRAM
Retention time: 5.95 (putrescine), 6.68 (cadaverine), 12.87 (spermidine), 21.36 (spermine)
Limit of detection: 1.1-9.9 pg

KEY WORDS
serum

REFERENCE
Faulkner, A.J.; Veening, H.; Becker, H.-D. 2-(9-Anthryl)ethyl chloroformate: A precolumn derivatizing reagent for polyamines determined by liquid chromatography and fluorescence detection, *Anal.Chem.*, **1991**, *63*, 292–296.

Benoxaprofen Chloride

RELATED REFERENCE

Spahn, H.; Krauss, D.; Mutschler, E. Enantiospecific high-performance liquid chromatographic (HPLC) determination of baclofen and its fluoro analog in biological material. *Pharm.Res.* **1988**, *5*, 107-112.

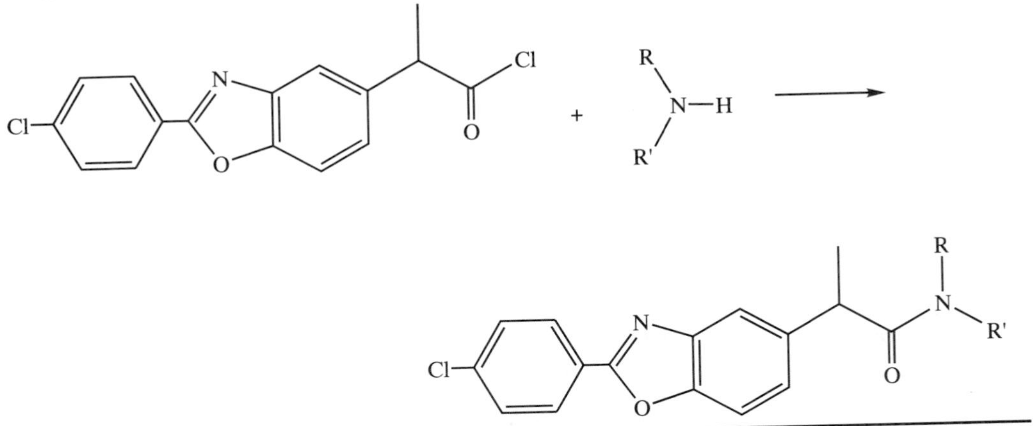

SAMPLE

Matrix: blood

Analyte: amines

Sample preparation: 1 mL Plasma + 1 mL 100 mM NaOH + 3 mL n-hexane, shake for 20 min, centrifuge for 10 min. Remove 2 mL of the organic layer and evaporate it to dryness using a vacuum centrifuge, reconstitute the residue in 500 µL 100 µg/mL (S)-(+)-benoxaprofen chloride in dried dichloromethane, let stand at room temperature for 30 min, inject a 10 µL aliquot. (Synthesis of benoxaprofen chloride is as follows. Dissolve 600 mg benoxaprofen in 50 mL toluene, slowly add 5 mL freshly-distilled thionyl chloride, reflux for 30 min, evaporate to dryness, recrystallize from dichloromethane to obtain benoxaprofen chloride (mp 91.5°).)

HPLC VARIABLES

Column: 250 × 4.6 7 µm Zorbax-Sil

Mobile phase: Cyclohexane:dichloromethane:THF 50:10:10

Flow rate: 1

Injection volume: 10

Detector: F ex 312 em 365

CHROMATOGRAM

Retention time: 6.7 ((R)-(+)-α-methylbenzylamine), 8.0 ((R)-(-)-amphetamine), 9.0 (S-(-)-tranylcypromine), 9.5 ((S)-(+)-amphetamine), 10.5 ((R)-(-)-methamphetamine), 10.7 (R-(+)-tranylcypromine), 10.8 ((S)-(-)-α-methylbenzylamine), 11.5 ((S)-(+)-methamphetamine)

KEY WORDS

plasma; normal phase; chiral

REFERENCE

Weber, H.; Spahn, H.; Mutschler, E.; Möhrke, W. Activated α-alkyl-α-arylacetic acid enantiomers for stereoselective thin-layer chromatographic and high-performance liquid chromatographic determination of chiral amines, *J.Chromatogr.*, **1984**, *307*, 145–153.

SAMPLE

Matrix: blood

Analyte: maprotiline

Sample preparation: 1 mL Plasma + 500 µL 10% potassium carbonate + 3 mL freshly-distilled n-hexane, shake for 30 min, centrifuge. Remove 2 mL of the organic layer and evaporate it to dryness under reduced pressure, reconstitute the residue in 500 μL 1 mg/mL benoxaprofen chloride in dry dichloromethane, heat at 50° for 30 min, inject a 10 μL aliquot. (Prepare benoxaprofen chloride as described above.)

HPLC VARIABLES

Column: 250 × 4.6 7 μm Zorbax-sil
Mobile phase: Cyclohexane:dichloromethane:THF 5:1:1
Flow rate: 1
Injection volume: 10
Detector: F ex 312 em 365

CHROMATOGRAM

Retention time: 8.6

OTHER SUBSTANCES

Simultaneously analyzed: amphetamine, benzylamine, methamphetamine, α-methylbenzylamine, phenylbutylamine, β-phenylethylamine, tolylethylamine, tranylcypromine
Interfering: procaine

KEY WORDS

normal phase; plasma

REFERENCE

Spahn, H.; Weber, H.; Mutschler, E.; Möhrke, W. α-Alkyl-α-arylacetic acid derivatives as fluorescence markers for thin-layer chromatographic and high-performance liquid chromatographic assay of amines and alcohols, *J.Chromatogr.*, **1984**, *310*, 167–178.

Benzoyl Chloride

RELATED REFERENCES

Asotra, S.; Mladenov, P.V.; Burke, R.D. Improved method for benzoyl chloride derivatization of polyamines for high-performance liquid chromatography. *J.Chromatogr.* **1987**, *408*, 227-233.

Hauschild, M.Z. Adsorption losses during extraction and derivatization efficiency by benzoylation of plant putrescine for high-performance liquid chromatographic analysis. *J.Chromatogr.* **1992**, *630*, 397-401.

Hwang, D.-F.; Chang, S.-H.; Shiua, C.-Y.; Chai, T. High-performance liquid chromatographic determination of biogenic amines in fish implicated in food poisoning. *J.Chromatogr.B* **1997**, *693*, 23-30.

Mei, Y.H. A sensitive and fast method for the determination of polyamines in biological samples. Benzoyl chloride pre-column derivatization high-performance liquid chromatography. *J.Liq.Chromatogr.* **1994**, *17*, 2413-2418.

Watanabe, S.; Saito, T.; Sato, S.; Nagase, S.; Ueda, S.; Tomita, M. Investigation of interfering products in the high-performance liquid chromatographic determination of polyamines as benzoyl derivatives. *J.Chromatogr.* **1990**, *518*, 264-267.

SAMPLE

Matrix: blood
Analyte: tromethamine
Sample preparation: 500 μL Serum + 50 μL 5.027 mg/mL 2,3-butanediol in water + 1 mL 4 M NaOH + 200 μL benzoyl chloride, rotate for 10 min, add 2 drops 13.1 mg/mL sodium glycine salt in water, let stand for 2-3 min, add 8 mL hexane, rotate for 5 min,

centrifuge. Remove the upper organic layer and evaporate it to dryness, reconstitute the residue in 300 µL MeOH, inject a 50 µL aliquot.

HPLC VARIABLES
Column: reverse-phase 10 C ODS
Mobile phase: MeCN:MeOH water 25:50:25
Flow rate: 2
Injection volume: 50
Detector: UV 237

CHROMATOGRAM
Retention time: 4
Internal standard: 2,3-butanediol (3)
Limit of detection: 20 µg/mL

KEY WORDS
serum

REFERENCE
Blanke, S.R.; Blanke, R.V. The Schotten-Baumann reaction as an aid to the analysis of polar compounds: application to the determination of tris(hydroxymethyl)aminomethane (THAM), *J.Anal.Toxicol.*, **1984**, *8*, 231–233.

SAMPLE
Matrix: blood, urine
Analyte: 4-aminopyridine
Sample preparation: Condition a 1 mL Bond Elut C18 SPE cartridge with one column volume 1 M HCl, two column volumes MeOH, and one column volume water, aspirate completely each time. 500 µL Serum or 250 µL urine + 50 (serum) or 100 (urine) µL 20 µg/mL N-propionylprocainamide in 100 mM HCl + 250 µL 4 M NaOH + 25 µL benzoyl chloride:MeCN 1:1 (freshly prepared), vortex vigorously, let stand at room temperature with occasional vortexing for 15 min, add 400 µL water, adjust pH to 5.5-6 with 4 M acetic acid, vortex, centrifuge at 1200 g for 2 min, add the supernatant to the SPE cartridge, wash with two column volumes of water, wash with two 400 µL aliquots of MeCN (make sure cartridge is drained completely after each wash), elute with 300 µL MeOH: 35% perchloric acid 99:1, drain cartridge completely by centrifuging at 800 g for 20 s, inject a 5 (urine) or 10 (serum) µL aliquot.

HPLC VARIABLES
Guard column: 15 × 3.2 7 µm RP-18 (Applied Biosystems)
Column: 150 × 4.6 5 µm Ultrasphere ODS
Mobile phase: MeCN:0.1% tetramethylammonium perchlorate 16:84, pH 4.2, containing 10% perchloric acid
Flow rate: 1.2
Injection volume: 5-10
Detector: UV 278

CHROMATOGRAM
Retention time: 10.5
Internal standard: N-propionylprocainamide (6)
Limit of quantitation: 3 ng/mL (20 µL injection)

KEY WORDS
serum; SPE

REFERENCE
Gupta, R.N.; Hansebout, R.R. Optimization of the determination of 4-aminopyridine in human serum and urine by column liquid chromatography, *J.Chromatogr.B*, **1996**, *677*, 183–189.

SAMPLE
Matrix: cell suspensions
Analyte: polyamines
Sample preparation: Add 2 mL cell suspension to 6 mL 4% albumin in water, centrifuge at 450 g for 5 min, replace the supernatant with 1 mL 6% perchloric acid, homogenize,

add 300 µL 167 µM 1,6-diaminohexane, vortex, centrifuge at 450 g for 10 min. Remove the supernatant and add it to 1 mL 2 M NaOH and 5 µL benzoyl chloride, vortex briefly, let stand for 20 min, add 2 mL saturated NaCl, add 2 mL chloroform, vortex for 1 min, centrifuge at 2000 g for 10 min. Remove a 1.5 mL aliquot of the lower organic layer and evaporate it to dryness under a stream of nitrogen, reconstitute the residue in 500 µL mobile phase, heat at 35° for 15 h (to degrade the interfering benzoic anhydride), add 200 µL 2 M NaOH, add 2 mL chloroform, mix for 1 min, centrifuge at 2000 g for 10 min. Remove a 1.5 mL aliquot of the organic layer and evaporate it to dryness under a stream of nitrogen, reconstitute the residue in 500 µL mobile phase, inject an aliquot.

HPLC VARIABLES
Column: 250 × 4.6 5 µm Ultrasphere C18
Mobile phase: MeOH:water 55:45
Flow rate: 1
Injection volume: 100
Detector: UV 234

CHROMATOGRAM
Retention time: 5.9 (N^1-acetylspermidine), 7.7 (putrescine), 8.5 (1,3-diaminopropane), 9.2 (cadaverine), 11.5 (N^1-acetylspermine), 16.9 (spermidine), 38.5 (spermine)
Internal standard: 1,6-diaminohexane (12.6)
Limit of quantitation: 16.5-40 pmole
Limit of detection: 2-12.5 pmole

REFERENCE
Schenkel, E.; Berlaimont, V.; Dubois, J.; Helson-Cambier, M.; Hanocq, M. Improved high-performance liquid chromatographic method for the determination of polyamines as their benzoylated derivatives: application to P388 cancer cells, *J.Chromatogr.B*, **1995**, *668*, 189–197.

SAMPLE
Matrix: perilymph
Analyte: antibiotics
Sample preparation: Lyophilize 6 µL perilymph, reconstitute with 90 µL pyridine, add 10 µL benzoyl chloride, heat at 80° for 30 min, evaporate to dryness under a stream of nitrogen, add 1 mL MeOH, heat at 80° for 10 min, add 50 mg solid sodium carbonate, add 1 mL MeOH saturated with sodium carbonate, wash 3 times with 2 mL portions of n-hexane, add 1 mL water, remove any hexane which separates, extract with 3 mL chloroform. Wash the chloroform layer 3 times with 1 mL portions of MeOH:water 50:50, evaporate the chloroform layer to dryness, reconstitute with 15 µL chloroform, inject a 5 µL aliquot.

HPLC VARIABLES
Column: 250 × 4.6 5-6 µm Zorbax SIL
Mobile phase: n-Hexane:THF 50:50
Flow rate: 2
Injection volume: 5
Detector: UV 230

CHROMATOGRAM
Retention time: 4.73 (streptomycin), 4.98 (dihydrostreptomycin), 5.13 (streptomycin), 5.46 (kanamycin), 5.76 (kanamycin), 5.86 (dihydrostreptomycin), 5.87 (streptomycin), 6.35 (dihydrostreptomycin), 8.02 (kanamycin), 8.11 (neomycin), 9.28 (neomycin), 10.08 (neomycin)
Limit of detection: 10 ng

KEY WORDS
normal phase; guinea pig

REFERENCE
Harada, T.; Iwamori, M.; Nagai, Y.; Nomura, Y. Analysis of aminoglycoside antibiotics as benzoyl derivatives by high-performance liquid chromatography and its application to the quantitation of neomycin in the perilymph, *J.Chromatogr.*, **1985**, *337*, 187–193.

SAMPLE

Matrix: polymers

Analyte: 4,4'-diaminodiphenylmethane

Sample preparation: Wash 12.5 g polymer with 250 mL water. Remove a 200 mL aliquot and extract it five times with 50 mL portions of ether. Combine the extracts and add 150 µL hexane:triethylamine 95:5, add 100 µL hexane:benzoyl chloride 95:5, let stand for 1 h, add 1 drop of morpholine, evaporate to dryness under reduced pressure at 50°, reconstitute with 1 mL 10 µg/mL 2-phenylnaphthalene, inject an aliquot.

HPLC VARIABLES

Column: 250 × 2.1 Sepralyte C18

Mobile phase: MeOH:THF:water 9:43:48

Column temperature: 35

Flow rate: 0.4

Injection volume: 10

Detector: UV 280

CHROMATOGRAM

Retention time: 4.7

Internal standard: 2-phenylnaphthalene (11.5)

Limit of detection: 70 ng/g

REFERENCE

Mazzu, A.L.; Smith, C.P. Determination of extractable methylene dianiline in thermoplastic polyurethanes by HPLC, *J.Biomed.Mater.Res.*, **1984**, *18*, 961–968.

SAMPLE

Matrix: semen

Analyte: polyamines

Sample preparation: Dilute 100-500 µL semen to 1.5 mL with water, add 20 µL benzoyl chloride, vortex, add 1 mL 2 M NaOH, rotate at 30 rpm for 20 min, add 2 mL saturated NaCl, mix well, add 3 mL diethyl ether, rotate at 30 rpm for 20 min, centrifuge at 2500 g for 3 min. Remove the organic layer and evaporate it to dryness under a stream of nitrogen at 37°, reconstitute the residue in 200 µL MeOH, sonicate for 30 s, inject a 20 µL aliquot.

HPLC VARIABLES

Column: 250 × 4 5 µm Bio-Sil ODS-5S (Bio-Rad)

Mobile phase: MeOH:water 60:40

Flow rate: 1

Injection volume: 20

Detector: UV 254

CHROMATOGRAM

Retention time: 4.5 (putrescine), 5 (cadaverine), 7.5 (spermidine), 14 (spermine)

Limit of detection: 5-50 ng

KEY WORDS

comparison with 3,5-dinitrobenzoyl chloride

REFERENCE

Wongyai, S.; Oefner, P.; Bonn, G. HPLC analysis of polyamines and their acetylated derivatives in the picomole range using benzoyl chloride and 3,5-dinitrobenzoyl chloride as derivatizing agent, *Biomed.Chromatogr.*, **1988**, *2*, 254–257.

SAMPLE

Matrix: semen

Analyte: polyamines

Sample preparation: Centrifuge semen at 1000 g for 3 h, dilute the supernatant with an equal volume of water, remove a 100-200 µL aliquot and add it to 20 µL benzoyl chloride, add 1 mL 2 M NaOH, rotate at 30 rpm for 20 min, add 2 mL saturated aqueous NaCl, add 3 mL diethyl ether, rotate at 30 rpm for 20 min, centrifuge at 1000 g for 3 min. Remove the organic layer and evaporate it to dryness under a stream of nitrogen at 37°, reconstitute the residue in 200 μL MeOH:water 60:40, sonicate for 10 s, vortex, inject an aliquot.

HPLC VARIABLES
Column: 250 × 4.0 5 μm Bio-Sil ODS-5S C18 (Bio-Rad)
Mobile phase: MeOH:water 60:40
Flow rate: 1
Injection volume: 20
Detector: UV 254

CHROMATOGRAM
Retention time: 5 (putrescine), 10 (spermidine), 17 (spermine)
Limit of detection: 57-124 pmole

REFERENCE
Oefner, P.J.; Wongyai, S.; Bonn, G. High-performance liquid chromatographic determination of free polyamines in human seminal plasma, *Clin.Chim.Acta*, **1992**, *205*, 11–18.

SAMPLE
Matrix: tissue
Analyte: polyamines
Sample preparation: Homogenize (Potter-Elvejhem) tissue with 5-10 volumes of cold 300 mM perchloric acid containing IS, centrifuge. Remove a 100-400 μL aliquot of the supernatant and add it to 2 mL 2 M NaOH, add 10 μL benzoyl chloride, mix vigorously, rotate gently at room temperature for 30 min, extract with 2 mL chloroform, centrifuge. Remove the organic layer and wash it with 2 mL water, centrifuge, evaporate to dryness under a stream of air, reconstitute the residue in 100-500 μL MeOH, inject a 2-10 μL aliquot.

HPLC VARIABLES
Guard column: reversed-phase
Column: two 100 × 3 5 μm ChromSpher C18 columns in series
Mobile phase: MeOH:water 65:35
Flow rate: 0.4
Injection volume: 2-10
Detector: UV 229

CHROMATOGRAM
Retention time: 2.96 (N^1-acetylputrescine), 3.63 (N^1-acetylspermidine), 3.64 (N^8-acetylspermidine), 4.18 (putrescine), 4.43 (1,3-propanediamine), 4.48 (N^1-acetylspermine), 4.54 (cadaverine), 4.78 ((2R,5R)-6-heptyne-2,5-diamine (MAP)), 5.80 (spermidine), 5.83 (S-adenosylmethionine), 5.94 (cyclohexylamine), 6.22 (1,7-heptanediamine), 8.10 (1,8-octanediamine), 8.36 (spermine)
Internal standard: 1,6-hexanediamine (5.18)
Limit of detection: 1 pmole

OTHER SUBSTANCES
Non-interfering: α-difluoromethylornithine, methylglyoxal bis(guanylhydrazone)

REFERENCE
Verkoelen, C.F.; Romijn, J.C.; Schroeder, F.H.; van Schalkwijk, W.P.; Splinter, T.A. Quantitation of polyamines in cultured cells and tissue homogenates by reversed-phase high-performance liquid chromatography of their benzoyl derivatives, *J.Chromatogr.*, **1988**, *426*, 41–54.

SAMPLE
Matrix: tissue
Analyte: polyamines
Sample preparation: Incubate chick embryo retinas in 4 mL medium at 37° for 4 h, centrifuge at 500 g for 1 min. For each 20 mg solid add 1 mL 20 μM 1,7-diaminoheptane in 200 mM perchloric acid, suspend, sonicate (Soniprep 150) with 10 μm amplitude in 10 s bursts, centrifuge at 20000 g for 15 min, neutralize with KOH, centrifuge. Remove a 1 mL aliquot and add it to 1 mL 2 M NaOH, add 5 μL benzoyl chloride, vortex briefly, let stand for 20 min, add 2 mL saturated NaCl, extract with 2 mL diethyl ether, centrifuge. Remove the upper organic phase and wash it with 2 mL 100 μM NaOH, dry the organic layer over a few mg anhydrous sodium sulfate, centrifuge. Remove the organic layer and

evaporate it to dryness under a stream of nitrogen, reconstitute the residue in 2 mL mobile phase, inject a 20 μL aliquot. (Medium was pH 7.4 serum- and glutamine-free Eagle's minimum essential medium containing 25 mM 4-(2-hydroxyethyl)-1-piperazine-ethane sulfonic acid (HEPES), 100 U/mL penicillin, and 100 μg/mL streptomycin.)

HPLC VARIABLES
Column: 150 × 4.6 3 μm Spherisorb ODS2
Mobile phase: MeOH:water 62:38
Flow rate: 1
Injection volume: 20
Detector: UV 254

CHROMATOGRAM
Retention time: 2 (N-acetylputrescine), 2.2 (N-acetylcadaverine), 2.7 (N^1-acetylspermi-dine), 3.2 (putrescine), 3.5 (cadaverine), 4 (N^1-acetylspermine), 5 (spermidine), 8.2 (spermine), 10.35 (histamine)
Internal standard: 1,7-diaminoheptane (5.4)
Limit of quantitation: 1.25 μM

KEY WORDS
retina; chicken

REFERENCE
Taibi, G.; Schiavo, M.R. Simple high-performance liquid chromatographic assay for polyamines and their monoacetyl derivatives, *J.Chromatogr.*, **1993**, *614*, 153–158.

SAMPLE
Matrix: urine
Analyte: amphetamine and methamphetamine
Sample preparation: 500 μL Urine + 100 μL 25 μg/mL N-n-propylaniline + 6 mL pH 10.0 carbonate buffer + 15 mL water, add mixture to an Extrelut SPE cartridge, let stand for 20 min, elute with 40 mL hexane:ethyl acetate 90:10. Add the eluate to 3 mL 100 mM sulfuric acid and 500 mg NaCl, stir for 20 min, centrifuge at 1000 g for 5 min. Remove the lower layer and add it to 3 mL 2.5 M NaOH and 20 μL benzoyl chloride, stir vigorously for 30 min. Extract the mixture with 1.5 mL chloroform. Wash the chloroform layer twice with 5 mL water and evaporate it to dryness at 40°, reconstitute the residue in 200 μL hexane:isopropanol 90:10, inject an aliquot.

HPLC VARIABLES
Column: 250 × 4.6 Chiralcel OB + 250 × 4.6 Chiralcel OJ
Mobile phase: Hexane:isopropanol 90:10
Column temperature: 48
Flow rate: 1-1.4
Detector: UV 220

CHROMATOGRAM
Retention time: 13 (d-methamphetamine), 15 (l-methamphetamine), 17 (l-amphetamine), 25 (d-amphetamine)
Internal standard: N-n-propylaniline (11)
Limit of detection: 25 ng

KEY WORDS
rat; SPE; chiral

REFERENCE
Nagai, T.; Kamiyama, S. Assay of the optical isomers of methamphetamine and amphetamine in rat urine using high-performance liquid chromatography with chiral cellulose-based columns, *J.Chromatogr.*, **1990**, *525*, 203–209.

SAMPLE
Matrix: urine
Analyte: amphetamine, methamphetamine, and metabolites
Sample preparation: 200-500 μL Rat urine + 200-500 μL pH 3.8 acetate buffer + 25 μL 40 μg/mL β-glucuronidase and 20 μg/mL arylsulfatase (Merck), heat at 37° for 24 h, add

100 μL 25 μg/mL 3-methoxytyramine in water, add 100 μL water, adjust pH to 9.0 with 1.9 M sodium carbonate, add to an Extrelut SPE cartridge, let stand for 20 min, elute with 6 mL ethyl acetate. Add the eluate to 1 mL 100 mM sulfuric acid, extract. Add the aqueous layer to 3 mL 2.5 M NaOH, add 25 μL benzoyl chloride, extract with 5 mL ethyl acetate. Wash the ethyl acetate layer with water, evaporate to dryness under a stream of nitrogen at 40°, reconstitute the residue in 50 μL EtOH, 3.5 mL 50 mM pH 8.0 Tris/HCl buffer, and 35 μL esterase (Type 1 porcine liver, Sigma). Heat at 25° for 45 min, add to an activated Sep-Pak C18 SPE cartridge, wash with 5 mL water, elute with 5 mL acetone. Evaporate the eluate to dryness under a stream of nitrogen at 40°, reconstitute the residue in 200 μL hexane:EtOH 89:11, inject a 20 μL aliquot.

HPLC VARIABLES
Column: 250 × 4.6 Chiralcel OB + 250 × 4.6 Chiralcel OJ
Mobile phase: n-Hexane:EtOH 89:11
Column temperature: 48
Flow rate: 1.4
Injection volume: 20
Detector: UV 220

CHROMATOGRAM
Retention time: 10 (D-methamphetamine), 11.5 (L-methamphetamine), 12.5 (L-amphetamine), 14.5 (D-amphetamine), 15 (D-p-hydroxymethamphetamine), 18 (L-p-hydroxymethamphetamine), 25 (L-p-hydroxyamphetamine), 36 (D-p-hydroxyamphetamine)
Internal standard: 3-methoxytyramine (54)

OTHER SUBSTANCES
Extracted: metabolites

KEY WORDS
rat; SPE

REFERENCE
Nagai, T.; Kamiyama, S. Simultaneous HPLC analysis of optical isomers of methamphetamine and its metabolites, and stereoselective metabolism of racemic methamphetamine in rat urine, *J.Anal.Toxicol.*, **1991**, *15*, 299–304.

SAMPLE
Matrix: urine
Analyte: amphetamine, ethylamphetamine, and methamphetamine
Sample preparation: 100-300 μL Urine + 100 μL 1.5 M NaOH + 5 μg IS, make up to 1 mL with water, add to an Extrelut 1 SPE cartridge, let stand for 20 min, elute with 6 mL benzene (Caution! Benzene is a carcinogen!). Add the eluate to 1 mL 100 mM sulfuric acid, extract. Remove the aqueous layer and add it to 3 mL 1.5 M NaOH, add 5 μL benzoyl chloride, vortex vigorously, extract twice with 3 mL portions of n-hexane. Combine the organic layers and wash them twice with 3 mL portions of water, evaporate the organic layer to dryness under a stream of nitrogen at 40°, reconstitute the residue in 200 μL mobile phase, inject a 20 μL aliquot.

HPLC VARIABLES
Column: Chiralcel OB-H
Mobile phase: n-Hexane:isopropanol 90:10
Column temperature: 55
Flow rate: 1
Injection volume: 20
Detector: UV 220

CHROMATOGRAM
Retention time: 5.5 (d-ethylamphetamine), 6.2 (l-ethylamphetamine), 7 (d-methamphetamine), 7.5 (l-amphetamine), 10 (l-methamphetamine), 18 (d-amphetamine)
Internal standard: l-p-methoxyamphetamine (12)
Limit of detection: 30 ng

KEY WORDS
rat; SPE; chiral

REFERENCE

Nagai, T.; Kamiyama, S.; Matsushima, K. Analysis of time-lapse changes of *d*- and *l*-enantiomers of racemic ethylamphetamine and stereoselective metabolism in rat urine by HPLC determination, *J.Anal.Toxicol.*, **1995**, *19*, 225–228.

4-(N-Chloroformylmethyl-N-methyl)amino-7-N,N-dimethylaminosulfonyl-2,1,3-benzoxadiazole

SAMPLE

Matrix: solutions

Analyte: amines

Sample preparation: Mix 10 μL 0.5 mM compound in anhydrous benzene containing 0.5 mM quinuclidine with 10 μL 25 mM DBD-COCl in anhydrous benzene (Caution! Benzene is a carcinogen!), heat at 60° for 15 min, add 980 μL MeCN:water:acetic acid 50:50:1, inject a 2 μL aliquot. (Purify quinuclidine by sublimation. DBD-COCl is 4-(N-chloroformylmethyl-N-methyl)amino-7-N,N-dimethylaminosulfonyl-2,1,3-benzoxadiazole. DBD-COCl is available from Tokyo Kasei (TCI America, Portland OR). Synthesis is as follows. Dissolve 0.5 g magnesium sulfate heptahydrate and 6 g NaOH in 60 mL water, throughout the reaction keep the flask at about 20° with cold water cooling, add 15 mL 30% hydrogen peroxide, add 75 mL MeOH, add 12.1 g powdered benzoyl peroxide in one go, stir for 10 min, pour into 150 mL 20% sulfuric acid, extract three times with 50 mL portions of chloroform, determine peroxybenzoic acid concentration by iodometric titration (Tetrahedron 1967, 23, 3327). Slowly add 110 mL 1 M peroxybenzoic acid in chloroform to 7 g 2,6-difluoroaniline dissolved in 100 mL chloroform, stir at room temperature, when reaction is complete (iodometric titration) wash with 2% sodium thiosulfate, wash with 5% sodium carbonate, wash with water, dry over anhydrous sodium sulfate, evaporate to dryness under reduced pressure, recrystallize 2,6-difluoronitrosobenzene from EtOH (mp 108.5-109.5). Stir 8.5 g 2,6-difluoronitrosobenzene in 85 mL DMSO at room temperature and add a solution of 3.91 g sodium azide in 85 mL DMSO dropwise, let stand for about 1 h, add to a large volume of water, extract with ether, dry the extracts over anhydrous sodium sulfate, evaporate to dryness under reduced pressure and distil to give 4-fluoro-2,1,3-benzoxadiazole as a colorless oil (bp 83°/12 mm Hg) (J.Chem.Soc.(C) 1970, 1433). Add 11 mL chlorosulfonic acid dropwise to 3 g 4-fluoro-2,1,3-benzoxadiazole in 10 mL chloroform at 0-10° (use a calcium chloride drying tube), stir at room temperature for 1 h, reflux for 2 h, cool, slowly pour into ice water, remove the organic layer, extract the aqueous layer with chloroform, combine the organic layers, wash, dry over anhydrous magnesium sulfate, evaporate under reduced pressure, take up the residue in 5 mL benzene (Caution! Benzene is a carcinogen!), chromatograph on a 150 × 30 column of silica gel (100-200 mesh Kanto Chemical) with n-hexane:benzene 50:50, evaporate the appropriate fractions to give 4-(chlorosulfonyl)-7-fluoro-2,1,3-benzoxadiazole (CBD-F) as pale yellow needles (mp 64-66°) (Anal. Chem. 1984. 56, 2461). Stir 0.76 g CBD-F in 70 mL

MeCN at 0-10° and add 1 g dimethylamine hydrochloride in 10 mL 100 mM pH 10 borax dropwise, adjust pH to 5 with 1 M HCl, concentrate to about 10 mL under reduced pressure, extract three times with 200 mL portions of diethyl ether, wash with water, dry over anhydrous magnesium sulfate, evaporate under reduced pressure, chromatograph on a 500 × 20 column of silica gel with chloroform, isolate the appropriate fraction and re-chromatograph on the same column with ethyl acetate:benzene 1:2 to give 4-(N,N-dimethylaminosulfonyl)-7-fluoro-2,1,3-benzoxadiazole (DBD-F) as white needles (mp 124-125°) (yield = 1% !) (Analyst 1989, 114, 413). On a Merck no. 5714 60F$_{254}$ tlc plate eluted with chloroform DBD-F has Rf 0.32 and lies between two other reaction products. DBD-F can also be purchased from Tokyo Kasei. Stir N-methylglycine and 2.3 g sodium carbonate in water at room temperature, add 880 mg DBD-F in 40 mL MeCN dropwise, stir for 1 h, evaporate to remove the MeCN, wash twice with 50 mL portions of ethyl acetate. Acidify the aqueous phase with HCl and extract it twice with 300 mL portions of ethyl acetate. Wash the organic layer twice with 100 mL portions of saturated aqueous NaCl, dry over anhydrous magnesium sulfate, evaporate to dryness under reduced pressure, recrystallize from ethyl acetate to give 4-(N-carboxymethyl-N-methyl)amino-7-dimethylaminosulfonyl-2,1,3-benzoxadiazole (DBD-COOH) as orange-yellow crystals (mp 209-210°). Add 3.5 mL oxalyl chloride and 24 μL DMF to 1 g DBD-COOH in anhydrous benzene, stir at room temperature for 30 min, reflux for 1 h, evaporate to dryness, add 20 mL dry benzene to the residue, filter, evaporate the filtrate to give 4-(N-chloroformylmethyl-N-methyl)amino-7-N,N-dimethylaminosulfonyl-2,1,3-benzoxadiazole (DBD-COCl) as yellow crystals (mp 102°).)

HPLC VARIABLES

Column: 150 × 4.6 5 μm Cosmosil 5C18
Mobile phase: Gradient. MeCN:water from 30:70 to 100:0 over 1 h.
Flow rate: 1
Injection volume: 2
Detector: F ex 445 em 555 (benzylamine); F ex 443 em 553 (phenetidine)

CHROMATOGRAM

Retention time: 15 (benzylamine), 22 (phenetidine)
Limit of detection: 56-89 fmole

REFERENCE

Imai, K.; Fukushima, T.; Yokosu, H. A novel electrophilic reagent, 4-(N-chloroformylmethyl-N-methyl)amino-7-N,N-dimethylaminosulphonyl-2,1,3-benzoxadiazole (DBD-COCl) for fluorometric detection of alcohols, phenols, amines and thiols, *Biomed.Chromatogr.*, **1994**, *8*, 107–113.

4-(2-Chloroformylpyrrolidin-1-yl)-7-(N,N-dimethylaminosulfonyl)-2,1,3-benzoxadiazole

SAMPLE
Matrix: bulk
Analyte: amines

Sample preparation: Dissolve 0.5 nmole amine and 5 nmole (S)-(-)-4-(2-chloroformylpyrrolidin-1-yl)-7-(N,N-dimethylaminosulfonyl)-2,1,3-benzoxadiazole in 1 mL anhydrous benzene:pyridine 99:1 (Caution! Benzene is a carcinogen!), heat at 50° for 1.5 h, cool in ice-water. Remove a 100 μL aliquot and make up to 1 mL with 1% methylamine in MeCN, inject a 5 μL aliquot. ((S)-(-)-4-(2-Chloroformylpyrrolidin-1-yl)-7-(N,N-dimethylaminosulfonyl)-2,1,3-benzoxadiazole ((S)-(-)-DBD-Pro-Cl) is available from Tokyo Kasei (TCI America, Portland OR). Synthesis is as follows. Dissolve 0.5 g magnesium sulfate heptahydrate and 6 g NaOH in 60 mL water, throughout the reaction keep the flask at about 20° with cold water cooling, add 15 mL 30% hydrogen peroxide, add 75 mL MeOH, add 12.1 g powdered benzoyl peroxide in one go, stir for 10 min, pour into 150 mL 20% sulfuric acid, extract three times with 50 mL portions of chloroform, determine peroxybenzoic acid concentration by iodometric titration (Tetrahedron 1967, 23, 3327). Slowly add 110 mL 1 M peroxybenzoic acid in chloroform to 7 g 2,6-difluoroaniline dissolved in 100 mL chloroform, stir at room temperature, when reaction is complete (iodometric titration) wash with 2% sodium thiosulfate, wash with 5% sodium carbonate, wash with water, dry over anhydrous sodium sulfate, evaporate to dryness under reduced pressure, recrystallize 2,6-difluoronitrosobenzene from EtOH (mp 108.5-109.5). Stir 8.5 g 2,6-difluoronitrosobenzene in 85 mL DMSO at room temperature and add a solution of 3.91 g sodium azide in 85 mL DMSO dropwise, let stand for about 1 h, add to a large volume of water, extract with ether, dry the extracts over anhydrous sodium sulfate, evaporate to dryness under reduced pressure and distil to give 4-fluoro-2,1,3-benzoxadiazole as a colorless oil (bp 83°/12 mm Hg) (J.Chem.Soc.(C) 1970, 1433). Add 11 mL chlorosulfonic acid dropwise to 3 g 4-fluoro-2,1,3-benzoxadiazole in 10 mL chloroform at 0-10° (use a calcium chloride drying tube), stir at room temperature for 1 h, reflux for 2 h, cool, slowly pour into ice water, remove the organic layer, extract the aqueous layer with chloroform, combine the organic layers, wash, dry over anhydrous magnesium sulfate, evaporate under reduced pressure, take up the residue in 5 mL benzene (Caution! Benzene is a carcinogen!), chromatograph on a 150 × 30 column of silica gel (100-200 mesh Kanto Chemical) with n-hexane:benzene 50:50, evaporate the appropriate fractions to give 4-(chlorosulfonyl)-7-fluoro-2,1,3-benzoxadiazole (CBD-F) as pale yellow needles (mp 64-66°) (Anal. Chem. 1984. 56, 2461). Stir 0.76 g CBD-F in 70 mL MeCN at 0-10° and add 1 g dimethylamine hydrochloride in 10 mL 100 mM pH 10 borax dropwise, adjust pH to 5 with 1 M HCl, concentrate to about 10 mL under reduced pressure, extract three times with 200 mL portions of diethyl ether, wash with water, dry over anhydrous magnesium sulfate, evaporate under reduced pressure, chromatograph on a 500 × 20 column of silica gel with chloroform, isolate the appropriate fraction and re-chromatograph on the same column with ethyl acetate:benzene 1:2 to give 4-(N,N-dimethylaminosulfonyl)-7-fluoro-2,1,3-benzoxadiazole (DBD-F) as white needles (mp 124-125°) (yield = 1% !). On a Merck no. 5714 60F₂₅₄ tlc plate eluted with chloroform DBD-F has Rf 0.32 and lies between two other reaction products (Analyst 1989, 114, 413). DBD-F can also be purchased from Tokyo Kasei. Add 100 mg DBD-F in 10 mL MeCN to 47 mg (S)-(-)proline in 20 mL 250 mM pH 11.5 sodium carbonate solution, stir at room temperature for 30 min, wash with ethyl acetate, adjust the pH of the aqueous layer to 1-2 with 2 M HCl, extract three times with 30 mL ethyl acetate. Combine the extracts and evaporate them under reduced pressure, recrystallize from benzene/ethyl acetate to give (S)-(-)-(N,N-dimethylaminosulfonyl)-7-(2-carboxypyrrolidin-1-yl)-2,1,3-benzoxadiazole (DBD-Pro) as yellow needles (mp 187-9° d) (Analyst 1989, 114, 1233). Suspend 55 mg (S)-(-)-DBD-Pro in 55 mL anhydrous diethyl ether at 0°, add 110 mg phosphorus pentachloride, stir at 5° for 1 h, filter quickly, evaporate to dryness under reduced pressure, dry under vacuum over phosphorus pentoxide for 12 h to give (S)-(-)-4-(2-chloroformylpyrrolidin-1-yl)-7-(N,N-dimethylaminosulfonyl)-2,1,3-benzoxadiazole (DBD-Pro-Cl) as yellow crystals (mp 116-17°) (Analyst 1993, 118, 759).)

HPLC VARIABLES

Column: 150 × 4.6 5 μm Inertsil ODS-80A
Mobile phase: MeCN:water 40:60
Column temperature: 40
Flow rate: 1
Injection volume: 5
Detector: F ex 450 em 560

CHROMATOGRAM
Retention time: 30.35 (S-1-cyclohexylethylamine), 33.64 (R-1-cyclohexylethylamine), 32.33 (S-1-(1-naphthyl)ethylamine), 33.46 (R-1-(1-naphthyl)ethylamine)

REFERENCE
Toyo'oka, T.; Liu, Y.M.; Jinno, H.; Hanioka, N.; Ando, M.; Imai, K. Chiral separation of amines by high-performance liquid chromatography after tagging with 4-(N,N-dimethylaminosulphonyl)-7-(2-chloroformylpyrrolidin-1-yl)-2,1,3-benzoxadiazole, *Biomed.Chromatogr.*, **1994**, *8*, 85–89.

SAMPLE
Matrix: solutions
Analyte: amines
Sample preparation: Mix a 500 µL aliquot of a solution in benzene:pyridine 98:2 with 500 µL 10 mM 4-(2-chloroformylpyrrolidin-1-yl)-7-(N,N-dimethylaminosulfonyl)-2,1,3-benzoxadiazole in anhydrous benzene (Caution! Benzene is a carcinogen!), heat at 50° for 30 min, cool in ice-water, dilute with MeCN:methylamine 99:1, inject a 10 µL aliquot. (Dry solvents over molecular sieve. 4-(2-Chloroformylpyrrolidin-1-yl)-7-(N,N-dimethylaminosulfonyl)-2,1,3-benzoxadiazolethe, DBD-Pro-COCl, is available from Tokyo Kasei (TCI America, Portland OR). Synthesis is as described above.)

HPLC VARIABLES
Column: 150 × 4.6 5 µm Inertsil ODS-2
Mobile phase: MeCN:water 50:50 (A) or 60:40 (B)
Column temperature: 40
Flow rate: 1
Injection volume: 10
Detector: F ex 450 em 560

CHROMATOGRAM
Retention time: 2.77 (methylamine (A)), 5.52 (diethylamine (A)), 6.11 (n-heptylamine (B)), 6.26 (aniline (A)), 8.29 (n-octylamine (B)), 11.19 ((R)-(-)-1-cyclohexylethylamine (A))
Limit of quantitation: 500 nM (laser F)

REFERENCE
Toyo'oka, T.; Liu, Y.-M.; Hanioka, N.; Jinno, H.; Ando, M. Determination of alcohols and amines, labeled with 4-(*N,N*-dimethylaminosulfonyl)-7-(2-chloroformylpyrrolidin-1-yl)-2,1,3-benzoxadiazole, by liquid chromatography with conventional and laser-induced fluorescence detection, *Anal.Chim.Acta*, **1994**, *285*, 343–351.

4-(2-Chloroformylpyrrolidin-1-yl)-7-nitro-2,1,3-benzoxadiazole

SAMPLE
Matrix: solutions
Analyte: amines
Sample preparation: Mix 50 µL 2 mM amine in anhydrous benzene:pyridine 98:2 with 50 µL 10 mM S-(-)-NBD-Pro-Cl in anhydrous benzene (Caution! Benzene is a carcinogen!), heat at 50° for 4 h, cool in ice water, add 900 µL MeCN:methylamine 99:1, inject a 5 µL aliquot. (S-(-)-NBD-Pro-Cl is available from Tokyo Kasei (TCI America, Portland OR).

Synthesis is as follows. Add a solution of 1 g 4-chloro-7-nitrobenzofurazan in 50 mL EtOH to a solution of 575 mg proline in 100 mL 5% sodium acetate solution, reflux for 10 min, acidify to pH 1.5 with 4 M HCl, evaporate to dryness. Dissolve the residue in water and extract with ethyl acetate. Evaporate the organic layer to dryness and recrystallize from benzene/ethyl acetate to give NBD-Pro as orange needles (mp 156-7°) (Anal.Chim.Acta 1983, 149, 305). Dissolve 2.6 g S-(-)-NBD-Pro in 200 mL anhydrous dichloromethane, add 10 mL oxalyl chloride and 200 µL DMF, stir at room temperature for 1 h, evaporate to dryness under reduced pressure, dissolve the residue in 100 mL anhydrous benzene, filter, evaporate the filtrate to dryness, dry over phosphorus pentoxide under vacuum to give S-(-)-NBD-Pro-Cl (S-(-)-4-(2-chloroformylpyrrolidin-1-yl)-7-nitrobenzofurazan) as red-orange crystals (mp 103-4° d).)

HPLC VARIABLES
Column: 150 × 4.6 5 µm Inertsil SIL
Mobile phase: n-Hexane:ethyl acetate 45:55
Column temperature: 40
Flow rate: 1
Injection volume: 5
Detector: F ex 470 em 540

CHROMATOGRAM
Retention time: 10.29 (S-1-phenylethylamine), 13.51 (R-1-phenylethylamine), 8.21 (S-1-(1-naphthyl)ethylamine), 10.12 (R-1-(1-naphthyl)ethylamine), 9.22 (S-1-cyclohexylethylamine), 11.45 (R-1-cyclohexylethylamine)

KEY WORDS
chiral; normal phase; can also be separated by reverse-phase with MeCN:water 45:55 but separation of enantiomers is less

REFERENCE
Toyo'oka, T.; Lui, Y.-M.; Hanioka, N.; Jinno, H.; Ando, M.; Imai, K. Resolution of enantiomers of alcohols and amines by high-performance liquid chromatography after derivatization with a novel fluorescent chiral reagent, *J.Chromatogr.A*, **1994**, *675*, 79–88.

3,4-Dihydro-6,7-dimethoxy-4-methyl-3-oxoquinoxaline-2-carboxyl Chloride

RELATED REFERENCES

Dave, K.J.; Riley, C.M.; Vander Velde, D.; Stobaugh, J.F. Improved preparation and structural confor-mation of the fluorescence labeling reagents 1,2-diamino-4,5-dimethoxybenzene and 3,4-dihydro-6,7-dimethoxy-4-methyl-3-oxoquinoxaline-2-carbonyl chloride. *J.Pharm.Biomed.Anal.* **1990**, *8*, 307-312.

Ishida, J.; Yamaguchi, M.; Iwata, T.; Nakamura, M. 3,4-Dihydro-6,7-dimethoxy-4-methyl-3-oxoquinoxa-line-2-carbonyl chloride as a sensitive fluorescence derivatization reagent for amines in liquid chro-matography. *Anal.Chim.Acta* **1989**, *223*, 319-326.

SAMPLE

Matrix: blood

Analyte: β-phenylethylamine

Sample preparation: Condition a Toyopak SP strong cation-exchange sulfopropyl SPE car-tridge (Tosoh) with two 1.8 mL portions of 2 M NaOH, two 5 mL portions of water, two 1.8 mL portions of MeCN:concentrated HCl 90:10, two 5 mL portions of water, and two 1.8 mL portions of 10 mM pH 5.0 sodium acetate/HCl buffer. 1 mL Plasma + 50 µL 2 µM p-methylbenzylamine in MeCN containing 2% Triton X-405 + 50 µL 70 mM HCl, add to the SPE cartridge, wash with two 5 mL portions of water, wash with two 1.8 mL portions of MeCN:water 40:60, elute with 3 mL MeCN:1 M NaCl 40:60. Add 600 µL 500 mM NaOH and 6 mL ethyl acetate to the eluate, shake for 10 min. Remove the organic layer and evaporate it to dryness under reduced pressure at 50°, reconstitute the residue in 200 µL 2% Triton X-405 in MeCN, add 100 µL 2 mM 3,4-dihydro-6,7-dime-thoxy-4-methyl-3-oxoquinoxaline-2-carboxyl chloride in MeCN, add 3 mg potassium car-bonate, mix, let stand at room temperature for about 1 min, inject a 20 µL aliquot of the supernatant. (3,4-Dihydro-6,7-dimethoxy-4-methyl-3-oxoquinoxaline-3-carboxyl chloride is available from Dojindo Molecular Technologies, Inc., 3 Bethesda Metro Center, Suite 700, Bethesda MD 20814; (301) 664-8448; www.dojindo.co.jp. Synthesis is as follows. Stir 483 g veratrole in 1.45 L acetic acid at 15°, add 683 g concentrated nitric acid (s.g. 1.05) over 1 h keeping the temperature below 40° (cool if necessary), add 2.127 L fuming nitric acid (s.g.1.50) over 1 h keeping the temperature below 30°, allow to stand for 2 h, pour into a large volume of cold water, filter, wash the solid until it is free from acid, recrys-tallize from EtOH to give 4,5-dinitroveratrole (mp 129.5-130.5°) (J. Am. Chem. Soc. 1946, 68, 1536). Reflux 5 g 4,5-dinitroveratrole in 200 mL benzene (Caution! Benzene is a car-cinogen!), add 100 g 60 mesh iron powder and 20 mL concentrated HCl in small portions over 1 h, reflux for 4 h, add 10 mL water, reflux for 2 h, cool, make alkaline with 2.5 M NaOH, extract several times with 200 mL portions of benzene. Combine the extracts and evaporate to dryness, add 10 mL concentrated HCl, recrystallize from EtOH to give 1,2-diamino-4,5-dimethoxybenzene hydrochloride as slightly pink needles (mp 240° d) (Anal. Chim. Acta 1982, 134, 39). Another synthesis of 1,2-diamino-4,5-dimethoxybenzene is as follows. Add 9.66 g veratrole in 9.5 mL glacial acetic acid dropwise to 26.6 mL fuming nitric acid over 30 min using an ice bath, stir at room temperature for 30 min, pour into 500 mL ice-cold water, filter to obtain 1,2-dimethoxy-4,5-dinitrobenzene, recrystallize repeatedly from EtOH. Dissolve 1 g 1,2-dimethoxy-4,5-dinitrobenzene in 50 mL EtOH, add 100 mg 10% palladium on charcoal, stir under hydrogen at atmospheric pressure and room temperature for 4 days, filter through Celite under an atmosphere of nitrogen, saturate the filtrate with hydrogen chloride gas, filter under nitrogen, dry the solid under vacuum to obtain 1,2-diamino-4,5-dimethoxybenzene hydrochloride (mp 240° d (Anal. Chim. Acta 1982, 134, 39)) (Anal. Chim. Acta 1992, 263, 137). 1,2-Diamino-4,5-dimethoxybenzene is available from Molecular Probes, Eugene OR or Dojindo. Dissolve 8 g 1,2-diamino-4,5-dimethoxybenzene monohydrochloride and 8 g α-ketomalonic acid in 20 mL 500 mM HCl, heat in a boiling water bath for 2 h, cool in ice, filter, wash the precipitate with water, recrystallize from dioxane:water 90:10 (Caution! Dioxane is a carcinogen!) to give 3,4-dihydro-6,7-dimethoxy-3-oxoquinoxaline-2-carboxylic acid as orange needles (mp 268°). Treat 5.5 g 3,4-dihydro-6,7-dimethoxy-3-oxoquinoxaline-3-carboxylic acid in 50 mL anhydrous MeOH with ethereal diazomethane, evaporate to dryness under reduced pressure, dissolve the residue in 30 mL chloroform, chromatograph on a 250 × 57 column of 130 g 70-230 mesh silica gel 60 (Merck) with n-hexane:ethyl acetate 50:50 to give methyl 3,4-dihydro-6,7-dimethoxy-4-methyl-3-oxoquinoxaline-2-carboxylate as yellow needles (mp 164°). Dissolve 2.5 g methyl 3,4-dihydro-6,7-dimethoxy-4-methyl-3-oxoquinoxaline-2-carboxylate in 200 mL 1 M NaOH, let stand at rom temperature for about 70 min, wash 5 times with 200 mL portions of ethyl acetate. Neutralize the aqueous layer with dilute HCl, filter, recrystallize the precipitate from dioxane:water 80:20 to give 3,4-dihydro-6,7-dimethoxy-4-methyl-3-oxoquinoxaline-2-carboxylic acid as yellow needles (mp 222°). Dissolve 1 g 3,4-dihydro-6,7-dimethoxy-4-methyl-3-oxoquinoxaline-2-carboxylic acid in 20 mL freshly distilled thionyl chloride, reflux for 1 h, cool, add 50 mL light petroleum (bp 30-60°), filter, recrystallize the precipitate from benzene:light petroleum 90:10 (Caution! Benzene is a carcinogen!) to give 3,4-dihydro-6,7-dimethoxy-4-methyl-3-oxoquinoxaline-2-carboxyl chloride as orange needles (mp 261°) (J. Chromatogr. 1986, 362, 209). Use MeCN solution within 3 h. Preparation of diazomethane is as follows. Caution! Diazomethane is toxic, explosive, and carcinogenic! A face shield and a safety screen should always be used and the preparation should only be carried out in a properly functioning chemical fume hood. Only smooth glass apparatus with rubber stoppers and plastic tubing should be used. Scratched glassware, ground glass joints, and sharp edges should be avoided (Org. Syn., Coll. Vol. VI; Wiley: New York, 1988, pp. 432-435). Procedures have been reported for the synthesis of diazomethane from N-methyl-N-nitrosourea (Org. Syn., Coll. Vol. II; Wiley: New York, 1943, pp. 165-167), N-nitroso-β-methylaminoisobutyl methyl ketone (Org. Syn., Coll. Vol. III; Wiley: New York, 1955, pp. 244-248), and N,N'-dimethyl-N,N'-dinitrosoterephthalamide (Org. Syn., Coll. Vol. V; Wiley: New York, 1973, pp. 351-355). Probably the most convenient starting material is N-methyl-N-nitroso-p-toluenesulfonamide (Diazald) (Aldrichimica Acta 1983, 16, 3-10; Org. Syn., Coll. Vol. IV, Wiley: New York, 1963, pp. 250-253). Add 10 mL 95% EtOH to a solution of 5 g KOH in 8 mL water, warm to 65° using a water bath, add a solution of 5 g N-methyl-N-nitroso-p-toluenesulfonamide in 45 mL ether over 20 min at such a rate as to keep the reaction volume constant. Collect the ether and diazomethane that distil in an ice-cooled receiving flask under a dry ice/acetone condenser. When all the N-methyl-N-nitroso-p-toluenesulfonamide has been used up, slowly add 10 mL ether to the reaction flask and continue distillation until the distillate is colorless. A purpose-built distillation apparatus can be purchased from Aldrich (Aldrichimica Acta 1983, 16, 3-10). Excess quantities of diazomethane can be destroyed by adding acetic acid until the yellow color of the diazomethane is discharged. The safe disposal of the nitroso compounds used to generate diazomethane has been discussed (Lunn,G.; Sansone,E.B. Destruction of Hazardous Chemicals in the Laboratory, Second Edition. Wiley: New York, 1994, pp. 277-289).)

HPLC VARIABLES

Column: 250 × 4.6 5 μm TSK gel ODS-120T (Tosoh)
Mobile phase: MeCN:50 mM ammonium acetate 33:67
Flow rate: 1
Injection volume: 20
Detector: F ex 406 em 485

CHROMATOGRAM
Retention time: 20
Internal standard: p-methylbenzylamine (27)
Limit of detection: 300 pM

KEY WORDS
plasma; SPE

REFERENCE
Ishida, J.; Yamaguchi, M.; Nakamura, M. High-performance liquid chromatographic determination of β-phenylethylamine in human plasma with fluorescence detection, *Anal.Biochem.*, **1990**, *184*, 86–89.

SAMPLE
Matrix: tissue
Analyte: 1,2,3,4-tetrahydroisoquinoline
Sample preparation: Homogenize rat brain with 10 mL 0.1% ascorbic acid in 400 mM perchloric acid containing 0.1% EDTA, centrifuge at 1000 g for 15 min, remove the supernatant, homogenize the pellet with 10 mL 0.1% ascorbic acid in 400 mM perchloric acid containing 0.1% EDTA, centrifuge at 1000 g for 15 min, Combine the supernatants, add 50 μL 2.5 μM 4-phenylpiperidine in MeCN, adjust the pH to 11.0 with 2 M NaOH and 100 mM pH 11.0 phosphate buffer, extract twice with 10 mL portions of dichloromethane. Combine the organic layers and add them to 10 mL 0.1% ascorbic acid in 100 mM HCl containing 0.1% EDTA, extract. Remove the aqueous phase and adjust the pH to 11.0 with 2 M NaOH and 100 mM pH 11.0 phosphate buffer, extract twice with 10 mL portions of dichloromethane. Combine the organic layers and dry them over 400 mg anhydrous potassium carbonate, evaporate to dryness under a stream of nitrogen, reconstitute with 200 μL MeCN, add 100 μL 2 mM 3,4-dihydro-6,7-dimethoxy-4-methyl-3-oxoquinoxaline-2-carboxyl chloride in MeCN, add 3 mg potassium carbonate, mix, let stand at room temperature for 3 min, inject a 20 μL aliquot of the supernatant. (3,4-Dihydro-6,7-dimethoxy-4-methyl-3-oxoquinoxaline-2-carboxyl chloride is available from Dojindo Molecular Technologies, Inc., 3 Bethesda Metro Center, Suite 700, Bethesda MD 20814; (301) 664-8448; www.dojindo.co.jp. Synthesis is as described above.)

HPLC VARIABLES
Column: 250 × 4.6 5 μm TSK gel ODS-120T
Mobile phase: MeOH:50 mM ammonium acetate 53:47
Flow rate: 1
Injection volume: 20
Detector: F ex 388 em 476

CHROMATOGRAM
Retention time: 15.5
Internal standard: 4-phenylpiperidine (31.0)
Limit of detection: 1 pmole/g

KEY WORDS
rat; brain

REFERENCE
Ishida, J.; Yamaguchi, M.; Nakamura, M. High-performance liquid chromatographic determination of 1,2,3,4-tetrahydroisoquinoline in rat brain with fluorescence detection, *Anal.Biochem.*, **1991**, *195*, 168–171.

4-(5',6'-Dimethoxybenzothiazolyl)benzoyl Fluoride

SAMPLE
Matrix: solutions
Analyte: amines
Sample preparation: Mix 100 μL of a solution in MeCN with 100 μL 10 mM quinuclidine in MeCN and 100 μL 3 mM 4-(5',6'-dimethoxybenzothiazolyl)benzoyl fluoride in MeCN, heat at 37° for 20 min, inject a 20 μL aliquot. (Preparation of 4-(5',6'-dimethoxybenzo-thiazolyl)benzoyl fluoride is as follows. Stir 20 g 4-bromoveratrole in 60 mL acetic acid, add 10 mL concentrated nitric acid dropwise while keeping the temperature at 10-30° with occasional cooling, pour into ice-water. Collect the precipitate and dissolve it in 500 mL hot EtOH, add activated charcoal, filter, add 40 mL water to the filtrate to obtain 4-bromo-5-nitroveratrole as light-yellow crystals (mp 121-122°). Dissolve 5 g 4-bromo-5-ni-troveratrole in 50 mL 95% EtOH, add sodium sulfide (prepare immediately before use by melting 5 g sodium sulfide nonahydrate and 700 mg sulfur together), reflux for 30 min, pour into ice-water. Collect the solid and recrystallize it from dichloromethane to obtain 2,2'-dithiobis(1-nitro-4,5-dimethoxybenzene) as yellow needles (mp 231-232°). Dissolve 2 g 2,2'-dithiobis(1-nitro-4,5-dimethoxybenzene) in 300 mL EtOH, add 8 g tin powder, add 30 mL concentrated HCl dropwise, make alkaline with 4 M NaOH, filter. Dilute the fil-trate with 200 mL water, extract twice with 100 mL portions of benzene (Caution! Ben-zene is a carcinogen!). Combine the extracts and evaporate them to dryness under reduced pressure, mix the residue with 10 mL benzene and 2 mL 10% hydrogen peroxide, stir for 30 min, recrystallize the precipitate from EtOH to give 2,2'-dithiobis(1-amino-4,5-dime-thoxybenzene) as colorless crystals (mp 155-156°) (Anal. Chim. Acta 1994, 291, 189). Re-flux 2 g 2,2'-dithiobis(1-amino-4,5-dimethoxybenzene) and 4 g 4-carboxybenzaldehyde in 50 mL EtOH containing 600 mg tri-n-butylphosphine and 800 mM HCl with stirring for 1 h, cool, concentrate to about 5 mL under reduced pressure, wash the solid with MeOH:water 50:50, dry under reduced pressure, chromatograph on a 200 × 35 column of about 100 g 70-230 mesh silica gel 60 (Merck) using chloroform:MeOH:water 20:2: 0.1, evaporate the main fraction to dryness under reduced pressure, recrystallize from EtOH to obtain 4-(5',6'-dimethoxybenzothiazolyl)benzoic acid as pale yellow needles (mp >300°). Add 100 mL pyridine and 60 mL (sic) cyanuric fluoride (Fluka) to a solution of 250 mg 4-(5',6'-dimethoxybenzothiazolyl)benzoic acid in 40 mL dichloromethane, stir at room temperature for 1 h, evaporate to dryness under reduced pressure, recrystallize from dichloromethane to obtain 4-(5',6'-dimethoxybenzothiazolyl)benzoyl fluoride as yellow nee-dles (mp 228-230°). Prepare solutions by dissolving the compound in 2-methoxyethanol then diluting with MeCN, use within 24 h.)

HPLC VARIABLES
Column: 250 × 4.6 5 μm TSK gel ODS-120T (Tosoh)
Mobile phase: MeOH:water 70:30
Flow rate: 1
Injection volume: 20
Detector: F ex 350 em 450

CHROMATOGRAM
Retention time: 30 (n-propylamine), 35 (n-heptylamine), 40 (N-methylhexylamine)
Limit of detection: 1.4-2 fmole

OTHER SUBSTANCES
Also analyzed: n-amylamine, benzylamine, n-butylamine, cyclohexylamine, dibenzylamine, di-n-butylamine, diethylamine, di-n-propylamine, N-ethylbenzylamine, n-hexylamine, histamine, 4-methylbenzylamine, n-nonylamine, n-octylamine, 2-phenylethylamine, tyramine

Non-interfering: acetoacetic acid, acetone, acetophenone, acetylacetone, N-acetylneuraminic acid, allantoin, alloxan, amino acids, aniline, ascorbic acid, benzil, bilirubin, n-butanol, cholesterol, citrulline, cortisone, cyclohexane, dehydroepiandrosterone, epiandrosterone, ethanol, D-fructose, D-glucose, glutathione, homogentisic acid, 3-hydroxybutyric acid, inositol, lactic acid, D-lactose, D-maltose, D-mannose, methanol, 4-methylcyclohexane, methylglyoxal, phenylpyruvic acid, pyruvic acid, thiamine, o-toluidine, urea, uric acid, D-xylose

REFERENCE
Hara, S.; Aoki, J.; Yoshikuni, K.-i.; Tatsuguchi, Y.; Yamaguchi, M. 4-(5',6'-Dimethoxybenzothiazolyl)benzoyl fluoride and 2-(5',6'-dimethoxybenzothiazolyl)benzenesulfonyl chloride as sensitive fluorescence derivatization reagents for amines in high-performance liquid chromatography, *Analyst*, **1997**, *122*, 475–479.

3,5-Dinitrobenzoyl Chloride

RELATED REFERENCES
Elrod, L., Jr.; White, L.B.; Spanton, S.G.; Stroz, D.G.; Cugier, P.J.; Luka, L.A. Determination of fortimicin A and 3-O-demethylfortimicin A as 3,5-dinitrobenzoyl derivatives by reverse-phase high-performance liquid chromatography. *Anal.Chem.* **1984**, *56*, 1786-1790.

Herráez-Hernández, R.; Campíns-Falcó, P.; Sevillano-Cabeza, A. Liquid chromatographic analysis of amphetamine and related compounds in urine using solid-phase extraction and 3,5-dinitrobenzoyl chloride for derivatization. *J.Chromatogr.Sci.* **1997**, *35*, 169-175.

Morley, J.; Elrod, L., Jr.; Linton, C.; Shaffer, D.; Krogh, S. Determination of residual amines used in bulk drug synthesis by pre-column derivatization with 3,5-dinitrobenzoyl chloride and high-performance liquid chromatography. *J.Chromatogr.A* **1997**, *766*, 77-83.

Pirkle, W.H.; Burke, J.A. Separation of the enantiomers of the 3,5-dinitrobenzamide derivatives of α-amino phosphonates on four chiral stationary phases. *J.Chromatogr.* **1992**, *598*, 159-167.

Yu, J.N.; Tian, S.J.; Ni, K.Y.; Cheng, G.X.; Li, Z.G. Studies on the analysis of ribostamycin by 3,5-dinitrobenzoyl chloride pre-column derivatization rapid RP-HPLC. *Chin.J.Pharm.Anal.(Yaowu Fenxi Zazhi)* **1995**, *15*, 28-32.

SAMPLE
Matrix: blood, tissue, urine
Analyte: hypotaurine and taurine
Sample preparation: Urine. Adjust urine to pH 7 with 2 M NaOH or 2 M HCl, dilute 5 (human) or 100 (rat) fold with water. Remove a 1 mL aliquot and add 10 μL triethylamine and 20 mg freshly powdered 3,5-dinitrobenzoyl chloride, shake vigorously on a mechanical shaker for 10 min, add 100 μL 2 M HCl, centrifuge at 1200 g for 5 min. Remove a 200 μL aliquot of the supernatant and dilute it to 5 mL with water, inject a 10-100 μL aliquot. Whole blood. Add 7 volumes of water to whole blood, mix, add 1 volume of 10% sodium

tungstate dihydrate solution, mix, add with shaking 1 volume of 333 mM sulfuric acid, shake vigorously (J. Biol. Chem. 1919, 38, 81), centrifuge at 1200 g for 10 min. Remove a 1 mL aliquot of the supernatant and add 10 μL triethylamine and 20 mg freshly powdered 3,5-dinitrobenzoyl chloride, shake vigorously on a mechanical shaker for 10 min, add 100 μL 2 M HCl, centrifuge at 1200 g for 5 min. Remove a 200 μL aliquot of the supernatant and dilute it to 5 mL with water, inject a 10-100 μL aliquot. Tissue. Liver or heart + 7 volumes of water + 1 volume 10% sodium tungstate + 1 volume 330 mM sulfuric acid, homogenize (Potter-Elvehjem, glass pestle), centrifuge at 1200 g for 10 min, dilute 2 (liver) or 5 (heart) fold with water. Remove a 1 mL aliquot and add 10 μL triethylamine and 20 mg freshly powdered 3,5-dinitrobenzoyl chloride, shake vigorously on a mechanical shaker for 10 min, add 100 μL 2 M HCl, centrifuge at 1200 g for 5 min. Remove a 200 μL aliquot of the supernatant and dilute it to 5 mL with water, inject a 10-100 μL aliquot.

HPLC VARIABLES
Column: 150 × 4.6 TSK gel ODS-80Ts C18 (Tosoh)
Mobile phase: MeCN:100 mM pH 3.7 ammonium acetate buffer 16:84
Flow rate: 0.8
Injection volume: 10-100
Detector: UV 254

CHROMATOGRAM
Retention time: 15 (hypotaurine), 18 (taurine)
Limit of detection: 500 nM

OTHER SUBSTANCES
Extracted: amino acids (some)

KEY WORDS
human; rat; whole blood; liver; heart

REFERENCE
Masuoka, N.; Yao, K.; Kinuta, M.; Ohta, J.; Wakimoto, M.; Ubuka, T. High-performance liquid chromatographic determination of taurine and hypotaurine using 3,5-dinitrobenzoyl chloride as derivatizing reagent, *J.Chromatogr.B*, **1994**, *660*, 31–35.

SAMPLE
Matrix: solutions
Analyte: polyamines
Sample preparation: Dilute 10 μL solution to 500 μL with water, add 1 mL 3,5-dinitrobenzoyl chloride, vortex, add 1 mL 2 M NaOH, rotate at 30 rpm for 20 min, add 2 mL saturated NaCl, mix well, add 3 mL diethyl ether, rotate at 30 rpm for 20 min, centrifuge at 2500 g for 3 min. Remove the organic layer and evaporate it to dryness under a stream of nitrogen at 37°, reconstitute the residue in 200 μL MeOH, sonicate for 30 s, inject a 20 μL aliquot.

HPLC VARIABLES
Column: 5 μm LiChrospher 100 RP-18 + 250 × 4 5 μm Bio-Sil ODS-5S (Bio-Rad) in series
Mobile phase: MeOH:water 60:40
Flow rate: 1
Injection volume: 20
Detector: UV 254

CHROMATOGRAM
Retention time: 8.8 (N^1-acetylspermidine), 9.5 (N^8-acetylspermidine), 12.8 (putrescine), 15.3 (cadaverine), 18.5 (N^1-acetylspermine), 25.4 (spermidine), 54.9 (spermine)
Limit of detection: 2-450 ng

KEY WORDS
comparison with benzoyl chloride

REFERENCE

Wongyai, S.; Oefner, P.; Bonn, G. HPLC analysis of polyamines and their acetylated derivatives in the picomole range using benzoyl chloride and 3,5-dinitrobenzoyl chloride as derivatizing agent, *Biomed.Chromatogr.*, **1988**, *2*, 254–257.

Ethyl Chloroformate

RELATED REFERENCE

Tinnerberg, H.; Spanne, M.; Dalene, M.; Skarping, G. Determination of complex mixtures of airborne isocyanates and amines. Part 3. Methylenediphenyl diisocyanate, methylenediphenylamino isocyanate and methylenediphenyldiamine and structural analogues after thermal degradation of polyurethane. *Analyst* **1997**, *122*, 275-278.

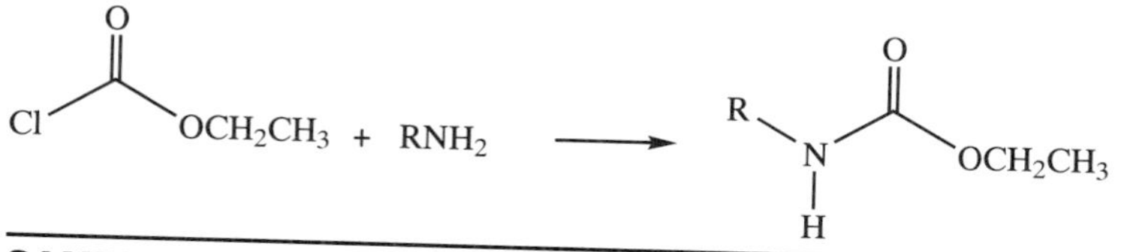

SAMPLE

Matrix: air

Analyte: isocyanates and amines

Sample preparation: Pull air through 10 mL 10 mM dibutylamine at 1 L/min, evaporate to dryness under reduced pressure, reconstitute with 1 mL toluene, add 1 mL 2 M pH 9.5 carbonate buffer, add 50 μL ethyl chloroformate, add 10 μL pyridine, shake for 5 min. Remove the organic layer and evaporate it to dryness, reconstitute the residue in 1 mL MeCN, inject a 10 μL aliquot. (Isocyanates react with dibutylamine to give ureas. Amines that are also collected are converted to urethanes by ethyl chloroformate.)

HPLC VARIABLES

Column: 150 × 2.1 5 μm Supelcosil LC18

Mobile phase: Gradient. MeCN:water from 46:54 to 77.5:22.5 over 14 min.

Flow rate: 0.5

Injection volume: 10

Detector: UV 240; MS, VG Organics Quattro quadrupole, atmospheric pressure chemical ionization or electrospray, positive ion monitoring, cone voltage 20 V (APCI) or 40 V (electrospray), ion source 150° (APCI) or 120° (electrospray), probe heater 500° (APCI)

CHROMATOGRAM

Retention time: 2.6 (2,4-toluenediamine), 6 (2-amino-6-isocyanatotoluene), 7 (2-amino-4-isocyanatotoluene), 7.3 (aminoisocyanatotoluene dimer), 9.9 (2,6-toluenediisocyanate), 10.3 (2,4-toluenediisocyanate), 12 (toluenediisocyanate dimer)

Limit of detection: 1-2 nM

REFERENCE

Tinnerberg, H.; Spanne, M.; Dalene, M.; Skarping, G. Determination of complex mixtures of airborne isocyanates and amines. Part 2. Toluene diisocyanate and aminoisocyanate and toluenediamine after thermal degradation of a toluene diisocyanate-polyurethane, *Analyst*, **1996**, *121*, 1101–1106.

(S)-(+)-Flunoxaprofen Chloride

RELATED REFERENCE
Spahn-Langguth, H.; Podkowik, B.; Stahl, E.; Martin, E.; Mutschler, E. Improved enantiospecific RP-
  HPLC assays for propranolol in plasma and urine with pronethalol as internal standard.
  J.Anal.Toxicol. **1991**, *15*, 209-213.

SAMPLE
Matrix: blood, urine
Analyte: tranylcypromine
Sample preparation: Plasma. 1 mL Plasma + 1 mL pH 11 sodium borate buffer, extract
  with 2.5 mL diisopropyl ether:EtOH 100:1.5 (Caution! Diisopropyl ether readily forms
  explosive peroxides!). Remove 2 mL of the organic layer and evaporate it to dryness under
  reduced pressure at room temperature, reconstitute the residue in 200 μL 10 μg/mL S-
  flunoxaprofen chloride in dichloromethane, let stand at room temperature for 1 h, add 20
  μL MeOH, evaporate to dryness, reconstitute with dichloromethane, inject a 5-50 μL
  aliquot. Urine. 1 mL Urine + 1 mL 50 mM NaOH, extract with 2.5 mL diisopropyl
  ether:EtOH 100:1.5. Remove 2 mL of the organic layer and evaporate it to dryness under
  reduced pressure at room temperature, reconstitute the residue in 200 μL 10 μg/mL S-
  flunoxaprofen chloride in dichloromethane, let stand at room temperature for 1 h, add 20
  μL MeOH, evaporate to dryness, reconstitute with dichloromethane, inject a 5-50 μL
  aliquot. (Prepare S-flunoxaprofen chloride as follows. Dissolve 1 mmole S-flunoxaprofen
  in 25 mL toluene, add a trace of DMF (J.Chromatogr. 1990, 528, 55), add 2.5 mL thionyl
  chloride, reflux for 30 min, remove solvent by evaporation, dry the residue under vacuum
  over KOH, recrystallize from dichloromethane (mp 73°).)

HPLC VARIABLES
Guard column: 4 × 4 LiChrosorb Si 60
Column: 250 × 4.6 7 μm Zorbax Sil
Mobile phase: Cyclohexane:dichloromethane:THF 70:10:10
Flow rate: 1.4
Injection volume: 5-50
Detector: F ex 305 em 355

CHROMATOGRAM
Retention time: 12 (S-(-)), 15 (R-(+))
Limit of detection: 2 ng/mL

KEY WORDS
normal phase; chiral; plasma

REFERENCE
Spahn, H. S-(+)-Flunoxaprofen chloride as chiral fluorescent reagent, *J.Chromatogr.*, **1988**, *427*, 131–
  137.

(+)-1-(9-Fluorenyl)ethyl Chloroformate

SAMPLE
Matrix: blood
Analyte: atenolol
Sample preparation: 100 μL Plasma + 5 μL 100 μg/mL racemic practolol in MeOH + 50 μL 0.5 M pH 12 glycine buffer + 50 μL 2 M NaOH + 1 mL saturated NaCl + 4 mL dichloromethane containing 3% (v/v) heptafluoro-1-butanol, extract for 10 min, centrifuge at 3015 g at 4° for 5 min, remove organic phase and evaporate to dryness at room temperature under a stream of nitrogen. Dissolve the residue in 40 μL 1 M pH 8.5 borate buffer and 50 μL 1 mM (+)-1-(9-fluorenyl)ethyl chloroformate (Flec) in acetone, let stand 30 min at room temperature, add 100 μL 30 mM hydroxyproline, after 2 min vortex mix with 300 μL n-pentane for 15 s, centrifuge at 3015 g at 4° for 10 min, discard n-pentane. Shake aqueous layer with 3 mL dichloromethane for 10 min, centrifuge at 3015 g at 4° for 10 min, remove organic phase and evaporate to dryness at room temperature under a stream of nitrogen. Dissolve the residue in 50 μL mobile phase, inject an aliquot.

HPLC VARIABLES
Column: 100 × 4.6 3μm Microspher C18 (Chrompack)
Mobile phase: MeCN:10 mM pH 7 sodium acetate buffer 50:50
Flow rate: 0.8
Injection volume: 10
Detector: F ex 227 em 310

CHROMATOGRAM
Retention time: 5.95 (S-(-)), 6.55 (R-(+))
Internal standard: practolol (8, 9)
Limit of quantitation: 10 ng/mL (each enantiomer)

KEY WORDS
plasma; chiral; rat

REFERENCE
Rosseel, M.T.; Vermeulen, A.M.; Belpaire, F.M. Reversed-phase high-performance liquid chromatographic analysis of atenolol enantiomers in plasma after chiral derivatization with (+)-1-(9-fluorenyl)ethyl chloroformate, *J.Chromatogr.*, **1991**, *568*, 239–245.

SAMPLE
Matrix: blood
Analyte: propranolol
Sample preparation: Condition a 1 mL 100 mg C18 Bond Elut SPE cartridge with 1.5 mL MeOH and 1.5 mL water. 1 mL Whole blood + 1 mL water + 100 μL 400 ng/mL methyl 4-propranolol hydrochloride in water + 3 mL 1 M sodium carbonate + 10 mL heptane:isopropanol 98:2, shake at 80-100 strokes/min for 15 min, centrifuge at 1500 g at 4° for 5 min. Remove 7 mL of the organic phase and evaporate it to dryness at 50° under a stream of nitrogen. Take up the residue in 150 μL acetone, vortex for 30 s, add

100 μL pH 7.85 borate buffer, add 50 μL 500 μg/mL (+)-1-(9-fluorenyl)ethyl chloroformate in acetone, mix for 30 s, let stand for 5 min at room temperature. Add 800 μL water and the reaction mixture to the SPE cartridge, wash with 1.5mL isooctane, elute with 500 μL dichloromethane. Evaporate the eluate to dryness and take up the residue in 35 μL MeCN, vortex vigorously for 15 s, add 75 μL water, inject a 100 μL aliquot.

HPLC VARIABLES
Column: 100 × 8 Nova-Pak C18 radial compression
Mobile phase: MeCN:water 75:25
Flow rate: 2
Injection volume: 100
Detector: F ex 260 em 340

CHROMATOGRAM
Retention time: 10.5 (S(-)), 11.3 (R(+))
Internal standard: methyl 4-propranolol hydrochloride (14.5 (S), 15.5 (R))
Limit of quantitation: 0.5 ng/mL

KEY WORDS
whole blood; chiral

REFERENCE
Roux, A.; Blanchot, G.; Baglin, A.; Flouvat, B. Liquid chromatographic analysis of propranolol enantiomers in human blood using precolumn derivatization with (+)-1-(9-fluorenyl)ethyl chloroformate, *J.Chromatogr.*, **1991**, *570*, 453–461.

SAMPLE
Matrix: blood
Analyte: S 12024-2 ((R,S)-1-methyl-8-[(morpholin-2-yl)methoxy]-1,2,3,4-tetrahydroquinoline, monomethanesulfonate salt)
Sample preparation: 1 mL Plasma + 50 μL 5 μg/mL IS in water + 50 μL 1 M NaOH + 5 mL diethyl ether:dichloromethane 70:30, shake vigorously for 3 min, centrifuge at 5° at 1500 g for 5 min. Remove the organic layer and evaporate it to dryness under a stream of nitrogen at 37°, reconstitute the residue in 250 μL water:0.1% triethylamine in MeCN 80:20, vortex for 1 min, add 200 μL 61 μg/mL (+)-1-(9-fluorenyl)ethyl chloroformate in MeCN, stir at room temperature for 1 h, add 10 mg/mL D,L-proline in water, stir for 1 min, inject an aliquot.

HPLC VARIABLES
Column: 150 × 4.6 3 μm Spherisorb cyanopropyl
Mobile phase: MeCN:buffer 25:75 (Buffer was 0.05% sulfuric acid adjusted to pH 3 with 10 M NaOH.)
Flow rate: 1.3
Injection volume: 20
Detector: F ex 260 em 310

CHROMATOGRAM
Retention time: 23.3 (R), 24.2 (S)
Internal standard: (+)-(S)-8-[(morpholin-2-yl)methoxy]-1,2,3,4-tetrahydroquinoline, monomethanesulfonate salt (S 12027) (18.5)
Limit of quantitation: 5 ng/mL

KEY WORDS
chiral; plasma; pharmacokinetics

REFERENCE
Boursier-Neyret, C.; Baune, A.; Klippert, P.; Castagne, I.; Sauveur, C. Determination of S-12024 enantiomers in human plasma by liquid chromatography after chiral pre-column derivatization, *J.Pharm.Biomed.Anal.*, **1993**, *11*, 1161–1166.

SAMPLE
Matrix: blood
Analyte: reboxetine

Sample preparation: 1 mL Plasma + 1 mL Tris buffer + 7 mL diethyl ether, vortex for 1 min, centrifuge at 1200 g for 5 min. Remove the organic phase and add it to 200 μL 10 mM phosphoric acid, vortex for 1 min, centrifuge at 1200 g for 5 min. Remove the aqueous phase and add it to 200 μL borate buffer and 200 μL reagent, shake, let stand at room temperature for 5 min, add 100 μL 100 mM L-proline in water, add 3 mL n-hexane, vortex for 1 min. Remove the organic layer and add it to 1 mL MeCN, extract, discard the upper n-hexane layer, wash the MeCN layer with 1 mL n-hexane. Evaporate the MeCN layer to dryness under a stream of nitrogen, reconstitute the residue in 500 μL mobile phase, inject a 200 μL aliquot. (Tris buffer was 25 mL 200 mM Tris solution and 5 mL 100mM HCl made up to 100 mL with water, pH 9.1. Borate buffer was 61.8 g boric acid in 900 mL water, adjust pH to 8.0 with 20% NaOH, make up to 1 L with water. Prepare reagent by diluting 1 mL 18 mM (+)-1-(9-fluorenyl)ethyl chloroformate in acetone to 50 mL with MeCN.)

HPLC VARIABLES
Guard column: 30-38 μm Survival pellicular ODS (Whatman)
Column: 250 × 4.6 4 μm Supersphere 60 RP-8 (end-capped) (Merck)
Mobile phase: THF:buffer 46.5:53.5 (Prepare buffer by dissolving 13.2 g $(NH_4)_2HPO_4$ in 900 mL water, adjust pH to 7.5 with 85% phosphoric acid, make up to 1 L with water.)
Flow rate: 0.5
Injection volume: 200
Detector: F ex 260 em 315

CHROMATOGRAM
Retention time: 57 (-), 59 (+)
Limit of quantitation: 1 ng/mL

KEY WORDS
plasma; chiral; pharmacokinetics

REFERENCE
Frigerio, E.; Pianezzola, E.; Strolin Benedetti, M. Sensitive procedure for the determination of reboxetine enantiomers in human plasma by reversed-phase high-performance liquid chromatography with fluorimetric detection after derivatization with (+)-1-(9-fluorenyl)ethyl chloroformate, *J.Chromatogr.A*, **1994**, *660*, 351–358.

SAMPLE
Matrix: bulk
Analyte: methamphetamine
Sample preparation: Prepare an aqueous solution. Adjust pH of 500 μL aqueous solution to 12 with 1 M NaOH, add 200 μL (+)-1-(9-fluorenyl)ethyl chloroformate:dichloromethane 1:100, let stand for 20 min, add 500 μL dichloromethane, shake for 30 min. Remove the organic phase and wash it twice with 500 μL aliquots of water, evaporate the organic phase to dryness under a stream of nitrogen, reconstitute the residue in 3 mL MeOH, filter (0.45 μm), inject a 25 μL aliquot.

HPLC VARIABLES
Column: 250 × 3.9 5 μm 5C18-AR (Waters)
Mobile phase: MeCN:50 mM pH 6.0 phosphate buffer 65:35
Flow rate: 1
Injection volume: 25
Detector: F ex 295 em 315

CHROMATOGRAM
Retention time: 44 ((S)-(+)-methamphetamine), 47 ((R)-(-)-methamphetamine)
Limit of quantitation: 16.7 ng/mL

OTHER SUBSTANCES
Extracted: ephedrine, pseudoephedrine

KEY WORDS
chiral

REFERENCE

Chen, Y.-P.; Hsu, M.-C.; Chien, C.S. Analysis of forensic samples using precolumn derivatization with (+)-1-(9-fluorenyl)ethyl chloroformate and liquid chromatography with fluorimetric detection, *J.Chromatogr.A*, **1994**, *672*, 135–140.

SAMPLE

Matrix: solutions

Analyte: propranolol

Sample preparation: Mix 20 μL of a solution of propranolol in buffer with 20 μL 4 mM (+)-1-(9-fluorenyl)ethyl chloroformate in dry acetone, let stand at room temperature for 10 min, inject a 10 μL aliquot. (Buffer was 100 mM boric acid/sodium bicarbonate buffer, adjusted to pH 8.5 with NaOH.)

HPLC VARIABLES

Column: 150 × 4 MicroPak SP C8

Mobile phase: MeCN:20 mM pH 4.0 sodium acetate 70:30

Flow rate: 2

Injection volume: 10

Detector: F ex 265 em 345

CHROMATOGRAM

Retention time: 6.7, 7.2 (enantiomers)

Limit of detection: 1 pmole

KEY WORDS

chiral

REFERENCE

Lai, F.; Mayer, A.; Sheehan, T. Chiral separation and detection enhancement of propranolol using automated pre-column derivatization, *J.Pharm.Biomed.Anal.*, **1993**, *11*, 117–120.

(-)-1-(9-Fluorenyl)ethyl Chloroformate

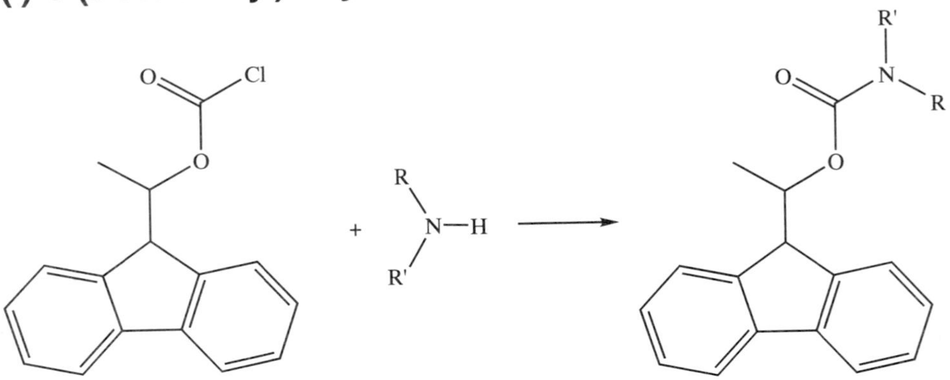

SAMPLE

Matrix: blood

Analyte: amphetamine and methamphetamine

Sample preparation: 100 μL Serum + 50 μL 100 ng/mL aniline sulfate in water + 200 μL 20 mM pH 10.6 carbonate buffer + 2 mL ethyl acetate, shake for 15 min, centrifuge at 1200 g for 5 min. Remove the organic layer and add it to 200 μL 50 mM HCl, shake for 15 min, centrifuge at 1200 g for 5 min. Remove the aqueous layer and add it to 40 μL 250 mM NaOH, add 50 μL 330 mM pH 7.8 phosphate buffer, add 250 μL MeCN, add 25 μL 1 mM (-)-1-(9-fluorenyl)ethyl chloroformate in acetone, let stand overnight at room temperature, add 30 μL 100 mM glycine in water, add 750 μL n-pentane, vortex for 2 min, centrifuge at 1200 g for 5 min. Remove the organic layer and evaporate it to dryness under vacuum, reconstitute the residue in MeCN:water 50:50, inject a 100 μL aliquot.

HPLC VARIABLES

Guard column: Direct-Connect column prefilter (Alltech)
Column: 150 × 4.6 3 μm Adsorbosphere HS C18 (Alltech)
Mobile phase: MeCN:THF:20 mM pH 3.6 acetate buffer 39:15:46
Flow rate: 1
Injection volume: 100
Detector: F ex 265 em 330

CHROMATOGRAM

Retention time: 22.6 (R-amphetamine), 23.6 (S-amphetamine), 27.7 (R-methamphetamine), 29.0 (S-methamphetamine)
Internal standard: aniline (21.0)
Limit of quantitation: 5 ng/mL

KEY WORDS

serum; rat; chiral

REFERENCE

Hutchaleelaha, A.; Walters, A.; Chow, H.-H.; Mayersohn, M. Sensitive enantiomer-specific high-performance liquid chromatographic analysis of methamphetamine and amphetamine from serum using precolumn fluorescent derivatization, *J.Chromatogr.B*, **1994**, *658*, 103–112.

SAMPLE

Matrix: blood
Analyte: mefloquine
Sample preparation: Plasma. 100 μL Plasma + 25 μL 100 mM zinc sulfate, vortex for 15 s, add 750 μL 1 μM IS in MeCN, vortex for 15 s, let stand for 15 min, centrifuge at 10000 g for 10 min. Remove the supernatant and add it to 4 mL 10 mM pH 10.0 borate buffer, add 5 mL MTBE, shake for 20 min, centrifuge at 3000 g for 5 min. Remove the upper organic layer and evaporate it to dryness under a stream of nitrogen at 65°, reconstitute the residue in 125 μL 400 μM (-)-1-(9-fluorenyl)ethyl chloroformate in MeCN, vortex, add 50 μL 15 mM pH 8.5 borate buffer, vortex, let stand at room temperature for 40 min, centrifuge at 3000 g for 10 min, inject a 100 μL aliquot. Dried blood. Cut paper containing 100 μL dried blood into small pieces and add them to 150 μL 5 μM IS in water, add 2 mL ammonia:water 90:10, let stand at room temperature for 30-60 min, sonicate at 37° for 30 min, add 5 mL MTBE, shake for 20 min, centrifuge at 3000 g for 5 min. Remove the upper organic layer and evaporate it to dryness under a stream of nitrogen at 65°, reconstitute the residue in 125 μL 400 μM (-)-1-(9-fluorenyl)ethyl chloroformate in MeCN, vortex, add 50 μL 15 mM pH 8.5 borate buffer, vortex, let stand at room temperature for 40 min, centrifuge at 3000 g for 10 min, inject a 100 μL aliquot.

HPLC VARIABLES

Column: 250 × 4.6 5 μm Ultrasphere octadecylsilica
Mobile phase: MeCN:water:acetic acid 82:18:0.07
Flow rate: 1
Injection volume: 100
Detector: UV 263; F ex 263 em 475

CHROMATOGRAM

Retention time: 12 (RS), 14 (SR)
Internal standard: D,L-erythro-α-(2-piperidyl)-2-trifluoromethyl-6,8-dichloro-4-quinoline-methanol (18, 21 (enantiomers))
Limit of detection: 250 nM (UV, plasma), 10-50 nM (F, plasma), 125 nM (F, dried blood)

OTHER SUBSTANCES

Non-interfering: chloroquine, pyrimethamine, quinine, sulfadoxine

KEY WORDS

plasma; dried blood; chiral; pharmacokinetics

REFERENCE

Bergqvist, Y.; Doverskog, M.; Al Kabbani, J. High-performance liquid chromatographic determination of (SR)- and (RS)-enantiomers of mefloquine in plasma and capillary blood sampled on paper after derivatization with (-)-1-(9-fluorenyl)ethyl chloroformate, *J.Chromatogr.B*, **1994**, *652*, 73–81.

SAMPLE

Matrix: urine

Analyte: amphetamine and methamphetamine

Sample preparation: Dilute urine 20-fold or more. 500 μL Diluted urine + 50 μL 500 ng/mL (+)-2,5-dimethoxyamphetamine hydrochloride + 100 μL 500 mM NaOH + 2 mL benzene (Caution! Benzene is a carcinogen!), shake for 15 min, centrifuge at 1200 g for 5 min. Remove 1.8 mL of the organic phase and add it to 220 μL 50 mM HCl, shake for 15 min, centrifuge at 1200 g for 5 min. Remove 200 μL of the aqueous phase and add it to 40 μL 250 mM NaOH, add 50 μL 330 mM pH 7.8 phosphate buffer, add 250 μL MeCN, add 25 μL 3 mM (-)-1-(9-fluorenyl)ethyl chloroformate, let stand at room temperature for 24 h, add 30 μL 100 mM glycine in water, let stand for 30 min, add 750 μL pentane, vortex for 2 min, centrifuge at 1200 g for 5 min. Remove the organic layer and evaporate it to dryness in a centrifugal evaporator at room temperature, reconstitute the residue in 300 μL MeCN:water 50:50, inject a 100 μL aliquot.

HPLC VARIABLES

Column: 150 × 4.6 3 μm Adsorbosphere HS C18

Mobile phase: MeCN:THF:20 mM pH 3.6 sodium acetate buffer 25:21:54

Flow rate: 1

Injection volume: 100

Detector: F ex 265 em 330

CHROMATOGRAM

Retention time: 37.5 (l-amphetamine), 40 (d-amphetamine), 42 (l-methamphetamine), 44 (d-methamphetamine)

Internal standard: (+)-2,5-dimethoxyamphetamine (33)

Limit of detection: 5 ng/mL

KEY WORDS

rat; chiral

REFERENCE

Sukbuntherng, J.; Hutchaleelaha, A.; Chow, H.-H.; Mayersohn, M. Separation and quantitation of the enantiomers of methamphetamine and its metabolites in urine by HPLC: Precolumn derivatization and fluorescence detection, *J.Anal.Toxicol.*, **1995**, *19*, 139–147.

9-Fluorenylmethyl Chloroformate

RELATED REFERENCES

Bartok, T.; Borcsok, G.; Sagi, F. RP-HPLC separation of polyamines after automatic FMOC-Cl derivatization and precolumn sample clean-up using column switching. *J.Liq.Chromatogr.* **1992**, *15*, 777-790.

Descombes, A.A.; Haerdi, W. HPLC separation of catecholamines after derivatization with 9-fluorenylmethyl chloroformate. *Chromatographia* **1992**, *33*, 83-86.

Harduf, Z.; Nir, T.; Juven, B.J. High-performance liquid chromatography of biogenic amines, derivatized with 9-fluorenylmethyl chloroformate. *J.Chromatogr.* **1988**, *437*, 379-386.

Huhn, G.; Mattusch, J.; Schulz, H. Determination of polyamines in biological materials by HPLC with 9-fluorenylmethyl chloroformate precolumn derivatization. *Fresenius' J.Anal.Chem.* **1995**, *351*, 563-566.

Kirschbaum, J.; Luckas, B.; Beinert, W.D. HPLC analysis of biogenic amines and amino acids in food after automatic pre-column derivatization with 9-fluorenylmethyl chloroformate. *Am.Lab.* **1994**, *26*, 28C-28F.

Maeder, G.; Pelletier, M.; Haerdi, W. Determination of amphetamines by high-performance liquid chromatography with ultraviolet detection. On-line pre-column derivatization with 9-fluorenylmethyl chloroformate and preconcentration. *J.Chromatogr.* **1992**, *593*, 9-14.

Mount, D.L.; Churchill, F.C.; Bergqvist, Y. Determination of mefloquine in blood, filter paper-absorbed blood and urine by 9-fluorenylmethyl chloroformate derivatization followed by liquid chromatography with fluorescence detection. *J.Chromatogr.* **1991**, *564*, 181-193.

Pfeiffer, P.; Radler, F. Determination of ethanolamine in wine by HPLC after derivatization with 9-fluorenylmethoxycarbonylchloride. *Am.J.Enol.Vitic.* **1992**, *43*, 315-317.

Price, J.R.; Metz, P.A.; Veening, H. HPLC of 9-fluorenylmethylchloroformate-polyamine derivatives with fluorescence detection. *Chromatographia* **1987**, *24*, 795-799.

Shah, J.A.; Weber, D.J. High-performance liquid chromatographic assay of pirlimycin in human serum and urine using 9-fluorenylmethylchloroformate. *J.Chromatogr.* **1984**, *309*, 95-105.

Teerlink, T.; Hennekes, M.W.T.; Mulder, C.; Brulez, H.F.H. Determination of dimethylamine in biological samples by high-performance liquid chromatography. *J.Chromatogr.B* **1997**, *691*, 269-276.

Wan, H.; Blomberg, L.G. Enantiomeric and diastereomeric separation of di- and tripeptides by capillary electrophoresis. *J.Chromatogr.A* **1997**, *758*, 303-311.

Wickstroem, K.; Betner, I. Analysis of polyamines and their acetylated forms with 9-fluorenylmethyl chloroformate and reversed phase HPLC. *J.Liq.Chromatogr.* **1991**, *14*, 675-697.

Zhang, R.; Zhang, Z.; Liu, G.; Hidaka, Y.; Shimonishi, Y. High-performance liquid chromatographic determination of neutral and amino monosaccharides by ultraviolet and fluorescence detection of sugar 9-fluorenylmethoxycarbonyl hydrazones and 9-fluorenylmethoxycarbonyl amino sugars at picomole and sub-picomole levels. *J.Chromatogr.* **1993**, *646*, 45-52.

SAMPLE
Matrix: blood
Analyte: selegiline metabolites
Sample preparation: 1 mL Plasma + 1 mL 500 mM pH 11 borate buffer, mix, add 2.5 mL diethyl ether, vortex for 5 min, centrifuge at 1200 g for 5 min, remove organic layer, repeat extraction. Combine the organic layers and add them to 200 μL 100 mM HCl, vortex for 2 min, centrifuge at 1200 g for 5 min. Remove the aqueous phase and add it to 150 μL 1 M pH 8 borate buffer and 100 μL 4 mM 9-fluorenylmethyl chloroformate in MeCN, shake, allow to react at 50° for 5 min, add 20 μL 20 mM proline in water, allow to react at 50° for 2 min, inject a 200 μL aliquot.

HPLC VARIABLES
Column: 150 × 3.9 4 μm Nova-Pak phenyl
Mobile phase: MeCN:50 mM pH 6.0 sodium phosphate buffer 50:50
Flow rate: 1
Injection volume: 200
Detector: F ex 260 em 315

CHROMATOGRAM
Retention time: 12 (amphetamine), 15 (methamphetamine), 19 (desmethyldeprenyl)
Limit of quantitation: 0.5 ng/mL

KEY WORDS
plasma

REFERENCE
La Croix, R.; Pianezzola, E.; Strolin Benedetti, M. Sensitive high-performance liquid chromatographic method for the determination of the three main metabolites of selegiline (L-deprenyl) in human plasma, *J.Chromatogr.B*, **1994**, *656*, 251–258.

SAMPLE
Matrix: blood
Analyte: 3'-amino-3'-deoxythymidine

Sample preparation: Condition a 3 mL 500 mg Bond Elut C18 SPE cartridge with three 1 mL portions of MeOH and two 3 mL portions of 50 mM pH 7.2 sodium phosphate buffer. 200 µL Plasma + 200 µL 50 mM pH 7.2 sodium phosphate buffer, mix, add to the SPE cartridge, let stand for 5 min, wash with 1 mL buffer, dry under vacuum for 5 min, elute with 1 mL MeOH. Evaporate the eluate to dryness under a stream of nitrogen, reconstitute the residue in 100 µL 200 mM pH 9.5 borate buffer, add 200 µL water, add 300 µL 2.5 mg/mL 9-fluorenylmethyl chloroformate in MeCN:acetone 50:50, vortex for 3 min, wash with 2 mL pentane. Remove the lower layer and dilute it with 5 volumes mobile phase, inject an aliquot.

HPLC VARIABLES
Column: 250 × 4.6 5 µm Hypersil ODS C18
Mobile phase: MeCN:10 mM pH 7 potassium phosphate buffer 32:68
Flow rate: 1.1
Detector: F ex 262 em 306

CHROMATOGRAM
Retention time: 27.8
Limit of quantitation: 900 pg/mL

OTHER SUBSTANCES
Non-interfering: amphotericin B, didanosine, ethambutol, isoniazid, pyrimethamine, rifampin, sulfadiazine, sulfamethoxazole, trimethoprim, zalcitabine, zidovudine

KEY WORDS
plasma; human; mouse; pharmacokinetics; SPE

REFERENCE
Nattouf, H.; Davrinche, C.; Sinet, M.; Farinotti, R. High-performance liquid chromatographic assay for 3'-amino-3'-deoxythymidine in plasma, with 9-fluorenyl methyl chloroformate as the derivatization agent, *J.Chromatogr.B*, **1995**, *664*, 365–371.

SAMPLE
Matrix: blood
Analyte: remikiren
Sample preparation: 500 µL Plasma + 5 µL 2 µg/mL IS in MeOH + 1 mL butyl acetate, extract for 2 min, centrifuge at 2000 g for 20 min. Remove the organic layer and evaporate it to dryness under a stream of helium at 60°, reconstitute the residue in 40 µL MeCN and 25 µL 1 M pH 6.3 borate buffer, vortex for 2 min, add 10 µL 25.85 mg/mL 9-fluorenylmethyl chloroformate in MeCN, vortex for 1 min, let stand for 1 h, add 20 µL 225 mM L-proline in water, inject an aliquot.

HPLC VARIABLES
Column: 125 × 4 3 µm Nucleosil C18
Mobile phase: A:B 92:8, after 15 min wash with MeCN for 5 min, re-equilibrate at initial conditions for 10 min. A was MeCN containing 350 µg/mL N-hexylmethylamine and 15 mM trifluoroacetic acid, adjusted to pH 3.0. B was 5 mM trifluoroacetic acid in water.
Flow rate: 0.75
Detector: F ex 261 em 308

CHROMATOGRAM
Retention time: 8.6
Internal standard: (S)-α--N-[1S,2R,3S-1-(cyclohexylmethyl)-2,3-dihydroxy-4-methylpentyl]imidazole-4-propionamide (Ro 42-4661) (10.8)
Limit of quantitation: 2 ng/mL

KEY WORDS
plasma; rat; dog; monkey; pharmacokinetics

REFERENCE
Coassolo, P.; Fischli, W.; Clozel, J.-P.; Chou, R.C. Pharmacokinetics of remikiren, a potent orally active inhibitor of human renin, in rat, dog and primates, *Xenobiotica*, **1996**, *26*, 333–345.

SAMPLE
Matrix: blood, broth

Analyte: gentamicin

Sample preparation: Condition a 3 mL 100 mg Isolute CBA-bonded (carboxypropyl) silica SPE cartridge (Jones Chromatography) with 1 mL MeOH and 1 mL 20 mM pH 7.4 phosphate buffer. Add 1 mL plasma or broth to the SPE cartridge, wash with 2 mL 20 mM pH 7.4 phosphate buffer, wash with 4 mL 200 mM pH 9.0 borate buffer, dry with 30 mL air, elute with 1 mL MeCN:200 mM pH 10.5 borate buffer 50:50, force out all the liquid with air. 1 mL Eluate + 200 μL 800 mM boric acid + 200 μL 2.5 mM 9-fluorenylmethyl chloroformate in MeCN, let stand at room temperature for 15 min, add 25 μL 100 mM glycine, let stand for 2 min, inject a 50 μL aliquot.

HPLC VARIABLES

Column: 200 $\times$ 4.6 3 μm ODS Hypersil
Mobile phase: MeCN:water 87:13
Flow rate: 1
Injection volume: 50
Detector: F ex 260 em 315

CHROMATOGRAM

Retention time: 18 (gentamicin C_{1a}), 20 (gentamicin C_2), 22 (gentamicin C_{2a}), 24 (gentamicin C_1)
Limit of detection: 10-50 ng/mL

KEY WORDS

plasma; SPE

REFERENCE

Stead, D.A.; Richards, R.M.E. Sensitive fluorimetric determination of gentamicin sulfate in biological matrices using solid-phase extraction, pre-column derivatization with 9-fluorenylmethyl chloroformate and reversed-phase high-performance liquid chromatography, *J.Chromatogr.B*, **1996**, *675*, 295–302.

SAMPLE

Matrix: blood, fermentation broth

Analyte: β-lactams

Sample preparation: Mix 1 mL filtered fermentation broth or serum with 100 μL 200 mM pH 7.7 sodium borate buffer and 2 mL acetone at 4°, centrifuge at 2600 g in a refrigerated centrifuge for 10 min, add 400 μL 15 mM 9-fluorenylmethyl chloroformate in dry acetone to the supernatant, wash twice with 3.4 mL portions of pentane, inject a 10 μL aliquot of the aqueous phase.

HPLC VARIABLES

Guard column: 10 $\times$ 4 3 μm Hypersil ODS
Column: 150 $\times$ 4 3 μm Hypersil ODS
Mobile phase: Gradient. MeCN:50 mM pH 5.0 acetate buffer from 18:82 to 20:80 over 20 min, to 23:77 over 25 min, to 100:0 over 5 min. (Place a 50 $\times$ 4 5 μm Spherisorb ODS column before the injector.)
Flow rate: 1
Injection volume: 10
Detector: F ex 260 em 311

CHROMATOGRAM

Retention time: 19 (deacetylcephalosporin C), 25 (deacetoxycephalosporin C), 31 (cephalosporin C), 41 (ampicillin)
Limit of detection: 250 ng/mL

KEY WORDS

serum

REFERENCE

Shah, A.J.; Adlard, M.W. Determination of β-lactams and their biosynthetic intermediates in fermentation media by pre-column derivatization followed by fluorescence detection, *J.Chromatogr.*, **1988**, *424*, 325–336.

SAMPLE
Matrix: eggs, milk, tissue
Analyte: erythromycin
Sample preparation: Homogenize (Ultra-Turrax) 30 g food with 100 mL MeCN and 1 mL saturated sodium carbonate solution for 1 min, rinse the shaft with 1-2 mL MeCN, centrifuge at 2000 g for 10 min. Remove the supernatant and add it to 3 g NaCl and 30 mL dichloromethane, shake for 1 min, let stand until the layers separate, discard the lower aqueous phase. Dry the clear organic layer over 2 g anhydrous sodium sulfate, filter through glass wool, rinse the filter with a few mL MeCN. Evaporate to dryness under reduced pressure, immediately dissolve the residue in 2 mL ethyl acetate, transfer to another tube. Rinse the flask with 2 mL ethyl acetate and 2.5 mL acetate buffer, combine all these samples, proceed rapidly. (Filter liver samples through glass wool and do not use acetate buffer for liver samples.) Vortex for 15 s, centrifuge on a tabletop centrifuge at medium speed for 20 s. Remove the lower aqueous phase and add it to 3 mL ethyl acetate and 200 μL saturated sodium carbonate solution, vortex briefly. Extract the ethyl acetate layer with 2 mL acetate buffer, neutralize the aqueous extract with 200 μL saturated sodium carbonate solution, repeat the extraction. Combine all the aqueous extracts, add 700 μL saturated sodium carbonate solution. Mix the ethyl acetate and acetate buffer layers, vortex, centrifuge. Remove the clear upper ethyl acetate layer, extract the alkaline aqueous layer twice with 2 mL portions of ethyl acetate. Combine the ethyl acetate layers and evaporate them to dryness under a stream of nitrogen at 35°, reconstitute with 500 μL mobile phase I, vortex for 1 min, centrifuge for 5 min, chromatograph the whole amount on a 4 × 4 LiChrospher RP-18 + 125 × 4 5 μm Nucleosil RP-18 column with mobile phase I at 1 mL/min at 30° with monitoring at 288 nm, collect a 5 mL fraction containing the erythromycin (retention time about 11 min). Add 800 μL 5% sodium carbonate solution to the column effluent, add 2 mL dichloromethane, vortex for 30 s. Repeat extraction twice more, combine the organic layers, centrifuge again, evaporate to dryness under a stream of nitrogen at 35°, reconstitute the residue in 100 μL MeCN, add 100 μL freshly-prepared 400 μg/mL 9-fluorenylmethylchloroformate in MeCN, add 100 μL pH 7.5 phosphate buffer, heat at 45° for 1 h, cool, make up to 500 μL with 200 μL MeCN:30 mM pH 7.0 phosphate buffer 50:50, centrifuge, inject a 30 μL aliquot. (Prepare acetate buffer by adjusting the pH of 100 mM sodium acetate to 4.0 with acetic acid. Prepare 150 mM pH 7.0 phosphate buffer by adjusting the pH of 150 mM K_2HPO_4 to 7.0 with 150 mM KH_2PO_4. Mobile phase I was MeCN:150 mM pH 7.0 phosphate buffer 30:70 adjusted to pH 6.9 with 150 mM phosphoric acid. Prepare pH 7.5 phosphate buffer by adjusting the pH of 100 mM K_2HPO_4 to 7.5 with 100 mM KH_2PO_4.)

HPLC VARIABLES
Guard column: 4 × 4 5 μm LiChrospher RP-18 endcapped
Column: 250 × 4 5 μm LiChrospher RP-18 endcapped
Mobile phase: MeCN:buffer 75:25, adjusted to pH 7.2 with 1 M phosphoric acid. (Prepare buffer by adjusting the pH of 30 mM K_2HPO_4 to 7.0 with 30 mM KH_2PO_4.)
Column temperature: 30
Flow rate: 1
Injection volume: 30
Detector: F ex 255 em 315

CHROMATOGRAM
Retention time: 13
Limit of detection: <10 ng/g

KEY WORDS
pig; muscle; kidney; liver

REFERENCE
Zierfels, G.; Petz, M. Fluorimetrische Bestimmung von Erythromycin-Rückständen in Lebensmitteln tierischer Herkunft nach Derivatisierung mit FMOC und HPLC-Trennung [Fluorimetric determination of erythromycin residues in foods of animal origin after derivatization with FMOC and HPLC separation], *Z.Lebensm.Unters.Forsch.*, **1994**, *198*, 307–312.

SAMPLE
Matrix: feed

Analyte: fumonisin B$_1$
Sample preparation: Condition a 3 mL Sep-Pak C18 SPE cartridge with 5 mL MeOH and 5 mL 1% KCl. Condition a Bond Elut strong anion-exchange (SAX) SPE cartridge with 8 mL MeOH:water 70:30. Blend 50 g feed with 200 mL MeCN:water 50:50 at medium speed for 2 min, centrifuge a 20 mL aliquot at 2000 rpm for 5 min. Add a 2 mL aliquot of the supernatant to 5 mL 1% KCl, mix, add to the C18 SPE cartridge, wash with 3 mL 1% KCl, wash with 2 mL MeCN:1% aqueous KCl 20:80, elute with 2 mL MeOH:water 70:30. Dilute the eluate to 4 mL with MeOH:water 70:30, add to the SAX SPE cartridge at 2 mL/min, wash with 8 mL MeOH:water 70:30, wash with 3 mL MeOH, elute with 14 mL MeOH:acetic acid 99:1, evaporate the eluate to dryness under a stream of nitrogen, reconstitute in 500 μL MeCN:water 50:50. Remove a 500 μL aliquot and add it to 100 μL buffer, add 100 μL 3 mg/mL (9-fluorenylmethyl) chloroformate in MeCN, let stand for 30 s, wash twice with 1 mL portions of pentane, inject a 10 μL aliquot. (Prepare buffer by adjusting the pH of 1 M boric acid to 7.5 with 6 M NaOH.)

HPLC VARIABLES

Guard column: present but not specified
Column: 250 × 4.6 5 μm Microsorb C8
Mobile phase: Gradient. A was MeCN:buffer 30:70. B was MeCN:buffer 70:30. A:B from 90:10 to 60:40 over 9 min, to 30:70 over 8 min, to 0:100 over 1 min, maintain at 0:100 for 17 min. (Buffer was 15 mM citric acid in water containing 10 mM tetramethylammonium chloride, pH adjusted to 4.7 with 6 M NaOH.)
Flow rate: 1.5
Injection volume: 10
Detector: F ex 263 em 313

CHROMATOGRAM

Retention time: 15 (fumonisin B$_1$), 19 (fumonisin B$_2$)
Limit of detection: 200 ppb

KEY WORDS

SPE

REFERENCE

Holcomb, M.; Thompson, H.C. Jr.; Hankins, L.J. Analysis of fumonisin B$_1$ in rodent feed by gradient elution HPLC using precolumn derivation with FMOC and fluorescence detection, *J.Agric.Food Chem.*, **1993**, *41*, 764–767.

SAMPLE

Matrix: fish, food, wine
Analyte: amines and amino acids
Sample preparation: Filter (0.45 μm) wine, fruit juice or vegetable juice, dilute 1:5 or 1:20. 50 μL Diluted wine, diluted juice, fish extract, or cheese extract (neutralize strongly acidic samples if necessary) + 200 μL 200 mM boric acid adjusted to pH 8.5 with 30% KOH + 200 μL 3 mM 9-fluorenylmethyl chloroformate in acetone, mix for 3 min at room temperature, add 50 μL reagent, mix for 3 min. Remove an 80 μL aliquot and add it to 320 μL initial mobile phase, inject a 20 μL aliquot. (Reagent was 3 mL heptylamine in 15 mL MeCN, adjusted to pH 7-8 with 175 mL 100 mM HCl. Extract fish as follows. Homogenize 20 g fish and 10 mL 100 mM HCl, add 40 mL 100 mM HCl, centrifuge, decant, extract residue with two 40 mL and one 20 mL portions of 100 mM HCl, filter, make up the extracts to 100 (?) mL with 100 mM HCl. Extract cheese as follows. Homogenize 25 g cheese and 18.75 mL 100 mM HCl, suspend with 40 mL 100 mM HCl, centrifuge, extract the residue twice with 20 mL portions of 100 mM HCl. Combine and filter the extracts, make up to 100 mL with 100 mM HCl.)

HPLC VARIABLES

Column: 250-4 Supersphere 60 RP-8 (Merck)
Mobile phase: Gradient. A was MeCN:100 mM pH 4.4 sodium acetate buffer 22:78. B was MeCN. A:B from 100:0 to 80:20 over 17 min, to 71:29 over 2 min, to 77:23 over 1 min, maintain at 77:23 for 7 min, to 64:36 over 3 min, maintain at 64:36 for 7 min, to 58:42 over 5 min, to 45:55 over 15 min, maintain at 45:55 for 6 min, 7:93 over 7 min,

to 0:100 over 3 min, maintain at 0:100 for 9 min, return to 100:0 over 1 min, re-equilibrate for 10 min.

Column temperature: 40
Flow rate: 1.25 (0.05 when not in use)
Injection volume: 20
Detector: F ex 265 em 315

CHROMATOGRAM
Retention time: 2.5 (cysteic acid), 8 (aspartic acid), 8 (serine), 8 (glutamic acid), 10 (threonine), 11.5 (glycine), 14.5 (alanine), 15 (tyrosine), 17 (proline), 17.5 (methionine), 20 (valine), 21 (phenylalanine), 24 (isoleucine), 25 (leucine), 32 (histidine), 34 (tyramine), 35 (lysine), 49 (phenylethylamine), 56 (putrescine), 57 (histamine), 59 (cadaverine), 63 (heptylamine), 70 (spermidine), 72 (spermine)

KEY WORDS
fruit juice; vegetable juice; cheese

REFERENCE
Kirschbaum, J.; Luckas, B.; Beinert, W.-D. Pre-column derivatization of biogenic amines and amino acids with 9-fluorenylmethyl chloroformate and heptylamine, *J.Chromatogr.A*, **1994**, *661*, 193–199.

SAMPLE
Matrix: fish, food, wine
Analyte: amines
Sample preparation: Filter (0.45 µm) wine, fruit juice or vegetable juice, dilute 1:5 or 1 :20. 50 µL Diluted wine, diluted juice, fish extract, or cheese extract (neutralize strongly acidic samples if necessary) + 200 µL 200 mM boric acid adjusted to pH 8.5 with 30% KOH + 200 µL 3 mM 9-fluorenylmethyl chloroformate in acetone, mix for 3 min at room temperature, add 50 µL reagent, mix for 3 min. Remove an 80 µL aliquot and add it to 320 µL initial mobile phase, inject a 20 µL aliquot. (Reagent was 3 mL heptylamine in 15 mL MeCN, adjusted to pH 7-8 with 175 mL 100 mM HCl. Extract fish as follows. Homogenize 20 g fish and 10 mL 100 mM HCl, add 40 mL 100 mM HCl, centrifuge, decant, extract residue with two 40 mL and one 20 mL portions of 100 mM HCl, filter, make up the extracts to 100 (?) mL with 100 mM HCl. Extract cheese as follows. Homogenize 25 g cheese and 18.75 mL 100 mM HCl, suspend with 40 mL 100 mM HCl, centrifuge, extract the residue twice with 20 mL portions of 100 mM HCl. Combine and filter the extracts, make up to 100 mL with 100 mM HCl.)

HPLC VARIABLES
Column: 250-4 Supersphere 60 RP-8 (Merck)
Mobile phase: Gradient. A was MeCN:100 mM pH 4.4 sodium acetate buffer 50:50. B was MeCN. A:B 100:0 for 7 min, to 90:10 over 5 min, to 70:30 over 15 min, maintain at 70 :30 for 6 min, to 10:90 over 7 min, to 0:100 over 3 min, maintain at 0:100 for 9 min, return to 100:0 over 1 min, re-equilibrate for 7 min.
Column temperature: 40
Flow rate: 1.2 (0.05 when not in use)
Injection volume: 20
Detector: F ex 265 em 315

CHROMATOGRAM
Retention time: 9.0 (tyramine (mono-derivative)), 21 (phenylethylamine), 27 (putrescine), 27.8 (histamine), 29 (cadaverine), 37.6 (tyramine (bis-derivative)), 40 (spermidine), 42.5 (spermine)
Limit of detection: <20 ng/mL

KEY WORDS
wine; fruit juice; vegetable juice; fish; cheese

REFERENCE
Kirschbaum, J.; Luckas, B.; Beinert, W.-D. Pre-column derivatization of biogenic amines and amino acids with 9-fluorenylmethyl chloroformate and heptylamine, *J.Chromatogr.A*, **1994**, *661*, 193–199.

SAMPLE
Matrix: formulations

Analyte: alendronate (4-amino-1-hydroxybutane-1,1-diphosphonic acid monosodium salt)

Sample preparation: Dilute injections to a concentration of 25 μg/mL with 100 mM sodium citrate. Shake 1 tablet or the contents of 1 capsule with 50 mL 100 mM sodium citrate for 30 min, sonicate for 5 min, make up to 100 mL with 100 mM sodium citrate, dilute to 25 μg/mL, centrifuge at 2000 rpm for 10 min. Mix 5 mL supernatant or diluted injection with 5 mL 100 mM pH 9.0 sodium borate, add 4 mL 500 μg/mL 9-fluorenylmethyl chloroformate in MeCN, vortex for 30 s, let stand for 30 min, add 25 mL dichloromethane, shake for 30-60 s, let stand for 5 min, centrifuge at 1000 rpm for 5 min, inject a 50 μL aliquot of the upper aqueous layer.

HPLC VARIABLES

Column: 250 × 4.1 10 μm PRP-1 (Hamilton)

Mobile phase: MeCN:MeOH:buffer 20:5:75 (Prepare buffer by dissolving 14.7 g sodium citrate dihydrate and 8.7 g K_2HPO_4 in 900 mL water, adjust pH to 8.0 with phosphoric acid.)

Column temperature: 35

Flow rate: 1

Injection volume: 50

Detector: UV 266

CHROMATOGRAM

Retention time: 7

OTHER SUBSTANCES

Simultaneously analyzed: impurities

KEY WORDS

injections; capsules; tablets

REFERENCE

De Marco, J.D.; Biffar, S.E.; Reed, D.G.; Brooks, M.A. The determination of 4-amino-1-hydroxybutane-1,1-diphosphonic acid monosodium salt trihydrate in pharmaceutical dosage forms by high-performance liquid chromatography, *J.Pharm.Biomed.Anal.*, **1989**, 7, 1719–1727.

SAMPLE

Matrix: groundwater

Analyte: dimethylamine

Sample preparation: Filter (0.22 μm) ground water. Remove a 400 μL aliquot and add it to 500 μL 15 mM 9-fluorenylmethylchloroformate in anhydrous MeCN and 100 μL 1 M pH 8.0 borate buffer, vortex for 2 min, add 5 μL glacial acetic acid, add 2 mL hexane, vortex for 30 s, inject a 20 μL aliquot of the aqueous phase.

HPLC VARIABLES

Column: 250 × 4 10 μm Spherisorb ODS-2

Mobile phase: MeCN:30 mM pH 4.0 acetate buffer 70:30

Flow rate: 1

Injection volume: 20

Detector: F ex 265 em 310

CHROMATOGRAM

Retention time: 5.70

Limit of detection: 450 nM

OTHER SUBSTANCES

Simultaneously analyzed: acetic acid, ammonia, aniline, benzylamine, n-butylamine, cadaverine, diethylamine, glycine, methylamine, proline, putrescine

REFERENCE

Rodríguez López, M.; González Alvarez, M.J.; Miranda Ordieres, A.J.; Tuñón Blanco, P. Determination of dimethylamine in groundwater by liquid chromatography and precolumn derivatization with 9-fluorenylmethylchloroformate, *J.Chromatogr.A*, **1996**, 721, 231–239.

SAMPLE

Matrix: gunshot residues

Analyte: 2,4-dinitrotoluene

Sample preparation: Extract swab with ethyl acetate, chromatograph by TLC (Silica gel 60, Merck) using toluene:ethyl acetate 15:1, remove the portion corresponding to 2,4-dinitrotoluene, elute with two 200 µL portions of ethyl acetate, evaporate the eluate to dryness, add 60 µL HCl, add 15 mg iron(II) ammonium sulfate, heat at 155° for 25 min, cool to room temperature, add 300 µL 3 M KOH, extract four times with 80 µL portions of ethyl acetate. Combine the organic layers and wash them twice with 180 µL portions of water, evaporate the organic layer nearly to dryness under a stream of nitrogen, reconstitute with 20 µL acetone, add 30 µL 258 µg/mL 9-fluorenylmethylchloroformate in MeCN, add 60 µL pH 8.0 borate buffer, heat at 52° for 20 min, cool to room temperature, add 10 µL 50 mM proline, let stand for 10 min, make up to 600 µL with mobile phase, inject a 20 µL aliquot. (2,4-Dinitrotoluene is reduced to 2,4-diaminotoluene which is then derivatized.)

HPLC VARIABLES

Column: 125 × 4.9 5 µm Exsil 80 ODS (Alltech)

Mobile phase: MeCN:buffer 78:22 (Buffer was citrate buffer adjusted to pH 6.6 with ammonia solution.)

Flow rate: 0.9

Injection volume: 20

Detector: F ex 266 em 302

CHROMATOGRAM

Retention time: 6.5

Limit of detection: 1 ng

REFERENCE

Meng, H.-h.; Caddy, B. High-performance liquid chromatographic analysis with fluorescence detection of ethyl centralite and 2,4-dinitrotoluene in gunshot residues after derivatization with 9-fluorenylmethylchloroformate, *J.Forensic Sci.*, **1996**, *41*, 213–220.

SAMPLE

Matrix: solutions

Analyte: angiotensin II

Sample preparation: Dry a 5 µL aliquot of a 0.1-4 mM aqueous solution, reconstitute with 100 µL pH 9.5 100 mM lithium carbonate/sodium bicarbonate buffer, add 100 µL 15 mM 9-fluorenylmethyl chloroformate in acetone, let stand at room temperature for 30 s, wash with five 500 µL portions of n-pentane. Remove a 10 µL aliquot of the aqueous phase and add it to 90 µL 10 mM NaH_2PO_4, mix, inject a 2 µL aliquot.

HPLC VARIABLES

Column: 125 × 2 5 µm LiChrospher RP-2

Mobile phase: MeCN:10 mM pH 4.9 NaH_2PO_4 54:46

Flow rate: 1

Injection volume: 2

Detector: F ex 260 em 310

CHROMATOGRAM

Retention time: 5.2

Limit of detection: 500 fmole

REFERENCE

Vogt, W.; Egeler, E.; Sommer, W.; Eisenbeiss, F.; Meyer, H.D. High-performance liquid chromatographic determination of hormonal peptides and their fluorenylmethoxycarbonyl derivatives, *J.Chromatogr.*, **1987**, *400*, 83–89.

SAMPLE

Matrix: solutions

Analyte: amphetamine

Sample preparation: 500 µL Solution + 250 µL buffer + 250 µL 20 mM 9-fluorenylmethyl chloroformate in MeCN, mix, add 1 mL MeCN, inject a 50 µL aliquot. (Prepare buffer by adjusting the pH of 4% sodium bicarbonate to 10 with 10% NaOH.)

HPLC VARIABLES
Column: 125 × 4 5 μm LiChrospher 100 RP 18
Mobile phase: Gradient. MeCN:water from 40:60 to 50:50 over 2.5 min, to 70:30 over 2.5 min, to 100:0 over 5 min.
Flow rate: 1.5
Injection volume: 50
Detector: F ex 264 em 313

CHROMATOGRAM
Retention time: 7.5 (amphetamine), 8.8 (methamphetamine)

REFERENCE
Herráez-Hernández, R.; Campins-Falcó, P.; Sevillano-Cabeza, A. On-line derivatization into precolumns for the determination of drugs by liquid chromatography and column switching: Determination of amphetamines in urine, *Anal.Chem.*, **1996**, *68*, 734–739.

SAMPLE
Matrix: solutions
Analyte: sulfamethazine
Sample preparation: 1 mL 1 mg/mL Solution + 1 mL 100 mM pH 8 sodium bicarbonate + 2 mL 10 mM 1-fluorenylmethyl chloroformate in acetone, let stand for 30 min, inject an aliquot.

HPLC VARIABLES
Column: 150 × 4 5 μm Micro Pak C18 TMS-capped
Mobile phase: Gradient. MeOH:buffer from 35:65 to 85:15 over 4 min (Waters medium concave gradient), maintain at 85:15. (Buffer was 50 mM NaH_2PO_4 adjusted to pH 3.5 with phosphoric acid.)
Flow rate: 1
Injection volume: 20
Detector: UV 264; F ex 264 em 307

CHROMATOGRAM
Retention time: 12.5

REFERENCE
Liang, G.S.; Zhang, Z.; Baker, W.L.; Cross, R.F. Formation and verification of the structure of the 1-fluorenylmethyl chloroformate derivative of sulfamethazine, *Anal.Chem.*, **1996**, *68*, 86–92.

SAMPLE
Matrix: tissue
Analyte: polyamines
Sample preparation: Sonicate (Heath Model W-220) dorsal root ganglia from 5 rats with 98 μL water and 14 μL 40% trichloroacetic acid at 0° for 10 s, centrifuge at 4° at 14000 rpm for 30 min. Remove a 50 μL aliquot and add it to 50 μL IS in water, add 1 mL 20 mM pH 9.6 borate buffer, add 1 mL acetone, add 100 μL 10 mM 9-fluorenylmethyl chloroformate in acetone (freshly prepared), vortex for 30 s, let stand at room temperature for 10 min, add 2 mL hexane:ethyl acetate 50:50, vortex for 30 s, repeat extraction. Combine the organic layers and evaporate them to dryness under a stream of nitrogen, reconstitute the residue in 2 mL acetone:water 50:50, inject a 20 μL aliquot.

HPLC VARIABLES
Column: Nova-Pak C18
Mobile phase: Gradient. MeCN:50 mM pH 3.65 sodium acetate buffer from 31:69 to 48:52 over 19 min, to 70:30 over 3 min (non-linear gradient), return to initial conditions over 8 min. (This gradient is given in J.Neurosci.Methods 1988, 26, 45 by these authors but it may not be valid for this separation; modification of these conditions should give acceptable results, however.)
Flow rate: 1
Injection volume: 20
Detector: F ex 254 em 313

CHROMATOGRAM
Retention time: 4.21 (putrescine), 10.09 (spermidine), 11.19 (spermine)
Internal standard: 1,6-hexanediamine (6.24)

KEY WORDS
rat; dorsal root ganglia

REFERENCE
Sabri, M.I.; Soiefer, A.I.; Kisby, G.E.; Spencer, P.S. Determination of polyamines by precolumn derivatization with 9-fluorenylmethyl chloroformate and reverse-phase high-performance liquid chromatography, *J.Neurosci.Methods*, **1989**, *29*, 27–31.

SAMPLE
Matrix: urine
Analyte: toluenediamines
Sample preparation: 500 μL Urine + 500 μL 15 μg/mL IS in MeOH:water 50:50 + 100 μL 6 M NaOH + 4 mL dichloromethane, shake for 4 min, centrifuge at 1500 g for 1.5 min, repeat extraction twice. Combine the organic layers and evaporate them to dryness under a stream of nitrogen, reconstitute the residue in 1 mL mobile phase. Remove a 400 μL aliquot and add it to 100 μL 1 M pH 7.7 sodium borate buffer, add 500 μL 15 mM fluorenylmethyl chloroformate in acetone, shake at room temperature for 3 min, shake at 50° for 3 min, inject an aliquot.

HPLC VARIABLES
Guard column: 4 × 4 5 μm LichroCART RP 18 (Merck)
Column: 250 × 4 5 μm LichroCART RP 18 Purospher (Merck)
Mobile phase: MeCN:water 80:20
Flow rate: 1
Injection volume: 20
Detector: F ex 260 em 310

CHROMATOGRAM
Retention time: 6.8 (2,6-toluenediamine), 8.0 (2,4-toluenediamine)
Internal standard: 4-chloro-1,2-phenylenediamine (10.3)
Limit of quantitation: 30 ng/mL
Limit of detection: 15 ng/mL

REFERENCE
De Lorenzi, E.; Massolini, G.; Macchia, M.; Caccialanza, G. HPLC determination of urinary 2,4- and 2,6-toluenediamines as potential degradation products of polyurethane breast implants, *Chromatographia*, **1995**, *41*, 661–664.

SAMPLE
Matrix: urine
Analyte: amines
Sample preparation: Inject 15 μL urine, inject a mixture of 5 μL 20 mM 9-fluorenylmethyl chloroformate in MeCN and 45 μL water, and inject 10 μL buffer on to column A and elute to waste with mobile phase A. After 2.8 min backflush the contents of column A on to column B with mobile phase B and start the gradient, monitor the effluent from column B. At the end of the run condition column A with 1 mL mobile phase A. (Buffer was 4% sodium bicarbonate adjusted to pH 10 with 10% NaOH.)

HPLC VARIABLES
Column: A 20 × 2.1 30 μm Hypersil ODS-C18; B 125 × 4 5 μm LiChrospher 100 PR-C18
Mobile phase: A water; B Gradient. MeCN:water from 40:60 to 70:30 over 15 min. to 100:0 over 5 min.
Flow rate: A 0.35; B 1
Injection volume: 15
Detector: F ex 264 em 313

CHROMATOGRAM
Retention time: 10.5 (norephedrine), 11.8 (ephedrine), 12.8 (pseudoephedrine), 15.3 (amphetamine), 15.7 (3-phenylpropylamine), 17.8 (methamphetamine)
Limit of detection: 5-25 ng/mL

KEY WORDS
column-switching; on-column derivatization

REFERENCE
Herráez-Hernández, R.; Campíns-Falcó, P.; Sevillano-Cabeza, A. Determination of amphetamine and related compounds in urine using on-line derivatization in octadecyl silica columns with 9-fluorenylmethyl chloroformate and liquid chromatography, *J.Chromatogr.B*, **1996**, *679*, 69–78.

SAMPLE
Matrix: urine
Analyte: amphetamine
Sample preparation: Inject 50 µL urine on to column A and elute to waste with mobile phase A, after 2 min inject a mixture of 15 µL 20 mM 9-fluorenylmethyl chloroformate in MeCN and 25 µL buffer on to column A, stop the flow of mobile phase A, backflush the contents of column A on to column B with mobile phase B and start the gradient, monitor the effluent from column B. After each run flush column A with 1 mL MeCN at 1 mL/min, re-equilibrate with mobile phase A. (Prepare buffer by adjusting the pH of 4% sodium bicarbonate to 10 with 10% NaOH.)

HPLC VARIABLES
Column: A 20 × 2.1 30 µm Hypersil ODS-C18; B 125 × 4 5 µm LiChrospher 100 RP 18
Mobile phase: A water; B Gradient. MeCN:water from 40:60 to 50:50 over 2.5 min, to 70:30 over 2.5 min, to 100:0 over 5 min.
Flow rate: A 1; B 1.5
Injection volume: 50
Detector: F ex 264 em 313

CHROMATOGRAM
Retention time: 7.5 (amphetamine), 8.8 (methamphetamine)
Limit of detection: 1 ng/mL

KEY WORDS
column-switching; on-column derivatization

REFERENCE
Herráez-Hernández, R.; Campins-Falcó, P.; Sevillano-Cabeza, A. On-line derivatization into precolumns for the determination of drugs by liquid chromatography and column switching: Determination of amphetamines in urine, *Anal.Chem.*, **1996**, *68*, 734–739.

SAMPLE
Matrix: water
Analyte: glyphosate and glufosinate
Sample preparation: 1.5 mL Water + 200 µL 125 mM pH 9 borate buffer + 1 mL 1 mg/mL 9-fluorenylmethyl chloroformate, mix, let stand at room temperature for 30 min, add 17.5 mL 25 mM pH 9 borate buffer, inject a 2 mL aliquot onto column A and elute to waste with mobile phase, after 2.21 min direct the effluent from column A onto column B, after 0.53 min remove column A from the circuit, elute column B with mobile phase, monitor the effluent from column B.

HPLC VARIABLES
Column: A 30 × 4.6 5 µm Spherisorb ODS-2; B 250 × 4.6 5 µm Adsorbosphere NH2 or 5 µm Spherisorb NH2
Mobile phase: MeCN:50 mM pH 5.5 phosphate buffer 35:65
Column temperature: 30 (column B only)
Flow rate: 1
Injection volume: 2000
Detector: F ex 263 em 317

CHROMATOGRAM
Retention time: 9 (aminomethylphosphonic acid), 14 (glufosinate), 19 (glyphosate)
Limit of detection: 1 ng/mL

KEY WORDS
column-switching

REFERENCE

Sancho, J.V.; Hernández, F.; López, F.J.; Hogendoorn, E.A.; Dijkman, E.; van Zoonen, P. Rapid determination of glufosinate, glyphosate and aminomethylphosphonic acid in environmental water samples using precolumn fluorogenic labeling and coupled-column liquid chromatography, *J.Chromatogr.A*, **1996**, *737*, 75−83.

SAMPLE

Matrix: wine

Analyte: amines

Sample preparation: 20 μL Wine + 50 μL buffer, mix, add 100 μL 8 mg/mL 9-fluorenylmethyl chloroformate in MeCN, mix, let stand for 3 min, add 50 μL 500 mM ammonia, mix, let stand for 3 min, add 300 μL MeCN:water:acetic acid 80:12:8, inject a 20 μL aliquot. (Prepare buffer by adjusting the pH of 200 mM boric acid to 8.5 with 5 M NaOH.)

HPLC VARIABLES

Column: 200 × 2.1 5 μm ODS Hypersil

Mobile phase: Gradient. A was MeCN:2-octanol 99:1. B was MeCN:water:phosphoric acid:dimethylcyclohexylamine 15:83.12:0.88:1, pH 2.7. A:B 15:85 for 18 min, 38:62 over 0.1 min, to 40:60 over 6.9 min, maintain at 40:60 for 5 min, to 42:58 over 37 min, to 85:15 over 36 min, maintain at 85:15 for 7 min.

Flow rate: 0.3

Injection volume: 2

Detector: F ex 263 em 313

CHROMATOGRAM

Retention time: 7 (histidine), 10 (arginine), 16 (agmatine), 33 (phenylalanine), 46 (phenylethylamine), 51 (ornithine), 76 (putrescine), 83 (cadaverine), 84 (tyrosine), 95 (tyramine), 100 (histamine), 102 (spermidine), 113 (spermine)

REFERENCE

Bauza, T.; Blaise, A.; Daumas, F.; Cabanis, J.C. Determination of biogenic amines and their precursor amino acids in wines of the Vallée du Rhône by high-performance liquid chromatography with precolumn derivatization and fluorimetric detection, *J.Chromatogr.A*, **1995**, *707*, 373−379.

Menthyl Chloroformate

RELATED REFERENCES

Witte, D.T.; de Zeeuw, R.A.; Drenth, B.F.H. Chiral derivatization of promethazine with (-)-menthyl chloroformate for enantiomeric separation by RP-HPLC. *J.High Resolut.Chromatogr.* **1990**, *13*, 569-570.

Witte, D.T.; Bosman, J.; de Boer, T.; Drenth, B.F.H.; Ensing, K.; de Zeeuw, R.A. Influence of chemical structure of tricyclic tertiary dimethylamines on chiral separation by reversed-phase high-performance liquid chromatography after derivatization with (-)-menthylchloroformate. *J.Chromatogr.* **1991**, *553*, 365-372.

SAMPLE

Matrix: blood

Analyte: atenolol

Sample preparation: 1 mL Whole blood + 1 mL 2.5 μg/mL methoxamine in 100 mM NaOH, vortex, add 5 mL ethyl acetate, shake at 230 oscillations/min on a reciprocating shaker for 15 min, centrifuge at 2000 g for 15 min, repeat the extraction with 3 mL ethyl acetate. Combine the organic layers and evaporate them to dryness under a stream of nitrogen at 37°, reconstitute the residue in 50 μL 100 mM NaOH, vortex briefly, add 200 μL 200 mM (-)-menthyl chloroformate in MeCN, vortex for 30 s, let stand at room temperature for 10 min, inject a 50 μL aliquot.

HPLC VARIABLES
Column: 150 × 4.6 5 μm Hypersil ODS
Mobile phase: MeCN:MeOH:water 43:25:32
Flow rate: 1.2
Injection volume: 50
Detector: F ex 230 em 305

CHROMATOGRAM
Retention time: 9.8 (S-(-)), 11.0 (R-(+))
Internal standard: methoxamine (14.0, 14.8 (enantiomers))
Limit of quantitation: 12.5 ng/mL

KEY WORDS
chiral; whole blood; pharmacokinetics

REFERENCE
Miller, R.B.; Guertin, Y. High-performance liquid chromatographic assay for the derivatized enantiomers of atenolol in whole blood, *J.Liq.Chromatogr.*, **1992**, *15*, 1289–1302.

SAMPLE
Matrix: blood
Analyte: tocainide
Sample preparation: Condition a 100 mg Bond Elut C18 SPE cartridge with 1 mL MeOH and 1 mL 20 mM pH 10 potassium phosphate buffer. 1 mL Plasma + 25 μL 1.2 mg/mL benzylamine hydrochloride in water + 100 mg solid sodium carbonate, vortex, add to the SPE cartridge, wash with 1 mL 20 mM pH 10 potassium phosphate buffer, elute with 500 μL MeOH. Add 20 μL 1 M (-)-menthyl chloroformate in MeCN to the eluate, shake briefly, let stand for 3 min, inject an aliquot.

HPLC VARIABLES
Column: 120 × 4.6 5 μm Spherisorb ODS-2
Mobile phase: MeOH:200 mM pH 5.0 potassium phosphate buffer:water 75:2.5:22.5
Flow rate: 1
Injection volume: 20
Detector: UV 262

CHROMATOGRAM
Retention time: 4.70 (R), 5.07 (S)
Internal standard: benzylamine (6.22)
Limit of quantitation: 1 μg/mL

KEY WORDS
chiral; SPE; plasma; rabbit; pharmacokinetics

REFERENCE
Christensen, E.B.; Hansen, S.H.; Rasmussen, S.N. Assay of tocainide enantiomers in plasma by solid-phase extraction and indirect chiral high-performance liquid chromatography after derivatization with (-)-menthyl chloroformate, *J.Chromatogr.B*, **1995**, *670*, 243–249.

SAMPLE
Matrix: blood, urine
Analyte: atenolol
Sample preparation: Plasma. 1 mL Plasma + 50 μL 10 μg/mL methoxamine in water + 100 μL 1 M NaOH + 4 mL ethyl acetate, vortex for 30 s, centrifuge at 3000 rpm for 10 min. Remove the organic layer and evaporate it to dryness under a stream of nitrogen at room temperature, reconstitute the residue in 200 μL saturated sodium carbonate and 200 μL 187 mM (-)-menthyl chloroformate in MeCN, vortex for 30 s, add 1 mL water, add 2 mL chloroform, vortex for 30 s, centrifuge for 3 min. Remove the organic layer and

evaporate it to dryness under a stream of nitrogen at room temperature, reconstitute the residue in 200 μL mobile phase, vortex for 5 s, centrifuge for 5 min, inject a 20-60 μL aliquot of the supernatant. Urine. Dilute urine 10 times with water. 100 μL Diluted urine + 50 μL 10 μg/mL methoxamine in water + 100 μL saturated sodium carbonate + 200 μL 187 mM (-)-menthyl chloroformate in MeCN, vortex for 30 s, add 1 mL water, add 2 mL chloroform, vortex for 30 s, centrifuge for 3 min. Remove the organic layer and evaporate it to dryness under a stream of nitrogen at room temperature, reconstitute the residue in 200 μL mobile phase, vortex for 5 s, centrifuge for 5 min, inject a 20-60 μL aliquot of the supernatant.

HPLC VARIABLES
Guard column: 50 mm long pellicular ODS (Whatman)
Column: 100 × 4.6 5 μm Partisil 5 ODS3
Mobile phase: MeCN:MeOH:water 35:22:43
Flow rate: 1.2
Injection volume: 20-60
Detector: F ex 195 em no emission filter

CHROMATOGRAM
Retention time: 13, 15 (enantiomers)
Internal standard: methoxamine (18 (-), 20 (+))
Limit of detection: 2.5 ng/mL

KEY WORDS
plasma; pharmacokinetics; chiral

REFERENCE
Mehvar, R. Liquid chromatographic analysis of atenolol enantiomers in human plasma and urine, *J.Pharm.Sci.*, **1989**, *78*, 1035-1039.

SAMPLE
Matrix: blood, urine
Analyte: sotalol
Sample preparation: 1 mL Plasma + 200 μL 2.5 μg/mL S-(-)atenolol + 660 μL 2 M perchloric acid, shake briefly, centrifuge at 2000 g for 5 min. Remove 1 mL of the supernatant and adjust the pH to 9 with 1 mL 2 M Tris-HCl buffer and 250 μL 2 M NaOH, add 5 mL chloroform:isopropanol 75:25, vortex for 1 min, centrifuge at 2000 g for 10 min, repeat the extraction. Combine the organic layers and dry them over about 3 g anhydrous sodium sulfate, evaporate to dryness under a stream of nitrogen, reconstitute with 200 μL saturated sodium carbonate solution, add 200 μL 40 μL/mL (-)-menthyl chloroformate in MeCN (prepare fresh daily), vortex for 30 s, add 1 mL water, add 2 mL chloroform, vortex for 1 min, centrifuge at 1500 g for 5 min. Remove the organic layer and evaporate it to dryness, reconstitute the residue in 100 μL mobile phase, centrifuge for 5 min, inject a 40 μL aliquot. Urine. Dilute 1 mL (?) urine 20 times with blank urine, add 200 μL 7.5 μg/mL S-(-)atenolol, adjust the pH to 9 with 1 mL 2 M Tris-HCl buffer, add 5 mL chloroform:isopropanol 75:25, vortex for 1 min, centrifuge at 2000 g for 10 min, repeat the extraction. Combine the organic layers and dry them over about 3 g anhydrous sodium sulfate, evaporate to dryness under a stream of nitrogen, reconstitute with 200 μL saturated sodium carbonate solution, add 200 μL 40 μL/mL (-)-menthylchloroformate in MeCN (prepare fresh daily), vortex for 30 s, add 1 mL water, add 2 mL chloroform, vortex for 1 min, centrifuge at 1500 g for 5 min. Remove the organic layer and evaporate it to dryness, reconstitute the residue in 100 μL mobile phase, centrifuge for 5 min, inject a 40 μL aliquot.

HPLC VARIABLES
Column: 250 × 4.6 5 μm C8 (Jones)
Mobile phase: MeCN:MeOH:water 15:50:35
Flow rate: 2
Injection volume: 40
Detector: F ex 235 em 300

CHROMATOGRAM
Retention time: 14 (l), 15 (d)

Internal standard: S-(-)-atenolol (12.3)
Limit of quantitation: 20 ng/mL
Limit of detection: 12.5 ng/mL

OTHER SUBSTANCES
Simultaneously analyzed: alprenolol, captopril, metoprolol, pindolol, propafenone
Non-interfering: acetaminophen, alginic acid, amiloride, amoxicillin, aspirin, diazoxide, digoxin, domperidone, flurazepam, furosemide, glyburide, heparin, hydralazine, hydrochlorothiazide, indapamide, lidocaine, lorazepam, lovastatin, mexiletine, nifedipine, nitroglycerin, omeprazole, oxazepam, procainamide, propranolol, propoxyphene, quinidine, ranitidine, timolol, triamterene, verapamil, warfarin

KEY WORDS
plasma; chiral

REFERENCE
Fiset, C.; Philippon, F.; Gilbert, M.; Turgeon, J. Stereoselective high-performance liquid chromatographic assay for the determination of sotalol enantiomers in biological fluids, *J.Chromatogr.*, **1993**, *612*, 231–237.

SAMPLE
Matrix: bulk
Analyte: encainide
Sample preparation: 0.01-5 µg Encainide + 50 µL MeCN:N,N-diisopropylethylamine 90:10 + 50 µL (-)-menthyl chloroformate:MeCN 10:90, heat at 60° for 2 h, evaporate to dryness, reconstitute in mobile phase, inject an aliquot.

HPLC VARIABLES
Column: 250 × 4.6 5 µm Ultrasphere silica
Mobile phase: Hexane:ethyl acetate:triethylamine 85:15:1
Flow rate: 1
Injection volume: 500
Detector: UV 261

CHROMATOGRAM
Retention time: 20 (+), 21 (-)
Limit of detection: 10 ng

KEY WORDS
chiral; normal phase

REFERENCE
Prakash, C.; Jajoo, H.K.; Blair, I.A.; Mayol, R.F. Resolution of enantiomers of the antiarrhythmic drug encainide and its major metabolites by chiral derivatization and high-performance liquid chromatography, *J.Chromatogr.*, **1989**, *493*, 325–335.

SAMPLE
Matrix: solutions
Analyte: adrenergic drugs
Sample preparation: Evaporate an aliquot of a solution in MeCN containing 62.5 (atenolol, metoprolol, propranolol, toliprolol) or 625 (metaproterenol, sotalol) ng drug to dryness under a stream of nitrogen at room temperature, add 200 µL saturated sodium carbonate, add 200 µL 4% (-)-menthyl chloroformate in MeCN, vortex for 30 s, add an excess amount of 4-hydroxy-L-proline, vortex for 30 s, centrifuge for 3 min, inject a 10-25 µL aliquot of the upper layer.

HPLC VARIABLES
Guard column: 50 × 4.6 Pellicular ODS (Whatman)
Column: 100 × 4.6 5 µm Partisil 5 ODS3
Mobile phase: MeCN:MeOH:water 35:22:43 (atenolol) or MeOH:water 60:40 (metaproterenol, sotalol) or MeCN:water 60:40 (metoprolol, propranolol, toliprolol)
Flow rate: 1.2 (atenolol) or 1 (others)
Injection volume: 10-25

Detector: F ex 200 (metoprolol, propranolol, toliprolol) or 232 (atenolol, metaproterenol, sotalol) em no emission filter

CHROMATOGRAM
Retention time: 19 ((-)-metoprolol), 20 ((+)-metoprolol), 24, 26 (toliprolol enantiomers), 32 ((-)-propranolol), 35 ((+)-propranolol), 19, 21 (metaproterenol enantiomers), 28 ((-)-sotalol), 31 ((+)-sotalol), 14, 16 (atenolol enantiomers)

KEY WORDS
chiral

REFERENCE
Mehvar, R. Stereospecific liquid chromatographic analysis of racemic adrenergic drugs utilizing precolumn derivatization with (-)-menthyl chloroformate, *J.Chromatogr.*, **1989**, *493*, 402–408.

SAMPLE
Matrix: solutions
Analyte: amlodipine
Sample preparation: 100 µL 0.1-1 mg/mL Amlodipine in MeCN:water 50:50 + 100 µL 1 M pH 6.8 sodium borate buffer, vortex, add 50 µL 100 mM (-)-(1R)-menthyl chloroformate in acetone, vortex, let stand for 5 min, add 1 mL water, add 2 mL dichloromethane, rotate for 10 min, centrifuge at 2000 rpm for 5 min. Remove the organic layer and evaporate it to dryness under a stream of nitrogen at 40°, reconstitute the residue in 200 µL MeCN, inject a 20 µL aliquot.

HPLC VARIABLES
Column: 100 × 10 Hypercarb-S porous graphitic carbon
Mobile phase: MeCN:dichloromethane:formic acid 20:50:30
Flow rate: 1.5
Injection volume: 20
Detector: UV 265

CHROMATOGRAM
Retention time: 6.5, 7.5 (enantiomers)

OTHER SUBSTANCES
Simultaneously analyzed: mexiletine, UK52.829
Interfering: propranolol

KEY WORDS
chiral

REFERENCE
Josefsson, M.; Carlsson, B.; Norlander, B. Fast chromatographic separation of (-)-menthyl chloroformate derivatives of some chiral drugs,with special reference to amlodipine, on porous graphitic carbon, *Chromatographia*, **1993**, *37*, 129–132.

SAMPLE
Matrix: urine
Analyte: metoprolol
Sample preparation: 2 mL Urine + 30 µL 1 mg/mL (±)-toliprolol in MeOH + 1 mL 2 M potassium carbonate + 1 g NaCl, extract twice with 5 mL ethyl acetate. Remove the organic layers and dry them over anhydrous sodium sulfate, evaporate to dryness under a stream of nitrogen at 50°, reconstitute the residue in 100 µL 0.4% triethylamine in MeCN:MeOH 50:50, add 100 µL 1% S-(-)-menthyl chloroformate in MeCN (prepare weekly in MeCN dried over anhydrous sodium sulfate), let stand at room temperature for 1 h, evaporate to dryness under a stream of nitrogen, reconstitute with 300 µL MeOH, inject a 10 µL aliquot.

HPLC VARIABLES
Guard column: 20 × 4 30 µm Hypersil HP ODS
Column: 250 × 4.6 Hypersil 5 C18
Mobile phase: Gradient. A was 13.8 g $NaH_2PO_4.H_2O$ and 1.59 g propylamine hydrochloride in 1 L water, pH adjusted to 3.2 with concentrated phosphoric acid. B was MeOH.

A:B from 25:75 to 15:85 over 15 min, maintain at 15:85 for 5 min, to 10:90 over 5 min, maintain at 10:90 for 3 min.
Flow rate: 1
Injection volume: 10
Detector: F ex 223 em 340 (cut-off filter)

CHROMATOGRAM
Retention time: 19.7 (-), 20.6 (+)
Internal standard: toliprolol (23.3 (-), 24.5 (+))
Limit of detection: 5 ng

OTHER SUBSTANCES
Extracted: metabolites

KEY WORDS
chiral; pharmacokinetics

REFERENCE
Li, F.; Cooper, S.F.; Côté, M. Determination of the enantiomers of metoprolol and its major acidic metabolite in human urine by high-performance liquid chromatography with fluorescence detection, *J.Chromatogr.B*, **1995**, *668*, 67–75.

7-Methoxycoumarin-3-carbonyl Fluoride

SAMPLE
Matrix: solutions
Analyte: amines
Sample preparation: Mix 80 μL of a 1.25 μM solution of amine in MeCN with 10 μL 10 mM triethylamine in MeCN, add 10 μL 1 mM 7-methoxycoumarin-3-carbonyl fluoride in MeCN, mix, let stand at room temperature for 30 s, inject a 10 μL aliquot. (Preparation of 7-methoxycoumarin-3-carbonyl fluoride is as follows. Add a solution of 1.9 g sodium cyanoacetate in 20 mL water to 1.5 g 2-hydroxy-4-methoxybenzaldehyde in 10 mL water and 10 mL 2 M NaOH, heat at 35° for 1 h, filter, dilute and acidify the filtrate. Collect the solid and wash it, dry to obtain trans-2-hydroxy-4-methoxybenzylidenecyanoacetic acid (mp 196°). Reflux 1.9 g trans-2-hydroxy-4-methoxybenzylidenecyanoacetic acid in 40 mL water for 5 min, filter, acidify and cool the filtrate. Recrystallize the precipitate from dilute EtOH to give 7-methoxycoumarin-3-carboxylic acid as faint greenish-yellow plates (mp 195°) (J. Chem. Soc. 1949, S12). Stir 242 mg 7-methoxycoumarin-3-carboxylic acid and 90 μL pyridine in 40 mL MeCN, add 60 μL cyanuric fluoride (Fluka), stir at room temperature for 2 h, evaporate to dryness under reduced pressure, recrystallize the residue from hexane to obtain 7-methoxycoumarin-3-carbonyl fluoride as pale-yellow needles (mp 158-160°).)

HPLC VARIABLES
Column: 250 × 4.6 5 μm TSK gel ODS-80TM (Tosoh)
Mobile phase: MeOH:water 100:30
Flow rate: 1
Injection volume: 10
Detector: F ex 350 em 405

CHROMATOGRAM
Retention time: 6 (propylamine), 8 (benzylamine), 10 (2-phenylethylamine), 14 (cyclohexylamine)
Limit of detection: 100 fmole

OTHER SUBSTANCES
Non-interfering: secondary amines

REFERENCE
Fujino, H.; Goya, S. 7-Methoxycoumarin-3-carbonyl fluoride as a fluorescent labeling reagent for primary amines in high-performance liquid chromatography, *Anal.Sci.*, **1990**, 6, 465–466.

6-Methoxy-2-methylsulfonylquinoline-4-carbonyl Chloride

SAMPLE
Matrix: solutions
Analyte: amines
Sample preparation: Mix a 200 μL aliquot of a solution in MeCN with 10 mg potassium carbonate and 200 μL 1 mM 6-methoxy-2-methylsulfonylquinoline-4-carbonyl chloride in MeCN, let stand at room temperature for 5 min. Remove a 50 μL aliquot and make it up to 1 mL with mobile phase, inject a 20 μL aliquot. (Preparation of 6-methoxy-2-methyl-sulfonylquinoline-4-carbonyl chloride is as follows. Dissolve 10 g p-anisidine and 15.6 g diethyl ketomalonate (ethyl mesoxalate) in 10 mL acetic acid, boil gently for 3 h, add 10 mL EtOH while warm, let stand in the refrigerator overnight, filter. (It may be necessary to use the hydrate of diethyl ketomalonate, diethyl dihydroxymalonate. Prepare this compound from equimolar amounts of diethyl ketomalonate and water, recrystallize from chloroform (mp 56-57°) (Org. Syn. 1932, I, 266).) Wash the solid with 15 mL EtOH and 15 mL ether, recrystallize several times from EtOH:water 75:25 with charcoal to yield ethyl 2,3-dihydro-3-hydroxy-5-methoxy-2-oxoindole-3-carboxylate as colorless crystals (mp 193-194°). Dissolve ethyl 2,3-dihydro-3-hydroxy-5-methoxy-2-oxoindole-3-carboxylate in 12 times as much 1 M KOH, heat on a water bath, filter. Acidify the filtrate with HCl, carbon dioxide is evolved. Reflux this weakly acid solution and add a small amount of a dilute solution of iron chloride, vigorously pull air through the mixture (?) until the solution turns deep red, cool, recrystallize the product from water to obtain 5-methoxyisatin as deep-brown-red needles (mp 201-202°). Gently heat 5-methoxyisatin with 2 equivalents of acetic anhydride for 30 min, while still warm add an equal volume of benzene (Caution! Benzene is a carcinogen!), pour into copious amounts of boiling benzene, extract the residue repeatedly with hot benzene. Treat the benzene solution with charcoal, filter, concentrate to a small volume, cool in ice, recrystallize from chloroform/benzene to obtain 1-acetyl-5-methoxyisatin as red crystals (mp 144-145°) (Ber. 1921, 54B, 3079; Chem. Abs. 1922, 16, 1771). Add 3 g 1-acetyl-5-methoxyisatin to a boiling mixture of 10 g caustic soda solution (d = 1.17) and 100 mL water, heat gently, cool, acidify with sulfuric acid, stir, filter. Wash the solid with acetone and ether, recrystallize from acetic acid to obtain 1,2-dihydro-6-methoxy-2-oxoquinoline-4-carboxylic acid (mp 326° d) (Ber. 1921, 54B, 3090;

Chem. Abs. 1922, 16, 1772). Reflux 1,2-dihydro-6-methoxy-2-oxoquinoline-4-carboxylic acid in phosphorus oxychloride to obtain 2-chloro-6-methoxyquinoline-4-carboxylic acid. Boil 2-chloro-6-methoxyquinoline-4-carboxylic acid with 1.2 equivalents thiourea in isopropanol for 15-30 min, dilute with water to obtain 1,2-dihydro-6-methoxy-2-thioxoquinoline-4-carboxylic acid (mp 271-273° d) (cf. Chem. Pharm. Bull. 1980, 28, 49). Dissolve 1 g 1,2-dihydro-6-methoxy-2-thioxoquinoline-4-carboxylic acid in 10 mL 1 M NaOH, add 2.3 g iodomethane, reflux for 1.5 h, cool, acidify with 10% HCl to obtain 6-methoxy-2-methylthioquinoline-4-carboxylic acid (mp 205-206°). React 6-methoxy-2-methylthioquinoline-4-carboxylic acid with 6.6% potassium permanganate solution (cf. Yakugaku Zasshi 1969, 89, 74) to obtain 6-methoxy-2-methylsulfonylquinoline-4-carboxylic acid (mp 215-216°) (Chem. Pharm. Bull. 1992, 40, 1322). Add 400 mg 6-methoxy-2-methylsulfonylquinoline-4-carboxylic acid to 5 mL thionyl chloride, reflux at 80° with stirring for 30 min, evaporate to dryness under a stream of nitrogen under reduced pressure, dissolve the residue in anhydrous benzene (Caution! Benzene is a carcinogen!), filter, add petroleum ether, collect the precipitate, recrystallize from benzene/petroleum ether to obtain 6-methoxy-2-methylsulfonylquinoline-4-carbonyl chloride as yellow needles (mp 175-176°) (Anal.Sci. 1992, 8, 355).)

HPLC VARIABLES

Column: 150 × 4.6 5 μm TSKgel ODS-80TM (Tosoh)
Mobile phase: MeCN:water 55:45 (A) or 40:60 (B)
Flow rate: 1
Injection volume: 20
Detector: F ex 342 em 448

CHROMATOGRAM

Retention time: 6.1 (pentylamine (A)), 6.8 (propylamine (B)), 8.7 (hexylamine (A)), 10.9 (butylamine (B)), 11.4 (dibutylamine (A)), 12.8 (heptylamine (A)), 13.7 (benzylamine (B)), 17 (cyclohexylamine (B)), 17.5 (dipropylamine (B)), 18 (2-phenylethylamine (B)), 19.4 (octylamine (A)), 21.4 (4-methylbenzylamine (B))
Limit of detection: 0.5-2 pmole (primary amines), 100 pmole (secondary amines)

OTHER SUBSTANCES

Non-interfering: alcohols

REFERENCE

Yoshida, T.; Moriyama, Y.; Nakamura, K.; Taniguchi, H. 6-Methoxy-2-methylsulfonylquinoline-4-carbonyl chloride as a fluorescence derivatization reagent for amines in liquid chromatography, *Analyst*, **1993**, *118*, 29–33.

(+)-4-(6-Methoxy-2-naphthyl)-2-butyl Chloroformate

SAMPLE
Matrix: blood
Analyte: metoprolol
Sample preparation: 1 mL Plasma + 50 μL 200 μg/mL (R)-(-)-flecainide in water + 1 mL
pH 11 borate buffer, mix, add 5 mL diisopropyl ether (Caution! Diisopropyl ether readily
forms explosive peroxides!), shake on a reciprocal shaker for 15 min, centrifuge at -10° at
4000 rpm, freeze in dry ice/acetone. Remove the organic layer and evaporate it to dryness
under reduced pressure, reconstitute the residue in 100 μL 10 mM HCl, adjust pH to 11
with 200 μL borate buffer, add 50 μL 100 μg/mL (+)-4-(6-methoxy-2-naphthyl)-2-butyl
chloroformate in MeCN, mix, let stand at room temperature for 1 h, add 3 mL dichloromethane, swirl mix for 30 s. Remove the organic layer and evaporate it to dryness under
reduced pressure, reconstitute the residue in 100 μL n-hexane:MeOH 100:6, inject a 20
μL aliquot. (Synthesis of (+)-4-(6-methoxy-2-naphthyl)-2-butyl chloroformate is as follows.
Slowly add 2 mmoles 4-(6-methoxy-2-naphthyl)-2-butanone (nabumetone) in 20 mL anhydrous diethyl ether to 0.55 mmoles lithium aluminum hydride in anhydrous diethyl
ether, when the reaction is complete (monitor by TLC) cautiously add water, add 6 M
HCl, remove the organic layer. Extract the aqueous layer 3 times with 10 mL portions of
diethyl ether. Combine the organic layers and dry them over anhydrous sodium sulfate,
evaporate to dryness to obtain racemic 4-(6-methoxy-2-naphthyl)-2-butanol. Dissolve 0.5
mmole 4-(6-methoxy-2-naphthyl)-2-butanol in 20 mL dry dichloromethane, add 0.55
mmole dry pyridine, stir, add 1 mmole (1S)-(-)-camphanoyl chloride ((1S)-(-)-camphanic
chloride) in small portions, stir at room temperature overnight, wash 3 times with 1 M
HCl, wash with water, wash with saturated sodium bicarbonate solution, wash with water
until neutral. Dry the organic layer over anhydrous magnesium sulfate, evaporate to
dryness to obtain the camphanoyl ester. Purify by preparative HPLC (250 × 50 (R)-
DNBPG (J.T. Baker), n-hexane:isopropanol 70:30, 65 mL/min, UV 220) to obtain diastereomerically pure camphanoyl esters. Add 5 mmole diastereomerically pure camphanoyl
ester dissolved in 30 mL dry diethyl ether dropwise under argon to 2.5 mmoles lithium
aluminum hydride stirred in diethyl ether at 0°, stir at room temperature for 1 h, stir in
an ice bath, cautiously add 10% aqueous ammonium chloride, remove the organic layer,
extract the aqueous layer three times with diethyl ether. Combine the organic layers and
dry them over anhydrous magnesium sulfate, evaporate to dryness. Dissolve 1 g in 10
mL dichloromethane, purify by flash chromatography on a 500 × 40 column filled with
200 g 40-63 μm silica gel 60 using dichloromethane:MeOH 99.5:0.5 at 45 mL/min to
obtain enantiomerically pure 4-(6-methoxy-2-naphthyl)-2-butanol. Dissolve 0.5 mmole
(+)-4-(6-methoxy-2-naphthyl)-2-butanol and 0.5 mmole triethylamine in 10 mL dry tolu-

ene, cool to 0°, stir, add 1 mL 20% phosgene in dry toluene, stir for 4 h, filter, evaporate under reduced pressure to obtain (+)-4-(6-methoxy-2-naphthyl)-2-butyl chloroformate as an oil.)

HPLC VARIABLES
Guard column: 20 × 4 6 μm Ultrasep ES 100 (Bischoff, Germany)
Column: 125 × 3 3 μm Nucleosil 120
Mobile phase: n-Hexane:MeOH 100:0.4
Flow rate: 1
Injection volume: 20
Detector: UV 230; F ex 270 em 350

CHROMATOGRAM
Retention time: 24.5 ((S)-(-)), 26.5 ((R)-(+))
Internal standard: (R)-(-)-flecainide (37.5)
Limit of quantitation: 2 ng/mL
Limit of detection: 0.9 ng/mL

OTHER SUBSTANCES
Also analyzed: acebutolol, alprenolol, flecainide, mexiletine, propafenone, propranolol, tocainide

KEY WORDS
plasma; chiral; pharmacokinetics; normal phase; protect from light

REFERENCE
Büschges, R.; Devant, R.; Mutschler, E.; Spahn-Langguth, H. 4-(6-Methoxy-2-naphthyl)-2-butyl chloroformate enantiomers: New reagents for the enantiospecific analysis of amino compounds in biogenic matrices, *J.Pharm.Biomed.Anal.*, **1996**, *15*, 201–220.

2-(6-Methoxy-2-naphthyl)-1-propyl Chloroformate

SAMPLE
Matrix: solutions
Analyte: drugs
Sample preparation: Mix 20 μL of a 1 mM solution in MeOH or water with 50 μL pH 8 borate buffer and 50 μL 18 mM 2-(6-methoxy-2-naphthyl)-1-propyl chloroformate in acetone, vortex, let stand at room temperature for 30 min, add 100 μL 10 mM trans-4-hydroxy-L-proline in water, mix, let stand for 2 min, add 2 mL dichloromethane, vortex for 30 s. Remove the organic layer and evaporate it to dryness under reduced pressure, reconstitute the residue in 100 μL mobile phase, inject an aliquot. Prepare 2-(6-methoxy-2-naphthyl)-1-propyl chloroformate as follows. Stir 1.5 mmoles lithium aluminum hydride in THF, slowly add 2 mmoles (S)-naproxen in 20 mL anhydrous THF, reflux for 1 h,

evaporate most of the solvent, cautiously add water with stirring, acidify with 6 N HCl, extract three times with diethyl ether. Combine the organic layers and dry them over anhydrous sodium sulfate, evaporate to dryness, chromatograph on silica gel with dichloromethane:MeOH 100:2 (flash chromatography), evaporate eluate to dryness, dry under vacuum over KOH to give 2-(6-methoxy-2-naphthyl)propanol as a white solid (mp 92-3°). Stir 0.5 mmoles 2-(6-methoxy-2-naphthyl)propanol and 0.5 mmoles triethylamine in 10 mL dry toluene at 0°, add 1 mL 20% phosgene in toluene (Caution! Phosgene is highly toxic, perform reaction in a chemical fume hood!) (Fluka), stir for 4 h, filter, evaporate to dryness under reduced pressure, dry under vacuum to give 2-(6-methoxy-2-naphthyl)-1-propyl chloroformate (mp 60°). Store under vacuum over phosphorus pentoxide at room temperature.)

HPLC VARIABLES
Column: 250 × 4 5 μm Zorbax-SIL
Mobile phase: n-Hexane:isopropanol 100:1 (A), 100:1.5 (B), 100:5 (C), or 100:0.25 (D)
Flow rate: 1.5
Injection volume: 100
Detector: UV 230; F ex 270 em 365

CHROMATOGRAM
Retention time: k' 4.6 (S-(-)-alprenolol, (A)), k' 5.1 (R-(+)-alprenolol, (A)); k' 7.4 (S-(-)-propranolol, (A)), k' 8.1 (R-(+)-propranolol, (A)); k' 12.6 (S-(-)-metoprolol, (B)), k' 13.5 (R-(+)-metoprolol, (B)); k' 20.4, 21.9 (acebutolol (enantiomers), (C)); k' 15.7 (S-(+)-tocainide, (B)), k' 16.9 (R-(-)-tocainide, (B)); k' 14.4 (R-(-)-flecainide, (B)), k' 15.2 (S-(+)-flecainide, (B)); k' 22.5 (S-(+)-mexiletine, (D)), k' 23.3 (R-(-)-mexiletine, (D)); k' 14.5 (S-(+)-propafenone, (B)), k' 15.4 (R-(-)-propafenone, (B))
Limit of detection: 0.2 ng/mL (metoprolol)

KEY WORDS
chiral; normal phase

REFERENCE
Büschges, R.; Linde, H.; Mutschler, E.; Spahn-Langguth, H. Chloroformates and isothiocyanates derived from 2-arylpropionic acids as chiral reagents: synthetic routes and chromatographic behaviour of the derivatives, *J.Chromatogr.A*, **1996**, *725*, 323–334.

(S)-(+)-α-Methoxyphenylacetyl Chloride

SAMPLE
Matrix: bulk
Analyte: phenylethylamine
Sample preparation: Dissolve 50 μL α-phenylethylamine in dioxane containing 70 μL triethylamine (Caution! Dioxane is a carcinogen!), add a 20% excess of (S)-(+)-α-methoxyphenylacetyl chloride. Prepare an ether/benzene solution of the reaction mixture and wash it with 1 M mineral acid and 1 M base (Caution! Benzene is a carcinogen!), evaporate to dryness, prepare a 10 mg/mL solution in mobile phase, inject a 10 μL aliquot. (Prepare (S)-(+)-α-methoxyphenylacetyl chloride by refluxing (S)-(+)-α-methoxyphenylacetic acid in thionyl chloride for 5 min.)

HPLC VARIABLES
Column: 200 × 3 5 μm Merckosorb SI 60 silica
Mobile phase: Isooctane:ethyl acetate 80:20
Flow rate: 3

Injection volume: 10
Detector: UV 255

CHROMATOGRAM
Retention time: 4.5 (S), 7.5 (R)
Limit of detection: 0.2% (minor enantiomer in presence of major enantiomer)

KEY WORDS
chiral; normal phase

REFERENCE
Helmchen, G.; Strubert, W. Determination of optical purity by high performance liquid chromatography, *Chromatographia*, **1974**, 7, 713–715.

(R)-(-)-α-Methoxyphenylacetyl Chloride

SAMPLE
Matrix: solutions
Analyte: ethambutol
Sample preparation: Mix a 2 mL aliquot of a 25 nM (?) solution of ethambutol in benzene (Caution! Benzene is a carcinogen!) with 5 mL 400 mM (R)-(-)-α-methoxyphenylacetyl chloride in benzene and 300 μL pyridine, stir vigorously at 40° for 30 min, pour into 5 mL ether and 5 mL water. Remove the ether layer and wash it with 5 mL 5% HCl, dry over anhydrous magnesium sulfate, filter, evaporate to dryness under reduced pressure, reconstitute with 500 μL chloroform, inject a 5 μL aliquot. (Synthesis of (R)-(-)-α-methoxyphenylacetyl chloride is as follows. Reflux (R)-(-)-α-methoxyphenylacetic acid with a 5 to 10-fold excess of thionyl chloride on a steam bath for 5 min, add benzene, evaporate under reduced pressure, repeat this procedure 3 or 4 times to remove excess thionyl chloride, distil at 70-75°/0.2 mm Hg to give (R)-(-)-α-methoxyphenylacetyl chloride (J. Org. Chem. 1968, 33, 1142).)

HPLC VARIABLES
Column: 250 × 4 5 μm LiChrosorb Si 60
Mobile phase: Chloroform:ethyl acetate 90:10
Flow rate: 0.7
Injection volume: 5
Detector: UV 254

CHROMATOGRAM
Retention time: 9 (-), 9.5 (+), 10.2 (meso)

OTHER SUBSTANCES
Simultaneously analyzed: 2-amino-1-butanol

KEY WORDS
normal phase; chiral

REFERENCE
Gamberini, G.; Ferioli, V. Determination of optical purity by high performance liquid chromatography of compounds of pharmaceutical interest, *Farmaco.[Prat].*, **1988**, 43, 357–363.

α-Methoxy-α-(trifluoromethyl)phenylacetyl Chloride

SAMPLE
Matrix: solutions
Analyte: 1-phenyl-2-aminopropanes
Sample preparation: 200 μL 0.5 mg/mL Compound in dichloromethane + 100 μL 500 mM (-)-α-methoxy-α-(trifluoromethyl)phenylacetyl chloride in dichloromethane + 50 μL pyridine, mix, heat at 70° for 30 min, cool in ice water, add 1 mL 1 M HCl, shake mechanically for 5 min, centrifuge at 1000 g for 15 min, discard the aqueous phase, add 500 μL 15% sodium carbonate, add 200 μL dichloromethane, shake for 5 min, centrifuge at 1000 g for 15 min. Remove a 40 μL aliquot of the organic layer and dilute it to 1 mL with MeOH, inject a 20 μL aliquot. ((-)-α-Methoxy-α-(trifluoromethyl)phenylacetyl chloride is available from Lancaster Synthesis, Windham NH or Tokyo Kasei (TCI America, Portland OR). Synthesis is as follows. Reflux 1 g (-)-α-methoxy-α-(trifluoromethyl)phenylacetic acid and 3 mL freshly distilled thionyl chloride for 50 h, remove excess thionyl chloride by evaporation under reduced pressure, dissolve the resulting oil in dichloromethane to give a 500 mM solution, store in the refrigerator (J. Pharm. Sci. 1977, 66, 169).)

HPLC VARIABLES
Column: 250 × 4.5 5 μm octadecyl (IBM)
Mobile phase: Gradient. MeOH:water 60:40 for 20 min, to 100:0 over 20 min.
Flow rate: 1-2
Injection volume: 20
Detector: UV 254

CHROMATOGRAM
Retention time: 9.7 ((R)-amphetamine), 10.4 ((S)-amphetamine), 13.9 ((R)-1-(4-chlorophenyl)-2-aminopropane), 14.9 ((S)-1-(4-chlorophenyl)-2-aminopropane), 22.1 ((R)-1-phenylethylamine), 22.8 ((S)-1-phenylethylamine), 33.0 ((R)-1-(2,5-dimethoxy-4-methylphenyl)-2-aminopropane), 33.5 ((S)-1-(2,5-dimethoxy-4-methylphenyl)-2-aminopropane), 32.5 ((R)-1-(2,4-dimethoxy-5-methylphenyl)-2-aminopropane), 38.9 ((S)-1-(2,4-dimethoxy-5-methylphenyl)-2-aminopropane), 30.8 ((R)-1-(2,5-dimethoxy-4-thiomethylphenyl)-2-aminopropane), 31.4 ((S)-1-(2,5-dimethoxy-4-thiomethylphenyl)-2-aminopropane)
Limit of quantitation: 100 ng

KEY WORDS
comparison with other derivatizing reagents; chiral

REFERENCE
Miller, K.J.; Gal, J.; Ames, M.M. High-performance liquid chromatographic resolution of enantiomers of 1-phenyl-2-aminopropanes (amphetamines) with four chiral reagents, *J.Chromatogr.*, **1984**, *307*, 335–342.

2-Naphthoxycarbonyl Chloride

SAMPLE
Matrix: food
Analyte: amines
Sample preparation: Dilute vinegar, wine, or juice 10-fold with water, centrifuge at 2000 g for 15 min. Remove a 20 μL aliquot and mix it with 150 μL buffer and 400 μL 5 mM 2-naphthoxycarbonyl chloride in MeCN, let stand for 3 min, add 50 μL 20 mM glycine in water, let stand for 3 min, add 500 μL MeCN:500 mM pH 4.4 sodium acetate buffer 75:25, mix, inject a 20 μL aliquot. (Prepare buffer by adjusting the pH of 33.43 g boric acid in 950 mL water to 9.0 with 20% KOH, make up to 1 L with water. Synthesis of 2-naphthoxycarbonyl chloride is as follows. Dissolve 30 mmoles 2-naphthol and 30 moles quinoline in 18 g toluene and 5 g dichloromethane, cool to 0°, add 47 mL 1.93 M phosgene in toluene, warm on a steam bath for 10 min, filter, evaporate to remove the solvent, distil the residue (bp 150-152°/9 mm Hg), take up the distillate in ether, crystallize by adding ligroin to obtain 2-naphthoxycarbonyl chloride (mp 66°) (J. Am. Chem. Soc. 1951, 73, 2080).)

HPLC VARIABLES
Guard column: present but not specified
Column: 250 × 4.6 4 μm Superspher 60 RP-18e
Mobile phase: Gradient. A was MeCN:100 mM pH 4.4 sodium acetate 40:60. B was MeCN. A:B 77:23 for 7 min, to 65:35 over 31 min, to 30:70 over 3 min, maintain at 30:70 for 5 min, to 0:100 over 1 min, maintain at 0:100 for 6 min, return to initial conditions over 1 min, re-equilibrate for 6 min.
Column temperature: 45
Flow rate: 1.1 for 46 min, to 1.5 over 1 min, maintain at 1.5 for 13 min
Injection volume: 20
Detector: F ex 274 em 335

CHROMATOGRAM
Retention time: 13.66 (2-phenylethylamine), 15.53 (putrescine), 17.15 (histamine), 19.38 (cadaverine), 37.53 (tyramine), 42.92 (spermidine), 47.69 (spermine)
Limit of detection: 49-747 ng/g

KEY WORDS
vinegar; wine; juice

REFERENCE
Kirschbaum, J.; Busch, I.; Brückner, H. Determination of biogenic amines in food by automated pre-column derivatization with 2-naphthyloxycarbonyl chloride (NOC-Cl), *Chromatographia*, **1997**, *45*, 263–268.

2-Naphthoyl Chloride

SAMPLE
Matrix: bulk, formulations
Analyte: amphetamine
Sample preparation: Bulk. Dissolve 10 mg bulk drug in 5 mL dichloromethane, add 5 mL 20% NaOH, add 10 mL 10 mM 2-naphthoyl chloride in dichloromethane, shake for 1 min. Remove the organic phase and wash the aqueous phase with 5 mL dichloromethane. Combine the organic layers and wash them with 5 mL 10 mM sulfuric acid. Filter the organic layer through a syringe containing a glass wool plug and anhydrous sodium sulfate, inject a 10 μL aliquot. Capsules. Sonicate 1 capsule in 5 mL 20% NaOH for 45 min or until dissolved, filter, add 5 mL dichloromethane to filtrate, add 10 mL 10 mM 2-naphthoyl chloride in dichloromethane, shake for 1 min. Remove the organic phase and wash the aqueous phase with 5 mL dichloromethane. Combine the organic layers and wash them with 5 mL 10 mM sulfuric acid. Filter the organic layer through a syringe containing a glass wool plug and anhydrous sodium sulfate, inject a 10 μL aliquot.

HPLC VARIABLES
Column: Bakerbond Chiral Phase (DNBPG)
Mobile phase: Hexane:isopropanol:MeCN 97:3:0.5
Column temperature: 20
Flow rate: 2
Injection volume: 10
Detector: UV 254

CHROMATOGRAM
Retention time: 28.5 (l), 31 (d)

KEY WORDS
capsules; chiral

REFERENCE
Alembik, M.C.; Wainer, I.W. Resolution and analysis of enantiomers of amphetamines by liquid chromatography on a chiral stationary phase: collaborative study, *J.Assoc.Off.Anal.Chem.*, **1988**, *71*, 530–533.

SAMPLE
Matrix: blood
Analyte: mexiletine
Sample preparation: 500 μL Plasma + 50 μL 300 mM NaOH, mix, add 4 mL diisopropyl ether (Caution! Diisopropyl ether readily forms explosive peroxides!), vortex for 2 min, centrifuge at 1800 g for 10 min, repeat the extraction. Combine the organic layers and add them to 50 μL 100 mM HCl in MeOH, evaporate to dryness under a stream of nitrogen, reconstitute with 100 μL aqueous 100 mM HCl, add 100 μL 2 M NaOH, add 200 μL water, vortex for 15 s, let stand for 15 min, add 100 μL 4 mg/mL 2-naphthoyl chloride in dichloromethane, vortex for 2 min, add 2 mL hexane:isopropanol 90:10, vortex for 2 min, centrifuge at 1800 g for 5 min. Remove the organic layer and evaporate it to dryness under a stream of nitrogen, reconstitute the residue in 200 μL hexane:isopropanol 90:10, inject an aliquot. (The 300 mM NaOH, 2 M NaOH, and aqueous 100 mM HCl solutions should be washed with diisopropyl ether before use.)

HPLC VARIABLES
Guard column: 50 mm long Chiralcel OJ
Column: 250 × 4.6 10 μm Chiralcel OJ

Mobile phase: Hexane:EtOH 71:29
Flow rate: 1
Injection volume: 10
Detector: F ex 230 em 340

CHROMATOGRAM
Retention time: 7.2 (R-(-)), 8.9 (S-(+))
Limit of quantitation: 50 ng/mL

OTHER SUBSTANCES
Extracted: metabolites
Non-interfering: amiodarone, captopril, clobazam, clonazepam, disopyramide, ergotamine, nitrazepam, nordisopyramide, procainamide, propafenone, sotalol

KEY WORDS
plasma; chiral

REFERENCE
Lanchote, V.L.; Bonato, P.S.; Dreossi, S.A.C.; Gonçalves, P.V.B.; Cesarino, E.J.; Bertucci, C. High-performance liquid chromatographic determination of mexiletine enantiomers in plasma using direct and indirect enantioselective separations, *J.Chromatogr.B*, **1996**, *685*, 281–289.

SAMPLE
Matrix: bulk, formulations
Analyte: amphetamine
Sample preparation: 10 mg Bulk drug, tablet, or capsule or 10 mL 1 mg/mL syrup + 5 mL 20% NaOH, sonicate until dissolved, add 10 mL 10 mM 2 naphthoyl chloride in dichloromethane, shake for 1 min. Remove the organic phase and extract the aqueous phase with 5 mL dichloromethane. Combine the organic layers and wash them with two 5 mL portions of 10 mM sulfuric acid, filter through a cotton plug, inject an aliquot.

HPLC VARIABLES
Guard column: 50 × 4.6 35-50 μm silica
Column: 250 × 4.6 Pirkle Covalent Phenylglycine (Regis)
Mobile phase: Hexane:isopropanol:MeCN 97:3:0.5
Column temperature: 20
Flow rate: 2
Injection volume: 20
Detector: UV 280

CHROMATOGRAM
Retention time: k' 20 (-), k' 22 (+)
Limit of detection: 1% (of major enantiomer)

KEY WORDS
tablets; capsules; syrup; chiral

REFERENCE
Wainer, I.W.; Doyle, T.D.; Adams, W.M. Liquid chromatographic chiral stationary phases in pharmaceutical analysis: determination of trace amounts of the (-)-enantiomer in (+)-amphetamine, *J.Pharm.Sci.*, **1984**, *73*, 1162–1164.

N-1-(2-Naphthylsulfonyl)-2-pyrrolidinecarbonyl Chloride

SAMPLE
Matrix: blood, urine
Analyte: temafloxacin
Sample preparation: 1 mL Serum or 500 μL urine + 500 μL 1 M ammonium acetate + 5 mL dichloromethane, shake vigorously for 10 min, centrifuge at 1500 g for 5 min. Remove the organic layer and evaporate it to dryness under a stream of nitrogen at 40°, reconstitute the residue in 500 μL MeOH:thionyl chloride 90:10, heat at 60° for 30 min, evaporate to dryness under a stream of nitrogen at 40°, add 500 μL 500 mM sulfuric acid, add 1 mL hexane, shake vigorously for 10 min, centrifuge for 5 min. Remove the aqueous layer and add it to 1 mL 2 M sodium carbonate and 3 mL ether, shake vigorously for 10 min, centrifuge for 5 min. Remove the organic layer and evaporate it to dryness under a stream of nitrogen at 40°, reconstitute the residue in 20 μL 2 mg/mL (S)-(-)-N-1-(2-naphthylsulfonyl)-2-pyrrolidinecarbonyl chloride in dichloromethane, add 5 μL triethylamine, let stand at room temperature for 10 min. Evaporate to dryness under a stream of nitrogen, reconstitute the residue in 500 μL dichloromethane, inject a 50 μL aliquot. (Prepare (S)-(-)-N-1-(2-naphthylsulfonyl)-2-pyrrolidinecarbonyl chloride as follows. Slowly add a solution of 24.5 g 2-naphthalenesulfonyl chloride in 186 mL diethyl ether to a solution of 10.35 g L-proline and 37.25 g potassium carbonate in 216 mL water stirred at 0°, stir for 2 days. Remove the aqueous layer and wash it with diethyl ether, acidify to pH 2 with 10% HCl, extract with 500 mL ethyl acetate. Wash the organic layer with saturated aqueous NaCl and dry it over anhydrous sodium sulfate, evaporate to dryness, recrystallize from benzene to give (S)-(-)-N-1-(2-naphthylsulfonyl)-2-pyrrolidinecarboxylic acid (mp 133-135°). Add 17.5 mL oxalyl chloride to a solution of 15.6 g (S)-(-)-N-1-(2-naphthylsulfonyl)-2-pyrrolidinecarboxylic acid in benzene, stir at 40-50° overnight, evaporate to dryness under reduced pressure, recrystallize from benzene/hexane to give (S)-(-)-N-1-(2-naphthylsulfonyl)-2-pyrrolidinecarbonyl chloride (mp 107-109°; $[\alpha]_D^{20}$ -81.6° (c = 1, chloroform)) (J. Chromatogr. 1986, 357, 119).)

HPLC VARIABLES
Column: 150 × 4.6 Zorbax SIL
Mobile phase: Hexane:methyl acetate:MeOH:aqueous ammonia 150:100:10:1
Flow rate: 0.8
Injection volume: 50
Detector: UV 280

CHROMATOGRAM
Retention time: 18.5 (S), 19.5 (R)
Limit of detection: 5 ng/mL

KEY WORDS
serum; chiral; pharmacokinetics; normal phase

REFERENCE
Matsuoka, M.; Banno, K.; Sato, T. Analytical chiral separation of a new quinolone compound in biological fluids by high-performance liquid chromatography, *J.Chromatogr.B*, **1996**, *676*, 117–124.

Naproxen Chloride

RELATED REFERENCE
Spahn, H.; Krauss, D.; Mutschler, E. Enantiospecific high-performance liquid chromatographic (HPLC) determination of baclofen and its fluoro analog in biological material. *Pharm.Res.* **1988**, *5*, 107–112.

SAMPLE
Matrix: blood
Analyte: tocainide
Sample preparation: 1 mL Plasma + 1 mL pH 11 sodium borate buffer + 2.5 mL diisopropyl ether:EtOH 100:1.5 (Caution! Diisopropyl ether readily forms explosive peroxides!), shake for 30 min, centrifuge, repeat the extraction. Combine the organic layers and evaporate them to dryness, reconstitute the residue in 100 μL toluene, evaporate to dryness, add 100 μL 400 μg/mL S-naproxen chloride in anhydrous dichloromethane, heat at 50° for 1 h, evaporate to dryness, reconstitute with mobile phase, inject an aliquot (cf Arch. Pharm. (Weinheim) 1990, 323, 465). (Synthesis of S-naproxen chloride is as follows. Protect all compounds from light. Dissolve 500 mg naproxen in 50 mL dry toluene, slowly add 5 mL thionyl chloride (freshly distilled from linseed oil), reflux for 1 h, evaporate to dryness under reduced pressure, dry over KOH under vacuum overnight to obtain S-naproxen chloride (mp 96°).)

HPLC VARIABLES
Column: 250 × 4.6 5 μm Zorbax Sil
Mobile phase: Cyclohexane:dichloromethane:THF 50:20:20
Detector: F ex 313 em 365

CHROMATOGRAM
Retention time: k' 4.0 (R), k' 5.2 (S)

KEY WORDS
protect from light; chiral; plasma; normal phase

REFERENCE
Spahn, H. S-(+)-Naproxen chloride as acylating agent for separating the enantiomers of chiral amines and alcohols, *Arch.Pharm.(Weinheim)*, **1988**, *321*, 847–850.

SAMPLE
Matrix: blood, urine
Analyte: carvedilol
Sample preparation: Plasma. 1 mL Plasma + 1 mL pH 9.8 buffer + 500 mg NaCl + 5 mL diisopropyl ether (Caution! Diisopropyl ether readily forms explosive peroxides!), shake for 30 min, centrifuge at 4000 g for 10 min. Remove a 4 mL aliquot of the organic layer and evaporate it to dryness, reconstitute the residue in 100 μL toluene, evaporate to dryness, add 100 μL 400 μg/mL S-naproxen chloride in anhydrous dichloromethane, heat at 50° for 1 h, evaporate to dryness, reconstitute with 120 μL mobile phase, inject

a 100 µL aliquot. Urine. 1 mL Urine (or plasma) + 1 mL pH 5 sodium citrate buffer + 10 µL (5.500 Fishman units) β-glucuronidase, heat at 37° for 4 h, add 200 µL 100 mM NaOH, add 1 mL buffer, add 500 mg NaCl, add 5 mL diisopropyl ether (Caution! Diisopropyl ether readily forms explosive peroxides!), shake for 30 min, centrifuge at 4000 g for 10 min. Remove a 4 mL aliquot of the organic layer and evaporate it to dryness, reconstitute the residue in 100 µL toluene, evaporate to dryness, add 100 µL 400 µg/mL S-naproxen chloride in anhydrous dichloromethane, heat at 50° for 1 h, evaporate to dryness, reconstitute with 120 µL mobile phase, inject a 100 µL aliquot. (Prepare buffer by mixing 42 mL 100 mM sodium carbonate with 58 mL 100 mM sodium bicarbonate, pH 9.8. Synthesis of S-naproxen chloride is described above.)

HPLC VARIABLES
Column: 250 × 4.6 5 µm Zorbax Sil
Mobile phase: n-Hexane:dichloromethane:EtOH 112:36:1.7
Flow rate: 1.5
Injection volume: 100
Detector: F ex 285 em 355

CHROMATOGRAM
Retention time: k' 13.3 (S-(-)), k' 15.7 (R-(+))
Limit of detection: 1 ng/mL

KEY WORDS
plasma; chiral; normal phase; pharmacokinetics

REFERENCE
Spahn, H.; Henke, W.; Langguth, P.; Schloos, J.; Mutschler, E. Measurement of carvedilol enantiomers in human plasma and urine using S-naproxen chloride for chiral derivatization, *Arch.Pharm.(Weinhein)*, **1990**, *323*, 465–469.

4-Nitrobenzoyl Chloride

SAMPLE
Matrix: solutions
Analyte: amines
Sample preparation: Mix 1 mL 20-300 µg/mL amine solution in water with 2 mL 50 mg/mL 4-nitrobenzoyl chloride in THF (freshly prepared) and 1 mL 1 M NaOH, heat at 65° for 1 h, cool, adjust pH to 12 with 1 M NaOH, extract with two 10 mL portions of chloroform. Combine the extracts and wash them with two 20 mL portions of 10% potassium carbonate, wash with water, dry over anhydrous magnesium sulfate. Evaporate to dryness under a stream of air, reconstitute the residue in MeOH, inject a 5 µL aliquot.

HPLC VARIABLES
Column: 300 × 3.9 µBondapak C18
Mobile phase: MeCN:water 35:65
Flow rate: 1.5
Injection volume: 5
Detector: UV 254

CHROMATOGRAM
Retention time: 9 (benzylamine), 11 (α-methylbenzylamine), 13 (amphetamine), 15 (methamphetamine), 33 (n-propylamphetamine)

REFERENCE

Clark, R.C.; Teague, J.D.; Wells, M.M.; Ellis, J.H. Gas and high-pressure liquid chromatographic properties of some 4-nitrobenzamides of amphetamines and related arylalkylamines, *Anal.Chem.*, **1977**, *49*, 912–915.

trans-4-Nitrocinnamoyl Chloride

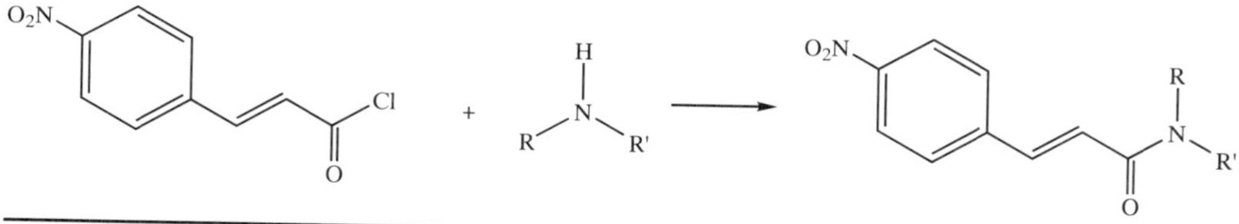

SAMPLE

Matrix: blood

Analyte: perhexiline

Sample preparation: 400 μL Plasma + 400 μL 1 M diammonium hydrogen phosphate + 50 μL 2.5 μg/mL di-n-hexylamine in water + 100 μL 1 mg/mL diisopropylethylamine + 5 mL isopentane:dichloromethane 60:40, vortex for 1.5 min, centrifuge at 1500 g for 5 min, freeze in dry ice/acetone. Remove the organic layer and evaporate it to dryness under reduced pressure at room temperature, reconstitute the residue in 100 μL 10 mg/mL trans-4-nitrocinnamoyl chloride in dry MeCN, vortex for 15 s, let stand at room temperature for 30 min, add 100 μL 25 mM sodium carbonate, vortex for 15 s, let stand at room temperature for 5 min, add 100 μL MeCN:50 mM ammonium acetate 50:50, vortex for 30 s, centrifuge at 1500 g for 5 min, inject a 130 μL aliquot of the supernatant.

HPLC VARIABLES

Column: 250 × 4.6 5 μm Spherisorb phenyl

Mobile phase: MeCN:water:glacial acetic acid 46:53.5:0.5

Flow rate: 2

Injection volume: 130

Detector: UV 340

CHROMATOGRAM

Retention time: 18.4

Internal standard: di-n-hexylamine (10.2)

Limit of detection: 30 ng/mL

OTHER SUBSTANCES

Extracted: metabolites

Simultaneously analyzed: dealkyldisopyramide, desethylamiodarone, desipramine, flecainide, mexiletine, nortriptyline, protriptyline, sotalol

KEY WORDS

plasma

REFERENCE

Grgurinovich, N. Method for the analysis of perhexiline and its hydroxy metabolite in plasma using high-performance liquid chromatography with precolumn derivatization, *J.Chromatogr.B*, **1997**, *696*, 75–80.

1-[(4-Nitrophenyl)sulfonyl]-L-prolyl Chloride

SAMPLE
Matrix: blood, urine
Analyte: flecainide
Sample preparation: Dilute urine 1:100. 1 mL Plasma or diluted urine + 100 μL 4 μg/mL IS in water + 1 mL water + 1 mL 1 M NaOH + 8 mL distilled diethyl ether, shake for 15 min, centrifuge at 1000 g for 5 min. Remove the organic layer and evaporate it to dryness under a stream of nitrogen, reconstitute the residue in 1 mL freshly prepared 49 mM 1-[(4-nitrophenyl)sulfonyl]-L-prolyl chloride in ethyl acetate, add 100 μL 0.016% triethylamine in ethyl acetate, vortex for 5 s, heat at 80° for 2 h, cool, wash with 3 mL 600 mM HCl, centrifuge for 5 min. Remove the organic layer and evaporate it to dryness under a stream of nitrogen, reconstitute the residue in 200 μL mobile phase, inject a 40-180 μL aliquot. (Synthesis of 1-[(4-nitrophenyl)sulfonyl]-L-prolyl chloride is as follows. Stir 40-45 mmoles L-(-)-proline in 40 mL THF and 200 mL 10% potassium carbonate, add 37-43 mmoles 4-nitrophenylsulfonyl chloride in 40 mL THF dropwise, heat at 50° for 3 h and maintain at pH 8 or above, cool, acidify to pH 2 and extract into chloroform. Extract with 10% potassium carbonate, acidify the aqueous layer, extract into chloroform. Dry the chloroform extract, evaporate, recrystallize from petroleum ether and benzene (Caution! Benzene is a carcinogen!). Stir 15 mmoles of the 4-nitrophenylsulfonylproline in 100 mL benzene under reflux condenser fitted with a calcium sulfate drying tube, add dropwise a five-fold molar excess of thionyl chloride in 50 mL benzene, heat at 35-40° until formation of the acid chloride is complete (about 48 h, monitor by IR spectroscopy), evaporate, recrystallize product from HPLC-grade heptane (mp 110-110.5°).)

HPLC VARIABLES
Guard column: 50 mm long octadecyl Pellicular ODS (Whatman)
Column: 300 × 3.9 Bondapak C18
Mobile phase: MeCN:water:triethylamine 45:55:0.2
Flow rate: 1
Injection volume: 40-180
Detector: UV 280

CHROMATOGRAM
Retention time: 29.8 (R), 31.9 (S)
Internal standard: (R,S)-N-(2-piperidylmethyl)-2,3-bis-(2,2,2-trifluoroethoxy)benzamide (25.5 (R), 27.9 (S))
Limit of quantitation: 50 ng/mL

KEY WORDS
plasma; pharmacokinetics; chiral

REFERENCE
Alessi-Severini, S.; Jamali, F.; Pasutto, F.M.; Coutts, R.T.; Gulamhusein, S. High-performance liquid chromatographic determination of the enantiomers of flecainide in human plasma and urine, *J.Pharm.Sci.*, **1990**, *79*, 257–260.

SAMPLE

Matrix: solutions
Analyte: amines
Sample preparation: Mix 40 mL of a 325-375 mM amine solution in THF with 150 mL 10% potassium carbonate, add dropwise 40 mL 275-325 mM reagent in THF, heat at 50° while maintaining at pH 8 or above for 3 h, cool, extract with chloroform. Evaporate the extracts to dryness, reconstitute, inject an aliquot. (Prepare reagent (1-[(4-nitro-phenyl)sulfonyl]-L-prolyl chloride) as described above.)

HPLC VARIABLES

Column: 150 × 4.6 5 μm Zorbax ODS
Mobile phase: MeOH:water 60:40
Flow rate: 1.5
Detector: UV 254

CHROMATOGRAM

Retention time: 5, 6 (ephedrine enantiomers), 7, 8.5 (pseudoephedrine enantiomers)

KEY WORDS

chiral

REFERENCE

Clark, C.R.; Barksdale, J.M. Synthesis and liquid chromatographic evaluation of some chiral derivatizing agents for resolution of amine enantiomers, *Anal.Chem.*, **1984**, *56*, 958–962.

SAMPLE

Matrix: solutions
Analyte: amphetamine
Sample preparation: 2 mL THF + 1 mL 33.5 mM reagent in THF (freshly prepared) + 1 mL 1 mg/mL amphetamine in water + 700 μL 10% sodium bicarbonate in water, heat at 65° for 1 h, cool, extract three times with 10 mL aliquots of chloroform. Combine the extracts and wash them with 10 mL water, dry over anhydrous magnesium sulfate, evaporate to dryness, reconstitute with 2.5 mL mobile phase, inject a 5 μL aliquot. (Prepare reagent (1-[(4-nitrophenyl)sulfonyl]-L-prolyl chloride) as described above.)

HPLC VARIABLES

Guard column: 70 × 2.1 30-38 μm HC Pellosil (Whatman)
Column: 150 × 4.6 5 μm Supelcosil LC-Si
Mobile phase: n-Heptane:chloroform 20:80
Flow rate: 1.4
Injection volume: 5
Detector: UV 254

CHROMATOGRAM

Retention time: 4 (R), 4.5 (S)

KEY WORDS

normal phase; chiral

REFERENCE

Barksdale, J.M.; Clark, C.R. Liquid chromatographic determination of the enantiomeric composition of amphetamine and related drugs by diastereomeric derivatization, *J.Chromatogr.Sci.*, **1985**, *23*, 176–180.

SAMPLE

Matrix: solutions
Analyte: methamphetamine
Sample preparation: 2 mL THF + 1 mL 33.5 mM reagent in THF (freshly prepared) + 1 mL 1 mg/mL amphetamine in water + 700 μL 10% sodium bicarbonate in water, heat at 65° for 1 h, cool, extract three times with 10 mL aliquots of chloroform. Combine the extracts and wash them with 10 mL water, dry over anhydrous magnesium sulfate, evaporate to dryness, reconstitute with 2.5 mL mobile phase, inject a 5 μL aliquot. (Prepare reagent (1-[(4-nitrophenyl)sulfonyl]prolyl chloride) as described above.)

HPLC VARIABLES
Column: 150 × 4.6 5 μm Zorbax ODS
Mobile phase: MeOH:water 60:40
Flow rate: 1.5
Injection volume: 5
Detector: UV 254

CHROMATOGRAM
Retention time: 12 (S), 14 (R)

KEY WORDS
chiral

REFERENCE
Barksdale, J.M.; Clark, C.R. Liquid chromatographic determination of the enantiomeric composition of amphetamine and related drugs by diastereomeric derivatization, *J.Chromatogr.Sci.*, **1985**, *23*, 176−180.

Pentafluorobenzoyl Chloride

SAMPLE
Matrix: solutions
Analyte: cytosine
Sample preparation: Evaporate 1 mL 1 mM cytosine in MeOH to dryness under reduced pressure at 4° for 30 min, add 1 mL 20 mM pentafluorobenzoyl chloride in MeCN containing 40 mM N-methylmorpholine, vortex, heat at 35° for 6 h while vortexing every hour, add 2.5 μL water, add 50 μL 250 mM dimethyl sulfate in MeCN containing 500 mM diisopropylethylamine, heat at 35° for 15 h, evaporate to dryness under reduced pressure at 44° for 45-60 min, add 500 μL MeCN, vortex, inject an aliquot.

HPLC VARIABLES
Column: 150 × 4.6 5 μm LC-8 (Supelco)
Mobile phase: Gradient. MeCN:pH 5.5 acetate buffer from 30:70 to 60:40 over 15 min, re-equilibrate at initial conditions for 5 min.
Flow rate: from 1 to 2 over 15 min
Injection volume: 50
Detector: UV 254

CHROMATOGRAM
Retention time: 8.5

REFERENCE
Fisher, D.H.; Adams, J.; Giese, R.W. Trace derivatization of cytosine with pentafluorobenzoyl chloride and dimethyl sulfate, *Environ.Health Perspect.*, **1985**, *62*, 67−71.

4-(2-Phthalimidyl)benzoyl Chloride

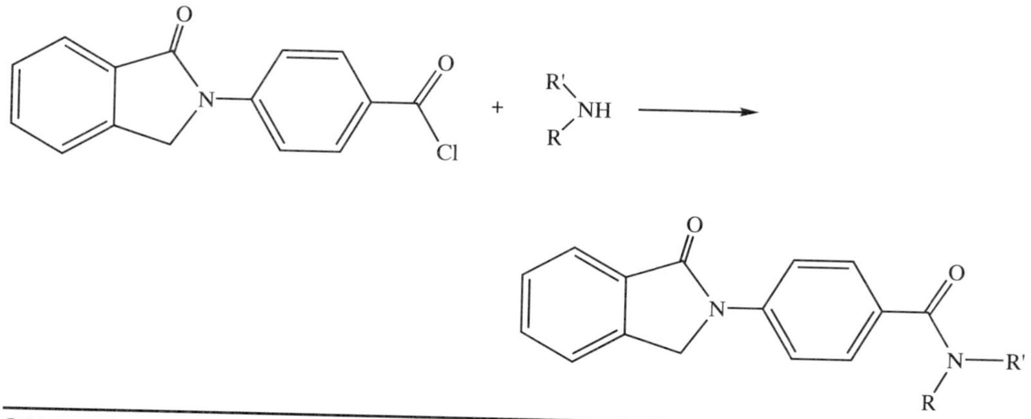

SAMPLE
Matrix: solutions

Analyte: amines derived from nitrosamines

Sample preparation: Mix 100 μL of a 0.64-200 μM nitrosamine solution with 70 μL HBr reagent, heat at 40° for 5 min, evaporate to dryness under a stream of nitrogen, reconstitute with 100 μL 400 mM sodium bicarbonate solution, add 100 μL 5 mM 4-(2-phthalimidyl)benzoyl chloride in MeCN, mix, let stand at room temperature for 1 min, inject a 10 μL aliquot. (Prepare the HBr reagent by diluting 5 mL 47% HBr in water to 26 mL with acetic anhydride. The nitrosamine is denitrosated by the HBr to give the corresponding amine which is then acylated with 4-(2-phthalimidyl)benzoyl chloride. Preparation of 4-(2-phthalimidyl)benzoyl chloride is as follows. Mix 1.34 g o-phthalaldehyde in 50 mL diethyl ether with 1.37 g 4-aminobenzoic acid in 150 mL diethyl ether, stir at room temperature overnight. Filter and dry the solid to obtain 4-(2-phthalimidyl)benzoic acid. Suspend the 4-(2-phthalimidyl)benzoic acid in 100 mL chloroform, add 6 mL thionyl chloride, reflux for 1 h, evaporate to dryness, wash the residue twice with 10 mL portions of chloroform, recrystallize from benzene (Caution! Benzene is a carcinogen!) to obtain 4-(2-phthalimidyl)benzoyl chloride as fine colorless needles (mp >230°) (Anal. Chim. Acta 1987, 192, 309).)

HPLC VARIABLES
Column: 125 × 4.6 5 μm Nucleosil C18

Mobile phase: MeCN:water 48:52

Flow rate: 0.8

Injection volume: 10

Detector: F ex 299 em 426

CHROMATOGRAM
Retention time: 3.5 (dimethylamine), 4 (pyrrolidine), 4.5 (diethylamine), 5 (piperidine), 8 (dipropylamine), 17 (dibutylamine)

Limit of detection: 0.4-1.6 pmole

REFERENCE
Zheng, M.; Fu, C.; Xu, H. High-performance liquid chromatographic detection of trace N-nitrosoamines by precolumn derivatization with 4-(2-phthalimidyl)benzoyl chloride, *Analyst*, **1993**, *118*, 269–271.

Propyl Chloroformate

SAMPLE
Matrix: blood
Analyte: 3-morpholinosydnonimine (active metabolite of molsidomine)
Sample preparation: 2 mL Plasma + 100 μL MeOH:concentrated HCl 100:0.1 + 25 μL 2 μg/mL IS + 100 μL propyl chloroformate + 15 mL 1,2-dichloroethane + 2 mL 500 mM pH 8.5 Tris HCl buffer, vortex vigorously for 2 min, centrifuge for 10 min. Remove the lower organic layer and evaporate it to dryness under a stream of nitrogen at 45°, reconstitute the residue in 3 mL diethyl ether, add 200 μL 10 mM HCl, vortex gently for 1 min, centrifuge for 2 min. Remove the lower aqueous layer and evaporate it to dryness under a stream of nitrogen at 45°, reconstitute the residue in 75-150 μL mobile phase, inject an aliquot.

HPLC VARIABLES
Guard column: 20 × 4.6 10 μm C18
Column: 150 × 4.6 5 μm Ultrasphere IP C18
Mobile phase: MeCN:10 mM pH 4.5 KH_2PO_4 18:82
Flow rate: 1
Injection volume: 50
Detector: UV 312

CHROMATOGRAM
Retention time: 11
Internal standard: ethoxycarbonyl derivative of 3-(N-methylsulfonylpiperazino)sydnonimine (7.4)
Limit of detection: 0.5 ng/mL

OTHER SUBSTANCES
Extracted: molsidomine (not derivatized)

KEY WORDS
plasma; protect from light; pharmacokinetics

REFERENCE
Dutot, C.; Moreau, J.; Cordonnier, P.; Spreux-Varoquaux, O.; Klein, C.; Ostrowski, J.; Advenier, C.; Gärtner, W.; Pays, M. Determination of the active metabolite of molsidomine in human plasma by reversed-phase high-performance liquid chromatography, *J.Chromatogr.*, **1990**, *528*, 435–446.

2-(1-Pyrenyl)ethyl Chloroformate

SAMPLE
Matrix: blood
Analyte: polyamines
Sample preparation: Reflux 5 mL serum with 5 mL 6 M HCl for 10 h, evaporate to dryness under reduced pressure at 60°, neutralize the residue with dilute KOH solution, evaporate to dryness, dilute to 10 mL, centrifuge at 1380 g for 30 min, filter (0.45 μm). Remove a 400 μL aliquot of the filtrate and add it to 300 μL 25 mM pH 9.0 buffer, add 300 μL 5.6 mM reagent in MeCN, heat at 36° for 5 min, dilute 400-fold with MeCN, inject an aliquot. (Prepare the reagent, 2-(1-pyrenyl)ethyl chloroformate (mp 71-73°) by reacting 2-(1-pyrenyl)ethanol with phosgene. For example, in an analogous reaction a solution of phosgene in toluene is added to a stirred solution of 2-(9-anthryl)ethanol in ether in the presence of pyridine (as an HCl scavenger) at 0° to give 2-(9-anthryl)ethyl chloroformate (Anal.Chem. 1991, 63, 292). Prepare 2-(1-pyrenyl)ethanol by adding 17 g ethylene oxide (Caution! Ethylene oxide is a carcinogen!) in 50 mL 1,1,2,2-tetrachloroethane to a solution of 50 g pyrene and 40 g aluminum trichloride in 200 mL 1,1,2,2-tetrachloroethane stirred at -5 to 0° under nitrogen, stir at 5-10° for 1.5 h, stir at 10-20° for 2.5 h, pour onto ice, add 30-40 mL concentrated HCl, remove the organic layer, extract the aqueous layer with ether. Combine the organic layers and evaporate them, distil (bp 230-233°/3 mm Hg) to give 2-(1-pyrenyl)ethanol (mp 57-59°) (J. Prakt. Chem. 1963, 22, 47 (Chem. Abstr. 1964, 60, 1669b)).)

HPLC VARIABLES
Column: 125 × 4 5 μm LiChrospher 100 RP-18
Mobile phase: Gradient. A was MeCN:water 90:10. B was MeCN:ethyl acetate 80:20. A:B 100:0 for 5 min, to 0:100 (step gradient)
Flow rate: 1
Injection volume: 20
Detector: F ex 275 em 389 (cut-off filter)

CHROMATOGRAM
Retention time: 5.8 (putrescine), 6.5 (cadaverine), 10.1 (spermidine), 12.1 (spermine)
Limit of detection: 0.9-2.9 pg

KEY WORDS
serum

REFERENCE
Cichy, M.A.; Stegmeier, D.L.; Veening, H.; Becker, H.-D. High-performance liquid chromatographic separation of biogenic polyamines using 2-(1-pyrenyl)ethyl chloroformate as a new fluorogenic derivatizing reagent, J.Chromatogr., **1993**, *613*, 15−21.

Salicyl Chloride

SAMPLE
Matrix: blood, urine
Analyte: amines
Sample preparation: Vortex plasma or urine for 30 s, centrifuge at 2000 g for 10 min. Remove a 500 μL aliquot of the supernatant and add it to 500 μL 100 mM pH 12.5 phosphate buffer, vortex, add to an ICT Bond-Elut C18 SPE cartridge, wash with two 1 mL portions of water, elute with two 500 μL portions of MeOH. Add 10 μL glacial acetic acid to the eluate, evaporate to dryness under a stream of nitrogen, reconstitute with 100 μL dry ethyl acetate, add 5 mg potassium carbonate, add 100 μL 0.5 mM salicyl chloride in toluene, heat at 70° for 50 min, cool. Remove a 100 μL aliquot and evaporate it to dryness under a stream of nitrogen, reconstitute the residue in 100 μL mobile phase, inject a 20 μL aliquot. (Prepare salicyl chloride by stirring 27.5 g freshly distilled thionyl chloride in 30 mL dry benzene at 0° (Caution! Benzene is a carcinogen!), protect the reaction with a calcium chloride drying tube and a nitrogen atmosphere, add 25 g sodium salicylate, stir at 0° for 1 h, remove solvent by vacuum distillation, take up the residue in 50 mL dry petroleum ether, stir for 15 min, centrifuge. Remove the petroleum ether layer and evaporate it to give salicyl chloride (Anal.Chim.Acta 1993, 273, 383).)

HPLC VARIABLES
Guard column: 5 × 4 30-40 μm Perisorb RP-18 (Merck)
Column: 125 × 4 3 μm Hypersil C18
Mobile phase: MeOH:water 42:58 containing 12 mM lithium perchlorate and 35 mM acetic acid, pH 3.5
Flow rate: 0.9
Injection volume: 20
Detector: E, Hewlett Packard Model 1049A, glassy carbon working electrode +1.2 V, Ag/AgCl reference electrode, carbon-filled PTFE auxiliary electrode; UV 210; UV 254

CHROMATOGRAM
Retention time: 15.5 (butylamine), 5 (diethylamine), 2 (dimethylamine), 19 (dipropylamine), 22.5 (2-pentylamine), 8 (propylamine)
Limit of detection: 1-5 pmole (E), 20-50 pmole (UV 210), 34-198 pmole (UV 254)

KEY WORDS
plasma; SPE

REFERENCE
Wintersteiger, R.; Barary, M.H.; El-Yazbi, F.A.; Sabry, S.M.; Wahbi, A.-A.M. Reversed phase-liquid chromatography of primary and secondary aliphatic amines after derivatization combined with electrochemical detection, *Anal.Chim.Acta*, **1995**, *306*, 273–283.

m-Toluoyl Chloride

SAMPLE
Matrix: bulk
Analyte: amines
Sample preparation: Prepare a 1% solution in pyridine, add 1,6-diaminohexane, add 3 equivalents of m-toluoyl chloride, mix, let stand at room temperature for 5 min, add 5 M HCl with stirring and cooling until the pH is about 2.5, cool to room temperature, extract with dichloromethane. wash the organic layer with 1% sodium carbonate in water, wash with 500 mM HCl, wash with water, dilute with MeCN, inject an aliquot. (Distil pyridine from phthalic anhydride just before use.)

HPLC VARIABLES
Column: 300 × 4 10 μm μBondapak C18
Mobile phase: MeCN:water 40:60
Flow rate: 0.9
Injection volume: 35
Detector: UV 254

CHROMATOGRAM
Retention time: 11 (2,5-dimethylmorpholine), 12 (2,6-dimethylmorpholine), 15 (1,2-diaminoethane), 18 (1,2-diaminopropane, 1,4-diaminobutane, piperazine), 22.5 (1,5-diaminopentane), 27 (2,5-dimethylpiperazine), 37 (ethanolamine), 40 (N-methylethanolamine)
Internal standard: 1,6-diaminohexane (32)
Limit of detection: 5 ng

REFERENCE
Wellons, S.L.; Carey, M.A. High-performance liquid chromatographic separation and quantitation of polyfunctional amines as their m-toluoyl derivatives, *J.Chromatogr.*, **1978**, *154*, 219–225.

Trichloroethyl Chloroformate

SAMPLE
Matrix: blood
Analyte: promethazine
Sample preparation: 2 mL Plasma or serum + 100 mg sodium carbonate, vortex for 3 s, add 10 mL 3 ng/mL triflupromazine in hexane:MeOH 99.7:0.3, shake for 15 min, centrifuge at 2000 rpm for 5 min. Remove 9.3 mL of the organic layer and evaporate it to dryness under a stream of air at 40°, reconstitute the residue in 50 μL ethyl acetate, add 25 μL trichloroethyl chloroformate, vortex, heat at 120° for 20 min, cool. Evaporate to dryness under a stream of air at 40°, reconstitute the residue in 100 μL MeOH, inject a 50 μL aliquot.

HPLC VARIABLES
Column: 300×4 MCH-10 reversed-phase (Varian)
Mobile phase: MeOH:water 84:16
Flow rate: 2
Injection volume: 50
Detector: UV 254

CHROMATOGRAM
Retention time: 4.2
Internal standard: triflupromazine (6)
Limit of detection: 1 ng/mL

OTHER SUBSTANCES
Simultaneously analyzed: chlorpromazine, chlorprothixene, phenothiazine-5-oxide, promazine, trimeprazine
Interfering: phenothiazine

KEY WORDS
plasma; serum

REFERENCE
Wallace, J.E.; Shimek, E.L. Jr.; Harris, S.C.; Stavchansky, S. Determination of promethazine in serum by liquid chromatography, *Clin.Chem.*, **1981**, *27*, 253–255.

ALDEHYDE

4-(Dimethylamino)benzaldehyde

SAMPLE
Matrix: eggs, milk, tissue
Analyte: sulfonamides
Sample preparation: Milk. Centrifuge at 2000 g and freeze at -20° to remove the cream. Mix a 5 mL aliquot with 5 mL saline solution and add 1 mL 1% sodium azide solution (Caution! Sodium azide is highly toxic! Do not discharge to the plumbing system!). Dialyze (24″ long cellulose acetate "Type C" membrane, Technicon, New York) an 8 mL aliquot against water pumped at 0.6 mL/min, pass the dialysate through column A to waste, after 10 min stop the dialysis and elute column A to waste with water, after 24 min backflush the contents of column A onto column B with mobile phase, after 5 min remove column A from the circuit, elute column B with mobile phase, monitor the effluent from column B. Meat. Blend 10 g homogenized meat with 20 mL saline, centrifuge, remove a 10 mL aliquot of the clear upper phase and add it to 1 mL 1% sodium azide (Caution! Sodium azide is highly toxic! Do not discharge to the plumbing system!). Dialyze (24″ long cellulose acetate "Type C" membrane, Technicon, New York) an 8 mL aliquot against water pumped at 0.6 mL/min, pass the dialysate through column A to waste, after 10 min stop the dialysis and elute column A to waste with water, after 24 min backflush the contents of column A onto column B with mobile phase, after 5 min remove column A from the circuit, elute column B with mobile phase, monitor the effluent from column B. Eggs. Dilute 10 g homogenized whole egg with 10 mL saline, add 3 mL 10% sodium azide solution (Caution! Sodium azide is highly toxic! Do not discharge to the plumbing system!). Dialyze (24″ long cellulose acetate "Type C" membrane, Technicon, New York) an 8 mL aliquot against water pumped at 0.6 mL/min, pass the dialysate through column A to waste, after 10 min stop the dialysis and elute column A to waste with water, after 24 min backflush

the contents of column A onto column B with mobile phase, after 5 min remove column A from the circuit, elute column B with mobile phase, monitor the effluent from column B.

HPLC VARIABLES
Column: A 60 × 4 50-100 μm XAD-4 (Rohm & Haas); B 250 × 4.6 7 μm Cp TM-Spher C18 (Chrompack)
Mobile phase: MeCN:50 mM pH 6.85 sodium acetate buffer 12.5:87.5
Detector: UV 450 following post-column reaction. The column effluent mixed with 1.5% 4-(dimethylamino)benzaldehyde in 17% phosphoric acid and the mixture flowed through a 7.5 m × 0.5 mm ID knitted PTFE coil to the detector.

CHROMATOGRAM
Retention time: k' 1.0 (sulfacetamide, sulfanilamide, sulfaguanidine), k' 2.0 (sulfadiazine), k' 2.5 (sulfachlorpyrazine, sulfamethoxazole), k' 2.6 (sulfadoxine), k' 3.5 (sulfatroxazole, sulfamerazine, sulfathiazole), k' 5.0 (sulfadimethoxine), k' 6.2 (sulfaquinoxaline), k' 6.3 (sulfamethazine), k' 11.0 (dapsone)
Limit of detection: 5-10 ng/g

OTHER SUBSTANCES
Non-interfering: chloramphenicol, trimethoprim

KEY WORDS
post-column reaction; meat; column-switching; dialysis

REFERENCE
Aerts, M.M.L.; Beek, W.M.J.; Brinkman, U.A.T. Monitoring of veterinary drug residues by a combination of continuous flow techniques and column-switching high-performance liquid chromatography. I. Sulphonamides in egg, meat and milk using post-column derivatization with dimethylaminobenzaldehyde, *J.Chromatogr.*, **1988**, *435*, 97–112.

SAMPLE
Matrix: feed, premix
Analyte: sulfonamides
Sample preparation: Shake premix or ground feed with 150 mM HCl in MeOH:water 25:75 for 1 h, dilute with 150 mM HCl in MeOH:water 25:75 to achieve a sulfonamide concentration of 5.5 μg/mL, filter (glass fiber), inject an aliquot.

HPLC VARIABLES
Guard column: 50 × 2 30-40 μm Perisorb RP-18
Column: 250 × 4.6 10 μm Partisil ODS-3
Mobile phase: MeOH:2% acetic acid 35:65 (sulfamethazine) or MeCN:2% acetic acid 18:82 (for sulfathiazole)
Flow rate: 1
Injection volume: 200
Detector: UV 450 following post-column reaction. The column effluent mixed with reagent pumped at 0.5 mL/min and the mixture flowed through a 3 m × 0.5 mm ID coil of PTFE tubing to the detector. (Prepare reagent by dissolving 3 g 4-(dimethylamino)benzaldehyde in 100 mL glacial acetic acid, add 60 mL MeOH, add 40 mL water, mix well.)

CHROMATOGRAM
Retention time: 6 (sulfamethazine), 8 (sulfathiazole)
Limit of quantitation: 1.65 μg/mL

KEY WORDS
post-column reaction

REFERENCE
Stringham, R.W.; Mundell, E.C.; Smallidge, R.L. Use of post-column derivatization in liquid chromatographic determination of sulfamethazine and sulfathiazole in feeds and feed premixes, *J.Assoc.Off.Anal.Chem.*, **1982**, *65*, 823–827.

SAMPLE
Matrix: feed, premix
Analyte: polyether antibiotics

Sample preparation: Feed. Shake 5 g feed with 15 mL MeOH for 2 h, filter, evaporate the filtrate to 3 mL and make up to 10 mL with MeOH, inject a 3 μL aliquot. Premix. Shake 0.5 g premix with 15 mL MeOH for 2 h, filter, make up the filtrate to 50 mL with MeOH, inject a 3 μL aliquot.

HPLC VARIABLES
Column: 150 × 1 5 μm Separon SGX C18 glass column (Tessek Prague)
Mobile phase: MeOH:water:glacial acetic acid 94:5.9:0.1
Flow rate: 0.02
Injection volume: 3
Detector: UV 592 following post-column reaction. The column effluent mixed with the reagent pumped at 0.015 mL/min and the mixture flowed through a 150 × 1 reactor containing 40-70 μm acid-washed glass beads at 75° to the detector. (The reagent was 500 mM 4-(dimethylamino)benzaldehyde in 1.2 M sulfuric acid in MeOH.)

CHROMATOGRAM
Retention time: 13 (monensin), 16 (salinomycin), 19 (narasin)
Limit of detection: 2.2 μg/mL

KEY WORDS
microbore; post-column reaction

REFERENCE
Fejglová, Z.; Dolezal, J.; Hrdlicka, A.; Frgalová, K. Microbore HPLC determination of polyether antibiotics using postcolumn derivatization with benzaldehyde reagents, *J.Liq.Chromatogr.*, **1994**, *17*, 359–372.

SAMPLE
Matrix: tissue
Analyte: sulfonamides
Sample preparation: Cut tissue into small pieces and homogenize in blender. 20 g Homogenized tissue + 200 μL 10 μg/mL methyl p-aminobenzoate in water + 60 mL acetone:chloroform 50:50, shake vigorously on a mechanical shaker for 10 min, centrifuge at 3000 g for 10 min, filter (Whatman No. 41 paper) supernatant, repeat extraction. Combine the extracts, if the extract is not clear centrifuge at 3000 g for 10 min and discard the aqueous layer, evaporate to an oily residue at 45° under reduced pressure, add 5 mL MeCN to flask, let stand for 10 min, remove MeCN layer, add 5 mL hexane and 5 mL MeCN, shake, centrifuge at 3000 g for 10 min, remove the MeCN layer, add 5 mL MeCN to the hexane layer, shake, centrifuge at 3000 g for 10 min, remove the MeCN layer. If hexane layer is not clear centrifuge at 3000 g for 10 min and remove the clear portion. Add 400 μL 15% trichloroacetic acid to the hexane layer, shake gently for 10 min, centrifuge at 3000 g for 10 min. Evaporate the MeCN layers, transfer the oily residue to a small flask with 3 mL hexane, add the aqueous trichloroacetic acid layer, shake gently for 10 min, centrifuge at 3000 g for 10 min. Discard the hexane layer, add 100 μL saturated aqueous sodium citrate solution to the aqueous layer, mix, inject a 50 μL aliquot.

HPLC VARIABLES
Guard column: 15 × 3.2 7 μm RP 18 (Brownlee)
Column: 300 × 3.9 10 μm μBondapak C18
Mobile phase: Gradient. A was 1% aqueous acetic acid. B was MeCN:water 80:20. A:B from 90:10 to 60:40 over 20 min, return to initial conditions over 5 min, re-equilibrate for 5 min.
Flow rate: 1.5
Injection volume: 50
Detector: UV 450 following post-column reaction. The column effluent mixed with the reagent pumped at 0.5 mL/min and the mixture flowed through a 7 m × 0.25 mm ID coil of stainless steel tubing to the detector. (Prepare reagent by dissolving 1 g 4-(dimethylamino)benzaldehyde in 30 mL MeCN, make up to 100 mL with 5% trichloroacetic acid in water.)

CHROMATOGRAM
Retention time: 9.1 (sulfadiazine), 11.1 (sulfapyridine), 12.1 (sulfamerazine), 14.4 (sulfamethazine (sulfadimidine)), 15.9 (sulfamethoxypyridazine), 26 (sulfaquinoxaline)

Internal standard: methyl p-aminobenzoate (18.6))
Limit of detection: 20 ng/g

KEY WORDS

chicken; liver; pig; kidney; sheep; cow; post-column reaction

REFERENCE

Bui, L.V. Liquid chromatographic determination of six sulfonamide residues in animal tissues using postcolumn derivatization, *J.AOAC Int.*, **1993**, *76*, 966–976.

Dimethylformamide Diethyl Acetal

$$(H_3C)_2N-\!\!\!\begin{array}{c} OCH_2CH_3 \\ \\ OCH_2CH_3 \end{array} \quad + \quad RNH_2 \quad \longrightarrow \quad R-N\!\!=\!\!\begin{array}{c} N(CH_3)_2 \end{array}$$

SAMPLE

Matrix: blood
Analyte: 5,6-dihydro-5-azacytidine
Sample preparation: Condition a 40 × 8 Affi-Gel 601 phenylboronate SPE cartridge (Bio-Rad) with 10 mL 250 mM pH 8.8 ammonium acetate buffer. Centrifuge plasma at 1000 g for 10 min. 1 mL Plasma + 80 μL 52.5 μg/mL IS, vortex briefly, filter (Amicon MPS-1 with YMP membrane) while centrifuging at 1145 g for 40 min. Mix 500 μL ultrafiltrate with 500 μL 250 mM pH 8.8 ammonium acetate, add to the SPE cartridge, wash with 10 mL 250 mM pH 8.8 ammonium acetate, elute with 10 mL 100 mM formic acid, add the eluate to a 15 × 8 Dowex 50W-X8 cation-exchange column (NH_4^+ form), wash with 15 mL water, elute with two 7.5 mL portions of 30 mM pH 10 ammonium hydroxide, evaporate eluate to dryness, dry under vacuum with warming for 10 min, add 100 μL DMF under argon, add 100 μL 2 M dimethylformamide diethyl acetal in pyridine under argon, stir briefly, let stand at room temperature for 30 min, evaporate to dryness under reduced pressure with gentle heating, heat for 3 min after residue appears dry, reconstitute with 500 μL water, inject an aliquot.

HPLC VARIABLES

Column: 5 μm Spherisorb S5 ODS II
Mobile phase: MeCN:water 20:80 containing 50 mM formic acid
Flow rate: 1.5
Detector: UV 264

CHROMATOGRAM

Retention time: 4
Internal standard: 5'-chloro-5'-deoxy-5,6-dihydro-5-azacytidine (Prepare by adding 200 mg 5,6-dihydro-5-azacytidine hydrochloride to 300 μL thionyl chloride in 2 mL HMPA, stir at room temperature for 16 h, let stand at 4° for 24 h, dilute with 18 mL water, add to a 60 × 50 column of 100-120 mesh Dowex AG 50W-X8 (H^+ form, Bio-Rad), wash with 1 L water, elute with 1 L 1 M ammonium hydroxide. Evaporate the eluate to dryness at 45°, purify by HPLC (300 × 7.8 10 μm μBondapak C18, MeOH:10 mM pH 5 ammonium phosphate 6:84, 2 mL/min), pass the collected fractions through a 70 × 8 cation-exchange column (NH_4^+ form), wash with 30 mL water, elute with 30 mL 30 mM pH 10 ammonium hydroxide, evaporate to give 5'-chloro-5'-deoxy-5,6-dihydro-5-azacytidine (mp 204°).) (7)
Limit of quantitation: 50 ng/mL
Limit of detection: 25 ng/mL

KEY WORDS

rat; human; plasma; SPE; ultrafiltrate; pharmacokinetics

REFERENCE

Huguenin, P.N.; Jayaram, H.N.; Kelley, J.A. Reverse phase HPLC determination of 5,6-dihydro-5-azacytidine in biological fluids, *J.Liq.Chromatogr.*, **1984**, *7*, 1433–1453.

5-(4-Pyridyl)-2-thiophenecarbaldehyde

SAMPLE
Matrix: solutions
Analyte: alkylamines
Sample preparation: Reflux a 10 mL aliquot of a 50-100 nM solution of amine in MeOH containing 170 equivalents of 5-(4-pyridyl)-2-thiophenecarbaldehyde for 5 h, dilute to 50 mL, inject a 10 μL aliquot. (Preparation of 5-(4-pyridyl)-2-thiophenecarbaldehyde is as follows. Dissolve 2 g 4-aminopyridine in 15 mL acetic acid, 1.5 mL concentrated sulfuric acid, and 2 mL water, stir at 0°, diazotize with 1 equivalent (?) sodium nitrite, add an excess of potassium iodide, keep at room temperature for 1 day, heat at 50° for 30 min, make alkaline, extract with ether. Evaporate the extract to dryness and recrystallize from EtOH:water to give 4-iodopyridine (yield = 15%; mp 100° d) (J. Chem. Soc. 1953, 3226). Heat 22.2 g iodine, 6.66 g periodic acid dihydrate, and 20 g 2-methylthiophene in 200 mL 80% acetic acid at 65° for 8 h, cool, remove the oil, basify the solution, extract with chloroform. Combine the oil and the residue from the extract, add to a NaOH solution containing sodium thiosulfate, steam distil. Distil the crude product to give 2-iodo-5-methylthiophene as a yellowish oil (bp 90°/4.1 kPa). Reflux 460 mg palladium amalgam, 14.7 g NaOH, 7.5 g 4-iodopyridine, 16.4 g 2-iodo-5-methylthiophene, and 2.34 g hydrazine hydrate in 47 mL water with stirring for 6 h (Caution! Hydrazine hydrate is a carcinogen!), filter, wash the precipitate with hot chloroform and hot water. Remove the organic layer from the filtrate and reduce its volume to 50 mL, extract with 15 mL 25% HCl. Remove the aqueous layer and basify it, extract with chloroform. Evaporate the chloroform layer to dryness, distil the residue at 150°/0.4 kPa, chromatograph the distillate on silica gel with acetone:chloroform 50:50 to give 4-(5-methyl-2-thienyl)pyridine as a white solid (mp 136.0-136.6°) in the first fraction. Reflux 1 g 4-(5-methyl-2-thienyl)pyridine and 957 mg selenium(IV) oxide in 35 mL m-xylene with stirring for 8 h, evaporate to dryness. Distil the residue at 200°/0.8 kPa and chromatograph the distillate on silica gel with ethyl acetate to give 5-(4-pyridyl)-2-thiophenecarbaldehyde as white plate crystals (mp 135-136°).)

HPLC VARIABLES
Column: 150 × 6 Shim-pack CLC-ODS (Shimadzu)
Mobile phase: MeOH:water 100:0 (A) or 90:10 (B)
Column temperature: 40
Flow rate: 1
Injection volume: 10
Detector: F ex 340 em 395 following post-column reaction. The column effluent mixed with 1 M nitric acid pumped at 0.2 mL/min and the mixture flowed through a 300 × 0.3 mm ID coil to the detector. (The original fluorescent 5-(4-pyridyl)-2-thiophenecarbaldehyde is regenerated by hydrolysis of the Schiff's base.)

CHROMATOGRAM
Retention time: 3.5 (1,3-propanediamine (A)), 3.7 (2-phenylethylamine (A)), 4.0 (benzylamine (B)), 4.0 (p-chloroaniline (A)), 4.3 (1,8-octanediamine (A)), 4.5 (1-phenylethylamine (B)), 4.6 (butylamine (B)), 4.7 (octylamine (A)), 5.0 (cyclohexylamine (B)), 5.1 (p-toluidine (B)), 5.5 (decylamine (A)), 6.1 (hexylamine (B)), 6.7 (dodecylamine (A)), 8.3 (tetradecylamine (A)), 10.8 (hexadecylamine (A)), 21.3 (octadecylamine (A))
Limit of detection: 0.1-0.15 pmole

REFERENCE
Nakajima, R.; Yamamoto, A.; Hara, T. Thienylpyridines as a new fluorescent reagent. I. Determination of primary alkylamines with 5-(4-pyridyl)-2-thiophenecarbaldehyde using HPLC, *Bull.Chem.Soc.Jpn.*, **1990**, *63*, 1968–1972.

Salicylaldehyde

RELATED REFERENCE
Prieto-Blanco, M.C.; López-Mahía, P.; Prada Rodríguez, D. Determination of dimethylaminopropylamine in alkylaminoamides by high-performance liquid chromatography. *J.Chromatogr.Sci.* **1997**, *35*, 265-269

SAMPLE
Matrix: polymers
Analyte: polyamines
Sample preparation: Dissolve 100-500 mg sample in 50 mL THF, sonicate until dissolved, dilute 20-fold with THF. Remove a 2 mL aliquot and add it to 5 mL mobile phase, add 2 mL reagent, add 5 mL mobile phase, shake gently at 70-75° for 2 min, cool, make up to 25 mL with mobile phase, filter (Gelbrand 030/70), inject an aliquot. (Prepare reagent by dissolving 250 mg salicylaldehyde and 400 mg diphenylboric anhydride in 50 mL MeCN, store in the dark at 4°, discard after 1 week.)

HPLC VARIABLES
Column: 250 × 4 7 μm LiChrosorb RP-18
Mobile phase: MeCN:20 mM pH 7.3 acetate buffer 70:30 (A) or 80:20 (B)
Flow rate: 1.5
Injection volume: 50
Detector: UV 234; F ex 385 em 470

CHROMATOGRAM
Retention time: 16 (ethylenediamine (A)), 7.5 (tetraethylenetetramine (B)), 8.5 (triethylenetetramine (B)), 15 (tetraethylpentamine (B))

REFERENCE
Winkler, E.; Hohaus, E.; Felber, E. Quantitative determination of free primary amines in polymeric amine hardeners. High-performance liquid chromatographic separation of ethylenediamine and triethylenetetramine after derivatization with salicylaldehyde-diphenylboron chelate and separation of 4,4'-diaminodiphenylmethane by high-performance liquid chromatography, *J.Chromatogr.*, **1988**, *436*, 447–454.

ALKENE

1-Phenylsulfonyl-3,3,3-trifluoropropene

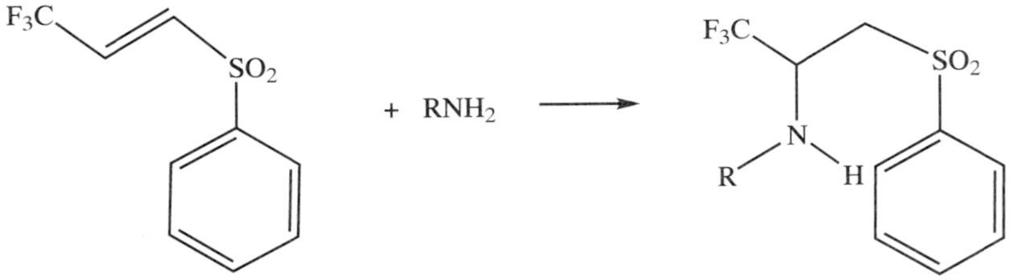

SAMPLE

Matrix: tissue

Analyte: polyamines

Sample preparation: Homogenize 200 mg tissue with 1 mL 30 μM IS in 100 mM perchloric acid, centrifuge at 3000 rpm for 20 min. Remove a 100 μL aliquot of the supernatant and add it to 400 μL 1% 1-phenylsulfonyl-3,3,3-trifluoropropene in EtOH, add 400 μL 12.5 mM sodium carbonate, heat at 35° for 4 h, filter (0.45 μm), inject a 90 μL aliquot of the filtrate. (Synthesis of 1-phenylsulfonyl-3,3,3-trifluoropropene is as follows. React 3,3,3-trifluoropropene with thiophenol and NaOH in EtOH at 90° for 4 days to give 1-phenylthio-3,3,3-trifluoropropane, reflux this product with sulfuryl chloride in carbon tetrachloride for 12 h then oxidize with 3-chloroperoxybenzoic acid (mCPBA) in dichloromethane, dehydrochlorinate with 1,8-diazabicyclo[5.4.0]undec-7-ene (DBU) in dichloromethane at room temperature to obtain 1-phenylsulfonyl-3,3,3-trifluoropropene (Chem. Pharm. Bull. 1985, 33, 4077).)

HPLC VARIABLES

Column: 150 × 6 5 μm Shim-pack CLC-ODS

Mobile phase: MeCN:MeOH:5 mM pH 6.4 phosphate buffer 21:42:27 containing 5 mM sodium 1-octanesulfonate

Column temperature: 30

Flow rate: 1.5

Injection volume: 90

Detector: UV 223

CHROMATOGRAM

Retention time: 6 (spermidine), 8 (putrescine), 10 (cadaverine), 23 (spermine)

Internal standard: 1,6-diaminohexane (15)

Limit of detection: 0.3-0.6 pmole

OTHER SUBSTANCES

Non-interfering: amino acids, catecholamines, diethylamine, diisopropylamine, histamine, indoleamines

KEY WORDS

rat; brain; kidney; liver; heart; testes

REFERENCE

Nakajima, M.; Wakabayashi, H.; Yamato, S.; Shimada, K. Determination of polyamines in rat tissues by high-performance liquid chromatography after derivatization with 1-phenylsulfonyl-3,3,3-trifluoropropene, *Anal.Sci.*, **1990**, *6*, 523–527.

ANHYDRIDE

Acetic Anhydride

SAMPLE
Matrix: solutions
Analyte: fluoroquinolone antibacterials
Sample preparation: Prepare a 450 µg/mL solution in MeCN:water 50:50. 5 mL Solution + 5 mL THF + 200 molar excess of acetic anhydride + 3 molar excess of 1 M NaOH, sonicate for 15 min, add 15 mL mobile phase, sonicate for 15 min, cool to room temperature, make up to 50 mL with mobile phase, inject a 50 µL aliquot.

HPLC VARIABLES
Column: 150 × 4.6 5 µm Nucleosil C18
Mobile phase: MeCN:buffer 35:65 (Buffer was prepared by mixing equal volumes of 20 mM citric acid and 20 mM sodium citrate, pH adjusted to 2.4 with perchloric acid.)
Flow rate: 1
Injection volume: 50
Detector: UV 280

CHROMATOGRAM
Retention time: 5.9 (norfloxacin), 6.4 (ciprofloxacin), 12 (sarafloxacin), 21.4 (temafloxacin)

REFERENCE
Morley, J.A.; Elrod, L. Jr. Determination of fluoroquinolone antibacterials as N-Acyl derivatives, *Chromatographia*, **1993**, 37, 295–299.

SAMPLE
Matrix: solutions
Analyte: acebutolol
Sample preparation: 100 µL 55 mM N-(Carbobenzyloxy)-L-phenylalanine (N-benzyloxy-carbonyl-L-phenylalanine, N-CBZ-L-Phe) in dichloromethane + 100 µL 14 mM N,N-di-methylaminopyrine (dimethylaminopyridine (?)) in dichloromethane + 100 µL 9-acetyl-anthracene in dichloromethane, cool in an ice bath, add 500 µL of a solution of acebutolol in dichloromethane, add 100 µL 240 mM dicyclohexylcarbodiimide in dichloromethane, shake mechanically at 0° for 30 min, add 100 µL 1.06 M acetic anhydride in dichloromethane, shake mechanically at 30° for 15 min, add 1 mL MeOH, mix, inject an aliquot. (N-CBZ-L-Phe derivatizes the alcohol and acetic anhydride derivatizes the secondary amine.)

HPLC VARIABLES
Guard column: 10 µm Nova-Pak C18 precolumn
Column: 150 × 3.9 4 µm Nova-Pak C18
Mobile phase: MeOH:water 60:40
Flow rate: 1.3
Detector: UV 254

CHROMATOGRAM
Retention time: 14.13 (S), 15.74 (R)
Internal standard: 9-acetylanthracene (6.52)

KEY WORDS
chiral

REFERENCE

Wen, Y.H.; Wu, S.S.; Wu, H.L. Chiral separation of acebutolol by derivatization and high-performance liquid chromatography, *J.Liq.Chromatogr.*, **1995**, *18*, 3329–3345.

SAMPLE

Matrix: tissue
Analyte: β-carbolines
Sample preparation: Homogenize (Polytron) rat lung with 5 volumes of ice-cold 400 mM perchloric acid, 1 mg semicarbazide hydrochloride, and 2 ng IS in an ice bath, centrifuge at 4° at 9000 g for 15 min, suspend the pellet in 2 mL cold 400 mM perchloric acid, centrifuge. Combine the supernatants, adjust the pH to 6-7 with 10 M NaOH, add 25 μL acetic anhydride, vortex, let stand for 5 min, adjust pH to 10, add 25 μL acetic anhydride, mix, let stand for 5 min, extract twice with 2 volumes of ethyl acetate. Combine the organic layers and evaporate them to dryness under a stream of nitrogen, reconstitute the residue in 10 mM perchloric acid, wash with 5 volumes of diethyl ether, inject an aliquot.

HPLC VARIABLES

Column: 250 × 4.6 5 μm Biosphere C18 (Bioanalytical Systems)
Mobile phase: MeOH:water 60:40 containing 0.5% triethylamine
Flow rate: 1.4
Injection volume: 50
Detector: F ex 252 em 430 (cut-off filter)

CHROMATOGRAM

Retention time: 4.2 (O-acetylnorharmol), 5.5 (O-acetylharmol), 6.5 (norharman), 7.6 (harman)
Internal standard: 1-propyl-9H-pyrido[3,4-b]indole (Synthesis is as follows. Heat 1 g l-tryptophan with 5 mL 500 mM sulfuric acid and 60 mL 7% butyraldehyde at 40° for 30 min, heat on a steam bath for 1.5 h, evaporate, add ammonia to obtain 3,4,5,6-tetrahydro-3-propyl-4-carboline-5-carboxylic acid. Dissolve 70 mg of this acid in 18 mL boiling water, add 3.5 mL 10% potassium dichromate, when a cloudiness appears add 700 μL acetic acid, heat for 1 min, cool, add dilute sodium sulfite to reduce the excess oxidizing agent, make alkaline with sodium carbonate, extract with ether, evaporate the ether layer to give 1-propyl-9H-pyrido[3,4-b]indole (cf J. Biol. Chem. 1936, 113, 759).) (10.2)
Limit of detection: 100-750 pg

KEY WORDS

rat; lung; use silanized glassware

REFERENCE

Bosin, T.R.; Faull, K.F. Measurement of β-carbolines by high-performance liquid chromatography with fluorescence detection, *J.Chromatogr.*, **1988**, *428*, 229–236.

Boc-L-Leu Anhydride

SAMPLE

Matrix: blood
Analyte: alprenolol and metoprolol
Sample preparation: 1 mL Plasma + 1 mL 1 M pH 9.9 carbonate buffer + 6 mL water-saturated diethyl ether, agitate for 15 min, centrifuge at 4200 g for 5 min. Remove 5 mL

of the organic layer and evaporate it to dryness under a stream of nitrogen at 35°, reconstitute the residue in 250 μL 142.4 mM triethylamine in dichloromethane, add 100 μL reagent, let stand for 15 min, evaporate to dryness under a stream of nitrogen at 35°, add 2 mL 100 mM NaOH, agitate for 10 min, add 6 mL ether, extract for 15 min, centrifuge. Remove 5 mL of the organic layer and evaporate it to dryness under a stream of nitrogen at 35°, cool in an ice bath, reconstitute the residue in 250 μL trifluoroacetic acid, let stand at 0° for 10 min, add 2 mL 2 M NaOH, extract with 6 mL ether. Remove a 5 mL aliquot of the organic layer and extract it with 100 μL 100 mM phosphoric acid, inject a 94 μL aliquot of the aqueous layer. (Purify triethylamine by drying it over NaOH pellets overnight, filter, add a volume of naphthylisocyanate equal to 2% of the volume of triethylamine, distil. Prepare reagent by dissolving 1 mmole N-tert-butoxycarbonyl-L-leucine (BOC-L-Leu) in 3 mL dichloromethane, add 2 mL 250 mM dicyclohexylcarbodiimide in dichloromethane, let stand at 0° for 1 h, filter, use the filtrate as the reagent.)

HPLC VARIABLES
Column: 100 × 3.2 10 μm μBondapak C18
Mobile phase: MeCN:100 mM pH 3.0 phosphate buffer 30:70 (metoprolol) or 35:65 (alprenolol)
Flow rate: 0.5
Injection volume: 94
Detector: F ex 193 em (no cutoff filter)

CHROMATOGRAM
Retention time: 5 (S-metoprolol), 6.5 (R-metoprolol), 7 (S-alprenolol), 9 (R-alprenolol)
Limit of detection: 0.5 ng/mL

KEY WORDS
pharmacokinetics; plasma; chiral

REFERENCE
Hermansson, J.; von Bahr, C. Determination of (R)- and (S)-alprenolol and (R)- and (S)-metoprolol as their diastereomeric derivatives in human plasma by reversed-phase liquid chromatography, *J.Chromatogr.*, **1982**, *227*, 113–127.

SAMPLE
Matrix: blood
Analyte: propranolol
Sample preparation: 100 μL Whole blood + 75 ng IS + 6 mL diethyl ether + 2 mL 1 M pH 9.9 sodium carbonate, shake horizontally for 25 min, centrifuge at 2000 g for 10 min. Remove the organic layer and evaporate it to dryness under a stream of nitrogen at 30°, reconstitute the residue in 250 μL 1.2 M triethylamine in dichloromethane, add 200 μL reagent, vortex briefly, let stand at room temperature for 1 h, evaporate under a stream of nitrogen at 30° for 17 min, reconstitute the residue in 2 mL 100 mM NaOH, agitate for 5 min, add 6 mL diethyl ether, shake for 15 min, centrifuge at 2000 g. Remove the organic layer and evaporate it to dryness under a stream of nitrogen at 30°, cool the tubes to 0°, reconstitute the residue in 250 μL trifluoroacetic acid, vortex briefly, let stand at 0°, after 7 min add 2 mL 2 M NaOH, add 6 mL diethyl ether. Remove the organic layer and add it to 150 μL 100 mM orthophosphoric acid, extract, inject a 20 μL aliquot of the aqueous phase. (Prepare the reagent, Boc-L-Leu anhydride, as described above.)

HPLC VARIABLES
Guard column: 30 × 2.1 5 μm Spheri-5-RP-18
Column: 250 × 2 5 μm Ultrasphere ODS
Mobile phase: MeOH:20 mM pH 2.8 (NH$_4$)H$_2$PO$_4$ 72:28
Flow rate: 0.2
Injection volume: 20
Detector: F ex 228 em 290 (cutoff filter)

CHROMATOGRAM
Retention time: 18 (S-(-)), 29 (S-(+))
Internal standard: cyclopentyldesisopropylpropranolol (Pierce) (33, 54 (enantiomers))
Limit of detection: 2.5 ng/mL

KEY WORDS
chiral; rat; whole blood; pharmacokinetics

REFERENCE
Guttendorf, R.J.; Kostenbauder, H.B.; Wedlund, P.J. Quantification of propranolol enantiomers in small blood samples from rats by reversed-phase high-performance liquid chromatography after chiral derivatization, *J.Chromatogr.*, **1989**, *489*, 333–343.

O,O-Dibenzoyltartaric Acid Anhydride

SAMPLE
Matrix: bulk

Analyte: 3-aminoquinuclidine

Sample preparation: Prepare the free base by dissolving the hydrochloride in MeOH, add 2 equivalents of sodium methoxide, after a few min evaporate to dryness under a stream of nitrogen, reconstitute with dry dioxane (Caution! Dioxane is a carcinogen!) at a concentration of 1-2 mg/mL, add a 10% molar excess of the reagent in dioxane, vortex, let stand at room temperature for 30 min, add mobile phase to form a homogeneous mixture, inject an aliquot. (Prepare the reagent, R,R-(+)-O,O-dibenzoyltartaric acid anhydride, by refluxing R,R-(+)-O,O-dibenzoyltartaric acid in acetic anhydride for 30 min, cool, filter, wash with light petroleum (bp 35-60°), recrystallize from xylene to give R,R-(+)-O,O-dibenzoyltartaric acid anhydride (mp 196°; $[\alpha]_D^{20} = +152°$).)

HPLC VARIABLES
Column: 150 × 4.6 5 μm Zorbax C8

Mobile phase: MeCN:MeOH:buffer 15:25:60 (Buffer was 0.15% acetic acid adjusted to pH 4.2 with triethylamine.)

Flow rate: 2

Injection volume: 10

Detector: UV 254

CHROMATOGRAM
Retention time: 1.6 (S), 2.2 (R)

Limit of detection: 0.1% (of major enantiomer)

KEY WORDS
chiral; comparison with other derivatizing reagents

REFERENCE
Demian, I.; Gripshover, D.F. Enantiomeric purity determination of 3-aminoquinuclidine by diastereomeric derivatization and high-performance liquid chromatographic separation, *J.Chromatogr.*, **1989**, *466*, 415–420.

N-α-(9-Fluorenylmethyloxycarbonyl)leucine-N-carboxyanhydride

SAMPLE
Matrix: bulk
Analyte: amines
Sample preparation: Dissolve 2-10 μmoles amines in 35 μL 28% diisopropylethylamine in DMF, shake at room temperature for 10 min, add 25 μL 200-1000 mM N-α-(9-fluorenylmethyloxycarbonyl)leucine-N-carboxyanhydride in DMF, mix, let stand at room temperature for 10 min, add 200 μL 250 mM pH 8 sodium glycinate, mix, let stand for 5 min, add 300 μL chloroform, extract, dilute 10000-50000-fold with n-hexane, inject a 20 μL aliquot.

HPLC VARIABLES
Guard column: 15 × 3.2 5 μm Kromasil silica
Column: 250 × 4.6 5 μm Kromasil silica
Mobile phase: n-Hexane:isopropanol 99:1
Flow rate: 0.8
Injection volume: 20
Detector: F ex 263 em 313

CHROMATOGRAM
Retention time: k' 2.51 (S-2-aminoheptane), k' 3.12 (R-2-aminoheptane), k' 3.41 (S-2-aminopentane), k' 4.14 (R-2-aminopentane), k' 4.19 (R-1-phenylethylamine), k' 4.66 (R-2-aminobutane), k' 5.32 (S-1-phenylethylamine), k' 5.32 (S-2-aminobutane), k' 5.73 (R-3-methylpiperidine), k' 6.39 (S-3-methylpiperidine), k' 6.83 (R-2-methylpiperidine), k' 7.68 (S-2-methylpiperidine)

KEY WORDS
normal phase; chiral; the use of related derivatization reagents is discussed

REFERENCE
Pugniere, M.; Mattras, H.; Castro, B.; Previero, A. Adsorption liquid chromatography on silica for the chiral separation of amino acids and asymmetric amines derivatized with optically active N-α-9-fluorenylmethyloxyxarbonyl-amino acid-N-carboxyanhydrides, *J.Chromatogr.A*, **1997**, *767*, 69–75.

Maleic Anhydride

SAMPLE
Matrix: solution
Analyte: phytotoxins
Sample preparation: Add at least a 10-fold molar excess of maleic anhydride (either crystals or a 1 M solution in dioxane (Caution! Dioxane is a carcinogen!)) to a solution of phytotoxin in 100 mM pH 9.2 sodium carbonate, if necessary maintain the pH above 9.0 with 100 mM NaOH, when reaction is complete (typically 5 min) adjust pH to 6-7 with HCl or mobile phase, inject an aliquot.

HPLC VARIABLES
Guard column: 10 μm Micro Pak MCH-10 C18
Column: 300 $\times$ 4 10 μm Micro Pak MCH-10 C18
Mobile phase: Gradient. MeOH:buffer from 50:50 to 70:30 over 1 h (Buffer was 50 mM potassium phosphate adjusted to pH 2.7 with 100 mM HCl.)
Flow rate: 0.5
Injection volume: 10
Detector: UV 250

CHROMATOGRAM
Retention time: 30 (TA), 47 (TB)
Limit of quantitation: 500 pmoles

REFERENCE
Siler, D.J.; Gilchrist, D.G. Determination of host-selective phytotoxins from Alternaria alternata f. sp. lycopersici as their maleyl derivatives by high-performance liquid chromatography, *J.Chromatogr.*, **1982**, *238*, 167–173.

N-Methylisatoic Anhydride

SAMPLE
Matrix: solutions
Analyte: peptides
Sample preparation: Mix 10 μL 50 μg/mL solution of peptides in water with 50 μL 100 mM pH 10 sodium carbonate buffer and 40 μL 5 mg/mL N-methylisatoic anhydride in MeCN, let stand at 37° with occasional vortexing for 16 h, add 400 μL 2% acetic acid in MeOH:water 50:50, inject a 10 μL aliquot. (The paper contains a full discussion of various reaction conditions.)

HPLC VARIABLES
Column: 300 × 3.9 μBondapak C18
Mobile phase: Gradient. A was 0.1% trifluoroacetic acid in water. B was 0.1% trifluoroacetic acid in MeCN:MeOH 90:10. A:B from 22:78 to 50:50 over 32 min.
Flow rate: 1
Injection volume: 10
Detector: F ex 360 em 420

CHROMATOGRAM
Retention time: 21 (bradykinin), 23 (lys-bradykinin), 24 (met-lys-bradykinin)
Limit of detection: 0.1-0.5 pmole

REFERENCE
Anumula, K.R.; Schulz, R.P.; Back, N. Fluorescent N-methylanthranilyl (Mantyl) tag for peptides: its application in subpicomole determination of kinins, *Peptides*, **1992**, *13*, 663–669.

Pentafluoropropionic Anhydride

$$\left[CF_3CF_2C\overset{\overset{O}{\|}}{}\!\!-\!\!O \right]_2 \;+\; \underset{R}{\overset{R'}{\diagup}}NH \;\longrightarrow\; \underset{R}{\overset{R'}{\diagdown}}N\!\!-\!\!\overset{\overset{O}{\|}}{C}\!\!-\!\!CF_2CF_3$$

SAMPLE
Matrix: urine
Analyte: 4,4'-methylenedianiline
Sample preparation: Add 2 mL 6 M HCl to 100 mL urine, store in the refrigerator. 800 μL Acidified urine + 1.2 mL 6 M HCl, heat at 98° for 2 h, cool to room temperature, add 3 mL toluene, add 4 mL saturated NaOH, shake for 10 min, centrifuge at 1500 g for 10 min. Remove a 2 mL aliquot of the organic phase and add it to 20 μL pentafluoropropionic anhydride, shake for 10 min. Remove a 1 mL aliquot and evaporate it to dryness under vacuum at 30°, reconstitute with 700 μL mobile phase, inject a 50 μL aliquot.

HPLC VARIABLES
Column: 250 × 4.6 Apex II octadecyl (Jones Chromatography)
Mobile phase: MeCN:water 67:33
Flow rate: 1
Injection volume: 50
Detector: UV 258

CHROMATOGRAM
Retention time: 8
Limit of detection: 8 ng/mL

REFERENCE
Tiljander, A.; Skarping, G. Determination of 4,4'-methylenedianiline in hydrolysed human urine using liquid chromatography with UV detection and peak identification by absorbance ratio, *J.Chromatogr.*, **1990**, *511*, 185–194.

Propionic Anhydride

SAMPLE
Matrix: tissue
Analyte: 5-aminosalicylic acid
Sample preparation: Freeze mucosal intestinal biopsy (ca. 5 mg) in liquid nitrogen and crush the sample, allow to warm to room temperature, add 20 μL propionic anhydride, add 100 μL 57.6 μg/mL N-propionyl-4-aminosalicylic acid (purified by HPLC) in mobile phase, add 500 μL 50 mM pH 7.4 phosphate buffer, wash grinding/mixing rod with 500 μL 50 mM pH 7.4 phosphate buffer, sonicate (80 W) by immersing microprobe tip in mixture for 1 min, vortex, let stand at 37° for 1 h, add 500 μL 10% NaCl, add 6 mL MeCN, extract, cool at 4° for 1 h, centrifuge at 1500 g for 10 min. Remove the organic layer and evaporate it to dryness under a stream of air, reconstitute the residue in 500 μL mobile phase, filter (0.45 μm), inject a 100 μL aliquot.

HPLC VARIABLES
Guard column: 10 × 4.6 5 μm Spherisorb ODS-2
Column: 150 × 4.6 3 μm Spherisorb ODS-2
Mobile phase: MeCN:100 mM acetic acid:triethylamine 8:92:0.2
Flow rate: 1.5
Injection volume: 100
Detector: F ex 315 em 430

CHROMATOGRAM
Retention time: 5.7
Internal standard: N-propionyl-4-aminosalicylic acid (7.3)
Limit of detection: 1 ng

OTHER SUBSTANCES
Extracted: N-acetyl-5-aminosalicylic acid

KEY WORDS
mucosal intestinal biopsy; intestine

REFERENCE
De Vos, M.; Verdievel, H.; Schoonjans, R.; Beke, R.; De Weerdt, G.A.; Barbier, F. High-performance liquid chromatographic assay for the determination of 5-aminosalicylic acid and acetyl-5-aminosalicylic acid concentrations in endoscopic intestinal biopsy in humans, *J.Chromatogr.*, **1991**, *564*, 296–302.

Trifluoroacetic Anhydride

RELATED REFERENCE
Ellis, J.D.; Hand, E.L.; Gilbert, J.D. Use of LC-MS/MS to cross-validate a radioimmunoassay for the fibrinogen receptor antagonist, Aggrastat (tirofiban hydrochloride) in human plasma. *J.Pharm.Biomed.Anal.* **1997**, *15*, 561-569

SAMPLE
Matrix: bulk
Analyte: nomifensine
Sample preparation: Mix 8 mg nomifensine maleate with 200 μL trifluoroacetic anhydride, stir at room temperature for 10 min, evaporate to dryness under a stream of nitrogen at 40°, reconstitute the residue in 2 mL mobile phase, inject a 5 μL aliquot.

HPLC VARIABLES
Column: 250 × 3.6 5 μm Spherisorb silica
Mobile phase: Chloroform:MeOH:water 100:3:0.15 containing 2 mM (+)-camphor-10-sulfonic acid
Flow rate: 1
Injection volume: 5
Detector: UV 254

CHROMATOGRAM
Retention time: 16 (d), 18 (l)
Limit of quantitation: 25 ng

KEY WORDS
chiral; normal phase

REFERENCE
Tsujiyama, T.; Tsuchiya, M.; Hamachi, Y.; Kuriki, T.; Fukunaga, T.; Suzuki, N. Ion-pair chromatographic separation of nomifensine maleate enantiomers, *Anal.Sci.*, **1989**, *5*, 285–288.

CARBOXYLIC ACID

Dansyl-L-proline

SAMPLE
Matrix: solutions
Analyte: amines
Sample preparation: Mix a 200 μL aliquot of a 10 μM solution in DMF with 200 μL 10 μM dansyl-L-proline in DMF, add 100 μL 22 μM diethyl cyanophosphonate (diethyl phosphorocyanidate) in DMF, add 100 μL 42 μM triethylamine in DMF, mix, let stand for 3 min, inject a 5 μL aliquot.

HPLC VARIABLES
Column: 250 × 4.6 5 μm Shim-pack ODS
Mobile phase: MeOH:water 70:30 (A) or MeOH:water 80:20 (B) or MeCN:water 50:50 (C) or MeOH:water 60:40 (D)
Column temperature: 45
Flow rate: 0.6 (A, D) or 0.7 (B, C)
Injection volume: 5
Detector: F ex 345 em 515

CHROMATOGRAM
Retention time: k' 2.39 (L-2-aminopropan-1-ol (C)), k' 2.84 (D-2-aminopropan-1-ol (C)), k' 2.88 (L-α-amino-gamma-butyrolactone (D)), k' 3.23 (L-2-aminobutan-1-ol (C)), k' 3.31 (L-N-methylphenylethanolamine (A)), k' 3.35 (D-2-aminobutan-1-ol (C)), k' 3.53 (D-N-methylphenylethanolamine (A)), k' 3.55 (L-α-methylbenzylamine (B)), k' 3.66 (D-α-methylbenzylamine (B)), k' 3.72 (L-1-aminopropan-2-ol (D)), k' 3.78 (D-1-aminopropan-2-ol (D)), k' 3.84 (D-α-amino-gamma-butyrolactone (D)), k' 6.98 (norephedrine (C)), k' 7.36 (norpseudoephedrine (C)), k' 7.70 (L-phenylethanolamine (D)), k' 7.80 (D-phenylethanolamine (D))
Limit of detection: 40-50 fmole

KEY WORDS
chiral

REFERENCE
Gamoh, K.; Sawamoto, H. Resolution of amine enantiomers using precolumn derivatization with dansyl-L-proline and reversed-phase liquid chromatography, *Anal.Chim.Acta*, **1991**, *243*, 251–257.

N-(9-Fluorenylmethoxycarbonyl)glycine

+ RNH₂ ⟶

SAMPLE
Matrix: solutions
Analyte: glycosylamines
Sample preparation: Mix 150 μL of a solution in DMSO:DMF 2:1 with 50 μL 68 mg/mL N-(9-fluorenylmethoxycarbonyl)glycine in DMF, 2 μL N-ethyldiisopropylamine, 100 μL 129 mg/mL O-benzotriazol-1-yl-N,N,N',N'-tetramethyluronium hexafluorophosphate in DMF, and 30 mg/mL 1-hydroxybenzotriazole in DMF, let stand at room temperature for 2 h, inject an aliquot.

HPLC VARIABLES
Column: 250 × 4.6 Spherisorb S5NH-250A (Hichrom)
Mobile phase: Gradient. MeCN:water 100:0 for 10 min, to 80:20 over 15 min, to 60:40 over 20 min, to 0:100 over 6 min.
Flow rate: 1
Detector: UV 266

CHROMATOGRAM
Retention time: 20-30 (depending on structure)

REFERENCE
Arsequell, G.; Dwek, R.A.; Wong, S.Y.C. 9-Fluorenylmethoxycarbonyl (Fmoc)-glycine coupling of saccharide β-glycosylamines for the fractionation of oligosaccharides and the formation of neoglycoconjugates, *Anal.Biochem.*, **1994**, *216*, 165–170.

N-[4-(6-Methoxy-2-benzoxazolyl)]benzoyl-L-phenylalanine

SAMPLE

Matrix: solutions

Analyte: amines

Sample preparation: Mix 100 μL 300 μg/mL BOX-L-Phe in dichloromethane with 100 μL 11 mg/mL 2,2'-dipyridyl disulfide in dichloromethane, add 100 μL 13 mg/mL triphenylphosphine in dichloromethane, add 100 μL of a 0.5-100 μM solution of the amine in dichloromethane, mix, let stand at room temperature for 5 min, evaporate to dryness under a stream of nitrogen, reconstitute the residue in 500 μL mobile phase, inject a 10 μL aliquot. (BOX-L-Phe is N-[4-(6-methoxy-2-benzoxazolyl)]benzoyl-L-phenylalanine. Synthesis is as follows. Hydrogenate 5-methoxy-2-nitrophenol in EtOH over platinum oxide to give 2-amino-5-methoxyphenol (J. Org. Chem. 1957, 22, 220). Alternatively 6.9 g 5-methoxy-2-nitrophenol can be reduced to 2-amino-5-methoxyphenol with 50 g sodium hydrosulfite in 100 mL boiling water. It should be possible to prepare ethyl 4-methoxycarbonylbenzimidate hydrochloride ($CH_3OCOC_6H_4C(=NH)OC_2H_5 \cdot HCl$) by passing dry hydrogen chloride into a mixture of methyl 4-cyanobenzoate and 1.2-1.5 equivalents EtOH in an inert solvent (e.g., benzene, chloroform, dioxane, ether, nitrobenzene (Caution! Benzene, chloroform, and dioxane are carcinogens!) at 0-5°, the benzimidate should crystallize from the mixture in 7-10 days (J. Chem. Soc. 1942, 103). Add 5.7 g 2-amino-5-methoxyphenol to a stirred suspension of 9.6 g ethyl 4-methoxycarbonylbenzimidate hydrochloride in 200 mL EtOH, reflux for 2 h, cool in an ice-water bath, filter, dry the precipitate under reduced pressure. Chromatograph on silica gel with dichloromethane to obtain methyl 4-(6-methoxy-2-benzoxazolyl)benzoate. Dissolve the product in 66 mL MeOH:water 90:10, add 3 mL 5 M NaOH, reflux for 3 h, cool in an ice-water bath, acidify with HCl, filter, dry the product under reduced pressure to obtain 4-(6-methoxy-2-benzoxazolyl)benzoic acid (mp 286°). Suspend 2 g 4-(6-methoxy-2-benzoxazolyl)benzoic acid in 50 mL dichloromethane, cool to 0°, add 1.3 g 1-(3-dimethylaminopropyl)-3-ethylcarbodiimide hydrochloride, stir for 30 min, add 1.7 g L-phenylalanine t-butyl ester hydrochloride (Fluka), add 1 mL triethylamine, stir at room temperature overnight, add 50 mL dichloromethane, wash with 500 mM HCl, wash with 10% sodium bicarbonate, wash with water. Dry the organic layer over anhydrous sodium sulfate and evaporate it to dryness under reduced pressure, chromatograph on silica gel with dichloromethane. Evaporate the main fraction to obtain N-[4-(6-methoxy-2-benzoxazolyl)]benzoyl-L-phenylalanine t-butyl ester, add the ester to 5 mL trifluoroacetic acid, stir at room temperature overnight, add dichloromethane, evaporate to dryness, add dichloromethane to the residue and repeat several times. Dissolve the residue in dichloromethane and wash with water, dry the organic layer over anhydrous sodium sulfate, evaporate to dryness under reduced pressure, recrystallize from MeCN:water 70:30 to obtain N-[4-(6-methoxy-2-benzoxazolyl)]benzoyl-L-phenylalanine as a white crystalline powder (mp 229°).)

HPLC VARIABLES

Column: 150 × 4.6 5 µm TSKgel Silica-60 (Tosoh)
Mobile phase: Hexane:ethyl acetate 65:35
Column temperature: 40
Flow rate: 1
Injection volume: 10
Detector: F ex 325 em 403

CHROMATOGRAM

Retention time: k' 2.98 (R-1-(1-naphthyl)ethylamine), k' 4.01 (S-1-(1-naphthyl)ethylamine), k' 3.70 (R-1-phenylethylamine), k' 4.87 (S-1-phenylethylamine), k' 3.13 (R-1-cyclohexylethylamine), k' 3.83 (S-1-cyclohexylethylamine), k' 4.83 (L-phenylalanine methyl ester), k' 6.00 (D-phenylalanine methyl ester), k' 9.25 (L-alanine methyl ester), k' 11.65 (D-alanine methyl ester)

KEY WORDS

normal phase; comparison with other derivatization reagents; little separation under reverse phase conditions; chiral

REFERENCE

Kondo, J.; Imaoka, T.; Suzuki, N.; Kawasaki, T.; Nakanishi, A.; Kawahara, Y. N-[4-(6-methoxy-2-benzoxazolyl)]benzoyl-L-amino acids as chiral derivatization reagents for amine enantiomers, *Anal.Sci.*, **1994**, *10*, 697–703.

N-[4-(6-Methoxy-2-benzoxazolyl)]benzoyl-L-proline

SAMPLE

Matrix: solutions
Analyte: amines
Sample preparation: Mix 100 µL 300 µg/mL BOX-L-Pro in dichloromethane with 100 µL 11 mg/mL 2,2'-dipyridyl disulfide in dichloromethane, add 100 µL 13 mg/mL triphenylphosphine in dichloromethane, add 100 µL of a 0.5-100 µM solution of the amine in dichloromethane, mix, let stand at room temperature for 5 min, evaporate to dryness

under a stream of nitrogen, reconstitute the residue in 500 μL mobile phase, inject a 10 μL aliquot. (BOX-L-Pro is N-[4-(6-methoxy-2-benzoxazolyl)]benzoyl-L-proline. Synthesis is as follows. Hydrogenate 5-methoxy-2-nitrophenol in EtOH over platinum oxide to give 2-amino-5-methoxyphenol (J. Org. Chem. 1957, 22, 220). Alternatively 6.9 g 5-methoxy-2-nitrophenol can be reduced to 2-amino-5-methoxyphenol with 50 g sodium hydrosulfite in 100 mL boiling water. It should be possible to prepare ethyl 4-methoxycarbonylbenzimidate hydrochloride ($CH_3OCOC_6H_4C(=NH)OC_2H_5.HCl$) by passing dry hydrogen chloride into a mixture of methyl 4-cyanobenzoate and 1.2-1.5 equivalents EtOH in an inert solvent (e.g., benzene, chloroform, dioxane, ether, nitrobenzene (Caution! Benzene, chloroform, and dioxane are carcinogens!) at 0-5°, the benzimidate should crystallize from the mixture in 7-10 days (J. Chem. Soc. 1942, 103). Add 5.7 g 2-amino-5-methoxyphenol to a stirred suspension of 9.6 g ethyl 4-methoxycarbonylbenzimidate hydrochloride in 200 mL EtOH, reflux for 2 h, cool in an ice-water bath, filter, dry the precipitate under reduced pressure. Chromatograph on silica gel with dichloromethane to obtain methyl 4-(6-methoxy-2-benzoxazolyl)benzoate. Dissolve the product in 66 mL MeOH:water 90:10, add 3 mL 5 M NaOH, reflux for 3 h, cool in an ice-water bath, acidify with HCl, filter, dry the product under reduced pressure to obtain 4-(6-methoxy-2-benzoxazolyl)benzoic acid (mp 286°). Suspend 3 g 4-(6-methoxy-2-benzoxazolyl)benzoic acid in 50 mL dichloromethane, cool to 0°, add 1.9 g 1-(3-dimethylaminopropyl)-3-ethylcarbodiimide hydrochloride, stir for 30 min, add 1.7 g L-proline methyl ester hydrochloride, add 2.8 mL triethylamine, stir at room temperature overnight, add 50 mL dichloromethane, wash with 500 mM HCl, wash with 10% sodium bicarbonate, wash with water. Dry the organic layer over anhydrous sodium sulfate and evaporate it to dryness under reduced pressure, chromatograph on silica gel with dichloromethane. Evaporate the main fraction to obtain N-[4-(6-methoxy-2-benzoxazolyl)]benzoyl-L-proline methyl ester, dissolve the ester in 50 mL MeOH:water 90:10, add 3 mL 5 M NaOH, let stand overnight, acidify with 5% HCl, extract with ethyl acetate. Wash the organic layer with water and dry it over anhydrous sodium sulfate, evaporate to dryness under reduced pressure to obtain N-[4-(6-methoxy-2-benzoxazolyl)]benzoyl-L-proline as a white crystalline powder (mp 182°).)

HPLC VARIABLES
Column: 150 × 4.6 5 μm TSKgel Silica-60 (Tosoh)
Mobile phase: Hexane:isopropanol 75:25
Column temperature: 40
Flow rate: 1
Injection volume: 10
Detector: F ex 325 em 403

CHROMATOGRAM
Retention time: k' 1.85 (R-1-(1-naphthyl)ethylamine), k' 1.09 (S-1-(1-naphthyl)ethylamine), k' 1.92 (R-1-phenylethylamine), k' 1.14 (S-1-phenylethylamine), k' 1.13 (R-1-cyclohexylethylamine), k' 0.80 (S-1-cyclohexylethylamine), k' 3.19 (L-phenylalanine methyl ester), k' 2.02 (D-phenylalanine methyl ester)
Limit of detection: 30 fmole

KEY WORDS
normal phase; comparison with other derivatization reagents; less separation under reverse phase conditions; chiral

REFERENCE
Kondo, J.; Imaoka, T.; Suzuki, N.; Kawasaki, T.; Nakanishi, A.; Kawahara, Y. N-[4-(6-methoxy-2-benzoxazolyl)]benzoyl-L-amino acids as chiral derivatization reagents for amine enantiomers, *Anal.Sci.*, **1994**, *10*, 697–703.

DIALDEHYDE

Anthracene-2,3-dicarboxaldehyde/Cyanide

SAMPLE
Matrix: solutions
Analyte: amphetamine
Sample preparation: Mix 500 μL of a solution in 20 mM pH 9.5 borate buffer with 100 μL 10 mM NaCN in 20 mM pH 9.5 borate buffer, add 500 μL 0.1 mM anthracene-2,3-dicarboxaldehyde (Molecular Probes, Eugene OR) in MeOH, mix, let stand at room temperature for 20 min, inject a 25 μL aliquot.

HPLC VARIABLES
Column: 150 × 3.1 5 μm LiChrosorb RP-18
Mobile phase: MeCN:2.5 mM pH 7.0 imidazole buffer 70:30
Flow rate: 0.5
Injection volume: 25
Detector: Chemiluminescence (418 nm cutoff filter) following post-column reaction. The column effluent mixed with 50 mM hydrogen peroxide in MeCN containing 5 mM bis(2-nitrophenyl)oxalate pumped at 0.2 mL/min and the mixture flowed into the detector.

CHROMATOGRAM
Retention time: 10
Limit of detection: 2 fmole

KEY WORDS
naphthalene-2,3-dicarboxaldehyde may be a better reagent

REFERENCE
Kwakman, P.J.M.; Koelewijn, H.; Kool, I.; Brinkman, U.A.T.; de Jong, G.J. Naphthalene- and anthracene-2,3-dialdehyde as precolumn labelling reagents for primary amines using reversed- and normal-phase liquid chromatography with peroxyoxalate chemiluminescence detection, *J.Chromatogr.*, **1990**, *511*, 155–166.

1-Methoxycarbonylindolizine-3,5-dicarbaldehyde

SAMPLE

Matrix: blood

Analyte: amikacin

Sample preparation: Evaporate 100 μL of a 1 mg/mL solution of 1-methoxycarbonylin-dolizine-3,5-dicarbaldehyde in ethyl acetate under reduced pressure into the bottom of a tube. Filter (Ultrafree C3LCC) 200 μL plasma while centrifuging at 4° at 2000 g for 20 min. Add 40 μL reaction buffer to the reaction tube, sonicate for 30 s, add 20 μL plasma ultrafiltrate, mix well, let stand in the dark at room temperature for 15 min, inject an aliquot. (Prepare phosphate-borate buffer by mixing equal volumes of 20 mM NaH_2PO_4 and 20 mM sodium tetraborate, adjust pH to 10 with 1 M NaOH. Prepare reaction buffer by mixing equal volumes of EtOH and phosphate-borate buffer. Prepare reagent (1-meth-oxycarbonylindolizine-3,5-dicarbaldehyde) as follows. Reflux 21.4 g 2-pyridinecarboxal-dehyde, 24 mL ethylene glycol, 10 g p-toluenesulfonic acid, and 300 mL benzene (Caution! Benzene is a carcinogen!) under a Dean-Stark separator for 64 h, pour into concentrated sodium carbonate solution. Remove the organic layer and extract the aqueous layer 4 times with benzene. Combine the organic layers and wash them with water, dry over anhydrous magnesium sulfate, evaporate, distil the residue to give 2-(1,3-dioxolan-2-yl)pyridine (bp 122°/4 mm Hg) (J.Org.Chem. 1963, 28, 83). Reflux 15.1 g 2-(1,3-dioxolan-2-yl)pyridine and 19.5 g tert-butyl bromoacetate in 100 mL dry MeCN for 7 h, let stand overnight at room temperature, filter, wash the precipitate with diethyl ether to give 1-(tert-butoxycarbonylmethyl)-2-(1,3-dioxolan-2-yl)pyridinium bromide (mp 110-2° from MeCN). Suspend 51.9 g of this compound in 1.5 L THF with stirring, add 62.1 g potassium carbonate, add 15.12 g methyl propiolate, stir at room temperature for 9 days, filter, evaporate the filtrate to dryness under reduced pressure, chromatograph the residue on silica gel with hexane:ethyl acetate 20:1-10:1, collect fractions and evaporate to dryness to give methyl 3-tert-butoxycarbonyl-5-(1,3-dioxolan-2-yl)indolizine-1-carboxylate (mp 138-9° from hexane). Reflux 20.82 g of this compound in 600 mL THF and 60 mL 10% HCl for 6 h, concentrate to one quarter of the original volume, add water, extract with chloroform. Wash the chloroform layer with water and dry it over anhydrous sodium sulfate, evaporate, chromatograph on silica gel with hexane:ethyl acetate 10:1 to give 1-methoxycarbonylindolizine-5-carbaldehyde (mp 135-7° from MeOH). Stir 12.18 g of this compound in 116 mL dry DMF at 0° under argon, add 17 mL phosphorus oxychloride, stir at room temperature for 1 h, pour into water, adjust pH to 9.0 with 5% potassium carbonate, extract with chloroform. Wash the organic layer with water and dry it over anhydrous sodium sulfate, concentrate until a precipitate forms, filter to obtain the prod-uct, concentrate the filtrate and chromatograph the residue on silica gel with hexane:ethyl acetate 5:1 to obtain more product. The product was 1-methoxycarbonylindolizine-3,5-dicarbaldehyde (mp 164-5° from methyl acetate) (J. Chromatogr. A 1996, 724, 169.)

CAPILLARY ELECTROPHORESIS

Capillary: 75 cm × 50 μm fused-silica

Running buffer: 20 mM pH 7 Phosphate-borate buffer containing 40 mM sodium dodecyl sulfate

Voltage/Current: 30 kV

Injection: Dynamic compression injection at 50 hPa (sic) for 6 s

Detector: F ex 414 em 482; UV 280

Model: Jasco Model CE-990
Migration time: 17.6
Limit of quantitation: 5 µg/mL

KEY WORDS
plasma; ultrafiltrate

REFERENCE
Oguri, S.; Miki, Y. Determination of amikacin in human plasma by high-performance capillary electro-phoresis with fluorescence detection, *J.Chromatogr.B*, **1996**, *686*, 205–210.

SAMPLE
Matrix: solutions
Analyte: antibiotics
Sample preparation: Inject an aliquot of a 100 µM solution in buffer. (Prepare buffer by mixing equal volumes of 20 mM NaH_2PO_4 and 20 mM sodium tetraborate and adjusting the pH to 10 with 1 M NaOH.)

CAPILLARY ELECTROPHORESIS
Capillary: 110 cm × 50 µm fused-silica (100 cm to detector)
Capillary preparation: Before injection rinse capillary with running buffer at 200 mbar with no voltage for 3 min and with running buffer at 4 mbar at 20 kV for 30 s.
Running buffer: 100 mM pH 10 phosphate borate buffer containing 860 µM 1-methoxy-carbonylindolizine-3,5-dicarbaldehyde (Prepare 1-methoxycarbonylindolizine-3,5-dicar-baldehyde as described above.)
Capillary temperature: 40
Voltage/Current: 20 kV
Injection: Pressure injection at 50 mbar for 6 s (with no applied voltage).
Detector: UV 409
Model: JASCO Model CE-990
Migration time: 21 (netilmicin), 22 (micromicin), 24.5 (arbekacin), 25.5 (amikacin)

KEY WORDS
on-column derivatization

REFERENCE
Oguri, S.; Fujiyoshi, T.; Miki, Y. In-capillary derivatization with 1-methoxycarbonylindolizine-3,5-dicar-baldehyde for high-performance capillary electrophoresis, *Analyst*, **1996**, *121*, 1683–1688.

Naphthalene-2,3-dicarboxaldehyde/N-Acetyl-D-penicillamine

SAMPLE
Matrix: blood
Analyte: cyclic heptapeptide
Sample preparation: Condition a cyclohexyl SPE cartridge (Analytichem) with 2 mL MeOH and 3 mL water. 1 mL Plasma + 1 mL 500 mM phosphoric acid, vortex, add to the SPE cartridge, wash with 3 mL water, elute with 500 μL eluant. Remove a 350 μL aliquot of the eluate and add it to 40 μL 1 M pH 10.6 carbonate buffer, add 20 μL 1 mg/mL N-acetyl-D-penicillamine in MeOH, add 10 μL 1 mg/mL naphthalene-2,3-dicarboxaldehyde in MeOH, mix, let stand for 1 min, inject a 50 μL aliquot onto column A and elute to waste with mobile phase A, after 27 min divert the effluent from column A onto column B, after 5 min divert the effluent from column A to waste and elute column B with mobile phase B, monitor the effluent from column B. (Eluant was MeCN:25 mM sodium citrate:25 mM pH 8.5 NaH$_2$PO$_4$ 30:35:35.)

HPLC VARIABLES
Column: A 150 × 4.6 5 μm 100 Å PLRP-S (Polymer Labs); B 150 × 4.6 5 μm 100 Å PLRP-S (Polymer Labs)
Mobile phase: A MeCN:buffer 1 25:75; B MeCN:buffer 2 32.5:67.5 (Buffer 1 was 25 mM sodium citrate containing 25 mM NaH$_2$PO$_4$, adjusted to pH 7.5 with phosphoric acid. Buffer 2 was 25 mM sodium citrate containing 25 mM NaH$_2$PO$_4$, adjusted to pH 5.1 with phosphoric acid.)
Injection volume: 50
Detector: F ex 436 em >440 (cut-off filter)

CHROMATOGRAM
Retention time: 37
Limit of quantitation: 1 ng/mL

KEY WORDS
column-switching; plasma; SPE

REFERENCE
Kline, W.F.; Matuszewski, B.K.; Hsieh, J.Y.-K. Determination of a cyclic heptapeptide, a novel fibrinogen receptor antagonist, in human plasma by high-performance liquid chromatography with automated

pre-column derivatization, column switching and fluorescence detection, *J.Chromatogr.*, **1992**, *578*, 31–37.

SAMPLE
Matrix: bone
Analyte: pamidronate
Sample preparation: Dissolve 25 mg ground (<20 μm) bone in 2 mL 200 mM HCl, add 5 μg IS, vortex, let stand overnight at room temperature, centrifuge. Remove a 500 μL aliquot and add it to 1 mL 10 mM NaOH and 50 μL 1 M NaOH, centrifuge at 1000 g for 10 min, wash the pellet with 1 mL water, centrifuge, discard the supernatant. Dissolve the pellet in 200 μL 200 mM phosphoric acid, add 250 μL 200 mM pH 10.3 EDTA in 200 mM NaOH, add 200 μL resin, vortex, centrifuge, filter (0.2 μm) a 550 μL aliquot of the supernatant, add 10 μL 10 M NaOH to the filtrate. Remove a 50 μL aliquot and add it to 50 μL 1 M pH 10.7 carbonate buffer, add 10 μL 1 mg/mL naphthalene-2,3-dicarboxaldehyde, add 10 μL 1 mg/mL N-acetyl-D-penicillamine, mix, let stand for 2 min, inject a 20-50 μL aliquot. (The resin was AG 50W-X8 (K$^+$ form) resin. Prepare the resin by adding 3 volumes of 1 M KOH to 200-400 mesh AG 50W-X8 cation-exchange resin H$^+$ form (Bio-Rad), stir for 30 s, decant the supernatant, repeat the procedure twice, wash five times with 3 volumes of water, store at 4°, before use wash 2 or 3 times with three volumes of water (J. Chromatogr. 1992, 584, 213).)

HPLC VARIABLES
Guard column: present but not specified
Column: 150 × 4.6 PLRP-S (Phenomenex)
Mobile phase: MeCN:25 mM pH 6.5 citrate-phosphate buffer 16:94
Flow rate: 1
Injection volume: 20-50
Detector: F ex 436 em 508

CHROMATOGRAM
Retention time: 5.0
Internal standard: 1-hydroxypentanylidene-1,1-bisphosphonate (Ciba-Geigy CGP 38146) (13.5)
Limit of quantitation: 7.5 μg/g

REFERENCE
King, L.E.; Vieth, R. Extraction and measurement of pamidronate from bone samples using automated pre-column derivatization, high-performance liquid chromatography and fluorescence detection, *J.Chromatogr.B*, **1996**, *678*, 325–330.

Naphthalene-2,3-dicarboxaldehyde/Cyanide

RELATED REFERENCES

de Montigny, P.; Stobaugh, J.F.; Givens, R.S.; Carlson, R.G.; Srinivasachar, K.; Sternson, L.A.; Higuchi, T. Naphthalene-2,3-dicarboxyaldehyde/cyanide ion: a rationally designed fluorogenic reagent for primary amines. *Anal.Chem.* **1987**, *59*, 1096-1101.

Eisenberg, E.J.; Bidgood, A.; Cundy, K.C. Penetration of GS4071, a novel influenza neuraminidase inhibitor, into rat bronchoalveolar lining fluid following oral administration of the prodrug GS4104. *Antimicrob.Agents Chemother.* **1997**, *41*, 1949-1952.

Kristjansson, F.; Thakur, A.; Stobaugh, J.F. Selective fluorogenic derivatization of Pro2-Lys peptides with naphthalene-2,3-dicarboxaldehyde/cyanide. *Anal.Chim.Acta* **1992**, *262*, 209-215.

Lunte, S.M.; Wong, O.S. Naphthalenedialdehyde-cyanide: a versatile fluorogenic reagent for the LC analysis of peptides and other primary amines. *LC-GC* **1989**, *7*, 908-914.

Lunte, S.M.; Wong, O.S. Precolumn derivatization with naphthalenedialdehyde/cyanide for fluorescence, chemiluminescence or electrochemical detection of primary amines and peptides. *Curr.Sep.* **1990**, *10*, 19-25.

Nerurkar, M.M.; Rose, M.J.; Stobaugh, J.F.; Borchardt, R.T. Selective fluorogenic derivatization of a peptide nucleic acid trimer with naphthalene-2,3-dicarboxaldehyde. *J.Pharm.Biomed.Anal.* **1997**, *15*, 945-950.

Pewnim, T.; Miles, C.; Seifert, J. A new method of L-kynurenine precolumn derivatization with naphthalene-2,3-dicarboxaldehyde and cyanide for high-performance liquid chromatography (HPLC). *Fresenius' J.Anal.Chem.* **1992**, *343*, 907–908.

SAMPLE
Matrix: CSF
Analyte: catecholamines
Sample preparation: 720 μL CSF + 80 μL 100 nM dihydroxybenzylamine in 175 mM perchloric acid + 160 μL 500 mM pH 8.6 borate buffer + 80 μL 22.5 mM NaCN in water + 40 μL 5 mM naphthalene-2,3-dicarboxaldehyde in MeOH, mix, let stand at room temperature for 5 min, inject an aliquot.

CAPILLARY ELECTROPHORESIS
Capillary: 43 cm × 25 μm fused-silica (23 cm to detector) (Polymicro Technologies)
Running buffer: 200 mM pH 7.05 ± 0.02 phosphate buffer
Capillary temperature: 25.5
Voltage/Current: 29 kV
Injection: Hydrodynamic injection by vacuum (300 mm Hg) to give 2.5 nL sample injection followed by 60 pL 200 mM orthophosphoric acid
Detector: F ex 442 (He-Cd laser) em 490 (bandpass filter)
Model: Zeta Technology IRIS 2000
Migration time: 1.319 (dopamine), 1.326 (norepinephrine)
Internal standard: dihydroxybenzylamine (1.346)
Limit of detection: 0.086 nM

OTHER SUBSTANCES
Non-interfering: amino acids

REFERENCE
Bert, L.; Robert, F.; Denoroy, L.; Renaud, B. High-speed separation of subnanomolar concentrations of noradrenaline and dopamine using capillary zone electrophoresis with laser-induced fluorescence detection, *Electrophoresis*, **1996**, *17*, 523–525.

SAMPLE
Matrix: blood
Analyte: gusperimus
Sample preparation: Filter (Amicon MPS-1 with 14 mm YM20 membrane) 1 mL plasma while centrifuging at 800 g for 1 h. Remove a 400 μL aliquot of the ultrafiltrate and add it to 50 μL 500 mM pH 6.8 phosphate buffer, add 50 μL 10 mM NaCN in water, add 100 μL 2 mM naphthalene-2,3-dicarboxaldehyde in MeCN, let stand at room temperature for 15 min, add 50 μL 500 mM pH 3.0 sodium acetate buffer, inject an aliquot.

HPLC VARIABLES
Column: 150 × 4.6 5 μm ODS Hypersil
Mobile phase: MeCN:100 mM KH$_2$PO$_4$:phosphoric acid 52:48:0.8 containing 18 mM sodium dodecyl sulfate
Column temperature: 40 ± 0.1°
Flow rate: 2
Injection volume: 50
Detector: F ex 420 em 490

CHROMATOGRAM
Retention time: 14
Limit of quantitation: 5 ng/mL

KEY WORDS
plasma; ultrafiltrate

REFERENCE
Sprancmanis, L.A.; Riley, C.M.; Stobaugh, J.F. Determination of the anticancer drug, 15-deoxyspergu-
alin, in plasma ultrafiltrate by liquid chromatography and precolumn derivatization with naphtha-
lene-2,3-dicarboxaldehyde/cyanide, *J.Pharm.Biomed.Anal.*, **1990**, *8*, 165–175.

SAMPLE
Matrix: blood
Analyte: peptides
Sample preparation: Condition a Baker-10 SPE-C8 SPE cartridge with 6 mL MeOH and
6 mL water. 4 mL Plasma + 750 µL 1% glycine in 1 M HCl + 250 µL 70% perchloric
acid, centrifuge at 2500 rpm for 15 min. Remove a 500 µL aliquot of the supernatant and
add it to the SPE cartridge, wash with 15 mL water, elute with 1.5 mL MeOH, evaporate
the eluate to dryness under a stream of nitrogen at 50°, reconstitute the residue in 200
µL buffer, add 25 µL 5 mM naphthalene-2,3-dicarboxaldehyde in MeCN (prepare fresh
each day), add 25 µL 10 mM NaCN in buffer (prepare fresh each day), mix, let stand at
room temperature for 3 min, inject a 50 µL aliquot. (Prepare buffer 0.5 moles KH_2PO_4
and 0.5 moles K_2HPO_4 in 1 L water, pH 6.8.)

HPLC VARIABLES
Guard column: 50 × 4.6 5 µm ODS Hypersil
Column: 250 × 4.6 3 µm ODS Hypersil
Mobile phase: MeCN:THF:buffer 45:4:51 (Prepare buffer by adjusting the pH of 50 mM
KH_2PO_4 to 3.0 with concentrated phosphoric acid.)
Flow rate: 1
Injection volume: 50
Detector: F ex 420 em 490

CHROMATOGRAM
Retention time: 7.5 (leucine enkephalinamide), 8.5 (leucine enkephalin)
Limit of quantitation: 310 nM

KEY WORDS
plasma; SPE

REFERENCE
de Montigny, P.; Riley, C.M.; Sternson, L.A.; Stobaugh, J.F. Fluorogenic derivatization of peptides with
naphthalene-2,3-dicarboxaldehyde/cyanide: optimization of yield and application in the determination of leucine-enkephalin spiked human plasma samples, *J.Pharm.Biomed.Anal.*, **1990**, *8*, 419–429.

SAMPLE
Matrix: blood
Analyte: histamine
Sample preparation: Mix plasma with pH 9.7 buffer containing 8.6 mM cyanide and 17.2
mM naphthalene-2,3-dicarboxaldehyde, let stand for 20 min, add 200 mM taurine, inject
an aliquot.

HPLC VARIABLES
Column: 250 × 4.6 5 µm Ultramex C8
Mobile phase: MeCN:pH 6.8 phosphate buffer 40:60
Detector: F

CHROMATOGRAM
Limit of quantitation: 75 pg

KEY WORDS
plasma

REFERENCE
James, J.; Lowe, D.; Karnes, H.T. Determination of histamine from plasma using derivatization with
naphthalene-2,3-dicarboxaldehyde and HPLC with fluorescence detection, *Pharm.Res.*, **1992**, *9*, S21.

SAMPLE
Matrix: blood, urine
Analyte: alendronate
Sample preparation: Urine. Condition an Analytichem 3 mL 500 mg diethylamine SPE cartridge with 3 mL water. 5 mL Urine + 50 μL 2.5 M calcium chloride, vortex, add 50 μL or more (as needed) 1 M NaOH to form a slight white precipitate, centrifuge. Discard supernatant and dissolve pellet in 50 μL 1 M HCl, add 5 mL water, precipitate by adding 50 μL 1 M NaOH, centrifuge. Discard the supernatant and dissolve the pellet in 50 μL 1 M HCl, add 1 mL 10 mM EDTA, add 2 mL 100 mM pH 4.0 sodium acetate buffer, add to the SPE cartridge, wash with 3 mL water, elute with 3 mL buffer. Remove 250 μL of the eluate and add it to 250 μL 100 mM pH 9.1 sodium borate buffer, vortex, add 10 μL 50 mM KCN, add 10 μL 2 mg/mL naphthalene-2,3-dicarboxaldehyde in MeOH, let stand for 15 min, inject a 100 μL aliquot. Plasma. Condition an Analytichem 3 mL 500 mg diethylamine SPE cartridge with 3 mL water. 1 mL Plasma + 1 mL 10% trichloroacetic acid, vortex, centrifuge at 5300 g for 10 min. Remove the supernatant and add it to 50 μL 100 mM sodium pyrophosphate in water, add 50 μL 2.5 M calcium chloride, vortex, add 50 μL or more (as needed) 1 M NaOH to form a slight white precipitate, centrifuge. Discard supernatant and dissolve pellet in 50 μL 1 M HCl, add 5 mL water, precipitate by adding 50 μL 1 M NaOH, centrifuge. Discard the supernatant and dissolve the pellet in 50 μL 1 M HCl, add 1 mL 10 mM EDTA, add 2 mL 100 mM pH 4.0 sodium acetate buffer, add to the SPE cartridge, wash with 3 mL water, elute with 3 mL buffer. Remove 250μL of the eluate and add it to 250 μL 100 mM pH 9.1 sodium borate buffer, vortex, add 10 μL 50 mM KCN, add 10 μL 2 mg/mL naphthalene-2,3-dicarboxaldehyde in MeOH, let stand for 15 min, inject a 100 μL aliquot. (Buffer was 50 mM sodium citrate : 50 mM pH 8.5 sodium phosphate buffer 1 : 1.)

HPLC VARIABLES
Guard column: 20 × 4.6 Hamilton PRP-1
Column: 150 × 4.6 5 μm 100 Å PLRP-S polymeric reversed-phase
Mobile phase: MeOH:buffer 40:60 (Buffer was 25 mM sodium citrate and 25 mM dihydrogenphosphate adjusted to pH 8.5 with 10 mM NaOH.)
Flow rate: 1
Injection volume: 100
Detector: F ex 420 em 490

CHROMATOGRAM
Retention time: 5
Limit of quantitation: 5 ng/mL (urine)

KEY WORDS
plasma

REFERENCE
Kline, W.F.; Matuszewski, B.K.; Bayne, W.F. Determination of 4-amino-1-hydroxybutane-1,1-bisphosphonic acid in urine by automated pre-column derivatization with 2,3-naphthalene dicarboxaldehyde and high-performance liquid chromatography with fluorescence detection, *J.Chromatogr.*, **1990**, *534*, 139–149.

SAMPLE
Matrix: dialysate
Analyte: eflornithine (α-difluoromethylornithine)
Sample preparation: 4 μL Dialysate + 2 μL 50 mM pH 10 borate buffer + 2 μL 15 mM NaCN in water + 2 μL 15 mM naphthalene-2,3-dicarboxaldehyde in MeCN, let stand for 10 min (color changes from colorless to bright yellow), inject an aliquot.

CAPILLARY ELECTROPHORESIS
Capillary: 75 cm × 50 μm fused-silica (50 cm to detector) (Polymicro Technologies)
Running buffer: 100 mM pH 6.5 Phosphate buffer
Voltage/Current: 15 kV/49 μA
Injection: Vacuum injection for 4 s (3.7 nL)
Detector: UV 254
Model: ISCO Model 3850

Migration time: 11
Limit of detection: 5 μM

KEY WORDS
rat; pharmacokinetics

REFERENCE
Hu, T.; Zuo, H.; Riley, C.M.; Stobaugh, J.F.; Lunte, S.M. Determination of α-difluoromethylornithine in blood by microdialysis sampling and capillary electrophoresis with UV detection, *J.Chromatogr.A*, **1995**, *716*, 381–388.

SAMPLE
Matrix: elastin
Analyte: desmosine and isodesmosine
Sample preparation: Dissolve 10 mg elastin in 1 mL 6 M HCl, heat in a sealed tube at 110° for 24 h, evaporate to dryness under a stream of nitrogen, reconstitute the residue in 5 mL buffer. Remove a 200 μL aliquot and add it to 100 μL 490 μg/mL NaCN in water, add 200 μL 6 mg/mL naphthalene-2,3-dicarboxaldehyde in MeCN, make up to 10 mL with buffer, let stand for 15 min, inject an aliquot. (Prepare buffer by adjusting the pH of 26 g/L sodium borate in water to 9.5 with NaOH.)

HPLC VARIABLES
Column: 150 × 4.6 3 μm LC-18 DB (Supelco)
Mobile phase: Gradient. A was MeOH:THF:5 mM sodium citrate 10:5:85 containing 50 mM sodium perchlorate. B was MeOH:5 mM citrate 90:10 containing 50 mM sodium perchlorate. A:B from 50:50 to 0:100 over 30 min.
Injection volume: 20
Detector: E, Bioanalytical Systems LC-4B, dual electrode +750 mV, Ag/AgCl reference electrode

CHROMATOGRAM
Retention time: 17 (isodesmosine), 19 (desmosine)
Limit of detection: 100 fmole

OTHER SUBSTANCES
Extracted: amino acids

REFERENCE
Lunte, S.M.; Mohabbat, T.; Wong, O.S.; Kuwana, T. Determination of desmosine, isodesmosine, and other amino acids by liquid chromatography with electrochemical detection following precolumn derivatization with naphthalenedialdehyde/cyanide, *Anal.Biochem.*, **1989**, *178*, 202–207.

SAMPLE
Matrix: solutions
Analyte: amines
Sample preparation: 100 μL 100 μM Solution + 300 μL 50 mM pH 9.0 sodium borate buffer + 400 μL MeCN + 100 μL 10 mM NaCN in water + 100 μL 2 mM naphthalene-2,3-dicarboxaldehyde in MeOH, mix, let stand in the dark for 2 h, dilute with MeCN:water 50:50, inject a 20 μL aliquot.

HPLC VARIABLES
Guard column: 30 × 4.6 Spheri-5 RP-18
Column: 250 × 4.6 Inertsil ODS-2
Mobile phase: MeCN:water 80:20 containing 1 mM imidazole, pH adjusted to 7.0 with nitric acid
Flow rate: 1
Injection volume: 20
Detector: Chemiluminescence following post-column reaction. The column effluent mixed with the reagent pumped at 1 mL/min and the mixture flowed through a 300 mm × 0.25 mm ID coil to the detector. (Prepare the reagent by dissolving 112 mg bis(2,4,6-trichlorophenyl) oxalate in 500 mL MeCN, add 8.6 mL 30% hydrogen peroxide, sonicate.); F ex 418 em 483

CHROMATOGRAM
Retention time: 6.5 (phenylpropanolamine), 7.5 (benzylamine), 8.5 (phenylethylamine), 11 (phenylpropylamine), 13 (phenylbutylamine)
Limit of detection: 0.2 fmole (chemiluminescence), 20-30 fmole (F)

KEY WORDS
post-column reaction; comparison with other derivatization reagents

REFERENCE
Hayakawa, K.; Hasegawa, K.; Imaizumi, N.; Wong, O.S.; Miyazaki, M. Determination of amphetamine-related compounds by high-performance liquid chromatography with chemiluminescence and fluorescence detections, *J.Chromatogr.*, **1989**, *464*, 343–352.

SAMPLE
Matrix: solutions
Analyte: amines
Sample preparation: Mix sample:50 (?) mM NaCN in 50 mM pH 9.3 borate buffer: 25 (?) mM naphthalene-2,3-dicarboxaldehyde in MeOH 3:1:1, let stand for 15 min, inject a 50 μL aliquot.

HPLC VARIABLES
Column: 200 × 3 5 μm Chromspher ODS-2 C18 (Chrompack)
Mobile phase: Gradient. A was THF:50 mM pH 6.8 potassium phosphate buffer 5:95. B was MeCN:MeOH:50 mM pH 6.8 potassium phosphate buffer 55:10:35. A:B from 70:30 to 0:100 over 1 h, maintain at 0:100 for 20 min.
Flow rate: 0.5
Injection volume: 50
Detector: F ex 420

CHROMATOGRAM
Retention time: 32 (baclofen), 66 (amphetamine), 69 (tranylcypromine)

REFERENCE
Koning, H.; Wolf, H.; Venema, K.; Korf, J. Automated precolumn derivatization of amino acids, small peptides, brain amines and drugs with primary amino groups for reversed-phase high-performance liquid chromatography using naphthalenedialdehyde as the fluorogenic label, *J.Chromatogr.*, **1990**, *533*, 171–178.

SAMPLE
Matrix: solutions
Analyte: angiotensin II
Sample preparation: Mix 20 μL of a 25 μM peptide solution in water with 50 μL 200 mM ascorbic acid in water, 100 μL 10 mM NaCN in water, 200 μL 5 mM naphthalene-2,3-dicarboxaldehyde in MeCN, and 540 μL 100 mM pH 7.0 phosphate buffer, mix, let stand at 0-4° for 20 min, add 50 μL 200 mM taurine in water, inject an aliquot.

HPLC VARIABLES
Column: 150 × 4.6 5 μm ODS Hypersil
Mobile phase: THF:100 mM pH 6.0 phosphate buffer 19:81 containing 30 mM sodium 1-octanesulfonate
Column temperature: 40 ± 0.1
Flow rate: 1-2
Detector: F ex 420 em 490

CHROMATOGRAM
Retention time: k' 11
Limit of detection: 50-100 fmole

OTHER SUBSTANCES
Simultaneously analyzed: similar peptides

REFERENCE
Patel, H.B.; Stobaugh, J.F.; Riley, C.M. Reversed-phase ion-pair liquid chromatography of the angiotensins, *J.Chromatogr.*, **1991**, *536*, 357–370.

SAMPLE
Matrix: solutions
Analyte: dipeptides
Sample preparation: 100 μL Solution + 700 μL 100 mM pH 9.0 borate buffer + 100 μL 10 mM NaCN + 100 μL 1 mM naphthalene-2,3-dialdehyde, vortex for 30 s, let stand at room temperature for 30 min, inject an aliquot.

CAPILLARY ELECTROPHORESIS
Capillary: 70 cm × 50 μm fused-silica (64 cm to detector) (Polymicro Technologies)
Capillary preparation: Before injection wash capillary with 100 mM NaOH for 15 min, with water for 15 min, and with running buffer for 30 min.
Running buffer: 100 mM Borate buffer containing 50 mM sodium dodecyl sulfate and 20 mM gamma-cyclodextrin
Voltage/Current: 20 kV
Injection: Electrokinetic injection
Detector: F ex 441.6 (He-Cd laser) em 497 (497 nm bandpass filter with a 475 nm longpass filter)
Migration time: 12.5 (L-Leu-L-Leu), 12.6 (D-Leu-D-Leu), 13.1 (L-Leu-D-Leu), 13.3 (D-Leu-L-Leu)

KEY WORDS
chiral

REFERENCE
DeSilva, K.; Jiang, Q.; Kuwana, T. Chiral separation of naphthalene-2,3-dialdehyde labelled peptides by cyclodextrin-modified electrokinetic chromatography, *Biomed.Chromatogr.*, **1995**, 9, 295–301.

SAMPLE
Matrix: tissue
Analyte: enkephalins
Sample preparation: Condition a Bond Elut C18 SPE cartridge with 3 mL water and 3 mL MeOH (in this order ?). Homogenize 200 mg tissue with 3 mL 100 mM HCl at 0-4°, add 25 μL 20 μM IS, add 1 mL 100 mM HCl, add 500 μL 2 M perchloric acid, mix, centrifuge at 800 g for 10 min, remove the supernatant, suspend the precipitate in 2 mL 200 mM perchloric acid, centrifuge. Combine the supernatants and adjust pH to 7-8 with about 2 mL 1 M sodium bicarbonate, add to the SPE cartridge, wash with 1 mL water, wash with 2 mL dichloromethane, wash with 1 mL water, wash with 3 mL 100 mM HCl, wash with 1 mL water, wash with 3 mL 100 mM pH 8.5 borate buffer, wash with 1 mL water, elute with 1 mL MeOH:water 90:10 (J. Chromatogr. 1988, 430, 271). Evaporate the eluate to dryness under vacuum at 30°, reconstitute with 540 μL 100 mM pH 6.8 phosphate buffer + 20 μL 25 μM IS + 50 μL 200 mM ascorbic acid + 100 μL 10 mM KCN in water + 200 μL 5 mM naphthalene-2,3-dicarboxaldehyde in MeCN (prepare fresh each week), shake by hand for 30 s, let stand in an ice bath for 20 min, add 50 μL 200 mM taurine, inject an aliquot onto column A and elute to waste with mobile phase, after 3.2 min divert the effluent from column A onto column B, after 1 min again divert the effluent from column A to waste, elute column B with mobile phase, monitor the effluent from column B.

HPLC VARIABLES
Column: A 150 × 4.6 5 μm Spherisorb phenyl; B 150 × 4.6 5 μm ODS Hypersil
Mobile phase: MeCN:26 mM pH 3.5 trifluoroacetic acid 45:55
Column temperature: 30
Flow rate: 1
Injection volume: 20
Detector: F ex 420 em 490

CHROMATOGRAM
Retention time: 10.5 (methionine enkephalin), 13 (leucine enkephalin)
Internal standard: Tyr-[D]-Ala-Gly-Phe-Met (11.5)
Limit of detection: 25 pmole/g

KEY WORDS
protect from light; rat; brain; column-switching; heart-cut; SPE

REFERENCE
Mifune, M.; Krehbiel, D.K.; Stobaugh, J.F.; Riley, C.M. Multi-dimensional high-performance liquid chromatography of opioid peptides following pre-column derivatization with naphthalene-2,3-dicarboxaldehyde in the presence of cyanide ion. Preliminary results on the determination of leucine- and methionine-enkephalin-like fluorescence in the striatum region of the rat brain, *J.Chromatogr.*, **1989**, *496*, 55–70.

SAMPLE
Matrix: tissue
Analyte: amines and amino acids
Sample preparation: Sonicate (Artix Sonic Dismembrator Model 150, power setting 30) 1-2 mg rat brain tissue and 10 μL EtOH:water 70:30 in an ice bath for 5-10 s, centrifuge at 16000 g for 10 min. 1 μL Supernatant + 1 μL 490 μM α-aminoadipic acid, mix, remove a 1 μL aliquot and add it to 5 μL 20 mM pH 9.0 sodium borate buffer, add 1.5 μL 20 mM NaCN in water, mix, add 1.5 μL 20 mM naphthalene-2,3-dicarboxaldehyde in MeCN, mix thoroughly, let stand at room temperature for 30 min (protect from light), inject an aliquot.

CAPILLARY ELECTROPHORESIS
Capillary: 115 cm × 50 μm fused-silica (95 cm to detector)
Running buffer: 20 mM pH 9.0 sodium borate buffer
Voltage/Current: 30 kV
Injection: Electrokinetic injection at 5 kV for 12 s
Detector: UV 420
Migration time: 13.6 (norepinephrine), 13.8 (dopamine), 14 (Trp), 14.3 (Phe), 14.5 (Tyr), 14.7 (Asn), 15 (gamma-aminobutyric acid), 15.1 (Ser), 15.4 (Ala), 15.8 (taurine), 16 (Gly), 19.6 (levodopa), 21 (phosphoethanolamine), 22.7 (Glu), 24.2 (Asp)
Internal standard: α-aminoadipic acid (21.6)
Limit of detection: 4.4 μM

KEY WORDS
rat; brain

REFERENCE
Weber, P.L.; O'Shea, T.J.; Lunte, S.M. Separation and quantitation of the amino acid neurotransmitters in rat brain by capillary electrophoresis, *J.Pharm.Biomed.Anal.*, **1994**, *12*, 319–324.

SAMPLE
Matrix: urine
Analyte: catecholamines
Sample preparation: 1 mL Urine + 2 mL water + 1 mL 500 μM EDTA in 1 mM HCl + 1 mL 2 M pH 8.5 phosphate buffer + 20 μL 100 μM 3,4-dihydroxybenzylamine + 50 mg alumina, vortex for 3 min, discard the supernatant, wash the alumina 3 times with 5 mL portions of water, elute by washing the alumina twice with 200 μL portions of 100 mM phosphoric acid for 30 s each time. Combine the acidic layers and add them to 560 μL 400 mM pH 9.0 borate buffer, add 20 μL 50 mM NaCN, add 20 μL 5 mM naphthalene-2,3-carboxaldehyde in MeOH, mix thoroughly, let stand at room temperature for 20 min, inject a 5 μL aliquot.

HPLC VARIABLES
Column: 150 × 4.6 5 μm ODS-120T (Toyo Soda)
Mobile phase: MeCN:THF:10 mM pH 2.5 phosphate buffer 38:6:56
Flow rate: 1
Injection volume: 5
Detector: F ex 420 em 483

CHROMATOGRAM
Retention time: 10 (dopamine), 18 (norepinephrine)
Internal standard: 3,4-dihydroxybenzylamine (15.5)
Limit of detection: 20 fmole

KEY WORDS
SPE

REFERENCE
Kawasaki, T.; Higuchi, T.; Imai, K.; Wong, O.S. Determination of dopamine, norepinephrine, and related trace amines by prechromatographic derivatization with naphthalene-2,3-dicarboxaldehyde, *Anal.Biochem.*, **1989**, *180*, 279–285.

SAMPLE
Matrix: urine
Analyte: cyanide
Sample preparation: Mix 20 mL urine with 100 µL 4 M NaOH, centrifuge at 1000 g for 15 min, dilute a 500 µL aliquot of the supernatant to 10 mL with water. Remove a 500 µL aliquot and add it to 100 µL 30 mM taurine in buffer and 100 µL reagent, let stand at room temperature for 2 h, inject a 20 µL aliquot. (Prepare buffer by mixing 53.5 mL 50 mM sodium borate solution with 46.5 mL KH_2PO_4, pH 8.0. Prepare reagent by mixing equal volumes of buffer and 4 mM 2,3-naphthalenedialdehyde in MeOH.)

HPLC VARIABLES
Guard column: LiChroCART RP-18 (Merck)
Column: 125 × 4 5 µm LiChrosorb RP-18
Mobile phase: MeOH:50 mM pH 6.8 potassium phosphate buffer 42:58
Flow rate: 1
Injection volume: 20
Detector: F ex 418 em 460

CHROMATOGRAM
Retention time: 13
Limit of detection: 30 nM

OTHER SUBSTANCES
Non-interfering: alanine, ammonia, arginine, cysteine, glutathione, glycine, histidine, sulfide, sulfite, thiocyanate, thiosulfate, threonine

REFERENCE
Sano, A.; Takezawa, M.; Takitani, S. High performance liquid chromatography determination of cyanide in urine by pre-column fluorescence derivatization, *Biomed.Chromatogr.*, **1989**, *3*, 209–212.

SAMPLE
Matrix: urine
Analyte: fluvoxamine
Sample preparation: 8 mL Urine + 2 mL 100 mM pH 9.5 borate buffer, mix, filter (0.2 µm). Remove a 500 µL aliquot of the filtrate and add it to 100 µL 10 mM NaCN in 20 mM pH 9.5 borate buffer, add 500 µL 1 mM naphthalene-2,3-dicarboxaldehyde in MeOH, mix, let stand at room temperature for 20 min, add 100 µL 100 mM glycine in 20 mM pH 9.5 borate buffer, mix, let stand for 10 min, add 500 µL hexane:toluene 50:50, extract. Remove a 400 µL aliquot of the organic layer and add it to 800 µL dichloromethane, mix, inject a 35 µL aliquot.

HPLC VARIABLES
Column: 250 × 3.1 5 µm LiChrosorb Si-60
Mobile phase: Dichloromethane:MeOH 99.8:0.2
Flow rate: 0.5
Injection volume: 35
Detector: Chemiluminescence (418 nm cutoff filter) following post-column reaction. The column effluent mixed with 50 mM hydrogen peroxide in MeCN:dichloromethane 50:50 containing 0.5 mM triethylamine pumped at 0.1 mL/min and with 5 mM bis(2,4,6-trichlorophenyl) oxalate in dichloromethane pumped at 0.1 mL/min and the mixture flowed into the detector.

CHROMATOGRAM
Retention time: 7
Limit of detection: 5 nM

KEY WORDS
normal phase

REFERENCE
Kwakman, P.J.; Koelewijn, H.; Kool, I.; Brinkman, U.A.; de Jong, G.J. Naphthalene- and anthracene-2,3-dialdehyde as precolumn labelling reagents for primary amines using reversed- and normal-phase liquid chromatography with peroxyoxalate chemiluminescence detection, *J.Chromatogr.*, **1990**, *511*, 155–166.

SAMPLE
Matrix: urine
Analyte: alendronate
Sample preparation: Condition an Analytichem 3 mL 200 mg diethylamine SPE cartridge with 3 mL water. 5 mL Urine + 50 μL 2.5 M calcium chloride, vortex, add 50 μL or more (as needed) 1 M NaOH to form a slight white precipitate, centrifuge. Discard supernatant and dissolve pellet in 50 μL 1 M HCl, add 5 mL water, precipitate by adding 50 μL 1 M NaOH, centrifuge. Discard the supernatant and dissolve the pellet in 50 μL 1 M HCl, add 1 mL 10 mM EDTA, add 2 mL 100 mM pH 4.0 sodium acetate buffer, add to the SPE cartridge, wash with 3 mL water, elute with 3 mL buffer. Remove 250 μL of the eluate and add it to 50 μL 1 M pH 10.7 carbonate buffer, vortex, add 10 μL 50 mM KCN, add 10 μL 2 mg/mL naphthalene-2,3-dicarboxaldehyde in MeOH, let stand for 15 min, inject a 100 μL aliquot. (Buffer was 50 mM sodium citrate:50 mM pH 8.5 sodium phosphate buffer 1:1.)

HPLC VARIABLES
Guard column: 20 × 4.6 Hamilton PRP-1
Column: 150 × 4.6 5 μm 100 Å PLRP-S polymeric reversed-phase
Mobile phase: MeOH:buffer 40:60 (Buffer was 25 mM sodium citrate and 25 mM dihydrogenphosphate adjusted to pH 8.5 with 10 mM NaOH.)
Flow rate: 1
Injection volume: 100
Detector: E, LC-4B (Bioanalytical Systems), LC-17A ED flow cell with glassy carbon electrode +0.65 V, RE-4 Ag/AgCl reference electrode

CHROMATOGRAM
Retention time: 5.4
Limit of quantitation: 2.5 ng/mL

REFERENCE
Kline, W.F.; Matuszewski, B.K. Improved determination of the bisphosphonate alendronate in human plasma and urine by automated precolumn derivatization and high-performance liquid chromatography with fluorescence and electrochemical detection, *J.Chromatogr.*, **1992**, *583*, 183–193.

Naphthalene-2,3-dicarboxaldehyde/Dithiothreitol

SAMPLE
Matrix: solutions
Analyte: proteins
Sample preparation: Mix 35 μL of a solution in 100 mM pH 9.0 sodium borate buffer with 2 μL 100 mM NaCN and 5 μL 4 mg/mL naphthalene-2,3-dicarboxaldehyde in MeOH, let stand for 3 min, add 5 μL 15 mM dithiothreitol, add 50 μL running buffer, centrifuge at 16000 g for 2 min, inject an aliquot.

CAPILLARY ELECTROPHORESIS
Capillary: 24 cm × 50 μm (19.4 cm to detector) (Polymicro Technologies)
Capillary preparation: Before each analysis purge capillary with 100 mM NaOH for 1.5 min, with 100 mM HCl for 1 min, and with running buffer for at least 2 min.
Running buffer: CE SDS sample buffer (Bio-Rad)
Capillary temperature: 20
Voltage/Current: -774 V/cm
Injection: Electrokinetic injection at -10 kV for 10 s
Detector: UV 280
Model: Bio-rad Biofocus 3000
Migration time: 5 (myoglobin), 8 (conalbumin)

KEY WORDS
comparison with other derivatizing reagents

REFERENCE
Gump, E.L.; Monnig, C.A. Pre-column derivatization of proteins to enhance detection sensitivity for sodium dodecyl sulfate non-gel sieving capillary electrophoresis, *J.Chromatogr.A*, **1995**, *715*, 167–177.

Naphthalene-2,3-dicarboxaldehyde/2-Mercaptoethanol

SAMPLE
Matrix: dialysate
Analyte: substance P
Sample preparation: Inject an aliquot of dialysate (artificial CSF).

CAPILLARY ELECTROPHORESIS
Capillary: 68 cm × 50 μm fused-silica (Polymicro Technologies)
Running buffer: 150 mM pH 7.0 Borate buffer containing 15 mM phytic acid and 5 mM
   sulfobutyl ether β-cyclodextrin (IV) (Cydex, Lenexa KS)
Voltage/Current: +17.5 kV
Injection: Hydrodynamic injection at 13.8 kPa for 5 s (20 nL).
Detector: F ex 442 (7-10 mW He-Cd laser) em 495 (long-pass) following post-column re-
   action. The capillary effluent mixed with reagent in a 48 cm long membrane reactor and
   passed to the detector (reactor construction details in paper). (Reagent was 70 mM pH
   9.5 sodium tetraborate containing 8 mM 2-mercaptoethanol, 1 mM naphthalene-2,3-di-
   carboxaldehyde, 5 mM sulfobutyl ether β-cyclodextrin (IV) (Cydex, Lenexa KS), and 30%
   MeOH.)
Model: laboratory-constructed
Migration time: 9
Limit of detection: 100 nM

OTHER SUBSTANCES
Extracted: metabolites

KEY WORDS
post-column reaction

REFERENCE
Kostel, K.L.; Lunte, S.M. Evaluation of capillary electrophoresis with post-column derivatization and
   laser-induced fluorescence detection for the determination of substance P and its metabolites,
   J.Chromatogr.B, **1997**, *695*, 27–38.

SAMPLE
Matrix: solutions
Analyte: enkephalins
Sample preparation: Inject a 60 nL aliquot.

HPLC VARIABLES
Column: 150 × 1 3 μm C18 (Isco)
Mobile phase: MeCN:20 mM pH 6.8 phosphate buffer 16:84
Flow rate: 0.05
Injection volume: 0.06
Detector: F ex 457.9 (200-800 mW argon ion laser) following post-column reaction. The
   column effluent mixed with the reagent pumped at 0.05 mL/min and the mixture flowed

through a 190 cm (?) × 0.125 mm ID PEEK coil (23 μL volume) to the detector. (Reagent was 11 mM naphthalene-2,3-dicarboxaldehyde in MeCN : 20 mM pH 6.8 phosphate buffer 50 : 50 containing 22 mM 2-mercaptoethanol.)

CHROMATOGRAM
Retention time: 9 (methionine enkephalin), 12.5 (leucine enkephalin)
Limit of detection: 600 nM

OTHER SUBSTANCES
Simultaneously analyzed: metabolites

KEY WORDS
post-column reaction; microbore

REFERENCE
Dave, K.J.; Stobaugh, J.F.; Rossi, T.M.; Riley, C.M. Reversed-phase liquid chromatography of the opioid peptides. 3. Development of a microanalytical system for opioid peptides involving microbore liquid chromatography, post-column derivatization and laser-induced fluorescence detection, *J.Pharm.Biomed.Anal.*, **1992**, *10*, 965–977.

Phthalaldehyde

RELATED REFERENCES
Allenmark, S.; Bergstrom, S.; Enerback, L. A selective postcolumn *o*-phthalaldehyde-derivatization system for the determination of histamine in biological material by high-performance liquid chromatography. *Anal.Biochem.* **1985**, *144*, 98-103.

Itoh, Y.; Oishi, R.; Adachi, N.; Saeki, K. A highly sensitive assay for histamine using ion-pair HPLC coupled with postcolumn fluorescent derivatization: its application to biological specimens. *J.Neurochem.* **1992**, *58*, 884-889.

Kondo, M.; Kimura, H.; Maekubo, T.; Tomita, T.; Senda, M.; Urata, G.; Kajiwara, M. Direct injection method for quantitation of δ-aminolevulinic acid in urine by high-performance liquid chromatography. *Chem.Pharm.Bull.* **1992**, *40*, 1948-1950.

Michelet, F.; Gueguen, R.; Leroy, P.; Wellman, M.; Nicolas, A.; Siest, G. Blood and plasma glutathione measured in healthy subjects by HPLC: relation to sex, aging, biological variables, and life habits. *Clin.Chem.* **1995**, *41*, 1509-1517.

Paroni, R.; De Vecchi, E.; Cighetti, G.; Arcelloni, C.; Fermo, I.; Grossi, A.; Bonini, P. HPLC with *o*-phthalaldehyde precolumn derivatization to measure total, oxidized, and protein-bound glutathione in blood, plasma, and tissue. *Clin.Chem.* **1995**, *41*, 448-454.

Schwedt, G.; Hauck, M. Neues HPLC-Verfahren für Cyclamat in Lebensmitteln mit prä-chromatographischer Derivatisierung [A new HPLC procedure for cyclamate in food with pre-chromatographic derivatization]. *Z.Lebensm.Unters.Forsch.* **1988**, *187*, 127-129.

Siegel, P.D.; Lewis, D.M.; Olenchock, S.A. High-performance liquid chromatographic method for the evaluation of possible interferences in basophil-histamine release measurements. *Anal.Biochem.* **1990**, *188*, 416-421.

Yamatodani, A.; Fukuda, H.; Wada, H.; Iwaeda, T.; Watanabe, T. High-performance liquid chromatographic determination of plasma and brain histamine without previous purification of biological samples: cation-exchange chromatography coupled with post-column derivatization fluorometry. *J.Chromatogr.* **1985**, *344*, 115-123.

Yan, C.C.; Huxtable, R.J. Fluorimetric determination of monobromobimane and *o*-phthalaldehyde adducts of gamma-glutamylcysteine and glutathione: application to assay of gamma-glutamylcysteinyl synthetase activity and glutathione concentration in liver. *J.Chromatogr.B* **1995**, *672*, 217-224.

SAMPLE
Analyte: histamine
Sample preparation: Blend 50 g tissue and 75 mL 5% trichloroacetic acid in a Waring Blendor at high speed for 2 min, centrifuge for 2 min, extract the solid twice more with 75 mL portions of 5% trichloroacetic acid, filter the supernatants through a glass wool plug, wash the funnel with 5% trichloroacetic acid, make up the filtrate to 250 mL with 5% trichloroacetic acid. Remove a 10 mL aliquot and add it to 4 g NaCl, 1 mL 50% NaOH, and 5 mL chloroform:n-butanol 50:50, shake vigorously for 2 min, centrifuge for 5 min, remove the upper organic layer, repeat the extraction twice more with 5 mL portions of chloroform:n-butanol 50:50. Combine the organic layers and add them to 15 mL n-heptane, extract three times with 1 mL portions of 200 mM HCl (J.Assoc.Off.Anal.Chem. 1978, 61, 139). Remove a 100 μL aliquot and add it to 900 μL 200 mM HCl, add 200 μL 2 M NaOH, add 100 μL 0.1% o-phthalaldehyde in MeOH, let stand in the dark for 4 min, add 200 μL 0.5 M sulfuric acid, let stand for 2 h, inject an aliquot.

HPLC VARIABLES
Column: 5 μm Lichrospher RP-18
Mobile phase: MeCN:50 mM NaH_2PO_4 30:70, pH 4.5
Flow rate: 0.5
Detector: F ex 305-395 (filter) em 430-470 (filter)

CHROMATOGRAM
Retention time: 2
Limit of detection: 1 ng

KEY WORDS
fish

REFERENCE
Kalligas, G.; Kaniou, I.; Zachariadis, G.; Tsoukali, H.; Epivatianos, P. Thin layer and high pressure liquid chromatographic determination of histamine in fish tissues, *J.Liq.Chromatogr.*, **1994**, *17*, 2457–2468.

Phthalaldehyde/Acetylcysteine

RELATED REFERENCES

Busto, O.; Miracle, M.; Guasch, J.; Borrull, F. Determination of biogenic amines in wines by high-performance liquid chromatography with on-column fluorescence derivatization. *J.Chromatogr.A* **1997**, *757*, 311-318.

Miyahara, M.; Akiyama, H.; Toyoda, M.; Saito, Y. New procedure for fumonisins B$_1$ and B$_2$ in corn and corn products by ion pair chromatography with o-phthaldialdehyde postcolumn derivatization and fluorometric detection. *J.Agric.Food Chem.* **1996**, *44*, 842-847.

Okuda, T.; Aoki, I.; Motohashi, M.; Yashiki, T. Sensitive high-performance liquid chromatographic determination of 2,2'-[(2-aminoethyl)imino]diethanol bis(butylcarbamate) and its metabolites in human serum following pre-column derivatization with o-phthalaldehyde and stabilization of the derivatives. *J.Chromatogr.* **1993**, *612*, 263-268.

Saito, K.; Horie, M.; Nose, N.; Nakagomi, K.; Nakazawa, H. High-performance liquid chromatography of histamine and 1-methylhistamine with on-column fluorescence derivatization. *J.Chromatogr.* **1992**, *595*, 163-168.

Wuis, E.W.; Beneken Kolmer, E.W.J.; Van Beijsterveldt, L.E.C.; Burgers, R.C.M.; Vree, T.B.; van der Kleyn, E. Enantioselective high-performance liquid chromatographic determination of baclofen after derivatization with a chiral adduct of o-phthaldialdehyde. *J.Chromatogr.* **1987**, *415*, 419-422.

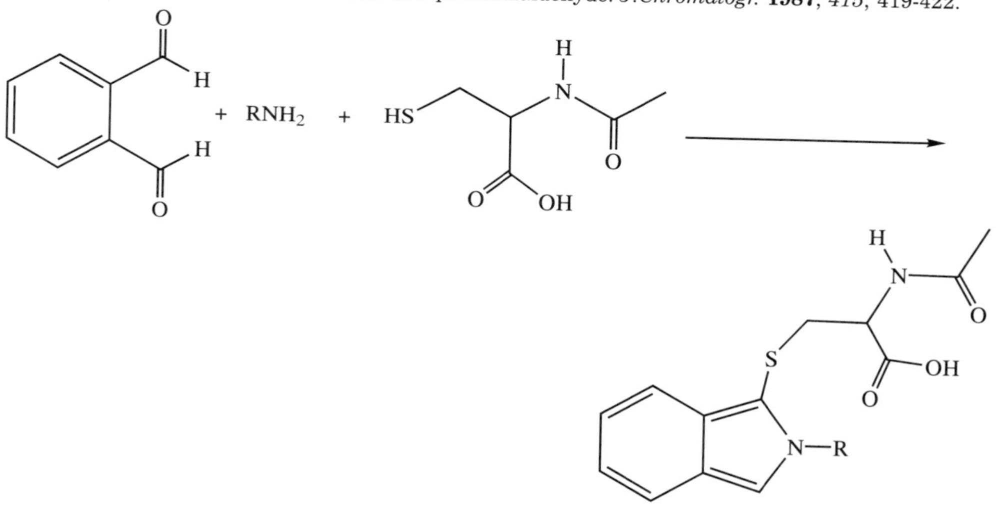

SAMPLE

Matrix: blood

Analyte: mexiletine

Sample preparation: 1 mL Plasma + 100 µL 0.4 µg/mL methylmexiletine hydrochloride in water + 1 mL 150 mM barium hydroxide + 1 mL 170 mM zinc sulfate, vortex, add 500 µL 2 M NaOH, extract twice with 5 mL portions of diethyl ether. Combine the organic layers and evaporate them to dryness at 45°, reconstitute the residue in 20 µL 30 mM HCl, add 60 µL 100 mM sodium borate, add 60-80 µL reagent, keep at 4°, inject an aliquot. (Reagent was 40 mg o-phthalaldehyde and 50 mg N-acetyl-L-cysteine in 5 mL MeOH, prepare daily.)

HPLC VARIABLES

Column: 250 × 4.6 5 µm Apex C18 (Jones Chromatography)

Mobile phase: MeOH:50 mM sodium acetate 65:35, pH 7.3

Flow rate: 1

Injection volume: 100

Detector: F ex 350 em 445

CHROMATOGRAM

Retention time: 14 (S-(+)), 15 (R-(-))

Internal standard: methylmexiletine (23, 24 (enantiomers))
Limit of detection: 1.5 ng/mL

OTHER SUBSTANCES
Non-interfering: metabolites

KEY WORDS
chiral; plasma; pharmacokinetics

REFERENCE
Abolfathi, Z.; Bélanger, P.-M.; Gilbert, M.; Rouleau, J.R.; Turgeon, J. Improved high-performance liquid chromatographic assay for the stereoselective determination of mexiletine in plasma, *J.Chromatogr.*, **1992**, *579*, 366–370.

SAMPLE
Matrix: blood
Analyte: mexiletine
Sample preparation: 500 μL Plasma + 50 μL 300 mM NaOH, mix, add 4 mL diisopropyl ether (Caution! Diisopropyl ether readily forms explosive peroxides!), vortex for 2 min, centrifuge at 1800 g for 10 min, repeat the extraction. Combine the organic layers and add them to 50 μL 100 mM HCl in MeOH, evaporate to dryness under a stream of nitrogen, reconstitute with 25 μL 30 mM aqueous HCl, add 50 μL 100 mM sodium borate, add 100 μL reagent, vortex for 30 s, inject an aliquot. (The 300 mM NaOH solution should be washed with diisopropyl ether before use. Reagent was 4 mg o-phthalaldehyde and 5 mg N-acetyl-L-cysteine in 500 μL MeOH.)

HPLC VARIABLES
Guard column: 4 × 4 5 μm LiChrospher 100 RP-18
Column: 150 × 4 5 μm LiChrospher 100 RP-18
Mobile phase: MeOH:50 mM pH 5.5 acetate buffer 65:35
Flow rate: 1
Injection volume: 20
Detector: F ex 350 em 455

CHROMATOGRAM
Retention time: 7.4 (S-(+)), 8.0 (R-(-))
Limit of quantitation: 1 ng/mL

OTHER SUBSTANCES
Extracted: metabolites
Simultaneously analyzed: quinidine
Non-interfering: aminopyrine, amiodarone, benzidamine, captopril, carbamazepine, clobazam, clomipramine, clonazepam, chlorpromazine, dapsone, digoxin, disopyramide, ethidocaine, lidocaine, metoclopramide, procainamide, propafenone, propranolol, sotalol, theophylline, trimipramine, verapamil

KEY WORDS
plasma; chiral; pharmacokinetics

REFERENCE
Lanchote, V.L.; Bonato, P.S.; Dreossi, S.A.C.; Gonçalves, P.V.B.; Cesarino, E.J.; Bertucci, C. High-performance liquid chromatographic determination of mexiletine enantiomers in plasma using direct and indirect enantioselective separations, *J.Chromatogr.B*, **1996**, *685*, 281–289.

SAMPLE
Matrix: blood, urine
Analyte: aminocaproic acid
Sample preparation: 100 μL Plasma or urine + 10 μL 10% zinc sulfate, mix, add 100 μL MeOH, vortex, centrifuge for 1 min. Remove a 50 μL aliquot and add it to 300 μL IS solution, add 50 μL reagent, mix, after 1 min inject a 50 μL aliquot. (Prepare IS solution by adding 500 μL 1 mg/mL DL-valine to 20 mL 1 M pH 9.8 borate buffer. Prepare reagent by dissolving 20 mg o-phthalaldehyde and 24 mg N-acetyl-L-cysteine in 6 mL MeOH: water 50:50.)

HPLC VARIABLES

Column: 150 × 4.2 Nucleosil 5-C18
Mobile phase: MeCN:buffer 10:90 (Buffer was 10 g/L (?) ammonium acetate containing 5 mM L-proline and 2.5 mM copper sulfate.)
Flow rate: 2
Injection volume: 50
Detector: F

CHROMATOGRAM

Retention time: 7
Internal standard: valine (9 (L), 13 (D))
Limit of detection: 50 ng/mL

OTHER SUBSTANCES

Simultaneously analyzed: amino acids

KEY WORDS

plasma

REFERENCE

Lam, S. High performance liquid chromatographic assay of Amicar, *epsilon*-aminocaproic acid, in plasma and urine after pre-column derivatization with *o*-phthalaldehyde for fluorescence detection, *Biomed.Chromatogr.*, **1990**, *4*, 175−177.

SAMPLE

Matrix: blood, urine
Analyte: tranylcypromine
Sample preparation: 1 mL Plasma or 500 μL urine + 1 mL 100 mM pH 11 sodium borate buffer + 20 μL 1 (plasma) or 10 (urine) μg/mL S-(+)-amphetamine in MeOH + 5 mL diethyl ether:EtOH 98.5:1.5, mix for 10 min, centrifuge at 1500 g for 10 min. Remove the organic layer and evaporate it to dryness, reconstitute the residue in 100 μL reagent, vortex briefly, let stand at room temperature for 5 min, inject an aliquot. (Reagent was 10 mg o-phthaldialdehyde, 500 μL EtOH, and 40 mg N-acetyl-L-cysteine in 5 mL buffer. Buffer was 14.75 g boric acid and 160 mL 1 M NaOH made up to 1 L with water, pH 10.)

HPLC VARIABLES

Column: 250 × 4.6 5 μm Zorbax ODS
Mobile phase: MeOH:THF:buffer 60:1:50 (Buffer was 653 mL 9.07 g/L KH_2PO_4 and 347 mL 11.87 g/L $Na_2HPO_4.2H_2O$)
Flow rate: 1.2
Detector: F ex 344 em 442

CHROMATOGRAM

Retention time: 32.5 (S-(-)), 35 (R-(+))
Internal standard: S-(+)-amphetamine (25)
Limit of detection: 0.5 ng/mL (plasma), 2 ng/mL (urine)

OTHER SUBSTANCES

Non-interfering: norephedrine, norepinephrine, norpseudoephedrine, tyramine

KEY WORDS

plasma; chiral; pharmacokinetics

REFERENCE

Spahn-Langguth, H.; Hahn, G.; Mutschler, E.; Möhrke, W.; Langguth, P. Enantiospecific high-performance liquid chromatographic assay with fluorescence detection for the monoamine oxidase inhibitor tranylcypromine and its applicability in pharmacokinetic studies, *J.Chromatogr.*, **1992**, *584*, 229−237.

Phthalaldehyde/t-Butanethiol

RELATED REFERENCES

Bourdelais, A.; Kalivas, P.W. High sensitivity HPLC assay for GABA in brain dialysis studies. *J.Neurosci.Methods* **1991**, *39*, 115-121.

Kehr, J.; Ungerstedt, U. Fast HPLC estimation of gamma-aminobutyric acid in microdialysis perfusates: effect of nipecotic and 3-mercaptopropionic acids. *J.Neurochem.* **1988**, *51*, 1308-1310.

Naini, A.B.; Vontzalidou, E.; Côté, L.J. Isocratic HPLC assay with electrochemical detection of free gamma-aminobutyric acid in cerebrospinal fluid. *Clin.Chem.* **1993**, *39*, 247-250.

Wall, G.M.; Baker, J.K. Determination of baclofen and α-baclofen in rat liver homogenate and human urine using solid-phase extraction, o-phthalaldehyde-*tert.*-butyl thiol derivatization and high-performance liquid chromatography with amperometric detection. *J.Chromatogr.* **1989**, *491*, 151-162.

SAMPLE

Matrix: blood

Analyte: baclofen

Sample preparation: Condition a 1 mL Bond Elut SCX strong cation exchange SPE cartridge with 2 mL hexane, 2 mL MeOH, 2 mL water, and 3 mL saturated NaCl solution. 1 mL Plasma + 100 µL water + 1 mL citrate buffer, mix, add to the SPE cartridge, wash with 4 mL water, wash with 1 mL saturated NaCl solution, dry the SPE cartridge, elute with 1.5 mL pH 10.4 borate buffer. Mix a 200 µL aliquot of the eluate and add it to 50 µL reagent, mix, inject a 20 µL aliquot. (Prepare citrate buffer by diluting 89 mL 100 mM citric acid to 100 mL with 200 mM Na_2HPO_4 pH 2.6. Prepare pH 10.4 borate buffer by mixing 54 mL 200 mM boric acid in 100 mM NaOH with 46 mL 100 mM NaOH. Prepare pH 9.3 borate buffer by mixing 87 mL 200 mM boric acid in 100 mM NaOH with 13 mL 100 mM NaOH. Prepare reagent daily by mixing 75 mg o-phthalaldehyde, 5 mL MeOH, 50 µL t-butanethiol, and 5 mL pH 9.3 borate buffer.)

HPLC VARIABLES

Column: 150 × 3.9 4 µm Novapak

Mobile phase: MeOH:buffer 74:36 (Prepare buffer by adjusting the pH of 60 mM Na_2HPO_4 to 7 with 60 mM KH_2PO_4.)

Flow rate: 0.8

Injection volume: 20

Detector: E, ESA Coulochem II, Model 5020 guard cell +1.2 V, Model 5011 glassy carbon working cell, screen electrode +0.2 V, quantifying electrode +0.7 V

CHROMATOGRAM

Retention time: 26

Limit of quantitation: 10 ng/mL

Limit of detection: 2.5 ng/mL

KEY WORDS

plasma; SPE; pharmacokinetics

REFERENCE

Millerioux, L.; Brault, M.; Gualano, V.; Mignot, A. High-performance liquid chromatographic determination of baclofen in human plasma, *J.Chromatogr.A*, **1996**, *729*, 309–314.

Phthalaldehyde/N,N-Dimethyl-2-mercaptoethylamine

SAMPLE

Matrix: blood

Analyte: SB 107647 ((S)-5-oxo-L-prolyl-L-α-glutamyl-L-α-aspartyl-N8-(5-amino-1-carboxy-pentyl)-8-oxo-N7-[N-[N-(5-oxo-L-prolyl)-L-α-glutamyl]-L-α-aspartyl]-L-threo-2,7,8-triami-nooctanoyl-L-lysine)

Sample preparation: Condition an Analytichem 1 mL 100 mg strong anion exchange SAX SPE cartridge with 1 mL MeOH and 1 mL water. 250 μL Plasma + 200 μL 50 mM acetic acid + 50 μL 5 μg/mL IS + 500 μL MeCN, mix, centrifuge at 2000 g for 5 min, add to the SPE cartridge, dry under vacuum, wash with 1 mL water, dry under vacuum, elute with 250 μL 100 mM pH 2.2 citric acid. Centrifuge the eluate at 2000 g for 5 min and inject a 25-130 μL aliquot of the supernatant.

HPLC VARIABLES

Guard column: 30 × 2.1 C8 (ABI Analytical)

Column: 150 × 2.1 5 μm Zorbax Rx octyl

Mobile phase: MeOH:buffer 20:80 (Prepare buffer as follows. Dissolve 28.3 g monochloroacetic acid, 9.8 g NaOH, and 1 g disodium EDTA in 1 L water (final pH 3.2). Dissolve 2.16 g sodium 1-octanesulfonate in 100 mL monochloroacetate buffer, add 900 mL water.)

Column temperature: 40

Flow rate: 0.3

Injection volume: 25-130

Detector: F ex 336 em 400 (cut-off filter) following post-column reaction. The column effluent mixed with the reagent pumped at 0.1 mL/min and the mixture flowed through a 1 mL coil (ABI Analytical) to the detector. (Prepare the reagent by dissolving 1 g NaOH in 500 mL water, filter (0.45 μm), sonicate for 10 min, sparge with helium for 10 min, add 2 mL 2 mg/mL o-phthalaldehyde in MeOH (freshly prepared), add 80 mg N,N-dimethyl-2-mercaptoethylamine hydrochloride (Thiofluor; Pickering Laboratories, Mountain View CA), swirl to mix, sparge continuously with helium, discard after 48 h.)

CHROMATOGRAM

Retention time: 14.5

Internal standard: SB 203285 (11.5)

Limit of quantitation: 20 ng/mL

Limit of detection: 10 ng/mL

KEY WORDS

post-column reaction; dog; rat; SPE; plasma

REFERENCE

Boppana, V.K.; Miller-Stein, C. Determination of a novel hematoregulatory peptide in dog plasma by reversed-phase high-performance liquid chromatography and an amine-selective o-phthaldialdehyde-thiol post-column reaction with fluorescence detection, *J.Chromatogr.A*, **1994**, *676*, 161–167.

SAMPLE

Matrix: plants

Analyte: aldicarb and metabolites

Sample preparation: Sonicate 1 g ground tobacco with 6 mL MeOH for 30 min, filter (0.45 μm), inject a 10 μL aliquot of the filtrate.

HPLC VARIABLES
Column: 200 × 4.6 Hypersil C8
Mobile phase: Gradient. A was 0.1% triethanolamine in MeCN:MeOH 20:80. B was 0.1% triethanolamine in water. A:B 10:90 for 4 min, to 30:70 over 7 min, to 90:10 over 5 min, maintain at 90:10 for 10 min.
Column temperature: 42
Flow rate: 1
Injection volume: 10
Detector: F ex 330 em 465 following post-column reaction. The column effluent mixed with 100 mM NaOH pumped at 0.3 mL/min and the mixture flowed through a coil at 100°. The effluent from this coil mixed with the reagent pumped at 0.3 mL/min and this mixture flowed through a coil at room temperature to the detector. (Prepare reagent by mixing 10 mL 10 mg/mL o-phthalaldehyde in MeOH with 1 L 2 g/L N,N-dimethyl-2-mercaptoethylamine hydrochloride (Thiofluor; Pickering Labs, Mountain View CA) in 50 mM pH 9.1 sodium tetraborate buffer.)

CHROMATOGRAM
Retention time: 10 (aldicarb sulfoxide), 11 (aldicarb sulfone), 18 (aldicarb)
Limit of detection: 0.5 ppm

KEY WORDS
post-column reaction; tobacco

REFERENCE
Yang, S.S.; Smetena, I. Determination of aldicarb, aldicarb sulfoxide and aldicarb sulfone in tobacco using high-performance liquid chromatography with dual post-column reaction and fluorescence detection, *J.Chromatogr.A*, **1994**, *664*, 289–294.

SAMPLE
Matrix: water
Analyte: pesticides
Sample preparation: Filter, inject a 400 μL aliquot of the filtrate.

HPLC VARIABLES
Guard column: C18
Column: 150 × 4.6 3 μm HS-3C18 (Perkin Elmer)
Mobile phase: Gradient. MeCN:water from 5:95 to 20:80 over 13 min, to 65:35 over 15 min, return to initial conditions over 2 min, re-equilibrate at initial conditions for 8 min.
Flow rate: 1
Injection volume: 400
Detector: F ex 340 em 460 following post-column reaction. The column effluent mixed with the reagent pumped at 0.1 mL/min and the mixture flowed through a 500 μL reaction coil at 95° to the detector. (Prepare the reagent by adding 1.25 mL 10 M NaOH to 100 mL water, add 10 mL 18 mg/mL N,N-dimethyl-2-mercaptoethylamine hydrochloride (Thiofluor; Pickering Laboratories, Mountain View CA), add 2.5 mL 10 mg/mL o-phthalaldehyde in MeOH, make up to 250 mL with water, filter (0.45 μm nylon), degas with helium for 10 min before use. Prepare fresh each day.)

CHROMATOGRAM
Retention time: 12.08 (aldicarb sulfoxide), 14.38 (aldicarb sulfone), 15.07 (oxamyl), 15.69 (methomyl), 21.23 (3-hydroxycarbofuran), 25.79 (aldicarb), 28.35 (propoxur), 28.71 (carbofuran), 29.78 (carbaryl), 32.92 (methiocarb)
Internal standard: 4-bromo-3,5-dimethylphenyl N-methylcarbamate (34)
Limit of detection: 0.4-1.2 ng/mL

KEY WORDS
post-column reaction

REFERENCE
Simon, V.A.; Pearson, K.S.; Taylor, A. Determination of N-methylcarbamates and N-methylcarbamoyloximes in water by high performance liquid chromatography with the use of fluorescence detection and a single o-phthalaldehyde post-column reaction, *J.Chromatogr.*, **1993**, *643*, 317–320.

Phthalaldehyde/Ethanethiol

RELATED REFERENCES

Leroy, P.; Nicolas, A.; Thioudellet, C.; Oster, T.; Wellman, M.; Siest, G. Rapid liquid chromatographic assay of glutathione in cultured cells. *Biomed.Chromatogr.* **1993**, *7*, 86-89.

Skaaden, T.; Greibrokk, T. Determination of polyamines by precolumn derivatization with o-phthalaldehyde and ethanethiol in combination with reversed-phase high-performance liquid chromatography. *J.Chromatogr.* **1982**, *247*, 111-122.

Todoriki, H.; Hayashi, T.; Naruse, H.; Hirakawa, A.Y. Sensitive high-performance liquid chromatographic determination of catecholamines in rat brain using a laser fluorimetric detection system. *J.Chromatogr.* **1983**, *276*, 45-54.

Xu, X.; L'Helgoualc'h, A.; Morier-Teissier, E.; Rips, R. Determination of gamma-aminobutyric acid in the mouse hypothalamus and hippocampus using liquid chromatography/electrochemistry. *J.Liq.Chromatogr.* **1986**, *9*, 2253-2267.

SAMPLE

Matrix: bacterial culture

Analyte: amines

Sample preparation: Prepare a column by filling a 1 mL pipette tip with 200 mg Extrelut, condition with 250 μL buffer. Add 10 μL bacterial culture and 10 μL IS solution to the column, elute with 1.3 mL ethyl methyl ketone:isopropanol 90:10 (prepared just prior to use) or with 1.3 mL ethyl methyl ketone. Add 200 μL reagent to the eluate, vortex for 15 s, let stand for 30 min, dilute 1 volume of the supernatant with 1 volume of MeOH and 2 volumes of water, inject an aliquot. (Prepare buffer by dissolving 50 mmoles ascorbic acid and 50 mmoles Na_2HPO_4 in 80 mL 1 M NaOH with stirring, adjust pH to 12.5 with 10 mM NaOH, make up to 100 mL with water, store in completely filled vials. Prepare the IS solution by adding 12 μL pentylamine to 100 mL100 mM HCl. Purify ethyl methyl ketone by passing 20 mL through 4 g acidic aluminum oxide (Merck). Prepare reagent each day by dissolving 10 mg o-phthaldialdehyde and 50 μL ethanethiol in 1 mL MeOH and adding 9 mL 400 mM pH 9.5 sodium borate buffer.)

HPLC VARIABLES

Column: 125 × 2 Superspher 100 RP-18 (Merck)

Mobile phase: MeOH:buffer 85:15 (Buffer was 7% triethylamine adjusted to pH 7.5 with acetic acid.)

Flow rate: 0.2

Injection volume: 1

Detector: F ex 340 em 450

CHROMATOGRAM

Retention time: 14.39 (cadaverine), 2.56 (histamine), 5.49 (isobutylamine), 6.07 (phenethylamine), 11.27 (putrescine), 4.51 (tryptamine), 3.16 (tyramine)

Internal standard: pentylamine (7.45)

KEY WORDS

SPE

REFERENCE

Bilic, N. Rapid identification of biogenic amine-producing bacterial cultures using isocratic high-performance liquid chromatography, *J.Chromatogr.A*, **1996**, *719*, 321–326.

SAMPLE
Matrix: tissue
Analyte: psychosine (galactosylsphingosine)
Sample preparation: Homogenize (all-glass Potter-Elvehjem) 100 mg tissue with 15 mL chloroform:MeOH 2:1, let stand at room temperature for 1 h, filter (paper), add 100 ng glucosylsphingosine, evaporate to dryness under reduced pressure. Suspend the residue in 3 mL MeOH, sonicate, add to a 1.5 mL column of AG 50WX8 (sodium form; Bio-Rad), rinse the flask with 3 mL MeOH, add the rinse to the column, wash with 15 mL MeOH, wash with 30 mL chloroform:MeOH 2:1, wash with 15 mL MeOH, elute with 15 mL MeOH:400 mM calcium chloride in water 75:25. Mix the eluate with 15 mL water and add it to a C18 Sep-Pak, wash with 100 mL water, elute with 5 mL MeOH, elute with 15 mL chloroform:MeOH 2:1, evaporate the eluate to dryness. Reconstitute the residue with 200 μL MeOH, 100 μL 400 mM pH 10.2 borate buffer, and 20 μL reagent, mix vigorously, let stand at room temperature for 10 min, add 2.5 mL chloroform:MeOH 2:1, add 450 μL water, mix vigorously, centrifuge at 500 g for 1 min, discard the upper layer, add 500 μL MeOH, add 500 μL water, mix, centrifuge, discard the upper layer. Evaporate the lower layer under a stream of nitrogen, reconstitute with 1 mL MeOH:water 90:10, add to a column of 1 mL AG 1X2 (acetate form; Bio-Rad) superimposed with 0.3 mL AG 50WX8 (sodium form; Bio-Rad) (previously washed with MeOH:water 90:10), elute with 9 mL MeOH:water 90:10. Evaporate the eluate to dryness, reconstitute the residue with 200 μL MeOH, inject an aliquot. (Prepare the reagent just prior to use by dissolving 10 mg o-phthalaldehyde in 1 mL MeOH and adding 100 μL 400 mM pH 10.2 borate buffer and 10 μL ethanethiol.)

HPLC VARIABLES
Column: 250 × 5.2 3 μm Nucleosil silica
Mobile phase: Gradient. Hexane:EtOH/water from 100:0 to 55:45 over 30 min (EtOH/water was EtOH:water 100:2.)
Flow rate: 1.5
Injection volume: 20
Detector: F ex 335 em 420

CHROMATOGRAM
Retention time: 18
Internal standard: glucosylsphingosine (17)

KEY WORDS
SPE; mouse; brain; normal phase

REFERENCE
Shinoda, H.; Kobayashi, T.; Katayama, M.; Goto, I.; Nagara, H. Accumulation of galactosylsphingosine (psychosine) in the twitcher mouse: determination by HPLC, *J.Neurochem.*, **1987**, *49*, 92–99.

SAMPLE
Matrix: tissue
Analyte: sphingoid bases
Sample preparation: Homogenize (all-glass Potter-Elvehjem) 50 mg tissue with 15 mL chloroform:MeOH 2:1 below 30°, add 88 pmole icosasphinganine, if sample is other than nervous tissue add 50 μg phosphatidylcholine and 250 μg cholesterol, let stand at room temperature with occasional shaking for 1 h, filter (paper), evaporate to dryness under a stream of nitrogen. Reconstitute the residue with 200 μL MeOH, 100 μL 400 mM pH 10.4 borate buffer, and 20 μL reagent, mix vigorously, let stand at room temperature for 10 min, add 2 mL MeOH:water 90:10, add to a 40 × 7 column of 200-400 mesh AG 1X2 (acetate form; Bio-Rad) superimposed with a 5 × 7 column of 200-400 mesh AG 50WX8 (hydrogen form; Bio-Rad) (previously washed with MeOH:water 90:10), elute with 4 mL MeOH. Evaporate the eluate to dryness, reconstitute the residue with 100 μL MeOH:water 90:10, filter (0.22 μm), inject a 20-40 μL aliquot. (Prepare the reagent just prior to use by dissolving 10 mg o-phthalaldehyde in 1 mL MeOH and adding 100 μL 400 mM pH 10.4 borate buffer and 10 μL ethanethiol.)

HPLC VARIABLES
Column: 150 × 4.6 Chemosorb 5-ODS-H (Chemo Co., Osaka)

Mobile phase: MeOH:THF:water 40:14:7
Flow rate: 0.9
Injection volume: 20-40
Detector: F ex 335 em 420

CHROMATOGRAM

Retention time: 7.7 (galactosylsphingenine), 8.1 (4-hydroxysphinganine), 8.8 (galactosyl-sphinganine), 11 (C_{18}-sphingenine), 12.9 (C_{18}-sphinganine), 16 (C_{20}-sphingenine)
Internal standard: icosasphinganine (19.1)
Limit of quantitation: 1 pmole

KEY WORDS

SPE; mouse; kidney; spinal cord

REFERENCE

Kobayashi, T.; Mitsuo, K.; Goto, I. Free sphingoid bases in normal murine tissues, *Eur.J.Biochem.*, **1988**, *172*, 747–752.

SAMPLE

Matrix: tissue
Analyte: lyosulfatide (sulfogalactosylsphingosine)
Sample preparation: Homogenize (all-glass Potter-Elvehjem) 1-10 mg tissue with 10 mL chloroform:MeOH 2:1, stir at room temperature for 1 h, filter (sintered glass), evaporate the filtrate to dryness under a stream of nitrogen. Reconstitute the residue with 10 mL MeOH, filter (sintered glass), evaporate to dryness, reconstitute with 5 mL MeOH:1 M KOH 90:10, heat at 37° for 2 h, add 5 mL water, add to a C18 Sep-Pak SPE cartridge, wash with 40 mL water, elute with 10 mL MeOH:water 80:20, add 333 pmole sphingen-ine to the eluate, evaporate to dryness. Reconstitute with 50 μL MeOH, 10 μL 400 mM pH 10.4 borate buffer, and 20 μL reagent, mix vigorously, let stand at room temperature for 10 min, inject an aliquot. (Prepare the reagent just prior to use by dissolving 10 mg o-phthalaldehyde in 1 mL MeOH and adding 100 μL 400 mM pH 10.4 borate buffer and 20 μL ethanethiol.)

HPLC VARIABLES

Column: 150 × 4.6 Chemcosorb 5-ODS-H (Chemo Co., Osaka)
Mobile phase: MeOH:water 90:10 containing 35 mM phosphoric acid
Flow rate: 1.5
Detector: F ex 335 em 420

CHROMATOGRAM

Retention time: 13 (galactosylsphingenine), 23.5 (sulfogalactosylsphingenine), 34 (sulfogalactosylsphinganine)
Internal standard: sphingenine (18)
Limit of quantitation: 50 pmole

KEY WORDS

SPE; white matter; grey matter; brain

REFERENCE

Toda, K.-I.; Kobayashi, T.; Goto, I.; Ohno, K.; Eto, Y.; Inui, K.; Okada, S. Lysosulfatide (sulfogalactosyl-sphingosine) accumulation in tissues from patients with metachromatic leukodystrophy, *J.Neurochem.*, **1990**, *55*, 1585–1591.

Phthalaldehyde/Mercaptoacetic Acid

SAMPLE
Matrix: bulk, formulations
Analyte: gentamicin
Sample preparation: Bulk. Weigh out amount equivalent to 100 mg gentamicin, dissolve in 100 mL water. Remove a 20 mL aliquot and add it to 10 mL reagent, make up to 50 mL with MeOH, heat at 90° for 15 min, cool for 5 min, inject a 20 μL aliquot. Injections. 500 μL of a 5% injection + 19.5 mL water + 10 mL reagent, heat at 90° for 15 min, cool for 5 min, inject a 20 μL aliquot. (Reagent was 400 mg o-phthalaldehyde in 4 mL MeOH, add 38 mL buffer, add 0.8 mL mercaptoacetic acid, adjust the pH to 10.4 with 45% KOH. Buffer was 6.18 g boric acid in 200 mL water, adjust pH to 10.4 with 45% KOH, make up to 250 mL with water.)

HPLC VARIABLES
Guard column: 45 × 4.6 Vydac reversed phase
Column: 150 × 4.6 Ultrasphere ODS
Mobile phase: Gradient. A was MeOH:water:acetic acid 700:250:50 containing 5 g/L sodium heptanesulfonate. B was MeOH. A:B 100:0 for 2 min, to 75:25 over 3 min, maintain at 75:25.
Flow rate: 1.5
Injection volume: 20
Detector: UV 330

CHROMATOGRAM
Retention time: 5.07 (C1), 11.06 (C1a), 12.67 (C2a), 13.77 (C2)

KEY WORDS
injections

REFERENCE
Albracht, J.H.; de Wit, M.S. Analysis of gentamicin in raw material and in pharmaceutical preparations by high-performance liquid chromatography, *J.Chromatogr.*, **1987**, *389*, 306–311.

SAMPLE
Matrix: reaction mixtures
Analyte: antibiotics
Sample preparation: 50 μL Buffered reaction mixture + 50 μL isopropanol + 50 μL reagent, heat at 60° for 10 min, centrifuge at 1000 g for 2 min, immediately inject a 50 μL aliquot of the supernatant. (Reagent was 80 mM o-phthalaldehyde and 250 mM mercaptoacetic acid in 1 M boric acid, pH adjusted to 10.4 with 40% KOH.)

HPLC VARIABLES
Column: 100 × 5 Hypersil ODS

Mobile phase: A was MeOH:water:acetic acid 50:45:5 containing 5 g/L heptanesulfonic acid. B was MeOH:water:acetic acid 75:20:5 containing 5 g/L heptanesulfonic acid.
Flow rate: 2
Injection volume: 50
Detector: UV 330

CHROMATOGRAM
Retention time: 18 (tobramycin (A:B 55:45)), 19 (netilmicin (A:B 10:90)), 19 (kanamycin A (A:B 60:40)), 20.5 (apramycin (A:B 70:30)), 21 (neomycin (A:B 45:55))

REFERENCE
Lovering, A.M.; White, L.O.; Reeves, D.S. Identification of aminoglycoside-acetylating enzymes by high-pressure liquid chromatographic determination of their reaction products, *Antimicrob.Agents Chemother.*, **1984**, *26*, 10–12.

Phthalaldehyde/2-Mercaptoethanol

RELATED REFERENCES
Agarwal, V.K. High performance liquid chromatographic determination of gentamycin in animal tissue. *J.Liq.Chromatogr.* **1989**, *12*, 613-628.

Agarwal, V.K. High performance liquid chromatographic determination of gentamicin in milk. *J.Liq.Chromatogr.* **1989**, *12*, 3265-3277.

Brückner, H.; Bosch, I.; Graser, T.; Fürst, P. Determination of α-alkyl-α-amino acids and alpha-amino alcohols by chiral-phase capillary gas chromatography and reverse-phase high-performance liquid chromatography. *J.Chromatogr.* **1987**, *395*, 569-590.

Buteau, C.; Duitschaever, C.L.; Ashton, G.C. High-performance liquid chromatographic detection and quantitation of amines in must and wine. *J.Chromatogr.* **1984**, *284*, 201-210.

Cabanes, A.; Cajal, Y.; Haro, I.; Garcia Anton, J.M.; Arboix, M; Reig, F. Gentamicin determination in biological fluids by HPLC, using tobramycin as internal standard. *J.Liq.Chromatogr.* **1991**, *14*, 1989-2010.

Chan-Yeung, M.; Chan, H.; Salari, H.; Wall, R.; Tse, K.S. Grain-dust extract induced direct release of mediators from human lung tissue. *J.Allerg.Clin.Immunol.* **1987**, *80*, 279-284.

Costa, C.A.; Trivelato, G.C.; Demasi, M.; Bechara, E.J.H. Determination of 5-aminolevulinic acid in blood plasma, tissues and cell cultures by high-performance liquid chromatography with electrochemical detection. *J.Chromatogr.B* **1997**, *695*, 245-250.

de Kok, A.; Hiemstra, M.; Brinkman, U.A.T. Low ng/L-level determination of twenty N-methylcarbamate pesticides and twelve of their polar metabolites in surface water via off-line solid-phase extraction and high-performance liquid chromatography with post-column reaction and fluorescence detection. *J.Chromatogr.* **1992**, *623*, 265-276.

Doko, M.B.; Visconti, A. Occurrence of fumonisins B1 and B2 in corn and corn-based human foodstuffs in Italy. *Food Addit.Contam.* **1994**, *11*, 433-439.

Dominguez, L.M.; Dunn, R.S. Analysis of OPA-derivatized amino sugars in tobacco by high-performance liquid chromatography with fluorimetric detection. *J.Chromatogr.Sci.* **1987**, *25*, 468-471.

Duff, R.; Murrill, E. Determination of L-buthionine-(S,R)-sulfoximine in plasma by high-performance liquid chromatography with o-phthalaldehyde derivatization and fluorometric detection. *J.Chromatogr.* **1987**, *385*, 275-282.

Dye, D.; East, T.; Bayne, W.F. High-performance liquid chromatographic method for post-column, in-line derivatization with o-phthalaldehyde and fluorometric detection of phenylpropanolamine in human urine. *J.Chromatogr.* **1984**, *284*, 457-461.

Edgell, K.W.; Biederman, L.A.; Longbottom, J.E. Direct aqueous injection-liquid chromatography with post-column derivatization for determination of N-methylcarbamoyloximes and N-methylcarbamates in finished drinking water: collaborative study. *J.Assoc.Off.Anal.Chem.* **1991**, *74*, 309-317.

Egger, D.; Reisbach, G.; Hültner, L. Simultaneous determination of histamine and serotonin in mast cells by high-performance liquid chromatography. *J.Chromatogr.B* **1994**, *662*, 103-107.

Endo, T.; Miyagi, M.; Ujie, A. Simultaneous determination of glycyl-L-histidyl-L-lysine and its metabolite, L-histidyl-L-lysine, in rat plasma by high-performance liquid chromatography with post-column derivatization. *J.Chromatogr.B* **1997**, *692*, 37-42.

Feng, Y.; Halaris, A.E.; Piletz, J.E. Determination of agmatine in brain and plasma using high-performance liquid chromatography with fluorescence detection. *J.Chromatogr.B* **1997**, *691*, 277-286.

Ferreira, I.M.P.L.V.O.; Nunes, M.V.; Mendes, E.; Remiao, F.; Ferreira, M.A. Development of an HPLC-UV method for determination of taurine in infant formulae and breast milk. *J.Liq.Chromatogr.Rel.Technol.* **1997**, *20*, 1269-1278.

Fukuda, H.; Yamatodani, A.; Imamura, I.; Maeyama, K.; Watanabe, T.; Wada, H. High-performance liquid chromatographic determination of histamine N-methyltransferase activity. *J.Chromatogr.* **1991**, *567*, 459-464.

Gamache, P.; Ryan, E.; Svendsen, C.; Murayama, K.; Acworth, I.N. Simultaneous measurement of monoamines, metabolites and amino acids in brain tissue and microdialysis perfusates. *J.Chromatogr.* **1993**, *614*, 213-220.

Guggisberg, D.; Koch, H. Methode zur quantitativen Bestimmung von Gentamicin in Fleisch. Leber und Niere mit HPLC und Nachsäulenderivatisation [Method for the quantitative determination of gentamicin in meat, liver and kidney by HPLC and post-column derivatization]. *Mitteilungen aus dem Gebiete der Lebensmitteluntersuchung und Hygiene* **1995**, *86*, 14-28.

Han, X.Q.; Vohra, M.M. A sensitive method for simultaneous determination of histamine and noradrenaline with high-performance liquid chromatography/electrochemistry. *J.Pharmacol.Methods* **1991**, *25*, 29-40.

Harsing, L.G., Jr.; Nagashima, H.; Vizi, E.S.; Duncalf, D. Electrochemical determination of histamine derivatized with o-phthalaldehyde and 2-mercaptoethanol. *J.Chromatogr.* **1986**, *383*, 19-26.

Heideman, R.L.; Fickling, K.B.; Walker, L.J. Free and total putrescine in cerebrospinal fluid quantified by reversed-phase liquid chromatography. *Clin.Chem.* **1984**, *30*, 1243-1245.

Huebert, N.D.; Schartz, J.-J.; Haegele, K.D. Analysis of 2-difluoromethyl-DL-ornithine in human plasma, cerebrospinal fluid and urine by cation-exchange high-performance liquid chromatography. *J.Chromatogr.A* **1997**, *762*, 293-298.

Hyvönen, T.; Keinanen, T.A.; Khomutov, A.R.; Khomutov, R.M.; Eloranta, T.O. Monitoring of the uptake and metabolism of aminooxy analogues of polyamines in cultured cells by high-performance liquid chromatography. *J.Chromatogr.* **1992**, *574*, 17-21.

Ide, T. Simple high-performance liquid chromatographic method for assaying cysteinesulfinic acid decarboxylase activity in rat tissue. *J.Chromatogr.B* **1997**, *694*, 325-332.

Izquierdo, A.; Tena, M.T.; Luque de Castro, M.D.; Valcárel, M. Supercritical fluid extraction of carbamate pesticides from soils and cereals. *Chromatographia* **1996**, *42*, 206-212.

Kijak, P.J.; Jackson, J.; Shaikh, B. Determination of gentamicin in bovine milk using liquid chromatography with post-column derivatization and fluorescence detection. *J.Chromatogr.B* **1997**, *691*, 377-382.

Körner, A.; Peter, A. Alternative methods for the determination of trace amounts of 4-aminomorpholine in molsidomine and linsidomine. *J.Chromatogr.A* **1995**, *689*, 235-245.

Krause, R.T. Liquid chromatographic determination of *N*-methylcarbamate insecticides and metabolites in crops. I. Collaborative study. *J.Assoc.Off.Anal.Chem.* **1985**, *68*, 726-733.

Löser, C.; Wunderlich, U.; Fölsch, U.R. Reversed-phase liquid chromatographic separation and simultaneous fluorimetric detection of polyamines and their monoacetyl derivatives in human and animal urine, serum and tissue samples: an improved, rapid and sensitive method for routine application. *J.Chromatogr.* **1988**, *430*, 249-262.

Mason, W.D.; Amick, E.N. High-pressure liquid chromatographic analysis of phenylpropanolamine in human plasma following derivatization with o-phthalaldehyde. *J.Pharm.Sci.* **1981**, *70*, 707-709.

Mimura, T.; Delmas, D. Rapid and sensitive method for muramic acid determination by high-performance liquid chromatography with precolumn fluorescence derivatization. *J.Chromatogr.* **1983**, *280*, 91-98.

Newsome, W.H.; Lau, B.P.-Y.; Ducharme, D.; Lewis, D. Comparison of liquid chromatography-atmospheric pressure chemical ionization/mass spectrometry and liquid chromatography-postcolumn fluorometry for determination of carbamates in food. *J.AOAC Int.* **1995**, *78*, 1312-1316.

Nozaki, O.; Ohba, Y. Determination of urinary free noradrenaline by reversed-phase high-performance liquid chromatography with on-line extraction and fluorescence derivatization. *J.Chromatogr.* **1990**, *515*, 621-627.

Olson, L.L.; Pick, J.; Ellis, W.Y.; Lim, P. A chemical assessment and HPLC assay validation of bulk paromomycin sulfate. *J.Pharm.Biomed.Anal.* **1997**, *15*, 783-793.

Patel, S.; Hazel, C.M.; Winterton, A.G.; Mortby, E. Survey of ethnic foods for mycotoxins. *Food Addit.Contam.* **1996**, *13*, 833-841.

Patel, B.M.; Moye, H.A.; Weinberger, R. Ultraviolet photolysis and ortho-phthal-aldehyde-mercaptoethanol derivatization of pharmaceuticals for enhanced fluorescence detection in HPLC. *Anal.Lett.* **1989**, *22*, 3057-3079.

Petridis, K.D.; Steinhart, H. Automatische Vorsäulenderivatisierung mit o-phthaldialdehyd (OPA). Eine neue RP-HPLC-Methode für die Bestimmung von biogenen Aminen in Lebensmitteln [Automated pre-column derivatization with o-phthaldialdehyde (OPA). A new RP-HPLC method for the determination of biogenic amines in food]. *Z.Lebensm.Unters.Forsch.* **1995**, *201*, 256-260.

Pfeffer, M.; Walenciak-Reddel, E. Determination of 6-amino-2,2-dimethyl-1,3-dioxepan-5-ol by postcolumn derivatization with o-phthalaldehyde/2-mercapto-ethanol after ion-pair chromatography. *Chromatographia* **1994**, *38*, 479-484.

Piñeiro, M.S.; Silva, G.E.; Scott, P.M.; Lawrence, G.A.; Stack, M.E. Fumonisin levels in Uruguayan corn products. *J.AOAC Int* **1997**, *80*, 825-828.

Prelusky, D.B.; Miller, J.D.; Trenholm, H.L. Disposition of ^{14}C-derived residues in tissues of pigs fed radiolabelled fumonisin B$_1$. *Food Addit.Contam.* **1996**, *13*, 155-162.

Puchala, R.; Piór, H.; Kulasek, G.W.; Shelford, J.A. Determination of diaminopimelic acid in biological materials using high-performance liquid chromatography. *J.Chromatogr.* **1992**, *623*, 63-67.

Rabouan-Guyon, S.; Courtois, P.; Barthes, D. Determination of acefylline heptaminol in pharmaceutical preparations by high performance liquid chromatography. *Farmaco* **1996**, *51*, 739-746.

Rice, L.G.; Ross, P.F.; Dejong, J.; Plattner, R.D.; Coats, J.R. Evaluation of a liquid chromatographic method for the determination of fumonisins in corn, poultry feed, and *Fusarium* culture material. *J.AOAC Int.* **1995**, *78*, 1002-1009.

Rouberty, F.; Fournier, J. Optimization of HPLC separation of carbamate insecticides (carbofuran, hydroxycarbofuran and aldicarb) by experimental design methodology. *J.Liq.Chrom.Rel.Technol.* **1996**, *19*, 37-55.

Scott, P.M.; Lawrence, G.A. Determination of hydrolysed fumonisin B$_1$ in alkali-processed corn foods. *Food Addit.Contam.* **1996**, *13*, 823-832.

Shephard, G.S.; Thiel, P.G.; Sydenham, E.W. Determination of fumonisin B$_1$ in plasma and urine by high-performance liquid chromatography.*J.Chromatogr.* **1992**, *574*, 299-304.

Shephard, G.S.; Thiel, P.G.; Marasas, W.F.O.; Sydenham, E.W.; Vleggaar, R. Isolation and determination of AAL phytotoxins from corn cultures of the fungus *Alternaria alternata* f. sp. *lycopersici*. *J.Chromatogr.* **1993**, *641*, 95-100.

Shi, R.J.Y.; Gee, W.L.; Williams, R.L.; Benet, L.Z.; Lin, E.T. Ion-pair liquid chromatographic analysis of phenylpropanolamine in plasma and urine by post-column derivatization with o-phthalaldehyde. *J.Liq.Chromatogr.* **1985**, *8*, 1489-1500.

Simmonds, R.J.; Wood, S.A.; Ackland, M.J. A sensitive high performance liquid chromatography assay for trospectomycin, an aminocyclitol antibiotic, in human plasma and serum. *J.Liq.Chromatogr.* **1990**, *13*, 1125-1142.

Simon, V.A.; Pearson, K.S.; Taylor, A. Determination of N-methylcarbamates and N-methylcarbamoyloximes in water by high performance liquid chromatography with the use of fluorescence detection and a single o-phthalaldehyde post-column reaction. *J.Chromatogr.* **1993**, *643*, 317-320.

Solfrizzo, M.; Avantaggiato, G.; Visconti, A. Rapid method to determine sphinganine/sphingosine in human and animal urine as a biomarker for fumonisin exposure. *J.Chromatogr.B* **1997**, *692*, 87-93.

Swanepoel, E.; de Villiers, M.M.; Du Preez, J.L. Fluorimetric method of analysis for D-norpseudoephedrine hydrochloride, glycine and L-glutamic acid by reversed-phase high-performance liquid chromatography. *J.Chromatogr.A* **1996**, *729*, 287-291.

Sydenham, E.W.; Shephard, G.S.; Thiel, P.G.; Stockenström, S.; Snijman, P.W.; Van Schalkwyk, D.J. Liquid chromatographic determination of fumonisins B$_1$, B$_2$, and B$_3$ in corn: AOAC-IUPAC collaborative study. *J.AOAC Int.* **1996**, *79*, 688-696.

Visconti, A.; Boenke, A.; Solfrizzo, M.; Pascale, M.; Doko, M.B. European intercomparison study for the determination of the fumonisins content in two maize materials. *Food Addit.Contam.* **1996**, *13*, 909-927.

Wagner, J.; Claverie, N.; Danzin, C. A rapid high-performance liquid chromatographic procedure for the simultaneous determination of methionine, ethionine, S-adenosylmethionine, S-adenosylethionine, and the natural polyamines in rat tissues. *Anal.Biochem.* **1984**, *140*, 108-116.

Wilson, J.M.; Cohen, R.I.; Kezer, E.A.; Smith, E.R. Analysis of dezocine in serum and urine by high performance liquid chromatography and pre-column derivatization. *J.Liq.Chromatogr.* **1994**, *17*, 4245-4257.

Zambonin, P.G.; Guerrieri, A.; Rotunno, T.; Palmisano, F. Simultaneous determination of gamma-aminobutyric acid and polyamines by o-phthalaldehyde-β-mercaptoethanol precolumn derivatization and gradient elution liquid chromatography with electrochemical detection. *Anal.Chim.Acta* **1991**, *251*, 101-107.

SAMPLE
Matrix: beverage
Analyte: amines

Sample preparation: Condition a 500 mg SAX SPE cartridge (Varian) with two 5 mL portions of MeOH and two 5 mL portions of water. Condition a 1000 mg C18 SPE cartridge with two 5 mL portions of MeOH and two 5 mL portions of 200 mM pH 4.5 sodium decanesulfonate. Adjust pH of 15 mL red wine to 8, pass through the SAX SPE cartridge, adjust the pH of the eluate to 4.5, add 100 μL 200 mM sodium decanesulfonate, add this solution to the C18 SPE cartridge, elute with 3 mL MeOH. Mix 10 μL reagent and 2 μL eluate for 1 min, inject the whole amount. (Prepare reagent by dissolving 45 mg o-phthalaldehyde and 200 μL 2-mercaptoethanol in 1 mL MeOH and making up to 10 mL with buffer. Prepare buffer by dissolving 3.81 g sodium tetraborate in 100 mL water and adjusting pH to 10.5 with 10 M NaOH.)

HPLC VARIABLES
Guard column: ODS Basic (Teknokroma, Barcelona)
Column: 250 × 4.6 5 μm ODS Basic (Teknokroma, Barcelona)
Mobile phase: Gradient. A was THF:50 mM pH 7 (?) sodium acetate 1:99. B was MeOH. A:B from 45:55 to 20:80 over 25 min, maintain at 20:80 for 3 min, return to initial conditions over 2 min.
Column temperature: 60
Flow rate: 1
Injection volume: 12
Detector: F ex 330 em 445

CHROMATOGRAM
Retention time: 3.5 (ethanolamine), 4.5 (histamine), 5 (methylamine), 6.5 (ethylamine), 6.7 (tyramine), 7.5 (isopropylamine), 8 (propylamine), 10.3 (tryptamine), 11 (butylamine), 12 (phenethylamine), 13 (putrescine), 13.5 (3-methylbutylamine), 14 (amylamine), 15 (cadaverine), 17 (hexylamine)
Limit of quantitation: 430 ng/mL

KEY WORDS
wine; SPE

REFERENCE
Busto, O.; Guasch, J.; Borrull, F. Improvement of a solid-phase extraction method for determining biogenic amines in wines, *J.Chromatogr.A*, **1995**, *718*, 309–317.

SAMPLE
Matrix: blood
Analyte: biogenic amines
Sample preparation: 20 mL Whole blood + 1 mL 20 mg/mL EDTA solution containing 10 mg/mL sodium metabisulfite, mix, centrifuge at 4° at 4000 g for 10 min. Remove the plasma and add concentrated perchloric acid until the concentration of perchloric acid is 400 mM, mix, let stand in the cold for 15 min, centrifuge at 4° at 20000 g for 20 min. Adjust pH of 2 mL supernatant to 7.0 ± 0.2 with 500 mM KOH, add 400 μL 875 μg/mL o-phthalaldehyde in pH 10.40 ± 0.02 buffer (containing 2-mercaptoethanol ?), add 2 g NaCl, add 2 mL ethyl acetate, shake for 1 min, centrifuge at 3400 g, repeat the extraction. Combine the organic layers and add them to 2 mL 35 mM pH 10.0 ± 0.1 Na_2HPO_4 buffer, shake for 1 min, centrifuge at 3400 g, discard the aqueous layer, wash the ethyl acetate layer again with phosphate buffer. Reduce the ethyl acetate volume to 100 μL under a stream of nitrogen, inject a 10-50 μL aliquot.

HPLC VARIABLES
Guard column: Co:Pell ODS
Column: 300 × 4 10 μm μBondapak phenyl
Mobile phase: Gradient. MeCN:25 mM pH 5.10 NaH_2PO_4 buffer 25:75 for 15 min then MeOH:25 mM pH 5.10 NaH_2PO_4 buffer 45:55 (step gradient).
Column temperature: 26
Flow rate: 1.5
Injection volume: 10-50
Detector: F ex 340 em 480

CHROMATOGRAM

Retention time: 7 (histamine), 13 (norepinephrine), 17 (octopamine), 29 (dopamine), 40 (serotonin)

Internal standard: tyramine (44)

Limit of detection: 0.5 ng/mL

KEY WORDS

plasma; whole blood; pig

REFERENCE

Davis, T.P.; Gehrke, C.W. Jr.; Williams, C.H.; Gehrke, C.W.; Gerhardt, K.O. Pre-column derivatization and high-performance liquid chromatography of biogenic amines in blood of normal and malignant hyperthermic pigs, *J.Chromatogr.*, **1982**, *228*, 113–122.

SAMPLE

Matrix: blood

Analyte: polyamines

Sample preparation: Add one volume of 50% trichloroacetic acid to 9 volumes of plasma, mix thoroughly for 3 min, let stand at 4° for 30 min, centrifuge at 2000 g for 20 min. Lyophilize the supernatant, reconstitute with 3 mL 6 M HCl, heat at 110° for 16 h, evaporate to dryness under reduced pressure, reconstitute with 500 μL 500 mM HCl, centrifuge, inject a 100 μL aliquot.

HPLC VARIABLES

Column: 75 × 2 CK-10S 11.5 μm cation-exchange (Mitsubishi Kasei)

Mobile phase: Gradient. 250 mM pH 6.5 Sodium citrate buffer containing x M NaCl. x = 0 for 16 min, x = 0.2 for 2 min, x = 0.5 for 4 min, x = 1.5 for 7 min, x = 2.0 for 8 min, x = 3.0 for 33 min (step gradient).

Column temperature: 60

Flow rate: 0.2

Injection volume: 100

Detector: F ex 340 em 455 following post-column reaction. The column effluent mixed with the reagent pumped at 0.2 mL/min and the mixture flowed through a reaction coil to the detector. (The reagent was 0.8 g/L o-phthalaldehyde and 2 mL/L 2-mercaptoethanol in 400 mM pH 10.5 potassium borate buffer.)

CHROMATOGRAM

Retention time: 43.8 (putrescine), 51.1 (cadaverine), 55.5 (spermidine), 63.7 (spermine)

Limit of detection: 6-10 pmole

KEY WORDS

plasma; post-column reaction

REFERENCE

Takagi, T.; Chung, T.G.; Saito, A. Determination of polyamines in hydrolysates of uremic plasma by high-performance cation-exchange column chromatography, *J.Chromatogr.*, **1983**, *272*, 279–285.

SAMPLE

Matrix: blood

Analyte: tranexamic acid

Sample preparation: 1 mL Plasma + 100 μL 250 μg/mL IS in water + 700 μL water, mix, add 200 μL 4 M perchloric acid, shake vigorously, let stand for 10 min, centrifuge at 3000 g for 5 min, inject an aliquot of the supernatant.

HPLC VARIABLES

Guard column: 30 × 4.6 10 μm cation-exchange (Brownlee)

Column: 250 × 4.6 10 μm Nucleosil SA

Mobile phase: MeOH:buffer 2:98 containing 100 μL/L caprylic acid (Buffer was 100 mM trisodium citrate adjusted to pH 4 with HCl.)

Column temperature: 26

Flow rate: 1.4

Injection volume: 100

Detector: F ex 410 em 450 following post-column reaction. The column effluent mixed with the reagent pumped at 0.7 mL/min and the mixture flowed through a 1 m × 0.3 mm ID coil of tubing to the detector. (Prepare reagent by adding 800 mg o-phthalaldehyde in 10 mL MeOH to 1 L 700 mM pH 9.5 potassium borate buffer containing 2 g EDTA and 2 mL 2-mercaptoethanol.)

CHROMATOGRAM
Retention time: 5.65
Internal standard: 4-aminomethyl bicyclo(2,2,2)octane-1-carboxylic acid (KabiVitrum, Uxbridge, UK) (6.65)
Limit of quantitation: 1 µg/mL

OTHER SUBSTANCES
Simultaneously analyzed: arginine, histidine
Non-interfering: amino acids

KEY WORDS
post-column reaction; plasma

REFERENCE
Elworthy, P.M.; Tsementzis, S.A.; Westhead, D.; Hitchcock, E.R. Determination of plasma tranexamic acid using cation-exchange high-performance liquid chromatography with fluorescence detection, *J.Chromatogr.*, **1985**, *343*, 109–117.

SAMPLE
Matrix: blood
Analyte: gentamicin
Sample preparation: Prepare a column of 150 mg dry silicic acid in a Pasteur pipette plugged with glass wool (10 mm height) and treat with 1 mL water. 500 µL Serum + 1.5 mL water, add to the column, rinse the tube with 1 mL water, add this water to the column, discard the eluate, add 500 µL reagent, let stand for 30 s, elute, discard the eluate, elute with 1.5 mL MeOH. Vortex the eluate, centrifuge, protect from light, inject a 20 µL aliquot. (Elution was performed under positive pressure from a rubber bulb. Reagent was prepared by dissolving 1 g of boric acid in 38 mL water, adjust pH to 10.4 with 450 g/L KOH, add 2 mL 100 mg/mL o-phthalaldehyde in MeOH, add 400 µL 2-mercaptoethanol. Prepare fresh each week.)

HPLC VARIABLES
Guard column: 23 × 3.9 µBondapak C18/Porasil B
Column: 300 × 3.9 10 µm µBondapak
Mobile phase: MeOH:buffer 21:79 (Buffer was 1% triethylamine adjusted to pH 6.2 ± 0.1 with phosphoric acid.)
Flow rate: 2
Injection volume: 25
Detector: F ex 260 em 418

CHROMATOGRAM
Retention time: 4.8 (C1), 6.8 (C1a), 8.6 (C2)
Limit of detection: 20 ng/mL (C1), 80 ng/mL (C1a, C2)

KEY WORDS
serum; SPE; pharmacokinetics

REFERENCE
Rumble, R.H.; Roberts, M.S. High-performance liquid chromatographic assay of the major components of gentamicin in serum, *J.Chromatogr.*, **1987**, *419*, 408–413.

SAMPLE
Matrix: blood
Analyte: eflornithine
Sample preparation: 100 µL Plasma + 20 µL 250 µg/mL 4-amino-3-hydroxybutyric acid in water + 400 µL MeOH, centrifuge at 800 g for 20 min. Remove the supernatant and add it to 200 µL 20 mM pH 7.5 phosphate buffer, mix an aliquot of this mixture with an equal volume of reagent, inject. (Reagent was 10 mg o-phthalaldehyde in 1 mL EtOH, add 100 µL 2-mercaptoethanol, add 10 mL 100 mM pH 7.5 phosphate buffer, store in the dark, freshly prepare every 3 days.)

HPLC VARIABLES
Guard column: 37-50 μm Bondapak C18/corasil
Column: 5 μm C18 Radial-Pak (Waters)
Mobile phase: Gradient. A was MeOH:isopropanol:100 mM pH 7.5 phosphate buffer 5:
    3:92. B was MeOH:MeCN:isopropanol:water 80:5:5:10. A:B 80:20 for 3 min, to 50:50
    over 3 min, maintain at 50:50 for 5 min, re-equilibrate at initial conditions for 7 min.
Flow rate: 0.2 for 1 min then 1.5
Detector: F ex 335 em 418

CHROMATOGRAM
Retention time: 19
Internal standard: 4-amino-3-hydroxybutyric acid (13)
Limit of quantitation: 500 ng/mL
Limit of detection: 250 ng/mL

KEY WORDS
plasma

REFERENCE
Smithers, J. A precolumn derivatization high-performance liquid chromatographic (HPLC) procedure
    for the quantitation of difluoromethylornithine in plasma, *Pharm.Res.*, **1988**, *5*, 684–686.

SAMPLE
Matrix: blood
Analyte: aminoglycoside antibiotics
Sample preparation: Plasma. Condition a 3 mL Baker cyanopropylsilane CN SPE car-
    tridge with 2 mL MeOH, 2 mL water, and 2 mL buffer. 1 mL Plasma + 100 μL 100 μg/mL
    dibekacin in water, vortex for 15 s, add 1 mL buffer, vortex for 15 s, centrifuge at 3100 g
    at 4° for 7 min, add to SPE cartridge, wash with 500 μL water, wash with 250 μL mobile
    phase, elute to dryness. Elute with 250 μL mobile phase, inject an aliquot of the eluate.
    Urine, dialysate. Dilute 1:100 with water, add 100 μL 100 μg/mL dibekacin per 1 mL of
    sample, mix well, inject a 100 μL aliquot. (Buffer was 0.94 g sodium hexanesulfonate in
    300 mL water, add 500 μL glacial acetic acid, dilute to 500 mL with water.)

HPLC VARIABLES
Guard column: 10 × 4.6 5 μm Hypersil C18
Column: 150 × 4.6 5 μm Hypersil C18
Mobile phase: MeOH:buffer 10:90 (Buffer was 3.76 g sodium hexanesulfonate + 28.4 g
    sodium sulfate in 2 L water, acidify to pH 3.4 with 2 mL glacial acetic acid.)
Column temperature: 25
Flow rate: 1.1
Injection volume: 100
Detector: F ex 338 em 418 (bandpass filter) following post-column reaction. The column
    effluent mixed with the reagent pumped at 0.4 mL/min and flowed through a 3 m × 0.05
    mm ID knitted PTFE reaction coil at 25° to the detector. (Derivatizing reagent was 0.4
    g o-phthalaldehyde in 3 mL MeOH added to 390 mL buffer, add 2 mL 2-mercaptoethanol,
    make up to 500 mL with water, store at 4°. Buffer was 1 M pH 10.4 borate from equal
    volumes of 1 M KOH and boric acid.)

CHROMATOGRAM
Retention time: 6.7 (isepamicin), 6 (kanamycin), 13 (tobramycin), 58 (netilmicin), 30, 48,
    50, 67 (gentamicin)
Internal standard: dibekacin (17)

KEY WORDS
plasma; dialysate; urine; post-column reaction; SPE

REFERENCE
Maloney, J.A.; Awni, W.M. High-performance liquid chromatographic determination of isepamicin in
    plasma, urine and dialysate, *J.Chromatogr.*, **1990**, *526*, 487–496.

SAMPLE
Matrix: blood

Analyte: isepamicin
Sample preparation: Dry blood on gauze, soak gauze in 500 μL 500 mM Na_2HPO_4 at 35° for 30 min, inject a 50 μL aliquot.

HPLC VARIABLES
Column: 300 × 3.9 5 μm μBondapak C18
Mobile phase: 16 mM Sodium sulfate containing 5 mM 1-heptanesulfonic acid (PIC B-7)
Flow rate: 0.6
Injection volume: 50
Detector: F ex 360 em 440 following post-column reaction. The column effluent mixed with reagent pumped at 0.3 mL/min and flowed through a 5 m × 0.25 mm ID coil of PTFE tubing to the detector. (Prepare reagent by dissolving 300 mg o-phthalaldehyde in 500 mL MeOH, add 1.25 mL 2-mercaptoethanol, add 500 mL 400 mM pH 10.4 potassium borate buffer.)

CHROMATOGRAM
Retention time: 12
Limit of detection: 100 ng/mL

KEY WORDS
post-column reaction; dried blood

REFERENCE
Shoshihara, M.; Kase, K.; Yoshizawa, E.; Takao, M.; Fujimoto, T. Column liquid chromatographic determination of isepamicin in nasal cavity using gauze, *J.Chromatogr.*, **1990**, *529*, 473–478.

SAMPLE
Matrix: blood
Analyte: vigabatrin
Sample preparation: 50 μL Serum + 50 μL 100 μg/mL in MeOH:water 10:90 + 1 mL MeOH, vortex, centrifuge at 200 g. Mix 2 volumes of supernatant with 1 volume of reagent, let stand for 1 min, inject a 10 μL aliquot. (Reagent was 10 mL 30 mg/mL o-phthaldialdehyde and 200 μL 2-mercaptoethanol made up to 50 mL with 400 mM pH 9.5 borate buffer.)

HPLC VARIABLES
Column: 150 × 4.6 5 μm Microsorb C18
Mobile phase: MeCN:MeOH:10 mM orthophosphoric acid 30:10:60
Flow rate: 2
Injection volume: 10
Detector: F ex 370 em 418-700 (filter)

CHROMATOGRAM
Retention time: 5.6
Internal standard: gamma-phenyl-gamma-aminobutyric acid (13.1)
Limit of quantitation: 540 ng/mL
Limit of detection: 80 ng/mL

OTHER SUBSTANCES
Non-interfering: carbamazepine, carbamazepine epoxide, ethosuximide, phenobarbital, phenytoin, primidone, valproic acid

KEY WORDS
serum

REFERENCE
Tsanaclis, L.M.; Wicks, J.; Williams, J.; Richens, A. Determination of vigabatrin in plasma by reversed-phase high-performance liquid chromatography, *Ther.Drug Monit.*, **1991**, *13*, 251–253.

SAMPLE
Matrix: blood
Analyte: tobramycin
Sample preparation: Condition a Bond Elut C18 SPE cartridge with 1 mL MeCN and 1 mL MeCN:20 mM pH 8 phosphate buffer 10:90. 100 μL Serum + 400 μL water + 500

µL 10% sulfosalicylic acid, vortex, centrifuge. Remove 60 µL supernatant and add it to 240 µL MeCN and 300 µL o-phthalaldehyde reagent, add to SPE cartridge, rinse in with 300 µL rinse + 100 µL MeCN:20 mM pH 8 phosphate buffer 10:90, wash with 500 µL MeCN:20 mM pH 8 phosphate buffer 10:90, elute with 440 µL MeCN, add 40 µL water to eluate, vortex, inject a 25 µL aliquot. (o-Phthalaldehyde reagent was 200 mg o-phthalaldehyde in 2 mL MeOH + 400 µL 2-mercaptoethanol. Mix with 1 g boric acid in 38 mL water adjusted to pH 10.4 with 450 g/L KOH, store under nitrogen at 4°.)

HPLC VARIABLES
Column: 150 × 4 MicroPak SP C8 (Varian)
Mobile phase: MeCN:20 mM pH 6.5 phosphate buffer 48:52
Flow rate: 2
Injection volume: 25
Detector: F ex 340 em 450

CHROMATOGRAM
Retention time: 8.4
Limit of detection: 10 pmol

KEY WORDS
serum; SPE

REFERENCE
Lai, F.; Sheehan, T. Enhancement of detection sensitivity and cleanup selectivity for tobramycin through pre-column derivatization, *J.Chromatogr.*, **1992**, *609*, 173–179.

SAMPLE
Matrix: blood
Analyte: netilmicin
Sample preparation: 100 µL Plasma + 100 µL trichloroacetic acid + 100 µL 25 µg/mL gentamicin, shake for 30 s, centrifuge at 3200 rpm for 5 min, add 100 µL 1 M NaOH to the supernatant, shake (?) for 30 s, add 1 mL pH 11 KH_2PO_4, add 2 mL dichloromethane, shake for 10 s, centrifuge at 3200 rpm for 5 min. Remove the aqueous layer and add it to 1 mL reagent, shake for 30 s, add 500 mg anhydrous sodium carbonate, shake for 30 s, add 500 µL isopropanol, shake, centrifuge for 5 min, inject an aliquot of the supernatant. (Reagent was 5 mg o-phthalaldehyde, 500 µL MeOH, 300 µL 2-mercaptoethanol, and 5 mL 100 mM pH 10.4 borate buffer. Prepare fresh each day.)

HPLC VARIABLES
Column: 300 × 4 10 µm RP-18
Mobile phase: Gradient. A was water:acetic acid:100 mM heptanesulfonic acid 80:10:10. B was MeCN. A:B from 60:40 to 50:50 over 10 min, re-equilibrate at initial conditions for 10 min.
Flow rate: 2
Detector: F ex 337 em 437

CHROMATOGRAM
Retention time: 6.5
Internal standard: gentamicin (11)
Limit of detection: 100 ng/mL

KEY WORDS
plasma

REFERENCE
Santos, M.; Garcia, E.; López, F.G.; Lanao, J.M.; Dominguez-Gil, A. Determination of netilmicin in plasma by HPLC, *J.Pharm.Biomed.Anal.*, **1995**, *13*, 1059–1062.

SAMPLE
Matrix: blood
Analyte: spectinomycin
Sample preparation: Condition a 3 mL High-hydrophobic C18 SPE cartridge (J.T. Baker) with 3 mL MeOH, 2 mL 20 mM sodium dioctyl sulfosuccinate (homogenize before use, pass slowly through cartridge), and 4 mL 20 mM pH 5.6 citric acid buffer, do not allow

to go dry. 2 mL Plasma + 8 mL water + 625 μL MeOH, adjust pH to 5.2-5.7 with ca. 60 μL 1 M HCl, vortex for 30 s, centrifuge at 2500 g for 10 min, remove the supernatant, rinse the residue with 3 mL citric acid buffer, centrifuge. Combine the rinse and the supernatant and add them to the SPE cartridge at 1-2 mL/min, wash with two 3 mL portions of citric acid buffer, allow to run dry. Centrifuge the SPE cartridge at 3000 g for 20 min then dry it in a stream of nitrogen for 20 min, elute with 4 mL MeOH, evaporate the eluate to dryness under a stream of nitrogen at room temperature, reconstitute with 1 mL MeOH:water 20:80, vortex, centrifuge at 2500 g for 10 min, inject a 50 μL aliquot of the supernatant.

HPLC VARIABLES
Guard column: 10 × 4.6 cation-exchange (Phase-Sep)
Column: 250 × 4.6 5 μm Spherisorb SCX
Mobile phase: MeCN:100 mM pH 2.6 sodium sulfate 20:80
Flow rate: 1.5
Injection volume: 50
Detector: F ex 340 em 460 following post-column reaction. The column effluent mixed with reagent A pumped at 0.25 mL/min and reagent B pumped at 0.75 mL/min and the mixture flowed through a 2.2 m × 0.5 mm ID coil at 70°. The effluent from the coil was mixed with reagent C pumped at 1 mL/min and this mixture flowed through a 1.7 m × 0.3 mm ID PTFE coil at 25° to the detector. Reagent A was prepared by adding 10 mL sodium hypochlorite solution (13% active chlorine) and 10 mL 100 mM pH 7.0 potassium phosphate buffer to 980 mL water. Reagent B was 400 mM pH 10.2 boric acid buffer prepared by dissolving 24.4 g boric acid and 20.0 g KOH in 1 L water. Reagent C was prepared by adding 10 mL 80 mg/mL o-phthalaldehyde in EtOH and 1 mL 2-mercaptoethanol to 990mL 300 mM pH 10.2 boric acid buffer. (The 300 mM pH 10.2 boric acid buffer was prepared by dissolving 18.55 g boric acid and 15.0 g KOH in 1 L water.)

CHROMATOGRAM
Retention time: 9
Limit of quantitation: 60 ng/mL

KEY WORDS
pig; cow; chicken; plasma; SPE; post-column reaction

REFERENCE
Haagsma, N.; Scherpenisse, P.; Simmonds, R.J.; Wood, S.A.; Rees, S.A. High-performance liquid chromatographic determination of spectinomycin in swine, calf and chicken plasma using post-column derivatization, *J.Chromatogr.B*, **1995**, *672*, 165–171.

SAMPLE
Matrix: blood
Analyte: anticonvulsants
Sample preparation: 500 μL Serum + 500 μL IS solution + 1 mL MeCN, vortex for 5 min, centrifuge for 15 min. Mix 6 μL buffer, 6 μL reagent, and 6 μL supernatant, let stand for 1 min, inject the whole amount. (Prepare IS solution by dissolving 100 mg gamma-phenyl-gamma-aminobutyric acid and 10 mg 1-(aminomethyl)cycloheptane acetic acid in 500 mL MeCN and 500 mL water. Prepare buffer by dissolving 15.5 mg boric acid in 500 mL water and adjusting to pH 9.5 with concentrated NaOH. Prepare reagent by mixing 100 mg o-phthalaldehyde, 9 mL MeOH, 1 mL buffer, and 100 μL 2-mercaptoethanol.)

HPLC VARIABLES
Column: 250 × 4 5 μm BANsil C18 (ASMT, Enger, Germany)
Mobile phase: Gradient. A was MeCN:MeOH:0.1% pH 2 phosphoric acid 10:10:80. B was MeCN:MeOH 50:50. A:B 90:10 for 1 min, to 30:70 over 25 min, maintain at 30:70 for 3 min, return to initial conditions over 0.1 min, re-equilibrate for 3.9 min.
Column temperature: 40
Flow rate: 1
Injection volume: 18
Detector: F ex 235 em 435

CHROMATOGRAM
Retention time: 19.9 (vigabatrin), 27.0 (gabapentin)
Internal standard: gamma-phenyl-gamma-aminobutyric acid (Marion Merrel Dow) (23.4), 1-(aminomethyl)cycloheptane acetic acid (Gö-3609, Parke Davis) (28.3)
Limit of quantitation: 500 ng/mL (gabapentin), 1 μg/mL (vigabatrin)
Limit of detection: 100 ng/mL

KEY WORDS
serum; degas mobile phase continuously with helium

REFERENCE
Juergens, U.H.; May, T.W.; Rambeck, B. Simultaneous HPLC determination of vigabatrin and gabapentin in serum with automated pre-injection derivatization, *J.Liq.Chromatogr.Rel.Technol.*, **1996**, *19*, 1459–1471.

SAMPLE
Matrix: blood, CSF, dialysate, tissue
Analyte: baclofen
Sample preparation: Dilute dialysate and CSF. Homogenize brain tissue with 4 volumes of MeOH:water in an ice bath, let stand at -20° for 8 h. 30 μL Plasma + 60 μL MeOH, mix, let stand at -20° for 1 h, centrifuge at 4° at 10000 rpm for 5 min, dilute the supernatant with water. Mix an aliquot of tissue homogenate, deproteinized plasma, diluted dialysate, or diluted CSF with an equal volume of the reagent, let stand for 1.5 min, inject a 30 μL aliquot. (Prepare reagent by mixing 50 mg o-phthalaldehyde, 900 μL MeOH, 100 μL 400 mM pH 9.2 borate buffer, and 50 μL 2-mercaptoethanol.)

HPLC VARIABLES
Guard column: μBondapak C18 Guard-Pak
Column: 250 × 4.6 5 μm Finepak SIL C18S ODS (Jasco)
Mobile phase: MeOH:THF:100 mM pH 6.95 acetate buffer 45.5:2:52.5 (dialysate, CSF, plasma) or 43:2:55 (tissue)
Flow rate: 1
Injection volume: 30
Detector: F ex 368 em 434

CHROMATOGRAM
Limit of detection: 100 nM

KEY WORDS
plasma; rat; brain; pharmacokinetics

REFERENCE
Deguchi, Y.; Inabe, K.; Tomiyasu, K.; Nozawa, K.; Yamada, S.; Kimura, R. Study on brain interstitial fluid distribution and blood-brain barrier transport of baclofen in rats by microdialysis, *Pharm.Res.*, **1995**, *12*, 1838–1844.

SAMPLE
Matrix: blood, tissue, urine
Analyte: taurine
Sample preparation: Urine. Centrifuge at 4° at 2500 g for 10 min, add 25-100 μL urine to the SPE column, discard the first 25-100 μL eluate, elute with 4 mL water, add 100 μL 100 μM homoserine to the eluate, add an equal volume of the reagent, let stand for 1.5 min, inject a 5-10 μL aliquot. Liver. Homogenize (polytron) 350-450 mg frozen liver in 4 mL 200 mM sulfosalicylic acid at 4°, centrifuge at 4° at 2500 g for 10 min, add a 25 μL aliquot to the SPE column, elute with 4 mL water, add 100 μL 100 μM homoserine to the eluate, add an equal volume of the reagent, let stand for 1.5 min, inject a 5-10 μL aliquot. Serum. 100 μL Serum + 100 μL 200 mM sulfosalicylic acid, mix at 4°, centrifuge at 4° at 11500 g for 2 min, add a 50 μL aliquot of the supernatant to the SPE column, discard the first 50 μL eluate, elute with 4 mL water, add 100 μL 100 μM homoserine to the eluate, add an equal volume of the reagent, let stand for 1.5 min, inject a 5-10 μL aliquot. Hepatocytes. 500 μL Cell suspension + 500 μL 200 mM sulfosalicylic acid, mix at 4°, centrifuge at 4° at 11500 g for 2 min, add a 200 μL aliquot of the supernatant (or 100 μL of the centrifuged medium) to the SPE column, elute with eight 500 μL aliquots

of water, add 100 μL 40 μM homoserine to the eluate, add an equal volume of the reagent, let stand for 1.5 min, inject a 5-10 μL aliquot. (Prepare SPE columns as follows. Wash 100 g 100-200 mesh Dowex 1-X4 (anion-exchange, Cl⁻ form) with three volumes of water to remove fines then with 250 mL 1 M HCl until pH was above 2.5. Wash 100 g 100-200 mesh Dowex 50W-X8 (cation-exchange, H+ form) with three volumes of water to remove fines then with 500 mL 4 M HCl in three washings then with 250 mL 1 M HCl. Prepare a column in a Pasteur pipette with 0.5 mL Dowex 1-X4 on top of 1.5 mL Dowex 50W-X8, wash with 12 mL water. After use regenerate with 12 mL 1 M HCl. Prepare reagent by adding 40 mg o-phthalaldehyde in 800 μL EtOH and 40 μL 2-mercaptoethanol to 10 mL buffer, dilute the mixture with an equal volume of water. Buffer was 3.1 g boric acid in 90 mL water adjusted to pH 10.4 with 5 M NaOH and made up to 100 mL.)

HPLC VARIABLES
Guard column: 4 × 4 LiChrospher
Column: 125 × 4 5 μm LiChrospher 100 RP-18
Mobile phase: MeOH:water 43:57 containing 50 mM NaH_2PO_4, pH 5.4
Flow rate: 2
Injection volume: 5-10
Detector: F ex 305-395 (filter) em 420-650 (filter)

CHROMATOGRAM
Retention time: 3.7
Internal standard: homoserine (3)
Limit of detection: 0.5 pmole

KEY WORDS
rat; liver; serum; dog; human; SPE; hepatocytes

REFERENCE
Waterfield, C.J. Determination of taurine in biological samples and isolated hepatocytes by high-performance liquid chromatography with fluorimetric detection, *J.Chromatogr.B*, **1994**, *657*, 37–45.

SAMPLE
Matrix: blood, urine
Analyte: amphetamines
Sample preparation: Adjust pH of 10 mL plasma or 20 mL urine to 11.4 with 5 M NaOH, add to a column containing 1.5 g Amberlite XAD-2, wash with 10 mL water, elute with 20 (plasma) or 40 (urine) mL chloroform:isopropanol 75:25, add 100 μL 6 M HCl in EtOH to the eluate, evaporate to dryness under reduced pressure, reconstitute with 1 mL EtOH, add 1 mL reagent, mix, filter, inject a 50 μL aliquot. (Prepare reagent by dissolving 200 mg o-phthalaldehyde in 2 mL MeOH, add 400 μL 2-mercaptoethanol, add to buffer, store in the dark in the refrigerator, discard after 5 days. Prepare buffer by dissolving 1 g boric acid in 38 mL water and adjusting pH to 10.4 with 4 M KOH.)

HPLC VARIABLES
Column: 200 × 5 10 μm Partisil ODS-2
Mobile phase: MeOH:water 73:27 containing 0.2% EDTA
Column temperature: 20
Flow rate: 1.8
Injection volume: 50
Detector: F ex 345 em 445

CHROMATOGRAM
Retention time: 3 (norephedrine), 4 (hydroxyamphetamine), 9 (amphetamine)
Internal standard: benzylamine (6)
Limit of quantitation: 500 ng/mL

KEY WORDS
plasma; comparison with other derivatization reagents and with ion-pair chromatography; SPE

REFERENCE
Farrell, B.M.; Jefferies, T.M. An investigation of high-performance liquid chromatographic methods for the analysis of amphetamines, *J.Chromatogr.*, **1983**, *272*, 111–128.

SAMPLE
Matrix: blood, urine
Analyte: cycloserine
Sample preparation: Plasma. 1 mL Plasma + 250 μL buffer + 30 μL 240 μg/mL 6-aminocaproic acid + 50 μL 6.5 mg/mL 5-methoxyindole-3-acetic acid, filter (Centriflo ultrafilter) while centrifuging at 723 g for 15 min, inject a 20-25 μL aliquot of the ultrafiltrate. Urine. 1 mL Urine + 30 μL 240 μg/mL 6-aminocaproic acid + 50 μL 3 mg/mL α-aminobutylhistidine + 1 mL 200 mg/mL sodium carbonate in water (prepare only 5-6 samples at a time), mix, add 2 mL isopropanol, vortex, centrifuge at 723 g for 5 min, inject a 25 μL aliquot of the top organic layer. (Buffer was 12.4 g boric acid + 100 mL 1 M NaOH diluted to 250 mL with water, pH 9.75.)

HPLC VARIABLES
Guard column: 30-38 μm Co:Pell ODS
Column: 240 × 5 10 μm ODS-Hypersil
Mobile phase: Isopropanol:water:glacial acetic acid:decanesulfonate 75:800:5:0.5 (plasma) or 65:800:5:0.5 (urine), pH adjusted to 4.4 with 1 M KOH
Flow rate: 2.3
Injection volume: 20-25
Detector: F ex 340 em 455 following post-column reaction. The column effluent mixed with reagent pumped at 1.2 mL/min and the mixture flowed through a 250 × 4.5 reactor packed with 50 μm glass beads to the detector. (Reagent was o-phthalaldehyde and 2-mercaptoethanol post-column derivatizing reagent (Fluoraldehyde, Pierce).)

CHROMATOGRAM
Retention time: 5 (plasma), 6 (urine)
Internal standard: 6-aminocaproic acid (9.5), 5-methoxyindole-3-acetic acid (11 (plasma), 8 (urine), detection at UV 313), α-aminobutylhistidine (15)
Limit of quantitation: 300 ng/mL

OTHER SUBSTANCES
Extracted: acetylacetonylcycloserine, metabolites

KEY WORDS
plasma; methoxyindoleacetic acid is IS for acetylacetonylcycloserine (UV 313 detection for both); post-column reaction

REFERENCE
Musson, D.G.; Maglietto, S.M.; Hwang, S.S.; Gravellese, D.; Bayne, W.F. Simultaneous quantification of cycloserine and its prodrug acetylacetonylcycloserine in plasma and urine by high-performance liquid chromatography using ultraviolet absorbance and fluorescence after post-column derivatization, *J.Chromatogr.*, **1987**, *414*, 121–129.

SAMPLE
Matrix: blood, urine
Analyte: gusperimus
Sample preparation: Plasma. 750 μL Plasma + 30 μL 70% perchloric acid, vortex, centrifuge at 15600 g for 5 min, filter (0.22 μm) the supernatant, inject a 10-200 μL aliquot of the filtrate. Urine. Dilute urine 10 to 25-fold with water, centrifuge, filter (0.22 μm) the supernatant, inject a 10-200 μL aliquot of the filtrate.

HPLC VARIABLES
Guard column: 15 × 3.2 7 μm Newguard RP 18
Column: two 100 × 4.6 5 μm RP 18 columns in series (Brownlee)
Mobile phase: Gradient. A was MeCN:100 mM pH 2.55 NaH_2PO_4 containing 8 mM octanesulfonic acid and 0.1 mM EDTA 2:98. B was MeCN:200 mM pH 3.1 NaH_2PO_4 containing 8 mM octanesulfonic acid 30:70. A:B from 55:45 to 25:75 over 20 min.
Flow rate: 1
Injection volume: 10-200
Detector: F ex 340 em 440 following post-column reaction. The column effluent mixed with the reagent pumped at 0.7 mL/min and flowed through a 2 m long reaction coil at 41° to

the detector. (Reagent was 500 mM pH 8.8 potassium borate buffer containing 0.8 g/L o-phthaldialdehyde, 1% MeOH, and 0.06% 2-mercaptoethanol.)

CHROMATOGRAM
Retention time: 22
Limit of quantitation: 100 nM

OTHER SUBSTANCES
Extracted: metabolites

KEY WORDS
use plasticware; plasma; post-column reaction; pharmacokinetics

REFERENCE
Muindi, J.F.; Lee, S.-J.; Baltzer, L.; Jakubowski, A.; Scher, H.I.; Sprancmanis, L.A.; Riley, C.M.; Vander Velde, D.; Young, C.W. Clinical pharmacology of deoxyspergualin in patients with advanced cancer, *Cancer Res.*, **1991**, *51*, 3096–3101.

SAMPLE
Matrix: blood, urine
Analyte: neomycin
Sample preparation: Plasma. 1 mL Plasma + 100 µL 30% trichloroacetic acid in water, vortex, centrifuge at 4° at 3600 g for 30 min. Remove 450 µL of the supernatant and add it to 50 µL buffer, vortex, centrifuge at 4° at 3600 g for 30 min, inject a 5-25 µL aliquot of the supernatant. Urine. Centrifuge urine at 4° at 3600 g for 30 min. Remove a 180 µL aliquot and add it to 20 µL buffer, vortex, inject a 5-25 µL aliquot. (Buffer contained 110 mM sodium 1-pentanesulfonate and 70 mM acetic acid.)

HPLC VARIABLES
Guard column: 20 × 4.6 5 µm Supelguard LC-8-DB (Supelco)
Column: 150 × 4.6 5 µm Supelcosil LC-8-DB
Mobile phase: MeOH : buffer 1.5 : 98.5 (Buffer contained 11 mM sodium 1-pentanesulfonate, 56 mM sodium sulfate, and 7 mM acetic acid.)
Column temperature: 32.5
Injection volume: 5-30
Detector: F ex 340 em 455 following post-column derivatization. The column effluent mixed with an o-phthalaldehyde reagent (Pierce) and flowed through a reaction coil (PCR 520, Applied Biosystems) at 33° to the detector.

CHROMATOGRAM
Retention time: 16
Limit of quantitation: 250 ng/mL (plasma), 1 µg/mL (urine)

KEY WORDS
plasma; cow; pharmacokinetics; post-column reaction

REFERENCE
Shaikh, B.; Jackson, J.; Guyer, G.; Ravis, W.R. Determination of neomycin in plasma and urine by high-performance liquid chromatography. Application to a preliminary pharmacokinetic study, *J.Chromatogr.*, **1991**, *571*, 189–198.

SAMPLE
Matrix: blood, urine
Analyte: mexiletine
Sample preparation: Plasma. 1 mL Plasma + 50 µL 20 µg/mL 4-methylmexiletine in MeOH + 1 mL 200 mM pH 9 borate buffer + 7 mL diethyl ether, shake on a reciprocating shaker for 10 min, centrifuge at 3000 g for 5 min, freeze in MeOH/dry ice. Remove the organic layer and evaporate it to dryness under a stream of nitrogen at 40°, reconstitute the residue in 250 µL MeOH : 10 mM phosphoric acid 48 : 52 containing 300 mg/L sodium 1-octanesulfonate, inject a 100 µL aliquot. Urine. 100 µL Urine + 300 µL 1 M pH 5.5 sodium acetate buffer + 2500 U β-glucuronidase (Helix pomatia Type H-5, Sigma), vortex, heat at 37° for 4 h, add 100 µL 20 µg/mL 4-methylmexiletine in MeOH, add 1 mL 200 mM pH 10 borate buffer, add 7 mL diethyl ether, shake on a reciprocating shaker for 10 min, centrifuge at 3000 g for 5 min, freeze in MeOH/dry ice. Remove the organic layer

and evaporate it to dryness under a stream of nitrogen at 40°, reconstitute the residue in 250 μL MeOH:10 mM phosphoric acid 48:52 containing 300 mg/L sodium 1-octanesulfonate, inject a 100 μL aliquot.

HPLC VARIABLES
Guard column: 10 × 4.6 Wakosil ODS (Wako)
Column: 250 × 4.6 Wakosil ODS (Wako)
Mobile phase: Gradient. A was MeOH:10 mM phosphoric acid 16:84 containing 300 mg/L sodium 1-octanesulfonate. B was MeOH:10 mM phosphoric acid 80:20 containing 300 mg/L sodium 1-octanesulfonate. A:B from 65:35 to 20:80 over 10 min, maintain at 20:80 for 4 min, return to initial conditions over 1 min, re-equilibrate for 10 min.
Flow rate: 1
Injection volume: 100
Detector: F ex 345 em 445 following post-column reaction. The column effluent mixed with reagent pumped at 0.6 mL/min and the mixture flowed through a 5 m × 0.5 mm ID Tefzel tube to the detector. (Reagent was 300 mg o-phthalaldehyde in 100 mL EtOH, add 900 mL 200 mM pH 10 borate buffer, add 500 μL 2-mercaptoethanol.)

CHROMATOGRAM
Retention time: 12.5
Internal standard: 4-methylmexiletine (14)
Limit of detection: 2 ng/mL (plasma)

OTHER SUBSTANCES
Extracted: metabolites

KEY WORDS
plasma; post-column reaction; pharmacokinetics

REFERENCE
Tateishi, T.; Harada, K.; Ebihara, A. Fluorescence detection of mexiletine and its p-hydroxylated and hydroxymethylated metabolites in human plasma and urine by high-performance liquid chromatography using post-column derivatization with o-phthalaldehyde, *J.Liq.Chromatogr.*, **1994**, *17*, 659–671.

SAMPLE
Matrix: blood, urine
Analyte: isepamicin
Sample preparation: 1 mL Plasma or urine + 0.5-5 μg dibekacin + 2 mL EtOH, mix, add 7 mL dichloromethane, add 1 mL water, centrifuge, inject an aliquot of the aqueous supernatant on to column A and elute to waste with mobile phase A, after 4 min elute the contents of column A on to column B with mobile phase B, monitor the effluent from column B.

HPLC VARIABLES
Column: A 3.9 × 4 10 μm Guard Pak with Cyano insert; B 150 × 4.6 5 μm Hypersil C18
Mobile phase: A 17 mM Acetic acid containing 10 mM hexanesulfonate; B MeOH:buffer 14.3:85.7 (Mobile B phase contained 100 mM sodium acetate, 17 mM acetic acid, 10 mM hexanesulfonate, and 3.53 M MeOH.)
Detector: F ex 338 (band-pass filter) em 418-700 and 450 cut-off filters following post-column reaction. The column effluent mixed with o-phthalaldehyde derivatizing reagent pumped at 0.2 mL/min and the mixture flowed to the detector.

CHROMATOGRAM
Internal standard: dibekacin
Limit of quantitation: 100 ng/mL

KEY WORDS
plasma; pharmacokinetics; post-column reaction; column-switching

REFERENCE
Lin, C.-C.; Radwanski, E.; Korduba, C.; Cayen, M.; Affrime, M. Pharmacokinetics of intravenously administered isepamicin in men, *Antimicrob.Agents Chemother.*, **1995**, *39*, 2774–2778.

SAMPLE
Matrix: blood, urine
Analyte: ranitidine
Sample preparation: Plasma. 500 μL Plasma + 100 μL 100 mM NaOH, mix, add 3 mL dichloromethane, shake for 10 min, centrifuge at 2000 g for 10 min, repeat the extraction. Combine the organic layers and evaporate them to dryness under a stream of argon, reconstitute with 500 μL mobile phase, inject a 100 μL aliquot. Urine. Dilute 1 mL urine with 25 mL water, filter (0.2 μm), inject a 100 μL aliquot.

HPLC VARIABLES
Guard column: 5 μm Spherisorb ODS-2
Column: 150 × 4 5 μm Spherisorb ODS-2
Mobile phase: Gradient. MeCN:7.5 mM pH 6 phosphate buffer 7:93 for 8 min, to 25:75 over 1 min, maintain at 25:75 for 6 min, return to initial conditions over 1 min, re-equilibrate for 10 min. (At the end of each day wash column with MeCN then water then re-equilibrate with mobile phase.)
Flow rate: 1
Injection volume: 100
Detector: F ex 350 em 450 following post-column reaction. The column effluent mixed with 5 mM sodium hypochlorite in 50 mM pH 4.5 sodium acetate buffer pumped at 1 mL/min and this mixture flowed through a 0.8 m × 0.5 mm ID PTFE coil at 25°. The effluent from this coil mixed with 20 mM o-phthalaldehyde in EtOH:500 mM pH 10.5 borate buffer 2:98 pumped at 1 mL/min and 100 mM 2-mercaptoethanol in 500 mM pH 10.5 borate buffer pumped at 1 mL/min and this mixture flowed through a 2.5 m × 0.5 mm ID PTFE coil at 25° to the detector. (The hypochlorite oxidizes the secondary to a primary amine which then reacts with the o-phthalaldehyde and 2-mercaptoethanol.)

CHROMATOGRAM
Retention time: 13
Limit of quantitation: 106 ng/mL
Limit of detection: 32 ng/mL

OTHER SUBSTANCES
Extracted: metabolites

KEY WORDS
post-column reaction; plasma

REFERENCE
Viñas, P.; Campillo, N.; López-Erroz, C.; Hernández-Córdoba, M. Use of post-column fluorescence derivatization to develop a liquid chromatographic assay for ranitidine and its metabolites in biological fluids, *J.Chromatogr.B*, **1997**, *693*, 443–449.

SAMPLE
Matrix: cell suspensions
Analyte: histamine
Sample preparation: 500 μL Cell suspension in buffer A containing 0.1% bovine serum albumin + 1.5 mL ice cold buffer A, centrifuge at 4° at 250 g for 10 min, remove the supernatant, add 2 mL buffer A to the pellet, boil this mixture for 3 min, cool on ice, centrifuge at 4° at 500 g for 10 min, remove the supernatant. Dry a 50-200 μL aliquot of each supernatant in a vacuum centrifuge and reconstitute the residue with 100 μL buffer B, add 20 μL reagent, inject an aliquot. (Buffer A contained 150 mM NaCl, 4 mM KCl, 1 mM calcium chloride, 1.2 mM magnesium sulfate, 2.46 mM Na_2HPO_4 0.615 mM KH_2PO_4, 5.6 mM glucose, and 10 mM HEPES, pH 7.4. Buffer B was THF:100 mM pH 9.5 sodium tetraborate buffer 30:70. The reagent was prepared by mixing equal volumes of 3.8 mM o-phthalaldehyde in MeOH (freshly prepared) and 2.5 mL/L 2-mercaptoethanol in MeOH.)

HPLC VARIABLES
Column: 100 × 3.2 3 μm Phase-II ODS (Bioanalytical Systems)
Mobile phase: MeCN:MeOH:buffer 14:16:70 containing 1 mM disodium EDTA (Buffer was 100 mM sodium phosphate buffer containing 0.4% triethylamine, pH 6.4.)
Flow rate: 0.6

Detector: E, Bioanalytical Systems Model LC4B, LC17A thin-layer electrochemical cell, glassy carbon working electrode +0.5 V, Ag/AgCl reference electrode

CHROMATOGRAM
Retention time: 10
Limit of detection: <0.1 pmole

KEY WORDS
rat; mast cells

REFERENCE
Jensen, T.B.; Marley, P.D. Development of an assay for histamine using automated high-performance liquid chromatography with electrochemical detection, *J.Chromatogr.B*, **1995**, *670*, 199–207.

SAMPLE
Matrix: eggs, milk, tissue
Analyte: sulfonamides
Sample preparation: Milk. Homogenize 3 g milk and 500 μL 30% trichloroacetic acid, centrifuge at 5000 rpm for 5 min. Remove the aqueous phase and extract the residue with 4 mL 3% trichloroacetic acid. Combine the aqueous layers and make up to 10 mL with trichloroacetic acid, filter (0.45 μm), inject a 50 μL aliquot. Fish, eggs. Homogenize (Ultra-Turrax) 3 g fish or 4 g eggs with 4 mL 3% trichloroacetic acid, centrifuge at 5000 rpm for 5 min. Remove the aqueous phase and extract the residue with 4 mL 3% trichloroacetic acid. Combine the aqueous layers and make up to 10 mL with trichloroacetic acid, filter (0.45 μm), inject a 50 μL aliquot.

HPLC VARIABLES
Guard column: 5 μm Spherisorb ODS-2
Column: 150 × 4.6 5 μm Spherisorb ODS-2
Mobile phase: MeCN:water 3:97 for 5 min, to 40:60 over 15 min, return to initial conditions over 1 min, re-equilibrate for 10 min. (At the end of each day wash with MeCN: ethyl acetate 5:95.)
Flow rate: 0.5
Injection volume: 50
Detector: F ex 302 em 412 following post-column reaction. The column effluent mixed with reagent 1 pumped at 0.25 mL/min and with reagent 2 pumped at 0.25 mL/min and this mixture flowed through a 2.5 m × 0.8 mm i.d. PTFE coil at 40° to the detector. (Reagent 1 was 10 mM o-phthalaldehyde in EtOH:700 mM phosphoric acid 2:98. Reagent 2 was 20 mM 2-mercaptoethanol in EtOH:700 mM phosphoric acid 2:98.)

CHROMATOGRAM
Retention time: 10 (sulfanilamide), 12 (sulfaguanidine), 20 (sulfadiazine), 21.5 (sulfapyridine), 26 (sulfamethoxazole)
Limit of detection: 11-19 ng/mL

OTHER SUBSTANCES
Non-interfering: sulfathiazole

KEY WORDS
post-column reaction

REFERENCE
Viñas, P. ; Erroz, C.L. ; Campillo, N. ; Hernández-Córdoba, M. Determination of sulphonamides in foods by liquid chromatography with postcolumn fluorescence derivatization, *J.Chromatogr.A* , **1996** , *726*, 125 –131.

SAMPLE
Matrix: erythrocytes
Analyte: carbonic anhydrase I
Sample preparation: Inject an aliquot of a solution in running buffer.

CAPILLARY ELECTROPHORESIS
Capillary: 70 cm × 15 μm
Running buffer: 20 mM Sodium tetraborate adjusted to pH 9.5 with NaOH

Voltage/Current: 300 V/cm
Injection: Electrokinetic for 5 s
Detector: F ex 325 (He-Cd laser) em 456 (filter) following post-column reaction. The effluent from the capillary mixed with reagent and this mixture moved through a 12 cm $\times$ 30 μm capillary under the influence of a 150 V/cm electric field to the detector. (Construction details for the post-column reactor are given in the paper. Prepare reagent by dissolving 2 mg o-phthalaldehyde in 3.9 mL buffer, 80 μL EtOH, and 20 μL 2-mercaptoethanol, store overnight at 4° before use, discard after 48 h. Prepare buffer by adjusting 20 mM sodium tetraborate solution to pH 9.5 with NaOH.)
Model: laboratory-constructed
Migration time: 10.2
Limit of detection: 3.8 amole

KEY WORDS
post-column reaction

REFERENCE
Zhang, L.; Yeung, E.S. Postcolumn reactor in capillary electrophoresis for laser-induced fluorescence detection, *J.Chromatogr.A*, **1996**, *734*, 331–337.

SAMPLE
Matrix: milk
Analyte: neomycin
Sample preparation: 1 mL Skim milk + 100 μL 20% trichloroacetic acid, vortex, centrifuge at 4° at 4000 rpm for 30 min. Remove a 180 μL aliquot and add it to 20 μL 100 mM sodium 1-pentanesulfonate containing 70 mM acetic acid, vortex, inject a 25 μL aliquot.

HPLC VARIABLES
Guard column: 20 $\times$ 4.6 5 μm Supelguard LC-8-DB
Column: 150 $\times$ 4.6 5 μm Supelcosil LC-8-DB
Mobile phase: MeOH:buffer 1.5:98.5 (Buffer contained 10 mM pentanesulfonic acid, 56 mM sodium sulfate, and 7 mM acetic acid.)
Column temperature: 32.5
Injection volume: 25
Detector: F ex 340 em 455 following post-column reaction. The column effluent mixed with o-phthalaldehyde solution (Pierce) and flowed through a reaction coil at 33° to the detector.

CHROMATOGRAM
Retention time: 20
Limit of detection: 150 ng/mL

KEY WORDS
cow; post-column reaction

REFERENCE
Shaikh, B.; Jackson, J. Determination of neomycin in milk by reversed phase ion-pairing liquid chromatography, *J.Liq.Chromatogr.*, **1989**, *12*, 1497–1515.

SAMPLE
Matrix: milk
Analyte: neomycin
Sample preparation: Prepare an SPE column as follows. Shake Amberlite CG 50 resin in buffer, equilibrate for 2 h, fill a plugged Pasteur pipette to a height of 35 mm with the slurry, wash with water until the eluent is neutral. 40 mL Milk + 2 g NaCl, shake well, add a 10 mL aliquot to the SPE column, wash with 8 mL water, add 600 μL reagent, let stand for 2 min, elute with 3 mL MeOH:buffer 80:20, make up the eluate to 4 mL with MeOH, store at -8 to -10°, after 15 min inject a 10-20 μL aliquot. (Prepare buffer by dissolving 76 g potassium tetraborate in 400 mL water, adjusting pH to 11 with KOH, and making up to 500 mL with water. Prepare reagent by dissolving 100 mg o-phthalaldehyde in 10 mL MeOH and adding 200 μL 2-mercaptoethanol and 10 mL borate buffer. Prepare borate buffer by dissolving 3.1 g boric acid in 100 mL water and adjusting the pH to 10.5 with 50% KOH, make up to 125 mL with water.)

HPLC VARIABLES

Column: 150 × 4.6 5 μm HISEP (Supelco)

Mobile phase: Gradient. A was 2 g/L tripotassium EDTA in MeOH:water 70:30. B was MeOH. A:B 100:0 to 40:60 over 15 min (LDC concave curve 4). (This curve holds the initial conditions for about 6 min.)

Flow rate: 1.7

Injection volume: 10-20

Detector: F ex 340 em KV418 filter

CHROMATOGRAM

Retention time: 8.2, 19.8

Limit of detection: 50 ppb

KEY WORDS

cow; SPE

REFERENCE

Agarwal, V.K. High performance liquid chromatographic determination of neomycin in milk using a HISEP column, *J.Liq.Chromatogr.*, **1990**, *13*, 2475–2487.

SAMPLE

Matrix: solutions

Analyte: spectinomycin and impurities

Sample preparation: Prepare an aqueous solution, inject an aliquot.

HPLC VARIABLES

Guard column: 43 × 4.2 Perisorb RP-8 (Merck)

Column: 250 × 4.6 LiChrosorb RP-8

Mobile phase: 0.1% Acetic acid containing 20 mM sodium heptanesulfonate and 200 mM sodium sulfate

Flow rate: 2

Injection volume: 15

Detector: F ex 350 em 450 following post-column reaction. The column effluent mixed with 10 mM sodium hypochlorite in 400 mM pH 10.4 potassium borate buffer pumped at 0.5 mL/min and the mixture flowed through a 2 m × 0.5 mm ID PTFE coil at 100°. The effluent from this coil mixed with the reagent pumped at 0.5 mL/min and this mixture passed through a 2 m × 0.5 mm ID PTFE coil at ambient temperature to the detector. (The reagent was prepared by adding 10 mL 80 mg/mL o-phthalaldehyde in 95% EtOH to 1 L 400 mM boric acid solution containing 2 mL 2-mercaptoethanol adjusted to pH 9.7 with KOH, then adding 1 g/L Brij. (Proc. Nat. Acad. Sci. USA 1975, 72, 619).)

CHROMATOGRAM

Retention time: 3 (actinamine), 4 (actinospectinoic acid), 7.5 (spectinomycin)

OTHER SUBSTANCES

Simultaneously analyzed: dihydrospectinomycin

KEY WORDS

post-column reaction

REFERENCE

Myers, H.N.; Rindler, J.V. Determination of spectinomycin by high-performance liquid chromatography with fluorimetric detection, *J.Chromatogr.*, **1979**, *176*, 103–108.

SAMPLE

Matrix: solutions

Analyte: eflornithine

HPLC VARIABLES

Column: 150 × 4.6 5 μm Ultrasphere ion pair

Mobile phase: Buffer prepared from 0.92 g L-proline and 1 g $CuSO_4.5H_2O$ in 1 L water, pH adjusted to 5.5 with 1 mL 5 M NaOH.

Column temperature: 30

Flow rate: 0.5

Detector: F ex 340 em 455 following post-column reaction. The column effluent mixed with reagent pumped at 0.35 mL/min and flowed through a 3 m × 0.3 mm ID PTFE coil at 50° to the detector. (Reagent was 800 mg o-phthalaldehyde in 10 mL EtOH + 500 mM pH 10.5 potassium borate buffer, filter (0.45 μm), add 3 mL Brij 35, add 2.5 mL 2-mercaptoethanol, store in the dark.)

CHROMATOGRAM
Retention time: 6.4 (+), 13.9 (−)

KEY WORDS
post-column reaction

REFERENCE
Wagner, J.; Gaget, C.; Heintzelmann, B.; Wolf, E. Resolution of the enantiomers of various α-substituted ornithine and lysine analogs by high-performance liquid chromatography with chiral eluant and by gas chromatography on Chirasil-Val, *Anal.Biochem.*, **1987**, *164*, 102–116.

SAMPLE
Matrix: solutions
Analyte: amphetamine
Sample preparation: 500 μL Solution + 250 μL water + 250 μL reagent, mix, inject a 50 μL aliquot. (Prepare reagent by dissolving 50 mg o-phthalaldehyde in 500 μL MeOH, add 100 μL 2-mercaptoethanol, dilute to 10 mL with buffer. Protect from light, discard after 3 days. Buffer was 15 mM boric acid adjusted to pH 10 with 10% NaOH.)

HPLC VARIABLES
Column: 125 × 4 5 μm LiChrospher 100 RP 18
Mobile phase: Gradient. MeCN:0.5% propylamine hydrochloride in water from 40:60 to 50:50 over 2.5 min, to 70:30 over 2.5 min, maintain at 70:30 for 5 min.
Flow rate: 0.8
Injection volume: 50
Detector: F ex 345 em 445

CHROMATOGRAM
Retention time: 7.5

REFERENCE
Herráez-Hernández, R.; Campins-Falcó, P.; Sevillano-Cabeza, A. On-line derivatization into precolumns for the determination of drugs by liquid chromatography and column switching: Determination of amphetamines in urine, *Anal.Chem.*, **1996**, *68*, 734–739.

SAMPLE
Matrix: solutions
Analyte: neomycin
Sample preparation: Inject a 20 μL aliquot of a solution in mobile phase.

HPLC VARIABLES
Column: 150 × 4.6 5 μm Supelcosil LC-8
Mobile phase: THF:buffer 3:97 (Buffer was 7 mM acetic acid containing 5.6 mM sodium sulfate and 10 mM sodium pentanesulfonate.)
Flow rate: 1.75
Injection volume: 20
Detector: F ex 365 em 418 following post-column reaction. The column effluent mixed with the reagent pumped at 0.4 mL/min and the mixture flowed through a 50 × 4.6 column packed with 75 μm glass beads and a 3.05 m × 0.5 mm knitted PTFE coil at 40° to the detector. (The reagent was 400 mM boric acid containing 380 mM KOH, 6 mL/L 40% Brij-35, 4 mL/L 2-mercaptoethanol, and 0.8 g/L o-phthalaldehyde.)

CHROMATOGRAM
Retention time: 5

KEY WORDS
post-column reaction

REFERENCE
Supelco Chromatography Products, Supelco, Inc. Bellefonte PA, **1996**, A29.

SAMPLE
Matrix: solutions
Analyte: N-methylcarbamate insecticides
Sample preparation: Inject a 20 μL aliquot.

HPLC VARIABLES
Guard column: 10 × 3 5 μm Kromasil-100-C18 (Akzo Nobel)
Column: 150 × 4 5 μm Kromasil-100-C18 (Akzo Nobel)
Mobile phase: MeCN:buffer 28:72 (Buffer was 820 mg/L sodium tetraborate decahydrate containing 50 μg/mL phthalaldehyde and 0.06 μL/mL 2-mercaptoethanol, adjusted to pH 8.5 with 100 mM HCl.)
Flow rate: 1
Injection volume: 20
Detector: F ex 340 em 460 following post-column reaction. The column effluent flowed through a 3 m × 0.51 mm ID stainless steel tube at 140° to the detector. (Although the reagents are in the mobile phase the derivatization reaction does not take place until the post-column reactor where the insecticides are hydrolyzed to methylamine that is then derivatized. This procedure avoids the use of a second pump for the post-column reagent.)

CHROMATOGRAM
Retention time: 2.5 (methomyl), 4.5 (dioxacarb), 6 (butocarboxim), 7 (aldicarb), 13 (propoxur), 14 (carbofuran), 19 (carbaryl)
Limit of detection: 200-700 pg

KEY WORDS
post-column reaction

REFERENCE
Sabala, A.; Portillo, J.L.; Broto-Puig, F.; Comellas, L. Development of a new high-performance liquid chromatography method to analyse N-methylcarbamate insecticides by a simple post-column derivatization system and fluorescence detection, *J.Chromatogr.A,* **1997**, *778*, 103–110.

SAMPLE
Matrix: tissue
Analyte: taurine
Sample preparation: Homogenize (Kontes micro-ultrasonic cell disrupter) 1-5 mg rat pineal tissue in 200 μL water, add 200 μL 2% picric acid in water (Caution! Picric acid is toxic and explosive when dry!), let stand at room temperature for 5 min, add to the column, rinse the container with three 200 μL portions of water, add the rinses to the column, add 1 mL water to the column, lyophilize the eluate, reconstitute with 100 μL water. Mix an aliquot of this solution with an equal volume of the reagent, let stand for 1 min, inject a 2-6 μL aliquot. (Prepare the column by making a 25 mm layer of Bio-Rad 200-400 mesh 50W-X8 (hydrogen form) ion-exchange resin in a 5 mm diameter column, add a 25 mm layer of 100-200 mesh Bio-Rad AG 1-X8 (chloride form) ion-exchange resin on top of this, wash with 10 mL water just before use. Prepare the reagent by dissolving 20 mg o-phthalaldehyde in 400 μL EtOH, adding 20 μL 2-mercaptoethanol, and adding 10 mL 500 mM pH 10.3 borate buffer. Dilute 1:10 with water before use.)

HPLC VARIABLES
Column: 300 × 3.9 μBondapak alkyl phenyl
Mobile phase: MeOH:buffer 42.75:57.25 (Prepare by mixing A:B in the ratio 43:57. A was prepared by dissolving 6.9 g $NaH_2PO_4.H_2O$ in water, adjusting pH to 5.3 with 5 M NaOH, and making up to 1 L with water. B was prepared by dissolving 6.9 g NaH_2PO_4 .H_2O in 250 mL water and making up to 1 L with MeOH.)
Flow rate: 2
Injection volume: 2-6
Detector: F ex 360 em 455

CHROMATOGRAM
Retention time: 4
Limit of quantitation: 600 pg

OTHER SUBSTANCES
Non-interfering: cysteic acid, hypotaurine, phosphoethanolamine

KEY WORDS
rat; pineal; SPE

REFERENCE
Larsen, B.R.; Grosso, D.S.; Chang, S.Y. A rapid method for taurine quantitation using high performance liquid chromatography, *J.Chromatogr.Sci.*, **1980**, *18*, 233–236.

SAMPLE
Matrix: tissue
Analyte: antibiotics
Sample preparation: Homogenize (Tissumizer) 1 g Ground tissue + 4 mL buffer at medium speed for 1 min, centrifuge at 3600 g for 20 min, remove the supernatant, re-homogenize pellet in 4 mL buffer for 10 min, centrifuge. Combine the supernatants, heat in a boiling water bath with occasional mixing for 5 min, centrifuge at 2000 g for 20 min, remove the supernatant, vortex the precipitate with 2 mL buffer for 30 s, centrifuge at 2000 g for 10 min. Combine the supernatants, acidify to pH 3.5-4 with 50-60 μL sulfuric acid, centrifuge at 2000 g for 10 min, inject an aliquot of the supernatant. (Buffer was 33.46 g K_2HPO_4 and 1.046 g KH_2PO_4 in 1 L water, pH 8.0.)

HPLC VARIABLES
Guard column: 10 μm RP-18
Column: 150 × 4.6 5 μm Supelcosil LC-8-DB
Mobile phase: MeOH:buffer 1.5:98.5 (Buffer was 10 mM sodium 1-pentanesulfonate, 56 mM sodium sulfate, and 7 mM acetic acid.)
Flow rate: 1.5
Detector: F ex 340 em 455 following post-column reaction with derivatization reagent pumped at 0.9 mL/min. (Derivatization reagent was commercially available (Pierce) or prepared by adding 2.5 mL 2-mercaptoethanol and 2.5 mL Brij-35 to 850 mg o-phthalaldehyde in 10 mL MeOH, mix until decolorization is complete, add 1 L buffer, filter (0.45 μm), and refrigerate until used. Buffer was prepared by adjusting pH of 250 mM boric acid to 9.5 with 5 M KOH.)

CHROMATOGRAM
Retention time: 6 (streptomycin), 9 (dihydrostreptomycin), 19 (paromomycin), 22 (neomycin)
Limit of detection: 3.5 ng

KEY WORDS
kidney; muscle; cow; pig; extraction validated only for paromomycin and neomycin; post-column reaction

REFERENCE
Shaikh, B.; Allen, E.H.; Gridley, J.C. Determination of neomycin in animal tissues, using ion-pair liquid chromatography with fluorometric detection, *J.Assoc.Off.Anal.Chem.*, **1985**, *68*, 29–36.

SAMPLE
Matrix: tissue
Analyte: amines
Sample preparation: 10 g Homogenized fish + 15 mL 600 mM perchloric acid, stir magnetically for 10 min, centrifuge at 3000 rpm for 10 min, remove the supernatant, add 10 mL 600 mM perchloric acid to the residue, stir magnetically for 10 min, centrifuge at 3000 rpm for 10 min. Combine the supernatants, make up to 25 mL with 600 mM perchloric acid, filter (0.45 μm), inject a 20 μL aliquot of the filtrate.

HPLC VARIABLES
Column: 150 × 3.9 4 μm Nova Pak C18
Mobile phase: Gradient. A was 100 mM sodium acetate containing 10 mM sodium octanesulfonate, adjusted to pH 5.20 with acetic acid. B was MeCN:buffer 66:34 (Buffer was 200 mM sodium acetate containing 10 mM sodium octanesulfonate, adjusted to pH 4.50

with acetic acid.) A:B from 80:20 to 20:80 over 50 min, maintain at 20:80 for 2 min, return to initial conditions over 2 min, re-equilibrate for 10 min.

Flow rate: 1

Injection volume: 20

Detector: F ex 340 em 445 following post-column reaction. The column effluent mixed with the reagent pumped at 0.5 mL/min, the mixture flowed through a 200 cm × 0.25 mm ID coil of stainless steel tubing to the detector. (Prepare reagent by dissolving 15.5 g boric acid and 13.1 g KOH in 500 mL water, adjust pH to 10.5-11 with 30% KOH (if necessary), add 1.5 mL 30% Brij-35, add 1.5 mL 2-mercaptoethanol, add 2.5 mL 40 μg/mL o-phthalaldehyde in MeOH, mix. Protect from light, prepare fresh daily.)

CHROMATOGRAM

Retention time: 24 (putrescine), 26 (tyramine), 28 (cadaverine), 32.5 (serotonin), 35 (histamine), 44 (agmatine), 50 (β-phenylethylamine), 52 (spermidine), 54 (tryptamine), 57 (spermine)

Limit of detection: 250 ng/g

KEY WORDS

post-column reaction; fish

REFERENCE

Veciana-Nogues, M.T.; Hernandez-Jover, T.; Marine-Font, A.; Vidal-Carou, M.C. Liquid chromatographic method for determination of biogenic amines in fish and fish products, *J.AOAC Int.*, **1995**, *78*, 1045–1050.

SAMPLE

Matrix: urine

Analyte: amphetamine

Sample preparation: Inject 50 μL urine on to column A and elute to waste with mobile phase A, after 2 min inject a mixture of 15 μL reagent on to column A, stop the flow of mobile phase A, backflush the contents of column A on to column B with mobile phase B and start the gradient, monitor the effluent from column B. After each run flush column A with ethyl acetate for 1 min, n-hexane for 1 min, and ethyl acetate for 1 min at 1 mL/min, re-equilibrate with mobile phase A. (Prepare reagent by dissolving 50 mg o-phthalaldehyde in 500 μL MeOH, add 100 μL 2-mercaptoethanol, dilute to 10 mL with buffer. Protect from light, discard after 3 days. Buffer was 15 mM boric acid adjusted to pH 10 with 10% NaOH.)

HPLC VARIABLES

Column: A 20 × 2.1 30 μm Hypersil ODS-C18; B 125 × 4 5 μm LiChrospher 100 RP 18

Mobile phase: A water; B Gradient. MeCN:0.5% propylamine hydrochloride in water from 40:60 to 50:50 over 2.5 min, to 70:30 over 2.5 min, maintain at 70:30 for 5 min.

Flow rate: A 1; B 0.8

Injection volume: 50

Detector: F ex 345 em 445

CHROMATOGRAM

Retention time: 7.5

Limit of detection: 10 ng/mL

KEY WORDS

column-switching; on-column derivatization

REFERENCE

Herráez-Hernández, R.; Campins-Falcó, P.; Sevillano-Cabeza, A. On-line derivatization into precolumns for the determination of drugs by liquid chromatography and column switching: Determination of amphetamines in urine, *Anal.Chem.*, **1996**, *68*, 734–739.

SAMPLE

Matrix: wine

Analyte: histamine

Sample preparation: Wash 5 g Amberlite CG-50 with water, add 5 mL 10 M NaOH, let stand for 30 min, rinse with water 3 times, add 25 mL 5 M HCl to a pH of 2, wash, mix with 5 mL 10 M NaOH, wash with 1 volume pH 7 buffer. Filter (0.45 μm) wine, add 10 mL to the resin in a 40 × 10 column, wash with water, elute with 10 mL 1 M HCl. Evaporate the eluate to 1 mL under reduced pressure, add heptylamine, derivatize with reagent, inject an aliquot. (Reagent was 45 mg o-phthalaldehyde in 1 mL MeOH, add 200 μL 2-mercaptoethanol, make up to 10 mL with buffer, prepare daily. Buffer was 3.81 g sodium tetraborate in 100 mL water, adjust to pH 10.5 with 10 M NaOH.)

HPLC VARIABLES
Column: 250 × 4.6 5 μm Spherisorb ODS-2
Mobile phase: Gradient. A was THF:water 1:99 containing 0.03% triethanolamine. B was MeOH. A:B from 40:60 to 20:80 over 25 min, re-equilibrate for 3 min. (Every 10 analyses flush with MeCN:80.8 mM acetic acid 30:70 for 1 h.)
Column temperature: 60
Flow rate: 1
Detector: F ex 330 em 445

CHROMATOGRAM
Retention time: 4.5
Internal standard: heptylamine (21.5)
Limit of quantitation: 100 ng/mL

OTHER SUBSTANCES
Simultaneously analyzed: biogenic amines

KEY WORDS
SPE

REFERENCE
Busto, O.; Mestres, M.; Guasch, J.; Borrull, F. Determination of biogenic amines in wine after clean-up by solid-phase extraction, *Chromatographia*, **1995**, *40*, 404–410.

Phthalaldehyde/3-Mercaptopropionic Acid

RELATED REFERENCES
Fujimoto, T.; Tawa, R.; Hirose, S. Fluorometric determination of sisomicin, an aminoglycoside antibiotic, in dried blood spots on filter paper by reversed-phase high-performance liquid chromatography with pre-column derivatization. *Chem.Pharm.Bull.* **1988**, *36*, 1571-1574.

Fujimoto, T.; Tawa, R.; Hirose, S. Determination of sisomicin in eluate from dried blood spot on filter paper disc for monitoring of blood level in rat, by reversed-phase high-performance liquid chromatography after pre-column fluorimetric derivatization. *Chem.Pharm.Bull.* **1989**, *37*, 174-176.

Kadin, H.; Brittain, H.G.; Ivashkiv, E.; Cohen, A.I. Body fluid analysis of a phosphonic acid angiotensin-converting enzyme inhibitor using high-performance liquid chromatography and post-column derivatization with o-phthalaldehyde. *J.Chromatogr.* **1989**, *487*, 135-141.

Koshide, K.; Tawa, R.; Hirose, S.; Fujimoto, T. Liquid-chromatographic determination of sisomicin in plasma, with fluorometric pre-column drivatization. *Clin.Chem.* **1985**, *31*, 1921-1922.

Miller-Stein, C.; Boppana, V.K.; Rhodes, G.R. High-performance liquid chromatographic determination of the N-ethyl tricarbamate ester pro-drug of fenoldopam utilizing simultaneous post-column hydrolysis and fluorescence derivatization. *J.Chromatogr.B* **1994**, *661*, 291-297.

Tawa, R.; Koshide, K.; Hirose, S.; Fujimoto, T. Pre-column derivatization of sisomicin with o-phthalaldehyde-β-mercaptopropionic acid and its application to sensitive high-performance liquid chromatographic determination with fluorimetric detection. *J.Chromatogr.* **1988**, *425*, 143-152.

Tsumura, Y.; Ujita, K.; Nakamura, Y.; Tonogai, Y.; Ito, Y. Simultaneous determination of aldicarb, ethiofencarb, methiocarb and their oxidized metabolites in grains, fruits and vegetables by high performance liquid chromatography. *J.Food Protect.* **1994**, *57*, 1001-1007.

Tsumura, Y.; Ujita, K.; Nakamura, Y.; Tonogai, Y.; Ito, Y. Simultaneous determination of aldicarb, ethiofencarb, methiocarb and their oxidized metabolites in grains, fruits and vegetables by high performance liquid chromatography. *J.Food Protect.* **1995**, *58*, 217-222.

van Eijk, H.M.H.; Rooyakkers, D.R.; Deutz, N.E.P. Automated determination of polyamines by high-performance liquid chromatography with simple sample preparation. *J.Chromatogr.A* **1996**, *730*, 115-120.

SAMPLE
Matrix: blood
Analyte: gabapentin
Sample preparation: 100 μL Plasma + 100 μL 2 M perchloric acid, vortex for 10 s, centrifuge at 15000 g for 3 min. Remove a 50 μL aliquot and add it to 200 μL MeOH, add 200 μL buffer, add 50 μL reagent, mix, let stand at room temperature for 5 min, inject a 20 μL aliquot. (Prepare buffer weekly by adjusting the pH of 500 mM boric acid to 9.5 with 1 M NaOH. Prepare reagent weekly by dissolving 50 mg o-phthalaldehyde in 4.5 mL MeOH, add 500 μL buffer, add 50 μL 3-mercaptopropionic acid.)

HPLC VARIABLES
Column: 250 × 4.6 5 μm Ultrasphere octadecyl
Mobile phase: MeCN:MeOH:buffer 30:30:40 (Prepare the buffer by diluting 7.5 mL glacial acetic acid to 400 mL with water, adding 40 mg EDTA, and adjusting the pH to 3.7 with 3 M NaOH.)
Flow rate: 1.5
Injection volume: 20
Detector: F ex 330 em 440

CHROMATOGRAM
Retention time: 11.5
Internal standard: 1-(aminomethyl)cycloheptaneacetic acid (Parke-Davis) (15)
Limit of quantitation: 500 ng/mL

OTHER SUBSTANCES
Non-interfering: alanine, arginine, aspartic acid, carbamazepine, clobazam, clonazepam, cysteine, felbamate, glutamic acid, glycine, histidine, isoleucine, lamotrigine, leucine, lysine, methionine, oxcarbazepine, phenobarbital, phenylalanine, phenytoin, primidone, proline, remacemide, serine, threonine, tiagabine, tyrosine, valine, valproic acid, vigabatrin

KEY WORDS
plasma

REFERENCE
Forrest, G.; Sills, G.J.; Leach, J.P.; Brodie, M.J. Determination of gabapentin in plasma by high-performance liquid chromatography, *J.Chromatogr.B*, **1996**, *681*, 421–425.

SAMPLE
Matrix: blood, dialysate, tissue
Analyte: amphetamine
Sample preparation: Plasma. 45 μL Plasma + 5 μL 40 μM tryptamine + 100 μL 100 mM borate buffer adjusted to pH 10.6 with NaOH + 200 μL ethyl acetate, vortex for 2 min, let sit on ice for 10 min, add 200 μL ethyl acetate, add 200 μL water, vortex briefly, centrifuge at 4° at 18000 g for 10 min. Remove 200 μL of the organic supernatant and evaporate it to dryness under a stream of nitrogen at 45°, reconstitute the residue in 80 μL 50 mM KH$_2$PO$_4$ adjusted to pH 2.6 with phosphoric acid, add 20 μL borate buffer

adjusted to pH 11.5 with NaOH, add 20 μL reagent, let stand for at least 2 min, keep at 4°, inject a 75 μL aliquot. Tissue. Sonicate brain tissue with 9 volumes of 8 μM tryptamine in 100 mM pH 10.6 borate buffer for 20 s, centrifuge at 4° at 18000 g for 10 min, remove a 100 μL aliquot of the supernatant and add 200 μL ethyl acetate, vortex for 2 min, let sit on ice for 10 min, add 200 μL ethyl acetate, add 200 μL water, vortex briefly, centrifuge at 4° at 18000 g for 10 min. Remove 200 μL of the organic supernatant and evaporate it to dryness under a stream of nitrogen at 45°, reconstitute the residue in 80 μL 50 mM KH_2PO_4 adjusted to pH 2.6 with phosphoric acid, add 20 μL borate buffer adjusted to pH 11.5 with NaOH, add 20 μL reagent, let stand for at least 2 min, keep at 4°, inject a 75 μL aliquot. Dialysate. 100 μL Dialysate + 20 μL reagent, let stand for at least 2 min, keep at 4°, inject a 75 μL aliquot. (Reagent was 27 mg o-phthaldialdehyde in 500 μL EtOH, add 5 mL 100 mM pH 9.6 borate buffer, add 40 μL 3-mercaptopropionic acid, refrigerate, use for up to 4 days.)

HPLC VARIABLES
Guard column: 20 × 2 37-50 μm Bondapak C18/Corasil
Column: 250 × 4.6 5 μm LC-18 (Supelco)
Mobile phase: Gradient. MeOH:buffer 35:65 for 3 min, to 65:35 over 1 min, maintain at 65:35 for 14 min, return to initial conditions over 1 min, re-equilibrate for 6 min. (Buffer was 50 mM KH_2PO_4 adjusted to pH 5.5 with KOH.)
Flow rate: 1.5
Injection volume: 75
Detector: F ex 340 em 440

CHROMATOGRAM
Retention time: 14.5
Internal standard: tryptamine (13)
Limit of quantitation: 100 pg (dialysate)
Limit of detection: 50 pg (dialysate), 200 pg (plasma, tissue)

OTHER SUBSTANCES
Extracted: metabolites, p-hydroxyamphetamine

KEY WORDS
rat; brain; plasma

REFERENCE
Bowyer, J.F.; Clausing, P.; Newport, G.D. Determination of d-amphetamine in biological samples using high-performance liquid chromatography after precolumn derivatization with o-phthaldialdehyde and 3-mercaptopropionic acid, *J.Chromatogr.B*, **1995**, *666*, 241–250.

Matrix: solutions
Analyte: α-methylbenzylamine
Sample preparation: Mix 70 μL of an aqueous solution with 300 μL phthalaldehyde solution and 30 μL thiol solution for 2 min, inject an aliquot. (Prepare the phthalaldehyde solution by dissolving 60 mg o-phthaldialdehyde in 3 mL MeOH and 15 mL 400 mM pH 9.4 sodium borate buffer. Prepare the thiol solution by dissolving 6.5 mg D-S-acetyl-3-mercapto-2-methylpropionic acid (Novabiochem) in 1 mL 1 M NaOH, stir at room temperature for 10 min, adjust the pH to 7.0 with phosphoric acid. The solution contains D-3-mercapto-2-methylpropionic acid.)

HPLC VARIABLES
Column: 250 × 4 3 μm Nucleosil-120-C18
Mobile phase: MeOH:50 mM pH 6.0 sodium acetate buffer 45:55
Column temperature: 40
Flow rate: 1
Injection volume: 20
Detector: F ex 338 em 415 (long-pass filter)

CHROMATOGRAM
Retention time: 46 (R), 50 (S)

KEY WORDS
chiral

REFERENCE
Duchateau, A.L.L.; Knuts, H.; Boesten, J.M.M.; Guns, J.J. Enantioseparation of amino compounds by derivatization with o-phthalaldehyde and D-3-mercapto-2-methylpropionic acid, *J.Chromatogr.*, **1992**, *623*, 237–245.

Matrix: food

Analyte: cyclamate

Sample preparation: Fruit juice. Centrifuge at 3000 g for 20 min, filter (0.45 μm) the supernatant. Remove a 2 mL aliquot of the filtrate and add it to 1 mL 30% hydrogen peroxide and 300 μL 37% HCl, heat at 100° under a reflux condenser for 1 h, cool to room temperature, neutralize with 400 μL 40% NaOH, make up to 25 mL with pH 10 borate buffer. Remove a 1 mL aliquot and make up to 10 mL with reagent, after 5 min inject an aliquot. Marmalade, preserves. Stir 10 g marmalade or preserve with 50 mL water for 30 min, centrifuge at 3000 g for 20 min, remove the supernatant, suspend the residue in 10 mL water, centrifuge. Combine the supernatants and make up to 100 mL with water. Remove a 2 mL aliquot and add it to 1 mL 30% hydrogen peroxide and 300 μL 37% HCl, heat at 100° under a reflux condenser for 1 h, cool to room temperature, neutralize with 400 μL 40% NaOH, make up to 25 mL with pH 10 borate buffer. Remove a 1 mL aliquot and make up to 10 mL with reagent, after 5 min inject an aliquot. (Prepare borate buffer by dissolving 25 g boric acid in 900 mL water, adjust pH to 10 with KOH. Prepare reagent by dissolving 200 mg o-phthaldialdehyde in 5 mL EtOH, add 1 mL 3-mercaptopropionic acid, make up to 100 mL with borate buffer. Cyclamate is hydrolyzed to cyclohexylamine which is then derivatized.)

HPLC VARIABLES
Column: 250 × 4 5 μm Lichrospher 100 RP-18

Mobile phase: MeCN:buffer 64:36 (Prepare buffer by dissolving 3 g $Na_2HPO_4.12H_2O$ and 3 g $NaH_2PO_4.H_2O$ in 1 L water.)

Column temperature: 40

Flow rate: 1

Injection volume: 10

Detector: F ex 350 em 440-650

CHROMATOGRAM
Retention time: 12

Limit of detection: 500 ng/g

KEY WORDS
fruit juice; marmalade; preserves

REFERENCE
Rüter, J.; Raczek, D.I.U. Empfindliches und selektives HPLC-Verfahren mit prächromatographischer Derivatisierung zur Bestimmung von Cyclamat in Lebensmitteln [Sensitive and selective HPLC procedure with prechromatographical derivatization for the determination of cyclamate in foods], *Z.Lebensm.Unters.Forsch.*, **1992**, *194*, 520–523.

Phthalaldehyde/Sodium Sulfite

RELATED REFERENCE
Jacobs, W.A. o-Phthalaldehyde-sulfite derivatization of primary amines for liquid chromatography-electrochemistry. *J.Chromatogr.* **1987**, *392*, 435-441.

SAMPLE
Matrix: dialysate
Analyte: gamma-aminobutyric acid
Sample preparation: 20 µL Dialysate + 2 µL reagent, mix, let stand at room temperature for 25 min, inject a 20 µL aliquot. (Prepare reagent by mixing 22 mg o-phthalaldehyde, 500 µL 1 M sodium sulfite, 500 µL EtOH, and 9 mL buffer. Prepare fresh each day. Prepare buffer by adjusting the pH of 400 mM boric acid to 10.4 with 5 M NaOH.)

HPLC VARIABLES
Guard column: 15 × 4.6 5 µm Microsorb C18
Column: 250 × 4.6 5 µm Microsorb C18
Mobile phase: MeOH:buffer 40:60, pH adjusted to 4.4 (Buffer was 100 mM NaH_2PO_4 containing 100 µM EDTA.)
Flow rate: 1
Injection volume: 20
Detector: E, Bioanalytical Systems LC-4/4A, glassy carbon working electrode, +0.85 V, Ag/AgCl reference electrode

CHROMATOGRAM
Retention time: 6
Limit of detection: 25-50 fmole

REFERENCE
Smith, S.; Sharp, T. Measurement of GABA in rat brain microdialysates using o-phthaldialdehyde-sulphite derivatization and high-performance liquid chromatography with electrochemical detection, *J.Chromatogr.B*, **1994**, *652*, 228–233.

Phthalaldehyde/Thioglucose Tetraacetate

SAMPLE
Matrix: solutions
Analyte: rimantadine
Sample preparation: 50 µL 5 mg/mL Rimantadine in 100 mM HCl + 50 µL buffer + 100 µL reagent, swirl for 1 min, place on ice for 5 min, add 2 mL mobile phase, inject a 5 µL aliquot. (Buffer was 100 mM sodium borate adjusted to pH 9.50 with 2 M NaOH. Reagent was 13.40 g o-phthaldialdehyde and 36.4 mg 1-thio-β-D-glucose tetraacetate (2,3,4,6-tetra-O-acetyl-1-thio-β-D-glucopyranoside) in 1 mL MeOH, protect from light, keep on ice.)

HPLC VARIABLES
Column: 150 × 3.9 4 µm Nova-Pak C18
Mobile phase: MeOH:buffer 85:15 (Buffer was 3 mL/L glacial acetic acid in water, pH adjusted to 7.20 with 2 M NaOH.)
Flow rate: 1
Injection volume: 5
Detector: F ex 338 em 420; UV 254

CHROMATOGRAM
Retention time: 5.88, 7.11 (enantiomers)
Limit of detection: 6 ng (UV)

KEY WORDS
protect from light; chiral

REFERENCE
Desai, D.M.; Gal, J. Enantiospecific drug analysis via the *ortho*-phthalaldehyde/homochiral thiol derivatization method, *J.Chromatogr.*, **1993**, *629*, 215–228.

HALOALDEHYDE

Bromoacetaldehyde

SAMPLE
Matrix: whole blood
Analyte: adenine compounds
Sample preparation: Dilute whole blood 10-100 fold with 320 mM sucrose, inject 10 μL of the diluted solution onto column A and column B in series with mobile phase A, elute with mobile phase A for 4 min then elute the contents of column B onto column C with mobile phase B, monitor the effluent from column C.

HPLC VARIABLES
Column: A 30 × 4.6 44-88 μm Butyl-Toyopearl 650-M (Tosoh) B 10 × 4 3 μm Hitachi gel 3013-N C 50 × 4.6 3 μm Hitachi gel 3013-N
Mobile phase: A MeCN:water 15:85; B MeCN:50 mM bromoacetaldehyde + 150 mM NaCl 15:85, containing 25 mM citrate, pH 4.0 (Bromoacetaldehyde may be prepared as follows. Stir 1.64 g bromoacetaldehyde dimethyl acetal, 40 mL acetone, and 600 μL water, add 0.4 g Amberlyst-15, stir at room temperature for 24 h, filter to obtain a solution of bromoacetaldehyde (Org. Prep. Proc. Int. 1993, 25, 469). It should be possible to run the reaction in a more convenient solvent, e.g., MeCN, or alternatively remove the acetone by careful evaporation under reduced pressure. Bromoacetaldehyde distills at 107-112° (J. Am. Chem. Soc. 923, 45, 734) or about 50°/1 mm Hg (J. Org. Chem. 1983, 48, 2111).)
Column temperature: 45 (column C)
Flow rate: A 0.3; B 0.3
Injection volume: 10
Detector: F ex 254 em 400 following post-column reaction. The column effluent flowed through a 15 m × 0.25 mm ID reaction coil (Jasco RU-150F unit) at 115° to the detector.

CHROMATOGRAM
Retention time: 6.5 (adenosine monophosphate), 9 (cyclic adenosine monophosphate), 10 (adenosine diphosphate), 25 (adenosine triphosphate)

OTHER SUBSTANCES
Non-interfering: adenine, adenosine

KEY WORDS
hamster; rat; human; post-column reaction; column-switching

REFERENCE
Fujimori, H.; Sasaki, T.; Hibi, K.; Senda, M.; Yoshioka, M. Direct injection of blood samples into a high-performance liquid chromatographic adenine analyser to measure adenine, adenosine, and the adenine nucleotides with fluorescence detection, *J.Chromatogr.*, **1990**, *515*, 363–373.

Chloroacetaldehyde

RELATED REFERENCES

Fujimori, H.; Yamauchi, M.; Pan-Hou, H. Availability of chloroacetaldehyde as a fluorescent reagent for a determination of adenine compounds by high-performance liquid chromatography. *Chem.Express* **1991**, *6*, 715-718.

Matuszewski, B.K.; Bayne, W.F. Fluorogenic reaction between adenine derivatives and chloroacetaldehyde and its application to the determination of 9-(2-chloro-6-fluorobenzyl)adenine in human plasma. *Anal.Chim.Acta* **1989**, *227*, 189-202.

Shaw, J.-P.; Louie, M.S.; Krishnamurthy, V.V.; Arimilli, M.N.; Jones, R.J.; Bidgood, A.M.; Lee, W.A.; Cundy, K.C. Pharmacokinetics and metabolism of selected prodrugs of PMEA in rats. *Drug Metab.Dispos.* **1997**, *25*, 362-366.

Sonoki, S.; Tanaka, Y.; Hisamatsu, S.; Kobayashi, T. High-performance liquid chromatographic analysis of fluorescent derivatives of adenine and adenosine and its nucleotides. Optimization of derivatization with chloroacetaldehyde and chromatographic procedures. *J.Chromatogr.* **1989**, *475*, 311-319.

SAMPLE

Matrix: blood

Analyte: adenosine

Sample preparation: 360 µL Blood + 40 µL stopping solution, centrifuge at 14000 g for 1 min. 100 µL Plasma + 10 µL 50% trichloroacetic acid, centrifuge for 5 min. Remove 75 µL of the supernatant and add it to 10 µL 2.3 M KOH, add 50 µL 1 M zinc sulfate, add 100 µL saturated barium hydroxide, vortex for 10 s, centrifuge at 14000 g for 5 min. Remove 100 µL supernatant (pH 5.4) and add it to 10 µL chloroacetaldehyde (45% in water (Fluka)), heat at 80° for 1 h, inject an aliquot. (Stopping solution was 1 mM dialazep, 10 µM erythro-9-(2-hydroxy-3-nonyl)adenine, 2 µg/mL indomethacin (final concentrations).)

HPLC VARIABLES

Column: 150 × 3.9 10 µm µBondapak C18

Mobile phase: Gradient. A was MeOH:10 mM pH 3.5 KH_2PO_4 12:88. B was MeOH:10 mM pH 3.5 KH_2PO_4 50:50. A for 8 min then B for 8 min, re-equilibrate for 4 min.

Flow rate: 1.5

Detector: F ex 280 em 380

CHROMATOGRAM

Retention time: 4.8

Limit of detection: 0.2 pmole

KEY WORDS

plasma; cat; rat; dog; mouse; rabbit; guinea pig

REFERENCE

Zhang, Y.; Geiger, J.D.; Lautt, W.W. Improved high-pressure liquid chromatographic-fluorometric assay for measurement of adenosine in plasma, *Am.J.Physiol.*, **1991**, *260*, G658–G664.

SAMPLE

Matrix: blood

Analyte: diadenosine 5',5'''-p^1,p^4-tetraphosphate

Sample preparation: 1 mL Plasma + 500 µL 10% perchloric acid, mix, centrifuge at 7000 g for 1 min. Remove the supernatant and add it to 500 µL 2 M potassium bicarbonate, centrifuge at 7000 g for 1 min, add 250 μL 2 M pH 4.5 sodium acetate buffer, add 20 μL 40% chloroacetaldehyde, heat at 80° for 1 h, cool in ice water, inject a 10 μL aliquot.

HPLC VARIABLES
Column: 150 × 4.6 TSK-gel ODS-80TM (Toso, Tokyo)
Mobile phase: MeOH:100 mM pH 6.0 potassium phosphate buffer 10:90
Column temperature: 40
Flow rate: 0.7
Injection volume: 10
Detector: F ex 275 em 410

CHROMATOGRAM
Retention time: 9
Limit of detection: 200 ng/mL

KEY WORDS
plasma; human; rat; dog

REFERENCE
Iwata, K.; Haruki, S.; Kimura, T. High-performance liquid chromatographic determination of diadenosine 5',5'''-p^1,p^4-tetraphosphate with precolumn fluorescence derivatization and its application to metabolism study in whole blood, *J.Chromatogr.B*, **1995**, *667*, 339–343.

SAMPLE
Matrix: blood, urine
Analyte: adenosine compounds
Sample preparation: Centrifuge plasma or urine at 1000 g at 4° for 10 min. Add 1 mL supernatant to 1 mL 12% trichloroacetic acid (in an ice bath), mix thoroughly, let stand for 15 min, centrifuge at 4° at 1000 g for 30 min. Remove the supernatant and extract it four times with water-saturated diethyl ether, discard the ether extracts. Remove traces of ether from the aqueous layer by heating it at 65° with intermittent suction, cool. Remove 500 μL of the aqueous layer and add it to 200 μL 40 mM chloroacetaldehyde in 28 mM pH 5.1 sodium acetate buffer. Heat at 80° for 40 min, cool, extract four times with 1 mL portions of water-saturated diethyl ether, discard the ether extracts. Heat the aqueous layer at 65° with intermittent suction to remove traces of ether, cool. Dilute urine preparation 1:9, do not dilute plasma preparations, inject a 50 μL aliquot.

HPLC VARIABLES
Guard column: 20 × 4 μBondapak C18/Corasil
Column: 250 × 4.6 Spherisorb S5 ODSII octadecyl
Mobile phase: Gradient. A was 20 mM sodium tetraborate decahydrate adjusted to pH 7.7 with 4.4 N phosphoric acid. B was MeOH:20 mM sodium tetraborate decahydrate adjusted to pH 7.7 with 4.4 N phosphoric acid 30:70. A:B from 100:0 to 0:100 over 15 min, maintain at 0:100 for 5 min, return to initial conditions, re-equilibrate for 5 min.
Flow rate: 1
Injection volume: 50
Detector: F ex 315 em 415

CHROMATOGRAM
Retention time: 7 (Ara-AMP), 10 (ara-H (UV detection)), 13 (adenosine), 15.2 (vidarabine (ara-A))
Limit of detection: 1.5-350 ng/mL

KEY WORDS
plasma; pharmacokinetics

REFERENCE
McCann, W.P.; Hall, L.M.; Siler, W.; Barton, N.; Whitley, R.J. High-pressure liquid chromatographic methods for determining arabinosyladenine-5'-monophosphate, arabinosyladenine, and arabinosylhypoxanthine in plasma and urine, *Antimicrob.Agents Chemother.*, **1985**, *28*, 265–273.

SAMPLE
Matrix: blood, urine
Analyte: acyclic adenine nucleoside phosphonates

Sample preparation: Plasma, serum. Add trichloroacetic acid to 100 µL serum or plasma so that the final concentration of trichloroacetic acid is 100 mg/mL, shake vigorously for 10 min, centrifuge at 9000 g for 5 min, add the supernatant to an equal volume of tri-n-octylamine:Freon 20:80, shake vigorously for 30 min, discard the lower organic layer. Treat the aqueous layer with pH 4.7 ammonium acetate buffer (final concentration 160 mM) and chloroacetaldehyde (final concentration 40 mM), heat at 95° for 40 min, cool at 4°, inject an aliquot equivalent to 60 µL plasma or serum. Urine. Centrifuge urine, treat with pH 4.7 ammonium acetate buffer and chloroacetaldehyde, heat at 95° for 20 min (Antimicrob. Agents Chemother. 1996, 40, 22).

HPLC VARIABLES
Column: 125 × 4.6 4 µm Superspher 60 C8 (Merck)
Mobile phase: Gradient. A was MeCN:2.5 mM pH 5.0 $(NH_4)H_2PO_4$ containing 2 mM tetrabutylammonium hydrogen sulfate 5:95. B was MeCN:75 mM pH 5.0 $(NH_4)H_2PO_4$ containing 2 mM tetrabutylammonium hydrogen sulfate 15:85. A:B 100:0 for 4 min, to 0:100 over 2 min, maintain at 0:100 for 4 min, re-equilibrate at initial conditions for 8 min.
Flow rate: 1
Detector: F ex 254 (filter) em 425 (filter)

CHROMATOGRAM
Retention time: 6.1 (adenosine), 9.8 (adenine), 10.8 (adenosine monophosphate), 11.4 ((S)-9-(3-hydroxy-2-phosphonylmethoxypropyl)adenine (HPMPA)), 11.7 (cAMP), 12.0 (9-(2-phosphonylmethoxyethyl)adenine (PMEA)), 12.6 ((R,S)-9-(3-fluoro-2-phosphonylmethoxypropyl)adenine (FPMPA)), 12.6 (NAD^+)
Limit of quantitation: 250 nM

KEY WORDS
monkey; cat; human; plasma; serum

REFERENCE
Naesens, L.; Balzarini, J.; De Clercq, E. Acyclic adenine nucleoside phosphonates in plasma determined by high-performance liquid chromatography with fluorescence detection, *Clin.Chem.*, **1992**, *38*, 480–485.

SAMPLE
Matrix: enzyme incubations
Analyte: cyclic adenosine monophosphate (cAMP)
Sample preparation: 150 µL Enzyme incubation + 150 µL 250 mM zinc sulfate + 150 µL 250 mM barium hydroxide, mix, centrifuge at 12500 g for 10 min. Remove a 300 µL aliquot of the supernatant and add it to 50 µL 1.55 mM chloroacetaldehyde, heat in a boiling water bath for 15 min, inject an aliquot.

HPLC VARIABLES
Guard column: Micro-Guard ODS-10 (Bio-Rad)
Column: 250 × 4 Bio-Sil ODS-10 (Bio-Rad)
Mobile phase: MeCN:buffer 6:94 containing 26 µM cetylpyridinium bromide (Buffer was 50 mM citric acid containing 50 mM Na_2HPO_4, adjusted to pH 4.8.)
Flow rate: 3
Injection volume: 250
Detector: F ex 298 (bandpass filter) em 418 (cut-off filter)

CHROMATOGRAM
Retention time: 8
Limit of quantitation: 2.5 pmole

OTHER SUBSTANCES
Simultaneously analyzed: adenine, adenosine, adenosine diphosphate, adenosine monophosphate, NAD
Non-interfering: guanosine triphosphate, inosine, 5'-inosine monophosphate, adenosine triphosphate

REFERENCE

Wojcik, W.; Olianas, M.; Parenti, M.; Gentleman, S.; Neff, N.H. A simple fluorometric method for cAMP: application to studies of brain adenylate cyclase activity, *J.Cyclic.Nucleotide.Res.*, **1981**, *7*, 27–35.

SAMPLE

Matrix: perfusate

Analyte: adenosine

Sample preparation: Mix 700 µL perfusate (Earle's medium) with 750 µL 24 mM phosphoric acid to adjust pH to 5.4, add 10 µL chloroacetaldehyde, heat at 100° for 40 min, cool in ice, inject a 50 µL aliquot. (Earle's medium contains 116 mM NaCl, 22.6 mM sodium bicarbonate, 5.4 mM KCl, 1.8 mM calcium chloride, 0.8 mM magnesium sulfate, 1.0 mM NaH_2PO_4, 5.5 mM glucose, and 40 g/L dextran (MW 32000-48000), pH 8.5-8.6. Prepare chloroacetaldehyde by refluxing chloroacetaldehyde dimethyl acetal:1.5 M sulfuric acid 5:1 for 30 min, distil, collect the fraction boiling at 85-95°, use undiluted.)

HPLC VARIABLES

Guard column: 15 × 3.2 7 µm NewGuard RP-18

Column: 100 × 4.6 3 µm ODS-Hypersil C18

Mobile phase: Gradient. MeOH:10 mM pH 6.7 $(NH_4)H_2PO_4$ from 0:100 to 1:99 over 1 min, to 3:97 over 1 min, to 6:94 over 1 min, to 10:90 over 1 min, to 15:85 over 1 min, to 23:77 over 1 min, to 35:65 over 1 min, to 50:50 over 1 min, to 60:40 over 1 min, to 80:20 over 0.5 min, maintain at 80:20 over 0.5 min, return to initial conditions over 0.2 min, re-equilibrate for 8.8 min.

Column temperature: 40

Flow rate: 1

Injection volume: 50

Detector: F ex 275 em 415

CHROMATOGRAM

Retention time: 10

Limit of quantitation: 2 nM

REFERENCE

Slegel, P.; Kitagawa, H.; Maguire, M.H. Determination of adenosine in fetal perfusates of human placental cotyledons using fluorescence derivatization and reversed-phase high-performance liquid chromatography, *Anal.Biochem.*, **1988**, *171*, 124–134.

SAMPLE

Matrix: perfusate

Analyte: adenine nucleotides and nucleosides

Sample preparation: 1 mL Perfusate (Krebs solution) + 40 µL chloroacetaldehyde + 360 µL buffer + 100 µL 600 nM vidarabine in water, heat at 80° for 40 min, cool on ice, inject an aliquot. (Krebs solution contained 113 mM NaCl, 4.8 mM KCl, 2.5 mM calcium chloride, 1.2 mM KH_2PO_4, 1.2 mM magnesium sulfate, 25 mM sodium bicarbonate, and 5.5 mM glucose. Prepare buffer by mixing 400 mL 100 mM citric acid with 245 mL 200 mM Na_2HPO_4, pH 4.0. (Prepare chloroacetaldehyde as follows. Cautiously add 1 mL concentrated sulfuric acid to 9 mL water (using eye protection and other protective equipment), add to 10 mL chloroacetaldehyde dimethyl acetal, distil slowly and collect the fraction boiling at 80-85° which contains 1-1.15 M chloroacetaldehyde, store at 0° (Anal. Biochem. 1984, 137, 93).)

HPLC VARIABLES

Guard column: 10 × 4.6 10 µm Ultron N-phenyl (Shinwa, Kyoto)

Column: 150 × 4.6 5 µm Ultron N-phenyl (Shinwa, Kyoto)

Mobile phase: MeCN:buffer 1.5:98.5, adjusted to pH 4.5 with 2-diethylaminoethanol (Prepare buffer by mixing 400 mL 100 mM citric acid with 245 mL 200 mM Na_2HPO_4, pH 4.0.)

Flow rate: 1

Detector: F ex 305 em 420

CHROMATOGRAM
Retention time: 3.5 (adenosine triphosphate), 4.5 (adenosine diphosphate), 7 (adenosine monophosphate), 22 (adenosine)
Internal standard: vidarabine (17)
Limit of detection: 0.1 pmole

REFERENCE
Mohri, K.; Takeuchi, K.; Shinozuka, K.; Bjur, R.A.; Westfall, D.P. Simultaneous determination of nerve-induced adenine nucleotides and nucleosides released from rabbit pulmonary artery, *Anal.Biochem.*, **1993**, *210*, 262−267.

SAMPLE
Matrix: solutions
Analyte: adenine-containing compounds
Sample preparation: Prepare a 2 mM solution in 20 mM pH 4.5 sodium phosphate buffer, add a 20 µL aliquot to 20 µL chloroacetaldehyde solution, heat at 95-100° for 20 min, cool in an ice bath, inject an aliquot. (Chloroacetaldehyde solution was prepared by diluting a 40-45% solution of chloroacetaldehyde in water with 20 mM pH 8.8 sodium phosphate buffer so as to achieve a chloroacetaldehyde concentration of 150 mM, final pH 4.6.)

CAPILLARY ELECTROPHORESIS
Capillary: 60 cm × 25 µm fused-silica (40 cm to detector)
Capillary preparation: Flush with running buffer between runs. Flush with 100 mM NaOH and water before use.
Running buffer: 20 mM pH 8.8 Sodium phosphate buffer
Voltage/Current: 20 kV
Injection: Siphon at 20.5 cm for 10 s (0.5 nL)
Detector: F ex 325 (He-Cd laser) em 375 (longpass filter) and 400 (bandpass filter)
Migration time: 2.9 (adenine), 3 (adenosine), 4.5 (cyclic adenosine monophosphate), 6 (adenosine monophosphate), 8 (adenosine diphosphate), 9 (adenosine triphosphate)

REFERENCE
Tseng, H.C.; Dadoo, R.; Zare, R.N. Selective determination of adenine-containing compounds by capillary electrophoresis with laser-induced fluorescence detection, *Anal.Biochem.*, **1994**, *222*, 55−58.

SAMPLE
Matrix: tissue
Analyte: adenine nucleotides
Sample preparation: Homogenize 1 g frozen powdered tissue with 9 mL 2.5% perchloric acid. Neutralize a 200 µL aliquot with 2 M KOH containing 200 mM K_2HPO_4, centrifuge at 4° at 9500 g for 1 min, inject an aliquot.

HPLC VARIABLES
Guard column: 10 × 2 5 µm Hypersil
Column: 125 × 4.6 5 µm Hypersil 5ODS
Mobile phase: Gradient. A was 100 mM pH 6.0 potassium phosphate buffer containing 8 mM tetra-n-butylammonium hydrogen sulfate and 15 mM chloroacetaldehyde. B was MeCN:water 75:25. A:B from 97:3 to 65:35 over 6 min, return to initial conditions over 1 min.
Flow rate: 1.6
Detector: F ex 230 em 430 following post-column reaction. The column effluent flowed through a knitted 10 m × 0.5 mm ID PTFE coil at 100° to the detector.

CHROMATOGRAM
Retention time: 5.5 (adenosine monophosphate), 7 (adenosine diphosphate), 8.5 (adenosine triphosphate)
Limit of detection: 10 pmole

KEY WORDS
post-column reaction; mouse; liver

REFERENCE

Stratford, M.R.L.; Dennis, M.F. Determination of adenine nucleotides by fluorescence detection using high-performance liquid chromatography and post-column derivatization with chloroacetaldehyde, *J.Chromatogr.B*, **1994**, *662*, 15–20.

SAMPLE

Matrix: tissue, perfusate

Analyte: adenine nucleotides and nucleosides

Sample preparation: Homogenize (glass to glass) tissue with 1 mL ice-cold 400 mM perchloric acid, centrifuge at 10000 g for 10 min, dilute the supernatant 10-fold with Krebs solution. Mix 1 mL (?) perfusate (Krebs solution) or tissue homogenate with 25 (perfusate) or 50 (tissue) μL chloroacetaldehyde, heat at 80° for 40 min, cool on ice, inject an aliquot. (Krebs solution contained 113 mM NaCl, 4.8 mM KCl, 2.5 mM calcium chloride, 1.2 mM KH_2PO_4, 1.2 mM magnesium sulfate, 25 mM sodium bicarbonate, and 5.5 mM glucose. Prepare buffer by mixing 400 mL 100 mM citric acid with 245 mL 200 mM Na_2HPO_4, pH 4.0. (Prepare chloroacetaldehyde as follows. Cautiously add 1 mL concentrated sulfuric acid to 9 mL water (using eye protection and other protective equipment), add to 10 mL chloroacetaldehyde dimethyl acetal, distil slowly and collect the fraction boiling at 80-85° which contains 1-1.15 M chloroacetaldehyde, store at 0°.)

HPLC VARIABLES

Column: 5 μm Radial-Pak C18

Mobile phase: Gradient. A was 100 mM pH 6.0 phosphate buffer. B was MeOH:100 mM pH 6.0 phosphate buffer 25:75. A:B from 100:0 to 0:100 over 15 min (Waters concave curve 8), maintain at 0:100 for 5 min, re-equilibrate at initial conditions for 5 min.

Flow rate: 2

Detector: F ex 300 em 420

CHROMATOGRAM

Retention time: 12 (adenosine triphosphate), 13 (adenosine diphosphate), 16 (adenosine monophosphate), 20 (adenosine)

Limit of detection: 0.5-1 pmole

KEY WORDS

guinea pig; vas deferens

REFERENCE

Levitt, B.; Head, R.J.; Westfall, D.P. High-performance liquid chromatographic-fluorometric detection of adenosine and adenine nucleotides: Application to endogenous content and electrically induced release of adenyl purines in guinea pig vas deferens, *Anal.Biochem.*, **1984**, *137*, 93–100.

SAMPLE

Matrix: urine

Analyte: S-(5'-deoxy-5'-adenosyl)-3-methylthiopropylamine

Sample preparation: Collect 24 h human urine with 50 mL 4 M perchloric acid, filter (0.22 μm). Remove a 4 mL aliquot and add it to 1 mL 500 nM IS, add to a 50 × 12 column of 200-400 mesh Dowex AG 50W-X8, wash with 10 mL water, wash with 25 mL 2 M HCl, elute with 25 mL 6 M HCl. Evaporate the eluate to dryness under reduced pressure at <40°, reconstitute the residue in 1 mL 200 mM perchloric acid. Remove a 500 μL aliquot and add it to 50 μL 45% 2-chloroacetaldehyde in water and 50 μL 3 M sodium acetate, heat at 40° for 12-16 h, cool in ice, inject a 10-20 μL aliquot.

HPLC VARIABLES

Guard column: 70 × 2 37-53 μm pellicular ODS (Whatman)

Column: 250 × 4.6 5 μm Ultrasphere ion-pair

Mobile phase: Gradient. A was MeCN:100 mM NaH_2PO_4 containing 100 μM EDTA and 8 mM octanesulfonic acid 2:98, pH adjusted to 3.65 with 3 M phosphoric acid. B was MeCN:200 mM NaH_2PO_4 26:74 containing 8 mM octanesulfonic acid 2:98, pH adjusted to 4.3 with 3 M phosphoric acid. A:B from 75:25 to 24:76 over 30 min, return to initial conditions over 30 s, re-equilibrate for 10 min.

Column temperature: 40

Flow rate: 1.5

Injection volume: 10-20
Detector: F ex 270 em 410

CHROMATOGRAM
Retention time: 26.46
Internal standard: S-(5'-deoxy-5'-adenosyl)-3-ethylthiopropylamine (27.67)

OTHER SUBSTANCES
Extracted: metabolites

KEY WORDS
SPE

REFERENCE
Wagner, J.; Hirth, Y.; Claverie, N.; Danzin, C. A sensitive high-performance liquid chromatographic procedure with fluorometric detection for the analysis of decarboxylated S-adenosylmethionine and analogs in urine samples, *Anal.Biochem.*, **1986**, *154*, 604–617.

SAMPLE
Matrix: urine
Analyte: adenosine
Sample preparation: Prepare a SPE column by adding 800 µL immobilized boronic acid gel (Pierce, Affipak) to a disposable 3 mL syringe and saturating with 100 mM pH 10 carbonate buffer. Condition a Sep-Pak C18 SPE cartridge with 5 mL MeOH and 5 mL water. Add sodium azide to urine so that the final concentration is 0.03%, centrifuge. Remove a 25 mL aliquot of the supernatant and add it to 25 mL 100 mM pH 7.5 phosphate buffer, mix, cool on ice, add a 10 mL aliquot to the SPE column, wash with 25 mM pH 10 carbonate buffer, elute with 9 mL 100 mM HCl. Adjust the pH of the eluate to 7 with 500 mM NaOH, add to the SPE cartridge, wash with 1 mL water, elute with 1.5 mL MeOH:50 mM HCl 25:75. Adjust the pH of the eluate to 6 with concentrated NaOH, lyophilize, resuspend in 1 mL water, add 100 µL 2 M chloroacetaldehyde in water, heat at 90° for 50 min, cool in ice, dilute with an equal volume of 10 mM pH 10 borate buffer containing 0.1% hydroxyethylcellulose, inject a 490 nL aliquot.

CAPILLARY ELECTROPHORESIS
Capillary: 65 cm × 50 µm fused-silica (50 cm to detector) (Polymicro Technologies)
Capillary preparation: Treat new capillaries with 1 M KOH for 15 min, wash with water for 20 min, wash with MeOH for 20 min, pull air through capillary for 30 min, continuously pull 50% trimethylchlorosilane in toluene through the capillary under vacuum for 8 h, rinse with MeOH, wash with water for 20 min, flush with running buffer for 20 min.
Running buffer: 80 mM pH 10 Borate buffer containing 0.1% hydroxyethylcellulose (Terminating electrolyte was the same as running buffer. Leading electrolyte was 180 mM HCl containing 0.1% hydroxyethylcellulose adjusted to pH 7.5 with 1 M Tris.)
Voltage/Current: 15 kV
Injection: Hydrodynamic injection at 20 cm
Detector: F ex 326 (6 mW He-Cd laser) em 415
Model: laboratory constructed
Migration time: 20
Limit of detection: 0.98 nM

KEY WORDS
SPE

REFERENCE
Wang, C.-C.; McCann, W.P.; Beale, S.C. Measurement of adenosine by capillary zone electrophoresis with on-column isotachophoretic preconcentration, *J.Chromatogr.B*, **1996**, *676*, 19–28.

ISOCYANATE

3,5-Dinitrophenyl Isocyanate

SAMPLE
Matrix: blood
Analyte: fenfluramine
Sample preparation: 1 mL Plasma + 100 µL 1.25 µg/mL IS in water + 200 µL water, vortex, add 400 µL 1 M trichloroacetic acid, vortex, centrifuge at 2500 g for 10 min. Remove 900 µL of the supernatant and add it to 250 µL 1 M NaOH, add 2 mL reagent, vortex for 2 s, let stand for 90 min. Remove a 1 mL aliquot of the lower organic layer and evaporate it to dryness under a stream of nitrogen, reconstitute the residue in 200 µL dichloromethane, inject a 35 µL aliquot. (Reagent was 10 µM 3,5-dinitrophenyl isocyanate in dichloromethane. Prepare as follows. Stir 6.40 g 3,5-dinitrobenzoyl chloride in 100 mL glacial acetic acid, add 1.80 g sodium azide in small increments, stir for 1 h, add 300 mL cold water. Filter off the 3,5-dinitrobenzoyl azide precipitate and wash it with a small portion of water. Dry overnight in a vacuum desiccator. Reflux 25 mg 3,5-dinitrobenzoyl azide dissolved in 5 mL toluene for 10 min, cool to room temperature, make up to 50 mL with dichloromethane, dilute an aliquot 1:200 with dichloromethane to give a 10 µM solution of 3,5-dinitrophenyl isocyanate. Prepare fresh daily. CAUTION! 3,5-Dinitrobenzoyl azide may be explosive and 3,5-dinitrophenyl isocyanate may be toxic!)

HPLC VARIABLES
Column: 250 × 4.6 5 µm (R)-naphthylurea Chiral (Supelco)
Mobile phase: Hexane:isopropanol:MeCN 89:9:2
Flow rate: 1.2 for 15 min, 3.5 for 13 min, 1.2 for 7 min
Injection volume: 35
Detector: UV 235

CHROMATOGRAM
Retention time: 10 (d), 11.5 (l)
Internal standard: β-methylphenethylamine (20.7, first peak)
Limit of quantitation: 10 ng/mL

OTHER SUBSTANCES
Extracted: metabolites

KEY WORDS
plasma; conduct analyses under yellow light; chiral; pharmacokinetics

REFERENCE
Zeng, J.-N.; Dou, L.; Duda, M.; Stuting, H.H. New chiral high-performance liquid chromatographic methodology used for the pharmacokinetic evaluation of dexfenfluramine, *J.Chromatogr.B*, **1994**, *654*, 231–248.

SAMPLE
Matrix: formulations
Analyte: fenfluramine
Sample preparation: Dissolve the contents of a 150 mg capsule in 1 L water, stir for 1 h, let stand for 30 min, centrifuge an aliquot at 1250 g for 5 min. Dilute 1 mL of the supernatant to 50 mL with water. Remove a 1 mL aliquot and add it to 100 µL 1 µg/mL IS,

add 100 μL 100 mM NaOH, add 2 mL reagent, vortex, let stand for 15 min, inject a 30 μL aliquot of upper aqueous layer (sic). (Reagent was 20 μM 3,5-dinitrophenyl isocyanate in dichloromethane. Prepare as described above.)

HPLC VARIABLES
Column: 250 × 4.6 5 μm (R)-naphthylurea Chiral (Supelco)
Mobile phase: Hexane:dichloromethane:MeOH:MeCN 81:17.5:1:0.5 (For l-isomer use 85:13.5:1:0.5 and 200 μM reagent.) (The exact ratios are very important.)
Flow rate: 1.5
Injection volume: 30
Detector: UV 235

CHROMATOGRAM
Retention time: 10.3 (d)
Internal standard: β-methylphenethylamine (17.6)
Limit of quantitation: 50 ng/mL

OTHER SUBSTANCES
Simultaneously analyzed: impurities

KEY WORDS
conduct analyses under yellow light; chiral; capsules

REFERENCE
Dou, L.; Zeng, J.-N.; Gerochi, D.D.; Duda, M.P.; Stuting, H.H. Chiral high-performance liquid chromatography methodology for quality control monitoring of dexfenfluramine, *J.Chromatogr.A*, **1994**, *679*, 367–374.

SAMPLE
Matrix: solutions
Analyte: 1-phenylethylamine
Sample preparation: Reflux a 10% excess of 3,5-dinitrobenzoyl azide in toluene for 6-10 min, add the amine, cool, dilute, inject an aliquot. (Dissolve 3,5-dinitrobenzoyl chloride in the minimum amount of glacial acetic acid, add one equivalent of sodium azide in portions, after 30 min dilute with water, filter to obtain 3,5-dinitrobenzoyl azide. Refluxing in toluene causes the azide to form the isocyanate which then reacts with the amine to give the carbamate.)

HPLC VARIABLES
Column: Chiral stationary phase CSP 2 (J. Chromatogr. 1984, 316, 585)
Mobile phase: Hexane:isopropanol 20:80
Flow rate: 2

CHROMATOGRAM
Retention time: k' 14.2 (α = 1.31)

OTHER SUBSTANCES
Also analyzed: other chiral amines in homologous series

KEY WORDS
chiral

REFERENCE
Pirkle, W.H.; Mahler, G.; Hyun, M.H. Separation of the enantiomers of 3,5-dinitrophenyl carbamates and 3,5-dinitrophenyl ureas, *J.Liq.Chromatogr.*, **1986**, *9*, 443–453.

FLOPIC

SAMPLE
Matrix: blood
Analyte: propranolol
Sample preparation: 1 mL Plasma + 1 mL pH 10 boric acid/KCl buffer + 500 mg NaCl + 5 mL toluene, extract. Remove the organic layer and evaporate it to dryness under a stream of nitrogen, reconstitute the residue in 100 μL dichloromethane, add 50 μL 1% triethylamine in MeOH, add 20 μL 0.1% FLOPIC in dichloromethane, mix, let stand at room temperature for 30 min, add 50 μL 1% ethanolamine in MeOH, mix, let stand at room temperature for 15 min, evaporate to dryness, reconstitute with 1% acetic acid in mobile phase, inject an aliquot. (FLOPIC is (-)-(S)-flunoxaprofen isocyanate; synthesis is as follows. Dissolve 1 g (+)-(S)-flunoxaprofen in 30 mL acetone, cool to 0°, add a solution of 500 μL triethylamine in 2 mL acetone dropwise, add a solution of 370 μL ethyl chloroformate in 2 mL acetone dropwise, stir at 0° for 15 min, add a solution of 250 mg sodium azide in 1 mL water dropwise (Caution! Sodium azide is highly toxic!), stir for 1 h, pour into 60 mL ice water, stir for 10 min, filter, wash the solid with two 50 mL aliquots of ice-water, dry under reduced pressure to obtain flunoxaprofen azide. Dissolve 100 mg flunoxaprofen azide in 3 mL dry toluene, reflux for 10-15 min, cool to room temperature, filter. Evaporate the filtrate to dryness under reduced pressure and dry under reduced pressure to obtain FLOPIC as a crystalline solid (mp 93-94°), store in a desiccator under reduced pressure (Chirality 1989, 1, 223).)

HPLC VARIABLES
Column: 150 × 3.9 4 μm Nova Pak C18
Mobile phase: MeOH:water 75:25
Flow rate: 1
Detector: F ex 305 em 355

CHROMATOGRAM
Retention time: 18, 20 (enantiomers)
Internal standard: pronethalol (15, 17 (enantiomers))
Limit of detection: 1-2 ng/mL

KEY WORDS
plasma; chiral; comparison with other derivatization reagents

REFERENCE
Spahn-Langguth, H.; Podkowik, B.; Stahl, E.; Martin, E.; Mutschler, E. Improved enantiospecific RP-HPLC assays for propranolol in plasma and urine with pronethalol as internal standard, *J.Anal.Toxicol.*, **1991**, *15*, 209–213.

SAMPLE
Matrix: solutions
Analyte: drugs

Sample preparation: Mix a 50 μL aliquot of a solution in MeOH:triethylamine 99:1 with 20 μL 0.1% FLOPIC in dry toluene, vortex briefly, let stand at room temperature in the dark for 30 min, add 50 μL 1% ethanolamine in MeOH, let stand at room temperature for 15 min, evaporate to dryness under reduced pressure, reconstitute with 100 μL mobile phase, sonicate for 30 s, inject a 20 μL aliquot. (FLOPIC is (-)-(S)-flunoxaprofen isocyanate; synthesis is as described above.)

HPLC VARIABLES
Column: 150 × 3.9 4 μm Nova Pak C18 (A-D) or 250 × 4.6 7 μm Nucleosil phenyl (E) or 200 × 4.6 5 μm Nucleosil cyano (F-I) or 250 × 4.6 7 μm Zorbax Sil (J)
Mobile phase: MeOH:water 75:25 (A) or MeOH:water 70:30 (B) or MeOH:water:THF 62:35:3 (C) or MeOH:water:THF 55:40:5 (D) or MeOH:80 mM NaCl in water 68:32 (E) or n-hexane:isopropanol:diethylamine 95:5:0.05 (F) or n-hexane:isopropanol 88:12 (G) or n-hexane:isopropanol 95:5 (H) or n-hexane:THF:isopropanol 95:2:3 (I) or n-hexane:THF:isopropanol 83:12:5 (J)
Flow rate: 1
Injection volume: 20
Detector: F ex 296 em 356

CHROMATOGRAM
Retention time: 12.9 (R-propranolol (A)), 14.6 (S-propranolol (A)), 27.8 (R-propafenone (B)), 29.8 (S-propafenone (B)), 20.9 (S-mexiletine (C)), 22.7 (R-mexiletine (C)), 29.0 (R-flecainide (C)), 31.0 (S-flecainide (C)), 27.0 (R-metoprolol (D)), 31.2 (S-metoprolol (D)), 18.4 (R-cibenzoline (E)), 22.4 (S-cibenzoline (E)), 22.2 (R-flecainide (F)), 26.8 (S-flecainide (F)), 15.9 (R-cibenzoline (G)), 18.6 (S-cibenzoline (G)), 20.6 ((-)-diprafenone (H)), 22.6 ((+)-diprafenone (H)), 22.7 (R-metoprolol (I)), 27.3 (S-metoprolol (I)), 19.6 ((-)-tranylcypromine (J)), 21.6 ((+)-tranylcypromine (J))

KEY WORDS
chiral; normal phase; reverse phase

REFERENCE
Martin, E.; Quinke, K.; Spahn, H.; Mutschler, E. (-)-(S)-Flunoxaprofen and (-)-(S)-naproxen isocyanate: two new fluorescent chiral derivatizing agents for an enantiospecific determination of primary and secondary amines, *Chirality*, **1989**, *1*, 223–234.

(R)-(+)-α-Methylbenzyl Isocyanate

SAMPLE
Matrix: blood
Analyte: propranolol
Sample preparation: 100 μL Serum + 10 μL 600 ng/mL IS in water + 200 μL 10% sodium bicarbonate + 5 mL diethyl ether, shake at 20 rpm for 15 min, centrifuge at 400 g for 5 min. Remove the organic layer and evaporate it to dryness under a stream of nitrogen at room temperature. Add 50 μL 5 μL/mL (R)-(+)-α-methylbenzyl isocyanate ((R)-(+)-phenylethylisocyanate) in diethyl ether to the residue, vortex vigorously for 30 s, keep at 4° for 30 min, allow to warm to room temperature, vortex for 15 s, evaporate under nitrogen, reconstitute in 100 μL mobile phase, let stand at room temperature for 20 min, inject a 20 μL aliquot.

HPLC VARIABLES
Guard column: 25-37 μm Whatman Co:Pell ODS pellicular C18
Column: 250 × 4.6 5 μm Ultrasphere C8
Mobile phase: MeOH:isopropanol:dichloromethane:water 67:7.5:1:25.5
Flow rate: 0.7
Injection volume: 20
Detector: F ex 220 em 300 (cut-off filter)

CHROMATOGRAM
Retention time: 13.7 (S(-)), 14.9 (R(+))
Internal standard: (±)-N-cyclopentyldesisopropylpropranolol (19.3, 21.2)
Limit of detection: 2 ng/mL

OTHER SUBSTANCES
Simultaneously analyzed: metabolites

KEY WORDS
serum; human; rat; chiral

REFERENCE
Laganière, S.; Kwong, E.; Shen, D.D. Stereoselective high-performance liquid chromatographic assay for propranolol enantiomers in serum, *J.Chromatogr.*, **1989**, *488*, 407–416.

SAMPLE
Matrix: blood
Analyte: propranolol
Sample preparation: 500 μL Plasma + 1 mL 200 mM pH 10.5 phosphate buffer + 5 mg ascorbic acid + 4 mL ethyl acetate, shake vigorously for 10 min, centrifuge at 1500 g for 15 min. Remove 3 mL of the organic layer and evaporate it to dryness under a stream of nitrogen at 37°, reconstitute the residue in 500 μL chloroform, add 30 μL triethylamine, add 2 μL (R)-(+)-α-methylbenzyl isocyanate (R-(+)-phenylethylisocyanate), shake, let stand at room temperature for 30 min, evaporate to dryness under a stream of nitrogen at room temperature, reconstitute the residue in 300 μL mobile phase, inject a 50 μL aliquot.

HPLC VARIABLES
Column: 125 × 4.6 5 μm Partisil 5 ODS-3
Mobile phase: MeOH:water 60.5:39.5
Flow rate: 1.2
Injection volume: 50
Detector: F ex 228 em 340 (cutoff filter)

CHROMATOGRAM
Retention time: 27 (-), 31 (+)

OTHER SUBSTANCES
Extracted: metabolites

KEY WORDS
chiral; plasma; pharmacokinetics

REFERENCE
Schaefer, H.G.; Spahn, H.; Lopez, L.M.; Derendorf, H. Simultaneous determination of propranolol and 4-hydroxypropranolol enantiomers after chiral derivatization using reversed-phase high-performance liquid chromatography, *J.Chromatogr.*, **1990**, *527*, 351–359.

SAMPLE
Matrix: blood
Analyte: CHF   1255   (5,6-dimethoxy-2-[3'-(p-hydroxyphenyl)-3'-hydroxy-2'-propyl]amino-tetralin)
Sample preparation: 1 mL Plasma + 100 μL 1 μg/mL IS + 1.5 mL 500 mM pH 9.5 borate buffer + 5 mL diethyl ether, shake mechanically for 15 min, centrifuge at 690 g for 10 min, repeat extraction. Combine the organic layers and add them to 1.5 mL 10 mM HCl, shake mechanically for 15 min, centrifuge at 690 g for 10 min. Remove the aqueous layer

and add it to 2.5 mL 500 mM pH 9.5 borate buffer, extract with 7 mL diethyl ether: MeOH 90:10. Remove the organic layer and evaporate it to dryness under a stream of air at 37°, reconstitute the residue in 100 μL buffer, add 10 μL 0.4% (R)-(+)-α-methylbenzyl isocyanate in MeCN, let stand at room temperature for 16 h, add 100 μL 1 mM HCl, inject a 50 μL aliquot. (Prepare buffer by adjusting pH of MeCN:0.05% sodium bicarbonate 80:20 to 7.5 with 100 mM HCl.)

HPLC VARIABLES
Guard column: 25 × 2.4 37-50 μm Bondapak C18/Corasil
Column: 250 × 4.6 5 μm Spherisorb ODS 2
Mobile phase: MeCN:50 mM sodium acetate 46:54, adjusted to pH 7.5 with glacial acetic acid
Flow rate: 1.2
Injection volume: 50
Detector: E, ESA Model 5100, Model 5020 guard cell before injector 0.9 V, Model 5011 analytical cell, screen electrode 0.45 V, sample electrode 0.70 V

CHROMATOGRAM
Retention time: 14.8, 16.9, 18.2, 19.9 (stereoisomers)
Internal standard: (S)-(-)-5,6-dimethoxy-2-[2'-(p-hydroxyphenyl)-1'-ethyl]aminotetralin (23.3)
Limit of quantitation: 0.59-1.50 ng/mL

KEY WORDS
chiral; plasma; human; rat

REFERENCE
Rondelli, I.; Mariotti, F.; Acerbi, D.; Redenti, E.; Amari, G.; Ventura, P. Selective method for plasma quantitation of the stereoisomers of a new aminotetralin by high-performance liquid chromatography with electrochemical detection, *J.Chromatogr.*, **1993**, *612*, 95–103.

SAMPLE
Matrix: blood, urine
Analyte: propranolol
Sample preparation: 500 μL Plasma or 1 mL urine + 50 μL 40 μg/mL pronethalol in MeOH + 1 mL 200 mM pH 9.8 carbonate buffer + 0.5 g NaCl (plasma samples only) + 5 mL toluene, shake horizontally for 30 min, centrifuge at 1500 g at 10° for 15 min. Remove 4 mL of the organic layer and evaporate it under reduced pressure. Reconstitute the residue in 200 μL MeOH, add 50 μL 1% triethylamine in MeOH, add 50 μL 2% (R)-(+)-α-methylbenzyl isocyanate ((R)-(+)-phenylethyl isocyanate) in dichloromethane, vortex briefly, heat at 30° for 35 min, evaporate under reduced pressure, reconstitute the residue in 200 μL mobile phase, inject a 20 μL aliquot.

HPLC VARIABLES
Column: 150 × 4.6 5 μm Zorbax ODS
Mobile phase: MeOH:water:acetic acid 70:30:0.1
Column temperature: 28
Flow rate: 1.2
Injection volume: 20
Detector: F ex 295 em 345

CHROMATOGRAM
Retention time: 18 (R), 21 (S)
Internal standard: pronethalol (14 (R), 16 (S))
Limit of detection: 1 ng/mL

KEY WORDS
plasma

REFERENCE
Spahn-Langguth, H.; Podkowik, B.; Stahl, E.; Martin, E.; Mutschler, E. Improved enantiospecific RP-HPLC assays for propranolol in plasma and urine with pronethalol as internal standard, *J.Anal.Toxicol.*, **1991**, *15*, 327–331.

SAMPLE

Matrix: blood, urine

Analyte: propranolol

Sample preparation: 100 μL Plasma or urine + 100 μL IS in MeOH + 100 μL 25% ammonium hydroxide + 2 mL MeOH:diethyl ether 10:90, vortex for 1.5 min, centrifuge at 1500 g for 5 min. Remove the organic layer and evaporate it to dryness under a stream of nitrogen at room temperature, reconstitute the residue in 100 μL 5 μL/mL (R)-(+)-α-methylbenzyl isocyanate (R-(+)-1-phenylethylisocyanate) in diethyl ether, vortex for 30 s, let stand at room temperature for 30 min, evaporate to dryness under a stream of nitrogen, add 100 μL mobile phase, vortex for 30 s, centrifuge at 3000 g for 7 min (plasma only), inject a 50 μL aliquot.

HPLC VARIABLES

Guard column: 15 × 3.2 7 μm C18 (Brownlee)

Column: 150 × 3.9 4 μm Novapak C18

Mobile phase: MeOH:water 72.5:27.5

Flow rate: 1.6

Injection volume: 50

Detector: F ex 232 em 340

CHROMATOGRAM

Retention time: 5.5 (S), 6.2 (R)

Internal standard: 4-methylpropranolol (Cambridge Research Biochemicals) (8.8 (S), 10.1 (R))

Limit of detection: 1 ng/mL

KEY WORDS

plasma; chiral; pharmacokinetics

REFERENCE

Pham-Huy, C.; Sahui-Gnassi, A.; Saada, V.; Gramond, J.P.; Galons, H.; Ellouk-Achard, S.; Levresse, V.; Fompeydie, D.; Claude, J.R. Microassay of propranolol enantiomers and conjugates in human plasma and urine by high-performance liquid chromatography after chiral derivatization for pharmacokinetic study, *J.Pharm.Biomed.Anal.*, **1994**, *12*, 1189–1198.

SAMPLE

Matrix: bulk

Analyte: propranolol and propylhexedrine

Sample preparation: Dissolve 1.5 mg compound in 1 mL reagent, add 3 μL triethylamine, sonicate for 20 min, add 3 μL diethylamine, let stand for 15 min, inject an aliquot. (Reagent was 2 mg/mL (R)-(+)-1-phenylethyl isocyanate ((R)-(+)-α-methylbenzyl isocyanate) solution in dry chloroform:DMF 80:20.)

HPLC VARIABLES

Column: 200 × 4.6 Silica 100 RP 18

Mobile phase: MeOH:water 70:30

Flow rate: 1

Detector: UV 254

CHROMATOGRAM

Retention time: k' 3.17, k' 3.66 (propranolol enantiomers), k' 4.36, k' 4.73 (propylhexedrine enantiomers)

KEY WORDS

chiral

REFERENCE

Jira, T.; Toll, C.; Vogt, C.; Beyrich, T. Zur Trennung einiger racemischer β-Blocker und α-Sympathikomimetika durch HPLC nach Derivatisierung [The separation of some racemic β-blockers and α-sympathomimetics with HPLC following derivatization], *Pharmazie*, **1991**, *46*, 432–434.

SAMPLE

Matrix: microsomal incubations

Analyte: bufuralol metabolites

Sample preparation: Cool 3 mL microsomal incubation in ice, add IS, add 500 μL saturated sodium carbonate solution, add 5 mL ethyl ether:dichloromethane 60:40, extract, centrifuge. Remove the organic layer and evaporate it to dryness under a stream of nitrogen, reconstitute the residue in 200 μL dichloromethane (dried over 3 Å molecular sieve), add 5 μL (R)-(+)-α-methylbenzyl isocyanate ((+)-(R)-1-phenethyl isocyanate), let stand for 1 h, evaporate to dryness, reconstitute with 500 μL MeOH, filter (0.45 μm), inject a 10 μL aliquot of the filtrate.

HPLC VARIABLES
Guard column: Hypersil C18
Column: 250×4.6 5 μm Ultrasphere ODS
Mobile phase: MeOH:water 69:31 (Wash column with MeOH:water 90:10 for 15 min after each run, re-equilibrate with mobile phase for 15 min.)
Flow rate: 1
Injection volume: 10
Detector: UV 248

CHROMATOGRAM
Retention time: 18.01, 19.18, 16.61, 17.89 (hydroxylated metabolites of bufuralol)
Internal standard: (1'S)-1''-oxobufuralol (24.51)
Limit of detection: 2 ng

KEY WORDS
human; liver

REFERENCE
Weerawarna, S.A.; Geisshusler, S.M.; Murthy, S.S.; Nelson, W.L. Enantioselective and diastereoselective hydroxylation of bufuralol. Absolute configuration of the 7-(1-hydroxyethyl)-2-[1-hydroxy-2-(tert-butylamino)ethyl]benzofurans, the benzylic hydroxylation metabolites, *J.Med.Chem.*, **1991**, *34*, 3091–3097.

SAMPLE
Matrix: solutions
Analyte: 1-phenyl-2-aminopropanes
Sample preparation: 100 μL 1 mg/mL Compound in dichloromethane + 100 μL 6.8 mM (R)-(+)-α-methylbenzyl isocyanate ((R)-(+)-1-phenylethyl isocyanate) in dichloromethane, mix, let stand at room temperature for 1 h, evaporate to dryness under a stream of nitrogen, add 1 mL 100 mM NaOH, vortex for 15 min, add 1 mL 20% NaOH, add 3 mL dichloromethane, shake mechanically for 15 min, centrifuge at 1000 g for 15 min. Remove the organic layer and wash it with 2 mL 100 mM HCl. Remove a 100 μL aliquot of the organic layer and dilute it to 1 mL with MeOH, inject a 20 μL aliquot.

HPLC VARIABLES
Column: 250×4.5 5 μm octadecyl (IBM)
Mobile phase: MeOH:water 60:40
Flow rate: 1-2
Injection volume: 20
Detector: UV 254

CHROMATOGRAM
Retention time: 13.5 ((R)-amphetamine), 14.0 ((S)-amphetamine), 23.6 ((R)-1-(4-chlorophenyl)-2-aminopropane), 24.2 ((S)-1-(4-chlorophenyl)-2-aminopropane), 13.0 ((R)-1-phenylethylamine), 13.0 ((S)-1-phenylethylamine), 26.3 ((R)-1-(2,5-dimethoxy-4-methylphenyl)-2-aminopropane), 28.0 ((S)-1-(2,5-dimethoxy-4-methylphenyl)-2-aminopropane), 26.3 ((R)-1-(2,4-dimethoxy-5-methylphenyl)-2-aminopropane), 27.2 ((S)-1-(2,4-dimethoxy-5-methylphenyl)-2-aminopropane)
Limit of quantitation: 100 ng

KEY WORDS
comparison with other derivatizing reagents; chiral

REFERENCE

Miller, K.J.; Gal, J.; Ames, M.M. High-performance liquid chromatographic resolution of enantiomers of 1-phenyl-2-aminopropanes (amphetamines) with four chiral reagents, *J.Chromatogr.*, **1984**, *307*, 335–342.

SAMPLE

Matrix: solutions
Analyte: propranolol
Sample preparation: Mix a 500 μL aliquot of a 20-120 μg/mL solution in chloroform with 5 μL (R)-(+)-α-methylbenzyl isocyanate (R-(+)-1-phenylethyl isocyanate), let stand at room temperature for 15 min, add 10 mL 100 mM HCl, shake on a reciprocating shaker for 10 min, centrifuge. Remove the organic layer and evaporate it to dryness under a stream of nitrogen, reconstitute the residue in 50 μL mobile phase, inject a 15 μL aliquot.

HPLC VARIABLES

Column: 250 × 4.6 10 μm silica (Alltech)
Mobile phase: Chloroform:MeOH 100:1.2
Flow rate: 1
Injection volume: 15
Detector: UV 313

CHROMATOGRAM

Retention time: 4 ((-)-propranolol), 4.5 ((+)-propranolol), 8 ((-)-4-hydroxypropranolol), 9 ((+)-4-hydroxypropranolol)

KEY WORDS

normal phase; chiral

REFERENCE

Wilson, M.J.; Walle, T. Silica gel high-performance liquid chromatography for the simultaneous determination of propranolol and 4-hydroxypropranolol enantiomers after chiral derivatization, *J.Chromatogr.*, **1984**, *310*, 424–430.

SAMPLE

Matrix: solutions
Analyte: ethambutol
Sample preparation: Dissolve 1 mg ethambutol in 200 μL MeCN (add 20 μL triethylamine if the compound is a hydrochloride), add 50 μL 60 mg/mL (R)-(+)-α-methylbenzyl isocyanate ((R)-(+)-phenylethyl isocyanate) in MeCN, mix briefly, let stand at room temperature for 30 min, add 500 μL MeCN, inject a 3-5 μL aliquot.

HPLC VARIABLES

Column: 100 × 2.6 3 μm octadecyl
Mobile phase: MeOH:water 65:35
Flow rate: 1
Injection volume: 3-5
Detector: UV 254

CHROMATOGRAM

Retention time: 8.5 (-), 9.7 (+), 12.7 (meso)

KEY WORDS

chiral

REFERENCE

Gamberini, G.; Ferioli, V. Determination of optical purity by high performance liquid chromatography of compounds of pharmaceutical interest, *Farmaco.[Prat].*, **1988**, *43*, 357–363.

(S)-(-)-α-Methylbenzyl Isocyanate

RELATED REFERENCES

Pflugmann, G.; Spahn, H.; Mutschler, E. Determination of metoprolol enantiomers in plasma and urine using (S)-(-)-phenylethyl isocyanate as a chiral reagent. *J.Chromatogr.* **1987**, *421*, 161-164.

Spahn-Langguth, H.; Podkowik, B.; Stahl, E.; Martin, E.; Mutschler,E. Improved enantiospecific RP-HPLC assays for propranolol in plasma and urine with pronethalol as internal standard. *J.Anal.Toxicol.* **1991**, *15*, 209-213.

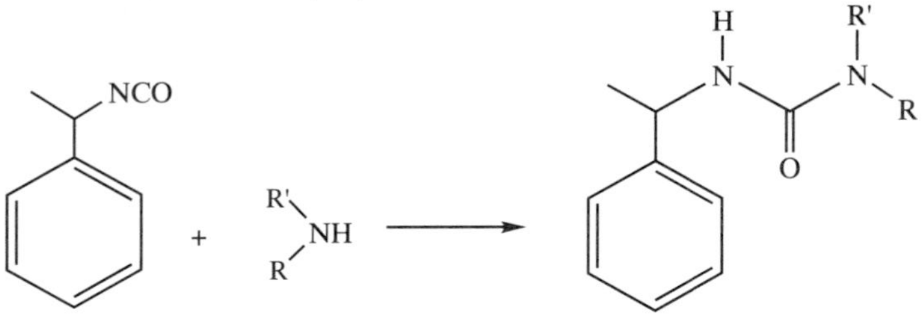

SAMPLE

Matrix: bulk

Analyte: 3-aminoquinuclidine

Sample preparation: Prepare the free base by dissolving the hydrochloride in MeOH, add 2 equivalents of sodium methoxide, after a few min evaporate to dryness under a stream of nitrogen, reconstitute with dry DMF at a concentration of 1-2 mg/mL, add a 10% molar excess of (S)-(-)-α-methylbenzyl isocyanate (S-(-)-1-phenylethyl isocyanate), vortex, let stand at room temperature for 30 min, dilute with mobile phase, inject an aliquot.

HPLC VARIABLES

Column: 150 × 4.6 5 μm Zorbax Sil

Mobile phase: Ethyl acetate:MeOH:concentrated ammonium hydroxide 85:10:5

Flow rate: 2

Injection volume: 10

Detector: UV 254

CHROMATOGRAM

Retention time: k' 1.77 (S), k' 2.04 (R)

KEY WORDS

chiral; normal phase; comparison with other derivatizing reagents

REFERENCE

Demian, I.; Gripshover, D.F. Enantiomeric purity determination of 3-aminoquinuclidine by diastereomeric derivatization and high-performance liquid chromatographic separation, *J.Chromatogr.*, **1989**, *466*, 415–420.

SAMPLE

Matrix: bulk

Analyte: dorzolamide

Sample preparation: Derivatize with (S)-(-)-α-methylbenzyl isocyanate, prepare a 2 mg/mL solution, inject a 10 μL aliquot.

HPLC VARIABLES

Column: 250 × 4.6 Zorbax SIL

Mobile phase: Heptane:MTBE:MeCN:water 35:62.855:1.95:0.195

Flow rate: 2

Injection volume: 10

Detector: UV 254

CHROMATOGRAM

Retention time: 12 (4S,6R(S) derivative), 16 (4R,6S(S) derivative)

KEY WORDS
chiral; rugged

REFERENCE
Dovletoglou, A.; Thomas, S.M.; Berwick, L.; Ellison, D.K.; Tway, P.C. Development of practical HPLC methods for analysis and quality assessment of the novel carbonic anhydrase inhibitor MK-0507 and the acetamidosulfonamide intermediate, *J.Liq.Chromatogr.*, **1995**, *18*, 2337–2352.

(R)-(-)-1-(1-Naphthyl)ethyl Isocyanate

SAMPLE
Matrix: blood
Analyte: propafenone
Sample preparation: 1 mL Plasma + 100 μL 1 μg/mL (-)-ephedrine in water + 200 μL saturated sodium carbonate + 4 mL hexane:isopropanol:heptafluoro-1-butanol 95:5:1.25, vortex for 30 s, centrifuge at 1800 g for 5 min. Remove the organic layer and evaporate it to dryness under a stream of nitrogen, reconstitute the residue in 100 μL reagent, vortex for 5 s, let stand at room temperature for 3 min, add 100 μL bupranolol solution, evaporate to dryness under a stream of nitrogen, reconstitute the residue in 400 μL hexane, add 200 μL 100 mM HCl, vortex for 15 s, centrifuge at 1800 g for 5 min, inject a 100 μL aliquot of the upper organic layer. (Prepare reagent by diluting commercially available R-(-)-1-(1-naphthyl)ethyl isocyanate 10-fold with hexane, pass through a 50 mm column of silica, dilute the eluate with hexane to a concentration of 0.1%, store in amber containers at -30°. Prepare working reagent immediately before use by diluting to 0.005% with hexane:isopropanol 95:5. Prepare bupranolol solution by dissolving bupranolol hydrochloride in 1 mL water, add 500 μL saturated sodium carbonate solution, extract with 25 mL hexane, use the hexane solution. Alprenolol, bupranolol, methoxamine, mexiletine, pindolol, propranolol, and tocainide can be derivatized and chromatographed to separate the enantiomers. Extraction from plasma has not been validated, however.)

HPLC VARIABLES
Guard column: 50 × 4.6 pellicular silica (Whatman)
Column: 100 × 4.6 5 μm Partisil 5 silica
Mobile phase: Hexane:isopropanol:isobutanol 96:2:2
Flow rate: 1.5
Injection volume: 100
Detector: UV 220

CHROMATOGRAM
Retention time: 8.3 (-), 10.1 (+)
Internal standard: (-)-ephedrine (16.4)
Limit of quantitation: 6.25 ng/mL

OTHER SUBSTANCES
Simultaneously analyzed: alprenolol, bupranolol, methoxamine, mexiletine, pindolol, propranolol, tocainide

KEY WORDS
plasma; normal phase; pharmacokinetics; chiral

REFERENCE

Mehvar, R. Liquid chromatographic analysis of propafenone enantiomers in human plasma, *J.Chromatogr.*, **1990**, *527*, 79–89.

SAMPLE

Matrix: blood

Analyte: nadolol

Sample preparation: 500 µL Plasma + 100 µL 1 M NaOH, vortex gently for 1 min, add to an Extrelut-1 SPE cartridge, rinse out the tube with 300 µL water, add the rinse to the SPE cartridge, let stand for 15 min, elute with 10 mL diethyl ether. Evaporate the eluate to dryness under a stream of nitrogen at 40°, reconstitute the residue in 100 µL water and 400 µL MeOH, evaporate to dryness under a stream of nitrogen at 50°, reconstitute with 50 µL MeOH, add 10 µL 5 mg/mL (R)-(-)-1-(1-naphthyl)ethylisocyanate in MeCN, vortex for 30 s, heat at 45° for 5 min, evaporate to dryness under a stream of nitrogen at room temperature, reconstitute with 100 µL MeOH, inject a 20 µL aliquot.

HPLC VARIABLES

Column: 250 × 4.6 5 µm YMC-AM-303 ODS (YMC)

Mobile phase: MeCN:water 40:60

Column temperature: 40

Flow rate: 1

Injection volume: 20

Detector: F ex 285 em 340

CHROMATOGRAM

Retention time: 31 (SR), 33 (RS), 35 (RR), 37 (SS)

Limit of quantitation: 2.5 ng/mL

KEY WORDS

pharmacokinetics; dog; plasma; chiral; SPE

REFERENCE

Hoshino, M.; Yajima, K.; Suzuki, Y.; Okahira, A. Determination of nadolol diastereomers in dog plasma using chiral derivatization and reversed-phase high-performance liquid chromatography with fluorescence detection, *J.Chromatogr.B*, **1994**, *661*, 281–289.

SAMPLE

Matrix: blood

Analyte: nadolol

Sample preparation: Condition a 500 mg non-end-capped cyanopropyl SPE cartridge (Varian) with 1 mL MeOH and 1 mL water. Condition a 100 mg phenylboronic acid SPE cartridge (Varian) with 1 mL MeOH, 2 mL 100 mM HCl, 4 mL 0.3% ammonium hydroxide, and 2 mL 100 mM pH 8.5 ammonium sulfate. 1 mL Plasma + 80 ng IS, add to the cyanopropyl SPE cartridge, wash with 3 mL water, wash with 3 mL MeCN, elute with 2.5 mL alkaline MeOH. Evaporate the eluate to dryness under a stream of nitrogen, reconstitute the residue in 1 mL MeOH, evaporate to dryness under a stream of nitrogen, reconstitute the residue in 50 µL distilled 1,2-dimethoxyethane, add 20 µL reagent, let stand at room temperature for 1 h, add 300 µL water, extract with 1 mL dichloromethane. Wash the organic layer with 300 µL water, evaporate the organic layer to dryness under a stream of nitrogen, reconstitute the residue in 1 mL MeOH:water 40:60. Add this solution to the phenylboronic SPE column, wash with 2 mL water, wash with 100 µL MeOH, wash with 1 mL hexane:ethyl acetate 80:20, wash with 1 mL hexane, elute with 5 mL MeOH:triethylamine 95:5. Evaporate the eluate to dryness under a stream of nitrogen, reconstitute the residue in 100 µL MeOH:2% tetramethylethylenediamine 50:50, inject an 80 µL aliquot. (Reagent was 0.1% R(-)-1-(naphthyl)ethylisocyanate in 1,2-dimethoxyethane.)

HPLC VARIABLES

Column: 250 × 4.6 5 µm C18 (Beckman)

Mobile phase: MeOH:THF:buffer 52:7:41 (Buffer was 0.4% tetramethylethylenediamine adjusted to pH 3.0 with trifluoroacetic acid.)

Column temperature: 28

Flow rate: 1
Injection volume: 80
Detector: F ex 230 em 330

CHROMATOGRAM
Retention time: 20.6 (SRS), 21.8 (RSR), 26.2 (SSR), 28.2 (RRS)
Internal standard: 5-(3-[methylethylamino]-2-hydroxypropoxy)-1,2,3,4-tetrahydro-2,3-naphthalenediol (SQ 11559, Bristol-Myers Squibb) (Racemate A was separated by HPLC using the above system with MeOH:100 mM ammonium acetate 18:82 mobile phase and UV 230.) (13.5, 15.5 (enantiomers))
Limit of quantitation: 2.5 ng/mL
Limit of detection: 1 ng/mL

KEY WORDS
chiral; plasma; SPE; pharmacokinetics

REFERENCE
Belas, F.J.; Phillips, M.A.; Srinivas, N.R.; Barbhaiya, R.H.; Blair, I.A. Simultaneous determination of nadolol enantiomers in human plasma by high-performance liquid chromatography using fluorescence detection, *Biomed.Chromatogr.*, **1995**, *9*, 140–145.

SAMPLE
Matrix: blood
Analyte: nadolol
Sample preparation: 1 mL Plasma + 50 μL 2 μg/mL desmethylnadolol + 750 μL 1 M NaOH, vortex for 10 s, add 5 mL dichloromethane, rotate for 15 min, centrifuge at 2600 rpm for 10 min. Remove the organic layer and evaporate it to dryness under a stream of nitrogen at 40°, reconstitute the residue in 50 μL 0.1% R-(-)-naphthylethylisocyanate in dichloromethane, let stand for 1 h, reconstitute with 150 μL mobile phase, inject a 125 μL aliquot.

HPLC VARIABLES
Column: 250 × 4.6 5 μm C18 ODS (Beckman)
Mobile phase: MeOH:THF:water:phosphoric acid 52:7:41:0.001 adjusted to pH 3 with tetramethylene diamine
Flow rate: 1.4
Injection volume: 125
Detector: F ex 230 em 330

CHROMATOGRAM
Retention time: 21 (SRS), 23 (RSR), 28 (SSR), 30 (RRS)
Internal standard: desmethylnadolol (13, 15 (enantiomers))
Limit of quantitation: 2 ng/mL

KEY WORDS
plasma; chiral

REFERENCE
Srinivas, N.R.; Shyu, W.C.; Dhah, V.R.; Campbell, D.A.; Barbhaiya, R.H. Stereoselective analysis of nadolol in human plasma, *Biomed.Chromatogr.*, **1995**, *9*, 226–228.

SAMPLE
Matrix: blood
Analyte: propafenone
Sample preparation: 500 μL Plasma + 100 μL saturated sodium carbonate + 2 mL hexane:propanol:heptafluorobutanol 95:5:1.25, shake for 30 min, centrifuge at 1800 g for 5 min. Remove the organic layer and evaporate it to dryness under a stream of nitrogen, reconstitute the residue in 100 μL 0.005% R-(-)-1-(1-naphthyl)ethylisocyanide in hexane, vortex, let stand at room temperature for a short time, add 100 μL bupranolol solution, evaporate to dryness, reconstitute in 200 μL hexane, add 200 μL 100 mM HCl, mix, centrifuge, inject a 100 μL aliquot of the upper organic phase. (The bupranolol solution was obtained by dissolving 5 mg bupranolol hydrochloride in 1 mL water, adding 500 μL saturated sodium carbonate, and extracting with 25 mL hexane. Use the hexane layer.)

HPLC VARIABLES
Column: 150 × 3.9 4 μm 60 Å Nova-Pak silica
Mobile phase: Hexane:isopropanol:isobutanol 96:2:2
Flow rate: 1.5
Injection volume: 100
Detector: UV 220

CHROMATOGRAM
Retention time: 7.5 (R(-)), 9.5 (S(+))
Internal standard: bupranolol (16)
Limit of detection: 20 ng/mL

KEY WORDS
plasma; chiral; normal phase; pharmacokinetics

REFERENCE
Volz, M.; Mitrovic, V.; Thiemer, J.; Schlepper, M. Steady-state plasma kinetics of slow-release propafenone, its two isomers and its main metabolites, *Arzneimittelforschung*, **1995**, *45*, 246–249.

SAMPLE
Matrix: blood
Analyte: sotalol
Sample preparation: Evaporate 100 (?) μL 10 μg/mL atenolol in MeOH in to the bottom of a tube, add 500 μL plasma, add 200 μL 1 M pH 9.3 carbonate buffer, add 6 mL ethyl acetate, shake for 4 min, centrifuge at 1000 g for 4 min. Remove the organic layer and evaporate it to dryness under reduced pressure, reconstitute the residue with 2 μL 1 M pH 9.3 carbonate buffer and 200 μL 0.005% R-(-)-1-(1-naphthyl)ethyl isocyanate in chloroform (prepared fresh daily), vortex for 15 s, let stand at room temperature for 1 h, evaporate to dryness with a stream of air, add 200 μL MeCN:water 39:61, vortex for 10 s, let stand at room temperature for 1 h, inject a 10 μL aliquot on to column A and column B in series and elute with mobile phase. Column A is subsequently removed from the circuit and backflushed with MeCN:water 50:50 for 3 min, with MeCN for 12 min, with MeCN:water 50:50 for 7.5 min, and with mobile phase for 7.5 min (this is very unclear in the paper).

HPLC VARIABLES
Column: A 75 × 3.9 4 μm Nova-Pak C18; B 100 × 8 4 μm Nova-Pak C18 Radial Compression
Mobile phase: MeCN:water 39:61
Flow rate: 1.5
Injection volume: 10
Detector: F ex 280 em 320

CHROMATOGRAM
Retention time: 23 (+), 26 (-)
Internal standard: atenolol (13)
Limit of quantitation: 10 ng/mL

OTHER SUBSTANCES
Non-interfering: metabolites, N-desisopropylpropranolol, 4-hydroxypropranolol, labetalol, 4-methylpropranolol, metoprolol, oxprenolol, pindolol, practolol, propranolol, propranolol glycol, timolol

KEY WORDS
plasma; chiral; pharmacokinetics; column-switching

REFERENCE
Hooper, W.D.; Baker, P.V. Enantioselective analysis of sotalol in plasma by reversed-phase high-performance liquid chromatography using diastereomeric derivatives, *J.Chromatogr.B*, **1995**, *672*, 89–96.

SAMPLE
Matrix: blood, tissue
Analyte: fluoxetine

Sample preparation: Homogenize (Brinkman Polytron) tissue with 5-10 volumes of water. 500 μL Plasma or tissue homogenate + 500 μL water + 50 μL 2 μg/mL IS in water + 100 μL 1 M NaOH, vortex gently, add 5 mL hexane:butanol 99.7:0.3, shake mechanically at 125-150 cycles/min for 30 min, centrifuge at 2000 g for 15 min. Remove the organic layer and mix it with 100 μL 200 μM R-(-)-1-(1-naphthyl)ethyl isocyanate in hexane, evaporate to dryness at 50-55° over 20-30 min, when all the hexane is gone dry more vigorously, reconstitute with 200 μL mobile phase, inject a 75 μL aliquot.

HPLC VARIABLES
Column: 250 × 4.6 5 μm Apex silica (Jones Chromatography)
Mobile phase: Isooctane:THF 70:30
Column temperature: 35
Flow rate: 1
Injection volume: 75
Detector: F ex 218 em 333

CHROMATOGRAM
Retention time: 8.3 (S), 9.3 (R)
Internal standard: S-nornisoxetine (15)
Limit of detection: 5 ng/mL (plasma), 25 ng/g (tissue)

OTHER SUBSTANCES
Extracted: metabolites, norfluoxetine

KEY WORDS
chiral; normal phase; plasma; silylate all glassware

REFERENCE
Potts, B.D.; Parli, C.J. Analysis of the enantiomers of fluoxetine and norfluoxetine in plasma and tissue using chiral derivatization and normal-phase liquid chromatography, *J.Liq.Chromatogr.*, **1992**, *15*, 665–681.

SAMPLE
Matrix: bulk
Analyte: fluoxetine
Sample preparation: Reflux 1.23 g fluoxetine with 788 mg (R)-(-)-1-(1-naphthyl)ethyl isocyanate in 25 mL toluene for 2 h, evaporate to dryness under reduced pressure, reconstitute, inject an aliquot.

HPLC VARIABLES
Column: 250 × 4.6 silica (IBM)
Mobile phase: Dichloromethane:MeOH 99.75:0.25
Flow rate: 2
Detector: UV 254

CHROMATOGRAM
Retention time: 6.69 (S), 7.52 (R)

KEY WORDS
chiral; normal phase

REFERENCE
Robertson, D.W.; Krushinski, J.H.; Fuller, R.W.; Leander, J.D. Absolute configurations and pharmacological activities of the optical isomers of fluoxetine, a selective serotonin-uptake inhibitor, *J.Med.Chem.*, **1988**, *31*, 1412–1417.

SAMPLE
Matrix: bulk
Analyte: 3-aminoquinuclidine
Sample preparation: Prepare the free base by dissolving the hydrochloride in MeOH, add 2 equivalents of sodium methoxide, after a few min evaporate to dryness under a stream of nitrogen, reconstitute with dry DMF at a concentration of 1-2 mg/mL, add a 10% molar excess of R-(-)-1-naphthylethyl isocyanate, vortex, let stand at room temperature for 30 min, dilute with mobile phase, inject an aliquot.

HPLC VARIABLES
Column: 150 × 4.6 5 μm Zorbax Sil
Mobile phase: Ethyl acetate:MeOH:concentrated ammonium hydroxide 85:10:5
Flow rate: 2
Injection volume: 10
Detector: UV 254

CHROMATOGRAM
Retention time: k' 1.60 (S), k' 1.98 (R)

KEY WORDS
chiral; normal phase; comparison with other derivatizing reagents

REFERENCE
Demian, I.; Gripshover, D.F. Enantiomeric purity determination of 3-aminoquinuclidine by diastereo-meric derivatization and high-performance liquid chromatographic separation, *J.Chromatogr.*, **1989**, *466*, 415–420.

SAMPLE
Matrix: bulk
Analyte: β-Blockers and sympathicomimetics
Sample preparation: Dissolve 1.5 mg compound in 1 mL reagent, add 3 μL triethylamine, sonicate for 20 min, add 3 μL diethylamine, let stand for 15 min, inject an aliquot. (Reagent was 2 mg/mL (R)-(-)-(naphth-1-yl)ethylisocyanate solution in dry chloroform:DMF 80:20.)

HPLC VARIABLES
Column: 200 × 4.6 Silica 100 RP 18
Mobile phase: MeOH:water 70:30
Flow rate: 1.5
Detector: UV 254

CHROMATOGRAM
Retention time: k' 0.68, k' 1.22 (atenolol enantiomers), k' 5.27, k' 6.21 (methylphenidate enantiomers), k' 6.98, k' 8.37 (metipranolol enantiomers), k' 2.50, k' 2.87 (pindolol enantiomers), k' 9.41 (D), k' 11.21 (L) (propranolol enantiomers), k' 9.26, k' 10.19 (propylhexedrine enantiomers), k' 1.66, k' 2.36 (talinolol enantiomers)

KEY WORDS
chiral

REFERENCE
Jira, T.; Toll, C.; Vogt, C.; Beyrich, T. Zur Trennung einiger racemischer β-Blocker und α-Sympathikom-imetika durch HPLC nach Derivatisierung [The separation of some racemic β-blockers and α-sym-pathomimetics with HPLC following derivatization], *Pharmazie*, **1991**, *46*, 432–434.

(S)-(+)-1-(1-Naphthyl)ethyl Isocyanate

RELATED REFERENCES

Darmon, A.; Thenot, J.P. Determination of betaxolol enantiomers by high-performance liquid chromatography. Application to pharmacokinetic studies. *J.Chromatogr.* **1986**, *374*, 321-328.

Gietl, Y.; Spahn, H.; Mutschler, E. Simultaneous determination of R- and S-prenylamine in plasma and urine by reversed-phase high-performance liquid chromatography. *J.Chromatogr.* **1988**, *426*, 305-314.

Peyton, A.L.; Carpenter, R.; Rutkowski, K. The stereospecific determination of fluoxetine and norfluoxetine enantiomers in human plasma by high-pressure liquid chromatography (HPLC) with fluorescence detection. *Pharm.Res.* **1991**, *8*, 1528-1532.

Piquette-Miller, M.; Jamali, F. Selective effect of adjuvant arthritis on the disposition of propranolol enantiomers in rats detected using a stereospecific HPLC assay. *Pharm.Res.* **1993**, *10*, 294-299.

SAMPLE

Matrix: blood

Analyte: tocainide

Sample preparation: 500 µL Plasma + 25 µL 10 µg/mL acebutolol in water + 200 µL 1 M NaOH + 5 mL chloroform, vortex for 30 s, centrifuge at 1800 g for 5 min. Remove the organic layer and evaporate it to dryness under reduced pressure, reconstitute the residue in 200 µL 0.05% S-(+)-1-(1-naphthyl)ethylisocyanate in chloroform, vortex for 30 s, evaporate to dryness under reduced pressure, reconstitute with 200 µL chloroform, inject a 50-175 µL aliquot.

HPLC VARIABLES

Column: 250 × 4.6 Partisil 5 silica

Mobile phase: Hexane:chloroform:MeOH 60:38:2

Flow rate: 2

Injection volume: 50-175

Detector: F ex 220 em 345

CHROMATOGRAM

Retention time: 3.5 (S), 4.5 (R) (tentative assignment)

Internal standard: (±)-acebutolol (14.9 (R), 16.5 (S))

Limit of quantitation: 250 ng/mL

Limit of detection: 25 ng/mL

KEY WORDS

plasma; chiral; pharmacokinetics

REFERENCE

Carr, R.A.; Foster, R.T.; Freitag, D.; Pasutto, F.M. Stereospecific high-performance liquid chromatographic determination of tocainide, *J.Chromatogr.*, **1991**, *566*, 155–162.

SAMPLE

Matrix: blood

Analyte: metoprolol

Sample preparation: 1 mL Plasma + 1 mL 5 µg/mL R-propranolol in water + 50 µL 100 mM NaOH + 4 mL chloroform, vortex for 30 s, centrifuge at 1800 g for 5 min. Remove organic layer and evaporate it to dryness in vacuum, reconstitute in 200 µL 0.05% S-(+)-1-(1-naphthyl)ethyl isocyanate in chloroform, vortex for 30 s, inject a 75-150 µL aliquot.

HPLC VARIABLES
Column: 250 mm long Whatman 5 µm silica
Mobile phase: Hexane:chloroform:MeOH 85:14:1
Flow rate: 2
Injection volume: 75-150
Detector: F ex 220 no emission filter

CHROMATOGRAM
Retention time: 14.1 (R-(+)), 16.2 (S-(-))
Internal standard: R-propranolol (8.5)
Limit of quantitation: 5 ng/mL

KEY WORDS
plasma; normal phase; chiral

REFERENCE
Bhatti, M.M.; Foster, R.T. Stereospecific high-performance liquid chromatographic assay of metoprolol, *J.Chromatogr.*, **1992**, *579*, 361–365.

SAMPLE
Matrix: blood
Analyte: oxprenolol
Sample preparation: 100 µL Plasma + 25 µL 25 ng/mL propranolol + 100 µL 1 M NaOH + 5 mL dichloromethane, shake, centrifuge at 1000 g for 10 min. Remove the organic layer and evaporate it to dryness under a stream of nitrogen at room temperature, reconstitute the residue in 100 µL dichloromethane, vortex for 5 s, add 10 µL 0.01% S-(+)-1-(1-naphthyl)ethyl isocyanate, heat at 37° for 2 h, add 20 µL tert-butylamine, evaporate under a stream of nitrogen, reconstitute with 50 µL mobile phase, inject a 20 µL aliquot.

HPLC VARIABLES
Column: 240 × 4.6 5 µm Spherisorb C18 ODS
Mobile phase: MeOH:THF:200 mM pH 3.6 acetate buffer 51:14:35
Column temperature: 30
Flow rate: 1
Injection volume: 20
Detector: F ex 226 em 333

CHROMATOGRAM
Retention time: 15.8 (S-(-)), 17.8 (R-(+))
Internal standard: propranolol (25.2 (R-(+)), 28.3 (S-(-)))
Limit of detection: 2.5 ng/mL

OTHER SUBSTANCES
Non-interfering: metabolites

KEY WORDS
plasma; chiral; pharmacokinetics

REFERENCE
Laethem, M.E.; Rosseel, M.T.; Wijnant, P.; Belpaire, F.M. Chiral high-performance liquid chromatographic determination of oxprenolol in plasma, *J.Chromatogr.*, **1993**, *621*, 225–229.

SAMPLE
Matrix: blood
Analyte: lomefloxacin
Sample preparation: 500 µL Plasma + 25 µL 100 µg/mL acebutolol in MeOH:water 10:90 + 100 µL 70 mM pH 7 phosphate buffer + 4 mL chloroform:isopentyl alcohol:diethyl ether 71.25:3.75:25, vortex for 30 s, centrifuge at 1800 g for 5 min. Remove the organic layer and evaporate it to dryness under reduced pressure, reconstitute the residue in 100 µL chloroform:triethylamine 100:1, add 100 µL 1% (S)-(+)-1-(1-naphthyl)ethyl isocyanate in chloroform, after 1 min add 50 µL 2% ethylchloroformate in chloroform, after 30 s add 50 µL 2.5% ethanolamine in chloroform, inject a 20-125 µL aliquot.

HPLC VARIABLES
Column: 100 × 8 4 µm Nova-Pak silica Radial Pak
Mobile phase: Hexane:chloroform:MeOH 64.5:33:2.5
Flow rate: 2
Injection volume: 20-125
Detector: F ex 245 em 420 for 12 min then ex 280 em 470

CHROMATOGRAM
Retention time: 21 ((S)-(-)), 22 ((R)-(+))
Internal standard: acebutolol (8.5, 9.5 enantiomers)
Limit of quantitation: 10 ng/mL

KEY WORDS
plasma; chiral; normal phase

REFERENCE
Foster, R.T.; Carr, R.A.; Pasutto, F.M.; Longstreth, J.A. Stereospecific high-performance liquid chromatographic assay of lomefloxacin in human plasma, *J.Pharm.Biomed.Anal.*, **1995**, *13*, 1243–1248.

SAMPLE
Matrix: blood
Analyte: DU-124884   (3-methylaminomethyl-3,4,5,6-tetrahydro-6-oxo-1H-azepino[5,4,3-cd] indole)
Sample preparation: 500 µL Plasma + 100 µL 2 µg/mL acebutolol in MeOH + 250 µL 0.5% triethylamine in MeOH, mix, let stand for 10 min, add 1 mL MeCN, mix, centrifuge. Remove the supernatant and evaporate it to dryness under a stream of nitrogen, reconstitute the residue in 500 µL EtOH, evaporate to dryness, reconstitute with 500 µL MeOH, add 25 µL 1% (S)-(+)-1-(1-naphthyl)ethyl isocyanate in chloroform, mix, evaporate to dryness under a stream of nitrogen, add 500 µL water, add 4 mL diethyl ether, extract. Remove the organic layer and evaporate it to dryness under a stream of nitrogen, reconstitute the residue in 200 µL chloroform, inject a 50 µL aliquot.

HPLC VARIABLES
Guard column: 30 × 4.6 5 µm Microsorb silica
Column: 250 × 4.6 5 µm Microsorb silica
Mobile phase: Hexane:chloroform:MeOH 80:12:8
Flow rate: 1
Injection volume: 50
Detector: F ex 320 em 440

CHROMATOGRAM
Retention time: 19 (-), 21 (+)
Internal standard: acebutolol (12)
Limit of quantitation: 0.1 ng/mL
Limit of detection: 0.02 ng/mL

OTHER SUBSTANCES
Extracted: metabolites

KEY WORDS
plasma; chiral; normal phase; use polypropylene tubes

REFERENCE
Naidong, W.; Pullen, R.H.; Arrendale, R.F.; Brennan, J.J.; Hulse, J.D.; Lee, J.W. Stereospecific determinations of (±)-DU-124884 and its metabolites (±)-KC-9048 in human plasma by liquid chromatography, *J.Pharm.Biomed.Anal.*, **1996**, *14*, 325–337.

SAMPLE
Matrix: blood, urine
Analyte: acebutolol
Sample preparation: Plasma. 1 mL Plasma + 100 µL 50 µg/mL pindolol in MeOH + 150 µL 1 M NaOH + 5 mL diethyl ether, vortex for 30 s, centrifuge at 1800 g for 5 min. Remove the organic layer and evaporate it to dryness under vacuum, add 200 µL 0.1%

S-(+)-1-(1-naphthyl)ethylisocyanate in chloroform, mix for 30 s, inject a 15-200 μL aliquot. Urine. Dilute 100 fold with water, proceed as above.

HPLC VARIABLES
Column: 250 mm long 5 μm Partisil silica
Mobile phase: Hexane:chloroform:MeOH 63:35:2
Flow rate: 2
Injection volume: 15-200
Detector: F ex 220 em 389

CHROMATOGRAM
Retention time: 12 (R), 13 (S)
Internal standard: pindolol (6,7 (enantiomers))
Limit of detection: 1 ng/mL

OTHER SUBSTANCES
Also analyzed: atenolol, nadolol, propranolol, sotalol, toliprolol, tocainide

KEY WORDS
plasma; chiral; normal phase

REFERENCE
Piquette-Miller, M.; Foster, R.T.; Pasutto, F.M.; Jamali, F. Stereospecific high-performance liquid chromatographic assay of acebutolol in human plasma and urine, *J.Chromatogr.*, **1990**, *526*, 129–137.

SAMPLE
Matrix: blood, urine
Analyte: tertatolol
Sample preparation: Evaporate 25 μL 1 μg/mL (-)-alprenolol hydrochloride in EtOH into the bottom of a tube, add 1 mL plasma or urine, add 100 μL 1 M NaOH, vortex for 1 min, add to an Extrelut SPE cartridge, elute with two 4 mL portions of diethyl ether. Evaporate the eluate to dryness under a stream of nitrogen, reconstitute the residue in 100 μL dichloromethane, add 10 μL 0.1% S-(+)-naphthylethylisocyanate in dichloromethane, shake for 1 min, let stand at room temperature for 12 h, add 10 μL tert-butylamine, evaporate to dryness under a stream of nitrogen, reconstitute the residue in 20 μL MeCN, vortex for 1 min, inject a 20 μL aliquot.

HPLC VARIABLES
Column: 70 × 4.6 3 μm Ultrasphere XLODS
Mobile phase: MeCN:water 40:60
Flow rate: 2
Injection volume: 20
Detector: F ex 220 em 320

CHROMATOGRAM
Retention time: 15 (-), 17 (+)
Internal standard: (-)-alprenolol (13)
Limit of quantitation: 6 ng/mL

KEY WORDS
SPE; chiral; plasma

REFERENCE
Lave, T.; Efthymiopoulos, C.; Koffel, J.C.; Jung, L. Determination of tertatolol enantiomers in biological fluids by high-performance liquid chromatography, *J.Chromatogr.*, **1991**, *572*, 203–210.

SAMPLE
Matrix: blood, urine
Analyte: dorzolamide
Sample preparation: 1 mL Whole blood + 600 μL 10% trichloroacetic acid, vortex for 1 min. Mix the acidified whole blood or 1 mL urine with 7 (blood) or 1 (urine) mL 200 mM pH 8.0 phosphate buffer, vortex for 10 s, add 10 mL toluene:ethyl acetate:isopropanol 49:50:1, shake mechanically at 60 strokes/min for 20 min, centrifuge for 5 min. Remove an 8 mL aliquot of the upper organic layer and evaporate it to dryness under a stream

of nitrogen at 70°, reconstitute the residue in 300 μL 5 μL/mL (S)-(+)-1-(1-naphthyl)ethyl isocyanate in dry dichloromethane, vortex for 2 min, let stand at room temperature overnight. Evaporate to dryness under a stream of nitrogen and reconstitute the residue in 300 μL 0.085% phosphoric acid, vortex for 2 min, add 2 mL hexane, vortex for 5 min, centrifuge. Remove a 250 μL aliquot of the lower aqueous layer and add it to 100 μL MeCN:MeOH 20:80, vortex, inject a 200 μL aliquot.

HPLC VARIABLES

Column: 250 × 4.6 5 μm C18 (Baker)
Mobile phase: MeCN:MeOH:0.085% phosphoric acid 8:34:58
Flow rate: 1
Injection volume: 200
Detector: UV 252

CHROMATOGRAM

Retention time: 58.2 (4S,6S), 55.4 (4R,6R), 65.2 (4S,6R), 74.4 (4R,6S)

KEY WORDS

chiral; whole blood

REFERENCE

Matuszewski, B.K.; Constanzer, M.L. Indirect chiral separation and analyses in human biological fluids of the stereoisomers of a thienothiopyran-2-sulfonamide (TRUSOPT), a novel carbonic anhydrase inhibitor with two chiral centers in the molecule, *Chirality*, **1992**, *4*, 515–519.

SAMPLE

Matrix: blood, urine
Analyte: oxprenolol
Sample preparation: Add propranolol to plasma or urine, make alkaline, extract with dichloromethane. Remove the organic layer and evaporate it to dryness, reconstitute the residue in 50 μL 0.02% S(+)-1-(1-naphthyl)ethylisocyanate in hexane:chloroform 50:50, vortex for 30 min, evaporate, reconstitute with 50 μL mobile phase, inject a 35 μL aliquot.

HPLC VARIABLES

Column: Lichrosorb DIOL
Mobile phase: Hexane:chloroform:MeOH 90:10:0.38
Injection volume: 35
Detector: F ex 226 em 340

CHROMATOGRAM

Internal standard: propranolol
Limit of quantitation: 10 ng/mL

KEY WORDS

plasma; chiral; pharmacokinetics

REFERENCE

Laethem, M.E.; Lefebvre, R.A.; Belpaire, F.M.; Vanhoe, H.L.; Bogaert, M.G. Stereoselective pharmacokinetics of oxprenolol and its glucuronides in humans, *Clin.Pharmacol.Ther.*, **1995**, *57*, 419–424.

SAMPLE

Matrix: microbial broth
Analyte: mexiletine
Sample preparation: Dilute microbial broth 1:20. 250 μL Diluted microbial broth + 5 μg IS + 100 μL 200 mM sodium carbonate, add 2 mL diethyl ether, vortex for 15 s, centrifuge at 1800 g for 4 min, remove 1.5 mL of the ether layer, repeat the extraction. Combine the organic layers and evaporate them to dryness under a stream of nitrogen, reconstitute the residue in 300 μL chloroform, add 75 μL 0.1% S-(+)-1-(1-naphthyl)ethyl isocyanate in chloroform, vortex for 10 s, evaporate to dryness under a stream of nitrogen. Reconstitute the residue in 220 μL chloroform, add 30 μL 0.33% n-butylamine in chloroform, inject a 10-65 μL aliquot.

HPLC VARIABLES

Column: 250 × 4 Partisil 5

Mobile phase: Hexane:chloroform:MeOH 65:34:1
Flow rate: 0.8 for 12 min then 2.5 for 23 min
Injection volume: 10-65
Detector: F ex 280 em 340

CHROMATOGRAM
Retention time: 7.0 (S-(+)), 7.6 (R-(-))
Internal standard: prenalterol ((±)-1-(4-hydroxyphenoxy)-3-isopropylaminopropan-2-ol) (27.7, 32.4 (enantiomers))
Limit of quantitation: 400 ng/mL

OTHER SUBSTANCES
Extracted: metabolites

KEY WORDS
chiral; normal phase

REFERENCE
Freitag, D.G.; Foster, R.T.; Coutts, R.T.; Pasutto, F.M. High-performance liquid chromatographic method for resolving the enantiomers of mexiletine and two major metabolites isolated from microbial fermentation media, *J.Chromatogr.*, **1993**, *616*, 253–259.

1-Naphthyl Isocyanate

SAMPLE
Matrix: blood
Analyte: ibutilide
Sample preparation: Basify 1 mL serum with 500 mM pH 10 carbonate buffer, extract with 1-chlorobutane. Extract the organic layer with 37.5 mM sulfuric acid. Basify the sulfuric acid layer (?) and extract it with 1-chlorobutane. Evaporate the organic layer to dryness, add 50 μL 130 ng/mL IS, add 50 μL 0.1% triethylamine in MeCN, evaporate to dryness, add 10 μL 0.1% acetic acid in MeCN, add 100 μL 220 μL/L naphthylisocyanate in MeCN, heat at 30° for 10 min, evaporate to dryness, reconstitute with 100 μL MeCN, use a column-switching technique during the first 6 min.

HPLC VARIABLES
Column: 100 × 4.6 3 μm 100 Å D-phenylglycine Pirkle column
Mobile phase: MeOH:triethylamine:trifluoroacetic acid 100:0.038:0.038
Flow rate: 1
Detector: F ex 224 em 340

CHROMATOGRAM
Retention time: 10.5 (+), 11.6 (-)
Internal standard: U-71175
Limit of quantitation: 0.5 ng/mL

KEY WORDS
chiral; serum

REFERENCE

Laganière, S.; Goernert, L.; Pereira, C. Stereoselective assay for ibutilide in human serum by derivatization and HPLC with fluorescence detection (Abstract APQ 1119), *Pharm.Res.*, **1996**, *13*, S32–S32.

SAMPLE

Matrix: solutions
Analyte: nadolol
Sample preparation: Prepare a 0.5-2 mg/mL solution in mobile phase, add a 5-fold molar excess of 1-naphthyl isocyanate, mix, let stand for 15 min, inject a 20 μL aliquot.

HPLC VARIABLES

Column: 250 × 4.6 5 μm CHI-1-LEU (R)-N-(3,5-dinitrobenzoyl)-L-leucine covalently bound to aminopropyl silica (Hichrom)
Mobile phase: n-Hexane:EtOH:MeCN 90:10:2
Flow rate: 2
Injection volume: 20
Detector: UV 239

CHROMATOGRAM

Retention time: 26 (SS), 28 (RS), 32 (SR), 36 (RR)

KEY WORDS

chiral

REFERENCE

Dyas, A.M.; Robinson, M.L.; Fell, A.F. Direct separation of nadolol enantiomers on a Pirkle-type chiral stationary phase, *J.Chromatogr.*, **1991**, *586*, 351–355.

SAMPLE

Matrix: solutions
Analyte: nadolol
Sample preparation: Prepare a 1 mg/mL solution in mobile phase, add 1 μL 1-naphthyl isocyanate for each mL of solution, let stand for 20 min, inject a 20 μL aliquot.

HPLC VARIABLES

Column: 150 × 4.6 3 μm aminopropylsilica (HPLC Technology) (Prepare column by passing through a solution of 5 g (R)-N-(3,5)-dinitrobenzoyl-L-leucine and 5 g 2-ethoxy-1-ethoxycarbonyl-1,2-dihydroquinoline in THF, wash with THF, wash with dichloromethane, wash with 100 mL dichloromethane containing 10 g trifluoroacetic anhydride, wash with dichloromethane, equilibrate with mobile phase (J. Chromatogr. 1991, 586, 351).)
Mobile phase: n-Hexane:isopropanol 80:20
Flow rate: 2
Injection volume: 20
Detector: UV 239

CHROMATOGRAM

Retention time: 26, 28, 31, 36 (enantiomers)

KEY WORDS

chiral

REFERENCE

Dyas, A.M.; Robinson, M.L.; Fell, A.F. Influence of the structure of the alcoholic modifier on the enantioselective separation of nadolol, *J.Chromatogr.A*, **1994**, *660*, 249–253.

NAPIC

SAMPLE
Matrix: solutions
Analyte: drugs
Sample preparation: Mix a 50 μL aliquot of a solution in MeOH:triethylamine 99:1 with 20 μL 0.1% NAPIC in dry toluene, vortex briefly, let stand at room temperature in the dark for 30 min, add 50 μL 1% ethanolamine in MeOH, let stand at room temperature for 15 min, evaporate to dryness under reduced pressure, reconstitute with 100 μL mobile phase, sonicate for 30 s, inject a 20 μL aliquot. (NAPIC is (-)-(S)-naproxen isocyanate; synthesis is as follows (protect from light). Dissolve 1 g (+)-(S)-naproxen in 30 mL acetone, cool to 0°, add a solution of 700 μL triethylamine in 2 mL acetone dropwise, add a solution of 450 μL ethyl chloroformate in 2 mL acetone dropwise, stir at 0° for 15 min, add a solution of 310 mg sodium azide in 1 mL water dropwise (Caution! Sodium azide is highly toxic!), stir for 1 h, pour into 60 mL ice water, stir for 10 min, filter, wash the solid with two 50 mL aliquots of ice-water, dry under reduced pressure to obtain naproxen azide. Dissolve 100 mg naproxen azide in 3 mL dry toluene, reflux for 10-15 min, cool to room temperature, filter. Evaporate the filtrate to dryness under reduced pressure and dry under reduced pressure to obtain NAPIC as an oil that crystallized in the desiccator (mp 48°), store in a desiccator under reduced pressure.)

HPLC VARIABLES
Column: 150 × 3.9 4 μm Nova Pak C18 (A-C) or 250 × 4.6 7 μm Nucleosil phenyl (D)
Mobile phase: MeOH:water 70:30 (A) or MeOH:water 65:35 (B) or MeOH:water:THF 74:25:1 (C) or MeOH:water:diethylamine 63:37:0.05 (D)
Flow rate: 1
Injection volume: 20
Detector: F ex 276 em 356

CHROMATOGRAM
Retention time: 18.6 (R-propranolol (A)), 20.3 (S-propranolol), 15.7 (R-metoprolol (B)), 17.1 (S-metoprolol (B)), 29.4 (S-prenylamine (C)), 30.8 (R-prenylamine (C)), 15.1 (S-cibenzoline (D)), 20.6 (R-cibenzoline (D))

KEY WORDS
chiral

REFERENCE
Martin, E.; Quinke, K.; Spahn, H.; Mutschler, E. (-)-(S)-Flunoxaprofen and (-)-(S)-naproxen isocyanate: two new fluorescent chiral derivatizing agents for an enantiospecific determination of primary and secondary amines, *Chirality*, **1989**, *1*, 223–234.

N-(p-Toluenesulfonyl)pyrrolidinyl Isocyanate

SAMPLE
Matrix: bulk
Analyte: drugs
Sample preparation: Reflux a 10% excess of reagent in toluene for 10 min, add the drug, let stand at room temperature for 10 min, cool, dilute, inject an aliquot. (The reagent was N-(p-toluenesulfonyl)prolyl azide and was prepared as follows. Mix 40-45 mmoles L-(-)-proline, 40 mL THF, and 200 mL 10% potassium carbonate, add 37-43 mmoles p-toluenesulfonyl chloride in 40 mL THF dropwise, heat at 50° and maintain at pH 8 or above for 3 h, cool, acidify to pH 2, extract with chloroform. Extract the organic layers with potassium carbonate in water. Acidify the aqueous layer and extract it with chloroform. Dry the chloroform layer and evaporate it to dryness, recrystallize the resulting 1-[(p-toluene)sulfonyl]proline from petroleum ether and benzene (Caution! Benzene is a carcinogen!) (Anal.Chem. 1984, 56, 958). Suspend 86 mmoles 1-[(p-toluene)sulfonyl]proline in 15 mL water and add sufficient acetone to give a clear solution, cool to 0°, add 10.2 g triethylamine in 175 mL acetone, slowly add 12.5 g ethyl chloroformate in 45 mL acetone while maintaining the temperature at 0°, stir at 0° for 30 min, add dropwise 8.6 g sodium azide in 30 mL water, stir at 0° for 1 h, pour into ice water, extract with ether, dry over anhydrous magnesium sulfate, evaporate under reduced pressure at room temperature to give N-(p-toluenesulfonyl)prolyl azide (cf J.Org.Chem. 1961, 26, 3511). Heating N-(p-toluenesulfonyl)prolyl azide in refluxing toluene produces N-(p-toluenesulfonyl) pyrrolidinyl isocyanate in situ.)

HPLC VARIABLES
Column: 300 × 4 7-9 μm silica gel
Mobile phase: Petroleum ether:isopropanol 97:3 (A) or 96.5:3.5 (B)
Flow rate: 1.5
Detector: UV 254

CHROMATOGRAM
Retention time: 13, 16 (amphetamine enantiomers (B)), 29.8, 44.3 (mexiletine enantiomers (A))

KEY WORDS
chiral; normal phase

REFERENCE
Zhou, Y.; Sun, Z.P.; Lin, D.K. Liquid chromatographic evaluation of a new chiral derivatizing agent for enantiomeric resolution of amine and alcohol drugs, *J.Liq.Chromatogr.*, **1990**, *13*, 875−885.

ISOTHIOCYANATE

Benzyl Isothiocyanate

SAMPLE
Matrix: bulk
Analyte: 2-amino-1-(p-nitrophenyl)-1,3-propanediol
Sample preparation: Dissolve 0.5 mg 2-amino-1-(p-nitrophenyl)-1,3-propanediol in 5 μL DMSO, add 50 μL 20 mg/mL benzyl isothiocyanate in MeCN, let stand at room temperature for 30 min, add 2 μL 2-aminoethanol, let stand for 10 min, add 50 μL 300 mM pH 3 ammonium phosphate buffer, inject a 2-5 μL aliquot.

HPLC VARIABLES
Column: 150 × 3.9 Nova-pak ODS
Mobile phase: MeOH:water 40:60
Flow rate: 1
Injection volume: 2-5
Detector: UV 254

CHROMATOGRAM
Retention time: 18 (1R,2R and 1S,2S), 20 (1R,2S and 1S,2R)

KEY WORDS
compound is a chloramphenicol precursor

REFERENCE
Gal, J.; Meyer-Lehnert, S. Reversed-phase liquid chromatographic separation of enantiomeric and diastereomeric bases related to chloramphenicol and thiamphenicol, *J.Pharm.Sci.*, **1988**, 77, 1062–1065.

SAMPLE
Matrix: bulk
Analyte: labetalol
Sample preparation: 1 mg Labetalol + 3 mg benzyl isothiocyanate + 100 μL MeCN + 50 μL water + 3 μL triethylamine, vortex, heat at 60° for 1 h. Remove a 150 μL aliquot and add it to 400 μL MeCN, vortex, inject a 5 μL aliquot.

HPLC VARIABLES
Column: 150 × 3.9 4 μm Nova-Pak C18
Mobile phase: MeOH:20 mM pH 4.60 $(NH_4)H_2PO_4$ 70:30
Flow rate: 1
Injection volume: 5
Detector: UV 254

CHROMATOGRAM
Retention time: 5.21 (RS,SR), 6.91 (SS,RR)

KEY WORDS
comparison with other derivatization reagents

REFERENCE
Desai, D.M.; Gal, J. Reversed-phase high-performance liquid chromatographic separation of the stereoisomers of labetalol via derivatization with chiral and non-chiral isothiocyanate reagents, *J.Chromatogr.*, **1992**, *579*, 165–171.

(4S-cis)-2,2-Dimethyl-5-isothiocyanato-4-phenyl-1,3-dioxane

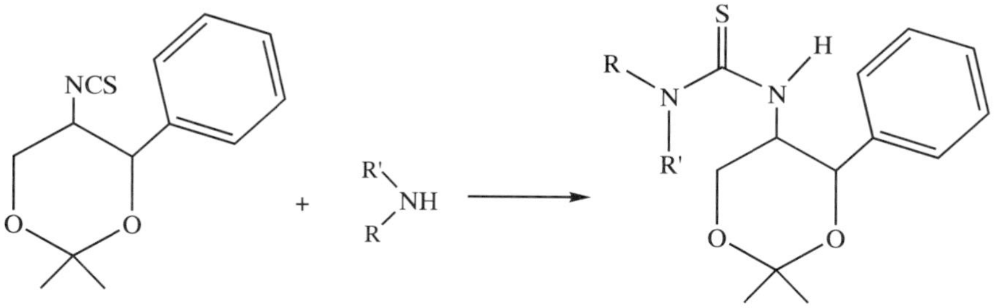

SAMPLE
Matrix: bulk
Analyte: labetalol
Sample preparation: 1 mg Labetalol + 3 mg reagent + 100 μL MeCN + 50 μL water + 3 μL triethylamine, vortex, heat at 60° for 1 h. Remove a 150 μL aliquot and add it to 400 μL MeCN, vortex, inject a 5 μL aliquot. (Prepare the reagent, (4S-cis)-2,2-dimethyl-5-isothiocyanato-4-phenyl-1,3-dioxane (PHEDIT), as follows. Add dropwise 5 g (4S,5S)-(+)-5-amino-2,2-dimethyl-4-phenyl-1,3-dioxane in 150 mL dichloromethane to 5 g 1,1'-thiocarbonyldiimidazole stirred in 150 mL dichloromethane, stir at room temperature for 3 h, wash the reaction mixture three times with 300 mL portions of 5% sodium bicarbonate, wash three times with 300 mL portions of water. Dry the organic layer over anhydrous sodium sulfate for 20 min, evaporate to dryness under reduced pressure, recrystallize from EtOH to give (4S-cis)-2,2-dimethyl-5-isothiocyanato-4-phenyl-1,3-dioxane as a pale yellow solid (mp 105-6°).)

HPLC VARIABLES
Column: 150 × 3.9 4 μm Nova-Pak C18
Mobile phase: MeOH:20 mM pH 4.60 $(NH_4)H_2PO_4$ 63:37
Flow rate: 1
Injection volume: 5
Detector: UV 254

CHROMATOGRAM
Retention time: 19.7 (RS or SR), 21.4 (RS or SR), 23.8 (SS), 30.0 (RR)

KEY WORDS
chiral; comparison with other derivatization reagents

REFERENCE
Desai, D.M.; Gal, J. Reversed-phase high-performance liquid chromatographic separation of the stereoisomers of labetalol via derivatization with chiral and non-chiral isothiocyanate reagents, *J.Chromatogr.*, **1992**, *579*, 165–171.

(R,R)-N-(3,5-Dinitrobenzoyl)-2-aminocyclohexyl Isothiocyanate

RELATED REFERENCE
Kleidernigg, O.P.; Lidner, W. Indirect resolution of chiral thiols by reversed-phase HPLC using (R,R)-dinitrobenzoyldiaminocyclohexyl isothiocyanate as chiral derivatizing agent (CDA). *GIT Spez.Chromatogr.* **1995**, *15*, 42-44.

SAMPLE
Matrix: bulk
Analyte: amines
Sample preparation: Dissolve 10 μmole compound (as free base or hydrochloride) in 500 μL MeCN, add 250 μL 5% sodium carbonate (for hydrochlorides only), add 500 μL 100 mM reagent in MeCN, vortex for 1 min, heat at 60° for 2 h, add 100 μmole L-proline, heat at 60° for 30 min. Remove a 100 μL aliquot and dilute it with mobile phase, neutralize with acetic acid, inject a 10 μL aliquot. Prepare the reagent ((R,R)-N-(3,5-dinitrobenzoyl)-2-aminocyclohexyl isothiocyanate) as follows. Add 0.7 mL carbon disulfide to 6 mL (1R,2R)-(-)-1,2-diaminocyclohexane, 12 mL water, and 12 mL EtOH, heat the oil bath to 80°, add 2.8 mL carbon disulfide dropwise (making sure that the product does not start to precipitate), when addition is complete reflux for 1 h, acidify with 500 μL 5 M HCl, reflux for 12 h, cool, filter, wash the solid with a little cold EtOH to give trans-4,5-tetramethyleneimidazolidine-2-thione as a white fluffy solid (mp 148-150°) (Tetrahedron 1993, 49, 4419). Stir 7.97 g 3,5-dinitrobenzoyl chloride in 30 mL dichloroethane at 50°, add a solution of 6 g trans-4,5-tetramethyleneimidazolidine-2-thione in 120 mL dichloroethane containing a catalytic amount of 4-(dimethylamino)pyridine over 15 min, reflux for 2 h, remove the crystals of (R,R)-N-(3,5-dinitrobenzoyl)-2-aminocyclohexylisothiocyanate by filtration, evaporate the filtrate to dryness and dissolve the residue in 60 mL dichloroethane, reflux for 16 h to obtain more (R,R)-N-(3,5-dinitrobenzoyl)-2-aminocyclohexyl isothiocyanate (mp >250°, $[\alpha]_{546}$ = -133° (c = 1) in MeCN).

HPLC VARIABLES
Column: 125 × 4 5 μm Lichrospher 60 RP Select B
Mobile phase: MeCN:20 mM ammonium acetate 55:45
Flow rate: 1
Injection volume: 10
Detector: UV 254

CHROMATOGRAM
Retention time: k' 3.10, k' 4.20 (acebutolol enantiomers), k' 9.56, k' 15.44 (alprenolol enantiomers), k' 11.86, k' 12.48 (amphetamine enantiomers (MeCN:20 mM ammonium acetate 45:55)), k' 1.63, k' 2.24 (atenolol enantiomers), k' 5.93, k' 8.59 (carazolol enantiomers), k' 6.11, k' 7.35 (carvedilol enantiomers), k' 2.52, k' 3.69 (formoterol enantiomers), k' 14.33, k' 15.90 (methamphetamine enantiomers), k' 7.62, k' 11.89 (metipranolol enantiomers), k' 5.33, k' 7.83 (metoprolol enantiomers), k' 4.43, k' 6.61 (nifenanol enantiomers), k' 3.73, k' 5.21 (nitrilo atenolol enantiomers), k' 4.47, k' 4.69 (normetoprolol (MeCN:20 mM ammonium acetate 45:55) enantiomers), k' 7.15, k' 10.81(oxprenolol enantiomers), k' 3.91, k' 5.59 (pindolol enantiomers), k' 5.07, k' 5.98 (propafenone (125 × 4 5 μm Hypersil ODS/MeCN:20 mM ammonium acetate 70:30) enantiomers), k' 8.48, k' 12.17 (propranolol enantiomers), k' 1.13, k' 2.27 (xamoterol enantiomers)

KEY WORDS
chiral

REFERENCE
Kleidernigg, O.P.; Posch, K.; Lindner, W. Synthesis and application of a new isothiocyanate as a chiral derivatizing agent for the indirect resolution of chiral amino alcohols and amines, *J.Chromatogr.A*, **1996**, *729*, 33–42.

Eu^{3+} Chelate of N^1-(p-Isothiocyanatobenzyl) diethylenetriaminetetraacetic Acid

SAMPLE
Matrix: solutions
Analyte: amines
Sample preparation: Mix a 100 µL aliquot of a 0.1-1 µM solution of amines with 50 µL 100-500 µM reagent and 50 µL 100 mM pH 9.8 carbonate buffer, heat at 40° for 8 h, inject a 20 µL aliquot. (The reagent was the Eu^{3+} chelate of N^1-(p-isothiocyanatobenzyl)diethylenetriaminetetraacetic acid; synthesis is as follows. Dissolve 100 g diethylenetriamine in toluene, add 48 g 4-nitrobenzyl bromide, stir at room temperature for 5 h, filter, extract the filtrate with water. Make the aqueous base alkaline and saturate it with NaCl, extract with chloroform, evaporate to dryness to give a mixture of N-(p-nitrobenzyl)diethylenetriamines. Dissolve this mixture in toluene, stir at 0°, slowly add 55 g salicylaldehyde (2-hydroxybenzaldehyde), reflux under a Dean and Stark trap to remove water, evaporate to remove the toluene, dissolve the residue in hot EtOH, allow to cool slowly to room temperature with vigorous stirring, filter. Dissolve the solid in warm EtOH, add HCl, add water, wash with diethyl ether. Make the aqueous layer alkaline and saturate it with NaCl, extract with chloroform, evaporate to dryness to obtain N^1-(p-nitrobenzyl)diethylenetriamine. Dissolve 53 g N^1-(p-nitrobenzyl)diethylenetriamine in water, adjust pH to 9-11, slowly add a solution of 160 g bromoacetic acid in water while adding 7 M KOH to maintain the pH at 9-11, stir at pH 9-11 for 24 h, acidify, concentrate by evaporation, add acetone to precipitate most of the salts, evaporate the precipitate to dryness to obtain crude N^1-(p-nitrobenzyl)diethylenetriaminetetraacetic acid. Dissolve N^1-(p-nitrobenzyl)diethylenetriaminetetraacetic acid in water, adjust pH to 5-6, add an equimolar amount of Eu^{3+} in water, maintain pH at 5-6, stir at room temperature for 30 min,

raise the pH to 9, filter, concentrate the filtrate by evaporation, purify by preparative HPLC using water as eluent to obtain the Eu^{3+} chelate of N^1-(p-nitrobenzyl)diethylenetriaminetetraacetic acid. Dissolve 2 g of the chelate in water, add 200 mg 5% palladium on activated charcoal, hydrogenate at 0-5° under 1.3 MPa hydrogen, monitor the reaction by UV spectroscopy. When the reaction is complete filter and evaporate the filtrate to dryness to obtain the Eu^{3+} chelate of N^1-(p-aminobenzyl) diethylenetriaminetetraacetic acid. Add a solution of 2.3 g of the Eu^{3+} chelate of N^1-(p-aminobenzyl)diethylenetriaminetetraacetic acid in water to a mixture of 1.7 g thiophosgene, 1 g sodium bicarbonate, and 15 mL chloroform, stir vigorously for 30 min. Wash the aqueous phase twice with chloroform and evaporate it to dryness, wash the product with EtOH to obtain the Eu^{3+} chelate of N^1-(p-isothiocyanatobenzyl)diethylenetriaminetetraacetic acid (Anal. Biochem. 1989, 176, 319).)

HPLC VARIABLES
Column: 250 × 4.6 5 μm Capcellpak SG-120 C18 (Shiseido)
Mobile phase: MeCN:20 mM pH 7.0 phosphate buffer 23:77
Flow rate: 1
Injection volume: 20
Detector: F ex 344 em 617 following post-column reaction. The column effluent mixed with the reagent pumped at 0.2 mL/min and the mixture flowed through a 2 m coil to the detector. (Prepare reagent by dissolving 15 g Triton X-100 in 500 mL water, add 58 mL acetic acid, add 20 mL 50 mM tri-n-octylphosphine oxide in MeCN, add 1 mL 100 mM 2-naphthoyltrifluoroacetone in MeCN, make up to 1 L with water. 2-Naphthoyltrifluoroacetone (4,4,4-trifluoro-1-(2-naphthyl)-1,3-butanedione) is available from Tokyo Kasei (TCI America, Portland OR). Prepare by adding 1 mole ethyl trifluoroacetate dropwise with stirring to 1.05 moles sodium methoxide in 100 mL anhydrous ether, add 1 mole 2'-acetonaphthone, let stand overnight, evaporate to dryness under reduced pressure, heat at 90° under reduced pressure to remove EtOH, add 1 mole 10% sulfuric acid, recrystallize from EtOH/water to obtain 2-naphthoyltrifluoroacetone (mp 70.1-71.1°) (J. Am. Chem. Soc. 1950, 72, 2948).); UV 254

CHROMATOGRAM
Retention time: 6 (phenethylamine), 10 (phenylpropylamine), 16 (phenylbutylamine)

REFERENCE
Okabayashi, Y.; Kitagawa, T. High-performance liquid chromatography for amino compounds and thiol compounds derivatized with europium chelate, *Anal.Chem.*, **1994**, *66*, 1448–1453.

Eu^{3+} Chelate of 1-(4-Isothiocyanobenzyl)ethylenediamine-N,N,N'N'-tetraacetic Acid

SAMPLE
Matrix: solutions
Analyte: 3-phenyl-1-propylamine
Sample preparation: Mix 10 μL of a 1-1000 μM solution with 45 μL reagent and 45 μL 100 mM pH 9.8 sodium carbonate buffer, heat at 40° for 8 h, inject an aliquot. (Prepare the reagent, Eu^{3+} chelate of 1-(4-isothiocyanobenzyl)ethylenediamine-N,N,N'N'-tetraacetic acid, by adding 1 mg isothiocyanobenzyl-EDTA (Dojindo Molecular Technologies,

Inc., 3 Bethesda Metro Center, Suite 700, Bethesda MD 20814; (301) 664-8448; www.dojindo.co.jp) to 5 mL 500 μM europium acetate solution.)

HPLC VARIABLES
Column: 50 × 0.15 Crest C18T-5
Mobile phase: MeCN:20 mM pH 7.0 sodium phosphate buffer 20:80
Flow rate: 0.1
Injection volume: 10
Detector: F ex 290 em 615 following post-column reaction. The column effluent mixed with reagent pumped at 0.02 mL/min and the mixture flowed through a 4 m × 0.25 mm ID tefzel coil at 40° to the detector. (Prepare reagent as follows. Mix 200 mL 5% Triton X-100, 100 mL 2 M pH 6.5 sodium acetate buffer, 10 mL 50 mM tri-n-octylphosphine oxide, and 32 mL 2.5 mM thienoyltrifluoroacetone (Dojindo Molecular Technologies, Inc., 3 Bethesda Metro Center, Suite 700, Bethesda MD 20814; (301) 664-8448; www.dojindo.co.jp), make up to 1 L with water. The construction of the detector is discussed.)

CHROMATOGRAM
Retention time: 8
Limit of quantitation: 100 nM

KEY WORDS
microbore

REFERENCE
Iwata, T.; Senda, M.; Kurosu, Y.; Tsuji, A.; Maeda, M. Construction of time-resolved fluorescence detector for amino compounds after high-performance liquid chromatography using europium chelate, *Anal.Chem.*, **1997**, *69*, 1861–1865.

Ferrocenyl Isothiocyanate

SAMPLE
Matrix: tissue
Analyte: 4-aminobutyric acid
Sample preparation: Homogenize 330 mg brain tissue with 4.8 mL ice-cold 10% perchloric acid. Remove a 500 μL aliquot of the homogenate and add it to 500 μL 10 μg/mL 5-aminovaleric acid in 10% perchloric acid, centrifuge at 1500 g for 15 min. Remove a 100 μL aliquot of the supernatant, adjust pH to 8.0 with triethylamine, add 100 μL 900 μg/mL (?) reagent in MeCN, heat at 70° for 30 min, cool, wash with ether, inject an aliquot of the aqueous layer. (Prepare reagent, ferrocenyl isothiocyanate, as follows. Stir 13 g ferrocene in 200 mL anhydrous THF at -30° under nitrogen, add 210 mmoles butyllithium in 160 mL ether dropwise over 25 min, stir at 0° for 2 h, and room temperature for 4 h, add 10.3 g methoxylamine in 75 mL anhydrous ether dropwise at -20° over 30 min, allow to warm gradually to room temperature, stir for 4 h, slowly add 10% HCl until pH of aqueous layer is 2, remove the ether layer and discard it. Make the aqueous layer strongly basic with KOH solution, extract with ether. Extract the ether layers with 2 M HCl. Make the aqueous layer strongly basic with KOH solution and extract it with ether. Dry the extracts over anhydrous magnesium sulfate and evaporate to dryness, recrystallize from ether/petroleum ether to give ferrocenylamine (mp 140-5°) (J. Org. Chem. 1959, 24, 1487).) Suspend 300 mg ferrocenylamine in 1.5 mL carbon disulfide:water 40:60, cool to 0°, stir, add 450 μL 83% KOH, heat at 100° for 25 min in a sealed tube, cool in ice, add 480 μL

ethyl chloroformate, stir at room temperature for 1 h, add 450 μL 67% KOH, stir for 1 h, extract with ether. Wash the organic layer with water and dry it over anhydrous sodium sulfate, evaporate to dryness, chromatograph on a 100 × 11 column of silica gel 60 (Merck) with benzene (Caution! Benzene is a carcinogen!) to give ferrocenyl isothiocyanate as a dark red oil.)

HPLC VARIABLES
Column: 150 × 4 5 μm TSKgel ODS-80TM
Mobile phase: MeCN:0.5% pH 4.0 sodium acetate 40:60
Flow rate: 1
Detector: E, EICOM ECD-100, glassy carbon electrode +0.50 V, Ag/AgCl reference electrode

CHROMATOGRAM
Retention time: 6.2
Internal standard: 5-aminovaleric acid (8)
Limit of detection: 0.05 pmole

KEY WORDS
rat; guinea pig; brain

REFERENCE
Shimada, K.; Kawai, Y.; Oe, T.; Nambara, T. Determination of amino acids by high-performance liquid chromatography with electrochemical detection using ferrocene derivatization reagents, *J.Liq.Chromatogr.*, **1989**, *12*, 359–371.

Fluorescein Isothiocyanate, Isomer I

RELATED REFERENCES
Reynolds, D.L.; Pachla, L.A. Analysis of 3-(2-(ethylamino)propyl)-1,2,3,4-tetrahydro-5H(1)benzo-pyrano(3,4-c)pyridin-5-one in plasma by liquid chromatographic column switching after derivatizing the secondary amine with fluorescein-6-isothiocyanate. *J.Pharm.Sci.* **1985**, *74*, 1091-1094.

Rodriguez, I.; Lee, H.K.; Li, S.F.Y. Separation of biogenic amines by micellar electrokinetic chromatography. *J.Chromatogr.A* **1996**, *745*, 255-262.

SAMPLE
Matrix: blood
Analyte: fumonisin
Sample preparation: Dilute 1.5 mL serum to 10 mL with water, add to a FumoniTest SPE affinity column (Vicam, Watertown MA) at 1.5 mL/min, wash with 5 mL 10 mM pH 7.5 phosphate buffer, wash with 5 mL water, elute with 1.6 mL MeOH. Evaporate the eluate to dryness under a stream of nitrogen at 60°, reconstitute with 120 μL DMSO:50 mM pH 9.5 borate buffer 90:10, add 30 μL 1.3 mM fluorescein 5-isothiocyanate (isomer I) in acetone, let stand at room temperature for 3 h, add 4.425 mL MeOH:50 mM pH 9.5 borate buffer 60:40, mix, inject an aliquot.

CAPILLARY ELECTROPHORESIS
Capillary: 50 cm $\times$ 75 μm
Capillary preparation: Between runs rinse capillary with 100 mM NaOH for 2 min, with water for 2 min, and with running buffer for 5 min.
Running buffer: MeOH:60 mM pH 9.5 borate buffer 7:93
Capillary temperature: 35
Voltage/Current: 17-30 kV/250 μA (fixed)
Injection: Hydrodynamic injection at 0.5 psi for 5 s (30 nL)
Detector: F ex 488 (argon ion laser)
Model: Beckman P/ACE 5000
Migration time: 8 (hydrolyzed fumonisin B_1), 20 (fumonisin B_1), 22 (fumonisin B_2)
Limit of detection: 25 ng/mL

KEY WORDS
SPE; horse; serum

REFERENCE
Maragos, C.M. Capillary zone electrophoresis and hplc for the analysis of fluorescein isothiocyanate-labeled fumonisin B1, *J.Agric.Food Chem.*, **1995**, *43*, 390–394.

SAMPLE
Matrix: cheese
Analyte: amines
Sample preparation: Homogenize 4 parts cheese with 3 parts 100 mM HCl, suspend 230 mg paste in 5 mL 100 mM HCl, centrifuge, extract the residue twice with 5 mL portions of 100 mM HCl. Filter the supernatants and neutralize them with 200 mM sodium carbonate solution. Remove a 1 mL aliquot and add it to 1 mL 83 μg/mL fluorescein isothiocyanate in acetone, let stand in the dark for 4 h (J.Chromatogr. 1991, 559, 183), dilute 100 000-fold, inject an aliquot.

CAPILLARY ELECTROPHORESIS
Capillary: 80 cm $\times$ 50 μm fused-silica (50 cm to detector) (Polymicro Technologies)
Capillary preparation: Rinse with 100 mM NaOH for 3 min, with water for 3 min, and with running buffer for 3 min.
Running buffer: 100 mM Boric acid containing 100 mM sodium dodecyl sulfate, pH adjusted to 9.2 with NaOH
Voltage/Current: 24 kV
Injection: Hydrodynamic injection for 2 s (17.5 nL)
Detector: F ex 488 (7 mW argon ion laser)
Model: TSP Spectra-Phoresis 100
Migration time: 9.65 (histamine), 11.3 (ornithine), 11.9 (ammonia), 12.6 (putrescine), 13.48 (tryptamine), 14 (β-phenylethylamine), 14.6 (cadaverine)
Limit of detection: 0.05-0.15 nM

OTHER SUBSTANCES
Extracted: amino acids

REFERENCE
Nouadje, G.; Nertz, M.; Verdeguer, P.; Couderc, F. Ball-lens laser-induced fluorescence detector as an easy-to-use highly sensitive detector for capillary electrophoresis. Application to the identification of biogenic amines in dairy products, *J.Chromatogr.A*, **1995**, *717*, 335–343.

SAMPLE
Matrix: dialysate
Analyte: amphetamine
Sample preparation: Mix 30 μL dialysate (artificial CSF) with 75 μL buffer and 15 μL 100 μM fluorescein isothiocyanate (isomer I) in acetone, let stand in the dark for 18 h, inject an aliquot.

CAPILLARY ELECTROPHORESIS
Capillary: 30 cm $\times$ 20-25 μm fused silica (20 cm to detector) (Polymicro Technologies)
Running buffer: 20 mM Carbonate
Voltage/Current: 20 kV

Injection: Vacuum injection at 19 psi for 0.3 s (0.3 nL)
Detector: F ex 488 (Ar laser) em 520 high-pass filter (with 488 nm notch filter))
Migration time: 3
Limit of detection: 3 nM

KEY WORDS
rat; brain

REFERENCE
Páez, X.; Rada, P.; Tucci, S.; Rodriguez, N.; Hernández, L. Capillary electrophoresis-laser-induced fluorescence detection of amphetamine in the brain, *J.Chromatogr.A*, **1996**, *735*, 263–269.

SAMPLE
Matrix: solutions
Analyte: formoterol
Sample preparation: Add 15 mL 8.57 mg/mL fluorescein isothiocyanate in EtOH dropwise to 103 mg formoterol and 500 μL triethylamine in 15 mL EtOH, stir at 40° for 4 h, evaporate, purify by TLC on silica gel using chloroform:ethyl acetate:MeOH 25:37.5:37.5, inject an aliquot of a 0.1 ng/mL solution.

CAPILLARY ELECTROPHORESIS
Capillary: 57 cm × 75 μm fused-silica (50 cm to detector) (Supelco)
Capillary preparation: Before analysis rinse capillary with 100 mM NaOH for 2 min, fill with running buffer.
Running buffer: MeOH:67 mM pH 8 phosphate buffer 10:90 containing 20 mM heptakis(2,3,6-tri-O-methyl-β-cyclodextrin) and 40 mM octanesulfonic acid
Capillary temperature: 25
Injection: Pressure injection at 0.3 psi for 5 s (5 nL)
Detector: F ex 488 (4 mW argon-ion laser) em 520 (band-pass filter)
Model: Beckman PACE 2100
Migration time: 27, 28 (enantiomers)
Limit of detection: 0.01 ng/mL

KEY WORDS
chiral

REFERENCE
Cherkaoui, S.; Faupel, M.; Francotte, E. Separation of formoterol enantiomers and detection of zeptomolar amounts by capillary electrophoresis using laser-induced fluorescence, *J.Chromatogr.A*, **1995**, *715*, 159–165.

SAMPLE
Matrix: urine
Analyte: amphetamine and methamphetamine
Sample preparation: 2 mL Urine + 20 μL 100 μM 1-phenylethylamine in water + 400 μL concentrated HCl, heat at 80° for 1 h, cool, neutralize with 600 μL 25% ammonia, add 5 mL 10% sodium carbonate solution, add 2 mL 500 mM pH 10.5 sodium borate buffer, add 2 mL chloroform:isopropanol 75:25, vortex for 1 min, centrifuge at 12.5° at 1500 g for 10 min, repeat the extraction. Combine the organic layers and remove a 100 μL aliquot, add 10 μL acetic acid to the aliquot, evaporate it to dryness under a stream of nitrogen at room temperature, reconstitute the residue in 50 μL carbonate buffer, add 50 μL 10 mM fluorescein isothiocyanate, isomer I (fluorescein-4-isothiocyanate) in EtOH, mix, heat in the dark at 80° for 15 min, inject a 20 μL aliquot. (Prepare carbonate buffer by adjusting the pH of 200 mM sodium bicarbonate to 9.0 with 200 mM sodium carbonate.)

HPLC VARIABLES
Column: 250 × 4.6 5 μm Daisopak SP-120-5-ODS (Daiso, Osaka)
Mobile phase: Gradient. MeCN:20 mM pH 7.9 sodium phosphate buffer 20:80 for 16 min then 24:76 (step-gradient).
Flow rate: 0.8
Injection volume: 20
Detector: F ex 496 em 518

CHROMATOGRAM
Retention time: 32 (amphetamine), 35.2 (methamphetamine)
Internal standard: 1-phenylethylamine (26.6)
Limit of detection: 5.5 nM

OTHER SUBSTANCES
Extracted: metabolites, norepinephrine

REFERENCE
Al-Dirbashi, O.; Kuroda, N.; Akiyama, S.; Nakashima, K. High-performance liquid chromatography of methamphetamine and its related compounds in human urine following derivatization with fluorescein isothiocyanate, *J.Chromatogr.B*, **1997**, *695*, 251−258.

SAMPLE
Matrix: water
Analyte: amines
Sample preparation: Mix 0.1-1 mL of a 30 ppb-10 ppm solution in water with 100 μL reagent, stir in the dark at room temperature for 16-48 h, make up to 10 mL with water. Mix a 1 mL aliquot with 1 mL 1 μM fluorescein, 1 mL MeCN, 1 mL 40 mM pH 7.0 phosphate buffer, and 1 mL water, filter 0.2 μm, inject an aliquot. (Reagent was 7-10 mg fluorescein isothiocyanate, isomer I and 300 mg sodium carbonate in 10 mL water.)

CAPILLARY ELECTROPHORESIS
Capillary: 57 cm × 75 μm (50 cm to detector)
Capillary preparation: Between runs wash with 100 mM NaOH then equilibrate with running buffer.
Running buffer: MeOH:40 mM pH 7.0 phosphate buffer 25:75
Capillary temperature: 25
Voltage/Current: 20-25 kV
Injection: Pressure injection for 1 s (5 nL).
Detector: F ex 488 (argon ion laser)
Model: Beckman P/ACE 2050
Migration time: 8.6 (1,5-pentanediamine), 12.5 (2-(2-aminoethyl)pyridine), 13.2 (diethylamine), 13.8 (propylamine)
Internal standard: fluorescein (17.5)
Limit of detection: 30 ppb

KEY WORDS
ground water

REFERENCE
Brumley, W.C.; Kelliher, V. Determination of aliphatic amines in water using derivatization with fluorescein isothiocyanate and capillary electrophoresis/laser-induced fluorescence detection, *J.Liq.Chromatogr.Rel.Technol.*, **1997**, *20*, 2193−2205.

6-Isothiocyanatobenzo[g]phthalazine-1,4(2H,3H)-dione

SAMPLE

Matrix: blood

Analyte: maprotiline

Sample preparation: 100 μL Plasma + 100 μL 100 ng/mL desipramine in 2-methoxy-ethanol containing 75 mM triethylamine + 500 μL 6 M NaOH + 3 mL n-hexane:isoamyl alcohol 95:5, vortex for 5 min, centrifuge at 1000 g for 5 min. Remove a 2 mL aliquot of the organic layer and evaporate it to dryness under a stream of nitrogen, reconstitute the residue in 100 μL 75 mM triethylamine in 2-methoxyethanol, add 100 μL 1 mM 6-iso-thiocyanatobenzo[g]phthalazine-1,4(2H,3H)-dione in DMSO, add 10 μL 75 mM triethyl-amine in 2-methoxyethanol, heat at 80° for 10 min, inject a 20 μL aliquot. (Prepare 6-isothiocyanatobenzo[g]phthalazine-1,4(2H,3H)-dione as follows. Dissolve 2.2 g 2,3-naphthalenedicarboxylic acid in 6.6 mL MeOH:concentrated sulfuric acid 10:1, reflux for 3 h, pour into ice-water, filter. Dissolve the precipitate in 20 mL ethyl acetate and purify on a 250 × 35 column containing 120 g 70-230 mesh silica gel 60 (Merck) using n-hexane:ethyl acetate 50:50 to give dimethyl 2,3-naphthalenedicarboxylate as a brownish crystalline solid (mp 46-49°). Stir 8 mL concentrated nitric acid:concentrated sulfuric acid 3:5 at 4°, add 2.5 g dimethyl 2,3-naphthalenedicarboxylate, stir at 4° for 4-6 h until the reaction is complete. (Check by TLC using Merck Kieselgel 60 F254 with n-hexane:ethyl acetate 50:50.) Pour the reaction mixture into ice-water and extract with 350 mL di-chloromethane. Wash the organic layer with water, with 5% sodium bicarbonate, and with water. Evaporate to dryness under reduced pressure, recrystallize from EtOH to give dimethyl 5-nitro-2,3-naphthalenedicarboxylate as yellow needles (mp 146-148°). Dissolve 1.25 g dimethyl 5-nitro-2,3-naphthalenedicarboxylate in 60 mL MeOH, add 500 mg 5% platinum on activated carbon, hydrogenate, filter, evaporate to dryness under reduced pressure, recrystallize from n-hexane/ethyl acetate to give dimethyl 5-amino-2,3-naph-thalenedicarboxylate as pale yellow-green crystals (mp 129-131°). Add 2.7 mL triethyl-amine and 2.7 mL hydrazine hydrate (Caution! Hydrazine hydrate is a carcinogen!) to 1.2 g dimethyl 5-amino-2,3-naphthalenedicarboxylate dissolved in 18 mL MeOH, reflux for 3 h, evaporate to remove the solvent, rinse the residue with EtOH, dry to give 6-aminobenzo[g]phthalazine-1,4(2H,3H)-dione as a yellow powder (mp 280° d). Add 1 mL thiophosgene dropwise to a solution of 1 g 6-aminobenzo[g]phthalazine-1,4(2H,3H)-dione in 20 mL water stirred at room temperature, stir for 1 h, filter, wash the precipitate with MeCN, dry to give 6-isothiocyanatobenzo[g]phthalazine-1,4(2H,3H)-dione as a pale brownish-yellow powder (mp 360° d) (Anal. Chim. Acta 1995, 302, 61).)

HPLC VARIABLES

Column: 150 × 4.6 5 μm TSKgel ODS-80 (Tosoh)

Mobile phase: MeCN:100 mM pH 3.2 acetate buffer 40:60

Flow rate: 1

Injection volume: 20

Detector: Chemiluminescence following post-column reaction. The column effluent mixed with 20 mM hydrogen peroxide in water pumped at 1 mL/min and with 40 mM potassium ferricyanide in 1.5 M NaOH pumped at 2 mL/min and the mixture flowed to the detector.

CHROMATOGRAM

Retention time: 59.5

Internal standard: desipramine (39.5)
Limit of detection: 0.1 ng/mL

OTHER SUBSTANCES

Simultaneously analyzed: amoxapine, metoprolol, nortriptyline, propranolol
Non-interfering: betamethasone, diazepam, phenobarbital, phenytoin, prednisolone, triazolam

KEY WORDS

plasma; pharmacokinetics

REFERENCE

Ishida, J.; Horike, N.; Yamaguchi, M. Determination of maprotiline in plasma by high-performance liquid chromatography with chemiluminescence detection, *J.Chromatogr.B*, **1995**, *669*, 390–396.

SAMPLE

Matrix: solutions
Analyte: amines
Sample preparation: Mix a 100 μL aliquot a solution of the amine in 2-methoxyethanol: Tween 80 96:4 with 100 μL 4 mM 6-isothiocyanatobenzo[g]phthalazine-1,4(2H,3H)-dione in DMSO, add 10 μL 750 mM triethylamine in 2-methoxyethanol, heat at 80° for 10 min, inject a 20 μL aliquot. (Prepare 6-isothiocyanatobenzo[g]phthalazine-1,4(2H,3H)-dione as described above.)

HPLC VARIABLES

Column: 150 × 4.6 5 μm TSKgel ODS-120T (Tosoh)
Mobile phase: MeCN:10 mM ammonium acetate 40:60
Flow rate: 1
Injection volume: 20
Detector: Chemiluminescence following post-column reaction. The column effluent mixed with 100 mM hydrogen peroxide in water pumped at 1 mL/min and with 40 mM potassium ferricyanide in 2 M NaOH pumped at 2 mL/min and the mixture flowed to the detector.

CHROMATOGRAM

Retention time: 4.6 (n-hexylamine), 6.4 (n-heptylamine), 7.4 (di-n-butylamine), 9.2 (dibenzylamine), 9.8 (n-octylamine), 10.5 (N-methyloctylamine), 11.7 (n-nonylamine), 12.4 (di-n-amylamine), 58.0 (di-n-hexylamine)
Limit of detection: 0.8-120 fmole

OTHER SUBSTANCES

Non-interfering: acetone, acetophenone, N-acetylneuraminic acid, 4-anisaldehyde, ascorbic acid, benzaldehyde, benzyl alcohol, caproic acid, cholesterol, cortisone, cyclohexanol, decanal, dehydroepiandrosterone, dextrose, epiandrosterone, fructose, glutathione, inositol, α-ketoisocaproic acid, α-ketoisovaleric acid, α-ketovaleric acid, lactic acid, lactose, maltose, mannose, methylglyoxal, mylistic acid, nonanol, palmitic acid, phenylpyruvic acid, pyruvic acid, thiamine, triethylamine, triphenylamine, urea, uric acid, xylose

REFERENCE

Ishida, J.; Horike, N.; Yamaguchi, M. 6-Isothiocyanatobenzo[*g*]phthalazine-1,4-(2*H*,3*H*)-dione as a highly sensitive chemiluminescence derivatization reagent for amines in liquid chromatography, *Anal.Chim.Acta*, **1995**, *302*, 61–67.

(R)-(-)-4-(3-Isothiocyanatopyrrolidin-1-yl)-7-(N,N-dimethylaminosulfonyl)-2,1,3-benzoxadiazole

SAMPLE

Matrix: blood, saliva

Analyte: propranolol

Sample preparation: 50 μL Plasma or saliva + 350 μL 1 μM oxprenolol hydrochloride in water + 100 μL 4 M NaOH, sonicate for 10 min, extract with 3 mL ethyl acetate, centrifuge at 3000 rpm for 3 min. Remove the organic layer and wash it with saturated NaCl solution, dry over anhydrous sodium sulfate, evaporate to dryness under reduced pressure, reconstitute the residue in 100 μL MeCN:water:triethylamine 50:50:0.1, add 100 μL 1 mM (R)-(-)-4-(3-isothiocyanatopyrrolidin-1-yl)-7-(N,N-dimethylaminosulfonyl)-2,1,3-benzoxadiazole in MeCN, heat in the dark at 65° for 1.5 h, inject a 50 μL aliquot. ((R)-(-)-4-(3-Isothiocyanatopyrrolidin-1-yl)-7-(N,N-dimethylaminosulfonyl)-2,1,3-benzoxadiazole ((R)-DBD-Py-NCS) is available from Tokyo Kasei (TCI America, Portland OR). Synthesis is as follows. Dissolve 0.5 g magnesium sulfate heptahydrate and 6 g NaOH in 60 mL water, throughout the reaction keep the flask at about 20° with cold water cooling, add 15 mL 30% hydrogen peroxide, add 75 mL MeOH, add 12.1 g powdered benzoyl peroxide in one go, stir for 10 min, pour into 150 mL 20% sulfuric acid, extract three times with 50 mL portions of chloroform, determine peroxybenzoic acid concentration by iodometric titration (Tetrahedron 1967, 23, 3327). Slowly add 110 mL 1 M peroxybenzoic acid in chloroform to 7 g 2,6-difluoroaniline dissolved in 100 mL chloroform, stir at room temperature, when reaction is complete (iodometric titration) wash with 2% sodium thiosulfate, wash with 5% sodium carbonate, wash with water, dry over anhydrous sodium sulfate, evaporate to dryness under reduced pressure, recrystallize 2,6-difluoronitroso-benzene from EtOH (mp 108.5-109.5). Stir 8.5 g 2,6-difluoronitrosobenzene in 85 mL DMSO at room temperature and add a solution of 3.91 g sodium azide in 85 mL DMSO dropwise, let stand for about 1 h, add to a large volume of water, extract with ether, dry the extracts over anhydrous sodium sulfate, evaporate to dryness under reduced pressure and distil to give 4-fluoro-2,1,3-benzoxadiazole as a colorless oil (bp 83°/12 mm Hg) (J.Chem.Soc.(C) 1970, 1433). Add 11 mL chlorosulfonic acid dropwise to 3 g 4-fluoro-2,1,3-benzoxadiazole in 10 mL chloroform at 0-10° (use a calcium chloride drying tube), stir at room temperature for 1 h, reflux for 2 h, cool, slowly pour into ice water, remove the organic layer, extract the aqueous layer with chloroform, combine the organic layers, wash, dry over anhydrous magnesium sulfate, evaporate under reduced pressure, take up the residue in 5 mL benzene (Caution! Benzene is a carcinogen!), chromatograph on a 150 × 30 column of silica gel (100-200 mesh Kanto Chemical) with n-hexane:benzene 50:50, evaporate the appropriate fractions to give 4-(chlorosulfonyl)-7-fluoro-2,1,3-benzoxadiazole (CBD-F) as pale yellow needles (mp 64-66°) (Anal. Chem. 1984. 56, 2461). Stir 0.76 g CBD-F in 70 mL MeCN at 0-10° and add 1 g dimethylamine hydrochloride in 10 mL 100 mM pH 10 borax dropwise, adjust pH to 5 with 1 M HCl, concentrate to about 10 mL under reduced pressure, extract three times with 200 mL portions of diethyl ether, wash with water, dry over anhydrous magnesium sulfate, evaporate under reduced pressure, chromatograph on a 500 × 20 column of silica gel with chloroform, isolate the appropriate fraction and re-chromatograph on the same column with ethyl acetate:benzene 1:2 to give 4-(N,N-dimethylaminosulfonyl)-7-fluoro-2,1,3-benzoxadiazole (DBD-F) as white needles (mp 124-125°) (yield = 1% !). On a Merck no. 5714 60F$_{254}$ TLC plate eluted with chloroform DBD-F has Rf 0.32 and lies between two other reaction products (Analyst 1989, 114, 413). DBD-F can also be purchased from Tokyo Kasei. Cool a solution of 16.4 g (S)-(-)-1-benzyl-3-pyrrolidinol in 164 mL pyridine to +5°, add 19.35 g p-toluenesulfonyl chloride, stir at +10° for 48 h, evaporate to dryness, chromatograph using dichloromethane:acetone 95:5 to obtain (3S)-3-[(4-tolylsulfonyl)oxy]-1-(phenylmethyl)-pyrrolidine (mp 68°). Heat a solution of (3S)-3-[(4-tolylsulfonyl)oxy]-1-(phenylmethyl)-pyrrolidine in 200 mL anhydrous DMF to 65°, add 33.5 g sodium azide (Caution! Sodium azide is highly toxic!), stir at 60° for 7 h, filter, evaporate the filtrate to dryness under reduced pressure, dissolve the residue in ethyl acetate, wash twice with water, dry over anhydrous magnesium sulfate, evaporate to obtain (3R)-3-azido-1-(phenylmethyl)pyrrolidine as an oil. Add 3.5 g 10% palladium on carbon under nitrogen to a solution of 7.05 g (3R)-3-azido-1-(phenylmethyl)pyrrolidine in 34.8 mL 1 M HCl in water and 245 mL EtOH, hydrogenate at atmospheric pressure for 30 min, add 3.5 g catalyst, hydrogenate for 2 h, filter, add 34.8 mL 1 M HCl to the filtrate, evaporate to dryness under reduced pressure, take up the residue in 70 mL EtOH, filter, evaporate the filtrate to dryness under reduced pressure, repeat this operation twice, crystallize with the minimum amount of EtOH to obtain (3R)-3-aminopyrrolidine dihydrochloride (J. Med. Chem. 1992, 35, 4205). 3R-(+)-Aminopyrrolidine is also available from Tokyo Kasei. Add 100 mg 4-(N,N-dimethylaminosulfonyl)-7-fluoro-2,1,3-benzoxadiazole in 20 mL MeCN dropwise to a stirred solution of 200 mg 3R-(+)-aminopyrrolidine in 20 mL MeCN at 0-10°, stir at room temperature for 30 min, remove the MeCN by evaporation under reduced pressure, dissolve the residue in 50 mL 5% HCl, wash 3 times with 50 mL portions of ethyl acetate, adjust the pH of the aqueous solution to 13-14 with 5% NaOH, extract 6 times with 50 mL portions of ethyl acetate. Combine the organic layers and wash them with 20 mL water, dry over anhydrous sodium sulfate, evaporate to dryness under reduced pressure, recrystallize from hexane to obtain (R)-(-)-4-(3-aminopyrrolidin-1-yl)-7-(N,N-dimethylaminosulfonyl)-2,1,3-benzoxadiazole as orange crystals (mp 96-98°) (Analyst 1992, 117, 727). Add 100 μL thiophosgene in 10 mL benzene (Caution! Benzene is a carcinogen!) to 100 mg (R)-(-)-4-(3-aminopyrrolidin-1-yl)-7-(N,N-dimethylaminosulfonyl)-2,1,3-benzoxadiazole in 100 mL acetone, reflux for 1 h, remove the solvent by evaporation under reduced pressure, suspend the residue in 100 mL water, extract 4 times with 25 mL portions of benzene. Combine the extracts and wash them with 20 mL water, dry over anhydrous sodium sulfate, evaporate to dryness under reduced pressure, recrystallize from hexane:benzene 1:2 to obtain (R)-(-)-4-(3-isothiocyanatopyrrolidin-1-yl)-7-(N,N-dimethylaminosulfonyl)-2,1,3-benzoxadiazole as yellow crystals (mp 160-170° d) (Analyst 1995, 120, 385).)

HPLC VARIABLES
Column: 150 × 4.6 5 μm Inertsil ODS-80A
Mobile phase: MeCN:water:trifluoroacetic acid 56:44:0.1
Column temperature: 40
Flow rate: 1
Injection volume: 50
Detector: F ex 460 em 550

CHROMATOGRAM
Retention time: 24, 31 (enantiomers)
Internal standard: oxprenolol
Limit of detection: 25-29 fmole

KEY WORDS
chiral; rat; plasma; pharmacokinetics

REFERENCE
Toyo'oka, T.; Toriumi, M.; Ishii, Y. Enantioseparation of β-blockers labelled with a chiral fluorescent reagent, R(-)-DBD-PyNCS, by reversed-phase liquid chromatography, *J.Pharm.Biomed.Anal.*, **1997**, *15*, 1467–1476.

SAMPLE
Matrix: solutions
Analyte: amines
Sample preparation: Mix 10 μL of a 1 mM amine solution in MeCN : water : triethylamine 50 : 50 : 2 with 10 μL 5 mM (R)-(-)-4-(3-isothiocyanatopyrrolidin-1-yl)-7-(N,N-dimethylaminosulfonyl)-2,1,3-benzoxadiazole in MeCN, heat at 55° for 10 min, add 480 μL 1 M acetic acid in MeCN : water 50 : 50, dilute 10-fold with MeCN, inject a 5 μL aliquot. ((R)-(-)-4-(3-isothiocyanatopyrrolidin-1-yl)-7-(N,N-dimethylaminosulfonyl)-2,1,3-benzoxadiazole, (R)-(-)-DBD-PyNCS, is available from Tokyo Kasei (TCI America, Portland OR). Synthesis is as described above.)

HPLC VARIABLES
Column: 150 × 4.6 5 μm Inertsil ODS-80A
Mobile phase: MeCN : water : trifluoroacetic acid 60 : 40 : 0.05
Column temperature: 40
Flow rate: 1
Injection volume: 5
Detector: F ex 460 em 550

CHROMATOGRAM
Retention time: k' 9.91 ((R)-(+)-alprenolol), k' 11.56 ((R)-(+)-propranolol), 12.91 ((S)-(-)-alprenolol), 14.56 ((S)-(-)-propranolol)

KEY WORDS
chiral

REFERENCE
Toyo'oka, T.; Liu, Y.-M. Development of optically active fluorescent Edman-type reagents, *Analyst*, **1995**, *120*, 385–390.

SAMPLE
Matrix: solutions
Analyte: peptides
Sample preparation: Mix 5 μL of a 100 μg/mL solution in water with 25 μL 0.2% triethylamine in water, add 20 μL 10 mM (R)-(-)-DBD-PyNCS in MeCN, vortex for several s, heat at 60° for 10 min, cool to room temperature, add 50 μL 10% acetic acid in water, store at 5°, inject a 5 μL aliquot. ((R)-(-)-DBD-PyNCS, (R)-(-)-4-(3-isothiocyanatopyrrolidin-1-yl)-7-(N,N-dimethylaminosulfonyl)-2,1,3-benzoxadiazole, is available from Tokyo Kasei (TCI America, Portland OR). Synthesis is as described above.)

HPLC VARIABLES
Column: 150 × 4.6 5 μm Inertsil ODS-80A
Mobile phase: MeCN : MeOH : water : trifluoroacetic acid 25 : 5 : 70 : 0.05
Column temperature: 40
Flow rate: 1
Injection volume: 5
Detector: F ex 450 em 560

CHROMATOGRAM
Retention time: k' 14.6 (D-Ala-D-Ala), k' 15.4 (L-Ala-L-Ala), k' 16.2 (D-Ala-L-Ala), k' 17.2 (L-Ala-D-Ala)

KEY WORDS
chiral; also for Ala-Leu and Ala-Ala-Gly with different mobile phases

REFERENCE
Liu, Y.-M.; Miao, J.-R.; Toyo'oka, T. Enantiomeric separation of di- and tri-peptides with chiral fluorescence labeling reagents by liquid chromatography, *Anal.Chim.Acta*, **1995**, *314*, 169–173.

SAMPLE
Matrix: solutions
Analyte: β-blockers
Sample preparation: Mix 100 μL of a 10 μM solution in MeCN : water : triethylamine 50 : 50 : 0.1 with 100 μL 1 mM (R)-(-)-4-(3-isothiocyanatopyrrolidin-1-yl)-7-(N,N-dimethylami-

nosulfonyl)-2,1,3-benzoxadiazole in MeCN, heat in the dark at 65° for 1.5 h, inject an aliquot. ((R)-(-)-4-(3-Isothiocyanatopyrrolidin-1-yl)-7-(N,N-dimethylaminosulfonyl)-2,1,3-benzoxadiazole ((R)-DBD-Py-NCS) is available from Tokyo Kasei (TCI America, Portland OR). Synthesis is as described above.)

HPLC VARIABLES
Column: 150 × 4.6 5 μm Inertsil ODS-80A
Mobile phase: MeCN:water:trifluoroacetic acid 35:65:0.1 (A) or 44:56:0.1 (B) or 62:38:0.1 (C) or 37:63:0.1 (D) or 50:50:0.1 (E)
Column temperature: 40
Flow rate: 1
Detector: F ex 460 em 550

CHROMATOGRAM
Retention time: 10.2, 12.0 (oxprenolol enantiomers) (C), 12.4, 15.9 (alprenolol enantiomers) (C), 23.4, 28.3 (carteolol enantiomers) (A), 26.7, 32.8 (atenolol enantiomers) (A), 28.8, 29.9 (timolol enantiomers) (A), 29.4, 36.5 (bupranolol enantiomers) (E), 30.9, 39.4 (pindolol enantiomers) (B), 37.2, 40.0 (bucumolol enantiomers) (D)
Limit of detection: 0.00125-320 fmole

KEY WORDS
chiral

REFERENCE
Toyo'oka, T.; Toriumi, M.; Ishii, Y. Enantioseparation of β-blockers labelled with a chiral fluorescent reagent, R(-)-DBD-PyNCS, by reversed-phase liquid chromatography, *J.Pharm.Biomed.Anal.*, **1997**, *15*, 1467–1476.

(R)-(-)-4-(3-Isothiocyanatopyrrolidin-1-yl)-7-nitro-2,1,3-benzoxadiazole

SAMPLE
Matrix: solutions
Analyte: amines
Sample preparation: Mix 10 μL of a 1 mM amine solution in MeCN:water:triethylamine 50:50:2 with 10 μL 5 mM (R)-(-)-4-(3-isothiocyanatopyrrolidin-1-yl)-7-nitro-2,1,3-benzoxadiazole in MeCN, heat at 55° for 10 min, add 480 μL 1 M acetic acid in MeCN:water 50:50, dilute 10-fold with MeCN, inject a 5 μL aliquot. (Synthesis of (R)-(-)-4-(3-isothiocyanatopyrrolidin-1-yl)-7-nitro-2,1,3-benzoxadiazole, (R)-(-)-NBD-PyNCS, is as follows. Cool a solution of 16.4 g (S)-(-)-1-benzyl-3-pyrrolidinol in 164 mL pyridine to +5°, add 19.35 g p-toluenesulfonyl chloride, stir at +10° for 48 h, evaporate to dryness, chromatograph using dichloromethane:acetone 95:5 to obtain (3S)-3-[(4-tolylsulfonyl)oxy]-1-(phenylmethyl)pyrrolidine (mp 68°). Heat a solution of (3S)-3-[(4-tolylsulfonyl)oxy]-1-

(phenylmethyl)pyrrolidine in 200 mL anhydrous DMF to 65°, add 33.5 g sodium azide (Caution! Sodium azide is highly toxic!), stir at 60° for 7 h, filter, evaporate the filtrate to dryness under reduced pressure, dissolve the residue in ethyl acetate, wash twice with water, dry over anhydrous magnesium sulfate, evaporate to obtain (3R)-3-azido-1-(phenylmethyl)pyrrolidine as an oil. Add 3.5 g 10% palladium on carbon under nitrogen to a solution of 7.05 g (3R)-3-azido-1-(phenylmethyl)pyrrolidine in 34.8 mL 1 M HCl in water and 245 mL EtOH, hydrogenate at atmospheric pressure for 30 min, add 3.5 g catalyst, hydrogenate for 2 h, filter, add 34.8 mL 1 M HCl to the filtrate, evaporate to dryness under reduced pressure, take up the residue in 70 mL EtOH, filter, evaporate the filtrate to dryness under reduced pressure, repeat this operation twice, crystallize with the minimum amount of EtOH to obtain (3R)-3-aminopyrrolidine dihydrochloride (J. Med. Chem. 1992, 35, 4205). 3R-(+)-Aminopyrrolidine is also available from Tokyo Kasei (TCI America, 911 North Harborgate St., Portland, OR 97203; 800-423-8616, 503-283-1681, (fax) 503-283-1987; www.tciamerica.com). Add 100 mg 4-fluoro-7-nitro-2,1,3-benzoxadiazole in 20 mL MeCN dropwise to a stirred solution of 200 mg 3R-(+)-aminopyrrolidine in 20 mL MeCN at 0-10°, stir at room temperature for 30 min, remove the MeCN by evaporation under reduced pressure, dissolve the residue in 50 mL water, extract 4 times with 80 mL portions of ethyl acetate. Combine the organic layers and wash them with 20 mL water, dry over anhydrous sodium sulfate, evaporate to dryness under reduced pressure, recrystallize from hexane to obtain (R)-(-)-4-(3-aminopyrrolidin-1-yl)-7-nitro-2,1,3-benzoxadiazole as dark red crystals (mp 178-181°) (Analyst 1992, 117, 727). Add 100 μL thiophosgene in 10 mL benzene (Caution! Benzene is a carcinogen!) to 100 mg (R)-(-)-4-(3-aminopyrrolidin-1-yl)-7-nitro-2,1,3-benzoxadiazole in 100 mL acetone, reflux for 1 h, remove the solvent by evaporation under reduced pressure, suspend the residue in 100 mL water, extract 4 times with 25 mL portions of benzene. Combine the extracts and wash them with 20 mL water, dry over anhydrous sodium sulfate, evaporate to dryness under reduced pressure, recrystallize from hexane:benzene 1:2 to obtain (R)-(-)-4-(3-isothiocyanatopyrrolidin-1-yl)-7-nitro-2,1,3-benzoxadiazole as red crystals (mp 165-170°).)

HPLC VARIABLES
Column: 150 × 4.6 5 μm Inertsil ODS-80A
Mobile phase: MeCN:water:trifluoroacetic acid 40:60:0.05
Column temperature: 40
Flow rate: 1
Injection volume: 5
Detector: F ex 490 em 530

CHROMATOGRAM
Retention time: 10.5 ((R)-(+)-propranolol), 11.5 ((S)-(-)-propranolol), 12.5 ((R)-(+)-alprenolol), 14.5 ((S)-(-)-alprenolol)

KEY WORDS
chiral

REFERENCE
Toyo'oka, T.; Liu, Y.-M. Development of optically active fluorescent Edman-type reagents, *Analyst*, **1995**, *120*, 385-390.

1-(6-Methoxy-2-naphthyl)ethyl Isothiocyanate

SAMPLE
Matrix: solutions
Analyte: drugs
Sample preparation: Mix 300 μL of a 30 μM solution in dichloromethane with 10 μL 20 mM 1-(6-methoxy-2-naphthyl)ethyl isothiocyanate in anhydrous dichloromethane and 50 μL 0.1% triethylamine in dichloromethane, vortex thoroughly, heat at 50° for 1.5 h, inject an aliquot. (Synthesize 1-(6-methoxy-2-naphthyl)ethyl isothiocyanate as follows (protect from light). Dissolve 500 mg (S)-(+)-naproxen in 50 mL dry toluene, slowly add 5 mL freshly distilled thionyl chloride, reflux for 1 h, evaporate to dryness under vacuum, dry the acyl chloride (mp 87.5°) under vacuum over KOH for 2 days. Dissolve 0.5 mmoles acyl chloride in 5 mL acetone, stir at 0°, add 0.6 mmoles sodium azide dissolved in ice water, stir at 0° for 30 min, add 10 mL ice-cold water, filter, dry solid in a desiccator under vacuum. Dissolve the solid in 1 mL toluene or dichloromethane (dried over 3 Å molecular sieve), reflux for 10 min, evaporate, store resulting isocyanate (mp 51°) under vacuum over a desiccant. Dissolve 0.5 mmole isocyanate in 5 mL acetone, add 20 mL 8.5% phosphoric acid, heat to 80° for 1.5 h, adjust to pH 13, extract with diethyl ether:dichloromethane 4:1. Wash the organic layer twice with water, dry over anhydrous sodium sulfate, evaporate to dryness, dissolve in 1 mL toluene, evaporate to give the amine from naproxen as crystals (mp 53°) (Pharm.Res. 1990, 7, 1262). Dissolve 1 mmole 1,1-thiocarbonyldiimidazole in 15 mL ice-cold chloroform, stir at 0°, add dropwise 1 mmole of the amine dissolved in 10 mL chloroform, stir at room temperature for 1.5 h, evaporate to dryness, reconstitute with carbon tetrachloride (Caution! Carbon tetrachloride is a carcinogen!), filter, evaporate the filtrate to dryness, store the resulting oil in a desiccator, purify on a short silica gel column with dichloromethane:light petroleum 50:50 to give 1-(6-methoxy-2-naphthyl)ethyl isothiocyanate as a slightly yellow liquid (store in the freezer under argon).)

HPLC VARIABLES
Column: 250 × 4 5 μm Zorbax ODS
Mobile phase: MeCN:water 50:50 (A), MeCN:water 55:45 (B), MeCN:water 70:30 (C, D), or MeOH:water 70:30 (E)
Flow rate: 1 (A-C), 0.8 (D, E)
Injection volume: 100
Detector: UV 230; F ex 270 em 350

CHROMATOGRAM
Retention time: k' 5.2 (S-(-)-atenolol, (A)), k' 6.1 (R-(+)-atenolol, (A)); k' 6.7 (S-(-)-diacetolol, (A)), k' 8.0 (R-(+)-diacetolol, (A)); k' 20.4 (S-(-)-metoprolol, (B)), k' 25.4 (R-(+)-metoprolol, (B)); k' 7.1 (S-(-)-carvedilol, (C)), k' 8.0 (R-(+)-carvedilol, (C)); k' 8.7 (S-(-)-propranolol, (D)), k' 10.7 (R-(+)-propranolol, (D)); k' 9.4 (S-(-)-alprenolol, (D)), k' 11.8 (R-(+)-alprenolol, (D)); k' 9.5 (R-(-)-propafenone, (C)), k' 10.2 (S-(+)-propafenone, (C)); k' 10.1 (S-(+)-tocainide, (E)), k' 11.1 (R-(-)-tocainide, (E)); k' 12.3 (flecainide, no enantiomeric separation(C))

KEY WORDS
chiral; F not much more sensitive than UV

REFERENCE
Büschges, R.; Linde, H.; Mutschler, E.; Spahn-Langguth, H. Chloroformates and isothiocyanates derived from 2-arylpropionic acids as chiral reagents: synthetic routes and chromatographic behaviour of the derivatives, *J.Chromatogr.A*, **1996**, *725*, 323–334.

D-α-Methylbenzyl Isothiocyanate

RELATED REFERENCE
Gal, J.; Sedman, A.J. R-α-methylbenzyl isothiocyanate, a new and convenient chiral derivatizing agent for the separation of enantiomeric amino compounds by high-performance liquid chromatography. *J.Chromatogr.* **1984**, *314*, 275-281.

SAMPLE
Matrix: bulk
Analyte: 2-amino-1-(p-methanesulfonylphenyl)-1,3-propanediol
Sample preparation: Dissolve 0.5 mg 2-amino-1-(p-methanesulfonylphenyl)-1,3-propanediol in 5 μL DMSO, add 100 μL 20 mg/mL D-α-methylbenzyl isothiocyanate (Trans World Chemicals, Kensington MD) in MeCN, let stand at room temperature for 30 min, add 2 μL 2-aminoethanol, let stand for 10 min, add 50 μL 300 mM pH 3 ammonium phosphate buffer, inject a 2-5 μL aliquot.

HPLC VARIABLES
Column: 150 × 3.9 Nova-pak ODS
Mobile phase: MeOH:water 30:70
Flow rate: 1
Injection volume: 2-5
Detector: UV 254

CHROMATOGRAM
Retention time: 16 (1R,2R), 21 (1S,2S), 24 (1R,2S and 1S,2R)

KEY WORDS
compound is a thiamphenicol precursor; chiral

REFERENCE
Gal, J.; Meyer-Lehnert, S. Reversed-phase liquid chromatographic separation of enantiomeric and diastereomeric bases related to chloramphenicol and thiamphenicol, *J.Pharm.Sci.*, **1988**, *77*, 1062–1065.

1-Naphthalenemethyl Isothiocyanate

SAMPLE
Matrix: bulk
Analyte: labetalol
Sample preparation: 1 mg Labetalol + 3 mg 1-naphthalenemethyl isothiocyanate (Trans World Chemicals, Chevy Chase MD) + 100 μL MeCN + 50 μL water + 3 μL triethylamine, vortex, heat at 60° for 1 h. Remove a 150 μL aliquot and add it to 400 μL MeCN, vortex, inject a 5 μL aliquot.

HPLC VARIABLES
Column: 150 × 3.9 4 μm Nova-Pak C18
Mobile phase: MeOH:20 mM pH 4.60 $(NH_4)H_2PO_4$ 70:30
Flow rate: 1
Injection volume: 5
Detector: UV 254

CHROMATOGRAM
Retention time: 10.42 (RS/SR), 14.73 (SS,RR)

KEY WORDS
comparison with other derivatization reagents

REFERENCE
Desai, D.M.; Gal, J. Reversed-phase high-performance liquid chromatographic separation of the stereoisomers of labetalol via derivatization with chiral and non-chiral isothiocyanate reagents, *J.Chromatogr.*, **1992**, *579*, 165–171.

1-Naphthyl Isothiocyanate

SAMPLE
Matrix: urine
Analyte: pamidronate
Sample preparation: Condition a 3 mL 500 mg Bakerbond quaternary amine SPE cartridge with 2.5 mL water, 1 mL 1 mg/mL etidronate, 2.5 mL 100 mM nitric acid, and two 2.5 mL portions of water. 2.5 mL Urine + 100 μL 1 mg/mL etidronate + 100 μL 2.16 μg/mL IS, mix, add 30 μL 1 M calcium chloride, vortex, add 25 μL portions of 1 M NaOH

(vortexing after each addition) until a precipitate is clearly present, add 25 μL 1 M NaOH, vortex, centrifuge at 3900 g for 2 min. Remove the pellet and dissolve it in 50 μL 1 M HCl, add 2.5 mL water, add 50 μL 1 M NaOH, centrifuge. Remove the pellet and dissolve it in the minimum amount of 1 M HCl (30-50 μL), add 2.5 mL water, add 30 μL 1 M NaOH, centrifuge. Remove the pellet and dissolve it in the minimum amount of 1M HCl (30-50 μL), add 2.5 mL water, add to the SPE cartridge, wash with two 2.5 mL portions of water, wash with 2.5 mL 10 mM nitric acid, elute with 2.5 mL 100 mM nitric acid. Evaporate the eluate to dryness under 0.8 bar nitrogen at 60° for 1.65-2 h, add 500 μL water, vortex. Remove a 250 μL aliquot and add it to 25 μL 1 mg/mL etidronate, add 40 μL triethylamine, add 250 μL 20 mg/mL 1-naphthyl isothiocyanate in pyridine, vortex, heat at 80° for 15 min, cool, add 2 mL 10 mg/mL tetrabutylammonium bromide (?) in chloroform, vortex, centrifuge, discard the lower organic phase, repeat the procedure. Remove a 250 μL aliquot of the aqueous phase and add it to 75 μL 3% hydrogen peroxide (to convert the thiourea derivative to the corresponding urea), mix, heat at 80° for 5 min, inject a 100 μL aliquot.

HPLC VARIABLES
Guard column: 10 × 2 R2 (Chrompack)
Column: 100 × 4.6 3 μm Microspher C18 (Chrompack)
Mobile phase: MeCN:buffer 65:35 (Buffer was 10 mM phosphate containing 10 mM tetraoctylammonium bromide and 2 mM etidronate (monosodium (1-hydroxyethylidene) bisphosphonate) (as adsorption suppressor), pH 7.6-7.9.)
Column temperature: 30
Flow rate: 0.8
Injection volume: 100
Detector: F ex 285 em 390

CHROMATOGRAM
Retention time: 10
Internal standard: (3-amino-3-phenyl-1-hydroxypropylidene)bisphosphonic acid (17)
Limit of quantitation: 3 ng/mL
Limit of detection: 1 ng/mL

KEY WORDS
SPE

REFERENCE
Sparidans, R.W.; Den Hartigh, J.; Beijnen, J.H.; Vermeij, P. Determination of pamidronate in urine by ion-pair liquid chromatography after derivatization with 1-naphthylisothiocyanate, *J.Chromatogr.B*, **1997**, *696*, 137–144.

Phenyl Isothiocyanate

RELATED REFERENCES
Anumula, K.R.; Schulz, R.P.; Back, N. Fluorescent N-methylanthranilyl (Mantyl) tag for peptides: its application in subpicomole determination of kinins. *Peptides (Pergamon)* **1992**, *13*, 663-669.

Anumula, K.R.; Taylor, P.B. Quantitative determination of phenyl isothiocyanate-derivatized amino sugars and amino sugar alcohols by high-performance liquid chromatography. *Anal.Biochem.* **1991**, *197*, 113-120.

Bogusz, M.J.; Kala, M.; Maier, R.-D. Determination of phenylisothiocyanate derivatives of amphetamine and its analogues in biological fluids by HPLC-APCI-MS or DAD. *J.Anal.Toxicol.* **1997**, *21*, 59-69.

Hagen, S.R. High-performance liquid chromatographic quantitation of phenylthiocarbamyl muramic acid and glucosamine from bacterial cell walls. *J.Chromatogr.* **1993**, *632*, 63-68.

Kariya, M.; Namiki, H. HPLC of phenylthiocarbamoyl labelled uridine-diphosphate-hexosamine. *Chromatographia* **1997**, *46*, 5-11.

Lippincott, S.E.; Friedman, A.L.; Siegel, F.L.; Pityer, R.M.; Chesney, R.W. HPLC analysis of the phenylisothiocyanate (PITC) derivatives of taurine from physiologic samples. *J.Am.Coll.Nutr.* **1988**, *7*, 491-497.

Osswald, W.F.; Jehle, J.; Firl, J. Quantification of fungal infection in plant tissues by determining the glucosamine phenyl isothiocyanate derivative using HPLC techniques. *J.Plant Physiol.* **1995**, *145*, 393-397.

SAMPLE
Matrix: CSF
Analyte: baclofen
Sample preparation: Prepare a 100 × 7 column of Dowex 50X4-400, treat with an excess of aqueous ammonia, wash with water until the washings are neutral, wash with an excess of 4 M HCl, wash with water until the washings are neutral, rinse with 8 mL water. Add 200 µL CSF to the column, wash with 8 mL water, elute with 2 mL 10% ammonia. Lyophilize the eluate, reconstitute the residue with 1 mL pH 7.4 phosphate buffer, extract twice with 1 mL portions of 1-butanol. Combine the organic layers and evaporate them to dryness, add 10-50 µL EtOH:water:triethylamine 50:25:25, evaporate to dryness, add 20 µL reagent, let stand at room temperature for 20 min, evaporate to dryness, reconstitute with pH 7.4 sodium phosphate buffer, inject a 5-25 µL aliquot. (Reagent was EtOH:triethylamine:water:phenyl isothiocyanate 70:10:10:10, store at -20° (Anal.Biochem. 1989, 176, 269).)

HPLC VARIABLES
Column: 150 × 3.9 Pico-Tag
Mobile phase: Gradient. A was sodium acetate adjusted to pH 6.4 with glacial acetic acid. B was MeCN:water 60:40. A:B from 100:0 to 60:40 over 10 min, to 0:100 over 1 min, maintain at 0:100 over 3 min.
Column temperature: 38
Injection volume: 5-25
Detector: UV 254

CHROMATOGRAM
Retention time: 13.2
Limit of detection: 5-10 ng/mL

KEY WORDS
SPE; pharmacokinetics

REFERENCE
Sallerin-Caute, B.; Monsarrat, B.; Lazorthes, Y.; Cros, J.; Bastide, R. A sensitive method for the determination of baclofen in human CSF by high performance liquid chromatography, *J.Liq.Chromatogr.*, **1988**, *11*, 1753–1761.

SAMPLE
Matrix: blood
Analyte: tranexamic acid
Sample preparation: 500 µL Serum + 5 µL 58.4 µg/mL IS, mix by swirling, add 2 mL EtOH, vortex, centrifuge at 1500 g for 10 min. Remove the supernatant and add it to 1 mL 10 mM pH 9.2 borax solution, add 13 µL phenyl isothiocyanate, heat at 40° for 30 min, add 2 mL xylene, agitate, centrifuge at 1500 g for 10 min, wash twice more. Acidify the aqueous layer with 1 mL concentrated HCl, heat at 80° for 10 min, evaporate to dryness under reduced pressure, reconstitute with 1 mL 100 mM borax solution, extract twice with 2 mL portions of benzene (Caution! Benzene is a carcinogen!). Combine the organic layers and evaporate them to dryness under a stream of nitrogen, reconstitute the residue in 500 µL mobile phase, inject a 10 µL aliquot.

HPLC VARIABLES
Column: 150 × 4.6 5 μm Cosmosil 5C8 (Nakarai Chemicals)
Mobile phase: EtOH:20 mM pH 7.0 phosphate buffer 10:90
Flow rate: 1.8
Injection volume: 10
Detector: UV 254

CHROMATOGRAM
Retention time: 10.8
Internal standard: 3-aminocyclohexanecarboxylic acid (14.1)
Limit of detection: 200 ng/mL

KEY WORDS
serum

REFERENCE
Matsubayashi, K.; Kojima, C.; Tachizawa, H. Determination of tranexamic acid in human serum by high-performance liquid chromatography using selective pre-column derivatization with phenyl isothiocyanate, *J.Chromatogr.*, **1988**, *433*, 225–234.

SAMPLE
Matrix: blood
Analyte: bradykinin
Sample preparation: 50 μL Plasma + 50 μL 1 M NaCl + 900 μL 0.5% HCl in isopropanol, boil for 10 min, centrifuge for 2 min. Suspend the pellet in 250 μL buffer (final pH 8.0), add 50 μL 10 mg/mL TPCK-treated trypsin in buffer (at pH 8.0), heat at 37° for 1 h, boil for 10 min, evaporate to dryness, reconstitute with water, remove a 10 μL aliquot and add it to 90 μL freshly-prepared MeOH:triethylamine:phenyl isothiocyanate 70:10:5, heat at 37° for 30 min with occasional vortexing, add 500 μL 2% phosphoric acid in MeOH:water 50:50, add 800 μL chloroform, mix vigorously, centrifuge, repeat extraction. (Buffer was 50 mM pH 11 sodium phosphate buffer containing 150 mM NaCl, 5 mM EDTA, and 2 mM 1,10-phenanthroline.)

HPLC VARIABLES
Column: 300 × 3.9 μBondapak C18
Mobile phase: Gradient. A was 0.1% trifluoroacetic acid in water. B was 0.05% trifluoroacetic acid in MeCN. A:B from 75:25 to 40:60 over 20 min.
Flow rate: 1
Detector: UV 254

CHROMATOGRAM
Retention time: 25

OTHER SUBSTANCES
Extracted: analogs

KEY WORDS
plasma; rat; human

REFERENCE
Anumula, K.R.; Schulz, R.; Back, N. Quantitative determination of kinins released by trypsin using enzyme-linked immunosorbent assay (ELISA) and identification by high-performance liquid chromatography (HPLC), *Biochem.Pharmacol.*, **1989**, *38*, 2421–2427.

SAMPLE
Matrix: bulk
Analyte: amines
Sample preparation: Prepare a 10 mg/mL solution in 500 mM sodium bicarbonate solutions, extract a 10 mL aliquot twice with 15 mL portions of dichloromethane. Combine the extracts and add 10 μL phenyl isothiocyanate, evaporate to dryness under a stream of air, reconstitute with 10 mL MeOH, inject a 10 μL aliquot.

HPLC VARIABLES
Guard column: 70 × 2.1 CO:PELL ODS

Column: 300 × 3.9 μBondapak C18
Mobile phase: MeOH:water:acetic acid 45:54:1
Flow rate: 2
Injection volume: 10
Detector: UV 254

CHROMATOGRAM
Retention time: 2 (lidocaine), 7 (ephedrine), 10 (phenylpropanolamine), 12 (pseudoephedrine)

REFERENCE
Noggle, F.T. Jr.; Clark, C.R. Liquid chromatographic analysis of samples containing cocaine, local anesthetics, and other amines, *J.Assoc.Off.Anal.Chem.*, **1983**, *66*, 151–157.

SAMPLE
Matrix: bulk
Analyte: clobenzorex (N-2-chlorobenzyl)-α-methylphenethylamine)
Sample preparation: Dissolve 2 mg of the hydrochloride in 30 mL 450 mM NaOH, extract with 30 mL chloroform. Add 10 μL phenyl isothiocyanate to the chloroform solution, evaporate to dryness under a stream of air, reconstitute with 1 mL MeOH, inject a 5 μL aliquot.

HPLC VARIABLES
Guard column: 70 × 2.1 CO:Pell ODS
Column: 300 × 3.9 μBondapak C18
Mobile phase: MeOH:water:acetic acid 65:34:1
Flow rate: 1.5
Injection volume: 5
Detector: UV 254

CHROMATOGRAM
Retention time: 20

OTHER SUBSTANCES
Simultaneously analyzed: 3- and 4-chloro analogs

REFERENCE
Noggle, F.T. Jr.; Clark, C.R.; Andurkar, S.V.; DeRuiter, J. Liquid chromatographic analysis of regioisomers and enantiomers of N-(chlorobenzyl)-α-methylphenethylamines: Analogues of clobenzorex, *J.Liq.Chromatogr.*, **1990**, *13*, 763–777.

SAMPLE
Matrix: peptide digests
Analyte: peptides
Sample preparation: Dry 5-3000 pmole tryptic digest, add 15 μL MeOH:water:triethylamine 40:40:20, evaporate to dryness under high vacuum for 30 min, repeat this step 3 times, dissolve in 20 μL MeOH:triethylamine:water:phenyl isothiocyanate 80:10:10:1, let stand at room temperature for 30 min, evaporate to dryness under high vacuum for 40-50 min, reconstitute with 6 M guanidinium hydrochloride containing 0.1% trifluoroacetic acid, inject an aliquot.

HPLC VARIABLES
Guard column: μBondapak C18/Corasil
Column: 150 × 3.9 Nova-Pak
Mobile phase: Gradient. MeCN:0.1% trifluoroacetic acid from 30:70 to 35:65 over 120 min, to 50:50 over 8 min
Flow rate: 0.5
Detector: UV 254

CHROMATOGRAM
Retention time: 40-120 (depending on peptide structure)

REFERENCE
Colilla, F.J.; Yadav, S.P.; Brew, K.; Mendez, E. Peptide maps at picomolar levels obtained by reversed-phase high-performance liquid chromatography and pre-column derivatization with phenyl isothiocyanate. Microsequencing of Phenylthiocarbamyl Peptides, *J.Chromatogr.*, **1991**, *548*, 303–310.

SAMPLE
Matrix: solutions
Analyte: hexosamines and hexosaminitol
Sample preparation: Condition a Chrom Prep mixed bed SPE cartridge (Hamilton) with 1 mL water. Evaporate ≤50 μL solution to dryness under reduced pressure, heat under vacuum in 4 M HCl at 100° for 4 h, evaporate to dryness, add 10 μL 1 mM p-aminophenyl-β-D-galactopyranoside, evaporate to dryness, add 10 μL EtOH:water:triethylamine 50:25:25, evaporate to dryness, add 20 μL EtOH:triethylamine:water:phenyl isothiocyanate 70:10:10:10, evaporate to dryness, reconstitute with 70 μL MeCN:5 mM pH 7.4 sodium phosphate buffer 5:95, add the sample to the SPE cartridge, elute with 130 μL water, inject all the eluate.

HPLC VARIABLES
Column: 150 × 4.6 PICO-TAG (Waters)
Mobile phase: Gradient. A was 140 mM pH 5.7 sodium acetate buffer containing 200 mM boric acid and 0.7 mL/L triethylamine. B was MeCN:water 60:40. A:B from 98:2 to 93:7 over 5 min, to 50:50 over 5 min, maintain at 50:50 for 0.5 min, return to initial conditions over 4.5 min.
Column temperature: 50
Flow rate: 2
Detector: UV 254

CHROMATOGRAM
Retention time: 3.1 (3-allosamine), 3.4 (3-glucosamine), 3.9 (galactosamine), 4.3 (glucosamine), 4.8 (allosamine), 6.4 (3-allosaminitol), 6.9 (glucosaminitol), 7.4 (mannosaminitol), 7.8 (allosaminitol), 8.1 (galactosaminitol)
Internal standard: p-aminophenyl-β-D-galactopyranoside (10.1)
Limit of detection: 1 nmole

KEY WORDS
SPE

REFERENCE
Cheng, P.-W. High-performance liquid chromatographic analysis of galactosamine, glucosamine, glucosaminitol, and galactosaminitol, *Anal.Biochem.*, **1987**, *167*, 265–269.

2,3,4,6-Tetra-O-acetyl-β-D-glucopyranosyl Isothiocyanate

RELATED REFERENCES
Miyazawa, T.; Iwanaga, H.; Yamada, T.; Kuwata, S. Resolution of cyclic imino acid and β-amino acid enantiomers by derivatization with 2,3,4,6-tetra-O-acetyl-β-D-glucopyranosyl isothiocyanate followed by reversed-phase HPLC analysis. *Anal.Lett.* **1992**, *26*, 367-378.
Nishi, H.; Fujimura, N.; Yamaguchi, H.; Fukuyama, T. Reversed-phase high-performance liquid chromatographic separation of the enantiomers of trimetoquinol hydrochloride by derivatization with 2,3,4,6-tetra-O-acetyl-β-D-glucopyranosyl isothiocyanate and application to the optical purity testing of drugs. *J.Chromatogr.* **1991**, *539*, 71-81.
Wu, S.T.; Chang, Y.P.; Gee, W.L.; Benet, L.Z.; Lin, E.T. Stereoselective high-performance liquid chromatography determination of propranolol and 4-hydroxypropranolol in human plasma after pre-column derivatization. *J.Chromatogr.B* **1997**, *692*, 133-140

SAMPLE
Matrix: blood
Analyte: carvedilol
Sample preparation: Condition a PrepSep RC18 SPE cartridge (Fisher) with 10 mL elution solvent and 2 mL MeCN:water 35:65. 1 mL Plasma + 1 mL 8 M guanidine hydrochloride, mix, add to SPE cartridge, wash with 2 mL MeCN:water 35:65, elute with two 200 μL portions of elution solvent. Add 50 μL 100 mM triethylamine in MeCN and 10 μL 20 mg/mL 2,3,4,6-tetra-O-acetyl-β-glucopyranosyl isothiocyanate in MeCN to the eluate, vortex briefly, allow to stand at room temperature for 30 min, add 150 μL 0.5% phosphoric acid, inject a 50-100 μL aliquot. (Elution solvent was MeCN:water:1 M triethylamine in water adjusted to pH 2.5 with phosphoric acid 80:17:3.)

HPLC VARIABLES
Column: 75 × 4.6 3 μm Ultrasphere ODS
Mobile phase: MeCN:MeOH:water:1 M triethylamine in water adjusted to pH 2.5 with phosphoric acid 29:29:41.5:0.5
Flow rate: 1.6
Injection volume: 50-100
Detector: F ex 285 em 360

CHROMATOGRAM
Retention time: 9.5 (S-(-)), 11.5 (R-(+))
Internal standard: N,N-bis-carvedilol (4)
Limit of quantitation: 2 ng/mL
Limit of detection: 0.6 ng/mL

OTHER SUBSTANCES
Extracted: metabolites

KEY WORDS
plasma; chiral; SPE

REFERENCE
Eisenberg, E.J.; Patterson, W.R.; Kahn, G.C. High-performance liquid chromatographic method for the simultaneous determination of the enantiomers of carvedilol and its O-desmethyl metabolite in human plasma after chiral derivatization, *J.Chromatogr.*, **1989**, *493*, 105–115.

SAMPLE
Matrix: blood
Analyte: carvedilol
Sample preparation: 1 mL Plasma + 100 ng IS in MeOH + 1 mL 100 mM pH 8 Britton-Robinson buffer + 5 mL ether, shake for 10 min at ca. 60 strokes/min, centrifuge at 1500 g for 10 min. Remove the organic layer and add it to 300 μL 50 mM sulfuric acid, shake, centrifuge. Remove the aqueous layer and add it to 1 mL 100 mM pH 8 Britton-Robinson buffer, add 5 mL ether, shake for 10 min, centrifuge. Remove the organic layer and dry it over 300 mg $MgSO_4.7H_2O$ (sic) by vortexing for a few s and standing for 10 min. Evaporate the ether solution to dryness under a stream of nitrogen. Take up the residue in 200 μL MeCN:triethylamine 99.6:0.4 and add 5 μL 25.2 mM 2,3,4,6-tetra-O-acetyl-β-D-glucopyranosyl isothiocyanate, let stand at room temperature for 10 min, evaporate under a stream of nitrogen, reconstitute the residue in 200 μL mobile phase, inject a 50 μL aliquot.

HPLC VARIABLES
Guard column: 15 × 3.2 7 μm ODS (Brownlee)
Column: 150 × 4.6 5 μm ODS 80TM (Tosoh)
Mobile phase: MeOH:EtOH:2 mM pH 7.0 $(NH_4)_2HPO_4$ 60:4:36
Column temperature: 50
Flow rate: 2.5
Injection volume: 50
Detector: F ex 285 em 355

CHROMATOGRAM
Retention time: 5.4 (R-(+)), 6.3 (S-(-))

Internal standard: 1-(2-methoxyphenyl)-4-[3-(naphthyloxy)-2-hydroxypropyl]piperazine (7.8)
Limit of detection: 1.55 ng/mL

KEY WORDS
plasma; human; monkey; pharmacokinetics

REFERENCE
Fujimaki, M.; Murakoshi, Y.; Hakusui, H. Assay and disposition of carvedilol enantiomers in humans and monkeys: evidence of stereoselective presystemic metabolism, *J.Pharm.Sci.*, **1990**, *79*, 568–572.

SAMPLE
Matrix: blood
Analyte: bevantolol
Sample preparation: Condition a 3 mL Bond-Elut C18 SPE cartridge with 1 volume MeCN and 1 volume 25 mM ammonium hydroxide, do not allow to dry. 500 μL Plasma + 80 μL water + 1 mL MeCN, vortex for 10 s, centrifuge at 2000 rpm for 10 min, remove the supernatant and add it to 1 mL 25 mM pH 10.5 ammonium hydroxide, vortex, add 100 μL 250 μg/mL 2,3,4,6-tetra-O-acetyl-β-D-glucopyranosyl isothiocyanate in MeCN (prepare fresh daily) with vortexing, let stand at room temperature for 10 min, add to SPE cartridge, wash with 1 volume 25 mM ammonium hydroxide, wash with 1 volume water, elute with 1.25 mL MeOH. Evaporate the eluent to dryness under a stream of nitrogen at 45°, reconstitute the residue in 500 μL mobile phase, inject a 75 μL aliquot.

HPLC VARIABLES
Column: 100 × 4.6 5 μm Partisil 5 RAC II
Mobile phase: MeCN:75 mM $(NH_4)_2HPO_4$ adjusted to pH 3.5 with phosphoric acid 50:50
Column temperature: 45
Flow rate: 2
Injection volume: 75
Detector: UV 220

CHROMATOGRAM
Retention time: 6 (+), 7 (-)
Limit of quantitation: 40 ng/mL
Limit of detection: 20 ng/mL

KEY WORDS
plasma; chiral; SPE

REFERENCE
Rose, S.E.; Randinitis, E.J. A high-performance liquid chromatographic assay for the enantiomers of bevantolol in human plasma, *Pharm.Res.*, **1991**, *8*, 758–762.

SAMPLE
Matrix: blood
Analyte: albuterol
Sample preparation: Condition a 1 mL Bond-Elut octadecylsilane SPE cartridge with 2 mL MeOH, 2 mL water, and 2 mL buffer (do not allow to dry). 1 mL Serum + 10 μL 100 μg/mL bamethan in water + 2 mL buffer, mix, add to SPE cartridge, wash with 1 mL buffer, 2 mL water, 2 mL MeOH:MeCN 15:85. Dry under full vacuum for 10 min and elute with 1 mL MeOH. Evaporate the eluate to dryness under a stream of nitrogen at room temperature. Dissolve the residue in 200 μL MeCN:triethylamine 199:1, heat at 45° for 20 min, add 10 μL 5 mg/mL 2,3,4,6-tetra-O-acetyl-β-D-glucopyranosyl isothiocyanate (TAGIT) in MeCN, heat at 45° for 2 h. Evaporate at room temperature under a stream of nitrogen, reconstitute in 250 μL mobile phase, inject a 100 μL aliquot. (Buffer was 100 mM Na_2HPO_4 adjusted to pH 7.3 with concentrated phosphoric acid. Prepare solutions of TAGIT in MeCN weekly.)

HPLC VARIABLES
Column: 100 × 4.6 5 μm Brownlee octadecylsilyl
Mobile phase: MeCN:water 29:71 containing 0.1% triethylamine (pH adjusted to 4.0 with concentrated phosphoric acid)

Flow rate: 0.8
Injection volume: 100
Detector: F ex 223 no emission filter

CHROMATOGRAM
Retention time: 5.77 (R(-)), 6.83 (S(+))
Internal standard: bamethan (9, 10)
Limit of detection: 1 ng/mL (each enantiomer)

KEY WORDS
serum; SPE; chiral

REFERENCE
He, L.; Stewart, J.T. A high performance liquid chromatographic method for the determination of albuterol enantiomers in human serum using solid phase extraction and chemical derivatization, *Bio-
med.Chromatogr.*, **1992**, *6*, 291–294.

SAMPLE
Matrix: blood
Analyte: fenoldopam
Sample preparation: Add 4.75 mL plasma to 250 μL 10% ascorbic acid before storage at
-20°. 1 mL Plasma + 50 μL 500 ng/mL IS in 50 mM acetic acid + 5 mL ethyl acetate +
500 μL 0.5 mM Na_2HPO_4 shake on a reciprocal shaker at 60 cycles/min for 10 min, centrifuge at 2000 g for 10 min. Remove 4.5 mL of the organic layer and evaporate it to
dryness under a stream of nitrogen at 40°, reconstitute the residue in 100 μL 200 μg/mL
2,3,4,6-tetra-O-acetyl-β-D-glucopyranosyl isothiocyanate in ethyl acetate (freshly prepared), add 50 μL 1% triethylamine in ethyl acetate (freshly prepared), let stand at room
temperature for 1 h, evaporate to dryness under a stream of nitrogen at 40°, reconstitute
the residue in 200 μL MeCN:50 mM acetic acid 30:70, inject a 5-50 μL aliquot.

HPLC VARIABLES
Guard column: 30 × 2.1 butylsilica (Pierce)
Column: 220 × 2.1 7 μm Aquapore butylsilica (Pierce)
Mobile phase: MeCN:MeOH:buffer:water 17:15:28:42 (Prepare buffer by dissolving 22
g sodium acetate trihydrate, 21 g citric acid monohydrate, 9.8 g NaOH, and 0.63 g diso-
dium EDTA in 2 L water, pH 5.6. Recycle mobile phase.)
Column temperature: 37
Flow rate: 0.3
Injection volume: 5-50
Detector: E, ESA, guard electrode +0.2 V, working electrode 1 -0.20 V, working electrode
2 +0.20 V

CHROMATOGRAM
Retention time: 3 (S), 6 (R)
Internal standard: 2,3,4,5-tetrahydro-1-phenyl-1H-3-benzazepine-7,8-diol (SK&F 38393-
A) (25, 30 (enantiomers))
Limit of quantitation: 0.5 ng/mL
Limit of detection: 0.25 ng/mL

KEY WORDS
chiral; plasma

REFERENCE
Boppana, V.K.; Geschwindt, L.; Cyronak, M.J.; Rhodes, G. Determination of the enantiomers of fenoldopam in human plasma by reversed-phase high-performance liquid chromatography after chiral
derivatization, *J.Chromatogr.*, **1992**, *592*, 317–322.

SAMPLE
Matrix: blood
Analyte: propranolol
Sample preparation: 500 μL + NaOH + 3 mL MTBE, extract. Remove the organic layer
and evaporate it to dryness, reconstitute the residue in equal parts of MeCN:triethylam-
ine 99.6:0.4 and 0.025% 2,3,4,6-tetra-O-acetyl-β-D-glucopyranosyl isothiocyanate (name

given in paper as 2,3,4,5-tetra-O-acetyl-α-d-glucopyranosyl isothiocyanate) in MeCN. Evaporate to dryness, reconstitute in 500 μL mobile phase, inject a 100 μL aliquot.

HPLC VARIABLES
Column: C18
Mobile phase: MeCN:75 mM pH 3 ammonium phosphate 50:50
Flow rate: 1.4
Injection volume: 100
Detector: F ex 216 em 340

CHROMATOGRAM
Limit of quantitation: 2.5 ng/mL

KEY WORDS
plasma; chiral; pharmacokinetics

REFERENCE
Bleske, B.E.; Welage, L.S.; Rose, S.; Amidon, G.L.; Shea, M.J. The effect of dosage release formulations on the pharmacokinetics of propranolol stereoisomers in humans, *J.Clin.Pharmacol.*, **1995**, *35*, 374–378.

SAMPLE
Matrix: blood, urine
Analyte: sotalol
Sample preparation: Condition a 500 mg Sep-Pak C18 SPE cartridge with 3 mL MeOH and 3 mL 200 mM sodium tetraborate. Dilute urine 10 times with 200 mM sodium tetraborate. Briefly vortex 1 mL plasma with 500 μL saturated pH 9.3 sodium tetraborate. Add 1 mL diluted urine or 1.5 mL diluted plasma to the SPE cartridge, wash with 2 mL 20 mM sodium tetraborate, wash with 2 mL water, wash with 1 mL dichloromethane, elute with 5 mL isopropanol. Add 100 μL 100 μg/mL isoamyl p-hydroxybenzoate in MeCN:water 10:90, 100 μL 4 mg/mL 2,3,4,6-tetra-O-acetyl-β-D-glucopyranosyl isothiocyanate in MeCN, and 100 μL 10 mM pH 8.0 $(NH_4)_2HPO_4$ to the eluate, heat at 50° for 3 h, evaporate to dryness under reduced pressure, reconstitute with 200 μL mobile phase, inject a 40 μL aliquot.

HPLC VARIABLES
Guard column: Guard-Pak Resolve C18 (Waters)
Column: 150 × 4.6 5 μm STR ODS II (Shimadzu)
Mobile phase: MeCN:20 mM pH 4.6 $(NH_4)H_2PO_4$ 40:60
Column temperature: 25
Flow rate: 1
Detector: UV 225

CHROMATOGRAM
Retention time: 11.4 ((-)-R), 14.3 ((+)-S)
Internal standard: isoamyl p-hydroxybenzoate (25.5)
Limit of quantitation: 22 ng/mL (plasma), 220 ng/mL (urine)

OTHER SUBSTANCES
Interfering: 40

KEY WORDS
chiral; rat; human; mouse; rabbit; plasma; SPE; pharmacokinetics

REFERENCE
Shimizu, T.; Hiraoka, M.; Nakanomyo, H. Enantioselective determination of sotalol enantiomers in biological fluids using high-performance liquid chromatography, *J.Chromatogr.B*, **1995**, *674*, 77–83.

SAMPLE
Matrix: bulk
Analyte: 3-aminoquinuclidine
Sample preparation: Prepare the free base by dissolving the hydrochloride in MeOH, add 2 equivalents of sodium methoxide, after a few min evaporate to dryness under a stream of nitrogen, reconstitute with dry DMF at a concentration of 1-2 mg/mL, add a 10% molar

excess of 2,3,4,6-tetra-O-acetyl-β-D-glucopyranosyl isothiocyanate, vortex, let stand at room temperature for 30 min, dilute with mobile phase, inject an aliquot.

HPLC VARIABLES
Column: 150 × 4.6 5 μm Zorbax C8
Mobile phase: MeCN:MeOH:buffer 15:25:60 (Buffer was 0.15% acetic acid containing 0.1% triethanolamine.)
Flow rate: 2
Injection volume: 10
Detector: UV 254

CHROMATOGRAM
Retention time: k' 9.24 (S), k' 10.65 (R)

KEY WORDS
chiral; comparison with other derivatizing reagents

REFERENCE
Demian, I.; Gripshover, D.F. Enantiomeric purity determination of 3-aminoquinuclidine by diastereomeric derivatization and high-performance liquid chromatographic separation, *J.Chromatogr.*, **1989**, *466*, 415–420.

SAMPLE
Matrix: bulk
Analyte: clobenzorex (N-(2-chlorobenzyl)-α-methylphenethylamine)
Sample preparation: Dissolve 2 mg of the hydrochloride in 30 mL 450 mM NaOH, extract with 30 mL chloroform. Add a 10% molar excess of 2,3,4,6-tetra-O-acetyl-β-D-glucopyranosyl isothiocyanate to the chloroform solution, let stand at room temperature for 10 min, evaporate to dryness under a stream of air, reconstitute with 1 mL chloroform, inject a 5 μL aliquot.

HPLC VARIABLES
Column: 300 × 3.9 μPorasil
Mobile phase: n-Hexane:chloroform:isopropanol 100:5:0.6
Flow rate: 1.5
Injection volume: 5
Detector: UV 254

CHROMATOGRAM
Retention time: 18 (+), 20 (-)

OTHER SUBSTANCES
Also analyzed: 3- and 4-chloro analogs

KEY WORDS
chiral; normal phase

REFERENCE
Noggle, F.T. Jr.; Clark, C.R.; Andurkar, S.V.; DeRuiter, J. Liquid chromatographic analysis of regioisomers and enantiomers of N-(chlorobenzyl)-α-methylphenethylamines: Analogues of clobenzorex, *J.Liq.Chromatogr.*, **1990**, *13*, 763–777.

SAMPLE
Matrix: formulations
Analyte: chloramphenicol
Sample preparation: Add capsule contents to water, extract with three 20 mL portions of ethyl acetate. Dry the extracts over potassium carbonate and evaporate them to dryness under reduced pressure. Add 120 mg of this product to 2.5 mL 5% HCl, heat at 95° for 2.5 h, cool to room temperature, adjust pH to 11 with concentrated NaOH, extract five times with 7 mL portions of ethyl acetate. Combine the extracts and dry them over potassium carbonate, evaporate to dryness, prepare a 10 mg/mL solution in MeOH. Add a 50 μL aliquot to 150 μL 10 mg/mL 2,3,4,6-tetra-O-acetyl-β-D-glucopyranosyl isothiocyanate (TAGIT) in MeCN, let stand at room temperature for 20 min, add 2 μL 2-aminoethanol, let stand for 10 min, add 50 μL 300 mM pH 3 ammonium phosphate, inject a

2-5 μL aliquot. (Chloramphenicol is de-acylated to 2-amino-1-(p-nitrophenyl)-1,3-pro-panediol which is then derivatized. The diastereomer obtained from chloramphen-icol (1R,2R) can be resolved from the other possible diastereomers.)

HPLC VARIABLES
Column: 150 × 3.9 Nova-pak ODS
Mobile phase: MeOH:water 42:58
Flow rate: 1
Injection volume: 2-5
Detector: UV 254

CHROMATOGRAM
Retention time: 12 (1R,2R), 18 (1S,2S), 22 (1R,2S or 1S,2R) , 25 (1R,2S or 1S,2R)

KEY WORDS
capsules; chiral

REFERENCE
Gal, J.; Meyer-Lehnert, S. Reversed-phase liquid chromatographic separation of enantiomeric and dias-tereomeric bases related to chloramphenicol and thiamphenicol, *J.Pharm.Sci.*, **1988**, 77, 1062–1065.

SAMPLE
Matrix: solutions
Analyte: 1-phenyl-2-aminopropanes
Sample preparation: 50 μL 1 mg/mL Compound in dichloromethane + 50 μL 7.6 mM 2,3,4,6-tetra-O-acetyl-β-D-glucopyranosyl isothiocyanate in MeCN or dichloromethane: triethylamine 99.8:0.2, mix, let stand at room temperature for 1 h, add 1 mL 1 M HCl, shake mechanically for 5 min, centrifuge at 1000 g for 15 min, discard the aqueous phase, add 1 mL 1 M NaOH, shake for 5 min, centrifuge at 1000 g for 15 min. Remove a 10 μL aliquot of the organic layer and dilute it to 1 mL with MeOH, inject a 20-50 μL aliquot.

HPLC VARIABLES
Column: 250 × 4.5 5 μm octadecyl (IBM)
Mobile phase: MeOH:water 55:45
Flow rate: 1-2
Injection volume: 20-50
Detector: UV 254

CHROMATOGRAM
Retention time: 26.4 ((R)-amphetamine), 28.5 ((S)-amphetamine), 49.2 ((R)-1-(4-chloro-phenyl)-2-aminopropane), 53.1 ((S)-1-(4-chlorophenyl)-2-aminopropane), 15.3 ((R)-1-phen-ylethylamine), 17.2 ((S)-1-phenylethylamine), 46.4 ((R)-1-(2,5-dimethoxy-4-methylphenyl)-2-aminopropane), 51.0 ((S)-1-(2,5-dimethoxy-4-methylphenyl)-2-aminopropane), 43.2 ((R)-1-(2,4-dimethoxy-5-methylphenyl)-2-aminopropane), 47.0 ((S)-1-(2,4-dimethoxy-5-methylphenyl)-2-aminopropane), 35.6 ((R)-1-(2,5-dimethoxy-4-thiomethylphenyl)-2-ami-nopropane), 38.0 ((S)-1-(2,5-dimethoxy-4-thiomethylphenyl)-2-aminopropane)
Limit of quantitation: 100 ng

KEY WORDS
comparison with other derivatizing reagents; chiral

REFERENCE
Miller, K.J.; Gal, J.; Ames, M.M. High-performance liquid chromatographic resolution of enantiomers of 1-phenyl-2-aminopropanes (amphetamines) with four chiral reagents, *J.Chromatogr.*, **1984**, 307, 335–342.

SAMPLE
Matrix: solutions
Analyte: 4'-hydroxypropranolol sulfate
Sample preparation: Evaporate a solution of 0.1-5 μg compound in MeOH to dryness under a stream of nitrogen, add 50 μL 0.4% triethylamine in MeCN:water 50:50, add 50 μL 60 mM 2,3,4,6-tetra-O-acetyl-β-D-glucopyranosyl isothiocyanate in MeCN, mix briefly, let stand at room temperature for 5-10 min, inject an aliquot.

HPLC VARIABLES
Column: 250 × 4.6 5 μm Spherisorb ODS
Mobile phase: MeCN:MeOH:water:acetic acid 35:5:59:1 containing 50 mM ammonium acetate, pH 4
Flow rate: 1
Detector: UV 313

CHROMATOGRAM
Retention time: 9 (-), 11 (+)
Limit of detection: 20 ng

KEY WORDS
plasma; dog; chiral

REFERENCE
Walle, T.; Christ, D.D.; Walle, U.K.; Wilson, M.J. Separation of the enantiomers of intact sulfate conjugates of adrenergic drugs by high-performance liquid chromatography after chiral derivatization, *J.Chromatogr.*, **1985**, *341*, 213–216.

SAMPLE
Matrix: solutions
Analyte: amphetamine
Sample preparation: Dissolve 2 mg in 15 mL 0.45 M NaOH, extract twice with 30 mL chloroform. Combine the organic layers and add a 10% molar excess of 2,3,4,6-tetra-O-acetyl-β-D-glucopyranosyl isothiocyanate in chloroform, let stand for 10 min, evaporate the chloroform under a stream of air, dissolve the residue in 1 mL THF or MeOH, inject a 5 μL aliquot.

HPLC VARIABLES
Guard column: 70 × 2.1 Co:Pell ODS
Column: 300 × 3.9 μBondapak C18
Mobile phase: MeOH:water:acetic acid 50:49:1
Flow rate: 1.5
Injection volume: 5
Detector: UV 254

CHROMATOGRAM
Retention time: 11 (R,R-norpseudoephedrine), 13 (R,S-norephedrine), 18 (S,R-norephedrine), 25 (R-amphetamine), 28 (S-amphetamine)

KEY WORDS
chiral

REFERENCE
Noggle, F.T. Jr.; DeRuiter, J.; Clark, C.R. Liquid chromatographic determination of the enantiomeric composition of amphetamine prepared from norephedrine and norpseudoephedrine, *J.Chromatogr.Sci.*, **1987**, *25*, 38–42.

SAMPLE
Matrix: solutions
Analyte: aminocyclopentanecarboxylic acid
Sample preparation: Prepare a 1 mg/mL solution of aminocyclopentanecarboxylic acid in MeCN:triethylamine 99.8:0.2. Remove a 50 μL aliquot and add it to 50 μL 0.5% 2,3,4,6-tetra-O-acetyl-β-D-glucopyranosyl isothiocyanate in MeCN, let stand at room temperature for 1 h, inject an aliquot (J. Chromatogr. 1980, 202, 375).

HPLC VARIABLES
Column: 125 × 4 5 μm Lichrospher 100 RP18
Mobile phase: MeOH:10 mM pH 3 sodium acetate 42.5:57.5
Flow rate: 0.8
Detector: UV 250

CHROMATOGRAM
Retention time: 10 (trans-(1S,2S)), 13 (trans-(1R,2R)), 16 (cis-(1S,2R)), 19 (cis-(1S,2R))
Limit of detection: 0.1% of major isomer

KEY WORDS
chiral

REFERENCE
Péter, A.; Fülöp, F. High-performance liquid chromatographic method for the separation of isomers of cis- and trans-2-amino-cyclopentane-1-carboxylic acid, *J.Chromatogr.A*, **1995**, *715*, 219–226 .

2,3,4,6-Tetra-O-benzoyl-β-D-glucopyranosyl Isothiocyanate

SAMPLE
Matrix: bulk
Analyte: amino acids and β-blockers
Sample preparation: Dissolve 0.1 mmole compound in 10 mL MeCN:water:triethylamine 50:50:0.55. Remove a 50 µL aliquot and add it to 50 µL 0.66% 2,3,4,6-tetra-O-benzoyl-β-D-glucopyranosyl isothiocyanate (Fluka) in MeCN, shake mechanically for 30 min, add 10 µL 0.26% ethanolamine in MeCN, shake for 10 min, make up to 1 mL with MeCN, inject a 10 µL aliquot.

HPLC VARIABLES
Column: 25 × 4 (sic) 5 µm LiChrospher 100 RP-18
Mobile phase: MeOH:water:67 mM pH 7.0 phosphate buffer 65:27:8 (A) or 70:25:5 (B) or 80:15:5 (C) or 70:30:0 (D) or 80:20:0 (E) or 85:15:0 (F) or 90:10:0 (G)
Flow rate: 0.42 (A) or 0.45 (B) or 0.50 (C, F, G) or 1 (D, E)
Injection volume: 10
Detector: UV 231

CHROMATOGRAM
Retention time: k' 2.65, 3.65 (propranolol enantiomers (G)), k' 3.00, 3.71 (pindolol enantiomers (F)), k' 5.01, 6.53 (sotalol enantiomers (E)), k' 5.19 (D-proline (B)), k' 5.35 (L-threonine (B)), k' 5.68, 6.66 (atenolol enantiomers (E)), k' 6.22 (L-tyrosine (B)), k' 6.24 (L-2-aminobutyric acid (B)), k' 6.24 (D-threonine (B)), k' 6.38 (L-phenylglycine (B)), k' 6.41 (L-proline (B)), k' 7.22 (L-valine (B)), k' 7.37 (L-penicillamine (B)), k' 7.41 (D-tyrosine (B)), k' 7.57 (D-2-aminobutyric acid (B)), k' 7.86 (D-phenylglycine (B)), k' 8.08 (L-methionine (B)), k' 8.86, 11.66 (5-(2-N-benzylamino-2-hydroxyethyl)salicylamide enantiomers (E)), k'9.16 (D-valine (B)), k' 9.27 (L-isoleucine (B)), k' 9.43 (L-tryptophan (B)), k' 9.51 (L-leucine (B)), k' 10.05 (D-penicillamine (B)), k' 10.24 (D-methionine (B)), k' 10.54 (L-phenylalanine (B)), k' 11.94 (L-ornithine (C)), k' 12.03 (D-tryptophan (B)), k' 12.35 (D-isoleucine (B)), k' 12.65 (D-leucine (B)), k' 12.89 (L-norleucine (B)), k' 13.48 (L-lysine (C)), k' 13.81 (D-phenylalanine (B)), k' 13.90 (D-ornithine (C)), k' 15.32 (D-lysine (C)), k' 16.81 (D-norleucine (B)), k' 16.88 (L-3-aminobutyric acid (A)), k' 18.00 (L-alanine (A)), k' 18.95 (D-3-aminobutyric acid (A)), k' 20.85 (D-alanine (A)), k' 22.61,25.67 (epinephrine enantiomers (D)), k' 33.96, 40.07 (phenylephrine enantiomers (D))

OTHER SUBSTANCES
Simultaneously analyzed: amino acids

KEY WORDS
chiral

REFERENCE

Lobell, M.; Schneider, M.P. 2,3,4,6-Tetra-O-benzoyl-β-D-glucopyranosyl isothiocyanate: an efficient reagent for the determination of enantiomeric purities of amino acids, β-adrenergic blockers and alkyloxiranes by high-performance liquid chromatography using standard reversed-phase columns, *J.Chromatogr.*, **1993**, *633*, 287–294.

2,3,4-Tri-O-acetyl-α-D-arabinopyranosyl Isothiocyanate

RELATED REFERENCE

Kinoshita, T.; Kasahara, Y.; Nimura, N. Reversed-phase high-performance liquid chromatographic resolution of nonesterified enantiomeric amino acids by derivatization with 2,3,4,6-tetra-O-acetyl-β-D-glucopyranosyl isothiocyanate and 2,3,4-tri-O-acetyl-α-D-arabinopyranosyl isothiocyanate. *J.Chromatogr.* **1981**, *210*, 77-81.

SAMPLE

Matrix: solutions

Analyte: 1-phenyl-2-aminopropanes

Sample preparation: 50 μL 1 mg/mL Compound in dichloromethane + 50 μL 7.6 mM 2,3,4-tri-O-acetyl-α-D-arabinopyranosyl isothiocyanate (Fluka) in MeCN or dichloromethane:triethylamine 99.8:0.2, mix, let stand at room temperature for 1 h, add 1 mL 1 M HCl, shake mechanically for 5 min, centrifuge at 1000 g for 15 min, discard the aqueous phase, add 1 mL 1 M NaOH, shake for 5 min, centrifuge at 1000 g for 15 min. Remove a 10 μL aliquot of the organic layer and dilute it to 1 mL with MeOH, inject a 20-50 μL aliquot.

HPLC VARIABLES

Column: 250 × 4.5 5 μm octadecyl (IBM)

Mobile phase: MeOH:water 55:45

Flow rate: 1-2

Injection volume: 20-50

Detector: UV 254

CHROMATOGRAM

Retention time: 18.2 ((R)-amphetamine), 16.8 ((S)-amphetamine), 35.4 ((R)-1-(4-chlorophenyl)-2-aminopropane), 32.3 ((S)-1-(4-chlorophenyl)-2-aminopropane), 11.4 ((R)-1-phenylethylamine), 10.4 ((S)-1-phenylethylamine), 33.1 ((R)-1-(2,5-dimethoxy-4-methylphenyl)-2-aminopropane), 30.2 ((S)-1-(2,5-dimethoxy-4-methylphenyl)-2-aminopropane), 31.2 ((R)-1-(2,4-dimethoxy-5-methylphenyl)-2-aminopropane), 28.7 ((S)-1-(2,4-dimethoxy-5-methylphenyl)-2-aminopropane), 24.8 ((R)-1-(2,5-dimethoxy-4-thiomethylphenyl)-2-aminopropane), 23.5 ((S)-1-(2,5-dimethoxy-4-thiomethylphenyl)-2-aminopropane)

Limit of quantitation: 100 ng

KEY WORDS

comparison with other derivatizing reagents; chiral

REFERENCE

Miller, K.J.; Gal, J.; Ames, M.M. High-performance liquid chromatographic resolution of enantiomers of 1-phenyl-2-aminopropanes (amphetamines) with four chiral reagents, *J.Chromatogr.*, **1984**, *307*, 335–342.

KETOALDEHYDE

3-(4-Carboxybenzoyl)-2-quinolinecarboxaldehyde

SAMPLE

Matrix: polysaccharides

Analyte: amino sugars

Sample preparation: Dissolve 50 mg chitosan polysaccharide in 2 mL water, add 500 μL 1 M HCl, heat at 100°, add 2.5 mL concentrated HCl, heat at 100° for 34 h. Remove a 50 μL aliquot and evaporate it to dryness, reconstitute with 50 μL water, add 10-20 μL 20 mM KCN in water, add 5-10 μL 3 mg/mL reagent in MeOH, let stand at room temperature for 1 h, inject an aliquot. (Reagent was 3-(4-carboxybenzoyl)-2-quinolinecarboxaldehyde, available from Molecular Probes, Eugene OR. Synthesis is as follows. Mix 9.69 g p-toluidine, 10.89 g o-nitrobenzaldehyde, and 25 mL EtOH and allow to react for 5 min, filter, wash the solid with EtOH to give 2-nitro-N-(p-tolyl)benzaldimine (Talanta 1989, 36, 321). (More 2-nitro-N-(p-tolyl)benzaldimine can be obtained by adding water to the filtrate.) Add, in portions in a thin stream, a hot solution of 46 g sodium sulfide in 23 mL water and 23 mL EtOH to a stirred refluxing solution of 100 mmole 2-nitro-N-(p-tolyl)benzaldimine in 50 mL EtOH. After a vigorous reaction 2-amino-N-(p-tolyl)benzaldimine crystallizes from the cooling solution (mp 102-103° after recrystallization from dilute MeOH) (Ber. 1943, 76, 1099). Wash 950 mg of a commercial 50% slurry of sodium hydride with pentane, add 7 mL dry THF (distilled from lithium aluminum hydride), add 1.51 g methyl 4-cyanobenzoate in 10 mL THF, add dropwise 1.38 mL acetone (distilled from calcium chloride), reflux for 1.5 h, cool, acidify with 3 M HCl. Remove the organic layer and wash it with brine and sodium bicarbonate solution, dry over anhydrous magnesium sulfate, evaporate to give (4-cyanobenzoyl)acetone. Reflux 433 mg (4-cyanobenzoyl)acetone, 486 mg 2-amino-N-(p-tolyl)benzaldimine, 69 mL piperidine, and 9 mL EtOH:water 95:5 for 18 h, remove volatiles by steam distillation, add the residue to water and dichloromethane. Remove the organic layer and dry it, evaporate to give 3-(4-cyanobenzoyl)-2-methylquinoline. Suspend 547 mg 3-(4-cyanobenzoyl)-2-methylquinoline in 13 mL EtOH:water 95:5, add 500 mg KOH, reflux for 6 h, cool, concentrate, add the residue to ether and water. Remove the aqueous layer and adjust the pH to 5 with tartaric acid, let stand for 15 min, filter, wash the precipitate with water, dry under vacuum to give 3-(4-carboxybenzoyl)-2-methylquinoline. Dissolve 266 mg 3-(4-carboxybenzoyl)-2-methylquinoline in 6 mL acetic acid, add 112 mg selenium dioxide, stir at 80° for 2 h, filter through Celite, wash the precipitate with several volumes of hot MeOH, dilute the filtrate with water, allow to stand, filter, wash the solid with water, dry under vacuum to give 3-(4-carboxybenzoyl)-2-quinolinecarboxaldehyde (Anal. Chem. 1991, 63, 408).)

CAPILLARY ELECTROPHORESIS

Capillary: 85 cm × 50 μm (55 cm to detector)

Running buffer: MeOH:buffer 30:70, pH 9.40 (Buffer was 10 mM Na_2HPO_4 containing 30 mM sodium borate and 50 mM sodium dodecyl sulfate.)
Voltage/Current: 20 kV/15 μA
Injection: Hydrodynamic or electromigration for 10 s
Detector: F ex 442 (50 mW He/Cd laser) em 550
Model: laboratory constructed
Migration time: 18-27 (depending on degree of polymerization)

REFERENCE
Liu, J.; Shirota, O.; Novotny, M. Capillary electrophoresis of amino sugars with laser-induced fluorescence detection, *Anal.Chem.*, **1991**, *63*, 413–417.

SAMPLE
Matrix: solutions
Analyte: saccharides
Sample preparation: Dissolve saccharides in water, add excess of 2 M ammonium sulfate or 4 M ammonium chloride, add excess 400 mM sodium cyanoborohydride, mix well, heat at 100° for 100-120 min, cool in an ice bath. Mix an aliquot with 10-20 μL 20 mM KCN in water, add 5-10 μL 10 mM reagent in MeOH, let stand at room temperature for 1 h, inject an aliquot. (Reagent was 3-(4-carboxybenzoyl)-2-quinolinecarboxaldehyde, available from Molecular Probes, Eugene OR. Synthesis is as described above.)

CAPILLARY ELECTROPHORESIS
Capillary: 88 cm × 50 μm (58 cm to detector) (Polymicro Technologies)
Running buffer: 10 mM pH 9.40 Na_2HPO_4 containing 10 mM sodium tetraborate
Voltage/Current: 20 kV/12 μA
Injection: Hydrodynamic injection for 5 s
Detector: F ex 457 (argon laser) em 552
Model: laboratory constructed
Migration time: 10.6 (glucosamine), 11.2 (galactosamine), 11.6 (erythrose), 12.4 (ribose), 13.2 (talose), 13.4 (mannose), 14 (dextrose), 14.2 (galactose), 17.2 (galacturonic acid), 18.2 (glucuronic acid), 18.8 (glucosaminic acid), 21 (glucose 6-phosphate), 11-14 (oligosaccharides, depending on degree of polymerization)
Limit of detection: 0.5-2.3 amole

REFERENCE
Liu, J.; Shirota, O.; Wiesler, D.; Novotny, M. Ultrasensitive fluorometric detection of carbohydrates as derivatives in mixtures separated by capillary electrophoresis, *Proc.Nat.Acad.Sci.USA*, **1991**, *88*, 2302–2306.

SAMPLE
Matrix: solutions
Analyte: TSB-51 protein
Sample preparation: Mix a 26-47 μg/mL solution of the protein in 20 mM pH 8.0 Tris-HCl buffer containing 2.5 mM EDTA with one volume of 10 mM 3-(4-carboxybenzoyl)-2-quinolinecarboxaldehyde (Molecular Probes, Eugene OR (see also synthesis above)) and two volumes of 20 mM KCN, let stand at room temperature for 4 h, inject an aliquot.

CAPILLARY ELECTROPHORESIS
Capillary: 57 cm × 75 μm fused-silica
Running buffer: 40 mM pH 9.0 Sodium borate buffer
Capillary temperature: 25
Voltage/Current: 25 kV
Injection: Pressure injection for 4 s.
Detector: F ex 488 (3 mW argon laser) em 520
Model: Beckman PACE 2100
Migration time: 8.23

REFERENCE
Asermely, K.E.; Broomfield, C.A.; Nowakowski, J.; Courtney, B.C.; Adler, M. Identification of a recombinant synatobrevin-thioredoxin fusion protein by capillary zone electrophoresis using laser-induced fluorescence detection, *J.Chromatogr.B*, **1997**, *695*, 67–75.

3,4-Dimethoxyphenylglyoxal

SAMPLE
Matrix: urine
Analyte: cGMP
Sample preparation: Condition a Toyopak DEAE S (anion-exchange, diethylaminoethyl resin; Tosoh) SPE cartridge with 2 mL water. Mix 200 μL filtered urine with 200 μL 20 mM HCl, add to the SPE cartridge, wash with 600 μL water, elute with 200 μL 500 mM NaCl. Remove a 50 μL aliquot of the eluate and add it to 50 μL 30 mM pH 7.0 sodium phosphate buffer, add 50 μL 100 mM 3,4-dimethoxyphenylglyoxal in DMSO:water 40:60, add 50 μL water, heat at 37° for 5 min, inject a 20 μL aliquot.

HPLC VARIABLES
Column: 150 × 4.6 5 μm TSK-ODS 120T (Tosoh)
Mobile phase: Gradient. MeCN:THF:50 mM pH 6.0 sodium phosphate buffer:water from 2:3:25:69 to 25:3:25:47 (?) over 30 min.
Flow rate: 1
Injection volume: 20
Detector: F ex 400 em 510

CHROMATOGRAM
Retention time: 4 (guanosine diphosphate), 5.5 (deoxyguanosine diphosphate), 7.5 (guanosine monophosphate), 9 (deoxyguanosine monophosphate), 12.5 (guanosine 3',5'-cyclic monophosphate)
Limit of detection: 4-14 pmole

KEY WORDS
SPE

REFERENCE

Ohba, Y.; Kai, M.; Zaitsu, K. High-performance liquid chromatographic determination of guanosine 3',5'-cyclic monophosphate in human urine with fluorescence detection using 3,4-dimethoxyphenylglyoxal, *Anal.Sci.*, **1997**, *13*, 469−472.

3-Furoylquinoline-2-carboxaldehyde

+ RNH$_2$ + CN$^\ominus$ ⟶

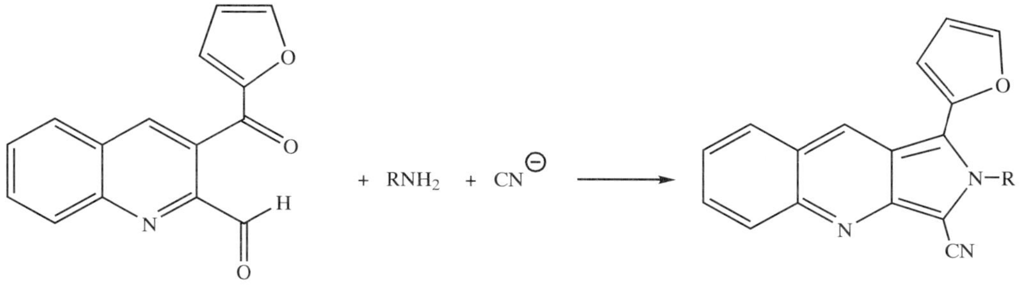

SAMPLE

Matrix: solutions

Analyte: ovalbumin

Sample preparation: Place a 10 μL aliquot of a 100 mM solution of reagent in MeOH in a tube and evaporate to dryness under reduced pressure, add 9 μL of a protein solution in 2.5 mM pH 9.4 borax, add 1 μL 25 mM KCN in 2.5 mM pH 9.4 borax, let stand for 30 min, inject an aliquot. (The reagent is described as 5-furoylquinoline-3-carboxaldehyde but it seems more likely that it is 3-furoylquinoline-2-carboxaldehyde, ATTO-TAG FQ supplied by Molecular Probes, Eugene OR).

CAPILLARY ELECTROPHORESIS

Capillary: 50 cm × 50 μm fused-silica (Polymicro Technologies)

Running buffer: 2.5 mM pH 9.4 borax containing 5 mM sodium dodecyl sulfate

Voltage/Current: 20 kV

Injection: Electrokinetic injection at 5 kV for 5 s.

Detector: F ex 488 (12 mW argon ion laser)

Model: laboratory-constructed

Migration time: 4.5

Limit of detection: 0.01 nM

REFERENCE

Pinto, D.M.; Arriaga, E.A.; Craig, D.; Angelova, J.; Sharma, N.; Ahmadzadeh, H.; Dovichi, N.J. Picomolar assay of native proteins by capillary electrophoresis precolumn labeling, submicellar separation, and laser-induced fluorescence detection, *Anal.Chem.*, **1997**, *69*, 3015−3021.

Glyoxal

SAMPLE
Matrix: solutions
Analyte: peptides
Sample preparation: Mix 100 μL of a 100 μM solution of peptide in water with 100 μL 200 mM pH 4.5 succinate buffer and 200 mM glyoxal in water, heat at 100° for 30 min, inject a 100 μL aliquot.

HPLC VARIABLES
Column: 150 × 4.6 5 μm TSKgel ODS-80TM (Tosoh)
Mobile phase: MeCN:MeOH:20 mM pH 6.0 phosphate buffer 23:17:60
Flow rate: 0.8
Injection volume: 100
Detector: F ex 275 em 465

CHROMATOGRAM
Retention time: 3.7 (Trp-Gly-Gly), 4.0 (Trp-Gly), 4.8 (Trp-Ala), 7.4 (Trp-NH$_2$), 7.8 (tryptophan), 7.8 (tryptamine), 8.0 (Trp-Trp), 9.2 (Trp-Leu), 18.2 (Trp-Met-Asp-Phe-NH$_2$)
Limit of detection: 55-4166 fmole

OTHER SUBSTANCES
Non-interfering: N-acetylgalactosamine, adenine, Ala-Trp, amino acids, cytosine, epinephrine, galactosamine, guanine, histamine, 3-hydroxykynurenine, 5-hydroxytryptophan, kynurenine, Lys-Trp-Lys, melatonin, norepinephrine, serotonin, spermidine, spermine, thymine, uracil

KEY WORDS
Method is specific for N-terminal tryptophan-containing peptides.

REFERENCE
Kai, M.; Kojima, E.; Ohkura, Y.; Iwasaki, M. High-performance liquid chromatography of N-terminal tryptophan-containing peptides with precolumn fluorescence derivatization with glyoxal, *J.Chromatogr.A,* **1993**, *653*, 235–240.

Phenylglyoxal

SAMPLE
Matrix: blood
Analyte: tryptophan
Sample preparation: 10 μL Serum + 30 μL water + 200 μL 300 mM perchloric acid, mix, centrifuge at 1000 g for 5 min. Remove a 100 μL aliquot of the supernatant and add it to 50 μL 60 mM phenylglyoxal hydrate in DMSO, heat at 100° for 15 min, inject a 100 μL aliquot.

HPLC VARIABLES
Column: 150 × 4.6 5 μm TSK gel ODS-80Tm (Tosoh)
Mobile phase: MeCN:50 mM pH 2.0 phosphate buffer:water 33:35:32
Flow rate: 1
Injection volume: 100
Detector: F ex 385 em 460

CHROMATOGRAM
Retention time: 4.6
Limit of detection: 72 nM

KEY WORDS
serum

REFERENCE
Kojima, E.; Kai, M.; Ohkura, Y. Determination of tryptophan in human serum by high-performance liquid chromatography with pre-column fluorescence derivatization using phenylglyoxal, *J.Chromatogr.*, **1993**, *612*, 187–190.

SAMPLE
Matrix: solutions
Analyte: guanine nucleosides and nucleotides
Sample preparation: 200 μL Solution + 100 μL 100 mM phenylglyoxal in DMSO + 100 μL 50 mM pH 6.0 phosphate buffer, heat at 37° for 15 min, inject a 100 μL aliquot.

HPLC VARIABLES
Column: 150 × 4.6 5 μm TSKgel ODS-120T
Mobile phase: Gradient. MeCN:5 mM pH 6.5 phosphate buffer from 5:95 to 22:78 over 28 min.
Flow rate: 1
Injection volume: 100
Detector: F ex 360 em 510

CHROMATOGRAM
Retention time: 8 (GTP), 10 (GDP, dGTP), 11 (dGDP), 13 (GMP), 16 (dGMP), 20 (cGMP), 23 (guanine), 24 (guanosine), 25 (deoxyguanosine)
Limit of detection: 140-720 fmole

OTHER SUBSTANCES
Non-interfering: adenine, adenosine, ADP, AMP, ATP, CDP, CMP, CTP, cytidine, cytosine, dADP, dATP, dATP, hypoxanthine, thymidine, thymine, uracil, uridine, UTP, xanthine

REFERENCE
Yonekura, S.; Iwasaki, M.; Kai, M.; Ohkura, Y. High-performance liquid chromatography of guanine and its nucleosides and nucleotides by pre-column fluorescence derivatization with phenylglyoxal reagent, *J.Chromatogr.*, **1993**, *641*, 235–239.

LACTONE

Fluorescamine

RELATED REFERENCES
Andarelli, M.A.; Farinotti, R.; Dauphin, A. HPLC determination of mexiletine in plasma after derivatization with fluorescamine. *J.Pharm.Clin.* **1982**, *1*, 113-122.

Baeyens, W.R.G.; Van Der Weken, G.; Van de Voorde, I.; Dewaele, C. Micro-bore LC of nonapeptides and some primary amine drugs on a polymer column applying postcolumn fluorescamine labelling. *Biomed.Chromatogr.* **1997**, *11*, 111-112.

Blanchin, M.D.; Fabre, H.; Mandrou, B. Fluorescamine post-column derivatization for the HPLC determination of cephalosporins in plasma and urine. *J.Liq.Chromatogr.* **1988**, *11*, 2993-3010.

Flesch, G.; Mann, C.; Degen, P.H. Quantitative determination of CGP 61755, a protease inhibitor, in plasma and urine by high-performance liquid chromatography and fluorescence detection. *J.Chromatogr.B* **1997**, *696*, 123-130.

García Sánchez, F.; Navaz Díaz, A.; García Pareja, A.; Bracho, V. Liquid chromatographic determination of asulam and amitrole with pre-column derivatization. *J.Liq.Chromatogr.Rel.Technol.* **1997**, *20*, 603-615.

Groeningsson, K.; Widahl-Naesman, M. Liquid chromatographic determination of felypressin after pre-column derivatization with fluorescamine. *J.Chromatogr.* **1984**, *291*, 185-194.

Ingles, D.L.; Gallimore, D. High-performance liquid chromatography of fluorescamine-labeled amines in acid solvents. *J.Chromatogr.* **1985**, *325*, 346-352.

Laloux, J.; Romnee, J.M.; Marin, C.; Vanwijnsberghe, D. Determination of sulfonamides in milk using fluorescamine derivatization and HPLC analysis. *Milchwissenschaft* **1996**, *51*, 517-520.

Rose, M.D.; Farrington, W.H.; Shearer, G. The effect of cooking on veterinary drug residues in food: 3. Sulphamethazine (sulphadimidine). *Food Addit.Contam.* **1995**, *12*, 739-750.

Scholten, A.H.M.T.; Brinkman, U.A.T.; Frei, R.W. Fluorescence detection of chloroanilines in liquid chromatography using a post-column reaction with fluorescamine. Comparison of reactor types and mixing tees. *J.Chromatogr.* **1981**, *218*, 3-13.

Sedman, A.J.; Gal, J. Pre-column derivatization with fluorescamine and high-performance liquid chromatographic analysis of drugs. Application to tocainide. *J.Chromatogr.* **1982**, *232*, 315-326.

Shibata, N.; Akabane, M.; Minouchi, T.; Ono, T.; Shimakawa, H. Fluorimetric determination of mexiletine in serum by high-performance liquid chromatography using pre-column derivatization with fluorescamine. *J.Chromatogr.* **1991**, *566*, 187-194.

Takeda, N.; Akiyama, Y. Rapid determination of sulfonamides in milk using liquid chromatographic separation and fluorescamine derivatization. *J.Chromatogr.* **1992**, *607*, 31-35

SAMPLE
Matrix: blood
Analyte: amifostine (WR 2721)
Sample preparation: 90 µL Plasma + 50 µL 1.4 µg/mL IS + 160 µL 50 mM pH 10 sodium borate/KCl buffer, add 200 µL 5 mg/mL fluorescamine in acetone (dried over 4 Å molecular sieve) while vortexing, vortex for 1 min, add 200 µL 5 mg/mL fluorescamine in acetone (dried over 4 Å molecular sieve), vortex for 20-30 s, centrifuge at 1500 rpm for 3 min, inject a 20-50 µL aliquot of the supernatant.

HPLC VARIABLES
Guard column: CoPell C18
Column: 100 × 8 5 µm spherical C18 RCM-100 (Waters)
Mobile phase: MeCN:water 22:78 containing 10 mM dibutylammonium phosphate, pH 3 (Prepare solutions of dibutylammonium phosphate by adjusting pH of dibutylamine solution to 2.5 with phosphoric acid. Flush column with MeOH:water 70:30 at the end of each day.)
Flow rate: 2
Injection volume: 20-50
Detector: F ex 395 em >460

CHROMATOGRAM
Retention time: 13.1
Internal standard: ^{14}C-labeled amifostine (Collect fraction containing amifostine and count it.)
Limit of quantitation: 2 µg/mL

OTHER SUBSTANCES
Non-interfering: metabolites

KEY WORDS
plasma; dog; pharmacokinetics

REFERENCE
Swynnerton, N.F.; McGovern, E.P.; Mangold, D.J.; Niño, J.A.; Gause, E.M.; Fleckenstein, L. HPLC assay for S-2-(3-aminopropylamino)ethyl phosphorothioate (WR 2721) in plasma, *J.Liq.Chromatogr.*, **1983**, *6*, 1523–1534.

SAMPLE
Matrix: blood
Analyte: amifostine
Sample preparation: 150 µL Plasma + 150 µL IS in 50 mM pH 10 sodium borate/KCl buffer + 200 µL 50 mM pH 7.6 sodium borate/KCl buffer, vortex, add 250 µL reagent, mix for 1 min, add 250 µL reagent, mix for 20-30 s, centrifuge at 1500 rpm for 3 min. Filter (0.45 µm) the supernatant and inject an aliquot of the supernatant. (Reagent was 5 mg/mL fluorescamine in acetone that had been previously dried over 4Å molecular sieve. Prepare fresh each week.)

HPLC VARIABLES
Guard column: CoPell C18
Column: 100 × 8 5 µm spherical C18 RCM-100 radial compression module (Waters)
Mobile phase: MeCN:water 22:78 containing 10 mM tetrabutylammonium phosphate, pH 3 or MeCN:EtOH:water 16:7:77 containing 10 mM tetrabutylammonium phosphate, pH 3 (Prepare 1 M solutions of tetrabutylammonium phosphate by adjusting the pH of 40% tetrabutylammonium hydroxide to pH 2.5 with phosphoric acid and adjusting to an appropriate volume.)
Flow rate: 2
Injection volume: 50
Detector: F ex 370 (bandpass filter) em 418-700 (cutoff filter)

CHROMATOGRAM
Internal standard: S-3-(4-aminobutylamino)propyl phosphorothioate (WR 80855)
Limit of quantitation: 500 ng/mL

OTHER SUBSTANCES
Non-interfering: metabolites

KEY WORDS
plasma; dog; pharmacokinetics

REFERENCE
Swynnerton, N.F.; McGovern, E.P.; Niño, J.A.; Mangold, D.J. An improved HPLC assay for S-2-(3-aminopropylamino)ethyl phosphorothioate (WR-2721) in plasma, *Int.J.Radiat.Oncol.Biol.Phys.*, **1984**, *10*, 1521–1524.

SAMPLE
Matrix: blood
Analyte: bacampicillin
Sample preparation: Add 100000 fold excess of ampicillin to eliminate adsorption of bacampicillin, extract into butyl acetate, re-extract into pH 2 buffer. Wash the aqueous phase with n-hexane, inject an aliquot.

HPLC VARIABLES
Column: 100 × 2.9 5 μm Nucleosil C18
Mobile phase: MeCN:pH 7.4 phosphate buffer 41:49
Injection volume: 20
Detector: F with post-column reaction. The column effluent mixed with two volumes of sodium borate buffer then with one volume of 150 μg/mL fluorescamine in acetone (Science 1972, 178, 871).

CHROMATOGRAM
Limit of detection: 800 pg/mL

KEY WORDS
post-column reaction

REFERENCE
Sjövall, J.; Westerlund, D.; Alván, G.; Magni, L.; Nord, C.E.; Sörstad, J. Rectal bioavailability of bacampicillin hydrochloride in man as determined by reversed-phase liquid chromatography, *Chemotherapy*, **1984**, *30*, 137–147.

SAMPLE
Matrix: blood
Analyte: epsilon-aminocaproic acid
Sample preparation: 20 μL Serum + 2 μL 3 mg/mL trans-4-aminomethylcyclohexanecarboxylic acid in water + 20 μL MeCN, mix, centrifuge at 10000 g for 3 min. Remove 5 μL of the supernatant and add it to 100 μL 25 mM pH 8 phosphate buffer, add 100 μL 300 μg/mL fluorescamine in acetone, vortex, inject a 20 μL aliquot.

HPLC VARIABLES
Column: 250 × 4.6 10 μm LiChrosorb RP 18
Mobile phase: MeCN:water:acetic acid:THF 30:69:0.5:0.5, containing 40 mM sodium acetate
Flow rate: 2
Injection volume: 20
Detector: F ex 390 em 475

CHROMATOGRAM
Retention time: 4
Internal standard: tranexamic acid (trans-4-aminomethylcyclohexanecarboxylic acid) (5)
Limit of detection: 6 μg/mL

KEY WORDS
serum

REFERENCE
Lacroix, C.; Levert, P.; Laine, G.; Goulle, J.P. Microdosage de deux antifibrinolytiques (acide epsilon-aminocaproque et acide tranexamique) par chromatographie liquide et détection fluorimétrique [Microanalysis of two antifibrinolytics (epsilon-aminocaproic acid and tranexamic acid) by liquid chromatography and fluorometry], *J.Chromatogr.*, **1984**, *309*, 183–186.

SAMPLE
Matrix: blood
Analyte: KABI 2161 (1-(ethoxycarbonyloxy)ethyl trans-4-(aminomethyl)cyclohexanecarboxylate)
Sample preparation: 500 µL Whole blood + 3 mL dichloromethane + 2 mL 32 ng/mL IS in bromothymol blue solution + 100 µL 3.8% (dog) or 10% (rat) sodium citrate, gently rotate by hand for 15 s, remove stopper after first turns, centrifuge at 1650 g. Remove the organic phase and add it to 1 mL 50 mM tetrabutylammonium hydrogen sulfate in 500 mL pH 3 phosphate buffer (µ = 0.1), shake mechanically for 30 min, centrifuge. Remove the upper aqueous phase and add it to 1 mL pH 7.5 phosphate buffer (µ = 0.1), rapidly add 500 µL 0.4% fluorescamine in MeCN, vortex, inject a 100-200 µL aliquot. (Prepare bromothymol blue solution by dissolving 6.46 g bromothymol blue sodium salt in 8 mL 1 M NaOH and making up to 1 L with water, discard after 1 month.)

HPLC VARIABLES
Guard column: 40 × 4.6 Spheri-5 RP8
Column: 150 × 4.6 5 µm Nucleosil C8
Mobile phase: MeCN:pH 2.5 phosphate buffer (µ = 0.1) 55:45
Flow rate: 1
Injection volume: 100-200
Detector: F ex 280 em 470 (filter)

CHROMATOGRAM
Retention time: 7.5
Internal standard: ethyl trans-4-(aminomethyl)cyclohexanecarboxylate (OAF 71901) (KabiVitrum) (5.5)
Limit of quantitation: 10 ng/mL
Limit of detection: 1-3 ng/mL

KEY WORDS
whole blood; human; dog; rat

REFERENCE
Abrahamsson, M. Determination of a prodrug of tranexamic acid in whole blood by reversed-phase liquid chromatography after pre-column derivatization with fluorescamine, *J.Pharm.Biomed.Anal.*, **1986**, *4*, 399–406.

SAMPLE
Matrix: blood
Analyte: amines
Sample preparation: Equilibrate a 95 × 3 column of Amberlite CG-50 with 200 mM pH 6.5 sodium phosphate buffer and wash with 10 mL 5 mM disodium EDTA. Condition two Sep-Pak C18 SPE cartridges connected in series with 5 mL MeOH and 5 mL water. 5 mL Heparinized whole blood + 50 µL 200 mM disodium EDTA + 50 µL 200 mM sodium metabisulfite, centrifuge at 3000 g for 10 min. Remove a 2 mL aliquot of the plasma and add it to 40 pmole benzylamine, add 2 mL 2.5% perchloric acid, centrifuge (?), adjust the pH of the supernatant to 6.5 with 2 M KOH, centrifuge at 3500 g for 15 min, add the supernatant to the Amberlite column, wash with 2 mL 800 mM lithium borate (to elute catecholamines), wash with 3 mL 100 mM pH 6.9 sodium phosphate buffer containing 5 mM disodium EDTA (to elute basic amino acids), elute with 7.5 mL 1 M pH 8.5 sodium phosphate buffer containing 5 mM disodium EDTA. Adjust the pH of the eluate to 8.5 with 4 M KOH, add to the two C18 SPE cartridges in series, wash with 2 mL water, wash with 1 mL MeOH:water 90:10, elute with 4 mL MeOH:water 90:10. Add 1 mL 100 mM pH 8.5 sodium phosphate buffer to the eluate and evaporate it to 1 mL under a stream of nitrogen at 38°, add 600 µL 30 µg/mL fluorescamine in acetone while vortexing for 1 min, evaporate to 1 mL under a stream of nitrogen at 38°, add 50 µL 4 M pH 5.5 sodium acetate buffer, add 3 mL ethyl acetate, vortex for 1 min, centrifuge briefly. Remove the organic layer and add it to 40 µL water, evaporate to remove the ethyl acetate under a stream of nitrogen at 38°, mix the residue with 40 µL mobile phase, stir briefly, inject a 50 µL aliquot.

HPLC VARIABLES
Column: 250 × 4.5 ODS-120T (Toyo Soda)
Mobile phase: MeOH:ethyl acetate:50 mM pH 2.5 citrate buffer 40:18:42
Flow rate: 0.8
Injection volume: 50
Detector: F ex 390 em 475

CHROMATOGRAM
Retention time: 9 (octopamine), 10 (tyramine), 12.5 (phenylethanolamine), 16 (phenethylamine)
Internal standard: benzylamine (14)
Limit of detection: 200 pM

KEY WORDS
plasma; whole blood; SPE

REFERENCE
Yonekura, T.; Kamata, S.; Wasa, M.; Okada, A.; Yamatodani, A.; Watanabe, T.; Wada, H. Simultaneous determination of plasma phenethylamine, phenylethanolamine, tyramine and octopamine by high-performance liquid chromatography using derivatization with fluorescamine, *J.Chromatogr.*, **1988**, *427*, 320–325.

SAMPLE
Matrix: blood
Analyte: Ro 19-6327 (N-(2-aminoethyl)-5-chloropyridine-2-carboxamide)
Sample preparation: 1 mL Plasma + 50 µL 400 mM NaOH, vortex, add 10 mL MTBE:1-butanol 80:20, shake at 30 rpm on a rotating shaker for 20 min, centrifuge at 2000 g for 5 min. Remove 9 mL of the organic layer and add it to 500 µL 0.17% phosphoric acid, extract for 20 min, centrifuge for 5 min. Remove the aqueous layer and add it to 500 µL buffer, add 500 µL 50 µg/mL fluorescamine in MeCN (prepare fresh each day) with constant vortexing, after 10 min remove the MeCN by evaporation under reduced pressure for exactly 10 min, vortex, inject a 100 µL aliquot. (Prepare buffer by dissolving 30 g Na_2HPO_4 in water, add 24 mL 1 M NaOH, make up to 1 L with water.)

HPLC VARIABLES
Column: 125 × 4 LiChroCART Superspher 60 RP-8e (Merck)
Mobile phase: MeCN:water 32:68 containing 100 mM NaH_2PO_4 and 5 mM Na_2HPO_4, pH 5.9 ± 0.1
Flow rate: 1
Injection volume: 100
Detector: F ex 370 em 485

CHROMATOGRAM
Retention time: 3.7
Limit of quantitation: 1 ng/mL
Limit of detection: 0.5 ng/mL

OTHER SUBSTANCES
Non-interfering: benserazide, dopamine, levodopa

KEY WORDS
plasma; rat; human

REFERENCE
Wyss, R.; Philipp, W. Determination of the monoamine oxidase B inhibitor Ro 19-6327 in plasma by high-performance liquid chromatography using precolumn derivatization with fluorescamine and fluorescence detection, *J.Chromatogr.*, **1990**, *507*, 187–198.

SAMPLE
Matrix: blood
Analyte: trientine
Sample preparation: Condition a Bond Elut SCX SPE cartridge with 5 mL water. 500 µL Plasma + 100 µL water + 1 mL MeCN, vortex briefly, centrifuge at 1000 g for 5 min. Add 1.2 mL of the supernatant to the SPE cartridge, wash with 3 mL water, wash with

2 mL 1 M KCl, wash with 3 mL 2 M KCl, elute with 1 mL 4 M KCl. Remove a 200 μL aliquot of the eluate and add it to 600 μL 100 mM pH 9.5 sodium phosphate buffer and 100 μL 0.15 mM trisodium EDTA in 100 mM pH 9.5 sodium phosphate buffer, mix, add 100 μL 10 mM fluorescamine in MeCN, vortex vigorously for 1 min, let stand for 20 min, add 50 μL 0.25 mM α-naphthylamine (Caution! α-Naphthylamine is a carcinogen!) in MeOH, inject a 20-50 μL aliquot.

HPLC VARIABLES
Column: 250 × 4.6 5 μm Nucleosil 5-CN
Mobile phase: MeCN:buffer 27:73 , pH adjusted to 6.0 with 2 M NaOH (Buffer was 140 mM ammonium chloride containing 48 mM sodium benzenesulfonate and 9.2 mM acetic acid.)
Column temperature: 40
Flow rate: 0.5
Injection volume: 20-50
Detector: F ex 380 em 485

CHROMATOGRAM
Retention time: 9.5
Internal standard: α-naphthylamine (13)
Limit of detection: 100 ng/mL

KEY WORDS
plasma; SPE; rat; human; pharmacokinetics

REFERENCE
Miyazaki, K.; Kishino, S.; Kobayashi, M.; Arashima, S.; Matsumoto, S.; Arita, T. Determination of triethylenetetramine in plasma of patients by high-performance liquid chromatography, *Chem.Pharm.Bull.*, **1990**, *38*, 1035−1038.

SAMPLE
Matrix: blood
Analyte: SK&F 110679 (NH_2-L-His-D-Trp-L-Ala-L-Trp-D-Phe-L-Lys-NH_2)
Sample preparation: Condition a 1 mL weak cation-exchange (CBA) SPE cartridge (Analytichem) with 1 mL MeOH and 1 mL water. 500 μL Plasma + 50 μL 1 μg/mL IS, mix, add to the SPE cartridge, rinse the tube with water, add the rinse to the SPE cartridge, wash with 1 mL 1% trifluoroacetic acid in water, wash with 3 mL water, wash with 1 mL MeOH:water 50:50, elute with 2 mL 2% ammonium acetate in MeOH. Evaporate the eluate to dryness under a stream of nitrogen, reconstitute the residue in 100 μL MeCN: 100 mM pH 4.0 citrate buffer 25:75. Remove a 20 μL aliquot and add it to 20 μL 500 mM pH 9.0 potassium phosphate buffer and 30 μL 2 mg/mL fluorescamine, mix, let stand at room temperature for 5 min, inject a 10-50μL aliquot.

HPLC VARIABLES
Column: 250 × 2 5 μm Ultrasphere octylsilica
Mobile phase: Gradient. MeOH:buffer from 45:55 to 65:35 over 10 min, maintain at 65:35 for 7 min. (Prepare buffer by dissolving 1.21 g Tris and 2.8 mL triethylamine in 1 L water, adjust pH to 7.0 with phosphoric acid.)
Column temperature: 50
Flow rate: 0.3
Injection volume: 10-50
Detector: F ex 390 em 470 (cutoff filter)

CHROMATOGRAM
Retention time: 16.4
Internal standard: SK&F 110910 (NH_2-His-D-Phe-Ala-Phe-D-Phe-Lys-NH_2) (18.2)
Limit of detection: 5 nM

KEY WORDS
plasma; SPE

REFERENCE

Boppana, V.K.; Miller-Stein, C.; Politowski, J.F.; Rhodes, G.R. High-performance liquid chromatographic determination of peptides in biological fluids by automated pre-column fluorescence derivatization with fluorescamine, *J.Chromatogr.*, **1991**, *548*, 319–327.

SAMPLE

Matrix: blood
Analyte: 3'-amino-3'-deoxythymidine
Sample preparation: 200 μL Plasma + 50 μL water, vortex, add 5 mL MTBE:1-butanol 60:40, shake for 15 min, centrifuge at 2000 g for 10 min. Remove the liquid phase and add it to 100 μL 100 mM KOH containing 5% NaCl, shake for 20 min, centrifuge at 2000 g for 10 min. Remove the lower aqueous phase and add it to 100 μL buffer, add 100 μL 200 μg/mL fluorescamine in anhydrous acetone (freshly prepared), vortex, evaporate the acetone under a stream of air for 2 min, inject a 170 μL aliquot of the remaining solution. (Buffer was 34.02 g/L KH_2PO_4, pH 4.3.)

HPLC VARIABLES

Column: 250 × 4 5 μm Hypersil octadecylsilane
Mobile phase: MeCN:MeOH:buffer 5:48:47 (Buffer was 80 mM KH_2PO_4 adjusted to pH 7.8 with 5 M KOH.)
Flow rate: 1
Injection volume: 170
Detector: F ex 265 em 475

CHROMATOGRAM

Retention time: 5.4
Limit of quantitation: 3 ng/mL
Limit of detection: <2 ng/mL

KEY WORDS

plasma; pharmacokinetics

REFERENCE

Zhou, X.-J.; Sommadossi, J.-P. Quantification of 3'-amino-3'-deoxythymidine, a toxic catabolite of 3'-azido-3'-deoxythymidine (zidovudine) in human plasma by high-performance liquid chromatography using precolumn derivatization with fluorescamine and fluorescence detection, *J.Chromatogr.B*, **1994**, *656*, 389–396.

SAMPLE

Matrix: blood
Analyte: CGP 53 437
Sample preparation: 1 mL Plasma + IS + 1 mL MeCN, mix, centrifuge at 2000 g for 5 min. Remove the supernatant and evaporate it to dryness, reconstitute the residue in 1 mL 71 mM pH 3 phthalate buffer and 100 μL 5 M NaCl, add 7 mL diisopropyl ether (Caution! Diisopropyl ether readily forms explosive peroxides!), shake horizontally at 300 rpm for 10 min, centrifuge at 1500 g for 5 min, freeze in dry ice/EtOH. Remove the supernatant and evaporate it to dryness under a stream of nitrogen at 40°, reconstitute the residue in 500 μL dichloromethane, add 200 μL trifluoroacetic acid, let stand at room temperature for 1 h, evaporate to dryness under a stream of nitrogen, reconstitute with 50 μL MeCN and 150 μL 60 mM pH 9 borax buffer, add 100 μL 3 mg/mL fluorescamine in MeCN with shaking, inject a 20 μL aliquot of this mixture.

HPLC VARIABLES

Column: 250 × 4 5 μm Lichrospher 100RP-18 endcapped
Mobile phase: MeCN:60 mM pH 9 borax buffer 42:58
Column temperature: 40
Flow rate: 1
Injection volume: 20
Detector: F ex 395 em 480

CHROMATOGRAM

Retention time: 13
Internal standard: CGP 54 451 (16)

Limit of quantitation: 5 nM
Limit of detection: 1 nM

KEY WORDS
plasma; dog; marmoset; rat; human; pharmacokinetics

REFERENCE
Flesch, G.; Mann, C.; Boss, E.; Lang, M.; Degen, P.H.; Dieterle, W. Quantitative determination of CGP 53 437, a new HIV protease inhibitor, in plasma by high-performance liquid chromatography and fluorescence detection, *J.Chromatogr.B*, **1994**, *657*, 155–161.

SAMPLE
Matrix: blood
Analyte: histamine
Sample preparation: Condition a CBA (carboxylic acid) SPE cartridge (Analytichem) with 1 mL MeOH and 2 mL 10 mM pH 7 phosphate buffer. 500 µL Plasma + 100 µL 1 mg/mL betazole + 2 mL ice-cold 10 mM pH 7 phosphate buffer, add to the SPE cartridge at 0.2-0.3 mL/min, dry for 30 s, wash with 1 mL hexane, elute with 1 mL MeOH:100 mM HCl 60:40. Evaporate the eluate to dryness under a stream of nitrogen at 60°, reconstitute the residue in 100 µL 200 mM pH 9 sodium borate buffer, add 50 µL 20 µg/mL fluorescamine in MeCN, vortex, store at 4°, inject a 50 µL aliquot.

HPLC VARIABLES
Guard column: Pelliguard LC-8 (Supelco)
Column: 250 × 4.6 5 µm Ultramex C8
Mobile phase: MeCN:500 mM pH 7 imidazole buffer 20:80
Flow rate: 1
Injection volume: 50
Detector: F ex 366 em 440

CHROMATOGRAM
Retention time: 15.2
Internal standard: betazole (26.9)
Limit of quantitation: 1 ng/mL

KEY WORDS
plasma; SPE

REFERENCE
Lowe, D.R.; March, C.; James, J.E.; Karnes, H.T. A high-performance liquid chromatographic method for histamine in plasma using solid phase extraction and fluorescamine derivatization, *J.Liq.Chromatogr.*, **1994**, *17*, 3563–3570.

SAMPLE
Matrix: blood
Analyte: histamine
Sample preparation: Condition a CBA SPE cartridge with 1 mL MeOH and 2 mL 10 mM pH 7 phosphate buffer. 500 µL Plasma + 2 mL chilled 10 mM pH 7 phosphate buffer, mix, add to the SPE cartridge, dry for 30 min, wash with 1 mL hexane, elute with 1 mL MeOH:100 mM HCl 40:60. Evaporate to dryness under a stream of nitrogen at 60°, reconstitute the residue in 100 µL 200 mM pH 9.7 borate buffer. Remove a 10 µL aliquot and add it to 40 µL 20 µg/mL fluorescamine in 200 mM pH 9.7 sodium borate buffer and 50 µL 200 mM pH 9.7 sodium borate buffer, inject an aliquot.

HPLC VARIABLES
Column: 250 × 4.6 5 µm Ultramex C-8
Mobile phase: MeCN:500 mM pH 7 imidazole nitrate buffer 20:80
Flow rate: 1
Injection volume: 50
Detector: F ex 366 em 440 (cut-off filter)

CHROMATOGRAM
Limit of quantitation: 166 pg
Limit of detection: 13 pg

KEY WORDS
plasma; SPE; details of chemiluminescence detection are also given in paper

REFERENCE
Walters, D.L.; James, J.E.; Vest, F.B.; Karnes, H.T. A comparison of fluorescence versus chemilumines-cence detection for analysis of the fluorescamine derivative of histamine by HPLC, *Biomed.Chromatogr.*, **1994**, *8*, 207–211.

SAMPLE
Matrix: blood
Analyte: gabapentin
Sample preparation: 50 μL Serum + 200 μL 2.5 mg/mL fluorescamine in MeCN (prepare fresh daily), vortex for 1 min, centrifuge at 14000 g for 30 s, inject an aliquot of the supernatant.

CAPILLARY ELECTROPHORESIS
Capillary: 600 mm × 50 μm
Capillary preparation: After each run wash capillary with 2 M NaOH for 2 min then with buffer. Condition capillary each morning with 2 M NaOH for 7 min and buffer for 7 min.
Running buffer: 200 mM Boric acid containing 26 mM K_2HPO_4
Capillary temperature: 24
Voltage/Current: 35 μA
Injection: Pressure injection for 15 s
Detector: UV 200
Model: Beckman Model 2000
Migration time: 11
Limit of detection: 1 μg/mL

OTHER SUBSTANCES
Non-interfering: acetaminophen, N-acetylprocainamide, amino acids, carbamazepine, disopyramide, GABA, gentamicin, lidocaine, phenytoin, primidone, procainamide, quinidine, salicylic acid, theophylline, tobramycin, valproic acid, vancomycin

KEY WORDS
serum

REFERENCE
Garcia, L.L.; Shihabi, Z.K.; Oles, K. Determination of gabapentin in serum by capillary electrophoresis, *J.Chromatogr.B*, **1995**, *669*, 157–162.

SAMPLE
Matrix: blood
Analyte: CGP 36 742 (3-aminopropyl-n-butylphosphinic acid)
Sample preparation: 500 μL Plasma + 5 nmole IS + 100 μL 1.5 M trichloroacetic acid, mix, centrifuge at 15000 g for 10 min. Adjust the pH of the supernatant to 5-6 with 1 M NaOH. Remove a 100 μL aliquot and add it to 100 μL buffer, while vortexing vigorously add 100 μL 1 mg/mL fluorescamine in MeCN over 5 s, add 100 μL dichloromethane, vortex for 10 s, centrifuge at 15000 g for 5 min, inject a 10 μL aliquot of the aqueous phase. Urine. 100 μL Urine + 50 nmole IS, make up to 5 mL with water. Remove a 100 μL aliquot and add it to 100 μL buffer, while vortexing vigorously add 100 μL 1 mg/mL fluorescamine in MeCN over 5 s, add 100 μL dichloromethane, vortex for 10 s, centrifuge at 15000 g for 5 min, inject a 10 μL aliquot of the aqueous phase. (Buffer was 43 mM disodium tetraborate containing 17 mM KH_2PO_4, pH 9.)

HPLC VARIABLES
Guard column: 10 × 4.6 25-40 μm LiChroprep RP-2 (Merck)
Column: 125 × 4.6 3 μm Hypersil ODS
Mobile phase: MeOH:buffer 30:70 (Buffer was 21.5 mM disodium tetraborate containing 8.5 mM KH_2PO_4, pH 8.8.)
Column temperature: 45 (urine), ambient (plasma)
Flow rate: 0.7
Injection volume: 10
Detector: F ex 395 em 480 (418 nm cut-off filter)

CHROMATOGRAM
Internal standard: CGP 44 962 (Ciba-Geigy)
Limit of quantitation: 1 μM (plasma), 60 μM (urine)

KEY WORDS
pharmacokinetics; plasma

REFERENCE
Gleiter, C.H.; Farger, G.; Möbius, H.-J. Pharmacokinetics of CGP 36 742, an orally active GABA$_B$ antagonist, in humans, *J.Clin.Pharmacol.*, **1996**, *36*, 428–438.

SAMPLE
Matrix: blood
Analyte: taurine
Sample preparation: 100 μL Plasma + 150 μL MeCN, vortex, centrifuge at 5800 g for 10 min. Remove the supernatant and add it to 50 μL buffer, add 50 μL 5 mM fluorescamine in MeCN, vortex, inject an aliquot. (Prepare buffer by adjusting the pH of 100 mM disodium tetraborate to 9.2 with 10 mM boric acid.)

HPLC VARIABLES
Guard column: C8
Column: 300 × 3.9 10 μm Bondclone C18
Mobile phase: MeCN:THF:buffer 24:4:72 (Prepare buffer by adjusting the pH of 15 mM KH$_2$PO$_4$ to 3.5 with phosphoric acid.)
Flow rate: 1
Injection volume: 20
Detector: UV 385

CHROMATOGRAM
Limit of quantitation: 5 μg/mL

KEY WORDS
plasma

REFERENCE
McMahon, G.P.; O'Kennedy, R.; Kelly, M.T. High-performance liquid chromatographic determination of taurine in human plasma using pre-column extraction and derivatization, *J.Pharm.Biomed.Anal.*, **1996**, *14*, 1287–1294.

SAMPLE
Matrix: blood
Analyte: CP-80,794
Sample preparation: 1 mL Plasma + 100 μL 2 μg/mL IS, mix, add 100 μL 1 M NaOH, mix, add 5 mL n-butyl chloride, shake at 100 rpm for 10 min, centrifuge. Remove the organic layer and add it to 2 mL MeCN, evaporate to dryness, reconstitute the residue in 20 μL reagent, vortex for 30 s, let stand at room temperature for 45 min, add 10 μL glycine solution, vortex for 30 s, let stand at room temperature for 15 min, add 10 μL chymotrypsin solution, vortex for 30 s, heat at 37° for 15 min, add 50 μL 90 nM fluorescamine in MeCN, inject a 75 μL aliquot. (Prepare reagent by dissolving 1.5 mg phthalaldehyde in 150 μL EtOH and adding 150 mg hydroxypropyl-β-cyclodextrin and 2.85 mL 100 mM pH 7.9 potassium phosphate buffer. Prepare glycine solution by dissolving 3 mg glycine and 50 mg hydroxypropyl-β-cyclodextrin in 1 mL 100 mM pH 7.9 potassium phosphate buffer. Prepare chymotrypsin solution by dissolving 1 mg chymotrypsin and 50 mg hydroxypropyl-β-cyclodextrin in 1 mL 100 mM pH 7.9 potassium phosphate buffer. Presumably the phthalaldehyde is added to derivatize endogenous amines and amino acids. An amide bond in CP-80-794 is then cleaved with chymotrypsin and the resulting amine is derivatized with fluorescamine. Presumably the endogenous phthalaldehyde compounds are not chromatographed under these conditions.)

HPLC VARIABLES
Guard column: 40 μm C18
Column: 150 × 3.9 5 μm Novapak C18
Mobile phase: MeCN:water 27:73

Flow rate: 1
Injection volume: 75
Detector: F ex 390 em 440

CHROMATOGRAM
Retention time: 4
Internal standard: CP-83,092 (6)
Limit of quantitation: 20 ng/mL (human), 50 ng/mL (dog)

KEY WORDS
plasma; dog; human; pharmacokinetics

REFERENCE
Allen, M.C.; Stafford, C.G.; Nocerini, M.R. Quantitation of 4-cyclohexyl-2-hydroxy-3-(3-methylsulfanyl-2-propionylamino)butyric acid isopropyl ester (CP-80,794), a renin inhibitor, and its hydrolytic cleavage metabolite 2-[(morpholine-4-carbonyl)amino]-3-phenylpropionic acid (CP-84,364) in dog and human plasma by high-performance liquid chromatography, *J.Chromatogr.B*, **1997**, *696*, 243–251.

SAMPLE
Matrix: blood, CSF, urine
Analyte: taurine
Sample preparation: Prepare a SPE column by placing a 2 cm layer of 200-400 mesh Dowex 2x8 Cl⁻ form in a column, on top of this place a 2 cm layer of 100-200 mesh Dowex 50 W-x8 H⁺ form. 1 mL Serum, CSF, or urine + 100 μL 4 M perchloric acid, centrifuge, add the supernatant to the SPE column, elute with three 1 mL portions of water, collect all the effluent from the column. Remove a 1 mL aliquot and add it to 500 μL 1 M pH 9.0 borate buffer, vortex while adding 500 μL 300 μg/mL fluorescamine in MeCN, let stand for 1 min, inject a 50 μL aliquot.

HPLC VARIABLES
Column: 250 mm long LiChrosorb RP-18-10A
Mobile phase: MeCN:15 mM pH 6.0 phosphate buffer 30:70
Flow rate: 1
Injection volume: 50
Detector: F ex 395 em 455

CHROMATOGRAM
Retention time: 3
Limit of detection: 0.25 pmole

OTHER SUBSTANCES
Non-interfering: amino acids, hypotaurine, phosphoethanolamine

KEY WORDS
serum; SPE

REFERENCE
Stabler, T.V.; Siegel, A.L. Rapid liquid-chromatographic/fluorometric method for taurine in biological fluids, involving pre-derivatization with fluorescamine, *Clin.Chem.*, **1981**, *27*, 1771–1771.

SAMPLE
Matrix: blood, eggs
Analyte: sulfamethazine
Sample preparation: 1 mL Serum or 1 g homogenized egg + 4 mL MeCN, vortex, centrifuge at 3000 rpm for 15 min. Remove the supernatant and evaporate it to dryness under a stream of nitrogen at 40°, reconstitute the residue in 50 μL water, mix vigorously, add 1 mL MeCN, centrifuge at 3000 rpm for 15 min. Remove the upper layer and evaporate it to dryness, reconstitute the residue in 1 mL 0.01% trichloroacetic acid, centrifuge at 3000 rpm for 15 min. Remove a 500 μL aliquot and add it to 100 μL 1 mg/mL fluorescamine in MeCN (freshly prepared), shake, let stand for 1 min, inject a 50 μL aliquot.

HPLC VARIABLES
Column: 300 × 3.9 10 μm Nova-Pak C18
Mobile phase: MeCN:10 mM KH₂PO₄ 30:70

Flow rate: 1
Injection volume: 50
Detector: Chemiluminescence following post-column reaction. The column effluent was mixed with reagent pumped at 0.5 mL/min and the mixture flowed to the detector. (Reagent was 1 mM bis[2-(3,6,9-trioxadecanyloxycarbonyl)-4-nitrophenyl] oxalate (Wako) and 300 mM hydrogen peroxide in MeCN.)

CHROMATOGRAM
Retention time: 8.1
Limit of detection: 1 ng/mL

KEY WORDS
chicken; serum

REFERENCE
Tsai, C.-E.; Kondo, F.; Ueyama, Y.; Azama, J. Determination of sulfamethazine residue in chicken serum and egg by high-performance liquid chromatography with chemiluminescence detection, *J.Chromatogr.Sci.*, **1995**, *33*, 365–369.

SAMPLE
Matrix: blood, milk
Analyte: sulfonamides
Sample preparation: 1 mL Serum or milk + 4 mL MeCN, vortex, centrifuge at 1000 g for 15 min. Remove the supernatant and evaporate it to dryness under a stream of nitrogen at 40°, reconstitute the residue in 50 μL water, mix vigorously, add 1 mL MeCN, centrifuge at 1000 g for 10 min. Remove the upper layer and evaporate it to dryness, reconstitute the residue in 1 mL 10 ng/mL p-aminobenzoic acid in 0.01% trichloroacetic acid, centrifuge at 1000 g for 10 min. Remove a 500 μL aliquot of the clear layer and add it to 100 μL 1 mg/mL fluorescamine in acetone (prepared fresh each day), mix for 1 min, inject a 50 μL aliquot.

HPLC VARIABLES
Column: 300 × 3.9 10 μm Nova-Pak C18
Mobile phase: MeCN:10 mM KH_2PO_4 30:70
Flow rate: 1
Injection volume: 50
Detector: F ex 390 em 475

CHROMATOGRAM
Retention time: 2.5 (sulfamethoxazole), 6.3 (sulfadiazine), 6.7 (sulfathiazole), 7.2 (sulfamethazine), 8.1 (sulfamonomethoxine), 16.5 (sulfadimethoxine)
Internal standard: p-aminobenzoic acid (5.6)
Limit of detection: 0.1 ng/mL

KEY WORDS
cow; serum

REFERENCE
Tsai, C.-E.; Kondo, F. Liquid chromatographic determination of fluorescent derivatives of six sulfonamides in bovine serum and milk, *J.AOAC Int.*, **1995**, *78*, 674–678.

SAMPLE
Matrix: blood, tissue
Analyte: sulfonamides
Sample preparation: 1 mL Serum or homogenized tissue + 4 mL MeCN, vortex, centrifuge at 1000 g for 15 min. Remove the supernatant and evaporate it to dryness under a stream of nitrogen at 40°, reconstitute the residue in 50 μL water, mix vigorously, add 1 mL MeCN, centrifuge at 1000 g for 15 min. Remove the upper layer and evaporate it to dryness, reconstitute the residue in 1 mL 10 ng/mL sulfadiazine in 0.01% trichloroacetic acid, shake, add 100 μL hexane, shake, centrifuge at 1000 g for 15 min. Remove a 500 μL aliquot of the clear aqueous layer and add it to 100 μL 1 mg/mL fluorescamine in MeCN (freshly prepared), shake by hand for 1 min, inject a 50 μL aliquot.

HPLC VARIABLES
Column: 300 × 3.9 10 μm Nova-Pack C18
Mobile phase: MeCN:10 mM KH_2PO_4 30:70
Flow rate: 1
Injection volume: 50
Detector: F ex 390 em 475

CHROMATOGRAM
Retention time: 7.9 (sulfamethazine), 9.1 (sulfamonomethoxine), 14.1 (sulfamethoxazole), 18.2 (sulfadimethoxine)
Internal standard: sulfadiazine (7.1)
Limit of detection: 0.1 ng/mL

KEY WORDS
serum; pig; kidney; muscle; liver

REFERENCE
Tsai, C.-E.; Kondo, F. A sensitive high-performance liquid chromatographic method for detecting sulfonamide residues in swine serum and tissues after fluorescamine derivatization, *J.Liq.Chromatogr.*, **1995**, *18*, 965–976.

SAMPLE
Matrix: blood, tissue, urine
Analyte: putrescine
Sample preparation: Tissue. Prepare a column of carboxymethyl cellulose, H^+ form (CM 22, Whatman) in a 130 × 5 glass column (0.75 mL bed volume), equilibrate with 1 M pyridine containing 1 M acetic acid, equilibrate with 10 mM pyridine containing 10 mM acetic acid. Homogenize 20 mg rat liver with 200 μL chilled 300 mM perchloric acid containing 1 nmole 1,6-hexanediamine and 20 nmole N-3-aminopropyl-1,6-hexanediamine at 4°, centrifuge at 3500 rpm for 15 min. Dilute the supernatant to 2 mL with 50 mM pyridine, add to the column, wash with 3 mL 10 mM pyridine containing 10 mM acetic acid, wash with 3 mL 50 mM pyridine containing 50 mM acetic acid, elute diamines with 200 mM pyridine containing 200 mM acetic acid, elute polyamines with 1 M pyridine containing 1 M acetic acid. Evaporate eluates to dryness at 3 mm Hg overnight in a desiccator that also contains 30% aqueous acetic acid. For polyamines reconstitute with 100 μL 100 mM pH 8 sodium borate buffer. Remove a 20 μL aliquot and add it rapidly to 10 μL 2 mg/mL fluorescamine in acetone, mix rapidly, add 5 μL 1 M pH 5.4 sodium acetate buffer, after 10 min inject an aliquot. For diamines reconstitute with 100 μL 10 mM HCl, evaporate to dryness under reduced pressure over KOH in a desiccator, reconstitute with 20 μL 100 mM pH 8 sodium borate buffer and add it rapidly to 10 μL 2 mg/mL fluorescamine in acetone, mix rapidly, add 5 μL 1 M pH 5.4 sodium acetate buffer, after 10 min inject an aliquot. Serum. Prepare a column of carboxymethyl cellulose, H^+ form (CM22, Whatman) in a 130 × 5 glass column (0.75 mL bed volume), equilibrate with 1 M pyridine containing 1 M acetic acid, equilibrate with 10 mM pyridine containing 10 mM acetic acid. Mix 500 μL human serum with 125 μL 1.5 M perchloric acid containing 0.5 nmole 1,6-hexanediamine and 0.5 nmole N-3-aminopropyl-1,6-hexanediamine, stir well, centrifuge. Dilute the supernatant to 3 mL with 50 mM pyridine, add to the column, wash with 3 mL 10 mM pyridine containing 10 mM acetic acid, wash with 3 mL 50 mM pyridine containing 50 mM acetic acid, elute diamines with 200 mM pyridine containing 200 mM acetic acid, elute polyamines with 1 M pyridine containing 1 M acetic acid. Evaporate eluates to dryness at 3 mm Hg overnight in a desiccator that also contains 30% aqueous acetic acid. Reconstitute with 100 μL 10 mM HCl, evaporate to dryness under reduced pressure over KOH in a desiccator, reconstitute with 20 μL 100 mM pH 8 sodium borate buffer and add it rapidly to 10 μL 2 mg/mL fluorescamine in acetone, mix rapidly, add 5 μL 1 M pH 5.4 sodium acetate buffer, after 10 min inject an aliquot. Urine. Prepare a column of carboxymethyl cellulose, H^+ form (CM 22, Whatman) in a 130 × 5 glass column (0.75 mL bed volume), equilibrate with 1 M pyridine containing 1 M acetic acid, equilibrate with 10 mM pyridine containing 10 mM acetic acid. Mix 500 μL human urine with 500 μL concentrated HCl containing 10 nmole 1,6-hexanediamine and 10 nmole N-3-aminopropyl-1,6-hexanediamine, hydrolyze, filter, evaporate to dryness, reconstitute with 5 mL 50 mM pyridine, add to the column, wash with 3 mL 10 mM pyridine

containing 10 mM acetic acid, wash with 3 mL 50 mM pyridine containing 50 mM acetic acid, elute diamines with 200 mM pyridine containing 200 mM acetic acid, elute polyamines with 1 M pyridine containing 1 M acetic acid. Evaporate eluates to dryness at 3 mm Hg overnight in a desiccator that also contains 30% aqueous acetic acid. Reconstitute with 100 μL 100 mM pH 8 sodium borate buffer. Remove a 20 μL aliquot and add it rapidly to 10 μL 2 mg/mL fluorescamine in acetone, mix rapidly, add 5 μL 1 M pH 5.4 sodium acetate buffer (for polyamines only), after 10 min inject an aliquot.

HPLC VARIABLES

Column: 500 × 3 Vydac reversed-phase glass column
Mobile phase: Gradient. MeOH:buffer from 10:90 to 40:60 over 20 min (Buffer was 100 mM boric acid adjusted to pH 8.0 with 6 M NaOH.)
Column temperature: 60
Flow rate: 1
Detector: F primary filter Corning No. 7-51, secondary filter Wratten No. 4

CHROMATOGRAM

Retention time: 8.8 (putrescine), 11.1 (cadaverine), 11.7 (spermidine), 8.4 (1,3-propanediamine), 15.4 (1,7-heptanediamine), 17.7 (1,8-octanediamine), 17 (spermine, using a gradient from 15:85 to 50:50 over 20 min)
Internal standard: 1,6-hexanediamine (13.4), N-3-aminopropyl-1,6-hexanediamine (17.7)
Limit of detection: 20 nM

KEY WORDS
rat; liver; human; serum

REFERENCE
Samejima, K.; Kawase, M.; Sakamoto, S.; Okada, M.; Endo, Y. A sensitive fluorometric method for the determination of aliphatic diamines and polyamines in biological materials by high-speed liquid chromatography, *Anal.Biochem.*, **1976**, *76*, 392–406.

SAMPLE
Matrix: blood, urine
Analyte: cephalosporins
Sample preparation: 1 mL Plasma + 1 mL 6% trichloroacetic acid, mix, centrifuge at 4000 rpm for 10 min, inject an aliquot of the supernatant. Inject an aliquot of urine directly.

HPLC VARIABLES
Guard column: 10 × 4 7 μm Lichrosorb RP 18
Column: 250 × 4 7 μm Lichrosorb RP 18
Mobile phase: MeCN:25 mM pH 7 phosphate buffer 10:90 (cefaclor (14 min), cephalexin, cephradine, cefroxadine (10)), 15:85 (cefaloglycine), or 5:95 (cefadroxil)
Flow rate: 1
Injection volume: 10
Detector: F ex 385 em 485 following post-column reaction. The column effluent mixed with 200 μg/mL fluorescamine in MeCN pumped at 0.25 mL/min and the mixture flowed through a 4.5 m × 0.25 mm ID coil of PTFE tubing to the detector.

CHROMATOGRAM
Limit of detection: 0.3-1.3 ng/mL

OTHER SUBSTANCES
Non-interfering: amidopyrin, aspirin, barbital, caffeine, cefmenoxime, cefotaxime, ceftizoxime, ceftriaxone, cetazidime, diazepam, dibekacin, gentamycin, kanamycin, lidocaine, netilmicin, tetracaine, theophylline, tobramycin

KEY WORDS
post-column reaction; plasma

REFERENCE
Blanchin, M.D.; Fabre, H.; Mandrou, B. Fluorescamine post-column derivatization for the HPLC determination of cephalosporins in plasma and urine, *J.Liq.Chromatogr.*, **1988**, *11*, 2993–3010.

SAMPLE
Matrix: blood, urine
Analyte: ampicillin
Sample preparation: Plasma. 100 μL Plasma + 200 μL water + 100 μL 70% perchloric acid:pH 5.4 buffer 25:75, vortex for 1 min, centrifuge at ca. 2400 g for 10 min. Remove 300 μL supernatant and add it to 75 μL 1 M NaOH, mix, inject a 100 μL aliquot on to column A and elute to waste with mobile phase A, after 2 min collect the effluent from column A in a 1 mL sample loop, after 1 min inject the contents of the sample loop on to column B with mobile phase B, elute column B with mobile phase B and monitor the effluent from column B. Urine. Dilute urine if necessary. 80 μL Urine + 720 μL pH 4.85 buffer, vortex, inject a 20 μL aliquot on to column A and elute to waste with mobile phase A, after 2 min collect the effluent from column A in a 1 mL sample loop, after 1 min inject the contents of the sample loop on to column B with mobile phase B, elute column B with mobile phase B and monitor the effluent from column B. (Prepare pH 5.4 buffer by dissolving 19.9 g Na_2HPO_4 and 8.4 g citric acid monohydrate in 250 mL water. Prepare pH 4.85 buffer by mixing 10 mL 500 mM Na_2HPO_4 and 350 mL water, adjust pH to 4.85 with 1 M citric acid, make up to 500 mL with water.)

HPLC VARIABLES
Column: A 15 × 3.2 7 μm New Guard RP18 + 3 × 3 3 μm Perkin-Elmer; B 100 × 4.6 3 μm Microspher C18 (Chrompack)
Mobile phase: A MeOH:buffer 17:83 containing 1 mM sodium hexylsulfate (Buffer was pH 7.4 phosphate buffer, ionic strength 0.05.); B MeOH:buffer 35:65 (plasma) or 30:70 (urine) (Buffer was pH 7.4 phosphate buffer, ionic strength 0.05.)
Flow rate: 1
Detector: F ex 372 em 470 following post-column reaction. The effluent from column B mixed with 160 μg/mL fluorescamine in MeCN pumped at 0.2 mL/min and the mixture flowed through a 5 m × 0.4 mm ID knitted PTFE tube to the detector.

CHROMATOGRAM
Retention time: 8
Limit of detection: 570 nM (urine), 14 nM (plasma)

OTHER SUBSTANCES
Interfering: 20-100

KEY WORDS
plasma; post-column reaction; column-switching; heart-cut

REFERENCE
Lanbeck-Vallén, K.; Carlqvist, J.; Nordgren, T. Determination of ampicillin in biological fluids by coupled-column liquid chromatography and post-column derivatization, *J.Chromatogr.*, **1991**, *567*, 121–128.

SAMPLE
Matrix: blood, urine
Analyte: pamidronate
Sample preparation: 2 mL Plasma or urine + 100 μL 25 μg/mL IS in 100 mM NaOH, adjust pH to 3 with concentrated HCl, filter (0.45 μm fluoro-membrane, Skan Model Acro LC 13). Add 0.5 mL 1.5 M trichloroacetic acid to the filtrate, centrifuge at 1500 g for 15 min. Remove the supernatant, add 20 μL 2.5 M calcium chloride solution, add 40 μL 500 mM NaH_2PO_4, adjust pH to 12.0 with 6.25 M NaOH then with 0.5 M NaOH, centrifuge at 1500 g for 15 min. Wash the precipitate with water, dissolve the precipitate in 200 μL 130 mM disodium EDTA adjusted to pH 10 with 6.25 M NaOH, add 100 μL 3 mg/mL fluorescamine in MeCN while vortexing vigorously, add 200 μL dichloromethane, extract, centrifuge at 1000 g for 3 min, remove the aqueous phase, inject a 10 μL aliquot of the aqueous phase.

HPLC VARIABLES
Column: 250 × 4 10 μm Nucleosil C18
Mobile phase: MeOH:1 mM disodium EDTA adjusted to pH 6.5 with 1 M NaOH 3:97
Column temperature: 40

Flow rate: 1
Injection volume: 10
Detector: F ex 395 em 480

CHROMATOGRAM
Retention time: 3.9
Internal standard: 6-amino-1-hydroxypentilidenebisphosphonate (CGP 33 637) (4.8)
Limit of quantitation: 700 nM (urine), 800 nM (plasma)
Limit of detection: 10 nM (plasma, urine)

KEY WORDS
pharmacokinetics; plasma

REFERENCE
Flesch, G.; Tominaga, N.; Degen, P. Improved determination of the bisphosphonate pamidronate disodium in plasma and urine by pre-column derivatization with fluorescamine, high-performance liquid
chromatography and fluorescence detection, *J.Chromatogr.*, **1991**, *568*, 261–266.

SAMPLE
Matrix: bulk
Analyte: interferon
Sample preparation: Prepare a solution in 100 mM pH 9.0 borax buffer. While continu-
ously and vigorously vortexing add 30 μL 3 mg/mL fluorescamine in acetone containing
20 μL/mL (?) pyridine to 70 μL sample, vortex for 2 min, inject an aliquot.

CAPILLARY ELECTROPHORESIS
Capillary: 57 cm $\times$ 75 μm (50 cm to the detector)
Capillary preparation: At the end of each run wash capillary with 2 M NaOH, 100 mM
NaOH, water, and running buffer.
Running buffer: 50 mM pH 7.0 Sodium phosphate buffer containing 50 mM LiCl
Capillary temperature: 25
Voltage/Current: 12 kV
Injection: Pressure injection at 0.5 psi for 4 s (24 nL)
Detector: UV 280
Model: Beckman P/ACE System 2000
Migration time: 23 (recombinant human leukocyte-A interferon)

OTHER SUBSTANCES
Simultaneously analyzed: acetone, ammonia

KEY WORDS
recombinant

REFERENCE
Guzman, N.A.; Moschera, J.; Bailey, C.A.; Iqbal, K.; Malick, A.W. Assay of protein drug substances
present in solution mixtures by fluorescamine derivatization and capillary electrophoresis,
J.Chromatogr., **1992**, *598*, 123–131.

SAMPLE
Matrix: enzyme incubations
Analyte: L-valyl-L-proline
Sample preparation: 500 μL Enzyme incubation + 500 μL 400 mM perchloric acid, centrifuge at 3000 g for 10 min. Remove the supernatant and add it to 4 mL 4 M KOH, heat
at 110° for 18 h (to hydrolyze the peptides), neutralize with perchloric acid, centrifuge at
3000 g for 10 min. Remove the supernatant and wash the precipitate twice with 500 μL
aliquots of water, combine the washes with the supernatant, filter (0.45 μm), inject a 20
μL aliquot.

HPLC VARIABLES
Column: 250 $\times$ 4.6 Partisil ODS-2
Mobile phase: Gradient. n-Propanol:50 mM pH 3.0 heptafluorobutyrate 0:100 for 5 min,
to 10:90 over 30 min, re-equilibrate at initial conditions for 25 min.
Flow rate: 1
Injection volume: 20

Detector: F ex 395 em 455 following post-column reaction. The column effluent mixed with 200 mM pH 8.6 borate buffer pumped at 0.05 mL/min and flowed through a reaction coil. The effluent from this coil mixed with 200 μg/mL fluorescamine in MeOH pumped at 0.075 mL/min and this mixture flowed through another reaction coil to the detector.

CHROMATOGRAM
Retention time: 17.33
Internal standard: glycyl-L-leucine (23.54)
Limit of quantitation: 200 ng

KEY WORDS
post-column reaction

REFERENCE
Stein, T.A.; Cohen, J.R.; Mandell, C.; Wise, L. Determination of elastase activity by reversed-phase high-performance liquid chromatography, *J.Chromatogr.*, **1989**, *461*, 267–270.

SAMPLE
Matrix: formulations
Analyte: lypressin
Sample preparation: Inject a 25 μL aliquot.

HPLC VARIABLES
Column: 250 × 3 10 μm RP 8 (Merck)
Mobile phase: MeCN:pH 7 phosphate buffer 15:80
Flow rate: 1.58
Injection volume: 25
Detector: F ex 390 em 470 following post-column reaction. The column effluent mixed with 300 μg/mL fluorescamine in MeCN pumped at 0.16 mL/min and the mixture flowed through a 4.4 m × 0.25 mm ID coil to the detector.

CHROMATOGRAM
Retention time: 8
Limit of detection: 6.4 ng

KEY WORDS
post-column reaction; injections

REFERENCE
Frei, R.W.; Michel, L.; Santi, W. Post-column fluorescence derivatization of peptides. Problems and potential in high-performance liquid chromatography, *J.Chromatogr.*, **1976**, *16*, 665–677.

SAMPLE
Matrix: formulations
Analyte: oxytocin
Sample preparation: Inject a 100 μL aliquot.

HPLC VARIABLES
Column: 250 × 3 10 μm RP 8 (Merck)
Mobile phase: MeCN:pH 7 phosphate buffer 20:80
Flow rate: 1.47
Injection volume: 100
Detector: F ex 390 em 470 following post-column reaction. The column effluent mixed with 300 μg/mL fluorescamine in MeCN pumped at 0.16 mL/min and the mixture flowed through a 4.4 m × 0.25 mm ID coil to the detector.

CHROMATOGRAM
Retention time: 6
Limit of detection: 9 ng

KEY WORDS
post-column reaction; injections

REFERENCE
Frei, R.W.; Michel, L.; Santi, W. Post-column fluorescence derivatization of peptides. Problems and potential in high-performance liquid chromatography, *J.Chromatogr.*, **1976**, *16*, 665–677.

SAMPLE
Matrix: formulations
Analyte: dexpanthenol
Sample preparation: Dissolve an amount of formulation containing 10-20 mg dexpanthenol in 10 mL 0.5 M HCl, heat at 85 ± 2° for 30 min to hydrolyze dexpanthenol to aminopropanol. Remove an aliquot containing 1-2 mg dexpanthenol and add it to 10 mL 0.4 mg/mL fluorescamine in MeCN, add 2 mL epsilon-aminocaproic acid (concentration 60% of that of dexpanthenol) in mobile phase, make up to 25 mL with mobile phase, inject a 20 μL aliquot.

HPLC VARIABLES
Column: 300 × 4.6 Chromegabond C18
Mobile phase: MeOH:100 mM borate buffer adjusted to pH 8.0 ± 0.1 with 2 M NaOH 30:70
Flow rate: 1
Injection volume: 20
Detector: F ex 390 em 475-490 or UV 390

CHROMATOGRAM
Retention time: 24
Internal standard: epsilon-aminocaproic acid (12)

REFERENCE
Umagat, H.; Tscherne, R. High performance liquid chromatographic determination of panthenol in bulk, premix, and multivitamin preparations, *Anal.Chem.*, **1980**, *52*, 1368–1370.

SAMPLE
Matrix: formulations
Analyte: dexpanthenol
Sample preparation: Weigh out amount containing about 1 mg dexpanthenol, add 50 mL 500 mM HCl, heat at 85° for 30 min, cool, centrifuge. Remove 1 mL of the supernatant and add it to 10 mL 400 μg/mL fluorescamine, dilute to 25 mL with 1% sodium borate, inject an aliquot.

HPLC VARIABLES
Column: μBondapak C18
Mobile phase: MeCN:100 mM ammonium acetate 18:82
Detector: F ex 390 em 475-490

CHROMATOGRAM
Retention time: 7

KEY WORDS
requires at least 100 μg/g dexpanthenol; liquid multivitamin product

REFERENCE
Hudson, T.S.; Subramanian, S.; Allen, R.J. Determination of pantothenic acid, biotin, and vitamin B12 in nutritional products, *J.Assoc.Off.Anal.Chem.*, **1984**, *67*, 994–998.

SAMPLE
Matrix: formulations
Analyte: oxytocin
Sample preparation: Inject a 6.5 mL aliquot of a solution in 5% dextrose or lactated Ringer's solution with 5% dextrose onto column A at 0.8 mL/min and elute to waste with 15 mM pH 3 sodium phosphate, after 12.5 min backflush the contents of column A onto column B with the mobile phase, after 2 min remove column A from the circuit, elute column B with mobile phase, monitor the effluent from column B.

HPLC VARIABLES
Column: A Alltech C18 guard cartridge; B Alltech C18 guard cartridge + 125 × 4.6 5 μm Partisphere C18
Mobile phase: MeCN:0.1% phosphoric acid 21:79
Flow rate: 1.3
Injection volume: 5000-7000

Detector: F ex 250 em 418 (cutoff filter). The column effluent mixed with 0.02% fluorescamine in MeCN pumped at 0.3 mL/min and buffer pumped at 0.7 mL/min and the mixture flowed through a 1 mL coil to the detector. (Prepare buffer by mixing 10 g KH_2PO_4, 100 mL MeCN, and 800 mL water, adjust pH to 8.1 with 10 M NaOH, make up to 1 L with water, add 3 mL 30% Brij-35.)

CHROMATOGRAM
Retention time: 12
Limit of detection: 2 ng/mL

OTHER SUBSTANCES
Simultaneously analyzed: degradation products

KEY WORDS
post-column reaction; column-switching; siliconize glassware with SurfaSil (Pierce); injections

REFERENCE
Brown, D.S.; Jenke, D.R. Determination of trace levels of oxytocin in pharmaceutical solutions by high-performance liquid chromatography, *J.Chromatogr.*, **1987**, *410*, 157–168.

SAMPLE
Matrix: formulations
Analyte: felypressin
Sample preparation: Inject a 20 μL aliquot of the undiluted formulation onto column A and elute to waste with mobile phase A, after 14 min elute the contents of column A onto column B with mobile phase B, monitor the effluent from column B, after 9 min reequilibrate column A with mobile phase A for 7 min.

HPLC VARIABLES
Column: A 25 × 4 4 μm Superspher 60 RP-8; B 50 × 4 4 μm Superspher 60 RP-8e
Mobile phase: A MeCN:50 mM pH 6.0 phosphate buffer 12:88; B MeCN:50 mM pH 6.0 phosphate buffer 20:80
Flow rate: 1
Injection volume: 20
Detector: F ex 390 em 470 following post-column reaction. The column effluent mixed with the reagent pumped at 0.25 mL/min and the mixture flowed through a 5 m × 0.5 mm ID knitted PTFE coil to the detector. (Reagent was 300 μg/mL fluorescamine in MeCN containing 0.1% Brij-35.)

CHROMATOGRAM
Retention time: 18
Limit of detection: 0.6 pmole

OTHER SUBSTANCES
Non-interfering: prilocaine

KEY WORDS
post-column reaction; injections; column-switching; Rinse all glass and plastic ware with 1 M acetic acid and then water to minimize adsorption.

REFERENCE
Svensson, M.; Gröningsson, K. Liquid chromatographic determination of felypressin using a column-switching technique and post-column derivatization, *J.Chromatogr.*, **1990**, *521*, 141–147.

SAMPLE
Matrix: protein
Analyte: 3-methylhistidine
Sample preparation: Precipitate 1 mg protein with 10% trichloroacetic acid, wash with EtOH, wash with diethyl ether, hydrolyse under vacuum with 6 M HCl at 110° for 22 h, evaporate to dryness under reduced pressure, dissolve in 200-400 μL citrate buffer, filter (0.22 μm). Add 100 μL of the filtrate to 250 μL borate buffer, add 250 μL 1.6 mg/mL fluorescamine in MeCN with vigorous agitation, add 35 μL concentrated perchloric acid, heat at 80° for 1 h, cool to room temperature, add 100 μL 500 mM morpholinopropane-

sulfonic acid in 3 M NaOH, inject a 20-30 μL aliquot. (Citrate buffer was 67 mM pH 2.2 sodium citrate containing 0.63 mM n-caprylic acid and 48 mM thiodiglycol. Borate buffer was 800 mM boric acid adjusted to pH 12.2 with NaOH.)

HPLC VARIABLES
Guard column: Upchurch
Column: 250 × 4.6 μBondapak C18
Mobile phase: MeCN:water 24:76
Flow rate: 1.5
Injection volume: 20-30
Detector: UV 365

CHROMATOGRAM
Retention time: 13
Limit of quantitation: 100 pmole

KEY WORDS
rat; rabbit; actin; myosin

REFERENCE
Dalla Libera, L. Determination of 3-methylhistidine in hydrolysed proteins by fluorescamine derivatization and high-performance liquid chromatography, *J.Chromatogr.*, **1991**, *536*, 283–288.

SAMPLE
Matrix: saliva
Analyte: sulfapyridine
Sample preparation: 1 mL Saliva + 148 ng sulfadiazine + 1 mL MeCN + 400 mg potassium carbonate, vortex for 1 min, centrifuge at ⩾1000 g for 10 min. Remove the upper MeCN layer and evaporate it to dryness under a stream of nitrogen at 60°, reconstitute the residue in 200 μL mobile phase, vortex, inject a 40 μL aliquot.

HPLC VARIABLES
Column: 125 × 4.6 5 μm RP-18 (Brownlee)
Mobile phase: MeOH:buffer 15:85 (Buffer was 50 mM NaHPO$_4$ (sic) containing 10 mM sodium 1-hexanesulfonate and 7.2 mM triethylamine, adjusted to pH 3.0 with phosphoric acid.)
Flow rate: 1
Injection volume: 40
Detector: F ex 395 em 470 following post-column reaction. The column effluent mixed with reagent pumped at 0.3 mL/min and the mixture flowed through a 4.8 m × 0.7 mm ID PTFE coil at 60° to the detector. (Prepare reagent by dissolving 400 mg fluorescamine in 250 mL MeOH, add 1 mL 2-mercaptoethanol, add 250 mL mobile phase.)

CHROMATOGRAM
Retention time: 7.32
Internal standard: sulfadiazine (5.66)
Limit of detection: 5 ng/mL

OTHER SUBSTANCES
Non-interfering: N-acetylsulfapyridine, 2-amino-3-phenyl-1-propanol, 5-aminosalicylic acid, amphetamine, furosemide, levallorphan, metoprolol, riboflavin, salicylic acid, sulfasalazine, viloxazine

KEY WORDS
post-column reaction

REFERENCE
Sista, H.S.; Dye, D.M.; Leonard, J. High-performance liquid chromatographic method for determination of sulfapyridine in human saliva using post-column, in-line derivatization with fluorescamine, *J.Chromatogr.*, **1983**, *273*, 464–468.

SAMPLE
Matrix: soil, water
Analyte: diethylenetriamine

Sample preparation: Soil. 2 g Soil + 4 mL 2 M calcium chloride solution, vortex vigorously for 3 min, centrifuge at 2000 rpm for 10 min. Add 1 mL supernatant to 20 μL 100 mM NaOH, rapidly add 400 μL 2 mg/mL fluorescamine in MeOH, mix rapidly, inject an aliquot. Water. Add 1 mL water to 20 μL 100 mM NaOH, rapidly add 100 μL 5 mg/mL fluorescamine in acetone, mix rapidly, inject an aliquot.

HPLC VARIABLES
Column: 250 × 4.6 10 μm Nucleosil C18
Mobile phase: MeOH:buffer 40:60 (soil) or 50:50 (water) (Prepare buffer by adjusting 100 mM boric acid to pH 8 with 100 mM NaOH.)
Flow rate: 1.8 (soil) or 1 (water)
Injection volume: 20
Detector: F ex 390 em 475

CHROMATOGRAM
Retention time: 3.7 (water), 5.8 (soil)
Limit of quantitation: 13.5 ng/mL (water), 130 ng/g (soil)

REFERENCE
Mao, J.; Dennis, M.W.; Foster, R.; Fackler, P.H. Analysis of diethylenetriamine in water and soil at ppb levels by high performance liquid chromatography with fluorescence detection, *J.Liq.Chromatogr.*, **1994**, *17*, 2485–2494.

SAMPLE
Matrix: solutions
Analyte: diamines
Sample preparation: Rapidly mix 1 volume of amine solution (<1 mM) with 1 volume buffer and 1 volume 2 mg/mL fluorescamine in acetone, inject an aliquot. (Buffer was 100 mM boric acid adjusted to pH 8.0 with 6 M NaOH.)

HPLC VARIABLES
Column: 500 × 3 Vydac octadecylsilane glass column
Mobile phase: Gradient. MeOH:buffer from 10:90 to 40:60 over 20 min (?). (Buffer was 100 mM boric acid adjusted to pH 8.0 with 6 M NaOH.)
Column temperature: 60
Flow rate: 1
Detector: F Corning No. 7-51 primary filter, Wratten No. 4 secondary filter

CHROMATOGRAM
Retention time: 7 (putrescine), 9 (cadaverine), 11 (1,6-hexanediamine), 12.5 (1,7-heptanediamine), 15 (1,8-octanediamine)

REFERENCE
Samejima, K. Separation of fluorescamine derivatives of aliphatic diamines and polyamines by high-speed liquid chromatography, *J.Chromatogr.*, **1974**, *96*, 250–254.

SAMPLE
Matrix: solutions
Analyte: catecholamines
Sample preparation: Dilute a few μL of a <1 mM solution to 20 μL with 50 mM pH 8.0 phosphate or borate buffer, add 10 μL 2 mg/mL (?) fluorescamine in acetone with vigorous shaking, inject an aliquot.

HPLC VARIABLES
Column: 500 × 3 Hitachi 3011 gel glass column
Mobile phase: MeOH:100 mM pH 8.0 Tris-HCl buffer 70:30
Flow rate: 0.72
Detector: F primary filter Corning No. 7-51, secondary filter No. 4-7116

CHROMATOGRAM
Retention time: 6 (normetanephrine), 8 (norepinephrine), 8 (3-methoxytyramine), 13 (dopamine)

REFERENCE
Imai, K. Fluorimetric assay of dopamine, norepinephrine and their 3-O-methyl metabolites by using fluorescamine, *J.Chromatogr.*, **1975**, *105*, 135–140.

SAMPLE
Matrix: solutions
Analyte: catecholamines
Sample preparation: 200 μL 5 μg/mL Amine solution + 300 μL 100 mM pH 8.0 phosphate buffer + 300 μL 200 μg/mL fluorescamine in acetone + 200 μL water, mix, saturate with NaCl, add 500 μL ethyl acetate, shake for 1 min, inject a 50 μL aliquot of the organic phase.

HPLC VARIABLES
Column: 250 × 2 10 μm LiChrosorb Si 60-10
Mobile phase: Benzene:dioxane:acetic acid 76:22:2 (Caution! Benzene and dioxane are carcinogens!)
Flow rate: 0.5
Detector: F ex 325-385 (filter) em 451

CHROMATOGRAM
Retention time: 4 (dopamine), 8 (norepinephrine)

OTHER SUBSTANCES
Interfering: 50

KEY WORDS
normal phase

REFERENCE
Schwedt, G. Hochdruck-Flüssigkeits-chromatographische Analyse der Katecholamine Dopamin und Noradrenalin als Fluorescaminderivate, *J.Chromatogr.*, **1976**, *118*, 429–432.

SAMPLE
Matrix: solutions
Analyte: kallidin and bradykinin
Sample preparation: Inject a 100 μL aliquot of a solution in 500 mM pH 8 phosphate buffer.

HPLC VARIABLES
Column: 250 × 4 10 μm RP-18 (Merck)
Mobile phase: Gradient. A was MeCN:buffer 5:95. B was MeCN:buffer 30:70. A:B from 50:50 to 10:90 over 7.5 min. (Buffer was 67 mM pH 7 phosphate buffer containing 25 g/L sodium sulfate. Maintain eluents at 40°.)
Column temperature: 40
Flow rate: 2
Injection volume: 100
Detector: F ex 390 em 476 following post-column reaction. The column effluent mixed with 1 mg/mL fluorescamine in MeCN pumped at 0.2 mL/min and the mixture flowed through a 2 m × 0.2 mm ID PTFE coil to the detector. For the first 2 min divert the column effluent to waste rather than through the post-column reaction system.

CHROMATOGRAM
Retention time: 4.9 (kallidin), 5.8 (bradykinin)
Limit of detection: 2.5-3 ng

KEY WORDS
post-column reaction; use plasticware

REFERENCE
Gau, W.; Haberland, G.L.; Ploschke, H.J.; Schmidt, K. A high performance liquid chromatographic method for the determination of kinin releasing activity of kallikrein, *Adv.Exp.Med.Biol.*, **1983**, *156*, 483–494.

SAMPLE
Matrix: solutions

Analyte: aromatic amines
Sample preparation: Condition a 1 mL 100 mg C18 SPE cartridge (J.T. Baker) with 1 mL MeOH and 1 mL water. Mix 5 mL 2% formic acid with 500 μL 0.05% fluorescamine in acetone and add it to the SPE cartridge. Filter (4-5.5 μm) water, add 50 mL filtrate to 1 mL 500 mM pH 5.5 citrate buffer, mix, add to the SPE cartridge at 10 mL/min, wash with 1 mL THF:water:formic acid 20:80:2, elute with 1 mL THF:water:formic acid 60 :40:2, inject a 20 μL aliquot.

HPLC VARIABLES
Column: 250 × 4.6 5 μm Spherisorb octyl C8
Mobile phase: THF:water:formic acid 42:56:2
Flow rate: 0.7
Injection volume: 20
Detector: F ex 390 em 495

CHROMATOGRAM
Retention time: 18 (aniline), 22 (p-toluidine), 28 (4-chloroaniline), 29 (4-bromoaniline)
Limit of detection: 6-30 ng/L

KEY WORDS
SPE; wastewater

REFERENCE
Djozan, D.; Faraj-Zadeh, M.A. Liquid chromatographic determination of aniline and derivatives in environmental waters at nanogram per litre levels using fluorescamine pre-column derivatization, *Chromatographia*, **1995**, *41*, 568–572.

SAMPLE
Matrix: solutions
Analyte: proteins
Sample preparation: Mix 85 μL of a solution in 100 mM pH 9.0 sodium borate buffer with 5 μL 5 mg/mL fluorescamine in acetone, let stand for 5 min, add 10 μL 300 mM dithiothreitol in 150 mM NaOH, add 100 μL CE SDS sample buffer (Bio-Rad), centrifuge at 16000 g for 2 min, inject an aliquot.

CAPILLARY ELECTROPHORESIS
Capillary: 37 cm × 50 μm fused-silica (25 cm to detector) (Polymicro Technologies)
Capillary preparation: At the beginning of each day purge capillary with 100 mM NaOH for 1.5 min, with 100 mM HCl for 1 min, and with CE SDS sample buffer (Bio-Rad) for at least 2 min.
Running buffer: CE SDS sample buffer (Bio-Rad)
Voltage/Current: -486 V/cm
Injection: Electrokinetic injection at -5 kV for 27 s
Detector: F ex 363.8 (20 mW argon ion laser) em 450 (bandpass filter)
Model: laboratory constructed
Migration time: 9.7 (myoglobin), 10.7 (α-chymotrypsinogen), 14.5 (conalbumin)
Limit of detection: 96-190 amole

KEY WORDS
comparison with other derivatizing reagents

REFERENCE
Gump, E.L.; Monnig, C.A. Pre-column derivatization of proteins to enhance detection sensitivity for sodium dodecyl sulfate non-gel sieving capillary electrophoresis, *J.Chromatogr.A*, **1995**, *715*, 167–177.

SAMPLE
Matrix: tears
Analyte: histamine
Sample preparation: 10 μL Tears + 40 μL 200 mM pH 9.1 sodium borate buffer + 50 μL 200 μg/mL fluorescamine in MeCN, mix vigorously, inject a 20 μL aliquot.

HPLC VARIABLES
Guard column: 20-40 μm LiChroprep RP-8

Column: 10 μm Nucleosil C8 or 10 μm RP-8 (Merck)
Mobile phase: MeCN:4 mM pH 3.5 KH$_2$PO$_4$ 65:35
Flow rate: 0.5
Injection volume: 20
Detector: F ex 390 em 480

CHROMATOGRAM
Retention time: 4
Limit of quantitation: 1 ng/mL

REFERENCE
Bettero, A.; Angi, M.R.; Moro, F.; Benassi, C.A. Histamine assay in tears by fluorescamine derivatization and high-performance liquid chromatography, *J.Chromatogr.*, **1984**, *310*, 390–395.

SAMPLE
Matrix: tears
Analyte: histamine
Sample preparation: 10 μL Tears + 40 μL 100 mM pH 9.1 borate buffer, mix, add 50 μL 200 μg/mL fluorescamine in MeCN with vigorous stirring, inject a 20 μL aliquot on to column A, divert the fraction containing the derivatized histamine on to column B then remove column A from the circuit (details are sketchy).

HPLC VARIABLES
Column: A 20-40 μm RP-8 (Merck); B 10 μm RP-8 (Merck)
Mobile phase: MeCN:4 mM pH 3.5 phosphate buffer 65:35
Flow rate: 0.5
Injection volume: 20
Detector: F ex 390 em 480

CHROMATOGRAM
Retention time: 4
Limit of detection: 0.1 ng/mL

KEY WORDS
column-switching; heart cut

REFERENCE
Bettero, A.; Galiano, F.; Benassi, C.A.; Angi, M.R. A rapid HPLC technique for determining levels of histamine in tears from normal and inflamed human eyes, *Food Chem.Toxicol.*, **1985**, *23*, 303–304.

SAMPLE
Matrix: tissue
Analyte: sulfonamides
Sample preparation: Blend 3 g meat with 30 mL chloroform for 2 min in a Polytron homogenizer, shake for 10 min, centrifuge at 1600 g for 5 min, filter (5A filter paper). Add 5 mL filtrate to 1 mL 3 M HCl, shake for 10 min, centrifuge at 1600 g for 5 min. 250 μL Aqueous layer + 250 μL 3.5 M sodium acetate solution, vortex, add 100 μL 0.2% fluorescamine in acetone, vortex, let stand for 20 min at room temperature, inject a 10 μL aliquot.

HPLC VARIABLES
Column: 150 × 4.6 5 μm Chemcosorb 5-ODS-H
Mobile phase: MeCN:2% acetic acid 5:3
Column temperature: 55
Flow rate: 1
Injection volume: 10
Detector: F ex 405 em 495

CHROMATOGRAM
Retention time: 4 (sulfisomidine), 6 (sulfadiazine), 7 (sulfamerazine), 7.5 (sulfamethazine (sulfadimidine)), 8 (sulfamonomethoxine), 12 (sulfamethoxazole), 14.5 (sulfadimethoxine), 16 (sulfaquinoxaline)
Limit of detection: 0.005 ng/g

KEY WORDS
cow; pig; chicken; ham; sausage; roast beef

REFERENCE
Takeda, N.; Akiyama, Y. Pre-column derivatization of sulfa drugs with fluorescamine and high-perfor-
mance liquid chromatographic determination at their residual levels in meat and meat products,
J.Chromatogr., **1991**, *558*, 175–180.

SAMPLE
Matrix: tissue
Analyte: sulfonamides
Sample preparation: Condition a 500 mg Chromabond SA cation-exchange SPE cartridge
(Macherey-Nagel) with 6 mL hexane, dry under vacuum for 10 min, condition with 6 mL
dichloromethane:acetone:acetic acid 50:50:2, do not allow to go dry. Homogenize (Polytron) 10 g sample with 60 mL dichloromethane:acetone 50:50 for 30 s, rinse the apparatus with 2-3 mL dichloromethane:acetone 50:50, centrifuge the mixture at 2500 rpm
for 10 min. Filter (cotton wool) the supernatant and wash it through with a little
dichloromethane:acetone 50:50, add 5 mL acetic acid to the filtrate, mix, remove one
tenth of this mixture and add it to the SPE cartridge at 2 mL/min, do not allow the SPE
cartridge to run dry, wash with 5 mL water, wash with 5 mL MeOH, dry under vacuum
for 10 min, pass gaseous ammonia through the SPE cartridge until the acid is neutralized
(when air is passed through the cartridge moist pH paper should turn blue), elute with
3 mL MeOH at 1-2 mL/min, carefully evaporate to dryness under reduced pressure (100
mbar) at 40°, reconstitute with 500 μL initial mobile phase, centrifuge, inject a 50 μL
aliquot of the supernatant.

HPLC VARIABLES
Column: 125 × 4 5 μm LiChrospher 100 RP-18
Mobile phase: Gradient. A was MeCN:20 mM pH 5 sodium acetate buffer 5.5:94.5. B was
MeCN:EtOH:20 mM pH 5 sodium acetate buffer 50:10:40. A:B from 100:0 to 0:100
over 32 min (concave gradient), return to initial conditions over 4 min, re-equilibrate at
initial conditions for 10 min.
Flow rate: 0.8
Injection volume: 50
Detector: UV 270; F ex 395 em 495 following post-column reaction. The column effluent
mixed with ice-cold reagent pumped at 0.3 mL/min and this mixture flowed through a
2.3 m × 0.5 mm ID coil in a cooled ultrasonic bath to the detector. (Prepare reagent by
dissolving 25 mg fluorescamine in 25 mL MeCN and adding 75 mL buffer and 200 μL
mercaptoethanol. Buffer was 20 mM NaH_2PO_4 adjusted to pH 3 with 1 M phosphoric
acid.)

CHROMATOGRAM
Retention time: 3 (sulfaguanidine), 3.7 (sulfanilamide), 10 (sulfadiazine), 15 (sulfathiazole), 15.7 (sulfapyridine), 16.5 (sulfamerazine), 18 (sulfamethizole), 20 (sulfamethazine
(sulfadimidine)), 21 (sulfamethoxypyridazine), 22 (sulfachlorpyridazine), 24 (sulfadoxine),
28 (sulfadimethoxine)
Limit of detection: 0.5-5 ppb

KEY WORDS
post-column reaction; muscle; kidney; SPE

REFERENCE
Pacciarelli, B.; Reber, S.; Douglas, C.; Dietrich, S.; Etter, R. Bestimmung von 12 Sulfonamiden in Fleisch
und Nieren mittels HPLC und Nachsäulenderivatisierung [Determination of 12 sulfonamides in
meat and kidney by HPLC and post column derivatization], *Mitt.geb.Lebensmittelunters.Hyg.*, **1991**,
82, 45–55.

SAMPLE
Matrix: tissue
Analyte: histamine and 3-methylhistamine
Sample preparation: Make a slurry of 40 μm Bakerbond carboxylic acid material in
MeOH and prepare SPE columns by adding an aliquot containing 150 mg material to a
Pasteur pipette plugged with glass fiber prefilter material (Sartorius). Wash column with

2 mL water, 2 mL 1 M HCl, 2 mL water, and 4 mL 200 mM pH 6.4 sodium phosphate buffer. Homogenize (Ultra-Turrax) rat heart and 500 ng 1-methylhistamine in ice-cold 50 mM pH 8.5 Tris-HCl buffer, sonicate (MSE sonifier, 12 W, 12 μm peak-peak) for 2 min with repeated intervals of 5 s, add perchloric acid to a final concentration of 300 mM, heat at 100° for 5 min, neutralize with KOH, cool to 0°, centrifuge at 4° at 2000 g for 5 min, adjust pH of supernatant to 6.4 with phosphate buffer, add 1-methylhistamine, add to the SPE column, wash with 4 mL 50 mM pH 6.4 disodium EDTA, wash with 4 mL water, elute with 1 mL 1 M HCl. Evaporate the eluate to dryness under a stream of nitrogen at 40° over about 15 min, reconstitute with 100 μL water, add 400 μL 50 mM pH 9.1 sodium borate, with continuous vigorous stirring add 500 μL 200 μg/mL fluorescamine in MeCN (freshly prepared), stir for 1 min, evaporate to dryness under a stream of nitrogen at 40°, reconstitute with 1 mL mobile phase, inject a 20 μL aliquot.

HPLC VARIABLES
Column: 200 × 3 5 μm Inertsil ODS-2 (see J. Chromatogr. B 1994, 657, 261)
Mobile phase: MeCN:MeOH:water:phosphoric acid 15:10:75:0.2 adjusted to pH 6.87 with ammonium hydroxide (see J. Chromatogr. B 1994, 657, 261)
Column temperature: 20
Flow rate: 0.4
Injection volume: 20
Detector: F ex 360 em 440

CHROMATOGRAM
Retention time: 15 (histamine), 20 (3-methylhistamine)
Internal standard: 1-methylhistamine (25)
Limit of detection: 20 pg

KEY WORDS
SPE; rat; heart

REFERENCE
van Haaster, C.M.C.J.; Engels, W.; Lemmens, P.J.M.R.; Hornstra, G.; van der Vusse, G.J. Rapid and highly sensitive high-performance liquid chromatographic method for the determination of histamine and 3-methylhistamine in biological samples using fluorescamine as the derivatizing agent, *J.Chromatogr.*, **1993**, *617*, 233–240.

SAMPLE
Matrix: tissue
Analyte: sulfadiazine
Sample preparation: Condition a 3 mL Bond Elut propylsulfonic acid strong cation exchange SPE cartridge with 4 mL MeCN, 16 mL 200 mM phosphoric acid, and 4 mL MeCN. Homogenize (Ultra-Turrax T-25 with S25N dispersing tool) 10 g chopped fish and 10 mL mobile phase at 16000 rpm for 1 min, add 90 mL MeCN, shake at low speed on shaker, centrifuge at 1500 rpm for 10 min, remove the supernatant, add 30 mL MeCN to the solid, shake, centrifuge, decant the supernatant. Combine the supernatants, add 100 mL water, add 2 mL diethylene glycol, add 60 mL dichloromethane, shake for 3 min, remove the organic layer, repeat the extraction with 40 mL dichloromethane. Combine the organic layers and evaporate in a rotary evaporator at 65° to 2-3 mL, add 3 mL dichloromethane, add to the SPE cartridge, rinse flask with MeCN:dichloromethane 60:40, add rinse to the SPE cartridge, rinse flask with 5 mL MeCN, add rinse to the SPE cartridge, wash with 2 mL MeCN:200 mM phosphoric acid 10:90, elute with 5 mL MeCN:200 mM phosphoric acid 10:90, inject a 10 μL aliquot of the eluate. (Do not allow SPE cartridge to go dry at any time.)

HPLC VARIABLES
Guard column: 20 × 2 pellicular C18
Column: 150 × 4.6 5 μm Inertsil ODS-2
Mobile phase: MeCN:2% acetic acid 10:90
Flow rate: 1
Injection volume: 10
Detector: F ex 400 em 495 following post-column reaction. The column effluent mixed with 500 μg/mL (?) fluorescamine in MeCN:2% acetic acid 55:45 pumped at 0.2 mL/min and

the mixture flowed through a 10.7 m $\times$ 0.4 mm ID coil of PTFE tubing at 70° to the detector.

CHROMATOGRAM
Retention time: 6.5
Limit of quantitation: 1 ng/g
Limit of detection: 0.2 ng/g

KEY WORDS
fish; salmon; post-column reaction; SPE

REFERENCE
Gehring, T.A.; Rushing, L.G.; Thompson, H.C. Jr. Liquid chromatographic determination of sulfadiazine in salmon by postcolumn derivatization and fluorescence detection, *J.AOAC Int.*, **1995**, *78*, 1161–1164.

SAMPLE
Matrix: tissue
Analyte: sulfonamides
Sample preparation: Homogenize (Ultra-Turrax) 3 g ground tissue with 30 mL chloroform for 2 min, centrifuge at 3000 g for 5 min, filter (paper). Remove a 10 mL aliquot of the filtrate and add it to 1 mL 3 M HCl, vortex for 1 min, centrifuge at 2000 g for 5 min. Remove a 250 µL aliquot of the aqueous layer and add it to 250 µL 3.8 M sodium acetate, add 100 µL 1 mg/mL fluorescamine in MeCN, vortex, let stand at room temperature for 20 min, inject a 20 µL aliquot. (Sodium acetate should be a highly pure grade.)

HPLC VARIABLES
Column: 250 $\times$ 4.6 5 µm Nucleosil 120 C18
Mobile phase: MeCN:20 mM pH 4 NaH_2PO_4 34:66 containing 20 mM sodium octanesulfonate
Column temperature: 30
Flow rate: 1.2
Injection volume: 20
Detector: F ex 405 em 495

CHROMATOGRAM
Retention time: 8 (sulfadiazine), 10 (sulfamethazine), 21 (sulfadimethoxine), 24 (sulfaquinoxaline)
Limit of detection: 3-40 ng/g

KEY WORDS
chicken; muscle

REFERENCE
Simeonidou, E.J.; Botsoglou, N.A.; Psomas, I.E.; Fletouris, D.J. Liquid chromatographic analysis of multiple sulfonamide residues in chicken muscle using pre-column derivatization and fluorescence detection, *J.Liq.Chromatogr.Rel.Technol.*, **1996**, *19*, 2349–2364.

SAMPLE
Matrix: tissue
Analyte: sulfonamides
Sample preparation: Homogenize (Ultra-Turrax T-25 with S25N dispersing tool) 10 g chopped fish and 10 mL mobile phase A at high speed for 30 s, add 90 mL MeCN, shake at low speed on shaker, centrifuge at 1500 rpm for 10 min, remove the supernatant, add 30 mL MeCN to the solid, shake, centrifuge, decant the supernatant. Combine the supernatants, add 100 mL water, add 2 mL diethylene glycol, add 60 mL dichloromethane, shake for 3 min, remove the organic layer, repeat the extraction with 40 mL dichloromethane. Combine the organic layers and evaporate in a rotary evaporator at 65° to ca. 2 mL, wash into a smaller tube with two 2 mL portions of MeOH, concentrate to about 1 mL with a stream of nitrogen at 65°, dilute to 4.5 mL with 200 mM phosphoric acid, add 5 mL hexane, vortex, centrifuge for 15 min, discard upper hexane layer. Dilute the lower aqueous layer to 5 mL with 200 mM phosphoric acid, inject a 20 µL aliquot.

HPLC VARIABLES
Guard column: C18
Column: 150 × 4.6 3.5 μm Symmetry C18 (Waters)
Mobile phase: Gradient. A was MeCN:MeOH:2% acetic acid in water 5:10:85. B was MeCN:MeOH:2% acetic acid in water 25:10:65. A:B from 100:0 to 0:100 over 25 min, maintain at 0:100 for 5 min.
Flow rate: 1
Injection volume: 20
Detector: F ex 400 em 495 following post-column reaction. The column effluent mixed with 500 μg/mL fluorescamine in MeCN:MeOH:2% acetic acid 52.5:5:42.5 pumped at 0.2 mL/min and the mixture flowed through a 10.7 m × 0.4 mm i.d. coil of PTFE tubing at 70° to the detector.

CHROMATOGRAM
Retention time: 4 (sulfanilamide), 6 (sulfadiazine), 6.5 (sulfathiazole), 7 (sulfapyridine), 7.5 (sulfamerazine), 9.5 (sulfamethazine), 10 (sulfamethizole), 11 (sulfamethoxypyridazine), 14 (sulfachloropyridazine), 14.5 (sulfamonomethoxine), 15 (sulfadoxine), 16 (sulfamethoxazole), 23.5 (sulfadimethoxine), 24.5 (sulfaquinoxaline)
Limit of quantitation: 1-5 ng/g

KEY WORDS
fish; salmon; post-column reaction

REFERENCE
Gehring, T.A.; Rushing, L.G.; Thompson, H.C. Jr. Determination of sulfonamides in edible salmon tissue by liquid chromatography with postcolumn derivatization and fluorescence detection, *J.AOAC Int*, **1997**, *80*, 751−755.

SAMPLE
Matrix: urine
Analyte: pamidronate
Sample preparation: Filter dog urine through filter paper. 2 mL Urine + 100 μL IS in water, adjust pH to 3 with concentrated HCl, filter (0.45 μm fluoro-membrane, Skan Model Acro LC 13). Add 0.5 mL 1.5 M trichloroacetic acid to the filtrate, centrifuge at 1500 g for 15 min. Remove the supernatant, add 20 μL 2.5 M calcium chloride solution, add 40 μL 500 mM NaH_2PO_4, adjust pH to 12.0 with 6.25 M NaOH then with 0.5 M NaOH, centrifuge at 1500 g for 15 min. Wash the precipitate with water, dissolve the precipitate in 200 μL 130 mM disodium EDTA adjusted to pH 9 with 6.25 M NaOH, add 100 μL 1 mg/mL fluorescamine in MeCN while vortexing vigorously, add 200 μL dichloromethane, extract, centrifuge at 1000 g for 3 min, remove the aqueous phase, inject a 10 μL aliquot of the aqueous phase.

HPLC VARIABLES
Column: 250 × 4 10 μm Nucleosil C18
Mobile phase: MeOH:1 mM disodium EDTA adjusted to pH 6.5 with 1 M NaOH 3:97
Column temperature: 40
Flow rate: 1
Injection volume: 10
Detector: F ex 395 em 480

CHROMATOGRAM
Retention time: 4
Internal standard: 6-amino-1-hydroxyhexylidenebisphosphonate (CGP 38 146) (8)
Limit of quantitation: 1000 nM
Limit of detection: 50 nM

KEY WORDS
human; dog

REFERENCE
Flesch, G.; Hauffe, S.A. Determination of the bisphosphonate pamidronate disodium in urine by precolumn derivatization with fluorescamine, high-performance liquid chromatography and fluorescence detection, *J.Chromatogr.*, **1989**, *489*, 446−451.

SAMPLE

Matrix: urine

Analyte: 3-methylhistidine

Sample preparation: 200 μL Urine + 50 μL perchloric acid, heat at 80° for 8-12 h or 100° for 1.5-4 h, add 750 μL water, invert tubes, vortex for 5 s, centrifuge at 3500 g for 5 min. Remove a 75 μL aliquot and add it to 250 μL borate buffer, add 250 μL 400 μg/mL fluorescamine in MeCN, vortex for 5 s, let stand for 5 min, add 35 μL perchloric acid, vortex, heat at 80° for 1 h, cool to room temperature, add 100 μL MOPS buffer, invert the tubes, vortex, inject a 100 μL aliquot. (Prepare borate buffer by dissolving 5.6 g boric oxide in 250 mL water and adjusting pH to 12.2 with 12 M NaOH. Prepare MOPS buffer by dissolving 20.93 g MOPS in 250 mL 2.4 M NaOH)

HPLC VARIABLES

Guard column: 10 μm Adsorbosphere C18

Column: 250 × 4.6 8 μm Dynamax C18 (Rainin)

Mobile phase: MeCN:buffer:diethylamine 20.5:79.5:0.025 (Buffer was 272 mg KH_2PO_4 and 1.136 g Na_2HPO_4 in 1 L water.)

Flow rate: 1.6

Injection volume: 100

Detector: F ex 365 em 460

CHROMATOGRAM

Retention time: 15

KEY WORDS

rat

REFERENCE

Kuhl, D.A.; Methvin, J.T.; Dickerson, R.N. Standardization of acid hydrolysis procedure for urinary 3-methylhistidine determination by high-performance liquid chromatography, *J.Chromatogr.B*, **1996**, *681*, 390–394.

2-Methoxy-2,4-diphenyl-3(2H)-furanone

SAMPLE
Matrix: solutions
Analyte: amines
Sample preparation: Mix a 100 μL aliquot of an aqueous solution with 100 μL 50 mM pH 9.6 sodium borate buffer, add 100 μL 10 mM 2-methoxy-2,4-diphenyl-3(2H)-furanone (Fluka) in MeCN while vortexing vigorously, vortex for 10 s, let stand at 20° for 30 min, inject an aliquot.

HPLC VARIABLES
Column: 300 × 4 5 μm TSK LS-410 K (Toyo Soda)
Mobile phase: MeOH:50 mM pH 7.0 phosphate buffer 70:30
Flow rate: 0.45
Injection volume: 10
Detector: F ex 390 em 480 following post-column reaction. The column effluent mixed with reagent pumped at 0.25 mL/min and the mixture flowed through a 3 m × 0.5 mm ID stainless steel coil at 60° and a 50 cm × 0.5 mm ID stainless steel coil at 0° to the detector. (The reagent was ethanolamine adjusted to pH 10.5 with concentrated HCl. If only primary amines are to be determined the post-column reaction system is not necessary.)

CHROMATOGRAM
Retention time: 14 (methylamine), 17 (diethylamine), 25 (di-n-propylamine), 30 (pentyl-amine), 42 (di-n-butylamine)
Limit of detection: 0.5-50 pmole

REFERENCE
Nakamura, H.; Takagi, K.; Tamura, Z.; Yoda, R.; Yamamoto, Y. Stepwise fluorometric determination of primary and secondary amines by liquid chromatography after derivatization with 2-methoxy-2,4-diphenyl-3(2H)-furanone, *Anal.Chem.*, **1984**, 56, 919-922.

2-Methyl-3-oxo-4-phenyl-2,3-dihydrofuran-2-yl Acetate

SAMPLE
Matrix: solutions
Analyte: peptides
Sample preparation: Mix 25 μL of a solution in pH 7.5 50 mM NaH_2PO_4 buffer containing 12.5 mM EDTA with 1 μL 1 mM 2-methyl-3-oxo-4-phenyl-2,3-dihydrofuran-2-yl acetate in DMSO, let stand at room temperature for 15 min, inject an aliquot. (Dry DMSO over molecular sieve before use. The reagent has a chiral center so each compound gives 2 peaks.)

HPLC VARIABLES
Column: 250 × 4.6 Biophase ODS C-18
Mobile phase: Gradient. MeOH:water from 5:95 to 45:55 over 35 min.
Flow rate: 1
Detector: F ex 390 em 470

CHROMATOGRAM
Retention time: 18, 22.5 (Val-Tyr-Val diastereomers), 19, 21 (methionine enkephalin diastereomers), 22, 24 (leucine enkephalin diastereomers)

KEY WORDS
comparison with capillary electrophoresis; comparison with fluorescamine derivatization

REFERENCE
Chen, P.; Novotny, M.V. 2-Methyl-3-oxo-4-phenyl-2,3-dihydrofuran-2-yl acetate: A fluorogenic reagent for detection and analysis of primary amines, *Anal.Chem.*, **1997**, *69*, 2806–2811.

SAMPLE
Matrix: solutions
Analyte: peptides
Sample preparation: Mix 4 μL of a solution in pH 9.0 buffer with 1 μL 1 mM 2-methyl-3-oxo-4-phenyl-2,3-dihydrofuran-2-yl acetate in DMSO, let stand at room temperature for 15 min, inject an aliquot. (Dry DMSO over molecular sieve before use.)

CAPILLARY ELECTROPHORESIS
Capillary: 60 cm × 50 μm coated capillary (45 cm to detector)
Capillary preparation: Coat the capillaries as follows. Treat capillary with 100 mM NaOH for 1 h, rinse with water, rinse with MeOH, fill capillary under nitrogen pressure with 10 μL (gamma-methacryloxypropyl)trimethoxysilane dissolved in dichloromethane containing 20 mM acetic acid, after 1 h rinse with MeOH, rinse with water, pass 4% acrylamide solution containing 1 μL/mL tetramethylethylenediamine and 1 mg/mL ammonium persulfate through the capillary under nitrogen pressure for 30 min (Caution! Acrylamide is a carcinogen!), rinse with water, dry under a stream of nitrogen (Anal. Chem. 1994, 66, 1134).
Running buffer: MeOH:buffer 10:90 (Buffer contained 150 mM Tris and 150 mM boronic acid, pH 8.3.)

Voltage/Current: 25 kV/ca. 5 μA

Injection: Hydrodynamic injection for 5 s.

Detector: F ex 325 (He-Cd laser) em 500

Model: laboratory-constructed

Migration time: 8 (Val-Tyr-Val), 9.5 (methionine enkephalin), 9.7 (leucine enkephalin), 15.5 (angiotensin II)

Limit of detection: <1 pmole

KEY WORDS

coated capillary; comparison with HPLC; comparison with fluorescamine derivatization

REFERENCE

Chen, P.; Novotny, M.V. 2-Methyl-3-oxo-4-phenyl-2,3-dihydrofuran-2-yl acetate: A fluorogenic reagent for detection and analysis of primary amines, *Anal.Chem.*, **1997**, *69*, 2806–2811.

MISCELLANEOUS REACTIONS

Diazo Coupling

$$\text{RNH}_2 \longrightarrow \text{RN}_2^{\oplus} \xrightarrow{\text{R'H}} \text{R-N=N-R'}$$

SAMPLE

Matrix: beverages

Analyte: aromatic amines

Sample preparation: Allow 200 mL soft drink to stand overnight, adjust pH to 2.1-2.6 with 20% sulfuric acid (if necessary), add 60 mg sodium dithionite (more if necessary), stir rapidly mechanically for 30-45 s, adjust pH to 8.5 with 10% NaOH, add 50 mg NaCl, stir for 1 min, add 25 mL chloroform, shake for 1 min, allow layers to separate, drain the chloroform layer through glass wool (previously washed with 25 mL chloroform), repeat the extraction twice more. Combine the chloroform layers and add them to 5 mL 5 mM sulfuric acid, evaporate the chloroform using a stream of nitrogen, add 1 mL 1 mg/mL sodium nitrite, swirl to mix, chill below 5° for 15 min, add 1 mL 5 mg/mL disodium 2-naphthol-3,6-disulfonate in 6 mg/mL sodium carbonate solution, swirl to mix, chill below 5° for 15 min, evaporate to dryness under reduced pressure at 45°, dry under a stream of air for 2 min (Anal. Chem. 1985, 57, 189), redissolve in 2 mL MeOH:water 25:75, add 20 μL 500 mM tetra-n-butylammonium phosphate, inject a 20 μL aliquot. (Purify disodium 2-naphthol-3,6-disulfonate (2-naphthol-3,6-disulfonic acid, disodium salt; Fluka) with decolorizing carbon and recrystallization from EtOH/water. Prepare tetra-n-butylammonium phosphate solution by adding 48 mL 40% tetra-n-butylammonium hydroxide to 100 mL 1.11 M KH_2PO_4 and filtering (0.45 μm nylon). Azo dyes are reduced to their parent amines which are then derivatized and chromatographed.)

HPLC VARIABLES

Column: 250 × 4.6 5 μm LC-18 (Supelco)

Mobile phase: Gradient. A was MeOH:water 60.5:39.5 containing 5 mM tetra-n-butylammonium phosphate. B was MeOH:water 69:31 containing 5 mM tetra-n-butylammonium phosphate. A:B 100:0 for 5 min, to 0:100 over 2 min, maintain at 0:100. (At the beginning of each day condition column with MeOH:water 55:45 for 30 min. At the end of each day flush with 100 mL water and 25 mL MeOH.)

Injection volume: 20

Detector: UV 512

CHROMATOGRAM

Retention time: 6 (aniline), 16.5 (2-naphthylamine), 17 (1-naphthylamine), 18.5 (2-aminobiphenyl), 19 (4-aminobiphenyl)

Limit of detection: 0.15-0.3 ng

KEY WORDS
soft drinks

REFERENCE
Lancaster, F.E.; Lawrence, J.F. Determination of total non-sulphonated aromatic amines in soft drinks and hard candies by reduction and derivatization followed by high-performance liquid chromatography, *Food Addit.Contam.*, **1992**, *9*, 171–182.

SAMPLE
Matrix: dyes
Analyte: aromatic amines
Sample preparation: Dissolve 1 g dye in 100 mL 10 mg/mL NaCl in 10 mM NaOH, add 25 mL chloroform, shake vigorously for 1 min, drain the chloroform layer through glass wool (previously washed with 25 mL chloroform), repeat the extraction twice more with 30 s shaking. Combine the chloroform layers and add them to 5 mL 5 mM sulfuric acid, evaporate the chloroform under reduced pressure at 45°, remove traces of chloroform under a stream of air for 2 min, chill in ice, add 1 mL 1 mg/mL sodium nitrite, swirl to mix, chill in ice for 15 min, add 1 mL 5 mg/mL disodium 2-naphthol-3,6-disulfonate (Fluka) in 6 mg/mL sodium carbonate solution, swirl to mix, chill in ice for 15 min, evaporate to dryness under reduced pressure at 45°, dry under a stream of air for 2 min, redissolve in 15 mg/mL NaH_2PO_4, inject a 250 μL aliquot.

HPLC VARIABLES
Column: 100 × 4.6 3 μm Microsorb C18
Mobile phase: Gradient. A was 15 mg/mL ammonium acetate in MeCN:water 0.5:99.5. B was MeCN. A:B from 100:0 to 60:40 over 30 min, to 0:100 (step gradient), maintain at 0:100 for 5 min, re-equilibrate at initial conditions for 10 min.
Flow rate: 1
Injection volume: 250
Detector: UV 510

CHROMATOGRAM
Retention time: 17 (aniline), 24.5 (2-aminobiphenyl), 24.5 (4-aminobiphenyl), 27 (4-aminoazobenzene)
Limit of detection: 13-1700 ppb

REFERENCE
Bailey, J.E. Jr. Determination of unsulfonated aromatic amines in D&C Red No. 33 by the diazotization and coupling procedure followed by reversed-phase liquid chromatographic analysis, *Anal.Chem.*, **1985**, *57*, 189–196.

SAMPLE
Matrix: food color
Analyte: naphthylamines
Sample preparation: Dissolve 1 g amaranth food color in 80 mL water, add 1.7 g sodium dithionite, stir for 15 min, adjust pH to 8.0 with about 4 mL 10% NaOH, add 1 g NaCl, stir for 5 min, extract three times with 25 mL portions of chloroform. Combine the chloroform layers and filter through glass wool, add 5 mL 5 mM sulfuric acid to the filtrate, evaporate the chloroform using a stream of nitrogen, chill below 5°, add 1 mL 1 mg/mL sodium nitrite, swirl to mix, chill below 5° for 15 min, add 1 mL 5 mg/mL disodium 2-naphthol-3,6-disulfonate in 6 mg/mL sodium carbonate solution, swirl to mix, chill below 5° for 15 min (Anal. Chem. 1985, 57, 189), evaporate to dryness under a stream of nitrogen, redissolve in 5 mL 5 mM tetrabutylammonium phosphate in MeOH:water 25:75, inject an aliquot. (Purify disodium 2-naphthol-3,6-disulfonate (2-naphthol-3,6-disulfonic acid, disodium salt; Fluka) with decolorizing carbon and recrystallization from EtOH/water. Prepare 500 mM tetra-n-butylammonium phosphate solution by adding 48 mL 40% tetra-n-butylammonium hydroxide to 100 mL 1.11 M KH_2PO_4 and filtering (0.45 μm nylon). Azo dyes are reduced to their parent amines which are then derivatized and chromatographed.)

HPLC VARIABLES
Column: 250 × 4.6 5 μm LC-18 (Supelco)

Mobile phase: MeOH:water 62.5:37.5 containing 5 mM tetra-n-butylammonium phosphate. (At the end of each day flush with 100 mL water and 25 mL MeOH.)
Detector: UV 522

CHROMATOGRAM
Retention time: 8 (2-naphthylamine), 9 (1-naphthylamine)
Limit of detection: 8 ng/g

KEY WORDS
amaranth

REFERENCE
Lancaster, F.E.; Lawrence, J.E. Determination of total non-sulphonated aromatic amines in the food colour amaranth by dithionite reduction followed by derivatization and high-performance liquid chromatography, *Food Addit.Contam.*, **1989**, *6*, 415–423.

SAMPLE
Matrix: formulations
Analyte: sulfonamides
Sample preparation: Powder tablets or pills. Weigh out an amount of powdered tablets or pills or capsule contents, dissolve in 5 mL EtOH, dilute with 150 mM HCl containing 40 mM sodium dodecyl sulfate. Dilute suspensions or drops with 150 mM HCl containing 40 mM sodium dodecyl sulfate. Filter solutions if necessary. 10 mL Solution in 150 mM HCl containing 40 mM sodium dodecyl sulfate + 1 mL 100 mM sodium nitrite, let stand for 5 min, add 1 mL 300 mM sulfamic acid, let stand for 10 min, add 500 µL 30 mM N-(1-naphthyl)ethylenediamine dihydrochloride, make up to 25 mL with water, inject an aliquot.

HPLC VARIABLES
Guard column: 35 × 4.6 C18 (Scharlau)
Column: 125 × 4.6 5 µm Spherisorb ODS-2 C18
Mobile phase: Pentanol:50 mM sodium dodecyl sulfate 2.4:97.6, pH adjusted to 7 with 100 mM phosphate buffer
Flow rate: 1
Injection volume: 20
Detector: UV 490

CHROMATOGRAM
Retention time: 4.9 (sulfacetamide), 5.2 (sulfamethizole), 7.5 (sulfamethoxazole), 7.8 (sulfaguanidine), 9.5 (sulfadiazine), 11 (sulfanilamide), 11.2 (sulfamerazine), 12 (sulfathiazole)
Limit of detection: 200 ng/mL

OTHER SUBSTANCES
Non-interfering: benzocaine

KEY WORDS
tablets; pills; capsules; suspensions; drops

REFERENCE
Garcia-Alvarez-Coque, M.C.; Simo-Alfonso, E.F.; Ramis-Ramos, G.; Esteve-Romero, J.S. High-performance micellar liquid chromatography determination of sulphonamides in pharmaceuticals after azodye precolumn derivatization, *J.Pharm.Biomed.Anal.*, **1995**, *13*, 237–245.

SAMPLE
Matrix: tissue
Analyte: sulfonamides
Sample preparation: Homogenize (Polytron) 10 g ground tissue with 40 mL acetone, centrifuge at 2800 g for 5 min, filter (paper) the supernatant. Homogenize (Polytron) the residue with 20 mL acetone for 1 min, centrifuge, filter. Combine the filtrates and add 60 mL 125 mM HCl, wash twice with 50 mL portions of n-hexane, add 10 mL 1 M pH 5.2 acetate buffer, adjust pH to 5.0-5.1 with 5 M NaOH, extract with 60 ml and 40 mL portions of ethyl acetate, combine the organic layers, evaporate to about 2 mL under reduced pressure at 45°C, add about 15 mL EtOH, evaporate to dryness under reduced

pressure at 50°, reconstitute immediately with 5-7 mL dichloromethane. Add to an 85 mm long column of silica gel made up in dichloromethane, rinse the flask twice with 1-2 mL portions of dichloromethane, add the rinses to the column, elute with 40 mL acetone:dichloromethane (60:40), elute to waste until the acetone front (visible against a dark background) is about 10 mm from the end of the column, collect the remaining eluate (Mitt. Gebiete. Lebensm. Hyg. 1990, 81, 522). Add 150 μL 10 μg/mL sulfabenzamide to the eluate, evaporate to dryness under reduced pressure at 45°, reconstitute the residue in 300 μL MeOH:water 50:50, filter (0.45 μm), inject a 20 μL aliquot.

HPLC VARIABLES
Guard column: 4 × 4 LiChrospher 5 μm 100 RP-18
Column: 250 × 4 5 μm Spherisorb ODS2
Mobile phase: MeCN:buffer 20:80 (Prepare buffer by dissolving 3.85 g ammonium acetate in 950 mL water, adjust pH to 4.00 with acetic acid, make up to 1 L with water.)
Column temperature: 35
Flow rate: 1
Injection volume: 20
Detector: UV 550 following post-column reaction. The column effluent mixed with ice-cold reagent pumped at 0.2 mL/min and the mixture flowed through a 25 cm × 0.33 mm ID coil. The effluent from this coil mixed with ice-cold 20 mg/mL ammonium sulfamate in water pumped at 0.2 mL/min and this mixture flowed through an ice-cooled 200 cm × 0.33 mm ID coil. The effluent from this coil mixed with ice-cold 4 mg/mL N-(1-naphthyl)ethylenediamine hydrochloride in water pumped at 0.2 mL/min and this mixture flowed through a 60 cm × 0.33 mm ID coil to the detector. (Reagent was 800 mg sodium nitrite dissolved in 150 mL water and 50 mL concentrated HCl.)

CHROMATOGRAM
Retention time: 3 (sulfanilamide), 4 (sulfadiazine, sulfacetamide), 4.5 (sulfathiazole, sulfapyridine), 5 (sulfamerazine), 6 (sulfamethoxypyridazine, sulfamethazine), 7 (sulfachloropyridazine), 8.5 (sulfamethoxazole), 15.5 (sulfadimethoxine), 17 (sulfaquinoxaline)
Internal standard: sulfabenzamide (8.8)
Limit of detection: 2 ppb

KEY WORDS
post-column reaction; muscle; liver; kidney; SPE

REFERENCE
Guggisberg, D.; Mooser, A.E.; Koch, H. Screening-Methode zur quantitativen Bestimmung von 12 Sulfonamiden in Fleisch, Leber und Niere mit HPLC und *on-line*-Nachsäulenderivatisation [Screening method for the quantitative determination of 12 sulfonamides in meat, liver and kidney by HPLC and on-line post-column derivatization], *Mitt.geb.Lebensmittelunters.Hyg.*, **1993**, *84*, 263–273.

SAMPLE
Matrix: tissue
Analyte: amines
Sample preparation: Homogenize (Polytron) 200 mg tissue with 5 volumes of 1 M perchloric acid, centrifuge at 10000 g for 20 min. Neutralize the supernatant with 10 M KOH, centrifuge at 10000 g for 20 min, remove a 20 μL aliquot of the supernatant and add it to 10 μL reagent, mix well, add 30 μL 10% sodium carbonate in EtOH:water 5:95, mix well, inject a 10 μL aliquot. (Prepare reagent prior to use by mixing equal volumes of 20 mM sulfanilic acid in 1 M HCl and 200 mM sodium nitrite solution.)

HPLC VARIABLES
Column: 250 × 4.6 5 μm TSK-ODS (Toso)
Mobile phase: Gradient. A was EtOH:150 mM pH 6.0 sodium acetate buffer 5:95. B was MeCN:water 60:40. A:B from 100:0 to 55:45 over 30 min.
Column temperature: 40
Flow rate: 1
Injection volume: 10
Detector: UV 420

CHROMATOGRAM
Retention time: 14.58 (histamine), 11.87 (histidine), 13.20 (carnosine), 15.31 (tyrosine)

Limit of detection: <7 nmole/g

OTHER SUBSTANCES
Non-interfering: anserine, 1-methylhistidine

KEY WORDS
fish; muscle

REFERENCE
Sato, M.; Nakano, T.; Takeuchi, M.; Kumagai, T.; Kanno, N.; Nagahisa, E.; Sato, Y. Specific determination of histamine in fish by high-performance liquid chromatography after diazo coupling, *Biosci.Biotechnol.Biochem.*, **1995**, *59*, 1208–1210.

SAMPLE
Matrix: urine
Analyte: amines
Sample preparation: 500 µL Urine diluted 1:10 + 50 µL MeOH + 100 µL 0.5 M HCl + 100 µL 0.1% sodium nitrite in water, vortex, let stand for 10 min, add 100 µL 2% ammonium sulfamate in water, let stand for 15 min, add 100 µL 0.05% 2-aminoanthracene in MeCN (Caution! 2-Aminoanthracene causes cancer in experimental animals!), let stand for 15 min in the dark, add 5 mL diethyl ether, shake for 5 min, centrifuge. Remove 4 mL of the organic layer and evaporate it to dryness under vacuum, reconstitute the residue in 300 µL MeOH, inject a 30 µL aliquot.

HPLC VARIABLES
Column: 150 × 6 5 µm YMC-Pack A-312 (YMC)
Mobile phase: MeOH:water:acetic acid 78:22:1
Flow rate: 1
Injection volume: 30
Detector: UV 279

CHROMATOGRAM
Retention time: 8 (sulfanilamide), 10 (4-aminobenzoyl-β-alanine), 20 (aminobenzoic acid), 27 (4-aminoacetophenone)
Limit of detection: 10 ng/mL

REFERENCE
Hayashi, T.; Amino, M.; Uchida, G.; Sato, M. High-performance liquid chromatographic determination of primary aromatic amines in urine after derivatization to an azo dye with 2-aminoanthracene, *J.Chromatogr.B*, **1995**, *665*, 209–212.

SAMPLE
Matrix: urine
Analyte: sulfonamides
Sample preparation: 2 mL Urine + 10 mL 150 mM HCl containing 40 mM sodium dodecyl sulfate + 1 mL 100 mM sodium nitrite, let stand for 5 min, add 1 mL 300 mM sulfamic acid, let stand for 10 min, add 500 µL 30 mM N-(1-naphthyl)ethylenediamine dihydrochloride, make up to 25 mL with water, inject a 20 µL aliquot.

HPLC VARIABLES
Guard column: 35 × 4.6 C18 (Scharlau)
Column: 125 × 4.6 5 µm Spherisorb ODS-2 C18
Mobile phase: Pentanol:50 mM sodium dodecyl sulfate 2.4:97.6, pH adjusted to 7 with 100 mM phosphate buffer
Flow rate: 1
Injection volume: 20
Detector: UV 490

CHROMATOGRAM
Retention time: 5.3 (sulfamethizole), 8.0 (sulfaguanidine), 8.5 (sulfamethoxazole), 8.8 (sulfadiazine), 10.5 (sulfathiazole)
Limit of detection: 100 ng/mL

REFERENCE
Simó-Alfonso, E.F.; Ramis-Ramos, G.; García-Alvarez-Coque, M.C.; Esteve-Romero, J.S. Determination of sulphonamides in human urine by azo dye precolumn derivatization and micellar liquid chromatography, *J.Chromatogr.B*, **1995**, *670*, 183–187.

Hantzch Reaction

SAMPLE
Matrix: urine
Analyte: aminolevulinic acid
Sample preparation: Centrifuge urine at 1000 g and store at -20°. 20 μL Urine + 5 mL acetylacetone:EtOH:4 g/L NaCl in water 15:10:75 + 450 μL 9.3 % formaldehyde solution, mix, boil for 15 min, cool with water, store sample in the dark at 15° until injection, inject a 50 μL aliquot.

HPLC VARIABLES
Column: 150 × 4.6 5 μm TSK-80 TM (Tosoh)
Mobile phase: Gradient. A was MeCN:MeOH:water:acetic acid 10:35:54:1. B was MeCN. A:B 100:0 for 7.5 min, to 50:50 over 1.5 min, return to initial conditions over 2 min, re-equilibrate for 2 min.
Column temperature: 40
Flow rate: 0.8
Injection volume: 50
Detector: F ex 246 em 458

CHROMATOGRAM
Retention time: 7.3
Limit of detection: 10 ng/mL

KEY WORDS
Improved version of A. Okayama et al. Clin. Chem. 1990, 36, 1494.; protect from light

REFERENCE
Endo, Y.; Okayama, A.; Endo, G.; Ueda, T.; Nakazono, N.; Horiguchi, S. Improvement of urinary δ-aminolevulinic acid determination by HPLC and fluorescence detection using condensing reaction with acetylacetone and formaldehyde, *Japan.J.Ind.Health*, **1994**, *36*, 49–56.

SAMPLE
Matrix: urine
Analyte: δ-aminolevulinic acid
Sample preparation: 1 mL Urine + 500 μL 10% trichloroacetic acid, mix, centrifuge at 12000 g for 5 min, inject a 10 μL aliquot of the supernatant.

HPLC VARIABLES
Column: 150 × 6 5 μm TSK-GEL SCX cation exchange (Toyo Soda)
Mobile phase: Gradient. A was 10 mM pH 3.0 sodium phosphate buffer. B was 10 mM Na_2HPO_4 containing 50 mM NaOH. A:B 100:0 for 3 min, to 80:20 (step gradient), maintain at 80:20 for 7 min, to 0:100 (step gradient), maintain at 0:100 for 5 min, return to initial conditions, re-equilibrate for 7 min.
Flow rate: 1
Injection volume: 10

Detector: F ex 363 em 473 following post-column reaction. The column effluent mixed with 18.5% formaldehyde in 2 M acetic acid pumped at 0.1 mL/min and the mixture flowed through a 10 m $\times$ 0.25 mm ID coil at 95°. The effluent from this coil mixed with EtOH: water:acetylacetone 25:25:50 pumped at 0.4 mL/min and this mixture flowed through a 20 m $\times$ 0.4 mm ID coil at 95° to the detector.

CHROMATOGRAM
Retention time: 14
Limit of detection: 100 pg

KEY WORDS
post-column reaction

REFERENCE
Okayama, A. Fluorimetric determination of urinary δ-aminolevulinic acid by high-performance liquid chromatography and post-column derivatization, *J.Chromatogr.*, **1988**, *426*, 365–369.

Oxidation

SAMPLE
Matrix: blood
Analyte: reserpine
Sample preparation: 2 mL Plasma + 2 mL saturated sodium borate in water + 3 mL benzene (Caution! Benzene is a carcinogen!), rotate at slow speed for 5 min, centrifuge for 5 min. Remove the organic layer and evaporate it to dryness under a stream of nitrogen at 50°, reconstitute the residue in 30 µL reagent, mix well, let stand for at least 10 min, inject a 20 µL aliquot. (Prepare reagent by adding 1 mL saturated vanadium pentoxide in concentrated phosphoric acid to 9 mL MeOH. Reserpine is oxidized to 3-dehydroreserpine.)

HPLC VARIABLES
Column: 300 $\times$ 4 µBondapak C18
Mobile phase: MeOH:10 mM sodium heptanesulfonate 65:35
Flow rate: 2.5
Injection volume: 20
Detector: F ex 390 em 470 (cut-off filter)

CHROMATOGRAM
Retention time: 6
Limit of detection: 100 pg/mL

OTHER SUBSTANCES
Interfering: rescinnamine

KEY WORDS
plasma; horse; comparison with TLC method

REFERENCE
Sams, R. Determination of reserpine in plasma using high-performance liquid chromatography with fluorescence detection, *Anal.Lett.*, **1978**, *B11*, 697–707.

SAMPLE
Matrix: blood
Analyte: reserpine
Sample preparation: 3 mL Plasma + 45 µL 100 ng/mL IS in 10 mM HCl + 1 mL 600 mM pH 9.5 carbonate buffer + 10 mL n-heptane:isoamyl alcohol 98.5:1.5, shake for 10 min, centrifuge for 10 min. Remove the organic layer and add it to 1.2 mL 100 mM HCl, shake for 10 min, centrifuge for 10 min. Remove the aqueous layer and add it to 500 µL 600 mM pH 9.5 carbonate buffer, add 500 µL MTBE, mix, centrifuge for 5 min. Remove the organic layer and evaporate it to dryness under reduced pressure, add 40 µL reagent, let stand for at least 10 min, inject the whole amount. (Reagent was a 1 in 10 mixture of

a saturated solution of vanadium pentoxide in MeOH. Reserpine is oxidized to 3-dehydroreserpine.)

HPLC VARIABLES
Column: 250 × 4.6 5 μm LC-1 trimethylsilyl (Supelco)
Mobile phase: MeCN:100 mM pH 4.2 acetate buffer containing 5 mM heptanesulfonate and 10 mM triethylamine
Flow rate: 1.6
Injection volume: 40
Detector: F ex 390 em 480

CHROMATOGRAM
Retention time: 5.2
Internal standard: methyl 18-triethoxybenzoylreserpate (7.8)
Limit of quantitation: 0.3 ng/mL
Limit of detection: 0.1 ng/mL

OTHER SUBSTANCES
Non-interfering: metabolites

KEY WORDS
plasma; pharmacokinetics

REFERENCE
Suckow, R.F.; Cooper, T.B.; Asnis, G.M. An improved method for the determination of reserpine in plasma using liquid chromatography with fluorescence detection, *J.Liq.Chromatogr.*, **1983**, *6*, 1111–1122.

SAMPLE
Matrix: blood
Analyte: reserpine
Sample preparation: 3 mL Plasma + 3 mL saturated sodium borate in water + 4.5 mL n-hexane:dichloromethane 50:50, mix for 20 min on a rotating tumbler, centrifuge at 2000 g for 10 min. Remove the organic layer and evaporate it to dryness under a stream of nitrogen at 50°, reconstitute the residue in 50 μL reagent, let stand for 25 min, inject a 20 μL aliquot. (Reagent was 1 mL of a saturated solution of vanadium pentoxide in concentrated phosphoric acid and 9 mL MeOH. Reserpine is oxidized to 3-dehydroreserpine.)

HPLC VARIABLES
Column: 300 × 3.9 μBondapak C18
Mobile phase: MeOH:10 mM sodium heptanesulfonate in water 65:35
Flow rate: 2
Injection volume: 20
Detector: F ex 390 em 470

CHROMATOGRAM
Retention time: 6.5
Limit of detection: 70 pg/mL

KEY WORDS
plasma; horse; pharmacokinetics

REFERENCE
Chapman, C.B.; Courage, P.; Huntington, P.J. Detection of reserpine in horses by high-performance liquid chromatography, *Aust.Vet.J.*, **1991**, *68*, 296–298.

SAMPLE
Matrix: blood
Analyte: lometrexol
Sample preparation: Condition a 10 mm Bond-Elut C8 SPE cartridge with 5 mL MeOH and 5 mL buffer. Dilute plasma with an equal volume of 100 ng/mL IS in buffer, vortex, centrifuge at 4° at 1000 g for 20 min, add a 2 mL aliquot of the supernatant to the SPE cartridge, wash with 5 mL buffer, elute with 1.5 mL MeCN:buffer 20:80. Centrifuge the eluate at 4° at 1000 g for 5 min, evaporate the supernatant to dryness under reduced

pressure, reconstitute with 20 μL 13% formic acid in water, add 100 μL of a freshly prepared 200 μg/mL suspension of activated manganese dioxide (Sigma) in water, vortex, heat at 37° for 1.5 h, cool to 4°, add 30 μL 5 M NaOH:1% pH 5 ammonium carbonate 50:50, cool on ice, centrifuge at room temperature at 13000 g for 10 min, inject a 100 μL aliquot of the supernatant. (Buffer was 1% aqueous formic acid adjusted to pH 3.7 with 5 M NaOH. The tetrahydropyridine ring is oxidized to a pyridine.)

HPLC VARIABLES
Column: 150 × 4.6 3 μm Apex II C18 (Jones Chromatography)
Mobile phase: MeCN:buffer 12:88 containing 171 μg/mL tetramethylammonium hydrogen sulfate (Buffer was 1% aqueous acetic acid adjusted to pH 5 with strong ammonia. Wash column with MeOH for 10 min between experiments.)
Flow rate: 1
Injection volume: 100
Detector: F ex 325 em 450

CHROMATOGRAM
Retention time: 4.2
Internal standard: C10-desmethylene lometrexol (5.3)
Limit of quantitation: 10 ng/mL

KEY WORDS
plasma; SPE; Analytical procedures that do not involve derivatization are also given in the paper; pharmacokinetics

REFERENCE
Wedge, S.R.; Laohavinij, S.; Taylor, G.A.; Newell, D.R. Measurement of 5,10-dideaza-5,6,7,8-tetrahydro-folate (lometrexol) in human plasma and urine by high-performance liquid chromatography, *J.Chromatogr.B*, **1995**, *663*, 327–335.

SAMPLE
Matrix: blood, urine
Analyte: taurine
Sample preparation: 500 μL Serum or 50 μL urine + 200 μL water + 200 μL 167 mM sulfuric acid + 300 μL 76 mM sodium tungstate(VI), shake mechanically, let stand for 10 min, centrifuge at 2000 g for 10 min, inject a 20 μL aliquot of the supernatant.

HPLC VARIABLES
Guard column: 50 × 8 Shodex Ionpac C-811P sulfonic acid-type polystyrene porous polymer (Showa Denko, Tokyo)
Column: 300 × 8 Shodex Ionpac KC-811 sulfonic acid-type polystyrene porous polymer (Showa Denko, Tokyo)
Mobile phase: 10 mM Phosphoric acid
Column temperature: 45
Flow rate: 0.8
Injection volume: 20
Detector: F ex 370 em 440 following post-column reaction. The column effluent mixed with the oxidant pumped at 0.3 mL/min and this mixture flowed through a 3 m × 0.5 mm ID PTFE coil at 65°. The effluent from this coil mixed with the reagent pumped at 0.3 mL/min and this mixture flowed through a 5 m × 0.5 mm ID PTFE coil at 65° and a 1 m × 0.5 mm ID PTFE cooling coil to the detector. (Prepare oxidant by adding 60 mL 1 M NaOH and 4 mL 25% Brij-35 to 6 mL 10% sodium hypochlorite solution, make up to 1 L with 100 mM pH 7.0 phosphate buffer, final pH 12.0. Prepare reagent by dissolving 22.8 g sodium nitrite and 200 mg thiamine hydrochloride in 100 mM pH 7.0 phosphate buffer, adjust to pH 7.0 with 100 mM Na_2HPO_4 and 100 mM NaH_2PO_4, make up to 1 L with 100 mM pH 7.0 phosphate buffer. Taurine is oxidized to N-chlorotaurine. N-Chlorotaurine oxidizes thiamine to thiochrome.)

CHROMATOGRAM
Retention time: 14
Limit of detection: 6 ng

KEY WORDS
post-column reaction; serum

REFERENCE
Yokoyama, T.; Kinoshita, T. High-performance liquid chromatographic determination of taurine in biological fluids by post-column fluorescence reaction with thiamine, *J.Chromatogr.*, **1991**, *568*, 212–218.

SAMPLE
Matrix: feed

Analyte: thiamine (vitamin B1)

Sample preparation: 1 g Ground feed + 20 mL 100 mM HCl, shake vigorously, heat on a boiling water bath for 30 min (shake every 5 min), cool in an ice bath for 5 min, centrifuge at 1000 rpm for 10 min. Remove a 5 mL aliquot and make it up to 50 mL with buffer, centrifuge at 1000 rpm for 5 min, inject a 10 μL aliquot. (Buffer was water adjusted to pH 4.0 with acetic acid.)

HPLC VARIABLES
Column: 250 × 4.6 SynChropack SCD-100 (SynChrome Inc.)

Mobile phase: MeOH:water 40:60 containing 50 mM sodium pentanesulfonate, pH adjusted to 4.0 with acetic acid

Flow rate: 1.5

Injection volume: 10

Detector: F ex 370 em 430 following post-column derivatization. The column effluent was mixed with 200 mM KOH and 0.01% potassium ferricyanide, each pumped at 0.5 mL/min. The mixture flowed in the dark through a 3 m × 0.8 mm ID knotted coil of PTFE tubing to the detector. (Thiamine is oxidized to thiochrome.)

CHROMATOGRAM
Retention time: 5

Limit of detection: 5 pg

KEY WORDS
post-column reaction

REFERENCE
Gehring, T.A.; Cooper, W.M.; Holder, C.L.; Thompson, H.C. Jr. Liquid chromatographic determination of thiamine in rodent feed by postcolumn derivatization and fluorescence detection, *J.AOAC Int.*, **1995**, *78*, 307–309.

SAMPLE
Matrix: rice

Analyte: vitamin B1 (thiamine)

Sample preparation: Grind rice to pass 30-mesh screen. Stir 3 g ground rice and 50 mL MeOH:100 mM HCl 40:60 with a glass rod until homogeneous, reflux for 30 min, vortex for 1 min, sonicate for 20 min, centrifuge at 3000 g for 20 min, filter (0.45 μm) the supernatant, inject a 5 μL aliquot of the filtrate.

HPLC VARIABLES
Column: 250 × 4.6 Zorbax TMS

Mobile phase: 10 mM NaH_2PO_4 containing 500 mM sodium perchlorate adjusted to pH 2.5 with 3 M perchloric acid

Column temperature: 55

Flow rate: 0.4

Injection volume: 5

Detector: F ex 375 em 435 following post-column reaction. The column effluent mixed with reagent pumped at 0.4 mL/min, the mixture flowed through a 300 × 0.8 stainless steel mixing coil to the detector. (Reagent was 0.1% potassium hexacyanoferrate(III) in 15% NaOH. Thiamine is oxidized to thiochrome.)

CHROMATOGRAM
Retention time: 8

Limit of quantitation: 1.5 μg/g

KEY WORDS
post-column reaction

REFERENCE
Ohta, H.; Maeda, M.; Nogata, Y.; Yoza, K.-I.; Takeda, Y.; Osajima, Y. A simple determination of thiamine in rice (*Oryza sativa* L.) by high-performance liquid chromatography with post-column derivatization, *J.Liq.Chromatogr.*, **1993**, *16*, 2617–2629.

SAMPLE
Matrix: tissue
Analyte: thiamine
Sample preparation: Homogenize (motor-driven glass homogenizer) 100 mg tissue with 300 µL 5% trichloroacetic acid at 4°, centrifuge at 4° at 5000 g for 1 h. Wash the supernatant with 3 volumes of water-saturated diethyl ether for 1 h. Remove an 80 µL aliquot of the aqueous phase and add it to 50 µL reagent, mix for 5-10 s, inject an aliquot 1 min after the addition of the reagent. (Prepare reagent by mixing 50 µL 10 mg/mL potassium ferricyanide with 2.5 mL 15% NaOH, store in the dark, discard after 1 day. Thiamine is oxidized to thiochrome.)

HPLC VARIABLES
Guard column: 4.2 × 3.2 30-44 µm 201RP (Vydac)
Column: 150 or 250 × 4.6 5 µm Ultrasphere ODS
Mobile phase: Gradient. MeOH:25 mM pH 8.4 phosphate buffer 10:90 for 1 min, to 100:0 over 3 min, maintain at 100:0 for 2 min, return to initial conditions over 2 min.
Flow rate: 1
Injection volume: 20
Detector: F ex 390 em 475

CHROMATOGRAM
Retention time: 4.2 (thiamine triphosphate), 4.5 (thiamine pyrophosphate), 5 (thiamine monophosphate), 6 (thiamine)
Limit of detection: 0.05 pmole

KEY WORDS
rat; nerve; heart

REFERENCE
Bontemps, J.; Philippe, P.; Bettendorff, L.; Lombet, J.; Dandrifosse, G.; Schoffeniels, E.; Crommen, J. Determination of thiamine and thiamine phosphates in excitable tissues as thiochrome derivatives by reversed-phase high-performance liquid chromatography on octadecyl silica, *J.Chromatogr.*, **1984**, *307*, 283–294.

PEROXIDE

Cerium(IV) Trihydroxyperoxide

SAMPLE
Matrix: blood
Analyte: methotrexate
Sample preparation: 500 µL Plasma + 500 µL acetone, vortex vigorously for 1 min, centrifuge at 3000 rpm for 4 min. Remove a 600 µL aliquot of the supernatant and add it to 600 µL butanol and 800 µL diethyl ether, mix vigorously, centrifuge at 2000 rpm for 2 min. Remove a 300 µL aliquot of the lower aqueous layer and evaporate it to dryness under reduced pressure, reconstitute the residue with 150 µL 0.9% NaCl in water, filter

(0.45 μm), inject a 50 μL aliquot onto column A and column B in series and elute with mobile phase A, after 5 min remove column A from the circuit, elute column B with mobile phase B, monitor the effluent from column B. (Methotrexate is oxidized on column A to the fluorescent compounds 2,4-diaminopteridine-6-carboxylic acid and 2,4-diaminopteridine-6-carboxaldehyde.)

HPLC VARIABLES

Column: A 70 × 4.5 cerium(IV) trihydroxyperoxide; B 250 × 4.6 5 μm ODS/TM silica (Preparation of cerium(IV) trihydroxyperoxide is as follows. Add 15 mL 35% ammonia solution with stirring to a cooled solution of 11.18 g cerium(III) chloride heptahydrate in 100 mL water, add 20 mL 30% hydrogen peroxide with stirring over 15 min, collect the orange-red precipitate by filtration, wash the solid several times with water, wash with acetone, dry under vacuum at room temperature overnight to obtain cerium(IV) trihydroxyperoxide, discard after 6 months. Suspend 1 g cerium(IV) trihydroxyperoxide in chloroform, degas under vacuum with stirring for 10 min, add the suspension to a 100 × 7 reservoir on top of the column, pump the slurry into the column using acetone as a purge solvent at 5 mL/min for 20 min, wash the column with acetone:pH 3.5 phosphate buffer 50:50 at 1 mL/min for 20 min, equilibrate with mobile phase A at 0.5 mL/min for 1 h (Analyst 1996, 121, 183).)

Mobile phase: A 40 mM pH 3.5 phosphate buffer; B MeCN:50 mM pH 6.6 phosphate buffer 10:90

Flow rate: A 0.2; B 1

Injection volume: 50

Detector: F ex 367 em 463

CHROMATOGRAM

Retention time: 5 (2,4-diaminopteridine-6-carboxylic acid), 6 (2,4-diaminopteridine-6-carboxaldehyde)

Limit of detection: 2.78 ng/mL

KEY WORDS

plasma; column-switching

REFERENCE

Emara, S.; Razee, S.; Khedr, A.; Masujima, T. On-line precolumn derivatization for HPLC determination of methotrexate using a column packed oxidant, *Biomed.Chromatogr.*, **1997**, *11*, 42−46.

SOLID PHASE REAGENT

Amberlite Benzoxazolesulfonate

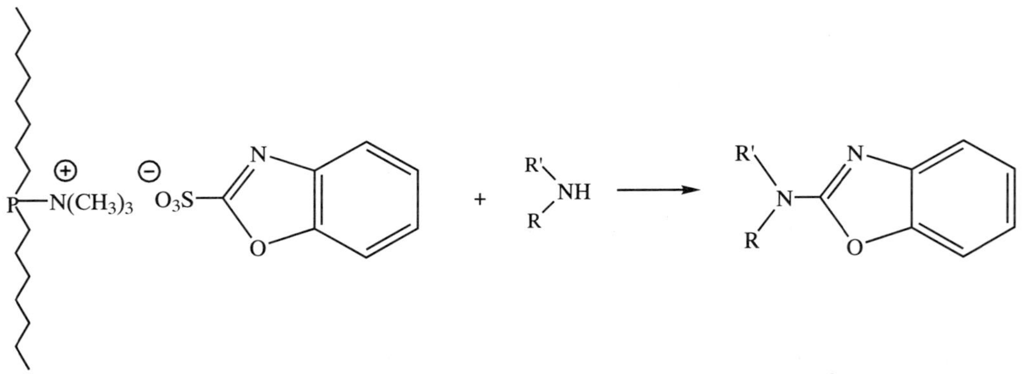

SAMPLE

Matrix: solutions

Analyte: amines
Sample preparation: Mix 100 μL of a solution of amine in MeOH with 1 g Amberlite benzoxazolesulfonate, heat at 60° for 5 min, inject an aliquot. (Prepare Amberlite benzoxazolesulfonate as follows. Heat 5 g 2-mercaptobenzoxazole with 15 g phosphorus pentachloride on a boiling water bath for 1 h then reflux for 1.5 h, cool to room temperature, add 50 mL 25% sodium sulfite, when the effervescence ceases add another 50 mL 25% sodium sulfite, reflux for 2 h, filter (paper) while hot, sodium benzoxazole-2-sulfonate precipitates as the solution cools. Stir 1 g air-dried Amberlite IRA 400 ion-exchange resin (chloride form) with 696 mg sodium benzoxazaole-2-sulfonate in 50 mL water at 75° for 30-45 min, wash resin several times with warm water by decantation, dry in air to give Amberlite benzoxazolesulfonate.)

HPLC VARIABLES
Column: 300 × 4.6 μBondapak C18
Mobile phase: MeCN:water 65:35
Flow rate: 2
Detector: UV 254

CHROMATOGRAM
Retention time: 2 (diethylamine), 2.8 (di-n-propylamine), 4 (di-n-butylamine)

KEY WORDS
derivative is fluorescent

REFERENCE
Idowu, O.R.; Adewuyi, G.O. Sodium benzoxazole-2-sulfonate: A derivatisation reagent for the analysis of amines and amino acids by HPLC with fluorescence or UV detection, *J.Liq.Chromatogr.*, **1993**, *16*, 3773–3791.

3,5-Dinitrobenzoyl Derivatized Silica

SAMPLE
Matrix: tissue
Analyte: histamine
Sample preparation: Sonicate 10 g tuna with 50 mL 1200 ppm triethylamine in MeOH for 10 min, filter, filter (0.45 μm) again, add 25 μL filtrate to 10 mg 3,5-dinitrobenzoyl derivatized silica, heat at 60° for 10 min, elute with 1 mL MeCN, inject a 20 μL aliquot of the eluate. (Prepare 3,5-dinitrobenzoyl derivatized silica as follows. Heat 4.7 g 4-hydroxy-3-nitrobenzoic acid, 7.5 mL thionyl chloride, 750 μL pyridine, and 60 mL benzene (Caution! Benzene is a carcinogen!) at 55° for 4 h, cool to room temperature, filter, evaporate under reduced pressure to obtain 4-hydroxy-3-nitrobenzoyl chloride. Store 10 μm LiChrosorb Si-100 silica in a desiccator over saturated aqueous LiCl solution for 2 weeks. 5 g Silica + 8.4 g 62% bis(2-hydroxyethyl)-3-aminopropyltriethoxysilane (Petrarch Systems, Bristol PA) in EtOH, add 40 mL EtOH:pyridine 99.5:0.5 (?), reflux with stirring

under nitrogen for 6 h, cool, filter, wash silica with three 25 mL portions of MeOH and four 25 mL portions of dichloromethane, dry under nitrogen at 40° overnight. Heat 5 g 4-hydroxy-3-nitrobenzoyl chloride, 300 μL pyridine, 5 g silica, and 60 mL benzene (previously dried over anhydrous sodium sulfate) at 70-75° for 2 h, filter, wash silica with three 75 mL portions of DMF and three 75 mL portions of dichloromethane, dry under vacuum at 40° for 12 h. Stir 400 mg silica, 600 mg 3,5-dinitrobenzoyl chloride, 25 mL benzene, and 300 μL pyridine at room temperature for 24 h, filter, wash with three 25 mL portions of dichloromethane to give 3,5-dinitrobenzoyl derivatized silica.)

HPLC VARIABLES
Column: 250 × 4 5 μm LiChrosphere C18
Mobile phase: MeCN:water 50:50 containing 0.05% ammonium hydroxide
Flow rate: 1.5
Injection volume: 20
Detector: UV 254

CHROMATOGRAM
Retention time: 2
Limit of detection: 152 ppb

KEY WORDS
tuna; fish

REFERENCE
Zhou.F.-X. ; Wahlberg, J.; Krull, I.S. Silica based 3,5-dinitrobenzoyl (DNB) reagent for off-line derivatization of amine nucleophiles in HPLC, *J.Liq.Chromatogr.*, **1991**, *14*, 1325–1350.

Fluorenylacetyl Derivatized Silica

SAMPLE
Matrix: solutions
Analyte: amines
Sample preparation: Mix 25 μL solution with 20 mg fluorenylacetyl derivatized silica, heat at 60° for 10 min, elute with 1 mL MeCN, inject a 20 μL aliquot of the eluate. (Prepare fluorenylacetyl derivatized silica as follows. Reflux 5 g 10 μm LiChrosorb Si 100, 8.4 g 62% bis(2-hydroxyethyl)-3-aminopropyltriethoxysilane in EtOH (Petrarch Systems, Bristol PA), 40 mL EtOH, and 200 μL pyridine under nitrogen for 8 h, wash the product with three 75 mL portions of MeOH, wash with three 75 mL portions of dichloromethane, dry under vacuum at 40° for 8 h to give functionalized silica (A). Heat 5.1 g 4-hydroxy-3-nitrobenzoic acid, 9 mL thionyl chloride, 700 μL pyridine, and 30 mL benzene (Caution! Benzene is a carcinogen!) at 55-60° for 4 h, filter, evaporate the filtrate under reduced pressure to obtain 4-hydroxy-3-nitrobenzoyl chloride. Heat 4 g functionalized silica (A), 5.7 g 4-hydroxy-3-nitrobenzoyl chloride, and 120 mL benzene:pyridine 99.5:0.5 at 70-75° for 2 h, wash the product with three 75 mL portions of DMF, wash with three 75 mL

portions of dichloromethane, dry under vacuum overnight to give functionalized silica (B). Heat 4 g 9-fluoreneacetic acid, 3.9 mL oxalyl chloride, 30 mL benzene, and 3 drops of triethylamine at 55° for 1 h, evaporate under reduced pressure to remove oxalyl chloride to obtain 9-fluoreneacetyl chloride (J. Chromatogr. 1992, 609, 103). Stir 1 g functionalized silica (B), 2.3 mmoles 9-fluoreneacetyl chloride, 1 mL pyridine, and 25 mL dichloromethane at room temperature for 1 h, filter, wash the product three times with 30 mL portions of DMF, wash with three 30 mL portions of dichloromethane, dry under vacuum at 40° overnight to obtain fluorenylacetyl derivatized silica. (Dry benzene over anhydrous sodium sulfate.))

HPLC VARIABLES
Column: 250 × 4 5 μm LiChrospher C18
Mobile phase: MeCN:water 70:30
Flow rate: 1.5
Injection volume: 20
Detector: F ex 254 em 313

CHROMATOGRAM
Retention time: 3.5 (n-propylamine), 4 (n-butylamine), 5 (n-amylamine), 6.5 (n-hexylamine), 9 (n-heptylamine), 13 (n-octylamine)
Limit of detection: 1 ppm

REFERENCE
Zhang, H.-M.; Zhou, F.-X.; Krull, I.S. Comparison of 9-fluorenylmethoxycarbonyl and 9-fluoreneacetyl-tagged silica-based derivatization reagents in high-performance liquid chromatography, *J.Pharm.Biomed.Anal.*, **1992**, *10*, 577–586.

SAMPLE
Matrix: urine
Analyte: cadaverine
Sample preparation: Mix 50 μL urine with 10 mg fluorenylacetyl-tagged silica, heat at 60° for 10 min, elute with 1 mL MeCN, inject a 20 μL aliquot of the eluate. (Prepare fluorenylacetyl derivatized silica as follows. Reflux 5 g 10 μm LiChrosorb Si 100, 8.4 g 62% bis(2-hydroxyethyl)-3-aminopropyltriethoxysilane in EtOH (Petrarch Systems, Bristol PA), 40 mL EtOH, and 200 μL pyridine under nitrogen for 8 h, wash the product with three 75 mL portions of MeOH, wash with three 75 mL portions of dichloromethane, dry under vacuum at 40° for 8 h to give functionalized silica (A). Heat 5.1 g 4-hydroxy-3-nitrobenzoic acid, 9 mL thionyl chloride, 700 μL pyridine, and 30 mL benzene (Caution! Benzene is a carcinogen!) at 55-60° for 4 h, filter, evaporate the filtrate under reduced pressure to obtain 4-hydroxy-3-nitrobenzoyl chloride. Heat 4 g functionalized silica (A), 5.7 g 4-hydroxy-3-nitrobenzoyl chloride, and 120 mL benzene:pyridine 99.5:0.5 at 70-75° for 2 h, wash the product with three 75 mL portions of DMF, wash with three 75 mL portions of dichloromethane, dry under vacuum overnight to give functionalized silica (B). Heat 4 g 9-fluoreneacetic acid, 3.9 mL oxalyl chloride, 30 mL benzene, and 3 drops of triethylamine at 55° for 1 h, evaporate under reduced pressure to remove oxalyl chloride to obtain 9-fluoreneacetyl chloride (J. Chromatogr. 1992, 609, 103). Stir 1 g functionalized silica (B), 2.3 mmoles 9-fluoreneacetyl chloride, 1 mL pyridine, and 25 mL dichloromethane at room temperature for 1 h, filter, wash the product three times with 30 mL portions of DMF, wash with three 30 mL portions of dichloromethane, dry under vacuum at 40° overnight to obtain fluorenylacetyl derivatized silica. (Dry benzene over anhydrous sodium sulfate.))

HPLC VARIABLES
Column: 250 × 4 5 μm LiChrospher C18
Mobile phase: MeCN:water 70:30
Flow rate: 1.5
Injection volume: 20
Detector: F ex 254 em 313

CHROMATOGRAM
Retention time: 6

REFERENCE
Zhang, H.-M.; Zhou, F.-X.; Krull, I.S. Comparison of 9-fluorenylmethoxycarbonyl and 9-fluoreneacetyl-tagged silica-based derivatization reagents in high-performance liquid chromatography, *J.Pharm.Biomed.Anal.*, **1992**, *10*, 577–586.

Polymer Bound Benzotriazole Acetylsalicylate

SAMPLE
Matrix: solutions
Analyte: amines
Sample preparation: Mix 60 μL of a 200 ppm solution in MeCN with 20 mg polymer bound benzotriazole acetylsalicylate, heat at 60° for 10 min, elute the resin with 1 mL MeCN, inject a 25 μL aliquot of the eluate. (Prepare polymer bound benzotriazole acetylsalicylate as follows. Add 10 g anhydrous aluminum chloride to 50 mL nitrobenzene, add 10 g 4-chloro-3-nitrobenzyl alcohol, add 10 g dried Porapak Q (Waters), shake (Fisher Tough-Mixer operated via a variable transformer (use of a vibrator is preferable to stirring)) at 65-75° for 3 days, filter, wash with three 50 mL portions of 1 M HCl in dioxane (Caution! Dioxane is a carcinogen!), wash with three 50 mL portions of MeOH, wash with three 50 mL portions of dichloromethane, dry under vacuum at 80°. Reflux the product in 60 mL hydrazine hydrate:2-ethoxyethanol 40:60 (Caution! Hydrazine hydrate is a carcinogen!) for 20 h, cool, filter, wash with water, dry in air. Suspend the product in 100 mL concentrated HCl:dioxane 50:50, reflux for 20 h, filter, wash with five 100 mL portions of water, wash with three 50 mL portions of MeOH, wash with three 50 mL portions of ether, dry under vacuum at 80° for 2 h. Stir the product in a mixture of 10 mL dichloromethane, 500 μL pyridine, and 810 mg o-acetylsalicyloyl chloride at room temperature for 30 min, filter, wash with three 20 mL portions of DMF, wash with three 20 mL portions of dichloromethane, wash with three 20 mL portions of dry ethyl ether, dry under vacuum at 80° for 3 h to give the product, polymer bound benzotriazole acetylsalicylate.)

HPLC VARIABLES
Column: 100 × 8 5 μm Resolve C18 Radial Pak (Waters)
Mobile phase: MeCN:water 70:30
Flow rate: 1.5
Injection volume: 25
Detector: UV 254

CHROMATOGRAM
Retention time: 2.5 (morpholine), 3 (n-propylamine), 3.3 (n-butylamine)

REFERENCE
Gao, C.-X.; Chou, T.-Y.; Colgan, S.T.; Krull, I.S.; Dorschel, C.; Bidlingmeyer, B. New polymeric benzotriazole reagents for off-line derivatizations of amines and polyamines in HPLC, *J.Chromatogr.Sci.*, **1988**, *26*, 449–457.

Polymer Bound Benzotriazole 6-Aminoquinoline

SAMPLE

Matrix: solutions

Analyte: amines

Sample preparation: Mix 50 μL 100 ppm amine solution, 50 μL 100 ppm triethylamine solution, and 15 mg polymer bound benzotriazole 6-aminoquinoline, heat at 70-75° for 10 min, wash with 1 mL MeCN, inject a 20 μL aliquot. (Prepare polymer bound benzotriazole 6-aminoquinoline as follows. React 10 g 4-chloro-3-nitrobenzyl alcohol, 10 g dried 16-20 μm styrene-divinylbenzene copolymer (12% cross-linked, 102 Å templated, Supelco), and 10 g anhydrous aluminum chloride in 50 mL nitrobenzene at 65-70° for 3 days, cool to room temperature, filter, wash with three 50 mL portions of 1 M HCl in dioxane (Caution! Dioxane is a carcinogen!), wash with three 50 mL portions of DMF, wash with three 50 mL portions of MeOH, wash with three 50 mL portions of dichloromethane, dry under reduced pressure at 100°. Reflux 19 g of this polymer in 60 mL hydrazine hydrate:2-ethoxyethanol 40:60 for 20 h, cool to room temperature, filter, wash thoroughly with water, suspend the polymer in 100 mL concentrated aqueous HCl:dioxane 50:50, reflux for 20 h, filter, wash with five 100 mL portions of water, wash with three 100 mL portions of MeOH, wash with three 50 mL portions of ether, dry under reduced pressure at 80° to obtain polymeric benzotriazole reagent (Eur. J. Biochem. 1975, 59, 55). Heat 1.6 g quinoline-6-carboxylic acid (ICN Biomedicals, Costa Mesa CA) with 2.7 mL thionyl chloride in 55 mL dry benzene at 65-75° for 1 h (Caution! Benzene is a carcinogen!), remove the benzene by evaporation under reduced pressure, add 21 mL acetic acid, stir strongly for several min, slowly add 600 mg sodium azide at room temperature, stir for 1 h, add 53 mL triethylamine, add 100 mL water, stir for 30 min, filter, wash ten times with 25 mL portions of water, dry under vacuum overnight to obtain quinoline-6-acyl azide. Heat 3.3 g polymeric benzotriazole reagent and 2 g quinoline-6-acyl azide in 150 mL toluene (dried over calcium hydride) at 60-65° for 2 h, wash with 600 mL warm (40°) dichloromethane, wash with 50 mL MeCN, dry under vacuum overnight to obtain polymer bound benzotriazole 6-aminoquinoline.)

HPLC VARIABLES

Column: 150 × 4.6 Supelcosil LC-C18-DB

Mobile phase: MeCN:water 50:50 containing 10 mM sodium dodecyl sulfate, 10 mM NaH_2PO_4, and 10 mM phosphoric acid

Flow rate: 1.5

Injection volume: 20

Detector: F ex 254 em 395

CHROMATOGRAM

Retention time: 4 (propylamine), 5 (butylamine), 12 (hexylamine), 20 (heptylamine)

REFERENCE

Yu, J.H.; Li, G.D.; Krull, I.S.; Cohen, S. Polymeric 6-aminoquinoline, an activated carbamate reagent for derivatization of amines and amino acids by high-performance liquid chromatography, *J.Chromatogr.B*, **1994**, *658*, 249–260.

Polymer Bound Benzotriazole 3,5-Dinitrobenzoate

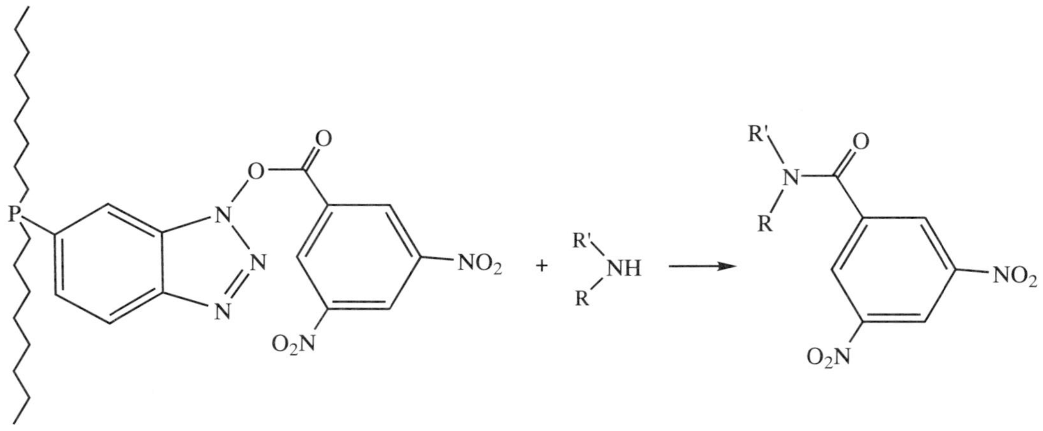

SAMPLE

Matrix: urine

Analyte: amphetamine

Sample preparation: Mix 100 μL urine with 60 mg polymer bound benzotriazole 3,5-di-nitrobenzoate, after 2 min elute with 500 μL MeCN, add 500 μL 50 mM NaOH to the eluate, mix, inject a 20 μL aliquot. (The reagent was polymer bound benzotriazole 3,5-dinitrobenzoate, synthesized as follows. (Caution! Chloroform, dichloromethane, dioxane, and hydrazine are carcinogenic in experimental animals! DMF may be carcinogenic! 3,5-Dinitrobenzoyl chloride and aluminum chloride are corrosive! Nitrobenzene is toxic!) 10 g Dried macroporous polystyrene (Xe-305, Rohm and Haas) + 10 g 3-nitro-4-chlorobenzyl alcohol + 10 g anhydrous aluminum chloride + 50 mL nitrobenzene, heat at 65-70° for 3 days, cool, filter, wash polymer with three 50 mL portions of 1 M HCl in dioxane, with three 50 mL portions of DMF, with three 50 mL portions of MeOH, and with three 50 mL portions of dichloromethane, dry under vacuum at 100°. Reflux 19 g of this polymer in 60 mL hydrazine hydrate:ethylene glycol monoethyl ether 40:60 for 20 h, cool to room temperature, filter off the polymer and wash it thoroughly with water. Suspend the polymer in 100 mL concentrated HCl:dioxane 50:50, reflux for 20 h, filter the polymer and wash it with five 100 mL portions of water, with three 100 mL portions of MeOH, and with three 50 mL portions of ether, dry under vacuum at 80°. Functionalization was 1.17 mmoles/g (Eur.J.Biochem. 1975, 59, 55). Add a portion of polymer to dry chloroform, add a three-fold excess of 3,5-dinitrobenzoyl chloride and pyridine, stir at 0-10° for 30 min, filter off polymer, wash with chloroform to give polymer bound benzotriazole 3,5-dinitrobenzoate (J.Org.Chem. 1984, 49, 922).)

HPLC VARIABLES

Column: 125 × 4 5 μm LiChrosorb C18

Mobile phase: MeCN:water 50:50 containing 0.05% ammonium hydroxide

Flow rate: 2

Injection volume: 20

Detector: UV

CHROMATOGRAM

Retention time: 4.1

REFERENCE

Bourque, A.J.; Krull, I.S. Solid-phase reagent containing the 3,5-dinitrophenyl tag for the improved derivatization of chiral and achiral amines, amino alcohols and amino acids in high-performance liquid chromatography with ultraviolet detection, *J.Chromatogr.*, **1991**, *537*, 123–152.

SAMPLE

Matrix: urine

Analyte: amphetamine

Sample preparation: Condition a 130 mg Bond Elut Certify SPE cartridge with 3 mL MeOH and 3 mL 100 mM pH 6.0 phosphate buffer, do not allow to go dry. 2 mL Urine + 800 μL 100 mM pH 6.0 phosphate buffer + 200 μL 10 μg/mL 1-methyl-3-phenylpropylamine in MeOH, if necessary adjust pH to 5-7 with 1 M KOH or 1 M HCl (with pH paper), add the mixture to the SPE cartridge, wash with 1 mL 1 M acetic acid, wash with 3 mL water, dry for 5 min under vacuum, wash with 6 mL MeOH, dry for 2 min under vacuum, elute with 2 mL ethyl acetate:30% ammonium hydroxide 98:2 at a flow rate of 4-6 drops/sec. Evaporate the eluate for 2 min under a stream of nitrogen, add 100 μL 1 M HCl in diethyl ether, evaporate to dryness under a stream of nitrogen at 40°, reconstitute the residue in 125 μL MeCN:water:triethylamine 90:8.6:1.4. Add the entire volume to 50 ± 5 mg polymer bound benzotriazole 3,5-dinitrobenzoate in a 1 mL plastic pipette tip plugged with a Kimwipe, react for 30 s, elute with 500 μL MeCN under pressure, inject a 10-20 μL aliquot. (Prepare polymer bound benzotriazole 3,5-dinitrobenzoate as described above.)

HPLC VARIABLES

Guard column: 20 mm long Microsorb octadecyldimethylsilyl silica (Rainin)

Column: 10 (sic)x 4.6 5 μm Microsorb octadecyldimethylsilyl silica (Rainin)

Mobile phase: MeCN:10 mM pH 2.5 phosphate buffer 45:55

Flow rate: 0.7

Injection volume: 10-20

Detector: UV 220

CHROMATOGRAM

Retention time: 16.3

Internal standard: 1-methyl-3-phenylpropylamine (23.0)

Limit of quantitation: 47 ng/mL

Limit of detection: 14 ng/mL

OTHER SUBSTANCES

Non-interfering: benzoylecgonine, cocaine, codeine, glutethimide, imipramine, meperidine, methadone, methamphetamine, methaqualone, morphine, nortriptyline, oxazepam, phencyclidine, propoxyphene, quinine

KEY WORDS

SPE

REFERENCE

Fisher, D.H.; Bourque, A.J. Quantification of amphetamine in urine: solid-phase extraction, polymeric reagent derivatization and reversed-phase high-performance liquid chromatography, *J.Chromatogr.*, **1993**, *614*, 142–147.

Polymer Bound Benzotriazole 3,5-Dinitrophenylcarbamate

SAMPLE
Matrix: solutions
Analyte: amphetamine and methamphetamine
Sample preparation: Mix 50 μL of a 200 ppm solution in MeCN:500 mM pH 9.0 borate buffer 50:50 with 25 mg polymer bound benzotriazole 3,5-dinitrophenylcarbamate, after 1 min elute with 1 mL hexane:THF 75:25, inject a 5 μL aliquot. (Synthesize polymer bound benzotriazole 3,5-dinitrophenylcarbamate as follows. (Caution! Chloroform, dichloromethane, dioxane, and hydrazine are carcinogenic in experimental animals! DMF may be carcinogenic! 3,5-Dinitrobenzoyl chloride and aluminum chloride are corrosive! Nitrobenzene is toxic!) 10 g Dried macroporous polystyrene (Xe-305, Rohm and Haas) + 10 g 3-nitro-4-chlorobenzyl alcohol + 10 g anhydrous aluminum chloride + 50 mL nitrobenzene, heat at 65-70° for 3 days, cool, filter, wash polymer with three 50 mL portions of 1 M HCl in dioxane, with three 50 mL portions of DMF, with three 50 mL portions of MeOH, and with three 50 mL portions of dichloromethane, dry under vacuum at 100°. Reflux 19 g of this polymer in 60 mL hydrazine hydrate:ethylene glycol monoethyl ether 40:60 for 20 h, cool to room temperature, filter off the polymer and wash it thoroughly with water. Suspend the polymer in 100 mL concentrated HCl:dioxane 50:50, reflux for 20 h, filter the polymer and wash it with five 100 mL portions of water, with three 100 mL portions of MeOH, and with three 50 mL portions of ether, dry under vacuum at 80°. Functionalization was 1.17 mmoles/g (Eur.J.Biochem. 1975, 59, 55). Dissolve 3,5-dinitrobenzoyl chloride in the minimum amount of glacial acetic acid, add an equimolar amount of sodium azide, stir for 30 min, dilute with water, filter to obtain 3,5-dinitrobenzoyl azide (Caution! Azides are toxic and potentially explosive!) (J. Liq. Chromatogr. 1986, 9, 443). Reflux (?) 71 mg 3,5-dinitrobenzoyl azide in 15 mL toluene (dried over calcium hydride) for 30 min (to form 3,5-dinitrophenylisocyanate), cool using an ice bath, add 200 mg polymer, allow to warm to room temperature with stirring for 1 h, filter, wash the polymer with four 10 mL portions of warm (40°) dichloromethane, dry under high vacuum for 1 h to obtain polymer bound benzotriazole 3,5-dinitrophenylcarbamate.)

HPLC VARIABLES
Column: 250 × 4.6 5 μm LC-(R)-naphthylurea (Supelco)
Mobile phase: Hexane:EtOH:MeCN 93:7:0.5
Flow rate: 2
Injection volume: 5
Detector: UV 254

CHROMATOGRAM
Retention time: 8.7, 10.7 (methamphetamine enantiomers), 12.2, 15.7 (amphetamine enantiomers)

KEY WORDS
chiral

REFERENCE
Bourque, A.J.; Krull, I.S. Immobilized isocyanates for derivatization of amines for chiral recognition in liquid chromatography with UV detection, *J.Pharm.Biomed.Anal.*, **1993**, *11*, 495–503.

Polymer Bound Benzotriazole 9-Fluorenylmethyl Carbonate

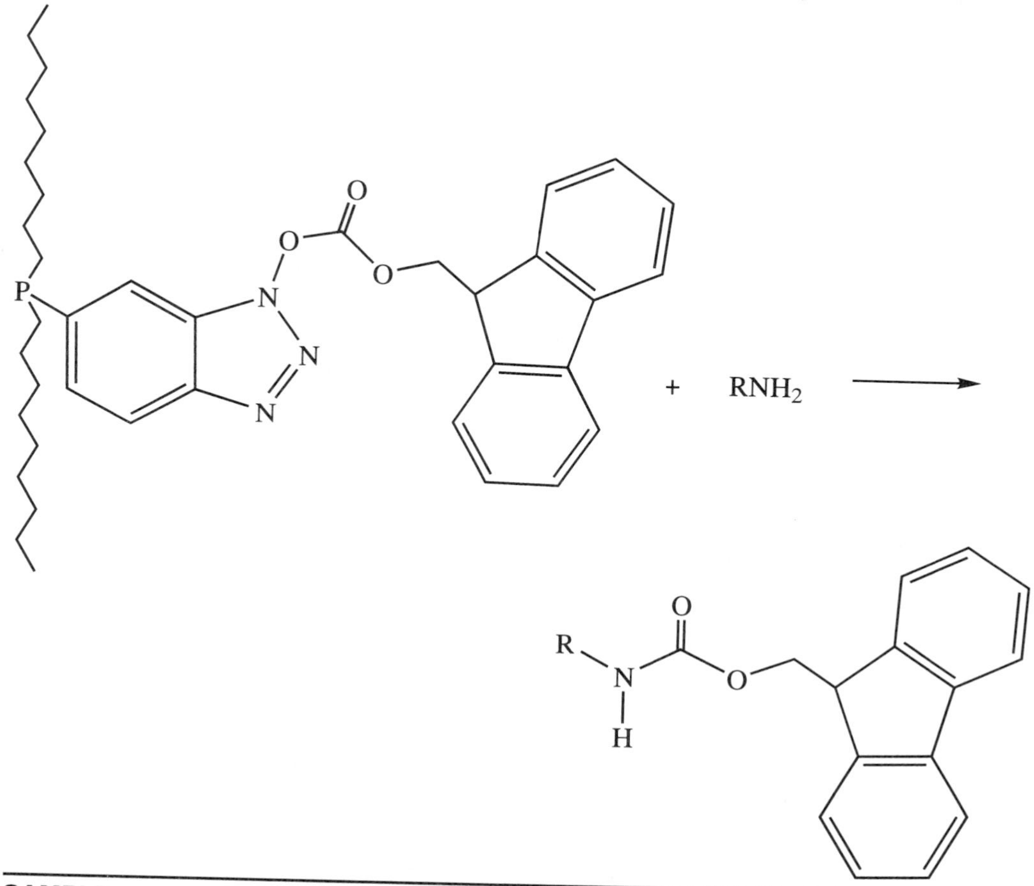

SAMPLE

Matrix: solutions

Analyte: polyamines

Sample preparation: Mix 60 μL of a 200 ppm solution in MeCN with 20 mg polymer bound benzotriazole 9-fluorenylmethyl carbonate, heat at 60° for 10 min, elute the resin with 1 mL MeCN, inject a 25 μL aliquot of the eluate. (Prepare polymer bound benzo-triazole 9-fluorenylmethyl carbonate as follows. Add 10 g anhydrous aluminum chloride to 50 mL nitrobenzene, add 10 g 4-chloro-3-nitrobenzyl alcohol, add 10 g dried Porapak Q (Waters), shake (Fisher Tough-Mixer operated via a variable transformer (use of a vibrator is preferable to stirring)) at 65-75° for 3 days, filter, wash with three 50 mL portions of 1 M HCl in dioxane (Caution! Dioxane is a carcinogen!), wash with three 50 mL portions of MeOH, wash with three 50 mL portions of dichloromethane, dry under vacuum at 80°. Reflux the product in 60 mL hydrazine hydrate:2-ethoxyethanol 40:60 (Caution! Hydrazine hydrate is a carcinogen!) for 20 h, cool, filter, wash with water, dry in air. Suspend the product in 100 mL concentrated HCl:dioxane 50:50, reflux for 20 h, filter, wash with five 100 mL portions of water, wash with three 50 mL portions of MeOH, wash with three 50 mL portions of ether, dry under vacuum at 80° for 2 h. Stir the product in a mixture of 10 mL dichloromethane, 500 µL pyridine, and 1.035 g 9-fluorenyl-methylchloroformate at room temperature for 30 min, filter, wash with three 20 mL portions of DMF, wash with three 20 mL portions of dichloromethane, wash with three 20 mL portions of dry ethyl ether, dry under vacuum at 80° for 3 h, to give the product, polymer bound benzotriazole 9-fluorenylmethyl carbonate.)

HPLC VARIABLES
Column: 100 × 8 5 µm Resolve C18 Radial Pak (Waters)
Mobile phase: MeCN:water 70:30
Flow rate: 1.5
Injection volume: 25
Detector: F ex 275 em 315

CHROMATOGRAM
Retention time: 7 (putrescine), 8 (cadaverine), 11 (1,7-diaminoheptane)
Limit of detection: 0.69-0.98 ppb

REFERENCE
Gao, C.-X.; Chou, T.-Y.; Colgan, S.T.; Krull, I.S.; Dorschel, C.; Bidlingmeyer, B. New polymeric benzo-triazole reagents for off-line derivatizations of amines and polyamines in HPLC, *J.Chromatogr.Sci.*, **1988**, *26*, 449–457.

SAMPLE
Matrix: urine
Analyte: amines
Sample preparation: 5 mL Urine + 1.2 mL concentrated HCl, heat at 110° for 16 h, adjust pH to 9 with 6 M NaOH, centrifuge at 4000 rpm for 10 min. Remove the supernatant and evaporate it to 1.5 mL. Remove a 50 µL aliquot and add it to 20 mg polymer bound benzotriazole 9-fluorenylmethyl carbonate, heat at 60° for 30 min, elute the resin with 1 mL MeCN, inject a 20 µL aliquot of the eluate. (Prepare polymer bound benzotriazole 9-fluorenylmethyl carbonate as follows. Add 10 g anhydrous aluminum chloride to 50 mL nitrobenzene, add 10 g dried Porapak Q (Waters), add 10 g 4-chloro-3-nitrobenzyl alcohol, shake (Fisher Tough-Mixer operated via a variable transformer (use of a vibrator is pref-erable to stirring)) at 65-75° for 3 days, filter, wash with three 50 mL portions of 1 M HCl in dioxane (Caution! Dioxane is a carcinogen!), wash with three 50 mL portions of MeOH, wash with three 50 mL portions of dichloromethane, dry under vacuum at 80°. Reflux the product in 60 mL hydrazine hydrate:2-ethoxyethanol 40:60 (Caution! Hydrazine hydrate is a carcinogen!) for 20 h, cool, filter, wash with water, dry in air. Suspend the product in 100 mL concentrated HCl:dioxane 50:50, reflux for 20 h, filter, wash with five 100 mL portions of water, wash with three 50 mL portions of MeOH, wash with three 50 mL portions of ether, dry under vacuum at 80° for 2 h. Stir the product in a mixture of 10 mL dichloromethane, 500 µL pyridine, and 1.035 g 9-fluorenylmethylchloroformate at room temperature for 30 min, filter, wash with three 20 mL portions of DMF, wash with three 20 mL portions of dichloromethane, wash with three 20 mL portions of dry ethyl ether, dry under vacuum at 80° for 3 h, to give the product, polymer bound benzotriazole 9-fluorenylmethyl carbonate (J.Chromatogr.Sci. 1988, 26, 449).)

HPLC VARIABLES
Column: 100 × 8 5 µm Resolve C18 Radial Pak (Waters)
Mobile phase: MeCN:water 70:30
Flow rate: 1.5
Injection volume: 20
Detector: UV 254; F ex 275 em 315

CHROMATOGRAM
Retention time: 7.5 (putrescine), 8 (cadaverine)
Internal standard: 1,7-diaminoheptane (12)
Limit of detection: 21-24 pmole

REFERENCE
Chou, T.-Y.; Gao, C.-X.; Colgan, S.T.; Krull, I.S.; Dorschel, C.; Bidlingmeyer, B. Polymeric benzotriazole reagent for the off-line high-performance liquid chromatographic derivatization of polyamines and related nucleophiles in biological fluids, *J.Chromatogr.*, **1988**, *454*, 169–183.

Polymer Bound o-Nitrobenzophenone 3,5-Dinitrobenzoate

SAMPLE

Matrix: solutions

Analyte: amines

Sample preparation: Mix 30 μL of a 20 ppm solution in MeCN with 70 mg polymer bound o-nitrobenzophenone 3,5-dinitrobenzoate, after 1 min add 200 μL 2 mM nitric acid in MeCN:water 80:20, elute with 1 mL MeCN, inject a 10 μL aliquot. (Reagent was polymer bound o-nitrobenzophenone 3,5-dinitrobenzoate, prepared as follows. (Caution! Dioxane is carcinogenic in experimental animals! DMF may be carcinogenic! 3,5-Dinitrobenzoyl chloride and aluminum chloride are corrosive! Nitrobenzene is toxic!) Soxhlet extract 200-400 mesh polystyrene cross-linked with 4% divinylbenzene (Fluka) with MeCN for 48 h. Add 25 g aluminum trichloride in 300 mL dry nitrobenzene to a mixture of 50 g of the polystyrene resin and 100 g 4-chloro-3-nitrobenzoyl chloride, stir the mixture mechanically at 60° for 5 h, pour into a mixture of 150 mL DMF, 100 mL concentrated HCl, and 150 g ice. Wash the beads with 300 mL portions of DMF:water 75:25 until the washings are colorless, wash with warm DMF (60°), wash with six 300 mL portions of dichloromethane:MeOH 2:1, dry. Add the polymer to 130 mL 40% benzyltrimethylammonium hydroxide in water, 130 mL water, and 260 mL dioxane, heat at 90° for 8 h, filter, repeat the process, filter, wash the beads with four portions of warm (60°) dioxane, add 30 mL acetic acid, stir for 15 min, wash with dioxane until the washings are neutral, wash with six 300 mL portions of dichloromethane:MeOH 2:1. Add a portion of polymer to dry chloroform, add a three-fold excess of 3,5-dinitrobenzoyl chloride and pyridine, stir at 0-10° for 30 min, filter off polymer, wash with chloroform to give the polymer bound o-nitrobenzophenone 3,5-dinitrobenzoate (J.Org.Chem. 1984, 49, 922).)

HPLC VARIABLES

Column: 125 × 4 5 μm LiChrosorb C18

Mobile phase: MeCN:water 50:50

Flow rate: 1.5

Injection volume: 10

Detector: UV

CHROMATOGRAM

Retention time: 1.9 (morpholine), 3.5 (n-propylamine), 5.3 (n-butylamine)

REFERENCE

Bourque, A.J.; Krull, I.S. Solid-phase reagent containing the 3,5-dinitrophenyl tag for the improved derivatization of chiral and achiral amines, amino alcohols and amino acids in high-performance liquid chromatography with ultraviolet detection, *J.Chromatogr.*, **1991**, *537*, 123–152.

SAMPLE

Matrix: solutions

Analyte: amphetamine

Sample preparation: Inject a 10 μL aliquot of a 100 ppm solution in MeCN: dichloromethane:triethylamine 50:50:0.05 into the mobile phase. The mobile phase flows through a 27 × 2.2 reactor packed with polymer bound o-nitrobenzophenone 3,5-dinitrobenzoate at 72° to the column. (Prepare polymer bound o-nitrobenzophenone 3,5-dinitrobenzoate as described above.)

HPLC VARIABLES
Column: 250 × 4.6 5 μm LC-(R)-Naphthyl Urea (Supelco)
Mobile phase: Hexane:dichloromethane:THF 70:27:3
Flow rate: 0.1 for 40 s, to 3.1 over 30 s, maintain at 3.1
Injection volume: 10
Detector: UV

CHROMATOGRAM
Retention time: 11.5 (d), 13 (l)

KEY WORDS
chiral

REFERENCE
Bourque, A.J.; Krull, I.S. Solid-phase reagent containing the 3,5-dinitrophenyl tag for the improved derivatization of chiral and achiral amines, amino alcohols and amino acids in high-performance liquid chromatography with ultraviolet detection, *J.Chromatogr.*, **1991**, *537*, 123–152.

Polymer-Bound o-Nitrobenzophenone 9-Fluoreneacetate

SAMPLE
Matrix: solutions
Analyte: amantadine
Sample preparation: Wash 70 mg polymeric reagent with 500 μL MeCN, heat at 75°, inject 50 μL of an amantadine solution in 50 mM NaOH, let stand at 75° for 3 min, flush contents of the reactor onto the column with mobile phase. Flush reactor with MeCN between runs. (Prepare polymeric reagent as follows. Prepare a porous rigid resin using a divinylbenzene:ethylstyrene:styrene 24:6:70 mixture with trimethylsilyl modified silica (102 Å average pore size, 1.08 mL/g pore volume, 366 m²/g surface area, 16-20 μm irregular particle shape, IMPAQ RG 1020 Si silica, PQ Co., Valley Forge PA). Further preparation details are not given but a typical procedure given in the cited reference is as follows. Aerate a mixture of 10 g modified silica in 100 mL water with nitrogen for 15 min, add 10 mL styrene:80% divinylbenzene:t-butyl peroxybenzoate 49:49:2 (remove preservative by passing through a butylcatechol remover (Scientific Polymer, Ontario NY), shake vigorously at room temperature for 4 h, add 150 mL 0.75% polyvinyl alcohol, shake for 4 h, heat at 120° for 24 h while shaking on a Parr instrument, cool to room temper-

ature, filter, wash with 100 mL water, wash with 50 mL MeOH. Add the solid to 500 mL 3 M NaOH in MeOH:water 40:60, shake at room temperature for 14 h (to dissolve the silica), filter, wash with water until the washings are neutral, wash with 100 mL MeOH, dry at 60°. The polymer has similar properties to the template silica (US Pat. 4 933 372 (1990)). Soxhlet extract the resin with dioxane for 8 h (Caution! Dioxane is a carcinogen!). Add 25 g aluminum trichloride in 300 mL dry nitrobenzene to 50 g resin and 100 g 4-chloro-3-nitrobenzoyl chloride, stir mechanically at 60° for 5 h, pour into a mixture of 150 mL DMF, 100 mL concentrated HCl, and 150 g ice, filter. Wash the solid with 300 mL portions of DMF:water 75:25 until the washings are colorless, wash with warm (60°) DMF, wash with six 300 mL portions of dichloromethane:MeOH 2:1. Stir the product in 130 mL 40% benzyltrimethylammonium hydroxide in water, 130 mL water, and 260 mL dioxane at 90° for 8 h, filter, repeat the process. Wash the product with four portions of warm (60°) dioxane. Stir the solid with 30 mL acetic acid for 15 min, filter. Wash the solid with dioxane until the washings are neutral, wash with six 300 mL portions of dichloromethane:MeOH 2:1 to give a nitrobenzophenol-substituted polymer (J. Org. Chem. 1984, 49, 924). Heat 4 g 9-fluoreneacetic acid, 3.9 mL oxalyl chloride, 30 mL benzene (dried over anhydrous sodium sulfate, Caution! Benzene is a carcinogen!), and 3 drops of triethylamine at 55° for 1 h, evaporate under reduced pressure to remove oxalyl chloride, dissolve the product in 35 mL dichloromethane to give a 120 mg/mL solution of 9-fluoreneacetyl chloride, dilute to obtain a 2 mM solution. Stir 1.3 g nitrobenzophenol-substituted polymer, 4.2 mL 2 mM 9-fluoreneacetyl chloride solution, 300 μL triethylamine, and 20 mL dichloromethane at room temperature for 1 h, filter, wash with three 20 mL portions of MeCN to obtain the reagent, polymer-bound nitrobenzophenone 9-fluoreneacetate (J. Chromatogr. 1992, 609, 103).)

HPLC VARIABLES
Column: 150 × 3.9 5 μm NovaPak C18
Mobile phase: MeCN:water 65:35
Flow rate: 1
Injection volume: 50
Detector: UV 254

CHROMATOGRAM
Retention time: 8

OTHER SUBSTANCES
Simultaneously analyzed: octylamine

REFERENCE
Szulc, M.; Swett, P.; Krull, I.S. Size-selective derivatizations with polymer immobilized reagents, *Biomed.Chromatogr.*, **1997**, *11*, 207–223.

SAMPLE
Matrix: blood, urine
Analyte: amantadine
Sample preparation: Add NaOH to plasma and urine so that the final NaOH concentration is 50 mM. Wash column A with 400 μL MeCN, with 400 μL 1 mM sodium dodecyl sulfate in MeCN:water 40:60, and with 400 μL 5 mM sodium dodecyl sulfate in water. Inject a 50 μL aliquot of plasma or urine at 0.36 mL/min, inject 50 μL 5 mM sodium dodecyl sulfate in water at 0.36 mL/min, after 3 min wash column A with 400 μL 5 mM sodium dodecyl sulfate in water, backflush the contents of column A onto column B with mobile phase and start the gradient, after 15 s remove column A from the circuit and wash it with 400 μL MeCN, elute column B with mobile phase, monitor the effluent from column B.

HPLC VARIABLES
Column: A 27 × 2.1 polymer bound o-nitrobenzophenone 9-fluoreneacetate; B reverse-phase (not otherwise specified) (Prepare polymer bound o-nitrobenzophenone 9-fluoreneacetate as described above.)
Mobile phase: Gradient. MeCN:water 55:45 for 1.5 min, to 85:15 over 4 min, maintain at 85:15 for 5 min, return to initial conditions over 1 min, re-equilibrate for 2 min.
Column temperature: 75 (column A only)

Flow rate: 1.5
Injection volume: 50
Detector: F ex 254 em 305-395

CHROMATOGRAM
Retention time: 8.7
Limit of detection: 0.74 ng (urine), 0.79 ng (plasma)

KEY WORDS
plasma; column-switching

REFERENCE
Zhou, F.-X.; Krull, I.S.; Feibush, B. Direct determination of adamantanamine in plasma and urine with automated solid phase derivatization, *J.Chromatogr.*, **1993**, *619*, 93–101.

SAMPLE
Matrix: blood, urine
Analyte: amphetamine and methamphetamine
Sample preparation: Serum. 1 mL Serum + 1 mL 120 mM sodium dodecyl sulfate in 100 mM NaOH, homogenize, filter (45 μm). Inject on to column A at 180 μL/min 200 μL 60 mM sodium dodecyl sulfate in MeCN:water 50:50, 25 μL filtrate, and 25 μL 60 mM sodium dodecyl sulfate in MeCN:water 10:90, wait for 1 min, inject 100 μL 60 mM sodium dodecyl sulfate in MeCN:water 10:90, inject 200 μL 10 mM sodium dodecyl sulfate in MeCN:water 10:90, backflush the contents of column A on to column B with mobile phase, after 18 s remove column A from the circuit, elute column B with mobile phase, monitor the effluent from column B. Wash column A with 300 μL 60 mM sodium dodecyl sulfate in MeCN:water 50:50 and 300 μL 10 mM sodium dodecyl sulfate in MeCN:water 80:20. (Only amphetamine is determined in serum). Urine. 5 mL Urine + 4.5 mL MeCN + 500 μL 1 M KOH, centrifuge at 2500 rpm for 10 min, filter (45 μm) the supernatant. Inject on to column A at 180 μL/min 25 μL 50 mM KOH in MeCN:water 20:80, 50 μL filtrate, 25 μL 50 mM KOH in MeCN:water 20:80, and 200 μL MeCN:water 20:80, backflush the contents of column A on to column B with mobile phase, after 18 s remove column A from the circuit, elute column B with mobile phase, monitor the effluent from column B. Wash column A with 400 μL MeCN.

HPLC VARIABLES
Column: A 20 × 2 polymer bound o-nitrobenzophenone 9-fluoreneacetate; B 250 × 4.6 5 μm Supelcosil LC-18-DB (with a guard column) (Prepare polymer bound o-nitrobenzophenone 9-fluoreneacetate as described above.)
Mobile phase: Gradient. MeCN:water 50:50 for 3.5 min, to 70:30 over 12 min, maintain at 70:30 for 2.5 min, return to initial conditions over 1 min. (Place a 100 × 4.6 30-40 μm silica column before the injector.)
Column temperature: 60 (column A only)
Injection volume: 25-50
Detector: F ex 254 em 305-395

CHROMATOGRAM
Retention time: 9 (amphetamine), 13.6 (methamphetamine)
Limit of quantitation: 25 ng/mL (urine), 600 ng/mL (serum)

KEY WORDS
serum

REFERENCE
Bourque, A.J.; Krull, I.S.; Feibush, B. Automated HPLC analyses of drugs of abuse via direct injection of biological fluids followed by simultaneous solid-phase extraction and derivatization with fluorescence detection, *Biomed.Chromatogr.*, **1994**, *8*, 53–62.

Polymer Bound o-Nitrobenzophenone Fluorenylmethoxycarbonyl Proline

SAMPLE
Matrix: solutions

Analyte: amines

Sample preparation: Add 20 µL of a solution in MeCN:water 50:50 to a cartridge containing 30 mg reagent, heat at 40° for 20 min, elute with 500 µL MeCN, inject a 20 µL aliquot of the eluate. Alternatively, add a 27 × 3 reactor filled with polymer bound o-nitrobenzophenone fluorenylmethoxycarbonyl proline as a pre-column to the analytical column, heat the reactor at 40°, inject 5 µL amine solution. (Prepare polymer bound o-nitrobenzophenone fluorenylmethoxycarbonyl proline as follows. Add 3 g aluminum trichloride to 5 g 60-90 µm macroporous polystyrene-divinylbenzene copolymer (Fluka) and 10 g 4-chloro-3-nitrobenzoyl chloride stirred in 100 mL dry nitrobenzene, stir at 60° for 5 h, pour into a mixture of 75 mL DMF, 50 mL concentrated HCl, and 50 g ice, filter. Wash the solid with three 50 mL portions of DMF:water 75:25, wash with 30 mL warm (60°) DMF, wash with five 50 mL portions of dichloromethane:MeOH 2:1. Stir the product in 15 mL 40% benzyltrimethylammonium hydroxide in water, 15 mL water, and 30 mL dioxane (Caution! Dioxane is a carcinogen!) at 90° for 8 h, filter. Wash the product with four 50 mL portions of warm (60°) dioxane. Stir the solid with 30 mL acetic acid for 15 min, filter. Wash the solid with three 50 mL portions of dioxane until the washings were neutral, wash with four 50 mL portions of dichloromethane:MeOH 2:1 (Anal.Chem. 1989, 61, 1538). Stir 5 g N-(9-fluorenylmethoxycarbonyl)-L-proline in 25 mL dichloromethane at 0°, add 14.8 mmole dicyclohexylcarbodiimide, stir at 0° for 30 min, filter, add 2 g polymer, add 2 mL pyridine, shake for 1 h at room temperature, filter. Wash the solid with three 200 mL portions of chloroform and three 200 mL portions of hexane, dry under vacuum to give the polymer bound o-nitrobenzophenone fluorenylmethoxycarbonyl proline.)

HPLC VARIABLES
Column: 250 × 4 5 µm LiChrospher Si-60

Mobile phase: Gradient. Hexane:isopropanol from 90:10 to 80:20 over 1 h.

Flow rate: 0.7
Injection volume: 5-20
Detector: F ex 275 em 315; UV 266

CHROMATOGRAM
Retention time: 20.2 (l-naphthylethylamine), 21.2 (d-naphthylethylamine), 21.7 (l-methylbenzylamine), 23.3 (d-methylbenzylamine)
Limit of detection: 10 ppb

KEY WORDS
chiral; normal phase

REFERENCE
Chou, T.Y.; Gao, C.-X.; Grinberg, N.; Krull, I.S. Chiral polymeric reagents for off-line and on-line derivatizations of enantiomers in high-performance liquid chromatography with ultraviolet and fluorescence detection: an enantiomer recognition approach, *Anal.Chem.*, **1989**, *61*, 1548–1558.

SAMPLE
Matrix: urine
Analyte: amphetamine
Sample preparation: Adjust pH of urine to 10 with 1 M NaOH, inject a 10 μL aliquot onto column A heated to 60°, after 6 s stop the flow through column A, after 5 min backflush the contents of column A onto column B with mobile phase, elute column B with mobile phase, monitor the effluent from column B.

HPLC VARIABLES
Column: A 27 × 30 packed with polymer bound o-nitrobenzophenone fluorenylmethoxycarbonyl proline; B 250 × 4 5 μm LiChrospher C18 (Prepare polymer bound o-nitrobenzophenone fluorenylmethoxycarbonyl proline as follows. Add 25 g aluminum trichloride in 300 mL dry nitrobenzene to 50 g 60-90 μm 96% styrene-4% divinylbenzene resin (Fluka) and 100 g 4-chloro-3-nitrobenzoyl chloride, stir mechanically at 60° for 5 h, pour into a mixture of 150 mL DMF, 100 mL concentrated HCl, and 150 g ice, filter. Wash the solid with 300 mL portions of DMF:water 75:25 until the washings are colorless, wash with warm (60°) DMF, wash with six 300 mL portions of dichloromethane:MeOH 2:1. Stir the product in 130 mL 40% benzyltrimethylammonium hydroxide in water, 130 mL water, and 260 mL dioxane at 90° for 8 h, filter, repeat the process. Wash the product with four portions of warm (60°) dioxane. Stir the solid with 30 mL acetic acid for 15 min, filter. Wash the solid with dioxane until the washings are neutral, wash with six 300 mL portions of dichloromethane:MeOH 2:1 to give a nitrobenzophenol-substituted polymer (J. Org. Chem. 1984, 49, 924). Dissolve 600 mg N-(9-fluorenylmethoxycarbonyl)-L-proline in 10 mL dichloromethane, cool to 0°, add 1.8 mmoles dicyclohexylcarbodiimide, stir at 0° for 30 min, filter. Add the filtrate to 1 g resin, add 500 μL pyridine, stir at room temperature for 1 h, filter, wash with three 50 mL portions of hexane, wash with three 50 mL portions of dichloromethane, wash with three 100 mL portions of MeCN to obtain polymer bound o-nitrobenzophenone fluorenylmethoxycarbonyl proline.)
Mobile phase: MeCN:water 48:52
Flow rate: 1.5
Injection volume: 10
Detector: UV 265; F ex 265 em 315

CHROMATOGRAM
Retention time: 28 (d), 30 (l)
Limit of detection: 50 ng/mL

KEY WORDS
chiral; column-switching

REFERENCE
Gao, C.-X.; Krull, I.S. Determination of enantiomeric drugs in physiological fluids using on-line solid phase derivatizations and reversed-phase liquid chromatography, *J.Pharm.Biomed.Anal.*, **1989**, *7*, 1183–1198.

Polymer Bound o-Nitrobenzophenone Fluorenylmethyl Carbonate

SAMPLE
Matrix: air
Analyte: amines
Sample preparation: Pull air through 1.4 g silica gel (SKC) at 400 mL/min for 4 h, sonicate with MeCN:500 mM sulfuric acid 50:50 for 20 min, remove a 500 μL aliquot and adjust the pH to 10 with 500 μL 1.1 M NaOH, inject a 5-10 μL aliquot on to column A, when the sample reaches column A stop the flow through column A, after 10 min elute the contents of column A on to column B with the mobile phase.

HPLC VARIABLES
Column: A 27 × 2 polymer bound o-nitrobenzophenone fluorenylmethyl carbonate; B 250 × 4.6 5 μm LiChrospher C18 (Prepare polymer bound o-nitrobenzophenone fluorenylmethyl carbonate as follows. Add 3 g aluminum trichloride to 5 g 60-90 μm macroporous polystyrene-divinylbenzene copolymer (Fluka) and 10 g 4-chloro-3-nitrobenzoyl chloride stirred in 100 mL dry nitrobenzene, stir at 60° for 5 h, pour into a mixture of 75 mL DMF, 50 mL concentrated HCl, and 50 g ice, filter. Wash the solid with three 50 mL portions of DMF:water 75:25, wash with 30 mL warm (60°) DMF, wash with five 50 mL portions of dichloromethane:MeOH 2:1. Stir the product in 15 mL 40% benzyltrimethylammonium hydroxide in water, 15 mL water, and 30 mL dioxane (Caution! Dioxane is a carcinogen!) at 90° for 8 h, filter. Wash the product with four 50 mL portions of warm (60°) dioxane. Stir the solid with 30 mL acetic acid for 15 min, filter. Wash the solid with three 50 mL portions of dioxane until the washings were neutral, wash with four 50 mL portions of dichloromethane:MeOH 2:1. Stir 1 g of the yellow product in 10 mL dichloromethane and 500 μL pyridine, add 1.24 g 9-fluorenylmethyl chloroformate, stir at room temperature for 30 min, filter. Wash the solid with three 20 mL portions of DMF, three 20 mL

portions of dichloromethane, and three 20 mL portions of dry diethyl ether, dry under vacuum to give the reagent (polymer bound o-nitrobenzophenone fluorenylmethyl carbonate). Store in the freezer, good for at least a year.)

Mobile phase: MeCN:water 50:50
Column temperature: 60 (column A only)
Flow rate: 1.5
Injection volume: 5-10
Detector: F ex 265 em 320; UV 265

CHROMATOGRAM
Retention time: 5 (methylamine), 10 (butylamine), 12 (diethylamine), 22 (cadaverine)
Limit of detection: 24-60 ppb

KEY WORDS
column-switching

REFERENCE
Gao, C.-X.; Krull, I.S.; Trainor, T. Determination of aliphatic amines in air by on-line solid-phase derivatization with HPLC-UV/FL, *J.Chromatogr.Sci.*, **1990**, *28*, 102–108.

SAMPLE
Matrix: urine
Analyte: amines
Sample preparation: Add 20 μL of a solution of amphetamine in MeCN:water 50:50 to a cartridge containing 30 mg polymer bound o-nitrobenzophenone fluorenylmethyl carbonate, heat at 60° for 5 min, elute with 500 μL MeCN, inject a 20 μL aliquot of the eluate. Alternatively, inject 10 μL of urine (adjusted to pH 10 with 100 mM NaOH and filtered) into a 27 × 3 reactor filled with polymer bound o-nitrobenzophenone fluorenylmethyl carbonate held at 60°, after 5 min elute the contents of the reactor on to the column with mobile phase, elute the column with mobile phase, monitor the effluent from the column. (Prepare polymer bound o-nitrobenzophenone fluorenylmethyl carbonate as follows. Add 3 g aluminum trichloride to 5 g 60-90 μm macroporous polystyrene-divinylbenzene copolymer (Fluka) and 10 g 4-chloro-3-nitrobenzoyl chloride stirred in 100 mL dry nitrobenzene, stir at 60° for 5 h, pour into a mixture of 75 mL DMF, 50 mL concentrated HCl, and 50 g ice, filter. Wash the solid with three 50 mL portions of DMF:water 75:25, wash with 30 mL warm (60°) DMF, wash with five 50 mL portions of dichloromethane:MeOH 2:1. Stir the product in 15 mL 40% benzyltrimethylammonium hydroxide in water, 15 mL water, and 30 mL dioxane (Caution! Dioxane is a carcinogen!) at 90° for 8 h, filter. Wash the product with four 50 mL portions of warm (60°) dioxane. Stir the solid with 30 mL acetic acid for 15 min, filter. Wash the solid with three 50 mL portions of dioxane until the washings are neutral, wash with four 50 mL portions of dichloromethane:MeOH 2:1. Stir 1 g of the yellow product in 10 mL dichloromethane and 500 μL pyridine, add 1.24 g 9-fluorenylmethyl chloroformate, stir at room temperature for 30 min, filter. Wash the solid with three 20 mL portions of DMF, three 20 mL portions of dichloromethane, and three 20 mL portions of dry diethyl ether, dry under vacuum to give polymer bound o-nitrobenzophenone fluorenylmethyl carbonate. Store in the freezer, good for at least a year.)

HPLC VARIABLES
Column: 250 × 4.6 5 μm LiChroSpher C18
Mobile phase: MeCN:water 50:50
Flow rate: 1.5
Injection volume: 20
Detector: F ex 265 em 320

CHROMATOGRAM
Retention time: 5 (amphetamine), 6 (morpholine), 7.5 (propylamine), 10 (butylamine), 12 (diethylamine)
Limit of detection: 100 ppb

KEY WORDS
solutions

REFERENCE

Gao, C.-X.; Chou, T.-Y.; Krull, I.S. Polymeric activated ester reagents for off-line and on-line derivatizations of amine nucleophiles in high-performance liquid chromatography with ultraviolet and fluorescence detection, *Anal.Chem.*, **1989**, *61*, 1538–1548.

SAMPLE

Matrix: urine

Analyte: amphetamine

Sample preparation: Inject a 10 μL aliquot into a 27 × 2 reactor packed with 65 mg polymeric reagents A and B in a 5:1 ratio heated at 60°, after 10 min flush the contents of the reactor onto the column with the mobile phase. (Prepare the polymeric reagents as follows. Stir mechanically 5 g 60-90 μm polystyrene-divinylbenzene copolymer (Fluka), 10 g 4-chloro-3-nitrobenzoyl chloride, and 3 g aluminum trichloride in 100 mL dry nitrobenzene at 60° for 5 h, pour into a mixture of 75 mL DMF, 50 mL concentrated HCl, and 50 g ice, filter. Wash the solid with three 50 mL portions of DMF:water 75:25, wash with warm (60°) DMF, wash with three 50 mL portions of dichloromethane:MeOH 2:1. Add the solid to 15 mL 40% benzyltrimethylammonium hydroxide in water, 15 mL water, and 30 mL dioxane (Caution! Dioxane is a carcinogen!), heat at 90° for 8 h, filter, wash with four 50 mL portions of warm (60°) dioxane. Add 30 mL acetic acid:water 50:50 and stir for 15 min, filter, wash with water until the washings are neutral, wash the three 50 mL portions of dioxane, wash with four 50 mL portions of dichloromethane:MeOH 2:1. Stir 1.12 g of this product with 248 mg 9-fluorenylmethyl chloroformate in 2 mL dichloromethane and 100 μL pyridine at room temperature for 30 min, filter, wash with three 4 mL portions of DMF, wash with three 4 mL portions of dichloromethane, wash with three 4 mL portions of dry ethyl ether, dry under vacuum to give polymer bound o-nitrobenzophenone fluorenylmethyl carbonate (A). Substitute 4-nitrobenzoyl chloride for 9-fluorenylmethyl chloroformate to obtain polymer bound o-nitrobenzophenone 4-nitrobenzoate (B).)

HPLC VARIABLES

Column: 250 × 4 5 μm LiChrospher C18

Mobile phase: MeCN:water 55:45

Flow rate: 1.5 for 6 min then 2.5

Injection volume: 10

Detector: UV 265; F ex 265 em 320

CHROMATOGRAM

Retention time: 6 (4-nitrobenzoyl derivative), 15 (fluorenylmethoxycarbonyl derivative)

Limit of quantitation: 10 ppm (F)

REFERENCE

Gao, C.X.; Schmalzing, D.; Krull, I.S. A mixed-bed, multi-derivatization approach using polymeric reagents for derivatizations of amines in high performance liquid chromatographic detection, *Biomed.Chromatogr.*, **1991**, *5*, 23–31.

Polymer Bound o-Nitrobenzophenone 4-Nitrobenzoate

SAMPLE
Matrix: blood, urine
Analyte: amantadine
Sample preparation: Urine. Adjust urine to pH 11 with NaOH, dilute with an equal volume of MeCN, filter (0.45 μm) add a 30 μL aliquot to 20 mg polymer bound o-nitrobenzophenone 4-nitrobenzoate, heat at 40° for 15 min, inject an aliquot of the reaction mixture. Plasma. Lyophilize plasma, reconstitute with 50 mM NaOH containing 5 mM sodium dodecyl sulfate, filter (0.45 μm), add a 50 μL aliquot to 30 mg polymer bound o-nitrobenzophenone 4-nitrobenzoate in a Pasteur pipette, heat at 75° for 10 min, wash with 5 mM sodium dodecyl sulfate, elute with 200 μL MeCN, dilute the MeCN with 400 μL running buffer, inject an aliquot. (Prepare polymer bound o-nitrobenzophenone 4-nitrobenzoate as follows. Soxhlet extract macroporous polystyrene cross linked with divinylbenzene with MeCN for 24 h. Heat 2.5 g resin and 4 g 4-chloro-3-nitrobenzoyl chloride in nitrobenzene in the presence of aluminum chloride at 60° for 24 h, pour into 75 mL DMF, 50 mL concentrated HCl, and 50 g ice. Wash the product with DMF/water, with DMF, and with dichloromethane. Add the product to 15 mL trimethylammonium hydroxide, 15 mL water, and 30 mL dioxane (Caution! Dioxane is a carcinogen!), heat at 90° for 16 h. React 1 g of this product with 0.5 g p-nitrobenzoyl chloride and 2.7 mmole pyridine in 20 mL dichloromethane at room temperature for 2 h, filter, wash with dichloromethane, with DMF, and with THF, dry under vacuum to obtain polymer bound o-nitrobenzophenone 4-nitrobenzoate.)

CAPILLARY ELECTROPHORESIS
Capillary: 60 cm × 75 μm (53 cm to detector)
Running buffer: MeCN:buffer 20:80 (Buffer was 25 mM pH 7.2 phosphate buffer.)
Voltage/Current: 20 kV/70 μA
Injection: Vacuum inject at 25 kPa
Detector: UV 254
Model: Isco Model 3140
Migration time: 9
Limit of quantitation: 10 ppm

KEY WORDS
plasma; cow

REFERENCE
Szulc, M.; Krull, I.S. Polymeric reagents for derivatizations in micellar electrokinetic chromatography, *Biomed.Chromatogr.*, **1994**, *8*, 212–218.

SAMPLE
Matrix: urine
Analyte: amphetamine
Sample preparation: Inject a 10 μL aliquot into a 27 × 2 reactor packed with 65 mg polymeric reagents A and B in a 5:1 ratio heated at 60°, after 10 min flush the contents of the reactor onto the column with the mobile phase. (Prepare the polymeric reagents

as follows. Stir mechanically 5 g 60-90 μm polystyrene-divinylbenzene copolymer (Fluka), 10 g 4-chloro-3-nitrobenzoyl chloride, and 3 g aluminum trichloride in 100 mL dry nitrobenzene at 60° for 5 h, pour into a mixture of 75 mL DMF, 50 mL concentrated HCl, and 50 g ice, filter. Wash the solid with three 50 mL portions of DMF:water 75:25, wash with warm (60°) DMF, wash with three 50 mL portions of dichloromethane:MeOH 2:1. Add the solid to 15 mL 40% benzyltrimethylammonium hydroxide in water, 15 mL water, and 30 mL dioxane (Caution! Dioxane is a carcinogen!), heat at 90° for 8 h, filter, wash with four 50 mL portions of warm (60°) dioxane. Add 30 mL acetic acid:water 50:50 and stir for 15 min, filter, wash with water until the washings are neutral, wash the three 50 mL portions of dioxane, wash with four 50 mL portions of dichloromethane:MeOH 2:1. Stir 1.12 g of this product with 248 mg 9-fluorenylmethyl chloroformate in 2 mL dichloromethane and 100 μL pyridine at room temperature for 30 min, filter, wash with three 4 mL portions of DMF, wash with three 4 mL portions of dichloromethane, wash with three 4 mL portions of dry ethyl ether, dry under vacuum to give polymer bound o-nitrobenzophenone fluorenylmethyl carbonate (A). Substitute 4-nitrobenzoyl chloride for 9-fluorenylmethyl chloroformate to obtain polymer bound o-nitrobenzophenone 4-nitrobenzoate (B).)

HPLC VARIABLES
Column: 250 × 4 5 μm LiChrospher C18
Mobile phase: MeCN:water 55:45
Flow rate: 1.5 for 6 min then 2.5
Injection volume: 10
Detector: UV 265; F ex 265 em 320

CHROMATOGRAM
Retention time: 6 (4-nitrobenzoyl derivative), 15 (fluorenylmethoxycarbonyl derivative)
Limit of quantitation: 10 ppm (F)

REFERENCE
Gao, C.X.; Schmalzing, D.; Krull, I.S. A mixed-bed, multi-derivatization approach using polymeric reagents for derivatizations of amines in high performance liquid chromatographic detection, *Biomed.Chromatogr.*, **1991**, *5*, 23–31.

SUCCINIMIDYL ESTER

6-Aminoquinolyl-N-hydroxysuccinimidyl Carbamate

RELATED REFERENCES
Chen, S.; Pawlowska, M.; Armstrong, D.W. HPLC enantioseparation of di- and tripeptides on cyclodextrin bonded stationary phases after derivatization with 6-aminoquinolyl-N-hydroxysuccinimidyl carbamate (AQC). *J.Liq.Chromatogr.* **1994**, *17*, 483-497.
Mascher, H.; Göd, B.; Kikuta, C. Determination of desmethylselegiline, methamphetamine and amphetamine - the main metabolites of selegiline in plasma by HPLC after derivatization. *J.Liq.Chromatogr.Rel.Technol.* **1997**, *20*, 797-809.
Velazquez, C.; van Bloemendal, C.; Sanchis, V.; Canela, R. Derivation of fumonisins B1 and B2 with 6-aminoquinolyl N-hydroxysuccinimidylcarbamate. *J.Agric.Food Chem.* **1995**, *43*, 1535-1537.

SAMPLE
Matrix: beverages
Analyte: amines

Sample preparation: Mix 10 μL wine with 6 μL pH 8.8 borate buffer (Waters), add 0.5 μL 10 mM 6-aminoquinolyl-N-hydroxysuccinimidyl carbamate in MeCN, mix, let stand for 5 min, inject a 10 μL aliquot. (6-Aminoquinolyl-N-hydroxysuccinimidyl carbamate can be purchased from Waters or synthesized as follows. Reflux 3 g N,N'-succinimidyl carbonate in 100 mL dry MeCN, add 1.5 g 6-aminoquinoline in 50 mL dry MeCN dropwise over 30 min, reflux for 30 min, evaporate to half volume under reduced pressure, cool for 24 h, filter, wash the solid with cold MeCN, recrystallize from MeCN to obtain 6-amino-quinolyl-N-hydroxysuccinimidyl carbamate as off-white crystals (mp 210-215° (d)) (Anal. Biochem. 1993, 211, 279).)

HPLC VARIABLES
Guard column: Spherisorb ODS-2
Column: 250 × 4.6 5 μm Spherisorb ODS-2
Mobile phase: Gradient. A was THF:50 mM sodium acetate. B was MeOH. A:B 75:25 for 5 min, to 20:80 over 20 min, to 0:100 (step gradient), maintain at 0:100 for 3 min, return to initial conditions, re-equilibrate for 2 min.
Column temperature: 65
Flow rate: 1
Injection volume: 10
Detector: F ex 250 em 395

CHROMATOGRAM
Retention time: 4 (ammonia), 5 (ethanolamine), 6.5 (methylamine), 9 (histamine), 10 (ethylamine), 12.5 (pyrrolidine), 13 (isopropylamine), 13.2 (propylamine), 13.5 (tyramine), 18 (butylamine), 18.5 (putrescine), 20.5 (cadaverine), 21.8 (phenethylamine), 22 (3-methyl-butylamine), 22.4 (amylamine), 24 (hexylamine)
Internal standard: heptylamine
Limit of detection: 100-500 ng/mL

KEY WORDS
wine

REFERENCE
Busto, O.; Guasch, J.; Borrull, F. Determination of biogenic amines in wine after precolumn derivatization with 6-aminoquinolyl-N-hydroxysuccinimidyl carbamate, *J.Chromatogr.A*, **1996**, *737*, 205–213.

SAMPLE
Matrix: blood
Analyte: polyamines
Sample preparation: Dilute whole blood 4-fold with buffer, centrifuge at 800 g for 5 min, wash the pellet three times by suspending it in buffer and centrifuging at 2000 g for 10 min. Suspend the cells in 2 volumes of buffer and sonicate (40 W, 70% duty cycle, Ultra-sonics Heat System) for 10 min, centrifuge at 2000 g for 5 min, heat the supernatant at 100° for 2 min, centrifuge at 2000 g for 5 min. Remove a 10-60 μL aliquot and add it to 10 μL 2 μM 1,7-diaminoheptane, make up to 90 μL with 200 mM pH 8.8 borate buffer containing 1 mM EDTA, add 10 μL reagent, mix, heat at 55° for 20 min, inject an aliquot. (Buffer was 58.5 mM Na_2HPO_4 containing 1.5 mM KH_2PO_4, 43.5 mM NaCl, 10 mM tri-sodium citrate, 10 mM dithiothreitol, and 2.7 mM KCl, pH 7.4. Reagent was N-hydroxy-succinimidyl-6-aminoquinoyl carbamate solution supplied by Waters. 6-Aminoquinolyl-N-hydroxysuccinimidyl carbamate can be purchased from Waters or synthesized as described above.)

HPLC VARIABLES
Column: 150 × 4.6 5 μm Microsorb-MV C8
Mobile phase: Gradient. MeCN:water:buffer from 0:0:100 to 1:0:99 over 1 min, to 5:0:95 over 17 min, to 9:0:91 over 1 min, to 17:0:83 over 10.5 min, to 40:0:60 over 3.5 min, to 60:40:0 over 9 min, return to initial conditions over 8 min. (Buffer was 140 mM pH 5.05 acetate containing 17 mM triethanolamine.)
Flow rate: 1
Detector: F ex 250 em 395

CHROMATOGRAM
Retention time: 37 (putrescine), 38 (spermine), 39 (spermidine)
Internal standard: 1,7-diaminoheptane (40)
Limit of detection: 660 fmole

KEY WORDS
mouse; whole blood

REFERENCE
Merali, S.; Clarkson, A.B. Jr. Polyamine analysis using N-hydroxysuccinimidyl-6-aminoquinoyl carbamate for pre-column derivatization, *J.Chromatogr.B*, **1996**, *675*, 321–326.

N-(tert-Butoxycarbonyl)-L-leucine N-Hydroxysuccinimide Ester

SAMPLE
Matrix: bulk
Analyte: vigabatrin
Sample preparation: 2.6 mg Vigabatrin + 250 μL 200 mM sodium bicarbonate, stir for 2 min, add 250 μL 65 mg/mL N-(tert-butoxycarbonyl)-L-leucine N-hydroxysuccinimide ester (N-t-Boc-L-leucine n-hydroxysuccinimide ester) in THF, stir for 30 min, evaporate to dryness under a stream of nitrogen, reconstitute with 200 μL trifluoroacetic acid, let stand at room temperature for 5 min, evaporate to dryness under a stream of nitrogen, reconstitute with 10 mL mobile phase, inject an aliquot.

HPLC VARIABLES
Column: 250 × 4 10 μm LiChrosorb RP-8
Mobile phase: MeCN:50 mM pH 7 phosphate buffer 4:96
Flow rate: 2
Injection volume: 100
Detector: UV 210

CHROMATOGRAM
Retention time: 4 (R-(-)), 17.5 (S-(+))
Limit of detection: 0.1% (of major enantiomer)

KEY WORDS
chiral

REFERENCE

Chen, T.-M.; Contario, J.J. High-performance liquid chromatographic resolution of enantiomers of gamma-vinyl-gamma-aminobutyric acid, *J.Chromatogr.*, **1984**, *314*, 495–498.

5-Carboxytetramethylrhodamine Succinimidyl Ester

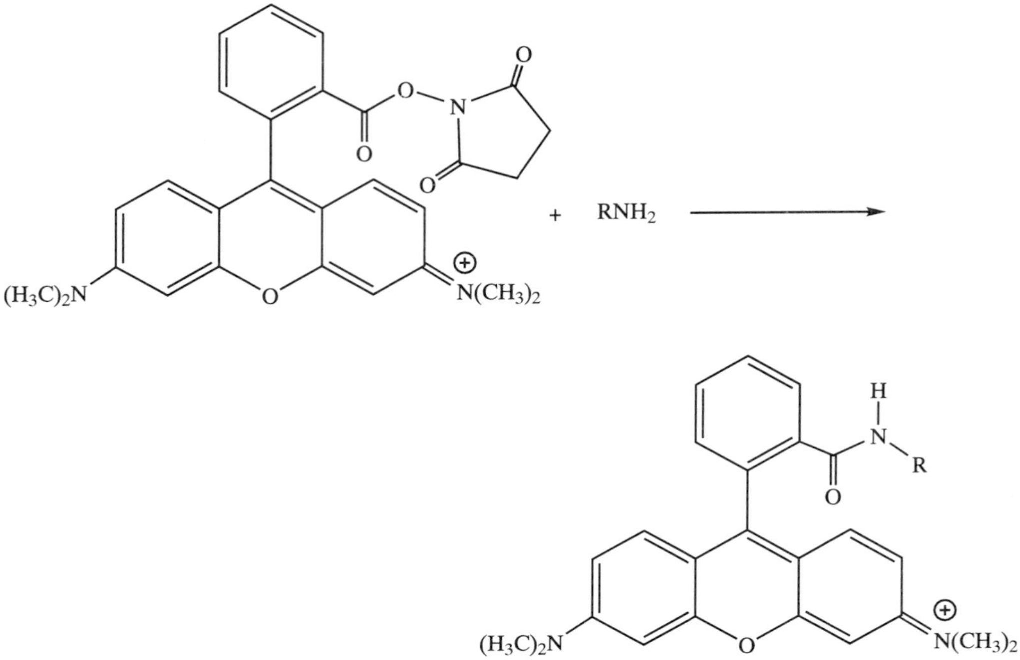

SAMPLE

Matrix: solutions

Analyte: aminated monosaccharides

Sample preparation: Mix 30 µL of a 20 mM solution in 185 mM pH 8.3 sodium carbonate buffer with 5 µL of a 1.28 mM solution of 5-carboxytetramethylrhodamine succinimidyl ester (Molecular Probes, Eugene OR) in DMF, vortex periodically, let stand in the dark at room temperature for 1 h

CAPILLARY ELECTROPHORESIS

Capillary: 74.4 cm × 10 µm fused-silica (Polymicro Technologies)

Running buffer: 10 mM pH 9.3 Na_2HPO_4 containing 10 mM borate, 18 mM phenyl boronic acid, and 10 mM sodium dodecyl sulfate

Voltage/Current: 29 kV

Injection: Electrokinetic injection at 2.5 kV for 5 s

Detector: F ex 543.5 (750 mW He-Ne laser, Melles Griot, Ontario) em 580 (filter)

Migration time: 19 (glucosamine), 19.2 (2-glucosamine), 19.8 (2-galactosamine), 20.2 (mannosamine), 20.4 (galactosamine), 21 (fucosamine)

REFERENCE

Zhao, J.Y.; Diedrich, P.; Zhang, Y.; Hindsgaul, O.; Dovichi, N.J. Separation of aminated monosaccharides by capillary zone electrophoresis with laser-induced fluorescence detection, *J.Chromatogr.B*, **1994**, *657*, 307–313.

SAMPLE

Matrix: solutions

Analyte: angiotensins

Sample preparation: Prepare a 1 mM solution in 100 mM pH 6 morpholinoethanesulfonic acid/NaOH buffer. Remove a 20 μL aliquot and add it to 5 μL 25 mM 5-carboxytetramethylrhodamine succinimidyl ester (Molecular Probes, Eugene OR) in DMF, let stand at room temperature overnight, add 50 μL 1 M pH 8 Tris-HCl buffer, inject a 5 μL aliquot.

HPLC VARIABLES
Guard column: reversed-phase
Column: reversed-phase
Mobile phase: Gradient. A was MeCN containing 0.1% trifluoroacetic acid. B was water containing 0.1% trifluoroacetic acid. A:B from 5:95 to 55:45 "applied just after application of a sample".
Flow rate: 1
Injection volume: 5
Detector: UV 280

CHROMATOGRAM
Retention time: 37 (angiotensin I), 38 (angiotensin II)

REFERENCE
Shimura, K.; Kasai, K.-I. Fluorescence-labeled peptides as isoelectric point (p*I*) markers in capillary isoelectric focusing with fluorescence detection, *Electrophoresis*, **1995**, *16*, 1479–1484.

N-(4-((2,5-Dioxo-1-pyrrolidinyl)oxycarboxyamino)butyl)-N-ethylisoluminol

RELATED REFERENCE
Nakashima, K.; Suetsugu, K.; Yoshida, K.; Imai, K.; Akiyama, S. High-performance liquid chromatography with chemiluminescence detection of methamphetamine and amphetamine in human urine. *Anal.Sci.* **1991**, *7*, 815-816

SAMPLE
Matrix: blood

Analyte: methamphetamine

Sample preparation: 100 μL Serum + 1 mL acetone, vortex for 2 min, centrifuge at 2000 g for 10 min. Remove a 500 μL aliquot of the supernatant and add it to 500 μL reagent, add 20 μL 0.5% triethylamine in MeOH, vortex, heat at 80° for 30 min, cool to room temperature, inject a 20 μL aliquot. (Prepare reagent, N-(4-((2,5-dioxo-1-pyrrolidinyl)oxycarboxyamino)butyl)-N-ethylisoluminol, by mixing 2 mL 2.5 mM N-(4-aminobutyl)-N-ethylisoluminol in MeOH with 2 mL 2.5 mM N,N'-disuccinimidyl carbonate in MeCN, let stand for 2 h.)

HPLC VARIABLES

Guard column: 30 × 4.6 TSKm Guardgel ODS-80TM (Toyo Soda)

Column: 150 × 6 5 μm Shimpack CLC C18 (Shimadzu)

Mobile phase: MeOH:water 54:46 containing 30 mM sodium 1-octanesulfonate

Flow rate: 1

Injection volume: 20

Detector: Chemiluminescence following post-column reaction. The column effluent mixed with 15 mM potassium ferricyanide in 2.5 M NaOH pumped at 1 mL/min and this mixture flowed through a 200 mm × 0.5 mm ID stainless steel coil. The effluent from this coil mixed with 300 mM hydrogen peroxide containing 10 mM β-cyclodextrin pumped at 1 mL/min and this mixture flowed through a 100 mm × 0.5 mm ID stainless steel coil to the detector.

CHROMATOGRAM

Retention time: 30

Limit of detection: 20 pM

KEY WORDS

serum

REFERENCE

Nakashima, K.; Suetsugu, K.; Akiyama, S.; Yoshida, M. High-performance liquid chromatography-chemiluminescence determination of methamphetamine in human serum using N-(4-aminobutyl)-N-ethylisoluminol as a chemiluminogen, *J.Chromatogr.*, **1990**, *530*, 154–159.

Luminarin 1

SAMPLE
Matrix: solutions
Analyte: pentylamine
Sample preparation: Mix 900 μL of a 10 mM solution of pentylamine in THF with 1 mL of a 10 mM solution of luminarin 1 in THF, heat at 60° for 20 min, evaporate to dryness, add 2 mL 100 mM pH 9 carbonate buffer, shake at 40° for 15 min, extract twice with 4 mL portions of chloroform. Combine the organic layers and evaporate them to dryness under a stream of nitrogen, reconstitute with MeCN, inject an aliquot. (Luminarin 1, 1-[[(2,3,6,7-tetrahydro-11-oxo-1H,5H,11H-[1]benzopyrano[6,7,8-ij]quinolizin-9-yl)acetyl]oxy]-2,5-pyrrolidinedione, may be obtained from Eurobio, Les Ulis, France. Synthesis is as follows. Reflux (with protection from moisture and with stirring) 2.12 g 8-hydroxyjulolidine, 2.22 g diethyl 1,3-acetonedicarboxylate (oxo-3-glutaric acid ethyl ester, Fluka), 1.71 g anhydrous zinc chloride, and 6 mL EtOH for 24 h, cool, add to 200 mL water, extract with 200 mL ethyl acetate, extract with 100 mL ethyl acetate. Combine the organic layers and wash them with water, dry over magnesium sulfate, evaporate to dryness, recrystallize from 5 parts ethyl acetate to give ethyl 2,3,6,7-tetrahydro-11-oxo-1H,5H,11H-[1]benzopyrano[6,7,8-ij]quinolizine-9-acetate. Heat 2 g of this compound with 42 mL 1.2% NaOH in water and 40 mL MeOH at 45° for 1 h, cool, wash with 50 mL chloroform, wash with 40 mL chloroform. Degas the aqueous phase and acidify it with 16 mL 3 M HCl, stir for 15 min, adjust pH to 6.5 with 13 mL 2.5 M NaOH, filter. Wash the precipitate with water and dry it to obtain 2,3,6,7-tetrahydro-11-oxo-1H,5H,11H-[1]benzopyrano[6,7,8-ij]-quinolizine-9-acetic acid. Stir 11.26 g of this compound, 10.62 g disuccinimidyl oxalate (dihydroxysuccinimide carbonate), 3.81 g anhydrous triethylamine, and 560 mL dry MeCN protected from moisture at room temperature for 1 h, stir at 35-40° for 1 h, filter. Concentrate the filtrate and chromatograph on silica gel with dichloromethane:THF 50:50 to give luminarin 1 (21%) (US Pat. 5 151 517 (Sept. 9, 1992)).

HPLC VARIABLES
Column: 250 × 4.6 5 μm Ultrasphere C18
Mobile phase: MeCN:10 mM pH 8 imidazole nitrate buffer 70:30
Flow rate: 1.2

Injection volume: 20
Detector: F ex 392 em 470

CHROMATOGRAM
Retention time: 4

REFERENCE
Tod, M.; Prevot, M.; Poulou, M.; Farinotti, R.; Chalom, J.; Mahuzier, G. Chromatographic and luminescence properties of a 7-aminocoumarin derivative with peroxylate chemiexcitation, *Anal.Chim.Acta*,
1989, *223*, 309–317.

SAMPLE
Matrix: solutions
Analyte: amines
Sample preparation: Primary amines. Mix 100 µL of a 1-10000 µM solution of the amine
in THF with 100 µL of a 1 mM solution of luminarin 1 in THF, heat at 50° for 20 min,
add 10 µL water, dilute with MeCN, inject a 20 µL aliquot. Secondary amines. Mix 100
µL of a solution of the amine in THF:DMSO 80:20 with 100 µL of a 1 mM solution of
luminarin 1 in THF, heat at 70° for 3 h min, add 10 µL water, dilute with MeCN, inject
a 20 µL aliquot. (Dry THF over molecular sieve. Luminarin 1, 1-[[(2,3,6,7-tetrahydro-
11-oxo-1H,5H,11H-[1]benzopyrano[6,7,8-ij]quinolizin-9-yl)acetyl]oxy] -2,5-pyrrolidinedione,
may be obtained from Eurobio, Les Ulis, France. Synthesis is as follows. Reflux (with
protection from moisture and with stirring) 2.12 g 8-hydroxyjulolidine, 2.22 g diethyl 1,3-
acetonedicarboxylate (oxo-3-glutaric acid ethyl ester, Fluka), 1.71 g anhydrous zinc chloride, and 6 mL EtOH for 24 h, cool, add to 200 mL water, extract with 200 mL ethyl
acetate, extract with 100 mL ethyl acetate. Combine the organic layers and wash them
with water, dry over magnesium sulfate, evaporate to dryness, recrystallize from 5 parts
ethyl acetate to give ethyl 2,3,6,7-tetrahydro-11-oxo-1H,5H,11H-[1]benzopyrano[6,7,8-
ij]quinolizine-9-acetate. Heat 2 g of this compound with 42 mL 1.2% NaOH in water and
40 mL MeOH at 45° for 1 h, cool, wash with 50 mL chloroform, wash with 40 mL chloroform. Degas the aqueous phase and acidify it with 16 mL 3 M HCl, stir for 15 min,
adjust pH to 6.5 with 13mL 2.5 M NaOH, filter. Wash the precipitate with water and dry
it to obtain 2,3,6,7-tetrahydro-11-oxo-1H,5H,11H-[1]benzopyrano[6,7,8-ij]quinolizine-9-
acetic acid. Stir 11.26 g of this compound, 10.62 g disuccinimidyl oxalate (dihydroxysuccinimide carbonate), 3.81 g anhydrous triethylamine, and 560 mL dry MeCN protected
from moisture at room temperature for 1 h, stir at 35-40° for 1 h, filter. Concentrate the
filtrate and chromatograph on silica gel with dichloromethane:THF 50:50 to give luminarin 1 (21%) (US Pat. 5 151 517 (Sept. 9, 1992)).

HPLC VARIABLES
Column: 250 × 4.6 5 µm Nucleosil C18
Mobile phase: MeCN:10 mM pH 7.8 imidazole nitrate buffer 60:40 (A) or 70:30 (B)
Column temperature: 40 (chemiluminescence only)
Flow rate: 1 (F), 1.1 (chemiluminescence)
Injection volume: 20
Detector: F ex 390 em 470; Chemiluminescence. The column effluent mixed with 5 mM
bis(2,4,6-trichlorophenyl) oxalate in methyl acetate pumped at 0.55 mL/min and with 400
mM hydrogen peroxide in THF pumped at 0.1 mL/min and the mixture flowed through
a 60 µL PTFE capillary containing glass beads at 40° to a Kratos FS970 detector.

CHROMATOGRAM
Retention time: 6.0 (pyrrolidine (A)), 9.0 (pentylamine (A)), 3.0 (proline (B)), 4.2 (tyramine
(B))
Limit of detection: 160-300 fmole (F), 15-100 fmole (chemiluminescence)

KEY WORDS
comparison with other derivatizing reagents

REFERENCE
Kouwatli, H.; Chalom, J.; Tod, M.; Farinotti, R.; Mahuzier, G. Precolumn derivatization of amines in
liquid chromatography using luminescent probes: comparison of several reagents with luminarin 1,
Anal.Chim.Acta, **1992**, *266*, 243–249.

Luminarin 2

SAMPLE

Matrix: bulk

Analyte: histamine

Sample preparation: Mix 0.5-5 nmole histamine with 50 nmole luminarin 2 in 500 μL acetone, add 100 μL 100 mM dimethylaminopyridine in acetone, evaporate to dryness, let stand in the dark for 1.5 h, reconstitute with 200 μL acetone, inject an aliquot. (Dry acetone over 0.4 nm molecular sieve. Luminarin 2, N-[6-[(2,5-dioxo-1-pyrrolidinyl)oxy]-6-oxohexyl]-2,3,6,7-tetrahydro-11-oxo-1H,5H,11H-[1]benzopyrano[6,7,8-ij]quinolizine-9-acetamide, may be obtained from Eurobio, Les Ulis, France. Synthesis is as follows. Reflux (with protection from moisture and with stirring) 2.12 g 8-hydroxyjulolidine, 2.22 g diethyl 1,3-acetonedicarboxylate (oxo-3-glutaric acid ethyl ester, Fluka), 1.71 g anhydrous zinc chloride, and 6 mL EtOH for 24 h, cool, add to 200 mL water, extract with 200 mL ethyl acetate, extract with 100 mL ethyl acetate. Combine the organic layers and wash them with water, dry over magnesium sulfate, evaporate to dryness, recrystallize from 5 parts ethyl acetate to give ethyl 2,3,6,7-tetrahydro-11-oxo-1H,5H,11H-[1]benzopyrano[6,7,8-ij]quinolizine-9-acetate. Heat 2 g of this compound with 42 mL 1.2% NaOH in water and 40 mL MeOH at 45° for 1 h, cool, wash with 50 mL chloroform, wash with 40 mL chloroform. Degas the aqueous phase and acidify it with 16 mL 3 M HCl, stir for 15 min, adjust pH to 6.5 with 13 mL 2.5 M NaOH, filter. Wash the precipitate with water and dry it to obtain 2,3,6,7-tetrahydro-11-oxo-1H,5H,11H-[1]benzopyrano[6,7,8-ij]quinolizine-9-acetic acid. React 30 g potassium carbonate with 30 g methyl 6-aminohexanoate hydrochloride (Fluka) for 30 h, filter. Stir 11.26 g of 2,3,6,7-tetrahydro-11-oxo-1H,5H,11H-[1]benzopyrano[6,7,8-ij]quinolizine-9-acetic acid, 10.62 g disuccinimidyl oxalate (dihydroxysuccinimide carbonate), 3.81 g anhydrous triethylamine, and 560 mL dry MeCN protected from moisture at room temperature for 1 h, stir at 35-40° for 1 h, add the methyl 6-aminohexanoate filtrate, stir for 8 h, add 20 g ethanolamine, stir for 30 min, filter, wash with water, remove the solvent, chromatograph on a silica gel column with dichloromethane, dichloromethane:THF 85:15, dichloromethane:THF 75:25, recrystallize from ethyl acetate to give methyl N-(2,3,6,7-tetrahydro-11-oxo-1H,5H,11H-[1]benzopyrano[6,7,8-ij]quinolizine-9-yl)acetyl-6-aminohexanoate (yield 36%). Heat 1.25 g of this compound with 42 mL 1.2% NaOH in water and 40 mL MeOH at 45° for 1 h, cool, wash with 50 mL chloroform, wash with 40 mL chloroform. Degas the aqueous phase and acidify it with 16 mL 3 M HCl, stir for 15 min, adjust pH to 6.5 with 13 mL 2.5 M NaOH, filter. Wash the precipitate with water and dry it to obtain N-(2,3,6,7-tetrahydro-11-oxo-1H,5H,11H-[1]benzopyrano[6,7,8-ij]quinolizine-9-yl)acetyl-6-aminohexanoic acid. React 750 mg of this acid with 912 mg triethylamine and 512 mg disuccinimidyl oxalate (di-hydroxysuccinimide carbonate) in 27 mL MeCN for 6 h, filter, evaporate, chromatograph on a silica gel column with dichloromethane, dichloromethane:THF 85:15, and dichloromethane:THF 75:25 to obtain Luminarin 2 (yield 19%) (World Pat. 89 12,052; Chem. Abstr. 1990, 113, 23889n).)

HPLC VARIABLES
Column: 150 × 4.6 5 μm Nucleosil C18
Mobile phase: MeCN:5 mM ammonium acetate 26:74
Column temperature: 40 (chemiluminescence only)
Flow rate: 2 (F) or 1.5 (chemiluminescence)
Injection volume: 20
Detector: F ex 390 em 490; Chemiluminescence. 1.1 mM Bis(2,4,6-trichlorophenyl) oxalate in methyl acetate pumped at 0.3 mL/min and 400 mM hydrogen peroxide in THF pumped at 0.3 mL/min mixed in a 292 μL capillary tube and this mixture mixed with the column effluent. The resulting mixture flowed through a 60 μL PTFE capillary at 40° to a Kratos FS970 detector fitted with a 470 nm long-pass filter.

CHROMATOGRAM
Retention time: 15.4
Limit of detection: 100 fmole (F), 50 fmole (chemiluminescence)

KEY WORDS
comparison with o-phthalaldehyde derivatization

REFERENCE
Tod, M.; Legendre, J.-Y.; Chalom, J.; Kouwatli, H.; Poulou, M.; Farinotti, R.; Mahuzier, G. Primary and secondary amine derivatization with luminarins 1 and 2: separation by liquid chromatography with peroxyoxalate chemiluminescence detection, *J.Chromatogr.*, **1992**, *594*, 386–391.

Succinimido 1-Naphthylcarbamate

SAMPLE
Matrix: solutions
Analyte: amines
Sample preparation: 20 μL Amine solution in MeCN + 10 μL 5 mM succinimido 1-naph-thylcarbamate in MeCN, let stand for 2 min, add 10 μL 1% ethanolamine in MeCN, mix, let stand for 2 min, inject a 5-10 μL aliquot. (Prepare succinimido 1-naphthylcarbamate as follows. Add 40 mmoles 1-naphthylamine in 40 mL MeCN dropwise with stirring to 50 mmoles N,N'-disuccinimidyl carbonate in 60 mL MeCN over 3-4 h, stir for 1 h, remove the MeCN by evaporation. Dissolve the residue in 100 mL ethyl acetate, wash with 1 M HCl, wash with 4% sodium bicarbonate, wash with water, dry over anhydrous sodium carbonate, evaporate to obtain succinimido 1-naphthylcarbamate, recrystallize from ben-zene/acetone (Caution! Benzene is a carcinogen!)

HPLC VARIABLES
Column: 150 × 4.6 Develosil ODS-5 (Nomura Chemical)
Mobile phase: MeOH:water 55:45
Flow rate: 1.5
Injection volume: 5-10
Detector: F ex 305 em 378

CHROMATOGRAM
Retention time: 5 (n-amylamine), 4 (n-butylamine), 10.3 (n-heptylamine), 7 (n-hexylamine), 15.3 (n-octylamine), 3.3 (n-propylamine)
Limit of detection: 3-8 pg

REFERENCE
Nimura, N.; Iwaki, K.; Kinoshita, T.; Takeda, K.; Ogura, H. Activated carbamate reagent as derivatizing agent for amino compounds in high-performance liquid chromatography, *Anal.Chem.*, **1986**, *48*, 2372–2375.

Succinimido Phenyl Carbamate

SAMPLE
Matrix: solutions
Analyte: amines
Sample preparation: 20 μL Amine solution in MeCN + 10 μL 5 mM succinimido phenyl carbamate in MeCN, let stand for 2 min, add 10 μL 1% ethanolamine in MeCN, mix, let stand for 2 min, inject a 5-10 μL aliquot. (Prepare succinimido phenyl carbamate as follows. Add 40 mmoles aniline in 40 mL MeCN dropwise with stirring to 50 mmoles N,N'-disuccinimidyl carbonate in 60 mL MeCN over 3-4 h, stir for 1 h, remove the MeCN by evaporation. Dissolve the residue in 100 mL ethyl acetate, wash with 1 M HCl, wash with 4% sodium bicarbonate, wash with water, dry over anhydrous sodium carbonate, evaporate to obtain succinimido phenyl carbamate, recrystallize from benzene/acetone (Caution! Benzene is a carcinogen!).)

HPLC VARIABLES
Column: 150 × 4.6 Develosil ODS-5 (Nomura Chemical)
Mobile phase: MeOH:water 50:50
Flow rate: 1
Injection volume: 5-10
Detector: UV 240

CHROMATOGRAM
Retention time: 5 (n-amylamine), 4 (n-butylamine), 10 (n-heptylamine), 7 (n-hexylamine), 15 (n-octylamine), 3.5 (n-propylamine)
Limit of detection: <1 ng

REFERENCE
Nimura, N.; Iwaki, K.; Kinoshita, T.; Takeda, K.; Ogura, H. Activated carbamate reagent as derivatizing agent for amino compounds in high-performance liquid chromatography, *Anal.Chem.*, **1986**, *48*, 2372–2375.

N-Succinimidyl 3-Ferrocenylpropionate

SAMPLE
Matrix: enzyme incubations
Analyte: putrescine
Sample preparation: 600 μL Enzyme incubation + 2.65 nmole 1,6-diaminohexane + 5 mL 5 mM sodium octanesulfonate in 100 mM acetic acid, mix, add to a Sep-Pak C18 SPE cartridge equilibrated with 5 mM sodium octanesulfonate in 100 mM acetic acid, wash with 6 mL MeCN:5 mM sodium octanesulfonate in 100 mM acetic acid 15:85, elute with 6 mL MeCN:water 2:1. Evaporate the eluate to dryness and reconstitute the residue with 200 μL 100 mM pH 8.5 borate buffer, add 200 μL 1 mg/mL reagent in pyridine, let stand at room temperature for 1 h, inject an aliquot. (Prepare reagent, N-succinimidyl-3-ferrocenylpropionate, as follows. Dissolve 650 mg ferrocenecarboxaldehyde and 310 mg malonic acid in 1 mL dry pyridine, add 2 drops piperidine, heat on a water bath for 1 h, remove the excess pyridine by evaporation under reduced pressure, dissolve the residue in EtOH, mix with water and some HCl, let stand in ice, filter off the precipitate, suck dry, recrystallize from EtOH/water to obtain ferrocenylacrylic acid (mp 183-185°). Hydrogenate ferrocenylacrylic acid in EtOH over 10% PdC for 15 min, recrystallize from benzene/petroleum ether to give 3-ferrocenylpropionic acid (mp 115-118°) (Monatsh.Chem. 1957, 88, 601). Stir 50 mg 3-ferrocenylpropionic acid, 25 mg N-hydroxysuccinimide, and 50 mg 1-(3-dimethylaminopropyl)-3-ethylcarbodiimide hydrochloride in 4 mL dioxane at room temperature for 5 h (Caution! Dioxane is a carcinogen!), extract with ethyl acetate, dry the organic layer over anhydrous sodium sulfate. Evaporate the extract to dryness and chromatograph the residue on a 150 × 12 column of silica gel 60 using ethyl acetate:hexane 1:2. Evaporate the eluate to dryness and recrystallize the residue from ethyl acetate/hexane to give N-succinimidyl-3-ferrocenylpropionate as yellow needles (mp 141-142°).)

HPLC VARIABLES
Column: 150 × 4 5 μm YMC-GEL C8 (Yamamura)
Mobile phase: MeCN:MeOH:water 60:20:50 containing 50 mM sodium perchlorate
Flow rate: 1
Detector: E, Yanagimoto VMD 501, +0.45 V, Ag/AgCl reference electrode

CHROMATOGRAM
Retention time: 10
Internal standard: 1,6-diaminohexane (14)
Limit of detection: 3 pmole

KEY WORDS
SPE

REFERENCE
Shimada, K.; Oe, T.; Tanaka, M.; Nambara, T. Sensitive ferrocene reagents for derivatization of amines for high-performance liquid chromatography with electrochemical detection, *J.Chromatogr.*, **1989**, *487*, 247–255.

(S)-N-Succinimidyl α-Methoxyphenylacetate

SAMPLE
Matrix: bulk
Analyte: 1-(phenylthio)-2-aminopropane
Sample preparation: Dissolve 1 mg amine in 1 mL THF and add 1 mL water, add 2-10 mg reagent, heat on a steam bath for 15 min, inject a 10 µL aliquot. (Synthesize the reagent, (S)-N-succinimidyl α-methoxyphenylacetate, as follows. Add 5.46 g dicyclohexylcarbodiimide in 30 mL dry THF dropwise to 4.0 g (S)-(+)-α-methoxyphenylacetic acid (2-phenyl-2-(S)-methoxyacetic acid) and 2.77 g N-hydroxysuccinimide in 20 mL dry THF stirred at room temperature under an argon atmosphere, stir for 3 h, filter, evaporate to dryness, dissolve the residue in 50 mL ethyl acetate. Wash the organic layer five times with 10 mL portions of water, wash five times with 10 mL portions of brine, quickly wash twice with 10 mL portions of ice-cold saturated sodium bicarbonate, evaporate to dryness under reduced pressure, recrystallize from hot EtOH to give (S)-N-succinimidyl α-methoxyphenylacetate (mp 95.7-96.3°) ($[\alpha]^{25}$ = +125° (c = 3, EtOH)).)

HPLC VARIABLES
Column: 250 × 4.6 5 µm silica gel (Rainin)
Mobile phase: Hexane:ethyl acetate 80:20
Flow rate: 1
Injection volume: 10
Detector: UV 254

CHROMATOGRAM
Retention time: 12 (S), 19 (R)

KEY WORDS
normal phase; chiral

REFERENCE
Husain, P.A.; Colbert, J.E.; Sirimanne, S.R.; VanDerveer, D.G.; Herman, H.H.; May, S.W. *N*-succinimidyl methoxyphenylacetic acid ester, an amine-directed chiral derivatizing reagent suitable for enzymatic scale resolutions, *Anal.Biochem.*, **1989**, *178*, 177–183.

SAMPLE
Matrix: bulk
Analyte: amines
Sample preparation: Add 30 mmole amine hydrochloride to 800 µL THF and 200 µL water, add 15.6 mmole (S)-N-succinimidyl α-methoxyphenylacetate, add 4.1 mL triethylamine, heat on a steam bath for 15 min, add 500 µL saturated sodium bicarbonate. (Synthesis of (S)-N-succinimidyl α-methoxyphenylacetate is as described above.)

HPLC VARIABLES
Column: 250 × 4.6 5 µm silica (Rainin)
Mobile phase: Hexane:ethyl acetate 75:25
Flow rate: 1
Detector: UV 254

CHROMATOGRAM
Retention time: k' 2.3 ((S)-1-methylpropylamine), k' 3.0 ((R)-1-methylpropylamine)

KEY WORDS

chiral; normal phase; numerous other examples

REFERENCE

Husain, P.A.; Debnath, J.; May, S.W. HPLC-based method for determination of absolute configuration of α-chiral amines, *Anal.Chem.*, **1993**, *65*, 1456–1461.

N-Succinimidyl Propionate

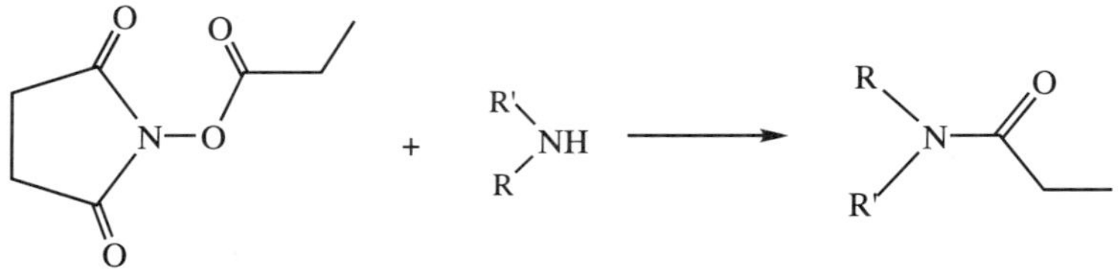

SAMPLE

Matrix: CSF

Analyte: serotonin

Sample preparation: Adjust pH of CSF to 11. 1 mL CSF + 60 μmoles N-succinimidylpropionate, heat at 70° for 2 h, add excess glycine, extract with ethyl acetate, evaporate the organic layer to dryness, reconstitute, inject an aliquot. (Prepare N-succinimidylpropionate as follows. Dissolve 1 mmole N-hydroxysuccinimide, 1 mmole triphenylphosphine, and 1 mmole propionic acid in 10 mL THF, add 1.1 mmole diethyl azodicarboxylate, let stand at room temperature overnight, evaporate to dryness under reduced pressure, dissolve the residue in 50 mL benzene (Caution! Benzene is a carcinogen!), wash with 10 mL aqueous sodium bicarbonate, wash twice with 10 mL portions of water, dry over anhydrous magnesium sulfate, evaporate to dryness, chromatograph on a silica gel column with benzene, recrystallize from n-hexane/benzene to obtain N-succinimidylpropionate (cf Synthesis 1977, 277).)

HPLC VARIABLES

Column: 100 mm long 3 μm Axxion C18

Mobile phase: MeCN:MeOH:pH 4.8 sodium acetate 6.3:2.3:91.4

Flow rate: 1.5

Detector: E, ESA 5011A, +0.30 V

CHROMATOGRAM

Retention time: 3 (N-acetylserotonin), 7 (serotonin)

REFERENCE

Jacobson, K.A.; Marshall, T.; Mine, K.; Kirk, K.L.; Linnoila, M. Electrochemical detection of biogenic amines following acylation by N-hydroxysuccinimide esters, *FEBS Lett.*, **1985**, *188*, 307–311.

N-Succinimidyl Tetrathiafulvalene-2-carboxylate

SAMPLE
Matrix: solutions
Analyte: β-phenethylamine
Sample preparation: Prepare a 10 μM solution of β-phenethylamine in MeCN:pyridine: water 50:10:40 containing 500 μM N-succinimidyl tetrathiafulvalene-2-carboxylate, let stand at room temperature for 1 h, inject an aliquot. (Prepare N-succinimidyl tetrathiafulvalene-2-carboxylate as follows. Stir 1.02 g tetrathiafulvalene in 100 mL anhydrous ether under argon or nitrogen at -60 to -70°, add a solution of 535 mg lithium diisopropylamide in 20 mL ether over 15 min (prepare lithium diisopropylamide by adding n-butyllithium to diisopropylamine in anhydrous ether solution), stir at -70° for 15 min, add this slurry at -70° to a flask containing anhydrous ether and solid carbon dioxide, warm to 15°, evaporate to dryness, add 25 mL water, filter. Acidify the filtrate with 1 M HCl, collect the red precipitate, recrystallize from benzene (Caution! Benzene is a carcinogen!) to give tetrathiafulvalene-2-carboxylic acid as red needles (mp 182-184°) (J. Org. Chem. 1979, 44, 1476). React tetrathiafulvalene-2-carboxylic acid with N-hydroxysuccinimide in the presence of dicyclohexylcarbodiimide, recrystallize from benzene to give N-succinimidyl tetrathiafulvalene-2-carboxylate as dark-purple crystals (mp 172.5-173.5°).)

HPLC VARIABLES
Column: 150 × 4.6 5 μm Chemcosorb 5-ODS-H
Mobile phase: MeCN:water 65:35 containing 50 mM sodium perchlorate
Flow rate: 1
Detector: E, EICOM ECD-100, glassy carbon working electrode +700 mV, Ag/AgCl reference electrode

CHROMATOGRAM
Retention time: k' 5
Limit of detection: 21 fmole

REFERENCE
Shimada, K.; Oe, T. Novel derivatization reagent with tetrathiafulvalene as an electrophore for precolumn labeling of amines in high-performance liquid chromatography, *Chem.Pharm.Bull.*, **1991**, *39*, 1897–1898.

Sulfosuccinimidyl-3-(4-hydroxyphenyl)propionate

SAMPLE
Matrix: tissue
Analyte: histamine
Sample preparation: Prepare SPE (A) column by washing 100-200 mesh Dowex 50 W with excess 2 M HCl, with water, with 2 M NaOH, and with water. Equilibrate the resin with 200 mM pH 6.5 sodium phosphate buffer and pack in a 16 × 5 glass column. Prepare SPE (B) column by washing cellulose-phosphate fibrous cation-exchanger (Sigma) in a 13 × 17 glass column with excess 100 mM NaOH, with water, with 100 mM HCl, and with water until the pH reaches 5-6. Homogenize (glass homogenizer) rat brain hypothalamus with 500 μL ice-cold 3% perchloric acid, rinse homogenizer 3 times with 500 μL portions of 3% perchloric acid. Combine the homogenate and rinses, add IS, centrifuge at 4° at 10000 g for 30 min, add the supernatant to SPE column (A), wash with 5 mL water, wash with 4 mL 2 M HCl, elute with 2.5 mL 3.5 M HCl. Evaporate the eluate to dryness, reconstitute the residue in 800 μL water, add 150 μL buffer, add 100 μL 20 mM sulfo-succinimidyl-3-(4-hydroxyphenyl)propionate (sulfo B-H, Pierce) in water, vortex for 30 s, adjust pH to 5.5-6.0 with 100 mM HCl, add to SPE column (B), wash with 5 mL water, wash with 6 mL 1 mM HCl, elute with 2.5 mL 100 mM HCl. Discard the first 500 μL eluate, evaporate the next 200 μL to dryness, reconstitute with 1 volume water, add 9 volumes water to a total volume of 0.11-1 mL, inject a 100 μL aliquot. (Prepare buffer by mixing 100 mM sodium carbonate and 100 mM sodium bicarbonate in a 10:1 ratio.)

HPLC VARIABLES
Guard column: μBondapak C18/Corasil
Column: 250 × 4.6 Ultrasphere ODS
Mobile phase: MeCN:140 mM sodium acetate 73:17 containing 3.89 M (sic) 1-octanesulfonic acid and 56 mg/L (?) EDTA, adjusted to pH 3.48 with glacial acetic acid
Flow rate: 1
Injection volume: 100
Detector: E, ESA Coulochem 5100A, Model 5011 analytical cell, cell 1 0.47 V, cell 2 (monitored) 0.56 V, oxidative screen mode

CHROMATOGRAM
Retention time: 8.4
Internal standard: N^α-methylhistamine (11.6)
Limit of detection: 0.1 pmole

OTHER SUBSTANCES
Extracted: N^{tau}-methylhistamine
Simultaneously analyzed: histidine, spermidine

KEY WORDS
rat; brain; SPE

REFERENCE

Mine, K.; Jacobson, K.A.; Kirk, K.L.; Kitajima, Y.; Linnoila, M. Simultaneous determination of histamine and N tau-methylhistamine with high-performance liquid chromatography using electrochemical detection, *Anal.Biochem.*, **1986**, *152*, 127–135.

SULFONIC ACID

Sodium 1,2-Naphthoquinone-4-sulfonate

RELATED REFERENCES

Falco, P.C.; Legua, C.M.; Hernandez, R.H.; Cabeza, A.S. Improved amphetamine and methamphetamine determination in urine by normal-phase high-performance liquid chromatography with sodium 1,2-naphthoquinone 4-sulfonate as derivatizing agent and solid-phase extraction for sample clean-up. *J.Chromatogr.B* **1995**, *663*, 235-245.

Smith, J.R.L.; Smart, A.U.; Hancock, F.E.; Twigg, M.V. High-performance liquid chromatographic determination of low levels of primary and secondary amines in aqueous media via derivatization with 1,2-naphthoquinone-4-sulfonate. *J.Chromatogr.* **1989**, *483*, 341-348.

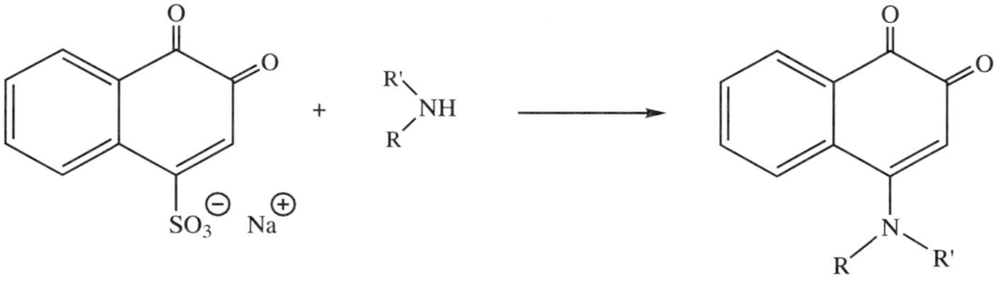

SAMPLE

Matrix: blood, urine

Analyte: amphetamines

Sample preparation: Adjust pH of 10 mL plasma or 20 mL urine to 11.4 with 5 M NaOH, add to a column containing 1.5 g Amberlite XAD-2, wash with 10 mL water, elute with 20 (plasma) or 40 (urine) mL chloroform:isopropanol 75:25, add 100 μL 6 M HCl in EtOH to the eluate, evaporate to dryness under reduced pressure, reconstitute with 1 mL 8% sodium bicarbonate, add 1 mL 0.5% sodium 1,2-naphthoquinone-4-sulfonate, heat at 70° for 20 min, add an equal volume of chloroform, vortex for 1 min, inject a 50 μL aliquot of the organic layer.

HPLC VARIABLES

Column: 150 × 5 Partisil 5

Mobile phase: Hexane:chloroform:ethyl acetate:EtOH 50:25:35:1

Column temperature: 20

Flow rate: 2.5

Injection volume: 50

Detector: UV 248

CHROMATOGRAM

Retention time: 3 (methamphetamine), 5 (amphetamine), 10 (norephedrine), 15.5 (hydroxyamphetamine)

Internal standard: phenylethylamine (6)

Limit of detection: 2 ng

KEY WORDS

plasma; normal phase; comparison with other derivatization reagents and with ion-pair chromatography; SPE

REFERENCE

Farrell, B.M.; Jefferies, T.M. An investigation of high-performance liquid chromatographic methods for the analysis of amphetamines, *J.Chromatogr.*, **1983**, *272*, 111–128.

SAMPLE

Matrix: formulations

Analyte: aminoglutethimide

Sample preparation: Weigh out amount of finely powdered tablet corresponding to 20.83 mg aminoglutethimide, add 1 mL EtOH, add 5 mL water, sonicate for 30 min, make up to 10 mL with water, filter, discard the first 2 mL filtrate. Remove a 40 μL aliquot of the filtrate and add it to 500 μL pH 4.5 phosphate buffer, add 300 μL 2.6 mg/mL 1,2-naphthoquinone-4-sulfonic acid in water (prepare fresh each day), heat at 60° for 15 min, cool, add 4 mL n-butanol:ethyl acetate 2:1, shake vigorously for 30 s, centrifuge at 1750 rpm for 10 min. Remove a 100 μL aliquot of the organic layer and add it to 250 μL MeOH, inject a 20 μL aliquot.

HPLC VARIABLES

Column: 300 × 3.9 10 μm μBondapak C18

Mobile phase: MeOH:water 90:10

Flow rate: 0.5

Injection volume: 20

Detector: UV 254

CHROMATOGRAM

Retention time: 5.53

Limit of detection: 50 ng

KEY WORDS

tablets

REFERENCE

Ozkirimli, S.; Sevingil, M. High pressure liquid chromatographic determination of aminoglutethimide, *Acta Pharm.Turc.*, **1989**, *31*, 57–60.

SAMPLE

Matrix: solutions

Analyte: amphetamine

Sample preparation: 1 mL Solution + 500 μL 0.5% sodium 1,2-naphthoquinone-4-sulfonate in water + 500 μL buffer, let stand for 10 min, extract three times with 2 mL aliquots of n-hexane:ethyl acetate 50:50. Combine the organic layers and evaporate them to dryness at 80°, reconstitute with 2 mL MeCN:water 50:50, inject a 50 μL aliquot. (Buffer was 4% sodium bicarbonate adjusted to pH 10 with 10% NaOH.)

HPLC VARIABLES

Column: 250 × 4 5 μm Hypersil ODS C18

Mobile phase: Gradient. MeCN:0.5% propylamine hydrochloride in water from 40:60 to 50:50 over 2.5 min, to 70:30 over 1 min, maintain at 70:30 for 4.5 min.

Flow rate: 1

Injection volume: 50

Detector: UV 280

CHROMATOGRAM

Retention time: 5 (amphetamine), 7 (methamphetamine)

REFERENCE

Herráez-Hernández, R.; Campins-Falcó, P.; Sevillano-Cabeza, A. On-line derivatization into precolumns for the determination of drugs by liquid chromatography and column switching: Determination of amphetamines in urine, *Anal.Chem.*, **1996**, *68*, 734–739.

SAMPLE

Matrix: urine

Analyte: amphetamine and methamphetamine

Sample preparation: Condition a 100 mg Bond-Elut C18 SPE cartridge with 500 μL MeOH and 500 μL water. Adjust pH of urine to 10, centrifuge at 1500 g. 2 mL Supernatant + 100 μL 75 μg/mL β-phenylethylamine hydrochloride in water, add to the SPE cartridge, wash with 2.5 mL water, elute with 2 mL MeOH, evaporate the eluate to dryness. reconstitute in water, add 500 μL 8% sodium bicarbonate, add 500 μL 0.5% sodium 1,2-naphthoquinone-4-sulfonate, make up to 1.5 mL with water, heat at 70° for 20 min, cool, add an equal volume of chloroform, shake for 2 min, centrifuge at 1500 g for 5 min. Remove the organic layer and dry it over anhydrous sodium sulfate, filter (0.45 μm), inject a 25 μL aliquot of the filtrate.

HPLC VARIABLES
Column: 125 × 4 5 μm LiChrospher Si-60
Mobile phase: EtOH:chloroform:ethyl acetate:n-hexane 1:22:32:45
Flow rate: 2
Injection volume: 25
Detector: UV 280

CHROMATOGRAM
Retention time: 2.6 (methamphetamine), 3.7 (amphetamine)
Internal standard: β-phenylethylamine hydrochloride (4.9)

KEY WORDS
SPE; normal phase

REFERENCE
Campins Falcó, P.; Molins Legua, C.; Herráez Hernandez, R.; Sevillano Cabeza, A. Improved amphetamine and methamphetamine determination in urine by normal-phase high-performance liquid chromatography with sodium 1,2-naphthoquinone 4-sulphonate as derivatizing agent and solid-phase extraction for sample clean-up, *J.Chromatogr.B*, **1995**, *663*, 235–245.

SAMPLE
Matrix: urine
Analyte: amphetamine and methamphetamine
Sample preparation: Condition an Extra-Sep C18 SPE cartridge (Teknokroma) with 1 mL MeOH and 1 mL buffer. Adjust pH of 2 mL urine to ca. 10 with 100 μL concentrated ammonium hydroxide, add 5 μg β-phenylethylamine, add to the SPE cartridge, wash with 5 mL water, wash with 1 mL MeCN, elute with 2 mL MeOH. Add 100 μL EtOH:concentrated HCl 6:1 to the eluate, evaporate to dryness. Reconstitute with 1 mL buffer and 1 mL 0.5% sodium 1,2-naphthoquinone-4-sulfonate, let stand at room temperature for 10 min, add 2 mL n-hexane:ethyl acetate 50:50, shake for 2 min, centrifuge at 1500 g for 5 min. Remove the organic layer and evaporate it to dryness, reconstitute the residue in 500 μL MeCN:water 50:50, inject a 50 μL aliquot. (Buffer was 1% aqueous sodium bicarbonate adjusted to pH 10 with 5 M NaOH.)

HPLC VARIABLES
Column: 250 × 4 5 μm Hypersil ODS-C18
Mobile phase: Gradient. MeCN:0.5% propylamine in water from 40:60 to 50:50 over 2.5 min, to 70:30 over 1 min, maintain at 70:30.
Flow rate: 1
Injection volume: 50
Detector: UV 280

CHROMATOGRAM
Retention time: 4.7 (amphetamine), 6.3 (methamphetamine)
Internal standard: β-phenylethylamine (4.1)
Limit of quantitation: 10 ng/mL
Limit of detection: 4 ng/mL

KEY WORDS
SPE

REFERENCE
Molíns Legua, C.; Campíns Falcó, P.; Sevillano Cabeza, A. Amphetamine and methamphetamine determination in urine by reversed-phase high-performance liquid chromatography with sodium 1,2-naphthoquinone 4-sulfonate as derivatizing agent and solid-phase extraction for sample clean-up, *J.Chromatogr.B*, **1995**, *672*, 81–88.

SAMPLE
Matrix: urine

Analyte: amphetamine

Sample preparation: Inject 50 μL urine on to column A and elute to waste with mobile phase A, after 2 min inject a mixture of 25 μL 0.5% sodium 1,2-naphthoquinone-4-sulfonate in water and 25 μL buffer on to column A, stop the flow of mobile phase A, after 10 min start pump A, after 5 min backflush the contents of column A on to column B with mobile phase B and start the gradient, monitor the effluent from column B. After each run flush column A with ethyl acetate for 1 min, n-hexane for 1 min, and ethyl acetate for 1 min, re-equilibrate with mobile phase A. (Buffer was 4% sodium bicarbonate adjusted to pH 10 with 10% NaOH.)

HPLC VARIABLES
Column: A 20 × 2.1 30 μm Hypersil ODS-C18; B 250 × 4 5 μm Hypersil ODS C18

Mobile phase: A water; B Gradient. MeCN:0.5% propylamine hydrochloride in water from 40:60 to 50:50 over 2.5 min, to 70:30 over 1 min, maintain at 70:30 for 4.5 min.

Flow rate: 1

Injection volume: 50

Detector: UV 280

CHROMATOGRAM
Retention time: 5 (amphetamine), 7 (methamphetamine)

Limit of detection: 25 ng/mL

KEY WORDS
column-switching; on-column derivatization

REFERENCE
Herráez-Hernández, R.; Campins-Falcó, P.; Sevillano-Cabeza, A. On-line derivatization into precolumns for the determination of drugs by liquid chromatography and column switching: Determination of amphetamines in urine, *Anal.Chem.*, **1996**, *68*, 734–739.

SAMPLE
Matrix: urine

Analyte: amphetamine and methamphetamine

Sample preparation: Condition a 200 mg Bond Elut C18 SPE cartridge with 1 mL MeOH and 1 mL 1% pH 10 sodium bicarbonate buffer. 2 mL Urine + 400 μL 8% pH 10 sodium bicarbonate buffer, mix, centrifuge at 1500 g for 2 min, add a 2 mL aliquot of the supernatant to the SPE cartridge, wash with 3 mL water, pass 500 μL 2% sodium 1,2-naphthoquinone-4-sulfonate through the cartridge, pass 500 μL 1% pH 10 sodium bicarbonate buffer through the cartridge, let stand at room temperature for 15 min, wash with 3 mL water, elute with 1 mL MeCN:water 50:50, inject a 20 μL aliquot of the eluate.

HPLC VARIABLES
Column: 250 × 4 5 μm Hypersil ODS

Mobile phase: Gradient. MeCN:buffer from 40:60 to 50:50 over 2.5 min, to 70:30 over 0.5 min, maintain at 70:30 for 1.5 min, to 85:15 over 1 min, maintain at 85:15 for 1.5 min. (Buffer was 5 mL/L propylamine in water.)

Flow rate: 1

Injection volume: 20

Detector: UV 280

CHROMATOGRAM
Retention time: 4.1 (amphetamine), 5.6 (methamphetamine)

Internal standard: β-phenylethylamine (3.6)

Limit of detection: 100-400 ng/mL

KEY WORDS
SPE

REFERENCE
Campíns-Falcó, P.; Sevillano-Cabeza, A.; Molíns-Legua, C.; Kohlmann, M. Amphetamine and methamphetamine determination in urine by reversed-phase high-performance liquid chromatography with

simultaneous sample clean-up and derivatization with 1,2-naphthoquinone 4-sulphonate on solid-phase cartridges, *J.Chromatogr.B*, **1996**, *687*, 239–246.

2,4,6-Trinitrobenzenesulfonic Acid

RELATED REFERENCES
Casals, I.; Reixach, M.; Amat, J.; Fuentes, M.; Serra-Majem, L. Quantification of cyclamate and cyclo-hexylamine in urine samples using high-performance liquid chromatography with trinitrobenzene-sulfonic acid pre-column derivatization. *J.Chromatogr.A* **1996**, *750*, 397-402.

Caudill, W.L.; Wightman, R.M. Trinitrobenzenesulfonic acid: a chromophore, electrophore and precol-umn derivatizing agent for high performance liquid chromatography of alkylamines. *Anal.Chim.Acta* **1982**, *141*, 269-278.

Tomarelli, R.M.; Yuhas, R.J.; Fisher, A.; Weaber, J.R. An HPLC method for the determination of reactive (available) lysine in milk and infant formulas. *J.Agric.Food Chem.* **1985**, *33*, 316-318

SAMPLE
Matrix: blood

Analyte: tobramycin

Sample preparation: Condition a Bond-Elut C18 SPE cartridge with 2 column volumes of MeOH and 2 column volumes of water. 50 μL Serum + 25 μL 2 M pH 10.3 Tris buffer + 100 μL 10 μg/mL sisomicin in MeCN, vortex, centrifuge at 15000 g for 1 min. Remove the supernatant and add it to 30 μL 250 mg/mL 2,4,6-trinitrobenzenesulfonic acid in MeCN, vortex, heat at 70° for 30 min. Add 700 μL wash solution then 200 μL sample to the SPE cartridge, wash with 3 column volumes of wash solution, elute with 300 μL MeCN, inject a 50 μL aliquot of the eluate. (Prepare wash solution by adding 10 mL 1 M K_2HPO_4 to 90 mL water, add 100 mL MeOH, adjust pH to 8.5 with phosphoric acid.)

HPLC VARIABLES
Column: 250 × 4.6 Ultrasphere octyl

Mobile phase: MeCN:buffer 70:30 (Buffer was 6.8 g/L KH_2PO_4 adjusted to pH 3.5 with phosphoric acid.)

Column temperature: 50

Flow rate: 3

Injection volume: 50

Detector: UV 340

CHROMATOGRAM
Retention time: 3

Internal standard: sisomicin (4)

Limit of detection: 1200 ng/mL

OTHER SUBSTANCES
Simultaneously analyzed: gentamicin

Non-interfering: acetaminophen, acetazolamide, N-acetylprocainamide, amikacin, amo-barbital, ampicillin, amitriptyline, caffeine, cefamandole, cefoxime, cefoxitin, cephalothin, clindamycin, chloramphenicol, chlordiazepoxide, diazepam, erythromycin, ethosuximide, glutethimide, imipramine, kanamycin, methaqualone, moxalactam, nafcillin, nitrofuran-toin, penicillin G, pentobarbital, phenobarbital, phenytoin, primidone, procainamide, quinidine, salicylic acid, secobarbital, tetracycline, theophylline, vancomycin

KEY WORDS
serum; SPE

REFERENCE

Kabra, P.M.; Bhatnagar, P.K.; Nelson, M.A.; Wall, J.H.; Marton, L.J. Liquid-chromatographic determination of tobramycin in serum with spectrophotometric detection, *Clin.Chem.*, **1983**, *29*, 672–674.

SAMPLE

Matrix: blood
Analyte: amikacin
Sample preparation: Condition a Bond-Elut C18 SPE cartridge with 2 column volumes of MeOH and 2 column volumes of water. 50 μL Serum + 25 μL Tris buffer + 100 μL 16 μg/mL kanamycin in MeCN, vortex, centrifuge at 15000 g for 1 min. Remove the supernatant and add it to 30 μL 250 mg/mL 2,4,6-trinitrobenzenesulfonic acid in MeCN, vortex, heat at 70° for 30 min. Add 700 μL wash solution then 250 μL sample solution to the SPE cartridge, wash with 3 column volumes of wash solution, elute with 300 μL MeCN, mix the eluate, inject a 50 μL aliquot of the eluate. (Tris buffer was 2 M pH 10.3 and was prepared by dissolving 24.2 g Trizma base in 100 mL water. Wash solution (10 mM) was 1.82 g potassium hydrogen phosphate in 1 L water adjusted to pH 8.6 with phosphoric acid.)

HPLC VARIABLES

Column: 250×4.6 5 μm Ultrasphere octyl
Mobile phase: MeCN:20 mM buffer 52:48 (Buffer was 2.68 g KH_2PO_4 in 1 L water adjusted to pH 3.0 with phosphoric acid.)
Column temperature: 50
Flow rate: 2
Injection volume: 50
Detector: UV 340

CHROMATOGRAM

Retention time: 8
Internal standard: kanamycin (12)
Limit of detection: 500 ng/mL

OTHER SUBSTANCES

Non-interfering: acetaminophen, acetazolamide, N-acetylprocainamide, amobarbital, ampicillin, amitriptyline, caffeine, cefamandole, cefoxime, cefoxitin, cephalothin, clindamycin, chloramphenicol, chlordiazepoxide, diazepam, erythromycin, ethosuximide, gentamicin, nitrofurantoin, penicillin G, pentobarbital, phenobarbital, phenytoin, primidone, procainamide, quinidine, salicylic acid, secobarbital, tetracycline, theophylline, tobramycin, vancomycin

KEY WORDS

serum

REFERENCE

Kabra, P.M.; Bhatnager, P.K.; Nelson, M.A. Liquid chromatographic determination of amikacin in serum with spectrophotometric detection, *J.Chromatogr.*, **1984**, *307*, 224–229.

SAMPLE

Matrix: blood
Analyte: copovithane
Sample preparation: Cool 2 mL plasma on ice for 5 min, add 200 μL 10 M perchloric acid, vortex vigorously, let stand on ice for 5 min, centrifuge at 17000 g for 15 min. Remove the supernatant and add it to 200 μL 10 M KOH, cool on ice for 5 min, centrifuge for 3 min. Remove the supernatant, add 2 mL hot (85°) saturated NaCl solution, add 3 mL chloroform, vortex vigorously, centrifuge at 17000 g for 15 min, repeat extract twice more. Combine the organic layers and evaporate them to dryness under a stream of nitrogen, reconstitute the residue in 1 mL 5 M HCl, heat at 160° for 16 h, cool on ice, add 1 mL 5 M NaOH, add 3 mL 1 M pH 8.0 sodium bicarbonate buffer, add 750 μL 0.5% trinitrobenzenesulfonic acid in acetone, let stand in the dark for 2.5 h, extract three times with 3 mL portions of ethyl acetate. Combine the organic layers and evaporate them to dryness under a stream of nitrogen, reconstitute the residue in 250 μL 200 mM pH 6.4 Na_2HPO_4

in MeCN, inject an aliquot. (Copovithane is hydrolyzed to methylamine and the methylamine is derivatized.)

HPLC VARIABLES
Column: 300 × 3.9 10 μm C18 (Waters)
Mobile phase: MeCN:water 30:70
Flow rate: 2
Detector: UV 340

CHROMATOGRAM
Retention time: 15
Limit of detection: 15 μg/mL

KEY WORDS
plasma; pharmacokinetics

REFERENCE
Rosenblum, M.G.; Hortobagyi, G.N.; Wingender, W.; Hersh, E.M. Analysis of the antitumor agent Bay i 7433 (copovithane) in plasma and urine by high performance liquid chromatography, *J.Liq.Chromatogr.*, **1984**, *7*, 159–166.

SAMPLE
Matrix: blood
Analyte: gabapentin
Sample preparation: Condition a 100 mg Bond Elut C18 SPE cartridge with 1 mL MeOH and 1 mL buffer, do not allow to go dry. Condition an Empore C18 SPE membrane by adding 500 μL MeOH, force through three drops, discard MeOH remaining in reservoir, add 500 μL water, force through three drops, discard water remaining in reservoir. Add 200 μL 3 μg/mL IS in buffer to the SPE cartridge, add 200 μL serum and force through at 1 drop/s, add 200 μL buffer, force all liquid through, elute with 500 μL MeOH. Add 100 μL saturated sodium tetraborate solution and 50 μL 5% 2,4,6-trinitrobenzenesulfonic acid in water to the eluate, mix, heat at 50° for 10 min, add 500 μL 250 mM acetic acid, centrifuge at 12500 g for 2 min, add the supernatant to the SPE membrane, force through using a syringe or by centrifuging at 100-120 g for 5 min, wash with 500 μL MeCN:water 20:80, elute with 75 μL MeCN then 125 μL water, mix the eluates, inject a 50 μL aliquot. (Buffer was saturated sodium tetraborate solution diluted with three volumes of water.)

HPLC VARIABLES
Guard column: 20 × 2 30 μm Permaphase ETH (DuPont)
Column: 250 × 4.6 Ultrasphere C18
Mobile phase: MeCN:water:acetic acid:n-butylamine 52:48:0.1:0.01 (pH should not exceed 4.5) (Connect a 150 × 4.6 37-53 μm silica (Whatman) column between pump and injector.)
Column temperature: 50
Flow rate: 1.2
Injection volume: 50
Detector: UV 340

CHROMATOGRAM
Retention time: 10
Internal standard: 1-(aminomethyl)cycloheptaneacetic acid (13)
Limit of detection: 50 ng/mL

OTHER SUBSTANCES
Non-interfering: acetaminophen, N-acetylprocainamide, amikacin, caffeine, carbamazepine epoxide, carbamazepine, chlordiazepoxide, demoxepam, desalkylflurazepam, desmethylchlordiazepoxide, desmethyldiazepam, diazepam, disopyramide, ethosuximide, flurazepam, gentamicin, lidocaine, phenobarbital, phenytoin, primidone, procainamide, quinidine, theophylline, tobramycin, valproic acid, vancomycin

KEY WORDS
SPE; pharmacokinetics

REFERENCE

Lensmeyer, G.L.; Kempf, T.; Gidal, B.E.; Wiebe, D.A. Optimized method for determination of gabapentin in serum by high-performance liquid chromatography, *Ther.Drug Monit.*, **1995**, *17*, 251–258.

SAMPLE

Matrix: blood

Analyte: gabapentin

Sample preparation: 500 μL Plasma + 50 μL IS solution + 200 μL 1.2 M perchloric acid, vortex, centrifuge at 3000 rpm for 5 min. Remove the supernatant and add it to 50 μL 5% 2,4,6-trinitrobenzenesulfonic acid in water, add 50 μL 50% NaOH, vortex, let stand at room temperature for 30 min, add 200 μL 6 M HCl, add 100 μL saturated NaCl, add 6 mL cyclohexane, shake for 10 min, centrifuge at 3000 rpm for 10 min. Remove the organic layer and evaporate it to dryness under a stream of nitrogen at 55°, reconstitute the residue in 200 μL EtOH:mobile phase 10:90, inject a 75 μL aliquot.

HPLC VARIABLES

Column: 5 μm Spherisorb ODSIII

Mobile phase: MeCN:100 mM pH 4 ammonium acetate 54:46

Column temperature: 40

Flow rate: 1

Injection volume: 75

Detector: UV 350

CHROMATOGRAM

Retention time: 5.2

Internal standard: 1-(aminomethyl)heptaneacetic acid (Parke-Davis) (13.3)

Limit of detection: 25 ng/mL

KEY WORDS

plasma; dog; pharmacokinetics

REFERENCE

Stevenson, C.M.; Radulovic, L.L.; Bockbrader, H.N.; Fleisher, D. Contrasting nutrient effects on the plasma levels of an amino acid-like antiepileptic agent from jejeunal administration in dogs, *J.Pharm.Sci.*, **1997**, *86*, 953–957.

SAMPLE

Matrix: blood, milk

Analyte: (S)-3-(aminomethyl)-5-methylhexanoic acid

Sample preparation: 100 μL Plasma or milk + 100 μL water + 100 μL 20.3 μg/ mL IS in water + 200 μL 2 M perchloric acid, vortex vigorously for 30 s, centrifuge at 11000 g for 10 min. Remove the aqueous phase and add it to 75 μL 5% 2,4,6-trinitrobenzenesulfonic acid in water and 25 μL 50% NaOH, vortex, let stand for 30 min, add 100 μL 6 M HCl, add 100 μL saturated NaCl solution, add 4 mL cyclohexane, shake horizontally for 10 min, centrifuge at 1600 g for 10 min. Remove the organic layer and evaporate it to dryness under reduced pressure at 50°, reconstitute the residue in 200 μL mobile phase:EtOH 90:10, inject a 25 μL aliquot.

HPLC VARIABLES

Column: 250 × 4.6 5 μm Spherisorb ODSII

Mobile phase: MeCN:100 mM pH 4.0 ammonium acetate buffer 57:43

Column temperature: 40

Flow rate: 1

Detector: UV 350

CHROMATOGRAM

Retention time: 5.8

Internal standard: 1-(aminomethyl)cycloheptaneacetic acid (PD 403609) (9.7)

Limit of quantitation: 1 μg/mL

OTHER SUBSTANCES

Interfering: 25

KEY WORDS
plasma; rat; pharmacokinetics

REFERENCE
Windsor, B.L.; Radulovic, L.L. Measurement of a new anticonvulsant, (S)-3-(aminomethyl)-5-methyl-hexanoic acid, in plasma and milk by high-performance liquid chromatography, *J.Chromatogr.B*, **1995**, *674*, 143–148.

SAMPLE
Matrix: blood, urine
Analyte: gabapentin
Sample preparation: Plasma. 500 μL Plasma + 10 μL IS in water + 5 drops 2 M perchloric acid, vortex vigorously for a few s, centrifuge at 15000 g for 2 min. Remove the supernatant and add it to 500 μL 1 M sodium bicarbonate, add 50 μL 2 M 2,4,6-trinitro-benzenesulfonic acid in water, adjust pH to 8.5 with 100 mM NaOH, let stand for 30 min, add 2 drops of 25% HCl, add 3 mL toluene, shake for 10 min, centrifuge at 5000 g for 2 min. Remove the upper organic layer and evaporate it to dryness under reduced pressure at 40°, reconstitute the residue in 100 μL 200 mM pH 8.5 sodium borate buffer, wash with 1 mL cyclohexane:toluene 90:10 by vortexing for 1 min, inject a 10-50 μL aliquot of the aqueous phase. Urine. 10-100 μL Urine + 10 μL 200μg/mL IS in water, add 500 μL 1 M sodium bicarbonate, add 50 μL 2 M 2,4,6-trinitrobenzenesulfonic acid in water, adjust pH to 8.5 with 100 mM NaOH, let stand for 30 min, add 2 drops of 25% HCl, add 3 mL toluene, shake for 10 min, centrifuge at 5000 g for 2 min. Remove the upper organic layer and evaporate it to dryness under reduced pressure at 40°, reconstitute the residue in 100 μL 200 mM pH 8.5 sodium borate buffer, wash with 1 mL cyclohexane:toluene 90:10 by vortexing for 1 min, inject a 10-50 μL aliquot of the aqueous phase.

HPLC VARIABLES
Column: 250 × 4 10 μm LiChrosorb RP-18
Mobile phase: MeCN:0.5% acetic acid 58:42
Flow rate: 1
Injection volume: 10-50
Detector: UV

CHROMATOGRAM
Retention time: 10.3
Internal standard: 1-(aminomethyl)cycloheptaneacetic acid (13.2)
Limit of quantitation: 20 ng/mL
Limit of detection: 10 ng/mL

KEY WORDS
plasma; pharmacokinetics

REFERENCE
Hengy, H.; Kölle, E.U. Determination of gabapentin in plasma and urine by high-performance liquid chromatography and pre-column labelling for ultraviolet detection, *J.Chromatogr.*, **1985**, *341*, 473–478.

SAMPLE
Matrix: bulk, formulations
Analyte: tobramycin
Sample preparation: Dilute 500 μL injection to 25 mL with Sorensen's phosphate buffer, dilute a 1 mL aliquot to 100 mL with Sorensen's phosphate buffer. Dissolve tobramycin-containing polymethylmethacrylate beads in 2 mL chloroform, extract 3 times with 5 mL aliquots of 50 mM KH_2PO_4, combine and dilute the extracts. Dilute bulk samples with Sorensen's phosphate buffer. 50 μL Solution + 25 μL 242 mg/mL pH 10.4 Tris buffer + 100 μL 6 μg/mL kanamycin in MeCN:water 50:50 + 30 μL 250 mg/mL 2,4,6-trinitro-benzenesulfonic acid in MeCN:water 80:20, vortex for 10 s, heat at 70° for 15 min, add 2 mL chloroform, shake horizontally at 180 cycles/min for 5 min, centrifuge at 750 g for 5 min. Remove the lower organic layer and evaporate it to dryness under a stream of nitrogen, reconstitute the residue in 200 μL MeCN, vortex, inject a 20 μL aliquot. (Sorensen's phosphate buffer was 197 mL 9.08 g/L KH_2PO_4 and 1803 mL 11.88 g/L $Na_2HPO_4.2H_2O$, pH 7.4.)

HPLC VARIABLES
Column: 250 × 4.6 5 μm Ultrasphere octyl
Mobile phase: MeCN:50 mM KH_2PO_4 62:38, pH adjusted to 3.5 with phosphoric acid
Flow rate: 2.5
Injection volume: 20
Detector: UV 340

CHROMATOGRAM
Retention time: 7.7
Internal standard: kanamycin (6.4)
Limit of quantitation: 780 ng/mL
Limit of detection: 160 ng/mL

KEY WORDS
injections; beads

REFERENCE
Dash, A.K.; Suryanarayanan, R. A liquid-chromatographic method for the determination of tobramycin, *J.Pharm.Biomed.Anal.*, **1991**, *9*, 237–245.

SAMPLE
Matrix: perfusate
Analyte: gamma-aminobutyric acid
Sample preparation: 2.4 mL Perfusate (physiological buffer) + 600 μL 167 ng/mL δ-aminovaleric acid in 100 mM disodium tetraborate + 100 μL 200 mM 2,4,6-trinitrobenzenesulfonic acid, mix, let stand at room temperature for 20 min, add 100 μL 3 M perchloric acid, extract with 2 mL toluene. Remove 1.6 mL of the organic layer and evaporate it to dryness under a stream of nitrogen, reconstitute the residue in 200 μL toluene, extract with 200 μL 100 mM disodium tetraborate. Remove the aqueous layer and acidify it with 20 μL 2 M perchloric acid, add 500 μL mobile phase, inject a 200 μL aliquot.

HPLC VARIABLES
Column: 100 × 6 ERC-ODS-1161 (ERMA Optical works, Tokyo)
Mobile phase: MeCN:buffer 25:75 (Buffer was 50 mM pH 5.0 acetate buffer containing 80 μg/mL tetramethylammonium chloride.)
Flow rate: 1.5
Injection volume: 200
Detector: E, Bioanalytical Systems LC4B, glassy carbon electrode, detector 1 -0.9 V, detector 2 +0.2 V (monitored)

CHROMATOGRAM
Retention time: 10
Internal standard: δ-aminovaleric acid (20)
Limit of detection: 0.83 ng/mL

OTHER SUBSTANCES
Simultaneously analyzed: alanine, β-alanine, glycine, tryptophan, valine
Non-interfering: ammonia, arginine, aspartic acid, ethanolamine, glutamic acid, glutamine, homocysteic acid, lysine, serine, taurine, threonine

REFERENCE
Waldmeier, P.C.; Wicki, P.; Feldtrauer, J.J.; Baumann, P.A. The measurement of the release of endogenous GABA from rat brain slices by liquid chromatography with electrochemical detection, *Naunyn-Schmiedeberg's Arch.Pharmacol.*, **1988**, *337*, 284–288.

SAMPLE
Matrix: solutions
Analyte: lysine salt of cephalexin
Sample preparation: Mix 2 mL of a 0.5-2 mM solution of the lysine salt of cephalexin in water with 5 mL 0.4% 2,4,6-trinitrobenzenesulfonic acid in water, shake, let stand in the dark for 1 h, add 20 mL pH 4.8 acetate buffer, add 2 mL 0.1% o-nitrophenol in water, make up to 100 mL with water, mix, inject a 1 μL aliquot.

HPLC VARIABLES
Column: 250 × 4.6 13 μm Amino Sil-X-1 alkylamine
Mobile phase: MeOH:water 10:80 containing 1% citric acid
Column temperature: 50
Flow rate: 0.75
Injection volume: 1
Detector: UV 425

CHROMATOGRAM
Retention time: 3.9 (lysine), 11.6 (cephalexin)

OTHER SUBSTANCES
Simultaneously analyzed: degradation products

REFERENCE
Fabregas, J.L.; Beneyto, J.E. Simultaneous determination of cephalexin and lysine in their salt using high-performance liquid chromatography of derivatives, *J.Pharm.Sci.*, **1980**, *69*, 1378–1380.

SAMPLE
Matrix: solutions
Analyte: insulin
Sample preparation: Dissolve 70 mg 2,4,6-trinitrobenzenesulfonic acid in 1 mL 100 mM pH 8.2 sodium bicarbonate and immediately add an aliquot to 50 volumes of 10 mg/mL insulin in 100 mM pH 8.2 sodium bicarbonate, let stand in the dark at room temperature for 2 h, add to a 150 × 60 column of Sephadex G25 made up in 100 mM pH 8.2 sodium bicarbonate, collect the major colored band and lyophilize it. Reconstitute, inject an aliquot.

HPLC VARIABLES
Column: μBondapak C18
Mobile phase: Gradient. MeCN:100 mM pH 3.6 sodium phosphate 25:75 for 10 min, to 45:55 over 1 h
Flow rate: 1
Detector: UV 280

CHROMATOGRAM
Retention time: 50

REFERENCE
Wallace, G.R.; McLeod, A.; Chain, B.M. Chromatographic analysis of the trinitrophenyl derivatives of insulin, *J.Chromatogr.*, **1988**, *427*, 239–246.

SULFONYL HALIDE

(1S)-(+)-10-Camphorsulfonyl Chloride

SAMPLE
Matrix: bulk
Analyte: amines
Sample preparation: Mix 200 μmole amine with 500 μmole (1S)-(+)-10-camphorsulfonyl chloride, 10 mL diethyl ether, and 10 mL 1 M NaOH, stir vigorously for 1 h, acidify with concentrated HCl, extract three times with diethyl ether. Combine the organic layers and wash them three times with water, evaporate to dryness, reconstitute with 1 mL MeOH, inject a 10 μL aliquot.

HPLC VARIABLES
Column: 200 × 4.6 5 μm Silica 100-RP 18
Mobile phase: MeOH:water 50:50
Column temperature: 40
Flow rate: 1.5
Injection volume: 10
Detector: UV 254

CHROMATOGRAM
Retention time: k' 15.55, 18.13 (amphetamine enantiomers), k' 9.46, 11.00 (bamethan enantiomers), k' 7.16, 8.26 (ephedrine enantiomers), k' 5.65, 6.72 (norpseudoephedrine enantiomers), k' 9.08, 11.86 (1-phenylethylamine enantiomers)

KEY WORDS
chiral

REFERENCE
Vogt, C.; Jira, T.; Beyrich, T. HPLC-Trennung racemischer Amine nach Derivatisierung mit (1S)-(+)-Campher-10-sulfonylchlorid, *Pharmazie*, **1990**, *45*, 691.

Dabsyl Chloride

RELATED REFERENCES

Bockhardt, A.; Krause, I.; Klostermeyer, H. Determination of biogenic amines by RP-HPLC of the dabsyl derivates. *Z.Lebensm.-Unters.Forsch.* **1996**, *203*, 65-70.

Lin, J-K.; Lai, C-C. High performance liquid chromatographic determination of naturally occurring primary and secondary amines with dabsyl chloride. *Anal.Chem.* **1980**, *52*, 630-635.

Sormiachi, K.; Ikeda, M.; Akimoto, K.; Niwa, A. Rapid determination of dabsylated hydroxyproline from cultured cells by reversed-phase high-performance liquid chromatography. *J.Chromatogr.B* **1995**, *664*, 435-439.

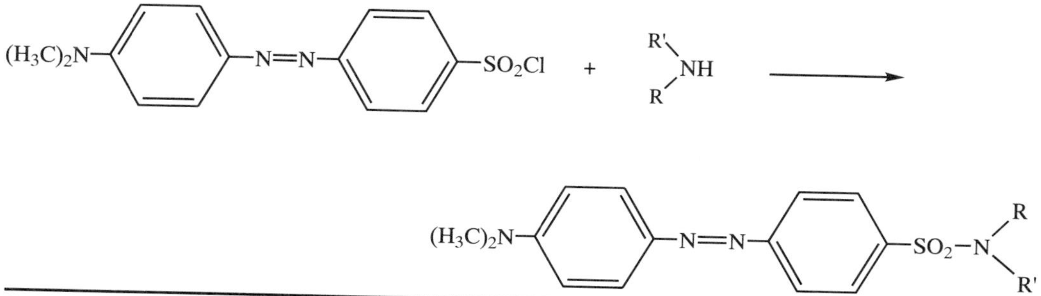

SAMPLE

Matrix: blood

Analyte: taurine

Sample preparation: 1 mL Plasma + 500 μL 5% perchloric acid, centrifuge at 3000 rpm for 10 min. Remove the supernatant and neutralize it with 3 M potassium carbonate, centrifuge at 3000 rpm for 5 min, adjust the volume to 2 mL. (Alternatively, filter (Amicon CF-50) 4 mL plasma while centrifuging at 2500 rpm for 10 min.) 10 μL Perchloric acid extract or ultrafiltrate + 10 μL 100 mM pH 9.0 sodium bicarbonate + 40 μL freshly prepared 4 mM dabsyl chloride in MeCN, heat at 70° for 10 min, cool, make up to 500 μL with EtOH:water 70:30, centrifuge at 14000 rpm for 3 min, inject a 5 μL aliquot.

HPLC VARIABLES

Guard column: 20 × 4.6 5 μm Supelcosil LC-18 T

Column: 150 × 4.6 3 μm Supelcosil LC-18 T

Mobile phase: Gradient. A was 25 mM pH 6.8 KH_2PO_4. B was MeCN:isopropanol 80:20. A:B 80:20 for 1 min, to 77:23 over 4 min, maintain at 77:23 for 7 min, to 73:27 over 11 min, to 70:30 over 7 min, to 40:60 over 9 min, to 30:70 over 1 min, maintain at 30:70 for 5 min, return to initial conditions over 1 min, re-equilibrate for 6 min.

Flow rate: 1.5

Injection volume: 5

Detector: UV 436

CHROMATOGRAM

Retention time: 25

OTHER SUBSTANCES

Extracted: amino acids

KEY WORDS

plasma; ultrafiltrate

REFERENCE

Stocchi, V.; Palma, F.; Piccoli, G.; Biagarelli, B.; Cucchiarini, L.; Magnani, M. HPLC analysis of taurine in human plasma sample using the DABS-Cl reagent with sensitivity at picomole level, *J.Liq.Chromatogr.*, **1994**, *17*, 347–357.

SAMPLE

Matrix: bulk

Analyte: sphingosine

Sample preparation: Reconstitute sphingosine with 20 μL 200 mM pH 8.75 sodium bicarbonate/NaOH buffer and 40 μL 5 mM dabsyl chloride in acetone, heat at 70° with constant shaking for 10 min, dilute with MeOH (if necessary), inject a 5-10 μL aliquot.

HPLC VARIABLES
Column: 250 × 4.6 10 μm Zorbax CN
Mobile phase: Gradient. MeCN : 17.5 mM pH 6.0 sodium acetate 60 : 40 for 10 min, to 90 : 10 over 5 min, return to initial conditions over 5 min, re-equilibrate for 10 min.
Flow rate: 1
Injection volume: 5-10
Detector: UV 430

CHROMATOGRAM
Retention time: 7
Limit of detection: 5 pmole

OTHER SUBSTANCES
Simultaneously analyzed: related compounds

REFERENCE
Rosenfelder, G.; Chang, J.-Y.; Braun, D.G. Sphingosine determination at the picomole level using dimethylaminoazobenzene sulphonyl chloride, *J.Chromatogr.*, **1983**, *272*, 21–27.

SAMPLE
Matrix: glycopeptides
Analyte: amino sugars
Sample preparation: Add 700 μL 6 M HCl to a lyophilized sample, flush with argon, evacuate to 0.1 mbar for 1-2 min, heat at 110° for 1-10 h, evaporate to dryness, add 20 μL pH 8.3 sodium bicarbonate buffer (Beckman), add 40 μL 1 mg/mL dabsyl chloride in MeCN, heat at 70° for 12-15 min, add 440 μL EtOH : water 50 : 50, inject a 30-60 μL aliquot.

HPLC VARIABLES
Column: 250 × 4.6 Ultrasphere-DABS C18
Mobile phase: Gradient. A was buffer. B was MeCN : DMF : buffer 67.2 : 2.8 : 30. Equilibrate column with A : B 71 : 29, to 49 : 51 at start, to 14 : 86 then 0 : 100 (timings are unclear). (Prepare buffer by mixing 115 mL 110 mM pH 6.51 sodium citrate, 845 mL water, and 40 mL DMF, final pH 6.50-6.52.)
Flow rate: 1.4
Injection volume: 30-60
Detector: UV 436

CHROMATOGRAM
Retention time: 14 (galactosamine), 20 (glucosamine)
Limit of detection: 0.2 nmole

OTHER SUBSTANCES
Simultaneously analyzed: amino acids

REFERENCE
Gorbics, L.; Urge, L.; Otvos-Papp, E.; Otvos, L. Jr. Determination of amino sugars in synthetic glycopeptides during the conditions of amino acid analysis utilizing precolumn derivatization and high-performance liquid chromatographic analysis, *J.Chromatogr.*, **1993**, *637*, 43–53.

SAMPLE
Matrix: urine
Analyte: hypotaurine and taurine
Sample preparation: 10 μL Urine + 50 μL 5 mM S-carboxymethyl-L-cysteine in water + 500 μL 100 mM pH 9.0 sodium bicarbonate buffer + 50 μL acetone + 1 mL 650 μg/mL dabsyl chloride in acetone, heat at 40° for 30 min, add 1.5 mL EtOH, let stand for 30 min, centrifuge, filter (0.2 μm) the supernatant, inject a 10 μL aliquot of the filtrate.

HPLC VARIABLES
Guard column: 15 × 3.2 5 μm TSKguardgel ODS-80Ts
Column: 150 × 4.6 5 μm TSKgel ODS-80Ts

Mobile phase: Gradient. MeCN:50 mM pH 4.00 sodium acetate buffer from 28:72 to 31: 69 over 20 min, to 32:68 over 20 min, wash with 95:5 for 5 min, re-equilibrate at initial conditions for 20 min.
Column temperature: 16
Injection volume: 10
Detector: UV 430

CHROMATOGRAM
Retention time: 23 (hypotaurine), 27 (taurine)
Internal standard: S-carboxymethyl-L-cysteine (23)
Limit of detection: 4 pmole

KEY WORDS
rat

REFERENCE
Futani, S.; Ubuka, T.; Abe, T. High-performance liquid chromatographic determination of hypotaurine and taurine after conversion to 4-dimethylaminoazobenzene-4'-sulfonyl derivatives and its application to the urine of cysteine-administered rats, *J.Chromatogr.B*, **1994**, *660*, 164–169.

Dansyl Chloride

Primary amines, secondary amines, and phenolic hydroxyls are all dansylated at about the same rate. Under these conditions tertiary amines and alkyl hydroxyls are not dansylated.

RELATED REFERENCES
Biondi, P.A.; Negri, A.; Ioppolo, A. High-performance liquid chromatographic determination of taurine in formulations as the dansyl derivative. *J.Chromatogr.* **1986**, *369*, 431-434.

Bontemps, J.; Etienne, A.; Kadri, M.; Van Cutsem, J.L.; Dandrifosse, G.; Forget, P.P. High-speed analysis of dansyl derivatives of polyamines. *Chromatographia* **1984**, *18*, 525-527.

Bringmann, G.; Feineis, D.; Hesselmann, C. Endogenous alkaloids in man. 12. Determination of 1,3-thiazolidinecarboxylic acids in urine by reversed-phase HPLC after fluorescence labeling with dansyl chloride. *Anal.Lett.* **1992**, *25*, 497-512.

Clausing, P.; Rushing, L.G.; Newport, G.D.; Bowyer, J.F. Determination of D-fenfluramine, D-norfenfluramine and fluoxetine in plasma, brain tissue and brain microdialysate using high-performance liquid chromatography after precolumn derivatization with dansyl chloride. *J.Chromatogr.B* **1997**, *692*, 419-426.

Fuh, M.S.; Wang, T.K.; Pan, W.H.T. Determination of free-form amphetamine in rat's brain by in vivo microdialysis and liquid chromatography with fluorescence detection. *J.Liq.Chromatogr.Rel.Technol.* **1997**, *20*, 1605-1615.

Gennaro, M.C.; Mentasti, E.; Sarzanini, C.; Porta, V. Aliphatic monoamines, diamines, polyamines: an HPLC method for their identification and separation by a dansylation reaction; the study of separation factors in homologous series. *Chromatographia* **1988**, *25*, 117-124.

Heimbecher, S.; Lee, Y.-C.; Tabibi, E.S.; Yalkowsky, S.H. Derivatization and high-performance liquid chromatographic analysis of pentaazapentacosane pentahydrochloride. *J.Chromatogr.B* **1997**, *691*, 173-178.

Lawrence, J.F.; Leduc, R. High-pressure liquid chromatographic analysis of carbofuran and two non-conjugated metabolites in crops as fluorescent dansyl derivatives. *J.Chromatogr.* **1978**, *152*, 507-513.

Nishitani, A.; Kanda, S.; Imai, K. A sensitive determination method for mexiletine derivatized with dansyl chloride in rat plasma utilizing a HPLC peroxyoxalate chemiluminescence detection system. *Biomed.Chromatogr.* **1992**, *6*, 124-127.

Outinen, K.; Lehtola, V.-M.; Vuorela, H. Behaviour of resolution by changing solvent strength and selectivity in the 'PRISMA' model using reversed-phase HPLC for biogenic amines. *J.Pharm.Biomed.Anal.* **1997**, *15*, 819-829.

Penmetsa, K.V.; Leidy, R.B.; Shea, D. Herbicide analysis by micellar electrokinetic capillary chromatography. *J.Chromatogr.A* **1996**, *745*, 201-208.

Price, N.P.J.; Firmin, J.L.; Gray, D.O. Screening for amines by dansylation and automated high-performance liquid chromatography. *J.Chromatogr.* **1992**, *598*, 51-57.

Price, N.P.J.; Firmin, J.L.; Robins, R.J.; Gray, D.O. High-performance liquid chromatography of the alkaloid perivine from Catharanthus roseus after derivatization with dansyl chloride. *J.Chromatogr.* **1993**, *653*, 161-166.

Roos, R.W.; Lau-Cam, C.A. Utility of dansyl derivatization to the high-performance liquid chromato-graphic analysis of 2-phenylethylamine drugs. *J.Chromatogr.* **1991**, *555*, 278-284.

Vallé, M.; Malle, P.; Bouquelet, S. Liquid chromatographic determination of fish decomposition indexes from analyses of plaice, whiting, and herring. *J.AOAC Int.* **1996**, *79*, 1134-1140.

Vallé, M.; Malle, P.; Bouquelet, S. Optimization of a liquid chromatographic method for determination of amines in fish. *J.AOAC Int.* **1997**, *80*, 49-56.

Wang, T.K.; Fuh, M.S. Determination of amphetamine in human urine by dansyl derivatization and high-performance liquid chromatography with fluorescence detection. *J.Chromatogr.B* **1996**, *686*, 285-290.

Wang, Z.; Xu, H.; Fu, C. Sensitive fluorescence detection of some nitrosamines by precolumn derivati-zation with dansyl chloride and high-performance liquid chromatography. *J.Chromatogr.* **1992**, *589*, 349-352.

SAMPLE
Matrix: formula, milk
Analyte: taurine
Sample preparation: 3 g Milk or formula + 80 mL water, heat at 50-60° with periodic agitation for 10 min, cool to room temperature, add 1 mL 150 mg/mL potassium ferro-cyanide trihydrate in water, swirl, add 1 mL 300 mg/mL zinc acetate dihydrate in water, mix, let stand with periodic inversion for 20 min, make up to 100 mL with water, mix thoroughly, filter, discard the first 3-5 mL. Mix 1 mL filtrate with 1 mL buffer, add 1 mL 1.5 mg/mL dansyl chloride in MeCN, mix by inversion, let stand in the dark (with mixing after 1 h) at room temperature for 2 h, add 100 μL 20 mg/mL methylamine hydrochloride in water, vortex, filter (0.45 μm), inject a 20 μL aliquot of the filtrate. (Buffer was 80 mM sodium carbonate adjusted to pH 9.5 with 1 M HCl.)

HPLC VARIABLES
Guard column: C18
Column: 5 μm Resolve (Waters)
Mobile phase: MeCN:buffer 16:84
Flow rate: 1
Injection volume: 20
Detector: UV 254; F ex 330 em 530

CHROMATOGRAM
Retention time: 4-6
Limit of detection: 50 μg/g

REFERENCE
Woollard, D.C.; Indyk, H.E. Taurine analysis in milk and infant formulae by liquid chromatography: Collaborative study, *J.AOAC Int*, **1997**, *80*, 860–865.

SAMPLE
Matrix: CSF, tissue culture, urine
Analyte: polyamines
Sample preparation: Urine, CSF. Condition a Bond-Elut SCX strong cation-exchange SPE cartridge with 2 column volumes MeOH, 2 column volumes water, and 2 column volumes 10 mM buffer. Condition a Bond-Elut C18 SPE cartridge with 2 column volumes of MeOH and 2 column volumes of water. 300 μL Urine or CSF + 100 μL 4 μM 1,7-diaminoheptane

+ 500 µL 20 mM buffer, add to the SCX SPE cartridge, wash with 2 column volumes water, wash with 2 column volumes 10 mM buffer, elute with two 300 µL portions of acetone:2.57 M buffer 20:80. Add 200 µL saturated sodium carbonate and 200 µL 10 mg/mL dansyl chloride in acetone to the eluate, vortex for 15 s, heat at 70° for 10 min, cool, add to the C18 SPE cartridge,wash with 2 column volumes of water, elute with 500 µL MeOH, inject a 10 (urine) or 50-100 (CSF) µL aliquot. Tissue culture. Condition a Bond-Elut C18 SPE cartridge with 2 column volumes of MeOH and 2 column volumes of water. Wash cells with phosphate-buffered saline, centrifuge, discard the supernatant. Add 200-250 µL 8% 5-sulfosalicylic acid for every million cells, sonicate for 20-30 s, let stand on ice for 30 min, mix, let stand on ice for 30 min, centrifuge at 12000 g for 5 min. Remove a 50 µL aliquot of the supernatant and add it to 50 µL 4 µM 1,6-diaminohexane, add 200 µL saturated sodium carbonate, add 200 µL 10 mg/mL dansyl chloride in acetone, vortex for 15 s, heat at 70° for 10 min, cool, add to the C18 SPE cartridge, wash with 2 column volumes of water, elute with 500µL MeOH, inject a 10 µL aliquot. (Prepare 2.57 M buffer by dissolving 178.75 g KCl, 28.75 g potassium citrate, and 3 mL 12 M HCl in 1 L water, pH 5.56. Dilute the 2.57 M buffer to obtain the 10 and 20 mM buffers.)

HPLC VARIABLES
Guard column: RP-18 spheri-guard (Rainin)
Column: 150 × 4.6 5 µm Ultrasphere ODS
Mobile phase: Gradient. MeCN:10 mM pH 4.4 phosphate buffer from 45:55 to 80:20 over 14 min, to 90:10 over 1 min, maintain at 90:10 for 5 min, re-equilibrate at initial conditions for 6 min.
Column temperature: 50
Flow rate: 2
Injection volume: 10-100
Detector: F ex 340 em 515

CHROMATOGRAM
Retention time: 12.5 (cadaverine), 11.7 (putrescine), 17.2 (spermidine), 19.8 (spermine)
Internal standard: 1,6-diaminohexane (13.5), 1,7-diaminoheptane (17)
Limit of detection: <20 pg

OTHER SUBSTANCES
Simultaneously analyzed: agmatine, amphetamine, 1,3-diaminopropane, gentamicin, histamine, 5-hydroxytryptamine, methamphetamine, norepinephrine, normetanephrine, tobramycin
Non-interfering: amikacin, deoxyepinephrine, dopamine

KEY WORDS
SPE

REFERENCE
Kabra, P.M.; Lee, H.K.; Lubich, W.P.; Marton, L.J. Solid-phase extraction and determination of dansyl derivatives of unconjugated and acetylated polyamines by reversed-phase liquid chromatography: improved separation systems for polyamines in cerebrospinal fluid, urine and tissue, *J.Chromatogr.*, **1986**, *380*, 19–32.

SAMPLE
Matrix: blood
Analyte: tocainide
Sample preparation: 0.05-2 mL Plasma or whole blood + 50 µL 13.5 µg/mL IS in 50 mM HCl + 200 µL 2 M NaOH + 3 mL ethyl acetate, mix with gentle tilting for 5 min, centrifuge at 2500 rpm for 10 min. Remove 2 mL of the organic layer and add it to 50 µL 50 mM sulfuric acid, vortex for 1 min, centrifuge for 5 min, discard the organic phase. Wash the aqueous phase with 1 mL hexane, freeze in dry ice/acetone, discard the organic layer. Thaw the aqueous layer and add 50 µL 1 M sodium bicarbonate and 200 µL reagent, heat at 40° for 30 min, add 100 µL 500 mM NaOH, evaporate under reduced pressure at 40° for 3 min, add 1 mL buffer and 200 µL carbon tetrachloride (Caution! Carbon tetrachloride is a carcinogen!), mix for 2 min, centrifuge for 5 min, inject a 2-5µL aliquot of the organic layer. (Reagent was 1 mg/mL dansyl chloride in acetone, store in the dark, prepare fresh each week. Buffer was a half-saturated solution of disodium citrate in water, adjusted to pH 6.0 with 85% phosphoric acid.)

HPLC VARIABLES
Column: 300 × 4 μBondapak NH2
Mobile phase: Hexane:dichloromethane:MeOH 50:50:1
Flow rate: 5
Injection volume: 2-5
Detector: F ex 360 (Corning 7-51 filter) em 490 (Wratten 8 filter)

CHROMATOGRAM
Retention time: 2.4
Internal standard: 2-amino-N-(2,6-dimethylphenyl)butanamide hydrochloride (Astra Pharmaceuticals, Worcester, MA) (1.9)
Limit of quantitation: 100 ng/mL

KEY WORDS
plasma; whole blood; pharmacokinetics

REFERENCE
Meffin, P.J.; Harapat, S.R.; Harrison, D.C. High-pressure liquid chromatographic analysis of drugs in biological fluids II: Determination of an antiarrhythmic drug, tocainide, as its dansyl derivative using a fluorescence detector, *J.Pharm.Sci.*, **1977**, *66*, 583–586.

SAMPLE
Matrix: blood
Analyte: indeloxazine
Sample preparation: 1 mL Plasma + 1 mL 300 ng/mL viloxazine in water + 500 μL 200 mM ammonium hydroxide + 4 mL ether, extract, centrifuge. Remove the organic layer and evaporate it to dryness at 45°, reconstitute the residue in 100 μL 3 mg/mL sodium bicarbonate and 200 μL 125 μg/mL dansyl chloride in acetone, heat at 45° for 20 min, cool, add 4 mL ether. Wash the ether solution twice with 3 mL water and evaporate it to dryness at 45°. Take up the residue in 100 μL n-heptane, inject a 3-5 μL aliquot.

HPLC VARIABLES
Column: 150 × 4 5 μm LiChrosorb SI-60
Mobile phase: n-Heptane:ethyl acetate 20:3
Flow rate: 1.5
Injection volume: 3-5
Detector: F ex 365 em 505

CHROMATOGRAM
Retention time: 3
Internal standard: viloxazine (4)
Limit of detection: 5 ng/mL

KEY WORDS
plasma; normal phase

REFERENCE
Kamimura, H.; Sasaki, H.; Yokoi, K.; Kawamura, S. Determination of indeloxazine in plasma by liquid chromatography and gas chromatography-mass spectrometry, *J.Pharm.Sci.*, **1985**, *74*, 559–561.

SAMPLE
Matrix: blood
Analyte: cyanamide
Sample preparation: 500 μL Plasma + 100 μL water + 2 mL ethyl acetate, vortex for 30 s, centrifuge at 2000 g for 5 min, remove 1.5 mL of the organic layer, add 2 mL ethyl acetate to the aqueous layer, repeat extraction, remove a 2 mL aliquot of the organic layer. Combine the organic layers and evaporate them to dryness under a stream of nitrogen at 40°, reconstitute the residue in 100 μL 200 mM pH 9.0 sodium carbonate/sodium bicarbonate buffer, add 100 μL 1 mg/mL dansyl chloride in acetone, vortex, heat at 40° for 1 h, cool to room temperature, inject a 50 μL aliquot.

HPLC VARIABLES
Guard column: μBondapak C18 Guard-Pak
Column: 100 × 8 10 μm μBondapak C18 Radial-Pak radial compression

Mobile phase: Gradient. A was 10 mM pH 7.0 potassium phosphate buffer. B was MeCN:
10 mM pH 7.0 potassium phosphate buffer 55:45. A:B from 70:30 to 40:60 over 7 min
(Waters curve 6), to 0:100 over 3 min (Waters curve 1), return to initial conditions over
2 min (Waters curve 1).
Flow rate: 4
Injection volume: 50
Detector: F ex 360 em 495 (longpass filter)

CHROMATOGRAM
Retention time: 5
Limit of detection: 4 ng/mL

KEY WORDS
human; rat; plasma; pharmacokinetics

REFERENCE
Pruñonosa, J.; Obach, R.; Vallès, J.M. Determination of cyanamide in plasma by high-performance liquid
chromatography, *J.Chromatogr.*, **1986**, *377*, 253–260.

SAMPLE
Matrix: blood
Analyte: paroxetine
Sample preparation: 1 mL Plasma + 250 μL buffer + 200 μL water + 100 μL 25 ng/mL
maprotiline in water + 4 mL toluene, extract on a tumble mixer for 15 min, centrifuge
at 1500 g for 5 min. Remove the organic layer and evaporate it to dryness under a stream
of nitrogen at 55°, reconstitute the residue in 50 μL acetone, add 25 μL 100 mM sodium
bicarbonate, add 10 μL 1 mg/mL dansyl chloride in acetone (prepare fresh daily), vortex
for 15 s, heat at 55° for 1 min, centrifuge for 1 min, let stand at room temperature for 30
min, add 25 μL 25 mg/mL L-proline in water (prepare fresh daily), vortex briefly, centrifuge for 1 min, let stand at room temperature for 5 min, add 500 μL water, add 2 mL
toluene, agitate on a tumble mixer for 10 min, centrifuge at 1500 g for 5 min. Remove
the organic layer and evaporate it to dryness under a stream of nitrogen at 55°, reconstitute the residue in 100 μL mobile phase, vortex for 30 s, inject an aliquot. (Buffer was
8.6 g $Na_2HPO_4.12H_2O$ in 100 mL water, adjusted to pH 12.0 with 4 M NaOH, made up
to 200 mL with water.)

HPLC VARIABLES
Guard column: 30 mm long 5 μm Spherisorb ODS
Column: 200 × 4 5 μm Spherisorb ODS
Mobile phase: MeOH:50 mM pH 4.5 sodium acetate buffer 84:16 (At the end of the day
wash column with 95:5.)
Flow rate: 1.5
Injection volume: 100
Detector: F ex 340 em 520

CHROMATOGRAM
Retention time: 5.8
Internal standard: maprotiline (8.2)
Limit of quantitation: 0.5-1 ng/mL
Limit of detection: 0.2 ng/mL

OTHER SUBSTANCES
Non-interfering: cimetidine, digoxin, methyldopa, phenobarbital, phenytoin, procyclidine,
tranylcypromine

KEY WORDS
plasma; pharmacokinetics

REFERENCE
Brett, M.A.; Dierdorf, H.-D.; Zussman, B.D.; Coates, P.E. Determination of paroxetine in human plasma,
using high-performance liquid chromatography with fluorescence detection, *J.Chromatogr.*, **1987**,
419, 438–444.

SAMPLE
Matrix: blood
Analyte: tetrafluoroputrescine
Sample preparation: Hemolyze red blood cells by vigorously vortexing cells from 1 mL blood with 1 mL water for 1 min. Add cold 10% perchloric acid to plasma or hemolyzed red blood cells, vortex, centrifuge, adjust the pH of the supernatant to 6.8. Remove a 1 mL aliquot and add it to 5-20 equivalents dansyl chloride in 1 mL acetone, heat at 60° in the dark for 1 h, adjust pH to 8 with saturated sodium carbonate, add a second aliquot of dansyl chloride, heat at 60° for 30 min, add 100 μL 200 mg/mL proline in water, add 700 pmoles dansyl diaminohexane, extract twice with 4 mL portions of ethyl acetate. Combine the organic layers and evaporate them to dryness under reduced pressure, reconstitute the residue in 6 mL ethyl acetate, inject an aliquot.

HPLC VARIABLES
Column: 250 × 4.6 spherical C18 (ASTEC)
Mobile phase: Gradient. MeOH:water 75:25 for 16 min, to 100:0 (step gradient), maintain at 100:0
Flow rate: 1
Injection volume: 20
Detector: UV 254

CHROMATOGRAM
Retention time: 12.6
Internal standard: dansyl diaminohexane (15.4)

OTHER SUBSTANCES
Extracted: putrescine, spermidine, spermine

KEY WORDS
plasma; red blood cells; rat

REFERENCE
Hawi, A.A.; Yip, H.; Sullivan, T.S.; Digenis, G.A. Development of an HPLC assay for the analysis of tetrafluoroputrescine—a putrescine analog, *Anal.Biochem.*, **1988**, *172*, 235–240.

SAMPLE
Matrix: blood
Analyte: eflornithine
Sample preparation: 1 mL Serum + 100 μL 1 mg/mL norvaline in water + 2 mL ice-cold 80% EtOH, vortex for 30 s, centrifuge at 3000 rpm for 5 min, remove the supernatant, extract the residue twice more with 2 mL portions of ice-cold 80% EtOH. Combine the supernatants and dry them under vacuum at 50°, dissolve in 2 mL water and 2 mL buffer. 100 μL Mixture + 100 μL 6 mg/mL dansyl chloride in acetone, let stand at room temperature in the dark for 4 h, add 800 μL water adjusted to pH 8.5 with NaOH, vortex for 15 s, centrifuge at 3000 rpm, inject a 25-50 μL aliquot. (Buffer was 500 mM sodium bicarbonate adjusted to pH 8.5 with 1 M NaOH.)

HPLC VARIABLES
Guard column: 15 × 4.6 5 μm C8 (Rainin)
Column: 150 × 4.6 5 μm C8 (Rainin)
Mobile phase: Gradient. A was THF:10 mM sodium acetate adjusted to pH 4.18 with 1 M acetic acid 5:95. B was MeCN:THF 90:10. A:B from 100:0 to 63.3:36.7 over 19 min, to 0:100 over 2 min, re-equilibrate at 100:0 for 7 min.
Flow rate: 1.5 for 21 min, 2.0 for 2 min, 1.75 for 3 min, 1.5 for 2 min
Injection volume: 25-50
Detector: UV 330

CHROMATOGRAM
Retention time: 14.73
Internal standard: norvaline (17.66)
Limit of detection: 10 μg/mL

KEY WORDS
serum; pharmacokinetics

REFERENCE
Cohen, J.L.; Ko, R.J.; Lo, A.T.; Shields, M.D.; Gilman, T.M. High-pressure liquid chromatographic analysis of eflornithine in serum, *J.Pharm.Sci.*, **1989**, *78*, 114–116.

SAMPLE
Matrix: blood
Analyte: fluvoxamine
Sample preparation: 2 mL Plasma + 200 μL 1 μg/mL metapramine in MeOH + 2 mL 1 M pH 10.0 phosphate buffer + 6 mL diethyl ether:hexane 50:50, shake for 15 min, centrifuge at 4000 g for 5 min. Remove the organic layer and add it to 2 mL 62.5 mM sulfuric acid, vortex for 5 min, centrifuge at 4000 g for 5 min. Remove the aqueous phase and add it to 1 mL 500 mM NaOH, vortex, add 6 mL hexane:diethyl ether 50:50, shake for 10 min, centrifuge at 4000 g for 5 min. Remove the organic layer and evaporate it to dryness under a stream of nitrogen at 50°, reconstitute the residue in 100 mM sodium carbonate, add 10 μL 10 mg/mL dansyl chloride in acetone, vortex for 1 min, heat at 45° for 30 min, evaporate under a stream of nitrogen at 50°. Reconstitute the residue in 200 μL MeCN:water 45:55, inject a 100 μL aliquot.

HPLC VARIABLES
Column: 125 × 4.6 5 μm Hypersil ODS
Mobile phase: Gradient. MeCN:water from 45:55 to 65:35 over 10 min, maintain at 65:35 for 20 min
Column temperature: 30
Flow rate: 1.5
Injection volume: 100
Detector: F ex Fluorichrom 7.54 and 7.60 filters em 3.71 and 4.76 filters

CHROMATOGRAM
Retention time: 19
Internal standard: metapramine (28)
Limit of detection: 1.5 ng/mL

OTHER SUBSTANCES
Non-interfering: alimemazine, alprazolam, amineptine, amitriptyline, caffeine, clobazam, clomipramine, clorazepate, cyamemazine, diazepam, demethyldiazepam, flunitrazepam, levomepromazine, loprazolam, lorazepam, meprobamate, nitrazepam, oxazepam, triazolam, viloxazine

KEY WORDS
plasma; protect from light

REFERENCE
Pommery, J.; Lhermitte, M. High performance liquid chromatographic determination of fluvoxamine in human plasma, *Biomed.Chromatogr.*, **1989**, *3*, 177–179.

SAMPLE
Matrix: blood
Analyte: taurine
Sample preparation: Subject whole blood to two freeze-thaw cycles. 50 μL Plasma or 10 μL whole blood + 200 μL MeCN:MeOH:triethylamine:water 25:22:3:50, vortex for 15 s, filter (Centricon-10 10000 MW exclusion filter) while centrifuging at 2677 g for 15 min. Remove a 20 μL aliquot of the ultrafiltrate and add it to 10 μL 10 mg/mL dansyl chloride in MeCN (prepare fresh daily), vortex, let stand in the dark for 30 min, add 10 μL water:ethylamine 96.5:3.5, inject a 10 μL aliquot.

HPLC VARIABLES
Column: 220 × 4.6 Spheri-5 ODS
Mobile phase: MeOH:water:acetic acid:triethylamine 33:66.5:0.5:0.025, after 6 min purge column with MeOH:water 90:10 for 4 min, re-equilibrate for 5 min
Column temperature: 50

Flow rate: 1.5
Injection volume: 10
Detector: F ex 329 em 530

CHROMATOGRAM
Retention time: 7
Limit of detection: 2.5 ng

KEY WORDS
cat; plasma; whole blood; ultrafiltrate

REFERENCE
Amiss, T.J.; Tyczkowska, K.L.; Aucoin, D.P. Analysis of taurine in feline plasma and whole blood by liquid chromatography with fluorimetric detection and confirmation by thermospray mass spectrometry, *J.Chromatogr.*, **1990**, *526*, 375–382.

SAMPLE
Matrix: blood
Analyte: fluoxetine
Sample preparation: 1 mL Plasma + 1 mL 0.6 M pH 9.8 carbonate buffer + 40 μL 5 μg/mL maprotiline in 10 mM HCl + 5 mL 200 g/L ethyl acetate in n-heptane, mix by rocking for 10 min, centrifuge at 1500 g for 10 min. Remove organic layer and add it to 150 μL 100 mM HCl, mix 10 min, centrifuge at 1500 g for 10 min. Discard organic layer and evaporate aqueous layer at 45° in a vacuum centrifuge for 1 h. Take up residue in 50 μL 1 M pH 10.3 carbonate buffer and 25 μL 10 mg/mL dansyl chloride in MeCN, vortex, allow to react at room temperature for 45 min, evaporate at 45° in a vacuum centrifuge for 20 min, reconstitute in 125 μL MeCN:water 75:25, vortex, centrifuge for 3-5 min, inject a 25-40 μL aliquot.

HPLC VARIABLES
Column: 250 × 4.6 5 μm Supelcosil LC-18
Mobile phase: MeCN:25 mM KH_2PO_4 75:25 containing 500 μL/L orthophosphoric acid and 600 μL/L n-butylamine
Flow rate: 2
Injection volume: 25-40
Detector: F ex 235 em 470 (cut-off)

CHROMATOGRAM
Retention time: 10.4
Internal standard: maprotiline (12.8)

OTHER SUBSTANCES
Simultaneously analyzed: amoxapine, clovoxamine, desipramine, fenfluramine, fluvoxamine, norfluoxetine, nortriptyline, propranolol, protriptyline, sertraline
Non-interfering: amitriptyline, atenolol, bupropion, carbamazepine, chlordiazepoxide, citalopram, clomipramine, clozapine, cyclobenzaprine, doxepin, imipramine, loxapine, metoprolol, mianserin, moclobemide, nomifensine, pindolol, thioridazine, tranylcypromine, trazodone, trimipramine

KEY WORDS
plasma; LOQ 3 ng/mL

REFERENCE
Suckow, R.F.; Zhang, M.F.; Cooper, T.B. Sensitive and selective liquid-chromatographic assay of fluoxetine and norfluoxetine in plasma with fluorescence detection after precolumn derivatization, *Clin.Chem.*, **1992**, *38*, 1756–1761.

SAMPLE
Matrix: blood
Analyte: fluvoxamine
Sample preparation: 1 mL Plasma + 10 μL 100 μg/mL nortriptyline in water + 100 μL 1 M pH 7.6 K_2HPO_4, vortex for 5 s, add 6 mL ethyl acetate, vortex for 1 min, centrifuge at 1900 g for 5 min. Remove the organic layer and evaporate it to dryness under a stream of nitrogen at 45°, reconstitute the residue in 250 μL 40 mM sodium bicarbonate, add 20

µL 10 mg/mL dansyl chloride in acetone, add 750 µL acetone, vortex for 30 s, let stand at room temperature at 22° for 15 min. Evaporate under a stream of nitrogen at 45°, reconstitute in 1 mL mobile phase, vortex for 1 min, centrifuge at 1900 g for 10 min, inject a 100 µL aliquot of the supernatant.

HPLC VARIABLES
Guard column: 15 × 3.2 7 µm NewGuard RP-8 (Brownlee)
Column: 250 × 4.6 5 µm Supelcosil LC-18-DB
Mobile phase: MeCN:10 mM pH 7.2 potassium phosphate 85:15
Flow rate: 1.5
Injection volume: 100
Detector: F (wavelengths not given)

CHROMATOGRAM
Retention time: 5.3 (fluvoxamine), 8.8 (desipramine)
Internal standard: nortriptyline (9.7)
Limit of quantitation: 10 ng/mL

KEY WORDS
plasma

REFERENCE
Pullen, R.H.; Fatmi, A.A. Determination of fluvoxamine in human plasma by high-performance liquid chromatography with fluorescence detection, *J.Chromatogr.*, **1992**, *574*, 101–107.

SAMPLE
Matrix: blood
Analyte: perhexiline
Sample preparation: 500 µL Plasma + 200 µL 1.4 µg/mL hexadiline hydrochloride in 100 mM HCl + 50 µL 2 M pH 8.75 Tris-HCl buffer + 4 mL n-hexane, shake horizontally at 100 oscillations/min for 15 min, centrifuge at 2500 rpm for 10 min. Remove the organic layer and evaporate it to dryness under a stream of nitrogen at 60°, reconstitute the residue in 100 µL 100 mM pH 10 sodium bicarbonate buffer, add 100 µL 5 mM dansyl chloride in acetone (freshly prepared), vortex, heat at 37° for 20 min, add 1.5 mL n-hexane, vortex, centrifuge at 10° at 2500 rpm for 3 min, freeze in dry ice/EtOH. Remove the organic layer and evaporate it to dryness at 60°, reconstitute the residue in 100 µL mobile phase, inject a 25 µL aliquot.

HPLC VARIABLES
Column: 100 × 3.2 3 µm Velosep (Brownlee)
Mobile phase: MeOH:water 86:14
Column temperature: 45
Flow rate: 0.5
Injection volume: 25
Detector: F ex 366 em 470

CHROMATOGRAM
Retention time: 15.9, 16.8 (isomers)
Internal standard: hexadiline (19.5)
Limit of quantitation: 150 ng/mL

OTHER SUBSTANCES
Non-interfering: metabolites

KEY WORDS
plasma; pharmacokinetics

REFERENCE
Morris, R.G.; Sallustio, B.C.; Saccoia, N.C.; Mangas, S.; Fergusson, L.K.; Kassapidis, C. Application of an improved HPLC perhexiline assay to human plasma specimens, *J.Liq.Chromatogr.*, **1992**, *15*, 3219–3232.

SAMPLE
Matrix: blood

Analyte: ephedrine and pseudoephedrine
Sample preparation: Condition a Sep-pak C18 SPE cartridge with EtOH, 5% aqueous bovine serum albumin, and water. 100-500 μL Plasma + 100 ng (l)-norephedrine + 2 mL 500 mM pH 7.0 phosphate buffer, add to the SPE cartridge, wash with 5 mL water, wash with 3 mL EtOH:water 20:80, elute with 8 mL EtOH. Evaporate the eluate to dryness, reconstitute the residue in 100 μL MeCN, add 100 μL 6 mg/mL dansyl chloride in MeCN containing 0.03% triethylamine, heat at 50° for 20 min, evaporate to dryness under a stream of nitrogen, reconstitute with 1 mL EtOH:water 90:10, add to an 18 × 6 column containing 80 mg carboxymethyl Sephadex LH-20 (0.95 meq/g), wash with EtOH:water 90:10, elute with 6 mL 50mM methylamine in EtOH:water 90:10 at 0.1 mL/min. Evaporate the eluate to dryness, reconstitute with 50-100 μL mobile phase, inject an aliquot.

HPLC VARIABLES
Guard column: 10 × 4 5 μm Shimpack G-ODS guard column (Shimadzu)
Column: 150 × 6 5 μm Shimpack CLC-ODS (Shimadzu)
Mobile phase: MeOH:0.6% pH 6.5 phosphate buffer 80:30
Flow rate: 1.3
Detector: F ex 316 em 486

CHROMATOGRAM
Retention time: 12.5 (l-ephedrine), 15 (d-pseudoephedrine)
Internal standard: (l)-norephedrine (10)

KEY WORDS
plasma; SPE; guinea pig; human; pharmacokinetics

REFERENCE
Shao, G.; Wang, D.-S.; Wu, F.; Chen, S.-J.; Luo, X. Separation and determination of (l)-ephedrine and (d)-pseudoephedrine in plasma by high-performance liquid chromatography with fluorescence detection, *J.Liq.Chromatogr.*, **1995**, *18*, 2133–2145.

SAMPLE
Matrix: blood
Analyte: L-buthionin (SR)-sulfoximine
Sample preparation: 100 μL Plasma + 50 μL 100 μg/mL L-norvaline, add 10% sulfosalicylic acid, mix, centrifuge at 13000 rpm for 5 min. Remove the supernatant and evaporate it to dryness under a stream of nitrogen, reconstitute the residue in 50 μL 400 mM pH 9.53 phosphate buffer, add 50 μL 3 mg/mL dansyl chloride in acetone, heat at 40° for 10 min, inject a 5 μL aliquot.

HPLC VARIABLES
Column: 150 × 4.6 Zorbax ODS
Mobile phase: MeCN:MeOH:10 mM pH 2.1 phosphate buffer 3:1:8
Column temperature: 40
Flow rate: 1
Injection volume: 5
Detector: F ex 335 em 525

CHROMATOGRAM
Retention time: 5
Internal standard: norvaline (10)
Limit of detection: 10 μg/mL

KEY WORDS
rat; plasma; protect from light

REFERENCE
Koyama, H.; Sugioka, N.; Hirata, I.; Ohta, T.; Kishimoto, H. Determination of L-buthionin (SR)-sulfoximine, gamma-glutamylcysteine synthetase inhibitor in rat plasma with HPLC after prelabeling with dansyl chloride, *J.Chromatogr.Sci.*, **1996**, *34*, 326–329.

SAMPLE
Matrix: blood
Analyte: duloxetine

Sample preparation: 1 mL Plasma + 1 mL 1 M pH 10 sodium carbonate buffer + 200 μL 100 ng/mL IS + 6 mL hexane:isopropanol 98:2, shake mechanically for 45 min, centrifuge at 825 g for 10 min, freeze in MeOH/dry ice. Remove the organic layer and evaporate it to dryness at 42°, reconstitute the residue in 2 mL acetone, add 150 μL 100 mM pH 10 sodium carbonate buffer, add 75 μL 500 μg/mL dansyl chloride, heat at 55° for 30 min. Evaporate to dryness under a stream of nitrogen at 42°, reconstitute the residue in 200 μL mobile phase, filter, inject a 100 μL aliquot.

HPLC VARIABLES
Guard column: 12.5 × 4 5 μm Zorbax ODS
Column: 250 × 4.6 5 μm Primesphere 5 C18 (Phenomenex)
Mobile phase: MeCN:50 mM ammonium acetate 79:21
Flow rate: 2
Injection volume: 100
Detector: F ex 285 em 525

CHROMATOGRAM
Retention time: 9.3
Internal standard: N-methyl-3-(1-naphthalenyloxy)-3-phenylpropanamine (Eli Lilly 113821) (10.9)
Limit of quantitation: 5 ng/mL

OTHER SUBSTANCES
Extracted: metabolites, desmethyl duloxetine

KEY WORDS
plasma; silanize glassware; pharmacokinetics

REFERENCE
Johnson, J.T.; Oldham, S.W.; Lantz, R.J.; Delong, A.F. High performance liquid chromatographic method for the determination of duloxetine and desmethyl duloxetine in human plasma, *J.Liq.Chromatogr.Rel.Technol.*, **1996**, *19*, 1631–1641.

SAMPLE
Matrix: blood, urine
Analyte: metapramine
Sample preparation: Plasma. 0.5-2 mL Plasma + 10-1000 ng maprotiline in water + 2 mL pH 10 phosphate buffer + 6 mL diethyl ether:hexane 50:50, shake for 15 min, centrifuge at 3750 g for 5 min. Remove the organic phase and add it to 2 mL 250 mM sulfuric acid, shake for 10 min, centrifuge for 5 min, discard the organic layer. Adjust the pH of the aqueous phase to 9.5-10.5 with 500 mM NaOH containing 1 M K_2HPO_4, add 6 mL diethyl ether:hexane 50:50, shake for 10 min, centrifuge for 10 min. Remove the organic layer and evaporate it to dryness under vacuum and a stream of nitrogen at 45°, reconstitute the residue in 100 μL 100 mM sodium carbonate, add 10 μL 1% dansyl chloride in acetone, vortex for 20-30 s, heat at 45° for 30 min. Evaporate the solvent, reconstitute in 100 μL mobile phase, inject a 10-50 μL aliquot. Urine. 0.5-2 mL Urine + 10-1000 ng maprotiline in water + 2 mL pH 10 phosphate buffer + 6 mL diethyl ether:hexane 50:50, shake for 15 min, centrifuge at 3750 g for 5 min. Remove the organic phase and evaporate it to dryness under vacuum and a stream of nitrogen at 45°, reconstitute the residue in 100 μL 100 mM sodium carbonate, add 10 μL 1% dansyl chloride in acetone, vortex for 20-30 s, heat at 45° for 30 min. Evaporate the solvent, reconstitute in 100 μL mobile phase, inject a 10-50 μL aliquot.

HPLC VARIABLES
Column: 125 × 4 5 μm LiChrosorb RP-18
Mobile phase: MeCN:water 65:35
Flow rate: 2
Injection volume: 10-50
Detector: F ex 248 em 470

CHROMATOGRAM
Retention time: 16.2
Internal standard: maprotiline (18.5)

Limit of detection: 1 ng/mL

OTHER SUBSTANCES
Extracted: metabolites

KEY WORDS
plasma; pharmacokinetics

REFERENCE
Sommadossi, J.P.; Lemar, M.; Necciari, J.; Sumirtapura, Y.; Cano, J.P.; Gaillot, J. High-performance liquid chromatographic method for the determination of plasma and urine metapramine after dansylation, *J.Chromatogr.*, **1982**, *228*, 205–213.

SAMPLE
Matrix: blood, urine
Analyte: vigabatrin
Sample preparation: Plasma. 100 µL Plasma + 20 µL 400 µg/mL IS in water + 200 µL MeCN + 100 µL 30 µM copper chloride, centrifuge at 800 g for 15 min. Remove the supernatant and add it to 200 µL buffer 1 and 200 µL 2 mg/mL dansyl chloride in MeCN, vortex for 5 s, heat at 50° for 15 min, cool to room temperature, wash with 1 mL diethyl ether. Extract the aqueous phase with ethyl acetate. Wash the ethyl acetate layer with 1 mL water. Evaporate the ethyl acetate layer to dryness under a stream of nitrogen at 35°, reconstitute the residue in 2-4 mL mobile phase, inject a 50-100 µL aliquot. Urine. 10 µL Urine + 100 µL water + 20 µL 400 µg/mL IS in water + 200 µL MeCN + 100 µL 15 µM copper chloride, vortex for 5 s, add 200 µL buffer 2, add 200 µL 2 mg/mL dansyl chloride in MeCN, vortex for 5 s, heat at 50° for 15 min, cool to room temperature, wash with 1 mL diethyl ether. Extract the aqueous phase with ethyl acetate. Wash the ethyl acetate layer with 1 mL water. Evaporate the ethyl acetate layer to dryness under a stream of nitrogen at 35°, reconstitute the residue in 2-4 mL mobile phase, inject a 50-100 µL aliquot. (Buffer 1 was 25 mL 200 mM boric acid and 20 mL 50 mM sodium borate made up to 100 mL, pH 8.45 ± 0.05. Buffer 2 was 50 mL 400 mM boric acid and 20 mL 125 mM sodium borate made up to 100 mL, pH 8.05 ± 0.05.)

HPLC VARIABLES
Column: 250 × 4.6 6 µm Zorbax C8
Mobile phase: MeCN:dioxane:500 mM orthophosphoric acid 35:15:50 (Caution! Dioxane is a carcinogen!)
Flow rate: 1
Injection volume: 50-100
Detector: F ex 345 em 418 (cut-off filter)

CHROMATOGRAM
Retention time: 8.8
Internal standard: gamma-aminobenzenebutanoic acid (13.5)
Limit of detection: 500 ng/mL (plasma), 10 µg/mL (urine)

KEY WORDS
plasma; dog; pharmacokinetics

REFERENCE
Smithers, J.A.; Lang, J.F.; Okerholm, R.A. Quantitative analysis of vigabatrin in plasma and urine by reversed-phase high-performance liquid chromatography, *J.Chromatogr.*, **1985**, *341*, 232–238.

SAMPLE
Matrix: cell suspensions, tissue
Analyte: diamines
Sample preparation: Tissue. Homogenize (glass tissue grinder) tissue with 3 volumes 400 mM perchloric acid, centrifuge at 2° at 25000 g for 30 min. Make up 5 (prostate), 25 (liver), 15 (spleen), or 53 (brain) µL to 100 µL with 300 mM perchloric acid, add 100 µL water, add 100 µL saturated sodium carbonate, add 20 µL 200 µM 1,2-diaminoethane, add 600 µL 7 mg/mL dansyl chloride in acetone, shake gently overnight in the dark. Evaporate to dryness under a stream of nitrogen at 40°, add 500 µL water, add 5 mL benzene (Caution! Benzene is a carcinogen!), vortex vigorously for 20 s, centrifuge at 1000 g for 2 min. Remove the organic layer and evaporate it to dryness under a stream of

nitrogen at 40°, reconstitute the residue in 50 µL dichloromethane, inject a 5-10 µL aliquot. Cell suspensions. Centrifuge cell suspension containing 1-2 million cells at 1000 g for 5 min, resuspend the pellet in 300 µL 300 mM perchloric acid, centrifuge. 100 µL Supernatant + 100 µL water + 100 µL saturated sodium carbonate + 20 µL 200 µM 1,2-diaminoethane + 600 µL 7 mg/mL dansyl chloride in acetone, shake gently overnight in the dark. Evaporate to dryness under a stream of nitrogen at 40°, add 500 µL water, add 5 mL benzene (Caution! Benzene is a carcinogen!), vortex vigorously for 20 s, centrifuge at 1000 g for 2 min. Remove the organic layer and evaporate it to dryness under a stream of nitrogen at 40°, reconstitute the residue in 50 µL dichloromethane, inject a 5-10 µL aliquot.

HPLC VARIABLES

Column: 250 × 2.5 Micropak CN-10

Mobile phase: Gradient. A was cyclohexane:isopropanol 98:2. B was cyclohexane:dichloromethane:isopropanol 84:12:4. A:B 100:0 for 5 min, to 0:100 over 15 min (Waters concave gradient curve no. 7), maintain at 0:100 for 2 min, re-equilibrate at initial conditions for 3 min.

Flow rate: 3

Detector: F (wavelengths not given)

CHROMATOGRAM

Retention time: 17 (putrescine), 20 (spermidine), 21 (spermine)

Internal standard: 1,2-diaminoethane (12)

OTHER SUBSTANCES

Interfering: 5-10

KEY WORDS

rat; prostate; liver; spleen; brain

REFERENCE

Newton, N.E.; Ohno, K.; Abdel-Monem, M.M. Determination of diamines and polyamines in tissues by high-pressure liquid chromatography, *J.Chromatogr.*, **1976**, *124*, 277–285.

SAMPLE

Matrix: cheese, tissue

Analyte: amines

Sample preparation: Add 1,7-diaminopentane as IS. Homogenize (Polytron) 10 g cheese with two 20 mL portions of 100 mM HCl. Homogenize (Polytron) 10 g tissue with three 15 mL portions of 5% trichloroacetic acid. Saturate the extracts with NaCl, adjust the pH to 11.5. Remove a 5 mL aliquot and add it to 5 mL butanol (cheese) or butanol:chloroform 50:50 (tissue) vortex for 5 min, repeat extraction twice more. Combine the organic extracts and remove a 1 mL aliquot, add 2 drops 1 M HCl, evaporate to dryness under reduced pressure, reconstitute the residue in 1 mL 100 mM HCl, add 0.5 µL saturated sodium bicarbonate, add 1 mL 5 mg/mL dansyl chloride, heat at 40° for 1 h, evaporate to dryness under reduced pressure, reconstitute with MeCN, inject an aliquot.

HPLC VARIABLES

Column: 150 × 1.6 3 µm Spherisorb 3S TG

Mobile phase: Gradient. MeCN:water 65:35 for 1 min, to 80:20 over 4 min, to 90:10 over 1 min.

Flow rate: 0.8

Injection volume: 10

Detector: UV 254

CHROMATOGRAM

Retention time: 2.5 (tryptamine), 3 (2-phenylethylamine), 3.2 (putrescine), 3.5 (cadaverine), 3.7 (histamine), 4.7 (tyramine), 4.9 (spermidine), 6 (spermine)

Internal standard: 1,7-diaminopentane (4.2)

KEY WORDS

fish; salmon; tuna; salami

REFERENCE
Moret, S.; Conte, L.S. High-performance liquid chromatographic evaluation of biogenic amines in foods. An analysis of different methods of sample preparation in relation to food characteristics, *J.Chromatogr.A*, **1996**, *729*, 363–369.

SAMPLE
Matrix: culture medium
Analyte: polyamines
Sample preparation: Prepare an SPE column by placing 0.7 mL 200-400 mesh Biorex-70 (Na$^+$) (Bio-Rad) in the barrel of a 12 mL syringe, wash with 10 mL water. Add 8-10 mL culture medium containing IS to the SPE column, wash with 1 mL water, wash with 3 mL 150 mM ammonium hydroxide, elute with 2.5 mL 4 M ammonium hydroxide. Add the eluate to 50 μL 2 M HCl, evaporate to dryness under reduced pressure, suspend the residue in 100 μL 400 mM HCl, add 35 μL saturated sodium carbonate, vortex. Add the supernatant to 200 μL 25 mg/mL dansyl chloride in acetone, heat at 70° for 15 min, allow to cool at room temperature for 1 h, make up to 1.5 mL with MeOH, let stand at -20° for at least 1 h, centrifuge at 1500 g for 5 min, inject a 50 μL aliquot of the supernatant.

HPLC VARIABLES
Guard column: 10 × 4.6 7 μm Adsorbosphere nucleotide guard column
Column: 250 × 4.6 5 μm Ultrasphere ODS-DABS
Mobile phase: Gradient, A was MeCN:MeOH:water 30:20:50. B was MeCN:MeOH 60:40. A:B from 90:10 to 72:28 over 8 min, to 65:35 over 3 min, to 25:75 over 12 min, to 10:90 over 4 min, to 5:95 over 2 min, maintain at 5:95 for 8 min, to 25:75 over 1 min, return to initial conditions over 1 min, re-equilibrate for 7 min.
Flow rate: 1
Injection volume: 50
Detector: F ex 330 em 510

CHROMATOGRAM
Retention time: 23 (putrescine), 31 (spermidine), 34 (spermine)
Internal standard: 1,7-diaminoheptane (28)

KEY WORDS
SPE

REFERENCE
Gilbert, R.S.; Gonzalez, G.G.; Hawel, L. III; Byus, C.V. An ion-exchange chromatography procedure for the isolation and concentration of basic amino acids and polyamines from complex biological samples prior to high-performance liquid chromatography, *Anal.Biochem.*, **1991**, *199*, 86–92.

SAMPLE
Matrix: enzyme incubations
Analyte: ammonia and methylamine
Sample preparation: Add 100 μL 1 M HCl in saturated potassium iodate, centrifuge at 2000 rpm for 15-30 min. Remove a 400 μL aliquot of the supernatant and add it to 400 μL buffer, add 400 μL 1.67 mM dansyl chloride in acetone, let stand at room temperature for 1.5 h or heat at 40° for 30 min, inject a 20 μL aliquot. (Prepare buffer by passing 164 mM pH 10.0 sodium borate buffer through a Dowex-50 cation-exchange column, store at room temperature, use within 1 month.)

HPLC VARIABLES
Guard column: 4 × 4.6 Anion-SA, Aminex resin based, sulfate form, anion-exchange (Bio-Rad) (After about 45 injections column can be regenerated by flushing with 1 M sodium sulfate at 0.2-0.4 mL/min for 4 h and with water for 4 h.)
Column: 150 × 4.6 5 μm Ultrasphere ODS
Mobile phase: MeCN:MeOH:water 20:43:36
Flow rate: 2
Injection volume: 20
Detector: F ex 368 em 500

CHROMATOGRAM
Retention time: 2.26 (ammonia), 2.99 (methylamine)

Limit of detection: 500 nM (ammonia), 25 nM (methylamine)

OTHER SUBSTANCES
Simultaneously analyzed: dimethylamine
Non-interfering: azide, cyanide, hydrazine, methylhydrazine, pyridine, valine

REFERENCE
Bravo, M.; Eran, H.; Zhang, F.X.; McKenna, C.E. A rapid, sensitive high-performance liquid chromatography analysis of ammonia and methylamine for nitrogenase assays, *Anal.Biochem.*, **1988**, *175*, 482–491.

SAMPLE
Matrix: formulations
Analyte: alkaloids
Sample preparation: Syrup. Dilute syrup with an equal volume of water. 2 mL Diluted syrup + 2 mL 1% dansyl chloride in acetone + 200 μL 1.5 M sodium carbonate, heat at 45 ± 2° in the dark for 20 min, cool, add 3 mL water, add 500 μL benzene (Caution! Benzene is a carcinogen!), shake. Remove the organic layer and evaporate it to dryness, reconstitute the residue in mobile phase, inject a 10 μL aliquot. Capsules. Sonicate the contents of a capsule in 20 mL water for 10 min, centrifuge at 2000 g for 10 min. 2 mL Supernatant + 2 mL 1% dansyl chloride in acetone + 200 μL 1.5 M sodium carbonate, heat at 45 ± 2° in the dark for 20 min, cool, add 3 mL water, add 500 μL benzene (Caution! Benzene is a carcinogen!), shake. Remove the organic layer and evaporate it to dryness, reconstitute the residue in mobile phase, inject a 10 μL aliquot.

HPLC VARIABLES
Column: 250 × 2.8 10 μm silica gel SI 100 (Merck)
Mobile phase: Diisopropyl ether:isopropanol:concentrated ammonia 48:2:0.3 (Caution! Diisopropyl ether readily forms explosive peroxides!)
Detector: UV 254; F ex 358 em 492 (cephaeline) or ex 356 em 481 (emetine)

CHROMATOGRAM
Retention time: 1.7 (ephedrine), 2 (narcotine), 2.5 (cephaeline), 3.5 (emetine)

OTHER SUBSTANCES
Simultaneously analyzed: codeine (not derivatized, detect at UV 254 only)
Interfering: 10

KEY WORDS
ipecac; syrup; capsules; normal phase

REFERENCE
Frei, R.W.; Santi, W.; Thomas, M. Liquid chromatography of dansyl derivatives of some alkaloids and the application to the analysis of pharmaceuticals, *J.Chromatogr.*, **1976**, *116*, 365–377.

SAMPLE
Matrix: formulations
Analyte: aminophylline
Sample preparation: Tablets. Weigh out powdered tablets containing 100 mg aminophylline, add 50 mL water, sonicate for 15 min, make up to 100 mL with water, mix, filter. Remove a 5 mL aliquot of the filtrate and add it to 10 mL 5 mg/mL dansyl chloride in acetone and 5 mL buffer, mix gently, let stand in the dark for 12 h, make up to 50 mL with acetone:water 50:50, mix, inject an aliquot. Injections, oral liquids. Measure out an amount containing 100 mg aminophylline, make up to 100 mL with water, mix. Remove a 5 mL aliquot and add it to 10 mL 5 mg/mL dansyl chloride in acetone and 5 mL buffer, mix gently, let stand in the dark for 12 h, make up to 50 mL with acetone:water 50:50, mix, inject an aliquot. (Prepare buffer by dissolving 550 mg anhydrous sodium carbonate in 300 mL water, add 300 mL acetone, mix.)

HPLC VARIABLES
Guard column: 70 × 2.1 Co:Pell ODS
Column: 300 × 3.9 10 μm μBondapak C18
Mobile phase: MeOH:water:acetic acid:triethylamine 60:38:1.5:0.5 (A) or 65:33:1.5:0.5 (B)

Flow rate: 1.5
Injection volume: 20
Detector: UV 254

CHROMATOGRAM
Retention time: 6.0 (theophylline (mobile phase B)), 7.3 (ethylenediamine (mobile phase B)), 7.45 (theophylline (mobile phase A)), 12.00 (ethylenediamine (mobile phase A))

KEY WORDS
tablets; injections; oral solutions

REFERENCE
Lau-Cam, C.A.; Roos, R.W. Simultaneous high performance liquid chromatographic determination of theophylline and ethylenediamine in aminophylline dosage forms as their dansyl derivatives, *J.Liq.Chromatogr.*, **1991**, *14*, 1939–1956.

SAMPLE
Matrix: formulations
Analyte: calcium cyanamide
Sample preparation: Grind tablets, weigh out amount corresponding to 25 mg calcium cyanamide, add 20 mL buffer, sonicate for 10 min, make up to 100 mL with 200 mM sodium carbonate, let stand at room temperature for 15 min. Remove a 200 µL aliquot of the clear supernatant and add it to 200 µL 10 mg/mL dansyl chloride in acetone, mix well, heat at 40° for 30 min, cool to room temperature, add 4 mL mobile phase, mix, inject a 10 µL aliquot. (Prepare buffer by dissolving 12 g anhydrous sodium acetate and 20 mL glacial acetic acid in 1 L water.)

HPLC VARIABLES
Column: 150 × 3.9 10 µm µBondapak C18
Mobile phase: MeCN:10 mM pH 6.3 sodium phosphate 25:75
Flow rate: 1
Injection volume: 10
Detector: UV 254

CHROMATOGRAM
Retention time: 5.3

KEY WORDS
tablets; stability-indicating

REFERENCE
Chen, S.; Ocampo, A.P.; Kucera, P.J. Liquid chromatographic method for the determination of calcium cyanamide using pre-column derivatization, *J.Chromatogr.*, **1991**, *558*, 141–146.

SAMPLE
Matrix: formulations
Analyte: aminocaproic acid
Sample preparation: Tablets. Weigh out powdered tablet containing aminocaproic acid, dissolve in 100 mL water, filter (0.45 µm). Mix a 5 mL aliquot of the filtrate with 10 mL 5 mg/mL dansyl chloride in acetone and 10 mL 400 µg/mL tranexamic acid in buffer, let stand in the dark at room temperature for 30 min, add 2 drops ethanolamine, mix, let stand at room temperature for 15 min, make up to 50 mL with acetone:water 50:50, mix, inject an aliquot. Injections, syrup. Weigh out amount of injection or syrup containing 250 mg aminocaproic acid, dilute with 100 mL water, dilute an aliquot 5-fold with water. Mix a 5 mL aliquot with 10 mL 5 mg/mL dansyl chloride in acetone and 10 mL 400 µg/mL tranexamic acid in buffer, let stand in the dark at room temperature for 30 min, add 2 drops ethanolamine, mix, let stand at room temperature for 15 min, make up to 50 mL with acetone:water 50:50, mix, inject an aliquot. (Prepare buffer by dissolving 550 mg anhydrous sodium carbonate in 300 mL water, add 300 mL acetone, mix.)

HPLC VARIABLES
Guard column: C18 (Alltech)
Column: 150 × 4.6 5 µm Econosphere C18
Mobile phase: MeOH:water:acetic acid:triethylamine 60:38:1.5:0.5

Flow rate: 1.5
Injection volume: 20
Detector: UV 335

CHROMATOGRAM
Retention time: 4.5
Internal standard: tranexamic acid (6.5)

KEY WORDS
tablets; injections; syrup

REFERENCE
Lau-Cam, C.A.; Roos, R.W. Assay of aminocaproic acid in dosage forms by reversed phase high performance liquid chromatography with dansylation, *J.Liq.Chromatogr.*, **1993**, *16*, 403–419.

SAMPLE
Matrix: formulations
Analyte: piperazine
Sample preparation: Tablets. Grind tablets to a fine powder, weigh out amount equivalent to about 200 mg piperazine citrate, dissolve in 50 mL water, sonicate for 15 min, make up to 100 mL with water, filter (0.45 μm). Remove a 3 mL aliquot and add it to 2 mL 4 mg/mL 1-benzylpiperazine in acetone:water 50:50, add 10 mL of a filtered 5 mg/mL solution of dansyl chloride in acetone, add 10 mL base solution, mix, sonicate for 10 min, let stand in the dark for 30 min, add 20 mL water, add 20 mL chloroform, shake vigorously for 1 min, filter the chloroform layer through anhydrous sodium sulfate, dilute 2 mL of the filtrate to 5 mL with mobile phase, inject an aliquot. Syrup. Dilute 5 mL syrup to 100 mL with water, mix, measure out an aliquot equivalent to about 200 mg piperazine citrate, make up to 100 mL with water, mix. Remove a 3 mL aliquot and add it to 2 mL 4 mg/mL 1-benzylpiperazine in acetone:water 50:50, add 10 mL of a filtered 5 mg/mL solution of dansyl chloride in acetone, add 10 mL base solution, mix, sonicate for 10 min, let stand in the dark for 30 min, add 20 mL water, add 20 mL chloroform, shake vigorously for 1 min, filter the chloroform layer through anhydrous sodium sulfate, dilute 2 mL of the filtrate to 5 mL with mobile phase, inject an aliquot. Granules, powders (effervescent). Weigh out amount equivalent to about 200 mg piperazine citrate, slowly add 50 mL water with swirling, sonicate for 10 min, make up to 100 mL with water, filter (0.45 μm). Remove a 3 mL aliquot and add it to 2 mL 4 mg/mL 1-benzylpiperazine in acetone:water 50:50, add 10 mL of a filtered 5 mg/mL solution of dansyl chloride in acetone, add 10 mL base solution, mix, sonicate for 10 min, let stand in the dark for 30 min, add 20 mL water, add 20 mL chloroform, shake vigorously for 1 min, filter the chloroform layer through anhydrous sodium sulfate, dilute 2 mL of the filtrate to 5 mL with mobile phase, inject an aliquot. (Prepare base solution by dissolving 550 mg anhydrous sodium carbonate in 300 mL water, add 300 mL acetone, mix.)

HPLC VARIABLES
Column: 250 × 4.6 5 μm CN5 SG cyanopropyl (Burdick & Jackson)
Mobile phase: Hexane:isopropanol 85:15
Flow rate: 1.5
Injection volume: 20
Detector: UV 335

CHROMATOGRAM
Retention time: 8.5
Internal standard: 1-benzylpiperazine (4.0)

KEY WORDS
tablets; syrup; granules; powders

REFERENCE
Lau-Cam, C.A.; Roos, R.W. Normal-phase high performance liquid chromatographic method with dansylation for the assay of piperazine citrate in dosage forms, *J.Liq.Chromatogr.*, **1995**, *18*, 3347–3357.

SAMPLE
Matrix: solutions

Analyte: diamines
Sample preparation: Prepare dansyl derivative, inject an aliquot.

HPLC VARIABLES
Column: 250 × 2.2 Micropak A1-5
Mobile phase: Chloroform:isopropanol 100:1
Flow rate: 1
Detector: UV 280

CHROMATOGRAM
Retention time: 0.5 (spermine), 1 (spermidine), 1.5 (cadaverine), 2.5 (putrescine), 4 (1,3-diaminopropane), 6 (1,2-diaminoethane), 11.5 (ammonia)

KEY WORDS
normal phase

REFERENCE
Abdel-Monem, M.M.; Ohno, K. Separation of the Dns derivatives of polyamines and related compounds by thin-layer and high-pressure liquid chromatography, *J.Chromatogr.*, **1975**, *107*, 416–419.

SAMPLE
Matrix: solutions
Analyte: amines
Sample preparation: 100 μL 100 μM Solution + 300 μL buffer + 500 μL 1 mM dansyl chloride in acetone, mix, heat at 45° in the dark for 1 h, dilute with MeCN:water 50:50, inject a 20 μL aliquot. (Prepare buffer by adjusting 10 mM sodium bicarbonate to pH 9.0 with NaOH.)

HPLC VARIABLES
Guard column: 30 × 4.6 Spheri-5 RP-18
Column: 250 × 4.6 Inertsil ODS-2
Mobile phase: MeCN:water 70:30 containing 1 mM imidazole, pH adjusted to 7.0 with nitric acid
Flow rate: 1
Injection volume: 20
Detector: Chemiluminescence following post-column reaction. The column effluent mixed with the reagent pumped at 1 mL/min and the mixture flowed through a 300 mm × 0.25 mm ID coil to the detector. (Prepare the reagent by dissolving 112 mg bis(2,4,6-trichlorophenyl) oxalate in 500 mL MeCN, add 8.6 mL 30% hydrogen peroxide, sonicate.); F ex 343 em 530

CHROMATOGRAM
Retention time: 7 (phenylpropanolamine), 8 (benzylamine), 9 (ephedrine), 10 (phenylethylamine), 11 (phenylpropylamine), 15 (phenylbutylamine), 16 (N-methylphenethylamine), 18 (methamphetamine), 20.5 (N-isopropylbenzylamine)
Limit of detection: 3-4 fmole (chemiluminescence), 40-50 fmole (F)

KEY WORDS
post-column reaction; comparison with other derivatization reagents

REFERENCE
Hayakawa, K.; Hasegawa, K.; Imaizumi, N.; Wong, O.S.; Miyazaki, M. Determination of amphetamine-related compounds by high-performance liquid chromatography with chemiluminescence and fluorescence detections, *J.Chromatogr.*, **1989**, *464*, 343–352.

SAMPLE
Matrix: solutions
Analyte: 2-methyltaurine
Sample preparation: Mix 2 mL of a 300 μg/mL solution of 2-methyltaurine in buffer with 1 mL 1 mg/mL dansyl chloride in MeCN, stir for 2 min, let stand at room temperature for 1 h, inject an aliquot. (Prepare buffer by adjusting the pH of 40 mM sodium carbonate to 9.5 with 2 M HCl.)

CAPILLARY ELECTROPHORESIS
Capillary: 80 cm × 75 μm fused-silica (71 cm to detector) (Supelco)

Capillary preparation: Wash with 100 mM NaOH for 10 min, with water for 10 min, and with running buffer for 10 min. At the end of each day wash with 100 mM NaOH for 10 min and with water for 10 min.

Running buffer: MeOH:buffer 20:80 containing 100 mM sodium dodecyl sulfate, 60 mM β-cyclodextrin and 10 mM gamma-cyclodextrin, pH 8.6 (Prepare by dissolving 618 mg boric acid in 100 mL MeOH:water 20:80, adjust pH to 8.6 with 2 M NaOH, add the other compounds.)

Voltage/Current: 20 kV/40 μA

Injection: Hydrodynamic injection at 5 kPa for 6 s

Detector: UV 254

Model: Prince (Lauerlabs, Netherlands)

Migration time: 22.56, 23.12 (enantiomers)

KEY WORDS
chiral

REFERENCE
Anselmi, S.; Braghiroli, D.; Di Bella, M.; Schmid, M.G.; Wintersteiger, R.; Gübitz, G. Chiral resolution of the dansyl derivative of 2-methyltaurine by capillary electrophoresis, *J.Chromatogr.B*, **1996**, *681*, 83–86.

SAMPLE
Matrix: tissue

Analyte: amines

Sample preparation: Homogenize (glass Potter) tissue with 4 volumes 300 mM perchloric acid, centrifuge at 10000 g for 30 min. Remove a 200 μL aliquot of the supernatant and add it to 20 μL 200 μM 1,8-diaminooctane, add 15 μL 3 M KOH, add 100 μL buffer, add 500 μL 10 mg/mL dansyl chloride in acetone, vortex, let stand in the dark overnight, add 200 μL 3 M KOH, vortex vigorously, extract with benzene (Caution! Benzene is a carcinogen!). Remove the organic layer and evaporate it to dryness under reduced pressure, reconstitute the residue in 20 μL isopropanol, inject a 2-10 μL aliquot. (Prepare buffer by adjusting pH of 1.5 M sodium carbonate to 10 with saturated sodium bicarbonate solution.)

HPLC VARIABLES
Guard column: 25 × 3.9 pellicular ODS

Column: 150 × 4.6 5 μm Ultrasphere ODS

Mobile phase: Gradient. A was MeCN:MeOH:water 30:20:50. B was MeCN:MeOH 60:40. A:B from 72:28 to 10:90 over 19 min (Waters concave curve no. 8), maintain at 10:90 for 7 min, return to initial conditions over 1 min, re-equilibrate for 4 min.

Flow rate: 1

Injection volume: 2-10

Detector: F ex 338 (filter) em 425 (cutoff filter)

CHROMATOGRAM
Retention time: 15 (N^8-acetylspermidine), 16 (N^1-acetylspermidine), 19 (putrescine), 20 (cadaverine), 24 (spermidine), 25 (spermine)

Internal standard: 1,8-diaminooctane (22.5)

Limit of detection: 1-3 pmole

KEY WORDS
rat; lung; spleen; liver

REFERENCE
Stefanelli, C.; Carati, D.; Rossoni, C. Separation of N^1- and N^8-acetylspermidine isomers by reversed-phase column liquid chromatography after derivatization with dansyl chloride, *J.Chromatogr.*, **1986**, *375*, 49–55.

SAMPLE
Matrix: tissue

Analyte: amines

Sample preparation: 10 g Homogenized fish + 30 mL 5% trichloroacetic acid in water, stir vigorously with a glass rod at 89-90° for 2 min, filter, dilute the filtrate 10-fold with

5% trichloroacetic acid in water. Remove a 1 mL aliquot and add it to 1 drop 4 M NaOH, add 1 mL pH 9 phosphate buffer (Titrisol, Merck), add 2 mL 5 mg/mL dansyl chloride in acetone (prepare fresh each day), stir for 30 s, heat in the dark at 55° for 1 h, cool, inject a 20 µL aliquot.

HPLC VARIABLES
Column: 250 × 3 10 µm Lichrosorb RP-8
Mobile phase: Gradient. MeOH:water from 70:30 to 80:20 over 7 min, to 100:0 over 4.5 min, maintain at 100:0 for 7.5 min, return to initial conditions over 0.1 min, re-equilibrate for 10 min.
Column temperature: 40
Flow rate: 1.28
Injection volume: 20
Detector: UV 254

CHROMATOGRAM
Retention time: 6 (putrescine), 7 (cadaverine), 9 (histamine), 11 (spermidine), 13 (spermine)

KEY WORDS
fish

REFERENCE
Rosier, J.; Van Peteghem, C. A screening method for the simultaneous determination of putrescine, cadaverine, histamine, spermidine and spermine in fish by means of high pressure liquid chromatography of their 5-dimethylaminonaphthalene-1-sulphonyl derivatives, *Z.Lebensm.Unters.Forsch.*, **1988**, *186*, 25–28.

SAMPLE
Matrix: tissue
Analyte: amines
Sample preparation: Homogenize (Ultraturrax) 5 g fish, 10 mL 200 mM perchloric acid, and 100 µL 800 µg/mL 1,3-diaminopropane dihydrochloride in water at -20° at 20000 rpm and centrifuge at 2° at 2500 g for 20 min. Remove a 100 µL aliquot of the supernatant and add it to 200 µL saturated sodium bicarbonate solution, add 400 µL 7.5 mg/mL dansyl chloride in acetone, agitate, heat at 60° in the dark for ?, add 100 µL 100 mg/mL L-proline in water, agitate, let stand in the dark at room temperature for 30 min, add 500 µL toluene, agitate. Remove the organic layer and evaporate it to dryness under a stream of nitrogen, reconstitute the residue in 300 µL MeCN, filter, inject an aliquot.

HPLC VARIABLES
Guard column: 30 × 4.6 5 µm Brownlee C18
Column: 250 × 4.6 5 µm Kromasil C18
Mobile phase: Gradient. MeCN:water from 60:40 to 75:25 over 6 min, maintain at 75:25 for 2 min, to 95:5 over 5 min, maintain at 95:5 for 7 min, re-equilibrate at initial conditions for 10 min.
Column temperature: 25
Flow rate: 1
Detector: UV 254

CHROMATOGRAM
Retention time: 13.1 (putrescine), 14.1 (cadaverine), 14.7 (histamine), 18.4 (spermidine), 21.7 (spermine)
Internal standard: 1,3-diaminopropane (12.5)

KEY WORDS
fish

REFERENCE
Malle, P.; Vallé, M.; Bouquelet, S. Assay of biogenic amines involved in fish decomposition, *J.AOAC Int.*, **1996**, *79*, 43–49.

SAMPLE
Matrix: urine

Analyte: polyamines
Sample preparation: 1 mL Urine + 1 mL 6 M HCl, heat at 110° for 20 h, cool, filter (Tessek cellulose microfilter), evaporate the filtrate to dryness, reconstitute with 300 μL water. Remove a 50 μL aliquot and add it to 100 μL 10 mg/mL dansyl chloride in MeCN, add 30 μL saturated sodium carbonate, heat at 55° for 5 min, evaporate to dryness, extract the residue twice with 1 mL portions of benzene (Caution! Benzene is a carcinogen!). Combine the extracts and evaporate them to dryness under reduced pressure, reconstitute with 50 μL MeOH, inject a 20 μL aliquot.

HPLC VARIABLES
Column: 150 × 3.2 7 μm Separon C-18 (Tessek, Prague)
Mobile phase: Gradient. MeOH:water from 75:25 to 95:5 over 6 min, maintain at 95:5 for 9 min, return to initial conditions over 5 min.
Flow rate: 0.5
Injection volume: 20
Detector: UV 250

CHROMATOGRAM
Retention time: 6.5 (putrescine), 10 (spermidine), 12 (spermine)
Limit of detection: 20-40 pmoles

REFERENCE
Brandsteterová, E.; Hatrik, S.; Blanárik, I.; Marcincinová, K. HPLC determination of polyamines in urine, *Neoplasma*, **1991**, *38*, 165–174.

SAMPLE
Matrix: urine
Analyte: hydroxylysine glycosides
Sample preparation: 100 μL Urine + 100 μL 300 mM pH 9.5 carbonate buffer + 200 μL freshly-prepared 40 mM dansyl chloride in MeCN, heat at 60° for 20 min, inject a 10 μL aliquot.

HPLC VARIABLES
Column: 250 × 4.6 Inertsil ODS-2
Mobile phase: Gradient. A was MeCN:25 mM pH 6.5 sodium acetate buffer 20:80. B was MeCN:25 mM pH 6.5 sodium acetate buffer 80:20. A:B from 0:100 to 80:20 over 30 min, maintain at 80:20 for 20 min, to 0:100 over 10 min.
Flow rate: 1
Injection volume: 10
Detector: F ex 325 em 535

CHROMATOGRAM
Retention time: 41.22 (2-O-β-D-glucopyranosyl-O-β-D-galactopyranosylhydroxylysine), 46.45 (β-D-galactopyranosylhydroxylysine)

REFERENCE
Yoshihara, K.; Mochidome, N.; Shida, Y.; Hayakawa, Y.; Nagata, M. Pre-column derivatization and its optimum conditions for quantitative determination of urinary hydroxylysine glycosides by high-performance liquid chromatography, *Biol.Pharm.Bull.*, **1993**, *16*, 604–607.

SAMPLE
Matrix: water
Analyte: carbamate insecticides
Sample preparation: Extract 500 mL water with two 25 mL portions of dichloromethane, combine the extracts and dry them over anhydrous sodium sulfate for 10 min, evaporate to dryness under a stream of air, reconstitute with 40 μL acetone, add 300 μL 100 mM sodium carbonate, heat at 45-50° for 30-40 min, cool, add 300 μL acetone, add 100 μL 0.2% dansyl chloride in acetone, mix well, heat at 45° for 20 min, cool, evaporate the acetone under a stream of air, extract with 300 μL benzene (Caution! Benzene is a carcinogen!). Remove the organic layer and dry it over anhydrous sodium sulfate, inject a 1-10 μL aliquot.

HPLC VARIABLES
Column: 1000 × 2.4 Zipax coated with 0.5% β,β'-oxydipropionitrile

Mobile phase: Hexane:EtOH 95:5
Injection volume: 1-10
Detector: F primary filter Turner 810, secondary filter Turner 827

CHROMATOGRAM
Retention time: k' 0.57 (carbofuran), k' 0.59 (carbaryl (Sevin)), k' 0.67 (propoxur (Baygon)), k' 1.03 (Carzol), k' 1.06 (dimethylamine), k' 1.25 (Mobam), k' 14.24 (aldicarb (Temik)), k' 3.87 (methylamine), k' 8.13 (methomyl)

KEY WORDS
lake water; normal phase

REFERENCE
Frei, R.W.; Lawrence, J.F.; Hope, J.; Cassidy, R.M. Analysis of carbamate insecticides by fluorigenic labelling and high-speed liquid chromatography, *J.Chromatogr.Sci.*, **1974**, *12*, 40–44.

5-Di-n-butylaminonaphthalene-1-sulfonyl Chloride

SAMPLE
Matrix: blood
Analyte: amosulalol (YM 09538)
Sample preparation: 1 mL Plasma + 2 mL 0.5 µg/mL IS in water + 0.5 g sodium bicarbonate + 4 mL ethyl acetate, extract, centrifuge. Remove the organic layer and evaporate it to dryness under reduced pressure, reconstitute the residue in 100 µL 3 mg/mL sodium bicarbonate in water and 200 µL 500 µg/mL 5-di-n-butylaminonaphthalene-1-sulfonyl chloride in acetone, heat at 45° for 90 min, cool, add 4 mL diethyl ether, wash mixture with 3 mL water for 10 s. Remove the organic layer and evaporate it to dryness at 40-50°, reconstitute the residue in 100 µL benzene (Caution! Benzene is a carcinogen!), inject a 5-10 µL aliquot. (Derivatization can also be performed with dansyl chloride, F ex 250 em 505, retention time of derivative 4.1 min. 5-Di-n-butylaminonaphthalene-1-sulfonyl chloride (Bans-Cl) is available from Tokyo Kasei (TCI America, Portland OR). Synthesis is as follows. Heat 5-amino-1-naphthalenesulfonic acid (1-naphthylamine-5-sulfonic acid; Fluka), n-butyl iodide, and anhydrous KF in DMF at 130-140° for 18 h with vigorous agitation to obtain 5-di-n-butylaminonaphthalene-1-sulfonic acid (J. Chromatogr. 1980, 192, 341). Grind 2.5 g 5-di-n-butylaminonaphthalene-1-sulfonic acid with 3.5 g phosphorus pentachloride in a mortar, pour into water (cf Biochem. J. 1952, 51, 155), extract with ethyl acetate, wash the organic layer several times with water, dry over sodium sulfate, evaporate to dryness under reduced pressure, chromatograph on a column of silica gel H (Merck) with toluene to obtain 5-di-n-butylaminonaphthalene-1-sulfonyl chloride as an oily orange-red product that slowly crystallizes (J. Chromatogr. 1973, 84, 95).)

HPLC VARIABLES
Column: 150 × 4 5 µm LiChrosorb SI-60
Mobile phase: Benzene:MeOH 50:1 (Caution! Benzene is a carcinogen)
Flow rate: 2

Injection volume: 5-10
Detector: F ex 356 em 500

CHROMATOGRAM
Retention time: 2.9
Internal   standard: 5-{1-hydroxy-2-[2-(o-methoxyphenoxy)ethylamino]ethyl}-2-methoxy-benzenesulfonamide (4.0)
Limit of detection: 20 ng/mL

KEY WORDS
plasma; normal phase; dog; pharmacokinetics

REFERENCE
Kamimura, H.; Sasaki, H.; Kawamura, S. Determination of the α,β-adrenoceptor blocker YM-09538 in plasma by high-performance liquid chromatography with fluorescence detection, *J.Chromatogr.*, **1981**, *225*, 115–121.

SAMPLE
Matrix: solutions
Analyte: melanotropin inhibiting factor (MIF)
Sample preparation: Mix 5 μL of a 3.6 mM solution with 5 μL 500 mM sodium bicarbonate, add 5 μL 1 mg/mL 5-di-n-butylaminonaphthalene-1-sulfonyl chloride in acetone, shake vigorously, heat at 37° in the dark for 1 h, add 100 μL ethyl acetate, inject a 1-20 μL aliquot of the organic layer. (Prepare 5-di-n-butylaminonaphthalene-1-sulfonyl chloride (Bans-Cl) as described above.)

HPLC VARIABLES
Column: μBondapak C18
Mobile phase: MeCN:10 mM pH 7.0 sodium sulfate buffer 45:55
Flow rate: 3
Injection volume: 1-20
Detector: UV 254; F ex 360 em 487

CHROMATOGRAM
Retention time: 8
Limit of detection: 3.388 nmoles

OTHER SUBSTANCES
Simultaneously analyzed: metabolites

KEY WORDS
comparison with other derivatizing reagents; other chromatographic systems described

REFERENCE
Hui, K-S.; Salschutz, M.; Davis, B.A.; Lajtha, A. Separation of alkylaminonaphthylenesulfonyl peptides and amino acids by high-performance liquid chromatography. Methods for measuring melanotropin inhibiting factor breakdown, *J.Chromatogr.*, **1980**, *192*, 341–350.

2-(5',6'-Dimethoxybenzothiazolyl)benzenesulfonyl Chloride

SAMPLE
Matrix: solutions
Analyte: amines
Sample preparation: Mix 100 μL of a solution in MeCN with 100 μL 10 mM triethylamine in MeCN and 100 μL 1 mM 2-(5',6'-dimethoxybenzothiazolyl)benzenesulfonyl chloride in MeCN, let stand at room temperature for 2-3 min, inject a 20 μL aliquot. (Preparation of 2-(5',6'-dimethoxybenzothiazolyl)benzenesulfonyl chloride is as follows. Stir 20 g 4-bromoveratrole in 60 mL acetic acid, add 10 mL concentrated nitric acid dropwise while keeping the temperature at 10-30° with occasional cooling, pour into ice-water. Collect the precipitate and dissolve it in 500 mL hot EtOH, add activated charcoal, filter, add 40 mL water to the filtrate to obtain 4-bromo-5-nitroveratrole as light-yellow crystals (mp 121-122°). Dissolve 5 g 4-bromo-5-nitroveratrole in 50 mL 95% EtOH, add sodium sulfide (prepare immediately before use by melting 5 g sodium sulfide nonahydrate and 700 mg sulfur together), reflux for 30 min, pour into ice-water. Collect the solid and recrystallize it from dichloromethane to obtain 2,2'-dithiobis(1-nitro-4,5-dimethoxybenzene) as yellow needles (mp 231-232°). Dissolve 2 g 2,2'-dithiobis(1-nitro-4,5-dimethoxybenzene) in 300 mL EtOH, add 8 g tin powder, add 30 mL concentrated HCl dropwise, make alkaline with 4 M NaOH, filter. Dilute the filtrate with 200 mL water, extract twice with 100 mL portions of benzene (Caution! Benzene is a carcinogen!). Combine the extracts and evaporate them to dryness under reduced pressure, mix the residue with 10 mL benzene and 2 mL 10% hydrogen peroxide, stir for 30 min, recrystallize the precipitate from EtOH to give 2,2'-dithiobis(1-amino-4,5-dimethoxybenzene) as colorless crystals (mp 155-156°) (Anal. Chim. Acta 1994, 291, 189). Add a solution of 2 g 2,2'-dithiobis(1-amino-4,5-dimethoxybenzene) in 50 mL EtOH containing 100 mg tri-n-butylphosphine and 2 M HCl to a solution of 2 g sodium 2-formylbenzenesulfonate in 30 mL water, reflux for 1 h, filter, wash the solid with water, dry under reduced pressure, recrystallize from EtOH to obtain 2-(5',6'-dimethoxybenzothiazolyl)benzenesulfonic acid as colorless needles (mp >300°). Dissolve 600 mg 2-(5',6'-dimethoxybenzothiazolyl)benzenesulfonic acid in 20 mL freshly distilled thionyl chloride, reflux for 1 h, cool, add 50 mL light petroleum (bp 30-60°), filter, recrystallize the solid from benzene:petroleum 50:50 to obtain 2-(5',6'-dimethoxybenzothiazolyl)benzenesulfonyl chloride as pale yellow needles (mp >300°). Prepare solutions by dissolving the compound in 2-methoxyethanol then diluting with MeCN, use within 24 h.)

HPLC VARIABLES
Column: 250 × 4.6 5 μm TSK gel ODS-120T (Tosoh)
Mobile phase: MeOH:water 70:30
Flow rate: 1
Injection volume: 20
Detector: F ex 330 em 450

CHROMATOGRAM
Retention time: 19 (n-propylamine), 25 (n-heptylamine), 41 (N-methylhexylamine)
Limit of detection: 10.2-13.1 fmole

OTHER SUBSTANCES
Also analyzed: n-amylamine, benzylamine, n-butylamine, cyclohexylamine, dibenzylamine, di-n-butylamine, diethylamine, di-n-propylamine, N-ethylbenzylamine, n-hexylamine, histamine, 4-methylbenzylamine, n-nonylamine, n-octylamine, 2-phenylethylamine, tyramine

Non-interfering: acetoacetic acid, acetone, acetophenone, acetylacetone, N-acetylneuraminic acid, allantoin, alloxan, amino acids, aniline, ascorbic acid, benzil, bilirubin, n-butanol, cholesterol, citrulline, cortisone, cyclohexane, dehydroepiandrosterone, epiandrosterone, ethanol, D-fructose, D-glucose, glutathione, homogentisic acid, 3-hydroxybutyric acid, inositol, lactic acid, D-lactose, D-maltose, D-mannose, methanol, 4-methylcyclohexane, methylglyoxal, phenylpyruvic acid, pyruvic acid, thiamine, o-toluidine, urea, uric acid, D-xylose

REFERENCE
Hara, S.; Aoki, J.; Yoshikuni, K.-i.; Tatsuguchi, Y.; Yamaguchi, M. 4-(5',6'-Dimethoxybenzothiazolyl)benzoyl fluoride and 2-(5',6'-dimethoxybenzothiazolyl)benzenesulfonyl chloride as sensitive fluorescence derivatization reagents for amines in high-performance liquid chromatography, *Analyst*, **1997**, *122*, 475–479.

2-Naphthalenesulfonyl Chloride

SAMPLE
Matrix: blood
Analyte: spectinomycin
Sample preparation: Condition a 3 mL C18 SPE cartridge (J.T. Baker) with 3 mL MeOH, 2 mL 20 mM sodium dioctylsulfosuccinate (slowly!), and 3 mL 20 mM pH 5.6 citric acid, do not allow to go dry. 2 mL Plasma + 8 mL water + 625 µL MeOH, adjust pH to 5.2-5.7 with about 60 µL 1 M HCl, vortex for 30 s, centrifuge at 2500 g for 10 min, remove the supernatant, resuspend the solid in 3 mL 20 mM pH 5.6 citric acid, centrifuge. Combine the supernatants and add them to the SPE cartridge at 1-2 mL/min, wash with two 3 mL portions of 20 mM pH 5.6 citric acid, allow to go dry. Centrifuge the SPE cartridge at 3000 g for 15 min, dry with a stream of nitrogen for 20 min, elute with 3 mL MeOH at 2 drops/s. Evaporate the eluate to dryness under a stream of nitrogen at room temperature, reconstitute the residue in 2 mL 10 g/L sodium bicarbonate containing 100 µL/L 1-methylpyrrole, vortex for 30 s, add 3 mL 12 mg/mL 2-naphthalenesulfonyl chloride in MeCN (prepare just before derivatization), vortex for 30 s, heat at 100° for 15 min, cool to room temperature, add 1 mL n-butyl chloride, vortex twice for 20 s, centrifuge at 1500 g for 5 min. Remove 2 mL of the organic layer and evaporate it to dryness under a stream of nitrogen at room temperature, reconstitute the residue in 1 mL n-butyl chloride, vortex for 30 s, centrifuge at 2500 g for 10 min, inject a 50 µL aliquot onto column A and elute to waste with mobile phase A, after 8 min divert the effluent from column A containing spectinomycin onto column B, after 2 min remove column A from the circuit, elute column B with mobile phase B, monitor the effluent from column B. Backflush column A with mobile phase A for 15 min.

HPLC VARIABLES
Column: A 10 × 2.1 40 μm pellicular silica + 100 × 3 5 μm Chromspher silica glass cartridge (Chrompack); B two 100 × 3 5 μm Chromspher silica glass cartridges in series (Chrompack)

Mobile phase: A Dichloromethane:MeCN:ethyl acetate:acetic acid 75:7.5:1.8:0.425; B Dichloromethane:MeCN:ethyl acetate:acetic acid 100:20:3.6:0.85

Flow rate: A 0.4; B 0.6

Injection volume: 50

Detector: UV 250

CHROMATOGRAM
Retention time: 14

Limit of quantitation: 40 ng/mL

KEY WORDS
cow; calf; chicken; pig; plasma; SPE; use glass not plastic tubes; column-switching; heart-cut; normal phase

REFERENCE
Haagsma, N.; Keegstra, J.R.; Scherpenisse, P. High-performance liquid chromatographic determination of spectinomycin in swine, calf and chicken plasma, *J.Chromatogr.*, **1993**, *615*, 289–295.

SAMPLE
Matrix: bulk

Analyte: spectinomycin

Sample preparation: 10 mL 250 μg/mL Spectinomycin solution in buffer + 10 mL 10 mg/mL 2-naphthalenesulfonyl chloride in MeCN containing 200 μg/mL methylprednisolone acetate (prepare fresh daily), shake, heat at 100° for 10 min, cool to room temperature, make up to 50 mL with mobile phase, shake vigorously for 10 min, centrifuge at <300 g for 3-5 min, inject a 40 μL aliquot of the organic layer. (Buffer was 4.2 g/L sodium bicarbonate containing 100 μL/L 1-methylpyrrole.)

HPLC VARIABLES
Column: 250 × 4.6 5 μm LiChrosorb SI-60

Mobile phase: Butyl chloride:THF:ethyl acetate:isopropanol:acetic acid 86:3.7:3:2.5:5 (Butyl chloride was 50% water-saturated.)

Flow rate: 1

Injection volume: 40

Detector: UV 254

CHROMATOGRAM
Retention time: 10

Internal standard: methylprednisolone acetate (18)

Limit of detection: 4.2 ng

OTHER SUBSTANCES
Simultaneously analyzed: actinosapectinoic acid

KEY WORDS
normal phase

REFERENCE
Tsuji, K.; Jenkins, K.M. Derivatization of secondary amines with 2-naphthalene-sulfonyl chloride for high-performance liquid chromatographic analysis of spectinomycin, *J.Chromatogr.*, **1985**, *333*, 365–380.

SAMPLE
Matrix: bulk

Analyte: neomycin

Sample preparation: Dissolve 20 mg neomycin sulfate powder in 100 mL 100 mM pH 8.0 sodium phosphate buffer. Remove a 10 mL aliquot and add it to 10 mL 40 mg/mL 2-naphthalenesulfonyl chloride in MeCN (prepare fresh daily), shake briefly, heat at 100-105° for 10 min, cool to room temperature, add 15 mL IS solution, shake vigorously for 10 min, centrifuge at <300 g for 3-5 min, inject a 50 μL aliquot of the lower organic layer.

(Prepare IS solution by dissolving 2 mg prednisolone in a small amount of THF, make up to 100 mL with chloroform.)

HPLC VARIABLES
Column: 125 × 4.6 5 μm P-E HS-5 silica (Perkin-Elmer)
Mobile phase: Chloroform:MeOH:acetic acid 95:2.3:2.5
Flow rate: 1.7
Injection volume: 50
Detector: UV 254

CHROMATOGRAM
Retention time: 6.2 (neomycin B), 8.3 (neomycin C)
Internal standard: prednisolone (12.3)

OTHER SUBSTANCES
Simultaneously analyzed: impurities, neamine

KEY WORDS
normal phase

REFERENCE
Tsuji, K.; Jenkins, K.M. Derivatization of primary amines by 2-naphthalenesulfonyl chloride for high-performance liquid chromatographic assay of neomycin sulfate, *J.Chromatogr.*, **1986**, *369*, 105–115.

8-Quinolinesulfonyl chloride

SAMPLE
Matrix: bulk
Analyte: amines
Sample preparation: Mix a 1 mg/mL solution in 1 M sodium carbonate with 2 mL 5 mg/mL 8-quinolinesulfonyl chloride in acetone, heat at 65° for 20 min, cool, extract twice with 30 mL portions of chloroform. Combine the extracts and dry them over anhydrous magnesium sulfate, evaporate to dryness under a stream of air, reconstitute, inject an aliquot.

HPLC VARIABLES
Guard column: 70 × 2.1 Co:Pell ODS
Column: 300 × 3.9 μBondapak C18
Mobile phase: MeCN:water:acetic acid 40:59:1
Flow rate: 1.5
Detector: UV 254; UV 280

CHROMATOGRAM
Retention time: 7.5 (phenylpropanolamine), 8 (ephedrine), 10 (pseudoephedrine), 16 (phenmetrazine), 19 (amphetamine), 25 (methamphetamine), 30 (phentermine)

REFERENCE
Noggle, F.T. Jr.; Clark, C.R. Liquid chromatographic determination of primary and secondary amines as 8-quinolinesulfonyl chloride derivatives, *J.Assoc.Off.Anal.Chem.*, **1984**, *67*, 687–691.

SAMPLE
Matrix: solutions
Analyte: amines
Sample preparation: Mix 2 mL of a solution of amines in 500 mM sodium carbonate with 10 mL 250 µg/mL 8-quinolinesulfonyl chloride in MeCN, heat at 65° for 20 min, cool, extract twice with 30 mL portions of chloroform. Combine the organic layers and dry them over anhydrous magnesium sulfate, evaporate to dryness under a stream of nitrogen, reconstitute the residue in 1 mL chloroform, inject an aliquot.

HPLC VARIABLES
Guard column: 20 × 4.6 Supelguard (Supelco)
Column: 250 × 4.6 Spherisorb hexyl
Mobile phase: MeCN:buffer:triethylamine 50:50:0.01 (Prepare the buffer by mixing equal volumes of 10 mM sodium acetate and 10 mM acetic acid.)
Flow rate: 1.5
Injection volume: 10
Detector: UV 254

CHROMATOGRAM
Retention time: 3 (methylamine), 3.5 (dimethylamine), 4 (ethylamine), 5 (propylamine), 5.5 (diethylamine), 6 (butylamine), 8 (pentylamine), 9 (dipropylamine), 14 (heptylamine), 16 (dibutylamine), 18 (octylamine)

REFERENCE
Saleh, M.I.; Pok, F.W. Separation of primary and secondary amines as their sulfonamide derivatives by reversed-phase high-performance liquid chromatography, *J.Chromatogr.A*, **1997**, *763*, 173–178.

p-Toluenesulfonyl Chloride

SAMPLE
Matrix: solutions
Analyte: glyphosate
Sample preparation: Mix 1 mL solution with 500 µL buffer, add 200 µL 10 mg/mL p-toluenesulfonyl chloride in MeCN, shake at 50° for 15 s, heat at 50° for 5 min, inject a 10 µL aliquot. (Prepare buffer by mixing 400 mM Na_2HPO_4, phosphoric acid and 400 mM Na_3PO_4, pH 11.0.)

HPLC VARIABLES
Column: 250 × 4.6 Develosil ODS-5 (Nomura Chemical, Aichi, Japan)
Mobile phase: MeCN:buffer 15:85 (Prepare buffer by mixing 200 mM phosphoric acid and 200 mM NaH_2PO_4, pH 2.30.)
Flow rate: 1
Injection volume: 10
Detector: UV 240

CHROMATOGRAM
Retention time: 10.5 (aminomethylphosphonic acid), 14 (glyphosate)
Limit of detection: 10 ng/mL

REFERENCE
Kawai, S.; Uno, B.; Tomita, M. Determination of glyphosate and its major metabolite aminomethylphosphonic acid by high-performance liquid chromatography after derivatization with p-toluenesulfonyl chloride, *J.Chromatogr.*, **1991**, *540*, 411–415.

AMINO ACID

ACTIVATED HALIDE

N-[4-((S)-1-Carbamoyl-2-methylpropylamino)-6-chloro-[1,3,5]triazin-2-yl]-L-phenylalanine Amide

SAMPLE
Matrix: solutions
Analyte: amino acids
Sample preparation: 30 μL 100 mM Amino acid in 1 M HCl + 45 μL 1 M sodium bicarbonate + 500 μL 10 mM reagent in DMSO, heat at 100° for 1 h, add 1.425 mL DMSO, inject a 1-5 μL aliquot. (Prepare reagent as follows. Add 2.14 g L-valinamide hydrochloride in 10 mL water to 2.58 g 2,4,6-trichloro-1,3,5-triazine (cyanuric chloride) in 35 mL acetone with stirring while maintaining the temperature at 5-10°, add 8 mL 2 M sodium carbonate, stir at 20° for 1 h, evaporate to dryness under reduced pressure, take up the residue in 350 mL ethyl acetate, wash with two 175 mL portions of water, add 10 mL toluene, evaporate to dryness to give N-(4,6-dichloro-[1,3,5]triazin-2-yl)-L-valine amide as a white solid (mp 151-152°). Add 1.64 g L-phenylalaninamide in 10 mL water to 2.64 g N-(4,6-dichloro-[1,3,5]triazin-2-yl)-L-valine amide in 35 mL acetone with stirring at room temperature, add 10 mL 1 M sodium bicarbonate, stir at room temperature for 3 h, evaporate to dryness under reduced pressure, take up the residue in 350 mL ethyl acetate, wash with two 175 mL portions of water, add 10 mL toluene, evaporate to dryness to give N-[4-((S)-1-carbamoyl-2-methylpropylamino)-6-chloro-[1,3,5]triazin-2-yl]-L-phenylalanine amide (mp 128-130°).)

HPLC VARIABLES
Column: 250 × 4 5 μm Nucleosil 100 C18
Mobile phase: MeCN:10 mM pH 4 sodium acetate 20:80
Flow rate: 1
Injection volume: 1-5
Detector: UV 254

CHROMATOGRAM
Retention time: 4.90 (L-Glu), 5.89 (D-Glu), 8.26 (L-Pro), 12.76 (D-Pro), 38.28 (L-Phe), 78.14 (D-Phe)

KEY WORDS
chiral

REFERENCE
Brückner, H.; Wachsmann, M. Liquid chromatographic separation of amino acid enantiomers on a silica-bonded chiral s-triazine column, *J.Chromatogr.A*, **1996**, *728*, 447–454.

4-Chloro-7-nitro-2,1,3-benzoxadiazole

RELATED REFERENCES
Ahnoff, M.; Grundevik, I.; Arfwidsson, A.; Fonselius, J.; Persson, B-A. Derivatization with 4-chloro-7-nitrobenzofurazan for liquid chromatographic determination of hydroxyproline in collagen hydrolysate. *Anal.Chem.* **1981**, *53*, 485-489.

Welch, R.W.; Acworth, I.; Levine, M. Coulometric electrochemical detection of hydroxyproline using 7-chloro-4-nitrobenzo-2-oxa-1,3-diazole. *Anal.Biochem.* **1993**, *210*, 199-205.

Yoshida, H.; Sumida, T.; Masujima, T.; Imai, H. Post-column fluorometric detection of amino acids with 7-chloro-4-nitrobenzo-2-oxa-1,3-diazole (NBD-Cl) - a sensitive method especially suitable for proline and hydroxyproline. *HRC CC,J.High Resolut.Chromatogr.Chromatogr.Commun.* **1982**, *5*, 509-511.

SAMPLE
Matrix: fruit juices, protein hydrolysate
Analyte: proline
Sample preparation: Dilute 500 μL fruit juice or 2.5 mg protein hydrolysate and 200 μL 6.55 mg/mL hydroxyproline in water containing 0.2% sodium azide to 20 mL with buffer, mix a 5 μL aliquot with 10 μL 5 mg/mL 4-chloro-7-nitrobenzofurazan in MeOH, inject the whole amount into a 25 × 4 reactor filled with 600 μm glass beads heated to 65° and a 6 m × 0.2 mm ID stainless steel coil heated to 65°. The reaction mixture flowed through the reactor and the coil to the analytical column. (Prepare buffer by dissolving 24.7 g boric acid in 900 mL water, adjusting pH to 10 with 50% NaOH, and making up to 1 L with water.)

HPLC VARIABLES
Column: 150 × 4 5 μm Nova-Pak C18
Mobile phase: Gradient. A was prepared by dissolving 770 mg Na_2HPO_4, 740 mg sodium acetate trihydrate, 5 mL THF, and 13 mL MeOH in 800 mL water, adjusting pH to 6.5 with 50% acetic acid, and making up to 1 L with water. B was MeOH:water 65:35. A:B 100:0 for 14 min, to 16:84 over 11 min, to 0:100 over 1 min, re-equilibrate at initial conditions for 8 min.
Column temperature: 21
Flow rate: 0.08 for 11.5 min, to 1 over 0.5 min, maintain at 1
Injection volume: 15
Detector: F ex 450 em 530

CHROMATOGRAM
Retention time: 23
Internal standard: hydroxyproline (19)
Limit of detection: 20 μM

KEY WORDS
grapefruit

REFERENCE
Carisano, A. Rapid and sensitive method for the determination of proline by reversed-phase high-performance liquid chromatography with automated pre-column fluorescence derivatization, *J.Chromatogr.*, **1985**, *318*, 132–138.

SAMPLE
Matrix: protein
Analyte: hydroxyproline
Sample preparation: Evaporate protein hydrolysate and make sure that residual HCl is removed, reconstitute with 200-500 μL water. Mix ≤100 μL with 100 μL 400 mM pH 9.50 potassium tetraborate buffer, add 2 μL 5 mM IS in water, add 100 μL 2 mM 4-chloro-7-nitro-2,1,3-benzoxadiazole in MeOH, heat in the dark at 37° for 20 min, add 50 μL 1 M HCl, add 150 μL MeOH:water 50:50, inject a 10-50 μL aliquot.

HPLC VARIABLES
Column: 250 × 4.6 Zorbax CN + 250 × 4.6 Spherisorb 10 ODS
Mobile phase: Gradient. MeOH:1% pH 2.73 acetic acid 5:95 for 5 min, to 30:70 over 25 min, to 50:50 over 10 min, return to initial conditions over 5 min.
Column temperature: 50
Flow rate: 1.5
Injection volume: 10-50
Detector: UV 495

CHROMATOGRAM
Retention time: 9 (trans-3-hydroxy-L-proline), 9.8 (trans-4-hydroxy-L-proline), 19 (cis-4-hydroxy-L-proline), 26.7 (proline)
Internal standard: trans-2,3-cis-3,4-dihydroxy-L-proline (5)
Limit of detection: 40 pmole

KEY WORDS
use siliconized glass

REFERENCE
Lindblad, W.J.; Diegelmann, R.F. Quantitation of hydroxyproline isomers in acid hydrolysates by high-performance liquid chromatography, *Anal.Biochem.*, **1984**, *138*, 390–395.

SAMPLE
Matrix: shrimp
Analyte: amino acids
Sample preparation: Blend (Waring) 1 g shrimp and 2 mL trichloroacetic acid, centrifuge, mix an aliquot of the supernatant with equal volumes of 400 mM borate buffer and 2 mg/mL 4-chloro-7-nitro-2,1,3-benzoxadiazole in MeOH, heat at 60° for 5 min, inject a 10 μL aliquot.

HPLC VARIABLES
Column: 300 × 4.6 5 μm Lichrosorb RP-C18
Mobile phase: Gradient. A was THF:100 mM sodium acetate buffer 1:99, pH 6.2. B was MeOH. A:B from 80:20 to 70:30 over 5 min, maintain at 70:30 for 3 min, to 50:50 over 1.5 min, maintain at 50:50 for 3 min, return to initial conditions over 2 min.
Flow rate: 1.2
Injection volume: 10
Detector: F ex 220 em 370 (filter)

CHROMATOGRAM
Retention time: 4.5 (hydroxyproline), 8.9 (proline)

REFERENCE
Vázquez-Ortiz, F.A.; Caire, G.; Higuera-Ciapara, I.; Hernández, G. High performance liquid chromatographic determination of free amino acids in shrimp, *J.Liq.Chromatogr.*, **1995**, *18*, 2059–2068.

N,N-Diethyl-2,4-dinitro-5-fluoroaniline

SAMPLE
Matrix: blood
Analyte: amino acids
Sample preparation: 100 μL Plasma + 300 μL MeCN, mix, centrifuge, lyophilize a 100 μL aliquot of the supernatant. Reconstitute with 50 μL 20 mM sodium bicarbonate solution, add 50 μL 5.1 mg/mL N,N-diethyl-2,4-dinitro-5-fluoroaniline (Fluka) in MeCN, heat at 100° for 15 min, evaporate to dryness under reduced pressure, reconstitute with 150 μL MeCN:buffer 50:50, inject a 15 μL aliquot. (Prepare buffer by adding 4 mL triethylamine and 3 mL acetic acid to water, make up to 1 L with water, adjust to pH 4.3 with 5 M NaOH.)

HPLC VARIABLES
Guard column: ChromSpher C18 (Chrompack)
Column: 250 × 4.6 ChromSpher C18 (Chrompack)
Mobile phase: Gradient. MeCN:buffer 20:80 for 15 min, to 35:65 over 20 min, maintain at 35:65 for 15 min, to 45:55 over 20 min. (Prepare buffer by adding 4 mL triethylamine and 3 mL acetic acid to water, make up to 1 L with water, adjusted to pH 4.3 with 5 M NaOH.)
Flow rate: 2
Injection volume: 15
Detector: UV 360

CHROMATOGRAM
Retention time: 5 (Cya), 9 (hydroxyproline), 15.5 (Asn), 16 (Gln), 18 (Ser), 21.5 (Asp), 23.5 (Glu), 25 (Thr), 29 (Gly), 30 (Pro), 32 (Ala), 40 (Met), 42 (Val), 45.5 (Trp), 47 (Phe), 49 (Leu), 51 (Ile), 65 (Orn), 72 (Lys), 76 (Tyr)
Internal standard: 6-aminocaproic acid (60)
Limit of detection: 1 pmole

KEY WORDS
plasma

REFERENCE
Fermo, I.; Rubino, F.M.; Bolzacchini, E.; Arcelloni, C.; Paroni, R.; Bonini, P.A. Pre-column derivatization of amino acids with N,N-diethyl-2,4-dinitro-5-fluoroaniline and reversed-phase liquid chromatographic separation, *J.Chromatogr.*, **1988**, *433*, 53–62.

SAMPLE
Matrix: urine
Analyte: hydroxyproline
Sample preparation: 50 μL Urine + 50 μL water + 100 μL concentrated HCl, heat at 110° for 16 h, evaporate under reduced pressure, reconstitute with 200 μL acetone, evaporate to dryness, reconstitute with 250 μL 40 μg/mL cysteic acid in 2 M sodium bicarbonate (prepared immediately before use), mix thoroughly. Remove a 50 μL aliquot and add it to 50 μL 5 mg/mL N,N-diethyl-2,4-dinitro-5-fluoroaniline (Fluka) in MeCN, heat

at 100° for 20 min, evaporate to dryness, reconstitute with 500 μL mobile phase (with heating if necessary), inject a 20 μL aliquot.

HPLC VARIABLES
Guard column: 30 × 4.6 5 μm Spheri 5 ODS
Column: 150 × 4.6 5 μm Ultrasphere ODS
Mobile phase: Gradient. MeCN:buffer 20:80 for 8 min, to 90:10 over 1 min, maintain at 90:10 for 5 min, return to initial conditions over 2 min, re-equilibrate for 3 min. (Buffer was 3 mL glacial acetic acid and 6 mL triethylamine in 1 L water, pH adjusted to 4.3 with 5 M HCl.)
Flow rate: 1.3
Injection volume: 20
Detector: UV 360

CHROMATOGRAM
Retention time: 7.3
Internal standard: cysteic acid (4.2)
Limit of detection: 2 μg/mL

REFERENCE
Paroni, R.; De Vecchi, E.; Fermo, I.; Arcelloni, C.; Diomede, L.; Magni, F.; Bonini, P.A. Total urinary hydroxyproline determined with rapid and simple high-performance liquid chromatography, *Clin.Chem.*, **1992**, *38*, 407−411.

4-(N,N-Dimethylaminosulfonyl)-7-fluoro-2,1,3-benzoxadiazole

SAMPLE
Matrix: solutions
Analyte: amino acids
Sample preparation: Mix 500 μL of a 10 μM solution of amino acids in 100 mM pH 9.3 borax buffer containing 1 mM disodium EDTA with 500 μL 20 mM DBD-F in MeCN, heat at 50° for 30 min, cool in ice, inject a 1-10 μL aliquot. (DBD-F can be purchased from Tokyo Kasei (TCI America, 911 North Harborgate St., Portland, OR 97203; 800-423-8616, 503-283-1681, (fax) 503-283-1987; www.tciamerica.com). Synthesis of DBD-F is as follows. Dissolve 0.5 g magnesium sulfate heptahydrate and 6 g NaOH in 60 mL water, throughout the reaction keep the flask at about 20° with cold water cooling, add 15 mL 30% hydrogen peroxide, add 75 mL MeOH, add 12.1 g powdered benzoyl peroxide in one go, stir for 10 min, pour into 150 mL 20% sulfuric acid, extract three times with 50 mL portions of chloroform, determine peroxybenzoic acid concentration by iodometric titration (Tetrahedron 1967, 23, 3327). Slowly add 110 mL 1 M peroxybenzoic acid in chloroform to 7 g 2,6-difluoroaniline dissolved in 100 mL chloroform, stir at room temperature, when reaction is complete (iodometric titration) wash with 2% sodium thiosulfate, wash with 5% sodium carbonate, wash with water, dry over anhydrous sodium sulfate, evaporate to dryness under reduced pressure, recrystallize 2,6-difluoronitrosobenzene from EtOH (mp 108.5-109.5). Stir 8.5 g 2,6-difluoronitrosobenzene in 85 mL DMSO at room temperature and add a solution of 3.91 g sodium azide in 85 mL DMSO dropwise, let stand for about

1 h, add to a large volume of water, extract with ether, dry the extracts over anhydrous sodium sulfate, evaporate to dryness under reduced pressure and distil to give 4-fluoro-2,1,3-benzoxadiazole as a colorless oil (bp 83°/12 mm Hg) (J.Chem.Soc.(C) 1970, 1433). Add 11 mL chlorosulfonic acid dropwise to 3 g 4-fluoro-2,1,3-benzoxadiazole in 10 mL chloroform at 0-10° (use a calcium chloride drying tube), stir at room temperature for 1 h, reflux for 2 h, cool, slowly pour into ice water, remove the organic layer, extract the aqueous layer with chloroform, combine the organic layers, wash, dry over anhydrous magnesium sulfate, evaporate under reduced pressure, take up the residue in 5 mL benzene (Caution! Benzene is a carcinogen!), chromatograph on a 150 × 30 column of silica gel (100-200 mesh Kanto Chemical) with n-hexane:benzene 50:50, evaporate the appropriate fractions to give 4-(chlorosulfonyl)-7-fluoro-2,1,3-benzoxadiazole (CBD-F) as pale yellow needles (mp 64-66°) (Anal. Chem. 1984. 56, 2461). Stir 0.76 g CBD-F in 70 mL MeCN at 0-10° and add 1 g dimethylamine hydrochloride in 10 mL 100 mM pH 10 borax dropwise, adjust pH to 5 with 1 M HCl, concentrate to about 10 mL under reduced pressure, extract three times with 200 mL portions of diethyl ether, wash with water, dry over anhydrous magnesium sulfate, evaporate under reduced pressure, chromatograph on a 500 × 20 column of silica gel with chloroform, isolate the appropriate fraction and re-chromatograph on the same column with ethyl acetate:benzene 1:2 to give 4-(N,N-dimethylaminosulfonyl)-7-fluoro-2,1,3-benzoxadiazole (DBD-F) as white needles (mp 124-125°) (yield = 1% !). On a Merck no. 5714 60F$_{254}$ tlc plate eluted with chloroform DBD-F has Rf 0.32 and lies between two other reaction products (Analyst 1989, 114, 413).)

HPLC VARIABLES
Guard column: 20 × 3.9 37-50 μm Bondapak C18/Corasil
Column: 300 × 3.9 10 μm μBondapak C18
Mobile phase: Gradient. A was 0.1% trifluoroacetic acid in water. B was 0.1% trifluoroacetic acid in MeCN. A:B from 90:10 to 30:70 over 1 h.
Column temperature: 40
Flow rate: 1
Injection volume: 1-10
Detector: F ex 450 em 590

CHROMATOGRAM
Retention time: 10 (hydroxyproline), 12 (glycine), 15 (alanine), 19 (proline), 23 (valine), 27 (isoleucine), 28 (phenylalanine), 30 (lysine), 35 (tyrosine)
Limit of detection: 0.11-0.79 pmole

REFERENCE
Toyo'oka, T.; Suzuki, T.; Saito, Y.; Uzu, S.; Imai, K. Evaluation of benzofurazan derivatives as fluorigenic reagents for thiols and amines using high-performance liquid chromatography, *Analyst*, **1989**, *114*, 1233–1240.

SAMPLE
Matrix: solutions
Analyte: amino acids
Sample preparation: Mix 10 μL of a 100 μM solution in 200 mM pH 9.3 borate buffer containing 4 mM disodium EDTA with 100 μL 50 mM DBD-F in MeCN, heat at 60° for 30 min, add 890 μL MeOH:acetic acid 99:1, inject a 10 μL aliquot. (DBD-F can be purchased from Tokyo Kasei (TCI America, 911 North Harborgate St., Portland, OR 97203; 800-423-8616, 503-283-1681, (fax) 503-283-1987; www.tciamerica.com). Synthesis of DBD-F is as described above.)

HPLC VARIABLES
Column: 250 × 4.6 5 μm Sumichiral OA 2500(S) ((S)-1-naphthylglycyl-3,5-dinitrophenylamide silica gel) (Sumika Analytical, Osaka)
Mobile phase: MeOH containing 20 mM ammonium acetate
Flow rate: 1
Injection volume: 10
Detector: F ex 450 em 590

CHROMATOGRAM
Retention time: 13 (D-Leu), 15 (L-Leu)

KEY WORDS
chiral; comparison with other derivatizing reagents

REFERENCE
Imai, K.; Fukushima, T. Derivatization with fluorogenic benzofurazan reagents of amino acid enantiomers and their separation on a Pirkle type column, *Biomed.Chromatogr.*, **1993**, *7*, 275–276.

2,4-Dinitrofluorobenzene

RELATED REFERENCES
Albalá-Hurtado, S.; Bover-Cid, S.; Izquierdo-Pulido, M.; Veciana-Nogués, M.; Vidal-Carou, M.C. Determination of available lysine in infant milk formulae by high-performance liquid chromatography. *J.Chromatogr.A* **1997**, *778*, 235-241.

Lin, C-E.; Li, F-K. Enantioseparation and recognition mechanisms of dinitrobenzoyl-derivatized amino acids and amino alcohols on chiral stationary phases consisting of cyanuric chloride with (S or R)-phenylalanyl-(S or R)-1-(1-naphthyl)ethylamide substituent. *J.Chromatogr.A* **1996**, *722*, 199-209.

Morton, R.C.; Gerber, G.E. Amino acid analysis by dinitrophenylation and reverse-phase high-pressure liquid chromatography. *Anal.Biochem.* **1988**, *170*, 220-227.

Reddy, K.S.; Menary, R.C. Separation of dinitrophenyl-derivatized amino acids from plant tissue by high-performance liquid chromatography. *Commun.Soil Sci.Plant Anal.* **1991**, *22*, 1207-1213.

Yik, Y.F.; Li, S.F.Y. Resolution optimization in micellar electrokinetic chromatography: use of overlapping resolution mapping scheme in the analysis of dinitrophenyl derivatized amino acids. *Chromatographia* **1993**, *35*, 560-566.

SAMPLE
Matrix: blood

Analyte: glutathione

Sample preparation: Mix 3 mL blood with 30 mg powdered 1,10-phenanthroline, centrifuge a 500 µL aliquot at 0-5° at 7000 g for 10 min, discard the supernatant. Add 400 µL 20 mM 1,10-phenanthroline to the remaining erythrocytes, add 300 µL 10% metaphosphoric acid, vortex, centrifuge at 15000 g for 15 min. Remove a 400 µL aliquot of the supernatant and add it to 100 µL 400 mM iodoacetic acid in water (freshly prepared), neutralize with 20 mg sodium bicarbonate, shake in the dark at room temperature for 1 h, add 500 µL 5% 2,4-dinitrofluorobenzene in EtOH, let stand in the dark at room temperature for 20 h, centrifuge at 15000 g for 15 min, inject a 20 µL aliquot.

HPLC VARIABLES
Guard column: 4 × 4 5 µm LiChrospher 100 NH2

Column: 250 × 4 5 µm LiChrospher 100 NH2

Mobile phase: Gradient. A was MeOH:water 80:20. B was MeOH:2 M pH 4.6 sodium acetate 64:36. A:B from 75:25 to 5:95 over 30 min, maintain at 5:95 for 15 min, reequilibrate at initial conditions for 15 min.

Column temperature: 40

Flow rate: 1.2

Injection volume: 20

Detector: UV 365

CHROMATOGRAM
Retention time: 31.7 (glutathione), 40.6 (oxidized glutathione)

KEY WORDS
erythrocytes

REFERENCE
Yoshida, T. Determination of reduced and oxidized glutathione in erythrocytes by high-performance liquid chromatography with ultraviolet absorbance detection, *J.Chromatogr.B*, **1996**, *678*, 157–164.

SAMPLE
Matrix: protein
Analyte: amino acids
Sample preparation: Heat 5 mg albumin with 1 mL 6 M HCl, 10 μL mercaptoethanol and 10 μL octanol under vacuum at 110° for 24 h, evaporate to dryness, take up in 1 M HCl, add to a 100 × 5 Dowex 50 ion-exchange column, elute with 1 M HCl, collect fractions. Evaporate fractions to dryness, reconstitute with 2 mL 3% sodium bicarbonate solution, add 4 mL 25 μL/mL 2,4-dinitrofluorobenzene, let stand in the dark at room temperature overnight, wash with ether, add 2 drops 6 M HCl, extract with ether. Evaporate the ether layer to dryness, reconstitute with EtOH, inject an aliquot.

HPLC VARIABLES
Column: 150 × 4.6 5 μm Ultrasphere ODS
Mobile phase: Gradient. MeOH:30 mM pH 5 KH_2PO_4 from 40:60 to 55:45 over 30 min, maintain at 55:45 for 5 min, return to initial conditions over 5 min, re-equilibrate for 5 min.
Flow rate: 1
Detector: UV 365

CHROMATOGRAM
Retention time: 5 (proline), 10 (tyrosine), 11 (s-methylcysteine), 18 (methionine), 20 (valine), 21 (cystine), 29 (phenylalanine), 31 (isoleucine, leucine)

REFERENCE
Seago, A.; Shuker, D.E.G.; Paine, A.J. Interaction of [^{14}C]dimethylnitrosamine with albumin produced by rat hepatocytes in culture, *Toxicol.Lett.*, **1986**, *30*, 41–48.

SAMPLE
Matrix: solutions
Analyte: amino acids
Sample preparation: Evaporate a 500 μL aliquot of an aqueous solution containing 25 μmoles amino acids to dryness under a stream of air at 37°, reconstitute with 250 μL 1 M sodium bicarbonate solution, add 500 μL 1% 2,4-dinitrofluorobenzene in acetone, mix, heat at 50° for 1 h, cool to room temperature, add 200 μL 2 M HCl, dry in a vacuum desiccator over NaOH, reconstitute in 1 mL MeOH, pass dry HCl gas through the solution for 5 min, cap the vial, let stand for 3 h, evaporate to dryness, dry in a vacuum desiccator over NaOH, reconstitute in 400 μL DMSO. Remove a 50 μL aliquot and add it to 200 μL 50 mM pH 7.5 HEPES buffer, add 50 μL 10.6 mg/mL carboxypeptidase-Y (Carlsberg Biotechnology) in water, mix, let stand at room temperature for 3 h, add 300 μL DMSO, inject a 25 μL aliquot. (All amino acids are converted to their dinitrophenyl methyl ester derivatives. The enzyme hydrolyses only L-amino acid methyl esters (at the α-position) and so D-amino acids (methyl ester at the α-position) can be distinguished from L-amino acids (carboxylic acid at the α-position). Methyl esters of carboxylic acid groups other than α are not affected. No racemization occurs during derivatization and so the procedure can be used to assess racemization reactions.)

HPLC VARIABLES
Column: 100 × 8 radial compression C18 (Waters)
Mobile phase: Gradient. MeCN:50 mM pH 3.0 triethylammonium phosphate buffer from 25:75 to 50:50 over 40 min.
Flow rate: 2
Injection volume: 25
Detector: UV 350

CHROMATOGRAM
Retention time: 8.10 (β-methoxy-L-Asp), 8.76 (L-Ala), 9.93 (gamma-methoxy-L-Glu), 15.93 (L-Met), 20.91 (dimethoxy-D-Asp), 22.75 (methoxy-D-Ala), 23.78 (L-Phe), 24.68 (dimethoxy-D-Glu), 32.58 (methoxy-D-Met), 39.59 (methoxy-D-Phe)

KEY WORDS
chiral

REFERENCE
Marfey, P.; Ottesen, M. Determination of D-amino acids. I. Hydrolysis of DNP-L-amino acid methyl esters with carboxypeptidase-Y, *Carlsberg Res.Commun.*, **1984**, *49*, 585–590.

SAMPLE
Matrix: solutions
Analyte: amino acids
Sample preparation: Mix a 100 µL aliquot of a 10-25 µg/mL solution of amino acids in water with 100 µL buffer, add 50 µL 5% 2,4-dinitrofluorobenzene in MeCN, vortex, heat at 50° for 1 h, add 250 µL 1% acetic acid, add 500 µL MeCN:water 50:50, inject a 20 µL aliquot. (Buffer was 100 mM sodium bicarbonate adjusted to pH 9.5 with 100 mM sodium carbonate.)

HPLC VARIABLES
Column: 150 × 4.6 5 µm 100 Å Kromasil (EKA Nobel) derivatized with a chiral quinine carbamate (J. Chromatogr.A 1996, 741, 33)
Mobile phase: MeCN:THF:110 mM ammonium acetate 44.55:0.45:55, adjusted to pH 5.5 with glacial acetic acid
Column temperature: 25
Flow rate: 1
Injection volume: 20
Detector: UV 390

CHROMATOGRAM
Retention time: 13.11 (L-Thr), 15.64 (L-Pro), 18.29 (D-Thr), 18.90 (L-Ala), 19.93 (L-Val), 21.47 (D-Pro), 22.43 (D-Ala), 23.79 (L-Leu), 26.04 (D-Val), 30.37 (D-Leu), 33.47 (L-Phe), 40.93 (D-Phe), 48.00 (L-Trp), 66.03 (D-Trp)

KEY WORDS
chiral

REFERENCE
Lämmerhofer, M.; Di Eugenio, P.; Molnar, I.; Lindner, W. Computerized optimization of the high-performance liquid chromatographic enantioseparation of a mixture of 4-dinitrophenyl amino acids on a quinine carbamate-type chiral stationary phase using DRYLAB, *J.Chromatogr.B*, **1997**, *689*, 123–135.

SAMPLE
Matrix: tissue
Analyte: glutathione
Sample preparation: Homogenize 1 g rat liver with 10 mL 10% perchloric acid containing 1 mM bathophenanthrolinedisulfonic acid, centrifuge at 4° at 15000 g for 3 min. Remove a 500 µL aliquot of the supernatant and add it to 50 µL 15 mM gamma-glutamylglutamate in 0.3% perchloric acid, add 50 µL 100 mM iodoacetic acid in 200 µM m-cresol purple solution, adjust pH to 8.0 with 2 M KOH containing 2.4 M potassium bicarbonate, let stand in the dark at room temperature for 10 min, add 1 mL 1% 2,4-dinitrofluorobenzene in EtOH, mix, inject a 100 µL aliquot.

HPLC VARIABLES
Guard column: 20 × 4.6 Spherisorb NH2
Column: 250 × 4.6 5 µm Spherisorb NH2
Mobile phase: Gradient. A as MeOH:water 80:20. B was MeOH:water 64:36 containing 500 mM sodium acetate. A:B 80:20 for 5 min, to 1:99 over 10 min, maintain at 1:99.
Flow rate: 1
Injection volume: 100
Detector: UV 365

CHROMATOGRAM
Retention time: 16
Internal standard: gamma-glutamylglutamate (15)

KEY WORDS
rat; liver

REFERENCE
Santori, G.; Domenicotti, C.; Bellocchio, A.; Pronzato, M.A.; Marinari, U.M.; Cottalasso, D. Different efficacy of iodoacetic acid and N-ethylmaleimide in high-performance liquid chromatographic measurement of liver glutathione, *J.Chromatogr.B*, **1997**, *695*, 427–433.

2,4-Dinitrophenyl-1-fluoro-5-L-alanine

SAMPLE
Matrix: peptides
Analyte: amino acids
Sample preparation: Dry 0.5-1 nmole peptide in a tube. add 500 µL 6 M HCl, evacuate, flush with nitrogen, evacuate, seal, heat in the vapor phase at 110° for 20-24 h, evaporate to dryness under reduced pressure, reconstitute with 5 µL 200 mM sodium bicarbonate, dry under reduced pressure, add 5 µL water, add a 2- to 3-fold molar excess of reagent in 10 µL acetone, heat in the dark with gentle shaking at 50° for 1 h, evaporate to dryness under reduced pressure, reconstitute, inject an aliquot. (Synthesize the reagent, 2,4-dinitrophenyl-1-fluoro-5-L-alanine, as follows. Add 10 mL 200 mM 1,5-difluoro-2,4-dinitrobenzene in acetone dropwise with stirring to 10 mM 200 mM L-alanine in 5 mM sodium bicarbonate, stir in the dark at room temperature at pH 8 for 4 h, acidify to pH 3 with 2 M HCl, evaporate to dryness, take up in 100 mM sodium bicarbonate, extract with diethyl ether, crystallize by the addition of hot 2 M HCl, filter, wash with acidified water, dry in the air in the dark to give 2,4-dinitrophenyl-1-fluoro-5-L-alanine as golden-yellow scales (mp 136-137°).)

HPLC VARIABLES
Column: 250 × 2 5 µm Ultrasphere octyl
Mobile phase: Gradient. A was 40 mM pH 2.2 triethylammonium phosphate. B was MeCN:isopropanol 80:20. A:B from 76:24 to 61:39 over 38 min, to 35:65 over 32 min, to 5:95 over 0.2 min, maintain at 5:95 for 3.8 min, return to initial conditions over 0.2 min.
Column temperature: 25
Flow rate: 0.2
Injection volume: 20
Detector: UV 340

CHROMATOGRAM
Retention time: 7 (D-His), 8 (L-His), 9.5 (D-Arg), 10 (L-Arg), 17 (L-Ser), 18 (D-Ser), 19 (L-Thr), 20.5 (L-Asp), 21.5 (D-Asp), 23.5 (L-Glu), 24.5 (D-Glu), 25 (D-Thr), 25.5 (Gly), 29.5 (ammonia), 30 (L-Ala), 37 (D-Ala), 40.5 (L-Met), 43.5 (L-Val), 44.5 (L-norvaline), 49 (D-Met), 50.5 (L-Phe), 51 (L-Ile), 51.5 (L-Leu), 52 (L-Lys, D-Val), 55 (D-Lys), 56 (D-Phe), 57.5 (D-Ile), 58.5 (D-Leu), 60 (L-Tyr), 65 (D-Tyr)

KEY WORDS
chiral

REFERENCE

Scaloni, A.; Simmaco, M.; Bossa, F. Determination of the chirality of amino acid residues in the course of subtractive Edman degradation of peptides, *Anal.Biochem.*, **1991**, *197*, 305–310.

2-Fluoro-4,5-diphenyloxazole

SAMPLE

Matrix: solutions

Analyte: amino acids

Sample preparation: Mix 250 μL of a solution of amino acids in 50 mM pH 9.3 borate buffer containing 1 mM disodium EDTA with 250 μL 1 mM 2-fluoro-4,5-diphenyloxazole (DIFOX) in MeCN, let stand in the dark at room temperature for 1 h, add 500 μL MeCN:1 M HCl 50:50, inject an aliquot. (Prepare DIFOX as follows. Gently reflux 21 g benzoin and 45 g urethane (Caution! Urethane is a carcinogen!) in 300 mL DMF for 6 h, cool, pour into water, filter, recrystallize to give 4,5-diphenyl-2-oxazolone (mp 211°), treat with phosphorus oxychloride to give 2-chloro-4,5-diphenyloxazole (Ber. 1956, 89, 1749). Add 3 g anhydrous KF to 1.62 g 2-chloro-4,5-diphenyloxazole in 60 mL MeCN, add 1.5 g 18-crown-6/MeCN complex, reflux for 24 h, cool, filter. Concentrate the filtrate and add 50 mL hexane, stir for 10 min, filter rapidly, repeat the extraction. Combine the filtrates and concentrate them to give a yellow oil, distil to give DIFOX (2-fluoro-4,5-diphenyloxazole) (bp 130°/0.02 mm Hg) (Analyst 1993, 118, 257). Prepare 18-crown-6/MeCN complex by heating 50 g 18-crown-6 in 125 mL MeCN until a homogeneous solution is obtained (use a calcium sulfate drying tube), stir the solution vigorously as it cools to room temperature, cool in a dry ice/acetone bath, filter rapidly, dry under high vacuum at ≤40° over 2-3 h to give 18-crown-6/MeCN complex (mp 36.5-38°) (Caution! The complex is hygroscopic!) (J. Org. Chem. 1974, 39, 2445). No experimental details are given for the phosphorus oxychloride reaction above but a procedure for the p-N,N-dimethylaminosulfonyl analogue proceeds as follows. Suspend 2 g dried 4,5-bis(p-N,N-dimethylaminosulfonylphenyl)-2-oxazolone in 30 mL phosphorus oxychloride, stir at 0°, add 610 μL triethylamine dropwise, heat at 100° for 7 h, remove the excess phosphorus oxychloride on a rotary evaporator. Dissolve the residue in dichloromethane and wash with cold saturated sodium bicarbonate, dry the organic layer over anhydrous magnesium sulfate, evaporate to dryness, chromatograph on silica gel to give 2-chloro-4,5-bis(p-N,N-dimethylaminosulfonylphenyl)oxazole (Analyst 1993, 118, 257).)

HPLC VARIABLES

Column: 250 × 4.6 5 μm LC-8 (Supelco)

Mobile phase: Gradient. A was MeCN:50 mM pH 7.0 phosphate buffer 25:75. B was MeCN:50 mM pH 7.0 phosphate buffer 50:50. A:B from 100:0 to 0:100 over 30 min, maintain at 0:100 for 30 min.

Flow rate: 1

Detector: F ex 320 em 420

CHROMATOGRAM

Retention time: 6 (aspartic acid), 6.5 (glutamic acid), 9.5 (hydroxyproline), 13.5 (arginine), 15 (proline), 15.5 (glycine), 16 (alanine), 17.5 (tyrosine), 20 (valine), 20.5 (methionine), 21.5 (leucine, isoleucine), 23.5 (tryptophan, phenylalanine), 35 (lysine)

Limit of detection: 19-64 fmole

REFERENCE

Toyo'oka, T.; Chokshi, H.P.; Givens, R.S.; Carlson, R.G.; Lunte, S.M.; Kuwana, T. Fluorescence and chemiluminescence detection of oxazole-labelled amines and thiols, *Biomed.Chromatogr.*, **1993**, *7*, 208−216.

SAMPLE

Matrix: solutions
Analyte: amino acids and peptides
Sample preparation: Mix 250 µL of a solution of amino acids or peptides in 50 mM pH 9.3 borate buffer containing 1 mM disodium EDTA with 250 µL 1 mM 2-fluoro-4,5-diphenyloxazole (DIFOX) in MeCN, let stand in the dark at room temperature for 1 h, add 500 µL MeCN:1 M HCl 50:50, inject an aliquot. (Prepare DIFOX as follows. Gently reflux 21 g benzoin and 45 g urethane (Caution! Urethane is a carcinogen!) in 300 mL DMF for 6 h, cool, pour into water, filter, recrystallize to give 4,5-diphenyl-2-oxazolone (mp 211°), treat with phosphorus oxychloride to give 2-chloro-4,5-diphenyloxazole (Ber. 1956, 89, 1749). Add 3 g anhydrous KF to 1.62 g 2-chloro-4,5-diphenyloxazole in 60 mL MeCN, add 1.5 g 18-crown-6/MeCN complex, reflux for 24 h, cool, filter. Concentrate the filtrate and add 50 mL hexane, stir for 10 min, filter rapidly, repeat the extraction. Combine the filtrates and concentrate them to give a yellow oil, distil to give 2-fluoro-4,5-diphenyloxazole (DIFOX) (bp 130°/0.02 mm Hg) (Analyst 1993, 118, 257). Prepare 18-crown-6/MeCN complex by heating 50 g 18-crown-6 in 125 mL MeCN until a homogeneous solution is obtained (use a calcium sulfate drying tube), stir the solution vigorously as it cools to room temperature, cool in a dry ice/acetone bath, filter rapidly, dry under high vacuum at ≤40° over 2-3 h to give 18-crown-6/MeCN complex (mp 36.5-38°) (Caution! The complex is hygroscopic!) (J. Org. Chem. 1974, 39, 2445). No experimental details are given for the phosphorus oxychloride reaction above but a procedure for the p-N,N-dimethylaminosulfonyl analogue proceeds as follows. Suspend 2 g dried 4,5-bis(p-N,N-dimethylaminosulfonylphenyl)-2-oxazolone in 30 mL phosphorus oxychloride, stir at 0°, add 610 µL triethylamine dropwise, heat at 100° for 7 h, remove the excess phosphorus oxychloride on a rotary evaporator. Dissolve the residue in dichloromethane and wash with cold saturated sodium bicarbonate, dry the organic layer over anhydrous magnesium sulfate, evaporate to dryness, chromatograph on silica gel to give 2-chloro-4,5-bis(p-N,N-dimethylaminosulfonylphenyl)oxazole.)

HPLC VARIABLES

Column: 250 × 4.6 5 µm LC-8 (Supelco)
Mobile phase: MeCN:50 mM pH 7 phosphate buffer 30:70
Flow rate: 1
Detector: F ex 320 em 420

CHROMATOGRAM

Retention time: 6 (hydroxy-L-proline), 8.6 (proline), 10 (L-prolyl-glycyl-glycine), 11.6 (alanine), 27 (L-prolyl-L-leucine)
Limit of detection: 8.6-11.6 fmole

REFERENCE

Toyo'oka, T.; Chokshi, H.P.; Carlson, R.G.; Givens, R.S.; Lunte, S.M. Oxazole-based tagging reagents for analysis of secondary amines and thiols by liquid chromatography with fluorescence detection, *Analyst*, **1993**, *118*, 257−263.

4-Fluoro-7-nitro-2,1,3-benzoxadiazole

RELATED REFERENCES

Imai, K.; Watanabe, Y.; Toyooka, T. Fluorometric assay of amino acids and amines by use of 7-fluoro-4-nitrobenzo-2-oxa-1,3-diazole (NBD-F) in high-performance liquid chromatography. *Chromatographia* **1982**, *16*, 214-215.

Kotaniguchi, H.; Kawakatsu, M.; Toyooka, T.; Imai, K. Automatic amino acid analysis utilizing 4-fluoro-7-nitrobenzo-2-oxa-1,3-diazole. *J.Chromatogr.* **1987**, *420*, 141-145.

Miyano, H.; Toyooka, T.; Imai, K. Further studies on the reaction of amines and proteins with 4-fluoro-7-nitrobenzo-2-oxa-1,3-diazole. *Anal.Chim.Acta* **1985**, *170*, 81-87.

Toyooka, T.; Miyano, H.; Imai, K. Amino acid composition analysis of minute amounts of cysteine-containing proteins using 4-(aminosulfonyl)-7-fluoro-2,1,3-benzoxadiazole and 4-fluoro-7-nitro-2,1,3-benzoxadiazole in combination with HPLC. *Biomed.Chromatogr.* **1986**, *1*, 15-20.

Watanabe, N.; Toyooka, T.; Imai, K. HPLC electrochemical fluorometric detection of amino acids including tryptophan using 4-fluoro-7-nitrobenzo-2-oxa-1,3-diazole. *Biomed.Chromatogr.* **1987**, *2*, 99-103.

Watanabe, Y.; Imai, K. High-performance liquid chromatography and sensitive detection of amino acids derivatized with 7-fluoro-4-nitrobenzo-2-oxa-1,3-diazole. *Anal.Biochem.* **1981**, *116*, 471-472.

Watanabe, Y.; Imai, K. Precolumn labeling for high-performance liquid chromatography of amino acids with 7-fluoro-4-nitrobenzo-2-oxa-1,3-diazole and its application to protein hydrolysates. *J.Chromatogr.* **1982**, *239*, 723-732.

SAMPLE
Matrix: beverages
Analyte: amino acids
Sample preparation: Filter (0.5 μm) wine, add 10 μL filtrate to 90 μL 200 mM pH 8.0 borate buffer containing 4 mM disodium EDTA, add 50 μL 50 mM 4-fluoro-7-nitro-2,1,3-benzoxadiazole in MeCN, heat at 60° for 5 min, add 850 μL MeOH:acetic acid 99:1, inject a 50 μL aliquot.

HPLC VARIABLES
Guard column: C18 (Tosoh)
Column: 250 × 4.6 5 μm Sumichiral OA-2500(R) or OA-2500(S) (1-naphthylglycyl-3,5-dinitrophenylamide silica) (Sumika)
Mobile phase: MeOH containing 4-5 mM citric acid
Flow rate: 0.8
Injection volume: 50
Detector: F ex 470 em 530

CHROMATOGRAM
Limit of detection: 2 nM

OTHER SUBSTANCES
Simultaneously analyzed: alanine, asparagine, aspartic acid, glutamic acid, glutamine, isoleucine, leucine, lysine

KEY WORDS
wine; chiral

REFERENCE
Kato, M.; Fukushima, T.; Santa, T.; Homma, H.; Imai, K. Determination of D-amino acids, derivatized with 4-fluoro-7-nitro-2,1,3-benzoxadiazole, in wine samples by high-performance liquid chromatography, *Biomed.Chromatogr.*, **1995**, *9*, 193–194.

SAMPLE
Matrix: blood, tissue
Analyte: amino acids
Sample preparation: Serum. 10 μL Serum + 90 μL MeOH, mix vigorously, centrifuge at 1000 g for 5 min. Remove a 10 μL aliquot of the supernatant and add it to 10 μL 200 mM pH 8.0 borate buffer containing 4 mM tetrasodium EDTA, add 30 μL 50 mM 4-fluoro-7-nitrobenzofurazan (4-fluoro-7-nitro-2,1,3-benzoxadiazole) in MeCN, heat at 60° for 5

min, cool, add 250 μL MeOH:acetic acid 99:1, filter (0.5 μm), inject a 10 μL aliquot of the filtrate. Rat brain. Homogenize (glass-Potter) rat brain tissue in 10 volumes MeOH at 4°, centrifuge at 1000 g for 10 min. Remove a 10 μL aliquot of the supernatant and add it to 10 μL 200 mM pH 8.0 borate buffer containing 4 mM disodium EDTA, add 30 μL 50 mM 4-fluoro-7-nitrobenzofurazan in MeCN, heat at 60° for 5 min, cool, add 250 μL MeOH:acetic acid 99:1, filter (0.5 μm), inject a 10 μL aliquot of the filtrate. Cow brain. Homogenize (glass-Potter) cow brain tissue in 10 volumes buffer at 4°, filter through gauze. Vigorously mix 10 μL filtrate with 90 μL MeOH, centrifuge at 1000 g for 5 min. Remove a 10 μL aliquot of the supernatant and add it to 10 μL 200 mM pH 8.0 borate buffer containing 4 mM disodium EDTA, add 30 μL 50 mM 4-fluoro-7-nitrobenzofurazan in MeCN, heat at 60° for 5 min, cool, add 250 μL MeOH:acetic acid 99:1, filter (0.5 μm), inject a 10 μL aliquot of the filtrate. (Buffer was 10 mM pH 7.4 N-2-hydroxyethylpiperazine-N'-2-ethanesulfonic acid (HEPES) containing 122 mM NaCl, 3 mM KCl, 25 mM sodium bicarbonate, 1.4 mM calcium chloride, 1.2 mM magnesium sulfate, and 0.4 mM glucose.)

HPLC VARIABLES
Guard column: Resolve C18 guard column (Waters)
Column: 250 × 4.6 Sumichiral OA-3100(S) (Sumika)
Mobile phase: MeOH containing 5 mM citric acid
Flow rate: 1 (standards) or 0.5 (biological samples)
Injection volume: 10
Detector: F ex 470 em 530

CHROMATOGRAM
Retention time: k' 1.01 (first (D) enantiomer), α = 1.52 (Leu); k' 1.18 (first (D) enantiomer), α = 1.25 (Ile); k' 1.29 (first (D) enantiomer), α = 1.22 (Val); k' 2.05 (first (D) enantiomer), α = 1.31 (Ala); k' 5.48 (first (D) enantiomer), α = 1.06 (Pro); k' 2.13 (first (D) enantiomer), α = 1.35 (Thr); k' 3.10 (first (D) enantiomer), α = 1.33 (Ser); k' 2.16 (first (D) enantiomer), α = 1.36 (Phe); k' 2.08 (first (D) enantiomer), α = 1.40 (Met); k' 3.00 (first (D) enantiomer), α = 1.27 (Gln); k' 4.96 (first (D) enantiomer), α = 1.57 (Lys)

KEY WORDS
serum; human; rat; cow; brain; chiral; detailed discussion of use of other chiral columns; detailed discussion of use of other derivatizing reagents

REFERENCE
Fukushima, T.; Kato, M.; Santa, T.; Imai, K. Enantiomeric separation and sensitive determination of D,L-amino acids derivatized with fluorogenic benzofurazan reagents on Pirkle type stationary phases, *Biomed.Chromatogr.*, **1995**, *9*, 10–17.

SAMPLE
Matrix: solutions
Analyte: amino acids
Sample preparation: Mix 10 μL of a 100 μM solution in 200 mM pH 8.0 borate buffer containing 4 mM disodium EDTA with 30 μL 50 mM 4-fluoro-7-nitro-2,1,3-benzoxadiazole in MeCN, heat at 60° for 2 min, add 960 μL MeOH:acetic acid 99:1, dilute 5 times with MeOH, inject a 10 μL aliquot.

HPLC VARIABLES
Column: 250 × 4.6 5 μm Sumichiral OA 2500(S) ((S)-1-naphthylglycyl-3,5-dinitrophenylamide silica gel) (Sumika Analytical, Osaka)
Mobile phase: MeOH containing 20 mM ammonium acetate
Flow rate: 1
Injection volume: 10
Detector: F ex 470 em 530

CHROMATOGRAM
Retention time: 18.57 (D-Leu), 20.8 (L-Leu), 25.3 (D-Phe), 30.6 (L-Phe)
Limit of detection: 25-34 fmole

KEY WORDS
chiral

REFERENCE
Imai, K.; Fukushima, T.; Uzu, S. Sensitive determination of enantiomers of amino acids derivatized with the fluorogenic reagent, 4-fluoro-7-nitro-2,1,3-benzoxadiazole, separated on a Pirkle-type column, Sumichiral OA 2500(S), *Biomed.Chromatogr.*, **1993**, 7, 177–178.

SAMPLE
Matrix: solutions
Analyte: amino acids and peptides
Sample preparation: Mix 20 µL of a solution in 100 mM ph 7.6 phosphate buffer containing 1 mM EDTA and 10 mM 2-mercaptoethanol with 20 µL 50 mM 4-fluoro-7-nitrobenzo-2-oxa-1,3-diazole in EtOH, heat at 65° for 5 min, add 160 µL 200 mM HCl, cool in an ice bath, inject a 20 µL aliquot.

HPLC VARIABLES
Column: 150 × 4.6 Cosmosil 5C18-AR (Nacalai Tesque)
Mobile phase: MeOH:0.1% phosphoric acid 50:50 (Arg, Arg-Gly-NH$_2$;) or 90:10 (Arg-Pro-Lys-Pro)
Flow rate: 1
Injection volume: 20
Detector: F ex 470 em 530

CHROMATOGRAM
Retention time: 5 (Arg-Gly-NH$_2$;), 7.5 (Arg-Pro-Lys-Pro)
Limit of detection: 1 pmole

KEY WORDS
comparison with other derivatization procedures

REFERENCE
Yoshinaga, K.; Kobayashi, N.; Nagatani, Y.; Tanaka, Y.; Ikeda, Y. A sensitive detection method for peptide using 4-fluoro-7-nitrobenzo-2-oxa-1,3-diazole and its application to measure prolyl endopeptidase activity, *Biomed.Chromatogr.*, **1994**, 8, 297–300.

Marfey's Reagent

SAMPLE
Matrix: bacteria
Analyte: 2,6-diaminopimelic acid
Sample preparation: Heat 4 mg freeze-dried powdered goat rumen bacteria with 4 mL 6 M HCl in a sealed tube at 110° for 20 h, cool, filter (Whatman No.2 paper), wash solid 3 times with water. Evaporate the filtrate to dryness under reduced pressure, add water to residue, evaporate, repeat process twice more, reconstitute residue with 1 mL water, filter

(0.45 μm). Remove a 100 μL aliquot of the filtrate and add it to 100 μL 1% Nα-(2,4-dinitro-5-fluorophenyl)-L-alaninamide (Marfey's reagent) in acetone and 40 μL 1 M sodium bicarbonate, heat at 40° for 1 h, add 20 μL 2 M HCl, inject a 10-50 μL aliquot.

HPLC VARIABLES
Column: 250 × 4 5 μm LiChrospher 100 RP-18
Mobile phase: MeCN:50 mM pH 3.0 triethylamine phosphate 28:72 (A) or 21.5:78.5 (B)
Column temperature: 40
Flow rate: 1
Injection volume: 10-50
Detector: UV 325

CHROMATOGRAM
Retention time: 17.5 (A) or 39.4 (meso), 63.1 (LL), 77.7 (DD) (B)
Limit of detection: 2.5 μM

REFERENCE
El-Waziry, A.M.; Tomita, Y.; Ling, J.R.; Onodera, R. Measurement of total and separate stereoisomers of diaminopimelic acid in rumen bacteria by high-performance liquid chromatography, *J.Chromatogr.B*, **1996**, *677*, 53–59.

SAMPLE
Matrix: bulk
Analyte: peptides
Sample preparation: Purge 100 μg peptide in deuterated HCl (20%):deuterated acetic acid 50:50 with nitrogen, heat at 130° for 18 h, cool, add 400 μL water, lyophilize. Reconstitute with 100 μL 1 M sodium bicarbonate, add 200 μL 38.7 mM 1-fluoro-2,4-dinitrophenyl-5-L-alanine amide (Marfey's reagent) in acetone, vortex, heat at 40° for 1 h, add 50 μL 2 M HCl, dilute 10-fold with mobile phase A, inject an 80 μL aliquot.

HPLC VARIABLES
Column: 250 × 4.6 Basic B-03-5 (YMC) (UV) or 250 × 2 Basic MCB-03-5 (YMC) (MS)
Mobile phase: Gradient. A was MeCN:MeOH:10 mM pH 5.2 ammonium formate 5:1:94. B was MeCN:MeOH:10 mM pH 5.2 ammonium formate 60:1:39. A:B from 100:0 to 0:100 over 45 min, re-equilibrate at initial conditions for 15 min (UV) or A:B from 80:20 to 20:80 over 20 min (MS).
Flow rate: 1 (UV) or 0.25 (MS)
Injection volume: 80
Detector: UV 340; MS, Finnigan MAT TSQ 700, electrospray +3.0 kV, column effluent was split so that 0.083 mL/min entered the MS, sheath liquid MeOH at 0.1 mL/min, auxiliary nitrogen 103.4 MPa, sheath nitrogen 179.3 MPa (Racemized amino acids are labeled with one deuterium on the α-carbon and this can be distinguished by MS.)

CHROMATOGRAM
Retention time: 7.88 (L-Asn), 7.98 (L-Asp), 8.48 (L-Ser), 8.97 (D-Gln), 9.01 (L-Thr), 9.08 (L-His), 9.23 (D-His), 9.42 (L-Glu), 9.63 (D-Asn), 9.67 (L-Lys), 9.85 (D-Asp), 9.93 (D-Ser), 10.03 (D-Gln), 10.29 (L-Arg), 10.57 (D-Lys), 10.89 (L-Cys), 10.93 (L-Ala), 11.03 (Gly), 11.17 (D-Arg), 11.35 (D-Glu), 11.57 (L-Pro), 12.28 (D-Thr), 12.56 (D-Cys), 13.34 (L-Tyr), 13.66 (D-Ala), 13.82 (D-Pro), 14.29 (L-Val), 14.54 (L-Met), 15.67 (D-Tyr), 17.01 (L-Ile), 17.58 (L-Leu), 17.98 (D-Met), 18.22 (D-Val), 18.33 (L-Phe), 18.37 (L-Trp), 20.52 (D-Trp), 20.94 (D-Ile), 21.23 (D-Phe), 21.28 (D-Leu) (Using UV detection)

KEY WORDS
chiral

REFERENCE
Goodlett, D.R.; Abuaf, P.A.; Savage, P.A.; Kowalski, K.A.; Mukherjee, T.K.; Tolan, J.W.; Corkum, N.; Goldstein, G.; Crowther, J.B. Peptide chiral purity determination: hydrolysis in deuterated acid, derivatization with Marfey's reagent and analysis using high-performance liquid chromatography-electrospray ionization-mass spectrometry, *J.Chromatogr.A*, **1995**, *707*, 233–244.

SAMPLE
Matrix: solutions
Analyte: amino acids

Sample preparation: Mix 50 µL of a 50 mM aqueous solution with 20 µL 1 M sodium bicarbonate, add 100 µL 1% Marfey's reagent in acetone, vortex, heat at 37° for 1 h, add 20 µL 1 M HCl, add 810 µL MeCN, inject a 1 µL aliquot.

HPLC VARIABLES

Column: 150 × 4.6 Cosmosil 5C18-AR (Nacalai Tesque)
Mobile phase: Gradient. MeCN:buffer from 15:85 to 45:55 over 5 min, re-equilibrate at initial conditions for 15 min. (Buffer was 100 mM ammonium acetate in water, adjusted to pH 3 with trifluoroacetic acid.)
Column temperature: 40
Flow rate: 1
Injection volume: 1
Detector: UV 340

CHROMATOGRAM

Retention time: 3.7 (D-β-threo-hydroxyaspartic acid), 3.8 (L-β-threo-hydroxyaspartic acid), 6.0 (D-histidine (mono-derivative)), 6.1 (D-β-erythro-hydroxyaspartic acid), 6.7 (L-asparagine), 6.7 (L-histidine (mono-derivative)), 6.9 (L-β-erythro-hydroxyaspartic acid), 7.3 (D-asparagine), 8.2 (L-aspartic acid), 8.5 (D-ornithine (mono-derivative)), 8.8 (L-glutamine), 9.0 (L-homoserine), 9.2 (L-serine), 9.3 (L-arginine), 9.4 (L-ornithine (mono-derivative)), 9.6 (D-glutamine), 9.6 (D-arginine), 9.8 (D-aspartic acid), 9.9 (D-serine), 10.0 (D-citrulline), 10.4 (L-allo-threonine), 10.4 (L-threonine), 10.5 (D-homoserine), 10.6 (L-lysine (mono-derivative)), 10.7 (L-citrulline),10.8 (D-lysine (mono-derivative)), 10.8 (L-glutamic acid), 12.3 (D-allo-threonine), 12.3 (L-O-methylserine), 13.1 (D-glutamic acid), 14.8 (D-threonine), 14.8 (L-alanine), 15.7 (L-proline), 17.2 (D-O-methylserine), 17.9 (L-2-amino-n-butyric acid), 18.2 (D-proline), 19.1 (D-alanine), 20.2 (L-methionine), 21.9 (L-histidine (bis-derivative)), 23.2 (L-valine), 23.8 (D-2-amino-n-butyric acid), 24.0 (L-norvaline), 25.3 (D-histidine (bis-derivative)), 26.4 (D-methionine), 27.7 (L-phenylalanine), 28.7 (L-isoleucine), 29.6 (L-leucine), 29.7 (D-valine), 29.9 (L-norleucine), 30.4(D-norvaline), 31.4 (L-ornithine (bis-derivative)), 31.4 (L-lysine (bis-derivative)),33.2 (D-phenylalanine), 34.5 (D-lysine (bis-derivative)), 34.5 (D-ornithine (bis-derivative)), 35.3 (D-isoleucine), 35.9 (D-leucine), 36.5 (D-norleucine), 37.8 (L-tyrosine (bis-derivative)), 42.6 (D-tyrosine (bis-derivative))

KEY WORDS

chiral

REFERENCE

Fujii, K.; Ikai, Y.; Mayumi, T.; Oka, H.; Suzuki, M.; Harada, K.-I. A nonempirical method using LC/MS for determination of the absolute configuration of constituent amino acids in a peptide: Elucidation of limitations of Marfey's method and of its separation mechanism, *Anal.Chem.*, **1997**, *69*, 3346–3352.

SAMPLE

Matrix: bulk
Analyte: leucine
Sample preparation: Treat 700 µg β-leucine with 20 µL 1% Nα-(2,4-dinitro-5-fluoro-phenyl)-L-alaninamide (Marfey's reagent) in acetone and 40 µL 1 M sodium bicarbonate, heat at 40° for 1 h, cool to 25°, add 20 µL 2 M HCl, filter (Gelman Acro LC-13), inject a 10 µL aliquot of the filtrate.

HPLC VARIABLES

Column: 150 × 3.9 Nova-Pak C18
Mobile phase: MeCN:50 mM pH 3.0 triethylamine phosphate 30:70
Flow rate: 2
Injection volume: 10
Detector: UV 340

CHROMATOGRAM

Retention time: 12.5 ((3R)-β-leucine), 15 ((2S)-α-leucine (L)), 32 ((3S)-β-leucine), 39 ((2R)-α-leucine (D))

KEY WORDS
chiral

REFERENCE
Aberhart, D.J.; Cotting, J.-A.; Lin, H.-J. Separation by high-performance liquid chromatography of (3R)- and (3S)-β-leucine as diastereomeric derivatives, *Anal.Biochem.*, **1985**, *151*, 88–91.

SAMPLE
Matrix: protein
Analyte: amino acids
Sample preparation: Hydrolyze 50-100 pmole protein with 200 µL 6 M HCl in a sealed argon-flushed vial at 110° for 24 h, evaporate to dryness, add 50 µL triethylamine:water 25:75, evaporate to dryness under reduced pressure, repeat this step, add 100 µL triethylamine:water 25:75, add 100 µL 1 mM Nα-(2,4-dinitro-5-fluorophenyl)-L-alaninamide (Marfey's reagent) in acetone, mix, shake gently in the dark at 40° for 1 h, add 20 µL 2 M HCl, dry under reduced pressure, dissolve in DMSO:water 50:50, inject an aliquot.

HPLC VARIABLES
Guard column: 20 × 4.6 5 µm Nucleosil C8
Column: 250 × 4.6 5 µm Aquapore RP-300 C8
Mobile phase: Gradient. A was THF:13 mM trifluoroacetic acid 4:96. B was MeCN:THF:13 mM trifluoroacetic acid 50:2:48. A:B from 100:0 to 65:35 over 50 min, to 60:40 (step gradient), maintain at 60:40 for 5 min, to 15:85 over 55 min.
Flow rate: 1
Injection volume: 50
Detector: UV 340

CHROMATOGRAM
Retention time: 22.5 (cysteic acid), 35.5 (S-carboxymethylcysteine), 40 (serine), 41.5 (threonine), 43.5 (arginine), 46 (glycine), 48 (aspartic acid), 54 (glutamic acid), 54.5 (alanine), 56 (proline), 68 (histidine), 68.5 (methionine), 70 (valine), 78.5 (isoleucine), 80.5 (leucine), 82.5 (tryptophan), 84 (phenylalanine), 87 (lysine), 88.5 (cysteine), 104.5 (tyrosine)
Limit of detection: 50 pmole

KEY WORDS
comparison with other derivatizing reagents

REFERENCE
Kochhar, S.; Christen, P. Amino acid analysis by high-performance liquid chromatography after derivatization with 1-fluoro-2,4-dinitrophenyl-5-L-alanine amide, *Anal.Biochem.*, **1989**, *178*, 17–21.

SAMPLE
Matrix: solutions
Analyte: amino acids
Sample preparation: Add 50 µL of a 50 mM solution in water to 100 µL 1% Nα-(2,4-dinitro-5-fluorophenyl)-L-alaninamide (Marfey's Reagent) in acetone, mix, add 20 µL 1 M sodium bicarbonate, mix at 30-40° for 1 h, cool to room temperature, add 10 µL 2 M HCl, mix, dry in a vacuum desiccator over NaOH, reconstitute with 500 µL DMSO, inject a 5-10 µL aliquot.

HPLC VARIABLES
Column: 100 × 8 C18 (Waters)
Mobile phase: Gradient. MeCN:50 mM pH 3.0 triethylammonium phosphate from 10:90 to 50:50 over 1 h.
Flow rate: 2
Injection volume: 5-10
Detector: UV 340

CHROMATOGRAM
Retention time: 17.68 (L-Asp), 19.40 (L-Glu), 20.28 (D-Asp), 21.40 (L-Ala), 22.71 (D-Glu), 26.72 (D-Ala), 28.21 (L-Met), 34.66 (D-Met), 35.82 (L-Phe), 41.22 (D-Phe)

KEY WORDS
chiral

REFERENCE
Marfey, P. Determination of D-amino acids. II. Use of a bifunctional reagent, 1,5-difluoro-2,4-dinitrobenzene, *Carlsberg Res. Commun.*, **1984**, *49*, 591–596.

SAMPLE
Matrix: solutions
Analyte: amino acids
Sample preparation: Mix 200 μL of a 10 mM solution of amino acid in 100 mM sodium bicarbonate with 200 μL 10 mM Nα-(2,4-dinitro-5-fluorophenyl)-L-alaninamide (Marfey's reagent) in acetone (freshly prepared), heat at 40° with frequent mixing for 1 h, cool, add 100 μL 200 mM HCl, degas, filter, inject an aliquot.

HPLC VARIABLES
Column: 100 × 8 10 μm μBondapak
Mobile phase: Gradient. X was MeCN containing 0.1% trifluoroacetic acid. Y was water containing 0.1% trifluoroacetic acid. X:Y from 10:90 to 60:40 over 50 min (mobile phase A) or from 5:95 to 20:80 over 70 min (Mobile Phase B) or isocratic MeCN:20 mM pH 4.0 sodium acetate buffer 8:92
Flow rate: 2
Detector: UV 340

CHROMATOGRAM
Retention time: k' 2.96 (L-His) (Mobile Phase A), k' 1.80 (D-His) (Mobile Phase A), k' 4.18 (L-Asp) (Mobile Phase A), k' 4.66 (D-Asp) (Mobile Phase A), k' 4.80 (L-Thr) (Mobile Phase A), k' 6.11 (D-Thr) (Mobile Phase A), k' 5.70 (L-Glu) (Mobile Phase A), k' 6.37 (D-Glu) (Mobile Phase A), k' 6.73 (L-Ala) (Mobile Phase A), k' 7.72 (D-Ala) (Mobile Phase A), k' 7.55 (L-Val) (Mobile Phase A), k' 9.08 (D-Val) (Mobile Phase A), k' 7.95 (L-Leu-NH2) (Mobile Phase A), k' 9.73 (D-Leu-NH2) (Mobile Phase A), k' 8.14 (L-Tyr) (Mobile Phase A), k' 9.05 (D-Tyr) (Mobile Phase A), k' 8.12 (L-Trp-NH2) (Mobile Phase A), k' 9.40 (D-Trp-NH2) (Mobile Phase A), k' 8.17 (L-Met) (Mobile Phase A), k' 9.85 (D-Met) (Mobile Phase A), k' 8.92 (L-Phe-NH2) (Mobile Phase A), k' 10.57 (D-Phe-NH2) (Mobile Phase A), k' 9.02 (L-Leu) (Mobile Phase A), k' 10.65 (D-Leu) (Mobile Phase A), k' 9.51 (L-Ile) (Mobile Phase A), k' 11.45 (D-Ile) (Mobile Phase A), k' 9.83 (L-Trp) (Mobile Phase A), k' 11.25 (D-Trp) (Mobile Phase A), k' 9.96 (L-Nle) (Mobile Phase A), k' 11.87 (D-Nle) (Mobile Phase A), k' 10.18 (L-allylglycine) (Mobile Phase A), k' 11.69 (D-allylglycine) (Mobile Phase A), k' 10.44 (L-Cys (tert-butyl)) (Mobile Phase A), k' 12.01 (D-Cys (tert-butyl)) (Mobile Phase A), k' 10.80 (L-Cystine) (Mobile Phase A), k' 11.47 (D-Cystine) (Mobile Phase A), k' 11.04 (L-Phe) (Mobile Phase A), k' 12.40 (D-Phe) (Mobile Phase A), k' 11.89 (L-Lys (bis derivative)) (Mobile Phase A), k' 12.58 (D-Lys (bis derivative)) (Mobile Phase A), k' 13.54 (L-His (triphenylmethyl)) (Mobile Phase A), k' 13.89 (D-His (triphenylmethyl)) (Mobile Phase A), k' 16.29 (L-Cys (triphenylmethyl)) (Mobile Phase A), k' 17.21 (D-Cys (triphenylmethyl)) (Mobile Phase A), k' 17.80 (L-Ser) (Mobile Phase B), k' 16.63 (D-Ser) (Mobile Phase B), k' 20.48 (L-Cys (S-acetamidomethyl)) (Mobile Phase B), k' 22.67 (D-Cys (S-acetamidomethyl)) (Mobile Phase B), k' 25.08 (L-Pro) (Mobile Phase B), k' 26.68 (D-Pro) (Mobile Phase B), k' 26.2 (L-Asn) (Mobile Phase C), k' 44.7 (D-Asn) (Mobile Phase C), k' 26.6 (L-Gln) (Mobile Phase C), k' 45.7 (D-Gln) (Mobile Phase C), k' 51.7 (L-Arg) (Mobile Phase C), k' 103.3 (D-Arg) (Mobile Phase C)

KEY WORDS
chiral

REFERENCE
Adamson, J.G.; Hoang, T.; Crivici, A.; Lajoie, G.A. Use of Marfey's reagent to quantitate racemization upon anchoring of amino acids to solid supports for peptide synthesis, *Anal.Biochem.*, **1992**, *202*, 210–214.

SAMPLE
Matrix: tissue
Analyte: phosphoserine

Sample preparation: Homogenize (Kontes microultrasonic cell disrupter) brain tissue with 0.5-1 mL ice-cold 1 M HCl, centrifuge at 40000 g for 45 min. Remove a 100 μL aliquot of the supernatant and add it to 20 μL 1 mg/mL 2-amino-5-phosphopentanoic acid in water, add 200 μL 1 M sodium bicarbonate, add 100 μL 7.5 mg/mL Nα-(2,4-dinitro-5-fluorophenyl)-L-alaninamide (Marfey's reagent) in acetone, heat at 50° for 2 h, evaporate to dryness under a stream of nitrogen at 60°, add 250 μL 50 mM Tris, vortex, centrifuge for 10 min, inject a 20 μL aliquot of the supernatant.

HPLC VARIABLES
Column: 250 × 4.6 Ultrasphere ODS
Mobile phase: Gradient. MeCN: 50 mM pH 8.0 Tris buffer from 0:100 to 10:90 over 30 min, to 70:30 over 5 min, maintain at 70:30 for 10 min, return to initial conditions over 5 min, re-equilibrate for 5 min.
Flow rate: 1
Injection volume: 20
Detector: UV 340

CHROMATOGRAM
Retention time: 16.5 (D), 20.5 (L)
Internal standard: 2-amino-5-phosphopentanoic acid (Tocris Cookson, Bristol UK) (24.5 (D enantiomer))
Limit of detection: 11 pmole

KEY WORDS
rat; brain; chiral

REFERENCE
Goodnough, D.B.; Lutz, M.P.; Wood, P.L. Separation and quantification of D- and L-phosphoserine in rat brain using Nα-(2,4-dinitro-5-fluorophenyl)-L-alaninamide (Marfey's reagent) by high-performance liquid chromatography with ultraviolet detection, *J.Chromatogr.B*, **1995**, *672*, 290−294.

4-Nitrobenzyl Bromide

SAMPLE
Matrix: bulk

Analyte: amino acids

Sample preparation: Dissolve 0.5 mmole amino acid and 0.5 mmole (1S)-(+)-10-camphorsulfonyl chloride in 10 mL ether, add 10 mL 1 M NaOH, stir vigorously for 1 h, acidify with 1 M HCl, extract with ether. Evaporate the extract to dryness and reconstitute with chloroform containing 0.5 mmole 4-nitrobenzyl bromide, reflux for 30 min, wash with water, dry over anhydrous sodium sulfate, evaporate to dryness.

HPLC VARIABLES

Column: 250 × 2 MicroPak Si-5
Mobile phase: Isooctane:isopropanol 98.5:1.5
Flow rate: 0.5
Detector: UV 254

CHROMATOGRAM

Retention time: 7.0 (L-isoleucine), 7.6 (D-isoleucine, L-leucine), 8.6 (D-leucine), 11.0 (L-phenylalanine), 12.4 (D-phenylalanine), 17.0 (L-alanine), 22.3 (D-alanine)

KEY WORDS

normal phase; chiral

REFERENCE

Furukawa, H.; Sakakibara, E.; Kamei, A.; Ito, K. Separation of L- and D-amino acids as diastereomeric derivatives by high performance liquid chromatography, *Chem.Pharm.Bull.*, **1975**, *23*, 1625–1626.

SAMPLE

Matrix: bulk

Analyte: amino acids

Sample preparation: Dissolve 1 mmole amino acid in 10 mL diethyl ether, add 20 mL 1 M NaOH, stir vigorously at 0°, add 2 mmole (1S)-(+)-10-camphorsulfonyl chloride in 30 mL ether dropwise, stir at room temperature for 1 h. Remove the aqueous layer and wash it twice with diethyl ether, acidify the aqueous layer with concentrated HCl, extract with diethyl ether. Dry the organic extract over anhydrous sodium sulfate, evaporate to dryness, reconstitute with 10 mL DMF, add 1 drop trimethylamine, add 1.1 mmole 4-nitrobenzyl bromide, heat at 55° for 2 h, dilute with 40 mL chloroform, wash with water, dry over anhydrous sodium sulfate, evaporate to dryness, reconstitute with chloroform, inject an aliquot.

HPLC VARIABLES

Column: 250 × 2.2 10 μm MicroPak-NH2
Mobile phase: Isooctane:dichloromethane:isopropanol 79:16:5
Flow rate: 0.4
Detector: UV 254

CHROMATOGRAM

Retention time: 3.9 (L-leucine), 4.4 (L-isoleucine, D-leucine), 5.0 (D-isoleucine), 6.2 (L-phenylalanine), 7.2 (L-alanine), 7.4 (L-methionine), 9.3 (D-alanine), 8.5 (D-phenylalanine), 10.0 (D-methionine), 12.8 (L-glutamic acid), 16.8 (D-glutamic acid), 29.2 (L-tryptophan), 33.2 (L-tyrosine), 47.2 (D-tyrosine), 49.6 (D-tryptophan)

KEY WORDS

normal phase; chiral

REFERENCE

Furukawa, H.; Mori, Y.; Takeuchi, Y.; Ito, K. Separation of L- and D-amino acids as diastereomeric derivatives by high-performance liquid chromatography, *J.Chromatogr.*, **1977**, *136*, 428–431.

ACYL HALIDE

2-(9-Anthryl)ethyl Chloroformate

SAMPLE
Matrix: solutions
Analyte: amino acids
Sample preparation: 400 µL Amino acid solution + 100 µL 1 M pH 8.9 borate buffer + 500 µL 10 mM 2-(9-anthryl)ethyl chloroformate in MeCN, mix, let stand for 5 min, add 1 mL pentane, mix, discard the pentane layer, inject an aliquot of the reaction mixture. (Prepare 2-(9-anthryl)ethyl chloroformate as follows. Stir a solution of 3 g of 9-bromoanthracene in 100 mL ether at 0° under argon or nitrogen, add 9 mL 1.6 M n-butyllithium over 5 min, stir for 30 min, add an ice-cold solution of 3 g ethylene oxide (Caution! Ethylene oxide is a carcinogen!) in 16 mL ether, stir for 1 h, add 70 mL water, add 50 mL ether, remove the organic layer, extract the aqueous layer with 100 mL dichloromethane. Combine the organic layers and wash them with water, dry over anhydrous sodium sulfate, evaporate to dryness, chromatograph on silica gel with dichloromethane to give 2-(9-anthryl)ethanol as pale yellow crystals (mp 106-8°) (J. Org. Chem. 1986, 51, 2956). Stir a solution of 2-(9-anthryl)ethanol in ether in the presence of pyridine (as an HCl scavenger) at 0°, add a solution of phosgene in toluene. 2-(9-anthryl)ethyl chloroformate is obtained as colorless crystals (mp 86-87° from pentane). Protect stock solutions from light and store them in the refrigerator (Anal. Chem. 1991, 63, 292). 2-(9-Anthryl)ethyl chloroformate) is also available from Eka Nobel.)

HPLC VARIABLES
Column: 650 × 0.25 5 µm Kromasil C8 (Eka Nobel)
Mobile phase: Gradient. MeCN:100 mM pH 4.1 acetate buffer from 40:60 to 85:15 over 45 min
Flow rate: 0.005-0.01
Injection volume: 0.06-1
Detector: F ex 351 (16 mW Ar laser) em 412 (bandpass filter); UV 256

CHROMATOGRAM
Retention time: 8.5 (Arg), 11 (Asp, Ser), 12 (Glu), 13 (Thr), 15 (Gly), 17.5 (Ala), 20.5 (Pro), 22 (Met), 24 (Phe), 24.5 (Val), 25.5 (Cys), 26.5 (Cystine), 27.5 (Ile), 28 (Leu), 33.5 (His), 36 (Lys), 39 (Tyr)
Limit of detection: 0.30 nM (F), 400 nM (UV)

KEY WORDS
capillary HPLC; comparison with capillary electrophoresis

REFERENCE
Engström, A.; Andersson, P.E.; Josefsson, B.; Pfeffer, W.D. Determination of 2-(9-anthryl)ethyl chloroformate-labeled amino acids by capillary electrophoresis and liquid chromatography with absorbance or fluorescence detection, *Anal.Chem.*, **1995**, *67*, 3018–3022.

SAMPLE
Matrix: solutions
Analyte: amino acids
Sample preparation: Mix 200 µL of a 2 mM amino acid solution in 200 mM pH 9.0 borate buffer with 200 µL 10 mM 2-(9-anthryl)ethyl chloroformate, let stand for 2 min, wash with 500 µL pentane, dilute 10-fold with water, inject an aliquot. (Prepare 2-(9-anthryl)ethyl chloroformate as described above.)

CAPILLARY ELECTROPHORESIS
Capillary: 67 cm × 25 µm (46 cm to detector) (Polymicro Technologies)
Capillary preparation: Before each run flush with MeOH:200 mM NaOH 10:90 for 10 min and with water for 5 min then equilibrate with running buffer for 10 min. Rinse new capillaries with 200 mM NaOH for 2 h.
Running buffer: Isopropanol:buffer 15:85 (Buffer was 50 mM pH 7.50 phosphate buffer containing 40 mM sodium dodecyl sulfate and 10 mM gamma-cyclodextrin
Capillary temperature: 25
Voltage/Current: 30 kV/12 µA
Injection: Pressure injection at 50 mbar for 15 s
Detector: UV 256
Model: Lauerlabs Prince
Migration time: 14.19 (Asp, α = 1.031), 14.23 (Glu, α = 1.030), 15.51 (Ser, α = 1.030), 16.16 (Thr, α = 1.039), 16.41 (Ala, α = 1.022), 19.90 (Val, α = 1.020), 23.70 (Ile, α = 1.053), 26.07 (Leu, α = 1.031), 31.04 (Phe, α = 1.048) [Retention time for the first eluting (L) isomer.)

KEY WORDS
chiral

REFERENCE
Wan, E.; Engström, A.; Blomberg, L.G. Direct chiral separation of amino acids derivatized with 2-(9-anthryl)ethyl chloroformate by capillary electrophoresis using cyclodextrins as chiral selectors. Effect of organic modifiers on resolution and enantiomeric elution order, *J.Chromatogr.A*, **1996**, *731*, 283–292.

Benzoyl Chloride

RELATED REFERENCE

Chen, S. The enantioresolution of N-benzoyl and its analogs derivatized amino acids on cyclodextrin bonded chiral stationary phases using a nonaqueous acetonitrile-based mobile phase. *J.Chin.Chem.Soc.(Taipei)* **1996**, *43*, 45-51.

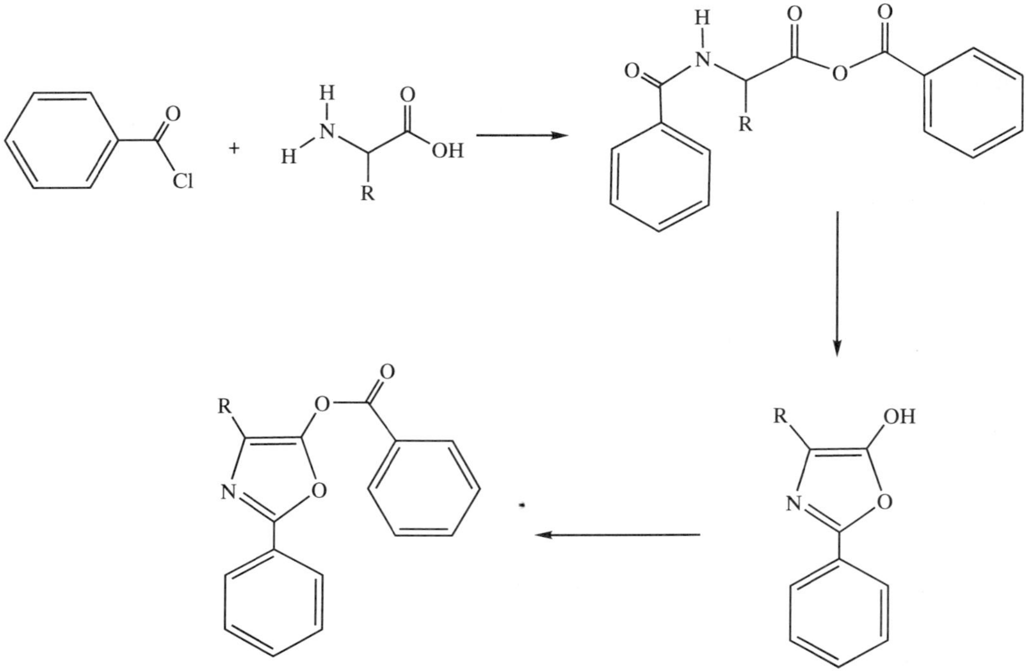

SAMPLE

Matrix: blood

Analyte: amino acids

Sample preparation: 90 μL Serum + 10 μL 2.5 mM cyclohexylalanine + 200 μL ethyl acetate + 2 μL benzoyl chloride + 6 μL triethylamine, vortex at 2500 vibrations/min for 2 min. Remove 150 μL of the ethyl acetate phase and evaporate it to dryness, dissolve the residue in 100 μL MeCN:water 70:30, inject an aliquot.

HPLC VARIABLES

Guard column: 5 μm Kromasil 100 C18

Column: 250 × 4 5 μm Kromasil 100 C18

Mobile phase: Gradient. MeCN:water from 70:30 to 95:5 over 30 min.

Flow rate: 1

Injection volume: 50

Detector: UV 274; MS, electrospray, Finnigan MAT, TSQ 700, flow rate 1 μL/min, 2.8 kV, drying gas 140°

CHROMATOGRAM

Retention time: 6.4 (lysine), 7.1 (glycine), 7.5 (alanine), 10.4 (glutamate), 8.5 (tryptophan), 8.5 (cystine), 9.4 (methionine), 11.8 (phenylalanine), 12.1 (valine), 13.3 (cysteine), 13.9 (leucine), 15.1 (isoleucine), 15.2 (naphthylalanine), 15.5 (tyrosine)

Internal standard: cyclohexylalanine (21.0)

Limit of quantitation: 10 pmol

KEY WORDS

serum; fetal bovine serum

REFERENCE
Oehlke, J.; Brudel, M.; Blasig, I.E. Benzoylation of sugars, polyols and amino acids in biological fluids for high-performance liquid chromatographic analysis, *J.Chromatogr.B*, **1994**, *655*, 105–111.

Benzyl Chloroformate

SAMPLE
Matrix: proteins
Analyte: amino acids
Sample preparation: Heat 300 mg protein and 1.5 mL 6 M HCl at 110° in a sealed tube for 24 h, evaporate to dryness. Dissolve an amount of protein hydrolysate containing 4 mmole amino acids in 2 mL 2 M NaOH, stir at 0°, add 1.36 g benzyl chloroformate and 250 μL 4 M NaOH simultaneously over 40 min so as to maintain the pH between 10 and 12, wash the reaction mixture with 250 μL pentane. Cool the aqueous layer and acidify to Congo red with 4 M HCl, extract with three 200 μL portions of ethyl acetate. Combine the extracts and dry them over anhydrous sodium sulfate, evaporate to dryness under reduced pressure, reconstitute with MeOH, inject a 10 μL MeOH.

HPLC VARIABLES
Column: 250 × 4.6 5 μm Econosil C18
Mobile phase: Gradient. A was 0.5% trifluoroacetic acid in water. B was 0.5% trifluoroacetic acid in MeCN. A:B from 100:0 to 40:60 over 1 h, to 0:100 over 5 min, maintain at 0:100 for 5 min, return to initial conditions over 5 min, re-equilibrate for 10 min.
Flow rate: 1.5
Injection volume: 10
Detector: UV 254

CHROMATOGRAM
Retention time: 17 (Asn), 18 (Gln), 18.5 (His), 19 (Ser), 20 (Thr), 22 (Asp), 23 (Glu), 24 (Gly), 28 (Ala), 33 (Pro), 34 (Tyr), 39 (Val), 42 (Met), 45.5 (Leu), 45.5 (Ile), 45.5 (Trp), 46 (Phe), 49.5 (Lys), 56 (Cys), 58 (Arg)

REFERENCE
Egorova, T.A.; Eremin, S.V.; Mitsner, B.I.; Zvonkova, E.N.; Shvets, V.I. Isolation of individual amino acids from various microbiological sources using reversed-phase high-performance liquid chromatography, *J.Chromatogr.B*, **1995**, *665*, 53–62.

3,5-Dinitrobenzoyl Chloride

SAMPLE
Matrix: solutions
Analyte: amino acids
Sample preparation: Shake a slurry of 100 μmoles amino acid, 100 μmole 3,5-dinitroben-
zoyl chloride, and 300 μmoles propylene oxide (Caution! Propylene oxide is a carcinogen!)
in 5 mL dry THF at room temperature for 7 days, filter, evaporate to dryness under
reduced pressure, dissolve the residue in 30 mL MeOH, add Amberlite IR-120, reflux for
1 h, evaporate to dryness, reconstitute with dichloromethane, chromatograph on a small
column of Kieselgel 60 (Merck) with hexane:ethyl acetate 90:10, evaporate to dryness,
inject an aliquot of a solution in the mobile phase.

HPLC VARIABLES
Column: 300 × 4 CSP 1 chiral column (Details of column preparation are in paper.)
Mobile phase: Hexane:isopropanol 90:10
Flow rate: 1
Detector: UV 254

CHROMATOGRAM
Retention time: k' 6.26 (first enantiomer), α = 1.18 (alanine); k' 4.48 (first enantiomer), α
= 1.16 (2-aminobutyric acid); k' 3.65 (first enantiomer), α = 1.11 (norvaline); k' 3.21 (first
enantiomer), α = 1.14 (valine); k' 3.26 (first enantiomer), α = 1.08 (norleucine); k' 3.07
(first enantiomer), α = 1.09 (leucine); k' 2.84 (first enantiomer), α = 1.09 (isoleucine); k'
6.27 (first enantiomer), α = 1.02 (phenylglycine); k' 5.53 (first enantiomer), α = 1.11 (phenylalanine); k' 7.11 (first enantiomer), α = 1.16 (methionine); k' 5.61 (first enantiomer), α
= 1.00 (proline); k' 3.27 (first enantiomer), α = 1.02 (N-methylalanine)

KEY WORDS
chiral

REFERENCE
Chen, C.-C.; Lin, C.-E. HPLC separation of enantiomers of amino acids and amino alcohols on ionically
bonded chiral stationary phases consisting of cyanuric chloride with amino acid and dialkylamine
substituents, *J.Chromatogr.Sci.*, **1995**, *33*, 229–235.

(S)-(+)-Flunoxaprofen Chloride

SAMPLE
Matrix: peptides
Analyte: amino acids
Sample preparation: Mix ≤5 mg peptide and 200 μL 6 M HCl, heat at 100° for 6 h, evaporate to dryness under a stream of nitrogen at 80-100°, reconstitute with 3 M HCl in isopropanol, heat at 105° for 20 min, evaporate to dryness under a stream of nitrogen at 105°, reconstitute with 1 mL 5 mg/mL sodium heptanesulfonate in water, extract with 5 mL dichloromethane. Evaporate the organic layer to dryness under a stream of nitrogen at 50°, add 20 mg anhydrous sodium carbonate, add a 3-fold molar excess of 1 mg/mL S-flunoxaprofen chloride in dichloromethane or ethyl acetate, heat at 40° for 1 h (or agitate moderately overnight), evaporate to dryness, reconstitute with 200 μL mobile phase, inject a 10-20 μL aliquot. (Prepare S-flunoxaprofen chloride as follows. Dissolve 1 mmole S-flunoxaprofen in 25 mL toluene, add a trace of DMF (this paper), add 2.5 mL thionyl chloride, reflux for 30 min, remove solvent by evaporation, dry the residue under vacuum over KOH, recrystallize from dichloromethane (mp 73°) (J.Chromatogr. 1988, 427, 131).)

HPLC VARIABLES
Column: 250 × 4.6 5 μm Zorbax-Sil
Mobile phase: n-Hexane:chloroform:EtOH 100:10:1
Flow rate: 2
Injection volume: 10-20
Detector: UV 305; F ex 305 em 355

CHROMATOGRAM
Retention time: k' 11.2 (D-alanine), k' 15.2 (L-alanine), k' 19.5 (glycine), k' 4.8 (D-isoleucine), k' 4.5 (L-isoleucine), k' 7.1 (D-leucine), k' 6.0 (L-leucine), k' 11.9 (D-methionine), k' 10.5 (L-methionine), k' 6.6 (D-phenylalanine), k' 6.4 (L-phenylalanine), k' 15.0 (D-proline), k' 12.3 (L-proline), k' 22.6 (D-tyrosine), k' 13.1 (L-tyrosine)
Limit of detection: 0.1-0.5 ng

KEY WORDS
normal phase; chiral

REFERENCE
Langguth, P.; Spahn, H.; Merkle, H.P. Fluorescence assay for small peptides and amino acids: high-performance liquid chromatographic determination of selected substrates using activated S-flunoxaprofen as a chiral derivatizing agent, *J.Chromatogr.*, **1990**, *528*, 55–64.

(+)-1-(9-Fluorenyl)ethyl Chloroformate

RELATED REFERENCE
Einarsson, S.; Josefsson, B.; Möller, P.; Sanchez, D. Separation of amino acid enantiomers and chiral amines using precolumn derivatization with (+)-1-(9-fluorenyl)ethyl chloroformate and reversed-phase liquid chromatography. *Anal.Chem.* **1987**, *59*, 1191-1195.

SAMPLE
Matrix: eye, tissue
Analyte: amino acids
Sample preparation: Homogenize eyes or nervous tissue with a 9-fold excess of 8% perchloric acid, centrifuge at 0° at 20000 g for 10 min. Neutralize the supernatant with solid potassium bicarbonate, centrifuge. Mix a 100 μL aliquot of the supernatant with 400 μL 300 mM pH 11.0 borate/NaOH buffer and 500 μL 18 mM (+)-1-(9-fluorenyl)ethyl chloroformate in acetone, let stand at room temperature for 45 s, wash twice with 2 mL portions of pentane, filter (0.45 μm) the aqueous phase, inject a 20 μL aliquot of the filtrate.

HPLC VARIABLES
Guard column: 10 × 4.6 Nucleosil C18
Column: 250 × 4.6 Shim-pack CLC-ODS (Shimadzu)
Mobile phase: Gradient. A was MeCN:THF:15 mM citric acid containing 10 mM tetramethylammonium chloride 10:75:15, adjusted to pH 2.0. B was MeCN:THF:15 mM citric acid containing 10 mM tetramethylammonium chloride 20:10:70, adjusted to pH 5.3. C was MeCN:THF:15 mM citric acid containing 10 mM tetramethylammonium chloride 60:10:30, adjusted to pH 6.2. A:B:C from 100:0:0 to 15:85:0 over 3 min, to 13:87:0 over 12 min, to 0:30:70 over 70 min, to 0:0:100 over 0.1 min, maintain at 0:0:100 for 9.9 min. (Adjust pH with 6 M HCl or 6 M NaOH.)
Column temperature: 55
Flow rate: 0.7
Injection volume: 20
Detector: F ex 260 em 310

CHROMATOGRAM
Retention time: 18.5 (D-Arg), 19.5 (L-Arg), 22.5 (taurine), 25.4 (D-Asn), 26 (L-Asn), 26.5 (D-Gln), 27.5 (L-Gln), 29 (D-citrulline), 29.3 (L-citrulline), 31.5 (D-Hyp, L-Ser, D-Ser), 32 (L-Hyp), 32.5 (D-Asp), 33.5 (L-Asp), 36 (D-Glu), 36.5 (L-Glu), 37 (D-Thr), 37.5 (L-Thr), 38.5 (Gly), 41.5 (L-α-aminoadipic acid), 45 (L-Pro), 48 (D-Ala, L-Ala), 50.5 (L-Pro, D-Pro), 52 (gamma-aminobutyric acid), 55 (D-β-aminobutyric acid, L-β-aminobutyric acid, L-α-aminobutyric acid), 57 (D-Met), 58 (L-Met), 62.5 (D-Val), 63.5 (L-Val), 65.5 (D-Phe), 66.5 (L-Phe), 71 (D-Ile), 71.5 (L-Ile), 72 (L-Ile), 72.5 (L-Leu), 75 (D-cystine), 76.5 (cystathionine),77 (L-cystine), 80.5 (D-Hyl, L-Hyl), 88.5 (D-ornithine), 90 (L-ornithine), 91.5 (D-Lys), 92.5 (L-Lys)
Limit of detection: <1 pmole

KEY WORDS
chiral; Compounds which co-elute under these conditions can be resolved by altering mobile phase conditions.; crab; lobster; prawn; crayfish; nervous tissue

REFERENCE

Okuma, E.; Abe, H. Simultaneous determination of D- and L-amino acids in the nervous tissues of crustaceans using precolumn derivatization with (+)-1-(9-fluorenyl)ethyl chloroformate and reversed-phase ion-pair high-performance liquid chromatography, *J.Chromatogr.B*, **1994**, *660*, 243–250.

SAMPLE

Matrix: solutions

Analyte: amino acids

Sample preparation: 10 μL Amino acid mixture + 10 μL 200 mM pH 8.0 borate buffer + 10 μL 4 mM (+)-1-(9-fluorenyl)ethyl chloroformate in acetone, mix for 4 min, add hydroxyproline, dilute with water, inject an aliquot.

CAPILLARY ELECTROPHORESIS

Capillary: 67 cm × 50 μm fused-silica (60 cm to detector) (Polymicro Technologies)

Capillary preparation: Flush capillary with running buffer for 2 min between runs. Wash new capillaries with 100 mM NaOH for 15 min and with water for 15 min.

Running buffer: MeCN:buffer 15:85 (Buffer was 10 mM pH 6.8 sodium phosphate buffer containing 25 mM sodium dodecyl sulfate.)

Capillary temperature: 20

Voltage/Current: 10 or 20 kV (?)

Injection: Pressure injection at 0.5 psi for 5 s.

Detector: UV 200; F ex 248 (0.5 mW KrF laser) em 310 (interference filter)

Model: Beckman P/ACE 2050

Migration time: 12.0 (L-Ser), 12.2 (D-Ser), 12.4 (L-Ala), 12.7 (D-Ala), 13.3 (L-Val), 14.0 (D-Val), 14.3 (L-Met), 14.7 (D-Met), 15.8 (L-Leu), 16.3 (D-Leu), 17.6 (L-Phe), 18.5 (D-Phe), 19.9 (L-Trp), 20.0(D-Trp)

Limit of detection: 300 nM (F)

KEY WORDS

chiral

REFERENCE

Chan, K.C.; Muschik, G.M.; Issaq, H.J. Enantiomeric separation of amino acids using micellar electrokinetic chromatography after pre-column derivatization with the chiral reagent 1-(9-fluorenyl)-ethyl chloroformate, *Electrophoresis*, **1995**, *16*, 504–509.

Fluorenylmethoxycarbonyl Glycyl Chloride

SAMPLE
Matrix: bulk
Analyte: amino acids
Sample preparation: Dissolve 1 mg amino acid in 250 µL water and 250 µL 40 mM pH 7.7 borate buffer, add 400 µL MeCN, add 3 mg reagent, vortex, let stand for about 30 min, extract with 3 mL ethyl acetate, inject an aliquot of the organic layer. (Synthesize the reagent, fluorenylmethoxycarbonyl glycyl chloride (FMOC-glycyl-Cl), as follows. Dissolve 1.0 g fluorenylmethoxycarbonyl glycine (FMOC glycine) in 15 mL dichloromethane, add 3.4 mL thionyl chloride, reflux for 4 h, evaporate to dryness under reduced pressure, dissolve the residue in 1 mL dichloromethane, add 10 mL hexane. Filter the precipitate and dry it under vacuum at room temperature for 2 h to obtain fluorenylmethoxycarbonyl glycyl chloride.)

HPLC VARIABLES
Column: 250 × 4.6 5 µm gamma-cyclodextrin Cyclobond II
Mobile phase: MeCN:triethylamine:acetic acid 100:1.2:0.3
Flow rate: 1
Detector: UV 265

CHROMATOGRAM
Retention time: k' 2.9 (methionine, α = 1.10), k' 3.0 (valine, α = 1.00), k' 3.1 (norleucine, α = 1.07), k' 3.1 (O-methyltyrosine, α = 1.10), k' 3.3 (norvaline, α = 1.09), k' 3.4 (phenylalanine, α = 1.11), k' 3.4 (alanine, α = 1.07), k' 3.6 (leucine, α = 1.15), k' 4.1 (homophenylalanine, α = 1.11), k' 8.3 (asparagine, α = 1.17), k' 8.66 (glutamine, α = 1.26), k' 12.1 (tryptophan, α = 1.29), k' 13.0 (aspartic acid, α = 1.06), k' 37.4 (glutamic acid, α = 1.21) (k' is the capacity factor of the first eluted enantiomer which is L except for aspartic acid, phenylalanine, and tryptophan.)

KEY WORDS
chiral; comparison with other derivatizing reagents; details of chromatography with other mobile phases and with a β-cyclodextrin column are also given in the paper

REFERENCE

Tang, Y.; Zukowski, J.; Armstrong, D.W. Investigation on enantiomeric separations of fluorenylmethox-ycarbonyl amino acids and peptides by high-performance liquid chromatography using native cy-clodextrins as chiral stationary phases, *J.Chromatogr.A*, **1996**, *743*, 261–271.

9-Fluorenylmethyl Chloroformate

RELATED REFERENCES

Bank, R.A.; Jansen, E.J.; Beekman, B.; te Koppele, J.M. Amino acid analysis by reverse-phase high-performance liquid chromatography: improved derivatization and detection conditions with 9-flu-orenylmethyl chloroformate. *Anal.Biochem.* **1996**, *240*, 167-176.

Betner, I.; Foeldi, P. New automated amino acid analysis by HPLC precolumns derivatization with fluorenylmethyloxycarbonylchloride. *Chromatographia* **1986**, *22*, 381-387.

Betner, I.; Foeldi, P. The FMOC-ADAM approach to amino acid analysis. *LC-GC* **1988**, *6*, 832-840.

Buetikofer, U.; Fuchs, D.; Bosset, J.O.; Gmuer, W. Automated HPLC-amino acid determination of protein hydrolyzates by precolumn derivatization with o-phthalaldehyde and 9-fluorenylmethylchloroformate and comparison with classical ion exchange chromatography. *Chromatographia* **1991**, *31*, 441-447.

Clapp, C.H.; Swan, J.S.; Poechmann, J.L. Identification of amino acids in unknown dipeptides: a deri-vatization with 9-fluorenylmethyl chloroformate and HPLC. *J.Chem.Educ.* **1992**, *69*, A122-A126.

Einarsson, S. Selective determination of secondary amino acids using precolumn derivatization with 9-fluorenylmethylchloroformate and reversed-phase high-performance liquid chromatography. *J.Chromatogr.* **1985**, *348*, 213-220.

Fernandez-Trapiella, A.C. Quantitative analysis of methionine, cysteine, and lysine in feeds by reverse-phase liquid chromatography using precolumn derivatization with 9-fluorenylmethyl chlorofor-mate: preliminary study. *J.Assoc.Off.Anal.Chem.* **1990**, *73*, 935-939.

Grzywacz, C.M. Identification of proteinaceous binding media in paintings by amino acid analysis using 9-fluorenylmethyl chloroformate derivatization and reversed-phase high-performance liquid chro-matography. *J.Chromatogr.A* **1994**, *676*, 177-183.

Gustavsson, B.; Betner, I. Fully automated amino acid analysis for protein and peptide hydrolyzates by precolumn derivatization with 9-fluorenyl methylchloroformate and 1-aminoadamantane. *J.Chromatogr.* **1990**, *507*, 67-77.

Kirschbaum, J.; Luckas, B.; Beinert, W.D. HPLC analysis of biogenic amines and amino acids in food after automatic pre-column derivatization with 9-fluorenylmethyl chloroformate. *Am.Lab.* **1994**, *26*, 28C-28F.

Miles, C.J.; Wallace, L.R.; Moye, H.A. Determination of glyphosate herbicide and (aminomethyl)-phosphonic acid in natural waters by liquid chromatography using pre-column fluorogenic labeling with 9-fluorenylmethyl chloroformate. *J.Assoc.Off.Anal.Chem.* **1986**, *69*, 458-461.

Niedbalski, J.S.; Ringer, D.P. Separation and quantitative analysis of O-linked phosphoamino acids by isocratic high-performance liquid chromatography of the 9-fluorenylmethyl chloroformate deriva-tives. *Anal.Biochem.* **1986**, *158*, 138-145.

Pecavar, A.; Golc-Wondra, A.; Prosek, M.; Skocir, E. Quantitative evaluation of amino acids, derivatizated with 9-fluorenylmethyl chloroformate. *Vestn.Slov.Kem.Drus.* **1991**, *38*, 183-194.

Pocklington, R.; Milley, J.E.; Bates, S.S.; Bird, C.J.; De Freitas, A.S.W.; Quilliam, M.A. Trace determi-nation of domoic acid in seawater and phytoplankton by high-performance liquid chromatography of the fluorenylmethoxycarbonyl (FMOC) derivative. *Int.J.Environ.Anal.Chem.* **1990**, *38*, 351-368.

Sancho, J.V.; Lopez, F.J.; Hernandez, F.; Hogendoorn, E.A.; van Zoonen, P. Rapid determination of glu-fosinate in environmental water samples using 9-fluorenylmethoxycarbonyl precolumn derivatiza-tion, large-volume injection and coupled-column liquid chromatography. *J.Chromatogr.A* **1994**, *678*, 59-67.

Smith, S.H.; Judge, M.D. An HPLC method using FMOC-ADAM for determination of hydroxyproline in muscle. *Meat Sci.* **1991**, *30*, 351-357.

Tan, H.S.I.; Zhang, H. Precolumn derivatization RP-HPLC assay of monosodium glutamate (MSG) in selected food products using UV detection (Abstract APQ 1061). *Pharm.Res.* **1996**, *13*, S18.

van Eijk, H.M.H.; Rooyakkers, D.R.; Wagenmakers, A.J.M.; Soeters, P.B.; Deutz, N.E.P. Isolation and quantitation of isotopically labeled amino acids from biological samples. *J.Chromatogr.B* **1997**, *691*, 287-296

SAMPLE
Matrix: blood
Analyte: amino acids
Sample preparation: Filter (Amicon Centrifree) plasma while centrifuging at 1500 g for 15 min, mix 2.5 μL ultrafiltrate with 2.5 μL reagent, add 1 μL 1 mg/mL 9-fluorenylmethyl chloroformate in MeCN, mix, let stand for 2.5 min, inject the whole amount. (Prepare reagent by dissolving 3 mg o-phthalaldehyde in 50 μL MeOH, add 450 μL 0.5 M pH 10.2 sodium borate buffer, add 5 μL 3-mercaptopropionic acid. Derivatization was performed automatically and took 5 min. o-Phthalaldehyde derivatized primary amino acids and 9-fluorenylmethyl chloroformate derivatized secondary amino acids (proline, sarcosine, and hydroxyproline).)

HPLC VARIABLES
Guard column: 20 × 2.1 5 μm Hypersil ODS
Column: two 100 × 2.1 5 μm Hypersil ODS columns in series
Mobile phase: Gradient. 15 mM pH 6.8 Sodium acetate:MeOH:10 mM pH 6.8 sodium acetate from 0:0:100 to 100:0:0 over 0.05 min, to 60:40:0 over 15 min, to 57.5:42.5:0 over 3.5 min, to 45:55:0 over 3.5 min, to 0:0:100 over 3 min, maintain at 0:0:100 for 5 min.
Column temperature: 40
Flow rate: 0.3
Injection volume: 6
Detector: F ex 230 em 450, after 20 min F ex 260 em 315

CHROMATOGRAM
Retention time: 1.8 (O-phospho-L-serine), 2 (aspartic acid), 2.5 (glutamic acid), 5 (glutathione (reduced)), 6 (asparagine), 6.2 (serine), 7.5 (glutamine), 8 (glycine), 8.5 (threonine), 8.8 (histidine), 9.2 (cystine), 9.5 (citrulline), 10.2 (taurine), 10.5 (alanine), 11.5 (arginine), 12.3 (tyrosine), 13.2 (α-amino-N-butyric acid), 15.3 (methionine), 15.5 (valine), 16 (norvaline), 16.2 (tryptophan), 16.5 (phenylalanine), 17.8 (isoleucine), 18.2 (ornithine), 18.5 (leucine), 19.5 (lysine), 20.5 (hydroxyproline), 22.3 (sarcosine), 24.5 (proline)
Limit of detection: 5 pmole

KEY WORDS
plasma; ultrafiltrate

REFERENCE
Worthen, H.G.; Liu, H. Automatic pre-column derivatization and reversed-phase high performance liquid chromatography of primary and secondary amino acids in plasma with photo-diode array and fluorescence detection, *J.Liq.Chromatogr.*, **1992**, *15*, 3323–3341.

SAMPLE
Matrix: blood, CSF, plant, tissue
Analyte: amino acids
Sample preparation: Seeds, serum, CSF. Homogenize seeds in 100 mL ice-cold water, centrifuge at 4° at 25000 g for 30 min. Mix seed homogenate, serum, or CSF with an equal volume of 10% trichloroacetic acid, centrifuge at 4° at 14000 g for 30 min, filter (0.2 μm) the supernatant. Dilute the supernatant with an equal volume of water. Remove a 20 (seed) or 30 (serum, CSF) μL aliquot and add it to 1 mL 25 mM pH 9.6 borate buffer, add 1 mL acetone, add 15 μL 100 μg/mL D-aminovaleric acid, add 100 μL 10 mM 9-fluorenylmethyl chloroformate in acetone (freshly prepared), vortex for 30 s, let stand at room temperature for 10 min, add 2 mL hexane:ethyl acetate 50:50, vortex for 30 s,

inject a 20 μL aliquot of the aqueous layer. Tissue. Sonicate rat sensorimotor cortex with 2 mL ice-cold PBS with two 10 s bursts at 0°. Remove a 200 μL aliquot of the homogenate and add it to 100 μL 10% trichloroacetic acid, mix, centrifuge at 4° at 25000 g for 30 min. Remove a 30 μL aliquot and add it to 1 mL 25 mM pH 9.6 borate buffer, add 1 mL acetone, add 15 μL 100 μg/mL D-aminovaleric acid, add 100 μL 10 mM 9-fluorenylmethyl chloroformate in acetone (freshly prepared), vortex for 30 s, let stand at room temperature for 10 min, add 2 mL hexane:ethyl acetate 50:50, vortex for 30 s, inject a 20 μL aliquot of the aqueous layer.

HPLC VARIABLES
Column: Nova-Pak C18
Mobile phase: Gradient. MeCN:50 mM pH 3.65 sodium acetate buffer from 31:69 to 48: 52 over 19 min, to 70:30 over 3 min (non-linear gradient), return to initial conditions over 8 min.
Flow rate: 1
Injection volume: 20
Detector: F ex 254 em 315

CHROMATOGRAM
Retention time: 2.5 (arginine), 3.5 (aspartate, serine), 4.5 (glutamate), 5 (threonine), 6.2 (glycine), 8 (alanine), 8.5 (tyrosine), 9.5 (gamma-aminobutyric acid), 10 (proline), 11.5 (methionine), 12.5 (valine), 14.5 (phenylalanine), 15.4 (isoleucine), 15.7 (leucine), 17.5 (β-N-methylamino-L-alanine), 18 (histidine), 18.5 (lysine, cystine)
Internal standard: D-aminovaleric acid (11)
Limit of detection: 0.6 pmole

KEY WORDS
rat; monkey; brain; serum; seeds

REFERENCE
Kisby, G.E.; Roy, D.N.; Spencer, P.S. Determination of β-N-methylamino-L-alanine (BMAA) in plant (*Cycas circinalis* L.) and animal tissue by precolumn derivatization with 9-fluorenylmethyl chloroformate (FMOC) and reversed-phase high-performance liquid chromatography, *J.Neurosci.Methods*, **1988**, *26*, 45–54.

SAMPLE
Matrix: blood, protein
Analyte: amino acids
Sample preparation: Plasma. Mix plasma vigorously with 3 volumes MeCN, centrifuge at 12000 g for 3 min, evaporate an aliquot of the supernatant to dryness under reduced pressure, reconstitute with buffer. Remove a 5 μL aliquot and add it to 5 μL 4.16 mg/mL 9-fluorenylmethyl chloroformate in MeCN, mix, let stand for 1.5 min, add 5 μL reagent, mix, let stand for 3.5 min, add 5 μL MeCN:water:acetic acid 80:12:8, mix, inject an aliquot. Protein. Hydrolyse protein with 500 μL 6 M HCl at 110° for 24 h, evaporate to dryness under reduced pressure, add 10 μL triethylamine:EtOH:water 40:40:20, evaporate to dryness, reconstitute with 5 μL buffer, add 5 μL 4.16 mg/mL 9-fluorenylmethyl chloroformate in MeCN, mix, let stand for 1.5 min, add 5 μL reagent, mix, let stand for 3.5 min, add 5 μL MeCN:water:acetic acid 80:12:8, mix, inject an aliquot. (Buffer was 200 mM boric acid adjusted to pH 8.5 with 5 M NaOH. Prepare reagent by mixing 170 μL 850 mM NaOH, 75 μL 500 mM hydroxylamine hydrochloride, and 5 μL 2-(methylthio)ethanol.)

HPLC VARIABLES
Guard column: 15 × 3.2 7 μm Newguard ODS
Column: 150 × 4.6 3 μm Spherisorb ODS-2
Mobile phase: Gradient. A was MeOH:20 mM pH 6.5 $(NH_4)H_2PO_4$ 15:85. B was MeCN: water 90:10. A:B 82:18 for 2 min, to 77:23 over 1 min, maintain at 77:23 for 7 min, to 64:36 over 10 min, to 52:48 over 1 min, maintain at 52:48 for 5 min, to 45:55 over 2 min, to 1:99 over 1 min (plasma). A:B 82:18 for 2 min, to 77:23 over 1 min, maintain at 77:23 for 3 min, to 60:40 over 10 min, to 55:45 over 1 min, maintain at 55:45 for 3 min, to 45:55 over 2 min, to 1:99 over 1 min (protein hydrolysate).
Column temperature: 35

Flow rate: 1
Detector: F ex 263 em 313; UV 263

CHROMATOGRAM

Retention time: 3 (phosphoserine), 3.5 (aspartic acid), 3.7 (glutamic acid), 4 (α-aminoadipic acid), 4.2 (S-carboxymethylcysteine), 6.7 (hydroxyproline), 7.5 (asparagine), 8.5 (glutamine), 8.7 (citrulline), 9 (serine), 9.5 (histidine), 10 (glycine), 10.6 (threonine), 11.2 (β-alanine), 11.7 (alanine), 12.3 (taurine), 13 (proline), 14.5 (tyrosine), 15 (α-aminobutyric acid), 15.8 (arginine), 17 (homoarginine), 17.7 (valine), 18.3 (methionine), 20.5 (isoleucine), 21 (leucine), 21.5 (norleucine), 22 (phenylalanine), 23 (cystathionine), 26.3 (ornithine), 27 (lysine) (Using plasma gradient.)
Limit of detection: 50 fmole

KEY WORDS
plasma

REFERENCE
Haynes, P.A.; Sheumack, D.; Greig, L.G.; Kibby, J.; Redmond, J.W. Applications of automated amino acid analysis using 9-fluorenylmethyl chloroformate, *J.Chromatogr.*, **1991**, *588*, 107–114.

SAMPLE
Matrix: blood, urine
Analyte: S-carboxymethyl-L-cysteine
Sample preparation: 50 µL Serum or urine + 5 ng (serum) or 5 µg (urine) carboxyethylcysteine + 400 µL 500 mM pH 7.7 sodium borate buffer + 400 µL 10 mM 9-fluorenylmethyl chloroformate in acetone, let stand at room temperature for 1 min, add 50 µL 5 M pH 7.7 ammonium acetate, inject an aliquot.

HPLC VARIABLES
Column: 250 × 4.6 5 µm Hypersil octadecylsilane
Mobile phase: Gradient. MeCN:buffer from 10:90 to 35:65 over 1 h. Buffer was DMF:25 mM pH 3.8 ammonium acetate 4:96.)
Flow rate: 0.8
Detector: F ex 266 em 305

CHROMATOGRAM
Retention time: 50
Internal standard: carboxyethylcysteine (55)
Limit of quantitation: 200 ng/mL

OTHER SUBSTANCES
Extracted: metabolites

KEY WORDS
serum; pharmacokinetics

REFERENCE
Brockmöller, J.; Staffeldt, B.; Roots, I. Evaluation of proposed sulphoxidation pathways of carbocysteine in man by HPLC quantification, *Eur.J.Clin.Pharmacol.*, **1991**, *40*, 387–392.

SAMPLE
Matrix: blood, urine
Analyte: S-carboxymethyl-L-cysteine and metabolites
Sample preparation: 25 µL Plasma or urine + 1.25 µg N-acetylalanine, evaporate to dryness under reduced pressure, reconstitute with 150 µL buffer A, add 150 µL 20 µM 1-octanesulfonic acid in isopropanol, add 150 µL 15 mM 1-pyrenyldiazomethane (Molecular Probes, Eugene OR) in ethyl acetate, heat at 40° for 30 min, add 150 µL isopropanol, add 400 µL buffer B, add 400 µL 10 mM 9-fluorenylmethyl chloroformate in acetone, let stand at room temperature for 50 s, add 35 µL 500 mM ammonium acetate solution, inject a 10 µL aliquot. (Buffer A was isopropanol:water 50:50 containing 100 mM phosphoric acid, adjusted to pH 4.0 with NaOH. Buffer B was 30.9 g/L boric acid adjusted to pH 7.7 with 30% NaOH. The pyrenyldiazomethane reacts with carboxylic acids and the fluorenylmethyl chloroformate reacts with amines. During the second derivatization the pyrenyl group is removed from amino acids and so amino acids exist only as fluorenyl derivatives.)

HPLC VARIABLES
Column: 150 × 4.6 5 μm Ultracarb ODS (Phenomenex)
Mobile phase: Gradient. MeCN:buffer from 25:75 to 50:50 over 80 min. (Buffer was MeOH:THF:40 mM triethylamine 10:5:85, adjusted to pH 3.2 with phosphoric acid.)
Column temperature: 60
Flow rate: 0.8
Injection volume: 10
Detector: F ex 340 em 380 (pyrenyldiazomethane derivatives); F ex 260 em 305 (fluorenylmethyl chloroformate derivatives)

CHROMATOGRAM
Retention time: 6 (S-carboxymethyl-L-cysteine sulfoxide*), 11 (S-methyl-L-cysteine sulfoxide*), 17 (N-acetyl-S-carboxymethyl-L-cysteine sulfoxide), 20 (S-carboxymethyl-L-cysteine*), 22 (thiodiglycolic acid sulfoxide), 25 (N-acetyl-S-methyl-L-cysteine sulfoxide), 28 (thiodiglycolic acid sulfone), 37 (N-acetyl-S-carboxymethyl-L-cysteine), 38 (methylcysteine*), 58 (thiodiglycolic acid), 64 (N-acetyl-S-methyl-L-cysteine) [* = fluorenyl derivatives, others pyrenyl derivatives]
Internal standard: N-acetylalanine (49)

KEY WORDS
plasma

REFERENCE
Staffeldt, B.; Brockmöller, J.; Roots, I. Determination of S-carboxymethyl-L-cysteine and some of its metabolites in urine and serum by high-performance liquid chromatography using fluorescent precolumn labelling, *J.Chromatogr.*, **1991**, *571*, 133–147.

SAMPLE
Matrix: blood, urine
Analyte: S-carboxymethyl-L-cysteine and metabolites
Sample preparation: 50 μL Plasma or urine + 50 μL 100 μg/mL carboxyethylcysteine in water, evaporate to dryness under reduced pressure, reconstitute with 400 μL buffer, add 400 μL 10 mM 9-fluorenylmethyl chloroformate in acetone, let stand at room temperature for 50 s, add 35 μL 500 mM ammonium acetate solution, inject an aliquot. (Buffer was 30.9 g/L boric acid adjusted to pH 7.7 with 30% NaOH.)

HPLC VARIABLES
Column: 250 × 4.6 5 μm Hypersil ODS
Mobile phase: Gradient. MeCN:buffer from 10:90 to 35:65 over 65 min. (Buffer was DMF:25% ammonia:water 0.4:0.2:99.4, pH adjusted to 3.8.)
Column temperature: 40
Flow rate: 0.8
Detector: F ex 260 em 305

CHROMATOGRAM
Retention time: 30 (S-carboxymethyl-L-cysteine sulfoxide), 31 (R-carboxymethyl-L-cysteine sulfoxide), 52 (S-carboxymethyl-L-cysteine)
Internal standard: carboxyethylcysteine (60)
Limit of quantitation: 100 ng/mL

KEY WORDS
plasma

REFERENCE
Staffeldt, B.; Brockmöller, J.; Roots, I. Determination of S-carboxymethyl-L-cysteine and some of its metabolites in urine and serum by high-performance liquid chromatography using fluorescent precolumn labelling, *J.Chromatogr.*, **1991**, *571*, 133–147.

SAMPLE
Matrix: cell suspensions
Analyte: amino acids
Sample preparation: Centrifuge 350 μL cell suspensions at 1470 g for 1 min, add 20 μL 16% trichloroacetic acid to the supernatant, freeze. Thaw the supernatant and adjust its

pH to 7.0 with 1 M NaOH, remove a 300 μL aliquot and make up to 400 μL with water, add 100 μL 1 M pH 6.2 boric acid, add 40 μL 10 μM dihydrokainic acid in MeCN:water 10:90, vortex for 10 s, add 500 μL 15 mM 9-fluorenylmethyl chloroformate, mix for 45 s, add 1 mL ethyl acetate, mix for 10 s, centrifuge at 10500 g for 2 min, repeat the ethyl acetate wash, inject an aliquot of the aqueous layer.

HPLC VARIABLES

Column: 250 × 4.6 C18 (Phenomenex)

Mobile phase: Gradient. A was 0.1% trifluoroacetic acid in MeCN. B was 0.1% trifluoroacetic acid in water. A:B from 30:70 to 50:50 over 15 min, to 100:0 over 2 min, maintain at 100:0 for 5 min, return to initial conditions over 2 min, re-equilibrate for 12 min (cf. Int. J. Environ. Anal. Chem. 1990, 38, 351).

Column temperature: 55

Flow rate: 1

Injection volume: 25

Detector: F ex 264 em 313

CHROMATOGRAM

Retention time: 3.5 (taurine), 11 (glutamine), 12.5 (glutamate), 13 (aspartate)

Internal standard: dihydrokainic acid (21)

KEY WORDS

derivatization

REFERENCE

Brown, J.A.; Nijjar, M.S. The release of glutamate and aspartate from rat brain synaptosomes in response to domoic acid (amnesic shellfish toxin) and kainic acid, *Mol.Cell Biochem.*, **1995**, *151*, 49–54.

SAMPLE

Matrix: cheese

Analyte: amino acids

Sample preparation: Homogenize (Ultra-Turrax T25) 16 g cheese with 30 mL water at room temperature for 2 min, centrifuge at 4° at 8650 g for 20 min, remove the water and fat layers. Homogenize the pellet in 30 mL water, centrifuge, remove the water and fat layers, repeat this process. Combine the fat layers and homogenize them with 20 mL water, centrifuge, remove the water layer. Combine all the water layers, filter (Whatman No. 42 paper) at 4°, filter (0.45 μm) at 4°, purify on a 200 × 4 Sephadex G-25 column with water at 84 mL/min with detection at UV 206, collect 84 mL fractions. Evaporate a 200 μL aliquot to dryness under reduced pressure, reconstitute with 200 μL 500 mM pH 7.8 borate buffer, add 200 μL 5.8 mM 9-fluorenylmethyl chloroformate in acetone, vortex for 45 s, wash with 400 μL pentane:ethyl acetate 80:20, inject an aliquot of the aqueous phase.

HPLC VARIABLES

Column: 250 × 4.6 5 μm Nucleosil C18

Mobile phase: Gradient. A was MeCN:100 mM pH 3.8 ammonium acetate buffer 20:80. B was MeCN:100 mM pH 4.2 ammonium acetate buffer 80:20. A:B 70:30 for 10 min, to 20:80 over 50 min, to 0:100 (step gradient), maintain at 0:100 for 10 min.

Column temperature: 40

Flow rate: 1

Detector: UV 214

CHROMATOGRAM

Retention time: 20 (arginine), 22.5 (serine, aspartic acid), 25 (glutamic acid), 27 (threonine), 30 (glycine), 35 (alanine), 40 (proline), 44 (valine), 46 (methionine), 51 (phenylalanine), 52.5 (isoleucine, leucine), 65 (histidine), 67.5 (lysine)

OTHER SUBSTANCES

Also analyzed: dipeptides, tripeptides

REFERENCE

Roturier, J.M.; Le Bars, D.; Gripon, J.C. Separation and identification of hydrophilic peptides in dairy products using FMOC derivatization, *J.Chromatogr.A*, **1995**, *696*, 209–217.

SAMPLE
Matrix: collagen
Analyte: amino acids
Sample preparation: Hydrolyze protein with 6 M HCl under nitrogen at 108° for 24 h, evaporate to dryness under reduced pressure, reconstitute with 100 mM sodium bicarbonate to give a hydrolysate concentration of 1 μg/mL. Remove a 5 μL aliquot and dilute it to 200 μL with 100 mM sodium bicarbonate, add 200 μL 4 mM 9-fluorenylmethyl chloroformate in dry acetone, shake quickly, let stand at room temperature for 10 min, wash twice with 600 μL portions of pentane:ethyl acetate 90:10, inject a 20 μL aliquot of the lower aqueous phase.

HPLC VARIABLES
Column: 150 × 4.6 MicroPak ODS-80TM
Mobile phase: Gradient. A was buffer. B was MeOH:buffer adjusted to pH 4.5 with phosphoric acid 20:80. C was MeCN. A:B:C 72.5:0:27.5 for 4.1 min, to 60:0:40 over 11.5 min, to 0:64:36 over 0.1 min, to 0:62:38 over 6.3 min, to 0:30:70 over 7 min, to 0:25:75 over 5 min, return to initial conditions, re-equilibrate for 11 min. (Buffer was 20 mM sodium citrate containing 5 mM tetramethylammonium chloride, pH 2.85.)
Column temperature: 30
Flow rate: 1.4
Injection volume: 20
Detector: F ex 254 em 340

CHROMATOGRAM
Retention time: 3.50 (histidine (mono-derivative)), 4.24 (cysteic acid), 4.78 (arginine), 7.60 (methionine sulfoxide), 8.92 (4-hydroxyproline), 9.80 (serine), 10.22 (3-hydroxyproline, homoserine), 10.80 (aspartic acid), 11.68 (glutamic acid), 12.48 (threonine), 13.05 (S-carboxymethylcysteine), 13.96 (glycine), 16.21 (ammonia), 16.60 (alanine), 17.48 (tyrosine (mono-derivative)), 18.30 (proline), 19.90 (methionine), 21.00 (valine), 23.27 (phenylalanine), 23.66 (isoleucine), 24.02 (leucine), 26.90 (cystine (bis-derivative)), 28.17 (hydroxylysine (bis-derivative), 28.81 (histidine (bis-derivative)), 30.00 (lysine (bis-derivative)), 31.85 (tyrosine (bis-derivative))
Limit of quantitation: 1 pmole

REFERENCE
Miller, E.J.; Narkates, A.J.; Niemann, M.A. Amino acid analysis of collagen hydrolysates by reverse-phase high-performance liquid chromatography of 9-fluorenylmethyl chloroformate derivatives, *Anal.Biochem.*, **1990**, *190*, 92–97.

SAMPLE
Matrix: collagen
Analyte: hydroxyproline and proline
Sample preparation: Hydrolyse collagen with 200 μL constant-boiling HCl and 1 μL liquified phenol in the gas phase at 150° for 1 h, dry under vacuum, reconstitute with 400 mM pH 10.4 borate buffer. Remove a 1 μL aliquot and mix it with 5 μL MeCN:buffer 40:100 and 1 μL reagent for 75 s, add 1 μL 2.5 mg/mL 9-fluorenylmethyl chloroformate in MeCN, mix for 50 s, inject the whole amount. (Buffer was 400 mM pH 10.4 borate buffer. Reagent was 10 mg/mL o-phthalaldehyde in buffer containing 3-mercaptopropionic acid. Primary amino acids are derivatized with o-phthalaldehyde then proline and hydroxyproline are derivatized with 9-fluorenylmethyl chloroformate and chromatographed.)

HPLC VARIABLES
Guard column: 4 × 4 5 μm LiChrospher 100 RP-8
Column: 125 × 4 5 μm C18 ODS Hypersil
Mobile phase: Gradient. A was THF:20 mM pH 7.2 sodium acetate containing 100 mM EDTA and 0.015% triethylamine 0.5:99.5. B was MeCN:100 mM pH 7.2 sodium acetate containing 100 mM EDTA 80:20. A:B from 75:25 to 70:30 over 1 min, to 60:40 over 2 min, to 50:50 over 1 min, to 0:100 over 0.25 min, maintain at 0:100 for 2.75 min, return to initial conditions over 0.25 min, re-equilibrate for 3.5 min.
Column temperature: 40

Flow rate: 1.5 for 1 min, to 2.5 over 2 min, maintain at 2.5 for 1 min, to 3.5 over 0.25 min, maintain at 2.5 for 2.25 min, to 2.5 over 0.25 min, to 1.5 over 0.25 min, maintain at 1.5
Injection volume: 8
Detector: F ex 340 em 450 for 1.5 min then F ex 266 em 305

CHROMATOGRAM
Retention time: 2.0 (hydroxyproline), 2.8 (proline)
Limit of quantitation: 10 pM

REFERENCE
Nathans, G.R.; Gere, D.R. Rapid robust separation of hydroxyproline and proline, *Anal.Biochem.*, **1992**, *202*, 262–267.

SAMPLE
Matrix: feces, insects, plants
Analyte: amino acids
Sample preparation: Grind black cherry leaves, insect larvae, or feces to pass 40 mesh. Weigh out an amount containing 0.2-2 mg protein, add 2 mL 4 M methanesulfonic acid in water containing 0.2% 3-(2-aminoethyl)indole, freeze in dry ice-acetone, evacuate to 50 µTorr, flush with nitrogen, evacuate to 50 µTorr, flush with nitrogen, evacuate to 50 µTorr, hydrolyze at 115° for 22 h, add c-allylglycine, buffer with 100 mM pH 8.0 sodium borate, adjust to pH 8.0 with NaOH, make up to 12 mL, centrifuge. Remove a 1 mL aliquot and add it to 1 mL 6 mM 9-fluorenylmethylchloroformate in acetone, let stand for 15 min at room temperature, extract twice with 2 mL portions of n-pentane, filter (0.45 µm), discard the first few drops, inject a 20 µL aliquot.

HPLC VARIABLES
Column: 150 × 4.6 end-capped ODS-80TM Aminotag (Varian)
Mobile phase: Gradient. A was MeCN:15 mM citric acid + 10 mM tetramethylammonium chloride adjusted to pH 1.85 with NaOH 27:73. B was MeCN:THF:15 mM citric acid + 10 mM tetramethylammonium chloride adjusted to pH 4.50 with NaOH 35:5:60. C was MeCN:THF:15 mM citric acid + 10 mM tetramethylammonium chloride adjusted to pH 4.50 with NaOH 62:13:25. A:B:C from 100:0:0 to 50:50:0 over 3 min, to 0:100:0 over 14 min (Waters concave gradient 7), to 0:96:4 over 6 min (Waters concave gradient 7), to 0:85:15 over 4 min, to 0:50:50 over 4 min, to 0:35:65 over 4 min, to 0:0:100 over 5 min. (Linear gradients except where shown).
Column temperature: 30
Flow rate: 1.4
Injection volume: 20
Detector: UV 264

CHROMATOGRAM
Retention time: 7.5 (His), 9.5 (Arg), 11.5 (Asn), 11.5 (Gln), 13 (Ser), 14 (Asp), 15 (Glu), 15.5 (Thr), 17 (Gly), 20.5 (Ala), 22.5 (Tyr), 23 (Pro), 26 (Met), 27 (Val), 29.5 (Phe), 30.5 (Trp), 31.5 (Ile), 32 (Leu), 33 (cystine), 35.5 (di-His), 36.5 (Cys), 37.5 (Lys), 40 (di-Tyr)
Internal standard: c-allylglycine (25)
Limit of quantitation: 30 µM

KEY WORDS
protein; cherry; leaves; larvae; ASN and GLN co-elute

REFERENCE
Malmer, M.F.; Schroeder, L.A. Amino acid analysis by high-performance liquid chromatography with methanesulfonic acid hydrolysis and 9-fluorenylmethylchloroformate derivatization, *J.Chromatogr.*, **1990**, *514*, 227–239.

SAMPLE
Matrix: protein
Analyte: S-methylcysteine
Sample preparation: Heat 3 mg protein with 500 µL 6 M HCl in an evacuated tube at 110° for 24 h, evaporate to dryness under reduced pressure, reconstitute with 1 mL water. Remove a 100 µL aliquot and add it to 100 µL 500 mM pH 8.3 Tris-HCl buffer, add 100 µL 10 mM 9-fluorenylmethyl chloroformate in acetone, mix, let stand at room tempera-

ture for 1.5 min, add 50 μL 25 mM 1-aminoadamantane in acetone:water 50:50, let stand at room temperature for 1.5 min, inject a 75 μL aliquot.

HPLC VARIABLES
Column: 4 μm RP-8
Mobile phase: MeCN:MeOH:20 mM pH 9.75 glycine buffer 8.5:17:74.5 (Place a column filled with RP-8 before the injector to saturate the mobile phase with silica gel.)
Flow rate: 1
Injection volume: 75
Detector: F ex 263 em 313

CHROMATOGRAM
Retention time: 58
Limit of detection: 200 fmole

KEY WORDS
comparison with other derivatizing reagents

REFERENCE
Müller, A.M.F.; Hallier, E.; Westphal, G.; Schröder, K.R.; Bolt, H.M. Determination of methylated globin and albumin for biomonitoring of exposure to methylating agents using HPLC with precolumn fluorescent derivatization, *Fresenius' J.Anal.Chem.*, **1994**, *350*, 712–715.

SAMPLE
Matrix: solutions
Analyte: amino acids
Sample preparation: Mix 400 μL aqueous sample solution with 100 μL buffer and 500 μL 15 mM 9-fluorenylmethyl chloroformate in acetone, let stand for 40 s, wash three times with 2 mL portions of pentane, inject an aliquot of the aqueous layer. (Prepare buffer by adjusting the pH of 1 M boric acid to 6.2 with NaOH.)

HPLC VARIABLES
Column: 125 × 4.6 3 μm ODS Hypersil
Mobile phase: Gradient. MeCN:MeOH:buffer 10:40:50 for 3 min, to 50:0:50 over 9 min.
Flow rate: 1.3 for 12 min, to 2 over 0.5 min
Injection volume: 10
Detector: F ex 250 (filter) em 320 (filter)

CHROMATOGRAM
Retention time: 4.5 (Asn), 4.8 (Gln), 5 (Asp), 5.5 (Ser), 6 (Glu), 7.5 (Gly), 8 (Thr), 8.7 (Arg), 10 (Ala), 10.5 (Tyr), 12.5 (Pro), 13 (Met), 14 (Val), 14.5 (Phe), 14.8 (Ile), 15 (Leu), 17.5 (His), 18 (Orn), 19.5 (Lys)
Limit of quantitation: 100 nM

REFERENCE
Einarsson, S.; Josefsson, B.; Lagerkvist, S. Determination of amino acids with 9-fluorenylmethyl chloroformate and reversed-phase high-performance liquid chromatography, *J.Chromatogr.*, **1983**, *282*, 609–618.

SAMPLE
Matrix: solutions
Analyte: iodinated thyronines
Sample preparation: 100 μL Solution + 100 μL 500 mM pH 7.7 borate buffer + 100 μL 2.5 mM 9-fluorenylmethyl chloroformate in dry acetone, mix, let stand at room temperature for 45 s, add 200 μL 12 mM 1-adamantamine in MeCN, inject an aliquot.

HPLC VARIABLES
Guard column: 10 × 3 30 μm Chromspher C18 (Chrompack)
Column: 100 × 3 5 μm Chromspher C18 (Chrompack)
Mobile phase: Gradient. A was MeOH:50 mM pH 4.2 sodium acetate buffer 40:60. B was MeCN:MeOH:50 mM pH 4.2 sodium acetate buffer 20:60:20. A:B from 100:0 to 0:100 over 40 min.
Column temperature: 35

Flow rate: 0.7
Detector: UV 260

CHROMATOGRAM

Retention time: 17 (3-monoiodotyrosine), 20 (thyronine), 21 (3,5-diiodotyrosine), 25 (3,5-diiodothyronine), 30 (liothyronine), 32 (3,3',5-triiodothyronine), 33 (levothyroxine)
Limit of detection: 1.6-3.5 pmole

KEY WORDS

comparison with other derivatization procedures

REFERENCE

Doorn, L.; Jansen, E.H.J.M.; Van Leeuwen, F.X.R. Comparison of high-performance liquid chromatographic detection methods for thyronine and tyrosine residues in toxicological studies of the thyroid, *J.Chromatogr.*, **1991**, *553*, 135–142.

SAMPLE

Matrix: solutions
Analyte: imino acids
Sample preparation: Mix 400 µL aqueous sample solution with 100 µL buffer and 500 µL 15 mM fluorenylmethyl chloroformate in acetone, let stand for 40 s, wash three times with 2 mL portions of pentane, inject an aliquot of the aqueous layer (J. Chromatogr. 1983, 282, 609). (Prepare buffer by adjusting the pH of 1 M boric acid to 6.2 with NaOH.)

HPLC VARIABLES

Column: 250 × 4.6 RN-β-CD (Astec)
Mobile phase: MeCN:triethylamine:acetic acid 100:0.6:0.4
Flow rate: 1
Injection volume: 0.2
Detector: F ex 250 (filter) em 320 (filter)

CHROMATOGRAM

Retention time: k' 2.9 (D-proline), k' 4.4 (L-proline), k' 7.3 (D-trans-4-hydroxyproline), k' 9.6 (L-trans-4-hydroxyproline), k' 1.8 (D-cis-4-hydroxyproline), k' 2.2 (L-cis-4-hydroxyproline), k' 1.9 (D-pyroglutamic acid), k' 2.4 (L-pyroglutamic acid), k' 2.6 (D-3,4-dehydroproline), k' 3.6 (L-3,4-dehydroproline), k' 1.9 (D-thiaproline), k' 2.2 (L-thiaproline), k' 1.2 (D-penicillamine acetone adduct), k' 1.4 (L-penicillamine acetone adduct), k' 2.2 (D-pipecolic acid), k' 2.3 (L-pipecolic acid)
Limit of detection: 0.0001%

KEY WORDS

chiral

REFERENCE

Zukowski, J.; Pawlowska, M.; Armstrong, D.W. Efficient enantioselective separation and determination of trace impurities in secondary amino acids (i.e. imino acids), *J.Chromatogr.*, **1992**, *623*, 33–41.

SAMPLE

Matrix: solutions
Analyte: amino acids
Sample preparation: 50 µL Amino acid solution in 500 mM pH 8 sodium borate buffer + 50 µL 20 mM norvaline in 500 mM pH 8 sodium borate buffer + 200 µL 30 mM 9-fluorenylmethyl chloroformate in dry acetone, shake, allow to stand at room temperature for 10 min, add 200 µL 25 mM 1-aminoamantadine in MeOH, let stand for 2 min, inject an aliquot.

HPLC VARIABLES

Column: 150 × 4.6 5 µm Aminotag (Varian)
Mobile phase: Gradient. A was MeCN:isopropanol 90:10. B was THF:50 mM sodium acetate buffer 4:96, adjusted to pH 4.03 with glacial acetic acid. A:B from 20:80 to 25:75 over 25 min, to 30:70 over 10 min, to 40:60 over 10 min, to 50:50 over 10 min, to 80:20 over 15 min, to 100:0 over 5 min, re-equilibrate at initial conditions for 15 min.
Column temperature: 32
Flow rate: 1.5

Injection volume: 20
Detector: UV 265; F ex 265 em 340

CHROMATOGRAM
Retention time: 12 (arginine), 15 (taurine), 16 (serine), 17 (aspartic acid), 20 (glutamic acid), 21 (threonine), 22 (glycine), 31 (proline), 37 (alanine), 41.5 (methionine), 43.5 (valine), 48 (phenylalanine), 48.5 (tryptophan), 49 (isoleucine), 50 (leucine), 59 (histidine), 60.5 (lysine), 64 (tyrosine)
Internal standard: norvaline (45)

REFERENCE
Carratú, B.; Boniglia, C.; Bellomonte, G. Optimization of the determination of amino acids in parenteral solutions by high-performance liquid chromatography with precolumn derivatization using 9-fluorenylmethyl chloroformate, *J.Chromatogr.A*, **1995**, *708*, 203–208.

SAMPLE
Matrix: solutions
Analyte: amino acids
Sample preparation: Mix 20 μL of an amino acid solution with 100 μL 400 mM pH 8.0 sodium borate buffer and 100 μL 3 mM 9-fluorenylmethyl chloroformate in acetone (MeCN if UV detection is used), let stand for 2 min, add 100 μL 40 mM 1-aminoadamantane in acetone:water 75:25, let stand for 2 min. Remove an 20 μL aliquot and add it to 380 μL MeCN:500 mM pH 4.0 sodium acetate buffer 50:50, inject a 20 μL aliquot.

HPLC VARIABLES
Guard column: 4 × 4 5 μm LiChrospher 100 RP-8
Column: 250 × 4 4 μm Superspher 60 RP-8
Mobile phase: Gradient. A was THF:DMF:100 mM pH 4.6 sodium acetate buffer 5:5:90. B was MeCN. A:B from 93:7 to 85:15 over 10 min, to 50:50 over 25 min, to 0:100 over 5 min, maintain at 0:100 for 5 min, re-equilibrate at initial conditions for 10 min. (Reflux DMF in the presence of ninhydrin then distil.)
Column temperature: 45
Flow rate: 1.25
Injection volume: 20
Detector: F ex 263 em 313

CHROMATOGRAM
Retention time: 8 (CyA), 15.2 (Asn), 16 (Gln), 17 (Asp), 17.5 (Ser), 18.5 (Glu), 19.5 (Thr), 20 (Arg), 20.5 (Gly), 21 (homo-Arg), 23 (Ala), 24 (Tyr (mono-derivative)), 25 (Pro), 26 (Met), 28 (Val, GABA), 29 (Phe), 30.8 (Ile), 32 (Leu), 32.5 (Hyl), 34 (His), 35 (Orn), 36.5 (Lys), 38 (Tyr (bis-derivative))
Limit of detection: 5-150 fmole

KEY WORDS
comparison with other derivatizing reagents

REFERENCE
Brückner, H.; Lüpke, M. Use of chromogenic and fluorescent oxycarbonyl chlorides as reagents for amino acid analysis by high-performance liquid chromatography, *J.Chromatogr.A*, **1995**, *697*, 295–307.

SAMPLE
Matrix: solutions
Analyte: amino acids
Sample preparation: 10 μL Amino acid mixture + 10 μL 200 mM pH 8.0 borate buffer + 10 μL 5 mM 9-fluorenylmethyl chloroformate in MeCN, mix for 1 min, dilute with water, inject an aliquot.

CAPILLARY ELECTROPHORESIS
Capillary: 67 cm × 50 μm fused-silica (60 cm to detector) (Polymicro Technologies)
Capillary preparation: Flush capillary with running buffer for 2 min between runs. Wash new capillaries with 100 mM NaOH for 15 min and with water for 15 min.
Running buffer: 5 mM pH 9.2 Sodium borate buffer containing 150 mM sodium dodecyl sulfate and 40 mM gamma-cyclodextrin
Capillary temperature: 20

Voltage/Current: 10 or 20 kV (?)
Injection: Pressure injection at 0.5 psi for 5 s.
Detector: UV 200
Model: Beckman P/ACE 2050
Migration time: 17.7 (D,L-Ser), 18 (D,L-Ala), 19.5 (D-Val), 19.6 (L-Val), 21.0 (D-Met), 21.1 (L-Met), 22.5 (D-Leu), 22.6 (L-Leu), 23.4 (D-Phe), L-Phe (23.7), 25.0 (D-Trp), 25.2 (L-Trp)

KEY WORDS
chiral

REFERENCE
Chan, K.C.; Muschik, G.M.; Issaq, H.J. Enantiomeric separation of amino acids using micellar electrokinetic chromatography after pre-column derivatization with the chiral reagent 1-(9-fluorenyl)-ethyl chloroformate, *Electrophoresis*, **1995**, *16*, 504–509.

SAMPLE
Matrix: solutions
Analyte: amino acids
Sample preparation: 10 μL Amino acid solution in 250 mM pH 8.8 borate buffer + 10 μL 4 mg/mL 9-fluorenylmethyl chloroformate in MeCN, mix, let stand for 1.5 min, add 10 μL reagent, mix, let stand for 3.5 min, add 10 mL (?) MeCN:glacial acetic acid 80:20, mix, inject an aliquot. (Reagent was 150 mL 500 mM hydroxylamine hydrochloride containing 340 μL 850 mM NaOH and 10 μL 2-(methylthio)ethanol.)

HPLC VARIABLES
Column: 150 × 4.6 5 μm ODS-Hypersil
Mobile phase: Gradient. A was MeOH:water 15:85 containing 30 mM $(NH_4)H_2PO_4$, pH 6.5. B was MeOH:water 15:85. C was MeCN:water 90:10. A:B:C 17:68:15 for 1 min, to 43.2:46:12 over 31 min, to 0:0:100 over 0.05 min, maintain at 0:0:100 for 1.95 min.
Column temperature: 38
Flow rate: 1
Detector: F ex 270 em 316

CHROMATOGRAM
Retention time: 3 (aspartic acid), 3.5 (glutamic acid), 7.5 (hydroxyproline), 9.7 (serine), 10.3 (histidine), 10.6 (glycine), 11.2 (threonine), 11.8 (alanine), 12.5 (proline), 13.5 (tyrosine), 15 (arginine), 16 (valine), 16.5 (methionine), 18.5 (isoleucine), 19 (leucine), 20 (phenylalanine)

REFERENCE
Ou, K.; Wilkins, M.R.; Yan, J.X.; Gooley, A.A.; Fung, Y.; Sheumack, D.; Williams, K.L. Improved high-performance liquid chromatography of amino acids derivatized with 9-fluorenylmethyl chloroformate, *J.Chromatogr.A*, **1996**, *723*, 219–225.

SAMPLE
Matrix: tissue
Analyte: basic amino acids
Sample preparation: Prepare an SPE column by placing 1 mL 200-400 mesh Biorex-70 (Na$^+$) (Bio-Rad) in the barrel of a 12 mL syringe, wash with 3 mL 1 M HCl, wash with 20 mL water until the pH of the effluent is 6-7. Homogenize 1 g tissue with 3 mL 200 mM perchloric acid, let stand on ice for 1 h, centrifuge at 1500 g for 20 min, centrifuge at 1500 g for 20 min. Remove a 125 μL aliquot of the supernatant and neutralize it with potassium bicarbonate, let stand on ice for 1 h, centrifuge, add to the SPE column, wash with 1 mL water, wash with 8 mL 150 mM ammonium hydroxide, elute with 2 mL 500 mM ammonium hydroxide. Evaporate the eluate to dryness under reduced pressure, suspend the residue in 1 mL 100 mM pH 7.9 sodium bicarbonate. Remove a 300 μL aliquot and add it to 300 μL 4 mM 9-fluorenylmethyl chloroformate in acetone, mix gently, let stand for 10 min, wash twice with 600 μL aliquots of pentane, inject a 25 μL aliquot of the aqueous layer.

HPLC VARIABLES
Guard column: 10 × 4.6 7 μm Adsorbosphere nucleotide guard column
Column: 250 × 4.6 5 μm Ultrasphere ODS-DABS

Mobile phase: Gradient, A was buffer. B was MeOH:buffer 20:80, adjusted to pH 4.5 with phosphoric acid. C was MeCN. A:B:C from 75:0:25 to 60:0:40 over 20 min, to 62:38:0 over 3 min, to 0:30:70 over 15 min, to 0:25:75 over 10 min, return to initial conditions over 1 min, re-equilibrate for 10 min. (Buffer was 20 mM sodium citrate containing 5 mM tetramethylammonium chloride, pH adjusted to 2.85 with phosphoric acid.)
Flow rate: 1.4
Injection volume: 25
Detector: F ex 269 em 340

CHROMATOGRAM
Retention time: 11 (His), 14 (Arg), 35 (Orn), 37 (Lys)

KEY WORDS
SPE; rat; liver; kidney

REFERENCE
Gilbert, R.S.; Gonzalez, G.G.; Hawel, L. III; Byus, C.V. An ion-exchange chromatography procedure for the isolation and concentration of basic amino acids and polyamines from complex biological samples prior to high-performance liquid chromatography, *Anal.Biochem.*, **1991**, *199*, 86–92.

SAMPLE
Matrix: urine
Analyte: 3-methylhistidine
Sample preparation: 400 µL Urine + 400 µL 1 M perchloric acid, mix, let stand for 10 min, centrifuge at 1500 g for 10 min, dilute an aliquot of the supernatant 10-fold with water. Remove a 100 µL aliquot and add it to 300 µL water, add 100 µL 800 mM pH 8.5 borate buffer, add 500 µL 15 mM 9-fluorenylmethyl chloroformate in MeCN, mix, let stand for at least 1 min at room temperature, wash twice with 2 mL portions of diethyl ether. Dilute the lower aqueous phase 5-fold with mobile phase, inject an aliquot.

HPLC VARIABLES
Column: 200 × 3 5 µm Spherisorb ODS
Mobile phase: MeCN:buffer 50:50 (Buffer was 30 mM glycine adjusted to pH 3.0 with HCl.)
Flow rate: 0.7
Detector: F ex 260 em 320

CHROMATOGRAM
Retention time: 13.2
Limit of detection: 0.3 pmole

OTHER SUBSTANCES
Simultaneously analyzed: arginine, 1-methylhistidine

REFERENCE
Teerlink, T.; de Boer, E. Determination of 3-methylhistidine in urine by high-performance liquid chromatography using pre-column derivatization with 9-fluorenylmethyl chloroformate, *J.Chromatogr.*, **1989**, *491*, 418–423.

SAMPLE
Matrix: urine
Analyte: secondary amino acids
Sample preparation: 900 µL Urine + 100 µL 1.7 mM 3,4-dehydroproline, add 1 mL concentrated HCl, heat at 110° for 16 h. Remove a 1 mL aliquot and add it to 800 µL 630 mM NaOH, add 150 µL 800 mM pH 9.5 sodium borate buffer, add 100 µL 370 mM o-phthalaldehyde in MeCN containing 370 mM 2-mercaptoethanol, after 30 s add 100 µL 750 mM iodoacetamide in MeCN, after 30 s add 300 µL 5 mM 9-fluorenylmethyl chloroformate in acetone, wash twice with 2 mL portions of diethyl ether. Add 1 mL water to the lower aqueous phase, inject a 30 µL aliquot.

HPLC VARIABLES
Column: 200 mm long 5 µm Spherisorb ODS
Mobile phase: MeCN:buffer 34:66 (Buffer was 50 mM acetic acid adjusted to pH 4.3 with 1 M NaOH.)
Flow rate: 1

Injection volume: 30
Detector: F ex 260 em 330

CHROMATOGRAM
Retention time: 2.8 (hydroxyproline), 6.0 (sarcosine), 9.2 (proline)
Internal standard: dehydroproline (5.4)
Limit of quantitation: 1 pmole

KEY WORDS
Primary amino acids are derivatized with o-phthalaldehyde (derivatives not chromatographed under these conditions) then secondary amino acids are derivatized with 9-fluorenylmethyl chloroformate.

REFERENCE
Teerlink, T.; Tavenier, P.; Netelenbos, J.C. Selective determination of hydroxyproline in urine by high-performance liquid chromatography using precolumn derivatization, *Clin.Chim.Acta*, **1989**, *183*, 309–315.

SAMPLE
Matrix: urine
Analyte: amino acids
Sample preparation: Filter urine, dilute 10-fold with water, 500 μL diluted urine + 100 μL 400 mM pH 9.3 sodium borate buffer + 400 μL 5 mM 9-fluorenylmethyl chloroformate in MeCN, mix for 1 min, wash twice with 2 mL pentane, dilute aqueous layer 100-fold with water, inject an aliquot.

CAPILLARY ELECTROPHORESIS
Capillary: 70 cm × 50 μm fused-silica (60 cm to detector) (Polymicro Technologies)
Running buffer: 20 mM pH 9.2 borate buffer containing 25 mM sodium dodecyl sulfate
Injection: Gravity injection at 10 cm for 20 s
Detector: F ex 248 (laser) em 315 (filters ARC No. 310-B-1D (FWHM = 60 nm) bandpass and Melles Griot WG-306 cut-off)
Migration time: 9.7 (alanine), 15.8 (arginine), 11.5 (aspartic acid), 15.6 (cystine), 11.1 (glutamic acid), 9.9 (glycine), 16.1 (histidine), 8.7 (hydroxyproline), 11.8 (isoleucine), 12.4 (leucine), 16.1 (lysine), 11.3 (methionine), 13.5 (phenylalanine), 10.4 (proline), 9.3 (serine), 9.5 (threonine), 10.8 (tyrosine), 10.5 (valine)
Limit of detection: 0.5 nM (S/N 2)

REFERENCE
Chan, K.C.; Janini, G.M.; Muschik, G.M.; Issaq, H.J. Laser-induced fluorescence detection of 9-fluorenylmethyl chloroformate derivatized amino acids in capillary electrophoresis, *J.Chromatogr.A*, **1993**, *653*, 93–97.

SAMPLE
Matrix: urine
Analyte: amino acids
Sample preparation: 80 μL Urine + 20 μL 10 mM N-methylalanine containing 1 mM norvaline, mix, filter (Ultrafree-MC) while centrifuging at 5000 g for 30 min. Remove a 6 μL aliquot of the ultrafiltrate and add it to 5 μL 0.5% 3-mercaptopropionic acid in 1 M pH 10.4 borate buffer, mix, add 1.5 μL 120 mM iodoacetic acid in 140 mM NaOH, mix, add 5 μL reagent, mix, add 2 μL 9-fluorenylmethyl chloroformate, mix, add 2.5 μL 1 M acetic acid, mix, inject the whole amount. (Reagent was 20 mg/mL o-phthalaldehyde in MeOH:500 mM pH 10.4 borate buffer:3-mercaptopropionic acid 10:88:2. The primary amino acids are derivatized with phthalaldehyde and the secondary amino acids (sarcosine and proline) with 9-fluorenylmethyl chloroformate.)

HPLC VARIABLES
Guard column: 20 × 4 5 μm ODS Hypersil
Column: 300 × 3.9 4 μm Nova-Pak C18
Mobile phase: Gradient. A was 60 mM pH 6.86 sodium acetate buffer containing 0.044% triethylamine. B was MeCN:MeOH:100 mM pH 5.45 sodium acetate buffer 74.5:4.5:21. A:B from 100:0 to 94.4:5.6 over 1 min, to 93.8:6.2 over 6 min, maintain at 93.8:6.2 for 2 min, to 92.3:7.7 over 12 min, maintain at 92.3:7.7 for 7 min, to 92:8 over 7 min, to

90.8:9.2 over 4 min, to 90.5:9.5 over 3 min, to 84:16 over 6 min, maintain at 84:16 for 1 min, to 82:18 over 1 min, to 78:22 over 20 min, to 72:28 over 7 min, to 68:32 over 8 min, to 0:100 over 9 min, maintain at 0:100 for 5 min, return to initial conditions over 1 min.

Column temperature: 40
Flow rate: 0.8
Injection volume: 22
Detector: F ex 340 em 450 for 79.5 min, F ex 260 em 315 for 1.5 min, F ex 340 em 450 for 6 min, F ex 260 em 315 for 13 min

CHROMATOGRAM

Retention time: 5.5 (Asp), 6.5 (Glu), 8 (Cys (S-carboxymethylated)), 9 (Aad), 10 (Asn), 11 (Ser), 11.5 (Homocysteine (S-carboxymethylated)), 15 (Gln), 17.5 (His), 18 (Gly), 20 (Thr), 25 (Cit), 26 (1-Methylhistidine), 28 (β-Alanine), 29.5 (Arg), 30.5 (3-Methylhistidine), 33 (Ala), 35 (Tau), 38.5 (Ans), 39 (Car), 40 (β-aminoisobutyric acid), 41 (gamma-aminobutyric acid), 52 (Tyr), 52.5 (Abu), 57 (Etn), 60 (Val), 61.5 (Met), 65 (Cysta), 70 (Trp), 71.5 (Ile), 72.5 (Phe), 75 (Hyl), 76.5 (Leu), 80 (Hyp), 83.5 (Lys), 85 (Orn), 90 (Sar), 91 (Pro)
Internal standard: norvaline (62.5), N-methylalanine (91.5)
Limit of quantitation: 10 μM
Limit of detection: 50-500 fmole

KEY WORDS

ultrafiltrate

REFERENCE

Carducci, C.; Birarelli, M.; Leuzzi, V.; Santagata, G.; Serafini, P.; Antonozzi, I. Automated method for the measurement of amino acids in urine by high-performance liquid chromatography, *J.Chromatogr.A*, **1996**, *729*, 173–180.

SAMPLE

Matrix: urine
Analyte: hydroxyproline peptide
Sample preparation: 20 μL Urine + 100 μL 60 μM 3,4-dehydroproline in 50 mM HCl + 200 μL 130 mM sodium carbonate, mix, add 50 μL 370 mM mercaptoethanol in MeCN, add 50 μL 370 mM o-phthalaldehyde in MeCN, let stand for 1 min, add 50 μL 750 mM iodoacetamide in MeCN, let stand for 30 s, add 100 μL 2 mM 9-fluorenylmethyl chloroformate in acetone, mix, add 4 mL water, inject a 10 μL aliquot.

HPLC VARIABLES

Guard column: AX guard system (Whatman)
Column: 110 × 4.7 5 μm Partisphere 5 SAX
Mobile phase: MeCN:isopropanol:20 mM pH 3.7 ammonium formate buffer 20:10:70
Flow rate: 1.2
Injection volume: 10
Detector: F ex 260 em 330

CHROMATOGRAM

Retention time: 5.0 (hydroxyproline peptide), 5.6 (hydroxyproline)
Internal standard: 3,4-dehydroproline (6.6)

KEY WORDS

Primary amino acids are derivatized with o-phthalaldehyde (derivatives not chromatographed under these conditions) then secondary amino acids are derivatized with 9-fluorenylmethyl chloroformate.

REFERENCE

Mazzi, G.; Fioravanzo, F.; Burti, E. New marker of bone resorption: hydroxyproline-containing peptide. High-performance liquid chromatographic assay without hydrolysis as an alternative to hydroxyproline determination: a preliminary report, *J.Chromatogr.B*, **1996**, *678*, 165–172.

SAMPLE

Matrix: urine
Analyte: selfotel (cis-4-(phosphonomethyl)-2-piperidinecarboxylic acid)

Sample preparation: 50 μL Urine + 25 μL 100 μg/mL IS in water + 1 mL buffer, vortex for 5 s, add 100 μL reagent, vortex for 10 s, let stand for 1.5 min, add 200 μL 100 mg/mL iodoacetamide in MeCN, vortex for 15 s, let stand for 1.5 min, add 500 μL 2 mg/mL 9-fluorenylmethyl chloroformate in acetone, vortex for 20 s, let stand for 2 min, add 5 mL ether, vortex for 1 min, repeat the ether wash. Remove traces of organic solvent from the aqueous layer with a stream of nitrogen, vortex the aqueous layer for 2 s. Remove a 50 μL aliquot and add it to 500 μL mobile phase, inject a 50 μL aliquot onto column A and elute to waste with mobile phase A, divert the fraction containing selfotel and the IS onto column B, after 2 min remove column A from the circuit, elute column B with mobile phase B, monitor the effluent from column B. (Prepare buffer by dissolving 15.5 g boric acid in 500 mL water, adjust pH to 9.5 with NaOH. Prepare reagent by adding 32 μL 3-mercaptopropionic acid to 1 mL 50 mg/mL o-phthalaldehyde in MeCN. Primary amino acids are removed by derivatization with o-phthalaldehyde/3-mercaptopropionic acid.)

HPLC VARIABLES
Column: A 75 × 4.6 5 μm Inertsil ODS-2; B 250 × 4.6 10 μm Chiralcel OD-R
Mobile phase: A MeCN:100 mM pH 2.50 phosphate buffer 35:65; B MeCN:100 mM pH 2.00 phosphate buffer 35:65
Column temperature: 30
Flow rate: A 0.5 for 16 min, to 2.0 over 0.3 min, maintain at 2.0 for 18.4 min, return to 0.5 over 0.3 min; B 0.5
Injection volume: 50
Detector: F ex 262 em 314

CHROMATOGRAM
Retention time: 19.2, 22.0
Internal standard: cis-4-(phosphonoethyl)-2-piperidinecarboxylic acid (CGS 20005) (36 (first eluting peak))
Limit of quantitation: 250 ng/mL

KEY WORDS
chiral; column-switching

REFERENCE
Knoche, B.; Milosavljev, S.; Gropper, S.; Brunner, L.A.; Powell, M.L. Enantioselective determination of selfotel in human urine by high-performance liquid chromatography on a chiral stationary phase after derivatization with 9-fluorenylmethyl chloroformate, *J.Chromatogr.B*, **1997**, *695*, 355–363.

SAMPLE
Matrix: vegetables
Analyte: amino acids
Sample preparation: Extract 100 g potato tubers with 100 mL boiling water for 2 h, centrifuge at 5000 g for 10 min, filter (0.45 μm) the supernatant, dilute 100 μL of the filtrate with 900 μL 111 ng/mL norvaline in water, add a 1 μL aliquot to 5 μL 0.4 N pH 10.4 potassium borate buffer and 1 μL reagent, mix, add 1 μL 2.5 mg/mL 9-fluorenyl-methyl chloroformate in anhydrous MeCN, mix, inject the whole amount. (Prepare reagent by dissolving 10 mg o-phthalaldehyde in 100 μL MeOH, make up to 1 mL with 0.4 N pH 10.4 borate buffer, add 20 μL 3-mercaptopropionic acid. Derivatization was performed automatically and took 5 min. o-Phthalaldehyde derivatized primary amino acids and 9-fluorenylmethyl chloroformate derivatized secondary amino acids (proline and hydroxyproline).)

HPLC VARIABLES
Guard column: present but not specified
Column: 100 × 4 3 μm Hypersil ODS
Mobile phase: Gradient. A was THF:18 mM sodium acetate containing 0.02% triethylamine (adjusted to pH 7.2 with 1% acetic acid) 0.3:99.7. B was MeCN:MeOH:100 mM pH 7.2 sodium acetate 40:40:20. A:B from 100:0 to 94:6 over 0.5 min, to 80:20 over 2.5 min, to 50:50 over 3.5 min, to 25:75 over 1.5 min, to 0:100 over 0.5 min, maintain at 0:100 for 4 min, return to initial conditions over 1 min, re-equilibrate for 5 min.
Column temperature: 40
Flow rate: 1.4

Injection volume: 8
Detector: F ex 340 em 450, after 6.8 min F ex 264 em 313

CHROMATOGRAM
Retention time: 0.709 (aspartic acid), 0.831 (glutamic acid), 1.747 (asparagine), 1.871 (serine), 2.160 (glutamine), 2.286 (histidine), 2.439 (glycine), 2.602 (threonine), 3.305 (alanine), 3.446 (arginine), 4.185 (tyrosine), 5.113 (valine), 5.221 (methionine), 5.657 (tryptophan), 5.835 (phenylalanine), 5.928 (isoleucine), 6.224 (leucine), 6.435 (lysine), 6.903 (hydroxyproline), 7.982 (proline)
Internal standard: norvaline (5.394)
Limit of quantitation: 20 pmole

KEY WORDS
potato tubers

REFERENCE
Bartók, T.; Szalai, G.; Lorincz, Z.; Börcsök, G.; Sági, F. High-speed RP-HPLC/FL analysis of amino acids after automated two-step derivatization with o-phthaldialdehyde/3-mercaptopropionic acid and 9-fluorenylmethyl chloroformate, *J.Liq.Chromatogr.*, **1994**, *17*, 4391−4403.

2-Naphthoxycarbonyl Chloride

SAMPLE
Matrix: solutions
Analyte: amino acids
Sample preparation: Mix 20 μL of an amino acid solution with 100 μL 400 mM pH 9.0 sodium borate buffer and 100 μL 10 mM 2-naphthoxycarbonyl chloride in MeCN, let stand for 2 min, add 250 μL 40 mM 1-aminoadamantane in acetone:water 75:25, let stand for 2 min. Remove an 40 μL aliquot and add it to 360 μL MeCN:500 mM pH 4.0 sodium acetate buffer 50:50, inject a 20 μL aliquot. (Synthesis of 2-naphthoxycarbonyl chloride is as follows. Dissolve 30 mmoles 2-naphthol and 30 moles quinoline in 18 g toluene and 5 g dichloromethane, cool to 0°, add 47 mL 1.93 M phosgene in toluene, warm on a steam bath for 10 min, filter, evaporate to remove the solvent, distil the residue (bp 150-152°/9 mm Hg, take up the distillate in ether, crystallize by adding ligroin to obtain 2-naphthoxycarbonyl chloride (mp 66°) (J. Am. Chem. Soc.1951, 73, 2080).)

HPLC VARIABLES
Guard column: 4 × 4 5 μm LiChrospher 100 RP-8
Column: 250 × 4 4 μm Superspher 100 RP-18 end-capped
Mobile phase: Gradient. 100 mM pH 7.0 sodium acetate buffer:100 mM pH 4.6 sodium acetate:MeCN buffer from 75:15:10 to 0:70:30 over 15 min, to 0:40:60 over 15 min, to 0:0:100 over 5 min, maintain at 0:0:100 for 8 min, re-equilibrate at initial conditions at 1.25 mL/min for 13 min.
Column temperature: 45
Flow rate: 1

Injection volume: 20
Detector: F ex 274 em 336

CHROMATOGRAM
Retention time: 10 (Cya), 12.5 (Asp), 13.5 (Glu), 14.5 (Asn), 15 (Ser, Gln), 15.5 (His (mono-derivative)), 16.5 (Gly), 16.8 (Arg), 17 (Thr), 17.5 (homo-Arg), 18.5 (Ala), 19 (Pro), 20 (Tyr (mono-derivative), His (mono-derivative)), 22.5 (Met), 23 (Val), 23.5 (GABA), 24.5 (cystine (matrix peak interferes)), 25 (Phe), 25.3 (Ile), 25.8 (Leu), 27 (Hyl), 29 (Orn), 29.5 (His (bis-derivative)), 30 (Lys), 33 (Tyr (bis-derivative))
Limit of detection: 25-500 fmole

KEY WORDS
comparison with other derivatizing reagents

REFERENCE
Brückner, H.; Lüpke, M. Use of chromogenic and fluorescent oxycarbonyl chlorides as reagents for amino acid analysis by high-performance liquid chromatography, *J.Chromatogr.A*, **1995**, *697*, 295–307.

2-(Naphthylmethyl)oxycarbonyl Chloride

SAMPLE
Matrix: solutions
Analyte: amino acids
Sample preparation: Mix 20 μL of an amino acid solution with 100 μL 400 mM pH 9.0 sodium borate buffer and 100 μL 10 mM 2-(naphthylmethyl)oxycarbonyl chloride in MeCN, let stand for 2 min, add 250 μL 40 mM 1-aminoadamantane in acetone:water 75:25, let stand for 2 min. Remove an 40 μL aliquot and add it to 360 μL MeCN:500 mM pH 4.0 sodium acetate buffer 50:50, inject a 20 μL aliquot. (Synthesis of 2-(naphthylmethyl)carbonyl chloride is as follows. Dissolve 10 mmoles 2-naphthalenemethanol in 15 mL toluene, cool to 0°, add 16 mL 1.93 M phosgene in toluene, warm on a steam bath for 10 min, filter, evaporate to remove the solvent, distil the residue under reduced pressure, take up the distillate in ether, crystallize by adding ligroin to obtain 2-(naphthylmethyl)carbonyl chloride (mp 59-62°) (cf J. Am. Chem. Soc.1951, 73, 2080).)

HPLC VARIABLES
Guard column: 4 × 4 5 μm LiChrospher 100 RP-8
Column: 250 × 4 4 μm Superspher 100 RP-18 end-capped
Mobile phase: Gradient. 100 mM pH 7.0 sodium acetate buffer:100 mM pH 4.6 sodium acetate:MeCN buffer from 75:15:10 to 0:70:30 over 15 min, to 0:40:60 over 15 min, to 0:0:100 over 5 min, maintain at 0:0:100 for 8 min, re-equilibrate at initial conditions at 1.25 mL/min for 13 min.
Column temperature: 45
Flow rate: 1
Injection volume: 20
Detector: F ex 274 em 336

CHROMATOGRAM

Retention time: 15.5 (Asp), 16 (Glu), 17.5 (Ser, His (mono-derivative)), 18.5 (Arg), 19 (Gly), 19.5 (Thr), 20.5 (Ala), 22.5 (Pro, Tyr (mono-derivative)), 24 (Met), 25 (Val (matrix peak interferes)), 26.3 (Phe, cystine (matrix peak interferes)), 27 (Ile), 27.5 (Leu), 30.5 (His (bis-derivative)), 32 (Lys), 40 (Tyr (bis-derivative) (matrix peak interferes))

KEY WORDS

comparison with other derivatizing reagents

REFERENCE

Brückner, H.; Lüpke, M. Use of chromogenic and fluorescent oxycarbonyl chlorides as reagents for amino acid analysis by high-performance liquid chromatography, *J.Chromatogr.A*, **1995**, *697*, 295–307.

2-Nitrobenzyloxycarbonyl Chloride

SAMPLE

Matrix: solutions

Analyte: amino acids

Sample preparation: Mix 30 µL of an amino acid solution in 100 mM HCl with 150 µL 500 mM pH 10.5 sodium borate buffer and 300 µL 50 mM 2-nitrobenzyloxycarbonyl chloride in MeCN, let stand for 2 min, add 300 µL 40 mM 1-aminoadamantane in acetone: water 75:25, let stand for 2 min. Remove an 80 µL aliquot and add it to 320 µL MeCN: 500 mM pH 4.0 sodium acetate buffer 50:50, inject a 20 µL aliquot. (2-Nitrobenzyloxy-carbonyl chloride can be purchased from Bachem, Bubendorf, Switzerland. Synthesis is as follows. Dissolve 10 mmoles 2-nitrobenzyl alcohol in 4 mL dioxane (Caution! Dioxane is a carcinogen!), add 15.5 mL 1.93 M phosgene in toluene, let stand overnight at room temperature, evaporate to obtain 2-nitrobenzyloxycarbonyl chloride as a yellow oil.)

HPLC VARIABLES

Guard column: 4 × 4 5 µm LiChrospher 100 RP-8

Column: 250 × 4 4 µm Superspher 60 RP-8

Mobile phase: Gradient. MeCN:100 mM pH 4.4 sodium acetate buffer from 10:90 to 25: 75 over 10 min, to 30:70 over 10 min, to 45:55 over 10 min, to 100:0 over 5 min, maintain at 100:0 for 5 min, re-equilibrate at initial conditions for 14 min.

Column temperature: 45

Flow rate: 1.25

Injection volume: 20

Detector: UV 265

CHROMATOGRAM

Retention time: 6.5 (Asp), 7 (Ser), 8 (Glu), 8.5 (Gly), 9 (Thr), 10.5 (Arg), 11 (Ala), 12.5 (Pro), 14 (cystine (matrix peak interferes)), 15 (Met), 16 (Val (matrix peak interferes)), 18.5 (Phe), 19 (Ile), 19.5 (Leu), 22.5 (His (bis-derivative)), 26 (Lys), 30 (Tyr (bis-derivative) (matrix peak interferes))

KEY WORDS

comparison with other derivatizing reagents

REFERENCE

Brückner, H.; Lüpke, M. Use of chromogenic and fluorescent oxycarbonyl chlorides as reagents for amino acid analysis by high-performance liquid chromatography, *J.Chromatogr.A*, **1995**, *697*, 295–307.

4-Nitrobenzyloxycarbonyl Chloride

SAMPLE
Matrix: solutions
Analyte: amino acids
Sample preparation: Mix 30 μL of an amino acid solution in 100 mM HCl with 150 μL 500 mM pH 10.5 sodium borate buffer and 300 μL 50 mM 4-nitrobenzyloxycarbonyl chloride in MeCN, let stand for 2 min. Remove an 80 μL aliquot and add it to 320 μL MeCN:500 mM pH 4.0 sodium acetate buffer 50:50, inject a 20 μL aliquot. (4-Nitrobenzyloxycarbonyl chloride can be purchased from Bachem, Bubendorf, Switzerland. Synthesis is as follows. Bubble phosgene gas into 180 mL cold dioxane until 174 g is absorbed (Caution! Dioxane is a carcinogen!). Dissolve 60 g 4-nitrobenzyl alcohol in 75 mL dioxane with slight warming and add this solution to the phosgene solution, stopper the flask with a cork, let stand at room temperature overnight, remove the excess phosgene and dioxane by distillation under reduced pressure, add more dioxane, repeat the distillation, repeat this process several times to remove traces of phosgene. Dissolve the residue in 120 mL toluene, cool to 0°, add 150 mL petroleum ether (bp 30-60°), scratch the container to induce crystallization, add 400 mL petroleum ether, cool to -50°, filter, wash the crystals with toluene/petroleum ether, wash with petroleum ether, dry under vacuum over phosphorus pentoxide to give 4-nitrobenzyloxycarbonyl chloride (mp 33.5-34°) (J. Am. Chem. Soc. 1952, 74, 3818).)

HPLC VARIABLES
Guard column: 4 × 4 5 μm LiChrospher 100 RP-8
Column: 250 × 4 4 μm Superspher 60 RP-8
Mobile phase: Gradient. MeCN:100 mM pH 4.4 sodium acetate buffer from 10:90 to 25:75 over 10 min, to 30:70 over 10 min, to 45:55 over 10 min, to 100:0 over 5 min, maintain at 100:0 for 5 min, re-equilibrate at initial conditions for 14 min.
Column temperature: 45
Flow rate: 1.25
Injection volume: 20
Detector: UV 265

CHROMATOGRAM
Retention time: 4.5 (Asp), 5 (Ser), 7.5 (Glu), 9 (Gly), 9.5 (Thr), 11 (Arg), 11.5 (Ala), 13.5 (Pro), 15 (cystine), 15.5 (Met), 16.5 (Val), 20 (Phe), 21 (Ile), 21.5 (Leu), 23 (His (bis-derivative)), 27 (Lys), 30.5 (Tyr)

KEY WORDS
comparison with other derivatizing reagents

REFERENCE
Brückner, H.; Lüpke, M. Use of chromogenic and fluorescent oxycarbonyl chlorides as reagents for amino acid analysis by high-performance liquid chromatography, *J.Chromatogr.A*, **1995**, *697*, 295–307.

4-Phenylazobenzyloxycarbonyl Chloride

SAMPLE
Matrix: solutions
Analyte: amino acids
Sample preparation: Mix 20 μL of an amino acid solution with 100 μL 500 mM pH 9.0 sodium borate buffer and 250 μL 5 mM 4-phenylazobenzyloxycarbonyl chloride in MeCN, let stand for 5 min, add 250 μL 40 mM 1-aminoadamantane in acetone:water 75:25, mix, let stand for 5 min. Remove an 80 μL aliquot and add it to 320 μL MeCN:500 mM pH 4.0 sodium acetate buffer 50:50, inject a 20 μL aliquot. (4-Phenylazobenzyloxycarbonyl chloride can be purchased from Bachem, Bubendorf, Switzerland. Synthesis is as follows. Dissolve 10 g 4-nitrobenzylalcohol in 100 mL MeOH, add 1 mL triethylamine, add 2 g Raney nickel, hydrogenate at room temperature and atmospheric pressure, filter. Evaporate the filtrate to dryness, add benzene (Caution! Benzene is a carcinogen!), evaporate to dryness to remove water, repeat this step to obtain 4-aminobenzyl alcohol as crystals. Mix 7.7 g 4-aminobenzyl alcohol with a solution of 7.4 g nitrosobenzene in 40 mL acetic acid with stirring at 0°, after 3 h filter, wash the solid with dilute acetic acid to obtain 4-phenylazobenzylalcohol (mp 142.5-143°). Dilute the filtrate with a lot of water, filter, extract with hot carbon tetrachloride, crystallize from carbon tetrachloride to obtain more p-phenylazobenzyl alcohol. Dissolve 10.9 g phosgene in 40 mL dioxane, add 5 g p-phenylazobenzylalcohol at 0°, stir at 0° for 15 min, let stand at room temperature for 3 h, filter, evaporate the filtrate to dryness under reduced pressure. Recrystallize the residue from petroleum ether to give 4-phenylazobenzyloxycarbonyl chloride (mp 82-83°) (Helv. Chim. Acta 1958, 41, 491). Alternatively, recrystallize in the cold from n-hexane/ethyl acetate.)

HPLC VARIABLES
Guard column: 4 × 4 5 μm LiChrospher 100 RP-18
Column: 250 × 4 5 μm LiChrospher 100 RP-18
Mobile phase: Gradient. MeCN:100 mM pH 7.0 sodium acetate from 22:78 to 50:50 over 40 min, to 80:20 over 5 min, to 100:0 over 1 min, maintain at 100:0 for 9 min, re-equilibrate at initial conditions for 12 min.
Column temperature: 45
Flow rate: 1.25
Injection volume: 20
Detector: UV 320

CHROMATOGRAM
Retention time: 7 (Asp), 8 (Glu), 14 (Ser), 15.5 (Gly), 16 (Thr), 16.5 (Pro, Arg), 17 (Ala), 19 (Tyr (mono-derivative)), 22 (Val), 22.5 (Met), 25 (Ile), 25.5 (Leu), 26.5 (Phe), 31.5 (cystine), 42.5 (His), 43 (Lys), 47 (Tyr (bis-derivative))
Limit of detection: 1-10 pmole

KEY WORDS
comparison with other derivatizing reagents

REFERENCE
Brückner, H.; Lüpke, M. Use of chromogenic and fluorescent oxycarbonyl chlorides as reagents for amino acid analysis by high-performance liquid chromatography, *J.Chromatogr.A*, **1995**, *697*, 295–307.

SAMPLE
Matrix: solutions
Analyte: amino acids
Sample preparation: Mix a 40 μL aliquot of a solution of amino acids in 100 mM HCl with 100 μL 500 mM pH 9.0 potassium borate buffer and 250 μL 0.5 mM 4-phenylazo-benzyloxycarbonyl chloride in MeCN, let stand for 5 min, add 100 μL reagent, mix, let stand for 5 min. Remove an 80 μL aliquot and add it to 320 μL MeCN:500 mM pH 4.0 sodium acetate buffer 20:80, mix, inject a 20 μL aliquot. (Prepare reagent by mixing 3 mL n-heptylamine, 15 mL MeCN, and 175 mL 100 mM HCl, pH 7-8. 4-Phenylazobenzyloxycarbonyl chloride can be purchased from Bachem, Bubendorf, Switzerland. Synthesis is as described above.)

HPLC VARIABLES
Column: 250 × 4 5 μm Grom-Sil 120 ODS-3 CP porous encapsulated polymer-coated spherical particles (Grom, Herrenberg, Germany)
Mobile phase: Gradient. A was MeCN:THF:100 mM pH 6.7 sodium acetate buffer 20:2:78. B was MeCN:THF 98:2. A:B from 100:0 to 95:5 over 15 min, to 85:15 over 18 min, to 75:25 over 24 min, to 50:50 over 10 min, to 0:100 over 1 min, maintain at 0:100 for 7 min, return to initial conditions over 0.1 min, re-equilibrate for 10 min.
Column temperature: 45
Flow rate: 1 for 5 min, to 1.25 over 0.1 min, maintain at 1.25
Injection volume: 20
Detector: UV 320

CHROMATOGRAM
Retention time: 7.75 (Asp), 8.95 (Glu), 18.05 (Ser), 19.56 (Arg), 19.94 (Gly), 20.84 (Thr), 21.45 (Pro), 22.37 (Ala), 26.77 (Val), 29.83 (Tyr), 31.21 (Met), 34.65 (Ile), 35.43 (Leu), 38.11 (Phe), 63.99 (His), 64.42 (Lys)
Limit of detection: 0.5 pmole

REFERENCE
Kirschbaum, J.; Brückner, H. Amino acid analysis by derivatization with chromogenic 4-phenylazoben-zyloxycarbonyl chloride (PAZ-Cl): Comparison of reversed phases, *Chromatographia*, **1996**, *43*, 275–278.

ALDEHYDE

4-Nitrobenzaldehyde

SAMPLE
Matrix: protein
Analyte: N-acylglycines
Sample preparation: Heat protein in 100 µL 50 mM HCl at 100° for 3-6 h, evaporate to dryness, add 50 µL water, vortex, evaporate to dryness. Add 10 µmoles of anhydrous sodium acetate in MeOH, evaporate to dryness under a stream of nitrogen, add 250 µL 40 mM p-nitrobenzaldehyde in acetic anhydride, heat at 100° for 30 min, cool, evaporate to dryness under a stream of nitrogen, add a small volume of MeCN, vortex vigorously, inject an aliquot.

HPLC VARIABLES
Guard column: 30 × 1.5 CO3-032 reverse-phase (Brownlee)
Column: 250 × 4.6 Vydac C18 protein/peptide column
Mobile phase: Gradient. MeCN:water 30:70 for 4 min, to 100:0 over 5 min, maintain at 100:0.
Flow rate: 1.5
Detector: UV 350

CHROMATOGRAM
Retention time: 10.5 (N-acetylglycine), 14 (N-decanoylglycine), 15 (N-lauroylglycine), 16 (N-myristoylglycine), 17.5 (N-palmitoylglycine)
Limit of detection: 5-10 pmole

REFERENCE
Goddard, C.; Felsted, R.L. Identification of N-myristoylated proteins by reverse-phase high performance liquid chromatography of an azlactone derivative of N-myristoylglycine, *Biochem.J.*, **1988**, *253*, 839–843.

SAMPLE
Matrix: protein
Analyte: N-myristylglycine
Sample preparation: Precipitate 1 mg protein with ice-cold 10% trichloroacetic acid, wash with 300 µL MeCN, dry under reduced pressure, suspend in 250 µL 50 mM HCl (constant boiling, Pierce), heat at 100° for 4-6 h, evaporate to dryness under a stream of nitrogen, suspend in 250 µL 40 mM sodium acetate in MeOH, vortex, dry under a stream of nitrogen, add 250 µL 40 mM p-nitrobenzaldehyde in acetic anhydride, heat at 100° for 30 min, cool, evaporate to dryness under a stream of nitrogen, suspend the residue in 120 µL MeCN, filter (0.45 µm), inject an aliquot of the filtrate.

HPLC VARIABLES
Column: 300 × 8 µBondapak C18
Mobile phase: Gradient. MeCN:water 30:70 for 4 min, to 100:0 over 5 min, maintain at 100:0 for 13 min.
Flow rate: 1
Detector: UV 350

CHROMATOGRAM
Retention time: 17

REFERENCE
Risinger, M.A.; Dotimas, E.M.; Cohen, C.M. Human erythrocyte protein 4.2, a high copy number membrane protein, is N-myristylated, *J.Biol.Chem.*, **1992**, *267*, 5680–5685.

ANHYDRIDE

O,O-Dibenzoyltartaric acid anhydride

SAMPLE
Matrix: solutions
Analyte: amino acids
Sample preparation: Dissolve 0.25 mmoles leucine in 25 mL MeCN, place under aspirator pressure for 2-3 min to remove traces of water, add 0.5 mmole (+)-O,O'-dibenzoyl-L-tartaric anhydride, stir at 50° for 20 h, evaporate to dryness under aspirator pressure, reconstitute with 3% aqueous ammonia, evaporate to dryness, inject an aliquot of a 100 μM solution in water. (Synthesize (+)-O,O'-dibenzoyl-L-tartaric anhydride as follows. Heat 1 mole L-(+)-tartaric acid and 2.7 mole benzoyl chloride at 150° for 4 h, cool, wash with ligroin, boil with xylene, remove the organic phase, repeat several times, filter, dry in a desiccator.)

CAPILLARY ELECTROPHORESIS
Capillary: 56 cm × 100 μm coated fused-silica (39 cm to detector) (Polymicro Technologies)
Capillary preparation: Coat capillary as follows. Suck 1% 3-(trimethoxysilyl)propylmethacrylate in acetone:water 50:50 into the capillary, after 1 h remove solution, fill capillary with 3% acrylamide (Caution! Acrylamide is a carcinogen!) solution in water containing 0.04% tetramethylethylenediamine and 0.05% ammonium persulfate, after 15-20 min rinse with water, dry by aspiration (Electrophoresis 1989, 10, 23).
Running buffer: 25 mM pH 5.8 phosphate buffer containing x% polyvinylpyrrolidone
Capillary temperature: 24
Voltage/Current: 12 kV/22 μA
Injection: Hydrodynamic injection at 10 cm for 5 s
Detector: UV 233
Model: Laboratory constructed
Migration time: 10.4, 10.7 (Leu, x = 0, enantiomers), 8.5, 8.6 (Val, x = 0, enantiomers), 9.1, 9.3 (Gln, x = 0, enantiomers), 9.2 (D-Thr, x = 3), 9.4 (L-Thr, x = 3), 11.5 (L-Phe, x = 3), 12 (D-Phe, x = 3), 14 (L-Trp, x = 2), 15.7 (D-Trp, x = 2), 11.8 (L-mandelic acid, x = 2), 12.5 (D-mandelic acid, x = 2)

KEY WORDS
chiral; anode at detector

REFERENCE
Schützner, W.; Caponecchi, G.; Fanali, S.; Rizzi, A.; Kenndler, E. Improved separation of diastereomeric derivatives of enantiomers by a physical network of linear polyvinylpyrrolidone applied as pseudophase in capillary zone electrophoresis, *Electrophoresis*, **1994**, *15*, 769–773.

SAMPLE

Matrix: solutions

Analyte: amino acids

Sample preparation: Prepare derivatives with reaction with a 3-fold excess of (+)-O,O'-dibenzoyl-L-tartaric anhydride in dichloroethane, THF, or acetone in the presence of an excess of trichloroacetic acid. Heat at 50° for several hours (J.Chromatogr. 1984, 316, 605). (Synthesize (+)-O,O'-dibenzoyl-L-tartaric anhydride as described above.)

CAPILLARY ELECTROPHORESIS

Capillary: 56 cm × 100 μm coated fused-silica (39 cm to detector) (Polymicro Technologies)

Capillary preparation: Coat capillary by filling with a 0.5% solution of gamma-methacryloxypropyltrimethoxysilane in acetone:water 50:50, after 1 h remove solution and fill capillary with 4% acrylamide (Caution! Acrylamide is a carcinogen!) solution containing 0.4 μL/mL tetramethylethylenediamine and 0.5 mg/mL ammonium persulfate, after 15-20 min rinse capillary with water, dry under vacuum (Electrophoresis 1989, 10, 23).

Running buffer: 30 mM NaH_2PO_4 containing 2.5% poly(vinylpyrrolidone) adjusted to pH 5.8 with NaOH

Capillary temperature: ambient

Voltage/Current: 12 kV

Injection: Hydrodynamic injection at 10 cm for 5 s

Detector: UV 233

Model: laboratory constructed

Migration time: 9.4 (D-Ser), 9.5 (L-Ser), 9.6 (D-Gln), 9.9 (L-Gln), 10.0 (D-Leu), 10.2 (L-Leu), 11.1 (L-Phe), 11.5 (D-Phe), 14 (L-Trp), 15.5 (D-Trp)

KEY WORDS

chiral

REFERENCE

Schnützer, W.; Fanali, S.; Rizzi, A.; Kenndler, E. Separation of diastereomers by capillary zone electrophoresis with polymer additives: Effect of polymer type and chain length, *Anal.Chem.*, **1995**, *67*, 3866–3870.

SAMPLE

Matrix: solutions

Analyte: leucine

Sample preparation: Dissolve 0.25 mmoles leucine in 25 mL MeCN, place under aspirator pressure for 2-3 min to remove traces of water, add 0.5 mmole (+)-O,O'-dibenzoyl-L-tartaric anhydride, stir at 50° for 20 h, evaporate to dryness under aspirator pressure, reconstitute with 3% aqueous ammonia, evaporate to dryness. (Synthesize (+)-O,O'-dibenzoyl-L-tartaric anhydride as described above.)

CAPILLARY ELECTROPHORESIS

Capillary: 56 cm × 100 μm coated fused-silica (39 cm to detector) (Polymicro Technologies)

Capillary preparation: Coat capillary as follows. Suck 1% 3-(trimethoxysilyl)propylmethacrylate in EtOH into the capillary, after 1 h remove solution, fill capillary with 4% acrylamide (Caution! Acrylamide is a carcinogen!) solution in water containing 0.1% tetramethylethylenediamine and 0.1% ammonium persulfate, after 15-20 min rinse with water, dry by aspiration (Electrophoresis 1989, 10, 23).

Running buffer: 30 mM pH 5.8 phosphate buffer

Injection: Hydrodynamic injection at 10 cm for 5 s

Migration time: 7.67 (D), 7.81 (L)

Detector: UV 233

Model: Laboratory constructed

Capillary temperature: 24

Voltage/Current: 12 kV/22 μA

KEY WORDS

chiral

REFERENCE

Schützner, W.; Fanali, S.; Rizzi, A.; Kenndler, E. Separation of diastereomers by capillary zone electrophoresis in free solution with polymer additive and organic solvent component. Effect of pH and solvent composition, *J.Chromatogr.A*, **1996**, *719*, 411–420.

N-α-(9-Fluorenylmethyloxycarbonyl)leucine-N-carboxyanhydride

SAMPLE
Matrix: bulk
Analyte: amino acids
Sample preparation: Treat 0.2-10 μmoles amino acids with 50-100 μL MeOH:thionyl chloride 95:5 at 60° for 2 h, evaporate to dryness, reconstitute with 35 μL 28% diisopropylethylamine in DMF, shake at room temperature for 10 min, add 25 μL 200-1000 mM N-α-(9-fluorenylmethyloxycarbonyl)leucine-N-carboxyanhydride in DMF, mix, let stand at room temperature for 10 min, add 200 μL 250 mM pH 8 sodium glycinate, mix, let stand for 5 min, add 300 μL chloroform, extract, dilute 10000-50000-fold with n-hexane, inject a 20 μL aliquot. (Synthesis of N-α-(9-fluorenylmethyloxycarbonyl)leucine-N-carboxyanhydride is as follows. Dry all solvents over 4 Å molecular sieve. Stir 20 mmoles L-leucine in 67 mL THF under nitrogen at 2° in an ice bath, add 22 mmoles 9-fluorenylmethyl chloroformate at once, slowly add 29 mmoles dry 4-methylmorpholine (N-methylmorpholine), stir at 2-5° for 2 h, slowly add 4 M HCl in dioxane (Caution! Dioxane is a carcinogen!) until the pH of a sample diluted with water reaches 4-5, filter, wash the solid with dry THF, concentrate the filtrate under reduced pressure, dissolve the resulting oil in the minimum volume of dry diisopropyl ether (Caution! Diisopropyl ether readily forms explosive peroxides!), add dry hexane until the solution just turns cloudy, let stand at -20° overnight, filter. Wash the solid with dry hexane and dry it under vacuum to obtain N-α-(9-fluorenylmethyloxycarbonyl)leucine-N-carboxyanhydride (mp 118-120°; $[\alpha]^{25}_D$ = +38.0°) (J. Am. Chem. Soc. 1990, 112, 7414).)

HPLC VARIABLES
Guard column: 15 × 3.2 5 μm Kromasil silica
Column: 250 × 4.6 5 μm Kromasil silica
Mobile phase: n-Hexane:isopropanol 98:2 (A) or 97:3 (B) or 95:5 (C) or 90:10 (D)
Flow rate: 0.8
Injection volume: 20
Detector: F ex 263 em 313

CHROMATOGRAM

Retention time: k' 1.35 (L-tyrosine (D)), k' 1.42 (L-tryptophan (D)), k' 1.66 (D-tyrosine (D)), k' 1.70 (L-threonine (D)), k' 1.72 (D-tryptophan (D)), k' 1.94 (D-threonine (D)), k' 1.97 (L-lysine (C)), k' 1.98 (L-leucine (A)), k' 2.11 (L-isoleucine (A)), k' 2.20 (L-valine (A)), k' 2.24 (L-2-aminohexanoic acid (A)), k' 2.37 (D-leucine (A)), k' 2.50 (L-β-(1-naphthyl)alanine (UV detection) (A)), k' 2.55 (L-phenylalanine (A)), k' 2.58 (L-2-amino-4-phenylbutyric acid (A)), k' 2.60 (D-isoleucine (A)), k' 2.60 (D-valine (A)), k' 2.60 (L-2-aminopentanoic acid (A)), k' 2.75 (D-2-aminohexanoic acid (A)), k' 2.87 (L-β-(2-naphthyl)alanine (UV detection) (A)), k' 2.88 (L-α-aminophenylacetic acid (A)), k' 2.94 (L-β-(2-thienyl)alanine (A)), k' 2.98 (L-β-(p-chlorophenyl)alanine (A)), k' 3.03 (D-lysine (C)), k' 3.12 (L-β-(3,4-dichlorophenyl)alanine (A)), k' 3.18 (D-2-aminopentanoic acid (A)), k' 3.27 (L-serine (C)), k' 3.35 (L-methionine (A)), k' 3.35 (D-2-amino-4-phenylbutyric acid (A)), k' 3.38 (L-1,2,3,4-tetrahydro-3-isoquinolinecarboxylic acid (A)), k' 3.42 (D-α-aminophenylacetic acid (A)), k' 3.51 (D-β-(1-naphthyl)alanine (UV detection) (A)), k' 3.59 (D-serine (C)), k' 3.61 (L-α-aminobutyric acid (A)), k' 4.11 (D-1,2,3,4-tetrahydro-3-isoquinolinecarboxylic acid (A)), k' 4.14 (D-β-(2-thienyl)alanine (A)), k' 4.34 (D-α-aminobutyric acid (A)), k' 4.42 (D-β-(2-naphthyl)alanine (UV detection) (A)), k' 4.71 (D-β-(p-chlorophenyl)alanine (A)), k' 4.88 (D-β-(3,4-dichlorophenyl)alanine (A)), k' 4.91 (D-methionine (A)), k' 4.91 (D-phenylalanine (A)), k' 5.04 (L-glutamic acid (B)), k' 5.37 (L-alanine (A)), k' 5.97 (D-alanine (A)), k' 6.24 (L-aspartic acid (B)), k' 7.35 (D-aspartic acid (B)), k' 7.8 (D-glutamic acid (B))

KEY WORDS

normal phase; the use of related derivatization reagents is discussed; chiral

REFERENCE

Pugniere, M.; Mattras, H.; Castro, B.; Previero, A. Adsorption liquid chromatography on silica for the chiral separation of amino acids and asymmetric amines derivatized with optically active N-α-9-fluorenylmethyloxyxarbonyl-amino acid-N-carboxyanhydrides, *J.Chromatogr.A*, **1997**, *767*, 69–75.

Trimethylacetic Anhydride

SAMPLE

Matrix: bulk

Analyte: liothyronine

Sample preparation: Dissolve 1 mg compound in 1 mL EtOH and 10 μL 1 M KOH in EtOH:water 50:50 (freshly prepared). Remove a 10 μL aliquot and evaporate it to dryness at <0.05 Torr at 45° for 30 min, add 50 μL 80 mM 4-dimethylaminopyridine in dry MeCN, add 5 μL EtOH, vortex thoroughly, add 50 μL trimethylacetic anhydride, vortex for 10 s, heat at 65-70° for 50 min, add 100 μL EtOH, heat at 65-70° for 10 min, evaporate to dryness, add 100 μL toluene, add 100 μL 100 mM pH 6 phosphate buffer, vortex, centrifuge. Remove the organic layer and evaporate it to dryness, add 100 μL MeCN, sonicate for 2 min, add 100 μL pH 2.1 phosphate buffer, mix, inject an aliquot. (Pass MeCN through an aluminum oxide column before use.)

HPLC VARIABLES

Column: 150 × 4.6 Supelcosil LC-8

Mobile phase: Gradient. MeCN:10 mM KH_2; PO_4; adjusted to pH 2.1 with phosphoric acid from 30:70 to 87:13 over 10 min.

Flow rate: 2

Detector: UV 214

CHROMATOGRAM

Retention time: 10.7

OTHER SUBSTANCES
Simultaneously analyzed: 3,5-diiodothyronine, thyroxine

REFERENCE
Joppich, M.; Joppich-Kuhn, R.; Sentissi, A.; Giese, R.W. Single-step, quantitative derivatization of amino, carboxyl, and hydroxyl groups in iodothyronine amino acids with ethanolic pivalic anhydride containing 4-dimethylaminopyridine, *Anal.Biochem.*, **1986**, *153*, 159–165.

BORINIC ACID

1,1-Diphenylborinic Acid

SAMPLE
Matrix: solutions
Analyte: amino acids
Sample preparation: Add a 10-fold excess of 1,1-diphenylborinic acid to a solution of amino acids in 1 M acetic acid in isopropanol:water 50:50, heat at 65-70° for 15-20 min, dry under reduced pressure, reconstitute with mobile phase, inject an aliquot. (Prepare 1,1-diphenylborinic acid as follows. Stir diphenylborinic acid ethanolamine ester in 1 M HCl under nitrogen for 30 min, extract with dichloromethane. Wash the organic layer twice with water, wash with brine, dry over anhydrous sodium sulfate, evaporate under reduced pressure, take up the oil in warm hexane, cool, collect and wash the crystals, recrystallize from hexane to obtain 1,1-diphenylborinic acid (mp 128-130°) (Biomed. Mass Spec. 1984, 11, 611).)

HPLC VARIABLES
Column: 250 × 4 Zorbax PTH
Mobile phase: MeCN:THF:6 mM pH 3.30 phosphoric acid 18:16:66
Detector: UV 254

CHROMATOGRAM
Retention time: 8.56 (phosphorylated serine), 10.78 (phosphorylated threonine), 11.33 (His), 13.77 (Asn), 14.22 (Gln), 14.63 (Gla), 15.53 (Ser), 15.63 (phosphorylated tyrosine), 16.13 (Asp), 16.63 (Gly), 16.94 (Thr), 17.12 (Glu), 18.30 (N-methylglycine), 18.75 (Ala), 19.12 (Pro), 20.70 (N,N-dimethylglycine), 21.97 (biocytin), 25.96 (Cys), 27.54 (gamma-methylglutamic acid), 28.49 (Val), 29.40 (glutathione), 29.99 (Tyr), 30.76 (Lys), 32.43 (Arg), 33.16 (diphthine), 34.52 (Met), 37.64 (Ile), 40.18 (Leu), 45.3 (Phe), 48.43 (Trp), 75.97 (S-(p-nitrobenzyl)glutathione), 89.69 (2,6-diaminopimelic acid), 90.60 (oxidized glutathione), 102.51 (glutamyllysine)

OTHER SUBSTANCES
Non-interfering: amines, peptides

REFERENCE
Strang, C.J.; Henson, E.; Okamoto, Y.; Paz, M.A.; Gallop, P.M. Separation and determination of α-amino acids by boroxazolidone formation, *Anal.Biochem.*, **1989**, *178*, 276–286.

CARBOXYLIC ACID

Dansyl-L-proline

SAMPLE
Matrix: solutions
Analyte: amino acid methyl esters
Sample preparation: Mix a 200 μL aliquot of a 10 μM solution in DMF with 200 μL 10 μM dansyl-L-proline in DMF, add 100 μL 22 μM diethyl cyanophosphonate (diethyl phosphorocyanidate) in DMF, add 100 μL 42 μM triethylamine in DMF, mix, let stand for 3 min, inject a 5 μL aliquot.

HPLC VARIABLES
Column: 250 × 4.6 5 μm Shim-pack ODS
Mobile phase: MeOH:water 70:30 (A) or MeOH:water 80:20 (B)
Column temperature: 45
Flow rate: 0.6 (A) or 0.7 (B)
Injection volume: 5
Detector: F ex 345 em 515

CHROMATOGRAM
Retention time: k' 2.68 (L-phenylalanine methyl ester (B)), k' 2.84 (D-phenylalanine methyl ester (B)), k' 3.43 (L-alanine methyl ester (A)), k' 3.84 (D-alanine methyl ester (A))
Limit of detection: 40-50 fmole

KEY WORDS
chiral

REFERENCE
Gamoh, K.; Sawamoto, H. Resolution of amine enantiomers using precolumn derivatization with dansyl-L-proline and reversed-phase liquid chromatography, *Anal.Chim.Acta*, **1991**, *243*, 251–257.

(-)-α-Methoxy-α-methyl-1-naphthaleneacetic Acid

SAMPLE
Matrix: bulk
Analyte: amino acid methyl esters
Sample preparation: Dissolve 100 μg amino acid methyl ester in 200 μL pyridine, add 2 mg (-)-α-methoxy-α-methyl-1-naphthaleneacetic acid, add 2 mg N,N'-dicyclohexylcarbodiimide, let stand at room temperature for 30 min, add 500 μL ethyl acetate, wash with 5% HCl, wash with 5% sodium bicarbonate, wash with water. Dry the organic layer over anhydrous sodium sulfate, inject a 5 μL aliquot. (Synthesis of (-)-α-methoxy-α-methyl-1-naphthaleneacetic acid is as follows. Stir 200 mg magnesium metal in 8 mL anhydrous ether, add 2 g 1-bromonaphthalene in 2 mL anhydrous ether, add a trace of iodomethane, stir under nitrogen at room temperature for 30 min, slowly add the resulting solution of 1-naphthylmagnesium bromide to a solution of 450 mg pyruvic acid in 3 mL ether, stir at room temperature for 30 min, acidify with 3% sulfuric acid, extract with ethyl acetate. Wash the organic layer with water, extract with 5% sodium carbonate solution. Acidify the aqueous layer with 5% HCl, extract with ethyl acetate. Wash the organic layer with water and dry it over anhydrous sodium sulfate, evaporate to dryness, recrystallize from hexane/acetone to give α-hydroxy-α-methyl-1-naphthaleneacetic acid as colorless plates (mp 110-112°). Add 620 mg sodium hydride to a solution of 1 g α-hydroxy-α-methyl-1-naphthaleneacetic acid in 4 mL anhydrous DMF at 0°, stir for 30 min, add 5 mL iodomethane, stir at room temperature for 45 min, pour into ice-water, extract with ethyl acetate. Wash the organic layer with water and dry it over anhydrous sodium sulfate, evaporate to dryness. Dissolve the residue in 20 mL 5% KOH in MeOH, heat at 50° for 1.5 h, concentrate under reduced pressure, acidify with 5% HCl, extract with ethyl acetate. Wash the organic layer with water, dry over anhydrous sodium sulfate, evaporate to dryness, recrystallize from hexane/ethyl acetate to give α-methoxy-α-methyl-1-naphthaleneacetic acid as colorless plates (mp 161-162°). Mix a solution of 1.8 g of α-methoxy-α-methyl-1-naphthaleneacetic acid in 5.5 mL EtOH with a solution of 760 mg (R)-(+)-α-methylbenzylamine in 5.5 mL EtOH, heat on a steam bath to dissolve the salt, let stand at room temperature for 20 h. Collect the precipitate and fractionally crystallize it from EtOH twice. Decompose the salt with 5% HCl, extract with ethyl acetate. Wash the organic layer with water and dry it over anhydrous sodium sulfate, evaporate to dryness, recrystallize from hexane/ether to give (-)-α-methoxy-α-methyl-1-naphthaleneacetic acid (mp 111-112°, $[\alpha]_D;^{13}$ -106.3° (c = 0.16, chloroform).)

HPLC VARIABLES
Column: 305 × 6.4 μPorasil
Mobile phase: Cyclohexane:ethyl acetate 80:20 (A), 2:1 (B), 50:50 (C), or 40:60 (D)
Flow rate: 0.7
Injection volume: 5
Detector: UV 280

CHROMATOGRAM
Retention time: k' 1.91 (D-alanine methyl ester (A)), k' 3.10 (L-alanine methyl ester (A)), k' 1.76 (D-valine methyl ester (A)), k' 1.96 (L-valine methyl ester (A)), k' 2.17 (D-norvaline methyl ester (A)), k' 1.22 (L-norvaline methyl ester (A)), k' 1.83 (D-leucine methyl ester (A)), k' 0.78 (L-leucine methyl ester (A)), k' 1.92 (D-norleucine methyl ester (A)), k' 0.96

(L-norleucine methyl ester (A)), k' 1.53 (D-isoleucine methyl ester (A)), k' 0.88 (L-isoleucine methyl ester (A)), k' 1.42 (D-proline methyl ester (A)), k' 2.38 (L-proline methyl ester (A)), k' 2.56 (D-phenylglycine methyl ester (A)), k' 1.32 (L-phenylglycine methyl ester (A)), k' 2.41 (D-phenylalanine methyl ester (A)), k' 1.52 (L-phenylalanine methyl ester (A)), k' 2.52 (D-serine methyl ester (C)), k' 7.32 (L-serine methyl ester (C)), k' 1.20 (D-threonine methyl ester (D)), k' 1.85 (L-threonine methyl ester (D)), k' 2.02 (D-tyrosine methyl ester (B)), k' 1.74 (L-tyrosine methyl ester (B)), k' 2.32 (D-DOPA methyl ester (C)), k' 2.04 (L-DOPA methyl ester (C)), k' 2.08 (D-cysteine methyl ester (C)), k' 0.98 (L-cysteine methyl ester (C)), k' 1.60 (D-methionine methyl ester (B)), k' 1.24 (L-methionine methyl ester (B)), k' 5.61 (D-aspartic acid methyl ester (A)), k' 7.00 (L-aspartic acid methyl ester (A)), k' 6.46 (D-glutamic acid methyl ester (A)), k' 6.81 (L-glutamic acid methyl ester (A)), k' 2.00 (D-ornithine methyl ester (D)), k' 3.52 (L-ornithine methyl ester (D)), k' 1.60 (D-lysine methyl ester (D)), k' 2.20 (L-lysine methyl ester (D)), k' 1.36 (D-histidine methyl ester (D)), k' 1.96 (L-histidine methyl ester (D))

KEY WORDS
chiral; normal phase; separation is not as good using reverse phase

REFERENCE
Goto, J.; Hasegawa, M.; Nakamura, S.; Shimada, K.; Nambara, T. New derivatization reagents for the resolution of amino acid enantiomers by high-performance liquid chromatography, *J.Chromatogr.*, **1978**, *152*, 413–419.

DIALDEHYDE

1-Methoxycarbonylindolizine-3,5-dicarbaldehyde

SAMPLE
Matrix: hair, protein
Analyte: amino acids
Sample preparation: Suspend 10 mg protein or hair in 10 mL 6 M HCl, seal tube under vacuum, heat at 110° for 20 h. Remove a 1 mL aliquot and evaporate it to dryness, dissolve the residue in 1 mL 100 mM HCl, make up to 10 mL with water, filter (Adovantex DISMIC-13cp), dilute the filtrate 10-fold with water. Place a 100 μL aliquot of 100 μg/mL solution of 1-methoxycarbonylindolizine-3,5-dicarbaldehyde in ethyl acetate in the bottom of a tube, evaporate to dryness under reduced pressure, add 100 μL reaction buffer, sonicate for 30 s, add 20 μL protein hydrolysate, mix well, let stand for 20 min in the dark, inject a 10 μL aliquot. (Prepare phosphate-borate buffer by mixing equal volumes of 20 mM NaH_2PO_4 and 20 mM sodium tetraborate, adjust pH to 10 with 1 M NaOH. Prepare reaction buffer by mixing equal volumes of EtOH and phosphate-borate buffer. Prepare reagent (1-methoxycarbonylindolizine-3,5-dicarbaldehyde) as follows. Reflux 21.4 g 2-pyridinecarboxaldehyde, 24 mL ethylene glycol, 10 g p-toluenesulfonic acid, and 300 mL benzene (Caution! Benzene is a carcinogen!) under a Dean-Stark separator for 64 h, pour into concentrated sodium carbonate solution. Remove the organic layer and extract the aqueous layer 4 times with benzene. Combine the organic layers and wash them with water, dry over anhydrous magnesium sulfate, evaporate, distil the residue to give 2-(1,3-

dioxolan-2-yl)pyridine (bp 122°/4 mm Hg) (J.Org.Chem. 1963, 28, 83). Reflux 15.1 g 2-(1,3-dioxolan-2-yl)pyridine and 19.5 g tert-butyl bromoacetate in 100 mL dry MeCN for 7 h, let stand overnight at room temperature, filter, wash the precipitate with diethyl ether to give 1-(tert-butoxycarbonylmethyl)-2-(1,3-dioxolan-2-yl)pyridinium bromide (mp 110-2° from MeCN). Suspend 51.9 g of this compound in 1.5 L THF with stirring, add 62.1 g potassium carbonate, add 15.12 g methyl propiolate, stir at room temperature for 9 days, filter, evaporate the filtrate to dryness under reduced pressure, chromatograph the residue on silica gel with hexane:ethyl acetate 20:1-10:1, collect fractions and evaporate to dryness to give methyl 3-tert-butoxycarbonyl-5-(1,3-dioxolan-2-yl)indolizine-1-carboxylate (mp 138-9° from hexane). Reflux 20.82 g of this compound in 600 mL THF and 60 mL 10% HCl for 6 h, concentrate to one quarter of the original volume, add water, extract with chloroform. Wash the chloroform layer with water and dry it over anhydrous sodium sulfate, evaporate, chromatograph on silica gel with hexane:ethyl acetate 10:1 to give 1-methoxycarbonylindolizine-5-carbaldehyde (mp 135-7° from MeOH). Stir 12.18 g of this compound in 116 mL dry DMF at 0° under argon, add 17 mL phosphorus oxychloride, stir at room temperature for 1 h, pour into water, adjust pH to 9.0 with 5% potassium carbonate, extract with chloroform. Wash the organic layer with water and dry it over anhydrous sodium sulfate, concentrate until a precipitate forms, filter to obtain the product, concentrate the filtrate and chromatograph the residue on silica gel with hexane:ethyl acetate 5:1 to obtain more product. The product was 1-methoxycarbonylindolizine-3,5-dicarbaldehyde (mp 164-5° from methyl acetate).)

HPLC VARIABLES
Column: 150 × 6 Asahipak ODP-50 (Asahi)
Mobile phase: Gradient. MeCN:buffer from 7:93 to 20:80 over 40 min, to 30:70 over 10 min, to 50:50 over 5 min, return to initial conditions over 5 min.) (Prepare buffer by adjusting the pH of 20 mM $(NH_4)H_2PO_4$ containing 10 mM sodium 1-octanesulfonate to 2.6 with phosphoric acid.)
Column temperature: 40
Flow rate: 1
Injection volume: 10
Detector: F ex 414 em 482

CHROMATOGRAM
Retention time: 13.8 (Asp), 15.5 (Ser), 17 (Gly), 18.8 (Glu), 19.2 (Cys), 20 (Thr), 21.5 (Ala), 30 (His), 33.8 (Val), 34.6 (ammonia), 35.8 (Met), 38.3 (Arg), 39.2 (Lys), 42.5 (Ile), 44.2 (Leu), 46.7 (Phe)
Limit of detection: 0.2-200 fmole

KEY WORDS
soybean

REFERENCE
Oguri, S.; Uchida, C.; Mishina, M.; Miki, Y.; Kakehi, K. Determination of amino acids by pre-column fluorescence derivatization with 1-methoxycarbonylindolizine-3,5-dicarbaldehyde, *J.Chromatogr.A*, **1996**, *724*, 169–177.

SAMPLE
Matrix: solutions
Analyte: amino acids
Sample preparation: Evaporate 100 μL of a 1 mg/mL solution of reagent in ethyl acetate into the bottom of a tube using reduced pressure in a desiccator, add 40 μL reaction buffer, add 20 μL of amino acid solution in water containing 2.5 mg/mL 3-nitrophenol, let stand for 15 min in the dark, inject an aliquot. (Prepare reaction buffer by mixing equal volumes of EtOH and phosphate/borate buffer. Prepare phosphate/borate buffer by mixing equal volumes of 20 mM NaH_2PO_4 and 20 mM sodium tetraborate and adjusting the pH to 10 with 1 M NaOH. Prepare reagent (1-methoxycarbonylindolizine-3,5-dicarbaldehyde) as described above.)

CAPILLARY ELECTROPHORESIS
Capillary: 60 cm × 50 μm fused-silica (50 cm to detector)

Running buffer: MeOH:buffer 3:97 (Buffer was 40 mM pH 7.0 borate/phosphate buffer containing 20 mM sodium dodecyl sulfate.)
Voltage/Current: 20 kV
Injection: Hydrostatic injection at 15 cm for 5 s
Detector: UV 280; UV 409; F ex 282 (or 414) em 482
Model: JASCO CE-800
Migration time: 6.7 (Ser), 6.9 (Cys), 7.1 (Glu), 7.3 (Asp), 7.5 (Gly), 7.7 (Ala), 10.3 (Val), 12.5 (His), 13.2 (Met), 13.8 (Ile), 14.2 (Leu), 15.7 (ammonia), 16.3 (Phe), 17.0 (Lys), 18.0 (Arg), 18.3 (Tyr)
Limit of detection: 5 nM

REFERENCE

Oguri, S.; Uchida, C.; Miyake, Y.; Miki, Y.; Kakehi, K. 1-Methoxycarbonylindolizine-3,5-dicarbaldehyde as a derivatization reagent for amino compounds in high-performance capillary electrophoresis, *Analyst*, **1995**, *120*, 63–68.

Naphthalene-2,3-dicarboxaldehyde

SAMPLE
Matrix: solutions
Analyte: glutathione
Sample preparation: 100 μL Solution + 100 μL running buffer + 100 μL 25 mM 2,3-naphthalenedicarboxaldehyde in MeCN, let stand for 10 min, inject an aliquot.

CAPILLARY ELECTROPHORESIS
Capillary: 80 cm × 27 μm fused-silica (Polymicro)
Running buffer: 11 mM pH 9.1 borate buffer
Voltage/Current: 25 kV
Detector: F ex 458 (Ar laser) em (532 nm notch filter, 480 nm long-pass filter)
Migration time: 10

OTHER SUBSTANCES
Simultaneously analyzed: glutamylcysteine

REFERENCE

Orwar, O.; Fishman, H.A.; Ziv, N.E.; Scheller, R.H.; Zare, R.N. Use of 2,3-naphthalenedicarboxaldehyde derivatization for single-cell analysis of glutathione by capillary electrophoresis and histochemical localization by fluorescence microscopy, *Anal.Chem.*, **1995**, *67*, 4261–4268.

Naphthalene-2,3-dicarboxaldehyde/Cyanide

RELATED REFERENCE
DeSilva, K.; Kuwana, T. Separation of chiral amino acids by micellar electrokinetic chromatography with derivatized cyclodextrins. *Biomed.Chromatogr.* **1997**, *11*, 230-235

SAMPLE
Matrix: blood
Analyte: arginine
Sample preparation: 125 μL Plasma + 10 μL 10 mM l-norleucine in water + 125 μL 10% perchloric acid, vortex, centrifuge at 15600 g for 10 min. Remove a 230 μL aliquot of the supernatant and add it to 63 μL 2 M NaOH, mix, add 187 μL 100 mM pH 9.2 sodium borate buffer, add 10 μL 100 mM NaCN in 100 mM pH 9.2 sodium borate buffer, add 10 μL 100 mM 2,3-naphthalenedicarboxaldehyde in MeOH, mix, let stand at room temperature for 30 min. Remove a 40 μL aliquot and add it to 60 μL MeCN:buffer 30:70, mix, inject the whole amount. (Buffer was 20 mM sodium citrate containing 5 mM tetramethylammonium chloride, pH adjusted to 2.8 with phosphoric acid.)

HPLC VARIABLES
Guard column: 4 μm Nova-Pak C18
Column: 250 × 4.6 4 μm Nova-Pak C18
Mobile phase: Gradient. A was buffer. B was MeCN:buffer (?) 60:40. A:B 100:0 for 3 min, to 70:30 over 5 min, to 40:60 over 16 min, to 20:80 over 6 min, to 10:90 over 5 min, to 0:100 over 3 min, maintain at 0:100 for 6 min, return to initial conditions over 2 min, re-equilibrate for 6 min. (Buffer was 20 mM sodium citrate containing 5 mM tetramethylammonium chloride, pH adjusted to 2.8 with phosphoric acid.)
Flow rate: 1
Injection volume: 100
Detector: UV 260; F ex 420 em 490

CHROMATOGRAM
Retention time: 21.4
Internal standard: l-norleucine (40.8)

KEY WORDS
plasma; pharmacokinetics

REFERENCE
Gopalakrishnan, V.; Burton, P.J.; Blaschke, T.F. High-performance liquid chromatographic assay for the quantitation of L-arginine in human plasma, *Anal.Chem.*, **1996**, *68*, 3520–3523.

SAMPLE
Matrix: dialysate
Analyte: amino acids
Sample preparation: Pump dialysate, 7 mM naphthalene-2,3-dicarboxaldehyde in 54:46 MeCN:water, 10 mM fluorescein, and 10 mM NaCN in 50 mM pH 9.3 borate buffer at 1

μL/min each through a 42 cm $\times$ 240 μm PEEK capillary, mix this mixture with running buffer pumped at 2-7 μL/min, inject an aliquot electrokinetically.

CAPILLARY ELECTROPHORESIS
Capillary: 30 cm $\times$ 25 μm (14 cm to detector) (Polymicro Technologies)
Capillary preparation: Flush with 100 mM NaOH before use.
Running buffer: 100 mM Tris pH 8.65
Voltage/Current: 25 kV
Injection: Electrokinetic injection
Detector: F ex 442 nm (5-7 mW He-Cd laser)
Migration time: 1.07 (glutamate), 1.17 (aspartate)
Internal standard: fluorescein (0.86)
Limit of detection: 30 nM

KEY WORDS
rat; on-line derivatization

REFERENCE
Zhou, S.Y.; Zuo, H.; Stobaugh, J.F.; Lunte, C.E.; Lunte, S.M. Continuous in vivo monitoring of amino acid neurotransmitters by microdialysis sampling with on-line derivatization and capillary electrophoresis separation, *Anal.Chem.*, **1995**, *67*, 594–599.

SAMPLE
Matrix: dialysate
Analyte: glutamate and aspartate
Sample preparation: 1.5 μL Dialysate (artificial CSF) + 2.9 μL 50 mM pH 9.5 borate buffer containing 20 mM NaCN + 0.1 μL 30 mM naphthalene-2,3-dicarboxaldehyde in MeOH, mix, let stand at room temperature for 3 min, inject an aliquot.

CAPILLARY ELECTROPHORESIS
Capillary: 37 cm $\times$ 75 μm fused-silica (Composite Metal Services, Hallow, UK)
Capillary preparation: Between analyses rinse with 100 mM NaOH for 1.5 min and with running buffer for 1.5 min.
Running buffer: 30 mM Boric acid, adjusted to pH 9.5 with 1 M NaOH
Voltage/Current: 30 kV
Injection: High pressure injection (5 nL).
Detector: F ex 442 (He/Cd laser)
Model: Beckman P/ACE 2050
Migration time: 1.48 (glutamate), 1.56 (aspartate)
Limit of detection: 1 nM

REFERENCE
Dawson, L.A.; Stow, J.M.; Palmer, A.M. Improved method for the measurement of glutamate and aspartate using capillary electrophoresis with laser induced fluorescence detection and its application to brain microdialysis, *J.Chromatogr.B*, **1997**, *694*, 455–460.

SAMPLE
Matrix: solutions
Analyte: amino acids
Sample preparation: Mix sample:50 (?) mM NaCN in 50 mM pH 9.3 borate buffer:25 (?) mM naphthalene-2,3-dicarboxaldehyde in MeOH 3:1:1, let stand for 15 min, inject a 50 μL aliquot.

HPLC VARIABLES
Column: 200 $\times$ 3 5 μm Chromspher ODS-2 C18 (Chrompack)
Mobile phase: Gradient. A was 50 mM pH 7.0 sodium phosphate buffer. B was MeOH: THF:water 50:20:30. A:B from 25:75 to 0:100 over 75 min.
Flow rate: 0.5
Injection volume: 50
Detector: F ex 420

CHROMATOGRAM
Retention time: 15 (aspartic acid), 20 (glutamic acid), 27.5 (asparagine), 30 (glutamine), 31.5 (serine), 32.5 (homoserine), 35.5 (glycine), 40 (taurine), 42 (alanine), 44 (β-alanine), 52 (norvaline), 54 (valine), 60 (leucine)

REFERENCE
Koning, H.; Wolf, H.; Venema, K.; Korf, J. Automated precolumn derivatization of amino acids, small peptides, brain amines and drugs with primary amino groups for reversed-phase high-performance liquid chromatography using naphthalenedialdehyde as the fluorogenic label, *J.Chromatogr.*, **1990**, *533*, 171–178.

SAMPLE
Matrix: solutions
Analyte: amino acids
Sample preparation: 32 μL Amino acid solution in 50 mM pH 9.0 borate buffer + 69 μL 1.96 mM cyanide, mix, add 6 μL 2.25 mM naphthalene-2,3-dicarboxaldehyde mixed with 13 μL 50 mM pH 9.0 borate buffer, mix, let stand for 30 min, inject a 20 μL aliquot.

HPLC VARIABLES
Column: 150 × 4 MicroPak SP C18
Mobile phase: Gradient. A was THF:50 mM pH 6.8 potassium phosphate buffer 10:90. B was MeCN:MeOH:THF:50 mM pH 6.8 potassium phosphate buffer 55:10:3.5:31.5. A: B from 90:10 to 45:55 over 45 min, to 20:80 over 1 min, maintain at 20:80 for 3 min, return to initial conditions over 2 min, re-equilibrate for 3 min.
Flow rate: 1
Injection volume: 20
Detector: F ex 420 em 490

CHROMATOGRAM
Retention time: 7 (aspartic acid), 10 (glutamic acid), 18 (histidine), 20 (serine), 22 (arginine), 22.5 (glycine), 24 (threonine), 26 (alanine), 29 (tyrosine), 31 (α-aminobutyric acid), 35.5 (valine), 36 (methionine), 40.5 (isoleucine), 41 (phenylalanine), 42 (leucine)
Limit of detection: 200 fmole

KEY WORDS
discussion of matrix interference in paper

REFERENCE
Lai, F.; Sheehan, T. Matrix effects in the derivatization of amino acids with naphthalene dicarboxaldehyde, 9-fluorenylmethyl chloroformate and phenylisothiocyanate, *BioTechniques*, **1993**, *14*, 642–649.

Phthalaldehyde

RELATED REFERENCES
August, T.F.; Musson, D.G.; Hwang, S.S.; Duggan, D.E.; Hooke, K.F.; Roman, I.J.; Ferguson, R.J.; Bayne, W.F. Bioanalysis and disposition of alpha-fluoromethylhistidine, a new histidine decarboxylase inhibitor. *J.Pharm.Sci.* **1985**, *74*, 871-875.
Scully, F.E., Jr.; Mazina, K.; Sonenshine, D.; Kopfler, F. Quantitation and identification of organic N-chloramines formed in stomach fluid on ingestion of aqueous hypochlorite. *Environ.Health Perspect.* **1986**, *69*, 259-265.

SAMPLE
Matrix: solutions
Analyte: amino acids
Sample preparation: Mix 2 μL of an amino acid solution in 50 mM pH 10.0 borate buffer with 12 μL 50 mM o-phthalaldehyde in 50 mM 50 mM pH 10.0 borate buffer containing 1.5 mM cinnamic acid, let stand for 20 min, inject an aliquot. (The two solutions can also be mixed in the inlet of the capillary and allowed to stand for 20 min before commencing the run. Inject reagent for 3 s, sample for 1 s, and reagent for 3 s.)

CAPILLARY ELECTROPHORESIS
Capillary: 72 cm × 50 μm fused-silica (50 cm to detector) (Polymicro Technologies)
Running buffer: 50 mM pH 10.0 Borate buffer
Capillary temperature: 30
Voltage/Current: 30 kV
Detector: UV 230
Model: Perkin Elmer ABI-270A
Migration time: 4.4 (Phe), 4.6 (Met), 5 (Ala), 7.5 (Glu)
Internal standard: cinnamic acid (5.5)

REFERENCE
Taga, A.; Honda, S. Derivatization at capillary inlet in high-performance capillary electrophoresis. Its reliability in quantification, *J.Chromatogr.A*, **1996**, *742*, 243–250.

Phthalaldehyde/Acetylcysteine

RELATED REFERENCES
Boomsma, F.; van der Hoorn, F.A.J.; Van in't Veld, A.J.; Schalekamp, M.A.D.H. Determination of D,L-*threo*-3,4-dihydroxyphenylserine and of the D- and L-enantiomers in human plasma and urine. *J.Chromatogr.* **1988**, *427*, 219-227.
Catala-Icardo, M.; Medina-Hernandez, M.J.; Garcia Alvarez-Coque, M.C. Determination of amino acids by micellar high-performance liquid chromatography and pre-column derivatization with o-phthalaldehyde and N-acetyl-L-cysteine. *J.Liq.Chromatogr.* **1995**, *18*, 2827-2841.
Florance, J.; Galdes, A.; Konteatis, Z.; Kosarych, Z.; Langer, K.; Martucci, C. High-performance liquid chromatographic separation of peptide and amino acid stereoisomers. *J.Chromatogr.* **1987**, *414*, 313-322.
Hisaka, A.; Kasamatsu, S.; Takenaga, N.; Ohtawa, M. Quantification of L-3-(3-hydroxy-4-pivaloyloxyphenyl)alanine (NB-355) by high-performance liquid chromatography using o-phthalaldehyde/N-acetyl-L-cysteine derivatization. *J.Chromatogr.* **1989**, *494*, 183-189.
Lovell, G.; Corran, P.H. Determination of L-thyroxine in reference serum preparations as the o-phthalaldehyde-N-acetylcysteine derivative by reversed-phase liquid chromatography with electrochemical detection. *J.Chromatogr.* **1990**, *525*, 287-296.
Maurs, M.; Trigalo, F.; Azerad, R. Resolution of α-substituted amino acid enantiomers by high-performance liquid chromatography after derivatization with a chiral adduct of *o*-phthalaldehyde. Application to glutamic acid analogues. *J.Chromatogr.* **1988**, *440*, 209-215.
Nimura, N.; Kinoshita, T. *o*-Phthalaldehyde-N-acetyl-L-cysteine as a chiral derivatization reagent for liquid chromatographic optical resolution of amino acid enantiomers and its application to conventional amino acid analysis. *J.Chromatogr.* **1986**, *352*, 169-177.
Soto-Otero, R.; Méndez-Alvarez, E.; Galán-Valiente, J.; Aguilar-Veiga, E.; Sierra-Marcuño, G. Quantitative analysis of neuroactive amino acids in brain tissue by liquid chromatography using fluorescent pre-column labelling with *o*-phthalaldehyde and *N*-acetyl-L-cysteine. *Biomed.Chromatogr.* **1994**, *8*, 114-118.

Takeuchi, T.; Niwa, T.; Ishii, D. Application of micro HPLC to the determination of amino acids after precolumn derivatization with o-phthalaldehyde and N-acetyl-L-cysteine. *HRC CC,J.High Resolut.Chromatogr.Chromatogr.Commun.* **1988**, *11*, 343-346.

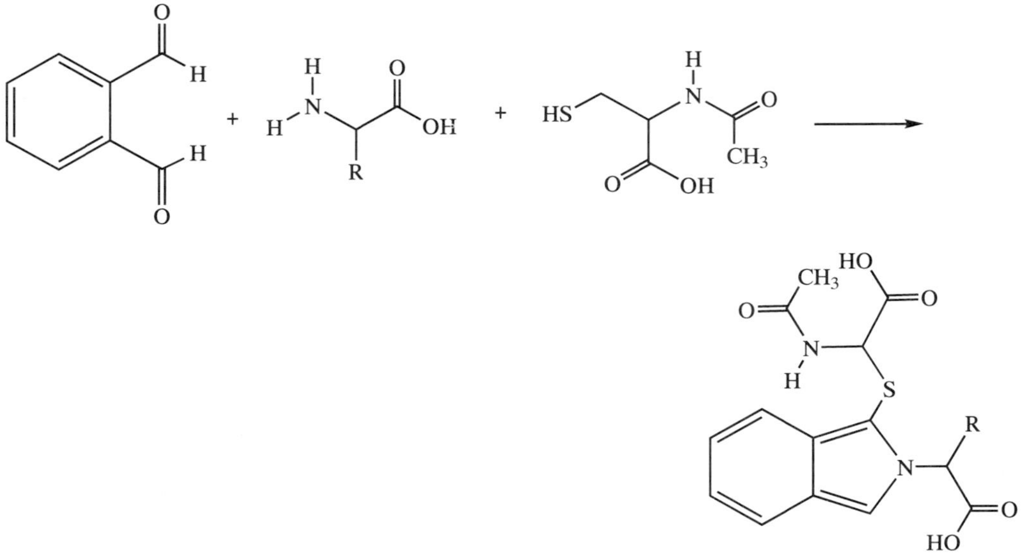

SAMPLE

Matrix: fermentation broth

Analyte: amino acids

Sample preparation: Wash 7.5 g broth on a 150 μm nylon mesh with ice-cold 0.8% NaCl, float cells in saline, wash cells with ice water, cool cells in ice, add 5 mL water, add 500 μL 1 M 4-morpholinepropanesulfonic acid buffer, sonicate for 15 s, shake for 30 s, sonicate for 15 s, filter 0.45 μm, filter (10 000 MW cutoff filter) while centrifuging. Add a 200 μL aliquot of the filtrate to a 15 × 8 column of quaternary ion-exchange resin, wash with 10 mL water, elute with 10 mL 1 M acetic acid, freeze dry, reconstitute with 200 μL water. Mix 20 μL solution with 5 μL reagent, let stand for 2-3 min, add 475 μL 50 mM pH 5.2 sodium acetate (Anal.Biochem. 1984, 137, 405), filter (0.45 μm), inject an aliquot of the filtrate. (Prepare reagent by dissolving 4 mg o-phthaldialdehyde in 300 μL MeOH, add 250 μL 400 mM pH 8.0 borate buffer, add 390 μL water, add 60 μL 1 M N-acetyl-L-cysteine (adjusted to pH 5.0-6.0 with NaOH). Store at 4°, discard after 3 weeks (Anal.Biochem. 1984, 137, 405).)

HPLC VARIABLES

Column: 150 × 5 3 μm Hypersil ODS

Mobile phase: Gradient. A was 50 mM pH 5.9 sodium acetate. B was MeOH:50 mM pH 5.9 sodium acetate 80:20. A:B from 100:0 to 0:100 over 50 min (?).

Flow rate: 1

Detector: F ex 330-375 em 418 (cut-off filter)

CHROMATOGRAM

Retention time: 3.5 (glutathione), 4 (L-Asp), 4.5 (L-Glu), 5.8 (L-Ser), 9.5 (L-α-aminoadipic acid), 10.5 (D-α-aminoadipic acid), 11 (L-His), 11.5 (L-Thr), 12 (Gly), 19 (L-Arg), 22.5 (A-Ala), 27 (cephalosporin C), 28 (L-Tyr), 28 (penicillin N), 28.5 (isopenicillin N), 32.5 (L-Val), 34.5 (L-Met), 35 (D-Val), 39 (L-Phe), 40 (L-Ile), 47.5 (L-Leu), 49 (L-Lys)

OTHER SUBSTANCES

Non-interfering: L-cysteine

KEY WORDS

SPE; chiral

REFERENCE

Usher, J.J.; Lewis, M.; Hughes, D.W. Determination by high-performance liquid chromatography of some compounds involved in the biosynthesis of penicillin and cephalosporin, *Anal.Biochem.*, **1985**, *149*, 105–110.

SAMPLE

Matrix: formulations
Analyte: amino acids
Sample preparation: Sonicate pills, capsules, powders, or drops in 50 mM sodium dodecyl sulfate, mix an aliquot with a 10-fold molar excess of reagent, let stand for 1 min, inject an aliquot. (Reagent was 2 mM o-phthalaldehyde and 2 mM N-acetyl-L-cysteine in 100 mM pH 9.5 sodium borate buffer.)

HPLC VARIABLES

Guard column: 35 × 4.6 5 μm Spherisorb ODS-2
Column: 120 × 4.6 5 μm Spherisorb ODS-2
Mobile phase: Propanol:50 mM sodium dodecyl sulfate 3:97, pH 3
Flow rate: 1
Injection volume: 20
Detector: UV 336

CHROMATOGRAM

Retention time: 2.5 (threonine), 3, 8 (lysine (different derivatives), 3 (glycine), 6 (methionine)

KEY WORDS

pills; capsules; powders; drops

REFERENCE

Catalá-Icardo, M.; Medina-Hernández, M.J.; García Alvarez-Coque, M.C. Determination of amino acids by micellar high-performance liquid chromatography and pre-column derivatization with o-phthalaldehyde and N-acetyl-L-cysteine, *J.Liq.Chromatogr.*, **1995**, *18*, 2827–2841.

SAMPLE

Matrix: proteins
Analyte: amino acids
Sample preparation: Grind soybeans to a fine powder, add 6 M HCl, heat at 110° for 22 h, evaporate to dryness in a desiccator over NaOH pellets, reconstitute with EtOH:0.2 N pH 3.2 sodium citrate 7:93, inject an aliquot.

HPLC VARIABLES

Column: 150 × 4 ISC-07/S1504 Na-type (strongly acidic cation-exchange resin of styrene-divinylbenzene copolymer with 10% crosslinkage) (Shimadzu)
Mobile phase: Gradient. A was EtOH:0.2 N pH 3.2 sodium citrate 7:93. B was 0.6 N pH 10.0 sodium citrate. C was 200 mM NaOH. A:B:C from 100:0:0 to 88:12:0 over 20 min, to 40:60:0 (step gradient), to 0:100:0 over 15 min, maintain at 0:100:0 for 5 min, to 0:0:100 (step gradient), maintain at 0:0:100 for 5 min, re-equilibrate at initial conditions for 10 min. (Parameters are approximate.)
Column temperature: 50
Flow rate: 0.3
Detector: F ex 348 em 450 following post-column reaction. The column effluent mixed with the reagent solution pumped at 0.2 mL/min and the mixture flowed through a 200 × 0.5 stainless steel or PTFE coil at 55°. The effluent from the coil mixed with the fluorescence solution pumped at 0.2 mL/min and flowed through a 2 m × 0.5 mm stainless steel or PTFE coil at 55° to the detector. (Prepare reagent solution by adding 400 μL NaOCl solution (chlorine concentration 10%) to 1 L buffer, discard after 2 weeks. Prepare fluorescence solution by adding 15 mL EtOH containing 1.6 g o-phthalaldehyde and 2.0 g N-acetyl-L-cysteine and 4 mL 10% Brij 35 in water to 980 mL buffer, discard after 1 month. Buffer contained 384 mM sodium carbonate, 216 mM boric acid, and108 mM potassium sulfate, pH 10.0.)

CHROMATOGRAM
Retention time: 8 (Asp), 10 (Thr), 11 (Ser), 13 (Glu), 14 (Pro), 19 (Gly), 20 (Ala), 22 (Cys), 25 (Val), 28 (Met), 30 (Ile), 31 (Leu), 33 (Tyr), 35 (Phe), 40 (His), 45 (Lys), 48 (ammonia), 54 (Arg)
Limit of quantitation: 10 pmole

KEY WORDS
post-column reaction; soybeans

REFERENCE
Fujiwara, M.; Ishida, Y.; Nimura, N.; Toyama, A.; Kinoshita, T. Postcolumn fluorometric detection system for liquid chromatographic analysis of amino and imino acids using o-phthalaldehyde/N-acetyl-L-cysteine reagent, *Anal.Biochem.*, **1987**, *166*, 72–78.

SAMPLE
Matrix: solutions
Analyte: amino acids
Sample preparation: 35 μL Solution of amino acids in water + 35 μL 3.3 mg/mL o-phthalaldehyde in MeOH:water 50:50 + 35 μL 4 mg/mL N-acetyl-L-cysteine in MeOH:water 50:50 + 175 μL 400 mM pH 9.4 potassium borate buffer, mix well, let stand at room temperature for at least 2 min, neutralize with 140 μL 1 M pH 3.5 sodium phosphate buffer, inject an aliquot.

HPLC VARIABLES
Column: 250 × 4 5 μm Nucleosil-120-C18 or 125 × 4 3 μm Nucleosil-120-C18
Mobile phase: MeOH:buffer 40:60 (Buffer was 2.5 mM copper(II) acetate containing 5 mM L-proline adjusted to pH 6.0 with ammonium acetate.)
Column temperature: 40
Flow rate: 1
Injection volume: 20
Detector: F ex 338 (bandpass filter) em 415 (longpass filter)

CHROMATOGRAM
Retention time: k' 1.8 (L-valine), k' 2.4 (D-valine), k' 2.8 (L-leucine), k' 3.2 (D-leucine)

OTHER SUBSTANCES
Also analyzed: other amino acids, α-alkylamino acids, amino acid amides

KEY WORDS
chiral

REFERENCE
Duchateau, A.; Crombach, M.; Kamphuis, J.; Boesten, W.H.J.; Schoemaker, H.E.; Meojer, E.M. Determination of the enantiomers of α-H-α-amino acids, α-alkyl-α-amino acids and the corresponding acid amides by high-performance liquid chromatography, *J.Chromatogr.*, **1989**, *471*, 263–270.

SAMPLE
Matrix: solutions
Analyte: amino acids
Sample preparation: 35 μL Solution of amino acids in water + 35 μL 3.3 mg/mL o-phthalaldehyde in MeOH:water 50:50 + 35 μL 4 mg/mL N-acetyl-L-cysteine in MeOH:water 50:50 + 175 μL 400 mM pH 9.4 potassium borate buffer, mix well, let stand at room temperature for at least 2 min (J.Chromatogr. 1989, 471, 263), neutralize with trifluoroacetic acid, inject an aliquot.

HPLC VARIABLES
Column: 250 × 4 5 μm Nucleosil 120-C18
Mobile phase: MeOH:buffer 40:60 (Buffer 50 mM ammonium acetate adjusted to pH 6.0 with acetic acid.)
Flow rate: 1
Injection volume: 20
Detector: MS, Finnigan MAT TSQ 70 triple quadrupole, thermospray, first and second quadrupoles in rf-only mode, third quadrupole in scan and mass selective mode, the column effluent was mixed with 1% trifluoroacetic acid in water pumped at 0.25 mL/min,

first 5 mL of column effluent was diverted from the detector, source 200°, repeller 120 V, vaporizer 90°

KEY WORDS
chiral

REFERENCE
van Leuken, R.G.J.; Duchateau, A.L.L.; Kwakkenbos, G.T.C. Thermospray liquid chromatography/mass spectrometry study of diastereomeric isoindole derivatives of amino acids and amino acid amides, *J.Pharm.Biomed.Anal.*, **1995**, *13*, 1459–1464.

SAMPLE
Matrix: yogurt
Analyte: amino acids
Sample preparation: 15 g Yogurt + 45 mL MeOH:water 80:20, stir for 10 min, centrifuge at 1630 g. Remove the supernatant and evaporate it to about 10 mL under reduced pressure, add 10 mL of a saturated solution of picric acid, mix, centrifuge at 1630 g. Remove the supernatant and wash it twice with 20 mL portions of light petroleum (bp 40-60°): diethyl ether 50:50, add the aqueous phase to a 50 × 10 column of Dowex 50W-X8 cation-exchange resin, wash with water, elute with 30 mL 2 M aqueous ammonia, evaporate the eluate to dryness, reconstitute with 500 μL 133 mM pH 10.4 borate buffer. Remove a 4 μL aliquot and add it to 2 μL 5 mg/mL o-phthaldialdehyde in 133 mM pH 10.4 borate buffer, add 2 μL 8 mg/mL N-acetyl-L-cysteine in 133 mM pH 10.4 borate buffer, mix for 3 min, inject an aliquot.

HPLC VARIABLES
Guard column: 20 × 4.6 3 μm Spherisorb ODS II
Column: 125 × 4.6 3 μm Spherisorb ODS II
Mobile phase: Gradient. A was 40 mM pH 6.5 sodium acetate. B was MeCN. A:B from 100:0 to 82:18 over 1 h.
Column temperature: 26
Flow rate: 0.9
Detector: UV 338

CHROMATOGRAM
Retention time: 3.8 (D-Asp), 4.1 (L-Asp), 7.4 (L-Glu), 8.0 (D-Glu), 12.3 (L-Ser), 12.7 (L-Asn), 12.9 (D-Ser), 13.7 (D-Asn), 16.5 (L-Gln), 18.0 (D-Gln), 18.2 (D,L-Thr), 18.4 (Gly), 21.0 (L-His), 21.6 (D-His), 25.4 (L-Ala), 26.0 (D-Ala), 27.2 (L-Arg), 28.8 (D-Arg), 36.5 (L-Tyr), 38.9 (D-Tyr), 39.5 (L-Val), 43.6 (D-Val), 44.1 (L-Met), 45.1 (D-Met), 48.8 (L-Ile), 51.5 (L-Trp), 52.6 (D-Ile), 53.2 (D-Trp), 53.6 (D-Phe), 54.1 (L-Phe), 55.1 (L-Leu), 55.7 (D-Leu), 59.3 (L-Lys), 59.7 (D-Lys)

KEY WORDS
chiral; SPE; comparison with derivatization procedures using other thiols

REFERENCE
Brückner, H.; Wittner, R.; Godel, H. Automated enantioseparation of amino acids by derivatization with o-phthaldialdehyde and N-acylated cysteines, *J.Chromatogr.*, **1989**, *476*, 73–82.

Phthalaldehyde/Acetylpenicillamine

RELATED REFERENCE
Euerby, M.R.; Nunn, P.B.; Partridge, L.Z. Resolution of neuroexcitatory nonprotein amino acid enanti-omers by high-performance liquid chromatography utilizing precolumn derivatization with o-phthal-dialdehyde chiral thiols. Application to omega-N-oxalyl diamino acids. *J.Chromatogr.* **1989**, *466*, 407-414.

SAMPLE
Matrix: peptides
Analyte: amino acids
Sample preparation: Heat 0.5-5 mg peptide with 1 mL 1% indolylpropionic acid in 6 M HCl under vacuum in a sealed tube at 115° for 4 h, evaporate to dryness, reconstitute with 10-100 mL water, remove a 100 μL aliquot, add 400 μL reagent, mix, let stand for 10 min, inject a 25 μL aliquot. (Iodomethane is used to protect cysteine as its S-methyl derivative. Prepare reagent by dissolving 30 mg o-phthaldialdehyde in 1 mL EtOH, add 22 mL 400 mM pH 10 sodium borate buffer, add 30 mg N-acetyl-D-penicillamine.)

HPLC VARIABLES
Column: 120 × 4 3 μm Hypersil ODS
Mobile phase: Gradient. A was 50 mM pH 7.0 phosphate buffer. B was MeOH:THF:50 mM pH 7.0 phosphate buffer 65:1:35. A:B from 70:30 to 0:100 over 48 min.
Flow rate: 1.4
Injection volume: 25
Detector: F ex 344 em 443

CHROMATOGRAM
Retention time: 6.8 (L-Ser), 7.5 (D-Ser), 9 (L-Gln), 10 (D-Gln), 15 (D-Ala), 16 (L-Ala), 21 (D-Tyr), 22.5 (L-Tyr), 35 (D-Val), 36 (L-Val), 38 (D-Trp), 39 (L-Trp, D-Ile), 30 (L-Ile)

KEY WORDS
chiral; Comparison with the results obtained with other thiols in the derivatization reagent.

REFERENCE
Buck, R.H.; Krummen, K. High-performance liquid chromatographic determination of enantiomeric amino acids and amino alcohols after derivatization with o-phthaldialdehyde and various chiral mer-captans. Application to peptide hydrolysates, *J.Chromatogr.*, **1987**, *387*, 255–265.

Phthalaldehyde/Boc-L-cysteine

RELATED REFERENCES

Buck, R.H.; Krummen, K. Resolution of amino acid enantiomers by high-performance liquid chromatography using automated pre-column derivatization with a chiral reagent. *J.Chromatogr.* **1984**, *315*, 279-285.

Euerby, M.R.; Partridge, L.Z.; Rajani, P. Resolution of lombricine enantiomers by high-performance liquid chromatography utilizing pre-column derivatization with o-phthaldialdehyde-chiral thiols. *J.Chromatogr.* **1988**, *447*, 392-397.

Euerby, M.R.; Nunn, P.B.; Partridge, L.Z. Resolution of neuroexcitatory nonprotein amino acid enantiomers by high-performance liquid chromatography utilizing precolumn derivatization with o-phthaldialdehyde chiral thiols. Application to omega-N-oxalyl diamino acids. *J.Chromatogr.* **1989**, *466*, 407-414.

Hashimoto, A.; Nishikawa, T.; Oka, T.; Takahashi, K.; Hayashi, T. Determination of free amino acid enantiomers in rat brain and serum by high-performance liquid chromatography after derivatization with N-*tert.*-butyloxycarbonyl-L-cysteine and o-phthaldialdehyde. *J.Chromatogr.* **1992**, *582*, 41-48.

Hashimoto, A.; Kumashiro, S.; Nishikawa, T.; Oka, T.; Takahashi, K.; Mito, T.; Takashima, S.; Doi, N.; Mizutani, Y.; Yamazaki, T. Embryonic development and postnatal changes in free D-aspartate and D-serine in the human prefrontal cortex. *J.Neurochem.* **1993**, *61*, 348-351.

SAMPLE

Matrix: peptides

Analyte: amino acids

Sample preparation: Heat 0.5-5 mg peptide with 1 mL 1% indolylpropionic acid in 6 M HCl under vacuum in a sealed tube at 115° for 4 h, evaporate to dryness, reconstitute with 10-100 mL water, remove a 1 mL aliquot, add 100 μL 20 mM dithioerythritol, add 100 μL 400 mM iodomethane in MeOH:water 50:50, add 200 μL 3 M NaOH, let stand for 10 min, add 200 μL 3 M HCl, mix briefly, add 400 μL 400 mM pH 10 sodium borate buffer, mix, check that pH is about 10. Remove a 100 μL aliquot, add 400 μL reagent, mix, let stand for 10 min, inject a 25 μL aliquot. (Iodomethane is used to protect cysteine as its S-methyl derivative. Prepare reagent by dissolving 30 mg o-phthaldialdehyde in 1 mL EtOH, add 22 mL 400 mM pH 10 sodium borate buffer, add 30 mg Boc-L-cysteine. Prepare Boc-L-cysteine by adding Boc-S-benzyl-L-cysteine to liquid ammonia, add metallic sodium until a blue color persists for 15 min, add ammonium chloride until all the excess sodium is destroyed, allow the ammonia to evaporate, add ice and water, extract with ethyl acetate.)

HPLC VARIABLES
Column: 120 × 4 3 μm Hypersil ODS
Mobile phase: Gradient. A was 50 mM pH 7.0 phosphate buffer. B was MeOH:THF:50
mM pH 7.0 phosphate buffer 65:1:35. A:B from 70:30 to 0:100 over 48 min.
Flow rate: 1.4
Injection volume: 25
Detector: F ex 344 em 443

CHROMATOGRAM
Retention time: 20.5 (L-Thr), 22.5 (D-Thr), 28.5 (S-methyl-L-Cys), 31 (L-threoninol), 31.5
(S-methyl-D-Cys), 32 (D-threoninol), 34 (L-Trp), 36.5 (L-Phe), 37 (D-Trp), 38 (D-Phe), 47
(L-Lys), 48 (D-Lys)

KEY WORDS
chiral; Comparison with the results obtained with other thiols in the derivatization reagent.

REFERENCE
Buck, R.H.; Krummen, K. High-performance liquid chromatographic determination of enantiomeric
amino acids and amino alcohols after derivatization with o-phthaldialdehyde and various chiral mer-
captans. Application to peptide hydrolysates, *J.Chromatogr.*, **1987**, *387*, 255–265.

Phthalaldehyde/N,N-Dimethyl-2-mercaptoethylamine

SAMPLE
Matrix: tissue
Analyte: glutamic acid
Sample preparation: 10 g Homogenized meat + 2 mL 5 mg/mL homocysteic acid in water
+ 100 mL hot water, mix for 2 min, heat for 30 min, cool to room temperature, make up
to 200 mL with water, mix, filter. Remove a 1 mL aliquot of the filtrate and add it to 6
mL MeOH, mix, degas, add 1 mL buffer, add 500 μL reagent, mix, let stand for 2 min,
inject a 10 μL aliquot. (Prepare buffer by dissolving 12.4 g boric acid in 800 mL water,
adjust pH to 9.5 with 50% NaOH, make up to 1 mL with water, mix. Prepare reagent by
dissolving 100 mg o-phthalaldehyde and 100 mg N,N-dimethyl-2-mercaptoethylamine hydrochloride (N,N-dimethyl-2-mercaptoethylammonium chloride; Thiofluor; E. Merck;
Pickering Laboratories, Mountain View CA) in 4.5 mL MeOH, add 500 μL buffer. Prepare
fresh each day.)

HPLC VARIABLES
Guard column: 10 × 4.6 5 μm Lichrospher 100
Column: 125 × 4.6 5 μm Lichrospher 100

Mobile phase: MeCN:buffer:water 8:18:74 (Prepare buffer by dissolving 3.54 g KH_2PO_4 and 5.66 g Na_2HPO_4 in 800 mL water, adjust pH to 7.0 with 50% NaOH, make up to 1 L with water (pH 7.0).)
Flow rate: 1
Injection volume: 10
Detector: F ex 340 em 389 or 440

CHROMATOGRAM
Retention time: 6
Internal standard: homocysteic acid (8.5)

REFERENCE
Beljaars, P.R.; van Dijk, R.; Bisschop, E.; Spiegelenberg, W.M. Liquid chromatographic determination of free glutamic acid in soup, meat product, and Chinese food: Interlaboratory study, *J.AOAC Int.*, **1996**, *79*, 697–702.

Phthalaldehyde/Ethanethiol

RELATED REFERENCES
Beninati, S.; Martinet, N.; Folk, J.E. High-performance liquid chromatographic method for the determination of epsilon-(gamma-glutamyl)lysine and mono- and bis-gamma-glutamyl derivatives of putrescine and spermidine. *J.Chromatogr.* **1988**, *443*, 329-335.
Dubruc, C.; Hermann, P.; Haddouche, A.; Badarani, M.M.; Thénot, J.P. Determination of S-carboxymethyl-L-cysteine in plasma by high-performance liquid chromatography with column switching following precolumn derivatization. *J.Chromatogr.* **1987**, *417*, 208-215.
Eslami, M.; Stuart, J.D.; Hill, D.W. Separation of orthophthalaldehyde/ethanethiol derivatives of taurine and closely eluting amino acids by high performance liquid chromatography. *J.Liq.Chromatogr.* **1984**, *7*, 1117-1131.
Eslami, M.; Stuart, J.D.; Dean, R.W. Analysis of taurine in blood plasma of epileptic patients using an improved isocratic HPLC method for amino acids. *J.Liq.Chromatogr.* **1987**, *10*, 977-995.
Euerby, M.R.; Partridge, L.Z.; Gibbons, W.A. High-performance liquid chromatographic determination of lombricine and N-phosphoryllombricine in the earthworm by pre-column fluorescence derivatization with o-phthaldialdehyde-ethanethiol. *J.Chromatogr.* **1988**, *445*, 433-440.
Fleury, M.O.; Ashley, D.V. High-performance liquid chromatographic analysis of amino acids in physiological fluids: on-line precolumn derivatization with o-phthaldialdehyde. *Anal.Biochem.* **1983**, *133*, 330-335.
Fournier, B.; Gineyts, E.; Delmas, P.D. Evidence that free gamma carboxyglutamic acid circulates in serum. *Clin.Chim.Acta* **1989**, *182*, 173-181.
Hill, D.; Burnworth, L.; Skea, W.; Pfeifer, R. Quantitative HPLC analysis of plasma amino acids as ortho-phthaldialdehyde/ethanethiol derivatives. *J.Liq.Chromatogr.* **1982**, *5*, 2369-2393.
Krishnamurti, C.R.; Heindze, A.M.; Galzy, G. Application of reversed-phase high-performance liquid chromatography using pre-column derivatization with o-phthaldialdehyde for the quantitative analysis of amino acids in adult and fetal sheep plasma, animal feeds and tissues. *J.Chromatogr.* **1984**, *315*, 321-331.
Weissberger, L.E.; Armstrong, M.K. Canavanine analysis of alfalfa extracts by high performance liquid chromatography using pre-column derivatization. *J.Chromatogr.Sci.* **1984**, *22*, 438-440.

SAMPLE
Matrix: protein
Analyte: amino acids
Sample preparation: Mix 20 µL 1 mg/mL protein solution and 500 µL concentrated HCl, seal in a tube under vacuum, heat at 110° for 24 h, lyophilize, dissolve the residue in 50 µL water. Add a 2-5 µL aliquot to 2-aminoethanol, add buffer, add 2-5 µL reagent, let

stand at room temperature for 1-2 min, inject an aliquot. (Prepare reagent by dissolving 5 mg o-phthalaldehyde in 40 μL MeOH, add 5 μL ethanethiol, add 50 μL buffer. Protect from light, prepare fresh for each use. Buffer was 400 mM boric acid adjusted to pH 9.50 with 1 M KOH.)

CAPILLARY ELECTROPHORESIS
Capillary: 86 cm × 50 μm fused-silica (53 cm to detector) (Polymicro Technologies)
Capillary preparation: Flush with water for 30 min, flush with 1 M KOH for 30 min, flush with 100 mM KOH for 30 min, rinse thoroughly with water, equilibrate with running buffer.
Running buffer: MeOH:THF:buffer 15:2:83 (Buffer was 50 mM pH 9.50 borate buffer containing 50 mM sodium dodecyl sulfate.)
Voltage/Current: 23 kV
Injection: Hydrodynamic injection for 10 s
Detector: F ex 365 em 418 (cut-off filter)
Migration time: 16.9 (glutamine), 17.3 (threonine), 17.6 (serine), 18.6 (histidine), 19.4 (alanine), 19.7 (glycine), 20.4 (valine), 20.8 (gamma-aminobutyric acid), 21.4 (methionine), 21.8 (taurine), 23.0 (isoleucine), 23.9 (tryptophan), 25.2 (leucine), 27.6 (lysine), 30.2 (glutamic acid), 37.4 (arginine)
Internal standard: 2-aminoethanol (39.8)

REFERENCE
Liu, J.P.; Cobb, K.A.; Novotny, M. Separation of precolumn *ortho*-phthalaldehyde-derivatized amino acids by capillary zone electrophoresis with normal and micellar solutions in the presence of organic modifiers, *J.Chromatogr.*, **1989**, *468*, 55–65.

Phthalaldehyde/N-Isobutyryl-L-cysteine

RELATED REFERENCES
Brückner, H.; Langer, M.; Lüpke, M.; Westhauser, T.; Godel, H. Liquid chromatographic determination of amino acid enantiomers by derivatization with o-phthaldialdehyde and chiral thiols. Application to food science. *J.Chromatogr.A* **1995**, *697*, 229-245.
Carolan, V.A.; Gardner, M.L.G.; Lucy, D.; Pollard, A.M. Some considerations regarding the use of amino acid racemization in human dentine as an indicator of age at death. *J.Forensic Sci.* **1997**, *42*, 10-16.

SAMPLE
Matrix: beverages, food
Analyte: amino acids
Sample preparation: Dilute orange juice or soy sauce with water, filter (0.2 μm) if necessary. 2.2 μL Solution + 5 μL 0.4 N pH 10.4 sodium borate buffer + 1 μL reagent, mix for 2 min, inject a 7 μL aliquot. (Reagent was 260 mM N-isobutyryl-L-cysteine and 170 mM o-phthaldialdehyde in 1 M pH 10.4 potassium borate buffer (Pierce fluoraldehyde diluent).

HPLC VARIABLES
Guard column: 20 × 4 5 μm Hypersil ODS
Column: 250 × 4 5 μm Hypersil ODS
Mobile phase: Gradient. A was 23 mM sodium acetate adjusted to pH 6.00 with 10% acetic acid. B was MeOH:MeCN 60:5. A:B from 100:0 to 46.5:53.5 over 75 min, to 0:100 (step gradient), maintain at 0:100 for 10 min, re-equilibrate at initial conditions for 5 min.
Column temperature: 25
Flow rate: 1
Injection volume: 7
Detector: F ex 230 em 445

CHROMATOGRAM
Retention time: 20 (L-Asp), 21 (D-Asp), 27 (L-Glu), 28 (L-Asn), 28.5 (D-Glu), 29 (L-Ser), 30 (D-Asn), 32 (D-Ser), 33.5 (L-Gln), 35 (D-Gln), 36 (L-Thr), 37.5 (Gly), 38 (D-Thr), 39 (L-His), 40 (D-His), 43 (L-Ala), 44 (L-Arg), 45.5 (D-Arg), 46.5 (D-Ala), 48 (L-homo-Arg), 50 (L-Tyr), 52 (D-Tyr), 56.5 (L-Val), 57.5 (L-Met), 60 (L-Trp), 61 (D-Met), 62 (D-Val), 63 (L-Ile), 63.5 (L-Phe), 64.5 (D-Trp), 65.5 (D-Phe), 68 (L-Leu), 68.5 (D-Ile), 71 (D-Leu), 72.5 (L-Lys), 73.5 (D-Lys)
Limit of detection: 1 pmole

KEY WORDS
orange juice; soy sauce; chiral

REFERENCE
Brückner, H.; Wittner, R.; Godel, H. Fully automated high-performance liquid chromatographic separation of DL-amino acids derivatized with o-phthaldialdehyde together with N-isobutyryl-cysteine. Application to food samples, *Chromatographia*, **1991**, *32*, 383–388.

SAMPLE
Matrix: blood, juice, urine
Analyte: amino acids
Sample preparation: Serum, urine. 400 μL Serum or urine + 50 μL 30% 5-sulfosalicylic acid + 50 μL 0.52 mM L-homo-Arg, centrifuge at 6000 g. Remove a 2 μL aliquot and add it to 5 μL 0.4 N pH 10.4 sodium borate buffer and 1 μL reagent, mix for 2 min, inject a 7 μL aliquot. Juice. Centrifuge filtered (paper) apple juice at 1650 g, remove a 1 mL aliquot and add it to 31.3 μL 1.6 mM L-homo-Arg in 100 mM HCl, adjust pH to 2.0 with 2 M HCl, add to a 50 × 10 column of Dowex 50W-X8 cation-exchanger, wash with water, elute with 30 mL 4 M aqueous ammonia, evaporate eluate to dryness, dissolve the residue in 2 mL 100 mM HCl. Remove a 2 μL aliquot and add it to 5 μL 0.4 N pH 10.4 sodium borate buffer and 1 μL reagent, mix for 2 min, inject a 7 μL aliquot. (Reagent was 260 mM N-isobutyryl–L-cysteine and 170 mM o-phthaldialdehyde in 1 M pH 10.4 potassium borate buffer (fluoraldehyde, Pierce) (Chromatographia 1991, 32, 383).)

HPLC VARIABLES
Guard column: 20 × 2.1 5 μm Hypersil ODS
Column: 250 × 4 5 μm Hypersil ODS
Mobile phase: Gradient. A was 3.13 g sodium acetate trihydrate in 990 mL water adjusted to pH 5.95 with 10% acetic acid, make up to 1 L. B was MeCN:MeOH 50:600. A:B from 100:0 to 46.5:53.5 over 75 min, re-equilibrate at 100:0 for 10 min.
Column temperature: 25
Flow rate: 1
Injection volume: 7
Detector: F ex 230 em 445 (280 nm cut-off filter)

CHROMATOGRAM
Retention time: 18.89 (L-Asp), 20.09 (D-Asp), 25.87 (L-Glu), 27.54 (D-Glu), 26.95 (L-Asn), 29.30 (D-Asn), 28.46 (L-Ser), 30.61 (D-Ser), 32.41 (L-Gln), 34.01 (D-Gln), 34.59 (L-Thr), 36.78 (D-Thr), 36.21 (Gly), 37.76 (L-His), 39.15 (D-His), 41.95 (L-Ala), 45.27 (D-Ala), 43.29 (L-Arg), 44.66 (D-Arg), 48.29 (L-Tyr), 50.87 (D-Tyr), 55.36 (L-Val), 60.71 (D-Val), 56.30 (L-Met), 59.65 (D-Met), 58.53 (L-Trp), 63.08 (D-Trp), 61.80 (L-Phe), 64.39 (D-Phe), 62.33 (L-Ile), 67.56 (D-Ile), 66.45 (L-Leu), 69.87 (D-Leu), 71.41 (L-Lys), 72.69 (D-Lys)
Internal standard: L-homo-Arg (47.5)

KEY WORDS
serum; human; dog; apple; chiral

REFERENCE
Brückner, H.; Haasmann, S.; Langer, M.; Westhauser, T.; Wittner, R.; Godel, H. Liquid chromatographic determination of D- and L-amino acids by derivatization with o-phthaldialdehyde and chiral thiols Applications with reference to biosciences, *J.Chromatogr.A*, **1994**, *666*, 259–273.

SAMPLE
Matrix: formulations
Analyte: amino acids
Sample preparation: Dissolve formulations in 10-100 mM HCl, filter, analyze an aliquot. Peptides not containing Cys or Trp. Mix 500 μg peptide with 500 μL 6 M HCl, flush tube with nitrogen, heat at 110 ± 1° for 24 h, evaporate to dryness under a stream of nitrogen, reconstitute with 1 mL 100 mM HCl. Add a 2 μL aliquot to 5 μL 200 mM pH 10.4 borate buffer and 1 μL reagent, mix for 2 min, inject the whole amount. Peptides containing Trp. Hydrolyze 500 μg peptide with 500 μL 4 M methanesulfonic acid at 110° for 24 h, cool, add 500 μL water. Add a 2 μL aliquot to 5 μL 200 mM pH 10.4 borate buffer and 1 μL reagent, mix for 2 min, inject the whole amount. Peptides containing Cys. Dissolve 500 μg peptide in 1 mL formic acid, add 500 μL MeOH, cool to -10°, add 2.5 mL performic acid, let stand at -10° for 2.5 h, evaporate to dryness under reduced pressure, reconstitute with 500 μL 6 M HCl, flush tube with nitrogen, heat at 110 ± 1° for 24 h, evaporate to dryness under a stream of nitrogen, reconstitute with 1 mL 100 mM HCl. Add a 2 μL aliquot to 5 μL 200 mM pH 10.4 borate buffer and 1 μL reagent, mix for 2 min, inject the whole amount. (Prepare performic acid immediately before use by mixing 98% formic acid and 30% hydrogen peroxide in a 98:2 ratio, let stand for 2 h. Reagent was 260 mM N-isobutyryl-L-cysteine and 170 mM o-phthaldialdehyde in 1 M potassium borate buffer (Pierce fluoraldehyde diluent).)

HPLC VARIABLES
Guard column: 20 × 2.1 5 μm Hypersil ODS
Column: 250 × 4 5 μm Hypersil ODS
Mobile phase: Gradient. A was 23 mM pH 6.0 sodium acetate buffer. B was MeOH:MeCN 60:5. A:B from 100:0 to 46.5:53.5 over 75 min.
Column temperature: 25
Flow rate: 1
Injection volume: 8
Detector: F ex 230 em 445

CHROMATOGRAM
Retention time: 19 (L-Asp), 20 (D-Asp), 26 (L-Glu), 28 (D-Glu), 35 (L-Thr), 37 (Gly), 38 (L-His), 42 (L-Ala), 44 (L-Arg), 45 (D-Arg), 45.5 (D-Ala), 56 (L-Val), 57 (L-Met), 59 (L-Trp), 61 (D-Val), 62 (L-Phe), 63 (L-Ile), 65 (D-Phe), 67 (L-Leu), 70 (D-Leu), 72 (L-Lys)

KEY WORDS
capsules; tablets; pills; dragees; granulates; chiral

REFERENCE
Brückner, H.; Westhauser, T.; Godel, H. Liquid chromatographic determination of D- and L-amino acids by derivatization with o-phthaldialdehyde and N-isobutyryl-L-cysteine. Applications with reference to the analysis of peptidic antibiotics, toxins, drugs and pharmaceutically used amino acids, *J.Chromatogr.A*, **1995**, *711*, 201–215.

SAMPLE
Matrix: solutions

Analyte: amino acids

Sample preparation: Inject a solution of amino acids in 400 mM buffer for 10 s, inject reagent for 16 s, inject 40 mM buffer for 24 s. (Reagent was 60 mM phthalaldehyde in MeCN containing 120 mM N-isobutyryl-L-cysteine. Prepare 400 mM buffer by adjusting the pH of 400 mM boric acid to 9.5 with 10 M NaOH. Prepare 40 mM buffer by mixing 37.5 mL 40 mM borax solution with 10.6 mL 100 mM NaOH and making up to 100 mL with water, pH 9.50-9.55.)

CAPILLARY ELECTROPHORESIS

Capillary: 56 cm × 50 μm fused-silica (39 cm to detector) (Polymicro Technologies)

Capillary preparation: Between runs rinse capillary with 100 mM NaOH for 1 min, with water for 1 min, and with running buffer for 1 min. Equilibrate with voltage on for 2 min before injection. Rinse new capillaries with 1 M NaOH for 10 min and then with water.

Running buffer: MeCN:buffer 4:96 (Prepare running buffer by dissolving 1.12 moles sodium dodecyl sulfate in a small amount of borate buffer, adding 1 mL MeCN, and making up to 25 mL with borate buffer. Prepare borate buffer by mixing 37.5 mL 40 mM borax solution with 10.6 mL 100 mM NaOH and making up to 100 mL with water, pH 9.50-9.55.)

Voltage/Current: 15 kV/ca. 30 μA

Injection: Hydrodynamic injection at 12 mbar of a solution of amino acids in 400 mM buffer for 10 s, reagent for 16 s, and 40 mM buffer for 24 s.

Detector: UV 235; UV 340

Model: laboratory-constructed

Migration time: 9.4 (L-Ala), 9.7 (D-Thr), 10 (L-Met), 10.5 (L-Leu), 11 (D-Leu), 11.3 (L-Lys)

Limit of detection: 1-2 μM (UV 235), 7-8 μM (UV 340)

KEY WORDS

chiral; on-column derivatization; comparison with other reagents

REFERENCE

Tivesten, A.; Folestad, S. Chiral *o*-phthaldialdehyde reagents for fluorogenic on-column labeling of D- and L-amino acids in micellar electrokinetic chromatography, *Electrophoresis*, **1997**, *18*, 970−977.

Phthalaldehyde/2-Mercaptoethanol

RELATED REFERENCES

Alonso, E.; Rubio, V. Determination of *N*-acetyl-L-glutamate using high-performance liquid chromatography. *Anal.Biochem.* **1985**, *146*, 252-259.

Altmann, F. Determination of amino sugars and amino acids in glycoconjugates using precolumn derivatization with *o*-phthalaldehyde. *Anal.Biochem.* **1992**, *204*, 215-219.

Archer, T.E.; Stokes, J.D. Residue analysis of glyphosate in blackberries by high-performance liquid chromatography and postcolumn reaction detection. *J.Agric.Food Chem.* **1984**, *32*, 586-588.

Bobbin, R.P.; Fallon, M.; LeBlanc, C.; Baber, A. Evidence that glutathione is the unidentified amine (Unk 2.5) released by high potassium into cochlear fluids. *Hear.Res.* **1995**, *87*, 49-54.

Brückner, H.; Bosch, I.; Graser, T.; Fürst, P. Determination of α-alkyl-α-amino acids and alpha-amino alcohols by chiral-phase capillary gas chromatography and reverse-phase high-performance liquid chromatography. *J.Chromatogr.* **1987**, *395*, 569-590.

Canevari, L.; Vieira, R.; Aldegunde, M.; Dagani, F. High-performance liquid chromatographic separation with electrochemical detection of amino acids focusing on neurochemical application. *Anal.Biochem.* **1992**, *205*, 137-142.

Chen, B.-M.; Xia, L.-W.; Zhao, R.-Q. Determination of NG,NG-dimethylarginine in human plasma by high-performance liquid chromatography. *J.Chromatogr.B* **1997**, *692*, 467-471.

Colette, C.; Benmbarek, A.; Boniface, H.; Astre, C.; Pares-Herbute, N.; Monnier, L.; Guitter, J. Determination of protein-bound urinary gamma-carboxyglutamic acid in calcium nephrolithiasis. *Clin.Chim.Acta* **1991**, *204*, 43-50.

Cooper, J.D.H.; Shearsby, N.; Fook Sheung, C.T.C. Use of the ASTED system to determine L-N^G-monomethylarginine (546C88) in human plasma by pre-column *o*-phthalaldehyde derivatization and high-performance liquid chromatography. *J.Chromatogr.B* **1997**, *696*, 117-122.

Cooper, J.D.H.; Ogden, G.; McIntosh, J.; Turnell, D.C. The stability of the *o*-phthalaldehyde/2-mercaptoethanol derivatives of amino acids: an investigation using high-pressure liquid chromatography with a precolumn derivatization technique. *Anal.Biochem.* **1984**, *142*, 98-102.

Costa, M.; Pecci, I.; Pensa, B.; Fontana, M.; Cavallini, D. High-performance liquid chromatography of cystathionine, lanthionine and aminoethylcysteine using o-phthaldialdehyde precolumn derivatization. *J.Chromatogr.* **1989**, *490*, 404-410.

Farrant, M.; Zia-Gharib, F.; Webster, R.A. Automated pre-column derivatization with *o*-phthalaldehyde for the determination of neurotransmitter amino acids using reversed-phase liquid chromatography. *J.Chromatogr.* **1987**, *417*, 385-390.

Fekkes, D.; van Dalen, A.; Edelman, M.; Voskuilen, A. Validation of the determination of amino acids in plasma by high-performance liquid chromatography using automated pre-column derivatization with *o*-phthaldialdehyde. *J.Chromatogr.B* **1995**, *669*, 177-186.

Fermo, I.; De Vecchi, E.; Diomede, L.; Paroni, R. Serum amino acid analysis with pre-column derivatization: comparison of the o-phthaldialdehyde and N,N-diethyl-2,4-dinitro-5-fluoroaniline methods. *J.Chromatogr.* **1990**, *534*, 23-35.

Fermo, I.; Arcelloni, C.; De Vecchi, E.; Vigano, S.; Paroni, R. High-performance liquid chromatographic method with fluorescence detection for the determination of total homocyst(e)ine in plasma. *J.Chromatogr.* **1992**, *593*, 171-176.

Fiedler, H.-P.; Plaga, A. Separation of amino acids and antibiotics by narrow-bore and normal-bore high-performance liquid chromatography with pre-column derivatization. *J.Chromatogr.* **1987**, *386*, 229-241.

Fiore, G.B.; Nicchitta, C.V.; Ellington, W.R. High-performance liquid chromatographic separation and quantification of alanopine and strombine in crude tissue extracts. *Anal.Biochem.* **1984**, *139*, 413-417.

Fried, V.A.; Ando, M.E.; Bell, A.J. Protein quantitation at the picomole level: an o-phthaldialdehyde-preTSK column-derivatization assay. *Anal.Biochem.* **1985**, *146*, 271-276.

Gamache, P.; Ryan, E.; Svendsen, C.; Murayama, K.; Acworth, I.N. Simultaneous measurement of monoamines, metabolites and amino acids in brain tissue and microdialysis perfusates. *J.Chromatogr.* **1993**, *614*, 213-220.

Garras, A.; Djurhuus, R.; Christensen, B.; Lillehaug, J.R.; Ueland, P.M. A nonradioactive assay for N^5-methyltetrahydrofolate-homocysteine methyltransferase (methionine synthase) based on *o*-phthaldialdehyde derivatization of methionine and fluorescence detection. *Anal.Biochem.* **1991**, *199*, 112-118.

Georgi, G.; Pietsch, C.; Sawatzki, G. High-performance liquid chromatographic determination of amino acids in protein hydrolysates and in plasma using automated pre-column derivatization with *o*-phthaldialdehyde/2-mercaptoethanol. *J.Chromatogr.* **1993**, *613*, 35-42.

Goldsmith, R.F.; Earl, J.W.; Cunningham, A.M. Determination of δ-aminobutyric acid and other amino acids in cerebrospinal fluid of pediatric patients by reversed-phase liquid chromatography. *Clin.Chem.* **1987**, *33*, 1736-1740.

Griffin, M.; Leah, J.; Mould, N.; Compton, G. Construction of an ion-exchange amino acid analyser kit for use with high-performance liquid chromatography apparatus. *J.Chromatogr.* **1988**, *431*, 285-295.

Haginaka, J.; Wakai, J. Fluorimetric determination of amino acids by high-performance liquid chromatography using a hollow-fibre membrane reactor. *J.Chromatogr.* **1987**, *396*, 297-305.

Haginaka, J.; Wakai, J. Liquid chromatographic determination of amino acids using a hollow-fiber membrane reactor. *Anal.Biochem.* **1988**, *171*, 398-403.

Halawa, I.; Baig, S.; Qureshi, G.A. Use of high performance liquid chromatography in defining the abnormalities in the free amino acid patterns in the cerebrospinal fluid of patients with aseptic meningitis. *Biomed.Chromatogr.* **1991**, *5*, 216-220.

Haroon, Y. Rapid assay for gamma-carboxyglutamic acid in urine and bone by precolumn derivatization and reversed-phase liquid chromatography. *Anal.Biochem.* **1984**, *140*, 343-348.

Harsing, L.G., Jr.; Lajtha, A.; Vizi, E.S. A high performance liquid chromatography/electrochemical assay for glutamatergic neurotransmitters in the rat brain. *Biomed.Chromatogr.* **1989**, *3*, 183-185.

Hayashi, K.; Maeda, Y.; Toyomizu, M.; Tomita, Y. High-performance liquid chromatographic method for the analysis of N^{tau}-methylhistidine in food, chicken excreta, and rat urine. *J.Nutr.Sci.Vitaminol.(Tokyo)* **1987**, *33*, 151-156.

Ho, J.W. Micro assay for urinary δ-aminolevulinic acid and porphobilinogen by high-performance liquid chromatography with pre-column derivatization. *J.Chromatogr.* **1990**, *527*, 134-139.

Huebert, N.D.; Schartz, J.-J.; Haegele, K.D. Analysis of 2-difluoromethyl-DL-ornithine in human plasma, cerebrospinal fluid and urine by cation-exchange high-performance liquid chromatography. *J.Chromatogr.A* **1997**, *762*, 293-298.

Hyland, K.; Bottiglieri, T. Measurement of total plasma and cerebrospinal fluid homocysteine by fluorescence following high-performance liquid chromatography and precolumn derivatization with *o*-phthaldialdehyde. *J.Chromatogr.* **1992**, *579*, 55-62.

Ida, S.; Tanaka, Y.; Ohkuma, S.; Kuriyama, K. Determination of cystamine by high-performance liquid chromatography. *Anal.Biochem.* **1984**, *136*, 352-356.

Jones, B.N.; Gilligan, J.P. *o*-Phthaldialdehyde precolumn derivatization and reversed-phase high-performance liquid chromatography of polypeptide hydrolysates and physiological fluids. *J.Chromatogr.* **1983**, *266*, 471-482.

Joseph, M.H.; Davies, P. Electrochemical activity of o-phthalaldehyde-mercaptoethanol derivatives of amino acids. Application to high-performance liquid chromatographic determination of amino acids in plasma and other biological materials. *J.Chromatogr.* **1983**, *277*, 125-136.

Lada, M.W.; Kennedy, R.T. Quantitative in vivo monitoring of primary amines in rat caudate nucleus using microdialysis coupled by a flow-gated interface to capillary electrophoresis with laser-induced fluorescence detection. *Anal.Chem.* **1996**, *68*, 2790-2797.

Langguth, P.; Merkle, H.P.; Amidon, G.L. Oral absorption of peptides: the effect of absorption site and enzyme inhibition on the systemic availability of metkephamid. *Pharm.Res.* **1994**, *11*, 528-535.

Lee, B.L.; Chua, S.C.; Ong, H.Y.; Lee, H.P.; Ong, C.N. Determination of beta-amino-isobutyric acid in urine and serum using pre-column derivatization technique. *J.Liq.Chromatogr.* **1992**, *15*, 1351-1360.

Lewisch, S.A.; Levine, R.L. Determination of 2-oxohistidine by amino acid analysis. *Anal.Biochem.* **1995**, *231*, 440-446.

Maestri, L.; Ghittori, S.; Grignani, E.; Fiorentino, M.L.; Imbriani, M. Dosaggio di un metabolita del benzene l'acido S-fenilmercapturico urinario (S-PMA), nell'uomo, mediante HPLC [The measurement of a benzene metabolite, urinary S-phenylmercapturic acid (S-PMA), in man by HPLC]. *Med.Lav.* **1993**, *84*, 55-65.

Maestri, L.; Ghittori, S.; Imbriani, M. Determination of urinary mercapturic acids of styrene in man by high-performance liquid chromatography with fluorescence detection. *J.Chromatogr.B* **1996**, *687*, 387-394.

Marchand, D.H.; Reed, D.J. Identification of the reactive glutathione conjugate S-(2-chloro-ethyl)glutathione in the bile of 1-bromo-2-chloroethane-treated rats by high-pressure liquid chromatography and precolumn derivatization with o-phthalaldehyde. *Chem.Res.Toxicol.* **1989**, *2*, 449-454.

McClung, G.; Frankenberger, W.T., Jr. Comparison of reverse-phase high-performance liquid chromatographic methods for precolumn-derivatized amino acids. *J.Liq.Chromatogr.* **1988**, *11*, 613-646.

Merino Merino, I.; Blanco Gonzalez, E.; Sanz-Medel, A. Liquid chromatographic separation of penicillamine enantiomers derivatized with OPA/2-ME on a β-cyclodextrin bonded phase. *Mikrochim.Acta* **1992**, *107*, 73-80.

Min, S.; Yisheng, Y.; Lu, Y. Determination of urinary 3-methylhistidine by high-performance liquid chromatography with o-phthaldialdehyde precolumn derivatization. *J.Chromatogr.* **1992**,*581*, 272-276.

Minkler, P.E.; Ingalls, S.T.; Griffin, R.L.; Hoppel, C.L. Rapid high-performance liquid chromatography of 3-methylhistidine in human urine. *J.Chromatogr.* **1987**, *413*, 33-42.

Moller, S.E. Quantification of physiological amino acids by gradient ion-exchange high-performance liquid chromatography. *J.Chromatogr.* **1993**, *613*, 223-230.

Musson, D.G.; Maglietto, S.M.; Bayne, W.F. Determination of the antibiotic fludalanine in plasma and urine by high-performance liquid chromatography using a packed-bed, post-column reactor with o-phthalaldehyde and 2-mercaptoethanol. *J.Chromatogr.* **1985**, *338*, 357-367.

Nakazawa, H. Fluorimetric determination of rat growth hormone in pituitary cell culture by high-performance liquid chromatography with post-column derivatization. *J.Chromatogr.* **1987**, *417*, 409-413.

Orwar, O.; Folestad, S.; Einarsson, S.; Andine, P.; Sandberg, M. Automated determination of neuroactive acidic sulphur-containing amino acids and gamma-glutamyl peptides using liquid chromatography with fluorescence and electrochemical detection. *J.Chromatogr.* **1991**, *566*, 39-55.

Patrizio, M.; Gallo, V.; Levi, G. Measurement of amino acid release from cultured cerebellar granule cells by an improved high performance liquid chromatography procedure. *Neurochem.Res.* **1989**, *14*, 627-633.

Pettersson, A.; Uggla, L.; Backman, V. Determination of dimethylated arginines in human plasma by high-performance liquid chromatography. *J.Chromatogr.B* **1997**, *692*, 257-262.

Pronce, T.; Tilquin, B. Trace analysis in chiral separation of selected amino enantiomers. *J.Pharm.Biomed.Anal.* **1996**, *14*, 1175-1184.

Qureshi, G.A.; Van den Berg, S.; Gutierrez, A.; Bergström, J. Determination of histidine and 3-methylhistidine in physiological fluids by high-performance liquid chromatography. *J.Chromatogr.* **1984**, *297*, 83-89.

Qureshi, G.A.; Qureshi, A.R. Determination of free amino acids in biological samples: problems of quantitation. *J.Chromatogr.* **1989**, *491*, 281-289.

Qureshi, G.A.; Qureshi, A.R.; Bergström, J. Quantitation of free amino acids in plasma and muscle samples in healthy subjects and uremic patients by high-performance liquid chromatography and fluorescence detection. *J.Pharm.Biomed.Anal.* **1989**, *7*, 377-384.

Rajendra, W. High performance liquid chromatographic determination of amino acids in biological samples by precolumn derivatization with o-phthaldialdehyde. *J.Liq.Chromatogr.* **1987**, *10*, 941-955.

Rama Sastry, B.V.; Janson, V.E.; Horst, M.; Stephan, C.C. HPLC analysis of amino acids with ion exchange chromatography and o-phthalaldehyde post-column derivatization: Applications to the assay of endogenous free amino acids and their transport in human placental villus. *J.Liq.Chromatogr.* **1986**, *9*, 1689-1710.

Rigas, P.G.; Arvanitis, S.J.; Pietrzyk, D.J. Ion interaction chromatographic separation of amino acids using a basic-tetraalkylammonium salt mobile phase. *J.Liq.Chromatogr.* **1987**, *10*, 2891-2910.

Rizzo, V.; Anesi, A.; Montalbetti, L.; Bellantoni, G.; Trotti, R.; Melzi d'Eril, G.V. Reference values of neuroactive amino acids in the cerebrospinal fluid by high-performance liquid chromatography with electrochemical and fluorescence detection. *J.Chromatogr.A* **1996**, *729*, 181-188.

Roesel, R.A.; Blankenship, P.R.; Hommes, F.A. HPLC assay of phenylalanine and tyrosine in blood spots on filter paper. *Clin.Chim.Acta* **1986**, *156*, 91-96.

Schmidt, J.; McClain, C.J. Separation strategies for o-phthalaldehyde-mercaptoethanol derivatives of amino acids for reversed-phase high-performance liquid chromatography. *J.Chromatogr.* **1987**, *419*, 1-16.

Spann, K.P.; Hargreaves, P.A. The determination of glyphosate in soils with moderate to high clay content. *Pestic.Sci.* **1994**, *40*, 41-48.

Spink, D.C.; Swann, J.W.; Snead, O.C.; Waniewski, R.A.; Martin, D.L. Analysis of aspartate and glutamate in human cerebrospinal fluid by high-performance liquid chromatography with automated pre-column derivatization. *Anal.Biochem.* **1986**, *158*, 79-86.

Suh, B.; Lee, H.W.; Hong, S.Y.; Kim, S.; Eshraghi, J.; Paik, W.K. A new HPLC analytical method for o-hydroxyhippuric acid in uremic serum. *J.Biochem.Biophys.Methods* **1986**, *13*, 211-220.

Sundaram, K.M.S.; Curry, J. A comparison of UV and fluorescence detectors in the liquid chromatographic analysis of glyphosate deposits after post-column derivatization. *J.Liq.Chromatogr.Rel.Technol.* **1997**, *20*, 511-524.

Swanepoel, E.; de Villiers, M.M.; Du Preez, J.L. Fluorimetric method of analysis for D-norpseudoephedrine hydrochloride, glycine and L-glutamic acid by reversed-phase high-performance liquid chromatography. *J.Chromatogr.A* **1996**, *729*, 287-291.

Turnell, D.C.; Cooper, J.D. Rapid assay for amino acids in serum or urine by pre-column derivatization and reversed-phase liquid chromatography. *Clin.Chem.* **1982**, *28*, 527-531.

Unnithan, S.; Moraga, D.A.; Schuster, S.M. A high-performance liquid chromatography assay for asparagine synthetase. *Anal.Biochem.* **1984**, *136*, 195-201.

van den Oetelaar, P.J.M.; van Beckhoven, J.R.C.M.; Hoenders, H.J. Analysis of aspartic acid racemization. Evaluation of a chiral capillary gas chromatographic and a diastereomeric high-performance-liquid chromatographic method. *J.Chromatogr.* **1987**, *388*, 441-447.

van Eijk, H.M.H.; Rooyakkers, D.R.; Wagenmakers, A.J.M.; Soeters, P.B.; Deutz, N.E.P. Isolation and quantitation of isotopically labeled amino acids from biological samples. *J.Chromatogr.B* **1997**, *691*, 287-296.

Vázquez-Ortiz, F.A.; Caire, G.; Higuera-Ciapara, I.; Hernández, G. High performance liquid chromatographic determination of free amino acids in shrimp. *J.Liq.Chromatogr.* **1995**, *18*, 2059-2068.

Venema, K.; Leever, W.; Bakker, J.O.; Haayer, G.; Korf, J. Automated precolumn derivatization device to determine neurotransmitter and other amino acids by reversed-phase high-performance liquid chromatography. *J.Chromatogr.* **1983**, *260*, 371-376.

Venkatakrishnan, A.; Abel, M.J.; Campbell, R.A.; Donahue, E.P.; Uselton, T.C.; Flakoll, P.J. Whole blood analysis of gluconeogenic amino acids for estimation of de novo gluconeogenesis using pre-column o-phthalaldehyde-derivatization and high-performance liquid chromatography. *J.Chromatogr.B* **1996**, *676*, 1-6.

Wagner, J.; Claverie, N.; Danzin, C. A rapid high-performance liquid chromatographic procedure for the simultaneous determination of methionine, ethionine, S-adenosylmethionine, S-adenosylethionine, and the natural polyamines in rat tissues. *Anal.Biochem.* **1984**, *140*, 108-116.

Walker, T.A.; Pietrzyk, D.J. Separation of free amino acids on reverse stationary phases using an alkyl sulfonate salt as a mobile phase additive. *J.Liq.Chromatogr.* **1985**, *8*, 2047-2079.

Walker, T.A.; Pietrzyk, D.J. Ion-interaction chromatographic separation of free amino acids. *J.Liq.Chromatogr.* **1987**, *10*, 161-174.

White, R.L.; DeMarco, A.C.; Shapiro, S.; Vining, L.C.; Wolfe, S. Measurement of δ-(L-α-aminoadipyl)-L-cysteinyl-D-valine synthetase activity in *Streptomyces clavuligerus* by high-performance liquid chromatography after precolumn derivatization with o-phthaldialdehyde. *Anal.Biochem.* **1989**, *178*, 399-403.

Yokoyama, Y.; Ozaki, O.; Sato, H. Separation and determination of amino acids, creatinine, bioactive amines and nucleic acid bases by dual-mode gradient ion-pair chromatography. *J.Chromatogr.A* **1996**, *739*, 333-342.

Yoshinaga, K.; Kobayashi, N.; Nagatani, Y.; Tanaka, Y.; Ikeda, Y. A sensitive detection method for peptide using 4-fluoro-7-nitrobenzo-2-oxa-1,3-diazole and its application to measure prolyl endopeptidase activity. *Biomed.Chromatogr.* **1994**, *8*, 297-300.

Zängerle, L.; Cuénod, M.; Winterhalter, K.H.; Do, K.Q. Screening of thiol compounds: depolarization-induced release of glutathione and cysteine from rat brain slices. *J.Neurochem.* **1992**, *59*, 181-189.

Zhao, Q.; Wada, H. On-line precolumn derivatization method for determination of amino acids with o-phthalaldehyde and 2-mercaptoethanol. *Anal.Sci.* **1989**, *5*, 487-488.

Zheng, L.; Chen, J.; Zhu, Y.; Yang, H.; Elmquist, W.; Hu, M. Comparison of the transport characteristics of D- and L-methionine in a human intestinal epithelial model (Caco-2) and in a perfused ratintestinal model. *Pharm.Res.* **1994**, *11*, 1771-1776.

SAMPLE
Matrix: amniotic fluid
Analyte: amino acids
Sample preparation: 200 µL Amniotic fluid + 800 µL MeOH, mix, centrifuge. 200 µL Supernatant + 80 µL pH 9.5 sodium borate + 60 µL reagent, mix, let stand for 3.5 min, add 25 µL 0.5 M HCl, mix, dilute 1:4 with 50 mM pH 7.0 sodium acetate buffer, inject a 20 µL aliquot. (Prepare reagent by dissolving 50 mg o-phthaldialdehyde in 4.5 mL MeOH, add 500 µL pH 9.5 sodium borate, add 50 µL 2-mercaptoethanol.)

HPLC VARIABLES
Guard column: 10-20 × 4 C18
Column: 300 × 3.9 5 µm NovaPak C18
Mobile phase: Gradient. MeOH:50 mM pH 7.0 sodium acetate buffer from 15:85 to 20:80 over 30 min, to 35:65 over 15 min, to 75:25 over 25 min, maintain at 75:25 for 5 min, return to initial conditions over 5 min.
Column temperature: 40
Flow rate: 1
Injection volume: 20
Detector: F ex 330 em 450

CHROMATOGRAM
Retention time: 3 (glutathione), 5 (aspartic acid), 8.5 (glutatic acid), 14 (saccharopine), 17 (2-aminoadipic acid), 20 (asparagine), 23 (serine), 31 (glutamine), 35 (histidine), 37 (homoserine), 40 (glycine), 42.5 (threonine), 44 (citruline), 46 (arginine), 50 (taurine), 52 (alanine), 53.5 (4-aminobutyric acid), 54.5 (tyrosine), 57 (3-aminobutyric acid), 59 (2-aminobutyric acid), 61.5 (tryptophan), 62 (methionine), 63 (valine), 64 (phenylalanine), 66 (isoleucine), 67 (leucine), 68 (5-hydroxylysine), 70 (ornithine), 72 (lysine)

REFERENCE
Klein, B.H.; Dudenhausen, J.W. Ion-exchange chromatography and ion-pair chromatography. Complementation of HPLC analysis of amino acids in body fluids by pre-column derivatization using ortho-phthaldialdehyde, *J.Liq.Chromatogr.*, **1995**, *18*, 4007–4028.

SAMPLE
Matrix: blood
Analyte: amino acids
Sample preparation: 100 µL Plasma + 5 µL 2.5 M homoserine, make up to 500 µL with 2 µL/mL 2-mercaptoethanol in MeCN, vortex, centrifuge for 4 min (Beckman microfuge). Remove a 40 µL aliquot of the supernatant and add it to 40 µL reagent and 20 µL 3.7% iodoacetic acid in 400 mM pH 9.5 sodium borate buffer, mix, let stand for 1 min, make up to 200 µL with 100 mM pH 4 potassium phosphate buffer, mix, inject a 20 µL aliquot. (Reagent was 50 mg o-phthaldialdehyde in 1 mL MeOH added to 11 mL 400 mM pH 9.5 sodium borate buffer, 50 µL 2-mercaptoethanol, and 10 mg nitrilotriacetic acid. Filter (0.2

µm), store in the dark at 4°, add 20 µL 2-mercaptoethanol each week to maintain the level of this reagent.)

HPLC VARIABLES
Guard column: 5 µm LiChrospher 100 RP-18
Column: 150 × 4.6 5 µm Dynamax Microsorb C18 (Rainin)
Mobile phase: Gradient. A was MeOH:100 mM pH 6.8 sodium acetate buffer 95:5. B was MeOH:100 mM pH 6.8 sodium acetate buffer 5:95. A:B from 15:85 to 30:70 over 15.5 min, to 55:45 over 9 min, to 60:40 over 2 min, to 100:0 over 8 min, maintain at 100:0 for 3 min, to 0:100 over 4 min, maintain at 0:100 for 3 min, to 15:85 over 1 min, stay at 15:85 for 2 min.
Column temperature: 35
Flow rate: 1.5 for 37.5 min, 1 for 10 min
Injection volume: 20
Detector: F ex 338 em 425

CHROMATOGRAM
Retention time: 6.37 (aspartic acid), 7.72 (glutamic acid), 10.81 (asparagine), 12.25 (serine), 14.20 (glutamine), 15.11 (histidine), 17.75 (glycine), 18.56 (threonine), 20.83 (arginine), 22.60 (taurine), 24.58 (alanine), 25.88 (tyrosine), 31.25 (tryptophan), 31.41 (methionine), 31.90 (valine), 32.75 (phenylalanine), 34.31 (isoleucine), 34.88 (leucine), 36.92 (lysine)
Internal standard: homoserine (16.31)
Limit of quantitation: 31 µM

KEY WORDS
plasma

REFERENCE
Uhe, A.M.; Collier, G.R.; McLennan, E.A.; Tucker, D.J.; O'Dea, K. Quantitation of tryptophan and other plasma amino acids by automated pre-column o-phthaldialdehyde derivatization high-performance liquid chromatography: improved sample preparation, *J.Chromatogr.*, **1991**, *564*, 81–91.

SAMPLE
Matrix: blood
Analyte: N^G-monomethyl-L-arginine
Sample preparation: 50 µL Serum + 20 µL 3% 5-sulfosalicylic acid + 250 µL MeOH, vortex vigorously, centrifuge at 4000 rpm for 10 min. Remove a 20 µL aliquot of the supernatant and add it to 100 µL reagent, mix for 3 min, inject a 25 µL aliquot. (Prepare reagent by mixing 5 mL o-phthalaldehyde and 200 µL 2-mercaptoethanol and making up to 100 mL with 10 mM pH 9.2 K_2HPO_4 buffer.)

HPLC VARIABLES
Column: 75 × 4.6 3 µm Ultrasphere C18
Mobile phase: Gradient. A was MeOH:10 mM K_2HPO_4, pH adjusted to 7.9 with phosphoric acid. B was 10 mM K_2HPO_4 adjusted to pH 6.9 with phosphoric acid.) A:B 70:30 for 15 min, to 10:90 over 1 min, maintain at 10:90 for 5 min, to 70:30 over 1 min, maintain at 70:30 for 3 min.
Flow rate: 1.8
Injection volume: 25
Detector: F ex 340 em 450

CHROMATOGRAM
Retention time: 9
Limit of quantitation: 50 ng/mL

KEY WORDS
serum; human; dog; pharmacokinetics

REFERENCE
Alak, A.M.; Moy, S. Determination of N^G-monomethyl-L-arginine in human and dog serum using pre-column o-phthaldialdehyde derivatization and high performance liquid chromatography, *J.Liq.Chromatogr.Rel.Technol.*, **1997**, *20*, 1839–1848.

SAMPLE
Matrix: blood, CSF

Analyte: amino acids
Sample preparation: Plasma. For each volume of plasma add 4 volumes of MeOH, centrifuge at 11600 g for 5 min. Remove a 10 μL aliquot and add it to 5 μL phthaldialdehyde/2-mercaptoethanol derivatizing reagent (Fluoraldehyde, Pierce) (use fresh reagent), allow to react at room temperature for 1 min, add 100 μL THF:100 mM sodium acetate 5:95 adjusted to pH 7.2 with glacial acetic acid, inject a 20 μL aliquot. CSF. Add an equal volume of MeOH to the CSF, centrifuge at 11600 g for 5 min. Remove a 10 μL aliquot and add it to 5 μL phthaldialdehyde, allow to react at room temperature for 1 min, add 100 μL THF:100 mM sodium acetate 5:95 adjusted to pH 7.2 with glacial acetic acid, inject a 20 μL aliquot.

HPLC VARIABLES
Guard column: 10 × 3 Spherisorb 5 ODS
Column: 50 × 4.6 Spherisorb 5 ODS
Mobile phase: Gradient. A was THF:100 mM sodium acetate 5:95 adjusted to pH 7.2 with glacial acetic acid. B was MeOH:THF 95:5. A:B from 10:90 to 0:100 over 13 min (sic), maintain at 0:100 for 4 min, return to initial conditions over 1 min.
Column temperature: 43
Flow rate: 1.5
Injection volume: 20
Detector: F (wavelengths not specified)

CHROMATOGRAM
Retention time: 0.5 (Asp), 0.7 (Glu), 1.2 (Tau), 1.5 (Ser), 1.8 (Gln), 2.2 (His), 2.3 (Gly), 3.0 (Thr), 3.1 (Asn), 3.2 (Ala), 3.3 (Arg), 3.8 (Tyr), 5.2 (Met), 5.4 (Val), 5.8 (Trp), 6.0 (Phe), 5.5 (Ile), 5.7 (Leu), 8.3 (Orn), 8.5 (Lys)
Limit of detection: 10 nM

KEY WORDS
plasma; Thr, Gly, His co-elute

REFERENCE
Begley, D.J.; Reichel, A.; Ermisch, A. Simple high-performance liquid chromatographic analysis of free primary amino acid concentrations in rat plasma and cisternal cerebrospinal fluid, *J.Chromatogr.B*, **1994**, *657*, 185–191.

SAMPLE
Matrix: blood, tissue
Analyte: amino acids
Sample preparation: Plasma. Mix 9 volumes of plasma with 1 volume of 35% 5-sulfosalicylic acid, centrifuge at 2000 g for 10 min. Neutralize the supernatant with 10 M KOH, dilute with 2 volumes of water. Mix an aliquot with an equal volume of reagent, inject a 20 μL aliquot within 1 min. Tissue. Homogenize tissue with four volumes 5% 5-sulfosalicylic acid, centrifuge at 5000 g for 10 min, neutralize the supernatant with 10 M KOH. Mix an aliquot with an equal volume of reagent, inject a 20 μL aliquot within 1 min. (Prepare reagent each day by dissolving 35 mg o-phthalaldehyde in 500 μL 95% EtOH and adding this mixture to 50 mL 100 mM pH 10.4 borate buffer, add 100 μL 2-mercaptoethanol.)

HPLC VARIABLES
Guard column: 37-50 μm Bondapak C18/Corasil
Column: 150 × 3.9 4 μm Nova-Pak C18
Mobile phase: Gradient. A was THF:water 3:97 containing 100 mM potassium phosphate, pH 7.0. B was THF:MeCN:water 3:40:57 containing 100 mM potassium phosphate, pH 7.0. A:B 97:3 for 1.5 min, to 68:32 over 17 min (Waters curve profile 3), to 0:100 over 2 min, maintain at 0:100 for 4.5 min, return to initial conditions over 2 min, re-equilibrate for 8 min.
Column temperature: 41
Flow rate: 1
Injection volume: 20
Detector: F ex 360 em 455

CHROMATOGRAM
Retention time: 3.5 (aspartate, cysteate), 5 (cysteinesulfinate), 6 (glutamate), 10.5 (serine), 11 (glutamine), 12.5 (arginine), 13.5 (glycine), 15 (threonine, glycerophosphorylethanolamine), 15.3 (O-phosphorylethanolamine), 18.5 (alanine), 20.5 (hypotaurine), 21.5 (taurine), 22 (β- and gamma-aminobutyrate), 25 (tyrosine), 25.5 (α-aminobutyrate), 26 (methionine), 29 (ethanolamine)

KEY WORDS
plasma; human; rat; liver; kidney; heart; brain

REFERENCE
Hirschberger, L.L.; De La Rosa, J.; Stipanuk, M.H. Determination of cysteinesulfinate, hypotaurine and taurine in physiological samples by reversed-phase high-performance liquid chromatography, *J.Chromatogr.*, **1985**, *343*, 303–313.

SAMPLE
Matrix: blood, urine
Analyte: cilastatin
Sample preparation: Plasma. 500 μL Plasma + 75 μL 30 μg/mL IS + 2 mL 500 mM pH 3 KH_2PO_4, vortex, add to an activated Sep-Pak C18 SPE cartridge, wash with 20 mL 1 mM orthophosphoric acid, elute with 1.5 mL MeOH, add the eluate to 1 mL water, vortex, inject a 50-200 μL aliquot. Urine. Stabilize urine by mixing with an equal volume of 1 M pH 6.8 MOPS buffer:ethylene glycol 50:50. 1 mL Stabilized urine + 50 μL 100 μg/mL IS + 2.5 mL 20 mM orthophosphoric acid, vortex, add to an activated Sep-Pak C18 SPE cartridge, wash with 20 mL 1 mM orthophosphoric acid, elute with 1.5 mL MeOH, add the eluate to 1 mL water, vortex, inject a 25-75 μL aliquot.

HPLC VARIABLES
Guard column: 40 × 4.6 ODS-10 (Bio-Rad)
Column: 250 × 4.6 Bio-Sil ODS-10 (Bio-Rad)
Mobile phase: Isopropanol:0.2% orthophosphoric acid 10.9:89.1, pH 3 (plasma) or isopropanol:0.2% orthophosphoric acid 6:94, pH 3 (urine)
Flow rate: 2
Injection volume: 25-200
Detector: F ex 335 em 455 following post-column reaction with o-phthalaldehyde reagent solution (Pierce) pumped at 1 mL/min. The mixture flowed through a 250 × 4.6 column packed with 40 μm glass beads (Whatman) to the detector.

CHROMATOGRAM
Retention time: 5.04 (plasma), 7.95 (urine)
Internal standard: S-(p-methylbenzyl)-L-cysteine (8.04 (plasma), 12.01 (urine))
Limit of detection: 750 ng/mL (plasma), 2.5 μg/mL (urine)

OTHER SUBSTANCES
Non-interfering: metabolites, imipenem

KEY WORDS
plasma; post-column reaction; SPE

REFERENCE
Demetriades, J.L.; Souder, P.R.; Entwistle, L.A.; Vincek, W.C.; Musson, D.G.; Bayne, W.F. High-performance liquid chromatographic determination of cilastatin in biological fluids, *J.Chromatogr.*, **1986**, *382*, 225–231.

SAMPLE
Matrix: bulk
Analyte: amino acids
Sample preparation: Prepare a 5 mg/mL solution in 1 M HCl or MeOH:water 25:75 depending on solubility, inject a 20 μL aliquot.

HPLC VARIABLES
Column: 300 mm long μBondapak C18
Mobile phase: 8 mM (S)-Proline containing 4 mM cupric acetate, adjusted to pH 5 with NaOH
Flow rate: 3

Injection volume: 20
Detector: F ex 365 em 455 following post-column reaction. The column effluent mixed with the reagent pumped at 1.5 mL/min and this mixture flowed to the detector. (Prepare the reagent by adding 7.5 mL 1% o-phthalaldehyde in EtOH:water 95:5, 1.8 mL Brij-35 surfactant, 200 μL 2-mercaptoethanol, and 570 mg EDTA to 285 mL 50 mM pH 9.5 boric acid buffer, adjust pH to 10-11 with NaOH.)

CHROMATOGRAM
Retention time: 4.54 (R-valine), 10.03 (S-valine), 11.00 (R-tyrosine), 20.24 (S-tyrosine)
Limit of detection: 0.01% (of major enantiomer)

KEY WORDS
chiral; post-column reaction

REFERENCE
Cotter, M.L.; Naldi, R.; Shaw, C.; Park, S.; Heavner, G.A. Detection and quantitation of low levels of protected and unprotected (*R*)-amino acids in the synthesis of thymopentin, an immunoregulatory peptide, *J.Pharm.Sci.*, **1985**, *74*, 489−491.

SAMPLE
Matrix: dialysate, tissue
Analyte: amino acids
Sample preparation: Homogenize (Kontes micro-ultrasonic cell disrupter) rat brain with 100 μL 50 mM ice-cold perchloric acid and 10 ng homoserine for 5 s, centrifuge at 4° at 13000 g for 5 min, filter (0.2 μm) the supernatant. Mix 25 μL of the filtrate from the tissue or dialysate (Ringer's) with 50 (tissue) or 12.5 (dialysate) μL working reagent, let stand for 2 min, inject an aliquot. (Prepare the reagent stock solution by dissolving 27 mg o-phthalaldehyde in 1 mL MeOH, add 5 μL 2-mercaptoethanol, add 9 mL 100 mM pH 9.3 sodium tetraborate, discard after 5 days. Prepare the working reagent by diluting 1 mL stock solution with 3 mL 100 mM sodium tetraborate, let stand for 24 h before use.)

HPLC VARIABLES
Column: 80 × 4.6 3 μm C18 HR-80 (ESA)
Mobile phase: MeOH:water 28:72 containing 100 mM Na_2HPO_4 and 0.13 mM disodium EDTA adjusted to pH 6.00 (tissue) or pH 6.40 (dialysate) with phosphoric acid. (Prepare by dissolving 14.2 g Na_2HPO_4 and 50 mg disodium EDTA in 720 mL water, add 280 mL MeOH, adjust pH. Recycle mobile phase.)
Flow rate: 1.2
Injection volume: 20
Detector: E, ESA Model 5100A coulometric, model 5011 dual electrode analytical cell preceded by a 0.2 μm carbon filter at -0.4 V and +0.6 V

CHROMATOGRAM
Retention time: 1.5 (Asp), 2 (Glu), 3 (Ser), 4 (Gln), 5.5 (Gly, Thr), 7 (phenylethanolamine), 9 (taurine), 10.5 (Ala), 14 (Tyr), 15 (GABA)
Internal standard: homoserine (3.5)
Limit of detection: 100-200 pg

KEY WORDS
rat; brain

REFERENCE
Donzanti, B.A.; Yamamoto, B.K. An improved and rapid HPLC-EC method for the isocratic separation of amino acid neurotransmitters from brain tissue and microdialysis perfusates, *Life Sci.*, **1988**, *43*, 913−922.

SAMPLE
Matrix: perfusate
Analyte: amino acids
Sample preparation: 30 μL Perfusate (artificial CSF) + 10 μL 200 mM perchloric acid. Mix a 25 μL aliquot with 12.5 μL reagent, let stand for 2 min, inject an aliquot. (Prepare a stock solution by dissolving 27 mg o-phthalaldehyde in 1 mL MeOH, add 5 μL 2-mercaptoethanol, add 9 mL 100 mM pH 9.3 sodium tetraborate containing 10 μM EDTA. This solution is good for 5 days in a sealed amber bottle at room temperature. Prepare

the working reagent by diluting 1 mL of the stock solution with 3 mL 100 mM pH 9.3 sodium tetraborate containing 10 μM EDTA, allow to stand for 24 h before use.)

HPLC VARIABLES
Column: two columns 150 × 4.6 5 μm M.S. Gel C18 (ESA)
Mobile phase: MeCN:MeOH:139 mM Na$_2$HPO$_4$ 3.1:25:71.9 adjusted to pH 6.8 with phosphoric acid
Column temperature: 33
Flow rate: 1.2
Detector: E, ESA Coulochem Electrode Array System Model 5500, detector temp 33°, oxidation potential 450 mV

CHROMATOGRAM
Retention time: 5.56 (Asp), 6.28 (Glu), 7.50 (Asn), 8.48 (His), 8.76 (Ser), 9.22 (Gln), 10.70 (Arg), 12.75 (Gly), 13.40 (Thr), 18.99 (Tau), 21.69 (Ala), 24.12 (Gaba), 25.02 (Tyr)
Limit of detection: 0.75 ng/mL

KEY WORDS
rat; pharmacokinetics

REFERENCE
Acworth, I.N.; Yu, J.; Ryan, E.; Gariepy, K.C.; Gamache, P.; Hull, K.; Maher, T. Simultaneous measurement of monoamine, amino acid, and drug levels, using high performance liquid chromatography and coulometric array technology: application to in vivo microdialysis perfusate analysis, *J.Liq.Chromatogr.*, **1994**, *17*, 685–705.

SAMPLE
Matrix: plant exudate
Analyte: amino acids
Sample preparation: 1 μL Plant exudate + 89 μL 100 mM sodium tetraborate + 10 μL reagent, vortex, inject an aliquot within 2 min. (Reagent was 60 mM o-phthalaldehyde in MeCN containing 120 mM 2-mercaptoethanol.)

CAPILLARY ELECTROPHORESIS
Capillary: 95 cm × 50 μm fused-silica (80 cm to detector) (CS-Chromatographie, Langerwehe, Germany)
Running buffer: MeCN:buffer 3:97 (Buffer was 27 mM pH 9.4 borate buffer containing 44 mM sodium dodecyl sulfate.)
Injection: Hydrodynamic injection for 1 s (2.4 nL).
Detector: F ex 325 (3.5 mW He-Cd laser) em >389
Model: SpectraPhoresis 100
Migration time: 17 (Gln), 17.5 (Asn, Thr), 17.8 (Ser), 18.5 (Ala), 19.2 (Val), 19.5 (Met), 21.5 (Ile), 23 (Leu), 27 (Glu), 28 (Asp), 38.5 (Lys)

REFERENCE
Bazzanella, A.; Lochmann, H.; Mainka, A.; Bächmann, K. Determination of inorganic anions, carboxylic acids and amino acids in plant matrices by capillary zone electrophoresis, *Chromatographia*, **1997**, *45*, 59–62.

SAMPLE
Matrix: tissue culture
Analyte: amino acids
Sample preparation: Dilute tissue culture 50-fold with initial mobile phase, filter (0.45 μm), remove a 10 μL aliquot of the filtrate and add it to 20 μL reagent, mix, inject. (Prepare reagent by mixing 12 mL Fluoraldehyde (Pierce) with 10 μL 2-mercaptoethanol. Fluoraldehyde contains o-phthalaldehyde, mercaptoethanol, and Brij-35.)

HPLC VARIABLES
Guard column: C18
Column: Resolve C18 (Waters)
Mobile phase: Gradient. A was MeOH:THF:50 mM pH 7.5 sodium acetate buffer containing 50 mM sodium phosphate 2:2:96. B was MeOH:water 65:35. A:B from 100:0 to 0:100 over 47 min, maintain at 0:100 for 15 min, re-equilibrate at initial conditions for 30 min.

Flow rate: 1.5
Detector: F ex 334 (filter) em 425 (filter)

CHROMATOGRAM
Retention time: 4.6 (Asp), 8.3 (Glu), 17.6 (Ser), 20.3 (Gln), 20.5 (His), 23.3 (Gly), 24.6 (Thr), 26.0 (Arg), 29.5 (Ala), 32.9 (Tyr), 40.0 (Met), 40.5 (Val), 42.5 (Phe), 44.7 (Ile), 45.8 (Leu), 52.0 (Lys)

KEY WORDS
paper contains discussion of ways to increase column life

REFERENCE
Krok, K.A.; Seaver, S.S. Realities of automating OPA HPLC amino acid analyses, *BioTechniques*, **1991**, *10*, 664–670.

SAMPLE
Matrix: urine
Analyte: amino acids
Sample preparation: Dilute urine 1:10 with water, inject an aliquot.

CAPILLARY ELECTROPHORESIS
Capillary: 90 cm × 75 μm fused-silica (Polymicro Technologies)
Running buffer: 15 mM pH 9.7 Borate buffer containing 10 mM sodium dodecyl sulfate
Voltage/Current: 20 kV
Injection: Hydrodynamic injection
Detector: F ex 345 em 455 following post-column reaction. The effluent from the column mixed with the reagent (under 180 mbar pressure) and the mixture flowed through a 22 cm × 50 μm capillary to the detector. (Reagent was 1.5 mL 25 mg/mL o-phthalaldehyde in MeOH, 18 μL 2-mercaptoethanol, 12.5 mL 100 mM pH 10 borax buffer, and 5 mL MeOH made up to 25 mL with water.)
Model: Lauer labs PRINCE
Migration time: 6.7 (Lys), 7 (Arg), 8.2 (Ala), 8.3 (Leu), 8.4 (Val), 8.5 (Trp), 9 (Gly), 9.2 (Phe), 9.4 (Met), 10 (Thr), 10.2 (Ser), 13.5 (Glu), 14 (Asp)
Limit of detection: 2-4 μM

KEY WORDS
post-column reaction

REFERENCE
Zhu, R.; Kok, W.T. Post-column reaction system for fluorescence detection in capillary electrophoresis, *J.Chromatogr.A*, **1995**, *716*, 123–133.

Phthalaldehyde/3-Mercaptopropionic Acid

RELATED REFERENCES
Bartók, T.; Szalai, G.; Lorincz, Z.; Börcsök, G.; Sági, F. High-speed RP-HPLC/FL analysis of amino acids after automated two-step derivatization with o-phthaldialdehyde/3-mercaptopropionic acid and 9-fluorenylmethyl chloroformate. *J.Liq.Chromatogr.* **1994**, *17*, 4391-4403.

Bertini, J.; Mannucci, C.; Noferini, R.; Perico, A.; Rovero, P. Rapid simultaneous determination of tryptophan and tyrosine in synthetic peptides by derivative spectroscopy. *J.Pharm.Sci.* **1993**, *82*, 179-182.

Birwé, H.; Hesse, A. High-performance liquid chromatographic determination of urinary cysteine and cystine. *Clin.Chim.Acta* **1991**, *199*, 33-42.

Durkin, T.A.; Anderson, G.M.; Cohen, D.J. High-performance liquid chromatographic analysis of neurotransmitter amino acids in brain. *J.Chromatogr.* **1988**, *428*, 9-15.

Fiorino, A.; Frigo, G.; Cucchetti, E. Liquid chromatographic analysis of amino and imino acids in protein hydrolysates by post-column derivatization with o-phthalaldehyde and 3-mercaptopropionic acid. *J.Chromatogr.* **1989**, *476*, 83-92.

Godel, H.; Graser, T.; Földi, P.; Pfaender, P.; Fürst, P. Measurement of free amino acids in human biological fluids by high-performance liquid chromatography. *J.Chromatogr.* **1984**, *297*, 49-61.

Graser, T.A.; Godel, H.G.; Albers, S.; Földi, P.; Fürst, P. An ultra rapid and sensitive high-performance liquid chromatographic method for determination of tissue and plasma free amino acids. *Anal.Biochem.* **1985**, *151*, 142-152.

Hashimoto, A.; Yamasaki, K.; Kokusenya, Y.; Miyamoto, T.; Sato, T. Investigation of "signal" constituents for the evaluation of animal crude drugs. I. Free amino acids and total amino acids. *Chem.Pharm.Bull.* **1994**, *42*, 1636-1641.

Moretti, F.; Birarelli, M.; Carducci, C.; Pontecorvi, A.; Antonozzi, I. Simultaneous high-performance liquid chromatographic determination of amino acids in a dried blood spot as a neonatal screening test. *J.Chromatogr.* **1990**, *511*, 131-136.

van Eijk, H.M.H.; van der Heijden, M.A.H.; van Berlo, C.L.H.; Soeters, P.B. Fully automated liquid-chromatographic determination of amino acids. *Clin.Chem.* **1988**, *34*, 2510-2513.

van Eijk, H.M.H.; Deutz, N.E.P.; Wagenmakers, A.J.M.; Soeters, P.B. 3-Methylhistidine determined in plasma by high-performance lipidchromatography. *Clin.Chem.* **1990**, *36*, 556-559.

van Eijk, H.M.H.; Huinck, M.P.L.; Rooyakkers, D.R.; Deutz, N.E.P. Automated simultaneous isolation and quantitation of labeled amino acid fractions from plasma and tissue by ion-exchange chromatography. *J.Chromatogr.B* **1994**, *660*, 251-257.

SAMPLE

Matrix: blood
Analyte: amino acids
Sample preparation: Filter (Amicon Centrifree) while centrifuging at 1500 g for 15 min, mix 2.5 μL ultrafiltrate with 2.5 μL reagent, add 1 μL 1 mg/mL 9-fluorenylmethyl chloroformate in MeCN, mix, let stand for 2.5 min, inject the whole amount. (Prepare reagent by dissolving 3 mg o-phthalaldehyde in 50 μL MeOH, add 450 μL 0.5 M pH 10.2 sodium borate buffer, add 5 μL 3-mercaptopropionic acid. Derivatization was performed automatically and took 5 min. o-Phthalaldehyde derivatized primary amino acids and 9-fluorenylmethyl chloroformate derivatized secondary amino acids (proline, sarcosine, and hydroxyproline).)

HPLC VARIABLES

Guard column: 20 × 2.1 5 μm Hypersil ODS
Column: two 100 × 2.1 5 μm Hypersil ODS columns in series
Mobile phase: Gradient. 15 mM pH 6.8 Sodium acetate:MeOH:10 mM pH 6.8 sodium acetate from 0:0:100 to 100:0:0 over 0.05 min, to 60:40:0 over 15 min, to 57.5:42.5:0 over 3.5 min, to 45:55:0 over 3.5 min, to 0:0:100 over 3 min, maintain at 0:0:100 for 5 min.
Column temperature: 40
Flow rate: 0.3
Injection volume: 6
Detector: F ex 230 em 450, after 20 min F ex 260 em 315

CHROMATOGRAM

Retention time: 1.8 (O-phospho-L-serine), 2 (aspartic acid), 2.5 (glutamic acid), 5 (gluta-thione (reduced)), 6 (asparagine), 6.2 (serine), 7.5 (glutamine), 8 (glycine), 8.5 (threonine), 8.8 (histidine), 9.2 (cystine), 9.5 (citrulline), 10.2 (taurine), 10.5 (alanine), 11.5 (arginine), 12.3 (tyrosine), 13.2 (α-amino-N-butyric acid), 15.3 (methionine), 15.5 (valine), 16 (nor-valine), 16.2 (tryptophan), 16.5 (phenylalanine), 17.8 (isoleucine), 18.2 (ornithine), 18.5 (leucine), 19.5 (lysine), 20.5 (hydroxyproline), 22.3 (sarcosine), 24.5 (proline)

Limit of detection: 5 pmole

KEY WORDS

plasma; ultrafiltrate

REFERENCE

Worthen, H.G.; Liu, H. Automatic pre-column derivatization and reversed-phase high performance liquid chromatography of primary and secondary amino acids in plasma with photo-diode array and fluorescence detection, *J.Liq.Chromatogr.*, **1992**, *15*, 3323–3341.

SAMPLE

Matrix: solutions

Analyte: amino acids

Sample preparation: Mix 70 µL of an aqueous amino acid solution with 300 µL phthal-aldehyde solution and 30 µL thiol solution for 2 min, inject an aliquot. (Prepare the phthalaldehyde solution by dissolving 60 mg o-phthalaldehyde in 3 mL MeOH and 15 mL 400 mM pH 9.4 sodium borate buffer. Prepare the thiol solution by dissolving 6.5 mg D-S-acetyl-3-mercapto-2-methylpropionic acid (Novabiochem) in 1 mL 1 M NaOH, stir at room temperature for 10 min, adjust the pH to 7.0 with phosphoric acid. The solution contains D-3-mercapto-2-methylpropionic acid.)

HPLC VARIABLES

Column: 250 × 4 3 µm Nucleosil-120-C18

Mobile phase: Gradient. MeCN:50 mM pH 6.0 sodium acetate buffer 0:100 for 10 min, to 62.5:37.5 (?) over 100 min.

Column temperature: 40

Flow rate: 1

Injection volume: 20

Detector: F ex 338 em 415 (long-pass filter)

CHROMATOGRAM

Retention time: 21 (D-Asp), 25 (L-Asp), 28.5 (D-Glu), 32 (L-Glu), 33 (D-Asn), 34 (D-Ser), 36 (L-Asn), 37 (L-Ser), 37.5 (D-Gln), 41 (D-Thr), 42 (L-Gln), 42.5 (Gly), 43 (D-His), 43.5 (L-His), 45 (D-Arg), 46 (L-Thr), 48 (D-Ala), 49 (L-Arg), 51.5 (L-Ala), 54 (D-Tyr), 57 (L-Tyr), 61 (D-Val), 64 (D-Met), 68 (L-Met), 69 (D-Ile, L-Val), 70 (D-Trp), 71 (D-Phe), 72 (D-Leu), 74 (L-Trp), 75 (L-Phe), 78.5 (L-Ile), 80 (L-Leu), 91 (D-Lys), 93 (L-Lys)

Limit of detection: 2 pmole

KEY WORDS

comparison with other thiols; chiral

REFERENCE

Duchateau, A.L.L.; Knuts, H.; Boesten, J.M.M.; Guns, J.J. Enantioseparation of amino compounds by derivatization with o-phthalaldehyde and D-3-mercapto-2-methylpropionic acid, *J.Chromatogr.*, **1992**, *623*, 237–245.

SAMPLE

Matrix: blood

Analyte: amino acids

Sample preparation: Add 100 µL 200 mg/mL 5-sulfosalicylic acid in EtOH to a 1 mL tube, evaporate EtOH at 50° overnight, add 200-500 µL plasma, vortex, freeze in liquid nitro-gen, store at -70°, thaw, centrifuge at 4° at 3000 g. 5 µL Supernatant + 20 µL water + 5 µL 1 mM norvaline in water + 90 µL reagent, mix thoroughly, incubate at room tem-perature for 3 min, add 50 µL neutralizing buffer, inject a 3 µL aliquot. (Prepare reagent stock solution by dissolving 25 mg o-phthalaldehyde in 500 µL MeOH, add 4.5 mL 100 mM pH 10.0 borate buffer, add 25 µL 3-mercaptopropionic acid. At the start of each day prepare reagent by diluting 1 part of stock solution with 20 parts 100 mM pH 10.0 borate buffer. Neutralizing buffer was 400 mM KH_2PO_4 containing 10mL/L triethylamine.

HPLC VARIABLES
Guard column: 10 × 2 Chrompack reverse phase
Column: 100 × 4.6 3 μm Microsphere C18 (Chrompack)
Mobile phase: Gradient. A was buffer:water:THF 50:50:0.2. B was MeOH:MeCN:buffer 35:15:50. A:B from 98:2 to 75:25 over 3.5 min, to 56:44 over 1.7 min, to 48:52 over 1.7 min, to 0:100 over 3.1 min, reset to initial conditions over 1 min.
Flow rate: 1.5
Injection volume: 3
Detector: F ex 230 em 389 (cut-off filter)

CHROMATOGRAM
Retention time: 2.3 (Asp), 4.1 (Glu), 4.8 (Asn), 5.1 (Ser), 5.8 (Gln), 6.2 (Gly), 6.4 (Thr), 6.6 (His), 6.8 (Cit), 7.0 (1-methylhistidine), 7.2 (3-methylhistidine), 7.4 (Ala), 7.6 (Tau), 7.7 (Arg), 8.5 (Tyr), 8.7 (α-aminobutyric acid), 9.8 (Val), 10.0 (Met), 10.5 (Trp), 10.7 (Phe), 11.0 (Ile), 11.2 (Orn), 11.4 (Leu), 11.6 (Lys)
Internal standard: norvaline (10.3)
Limit of quantitation: 5 μM

KEY WORDS
plasma

REFERENCE
Teerlink, T.; Van Leeuwen, P.A.M.; Houdijk, A. Plasma amino acids determined by liquid chromatography within 17 minutes, *Clin.Chem.*, **1994**, *40*, 245–249.

SAMPLE
Matrix: urine
Analyte: amino acids
Sample preparation: 80 μL Urine + 20 μL 10 mM N-methylalanine containing 1 mM norvaline, mix, filter (Ultrafree-MC) while centrifuging at 5000 g for 30 min. Remove a 6 μL aliquot of the ultrafiltrate and add it to 5 μL 0.5% 3-mercaptopropionic acid in 1 M pH 10.4 borate buffer, mix, add 1.5 μL 120 mM iodoacetic acid in 140 mM NaOH, mix, add 5 μL reagent, mix, add 2 μL 9-fluorenylmethyl chloroformate, mix, add 2.5 μL 1 M acetic acid, mix, inject the whole amount. (Reagent was 20 mg/mL o-phthalaldehyde in MeOH:500 mM pH 10.4 borate buffer:3-mercaptopropionic acid 10:88:2. The primary amino acids are derivatized with phthalaldehyde and the secondary amino acids (sarcosine and proline) with 9-fluorenylmethyl chloroformate.)

HPLC VARIABLES
Guard column: 20 × 4 5 μm ODS Hypersil
Column: 300 × 3.9 4 μm Nova-Pak C18
Mobile phase: Gradient. A was 60 mM pH 6.86 sodium acetate buffer containing 0.044% triethylamine. B was MeCN:MeOH:100 mM pH 5.45 sodium acetate buffer 74.5:4.5:21. A:B from 100:0 to 94.4:5.6 over 1 min, to 93.8:6.2 over 6 min, maintain at 93.8:6.2 for 2 min, to 92.3:7.7 over 12 min, maintain at 92.3:7.7 for 7 min, to 92:8 over 7 min, to 90.8:9.2 over 4 min, to 90.5:9.5 over 3 min, to 84:16 over 6 min, maintain at 84:16 for 1 min, to 82:18 over 1 min, to 78:22 over 20 min, to 72:28 over 7 min, to 68:32 over 8 min, to 0:100 over 9 min, maintain at 0:100 for 5 min, return to initial conditions over 1 min.
Column temperature: 40
Flow rate: 0.8
Injection volume: 22
Detector: F ex 340 em 450 for 79.5 min, F ex 260 em 315 for 1.5 min, F ex 340 em 450 for 6 min, F ex 260 em 315 for 13 min

CHROMATOGRAM

Retention time: 5.5 (Asp), 6.5 (Glu), 8 (Cys (S-carboxymethylated)), 9 (Aad), 10 (Asn), 11 (Ser), 11.5 (Homocysteine (S-carboxymethylated)), 15 (Gln), 17.5 (His), 18 (Gly), 20 (Thr), 25 (Cit), 26 (1-Methylhistidine), 28 (β-Alanine), 29.5 (Arg), 30.5 (3-Methylhistidine), 33 (Ala), 35 (Tau), 38.5 (Ans), 39 (Car), 40 (β-aminoisobutyric acid), 41 (gamma-aminobutyric acid), 52 (Tyr), 52.5 (Abu), 57 (Etn), 60 (Val), 61.5 (Met), 65 (Cysta), 70 (Trp), 71.5 (Ile), 72.5 (Phe), 75 (Hyl), 76.5 (Leu), 80 (Hyp), 83.5 (Lys), 85 (Orn), 90 (Sar), 91 (Pro)

Internal standard: norvaline (62.5), N-methylalanine (91.5)

Limit of quantitation: 10 μM

Limit of detection: 50-500 fmole

KEY WORDS

ultrafiltrate

REFERENCE

Carducci, C.; Birarelli, M.; Leuzzi, V.; Santagata, G.; Serafini, P.; Antonozzi, I. Automated method for the measurement of amino acids in urine by high-performance liquid chromatography, *J.Chromatogr.A*, **1996**, *729*, 173–180.

SAMPLE

Matrix: vegetables

Analyte: amino acids

Sample preparation: Extract 100 g potato tubers with 100 mL boiling water for 2 h, centrifuge at 5000 g for 10 min, filter (0.45 μm) the supernatant, dilute 100 μL of the filtrate with 900 μL 111 ng/mL norvaline in water, add a 1 μL aliquot to 5 μL 0.4 N pH 10.4 potassium borate buffer and 1 μL reagent, mix, add 1 μL 2.5 mg/mL 9-fluorenyl-methyl chloroformate in anhydrous MeCN, mix, inject the whole amount. (Prepare reagent by dissolving 10 mg o-phthalaldehyde in 100 μL MeOH, make up to 1 mL with 0.4 N pH 10.4 borate buffer, add 20 μL 3-mercaptopropionic acid. Derivatization was performed automatically and took 5 min. o-Phthalaldehyde derivatized primary amino acids and 9-fluorenylmethyl chloroformate derivatized secondary amino acids (proline and hydroxyproline).)

HPLC VARIABLES

Guard column: present but not specified

Column: 100 × 4 3 μm Hypersil ODS

Mobile phase: Gradient. A was THF:18 mM sodium acetate containing 0.02% triethylamine (adjusted to pH 7.2 with 1% acetic acid) 0.3:99.7. B was MeCN:MeOH:100 mM pH 7.2 sodium acetate 40:40:20. A:B from 100:0 to 94:6 over 0.5 min, to 80:20 over 2.5 min, to 50:50 over 3.5 min, to 25:75 over 1.5 min, to 0:100 over 0.5 min, maintain at 0:100 for 4 min, return to initial conditions over 1 min, re-equilibrate for 5 min.

Column temperature: 40

Flow rate: 1.4

Injection volume: 8

Detector: F ex 340 em 450, after 6.8 min F ex 264 em 313

CHROMATOGRAM

Retention time: 0.709 (aspartic acid), 0.831 (glutamic acid), 1.747 (asparagine), 1.871 (serine), 2.160 (glutamine), 2.286 (histidine), 2.439 (glycine), 2.602 (threonine), 3.305 (alanine), 3.446 (arginine), 4.185 (tyrosine), 5.113 (valine), 5.221 (methionine), 5.657 (tryptophan), 5.835 (phenylalanine), 5.928 (isoleucine), 6.224 (leucine), 6.435 (lysine), 6.903 (hydroxyproline), 7.982 (proline)

Internal standard: norvaline (5.394)

Limit of quantitation: 20 pmole

KEY WORDS

potato tubers

REFERENCE

Bartók, T.; Szalai, G.; Lorincz, Z.; Börcsök, G.; Sági, F. High-speed RP-HPLC/FL analysis of amino acids after automated two-step derivatization with o-phthaldialdehyde/3-mercaptopropionic acid and 9-fluorenylmethyl chloroformate, *J.Liq.Chromatogr.*, **1994**, *17*, 4391–4403.

Phthalaldehyde/Sodium Sulfite

RELATED REFERENCE
Turiák, G.; Volicer, L. Stability of *o*-phthalaldehyde-sulfite derivatives of amino acids and their methyl esters: electrochemical and chromatographic properties. *J.Chromatogr.A* **1994**, *668*, 323-329.

SAMPLE
Matrix: CSF
Analyte: amino acids
Sample preparation: 20 µL CSF + 0.4 µL reagent, mix, let stand for 10 min, inject an aliquot. (Prepare reagent by mixing 22 mg o-phthalaldehyde, 500 µL 1 M sodium sulfite, 500 µL EtOH, and 900 µL buffer. Prepare fresh each day, protect from light. The most satisfactory o-phthalaldehyde was obtained from Aldrich. Prepare buffer by adjusting the pH of 100 mM sodium tetraborate to 10.4 with 5 M NaOH.)

HPLC VARIABLES
Column: 250 × 4.6 5 µm Dynamax C18 (Rainin)
Mobile phase: MeOH:buffer 25:75, adjusted to pH 4.5 with 1 M phosphoric acid (Buffer was 100 mM NaH_2PO_4 containing 500 µM EDTA.)
Flow rate: 0.7
Injection volume: 20
Detector: E, Antec VT-03, glassy carbon working electrode +0.85 V, Ag/AgCl reference electrode

CHROMATOGRAM
Retention time: 5 (serine), 6.5 (glycine), 7 (taurine), 8 (glutamic acid), 9 (arginine), 10 (alanine), 24.5 (gamma-aminobutyric acid)
Limit of detection: 5-10 fmole

OTHER SUBSTANCES
Non-interfering: aspartic acid, cysteine, leucine, methionine, phenylalanine, tryptophan, tyrosine

KEY WORDS
rat

REFERENCE
Rowley, H.L.; Martin, K.F.; Marsden, C.A. Determination of in vivo amino acid neurotransmitters by high-performance liquid chromatography with *o*-phthalaldehyde-sulphite derivatisation, *J.Neurosci.Methods*, **1995**, *57*, 93–99.

Phthalaldehyde/1-Thio-β-D-galactopyranose

SAMPLE
Matrix: solutions
Analyte: amino acids
Sample preparation: Mix 10 μL 4 mM amino acids in water with 100 μL buffer, 50 μL 50 mg/mL 1-thio-β-D-galactopyranose sodium salt in water, and 50 μL 40 mg/mL o-phthalaldehyde in MeOH, stir thoroughly for 1 min, inject a 10 μL aliquot. (Buffer was 500 mg boric acid in 19 mL water, pH adjusted to 9.30 with 45% KOH.)

HPLC VARIABLES
Column: 250 × 4 5 μm LiChrosorb RP-8
Mobile phase: Gradient. A was 50 mM sodium acetate adjusted to pH 6.10 with acetic acid. B was MeOH:100 mM pH 7.60 sodium acetate buffer 90:10. A:B 100:0 for 8 min, to 45:55 over 47 min, to 0:100 over 5 min.
Column temperature: 35
Flow rate: 1.2
Injection volume: 10
Detector: F ex 360 em 420

CHROMATOGRAM
Retention time: 4.6 (L-Asp, D-Asp), 8.9 (L-Glu, D-Glu), 12.1 (L-Ser), 15.2 (D-Ser), 20.7 (L-Tre), 24.2 (D-Tre), 27.7 (L-Ala), 29.9 (L-Arg), 30.1 (D-Ala), 30.3 (D-Arg), 37.1 (L-Tyr), 37.8 (D-Tyr), 43.0 (D-Val), 44.9 (L-norvaline), 45.1 (L-Trp), 45.4 (L-Val), 45.9 (D-norvaline), 46.8 (D-Trp), 47.6 (L-Phe), 49.6 (D-Phe), 51.8 (L-Leu), 51.8 (D-Ile), 52.2 (L-norleucine), 52.2 (L-Ile), 53.1 (D-norleucine), 53.1 (D-Leu), 54.2 (D-Lys), 55.5 (L-Lys)
Limit of detection: <10 pmole

KEY WORDS
chiral

REFERENCE
Jegorov, A.; Triska, J.; Trnka, T. 1-Thio-β-D-galactose as a chiral derivatization agent for the resolution of D,L-amino acid enantiomers, *J.Chromatogr.A*, **1994**, *673*, 286–290.

Phthalaldehyde/1-Thio-β-D-glucose

RELATED REFERENCE
Jegorov, A.; Triska, J.; Trnka, T. 1-Thio-β-D-galactose as a chiral derivatization agent for the resolution of D,L-amino acid enantiomers. *J.Chromatogr.A* **1994**, *673*, 286-290.

SAMPLE
Matrix: solutions
Analyte: amino acids
Sample preparation: Mix 10 μL of an amino acid solution in water with 30 μL buffer, add 20 μL 50 mg/mL sodium 1-thio-β-D-glucose, add 20 μL 40 mg/mL o-phthalaldehyde in MeOH, stir thoroughly for 1 min, add 120 μL 50 mM pH 6.05 sodium acetate, inject a 10 μL aliquot. (Prepare buffer by dissolving 500 mg boric acid in 19 mL water and adjusting the pH to 10.40 with KOH solution (45 g KOH in 100 mL water).)

HPLC VARIABLES
Column: 250 × 4 5 μm LiChrosorb RP-8
Mobile phase: Gradient. A was 50 mM sodium acetate adjusted to pH 6.05 with acetic acid. B was MeOH:100 mM pH 7.60 sodium acetate 90:10. A:B 100:0 for 8 min, to 45:55 over 47 min, wash with 0:100 for 5 min.
Flow rate: 1.2
Injection volume: 10
Detector: F ex 360 em 420 (cut-off filter)

CHROMATOGRAM
Retention time: 6.4 (L-Asp), 5.2 (D-Asp), 14.4 (L-Glu), 15.6 (D-Glu), 18.2 (L-Ser), 20.1 (D-Ser), 25.0 (L-Thr), 28.1 (D-Thr), 29.0 (L-Arg), 30.3 (D-Arg), 31.6 (L-Ala), 34.0 (D-Ala), 37.0 (L-Tyr), 37.3 (D-Tyr), 48.0 (L-Val), 46.4 (D-Val), 46.1 (L-norvaline), 47.2 (D-norvaline), 47.9 (L-Trp), 48.7 (D-Trp), 51.5 (L-Phe), 50.0 (D-Phe), 53.9 (L-Ileu), 52.8 (D-Ileu), 54.0 (L-Leu), 54.9 (D-Leu), 54.3 (L-norleucine), 55.2 (D-norleucine)
Limit of detection: <1 pmole

KEY WORDS
chiral

REFERENCE
Jegorov, A.; Triska, J.; Trnka, T.; Cerny, M. Separation of α-amino acid enantiomers by reversed-phase high-performance liquid chromatography after derivatization with o-phthaldialdehyde and a sodium salt of 1-thio-β-D-glucose, *J.Chromatogr.*, **1988**, *434*, 417–422.

Phthalaldehyde/Thioglucose Tetraacetate

RELATED REFERENCE
Einarsson, S.; Folestad, S.; Josefsson, B. Separation of amino acid enantiomers using precolumn derivatization with o-phthalaldehyde and 2,3,4,6-tetra-O-acetyl-1-thio-β-glucopyranoside. *J.Liq.Chromatogr.* **1987**, *10*, 1589-1601.

SAMPLE
Matrix: blood, urine
Analyte: aspartic acid and glutamic acid
Sample preparation: Centrifuge urine at 4500 rpm for 15 min, dilute 50 times with water. Mix 400 μL serum with 600 μL MeCN, vortex for 5 min, centrifuge at 4500 rpm for 15 min. Mix 225 μL of the supernatant with 45 μL buffer and 45 μL reagent, let stand for 6 min, inject an aliquot. (Prepare buffer by adjusting the pH of 400 mM boric acid to 9.5 with 10 M NaOH. The reagent was 60 mM phthalaldehyde in MeCN containing 120 mM 2,3,4,6-tetra-O-acetyl-1-thio-β-D-glucopyranose.)

CAPILLARY ELECTROPHORESIS
Capillary: 55 cm × 50 μm fused-silica (40 cm to detector) (Polymicro Technologies)
Capillary preparation: Between runs rinse with 100 mM NaOH for 5 min, with water for 5 min, and with running buffer for 5 min then equilibrate at 15 kV for 5 min. Flush new capillaries with 1 M NaOH at 1 bar for 10 min then with water and with running buffer.
Running buffer: MeCN:pH 6.5 buffer 5:95 containing 10 mM octyl-β-D-glucopyranoside
Voltage/Current: +15 kV/16 μA
Injection: Hydrodynamic injection at 20 mbar for 8 s (3.7 nL).
Detector: F ex 325 (5.6 mW He-Cd laser) em 389 (cut-off filter); UV 230
Model: laboratory-constructed
Migration time: 6.8 (D-Glu), 7 (L-Glu), 7.3 (D-Asp), 7.5 (L-Asp)
Limit of detection: 1.1 μM (UV), 98 nM (F)

KEY WORDS
chiral; serum

REFERENCE
Tivesten, A.; Lundqvist, A.; Folestad, S. Selective chiral determination of aspartic and glutamic acid in biological samples by capillary electrophoresis, *Chromatographia*, **1997**, *44*, 623–633.

SAMPLE
Matrix: solutions
Analyte: amino acids
Sample preparation: Mix 100 μL of a 75 μM amino acid solution with 15 μL buffer and 15 μL reagent, let stand for 6 min, inject an aliquot. (Prepare reagent by dissolving 8 mg o-phthaldialdehyde and 44 mg 2,3,4,6-tetra-O-acetyl-1-thio-β-D-glucopyranose in 1 mL MeOH. Prepare buffer by adjusting 400 mM boric acid to pH 9.5 with ca. 1.2 mL 10 M NaOH.)

CAPILLARY ELECTROPHORESIS
Capillary: 27 cm × 50 μm fused-silica (19 cm to detector) (Polymicro Technologies)
Capillary preparation: Between each run rinse capillary with 100 mM NaOH for 1 min, with water for 1 min, with running buffer for 1 min, and then equilibrate with running buffer by electrokinetic pumping for 2 min. Flush new capillaries with 1 M NaOH for 10 min, rinse with water, and rinse with running buffer.
Running buffer: MeCN:pH 9.55 sodium borate buffer (I = 40 mM) containing 45 mM sodium dodecyl sulfate
Voltage/Current: 405 V/cm/ca. 52 μA
Injection: Hydrodynamic injection at 0.5 psi for 2 s.
Detector: UV 340; F ex 350 em 415
Migration time: 1.83 (L-Ser, L-His), 1.90 (L-Tyr, L-Thr), 1.93 (D-Ser, L-Ala), 2.0 (D-His), 2.12 (Gly), 2.16 (L-Glu, L-Asp, D-Glu, D-Asp), 2.24 (L-Trp, L-Val, D-Ala, D-Thr), 2.28 (D-Tyr), 2.35 (L-Met), 2.48 (L-Phe), 2.66 (L-Ile), 2.83 (D-Val, D-Met), 2.90 (D-Trp), 3.0 (L-Leu), 3.29 (D-Phe), 3.34 (D-Ile), 3.41 (D-Leu), 3.97 (L-Arg), 4.12 (ammonia), 4.40 (D-Arg), 4.48 (L-Lys), 4.59 (L-Orn), 4.61 (D-Orn), 4.69 (D-Lys)

KEY WORDS
chiral

REFERENCE
Tivesten, A.; Folestad, S. Separation of precolumn-labelled D- and L-amino acids by micellar electrokinetic chromatography with UV and fluorescence detection, *J.Chromatogr.A*, **1995**, *708*, 323–337.

SAMPLE
Matrix: solutions
Analyte: amino acids
Sample preparation: Mix 180 μL of an amino acid solution in buffer with 30 μL 60 mM phthalaldehyde in MeCN containing 120 mM 2,3,4,6-tetra-O-acetyl-1-thio-β-D-glucose, let stand for 6 min, inject an aliquot. (Prepare buffer by adjusting the pH of 400 mM boric acid to 9.5 with 10 M NaOH.)

CAPILLARY ELECTROPHORESIS
Capillary: 56 cm × 50 μm fused-silica (39 cm to detector) (Polymicro Technologies)
Capillary preparation: Between runs rinse capillary with 100 mM NaOH for 1 min, with water for 1 min, and with running buffer for 1 min. Equilibrate with voltage on for 2 min before injection. Rinse new capillaries with 1 M NaOH for 10 min and then with water.
Running buffer: MeCN:buffer 4:96 (Prepare running buffer by dissolving 1.12 moles sodium dodecyl sulfate in a small amount of borate buffer, adding 1 mL MeCN, and making up to 25 mL with borate buffer. Prepare borate buffer by mixing 37.5 mL 40 mM borax solution with 10.6 mL 100 mM NaOH and making up to 100 mL with water, pH 9.50-9.55.)
Voltage/Current: 15 kV/ca. 30 μA
Injection: Hydrodynamic injection at 80 mbar for 5 s.
Detector: UV 235; UV 340
Model: laboratory-constructed
Migration time: 7 (L-Ala), 8.5 (D-Thr), 9 (L-Met), 11.5 (L-Leu), 13.5 (D-Leu), 15.5 (L-Lys)
Limit of detection: 1-2 μM (UV 235), 7-8 μM (UV 340)

KEY WORDS
chiral; comparison with other reagents

REFERENCE
Tivesten, A.; Folestad, S. Chiral *o*-phthaldialdehyde reagents for fluorogenic on-column labeling of D- and L-amino acids in micellar electrokinetic chromatography, *Electrophoresis*, **1997**, *18*, 970–977.

DIAZO COMPOUND

1-Pyrenyldiazomethane

SAMPLE
Matrix: blood, urine
Analyte: thiodiglycolic acid
Sample preparation: 50 μL Plasma or urine + 5 μg N-acetyl-L-alanine + 400 μL isopropanol:50 mM pH 4 phosphate buffer 50:50 + 400 μL 20 mM octadecylsulfonic acid in isopropanol + 400 μL 1-pyrenyldiazomethane (Molecular Probes, Eugene OR) in ethyl acetate, heat at 40° for 2 h, inject an aliquot.

HPLC VARIABLES
Column: 160 × 4.6 3 μm Hypersil reverse phase
Mobile phase: Gradient. MeCN:buffer from 20:80 to 35:65 over 70 min. Buffer was MeOH:THF:15 mM triethanolamine adjusted to pH 2.4 with phosphoric acid 20:5:75.)
Flow rate: 0.8
Detector: F ex 340 em 380

CHROMATOGRAM
Retention time: 50
Internal standard: N-acetyl-L-alanine

KEY WORDS
plasma; pharmacokinetics

REFERENCE
Brockmöller, J.; Staffeldt, B.; Roots, I. Evaluation of proposed sulphoxidation pathways of carbocysteine in man by HPLC quantification, *Eur.J.Clin.Pharmacol.*, **1991**, *40*, 387–392.

ETHOXYMETHYLENE COMPOUND

Diethyl Ethoxymethylenemalonate

SAMPLE
Matrix: enzyme incubations
Analyte: amino acids
Sample preparation: 1 mL Enzyme incubation + 11 mL 155 μM DL-α-aminobutyric acid in buffer, mix, add a 2 mL aliquot to an 8 mL column of Sephadex G-25 (Pharmacia PD-10), elute with buffer, discard the first 6 mL eluate, collect the next 4 mL eluate. Add 3.2 μL diethyl ethoxymethylenemalonate to this fraction, shake at 50° for 50 min, inject a 15 μL aliquot. (Buffer was 1 M pH 9 sodium borate buffer containing 0.02% sodium azide.)

HPLC VARIABLES
Column: 300 × 3.9 4 μm Nova-Pak C18
Mobile phase: Gradient. MeCN:25 mM pH 6 sodium acetate containing 0.02% sodium azide from 9:91 to 14:86 over 3 min, maintain at 14:86 for 10 min, to 31:69 over 17 min, maintain at 31:69 for 20 min. (Caution! Sodium azide is highly toxic! Do not discharge to the plumbing system!)
Flow rate: 0.9
Injection volume: 15
Detector: UV 280

CHROMATOGRAM
Retention time: 17.7 (N-epsilon-(2-propenal)lysine), 35 (lysine), 44.8 (lysine dipeptide)
Internal standard: DL-α-aminobutyric acid (19)
Limit of detection: <1 pmole

REFERENCE
Girón, J.; Alaiz, M.; Vioque, E. High-performance liquid chromatographic determination of *N*-epsilon-(2-propenal)lysine in biological samples after derivatization with diethylethoxymethylenemalonate, *Anal.Biochem.*, **1992**, *206*, 155–160.

SAMPLE
Matrix: protein hydrolysate
Analyte: amino acids
Sample preparation: 2-200 μg Protein hydrolysate + 1 mL 1 M pH 9.0 sodium borate buffer containing 0.02% sodium azide + 0.8 μL diethyl ethoxymethylenemalonate (Fluka), heat at 50° with vigorous shaking for 50 min, cool to room temperature, inject a 15 μL aliquot.

HPLC VARIABLES
Column: 300 × 3.9 4 μm Nova-Pak C18

Mobile phase: Gradient. MeCN:buffer from 9:91 to 14:86 over 3 min, maintain at 14:86 for 10 min, to 31:69 over 17 min, maintain at 31:69 for 5 min. (Buffer was 25 mM pH 6.0 sodium acetate containing 0.02% sodium azide. Caution! Sodium azide is highly toxic! Do not discharge solutions containing sodium azide to the plumbing system!)
Column temperature: 18
Flow rate: 0.9
Injection volume: 15
Detector: UV 280

CHROMATOGRAM

Retention time: 3.10 (aspartic acid), 3.31 (glutamic acid), 7.48 (serine), 9.35 (histidine), 10.25 (glycine), 11.02 (threonine), 12.66 (arginine), 13.45 (alanine), 15.25 (proline), 20.05 (tyrosine), 24.21 (ammonia), 25.84 (valine), 27.22 (methionine), 28.52 (cystine), 30.19 (isoleucine), 31.02 (leucine), 31.95 (phenylalanine), 34.73 (lysine)
Internal standard: α-aminobutyric acid (18.15)
Limit of detection: 3 pmole

REFERENCE

Alaiz, M.; Navarro, J.L.; Girón, J.; Vioque, E. Amino acid analysis by high-performance liquid chromatography after derivatization with diethyl ethoxymethylenemalonate, *J.Chromatogr.*, **1992**, *591*, 181–186.

ISOCYANATE

(S)-(+)-1-(1-Naphthyl)ethyl Isocyanate

SAMPLE

Matrix: solutions
Analyte: amino acids
Sample preparation: Mix a 50 µL 5 mM amino acid solution with 50 µL 400 mM pH 9.0 sodium borate buffer, add 100 µL 16 mg/mL (S)-(+)-1-(1-naphthyl)ethyl isocyanate in dry acetone while vortexing, vortex for 30 s, let stand at room temperature for 5 min, centrifuge at 4000 g for 2 min. Remove a 150 µL aliquot of the supernatant and wash it three times with 1 mL portions of wash solution, inject an aliquot. (Prepare wash solution by mixing equal volumes of cyclohexane saturated with 400 mM pH 9.0 borate buffer and ether. Wash ether with ferrous sulfate to remove peroxides if methionine is present.)

HPLC VARIABLES

Column: 250 × 4.6 5 µm octadecyl Si100 Polyol (Serva)
Mobile phase: MeCN:50 mM pH 6.2 ammonium acetate 15:85 (A) or 20:80 (B) or 25:75 (C) or 27.5:72.5 (D) or 35:65 (E) or 40:60 (F) or MeCN:100 mM acetic acid 20:80 (G)
Flow rate: 1
Detector: F ex 214 em 320 (cut-off filter)

CHROMATOGRAM

Retention time: 22.6 (D-Asp (G)), 24.4 (L-Asp (G)), 26.6 (D-Glu (G)), 28.2 (L-Glu (G)), 20.8 (D-Asn (A)), 23.2 (L-Asn (A)), 19.3 (D-Ser (A)), 20.8 (L-Ser (A)), 27.6 (D-His (A)), 31.5 (L-His (A)), 9.9 (D-Thr (B)), 11.9 (L-Thr (B)), 11.0 (D-Pro (B)), 12.2 (L-Pro (B)), 11.7 (D-Ala (B)), 13.2 (L-Ala (B)), 6.9 (D-Arg (C)), 7.7 (L-Arg (C)), 10.4 (D-Val (C)), 13.2 (L-Val (C)), 5.8 (D-Met (D)), 7.4 (L-Met (D)), 9.7 (D-Ile (D)), 12.1 (L-Ile (D)), 11.0 (D-Leu (D)), 13.5 (L-Leu (D)), 16.0 (D-Trp (D) (UV 215)), 18.3 (L-Trp (D) (UV 215)), 17.3 (D-Phe (D)), 20.6(L-Phe (D)), 15.5 (D-Lys (E)), 17.8 (L-Lys (E)), 17.3 (D-Tyr (F)), 19.0 (L-Tyr (F))

Limit of detection: <1 pmole

KEY WORDS
chiral

REFERENCE
Dunlop, D.S.; Neidle, A. The separation of D/L amino acid pairs by high-performance liquid chromatography after precolumn derivatization with optically active naphthylethyl isocyanate, *Anal.Biochem.*, **1987**, *165*, 38−44.

1-Naphthyl Isocyanate

SAMPLE

Matrix: blood, CSF, tissue

Analyte: amino acids

Sample preparation: Blood. Mix plasma or whole blood with an equal volume of 0.5% sodium dodecyl sulfate, let stand for 15 min, add 2 volumes of 6% perchloric acid, add norleucine, centrifuge. Remove a 100 μL aliquot of the supernatant and add it to 25 μL buffer, add 125 μL freshly-prepared 2.5 μL/mL 1-naphthyl isocyanate in acetone, mix, let stand for 45 s, add 1 mL cyclohexane, vortex vigorously, discard the cyclohexane layer, repeat the wash twice more, centrifuge the aqueous layer, inject a 20 μL aliquot of the supernatant. CSF. Mix 1 volume of 60% perchloric acid with 20 volumes of CSF, centrifuge. Remove a 100 μL aliquot of the supernatant and add it to 25 μL buffer, add 125 μL freshly-prepared 2.5 μL/mL 1-naphthyl isocyanate in acetone, mix, let stand for 45 s, add 1 mL cyclohexane, vortex vigorously, discard the cyclohexane layer, repeat the wash twice more, centrifuge the aqueous layer, inject a 20 μL aliquot of the supernatant. Tissue. Homogenize brain tissue with 9 volumes 3% perchloric acid, centrifuge at 9000 g for 3 min. Remove the supernatant and neutralize it with NaOH, add 3 volumes of water, add norleucine. Remove a 100 μL aliquot and add it to 25 μL buffer, add 125 μL freshly-prepared 2.5 μL/mL 1-naphthyl isocyanate in acetone, mix, let stand for 45 s, add 1 mL cyclohexane, vortex vigorously, discard the cyclohexane layer, repeat the wash twice more, centrifuge the aqueous layer, inject a 5 μL aliquot of the supernatant. (Prepare buffer by adjusting the pH of 1 M boric acid to6.25 with NaOH. Dry acetone over anhydrous sodium sulfate.)

HPLC VARIABLES

Column: 250 × 4.6 5 μm Absorbosphere HS C18

Mobile phase: Gradient. A was MeCN:MeOH:buffer:water 2.5:12.5:4.5:80.5. B was MeCN:buffer:water 45:4:51. C was MeCN:water 70:30. A:B:C 100:0:0 for 45 min, to 0:100:0 over 150 min, to 0:0:100 (step gradient), maintain at 0:0:100 for 10 min, re-

equilibrate at initial conditions for 30 min. (Buffer was 100 mM NaH_2PO_4 containing 100 mM sodium acetate, adjusted to pH 5.4 with phosphoric acid.)
Flow rate: 1
Injection volume: 5-20
Detector: F ex 228 em 320 (cut-off filter); UV 225

CHROMATOGRAM
Retention time: 27 (Asp), 34 (Glu), 44 (Asn), 50 (Ser), 57 (Gly), 60 (Gln), 65 (Tau), 69 (His), 78 (Thr), 81 (Pro), 83 (Ala), 88 (Arg), 90 (GABA), 95 (ammonia), 105 (glutathione), 109 (Tyr), 116 (Val), 118 (Met), 130 (Ile), 131 (Leu (some interference from matrix)), 136 (Phe), 139 (Trp), 140 (Cys), 154 (Orn), 158 (Lys)
Internal standard: norleucine (133)

KEY WORDS
plasma; whole blood; brain; human; mouse

REFERENCE
Neidle, A.; Banay-Schwartz, M.; Sacks, S.; Dunlop, D.S. Amino acid analysis using 1-naphthylisocyanate as a precolumn high performance liquid chromatography derivatization reagent, *Anal.Biochem.*, **1989**, *180*, 291–297.

ISOTHIOCYANATE

Benzyl Isothiocyanate

RELATED REFERENCE
Woo, K.-L.; Ahan, Y-K. Determination of protein amino acids as benzylthiocarbamyl derivatives compared with phenylthiocarbamyl derivatives by reversed-phase high-performance liquid chromatography. *J.Chromatogr.A* **1996**, *738*, 285-289.

SAMPLE
Matrix: soybean meal
Analyte: amino acids
Sample preparation: Mix 200 mg soybean meal with 15 mL 0.1% phenol in 6 M HCl, pump out the tube then flush with nitrogen 3 times, heat at 145° for 4 h, filter, evaporate to dryness under reduced pressure at 50°, reconstitute with 50 mL 10 mM HCl. Add a 5 mL aliquot to a 100 × 13 Dowex 5X8 cation exchange column, elute with 4 M ammonia. Evaporate the eluate to dryness under reduced pressure at 50°, reconstitute with 50 mL 10 mM HCl. Evaporate a 500 μL aliquot to dryness under a stream of nitrogen at 50°, add 30 μL MeCN, evaporate to dryness under a stream of nitrogen at 50°, reconstitute the residue in 50 μL 2.5 mM L-norleucine in MeCN:MeOH:triethylamine 10:5:2, add 3 μL benzyl isothiocyanate, heat at 50° for 30min, evaporate to dryness under a stream of nitrogen with simultaneous vacuum pump evacuation at room temperature, reconstitute the residue in 50 μL MeCN, evaporate to dryness as before, reconstitute the residue in 1 mL initial mobile phase, filter (0.2 μm), inject a 10 μL aliquot of the filtrate.

HPLC VARIABLES
Column: 300 × 3.9 4 μm Nova-Pak C18

Mobile phase: Gradient. A was MeOH:THF:20 mM NaH_2PO_4 5:1.5:93.5, adjusted to pH 6.8 with phosphoric acid. B was MeCN:MeOH:THF:20 mM NaH_2PO_4 50:2.5:0.75:46.75, adjusted to pH 6.8 with phosphoric acid. C was MeCN:water 70:30. A:B:C from 100:0:0 to 80:20:0 over 10 min, to 76:20:4 over 5 min, to 70:20:10 over 5 min, to 50:30:20 over 10 min, to 30:35:35 over 10 min, to 0:0:100 (step gradient), maintain at 0:0:100 for 20 min.

Column temperature: 40

Flow rate: 1.2

Injection volume: 10

Detector: UV 246

CHROMATOGRAM

Retention time: 4 (Asp), 4.5 (Glu), 7 (Hyp), 9.5 (Asn), 10 (Gln), 10.5 (Ser), 11 (Gly), 11.5 (His), 12 (Pro), 12.5 (Arg), 13 (Thr), 13.5 (Ala), 17.5 (ammonia), 19 (Asp), 20 (Val), 20.5 (Tyr), 22 (Met), 25 (Ile), 25.5 (Leu), 27.5 (cystine), 28 (Phe), 29.5 (Trp), 32 (Lys), 34.5 (cysteine)

Internal standard: L-norleucine (26)

Limit of detection: 3.9 pmole

KEY WORDS

SPE; comparison with phenyl isothiocyanate derivatization

REFERENCE

Woo, K.-L.; Ahan, Y.-K. Determination of protein amino acids as benzylthiocarbamyl derivatives compared with phenylthiocarbamyl derivatives by reversed-phase high-performance liquid chromatography, ultraviolet detection and precolumn derivatization, *J.Chromatogr.A*, **1996**, *740*, 41–50.

Butyl Isothiocyanate

RELATED REFERENCE

Lawrence, J.F.; Menard, C. Confirmation of domoic acid in shellfish using butyl isothiocyanate and reversed-phase liquid chromatography. *J.Chromatogr.* **1991**, *550*, 595-601.

SAMPLE

Matrix: solutions

Analyte: amino acids

Sample preparation: Evaporate a solution of amino acids in 10 mM HCl under a stream of nitrogen at 50°, add 50 μL MeCN, evaporate under a stream of nitrogen, add 50 μL MeCN:MeOH:triethylamine 10:5:2, sonicate for 1 min, add 3 μL butyl isothiocyanate, cap the vial with a septum, heat at 40° for 30 min. Pass nitrogen into the vial with one needle and apply a vacuum with another needle until solvent is removed (ca. 15 min), add 100 μL MeCN, remove MeCN in a similar fashion. Dissolve the residue in 500 μL 200 mM ammonium acetate, filter (0.20 μm), inject a 10 μL aliquot of the filtrate.

HPLC VARIABLES

Column: 300 × 3.9 4 μm Nova-Pak C18

Mobile phase: Gradient. A was 50 mM ammonium acetate adjusted to pH 6.7 with phosphoric acid. B was MeCN:50 mM ammonium acetate adjusted to pH 6.7 with phosphoric acid 50:50. C was MeCN:water 70:30. A:B:C from 100:0:0 to 85:15:0 over 8 min, to 70:20:10 over 6 min, to 60:20:20 over 6 min, to 20:0:80 over 5 min, to 0:0:100 over 5 min, maintain at 0:0:100 for 5 min.

Column temperature: 40

Flow rate: 1

Injection volume: 10
Detector: UV 250

CHROMATOGRAM
Retention time: 8.05 (Asp), 9.00 (Glu), 11.27 (Hyp), 13.75 (Asn), 13.75 (Ser), 14.39 (Gly), 14.63 (Gln), 15.71 (His), 16.13 (Thr), 16.29 (Ala), 16.45 (Arg), 16.84 (Pro), 17.76 (Cyt), 20.27 (Tyr), 20.51 (Val), 21.73 (Met), 23.29 (Ile), 23.67 (Leu), 25.18 (Phe), 25.53 (Trp), 26.16 (Lys), 26.84 (Cys)
Limit of quantitation: 0.5 nmole

KEY WORDS
Asp and Ser not resolved

REFERENCE
Woo, K.-L.; Lee, S.-H. Determination of protein amino acids as butylthiocarbamyl derivatives by reversed-phase high-performance liquid chromatography with precolumn derivatization and UV detection, *J.Chromatogr.A*, **1994**, *667*, 105–111.

SAMPLE
Matrix: solutions
Analyte: amino acids
Sample preparation: Evaporate 25 μL of an amino acid solution to dryness under a stream of nitrogen at 50°, add MeCN, evaporate to dryness under a stream of nitrogen at 50°, reconstitute the residue in 50 μL MeCN:MeOH:triethylamine 10:5:2, add 3 μL butylisothiocyanate, sonicate for 1 min, heat at 40° for 30 min, evaporate to dryness under a stream of nitrogen at room temperature, reconstitute the residue in 100 μL MeCN, evaporate to dryness under a stream of nitrogen, reconstitute the residue in 1 mL 20 mM ammonium acetate, filter (0.25 μm), inject a 10 μL aliquot of the filtrate.

HPLC VARIABLES
Column: 300 × 3.9 4 μm Nova-Pak C18
Mobile phase: Gradient. A was 50 mM ammonium acetate adjusted to pH 6.7 with phosphoric acid. B was MeCN:MeOH:THF:20 mM Na_2HPO_4 50:2.5:0.75:46.75. C was MeCN:water 70:30. A:B:C from 100:0:0 to 85:15:0 over 8 min, to 70:20:10 over 6 min, to 60:20:20 over 6 min, to 30:20:50 over 5 min, to 10:20:70 over 5 min, to 0:0:100 (step gradient), maintain at 0:0:100 for 20 min.
Column temperature: 40
Flow rate: 1
Injection volume: 10
Detector: UV 240

CHROMATOGRAM
Retention time: 8 (Asp), 9 (Glu), 12 (Hyp), 14 (Asn, Ser), 14.3 (Gly), 14.6 (Gln), 15.6 (His), 15.9 (Thr), 16.2 (Ala), 16.6 (Arg), 17 (Pro), 20.3 (Tyr), 20.6 (Val), 22 (Met), 23 (Ile), 23.5 (Leu), 24 (cystine), 25.3 (Phe), 26 (Trp), 26.5 (Lys), 27.7 (cysteine)
Internal standard: norleucine (24.5)
Limit of detection: 3.9 pmole

KEY WORDS
comparison with phenyl isothiocyanate derivatization

REFERENCE
Woo, K.-L.; Hwang, Q.-C.; Kim, H.-S. Determination of amino acids in the foods by reversed-phase high-performance liquid chromatography with a new precolumn derivative, butylthiocarbamyl amino acid compared to the conventional phenylthiocarbamyl derivatives and ion-exchange chromatography, *J.Chromatogr.A*, **1996**, *740*, 31–40.

4-N,N-Dimethylaminoazobenzene-4'-isothiocyanate

RELATED REFERENCES

Nair, P.P.; Kessie, G.; Patnaik, R.; Guidry, C. Isolation and HPLC of N-epsilon-lithocholyl lysine as its fluorescamine and dimethylaminoazobenzene isothiocyanate derivatives. *Steroids* **1994**, *59*, 212-216.

Winkler, G.; Heinz, F.X.; Kunz, C. Exclusive use of high-performance liquid chromatographic techniques for the isolation, 4-dimethylaminoazobenzene-4'-sulfonyl chloride amino acid analysis and 4-N,N-dimethylaminoazobenzene-4'-isothiocyanate-phenyl isothiocyanate sequencing of a viral membrane protein. *J.Chromatogr.* **1984**, *297*, 63-73.

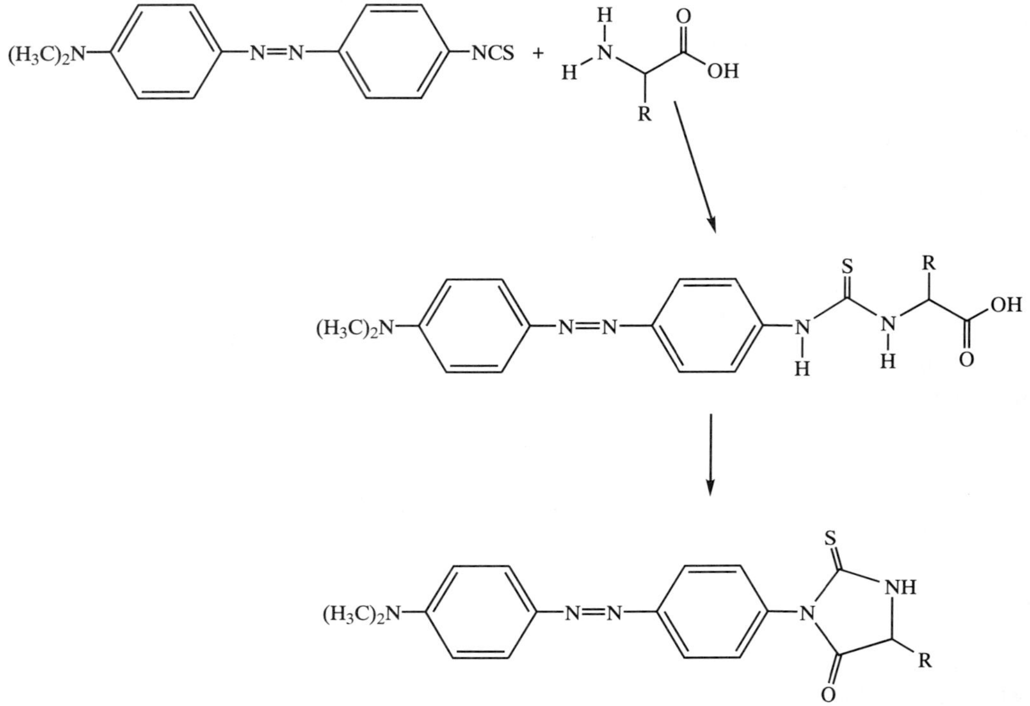

SAMPLE

Matrix: bulk

Analyte: amino acids

Sample preparation: Dissolve 500 μg amino acids in 1 mL pH 10.65 triethylamine/acetic acid buffer, remove a 100 μL aliquot and add it to 50 μL 4 mM 4-N,N-dimethylaminoazobenzene-4'-isothiocyanate in acetone, heat at 54° for 1 h, dry under vacuum, dissolve the residue in 100 μL 50% trifluoroacetic acid, heat at 54° for 45 min, dry under vacuum, reconstitute with EtOH:water, inject a 20 μL aliquot. (Purify 4-N,N-dimethylaminoazobenzene-4'-isothiocyanate by dissolving 500 mg in 50 mL boiling acetone, filter (sintered glass), store the filtrate at -20° overnight to obtain 4-N,N-dimethylaminoazobenzene-4'-isothiocyanate as needle-shaped crystals.)

HPLC VARIABLES

Guard column: 20 × 4.6 40 μm Pellicular Packing LC-18

Column: 250 × 4.6 5 μm Supelcosil LC-18

Mobile phase: Gradient. MeCN:35 mM pH 5.1 sodium acetate buffer 39:61 for 8 min, to 53:47 over 4 min, maintain at 53:47 for 28 min, re-equilibrate at initial conditions for 10 min.

Flow rate: 1

Injection volume: 20

Detector: UV 436

CHROMATOGRAM
Retention time: 5 (cysteic acid), 6 (aspartic acid), 7 (carboxymethylcysteine), 5 (glutamic acid), 10.8 (glutamine), 12 (glutamine), 12.8 (asparagine), 13.2 (serine), 14 (threonine), 16.4 (glycine), 17.2 (histidine), 18.8 (tyrosine), 19.6 (alanine), 17.2 (methionine), 18 (tryptophan), 18.8 (valine), 19.6 (proline), 32.4 (phenylalanine), 37.2 (isoleucine), 38.4 (leucine)
Limit of detection: <1 pmole

REFERENCE
Stocchi, V.; Cucchiarini, L.; Piccoli, G.; Magnani, M. Complete high-performance liquid chromatographic separation of 4-N,N-dimethylaminoazobenzene-4'-thiohydantoin and 4-dimethylaminoazobenzene-4'-sulphonyl chloride amino acids utilizing the same reversed-phase column at room temperature, *J.Chromatogr.*, **1985**, *349*, 77–82.

SAMPLE
Matrix: bulk
Analyte: amino acids
Sample preparation: Dissolve a few mg amino acids in 200 μL reagent, let stand at room temperature in the dark for 5-10 min, evaporate to dryness under reduced pressure, reconstitute with mobile phase, inject an aliquot. (Prepare reagent by mixing 5 mL MeOH, 2.5 mL pyridine, 1 mL triethylamine, 1.5 mL water, and 500 μL 100 mM 4-dimethylaminoazobenzene-4'-isothiocyanate in MeCN.)

HPLC VARIABLES
Column: 250 × 4 5 μm Chiradex (immobilized β-cyclodextrin) (Merck)
Mobile phase: MeOH:100 mM pH 6.5 ammonium acetate buffer containing 0.1% triethylamine 40:60
Column temperature: 20 ± 0.1
Flow rate: 0.5
Injection volume: 20
Detector: UV 254; F ex 240 em 480

CHROMATOGRAM
Retention time: k' 3.18 (second enantiomer), α = 1.00 (Ala); k' 2.72 (second enantiomer), α = 1.00 (Val); k' 6.86 (second enantiomer), α = 1.43 (Leu); k' 4.19 (second enantiomer), α = 1.10 (Ile); k' 6.92 (second enantiomer), α = 1.44 (Pro); k' 4.89 (second enantiomer), α = 1.19 (Met); k' 5.58 (second enantiomer), α = 1.33 (Cys); k' 3.39 (second enantiomer), α = 1.00 (Ser); k' 3.00 (second enantiomer), α = 1.00 (Thr); k' 1.86 (second enantiomer), α = 1.12 (Lys); k' 1.85 (second enantiomer), α = 1.14 (Arg); k' 4.65 (second enantiomer), α = 1.18 (Asn); k' 3.39 (second enantiomer), α = 1.07 (Gln); k' 20.44 (second enantiomer), α = 1.05 (Glu); k' 7.07 (second enantiomer), α = 1.16 (Phe); k' 7.75 (second enantiomer), α = 1.00 (Trp); k' 3.64 (second enantiomer), α = 1.09 (His); k' 4.37 (second enantiomer), α = 1.11 (Tyr)

KEY WORDS
chiral; α = k' (second enantiomer)/k' (first enantiomer); detailed comparison with other derivatizing reagents

REFERENCE
Rizzi, A.M.; Cladrowa-Runge, S.; Jonsson, H.; Osla, S. Enantiomeric resolution of derivatized DL-amino acids by high-performance liquid chromatography using a β-cyclodextrin chiral stationary phase: A comparison between derivatization labels, *J.Chromatogr.A*, **1995**, *710*, 287–295.

SAMPLE
Matrix: perfusate
Analyte: homocysteic acid
Sample preparation: Lyophilize perfusate, reconstitute with 1 mL water, add 300 μL 1 μg/mL IS in water, add 1.5 mL 300 μg/mL 4-[(4-isothiocyanatophenyl)azo]-N,N-dimethylaniline (4-(N,N-dimethylamino)azobenzene-4'-isothiocyanate) in pyridine (freshly prepared), heat at 70° for 90 min, wash with heptane:ethyl acetate 2:1. Evaporate the aqueous phase to dryness, reconstitute with 1 mL EtOH:water 50:50, filter, inject an aliquot.

HPLC VARIABLES
Guard column: present but not specified
Column: 250 × 4.6 5 μm Spherisorb ODS II
Mobile phase: Gradient. MeCN:35 mM pH 5.5 acetate buffer 20:80 for 60-80 min, to 80:20 (step gradient).
Flow rate: 1
Detector: UV 455

CHROMATOGRAM
Retention time: 15 (L-cysteine sulfinic acid), 16 (L-cysteic acid), 19 (L-homocysteic acid)
Internal standard: (D,L)-2-amino-7-sulfonoheptanoic acid (AS-7) (21)

REFERENCE
Do, K.Q.; Herrling, P.L.; Streit, P.; Turski, W.A.; Cuenod, M. *In vitro* release and electrophysiological effects *in situ* of homocysteic acid, an endogenous *N*-methyl-(D)-aspartic acid agonist, in the mammalian striatum, *J.Neurosci.*, **1986**, *6*, 2226–2234.

SAMPLE
Matrix: solutions
Analyte: amino acids
Sample preparation: 100 μL Solution + 50 μL 4 mM dimethylaminoazobenzene isothiocyanate in acetone, mix, heat at 52 ± 2° for 1 h, evaporate to dryness under reduced pressure at 30°, reconstitute the residue with 40 μL water and 80 μL 6 M HCl:glacial acetic acid 1:2, heat for 50 min, evaporate to dryness under reduced pressure, reconstitute with 500 μL 10 mM pH 2.5 phosphate buffer, inject an aliquot.

CAPILLARY ELECTROPHORESIS
Capillary: 110 cm × 50 μm fused-silica (104.5 cm to detector) (Polymicro Technologies)
Running buffer: MeCN:10 mM pH 2.5 phosphate buffer 40:60 containing 10 mM sodium dodecyl sulfate
Injection: Electrokinetic injection at 20 kV for 5 s
Detector: Thermooptical, pump 442 nm (15 mW He-Cd laser), probe 632.8 nm (2 mW He-Ne laser)
Migration time: 35 (Cis), 36 (His), 36.5 (Lys), 38 (Arg), 42.5 (Lys), 43.5 (Cys), 62 (Asn), 63.5 (Gly), 64 (Thr), 65 (Asp), 66 (Ala), 66.5 (Gln), 67 (Glu, Pro), 68.5 (Val), 70 (Tyr), 71.5 (Met), 73 (Phe), 74 (Trp)

KEY WORDS
laser

REFERENCE
Waldron, K.C.; Wu, S.L.; Earle, C.W.; Harke, H.R.; Dovichi, N.J. Capillary zone electrophoresis separation and laser-based detection of both fluorescein thiohydantoin and dimethylaminoazobenzene thiohydantoin derivatives of amino acids, *Electrophoresis*, **1990**, *11*, 777–780.

4-N,N-Dimethylamino-1-naphthyl Isothiocyanate

RELATED REFERENCE
Lee, Y.-M.; Nakamura, H.; Nakajima, T. Identification of 4-N,N-dimethylamino-1-naphthylthiohydantoin amino acids by high-performance liquid chromatography with enhanced fluorescence detection. *Anal.Sci.* **1989**, *5*, 281-284.

SAMPLE
Matrix: bulk
Analyte: amino acids
Sample preparation: Dissolve 250 nmoles of the amino acid in 500 μL 50 mM pH 10.0 sodium borate buffer, add 500 μL 6 mM 4-N,N-dimethylamino-1-naphthyl isothiocyanate in dioxane (Caution! Dioxane is a carcinogen!), purge with nitrogen, heat at 40° for 1.5 min, concentrate under a stream of nitrogen, add 500 μL water, add 1 mL hexane, vortex, centrifuge, discard the hexane layer, repeat the hexane wash twice more. Dry the aqueous layer under a stream of nitrogen, reconstitute with 500 μL trifluoroacetic acid:water 50:50, purge with nitrogen, heat at 80° for 10 min, dry under a stream of nitrogen, reconstitute with 1 mL MeCN, inject a 10 μL aliquot.

HPLC VARIABLES
Column: 150 × 4.6 5 μm TSK-gel ODS-120A (Toyo Soda, Tokyo)
Mobile phase: Gradient. A was MeCN:50 mM pH 7.0 imidazole nitrate buffer 15:85. B was MeCN:water 60:40. A:B from 100:0 to 0:100 over 25 min.
Flow rate: 1
Injection volume: 10
Detector: F ex 345 em 435 following post-column reaction. The column effluent mixed with MeCN:150 mM NaOH 60:40 pumped at 1 mL/min and the mixture flowed through a 50 cm × 0.5 mm ID stainless steel coil to the detector.

CHROMATOGRAM
Retention time: 12.1, 12.4 (Asp), 13.6 (Glu), 13.6, 14.1 (carboxymethylcysteine), 17.4 (Asn), 17.6, 18.7 (His), 18.0, 18.4 (Ser), 18.5 (Gln), 18.6, 19.0 (Thr), 20.3, 20.9 (Arg), 20.6 (Gly), 21.8, 23.0 (Tyr), 22.7 (Ala), 25.9, 27.2 (Trp), 26.2, 26.6 (Met), 26.7 (Val), 27.1, 31.5 (Lys), 27.8, 28.5 (Phe), 28.7 (Ile), 29.0 (Leu)
Limit of detection: 0.2 pmole

KEY WORDS
post-column reaction; stereoisomers give rise to two peaks for some amino acids

REFERENCE
Miyano, H.; Nakajima, T.; Imai, K. Micro-scale sequence analysis from the N-terminus of peptides using the fluorogenic Edman reagent 4-N,N-dimethylamino-1-naphthyl isothiocyanate, *Biomed.Chromatogr.*, **1987**, 2, 139–144.

SAMPLE
Matrix: bulk
Analyte: amino acids
Sample preparation: Dissolve a few mg amino acids in 200 μL reagent, let stand at room temperature in the dark for 5-10 min, evaporate to dryness under reduced pressure, reconstitute with mobile phase, inject an aliquot. (Prepare reagent by mixing 5 mL MeOH, 2.5 mL pyridine, 1 mL triethylamine, 1.5 mL water, and 500 μL 100 mM 4-dimethylamino-1-naphthylisothiocyanate in MeCN.)

HPLC VARIABLES
Column: 250 × 4 5 μm Chiradex (immobilized β-cyclodextrin) (Merck)
Mobile phase: MeOH:100 mM pH 6.5 ammonium acetate buffer containing 0.1% triethylamine 40:60

Column temperature: 20 ± 0.1
Flow rate: 0.5
Injection volume: 20
Detector: UV 254; F ex 240 em 490

CHROMATOGRAM

Retention time: k' 1.98 (second enantiomer), α = 1.00 (Ala); k' 2.19 (second enantiomer), α = 1.00 (Val); k' 4.58 (second enantiomer), α = 1.40 (Leu); k' 2.84 (second enantiomer), α = 1.08 (Ile); k' 3.79 (second enantiomer), α = 1.10 (Pro); k' 3.32 (second enantiomer), α = 1.18 (Met); k' 3.58 (second enantiomer), α = 1.00 (Cys); k' 2.28 (second enantiomer), α = 1.00 (Ser); k' 2.14 (second enantiomer), α = 1.00 (Thr); k' 1.63 (second enantiomer), α = 1.11 (Lys); k' 4.97 (second enantiomer), α = 1.32 (Lys; bis derivative); k' 1.60 (second enantiomer), α = 1.13 (Arg); k' 3.06 (second enantiomer), α = 1.17 (Asn); k' 19.26 (second enantiomer), α = 1.10 (Asp); k' 11.66 (second enantiomer), α = 1.00 (Glu); k' 4.66 (second enantiomer), α = 1.14 (Phe); k' 5.10 (second enantiomer), α = 1.00 (Trp); k' 2.96 (second enantiomer), α = 1.08 (His); k' 2.95 (second enantiomer), α = 1.10 (Tyr)

KEY WORDS

chiral; α = k' (second enantiomer)/k' (first enantiomer); detailed comparison with other derivatizing reagents

REFERENCE

Rizzi, A.M.; Cladrowa-Runge, S.; Jonsson, H.; Osla, S. Enantiomeric resolution of derivatized DL-amino acids by high-performance liquid chromatography using a β-cyclodextrin chiral stationary phase: A comparison between derivatization labels, *J.Chromatogr.A*, **1995**, *710*, 287–295.

p-N,N-Dimethylaminophenylisothiocyanate

SAMPLE

Matrix: solutions
Analyte: amino acids
Sample preparation: Mix amino acid solution with 2 mL freshly-prepared buffer, add 5 mL 10 ppm p-N,N-dimethylaminophenylisothiocyanate in acetone, purge tube with nitro-

gen, heat at 50° for 20 min, add 5 mL acid solution, purge with nitrogen, heat at 50° for 20 min, let stand for 20 min, evaporate to dryness under reduced pressure, wash the residue several times with isooctane, reconstitute with 100 mM pH 2 phosphate buffer, inject a 10 μL aliquot. (Prepare the buffer by mixing 50 mL water, 50 mL acetone, 5 mL 200 mM acetic acid, 1 mL triethylamine, and enough pyridine to adjust the pH to 9.5. Prepare the acid solution by mixing equal volumes of water and glacial acetic acid saturated with HCl. Synthesis of p-N,N-dimethylaminophenylisothiocyanate is as follows. Add 150 mL water containing 33 g sodium carbonate and 24 g thiophosgene to 60 mmoles N,N-dimethyl-1,4-phenylenediamine (p-N,N-dimethylaminoaniline) in 100 mL EtOH:water 50:50 with vigorous stirring, stir at room temperature for 2 h, extract with dichloromethane. Dry the organic layer and evaporate it to dryness under reduced pressure to obtain p-N,N-dimethylaminophenylisothiocyanate as an oil.)

HPLC VARIABLES

Column: 250 × 4.6 10 μm RP-8 (Hewlett-Packard)

Mobile phase: MeCN:18.13 g/L KH_2PO_4 25:75, adjusted to pH 2 with sulfuric acid or KOH

Flow rate: 2.5

Injection volume: 10

Detector: E, Bioanalytical Systems LC-4, TL-5 glassy carbon flow-cell 0.85 V, Ag/AgCl reference electrode

CHROMATOGRAM

Retention time: 4 (Cys), 5 (Asp), 6 (CyssCy), 7 (Asn), 9.5 (Ser), 10 (Glu), 10.5 (Gln), 15 (Thr), 18 (Gly), 19 (His), 20 (Ala), 22 (Arg), 25 (Met), 29 (Pro), 34 (Val), 42 (Trp), 49 (Phe), 53 (Tyr), 61 (Lys), 70 (Ile), 74 (Leu)

Limit of detection: 0.2-0.6 ng

REFERENCE

Mahachi, T.J.; Carlson, R.M.; Poe, D.P. p-N,N-Dimethylaminophenylisothiocyanate as an electrochemical label for high-performance liquid chromatographic determination of amino acids, *J.Chromatogr.*, **1984**, *298*, 279–288.

4-(N,N-Dimethylaminosulfonyl)-7-isothiocyanato-2,1,3-benzoxadiazole

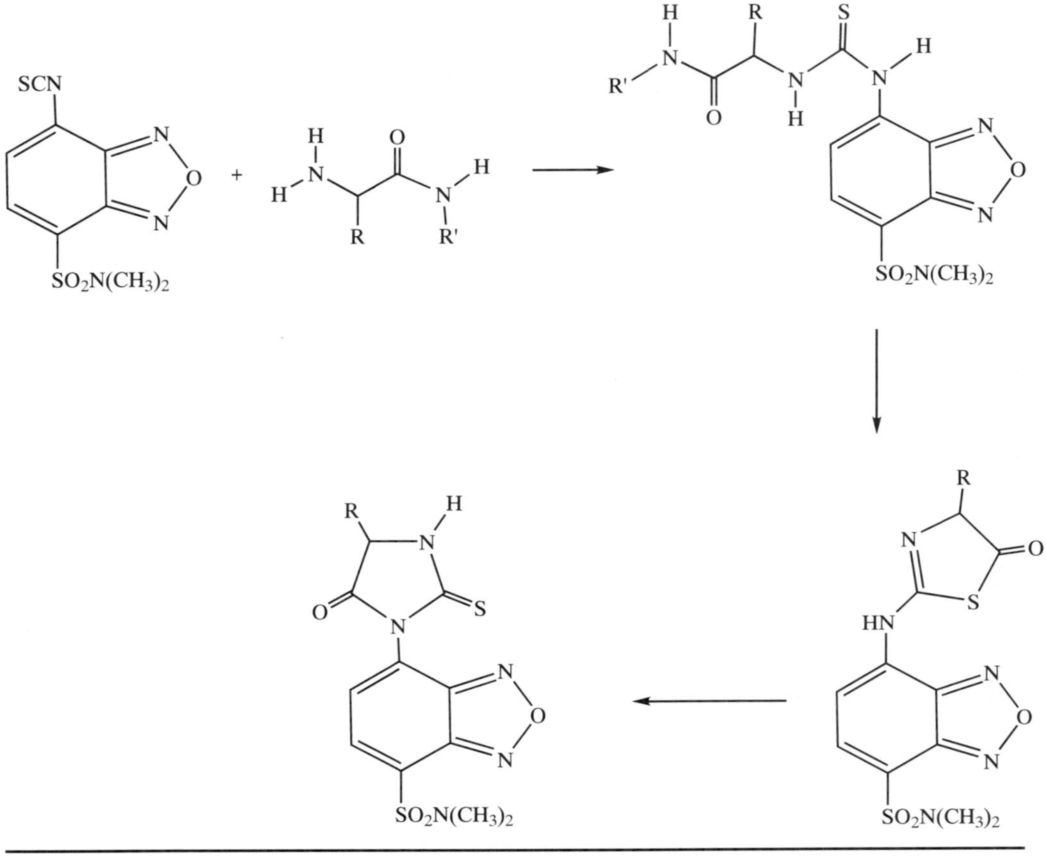

SAMPLE

Matrix: peptides

Analyte: amino acids

Sample preparation: Vortex 10 μL of a 50-100 μM solution of a dipeptide in pyridine:water 50:50 with 10 μL 20 mM 4-(N,N-dimethylaminosulfonyl)-7-isothiocyanato-2,1,3-benzoxadiazole in pyridine:water 50:50, heat at 50° for 15 min, wash 3 times with 100 μL portions of n-heptane:dichloromethane 80:20. Evaporate the aqueous layer to dryness at 50° for 15 min, add 30 μL 1% boron trifluoride in dichloromethane containing 0.02% ethanethiol, heat at 50° for 5 min, evaporate to dryness under a stream of nitrogen, add 20 μL water, add 100 μL n-heptane:dichloromethane 70:30, mix, centrifuge at 1000 g for 5 min, repeat the extraction 3 times. Combine the organic layers and evaporate them to dryness under a stream of nitrogen, reconstitute the residue in MeCN, inject an aliquot. (Synthesis of 4-(N,N-dimethylaminosulfonyl)-7-isothiocyanato-2,1,3-benzoxadiazole is as follows. Dissolve 0.5 g magnesium sulfate heptahydrate and 6 g NaOH in 60 mL water, throughout the reaction keep the flask at about 20° with cold water cooling, add 15 mL 30% hydrogen peroxide, add 75 mL MeOH, add 12.1 g powdered benzoyl peroxide in one go, stir for 10 min, pour into 150 mL 20% sulfuric acid, extract three times with 50 mL portions of chloroform, determine peroxybenzoic acid concentration by iodometric titration (Tetrahedron 1967, 23, 3327). Slowly add 110 mL 1 M peroxybenzoic acid in chloroform to 7 g 2,6-difluoroaniline dissolved in 100 mL chloroform, stir at room temperature, when reaction is complete (iodometric titration) wash with 2% sodium thiosulfate, wash with 5% sodium carbonate, wash with water, dry over anhydrous sodium sulfate, evaporate to dryness under reduced pressure, recrystallize 2,6-difluoronitrosobenzene from EtOH (mp 108.5-109.5). Stir 8.5 g 2,6-difluoronitrosobenzene in 85 mL DMSO at room temperature and add a solution of 3.91 g sodium azide in 85 mL DMSO dropwise, let stand for about 1 h, add to a large volume of water, extract with ether, dry the extracts over anhydrous sodium sulfate, evaporate to dryness under reduced pressure and distil to give 4-fluoro-2,1,3-benzoxadiazole as a colorless oil (bp 83°/12 mm Hg) (J.Chem.Soc.(C) 1970, 1433). Add 11 mL chlorosulfonic acid dropwise to 3 g 4-fluoro-2,1,3-benzoxadiazole in 10 mL chloroform at 0-10° (use a calcium chloride drying tube), stir at room temperature for 1 h, reflux for 2 h, cool, slowly pour into ice water, remove the organic layer, extract the aqueous layer with chloroform, combine the organic layers, wash, dry over anhydrous magnesium sulfate, evaporate under reduced pressure, take up the residue in 5 mL benzene (Caution! Benzene is a carcinogen!), chromatograph on a 150×30 column of silica gel (100-200 mesh Kanto Chemical) with n-hexane:benzene 50:50, evaporate the appropriate fractions to give 4-(chlorosulfonyl)-7-fluoro-2,1,3-benzoxadiazole (CBD-F) as pale yellow needles (mp 64-66°) (Anal. Chem. 1984. 56, 2461). Stir 0.76 g CBD-F in 70 mL MeCN at 0-10° and add 1 g dimethylamine hydrochloride in 10 mL 100 mM pH 10 borax dropwise, adjust pH to 5 with 1 M HCl, concentrate to about 10 mL under reduced pressure, extract three times with 200 mL portions of diethyl ether, wash with water, dry over anhydrous magnesium sulfate, evaporate under reduced pressure, chromatograph on a 500×20 column of silica gel with chloroform, isolate the appropriate fraction and re-chromatograph on the same column with ethyl acetate:benzene 1:2 to give 4-(N,N-dimethylaminosulfonyl)-7-fluoro-2,1,3-benzoxadiazole (DBD-F) as white needles (mp 124-125°) (yield = 1% !). On a Merck no. 5714 $60F_{254}$ tlc plate eluted with chloroform DBD-F has Rf 0.32 and lies between two other reaction products (Analyst 1989, 114, 413). DBD-F can also be purchased from Tokyo Kasei (TCI America, 911 North Harborgate St., Portland, OR 97203; 800-423-8616, 503-283-1681, (fax) 503-283-1987; www.tciamerica.com). Add 100 µL 28% ammonia in water to 50 mg 4-(N,N-dimethylaminosulfonyl)-7-fluoro-2,1,3-benzoxadiazole in 15 mL MeCN, stir at room temperature overnight, filter, evaporate the filtrate to dryness under reduced pressure, recrystallize from MeCN to give 4-amino-7-(N,N-dimethylaminosulfonyl)-2,1,3-benzoxadiazole as pale yellow needles (mp 214-217°). Add 1 mL 30% thiophosgene in benzene dropwise to 200 mg 4-amino-7-(N,N-dimethylaminosulfonyl)-2,1,3-benzoxadiazole in 15 mL MeCN, reflux for 5 h, concentrate under reduced pressure, extract the residue twice with 20 mL portions of chloroform. Combine the chloroform layers, filter, evaporate the filtrate to dryness, chromatograph the residue on 15 g silica gel G-200 with chloroform. Collect the fraction containing the product and evaporate it to dryness under reduced pressure, recrystallize from benzene/n-hexane to obtain 4-(N,N-dimethylaminosulfonyl)-7-isothiocyanato-2,1,3-benzoxadiazole as pale yellow-white crystals (mp 122-124°; yield 26%) (Biomed. Chromatogr. 1993, 7, 56).)

HPLC VARIABLES
Column: 150×6 5 µm ES-1/4phCD phenylcarbamoylated β-cyclodextrin (Shinwa, Kyoto)
Mobile phase: MeCN:MeOH:water 15:40:45 containing 10 mM acetic acid
Injection volume: 20
Detector: F ex 387 em 524

CHROMATOGRAM
Retention time: 21 (D-Pro), 23 (L-Pro), 27 (L-Leu), 30 (D-Leu)

KEY WORDS
chiral

REFERENCE
Matsunaga, H.; Iida, T.; Fukushima, T.; Santa, .; Homma, H.; Imai, K. Boron-trifluoride etherate (Lewis acid) as an efficient acid at cyclization/cleavage reaction of D/L-amino acids affording the retention of their original configuration in the Edman sequencing method of peptides, *Biomed.Chromatogr.*, **1996**, *10*, 95–96.

SAMPLE
Matrix: peptides
Analyte: amino acids
Sample preparation: Mix 10 µL of a 50 µM peptide solution in pyridine:water 50:50 with 10 µL 20 mM 4-(N,N-dimethylaminosulfonyl)-7-isothiocyanato-2,1,3-benzoxadiazole in

pyridine:water 50:50, vortex, heat at 50° for 15 min, wash three times with 100 μL portions of n-heptane:dichloromethane 80:20, dry at 50° under reduced pressure for 15 min, add 30 μL 1% boron trifluoride etherate in dichloroethane containing 0.1% ethanethiol, heat at 50° for 5 min, evaporate to dryness under a stream of nitrogen, add 20 μL water, add 100 μL ethyl acetate:benzene 75:25 (Caution! Benzene is a carcinogen!), mix, centrifuge at 1000 g for 1 min, repeat the extraction 3 times. Combine the organic layers and evaporate them to dryness under a stream of nitrogen, reconstitute the residue in mobile phase, inject an aliquot. (Evaporate the aqueous phase to dryness and continue the cycle. Boron trifluoride gives less racemization than trifluoroacetic acid. Synthesis of 4-(N,N-dimethylaminosulfonyl)-7-isothiocyanato-2,1,3-benzoxadiazole is as described above.)

HPLC VARIABLES
Column: 150 × 6 5 μm ES-1/4phCD 25% phenylcarbamoylated β-cyclodextrin (Shinwa)
Mobile phase: MeOH:water 70:30 containing 10 mM acetic acid
Column temperature: 5
Flow rate: 0.5
Injection volume: 20
Detector: F ex 387 em 524

CHROMATOGRAM
Retention time: 19 (L-Arg), 21 (D-Arg), 30 (L-Phe), 33 (D-Phe), 35 (L-Met), 38 (D-Met)

KEY WORDS
chiral

REFERENCE
Matsunaga, H.; Santa, T.; Iida, T.; Fukushima, T.; Homma, H.; Imai, K. Proton: A major factor for the racemization and the dehydration at the cyclization/cleavage stage in the Edman sequencing method, *Anal.Chem.*, **1996**, *68*, 2850–2856.

SAMPLE
Matrix: solutions
Analyte: amino acids
Sample preparation: Mix 10 μL of a 50 μM amino acid solution in 200 mM pH 10.0 borate buffer with 10 μL 10 mM 4-(N,N-dimethylaminosulfonyl)-7-isothiocyanato-2,1,3-benzoxadiazole in MeCN, heat at 55° for 2 min, add 180 μL 1 M acetic acid, inject a 20 μL aliquot. (Synthesis of 4-(N,N-dimethylaminosulfonyl)-7-isothiocyanato-2,1,3-benzoxadiazole is as described above.)

HPLC VARIABLES
Column: 150 × 4.6 5 μm TSK-ODS 80TM (Tosoh)
Mobile phase: Gradient. A was MeCN:water 23:77 containing 0.05% trifluoroacetic acid. B was MeCN:water 52:48 containing 0.05% trifluoroacetic acid. A:B from 100:0 to 0:100 over 30 min.
Flow rate: 1
Injection volume: 20
Detector: F ex 385 em 505

CHROMATOGRAM
Retention time: 11 (Asp), 12.5 (ammonia), 15 (Ala), 20 (Tyr), 25 (Ile), 27 (Phe)

REFERENCE
Imai, K.; Uzu, S.; Nakashima, K.; Akiyama, S. Synthesis of novel fluorogenic Edman reagents, 7-N,N-dimethylaminosulphonyl-4-(2,1,3-benzoxadiazolyl)isothiocyanate (DBD-NCS) and 7-aminosulphonyl-4-(2,1,3-benzoxadiazolyl)isothiocyanate, *Biomed.Chromatogr.*, **1993**, *7*, 56–57.

SAMPLE
Matrix: solutions
Analyte: amino acids
Sample preparation: Vortex 10 μL peptide solution in pyridine:water 50:50 and 10 μL 20 mM 4-(N,N-dimethylaminosulfonyl)-7-isothiocyanato-2,1,3-benzoxadiazole in pyridine:water 50:50, heat at 50° for 15 min, wash 3 times with 100 μL porions of n-heptane:dichloromethane 90:10. Remove the aqueous phase and dry it in a centrifugal evaporator

at 50° for 15 min, add 30 μL trifluoroacetic acid to the residue, heat at 50° for 10 min, evaporate to dryness under a stream of nitrogen, reconstitute with 20 μL water, extract 3 times with n-heptane:dichloromethane 80:20 (retain the aqueous phase for the next cycle). Combine the extracts and evaporate them to dryness under a stream of nitrogen, reconstitute in MeCN, inject an aliquot. (Perform all reactions under nitrogen. N-Terminal amino acid is derivatized and sequenced. Reagent was 4-(N,N-dimethylaminosulfonyl)-7-isothiocyanato-2,1,3-benzoxadiazole and it was prepared as described above.)

HPLC VARIABLES
Column: 250 × 4.6 5 μm YMC J'sphere ODS H-80 (YMC) + 250 × 4.6 5 μm YMC-Pack Ph phenyl (YMC) in series
Mobile phase: MeCN:water 60:40 containing 10 mM formic acid
Flow rate: 0.5
Injection volume: 20
Detector: F ex 387 em 524

CHROMATOGRAM
Retention time: 14 (Ser), 17 (Arg), 17.5 (Thr), 21 (Gly), 23 (Tyr), 24 (Ala), 26 (Asn), 28 (His), 29 (Pro), 30.5 (Asp), 31 (Glu), 32 (Met), 33, 45 (Lys), 36 (Val), 41 (Phe), 46 (Ile), 48 (Leu)
Limit of detection: 200 pmole

REFERENCE
Matsunaga, H.; Santa, T.; Hagiwara, K.; Homma, H.; Imai, K.; Uzu, S.; Nakashima, K.; Akiyama, S. Development of an efficient amino acid sequencing method using fluorescent Edman reagent 7-[(N,N-dimethylamino)sulfonyl]-2,1,3-benzoxadiazol-4-yl isothiocyanate, *Anal.Chem.*, **1995**, *67*, 4276−4282.

Diphenyl Phosphoroisothiocyanatidate

SAMPLE
Matrix: peptides
Analyte: amino acids

Sample preparation: Heat 60 nmole peptide with 100 μL acetic anhydride at 50° for 10 min, evaporate to dryness under reduced pressure, reconstitute with 100 μL water:triethylamine 98:2, let stand at 25° for 5 min, evaporate to dryness under reduced pressure, reconstitute with 40 μL anhydrous MeCN, add 10 μL pyridine, add 60 μL 1 mM diphenyl phosphoroisothiocyanate in MeCN, heat at 50° for 30 min, evaporate to dryness under reduced pressure, reconstitute with 100 μL 0.1% trifluoroacetic acid in water, inject an aliquot. (Diphenyl phosphoroisothiocyanatidate is available from Lancaster Synthesis, Windham NH. Synthesis is as follows. Add 50.1 g diphenyl chlorophosphate to a solution of 19.3 g potassium thiocyanate in 200 mL MeCN, shake for 3 h, let stand for 3 h, dilute with 300 mL dry benzene (Caution! Benzene is a carcinogen!), filter, evaporate the filtrate, distil the residue to give diphenyl phosphoroisothiocyanatidate (bp 105°/0.1 mm Hg) (J. Chem. Soc. 1953, 673).)

HPLC VARIABLES
Column: 250 × 2.0 3 μm Reliasil C18 (Column Engineering, Ontario CA)
Mobile phase: Gradient. A was 0.1% trifluoroacetic acid in water. B was MeCN:MeOH: water 80:10:10. A:B 100:0 for 2 min, to 96:4 over 3 min, to 65:35 over 35 min, to 50: 50 over 10 min, return to initial conditions over 2 min.
Detector: UV 265

CHROMATOGRAM
Retention time: 7 (asparagine), 8 (glycine), 10 (aspartic acid), 12.5 (glutamine), 13.5 (histidine), 14 (alanine), 15.5 (glutamic acid), 17 (serine), 19 (arginine), 21 (cysteine), 23 (lysine), 23.5 (threonine), 26 (valine), 28 (tyrosine), 28.5 (methionine), 35.5 (isoleucine), 38 (leucine), 41 (phenylalanine), 44 (tryptophan)

REFERENCE
Bailey, J.M.; Nikfarjam, F.; Shenoy, N.R.; Shively, J.E. Automated carboxy-terminal sequence analysis of peptides and proteins using diphenyl phosphoroisothiocyanatidate, *Protein Sci.*, **1992**, *1*, 1622–1633.

Fluorescein Isothiocyanate, Isomer I

SAMPLE
Matrix: CSF
Analyte: amino acids
Sample preparation: 50 μL CSF + 50 μL 210 μM fluorescein isothiocyanate isomer I in acetone, mix, let stand in the dark for 2 h, dilute 100-1000 times, inject an aliquot.

CAPILLARY ELECTROPHORESIS
Capillary: 75 cm × 50 μm fused-silica (42 cm to detector) (Polymicro Technologies)
Capillary preparation: Rinse with 100 mM NaOH for 3 min, with water for 2 min, and with running buffer for 3 min.
Running buffer: 100 mM Boric acid containing 100 mM sodium dodecyl sulfate, adjusted to pH 9.3 with NaOH

Voltage/Current: +20 kV/42 μA
Injection: Hydrodynamic injection for 2 s (15 nL)
Detector: F ex 488 (laser)
Model: TSP Spectraphoresis 100
Migration time: 8.4 (Lys), 8.6 (Arg), 9.1 (ornithine), 9.9 (ammonia), 10.0 (tyramine), 7.2 (Leu), 7.3 (Gln), 7.4 (Tyr), 7.6 (Val), 7.7 (Thr), 7.8 (Phe), 8.2 (Ser), 8.4 (Ala), 8.9 (taurine), 9.2 (Gly), 18.4 (Glu), 19.6 (Asp)
Limit of detection: <0.2 nM

REFERENCE

Nouadje, G.; Rubie, H.; Chatelut, E.; Canal, P.; Nertz, M.; Puig, P.; Couderc, F. Child cerebrospinal fluid analysis by capillary electrophoresis and laser-induced fluorescence detection, *J.Chromatogr.A*, **1995**, *717*, 293–298.

SAMPLE

Matrix: cheese
Analyte: amino acids
Sample preparation: Homogenize 4 parts cheese with 3 parts 100 mM HCl, suspend 230 mg paste in 5 mL 100 mM HCl, centrifuge, extract the residue twice with 5 mL portions of 100 mM HCl. Filter the supernatants and neutralize them with 200 mM sodium carbonate solution. Remove a 1 mL aliquot and add it to 1 mL 83 μg/mL fluorescein isothiocyanate in acetone, let stand in the dark for 4 h (J.Chromatogr. 1991, 559, 183), dilute 100 000-fold, inject an aliquot.

CAPILLARY ELECTROPHORESIS

Capillary: 80 cm × 50 μm fused-silica (50 cm to detector) (Polymicro Technologies)
Capillary preparation: Rinse with 100 mM NaOH for 3 min, with water for 3 min, and with running buffer for 3 min.
Running buffer: 100 mM Boric acid containing 100 mM sodium dodecyl sulfate, pH adjusted to 9.2 with NaOH
Voltage/Current: 24 kV
Injection: Hydrodynamic injection for 2 s (17.5 nL)
Detector: F ex 488 (7 mW argon ion laser)
Model: TSP Spectra-Phoresis 100
Migration time: 10.5 (Lys), 10.7 (Arg), 14.8 (Tyr), 15.4 (Phe), 15.8 (Ser), 16.1 (Ala), 16.9 (Gly), 21 (Asp)
Limit of detection: 0.05-0.15 nM

OTHER SUBSTANCES

Extracted: ammonia, cadaverine, histamine, ornithine, β-phenylethylamine, putrescine, tryptamine

REFERENCE

Nouadje, G.; Nertz, M.; Verdeguer, P.; Couderc, F. Ball-lens laser-induced fluorescence detector as an easy-to-use highly sensitive detector for capillary electrophoresis. Application to the identification of biogenic amines in dairy products, *J.Chromatogr.A*, **1995**, *717*, 335–343.

SAMPLE

Matrix: dialysate
Analyte: glutamate
Sample preparation: Mix dialysate (artificial CSF) with 20 mM pH 9.4 carbonate buffer and 400 μM fluorescein isothiocyanate, isomer I in acetone in the ratio 5:1:1, let stand for 16 h, dilute 10-fold with 20 mM pH 9.4 carbonate buffer, inject an aliquot.

CAPILLARY ELECTROPHORESIS

Capillary: 30 cm × 20 μm fused-silica (20 cm to detector) (Polymicro Technologies)
Capillary preparation: After each run rinse with 100 mM NaOH for 2 min, rinse with water for 2 min, rinse with running buffer for 3 min.
Running buffer: 20 mM pH 9.4 Carbonate buffer
Voltage/Current: 20 kV
Injection: Pressure injection by 19 psi for 0.3 s.
Detector: F ex 488 (3 mW argon ion laser) em 510
Migration time: 3.5

REFERENCE
Tucci, S.; Rada, P.; Sepúlveda, M.J.; Hernandez, L. Glutamate measured by 6-s resolution brain microdialysis: capillary electrophoretic and laser-induced fluorescence detection application, *J.Chromatogr.B*, **1997**, *694*, 343–349.

SAMPLE
Matrix: solutions
Analyte: amino acids
Sample preparation: Prepare a solution in 5 mM pH 10 carbonate buffer. Add a 2-5 mL aliquot to 20 µL reagent, let stand for 2-4 h in the dark, inject an aliquot. (Reagent was 550 µM fluorescein isothiocyanate, isomer 1 in acetone containing a trace of pyridine.)

CAPILLARY ELECTROPHORESIS
Capillary: 99 cm × 50 µm fused-silica
Running buffer: 5 mM pH 10 carbonate buffer
Voltage/Current: 25 kV
Injection: Electromigration at 2 kV for 10 s
Detector: F ex 488 (1 W Ar laser) em 495 (long pass filter) - 560 (short pass filter) (details of detector given in paper)
Migration time: 13.2 (Arg), 14.8 (Lys), 17.7 (Leu), 17.8 (Ile, Trp), 18.0 (Met), 18.2 (Phe, Val, His, Pro), 18.5 (Thr), 18.9 (Ser), 19.0 (Cys), 19.15 (Ala), 19.7 (Gly), 20.5 (Tyr), 24.0 (Glu), 25.0 (Asp)
Limit of detection: 0.005-0.086 nM

KEY WORDS
laser

REFERENCE
Cheng, Y.-F.; Dovichi, N.J. Subattomole amino acid analysis by capillary zone electrophoresis and laser-induced fluorescence, *Science*, **1988**, *242*, 562–564.

SAMPLE
Matrix: solutions
Analyte: amino acids
Sample preparation: 5 µL 200 µM Amino acid in 200 mM pH 9.0 carbonate buffer + 20 µL 200 mM pH 9.0 carbonate buffer + 24 µL acetone + 1 µL 5 mM fluorescein isothiocyanate isomer 1 (Sigma) in acetone, heat at 40° for 4 h, dilute with water, inject an aliquot.

CAPILLARY ELECTROPHORESIS
Capillary: 57 cm × 50 µm fused-silica (50 cm to detector) (Polymicro Technologies)
Capillary preparation: Before each injection rinse with running buffer for 2 min.
Running buffer: 50 mM pH 7.5 Phosphate-borate buffer containing 75 mM sodium dodecyl sulfate
Capillary temperature: 35
Voltage/Current: 30 kV
Injection: Pneumatic injection of sample for 2 s then 0.1 s injection of running buffer
Detector: F ex 488 (?) (argon ion laser)
Model: Beckman P/ACE System 2100
Migration time: 4.0, 4.2 (Lys), 4.5 (Arg), 5.5 (His), 5.7 (Hyp), 5.78 (Tyr), 5.82 (Gln), 5.89 (Pro), 5.91 (Leu, Ile), 5.95 (Met), 6.0 (Val), 6.05 (Asn), 6.1 (Phe), 6.15 (Ser), 6.35 (Ala), 6.52 (Gly), 8.7 (Glu), 9.2 (Asp)

REFERENCE
Lalljie, S.P.D.; Sandra, P. MEKC analysis of FITC and DTAF amino acid derivatives with LIF detection, *Chromatographia*, **1995**, *40*, 513–518.

9-Isothiocyanatoacridine

SAMPLE
Matrix: protein
Analyte: amino acids
Sample preparation: Heat 2 mg protein and 1 mL 6 M HCl in a sealed tube at 110° for 20 h, reconstitute with 5 mL buffer. Remove a 400 μL aliquot and add it to 600 μL buffer and 400 μL 2 mM 9-isothiocyanatoacridine in MeCN, heat at 60° with occasional shaking for 1.5 h, cool, inject a 20 μL aliquot. (Prepare buffer by mixing 500 mL 100 mM boric acid containing 100 mM KCl with 408 mL 100 mM NaOH, make up to 1 L with water, pH 9.8. Prepare 9-isothiocyanatoacridine as follows. Reflux 15 mmoles 9-chloroacridine (Eastman) and 16 mmoles silver thiocyanate (Aldrich) in 150 mL anhydrous toluene, recrystallize 9-isothiocyanatoacridine from anhydrous acetone, mp 131-2° (Chem.Abs. 1970, 72, 21584v).)

HPLC VARIABLES
Column: 150 × 3.3 5 μm Separon C-18 SGX (Tessek, Prague)
Mobile phase: MeCN:23 mM pH 6 ammonium formate 20:80
Flow rate: 0.2
Injection volume: 20
Detector: UV 280

CHROMATOGRAM
Retention time: 4.62 (Glu), 4.85 (Asp), 5.88 (Ala), 6.44 (Asn), 6.83 (Gly), 7.10 (Ser), 7.90 (Thr), 8.37 (His), 10.73 (Gln), 11.63 (Val), 12.27 (Tyr), 13.37 (Pro, Lys), 15.20 (Arg), 16.97 (Met), 26.07 (Ile), 26.60 (Leu), 36.93 (Phe), 38.97 (Trp)
Limit of detection: 100-400 pmole

REFERENCE
Oravec, P.; Podhradsky, D. High-performance liquid chromatography of amino acids after derivatization with 9-isothiocyanatoacridine, *J.Biochem.Biophys.Methods*, **1995**, *30*, 145−152.

(R)-(-)-4-(3-Isothiocyanatopyrrolidin-1-yl)-7-(N,N-dimethylaminosulfonyl)-2,1,3-benzoxadiazole

RELATED REFERENCE
Toyo'oka, T.; Liu, Y.-M. Determination of D- and L-amino acid residues in peptides with fluorescent chiral tagging reagents by high-performance liquid chromatography. *Chromatographia* **1995**, *40*, 645-651.

SAMPLE
Matrix: solutions

Analyte: amino acids

Sample preparation: Mix 10 μL of an amino acid solution in MeCN:water:triethylamine 50:50:2 with 10 μL 5 mM (R)-(-)-4-(3-isothiocyanatopyrrolidin-1-yl)-7-(N,N-dimethylaminosulfonyl)-2,1,3-benzoxadiazole in MeCN, heat at 55° for 10 min, add 480 μL 1 M acetic acid in MeCN:water 50:50, dilute 10-fold with MeCN, inject a 5 μL aliquot. ((R)-(-)-4-(3-Isothiocyanatopyrrolidin-1-yl)-7-(N,N-dimethylaminosulfonyl)-2,1,3-benzoxadiazole ((R)-DBD-Py-NCS) is available from Tokyo Kasei (TCI America, Portland OR). Synthesis is as follows. Dissolve 0.5 g magnesium sulfate heptahydrate and 6 g NaOH in 60 mL water, throughout the reaction keep the flask at about 20° with cold water cooling, add 15 mL 30% hydrogen peroxide, add 75 mL MeOH, add 12.1 g powdered benzoyl peroxide in one go, stir for 10 min, pour into 150 mL 20% sulfuric acid, extract three times with 50 mL portions of chloroform, determine peroxybenzoic acid concentration by iodometric titration (Tetrahedron 1967, 23, 3327). Slowly add 110 mL 1 M peroxybenzoic acid in chloroform to 7 g 2,6-difluoroaniline dissolved in 100 mL chloroform, stir at room temperature, when reaction is complete (iodometric titration) wash with 2% sodium thiosulfate, wash with 5% sodium carbonate, wash with water, dry over anhydrous sodium sulfate, evaporate to dryness under reduced pressure, recrystallize 2,6-difluoronitrosobenzene from EtOH (mp 108.5-109.5). Stir 8.5 g 2,6-difluoronitrosobenzene in 85 mL DMSO at room temperature and add a solution of 3.91 g sodium azide in 85 mL DMSO dropwise, let stand for about

1 h, add to a large volume of water, extract with ether, dry the extracts over anhydrous sodium sulfate, evaporate to dryness under reduced pressure and distil to give 4-fluoro-2,1,3-benzoxadiazole as a colorless oil (bp 83°/12 mm Hg) (J.Chem.Soc.(C) 1970, 1433). Add 11 mL chlorosulfonic acid dropwise to 3 g 4-fluoro-2,1,3-benzoxadiazole in 10 mL chloroform at 0-10° (use a calcium chloride drying tube), stir at room temperature for 1 h, reflux for 2 h, cool, slowly pour into ice water, remove the organic layer, extract the aqueous layer with chloroform, combine the organic layers, wash, dry over anhydrous magnesium sulfate, evaporate under reduced pressure, take up the residue in 5 mL benzene (Caution! Benzene is a carcinogen!), chromatograph on a 150 × 30 column of silica gel (100-200 mesh Kanto Chemical) with n-hexane:benzene 50:50, evaporate the appropriate fractions to give 4-(chlorosulfonyl)-7-fluoro-2,1,3-benzoxadiazole (CBD-F) as pale yellow needles (mp 64-66°) (Anal. Chem. 1984. 56, 2461). Stir 0.76 g CBD-F in 70 mL MeCN at 0-10° and add 1 g dimethylamine hydrochloride in 10 mL 100 mM pH 10 borax dropwise, adjust pH to 5 with 1 M HCl, concentrate to about 10 mL under reduced pressure, extract three times with 200 mL portions of diethyl ether, wash with water, dry over anhydrous magnesium sulfate, evaporate under reduced pressure, chromatograph on a 500 × 20 column of silica gel with chloroform, isolate the appropriate fraction and re-chromatograph on the same column with ethyl acetate:benzene 1:2 to give 4-(N,N-dimethylaminosulfonyl)-7-fluoro-2,1,3-benzoxadiazole (DBD-F) as white needles (mp 124-125°) (yield = 1% !). On a Merck no. 5714 60F$_{254}$ TLC plate eluted with chloroform DBD-F has Rf 0.32 and lies between two other reaction products (Analyst 1989, 114, 413). DBD-F can also be purchased from Tokyo Kasei. Cool a solution of 16.4 g (S)-(-)-1-benzyl-3-pyrrolidinol in 164 mL pyridine to +5°, add 19.35 g p-toluenesulfonyl chloride, stir at +10° for 48 h, evaporate to dryness, chromatograph using dichloromethane:acetone 95:5 to obtain (3S)-3-[(4-tolylsulfonyl)oxy]-1-(phenylmethyl)pyrrolidine (mp 68°). Heat a solution of (3S)-3-[(4-tolylsulfonyl)oxy]-1-(phenylmethyl)pyrrolidine in 200 mL anhydrous DMF to 65°, add 33.5 g sodium azide (Caution! Sodium azide is highly toxic!), stir at 60° for 7 h, filter, evaporate the filtrate to dryness under reduced pressure, dissolve the residue in ethyl acetate, wash twice with water, dry over anhydrous magnesium sulfate, evaporate to obtain (3R)-3-azido-1-(phenylmethyl)pyrrolidine as an oil. Add 3.5 g 10% palladium on carbon under nitrogen to a solution of 7.05 g (3R)-3-azido-1-(phenylmethyl)pyrrolidine in 34.8 mL 1 M HCl in water and 245 mL EtOH, hydrogenate at atmospheric pressure for 30 min, add 3.5 g catalyst, hydrogenate for 2 h, filter, add 34.8 mL 1 M HCl to the filtrate, evaporate to dryness under reduced pressure, take up the residue in 70 mL EtOH, filter, evaporate the filtrate to dryness under reduced pressure, repeat this operation twice, crystallize with the minimum amount of EtOH to obtain (3R)-3-aminopyrrolidine dihydrochloride (J. Med. Chem. 1992, 35, 4205). 3R-(+)-Aminopyrrolidine is also available from Tokyo Kasei. Add 100 mg 4-(N,N-dimethylaminosulfonyl)-7-fluoro-2,1,3-benzoxadiazole in 20 mL MeCN dropwise to a stirred solution of 200 mg 3R-(+)-aminopyrrolidine in 20 mL MeCN at 0-10°, stir at room temperature for 30 min, remove the MeCN by evaporation under reduced pressure, dissolve the residue in 50 mL 5% HCl, wash 3 times with 50 mL portions of ethyl acetate, adjust the pH of the aqueous solution to 13-14 with 5% NaOH, extract 6 times with 50 mL portions of ethyl acetate. Combine the organic layers and wash them with 20 mL water, dry over anhydrous sodium sulfate, evaporate to dryness under reduced pressure, recrystallize from hexane to obtain (R)-(-)-4-(3-aminopyrrolidin-1-yl)-7-(N,N-dimethylaminosulfonyl)-2,1,3-benzoxadiazole as orange crystals (mp 96-98°) (Analyst 1992, 117, 727). Add 100 μL thiophosgene in 10 mL benzene (Caution! Benzene is a carcinogen!) to 100 mg (R)-(-)-4-(3-aminopyrrolidin-1-yl)-7-(N,N-dimethylaminosulfonyl)-2,1,3-benzoxadiazole in 100 mL acetone, reflux for 1 h, remove the solvent by evaporation under reduced pressure, suspend the residue in 100 mL water, extract 4 times with 25 mL portions of benzene. Combine the extracts and wash them with 20 mL water, dry over anhydrous sodium sulfate, evaporate to dryness under reduced pressure, recrystallize from hexane:benzene 1:2 to obtain (R)-(-)-4-(3-isothiocyanatopyrrolidin-1-yl)-7-(N,N-dimethylaminosulfonyl)-2,1,3-benzoxadiazole as yellow crystals (mp 160-170° d) (Analyst 1995, 120, 385).

HPLC VARIABLES
Column: 150 × 4.6 5 μm Inertsil ODS-80A
Mobile phase: Gradient. A was 0.05% trifluoroacetic acid in water. B was 0.05% trifluoroacetic acid in MeCN. A:B 75:25 for 5 min, to 60:40 over 20 min, maintain at 60:40 for 25 min.

Column temperature: 40
Flow rate: 1
Injection volume: 5
Detector: F ex 460 em 540

CHROMATOGRAM

Retention time: 7.5 (D-histidine), 8 (L-histidine), 16 (glycine), 16.5 (D-threonine), 17 (L-threonine), 20 (D-proline), 21 (L-proline), 26.5 (D-valine), 28 (L-valine), 31 (D-leucine), 32 (L-leucine), 33 (D-tryptophan), 35 (L-tryptophan)

KEY WORDS

chiral

REFERENCE

Toyo'oka, T.; Liu, Y.-M. High-performance liquid chromatographic resolution of amino acid enantiomers derivatized with fluorescent chiral Edman reagents, *J.Chromatogr.A*, **1995**, *689*, 23–30.

(R)-(-)-4-(3-Isothiocyanatopyrrolidin-1-yl)-7-nitro-2,1,3-benzoxadiazole

RELATED REFERENCE

Toyo'oka, T.; Liu, Y.-M. Determination of D- and L-amino acid residues in peptides with fluorescent chiral tagging reagents by high-performance liquid chromatography. *Chromatographia* **1995**, *40*, 645-651.

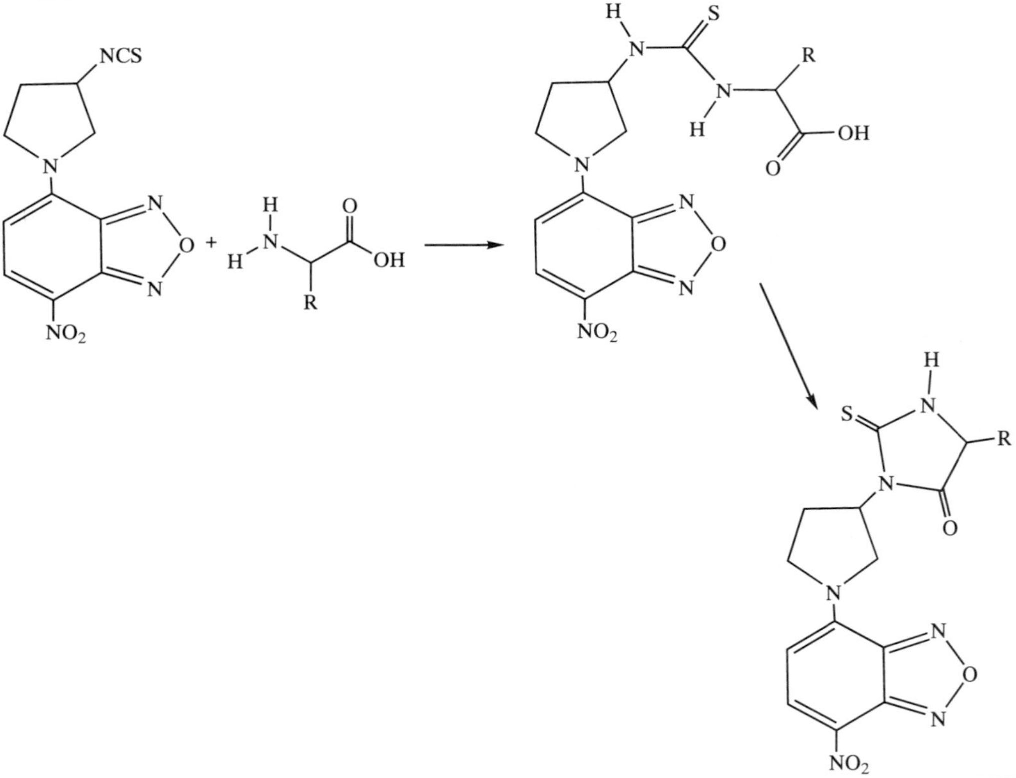

SAMPLE

Matrix: solutions
Analyte: amino acids
Sample preparation: Mix 10 µL of an amino acid solution in MeCN:water:triethylamine 50:50:2 with 10 µL 5 mM (R)-(-)-4-(3-isothiocyanatopyrrolidin-1-yl)-7-nitro-2,1,3-benzox-

adiazole in MeCN, heat at 55° for 10 min, add 480 μL 1 M acetic acid in MeCN:water 50:50, dilute 10-fold with MeCN, inject a 5 μL aliquot. (Synthesis of (R)-(-)-4-(3-isothiocyanatopyrrolidin-1-yl)-7-nitro-2,1,3-benzoxadiazole is as follows. Cool a solution of 16.4 g (S)-(-)-1-benzyl-3-pyrrolidinol in 164 mL pyridine to +5°, add 19.35 g p-toluenesulfonyl chloride, stir at +10° for 48 h, evaporate to dryness, chromatograph using dichloromethane:acetone 95:5 to obtain (3S)-3-[(4-tolylsulfonyl)oxy]-1-(phenylmethyl)-pyrrolidine (mp 68°). Heat a solution of (3S)-3-[(4-tolylsulfonyl)oxy]-1-(phenylmethyl)-pyrrolidine in 200 mL anhydrous DMF to 65°, add 33.5 g sodium azide (Caution! Sodium azide is highly toxic!), stir at 60° for 7 h, filter, evaporate the filtrate to dryness under reduced pressure, dissolve the residue in ethyl acetate, wash twice with water, dry over anhydrous magnesium sulfate, evaporate to obtain (3R)-3-azido-1-(phenylmethyl)pyrrolidine as an oil. Add 3.5 g 10% palladium on carbon under nitrogen to a solution of 7.05 g (3R)-3-azido-1-(phenylmethyl)pyrrolidine in 34.8 mL 1 M HCl in water and 245 mL EtOH, hydrogenate at atmospheric pressure for 30 min, add 3.5g catalyst, hydrogenate for 2 h, filter, add 34.8 mL 1 M HCl to the filtrate, evaporate to dryness under reduced pressure, take up the residue in 70 mL EtOH, filter, evaporate the filtrate to dryness under reduced pressure, repeat this operation twice, crystallize with the minimum amount of EtOH to obtain (3R)-3-aminopyrrolidine dihydrochloride (J. Med. Chem. 1992, 35, 4205). 3R-(+)-Aminopyrrolidine is also available from Tokyo Kasei (TCI America, 911 North Harborgate St., Portland, OR 97203; 800-423-8616, 503-283-1681, (fax) 503-283-1987; www.tciamerica.com). Add 100 mg 4-fluoro-7-nitro-2,1,3-benzoxadiazole in 20 mL MeCN dropwise to a stirred solution of 200 mg 3R-(+)-aminopyrrolidine in 20 mL MeCN at 0-10°, stir at room temperature for 30 min, remove the MeCN by evaporation under reduced pressure, dissolve the residue in 50 mL water, extract 4 times with 80 mL portions of ethyl acetate. Combine the organic layers and wash them with 20 mL water, dry over anhydrous sodium sulfate, evaporate to dryness under reduced pressure, recrystallize from hexane to obtain (R)-(-)-4-(3-aminopyrrolidin-1-yl)-7-nitro-2,1,3-benzoxadiazole as dark red crystals (mp 178-181°) (Analyst 1992, 117, 727). Add 100 μL thiophosgene in 10 mL benzene (Caution! Benzene is a carcinogen!) to 100 mg (R)-(-)-4-(3-aminopyrrolidin-1-yl)-7-nitro-2,1,3-benzoxadiazole in 100 mL acetone, reflux for 1 h, remove the solvent by evaporation under reduced pressure, suspend the residue in 100 mL water, extract 4 times with 25 mL portions of benzene. Combine the extracts and wash them with 20 mL water, dry over anhydrous sodium sulfate, evaporate to dryness under reduced pressure, recrystallize from hexane:benzene 1:2 to obtain (R)-(-)-4-(3-isothiocyanatopyrrolidin-1-yl)-7-nitro-2,1,3-benzoxadiazole as red crystals (mp 165-170°) (Analyst 1995, 120, 385).)

HPLC VARIABLES
Column: 150 × 4.6 5 μm Inertsil ODS-80A
Mobile phase: MeCN:water:trifluoroacetic acid 25:75:0.05 (A) or 30:70:0.05 (B) or 35:65:0.05 (C) or 40:60:0.05 (D) or 45:55:0.05 (E)
Column temperature: 40
Flow rate: 1
Injection volume: 5
Detector: F ex 490 em 530

CHROMATOGRAM
Retention time: k' 10.71 (D-alanine (A)), k' 11.19 (L-alanine (A)), k' 3.34 (D-cystine (E)), k' 4.03 (L-cystine (E)), k' 10.94 (D-isoleucine (C)), k' 12.30 (L-isoleucine (C)), k' 5.61 (D-leucine (D)), k' 6.29 (L-leucine (D)), k' 7.91 (D-lysine (D)), k' 9.13 (L-lysine (D)), k' 6.37 (D-methionine (C)), k' 6.92 (L-methionine (C)), k' 12.17 (D-norleucine (C)), k' 13.75 (L-norleucine (C)), k' 4.87 (D-proline (B)), k' 6.06 (L-proline (B)), k' 12.52 (D-phenylalanine (C)), k' 14.90 (L-phenylalanine (C)), k' 7.77 (D-threonine (A)), k' 8.25 (L-threonine (A)), k' 5.95 (D-tryptophan (D)), k' 7.05 (L-tryptophan (D)), k' 9.20 (D-tyrosine (B)), k' 10.00 (L-tyrosine (B)), k' 6.11 (D-valine (C)), k' 6.81 (L-valine (C))

KEY WORDS
chiral

REFERENCE
Toyo'oka, T.; Liu, Y.-M. High-performance liquid chromatographic resolution of amino acid enantiomers derivatized with fluorescent chiral Edman reagents, *J.Chromatogr.A*, **1995**, *689*, 23–30.

1-Naphthyl Isothiocyanate

RELATED REFERENCE

Levin, J.O.; Andersson, K.; Faengmark, I.; Hallgren, C. Determination of gaseous and particulate polyamines in air using sorbent or filter coated with naphthyl isothiocyanate. *Appl.Ind.Hyg.* **1989**, *4*, 98-100.

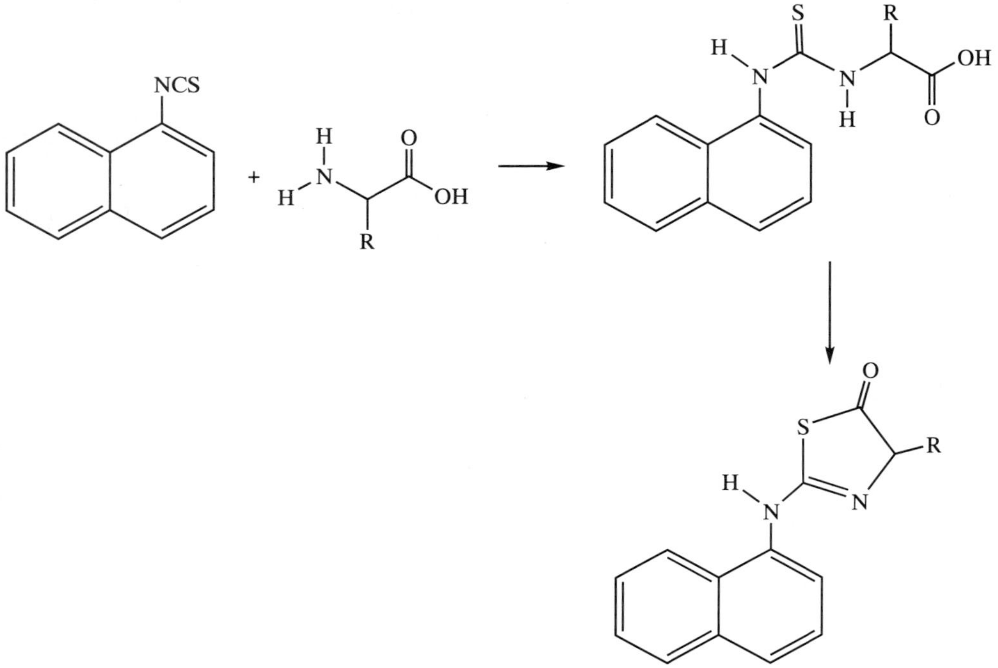

SAMPLE

Matrix: bulk

Analyte: amino acids

Sample preparation: Dissolve 500 nmoles of the amino acid in 500 μL 50 mM pH 10.0 sodium borate buffer, add 500 μL 9.5 mM 1-naphthyl isothiocyanate in dioxane (Caution! Dioxane is a carcinogen!), purge with nitrogen, heat at 40° for 1.5 min, concentrate under a stream of nitrogen, add 500 μL water, add 1 mL hexane, vortex, centrifuge, discard the hexane layer, repeat the hexane wash twice more. Dry the aqueous layer under a stream of nitrogen, reconstitute with 500 μL trifluoroacetic acid:water 50:50, purge with nitrogen, heat at 80° for 10 min, dry under a stream of nitrogen, reconstitute with 1 mL MeCN, inject a 10 μL aliquot.

HPLC VARIABLES

Column: 150 × 4.6 5 μm TSK-gel ODS-120A (Toyo Soda, Tokyo)

Mobile phase: Gradient. A was MeCN:50 mM pH 7.0 imidazole nitrate buffer 15:85. B was MeCN:water 60:40. A:B from 100:0 to 0:100 over 25 min.

Flow rate: 1

Injection volume: 10

Detector: UV 270

KEY WORDS

for Ala and Phe; stereoisomers give rise to two peaks for Phe

REFERENCE

Miyano, H.; Nakajima, T.; Imai, K. Micro-scale sequence analysis from the N-terminus of peptides using the fluorogenic Edman reagent 4-*N,N*-dimethylamino-1-naphthyl isothiocyanate, *Biomed.Chromatogr.*, **1987**, *2*, 139–144.

SAMPLE

Matrix: bulk

Analyte: amino acids

Sample preparation: Dissolve a few mg amino acids in 200 μL reagent, let stand at room temperature in the dark for 5-10 min, evaporate to dryness under reduced pressure, reconstitute with mobile phase, inject an aliquot. (Prepare reagent by mixing 5 mL MeOH, 2.5 mL pyridine, 1 mL triethylamine, 1.5 mL water, and 500 μL 100 mM 1-naphthyl isothiocyanate in MeCN.)

HPLC VARIABLES

Column: 250 × 4 5 μm Chiradex (immobilized β-cyclodextrin) (Merck)

Mobile phase: MeOH:100 mM pH 6.5 ammonium acetate buffer containing 0.1% triethylamine 40:60

Column temperature: 20 ± 0.1

Flow rate: 0.5

Injection volume: 20

Detector: UV 254; F ex 250 em 410

CHROMATOGRAM

Retention time: k' 2.93 (second enantiomer), α = 1.06 (Ala); k' 2.79 (second enantiomer), α = 1.10 (Val); k' 4.90 (second enantiomer), α = 1.14 (Leu); k' 3.32 (second enantiomer), α = 1.00 (Ile); k' 3.81 (second enantiomer), α = 1.22 (Pro); k' 3.94 (second enantiomer), α = 1.08 (Met); k' 4.91 (second enantiomer), α = 1.07 (Cys); k' 3.77 (second enantiomer), α = 1.07 (Ser); k' 2.77 (second enantiomer), α = 1.06 (Thr); k' 1.48 (second enantiomer), α = 1.14 (Lys); k' 1.42 (second enantiomer), α = 1.13 (Arg); k' 15.82 (second enantiomer), α = 1.04 (Glu); k' 8.54 (second enantiomer), α = 1.13 (Phe); k' 10.34 (second enantiomer), α = 1.35 (Trp); k' 3.12 (second enantiomer), α = 1.12 (His); k' 4.94 (second enantiomer), α = 1.00 (Tyr)

KEY WORDS

chiral; α = k' (second enantiomer)/k' (first enantiomer); detailed comparison with other derivatizing reagents

REFERENCE

Rizzi, A.M.; Cladrowa-Runge, S.; Jonsson, H.; Osla, S. Enantiomeric resolution of derivatized DL-amino acids by high-performance liquid chromatography using a β-cyclodextrin chiral stationary phase: A comparison between derivatization labels, *J.Chromatogr.A*, **1995**, *710*, 287–295.

4-Nitrophenyl Isothiocyanate

SAMPLE

Matrix: solutions

Analyte: amino acids

Sample preparation: Dry an aqueous solution of amino acids into a tube, add 10 μL reagent, vortex vigorously, let stand for 10 min, add 40 μL water, wash with 50 μL toluene, inject a 1-10 μL aliquot of he aqueous layer. (Prepare reagent by mixing 490 μL 50 mM 4-nitrophenyl isothiocyanate in MeCN, 50 μL 10% triethylamine in MeCN, and 50 μL water. Purify commercial 4-nitrophenyl isothiocyanate by sublimation at 96° and 13 mm Hg to give a pale yellow powder (mp 106.5°).)

HPLC VARIABLES
Column: 300 × 3.9 Pico-Tag free amino acid analysis column (Waters)
Mobile phase: Gradient. A was MeCN:140 mM sodium acetate:triethylamine 6:94:0.05, pH adjusted to 6.4 with acetic acid. B was MeCN:water 60:40. A:B from 85:15 to 40:60 over 20 min, to 0:100 (step gradient), maintain at 0:100 for 5 min, re-equilibrate at initial conditions for 12 min.
Column temperature: 46
Flow rate: 1
Injection volume: 1-10
Detector: UV 254

CHROMATOGRAM
Retention time: 3.3 (Asp), 3.5 (Glu), 7.2 (Ser), 8 (Gly), 8.3 (His), 9 (Arg), 9.8 (Thr), 10 (Ala), 10.7 (Pro), 12.2 (Tyr), 13.3 (Val), 14 (Met), 15.5 (Ile), 15.7 (Leu), 16 (cystine), 16.8 (Phe), 21.7 (Lys)
Limit of detection: 0.5-1 pmole

REFERENCE
Cohen, S.A. Analysis of amino acids by liquid chromatography after precolumn derivatization with 4-nitrophenylisothiocyanate, *J.Chromatogr.*, **1990**, *512*, 283–290.

Phenyl Isothiocyanate

The initially formed thiourea can cyclize to the corresponding phenylthiohydantoin under acid conditions.

RELATED REFERENCES

Acharya, A.S.; Sussman, L.G.; Manjula, B.N. Application of reductive dihydroxypropylation of amino groups of proteins in primary structural studies: identification of phenylthiohydantoin derivative of epsilon-dihydroxypropyl-lysine residues by high-performance liquid chromatography. *J.Chromatogr.* **1984**, *297*, 37-48.

Beaver, R.W.; Wilson, D.M.; Jones, H.M.; Haydon, K.D. Amino acid analysis in feeds and feedstuffs using precolumn phenyl isothiocyanate derivatization and liquid chromatography - preliminary study. *J.Assoc.Off.Anal.Chem.* **1987**, *70*, 425-428.

Calull, M.; Fabregas, J.; marce, R.M.; Borrull, F. Determination of free amino acids by precolumn derivatization with phenylisothiocyanate. Application to wine samples. *Chromatographia* **1991**, *31*, 272-276.

Chang, K.C.; Skauge, L.H.; Satterlee, L.D. Analysis of amino acids in soy isolates and navy beans using precolumn derivatization with phenylisothiocyanate and reversed-phase high performance liquid chromatography. *J.Food Sci.* **1989**, *54*, 756-759.

Chauvaux, N.; Van Dongen, W.; Esmans, E.L.; Van Onckelen, H.A. Quantitative analysis of 1-aminocyclopropane-1-carboxylic acid by liquid chromatography coupled to electrospray tandem mass spectrometry. *J.Chromatogr.A* **1997**, *775*, 143-150.

Christie, D.L.; Hill, R.M.; Isakow, K.; Barling, P.M. Identification of tyrosine O-sulfate in proteins by reverse-phase high-performance liquid chromatography: use of base hydrolysis combined with precolumn derivatization using phenyl isothiocyanate. *Anal.Biochem.* **1986**, *154*, 92-99.

Cohen, K.A.; Dolan, J.W.; Grillo, S.A. Examination of theoretical principles of gradient elution as applied to reversed-phase high-performance liquid chromatography separation of phenylthiohydantoin amino acids. *J.Chromatogr.* **1984**, *316*, 359-372.

Cohen, S.A.; Tarvin, T.L.; Bidlingmeyer, B.A. Analysis of amino acids using precolumn derivatization with phenylisothiocyanate. *Am.Lab.* **1984**, *16*, 48-59.

Cohen, S.A.; Bidlingmeyer, B.A.; Tarvin, T.L. PITC derivatives in amino acid analysis. *Nature* **1986**, *320*, 769-770.

Crimmins, D.L.; Thoma, R.S. Semi-automated chromatographic procedure for the isolation of acetylated N-terminal fragments from protein digests. *J.Chromatogr.* **1993**, *634*, 241-250.

Daniels, D.H.; Joe, F.L., Jr.; Diachenko, G.W. Determination of free glutamic acid in a variety of foods by high-performance liquidchromatography. *Food Addit.Contam.* **1995**, *12*, 21-29.

Dunphy, M.J.; Bhide, M.V.; Smith, D.J. Determination of hydroxyproline in tissue collagen hydrolysate by derivatization and isocratic reversed-phase high-performance liquid chromatography. *J.Chromatogr.* **1987**, *420*, 394-397.

Elkin, R.G.; Wasynczuk, A.M. Amino acid analysis of feedstuff hydrolyzates by precolumn derivatization with phenyl isothiocyanate and reversed-phase high-performance liquid chromatography. *Cereal Chem.* **1987**, *64*, 226-229.

Fields, C.G.; Loffet, A.; Kates, S.A.; Fields, G.B. The development of high-performance liquid chromatographic analysis of allyl and allyloxycarbonyl side-chain-protected phenylthiohydantoin amino acids. *Anal.Biochem.* **1992**, *203*, 245-251.

Gamcsik, M.P. Determination of low levels of 4-fluorophenylalanine incorporation into proteins. *Anal.Biochem.* **1986**, *154*, 311-315.

Gimenez-Gallego, G.; Thomas, K.A. High-performance liquid chromatography of phenylthiocarbamyl-amino acids. Application to carboxyl-terminal sequencing of proteins. *J.Chromatogr.* **1987**, *409*, 299-304.

González-Castro, M.J.; López-Hernández, J.; Simal-Lozano, J.; Oruña-Concha, M.J. Determination of amino acids in green beans by derivatization with phenylisothiocianate and high-performance liquid chromatography with ultraviolet detection. *J.Chromatogr.Sci.* **1997**, *35*, 181-185.

Granberg, R.R. High-resolution analysis of PITC-derivatized amino acids with UV and electrochemical detection. *LC,Liq.Chromatogr.HPLC Mag.* **1984**, *2*, 776-778.

Guitart, A.; Hernandez Orte, P.; Cacho, J. Stability of phenyl(thiocarbamoyl)amino acids and optimization of their separation by high-performance liquid chromatography. *Analyst* **1991**, *116*, 399-403.

Gunawan, S.; Walton, N.Y.; Treiman, D.M. High-performance liquid chromatographic determination of selected amino acids in rat brain by precolumn derivatization with phenylisothiocyanate. *J.Chromatogr.* **1990**, *503*, 177-187.

Hagen, S.R.; Frost, B.; Augustin, J. Precolumn phenyl isothiocyanate derivatization and liquid chromatography of amino acids in food. *J.Assoc.Off.Anal.Chem.* **1989**, *72*, 912-916.

Hagen, S.R.; Augustin, J.; Grings, E.; Tassinari, P. Precolumn phenylisothiocyanate derivatization and liquid chromatography of free amino acids in biological samples. *Food Chem.* **1993**, *46*, 319-323.

Hanis, T.; Deyl, Z.; Struzinsky, R.; Miksik, I. Separation of elastin cross-links as phenylisothiocyanate derivatives. *J.Chromatogr.* **1991**, *553*, 93-99.

Hariharan, M.; Naga, S.; VanNoord, T. Systematic approach to the development of plasma amino acid analysis by high-performance liquid chromatography with ultraviolet detection with precolumn derivatization using phenyl isothiocyanate. *J.Chromatogr.* **1993**, *621*, 15-22.

Heinrikson, R.L.; Meredith, S.C. Amino acid analysis by reverse-phase high-performance liquid chromatography: precolumn derivatization with phenylisothiocyanate. *Anal.Biochem.* **1984**, *136*, 65-74.

Henderson, L.E.; Copeland, T.D.; Oroszlan, S. Separation of amino acid phenylthiohydantoins by high-performance liquid chromatography on phenylalkyl support. *Anal.Biochem.* **1980**, *102*, 1-7.

Hoogerheide, J.G.; Campbell, C.M. Determination of cysteine plus half-cystine in protein and peptide hydrolysates: use of dithiodiglycolic acid and phenylisothiocyanate derivatization. *Anal.Biochem.* **1992**, *201*, 146-151.

Jansen, E.H.J.M.; Both-Miedema, R. Instability of phenylthiohydantoin amino acids. *J.Chromatogr.* **1988**, *435*, 363-367.

Janssen, P.S.L.; van Nispen, J.W.; Melgers, P.A.T.A.; Van den Bogaart, H.W.M.; Van Aalst, G.W.M.; Goverde, B.C. HPLC analysis of phenylthiocarbamyl (PTC) amino acids. II. Application in the analysis of (poly)peptides. *Chromatographia* **1986**, *22*, 351-357.

Karnaukhova, E.; Niessen, W.M.A.; Tjaden, U.R.; Raap, J.; Lugtenburg, J.; van der Greef, J. Determination of deuterium-labeled tryptophan in proteins by means of high-performance liquid chromatography and thermospray mass spectrometry. *Anal.Biochem.* **1989**, *181*, 271-275.

Khan, J.K.; Kebede, N.; Kuo, Y.H.; Lambein, F.; De Bruyn, A. Analysis of the neurotoxin β-ODAP and its α-isomer by precolumn derivatization with phenylisothiocyanate. *Anal.Biochem.* **1993**, *208*, 237-240.

Khan, J.K.; Kuo, Y.-H.; Kebede, N.; Lambein, F. Determination of non-protein amino acids and toxins in *Lathyrus* by high-performance liquid chromatography with precolumn phenyl isothiocyanate derivatization. *J.Chromatogr.A* **1994**, *687*, 113-119.

Klotz, A.V.; Thomas, B.A.; Glazer, A.N.; Blacher, R.W. Detection of methylated asparagine and glutamine residues in polypeptides. *Anal.Biochem.* **1990**, *186*, 95-100.

L'Italien, J.J.; Kent, S.B.H. Protein microsequencing with postcolumn fluorescent phenylisothiocyanate analogs. *J.Chromatogr.* **1984**, *283*, 149-156.

Lanneluc-Sanson, D.; Phan, C.T.; Granger, R.L. Analysis by reverse-phase high-pressure liquid chromatography of phenylisothiocyanate-derivatized 1-aminocyclopropane-1-carboxylic acid in apple extracts. *Anal.Biochem.* **1986**, *155*, 322-327.

Lemaire, S.; Dumont, M.; Nolet, S. Sensitive method of detection, quantitation and purification of peptides using pre-column derivatization with phenyl isothiocyanate. *J.Chromatogr.* **1988**, *425*, 77-86.

Marce, R.M.; Calull, M.; Guasch, J.; Borrull, F. Determination of free amino acids in wine by HPLC using precolumn derivatization with phenylisothiocyanate. *Am.J.Enol.Vitic.* **1989**, *40*, 194-198.

McBroom, T.; Stubbs, H.J.; Wadhwa, M.S.; Thomas, V.H.; Rice, K.G. Reversal of tyrosinamide-oligosaccharide derivatization by Edman degradation. *Anal.Biochem.* **1994**, *222*, 243-250.

Meyer, H.E.; Heber, M.; Eisermann, B.; Korte, H.; Metzger, J.W.; Jung, G. Sequence analysis of lantibiotics: chemical derivatization procedures allow a fast access to complete Edman degradation. *Anal.Biochem.* **1994**, *223*, 185-190.

Mora, R.; Berndt, K.D.; Tsai, H.; Meredith, S.C. Quantitation of aspartate and glutamate in HPLC analysis of phenylthiocarbamyl amino acids. *Anal.Biochem.* **1988**, *172*, 368-376.

Mueller, B.; Burgstaller, W.; Strasser, H.; Zanella, A.; Schinner, F. Leaching of zinc from an industrial filter dust with *Penicillum, Pseudomonas* and *Corynebacterium*: Citric acid is the leaching agent rather than amino acids. *J.Ind.Microbiol.* **1995**, *14*, 208-212.

Murthy, L.R.; Iqbal, K. Measurement of picomoles of phosphoamino acids by high-performance liquid chromatography. *Anal.Biochem.* **1991**, *193*, 299-305.

O'Hare, M.M.T.; Tortora, O.; Gether, U.; Nielsen, H.V.; Schwartz, T.W. High-performance liquid chromatography of phenylthiocarbamyl derivatives of amino acids and side-chain derivatized amino acids. *J.Chromatogr.* **1987**, *389*, 379-388.

Puig-Deu, M.; Buxaderas, S. Analytical conditions for the determination of 23 phenylthiocarbamyl amino acids and ethanolamine in musts and wines by high-performance liquid chromatography. *J.Chromatogr.A* **1994**, *685*, 21-30.

Rosenlund, B.L. Time-saving method for the reversed-phase high-performance liquid chromatography of phenylthiocarbamylamino acid derivatives of free amino acids in plasma. *J.Chromatogr.* **1990**, *529*, 258-262.

Rossetti, V.; Lombard, A. Determination of glutamate decarboxylase by high-performance liquid chromatography. *J.Chromatogr.B* **1996**, *681*, 63-67.

Santucci, A.; Lozzi, L.; Bracci, L.; Petreni, S.; Rustici, M.; Soldani, P.; Neri, P. Micro-determination of amino acid composition of proteins electroblotted onto polyvinylidene difluoride membranes. *Ital.J.Biochem.* **1989**, *38*, 349-359.

Sarwar, G.; Botting, H.G.; Peace, R.W. Complete amino acid analysis in hydrolyxates of foods and feces by liquid chromatography of precolumn phenylisothiocyanate derivatives. *J.Assoc.Off.Anal.Chem.* **1988**, *71*, 1172-1175.

Sarwar, G.; Botting, H.G. Rapid analysis of nutritionally important free amino acids in serum and organs (liver, brain, and heart) by liquid chromatography of precolumn phenylisothiocyanate derivatives. *J.Assoc.Off.Anal.Chem.* **1990**, *73*, 470-475.

Sato, K.; Tsukamasa, Y.; Imai, C.; Ohtsuki, K.; Shimizu, Y.; Kawabata, M. Improved method for identification and determination of epsilon-(gamma-glutamyl)lysine cross-link in protein using proteolytic digestion and derivatization with phenyl isothiocyanate followed by high-performance liquid chromatography separation. *J.Agric.Food Chem.* **1992**, *40*, 806-810.

Sato, M.; Suzuki, S.; Yasuda, Y.; Kawauchi, H.; Kanno, N.; Sato, Y. Quantitative HPLC analysis of acidic opines by phenylthiocarbamyl derivatization. *Anal.Biochem.* **1988**, *174*, 623-627.

Saunders, J.A.; Saunders, J.M.; Morris, S.; Wynne, S.A., II Amino acid analysis of subcellular fractions by PITC and OPA. *Chromatogram* **1988**, *9*, 2-4.

Schmeer, K.; Khalifa, M.; Császár, J.; Farkas, G.; Bayer, E.; Molnár-Perl, I. Compositional analysis of the phenylthiocarbamyl amino acids by liquid chromatography - atmospheric pressure ionization mass spectrometry with particular attention to the cyst(e)ine derivatives. *J.Chromatogr.A* **1995**, *691*, 285-299.

Scholze, H. Determination of phenylthiocarbamyl amino acids by reversed-phase high-performance liquid chromatography. *J.Chromatogr.* **1985**, *350*, 453-460.

Seferiadis, K.; Frillingos, S.; Tsolas, O. Amino acid analysis by PITC derivatization. *Chromatogram* **1987**, *8*, 2.

Semensi, V.; Sugumaran, M. Simultaneous determination of trans-hydroxyproline and other amino acids in protein hydrolysates at picomolar levels by HPLC using precolumn derivatization with phenylisothiocyanate. *LC-GC* **1986**, *4*, 1108-1110.

Senden, M.H.M.N.; Van der Meer, A.J.G.M.; Limborgh, J.; Wolterbeek, H.T. Analysis of major tomato xylem organic acids and PITC-derivatives of amino acids by RP-HPLC and UV detection. *Plant Soil* **1992**, *142*, 81-89.

Shang, S.F.; Wang, H. Sensitive determination of amino acids in kelp by reversed phase high performance liquid chromatography with precolumn derivatization using phenylisothiocyanate. *Chromatographia* **1996**, *43*, 309-312.

Siebert, J.; Palmer, R.J., Jr.; Hirsch, P. Analysis of free amino acids in microbially colonized sandstone by precolumn phenyl isothiocyanate derivatization and high-performance liquid chromatography. *Appl.Environ.Microbiol.* **1991**, *57*, 879-881.

Simmons, J.; Schlesinger, D.H. High-performance liquid chromatography of side-chain-protected amino acid phenylthiohydantoins. *Anal.Biochem.* **1980**, *104*, 254-258.

Smalley, D.M.; Preusch, P.C. Analysis of gamma-carboxyglutamic acid by reverse phase HPLC of its phenylthiocarbamyl derivative. *Anal.Biochem.* **1988**, *172*, 241-247.

Suzuki, E.Y.; Early, R.J. High-performance liquid chromatographic separation of phenylthiocarbamyl derivatives of amino acids from protein hydrolyzates using a Partisphere C18 column. *J.Chromatogr.* **1993**, *657*, 204-207.

Tarcsa, E.; Fesus, L. Determination of epsilon(gamma-glutamyl)lysine crosslink in proteins using phenylisothiocyanate derivatization and high-pressure liquid chromatographic separation. *Anal.Biochem.* **1990**, *186*, 135-140.

Tedesco, J.L.; Schafer, R. Increased yield of the phenylthiocarbamyl-amino acids tyrosine, cysteine and phenylalanine using diethyl ether-based liquid-liquid extraction. *J.Chromatogr.* **1987**, *403*, 299-306.

van der Leij, F.R.; Welling, G.W. Determination of β-aspartylpeptidase activity in human faeces by high-performance liquid chromatography using pre-column derivatization with phenyl isothiocyanate. *J.Chromatogr.* **1986**, *383*, 35-42.

Van Poelje, P.D.; Snell, E.E. Use of p-aminobenzoic acid and tritiated cyanoborohydride for the detection of pyruvoyl residues in proteins. *Anal.Biochem.* **1987**, *161*, 420-424.

Waldmann, G.; Podschun, B. Assay for β-ureidopropionase by high-performance liquid chromatography. *Anal.Biochem.* **1990**, *188*, 233-236.

Winkler, G.; Heinz, F.X.; Kunz, C. Exclusive use of high-performance liquid chromatographic techniques for the isolation, 4-dimethylaminoazobenzene-4'-sulfonyl chloride amino acid analysis and 4-N,N-dimethylaminoazobenzene-4'-isothiocyanate-phenyl isothiocyanate sequencing of a viral membrane protein. *J.Chromatogr.* **1984**, *297*, 63-73.

Woo, K.-L.; Ahan, Y.-K. Determination of protein amino acids as benzylthiocarbamyl derivatives compared with phenylthiocarbamyl derivatives by reversed-phase high-performance liquid chromatography, ultraviolet detection and precolumn derivatization. *J.Chromatogr.A* **1996**, *740*, 41-50.

Woo, K.-L.; Hwang, Q.-C.; Kim, H.-S. Determination of amino acids in the foods by reversed-phase high-performance liquid chromatography with a new precolumn derivative, butylthiocarbamyl amino acid compared to the conventional phenylthiocarbamyl derivatives and ion-exchange chromatography. *J.Chromatogr.A* **1996**, *740*, 31-40.

Yaegaki, K.; Tonzetich, J.; Ng, A.S. Improved high-performance liquid chromatography method for quantitation of proline and hydroxyproline in biological materials. *J.Chromatogr.* **1986**, *356*, 163-170.

Yamaya, T.; Matsumoto, H. Analysis of phenylthiocarbamyl-amino acids at pico-mole level by high performance liquid chromatography and application to plant materials. *Soil Sci.Plant Nutr.(Tokyo)* **1988**, *34*, 297-302.

Yang, C.Y.; Sepulveda, F.I. Separation of phenylthiocarbamyl amino acids by high-performance liquid chromatography on Spherisorb octadecylsilane columns. *J.Chromatogr.* **1985**, *346*, 413-416.

SAMPLE
Matrix: ascitic fluid, blood, tissue, urine
Analyte: amino acids
Sample preparation: Homogenize liver with 5 volumes ice-cold 3% sulfosalicylic acid, centrifuge at 2000 g for 15 min. Add an equal volume of cold 6% sulfosalicylic acid to plasma, urine, or ascitic fluid, mix, centrifuge at 4° at 1800 g for 12 min. Dilute the supernatant or liver homogenate with an equal volume of 400 μM methionine sulfone in 100 mM HCl, evaporate an aliquot to dryness under reduced pressure, add 20 μL EtOH:water:triethylamine 40:40:20, evaporate to dryness under reduced pressure, reconstitute with 20 μL EtOH:triethylamine:water:phenyl isothiocyanate 70:10:10:10 (freshly prepared), let stand at room temperature for 20 min, evaporate to dryness under reduced pressure (70 mTorr) for 1.5-2 h, reconstitute with 250 μL pH 7.4 phosphate buffer, inject an aliquot.

HPLC VARIABLES
Column: 150 × 3.9 Pico-Tag C18 (Waters)
Mobile phase: Gradient. A was MeCN:buffer 6:94. B was MeCN:water 60:40. A:B from 98:2 to 54:46 over 10 min (Waters convex curve No. 5), to 0:100 over 0.5 min, maintain at 0:100 for 5 min, re-equilibrate at initial conditions for 7 min. (Buffer was 140 mM sodium acetate in water containing 500 μL/L triethylamine, pH adjusted to 6.40 with glacial acetic acid.)
Column temperature: 38

Flow rate: 1
Detector: UV 254

CHROMATOGRAM
Retention time: 1.7 (Asp), 1.9 (Glu), 2.7 (Hpro), 3.45 (Asn), 3.65 (Ser), 3.8 (Gln), 4.0 (Gly), 4.6 (His), 5.25 (Tau), 5.5 (Arg), 5.7 (Thr), 5.9 (Ala), 6.3 (Pro), 7.2 (AAB), 7.8 (Tyr), 8.4 (Val), 8.8 (Met), 9.5 (Cys), 9.75 (Ile), 9.9 (Leu), 10.9 (Phe), 11.3 (Trp), 12 (Lys)
Internal standard: methionine sulfone (6.5)
Limit of detection: 3 μM

KEY WORDS
plasma; liver

REFERENCE
Fierabracci, V.; Masiello, P.; Novelli, M.; Bergamini, E. Application of amino acid analysis by high-performance liquid chromatography with phenyl isothiocyanate derivatization to the rapid determination of free amino acids in biological samples, *J.Chromatogr.*, **1991**, *570*, 285–291.

SAMPLE
Matrix: blood
Analyte: amino acids
Sample preparation: 1 mL Plasma + 1 mL acetone, shake thoroughly, centrifuge at 5000 rpm for 10 min. Remove the supernatant and add it to 1 mL buffer and 5 μL phenyl isothiocyanate, heat at 40° for 1 h, add 2 mL benzene (Caution! Benzene is a carcinogen!), shake, centrifuge at 3000 rpm for 3 min, discard the organic phase. Add 1 mL 1 M hydrogen chloride in acetic acid to the aqueous phase and flush the tube with nitrogen, heat at 80° for 50 min, cool, extract with two 10 mL aliquots of dichloromethane. Combine the extracts and evaporate them to dryness, reconstitute the residue with 5 mL dichloromethane and evaporate to dryness with a stream of nitrogen, reconstitute with 270 μL mobile phase, shake thoroughly, centrifuge at 3000 rpm for 3 min, inject a 200 μL aliquot of the supernatant. (Prepare the buffer by adding 2 mL 2 M acetic acid to 1.2 mL triethylamine and diluting with 96.8 mL acetone:water 50:50.)

HPLC VARIABLES
Column: 500 × 6 5 μm Lichrosorb Si 100
Mobile phase: n-Hexane:dichloromethane:saturated dichloromethane 25:55:20 (Prepare saturated dichloromethane by stirring 885 mL dichloromethane, 75 mL EtOH, and 40 mL water for 2 h, let stand for 2 h. The organic phase is saturated dichloromethane.)
Flow rate: 2.5
Injection volume: 200
Detector: UV 268

CHROMATOGRAM
Retention time: 4.8 (proline), 5.2 (leucine), 6.0 (valine), 6.2 (phenylalanine), 8.4 (alanine), 9.5 (tryptophan), 11.2 (glycine), 22.0 (tyrosine)

KEY WORDS
plasma; normal phase; pharmacokinetics

REFERENCE
Trefz, F.K.; Byrd, D.J.; Blaskovics, M.E.; Kochen, W.; Lutz, P. Determination of deuterium-labeled phenylalanine and tyrosine in human plasma with high pressure liquid chromatography and mass spectrometry, *Clin.Chim.Acta*, **1976**, *73*, 431–438.

SAMPLE
Matrix: blood
Analyte: amino acids
Sample preparation: 50 μL Serum or plasma + 450 μL cold MeOH + 10 μL 1 mM L-α-aminoadipate, vortex, centrifuge at 8700 g for 1 min. Remove the supernatant and evaporate it to dryness under reduced pressure at 45° over 1 h, reconstitute the residue in 25 μL water, 25 μL triethylamine, and 50 μL EtOH, evaporate to dryness under high vacuum over 30 min. Dissolve the residue in 70 μL EtOH, 10 μL water, 10 μL triethylamine, and 10 μL phenyl isothiocyanate, let stand at room temperature for 20 min, evaporate to dryness under high vacuum over 1 h, reconstitute with 50 μL buffer, inject a 10 μL

aliquot. (Buffer was MeCN:7 mM Na_2HPO_4 5:95, adjusted to pH 7.4 with phosphoric acid.)

HPLC VARIABLES
Column: 250 × 4 Bio-sil ODS-5S (Bio-Rad)
Mobile phase: Gradient. A was 50 mM ammonium acetate adjusted to pH 6.8 with phosphoric acid. B was MeCN:100 mM pH 6.8 ammonium acetate 50:50. A:B from 100:0 to 92:8 over 40 min, to 40:60 over 40 min, wash with 0:100 for 15 min, re-equilibrate at initial conditions for 10 min.
Flow rate: 1
Injection volume: 10
Detector: UV 254

CHROMATOGRAM
Retention time: 6 (Asp), 7.5 (Glu), 12.5 (S-carboxmethyl-L-cysteine), 16.5 (OH-Pro), 20 (Ser), 21.5 (Gly), 23 (Asn), 27 (Glu), 35 (Thr), 37 (Ala), 40 (His), 45 (Pro), 49.5 (Arg), 59.5 (Tyr), 60 (Val), 62.5 (Met), 63 (Cys), 67 (Ile), 68 (Leu), 71 (Phe), 73 (Trp), 76 (Lys)
Internal standard: α-aminoadipate (14)
Limit of quantitation: 10 pmole

KEY WORDS
serum; plasma

REFERENCE
Lavi, L.E.; Holcenberg, J.S.; Cole, D.E.; Jolivet, J. Sensitive analysis of asparagine and glutamine in physiological fluids and cells by precolumn derivatization with phenylisothiocyanate and reversed-phase high-performance liquid chromatography, *J.Chromatogr.*, **1986**, *377*, 155–163.

SAMPLE
Matrix: blood
Analyte: amino acids
Sample preparation: Dried blood. Add paper containing about 20 μL dried blood to 200 μL 500 μM norvaline in MeCN:triethylamine:water 50:25:15, rotate at 60-100 rpm for 30-60 min. Add 20 μL phenyl isothiocyanate to the liquid, let stand at room temperature for 20-30 min, add 500 μL 100 mM pH 6.5 sodium acetate buffer, vortex for 30 s, add 2.5 mL dichloromethane, centrifuge, inject a 50 μL aliquot of the aqueous layer. Serum. 20 μL Serum + 200 μL 500 μM norvaline in MeCN:triethylamine:water 50:25:15, mix, centrifuge at 1500 g for 5 min. Add 20 μL phenyl isothiocyanate to the supernatant, let stand at room temperature for 20-30 min, add 500 μL 100 mM pH 6.5 sodium acetate buffer, vortex for 30 s, add 2.5 mL dichloromethane, centrifuge, inject a 50 μL aliquot of the aqueous layer.

HPLC VARIABLES
Column: 150 × 4.6 5 μm Zorbax C18
Mobile phase: Gradient. MeCN:100 mM pH 6.5 sodium acetate buffer 15:85 for 1 min, to 28:72 over 7 min, to 60:40 (step gradient), maintain at 60:40 for 5 min, return to initial conditions over 2 min.
Column temperature: 45
Injection volume: 50
Detector: UV 254

CHROMATOGRAM
Retention time: 3 (tyrosine), 8 (phenylalanine)
Internal standard: norvaline (4.5)
Limit of detection: <10 μM

OTHER SUBSTANCES
Non-interfering: other amino acids, bilirubin, lipids, carbamazepine, ethosuximide, phenobarbital, phenytoin, primidone

KEY WORDS
serum; dried blood

REFERENCE
Rudy, J.L.; Rutledge, J.C.; Lewis, S.L. Phenylalanine and tyrosine in serum and eluates from dried blood spots as determined by reversed-phase liquid chromatography, *Clin.Chem.*, **1987**, *33*, 1152–1154.

SAMPLE
Matrix: blood

Analyte: amino acids

Sample preparation: Mix whole blood immediately with an equal volume of 1 M perchloric acid, centrifuge at 4° at 1500 g for 20 min, filter (0.45 µm) the supernatant. Evaporate a 20 µL aliquot of the filtrate to dryness under vacuum (80 mTorr) for 30 min, reconstitute with 20 µL MeOH:1 M sodium acetate:triethylamine 40:40:20, evaporate to dryness under vacuum (80 mTorr) for 15 min, reconstitute with 20 µL MeOH:water: triethylamine:phenyl isothiocyanate 70:10:10:10. Let stand at room temperature for 20 min, evaporate to dryness under vacuum (60 mTorr) for 1.5 h, reconstitute with 200 µL pH 7.40 phosphate buffer, vortex, sonicate for a few s, inject a 2-10 µL aliquot.

HPLC VARIABLES
Column: 300 × 3.9 Pico Tag (Waters)

Mobile phase: Gradient. A was MeCN:70 mM sodium acetate 2.5:97.5 containing 1 ppm EDTA, adjusted to pH 6.50 with 10% acetic acid. B was MeCN:MeOH:water 45:15:40. A:B 100:0 for 13 min, to 97:3 (step gradient), maintain at 97:3 for 7 min, to 92:8 over 10 min (convex gradient), to 66:34 over 20 min, maintain at 66:34 for 14 min, to 0:100 over 1.5 min, maintain at 0:100 for 10 min, re-equilibrate at initial conditions for 20 min.

Column temperature: 46 ± 1

Flow rate: 1

Injection volume: 2-10

Detector: UV 254

CHROMATOGRAM
Retention time: 3.48 (aspartic acid), 4.02 (glutamic acid), 7.19 (hydroxyproline), 9.12 (serine), 9.67 (asparagine), 10.22 (glycine), 14.69 (taurine), 17.16 (histidine), 17.71 (GABA), 18.48 (citrulline), 20.04 (threonine), 20.93 (alanine), 23.78 (arginine), 27.25 (proline), 42.55 (tyrosine), 45.61 (valine), 47.20 (methionine), 53.26 (isoleucine), 53.92 (leucine), 57.34 (phenylalanine), 58.84 (ornithine), 63.52 (lysine)

KEY WORDS
whole blood; tryptophan, cysteine not detected; Glutamine and asparagine are converted to glutamate and aspartate over time.

REFERENCE
Buzzigoli, G.; Lanzone, L.; Ciociaro, D.; Frascerra, S.; Cerri, M.; Scandroglio, A.; Coldani, R.; Ferrannini, E. Characterization of a reversed-phase high-performance liquid chromatographic system for the determination of blood amino acids, *J.Chromatogr.*, **1990**, *507*, 85–93.

SAMPLE
Matrix: blood

Analyte: amino acids

Sample preparation: 100 µL Plasma + 10 µL carboxymethylcysteine solution + 400 µL EtOH, mix thoroughly, centrifuge at 3000 g for 5 min. Remove the supernatant and evaporate it to dryness under a stream of nitrogen at 45°, reconstitute the residue in 400 µL EtOH, evaporate to dryness under a stream of nitrogen, reconstitute with 200 µL MeOH:triethylamine 95:5, add 10 µL MeOH:phenyl isothiocyanate 87.5:12.5, mix, let stand for 5 min, evaporate to dryness under reduced pressure, reconstitute with 250 µL buffer, when dissolved add 100 µL dichloromethane, vortex for 1 min, centrifuge at 3000 g for 1-3 min, inject a 20 µL aliquot of the aqueous layer. (Prepare buffer by dissolving 1.36 g sodium acetate trihydrate in water, adjust pH to 6.4 ± 0.01 with 1 M orthophosphoric acid, make up to 1 L with water.)

HPLC VARIABLES
Column: 250 × 4.6 5 µm ODS-Hypersil

Mobile phase: Gradient. A was 1.36 g sodium acetate trihydrate in water, pH adjusted to 6.4 ± 0.01 with 1 M orthophosphoric acid, made up to 1 L with water. B was 1.36 g sodium acetate trihydrate in water, pH adjusted to 6.4 ± 0.01 with 1 M orthophosphoric acid, made up to 400 mL with water, mixed with 600 mL MeCN. A:B from 100:0 to 89:11 over 15 min, to 0:100 over 3 min, maintain at 0:100 for 2 min, return to initial conditions for 3 min, re-equilibrate for 2 min.

Flow rate: 1 for 18 min, to 2 over 2 min, maintain at 2

Injection volume: 20

Detector: E, EDT Research LCA 15 or Chromajet, glassy carbon electrode +1.10 V, Ag/AgCl reference electrode

CHROMATOGRAM

Retention time: 6.3 (glutamic acid), 13.1 (serine), 14 (glycine), 15.3 (alanine)

Internal standard: carboxymethylcysteine (8.1)

KEY WORDS

plasma

REFERENCE

Sherwood, R.A.; Bayliss, E.M.; Chappatte, O. Assay of plasma glycine by HPLC with electrochemical detection in patients undergoing glycine irrigation during gynaecological surgery, *Clin.Chim.Acta*, **1991**, *203*, 275–283.

SAMPLE

Matrix: blood, amniotic fluid, CSF, urine

Analyte: amino acids

Sample preparation: Plasma. Condition a 100 mg Bond Elut SCX (propylbenzenesulfonic acid, H^+ form) SPE cartridge with 1 mL 50 mM HCl, 1 mL MeOH, 2 mL water, and 1 mL 50 mM HCl. 100 µL Plasma + 100 µL 250 µM norleucine in 100 mM HCl + 10 mg solid sulfosalicylic acid + 800 µL acetone or MeOH, mix, centrifuge, add a 50 µL aliquot to the SPE cartridge, wash with 2 mL water, elute with two 500 µL portions of MeOH:water:triethylamine 40:40:20, dry the eluate under vacuum, add 10 µL MeOH:1 M sodium acetate:triethylamine 40:40:20, dry under vacuum at 70 mTorr, reconstitute with 20 µL MeOH:triethylamine:water:phenyl isothiocyanate 70:10:10:10, let stand at room temperature for 20 min, evaporate to dryness under vacuum, reconstitute with 100 µL MeCN:5 mM pH 7.4 sodium phosphate buffer 5:95, inject a 20 µL aliquot. Dried blood. Add 25 µL 250 µM norleucine in 100 mM HCl to a 6 mm filter paper disc containing dried blood, add 100 µL MeCN, let stand for 30 min, centrifuge, remove a 75 µL aliquot of the supernatant, evaporate to dryness under reduced pressure, add 10 µL MeOH:1 M sodium acetate:triethylamine 40:40:20, dry under vacuum at 70 mTorr, reconstitute with 20 µL MeOH:triethylamine:water:phenyl isothiocyanate 70:10:10:10, let stand at room temperature for 2 min, evaporate to dryness under vacuum, reconstitute with 50 µL MeCN:5 mM pH 7.4 sodium phosphate buffer 5:95, inject a 20 µL aliquot. Amniotic fluid, CSF. Mix amniotic fluid or CSF with an equal volume of 250 µM norleucine in 100 mM HCl, filter (Centrifree 10000 MW cutoff) while centrifuging at 2200 g. Evaporate a 50 µL aliquot of the ultrafiltrate to dryness under vacuum, add 10 µL MeOH:1 M sodium acetate:triethylamine 40:40:20, dry under vacuum at 70 mTorr, reconstitute with 20 µL MeOH:triethylamine:water:phenyl isothiocyanate 70:10:10:10, let stand at room temperature for 20 min, evaporate to dryness under vacuum, reconstitute with 50 (CSF) or 100 (amniotic fluid) µL MeCN:5 mM pH 7.4 sodium phosphate buffer 5:95, inject a 20 µL aliquot. Urine. Dilute urine with water to a creatinine concentration of 1 mM, mix an aliquot with an equal volume of 250 µM norleucine in 100 mM HCl, filter (Centrifree 10000 MW cutoff) while centrifuging at 2200 g. Evaporate a 50 µL aliquot of the ultrafiltrate to dryness under vacuum, add 10 µL MeOH:1 M sodium acetate:triethylamine 40:40:20, dry under vacuum at 70 mTorr, reconstitute with 20 µL MeOH:triethylamine:water:phenyl isothiocyanate 70:10:10:10, let stand at room temperature for 20 min, evaporate to dryness under vacuum, reconstitute with 100 µL MeCN:5 mM pH 7.4 sodium phosphate buffer 5:95, inject a 20 µL aliquot.

HPLC VARIABLES

Column: 300 × 3.9 Pico-Tag amino acid column (Waters)

Mobile phase: Gradient. A was MeCN:70 mM pH 6.55 sodium acetate 2.5:97.5. B was MeCN:MeOH:water 45:15:40. A:B 100:0 for 13.5 min, to 97:3 (step gradient), to 94:6 over 10.5 min (Waters curve 8 (slightly concave)), to 91:9 over 6 min (Waters curve 5

(slightly convex)), to 66:34 over 20 min, maintain at 66:34 for 12 min, to 0:100 over 0.5 min, maintain at 0:100 for 4 min, return to initial conditions over 0.5 min.
Column temperature: 46
Flow rate: 1
Injection volume: 20
Detector: UV 254

CHROMATOGRAM
Retention time: 2.97 (phosphoserine), 3.08 (cysteic acid), 3.36 (aspartic acid), 3.80 (glutamic acid), 5.23 (α-aminoadipic acid), 6.83 (hydroxyproline), 7.21 (phosphoethanolamine), 8.70 (serine), 8.98 (galactosamine), 9.03 (aspartylglucosamine), 9.09 (asparagine), 9.75 (glycine), 9.91 (glucosamine), 10.46 (glutamine), 11.84 (α-alanine), 12.12 (homoserine), 12.12 (sarcosine), 12.56 (glycylglycine), 13.82 (taurine), 15.92 (histidine), 16.63 (gamma-aminobutyric acid), 17.02 (ammonia), 17.18 (citrulline), 18.23 (glycylhistidine), 18.83 (threonine), 19.8 (alanine), 21.26 (β-aminoisobutyric acid), 22.14 (carnosine), 22.25 (β-amino-n-butyric acid), 23.07 (arginine), 23.40 (methionine sulfone), 26.10 (proline), 26.71 (σ-amino-n-valeric acid), 27.48 (1-methylhistidine), 27.98 (anserine), 28.31 (homocitrulline), 28.36 (3-methylhistidine), 29.35 (4-aminobenzoic acid), 31.00 (ethanolamine), 31.89 (homoarginine), 33.59 (cysteine), 34.03 (gamma-amino-n-butyric acid), 34.03 (glutathionine (oxidized)), 34.14 (levodopa), 34.47 (Tris), 38.55 (4-aminophenylacetic acid), 40.37 (4-aminohippuric acid), 40.70 (glycyltyrosine), 42.29 (tyrosine), 45.38 (valine), 46.97 (methionine), 47.47 (3-hydroxyanthranilic acid), 48.74 (cystathionine), 48.90 (3-hydroxykynurenine), 50.28 (ethylamine), 50.83 (cystine), 51.05 (α-aminophenylacetic acid),51.71 (glycylleucine), 51.77 (3-amino-3-phenylpropionic acid), 53.14 (isoleucine), 53.42 (alloisoleucine), 53.80 (leucine), 53.97 (cysteine-homocysteine (mixed disulfide)), 54.19 (ethionine), 54.79 (glycylphenylalanine), 56.12 (kynurenine), 56.12 (homocystine), 57.11 (phenylalanine), 57.99 (tryptophan), 58.43 (ornithine), 62.89 (lysine), 65.92 (serotonin)
Internal standard: norleucine (55.07)

OTHER SUBSTANCES
Non-interfering: cadaverine, 2-phenylethylamine

KEY WORDS
SPE; ultrafiltrate; plasma; dried blood

REFERENCE
Davey, J.F.; Ersser, R.S. Amino acid analysis of physiological fluids by high-performance liquid chromatography with phenylisothiocyanate derivatization and comparison with ion-exchange chromatography, *J.Chromatogr.*, **1990**, *528*, 9–23.

SAMPLE
Matrix: blood, urine
Analyte: amino acids
Sample preparation: Serum. 100 μL Serum + 400 μL EtOH, mix, centrifuge. Remove the supernatant and evaporate it to dryness under reduced pressure, reconstitute with 200 μL MeOH:triethylamine 95:5 (prepare fresh each day), add 10 μL phenyl isothiocyanate:MeOH 12.5:87.5 (prepare fresh each day), mix, let stand at room temperature for 5 min, evaporate to dryness under reduced pressure, reconstitute with 500 μL buffer, add 200 μL dichloromethane, vortex for 1 min, centrifuge at 1200 g for 1-3 min, inject an aliquot of the aqueous layer. Urine. Add 10 μL IS solution to 50/x μL urine (x = concentration of creatinine (mM)), add 200 μL EtOH, mix, centrifuge. Remove the supernatant and evaporate it to dryness under reduced pressure, reconstitute with 200 μL MeOH:triethylamine 95:5 (prepare fresh each day), add 10 μL phenyl isothiocyanate: MeOH 12.5:87.5 (prepare fresh each day), mix, let stand at room temperature for 5 min, evaporate to dryness under reduced pressure, reconstitute with 500 μL buffer, add 200 μL dichloromethane, vortex for 1 min, centrifuge at 1200 g for 1-3 min, inject an aliquot of the aqueous layer. (Prepare buffer by dissolving 1.36 g sodium acetate trihydrate in water, adjust pH to 6.40 ± 0.01 with 2% orthophosphoric acid, make up to 1 L with water.)

HPLC VARIABLES
Column: 250 × 4.6 5 μm Hypersil-ODS

Mobile phase: Gradient. A was 1.36 g/L sodium acetate trihydrate in water, pH adjusted to 6.40 ± 0.01 with 2% orthophosphoric acid. B was 1.36 g/L sodium acetate trihydrate in MeCN:water 60:40, pH adjusted to 6.40 ± 0.01 with 2% orthophosphoric acid. A:B from 100:0 to 87:13 over 20 min, to 45:55 over 45 min, to 0:100 over 2.5 min, maintain at 0:100 for 2.5 min, return to initial conditions over 5 min, re-equilibrate for 5 min.
Flow rate: 1 for 65 min then 2
Injection volume: 20
Detector: E, EDT Research LCA 15, glassy carbon electrode +1.10 V, Ag/AgCl reference electrode

CHROMATOGRAM
Retention time: 3.82 (phosphoserine), 5.44 (aspartic acid), 6.83 (glutamic acid), 10.35 (gamma-aminoadipic acid), 12.26 (hydroxyproline), 12.45 (phosphoethanolamine), 14.52 (serine), 15.52 (glycine), 15.54 (asparagine), 16.91 (sarcosine), 17.42 (β-alanine), 18.98 (taurine), 20.48 (gamma-aminobutyric acid), 20.63 (citrulline), 21.11 (threonine), 21.75 (alanine), 22.27 (β-aminoisobutyric acid), 23.48 (proline), 23.91 (histidine), 25.71 (carnosine), 28.22 (arginine), 28.39 (1-methylhistidine, 3-methylhistidine), 28.60 (α-aminobutyric acid), 29.06 (anserine), 35.02 (tyrosine), 35.76 (valine), 37.65 (ethanolamine), 38.10 (methionine), 38.19 (cystathionine), 40.88 (cystine), 42.80 (isoleucine), 43.47 (leucine), 47.78 (hydroxylysine), 48.20 (phenylalanine), 48.46 (hydroxylysine), 49.67 (ornithine), 50.16 (tryptophan), 52.79 (lysine)
Internal standard: norleucine (45.07)
Limit of quantitation: 1 μM

KEY WORDS
serum

REFERENCE
Sherwood, R.A.; Titheradge, A.C.; Richards, D.A. Measurement of plasma and urine amino acids by high-performance liquid chromatography with electrochemical detection using phenylisothiocyanate derivatization, *J.Chromatogr.*, **1990**, *528*, 293–303.

SAMPLE
Matrix: blood, urine
Analyte: buthionine sulfoximine
Sample preparation: Plasma. 200 μL Plasma + 200 μL 100 μg/mL L-norleucine in 100 mM HCl, vortex, filter (Millipore 10000 nominal molecular weight limit) while centrifuging at 5000 g for 15 min. Remove a 40 μL aliquot of the ultrafiltrate and add it to 80 μL freshly-prepared EtOH:phenyl isothiocyanate:triethylamine 40:1:1, vortex, let stand at room temperature for 15 min, add 10 μL 316 mg/mL L-serine, let stand for 15 min, evaporate to dryness under reduced pressure, reconstitute with 600 μL A, inject a 250 μL aliquot. Urine. Condition a Bond Elut C18 SPE cartridge with 3 mL MeOH and 3 mL water. 100 μL Urine + 100 μL 100 μg/mL L-norleucine in 100 mM HCl, vortex, add to the SPE cartridge, elute with 300 μL 100 mM HCl, elute with 500 μL MeCN:100 mM HCl 30:70, collect all the effluent from the cartridge, vortex thoroughly. Remove a 40 μL aliquot and add it to 80 μL freshly-prepared EtOH:phenyl isothiocyanate:triethylamine 40:1:1, vortex, let stand at room temperature for 15 min, add 10 μL 316 mg/mL L-serine, let stand for 15 min, evaporate to dryness under reduced pressure, reconstitute with 600 μL A, inject a 250 μL aliquot.

HPLC VARIABLES
Guard column: 10 × 4.6 5 μm Adsorbosphere C18
Column: 250 × 4.6 5 μm Adsorbosphere C18
Mobile phase: Gradient. A:B 91.5:8.5 for 30 min, to 55:45 over 10 min, to 0:100 over 2 min, maintain at 0:100 for 6 min, return to initial conditions over 2 min, re-equilibrate for 8 min. A was MeCN:140 mM pH 6.40 sodium acetate buffer containing 500 μL/L triethylamine and 200 μL/L 1 mg/mL EDTA 6:94. B was MeCN:water 60:40 containing 200 μL/L 1 mg/mL EDTA.
Flow rate: 1.25
Injection volume: 250
Detector: UV 254

CHROMATOGRAM
Retention time: 26.0 (S), 26.6 (R)
Internal standard: L-norleucine (39.3)
Limit of quantitation: 2 µg/mL (plasma), 10 µg/mL (urine), 6 µg/mL (urine)
Limit of detection: 1 µg/mL (plasma)

KEY WORDS
plasma; ultrafiltrate; SPE

REFERENCE
Brennan, J.M.; O'Dwyer, P.J.; Ozols, R.F.; LaCreta, F.P. High-performance liquid chromatographic determination of the S- and R-diastereoisomers of L-buthionine (SR)-sulfoximine in human plasma and urine, *J.Chromatogr.*, **1993**, *620*, 121–128.

SAMPLE
Matrix: bulk
Analyte: amino acids
Sample preparation: Dissolve a few mg amino acids in 200 µL reagent, let stand at room temperature in the dark for 5-10 min, evaporate to dryness under reduced pressure, reconstitute with mobile phase, inject an aliquot. (Prepare reagent by mixing 5 mL MeOH, 2.5 mL pyridine, 1 mL triethylamine, 1.5 mL water, and 500 µL 100 mM phenyl isothiocyanate in MeCN.)

HPLC VARIABLES
Column: 250 × 4 5 µm Chiradex (immobilized β-cyclodextrin) (Merck)
Mobile phase: MeOH:100 mM pH 6.5 ammonium acetate buffer containing 0.1% triethylamine 40:60
Column temperature: 20 ± 0.1
Flow rate: 0.5
Injection volume: 20
Detector: UV 254

CHROMATOGRAM
Retention time: k' 3.01 (second enantiomer), α = 1.17 (Ala); k' 3.34 (second enantiomer), α = 1.30 (Val); k' 5.00 (second enantiomer), α = 1.00 (Leu); k' 4.22 (second enantiomer), α = 1.21 (Ile); k' 3.13 (second enantiomer), α = 1.14 (Pro); k' 3.86 (second enantiomer), α = 1.06 (Met); k' 6.68 (second enantiomer), α = 1.82 (Cys); k' 3.11 (second enantiomer), α = 1.11 (Ser); k' 2.93 (second enantiomer), α = 1.25 (Thr); k' 1.61 (second enantiomer), α = 1.28 (Lys); k' 1.42 (second enantiomer), α = 1.24 (Arg); k' 3.68 (second enantiomer), α = 1.25 (Asn); k' 2.69 (second enantiomer), α = 1.16 (Gln); k' 29.69 (second enantiomer), α = 1.24 (Asp); k' 17.65 (second enantiomer), α = 1.14 (Glu); k' 8.11 (second enantiomer), α = 1.21 (Phe); k' 7.87 (second enantiomer), α = 1.24 (Trp); k' 3.00 (second enantiomer), α = 1.18 (His); k' 5.33 (second enantiomer), α = 1.06 (Tyr)

KEY WORDS
chiral; α = k' (second enantiomer)/k' (first enantiomer); detailed comparison with other derivatizing reagents

REFERENCE
Rizzi, A.M.; Cladrowa-Runge, S.; Jonsson, H.; Osla, S. Enantiomeric resolution of derivatized DL-amino acids by high-performance liquid chromatography using a β-cyclodextrin chiral stationary phase: A comparison between derivatization labels, *J.Chromatogr.A*, **1995**, *710*, 287–295.

SAMPLE
Matrix: bulk
Analyte: amino acids
Sample preparation: Mix 0.1-1 mg amino acids with 100 µL MeOH:water:triethylamine 70:10:10, add 5-10 µL phenyl isothiocyanate, vortex, heat at 55° for 30 min, evaporate to dryness at 55°, add 100 µL 12.5-25% trifluoroacetic acid in water, heat at 55° for 40 min, extract 3 times with 1 mL portions of ethyl acetate. Combine the extracts and evaporate them to dryness under a stream of nitrogen, wash the residue 3 times with 1 mL portions of n-heptane, reconstitute, inject an aliquot.

CAPILLARY ELECTROPHORESIS
Capillary: 50 cm × 50 μm (30 cm to detector) (GL Science, Tokyo)
Running buffer: 50 mM pH 3.0 Sodium phosphate buffer containing 25 mM β-escin, 25 mM digitonin, and 50 mM sodium dodecyl sulfate
Injection: Gravimetric injection at 5 cm for 5-40 s.
Model: JASCO CE-800
Migration time: 18.17, 18.35 (Phe enantiomers), 19.14, 19.29 (Leu enantiomers), 19.27, 19.71 (Ile enantiomers), 22.03, 22.31 (Met enantiomers), 22.53, 22.85 (Tyr enantiomers), 22.74, 23.12 (Val enantiomers), 31.03, 31.63 (Glu enantiomers), 31.14, 31.90 (Ala enantiomers), 33.70, 34.29 (Asp enantiomers), 35.00, 35.62 (Gln enantiomers), 42.87, 43.98 (Thr enantiomers), 47.96, 49.14 (Asn enantiomers), 50.68, 51.78 (Ser enantiomers)

KEY WORDS
chiral

REFERENCE
Kurosu, Y.; Murayama, K.; Shindo, N.; Shisa, Y.; Satou, Y.; Ishioka, N. Optical resolution of phenylthio-hydantoin-amino acids by capillary electrophoresis for protein sequencing, *J.Chromatogr.A*, **1997**, *771*, 311–317.

SAMPLE
Matrix: contact lenses
Analyte: amino acids
Sample preparation: Hydrolyze contact lens with 200 μL concentrated HCl at 145° for 1 h or with 200 μL 6 M HCl at 105° for 20 h, remove a 50-150 μL aliquot and evaporate it to dryness under reduced pressure, add 20 μL EtOH:water:triethylamine 40:40:20, evaporate to dryness under reduced pressure, add 30 μL EtOH:water:triethylamine:phenyl isothiocyanate 70:10:10:10, let stand at room temperature for 25 min, reconstitute with 30-200 μL buffer, inject a 10 μL aliquot. (Buffer was MeCN:5 mM Na_2HPO_4 adjusted to pH 6.8 with 2% phosphoric acid.)

HPLC VARIABLES
Column: 150 × 4.6 Ultrasphere ODS C18
Mobile phase: Gradient. A was 700 μL triethylamine in 1 L 100 mM sodium acetate, adjusted to pH 5.5 with glacial acetic acid. B was MeCN:A:water 315:250:185. A:B 90:10 for 2 min, to 62:38 over 6 min, to 55:45 over 0.5 min, to 40:60 over 7 min, to 10:90 over 3 min, to 0:100 over 1 min, return to initial conditions over 8 min, re-equilibrate for 6 min. (Mobile phase A maintained at 60°, mobile phase B maintained at 40°.)
Column temperature: 43
Flow rate: 1
Injection volume: 10
Detector: UV 254

REFERENCE
Yan, G.; Nyquist, G.; Caldwell, K.D.; Payor, R.; McCraw, E.C. Quantitation of total protein deposits on contact lenses by means of amino acid analysis, *Invest.Ophthalmol.Vis.Sci.*, **1993**, *34*, 1804–1813.

SAMPLE
Matrix: exopeptidase digest, protein hydrolysate
Analyte: amino acids
Sample preparation: 5-10 μL Protein hydrolysate or exopeptidase digest + 40 μL MeOH:triethylamine 8:1 + 5 μL phenyl isothiocyanate, mix, let stand at room temperature for 20 min, wash twice with 15 μL portions of heptane, dilute the aqueous phase with 40 μL 50 mM pH 5.4 sodium acetate buffer:acetic acid 100:3, let stand for 30 s, add 40 μL 50 mM pH 5.4 sodium acetate buffer:acetic acid 100:3, let stand for 30 s, inject a 90 μL aliquot.

HPLC VARIABLES
Mobile phase: Gradient. A was 50 mM pH 5.4 sodium acetate. B was MeCN:water 70:30. A:B from 93:7 to 70:30 over 10 min, to 42:58 over 10 min.
Column temperature: 37
Flow rate: 0.3

Injection volume: 90
Detector: UV 254

CHROMATOGRAM
Retention time: 5.5 (Asp), 6 (Glu), 6.7 (Ser), 7.2 (Gly), 7.5 (His), 8 (Arg), 8.5 (Thr), 9 (Ala), 9.3 (Pro), 11.6 (Tyr), 12.6 (Val), 13 (Met), 13.7 (Cystine), 14.8 (Ile), 15 (Leu), 15.4 (Norleucine), 15.8 (Phe), 17 (Lys)
Limit of quantitation: 50 pmole

REFERENCE
Thoma, R.S.; Crimmins, D.L. Automated phenylthiocarbamyl amino acid analysis of carboxypeptidase/aminopeptidase digests and acid hydrolysates, *J.Chromatogr.*, **1991**, *537*, 153–165.

SAMPLE
Matrix: food
Analyte: amino acids
Sample preparation: Homogenize (Sorvall) 5 g food with 50 mL 600 mM perchloric acid, centrifuge at 3500 rpm for 20 min, filter (0.45 μm) the supernatant, adjust the pH of the filtrate to 7.0 ± 0.2 with 30% KOH, place in the fridge for 5 min. Evaporate a 1 mL extract to dryness under reduced pressure at 37°, add 20 μL reagent, mix. Let stand at room temperature for 20 min, evaporate to dryness under reduced pressure at 37°, reconstitute with 200 μL buffer, inject a 20 μL aliquot. (Prepare reagent by mixing 70 μL EtOH, 10 μL water, 10 μL triethylamine, and 10 μL phenyl isothiocyanate just before use. Buffer was MeCN:water containing 710 mg Na_2HPO_4 adjust pH to 7.40 with phosphoric acid.)

HPLC VARIABLES
Column: 250 × 4.6 5 μm Ultrabase C18
Mobile phase: Gradient. A MeCN:buffer 6:76. A was MeCN:water 60:40. A:B from 100:0 to 54:46 over 14.5 min, to 0:100 over 0.5 min, return to initial conditions over 2 min, re-equilibrate at initial conditions for 3.5 min. (Buffer was 19 g/L sodium acetate trihydrate containing 500 μL/L triethylamine, adjust pH to 6.40 with glacial acetic acid.)
Column temperature: 38 ± 1
Flow rate: 1 for 15 min, to 1.5 over 2 min, maintain at 1.5 for 3 min, return to 1 over 0.5 min
Injection volume: 20
Detector: UV 254

CHROMATOGRAM
Retention time: 4.2 (Asp), 5 (Glu), 8.2 (Ser), 8.5 (Gly), 9 (His), 9.7 (Arg), 10.2 (Thr), 10.8 (Ala), 12.3 (Pro), 13.5 (Tyr), 14.4 (Val), 15.1 (Met), 15.6 (Cys), 16.5 (Ile), 16.7 (Leu), 17.5 (Phe), 17.7 (Trp), 17.9 (Lys)

REFERENCE
Alonso, M.L.; Alvarez, A.I.; Zapico, J. Rapid analysis of free amino acids in infant foods, *J.Liq.Chromatogr.*, **1994**, *17*, 4019–4030.

SAMPLE
Matrix: infant formula
Analyte: tryptophan
Sample preparation: Mix 500 mg infant formula with 4.83 g barium hydroxide and 5 mL boiling water, remove air with nitrogen and sonication for 30 s, heat at 120° for 8 h, adjust pH to 3-4 with HCl, filter (0.22 μm), make up to 25 mL with water. Mix 163 μL hydrolysate and 10 μL 4 mM norleucine, dry under high vacuum, for each 100 nmoles of amino acids add 2 μL EtOH:water:triethylamine 40:40:20, dry under high vacuum, for each 100 nmoles of amino acids add 5 μL EtOH:triethylamine:water:phenyl isothiocyanate 70:10:10:10 (freshly prepared), let stand at room temperature for 20 min, evaporate to dryness under high vacuum, reconstitute with 400 μL MeCN:5mM pH 7.4 sodium acetate buffer 5:95, sonicate for 1 min, filter (0.22 μm), inject an aliquot of the filtrate.

HPLC VARIABLES
Column: 250 × 4.6 5 μm Spherisorb ODS-2

Mobile phase: MeCN:buffer 16:84 (Buffer was 17 mM sodium acetate containing 325 μL/L triethylamine, pH adjusted to 6.8 with glacial acetic acid.)
Injection volume: 20
Detector: UV 254

CHROMATOGRAM
Retention time: 25
Internal standard: norleucine (12.5)
Limit of quantitation: 510 μg/g
Limit of detection: 180 μg/g

REFERENCE
Alegría, A.; Barberá, R.; Farré, R.; Ferrerés, M.; Lagarda, M.J.; López, J.C. Isocratic high-performance liquid chromatographic determination of tryptophan in infant formulas, *J.Chromatogr.A*, **1996**, *721*, 83−88.

SAMPLE
Matrix: peptides
Analyte: amino acids
Sample preparation: Mix 10 μL of a solution containing 50 pmole peptide with 35 μL reagent, heat at 50° for 10 min, evaporate to dryness at 50° in a centrifugal evaporator for 5 min, reconstitute with 10 μL water, wash with three 100 μL portions of n-heptane: dichloromethane 90:10, evaporate to dryness at 50° in a centrifugal evaporator for 15 min, reconstitute with 30 μL trifluoroacetic acid, heat at 50° for 5 min, evaporate to dryness under a stream of nitrogen. Reconstitute with 20 μL water, add 100 μL n-heptane:dichloromethane 70:30, mix, centrifuge at 1000 g for 5 min, repeat extraction 3 times. (The aqueous phase contains the residual peptide and can be subjected to the same procedure to determine the next amino acid.) Combine the organic layers and evaporate them to dryness under a stream of nitrogen, reconstitute with 20 μL water:trifluoroacetic acid 80:20, heat at 50° for 10 min, dry under a stream of nitrogen, reconstitute with mobile phase, inject an aliquot. (Reagent was phenyl isothiocyanate:EtOH:pyridine 1: 4:2.)

HPLC VARIABLES
Column: 150 × 6 5 μm ES-1/2phCD (β-cyclodextrin 50% modified with phenylcarbamoyl) (Shinwakakou, Kyoto) (Column A) or 150 × 6 5 μm Ultron ES-phCD (phenylcarbamoylated β-cyclodextrin) (Shinwakakou, Kyoto) (Column B)
Mobile phase: MeOH:water 25:75 containing 10 mM formic acid (Mobile Phase A) or MeCN:MeOH:water 10:45:45 containing 10 mM formic acid (Mobile Phase B)
Flow rate: 0.7
Detector: UV 269

CHROMATOGRAM
Retention time: 15 (Gly) (Column A, Mobile Phase A), 16 (L-Ala) (Column A, Mobile Phase A), 17 (D-Ala) (Column A, Mobile Phase A), 17 (D-Leu) (Column B, Mobile Phase B), 17.5 (L-Leu) (Column B, Mobile Phase B), 21 (D-Phe) (Column B, Mobile Phase B), 22.5 (L-Tyr) (Column A, Mobile Phase A), 23 (L-Phe) (Column B, Mobile Phase B), 24 (D-Tyr) (Column A, Mobile Phase A)

KEY WORDS
chiral

REFERENCE
Imai, K.; Matsunaga, H.; Santa, T.; Homma, H. Availability of phenylisothiocyanate for the amino acid sequence/configuration determination of peptides containing D/L-amino acids, *Biomed.Chromatogr.*, **1995**, *9*, 195−196.

SAMPLE
Matrix: perfusate
Analyte: amino acids
Sample preparation: Dry 120 μL perfusate under vacuum, add 50 μL EtOH:water:triethylamine 40:40:20, vortex, dry under vacuum, add 20 μL EtOH:water:triethylamine: phenyl isothiocyanate 70:10:10:10, vortex, let stand at room temperature for 20 min,

vacuum dry for at least 8 h (most of the derivatizing reagent must be removed in the first hour), reconstitute with 100 μL mobile phase A, inject a 70 μL aliquot.

HPLC VARIABLES
Guard column: 30-40 μm pellicular RP18
Column: 150 × 3.9 Novapak C18
Mobile phase: Gradient. A was MeCN:14 mM sodium acetate:triethylamine 4.5:95.5:0.05, adjusted to pH 6.6 with glacial acetic acid. B was MeCN:water 60:40. A:B 100:0 for 1 min, to 95:5 (step gradient), maintain at 95:5 for 9 min, to 94:6 (step gradient), to 0:100 over 4 min, maintain at 0:100 for 4 min, return to initial conditions, re-equilibrate for 10 min.
Column temperature: 37
Flow rate: 1.1 for 10 min, to 1.5 over 4 min, maintain at 1.5
Injection volume: 70
Detector: UV 254

CHROMATOGRAM
Retention time: 2 (aspartate), 2.4 (glutamate), 3.1 (adenosine), 3.4 (hydroxyproline), 5 (glutamine), 6.4 (taurine), 6.7 (gamma-aminobutyric acid), 9 (kainic acid)

REFERENCE
Rogers, K.L.; Philibert, R.A.; Allen, A.J.; Molitor, J.; Wilson, E.J.; Dutton, G.R. HPLC analysis of putative amino acid neurotransmitters released from primary cerebellar cultures, *J.Neurosci.Methods*, **1987**, *22*, 173–179.

SAMPLE
Matrix: proteins
Analyte: amino acids
Sample preparation: Add 200 μL concentrated HCL containing 1% phenol to 0.1-5 μg protein, seal tube under vacuum, heat at 150° for 1 h or 108° for 24 h, evaporate to dryness under reduced pressure, add 10-20 μL EtOH:water:triethylamine 40:40:20, evaporate to dryness under reduced pressure, add 20 μL reagent, let stand at room temperature for 20 min, evaporate to dryness under reduced pressure, reconstitute with PTC amino acid diluent (Waters), inject a 1-40 μL aliquot (J.Chromatogr. 1984, 336, 93). (Reagent was EtOH:triethylamine:water:phenyl isothiocyanate 70:10:10:10, store at -20°.)

HPLC VARIABLES
Column: 150 × 3.9 Nova Pak C18
Mobile phase: Gradient. A was MeCN:145 mM pH 6.2 ammonium acetate 6:94. B was MeCN:water 60:40. A:B from 100:0 to 40:60 over 15 min (Waters convex gradient 5), clean with 0:100 for 2 min.
Column temperature: 43
Flow rate: 0.7
Injection volume: 1-40
Detector: UV 254; MS, VG Masslab Model 30-250 quadrupole, VG Masslab thermospray interface, the column effluent was combined with 200 mM ammonium acetate pumped at 0.3 mL/min and this mixture flowed into the MS, thermospray nozzle 32°, thermospray chamber 340°

CHROMATOGRAM
Retention time: 1.6 (Asp), 1.9 (Glu), 4.7 (Ser), 5.0 (Gly), 6.1 (His), 8.1 (Arg), 8.7 (Thr), 9.3 (Ala), 10.1 (ammonia), 11.2 (Pro), 14.5 (Tyr), 15.4 (Val), 16.0 (Met), 16.8 (Cys), 17.3 (Ile), 17.5 (Leu), 18.6 (Phe), 20.2 (Lys)

REFERENCE
Pramanik, B.C.; Moomaw, C.R.; Evans, C.T.; Cohen, S.A.; Slaughter, C.A. Identification of phenylthiocarbamyl amino acids for compositional analysis by thermospray liquid chromatography/mass spectrometry, *Anal.Biochem.*, **1989**, *176*, 269–277.

SAMPLE
Matrix: solutions
Analyte: amino acids

Sample preparation: Evaporate 10 μL of a 2.5 mM solution in 100 mM HCl to dryness under reduced pressure, reconstitute with 100 μL MeCN:pyridine:triethylamine:water 50:25:10:15, evaporate to dryness, reconstitute with 100 μL MeCN:pyridine: triethylamine:water 50:25:10:15, add 5 μL phenyl isothiocyanate, mix, let stand at room temperature for 5 min, evaporate to dryness, reconstitute with 250 μL MeCN:water 20: 70, inject an aliquot.

HPLC VARIABLES
Guard column: ODS-5 (Altex)
Column: 250 × 4.6 5 μm Ultrasphere ODS
Mobile phase: Gradient. A was 115 mM pH 6.0 ammonium acetate. B was 230 mM pH 6.0 ammonium acetate in MeCN:MeOH:water 44:10:46. A:B from 100:0 to 85:15 over 15 min, to 50:50 over 15 min, to 100:0 over 4 min, maintain at 100:0 for 3 min, return to initial conditions over 13 min.
Flow rate: 1
Detector: UV 254

CHROMATOGRAM
Retention time: 9 (Glu), 11 (Asp), 12.5 (Ser), 16.3 (Hyp), 17.3 (Gly), 17.7 (Gln), 18.5 (Asn), 22 (His), 23 (Thr), 23.6 (Ala), 24.6 (Arg), 24.8 (Pro), 32.5 (Tyr), 33.8 (Val), 35 (Met), 35.8 (Cys), 37.4 (Ile), 37.6 (Leu), 38.6 (Phe), 39 (Trp), 39.4 (Lys)
Limit of detection: 50-500 pmole

KEY WORDS
comparison with other derivatization procedures

REFERENCE
McClung, G.; Frankenberger, W.T. Jr. Comparison of reverse-phase high-performance liquid chromatographic methods for precolumn-derivatized amino acids, *J.Liq.Chromatogr.*, **1988**, *11*, 613–646.

SAMPLE
Matrix: tissue
Analyte: desmosine and isodesmosine
Sample preparation: Prepare a column by making a slurry of 10 g CF$_1$ cellulose powder (Whatman) in 200 mL n-butanol:acetic acid:water 4:1:1, shake, sonicate for 2-4 min to remove air bubbles, fill a 7 mm ID column to a height of 45-50 mm with slurry, allow to drain, wash with 5 mL n-butanol:acetic acid:water 4:1:1. Lyophilize tissue, hydrolyse a 20 mg portion with 500 μL 6 M HCl under nitrogen in a sealed vial for 36 h. Open the vial and mix the contents with 500 μL acetic acid, 500 μL of a 50 mg/mL slurry of CF$_1$ cellulose in n-butanol:acetic acid:water 4:1:1, and 2 mL n-butanol, add the slurry to the column, rinse the vial with1.5 mL n-butanol:acetic acid:water 4:1:1, add the rinse to the column, wash with 15 mL n-butanol:acetic acid:water 4:1:1, elute with 5 mL water (discard any n-butanol phase) (J. Chromatogr. 1982, 229, 200). Evaporate the eluate under reduced pressure, reconstitute with EtOH:water:triethanolamine 40:40:20, dry under vacuum at 40°, repeat this neutralization step twice more, reconstitute with 20 μL EtOH:water:triethanolamine 80:10:10, add 5 μL phenyl isothiocyanate, heat at 40° in a sealed tube for 30 min, dry under vacuum at 40°, reconstitute with 500 μL mobile phase, inject a 20 μL aliquot.

HPLC VARIABLES
Guard column: 4 × 4 LiChrocart
Column: 250 × 4 5 μm Nucleosil 120-5C18
Mobile phase: MeCN:MeOH:buffer 34:16:50 (Buffer was 80 mM pH 6.5 phosphate buffer containing 10 mM tetrabutylammonium phosphate.)
Column temperature: 32
Flow rate: 1
Injection volume: 20
Detector: UV 254

CHROMATOGRAM
Retention time: 5.0 (isodesmosine), 5.6 (desmosine)
Limit of detection: 12.5 ng/g

KEY WORDS
rat; human; vein; lung; aorta; SPE

REFERENCE
Salomoni, M.; Muda, M.; Zuccato, E.; Mussini, E. High-performance liquid chromatographic determination of desmosine and isodesmosine after phenylisothiocyanate derivatization, *J.Chromatogr.*, **1991**, *572*, 312–316.

SAMPLE
Matrix: tissue
Analyte: amino acids
Sample preparation: Homogenize oocytes with 3 volumes of water, add 9 volumes ice-cold MeOH, centrifuge at 4° at 2000 g for 10 min. Evaporate the supernatant to dryness under reduced pressure, resuspend in 100 μL MeOH:water:triethylamine 40:40:20, evaporate to dryness under reduced pressure, add 1-10 μL MeOH:triethylamine:water:phenyl isothiocyanate 70:10:10:10 per oocyte, let stand at room temperature for 15 min, dry under vacuum, reconstitute with 50 mM pH 6.8 ammonium acetate buffer, inject an aliquot.

HPLC VARIABLES
Column: 100 × 5 Nova-Pak C18
Mobile phase: Gradient. A was 50 mM pH 6.8 ammonium acetate buffer. B was MeCN:100 mM ammonium acetate 50:50. A:B 100:0 for 5 min, to 90:10 over 1 min, maintain at 90:10 for 4 min, to 50:50 over 15 min, maintain at 50:50 for 5 min.
Detector: UV 254

CHROMATOGRAM
Retention time: 2.6 (Asp), 3.5 (Glu), 9 (Ser), 10 (Gly), 12.5 (Asn), 13.3 (Gln), 15 (Thr), 15.4 (Ala, His), 17 (Pro), 20.6 (Tyr), 20.7 (Val), 21.7 (Met), 23 (Ile), 23.3 (Leu), 25 (Phe), 26 (Trp), 27.5 (Lys)

KEY WORDS
oocytes

REFERENCE
O'Connor, C.M. Analysis of aspartic acid and asparagine metabolism in *Xenopus laevis* oocytes using a simple and sensitive HPLC method, *Mol.Reprod.Dev.*, **1994**, *39*, 392–396.

SAMPLE
Matrix: urine
Analyte: amino acids
Sample preparation: Freeze urine at -20°, thaw, stir thoroughly, allow to settle. 600 μL Supernatant + 600 μL concentrated HCl, heat at 110 ± 5° for 18 h, cool, filter, dilute (if necessary). Evaporate a 25-200 μL aliquot to dryness under reduced pressure, reconstitute with 30 μL water:EtOH:triethylamine 40:40:20, evaporate to dryness under reduced pressure, add 50 μL phenyl isothiocyanate:EtOH:triethylamine:water 10:70:10:10, vortex, let stand for 10 min, evaporate to dryness under reduced pressure, reconstitute with 500 μL buffer, filter, inject an aliquot. (Buffer was MeCN:10 mM NaH_2PO_4 containing 2 mM ethylenebis(oxyethylenenitrilo)tetraacetic acid (EGTA) 1.8:99.2, pH 6.0.)

HPLC VARIABLES
Guard column: Guard-Pak (Waters)
Column: 250 × 4 5 μm octadecyl (IBM)
Mobile phase: Gradient. A was MeCN:10 mM NaH_2PO_4 containing 2 mM ethylenebis(oxyethylenenitrilo)tetraacetic acid (EGTA) 1.8:99.2, pH 6.0. B was MeCN:water 60:40. A:B 100:0 for 7 min, to 0:100 over 1 min, maintain at 0:100 for 2 min, return to initial conditions over 1 min, re-equilibrate for 7 min
Column temperature: 34 or 43
Flow rate: 1.2
Detector: UV 254

CHROMATOGRAM
Retention time: 2.60 (O-phosphoserine), 2.77 (aspartic acid), 3.12 (glutamic acid), 4.13 (α-aminoadipic acid), 5.10 (hydroxyproline), 6.22 (phosphoethanolamine), 6.74 (asparagine),

6.90 (serine), 7.67 (glycine), 7.75 (glutamine), 8.5 (homoserine), 8.72 (sarcosine), 9.24 (β-alanine), 9.25 (anserine), 9.54 (glycerophosphoryl ethanolamine), 10.57 (taurine), 11.57 (citrulline), 12.50 (threonine), 12.84 (gamma-aminobutyric acid), 13.53 (alanine), 14.29 (β-aminoisobutyric acid), 15.06 (histidine), 16.31 (proline), 17.47 (carnosine)
Limit of quantitation: 250 pmoles

REFERENCE
Lippincott, S.; Chesney, R.W.; Friedman, A.; Pityer, R.; Barden, H.; Mazess, R.B. Rapid determination of total hydroxyproline (HYP) in human urine by HPLC analysis of the phenylisothiocyonate (PITC)-derivative, *Bone*, **1989**, *10*, 265–268.

SAMPLE
Matrix: urine
Analyte: 3-methylhistidine
Sample preparation: 100 μL Urine + 10 μL 10 μM N-methyllysine + 200 μL reagent + 10 μL triethylamine:MeOH 1:10, heat at 35° for 10 min, evaporate to dryness under reduced pressure, reconstitute with 500 μL water, add 200 μL dichloromethane, vortex for 1 min, centrifuge at 2500 g for 3 min, inject an aliquot of the aqueous phase. (Prepare reagent by mixing 500 μL phenyl isothiocyanate with 3.5 mL MeOH.)

HPLC VARIABLES
Column: 150 × 4.6 5 μm Apex phenyl RP (Jones Chromatography)
Mobile phase: MeCN:15 mM sodium acetate 8:92, adjusted to pH 6.2 with acetic acid
Flow rate: 1
Injection volume: 100
Detector: E, ESA Model 5100A Coulochem, Model 5011 analytical cell with two porous graphite working electrodes, palladium reference electrode, first electrode +0.40 V, second electrode +0.80 V

CHROMATOGRAM
Retention time: 10
Internal standard: N-methyllysine (15)
Limit of quantitation: 0.1 pmole

OTHER SUBSTANCES
Extracted: histidine

REFERENCE
Betto, P.; Ricciarello, G.; Pichini, S.; Dello Strologo, L.; Rizzoni, G. High-performance liquid chromatography-electrochemical detection of 3-methylhistidine in human urine, *J.Chromatogr.*, **1992**, *584*, 256–260.

2,3,4,6-Tetra-O-acetyl-β-D-glucopyranosyl Isothiocyanate

RELATED REFERENCES
Kinoshita, T.; Kasahara, Y.; Nimura, N. Reversed-phase high-performance liquid chromatographic resolution of nonesterified enantiomeric amino acids by derivatization with 2,3,4,6-tetra-O-acetyl-β-D-glucopyranosyl isothiocyanate and 2,3,4-tri-O-acetyl-α-D-arabinopyranosyl isothiocyanate. *J.Chromatogr.* **1981**, *210*, 77-81.
Miyazawa, T.; Iwanaga, H.; Yamada, T.; Kuwata, S. Resolution of cyclic imino acid and β-amino acid enantiomers by derivatization with 2,3,4,6-tetra-O-acetyl-β-D-glucopyranosyl isothiocyanate followed by reversed-phase HPLC analysis. *Anal.Lett.* **1992**, *26*, 367-378.

SAMPLE
Matrix: solutions
Analyte: amino acids
Sample preparation: Mix a 50 μL aliquot of a 1 mg/mL amino acid solution in MeCN: triethylamine 99.8:0.2 with 50 μL 0.5% 2,3,4,6-tetra-O-acetyl-β-D-glucopyranosyl isothiocyanate in MeCN, let stand at room temperature for 1 h, inject a 5 μL aliquot. (2,3,4,6-Tetra-O-acetyl-β-D-glucopyranosyl isothiocyanate may be purchased from Aldrich, synthesis details are also given in this paper.)

HPLC VARIABLES
Column: 250 × 4 5 μm LiChrosorb RP-18
Mobile phase: MeOH:water 50:50 (A) or 60:40 (B)
Flow rate: 0.4
Injection volume: 5
Detector: UV 250

CHROMATOGRAM
Retention time: 10 (D-Ser (A)), 11 (L-Ser (A)), 14 (L-Tyr (B)), 15 (L-Ala (A)), 16 (D-Ala (A), D-Tyr (B)), 19 (L-Val (B)), 21 (D-Val (B)), 22 (L-phenylglycine (B)), 23 (D-phenylglycine (B)), 25 (L-Asp (A)), 28 (L-Leu (B)), 28.5 (D-Asp (A), L-Glu (A)), 30 (D-Glu (A)), 32 (D-Leu (B), L-Phe (B)), 38 (D-Phe (B))
Limit of detection: 5 ng

KEY WORDS
chiral

REFERENCE
Nimura, N.; Ogura, H.; Kinoshita, T. Reversed-phase liquid chromatographic resolution of amino acid enantiomers by derivatization with 2,3,4,6-tetra-O-acetyl-β-D-glucopyranosyl isothiocyanate, *J.Chromatogr.*, **1980**, *202*, 375–379.

SAMPLE
Matrix: solutions
Analyte: amino acids
Sample preparation: 100 μL 20 mM Amino acid (10 mM for Lys) in MeCN:water 50:50 + 2 μL triethylamine + 20 μL 40 mg/mL 2,3,4,6-tetra-O-acetyl-β-D-glucopyranosyl isothiocyanate in MeCN, let stand at room temperature for 30 min, add 50 μL 1 M HCl, add 830 μL MeCN:water 25:75, inject a 5-20 μL aliquot.

HPLC VARIABLES
Column: 200 × 4.6 7 μm Hypercarb S
Mobile phase: Gradient. A was 0.1% trifluoroacetic acid in water. B was 0.1% trifluoroacetic acid in MeCN:water 90:10. A:B 70:30 for 15 min, to 64:36 over 10 min, maintain at 64:36 for 3 min, to 61:39 over 5 min, maintain at 61:39 for 2 min, to 55:45 over 10 min, maintain at 55:45 for 5 min, to 0:100 over 33 min, maintain at 0:100 for 10 min, return to initial conditions over 3 min.
Flow rate: 1.1
Injection volume: 5-20
Detector: UV 250

CHROMATOGRAM
Retention time: k' 7.15 (D-His), k' 4.38 (L-His), k' 10.54 (D-Arg), k' 8.92 (L-Arg), k' 11.89 (D-Ser), k' 11.60 (L-Ser), k' 12.43 (D-Pro), k' 12.78 (L-Pro), k' 16.67 (D-Thr), k' 13.71 (L-Thr), k' 17.07 (D-Ala), k' 14.49 (L-Ala), k' 17.97 (D-Asp), k' 16.03 (L-Asp), k' 17.97 (D-Glu), k' 16.03 (L-Glu), k' 25.52 (D-Val), k' 21.69 (L-Val), k' 28.02 (D-Leu), k' 25.17 (L-Leu), k' 29.9 (D-Ile), k' 27.41 (L-Ile), k' 41.0 (D-Phe), k' 34.27 (L-Phe), k' 37.17 (D-Lys), k' 36.65 (L-Lys), k' 45.04 (D-Tyr), k' 34.67 (L-Tyr)

KEY WORDS
chiral

REFERENCE
Chan, W.C.; Micklewright, R.; Barrett, D.A. Porous graphitic carbon for the chromatographic separation of O-tetraacetyl-β-D-glucopyranosyl isothiocyanate-derivatised amino acid enantiomers, *J.Chromatogr.A*, **1995**, *697*, 213–217.

2,3,4,6-Tetra-O-benzoyl-β-D-glucopyranosyl Isothiocyanate

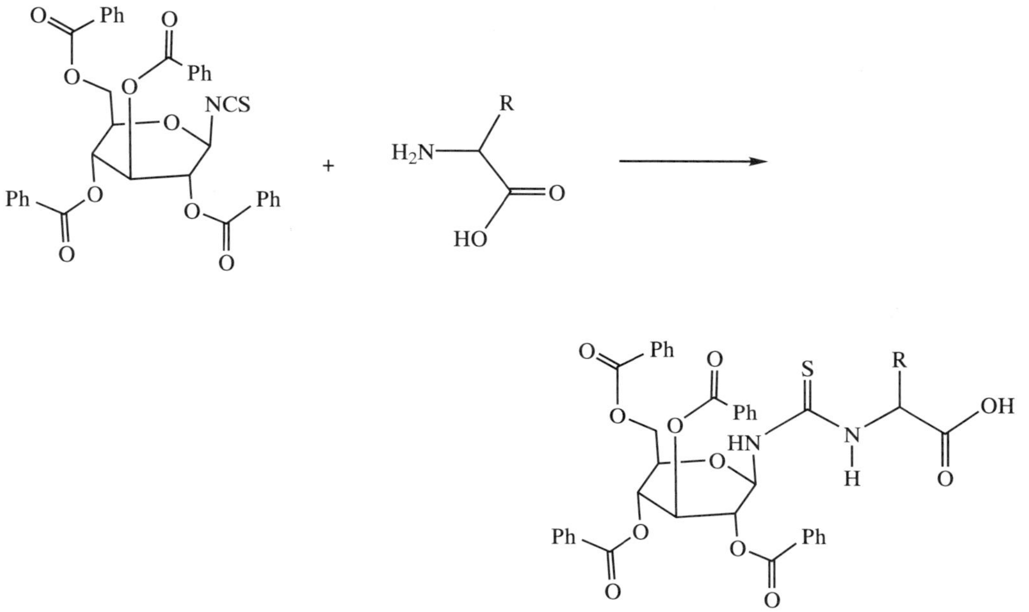

SAMPLE
Matrix: bulk

Analyte: amino acids and β-blockers

Sample preparation: Dissolve 5 mg amino acids in 10 mL MeCN:water:triethylamine 50:50:0.55. Remove a 50 μL aliquot and add it to 50 μL 0.66% 2,3,4,6-tetra-O-benzoyl-β-D-glucopyranosyl isothiocyanate (Fluka) in MeCN, shake mechanically for 30 min, add 10 μL 0.26% ethanolamine in MeCN, shake for 10 min, make up to 1 mL with MeCN, inject a 10 μL aliquot.

HPLC VARIABLES
Column: 25 × 4 (sic) 5 μm LiChrospher 100 RP-18

Mobile phase: MeOH:water:67 mM pH 7.0 phosphate buffer 65:27:8 (A) or 70:25:5 (B) or 80:15:5 (C) or 70:30:0 (D) or 80:20:0 (E) or 85:15:0 (F) or 90:10:0 (G)

Flow rate: 0.42 (A) or 0.45 (B) or 0.50 (C, F, G) or 1 (D, E)

Injection volume: 10

Detector: UV 231

CHROMATOGRAM
Retention time: k' 2.65, 3.65 (propranolol enantiomers (G)), k' 3.00, 3.71 (pindolol enantiomers (F)), k' 5.01, 6.53 (sotalol enantiomers (E)), k' 5.19 (D-proline (B)), k' 5.35 (L-threonine (B)), k' 5.68, 6.66 (atenolol enantiomers (E)), k' 6.22 (L-tyrosine (B)), k' 6.24 (L-2-aminobutyric acid (B)), k' 6.24 (D-threonine (B)), k' 6.38 (L-phenylglycine (B)), k' 6.41 (L-proline (B)), k' 7.22 (L-valine (B)), k' 7.37 (L-penicillamine (B)), k' 7.41 (D-tyrosine (B)), k' 7.57 (D-2-aminobutyric acid (B)), k' 7.86 (D-phenylglycine (B)), k' 8.08 (L-methionine (B)), k' 8.86, 11.66 (5-(2-N-benzylamino-2-hydroxyethyl)salicylamide enantiomers (E)), k' 9.16 (D-valine (B)), k' 9.27 (L-isoleucine (B)), k' 9.43 (L-tryptophan (B)), k' 9.51 (L-leucine (B)), k' 10.05 (D-penicillamine (B)), k' 10.24 (D-methionine (B)), k' 10.54 (L-phenylalanine

(B)), k' 11.94 (L-ornithine (C)), k' 12.03 (D-tryptophan (B)), k' 12.35 (D-isoleucine (B)), k' 12.65 (D-leucine (B)), k' 12.89 (L-norleucine (B)), k' 13.48 (L-lysine (C)), k' 13.81 (D-phenylalanine (B)), k' 13.90 (D-ornithine (C)), k' 15.32 (D-lysine (C)), k' 16.81 (D-norleucine (B)), k' 16.88 (L-3-aminobutyric acid (A)), k' 18.00 (L-alanine (A)), k' 18.95 (D-3-aminobutyric acid (A)), k' 20.85 (D-alanine (A)), k' 22.61, 25.67 (epinephrine enantiomers (D)), k' 33.96, 40.07 (phenylephrine enantiomers (D))

OTHER SUBSTANCES
Simultaneously analyzed: amines

KEY WORDS
chiral

REFERENCE
Lobell, M.; Schneider, M.P. 2,3,4,6-Tetra-O-benzoyl-β-D-glucopyranosyl isothiocyanate: an efficient reagent for the determination of enantiomeric purities of amino acids, β-adrenergic blockers and alkyloxiranes by high-performance liquid chromatography using standard reversed-phase columns, *J.Chromatogr.*, **1993**, *633*, 287–294.

KETOALDEHYDE

3-(4-Carboxybenzoyl)-2-quinolinecarboxaldehyde

SAMPLE
Matrix: solutions
Analyte: amino acids
Sample preparation: Mix 2-5 μL of a 1-100 μM solution or protein hydrolysate with 10-20 μL 10 mM KCN in water and 5-10 μL 3 mg/mL reagent in MeOH, let stand at room temperature for at least 1 h, inject an aliquot. (Reagent was 3-(4-carboxybenzoyl)-2-quinolinecarboxaldehyde, available from Molecular Probes, Eugene OR. Synthesis is as follows. Mix 9.69 g p-toluidine, 10.89 g o-nitrobenzaldehyde, and 25 mL EtOH and allow to react for 5 min, filter, wash the solid with EtOH to give 2-nitro-N-(p-tolyl)benzaldimine (Talanta 1989, 36, 321). (More 2-nitro-N-(p-tolyl)benzaldimine can be obtained by adding water to the filtrate.) Add, in portions in a thin stream, a hot solution of 46 g sodium sulfide in 23 mL water and 23 mL EtOH to a stirred refluxing solution of 100 mmole 2-nitro-N-(p-tolyl)benzaldimine in 50 mL EtOH. After a vigorous reaction 2-amino-N-(p-tolyl)benzaldimine crystallizes from the cooling solution (mp 102-103° after recrystallization from dilute MeOH) (Ber. 1943, 76, 1099). Wash 950 mg of a commercial 50% slurry of sodium hydride with pentane, add 7 mL dry THF (distilled from lithium aluminum hydride), add 1.51 g methyl 4-cyanobenzoate in 10 mL THF, add dropwise 1.38 mL acetone (distilled from calcium chloride), reflux for 1.5 h, cool, acidify with 3 M HCl. Remove the organic layer and wash it with brine and sodium bicarbonate solution, dry over an-

hydrous magnesium sulfate, evaporate to give (4-cyanobenzoyl)acetone. Reflux 433 mg (4-cyanobenzoyl)acetone, 486 mg 2-amino-N-(p-tolyl)benzaldimine, 69 mL piperidine, and 9 mL EtOH:water 95:5 for 18 h, remove volatiles by steam distillation, add the residue to water and dichloromethane. Remove the organic layer and dry it, evaporate to give 3-(4-cyanobenzoyl)-2-methylquinoline. Suspend 547 mg 3-(4-cyanobenzoyl)-2-methylquinoline in 13 mL EtOH:water 95:5, add 500 mg KOH, reflux for 6 h, cool, concentrate, add the residue to ether and water. Remove the aqueous layer and adjust the pH to 5 with tartaric acid, let stand for 15 min, filter, wash the precipitate with water, dry under vacuum to give 3-(4-carboxybenzoyl)-2-methylquinoline. Dissolve 266 mg 3-(4-carboxybenzoyl)-2-methylquinoline in 6 mL acetic acid, add 112 mg selenium dioxide, stir at 80° for 2 h, filter through Celite, wash the precipitate with several volumes of hot MeOH, dilute the filtrate with water, allow to stand, filter, wash the solid with water, dry under vacuum to give 3-(4-carboxybenzoyl)-2-quinolinecarboxaldehyde.)

CAPILLARY ELECTROPHORESIS
Capillary: 104 cm × 50 μm fused-silica (73 cm to detector)
Running buffer: 50 mM pH 7.02 2-[N-[tris(hydroxymethyl)methyl]amino]ethanesulfonic acid (TES) containing 50 mM sodium dodecyl sulfate
Voltage/Current: 25 kV/14 μA
Injection: Hydrodynamic or electromigration for 10 s
Detector: F ex 442 (50 mW He-Cd laser) em 560 (bandwith filter)
Migration time: 14 (Arg), 15.2 (Trp), 15.4 (Tyr), 15.6 (His), 15.8 (Met), 16 (Ile), 16.5 (Gln), 16.7 (Asn), 17 (Thr), 18.5 (Phe), 18.8 (Leu), 19.2 (Val), 20.5 (Ser), 21 (Ala), 22.5 (Gly), 24.5 (Glu), 30 (Asp)
Limit of detection: 4.6-13.8 amole

REFERENCE
Liu, J.; Hsieh, Y.-Z.; Wiesler, D.; Novotny, M. Design of 3-(4-carboxybenzoyl)-2-quinolinecarboxaldehyde as a reagent for ultrasensitive determination of primary amines by capillary electrophoresis using laser fluorescence detection, *Anal.Chem.*, **1991**, *63*, 408–412.

SAMPLE
Matrix: dialysate
Analyte: amino acids
Sample preparation: 10 μL Dialysate + 1 μL 198 μM somatostatin in water + 5 μL 3 mg/mL 3-(4-carboxybenzoyl)-2-quinolinecarboxaldehyde (CBQCA; Molecular Probes, Eugene OR) in MeOH + 2 μL 50 mM KCN in 50 mM pH 9.0 borate buffer, mix, let stand at room temperature for 2 h, inject an aliquot.

CAPILLARY ELECTROPHORESIS
Capillary: 57 cm × 75 μm fused-silica (50 cm to detector) (Polymicro Technologies)
Capillary preparation: Before each injection rinse capillary with 100 mM NaOH at high pressure for 2 min and with running buffer at high pressure for 3 min
Running buffer: DMSO:buffer 20:80 (Buffer was 50 mM pH 9.0 Sodium borate buffer containing 30 mM sodium dodecyl sulfate.)
Capillary temperature: 25
Voltage/Current: 20 kV/33 μA
Injection: Pressure injection for 1 s (5.9 nL).
Detector: F ex 488 (4 nW argon ion laser) em 560 ± 40
Model: P/ACE 2100
Migration time: 9.5 (arginine), 14 (glutamine), 14.3 (valine), 14.5 (GABA), 15 (alanine), 15.2 (glycine), 21.6 (glutamic acid), 22.3 (aspartic acid)
Internal standard: somatostatin (3-6) (H-Cys-Lys-Asn-Phe-OH) (10.7)
Limit of detection: 0.29-68 nM

OTHER SUBSTANCES
Non-interfering: ammonia, aspargine, citrulline, ethanolamine, histidine, isoleucine, leucine, lysine, phenylalanine, o-phosphoethanolamine, serine, taurine, threonine, tyrosine

KEY WORDS
rat

REFERENCE

Bergquist, J.; Vona, M.J.; Stiller, C.-O.; O'Connor, W.T.; Falkenberg, T.; Ekman, R. Capillary electrophoresis with laser-induced fluorescence detection: A sensitive method for monitoring extracellular
concentrations of amino acids in the periaqueductal gray matter, *J.Neurosci.Methods*, **1996**, *65*,
33–42.

LACTONE

Fluorescamine

RELATED REFERENCES

Clements, L.A.J.; Hilbish, T.J. Comparison of fluorescamine and o-phthalidialdehyde methods for measuring amino acid exchange in marine organisms. *Limnol.Oceanogr.* **1992**, *36*, 1463-1471.
Garcia Sanchez, F.; Aguilar Gallardo, A. Liquid chromatographic and spectrofluorimetric determination
of aspartame and glutamate in foodstuffs following fluorescamine fluorigenic labeling.
Anal.Chim.Acta **1992**, *270*, 45-53.
Nair, P.P.; Kessie, G.; Patnaik, R.; Guidry, C. Isolation and HPLC of N-epsilon-lithocholyl lysine as its
fluorescamine and dimethylaminoazobenzene isothiocyanate derivatives. *Steroids* **1994**, *59*, 212-216

SAMPLE

Matrix: solutions

Analyte: proline and hydroxyproline

Sample preparation: Add 30 μL reagent to 70 μL of a solution in 100 mM pH 9.0 sodium
tetraborate buffer while continuously and vigorously vortexing, after 2 min inject an aliquot. (Reagent was 3 mg/mL fluorescamine in acetone containing 20 μL/mL (?) pyridine.)

CAPILLARY ELECTROPHORESIS

Capillary: 107 cm × 75 μm (100 cm to detector)

Capillary preparation: At the end of each run wash capillary with 2 M NaOH, with 1 M NaOH, with water, and with running buffer.
Running buffer: 50 mM pH 8.3 Sodium tetraborate buffer containing 25 mM LiCl
Capillary temperature: 25
Voltage/Current: 12 kV
Injection: Pressure injection with nitrogen at 0.5 psi for 4 s (24 nL)
Detector: UV 214
Model: Beckman P/ACE System 2000
Migration time: 41 (4-hydroxyproline), 42 (proline)

REFERENCE
Guzman, N.A.; Moschera, J.; Iqbal, K.; Malick, A.W. A quantitative assay for the determination of proline and hydroxyproline by capillary electrophoresis, *J.Liq.Chromatogr.*, **1992**, *15*, 1163–1177.

MISCELLANEOUS REACTIONS

Cobalt/Hydroxylamine

RELATED REFERENCE
Kai, M.; Nakashima, A.; Ohkura, Y. High-performance liquid chromatographic separation of kyotorphin, a basic Tyr-Arg dipeptide, in rat brain tissue and quantification using fluorimetric detection. *J.Chromatogr.B* **1997**, *688*, 205-212.

SAMPLE
Matrix: enzyme incubations
Analyte: N-terminal tyrosine-containing peptides
Sample preparation: Heat 200 µL enzyme incubation at 100° for 1 min, centrifuge at 2000 g for 10 min, inject a 50 µL aliquot of the supernatant.

HPLC VARIABLES
Column: 200 × 4 5 µm TSKgel ODS-120T
Mobile phase: Gradient. A was MeCN:300 mM pH 2.3 sodium phosphate buffer:water 1: 20:79. B was MeCN:300 mM pH 2.3 sodium phosphate buffer:water 60:20:20. A:B 100:0 for 5 min, to 99:1 (step gradient), maintain at 99:1 for 5 min, to 60:40 (step gradient), maintain at 60:40 for 2 min, to 40:60 over 5 min, maintain at 40:60 for 6 min, return to initial conditions (step-gradient), re-equilibrate for 7 min.
Flow rate: 1
Injection volume: 50
Detector: F ex 335 em 435 following post-column reaction. The column effluent mixed with the reagent pumped at 0.5 mL/min and with buffer pumped at 0.5 mL/min and the mixture flowed through a 19 m × 0.5 mm ID PTFE coil at 75 ± 1° to the detector. (Prepare reagent by mixing equal volumes of 40 mM hydroxylamine hydrochloride and 0.1 mM cobalt(II) chloride hexahydrate, usable for at least 1 month when stored a refrigerator. Prepare buffer by dissolving 9.27 g boric acid in 400 mL water, adjusting pH to 11.0 with 300 mM NaOH, make up to 500 mL with water. Post-column reaction is specific for peptides containing an N-terminal tyrosine.)

CHROMATOGRAM
Retention time: 10.5 (tyrosine), 12.5 (Tyr-Gly-Gly), 14.5 (Tyr-Gly), 18.5 (Tyr-Tyr), 20 (Tyr-Gly-Gly-Phe), 22 (methionine enkephalin), 23.5 (leucine enkephalin)
Limit of detection: 40-200 nM

KEY WORDS
post-column reaction

REFERENCE
Ohno, M.; Kai, M.; Ohkura, Y. On-line post-column fluorescence detection for N-terminal tyrosine-containing peptides in high-performance liquid chromatography, *J.Chromatogr.*, **1987**, *421*, 245–256.

SAMPLE
Matrix: solutions
Analyte: oligopeptides
Sample preparation: 50 µL Oligopeptide solution + 50 µL reagent solution + 50 µL 400 mM pH 8.5 sodium borate buffer, heat in a boiling water bath for 3 min, add 25 µL 20 mM 2-mercaptoethanol, inject a 100 µL aliquot. (Prepare reagent solution by dissolving 20.5 mg hydroxylamine oxalate and 3.1 mg cobalt(II) acetate tetrahydrate in 10 mL water, make up to 25 mL with water. Prepare hydroxylamine oxalate as follows. Mechanically stir a solution of 140 g hydroxylamine hydrochloride in 400 mL water and add 126 g oxalic acid dihydrate in one portion, after a few min rapidly add a solution of 80 g NaOH in 200 mL water dropwise, when addition is complete stir for 30 min, cool to 15°, filter, wash the crystals with three 100 mL portions of water, recrystallize from water (about 8 volumes) (Inorg. Syn. 1950, 3, 81).)

HPLC VARIABLES
Column: 150 × 4 5 µm TSKgel ODS-120T (Toyo Soda)
Mobile phase: Gradient. MeCN:buffer from 2:98 to 20:80 over 10 min, to 26:74 over 15 min. (Buffer was 100 mM pH 8.5 sodium borate buffer:10 mM tetra-n-butylammonium chloride:water 45:15:40.)
Flow rate: 1
Injection volume: 100
Detector: F ex 330 em 440

CHROMATOGRAM
Retention time: 4.6 (Tyr-Arg), 6.1 (tyrosinamide), 13.6 (tyrosine), 13.6 (Tyr-Gly), 13.6 (Tyr-Gly-Gly), 14.5 (leucine-enkephalin-Arg), 15.2 (Tyr-Tyr), 16.2 (methionine-enkephalin-NH$_2$), 16.6 (Tyr-Gly-Gly-Phe), 16.6 (methionine-enkephalin-Arg-Gly-Leu), 18.4 (Tyr-Phe), 18.4 (leucine-enkephalin-NH$_2$), 18.6 (methionine-enkephalin-Arg-Phe), 19.8 (methionine-enkephalin), 21.2 (leucine-enkephalin), 23.6 ([Ala^2,Ala^2]methionine-enkephalin)
Limit of detection: 150-2220 fmole

KEY WORDS
Method is specific for N-terminal tyrosine-containing oligopeptides.

REFERENCE
Nakano, M.; Kai, M.; Ohno, M.; Ohkura, Y. High-performance liquid chromatography of N-terminal tyrosine-containing oligopeptides by pre-column fluorescence derivatization with hydroxylamine, cobalt (II) and borate reagents, *J.Chromatogr.*, **1987**, *411*, 305–311.

SAMPLE
Matrix: tissue
Analyte: enkephalins
Sample preparation: Homogenize 50 mg tissue with 500 µL 100 mM HCl at 0-4°, centrifuge at 350 g for 20 min, add 1 mL acetone to the supernatant, centrifuge at 1500 g for 10 min, suspend the precipitate in 350 µL water. Remove a 300 µL aliquot and add it to 50 µL 500 mM pH 7.8 Tris-HCl buffer containing 100 mM calcium chloride and 100 mM NaCl and 125 µL 24000 U/mL trypsin (EC 3.4.21.4 from bovine pancreas, TPCK treated, Sigma), mix, make up to 500 µL with water, heat at 37° for 1 h, add 200 µL 2 M perchloric acid, add 15 µL 5 µM IS, centrifuge at 1500 g for 5 min, add the supernatant to a 200 mg 40 µm Asahipregel TC18 SPE cartridge, wash twice with 2 mL portions of water, elute with 1.5 mL MeCN:water 90:10. Evaporate the eluate to dryness under reduced pressure at 30°, reconstitute the residue in 250 µL water. Remove a 60 µL aliquot and make up to 100 µL with water, inject the whole amount. (Procedure is specific for N-terminal tyrosine peptides.)

HPLC VARIABLES
Column: 150 × 6 Asahipak ODP-50 octadecylated polyvinylalcohol copolymer gel column (Ashai Chemical)
Mobile phase: Gradient. A was MeCN:50 mM pH 7.0 borate buffer:water 5:20:75. B was MeCN:50 mM pH 7.0 borate buffer:water 60:20:20. A:B from 93:7 to 55:45 over 40 min.
Flow rate: 1

Injection volume: 100

Detector: F ex 330 em 440 following post-column reaction. The column effluent mixed with 300 mM pH 9.0 sodium borate buffer pumped at 0.4 mL/min and 8 mM hydroxylamine oxalate containing 0.2 mM cobalt(II) acetate pumped at 0.4 mL/min and the mixture flowed through a 10 m $\times$ 0.5 mm ID stainless-steel coil at 100° to the detector.

CHROMATOGRAM
Retention time: 16 (methionine enkephalin), 18 (leucine enkephalin)
Internal standard: [D-Ala^2]ME (19)
Limit of detection: 0.7-2.8 pmole

OTHER SUBSTANCES
Extracted: numerous other peptides

KEY WORDS
post-column reaction; rat; brain; SPE

REFERENCE
Zhang, G.-Q.; Kai, M.; Ohkura, Y. High-performance liquid chromatographic determination of peptides released by tryptic degradation from opioid peptide precursors in rat brain, *Chem.Pharm.Bull.*, **1991**, *39*, 2369–2372.

SAMPLE
Matrix: tissue
Analyte: enkephalins
Sample preparation: Condition a 200 mg 9 μm Asahipregel T C18 SPE cartridge with 2 mL MeOH, 2 mL 100 mM pH 8.5 borate buffer, and 2 mL water. Homogenize 100-200 mg tissue with 3 mL 100 mM HCl at 0-4°, add 20 μL 5 μm IS, add 20 μL water, add 500 μL 2 M perchloric acid, centrifuge at 2450 g for 10 min. Adjust the pH of the supernatant to 7-8 with 1 M sodium bicarbonate solution (ca. 2 mL), add 2 mL 50 mM disodium EDTA, add to the SPE cartridge, wash with 1 mL water, wash with 3 mL 100 mM HCl, wash with 3 mL water, wash with 3 mL 100 mM pH 8.5 borate buffer, wash with 1 mL water, elute with 1.5 mL MeOH:200 mM pH 8.5 borate buffer 80:20. Evaporate the eluate to dryness under reduced pressure at 30°, reconstitute the residue in 300 μL water. Remove a 60 μL aliquot and add it to 25 μL 10 mM hydroxylamine hydrochloride containing 0.5 mM cobalt(II) acetate and 15 μL water, heat at 100° for 3 min, add 25 μL 20 mM 2-mercaptoethanol, inject an aliquot. (Method is specific for N-terminal tyrosine peptides.)

HPLC VARIABLES
Column: 150 $\times$ 6 Asahipak ODP-50 octadecylated polyvinylalcohol copolymer gel column (Ashai Chemical)
Mobile phase: Gradient. A was MeCN:100 mM pH 9.5 borate buffer:water 2:20:78. B was MeCN:100 mM pH 9.5 borate buffer:water 50:20:30. A:B from 100:0 to 73:27 over 10 min, maintain at 73:27 for 20 min, to 21:79 over 20 min, to 0:100 (step gradient), maintain at 0:100 for 6 min, return to initial conditions (step gradient).
Flow rate: 0.8 for 25 min then 1 for 30 min
Injection volume: 100
Detector: F ex 330 em 440

CHROMATOGRAM
Retention time: 14 (methionine enkephalin), 16 (leucine enkephalin)
Internal standard: [D-Ala^2]LE-NH$_2$; (30)
Limit of detection: 0.33-1.21 pmole

OTHER SUBSTANCES
Extracted: numerous other peptides

KEY WORDS
rat; brain; SPE

REFERENCE
Zhang, G.Q.; Kai, M.; Nakano, M.; Ohkura, Y. Pre-column fluorescence derivatization high-performance liquid chromatography of opioid peptides in rat brain and its use for enzymatic peptide characterization, *Chem.Pharm.Bull.*, **1991**, *39*, 126–129.

Cyclization

SAMPLE
Matrix: blood
Analyte: tryptophan
Sample preparation: Plasma. 10 μL Plasma + 1 mL ice-cold 80-200 nM IS in 10% trichloroacetic acid, centrifuge at 4° at 10000 g for 10 min. Remove a 900 μL aliquot of the supernatant and add it to 100 μL 10% formaldehyde, add 70 μL 5 M NaOH, add 50 μL 6 mM potassium ferricyanide, cap tightly, vortex, heat at 100-105° for 15 min, immediately after heating loosen caps and vortex in a fume hood to remove chloroform and carbon dioxide, inject a 200 μL aliquot.

HPLC VARIABLES
Column: 300 × 3.9 μBondapak C18
Mobile phase: DMF:buffer:water 25:20:55, pH 2.5 (Buffer was 8.5% phosphoric acid adjusted to pH 1.5 with NaOH.)
Flow rate: 1
Injection volume: 200
Detector: F ex 360 em 425 (cut-off)

CHROMATOGRAM
Retention time: 6
Internal standard: 5-methyl-DL-tryptophan (7.5)
Limit of quantitation: 300 nM

KEY WORDS
human; mouse; plasma

REFERENCE
Inoue, S.; Tokuyama, T.; Takai, K. Picomole analyses of tryptophan by derivatization to 9-hydroxymethyl-beta-carboline, *Anal.Biochem.*, **1983**, *132*, 468–480.

SOLID PHASE REAGENT

Polymer Bound Benzotriazole 6-Aminoquinoline

SAMPLE
Matrix: protein
Analyte: amino acids
Sample preparation: Heat 50 µg protein with 200 µL 6 M HCl and 1 crystal of phenol at 153° for 1 h, evaporate to dryness under reduced pressure, reconstitute with water. Mix 25 µL hydrolyzate, 25 µL saturated sodium borate, 25 µL 20 mM cetyltrimethylammonium bromide, 12.5 µL MeCN, and 15 mg polymer bound benzotriazole 6-aminoquinoline, heat at 70° for 10 min, elute with 1 mL MeCN:water 70:30, inject a 20 µL aliquot of the eluate. (Prepare polymer bound benzotriazole 6-aminoquinoline as follows. React 10 g 4-chloro-3-nitrobenzyl alcohol, 10 g dried 16-20 µm styrene-divinylbenzene copolymer (12% cross-linked, 102 Å templated, Supelco), and 10 g anhydrous aluminum chloride in 50 mL nitrobenzene at 65-70° for 3 days, cool to room temperature, filter, wash with three 50 mL portions of 1 M HCl in dioxane (Caution! Dioxane is a carcinogen!), wash with three 50 mL portions of DMF, wash with three 50 mL portions of MeOH, wash with three 50 mL portions of dichloromethane, dry under reduced pressure at 100°. Reflux 19 g of this polymer in 60 mL hydrazine hydrate:2-ethoxyethanol 40:60 for 20 h, cool to room temperature, filter, wash thoroughly with water, suspend the polymer in 100 mL concentrated aqueous HCl:dioxane 50:50, reflux for 20 h, filter, wash with five 100 mL portions of water, wash with three 100 mL portions of MeOH, wash with three 50 mL portions of ether, dry under reduced pressure at 80° to obtain polymeric benzotriazole reagent (Eur.J. Biochem. 1975, 59, 55). Heat 1.6 g quinoline-6-carboxylic acid (ICN Biomedicals, Costa Mesa CA) with 2.7 mL thionyl chloride in 55 mL dry benzene at 65-75° for 1 h (Caution! Benzene is a carcinogen!), remove the benzene by evaporation under reduced pressure, add 21 mL acetic acid, stir strongly for several min, slowly add 600 mg sodium azide at room temperature, stir for 1 h, add 53 mL triethylamine, add 100 mL water, stir for 30 min, filter, wash ten times with 25 mL portions of water, dry under vacuum overnight to obtain quinoline-6-acyl azide. Heat 3.3 g polymeric benzotriazole reagent and 2 g quinoline-6-acyl azide in 150 mL toluene (dried over calcium hydride) at 60-65° for 2 h, wash with 600 mL warm (40°) dichloromethane, wash with 50 mL MeCN, dry under vacuum overnight to obtain polymer bound benzotriazole 6-aminoquinoline.)

HPLC VARIABLES
Column: 150 × 3.9 4 µm Nova-Pak C18

Mobile phase: Gradient. A was 140 mM sodium acetate containing 17 mM triethylamine, adjusted to pH 5.05 with phosphoric acid. B was MeCN:water 60:40. A:B from 100:0 to 50:50 over 30 min, to 0:100 (step gradient), maintain at 0:100 for 5 min, return to initial conditions, re-equilibrate at initial conditions for 10 min.
Flow rate: 1.5
Injection volume: 20
Detector: F ex 254 em 395

CHROMATOGRAM
Retention time: 15-30 (depending on structure; peaks not identified)
Limit of detection: 1 ppm

REFERENCE
Yu, J.H.; Li, G.D.; Krull, I.S.; Cohen, S. Polymeric 6-aminoquinoline, an activated carbamate reagent for derivatization of amines and amino acids by high-performance liquid chromatography, *J.Chromatogr.B*, **1994**, *658*, 249–260.

Polymer Bound Benzotriazole 3,5-Dinitrobenzoate

SAMPLE
Matrix: solutions
Analyte: amino acids
Sample preparation: Mix 50 μL of a 200 ppm solution in MeCN:50 mM NaOH 80:20 with 30 mg polymer bound benzotriazole 3,5-dinitrobenzoate, after 2 min elute with 500 μL MeCN, add 500 μL water to the eluate, mix, inject a 20 μL aliquot. (The reagent was polymer bound benzotriazole 3,5-dinitrobenzoate, synthesized as follows. (Caution! Chloroform, dichloromethane, dioxane, and hydrazine are carcinogenic in experimental animals! DMF may be carcinogenic! 3,5-Dinitrobenzoyl chloride and aluminum chloride are corrosive! Nitrobenzene is toxic!) 10 g Dried macroporous polystyrene (Xe-305, Rohm and Haas) + 10 g 3-nitro-4-chlorobenzyl alcohol + 10 g anhydrous aluminum chloride + 50 mL nitrobenzene, heat at 65-70° for 3 days, cool, filter, wash polymer with three 50 mL portions of 1 M HCl in dioxane, with three 50 mL portions of DMF, with three 50 mL portions of MeOH, and with three 50 mL portions of dichloromethane, dry under vacuum at 100°. Reflux 19 g of this polymer in 60 mL hydrazine hydrate:ethylene glycol monoethyl ether 40:60 for 20 h, cool to room temperature, filter off the polymer and wash it thoroughly with water. Suspend the polymer in 100 mL concentrated HCl:dioxane 50:50, reflux for 20 h, filter the polymer and wash it with five 100 mL portions of water, with three 100 mL portions of MeOH, and with three 50 mL portions of ether, dry under vacuum at 80°. Functionalization was 1.17 mmoles/g (Eur.J.Biochem. 1975, 59, 55). Add

a portion of polymer to dry chloroform, add a three-fold excess of 3,5-dinitrobenzoyl chloride and pyridine, stir at 0-10° for 30 min, filter off polymer, wash with chloroform to give polymer bound benzotriazole 3,5-dinitrobenzoate (J.Org.Chem. 1984, 49, 922).)

HPLC VARIABLES
Column: 100 × 4.6 3 μm Spherisorb CN 100
Mobile phase: MeCN:water:trifluoroacetic acid 20:80:0.1
Injection volume: 20
Detector: UV

CHROMATOGRAM
Retention time: 2 (valine), 2.2 (methionine), 3.3 (phenylalanine), 4 (tryptophan)

REFERENCE
Bourque, A.J.; Krull, I.S. Solid-phase reagent containing the 3,5-dinitrophenyl tag for the improved derivatization of chiral and achiral amines, amino alcohols and amino acids in high-performance liquid chromatography with ultraviolet detection, *J.Chromatogr.*, **1991**, *537*, 123–152.

Polymer Bound o-Nitrobenzophenone 9-Fluoreneacetate

SAMPLE
Matrix: protein
Analyte: amino acids
Sample preparation: Dissolve 1 mg protein in 2 mL 6 M HCl, heat in a sealed tube at 90° for 12 h, evaporate to dryness under reduced pressure, reconstitute with 2 mL water. Mix 25 μL hydrolyzate, 25 μL pH 9.1 saturated sodium borate, 25 μL 20 mM cetyltrimethylammonium bromide, 12.5 μL MeCN, and 10 mg polymer bound o-nitrobenzophenone 9-fluoreneacetate, heat at 70° for 10 min, elute with 1 mL MeCN:water 70:30, inject a 20 μL aliquot of the eluate. (Prepare polymer bound o-nitrobenzophenone 9-fluoreneacetate as follows. Soxhlet extract styrene-divinylbenzene copolymer (12% cross-linked, 60 Å templated, 10-20 μm, Supelco) with dioxane for 8 h (Caution! Dioxane is a carcin-

ogen!). Add 25 g aluminum trichloride in 300 mL dry nitrobenzene to 50 g resin and 100 g 4-chloro-3-nitrobenzoyl chloride, stir mechanically at 60° for 5 h, pour into a mixture of 150 mL DMF, 100 mL concentrated HCl, and 150 g ice, filter. Wash the solid with 300 mL portions of DMF:water 75:25 until the washings are colorless, wash with warm (60°) DMF, wash with six 300 mL portions of dichloromethane:MeOH 2:1. Stir the product in 130 mL 40% benzyltrimethylammonium hydroxide in water, 130 mL water, and 260 mL dioxane at 90° for 8 h, filter, repeat the process. Wash the product with four portions of warm (60°) dioxane. Stir the solid with 30 mL acetic acid for 15 min, filter. Wash the solid with dioxane until the washings are neutral, wash with six 300 mL portions of dichloromethane:MeOH 2:1 to give a nitrobenzophenol-substituted polymer (J. Org. Chem. 1984, 49, 924). Heat 4 g 9-fluoreneacetic acid, 3.9 mL oxalyl chloride, 30 mL benzene (dried over anhydrous sodium sulfate, Caution! Benzene is a carcinogen!), and 3 drops of triethylamine at 55° for 1 h, evaporate under reduced pressure to remove oxalyl chloride, dissolve the product in 35 mL dichloromethane to give a 120 mg/mL solution of 9-fluoreneacetyl chloride, dilute to obtain a 2 mM solution. Stir 1.3 g nitrobenzophenol-substituted polymer, 4.2 mL 2 mM 9-fluoreneacetyl chloride solution, 300 μL triethylamine, and 20 mL dichloromethane at room temperature for 1 h, filter, wash with three 20 mL portions of MeCN to obtain the reagent, polymer bound o-nitrobenzophenone 9-fluoreneacetate (J. Chromatogr. 1992, 609, 103).)

HPLC VARIABLES

Column: 250 × 4.6 YMC AP-303 300 Å ODS (YMC)

Mobile phase: Gradient. A was 0.05% trifluoroacetic acid in water. B was 0.05% trifluoroacetic acid in MeCN. A:B from 70:30 to 30:70 over 16 min, maintain at 30:70 for 6 min, return to initial conditions over 30 s.

Flow rate: 1.5

Injection volume: 20

Detector: F ex 254 em 305-395

CHROMATOGRAM

Retention time: 5.8 (His), 6.3 (Asn, Gln), 7.2 (Ser), 7.4 (Glu), 7.8 (Thr), 9.1 (Ala), 10.6 (Pro), 11 (Met), 12.2 (Tyr), 12.6 (Ile), 12.8 (Leu), 13 (Phe), 14.6 (Lys)

REFERENCE

Zhou, F.-X.; Krull, I.S.; Feibush, B. Solid-phase derivatization of amino acids and peptides in high-performance liquid chromatography, *J.Chromatogr.*, **1993**, *648*, 357–365.

SUCCINIMIDYL ESTER

6-Aminoquinolyl-N-hydroxysuccinimidyl Carbamate

RELATED REFERENCES

Liu, H.J. Determination of amino acids by precolumn derivatization with 6-aminoquinolyl-N-hydroxy-succinimidyl carbamate and high-performance liquid chromatography with ultraviolet detection. *J.Chromatogr.A* **1994**, *670*, 59-66.

Pawlowska, M.; Chen, S.; Armstrong, D.W. Enantiomeric separation of fluorescent, 6-aminoquinolyl-N-hydroxysuccinimidyl carbamate, tagged amino acids. *J.Chromatogr.* **1993**, *641*, 257-265.

Reverter, M.; Lundh, T.; Lindberg, J.E. Determination of free amino acids in pig plasma by precolumn derivatization with 6-N-aminoquinolyl-N-hydroxysuccinimidyl carbamate and high-performance liquid chromatography. *J.Chromatogr.B* **1997**, *696*, 1-8.

van Wandelen, C.; Cohen, S.A. Using quaternary high-performance liquid chromatography eluent systems for separating 6-aminoquinolyl-N-hydroxysuccinimidyl carbamate-derivatized amino acid mixtures. *J.Chromatogr.A* **1997**, *763*, 11-22

SAMPLE

Matrix: blood

Analyte: amino acids

Sample preparation: 500 μL Plasma + 500 μL MeCN, vortex, let stand for 10 min, centrifuge at 12000 rpm for 15 min. Remove a 10 μL aliquot of the supernatant and add it to 70 μL 200 mM pH 8.8 borate buffer, vortex, add 20 μL 3 mg/mL 6-aminoquinolyl-N-hydroxysuccinimidyl carbamate in MeCN, vortex immediately, heat at 50° for 10 min, dilute (if necessary), inject an aliquot. (6-Aminoquinolyl-N-hydroxysuccinimidyl carbamate can be purchased from Waters or synthesized as follows. Reflux 3 g N,N'-succinimidyl carbonate in 100 mL dry MeCN, add 1.5 g 6-aminoquinoline in 50 mL dry MeCN dropwise over 30 min, reflux for 30 min, evaporate to half volume under reduced pressure, cool for 24 h, filter, wash the solid with cold MeCN, recrystallize from MeCN to obtain 6-aminoquinolyl-N-hydroxysuccinimidyl carbamate as off-white crystals (mp 210-215° (d)) (Anal. Biochem. 1993, 211, 279).)

HPLC VARIABLES

Column: 250 × 4.6 5 μm YMC AP303 300 Å ODS (YMC)

Mobile phase: Gradient. A was MeCN:water:buffer 2:96:2. B was MeCN:water:buffer 60:38:2. A:B 100:0 for 2 min, to 70:30 over 40 min, maintain at 70:30 for 10 min. (Prepare buffer by adjusting the pH of 1 M NaH_2PO_4 to 3.0 with phosphoric acid.)

Flow rate: 1.5

Detector: E, Bioanalytical Systems LC-4B, Model MF-1000 glassy carbon working electrode +1.1 V, stainless steel counter electrode, Ag/AgCl reference electrode

CHROMATOGRAM
Retention time: 15.2 (ammonia), 16.8 (His), 18.5 (Ser), 19.2 (Arg), 19.5 (Gly), 20.8 (Asp), 22 (Glu), 23.2 (Thr), 24.8 (Ala), 27.2 (Pro), 33 (Cys), 34.8 (Lys), 36.8 (Met), 36.8 (Tyr), 44.8 (Ile), 45.6 (Leu), 47.2 (Phe), 50.4 (Trp)
Limit of quantitation: 500 nM
Limit of detection: 250 nM

KEY WORDS
plasma; cow; human

REFERENCE
Li, G.-D.; Krull, I.S.; Cohen, S.A. Electrochemical activity of 6-aminoquinolyl urea derivatives of amino acids and peptides. Application to high-performance liquid chromatography with electrochemical detection, *J.Chromatogr.A*, **1996**, *724*, 147–157.

SAMPLE
Matrix: blood, dialysate
Analyte: (R)-4-oxo-5-phosphononorvaline (MDL 100,453)
Sample preparation: Plasma. Add an equal volume of MeCN to plasma, vortex, centrifuge at 1700 rpm for 5 min. Remove 10 μL of the supernatant and add it to 70 μL 200 mM pH 8.8 borate buffer, add 20 μL 3 mg/mL 6-aminoquinolyl-N-hydroxysuccinimidyl carbamate in MeCN, let stand for 2 min, inject a 10 μL aliquot. Dialysate. Evaporate dialysate (Dulbecco's PBS) to dryness under reduced pressure, reconstitute with 30 μL 200 mM pH 8.8 borate buffer, add 20 μL 3 mg/mL 6-aminoquinolyl-N-hydroxysuccinimidyl carbamate in MeCN, let stand for 2 min, inject a 10 μL aliquot. (6-Aminoquinolyl-N-hydroxysuccinimidyl carbamate can be purchased from Waters or synthesized as described above.)

HPLC VARIABLES
Column: 150 × 3.9 Nova-Pak C18
Mobile phase: Gradient. A was 190.4 g sodium acetate trihydrate, 10 mg EDTA, and 23.7 mL triethylamine in 1 L water, adjust pH to 6.0 with phosphoric acid, dilute 10-fold with water. B was MeCN. For plasma A:B from 100:0 to 96:4 over 0.5 min, to 95:5 over 17.5 min, to 60:40 (?) over 1 min, maintain at this level for 16 min. For dialysate A:B from 100:0 to 97:3 over 18 min, to 60:40 (?) over 1 min, maintain at this level for 16 min.
Column temperature: 32
Flow rate: 1
Injection volume: 10
Detector: F ex 250 em 395

CHROMATOGRAM
Retention time: 5.5 (plasma), 14 (dialysate)
Limit of detection: <10 μM

KEY WORDS
rat; brain; plasma; pharmacokinetics

REFERENCE
Cornelius, K.E.; Fadayel, G.M.; Baron, B.M.; Schmidt, C.J.; Haegele, K.D.; Chen, T.-M. Analysis of (R)-4-oxo-5-phosphononorvaline (MDL 100,453) in rat plasma and brain dialysate using liquid chromatography after derivatization with 6-aminoquinolyl-N-hydroxysuccinimidyl carbamate, *J.Pharm.Biomed.Anal.*, **1996**, *14*, 143–150.

SAMPLE
Matrix: bulk
Analyte: amino acids
Sample preparation: Dissolve amino acids in 70 μL AccQ.Fluor Borate Buffer (Waters), add 20 μL AccQ.Fluor Reagent (10 mM 6-aminoquinolyl-N-hydroxysuccinimidyl carbamate in MeCN), vortex, let stand for 1 h at room temperature, heat at 55° for 10 min, inject an aliquot. (6-Aminoquinolyl-N-hydroxysuccinimidyl carbamate can be purchased from Waters or synthesized as described above.)

HPLC VARIABLES
Column: 250 × 4 5 μm Chiradex (immobilized β-cyclodextrin) (Merck)

Mobile phase: MeOH:100 mM pH 6.5 ammonium acetate buffer containing 0.1% triethylamine 50:50
Column temperature: 20 ± 0.1
Flow rate: 0.5
Injection volume: 20
Detector: UV 254; F ex 250 em 395

CHROMATOGRAM
Retention time: k' 2.45 (second enantiomer), α = 1.08 (Ala); k' 2.15 (second enantiomer), α = 1.00 (Val); k' 4.89 (second enantiomer), α = 1.39 (Leu); k' 3.16 (second enantiomer), α = 1.13 (Ile); k' 2.51 (second enantiomer), α = 1.10 (Pro); k' 2.90 (second enantiomer), α = 1.08 (Met); k' 2.40 (second enantiomer), α = 1.08 (Cys); k' 2.50 (second enantiomer), α = 1.07 (Ser); k' 2.29 (second enantiomer), α = 1.07 (Thr); k' 4.86 (second enantiomer), α = 1.16 (Lys); k' 1.44 (second enantiomer), α = 1.13 (Arg); k' 2.46 (second enantiomer), α = 1.11 (Asn); k' 2.05 (second enantiomer), α = 1.06 (Gln); k' 16.64 (second enantiomer), α = 1.20 (Asp); k' 10.08 (second enantiomer), α = 1.00 (Glu); k' 7.06 (second enantiomer), α = 1.13 (Phe); k' 4.82 (second enantiomer), α = 1.04 (Trp); k' 2.50 (second enantiomer), α = 1.06 (His); k' 5.19 (second enantiomer), α = 1.25 (Tyr)

KEY WORDS
chiral; α = k' (second enantiomer)/k' (first enantiomer); detailed comparison with other derivatizing reagents

REFERENCE
Rizzi, A.M.; Cladrowa-Runge, S.; Jonsson, H.; Osla, S. Enantiomeric resolution of derivatized DL-amino acids by high-performance liquid chromatography using a β-cyclodextrin chiral stationary phase: A comparison between derivatization labels, *J.Chromatogr.A*, **1995**, *710*, 287–295.

SAMPLE
Matrix: bulk
Analyte: amino acids
Sample preparation: Dissolve 3 mg amino acids in 70 μL AccQ.Fluor borate buffer (Waters), add 20 μL 10 mM 6-aminoquinolyl-N-hydroxysuccinimidyl carbamate in MeCN (Waters (synthesis given above)), vortex for several s, let stand at room temperature for 1 min, heat at 55° for 10 min, inject an aliquot.

CAPILLARY ELECTROPHORESIS
Capillary: 48.5 cm × 50 μm fused-silica (40 cm to detector) (Hewlett-Packard)
Running buffer: 10 mM 1,3-Bis[tris(hydroxymethyl)ethylamino]propane (BTP) containing 5 mM heptakis(2,6-di-O-methyl)-β-cyclodextrin, adjusted to pH 7.0 with HCl
Capillary temperature: 20 ± 0.1
Voltage/Current: 30 kV
Detector: UV 245
Model: Hewlett-Packard HP-3D
Migration time: 14.358, 14.933 (alanine enantiomers), 12.740, 13.586 (valine enantiomers)

KEY WORDS
chiral; discussion of chiral running buffer modifiers

REFERENCE
Cladrowa-Runge, S.; Rizzi, A. Enantioseparation of 6-aminoquinolyl-N-hydroxysuccinimidyl carbamate-derivatized-amino acids by capillary zone electrophoresis using native and substituted β-cyclodextrin as chiral additives. I. Discussion of optimum separation conditions, *J.Chromatogr.A*, **1997**, *759*, 157–165.

SAMPLE
Matrix: chitin, protein soil
Analyte: amino acids
Sample preparation: Hydrolyse 2.5 g sample with 90 mL boiling 6 M HCl for 6 h, filter, dilute filtrate to 100 mL, add 2 mL L-norleucine solution. Evaporate a 10 μL aliquot to dryness under reduced pressure, reconstitute with 20 μL 20 mM HCl, add 60 μL 200 mM pH 8.8 borate buffer, add 20 μL 6-aminoquinolyl-N-hydroxysuccinimidyl carbamate in

MeCN, heat at 40° for 5 min, inject an aliquot. (6-Aminoquinolyl-N-hydroxysuccinimidyl carbamate can be purchased from Waters or synthesized as described above.)

HPLC VARIABLES
Column: 150 × 3.9 4 μm AccQ-Tag C18 (Waters)
Mobile phase: Gradient. MeCN:water:buffer from 0:0:100 to 1:0:99 over 0.5 min, to 5:0:95 over 17.5 min, to 9:0:91 over 1 min, to 17:0:83 over 10.5 min, to 60:40:0 over 3.5 min, return to initial conditions over 22 min. (Buffer was 140 mM sodium acetate containing 17 mM triethylamine, adjusted to pH 5.05 with phosphoric acid.)
Column temperature: 37
Flow rate: 1
Injection volume: 5
Detector: F ex 250 em 395

CHROMATOGRAM
Retention time: 7.8 (β-Galactosamine), 18.9 (α-Galactosamine), 9.1 (β-Glucosamine), 15.9 (α-Glucosamine), 17.0 (Asp), 19.3 (Ser), 19.6 (Glu), 21.3 (Gly), 22.2 (His), 23.3 (ammonia), 25.8 (Arg), 26.2 (Thr), 26.9 (Ala), 28.2 (Pro), 29.3 (α-aminobutyric acid), 31.3 (Tyr), 32.4 (Val), 32.8 (Met), 35.7 (Lys), 36.1 (Ile), 36.5 (Leu), 38.3 (Phe)
Internal standard: L-norleucine (37.4)
Limit of detection: 49-780 fmole

REFERENCE
Díaz, J.; Lliberia, J.L.; Comellas, L.; Broto-Puig, F. Amino acid and amino sugar determination by derivatization with 6-aminoquinolyl-N-hydroxysuccinimidyl carbamate followed by high-performance liquid chromatography and fluorescence detection, *J.Chromatogr.A*, **1996**, *719*, 171–179.

SAMPLE
Matrix: feed, food
Analyte: amino acids
Sample preparation: Add 2 mL chilled performic acid to 50-70 mg sample, let stand at 0° for 16 h, add 300 μL 48% HBr, let stand at 0° for 15 min, evaporate to dryness under reduced pressure at ≤60°, add 10 mL 6 M HCl, freeze in liquid nitrogen, seal tube under vacuum, heat at 110 ± 2° for 22 h, cool to room temperature, filter. Remove a 1-2 mL aliquot and evaporate it to dryness under reduced pressure at ≤50°, reconstitute with 2.5 mM IS solution and water so that the amino acid concentration is <13 mM and the IS concentration is 250 μM. Remove a 10 μL aliquot and add it to 70 μL 200 mM pH 8.8 borate buffer, vortex for 10 s, add 20 μL 3 mg/mL 6-aminoquinolyl-N-hydroxysuccinimidyl carbamate in MeCN, vortex immediately, heat at 50° for 10 min, inject a 4 μL aliquot. (6-Aminoquinolyl-N-hydroxysuccinimidyl carbamate can be purchased from Waters or synthesized as described above.)

HPLC VARIABLES
Column: 150 × 3.9 4 μm AccQ-Tag C18 (Waters)
Mobile phase: Gradient. A was 140 mM sodium acetate containing 17 mM triethylamine and 100 mg/L sodium azide (Caution! Sodium azide is carcinogenic and toxic! Do not discharge to the plumbing system!), pH adjusted to 4.95 with phosphoric acid. B MeCN:water:acetone 60:40:0.01. A:B from 100:0 to 92:8 over 17 min, to 83:17 over 4 min, to 73:27 over 11 min, to 50:50 over 2 min, maintain at 50:50 for 1 min, to 0:100 over 2 min, return to initial conditions over 1 min, re-equilibrate for 7 min.
Column temperature: 47
Flow rate: 1
Injection volume: 4
Detector: UV 248

CHROMATOGRAM
Retention time: 7.5 (cysteic acid), 12.5 (Asp), 13.5 (Ser), 14.5 (Glu), 15 (Gly), 15.5 (His), 19 (Arg), 19.5 (Thr), 20.7 (methionine sulfone), 21 (Ala), 24 (Pro), 28 (Val), 32 (Lys), 33 (Ile), 33.5 (Leu), 34.5 (Phe)
Internal standard: α-aminobutyric acid (25)

KEY WORDS
corn; shrimp

REFERENCE

Liu, H.J.; Chang, B.Y.; Yan, H.W.; Yu, F.H.; Liu, X.X. Determination of amino acids in food and feed by derivatization with 6-aminoquinolyl-*N*-hydroxysuccinimidyl carbamate and reversed-phase liquid chromatographic separation, *J.AOAC Int.*, **1995**, *78*, 736–744.

SAMPLE

Matrix: peptides

Analyte: amino acids

Sample preparation: Dry a solution containing 1 μg peptide under vacuum, add 200 μL 0.06% phenol in constant-boiling HCl, evacuate and purge with nitrogen 3 times, evacuate and seal, heat at 114° for 22 h, cool, evaporate to dryness under reduced pressure, reconstitute with 20 μL 20 mM HCl, add 60 μL 200 mM pH 8.8 sodium borate containing 5 mM EDTA, add 20 μL 3 mg/mL 6-aminoquinolyl-N-hydroxysuccinimidyl carbamate) in MeCN, inject an aliquot. (6-Aminoquinolyl-N-hydroxysuccinimidyl carbamate can be purchased from Waters or synthesized as described above.)

HPLC VARIABLES

Column: 150 × 3.9 Nova-Pak C18

Mobile phase: Gradient. MeCN:buffer:water 0:100:0 for 0.5 min, to 1:99:0 (step gradient), to 5:95:0 over 17.5 min, to 9:91:0 over 1 min, to 17:83:0 over 10.5 min, to 32:68:0 over 3.5 min, to 60:0:40 (step gradient), maintain at 60:0:40 for 3 min, re-equilibrate at initial conditions for 9 min. (Buffer was 140 mM sodium acetate containing 17 mM triethylamine and 3 mM EDTA, adjusted to pH 5.05 with phosphoric acid.)

Column temperature: 37

Flow rate: 1

Detector: F ex 250 em 395

CHROMATOGRAM

Retention time: 14 (Asp), 15 (Ser), 16 (Glu), 17 (Gly), 17.5 (His), 19 (ammonia), 21.8 (Arg), 22 (Thr), 23 (Ala), 24 (Pro), 26.8 (Cys), 27 (Tyr), 28.3 (Val), 28.7 (Met), 30.7 (Lys), 31.8 (Ile), 32.5 (Leu), 33.5 (Phe)

Limit of detection: 38-794 fmole

KEY WORDS

Procedure can also be used to derivatize peptides; changes to the gradient are required.

REFERENCE

De Antonis, K.M.; Brown, P.R.; Cohen, S.A. High-performance liquid chromatographic analysis of synthetic peptides using derivatization with 6-aminoquinolyl-*N*-hydroxysuccinimidyl carbamate, *Anal.Biochem.*, **1994**, *223*, 191–197.

SAMPLE

Matrix: solutions

Analyte: amino acids

Sample preparation: Mix 100 μL of a 250 μg/mL solution of amino acids in water with 700 μL 200 mM pH 8.8 sodium borate buffer, vortex for 20 s, add 100 μL 3 mg/mL 6-aminoquinoyl-N-hydroxysuccinimidyl carbamate in MeCN, vortex for 20 s, dilute with water or running buffer (if desired), inject an aliquot. (6-Aminoquinolyl-N-hydroxysuccinimidyl carbamate can be purchased from Waters or synthesized as described above.)

CAPILLARY ELECTROPHORESIS

Capillary: 60 cm × 50 μm (52.5 cm to detector) (Waters AccuSep)

Capillary preparation: Rinse with running buffer for 3 min between runs. Rinse new capillaries with 500 mM NaOH for 10 min, with water for 10 min, and with running buffer for 10 min.

Running buffer: 25 mM pH 9.0 Na_2HPO_4/sodium tetraborate containing 100 mM (R)-N-dodecoxycarbonylvaline (Prepare (R)-N-dodecoxycarbonylvaline as follows. Prepare dodecyl chloroformate by reacting 1-dodecanol with 0.33 equivalents of triphosgene in dichloromethane solution in the presence of pyridine. Add dodecyl chloroformate dropwise to (R)-valine in 1 M NaOH solution, filter, wash with hexane, recrystallize from ether/petroleum ether (J. Chromatogr. A, 1994, 680, 125).)

Capillary temperature: 30

Voltage/Current: 16 kV
Injection: Hydrostatic injection at 10 cm for 20 s
Detector: UV 254
Model: Waters Quanta 4000E
Migration time: 11.8 (d-Ile), 12 (l-Ile), 12.4 (d-Leu), 12.6 (l-Leu), 13.1 (d-Orn), 13.3 (l-Orn), 13.5 (d-Lys), 13.7 (l-Lys), 14.2 (d-Trp), 14.8 (l-Trp), 18.4 (d-Arg), 18.8 (l-Arg)
Limit of quantitation: 300 ng/mL
Limit of detection: 100 ng/mL

KEY WORDS
chiral

REFERENCE
Swartz, M.E.; Mazzeo, J.R.; Grover, E.R.; Brown, P.R. Separation of amino acid enantiomers by micellar electrokinetic capillary chromatography using synthetic chiral surfactants, *Anal.Biochem.*, **1995**, *231*, 65–71.

N-(tert-Butoxycarbonyl)-L-leucine N-Hydroxysuccinimide Ester

SAMPLE
Matrix: solutions
Analyte: thyroxines
Sample preparation: Take up 1.5 mg dextrothyroxine in 200 μL 100 mM sodium bicarbonate and 400 μL reagent, stir in an ice bath for 30 min, evaporate to dryness below 30°, add 100 μL trifluoroacetic acid to the dry residue, let stand for 30 min at room temperature, add 2 mL 1 M sodium bicarbonate, centrifuge. Remove the precipitate and

dissolve it in 600 μL MeOH:20 mM NaOH 50:50, inject a 15 μL aliquot. Reagent was 7 mg/mL N-(tert-butoxycarbonyl)-L-leucine N-hydroxysuccinimide ester (BOC-L-Leu-SU) in MeOH, prepared immediately before use.)

HPLC VARIABLES
Column: 150 × 3.2 7 μm LiChrosorb RP-18
Mobile phase: MeOH:water 60:40 containing 0.05% methanesulfonic acid
Flow rate: 1
Injection volume: 15
Detector: UV 230

CHROMATOGRAM
Retention time: 11.5 (dextrothyroxine), 13 (levothyroxine)
Limit of detection: 0.05% (of major enantiomer)

OTHER SUBSTANCES
Simultaneously analyzed: impurities

KEY WORDS
chiral

REFERENCE
Lankmayr, E.P.; Budna, K.W.; Nachtmann, F. Separation of enantiomeric iodinated thyronines by liquid chromatography of diastereomers, *J.Chromatogr.*, **1980**, *198*, 471–479.

N-(9-Fluorenylmethoxycarbonyloxy)succinimide

SAMPLE
Matrix: perfusate
Analyte: amino acids
Sample preparation: Lyophilize perfusate, reconstitute with water. Remove a 400 μL aliquot and add it to 100 μL 100 mM pH 6.2 borate buffer, adjust pH to 7.7 with 100 mM NaOH, add 500 μL 15.4 mM N-(9-fluorenylmethoxycarbonyloxy)succinimide (9-fluorenylmethyl N-succinimidyl carbonate) in acetone, let stand at room temperature for 1 min, extract twice with 2 mL portions of pentane. Evaporate the aqueous layer to dryness under reduced pressure, reconstitute with 200 μL water, filter (0.45 μm), inject a 10 μL aliquot of the filtrate.

HPLC VARIABLES
Column: 250 × 4.6 5 μm Spherisorb ODS II
Mobile phase: Gradient. MeCN:35 mM pH 5.5 ammonium acetate from 20:80 to 25:75 over 10 min, maintain at 25:75 for 5 min, to 50:50 over 25 min, to 80:20 over 10 min.
Flow rate: 1
Injection volume: 10
Detector: F ex 254 em 313

CHROMATOGRAM
Retention time: 14.5 (Asp), 15 (Glu), 20 (Asn), 21 (Gln), 22.5 (Ser), 26.5 (Thr, Gly), 27.5 (Tau), 29 (Ala), 32.5 (Arg, Pro), 33 (β-Ala), 36 (Met), 37 (Bal), 39 (Phe), 39.5 (Ile, Leu)

REFERENCE
Keller, H.J.; Do, K.Q.; Zollinger, M.; Winterhalter, K.H.; Cuénod, M. Cysteine: depolarization-induced release from rat brain in vitro, *J.Neurochem.*, **1989**, *52*, 1801–1806.

Succinimido p-Bromophenylcarbamate

SAMPLE
Matrix: solutions
Analyte: amino acids
Sample preparation: 20 μL Amino acid solution in 100 mM HCl + 30 μL 300 mM pH 9.5 borate buffer + 50 μL 5 mM succinimido p-bromophenylcarbamate in MeCN, let stand for 5 min, inject a 5 μL aliquot. (Prepare succinimido p-bromophenylcarbamate as follows. Add 40 mmoles p-bromoaniline in 40 mL MeCN dropwise with stirring to 50 mmoles N,N'-disuccinimidyl carbonate in 60 mL MeCN over 3-4 h, stir for 1 h, remove the MeCN by evaporation. Dissolve the residue in 100 mL ethyl acetate, wash with 1 M HCl, wash with 4% sodium bicarbonate, wash with water, dry over anhydrous sodium carbonate, evaporate to obtain succinimido p-bromophenylcarbamate, recrystallize from benzene/acetone (Caution! Benzene is a carcinogen!).

HPLC VARIABLES
Column: 150 × 4.6 Develosil ODS-5 (Nomura Chemical)
Mobile phase: MeOH:0.1% phosphoric acid 45:55
Flow rate: 1
Injection volume: 5
Detector: UV 250

CHROMATOGRAM
Retention time: 7 (His), 8 (Arg), 16 (Ser), 21.5 (Pro), 26 (Thr), 28 (Ala)
Limit of detection: 0.15-0.3 ng

OTHER SUBSTANCES

Also analyzed: tert-butylamine, diethylamine, diisobutylamine, diisopropylamine, isobutylamine (mobile phase MeCN:water 55:45)

REFERENCE

Nimura, N.; Iwaki, K.; Kinoshita, T.; Takeda, K.; Ogura, H. Activated carbamate reagent as derivatizing agent for amino compounds in high-performance liquid chromatography, *Anal.Chem.*, **1986**, *48*, 2372–2375.

Succinimido 1-Naphthylcarbamate

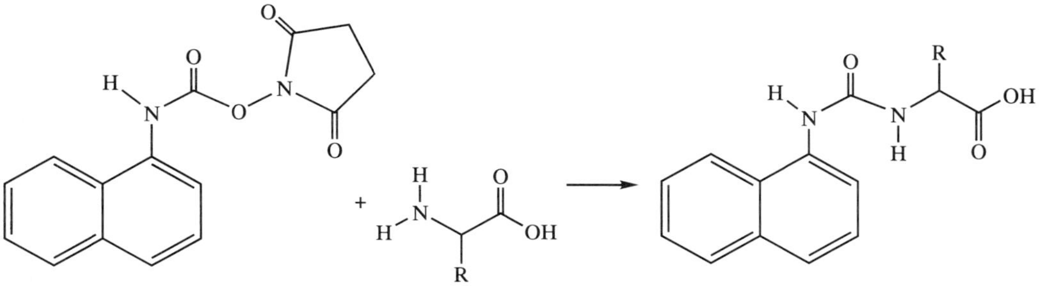

SAMPLE

Matrix: solutions

Analyte: amino acids

Sample preparation: Mix a 20 μL aliquot of a solution in 100 mM HCl with 20 μL 500 mM pH 9.5 borate buffer and 20 μL 1.5 mg/mL succinimido 1-naphthylcarbamate, mix for 1.5 min, let stand for 3 min, inject a 6 μL aliquot. (Prepare the reagent, succinimido 1-naphthylcarbamate, as follows. Add 40 mL 1 M 1-naphthylamine in MeCN dropwise to 60 mL MeCN containing 50 mmoles N,N'-disuccinimidyl carbonate with stirring at room temperature over 3-4 h, after addition is complete stir for 1 h at room temperature, evaporate to dryness, dissolve the residue in 100 mL ethyl acetate. Wash the solution with 1 M HCl, wash with 4% sodium bicarbonate, wash with water, dry over anhydrous sodium sulfate, evaporate to dryness, recrystallize from benzene/acetone to obtain succinimido 1-naphthylcarbamate (Caution! Benzene is a carcinogen!) (Anal.Chem. 1986, 58, 2372).)

HPLC VARIABLES

Column: 10 × 6 Develosil ODS-5

Mobile phase: Gradient. MeCN:100 mM pH 6.30 sodium acetate from 14:86 to 16:84 over 5 min, to 25:75 over 10 min, to 39:61 over 6 min, to 70:30 (step gradient), maintain at 70:30 for 6 min, re-equilibrate at initial conditions for 13 min.

Column temperature: 30

Flow rate: 1.2

Injection volume: 6

Detector: F ex 290 em 370

CHROMATOGRAM

Retention time: 3.5 (Asp), 4 (Glu), 8 (Ser), 8.4 (His), 8.8 (Pro), 9.2 (Gly), 9.6 (Arg), 10 (Thr), 11 (Ala), 15 (Tyr), 16 (ammonia), 16.7 (Val), 17.5 (Met), 19.5 (Ile), 20 (Leu), 21.5 (Phe), 22 (Cys), 24.5 (Lys)

Limit of detection: 75-1200 fmole

REFERENCE

Iwaki, K.; Nimura, N.; Hiraga, Y.; Kinoshita, T.; Takeda, K.; Ogura, H. Amino acid analysis by reversed-phase high-performance liquid chromatography. Automatic pre-column derivatization with activated carbamate reagent, *J.Chromatogr.*, **1987**, *407*, 273–279.

SULFONIC ACID

Sodium 1,2-Naphthoquinone-4-sulfonate

RELATED REFERENCE
Saurina, J.; Hernandez-Cassou, S. Determination of amino acids by ion-pair liquid chromatography with post-column derivatization using 1,2-naphthoquinone-4-sulfonate. *J.Chromatogr.A* **1994**, *676*, 311-319.

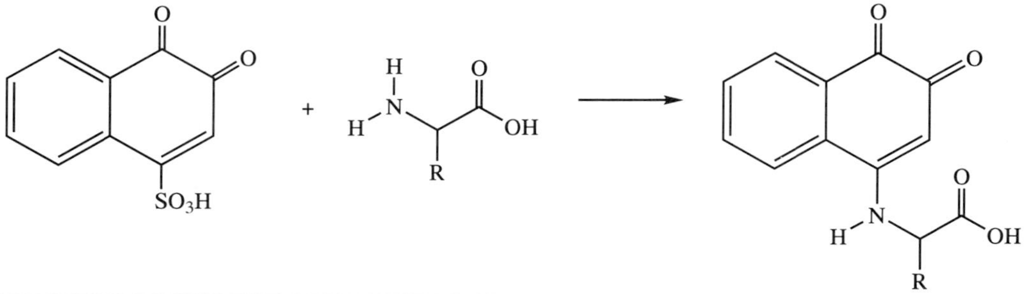

SAMPLE
Matrix: feed

Analyte: amino acids

Sample preparation: Heat 500 mg feed and 2 mL 6 M HCl containing 1% phenol at 105° for 24 h, neutralize with 100 mM NaOH, make up to 100 mL, filter (0.22 μm nylon). Mix 150 μL hydrolysate with 150 μL 30 mM sodium 1,2-naphthoquinone-4-sulfonate in 100 mM HCl and 150 μL 50 mM sodium borate containing 90 mM NaOH, heat at 65° for 5 min, add 60 μL 250 mM HCl, inject a 50 μL aliquot.

HPLC VARIABLES
Column: 150 × 4.6 3 μm Spherisorb ODS 2

Mobile phase: Gradient. A was buffer. B was MeCN:buffer 50:50. A:B 100:0 for 15 min, to 94:6 over 5 min, maintain at 94:6 for 10 min, to 86:14 over 1 min, to 83:17 over 14 min, to 73:27 over 1 min, to 71:29 over 14 min, to 0:100 over 2 min, maintain at 0:100 for 8 min, return to initial conditions over 2 min, re-equilibrate for 3 min. (Buffer was 50 mM acetic acid containing 50 mM sodium acetate, pH 4.75.)

Column temperature: 50

Flow rate: 0.8

Injection volume: 50

Detector: UV 305

CHROMATOGRAM
Retention time: 10 (Cys), 12 (Asp), 14.5 (Ser), 21 (Gly), 22 (Glu), 24 (His), 26 (Thr), 29.5 (Pro), 30.5 (Arg), 33.5 (Ala), 41 (Tyr), 45 (Orn), 47.5 (Val), 48 (Met), 42.5 (Lys), 43.5 (Ile), 46 (Leu), 47 (Phe)

Limit of detection: 40-100 pmole

REFERENCE
Saurina, J.; Hernández-Cassou, S. Chromatographic determination of amino acids by pre-column derivatization using 1,2-naphthoquinone-4-sulfonate as reagent, *J.Chromatogr.A*, **1996**, *740*, 21–30.

2,4,6-Trinitrobenzenesulfonic Acid

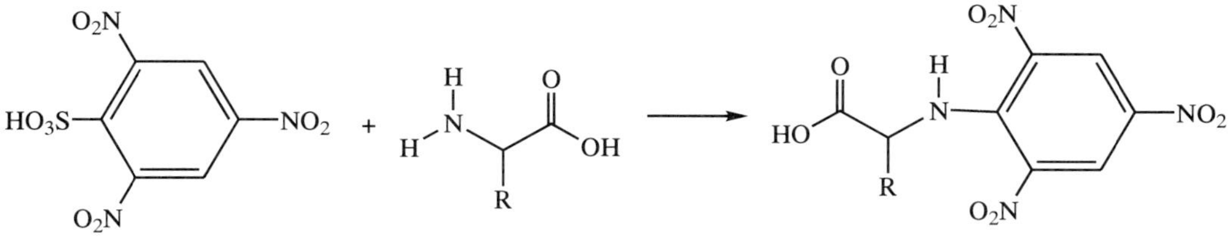

SAMPLE

Matrix: solutions

Analyte: amino acids

Sample preparation: 1 mL Solution + 3 mL 200 mM pH 9.0 borate buffer + 1 mL 30 mM 2,4,6-trinitrobenzenesulfonic acid in water, let stand at room temperature for 1 h, dilute, inject an aliquot. Alternatively, acidify reaction mixture with 1 drop 6 M HCl, add 2.5 mL hexane:ethyl acetate 20:80, vortex for 15 s. Remove 2 mL of the organic layer and evaporate it to dryness, reconstitute the residue in 500 μL 100 mM perchloric acid, dilute 1: 200, inject an aliquot.

HPLC VARIABLES

Column: 250 × 4.6 5 μm Biophase 6032 octyl (Bioanalytical Systems)

Mobile phase: n-Propanol:100 mM pH 5.0 sodium acetate 12.5:87.5

Flow rate: 1

Detector: E, Bioanalytical Systems LC-154, TL-5A glassy carbon electrode -0.85 V, Ag/AgCl reference electrode

CHROMATOGRAM

Retention time: 15.5 (glutamic acid), 16 (serine), 17 (threonine), 18 (glycine), 19.5 (alanine)

REFERENCE

Jacobs, W.A.; Kissinger, P.T. Nitroaromatic reagents for determination of amines and amino acids by liquid chromatography/electrochemistry, *J.Liq.Chromatogr.*, **1982**, *5*, 881–895.

SULFONYL HALIDE

(1S)-(+)-10-Camphorsulfonyl Chloride

SAMPLE
Matrix: bulk
Analyte: amino acids
Sample preparation: Dissolve 0.5 mmole amino acid and 0.5 mmole (1S)-(+)-10-camphorsulfonyl chloride in 10 mL ether, add 10 mL 1 M NaOH, stir vigorously for 1 h, acidify with 1 M HCl, extract with ether. Evaporate the extract to dryness and reconstitute with chloroform containing 0.5 mmole 4-nitrobenzyl bromide, reflux for 30 min, wash with water, dry over anhydrous sodium sulfate, evaporate to dryness.

HPLC VARIABLES
Column: 250 × 2 MicroPak Si-5
Mobile phase: Isooctane:isopropanol 98.5:1.5
Flow rate: 0.5
Detector: UV 254

CHROMATOGRAM
Retention time: 7.0 (L-isoleucine), 7.6 (D-isoleucine, L-leucine), 8.6 (D-leucine), 11.0 (L-phenylalanine), 12.4 (D-phenylalanine), 17.0 (L-alanine), 22.3 (D-alanine)

KEY WORDS
normal phase; chiral

REFERENCE
Furukawa, H.; Sakakibara, E.; Kamei, A.; Ito, K. Separation of L- and D-amino acids as diastereomeric derivatives by high performance liquid chromatography, *Chem.Pharm.Bull.*, **1975**, *23*, 1625–1626.

SAMPLE
Matrix: bulk
Analyte: amino acids

Sample preparation: Dissolve 1 mmole amino acid in 10 mL diethyl ether, add 20 mL 1 M NaOH, stir vigorously at 0°, add 2 mmole (1S)-(+)-10-camphorsulfonyl chloride in 30 mL ether dropwise, stir at room temperature for 1 h. Remove the aqueous layer and wash it twice with diethyl ether, acidify the aqueous layer with concentrated HCl, extract with diethyl ether. Dry the organic extract over anhydrous sodium sulfate, evaporate to dryness, reconstitute with 10 mL DMF, add 1 drop trimethylamine, add 1.1 mmole 4-nitrobenzyl bromide, heat at 55° for 2 h, dilute with 40 mL chloroform, wash with water, dry over anhydrous sodium sulfate, evaporate to dryness, reconstitute with chloroform, inject an aliquot.

HPLC VARIABLES
Column: 250 × 2.2 10 μm MicroPak-NH2
Mobile phase: Isooctane:dichloromethane:isopropanol 79:16:5
Flow rate: 0.4
Detector: UV 254

CHROMATOGRAM
Retention time: 3.9 (L-leucine), 4.4 (L-isoleucine, D-leucine), 5.0 (D-isoleucine), 6.2 (L-phenylalanine), 7.2 (L-alanine), 7.4 (L-methionine), 9.3 (D-alanine), 8.5 (D-phenylalanine), 10.0 (D-methionine), 12.8 (L-glutamic acid), 16.8 (D-glutamic acid), 29.2 (L-tryptophan), 33.2 (L-tyrosine), 47.2 (D-tyrosine), 49.6 (D-tryptophan)

KEY WORDS
normal phase; chiral

REFERENCE
Furukawa, H.; Mori, Y.; Takeuchi, Y.; Ito, K. Separation of L- and D-amino acids as diastereomeric derivatives by high-performance liquid chromatography, *J.Chromatogr.*, **1977**, *136*, 428–431.

Dabsyl Chloride

RELATED REFERENCES
Chang, J.Y. Analysis of phospho-amino acids and amino acid amides at the picomole level using 4'-dimethylaminoazobenzene-4-sulfonyl chloride. *J.Chromatogr.* **1984**, *295*, 193-200.

de Witte, P.A.; Cuveele, J.F.; Merlevede, W.J.; Vandenheede, J.R. Analysis of phosphorylhydroxyamino acids present in hydrolyzed cell extracts using dabsyl derivatization. *Anal.Biochem.* **1995**, *226*, 1-9.

Gorbics, L.; Urge, L.; Lang, E.; Szendrei, G.I.; Otvos, L., Jr. Successful and rapid verification of the presence of a phosphate group in synthetic phosphopeptides using the conditions of standard DABS-CL amino acid analysis. *J.Liq.Chromatogr.* **1994**, *17*, 175-189.

Gorbics, L.; Urge, L.; Otvos, L.J. Comparative and optimized dabsyl-amino acid analysis of synthetic phosphopeptides and glycopeptides. *J.Chromatogr.A* **1994**, *676*, 169-176.

Hughes, G.J.; Frutiger, S.; Fonck, C. Quantitative high-performance liquid chromatographic analysis of Dabsyl-amino acids within 14 min. *J.Chromatogr.* **1987**, *389*, 327-333.

Ikeda, M.; Sorimachi, K.; Akimoto, K.; Okazaki, M.; Sunagawa, M.; Niwa, A. Analysis of hydroxyproline in urine by high-performance liquid chromatography after dabsyl-chloride derivatization. *Amino Acids* **1995**, *8*, 401-407.

Lin, J-K.; Wang, C-H. Determination of urinary amino acids by liquid chromatography with "dabsyl chloride". *Clin.Chem.* **1980**, *26*, 579-583.

Lin, J.K.; Shiau, S.Y.L. Determination of amino acids in human specimens by concave and linear gradient liquid chromatography with dabsyl chloride. *J.Chin.Biochem.Soc.* **1983**, *12*, 47-60.

Reitsma, B.H.; Yeung, E.S. Optical activity and ultraviolet absorbance detection of dansyl-L-amino acids separated by gradient liquid chromatography. *Anal.Chem.* **1987**, *59*, 1059-1061.

Schneider, H.J. Amino acid determination using DABS-CL. *Chromatographia* **1989**, *28*, 45-48.

Vendrell, J.; Aviles, F.X. Complete amino acid analysis of proteins by dabsyl derivatization and reversed-phase liquid chromatography. *J.Chromatogr.* **1986**, *358*, 401-413.

Weiner, S.; Tishbee, A. Separation of Dns-amino acids using reversed-phase high-performance liquid chromatography: a sensitive method for determining N-terminuses of peptides and proteins. *J.Chromatogr.* **1981**, *213*, 501-506.

Winkler, G.; Heinz, F.X.; Kunz, C. Exclusive use of high-performance liquid chromatographic techniques for the isolation, 4-dimethylaminoazobenzene-4'-sulfonyl chloride amino acid analysis and 4-N,N-dimethylaminoazobenzene-4'-isothiocyanate-phenyl isothiocyanate sequencing of a viral membrane protein. *J.Chromatogr.* **1984**, *297*, 63-73.

Yang, C.Y.; Yang, T.; Yeh, B.K. Liquid chromatographic analysis of amino acids: Using dimethylaminoazobenzenesulfonyl chloride and Hypersil ODS column to analyze the composition of apo B peptides. *J.Protein Chem.* **1993**, *12*, 11-14.

SAMPLE
Matrix: blood
Analyte: amino acids
Sample preparation: 200 μL Serum + 400 μL MeCN, mix, centrifuge. Remove a 40 μL aliquot of the supernatant and add it to 40 μL 4 nM dabsyl chloride in MeCN, heat at 70° for 10 min, add 70 μL MeOH:50 mM pH 7.0 phosphate buffer 50:50, inject a 20 μL aliquot.

HPLC VARIABLES
Column: 150 × 4.6 5 μm Hypersil ODS
Mobile phase: Gradient. A was MeCN:MeOH:100 mM pH 6.5 sodium acetate 16:28:56. B was MeCN:25 mM pH 6.5 sodium acetate 29:71. A:B 100:0 for 16 min, to 0:100 over 1 min, maintain at 0:100 for 18 min.
Flow rate: 1
Injection volume: 20
Detector: UV 436

KEY WORDS
serum

REFERENCE
Jansen, E.H.J.M.; van den Berg, R.H.; Both-Miedema, R.; Doorn, L. Advantages and limitations of precolumn derivatization of amino acids with dabsyl chloride, *J.Chromatogr.*, **1991**, *553*, 123−133.

SAMPLE
Matrix: blood, food, peptides, plants, tissue
Analyte: amino acids
Sample preparation: Hydrolyze peptide with 6 M HCl containing 0.2% 3,3'-thiodipropionic acid at 110° for 24 h, evaporate to dryness, reconstitute with 50-200 μL 0.1% HCl containing 0.2% 3,3'-thiodipropionic acid. Homogenize (Ultra-Turrax) 0.1-1 g food, tissue, plant material, lyophilized plasma, or lyophilized tissue in 10 mL 250 nM IS in 100 mM HCl containing 0.2% 3,3'-thiodipropionic acid at 20000 rpm for 2 min, sonicate for ≤30 min, centrifuge at 5000 g for 20 min, discard fat layer, filter (Millipore ultrafiltration insert (MW cutoff 5000) prewashed with 200 μL 100 mM HCl containing 0.2% 3,3'-thiodipropionic acid) 3 mL supernatant while centrifuging at 3500 g for 1 h. Mix 20 μL deproteinized sample (or 10 μL peptide hydrolysate) with 180 μL buffer, vortex, add 200 μL reagent, mix, heat at 70° for 15 min with mixing at 1 min and 12 min, cool in an ice bath for 5 min, centrifuge at 10000 g for 10 s, add 400 μL diluent, mix thoroughly, centrifuge at 15000 g for 5 min, inject a 10 μL aliquot of the supernatant. (Prepare buffer by dissolving 630 mg sodium bicarbonate in 40 mL water, adjusting pH to 8.6 with NaOH, and making up to 50 mL with water. Prepare reagent by sonicating 40 mg dabsyl chloride in

10 mL acetone for 10 min, then filtering into brown vials and storing at -20°. Prepare diluent by mixing 50 mL MeCN, 25 mL EtOH, and 25 mL mobile phase A.)

HPLC VARIABLES

Guard column: present but not specified

Column: 150 × 3.9 4 μm Novapak C18

Mobile phase: Gradient. A was DMF:9 mM NaH_2; PO_4; containing 0.16% triethylamine, adjusted to pH 6.55 with phosphoric acid. B was MeCN:water 80:20. A:B 92:8 for 2 min, to 80:20 over 5 min (Waters convex curve 5), to 65:35 over 28 min (Waters concave curve 7), to 50:50 over 10 min, to 0:100 over 21 min, maintain at 0:100 for 11 min, return to initial conditions over 0.5 min, re-equilibrate for 12.5 min.

Column temperature: 50

Flow rate: 1

Injection volume: 10

Detector: UV 436

CHROMATOGRAM

Retention time: 13.95 (O-phosphoserine), 14.94 (aspartic acid), 15.15 (O-phosphothreonine), 15.91 (glutamic acid), 16.39 (carboxymethylcysteine), 16.89 (S-sulfocysteine), 17.10 (β-aminoadipic acid), 22.51 (hydroxyproline), 23.13 (asparagine), 24.72 (glutamine), 25.35 (citrulline), 26.17 (serine), 27.43 (phosphoethanolamine), 27.69, 28.15 (methionine sulfoxide (diastereomers)), 28.45 (threonine), 28.94 (glycine), 28.94 (1- and 3-methylhistidine), 29.38 (arginine), 30.65 (alanine), 30.92 (β-alanine), 31.52 (anserine), 32.31 (taurine), 32.58 (sarcosine), 32.75 (α-aminobutyric acid), 33.27 (gamma-aminobutyric acid), 33.99 (proline), 34.40 (β-aminoisobutyric acid), 35.55(valine), 37.60 (methionine), 39.35 (isoleucine), 40.03 (leucine), 40.51 (tryptophan), 41.57 (phenylalanine), 42.05 (ammonia), 43.62 (lanthionine), 44.00 (agmatine), 44.76 (2-aminoethanol), 44.76 (cystathionine), 45.65 (cysteine), 46.83 (homocysteine), 48.02 (1-amino-2-propanol), 50.04 (hydroxylysine), 51.02 (ornithine), 51.58 (lysine), 52.10 (histidine), 52.38 (carnosine), 52.59 (ethylamine), 54.30 (tyrosine), 57.22 (pyrrolidine), 57.22 (tryptamine), 57.79 (isobutylamine), 58.34 (3,4-dihydroxyphenylalanine), 59.20 (phenylethylamine), 59.94 (methylbutylamine), 62.16 (putrescine), 63.29 (cadaverine), 63.87 (histamine), 63.87 (cystamine), 65.00 (serotonin),67.31 (tyramine), 67.98 (spermidine), 68.49 (norepinephrine), 69.05 (dopamine), 70.61 (epinephrine), 71.94 (spermine)

Internal standard: norleucine (40.90), norvaline (35.06)

Limit of quantitation: 0.4-1.5 pmole

Limit of detection: 0.12-0.52 pmole

KEY WORDS

rinse glass and plasticware with 70% EtOH and water and dry before use; cheese; meat; sausage; fish; plasma

REFERENCE

Krause, I.; Bockhardt, A.; Neckermann, H.; Henle, T.; Klostermeyer, H. Simultaneous determination of amino acids and biogenic amines by reversed-phase high-performance liquid chromatography of the dabsyl derivatives, *J.Chromatogr.A*, **1995**, *715*, 67–79.

SAMPLE

Matrix: blood, tissue

Analyte: amino acids

Sample preparation: 250 μL Plasma + 5 μL 2 mM norleucine + 2 mL 10% trichloroacetic acid, mix, centrifuge at 10000 g for 10 min. Adjust the pH of the supernatant to 9.0 with KOH. Remove a 40 μL aliquot and add it to 40 μL 100 mM pH 8.3 sodium bicarbonate, add 80 μL 4 mM dabsyl chloride in MeCN, heat at 70° for 12 min (mix after 1 and 4 min), cool to room temperature for 5 min, add 440 μL EtOH:50 mM pH 7.0 sodium phosphate buffer 50:50, inject a 50 μL aliquot. Liver. Homogenize (Polytron) 200 mg liver with 5 volumes of 10% trichloroacetic acid, mix, centrifuge at 10000 g for 10 min. Adjust the pH of the supernatant to 9.0 with KOH. Remove a 40 μL aliquot and add it to 40 μL 100 mM pH 8.3 sodium bicarbonate, add 80 μL 4mM dabsyl chloride in MeCN, heat at 70° for 12 min (mix after 1 and 4 min), cool to room temperature for 5 min, add 440 μL EtOH:50 mM pH 7.0 sodium phosphate buffer 50:50, inject a 50 μL aliquot.

HPLC VARIABLES
Guard column: 5 μm Adsorbosphere C18
Column: 250 × 4.6 5 μm Econosphere
Mobile phase: Gradient. A was DMF:10 mM pH 6.5 citrate buffer 4:96. B was MeCN:10 mM pH 6.5 citrate buffer:DMF 67.2:28.8:4. A:B 83:17 for 5 min, to 74:26 over 18.6 min, to 63:37 over 16 min, to 19:81 over 34.7 min, to 0:100 over 0.3 min, maintain at 0:100 for 1.4 min, return to initial conditions, re-equilibrate for 14 min. (Condition column with DMF:100 mM pH 3.5 sodium citrate 20:80 at 1 mL/min for 1 h before use.)
Flow rate: 1.4
Injection volume: 50
Detector: UV 436

CHROMATOGRAM
Retention time: 28.7 (aspartate), 30.9 (cysteine), 32.7 (glutamate), 43.4 (glutamine), 43.8 (serine), 44.6 (threonine), 45.3 (arginine), 46.0 (glycine), 46.5 (alanine), 48.6 (proline), 49.5 (valine), 51.0 (methionine), 51.5 (leucine), 52.2 (isoleucine), 54 (phenylalanine), 66.4 (lysine), 67.6 (histidine), 71.5 (tyrosine)
Internal standard: norleucine (53)

KEY WORDS
rat; plasma; liver

REFERENCE
Drnevich, D.; Vary, T.C. Analysis of physiological amino acids using dabsyl derivatization and reversed-phase liquid chromatography, *J.Chromatogr.*, **1993**, *613*, 137−144.

SAMPLE
Matrix: bulk
Analyte: amino acids
Sample preparation: Dissolve 5 mg amino acids in ? μL 100 mM pH 7 phosphate buffer, add 100 μL 10 mg/mL dabsyl chloride in acetone, heat at 65° for 30 min, evaporate to dryness under reduced pressure, reconstitute with 500 μL acetone:1 M HCl 95:5, centrifuge for 5 min, evaporate the liquid to dryness under reduced pressure, reconstitute with mobile phase, inject an aliquot.

HPLC VARIABLES
Column: 250 × 4 5 μm Chiradex (immobilized β-cyclodextrin) (Merck)
Mobile phase: MeOH:100 mM pH 5.5 ammonium acetate buffer containing 0.1% triethylamine 90:10
Column temperature: 20 ± 0.1
Flow rate: 0.5
Injection volume: 20
Detector: UV 456

CHROMATOGRAM
Retention time: k' 9.57 (second enantiomer), α = 1.14 (Ala); k' 12.34 (second enantiomer), α = 1.64 (Val); k' 14.05 (second enantiomer), α = 1.19 (Leu); k' 18.21 (second enantiomer), α = 1.00 (Pro); k' 11.37 (second enantiomer), α = 1.31 (Met); k' 8.32 (second enantiomer), α = 1.00 (Cys); k' 13.39 (second enantiomer), α = 1.15 (Ser); k' 10.10 (second enantiomer), α = 1.13 (Thr); k' 5.23 (second enantiomer), α = 1.20 (Lys); k' 21.28 (second enantiomer), α = 1.25 (Lys bis-derivative); k' 3.68 (second enantiomer), α = 1.22 (Arg); k' 19.89 (second enantiomer), α = 1.10 (Asn); k' 12.84 (second enantiomer), α = 1.16 (Gln); k' 13.23 (second enantiomer), α = 1.13 (Phe); k' 14.65 (second enantiomer), α = 1.19 (Trp); k' 14.75 (second enantiomer), α = 1.14 (His); k' 15.30 (second enantiomer), α = 1.20 (Tyr)

KEY WORDS
chiral; α = k' (second enantiomer)/k' (first enantiomer); detailed comparison with other derivatizing reagents

REFERENCE
Rizzi, A.M.; Cladrowa-Runge, S.; Jonsson, H.; Osla, S. Enantiomeric resolution of derivatized DL-amino acids by high-performance liquid chromatography using a β-cyclodextrin chiral stationary phase: A comparison between derivatization labels, *J.Chromatogr.A*, **1995**, *710*, 287−295.

SAMPLE
Matrix: cells
Analyte: hydroxyproline
Sample preparation: Solubilize fibroblasts with 500 mM NaOH, wash three times with warm Hank's balanced salt solution, collect, hydrolyze with 3 M HCl at 130° overnight, adjust pH to 8.9 with 4 M potassium carbonate. Remove a 20 μL aliquot and add it to 20 μL 6.2 mM N-methyltaurine in 100 mM pH 8.9 sodium carbonate/sodium bicarbonate buffer, add 20 μL 3.3 mg/mL dabsyl chloride in acetone, heat at 70° with constant shaking for 6 min, inject a 10 μL aliquot.

HPLC VARIABLES
Column: 250 × 4.6 10 μm RP-8 (Brownlee)
Mobile phase: Gradient. A was MeCN containing 1.66 mM acetic acid and 6.99 mM orthophosphoric acid, pH 3.5. B was 10 mM sodium acetate containing 1.66 mM acetic acid, pH adjusted to 3.00 with orthophosphoric acid. A:B from 30:70 to 34:66 over 10 min, to 60:40 over 10 min (Perkin-Elmer convex curve 0.3), to 100:1 step gradient, maintain at 100:0 for 5 min, return to initial conditions, re-equilibrate for 15 min.
Column temperature: 40
Flow rate: 1.5
Injection volume: 10
Detector: UV 486

CHROMATOGRAM
Retention time: 15.2
Internal standard: N-methyltaurine (8)
Limit of detection: 458 pmole

OTHER SUBSTANCES
Extracted: arginine, glutamine, glutamic acid, glycine

KEY WORDS
fibroblasts

REFERENCE
Casini, A.; Martini, F.; Nieri, S.; Ramarli, D.; Franconi, F.; Surrenti, C. High-performance liquid chromatographic determination of hydroxyproline after derivatization with 4-dimethylaminoazobenzene-4'-sulphonyl chloride, *J.Chromatogr.*, **1982**, *249*, 187–192.

SAMPLE
Matrix: peptides
Analyte: amino acids
Sample preparation: Freeze-dry 0.1-1 μg peptide containing 1-10 nmole total amino acids, add 20 μL 6 M HCl, seal in a tube under reduced pressure, heat at 110° for 24 h, evaporate under vacuum, add 10 μL 100 mM pH 9.0 sodium bicarbonate buffer, add 20 μL 4 mM dabsyl chloride in acetone, heat at 70° for 10-15 min with occasional shaking, dilute to 100-500 μL with EtOH:water 70:30, inject a 10 μL aliquot. (Recrystallize dabsyl chloride from acetone.)

HPLC VARIABLES
Column: Zorbax ODS
Mobile phase: Gradient. G1-MeCN:buffer from 20:80 to 70:30 over 25 min, stay at 70:30 for 5 min, wash with 100:0 for 15 min, return to initial conditions over 5 min, re-equilibrate for 10 min. (Buffer was 5.44 g sodium acetate trihydrate and 7.7 mL acetic acid made up to 900 mL with water, pH 4.13.) (Using G2-MeCN:pH 7.2 phosphate buffer from 20:80 to 35:65 over 15 min, stay at 35:65 for 5 min, go to 65:35 over 5 min, stay at 65:35 for 5 min, Asp and Ser are separated but other separations are not as good.)
Flow rate: 1.2
Injection volume: 10
Detector: UV 436

CHROMATOGRAM
Retention time: 13.5 (8.5)(Asp), 13.5 (14.5) (Ser), 14 (9) (Glu), 14.5 (15) (Thr), 15 (15.5) (Gly), 16 (18) (Arg), 17 (16) (Ala), 19 (18.5) (Met), 20 (17) (Pro), 20.5 (17.5) (Val), 21 (21)

(Phe), 22.5 (19.2) (Leu), 23 (18.8) (Ile), 26 (27.3) (His), 27 (27) (Lys), 30 (28) (Tyr) (times for G1, G2 times in parentheses)
Limit of detection: 2.5 pmole

KEY WORDS
Asp and Ser co-elute

REFERENCE
Chang, J.-Y.; Knecht, R.; Braun, D.G. Amino acid analysis at the picomole level. Application to the C-terminal sequence analysis of polypeptides, *Biochem.J.*, **1981**, *199*, 547–555.

SAMPLE
Matrix: peptides
Analyte: amino acids
Sample preparation: Freeze dry 0.1-1 µg peptide in a tube, add 20 µL 6 M HCl, seal under reduced pressure, heat at 120° for 24 h, centrifuge, evaporate to dryness, reconstitute with 10 µL 200 mM pH 9.0 sodium bicarbonate buffer, add 20 µL 4 mM dabsyl chloride in acetone, heat at 70° with occasional shaking for 10-15 min, dilute to 100-500 µL with EtOH:water 70:30 or MeCN:water 50:50, inject a 10 µL aliquot.

HPLC VARIABLES
Column: Lichrosorb C-18
Mobile phase: Gradient. A was DMF:17 mM pH 6.5 phosphate buffer 2:98. B was MeCN:DMF 96:4. A:B from 85:15 to 55:45 over 25 min, to 30:70 over 10 min.
Column temperature: 50
Flow rate: 1
Injection volume: 10
Detector: UV 436

CHROMATOGRAM
Retention time: 11 (Asp), 12 (Glu), 18 (Ser), 18.7 (Thr), 19.3 (Gly), 20 (Ala), 21.5 (Pro), 21.7 (Val), 22.2 (Arg), 23 (Met), 24 (Ile), 24.5 (Leu), 25 (Phe), 32.5 (ammonia), 34.5 (lysine), 35 (His), 36.5 (Tyr)

REFERENCE
Chang, J.-Y.; Knecht, R.; Braun, D.G. Amino acid analysis in the picomole range by precolumn derivatization and high-performance liquid chromatography, *Methods Enzymol.*, **1983**, *91*, 41–48.

SAMPLE
Matrix: protein
Analyte: amino acids
Sample preparation: 0.1-1 µg protein + 20 µL 5.7 M HCl (twice distilled), seal under reduced pressure, heat at 110° for 18 h, dry under vacuum, add 10 µL 4 mM dabsyl chloride in acetone, heat at 70° for 10 min, dry under vacuum, reconstitute with 100-500 µL MeCN:water 50:50 over EtOH:water 70:30, inject a 10 µL aliquot.

HPLC VARIABLES
Column: 150 × 4.6 Cosmosil 5 octadecylsilane (Nakarai Chemicals, Kyoto)
Mobile phase: Gradient. A was DMF:17 mM pH 6.5 phosphate buffer 2:98. B was MeCN:DMF 96:4. A:B from 85:15 to 55:45 over 25 min, to 30:70 over 10 min.
Flow rate: 1
Injection volume: 10
Detector: UV 420

CHROMATOGRAM
Retention time: 13.5 (Asp), 14.5 (Glu), 20 (Ser), 20.8 (Thr), 21.5 (Gly), 21.8 (Ala), 22.2 (Arg), 22.5 (Pro), 23 (Val), 24.5 (Met), 25 (Ile), 25.5 (Leu), 27 (Phe), 35 (Lys), 35.5 (His), 38 (Tyr)

REFERENCE
Odani, S.; Kenmochi, N.; Ogata, K. A comparative study on 40S ribosomal proteins of Artemia salina and rat liver: micro analysis of amino acid composition by high-performance liquid chromatography, *J.Biochem.*, **1988**, *103*, 872–877.

SAMPLE
Matrix: proteins
Analyte: amino acids
Sample preparation: Hydrolyze 500 ng protein with 400 μL 6 M HCl at 110° for 24 h (flush tube with argon and evacuate to 0.1 mbar before sealing, alternatively hydrolyze in liquid phase with 25 μL 6 M HCl), evaporate to dryness under reduced pressure, reconstitute with 20 μL 50 mM pH 8.1 sodium bicarbonate, add 40 μL 4 mM dabsyl chloride in MeCN, heat at 70° for 10 min, dilute to 1 mL with EtOH:50 mM pH 7.0 sodium phosphate 50:50, inject a 20 μL aliquot.

HPLC VARIABLES
Column: 5 μm Lichrosphere 100 CH-18/2
Mobile phase: Gradient. A was DMF:25 mM sodium acetate 4:96, pH 2.5. B was MeCN. A:B from 85:15 to 60:40 over 20 min, to 30:70 over 12 min, maintain at 30:70 for 2 min, return to initial conditions over 2 min, re-equilibrate for 8 min.
Column temperature: 40
Flow rate: 1
Injection volume: 20
Detector: UV 436

CHROMATOGRAM
Retention time: 15 (aspartate), 16 (glutamate), 20.5 (serine), 21 (threonine), 22 (glycine), 22.5 (alanine), 23 (arginine), 23.5 (proline), 24 (valine), 25 (methionine), 20.5 (isoleucine), 21 (leucine), 22 (phenylalanine), 23.5 (cysteine), 32 (ammonia), 33.5 (lysine), 34 (histidine), 35.5 (tyrosine)
Limit of detection: <0.5 pmole

REFERENCE
Knecht, R.; Chang, J.Y. Liquid chromatographic determination of amino acids after gas-phase hydrolysis and derivatization with (dimethylamino)azobenzenesulfonyl chloride, *Anal.Chem.*, **1986**, *58*, 2375–2379.

SAMPLE
Matrix: proteins
Analyte: amino acids
Sample preparation: Hydrolyze protein with 80 μL 3% phenol in 6 M HCl in an evacuated tube at 166° for 25 min, evaporate to dryness under reduced pressure at 50°, reconstitute with 10 μL 200 mM pH 9.0 carbonate buffer, add 20 μL 4 mM dabsyl chloride in MeCN, heat at 70° with occasional shaking for 10 min, dilute to 200 μL with MeCN:water 66:34, centrifuge, inject a 10 μL aliquot of the supernatant.

HPLC VARIABLES
Column: 100 × 4.6 3 μm Hypersil ODS
Mobile phase: Gradient. A was acetone:10 mM pH 6.5 sodium phosphate buffer 7.5:92.5. B was acetone:10 mM pH 6.5 sodium phosphate buffer 50:50. A:B from 100:0 to 0:100 over 40 min. (Place a 50 × 4.6 column of 30 μm Wakogel LC ODS-30K before the injector.)
Column temperature: 60
Flow rate: 1
Injection volume: 10
Detector: UV 436

CHROMATOGRAM
Retention time: 11 (aspartic acid), 12 (glutamic acid), 18.5 (serine), 19.3 (threonine), 19.7 (glycine), 20.5 (alanine), 21.5 (proline), 23 (valine), 24 (arginine), 25 (methionine), 26 (isoleucine), 26.5 (leucine), 27 (tryptophan), 27.5 (phenylalanine), 30 (cysteine), 38.5 (lysine), 39.5 (histidine), 43 (tyrosine)

REFERENCE
Muramoto, K.; Kamiya, H. Recovery of tryptophan in peptides and proteins by high-temperature and short-term acid hydrolysis in the presence of phenol, *Anal.Biochem.*, **1990**, *189*, 223–230.

SAMPLE
Matrix: solutions

Analyte: amino acids

Sample preparation: 100 µL 2 mM Amino acids in 100 mM pH 9.0 sodium bicarbonate buffer + 1 mL 2 µM dabsyl chloride in acetone, mix, heat with occasional mixing at 70° for 10 min, evaporate to dryness under reduced pressure, reconstitute with 2 mL EtOH : water 70 : 30, inject an aliquot.

HPLC VARIABLES

Guard column: ODS-5 (Altex)

Column: 250 × 4.6 5 µm Ultrasphere ODS

Mobile phase: Gradient. MeCN : buffer from 20 : 80 to 70 : 30 over 25 min, maintain at 70 : 30 for 5 min. (Buffer was 5.44 g sodium acetate trihydrate and 7.7 mL acetic acid diluted to 900 mL, pH 4.13.)

Flow rate: 1.2

Detector: UV 436

CHROMATOGRAM

Retention time: 13 (Cys), 14 (Asn), 14.2 (Gln), 14.8 (Asp, Ser), 15 (Glu), 15.8 (Arg), 16 (Thr), 16.7 (Gly), 18 (Ala), 20 (Met), 21 (Pro), 21.5 (Val), 22.5 (Phe, Trp), 23.5 (Leu)

Limit of detection: 50-500 pmole

KEY WORDS

comparison with other derivatization procedures; His, Lys, Tyr, Cys give multiple peaks

REFERENCE

McClung, G.; Frankenberger, W.T. Jr. Comparison of reverse-phase high-performance liquid chromatographic methods for precolumn-derivatized amino acids, *J.Liq.Chromatogr.*, **1988**, *11*, 613–646.

SAMPLE

Matrix: solutions

Analyte: amino acids

Sample preparation: Dissolve amino acids in 20 µL 50 mM pH 9.0 sodium bicarbonate, add 40 µL 4 mM dabsyl chloride in MeCN, heat at 70° for 10 min, dry under vacuum, dissolve the residue in EtOH : water 70 : 30 (J. Chromatogr. 1985, 349 77), inject a 5 µL aliquot.

HPLC VARIABLES

Guard column: 20 × 4.6 5 µm Supelcosil LC-18

Column: 150 × 4.6 3 µm Supelcosil LC-18

Mobile phase: Gradient. A was 25 mM pH 6.8 KH_2PO_4. B was MeCN : isopropanol 75 : 25. A : B 80 : 20 for 1 min, to 77 : 23 over 3 min, maintain at 77 : 23 for 5 min, to 73 : 27 over 1 min, maintain at 73 : 27 for 4 min, to 65 : 35 over 5 min, to 40 : 60 over 6 min, to 70 : 30 over 1 min, maintain at 30 : 70 for 3 min, re-equilibrate at initial conditions for 6 min.

Flow rate: 2

Injection volume: 5

Detector: UV 436

CHROMATOGRAM

Retention time: 4 (aspartic acid), 4.5 (cysteamine), 5 (glutamic acid), 5.7 (carboxymethyl-cysteine), 6 (S-sulfocysteine), 8.5 (asparagine), 9.5 (glutamine), 10 (serine), 10.7 (threonine), 11.4 (glycine), 12 (alanine), 12.5 (arginine), 13.5 (taurine), 14 (proline), 14.5 (valine), 16.4 (methionine), 17.2 (isoleucine), 17.7 (leucine), 18.2 (tryptophan), 18.5 (norleucine), 19 (phenylalanine), 20 (ammonia), 20.5 (cystine), 22.4 (OH-Lys), 23 (lysine), 23.5 (histidine), 24.3 (tyrosine)

REFERENCE

Stocchi, V.; Piccoli, G.; Magnani, M.; Palma, F.; Biagiarelli, B.; Cucchiarini, L. Reversed-phase high-performance liquid chromatography separation of dimethylaminoazobenzene sulfonyl- and dimethyl-aminoazobenzene thiohydantoin-amino acid derivatives for amino acid analysis and microsequencing studies at the picomole level, *Anal.Biochem.*, **1989**, *178*, 107–117.

SAMPLE

Matrix: solutions

Analyte: amino acids

Sample preparation: 1 mL Amino acid solution in 50 mM pH 8.9 carbonate buffer (chloroform-saturated) + 1 mL 6 mM dabsyl chloride in acetone, heat at 75° with gentle stirring until the color changes from red to yellow-orange, evaporate to dryness under a stream of nitrogen, reconstitute with 1 mL water, dilute, inject an aliquot.

CAPILLARY ELECTROPHORESIS
Capillary: 115 cm × 50 μm fused-silica (105 cm to detector)
Running buffer: MeCN:20 mM pH 7 phosphate buffer 50:50 containing 5 mM sodium dodecyl sulfate
Voltage/Current: 27 kV
Injection: Electromigration at 5 kV for 5 s (0.7 nL)
Detector: Thermooptical detection, 458 nm (pump beam, 130 mW Ar laser), 632.8 nm (probe beam, 1 mW He-Ne laser)
Migration time: 22 (arginine), 33 (histidine), 34 (lysine), 38 (cysteine, tyrosine), 38.5 (tryptophan), 39 (proline), 40 (phenylalanine), 40.5 (leucine), 41 (methionine), 41.5 (isoleucine, valine), 42 (tyrosine, serine), 42.5 (alanine), 43.5 (glycine), 66.5 (glutamic acid), 68 (aspartic acid)
Limit of detection: 50-500 nM

KEY WORDS
laser

REFERENCE
Yu, M.; Dovichi, N.J. Attomole amino acid determination by capillary zone electrophoresis with thermooptical absorbance detection, *Anal.Chem.*, **1989**, *61*, 37−40.

SAMPLE
Matrix: solutions
Analyte: iodinated thyronines
Sample preparation: 20 μL Solution + 40 μL 50 mM pH 8.5 borate buffer + 40 μL 4 mM dabsyl chloride in MeCN, mix, heat at 70° for 15 min, add 100 μL 25 mM pH 6.5 sodium acetate buffer, inject an aliquot.

HPLC VARIABLES
Guard column: 10 × 3 30 μm Chromspher C18 (Chrompack)
Column: 100 × 3 5 μm Chromspher C18 (Chrompack)
Mobile phase: Gradient. A was MeOH:25 mM pH 6.5 sodium acetate buffer 70:30. B was MeOH:25 mM pH 6.5 sodium acetate buffer 90:10. A:B from 100:0 to 0:100 over 15 min, maintain at 0:100 for 5 min.
Column temperature: 35
Flow rate: 0.7
Detector: UV 436

CHROMATOGRAM
Retention time: 8 (tyrosine), 9 (3-monoiodotyrosine), 10 (3,5-diiodotyrosine), 11.5 (thyronine), 12.5 (3,5-diiodothyronine), 14 (liothyronine), 14.5 (levothyroxine), 15 (3,3',5-triiodothyronine)
Limit of detection: 0.35-1.1 pmole

KEY WORDS
comparison with other derivatization procedures

REFERENCE
Doorn, L.; Jansen, E.H.J.M.; Van Leeuwen, F.X.R. Comparison of high-performance liquid chromatographic detection methods for thyronine and tyrosine residues in toxicological studies of the thyroid, *J.Chromatogr.*, **1991**, *553*, 135−142.

SAMPLE
Matrix: solutions
Analyte: hypusine
Sample preparation: Take up 2 nmole in 10 μL 100 mM pH 8.4 sodium bicarbonate, add 20 μL 4 mM dabsyl chloride in MeCN, mix, heat at 70° for 10 min, cool to room temperature, dilute with EtOH:50 mM pH 7.0 phosphate buffer, inject an aliquot.

HPLC VARIABLES
Guard column: present but not specified
Column: 100×4.2 5 μm Spherisorb C18
Mobile phase: Gradient. MeCN:25 mM pH 6.5 sodium acetate buffer from 30:70 to 58:42 over 24 min.
Flow rate: 1
Detector: UV 436

CHROMATOGRAM
Retention time: 23
Limit of detection: 500 fmole

OTHER SUBSTANCES
Non-interfering: other amino acids

REFERENCE
Bartig, D.; Klink, F. Determination of the unusual amino acid hypusine at the lower picomole level by derivatization with 4-dimethylaminoazobenzene-4'-sulphonyl chloride and reversed-phase high-performance or medium-pressure liquid chromatography, *J.Chromatogr.*, **1992**, *606*, 43−48.

SAMPLE
Matrix: solutions
Analyte: proline and hydroxyproline
Sample preparation: Evaporate 20 μL of an amino acid solution to dryness, reconstitute with 20 μL reagent, let stand at room temperature for 5 min, add 20 μL 13 mg/mL dabsyl chloride in MeCN (freshly prepared), mix, heat at 70° for 20 min, add 60 μL MeCN:water 50:50, shake gently, inject a 10 μL aliquot. (Prepare reagent by dissolving 60 mg o-phthalaldehyde in 2 mL acetone and adding 30 μL 2-mercaptoethanol just before use. Immediately dilute 64-fold with 50 mM pH 9.0 carbonate/bicarbonate buffer. Recrystallize dabsyl chloride by dissolving 1 g in acetone, let stand at -20° overnight, filter. To prepare the dabsyl chloride solution dissolve 1.3 mg of the recrystallized material in 2 mL acetone, evaporate a 400 μL aliquot to dryness under reduced pressure, reconstitute with 200 μL MeCN, use a 20 μL aliquot. Primary (but not secondary) amino acids react with the o-phthalaldehyde and are not retained under these chromatographic conditions. Secondary amino acids are derivatized with the dabsyl chloride.)

HPLC VARIABLES
Column: 250×4.6 5 μm Wakosil 5C18 (Wako)
Mobile phase: Gradient. A was 40 mM NaH_2PO_4 adjusted to pH 3.0 with phosphoric acid. B was MeCN:water 76.8:19.2:4. A:B 55:45 for 3 min, to 40:60 over 5 min, to 33:67 over 4 min, to 0:100 over 1 min, maintain at 0:100 for 2 min, return to initial conditions over 0.5 min.
Flow rate: 1.5
Injection volume: 10
Detector: UV 436

CHROMATOGRAM
Retention time: 10.27 (hydroxyproline), 16.02 (proline)
Limit of quantitation: 20 pmole

REFERENCE
Ikeda, M.; Sorimachi, K.; Akimoto, K.; Yasumura, Y. Reversed-phase high-performance liquid chromatographic analysis of hydroxyproline and proline from collagen by derivatization with dabsyl chloride, *J.Chromatogr.*, **1993**, *621*, 133−138.

SAMPLE
Matrix: tissue
Analyte: levothyroxine (T4) and related compounds
Sample preparation: 100 μL Thyroid tissue + 200 μL MeCN, mix, centrifuge. Remove a 100 μL aliquot of the supernatant and add it to 100 μL 4 nM dabsyl chloride in MeCN, heat at 70° for 10 min, add 400 μL MeOH:50 mM pH 7.0 phosphate buffer 50:50, inject a 20 μL aliquot.

HPLC VARIABLES

Column: 150 × 4.6 5 μm Hypersil ODS

Mobile phase: Gradient. A was MeOH:25 mM pH 6.5 sodium acetate 56:44. B was MeOH. A:B from 80:20 to 35:65 over 15 min, maintain at 35:65 for 3 min, to 0:100 over 1 min, maintain at 0:100 for 2 min.

Flow rate: 1

Injection volume: 20

Detector: UV 436

CHROMATOGRAM

Retention time: 17.5 (levothyroxine (T4)), 17 (liothyronine (T3)), 16 (diiodothyronine (T2))

KEY WORDS

thyroid

REFERENCE

Jansen, E.H.J.M.; van den Berg, R.H.; Both-Miedema, R.; Doorn, L. Advantages and limitations of pre-column derivatization of amino acids with dabsyl chloride, *J.Chromatogr.*, **1991**, *553*, 123–133.

SAMPLE

Matrix: tissue

Analyte: amino acids

Sample preparation: Homogenize (glass/PTFE homogenizer) tissue in 320 mM sucrose, mix 60 μL homogenate with 100 μL 400 mM perchloric acid, centrifuge at 10000 g for 10 min. Remove a 100 μL aliquot of the supernatant and add it to 65 μL 2 M potassium bicarbonate, centrifuge at 10000 g for 10 min. Remove a 10 μL aliquot of the supernatant and mix it with 10 μL 1 mM α-aminoadipic acid, 20 μL 25 mM sodium bicarbonate, and 80 μL 4 mM dabsyl chloride in MeCN, vortex thoroughly, heat at 70° for 12 min, add 380 μL 20% acetic acid, inject a 20 μL aliquot.

HPLC VARIABLES

Column: 125 × 4 5 μm LiChrospher C18

Mobile phase: Gradient. A was DMF:25 mM pH 6.4 sodium acetate buffer 4:96. B was MeCN. A:B 85:15 for 2 min, to 70:30 over 28 min, to 59:41 over 5 min, to 46:54 over 2 min, to 43:57 over 2 min, to 33:67 over 2 min, to 10:90 over 1 min, return to initial conditions over 2 min, re-equilibrate for 11 min.

Column temperature: 40

Flow rate: 1

Injection volume: 20

Detector: UV 436

CHROMATOGRAM

Retention time: 10 (aspartate), 11 (glutamate), 19 (glutamine), 20 (serine), 21 (threonine), 22 (glycine), 23 (alanine), 25 (arginine, taurine), 26.5 (proline, gamma-aminobutyric acid), 27.5 (valine), 30 (methionine), 31.5 (isoleucine), 32.5 (leucine), 35 (phenylalanine), 37.5 (cystine), 38.5 (ammonia), 40.5 (lysine), 41 (histidine), 42 (tyrosine)

Internal standard: α-aminoadipic acid (12)

KEY WORDS

brain

REFERENCE

Watanabe, A.; Semba, J.; Kurumaji, A.; Kumashiro, S.; Toru, M. Measurement of glutamate, aspartate and glycine and its potential precursors in human brain using high-performance liquid chromatography by pre-column derivatization with dimethylaminoazobenzene sulphonyl chloride, *J.Chromatogr.*, **1992**, *583*, 241–245.

SAMPLE

Matrix: urine

Analyte: hydroxyproline

Sample preparation: 500 μL Urine + 1.5 mL pH 5.0 buffer + 200 mg 100-200 mesh Dowex 50W-X8(H⁺) resin, shake gently for 10 min, centrifuge, discard the supernatant, add 6 mL water, mix, centrifuge at 1000 g, discard the supernatant, heat the resin in a capped tube at 120° for 16 h, elute with 1 mL pH 6.0 buffer, add 40 μL 10 mM N-methyltaurine, mix. Remove a 50 μL aliquot and add it to 450 μL pH 9.2 buffer, mix. Remove a 50 μL aliquot and add it to 100 μL 4 mM dabsyl chloride in acetone, shake at 70° for 10 min, evaporate to dryness under a stream of nitrogen at 40°, reconstitute the residue in 500 μL mobile phase, inject a 10 μL aliquot. (Prepare pH 5.0 buffer by mixing 200 mM Na$_2$HPO$_4$ and 100 mM citric acid in the ratio 103:97. Prepare pH 6.0 buffer by dissolving 57.1 g sodium acetate trihydrate, 58.7 g trisodium citrate dihydrate, and 5.5 g citric acid in 800 mL water. Prepare pH 9.2 buffer by mixing sodium carbonate and sodium bicarbonate in the ratio 14:86.)

HPLC VARIABLES
Column: 125 × 4 5 μm Hibar RP-18 (Merck)
Mobile phase: MeCN:12 mM pH 4.14 sodium acetate 30:70
Column temperature: 40
Flow rate: 0.7
Injection volume: 10
Detector: UV 471

CHROMATOGRAM
Retention time: 9.54
Internal standard: N-methyltaurine (13.25)
Limit of quantitation: 5 μM

OTHER SUBSTANCES
Simultaneously analyzed: alanine, glutamine, glycine, proline

KEY WORDS
SPE

REFERENCE
Pang, C.-P.; Ho, K.-C.; Jones, M.G.; Cheung, C.-K. Analysis of total hydroxyproline in urine by high-performance liquid chromatography and pre-column derivatization, *J.Chromatogr.*, **1987**, *386*, 309–314.

SAMPLE
Matrix: urine
Analyte: hydroxyproline
Sample preparation: 1 mL Urine + 1 mL 37% HCl + 100 μL 65.5 μg/mL methyltaurine, heat at 100° for at least 15 h, dilute with 4 volumes of water. Remove a 100 μL aliquot and add it to 800 μL 500 mM NaOH and 800 μL 250 mM sodium carbonate. Remove a 500 μL aliquot and add it to 200 μL 5 mg/mL o-phthalaldehyde in MeCN. Remove a 150 μL aliquot and add it to 200 μL 5 mg/mL dabsyl chloride in MeCN, heat at 70° for 10 min, cool, add 500 μL mobile phase, inject an aliquot. (Primary amino acids are derivatized with o-phthalaldehyde then hydroxyproline (a secondary amino acid) is derivatized with dabsyl chloride.)

HPLC VARIABLES
Guard column: 30 × 4.6 10 μm Microguard RP (Bio-Rad)
Column: 150 × 4.6 5 μm ODS (Bio-Rad)
Mobile phase: MeCN:50 mM citric acid:10 mM sodium phosphate 29:33:38
Column temperature: 60
Flow rate: 1.5
Detector: UV 471

CHROMATOGRAM
Retention time: 5
Internal standard: methyltaurine (3.7)
Limit of quantitation: 6.7 mg/mL

REFERENCE
Bianchi, V.; Mazza, L. Rapid reversed-phase high-performance liquid chromatographic method with double derivatization for the assay of urinary hydroxyproline, *J.Chromatogr.B*, **1995**, *665*, 295–302.

Dansyl Chloride

RELATED REFERENCES

Badoud, R.; Pratz, G. Simple and rapid quantitative determination of lysinoalanine and protein hydrolyzate amino acids by high-performance liquid chromatography after derivatization with dansyl chloride. *Chromatographia* **1984**, *19*, 155-164.

De Jong, C.; Hughes, G.J.; Van Wieringen, E.; Wilson, K.J. Amino acid analyses by high-performance liquid chromatography. An evaluation of the usefulness of precolumn Dns derivatization. *J.Chromatogr.* **1982**, *241*, 345-359.

Fujimura, K.; Suzuki, S.; Hayashi, K.; Masuda, S. Retention behavior and chiral recognition mechanism of several cyclodextrin-bonded stationary phases for dansyl amino acids. *Anal.Chem.* **1990**, *62*, 2198-2205.

Herrndobler, A. Determination of amino acids in an infusion solution by reversed-phase HPLC and precolumn dansyl chloride derivatization. *HRC CC,J.High Resolut.Chromatogr. Chromatogr. Commun.* **1986**, *9*, 602-604.

Simmaco, M.; De Biase, D.; Barra, D.; Bossa, F. Automated amino acid analysis using precolumn derivatization with dansylchloride and reversed-phase high-performance liquid chromatography. *J.Chromatogr.* **1990**, *504*, 129-138.

Skocir, E.; Prosek, M.; Oskomic, M. Separation of di-dansyl amino acids by MEKC with a co-surfactant. *Chromatographia* **1997**, *44*, 267-273.

Takeuchi, T.; Yamazaki, M.; Ishii, D. Micro high-performance liquid chromatography of 5-dimethylaminonaphthalenesulfonylamino acids. *J.Chromatogr.* **1984**, *295*, 333-339.

Torsi, G.; Gioacchini, A.M.; Gandini, N. Absolute analysis method applied to the dansyl derivatives of amino acids in HPLC. Stability of amino acids in 6N HCl. *Ann.Chim.(Rome)* **1993**, *83*, 345-354.

Weinstein, S.; Weiner, S. Enantiomeric analysis of a mixture of the common protein amino acids as their Dns derivatives. Single-analysis reversed-phase high-performance liquid chromatographic procedure using a chiral mobile phase additive. *J.Chromatogr.* **1984**, *303*, 244-250.

Wiedmeier, V.T.; Porterfield, S.P.; Hendrich, C.E. Quantitation of Dns-amino acids from body tissues and fluids using high-performance liquid chromatography. *J.Chromatogr.* **1982**, *231*, 410-417.

SAMPLE

Matrix: ascitic fluid, blood

Analyte: amino acids

Sample preparation: 100 µL Plasma or ascitic fluid + 400 µL MeOH, shake gently, let stand at 4° for 10 min, centrifuge at 11600 g for 5 min, remove the supernatant, wash the precipitate with 100 µL MeOH:water 80:20, centrifuge at 11600 g for 5 min, remove the supernatant, wash the precipitate twice more. Combine the supernatants and evaporate them to 30-50 µL at 80°, make up to the original volume with water. Remove a 100 µL aliquot and add it to 100 µL 40 mM pH 9.5 lithium carbonate buffer, add 100 µL 20 mM dansyl chloride in MeCN, stir, let stand at room temperature for 1 h, inject a 10 µL aliquot.

HPLC VARIABLES

Guard column: 15 × 3.2 7 µm New Guard RP-18

Column: 150 × 4.6 5 µm Supelcosil LC-18

Mobile phase: Gradient. A was MeOH. B was 0.6% acetic acid containing 0.008% triethylamine in water. A:B 30:70 for 20 min, to 40:60 over 5 min, to 50:50 over 25 min, maintain at 50:50 for 5 min, to 65:35 over 10 min, to 75:25 over 5 min, maintain at 75:25 over 5 min.

Flow rate: 1 for 25 min, to 1.5 for 25 min, to 2 over 5 min
Injection volume: 10
Detector: UV 254

CHROMATOGRAM
Retention time: 3 (Cys), 3.5 (Cya), 4.5 (Tau), 12 (Asn), 15 (Gln), 17 (ammonia), 18.5 (Ser), 20 (Asp), 22 (Glu), 25 (Gly), 28 (Thr), 29 (β-alanine), 30 (Ala), 31.5 (GABA), 35.5 (α-ABA), 38 (Pro), 41 (Met), 43 (Val), 46 (Arg), 49 (Trp), 52 (Ile, Phe), 53.5 (Leu), 61 (Cis (bis-derivative)), 65 (Orn (bis-derivative)), 66 (Lys (bis-derivative)), 68 (His (bis-derivative)), 70 (Tyr (bis-derivative))

KEY WORDS
mouse; plasma; protect from light

REFERENCE
Márquez, F.J.; Quesada, A.R.; Sánchez-Jiménez, F.; Núñez de Castro, I. Determination of 27 dansyl amino acid derivatives in biological fluids by reversed-phase high-performance liquid chromatography, *J.Chromatogr.*, **1986**, *380*, 275–283.

SAMPLE
Matrix: blood
Analyte: imino acids
Sample preparation: 30 μL Plasma + 500 μL EtOH:water 70:30, shake vigorously, centrifuge at 7000 g for 15 min. Remove the supernatant and evaporate it to dryness, reconstitute the residue in 100 μL 100 mM pH 6.8 phosphate buffer containing 10% formaldehyde, sonicate for 2 min, add 50 μL 500 mM sodium bicarbonate, add 100 μL 2 mg/mL dansyl chloride in acetone, heat at 37° in the dark for 20 min, centrifuge, inject a 50 μL aliquot of the supernatant. (Amino acids react with the formaldehyde leaving the imino acids to react with the dansyl chloride.)

HPLC VARIABLES
Column: 250 × 4 5 μm LiChrosorb RP-18
Mobile phase: Gradient. A was MeCN:50 mM pH 4.0 acetate buffer 10:90. B was MeCN:water 70:30. A:B from 100:0 to 20:80 over 15 min (convex gradient).
Column temperature: 50
Flow rate: 1
Injection volume: 50
Detector: F ex 340 em 510

CHROMATOGRAM
Retention time: 10 (hydroxyproline), 12.5 (proline)
Limit of quantitation: 1 μg/mL

KEY WORDS
plasma

REFERENCE
Tsuchiya, H.; Hayashi, T.; Tatsumi, M.; Fukita, T.; Takagi, N. Simplified procedure for specific determination of imino acids in human blood plasma by high-performance liquid chromatography, *J.Chromatogr.*, **1985**, *339*, 59–65.

SAMPLE
Matrix: bulk
Analyte: amino acids
Sample preparation: Dissolve 5 mg amino acids in 460 μL 100 mM boric acid adjusted to pH 9 with NaOH, add 300 μL 100 mM dansyl chloride in acetone, let stand in dark for 2 h, evaporate to dryness under reduced pressure, reconstitute with 500 μL acetone:1 M HCl 95:5, centrifuge for 5 min, evaporate the liquid to dryness under reduced pressure, reconstitute with mobile phase, inject an aliquot.

HPLC VARIABLES
Column: 250 × 4 5 μm Chiradex (immobilized β-cyclodextrin) (Merck)
Mobile phase: MeOH:100 mM pH 5.5 ammonium acetate buffer containing 0.1% triethylamine 70:30

Column temperature: 20 ± 0.1
Flow rate: 0.5
Injection volume: 20
Detector: UV 254; F ex 325 em 350

CHROMATOGRAM
Retention time: k' 2.52 (second enantiomer), α = 1.08 (Ala); k' 2.03 (second enantiomer), α = 1.22 (Val); k' 2.54 (second enantiomer), α = 1.75 (Leu); k' 1.68 (second enantiomer), α = 1.32 (Ile); k' 1.40 (second enantiomer), α = 1.10 (Pro); k' 1.33 (second enantiomer), α = 1.24 (Met); k' 2.99 (second enantiomer), α = 1.15 (Cys); k' 1.37 (second enantiomer), α = 1.16 (Ser); k' 1.46 (second enantiomer), α = 1.29 (Thr); k' 1.22 (second enantiomer), α = 1.10 (Lys); k' 0.84 (second enantiomer), α = 1.16 (Arg); k' 1.68 (second enantiomer), α = 1.11 (Asn); k' 1.38 (second enantiomer), α = 1.14 (Gln); k' 18.03 (second enantiomer), α = 1.13 (Asp); k' 9.17 (second enantiomer), α = 1.15 (Glu); k' 2.42 (second enantiomer), α = 1.43 (Phe); k' 1.63 (second enantiomer), α = 1.00 (Trp); k' 1.66 (second enantiomer), α = 1.07 (His); k' 2.67 (second enantiomer), α = 1.25 (Tyr)

KEY WORDS
chiral; α = k' (second enantiomer)/k' (first enantiomer); detailed comparison with other derivatizing reagents

REFERENCE
Rizzi, A.M.; Cladrowa-Runge, S.; Jonsson, H.; Osla, S. Enantiomeric resolution of derivatized DL-amino acids by high-performance liquid chromatography using a β-cyclodextrin chiral stationary phase: A comparison between derivatization labels, *J.Chromatogr.A*, **1995**, *710*, 287–295.

SAMPLE
Matrix: enzyme incubations
Analyte: glutamine, glutamic acid, and gamma-aminobutyric acid
Sample preparation: Adjust the pH of 50 µL enzyme incubation 9-9.5 with 100 mM NaOH, centrifuge for 5 min. Lyophilize a 50 µL aliquot of the supernatant, reconstitute with 50 µL 100 mM pH 9.5 sodium bicarbonate buffer, add 100 µL 0.25% dansyl chloride in acetone, mix thoroughly, heat at 70° for 15 min, cool in ice for 5 min, inject a 20 µL aliquot.

HPLC VARIABLES
Column: 250 × 4.6 Ultrasphere ODS-5
Mobile phase: Gradient. MeCN:100 mM pH 2.1 KH_2PO_4 from 0:100 to 10:90 over 9 min, to 40:60 over 30 min
Flow rate: 1
Injection volume: 20
Detector: UV 206

CHROMATOGRAM
Retention time: 25 (glutamine), 27.5 (glutamic acid), 33.5 (gamma-aminobutyric acid)

REFERENCE
Pahuja, S.L.; Albert, J.; Reid, T.W. Use of reversed-phase high-performance liquid chromatography for simultaneous determination of glutamine synthetase and glutamic acid decarboxylase in crude extracts, *J.Chromatogr.*, **1981**, *225*, 37–45.

SAMPLE
Matrix: flour
Analyte: amino acids
Sample preparation: Hydrolyze 30-40 mg flour with 4 mL 6 M HCl at 120° for 24 h, evaporate to dryness under reduced pressure, reconstitute with 10 mL water. Remove a 2.5 mL aliquot and add it to 2.5 mL 500 mM sodium bicarbonate, add 2.5 mL 20 mM dansyl chloride in acetone, sonicate for 10 min, heat at 50° for 1 h, add 100 µL concentrated ammonia, heat at 50° for 15 min, add to a 30 × 11 glass column of Amberlite CG 50 1, elute twice with 2 mL portions of water, elute with three 5 mL portions of MeOH, combine all the eluates and evaporate them to dryness under reduced pressure, reconstitute with 1 mL MeOH:water 20:80, inject an aliquot.

CAPILLARY ELECTROPHORESIS
Capillary: 57.5 × 50 μm fused-silica (50 cm to detector)
Capillary preparation: Flush capillary with running buffer for 2 min before each run.
Running buffer: 20 mM Borax buffer containing 125 mM sodium dodecyl sulfate
Capillary temperature: 10
Voltage/Current: 25 kV
Injection: Hydrodynamic injection for 3 s
Detector: UV 214
Model: Thermo Separation Products SpectraPHORESIS 500
Migration time: 11.7 (Thr), 11.8 (Ser), 12.1 (Asn), 12.3 (Gln), 12.5 (Ala), 12.7 (Glu), 13.1 (α-ABA), 13.3 (Asp), 13.6 (Gly), 13.8 (Val), 14.3 (cysteic acid), 15.8 (Pro), 16.2 (Met), 17.7 (Ile), 18.5 (Leu), 20.7 (Phe), 21.2 (Trp), 23.3 (Cys, bis derivative), 24.8 (ammonia), 26 (Lys), 26.8 (Arg), 27.5 (Lys, bis derivative), 27.5 (His, bis derivative), 27.9 (Tyr, bis derivative)

KEY WORDS
corn; protect from light; SPE

REFERENCE
Skocir, E.; Prosek, M. Determination of amino acid ratios in natural products by micellar electrokinetic chromatography, *Chromatographia*, **1995**, *41*, 638–644.

SAMPLE
Matrix: fungal spore walls
Analyte: amino acids
Sample preparation: Hydrolyse fungal spore walls with 6 M HCl at 110° for 12 h, evaporate to dryness under reduced pressure, chromatograph on a DeltaPak C18 column (Waters) with 0.1% trifluoroacetic acid as mobile phase and detection at UV 214. Collect the unretained material and evaporate it to dryness under reduced pressure, dissolve 20 nmoles crude amino acids in 400 μL 100 mM pH 9 borate buffer, add 300 μL 10 mg/mL dansyl chloride in acetone, let stand in the dark for 2 h, evaporate to dryness under a stream of nitrogen, reconstitute with acetone:1 M HCl 95:5, centrifuge, evaporate the supernatant to dryness, repeat this extraction, reconstitute with mobile phase A, inject a 20 μL aliquot onto column A, elute to waste with mobile phase A, divert the fraction (200 μL) containing the amino acid from column A to column B (16.6-17 min for alanine; 24.5-24.8 min for glutamic acid), elute column B with mobile phase B, monitor the effluent from column B.

HPLC VARIABLES
Column: A 250 × 4 7 μm LiChrosorb RP-18; B 250 × 4 5 μm LiChrospher 100 RP-18
Mobile phase: A EtOH:25 mM pH 5.5 ammonium acetate containing 0.1% triethylamine 21:79 or 30:70; B EtOH:15 mM pH 5.5 ammonium acetate 20:80 containing 35 mM β-cyclodextrin and 1 M urea.
Flow rate: 0.5
Injection volume: 20
Detector: F ex 340 em 480

CHROMATOGRAM
Retention time: 12.5 (D-Glu (mobile phase A 21:79)), 13.3 (L-Glu (mobile phase A 21:79)), 38.6 (D-Ala (mobile phase A 30:70)), 41.5 (L-Ala (mobile phase A 30:70))

KEY WORDS
chiral; column-switching

REFERENCE
Rizzi, A.M.; Briza, P.; Breitenbach, M. Determination of D-alanine and D-glutamic acid in biological samples by coupled-column chromatography using β-cyclodextrin as mobile phase additive, *J.Chromatogr.*, **1992**, *582*, 35–40.

SAMPLE
Matrix: perfusate
Analyte: amino acids
Sample preparation: Add L-norvaline to a concentration of 750 μM, remove a 1 mL aliquot and add it to 75 μL ice-cold 60% perchloric acid, centrifuge, neutralize supernatant

with 200 mM potassium carbonate, add (?) 0.5 M potassium carbonate:potassium hydrogen carbonate 30:70, adjust pH to 9.40-9.50 with 5 M KOH. Remove a 200 µL aliquot and add it to 100 µL 1.25 mg/mL dansyl chloride in MeCN, let stand at room temperature in the dark for 1 h, add 6 µL 0.2% triethylamine in water, add acetic acid to a final concentration of 3%, inject a 20 µL aliquot.

HPLC VARIABLES
Guard column: 30 × 4.6 Biosil ODS-5S microguard refill cartridge (Biorad)
Column: 250 × 4 Biosil C18 ODS-5S (Biorad)
Mobile phase: Gradient. A was MeOH:water 15:85 containing 1% glacial acetic acid and 0.030% triethylamine. B was MeOH:MeCN 70:30 containing 3% glacial acetic acid and 0.030% triethylamine. A:B from 70:30 to 50:50 over 52 min, to 25:75 over 21 min, maintain at 25:75 for 5 min, reset to initial conditions over 7 min.
Flow rate: 1
Injection volume: 20
Detector: F ex 340 em 520

CHROMATOGRAM
Retention time: 10 (Asn), 11 (Gln), 13 (Ser), 14 (Glu), 15 (Asp), 17 (Gly), 18 (Thr), 22 (Ala), 24 (Arg), 32 (Pro), 35 (Met), 37 (Val), 45 (Try), 53 (Phe), 55 (Iso-leu), 57 (Leu), 73 (Orn), 75 (Lys), 79 (His), 83 (Tyr)
Internal standard: L-norvaline (41)

KEY WORDS
rat

REFERENCE
Zezza, F.; Kerner, J.; Pascale, M.R.; Giannini, R.; Arrigoni Martelli, E. Rapid determination of amino acids by high-performance liquid chromatography: release of amino acids by perfused rat liver, *J.Chromatogr.*, **1992**, *593*, 99–101.

SAMPLE
Matrix: protein
Analyte: amino acids
Sample preparation: Heat protein in 5.9 M HCl (twice distilled) in a sealed tube at 108° for 22 h, evaporate to dryness under reduced pressure, reconstitute with water, evaporate to dryness under reduced pressure, repeat reconstitution and evaporation step, dry over NaOH under reduced pressure overnight, reconstitute with water. Mix 100 µL of this solution (containing 10-40 µg free amino acids) with 100 µL 500 mM sodium bicarbonate solution and 100 µL freshly prepared 20 mM dansyl chloride, heat at 65° for 40 min (or let stand at room temperature for 24 h) in the dark, inject an aliquot.

HPLC VARIABLES
Column: 250 × 4.6 5 µm Ultrasphere ODS C18
Mobile phase: Gradient. A was MeCN:25 mM NaH_2PO_4 containing 25 mM acetic acid 14:86, adjusted to pH 7.00 with concentrated NaOH. B was MeCN. A:B 100:0 for 5 min, to 92:8 over 5 min, maintain at 92:8 for 15 min, to 55:45 over 30 min, to 42:58 over 5 min, re-equilibrate at initial conditions for 20 min. Every 10-15 runs wash column with 200 mM acetic acid for 40 min.
Flow rate: 1
Injection volume: 20
Detector: UV 254

CHROMATOGRAM
Retention time: 8.49 (Asp), 10.13 (Glu), 16.28 (4-hydroxyproline), 18.60 (Ser), 19.00 (His (α-derivative)), 19.70 (Arg), 20.14 (Thr), 20.59 (Gly), 22.06 (Ala), 25.18 (Pro), 31.09 (Val), 33.65 (Met), 35.49 (Ile), 35.87 (Leu), 37.85 (Phe), 38.95 (Cys (bis derivative)), 40.85 (ammonia), 44.05 (desmosine), 44.05 (isodesmosine), 44.46, 44.80 (5-hydroxylysine (bis derivative)), 47.34 (Lys (bis derivative)), 49.88 (His (bis derivative)), 52.82 (Tyr (bis derivative))
Limit of quantitation: 100 pmole

REFERENCE

Negro, A.; Garbisa, S.; Gotte, L.; Spina, M. The use of reverse-phase high-performance liquid chromatography and precolumn derivatization with dansyl chloride for quantitation of specific amino acids in collagen and elastin, *Anal.Biochem.*, **1987**, *160*, 39–46.

SAMPLE

Matrix: protein
Analyte: amino acids
Sample preparation: Hydrolyze protein with 1 mL 6 M HCl for each 1 mg protein, heat under nitrogen at 110° for 24 h, filter (paper), make up the filtrate to 250 mL, evaporate to dryness under reduced pressure at 40°, reconstitute the residue with water so as to give a protein concentration of 150-250 μg/mL. 1 mL Hydrolysate + 2 mL 40 mM pH 9.5 lithium carbonate + 1 mL 4 mg/mL dansyl chloride in MeCN, mix, heat at 60° for 30 min, add 50 μL 2% methylamine in water, inject an aliquot.

HPLC VARIABLES

Column: 300 × 3.9 10 μm Spherisorb ODS-2
Mobile phase: MeCN:10 mM pH 7.0 phosphate buffer 39:61
Column temperature: 40
Flow rate: 1.5
Injection volume: 20
Detector: UV 254

CHROMATOGRAM

Retention time: 4.2 (lysine, bis derivative), 4.7 (histidine, bis derivative), 6.7 (ammonia), 11.7 (tyrosine), 15 (methylamine)

KEY WORDS

casein; lysozyme; lentils; enteral solution

REFERENCE

Sanz, M.A.; Castillo, G.; Hernández, A. Isocratic high-performance liquid chromatographic method for quantitative determination of lysine, histidine and tyrosine in foods, *J.Chromatogr.A*, **1996**, *719*, 195–201.

SAMPLE

Matrix: proteins
Analyte: amino acids
Sample preparation: For N-terminal amino acid. Add 200 pmole protein and 200 pmole nor-leucine to a tube, dry under reduced pressure, dissolve in 40 μL pH 9.5 lithium carbonate buffer, add 20 μL 1.5 mg/mL dansyl chloride in MeCN, shake for 2 min, heat at 37° for 40 min, add 2 μL 8.9 M ethylamine hydrochloride, heat at 37° for 10 min, evaporate to dryness under reduced pressure at room temperature, reconstitute with 100 μL constant boiling HCl, heat at 110° for 18 h, evaporate to dryness over solid NaOH, reconstitute with 100 μL initial mobile phase, inject an aliquot. For amino acid mixtures. Add 20 nmole amino acid mixture to a tube, dry under reduced pressure, dissolve in 40 μL pH 9.5 lithium carbonate buffer, add 20 μL 1.5 mg/mL dansyl chloride in MeCN, shake for 2 min, heat at 37° for 40 min, add 2 μL 8.9 M ethylamine hydrochloride, heat at 37° for 10 min, evaporate to dryness under reduced pressure at room temperature, reconstitute with 100 μL initial mobile phase, inject an aliquot.

HPLC VARIABLES

Guard column: 50 mm long 37-50 μm C18/Corasil
Column: 150 × 3.9 4 μm NovaPak C18
Mobile phase: Gradient. A was MeOH:THF:100 mM pH 7.4 sodium phosphate buffer 5:6.5:88.5. B was MeOH:water 70:30. A:B from 100:0 to 0:100 over 30 min (Waters curve no.6), return to initial conditions over 10 min, re-equilibrate at initial conditions for 10 min. (At the end of each day wash column with 35 mL water and 20 mL MeOH:water 70:30.)
Column temperature: 25
Flow rate: 1
Detector: F ex 338 (bandpass filter) ex 455 (long-pass filter)

CHROMATOGRAM
Retention time: 4.5 (Asp), 5.5 (Glu), 12 (Ser), 12.5 (Arg), 13 (Thr), 14 (Gly), 14.7 (Ala), 15 (Pro), 18.5 (Val), 19 (ammonia), 20 (Met), 21.5 (Ile), 22 (Leu), 23.5 (Phe), 25 (ethylamine), 26 (cystine, bis-derivative), 29 (Lys, bis-derivative), 30.5 (His, bis-derivative), 31 (Tyr, bis-derivative)
Internal standard: norleucine (23)
Limit of detection: 2 pmole

REFERENCE
Martins, A.R.; Padovan, A.P. A practical approach to improve the resolution of dansyl-amino acids by high-performance liquid chromatography, *J.Liq.Chrom.Rel.Technol.*, **1996**, *19*, 467–476.

SAMPLE
Matrix: solutions
Analyte: amino acids
Sample preparation: 20-100 μL Amino acid solution in water + 20-100 μL pH 9.0 Titrisol buffer (Merck) + 20-100 μL 2.7 mg/mL dansyl chloride in water-free acetone, let stand in the dark at room temperature for 1 h, evaporate to dryness under reduced pressure, reconstitute with 100 μL acetone:1 M HCl 95:5, inject a 0.1-1 μL aliquot.

HPLC VARIABLES
Column: 500 × 3 5 μm LiChrosorb SI 60
Mobile phase: Gradient. A was benzene:pyridine:acetic acid 100:10:1. B was pyridine:acetic acid 90:9. A:B from 100:0 to 0:100 over 25 min in a complex fashion. The gradient was obtained by adding 15 mL B over 15 min to a reservoir containing 50 mL A, then adding 15 mL B over 5 min. While these additions were happening the mobile phase was pumped out of the reservoir at 1 mL/min. Re-equilibrate at initial conditions for 5 min before the next injection. (Caution! Benzene is a carcinogen!)
Column temperature: 65
Flow rate: 1
Injection volume: 0.1–1
Detector: F ex 340 em 510

CHROMATOGRAM
Retention time: 4 (isoleucine), 5 (leucine), 6 (valine), 7.5 (proline), 10 (phenylalanine), 12 (methionine), 13 (alanine), 14 (lysine, didansyl), 15 (tyrosine), 17 (glycine), 18 (tryptophan), 19 (glutamic acid), 20 (threonine), 22 (serine), 23 (aspartic acid), 2.5 (cystine, mobile phase benzene:pyridine:acetic acid:MeOH 100:50:0.5:50), 3 (histidine, mobile phase benzene:pyridine:acetic acid:MeOH 100:50:0.5:50), 5 (arginine, mobile phase benzene:pyridine:acetic acid:MeOH 100:50:0.5:50)

KEY WORDS
normal phase

REFERENCE
Bayer, E.; Grom, E.; Kaltenegger, B.; Uhmann, R. Separation of amino acids by high performance liquid chromatography, *Anal.Chem.*, **1976**, *48*, 1106–1109.

SAMPLE
Matrix: solutions
Analyte: amino acids
Sample preparation: Rapidly mix 2 mL 0.001-1 mM amino acid solution in buffer with 1 mL 1.5 mg/mL dansyl chloride in MeCN, shake gently for 2 min, let stand at room temperature for 35 min, add 100 μL 2% ethylamine hydrochloride, inject a 25 μL aliquot. (Buffer was 40 mM lithium carbonate adjusted to pH 9.5 with HCl. Distil MeCN from dansyl chloride to remove trace impurities. Protect reaction from light.)

HPLC VARIABLES
Column: 150 × 4.6 5 μm Supelcosil C8
Mobile phase: MeOH:buffer 42:58 (Buffer was 0.6% glacial acetic acid containing 0.008% triethylamine.)

Flow rate: 2
Injection volume: 25
Detector: F ex 250 em 470

CHROMATOGRAM
Retention time: 2.5 (Asn), 3.5 (ammonia), 4 (Asp), 5.5 (Thr), 7 (Ala), 17 (Met)
Limit of detection: 1 pmole

REFERENCE
Tapuhi, Y.; Schmidt, D.E.; Lindner, W.; Karger, B.L. Dansylation of amino acids for high-performance liquid chromatography analysis, *Anal.Biochem.*, **1981**, *115*, 123–129.

SAMPLE
Matrix: solutions
Analyte: amino acids
Sample preparation: 2 mL 1 mM Amino acids in 40 mM pH 9.5 lithium carbonate buffer + 1 mL 5.56 mM dansyl chloride in MeCN, shake gently for 2 min, let stand at room temperature for 35 min, add 100 μL 2% ethylamine, inject a 20 μL aliquot. (Purify MeCN by distillation from dansyl chloride.)

HPLC VARIABLES
Guard column: ODS-5 (Altex)
Column: 250 × 4.6 5 μm Ultrasphere ODS
Mobile phase: Gradient. MeCN:30 mM pH 7.6 sodium phosphate buffer 10:90 for 0.1 min, to 45:55 over 22.9 min, return to initial conditions over 7 min.
Flow rate: 2
Injection volume: 20
Detector: UV 250

CHROMATOGRAM
Retention time: 9.5 (Asp), 10.5 (Glu), 17 (Asn, Hyp), 18.5 (Ser), 14.8 (Thr), 15.2 (Gly), 15.5 (Ala), 16 (Pro), 17 (Val), 18 (Met), 18.5 (Leu, Ile), 19 (Phe), 19.3 (Trp), 19.5 (Cys), 24 (Lys), 26 (Thr)
Limit of detection: 50-500 pmole

KEY WORDS
comparison with other derivatization procedures; protect from light

REFERENCE
McClung, G.; Frankenberger, W.T. Jr. Comparison of reverse-phase high-performance liquid chromatographic methods for precolumn-derivatized amino acids, *J.Liq.Chromatogr.*, **1988**, *11*, 613–646.

SAMPLE
Matrix: solutions
Analyte: amino acids
Sample preparation: Prepare dansyl derivatives, inject an aliquot.

CAPILLARY ELECTROPHORESIS
Capillary: 34 cm × 25 μm fused-silica (19 cm to detector) (Polymicro Technologies)
Capillary preparation: Before each run rinse with MeOH, water, and separation buffer
Running buffer: 25 mM pH 2.40 sodium phosphate buffer containing 100 mM Tween 20
Voltage/Current: 16 kV
Detector: UV 214
Migration time: 6 (Arg), 10 (Asn), 11 (Gln), 11.5 (Ser), 12 (Thr), 13 (Hyp), 15.5 (Glu), 16 (Gly), 16.5 (Ala), 17.5 (Lys, didansyl), 20.5 (Aba), 23 (Pro), 26 (Tyr, O-dansyl derivative), 27 (Val), 32.5 (Nval, Met), 42 (Ile), 45 (Phe), 47 (Leu), 65 (Trp), 67 (Asp), 70 (Cys, didansylcystine)

REFERENCE
Matsubara, N.; Terabe, S. Separation of 24 dansylamino acids by capillary electrophoresis with a nonionic surfactant, *J.Chromatogr.A*, **1994**, *680*, 311–315.

SAMPLE
Matrix: solutions

Analyte: amino acids
Sample preparation: Dissolve dansylated amino acids in MeOH, dilute with water so that the MeOH concentration does not exceed 7.5%, inject an aliquot.

CAPILLARY ELECTROPHORESIS
Capillary: 57.5 cm × 50 μm fused-silica (50 cm to detector)
Capillary preparation: Rinse with running buffer for 2 min before each run.
Running buffer: 20 mM pH 9.2 borax buffer containing 102.5 mM sodium dodecyl sulfate
Capillary temperature: 10
Voltage/Current: 25 kV
Injection: Hydrodynamic injection for 3 s
Detector: UV 214
Model: SpectraPhoresis 500 (Therm Separation Products)
Migration time: 9.21 (Hyp), 10.01 (Thr), 10.21 (Ser), 10.30 (Asn), 10.46 (Gln), 10.61 (Ala), 11.05 (Aba), 11.46 (Gly), 11.72 (Glu), 11.90 (Val), 12.40 (Asp), 12.69 (cysteic acid), 13.10 (Pro), 13.48 (Met), 14.56 (Ile), 15.21 (Leu), 17.27 (Phe), 17.81 (Trp), 19.78 (Cys, dansyl), 23.02 (Lys), 23.62 (Arg), 24.48 (Lys, dansyl), 24.68 (Tyr, dansyl)

REFERENCE
Skocir, E.; Vindevogel, J.; Sandra, P. Separation of 23 dansylated amino acids by micellar electrokinetic chromatography at low temperature, *Chromatographia*, **1994**, *39*, 7–10.

SAMPLE
Matrix: solutions
Analyte: amino acids
Sample preparation: Rapidly mix 1 mL 10 mM dansyl chloride in MeCN with 2 mL amino acid solution in buffer, sonicate for 10 min in the dark, let stand at room temperature in the dark for 40 min, inject a 25 μL aliquot. (Buffer was 40 mM lithium carbonate adjusted to pH 9.5 with HCl.)

HPLC VARIABLES
Guard column: 70 × 2 Partisil ODS
Column: 250 × 4.6 5 μm Zorbax ODS
Mobile phase: MeCN:50 mM pH 7.5 phosphate buffer 17:83 containing 0.5 mM tris(2,2'-bipyridyl)ruthenium(II) chloride (Ru(bpy)$_3^{2+}$)
Flow rate: 1
Injection volume: 25
Detector: chemiluminescence following post-column electrochemical oxidation of Ru(bpy)$_3^{2+}$ to Ru(bpy)$_3^{3+}$. The column effluent flowed through a Bioanalytical Systems BAS 100A electrochemical detector with a Pt electrode at 1250 mV (relative to an Ag/AgCl reference electrode) to the chemiluminescence detector.

CHROMATOGRAM
Retention time: 13 (Glu), 23 (Asn), 32 (Ser), 40 (Thr), 46 (Gly), 52 (Ala)
Limit of detection: 100 nM

REFERENCE
Skotty, D.R.; Lee, W.-Y.; Nieman, T.A. Determination of dansyl amino acids and oxalate by HPLC with electrogenerated chemiluminescence detection using tris(2,2'-bipyridyl)ruthenium(II) in the mobile phase, *Anal.Chem.*, **1996**, *68*, 1530–1535.

SAMPLE
Matrix: solutions
Analyte: amino acids
Sample preparation: Mix 50 μL of a 1 mM amino acid solution in 100 mM pH 9.5 borate buffer with 100 μL 100 mM pH 9.5 borate buffer and 100 μL 8 mM dansyl chloride in MeCN, let stand at room temperature for 1 h, filter (0.45 μm PTFE), inject an aliquot.

CAPILLARY ELECTROPHORESIS
Capillary: 50 cm × 50 μm fused silica (41.5 cm to detector)
Running buffer: N-Methylformamide containing 10 mM NaCl and 100 mM β-cyclodextrin
Capillary temperature: 25
Voltage/Current: 30 kV

Injection: Pressure injection at 50 mbar for 5 s
Detector: UV 254
Model: Hewlett-Packard HP3D
Migration time: 12.45 (D-Pro), 23.74 (L-Pro), 12.13 (D-Ala), 12.90 (L-Ala), 12.53 (D-Ser), 13.36 (L-Ser), 11.10 (D-Lys), 11.34 (L-Lys), 12.64 (D-Asn), 15.00 (D-Leu), 15.77 (L-Leu), 12.23 (D-Nval), 12.84 (L-Nval), 14.66 (D-Asp), 15.54 (L-Asp), 12.87 (D-Met), 13.56 (L-Met), 12.99 (D-Thr), 14.17 (L-Thr), 12.38 (D-Val), 13.08 (L-Val)

KEY WORDS
chiral

REFERENCE
Valkó, I.E.; Sirén, H.; Riekkola, M.-L. Chiral separation of dansyl-amino acids in a non-aqueous medium by capillary electrophoresis, *J.Chromatogr.A*, **1996**, *737*, 263–272.

SAMPLE
Matrix: tissue
Analyte: hydroxyproline
Sample preparation: Hydrolyze 10-50 mg tissue with 6 M HCl in an evacuated sealed tube at 110° for 24 h, cool, adjust pH to 6.7 with NaOH, centrifuge at 27000 g for 10 min. Remove a 70 μL aliquot of the supernatant and add it to 20 μL 375 mM dansyl chloride in MeCN, add 60 μL 1.5 M pH 11.0 lithium carbonate, heat at 60° for 15 min, centrifuge at 27000 g for 10 min, inject an aliquot.

HPLC VARIABLES
Column: 250 × 4.5 C18 (IBM)
Mobile phase: MeOH:10 mM Tris-HCl 20:80:100 mM Tris-HCl 43.3:53.3:3.3, pH 7.75
Flow rate: 1
Detector: F (wavelengths not given)

CHROMATOGRAM
Retention time: 23
Internal standard: gamma-aminobutyric acid (37), Nleu (55)
Limit of quantitation: 100 pmole

REFERENCE
Takahashi, S.; Lee, M.-J. Quantitative study of tissue collagen metabolism, *Anal.Biochem.*, **1987**, *162*, 553–561.

SAMPLE
Matrix: tissue
Analyte: amino acids
Sample preparation: Homogenize (Polytron for >10 mg; Kontes micro-ultrasonic cell disrupter for <10 mg) tissue with 40 volumes 25 μg/mL β-aminoisobutyric acid in EtOH:water:glacial acetic acid 75:20:5, centrifuge at 4° at 25000 g for 20 min. Remove a 50 μL aliquot of the supernatant and evaporate it to dryness under reduced pressure, suspend the residue in 100 μL 100 mM sodium bicarbonate by sonicating or vortexing, add 200 μL 1.25 mg/mL dansyl chloride in acetone, vortex, heat at 90° for 30 min, centrifuge at 5000 g for 20 min, inject a 4 μL aliquot of the supernatant.

HPLC VARIABLES
Column: 75 × 4.6 3 μm Ultrasphere ODS
Mobile phase: MeCN:water:phosphoric acid 13:87:0.15
Flow rate: 1
Injection volume: 4
Detector: UV 254

CHROMATOGRAM
Retention time: k' 0.61 (cysteic acid), k' 0.93 (glutathione), k' 0.98 (ethanolamine), k' 1.24 (asparagine), k' 1.26 (taurine), k' 1.30 (methionine), k' 1.71 (ammonia), k' 1.58 (glutamine), k' 1.94 (cystathionine), k' 1.95 (leucine), k' 1.95 (lysine), k' 1.96 (isoleucine), k' 2.01 (cysteine), k' 2.14 (proline), k' 2.43 (homocarnosine), k' 2.45 (urea), k' 2.54 (arginine), k' 2.54 (hydroxyproline), k' 2.70 (glutamic acid), k' 2.86 (aspartic acid), k' 2.86 (serine), k' 3.65 (threonine), k' 4.01 (glycine), k' 4.55 (norvaline), k' 5.91 (alanine), k' 5.93 (valine), k' 6.48 (GABA), k' 8.58 (6-aminocaproic acid), k' 11.42 (α-aminobutyric acid), k' 16.22 (tryptophan), k'22.13 (tyrosine)

Internal standard: β-aminoisobutyric acid (k' 9.25)
Limit of quantitation: 10 pmole

KEY WORDS
rat; brain

REFERENCE
Saller, C.F.; Czupryna, M.J. gamma-Aminobutyric acid, glutamate, glycine and taurine analysis using reversed-phase high-performance liquid chromatography and ultraviolet detection of dansyl chloride derivatives, *J.Chromatogr.*, **1989**, *487*, 167–172.

SAMPLE
Matrix: tissue
Analyte: amino acids
Sample preparation: Homogenize mouse liver with chloroform:MeOH 2:1, extract the homogenate with water. Lyophilize the aqueous phase, reconstitute with water. Remove a 100 μL aliquot, add 10 μL 10 mM NaOH, add 100 μL 100 mM pH 9.0 borate buffer, add 100 μL 1 mM dansyl chloride in MeCN, vortex, heat at 40° for 45 min, cool to room temperature, inject a 75 μL aliquot.

HPLC VARIABLES
Column: 250 × 4.6 LiChrosorb RP-18
Mobile phase: Gradient. A was MeCN:10 mM pH 7.0 phosphate buffer 10:90. B was MeCN:10 mM pH 7.0 phosphate buffer 50:50. A:B from 75:25 to 30:70 over 20 min, maintain at 30:70 for 10 min, return to initial conditions over 0.1 min, re-equilibrate for 10 min.
Flow rate: 3
Injection volume: 75
Detector: F ex 330 em 565; UV 254

CHROMATOGRAM
Retention time: 8.24 (tyrosine), 11.9 (tryptophan)

KEY WORDS
mouse; liver

REFERENCE
Manwaring, J.D.; Csallany, A.S. Identification of vitamin E-dependent water soluble fluorescent compounds in mouse tissues, *Lipids*, **1990**, *25*, 22–26.

SAMPLE
Matrix: tissue
Analyte: glutathione and cysteine
Sample preparation: Homogenize (PTFE/glass homogenizer) 200 mg tissue with 10 mL buffer at 0-4°, centrifuge at 4° at 10000 g for 10 min. Remove the supernatant and add it to 4 nmoles IS, add 1 μmole iodoacetic acid, adjust pH to 8.0-8.5 with 2 M LiOH (indicator color change from yellow to mauve), mix, let stand at 25° in the dark for 30 min, add an equal volume of 1 mg/mL dansyl chloride in acetone, mix, let stand at 25° for 1 h, inject an aliquot. (To increase sensitivity add an equal volume of chloroform to this mixture, vortex, centrifuge at 3000 g for 5 min, inject an aliquot of the upper aqueous phase.) (Buffer was 5% perchloric acid containing 2 mM diethylenetriaminepentaacetic acid, 200 mM boric acid, and 5 μg/mL cresol red indicator.)

HPLC VARIABLES
Guard column: 4 × 4 5 μm LiChrospher aminopropyl
Column: 125 × 4 5 μm LiChrospher aminopropyl
Mobile phase: Gradient. A was MeOH:water 80:20. B was MeOH:water:buffer 72:18:10. (Buffer was 272 g sodium acetate trihydrate, 378 mL glacial acetic acid, and 122 mL water.) A:B from 100:0 to 0:100 over 20 min, maintain at 0:100 for 10 min.
Flow rate: 1.3
Detector: F ex 328 em 541

CHROMATOGRAM
Retention time: 14.6 (cystine), 15.7 (cysteine), 21.7 (glutathione), 26.0 (oxidized glutathione)

Internal standard: gamma-glutamylglutamine
Limit of detection: 1 pmole

KEY WORDS
rat; lung; protect from light

REFERENCE
Martin, J.; White, I.N. Fluorimetric determination of oxidised and reduced glutathione in cells and tissues by high-performance liquid chromatography following derivatization with dansyl chloride, *J.Chromatogr.*, **1991**, *568*, 219–225.

SAMPLE
Matrix: tissue
Analyte: amino acids
Sample preparation: Heat 10-30 mg tissue protein in 10 mL 6 M HCl at 120° for 20 h, dry with air, reconstitute with water, dry with air, reconstitute with 2 mL water, add 10 μmoles norvaline, add 5 μmoles 3,4-dimethoxyphenylammonium chloride. Remove a 50 μL aliquot and add it to 1 mL 25 mM dansyl chloride in MeCN and 1 mL 40 mM pH 9.5 lithium carbonate, let stand in the dark for 2 h, add 100 μL 4% ethylamine, dry with air, reconstitute with MeOH:water 20:80, centrifuge at 2000 g for 2 min, inject an aliquot.

CAPILLARY ELECTROPHORESIS
Capillary: 72 cm × 50 μm fused-silica (52 cm to detector) (J&W Scientific)
Capillary preparation: Before each run wash capillary with 1 M NaOH for 2 min, with water for 2 min, and with running buffer for 5 min.
Running buffer: 100 mM pH 8.3 Boric acid containing 150 mM sodium dodecyl sulfate
Capillary temperature: 25
Voltage/Current: 15 kV
Injection: Vacuum injection for 1 s (4.5 nL)
Detector: UV 216
Model: Applied Biosystems Model 270 A-HT
Migration time: 18.9 (Thr), 19.1 (Asn), 19.2 (Ser), 19.5 (Gln), 20.2 (Ala), 20.6 (Glu), 21.3 (Asp), 21.6 (Gly), 23.2 (Val), 25.5 (Pro), 26.0 (Met), 28.6 (Ile), 29.7 (Leu), 33.0 (Phe), 33.7 (Trp), 43.6 (Arg), 44.9 (Lys, didansyl), 45.3 (tryptamine), 46.0 (Tyr, didansyl)
Internal standard: norvaline (25.3)
Limit of detection: 0.1-1.3 μM

KEY WORDS
skin; mink

REFERENCE
Michaelsen, S.; Moller, P.; Sorensen, H. Analysis of dansyl amino acids in feedstuffs and skin by micellar electrokinetic capillary chromatography, *J.Chromatogr.A*, **1994**, *680*, 299–310.

4-(5,6-Dimethoxy-2-phthalimidinyl)phenylsulfonyl Chloride

SAMPLE
Matrix: blood
Analyte: proline and hydroxyproline
Sample preparation: 10 μL Serum + 20 μL 50 μM nipecotic acid + 270 μL 100 mM pH 8.5 borate buffer + 50 μL 4% o-phthalaldehyde in acetone, mix, let stand for 3 min, add 450 μL 1.2 mM 4-(5,6-dimethoxy-2-phthalimidinyl)phenylsulfonyl chloride in acetone, heat at 30° for 10 min, cool, add 800 μL dichloromethane, vortex, centrifuge at 500 g for 10 min, inject an aliquot of the aqueous layer. (o-Phthalaldehyde removes primary amino acids from the chromatogram but does not affect proline or hydroxyproline. 4-(5,6-Di-methoxy-2-phthalimidinyl)phenylsulfonyl chloride is available from Dojindo Molecular Technologies, Inc., 3 Bethesda Metro Center, Suite 700, Bethesda MD 20814; (301) 664-8448; www.dojindo.co.jp. Synthesis is as follows. Saturate 230 mL formaldehyde (37% solution) with HCl gas at 15-20°, add 32 g 3,4-dimethoxybenzoic acid, heat to 60-70° for 7 h and slowly bubble HCl gas through the mixture, cool, let stand overnight, concentrate under reduced pressure, add 100 mL water, neutralize with 2:3 dilute ammonium hydroxide. Collect the solid and wash it with water, dry, recrystallize from MeOH to give m-meconin (mp 154-5°). Reflux 7.5 g lithium aluminum hydride in 80 mL THF and add a warm solution of 14 g m-meconin in 150 mL anhydrous THF slowly, reflux for 4 h, cool in an ice bath, stir, add 7.5 mL water dropwise, add 7.5 mL 15% NaOH, add 22.5 mL water, extract the precipitate with hot THF. Evaporate the THF layers and extract the residue with chloroform. Evaporate the chloroform extract to give 4,5-dimethoxyphthalyl alcohol, recrystallize from benzene (mp 110-1°) (Caution! Benzene is a carcinogen!). Vigorously stir 20 g activated manganese dioxide in 200 mL chloroform, add 3 g 4,5-dimethoxyphthalyl alcohol, stir at room temperature for 3 days, filter, wash the precipitate with chloroform, evaporate the filtrate to give 4,5-dimethoxyphthalaldehyde, recrystallize from benzene (mp 168-9°) (J.Pharm.Sci. 1980, 69, 120). Add 0.96 g aniline in 20 mL dioxane to 2.0 g 4,5-dimethoxyphthalaldehyde stirred in 230 mL dioxane (Caution! Dioxane is a carcinogen!), stir at room temperature for 3 days, concentrate to about 5 mL under reduced pressure, add diethyl ether, filter the solid, recrystallize 5,6-dimethoxy-2-phenylphthalimide from EtOH (colorless plates, mp 198.7-199.3°). Dissolve 500 mg 5,6-dimethoxy-2-phenylphthalimide in portions in 1 mL chlorosulfonic acid cooled in an ice bath, stir at 50° for 5 min, cool to room temperature, pour dropwise onto 20 g crushed ice, add 200 mL water, extract three times with 150 mL portions of chloroform. Combine the extracts and wash them three times with 200 mL portions of water, dry over anhydrous sodium sulfate for 2 h, evaporate to dryness, chromatograph on a 120 × 40 column of silica gel (Wakogel C-200) with chloroform, collect the non-fluorescent fraction (second 200 mL) and evaporate it to dryness under reduced pressure. Recrystallize 4-(5,6-dimethoxy-2-phthalimidinyl)phenylsulfonyl chloride from MeCN as white needles (mp 214.4°).)

HPLC VARIABLES
Guard column: 15 × 3.2 TSK Guardgel ODS-80TM (Tosoh)
Column: 150 × 3.9 4 μm Nova-Pak C18

Mobile phase: Gradient. MeCN:1 mM pH 7 phosphate buffer from 90:10 to 78:22 over 12 min, to 20:80 (step gradient), maintain at 20:80 for 5 min, re-equilibrate at initial conditions for 5 min.
Column temperature: 25
Flow rate: 1
Injection volume: 10
Detector: F ex 315 em 385

CHROMATOGRAM
Retention time: 6.8 (hydroxyproline), 9.0 (proline)
Internal standard: nipecotic acid (12.0)
Limit of detection: 10 fmole

KEY WORDS
serum

REFERENCE
Inoue, H.; Moritani, K.; Date, Y.; Kohashi, K.; Tsuruta, Y. Determination of free hydroxyproline and proline in human serum by high-performance liquid chromatography using 4-(5,6-dimethoxy-2-phthalimidinyl)phenylsulfonyl chloride as a pre-column fluorescent labelling reagent, *Analyst*, **1995**, *120*, 1141–1145.

SAMPLE
Matrix: blood, urine
Analyte: secondary amino acids
Sample preparation: Condition a 1 mL 100 mg Bond Elut C18 SPE cartridge with 2 mL MeOH, 2 mL water, and 400 μL 100 mM pH 8.5 borate buffer. 10 μL Serum or urine + 100 μL concentrated HCl + 100 μL 2 (serum) or 0.05 (urine) mM nipecotic acid, heat at 120° for 16 h, cool, add 300 μL 2 M sodium carbonate. Remove a 100 μL aliquot and add it to 300 μL 100 mM pH 8.5 borate buffer and 100 μL 4% o-phthalaldehyde in MeCN: 100 mM pH 8.5 borate buffer 50:50, mix, let stand at room temperature for 3 min, add a 200 μL aliquot to the SPE cartridge, elute with two 400 μL portions of borate buffer, mix all the effluent from the SPE cartridge. Remove a 300 μL aliquot and add it to 500 μL 1.2 mM 4-(5,6-dimethoxy-2-phthalimidinyl)phenylsulfonyl chloride in acetone, heat at 30° for 10 min, cool, add 800 μL dichloromethane, vortex, centrifuge at 500 g for 10 min, inject a 20 μL aliquot of the aqueous layer. (o-Phthalaldehyde removes primary amino acids from the chromatogram but does not affect proline, hydroxyproline, or sarcosine. 4-(5,6-Dimethoxy-2-phthalimidinyl)phenylsulfonyl chloride is available from Dojindo Molecular Technologies, Inc., 3 Bethesda Metro Center, Suite 700, Bethesda MD 20814; (301) 664-8448; www.dojindo.co.jp. Synthesis is as described above.)

HPLC VARIABLES
Guard column: 15 × 3.2 TSK Guardgel ODS-80TM (Tosoh)
Column: 150 × 3.9 4 μm Nova-Pak C18
Mobile phase: Gradient. MeCN:1 mM pH 7 phosphate buffer from 12:88 to 20:80 over 15 min, to 80:20 (step gradient), maintain at 80:20 for 5 min, re-equilibrate at initial conditions for 5 min.
Column temperature: 25
Flow rate: 1
Injection volume: 20
Detector: F ex 315 em 385

CHROMATOGRAM
Retention time: 5.7 (hydroxyproline), 8.9 (proline), 9.7 (sarcosine)
Internal standard: nipecotic acid (14.2)
Limit of detection: 1 μM

KEY WORDS
serum; SPE

REFERENCE
Inoue, H.; Date, Y.; Kohashi, K.; Yoshitomi, H.; Tsuruta, Y. Determination of total hydroxyproline and proline in human serum and urine by HPLC with fluorescence detection, *Biol.Pharm.Bull.*, **1996**, *19*, 163–166.

4-(N-Phthalimidyl)benzenesulfonyl Chloride

SAMPLE
Matrix: solutions
Analyte: amino acids
Sample preparation: Mix 5 μL of a 0.05-10 mM solution with 20 μL 100 mM NaOH and 150 μL 5 mM 4-(N-phthalimidyl)benzenesulfonyl chloride in acetone, heat at 50° for 15 min, dilute 10-fold with mobile phase, inject a 20 μL aliquot. (Synthesis of 4-(N-phthalimidyl)benzenesulfonyl chloride (Phisyl-Cl) is as follows. Mix 2.68 g o-phthalaldehyde in 100 mL diethyl ether with 1.86 g aniline in 20 mL diethyl ether, stir at room temperature overnight, filter. Wash the solid with diethyl ether and recrystallize it from MeOH to yield N-phenylphthalimidine. Drop 6.6 g chlorosulfonic acid onto 2.09 g of crystals of N-phenylphthalimidine in an ice bath with vigorous stirring over 20 min, heat at 60° for 2 h, add 30 g crushed ice, recrystallize the precipitate from benzene to obtain 4-(N-phthalimidyl)benzenesulfonyl chloride as fine colorless needles (mp 186-187°) (Caution! Benzene is a carcinogen!).)

HPLC VARIABLES
Guard column: 25 × 4.6 5 μm ODS (Yamamura, Japan)
Column: 250 × 4.6 5 μm YMC AM-303 ODS (Yamamura)
Mobile phase: Gradient. A was MeCN:30 mM pH 6.5 Tris buffer 10:90. B was MeCN:30 mM pH 6.5 Tris buffer 75:25. A:B 100:0 for 8 min, 90:10 for 12 min, 80:20 for 18 min, 70:30 for 10 min, 55:45 for 5 min, 40:60 for 7 min, 0:100 for 5 min, re-equilibrate at initial conditions for 15 min (step gradient).
Flow rate: 0.6
Injection volume: 20
Detector: F ex 295 em 425

CHROMATOGRAM
Retention time: 18.5 (cysteine), 19 (aspartic acid), 20 (glutamic acid), 30 (hydroxyproline), 32.5 (asparagine), 35 (serine), 36 (methionine), 36.5 (threonine), 38 (glycine), 39 (alanine), 43 (proline), 47 (valine), 51 (isoleucine), 52.5 (leucine), 55 (phenylalanine), 56.5 (cystine), 59.8 (ornithine), 60.2 (lysine), 61 (histidine), 63 (tyrosine)
Limit of detection: <0.2 pmole

REFERENCE
Tsuruta, Y.; Date, Y.; Kohashi, K. Phthalimidylbenzenesulfonyl chlorides as fluorescence labeling reagents for amino acids in high-performance liquid chromatography, *J.Chromatogr.*, **1990**, *502*, 178–183.

THIOACID

Thioacetylthioglycolic Acid

SAMPLE

Matrix: protein

Analyte: amino acids

Sample preparation: Treat 10 nmoles protein immobilized on 5 mg p-phenylenediisothio-
cyanate glass with 20 mg thio-acetylated glass beads (Sigma), 500 μL 10% thioacetylthioglycolic acid in MeOH, and 200 μL 20% triethylamine in MeOH with constant agitation at 45° for 30 min, wash with 2 mL 20% triethylamine in MeOH, wash with 3 mL
MeOH, wash with 3 mL dichloromethane, dry under vacuum for 15 min, add 500 μL
trifluoroacetic acid, let stand at room temperature for 15 min. Remove the acid, wash the
solid three times with 200 μL portions of dichloromethane. Combine the acid and the
washes and add 100 μL 4-nitrobenzenesulfonyl chloride (2.5 eq) in dichloromethane.
Evaporate to dryness under a stream of nitrogen at 35°, reconstitute the residue in 500
μL dichloromethane, add 50 μL 20% triethylamine in dichloromethane, concentrate immediately under reduced pressure, reconstitute, inject a 20 μL aliquot. (Synthesis of
thioacetylthioglycolic acid is as follows. Reflux 8.8 g acetaldehyde, 30 mL piperidine, and
9.6 g sulfur for 1 h, cool, add water, acidify with concentrated HCl, crush the solid, wash
with water, recrystallize from 200 mL EtOH to give thioacetylpiperidine (mp 55-56°).
Dissolve 10 g thioacetylpiperidine in 50 mL dry benzene (Caution! Benzene is a carcinogen!), add 10.7 g bromoacetic acid, let stand at room temperature for 24 h, add 150 mL
dry ether, recrystallize from EtOH/ether to give S-carboxyethylthiomethylpiperidinium
bromide (mp 168-169°). Dissolve 10.1 g S-carboxyethylthiomethylpiperidinium bromide in
40 mL EtOH, cool in ice, pass hydrogen sulfide through the solution at 2-3 bubbles/s for
3-4 h (Caution! Hydrogen sulfide is highly toxic!), let stand overnight at 0°, remove EtOH
by evaporation under reduced pressure, repeatedly extract the residue with ether until
it is colorless. Evaporate the ether solution to dryness under reduced pressure, recrystallize from light petroleum (bp 60-100°) to obtain thioacetylthioglycolic acid (mp 80-81°)
(Acta Chem. Scand. 1961, 15, 1087). 4-Acetylmorpholine can also be used as the starting
reagent instead of acetaldehyde and piperidine (acetylpiperidine) (Anal. Biochem. 1989,
181, 113).)

HPLC VARIABLES

Column: 150 × 4.6 5 μm IBM C18

Mobile phase: Gradient. MeCN:MeOH:buffer from 35:0:65 to 42:0:58 over 4.9 min, to
36:9:55 over 0.1 min, to 60:15:25 over 10 min, maintain at 60:15:25 for 2 min, to 35:

0:65 over 0.1 min, maintain at 35:0:65 for 3.9 min. (Prepare buffer by dissolving 10 mL acetic acid, 4 mL triethylamine, and 2 mL trifluoroacetic acid, pH 3.1.)

Flow rate: 1.5

Injection volume: 20

Detector: UV 248

CHROMATOGRAM

Retention time: 3.2 (Asn), 3.7 (Gln), 4.6 (Asp), 5.0 (Arg), 5.9 (Glu), 8.3 (Gly), 9.3 (Ala), 10.0 (Lys), 12.6 (Met), 13.4 (Val), 13.9 (Trp), 14.7 (Phe), 15.0 (Leu, Ile), 17.0 (Tyr)

REFERENCE

Stolowitz, M.L.; Paape, B.A.; Dixit, V.M. Thioacetylation method of protein sequencing: derivatization of 2-methyl-5(4*H*)-thiazolones for high-performance liquid chromatographic detection, *Anal.Biochem.*, **1989**, *181*, 113–119.

AMINO ALCOHOL

ACYL HALIDE

Phosgene

SAMPLE
Matrix: blood
Analyte: propranolol
Sample preparation: 1 mL Plasma + 500 μL 100 mM KOH + 30 μL 2 μg/mL (R,S)-n-pentyl propranolol hydrochloride, vortex, add 7 mL n-hexane:n-butanol 99:1, extract. Remove 6 mL of the organic layer and evaporate it to dryness under a stream of nitrogen, reconstitute the residue in 500 μL 20% phosgene in toluene, add 1 mg 4-dimethylaminopyridine, heat at 40° for 3 h. Evaporate to dryness under a stream of nitrogen, reconstitute the residue in 100 μL mobile phase, inject a 20 μL aliquot.

HPLC VARIABLES
Column: 250 × 4 LiChrosorb Si 100 modified with (R,R)-DACH-DNB (see J. Chromatogr. 1991, 539, 25)
Mobile phase: Dichloromethane:MeOH 99.75:0.25
Flow rate: 1
Injection volume: 20
Detector: F ex 290 em 330

CHROMATOGRAM
Internal standard: (R,S)-n-pentyl propranolol hydrochloride
Limit of detection: 0.5-0.6 ng/mL

KEY WORDS
plasma; chiral

REFERENCE
Stoschitzky, K.; Kahr, S.; Donnerer, J.; Schumacher, M.; Luha, O.; Maier, R.; Klein, W.; Lindner, W. Stereoselective increase of plasma concentrations of the enantiomers of propranolol and atenolol during exercise, *Clin.Pharmacol.Ther.*, **1995**, *57*, 543–551.

AMINO KETONE

DIKETONE

Acetylacetone

RELATED REFERENCES

Endo, Y.; Okayama, A.; Endo, G.; Ueda, T.; Nakazono, N.; Horiguchi, S. Improvement of urinary δ-aminolevulinic acid determination by HPLC and fluorescence detection using condensing reaction with acetylacetone and formaldehyde. *Sangyo Igaku* **1994**, *36*, 49-56.

Nishikawa, Y. Liquid chromatographic determination of aliphatic diamines in water via derivatization with acetylacetone. *J.Chromatogr.* **1987**, *392*, 349-359.

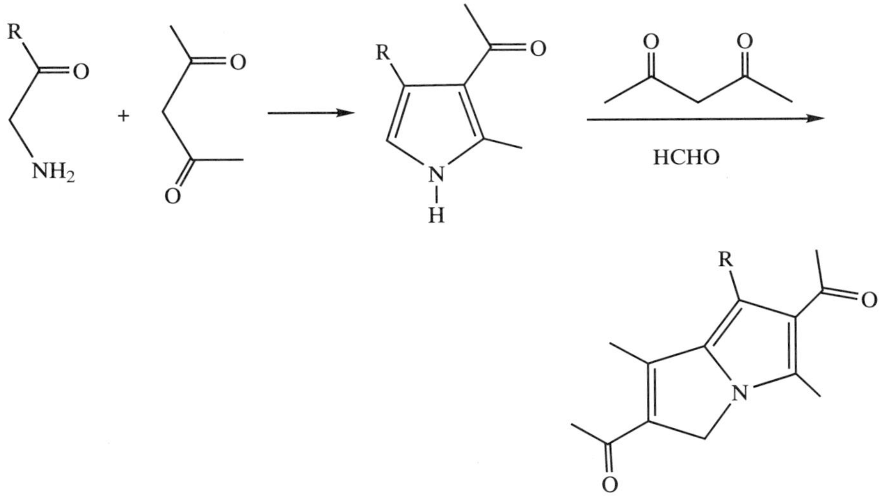

SAMPLE

Matrix: blood, urine

Analyte: delta-aminolevulinic acid

Sample preparation: 50 μL Plasma or urine + 3.5 mL reagent + 450 μL 10% formaldehyde, vortex for 3 s, heat at 100° for 10 min, cool in an ice bath, filter (0.8 μm, plasma samples only), inject a 10 (urine) or 20 (plasma) μL aliquot. (Prepare the reagent by mixing 15 mL acetylacetone, 10 mL EtOH, and 75 mL water.)

HPLC VARIABLES

Column: 150 × 4.6 Shim-pack CLC-ODS (Shimadzu)

Mobile phase: MeOH:water:acetic acid 50:50:1

Column temperature: 40

Flow rate: 0.7

Injection volume: 10-20

Detector: F ex 370 em 460

CHROMATOGRAM

Retention time: 6.1

Limit of detection: 3 ng/mL

KEY WORDS

plasma; protect from light

REFERENCE

Oishi, H.; Nomiyama, H.; Nomiyama, K.; Tomokuni, K. Fluorometric HPLC determination of delta-aminolevulinic acid (ALA) in the plasma and urine of lead workers: Biological indicators of lead exposure, *J.Anal.Toxicol.*, **1996**, *20*, 106–110.

ANHYDRIDE

MISCELLANEOUS REACTIONS

Hydrolysis

SAMPLE
Matrix: solutions
Analyte: anhydrides
Sample preparation: Mix 100 μL of a 20 μg/mL solution in ethyl acetate with 1 mL 100 mM NaOH, let stand for 30 min, make up to 10 mL with water, inject an aliquot. (The anhydride is hydrolyzed to the corresponding acid.)

CAPILLARY ELECTROPHORESIS
Capillary: 64 cm × 50 μm fused-silica (55 cm to detector)
Running buffer: 12.5 mM Sodium borate containing 50 mM sodium dodecyl sulfate, pH adjusted to 9 with 500 mM phosphoric acid
Voltage/Current: 25 kV
Detector: UV 210
Model: Prince
Migration time: 10 (crotonic acid), 10.5 (methacrylic acid), 15 (dimethylmaleic acid), 17 (citraconic acid), 24 (maleic acid)
Limit of quantitation: 100-400 ppb

REFERENCE
He, Y.; Lee, H.K. Analysis of some anhydrides as their corresponding acids by capillary electrophoresis, *Chromatographia*, **1997**, *46*, 67–71.

AZIDE

ACYL HALIDE

Benzoyl Chloride

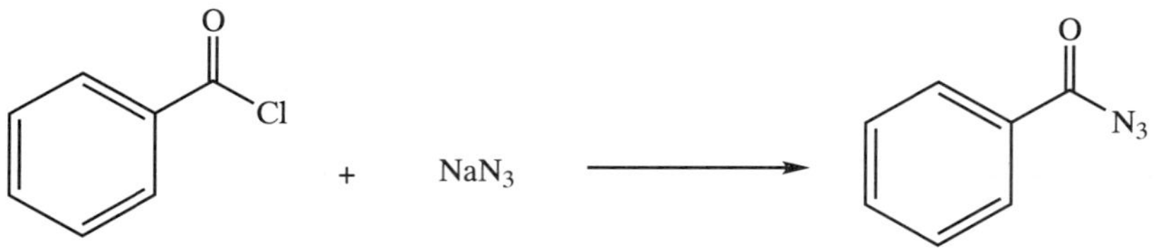

SAMPLE
Matrix: bile, blood, gastric contents, tissue
Analyte: sodium azide
Sample preparation: Homogenize 20 g tissue with 40 mL 1 M pH 7.4 Tris buffer, centrifuge. 1 mL Tissue homogenate, blood, bile, or gastric contents + 1 mL MeCN, shake on an oscillating agitator for 15 min, centrifuge at 900 g for 5 min. Remove the supernatant and add it to 50 μL 1 M pH 4.6 Tris buffer and 50 μL 100 mg/mL benzoyl chloride in MeOH, shake for 15 min, let stand at room temperature for 15 min, inject a 20 μL aliquot.

HPLC VARIABLES
Column: 150 × 4.6 5 μm Novapack C18
Mobile phase: MeCN:50 mM pH 4.6 KH_2PO_4/Tris buffer 45:55
Flow rate: 1.1
Injection volume: 20
Detector: UV 252

CHROMATOGRAM
Retention time: 10
Limit of quantitation: 500 ng/mL
Limit of detection: 200 ng/mL

KEY WORDS
whole blood; liver; brain; kidney; lung

REFERENCE
Marquet, P.; Clément, S.; Lotfi, H.; Dreyfuss, M.-F.; Debord, J.; Dumont, D.; Lachâtre, G. Analytical findings in a suicide involving sodium azide, *J.Anal.Toxicol.*, **1996**, *20*, 134–138.

3,5-Dinitrobenzoyl Chloride

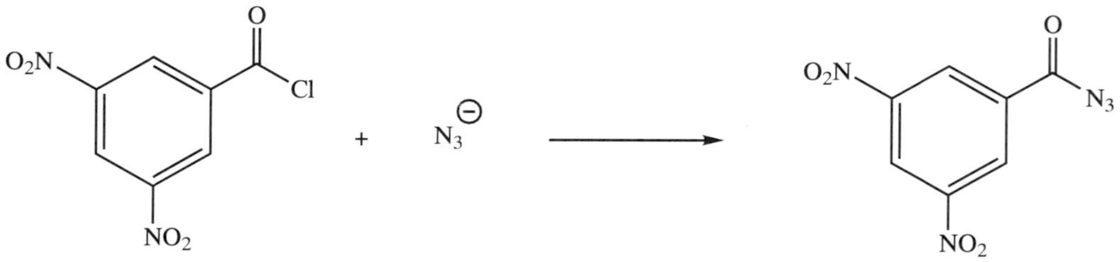

SAMPLE
Matrix: air

Analyte: sodium azide

Sample preparation: Pull air through 10 mM sodium carbonate solution. Remove a 5 mL aliquot and add it to 2 mL MeCN and 5 drops indicator, add 200 mM HCl dropwise until the color changes from blue to yellow, add 1 more drop of 200 mM HCl, add 50 μL freshly-prepared 100 mg/mL 3,5-dinitrobenzoyl chloride in MeCN, shake for several s, let stand for 3 min, inject an aliquot. (Prepare indicator by dissolving 100 mg bromothymol blue in 3.2 mL 200 mM NaOH, dilute to 100 mL.)

HPLC VARIABLES

Guard column: 20 × 3.2 RP-18 microparticulate (Altex)
Column: 250 × 4.6 Zorbax ODS
Mobile phase: MeCN:water 50:50
Flow rate: 1
Injection volume: 27
Detector: UV 254

CHROMATOGRAM

Retention time: 13
Limit of detection: 10 ng/mL

OTHER SUBSTANCES

Interfering: aniline (at high concentrations)
Non-interfering: acetic acid, ferric ions, iodide, MeOH, thiocyanate

REFERENCE

Swarin, S.J.; Waldo, R.A. Liquid chromatographic determination of azide as the 3,5-dinitrobenzoyl derivative, *J.Liq.Chromatogr.*, **1982**, 5, 597–604.

SAMPLE

Matrix: bile, blood, gastric contents, tissue
Analyte: sodium azide

Sample preparation: Gastric contents. 1 mL Gastric contents + 4 mL 10 mM potassium carbonate + 250 μL indicator, adjust pH to 4 by adding 200 mM HCl dropwise until the color changed from blue to yellow, add 50 μL 10 mg/mL 3,5-dinitrobenzoyl chloride in MeCN, shake, let stand for 10 min, inject a 50 μL aliquot. Blood, bile. 1 mL Blood or bile + 2 mL 10 mM potassium carbonate + 1 mL MeCN, vortex for 30 s, centrifuge at 3000 g for 5 min. Remove a 3 mL aliquot of the supernatant and add it to 3 mL 20 mM potassium carbonate and 1 mL MeCN, add 250 μL indicator, adjust pH to 4 by adding 200 mM HCl dropwise until the color changed from blue to yellow, add 50 μL 10 mg/mL 3,5-dinitrobenzoyl chloride in MeCN, shake, let stand for 10 min, inject a 50 μL aliquot. Tissue. Homogenize 1 g tissue with 3 mL 10 mM potassium carbonate, add 1 mL MeCN, vortex for 30 s, centrifuge at 3000 g for 5 min. Remove a 3 mL aliquot of the supernatant and add it to 3 mL 20 mM potassium carbonate and 1 mL MeCN, add 250 μL indicator, adjust pH to 4 by adding 200 mM HCl dropwise until the color changed from blue to yellow, add 50 μL 10 mg/mL 3,5-dinitrobenzoyl chloride in MeCN, shake, let stand for 10 min, inject a 50 μL aliquot. (Prepare indicator by dissolving 100 mg bromothymol blue in 3.2 mL 200 mM NaOH.)

HPLC VARIABLES

Column: 150 × 4.6 5 μm Ultrasphere ODS
Mobile phase: MeCN:water 50:50
Flow rate: 1
Injection volume: 50
Detector: UV 240; UV 254

CHROMATOGRAM

Retention time: 5.65
Limit of detection: 80 ng/mL

KEY WORDS

whole blood

REFERENCE

Lambert, W.E.; Piette, M.; Van Peteghem, C.; De Leenheer, A.P. Application of high-performance liquid chromatography to a fatality involving azide, *J.Anal.Toxicol.*, **1995**, *19*, 261–264.

SAMPLE

Matrix: reaction mixtures
Analyte: sodium azide
Sample preparation: Reaction mixtures containing nitrite. Remove a 5 mL aliquot of the reaction mixture and remove excess nitrite by adding at least 1 mL 20% sulfamic acid in water, check for complete removal of nitrite. (More sulfamic acid solution may be required for strongly basic reaction mixtures or those containing high concentrations of nitrite.) After standing for at least 3 min the analytical solution may be spiked, if desired, by adding ca. 10-20 μg/mL 100 μg/mL sodium azide in water. Add 1 drop of indicator, add 1 M KOH solution until it turns purple (typically, 3-10 mL are required), add 2 mL of MeCN (4 mL if more than 1 mL of sulfamic acid was used), add HCl dropwise until the mixture is acidic (yellow color) followed by 1 more drop. Prepare a 10% (w/v) solution of 3,5-dinitrobenzoyl chloride in MeCN and add 50 μL. Shake the mixture and allow it to stand for at least 3 min, inject an aliquot. Reaction mixtures containing ceric salts. Remove a 10 mL aliquot of the reaction mixture and dilute with 40 mL water. Add 5 mL of this solution to 3 mL of 1 *M* KOH. (If less than 3 mL of 1 *M* KOH is used precipitation of ceric salts is not complete.) Shake, centrifuge, remove 2 mL of the supernatant and add to 1 mL MeCN. At this point the analytical solution may be spiked, if desired, with 10-20 μL 100 μg/mL sodium azide in water. Add 1 drop of indicator and then add HCl dropwise until the mixture is acidic (yellow color) followed by 1 more drop. Prepare a 10% (w/v) solution of 3,5-dinitrobenzoyl chloride in MeCN and add 50 μL. Shake the mixture and allow it to stand for at least 3 min, inject an aliquot. (It is crucial to use freshly prepared 3,5-dinitrobenzoyl azide solution. Use within minutes of preparation. It is generally most convenient to prepare all the analytical samples (spiked and unspiked) with the fresh solution at the beginning of the day and then analyze them in the course of the day. Prepare indicator by dissolving 100 mg bromocresol purple in 18.5 mL 10 mM KOH, make up to 25 mL with water. Check for complete destruction of nitrite as follows. Boil 0.1 g α-naphthylamine in 20 mL of water until it dissolves. Pour this solution, while hot, into 150 mL of 15% v/v aqueous acetic acid. To this solution add a solution of 0.5 g of sulfanilic acid in 150 mL of 15% v/v aqueous acetic acid. Store the reagent in a brown bottle. Add 3 mL of the mixture to be tested to 1 mL of the reagent and allow to stand at room temperature for 6 min. Measure the absorbance at 520 nm against a suitable blank (Official Methods of Analysis, 12th ed., p 422). Limit of detection was 60 ng/mL of sodium nitrite. Note that at high pH the reaction between azide and nitrite is quite slow, and so the presence of excess nitrite does not mean that all the azide has been degraded. N-(1-Naphthyl)ethylenediamine dihydrochloride may be substituted for α-naphthylamine (Official Methods of Analysis, 15th ed., p 938).)

HPLC VARIABLES

Guard column: 15 × 4.6 5 μm Microsorb C8
Column: 250 × 4.6 5 μm Microsorb C8
Mobile phase: MeCN:water 50:50
Flow rate: 1
Injection volume: 20
Detector: UV 254

CHROMATOGRAM

Retention time: 9
Limit of detection: 200 ng/mL

REFERENCE

Lunn, G.; Sansone, E.B. *Destruction of Hazardous Chemicals in the Laboratory, 2nd edition; John Wiley & Sons, Inc. New York*, **1994**, pp. 59–61.

AZIRIDINE

SULFIDE

Sodium Sulfide

SAMPLE
Matrix: blood
Analyte: thiotepa and tepa
Sample preparation: 1 mL Plasma + 2.2 mL water, add 3 mL to an Extrelut 3 SPE cartridge, let stand for 15 min, elute with chloroform. Collect the first 8 mL of effluent, evaporate to dryness under a stream of nitrogen at 20°, reconstitute the residue in 500 μL 1-propanol. Remove a 100 μL aliquot and add 10 μL reagent, heat at 80° for 30 min, cool in an ice bath, add 400 μL taurine solution, add 400 μL OPA solution, let stand for 10 min, inject a 20 μL aliquot. (Reagent was prepared by mixing equal volumes of 80 mM sodium sulfide solution and 100 mM disodium EDTA, prepare fresh daily. Taurine solution was 0.2 mM taurine in 100 mM pH 8.0 phosphate buffer. OPA solution was 0.3 mM o-phthalaldehyde in 100 mM pH 8.0 phosphate buffer.)

HPLC VARIABLES
Column: 250 × 4 5 μm LiChrosorb RP-18
Mobile phase: MeCN:100 mM pH 5.7 phosphate buffer 28:72
Flow rate: 1
Injection volume: 20
Detector: F ex 340 em 440

CHROMATOGRAM
Retention time: 4.0 (TEPA), 17.0 (thiotepa)
Limit of detection: 10 mg/mL

OTHER SUBSTANCES
Extracted: metabolites

KEY WORDS
rabbit; plasma; SPE

REFERENCE
Sano, A.; Matsutani, S.; Takitani, S. High-performance liquid chromatography of the anti-tumour agent triethylenethiophosphoramide and its metabolite triethylenephosphoramide with sodium sulphide, taurine and o-phthalaldehyde as pre-column fluorescent derivatization reagents, *J.Chromatogr.*, **1988**, *458*, 295–301.

AZO COMPOUND

MISCELLANEOUS REACTIONS

Reduction

$$R-N{=}N-R' \xrightarrow{\text{sodium dithionite}} RNH_2 \ + \ R'NH_2$$

SAMPLE

Matrix: blood, feces, urine

Analyte: sulfasalazine

Sample preparation: Urine. 1 mL Urine + 300 μL 10% sodium dithionite (freshly prepared), heat at 30° for 5 min, add 500 μL 10 mM IS, add 20 μL propionic anhydride, let stand for a few min at room temperature, add 8.5 mL MeOH, vortex, let stand for 30 min, centrifuge. Remove a 500 μL aliquot and add it to 4.5 mL water, inject a 20 μL aliquot. Feces. Mix feces with 300 mL 0.2% mercury chloride, mix thoroughly. 1 mL Feces homogenate + 300 μL 10% sodium dithionite (freshly prepared), heat at 30° for 5 min, add 500 μL 10 mM IS, add 20 μL propionic anhydride, let stand for a few min at room temperature, add 8.5 mL MeOH, vortex, let stand for 30 min, centrifuge. Remove a 500 μL aliquot and add it to 4.5 mL water, inject a 20 μL aliquot. Serum. 500 μL Serum + 200 μL 10% sodium dithionite (freshly prepared), heat at 30° for 5 min, add 100 μL 500 μM IS, add 20 μL propionic anhydride, let stand for a few min at room temperature, add 4 mL 10 mM pH 8 phosphate buffer, centrifuge, add to a 3 mL Baker quaternary amine SPE cartridge (No. 7091) (pre-equilibrated with 10 mL 10 mM pH 8 phosphate buffer), wash with 3 mL 10 mM pH 8 phosphate buffer, elute with 2 mL 200 mM pH 5.7 phosphate buffer, inject a 20 μL aliquot of the eluate.

HPLC VARIABLES

Guard column: 75 × 12 (sic) Chrompack

Column: 150 × 3 5 μm LiChrosorb RP-18

Mobile phase: MeOH:10 mM pH 3 citric acid 20:80

Flow rate: 1.4

Injection volume: 20

Detector: F ex 305 em 396 (cut-off filter)

CHROMATOGRAM

Retention time: 6.5 (as the N-propionyl derivative of the metabolite and reduction product, 5-aminosalicylic acid)

Internal standard: N-propionyl-4-amino-2-hydroxybenzoic acid (12.5)

Limit of detection: 100 nM

OTHER SUBSTANCES

Extracted: metabolites

KEY WORDS

serum; SPE

REFERENCE

van Hogezand, R.A.; van Balen, H.C.J.G.; van Schaik, A.; Tangerman, A.; van Hees, P.A.M.; Zwanenburg, B.; van Tongeren, J.H.M. Determination of sodium azodisalicylate, salazosulphapyridine and their metabolites in serum, urine and faeces by high-performance liquid chromatography, *J.Chromatogr.*, **1984**, *305*, 470−476.

BORIC ACID

DIOL

4,5-Dihydroxynaphthalene-2,7-disulfonic Acid Disodium Salt

SAMPLE
Matrix: solutions
Analyte: boric acid
Sample preparation: 10 mL Solution + 500 µL reagent A + 500 µL reagent B, mix, let stand at room temperature for 40 min, inject a 10 µL aliquot. (Reagent A was 20 mM 4,5-dihydroxynaphthalene-2,7-disulfonic acid disodium salt (chromotropic acid) in 100 mM EDTA. Reagent B was 2 M octyltrimethylammonium chloride (bromide is available from Lancaster Synthesis, Windham NH) in 1 M pH 4.8 acetate buffer).

HPLC VARIABLES
Column: 50 × 4.6 TSK gel IC-Anion-PW anion-exchange (Tosoh) (cf. Analyst 1988, 113, 1631)
Mobile phase: 1 mM pH 9.3 Ammonium buffer containing 200 mM sodium perchlorate
Column temperature: 40
Flow rate: 1
Injection volume: 10
Detector: F ex 355 em 328

CHROMATOGRAM
Retention time: 3
Limit of detection: 1 ppb boron

REFERENCE
Oshima, M.; Motomizu, S.; Jun, Z. Fluorimetric determination of boric acid by high performance liquid chromatography after derivatization with chromotropic acid, *Anal.Sci.*, **1990**, *6*, 627–628.

CARBAMATE

MISCELLANEOUS REACTIONS

Hydrolysis

SAMPLE

Matrix: fruits, vegetables

Analyte: methyl carbamate insecticides

Sample preparation: Homogenize (Omni-Mixer) 100 g chopped sample with 250 mL MeOH at half-speed for 5 min, filter (Whatman No. 1 PS paper), make up filtrate to 500 mL with MeOH. Remove 100 mL filtrate and add it to 125 mL 4% aqueous sodium sulfate, shake well, extract mixture with 75, 50, and 50 mL portions of dichloromethane with 30 s shaking each time, drain organic layers through anhydrous sodium sulfate. Combine the organic layers and evaporate them to 1 mL under reduced pressure at 30°, transfer residue to a tube with two 2 mL rinses of dichloromethane:cyclohexane 50:50, make volume up to 10 mL with dichloromethane:cyclohexane 50:50, filter (0.45 μm), add 5 mL to a 600 × 25 tube containing 60 g 200-400 mesh BioBeads SX-3 resin (Analytical BioChemistry Laboratories), pump through at 5 mL/min with dichloromethane:cyclohexane 50:50 mobile phase, discard mobile phase for 24 min, collect fraction containing the compound for 12 min, evaporate under reduced pressure at 30° to low volume, add 15 mL MeOH, evaporate to about 1 mL, filter (0.45 μm), inject a 20 μL aliquot. Alternatively, run output from BioBeads column through a column containing 0.5 g of a mixture of Nuchar S-N (Fisher):Celite 545 1:4, at the end of the chromatography elute this column with 10 mL MeCN:toluene 75:25, evaporate the eluate under reduced pressure at 30° to low volume, add 15 mL MeOH, evaporate to about 1 mL, filter (0.45μm), inject a 20 μL aliquot.

HPLC VARIABLES

Guard column: 50 × 4.6 Pellicular ODS (Whatman)

Column: 250 × 4.6 5 μm Apex ODS (Jones Chromatography)

Mobile phase: Gradient. MeOH:water from 10:90 to 90:10 over 23 min, to 10:90 over 4 min, re-equilibrate at 10:90 for 10 min

Flow rate: 1

Injection volume: 20

Detector: F ex 340 em 455, following post-column reaction. The column effluent mixed with 200 mM NaOH pumped at 0.8 mL/min and the mixture flowed through a 1 mL coil at 95°. The effluent from this coil mixed with 500 mg/L o-phthalaldehyde in 50 mM sodium tetraborate containing 1 mL/L 2-mercaptoethanol pumped at 0.8 mL/min and this mixture flowed through a 0.5 mL coil at ambient temperature to the detector.

CHROMATOGRAM

Retention time: 14.7 (oxamyl), 15.5 (methomyl), 22.5 (aldicarb), 24 (carbofuran, propoxur), 25 (carbaryl), 28.5 (methiocarb)

Limit of detection: 5-10 ppb

KEY WORDS
apples; broccoli; cabbage; cauliflower; potatoes; post-column reaction

REFERENCE
Chaput, D. Simplified multiresidue method for liquid chromatographic determination of N-methyl carbamate insecticides in fruits and vegetables, *J.Assoc.Off.Anal.Chem.*, **1988**, 71, 542−546.

SAMPLE
Matrix: tissue
Analyte: methylcarbamate pesticides
Sample preparation: 21 g Liver + 60 g anhydrous sodium sulfate, mix with spatula, add 200 mL dichloromethane, mix with spatula, homogenize (VirTis 45) for 2 min at medium speed, filter through 5 g anhydrous sodium sulfate, re-extract tissue and sodium sulfate with 100 mL dichloromethane, filter, wash out flask with 25 mL dichloromethane, filter. Combine filtrates and filter them through 2 g anhydrous sodium sulfate, rinse flask with 20 mL dichloromethane, wash filter with 10 mL dichloromethane. Concentrate filtrate to 1-2 mL under reduced pressure at 30° (do not allow to go dry), transfer residue to a tube with 1-2 mL cyclohexane, wash in with dichloromethane:cyclohexane 50:50, make volume in tube 7.5 mL, filter (0.45 μm), add 5 mL to a 600 × 25 tube containing 60 g 200-400 mesh BioBeads SX-3 resin (Analytical BioChemistry Laboratories), pump through at 5 mL/min with dichloromethane:cyclohexane 50:50 mobile phase, collect fraction containing the compound, evaporate under reduced pressure at 30° to about 1 mL, make up to 2 mL with dichloromethane, add 1 mL to 1 mL 100 mg Bond Elut aminopropyl SPE cartridge (previously conditioned with 1 mL dichloromethane), elute with 3-5 mL dichloromethane:MeOH 98.5:1.5, evaporate eluate to dryness at 30° under reduced pressure (do not over dry), reconstitute in 200 μL MeOH, vortex for 5 s, filter (0.45 μm), inject a 20-30 μL aliquot.

HPLC VARIABLES
Guard column: Guard-PAK (Waters no. 88070)
Column: 250 × 4.6 5 μm Zorbax C8
Mobile phase: Gradient. MeCN:water from 12:88 to 70:30 over 30 min, to 80:20 over 1 min, maintain at 80:20 for 8 min, re-equilibrate at initial conditions for 10 min.
Flow rate: 1.5
Injection volume: 20-30
Detector: F ex 340 em 418, following post-column derivatization. The column effluent mixed with 50 mM NaOH pumped at 0.27 mL/min and the mixture flowed through a 1 mL coil at 80°. The effluent from this coil mixed with 140 μg/mL o-phthalaldehyde in 50 mM pH 10.5 potassium borate buffer containing 1 mL/L mercaptoethanol pumped at 0.27 mL/min and this mixture flowed through a 1 mL coil at 40° to the detector.)

CHROMATOGRAM
Retention time: 10.6 (methomyl), 18.5 (aldicarb), 22.3 (carbofuran), 23.1 (carbaryl), 24.6 (methiocarb), 27.0 (bufencarb)
Limit of quantitation: 5 ppb

KEY WORDS
liver; pig; cow; duck; SPE; post-column reaction

REFERENCE
Ali, M.S. Determination of N-methylcarbamate pesticides in liver by liquid chromatography, *J.Assoc.Off.Anal.Chem.*, **1989**, 72, 586−592.

SAMPLE
Matrix: tissue
Analyte: methylcarbamate pesticides
Sample preparation: 21 g Liver + 60 g anhydrous sodium sulfate, mix with spatula, add 200 mL dichloromethane, mix with spatula, homogenize (VirTis 45) for 2 min at medium speed, filter through 5 g anhydrous sodium sulfate, re-extract tissue and sodium sulfate with 100 mL dichloromethane, filter, wash out flask with 25 mL dichloromethane, filter. Combine filtrates and filter them through 2 g anhydrous sodium sulfate, rinse flask with

20 mL dichloromethane, wash filter with 10 mL dichloromethane. Concentrate filtrate to 1-2 mL under reduced pressure at 30° (do not allow to go dry), transfer residue to a tube with 1-2 mL cyclohexane, wash in with dichloromethane:cyclohexane 50:50, make volume in tube 7.5 mL, filter (0.45 μm), add 5 mL to a 600 ×25 tube containing 60 g 200-400 mesh BioBeads SX-3 resin (Analytical BioChemistry Laboratories), pump through at 5 mL/min with dichloromethane:cyclohexane 50:50 mobile phase, collect fraction containing the compound, evaporate under reduced pressure at 30° to about 1 mL, make up to 2 mL with dichloromethane, add 1 mL to a 1 mL 100 mg Bond Elut aminopropyl SPE cartridge (previously conditioned with 1 mL dichloromethane), elute with 3-5 mL dichloromethane:MeOH 98.5:1.5, evaporate eluate to dryness at 30° under reduced pressure (do not over dry), reconstitute in 200 μL MeOH, vortex for 5 s, filter (0.45 μm), inject a 20-30 μL aliquot.

HPLC VARIABLES
Guard column: Guard-PAK (Waters no. 88070)
Column: 250 × 4.6 5 μm Zorbax C8
Mobile phase: Gradient. MeCN:water from 12:88 to 70:30 over 30 min, to 80:20 over 1 min, maintain at 80:20 for 8 min, re-equilibrate at initial conditions for 10 min.
Flow rate: 1.5
Injection volume: 20-30
Detector: F ex 340 em 418, following post-column reaction. The column effluent mixed with 50 mM NaOH pumped at 0.27 mL/min and the mixture flowed through a 1 mL coil at 80°. The effluent from this coil mixed with 140 μg/mL o-phthalaldehyde in 50 mM pH 10.5 potassium borate buffer containing 1 mL/L mercaptoethanol pumped at 0.27 mL/min and flowed through a 1 mL coil at 40° to the detector.

CHROMATOGRAM
Retention time: 9.3 (oxamyl), 10.5 (methomyl), 15.3 (dioxacarb), 18.5 (aldicarb), 22.1 (propoxur), 22.5 (carbofuran), 22.7 (bendiocarb), 24 (carbaryl), 25.6 (isoprocarb), 28 (methiocarb), 29.7 (promecarb), 32 (bufencarb)
Limit of quantitation: 5 ppb

KEY WORDS
liver; pig; cow; duck; SPE; post-column reaction

REFERENCE
Ali, M.S.; White, J.D.; Bakowski, R.S.; Stapleton, N.K.; Williams, K.A.; Johnson, R.C.; Phillippo, T.; Woods, R.W.; Ellis, R.L. Extension of a liquid chromatographic method for N-methylcarbamate pesticides in cattle, swine, and poultry liver, *J.AOAC Int.*, **1993**, *76*, 907–910.

SAMPLE
Matrix: water
Analyte: N-methylcarbamate pesticides
Sample preparation: Condition a 10 × 4 55 mg 40 μm C18/OH Bondesil SPE cartridge (Varian/Analytichem) with 1 mL MeOH and 1 mL water, pass through 5 mL test water at 1 mL/min, pass through 500 μL pure water, elute the contents of the SPE cartridge onto the analytical column with mobile phase.

HPLC VARIABLES
Guard column: 10 × 4 4 μm Supersphere RP-8 (Merck)
Column: 250 × 4 4 μm Supersphere RP-8 (Merck)
Mobile phase: Gradient. A was MeCN:water 20:80 containing 2.5 mM sodium acetate. B was MeOH:water 20:80 containing 2.5 mM sodium acetate. C was MeCN:water 60:40 containing 2.5 mM sodium acetate. A:B:C 75:25:0 for 5 min, to 0:0:100 over 20 min, maintain at 0:0:100 for 5 min, re-equilibrate at initial conditions for 15 min.
Column temperature: 35
Flow rate: 0.75
Injection volume: 100
Detector: F ex 340 em 445 following post-column reaction. The column effluent flowed through a 50 × 4 Aminex A-27 (Bio-Rad) column at 120-140° and was mixed with reagent pumped at 1 mL/min, this mixture flowed through a 200 × 0.12 PTFE tube to the detector. (Reagent was prepared by adding 2 mL 25 mg/mL o-phthalaldehyde in MeCN and 100

µL 2-mercaptoethanol to 200 mL 5 mg/mL disodium tetraborate in water then making up to 250 mL with water.)

CHROMATOGRAM
Retention time: 9.44 (oxamyl), 11.70 (methomyl), 15.15 (tranid), 16.45 (dioxacarb), 19.30 (butocarboxim), 20.21 (aldicarb), 22.95 (cloethocarb), 23.09 (propoxur), 23.36 (carbofuran), 23.36 (bendiocarb), 24.53 (carbaryl), 25.00 (thiofanox), 25.06 (ethiofencarb), 26.09 (isoprocarb), 26.75 (carbanolate), 28.48 (methiocarb), 28.69 (fenobucarb), 29.60 (promecarb), 33.52 (bufencarb)

Internal standard: trimethacarb (26.12)

Limit of detection: 0.03-0.05 ng/mL

KEY WORDS
SPE; post-column reaction

REFERENCE
Hiemstra, M.; de Kok, A. Determination of N-methylcarbamate pesticides in environmental water samples using automated on-line trace enrichment with exchangeable cartridges and high-performance liquid chromatography, *J.Chromatogr.A*, **1994**, *667*, 155–166.

SAMPLE
Matrix: water

Analyte: pesticides

Sample preparation: Mix water sample with 2 mL 500 mM NaOH, make up to 40 mL with water, shake for a few s, let stand for 10 min, add 1 mL glacial acetic acid, make up to 50 mL with water, filter (0.45 µm), sonicate the filtrate, inject a 20 µL aliquot.

HPLC VARIABLES
Column: 33 × 4.6 3 µm Pecosphere 3x3 CR C18

Mobile phase: MeCN:water:glacial acetic acid 50:49.5:0.5 containing 10 mM sodium perchlorate

Flow rate: 1

Injection volume: 20

Detector: E, ESA Coulochem II, model 5021 conditioning cell 0 V, model 5011 dual analytical cell with porous graphite working electrodes at +0.1 V and +0.6 V (monitored), cells protected with 0.2 µm porous graphite filters

CHROMATOGRAM
Retention time: 1.00 (carbofuran), 1.21 (carbaryl), 2.17 (fenobucarb)

Limit of detection: 1-2 nM

KEY WORDS
river water

REFERENCE
Galeano Díaz, T.; Guiberteau, A.; Salinas, F.; Ortiz, J.M. Rapid and sensitive determination of carbaryl, carbofuran and fenobucarb by liquid chromatography with electrochemical detection, *J.Liq.Chromatogr.& Rel.Technol.*, **1996**, *19*, 2681–2690.

CARBOXYLIC ACID

ACTIVATED HALIDE

N-(9-Acridinyl)bromoacetamide

SAMPLE
Matrix: solutions
Analyte: carboxylic acids
Sample preparation: Mix a 100 μL aliquot of a 1-100 μM solution in 20 mM pH 7.1 phosphate buffer with 10 μL 10 mM tetrabutylammonium hydrogen sulfate in water and 100 μL 1 mM N-(9-acridinyl)bromoacetamide in chloroform, stir at 90° for 20-30 min, cool. Remove a 10 μL aliquot of the organic layer and evaporate it to dryness under reduced pressure, reconstitute with mobile phase, sonicate, inject an aliquot. (Synthesis of N-(9-acridinyl)bromoacetamide is as follows. Dissolve 2.49 g 9-aminoacridine hydrochloride hydrate in water, add dilute NaOH to precipitate the free base, extract with ethyl acetate, dry over anhydrous magnesium sulfate, filter, evaporate to give 9-aminoacridine as yellow needle-shaped crystals (mp 239-240°). Add 1.01 g bromoacetyl bromide in 20 mL diethyl ether dropwise with stirring to 970 mg 9-aminoacridine dissolved in 50 mL acetone containing 1.02 g triethylamine, filter, wash the solid with acetone. Evaporate the filtrate and chromatograph the residue on a 260 × 30 glass column of 70-230 mesh silica gel 60 (Merck) with chloroform:ethyl acetate 2:1. Collect the strong yellow band and evaporate it to dryness, recrystallize from MeOH to give N-(9-acridinyl)bromoacetamide as light yellow crystals (mp 180-182° d).)

HPLC VARIABLES
Column: 150 × 4.6 5 μm Nucleosil C18
Mobile phase: MeCN:water:phosphoric acid 40:60:0.2 (A) or 50:50:0.2 (B)
Flow rate: 1
Injection volume: 20
Detector: F ex 357.5 em 482

CHROMATOGRAM
Retention time: 5 (okadaic acid (B)), 10 (cholic acid (A)), 32 (deoxycholic acid (A)), 36 (chenodiol (A))
Limit of detection: 10 fmole

REFERENCE
Allenmark, S.; Chelminska-Bertilsson, M.; Thompson, R.A. *N*-(9-Acridinyl)-bromoacetamide - A powerful reagent for phase-transfer-catalyzed fluorescence labeling of carboxylic acids for liquid chromatography, *Anal.Biochem.*, **1990**, *185*, 279–285.

2-Bromo-2'-acetonaphthone

RELATED REFERENCE
Shiao, M.S.; Hao, Y.Y. Simultaneous derivatization of gibberellic, indolyl-3-acetic and abscisic acids by α-bromo-2'-acetonaphthone and separation of derivatives by high performance liquid chromatography. *Bot.Bull.Acad.Sin.* **1985**, *26*, 105-111.

SAMPLE
Matrix: blood
Analyte: valproic acid
Sample preparation: 500 μL Plasma + 500 μL 400 μM cyclohexanecarboxylic acid in MeCN + 50 μL 1 M HCl, vortex briefly, add about 300 mg KCl, vortex vigorously, centrifuge at 1500 g for at least 5 min. Remove a 200 μL aliquot of the clear MeCN supernatant and add it to 20 mg potassium carbonate, vortex for 1 h, add 200 μL reagent, mix, let stand at room temperature for 1 h, inject a 10 μL aliquot. (Reagent was 20 mM 2-bromo-2'-acetonaphthone in MeCN containing 1.5 mM 15-crown-5.)

HPLC VARIABLES
Column: 250 × 4 5 μm LiChrosorb RP-18
Mobile phase: MeCN:water 83:17
Flow rate: 1
Injection volume: 10
Detector: UV 280

CHROMATOGRAM
Retention time: 8.4
Internal standard: cyclohexanecarboxylic acid (6.2)

OTHER SUBSTANCES
Extracted: phenobarbital

KEY WORDS
plasma

REFERENCE
Alric, R.; Cociglio, M.; Blayac, J.P.; Puech, R. Performance evaluation of a reversed-phase, high-performance liquid chromatographic assay of valproic acid involving a "solvent demixing" extraction procedure and precolumn derivatization, *J.Chromatogr.*, **1981**, *224*, 289–299.

SAMPLE
Matrix: blood
Analyte: valproic acid
Sample preparation: 200 μL Plasma + 500 μL water, shake, add 300 μL 25 mM perchloric acid, add 3 mL cyclohexane, vortex for 3 min, centrifuge at 3500 rpm for 5 min, repeat extraction twice more. Combine the organic layers and dry them over anhydrous sodium sulfate, add 1 μmole sodium methoxide in MeOH, shake for 1 min, evaporate to dryness under a stream of nitrogen at room temperature, reconstitute with 200 μL MeCN, add 100 μL buffer, add 200 μL reagent, heat at 70° for 40 min, inject an aliquot. (Prepare the buffer by dissolving 3.8 g KH_2PO_4 and 5.96 g Na_2HPO_4 in 200 mL water, pH 7.4. The reagent was 17 mg/mL 2-bromo-2'-acetonaphthone (2-naphthacyl bromide) in MeCN containing 1 mg/mL dicyclohexane-18-crown-6.)

HPLC VARIABLES
Column: 125 mm long 3 μm HS C18 (Perkin-Elmer)
Mobile phase: MeOH:water 77:23
Flow rate: 1
Injection volume: 6
Detector: UV 280

CHROMATOGRAM
Retention time: 5.5
Limit of detection: 3.47 μg/mL

KEY WORDS
plasma; comparison with TLC

REFERENCE
Corti, P.; Cenni, A.; Corbini, G.; Dreassi, E.; Murratzu, C.; Caricchia, A.M. Thin-layer chromatography and densitometry in drug assay: comparison of methods for monitoring valproic acid in plasma, *J.Pharm.Biomed.Anal.*, **1990**, *8*, 431–436.

SAMPLE
Matrix: bulk
Analyte: fatty acids
Sample preparation: Dissolve 10 μmoles linoleic acid, 20 μmoles 2-bromo-2'-acetonaphthone (2-naphthacyl bromide), and 40 μmoles N,N-diisopropylethylamine in 1 mL DMF, heat at 60° for 10 min, inject an aliquot.

HPLC VARIABLES
Column: 914 × 1.8 Corasil C18
Mobile phase: MeOH:water 85:15
Flow rate: 0.2
Detector: UV 254

CHROMATOGRAM
Retention time: 40 (linolenic acid), 50 (arachidonic acid), 55 (linoleic acid), 70 (dihomolinolenic acid), 90 (oleic acid)

REFERENCE
Cooper, M.J.; Anders, M.W. Determination of long chain fatty acids as 2-naphthacyl esters by high pressure liquid chromatography and mass spectrometry, *Anal.Chem.*, **1974**, *46*, 1849–1852.

SAMPLE
Matrix: bulk
Analyte: fatty acids
Sample preparation: Mix 10 μmoles fatty acid, 40 μmoles N,N-diisopropylethylamine or lithium carbonate, and 10-20 μmoles 2-bromo-2'-acetonaphthone (2-naphthacyl bromide) in 1 mL DMF, heat at 65° for 15 min, inject an aliquot.

HPLC VARIABLES
Column: two 300 × 3.9 μBondapak C18 columns in series
Mobile phase: Gradient. MeCN:water from 40:60 to 100:0 (Waters convex curve 5) over 3 h.
Flow rate: 1
Detector: UV 254

CHROMATOGRAM
Retention time: 18 (lactic acid), 23 (acetic acid), 27 (propionic acid), 33 (butyric acid), 44 (caproic acid), 56 (caprylic acid), 70 (capric acid), 84 (lauric acid), 97 (linolenic acid), 100 (myristic acid), 103 (arachidonic acid, palmitoleic acid), 107 (linoleic acid), 115 (palmitic acid), 117 (oleic acid), 119 (elaidic acid, vaccenic acid), 128 (stearic acid), 142 (arachidic acid), 154 (behenic acid, nervonic acid), 163 (lignoceric acid)

REFERENCE
Jordi, H.C. Separation of long and short chain fatty acids as naphthacyl and substituted phenacyl esters by high performance liquid chromatography, *J.Liq.Chromatogr.*, **1978**, *1*, 215–230.

SAMPLE
Matrix: bulk
Analyte: pesticides
Sample preparation: Prepare a 100 μg/mL solution in acetone. 1 mL Solution + 1 mL 2-10 mg/mL 2-bromo-2'-acetonaphthone (2-naphthacyl bromide) in acetone + 5-10 mg cesium carbonate, heat in the dark at 35° for 45 min, inject an aliquot. Alternatively, add 100 μL glacial acetic acid to the reaction mixture, extract twice with 5 mL portions of light petroleum (bp 40-60°). Combine the extracts and evaporate them to dryness, reconstitute, inject an aliquot.

HPLC VARIABLES
Column: 150 × 4.6 Hypersil ODS
Mobile phase: MeOH:water 80:20
Flow rate: 1
Detector: UV 254

CHROMATOGRAM
Retention time: k' 4.35 (2,3,6-trichlorobenzoic acid), k' 3.32 (dicamba), k' 3.45 (2,4-dichlorophenoxyacetic acid), k' 5.23 (2-(2,4-dichlorophenoxy)propionic acid), k' 5.06 (mecoprop), k' 3.81 (4-(2-methyl-4-chlorophenoxy)butyric acid), k' 8.16 (phenoprop)
Limit of detection: 200-300 pg

KEY WORDS
comparison with derivatization with 4-bromomethyl-7-methoxycoumarin and with no derivatization

REFERENCE
Roseboom, H.; Herbold, H.A.; Berkhoff, C.J. Determination of phenoxy carboxylic acid pesticides by gas and liquid chromatography, *J.Chromatogr.*, **1982**, *249*, 323–331.

SAMPLE
Matrix: bulk
Analyte: fatty acids
Sample preparation: 100 μg Fatty acid mixture + 25 μL 12.5 mg/mL 2-bromo-2'-acetonaphthone (naphthacyl bromide) in acetone + 25 μL 10 mg/mL triethylamine in acetone, heat in a boiling water bath for 15 min, add 35 μL 2 mg/mL acetic acid in acetone, heat for 5 min, evaporate to dryness under a stream of nitrogen at 40°, reconstitute with 100 μL MeCN, inject a 5-10 μL aliquot.

HPLC VARIABLES
Guard column: octadecyl
Column: 250 × 4.5 5 μm octadecyl-bonded spherical silica (IBM)
Mobile phase: Gradient. MeCN:water 87:13 for 25 min, to 92:8 over 15 min
Flow rate: 2
Injection volume: 5-10

Detector: UV 242; UV 254

CHROMATOGRAM
Limit of quantitation: 10 ng

OTHER SUBSTANCES
Also analyzed: arachidonic acid, all-cis-delta^{8,11,14}-eicosatrienoic acid, cis-delta^{11}-eicosenoic acid, elaidic acid, heptadecanoic acid, trans-delta^9-hexadecenoic acid, lauric acid, linoleic acid, linolenic acid, myristic acid, myristoleic acid, oleic acid, palmitic acid, pentadecanoic acid, cis-delta^{10}-pentadecenoic acid, stearic acid, tridecanoic acid

REFERENCE
Wood, R.; Lee, T. High-performance liquid chromatography of fatty acids: Quantitative analysis of saturated, monoenoic, polyenoic and geometrical isomers, *J.Chromatogr.*, **1983**, *254*, 237–246.

SAMPLE
Matrix: bulk
Analyte: prostaglandins
Sample preparation: Make up a 20 μg/mL solution in dichloromethane, evaporate 1 mL under a stream of nitrogen, add 100 μL 20 mg/mL 2-bromo-2'-acetonaphthone (2-naphthacyl bromide) in MeCN, add 100 μL 15 μL/mL N,N-diisopropylethylamine, add 100 μL MeCN, heat at 45° for 1 h, evaporate under a stream of nitrogen, take up the residue in 5 mL dichloromethane. (Purify 2-bromo-2'-acetonaphthone in acetone solution with activated carbon, then recrystallize 2-bromo-2'-acetophenone from carbon tetrachloride.)

HPLC VARIABLES
Column: 250 × 4 Zorbax SIL
Mobile phase: Dichloromethane:MeCN:water:acetic acid:silver nitrate 87.5:12.5:0.4: 0.2:0.17 (v:v:v:v:w)
Flow rate: 1
Injection volume: 10
Detector: UV 254

CHROMATOGRAM
Retention time: 11.5 (5,6-trans-16,16-dimethylprostaglandin E$_2$), 12.1 (5,6-trans-15-(R)-methylprostaglandin E$_2$), 14.1 (16,16-dimethylprostaglandin E$_2$), 14.5 (arbaprostil), 15.4 (5,6-trans-9-deoxo-16,16-dimethyl-9-methyleneprostaglandin E$_2$), 17.3 (5,6-trans-prostaglandin E$_2$), 20.8 (meteneprost), 22 (dinoprostone)

KEY WORDS
normal phase

REFERENCE
Kissinger, L.D.; Robins, R.H. Silver-modified mobile phase for normal-phase liquid chromatographic determination of prostaglandins and their 5,6-*trans* isomers in prostaglandin bulk drugs and triacetin solutions, *J.Chromatogr.*, **1985**, *321*, 353–362.

SAMPLE
Matrix: bulk, formulations
Analyte: alprostadil
Sample preparation: Prepare a 500 μg/mL solution of the bulk drug in EtOH. Evaporate a 2 mL aliquot of the EtOH solution or an aliquot of the formulation containing 1 mg compound to dryness under a stream of nitrogen, add 200 μL 20 mg/mL 2-bromo-2'-acetonaphthone (α-bromoacetonaphthone), swirl, add 100 μL 10 μL/mL N,N-diisopropylethylamine, swirl, heat at 45° for 1 h with swirling every 15 min, evaporate to dryness under a stream of nitrogen, reconstitute with 10 mL 400 μg/mL methylprednisolone in dichloromethane, inject a 10 μL aliquot.

HPLC VARIABLES
Column: 300 × 3.9 10 μm μPorasil
Mobile phase: Dichloromethane:1,3-butanediol:water 99.5:0.5:0.05
Flow rate: 1.5
Injection volume: 10
Detector: UV 254

CHROMATOGRAM
Retention time: 15
Internal standard: methylprednisolone (25)

OTHER SUBSTANCES
Simultaneously analyzed: dinoprostone, 8-isoprostaglandin E_1, 8-isoprostaglandin E_2, 5,6-trans-prostaglandin E_2

KEY WORDS
injections; normal phase

REFERENCE
Zoutendam, P.H.; Bowman, P.B.; Ryan, T.M.; Rumph, J.L. Quantitative determination of alprostadil (PGE₁) in bulk drug and pharmaceutical formulations by high-performance liquid chromatography, *J.Chromatogr.*, **1984**, *283*, 273–280.

2-Bromoacetophenone

SAMPLE
Matrix: blood
Analyte: valproic acid
Sample preparation: 50 µL Serum + 1 mL 4 µg/mL IS in MeCN, vortex for 10 s, centrifuge. Remove an 800 µL aliquot of the supernatant and add it to 200 µL 3 mg/mL 2-bromoacetophenone (phenacyl bromide) in MeCN and 100 µL triethylamine, heat in an open tube at 80° for 30 min, cool, inject a 10 µL aliquot.

HPLC VARIABLES
Column: 250 × 4.6 5 µm Finepak Sil $C_{18.5}$ (Japan Spectroscopic, Tokyo)
Mobile phase: MeCN:water 60:40
Column temperature: 30
Flow rate: 1
Injection volume: 10
Detector: UV 245

CHROMATOGRAM
Retention time: 21
Internal standard: cyclohexane carboxylic acid (13)
Limit of quantitation: 500 ng/mL

OTHER SUBSTANCES
Extracted: hexobarbital, phenobarbital
Non-interfering: carbamazepine, phenytoin

KEY WORDS
serum

REFERENCE
Nakamura, M.; Kondo, K.; Nishioka, R.; Kawai, S. Improved procedure for the high-performance liquid chromatographic determination of valproic acid in serum as its phenacyl ester, *J.Chromatogr.*, **1984**, *310*, 450–454.

SAMPLE
Matrix: blood

Analyte: acetate, hydroxybutyrate, and lactate

Sample preparation: 1 mL Plasma + 50 μL 15000 dpm/mL [2-³H]acetate in 2 mM sodium carbonate, vortex, add 100 μL 1 M HCl, vortex, add 10 mL ether, shake mechanically for 1 h, centrifuge at 1000 g for 15 min, discard the organic layer. Evaporate the aqueous layer to dryness under reduced pressure (to prevent acetate transfer cover tube with paper soaked in 100 mM sodium carbonate and then dried), add 100 μL 40 mM 2-bromoacetophenone (phenacyl bromide) in acetone containing 80 mM 18-crown-6, sonicate for 10 s, vortex, heat at 100° for 15 min, cool for 5-10 min, add 50 μL 150 mM propionic acid in acetone, heat at 100° for 5 min. Evaporate to dryness under reduced pressure, reconstitute with 240 μL mobile phase, inject a 200 μL aliquot. (It is important to use fat extraction grade diethyl ether (Mallinckrodt). On the day of the assay wash 600 mL ether with 100 mM 50 mM sodium carbonate.)

HPLC VARIABLES

Column: 250 × 4.6 C18 (Jones Chromatography)
Mobile phase: MeCN:water 20:80
Column temperature: 50
Flow rate: 1.2
Injection volume: 200
Detector: UV 320

CHROMATOGRAM

Retention time: 11 (lactate), 14 (β-hydroxybutyrate), 17 (acetate)
Internal standard: [2-³H]acetate (Collect the acetate fraction and count it to provide IS data.)

OTHER SUBSTANCES

Simultaneously analyzed: 3-hydroxyisobutyrate

KEY WORDS

plasma; dog

REFERENCE

Persson, M.; Bleiberg, B.; Kiss, D.; Miles, J. Measurement of plasma acetate kinetics using high-performance liquid chromatography, *Anal.Biochem.*, **1991**, *198*, 149–153.

SAMPLE

Matrix: blood

Analyte: lactic acid

Sample preparation: 1 mL Plasma + 50 μL acetate + 100 μL 2 M HCl + 10 mL diethyl ether, shake for 1 h, centrifuge at 2500 g for 10 min. Remove the organic layer and add it to 1 mL 2 mM sodium carbonate, shake for 5 min, centrifuge for 5 min. Remove the aqueous phase and evaporate it to dryness under reduced pressure, add 50 μL 25 mg/mL 2-bromoacetophenone (phenacyl bromide) in acetone containing 33 mg/mL 18-crown-6, sonicate briefly, heat at 100° for 15 min, add 100 μL 50 mM propionic acid in acetone, heat at 100° for 5 min, evaporate to dryness under reduced pressure, suspend in 250 μL MeCN:water 30:70, inject a 240 μL aliquot.

HPLC VARIABLES

Column: 250 × 4.6 5 μm C18 (Beckman)
Mobile phase: MeCN:water 30:70
Flow rate: 1
Injection volume: 240
Detector: UV 320

CHROMATOGRAM

Retention time: 8.89 (lactate), 10.49 (β-hydroxybutyrate)
Internal standard: acetate (16.55)
Limit of detection: 36 pmole

KEY WORDS

plasma

REFERENCE

Bleiberg, B.; Steinberg, J.J.; Katz, S.D.; Wexler, J.; LeJemtel, T. Determination of plasma lactic acid concentration and specific activity using high-performance liquid chromatography, *J.Chromatogr.*, **1991**, *568*, 301–308.

SAMPLE

Matrix: bulk

Analyte: fatty acids

Sample preparation: 100 µg Fatty acid mixture + 25 µL 10 mg/mL 2-bromoacetophenone (phenacyl bromide) in acetone + 25 µL 10 mg/mL triethylamine in acetone, heat in a boiling water bath for 15 min, add 35 µL 2 mg/mL acetic acid in acetone, heat for 5 min, evaporate to dryness under a stream of nitrogen at 40°, reconstitute with 100 µL MeCN, inject a 5-10 µL aliquot.

HPLC VARIABLES

Guard column: octadecyl

Column: 250 × 4.5 5 µm octadecyl-bonded spherical silica (IBM)

Mobile phase: Gradient. MeCN:water 80:20 for 25 min, to 85:15 over 15 min

Flow rate: 2

Injection volume: 5-10

Detector: UV 242; UV 254

CHROMATOGRAM

Retention time: 8 (lauric acid), 9 (myristoleic acid), 12 (tridecanoic acid), 13 (cis-delta[10]-pentadecenoic acid), 13.5 (linolenic acid), 16 (myristic acid), 18 (arachidonic acid), 20 (trans-delta[9]-hexadecenoic acid), 21 (linoleic acid), 23.5 (pentadecanoic acid), 24.5 (all-cis-delta[8,11,14]-eicosatrienoic acid), 32 (palmitic acid), 34 (oleic acid), 36 (elaidic acid), 40.5 (heptadecanoic acid), 51 (stearic acid), 52.5 (cis-delta[11]-eicosenoic acid)

Limit of quantitation: 10 ng

REFERENCE

Wood, R.; Lee, T. High-performance liquid chromatography of fatty acids: Quantitative analysis of saturated, monoenoic, polyenoic and geometrical isomers, *J.Chromatogr.*, **1983**, *254*, 237–246.

SAMPLE

Matrix: fat

Analyte: fatty acids

Sample preparation: 50 mg Fat + 1 mL 25% KOH in 96% EtOH, heat in a boiling water bath for 1 h, cool, adjust pH to 2 with 3 M HCl, extract three times with 2 mL portions of n-hexane:diethyl ether 50:50. Combine the organic layers and evaporate them to dryness under a stream of nitrogen, reconstitute the residue with 25 µL 10 mg/mL 2-bromoacetophenone (phenacyl bromide) in acetone and 25 µL 10 mg/mL triethylamine in acetone, heat in a boiling water bath for 5 min, add 40 µL 2 mg/mL acetic acid in acetone, heat for 5 min, evaporate to dryness under a stream of nitrogen at 40°, reconstitute with 100 µL MeOH, inject a 10 µL aliquot.

HPLC VARIABLES

Guard column: 50 × 4 7 µm Separon SGX C18 (Tessek, Prague)

Column: 250 × 4 5 µm Separon SGX C18 (Tessek, Prague)

Mobile phase: Gradient. MeCN:MeOH:water from 40.5:40:19.5 to 0:81.5:18.5 over 25 min, to 0:90:10 over 45 min, to 0:100:0 over 20 min.

Column temperature: 40

Flow rate: 1

Injection volume: 10

Detector: UV 242

CHROMATOGRAM

Retention time: 6 (caproic acid), 8 (caprylic acid), 9 (capric acid), 14 (lauric acid), 15 (myristoleic acid), 22 (linolenic acid), 25 (myristic acid), 26 (docosahexaenoic acid), 28 (palmitoleic acid), 30 (arachidonic acid), 31 (linoleic acid), 34 (pentadecanoic acid), 38 (linoelaidic acid), 41 (eicosatrienoic acid), 45 (palmitic acid), 47 (oleic acid), 50 (elaidic acid), 52 (eicosadienoic acid), 53 (heptadecanoic acid), 61 (stearic acid), 62 (eicosenoic acid), 72 (erucic acid)

Limit of detection: 0.8-1.2 ng

KEY WORDS
rat

REFERENCE
Hanis, T.; Smrz, M.; Klir, P.; Macek, K.; Klima, J.; Base, J.; Deyl, Z. Determination of fatty acids as phenacyl esters in rat adipose tissue and blood vessel walls by high-performance liquid chromatography, *J.Chromatogr.*, **1988**, *452*, 443−457.

SAMPLE
Matrix: formulations
Analyte: valproic acid
Sample preparation: Capsules. Dissolve 10 broken capsules in 200 mL acetone. Remove a 20 mL aliquot and make it up to 100 mL with acetone:water 45:55. Remove a 1 mL aliquot and make it up to 10 mL with acetone, mix. Remove a 1 mL aliquot and add it to 1 mL IS solution, add 50 μL 12.8 mg/mL 2-bromoacetophenone (phenacyl bromide) in acetone, add 50 μL 10 mg/mL triethylamine in acetone, mix with gentle swirling, heat at 50° for 2 h, cool to room temperature, inject an aliquot. Syrup. Dilute 5 mL syrup to 100 mL with acetone:water 45:55. Remove a 1 mL aliquot and make it up to 10 mL with acetone, mix. Remove a 1 mL aliquot and add it to 1 mL IS solution, add 50 μL 12.8 mg/mL 2-bromoacetophenone (phenacyl bromide) in acetone, add 50 μL 10 mg/mL triethylamine in acetone, mix with gentle swirling, heat at 50° for 2 h, cool to room temperature, inject an aliquot. Tablets. Weigh out powdered tablets equivalent to 250 mg valproic acid, add 50 mL acetone:water 45:55, sonicate for 10 min, make up to 100 mL with acetone:water 45:55, sonicate for 10 min, mix, centrifuge an aliquot of the suspension at 4000 rpm for 5 min. Remove a 1 mL aliquot of the supernatant and make it up to 10 mL with acetone, mix. Remove a 1 mL aliquot and add it to 1 mL IS solution, add 50 μL 12.8 mg/mL 2-bromoacetophenone (phenacyl bromide) in acetone, add 50 μL 10 mg/mL triethylamine in acetone, mix with gentle swirling, heat at 50° for 2 h, cool to room temperature, inject an aliquot. (Prepare IS solution by diluting a 400 μg/mL solution of sodium caproate in acetone:water 45:55 with an equal volume of acetone.)

HPLC VARIABLES
Column: 250 × 4.6 5 μm Microsorb-MV C18
Mobile phase: MeCN:MeOH:water 50:20:30
Flow rate: 2
Injection volume: 50
Detector: UV 245

CHROMATOGRAM
Retention time: 8.5
Internal standard: caproic acid (4.5)

KEY WORDS
capsules; syrup; tablets

REFERENCE
Lau-Cam, C.A.; Roos, R.W. HPLC method with precolumn phenacylation for the assay of valproic acid and its salts in pharmaceutical dosage forms, *J.Liq.Chromatogr.Rel.Technol.*, **1997**, *20*, 2075−2087.

SAMPLE
Matrix: solutions
Analyte: dicarboxylic acids
Sample preparation: Prepare a solution in water or MeOH, neutralize to a phenolphthalein endpoint with KOH in MeOH, evaporate to dryness under reduced pressure, dry, for every 100 μmoles of acid add 1.1 mL reagent, reflux for 30 min, inject an aliquot. (Reagent contained 200 μmoles 2-bromoacetophenone (phenacyl bromide) and 150 pmoles 18-crown-6 in 1 mL MeCN.)

HPLC VARIABLES
Column: 250 × 4 Corasil II C9 (22% coverage)
Mobile phase: MeOH:water 32:68
Column temperature: 40 ± 0.02

Flow rate: 3.6
Detector: UV 254

CHROMATOGRAM
Retention time: 5 (malonic acid), 7 (succinic acid), 11 (glutaric acid), 18 (adipic acid)

REFERENCE
Grushka, E.; Durst, H.D.; Kikta, E.J. Jr. Liquid chromatographic separation and detection of nanogram quantities of biologically important dicarboxylic acids, *J.Chromatogr.*, **1975**, *112*, 673–678.

SAMPLE
Matrix: solutions
Analyte: fatty acids
Sample preparation: Mix 100 µg fatty acid, 10 µL 12 mg/mL 2-bromoacetophenone (phenacyl bromide) in acetone, and 10 µL 10 mg/mL triethylamine in acetone, let stand at room temperature overnight, inject an aliquot.

HPLC VARIABLES
Column: 900 × 6.4 10 µm µBondapak C18
Mobile phase: Gradient. MeCN:water 67:33 for 125 min, 74:26 for 50 min, 80:20 for 40 min, 97:3 for 35 min (step gradients).
Flow rate: 2
Detector: UV 254

CHROMATOGRAM
Retention time: 100 (lauric acid), 113 (myristoleic acid), 120 (linoleic acid), 168 (linolenic acid), 181 (myristic acid), 205 (palmitoleic acid), 210 (arachidonic acid), 250 (pentadecanoic acid), 265 (linolelaidic acid), 280 (eicosatrienoic acid), 305 (palmitic acid), 322 (oleic acid), 322 (vaccenic acid), 339 (petroselinic acid), 344 (elaidic acid), 356 (eicosadienoic acid), 372 (heptadecanoic acid), 418 (stearic acid), 426 (eicosaenoic acid), 442 (nonadecanoic acid), 450 (arachidic acid), 450 (erucic acid), 458 (heneicosanoic acid), 466 (nervonic acid), 466 (behenic acid), 484 (lignoceric acid)
Limit of detection: 100 ng

REFERENCE
Borch, R.F. Separation of long chain fatty acids as phenacyl esters by high pressure liquid chromatography, *Anal.Chem.*, **1975**, *47*, 2437–2439.

SAMPLE
Matrix: solutions
Analyte: bile acids
Sample preparation: Mix an aliquot of solution (or hydrolyzed bile) with a 50% molar excess of triethylamine in MeCN, warm briefly, add a 50% molar excess of 100 mM 2-bromoacetophenone in MeCN, heat at 80-90° for 45-60 min, evaporate to dryness, reconstitute with dioxane (Caution! Dioxane is a carcinogen!), filter (0.47 µm), inject an aliquot.

HPLC VARIABLES
Column: 250 × 4.6 Partisil 10/25 ODS
Mobile phase: Gradient. n-Heptane:dioxane 90:10 for 3 min then n-heptane:dioxane:isopropanol 70:25:5 (step gradient). (Caution! Dioxane is a carcinogen!)
Flow rate: 1.2
Detector: UV 254

CHROMATOGRAM
Retention time: 15 (lithocholic acid), 17 (deoxycholic acid), 18 (chenodiol), 19 (ursodiol), 24 (hyodeoxycholic acid), 26 (cholic acid)
Limit of quantitation: 5 pmole

REFERENCE
Stellaard, F.; Hachey, D.L.; Klein, P.D. Separation of bile acids as their phenacyl esters by high-pressure liquid chromatography, *Anal.Biochem.*, **1978**, *87*, 359–366.

SAMPLE
Matrix: wine

Analyte: carboxylic acids
Sample preparation: Adjust pH of wine to 7-8 with potassium bicarbonate. Remove a 1 mL aliquot and add it to 1 mL 170 mM 2-bromoacetophenone (phenacyl bromide) in acetone, add 1 mL 17 mM 18-crown-6 in acetone, add 1 mL acetone, heat in a boiling water bath for 75 min, cool, inject a 10 μL aliquot. (Recrystallize phenacyl bromide from n-heptane.)

HPLC VARIABLES
Guard column: 37-50 μm Bondapak C18/Corasil
Column: 250 × 4 7 μm RP-18 (Merck)
Mobile phase: Gradient. MeOH:water from 35:65 to 85:15 over 20 min.
Flow rate: 2
Injection volume: 10
Detector: UV 254

CHROMATOGRAM
Retention time: 2.5 (galacturonic acid), 3.9 (glycolic acid), 5.2 (glyoxylic acid), 5.9 (pyruvic acid), 6.0 (lactic acid), 6.9 (acetic acid), 9.6 (propionic acid), 10.5 (mandelic acid), 10.9 (tartaric acid), 11.1 (ascorbic acid), 11.7 (salicylic acid), 11.7 (p-hydroxybenzoic acid), 11.9 (vanillic acid), 12.2 (butyric acid), 12.4 (malic acid), 13.1 (α-ketoglutaric acid), 13.6 (citramalic acid), 13.7 (succinic acid), 14.2 (phenylacetic acid), 14.6 (cinnamic acid), 14.7 (benzoic acid), 14.8 (glutaric acid), 15.0 (valeric acid), 15.2 (sorbic acid), 15.4 (fumaric acid), 15.6 (anisic acid), 16.3 (gallic acid), 16.7 (isocitric acid), 16.9 (citric acid), 17.1 (benzilic acid), 17.9 (protocatechuic acid), 19.7 (enanthic acid), 21.1 (caprylic acid)

REFERENCE
Mentasti, E.; Gennaro, M.C.; Sarzanini, C.; Baiocchi, C.; Savigliano, M. Derivatization, identification and separation of carboxylic acids in wines and beverages by high-performance liquid chromatography, *J.Chromatogr.*, **1985**, *322*, 177–189.

5-Bromoacetyl Acenaphthene

SAMPLE
Matrix: blood
Analyte: ibuprofen
Sample preparation: 100 μL Plasma + 50 μL 2 μg/mL flurbiprofen + 25 μL 2 M HCl, vortex for 15 s, add 2 mL isooctane:isopropanol 85:15, rotate for 5 min, centrifuge at 3000 rpm for 10 min. Remove the upper organic layer and evaporate it to dryness under a stream of nitrogen at 45°, reconstitute the residue in 25 μL 5 mg/mL 5-bromoacetyl acenaphthene in MeCN, add 10 μL 3% triethylamine in MeCN, vortex for 30 s, heat at 75° for 5 min, evaporate to dryness under reduced pressure, reconstitute with 25 μL MeCN, inject a 20 μL aliquot. (Prepare 5-bromoacetyl acenaphthene as follows. Add 43 g bromoacetyl chloride to 43 g acenaphthene dissolved in 200 mL dichloroethane, cool to

-5° in an ice/salt bath, stir vigorously and add 38 g aluminum chloride in small portions over 90 min, do not allow temperature to go above 3°, place under reduced pressure for 30 min, add an excess of crushed ice. Separate the dichloroethane layer and wash it with two 100 mL portions of dilute HCl, wash with 100 mL 5% sodium carbonate solution. Dry the organic layer over anhydrous magnesium sulfate, remove the solvent under reduced pressure, allow the oily residue to solidify, remove liquid by blotting with filter paper. Purify the solid by chromatography on a 300 × 20 column of 60-120 mesh silica gel, elute with toluene, unreacted acenaphthene elutes first followed by 5-bromoacetyl acenaphthene (mp 87-90°).)

HPLC VARIABLES
Column: 250 × 4.6 5 μm Hypersil C18
Mobile phase: MeCN:water 90:10
Flow rate: 1
Injection volume: 20
Detector: F ex 250 em 450

CHROMATOGRAM
Internal standard: flurbiprofen
Limit of detection: 2.5 pmole

KEY WORDS
rat; plasma; protect from light; pharmacokinetics

REFERENCE
Gifford, L.A.; Owusu-Daaku, F.T.K.; Stevens, A.J. Acenaphthene fluorescence derivatization reagents for use in high-performance liquid chromatography, *J.Chromatogr.A*, **1995**, *715*, 201–212.

2-Bromoacetyl-6-methoxynaphthalene

SAMPLE
Matrix: bile, formulations
Analyte: bile acids
Sample preparation: Bile. Condition a 200 mg Bond Elut C18 SPE cartridge with 5 mL MeOH and 5 mL water. Condition a 500 mg Bond Elut SAX SPE cartridge with 5 mL MeOH, 5 mL water, and 5 mL MeOH. 50 μL Bile + 5 mL 50 mM pH 7.5 phosphate buffer, vortex, add to the C18 SPE cartridge, wash with 5 mL MeOH:40 mM pH 4.3 acetate buffer 40:60, wash with 10 mL water, elute with 2 mL MeOH. Add the eluate to the SAX SPE cartridge, elute with 3.5 mL MeOH, collect all the effluent from the cartridge (J. Pharm. Biomed. Anal. 1990, 8, 235). Evaporate to dryness under a stream of nitrogen, reconstitute with 2 mL MeOH, sonicate at 40° for 3 min, filter (0.2 μm). Add a 500 μL aliquot of the filtrate to 50 μL 0.01% KOH in MeOH, evaporate to dryness, reconstitute with 200 μL MeOH:water 10:90, sonicate at 40° for 3 min, add 300 μL 20 mM tetra-hexylammonium bromide in 100 mM pH 7.0 phosphate buffer, add 50 μL 2.1 mg/mL 2-bromoacetyl-6-methoxynaphthalene in acetone, sonicate at 40° for 10 min, add 50 μL 43.6

µg/mL IS in MeOH:water 75:25, add 300 µL MeCN, sonicate at room temperature for 1 min, inject a 50 µL aliquot. Formulations. Powder capsule contents, weigh out amount containing about 25 mg compound, add 100 mL MeOH (water for bile acid salts), stir for 10 min, filter, dilute the filtrate 10-fold with water (or MeOH:water 10:90 for bile acid salts). Evaporate 50 µL 0.01% KOH in MeOH into a tube, add a 200 µL aliquot of the diluted filtrate, add 300 µL 20 mM tetrahexylammonium bromide in 100 mM pH 7.0 phosphate buffer, add 50 µL 2.1 mg/mL 2-bromoacetyl-6-methoxynaphthalene in acetone, sonicate at 40° for 10 min, add 50 µL 43.6 µg/mL IS in MeOH:water 75:25, add 300 µL MeCN, sonicate at room temperature for 1 min, inject a 50 µL aliquot. (Prepare 2-bromoacetyl-6-methoxynaphthalene as follows. Stir equimolar amounts of 2-acetyl-6-methoxynaphthalene (6'-methoxy-2'-acetonaphthone, Aldrich) and methyltriphenylphosphonium tribromide in anhydrous THF at room temperature under nitrogen for 1 h, dilute the reaction mixture with ether, wash with sodium bisulfite solution, wash with water (Phosphorus and Sulfur 1985, 25, 357). Purify by column chromatography on silica gel with chloroform:petroleum ether 50:50 to obtain 2-bromoacetyl-6-methoxynaphthalene (mp 109-112°) (Chromatographia 1992, 33, 13). (By analogy with a related compound it should be possible to prepare methyltriphenylphosphonium tribromide by slowly adding ca. 2 equivalents bromine to a stirred solution of methyltriphenylphosphonium bromide in acetic acid at 0°, filter to obtain the product (cf. Tet.Lett. 1975, 373). Bromination can also be achieved with phenyltrimethylammonium tribromide over 3 h but the reaction is less selective.))

HPLC VARIABLES
Column: 250 × 4.6 5 µm Hypersil RP-18
Mobile phase: Gradient. For bile use MeCN:water 60:40 for 10 min, to 80:20 over 10 min, maintain at 80:20 for 25 min, return to initial conditions over 5 min. For formulations use isocratic MeCN:water 78:22.
Flow rate: 1
Injection volume: 50
Detector: F ex 300 em 460

CHROMATOGRAM
Retention time: 6.5 (glycocholic acid), 7.5 (glycoursodeoxycholic acid), 12.5 (glycochenodeoxycholic acid), 14 (cholic acid), 18 (ursodiol), 21.5 (glycolithocholic acid), 26.5 (chenodiol), 30 (deoxycholic acid), 31 (lithocholic acid) (gradient elution), 7 (ursodiol), 12 (chenodiol) (isocratic elution)
Internal standard: 6-methoxynaphthacyl ester of valproic acid (23 (gradient), 10.5 (isocratic))
Limit of detection: 1-2 pmole

KEY WORDS
capsules; SPE

REFERENCE
Cavrini, V.; Gatti, R.; Roda, A.; Cerrè, C.; Roveri, P. HPLC-fluorescence determination of bile acids in pharmaceuticals and bile after derivatization with 2-bromoacetyl-6-methoxynaphthalene, *J.Pharm.Biomed.Anal.*, **1993**, *11*, 761–770.

SAMPLE
Matrix: blood
Analyte: enprostil acid
Sample preparation: Prepare an SPE cartridge by adding 150 mg 40 µm Bondesil phenyl (Analytichem) to a 5 mL column. Condition SPE cartridge with 2 mL MeOH, 2 mL water, and 1 mL 20 mM pH 3.0 sodium acetate buffer. 2 mL Plasma + IS (500-1000 cpm), filter (20 µm), add to SPE cartridge, wash with 300 µL pH 3 buffer, wash with 1 mL MeOH:water 40:60, wash with 300 µL 0.3% acetic acid, wash with 1 mL water, elute with 1.5 mL MeOH:water 60:40. Evaporate the eluate under a stream of nitrogen, reconstitute in 125 µL 1 mg/mL 2-bromoacetyl-6-methoxynaphthalene in MeCN and 100 µL 34.6 mg/mL 18-crown-6 in MeCN, add about 2 mg anhydrous sodium sulfate, add about 2 mg potassium carbonate, shake gently for 1 h, evaporate under nitrogen, dissolve the residue in 250 µL dichloromethane. Inject a 200 µL aliquot onto a 100 × 4.6 5 µm Spheri-5 silica column, elute to waste with dichloromethane:MeCN 20:70, elute a 1 min fraction con-

taining the analyte onto a 220 × 4.6 5 μm Spheri-5 silica column, elute this column with the same mobile phase, collect a fraction containing the analyte. Evaporate the fraction to dryness and reconstitute it in 1 mL dichloromethane, add it to an AASP silica SPE cartridge (Analytichem), elute the contents of the cartridge onto column A with mobile phase A for 1 min, elute column A to waste with mobile phase A, elute a 1 min fraction containing the analyte and mix it with water pumped at 2 mL/min, the combined effluent flows onto column B. At the end of this time elute column B with mobile phase B onto column C, elute a 1 min fraction containing the analyte and mix it with water pumped at 0.15 mL/min, store the diluted column effluent in a 600 μL sample loop. Flush the contents of the sample loop onto column D with mobile phase C for 8 min, elute the contents of column D with mobile phase D onto column E, monitor the effluent from column E. (Prepare 2-bromoacetyl-6-methoxynaphthalene as described above.)

HPLC VARIABLES
Column: A 250 × 4.6 7 μm Chemcosorb 7CN (Dychrom); B 250 × 4.6 5 μm Spheri-5 C18 C 250 × 1 5 μm Hypersil C18; D 30 × 1 5 μm Hypersil C18; E 150 × 1 3 μm Hypersil C18
Mobile phase: A MeOH:water:acetic acid 40:60:0.1; B MeOH:water:acetic acid 50:50:0.1; C water; D MeOH:water 40:60
Flow rate: A 1; B 0.05; C 0.1; D 0.05
Injection volume: 200
Detector: F ex 325 em 450 (Corrion S40-450 and LL-400 filters) (laser fluorescence, specially constructed apparatus)

CHROMATOGRAM
Retention time: 20
Internal standard: tritiated enprostil acid
Limit of quantitation: 0.005 ng/mL

KEY WORDS
plasma; SPE; column-switching; normal phase; reverse phase; heart cut; microbore; laser

REFERENCE
Kiang, C.H.; Nolan, T.; Huang, B.L.; Lee, C.P. Determination of femtomole/milliliter concentrations of enprostil acid in human plasma using high-performance liquid chromatography-laser-induced fluorescence detection, *J.Chromatogr.*, **1991**, *567*, 195–212.

SAMPLE
Matrix: formulations
Analyte: azelaic acid
Sample preparation: Ointment, lotion. Weigh out ointment or lotion equivalent to about 15 mg azelaic acid, dissolve in 100 mL MeOH, dilute an aliquot 1:5 with water. 200 μL Sample + 150 μL 20 mM tetrahexylammonium bromide in 100 mM pH 7.0 phosphate buffer + 100 μL 4.2 mg/mL 2-bromoacetyl-6-methoxynaphthalene in acetone, stir for 33 min at 70°, add 150 μL 20 μg/mL IS in MeCN, sonicate for 1 min, inject a 50 μL aliquot into mobile phase A. Ointment. Dissolve in MeCN to give a concentration of 18 μg/mL. 100-200 μL Sample + 100 μL 4.2 mg/mL 2-bromoacetyl-6-methoxynaphthalene in MeCN + 100 μL 1% triethylamine in MeCN, heat at 40° for 40 min, reconstitute in 150 μL 40 μg/mL IS in Mobile Phase B and 450 μL mobile phase B, sonicate for 1 min, inject a 50 μL aliquot into mobile phase B. Powder. Weigh out powder equivalent to about 15 mg azelaic acid, dissolve in 100 mL MeOH, sonicate for 10 min, centrifuge at 4000 rpm for 20 min, filter the supernatant, dilute an aliquot of the filtrate 1:5 with water. 200 μL Sample + 150 μL 20 mM tetrahexylammonium bromide in 100 mM pH 7.0 phosphate buffer + 100 μL 4.2 mg/mL 2-bromoacetyl-6-methoxynaphthalene in acetone, stir for 33 min at 70°, add 150 μL 20 μg/mL IS in MeCN, sonicate for 1 min, inject a 50 μL aliquot into mobile phase A. (Synthesis of 2-bromoacetyl-6-methoxynaphthalene is as described above.)

HPLC VARIABLES
Column: 250 × 4.6 Hypersil 5 ODS
Mobile phase: MeCN:MeOH:THF:water 38.5:28:3.5:30 (A) or 37.4:27.2:3.4:32 (B)
Column temperature: 35

Flow rate: 1.2 (A), 1.6 (B)
Injection volume: 50
Detector: F ex 300 em 460

CHROMATOGRAM
Retention time: 18
Internal standard: valproic acid 6-methoxynaphthacylester (15.5) [Prepare by dissolving 2 mmoles valproic acid and 1 mmole 2-bromoacetyl-6-methoxynaphthalene in 10 mL anhydrous MeCN, add 0.5 mL triethylamine, heat at 60° for 30 min, cool, dilute with 30 mL water, extract three times with 10 mL portions of diethyl ether. Combine the extracts, wash with 5% sodium bicarbonate, wash three times with 10 mL portions of water, dry over anhydrous sodium sulfate, evaporate under reduced pressure, recrystallize from MeOH/water to give white crystals, mp 56-7° (Chromatographia 1992, 33, 13).]

KEY WORDS
ointment; lotion; powder

REFERENCE
Gatti, R.; Andrisano, V.; Di Pietra, A.M.; Cavrini, V. Analysis of aliphatic dicarboxylic acids in pharmaceuticals and cosmetics by liquid chromatography (HPLC) with fluorescence detection, *J.Pharm.Biomed.Anal.*, **1995**, *13*, 589–595.

SAMPLE
Matrix: formulations
Analyte: sodium fusidate
Sample preparation: Weigh out ointment containing 2.14 mg sodium fusidate, take up in 100 mL MeOH:water 20:80. Remove a 200 µL aliquot and add it to 150 µL 20 mM tetrahexylammonium bromide in 100 mM pH 7.0 phosphate buffer and 100 µL 4.2 mg/mL 2-bromoacetyl-6-methoxynaphthalene in acetone, mix, let stand at room temperature for 5 min, add 150 µL 8.9 µg/mL IS in MeCN, sonicate at room temperature for 1 min, inject a 50 µL aliquot. (Prepare 2-bromoacetyl-6-methoxynaphthalene as described above.)

HPLC VARIABLES
Column: 250 × 4.6 5 µm Hypersil ODS
Mobile phase: MeCN:MeOH:water 51:34:15
Column temperature: 35
Flow rate: 1.6
Injection volume: 50
Detector: F ex 300 em 460

CHROMATOGRAM
Retention time: 10
Internal standard: nonanoic acid naphthacyl ester (Prepare as follows. Dissolve 2 mmoles nonanoic acid and 1 mmole 2-bromoacetyl-6-methoxynaphthalene in 10 mL anhydrous MeCN, add 500 µL triethylamine, heat to 60° for 30 min, cool, add 30 mL water, extract three times with 10 mL portions of ether. Combine the extracts and wash them with 5% sodium bicarbonate solution and with three 10 mL portions of water, dry over anhydrous sodium sulfate, evaporate to dryness under reduced pressure, recrystallize from MeOH/water (mp 66-8°) (Chromatographia 1992, 33, 13).) (5.5)
Limit of detection: 1 pmole

KEY WORDS
ointment

REFERENCE
Gatti, R.; Gotti, R.; Bonazzi, D.; Cavrini, V. A comparative evaluation of three detectors in the HPLC analysis of sodium fusidate, *Farmaco*, **1996**, *51*, 115–119.

SAMPLE
Matrix: solutions
Analyte: bile acids
Sample preparation: Mix 100-500 µL of a solution in MeCN with 300 µL 1.28 mg/mL 2-bromoacetyl-6-methoxynaphthalene in MeCN, add 50 µL 3% triethylamine in MeCN, heat

at 70° for 30 min, cool, inject a 50 μL aliquot. (Prepare 2-bromoacetyl-6-methoxy-naphthalene as described above.)

HPLC VARIABLES
Column: 150 × 4.6 5 μm Hypersil RP-18
Mobile phase: Gradient. MeCN:water from 55:45 to 80:20 over 20 min, maintain at 80:20 for 10 min, return to initial conditions over 10 min.
Flow rate: 1
Injection volume: 50
Detector: F ex 300 em 460

CHROMATOGRAM
Retention time: 5 (glycocholic acid), 8.5 (glycochenodeoxycholic acid), 11 (cholic acid), 18 (chenodiol), 19.5 (deoxycholic acid), 29 (lithocholic acid)
Limit of quantitation: 2-3 pmole

REFERENCE
Gatti, R.; Cavrini, V.; Roveri, P. 2-Bromoacetyl-6-methoxynaphthalene: A useful fluorescent labelling reagent for HPLC analysis of carboxylic acids, *Chromatographia*, **1992**, *33*, 13–18.

SAMPLE
Matrix: solutions
Analyte: 4-hydroxybutyric acid
Sample preparation: Mix 200 μL of a solution in MeOH:water 10:90 with 150 μL 0.5 mM tetrakis(decyl)ammonium bromide in 5 mM pH 7.0 phosphate buffer and 100 μL 4.2 mg/mL 2-bromoacetyl-6-methoxynaphthalene in acetone, stir at 70° for 65 min, add 150 μL 80 μg/mL IS in MeCN, sonicate at room temperature for 1 min, inject a 50 μL aliquot. (Prepare 2-bromoacetyl-6-methoxynaphthalene as described above.)

HPLC VARIABLES
Column: 250 × 4.5 Hypersil 5 ODS
Mobile phase: MeCN:MeOH:25 mM pH 4.5 phosphate buffer 18.7:15.3:66
Column temperature: 35
Flow rate: 1.7
Injection volume: 50
Detector: F ex 300 em 460

CHROMATOGRAM
Retention time: 33
Internal standard: 4-(6-methoxy-2-naphthyl)-4-oxobutanoic acid (Synthesis of 4-(6-methoxy-2-naphthyl)-4-oxobutanoic acid is as follows. Dissolve 43 g anhydrous aluminum chloride in 200 mL nitrobenzene (dried over calcium chloride) with mechanical stirring, add 39.5 g finely-ground 2-methoxynaphthalene, cool to 5° with an ice bath, add 23 mL re-distilled acetyl chloride dropwise over 15-20 min so as to maintain the temperature between 10.5° and 13°. After addition is complete cool in ice-water with stirring for 2 h, let stand at room temperature for 12 h, cool in an ice bath, add with stirring to 200 g crushed ice, add 100 mL concentrated HCl, add 50 mL chloroform. Remove the organic layer and wash it three times with 100 mL portions of water, steam distil while heating the reaction flask to 120°. After 3-4 L water has distilled (3 h) allow the residue in the flask to cool. Remove residual water and extract it with chloroform, dissolve the residue in 100 mL chloroform, combine the chloroform layers, dry over anhydrous magnesium sulfate, evaporate under reduced pressure, distil at 150-165°/0.02 mm Hg with the receiving flask cooled in an ice bath, recrystallize 40 g of the distillate from 75 mL MeOH to obtain 2-acetyl-6-methoxynaphthalene as white crystals (mp 106.5-108°) (do not cool the MeOH solution below 0°) (Org. Syn. 1988, Coll. Vol. VI, 34). 2-Acetyl-6-methoxynaphthalene may also be obtained from Janssen Chimica, Belgium. Dissolve 5 g 2-acetyl-6-methoxynaphthalene in the minimum amount of warm glacial acetic acid, add 2.5 g glyoxylic acid, reflux for 24 h, evaporate to dryness under reduced pressure, dissolve the residue in chloroform, extract three times with 5% sodium carbonate solution. Combine the extracts and acidify them with concentrated HCl, filter to recover the product, recrystallize from MeOH/water to obtain 4-(6-methoxy-2-naphthyl)-4-oxo-2-butenoic acid (mp 165-168°). Dissolve 4 g 4-(6-methoxy-2-naphthyl)-4-oxo-2-butenoic acid in the minimum amount of THF,

add palladium on charcoal, hydrogenate until 450 mL hydrogen are absorbed, filter, evaporate to dryness under reduced pressure, recrystallize from acetic acid to obtain 4-(6-methoxy-2-naphthyl)-4-oxobutanoic acid as a white solid (mp 148°) (Farmaco Ed. Sci. 1982, 37,171).) (22)

REFERENCE
Gatti, R.; Bousquet, E.; Bonazzi, D.; Cavrini, V. Determination of carboxylic acid salts in pharmaceuticals by high-performance liquid chromatography after pre-column fluorogenic labelling, *Biomed.Chromatogr.*, **1996**, *10*, 19–24.

SAMPLE
Matrix: solutions
Analyte: pyroglutamic acid (2-pyrrolidone-5-carboxylic acid)
Sample preparation: Mix 100 μL of a solution in MeCN with 100 μL 4.2 mg/mL 2-bromoacetyl-6-methoxynaphthalene in MeCN and 100 μL 1% triethylamine in MeCN, stir at 40° for 22 min, evaporate to dryness under a stream of nitrogen, reconstitute with 150 μL 67 μg/mL griseofulvin in mobile phase, add 500 μL mobile phase, sonicate at room temperature for 1 min, inject a 50 μL aliquot. (Prepare 2-bromoacetyl-6-methoxynaphthalene as described above.)

HPLC VARIABLES
Column: 250 × 4.5 Hypersil 5 ODS
Mobile phase: MeCN:MeOH:water 25.8:17.2:57
Column temperature: 35
Flow rate: 1.2
Injection volume: 50
Detector: F ex 300 em 460

CHROMATOGRAM
Retention time: 12
Internal standard: griseofulvin (15)

REFERENCE
Gatti, R.; Bousquet, E.; Bonazzi, D.; Cavrini, V. Determination of carboxylic acid salts in pharmaceuticals by high-performance liquid chromatography after pre-column fluorogenic labelling, *Biomed.Chromatogr.*, **1996**, *10*, 19–24.

SAMPLE
Matrix: solutions
Analyte: thioctic acid
Sample preparation: Mix 200 μL of a solution in MeOH:water 10:90 with 150 μL 20 mM tetrahexylammonium bromide in 100 mM pH 7.0 phosphate buffer and 100 μL 4.2 mg/mL 2-bromoacetyl-6-methoxynaphthalene in acetone, sonicate for 3 min, let stand for 22 min, add 150 μL 4.5 μg/mL IS in MeCN, sonicate at room temperature for 1 min, inject a 50 μL aliquot. (Prepare 2-bromoacetyl-6-methoxynaphthalene as described above.)

HPLC VARIABLES
Column: 250 × 4.5 Hypersil 5 ODS
Mobile phase: MeCN:MeOH:water 37.4:30.6:32
Column temperature: 35
Flow rate: 1.1
Injection volume: 50
Detector: F ex 300 em 460

CHROMATOGRAM
Retention time: 16.5
Internal standard: n-hexanoic acid 6-methoxynaphthacyl ester (Dissolve 2 mmole n-hexanoic acid and 1 mmole 2-bromoacetyl-6-methoxynaphthalene in 10 mL anhydrous MeCN, add 500 μL triethylamine, heat to 60° for 30 min, cool, add 30 mL water, extract three times with 10 mL portions of diethyl ether. Combine the organic layers and wash them with 5% sodium bicarbonate solution, wash three times with 10 mL portions of water, dry over anhydrous sodium sulfate, evaporate to dryness under reduced pressure, recrys-

tallize from MeOH/water to give 6-methoxynaphthacyl ester of n-hexanoic acid (mp 79-80°) (J. Pharm. Biomed. Anal. 1993, 11, 761) (19)

REFERENCE
Gatti, R.; Bousquet, E.; Bonazzi, D.; Cavrini, V. Determination of carboxylic acid salts in pharmaceuticals by high-performance liquid chromatography after pre-column fluorogenic labelling, *Biomed.Chromatogr.*, **1996**, *10*, 19–24.

SAMPLE
Matrix: solutions
Analyte: bile acids
Sample preparation: Mix 200 μL of a solution of bile acids with 50 μL 2.1 mg/mL 2-bromoacetyl-6-methoxynaphthalene in acetone, add 300 μL 10 mM tetrakis (decyl)ammonium bromide in 100 mM pH 7.0 phosphate buffer, heat at 40° for with sonication 10 min, add 300 μL 5.1 μM IS in MeCN, sonicate at room temperature for 1 min, inject a 50 μL aliquot. (Prepare 2-bromoacetyl-6-methoxynaphthalene as described above.)

HPLC VARIABLES
Column: 250 × 4.6 Ultracarb 5 ODS
Mobile phase: Gradient. A was water. B was MeCN:MeOH 60:40. A:B 55:45 for 20 min, to 30:70 over 10 min, maintain at 30:70 for 25 min, return to initial conditions over 5 min.
Column temperature: 35
Flow rate: 1.2
Injection volume: 50
Detector: F ex 300 em 460

CHROMATOGRAM
Retention time: 13 (ursodiol), 16 (cholic acid), 29 (chenodiol), 31 (deoxycholic acid), 40 (lithocholic acid)
Internal standard: 6-methoxynaphthacyl ester of lauric acid (36)
Limit of detection: 1-2 pmole

REFERENCE
Gatti, R.; Roda, A.; Cerre, C.; Bonazzi, D.; Cavrini, V. HPLC-fluorescence determination of individual free and conjugated bile acids in human serum, *Biomed.Chromatogr.*, **1997**, *11*, 11–15.

1-(Bromoacetyl)pyrene

RELATED REFERENCES
Comesana-Losada, M.; Gago-Martínez, A.; Leao-Martins, J.M.; Rodríguez-Vázquez, J.A. High-performance liquid chromatographic methods for determination of marine biotoxins. *Analyst* **1996**, *121*, 1665-1670.

Kelly, S.S.; Bishop, A.G.; Carmody, E.P.; James, K.J. Isolation of dinophysistoxin-2 and the high-performance liquid chromatographic analysis of diarrhetic shellfish toxins using derivatisation with 1-bromoacetylpyrene. *J.Chromatogr.A* **1996**, *749*, 33-40.

SAMPLE
Matrix: bile, blood
Analyte: bile acids
Sample preparation: Serum. 100-200 μL Serum + 1 mL MeOH, mix, sonicate for 15 min. Remove a 600 μL aliquot of the supernatant and evaporate it to dryness under a stream of nitrogen, reconstitute with 1 mL 50 mM pH 7.0 phosphate buffer, add to a Sep-Pak C18 SPE cartridge, wash with 2 mL MeOH:water 20:80, elute with 4 mL MeOH:water 80:20. Evaporate the eluate to dryness under reduced pressure at 40°, reconstitute with 1 mL MeOH. Remove a 500 μL aliquot and add it to 50 μL 100 μM lauric acid in MeOH, add 50 μL 0.1 mg/mL KOH on MeOH, evaporate to dryness under a stream of nitrogen, add 100 μL 1 mg/mL dicyclohexyl-18-crown-6 in MeCN, add 100 μL 25 mM 1-(bromoace-tyl)pyrene in MeCN, mix, heat at 40° for 30 min, cool, inject an 8 μL aliquot. Bile. Mix 10 μL bile with 10 mL 50 mM pH 7.0 phosphate buffer, add a 1 mL aliquot to a Sep-Pak C18 SPE cartridge, wash with 2 mL MeOH:water 20:80, elute with 4 mL MeOH:water 80:20. Evaporate the eluate to dryness under reduced pressure at 40°, reconstitute with 1 mL MeOH. Remove a 500 μL aliquot and add it to 50 μL 100 μM lauric acid in MeOH, add 50 μL 0.1 mg/mL KOH on MeOH, evaporate to dryness under a stream of nitrogen, add 100 μL 1 mg/mL dicyclohexyl-18-crown-6 in MeCN, add 100 μL 25 mM 1-(bromoace-tyl)pyrene in MeCN, mix, heat at 40° for 30 min, cool, inject an 8 μL aliquot.

HPLC VARIABLES
Column: 100 × 8 10 μm Model RCM-100 Radial-Pak A (Waters)
Mobile phase: Gradient. MeCN:MeOH:water 100:50:40 for 30 min then 100:50:20 (step gradient).
Flow rate: 2
Injection volume: 8
Detector: F ex 370 em 440

CHROMATOGRAM
Retention time: 8 (glycoursodeoxycholic acid), 11 (glycocholic acid), 20 (glycochenodeoxy-cholic acid), 23 (glycodeoxycholic acid), 26 (ursodiol), 32 (cholic acid), 37 (glycolithocholic acid), 44 (chenodiol), 45 (deoxycholic acid), 66 (lithocholic acid)
Internal standard: lauric acid (56)
Limit of quantitation: 50 pmole
Limit of detection: 10 pmole

KEY WORDS
serum; SPE

REFERENCE
Kamada, S.; Maeda, M.; Tsuji, A. Fluorescence high-performance liquid chromatographic determination of free and conjugated bile acids in serum and bile using 1-bromoacetylpyrene as a pre-labeling reagent, *J.Chromatogr.*, **1983**, *272*, 29–41.

SAMPLE
Matrix: blood
Analyte: simvastatin
Sample preparation: Condition a 2.8 mL 500 mg Bond Elut C8 SPE cartridge with 2 mL MeOH and 2.5 mL water. Condition a 3 mL 200 mg Bond Elut C18 SPE cartridge with 1.5 mL MeCN and 2 mL water. 1 mL Plasma + 2.5 ng IS, mix, add to the C8 SPE cartridge, wash with 2 mL MeCN:water 10:90, wash with 1 mL MeOH:water 30:70, wash with 2 mL MeOH:water 60:40 (the ring-opened active metabolite elutes in this fraction), elute with 2 mL MeCN, add 100 μL 20 mM potassium carbonate to the eluate, evaporate to dryness under reduced pressure at 40° (this hydrolyses the lactone), dry under vacuum for more than 30 min, reconstitute with 100 μL 10 mM 1-(bromoace-tyl)pyrene in DMF, add 100 μL 10 mM 18-crown-6 in DMF, mix, let stand at room temperature for 30 min, add 2 mL MeCN:triethylamine 90:10, add to a 10 mL 100 mg Bond Elut LRC PBA SPE cartridge, wash with 4 mL MeOH, wash with 2 mL MeCN, elute with 2 mL MeCN:propylene glycol 60:40, dilute the eluate with 1 mL water, add to the C18 SPE cartridge, wash with 2 mL MeCN:water 70:30, elute with 3 mL MeCN. Evaporate the eluate to dryness, reconstitute the residue in 300 μL MeCN:water 70:30, inject a 150

µL aliquot onto column A and elute to waste with mobile phase A, after 12.5 min elute column A onto column B with mobile phase A, after another 5 min remove column A from the circuit and elute column B with mobile phase B, monitor the effluent from column B. Flush column A with MeOH then re-equilibrate with mobile phase A for 6 min. (The procedure can also be modified to determine the active metabolite.)

HPLC VARIABLES
Column: A 150 × 4.6 5 µm Bondesil CH (Varian); B 150 × 4.6 5 µm Capcell Pak C18 UG 120 (Shiseido)
Mobile phase: A MeOH:water 80:20; B MeCN:water 80:20
Column temperature: 40
Flow rate: 1
Injection volume: 150
Detector: F ex 360 em 430

CHROMATOGRAM
Retention time: 27.5
Internal standard: 2-ethyl-2-methylbutanoate ester analog of simvastatin (31)
Limit of detection: 20 pg/mL

KEY WORDS
plasma; column-switching; SPE; heart-cut; pharmacokinetics

REFERENCE
Ochiai, H.; Uchiyama, N.; Imagaki, K.; Hata, S.; Kamei, T. Determination of simvastatin and its active metabolite in human plasma by column-switching high-performance liquid chromatography with fluorescence detection after derivatization with 1-bromoacetylpyrene, *J.Chromatogr.B*, **1997**, *694*, 211–217.

SAMPLE
Matrix: solutions
Analyte: polyether antibiotics
Sample preparation: Condition a Mega Bond Elut silica gel SPE cartridge with benzene (Caution! Benzene is a carcinogen!). Evaporate a solution in MeOH to dryness, add 5 mL 5.28 mg/mL 1-bromoacetylpyrene in MeCN, add 5 mL 1.28 mg/mL Kryptofix 222 in MeCN, heat at 50° for 1.5 h, cool. Either inject this solution directly or evaporate it to dryness, dissolve the residue in 5 mL benzene:chloroform 50:50, rinse out the flask with two 5 mL portions of benzene:chloroform 50:50, filter, add the filtrate to the SPE cartridge, elute with two 5 mL portions of benzene:acetone 70:30. Evaporate the eluate to dryness, reconstitute the residue in 10 mL MeCN, inject an aliquot.

HPLC VARIABLES
Column: 250 × 4.6 5 µm Develosil 5C18
Mobile phase: MeOH:water 97:3
Flow rate: 1
Detector: F ex 360 em 420

CHROMATOGRAM
Retention time: 13.5 (monensin), 14.5 (lasalocid), 21.5 (salinomycin), 27 (narasin)
Internal standard: 18,19-dihydrosalinomycin (25), 18,19-dihydro-20-ketosalinomycin (16.5)
Limit of quantitation: 200 ng/mL

KEY WORDS
SPE

REFERENCE
Asukabe, H.; Murata, H.; Harada, K.-I.; Suzuki, M.; Oka, H.; Ikai, Y. Improvement of chemical analysis of antibiotics. XX. Basic study on high-performance liquid chromatographic determination of four polyether antibiotics pre-derivatized with 1-bromoacetylpyrene, *J.Chromatogr.A*, **1993**, *657*, 349–356.

2-Bromo-4'-chloroacetophenone

SAMPLE
Matrix: bulk
Analyte: fatty acids
Sample preparation: Mix 10 μmoles fatty acid, 40 μmoles N,N-diisopropylethylamine or lithium carbonate, and 10-20 μmoles 2-bromo-4'-chloroacetophenone (p-chlorophenacyl bromide) in 1 mL DMF, heat at 65° for 15 min, inject an aliquot.

HPLC VARIABLES
Column: two 300 × 3.9 μBondapak C18 columns in series
Mobile phase: Gradient. MeCN:water from 40:60 to 100:0 (Waters convex curve 5) over 3 h.
Flow rate: 1
Detector: UV 254

CHROMATOGRAM
Retention time: 14 (lactic acid), 18 (acetic acid), 22 (propionic acid), 27 (butyric acid), 26 (caproic acid), 49 (caprylic acid), 60 (capric acid), 75 (lauric acid), 86 (linolenic acid), 88 (myristic acid), 90 (arachidonic acid), 92 (palmitoleic acid), 95 (linoleic acid), 103 (palmitic acid), 105 (oleic acid), 107 (elaidic acid, vaccenic acid), 117 (stearic acid), 130 (arachidic acid), 142 (behenic acid, nervonic acid), 153 (lignoceric acid)

REFERENCE
Jordi, H.C. Separation of long and short chain fatty acids as naphthacyl and substituted phenacyl esters by high performance liquid chromatography, *J.Liq.Chromatogr.*, **1978**, *1*, 215–230.

4-Bromomethyl-7-acetoxycoumarin

SAMPLE
Matrix: blood
Analyte: fatty acids
Sample preparation: 10 μL Plasma + 200 μL 500 mM pH 6.5 phosphate buffer + 50 μL 20 μM IS in MeOH + 2 mL n-heptane:chloroform 50:50, vortex for 2 min, centrifuge at 1000 g for 10 min. Remove the lower organic layer and evaporate it to dryness, reconsti-

tute the residue in two 100 μL aliquots of acetone. Evaporate the acetone solution to dryness, add 2-3 mg finely powdered potassium bicarbonate:sodium sulfate 50:50, add 50 μL 800 μM dibenzo-18-crown-6 in acetone, add 50 μL 2 mM 4-bromomethyl-7-acetoxycoumarin in acetone, heat at 50° in the dark for 30 min, inject a 50 μL aliquot. (4-Bromomethyl-7-acetoxycoumarin (7-acetoxy-4-bromomethylcoumarin) is available from Tokyo Kasei (TCI America, Portland OR). Synthesis is as follows. Reflux 50 g 7-hydroxy-4-methylcoumarin (β-methylumbelliferone) and 100 mL acetic anhydride for 1 h, cool, pour into 500 mL cold water, filter, dry the solid, recrystallize from EtOH to give 4-methyl-7-acetoxycoumarin. Reflux 10 g 4-methyl-7-acetoxycoumarin, 9 g N-bromosuccinimide, a little 2,2'-(azobis(2-methylpropionitrile) (α,α'-azobisisobutyronitrile, Eastman), and 100 mL carbon tetrachloride for 20 h, cool, evaporate under reduced pressure to remove the solvent, wash the residue with water, filter, dry, recrystallize from ethyl acetate/cyclohexane to give 4-bromomethyl-7-acetoxycoumarin (mp 184-185°) (J. Chromatogr. 1982, 234, 121).)

HPLC VARIABLES

Column: 250 × 4 5 μm LiChrosorb RP-18
Mobile phase: Gradient. MeCN:MeOH:water from 35:35:30 to 0:90:10 over 70 min (convex gradient).
Column temperature: 40
Flow rate: 1.2
Injection volume: 50
Detector: F ex 365 em 460 following post-column reaction. The column effluent mixed with 200 mM NaOH in MeOH:water 80:20 pumped at 0.4 mL/min and the mixture flowed through a 3.5 m × 0.5 mm ID stainless steel coil at 50° to the detector.

CHROMATOGRAM

Retention time: 7 (caproic acid), 9 (heptanoic acid), 12 (caprylic acid), 14 (nonanoic acid), 17 (capric acid), 21 (undecanoic acid), 26 (lauric acid), 28 (myristoleic acid), 31 (tridecanoic acid), 35 (linolenic acid), 36 (myristic acid), 38 (arachidonic acid), 39 (palmitoleic acid), 42 (linoleic acid), 50 (palmitic acid), 51 (oleic acid), 66 (stearic acid)
Internal standard: margaric acid (57)
Limit of quantitation: 5 pmole

KEY WORDS

post-column reaction; plasma

REFERENCE

Tsuchiya, H.; Hayashi, T.; Sato, M.; Tatsumi, M.; Takagi, N. Simultaneous separation and sensitive determination of free fatty acids in blood plasma by high-performance liquid chromatography, *J.Chromatogr.*, **1984**, *309*, 43–52.

SAMPLE

Matrix: blood
Analyte: carboxylic acids
Sample preparation: 15 μL Plasma + 485 μL 6.2 μM IS in water, mix, add 1 mL MeOH, vortex, add 3 mL 50 mg/mL BHT in chloroform, vortex for 1 min, centrifuge at 2000 g for 10 min, remove the organic layer, extract the aqueous layer with 4 mL chloroform:MeOH 75:25. Combine the organic layers and evaporate them to dryness under a stream of nitrogen, reconstitute with 1 mL chloroform, add to a Bond Elut aminopropyl SPE cartridge, elute with diethyl ether:acetic acid 98:2. Evaporate the eluate to dryness under a stream of nitrogen, reconstitute with acetone, evaporate to dryness (?), add 50 μL 1 mg/mL 4-bromomethyl-7-acetoxycoumarin in acetone, add 50 μL 800 μM dibenzo-18-crown-6 in acetone, add 2-3 mg of a finely powdered mixture of potassium bicarbonate and sodium sulfate (50:50 ?), shake at 50° for 30 min, inject a 5-50 μL aliquot. (4-Bromomethyl-7-acetoxycoumarin (7-acetoxy-4-bromomethylcoumarin) is available from Tokyo Kasei (TCI America, Portland OR). Synthesis is as follows. Reflux 50 g 7-hydroxy-4-methylcoumarin (β-methylumbelliferone) and 100 mL acetic anhydride for 1 h, cool, pour into 500 mL cold water, filter, dry the solid, recrystallize from EtOH to give 4-methyl-7-acetoxycoumarin. Reflux 10 g 4-methyl-7-acetoxycoumarin, 9 g N-bromosuccinimide, a little 2,2'-(azobis(2-methylpropionitrile) (α,α'-azobisisobutyronitrile, Eastman), and 100 mL carbon tetrachloride for 20 h, cool, evaporate under reduced pressure to remove the

solvent, wash the residue with water, filter, dry, recrystallize from ethyl acetate/cyclohexane to give 4-bromomethyl-7-acetoxycoumarin (mp 184-185°) (J. Chromatogr. 1982, 234, 121).)

HPLC VARIABLES

Column: 100×5 5 μm Nova Pak radial compression

Mobile phase: Gradient. A was MeCN:MeOH:water 35:35:30. B was MeOH:water 90:10. A:B from 100:0 to 0:100 over 70 min (convex gradient).

Flow rate: 2

Injection volume: 5-50

Detector: F ex 365 em 460 following post-column reaction. The column effluent mixed with 200 mM NaOH in MeOH:water 80:20 pumped at 0.4 mL/min and the mixture flowed through a 9 m × 0.23 mm ID stainless steel coil at 80° to the detector.

CHROMATOGRAM

Retention time: 22 (lauric acid), 26 (myristoleic acid), 36 (linolenic acid), 39 (myristic acid), 41 (arachidonic acid), 44 (palmitoleic acid), 50 (linoleic acid), 54 (eicosatrienoic acid), 56 (palmitic acid), 57 (oleic acid), 69 (stearic acid)

Internal standard: margaric acid (62)

Limit of detection: 10 fmole

KEY WORDS

post-column reaction; rat; plasma; SPE

REFERENCE

Kelly, R.A.; O'Hara, D.S.; Kelley, V. High-performance liquid chromatographic separation of femtomolar quantities of endogenous carboxylic acids, including arachidonic acid metabolites, as 4-bromomethyl-7-acetoxycoumarin derivatives, *J.Chromatogr.*, **1987**, *416*, 247–254.

SAMPLE

Matrix: blood

Analyte: ciprostene

Sample preparation: Condition a 1 mL 100 mg Bond Elut C2 SPE cartridge with two 1 mL portions of MeCN, with 1 mL water, and with 1 mL 1% phosphoric acid, do not allow to dry. Condition a Bond Elut CN SPE cartridge with two 1 mL portions of n-hexane. 250 μL Plasma + 50 μL 375 ng/mL carbacyclin in MeOH + 750 μL 1% phosphoric acid, add to the C2 SPE cartridge, wash with two 1 mL portions of water, wash with two 1 mL portions of MeOH:water 40:60, dry under vacuum for 10 min, wash with two 1 mL portions of n-hexane:MTBE 75:25, elute with 1 mL n-hexane:MTBE 20:80. Evaporate the eluate to dryness under reduced pressure, reconstitute with 100 μL MeOH, evaporate to dryness under a stream of air, add 10 mg solid potassium bicarbonate:sodium sulfate 1:1, add 50 μL 200 μM dibenzo-18-crown-6 in acetone, add 50 μL 1 mM 4-bromomethyl-7-acetoxycoumarin in acetone, stir at 50° for 30 min, cool, add 900 μL n-hexane, vortex, add to the CN SPE cartridge, wash with two 1 mL portions of n-hexane, wash with two 1 mL portions of n-hexane:ethyl acetate 80:20, dry under vacuum for 1 min, elute slowly with 1 mL MeCN. Evaporate the eluate to dryness under a stream of air at 30°, reconstitute the residue in 50 μL MeCN:water:trifluoroacetic acid 50:50:0.1, inject an aliquot. (4-Bromomethyl-7-acetoxycoumarin (7-acetoxy-4-bromomethylcoumarin) is available from Tokyo Kasei (TCI America, Portland OR). Synthesis is as described above.)

HPLC VARIABLES

Guard column: 15×3.2 NewGuard RP18

Column: 250×4.6 5 μm Zorbax ODS

Mobile phase: MeCN:water:trifluoroacetic acid 55:44.9:0.1

Column temperature: 45

Flow rate: 1.5

Detector: F ex 370 em 466 following post-column reaction. The column effluent mixed with 100 mM NaOH pumped at 0.5 mL/min and the mixture flowed through a 4 mL knitted coil of PTFE tubing at 80° to the detector.

CHROMATOGRAM

Retention time: 31.3

Internal standard: carbacyclin (23.7)
Limit of quantitation: 5 ng/mL

KEY WORDS
plasma; post-column reaction; SPE

REFERENCE
James, C.A.; Simmonds, R.J.; Burton, N.K. An HPLC assay for a prostacyclin analogue, ciprostene calcium, in human plasma, *J.Liq.Chromatogr.*, **1990**, *13*, 1143–1158.

SAMPLE
Matrix: bulk
Analyte: carboxylic acids
Sample preparation: Dissolve 0.1-20 nmole carboxylic acid, a 2.5-fold excess of 4-bromomethyl-7-acetoxycoumarin, and an equimolar amount of dibenzo-18-crown-6 in 50 μL acetone, add 3 mg finely powdered potassium bicarbonate:sodium sulfate 50:50, heat at 50° in the dark for 30 min, cool, inject an aliquot. (4-Bromomethyl-7-acetoxycoumarin (7-acetoxy-4-bromomethylcoumarin) is available from Tokyo Kasei (TCI America, Portland OR). Synthesis is as described above.)

HPLC VARIABLES
Column: 250 × 2.1 10 μm ODS-6013 (Kyowa Seimitsu, Tokyo)
Mobile phase: Gradient. MeCN:water from 40:60 to 90:10 over 32 min (convex gradient).
Column temperature: 50
Flow rate: 0.8
Detector: F ex 365 em 460 following post-column reaction. The column effluent mixed with 100 mM pH 11.0 borate buffer pumped at 0.4 mL/min and the mixture flowed through a 10 m × 0.5 mm ID stainless steel coil at 50° to the detector.

CHROMATOGRAM
Retention time: 8 (butyric acid), 11 (caproic acid), 15 (caprylic acid), 19 (capric acid), 22 (lauric acid), 28 (palmitic acid), 32 (stearic acid), 36 (arachidic acid)
Limit of detection: 10 fmole

KEY WORDS
post-column reaction

REFERENCE
Tsuchiya, H.; Hayashi, T.; Naruse, H.; Takagi, N. High-performance liquid chromatography of carboxylic acids using 4-bromomethyl-7-acetoxycoumarin as fluorescence reagent, *J.Chromatogr.*, **1982**, *234*, 121–130.

SAMPLE
Matrix: seminal fluid
Analyte: prostaglandins
Sample preparation: 1-5 μL Seminal fluid + 100 μL 5 μM IS in MeOH, mix, add 3 mL water to the supernatant, acidify to pH 3-4 with 100 mM HCl, extract with 7 mL ethyl acetate. Remove the ethyl acetate layer and evaporate it to dryness, reconstitute with MeOH. Evaporate to dryness in a clean tube, add 10 mg finely-powdered potassium bicarbonate:sodium sulfate 50:50, add 50 μL 0.4-1 mM 4-bromomethyl-7-acetoxycoumarin in acetone, add 50 μL 200 μM dibenzo-18-crown-6 in acetone, heat in the dark at 80° for 1 h, cool, inject a 20-40 μL aliquot. (4-Bromomethyl-7-acetoxycoumarin (7-acetoxy-4-bromomethylcoumarin) is available from Tokyo Kasei (TCI America, Portland OR). Synthesis is as described above.)

HPLC VARIABLES
Column: 250 × 4 5 μm LiChrosorb RP-18
Mobile phase: Gradient. MeCN:water from 30:70 to 90:10 over 99 min (Concave 1 curve (64 min) using a Japan Spectroscopic Model GP-A30 solvent programmer).
Column temperature: 50
Flow rate: 1
Injection volume: 20-40
Detector: F ex 365 em 460 following post-column reaction. The effluent from the column mixed with 100 mM NaOH pumped at 0.4 mL/min and the mixture flowed through a 10

m × 0.5 mm ID stainless steel coil at 50° to the detector. (The prostaglandins are chromatographed as the coumarin derivatives then hydrolyzed in the post-column reactor to fluorescent 7-hydroxy-4-hydroxymethylcoumarin.)

CHROMATOGRAM
Retention time: 43 (dinoprost), 46 (dinoprostone), 47 (alprostadil)
Internal standard: 16-methylprostaglandin $F_{1\alpha}$ (49)
Limit of detection: 10 fmole

KEY WORDS
post-column reaction

REFERENCE
Tsuchiya, H.; Hayashi, T.; Naruse, H.; Takagi, N. Sensitive high-performance liquid chromatographic method for prostaglandins using a fluorescence reagent, 4-bromomethyl-7-acetoxycoumarin, *J.Chromatogr.*, **1982**, *231*, 247−254.

9-Bromomethylacridine

SAMPLE
Matrix: blood
Analyte: fatty acids
Sample preparation: 50 μL Plasma + 10 μL IS solution + 450 μL micelle solution, vortex for 10 s, add 25 μL 28 mg/mL 9-bromomethylacridine in acetone, mix. Remove a 50 μL aliquot and heat it to 60° for 6 min, inject the whole amount through a 2 μm stainless steel filter of 8 sq mm area onto column A, wash to waste with 400 μL mobile phase A, back flush the contents of column A onto column B with the mobile phase B, monitor the effluent from column B. After each injection backflush the stainless steel filter with 1 mL buffer. (Micelle solution was 25 mM Arkopal N-130 (a polyoxyethylene(13)nonylphenol, Hoechst Holland, Amsterdam) in 10 mM pH 7.0 phosphate buffer containing 6 mM tetrakis(decyl)ammonium bromide. Synthesize 9-bromomethylacridine as follows. Heat 10 g diphenylamine, 10 mL glacial acetic acid, and 40 g anhydrous zinc chloride to 220°, evaporate excess acetic acid with stirring, heat at 220-230° for 6 h, digest with hot 10% sulfuric acid, make strongly alkaline with 25% ammonia to dissolve the zinc chloride. Extract the insoluble residue with toluene. Extract the organic layer with 10% sulfuric acid, make the aqueous layer alkaline with aqueous ammonia. Collect the yellow precipitate that separates and recrystallize it twice from petroleum ether to give 9-methyl acridine as pale yellow needles (Chromatographia 1989, 28, 267). Reflux 560 mg 9-methylacridine, 445 mg N-bromosuccinimide, and 10 mg benzoyl peroxide in 30 mL carbon tetrachloride for more than 2 h, cool, chromatograph on silica gel with benzene:ethyl

acetate 30:1 (Caution! Benzene is a carcinogen!) to obtain 9-bromomethylacridine as yellow crystals (mp 147-151°) (Anal. Lett. 1987, 20, 1581).)

HPLC VARIABLES

Column: A 10 × 2.1 40 μm Chromsep C18 (Chrompack); B 100 × 3 5 μm Chromspher C18 (Chrompack)

Mobile phase: A 10 mM pH 7.0 phosphate buffer; B Gradient. MeOH:water 75:25 for 3 min, to 100:0 over 12 min (concave gradient).

Injection volume: 50

Detector: UV 254; F ex 362 em 418

CHROMATOGRAM

Retention time: 6.5 (linolenic acid), 7 (arachidonic acid), 7.5 (palmitoleic acid), 8.5 (myristic acid), 8.9 (linoleic acid), 10 (oleic acid), 10.3 (palmitic acid), 13 (stearic acid)

Internal standard: heptadecanoic acid (11.8)

Limit of detection: 300 nM

KEY WORDS

column-switching; plasma

REFERENCE

van der Horst, F.A.L.; Post, M.H.; Holthuis, J.J.M.; Brinkman, U.A.T. Automated high-performance liquid chromatographic determination of plasma free fatty acids using on-line derivatization with 9-bromomethylacridine based on micellar phase-transfer catalysis, *J.Chromatogr.*, **1990**, *500*, 443–452.

SAMPLE

Matrix: food

Analyte: fatty acids

Sample preparation: Dissolve 20-40 mg oil, butter, or margarine in 5 mL chloroform. Remove a 100 μL aliquot and add it to 1.5 mL 25 mM tetraethylammonium carbonate in MeOH, heat in a capped tube at 70° for 5 min, heat at 70° for 30 min in an open tube, evaporate to dryness under a stream of nitrogen, reconstitute with 1 mL DMF. Remove a 100 μL aliquot and add it to 200 μL 5 mM 9-bromomethylacridine in DMF, let stand at room temperature for at least 10 min, inject a 10 μL aliquot. (Synthesize 9-bromomethylacridine as described above. Prepare tetraethylammonium carbonate by adding dry ice to an aqueous solution of tetraethylammonium hydroxide, evaporate to dryness under reduced pressure over phosphorus pentoxide at 56° to obtain tetraethylammonium carbonate as a white hygroscopic powder (decomposes 284-288°).)

HPLC VARIABLES

Column: 150 × 4.6 TSK-gel ODS 120A (Toyo Soda)

Mobile phase: Gradient. MeOH:water from 90:10 to 97:3 over 32 min (JASCO concave curve 2), maintain at 97:3 for 30 min.

Flow rate: 0.8

Injection volume: 10

Detector: UV 252; F ex 365 em 425

CHROMATOGRAM

Retention time: 8.5 (caprylic acid), 14 (capric acid), 22 (lauric acid), 25 (eicosapentaenoic acid), 29 (linolenic acid), 30 (docosahexaenoic acid), 32 (arachidonic acid), 32.5 (palmitoleic acid), 35 (myristic acid), 36 (linoleic acid), 39 (eicosatrienoic acid), 42.5 (oleic acid), 45 (palmitic acid), 50 (heptadecanoic acid), 56 (stearic acid)

KEY WORDS

oil; butter; margarine

REFERENCE

Akasaka, K.; Suzuki, T.; Ohrui, H.; Meguro, H.; Shindo, Y.; Takahashi, H. 9-Bromomethylacridine a novel fluorescent labeling reagent of carboxylic group for HPLC, *Anal.Lett.*, **1987**, *20*, 1581–1594.

SAMPLE

Matrix: solutions

Analyte: carboxylic acids

Sample preparation: 10 μL Solution + 500 μL micelle solution + 25 μL 28 mg/mL 9-bromomethylacridine in acetone, mix, heat at 60° for 6 min, inject a 20 μL aliquot. (Micelle solution was 25 mM Arkopal N-130 (a polyoxyethylene(13)nonylphenol, Hoechst Holland, Amsterdam) in 10 mM pH 7.0 phosphate buffer containing 6 mM tetrakis(decyl) ammonium bromide. Synthesize 9-bromomethylacridine as described above.)

HPLC VARIABLES
Guard column: 10 × 2.1 40 μm Chromsep C18 (Chrompack)
Column: 100 × 3 5 μm Chromspher C18 (Chrompack)
Mobile phase: Gradient. MeOH:water from 20:80 to 100:0 over 13 min.
Injection volume: 20
Detector: UV 254; F ex 362 em 418

CHROMATOGRAM
Retention time: 6 (salicylic acid), 7 (ibuprofen), 10 (valproic acid), 12.5 (cholic acid)

REFERENCE
van der Horst, F.A.L.; Post, M.H.; Holthuis, J.J.M.; Brinkman, U.A.T. Derivatization of carboxylic acids with 9-bromomethylacridine using micellar phase-transfer catalysis, *Chromatographia*, **1989**, *28*, 267–273.

4-(Bromomethyl)-6,7-dimethoxycoumarin

SAMPLE
Matrix: microsomal incubations
Analyte: hydroxylauric acid
Sample preparation: 2 mL Microsomal incubation + 300 μL 3 M HCl, extract twice with 3 mL portions of diethyl ether. Combine the organic layers and evaporate them to dryness under a stream of nitrogen at 35°, reconstitute the residue in 1 mL 250 μg/mL 18-crown-6 in acetone, add 1 mg potassium carbonate (dried), add 500 μL 2 mg/mL 4-(bromo-methyl)-6,7-dimethoxycoumarin in acetone, heat at 70° for 45 min, evaporate to dryness under a stream of nitrogen at 40°, reconstitute with MeOH, inject an aliquot.

HPLC VARIABLES
Column: 150 × 4.6 5 μm Nucleosil C18
Mobile phase: Gradient. MeOH:water 67:33 for 25 min, to 97:3 over 10 min, maintain at 97:3 for 5 min
Column temperature: 30
Flow rate: 1
Injection volume: 20
Detector: F ex 340 em 420

CHROMATOGRAM
Retention time: 19.5 (11-hydroxylauric acid), 22.0 (12-hydroxylauric acid), 40.0 (lauric acid)

KEY WORDS
rat; liver

REFERENCE
Dirven, H.A.A.M.; de Bruijn, A.A.G.M.; Sessink, P.J.M.; Jongeneelen, F.J. Determination of the cytochrome P-450 IV marker, omega-hydroxylauric acid, by high-performance liquid chromatography and fluorimetric detection, *J.Chromatogr.*, **1991**, *564*, 266–271.

SAMPLE
Matrix: microsomal incubations
Analyte: lauric acid and its metabolites
Sample preparation: 2 mL Microsomal incubation + 800 µL 10% sulfuric acid, mix, extract twice with 6 mL portions of diethyl ether. Combine the organic layers and dry them over anhydrous sodium sulfate, evaporate to dryness under a stream of nitrogen, reconstitute the residue in 100 µL 2.5 mg/mL 18-crown-6 in MeCN, add 2 mg dried potassium carbonate, add 100 µL 10 mg/mL 4-bromomethyl-6,7-dimethoxycoumarin in acetone, shake vigorously, heat at 70° in the dark for 1 h, cool, dilute with MeCN, inject an aliquot.

HPLC VARIABLES
Column: 250 × 4.6 5 µm Nucleosil C18
Mobile phase: Gradient. MeCN:1% acetic acid 55:45 for 20 min, to 95:5 over 3 min, maintain at 95:5 for 15 min.
Flow rate: 1
Detector: F ex 300-400 em 417-700

CHROMATOGRAM
Retention time: 20.5 (11-hydroxylauric acid), 22 (12-hydroxylauric acid), 41 (lauric acid)
Limit of detection: 75 pg

KEY WORDS
human; rat; liver

REFERENCE
Amet, Y.; Berthou, F.; Menez, J.F. Simultaneous radiometric and fluorimetric detection of lauric acid metabolites using high-performance liquid chromatography following esterification with 4-bromomethyl-6,7-dimethoxycoumarin in human and rat liver microsomes, *J.Chromatogr.B*, **1996**, *681*, 233–239.

SAMPLE
Matrix: solutions
Analyte: carboxylic acids
Sample preparation: Neutralize 100 µL of a solution in anhydrous acetone with 10% KOH in MeOH (phenolphthalein end-point), add a 3-fold excess of 4-(bromomethyl)-6,7-dimethoxycoumarin in acetone, add 0.2 mole (sic) 18-crown-6 in MeCN, add 2 mg anhydrous potassium carbonate, heat at 70° for 30 min, cool, inject an aliquot.

HPLC VARIABLES
Column: 250 × 4.6 5 µm Ultrasphere ODS
Mobile phase: MeOH:water 50:50 (A) or 95:5 (B)
Flow rate: 2 (A) or 3 (B)
Injection volume: 20
Detector: F em 425

CHROMATOGRAM
Retention time: 3 (acetic acid (A)), 4.5 (propionic acid (A)), 9.5 (butyric acid (A)), 16.5 (isovaleric acid (A)), 19 (valeric acid (A)), 2 (myristic acid (B)), 2.5 (palmitic acid (B)), 3.5 (stearic acid (B)), 5 (arachidic acid (B))
Limit of detection: 0.5 pmole

KEY WORDS
protect from light

REFERENCE

Farinotti, R.; Siard, P.; Bourson, J.; Kirkiacharian, S.; Valeur, B.; Mahuzier, G. 4-Bromomethyl-6,7-dimethoxycoumarin as a fluorescent label for carboxylic acids in chromatographic detection, *J.Chromatogr.*, **1983**, *269*, 81–90.

3-Bromomethyl-6,7-dimethoxy-1-methyl-2(1*H*)-quinoxalinone

RELATED REFERENCE

Yamaguchi, M.; Matsunaga, R.; Fukuda, K.; Nakamura, M. Determination of free and total phenylacetic acid and p- and m-hydroxyphenylacetic acids in human urine by liquid chromatography with fluorimetric detection. *J.Chromatogr.* **1987**, *414*, 275-284.

SAMPLE

Matrix: blood, tissue

Analyte: E-64-c ((+)-(2S,3S)-3-[(S)-3-methyl-1-(4-guanidinobutylcarbamoyl)butylcarbamoyl]-2-oxiranecarboxylic acid)

Sample preparation: Serum. 10 μL Serum + 50 μL water + 200 μL acetone, mix, centrifuge at 1000 g for 5 min. Remove a 20 μL aliquot of the supernatant and add it to 10 μL 100 mM potassium bicarbonate in DMF:water 20:80, add 70 μL 4 mM bromomethyl-6,7-dimethoxy-1-methyl-2(1H)-quinoxalinone in DMF, heat at 100° in the dark for 20 min, inject a 50 μL aliquot. Tissue. Homogenize 20 mg tissue with 50 μL water, add 200 μL acetone, mix, centrifuge at 1000 g for 5 min. Remove a 20 μL aliquot of the supernatant and add it to 10 μL 100 mM potassium bicarbonate in DMF:water 20:80, add 70 μL 4 mM bromomethyl-6,7-dimethoxy-1-methyl-2(1H)-quinoxalinone in DMF, heat at 100° in the dark for 20 min, inject a 50μL aliquot. (3-Bromomethyl-6,7-dimethoxy-1-methyl-2(1H)-quinoxalinone is available from Molecular Probes, Eugene OR; Wako Chemicals, Richmond VA; and Dojindo Molecular Technologies, Inc., 3 Bethesda Metro Center, Suite 700, Bethesda MD 20814, (301) 664-8448, www.dojindo.co.jp. Synthesis is as follows. Stir 483 g veratrole in 1.45 L acetic acid at 15° for 1 h, add 683 g concentrated nitric acid (d 1.05) over 1 h (maintain the temperature below 40° by cooling and regulating the rate of addition of the nitric acid). Continue stirring and add 2.127 L fuming nitric acid (d 1.50) over 1 h while maintaining the temperature below 30°, let stand for 2 h, pour into a large volume of cold water, filter, wash the solid with water until the washings are neutral, recrystallize from EtOH to give 4,5-dinitroveratrole (mp 129.5-130.5°) (J. Am. Chem. Soc. 1946, 68, 1536). Reflux 5 g 4,5-dinitroveratrole in 200 mL benzene (Caution! Benzene is a carcinogen!), add 100 g 60 mesh iron powder and 20 mL concentrated HCl in small portions over 1 h, reflux for 4 h, add 10 mL water, reflux for 2 h, cool, make alkaline with 2.5 M NaOH, extract several times with 200 mL portions of benzene. Combine the organic layers and evaporate them to dryness, add 10 mL concentrated HCl, recrystallize from EtOH to give 1,2-diamino-4,5-dimethoxybenzene monohydrochloride as very slightly pink needles (mp 240°) (Anal. Chim. Acta 1982, 134, 39). Another synthesis of 1,2-diamino-4,5-dimethoxybenzene is as follows. Add 9.66g veratrole in 9.5 mL glacial acetic acid dropwise to 26.6 mL fuming nitric acid over 30 min using an ice bath, stir at room temperature for 30 min, pour into 500 mL ice-cold water, filter to obtain 1,2-dimethoxy-4,5-dinitrobenzene, recrystallize repeatedly from EtOH. Dissolve 1 g 1,2-dimethoxy-4,5-dinitrobenzene in 50 mL EtOH, add 100 mg 10% palladium on charcoal, stir under hydrogen at atmospheric pressure and room temperature for 4 days, filter through Celite under an atmosphere of nitrogen, saturate the filtrate with hydrogen chloride gas, filter under nitrogen, dry the solid under vacuum to obtain 1,2-diamino-4,5-dimethoxybenzene hydrochloride (mp 240° d (Anal. Chim. Acta 1982, 134, 39)) (Anal. Chim. Acta 1992, 263, 137). 1,2-Diamino-4,5-dimethoxybenzene is also available from Molecular Probes, Eugene OR or Dojindo Molecular Technologies. Heat 2.5 mmoles 1,2-diamino-4,5-dimethoxybenzene hydrochloride and 2.4 mmoles pyruvic acid in 30 mL 500 mM HCl on a boiling water bath for 2 h, cool with ice-water, filter. Wash the precipitate with water and dry it under vacuum, recrystallize from MeOH:water 90:10 to give 6,7-dimethoxy-3-methyl-2(1H)-quinoxalinone as yellow needles (mp 255°) (Chem. Pharm. Bull. 1985, 33, 3493). Treat 1 g 6,7-dimethoxy-3-methyl-2(1H)-quinoxalinone dissolved in 50 mL anhydrous MeOH with a solution of diazomethane in ether, evaporate to dryness under reduced pressure, dissolve the residue in 5 mL ethyl acetate, chromatograph on a 250 × 35 column filled with 130g 70-230 mesh silica gel 60 (Merck) using n-hexane:ethyl acetate 25:75 to give 6,7-dimethoxy-1,3-dimethyl-2(1H)-quinoxalinone as yellow needles (mp 170-171°). Dissolve 350 mg 6,7-dimethoxy-1,3-dimethyl-2(1H)-quinoxalinone in 3 mL acetic acid, add 350 mg anhydrous sodium acetate, add 2 mL 1.5 M bromine in acetic acid, heat at 100° for 15 min, cool, add 10 mL ether, filter, wash the solid 2 or 3 times with small portions of ether. Combine the filtrate and washings and evaporate them to dryness, dissolve the residue in 5 mL ethyl acetate, chromatograph on a 250 × 35 column filled with 130 g 70-230 mesh silica gel 60 (Merck) using ether, evaporate the main fraction to dryness, recrystallize the residue from n-hexane:ethyl acetate 50:50 to give 3-bromomethyl-6,7-dimethoxy-1-methyl-2(1H)-quinoxalinone as yellow needles (mp 161-163°) (J. Chromatogr. 1985, 346, 227). Preparation of diazomethane is as follows. Caution! Diazomethane is toxic, explosive, and carcinogenic! A face shield and a safety screen should always be used and the preparation should only be carried out in a properly functioning chemical fume hood. Only smooth glass apparatus with rubber stoppers and plastic tubing should be used. Scratched glassware, ground glass joints, and sharp edges should be avoided (Org. Syn., Coll. Vol. VI; Wiley: New York, 1988, pp. 432-435). Procedures have been reported for the synthesis of diazomethane from N-methyl-N-nitrosourea (Org. Syn., Coll. Vol. II; Wiley: New York, 1943, pp. 165-167), N-nitroso-β-methylaminoisobutyl methyl ketone (Org. Syn., Coll. Vol. III; Wiley: New York, 1955, pp. 244-248), and N,N'-dimethyl-N,N'-dinitrosoterephthalamide (Org. Syn., Coll. Vol. V; Wiley: New York, 1973, pp. 351-355). Probably the most convenient starting material is N-methyl-N-nitroso-p-toluenesulfonamide (Diazald) (Aldrichimica Acta 1983, 16, 3-10; Org. Syn., Coll. Vol. IV, Wiley: New York, 1963, pp. 250-253). Add 10 mL 95% EtOH to a solution of 5 g KOH in 8 mL water, warm to 65° using a water bath, add a solution of 5 g N-methyl-N-nitroso-p-toluenesulfonamide in 45 mL ether over 20 min at such a rate as to keep the reaction volume constant. Collect the ether and diazomethane that distil in an ice-cooled receiving flask under a dry ice/acetone condenser. When all the N-methyl-N-nitroso-p-toluenesulfonamide has been used up, slowly add 10 mL ether to the reaction flask and continue distillation until the distillate is colorless. A purpose-built distillation apparatus can be purchased from Aldrich (Aldrichimica Acta 1983, 16, 3-10). Excess quantities of diazomethane can be destroyed by adding acetic acid until the yellow color of the diazomethane is discharged. The safe disposal of the nitroso compounds used to generate diazomethane has been discussed (Lunn,G.; Sansone,E.B. Destruction of Hazardous Chemicals in the Laboratory, Second Edition. Wiley: New York, 1994, pp. 277-289).)

HPLC VARIABLES

Column: 150 × 4.6 5 μm TSK gel ODS-120T (Tosoh)

Mobile phase: Gradient. MeCN:water 40:60 for 25 min, to 95:5 (step gradient), maintain at 95:5 for 30 min, re-equilibrate at initial conditions for 10 min.

Flow rate: 0.8

Injection volume: 20
Detector: F ex 380 em 460

CHROMATOGRAM
Retention time: 20
Limit of detection: 500 nM (serum), 300 pmole/g (muscle)

OTHER SUBSTANCES
Non-interfering: acetone, acetophenone, arginine, benzaldehyde, epinephrine, formaldehyde, fructose, glucose, histamine, p-hydroxyphenylpyruvic acid, lactic acid, maleic acid, oleic acid, palmitic acid, phenylpyruvic acid, tyramine, tyrosine, 3,5-xylenol

KEY WORDS
mouse; serum; muscle; pharmacokinetics

REFERENCE
Chao, W.-F.; Kai, M.; Ohkura, Y. High-performance liquid chromatographic determination of a cysteine protease inhibitor and its ethyl ester in mouse serum and muscle by pre-column fluorescence derivatization, *J.Chromatogr.*, **1990**, *526*, 77–86.

SAMPLE
Matrix: solutions
Analyte: carboxylic acids
Sample preparation: Add 500 µL of a solution in MeCN to 100 mg finely powdered potassium carbonate, add 250 µL 3.8 mM 18-crown-6 in MeCN, add 250 µL 0.8 mM 3-bromomethyl-6,7-dimethoxy-1-methyl-2(1*H*)-quinoxalinone in MeCN, heat at 80° in the dark for 20 min, cool, inject a 5 µL aliquot. (3-Bromomethyl-6,7-dimethoxy-1-methyl-2(1H)-quinoxalinone is available from Wako Chemicals, Richmond VA; Molecular Probes, Eugene OR; and Dojindo Molecular Technologies, Inc., 3 Bethesda Metro Center, Suite 700, Bethesda MD 20814, (301) 664-8448, www.dojindo.co.jp. Synthesis is as described above.)

HPLC VARIABLES
Column: 100 × 4 10 µm Radial-Pak C18 (Waters)
Mobile phase: Gradient. MeOH:water from 57:43 to 100:0 over 20 min, maintain at 100:0 for 12 min (A) or from 30:70 to 70:30 over 30 min (B)
Flow rate: 2
Injection volume: 5
Detector: F ex 370 em 450

CHROMATOGRAM
Retention time: 5 (propionic acid (A)), 7 (butyric acid (A)), 7.1 (salicylic acid (A)), 7.9 (1-methyl-4-imidazoleacetic acid (A)), 8.3 (imidazole-4-acetic acid (A)), 9 (valeric acid (A)), 9.4 (benzoic acid (A)), 10.2 (glucuronic acid (B)), 11 (caproic acid (A)), 13.7 (uridine (A)), 14.1 (deoxyuridine (A)), 15 (caprylic acid (A)), 15.2 (thymidine (A)), 17 (p-aminobenzoic acid (B)), 18 (capric acid (A)), 20.7 (myristoleic acid (A)), 21 (lauric acid (A)), 22.2 (palmitoleic acid (A)), 22.2 (linolenic acid (A)), 22.3 (arachidonic acid (A)), 22.4 (linoleic acid (A)), 23 (myristic acid (A)), 23.8 (oleic acid (A)),25 (palmitic acid (A)), 26 (margaric acid (A)), 28 (stearic acid (A)), 31 (arachidic acid (A))
Limit of detection: 0.3-1 fmole

REFERENCE
Yamaguchi, M.; Hara, S.; Matsunaga, R.; Nakamura, M.; Ohkura, Y. 3-Bromomethyl-6,7-dimethoxy-1-methyl-2(1*H*)-quinoxalinone as a new fluorescence derivatization reagent for carboxylic acids in high-performance liquid chromatography, *J.Chromatogr.*, **1985**, *346*, 227–236.

5-(Bromomethyl)fluorescein

RELATED REFERENCES

Mukherjee, P.S.; Karnes, H.T. Reaction of 5-bromomethylfluorescein (5-BMF) with cefuroxime and other carboxyl-containing analytes to form derivatives suitable for laser-induced fluorescence detection. *Analyst* **1996**, *121*, 1573-1579.

Mukherjee, P.S.; De Silva, K.H.; Karnes, H.T. An argon-laser fluorometer for optimum determination of 5-bromomethylfluorescein derivatized carboxylic acids. *Mikrochim.Acta* **1996**, *124*, 99-109.

SAMPLE

Matrix: solutions

Analyte: palmitic acid

Sample preparation: 188 μL 100 μg/mL Palmitic acid in acetone + 250 μL 125 μg/mL 5-bromomethyl fluorescein (Molecular Probes, Eugene OR) in MeCN + 20-25 mg anhydrous potassium carbonate + 48 μL 700 μg/mL 18-crown-6 in MeCN, mix for 10 s, reflux for 1 h, dilute with mobile phase, inject a 50 μL aliquot. (Dry all solvents over molecular sieve. The later eluting peak is formed by the reaction of palmitic acid with the phenolic group of 5-(bromomethyl)fluorescein and hydrolysis of the bromomethyl group to a bromohydroxy group.)

HPLC VARIABLES

Guard column: 10 × 4.6 50 μm Corasil

Column: 250 × 3.2 5 μm Spherisorb ODS-2

Mobile phase: MeCN:water 85:15

Flow rate: 1

Injection volume: 50

Detector: UV 250; F ex 488 (25 mW argon ion laser, Lexel model 95-5) em 523 (filter); F ex 501 em 523

CHROMATOGRAM

Retention time: 18, 28

Limit of detection: 6 nM (conventional F), 0.756 nM (laser F)

REFERENCE

Mukherjee, P.S.; DeSilva, K.H.; Karnes, H.T. 5-Bromomethyl fluorescein (5-BMF) for derivatization of carboxyl containing analytes for use with laser-induced fluorescence detection, *Pharm.Res.*, **1995**, *12*, 930−936.

SAMPLE

Matrix: tissue

Analyte: gamma-(cholesteryloxy)butyric acid

Sample preparation: Freeze tissue in liquid nitrogen and pulverize. 50 mg Pulverized tissue + 300 μL 100 mM sodium dodecyl sulfate in 1% ascorbic acid, vortex for a few s, add 450 μL EtOH, vortex, sonicate, add 800 μL hexane, vortex for 20 s, centrifuge at 10000 rpm for 1 min. Remove the hexane layer and evaporate it to dryness under a stream of nitrogen, reconstitute with 300 μL acetone. 50 μL Solution + 250 μL 125 μg/mL 5-(bromomethyl)fluorescein (Molecular Probes Inc., Eugene, OR) in MeCN + 200 μL 700 μg/mL 18-crown-6 in MeCN + 20-25 mg anhydrous potassium carbonate, stir magnetically at 87° for 1.5 h, inject a 50 μL aliquot of the reaction mixture. (Dry all solvents over molecular sieve. Store all reagents at 4° in amber bottles.)

HPLC VARIABLES

Guard column: 10 × 4.6 50 μm Corasil
Column: 150 × 4.6 5 μm Vydac 201-TP5415 C-18
Mobile phase: MeOH:water 96:4
Flow rate: 2
Injection volume: 50
Detector: F ex 488 em 520

CHROMATOGRAM

Retention time: 10.37
Limit of detection: 2.2 ng/mL (conventional F detector), 29.9 pg/mL (laser F detector)

KEY WORDS

rat; liver; heart; brain; lung; kidney; spleen

REFERENCE

Mukherjee, P.S.; Karnes, H.T. Analysis of gamma-(cholesteryloxy)butyric acid in biologic samples by derivatization with 5-(bromomethyl)fluorescein followed by high-performance liquid chromatography with laser-induced fluorescence detection, *Anal.Chem.*, **1996**, *68*, 327–332.

3-Bromomethyl-7-methoxy-1,4-benzoxazin-2-one

RELATED REFERENCE

Naganuma, H.; Nakanishi, A.; Kondo, J.; Watanabe, K.; Kawahara, Y. The synthesis and analytical application of 3-bromomethyl-7-methoxy- 1,4-benzoxazin-2-one (BrMB) as a highly sensitive fluorescence derivatization reagent for carboxylic acids in high-performance liquid chromatography. *Sankyo Kenkyusho Nenpo* **1988**, *40*, 51-56.

SAMPLE

Matrix: bile
Analyte: glycine-conjugated bile acids
Sample preparation: Condition a Sep-Pak C18 SPE cartridge with 5 mL EtOH and 5 mL water. Dilute 50 μL bile with 5 mL 500 mM pH 7.0 phosphate buffer. Remove a 100 μL aliquot of this solution and add it to 100 μL 20 μg/mL IS in MeOH and 2 mL 500 mM pH 7.0 phosphate buffer, add to the SPE cartridge, wash with 2 mL water, wash with 1 mL 1.5% EtOH, elute with 2 mL 90% EtOH, add the eluate to a 20 × 6 column of PHP-LH-20 Sephadex, wash with 1 mL 90% EtOH, wash with 5 mL 100 mM acetic acid in 90% EtOH, elute with 5 mL 200 mM formic acid in 90% EtOH (J. Chromatogr. 1983, 276, 289). Remove one fifth of the eluate containing the glycine-conjugated bile acids and evaporate it to dryness. Add 100 μL 12 mg/mL 3-bromomethyl-7-methoxy-1,4-benzoxazin-2-one in MeCN, 100μL 12 mg/mL 18-crown-6 in MeCN, and 1 mg KF to the residue, heat

at 40° for 1 h, dilute with 3 mL benzene (Caution! Benzene is a carcinogen!). Chromatograph on a 40 × 6 column of 70-230 mesh silica gel, wash with 2 mL benzene:ethyl acetate 100:20 to remove excess reagent, elute with 10 mL chloroform:MeOH 70:30. Evaporate the eluate to dryness, reconstitute with ethyl acetate, inject an aliquot. (3-Bromomethyl-7-methoxy-1,4-benzoxazin-2-one is available from Tokyo Kasei (TCI America, 911 North Harborgate St., Portland, OR 97203; 800-423-8616, 503-283-1681, (fax) 503-283-1987; www.tciamerica.com). Synthesis of 3-bromomethyl-7-methoxy-1,4-benzoxazin-2-one is as follows. Add 36 g sodium hydrosulfite to 5.07 g 5-methoxy-2-nitrophenol in 60 mL water, reflux under an inert gas for 30 min, cool in ice, filter to obtain 2-amino-5-methoxyphenol as amber prisms. Add 3.48 g ethyl pyruvate to a solution of 4.17 g 2-amino-5-methoxyphenol in 45 mL EtOH and 11 mL acetic acid, stir at room temperature in the dark overnight, filter, recrystallize the solid from EtOH to obtain 3-methyl-7-methoxy-1,4-benzoxazin-2-one as orange-red crystals (mp 126°). Add 3.76 g phenyltrimethylammonium tribromide to a solution of 1.09 g 3-methyl-7-methoxy-1,4-benzoxazin-2-one in 20 mL THF stirred at 0°, stir gently overnight in a refrigerator, dilute with 50 mL diethyl ether, wash with 25 mL 100 mM sodium bisulfite, wash with 25 mL 100 mM sodium bicarbonate, wash with water. Dry the organic layer over anhydrous sodium sulfate, evaporate to dryness under reduced pressure, chromatograph the residue on 50 g silica gel with n-hexane:dichloromethane 50:50, recrystallize the product from n-hexane:dichloromethane 50:50 to obtain 3-bromomethyl-7-methoxy-1,4-benzoxazin-2-one as faint yellow prisms (mp 146°) (J. Chromatogr. 1992, 591, 159). Preparation of PHP-LH-20 Sephadex is as follows. Suspend 75.7 g Sephadex LH-20 in 200 mL dichloromethane using a glass stirring rod (not a magnetic stirrer) for 30 min, add 19 mL boron trifluoride ethyl etherate, after 15 min add 50 mL 35% epichlorohydrin in dichloromethane at 1-2 mL/min (Caution! Epichlorohydrin is a carcinogen!), stir for another 30 min, filter, wash with EtOH, dry chlorohydroxypropyl Sephadex LH-20 at 50° (J.Chromatogr. 1971, 59, 45). Stir 27.2 g chlorohydroxypropyl Sephadex LH-20 in 100.5 mL piperidine at room temperature for 30 min, add 5.74 g KOH in 302 mL MeOH, heat at 50-60° for 3 h with occasional shaking, filter, wash with EtOH:water 50:50, wash with 200 mM acetic acid in EtOH:water 70:30, wash with EtOH:water 90:10 until washings become neutral, store in EtOH:water 90:10 (Clin. Chim. Acta 1978, 87, 141).)

HPLC VARIABLES
Column: 150 × 4.6 5 μm YMC.GEL C8 (YMC)
Mobile phase: MeCN:MeOH:water 30:35:40 containing 2.5 mM gamma-cyclodextrin
Flow rate: 1
Detector: F ex 355 em 430

CHROMATOGRAM
Retention time: 14 (glycocholic acid), 23 (glycochenodeoxycholic acid), 25 (glycolithocholic acid), 30 (glycodeoxycholic acid)
Internal standard: glycodeoxycholic acid 12-propionate (33)
Limit of detection: 20 fmole

KEY WORDS
SPE

REFERENCE
Shimada, K.; Mitamura, K.; Ishitoya, S.; Hirakata, K. High performance liquid chromatographic separation of sensitive fluorescent derivatives of bile acids with cyclodextrin-containing mobile phase, *J.Liq.Chromatogr.*, **1993**, 16, 3965–3976.

SAMPLE
Matrix: microsomal incubations
Analyte: hydroxylauric acids
Sample preparation: 1 mL Microsomal incubation + 250 μL 4 M HCl + 250 μL 2.5 μg/mL 10-hydroxycapric acid, mix, saturate with NaCl, extract with 4.5 mL diethyl ether. Remove the organic layer and wash it with 4 mL water, evaporate to dryness under reduced pressure, reconstitute the residue in 100 μL 10 mM 3-bromomethyl-7-methoxy-1,4-benzoxazin-2-one in MeCN, add 100 μL 15 mg/mL 18-crown-6 in MeCN saturated with finely powdered potassium carbonate, heat at 40° for 30 min, add 50 μL 2% acetic acid, inject a 2 μL aliquot. (3-Bromomethyl-7-methoxy-1,4-benzoxazin-2-one is also available from

Tokyo Kasei (TCI America, 911 North Harborgate St., Portland, OR 97203; 800-423-8616, 503-283-1681, (fax) 503-283-1987; www.tciamerica.com). Synthesis of 3-bromomethyl-7-methoxy-1,4-benzoxazin-2-one is as described above.)

HPLC VARIABLES
Guard column: 10 μm Resolve C18 (Waters)
Column: 150 × 6 5 μm YMC pack A-321 ODS (YMC)
Mobile phase: MeCN:1% acetic acid 58:42 (At the end of each run wash with THF for 3 min, re-equilibrate for 6 min.)
Flow rate: 1.5
Injection volume: 2
Detector: F ex 355 em 430

CHROMATOGRAM
Retention time: 14.40 (11-hydroxylauric acid), 15.44 (12-hydroxylauric acid)
Internal standard: 10-hydroxycapric acid (7.55)
Limit of detection: 50 fmole

KEY WORDS
rat; liver

REFERENCE
Yamada, J.; Sakuma, M.; Suga, T. Assay of fatty acid omega-hydroxylation using high-performance liquid chromatography with fluorescence labeling reagent, 3-bromomethyl-7-methoxy-1,4-benzoxazin-2-one (BrMB), *Anal.Biochem.*, **1991**, *199*, 132–136.

SAMPLE
Matrix: solutions
Analyte: carboxylic acids
Sample preparation: Mix 1 mL of a solution in MeCN with 50 μL 40 mM 3-bromomethyl-7-methoxy-1,4-benzoxazin-2-one in MeCN, 50 μL 40 mM 18-crown-6 in MeCN, and 10 mg potassium carbonate, heat at 40° for 10 min, inject a 10 μL aliquot. (3-Bromomethyl-7-methoxy-1,4-benzoxazin-2-one is available from Tokyo Kasei (TCI America, 911 North Harborgate St., Portland, OR 97203; 800-423-8616, 503-283-1681, (fax) 503-283-1987; www.tciamerica.com). Synthesis of 3-bromomethyl-7-methoxy-1,4-benzoxazin-2-one is as described above.)

HPLC VARIABLES
Column: 150 × 6 5 μm ODS YMC Pack A-312 (YMC)
Mobile phase: MeCN:water 80:20
Flow rate: 1.5
Detector: F ex 345 em 440

CHROMATOGRAM
Retention time: 4 (hexanoic acid), 5 (heptanoic acid), 6 (octanoic acid), 8 (nonanoic acid), 10 (decanoic acid), 14 (undecanoic acid), 20 (dodecanoic acid), 29 (tridecanoic acid)
Limit of quantitation: 2-10 fmole

REFERENCE
Nakanishi, A.; Naganuma, H.; Kondo, J.; Watanabe, K.; Hirano, K.; Kawasaki, T.; Kawahara, Y. 3-Bromomethyl-7-methoxy-1,4-benzoxazin-2-one as a highly sensitive fluorescence derivatization reagent for carboxylic acids in high-performance liquid chromatography, *J.Chromatogr.*, **1992**, *591*, 159–164.

4-(Bromomethyl)-7-methoxycoumarin

RELATED REFERENCES
Berger, K.; Petz, M. Fluorescence HPLC determination of penicillins with 4-bromomethyl-7-methoxycoumarin as a precolumn labeling agent. *Dtsch.Lebensm.-Rundsch.* **1991**, *87*, 137-141.
Cohen, H.; Boutin-Muma, B. High-performance liquid chromatographic determination of fusilade using a fluorescence reagent, 4-bromomethyl-7-methoxycoumarin. *J.Liq.Chromatogr.* **1991**, *14*, 313-326.

Elbert, W.; Breitenbach, S.; Neftel, A.; Hahn, J. 4-Methyl-7-methoxycoumarin as a fluorescent label for high-performance liquid chromatographic analysis of dicarboxylic acids. *J.Chromatogr.* **1985**, *328*, 111-120.

Ertel, K.D.; Carstensen, J.T. Quantitative determination of octanoic acid by high-performance liquid chromatography following derivatization with 4-bromomethyl-7-methoxycoumarin. *J.Chromatogr.* **1987**, *411*, 297-304.

Lam, S.; Grushka, E. Labeling of fatty acids with 4-bromomethyl-7-methoxycoumarin via crown ether catalyst for fluorimetric detection in high-performance liquid chromatography. *J.Chromatogr.* **1978**, *158*, 207-214.

Leroy, P.; Chakir, S.; Nicolas, A. Measurement of the formation of menthol glucuronide in vitro, by reversed-phase high-performance liquid chromatography after pre-column labeling with 4-bromo-methyl-7-methoxycoumarin. *J.Chromatogr.* **1986**, *351*, 267-274.

Mazur, H.; Kosakowska, A.; Pazdro, K. Determination of indole-3-acetic acid in the Gulf of Gdansk by high-performance liquid chromatography of its 4-methyl-7-methoxycoumarin derivative. *J.Chromatogr.A* **1997**, *766*, 261-266.

Penmetsa, K.V.; Leidy, R.B.; Shea, D. Herbicide analysis by micellar electrokinetic capillary chromatography. *J.Chromatogr.A* **1996**, *745*, 201-208.

Slebioda, M.; Pazdro, K.; Lewandowska, J.; Falkowski, L. Modification of 4-bromomethyl-7-methoxycoumarin derivatization method of fatty acids for HPLC analysis. *Chem.Anal.(Warsaw)* **1994**, *39*, 439-443.

Zelenski, S.G.; Huber, J.W..I. Application of 4-methyl-7-methoxycoumarin derivatives to the high pressure liquid chromatographic analysis of fatty acids. *Chromatographia* **1978**, *11*, 645-646.

SAMPLE
Matrix: bile, blood, feces, gastric juice, tissue
Analyte: bile acids
Sample preparation: Condition a Sep-Pak C18 cartridge with 2 mL 720 mM MeOH in water and 6 mL 100 mM pH 7.0 potassium phosphate buffer. Serum. 200 µL Serum + 1 mL MeCN, mix, sonicate for 10 min, centrifuge at 17000 g for 15 min. Remove a 600 µL aliquot of the supernatant and evaporate it to dryness under a stream of nitrogen at 75°, reconstitute with 5 mL 100 mM pH 7.0 potassium phosphate buffer. Add to the SPE cartridge at 0.5 mL/min, wash with 2 mL 40 mM MeOH in water, elute with 4 mL 720 mM MeOH in water, filter (0.45 µm), evaporate the filtrate to dryness, reconstitute with 50 µL 250 µM lauric acid in MeOH, add 50 µL 1.8 mM KOH in MeOH, evaporate to dryness under a stream of nitrogen at 75°, reconstitute with 100 µL 10 mM 4-(bromo-methyl)-7-methoxycoumarin in MeCN containing 5 mM dicyclohexyl-18-crown-6, let stand at room temperature for 35 min, inject an aliquot. Liver. Homogenize (glass homogenizer) liver in 1 mL 720 mM EtOH in water, add 2 mL 720 mM EtOH in water, heat at 75° for 15 min, centrifuge at 17000 g for 10 min, remove the supernatant, extract the residue twice more. Combine the supernatants and evaporate them to dryness at 75°, reconstitute with 5 mL 100 mM pH 7.0 potassium phosphate buffer. Add to the SPE cartridge at 0.5 mL/min, wash with 2 mL 40 mM MeOH in water, elute with 4 mL 720 mM MeOH in water, filter (0.45 µm), evaporate the filtrate to dryness, reconstitute with 50 µL 250 µM lauric acid in MeOH, add 50 µL 1.8 mM KOH in MeOH, evaporate to dryness under a stream of nitrogen at 75°, reconstitute with 100 µL 10 mM 4-(bromomethyl)-7-methoxy-

coumarin in MeCN containing 5 mM dicyclohexyl-18-crown-6, let stand at room temperature for 35 min, inject an aliquot. Bile. Dilute 20 μL bile with 10 mL 100 mM pH 7.0 potassium phosphate buffer. Add 1 mL to the SPE cartridge at 0.5 mL/min, wash with 2 mL 40 mM MeOH in water, elute with 4 mL 720 mM MeOH in water, filter (0.45 μm), evaporate the filtrate to dryness, reconstitute with 50 μL 250 μM lauric acid in MeOH, add 50 μL 1.8 mM KOH in MeOH, evaporate to dryness under a stream of nitrogen at 75°, reconstitute with 100 μL 10 mM 4-(bromomethyl)-7-methoxycoumarin in MeCN containing 5 mM dicyclohexyl-18-crown-6, let stand at room temperature for 35 min, inject an aliquot. Gastric juice. Dilute 1 mL gastric juice with 9 mL 100 mM pH 7.0 potassium phosphate buffer, sonicate for 10 min. Add 1 mL to the SPE cartridge at 0.5 mL/min, wash with 2 mL 40 mM MeOH in water, elute with 4 mL 720 mM MeOH in water, filter (0.45 μm), evaporate the filtrate to dryness, reconstitute with 50 μL 250 μM lauric acid in MeOH, add 50 μL 1.8 mM KOH in MeOH, evaporate to dryness under a stream of nitrogen at 75°, reconstitute with 100 μL 10 mM 4-(bromomethyl)-7-methoxycoumarin in MeCN containing 5 mM dicyclohexyl-18-crown-6, let stand at room temperature for 35 min, inject an aliquot. Feces. Dilute 1 g feces with 9 mL MeOH, mix thoroughly, sonicate for 10 min, centrifuge at 17000 g for 10 min. Remove a 1 mL aliquot of the supernatant and evaporate it to dryness, reconstitute with 5 mL 100 mM pH 7.0 potassium phosphate buffer. Add to the SPE cartridge at 0.5 mL/min, wash with 2 mL 40 mM MeOH in water, elute with 4 mL 720 mM MeOH in water, filter (0.45 μm), evaporate the filtrate to dryness, reconstitute with 50 μL 250 μM lauric acid in MeOH, add 50 μL 1.8 mM KOH in MeOH, evaporate to dryness under a stream of nitrogen at 75°, reconstitute with 100 μL 10 mM 4-(bromomethyl)-7-methoxycoumarin in MeCN containing 5 mM dicyclohexyl-18-crown-6, let stand at room temperature for 35 min, inject an aliquot.

HPLC VARIABLES

Column: 250 × 4.6 5 μm Ultrasphere I.P. C18

Mobile phase: Gradient. A was MeCN:MeOH:water 100:50:75. B was MeCN:MeOH 100:50. A:B 100:0 for 7 min, to 70:30 over 0.5 min, maintain at 70:30 for 5 min, to 50:50 over 0.5 min, maintain at 50:50 over 7 min, to 25:75 over 1 min, maintain at 25:75 for 7 min.

Column temperature: 35

Flow rate: 1.7

Injection volume: 100

Detector: F

CHROMATOGRAM

Retention time: 6.8 (glycineursodeoxycholic acid), 8.8 (glycinecholic acid), 12.5 (glycine-chenodeoxycholic acid), 13.5 (glycinedeoxycholic acid), 15 (ursodiol, ursodeoxycholic acid), 16 (cholic acid), 17.5 (glycinelithocholic acid), 21.3 (chenodiol, chenodeoxycholic acid), 22 (deoxycholic acid), 27 (lithocholic acid)

Internal standard: lauric acid (24.5)

Limit of detection: 0.5 pmole

KEY WORDS

SPE; liver; serum

REFERENCE

Güldütuna, S.; You, T.; Kurts, W.; Leuschner, U. High performance liquid chromatographic determination of free and conjugated bile acids in serum, liver biopsies, bile, gastric juice and feces by fluorescence labeling, *Clin.Chim.Acta*, **1993**, *214*, 195–207.

SAMPLE

Matrix: blood

Analyte: 2-phenylbutyric acid

Sample preparation: 500 μL Plasma + 100 μL 1 M HCl + 75 μL 10 μg/mL anisic acid in ether + 2 mL ether, vortex for 1 min, centrifuge at -9° at 2000 g for 10 min, repeat the extraction. Combine the ether layers and dry them over anhydrous sodium sulfate, add 10 μL 15 mM sodium methoxide solution, evaporate to dryness under a stream of nitrogen, reconstitute the residue in 500 μL 4-(bromomethyl)-7-methoxycoumarin, add 250 μL 18-crown-6, heat in the dark at 70° for 15 min, evaporate to dryness under a stream of nitrogen, reconstitute the residue in 100 μL MeOH, inject a 20 μL aliquot.

HPLC VARIABLES
Column: 150 × 4.6 5 μm Spherisorb ODS
Mobile phase: MeOH:water:perchloric acid 65:35:0.075
Flow rate: 2.5
Injection volume: 20
Detector: F ex 300 em 418

CHROMATOGRAM
Retention time: 6.5
Internal standard: anisic acid (9)
Limit of detection: 50 ng/mL

KEY WORDS
plasma; comparison with GC

REFERENCE
Tsamis, M.T.; Mange, A.M.; Farinotti, R.; Mahuzier, G. Dosage de l'acide phenyl-2 butyrique dans le plasma par chromatographie gazeuze et chromatograhie liquide [Assay of 2-phenylbutyric acid in plasma with gas and liquid chromatography], *J.Chromatogr.*, **1983**, *277*, 61−69.

SAMPLE
Matrix: blood
Analyte: valproic acid
Sample preparation: Perform all operations with the exclusion of light. Evaporate 240 μL derivatization solution into a vial, add 400 μL 50 mM pH 7.0 phosphate buffer, add 100 μL plasma, add 10 μL 46 μg/mL undecylenic acid in MeCN, vortex for 5 s, heat at 70° for 40 min, add 500 μL MeCN, centrifuge at 3000 g for 5 min, inject a 20 μL aliquot. (Derivatization solution was 1.65 g Arkopal N-130 (a non-ionic surfactant, nonylphenol/13 unit chain polyoxyethylene) + 650 mg tetrahexylammonium bromide + 60 mg 4-(bromomethyl)-7-methoxycoumarin in 20 mL acetone.)

HPLC VARIABLES
Guard column: 10 × 3 5-20 μm LiChroprep RP-8
Column: 100 × 3 5 μm Chromspher C18
Mobile phase: Gradient. MeOH:water 80:20 for 3 min, then to 100:0 over 6 min, then held at 100:0 for 4 min.
Injection volume: 20
Detector: F ex 330 em 395

CHROMATOGRAM
Retention time: 4
Internal standard: undecylenic acid (7)
Limit of detection: 1 μg/mL

OTHER SUBSTANCES
Extracted: metabolites
Non-interfering: carbamazepine, phenobarbital, phenytoin

KEY WORDS
plasma

REFERENCE
van der Horst, F.A.; Eikelboom, G.G.; Holthuis, J.J. High-performance liquid chromatographic determination of valproic acid in plasma using a micelle-mediated pre-column derivatization, *J.Chromatogr.*, **1988**, *456*, 191−199.

SAMPLE
Matrix: blood
Analyte: atropine
Sample preparation: 2 mL Plasma + 100 μL 1 M NaOH + 10 mL chloroform, shake for 30 s, remove a 9 mL aliquot of the organic phase and add it to 1 mL 100 mM HCl, extract. Remove a 900 μL aliquot of the aqueous layer and add it to 100 μL 5 M NaOH, heat at 38° for 3 h, acidify with 5 M HCl, add 8 mL dichloromethane, vortex for 2 min, centrifuge. Remove the organic layer and evaporate it to dryness under a stream of nitrogen at 40-

45°, add 3 mg solid potassium bicarbonate, add 100 μL 10 μg/mL mandelic acid in MeCN, add 300 μL 50 μg/mL 4-(bromomethyl)-7-methoxycoumarin in MeCN, add 100 μL 15 μg/mL 18-crown-6 in MeCN, vortex for 15 s, heat at 70° for 45 min, inject a 25 μL aliquot. (Atropine is determined after hydrolysis to tropic acid.)

HPLC VARIABLES
Column: 150 × 4.6 5 μm ODS (Dupont)
Mobile phase: MeCN:buffer 33:67 (Buffer was 10 mM $(NH_4)H_2PO_4$ adjusted to pH 5.0.)
Column temperature: 40
Flow rate: 2
Injection volume: 25
Detector: F ex 328 em 389 (cutoff filter)

CHROMATOGRAM
Retention time: 10
Internal standard: mandelic acid (7)
Limit of quantitation: 125 ng/mL
Limit of detection: 108 ng/mL

KEY WORDS
plasma

REFERENCE
Li, S.; Wahba Khalil, S.K. An HPLC method for determination of atropine in human plasma, *J.Liq.Chromatogr.*, **1990**, *13*, 1339–1350.

SAMPLE
Matrix: blood
Analyte: bile acids
Sample preparation: Condition a Sep-Pak C18 SPE cartridge with 5 mL MeOH and 5 mL water. Dilute 100-200 μL serum with 4 mL 400 mM sodium bicarbonate, add to the SPE cartridge, wash with 20 mL water, elute with 2 mL MeOH. Evaporate the eluate to dryness under a stream of nitrogen at 45°, reconstitute the residue in 100 μL 2 mg/mL 4-(bromomethyl)-7-methoxycoumarin in MeCN, add 400 μg sodium carbonate, add 50 μL 20 mg/mL 18-crown-6 in MeCN, heat at 40° for 1 h, make up to 500 μL with MeCN, inject a 10 μL aliquot.

HPLC VARIABLES
Column: 150 × 3.9 5 μm Nova-Pak ODS
Mobile phase: Gradient. A was MeCN:MeOH:water 15:13.8:71.2. B was MeCN. A:B from 100:0 to 37:63 over 47 min (Waters convex curve + 2), to 0:100 over 0.1 min (Waters curve +9), maintain at 0:100 for 7.9 min, re-equilibrate at initial conditions for 6 min.
Flow rate: 1 for 47 min then 1.5
Injection volume: 10
Detector: F ex 320 em 385

CHROMATOGRAM
Retention time: 25.94 (glycocholic acid), 31.77 (glycochenodeoxycholic acid), 33.27 (glyco-deoxycholic acid), 35.87 (cholic acid), 36.43 (ursodiol), 41.47 (glycolithocholic acid), 45.23 (chenodiol), 46.47 (deoxycholic acid)
Limit of detection: 50-80 nM

KEY WORDS
serum; SPE

REFERENCE
Wang, G.F.; Stacey, N.H.; Earl, J. Determination of individual bile acids in serum by high performance liquid chromatography, *Biomed.Chromatogr.*, **1990**, *4*, 136–140.

SAMPLE
Matrix: blood
Analyte: fatty acids
Sample preparation: 20 μL Whole blood + 1 mL margaric acid in MeCN, mix, centrifuge. Remove the supernatant and add it to an equal volume of suspension, add an equal (?)

volume of 1 mg/mL 4-(bromomethyl)-7-methoxycoumarin in MeCN to this mixture, inject a 20 μL aliquot. (Suspension was 100 mg potassium carbonate + 50 μL water + 5 mL 20 mM 18-crown-6 in MeCN, sonicate for 30 min, add 5 mL MeCN, separate the suspension from the precipitated potassium carbonate.)

HPLC VARIABLES
Column: 200 × 3 Chromspher C18
Mobile phase: Gradient. A was MeCN:water 80:20 containing 25 mM phosphoric acid. B was MeCN:MeOH 50:50 containing 25 mM phosphoric acid. A:B 100:0 to 0:100 over 45 min.
Flow rate: 0.5
Injection volume: 20
Detector: F ex 325 em 398 (cutoff filter)

CHROMATOGRAM
Retention time: 19 (palmitoleic acid), 21 (myristic acid), 32 (oleic acid), 34 (palmitic acid), 58 (stearic acid), 75 (arachidonic acid)
Internal standard: margaric acid (47)

KEY WORDS
rat; whole blood; Derivatization gives double peaks for each acid; use second peak for quantitation.

REFERENCE
Wolf, J.H.; Korf, J. Improved automated precolumn derivatization reaction of fatty acids with bromomethylmethoxycoumarin as label, *J.Chromatogr.*, **1990**, *502*, 423–430.

SAMPLE
Matrix: blood
Analyte: valproic acid
Sample preparation: Filter (Amicon Centrifree) serum at 550 g for 15 min. Remove a 25 μL aliquot of the ultrafiltrate and add it to 475 μL 1 μg/mL undecylenic acid in MeCN, centrifuge. Remove a 50 μL aliquot of the supernatant and add it to 100 μL 18-crown-6 suspension and 50 μL 1 mg/mL 4-(bromomethyl)-7-methoxycoumarin in MeCN, heat at 65° in the dark for 30 min, inject a 5 μL aliquot. (Prepare the 18-crown-6 suspension by dissolving 100 mg potassium carbonate in 50 μL water then adding this mixture to 5 mL 20 mM 18-crown-6 in MeCN, sonicate for 30 min, add 5 mL MeCN.)

HPLC VARIABLES
Column: 100 × 2.1 5 μm HP Hypersil-ODS
Mobile phase: MeOH:water 80:20
Column temperature: 40
Flow rate: 0.3
Injection volume: 5
Detector: UV 200; UV 320; F ex 322 em 695

CHROMATOGRAM
Retention time: 2.5
Internal standard: undecylenic acid (4.4)
Limit of detection: 1.25 ng/mL (S/N 12, fluorescence)

OTHER SUBSTANCES
Non-interfering: acetaminophen, carbamazepine, chlordiazepoxide, clonazepam, desmethyldiazepam, diazepam, digoxin, disopyramide, ethosuximide, gentamicin, lidocaine, lorazepam, methotrexate, nitrazepam, oxazepam, phenobarbital, phenytoin, prazepam, primidone, procainamide, quinidine, temazepam, theophylline, vancomycin

KEY WORDS
serum

REFERENCE
Liu, H.; Montoya, J.L.; Forman, L.J.; Eggers, C.M.; Barham, C.F.; Delgado, M. Determination of free valproic acid: evaluation of the Centrifree system and comparison between high-performance liquid chromatography and enzyme immunoassay, *Ther.Drug Monit.*, **1992**, *14*, 513–521.

SAMPLE
Matrix: blood
Analyte: valproic acid
Sample preparation: Prepare ultrafiltrate from serum with an Amicon Centifree unit by centrifuging at 700 g for 10 min. 25 μL Ultrafiltrate + 475 μL 10 μg/mL undecylenic acid in MeCN, centrifuge. Remove 50 μL supernatant, add 100 μL 18-crown-6 solution, add 50 μL 1 mg/mL 4-(bromomethyl)-7-methoxycoumarin in MeCN, let stand in the dark at 65° for 30 min, inject a 5 μL aliquot. (Prepare 18-crown-6 solution by dissolving 100 mg potassium carbonate in 50 μL water, add 5 mL 20 mM 18-crown-6 in MeCN, sonicate for 30 min, add 5 mL MeCN.)

HPLC VARIABLES
Column: 100 × 2.1 5 μm HP Hypersil-ODS
Mobile phase: MeOH:water 80:20
Column temperature: 40
Flow rate: 0.3
Injection volume: 5
Detector: F ex 322 em 695

CHROMATOGRAM
Retention time: 2.5
Internal standard: undecylenic acid (4.5)
Limit of quantitation: 6.25 μg/mL

OTHER SUBSTANCES
Non-interfering: carbamazepine, phenobarbital, phenytoin

KEY WORDS
serum

REFERENCE
Liu, H.; Forman, L.J.; Montoya, J.; Eggers, C.; Barham, C.; Delgado, M. Determination of valproic acid by high-performance liquid chromatography with photodiode-array and fluorescence detection, *J.Chromatogr.*, **1992**, *576*, 163–169.

SAMPLE
Matrix: blood, dialysate
Analyte: valproic acid
Sample preparation: Blood. 20 μL Plasma or whole blood + 1 mL 1 μg/mL nonanoic acid in MeCN, centrifuge. Remove 100 μL supernatant, add 100 μL suspension, 15 min before injection add 200 μL 0.5 mg/mL 4-(bromomethyl)-7-methoxycoumarin, mix, inject. Dialysate. Lyophilize, add 40 μL suspension with twice the amount of crown ether, add 40 μL 0.5 mg/mL bromomethylmethoxycoumarin, mix, inject after 15 min. (Suspension was 100 mg potassium carbonate + 50 μL water + 5 mL 20 mM 18-crown-6 in MeCN, separate the suspension from the precipitated potassium carbonate.)

HPLC VARIABLES
Column: 200 × 3 5 μm Chromspher ODS
Mobile phase: MeCN:2.5 M formic acid 75:25
Flow rate: 0.4
Detector: F ex 325 em 398 (cut-off filter)

CHROMATOGRAM
Retention time: 8, 11 (double peak caused by impurity in derivatizing reagent)
Internal standard: nonanoic acid (12, 16)
Limit of quantitation: 1 μg/mL

KEY WORDS
plasma; whole blood; human; rat

REFERENCE
Wolf, J.H.; Veenma-van der Duin, L.; Korf, J. Automated analysis procedure for valproic acid in blood, serum and brain dialysate by high-performance liquid chromatography with bromomethylmethoxy-coumarin as fluorescent label, *J.Chromatogr.*, **1989**, *487*, 496–502.

SAMPLE
Matrix: bulk
Analyte: pesticides
Sample preparation: Prepare a 100 µg/mL solution in acetone. 1 mL solution + 1 mL 2-10 mg/mL 4-(bromomethyl)-7-methoxycoumarin in acetone + 5-10 mg cesium carbonate, heat in the dark at 35° for 45 min, inject an aliquot. Alternatively, add 100 µL glacial acetic acid to the reaction mixture, extract twice with 5 mL portions of light petroleum (bp 40-60°). Combine the extracts and evaporate them to dryness, reconstitute, inject an aliquot.

HPLC VARIABLES
Column: 150 × 4.6 Hypersil ODS
Mobile phase: MeOH:water 80:20
Flow rate: 1
Detector: UV 340

CHROMATOGRAM
Retention time: k' 2.67 (2,3,6-trichlorobenzoic acid), k' 2.25 (dicamba), k' 1.83 (2,4-dichlorophenoxyacetic acid), k' 2.33 (2-(2,4-dichlorophenoxy)propionic acid), k' 2.83 (mecoprop), k' 3.58 (4-(2-methyl-4-chlorophenoxy)butyric acid), k' 3.42 (phenoprop)
Limit of detection: 500-700 pg

KEY WORDS
comparison with derivatization with 2-naphthacyl bromide and with no derivatization

REFERENCE
Roseboom, H.; Herbold, H.A.; Berkhoff, C.J. Determination of phenoxy carboxylic acid pesticides by gas and liquid chromatography, *J.Chromatogr.*, **1982**, *249*, 323–331.

SAMPLE
Matrix: formulations
Analyte: prostaglandins
Sample preparation: Dilute with acetone, add hexanoic acid, heptanoic acid, and octanoic acid, add mixture to 5 g of a powdered 1:1 mixture of sodium sulfate and potassium bicarbonate + 2.7 mg dibenzo-18-crown-6 + 2.7 mg 4-bromomethyl-7-methoxycoumarin, let stand in the dark at 37° for 6 h, inject an aliquot.

HPLC VARIABLES
Column: 1060 × 0.2 3 µm Micro-Pak SP-18 (Varian)
Mobile phase: MeOH:MeCN:water 47.6:23.8:28.6
Column temperature: 35
Flow rate: 0.0006
Injection volume: 0.011
Detector: F ex 325 em 430 (laser-fluorescence)

CHROMATOGRAM
Retention time: 148 (dinoprostone), 155 (prostaglandin D_2), 160 (dinoprost), 175 (alprostadil), 180 (prostaglandin $F_{1\alpha}$), 233 (prostaglandin A_2), 237 (prostaglandin B_2), 273 (prostaglandin B_1), 277 (prostaglandin A_1)
Internal standard: hexanoic acid (136), heptanoic acid (203), octanoic acid (313)
Limit of detection: 40-125 fmole

KEY WORDS
capillary HPLC; laser

REFERENCE
McGuffin, V.L.; Zare, R.N. Femtomole analysis of prostaglandin pharmaceuticals, *Proc.Natl.Acad.Sci.U.S.A.*, **1985**, *82*, 8315–8319.

SAMPLE
Matrix: microsomal incubations
Analyte: lauric acid metabolites
Sample preparation: 1 mL Microsomal incubation + 5 mL diethyl ether, extract, centrifuge at 2800 rpm for 5 min, repeat extraction. Combine the organic layers and evaporate

them to dryness under a stream of nitrogen, reconstitute the residue in 2 mL MeCN. Remove a 550 μL aliquot of this solution and mix it with 1 mg potassium carbonate, 200 μL 850 μg/mL 4-(bromomethyl)-7-methoxycoumarin, 100 μL 330 μg/mL 18-crown-6, and 50 μL 10 μg/mL octanoic acid, heat at 80° for 40 min, inject a 20 μL aliquot.

HPLC VARIABLES
Guard column: 10 × 3 Chrompack C18
Column: 250 × 4.6 Chromspher C18 (Chrompack)
Mobile phase: MeOH:water 67:33, after 44 min wash with MeOH:water 97:3
Column temperature: 30
Flow rate: 1
Injection volume: 20
Detector: F ex 330 em 396

CHROMATOGRAM
Retention time: 34 (11-hydroxylauric acid), 37 (12-hydroxylauric acid)
Internal standard: octanoic acid (44)
Limit of detection: 0.25 pmole (12-hydroxylauric acid)

KEY WORDS
rat; liver

REFERENCE
Jansen, E.H.J.M.; De Fluiter, P. Determination of lauric acid metabolites in peroxisome proliferation after derivatization and HPLC analysis with fluorimetric detection, *J.Liq.Chromatogr.*, **1992**, *15*, 2247–2260.

SAMPLE
Matrix: solutions
Analyte: glucuronides
Sample preparation: Condition a Sep-Pak C18 SPE cartridge with 5 mL MeOH:water 80:20. 1 mL Aqueous solution + 500 μL acetic acid:water 20:80, add to the SPE cartridge, wait for 15 min, wash with 3 mL water, flush with air, elute with 3 mL 0.1% triethylamine in MeOH. Evaporate the eluate to dryness under a stream of nitrogen, reconstitute the residue in 80 μL DMF and 500 μL acetone, add 200 μL 1.5 mg/mL 4-(bromomethyl)-7-methoxycoumarin, add 100 μL 1 mg/mL 18-crown-6 in acetone, add 20 mg potassium carbonate, heat at 70° for 30 min, cool in ice, inject an aliquot.

HPLC VARIABLES
Guard column: 4 × 4 5 μm LiChrospher 100 Diol
Column: 250 × 4 5 μm LiChrospher 100 Diol
Mobile phase: Hexane:EtOH 80:20 (A) or 70:30 (B)
Flow rate: 1.2
Injection volume: 10
Detector: UV 328

CHROMATOGRAM
Retention time: k' 6.0 (menthol glucuronide (A)), k' 7.0 (borneol glucuronide (A)), k' 6.6 (testosterone glucuronide (B)), k' 7.6 (estrone glucuronide (B)), k' 4.7 (phenol glucuronide (B))
Limit of detection: 10 pmole

KEY WORDS
SPE

REFERENCE
Chakir, S.; Leroy, P.; Nicolas, A.; Ziegler, J.M.; Labory, P. High-performance liquid chromatographic analysis of glucuronic acid conjugates after derivatization with 4-bromomethyl-7-methoxycoumarin, *J.Chromatogr.*, **1987**, *395*, 553–561.

4-Bromomethyl-6,7-methylenedioxycoumarin

RELATED REFERENCE
Naganuma, H.; Kawahara, Y. Sensitive fluorescence labelling for determination of carboxylic acids with 4-bromomethyl-6,7-methylenedioxycoumarin. *J.Chromatogr.* **1989**, *478*, 149-158.

SAMPLE
Matrix: blood, urine
Analyte: loxoprofen
Sample preparation: Dilute urine nine-fold with water. 500 μL Plasma or diluted urine + 200 μL 2 M HCl + 500 μL 2 μg/mL 1-naphthoic acid, extract with 6 mL benzene (Caution! Benzene is a carcinogen!), centrifuge at 900 g for 10 min, repeat extraction. Combine the upper organic layers and evaporate them to dryness under reduced pressure, reconstitute the residue in 500 μL freshly prepared 250 μg/mL 4-bromomethyl-6,7-methylenedioxycoumarin in MeCN and 500 μL freshly prepared 1.5 mg/mL 18-crown-6 in MeCN (saturated with finely powdered potassium carbonate), sonicate briefly, heat at 40° for 30 min, add 100 μL MeCN:acetic acid 90:10, inject a 5 μL aliquot. (To hydrolyze glucuronides add 500 μL 500 mM NaOH to 500 μL diluted urine, let stand for 30 min, neutralize with 500 mM HCl. Prepare 4-bromomethyl-6,7-methylenedioxycoumarin as follows. Add 250 mmole finely powdered crystalline citric acid stepwise to 67.5 mL concentrated sulfuric acid at 70° (Caution! Carbon monoxide is evolved!), cool in ice, add 250 mmoles finely ground sesamol (3,4-methylenedioxyphenol) keeping the temperature below 5°, add 29 mL concentrated sulfuric acid, stir gently overnight, dilute with 50 mL chilled water, filter. Wash the brown solid with 10 mM sulfuric acid, with 2 M NaOH, and with 2 M HCl. Recrystallize from MeCN to give white prisms, mp 170°. Suspend 10 mL of this solid in 7.5 mL acetic acid, slowly add 7.5 mL acetic acid containing 10 mmoles bromine, reflux for 1 h, cool, chromatograph the solid on a silica gel column, elute with acetone, recrystallize 4-bromomethyl-6,7-methylenedioxycoumarin from MeOH to give yellow prisms, mp 241° (J.Chromatogr. 1989, 478, 149).)

HPLC VARIABLES
Column: 250 × 4.6 Zorbax ODS
Mobile phase: MeCN:water:acetic acid 55:45:1
Flow rate: 1.2 (plasma), 1.5 (urine)
Injection volume: 5
Detector: F ex 355 em 435

CHROMATOGRAM
Retention time: 15 (plasma), 12 (urine)
Internal standard: 1-naphthoic acid (20 (plasma), 14 (urine))
Limit of quantitation: 10 ng/mL (plasma), 50 ng/mL (urine)

OTHER SUBSTANCES
Extracted: metabolites

KEY WORDS
plasma; pharmacokinetics

REFERENCE
Naganuma, H.; Kawahara, Y. High-performance liquid chromatographic determination of loxoprofen and its diastereomeric alcohol metabolites in biological fluids by fluorescence labelling with 4-bromomethyl-6,7-methylenedioxycoumarin, *J.Chromatogr.*, **1990**, *530*, 387–396.

3-Bromomethyl-6,7-methylenedioxy-1-methyl-2(1H)-quinoxalinone

SAMPLE

Matrix: solutions

Analyte: fatty acids

Sample preparation: Mix a 200 μL aliquot of a fatty acid solution in MeCN with 10 mg finely-powdered potassium carbonate, 100 μL 3.8 mM 18-crown-6 in MeCN, and 100 μL 3 mM 3-bromomethyl-6,7-methylenedioxy-1-methyl-2(1H)-quinoxalinone in MeCN, heat in the dark at 80° for 15 min, cool, inject a 10 μL aliquot. (Preparation of 3-bromomethyl-6,7-methylenedioxy-1-methyl-2(1H)-quinoxalinone is as follows. Add 50 g piperonal in small portions to 250 mL sulfuric acid (d 1.48), 250 mL nitric acid (d 1.40), and 11.8 g mercury(I) nitrate at 0°, neutralize with KOH (Caution! Exothermic!), add sodium bisulfite solution to remove aldehydes, steam distil to remove volatiles, recrystallize to obtain 1,2-dinitro-4,5-methylenedioxybenzene (J. Am. Chem. Soc. 1928,50, 2711). Reflux 5 g 1,2-dinitro-4,5-methylenedioxybenzene in 200 mL benzene (Caution! Benzene is a carcinogen!), add 100 g 80 mesh iron powder and 20 mL concentrated HCl in portions over 1 h, reflux for 4 h, add 10 mL water, reflux for 2 h, cool, make alkaline with 2.6 M NaOH, extract 3 times with 200 mL portions of benzene. Combine the extracts and evaporate them to dryness, add 10 mL concentrated HCl, recrystallize from EtOH to give 1,2-diamino-4,5-methylenedioxybenzene dihydrochloride as colorless needles (mp 176-179°) (Chem. Pharm. Bull. 1987, 35, 687). Purify 1,2-diamino-4,5-methylenedioxybenzene dihydrochloride by dissolving 1 g in 28 mL MeOH and adding 500 μL 1 M 2-mercaptoethanol in 400 mM HCl containing 28 mM sodium hydrosulfite, mix slowly with 120 mL MeCN. Wash the resulting needles twice with 2 mL portions of MeOH:MeCN 1:5, dry under vacuum overnight. 1,2-Diamino-4,5-methylenedioxybenzene dihydrochloride is available from Dojindo, Japan (Dojindo Molecular Technologies, Inc., 3 Bethesda Metro Center, Suite 700, Bethesda MD 20814; (301) 664-8448; www.dojindo.co.jp). Heat 560 mg 1,2-diamino-4,5-methylenedioxybenzene dihydrochloride and 250 mg pyruvic acid in 30 mL 500 mM HCl on a boiling water bath for 50 min, cool with ice-water, filter. Wash the precipitate with water and dry it under vacuum, recrystallize from MeOH to give 6,7-methylenedioxy-3-methyl-2(1H)-quinoxalinone as pale-yellow needles (mp 273-274° d) (Chem. Pharm. Bull. 1987, 35, 687). Treat 1 g 6,7-methylenedioxy-3-methyl-2(1H)-quinoxalinone dissolved in 50 mL anhydrous MeOH with a solution of diazomethane in ether, evaporate to dryness under reduced pressure, dissolve the residue in 5 mL ethyl acetate, chromatograph on a 250 × 35 column filled with 130 g 70-230 mesh silica gel 60 (Merck) using n-hexane:ethyl acetate 25:75 to give 1,3-dimethyl-6,7-methylenedioxy-2(1H)-quinoxalinone as pale-yellow needles (mp 223-225°). Dissolve 350 mg 1,3-dimethyl-6,7-methylenedioxy-2(1H)-quinoxalinone in 3 mL acetic acid, add 350 mg anhydrous sodium acetate, add 2 mL 1.5 M bromine in acetic acid, heat at 100° for 15 min, cool, add 10 mL ether, filter, wash the solid 2 or 3 times with small portions of ether. Combine the filtrate and washings and evaporate them to dryness, dissolve the residue in 5 mL ethyl acetate, chromatograph on a 250 × 35 column filled with 130 g 70-230 mesh silica gel 60 (Merck) using ether, evaporate the main fraction to dryness, recrystallize the residue from n-hexane:ethyl acetate 50:50 to give 3-bromomethyl-6,7-methylenedioxy-1-methyl-2(1H)-quinoxalinone as yellow needles (mp 170°) (cf. J. Chromatogr. 1985, 346, 227). Preparation of diazomethane is as follows. Caution! Diazomethane is toxic, explosive, and carcinogenic! A face shield and a safety screen should always be used and the preparation should only be carried out in a properly functioning chemical fume hood. Only smooth glass apparatus with rubber stoppers and plastic tubing should be used. Scratched glassware, ground glass joints, and sharp edges should be avoided (Org. Syn., Coll. Vol. VI; Wiley: New York, 1988, pp. 432-435). Procedures have been reported for the synthesis of diazomethane from N-methyl-N-nitrosourea (Org. Syn., Coll. Vol. II; Wiley: New York, 1943, pp. 165-167), N-nitroso-β-methylaminoisobutyl methyl ketone (Org. Syn., Coll. Vol. III; Wiley: New York, 1955, pp. 244-248), and N,N'-dimethyl-N,N'-dinitrosoterephthalamide (Org. Syn., Coll. Vol. V; Wiley: New York, 1973, pp. 351-355). Probably the most convenient starting material is N-methyl-N-nitroso-p-toluenesulfonamide (Diazald) (Aldrichimica Acta 1983, 16, 3-10; Org. Syn., Coll. Vol. IV, Wiley:New York, 1963, pp. 250-253). Add 10 mL 95% EtOH to a solution of 5 g KOH in 8 mL water, warm to 65° using a water bath, add a solution of 5 g N-methyl-N-nitroso-p-toluenesulfonamide in 45 mL ether over 20 min at such a rate as to keep the reaction volume constant. Collect the ether and diazomethane that distil in an ice-cooled receiving flask under a dry ice/acetone condenser. When all the N-methyl-N-nitroso-p-toluenesulfonamide has been used up, slowly add 10 mL ether to the reaction flask and continue distillation until the distillate is colorless. A purpose-built distillation apparatus can be purchased from Aldrich (Aldrichimica Acta 1983, 16, 3-10). Excess quantities of diazomethane can be destroyed by adding acetic acid until the yellow color of the diazomethane is discharged. The safe disposal of the nitroso compounds used to generate diazomethane has been discussed (Lunn,G.; Sansone,E.B. Destruction of Hazardous Chemicals in the Laboratory, Second Edition. Wiley: New York, 1994, pp. 277-289).)

HPLC VARIABLES
Column: 100 × 8 5 μm Radial Pak C18
Mobile phase: Gradient. MeOH:water 35:65 for 18 min, to 87:13 over 22 min, to 100:0 over 20 min.
Flow rate: 2
Injection volume: 10
Detector: F ex 363 em 437

CHROMATOGRAM
Retention time: 20 (propionic acid), 22 (butyric acid), 24 (valeric acid), 27 (caproic acid), 32 (caprylic acid), 37 (capric acid), 42 (lauric acid), 45 (myristic acid), 50 (palmitic acid), 51 (margaric acid), 52 (stearic acid), 56 (arachidic acid)
Limit of detection: 0.2-0.8 fmole

REFERENCE
Yamaguchi, M.; Takehiro, O.; Hara, S.; Nakamura, M.; Ohkura, Y. 3-Bromomethyl-6,7-methylenedioxy-1-methyl-2(1H)-quinoxalinone as a highly sensitive fluorescence derivatization reagent for carboxylic acids in high-performance liquid chromatography, Chem.Pharm.Bull., 1988, 36, 2263–2266.

3-[4-(Bromomethyl)phenyl]-7-(diethylamino)-2H-1-benzopyran-2-one

SAMPLE

Matrix: solutions

Analyte: fatty acids

Sample preparation: Heat a 100 μL aliquot of a 100 μM solution in MeCN with 100 μL 1 mM 3-[4-(bromomethyl)phenyl]-7-(diethylamino)-2H-1-benzopyran-2-one in MeCN, 100 μL 1 mM 18-crown-6, 100 μL 100 μM IS in MeCN, and 10 mg potassium bicarbonate at 60° for 30 min, make up to 5 mL with MeOH, inject a 10 μL aliquot. (Synthesize the reagent, 3-[4-(bromomethyl)phenyl]-7-(diethylamino)-2H-1-benzopyran-2-one, as follows. Add 18.8 g aluminum trichloride to a solution of 94 mmoles m-diethylaminophenol and 84 g triethyl orthoformate in 185 mL chloroform at room temperature, mix for 10 min, add 50 mL 10% HCl, stir, neutralize with 10% NaOH, filter through a short column of Celite, wash through with chloroform. Wash the organic layer with saturated NaCl and dry it over anhydrous magnesium sulfate, evaporate to dryness under reduced pressure, recrystallize the product from chloroform to give 4-(diethylamino)-2-hydroxybenzaldehyde (Bull.Chem.Soc.Jpn. 1985, 58, 2192). Reflux 90 g α-bromo-p-toluic acid and 25 mL concentrated sulfuric acid in 1 L MeOH for 75 min, concentrate to 200 mL under reduced pressure, dilute with 3 mL water, recrystallize from MeOH to give methyl α-bromo-p-toluate as colorless needles (mp 54-55°). Add 35 g NaCN in 85 mL water over 10 min to a stirred solution of 70 g methyl α-bromo-p-toluate in 350 mL MeOH, heat at 50° for 30 min, dilute to 1 L with cold water, filter, recrystallize from light petroleum:benzene 2:1 (Caution! Benzene is a carcinogen!) with the aid of charcoal to give methyl α-cyano-p-toluate as colorless needles (mp 63-64°). Reflux 500 mg methyl α-cyano-p-toluate in 12 mL 10% NaOH in water for 30 min, filter, acidify the filtrate. Collect the precipitate and recrystallize it from glacial acetic acid to give 4-carboxyphenylacetic acid (homoterephthalic acid) (mp 239-241°) (J. Org. Chem. 1952, 17, 1035). Reflux 4-carboxyphenylacetic acid in MeOH in the presence of sulfuric acid to obtain methyl 4-carbomethoxyphenyl acetate (bp 172-175°/20 mm Hg) (J. Indian Chem. Soc. 1987, 64, 34). Heat 8.69 g 4-(diethylamino)-2-hydroxybenzaldehyde, 15.6 g methyl 4-carbomethoxyphenyl acetate, 2 mL piperidine, and 68 mL pyridine at 100° for 15 h, evaporate to dryness, recrystallize from ethyl acetate/hexane to give methyl 4-[7-(diethylamino)-2-oxo-2H-1-benzopyran-3-yl]benzoate as reddish-yellow prisms (mp 179-180.5). (A second crop can be obtained from the mother liquor by chromatography on silica gel using chloroform:hexane:acetone 20:18:1.) Suspend 7.1 g methyl 4-[7-(diethylamino)-2-oxo-2H-1-benzopyran-3-yl]benzoate in 240 mL 6 M HCl, reflux for 19 h, evaporate most of the solvent under reduced pressure, neutralize the residue with a saturated aqueous solution of sodium bicarbonate, filter, recrystallize from acetone to give 4-[7-(diethylamino)-2-oxo-2H-1-benzopyran-3-yl]benzoic acid as reddish-yellow prisms (mp 282-283.5°). Add 2.2 mL ethyl chloroformate and 3.2 mL triethylamine to a solution of 1.69 g 4-[7-(diethylamino)-2-oxo-2H-1-benzopyran-3-yl]benzoic acid in 200 mL THF, stir at room temperature for 1 h, add a solution of 2.27 g sodium borohydride in 4.8 mL water over 30 min, stir for 1 h, acidify with acetic acid, evaporate the THF under reduced pressure, add chloroform and water to the residue. Remove the organic layer and wash it with water, dry over anhydrous magnesium sulfate, evaporate to dryness, chromatograph on silica gel with chloroform, recrystallize from ethyl acetate/hexane to give 7-(diethylamino)-3-[4-(hydroxymethyl)phenyl]-2H-1-benzopyran-2-one as yellow needles (mp 153-154°). Stir 3.23 g 7-(diethylamino)-3-[4-(hydroxymethyl)phenyl]-2H-1-benzopyran-2-one and 86 mL phosphorus tribromide at 40-50° for 3 days, pour into ice-water, filter. Dissolve the solid in chloroform and wash with saturated aqueous sodium bicarbonate, wash with water, dry over anhydrous magnesium sulfate, evaporate to dryness, recrystallize from ethyl acetate/hexane to give 3-[4-(bromomethyl)phenyl]-7-(diethylamino)-2H-1-benzopyran-2-one as yellow needles (mp 166-167°).)

HPLC VARIABLES
Column: 150 × 4.6 5 μm Inertsil ODS-2
Mobile phase: Gradient. MeOH:water over 35 min
Flow rate: 1
Injection volume: 10
Detector: F ex 403 em 474

CHROMATOGRAM
Retention time: 8 (butyric acid), 11 (valeric acid), 14 (caproic acid), 21 (caprylic acid), 25 (capric acid), 27 (lauric acid), 32 (palmitic acid), 35 (stearic acid)
Internal standard: myristic acid (29)
Limit of detection: 1.5 nM

REFERENCE
Takechi, H.; Kamada, S.; Machida, M. 3-[4-(Bromomethyl)phenyl]-7-(diethylamino)-2H-1-benzopyran-2-one (MPAC-Br): A highly sensitive fluorescent derivatization reagent for carboxylic acids in high-performance liquid chromatography, *Chem.Pharm.Bull.*, **1996**, *44*, 793–799.

SAMPLE
Matrix: solutions
Analyte: chenodiol
Sample preparation: Mix 100 μL of a solution in MeOH with 100 μL 2.5 μg/mL IS in MeOH, evaporate to dryness, add 150 μL 0.2% 3-[4-(bromomethyl)phenyl]-7-(diethylamino)-2H-1-benzopyran-2-one (3-(4-bromomethylphenyl)-7-diethylaminocoumarin) in MeCN, add crown ether solution, heat at 60° for 20 min, evaporate to dryness at room temperature, reconstitute with 500 μL chloroform, add to a 55 × 6 column of silica gel, wash with 6 mL chloroform:MeOH 200:1, elute with 5 mL chloroform:MeOH 8:1, evaporate to dryness, reconstitute with 200 μL MeOH, inject a 5 μL aliquot. (Prepare crown ether solution by adding a large excess of solid potassium bicarbonate to a 0.4% 18-crown-6 solution in MeCN, sonicate at room temperature for 10 min, centrifuge at 1000 g for 10 min, use the supernatant. Synthesize 3-[4-(bromomethyl)phenyl]-7-(diethylamino)-2H-1-benzopyran-2-one as described above.)

HPLC VARIABLES
Column: 250 × 4.6 Inertsil C8
Mobile phase: Gradient. A was MeOH:20 mM pH 7.5 Tris-acetate buffer 78:22. B was MeOH:20 mM pH 7.5 Tris-acetate buffer 90:10. A:B 100:0 for 20 min, to 65:35 over 30 min, to 35:65 (step gradient), to 10:90 over 20 min.
Flow rate: 1
Injection volume: 5
Detector: F ex 400 em 475

CHROMATOGRAM
Retention time: 52
Internal standard: 3α,7α,12α-trihydroxy-26a,26b-dihomo-27-nor-5β-cholestan-26b-oic acid (60)

Limit of detection: 15 fmole

OTHER SUBSTANCES
Simultaneously analyzed: numerous other bile acids, cholic acid

REFERENCE
Kurosawa, T.; Sato, H.; Sato, M.; Takechi, H.; Machida, M.; Tohma, M. Analysis of stereoisomeric C$_{27}$-bile acids by high-performance liquid chromatography with fluorescence detection, *J.Pharm.Biomed.Anal.*, **1997**, *15*, 1375–1382.

2-Bromo-4'-nitroacetophenone

SAMPLE
Matrix: bulk
Analyte: prostaglandins
Sample preparation: Dissolve 0.4-3.1 mg dinoprostone in 1 mL MeCN, add 9.60 mg 2-bromo-4'-nitroacetophenone (p-nitrophenacyl bromide), add 2.9 µL N,N-diisopropylethylamine, let stand at room temperature for 15 min, inject an aliquot.

HPLC VARIABLES
Column: two 250 × 2.1 Zorbax-Sil columns in series
Mobile phase: Dichloromethane:MeCN:DMF 80:20:0.5 (A) or dichloromethane:hexane:MeOH 55:5:5 (B)
Flow rate: 0.28 (A) or 0.3 (B)
Detector: UV 254

CHROMATOGRAM
Retention time: 10.95 (dinoprost (B)), 30.05 (dinoprostone (A)), 34.25 (alprostadil (A))

KEY WORDS
numerous other prostaglandins determined; normal phase

REFERENCE
Morozowich, W.; Douglas, S.L. Resolution of prostaglandin p-nitrophenacyl esters by liquid chromatography and conditions for rapid, quantitative p-nitrophenacylation, *Prostaglandins*, **1975**, *10*, 19–40.

SAMPLE
Matrix: bulk
Analyte: dinoprostone
Sample preparation: Dissolve 10 mg sample in 1 mL 15 mg/mL 2-bromo-4'-nitroacetophenone in MeCN, add 5 µL N,N-diisopropylethylamine, mix, let stand for 2 h at room temperature. Evaporate to dryness under a stream of nitrogen, reconstitute the residue in 1 mL chloroform, add 500 µL 200 mg/mL silver nitrate in water, mix thoroughly, centrifuge, filter (0.2 µm) the chloroform layer, inject a 4 µL aliquot of the filtrate.

HPLC VARIABLES
Column: 1000 × 2.1 30-44 µm Vydac strong cation exchange resin containing silver (Prepare as follows. Equilibrate 6 g Vydac strong cation exchange resin with 600 mM silver nitrate in water, wash with water until no silver ion is detected, wash with 120 mL EtOH, wash with 120 mL acetone, wash with 120 mL ethyl acetate, wash with 120 mL 1,1,1-trichloroethane, wash with 120 mL hexane, dry under vacuum at 35-40° for 4 h, pack column.)
Mobile phase: Hexane:chloroform:MeCN 31.08:62.17:6.75

Column temperature: 26
Flow rate: 0.5
Detector: UV 254

CHROMATOGRAM
Retention time: 10 (dinoprostone), 6 (trans-dinoprostone)
Limit of detection: 0.2%

OTHER SUBSTANCES
Simultaneously analyzed: impurities
Interfering: 4

REFERENCE
Merritt, M.V.; Bronson, G.E. Determination of 5-*trans*-prostaglandin E$_2$ in prostaglandin E$_2$ *via* high performance liquid chromatography of their *p*-nitrophenacyl esters on a silver ion-loaded cation exchange resin, *Anal.Chem.*, **1976**, *48*, 1851–1853.

SAMPLE
Matrix: bulk
Analyte: prostaglandins
Sample preparation: Add 10 mg prostaglandin to 1 mL 15 mg/mL 2-bromo-4'-nitroacetophenone in MeCN, add 5 μL N,N-diisopropylethylamine, mix, let stand at room temperature for at least 2 h. Evaporate to dryness under a stream of nitrogen, reconstitute the residue in 1 mL chloroform, add 500 μL 200 mg/mL silver nitrate in water, mix thoroughly, centrifuge. Filter (0.2 μm) the chloroform layer and inject an aliquot of the filtrate.

HPLC VARIABLES
Column: 250 × 4.6 10 μm Partisil SCX impregnated with silver ion (Prepare the column by pumping 80 mL 1 M silver nitrate in water through the column, wash the column with water until a negative test for silver ion is obtained, wash with 50 mL EtOH, wash with 50 mL acetone, wash with 50 mL ethyl acetate, wash with 50 mL trichloroethane, and wash with 50 mL hexane.)
Mobile phase: Dioxane:MeCN 99.94:0.06 (Caution! Dioxane is a carcinogen!)
Detector: UV 254

CHROMATOGRAM
Retention time: k' 1.2 (alprostadil), k' 1.9 (dinoprostone), k' 2.9 (prostaglandin F$_{1\alpha}$), k' 5.8 (dinoprost)

OTHER SUBSTANCES
Simultaneously analyzed: degradation products

REFERENCE
Merritt, M.V.; Bronson, G.E. High-performance liquid chromatography of p-nitrophenacyl esters of selected prostaglandins on silver ion-loaded microparticulate cation-exchange resin, *Anal.Biochem.*, **1977**, *80*, 392–400.

SAMPLE
Matrix: formulations
Analyte: arbaprostil
Sample preparation: 1 mL Formulation + IS + 500 μL 2% phosphoric acid, mix, add 10 mL ethyl ether:chloroform 80:20, extract, centrifuge. Remove 8 mL of the organic layer and evaporate it to dryness under a stream of nitrogen at 40°, reconstitute the residue in 1 mL 2.5 mg/mL 2-bromo-4'-nitroacetophenone (p-nitrophenacyl bromide) in MeCN, add 500 μL 12.5 μL/mL N,N-diisopropylethylamine in MeCN, vortex briefly, heat at 40° for 30 min, evaporate to dryness under a stream of nitrogen at 40°, reconstitute the residue in 2 mL mobile phase, vortex, inject a 5-25 μL aliquot.

HPLC VARIABLES
Column: μPorasil silica gel
Mobile phase: MeCN:dichloromethane:water 30:70:0.5
Flow rate: 1.5
Injection volume: 5-25
Detector: UV 254

CHROMATOGRAM
Retention time: 6.8
Internal standard: 17β-hydroxy-17-methyl-4-androstene-3,11-dione (5.8)

OTHER SUBSTANCES
Simultaneously analyzed: s-epimer

KEY WORDS
injections; normal phase; siliconise glassware with Surfasil (Pierce)

REFERENCE
Peng, G.W.; Sood, V.K. Liquid chromatographic assay of arbaprostil, *J.Liq.Chromatogr.*, **1983**, *6*, 1499–1511.

2-Bromo-4'-phenylazoacetophenone

SAMPLE
Matrix: bulk
Analyte: fatty acids
Sample preparation: Mix 5-10 mg fatty acid with 1 mL 5% triethylamine in acetone and 1.5 mL acetone containing 1 equivalent of 2-bromo-4'-phenylazoacetophenone (p-phenylazophenacyl bromide), shake for 30 min, purify by TLC on silica gel G with benzene (Caution! Benzene is a carcinogen!), scrape off the appropriate band and elute it with benzene, evaporate the eluate to dryness under a stream of nitrogen, reconstitute with mobile phase, inject an aliquot. (p-Phenylazophenacyl bromide is available from Tokyo Kasei (TCI America, Portland OR). Synthesis is as follows. React equimolar quantities of p-aminoacetophenone and nitrosobenzene in glacial acetic acid at room temperature, recrystallize the crude product from EtOH to give p-phenylazoacetophenone as orange crystals (mp 114.5-116°) (Anal. Chem. 1954, 26, 1228). React 3.5 g bromine in 25 mL chloroform with 5 g p-phenylazoacetophenone in 50 mL chloroform at room temperature to obtain p-phenylazophenacyl bromide as orange red prisms after recrystallization from petroleum ether (mp 103°) (J. Chem. Soc. Japan, Pure Chem. Sect. 1951, 72, 152; Chem. Abs. 1952, 46, 3447i).)

HPLC VARIABLES
Guard column: ODS-5S (Bio-Rad)
Column: 250 × 4 5 μm Bio-Sil ODS (Bio-Rad)
Mobile phase: MeCN:water 95:5 (A) or 98:2 (B) or 99:1 (C)
Flow rate: 1 (A, B) or 1.5 (C)
Injection volume: 20
Detector: UV 330

CHROMATOGRAM
Retention time: 2.2 (formic acid (A)), 2.8 (acetic acid (A)), 3 (propionic acid (A)), 3.2 (butyric acid (A)), 3.4 (valeric acid (A)), 3.8 (caproic acid (A)), 4.2 (heptanoic acid (A)), 4.8 (caprylic acid (A)), 5.5 (nonanoic acid (A)), 6.6 (capric acid (A)), 7.7 (undecanoic acid (A)), 3 (capric

acid (B)), 3.5 (undecanoic acid (B)), 4 (lauric acid (B)), 4.6 (tridecanoic acid (B)), 5.5 (myristic acid (B)), 6.5 (pentadecanoic acid (B)), 8.2 (palmitic acid (B)), 10 (margaric acid (B)), 12 (stearic acid (B)), 20 (arachidic acid (B)), 4.9 (linolenic acid (C)), 6.5 (linoleic acid (C)), 9.2 (oleic acid (C))

REFERENCE

Vioque, E.; Maza, M.P.; Millán, F. High-performance liquid chromatography of fatty acids as their p-phenylazophenacyl esters, *J.Chromatogr.*, **1985**, *331*, 187–192.

9-(Chloromethyl)anthracene

SAMPLE

Matrix: solutions

Analyte: carboxylic acids

Sample preparation: Treat a solution in MeOH with a slight excess of tetramethylammonium hydroxide in MeOH, evaporate to dryness under a stream of nitrogen, reconstitute with MeCN, add a 2-10 fold excess of 9-(chloromethyl)anthracene in cyclohexane, heat at 75° for 15 min (very dilute solutions may require longer times), dilute with MeCN, inject an aliquot.

HPLC VARIABLES

Column: 300 mm long "Fatty Acid" reversed-phase (Waters)

Mobile phase: MeOH:water 88:12 (A) or 82:18 (B)

Flow rate: 0.75

Detector: UV 254

CHROMATOGRAM

Retention time: 11 (lauric acid (A)), 15 (myristic acid (A)), 19 (linolenic acid (A)), 21 (palmitic acid (A)), 24 (oleic acid (A)), 30 (stearic acid (A)), 10 (glycocholic acid (B)), 14 (glycochenodeoxycholic acid (BA)), 16 (glycodeoxycholic acid (B)), 20 (cholic acid (B)), 32 (deoxycholic acid (B)), 32 (chenodiol (B))

REFERENCE

Korte, W.D. 9-(Chloromethyl)anthracene: a useful derivatizing reagent for enhanced ultraviolet and fluorescence detection of carboxylic acids with liquid chromatography, *J.Chromatogr.*, **1982**, *243*, 153–157.

SAMPLE

Matrix: beverages

Analyte: fatty acids

Sample preparation: Heat 240 mL beer at 60°, cool. Remove a 200 mL aliquot and add it to 10 mL 3 M sulfuric acid, add 100 μL 100 μg/mL n-heptadecanoic acid in MeOH, add 200 mL ether:pentane 50:50, add 60 g NaCl, shake for 30 min, centrifuge at 5000 rpm for 5 min. Remove a 150 mL aliquot of the organic layer and wash it with 150 mL 5% sodium bicarbonate, evaporate to dryness under reduced pressure, reconstitute the residue in 1 mL 6 mM tetramethylammonium hydroxide in DMF, add 1 mL 7.5 mM 9-(chloromethyl)anthracene in DMF, heat at 75° for 20 min, inject a 50 μL aliquot.

HPLC VARIABLES
Column: 250 × 4.6 Kaseisorb LC ODS-60-5S C18 (Tokyo Kasei)
Mobile phase: Gradient. MeCN:water 90:10 for 5 min, to 95:5 over 40 min, maintain at 95:5 for 20 min, to 100:0 over 5 min, maintain at 100:0 for 15 min.
Column temperature: 40
Flow rate: 1.1
Injection volume: 50
Detector: F ex 365 em 412

CHROMATOGRAM
Retention time: 33 (lauric acid), 42 (linolenic acid), 49 (myristic acid), 51 (palmitoleic acid), 54 (linoleic acid), 71 (palmitic acid), 73 (oleic acid), 80 (n-heptadecanoic acid), 87 (stearic acid)
Limit of quantitation: 0.4 ng/mL
Limit of detection: 0.2 ng/mL

KEY WORDS
beer

REFERENCE
Kaneda, H.; Kano, Y.; Kamimura, M.; Osawa, T.; Kawakishi, S. Analysis of long-chain fatty acids in beer by HPLC-fluorescence detection method, *J.Agric.Food Chem.*, **1990**, *38*, 1363–1367.

SAMPLE
Matrix: tissue
Analyte: okadaic acid
Sample preparation: Condition a 690 mg Sep-Pak Classic SPE cartridge with hexane: dichloromethane 50:50. Extract 1 g mussel hepatopancreas with 4 mL MeOH:water 80:20, centrifuge. Remove a 2.5 mL aliquot of the supernatant and wash it twice with 2.5 mL portions of hexane. Add 1 mL water to the MeOH layer, extract twice with 4 mL portions of chloroform. Combine the chloroform layers and make up to 10 mL (Agric. Biol. Chem. 1987, 51, 877), evaporate a 1 mL aliquot to dryness under a stream of nitrogen, add 200 μL 0.1% 9-(chloromethyl)anthracene in DMF, add 100 μL 0.05% tetramethylammonium hydroxide in DMF, heat at 75° for 30 min, evaporate to dryness under a stream of nitrogen, reconstitute with three 3.5 mL portions of hexane:dichloromethane 50:50, add these portions to the SPE cartridge, wash with 7 mL hexane:dichloromethane 50:50, wash with 7 mL dichloromethane:acetone 90:10, elute with 17.5 mL MeCN:dichloromethane 50:50. Evaporate the eluate to dryness under a stream of nitrogen, reconstitute with 100 μL MeOH (cf Toxicon 1991, 29, 21), inject a 10 μL aliquot.

HPLC VARIABLES
Guard column: 40 × 4 15 μm pellicular C18 (Merck)
Column: 250 × 4 5 μm LiChrosorb RP-8
Mobile phase: Gradient. A was acetone. B was MeCN:water 50:50. A:B 55:45 for 6.5 min, to 95:5 over 20 min, to 98:2 (step gradient), maintain at 98:2 for 5 min, re-equilibrate at initial conditions for 10 min.
Flow rate: 1
Injection volume: 10
Detector: F ex 366 em 404

CHROMATOGRAM
Retention time: 6
Limit of detection: 1 μg/g

KEY WORDS
mussels; hepatopancreas

REFERENCE
Zonta, F.; Stancher, B.; Bogoni, P.; Masotti, P. High-performance liquid chromatography of okadaic acid and free fatty acids in mussels, *J.Chromatogr.*, **1992**, *594*, 137–144.

SAMPLE
Matrix: tissue

Analyte: okadaic acid

Sample preparation: Condition a 3 mL 500 mg silica gel SPE cartridge (Supelco) with 6 mL dichloromethane and 6 mL hexane:dichloromethane 50:50 (A), condition another one with 6 mL dichloromethane (B). Homogenize 1 g hepatopancreas with 6 mL MeOH:water 80:20, centrifuge, remove the supernatant, suspend the residue in 2 mL MeOH:water 80:20, centrifuge. Combine the supernatants and wash them three times with 15 mL portions of hexane:dichloromethane 85:15, add 5 mL water to the aqueous phase, extract three times with 15 mL portions of hexane:dichloromethane 50:50. Combine the organic layers and dry them over anhydrous sodium sulfate, evaporate to dryness under reduced pressure at 40°, reconstitute with 1 mL MeOH. Remove a 25-50 µL aliquot and evaporate it to dryness under a stream of nitrogen at room temperature, reconstitute the residue in 400 µL 800 µM tetramethylammonium hydroxide in MeCN, heat at 40° for 2 min, evaporate to dryness at 40°, add 400 µL 0.8 mM 9-(chloromethyl)anthracene in MeCN, heat at 90° for 1 h, cool, evaporate to dryness under a stream of nitrogen at room temperature, reconstitute with 300 µL hexane:dichloromethane 50:50, add to the SPE cartridge (A), rinse the sample vial with two 500 µL aliquots of hexane:dichloromethane 50:50, add the rinses to SPE cartridge (A), wash with 6 mL hexane:dichloromethane 50:50, wash with 7 mL dichloromethane:MeOH 99:1. Elute with 7 mL dichloromethane:MeOH 95:5 and pass the eluate directly in to the SPE cartridge (B). Collect all the effluent from cartridge (B) and evaporate it to dryness, reconstitute with 2 mL mobile phase, inject a 50 µL aliquot.

HPLC VARIABLES

Column: 150 × 4.6 5 µm reversed-phase C18 (Supelco)
Mobile phase: MeCN:water 75:25
Flow rate: 1
Injection volume: 50
Detector: F ex 365 em 412

CHROMATOGRAM

Retention time: 20
Limit of detection: 70 ng/g

KEY WORDS

nussel; hepatopancreas; SPE

REFERENCE

Lawrence, J.F.; Roussel, S.; Ménard, C. Liquid chromatographic determination of okadaic acid and dinophysistoxin-1 in shellfish after derivatization with 9-chloromethylanthracene, *J.Chromatogr.A*, **1996**, *721*, 359–364.

1-Chloromethylisatin

SAMPLE
Matrix: bulk
Analyte: carboxylic acids
Sample preparation: Dissolve 0.02-2 μmoles in DMF, add a 10-fold molar excess of 1-chloromethylisatin, add a 5-fold molar excess of dibenzo-18-crown-6, add a 50-fold molar excess of finely powdered potassium bicarbonate (final volume 50 μL), heat at 50° for 10 min, cool, add 100 μL water, extract with 100 μL chloroform, inject a 20 μL aliquot of the organic layer (Preparation of 1-chloromethylisatin is as follows. Dissolve 29.4 g isatin in 800 mL water with heating, add 24 mL 32% formaldehyde solution, heat until small yellow crystals are deposited on the container walls, filter while hot. 1-Hydroxymethylisatin crystallizes from the filtrate on cooling, concentrate the mother liquor to obtain a second crop. Recrystallize from MeOH or acetic acid to obtain 1-hydroxymethylisatin as shiny red crystals (mp 156-157°) (Ber. 1924, 57, 989). Reflux 10 g 1-hydroxymethylisatin with 35 mL thionyl chloride for 40 min, remove the excess thionyl chloride under reduced pressure, recrystallize the residue from cyclohexane/benzene to give 1-chloromethylisatin as orange-yellow needles (mp 100°) (Sci. Pharm. 1970, 38, 227) (Caution! Benzene is a carcinogen!).)

HPLC VARIABLES
Column: 250 × 4 10 μm Hibar RP-8 (Merck)
Mobile phase: Gradient. MeOH:water 50:50 for 5 min, to 100:0 over 50 min.
Flow rate: 1
Injection volume: 20
Detector: UV 240

CHROMATOGRAM
Retention time: 5 (formic acid), 6 (acetic acid), 8 (propionic acid), 10 (n-butyric acid), 13 (n-valeric acid), 17 (caproic acid), 22 (heptanoic acid), 26 (caprylic acid), 30 (capric acid), 34 (lauric acid), 37 (myristic acid), 40 (palmitic acid), 42 (stearic acid)
Limit of detection: 1-10 ng

REFERENCE
Gübitz, G. Derivatization of fatty acids with 1-chloromethylisatin for high-performance liquid chromatography, *J.Chromatogr.*, **1980**, *187*, 208–211.

N-(Chloromethyl)phthalimide

SAMPLE
Matrix: solutions
Analyte: carboxylic acids
Sample preparation: Evaporate a solution in water, MeOH, or diethyl ether to dryness, add a 3-fold molar excess of triethylamine, add 0.5-3 mL MeCN, add a 3-fold molar excess of N-(chloromethyl)phthalimide, heat at 60° for 1 h, inject an aliquot.

HPLC VARIABLES
Column: 7 μm LiChrosorb RP8
Mobile phase: MeCN:water 60:40
Flow rate: 1.5
Detector: UV 254

CHROMATOGRAM
Retention time: 2.5 (malonic acid), 3 (succinic acid), 4 (glutaric acid), 5.5 (adipic acid)

REFERENCE
Lindner, W.; Santi, W. N-Chloromethylphthalimides as derivatization reagents for high-performance liquid chromatography, *J.Chromatogr.*, **1979**, *176*, 55–64.

Dansyl-BAP

SAMPLE
Matrix: solutions
Analyte: carboxylic acids
Sample preparation: Mix a 100 μL aliquot of a 5-1000 nM solution of a carboxylic acid in MeCN with 100 μL 18-crown-6 solution, add 100 μL 100 μM dansyl-BAP in MeCN, mix, let stand at room temperature (aliphatic acids) or 55° (aromatic acids) for 30 min, add 100 μL 3 mM thymine in MeCN, add 5 mg potassium bicarbonate, vortex for 30 s, let stand for 30 min, evaporate to dryness under a stream of nitrogen. Reconstitute with dichloromethane, add to a Bond-Elut silica SPE cartridge, elute with 1.5 mL MeCN: dichloromethane 50:50. Evaporate the eluate to dryness and reconstitute the residue with 500 μL mobile phase, inject a 25 μL aliquot. (Prepare 18-crown-6 solution by sonicating a 1 mg/mL solution of 18-crown-6 in MeCN containing 1 mg/mL potassium bicarbonate for 20 min. Prepare dansyl-BAP (N-(bromoacetyl)-N'-[5-(dimethylamino)naphthalene-1-sulfonyl]piperazine) as follows. Slowly add a solution of 135 mg dansyl chloride in 30 mL acetone to a 10-fold molar excess of piperazine in acetone:water 75:25, stir at 50° for 1 h, evaporate the acetone. Acidify the remaining aqueous layer with concentrated nitric acid, wash 3 times with 15 mL portions of dichloromethane, adjust the pH of the aqueous layer to 11 with concentrated NaOH, extract three times with 15 mL portions of dichloromethane. Combine the extracts and dry them over anhydrous calcium chloride, concentrate to about 5 mL, chromatograph on a 400 × 25 column of 60-200 μm silica gel Si-60, wash with about 20 mL dichloromethane:MeOH 99:1 to remove a small fluorescent band, elute with about 30 mL dichloromethane:MeOH 95:5-94:6 to obtain a solution of dansylpiperazine, determine the concentration by UV absorption at 340 nm (extinction coefficient = 4300 in MeOH). Stir a solution of 700 mg bromoacetic acid and 1.1 g dicyclohexylcarbodiimide in 100 mL MeCN at room temperature for 1 min, slowly add 470 μmoles dansylpiperazine in MeCN, stir for 1 h, evaporate to dryness, reconstitute the

residue with 10 mL dichloromethane, filter. Chromatograph the filtrate on a 400 × 25 column of 40-60 µm Si-60 silica gel with dichloromethane, when the first-eluting, strongly-fluorescent yellow band reaches the outlet change the eluent to dichloromethane:MeOH 99:1, collect about 50 mL eluate to obtain dansyl-BAP.)

HPLC VARIABLES
Column: 150 × 3.1 5 µm LiChrosorb RP-18
Mobile phase: MeCN:water 60:40, containing 2.5 mM pH 7.0 imidazole buffer
Flow rate: 0.5
Injection volume: 25
Detector: F ex 246 em 490 (cut-off filter) following post-column reaction. The column effluent mixed with 50 mM hydrogen peroxide in MeCN containing 5 mM bis(2-nitrophenyl)oxalate pumped at 0.3 mL/min and the mixture flowed immediately to the detector. (Prepare bis(2-nitrophenyl)oxalate by dissolving 13.9 g 2-nitrophenol in 250 mL benzene (Caution! Benzene is a carcinogen!), remove 50 mL benzene by azeotropic distillation, cool to 10°, add 10.1 g freshly distilled triethylamine, add 7 g oxalyl chloride dropwise, allow to warm to room temperature, let stand overnight, evaporate to dryness under reduced pressure, recrystallize to give bis(2-nitrophenyl)oxalate (J. Chem. Educ. 1974, 51, 529).)

CHROMATOGRAM
Retention time: 6.5 (benzoic acid), 7 (2-methoxybenzoic acid), 9 (2,4-dichlorobenzoic acid), 10 (lipoic acid), 12 (naproxen), 16 (octanoic acid), 24 (ibuprofen)
Limit of detection: 0.8-1 pmole

KEY WORDS
SPE

REFERENCE
Kwakman, P.J.M.; Van Schaik, H.P.; Brinkman, U.A.T.; de Jong, G.J. N-(Bromoacetyl)-N'-[5-(dimethylamino)naphthalene-1-sulfonyl]piperazine as a sensitive labeling reagent for the determination of carboxylic acids by liquid chromatography with peroxyoxalate chemiluminescence and fluorescence detection, *Analyst*, **1991**, *116*, 1385–1391.

2,4'-Dibromoacetophenone

RELATED REFERENCES
Jordi, H.C. Separation of long and short chain fatty acids as naphthacyl and substituted phenacyl esters by high performance liquid chromatography. *J.Liq.Chromatogr.* **1978**, *1*, 215-230.

Mingrone, G.; Greco, A.V.; Passi, S. Reversed-phase high-performance liquid chromatographic separation and quantification of individual human bile acids. *J.Chromatogr.* **1980**, *183*, 277-286.

Minkler, P.E.; Ingalls, S.T.; Kormos, L.S.; Weir, D.E.; Hoppel, C.L. Determination of carnitine, butyrobetaine, and betaine as 4'-bromophenacyl ester derivatives by high-performance liquid chromatography. *J.Chromatogr.* **1984**, *336*, 271-283.

Nagels, L.; Debeuf, C.; Esmans, E. Quantitative determination of quinic acid and derivatives by high-performance liquid chromatography after derivatization with *p*-bromophenacyl bromide. *J.Chromatogr.* **1980**, *190*, 411-417.

Patience, R.L.; Thomas, J.D. Rapid concentration and analysis of short chain carboxylic acids: variation on a theme. *J.Chromatogr.* **1982**, *234*, 225-230.

SAMPLE
Matrix: blood

Analyte: valproic acid

Sample preparation: 250 μL Serum + 100 μL 2.5 mM heptanoic acid + 300 μL 2 M sulfuric acid + 1 mL petroleum ether (40-60°), shake for 5 min, centrifuge. Remove a 900 μL aliquot of the organic layer and add it to 10 μL tributylamine, evaporate to dryness at 70° in about 10 min, reconstitute the residue in 400 μL MeCN, add 50 μL 10 mg/mL 2,4'-dibromoacetophenone (p-bromophenacyl bromide) in MeOH, heat at 70° for 10 min, inject a 20 μL aliquot.

HPLC VARIABLES

Column: 250 × 4 10 μm Lichrosorb RP-2
Mobile phase: MeOH:50 mM pH 2.3-2.5 phosphate buffer 75:25
Column temperature: 50
Flow rate: 2
Injection volume: 20
Detector: UV 260

CHROMATOGRAM

Retention time: 3.14
Internal standard: heptanoic acid (2.74)

OTHER SUBSTANCES

Non-interfering: carbamazepine, clonazepam, ethosuximide, phenobarbital, phenytoin, primidone

KEY WORDS

serum

REFERENCE

Ehrenthal, W.; Rochel, M. Kontrolle der Valproinsäure-Therapie durch Serumspiegelbestimmungen mit HPLC oder EMIT und durch gleichzeltige Bestimmung klinisch-chemischer Parameter [Drug monitoring for valproic acid with HPLC or EMIT and concomitant measurements of clinical-chemical parameters], *Arzneimittelforschung*, **1982**, *32*, 449–452.

SAMPLE

Matrix: blood
Analyte: valproic acid

Sample preparation: 1 mL Serum + 50 μg IS + 100 μL concentrated phosphoric acid + 2 mL dichloromethane, shake for 30 min, centrifuge at 2500 rpm for 10 min. Remove a 500 μL aliquot of the organic layer and add it to 100 μL reagent and 0.5 mg sodium bicarbonate, heat at 75° for 30 min, evaporate to 0.5 mL under a stream of nitrogen, make up to 1 mL with MeCN, inject an aliquot. (Reagent was 20 mM 2,4'-dibromoacetophenone (p-bromophenacyl bromide) in MeCN containing 1 mM 18-crown-6, prepared by diluting Phenacyl-8 (Pierce) 5 times with MeCN.)

HPLC VARIABLES

Column: C18
Mobile phase: MeCN:water 65:35
Detector: UV 254

CHROMATOGRAM

Retention time: 10
Internal standard: cyclohexane carboxylic acid (6.5)
Limit of quantitation: 5 μg/mL

KEY WORDS

serum

REFERENCE

Kline, W.F.; Enagonio, D.P.; Reeder, D.J.; May, W.E. Liquid chromatographic determination of valproic acid in human serum, *J.Liq.Chromatogr.*, **1982**, *5*, 1697–1709.

SAMPLE

Matrix: blood
Analyte: valproic acid

Sample preparation: 100 μL Serum + 25 μL buffer + 250 μL 167 μg/mL nonanoic acid in MeCN, vortex for 10 s, centrifuge for 5 min. Remove a 200 μL aliquot of the supernatant and add it to 50 μL 20 mg/mL 2,4'-dibromoacetophenone (p-bromophenacyl bromide) in MeCN containing 500 μg/mL dicyclohexane-18-crown-6, heat at 70° for 15 min, cool, inject a 5 μL aliquot. (Prepare buffer by dissolving 19.04 g KH_2PO_4 and 37.4 g $Na_2HPO_4.2H_2O$ in 1 L water, pH 7.0.)

HPLC VARIABLES
Column: 100 × 5 5 μm Hypersil ODS
Mobile phase: MeCN:3 mM KH_2PO_4 70:30
Flow rate: 2
Injection volume: 5
Detector: UV 254

CHROMATOGRAM
Retention time: 4
Internal standard: nonanoic acid (6)
Limit of detection: 60 μM

OTHER SUBSTANCES
Non-interfering: acetaminophen, clonazepam, diazepam, phenobarbital, phenytoin, primidone, salicylic acid, theophylline

KEY WORDS
horse; serum

REFERENCE
Moody, J.P.; Allan, S.M. Measurement of valproic acid in serum as the 4-bromophenacyl ester by high performance liquid chromatography, *Clin.Chim.Acta*, **1983**, *127*, 263–269.

SAMPLE
Matrix: blood
Analyte: butyric acid
Sample preparation: Filter (Amicon YMT membrane) 1 mL plasma while centrifuging at 2000 g for 45 min. Mix 300 μL ultrafiltrate with 10 μL 1 M potassium bicarbonate, evaporate to dryness under a stream of nitrogen, add 100 μL reagent, add 400 μL MeCN, heat with shaking at 80° for 30 min, centrifuge at 1000 g for 5 min, inject a 10 μL aliquot of the supernatant. (Reagent was a 20:1 mixture of 200 mM 2,4'-dibromoacetophenone (p-bromophenacyl bromide) in MeCN and 200 mM dicyclohexyl-18-crown-6 in MeCN.)

HPLC VARIABLES
Column: 300 × 3.9 10 μm μBondapak C18
Mobile phase: MeCN:water 45:55
Flow rate: 2
Injection volume: 10
Detector: UV 254

CHROMATOGRAM
Retention time: 12
Limit of detection: 2 μm

KEY WORDS
plasma; ultrafiltrate; pharmacokinetics

REFERENCE
Miller, A.A.; Kurschel, E.; Osieka, R.; Schmidt, C.G. Clinical pharmacology of sodium butyrate in patients with acute leukemia, *Eur.J.Cancer Clin.Oncol.*, **1987**, *23*, 1283–1287.

SAMPLE
Matrix: blood
Analyte: fatty acids
Sample preparation: 100 μL Serum + 20 μL 5 mM IS in isopropanol, mix, add 500 μL isopropanol:n-heptane:2 M phosphoric acid 40:10:1, mix, let stand at room temperature for 5-10 min, add 200 μL n-heptane, add 300 μL water, vortex thoroughly, centrifuge at

1000 g for 5 min. Remove a 200 µL aliquot of the upper organic layer and evaporate it to dryness under a stream of nitrogen, reconstitute the residue in 6 µL reagent, 500 µL MeCN, and ca. 1 mg potassium bicarbonate, flush the tube with nitrogen, close the PTFE-lined cap tightly, heat at 85° with vigorous stirring for 45 min (weigh vial before and after heating to check for leakage), cool, remove stir bar, centrifuge, inject a 10-25 µL aliquot of the supernatant. (Reagent was 50 mM 2,4'-dibromoacetophenone (p-bromophenacyl bromide) in MeCN containing 5 mM 18-crown-6, store protected from light.)

HPLC VARIABLES
Guard column: 4 × 4 5 µm CN
Column: 25 × 4 3 µm Spherisorb C6
Mobile phase: MeCN:water 77:23
Column temperature: 30
Flow rate: 1.3
Injection volume: 10-25
Detector: UV 254

CHROMATOGRAM
Retention time: 6 (lauric acid), 6.6 (myristoleic acid), 7.5 (eicosapentaenoic acid), 8 (lino-lenic acid), 8.5 (myristic acid), 8.8 (docosahexaenoic acid), 9 (palmitoleic acid), 9.5 (arach-idonic acid), 10.3 (linoleic acid), 12 (palmitic acid), 13 (oleic acid), 14 (elaidic acid), 18 (stearic)
Internal standard: heptadecanoic acid (margaric acid) (14.7)
Limit of detection: 800 nM

KEY WORDS
serum

REFERENCE
Püttmann, M.; Krug, H.; von Ochsenstein, E.; Kattermann, R. Fast HPLC determination of serum free fatty acids in the picomole range, *Clin.Chem.*, **1993**, *39*, 825–832.

SAMPLE
Matrix: bulk
Analyte: fatty acids
Sample preparation: Add 3-5 equivalents of anhydrous potassium carbonate to the acid, add 0.5-1.5 mL MeCN containing an excess of 2,4'-dibromoacetophenone:dicyclohexyl-18-crown-6 10:1, reflux with vigorous stirring for 45 min, evaporate to dryness, take up in chloroform, filter, dilute the filtrate with chloroform, inject an aliquot.

HPLC VARIABLES
Column: 250 × 5 5 µm HI-EFF Micropart C18 (Applied Science Laboratories)
Mobile phase: MeOH:water 90:10
Flow rate: 1.5
Detector: UV 254

CHROMATOGRAM
Retention time: 14 (linolenic acid), 16 (palmitoleic acid), 17 (palmitelaidic acid), 18 (linoleic acid)), 19 (elaidic acid), 20 (linoleaidic acid), 24 (petroselinic acid), 26 (oleic acid), 45 (ste-aric acid)

REFERENCE
Pei, P.T.-S.; Kossa, W.C.; Ramachandran, S.; Henly, R.S. High pressure reverse phase liquid chromatog-raphy of fatty acid *p*-bromophenacyl esters, *Lipids*, **1976**, *11*, 814–816.

SAMPLE
Matrix: bulk
Analyte: gangliosides
Sample preparation: Dry ganglioside containing 1-15 µg sialic acid, add 20 µL 10 mg/mL 2,4'-dibromoacetophenone (p-bromophenacyl bromide) in DMF, heat at 60° for 1 h, inject a 5 µL aliquot.

HPLC VARIABLES
Column: 150 × 4.6 5-6 µm Zorbax C8

Mobile phase: MeCN:MeOH 80:20
Flow rate: 0.5
Injection volume: 5
Detector: UV 261

CHROMATOGRAM
Retention time: 16 (sphingosine GM1), 20.4 (eicosasphingosine GM1)

REFERENCE
Nakabayashi, H.; Iwamori, M.; Nagai, Y. Analysis and quantitation of gangliosides as p-bromophenacyl derivatives by high-performance liquid chromatography, *J.Biochem.*, **1984**, *96*, 977–984.

SAMPLE
Matrix: cells
Analyte: mycolic acids
Sample preparation: Vortex cells with 2 mL 25% KOH in MeOH:water 50:50 for 30 s, heat at 100° for 2 h, cool to room temperature, add 2 mL chloroform, add 1.5 mL 6 M HCl, vortex for 30 s. Remove the organic layer and evaporate it to dryness under a stream of air at 85°, reconstitute the residue in 100 μL 2% potassium bicarbonate in MeOH: water 50:50, evaporate to dryness under a stream of air at 85°, cool to room temperature, add 1 mL reagent:chloroform 5:95, vortex for 30 s, heat at 85° for 30 min, cool, add 1 mL 6 M HCl:MeOH 50:50, vortex for 30 s. Remove the lower organic layer and evaporate it to dryness under a stream of air at 85°, reconstitute the residue in 100 μL chloroform and 5 μL 50 μg/mL IS in chloroform, inject a 10 μL aliquot. (Reagent was 100 mM 2,4'-dibromoacetophenone (p-bromophenacyl bromide) and 5 mM crown ether in MeCN (p-bromophenacyl-8, Pierce).)

HPLC VARIABLES
Column: 70 × 4.6 3 μm Ultrasphere XL-ODS
Mobile phase: Gradient. MeOH:dichloromethane from 98:2 to 80:20 over 1 min, to 35:65 over 9 min, maintain at 35:65 for 1.5 min, return to initial conditions over 25 s, re-equilibrate for 1.5 min.
Column temperature: 25
Flow rate: 2 to 2.5 over 1 min, maintain at 2.5 for 10.5 min, return to 2 over 25 s, maintain at 2 for 1.5 min
Injection volume: 10
Detector: UV 254

CHROMATOGRAM
Retention time: 6-10
Internal standard: Ribi High-Molecular-Weight Internal Standard for HPLC (Ribi ImmunoChem research, Inc., Hamilton MT) (11)

KEY WORDS
used to identify bacterial strains

REFERENCE
Cage, G.D. High-performance liquid chromatography patterns of *Mycobacterium gordonae* mycolic acids, *J.Clin.Microbiol.*, **1992**, *30*, 2402–2407.

SAMPLE
Matrix: enzyme incubations
Analyte: prostaglandins
Sample preparation: Add 500 μL enzyme incubation to 1 mL MeOH, mix, add 4 mL 100 mM citric acid, add 500 mg anhydrous sodium sulfate, extract twice (alprostadil, dinoprostone) or 3 times (dinoprost) with 5 mL portions of dichloromethane. Pass the extracts through 1 g anhydrous sodium sulfate and evaporate them to dryness, reconstitute with 1 mL anhydrous MeCN containing a 3-fold molar excess of 2,4'-dibromoacetophenone (p-bromophenacyl bromide), add 2 μL diisopropylethylamine, let stand for 1 h, evaporate to dryness, reconstitute with 200 μL MeOH, inject a 10 μL aliquot.

HPLC VARIABLES
Column: μBondapak C18
Mobile phase: MeCN:water 50:50
Flow rate: 1.2
Injection volume: 10
Detector: UV 254

CHROMATOGRAM
Retention time: 11 (dinoprost), 14 (dinoprostone)
Limit of quantitation: 5 μM

OTHER SUBSTANCES
Extracted: metabolites, alprostadil

REFERENCE
Fitzpatrick, F.A. High-performance liquid chromatographic analysis of prostaglandins formed during *in vitro* incubations with prostaglandin 15-dehydrogenase, *J.Pharm.Sci.*, **1976**, *65*, 1609–1613.

SAMPLE
Matrix: enzyme incubations
Analyte: prostaglandins
Sample preparation: 2 mL Enzyme incubation + 2 mL MeOH, centrifuge. Remove the supernatant and add it to 2 mL 100 mM citric acid and 500 mg anhydrous sodium sulfate, extract twice with 5 mL portions of dichloromethane. Dry the extracts over anhydrous sodium sulfate and evaporate them to dryness under a stream of nitrogen, reconstitute with 100-200 μL 1.2 mg/mL 2,4'-dibromoacetophenone (p-bromophenacyl) bromide in MeCN, add 0.5 μL diisopropylethylamine, let stand at room temperature for 1 h, inject an aliquot.

HPLC VARIABLES
Column: 250 × 4 μBondapak C18
Mobile phase: MeCN:water 50:50
Flow rate: 1.2
Detector: UV 254

CHROMATOGRAM
Retention time: 12 (dinoprost), 15 (dinoprostone), 17 (prostaglandin D_2), 32 (prostaglandin A_2), 34 (prostaglandin B_2), 43 (15-methylprostaglandin B_2)
Limit of quantitation: <3 μg

REFERENCE
Fitzpatrick, F.A. High performance liquid chromatographic determination of prostaglandins $F_{2:ga}$ E_2 and D_2; from in vitro enzyme incubations, *Anal.Chem.*, **1976**, *48*, 499–502.

SAMPLE
Matrix: formulations
Analyte: azelaic acid
Sample preparation: Condition a 1 g 100 μm Bakerbond Florisil SPE cartridge with 5 mL THF:hexane 40:60. Condition a 500 mg 40 μm Bakerbond C18 SPE cartridge with MeCN and MeCN:water 65:35. 200 mg Cream + 3 mL THF:hexane 40:60, sonicate, centrifuge at 3000 rpm for 2 min, repeat extraction twice. Add the supernatants to the Florisil SPE cartridge, wash with 2 mL THF:hexane 40:60, dry under vacuum, elute with two 2 mL portions of hexane:isopropanol 70:30. Evaporate the eluate to dryness under a stream of nitrogen, reconstitute the residue in 10 mL MeOH, neutralize (phenolphthalein endpoint) with 0.01% KOH in MeOH, evaporate to dryness under a stream of nitrogen at room temperature, reconstitute with 15 mL MeCN, add 5 mL 2 mM 2,4'-dibromoacetophenone (p-bromophenacyl bromide) in MeCN containing 100 μM 18-crown-6, add 10 mL MeCN, stir at 80° for 30 min, cool to 4°, dilute 1:50 with MeCN:water 65:35, add a 1 mL aliquot to the C18 SPE cartridge, wash with two 3 mL aliquots of MeCN:water 75:25, elute with 10 mL MeCN. Remove a 1 mL aliquot of the eluate and add it to 100 μL 85.125 μg/mL sebacic acid, inject a 5 μL aliquot.

HPLC VARIABLES
Column: 125 × 4 5 μm LiChrospher 100-RP-18
Mobile phase: MeCN:water 75:25
Flow rate: 1
Injection volume: 5
Detector: UV 254

CHROMATOGRAM
Retention time: 6.06
Internal standard: sebacic acid (7.59)

KEY WORDS
cream; SPE

REFERENCE
Ferioli, V.; Rustichelli, C.; Vezzalini, F.; Gamberini, G. Determination of azelaic acid in pharmaceuticals and cosmetics by RP-HPLC after pre-column derivatization, *Farmaco*, **1994**, *49*, 421–425.

SAMPLE
Matrix: solutions
Analyte: fatty acids
Sample preparation: Neutralize a solution of the acid in MeOH or water with KOH in MeOH (phenolphthalein endpoint). Evaporate to dryness under reduced pressure, dissolve in 0.5-1.5 mL MeCN containing 2,4'-dibromoacetophenone:18-crown-6 10:1, stir at 80° for 15 min, cool, inject an aliquot.

HPLC VARIABLES
Column: 250 × 4 Corasil II C9
Mobile phase: MeOH:water 62.5:37.5, after 132 min 75:25
Column temperature: 40
Flow rate: 0.3
Detector: UV 254

CHROMATOGRAM
Retention time: 75 (linolenic acid), 100 (linoleic acid), 140 (oleic acid), 150 (stearic acid)

REFERENCE
Durst, H.D.; Milano, M.; Kikta, E.J. Jr.; Connelly, S.A.; Grushka, E. Phenacyl esters of fatty acids via crown ether catalysts for enhanced ultraviolet detection in liquid chromatography, *Anal.Chem.*, **1975**, *47*, 1797–1801.

SAMPLE
Matrix: solutions
Analyte: biotin and its analogs
Sample preparation: Evaporate biological samples, add 10 μmoles dibenzo-18-crown-6, add 100 μmoles 2,4'-dibromoacetophenone, add 25 mg anhydrous potassium carbonate, add 1.6 mL MeCN, reflux for 1 h. For bulk quantities take 55 mg biotin + 5 mL EtOH + 3 mL water, neutralize (phenolphthalein) with 20 mM KOH in MeOH, remove the solvent. Suspend in 5 mL MeCN, add 60 mg potassium carbonate, add 24 mg dibenzo-18-crown-6, add 166 mg 2,4'-dibromoacetophenone (p-bromophenacyl bromide), reflux for 1 h.

HPLC VARIABLES
Column: 330 × 4 10 μm μBondapak C18
Mobile phase: MeOH:water 60:40
Flow rate: 2
Detector: UV 254

CHROMATOGRAM
Retention time: 12.5 (biotin-l-sulfoxide), 13.2 (biotin-d-sulfoxide), 14.6 (biotin sulfone), 17.4 (biotin), 18.4 (dethiobiotin)
Limit of detection: 10 ng/mL

REFERENCE
Desbene, P.-L.; Coustal, S.; Frappier, F. Separation of biotin and its analogs by high-performance liquid chromatography: convenient labeling for ultraviolet or fluorimetric detection, *Anal.Biochem.*, **1983**, *128*, 359–362.

SAMPLE
Matrix: urine
Analyte: butyric acid
Sample preparation: 1 mL Urine + 200 µL 1 M HCl + 3 mL chloroform, mix vigorously for 5 min, centrifuge at 1000 g for 10 min. Remove the organic layer and add it to 1 mL water and 50 µL 1 M KOH, mix for 5 min, centrifuge at 1000 g for 10 min. Remove the aqueous layer and evaporate to it dryness under a stream of nitrogen, add 100 µL reagent, add 400 µL MeCN, heat with shaking at 80° for 30 min, centrifuge at 1000 g for 5 min, inject a 10 µL aliquot of the supernatant. (Reagent was a 20:1 mixture of 200 mM 2,4'-dibromoacetophenone (p-bromophenacyl bromide) in MeCN and 200 mM dicyclohexyl-18-crown-6 in MeCN.)

HPLC VARIABLES
Column: 150 × 3.9 5 µm Resolve C18
Mobile phase: MeCN:water 35:65
Flow rate: 2
Injection volume: 10
Detector: UV 254

CHROMATOGRAM
Retention time: 15
Limit of detection: 5 µm

KEY WORDS
pharmacokinetics

REFERENCE
Miller, A.A.; Kurschel, E.; Osieka, R.; Schmidt, C.G. Clinical pharmacology of sodium butyrate in patients with acute leukemia, *Eur.J.Cancer Clin.Oncol.*, **1987**, *23*, 1283–1287.

SAMPLE
Matrix: urine
Analyte: carnitine and acylcarnitines
Sample preparation: Condition a 500 mg silica SPE cartridge (Baker) with 5 mL MeOH. 500 µL Urine + 100 nmoles IS in water, evaporate to dryness under reduced pressure, reconstitute with 500 µL MeOH, centrifuge at 10000 g for 3 min, add the supernatant to the SPE cartridge, wash with 2 mL MeOH, elute with 3 mL MeOH:water:acetic acid 45:50:5. Collect the final 2.5 mL of eluate and evaporate it to dryness under reduced pressure, reconstitute with 500 µL MeOH, centrifuge. Evaporate the supernatant to dryness under a stream of nitrogen in a clean tube at 40°, add 300 µL 6.8 mM N,N-diisopropylethylamine in MeCN, vortex for 10 s, sonicate for 15 min, add 200 µL 25 mM 2,4'-dibromoacetophenone (p-bromophenacyl bromide) in MeCN, vortex for 10 s, heat at 37° for 30 min, filter (0.45 µm), inject a 20 µL aliquot.

HPLC VARIABLES
Guard column: 5 µm Hypersil BDS C8
Column: 200 × 4.6 5 µm Hypersil BDS C8
Mobile phase: Gradient. A was MeCN:water 70:30. B was MeCN:100 mM pH 5.0 triethylamine phosphate buffer 95:5. A:B 97:3 for 9 min, to 90:10 over 3 min, to 50:50 over 8 min, to 10:90 over 6 min, maintain at 10:90 for 3 min, return to initial conditions over 10 min, re-equilibrate for 5 min
Flow rate: 1.2 for 12 min, then 1.4
Injection volume: 20
Detector: UV 260

CHROMATOGRAM
Retention time: 14 (carnitine), 15 (acetylcarnitine), 16 (propionylcarnitine), 18 (isovalerylcarnitine), 19 (hexanoylcarnitine), 21 (octanoylcarnitine), 22 (nonanoylcarnitine), 23 (decanoylcarnitine), 29 (palmitoylcarnitine)

Internal standard: undecanoyl-L-carnitine (24)

KEY WORDS
SPE

REFERENCE
Poorthuis, B.J.H.M.; Jille-Vlcková, T.; Onkenhout, W. Determination of acylcarnitines in urine of patients with inborn errors of metabolism using high-performance liquid chromatography after derivatization with 4'-bromophenacylbromide, *Clin.Chim.Acta*, **1993**, *216*, 53–61.

2-Diethylaminoethyl Chloride

SAMPLE
Matrix: solutions
Analyte: prostaglandins
Sample preparation: Add 100 mg/mL 2-diethylaminoethyl chloride in MeCN:8% diethyl-isopropylamine in MeCN 1:10 to the sample, heat at 75° for 1 h.

HPLC VARIABLES
Column: 150 × 4.6 5 μm Econosphere C18
Mobile phase: Gradient. A was MeCN:water 30:70 containing 100 mM ammonium acetate. B was MeCN:water 70:30 containing 100 mM ammonium acetate. A:B from 100:0 to 0:100 over 20 min.
Flow rate: 1.2
Detector: MS, Finnigan MAT 4500, Vestec thermospray interface, positive ion mode, vaporizer 224 ± 10°, source 250°, filament 1000 eV, emission current 150 μA, SIM, ion evaporation

CHROMATOGRAM
Retention time: 10.5 (thromboxane B_2), 12.2 (dinoprost), 12.2 (dinoprostone), 13 (prostaglandin D_2), 13.9 (6-ketoprostaglandin $F_{1\alpha}$), 15.2 (prostaglandin A_2), 16.6 (prostaglandin A_1)
Limit of detection: 20-900 pg

REFERENCE
Voyksner, R.D.; Bush, E.D.; Brent, D. Derivatization to improve thermospray HPLC/MS sensitivity for the determination of prostaglandins and thromboxane B2, *Biomed.Environ.Mass.Spectrom.*, **1987**, *14*, 523–531.

1-(2,5-Dihydroxyphenyl)-2-bromoethanone

SAMPLE
Matrix: bulk
Analyte: bile acids
Sample preparation: Add 5 mL of a 39.2 mg/mL solution in dry MeCN to 120 mg 1-(2,5-dihydroxyphenyl)-2-bromoethanone and 100 μL triethylamine, heat at 70° for 2 h, dilute with 20 mL water, extract 3 times with diethyl ether. Combine the extracts and wash them with saturated sodium bicarbonate and water, dry over anhydrous sodium sulfate, evaporate, reconstitute, inject a 5 μL aliquot. (Preparation of 1-(2,5-dihydroxyphenyl)-2-bromoethanone is as follows. Slowly add 2.5 g phenyltrimethylammonium tribromide to a solution of 2',5'-dihydroxyacetophenone in 20 mL dry THF, stir at room temperature overnight (check by TLC with cyclohexane:ethyl acetate 70:30). Remove the precipitate by filtration and dry under reduced pressure, chromatograph using cyclohexane:ethyl acetate 70:30 to give 1-(2,5-dihydroxyphenyl)-2-bromoethanone. An alternative procedure is as follows. Stir 27.6 g 1,4-dimethoxybenzene and 28 mL bromoacetyl bromide at 0°, add 53.4 g aluminum bromide over 10 min (an exothermic reaction ensues), let stand at room temperature for 12 h, add 100 mL 48% HBr, add 100 g ice, stir for 1 h, extract twice with 200 mL portions of diethyl ether. Combine the extracts and wash them 3 times with 200 mL portions of water, dry over 40 g anhydrous magnesium sulfate, evaporate to dryness, recrystallize the product 3 times from EtOH to yield 1-(2,5-dihydroxyphenyl)-2-bromo-ethanone monobromoacetate (mp 105-107°). Dissolve 11 g 1-(2,5-dihydroxyphenyl)-2-bromoethanone monobromoacetate in 200 mL warm dry MeOH saturated with HBr, stir for 18 h, add 200 mL water, cool to -10°. Collect the yellow solid and dry it under vacuum at 50° for 48 h, recrystallize from toluene:heptane 50:50 then toluene to obtain 1-(2,5-dihydroxyphenyl)-2-bromoethanone as yellow needles (mp 117-119°) (J. Chromatogr. 1988, 442, 209).)

HPLC VARIABLES
Guard column: 4 × 4 5 μm 5 μm Hypersyl ODS RP-18
Column: 100 × 4.6 3 μm Adsorbosphere
Mobile phase: MeCN:MeOH:100 mM pH 6.5 sodium acetate buffer 20:60:20
Flow rate: 1
Injection volume: 5
Detector: E, ESA Coulochem Model 5100A, Model 5010 analytical cell, porous graphite electrodes +0.6 V

CHROMATOGRAM
Retention time: 4.07 (ursodiol), 8.71 (chenodiol)
Limit of detection: 0.78-0.88 nM

REFERENCE
Bousquet, E.; Santagati, N.A.; Tirendi, S. Determination of chenodeoxycholic acid in pharmaceutical preparations of ursodeoxycholic acid by high performance liquid chromatography with coulometric electrochemical detection, *J.Liq.Chromatogr.Rel.Technol.*, **1997**, *20*, 757–770.

SAMPLE
Matrix: solutions
Analyte: antibiotics

Sample preparation: React the antibiotic, triethylamine, and 1-(2,5-dihydroxyphenyl)-2-bromoethanone in a 1:2:4 molar ratio in DMF at 45° for 2 h (use 18-crown-6 for potassium salts and dibenzo-18-crown-6 for sodium salts), inject a 10 μL aliquot. (Preparation of 1-(2,5-dihydroxyphenyl)-2-bromoethanone is as described above.)

HPLC VARIABLES
Column: 250 × 4 7 μm RP-18 LiChrocart (Merck)
Mobile phase: MeOH:100 mM pH 6.5 sodium acetate 58:42 (A) or 45:55 (B)
Flow rate: 1
Injection volume: 10
Detector: E, Bioanalytical Systems Model LC4B, glassy carbon electrode 0.8 V, Ag/AgCl reference electrode

CHROMATOGRAM
Retention time: 6.3 (cephapirin sodium (A)), 7.8 (hetacillin potassium (A)), 10.3 (ampicillin (B)), 10.7 (methicillin sodium (A)), 12.8 (penicillin G sodium (A)), 13.7 (carbenicillin monosodium (A)), 16.7 (amoxicillin (B), 19 (oxacillin monosodium (A)), 19.3 (cephapirin sodium (B)), 21 (cloxacillin sodium (A)), 28.8 (dicloxacillin sodium (A)), 30.7 (nafcillin sodium (A))

REFERENCE
Munns, R.K.; Roybal, J.E.; Shimoda, W.; Hurlbut, J.A. 1-(4-Hydroxyphenyl)-, 1-(2,4-dihydroxyphenyl)- and 1-(2,5-dihydroxyphenyl)-2-bromoethanones: new labels for determination of carboxylic acids by high-performance liquid chromatography with electrochemical and ultraviolet detection, *J.Chromatogr.*, **1988**, *442*, 209–218.

SAMPLE
Matrix: solutions
Analyte: carboxylic acids
Sample preparation: React the carboxylic acid, triethylamine, and 1-(2,5-dihydroxy-phenyl)-2-bromoethanone in a 1:2:4 molar ratio in MeCN at 45° for 2 h, inject a 10 μL aliquot. (Preparation of 1-(2,5-dihydroxyphenyl)-2-bromoethanone is as described above.)

HPLC VARIABLES
Column: 250 × 4 7 μm RP-18 LiChrocart (Merck)
Mobile phase: MeOH:100 mM pH 6.5 sodium acetate 58:42
Flow rate: 1
Injection volume: 10
Detector: E, Bioanalytical Systems Model LC4B, glassy carbon electrode 0.6 V, Ag/AgCl reference electrode

CHROMATOGRAM
Retention time: 4 (quinoxaline-2-carboxylic acid), 5 (benzoic acid), 6 (salicylic acid)
Limit of detection: 1 pmole

REFERENCE
Munns, R.K.; Roybal, J.E.; Shimoda, W.; Hurlbut, J.A. 1-(4-Hydroxyphenyl)-, 1-(2,4-dihydroxyphenyl)- and 1-(2,5-dihydroxyphenyl)-2-bromoethanones: new labels for determination of carboxylic acids by high-performance liquid chromatography with electrochemical and ultraviolet detection, *J.Chromatogr.*, **1988**, *442*, 209–218.

Panacyl Bromide

RELATED REFERENCES
Engels, W.; Kamps, M.A.F.; Lemmens, P.J.M.R.; van der Vusse, G.J.; Reneman, R.S. Determination of prostaglandins and thromboxane in whole blood by high-performance liquid chromatography with fluorimetric detection. *J.Chromatogr.* **1988**, *427*, 209-218.

Hawkes, J.S.; James, M.J.; Cleland, L.G. Separation and quantification of PGE3 following derivatization with panacyl bromide by high pressure liquid chromatography with fluorometric detection. *Prostaglandins* **1991**, *42*, 355-368.

Pullen, R.H.; Howell, J.A.; Cox, J.W. High performance liquid chromatographic determination of thromboxane B2 in human serum as a methoxime-panacyl ester derivative. *Prostaglandins,Leukotrienes Med.* **1987**, *29*, 205-219.

Stein, J.; Milovic, V.; Zeuzem, S.; Caspary, W.F. Fluorometric high-performance liquid chromatography of free fatty acids using panacyl bromide. *J.Liq.Chromatogr.* **1993**, *16*, 2915-2922.

SAMPLE

Matrix: blood

Analyte: arbaprostil

Sample preparation: Condition a 3 mL 200 mg and a 1 mL 100 mg Bond-Elut C18 SPE cartridge with 4 mL MeOH and 4 mL water. 3 mL Plasma + 12 μL 32.5 ng/mL IS in MeCN + 300 μL 5% formic acid, vortex, centrifuge at 4° at 1500 g for 15 min, add the supernatant to the 3 mL SPE cartridge, wash with two 2 mL portions of water, wash with two 2 mL portions of MeOH:water 10:90, wash with 2 mL toluene, elute with 1 mL ethyl acetate. Evaporate the eluate to dryness under a stream of nitrogen at 40°, reconstitute with 610 μL MeOH, vortex for 30 s, add 1.42 mL 0.01% formic acid, inject a 1.8 mL aliquot onto a 30 × 4.6 5 μm Brownlee RP-8 column and elute to waste with MeCN:water:formic acid 40:60:0.01 at 2 mL/min, after 2.5 min backflush the contents of this column onto a 250 × 4.6 5 μm Supelcosil LC-18 column eluted with MeCN:water:formic acid 40:60:0.01 at 2 mL/min, after about 6.5-7 min collect a fraction containing the prostaglandins. Dilute this fraction with an equal volume of water, add to the 1 mL SPE cartridge, wash with 1 mL hexane, elute with two 500 μL portions of ethyl acetate. Evaporate the eluate to dryness under a stream of nitrogen at 40°, reconstitute the residue in 250 μL 100 μg/mL panacyl bromide in THF:MeCN 20:80, vortex for 30 s, add 3 μL N,N-diisopropylethylamine, heat at 40° for 1 h (Anal. Chem. 1984, 56, 1866). Evaporate to dryness under a stream of nitrogen at 40°, reconstitute the residue in 230 μL isooctane:ethylene dichloride:isopropanol 70:30:1, sonicate for 10 min, inject a 200 μL aliquot onto column A and elute to waste with mobile phase A, after 3 min divert the effluent from column A onto column B and elute both to waste, after 1.5 min remove column A from the circuit, continue to elute column B to waste with mobile phase A, after 7.5 min collect the effluent from column B in a 2.2 mL sample loop, after 2 min inject the contents of this sample loop onto column C with mobile phase B, elute column C with mobile phase B, monitor the effluent from column C. (Panacyl bromide (p-(9-anthroyloxy)phenacyl bromide) is available from Molecular Probes, Eugene OR. Synthesis is as follows. Add 3.04 g benzyltrimethylammonium dichloroiodate to a solution of 500 mg 4'-hydroxyacetophenone in 50 mL dichloroethane and 20 mL MeOH, reflux for 10 h, remove the solvent by distillation, add 20 mL 5% sodium bisulfite to the residue, extract four times with 40 mL portions of ether, dry over anhydrous magnesium sulfate, evaporate to dryness under reduced pressure to give p-hydroxyphenacyl chloride (mp 151-152°) (Synthesis 1988, 545). Purify p-hydroxyphenacyl chloride by suspending 100 g in 1 L boiling toluene, filter, cool to obtain white crystals of p-hydroxyphenacyl chloride. Repeat this process a number of times to obtain more pure product. Reflux 10 g 9-anthracenecarboxylic acid in 150 mL redistilled thionyl chloride for 2 h, evaporate to dryness under reduced pressure at 30°, dissolve the residue in 150 mL dry toluene containing 11.5 g p-hydroxyphenacyl chloride, reflux for 2 h, evaporate to dryness under reduced pressure, recrystallize from 200 mL hot MeCN to give p-(9-anthroyloxy)phenacyl chloride as deep yellow crystals (mp 159.8-161.6°). Dissolve 2.5 g p-(9-anthroyloxy)phenacyl chloride in 25 mL THF:MeCN 20:80, add 8 g anhydrous LiBr, reflux briefly, cool to room temperature, filter, wash the solid with water to obtain p-(9-anthroyloxy)phenacyl bromide (panacyl bromide) as deep yellow crystals (mp 173.3-173.6°) (Anal. Biochem. 1987, 165, 220).)

HPLC VARIABLES
Column: A 10 × 4.6 Co:Pell PAC; B 150 × 4.6 6 μm Zorbax CN; C 240 × 4.6 6 μm Zorbax Sil
Mobile phase: A Hexane:dichloromethane:isopropanol 70:30:1; B Hexane:dichloromethane:THF:isopropanol 60:20:20:1
Flow rate: 1
Injection volume: 200
Detector: F ex 375 em 470

CHROMATOGRAM
Retention time: 36 (arbaprostil), 39.5 (15S epimer)
Internal standard: 5,6-trans-(15R)-15-methylprostaglandin E_2 (U-67205) (37.5)
Limit of quantitation: 10 pg/mL

KEY WORDS
SPE; plasma; column-switching; normal phase

REFERENCE
Pullen, R.H.; Cox, J.W. Determination of (15R)- and (15S)-15-methylprostaglandin E2 in human plasma with picogram per milliliter sensitivity by column-switching high-performance liquid chromatography, *J.Chromatogr.*, **1985**, *343*, 271–283.

SAMPLE
Matrix: enzyme incubations
Analyte: dinoprostone
Sample preparation: Condition a Sep-Pak silica SPE cartridge with dichloromethane. Centrifuge 1 mL enzyme incubation at 15000 g for 1 min, add 50 ng IS to the supernatant, add 1.5 mL cold (-20°) acetone, add 1.5 mL light petroleum, mix, discard the light petroleum layer, repeat the light petroleum wash. Adjust the pH of the aqueous phase to 3.5 with formic acid, extract twice with 1 mL portions of ethyl acetate. Combine the organic layers and evaporate them to dryness under a stream of nitrogen at 45°, reconstitute the residue in 1 mL 10 μg/mL panacyl bromide in MeCN:THF 80:20, add 1 μL triethylamine, mix, let stand at room temperature for 2 h, add to the SPE cartridge, wash with 20 mL dichloromethane, elute with 2 mL MeCN:MeOH 85:15. Evaporate the eluate to dryness under a stream of nitrogen, reconstitute the residue in 200 μL MeCN, inject a 20 μL aliquot. (Panacyl bromide (p-(9-anthroyloxy)phenacyl bromide) is available from Molecular Probes, Eugene OR. Synthesis is as described above.)

HPLC VARIABLES
Column: 250 × 4.6 Zorbax Sil
Mobile phase: Dichloromethane:MeCN:MeOH 90:9:1
Flow rate: 2.3
Injection volume: 20
Detector: F ex 280 em 400 (cutoff filter)

CHROMATOGRAM
Retention time: 2.8
Internal standard: 13,14-dihydro-15-keto-PGF$_{2\alpha}$ (2.1)

Limit of detection: 40 pg

KEY WORDS
SPE; normal phase; rat; gastric mucosa

REFERENCE
Stein, T.A.; Angus, L.; Borrero, E.; Auguste, L.J.; Wise, L. Picogram measurement of prostaglandin E_2 synthesis by gastric mucosa by high-performance liquid chromatography, *J.Chromatogr.*, **1987**, *385*, 377–382.

SAMPLE
Matrix: solutions
Analyte: prostaglandins
Sample preparation: Condition a Sep-Pak silica SPE cartridge with 5 mL THF:water 95:5, 5 mL MeCN, and 5 mL dichloromethane. Prepare a solution in EtOH, add a 500 ng/mL solution of panacyl bromide in THF:MeCN 20:80, for each 1 mL of reaction mixture add 3 µL triethylamine, heat at 37° for 3 h, add 0.5 mL of the reaction mixture to the SPE cartridge, wash with 10 mL dichloromethane, elute with 3 mL MeCN:MeOH 85:15, evaporate the eluate to dryness under a stream of nitrogen, reconstitute in MeCN, inject an aliquot. (Panacyl bromide (p-(9-anthroyloxy)phenacyl bromide) is available from Molecular Probes, Eugene OR. Synthesis is as described above.)

HPLC VARIABLES
Column: 300 × 3.9 fatty acid analysis column (Waters?)
Mobile phase: Gradient. MeCN:water:acetic acid from 56:44:0.1 to 65:35:0.1 over 15 min
Flow rate: 1.2
Detector: F ex 249 em 413 (cut-off filter); UV 254

CHROMATOGRAM
Retention time: 18.58 (6-keto-prostaglandin $F_{1\alpha}$), 21.71 (6-keto-prostaglandin E_1), 22.78 (dinoprost), 25.28 (dinoprostone), 26.41 (prostaglandin D_2), 27.68 (13,14-dihydro-15-keto-prostaglandin $F_{2\alpha}$)
Limit of detection: 50 pg (F), 280 pg (UV)

KEY WORDS
SPE

REFERENCE
Watkins, W.D.; Peterson, M.B. Fluorescent/ultraviolet absorbing ester derivative formation and analysis of eicosanoids by high-pressure liquid chromatography, *Anal.Biochem.*, **1982**, *125*, 30–40.

SAMPLE
Matrix: solutions
Analyte: prostaglandins
Sample preparation: Dissolve compound in 1 mL MeCN:THF 80:20, add 70 µg panacyl bromide, add 1.025 µL N,N-diisopropylethylamine, mix, let stand at room temperature for 3 h, inject an aliquot onto column A (pre-equilibrated with 10 mL dichloromethane) and elute to waste with 15 mL dichloromethane, elute the contents of column A onto column B with the mobile phase and start the gradient, monitor the effluent from column B. (Panacyl bromide (p-(9-anthroyloxy)phenacyl bromide) is available from Molecular Probes, Eugene OR. Synthesis is as described above.)

HPLC VARIABLES
Column: A Guard-Pak silica; B 250 × 4.6 5 µm Hibar Silica (Merck)
Mobile phase: Gradient. A was hexane:dichloromethane:THF:MeCN:MeOH 35:50:11:4:0.25. B was dichloromethane:MeOH 98:2. C was dichloromethane:MeOH:THF 92:7:1. A:B:C 100:0:0 for 35 min, to 0:100:0 over 10 min, maintain at 0:100:0 for 20 min, to 0:0:100 over 20 min, maintain at 0:0:100 for 15 min
Flow rate: 1
Injection volume: 20
Detector: F ex 253 em 445

CHROMATOGRAM
Retention time: 10 (13,14-dihydro-15-ketoprostaglandin E_2), 15 (prostaglandin A_2), 32 (prostaglandin D_2), 49 (8-isoprostaglandin E_2), 52 (11-epiprostaglandin E_2), 63 (dinoprostone), 71 (alprostadil), 83 (dinoprost), 94 (thromboxane B_2), 98 (6-ketoprostaglandin $F_{1\alpha}$)

Limit of detection: 30 pg

KEY WORDS
column-switching; normal phase

REFERENCE
Salari, H.; Yeung, M.; Douglas, S.; Morozowich, W. Detection of prostaglandins by high-performance liquid chromatography after conversion to *p*-(9-anthroyloxy)phenacyl esters, *Anal.Biochem.*, **1987**, *165*, 220−229.

SAMPLE
Matrix: solutions
Analyte: prostaglandins
Sample preparation: Condition a Sep-Pak silica SPE cartridge with dichloromethane. Dissolve 0.04-100 ng compound and 50 ng IS in 1 mL MeCN:THF 80:20, add 10 µg panacyl bromide, add 1 µL triethylamine, mix, let stand at room temperature for 2 h, add to the SPE cartridge, wash with 20 mL dichloromethane, elute with 2 mL MeCN:MeOH 85:15. Evaporate the eluate to dryness under a stream of nitrogen, reconstitute the residue in 200 µL MeCN, inject a 20 µL aliquot. (Panacyl bromide (p-(9-anthroyloxy)phenacyl bromide) is available from Molecular Probes, Eugene OR. Synthesis is as described above.)

HPLC VARIABLES
Column: 250 × 4.6 Zorbax Sil
Mobile phase: Dichloromethane:MeCN:MeOH 90:9:1
Flow rate: 1.5
Injection volume: 20
Detector: F ex 280 em 400 (cutoff filter)

CHROMATOGRAM
Retention time: 7.90 (dinoprostone), 10.05 (thromboxane B_2), 12.26 (6α-ketoprostaglandin $F_{1\alpha}$), 13.98 (dinoprost)
Internal standard: 13,14-dihydro-15-keto-PGF$_{2\alpha}$ (7.14)
Limit of quantitation: 40-400 pg

KEY WORDS
SPE; normal phase

REFERENCE
Stein, T.A.; Angus, L.; Borrero, E.; Auguste, L.J.; Wise, L. High-performance liquid-chromatographic assay for prostaglandins with the use of *p*-(9-anthroyloxy)phenacyl bromide, *J.Chromatogr.*, **1987**, *395*, 591−595.

SAMPLE
Matrix: tissue
Analyte: biotin
Sample preparation: Condition two Sep-Pak C18 SPE cartridges with 10 mL MeOH and 10 mL water. 2-3 g Gut tissue or liver + 5 mL 5% trichloroacetic acid + 5 nmole dethiobiotin, homogenize, centrifuge at 10000 g for 15 min, re-extract pellet with 5 mL 5% trichloroacetic acid twice. Combine the supernatants and add them to a SPE cartridge, wash with 10 mL MeCN:water 2:98, elute with 10 mL MeCN:water 15:85, add the eluate to a 70 × 8 column of 200-400 mesh Dowex 1x8 formate, wash with 10 mL water, wash with 10 mL 100 mM potassium formate, elute with 30 mL 100 mM potassium formate. Add the eluate to a SPE cartridge, wash with 10 mL water, elute with 10 mL methyl formate. Evaporate the eluate under a stream of nitrogen, dissolve the residue in 100 µL 2.5 mM panacyl bromide (p-(9-anthroyl)phenacyl bromide) and 0.5 mM dibenzo-18-crown-6 in acetone, add 20-30 mg potassium carbonate, heat at 57° for 3 h, inject an aliquot. (Panacyl bromide is available from Molecular Probes, Eugene OR. Synthesis is as described above.)

HPLC VARIABLES
Column: 150 × 4.6 3 μm Hypersil
Mobile phase: Dichloromethane:MeOH 95:5
Flow rate: 1.4
Injection volume: 100
Detector: F ex 380 em 470

CHROMATOGRAM
Retention time: 6.46
Internal standard: dethiobiotin (5.56)

KEY WORDS
SPE; gut; rat; liver; normal phase

REFERENCE
Stein, J.; Hahn, A.; Lembcke, B.; Rehner, G. High-performance liquid chromatographic determination of biotin in biological materials after crown ether-catalyzed fluorescence derivatization with panacyl bromide, *Anal.Biochem.*, **1992**, *200*, 89–94.

Pentafluorobenzyl Bromide

SAMPLE
Matrix: blood
Analyte: retinoic acid
Sample preparation: Condition a 100 mg methyl-C1 Accubond SPE cartridge (J&W) with three 1 mL portions of MeOH and three 1 mL portions of 1% ammonium acetate. 500 μL Plasma + 20 μL 1 μg/mL acitretin in MeCN containing 10 mM BHT + 1 mL 10 mM BHT in isopropanol, vortex, rotate for 15 min, centrifuge at 16000 g for 10 min. Remove the supernatant and add it to 11 mL 1% ammonium acetate, add to the SPE cartridge, wash with 1 mL 0.1% ammonium acetate, wash with 1 mL MeOH:0.1% ammonium acetate 50:50, dry under vacuum for 30 s, elute with 1.5 mL 10 mM BHT in MeCN. Add 10 μL pentafluorobenzyl bromide and 10 μL 10 mg/mL potassium carbonate in MeCN:water 50:50 to the eluate, vortex, let stand at room temperature for1 h, evaporate to dryness under reduced pressure for 2 h, reconstitute with 20-100 μL 10 mM BHT in MeCN, inject a 20 μL aliquot.

HPLC VARIABLES
Column: 150 × 3.9 Nova-Pak C18 + 75 × 3.9 Nova-Pak C18 (in series)
Mobile phase: Gradient. MeCN:buffer 80:20 for 10 min, to 90:10 (step gradient). (Buffer was 100 mM ammonium acetate adjusted to pH 5.0 with acetic acid.)
Column temperature: 40
Injection volume: 20
Detector: UV 369; MS Hewlett-Packard model 5988A, particle beam interface nebulizer 60°, helium 35 psi, m/z 299

CHROMATOGRAM
Retention time: 26 (isotretinoin), 27.3 (9-cis-retinoic acid), 28.2 (tretinoin)
Internal standard: acitretin (m/z 325) (16)
Limit of detection: 0.05 ng/mL

KEY WORDS
plasma; protect from light; SPE

REFERENCE
Lehman, P.A.; Franz, T.J. A sensitive high-pressure liquid chromatography/particle beam/mass spectrometry assay for the determination of all-trans-retinoic acid and 13-cis-retinoic acid in human plasma, *J.Pharm.Sci.*, **1996**, *85*, 287–290.

SAMPLE
Matrix: bulk
Analyte: fatty acids
Sample preparation: Dissolve up to 1 mg fatty acid in 1 mL dichloromethane, add 1 mL of a solution containing 100 µmoles tetrabutylammonium hydrogen sulfate and 200 µmoles NaOH, add 20 µL pentafluorobenzyl bromide, shake vigorously at room temperature for 30 min. Remove the dichloromethane layer and evaporate it to dryness, reconstitute with hexane, add to a 20 × 10 silica Sep-Pak SPE cartridge, elute with hexane:dichloromethane 85:15, inject an aliquot.

HPLC VARIABLES
Column: 300 × 7.8 µPorasil
Mobile phase: Hexane:dichloromethane 85:15, half saturated with water
Flow rate: 4
Detector: UV 254

CHROMATOGRAM
Retention time: 11 (stearic acid), 12.5 (oleic acid), 15 (linoleic acid), 17 (linolenic acid)

KEY WORDS
SPE; normal phase; semi-preparative

REFERENCE
Netting, A.G.; Duffield, A.M. Pentafluorobenzyl esters as derivatives for the semi-preparative high-performance liquid chromatography of fatty acids, *J.Chromatogr.*, **1983**, *257*, 174–179.

N-(1-Pyrenyl)bromoacetamide

SAMPLE
Matrix: solutions
Analyte: fatty acids
Sample preparation: Mix 50 µL of a 5-50 µM solution in 100 mM pH 6.8 phosphate buffer, 10 µL 10 mM tetrabutylammonium hydrogen sulfate in water, and 50 µL 1 mM N-(1-pyrenyl)bromoacetamide in 1,2-dichloroethane, heat at 80-90° with agitation for 15 min, cool. Remove a 5-10 µL aliquot of the organic layer and dilute to 50-500 µL with mobile phase, inject a 20 µL aliquot. (Preparation of N-(1-pyrenyl)bromoacetamide is as follows. Add a solution of 202 mg bromoacetyl bromide in 5 mL dry diethyl ether dropwise to a solution of 217 mg 1-aminopyrene and 300 µL triethylamine in 30 mL dry diethyl ether. Filter off the precipitate and wash it with MeOH:water 20:80, dry, chromatograph on a

300 × 20 glass column of 70-230 mesh silica gel 60 (Merck) with 1,2-dichloroethane:ethyl acetate 90:10, recrystallize from ethyl acetate to obtain N-(1-pyrenyl)bromoacetamide as a light-yellow solid (mp 222-224°).)

HPLC VARIABLES
Column: 150 × 4.6 5 μm Nucleosil C18
Mobile phase: MeCN:MeOH:water 80:10:10
Flow rate: 1.8
Injection volume: 20
Detector: F ex 344 em 386.5

CHROMATOGRAM
Retention time: 7.5 (lauric acid), 12 (myristic acid), 20 (arachidic acid)

REFERENCE
Allenmark, S.; Chelminska-Bertilsson, M. Precolumn fluorogenic labeling of carboxylic acids with the use of N-(1-pyrenyl)bromoacetamide and phase-transfer catalysis, *Chromatographia*, **1989**, *28*, 367–369.

ALCOHOL

9-Anthracenemethanol

SAMPLE
Matrix: blood
Analyte: ibuprofen
Sample preparation: 150 μL Plasma + 1 μg ibuprofen + 10 μL 4 M HCl, vortex, add 1 mL dichloromethane, extract, centrifuge at 2500 g for 5 min. Remove the organic layer and evaporate it to dryness under a stream of nitrogen, reconstitute the residue in 100 μL 50 μg/mL 1-hydroxybenzotriazole in chloroform, evaporate to dryness under a stream of nitrogen, add 30 μL 1 mg/mL 1-(3-dimethylaminopropyl)-3-ethylcarbodiimide hydrochloride in chloroform, add 5 μL triethylamine, mix, let stand at 0° for 30 min, add 100 μL 200 μg/mL 9-anthracenemethanol (9-(hydroxymethyl)anthracene) in chloroform, let stand at room temperature for 10 min, inject a 10 μL aliquot. (Phthalic acid, salicylic acid, and p-sulfamoylbenzoic acid can be determined using 1,1'-carbonyldiimidazole as a coupling reagent. Mix 10 μL of a chloroform solution with 3 mg 1,1'-carbonyldiimidazole, let stand at 25° for 5 min, add 1 μL 4 M HCl, heat at 60° for 30 s, add 50 μL 10 mg/mL 9-anthracenemethanol in chloroform, vortex for 15 s, heat at 60° for 20 min.)

HPLC VARIABLES
Column: 300 × 1.9 10 μm LiChrosorb RP-18
Mobile phase: MeOH:water 90:10
Flow rate: 1
Injection volume: 10
Detector: F ex 365 em 415

CHROMATOGRAM
Retention time: 15
Limit of detection: 100 fmole

OTHER SUBSTANCES
Also analyzed: acetic acid, aminocaproic acid, aspirin, benzoic acid, diclofenac, indomethacin, lauric acid, mandelic acid, myristic acid, nalidixic acid, nicotinic acid, phenylacetic
acid, probenecid, sorbic acid, stearic acid
Non-interfering: barbituric acid, mercaptopurine, phenol, phthalic acid, purine, salicylic
acid, p-sulfamoylbenzoic acid, sulfanilic acid

KEY WORDS
plasma

REFERENCE
Lingeman, H.; Hulshoff, A.; Underberg, W.J.M.; Offerman, F.B.J.M. Rapid, sensitive and specific derivatization methods with 9-(hydroxymethyl)anthracene for the fluorimetric detection of carboxylic
acids prior to reversed-phase high-performance liquid chromatographic separation, *J.Chromatogr.*,
1984, *290*, 215–222.

SAMPLE
Matrix: blood
Analyte: fatty acids
Sample preparation: 200 µL Plasma + 3 mL 3.9 µg/mL margaric acid in chloroform:n-
heptane:MeOH 28:21:1, vortex for 2 min, centrifuge at 2000 g for 20 min. Remove a 2
mL aliquot of the organic layer and evaporate it to dryness under a stream of nitrogen,
reconstitute with 50 µL 2 mg/mL 9-anthracenemethanol (9-(hydroxymethyl)anthracene)
in dichloromethane, add 50 µL reagent solution, add 10 µL triethylamine, sonicate for 15
s, vortex for 15 s, heat at 50° for 30 min, evaporate to dryness under a stream of nitrogen,
reconstitute with 1 mL mobile phase, inject an aliquot. (Prepare the reagent, 2-bromo-1-
methylpyridinium iodide, as follows. Reflux 5 mL 2-bromopyridine and 7 mL iodomethane
in 20 mL dry diethyl ether for 1 h, wash the precipitate of 2-bromo-1-methylpyridinium
iodide with ether. The reagent solution was a 20 mg/mL suspension of 2-bromo-1-meth-
ylpyridinium iodide in dichloromethane.)

HPLC VARIABLES
Column: 250 × 4.5 3 µm Spherisorb C8
Mobile phase: Gradient. MeCN:water 93:7 for 12 min, 86:14 for 5 min, 100:0 for 23 min
(step gradient?).
Flow rate: 1
Injection volume: 20
Detector: F ex 360 em 420

CHROMATOGRAM
Retention time: 9 (lauric acid), 10 (linolenic acid), 11.5 (arachidonic acid), 12.5 (myristic
acid), 13.5 (palmitoleic acid), 15 (linoleic acid), 21 (palmitic acid), 22 (oleic acid), 25 (stearic
acid), 28.5 (arachidic acid), 33 (behenic acid), 38 (lignoceric acid)
Internal standard: margaric acid (23)
Limit of detection: 50 ng

KEY WORDS
plasma

REFERENCE
Baty, J.D.; Pazouki, S.; Dolphin, J. Analysis of fatty acids as their anthrylmethyl esters by high-performance liquid chromatography with fluorescence detection, *J.Chromatogr.*, **1987**, *395*, 403–411.

Dansyl Ethanolamine

SAMPLE
Matrix: bulk
Analyte: fatty acids
Sample preparation: React 4 mg fatty acid, 3 mg dicyclohexylcarbodiimide, and 120 μL 40 mg/mL dansyl ethanolamine in chloroform in the dark at room temperature overnight, add water, filter, inject an aliquot of the organic layer of the filtrate. (Prepare dansyl ethanolamine by adding dansyl chloride to a large excess of stirred ethanolamine. Collect the precipitate of dansyl ethanolamine and recrystallize it from MeOH.)

HPLC VARIABLES
Column: 250 × 4.65 μm Ultrasphere ODS
Mobile phase: MeCN:MeOH:20 mM silver nitrate in water 45:45:10
Flow rate: 2
Detector: F ex 360 em 420

CHROMATOGRAM
Retention time: 9.3 (linolenic acid), 10.2 (palmitoleic acid), 13.2 (eicosadienoic acid), 15.0 (linoleic acid), 23.7 (palmitic acid), 24.9 (oleic acid), 27.0 (elaidic acid), 32.1 (margaric acid)

REFERENCE
Ryan, P.J.; Honeyman, T.W. Determination of fatty acids by high-performance liquid chromatography of Dns-ethanolamine derivatives, *J.Chromatogr.*, **1984**, *312*, 461–466.

Methanol

$$CH_3OH \quad + \quad \underset{R}{\overset{O}{\|}}\!\!-\!\!OH \quad \longrightarrow \quad \underset{R}{\overset{O}{\|}}\!\!-\!\!OCH_3$$

SAMPLE
Matrix: blood
Analyte: ochratoxin A
Sample preparation: 2.5 mL Serum + 5 mL chloroform + 10 mL 50 mM HCl containing 100 mM magnesium chloride, rotate at 20 rpm for 10 min, centrifuge at 10000 g for 10 min (Appl. Environ. Microbiol. 1979, 38, 772). Remove the organic layer and add 10 mL 14% boron trifluoride in MeOH, heat on a steam bath for 5 min, add 30 mL water, extract with three 10 mL portions of chloroform (J. Assoc. Off. Anal. Chem. 1973, 56, 817). Combine the extracts and wash them three times with 10 mL portions of water, purify on C18 TLC plates (Whatman Type KC 18F) with MeOH:water 70:30, elute from the plate with MeOH, inject an aliquot.

HPLC VARIABLES
Column: 250 × 4.6 5 μm Ultrasphere ODS
Mobile phase: MeOH:buffer 70:30 (Buffer was water adjusted to pH 2.1 with phosphoric acid.)
Column temperature: 50
Flow rate: 1.5
Injection volume: 50
Detector: F ex 333 em 418

CHROMATOGRAM
Retention time: 10.2
Limit of detection: 20 ng/mL

KEY WORDS
pig; serum

REFERENCE
Marquardt, R.R.; Frohlich, A.A.; Sreemannarayana, O.; Abramson, D.; Bernatsky, A. Ochratoxin A in blood from slaughter pigs in western Canada, *Can.J.Vet.Res.*, **1988**, *52*, 186–190.

SAMPLE
Matrix: bulk
Analyte: amino acids
Sample preparation: Treat 0.2-10 μmoles amino acids with 50-100 μL MeOH:thionyl chloride 95:5 at 60° for 2 h, evaporate to dryness, reconstitute with 35 μL 28% diisopropylethylamine in DMF, shake at room temperature for 10 min, add 25 μL 200-1000 mM N-α-(9-fluorenylmethyloxycarbonyl)leucine-N-carboxyanhydride in DMF, mix, let stand at room temperature for 10 min, add 200 μL 250 mM pH 8 sodium glycinate, mix, let stand for 5 min, add 300 μL chloroform, extract, dilute 10000-50000-fold with n-hexane, inject a 20 μL aliquot. (Synthesis of N-α-(9-fluorenylmethyloxycarbonyl)leucine-N-carboxyanhydride is as follows. Dry all solvents over 4 Å molecular sieve. Stir 20 mmoles L-leucine in 67 mL THF under nitrogen at 2° in an ice bath, add 22 mmoles 9-fluorenylmethyl chloroformate at once, slowly add 29 mmoles dry 4-methylmorpholine (N-methylmorpholine), stir at 2-5° for 2 h, slowly add 4 M HCl in dioxane (Caution! Dioxane is a carcinogen!) until the pH of a sample diluted with water reaches 4-5, filter, wash the solid with dry THF, concentrate the filtrate under reduced pressure, dissolve the resulting oil in the minimum volume of dry diisopropyl ether (Caution! Diisopropyl ether readily forms explosive peroxides!), add dry hexane until the solution just turns cloudy, let stand at -20° overnight, filter. Wash the solid with dry hexane and dry it under vacuum to obtain N-α-(9-fluorenylmethyloxycarbonyl)leucine-N-carboxyanhydride (mp 118-120°; [α]$^{25}_D$ = +38.0°) (J.Am. Chem. Soc. 1990, 112, 7414).)

HPLC VARIABLES
Guard column: 15 × 3.2 5 μm Kromasil silica
Column: 250 × 4.6 5 μm Kromasil silica
Mobile phase: n-Hexane:isopropanol 98:2 (A) or 97:3 (B) or 95:5 (C) or 90:10 (D)
Flow rate: 0.8
Injection volume: 20
Detector: F ex 263 em 313

CHROMATOGRAM
Retention time: k' 1.35 (L-tyrosine (D)), k' 1.42 (L-tryptophan (D)), k' 1.66 (D-tyrosine (D)), k' 1.70 (L-threonine (D)), k' 1.72 (D-tryptophan (D)), k' 1.94 (D-threonine (D)), k' 1.97 (L-lysine (C)), k' 1.98 (L-leucine (A)), k' 2.11 (L-isoleucine (A)), k' 2.20 (L-valine (A)), k' 2.24 (L-2-aminohexanoic acid (A)), k' 2.37 (D-leucine (A)), k' 2.50 (L-β-(1-naphthyl)alanine (UV detection) (A)), k' 2.55 (L-phenylalanine (A)), k' 2.58 (L-2-amino-4-phenylbutyric acid (A)), k' 2.60 (D-isoleucine (A)), k' 2.60 (D-valine (A)), k' 2.60 (L-2-aminopentanoic acid (A)), k' 2.75 (D-2-aminohexanoic acid (A)), k' 2.87 (L-β-(2-naphthyl)alanine (UV detection) (A)), k' 2.88 (L-α-aminophenylacetic acid (A)), k' 2.94 (L-β-(2-thienyl)alanine (A)), k' 2.98 (L-β-(p-chlorophenyl)alanine (A)), k' 3.03 (D-lysine (C)), k' 3.12 (L-β-(3,4-dichlorophenyl)alanine (A)), k' 3.18 (D-2-aminopentanoic acid (A)), k' 3.27 (L-serine (C)), k' 3.35 (L-methionine (A)), k' 3.35 (D-2-amino-4-phenylbutyric acid (A)), k' 3.38 (L-1,2,3,4-tetrahydro-3-isoquinolinecarboxylic acid (A)), k' 3.42 (D-α-aminophenylacetic acid (A)), k' 3.51 (D-β-(1-naphthyl)alanine (UV detection) (A)), k' 3.59 (D-serine (C)), k' 3.61 (L-α-aminobutyric acid (A)), k' 4.11 (D-1,2,3,4-tetrahydro-3-isoquinolinecarboxylic acid (A)), k' 4.14 (D-β-(2-thienyl)alanine (A)), k' 4.34 (D-α-aminobutyric acid (A)), k' 4.42 (D-β-(2-naphthyl)alanine (UV detection) (A)), k' 4.71 (D-β-(p-chlorophenyl)alanine (A)), k' 4.88 (D-β-(3,4-dichlorophenyl)alanine (A)), k' 4.91 (D-methionine (A)), k' 4.91 (D-phenylalanine (A)), k' 5.04 (L-glutamic acid (B)), k' 5.37 (L-alanine (A)), k' 5.97 (D-alanine (A)), k' 6.24 (L-aspartic acid (B)), k' 7.35 (D-aspartic acid (B)), k' 7.8 (D-glutamic acid (B))

KEY WORDS
normal phase; the use of related derivatization reagents is discussed; chiral

REFERENCE
Pugniere, M.; Mattras, H.; Castro, B.; Previero, A. Adsorption liquid chromatography on silica for the chiral separation of amino acids and asymmetric amines derivatized with optically active N-α-9-fluorenylmethyloxyxarbonyl-amino acid-N-carboxyanhydrides, *J.Chromatogr.A*, **1997**, *767*, 69–75.

SAMPLE
Matrix: sewage
Analyte: carboxylated poly(ethylene glycols)
Sample preparation: Prepare an SPE cartridge by packing 500 mg 120-400 mesh Carbograph 4 (surface area 210 m^2/g, Carbochimica Romana, Rome) in glass tubes with PTFE frits (Baker) above and below the sorbent and washing with 10 mL eluent, 2 mL MeOH, and 10 mL water acidified to pH 2 with HCl. Filter (Whatman GF/C) 50 mL raw sewage and pass it through the SPE cartridge, wash with 1.5 mL MeOH, wash with 8 mL dichloromethane:MeOH 80:20 at 5 mL/min, elute with 8 mL eluent. Evaporate the eluate to 200 μL under a stream of nitrogen at 40° (during this process the carboxylic acids were converted to their methyl esters), wash the walls of the vial with 100 μL 20 mM HCl in MeOH, evaporate to dryness, reconstitute the residue in 100 μL MeOH:water 20:80, inject a 20 μL aliquot. (Eluent was 20 mM HCl in dichloromethane:MeOH 80:20.)

HPLC VARIABLES
Column: 50 × 4.6 5 μm Alltima C18 (Alltech)
Mobile phase: Gradient. A was 10 μM NaCl in MeOH. B was 10 μM NaCl in water. A:B from 20:80 to 65:35 over 25 min. (NaCl was recrystallized.)
Flow rate: 1
Injection volume: 20
Detector: MS, Fisons VG Platform single quadrupole, electrospray, 40 μL/min of the mobile phase was introduced through a 40 cm × 75 μm ID PEEK capillary, 4 kV applied to capillary, positive ion mode, source 70°

CHROMATOGRAM
Retention time: 8-24 (depending on number of ethoxy units)
Limit of quantitation: 0.11-0.28 ng/L (SIM)

KEY WORDS
SPE

REFERENCE
Crescenzi, T.; Di Corcia, A.; Marcomini, A.; Samperi, R. Detection of poly(ethylene glycols) and related acidic forms in environmental waters by liquid chromatography/electrospray/mass spectrometry, *Environ.Sci.Technol.*, **1997**, *31*, 2679–2685.

(S)-(+)-2-Octanol

RELATED REFERENCES
Anelli, P.L.; Tomba, C.; Uggeri, F. Optical resolution of 2-chloro-3-phenylmethoxypropanoic acid after derivatization with (S)-2-octanol by high-performance liquid chromatography. *J.Chromatogr.* **1992**, *589*, 346-348.

Johnson, D.M.; Reuter, A.; Collins, J.M.; Thompson, G.F. Enantiomeric purity of naproxen by liquid chromatographic analysis of its diastereomeric octyl esters. *J.Pharm.Sci.* **1979**, *68*, 112-114.

SAMPLE
Matrix: solutions
Analyte: ketorolac
Sample preparation: Acidify 5 mL solution with concentrated HCl, extract with two 5 mL portions of dichloromethane. Combine the organic layers and evaporate them to dryness, reconstitute the residue in 1 mL d-2-octanol:toluene:sulfuric acid 2:100:0.1, heat at 40° for 19 h, neutralize with 1 mL 20 mM sodium bicarbonate. Remove the organic layer and dry it over anhydrous sodium sulfate. Remove a 200 µL aliquot and evaporate it to dryness, reconstitute with 10 mL mobile phase, inject an aliquot.

HPLC VARIABLES
Column: 250 × 4.6 5 µm Ultrasphere-Si
Mobile phase: Hexane:ethyl acetate 96:42
Detector: UV 325

KEY WORDS
chiral; normal phase

REFERENCE
Brandl, M.; Conley, D.; Johnson, D. Racemization of ketorolac in aqueous solution, *J.Pharm.Sci.*, **1995**, *84*, 1045–1048.

5-(4-Pyridyl)-2-thiophenemethanol

SAMPLE
Matrix: solutions
Analyte: carboxylic acids
Sample preparation: Mix 200 μL of a chloroform solution of carboxylic acids with 200 μL 4 mM 1-(3-dimethylaminopropyl)-3-isopropylcarbodiimide perchlorate in chloroform and 200 μL 2 mM 5-(4-pyridyl)-2-thiophenemethanol in chloroform containing 8 mM 4-dimethylaminopyridine, heat at 60° for 3 h, cool, inject a 1 μL aliquot. (Synthesis of 5-(4-pyridyl)-2-thiophenemethanol is as follows. Dissolve 2 g 4-aminopyridine in 15 mL acetic acid, 1.5 mL concentrated sulfuric acid, and 2 mL water, stir at 0°, diazotize with 1 equivalent (?) sodium nitrite, add an excess of potassium iodide, keep at room temperature for 1 day, heat at 50° for 30 min, make alkaline, extract with ether. Evaporate the extract to dryness and recrystallize from EtOH:water to give 4-iodopyridine (yield 15%; mp 100° d) (J. Chem. Soc. 1953, 3226). Heat 22.2 g iodine, 6.66 g periodic acid dihydrate, and 20 g 2-methylthiophene in 200 mL 80% acetic acid at 65° for 8 h, cool, remove the oil, basify the solution, extract with chloroform. Combine the oil and the residue from the extract, add to a NaOH solution containing sodium thiosulfate, steam distil. Distil the crude product to give 2-iodo-5-methylthiophene as a yellowish oil (bp 90°/4.1 kPa). Reflux 460 mg palladium amalgam, 14.7 g NaOH, 7.5 g 4-iodopyridine, 16.4 g 2-iodo-5-methylthiophene, and 2.34 g hydrazine hydrate in 47 mL water with stirring for 6 h (Caution! Hydrazine hydrate is a carcinogen!), filter, wash the precipitate with hot chloroform and hot water. Remove the organic layer from the filtrate and reduce its volume to 50 mL, extract with 15 mL 25% HCl. Remove the aqueous layer and basify it, extract with chloroform. Evaporate the chloroform layer to dryness, distil the residue at 150°/0.4 kPa, chromatograph the distillate on silica gel with acetone:chloroform 50:50 to give 4-(5-methyl-2-thienyl)pyridine as a white solid (mp 136.0-136.6°) in the first fraction. Reflux 1 g 4-(5-methyl-2-thienyl)pyridine and 957 mg selenium(IV) oxide in 35 mL m-xylene with stirring for 8 h, evaporate to dryness. Distil the residue at 200°/0.8 kPa and chromatograph the distillate on silica gel with ethyl acetate to give 5-(4-pyridyl)-2-thiophenecarbaldehyde as white plate crystals (mp 135-136°) (Bull. Chem. Soc. Japan 1990, 63, 968). Dissolve 2 mmoles 5-(4-pyridyl)-2-thiophenecarbaldehyde and a trace of Methyl Orange in 40 mL MeOH, stir, add 2.6 mmoles sodium cyanoborohydride, add MeOH:concentrated HCl 5:1 dropwise to maintain the red color, after 5 h evaporate to dryness under reduced pressure. Take up the residue in 20 mL 5% NaOH and extract three times with 20 mL portions of chloroform. Combine the chloroform layers and evaporate them to dryness under reduced pressure, chromatograph the residue on silica with acetone:ethyl acetate 50:50, recrystallize the crude product from chloroform to obtain 5-(4-pyridyl)-2-thiophenemethanol as colorless needles (mp 139.0-139.5°). Prepare 1-isopropyl-3-(3-dimethylaminopropyl)carbodiimide perchlorate as follows. Stir 1.41 moles isopropylisocyanate in 750 mL dichloromethane at 5°, add 144 g 3-dimethylaminopropylamine (N,N-dimethyl-1,3-propanediamine) in 250 mL dichloromethane at such a rate that the temperature does not exceed 10°, add 500 mL triethylamine, add 300 g p-toluenesulfonyl chloride in 300 mL dichloromethane at such a rate that the temperature does not exceed 10°, reflux for 3 h, add 400 g anhydrous sodium carbonate, add 3.5 L ice water, stir vigorously for 30 min, remove the organic phase. Extract the aqueous phase three times with 500 mL portions of dichloromethane. Combine the organic layers and dry them over anhydrous sodium sulfate, evaporate under reduced pressure, distil the residue to give 1-isopropyl-

3-(3-dimethylaminopropyl)carbodiimide (bp 91-92°/10 mm Hg (Ber. 1941, 74B, 1285)) (cf. Org. Syn. 1973, Coll. Vol. V, 555). Prepare pyridine perchlorate from pyridine and 20% perchloric acid, crystallize from EtOH (Ber. 1926, 59, 446). Add 18 g pyridine perchlorate in portions to 100 mmoles 1-isopropyl-3-(3-dimethylaminopropyl)carbodiimide stirred in 200 mL dichloromethane at 0°, let stand for 30 min, filter, add 200 mL anhydrous diethyl ether to the filtrate. Filter off the precipitate and recrystallize it from dichloromethane/diethyl ether to give 1-isopropyl-3-(3-dimethylaminopropyl)carbodiimide perchlorate (mp 88-90°) (Chem. Pharm. Bull. 1985, 33, 5375).)

HPLC VARIABLES
Column: 150 × 6 Shim-pack CLC-ODS (Shimadzu)
Mobile phase: Gradient. MeOH:water from 70:30 to 100:0 over 20 min (A) or isocratic MeOH:water 100:0 (B) or 95:5 (C)
Flow rate: 1
Injection volume: 1
Detector: F ex 300 em 360

CHROMATOGRAM
Retention time: 5.1 (o-anisic acid (C)), 6.7 (m-anisic acid (C)), 6.7 (o-bromobenzoic acid (C)), 6.9 (benzoic acid (C)), 7 (acetic acid (A)), 7 (lauric acid (B)), 7.7 (p-anisic acid (C)), 8.9 (p-toluic acid (C)), 9 (myristic acid (B)), 9.5 (p-bromobenzoic acid (C)), 10 (m-bromobenzoic acid (C)), 10 (propionic acid (A)), 12 (butyric acid (A)), 12 (palmitic acid (B)), 15 (valeric acid (A)), 16 (stearic acid (B)), 19 (heptanoic acid (A))
Limit of detection: 5-5800 fmole

REFERENCE
Nakajima, R.; Shimada, K.; Fujii, Y.; Yamamoto, A.; Hara, T. Thienylpyridines as a new fluorescent reagent. II. Determination of carboxylic acid with 5-(4-pyridyl)-2-thiophenemethanol using HPLC, *Bull.Chem.Soc.Jpn.*, **1991**, *64*, 3173–3175.

AMINE

L-Alanine β-Naphthylamide

SAMPLE
Matrix: bulk
Analyte: imidapril
Sample preparation: 5 mg Imidapril hydrochloride + 15 mg L-alanine β-naphthylamide hydrobromide + 100 μL chloroform + 50 μL pyridine + 2 mL 4.5 g/L N,N'-dicyclohexylcarbodiimide in chloroform, shake vigorously, let stand for 1 h, wash with 2 mL 1 M HCl,

wash with two 2 mL portions of water. Remove a 1 mL aliquot and dilute it to 5 mL with chloroform, inject a 20 μL aliquot.

HPLC VARIABLES
Column: 150 × 4.6 5 μm Zorbax Sil
Mobile phase: Chloroform:MeOH:EtOH:diethylamine 600:10:2:0.1
Column temperature: 40
Flow rate: 1
Injection volume: 20
Detector: UV 254

CHROMATOGRAM
Retention time: 5 (SRS), 5.5 (SRR), 7.5 (SSS, SSR), 15 (RSS), 17.5 (RRR), 19 (RRS), 21 (RSS)
Limit of detection: 0.05% of larger isomer

KEY WORDS
chiral; normal phase

REFERENCE
Nishi, H.; Yamasaki, K.; Kokusenya, Y.; Sato, T. Optical resolution of imidapril hydrochloride by high-performance liquid chromatography and application to the optical purity testing of drugs, *J.Chromatogr.A*, **1994**, *672*, 125–133.

1-Aminoanthracene

SAMPLE
Matrix: blood
Analyte: carnitine and acylcarnitines
Sample preparation: Condition a 100 mg SAX-Isolute SPE cartridge (Stepbio, Bologna) with 500 μL MeOH and 1 mL water. 100 μL Plasma + 30 μL 17.6 μg/mL methanesulfonyl-L-carnitine in water containing 640 ng/mL isobutyryl-L-carnitine + 370 μL water, add to the SPE cartridge, elute with 500 μL 10 mM pH 3.5 phosphate buffer. With continuous vortexing add 20 μL 1 M HCl, 100 μL 16 mg/mL 1-aminoanthracene in acetone, and 100 μL 160 mg/mL 1-(3-dimethylaminopropyl)-3-ethylcarbodiimide in 10 mM pH 3.5 NaH_2PO_4 buffer (in 20 μL aliquots) to the eluate, let stand at 25° for 20 min, wash with 5 mL diethyl ether. Remove a 300 μL aliquot of the aqueous phase and add it to 700 μL 10 mM pH 9.1 Na_2HPO_4 buffer, wash with 5 mL chloroform. Remove a 500 μL aliquot of the aqueous phase and add it to 500 μL 10 mM pH 3.5 NaH_2PO_4 buffer, inject a 20 μL aliquot.

HPLC VARIABLES
Column: 250 × 4.6 5 μm Kromasil C18
Mobile phase: MeCN:100 mM pH 3.5 ammonium acetate 30:70
Flow rate: 1.3
Injection volume: 20
Detector: F ex 248 em 418

CHROMATOGRAM
Retention time: 8 (L-carnitine), 12.5 (acetyl-L-carnitine), 18.5 (propionyl-L-carnitine)

Internal standard: methanesulfonyl-L-carnitine (14.5), isobutryl-L-carnitine (21)
Limit of quantitation: 0.25-5 μM

KEY WORDS
plasma; SPE

REFERENCE
Longo, A.; Bruno, G.; Curti, S.; Mancinelli, A.; Miotto, G. Determination of L-carnitine, acetyl-L-carnitine and propionyl-L-carnitine in human plasma by high-performance liquid chromatography after pre-column derivatization with 1-aminoanthracene, *J.Chromatogr.B*, **1996**, *686*, 129–139.

4-(2-Aminoethylamino)-7-nitro-2,1,3-benzoxadiazole

SAMPLE
Matrix: bulk
Analyte: carnitine and acylcarnitines
Sample preparation: React 50 μmoles carnitines, 2 mmoles 4-(2-aminoethylamino)-7-nitro-2,1,3-benzoxadiazole, and 35 mmoles 1-(3-dimethylaminopropyl)-3-ethylcarbodiimide in 100 μL pyridine:DMF 20:80, let stand at room temperature for 2 h, add 900 μL 10 mM HCl in MeOH:water 80:20, add to a 150 μL 0.06 mequiv Toyopak IC-SP S (Tosoh) SPE cartridge, wash with 5 mL 10 mM HCl:MeOH 50:50, elute with 3 mL 1 M pH 7.0 triethylamine acetate in MeOH. Evaporate the eluate to dryness, reconstitute with 1 mL 100 mM trifluoroacetic acid:DMF 20:80, inject an aliquot. (Preparation of 4-(2-amino-ethylamino)-7-nitro-2,1,3-benzoxadiazole is as follows. Add 10 mg 4-fluoro-7-nitro-2,1,3-benzoxadiazole in 10 mL MeCN dropwise to a stirred solution of 550 μmoles ethylene-diamine in MeCN over 30 min, stir at room temperature for 2 h, remove MeCN by evaporation under reduced pressure, acidify with 5% HCl, purify by reverse phase HPLC using a 10 mM HCl/MeCN gradient (no further details).)

HPLC VARIABLES
Column: 150 × 4.6 5 μm TSKgel ODS 80Ts (Tosoh)
Mobile phase: Gradient. A was 10 mM trifluoroacetic acid in water. B was 10 mM trifluoroacetic acid in MeCN:water 90:10. A:B from 100:0 to 85:15 over 1 min, to 75:25 over 9 min, to 65:35 over 10 min, to 0:100 over 10 min, maintain at 0:100 for 2 min, return to initial conditions over 1 min.
Injection volume: 100
Detector: F ex 485 em 540

CHROMATOGRAM
Retention time: 8 (carnitine), 10 (acetylcarnitine), 12 (propionylcarnitine), 15 (isobutyryl-carnitine), 16 (butyrylcarnitine), 19 (isovalerylcarnitine), 20 (valerylcarnitine), 24 (hex-anoylcarnitine), 26 (heptanoylcarnitine), 27 (valproylcarnitine), 28 (octanoylcarnitine), 29 (nonanoylcarnitine), 30 (decanoylcarnitine), 32 (lauroylcarnitine), 34 (myristoylcarnitine), 36 (palmitoylcarnitine), 38 (stearoylcarnitine)
Limit of detection: 10-100 fmole

KEY WORDS
SPE

REFERENCE
Matsumoto, K.; Ichitani, Y.; Ogasawara, N.; Yuki, H.; Imai, K. Precolumn fluorescence derivatization of carnitine and acylcarnitines with 4-(2-aminoethylamino)-7-nitro-2,1,3-benzoxadiazole prior to high-performance liquid chromatography, *J.Chromatogr.A*, **1994**, *678*, 241–247.

(-)-2-[4-(1-Aminoethyl)phenyl]-6-methoxybenzoxazole

SAMPLE
Matrix: blood
Analyte: ibuprofen
Sample preparation: Place 100 µL 100 µM diclofenac in dichloromethane in the bottom of a tube and evaporate it to dryness under a stream of nitrogen, add 100 µL plasma, add 25 µL 1 M HCl, mix, add to a dry Chem Elut diatomaceous earth SPE cartridge (Varian), let stand for 5 min, elute with 6 mL n-hexane:diethyl ether:isopropanol 50:50:1. Evaporate the eluate to dryness under a stream of nitrogen at 40°, reconstitute the residue in 200 µL 2 mM (-)-2-[4-(1-aminoethyl)phenyl]-6-methoxybenzoxazole in dichloromethane, add 100 µL 20 mM 2,2'-dipyridyl disulfide in dichloromethane, add 100 µL 20 mM triphenylphosphine in dichloromethane, mix, let stand at room temperature for 5 min. Evaporate to dryness under a stream of nitrogen at 40°, reconstitute the residue in 400 µL mobile phase, inject a 10 µL aliquot. (Synthesis of (-)-2-[4-(1-aminoethyl)phenyl]-6-methoxybenzoxazole ((-)-APMB) is as follows. Hydrogenate 5-methoxy-2-nitrophenol in EtOH over platinum oxide to give 2-amino-5-methoxyphenol (J. Org. Chem. 1957, 22, 220). It should be possible to prepare ethyl 4-acetylbenzimidate hydrochloride ($CH_3COC_6H_4C(=NH)OC_2H_5$.HCl) by passing dry hydrogen chloride into a mixture of 4-acetylbenzonitrile and 1.2-1.5 equivalents EtOH in an inert solvent (e.g., benzene, chloroform, dioxane, ether, nitrobenzene (Caution! Benzene, chloroform, and dioxane are carcinogens!) at 0-5°, the benzimidate should crystallize from the mixture in 7-10 days (J. Chem. Soc. 1942, 103). Add a solution of 5.5 g 2-amino-5-methoxyphenol in 200 mL MeOH to 9 g ethyl 4-acetylbenzimidate hydrochloride, stir at 60-70° for 4 h, evaporate to dryness under reduced pressure, recrystallize from EtOH to give 4-(6-methoxy-2-benzoxazolyl)acetophenone as fine orange-yellow crystals (mp 167°) (J. Chromatogr. 1990, 532, 65). Add 7.0 g hydroxylamine hydrochloride and 8.2 g sodium acetate to 10.1 g 4-(6-methoxy-2-benzoxazolyl)acetophenone in 500 mL EtOH:water 95:5, reflux for 1 h, pour into ice-water, filter, recrystallize from EtOH:water 90:10 to give 4-(6-methoxy-2-benzoxazolyl)acetophenone oxime as faint reddish needles (mp 212°). Dissolve 4.7 g 4-(6-methoxy-2-benzoxazolyl)acetophenone oxime in 300 mL MeOH, add 3 g 10% palladium on charcoal, add 10.5 g ammonium formate, reflux for 30 min, filter, evaporate the

filtrate to dryness under reduced pressure. Take up the residue in 100 mL 5% HCl and wash the aqueous phase with 100 mL ethyl acetate. Adjust the pH of the aqueous layer to 13-14 with 10% NaOH and extract with 200 mL ethyl acetate. Wash the organic layer with 100 mL water and dry it over anhydrous sodium sulfate, evaporate to dryness under reduced pressure to give racemic 2-[4-(1-aminoethyl)phenyl]-6-methoxybenzoxazole. Dissolve 3.6 g racemic 2-[4-(1-aminoethyl)phenyl]-6-methoxybenzoxazole in 50 mL EtOH and add 3.5 g (S)-(-)-α-methoxy-α-trifluoromethylphenylacetic acid, allow to stand overnight at 5°. Collect the precipitate and fractionally crystallize it from EtOH 4 times. Take up the final product in 5% NaOH and extract it with ethyl acetate, wash the organic layer with water, dry over anhydrous sodium sulfate, evaporate to dryness under reduced pressure, recrystallize from EtOH to give (-)-2-[4-(1-aminoethyl)phenyl]-6-methoxybenzoxazole as pale yellow crystals (mp 74°).)

HPLC VARIABLES
Column: 150 × 4.6 5 μm TSK gel ODS-80TS (Tosoh)
Mobile phase: MeCN:water:acetic acid 70:30:0.1
Column temperature: 40
Flow rate: 1
Injection volume: 10
Detector: F ex 320 em 380

CHROMATOGRAM
Retention time: 11.0 (S), 12.2 (R)
Internal standard: diclofenac (14.0)
Limit of quantitation: 200 ng/mL (S-ibuprofen), 400 ng/mL (R-ibuprofen)

KEY WORDS
rat; plasma; chiral; pharmacokinetics; SPE

REFERENCE
Kondo, J.; Suzuki, N.; Naganuma, H.; Imaoka, T.; Kawasaki, T.; Nakanishi, A.; Kawahara, Y. Enantiospecific determination of ibuprofen in rat plasma using chiral fluorescence derivatization reagent, (-)-2-[4-(1-aminoethyl)phenyl]-6-methoxybenzoxazole, *Biomed.Chromatogr.*, **1994**, *8*, 170–174.

SAMPLE
Matrix: solutions
Analyte: carboxylic acids
Sample preparation: Mix 100 μL of a 1-200 μM solution of carboxylic acid in dichloromethane with 100 μL 800 μM (-)-2-[4-(1-aminoethyl)phenyl]-6-methoxybenzoxazole in dichloromethane, 100 μL 1.6 mM 2,2'-dipyridyl disulfide in dichloromethane, and 100 μL 1.6 mM triphenylphosphine in dichloromethane, let stand at room temperature for 20 min. Evaporate to dryness under a stream of nitrogen, reconstitute the residue in 400 μL mobile phase, inject a 10 μL aliquot. (Synthesis of (-)-2-[4-(1-aminoethyl)phenyl]-6-methoxybenzoxazole ((-)-APMB) is as described above.)

HPLC VARIABLES
Column: 150 × 4.6 5 μm TSK gel ODS-80TM (Tosoh)
Mobile phase: MeCN:water:acetic acid 60:40:0.1
Flow rate: 1
Injection volume: 10
Detector: F ex 320 em 380

CHROMATOGRAM
Retention time: 7.5 ((S)-2-phenylpropionic acid), 8 ((R)-2-phenylpropionic acid), 11 ((S)-naproxen), 12 ((R)-naproxen), 16 ((S)-flurbiprofen), 18 ((R)-flurbiprofen), 21 ((S)-ibuprofen), 24 ((R)-ibuprofen)
Limit of detection: 10 fmole

KEY WORDS
chiral

REFERENCE

Kondo, J.; Imaoka, T.; Kawasaki, T.; Nakanishi, A.; Kawahara, Y. Fluorescence derivatization reagent for resolution of carboxylic enantiomers by high-performance liquid chromatography, *J.Chromatogr.*, **1993**, *645*, 75–81.

4-Aminomethyl-6,7-dimethoxycoumarin

SAMPLE

Matrix: blood

Analyte: fatty acids

Sample preparation: 50 μL Serum + 200 μL ethyl acetate + 100 μL dilute HCl (pH 3.0), vortex for 5 min. Remove the supernatant and add it to 50 μL 4 mM 4-aminomethyl-6,7-dimethoxycoumarin in DMF containing 25 mM 1-hydroxybenzotriazole, add 40 μL 500 mM pH 7.0 phosphate buffer, add 100 μL 2 M 1-(3-dimethylaminopropyl)-3-ethylcarbodiimide in water, add 10 μL 1 mM nonadecanoic acid in MeOH, vortex at 25° for 5 min, inject a 10 μL aliquot of the organic layer. (Preparation of 4-aminomethyl-6,7-dimethoxycoumarin is as follows. Heat 1 g 4-bromomethyl-6,7-dimethoxycoumarin and 1.3 g potassium phthalimide in 40 mL DMF at 80° for 1 h, cool, add 100 mL chloroform, wash with 50 mL 100 mM NaOH, wash with 100 mL water. Dry the organic layer over anhydrous sodium sulfate, concentrate to a small volume, add ether, collect 4-phthalimidylmethyl-6,7-dimethoxycoumarin as a crystalline solid (mp 240.5-242°). Reflux 950 mg 4-phthalimidylmethyl-6,7-dimethoxycoumarin and 520 mg hydrazine hydrate in 40 mL THF:MeOH 50:50 for 1.5 h (Caution! Hydrazine hydrate is a carcinogen and explodes on distillation in air!), filter, evaporate the filtrate to dryness, add 20 mL dilute HCl, evaporate to dryness, recrystallize from EtOH:ether 50:50 to give 4-aminomethyl-6,7-dimethoxycoumarin hydrochloride as slightly yellow crystals (mp 240-242.5°). Dissolve 370 mg 4-aminomethyl-6,7-dimethoxycoumarin hydrochloride in 20 mL 5% NaOH, extract 5 times with 20 mL portions of ethyl acetate, evaporate to dryness under reduced pressure, chromatograph on silica with chloroform:MeOH 98:2 to give 4-aminomethyl-6,7-dimethoxycoumarin as a slightly yellow powder (mp 137-147°).)

HPLC VARIABLES

Column: 250 × 4.6 5 μm Wakosil-II 5C18

Mobile phase: Gradient. MeCN:water from 40:60 to 100:0 over 35 min, maintain at 100:0.

Column temperature: 40

Flow rate: 1

Injection volume: 10

Detector: F ex 348 em 429

CHROMATOGRAM

Retention time: 13 (lauric acid), 15 (myristoleic acid), 21 (linolenic acid), 22 (myristic acid), 25 (palmitoleic acid), 26 (arachidonic acid), 27 (linoleic acid), 32 (palmitic acid), 34 (oleic acid), 37 (margaric acid), 40 (stearic acid)

Internal standard: nonadecanoic acid (42)

Limit of detection: 20-50 fmole

KEY WORDS
serum

REFERENCE
Sasamoto, K.; Ushijima, T.; Saito, M.; Ohkura, Y. Precolumn fluorescence derivatization of carboxylic acids using 4-aminomethyl-6,7-dimethoxycoumarin in a two-phase medium, *Anal.Sci.*, **1996**, *12*, 189–193.

1-Aminonaphthalene

RELATED REFERENCE
Ikeda, M.; Shimada, K.; Sakaguchi, T. Application of 1-naphthylamine to labeling of fatty acids for high-performance liquid chromatography. *Chem.Pharm.Bull.* **1982**, *30*, 2258-2261.

SAMPLE
Matrix: blood
Analyte: fatty acids
Sample preparation: 500 μL Serum + 100 μL 100 μg/mL margaric acid in MeOH + 1.4 mL 1/15 M pH 7.0 phosphate buffer, shake, add to a 45 × 12 glass column packed with 1 g Extrelut, let stand for 20 min, elute with 10 mL chloroform. Evaporate the eluate to dryness under reduced pressure and reconstitute the residue with 600 μL benzene (Caution! Benzene is a carcinogen!), add 600 μL 2% oxalyl chloride in benzene, heat at 70° for 30 min, evaporate to dryness under reduced pressure, add 100 μL 40 mM 1-aminonaphthalene (1-naphthylamine) in benzene (Caution! 1-Aminonaphthalene is a carcinogen!), add 10 μL 400 μM triethylamine in benzene, heat at 30° for 15 min, inject a 40 μL aliquot.

HPLC VARIABLES
Column: 300 × 4 8-10 μm μBondapak C18
Mobile phase: MeOH:water 81:19
Column temperature: 40
Flow rate: 2
Injection volume: 40
Detector: UV 280

CHROMATOGRAM
Retention time: 9 (myristic acid), 10 (palmitoleic acid), 12 (linoleic acid), 15 (palmitic acid), 18 (oleic acid), 30 (stearic acid)
Internal standard: margaric acid (21)
Limit of detection: 4 ng

KEY WORDS
serum

REFERENCE
Ikeda, M.; Shimada, K.; Sakaguchi, T. High-performance liquid chromatographic determination of free fatty acids with 1-naphthylamine, *J.Chromatogr.*, **1983**, *272*, 251–259.

SAMPLE
Matrix: blood
Analyte: Cremophor EL (polyoxyethyleneglycerol triricinoleate 35)

Sample preparation: 20 μL Plasma + 10 μL 2 mg/mL margaric acid in MeOH + 200 μL 500 mM KOH in EtOH, vortex for 30 s, centrifuge at 2500 g for 10 min. Remove the supernatant and heat it at 100° for 30 min, cool, add 200 μL 1 M HCl, add 2 mL chloroform, mix vigorously for 5 min, centrifuge at 460 g for 5 min. Remove the organic layer and evaporate it to dryness under reduced pressure at 43°, reconstitute the residue in 500 μL benzene (Caution! Benzene is a carcinogen!), sonicate for 1 min, add 500 μL 2% oxalyl chloride in benzene, heat at 70° for 30 min, evaporate to dryness under reduced pressure, reconstitute with 100 μL benzene, add 100 μL 40 mM 1-aminonaphthalene (1-naphthylamine) (Caution! 1-Aminonaphthalene is a carcinogen!) in benzene, vortex for 30 s, heat at 37° for 30 min, evaporate to dryness, reconstitute with MeOH:MeCN:water 72:13:15, vortex for 20 s, inject a 20 μL aliquot. (Cremophor EL is saponified to ricinoleic acid which is then derivatized with 1-naphthylamine.)

HPLC VARIABLES
Column: two 100 × 3 5 μm Spherisorb ODS-I glass columns in series
Mobile phase: MeCN:MeOH:10 mM pH 7.0 potassium phosphate buffer 13:72:15
Flow rate: 0.4
Injection volume: 20
Detector: UV 280

CHROMATOGRAM
Retention time: 10.0
Internal standard: margaric acid (27.8)
Limit of quantitation: 0.01%
Limit of detection: 0.005%

OTHER SUBSTANCES
Simultaneously analyzed: arachidonic acid, linoleic acid, linolenic acid, myristic acid, oleic acid, palmitic acid, palmitoleic acid, stearic acid

KEY WORDS
plasma; human; mouse

REFERENCE
Sparreboom, A.; van Tellingen, O.; Huizing, M.T.; Nooijen, W.J.; Beijnen, J.H. Determination of polyoxyethyleneglycerol triricinolate 35 (Cremophor EL) in plasma by pre-column derivatization and reversed-phase high-performance liquid chromatography, *J.Chromatogr.B*, **1996**, *681*, 355–362.

SAMPLE
Matrix: solutions
Analyte: 2-arylpropionic acids
Sample preparation: Mix 1 mL 100 μg/mL compound in dichloromethane with 300 μL 100 μg/mL 1-hydroxybenzotriazole in dichloromethane:pyridine 99:1, 300 μL 1.1 mg/mL 1-ethyl-3-dimethylaminopropylcarbodiimide hydrochloride in dichloromethane, and 300 μL 380 μg/mL 1-aminonaphthalene (1-naphthylamine) in dichloromethane (Caution! 1-Aminonaphthalene is a carcinogen!), vortex, let stand at room temperature for 1.5 h, evaporate to dryness under a stream of nitrogen, reconstitute the residue in 500 μL MeOH, inject an aliquot.

HPLC VARIABLES
Column: 150 × 4.6 10 μm EXP B101 tris(4-methylbenzoate) cellulose on silica (Bio-Rad)
Mobile phase: MeOH:buffer 80:20 (90:10 for benoxaprofen) (Prepare buffer solution by dissolving 14.05 g sodium perchlorate in water, adjust pH to 2.0, make up to 1 L with water.)
Flow rate: 1
Detector: UV 230

CHROMATOGRAM
Retention time: k' 8.61, k' 23.89 (benoxaprofen enantiomers), k' 17.74, k' 22.14 (carprofen enantiomers), k' 9.46, k' 11.40 (fenoprofen enantiomers), k' 14.08, k' 15.11 (flurbiprofen enantiomers), k' 2.18, k' 3.61 (ibuprofen enantiomers), k' 6.10, k' 7.69 (pirprofen enantiomers), k' 6.24, k' 9.14 (tiaprofenic acid enantiomers)

KEY WORDS
chiral; ketoprofen enantiomers are not resolved

REFERENCE
Van Overbeke, A.; Baeyens, W.; Van den Bossche, W.; Dewaele, C. Separation of 2-arylpropionic acids on a cellulose based chiral stationary phase by RP-HPLC, *J.Pharm.Biomed.Anal.*, **1994**, *12*, 901–909.

SAMPLE
Matrix: solutions
Analyte: ketoprofen
Sample preparation: 1 mL 1.23 mg/mL ketoprofen in dichloromethane + 300 µL 1 mg/mL hydroxybenzotriazole in dichloromethane:pyridine 99:1 + 300 µL 11 mg/mL 1-ethyl-3-dimethylaminopropylcarbodiimide in dichloromethane + 300 µL 3.47 mg/mL 1-aminonaphthalene (1-naphthylamine) (Caution! 1-Aminonaphthalene in a carcinogen!) in dichloromethane, vortex, let stand for 1 h, evaporate to dryness under a stream of nitrogen, reconstitute with 5 mL MeOH, inject an aliquot.

HPLC VARIABLES
Column: 150 × 2.1 Tolycellulose EXP B101 (tris(4-methylbenzoate)cellulose covalently bonded to 10 µm aminopropylsilica)
Mobile phase: MeOH:buffer 85:15 (Buffer was 14.05 g/L sodium perchlorate adjusted to pH 2.0.)
Flow rate: 0.21
Injection volume: 1
Detector: UV 220

CHROMATOGRAM
Retention time: 5.9, 8.6 (enantiomers)
Limit of detection: 100 pg

OTHER SUBSTANCES
Also analyzed: fenoprofen, flurbiprofen, ibuprofen, tiaprofenic acid

KEY WORDS
narrow-bore; chiral

REFERENCE
Van Overbeke, A.; Baeyens, W.; Van Der Weken, G.; Van de Voorde, I.; Dewaele, C. Comparative chromatographic study on the chiral separation of the 1-naphthylamine derivative of ketoprofen on cellulose-based columns of different sizes, *Biomed.Chromatogr.*, **1995**, *9*, 289–290.

7-Amino-1,3-naphthalenedisulfonic Acid

RELATED REFERENCE
Mechref, Y.; El Rassi, Z. Capillary electrophoresis of herbicides: IV. Evaluation of octylmaltopyranoside chiral surfactant in the enantiomeric separation of fluorescently labeled phenoxy acid herbicides and their laser-induced fluorescence detection. *Electrophoresis* **1997**, *18*, 220-226

SAMPLE
Matrix: bulk
Analyte: monosaccharides
Sample preparation: 1 mg Monosaccharide + 60 μL 100 mM pH 5.0 1-ethyl-3-(3-dimethylaminopropyl) carbodiimide in water + 100 μL 100 mM 7-amino-1,3-naphthalenedisulfonic acid in water, stir at room temperature for at least 2 h, inject an aliquot.

CAPILLARY ELECTROPHORESIS
Capillary: 80 cm × 50 μm fused-silica (50 cm to detector) (Polymicro Technologies)
Capillary preparation: Replace running buffer after each run.
Running buffer: 100 mM pH 10.0 Borate buffer
Voltage/Current: +20 kV
Injection: Hydrodynamic injection at 15 cm for 5 s (1.4 nL)
Detector: UV 247; F ex 315 em 400 (cutoff filter)
Model: laboratory constructed
Migration time: 15 (N-acetylneuraminic acid), 21 (gluconic acid), 24 (galacturonic acid), 25 (glyceric acid)

REFERENCE
Mechref, Y.; El Rassi, Z. Capillary zone electrophoresis of derivatized acidic monosaccharides, *Electrophoresis*, **1994**, *15*, 627–634.

SAMPLE
Matrix: solutions
Analyte: herbicides
Sample preparation: Mix 5 μL of a 50 μM solution in MeCN with 10 μL 50 μM 7-amino-1,3-naphthalenedisulfonic acid in 50 mM pH 3.0 phosphate buffer, add 5 μL 50 μM pH 5.0 1-ethyl-3-(3-dimethylaminopropyl) carbodiimide in water, stir at room temperature for 2 h, inject an aliquot.

CAPILLARY ELECTROPHORESIS
Capillary: 80 cm × 50 μm fused-silica (50 cm to detector) (Polymicro Technologies)
Capillary preparation: Between runs flush capillary with 100 mM NaOH for 1 min, with water for 1 min, and with running buffer for 1 min.
Running buffer: 600 mM pH 5.0 Borate buffer containing 25 mM sodium phosphate and 10 mM heptakis(2,3,6-tri-O-methyl)-β-cyclodextrin
Voltage/Current: 20 kV/10 μA
Injection: Hydrodynamic injection at 15 cm for 5 s (1.4 nL)
Detector: F ex 315 em 400 (cutoff filter)

Model: laboratory constructed

Migration time: 13 (silvex), 14, 15 (mecoprop enantiomers), 15.5, 16.2 (dichlorprop enantiomers), 15.8, 16.5 (2-(4-chlorophenoxy)propionic acid enantiomers), 18.5 (2-(3-chlorophenoxy)propionic acid), 21, 21.5 (2-(2-chlorophenoxy)propionic acid enantiomers), 23.5 (2-phenoxypropionic acid)

REFERENCE
Mechref, Y.; El Rassi, Z. Capillary electrophoresis of herbicides. 1. Precolumn derivatization of chiral and achiral phenoxy acid herbicides with a fluorescent tag for electrophoretic separation in the presence of cyclodextrins and micellar phases, *Anal.Chem.*, **1996**, *68*, 1771–1777.

SAMPLE
Matrix: solutions

Analyte: chrysanthenic acids

Sample preparation: Add 4 μL of 1 M 1,3-dicyclohexylcarbodiimide in dichloromethane to 2 mL of a 2 mM solution of the carboxylic acid in pyridine, add 10 equivalents of 7-amino-1,3-naphthalenedisulfonic acid, let stand overnight, evaporate to dryness under reduced pressure, reconstitute with 2 mL MeCN:water 50:50, inject an aliquot.

CAPILLARY ELECTROPHORESIS
Capillary: 57 cm × 50 μm fused-silica (50 cm to detector) (Polymicro Technologies)

Capillary preparation: Between runs rinse capillary with 100 mM NaOH for 2 min, water for 3 min, and running buffer for 1 min.

Running buffer: MeCN:100 mM pH 6.5 sodium phosphate buffer 10:90 containing 70 mM octyl-β-glucopyranoside

Capillary temperature: 20

Voltage/Current: +19 kV

Injection: Pressure injection at 3.5 kPa.

Detector: F ex 325 (8 mW He-Cd laser) em 380 ± 2 (band-pass)

Model: Beckman P/ACE 5510

Migration time: 11.5, 11.6 (cis-dichlorochrysanthenic acid enantiomers), 11.8, 11.9 (2-(4-chlorophenyl)-3-methylbutanoic acid enantiomers), 12.2, 12.3 (trans-dichlorochrysanthenic acid enantiomers), 13.5, 13.7 (cis-chrysanthenic acid enantiomers), 14.2, 14.4 (trans-chrysanthenic acid enantiomers), 17.5 (2,2,3,3-tetramethylcyclopropanecarboxylic acid enantiomers)

Internal standard: 1,1-binaphthyl-2,2'-dihydrogen phosphate

Limit of detection: 98-140 nM

KEY WORDS
chiral

REFERENCE
Karcher, A.; El Rassi, Z. Capillary electrophoresis of pesticides: V. Analysis of pyrethroid insecticides *via* their hydrolysis products labeled with a fluorescing and UV absorbing tag for laser-induced fluorescence and UV detection, *Electrophoresis*, **1997**, *18*, 1173–1179.

SAMPLE
Matrix: bulk

Analyte: gangliosides

Sample preparation: Add 50 μL 100 mM pH 5.0 1-ethyl-3-(3-dimethylaminopropyl) carbodiimide in water to <1 μg gangliosides, stir for 1 h, add 50 μL 100 mM 7-amino-1,3-naphthalenedisulfonic acid in water, stir for 1.5 h, inject an aliquot.

CAPILLARY ELECTROPHORESIS
Capillary: 80 cm × 50 μm fused-silica (50 cm to detector) (Polymicro Technologies)

Running buffer: MeCN:10 mM pH 10.0 sodium phosphate buffer 50:50

Voltage/Current: 20 kV

Injection: Hydrodynamic injection at 15 cm for 5 s (1.4 nL)

Detector: F ex 315 em 400 (cutoff filter); UV 247

Model: laboratory constructed

Migration time: 16-19 (depending on structure)

REFERENCE

Mechref, Y.; Ostrander, G.K.; El Rassi, Z. Capillary electrophoresis of carboxylated carbohydrates. I. Selective precolumn derivatization of gangliosides with UV absorbing and fluorescent tags, *J.Chromatogr.A*, **1995**, *695*, 83–95.

9-Aminophenanthrene

RELATED REFERENCE

Nakajima, M.; Sato, A.; Shimada, K. Determination of serum valproate by high-performance liquid chromatography using fluorescence labeling with 9-aminophenanthrene. *Anal.Sci.* **1988**, *4*, 385-388.

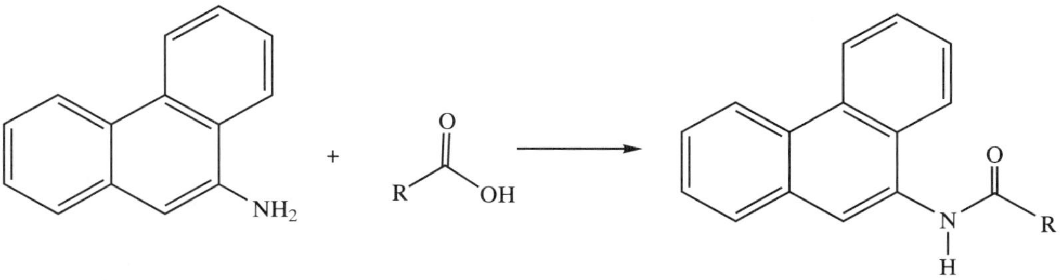

SAMPLE

Matrix: solutions

Analyte: fatty acids

Sample preparation: Shake 200-400 μL of a solution of fatty acids in benzene with an equal volume of 2% oxalyl chloride in benzene (Caution! Benzene is a carcinogen!), heat at 70° for 30 min, evaporate to dryness under a stream of nitrogen, add 100 μL 7.76 mM 9-aminophenanthrene in benzene, add 100 μL 0.1% triethylamine in benzene, heat at 70° for 45 min. (Purify 9-aminophenanthrene by dissolving 50 mg in 30 mL EtOH, filter, add hydrochloric acid saturated ether to the filtrate until precipitate no longer appears. Filter off the precipitate and wash it with ether, dry in a desiccator under vacuum, dissolve in hot water, basify with ammonia, filter to obtain pure 9-aminophenanthrene as white crystals (mp 134°).)

HPLC VARIABLES

Column: 300 × 4 8-10 μm μBondapak C18

Mobile phase: MeCN:MeOH:water 27:53:20

Column temperature: 40

Flow rate: 2

Detector: F ex 303 em 376

CHROMATOGRAM

Retention time: 11 (myristic acid), 12 (palmitoleic acid), 13.5 (arachidonic acid), 14.5 (linoleic acid), 19.5 (palmitic acid), 21 (oleic acid), 35.5 (stearic acid)

Internal standard: margaric acid (26)

Limit of detection: 10-15 pmole

KEY WORDS

protect from light; paper contains some details for analysis of fatty acids in serum

REFERENCE

Ikeda, M.; Shimada, K.; Sakaguchi, T.; Matsumoto, U. Fluorometric high-performance liquid chromatography of 9-aminophenanthrene-derivatized free fatty acids, *J.Chromatogr.*, **1984**, *305*, 261–270.

4-Aminophenol

RELATED REFERENCE

Khalaf, K.D.; Morales-Rubio, A.; De La Guardia, M.; Garcia, J.M.; Jimenez, F.; Arias, J.J. Simultaneous kinetic determination of carbamate pesticides after derivatization with p-aminophenol by using partial least squares. *Microchemical J.* **1996**, *53*, 461-471.

SAMPLE

Matrix: bile
Analyte: bile acids
Sample preparation: 200 μL Bile + 4 M NaOH:MeOH 50:50, heat at 80° for 16 h, adjust pH to 1.5 with 6 M HCl, extract three times with 10 mL portions of ethyl acetate. Combine the organic layers and evaporate them to dryness under a stream of nitrogen, reconstitute the residue in 1 mL dry dichloromethane, add 1 mg 4-aminophenol, add 150 μL triethylamine, add at least a 3-fold molar excess of 2-bromo-1-methylpyridinium iodide, heat at 60° for 30 min, cool, concentrate under a stream of nitrogen, add 1 mL 100 mM HCl, add 1 mL ethyl acetate, shake vigorously, centrifuge at 2500 rpm for 5 min, inject an aliquot of the supernatant. (Prepare 2-bromo-1-methylpyridinium iodide by analogy with the preparation of 2-chloro-1-methylpyridinium iodide. Add 15 g methyl iodide to 13.9 g 2-bromopyridine in 3 mL acetone at 0°, stir at room temperature for 3 days. Filter the precipitate and wash it with 50 mL dry ether, dry under reduced pressure to give 2-bromo-1-methylpyridinium iodide (Bull. Chem. Soc. Japan 1977, 50, 1863).)

HPLC VARIABLES

Column: 250 × 4.6 10 μm Nucleosil C-18
Mobile phase: MeOH:water:perchloric acid 75:25:0.1, containing 50 mM sodium perchlorate
Column temperature: 25 ± 0.1
Flow rate: 0.9
Detector: E, 0.75 v, Ag/AgCl reference electrode

CHROMATOGRAM

Retention time: 9.5 (cholic acid), 15.5 (chenodiol), 17 (deoxycholic acid), 30 (lithocholic acid)
Limit of detection: 2 ng

REFERENCE

Ikenoya, S.; Hiroshima, O.; Ohmae, M.; Kawabe, K. Electrochemical detector for high performance liquid chromatography. IV. Analysis of fatty acids, bile acids and prostaglandins by derivatization to an electrochemically active form, *Chem.Pharm.Bull.*, **1980**, *28*, 2941–2947.

SAMPLE

Matrix: blood
Analyte: fatty acids
Sample preparation: 500 μL Plasma + 20 mL chloroform:MeOH 2:1, vortex for 30 s, add 4 mL water, shake for 5 min, centrifuge at 2500 rpm for 5 min, extract the aqueous layer twice more with 10 mL portions of chloroform:MeOH 2:1. Combine the organic layers and dry them over anhydrous sodium sulfate, evaporate to dryness under a stream of nitrogen at 40°, reconstitute the residue in 1 mL dry dichloromethane, add 1 mg p-aminophenol, add 150 μL triethylamine, add at least a 3-fold molar excess of 2-bromo-1-methylpyridinium iodide, heat at 60° for 30 min, cool, concentrate under a stream of nitrogen, add 1 mL 100 mM HCl, add 1 mL ethyl acetate, shake vigorously, centrifuge at 2500 rpm for 5 min, inject an aliquot of the supernatant. (Prepare 2-bromo-1-methylpyridinium iodide by analogy with the preparation of 2-chloro-1-methylpyridinium iodide.

Add 15 g methyl iodide to 13.9 g 2-bromopyridine in 3 mL acetone at 0°, stir at room temperature for 3 days. Filter the precipitate and wash it with 50 mL dry ether, dry under reduced pressure to give 2-bromo-1-methylpyridinium iodide (Bull. Chem. Soc. Japan 1977, 50, 1863).)

HPLC VARIABLES
Column: 250 × 4.6 10 μm Nucleosil C-18
Mobile phase: MeOH:water:perchloric acid 88:12:0.1, containing 50 mM sodium perchlorate
Column temperature: 25 ± 0.1
Flow rate: 1.2
Detector: E, 0.75 V, Ag/AgCl reference electrode

CHROMATOGRAM
Retention time: 8 (linolenic acid), 9 (palmitoleic acid), 10.5 (linoleic acid), 12.5 (palmitic acid), 13.5 (oleic acid), 19 (stearic acid)
Limit of detection: 0.5 ng (stearic acid)

KEY WORDS
guinea pig; plasma

REFERENCE
Ikenoya, S.; Hiroshima, O.; Ohmae, M.; Kawabe, K. Electrochemical detector for high performance liquid chromatography. IV. Analysis of fatty acids, bile acids and prostaglandins by derivatization to an electrochemically active form, *Chem.Pharm.Bull.*, **1980**, *28*, 2941–2947.

SAMPLE
Matrix: bulk
Analyte: prostaglandins
Sample preparation: Mix 50-200 μg prostaglandins with 1 mL dry dichloromethane, 1 mg 4-aminophenol, and 150 μL triethylamine, sonicate at 25° under nitrogen until dissolved, add at least a 3-fold molar excess of 2-bromo-1-methylpyridinium iodide, sonicate at 25° until the solution becomes cloudy, concentrate under a stream of nitrogen, add 1 mL 100 mM HCl, add 1 mL ethyl acetate, shake vigorously, centrifuge at 2500 rpm for 5 min, inject an aliquot of the supernatant. (Prepare 2-bromo-1-methylpyridinium iodide by analogy with the preparation of 2-chloro-1-methylpyridinium iodide. Add 15 g methyl iodide to 13.9 g 2-bromopyridine in 3 mL acetone at 0°, stir at room temperature for 3 days. Filter the precipitate and wash it with 50 mL dry ether, dry under reduced pressure to give 2-bromo-1-methylpyridinium iodide (Bull. Chem. Soc. Japan 1977, 50, 1863).)

HPLC VARIABLES
Column: 250 × 4.6 10 μm Nucleosil C-18
Mobile phase: MeOH:water:perchloric acid 60:40:0.1, containing 50 mM sodium perchlorate
Column temperature: 25 ± 0.1
Flow rate: 1.1
Detector: E, 0.7 V, Ag/AgCl reference electrode

CHROMATOGRAM
Retention time: 9 (dinoprost), 10 (dinoprostone)
Limit of detection: 2 ng

KEY WORDS
guinea pig; plasma; human

REFERENCE
Ikenoya, S.; Hiroshima, O.; Ohmae, M.; Kawabe, K. Electrochemical detector for high performance liquid chromatography. IV. Analysis of fatty acids, bile acids and prostaglandins by derivatization to an electrochemically active form, *Chem.Pharm.Bull.*, **1980**, *28*, 2941–2947.

(S)-(+)-4-(3-Aminopyrrolidin-1-yl)-7-(aminosulfonyl)-2,1,3-benzoxadiazole

RELATED REFERENCES

Toyo'oka, T.; Ishibashi, M.; Terao, T. Resolution of carboxylic acid enantiomers by high-performance liquid chromatography with peroxyoxalate chemiluminescence detection. *J.Chromatogr.* **1992**, *627*, 75-86.

Toyo'oka, T.; Ishibashi, M.; Terao, T. Further studies on the resolution of carboxylic acid enantiomers by liquid chromatography with fluorescence and laser-induced fluorescence detection. *Anal.Chim.Acta* **1993**, *278*, 71-81.

SAMPLE

Matrix: solutions

Analyte: naproxen

Sample preparation: Prepare a 10 μM solution of the compound in MeCN containing 2 mM (+)-4-(3-aminopyrrolidin-1-yl)-7-(aminosulfonyl)-2,1,3-benzoxadiazole, 3 mM 2,2'-dipyridyl disulfide, and 3 mM triphenylphosphine, let stand at room temperature for 2 h, inject a 5 μL aliquot. (Synthesis of (+)-4-(3-aminopyrrolidin-1-yl)-7-(aminosulfonyl)-2,1,3-benzoxadiazole (D-ABD-APy), is as follows. Cool a solution of 16.4 g (R)-(+)-1-benzyl-3-pyrrolidinol in 164 mL pyridine to +5°, add 19.35 g p-toluenesulfonyl chloride, stir at +10° for 48 h, evaporate to dryness, chromatograph using dichloromethane:acetone 95: 5 to obtain (3R)-3-[(4-tolylsulfonyl)oxy]-1-(phenylmethyl)pyrrolidine. Heat a solution of (3R)-3-[(4-tolylsulfonyl)oxy]-1-(phenylmethyl)pyrrolidine in 200 mL anhydrous DMF to 65°, add 33.5 g sodium azide (Caution! Sodium azide is highly toxic!), stir at 60° for 7 h, filter, evaporate the filtrate to dryness under reduced pressure, dissolve the residue in ethyl acetate, wash twice with water, dry over anhydrous magnesium sulfate, evaporate to obtain (3S)-3-azido-1-(phenylmethyl)pyrrolidine as an oil. Add 3.5 g 10% palladium on carbon under nitrogen to a solution of 7.05 g (3S)-3-azido-1-(phenylmethyl)pyrrolidine in 34.8 mL 1 M HCl in water and 245 mL EtOH, hydrogenate at atmospheric pressure for 30 min, add 3.5 g catalyst, hydrogenate for 2 h, filter, add 34.8 mL 1 M HCl to the filtrate, evaporate to dryness under reduced pressure, take up the residue in 70 mL EtOH, filter, evaporate the filtrate to dryness under reduced pressure, repeat this operation twice, crystallize with the minimum amount of EtOH to obtain (3S)-3-aminopyrrolidine dihydrochloride (by analogy with J. Med. Chem. 1992, 35, 4205 for the other enantiomer). 3S-(-)-Aminopyrrolidine is also available from Tokyo Kasei (TCI America, 911 North Harborgate St., Portland, OR 97203; 800-423-8616, 503-283-1681, (fax) 503-283-1987; www.tciamerica.com). Dissolve 0.5 g magnesium sulfate heptahydrate and 6 g NaOH in 60 mL water, throughout the reaction keep the flask at about 20° with cold water cooling, add 15 mL 30% hydrogen peroxide, add 75 mL MeOH, add 12.1 g powdered benzoyl peroxide in one go, stir for 10 min, pour into 150 mL 20% sulfuric acid, extract three times with 50 mL portions of chloroform, determine peroxybenzoic acid concentration by iodometric titration (Tetrahedron 1967, 23, 3327). Slowly add 110 mL 1 M peroxybenzoic acid in chloroform to 7 g 2,6-difluoroaniline dissolved in 100 mL chloroform, stir at room temperature, when reaction is complete (iodometric titration) wash with 2% sodium thiosulfate, wash with 5% sodium carbonate, wash with water, dry over anhydrous sodium sulfate, evaporate to dryness under reduced pressure, recrystallize 2,6-difluoronitroso-benzene from EtOH (mp 108.5-109.5). Stir 8.5 g 2,6-difluoronitrosobenzene in 85 mL DMSO at room temperature and add a solution of 3.91 g sodium azide in 85 mL DMSO dropwise, let stand for about 1 h, add to a large volume of water, extract with ether, dry the extracts over anhydrous sodium sulfate, evaporate to dryness under reduced pressure and distil to give 4-fluoro-2,1,3-benzoxadiazole as a colorless oil (bp 83°/12 mm Hg) (J.Chem.Soc.(C) 1970, 1433). Add 11 mL chlorosulfonic acid dropwise to 3 g 4-fluoro-2,1,3-benzoxadiazole in 10 mL chloroform at 0-10° (use a calcium chloride drying tube), stir at room temperature for 1 h, reflux for 2 h, cool, slowly pour into ice water, remove the organic layer, extract the aqueous layer with chloroform, combine the organic layers, wash, dry over anhydrous magnesium sulfate, evaporate under reduced pressure, take up the residue in 5 mL benzene (Caution! Benzene is a carcinogen!), chromatograph on a 150 × 30 column of silica gel (100-200 mesh Kanto Chemical) with n-hexane:benzene 50:50, evaporate the appropriate fractions to give 4-(chlorosulfonyl)-7-fluoro-2,1,3-benzoxadiazole (CBD-F) as pale yellow needles (mp 64-66°). Add 1 g CBD-F dropwise to 100 mL 6% ammonium hydroxide, neutralize with 10% HCl, evaporate under reduced pressure, add 200 mL MeCN to the residue, filter. Evaporate the filtrate and chromatograph on a 300 × 20 column of 100-200 mesh silica with chloroform, collect the appropriate fractions and evaporate them to give ABD-F (4-(aminosulfonyl)-7-fluoro-2,1,3-benzoxadiazole) as white needles (mp 145-6°) after recrystallization from n-hexane/benzene (Caution! Benzene is a carcinogen! (Anal. Chem. 1984, 56, 2461).) ABD-F is also available from Wako Chemicals, Richmond VA. Add 100 mg ABD-F in 20 mL MeCN dropwise to a stirred solution of 200 mg 3S-(-)-aminopyrrolidine in 20 mL MeCN at 0-10°, stir at room temperature for 30 min, remove the MeCN by evaporation under reduced pressure, dissolve the residue in 50 mL water, extract 4 times with 80 mL portions of ethyl acetate. Combine the organic layers and wash them with 20 mL water, dry over anhydrous sodium sulfate, evaporate to dryness under reduced pressure, recrystallize from hexane to obtain (+)-4-(3-aminopyrrolidin-1-yl)-7-(aminosulfonyl)-2,1,3-benzoxadiazole (D-ABD-APy) as orange crystals (mp 199-201°).)

HPLC VARIABLES
Column: 150 × 4.6 5 μm Inertsil ODS-2
Mobile phase: MeCN:water 38:62
Column temperature: 40
Flow rate: 1
Injection volume: 5
Detector: F ex 470 em 585

CHROMATOGRAM
Retention time: k' 8.71 (D-naproxen), k' 11.34 (L-naproxen)
Limit of detection: 30 fmole

KEY WORDS
chiral

REFERENCE
Toyo'oka, T.; Ishibashi, M.; Terao, T. Fluorescent chiral derivatization reagents for carboxylic acid enantiomers in high-performance liquid chromatography, *Analyst*, **1992**, *117*, 727-733.

(R)-(-)-4-(3-Aminopyrrolidin-1-yl)-7-(N,N-dimethylaminosulfonyl)-2,1,3-benzoxadiazole

SAMPLE

Matrix: solutions

Analyte: carboxylic acids

Sample preparation: Prepare a 10 μM solution of the compound in MeCN containing 2 mM (R)-(-)-4-(3-aminopyrrolidin-1-yl)-7-(N,N-dimethylaminosulfonyl)-2,1,3-benzoxadiazole, 3 mM 2,2'-dipyridyl disulfide, and 3 mM triphenylphosphine, let stand at room temperature for 2 h, inject a 5 μL aliquot. (Synthesis of (R)-(-)-4-(3-aminopyrrolidin-1-yl)-7-(N,N-dimethylaminosulfonyl)-2,1,3-benzoxadiazole (L-DBD-APy) is as follows. Cool a solution of 16.4 g (S)-(-)-1-benzyl-3-pyrrolidinol in 164 mL pyridine to +5°, add 19.35 g p-toluenesulfonyl chloride, stir at +10° for 48 h, evaporate to dryness, chromatograph using dichloromethane:acetone 95:5 to obtain(3S)-3-[(4-tolylsulfonyl)oxy]-1-(phenylmethyl)-pyrrolidine (mp 68°). Heat a solution of (3S)-3-[(4-tolylsulfonyl)oxy]-1-(phenylmethyl)-pyrrolidine in 200 mL anhydrous DMF to 65°, add 33.5 g sodium azide (Caution! Sodium azide is highly toxic!), stir at 60° for 7 h, filter, evaporate the filtrate to dryness under reduced pressure, dissolve the residue in ethyl acetate, wash twice with water, dry over anhydrous magnesium sulfate, evaporate to obtain (3R)-3-azido-1-(phenylmethyl)-pyrrolidine as an oil. Add 3.5 g 10% palladium on carbon under nitrogen to a solution of 7.05 g (3R)-3-azido-1-(phenylmethyl)pyrrolidine in 34.8 mL 1 M HCl in water and 245 mL EtOH, hydrogenate at atmospheric pressure for 30 min, add 3.5 g catalyst, hydrogenate for 2 h, filter, add 34.8 mL 1 M HCl to the filtrate, evaporate to dryness under reduced pressure, take up the residue in 70 mL EtOH, filter, evaporate the filtrate to dryness under reduced pressure, repeat this operation twice, crystallize with the minimum amount of EtOH to obtain (3R)-3-aminopyrrolidine dihydrochloride (J. Med. Chem. 1992, 35, 4205). 3R-(+)-Aminopyrrolidine is also available from Tokyo Kasei (TCI America, 911 North Harborgate St., Portland, OR 97203; 800-423-8616, 503-283-1681, (fax) 503-283-1987; www.tciamerica.com). Dissolve 0.5 g magnesium sulfate heptahydrate and 6 g NaOH in 60 mL water, throughout the reaction keep the flask at about 20° with cold water cooling, add 15 mL 30% hydrogen peroxide, add 75 mL MeOH, add 12.1 g powdered benzoyl peroxide in one go, stir for 10 min, pour into 150 mL 20% sulfuric acid, extract three times with 50 mL portions of chloroform, determine peroxybenzoic acid concentration by iodometric titration (Tetrahedron 1967, 23, 3327). Slowly add 110 mL 1 M peroxybenzoic acid in chloroform to 7 g 2,6-difluoroaniline dissolved in 100 mL chloroform, stir at room temperature, when reaction is complete (iodometric titration) wash with 2% sodium thiosulfate, wash with 5% sodium carbonate, wash with water, dry over anhydrous sodium sulfate, evaporate to dryness under reduced pressure, recrystallize 2,6-difluoronitrosobenzene from EtOH (mp 108.5-109.5). Stir 8.5 g 2,6-difluoronitrosobenzene in 85 mL DMSO at room temperature and add a solution of 3.91 g sodium azide in 85 mL DMSO dropwise, let stand for about 1 h, add to a large volume of water, extract with ether, dry the extracts over anhydrous sodium sulfate, evaporate to dryness under reduced pressure and distil to give 4-fluoro-2,1,3-benzoxadiazole as a colorless oil (bp 83°/12 mm Hg) (J.Chem.Soc.(C) 1970, 1433). Add 11 mL chlorosulfonic acid dropwise to 3 g 4-fluoro-2,1,3-benzoxadiazole in 10 mL chloroform at 0-10° (use a calcium chloride drying tube), stir at room temperature for 1 h, reflux for 2 h, cool, slowly pour into ice water, remove the organic layer, extract the aqueous layer with chloroform, combine the organic layers, wash, dry over anhydrous magnesium sulfate, evaporate under reduced pressure, take up the residue in 5 mL benzene (Caution! Benzene is a carcinogen!), chromatograph on a 150 × 30 column of silica gel (100-200 mesh Kanto Chemical) with n-hexane:benzene 50:50, evaporate the appropriate fractions to give 4-(chlorosulfonyl)-7-fluoro-2,1,3-benzoxadiazole (CBD-F) as pale yellow needles (mp 64-66°) (Anal. Chem. 1984. 56, 2461). Stir 0.76 g CBD-F in 70 mL MeCN at 0-10° and add 1 g dimethylamine hydrochloride in 10 mL 100 mM pH 10 borax dropwise, adjust pH to 5 with 1 M HCl, concentrate to about 10 mL under reduced pressure, extract three times with 200 mL portions of diethyl ether, wash with water, dry over anhydrous magnesium sulfate, evaporate under reduced pressure, chromatograph on a 500 × 20 column of silica gel with chloroform, isolate the appropriate fraction and re-chromatograph on the same column with ethyl acetate:benzene 1:2 to give 4-(N,N-dimethylaminosulfonyl)-7-fluoro-2,1,3-benzoxadiazole (DBD-F) as white needles (mp 124-125°) (yield 1% !). On a Merck no. 5714 60F$_{254}$ TLC plate eluted with chloroform DBD-F has Rf 0.32 and lies between two other reaction products (Analyst 1989, 114, 413). DBD-F can also be purchased from Tokyo Kasei. Add 100 mg 4-(N,N-dimethylaminosulfonyl)-7-fluoro-2,1,3-benzoxadiazole in 20 mL MeCN dropwise to a stirred solution of 200 mg 3R-(+)-aminopyrrolidine in 20 mL MeCN at 0-10°, stir at room temperature for 30 min, remove the MeCN by evaporation under reduced pressure, dissolve the residue in 50 mL 5% HCl, wash 3 times with 50 mL portions of ethyl acetate, adjust the pH of the aqueous solution to 13-14 with 5% NaOH, extract 6 times with 50 mL portions of ethyl acetate. Combine the organic layers and wash them with 20 mL water, dry over anhydrous sodium sulfate, evaporate to dryness under reduced pressure, recrystallize from hexane to obtain (R)-(-)-4-(3-aminopyrrolidin-1-yl)-7-(N,N-dimethylaminosulfonyl)-2,1,3-benzoxadiazole as orange crystals (mp 96-98°).)

HPLC VARIABLES

Column: 150 × 4.6 5 μm Inertsil ODS-2
Mobile phase: MeCN:water:trifluoroacetic acid 55:45:0.1 (A) or 45:55:0.1 (B) or 40:60:0.1 (C)
Column temperature: 40
Flow rate: 1
Injection volume: 5
Detector: F ex 470 em 585

CHROMATOGRAM

Retention time: k' 8.27 (L-ibuprofen (A)), k' 10.04 (D-ibuprofen (A)), k' 10.29 (L-naproxen (B)), k' 12.75 (L-loxoprofen (C)), k' 12.82 (D-naproxen (B)), k' 15.42 (D-loxoprofen (C))
Limit of detection: 10 fmole

KEY WORDS

chiral

REFERENCE

Toyo'oka, T.; Ishibashi, M.; Terao, T. Fluorescent chiral derivatization reagents for carboxylic acid enantiomers in high-performance liquid chromatography, *Analyst*, **1992**, *117*, 727–733.

(S)-(+)-4-(3-Aminopyrrolidin-1-yl)-7-(N,N-dimethylaminosulfonyl)-2,1,3-benzoxadiazole

RELATED REFERENCES

Toyo'oka, T.; Ishibashi, M.; Terao, T. Resolution of carboxylic acid enantiomers by high-performance liquid chromatography with peroxyoxalate chemiluminescence detection. *J.Chromatogr.* **1992**, *627*, 75-86.

Toyo'oka, T.; Ishibashi, M.; Terao, T. Further studies on the resolution of carboxylic acid enantiomers by liquid chromatography with fluorescence and laser-induced fluorescence detection. *Anal.Chim.Acta* **1993**, *278*, 71-81.

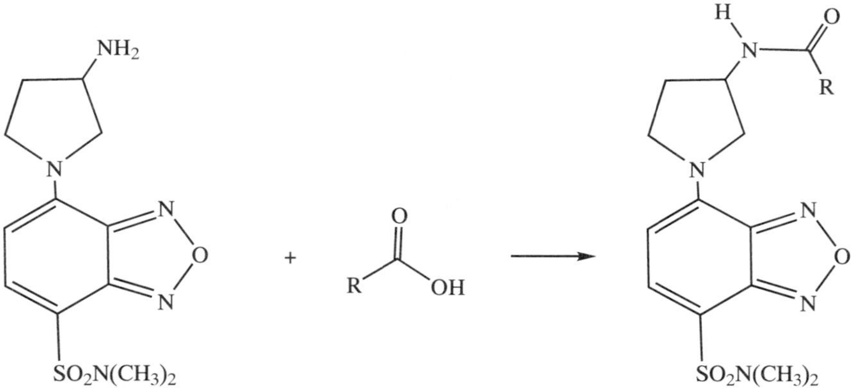

SAMPLE

Matrix: solutions

Analyte: carboxylic acids

Sample preparation: Prepare a 10 μM solution of the compound in MeCN containing 2 mM (S)-(+)-4-(3-aminopyrrolidin-1-yl)-7-(N,N-dimethylaminosulfonyl)-2,1,3-benzoxadiazole, 3 mM 2,2'-dipyridyl disulfide, and 3 mM triphenylphosphine, let stand at room temperature for 2 h, inject a 5 μL aliquot. (Synthesis of (S)-(+)-4-(3-aminopyrrolidin-1-yl)-7-(N,N-dimethylaminosulfonyl)-2,1,3-benzoxadiazole (D-DBD-APy) is as described above for the other enantiomer.)

HPLC VARIABLES

Column: 150 × 4.6 5 μm Inertsil ODS-2

Mobile phase: MeCN:water:trifluoroacetic acid 55:45:0.1 (A) or 45:55:0.1 (B) or 40:60:0.1 (C)

Column temperature: 40

Flow rate: 1

Injection volume: 5

Detector: F ex 470 em 585

CHROMATOGRAM

Retention time: k' 8.32 (D-ibuprofen (A)), k' 10.10 (L-ibuprofen (A)), k' 10.24 (D-naproxen (B)), k' 12.73 (L-naproxen (B)), k' 12.89 (D-loxoprofen (C)), k' 15.64 (L-loxoprofen (C))

Limit of detection: 10 fmole

KEY WORDS

chiral

REFERENCE

Toyo'oka, T.; Ishibashi, M.; Terao, T. Fluorescent chiral derivatization reagents for carboxylic acid enantiomers in high-performance liquid chromatography, *Analyst*, **1992**, *117*, 727–733.

(R)-(-)-4-(3-Aminopyrrolidin-1-yl)-7-nitro-2,1,3-benzoxadiazole

SAMPLE
Matrix: solutions
Analyte: carboxylic acids
Sample preparation: Prepare a 10 μM solution of the compound in MeCN containing 2 mM (R)-(-)-4-(3-aminopyrrolidin-1-yl)-7-nitro-2,1,3-benzoxadiazole, 3 mM 2,2'-dipyridyl disulfide, and 3 mM triphenylphosphine, let stand at room temperature for 2 h, inject a 5 μL aliquot. ((R)-(-)-4-(3-Aminopyrrolidin-1-yl)-7-nitro-2,1,3-benzoxadiazole (L-NBD-APy), is available from Tokyo Kasei (TCI America, Portland OR). Synthesis is as follows. Cool a solution of 16.4 g (S)-(-)-1-benzyl-3-pyrrolidinol in 164 mL pyridine to +5°, add 19.35 g p-toluenesulfonyl chloride, stir at +10° for 48 h, evaporate to dryness, chromatograph using dichloromethane:acetone 95:5 to obtain(3S)-3-[(4-tolylsulfonyl)oxy]-1-(phenylmethyl)pyrrolidine (mp 68°). Heat a solution of (3S)-3-[(4-tolylsulfonyl)oxy]-1-(phenylmethyl)pyrrolidine in 200 mL anhydrous DMF to 65°, add 33.5 g sodium azide (Caution! Sodium azide is highly toxic!), stir at 60° for 7 h, filter, evaporate the filtrate to dryness under reduced pressure, dissolve the residue in ethyl acetate, wash twice with water, dry over anhydrous magnesium sulfate, evaporate to obtain (3R)-3-azido-1-(phenylmethyl)pyrrolidine as an oil. Add 3.5 g 10% palladium on carbon under nitrogen to a solution of 7.05 g (3R)-3-azido-1-(phenylmethyl)pyrrolidine in 34.8 mL 1 M HCl in water and 245 mL EtOH, hydrogenate at atmospheric pressure for 30 min, add 3.5g catalyst, hydrogenate for 2 h, filter, add 34.8 mL 1 M HCl to the filtrate, evaporate to dryness under reduced pressure, take up the residue in 70 mL EtOH, filter, evaporate the filtrate to dryness under reduced pressure, repeat this operation twice, crystallize with the minimum amount of EtOH to obtain (3R)-3-aminopyrrolidine dihydrochloride (J. Med. Chem. 1992, 35, 4205). 3R-(+)-Aminopyrrolidine is also available from Tokyo Kasei. Add 100 mg 4-fluoro-7-nitro-2,1,3-benzoxadiazole in 20 mL MeCN dropwise to a stirred solution of 200 mg 3R-(+)-aminopyrrolidine in 20 mL MeCN at 0-10°, stir at room temperature for 30 min, remove the MeCN by evaporation under reduced pressure, dissolve the residue in 50 mL water, extract 4 times with 80 mL portions of ethyl acetate. Combine the organic layers and wash them with 20 mL water, dry over anhydrous sodium sulfate, evaporate to dryness under reduced pressure, recrystallize from hexane to obtain (R)-(-)-4-(3-aminopyrrolidin-1-yl)-7-nitro-2,1,3-benzoxadiazole as dark red crystals (mp 178-181°).)

HPLC VARIABLES
Column: 150 × 4.6 5 μm Inertsil ODS-2
Mobile phase: MeCN:water:trifluoroacetic acid 50:50:0.1 (A) or 45:55:0.1 (B) or 40:60:0.1 (C)
Column temperature: 40
Flow rate: 1
Injection volume: 5
Detector: F ex 470 em 540

CHROMATOGRAM
Retention time: k' 6.12 (L-naproxen (B)), k' 8.83 (L-loxoprofen (C)), k' 9.09 (D-naproxen (B)), k' 10.95 (D-loxoprofen (C)), k' 11.07 (L-ibuprofen (A)), k' 14.02 (D-ibuprofen (A))
Limit of detection: 15 fmole

KEY WORDS
chiral

REFERENCE
Toyo'oka, T.; Ishibashi, M.; Terao, T. Fluorescent chiral derivatization reagents for carboxylic acid enantiomers in high-performance liquid chromatography, *Analyst*, **1992**, *117*, 727–733.

(S)-(+)-4-(3-Aminopyrrolidin-1-yl)-7-nitro-2,1,3-benzoxadiazole

RELATED REFERENCES
Pecanac, D.; Baeyens, W.R.G.; Imai, K.; Van Overbeke, A.; Van Der Weken, G.; De Waele, C. Enantiomeric separation of some 2-arylpropionic acids with a chiral fluorescence labelling reagent by narrowbore liquid chromatography. *Biomed.Chromatogr.* **1997**, *11*, 83-84.
Toyo'oka, T.; Ishibashi, M.; Terao, T. Further studies on the resolution of carboxylic acid enantiomers by liquid chromatography with fluorescence and laser-induced fluorescence detection. *Anal.Chim.Acta* **1993**, *278*, 71-81.

SAMPLE
Matrix: solutions
Analyte: carboxylic acids
Sample preparation: Prepare a 10 μM solution of the compound in MeCN containing 2 mM (S)-(+)-4-(3-aminopyrrolidin-1-yl)-7-nitro-2,1,3-benzoxadiazole, 3 mM 2,2'-dipyridyl disulfide, and 3 mM triphenylphosphine, let stand at room temperature for 2 h, inject a 5 μL aliquot. ((S)-(+)-4-(3-Aminopyrrolidin-1-yl)-7-nitro-2,1,3-benzoxadiazole (D-NBD-APy) is available from Tokyo Kasei (TCI America, Portland OR). Synthesis is as follows. Cool a solution of 16.4 g (R)-(+)-1-benzyl-3-pyrrolidinol in 164 mL pyridine to +5°, add 19.35 g p-toluenesulfonyl chloride, stir at +10° for 48 h, evaporate to dryness, chromatograph using dichloromethane:acetone 95:5 to obtain (3R)-3-[(4-tolylsulfonyl)oxy]-1-(phenylmethyl)pyrrolidine. Heat a solution of (3R)-3-[(4-tolylsulfonyl)oxy]-1-(phenylmethyl)pyrrolidine in 200 mL anhydrous DMF to 65°, add 33.5 g sodium azide (Caution! Sodium azide is highly toxic!), stir at 60° for 7 h, filter, evaporate the filtrate to dryness under reduced pressure, dissolve the residue in ethyl acetate, wash twice with water, dry over anhydrous magnesium sulfate, evaporate to obtain (3S)-3-azido-1-(phenylmethyl)pyrrolidine as an oil. Add 3.5 g 10% palladium on carbon under nitrogen to a solution of 7.05 g (3S)-3-azido-1-(phenylmethyl)pyrrolidine in 34.8 mL 1 M HCl in water and 245 mL EtOH, hydrogenate at atmospheric pressure for 30 min, add 3.5 g catalyst,

hydrogenate for 2 h, filter, add 34.8 mL 1 M HCl to the filtrate, evaporate to dryness under reduced pressure, take up the residue in 70 mL EtOH, filter, evaporate the filtrate to dryness under reduced pressure, repeat this operation twice, crystallize with the minimum amount of EtOH to obtain (3S)-3-aminopyrrolidine dihydrochloride (by analogy with J. Med. Chem. 1992, 35, 4205 for the other enantiomer). 3S-(-)-Aminopyrrolidine is also available from Tokyo Kasei. Add 100 mg 4-fluoro-7-nitro-2,1,3-benzoxadiazole in 20 mL MeCN dropwise to a stirred solution of 200 mg 3S-(-)-aminopyrrolidine in 20 mL MeCN at 0-10°, stir at room temperature for 30 min, remove the MeCN by evaporation under reduced pressure, dissolve the residue in 50 mL water, extract 4 times with 80 mL portions of ethyl acetate. Combine the organic layers and wash them with 20 mL water, dry over anhydrous sodium sulfate, evaporate to dryness under reduced pressure, recrystallize from hexane to obtain (S)-(+)-4-(3-aminopyrrolidin-1-yl)-7-nitro-2,1,3-benzoxadiazole as dark red crystals (mp 177-180°).)

HPLC VARIABLES
Column: 150 × 4.6 5 μm Inertsil ODS-2
Mobile phase: MeCN:water:trifluoroacetic acid 50:50:0.1 (A) or 45:55:0.1 (B) or 40:60:0.1 (C)
Column temperature: 40
Flow rate: 1
Injection volume: 5
Detector: F ex 470 em 540

CHROMATOGRAM
Retention time: k' 6.10 (D-naproxen (B)), k' 8.78 (D-loxoprofen (C)), k' 8.95 (L-naproxen (B)), k' 10.87 (L-loxoprofen (C)), k' 11.04 (D-ibuprofen (A)), k' 13.98 (L-ibuprofen (A))
Limit of detection: 15 fmole

KEY WORDS
chiral

REFERENCE
Toyo'oka, T.; Ishibashi, M.; Terao, T. Fluorescent chiral derivatization reagents for carboxylic acid enantiomers in high-performance liquid chromatography, *Analyst*, **1992**, *117*, 727–733.

4-(Aminosulfonyl)-7-(2-aminoethylamino)-2,1,3-benzoxadiazole

SAMPLE
Matrix: solutions
Analyte: fatty acids
Sample preparation: Mix 200 μL of a 10 μM solution in DMF with 200 μL of 10 mM ABD-AE in DMF, add 5 μL diethyl cyanophosphonate (diethyl phosphorocyanidate), let stand at room temperature for 6 h, inject a 2 μL aliquot. (Synthesis of ABD-AE, 4-(aminosulfonyl)-7-(2-aminoethylamino)-2,1,3-benzoxadiazole, is as follows. Dissolve 0.5 g magnesium sulfate heptahydrate and 6 g NaOH in 60 mL water, throughout the reaction keep

the flask at about 20° with cold water cooling, add 15 mL 30% hydrogen peroxide, add 75 mL MeOH, add 12.1 g powdered benzoyl peroxide in one go, stir for 10 min, pour into 150 mL 20% sulfuric acid, extract three times with 50 mL portions of chloroform, determine peroxybenzoic acid concentration by iodometric titration (Tetrahedron 1967, 23,3327). Slowly add 110 mL 1 M peroxybenzoic acid in chloroform to 7 g 2,6-difluoroaniline dissolved in 100 mL chloroform, stir at room temperature, when reaction is complete (iodometric titration) wash with 2% sodium thiosulfate, wash with 5% sodium carbonate, wash with water, dry over anhydrous sodium sulfate, evaporate to dryness under reduced pressure, recrystallize 2,6-difluoronitrobenzene from EtOH (mp 108.5-109.5). Stir 8.5 g 2,6-difluoronitrobenzene in 85 mL DMSO at room temperature and add a solution of 3.91 g sodium azide in 85 mL DMSO dropwise, let stand for about 1 h, add to a large volume of water, extract with ether, dry the extracts over anhydrous sodium sulfate, evaporate to dryness under reduced pressure and distil to give 4-fluoro-2,1,3-benzoxadiazole as a colorless oil (bp 83°/12 mm Hg) (J.Chem.Soc.(C) 1970, 1433). Add 11 mL chlorosulfonic acid dropwise to 3 g 4-fluoro-2,1,3-benzoxadiazole in 10 mL chloroform at 0-10° (use a calcium chloride drying tube), stir at room temperature for 1 h, reflux for 2 h, cool, slowly pour into ice water, remove the organic layer, extract the aqueous layer with chloroform, combine the organic layers, wash, dry over anhydrous magnesium sulfate, evaporate under reduced pressure, take up the residue in 5 mL benzene (Caution! Benzene is a carcinogen!), chromatograph on a 150 × 30 column of silica gel (100-200 mesh Kanto Chemical) with n-hexane:benzene 50:50, evaporate the appropriate fractions to give 4-(chlorosulfonyl)-7-fluoro-2,1,3-benzoxadiazole (CBD-F) as pale yellow needles (mp 64-66°). Add 1 g CBD-F dropwise to 100 mL 6% ammonium hydroxide, neutralize with 10% HCl, evaporate under reduced pressure, add 200 mL MeCN to the residue, filter. Evaporate the filtrate and chromatograph on a 300 × 20 column of 100-200 mesh silica with chloroform, collect the appropriate fractions and evaporate them to give ABD-F (4-(aminosulfonyl)-7-fluoro-2,1,3-benzoxadiazole) as white needles (mp 145-6°) after recrystallization from n-hexane/benzene (Caution! Benzene is a carcinogen!). ABD-F is also available from Molecular Probes, Eugene OR; Wako Chemicals, Richmond VA; or Dojindo Molecular Technologies, Inc., 3 Bethesda Metro Center, Suite 700, Bethesda MD 20814, (301) 664-8448, www.dojindo.co.jp. Add 109 mg ABD-F in 20 mL MeCN dropwise at room temperature to a stirred solution of 90 mg ethylenediamine in 20 mL MeCN, stir at room temperature for 30 min, evaporate to dryness under reduced pressure, dissolve the residue in 50 mL 5% HCl, wash three times with 50 mL portions of ethyl acetate. Adjust the pH of the aqueous layer to 13-14 with 5% NaOH, extract 6 times with 50 mL portions of ethyl acetate. Combine the organic layers and wash them with 20 mL water, dry over anhydrous sodium sulfate, evaporate to dryness under reduced pressure to obtain 4-(aminosulfonyl)-7-(2-aminoethylamino)-2,1,3-benzoxadiazole as orange crystals (mp 144-146°).)

HPLC VARIABLES

Column: 150 × 4.6 5 μm Inertsil ODS-2
Mobile phase: Gradient. MeCN:water from 70:30 to 98:2 over 1 h, maintain at 98:2 for 10 min
Column temperature: 40
Flow rate: 1
Injection volume: 2
Detector: F ex 429 em 573

CHROMATOGRAM

Retention time: 5.12 (lauric acid), 8.76 (myristic acid), 15.12 (palmitic acid), 19.40 (margaric acid), 24.25 (stearic acid), 34.93 (arachidic acid), 45.59 (behenic acid), 55.16 (lignoceric acid)
Limit of detection: 23-50 fmole

REFERENCE

Toyo'oka, T.; Ishibashi, M.; Takeda, Y.; Imai, K. 4-(Aminosulphonyl)-2,1,3-benzoxadiazole derivatives as pre-column fluorogenic tagging reagents for carboxylic acids in high-performance liquid chromatography, *Analyst*, **1991**, *116*, 609–613.

4-(Aminosulfonyl)-7-(5-aminopentylamino)-2,1,3-benzoxadiazole

SAMPLE
Matrix: solutions
Analyte: fatty acids
Sample preparation: Mix 200 μL of a 10 μM solution in DMF with 200 μL of a 10 mM ABD-AP in DMF, add 5 μL diethyl cyanophosphonate (diethyl phosphorocyanidate), let stand at room temperature for 6 h, inject a 2 μL aliquot. (Synthesis of ABD-AP, 4-(aminosulfonyl)-7-(5-aminopentylamino)-2,1,3-benzoxadiazole, is as follows. Dissolve 0.5 g magnesium sulfate heptahydrate and 6 g NaOH in 60 mL water, throughout the reaction keep the flask at about 20° with cold water cooling, add 15 mL 30% hydrogen peroxide, add 75 mL MeOH, add 12.1 g powdered benzoyl peroxide in one go, stir for 10 min, pour into 150 mL 20% sulfuric acid, extract three times with 50 mL portions of chloroform, determine peroxybenzoic acid concentration by iodometric titration (Tetrahedron 1967, 23, 3327). Slowly add 110 mL 1 M peroxybenzoic acid in chloroform to 7 g 2,6-difluoroaniline dissolved in 100 mL chloroform, stir at room temperature, when reaction is complete (iodometric titration) wash with 2% sodium thiosulfate, wash with 5% sodium carbonate, wash with water, dry over anhydrous sodium sulfate, evaporate to dryness under reduced pressure, recrystallize 2,6-difluoronitrosobenzene from EtOH (mp 108.5-109.5). Stir 8.5 g 2,6-difluoronitrosobenzene in 85 mL DMSO at room temperature and add a solution of 3.91 g sodium azide in 85 mL DMSO dropwise, let stand for about 1 h, add to a large volume of water, extract with ether, dry the extracts over anhydrous sodium sulfate, evaporate to dryness under reduced pressure and distil to give 4-fluoro-2,1,3-benzoxadiazole as a colorless oil (bp 83°/12 mm Hg) (J.Chem.Soc.(C) 1970, 1433). Add 11 mL chlorosulfonic acid dropwise to 3 g 4-fluoro-2,1,3-benzoxadiazole in 10 mL chloroform at 0-10° (use a calcium chloride drying tube), stir at room temperature for 1 h, reflux for 2 h, cool, slowly pour into ice water, remove the organic layer, extract the aqueous layer with chloroform, combine the organic layers, wash, dry over anhydrous magnesium sulfate, evaporate under reduced pressure, take up the residue in 5 mL benzene (Caution! Benzene is a carcinogen!), chromatograph on a 150 × 30 column of silica gel (100-200 mesh Kanto Chemical) with n-hexane:benzene 50:50, evaporate the appropriate fractions to give 4-(chlorosulfonyl)-7-fluoro-2,1,3-benzoxadiazole (CBD-F) as pale yellow needles (mp 64-66°). Add 1 g CBD-F dropwise to 100 mL 6% ammonium hydroxide, neutralize with 10% HCl, evaporate under reduced pressure, add 200 mL MeCN to the residue, filter. Evaporate the filtrate and chromatograph on a 300 × 20 column of 100-200 mesh silica with chloroform, collect the appropriate fractions and evaporate them to give ABD-F (4-(aminosulfonyl)-7-fluoro-2,1,3-benzoxadiazole) as white needles (mp 145-6°) after recrystallization from n-hexane/benzene (Caution! Benzene is a carcinogen!). ABD-F is also available from Molecular Probes, Eugene OR; Wako Chemicals, Richmond VA; or Dojindo Molecular Technologies, Inc., 3 Bethesda Metro Center, Suite 700, Bethesda MD 20814, (301) 664-8448, www.dojindo.co.jp. Add 109 mg ABD-F in 20 mL MeCN dropwise at room temperature to a stirred solution of 153 mg cadaverine in 20 mL MeCN, stir at room temperature for 30 min, evaporate to dryness under reduced pressure, dissolve the residue in 50 mL 5% HCl, wash three times with 50 mL portions of ethyl acetate. Adjust the pH of the aqueous layer to 13-14 with 5% NaOH, extract 6 times with 50 mL portions of ethyl acetate. Combine the organic layers and wash them with 20 mL water, dry over anhydrous sodium sulfate, evaporate to dryness under reduced pressure to obtain 4-(aminosulfonyl)-7-(5-aminopentylamino)-2,1,3-benzoxadiazole as orange crystals (mp 148-150°).)

HPLC VARIABLES

Column: 150 × 4.6 5 μm Inertsil ODS-2
Mobile phase: Gradient. MeCN:water from 70:30 to 98:2 over 1 h, maintain at 98:2 for 10 min
Column temperature: 40
Flow rate: 1
Injection volume: 2
Detector: F ex 438 em 570

CHROMATOGRAM

Retention time: 5.97 (lauric acid), 10.25 (myristic acid), 17.32 (palmitic acid), 21.84 (margaric acid), 26.83 (stearic acid), 37.44 (arachidic acid), 47.75 (behenic acid), 56.92 (lignoceric acid)
Limit of detection: 11-45 fmole

REFERENCE

Toyo'oka, T.; Ishibashi, M.; Takeda, Y.; Imai, K. 4-(Aminosulphonyl)-2,1,3-benzoxadiazole derivatives as pre-column fluorogenic tagging reagents for carboxylic acids in high-performance liquid chromatography, *Analyst*, **1991**, *116*, 609–613.

4-(Aminosulfonyl)-7-(1-piperazinyl)-2,1,3-benzoxadiazole

SAMPLE

Matrix: solutions
Analyte: fatty acids
Sample preparation: Mix 200 μL of a 10 μM solution in DMF with 200 μL of 10 mM ABD-PZ in DMF, add 5 μL diethyl cyanophosphonate (diethyl phosphorocyanidate), let stand at room temperature for 6 h, inject a 2 μL aliquot. (Synthesis of ABD-PZ, 4-(aminosulfonyl)-7-(1-piperazinyl)-2,1,3-benzoxadiazole, is as follows. Dissolve 0.5 g magnesium sulfate heptahydrate and 6 g NaOH in 60 mL water, throughout the reaction keep the flask at about 20° with cold water cooling, add 15 mL 30% hydrogen peroxide, add 75 mL MeOH, add 12.1 g powdered benzoyl peroxide in one go, stir for 10 min, pour into 150 mL 20% sulfuric acid, extract three times with 50 mL portions of chloroform, determine peroxybenzoic acid concentration by iodometric titration (Tetrahedron 1967, 23, 3327). Slowly add 110 mL 1 M peroxybenzoic acid in chloroform to 7 g 2,6-difluoroaniline dissolved in 100 mL chloroform, stir at room temperature, when reaction is complete (iodometric titration) wash with 2% sodium thiosulfate, wash with 5% sodium carbonate, wash with water, dry over anhydrous sodium sulfate, evaporate to dryness under reduced pressure, recrystallize 2,6-difluoronitrosobenzene from EtOH (mp 108.5-109.5). Stir 8.5 g 2,6-difluoronitrosobenzene in 85 mL DMSO at room temperature and add a solution of 3.91 g sodium azide in 85 mL DMSO dropwise, let stand for about 1 h, add to a large

volume of water, extract with ether, dry the extracts over anhydrous sodium sulfate, evaporate to dryness under reduced pressure and distil to give 4-fluoro-2,1,3-benzoxadiazole as a colorless oil (bp 83°/12 mm Hg) (J.Chem.Soc.(C) 1970, 1433). Add 11 mL chlorosulfonic acid dropwise to 3 g 4-fluoro-2,1,3-benzoxadiazole in 10 mL chloroform at 0-10° (use a calcium chloride drying tube), stir at room temperature for 1 h, reflux for 2 h, cool, slowly pour into ice water, remove the organic layer, extract the aqueous layer with chloroform, combine the organic layers, wash, dry over anhydrous magnesium sulfate, evaporate under reduced pressure, take up the residue in 5 mL benzene (Caution! Benzene is a carcinogen!), chromatograph on a 150 × 30 column of silica gel (100-200 mesh Kanto Chemical) with n-hexane:benzene 50:50, evaporate the appropriate fractions to give 4-(chlorosulfonyl)-7-fluoro-2,1,3-benzoxadiazole (CBD-F) as pale yellow needles (mp 64-66°). Add 1 g CBD-F dropwise to 100 mL 6% ammonium hydroxide, neutralize with 10% HCl, evaporate under reduced pressure, add 200 mL MeCN to the residue, filter. Evaporate the filtrate and chromatograph on a 300 × 20 column of 100-200 mesh silica with chloroform, collect the appropriate fractions and evaporate them to give ABD-F (4-(aminosulfonyl)-7-fluoro-2,1,3-benzoxadiazole) as white needles (mp 145-6°) after recrystallization from n-hexane/benzene (Caution! Benzene is a carcinogen!). ABD-F is also available from Molecular Probes, Eugene OR; Wako Chemicals, Richmond VA; or Dojindo Molecular Technologies, Inc., 3 Bethesda Metro Center, Suite 700, Bethesda MD 20814, (301) 664-8448, www.dojindo.co.jp). Add 109 mg ABD-F in 20 mL MeCN dropwise at room temperature to a stirred solution of 129 mg piperazine in 20 mL MeCN, stir at room temperature for 30 min, evaporate to dryness under reduced pressure, dissolve the residue in 50 mL 5% HCl, wash three times with 50 mL portions of ethyl acetate. Adjust the pH of the aqueous layer to 13-14 with 5% NaOH, extract 6 times with 50 mL portions of ethyl acetate. Combine the organic layers and wash them with 20 mL water, dry over anhydrous sodium sulfate, evaporate to dryness under reduced pressure to obtain 4-(aminosulfonyl)-7-(1-piperazinyl)-2,1,3-benzoxadiazole as orange crystals (mp 178-180°).)

HPLC VARIABLES

Column: 150 × 4.6 5 μm Inertsil ODS-2
Mobile phase: Gradient. MeCN:water from 70:30 to 98:2 over 1 h, maintain at 98:2 for 10 min
Column temperature: 40
Flow rate: 1
Injection volume: 2
Detector: F ex 440 em 580

CHROMATOGRAM

Retention time: 7.86 (lauric acid), 13.63 (myristic acid), 22.29 (palmitic acid), 27.40 (margaric acid), 32.83 (stearic acid), 43.61 (arachidic acid), 53.45 (behenic acid), 61.83 (lignoceric acid)
Limit of detection: 10-20 fmole

REFERENCE

Toyo'oka, T.; Ishibashi, M.; Takeda, Y.; Imai, K. 4-(Aminosulphonyl)-2,1,3-benzoxadiazole derivatives as pre-column fluorogenic tagging reagents for carboxylic acids in high-performance liquid chromatography, *Analyst*, **1991**, *116*, 609–613.

p-Anisidine

SAMPLE .
Matrix: solutions
Analyte: fatty acids
Sample preparation: Dissolve sample in 2 mL carbon tetrachloride, add 1.4 g triphenyl-phosphine, polymer supported, heat at 80° with gentle shaking for 5 min, cool to room temperature, add 8 mL 62.5 mg/mL p-anisidine in ethyl acetate, heat at 80° for 10 min, inject a 2 μL aliquot.

HPLC VARIABLES
Column: 300 × 3.9 μBondapak C18
Mobile phase: Gradient. MeCN:water from 0:100 to 100:0 over 40 min (Waters curve 2).
Flow rate: 1
Injection volume: 2
Detector: UV 254

CHROMATOGRAM
Retention time: 12 (caproic acid), 14 (caprylic acid), 16 (capric acid), 19 (kauric acid), 22 (linolenic acid), 23 (myristic acid), 23 (4,7,10,13,16,19-docoshexaenoic acid), 24 (arachidonic acid), 24 (palmitoleic acid), 24.5 (linoleic acid), 25 (pentadecanoic acid), 28 (palmitic acid), 28.5 (oleic acid), 30 (margaric acid), 32 (stearic acid), 36 (erucic acid), 36 (arachidic acid), 41 (nervonic acid), 42 (docosanoic acid), 46 (tetracosanoic acid)

REFERENCE
Hoffman, N.E.; Liao, J.C. High pressure liquid chromatography of p-methoxyanilides of fatty acids, *Anal.Chem.*, **1976**, *48*, 1104–1106.

SAMPLE
Matrix: solutions
Analyte: prostaglandins
Sample preparation: Condition a Baker C18 SPE cartridge with 2 mL MeOH and 2 mL water. Mix 100 μL of a solution in MeOH with 12.5 μL 20 mM p-anisidine hydrochloride in water and 25 μL 125 mM 1-(3-dimethylaminopropyl)-3-ethylcarbodiimide hydrochloride in EtOH:pyridine 98.5:1.5, shake at 37° for 1 h, add 1.5 mL water (?), add to the SPE cartridge, wash with 2 mL buffer, wash with 1 mL MeOH:water 50:50, elute with 2 mL MeOH, evaporate the eluate to dryness under a stream of nitrogen, reconstitute the residue in 100 μL MeOH (J. Chromatogr. 1988, 442, 444), inject an aliquot. (Buffer was MeOH:water 50:50 adjusted to pH 2.75 with HCl.)

HPLC VARIABLES
Column: 250 × 4.6 5 μm Nucleosil C18
Mobile phase: MeCN:MeOH:water 35:22:43 containing 500 μg/mL lithium chlorate, adjusted to pH 4.1 with trifluoroacetic acid
Flow rate: 1
Detector: E, Waters 460, glassy-carbon working electrode 1.10 V, Ag/AgCl reference electrode

CHROMATOGRAM
Retention time: 11 (thromboxane B_2), 13.5 (dinoprost), 15 (dinoprostone), 16 (prostaglandin D_2)
Internal standard: 16,16-dimethylprostaglandin E_2 (30)
Limit of detection: 45-75 pg

KEY WORDS
SPE

REFERENCE
Knospe, J.; Herrmann, T.; Steinhilber, D.; Roth, H.J. Derivatization of prostaglandins to corresponding anilides and analysis by HPLC, *Adv.Prostaglandin Thromboxane Leukot.Res.*, **1989**, *19*, 692–695.

SAMPLE
Matrix: solutions
Analyte: ibuprofen
Sample preparation: Condition an Analytichem AASP propylbenzenesulfonic acid (SCX) SPE cartridge with isopropanol at 4 mL/min for 1.5 min and with mobile phase at 4 mL/min for 2 min. 100 µL Ibuprofen solution in dichloromethane + 100 µL 500 µg/mL 2-phenylpropionic acid in dichloromethane + 200 µL 100 mM triethylamine in dichloromethane + 100 µL 60 mM ethyl chloroformate, vortex for 15 s, let stand for 15 min, add 100 µL 500 mM p-anisidine in dichloromethane, vortex for 15 s, let stand for 5 min, add 600 µL isopropanol:hexane 10:90, vortex for 15 s, add a 25 µL aliquot to the SPE cartridge, elute the contents of the SPE cartridge on to the analytical column with mobile phase, after 2.43 min remove the SPE cartridge from the circuit, elute the analytical column with mobile phase and monitor the effluent from the column.

HPLC VARIABLES
Column: 250 × 4.6 5 µm Rexchrom Regis Pirkle D-phenylglycine (Regis)
Mobile phase: Hexane:isopropanol 90:10
Flow rate: 2
Injection volume: 25
Detector: UV 254

CHROMATOGRAM
Retention time: 10 (S), 12.5 (R)
Internal standard: 2-phenylpropionic acid (16 (S), 18 (R))
Limit of detection: 500 ng/mL

KEY WORDS
chiral; SPE

REFERENCE
Nicoll-Griffith, D.; Scartozzi, M.; Chiem, N. Automated derivatization and high-performance liquid chromatographic analysis of ibuprofen enantiomers, *J.Chromatogr.A*, **1993**, *653*, 253–259.

l-1-(1-Anthryl)ethylamine

SAMPLE
Matrix: solutions
Analyte: naproxen
Sample preparation: Mix 200 µL 1.25 µg/mL naproxen in dichloromethane with 5 µg 1-hydroxybenzotriazole, 100 µg 1-(3-dimethylaminopropyl)-3-ethylcarbodiimide hydrochloride, and 40 µg l-1-(1-anthryl)ethylamine, heat at 40°, dilute with ethyl acetate, wash with 5% HCl, wash with 5% sodium bicarbonate, wash with water. Remove the organic layer and evaporate it to dryness under a stream of nitrogen, reconstitute the residue in 500 µL ethyl acetate, inject a 10 µL aliquot. (Synthesis of l-1-(1-anthryl)ethylamine is as

follows. Dissolve 50 g benzanthrone in 500 mL concentrated sulfuric acid with gentle warming, pour this solution cautiously into 4 L hot water with vigorous stirring. Boil the suspension and slowly add 200 g chromium(VI) oxide (Caution! Chromium oxide is a carcinogen and highly corrosive!), after 6 h cool the mixture, filter, wash the precipitate with hot water. Dissolve the precipitate in dilute ammonia and precipitate with acid, crystallize from boiling concentrated nitric acid to give anthraquinone-1-carboxylic acid (Ber. 1924, 57, 1775). Warm, on a water bath, anthraquinone-1-carboxylic acid in dilute ammonia with twice the amount of zinc dust, when the reaction has ceased (30 min ?) filter the reaction mixture, add HCl to the filtrate to obtain anthracene-1-carboxylic acid as yellow needles, recrystallize from EtOH (mp 245°) (Ber 1897, 30, 1118). Anthracene-1-carboxylic acid is available from Tokyo Kasei (TCI America, Portland OR). Add 2 mL oxalyl chloride to a solution of 500 mg anthracene-1-carboxylic acid in 10 mL dichloromethane, reflux for 7 h, evaporate to dryness. Dissolve the residue in ether and add it dropwise to a solution of diazomethane in ether, stir at room temperature for 1 h, evaporate to dryness. Take up the residue in 5 mL chloroform and add it dropwise to 400 μL 55% hyroiodic acid, stir at room temperature for 2 h, wash with 5% sodium thiosulfate, wash with water, evaporate to dryness, chromatograph on 15 g silica gel with hexane:ethyl acetate 100:10, recrystallize from acetone:hexane to obtain 1-acetylanthracene as a yellow powder (mp 69.5-71°). Reflux 300 mg 1-acetylanthracene, 300 mg hydroxylamine hydrochloride, and 600 mg sodium acetate in 8 mL EtOH for 2 h, filter. Evaporate the filtrate to dryness and chromatograph the oily residue on a column of 10 g silica gel with hexane:ethyl acetate 100:10, recrystallize the product from acetone/hexane to obtain 1-acetylanthracene oxime as colorless needles (mp 170-173°). Stir 400 mg 1-acetylanthracene oxime in 20 mL EtOH and 20 mL 10% NaOH at 0° and add 1 g Raney nickel in portions, stir the suspension for 7 h, filter. Concentrate the filtrate and extract the residue with ethyl acetate. Wash the organic layer with water and dry it over anhydrous sodium sulfate, evaporate to dryness. Chromatograph the residue on a column of 12 g of silica gel with ethyl acetate:100 mM methylamine in EtOH 100:10 to obtain dl-1-(1-anthryl)ethylamine as an oil. Mix 2 mL of a 400 mg/mL solution of dl-1-(1-anthryl)ethylamine in EtOH with 4 mL of a 200 mg/mL solution of l-α-methoxy-α-methyl-1-naphthaleneacetic acid. Collect the resulting precipitate and fractionally crystallize it from EtOH several times. Add the salt to 10% NaOH and extract with ether, wash the organic layer with water, dry over anhydrous sodium sulfate, evaporate to dryness, take up the residue in ether, pass anhydrous HCl gas through the solution to obtain l-1-(1-anthryl)ethylamine hydrochloride as colorless needles (mp 236-238° (d), $[\alpha]_D^{25}$ -33.2° (c = 0.20, MeOH)). Synthesis of l-α-methoxy-α-methyl-1-naphthaleneacetic acid is as follows. Stir 200 mg magnesium metal in 8 mL diethyl ether under nitrogen, add a trace amount of iodomethane, cautiously add 2 g 1-bromonaphthalene in 2 mL anhydrous ether, stir at room temperature for 30 min, slowly add this solution of 1-naphthylmagnesium bromide to a solution of 450 mg pyruvic acid in 3 mL ether, stir at room temperature for 30 min, acidify with 3% sulfuric acid, extract with ethyl acetate. Wash the organic layer with water and extract it with 5% sodium carbonate. Acidify the aqueous layer with 5% HCl, extract with ethyl acetate. Wash the organic layer with water and dry it over anhydrous sodium sulfate, evaporate to dryness, recrystallize from acetone/hexane to give α-methoxy-α-methyl-1-naphthaleneacetic acid as colorless plates (mp 161-162°). Prepare the (R)-(+)-α-methylbenzylamine salt of α-methoxy-α-methyl-1-naphthaleneacetic acid in EtOH and fractionally recrystallize it twice from EtOH. Decompose the salt with 5% HCl, extract with ethyl acetate. Wash the organic layer with water and dry it over anhydrous sodium sulfate, evaporate to dryness, recrystallize the crude product from ether/hexane to obtain l-α-methoxy-α-methyl-1-naphthaleneacetic acid as colorless plates (mp 111-112°; $[\alpha]_D^{13}$ -106.3° (c = 0.16, chloroform) (J. Chromatogr. 1978, 152, 413). Preparation of diazomethane is as follows. Caution! Diazomethane is toxic, explosive, and carcinogenic! A face shield and a safety screen should always be used and the preparation should only be carried out in a properly functioning chemical fume hood. Only smooth glass apparatus with rubber stoppers and plastic tubing should be used. Scratched glassware, ground glass joints, and sharp edges should be avoided (Org. Syn., Coll. Vol. VI; Wiley: New York, 1988, pp. 432-435). Procedures have been reported for the synthesis of diazomethane from N-methyl-N-nitrosourea (Org. Syn., Coll. Vol. II; Wiley: New York, 1943, pp. 165-167), N-nitroso-β-methylaminoisobutyl methyl ketone (Org. Syn., Coll. Vol. III; Wiley: New York,

1955, pp. 244-248), and N,N'-dimethyl-N,N'-dinitrosoterephthalamide (Org. Syn., Coll. Vol. V; Wiley: New York, 1973, pp. 351-355). Probably the most convenient starting material is N-methyl-N-nitroso-p-toluenesulfonamide (Diazald) (Aldrichimica Acta 1983,16, 3-10; Org. Syn., Coll. Vol. IV, Wiley: New York, 1963, pp. 250-253). Add 10 mL 95% EtOH to a solution of 5 g KOH in 8 mL water, warm to 65° using a water bath, add a solution of 5 g N-methyl-N-nitroso-p-toluenesulfonamide in 45 mL ether over 20 min at such a rate as to keep the reaction volume constant. Collect the ether and diazomethane that distil in an ice-cooled receiving flask under a dry ice/acetone condenser. When all the N-methyl-N-nitroso-p-toluenesulfonamide has been used up, slowly add 10 mL ether to the reaction flask and continue distillation until the distillate is colorless. A purpose-built distillation apparatus can be purchased from Aldrich (Aldrichimica Acta 1983, 16, 3-10). Excess quantities of diazomethane can be destroyed by adding acetic acid until the yellow color of the diazomethane is discharged. The safe disposal of the nitroso compounds used to generate diazomethane has been discussed (Lunn,G.; Sansone,E.B. Destruction of Hazardous Chemicals in the Laboratory, Second Edition. Wiley: New York, 1994, pp. 277–289).)

HPLC VARIABLES
Column: 150 × 4.6 5 μm Resolve spherical silica
Mobile phase: Hexane:ethyl acetate 80:20
Flow rate: 1
Injection volume: 10
Detector: F ex 260 em 400

CHROMATOGRAM
Retention time: 6 (d), 10 (l)
Limit of detection: 100 fmole

KEY WORDS
chiral; normal phase

REFERENCE
Goto, J.; Ito, M.; Katsuki, S.; Saito, N.; Nambara, T. Sensitive derivatization reagents for optical resolution of carboxylic acids by high performance liquid chromatography with fluorescence detection, *J.Liq.Chromatogr.*, **1986**, 9, 683–694.

Benzylamine

SAMPLE
Matrix: solutions
Analyte: 2-arylpropionic acids
Sample preparation: Mix 1 mL 100 μg/mL compound in dichloromethane with 300 μL 1 mg/mL 1-hydroxybenzotriazole in dichloromethane:pyridine 99:1, 300 μL 1 mg/mL 1-ethyl-3-dimethylaminopropylcarbodiimide hydrochloride in dichloromethane, and 300 μL 1-1.5 mg/mL benzylamine in dichloromethane, vortex, let stand at room temperature for 1.5 h, wash with 1 mL 250 mM HCl, wash with 1 mL water. Remove the organic layer and evaporate it to dryness under a stream of nitrogen, reconstitute the residue in MeOH, inject an aliquot.

HPLC VARIABLES
Column: 10 μm EXP B101 4-methylbenzoate cellulose on silica gel (Bio-Rad)
Mobile phase: MeOH:50 mM pH 1.5 perchlorate buffer 80:20

Flow rate: 1
Detector: UV 230

CHROMATOGRAM
Retention time: 4, 4.5 (ketoprofen enantiomers), 7, 8 (flurbiprofen enantiomers)

KEY WORDS
chiral

REFERENCE
Van Overbeke, A.; Baeyens, W.; Van den Bossche, W.; Dewaele, C. Enantiomeric separation of amide derivatives of some 2-arylpropionic acids by HPLC on a cellulose-based chiral stationary phase, *J.Pharm.Biomed.Anal.*, **1994**, *12*, 911–916.

SAMPLE
Matrix: solutions
Analyte: 2-arylpropionic acids
Sample preparation: Mix 1 mL 100 μg/mL compound in dichloromethane with 300 μL 100 μg/mL 1-hydroxybenzotriazole in dichloromethane:pyridine 99:1, 300 μL 1.1 mg/mL 1-ethyl-3-dimethylaminopropylcarbodiimide hydrochloride in dichloromethane, and 300 μL 300 μg/mL benzylamine in dichloromethane, vortex, let stand at room temperature for 1.5 h, evaporate to dryness under a stream of nitrogen, reconstitute the residue in 500 μL MeOH, inject an aliquot.

HPLC VARIABLES
Column: 150 × 4.6 10 μm EXP B101 tris(4-methylbenzoate) cellulose on silica (Bio-Rad)
Mobile phase: MeOH:buffer 70:30 (80:20 for benoxaprofen) (Prepare buffer solution by dissolving 14.05 g sodium perchlorate in water, adjust pH to 2.0, make up to 1 L with water.)
Flow rate: 1
Detector: UV 230

CHROMATOGRAM
Retention time: k' 8.43, k' 25.11 (benoxaprofen enantiomers), k' 28.22, k' 36.55 (carprofen enantiomers), k' 6.82, k' 9.98 (fenoprofen enantiomers), k' 13.11, k' 16.60 (flurbiprofen enantiomers), k' 2.77, k' 3.75 (ibuprofen enantiomers), k' 4.06, k' 5.69 (ketoprofen enantiomers), k' 5.15, k' 6.62 (pirprofen enantiomers), k' 5.36, k' 6.45 (tiaprofenic acid enantiomers)

KEY WORDS
chiral

REFERENCE
Van Overbeke, A.; Baeyens, W.; Van den Bossche, W.; Dewaele, C. Separation of 2-arylpropionic acids on a cellulose based chiral stationary phase by RP-HPLC, *J.Pharm.Biomed.Anal.*, **1994**, *12*, 901–909.

4-N-Cadaverino-7-(N,N-dimethylaminosulfonyl)-2,1,3-benzoxadiazole

SAMPLE
Matrix: solutions
Analyte: fatty acids
Sample preparation: Mix 200 μL of a 10 μM solution of fatty acids in DMF containing 140 mM diethylphosphorocyanidate with 200 μL 10 mM 4-N-cadaverino-7-(N,N-dimethyl-aminosulfonyl)-2,1,3-benzoxadiazole in MeCN, let stand at room temperature for 6 h, inject a 1 μL aliquot. (Synthesis of 4-N-cadaverino-7-(N,N-dimethylaminosulfonyl)-2,1,3-benzoxadiazole (DBD-CD) is as follows. Dissolve 0.5 g magnesium sulfate heptahydrate and 6 g NaOH in 60 mL water, throughout the reaction keep the flask at about 20° with cold water cooling, add 15 mL 30% hydrogen peroxide, add 75 mL MeOH, add 12.1 g powdered benzoyl peroxide in one go, stir for 10 min, pour into 150 mL 20% sulfuric acid, extract three times with 50 mL portions of chloroform, determine peroxybenzoic acid concentration by iodometric titration (Tetrahedron 1967, 23, 3327). Slowly add 110 mL 1 M peroxybenzoic acid in chloroform to 7 g 2,6-difluoroaniline dissolved in 100 mL chloroform, stir at room temperature, when reaction is complete (iodometric titration) wash with 2% sodium thiosulfate, wash with 5% sodium carbonate, wash with water, dry over anhydrous sodium sulfate, evaporate to dryness under reduced pressure, recrystallize 2,6-difluoronitrosobenzene from EtOH (mp 108.5-109.5). Stir 8.5 g 2,6-difluoronitrosobenzene in 85 mL DMSO at room temperature and add a solution of 3.91 g sodium azide in 85 mL DMSO dropwise, let stand for about 1 h, add to a large volume of water, extract with ether, dry the extracts over anhydrous sodium sulfate, evaporate to dryness under reduced pressure and distil to give 4-fluoro-2,1,3-benzoxadiazole as a colorless oil (bp 83°/12 mm Hg) (J.Chem.Soc.(C) 1970, 1433). Add 11 mL chlorosulfonic acid dropwise to 3 g 4-fluoro-2,1,3-benzoxadiazole in 10 mL chloroform at 0-10° (use a calcium chloride drying tube), stir at room temperature for 1 h, reflux for 2 h, cool, slowly pour into ice water, remove the organic layer, extract the aqueous layer with chloroform, combine the organic layers, wash, dry over anhydrous magnesium sulfate, evaporate under reduced pressure, take up the residue in 5 mL benzene (Caution! Benzene is a carcinogen!), chromatograph on a 150 × 30 column of silica gel (100-200 mesh Kanto Chemical) with n-hexane:benzene 50:50, evaporate the appropriate fractions to give 4-(chlorosulfonyl)-7-fluoro-2,1,3-ben-zoxadiazole (CBD-F) as pale yellow needles (mp 64-66°) (Anal. Chem. 1984. 56, 2461). Stir 0.76 g CBD-F in 70 mL MeCN at 0-10° and add 1 g dimethylamine hydrochloride in 10 mL 100 mM pH 10 borax dropwise, adjust pH to 5 with 1 M HCl, concentrate to about 10 mL under reduced pressure, extract three times with 200 mL portions of diethyl ether, wash with water, dry over anhydrous magnesium sulfate, evaporate under reduced pressure, chromatograph on a 500 × 20 column of silica gel with chloroform, isolate the appropriate fraction and re-chromatograph on the same column with ethyl acetate:benzene 1:2 to give 4-(N,N-dimethylaminosulfonyl)-7-fluoro-2,1,3-benzoxadiazole (DBD-F) as white needles (mp 124-125°) (yield 1% !). On a Merck no. 5714 60F$_{254}$ tlc plate eluted with chloroform DBD-F has Rf 0.32 and lies between two other reaction products (Analyst 1989, 114, 413). DBD-F can also be purchased from Tokyo Kasei (TCI America, 911 North Harborgate St., Portland, OR 97203; 800-423-8616, 503-283-1681, (fax) 503-283-1987;

www.tciamerica.com). Add 123 mg 4-(N,N-dimethylaminosulfonyl)-7-fluoro-2,1,3-benzox-adiazole in 20 mL MeCN dropwise to 153 mg cadaverine in 20 mL MeCN at room temperature, stir for 30 min, evaporate under reduced pressure, dissolve residue in 50 mL 5% HCl, wash three times with 20 mL ethyl acetate, discard ethyl acetate extracts, adjust pH of aqueous solution to 13-14 with 5% NaOH, extract five times with 50 mL ethyl acetate, combine extracts, wash with 20 mL water, dry over anhydrous sodium sulfate, evaporate under reduced pressure to give 4-N-cadaverino-7-(N,N-dimethylaminosulfonyl)-2,1,3-benzoxadiazole as orange crystals (mp 100-101°; yield 30%).)

HPLC VARIABLES
Column: 150 × 4.6 5 μm Inertsil ODS-2
Mobile phase: Gradient. MeCN:water from 70:30 to 98:2 over 1 h, maintain at 98:2 for 20 min.
Column temperature: 40
Flow rate: 1
Injection volume: 1
Detector: F ex 437 em 561

CHROMATOGRAM
Retention time: 10 (lauric acid), 13 (tridecanoic acid), 17 (myristic acid), 21 (pentadecanoic acid), 27 (palmitic acid), 32 (margaric acid), 38 (stearic acid), 43 (nonadecanoic acid), 48 (arachidic acid), 53 (heneicosanoic acid), 57 (behenic acid), 62 (tricosanoic acid), 66 (lignoceric acid)

REFERENCE
Toyo'oka, T.; Ishibashi, M.; Takeda, Y.; Nakashima, K.; Akiyama, S.; Uzu, S.; Imai, K. Precolumn fluorescence tagging reagent for carboxylic acids in high-performance liquid chromatography: 4-substituted-7-aminoalkylamino-2,1,3-benzoxadiazoles, *J.Chromatogr.*, **1991**, *588*, 61−71.

(S)-1-(4-Dansylaminophenyl)ethylamine

SAMPLE
Matrix: solutions

Analyte: carboxylic acids

Sample preparation: Mix 100 μL of a solution of the carboxylic acid in MeCN with 100 μL 2 mM (S)-1-(4-dansylaminophenyl)ethylamine in MeCN, add 100 μL 3 mM 2,2'-dipyridyl disulfide (Aldrithiol-2) in MeCN, add 100 μL 3 mM triphenylphosphine in MeCN, vortex, let stand at room temperature for 3 h, inject a 5 μL aliquot. (Synthesis of (S)-1-(4-dansylaminophenyl)ethylamine is as follows. Add 2.2 g di-tert-butyl dicarbonate dropwise to a stirred solution of 2 g (S)-α-methyl-4-nitrobenzylamine hydrochloride ((S)-1-(4-nitrophenyl)ethylamine hydrochloride) and 1.1 g triethylamine in 20 mL MeCN at 0°, stir at room temperature for 1 h, evaporate to dryness under reduced pressure. Dissolve the residue in 50 mL ethyl acetate and wash with 10% aqueous citric acid, wash with water, dry over anhydrous sodium sulfate, evaporate to dryness under reduced pressure to give (S)-N-tert-butoxycarbonyl-1-(4-nitrophenyl)ethylamine as white crystals (mp 86-89°). Add 200 mg 5% PdC to a solution of 2 g (S)-N-tert-butoxycarbonyl-1-(4-nitrophenyl)ethylamine in 40 mL MeOH, stir, hydrogenate at room temperature for 3 h, filter, evaporate the filtrate to dryness to obtain (S)-N-tert-butoxycarbonyl-1-(4-aminophenyl)ethylamine. Stir 1.4 g (S)-N-tert-butoxycarbonyl-1-(4-aminophenyl)ethylamine in 10 mL MeCN and 50 mL 100 mM pH 9.0 sodium bicarbonate, add a solution of 1.9 g dansyl chloride in 30 mL MeCN dropwise while maintaining the pH of the solution at 9.0 with 1 M NaOH, stir at room temperature for 20 min, stir at 45° for 2 h, cool to room temperature, extract three times with 30 mL portions of ethyl acetate. Combine the organic layers and evaporate them to dryness under reduced pressure, recrystallize the residue from benzene/hexane to obtain (S)-N-tert-butoxycarbonyl-1-(4-dansylaminophenyl)ethylamine as pale-yellow crystals (mp 97-101°) (Caution! Benzene is a carcinogen!). Add 2 mL concentrated HCl to a solution of 1.5 g (S)-N-tert-butoxycarbonyl-1-(4-dansylaminophenyl)ethylamine in 10 mL MeOH, stir at room temperature for 30 min, evaporate to dryness under reduced pressure. Dissolve the residue in 30 mL water and adjust the pH to 8.0 with sodium bicarbonate, extract three times with 30 mL portions of ethyl acetate. Combine the organic layers and evaporate them to dryness under reduced pressure, recrystallize the residue from EtOH to obtain (S)-1-(4-dansylaminophenyl)ethylamine as pale-yellow crystals (mp 157-160°; $[\alpha]_D^{28}$ -11.1° (c = 0.2 in MeCN)). (The (R)-enantiomer can be prepared in an exactly analogous fashion.))

HPLC VARIABLES

Column: 150 × 4.6 5 μm ODS-80TM (Tosoh)

Mobile phase: MeCN:50 mM pH 6.5 sodium acetate buffer 55:45 (A) or 70:30 (B) or 65:35 (C) or 60:40 (D)

Flow rate: 1

Injection volume: 5

Detector: F ex 338 em 535

CHROMATOGRAM

Retention time: k' 8.45 (d-2-phenylpropionic acid (A)), k' 9.69 (l-2-phenylpropionic acid (A)), k' 7.44 (d-ibuprofen (B)), k' 8.90 (l-ibuprofen (B)), k' 8.71 (d-flurbiprofen (C)), k' 10.57 (l-flurbiprofen (C)), k' 6.76 (d-pranoprofen (A)), k' 8.01 (l-pranoprofen (A)), k' 8.29 (d-phenoprofen (C)), k' 9.76 (l-phenoprofen (C)), k' 4.52 (d-naproxen (D)), k' 5.45 (l-naproxen (D))

Limit of detection: 170 fmole

KEY WORDS

chiral

REFERENCE

Iwaki, K.; Bunrin, T.; Kameda, Y.; Yamazaki, M. Resolution and sensitive detection of carboxylic acid enantiomers using fluorescent chiral derivatization reagents by high-performance liquid chromatography, *J.Chromatogr.A*, **1994**, *662*, 87–93.

Dansyl Semipiperazide

SAMPLE
Matrix: blood
Analyte: fatty acids
Sample preparation: Condition a Sep-Pak silica SPE cartridge with chloroform. 500 μL Serum + 1 mL 20 μM tridecanoic acid in chloroform containing 9.53 μM 2-heptenoic acid + 100 μL 500 mM sulfuric acid + 5 mL chloroform + 20 mL MeCN, shake slowly, add 1.5 g anhydrous calcium chloride, invert slowly, let stand for 30 min, filter (No 5A paper) the upper layer. Add 500 μL 45 mM dansyl semipiperazide in chloroform and 150 mg dicyclohexylcarbodiimide to the filtrate, let stand for 30 min, remove the solvent by evaporation under reduced pressure, dissolve the residue in the minimum amount of chloroform, add to the SPE cartridge, elute with two 1 mL portions of chloroform. Evaporate the eluate to dryness under reduced pressure, reconstitute with the minimum amount of MeOH, inject an aliquot. (Prepare dansyl semipiperazide as follows. Add 8 g dansyl chloride to 50 g piperazine dissolved in 500 mL acetone with stirring at room temperature over 30 min, stir overnight, evaporate to dryness under reduced pressure. Dissolve the residue in 300 mL chloroform, filter (5A paper), wash the filtrate three times with 5% sodium bicarbonate, wash with water. Extract the organic layer three times with 100 mL portions of 1 M HCl, combine the extracts and wash them repeatedly with chloroform, make alkaline with a slight excess of powdered sodium carbonate, extract with benzene (Caution! Benzene is a carcinogen!). Wash the benzene layer twice with 5% sodium bicarbonate, dry over anhydrous sodium sulfate, evaporate to dryness under reduced pressure. Dissolve the residue in the minimum amount of chloroform, chromatograph on a 200 × 50 column of Wako Gel C-200 silica gel with chloroform. When the first eluting yellow band reaches the bottom change the eluent to MeOH, continue eluting until all the yellow band is collected. Evaporate the eluate to dryness to obtain dansyl semipiperazide, store in the dark at 5°. Determine the concentration of dansyl semipiperazide by spectrophotometry, extinction coefficient at 340 nm in MeOH or EtOH is 4300 $M^{-1}cm^{-1}$.)

HPLC VARIABLES
Column: 250 × 4.6 Hitachi C18 3056
Mobile phase: Gradient. MeCN:water 45:55 for 43 min, 60:40 for 32 min, 75:25 for 28 min, 85:15 for 100 min, 100:0 for rest of run (step gradient).
Flow rate: 0.8
Detector: F ex 350 em 530

CHROMATOGRAM
Retention time: 13 (lactic acid), 16 (acetic acid), 24 (propionic acid), 35 (isobutyric acid), 37 (n-butyric acid), 55 (isovaleric acid), 57 (n-valeric acid), 64 (isocaproic acid), 67 (n-caproic acid), 91 (caprylic acid), 100 (nonanoic acid), 111 (capric acid), 121 (undecanoic acid), 133 (lauric acid, myristoleic acid), 145 (linolenic acid), 150 (docosahexaenoic acid), 157 (myristic acid), 160 (palmitoleic acid, arachidonic acid), 168 (linoleic acid), 180 (pen-

tadecanoic acid, eicosatrienoic acid), 212 (palmitic acid, elaidic acid, vaccenic acid), 218 (petroselenic acid), 226 (margaric acid), 233 (eicosenoic acid), 236 (stearic acid), 252 (erucic acid), 257 (arachidic acid)

Internal standard: tridecanoic acid (139), 2-heptenoic acid (77)

KEY WORDS
serum; SPE

REFERENCE
Yanagisawa, I.; Yamane, M.; Urayama, T. Simultaneous separation and sensitive determination of free fatty acids in blood plasma by high-performance liquid chromatography, *J.Chromatogr.*, **1985**, *345*, 229–240.

2,4-Dimethoxyaniline

SAMPLE
Matrix: solutions
Analyte: prostaglandins
Sample preparation: Condition a Baker C18 SPE cartridge with 2 mL MeOH and 2 mL water. Mix 20 μL of a solution in MeOH with 2.5 μL 20 mM 2,4-dimethoxyaniline hydrochloride in water and 5 μL 125 mM 1-(3-dimethylaminopropyl)-3-ethylcarbodiimide hydrochloride in EtOH:pyridine 98.5:1.5, heat at 37° for 1 h, add 300 μL water, add to the SPE cartridge, wash with 2 mL buffer, wash with 1 mL MeOH:water 50:50, elute with 2 mL MeOH, evaporate the eluate to dryness under a stream of nitrogen at 37°, reconstitute the residue in 100 μL MeOH, inject a 10 μL aliquot. (Buffer was MeOH:water 50:50 adjusted to pH 2.75 with HCl.)

HPLC VARIABLES
Column: 250 × 4.6 5 μm Nucleosil C18
Mobile phase: MeCN:MeOH:water 35:22:43 containing 500 μg/mL lithium chlorate, adjusted to pH 4.1 with trifluoroacetic acid
Flow rate: 1
Detector: E, Waters 460, thin-layer glassy-carbon working electrode 1.10 V, Ag/AgCl reference electrode

CHROMATOGRAM
Retention time: 11 (thromboxane B_2), 14 (dinoprost), 15 (dinoprostone), 17 (prostaglandin D_2)
Internal standard: 16,16-dimethylprostaglandin E_2 (30)
Limit of detection: 40-70 pg

KEY WORDS
SPE

REFERENCE
Knospe, J.; Steinhilber, D.; Herrmann, T.; Roth, H.J. Picomole determination of 2,4-dimethoxyanilides of prostaglandins by high-performance liquid chromatography with electrochemical detection, *J.Chromatogr.*, **1988**, *442*, 444–450.

SAMPLE
Matrix: solutions
Analyte: prostaglandins

Sample preparation: Condition a Baker C18 SPE cartridge with 2 mL MeOH and 2 mL water. Mix 100 μL of a solution in MeOH with 12.5 μL 20 mM 4-methoxyaniline hydrochloride in water and 25 μL 125 mM 1-(3-dimethylaminopropyl)-3-ethylcarbodiimide hydrochloride in EtOH:pyridine 98.5:1.5, shake at 37° for 1 h, add 1.5 mL water (?), add to the SPE cartridge, wash with 2 mL buffer, wash with 1 mL MeOH:water 50:50, elute with 2 mL MeOH, evaporate the eluate to dryness under a stream of nitrogen, reconstitute the residue in 100 μL MeOH (J. Chromatogr. 1988, 442, 444), inject an aliquot. (Buffer was MeOH:water 50:50 adjusted to pH 2.75 with HCl.)

HPLC VARIABLES
Column: 250 × 4.6 5 μm Nucleosil C18
Mobile phase: MeCN:water 42:58, adjusted to pH 3.5 with trifluoroacetic acid
Flow rate: 1
Detector: UV 249

CHROMATOGRAM
Retention time: 7.5 (6-ketoprostaglandin $F_{1\alpha}$), 10.5 (thromboxane B_2), 14 (dinoprost), 17 (dinoprostone), 20 (prostaglandin D_2)
Internal standard: 16,16-dimethylprostaglandin E_2 (35.5)
Limit of detection: 1.2-2.2 ng

KEY WORDS
SPE

REFERENCE
Knospe, J.; Herrmann, T.; Steinhilber, D.; Roth, H.J. Derivatization of prostaglandins to corresponding anilides and analysis by HPLC, *Adv.Prostaglandin Thromboxane Leukot.Res.*, **1989**, *19*, 692–695.

1-(5-Dimethylamino-1-naphthalenesulfonyl)-(S)-3-aminopyrrolidine

SAMPLE
Matrix: solutions
Analyte: carboxylic acids
Sample preparation: Mix 50 μL of a 0.001-5 mM solution in MeCN with 50 μL 1 mM DNS-APy in MeCN containing 50 mM 2,2'-dipyridyl disulfide and 50 mM triphenylphosphine, let stand at room temperature for 30 min. Remove a 10 μL aliquot and dilute it to 100 μL with MeCN, inject a 2 μL aliquot. (Synthesis of DNS-APy, 1-(5-dimethylamino-1-naphthalenesulfonyl)-(S)-3-aminopyrrolidine, is as follows. Cool a solution of 16.4 g (R)-(+)-1-benzyl-3-pyrrolidinol in 164 mL pyridine to +5°, add 19.35 g p-toluenesulfonyl chloride, stir at +10° for 48 h, evaporate to dryness, chromatograph using dichloromethane:acetone 95:5 to obtain (3R)-3-[(4-tolylsulfonyl)oxy]-1-(phenylmethyl)pyrrolidine (mp 68°). Heat a solution of(3R)-3-[(4-tolylsulfonyl)oxy]-1-(phenylmethyl)pyrrolidine in 200 mL an-

hydrous DMF to 65°, add 33.5 g sodium azide (Caution! Sodium azide is highly toxic!), stir at 60° for 7 h, filter, evaporate the filtrate to dryness under reduced pressure, dissolve the residue in ethyl acetate, wash twice with water, dry over anhydrous magnesium sulfate, evaporate to obtain (3S)-3-azido-1-(phenylmethyl)pyrrolidine as an oil. Add 3.5 g 10% palladium on carbon under nitrogen to a solution of 7.05 g (3S)-3-azido-1-(phenylmethyl)pyrrolidine in 34.8 mL 1 M HCl in water and 245 mL EtOH, hydrogenate at atmospheric pressure for 30 min, add 3.5 g catalyst, hydrogenate for 2 h, filter, add 34.8 mL 1 M HCl to the filtrate, evaporate to dryness under reduced pressure, take up the residue in 70 mL EtOH, filter, evaporate the filtrate to dryness under reduced pressure, repeat this operation twice, crystallize with the minimum amount of EtOH to obtain (3S)-(+)-3-aminopyrrolidine dihydrochloride (J. Med. Chem. 1992, 35, 4205). (3S)-(+)-Aminopyrrolidine dihydrochloride is also available from Tokyo Kasei (TCI America, 911 North Harborgate St., Portland, OR 97203; 800-423-8616, 503-283-1681, (fax) 503-283-1987; www.tciamerica.com). Stir 800 mg (3S)-(+)-3-aminopyrrolidine dihydrochloride and 2 mL triethylamine in 800 mL MeCN at 0-10°, add a solution of 440 mg dansyl chloride in 80 mL MeCN dropwise, stir in the dark for 30 min, evaporate to dryness under reduced pressure, dissolve the residue in 200 mL 5% HCl, wash twice with 40 mL portions of dichloromethane. Adjust the pH of the organic layer to 13-14 with 5% NaOH, extract twice with 10 mL portions of dichloromethane. Combine the organic layers and wash them with 80 mL water. Dry the organic layer over anhydrous sodium sulfate, evaporate to dryness under reduced pressure, dissolve the residue in dichloromethane:MeOH 90:10, chromatograph on silica gel with dichloromethane:MeOH 90:10. Collect the greenish-yellow fluorescent band and evaporate it under reduced pressure to obtain DNS-APy as a greenish-yellow oil.)

HPLC VARIABLES
Column: 150 × 4.6 5 µm TSK gel ODS-80TM (Tosoh)
Mobile phase: MeCN:water 60:40 (A) or 50:50 (B) or 40:60 (C)
Flow rate: 1
Injection volume: 2
Detector: F ex 340 em 530

CHROMATOGRAM
Retention time: 30 ((S)-(+)-ibuprofen) (A), 32 ((R)-(-)-ibuprofen) (A), 23 ((S)-(+)-pranoprofen) (B), 26 ((R)-(-)-pranoprofen) (B), 37 ((S)-(+)-ketoprofen) (B), 40 ((R)-(-)-ketoprofen) (B), 17 ((S)-tropic acid) (C), 19 ((R)-tropic acid) (C)
Limit of detection: 0.1 pmole

KEY WORDS
chiral

REFERENCE
Al-Kindy, S.; Santa, T.; Fukushima, T.; Homma, H.; Imai, K. 1-(5-Dimethylamino-1-naphthalenesulphonyl)-(S)-3-aminopyrrolidine (DNS-Apy) as a fluorescence chiral labelling reagent for carboxylic acid enantiomers, *Biomed.Chromatogr.*, **1997**, *11*, 137−142.

(1R)-1-(4-Dimethylaminonaphthalen-1-yl)ethylamine

SAMPLE
Matrix: blood
Analyte: naproxen
Sample preparation: Condition a Sep-Pak C18 SPE cartridge with EtOH and water. 20-100 μL Serum + 2 mL 100 mM pH 2.0 phosphate buffer, mix, add to the SPE cartridge, wash with 10 mL water, wash with 500 μL EtOH:water 70:30, elute with 2 mL EtOH:water 70:30. Evaporate the eluate to dryness, reconstitute with 1 mL 50 mM pH 10 sodium carbonate buffer, wash three times with 2 mL portions of n-hexane. Acidify the aqueous layer with 200 μL 1 M HCl, extract three times with 2 mL portions of ethyl acetate. Add 5 μg 1-hydroxybenzotriazole and 300 ng IS to the combined organic layers and evaporate to dryness under a stream of nitrogen, add 20 μL 1.5 mg/mL (1R)-1-(4-dimethylaminonaphthalen-1-yl)ethylamine hydrochloride in pyridine to the residue, add 200 μL 500 μg/mL 1-(3-dimethylaminopropyl)-3-ethylcarbodiimide hydrochloride in dichloromethane, let stand at 4° for 45 min, evaporate to dryness under a stream of nitrogen, reconstitute with 500 μL n-hexane:ethyl acetate 75:25, wash with 5% sodium bicarbonate solution, add to a 30 × 4 silica gel column, wash with 700 μL n-hexane:ethyl acetate 75:25, elute with 2 mL ethyl acetate. Evaporate the eluate to dryness and reconstitute with 400 μL mobile phase, inject a 5-10 μL aliquot. (A synthesis of (1R)-1-(4-dimethylaminonaphthalen-1-yl)ethylamine (1 enantiomer) is given in Anal. Chim. Acta 1980, 120, 187 but the resolution of the enantiomers is problematic (Chem. Pharm.Bull. 1984, 32, 251). It should be possible to prepare (1R)-1-(4-dimethylaminonaphthalen-1-yl)ethylamine from (R)-(+)-1-(1-naphthyl)ethylamine by analogy with the synthesis of (1S)-1-(4-dimethylaminonaphthalen-1-yl)ethylamine. Synthesis of (1S)-1-(4-dimethylaminonaphthalen-1-yl)ethylamine is as follows. Mix 9.6 g (S)-(-)-1-(1-naphthyl)ethylamine ((1S)-1-(naphthalen-1-yl)ethylamine) with 37 mL acetic anhydride, let stand at room temperature for 30 min, cool to 12°, add 7.1 mL concentrated nitric acid dropwise, stir at room temperature for 1 h, adjust pH to 9 with 1 M NaOH, extract with ethyl acetate. Dry the organic layer over anhydrous sodium sulfate and evaporate to dryness under reduced pressure, chromatograph on a column of silica gel with ethyl acetate:benzene 10:1 (Caution! Benzene is a carcinogen!). Evaporate the eluate to dryness under reduced pressure to give (1S)-1-(4-nitronaphthalen-1-yl)-N-acetylethylamine as colorless needles. Stir 100 mg 10% PdC and 2.5 g (1S)-1-(4-nitronaphthalen-1-yl)-N-acetylethylamine in 15 mL EtOH, add 1.5 mL hydrazine hydrate in portions (Caution! Hydrazine hydrate is a carcinogen and explodes on distillation in air!), add 100 mg 10% PdC, reflux for 1 h, cool, filter. Evaporate the filtrate to dryness under reduced pressure and chromatograph the residue on a column of silica gel with ethyl acetate, evaporate the eluate to dryness under reduced pressure to give (1S)-1-(4-aminonaphthalen-1-yl)-N-acetylethylamine as a colorless powder. Stir 1.5 g (1S)-1-(4-aminonaphthalen-1-yl)-N-acetylethylamine and 1.7 g sodium bicarbonate in 5 mL water at 0°, add 2 mL dimethyl sulfate (Caution! Dimethyl sulfate is highly toxic!), stir at room temperature for 3 h, distil to remove excess reagent. Concentrate the aqueous layer under reduced pressure, purify the residue by TLC using ethyl acetate (R_f 0.30) to obtain (1S)-1-(4-dimethylaminonaphthalen-1-yl)-N-acetylethylamine as a colorless powder. Reflux 1.2 g (1S)-1-(4-dimethylaminonaphthalen-1-yl)-N-ace-

tylethylamine in 15 mL concentrated HCl for 35 h, adjust the pH to 9 with 2 M NaOH (Caution! Exothermic!), cool, extract with ethyl acetate. Dry the organic layer over anhydrous sodium sulfate and evaporate it to dryness under reduced pressure, purify by TLC using ethyl acetate to obtain (1S)-1-(4-dimethylaminonaphthalen-1-yl)ethylamine as a slightly yellow oil ($[\alpha]_D$ +19.75° (c = 2.0, EtOH) (Chem. Pharm. Bull. 1984, 32, 251).)

HPLC VARIABLES
Column: 305 × 6.4 Silica
Mobile phase: n-Hexane:THF 80:26
Flow rate: 0.6
Injection volume: 5-10
Detector: F ex 320 em 410

CHROMATOGRAM
Retention time: 11 (l), 13 (d)
Internal standard: 3,5-di-tert-butyl-4-hydroxybenzaldehyde O-(methoxycarbonylmethyl)oxime (12)
Limit of quantitation: 100 pg

KEY WORDS
serum; chiral; normal phase; SPE; rabbit; pharmacokinetics

REFERENCE
Goto, J.; Goto, N.; Nambara, T. Separation and determination of naproxen enantiomers in serum by high-performance liquid chromatography, *J.Chromatogr.*, **1982**, *239*, 559–564.

(1S)-1-(4-Dimethylaminonaphthalen-1-yl)ethylamine

SAMPLE
Matrix: blood
Analyte: loxoprofen
Sample preparation: 1 mL Plasma + 100 µL 1 M HCl, extract with 5 mL hexane:ethyl acetate 75:25, wash the organic layer with 1 mL 100 mM HCl. Remove a 4 mL aliquot of the organic layer and evaporate it to dryness under reduced pressure, reconstitute the residue in 100 µL 1 mg/mL (1S)-1-(4-dimethylaminonaphthalen-1-yl)ethylamine in dichloromethane:pyridine 10:1, add 20 µL 10 mg/mL 1-hydroxybenzotriazole in dichloromethane:pyridine 10:1, add 100 µL 2 mg/mL N,N'-dicyclohexylcarbodiimide, mix, let stand at room temperature for 1 h, evaporate to dryness, add 500 µL hexane:ethyl acetate 75:25, add 500 µL 100 mM HCl, vortex for 1 min. Remove the organic layer and dry it over anhydrous sodium sulfate, inject a 20 µL aliquot. (Synthesis of (1S)-1-(4-dimethylaminonaphthalen-1-yl)ethylamine is as follows. Mix 9.6 g (S)-(-)-1-(1-naphthyl)ethylamine ((1S)-1-(naphthalen-1-yl)ethylamine) with 37 mL acetic anhydride, let stand at room temperature for 30 min, cool to 12°, add 7.1 mL concentrated nitric acid dropwise, stir at room temperature for 1 h, adjust pH to 9 with 1 M NaOH, extract with ethyl acetate. Dry the organic layer over anhydrous sodium sulfate and evaporate to dryness under reduced pressure, chromatograph on a column of silica gel with ethyl

acetate:benzene 10:1 (Caution! Benzene is a carcinogen!). Evaporate the eluate to dryness under reduced pressure to give (1S)-1-(4-nitronaphthalen-1-yl)-N-acetylethylamine as colorless needles. Stir 100 mg 10% PdC and 2.5 g (1S)-1-(4-nitronaphthalen-1-yl)-N-acetylethylamine in 15 mL EtOH, add 1.5 mL hydrazine hydrate in portions (Caution! Hydrazine hydrate is a carcinogen and explodes on distillation in air!), add 100 mg 10% PdC, reflux for 1 h, cool, filter. Evaporate the filtrate to dryness under reduced pressure and chromatograph the residue on a column of silica gel with ethyl acetate, evaporate the eluate to dryness under reduced pressure to give (1S)-1-(4-aminonaphthalen-1-yl)-N-acetylethylamine as a colorless powder. Stir 1.5 g (1S)-1-(4-aminonaphthalen-1-yl)-N-acetylethylamine and 1.7 g sodium bicarbonate in 5 mL water at 0°, add 2 mL dimethyl sulfate (Caution! Dimethyl sulfate is highly toxic!), stir at room temperature for 3 h, distil to remove excess reagent. Concentrate the aqueous layer under reduced pressure, purify the residue by TLC using ethyl acetate (R_f 0.30) to obtain (1S)-1-(4-dimethylamino-naphthalen-1-yl)-N-acetylethylamine as a colorless powder. Reflux 1.2 g (1S)-1-(4-di-methylaminonaphthalen-1-yl)-N-acetylethylamine in 15 mL concentrated HCl for 35 h, adjust the pH to 9 with 2 M NaOH (Caution! Exothermic!), cool, extract with ethyl acetate. Dry the organic layer over anhydrous sodium sulfate and evaporate it to dryness under reduced pressure, purify by TLC using ethyl acetate to obtain (1S)-1-(4-dimethylamino-naphthalen-1-yl)ethylamine as a slightly yellow oil ($[\alpha]_D$ +19.75° (c = 2.0, EtOH).)

HPLC VARIABLES
Column: 300 × 3.9 μPorasil
Mobile phase: n-Hexane:ethyl acetate 68:32
Flow rate: 1.7
Injection volume: 20
Detector: F ex 313 em 420

CHROMATOGRAM
Retention time: 6 (S), 7 (R)
Limit of detection: 1 ng

OTHER SUBSTANCES
Extracted: metabolites

KEY WORDS
normal phase; chiral; rat; plasma; pharmacokinetics

REFERENCE
Nagashima, H.; Tanaka, Y.; Watanabe, H.; Hayashi, R.; Kawada, K. Optical inversion of (2R)- to (2S)-isomers of 2-[4-(2-oxocyclopentylmethyl)phenyl]propionic acid (loxoprofen), a new anti-inflammatory agent, and its monohydroxy metabolites in the rat, *Chem.Pharm.Bull.*, **1984**, *32*, 251–257.

SAMPLE
Matrix: solutions
Analyte: N-acetylaminoacids
Sample preparation: Mix 200 μL of a 2.5 mg/mL solution in pyridine or dichloromethane with 20 μL tri-n-butylamine, add 500 μg (1S)-1-(4-dimethylaminonaphthalen-1-yl)ethylamine, add 1 mg 1-(3-dimethylaminopropyl)-3-ethylcarbodiimide hydrochloride, let stand at room temperature for 3 h, extract with ethyl acetate, inject a 5 μL aliquot. (Synthesis of (1S)-1-(4-dimethylaminonaphthalen-1-yl)ethylamine is as described above.)

HPLC VARIABLES
Column: 305 × 6.4 μPorasil
Mobile phase: n-Hexane:ethyl acetate 25:75
Flow rate: 1 (A) or 3 (B)
Injection volume: 5
Detector: F ex 320 em 395

CHROMATOGRAM
Retention time: k' 11.20 (N-acetyl-L-alanine (B)), k' 13.40 (N-acetyl-D-alanine (B)), k' 6.93 (N-acetyl-L-α-aminobutyric acid (A)), k' 11.33 (N-acetyl-D-α-aminobutyric acid (A)), k' 3.20 (N-acetyl-L-valine (A)), k' 7.07 (N-acetyl-D-valine (A)), k' 4.00 (N-acetyl-L-norvaline (A)), k' 8.60 (N-acetyl-D-norvaline (A)), k' 2.73 (N-acetyl-L-leucine (A)), k' 7.00 (N-acetyl-D-

leucine (A)), k' 4.00 (N-acetyl-L-norleucine (A)), k' 7.73 (N-acetyl-D-norleucine (A)), k' 22.80 (N-acetyl-L-proline (B)), k' 34.00 (N-acetyl-D-proline (B)), k' 2.00 (N-acetyl-L-phenylglycine (A)), k' 4.63 (N-acetyl-D-phenylglycine (A)), k' 3.10 (N-acetyl-L-phenylalanine (A)), k'6.33 (N-acetyl-D-phenylalanine (A))

KEY WORDS
normal phase; chiral

REFERENCE
Goto, J.; Goto, N.; Hikichi, A.; Nishimaki, T.; Nambara, T. Sensitive derivatization reagents for the resolution of carboxylic acid enantiomers by high-performance liquid chromatography, *Anal.Chim.Acta*, **1980**, *120*, 187–192.

SAMPLE
Matrix: solutions
Analyte: carboxylic acids
Sample preparation: Mix 200 μL of a 2.5 mg/mL solution in pyridine or dichloromethane with 20 μL tri-n-butylamine, add 500 μg (1S)-1-(4-dimethylaminonaphthalen-1-yl)ethylamine, add 1 mg 1-(3-dimethylaminopropyl)-3-ethylcarbodiimide hydrochloride, let stand at room temperature for 3 h, extract with ethyl acetate, inject a 5 μL aliquot. (Synthesis of (1S)-1-(4-dimethylaminonaphthalen-1-yl)ethylamine is as described above.)

HPLC VARIABLES
Column: 305 × 6.4 μPorasil
Mobile phase: n-Hexane:ethyl acetate 100:30
Flow rate: 1
Injection volume: 5
Detector: F ex 320 em 395

CHROMATOGRAM
Retention time: k' 1.36 (D-ibuprofen), k' 2.05 (L-ibuprofen), k' 19.10 (D-indoprofen), k' 24.70 (L-indoprofen), k' 3.04 (D-naproxen), k' 4.37 (L-naproxen)

KEY WORDS
normal phase; chiral

REFERENCE
Goto, J.; Goto, N.; Hikichi, A.; Nishimaki, T.; Nambara, T. Sensitive derivatization reagents for the resolution of carboxylic acid enantiomers by high-performance liquid chromatography, *Anal.Chim.Acta*, **1980**, *120*, 187–192.

SAMPLE
Matrix: urine
Analyte: loxoprofen
Sample preparation: 1 mL Urine + 1 mL 1 M NaOH, mix, let stand at room temperature for 1 h, add 1.5 mL M HCl, add 10 mL mobile phase, shake for 5 min, centrifuge at 1000 g for 5 min. Remove a 1 mL aliquot of the organic layer and evaporate it to dryness under reduced pressure at 40°, reconstitute the residue in 100 μL 2 mg/mL (1S)-1-(4-dimethylaminonaphthalen-1-yl)ethylamine in dichloromethane:pyridine 10:1, add 20 μL 10 mg/mL 1-hydroxybenzotriazole in dichloromethane:pyridine 10:1, add 100 μL 2 mg/mL N,N'-dicyclohexylcarbodiimide, mix, let stand at room temperature for 1 h, evaporate to dryness, add 500 μL mobile phase, add 500 μL 100 mM HCl, vortex for 1 min. Remove the organic layer and dry it over anhydrous sodium sulfate, inject a 20 μL aliquot. (Synthesis of (1S)-1-(4-dimethylaminonaphthalen-1-yl)ethylamine is as described above.)

HPLC VARIABLES
Column: 300 × 3.9 10 μm μPorasil
Mobile phase: n-Hexane:ethyl acetate 68:32
Flow rate: 1.7
Injection volume: 20
Detector: F ex 313 em 420

CHROMATOGRAM
Retention time: 6 (S), 7 (R)

Limit of quantitation: 5 ng

OTHER SUBSTANCES
Extracted: metabolites

KEY WORDS
normal phase; chiral

REFERENCE
Nagashima, H.; Tanaka, Y.; Hayashi, R. Column liquid chromatography for the simultaneous determination of the enantiomers of loxoprofen sodium and its metabolites in human urine, *J.Chromatogr.*, **1985**, *345*, 373−379.

4-(N,N-Dimethylaminosulfonyl)-7-N-piperazino-2,1,3-benzoxadiazole

RELATED REFERENCES
Fukushima, T.; Santa, T.; Homma, H.; Al-kindy, S.M.; Imai, K. Enantiomeric separation and detection of 2-arylpropionic acids derivatized with [(N,N-dimethylamino)sulfonyl]benzofurazan reagents on a modified cellulose stationary phase by high-performance liquid chromatography. *Anal.Chem.* **1997**, *69*, 1793-1799.

Hasegawa, H.; Takahara, E.; Yamada, T.; Nagata, O. Sensitive analytical method for the novel H_1-receptor antagonist HSR-609 in human plasma and urine by high-performance liquid chromatography with pre-column fluorescent derivatization. *J.Chromatogr.B* **1996**, *687*, 419-425.

SAMPLE
Matrix: solutions
Analyte: carboxylic acids
Sample preparation: Sample + 400 μL 5 mM 4-(N,N-dimethylaminosulfonyl)-7-N-piperazino-2,1,3-benzoxadiazole + 70 mM diethylphosphorocyanidate in MeCN, react for 6 h, inject a 1 μL aliquot. (4-(N,N-Dimethylaminosulfonyl)-7-N-piperazino-2,1,3-benzoxadiazole (DBD-PZ) is available from Tokyo Kasei (TCI America, Portland OR). Synthesis is as follows. Dissolve 0.5 g magnesium sulfate heptahydrate and 6 g NaOH in 60 mL water, throughout the reaction keep the flask at about 20° with cold water cooling, add 15 mL 30% hydrogen peroxide, add 75 mL MeOH, add 12.1 g powdered benzoyl peroxide in one go, stir for 10 min, pour into 150 mL 20% sulfuric acid, extract three times with 50 mL portions of chloroform, determine peroxybenzoic acid concentration by iodometric titration (Tetrahedron 1967, 23, 3327). Slowly add 110 mL 1 M peroxybenzoic acid in chloroform to 7 g 2,6-difluoroaniline dissolved in 100 mL chloroform, stir at room temperature, when reaction is complete (iodometric titration) wash with 2% sodium thiosulfate, wash with 5% sodium carbonate, wash with water, dry over anhydrous sodium sulfate, evaporate to dryness under reduced pressure, recrystallize 2,6-difluoronitrosobenzene from EtOH (mp 108.5-109.5). Stir 8.5 g 2,6-difluoronitrosobenzene in 85 mL DMSO at room temperature

and add a solution of 3.91 g sodium azide in 85 mL DMSO dropwise, let stand for about 1 h, add to a large volume of water, extract with ether, dry the extracts over anhydrous sodium sulfate, evaporate to dryness under reduced pressure and distil to give 4-fluoro-2,1,3-benzoxadiazole as a colorless oil (bp 83°/12 mm Hg) (J.Chem.Soc.(C) 1970, 1433). Add 11 mL chlorosulfonic acid dropwise to 3 g 4-fluoro-2,1,3-benzoxadiazole in 10 mL chloroform at 0-10° (use a calcium chloride drying tube), stir at room temperature for 1 h, reflux for 2 h, cool, slowly pour into ice water, remove the organic layer, extract the aqueous layer with chloroform, combine the organic layers, wash, dry over anhydrous magnesium sulfate, evaporate under reduced pressure, take up the residue in 5 mL benzene (Caution! Benzene is a carcinogen!), chromatograph on a 150 × 30 column of silica gel (100-200 mesh Kanto Chemical) with n-hexane:benzene 50:50, evaporate the appropriate fractions to give 4-(chlorosulfonyl)-7-fluoro-2,1,3-benzoxadiazole (CBD-F) as pale yellow needles (mp 64-66°) (Anal. Chem. 1984. 56, 2461). Stir 0.76 g CBD-F in 70 mL MeCN at 0-10° and add 1 g dimethylamine hydrochloride in 10 mL 100 mM pH 10 borax dropwise, adjust pH to 5 with 1 M HCl, concentrate to about 10 mL under reduced pressure, extract three times with 200 mL portions of diethyl ether, wash with water, dry over anhydrous magnesium sulfate, evaporate under reduced pressure, chromatograph on a 500 × 20 column of silica gel with chloroform, isolate the appropriate fraction and re-chromatograph on the same column with ethyl acetate:benzene 1:2 to give 4-(N,N-dimethylaminosulfonyl)-7-fluoro-2,1,3-benzoxadiazole (DBD-F) as white needles (mp 124-125°) (yield 1% !). On a Merck no. 5714 60F$_{254}$ tlc plate eluted with chloroform DBD-F has Rf 0.32 and lies between two other reaction products (Analyst 1989, 114, 413). DBD-F can also be purchased from Tokyo Kasei. Add 123 mg 4-(N,N-dimethylaminosulfonyl)-7-fluoro-2,1,3-benzoxadiazole in 20 mL MeCN dropwise to 129 mg piperazine in 20 mL MeCN at room temperature, stir for 30 min, evaporate under reduced pressure, dissolve residue in 50 mL 5% HCl, wash three times with 20 mL ethyl acetate, discard ethyl acetate extracts, adjust pH of aqueous solution to 13-14 with 5% NaOH, extract five times with 50 mL ethyl acetate, combine extracts, wash with 20 mL water, dry over anhydrous sodium sulfate, evaporate under vacuum to give 4-(N,N-dimethylaminosulfonyl)-7-N-piperazino-2,1,3-benzoxadiazole as orange crystals (mp 121-2°).)

HPLC VARIABLES
Column: 150 × 4.6 5 μm Inertsil ODS-2
Mobile phase: MeCN:water 65:35 (A) or 50:50 (B) or 45:55 (C)
Column temperature: 40
Flow rate: 1
Injection volume: 1
Detector: F ex 437 em 561

CHROMATOGRAM
Retention time: 8 (indomethacin (A)), 10 (ibuprofen (A)), 10 (dehydrocholic acid (B)), 12 (prednisolone succinate (C)), 12 (dinoprost (C)), 13 (hydrocortisone succinate (C)), 6 (alprostadil (C)), 17 (ursodiol (B))
Limit of detection: 3.9-14 fmole

REFERENCE
Toyo'oka, T.; Ishibashi, M.; Takeda, Y.; Nakashima, K.; Akiyama, S.; Uzu, S.; Imai, K. Precolumn fluorescence tagging reagent for carboxylic acids in high-performance liquid chromatography: 4-substituted-7-aminoalkylamino-2,1,3-benzoxadiazoles, *J.Chromatogr.*, **1991**, *588*, 61–71.

SAMPLE
Matrix: solutions
Analyte: fatty acids
Sample preparation: Mix 200 μL of a 10 μM solution of fatty acids in DMF containing 140 mM diethylphosphorocyanidate with 200 μL 10 mM 4-(N,N-dimethylaminosulfonyl)-7-N-piperazino-2,1,3-benzoxadiazole in MeCN, let stand at room temperature for 6 h, inject a 1 μL aliquot. (4-(N,N-Dimethylaminosulfonyl)-7-N-piperazino-2,1,3-benzoxadiazole (DBD-PZ) is available from Tokyo Kasei (TCI America, Portland OR). Synthesis is as follows. Dissolve 0.5 g magnesium sulfate heptahydrate and 6 g NaOH in 60 mL water, throughout the reaction keep the flask at about 20° with cold water cooling, add 15 mL 30% hydrogen peroxide, add 75 mL MeOH, add 12.1 g powdered benzoyl peroxide in one

go, stir for 10 min, pour into 150 mL 20% sulfuric acid, extract three times with 50 mL portions of chloroform, determine peroxybenzoic acid concentration by iodometric titration (Tetrahedron 1967, 23, 3327). Slowly add 110 mL 1 M peroxybenzoic acid in chloroform to 7 g 2,6-difluoroaniline dissolved in 100 mL chloroform, stir at room temperature, when reaction is complete (iodometric titration) wash with 2% sodium thiosulfate, wash with 5% sodium carbonate, wash with water, dry over anhydrous sodium sulfate, evaporate to dryness under reduced pressure, recrystallize 2,6-difluoronitrosobenzene from EtOH (mp 108.5-109.5). Stir 8.5 g 2,6-difluoronitrosobenzene in 85 mL DMSO at room temperature and add a solution of 3.91 g sodium azide in 85 mL DMSO dropwise, let stand for about 1 h, add to a large volume of water, extract with ether, dry the extracts over anhydrous sodium sulfate, evaporate to dryness under reduced pressure and distil to give 4-fluoro-2,1,3-benzoxadiazole as a colorless oil (bp 83°/12 mm Hg) (J.Chem.Soc.(C) 1970, 1433). Add 11 mL chlorosulfonic acid dropwise to 3 g 4-fluoro-2,1,3-benzoxadiazole in 10 mL chloroform at 0-10° (use a calcium chloride drying tube), stir at room temperature for 1 h, reflux for 2 h, cool, slowly pour into ice water, remove the organic layer, extract the aqueous layer with chloroform, combine the organic layers, wash, dry over anhydrous magnesium sulfate, evaporate under reduced pressure, take up the residue in 5 mL benzene (Caution! Benzene is a carcinogen!), chromatograph on a 150 × 30 column of silica gel (100-200 mesh Kanto Chemical) with n-hexane:benzene 50:50, evaporate the appropriate fractions to give 4-(chlorosulfonyl)-7-fluoro-2,1,3-benzoxadiazole (CBD-F) as pale yellow needles (mp 64-66°) (Anal. Chem. 1984. 56, 2461). Stir 0.76 g CBD-F in 70 mL MeCN at 0-10° and add 1 g dimethylamine hydrochloride in 10 mL 100 mM pH 10 borax dropwise, adjust pH to 5 with 1 M HCl, concentrate to about 10 mL under reduced pressure, extract three times with 200 mL portions of diethyl ether, wash with water, dry over anhydrous magnesium sulfate, evaporate under reduced pressure, chromatograph on a 500 × 20 column of silica gel with chloroform, isolate the appropriate fraction and re-chromatograph on the same column with ethyl acetate:benzene 1:2 to give 4-(N,N-dimethylaminosulfonyl)-7-fluoro-2,1,3-benzoxadiazole (DBD-F) as white needles (mp 124-125°) (yield 1% !). On a Merck no. 5714 60F$_{254}$ tlc plate eluted with chloroform DBD-F has Rf 0.32 and lies between two other reaction products (Analyst 1989, 114, 413). DBD-F can also be purchased from Tokyo Kasei. Add 123 mg 4-(N,N-dimethylaminosulfonyl)-7-fluoro-2,1,3-benzoxadiazole in 20 mL MeCN dropwise to 129 mg piperazine in 20 mL MeCN at room temperature, stir for 30 min, evaporate under reduced pressure, dissolve residue in 50 mL 5% HCl, wash three times with 20 mL ethyl acetate, discard ethyl acetate extracts, adjust pH of aqueous solution to 13-14 with 5% NaOH, extract five times with 50 mL ethyl acetate, combine extracts,wash with 20 mL water, dry over anhydrous sodium sulfate, evaporate under vacuum to give 4-(N,N-dimethylaminosulfonyl)-7-N-piperazino-2,1,3-benzoxadiazole as orange crystals (mp 121-2°).)

HPLC VARIABLES

Column: 150 × 4.6 5 μm Inertsil ODS-2

Mobile phase: Gradient. MeCN:water from 70:30 to 98:2 over 1 h, maintain at 98:2 for 20 min.

Column temperature: 40

Flow rate: 1

Injection volume: 1

Detector: F ex 440 em 569

CHROMATOGRAM

Retention time: 15 (lauric acid), 19 (tridecanoic acid), 23 (myristic acid), 28 (pentadecanoic acid), 34 (palmitic acid), 40 (margaric acid), 45 (stearic acid), 50 (nonadecanoic acid), 55 (arachidic acid), 59 (heneicosanoic acid), 63 (behenic acid), 67 (tricosanoic acid), 71 (lignoceric acid)

Limit of detection: 3.2-4.7 fmole

REFERENCE

Toyo'oka, T.; Ishibashi, M.; Takeda, Y.; Nakashima, K.; Akiyama, S.; Uzu, S.; Imai, K. Precolumn fluorescence tagging reagent for carboxylic acids in high-performance liquid chromatography: 4-substituted-7-aminoalkylamino-2,1,3-benzoxadiazoles, *J.Chromatogr.*, **1991**, *588*, 61–71.

3,5-Dimethylaniline

SAMPLE
Matrix: bulk
Analyte: NSAIDs
Sample preparation: 10 mg Etodolac + 10 mg 1-(3-dimethylaminopropyl)-3-ethylcarbo-diimide hydrochloride + 2 drops 3,5-dimethylaniline + 1.5 mL dichloromethane, mix, after 30 min add 1 mL 1 M HCl, shake vigorously. Remove the lower organic layer and dry it over anhydrous magnesium sulfate, inject an aliquot.

HPLC VARIABLES
Column: 250 × 4.6 5 μm D N-(3,5-dinitrobenzoyl)phenylglycine (Regis)
Mobile phase: Hexane:isopropanol 80:20
Flow rate: 2
Injection volume: 20
Detector: UV 254; UV 280

CHROMATOGRAM
Retention time: k' 6.90, α = 1.09 (carprofen); k' 6.00, α = 1.22 (cicloprofen); k' 1.33, α = 1.35 (etodolac); k' 2.50, α = 1.25 (fenoprofen); k' 2.27, α = 1.26 (flurbiprofen); k' 1.23, α = 1.30 (ibuprofen); k' 4.97, α = 1.18 (ketoprofen); k' 9.10, α = 1.17 (naproxen); k' 4.30, α = 1.27 (pirprofen); k' 7.50, α = 1.13 (tiaprofenic acid) (k' for first enantiomer)

KEY WORDS
chiral

REFERENCE
Pirkle, W.H.; Murray, P.G. The separation of the enantiomers of a variety of non-steroidal anti-inflamatory drugs (NSAIDS) as their anilide derivatives using a chiral stationary phase, *J.Liq.Chromatogr.*, **1990**, *13*, 2123–2134.

Diphenylamine

SAMPLE
Matrix: bulk
Analyte: 2-(4-chloro-2-methylphenoxy)propanoic acid

Sample preparation: 500 mg 2-(4-Chloro-2-methylphenoxy)propanoic acid + 300 mg thionyl chloride, heat on steam bath for 1 h, evaporate to dryness under reduced pressure, reconstitute with chloroform, add 2.4 mmoles diphenylamine dissolved in 500 μL chloroform, shake regularly for 1 h, evaporate to dryness, dissolve in 5 mL EtOH:water 99.8:0.2, filter, evaporate to 1 mL, add 1 drop water, cool in the refrigerator to obtain the derivative.

HPLC VARIABLES
Column: 300 mm long N-(3,5-dinitrobenzoyl)-(R)-(-)-phenylglycine bonded to gamma-aminopropyl silica (Pirkle column)
Mobile phase: Hexane:isopropanol 90:10
Flow rate: 1
Detector: UV 260

CHROMATOGRAM
Retention time: 11 (+), 12 (-)

KEY WORDS
chiral

REFERENCE
Blessington, B.; Crabb, N.; O'Sullivan, J. Chiral high-performance liquid chromatographic studies of 2-(4-chloro-2-methylphenoxy)propanoic acid, *J.Chromatogr.*, **1987**, *396*, 177–182.

2-Ferrocenylethylamine

SAMPLE
Matrix: bulk
Analyte: estrogen glucuronides
Sample preparation: Dissolve 100 μg estrogen glucuronide in 50 μg pyridine and 400 μL chloroform, add 100 μL 5 mg/mL 2-ferrocenylethylamine in dichloromethane, add 400 μL 1 mg/mL 1-ethyl-3-(3-dimethylaminopropyl)carbodiimide in dichloromethane, add 20 μL 1 mg/mL 1-hydroxybenzotriazole in pyridine, heat at 37° overnight, extract with ethyl acetate. Wash the organic layer with 5% HCl, with water, with 5% sodium bicarbonate, and with water then evaporate it to dryness, reconstitute, inject an aliquot. (Prepare 2-ferrocenylethylamine as follows. Reflux 20 g lithium aluminum hydride in 1 L ether for 1 h, add a solution of 77 g ferroceneacetonitrile in 500 mL ether at such a rate as to produce gentle refluxing, reflux for 2 h, cool in ice, add 20 mL water, add 15 mL 20% NaOH in water, add 90 mL water, decant the organic layer, wash the solid several times with ether. Combine the ether layers and saturate them with hydrogen chloride, decant the ether under nitrogen, add the solid (soaked in ether) to 2 M NaOH, extract with ether, dry the ether layer over anhydrous sodium sulfate, evaporate under reduced pressure, distil at 118-120/0.5 mm Hg to obtain 2-ferrocenylethylamine as a dark brown oil (J. Org. Chem. 1958, 23, 653). The hydrochloride can also be isolated as red prisms (mp 200-205° d) after recrystallization from MeOH/acetone (J. Pharm. Biomed. Anal. 1987, 5, 361).)

HPLC VARIABLES
Column: 150 × 4 Develosil ODS-5
Mobile phase: MeCN:water 50:60 containing 50 mM sodium perchlorate, pH adjusted to 3.0 with phosphoric acid

Flow rate: 1
Injection volume: 2000
Detector: E, Yanagimoto Model VMD 101, +0.5 V, Ag/AgCl reference electrode

CHROMATOGRAM
Retention time: k' 7.00 (2-hydroxyestrone 3-glucuronide), k' 7.80 (4-hydroxyestrone 3-glucuronide), k' 8.40 (2-hydroxyestrone 2-glucuronide), k' 10.00 (4-hydroxyestrone 4-glucuronide)

REFERENCE
Shimada, K.; Xie, F.; Niwa, T.; Wakasawa, T.; Nambara, T. Studies on steroids. CCXXIX. Separation and characterization of catechol oestrogen glucuronides in urine of pregnant women by high-performance liquid chromatography, *J.Chromatogr.*, **1987**, *400*, 215–221.

SAMPLE
Matrix: urine
Analyte: estrogen glucuronides
Sample preparation: Dilute 100 µL urine with 5 mL water, add to a Sep-Pak C18 SPE cartridge, elute with 5 mL MeOH:water 50:50. Evaporate the eluate to dryness and reconstitute with 40 µL pyridine, add 50 µL 1 mg/mL 2-ferrocenylethylamine in chloroform, add 400 µL 1 mg/mL 1-ethyl-3-(3-dimethylaminopropyl)carbodiimide in chloroform, add 25 µL 400 µg/mL 1-hydroxybenzotriazole in pyridine:chloroform 40:60, heat at 37° for 2 h, extract with ethyl acetate. Wash the organic layer with 5% HCl, with 5% aqueous sodium bicarbonate, and with water. Evaporate the organic layer to dryness and reconstitute with 2 mL benzene:ethyl acetate:MeOH 2:1:0.1 (Caution! Benzene is a carcinogen!), add to a 30 × 5 silica gel column, wash with 3 mL benzene, wash with 3 mL ethyl acetate, elute with 3 mL ethyl acetate:MeOH 80:20, inject an aliquot of the eluate. (Prepare 2-ferrocenylethylamine as follows. Reflux 20 g lithium aluminum hydride in 1 L ether for 1 h, add a solution of 77 g ferroceneacetonitrile in 500 mL ether at such a rate as to produce gentle refluxing, reflux for 2 h, cool in ice, add 20 mL water, add 15 mL 20% NaOH in water, add 90 mL water, decant the organic layer, wash the solid several times with ether. Combine the ether layers and saturate them with hydrogen chloride, decant the ether under nitrogen, add the solid (soaked in ether) to 2 M NaOH, extract with ether, dry the ether layer over anhydrous sodium sulfate, evaporate under reduced pressure, distil at 118-120/0.5 mm Hg to obtain 2-ferrocenylethylamine as a dark brown oil (J. Org. Chem. 1958, 23, 653). The hydrochloride can also be isolated as red prisms (mp 200-205° d) after recrystallization from MeOH/acetone.)

HPLC VARIABLES
Column: 150 × 4 5 µm TSKgel ODS-80TM
Mobile phase: MeCN:MeOH:water 4:16:80 containing 50 mM sodium perchlorate
Flow rate: 1
Detector: E, Yanagimoto VMD-501, +0.45 V, Ag/AgCl reference electrode

CHROMATOGRAM
Retention time: 6.5 (estriol 16-glucuronide), 10 (estradiol 3-glucuronide), 11 (estrone 3-glucuronide), 16.5 (estradiol 17-glucuronide)
Internal standard: 16-epiestriol 17-glucuronide (7)
Limit of detection: 0.5 pmole

KEY WORDS
SPE

REFERENCE
Shimada, K.; Nagashima, E.; Orii, S.; Nambara, T. New derivatization method using ferrocene reagents for the determination of steroid glucuronides by high performance liquid chromatography with electrochemical detection, *J.Pharm.Biomed.Anal.*, **1987**, *5*, 361–368.

S-FLOPA

SAMPLE
Matrix: blood, urine
Analyte: beclobrate (as beclobric acid)
Sample preparation: 200 μL Plasma or 100 μL urine + 500 (plasma) or 300 (urine) mg
NaCl + 50 μL 1 μg/mL clofibric acid in MeOH + 1 (plasma) or 0.3 (urine) mL pH 4 buffer
+ 5 mL n-hexane:EtOH 90:10, shake horizontally for 10 min, centrifuge. Remove 4 mL
of the organic layer and evaporate it to dryness under a stream of nitrogen at 55°, add
50 μL toluene and evaporate it to remove traces of water. Reconstitute the residue in 500
μL dichloromethane, add 50 μL 1 mg/mL 1-hydroxybenzotriazole in dichloromethane:
pyridine 99:1, add 50 μL 1 mg/mL 1-(3-dimethylaminopropyl)-3-ethylcarbodiimide hydrochloride in dichloromethane, add 50 μL 1 mg/mL FLOPA, vortex, let stand at room temperature for 2 h, evaporate to dryness, reconstitute in 500 μL mobile phase, inject a 50
μL aliquot. (FLOPA ((S)-(+)-5-(1-aminoethyl)-2-(4-fluorophenyl)benzoxazole) is the corresponding amine hydrochloride from (+)-(S)-flunoxaprofen. Synthesis is as follows (protect
from light). Take up 500 mg (+)-(S)-flunoxaprofen in 50 mL dry toluene, slowly add 5 mL
freshly distilled thionyl chloride, reflux for 1 h, evaporate to dryness under vacuum, dry
the acyl chloride under vacuum over KOH for 2 days. Dissolve 0.5 mmoles acyl chloride
in 5 mL acetone, add 600 mg sodium azide dissolved in ice water with stirring, stir at 0°
for 30 min, add 10 mL ice-cold water, filter, dry solid in a desiccator under vacuum.
Dissolve the solid in 1 mL toluene or dichloromethane (dried over 3 Å molecular sieve),
reflux for 10 min, evaporate, store resulting isocyanate under vacuum over a desiccant.
Dissolve 0.5 mmole isocyanate in 5 mL acetone, add 20 mL 8.5% phosphoric acid, heat
to 80° for 1.5 h, adjust to pH 13, extract with diethyl ether:dichloromethane 4:1. Wash
the organic layer twice with water, dry over anhydrous sodium sulfate, evaporate to dryness, dissolve in 1 mL toluene, evaporate to give crystals (mp 91°). Dissolve in ether, add
0.5 M HCl in ether, filter, dissolve solid in a small volume of MeOH, precipitate with
ether, dry FLOPA over phosphorus pentoxide under vacuum (Pharm.Res. 1990, 7, 1262).)

HPLC VARIABLES
Column: 250 × 4.6 5 μm Zorbax Sil
Mobile phase: Gradient. A was n-hexane:chloroform:EtOH 100:10:0.75. B was n-hex-
ane:chloroform:EtOH 100:10:20. A:B 100:0 for 10 min, 50:50 for 5 min, 100:0 for 5
min (stepwise).
Flow rate: 2
Injection volume: 50
Detector: F ex 305 em 355

CHROMATOGRAM
Retention time: 6 (-), 6.5 (+)
Internal standard: clofibric acid (8)
Limit of detection: 25 ng/mL

KEY WORDS
plasma; pharmacokinetics; chiral; normal phase

REFERENCE
Mayer, S.; Mutschler, E.; Spahn-Langguth, H. Pharmacokinetic studies with the lipid-regulating agent beclobrate: enantiospecific assay for beclobric acid using a new fluorescent chiral coupling component (S-FLOPA), *Chirality*, **1991**, *3*, 35–42.

SAMPLE
Matrix: blood, urine

Analyte: α-phenylcyclopentylacetic acid

Sample preparation: Plasma. 4 mL Plasma + 500 mg NaCl + 50 μL 1 μg/mL (R)-(-)-ibuprofen in MeOH + 250 μL 1 M HCl + 5 mL n-hexane:EtOH 90:10, shake horizontally for 10 min, centrifuge. Remove 4 mL of the organic layer and evaporate it to dryness under a stream of nitrogen at 55°, add 50 μL toluene, evaporate, reconstitute the residue in 500 μL dichloromethane, add 50 μL 1 mg/mL 1-hydroxybenzotriazole in dichloromethane:pyridine 99:1, add 50 μL 1 mg/mL 1-(3-dimethylaminopropyl)-3-ethyl-carbodiimide hydrochloride in dichloromethane, add 50 μL 1 mg/mL S-FLOPA in dichloromethane, vortex, let stand at room temperature for 2 h, evaporate to dryness, reconstitute with 500 μL mobile phase, inject a 50 μL aliquot. Urine. 1 mL Urine + 2 mL pH 4 buffer + 50 ng (R)-(-)-ibuprofen + 500 mg NaCl + 5 mL n-hexane:EtOH 90:10, shake horizontally for 10 min, centrifuge. Remove 4 mL of the organic layer and evaporate it to dryness under a stream of nitrogen at 55°, add 50 μL toluene, evaporate, reconstitute the residue in 500 μL dichloromethane, add 50 μL 1 mg/mL 1-hydroxybenzotriazole in dichloromethane:pyridine 99:1, add 50 μL 1 mg/mL 1-(3-dimethylaminopropyl)-3-ethyl-carbodiimide hydrochloride in dichloromethane, add 50 μL 1 mg/mL S-FLOPA in dichloromethane, vortex, let stand at room temperature for 2 h, evaporate to dryness, reconstitute with 500 μL mobile phase, inject a 50 μL aliquot. (S-FLOPA is the corresponding amine hydrochloride from (S)-(+)-flunoxaprofen. Synthesis is as follows (protect from light). Take up 500 mg (S)-(+)-flunoxaprofen in 50 mL dry toluene, slowly add 5 mL freshly distilled thionyl chloride, reflux for 1 h, evaporate to dryness under vacuum, dry the acyl chloride under vacuum over KOH for 2 days. Dissolve 0.5 mmoles acyl chloride in 5 mL acetone, add 600 mg sodium azide dissolved in ice water with stirring, stir at 0° for 30 min, add 10 mL ice-cold water, filter, dry the solid in a desiccator under vacuum. Dissolve the solid in 1 mL toluene or dichloromethane (dried over 3 Å molecular sieve), reflux for 10 min, evaporate, store resulting isocyanate under vacuum over a desiccant. Dissolve 0.5 mmole isocyanate in 5 mL acetone, add 20 mL 8.5% phosphoric acid, heat to 80° for 1.5 h, adjust to pH 13, extract with diethyl ether:dichloromethane 4:1. Wash the organic layer twice with water, dry over anhydrous sodium sulfate, evaporate to dryness, dissolve in 1 mL toluene, evaporate to give crystals (mp 91°). Dissolve in ether, add 0.5 M HCl in ether, filter, dissolve solid in a small volume of MeOH, precipitate with ether, dry S-FLOPA over phosphorus pentoxide under vacuum (Pharm. Res. 1990, 7, 1262).)

HPLC VARIABLES
Column: 250 × 4.6 5 μm Zorbax Sil

Mobile phase: Gradient. A was n-hexane:chloroform:EtOH 100:10:1. B was n-hexane:chloroform:EtOH 100:10:20. A:B 100:0 for 25 min, 50:50 for 5 min (step gradient), 0:100 for 5 min.

Flow rate: 2

Injection volume: 50

Detector: F ex 305 em 355

CHROMATOGRAM
Retention time: 10 (-), 11 (+)

Internal standard: (R)-(-)-ibuprofen (16)

Limit of quantitation: 1 ng/mL (plasma), 10 ng/mL (urine)

KEY WORDS
chiral; normal phase; plasma; pharmacokinetics; α-phenylcyclopentylacetic acid is a metabolite of ciclotropium bromide

REFERENCE

Liebmann, B.; Mayer, S.; Mutschler, E.; Spahn-Langguth, H. Studies on the metabolic clearance of ciclotropium to α-phenylciclopentylacetic acid using a new enantiospecific metabolite assay, *Arzneimittelforschung*, **1992**, *42*, 1354–1358.

SAMPLE

Matrix: urine

Analyte: ibuprofen

Sample preparation: 100 μL Urine + 200 μL pH 5 citrate/NaOH buffer + 500 ng clofibric acid + 5 mL diethyl ether:dichloromethane 80:20, shake for 10 min, centrifuge. Remove the organic layer and evaporate it to dryness, add 200 μL toluene and evaporate it to remove traces of water. Reconstitute the residue in 500 μL dichloromethane, add 50 μL 1 mg/mL 1-hydroxybenzotriazole in dichloromethane:pyridine 99:1, add 50 μL 1 mg/mL 1-(3-dimethylaminopropyl)-3-ethylcarbodiimide hydrochloride in dichloromethane, add 50 μL 1 mg/mL S-FLOPA in dichloromethane, vortex, let stand at room temperature for 2 h, evaporate to dryness, reconstitute in mobile phase, inject a 10-20 μL aliquot. (To hydrolyse glucuronides add 100 μL 1 M NaOH to 100 μL urine, let stand for 1 h, add 100 μL 1 M HCl, proceed as above.) (S-FLOPA ((S)-(+)-5-(1-aminoethyl)-2-(4-fluorophenyl)benzoxazole) is the corresponding amine hydrochloride from (+)-(S)-flunoxaprofen. Synthesis is as follows (protect from light). Take up 500 mg (+)-(S)-flunoxaprofen in 50 mL dry toluene, slowly add 5 mL freshly distilled thionyl chloride, reflux for 1 h, evaporate to dryness under vacuum, dry the acyl chloride under vacuum over KOH for 2 days. Dissolve 0.5 mmoles acyl chloride in 5 mL acetone, add 600 mg sodium azide dissolved in ice water with stirring, stir at 0° for 30 min, add 10 mL ice-cold water, filter, dry solid in a desiccator under vacuum. Dissolve the solid in 1 mL toluene or dichloromethane (dried over 3 Å molecular sieve), reflux for 10 min, evaporate, store resulting isocyanate under vacuum over a desiccant. Dissolve 0.5 mmole isocyanate in 5 mL acetone, add 20 mL 8.5% phosphoric acid, heat to 80° for 1.5 h, adjust to pH 13, extract with diethyl ether:dichloromethane 4:1. Wash the organic layer twice with water, dry over anhydrous sodium sulfate, evaporate to dryness, dissolve in 1 mL toluene, evaporate to give crystals (mp 91°). Dissolve in ether, add 0.5 M HCl in ether, filter, dissolve solid in a small volume of MeOH, precipitate with ether, dry S-FLOPA over phosphorus pentoxide under vacuum.)

HPLC VARIABLES

Column: 250 × 4.6 5 μm Zorbax Sil

Mobile phase: n-Hexane:chloroform:EtOH 100:10:1.25

Flow rate: 2

Injection volume: 10-20

Detector: F ex 305 em 355

CHROMATOGRAM

Retention time: 6.5 (R-(-)), 13.5 (R-(+))

Internal standard: clofibric acid (5)

KEY WORDS

pharmacokinetics; chiral; normal phase

REFERENCE

Spahn, H.; Langguth, P. Chiral amines derived from 2-arylpropionic acids: novel reagents for the liquid chromatographic (LC) fluorescence assay of optically active carboxylic acid xenobiotics, *Pharm.Res.*, **1990**, *7*, 1262–1268.

L-Leucinamide

RELATED REFERENCES

Spahn, H. Formation of diastereomeric derivatives of 2-arylpropionic acids using L-leucinamide. *J.Chromatogr.* **1987**, *423*, 334-339.

Spahn, H.; Spahn, I.; Pflugmann, G.; Mutschler, E.; Benet, L.Z. Measurement of carprofen enantiomer concentrations in plasma and urine using L-leucinamide as the chiral coupling component. *J.Chromatogr.* **1988**, *433*, 331-338.

SAMPLE

Matrix: blood

Analyte: indobufen and indoprofen

Sample preparation: 0.5-1 mL Plasma + 500 µL water + 100 µL 600 mM sulfuric acid + 40 mg NaCl + 4 mL diethyl ether, extract on a rotamixer, centrifuge at 1200 g. Remove the organic layer and evaporate it to dryness under a stream of air at 30°, add 5 drops toluene, evaporate to dryness under a stream of air at 30°, reconstitute the residue in 200 µL 50 mM triethylamine in MeCN, add 100 µL 60 mM ethyl chloroformate in MeCN, after 30 s add 100 µL 1 M L-leucinamide hydrochloride in MeOH containing 1 M triethylamine, after 2 min add 500 µL 250 mM HCl, extract with 4 mL ethyl acetate. Evaporate the organic layer to dryness under a stream of air at 30°, reconstitute the residue with 500 µL MeOH, inject a 10 µL aliquot.

HPLC VARIABLES

Guard column: 30 × 4 Perisorb RP-18 (Merck)

Column: 250 × 4 7 µm LiChroCart RP-18 (Merck)

Mobile phase: MeCN:10 mM pH 6.5 phosphate buffer 38:62

Flow rate: 2

Injection volume: 10

Detector: UV 275

CHROMATOGRAM

Retention time: 4.5 ((-)-indoprofen), 5 ((+)-indoprofen), 6 ((-)-indobufen), 8 ((+)-indobufen)

KEY WORDS

plasma; chiral

REFERENCE

Björkman, S. Determination of the enantiomers of indoprofen in blood plasma by high-performance liquid chromatography after rapid derivatization by means of ethyl chloroformate, *J.Chromatogr.*, **1985**, *339*, 339–346.

SAMPLE

Matrix: blood

Analyte: ketoprofen

Sample preparation: 0.5-1 mL Plasma + 500 µL 4-8 µg/mL IS in water + 500 µL buffer + 4 mL dichloromethane:n-propanol 99:1, extract on a rotamixer, centrifuge at 1200 g. Remove the organic layer and evaporate it to dryness under a stream of air at 30°, add 5 drops toluene, evaporate to dryness under a stream of air at 30°, reconstitute the residue in 200 µL 50 mM triethylamine in MeCN, add 100 µL 60 mM ethyl chloroformate in MeCN, after 30 s add 100 µL 1 M L-leucinamide hydrochloride in MeOH containing 1 M triethylamine, after 2 min add 500 µL 250 mM HCl, extract with 4 mL ethyl acetate. Evaporate the organic layer to dryness under a stream of air at 30°, reconstitute the residue with 100 µL MeCN, add 400 µL 10 mM pH 6.5 phosphate buffer, inject a 60 µL aliquot. (Prepare buffer as follows. Neutralize a 1 M solution of tetrabutylammonium sulfate in water with NaOH, wash 5 times with dichloromethane, wash twice with heptane. Prepare a 100 mM pH 9.6 sodium carbonate buffer containing 0.5 M of the neutralized and washed tetrabutylammonium salt.)

HPLC VARIABLES
Guard column: 30 × 4 Perisorb RP-18 (Merck)
Column: 250 × 4 7 μm LiChroCart RP-18 (Merck)
Mobile phase: MeCN:10 mM pH 6.5 phosphate buffer 38:62
Flow rate: 2
Injection volume: 10
Detector: UV 260

CHROMATOGRAM
Retention time: 8.5 (-), 10 (+)
Internal standard: 2-(4-benzoylphenyl)butyric acid (12, 15 (enantiomers))
Limit of quantitation: 250 ng/mL

KEY WORDS
plasma; chiral; pharmacokinetics

REFERENCE
Björkman, S. Determination of the enantiomers of ketoprofen in blood plasma by ion-pair extraction and high-performance liquid chromatography of leucinamide derivatives, *J.Chromatogr.*, **1987**, *414*, 465−471.

SAMPLE
Matrix: blood
Analyte: indobufen
Sample preparation: 0.5-1 mL Plasma + 100 μL 600 mM sulfuric acid + 40 mg NaCl + 4 mL diethyl ether, extract by rotation, centrifuge at 1200 g. Remove the organic layer and evaporate it to dryness under a stream of air at 30°, add 5 drops of toluene, evaporate to dryness under a stream of air. Reconstitute the residue in 200 μL 50 mM triethylamine in MeCN, add 100 μL 60 mM ethyl chloroformate in MeCN, let stand for 30 s, add 100 μL 1 M L-leucinamide hydrochloride in MeOH containing 1 M triethylamine, let stand for 2 min, add 500 μL 250 mM HCl, extract with 4 mL ethyl acetate. Remove the organic layer and evaporate it to dryness under a stream of air at 30°, reconstitute in 500 μL MeOH, inject a 10 μL aliquot (J.Chromatogr. 1985, 339, 339).

HPLC VARIABLES
Column: 300 × 3.9 10 μm μBondapak C18
Mobile phase: MeCN:pH 6.4 phosphate buffer 40:60
Flow rate: 1.5
Injection volume: 10
Detector: UV 275

CHROMATOGRAM
Retention time: 6.977 (R), 8.848 (S)

KEY WORDS
chiral; plasma

REFERENCE
Perrone, G.; Farina, M. High-performance liquid chromatographic method for direct resolution of the indobufen enantiomeric components, *J.Chromatogr.*, **1990**, *520*, 373−378.

SAMPLE
Matrix: blood
Analyte: tiaprofenic acid
Sample preparation: 500 μL Plasma + 50 μL 100 μg/mL ketorolac + 100 μL 600 mM sulfuric acid + 3 mL isooctane:isopropanol 95:5, vortex for 30 s, centrifuge at 1800 g for 5 min. Remove the organic layer and evaporate it to dryness, reconstitute the residue in 100 μL 2 mg/mL 4-(dimethylamino)pyridine in MeCN, add 100 μL 60 mM trichloroethyl chloroformate in MeCN, add 1 M L-leucinamide in MeCN, let stand for 2 min, add 500

μL 250 mM HCl, extract with chloroform. Remove the organic layer and evaporate it to dryness, reconstitute the residue in mobile phase, inject a 10-100 μL aliquot. (A 7% conversion of S to R is observed during the derivatization procedure. No racemization is observed using a direct procedure with a chiral column.)

HPLC VARIABLES
Column: 100 × 4.6 5 μm Partisil 5 ODS-2
Mobile phase: MeCN:60 mM KH_2PO_4:triethylamine 30:70:0.02
Flow rate: 1
Injection volume: 10-100
Detector: UV 310

CHROMATOGRAM
Retention time: 17.2 (R), 19.5 (S)
Internal standard: ketorolac (10.7 (R), 19.5 (S))
Limit of quantitation: 50 ng/mL

KEY WORDS
chiral; plasma; comparison with a method involving a chiral column; pharmacokinetics

REFERENCE
Vakily, M.; Jamali, F. Pharmacokinetics of tiaprofenic acid in humans: Lack of stereoselectivity in plasma using both direct and precolumn derivatization methods, *J.Pharm.Sci.*, **1996**, *85*, 638–642.

SAMPLE
Matrix: blood, urine
Analyte: ketoprofen
Sample preparation: Plasma. 500 μL Plasma + 100 μL 100 μg/mL calcium fenoprofen in water + 100 μL 600 mM sulfuric acid + 3 mL isooctane:isopropanol 95:5, vortex for 30 s, centrifuge at 1800 RCF for 5 min. Remove the organic layer and add it to 3 mL water, vortex for 30 s, centrifuge for 3 min. Remove the aqueous layer and add it to 200 μL 600 mM sulfuric acid, add 3 mL chloroform, vortex for 30 s, centrifuge for 3 min. Remove the organic layer and evaporate it to dryness under reduced pressure, reconstitute with 100 μL 50 mM triethylamine in MeCN, add 50 μL 60 mM ethyl chloroformate in MeCN, after 30 s add 50 μL 1 M L-leucinamide hydrochloride in MeOH containing 1 M triethylamine, after 2 min add 50 μL water, inject a 10-40 μL aliquot. Urine. 100-500 μL Urine + 25-125 μL 1 M NaOH, mix, add a volume of 600 mM sulfuric acid equal to the volume of 1 M NaOH plus 100 μL, add 100 μL 100 μg/mL calcium fenoprofen in water, add 3 mL isooctane:isopropanol 95:5, vortex for 30 s, centrifuge at 1800 RCF for 5 min. Remove the organic layer and add it to 3 mL water, vortex for 30 s, centrifuge for 3 min. Remove the aqueous layer and add it to 200 μL 600 mM sulfuric acid, add 3 mL chloroform, vortex for 30 s, centrifuge for 3 min. Remove the organic layer and evaporate it to dryness under reduced pressure, reconstitute with 100 μL 50 mM triethylamine in MeCN, add 50 μL 60 mM ethyl chloroformate in MeCN, after 30 s add 50 μL 1 M L-leucinamide hydrochloride in MeOH containing 1 M triethylamine, after 2 min add 50 μL water, inject a 10-40 μL aliquot.

HPLC VARIABLES
Guard column: 50 mm long 37-53 μm C18
Column: 100 mm long Partisil 5 ODS-3
Mobile phase: MeCN:60 mM KH_2PO_4:triethylamine 64:36:0.02
Flow rate: 1
Injection volume: 10-40
Detector: UV 275

CHROMATOGRAM
Retention time: 9.8 (R-(-)), 11.3 (S-(+))
Internal standard: fenoprofen (17.7 (R-(-)), 19.9 (S-(+)))
Limit of quantitation: 50 ng/mL

OTHER SUBSTANCES
Simultaneously analyzed: flurbiprofen
Interfering: naproxen

KEY WORDS
plasma; chiral

REFERENCE
Foster, R.T.; Jamali, F. High-performance liquid chromatographic assay of ketoprofen enantiomers in human plasma and urine, *J.Chromatogr.*, **1987**, *416*, 388–393.

SAMPLE
Matrix: blood, urine

Analyte: fenoprofen

Sample preparation: Plasma. 500 μL Plasma + 50 μL 50 μg/mL ketoprofen in 10 mM NaOH + 100 μL 600 mM sulfuric acid + 3 mL isooctane:isopropanol 95:5, vortex for 30 s, centrifuge at 3000 rpm for 5 min. Remove the organic layer and add it to 2.5 mL water, vortex for 15 s, centrifuge for 3 min. Remove the aqueous layer and add it to 200 μL 600 mM sulfuric acid, add 2.5 mL chloroform, vortex for 15 s, centrifuge for 3 min. Remove the organic layer and evaporate it to dryness under reduced pressure, reconstitute with 100 μL 50 mM triethylamine in MeCN, add 50 μL 60 mM ethyl chloroformate in MeCN, after 30 s add 50 μL 1 M L-leucinamide in MeOH containing 1 M triethylamine, after 2 min add 50 μL water, inject a 10-40 μL aliquot. Urine. 500 μL Urine + 250 μL 1 M NaOH, mix, add 300 μL 600 mM sulfuric acid, add 50 μL 50 μg/mL ketoprofen in 10 mM NaOH, add 3 mL isooctane:isopropanol 95:5, vortex for 30 s, centrifuge at 3000 rpm for 5 min. Remove the organic layer and add it to 2.5 mL water, vortex for 15 s, centrifuge for 3 min. Remove the aqueous layer and add it to 200 μL 600 mM sulfuric acid, add 2.5 mL chloroform, vortex for 15 s, centrifuge for 3 min. Remove the organic layer and evaporate it to dryness under reduced pressure, reconstitute with 100 μL 50 mM triethylamine in MeCN, add 50 μL 60 mM ethyl chloroformate in MeCN, after 30 s add 50 μL 1 M L-leucinamide in MeOH containing 1 M triethylamine, after 2 min add 50 μL water, inject a 10-40 μL aliquot.

HPLC VARIABLES
Guard column: 20 mm long 37-53 μm reverse phase

Column: 100 × 4.6 5 μm Partisil 5 ODS-3

Mobile phase: MeCN:70 mM KH_2PO_4:triethylamine 65:35:0.02, pH 6.0

Flow rate: 1 (plasma), 1.2 (urine)

Injection volume: 10-40

Detector: UV 275 for 13 min then UV 232

CHROMATOGRAM
Retention time: 16.3 (R, urine), 19.1 (R, plasma or S, urine), 22.0 (S, plasma)

Internal standard: ketoprofen (9, 10 (enantiomers))

Limit of quantitation: 250 ng/mL

KEY WORDS
plasma; pharmacokinetics; chiral

REFERENCE
Mehvar, R.; Jamali, F. Stereospecific high-performance liquid chromatographic (HPLC) assay of fenoprofen enantiomers in plasma and urine, *Pharm.Res.*, **1988**, *5*, 53–56.

SAMPLE
Matrix: blood, urine

Analyte: flurbiprofen

Sample preparation: 500 μL Urine + 100 μL 1 M NaOH (to hydrolyze conjugates), vortex. 500 μL Plasma or 600 μL basified urine + 200 μL 600 mM sulfuric acid, vortex, add 50 μL 100 μg/mL ketoprofen in water + 3 mL isooctane:isopropanol 96:5, mix vigorously for 45 s, centrifuge at 3000 rpm for 5 min. Remove the top layer and add it to 3 mL water, mix vigorously, centrifuge for 5 min, discard the organic layer. Add the aqueous layer to 350 μL 600 mM sulfuric acid, add 3 mL chloroform, mix, centrifuge for 5 min. Remove the organic layer and evaporate it to dryness under reduced pressure, reconstitute the residue in 100 μL 50 mM triethylamine in MeCN, after 30 s add 50 μL 60 mM ethyl chloroformate in MeCN, after 30 s add 50 μL 1 M L-leucinamide, after 2 min add 50 μL water, inject a 10-50 μL aliquot.

HPLC VARIABLES
Guard column: 50 mm long 10 μm Partisil 5 ODS-3
Column: 100 × 4.6 5 μm Partisil 5 ODS-3
Mobile phase: MeCN:67 mM KH_2PO_4:triethylamine 35:65:0.02
Flow rate: 1
Injection volume: 10-50
Detector: UV 250

CHROMATOGRAM
Retention time: 17 (-), 21 (+)
Internal standard: ketoprofen (UV 275) (8 (-), 10 (+))
Limit of quantitation: 100 ng/mL (plasma), 250 ng/mL (urine)

KEY WORDS
plasma; pharmacokinetics; chiral

REFERENCE
Berry, B.W.; Jamali, F. Stereospecific high-performance liquid chromatographic (HPLC) assay of flurbiprofen in biological specimens, *Pharm.Res.*, **1988**, *5*, 123–125.

SAMPLE
Matrix: blood, urine
Analyte: ofloxacin
Sample preparation: 500 μL Plasma or serum or 200 μL urine + 1 mL pH 7 phosphate buffer, mix, add 2 mL dichloromethane, shake, centrifuge at 2500 g for 5 min. Remove a 1 mL aliquot of the organic layer and add it to 1 mL dichloromethane, add 20 μL diphenylphosphinic chloride (Fluka; Janssen, Belgium), add 20 μL triethylamine, vortex for 10 s, add 500 μL reagent, shake for 10 min, extract with 200 (plasma, serum) or 500 (urine) μL of 1 (plasma, serum) or 0.5 (urine) M HCl, inject a 100 μL aliquot of the aqueous supernatant. (Prepare reagent by adding 10 mL dichloromethane and 1 mL 5 M NaOH to 500 mg L-leucinamide hydrochloride, shake, retain the lower organic layer and store it over anhydrous sodium sulfate.)

HPLC VARIABLES
Column: 125 × 4.6 5 μm Nucleosil 120-5 C18
Mobile phase: MeCN:buffer 20:80 (Buffer was 200 mM phosphoric acid adjusted to pH 1.85 tetraethylammonium hydroxide.)
Column temperature: 40
Flow rate: 1.5
Injection volume: 100
Detector: F ex 298 em 458

CHROMATOGRAM
Retention time: 2.6 (-), 3.8 (+)
Limit of detection: 3 ng/mL (plasma), 80 ng/mL (urine)

KEY WORDS
plasma; serum; chiral; pharmacokinetics; comparison with use of chiral column

REFERENCE
Lehr, K.-H.; Damm, P. Quantification of the enantiomers of ofloxacin in biological fluids by high-performance liquid chromatography, *J.Chromatogr.*, **1988**, *425*, 153–161.

SAMPLE
Matrix: blood, urine
Analyte: fenoprofen
Sample preparation: 50 μL Plasma or urine + ketoprofen + flunoxaprofen + 10 μL 1 M NaOH, heat at 37° for 2 h, add 10 μL 1 M HCl, extract with 1 mL ethyl acetate. Remove the organic layer and evaporate it to dryness, reconstitute the residue in 100 μL 50 mM triethylamine in MeCN, add 50 μL 60 mM ethyl chloroformate in MeCN, let stand for 2 min, add 50 μL 1 M L-leucinamide in 1 M triethylamine in MeOH. Evaporate, take up the residue in 100 μL mobile phase, inject a 100 μL aliquot. (Hydrolysis of glucuronides may be omitted.)

HPLC VARIABLES
Column: 250 × 4.6 5 μm Ultrasphere ODS
Mobile phase: MeCN:60 mM pH 6 potassium phosphate buffer 40:60
Flow rate: 1
Injection volume: 100
Detector: UV 272

CHROMATOGRAM
Retention time: 26.5 (R), 30.0 (S)
Internal standard: ketoprofen (R, 13.6; S, 16.0), flunoxaprofen (R, 21.2; S, 23.8)
Limit of quantitation: 250 ng/mL
Limit of detection: 25 ng/mL

OTHER SUBSTANCES
Extracted: metabolites, glucuronides

KEY WORDS
plasma; chiral; pharmacokinetics

REFERENCE
Volland, C.; Sun, H.; Benet, L.Z. Stereoselective analysis of fenoprofen and its metabolites, *J.Chromatogr.*, **1990**, *534*, 127–138.

SAMPLE
Matrix: blood, urine
Analyte: ketoprofen and probenecid
Sample preparation: Plasma. 100 μL Plasma + 100 μL 100 μg/mL indoprofen in water + 100 μL 600 mM sulfuric acid + 5 mL isooctane:isopropanol 95:5, vortex 30 s, centrifuge at 1800 g for 5 min. Remove organic layer and add 5 mL water to it. Vortex for 30 s, centrifuge for 3 min. Remove organic layer and evaporate it to dryness on a Speed Vac concentrator. Reconstitute residue in 100 μL 50 mM triethylamine in MeCN, vortex 30 s, add 50 μL 60 mM ethyl chloroformate in MeCN, let stand 30 s, add 50 μL 1 M L-leucinamide hydrochloride and 1 M triethylamine in MeOH, let stand 2 min, add 50 μL water, inject 10-60 μL aliquots. Urine. 100 μL Urine + 25 μL 1 M NaOH, add 125 μL 600 mM sulfuric acid, proceed as for plasma.

HPLC VARIABLES
Guard column: 50 × 5 37-53 μm C18 material
Column: 100 × 4.6 5 μm Partisil 5 ODS-3
Mobile phase: MeCN:60 mM KH_2PO_4:triethylamine 35:65:0.1
Flow rate: 1
Injection volume: 10-60
Detector: UV 275

CHROMATOGRAM
Retention time: 10 (R-ketoprofen), 12 (S-ketoprofen), 19 (probenecid)
Internal standard: indoprofen (6 (R), 7 (S))
Limit of quantitation: 500 ng/mL

OTHER SUBSTANCES
Also analyzed: carprofen, cicloprofen, fenoprofen, flurbiprofen, indoprofen, pirprofen
Simultaneously analyzed: probenecid

KEY WORDS
plasma; rat; chiral

REFERENCE
Palylyk, E.L.; Jamali, F. Simultaneous determination of ketoprofen enantiomers and probenecid in plasma and urine by high-performance liquid chromatography, *J.Chromatogr.*, **1991**, *568*, 187–196.

SAMPLE
Matrix: urine
Analyte: indobufen

Sample preparation: 500 µL Urine + 50 µL 5000 U/mL β-glucuronidase (Sigma) + 2 mL 100 mM pH 5 acetate buffer, heat at 37° for 16 h, add S-indoprofen, add 1 mL 1 M HCl, extract with diethyl ether. Remove the organic phase and extract it with 500 µL 1 M NaOH. Remove the aqueous phase and add it to 1 mL 1 M HCl, extract with diethyl ether. Remove the organic layer and evaporate it to dryness under a stream of air at 30°, add 5 drops of toluene, evaporate to dryness under a stream of air. Reconstitute the residue in 200 µL 50 mM triethylamine in MeCN, add 100 µL 60 mM ethyl chloroformate in MeCN, let stand for 30 s, add 100 µL 1 M L-leucinamide hydrochloride in MeOH containing 1 M triethylamine, let stand for 2 min, add 500 µL 250 mM HCl, extract with 4 mL ethyl acetate. Remove the organic layer and evaporate it to dryness under a stream of air at 30°, reconstitute in 500 µL MeOH, inject a 10 µL aliquot (J.Chromatogr. 1985, 339, 339).

HPLC VARIABLES
Column: 125 × 4 4 µm Lichrocart Superspher (Merck)
Mobile phase: MeCN:10 mM pH 6.5 phosphate buffer 35:65
Injection volume: 10
Detector: UV 275

CHROMATOGRAM
Internal standard: indoprofen
Limit of detection: 5 µg/mL

KEY WORDS
chiral; pharmacokinetics

REFERENCE
Strolin Benedetti, M.; Frigerio, E.; Tamassia, V.; Noseda, G.; Caldwell, J. The dispositional enantiose-lectivity of indobufen in man, *Biochem.Pharmacol.*, **1992**, *43*, 2032–2034.

SAMPLE
Matrix: blood
Analyte: arylpropionic acids
Sample preparation: 500 µL Plasma + 100 µL 25 µg/mL indomethacin + 500 µL 600 mM sulfuric acid + 15 mL dichloromethane, mix for 20 min, centrifuge at 2000 g for 5 min. Remove the organic layer and evaporate it to dryness under a stream of nitrogen at room temperature, reconstitute the residue in 100 µL 50 mM triethylamine in MeCN + 50 µL 60 mM ethyl chloroformate in MeCN, vortex for 30 s, add 50 µL 100 mM L-leucinamide in MeOH:triethylamine 100:14, let stand for 2 min, add 50 µL water, inject a 10-50 µL aliquot of the reaction mixture.

HPLC VARIABLES
Column: 250 × 4.6 5 µm Ultrabase C18 (Shandon)
Mobile phase: MeCN:60 mM KH_2PO_4:triethylamine 49:51:0.1
Flow rate: 1.8
Injection volume: 10-50
Detector: UV 275; UV 225

CHROMATOGRAM
Retention time: 3.5 (R-(-)-flurbiprofen (UV 275)), 4.4 (S-(+)-flurbiprofen (UV 275)), 5.4 (R-(-)-ibuprofen (UV 225)), 5.8 (S-(+)-ibuprofen (UV 225))
Internal standard: naproxen (UV 225) (2.6), indomethacin (UV 275) (5.3)
Limit of detection: 100 ng/mL

OTHER SUBSTANCES
Extracted: ketoprofen

KEY WORDS
plasma; chiral

REFERENCE
Péhourcq, F.; Lagrange, F.; Labat, L.; Bannwarth, B. Simultaneous measurement of flurbiprofen, ibuprofen, and ketoprofen enantiomer concentrations in plasma using L-leucinamide as the chiral coupling component, *J.Liq.Chromatogr.*, **1995**, *18*, 3969–3979.

L-Leucine-(4-methyl-7-coumarinylamide)

SAMPLE
Matrix: blood
Analyte: carboxylic acids
Sample preparation: Condition a 500 mg Bond Elut NH2 SPE cartridge with 3 mL MeOH and 12 mL 100 mM pH 7 NaH_2PO_4 buffer. Add 1 mL plasma to the SPE cartridge, wash with 3 mL water, wash with 3 mL MeCN, dry under vacuum, elute with 1 mL 500 mM formic acid in MeCN. Evaporate the eluate to dryness under a stream of nitrogen, reconstitute with 100 μL 50 mM triethylamine in MeCN, vortex for 1 min, add 50 μL 60 mM ethyl chloroformate in MeCN, vortex for 1 min, add 200 μL 3 mM L-leucine-(4-methyl-7-coumarinylamide) in MeOH, vortex for 1 min, let stand for 4 min, evaporate to dryness under a stream of nitrogen, reconstitute with 200 μL MeCN:water 50:50, inject a 5-15 μL aliquot.

HPLC VARIABLES
Column: 250 × 4.6 5 μm Axxiom octyl (Richard Scientific, Novato)
Mobile phase: Gradient. A was 1 L water containing 2 mL 85% phosphoric acid. B was 1 L MeCN containing 2 mL 85% phosphoric acid. A:B 55:45 for 15 min, to 10:90 over 15 min.
Injection volume: 5-15
Detector: F ex 330 em 390

CHROMATOGRAM
Retention time: 25 (pimelic acid), 30 (azelaic acid), 36.5 (tetradecanedioic acid), 35 (dodecanedioic acid), 39.5 (hexadecanedioic acid)

OTHER SUBSTANCES
Also analyzed: enalaprilat

KEY WORDS
plasma; SPE; dog; human

REFERENCE
Lévai, F.; Liu, C.-M.; Tse, M.M.; Lin, E.T. Pre-column fluorescence derivatization using leucine-coumarinylamide for HPLC determination of mono- and dicarboxylic acids in plasma, *Acta Physiol.Hung.*, **1995**, *83*, 39–46.

Luminarin 4

SAMPLE
Matrix: solutions
Analyte: carboxylic acids
Sample preparation: Mix 50 μL of a solution in DMF with 100 μL 62.5 mM N-hydroxy-succinimide in DMF and 100 μL 75 mM dicyclohexylcarbodiimide in DMF, let stand at 20° for 12 h, add 50 μL 2 mM coumarin 102 in DMSO, add 500 μL 1 mM luminarin 4 in DMSO, heat at 70° for 1 h, dilute 10-fold (or more) with DMSO, inject an aliquot. (Luminarin 4, N-(4-aminobutyl)-2,3,6,7-tetrahydro-11-oxo-1H,5H,11H-[1]benzopyrano[6,7,8-ij]quinolizine-9-acetamide, may be obtained from Eurobio, Les Ulis, France. Synthesis is as follows. Reflux (with protection from moisture and with stirring) 2.12 g 8-hydroxyju-lolidine, 2.22 g diethyl 1,3-acetonedicarboxylate (oxo-3-glutaric acid ethyl ester, Fluka), 1.71 g anhydrous zinc chloride, and 6 mL EtOH for 24 h, cool, add to 200 mL water, extract with 200 mL ethyl acetate, extract with 100 mL ethyl acetate. Combine the organic layers and wash them with water, dry over magnesium sulfate, evaporate to dryness, recrystallize from 5 parts ethyl acetate to give ethyl 2,3,6,7-tetrahydro-11-oxo-1H,5H,11H-[1]benzopyrano[6,7,8-ij]quinolizine-9-acetate. Heat 2 g of this compound with 42 mL 1.2% NaOH in water and 40 mL MeOH at 45° for 1 h, cool, wash with 50 mL chloroform, wash with 40 mL chloroform. Degas the aqueous phase and acidify it with 16 mL 3 M HCl, stir for 15 min, adjust pH to 6.5 with 13 mL 2.5 M NaOH, filter. Wash the precipitate with water and dry it to obtain 2,3,6,7-tetrahydro-11-oxo-1H,5H,11H-[1]benzopyrano[6,7,8-ij]quinolizine-9-acetic acid. Stir 11.26 g of this compound, 10.62 g disuccinimidyl oxalate (dihydroxysuccinimide carbonate), 3.81 g anhydrous triethylamine, and 560 mL dry MeCN protected from moisture at room temperature for 1 h, stir at 35-40° for 1 h, filter. Concentrate the filtrate and chromatograph on silica gel with dichloromethane:THF 50:50 to give luminarin 1 (21%). Stir 16.5 g luminarin 1 and 18.3 g 1,4-diaminobutane in 400 mL dry THF for 24 h, filter, evaporate the filtrate to dryness. Dissolve the residue in 100 mL dichloromethane and wash five times with 100 mL por-

tions of water, evaporate to dryness, stir the residue with 25 mL dichloromethane to give luminarin 4 (US Pat. 5 151 517 (Sept. 9, 1992)).

HPLC VARIABLES

Column: 150 × 4.6 5 μm Spherisorb ODS-2
Mobile phase: MeCN:DMSO:5 (F) or 10 (chemiluminescence) mM pH 7 imidazole nitrate buffer 45:5:50 (F) or 45:0:55 (chemiluminescence)
Column temperature: 40 (chemiluminescence only)
Flow rate: 1.5 (F), 1.2 (chemiluminescence)
Injection volume: 20
Detector: F ex 390 em 470 (cut-off filter); Chemiluminescence. 1 mg/mL Bis(2,4,6-trichlorophenyl) oxalate in methyl acetate pumped at 0.25 mL/min and 400 mM hydrogen peroxide in THF pumped at 0.25 mL/min mixed in a 292 μL capillary tube at 40° and this mixture mixed with the column effluent. The resulting mixture flowed through a 60 μL PTFE capillary at 40° to a Kratos FS970 detector fitted with a 470 nm long-pass filter.

CHROMATOGRAM

Retention time: 4.5 (dinoprostone (F)), 10.5 (dinoprostone (chemiluminescence)), 12.6 (isovaleric acid (F)), 16.0 (nonanoic acid (chemiluminescence)), 17.0 (nonanoic acid (F))
Internal standard: coumarin 102 (Eastman) (9.0 (F), 15.0 (chemiluminescence))
Limit of detection: 300 fmole (F), 50 fmole (chemiluminescence)

REFERENCE

Tod, M.; Prevot, M.; Chalom, J.; Farinotti, R.; Mahuzier, G. Luminarin 4 as a labelling reagent for carboxylic acids in liquid chromatography with peroxylate chemiluminescence detection, *J.Chromatogr.*, **1991**, *542*, 295–306.

S-(-)-α-Methylbenzylamine

RELATED REFERENCE

Jiang, M.; Soderlund, D.M. Liquid chromatographic determination and resolution of the enantiomers of the acid moieties of pyrethroid insecticides as their (-)-1-(1-phenyl)ethylamide derivatives. *J.Chromatogr.* **1982**, *248*, 143-149.

SAMPLE

Matrix: blood
Analyte: ibuprofen
Sample preparation: Dialyze (Spectrum Medical Industries, Inc. Spectrophor 2, 12000-14000 molecular weight cutoff) 3.5 mL plasma with 3.5 mL buffer at 37° with one 6 cm oscillation per s for 16-17 h. Remove a 2 mL aliquot of the buffer and add it to 500 μL 2 M sulfuric acid and 10 mL heptane:isopropanol 95:5, vortex for 1 min, centrifuge at 1000 g for 5 min. Remove the organic layer and evaporate it to dryness under a stream of nitrogen at 50°, reconstitute the residue in 100 μL 1% thionyl chloride in dichloromethane (freshly prepared), vortex briefly, heat at 70° in a tube securely sealed with a PTFE-lined cap for 1 h, let cool for 15 min, add 500 μL 1% (S)-(-)-α-methylbenzylamine ((S)-(-)-1-phenylethylamine) in dichloromethane (freshly prepared), vortex briefly, let stand at room temperature for 20 min, add 500 μL 2 M sulfuric acid, add 5 mL heptane, vortex for 1 min, centrifuge at 1000 g for 5 min. Remove the organic layer and evaporate it to dryness under a stream of nitrogen at 50°, reconstitute the residue in 100 μL mobile phase, vortex

briefly, inject a 50-90 μL aliquot. (Buffer was isotonic pH 7.4 phosphate buffer prepared from 67 mM NaH_2PO_4, 67 mM Na_2HPO_4 and NaCl.)

HPLC VARIABLES
Column: 250 × 4 5 μm Hibar Lichrosorb Si60
Mobile phase: Heptane:isopropanol 97.5:2.5
Flow rate: 2
Injection volume: 50-90
Detector: UV 216

CHROMATOGRAM
Retention time: 3 (R-(-)), 7 (S-(+))

KEY WORDS
plasma; normal phase

REFERENCE
Evans, A.M.; Nation, R.L.; Sansom, L.N.; Bochner, F.; Somogyi, A.A. Stereoselective plasma protein binding of ibuprofen enantiomers, *Eur.J.Clin.Pharmacol.*, **1989**, *36*, 283–290.

SAMPLE
Matrix: blood
Analyte: 5-dimethylsulfamoyl-6,7-dichloro-2,3-dihydrobenzofuran-2-carboxylic acid
Sample preparation: Condition a 3 mL Bond Elut C18 SPE cartridge with five 2.5 mL portions of MeOH and one portion of MeOH:water:acetic acid 45:54.5:0.5. Make up 0.1-0.5 mL plasma to 1 mL with water, add 50 μL 20 μg/mL IS in MeCN, add 500 μL 1 M HCl, add 500 μL pH 1 buffer, add 7 mL diethyl ether (chloroform ?), vortex, centrifuge at 1300 g. Remove the organic layer and add it to 1.5 mL pH 7.5 buffer, vortex. Remove the aqueous layer and wash it with 6 mL diethyl ether, add 500 μL 1 M HCl and 500 μL pH 1 buffer to the aqueous layer, extract the aqueous layer with 6 mL diethyl ether (chloroform ?). Dry the organic layer and add it to 100 μL 0.2% 1,1'-carbonylbis(2-methylimidazole) in MeCN, let stand at room temperature for 30-60 min, add 50 μL 0.2% (S)-(-)-α-methylbenzylamine in MeCN, add 10 μL 1% acetic acid in MeCN, mix, evaporate to dryness, reconstitute with 2.7 mL MeOH:water:acetic acid 45:54.5:0.5, filter (0.5 μm), add the filtrate to the SPE cartridge, wash with two 2.5 mL portions of MeOH:water:acetic acid 45:54.5:0.5, elute with 2.5 mL MeCN:water 70:30, evaporate to dryness, reconstitute with 150 μL MeCN, inject a 15 μL aliquot. (pH 1 Buffer was 250 mM sodium acetate adjusted to pH 1 with HCl. pH 7.5 Buffer was 200 mM KH_2PO_4 adjusted to pH 7.5 with 200 mM Na_2HPO_4.)

HPLC VARIABLES
Column: 250 × 4.6 5 μm Nucleosil C18
Mobile phase: MeCN:MeOH:water 45:5:50
Flow rate: 1
Injection volume: 15
Detector: UV 223

CHROMATOGRAM
Retention time: 24 (R-(+)), 26 (S-(-))
Internal standard: 5-diethylsulfamoyl-7-chloro-2,3-dihydrobenzofuran-2-carboxylic acid (Shiongi Research Labs) (32, 34 (enantiomers))
Limit of detection: 10 ng/mL

OTHER SUBSTANCES
Extracted: metabolites

KEY WORDS
monkey; plasma; chiral; SPE

REFERENCE
Nakano, M.; Kawahara, S. High-performance liquid chromatographic method for simultaneous determination of enantiomers of 5-dimethylsulphamoyl-6,7-dichloro-2-dihydrobenzofuran-2, carboxylic acid and its N-monodemethyl metabolite in monkey plasma and urine after chiral derivatization, *J.Chromatogr.*, **1991**, *564*, 235–241.

SAMPLE
Matrix: blood
Analyte: ketoprofen
Sample preparation: 1 mL Plasma + 50 μL 200 μg/mL S-naproxen in MeOH + 500 μL 2 M sulfuric acid + 8 mL n-hexane:ethyl acetate 90:10, mix gently at 30 rpm for 10 min, centrifuge at 1500 g for 10 min. Remove organic layer and evaporate it to dryness at 45° under a stream of nitrogen. Reconstitute in 100 μL 1.5% thionyl chloride in n-hexane (freshly prepared), heat at 75° for 1 h in a capped tube, cool to room temperature, add 500 μL 2% (S)-(-)-α-methylbenzylamine ((S)-1-phenylethylamine) in dichloromethane (freshly prepared), let stand for 15 min, add 500 μL 2 M sulfuric acid + 5 mL n-hexane, mix gently at 30 rpm for 10 min, centrifuge at 1500 g for 10 min. Remove organic layer and evaporate it to dryness at 45° under a stream of nitrogen. Reconstitute in 250 μL mobile phase, inject 200 μL aliquot.

HPLC VARIABLES
Column: 250 × 4 5 μm SGE silica glass column
Mobile phase: n-Heptane:isopropanol 92:8
Flow rate: 1
Injection volume: 200
Detector: UV 254

CHROMATOGRAM
Retention time: 5.2(R), 6.6(S)
Internal standard: S-naproxen (5.9)
Limit of quantitation: 150 ng/mL

OTHER SUBSTANCES
Simultaneously analyzed: fenoprofen, ibuprofen, mefenamic acid, salicylic acid
Non-interfering: diazepam, digoxin, methylprednisolone, midazolam, nifedipine, penicillamine, ranitidine, theophylline

KEY WORDS
plasma; normal phase; chiral

REFERENCE
Hayball, P.J.; Nation, R.L.; Bochner, F.; Le Leu, R.K. Enantiospecific analysis of ketoprofen in plasma by high-performance liquid chromatography, *J.Chromatogr.*, **1991**, *570*, 446–452.

SAMPLE
Matrix: blood
Analyte: ketorolac
Sample preparation: 1 mL Plasma + 50 μL 15 μg/mL (S)-ketoprofen in MeOH:water 1:4 + 75 μL 2 M sulfuric acid + 100 μL MeOH:water 1:4 + 8 mL hexane:ethyl acetate 80:20, rotary mix for 10 min, centrifuge at 2000 g for 10 min. Remove the organic layer and evaporate it to dryness at 55° under a stream of nitrogen. Reconstitute in 100 μL 1.5% thionyl chloride in n-hexane (freshly prepared),, vortex, heat at 80° for 30 min, cool to room temperature, add 500 μL reagent, vortex, let stand at room temperature for 10 min, evaporate to dryness under a stream of nitrogen, add 1 mL 2 M sulfuric acid, add 5 mL ethyl acetate, mix, centrifuge. Remove the organic layer and evaporate it to dryness at 55° under a stream of nitrogen. Reconstitute in 125 μL mobile phase, inject a 100 μL aliquot. (Reagent was 3% (S)-(-)-α-methylbenzylamine ((S)-1-phenylethylamine) in dry dichloromethane prepared within 30 min of use.)

HPLC VARIABLES
Column: 100 × 8 4 μm Nova-Pak phenyl radially compressed bonded phase cartridge
Mobile phase: MeCN:20 mM sodium acetate buffer 50:50 containing 0.1% triethylamine, final pH 5.5
Flow rate: 2
Injection volume: 100
Detector: UV 310 for 8 min (derivatized ketorolac) then UV 254 (derivatized ketoprofen)

CHROMATOGRAM
Retention time: 6.5 (S), 7.2 (R)

Internal standard: (S)-ketoprofen (8.6)
Limit of quantitation: 50 ng/mL

KEY WORDS
plasma; chiral

REFERENCE
Hayball, P.J.; Tamblyn, J.G.; Holden, Y.; Wrobel, J. Stereoselective analysis of ketorolac in human plasma by high-performance liquid chromatography, *Chirality*, **1993**, *5*, 31–35.

SAMPLE
Matrix: blood
Analyte: flurbiprofen
Sample preparation: 500 μL Plasma + 1 mL water + ibuprofen (15 μg per 100 μL) + 2 mL 10% trichloroacetic acid, mix, add 5 mL hexane, mix for 15 min, centrifuge at 800 g for 5-10 min, repeat extraction. Combine hexane layers and evaporate them under a stream of nitrogen at 37-40°. Reconstitute with 300 μL chloroform, add 200 μL 65 mg/mL 1,1'-carbonyldiimidazole in chloroform, let stand 5-10 min at room temperature, add 10 μL glacial acetic acid, vortex briefly, let stand 5-10 min at room temperature, add 50 μL S-(-)-α-methylbenzylamine, mix briefly, let stand for 30 min at room temperature, add 3 mL 0.5 M ammonium hydroxide, add 5 mL hexane, mix gently for 15 min. Remove hexane and wash it with 3 mL 1 M HCl, 3 mL 0.5 M ammonium hydroxide, and 3 mL 1 M HCl (with 15 min mixing each time). Evaporate hexane under a stream of nitrogen at 37°, dissolve in 150 μL mobile phase, inject 25 μL aliquot.

HPLC VARIABLES
Guard column: Brownlee RP18
Column: 250 × 4.6 5 μm Ultrasphere ODS
Mobile phase: MeCN:water 62:38
Flow rate: 1
Injection volume: 25
Detector: UV 245

CHROMATOGRAM
Retention time: 12 (S), 14 (R)
Internal standard: ibuprofen (17 S, 19 R)
Limit of quantitation: 25 ng/mL
Limit of detection: 10 ng/mL

OTHER SUBSTANCES
Simultaneously analyzed: metabolites

KEY WORDS
plasma; chiral

REFERENCE
Knadler, M.P.; Hall, S.D. High-performance liquid chromatographic analysis of the enantiomers of flurbiprofen and its metabolites in plasma and urine, *J.Chromatogr.*, **1989**, *494*, 173–182.

SAMPLE
Matrix: cell suspensions
Analyte: 2-arylpropionic acids
Sample preparation: Centrifuge cell suspension at 2000 g for 4 min. Remove a 2 mL aliquot of the supernatant and add it to 200 μL IS in DMF, mix, add 200 μL 5 M HCl, extract twice with 3 mL portions of toluene. Combine the organic layers and evaporate them to dryness under a stream of nitrogen, reconstitute the residue in 1 mL dichloromethane, add 20 μL 10 mg/mL 1-hydroxybenzotriazole in dichloromethane:pyridine 99:1, add 300 μL 10 mg/mL 1-(3-dimethylaminopropyl)-3-ethylcarbodiimide hydrochloride in dichloromethane, add 300 μL 10 mg/mL (S)-(-)-α-methylbenzylamine in dichloromethane, let stand for 30 min, evaporate to dryness, reconstitute with 500 μL mobile phase, inject a 10 μL aliquot. (For ketoprofen increase the concentration of the derivatization reagents to 40 mg/mL because of the high IS concentration.)

HPLC VARIABLES

Guard column: 10 mm long Techsphere ODS (HPLC Technology, Macclesfield UK)
Column: 250 × 5 5 μm Techsphere ODS (HPLC Technology, Macclesfield UK)
Mobile phase: MeCN:7.5 mM NaH$_2$PO$_4$ X:Y, containing 5 mM sodium pentanesulfonate, pH adjusted to 2.8 with phosphoric acid
Flow rate: 1
Injection volume: 10
Detector: UV 254

CHROMATOGRAM

Retention time: k' 6.25, 7.10 (2-phenylpropionic acid enantiomers; X:Y 45:55), k' 6.63, 7.11 (ibuprofen enantiomers; X:Y 60:40), k' 6.60, 7.55 (ketoprofen enantiomers; X:Y 50:50), k' 4.75, 5.35 (flurbiprofen enantiomers; X:Y 60:40), k' 4.95, 5.42 (fenoprofen enantiomers; X:Y 55:45), k' 4.60, 5.10 (indoprofen enantiomers; X:Y 50:50), k' 5.65, 6.40 (suprofen enantiomers; X:Y 50:50), k' 5.67, 6.08 (etodolac enantiomers; X:Y 65:35)

Internal standard: phenylacetic acid (IS for 2-phenylpropionic acid; X:Y 45:55; 200 μg/mL; k' 4.25), (S)-naproxen (IS for ibuprofen; X:Y 60:40; 1 μg/mL; k' 3.05), phenylacetic acid (IS for ketoprofen; X:Y 50:50; 4 mg/mL; k' 2.50), (S)-naproxen (IS for flurbiprofen; X:Y 60:40; 200 μg/mL; k' 3.15), (S)-naproxen (IS for fenoprofen; X:Y 55:45; 100 μg/mL; k' 3.42), (S)-naproxen (IS for indoprofen; X:Y 50:50; 100 μg/mL; k' 7.45), (S)-naproxen (IS for suprofen; X:Y 50:50; 200 μg/mL; k' 7.45), (R)-ibuprofen (IS for etodolac; X:Y 65:35; 100 μg/mL; k' 5.21)

Limit of detection: 0.5-10 μg/mL

KEY WORDS

chiral

REFERENCE

Thomason, M.J.; Hung, Y.-F.; Rhys-Williams, W.; Hanlon, G.W.; Lloyd, A.W. Indirect enantiomeric separation of 2-arylpropionic acids and structurally related compounds by reversed phase HPLC, *J.Pharm.Biomed.Anal.*, **1997**, *15*, 1765–1774.

SAMPLE

Matrix: urine
Analyte: 5-dimethylsulfamoyl-6,7-dichloro-2,3-dihydrobenzofuran-2-carboxylic acid
Sample preparation: Condition a 3 mL Bond Elut C18 SPE cartridge with five 2.5 mL portions of MeOH and one portion of MeOH:water:acetic acid 45:54.5:0.5. Make up 0.1-1 mL urine to 1 mL with water, add 50 μL 200 μg/mL IS in MeCN, add 500 μL 1 M HCl, add 500 μL buffer, add 7 mL diethyl ether, vortex, centrifuge at 1300 g. Remove the organic layer and add it to 1.5 mL 1% pH 8.6 sodium bicarbonate solution, vortex. Remove the aqueous layer and wash it with 6 mL diethyl ether, add 500 μL 1 M HCl and 500 μL buffer to the aqueous layer, extract the aqueous layer with 6 mL chloroform. Dry the organic layer and add it to 500 μL 0.2% 1,1'-carbonylbis(2-methylimidazole) in MeCN, let stand at room temperature for 30-60 min, add 250 μL 0.2% (S)-(-)-α-methylbenzylamine in MeCN, add 50 μL 1% acetic acid in MeCN, mix, evaporate to dryness, reconstitute with 2.7 mL MeOH:water:acetic acid 45:54.5:0.5, filter (0.5 μm), add the filtrate to the SPE cartridge, wash with two 2.5 mL portions of MeOH:water:acetic acid 45:54.5:0.5, elute with 2.5 mL MeCN:water 70:30, evaporate to dryness, reconstitute with 150 μL MeCN, inject a 15 μL aliquot. (Buffer was 250 mM sodium acetate adjusted to pH 1 with HCl.)

HPLC VARIABLES

Column: 200 × 4.6 5 μm Nucleosil C18 + 150 × 4.6 5 μm Nucleosil C18 in series
Mobile phase: MeCN:isopropanol:water 40:5:55
Flow rate: 1
Injection volume: 15
Detector: UV 254

CHROMATOGRAM

Retention time: 48 (R-(+)), 52 (S-(-))
Internal standard: 5-diethylsulfamoyl-7-chloro-2,3-dihydrobenzofuran-2-carboxylic acid (Shiongi Research Labs) (62, 67 (enantiomers))

Limit of detection: 10 ng/mL

OTHER SUBSTANCES
Extracted: metabolites

KEY WORDS
monkey; SPE; chiral

REFERENCE
Nakano, M.; Kawahara, S. High-performance liquid chromatographic method for simultaneous determination of enantiomers of 5-dimethylsulphamoyl-6,7-dichloro-2-dihydrobenzofuran-2, carboxylic acid and its N-monodemethyl metabolite in monkey plasma and urine after chiral derivatization, *J.Chromatogr.*, **1991**, *564*, 235–241.

(R)-(+)-α-Methylbenzylamine

RELATED REFERENCE
Noe, C.R.; Freissmuth, J.; Rothley, D.; Lachmann, B.; Richter, P. Kapillarelektrophoretische Analytik komplexer Kohlenhydratgemische [Capillary zone electrophoresis of carbohydrate mixtures]. *Pharmazie* **1996**, *51*, 868-873.

SAMPLE
Matrix: blood, urine
Analyte: ibuprofen
Sample preparation: 100 μL Urine or rat plasma or 500 μL human plasma + 50 μL 25 μg/mL fenoprofen in MeOH:10 mM NaOH 10:90 + 200 μL 600 mM sulfuric acid + 3 mL isooctane:isopropanol 95:5, vortex for 30 s, centrifuge at 1800 g for 5 min. Remove the organic layer and evaporate it to dryness, reconstitute the residue in 200 μL 50 mM triethylamine in MeCN, add 50 μL 6 mM ethyl chloroformate in MeCN, vortex for 30 s, add 50 μL 500 mM (R)-(+)-α-methylbenzylamine ((R)-(+)-α-phenylethylamine) in MeCN:triethylamine 80:20, vortex briefly, let stand for 2 min, add 1 mL 250 mM HCl, add 3 mL chloroform, vortex for 30 s, centrifuge at 1800 g for 2 min. Remove the organic layer and evaporate it to dryness, reconstitute the residue in 200 μL mobile phase, inject a 10-150 μL aliquot.

HPLC VARIABLES
Guard column: 20 × 4.6 37-53 μm reversed-phase
Column: 100 × 4.6 5 μm C18 (Phenomenex)
Mobile phase: MeCN:water:acetic acid:triethylamine 46.5:53.5:0.1:0.03, pH 4.9
Flow rate: 1.6
Injection volume: 10-150
Detector: UV 225

CHROMATOGRAM
Retention time: 15.69 (R), 17.65 (S)
Internal standard: fenoprofen (11.70, 13.40 (enantiomers))
Limit of quantitation: 250 ng/mL

KEY WORDS
human; rat; plasma; chiral; pharmacokinetics

REFERENCE

Wright, M.R.; Sattari, S.; Brocks, D.R.; Jamali, F. Improved high-performance liquid chromatographic assay method for the enantiomers of ibuprofen, *J.Chromatogr.*, **1992**, *583*, 259–265.

2-[p-(5,6-Methylenedioxy-2H-benzotriazol-2-yl]phenethylamine

SAMPLE

Matrix: blood

Analyte: ibuprofen

Sample preparation: 1 mL Serum + 100 μL 1 M HCl, mix, add to a Bond Elut C18 SPE cartridge, wash with 10 mL water, wash with 6 mL MeOH:water 50:50, elute with 2 mL MeOH:water 85:15. Evaporate the eluate to dryness under a stream of nitrogen, reconstitute the residue in 500 μL MeCN, add 50 μL 3.7 mM 2-bromo-1-ethylpyridinium tetrafluoroborate in MeCN, add 50 μL 2.7 mM 9-methyl-3,4-dihydro-2H-pyrido [1,2-a]pyrimidin-2-one in MeCN, add 50 μL 3.7 mM 2-[p-(5,6-methylenedioxy-2H-benzo-triazol-2-yl]phenethylamine in MeCN, mix for 10 s, let stand at room temperature for 1.5 h. Remove a 500 μL aliquot and add it to a Bond Elut silica SPE cartridge, elute with 1.5 mL MeCN. Make up the eluate to 2.5 mL with MeCN, inject a 20μL aliquot. (Synthesis of 2-[p-(5,6-methylenedioxy-2H-benzotriazol-2-yl]phenethylamine is as follows. Stir 2.7 g N-(p-aminophenethyl)-2,2,2-trifluoroacetamide hydrochloride in 50 mL 10% HCl at 0°, add 700 mg sodium nitrite in 10 mL water dropwise, stir for 15 min, add 1.5 g ammonium sulfamate, adjust to pH 5 with sodium acetate, add 1.4 g 3,4-(methylenedioxyaniline), stir for 2 h, adjust to pH 9 with 1 M NaOH, extract with ethyl acetate (cf Chem. Pharm. Bull. 1989, 37, 1009), recrystallize the resulting azo compound from DMF/MeOH to give red needles (mp 165-166°). Dissolve 2.5 g of the azo compound in 40 mL pyridine, stir, add ammoniacal cupric sulfate, reflux for 1 h, add 200 mL water. Collect the precipitate and wash it with 1% ammonium hydroxide, dissolve in 30 mL pyridine, add 30 mL 1 M NaOH, reflux for 1 h, pour into 200 mL water. Collect the precipitate and recrystallize it from MeOH to give 2-[p-(5,6-methylenedioxy-2H-benzotriazol-2-yl]phenethylamine as colorless needles (mp 145-147°). Prepare ammoniacal cupric sulfate by dissolving 10 g copper sulfate pentahydrate and 5 g ammonium chloride in 80 mL water, neutralize with sodium bicarbonate. Prepare N-(p-aminophenethyl)-2,2,2-trifluoroacetamide hydrochloride by adding 6.7 g trifluoroacetic anhydride dropwise with stirring to a cooled solution of 5 g 4-nitrophenethylamine in 40 mL benzene (Caution! Benzene is a carcinogen!), reflux for 3 h, evaporate to dryness under reduced pressure. Wash the residue with water and dilute HCl, recrystallize from EtOH to give N-trifluoroacetyl-4-nitrophenethylamine (mp 97-98°).

Dissolve 2 g N-trifluoroacetyl-4-nitrophenethylamine in 15 mL THF, add 300 mg 10% palladium on charcoal, hydrogenate at 2-3 atmospheres, after the theoretical amount of hydrogen has been absorbed filter, evaporate to dryness, wash the residue with acetone to give N-(p-aminophenethyl)-2,2,2-trifluoroacetamide as a white solid (mp 83-86°) (J. Pharm. Sci. 1969, 58, 1558). Prepare 2-bromo-1-ethylpyridinium tetrafluoroborate by adding a solution of 2.88 g 2-bromopyridine in dichloromethane (?) to 4 g triethyloxonium tetrafluoroborate under argon at room temperature, heat at 50-60° for 1 h, cool in an ice-bath, filter, wash the precipitate with 15 mL dry ether, dry under reduced pressure at room temperature to obtain 2-bromo-1-ethylpyridinium tetrafluoroborate (Bull. Chem. Soc. Japan 1977, 50, 1863). Prepare 9-methyl-3,4-dihydro-2H-pyrido[1,2-a]pyrimidin-2-one by adding 1.6 g ethyl 3-chloropropionate to a solution of 1 g 2-amino-3-picoline (2-amino-3-methylpyridine) in anhydrous EtOH, concentrate under reduced pressure, add a few drops of concentrated HCl to the oil to give 9-methyl-3,4-dihydro-2H-pyrido[1,2-a]pyrimidin-2-one hydrochloride as colorless prisms (Chem. Pharm. Bull. 1971, 19, 764).)

HPLC VARIABLES
Guard column: 30 × 4 Nucleosil 5C18
Column: 150 × 3.9 Nova Pak C18
Mobile phase: MeCN:water 52:40
Flow rate: 1
Injection volume: 20
Detector: F ex 333 em 372

CHROMATOGRAM
Retention time: 23
Limit of detection: 1.5 pg

KEY WORDS
serum; SPE

REFERENCE
Narita, S.; Kitagawa, T. 2-[p-(5,6-Methylenedioxy-2H-benzotriazol-2-yl)]phenethylamine as fluorescence derivatization reagent for carboxylic acids in high-performance liquid chromatography, Anal.Sci., **1989**, 5, 31−34.

(R)-(+)-α-Methyl-4-nitrobenzylamine

SAMPLE
Matrix: solutions
Analyte: citronellic acid (3,7-dimethyl-6-octenoic acid)
Sample preparation: Reflux 0.5 mmole citronellic acid (3,7-dimethyl-6-octenoic acid) and 1.5 mmole oxalyl chloride in 5 mL benzene for 30 min (Caution! Benzene is a carcinogen!). Evaporate to dryness under reduced pressure at 45°, reconstitute with 5 mL ether, cool in an ice bath, add 3 mL 667 mM (R)-(+)-α-methyl-4-nitrobenzylamine in ether in small portions, stir at 0-5° for 1 h, dilute with 100 mL ether, wash with 1 M HCl, wash with saturated sodium bicarbonate solution, wash with water, dry over anhydrous magnesium sulfate, evaporate to dryness, prepare a 1% solution in ethyl acetate, inject a 6 μL aliquot.

HPLC VARIABLES
Column: two 500 × 4 10 μm Partisil columns in series

Mobile phase: n-Heptane:THF 80:20
Flow rate: 1.5
Injection volume: 6
Detector: UV 254

CHROMATOGRAM
Retention time: k' 7.2 ((R)-citronellic acid (3,7-dimethyl-6-octenoic acid)), 8.1 ((S)-citronellic acid (3,7-dimethyl-6-octenoic acid)); 6.3 ((3R)-3,7-dimethyloctanoic acid), 7.6 ((3S)-3,7-dimethyloctanoic acid); 5.1 ((RRR, SRR)-3,7,11-trimethyldodecanoic acid), 6.3 ((RSR, SSR)-3,7,11-trimethyldodecanoic acid)

KEY WORDS
normal phase; chiral

REFERENCE
Valentine, D. Jr.; Chan, K.K.; Scott, C.G.; Johnson, K.K.; Toth, K.; Saucy, G. Direct determination of R/S enantiomer ratios of citronellic acid and related substances by nuclear magnetic resonance spectroscopy and high pressure liquid chromatography, *J.Org.Chem.*, **1976**, *41*, 62–65.

(S)-(-)-α-Methyl-4-nitrobenzylamine

SAMPLE
Matrix: solutions
Analyte: 2-methylhexanoic acid
Sample preparation: Reflux 0.5 mmole 2-methylhexanoic acid and 1.5 mmole oxalyl chloride in 5 mL benzene for 30 min (Caution! Benzene is a carcinogen!). Evaporate to dryness under reduced pressure at 45°, reconstitute with 5 mL ether, cool in an ice bath, add 3 mL 667 mM (S)-(-)-α-methyl-4-nitrobenzylamine in ether in small portions, stir at 0-5° for 1 h, dilute with 100 mL ether, wash with 1 M HCl, wash with saturated sodium bicarbonate solution, wash with water, dry over anhydrous magnesium sulfate, evaporate to dryness, prepare a 1% solution in ethyl acetate, inject an aliquot. (J.Org.Chem. 1976, 41, 62)

HPLC VARIABLES
Column: 500 × 4 10 μm Partisil
Mobile phase: n-Heptane:THF 80:20
Flow rate: 1.5
Detector: UV 254

CHROMATOGRAM
Retention time: k' 3.13 (R), 6.92 (S)

OTHER SUBSTANCES
Also analyzed: other 3 and 4-methylcarboxylic acids
Interfering: 50

KEY WORDS
normal phase; chiral

REFERENCE
Scott, C.G.; Petrin, M.J.; McCorkle, T. The liquid chromatographic separation of some acyclic isoprenoid acid enantiomers via diastereomer derivatization, *J.Chromatogr.*, **1976**, *125*, 157–161.

Monodansyl Cadaverine

RELATED REFERENCES

Lee, Y.M.; Nakamura, H.; Nakajima, T. Fluorometric determination of carboxylic acids by high-performance liquid chromatography after derivatization with monodansylcadaverine. *Anal.Sci.* **1989**, *5*, 681-685.

Lee, Y.M.; Nakamura, H.; Nakajima, T. Monodansylcadaverine as a versatile reagent for fluorogenic precolumn derivatization of carboxylic acids. *Anal.Sci.* **1989**, *5*, 209-210.

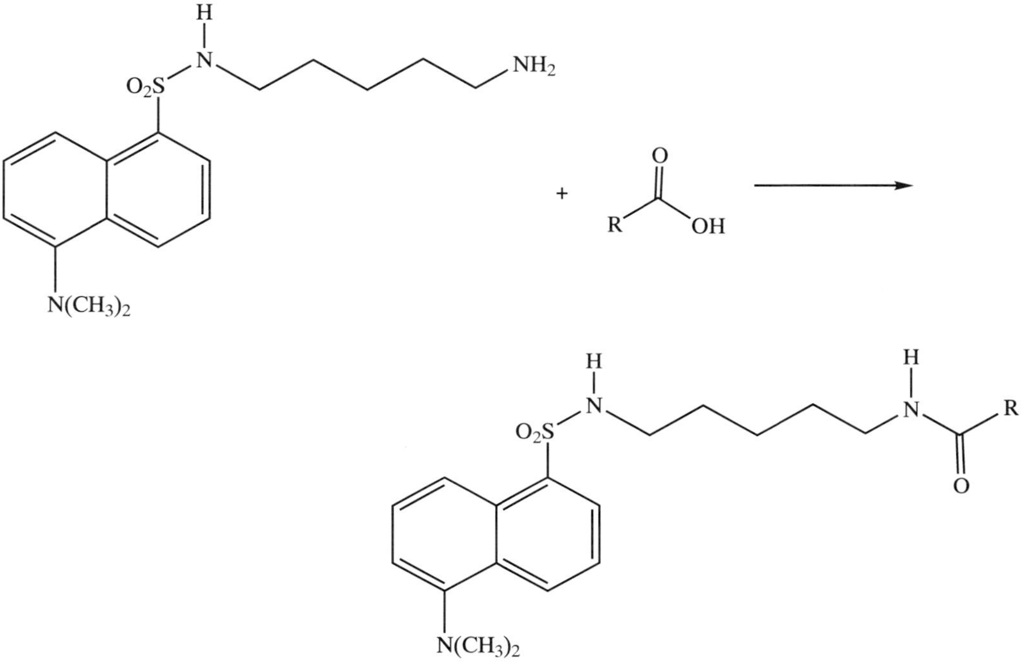

SAMPLE

Matrix: blood

Analyte: fatty acids

Sample preparation: Mix 1 mL rat platelets suspended in pH 7.2 Tyrode-HEPES buffer containing 2 mM calcium chloride and 2.5 U/mL thrombin with 50 μL 2 μM margaric acid in MeOH, filter (0.45 μm), add 10 μL 20% HCl to the filtrate, extract with 2 mL chloroform:MeOH 50:50, centrifuge at 880 g for 10 min, remove the organic layer, extract with 1 mL chloroform, centrifuge at 800 g for 10 min. Combine the organic layers and evaporate them to dryness under a stream of nitrogen, reconstitute the residue in 50 μL DMF, add 50 μL 12 mM monodansylcadaverine in DMF, add 2 μL diethyl cyanophosphonate (diethylphosphorocyanidate), stir for 10 s, let stand at room temperature for 15 min, inject an aliquot.

HPLC VARIABLES

Column: 250 × 4.6 TSKgel ODS-80TM (Tosoh)

Mobile phase: Gradient. A was MeOH:200 mM pH 7.8 Tris-HCl buffer 50:50. B was MeCN. A:B from 50:50 to 10:90 over 1 h.

Column temperature: 40

Flow rate: 1

Detector: F ex 340 em 518

CHROMATOGRAM

Retention time: 20 (lauric acid), 22 (myristoleic acid), 28 (linolenic acid), 29 (myristic acid), 30 (palmitoleic acid), 32 (linoleic acid), 38 (palmitic acid), 40 (oleic acid), 49 (stearic acid)

Internal standard: margaric acid (42)
Limit of detection: 0.1 pmole

KEY WORDS
rat; platelets

REFERENCE
Lee, Y.M.; Nakamura, H.; Nakajima, T. Rapid determination by high-performance liquid chromatography of free fatty acids released from rat platelets after derivatization with monodansylcadaverine, *J.Chromatogr.*, **1990**, *515*, 467–473.

SAMPLE
Matrix: bulk
Analyte: melittin
Sample preparation: Heat 100 µg melittin with 100 µL 700 µM monodansyl cadaverine (N-(5-aminopentyl)-5-dimethylamino-1-naphthalenesulfonamide) in 50 mM pH 8.0 Tris-HCl buffer containing 2 µM guinea pig liver transglutaminase (Sigma), 20 mM dithiothreitol, and 20 mM calcium chloride at 37° for 3 h, heat at 60° for 5 min, inject a 5-150 µL aliquot.

HPLC VARIABLES
Column: 300 × 7.8 µBondapak C18
Mobile phase: Gradient. MeCN:0.1% trifluoroacetic acid in water from 30:70 to 70:30 over 20 min. (Use a µBondapak C18 Guard-Pak between pump and injector.)
Flow rate: 1
Injection volume: 5-150
Detector: F ex 330 em 520

CHROMATOGRAM
Retention time: 17

KEY WORDS
derivatization is incomplete under these conditions; longer reaction times give more complete derivatization

REFERENCE
Perez-Paya, E.; Braco, L.; Abad, C.; Dufourcq, J. High-performance liquid chromatographic separation of modified and native melittin following transglutaminase-mediated derivatization with a dansyl fluorescent probe, *J.Chromatogr.*, **1991**, *548*, 351–359.

1-Naphthalenemethylamine

SAMPLE
Matrix: solutions
Analyte: 2-arylpropionic acids
Sample preparation: Mix 1 mL 100 µg/mL compound in dichloromethane with 300 µL 1 mg/mL 1-hydroxybenzotriazole in dichloromethane:pyridine 99:1, 300 µL 1 mg/mL 1-ethyl-3-dimethylaminopropylcarbodiimide hydrochloride in dichloromethane, and 300 µL 1-1.5 mg/mL 1-naphthalenemethylamine (1-naphthylmethylamine) in dichloromethane, vortex, let stand at room temperature for 1.5 h, wash with 1 mL 250 mM HCl, wash with

1 mL water. Remove the organic layer and evaporate it to dryness under a stream of nitrogen, reconstitute the residue in MeOH, inject an aliquot.

HPLC VARIABLES
Column: 10 μm EXP B101 4-methylbenzoate cellulose on silica gel (Bio-Rad)
Mobile phase: MeOH:50 mM pH 1.5 perchlorate buffer 80:20
Flow rate: 1
Detector: UV 230

CHROMATOGRAM
Retention time: 6, 8 (ibuprofen enantiomers), 13, 18 (tiaprofenic acid enantiomers)

KEY WORDS
chiral

REFERENCE
Van Overbeke, A.; Baeyens, W.; Van den Bossche, W.; Dewaele, C. Enantiomeric separation of amide derivatives of some 2-arylpropionic acids by HPLC on a cellulose-based chiral stationary phase, *J.Pharm.Biomed.Anal.*, **1994**, *12*, 911–916.

(R)-(+)-1-(1-Naphthyl)ethylamine

SAMPLE
Matrix: blood, urine
Analyte: ibuprofen
Sample preparation: Serum. Activate an SPE cartridge filled with 100 mg 40-63 μm silica gel (Merck) with 1 mL MeOH and dry in a hot air oven at 100° for 1 h, equilibrate with 1 mL dichloromethane before use. 500 μL Serum + 10 μL 100 μg/mL flurbiprofen in MeCN + 100 μL 1 M HCl, mix, add 1 mL 1 M pH 3.8 sodium phosphate buffer, mix, add 3 mL diethyl ether, rock for 20 min, centrifuge at 1000 g for 2 min. Remove the organic layer and evaporate it to dryness under a stream of nitrogen at 40°, reconstitute the residue in 100 μL 1 mg/mL 1-(3-dimethylaminopropyl)-3-ethylcarbodiimide hydrochloride in dichloromethane, add 100 μL 1 mg/mL 1-hydroxybenzotriazole in dichloromethane, add 100 μL 1 mg/mL (R)-(+)-1-(1-naphthyl)ethylamine in dichloromethane, vortex briefly, let stand at room temperature for 2 h, add to the SPE cartridge, elute with two 1 mL portions of dichloromethane:MeCN 90:10. Combine all the eluate and evaporate it to dryness under a stream of nitrogen at 40°, reconstitute the residue in 100 μL mobile phase, inject a 50 μL aliquot. Urine. Activate an SPE cartridge filled with 100 mg 40-63 μm silica gel (Merck) with 1 mL MeOH and dry in a hot air oven at 100° for 1 h, equilibrate with 1 mL dichloromethane before use. 500 μL Urine + 10 μL 100 μg/mL flurbiprofen in MeCN + 100 μL 1 M HCl, mix, add 1.5 mL 1 M pH 3.8 sodium phosphate buffer, mix, add 5 mL hexane:isopropanol 90:10, rock for 20 min, centrifuge at 1000 g for 5 min. Remove the organic layer and evaporate it to dryness under a stream of nitrogen at 40°, reconstitute the residue in 100 μL 1 mg/mL 1-(3-dimethylaminopropyl)-3-ethylcarbodiimide hydrochloride in dichloromethane, add 100 μL 1 mg/mL 1-hydroxybenzotriazole in dichloromethane, add 100 μL 1 mg/mL (R)-(+)-1-(1-naphthyl)ethylamine in dichloromethane, vortex briefly, let stand at room temperature for 2 h, add to the SPE cartridge, elute with two 1 mL portions of dichloromethane:MeCN 90:10. Combine all the eluate

and evaporate it to dryness under a stream of nitrogen at 40°, reconstitute the residue in 100 µL mobile phase, inject a 50 µL aliquot.

HPLC VARIABLES
Guard column: 10 × 2.1 40-63 µm pellicular C18 (Alltech)
Column: 150 × 3.9 5 µm Resolve C18 (Waters)
Mobile phase: MeCN:10 mM pH 3.5 phosphate buffer 50:50
Flow rate: 1.5
Injection volume: 50
Detector: F ex 290 em 330

CHROMATOGRAM
Retention time: 22.2 (R), 25.6 (S)
Internal standard: flurbiprofen (quantitation uses the peak for the (S)-enantiomer) (14.5 (R), 17.8 (S))
Limit of quantitation: 100 ng/mL

KEY WORDS
chiral; serum

REFERENCE
Tan, S.C.; Jackson, S.H.D.; Swift, C.G.; Hutt, A.J. Enantiospecific analysis of ibuprofen by high performance liquid chromatography: Determination of free and total drug enantiomer concentrations in serum and urine, *Chromatographia*, **1997**, *46*, 23–32.

SAMPLE
Matrix: solutions
Analyte: loxoprofen
Sample preparation: Add 500 ng IS, evaporate to dryness, add 100 µL 200 µg/mL (+)-(R)-1-(1-naphthyl)ethylamine in dichloromethane:pyridine 98:2 + 20 µL 2 mg/mL 1-hydroxybenzotriazole in dichloromethane:pyridine 98:2 + 100 µL 400 µg/mL N,N'-dicyclohexylcarbodiimide in dichloromethane:pyridine 98:2, let stand at room temperature for 1 h, evaporate to dryness, reconstitute in 500 µL ethyl acetate and 500 µL HCl, vortex for 1 min. Remove the organic phase and dry it over anhydrous sodium sulfate, inject a 50 µL aliquot.

HPLC VARIABLES
Column: 250 × 6 ERC-Silica-1282 (ERMA CR, Saitama, Japan)
Mobile phase: n-Hexane:ethyl acetate 59:41
Flow rate: 2
Injection volume: 50
Detector: F ex 283 em 330

CHROMATOGRAM
Retention time: 7 (S), 8.5 (R)
Internal standard: (+)-(S)-2-(2-(2-hydroxyprop-2-yl)indan-5-yl)propionic acid (16)

OTHER SUBSTANCES
Simultaneously analyzed: metabolites

KEY WORDS
chiral; normal phase; Plasma is purified on an immobilized antibody column (full preparative details given) before analysis.

REFERENCE
Takasaki, W.; Tanaka, Y. Application of antibody-mediated extraction for the stereoselective determination of the active metabolite of loxoprofen in human and rat plasma, *Chirality*, **1992**, *4*, 308–315.

(S)-(-)-1-(1-Naphthyl)ethylamine

RELATED REFERENCES

Avgerinos, A.; Hutt, A.J. Determination of the enantiomeric composition of ibuprofen in human plasma by high-performance liquid chromatography. *J.Chromatogr.* **1987**, *415*, 75-83.

Bergot, B.J.; Anderson, R.J.; Schooley, D.A.; Henrick, C.A. Liquid chromatographic analysis of enantiomeric purity of several terpenoid acids as their 1-(1-naphthyl)ethylamide derivatives. *J.Chromatogr.* **1978**, *155*, 97-105.

Robinett, R.S.R.; Hsieh, J.Y.-K. Stereoselective determination of *R*(-)- and *S*(+)-MK-571, a leukotriene D$_4$ antagonist, in human plasma by chiral high-performance liquid chromatography. *J.Chromatogr.* **1991**, *570*, 157-165.

Tsina, I.; Tam, Y.L.; Boyd, A.; Rocha, C.; Massey, I.; Tarnowski, T. An indirect (derivatization) and a direct HPLC method for the determination of the enantiomers of ketorolac in plasma. *J.Pharm.Biomed.Anal.* **1996**, *15*, 403-417.

SAMPLE

Matrix: blood

Analyte: ibuprofen

Sample preparation: 500 μL Plasma + 100 μL 100 μg/mL IS in 10 mM NaOH + 200 μL 600 mM sulfuric acid + 3 mL isooctane:isopropanol 95:5, vortex for 30 s, centrifuge at 1800 g for 5 min. Remove the organic layer and evaporate it to dryness, reconstitute the residue in 100 μL 50 mM triethylamine in MeCN, add 50 μL 6 mM ethyl chloroformate in MeCN, after 30 s add 25 μL 1 mL/L (S)-(-)-1-(1-naphthyl)ethylamine in MeCN, let stand for 3 min, add 500 μL 250 mM HCl, add 2 mL chloroform, vortex for 15 s, centrifuge at 1800 g for 2 min. Remove the organic layer and evaporate it to dryness under reduced pressure, reconstitute the residue in 200 μL mobile phase, inject a 10-50 μL aliquot.

HPLC VARIABLES

Guard column: 20 mm long 37-53 μm reversed-phase

Column: 100 × 4.6 5 μm Partisil 5 ODS-3

Mobile phase: MeCN:water:acetic acid:triethylamine 55:45:0.1:0.02, pH 4.9

Flow rate: 1

Injection volume: 10-50

Detector: UV 232

CHROMATOGRAM

Retention time: 18.5 (S), 21.0 (R)

Internal standard: (±)-2-(4-benzoylphenyl)butyric acid (11, 13 (enantiomers))

Limit of quantitation: 100 ng/mL

OTHER SUBSTANCES

Simultaneously analyzed: etodolac (enantiomers not resolved), flurbiprofen, ketoprofen, tiaprofenic acid (not derivatized)

KEY WORDS

plasma; chiral; pharmacokinetics

REFERENCE

Mehvar, R.; Jamali, F.; Pasutto, F.M. Liquid-chromatographic assay of ibuprofen enantiomers in plasma, *Clin.Chem.*, **1988**, *34*, 493–496.

SAMPLE
Matrix: blood
Analyte: ibuprofen
Sample preparation: 500 µL Plasma + 50 µL 200 µg/mL fenoprofen in MeOH:water 1:4 + 200 µL 1 M sulfuric acid + 3 mL isooctane:isopropanol 95:5, vortex 30 s, centrifuge at 1800 g for 5 min. Remove organic layer and evaporate it to dryness. Add 300 µL 50 mM triethylamine in MeCN and 50 µL 6 mM ethyl chloroformate in MeCN, wait 30 s, add 25 µL 0.1% (S)-(-)-1-(1-naphthyl)ethylamine in MeCN:triethylamine 98:2, after 3 min add 25 µL 2.5% ethanolamine in MeCN, inject 2-30 µL aliquot.

HPLC VARIABLES
Column: 100 × 4.6 5 µm Partisil ODS 3 RAC
Mobile phase: MeCN:water:acetic acid:triethylamine 60:40:0.1:0.02, final pH 5.0 (After every third injection flush with MeCN for 6 min at 1.6 mL/min, equilibrate with mobile phase for 9 min.)
Flow rate: 1.2
Injection volume: 2-30
Detector: F ex 280 em 320

CHROMATOGRAM
Retention time: 10.5 (S-(+)), 11.8 (R-(-))
Internal standard: fenoprofen (7.5 (S), 8.8 (R))
Limit of detection: 10 ng/mL

KEY WORDS
plasma; chiral; also UV 232 (Clin.Chem. 1988, 34, 493)

REFERENCE
Lemko, C.H.; Caillé, G.; Foster, R.T. Stereospecific high-performance liquid chromatographic assay of ibuprofen: improved sensitivity and sample processing efficiency, *J.Chromatogr.*, **1993**, *619*, 330–335.

SAMPLE
Matrix: blood
Analyte: ibuprofen
Sample preparation: 500 µL Plasma + 50 µL 50 µg/mL fenoprofen in water + 200 µL 20% sulfuric acid + 6 mL n-butyl chloride, vortex for 5 min, centrifuge at 950 g for 5 min, freeze in dry ice/acetone. Remove the organic layer and evaporate it to dryness under a stream of nitrogen, reconstitute the residue in 300 µL 50 mM triethylamine, sonicate for 1 min, vortex for 30 s, add 50 µL 6 mM ethyl chloroformate, let stand for 30 s, add 25 µL 10 mM S-(-)-1-(1-naphthyl)ethylamine, let stand for 3 min, add 25 µL MeCN:ethanolamine 40:1. Evaporate to dryness under a stream of nitrogen at 40°, reconstitute the residue in 250 µL mobile phase, inject a 25 µL aliquot.

HPLC VARIABLES
Guard column: 15 × 3.2 7 µm Newguard RP-18
Column: 150 × 4.6 5 µm Inertsil ODS-2
Mobile phase: MeCN:water (pH 3.0) 66.5:33.5
Column temperature: 27
Flow rate: 1.2
Injection volume: 25
Detector: F ex 280 em 320

CHROMATOGRAM
Retention time: 11.3 (S-(+)), 12.3 (R-(-))
Internal standard: fenoprofen (7.7 (S), 8.5 (R))
Limit of quantitation: 100 ng/mL

KEY WORDS
chiral; plasma

REFERENCE

Lau, Y.Y. Determination of ibuprofen enantiomers in human plasma by derivatization and high-performance liquid chromatography with fluorescence detection, *J.Liq.Chromatogr.Rel.Technol.*, **1996**, *19*, 2143–2153.

SAMPLE

Matrix: cell suspensions
Analyte: 2-arylpropionic acids
Sample preparation: Centrifuge cell suspension at 2000 g for 4 min. Remove a 2 mL aliquot of the supernatant and add it to 200 μL 200 μg/mL IS in DMF, mix, add 200 μL 5 M HCl, extract twice with 3 mL portions of toluene. Combine the organic layers and evaporate them to dryness under a stream of nitrogen, reconstitute the residue in 1 mL dichloromethane, add 20 μL 10 mg/mL 1-hydroxybenzotriazole in dichloromethane:pyridine 99:1, add 300 μL 10 mg/mL 1-(3-dimethylaminopropyl)-3-ethylcarbodiimide hydrochloride in dichloromethane, add 300 μL 10 mg/mL (S)-(-)-1-(1-naphthyl)ethylamine in dichloromethane, let stand for 30 min, evaporate to dryness, reconstitute with 500 μL mobile phase, inject a 10 μL aliquot.

HPLC VARIABLES

Guard column: 10 mm long Techsphere ODS (HPLC Technology, Macclesfield UK)
Column: 250 × 5 5 μm Techsphere ODS (HPLC Technology, Macclesfield UK)
Mobile phase: MeCN:7.5 mM NaH_2PO_4 X:Y, containing 5 mM sodium pentanesulfonate, pH adjusted to 2.8 with phosphoric acid
Flow rate: 1
Injection volume: 10
Detector: UV 254

CHROMATOGRAM

Retention time: k' 8.85, 9.35 (2-phenoxypropionic acid enantiomers; X:Y 50:50), k' 7.00, 7.45 (2-phenylbutyric acid enantiomers; X:Y 53:47), k' 3.67, 4.58 (α-methoxyphenylacetic acid enantiomers; X:Y 70:30), k' 3.17, 3.54 (mandelic acid enantiomers; X:Y 65:35), k' 5.13, 5.75 (2-hydroxy-2-phenylpropionic acid enantiomers; X:Y 65:35)
Internal standard: phenylacetic acid (IS for 2-phenoxypropionic acid; X:Y 50:50; k' 5.45), phenylacetic acid (IS for 2-phenylbutyric acid; X:Y 53:47; k' 3.85), phenylacetic acid (IS for α-methoxyphenylacetic acid; X:Y 70:30; k' 2.79), phenylacetic acid (IS for mandelic acid; X:Y 65:35; k' 4.63), p-toluic acid (IS for 2-hydroxy-2-phenylpropionic acid; X:Y 65:35; k' 7.04)
Limit of detection: 2.5-5 μg/mL

KEY WORDS

chiral

REFERENCE

Thomason, M.J.; Hung, Y.-F.; Rhys-Williams, W.; Hanlon, G.W.; Lloyd, A.W. Indirect enantiomeric separation of 2-arylpropionic acids and structurally related compounds by reversed phase HPLC, *J.Pharm.Biomed.Anal.*, **1997**, *15*, 1765–1774.

SAMPLE

Matrix: urine
Analyte: arylpropionic acids
Sample preparation: 500 μL Urine + 500 μL 1 M NaOH, mix, let stand at room temperature for 2 h (to hydrolyse glucuronides), add 100 μL 1 mg/mL 1-naphthylacetic acid, add 2 mL 1 M HCl, add 7 mL benzene (Caution! Benzene is a carcinogen!), centrifuge. Remove the organic layer and dry it over anhydrous sodium sulfate, evaporate to dryness under a stream of nitrogen, reconstitute the residue in 1 mL dichloromethane, add 10 μL 1 mg/mL 1-hydroxybenzotriazole in dichloromethane:pyridine 99:1, add 500 μL 1 mg/mL 1-ethyl-3-(dimethylaminopropyl)carbodiimide hydrochloride in dichloromethane, add 500 μL 1 mg/mL (S)-(-)-1-(1-naphthyl)ethylamine in dichloromethane, mix thoroughly, let stand at room temperature for 1.5 h, evaporate to dryness under reduced pressure, reconstitute with 500 μL mobile phase, inject a 10 μL aliquot. (Extraction from urine was only demonstrated for 2-phenylpropionic acid, other compounds were derivatized as solutions in dichloromethane.)

HPLC VARIABLES
Column: 100 × 5 10 μm Porasil Radial-Pak radial compression
Mobile phase: Hexane:ethyl acetate 80:20
Flow rate: 0.8
Injection volume: 10
Detector: UV 254

CHROMATOGRAM
Retention time: k' 1.26 ((R)-2-phenylpropionic acid), k' 2.32 ((S)-2-phenylpropionic acid), k' 0.84 ((R)-ibuprofen), k' 1.79 ((S)-ibuprofen), k' 8.84 ((R)-carprofen), k' 15.11 ((S)-carprofen), k' 1.75 ((-)-pirprofen), k' 2.80 ((+)-pirprofen)
Internal standard: 1-naphthylacetic acid

KEY WORDS
chiral; rat; normal phase

REFERENCE
Hutt, A.J.; Fournel, S.; Caldwell, J. Application of a radial compression column to the high-performance liquid chromatographic separation of the enantiomers of some 2-arylpropionic acids as their diastereoisomeric S-(-)-1-(naphthen-1-yl)ethylamines, *J.Chromatogr.*, **1986**, *378*, 409–418.

Nile Blue

SAMPLE
Matrix: blood
Analyte: phenylacetic acid
Sample preparation: 200 μL Plasma + 50 μL 4 M HCl, vortex for 10 s. 780 μL Dichloromethane + 20 μL 100 mM triethylamine + 100 μL 80 μM Nile blue + 100 μL 2-chloro-1-methylpyridinium iodide, vortex for 10 s. Combine the two solutions, vortex for 30 s, centrifuge at 10000 rpm for 5 min. Remove the organic layer and evaporate it to dryness under a stream of nitrogen, reconstitute with 1 mL MeCN:water 80:20, inject a 50 μL aliquot.

HPLC VARIABLES
Guard column: 20 × 4.6 5 μm 201TP (Vydac)

Column: 150 × 4.6 5 μm 201TP C18 (Vydac)
Mobile phase: MeCN:water:trifluoroacetic acid 40:59.9:0.1 (May not be correct.)
Flow rate: 2
Injection volume: 50
Detector: F ex 635 em 650 following post-column reaction. The column effluent mixed with 500 mM NaOH pumped at 0.035 mL/min and the mixture flowed to the detector. (A laser or conventional detector may be used.)

KEY WORDS
post-column reaction; plasma; cf J. Pharm. Biomed. Anal. 1996, 15, 83

REFERENCE
Rahavendran, S.V.; Karnes, H.T. RPHPLC-visible diode laser induced fluorescence detection (VDLIF) of phenyl acetic acid in plasma derivatized with nile blue using precolumn phase transfer catalysis (Abstract APQ 1090), *Pharm.Res.*, **1996**, *13*, S25–S25.

SAMPLE
Matrix: blood
Analyte: phenylacetic acid
Sample preparation: 200 μL Plasma or serum + 50 μL 4 M HCl, vortex for 10 s. In a separate vial mix 780 μL dichloromethane, 20 μL 100 mM triethylamine, 100 μL 11.74 mM 2-chloro-1-methylpyridinium iodide, and 100 μL 80 μM Nile blue perchlorate, vortex for 10 s, add the contents of this vial to the first vial, vortex for 30 s, centrifuge at 10000 rpm for 5 min. Remove the organic layer and evaporate it to dryness under a stream of nitrogen, reconstitute with 1 mL MeCN:water 50:50, inject a 50 μL aliquot.

HPLC VARIABLES
Guard column: 20 × 4.6 5 μm 201 TP (Vydac)
Column: 250 × 4.6 5 μm 218 TP C18 (Vydac)
Mobile phase: MeCN:water:trifluoroacetic acid 34:66:0.08
Column temperature: 36
Flow rate: 2
Injection volume: 50
Detector: F ex 630 em 700 following post-column reaction. The column effluent mixed with 1 M NaOH pumped at 50 μL/min (?) and the mixture flowed to the detector.

CHROMATOGRAM
Retention time: 17.5
Limit of detection: 1 nM

KEY WORDS
plasma; serum

REFERENCE
Rahavendran, S.V.; Karnes, H.T. Visible diode laser-induced fluorescence detection of phenylacetic acid in plasma derivatized with Nile blue and using precolumn phase transfer catalysis, *Anal.Chem.*, **1997**, *69*, 3022–3027.

4-Nitrobenzylamine

SAMPLE
Matrix: blood

Analyte: ibuprofen

Sample preparation: 500 μL Plasma + 100 μL 200 μg/mL tridecanoic acid in ethylene chloride + 200 μL 600 mM sulfuric acid + 3 mL isooctane:isopropanol 95:5, extract. Remove the organic layer and evaporate it to dryness, reconstitute the residue in 1 mL 2.4 mg/mL 2-ethoxy-1-ethoxycarbonyl-1,2-dihydroquinoline (EEDQ) in ethylene chloride, add 5 mL reagent, reflux for 10 min, dilute with 10 mL ethylene chloride, wash with an equal volume of 200 mM NaOH, wash with an equal volume of 1 M HCl, wash with an equal volume of water. Remove the organic layer and dry it over sodium sulfate, evaporate to dryness, reconstitute the residue in mobile phase, inject an aliquot. (Prepare reagent by dissolving 5 mg 4-nitrobenzylamine hydrochloride in 5 mL 200 mM NaOH, extract with 5 mL ethylene chloride, dry the organic layer over anhydrous sodium sulfate, use this solution as the reagent.)

HPLC VARIABLES

Column: 100 × 4.6 3 μm (R)-(-)-(1-naphthyl)ethylurea (Prepare by pumping 2 g (R)-(-)-1-(1-naphthyl)ethyl isocyanate in 100 mL dichloromethane through a 100 × 4.6 3 μm aminopropyl-silanized silica column (Regis) at 2 mL/min (without detector), after 12.5 min recycle the mobile phase, after 2 h wash the column with 300 mL dichloromethane at 2 mL/min, wash with hexane:isopropanol 80:20 until a steady baseline is achieved.)

Mobile phase: Hexane:isopropanol 87.5:12.5

Flow rate: 1.5

Detector: UV 235

CHROMATOGRAM

Retention time: 21 (S), 22.5 (R)

Internal standard: tridecanoic acid (19)

Limit of quantitation: 2.5 μg/mL

KEY WORDS

dog; chiral; plasma; pharmacokinetics

REFERENCE

Ahn, H.-Y.; Shiu, G.K.; Trafton, W.F.; Doyle, T.D. Resolution of the enantiomers of ibuprofen; comparison study of diastereomeric method and chiral stationary phase method, *J.Chromatogr.B*, **1994**, *653*, 163–169.

Sulfanilic Acid

SAMPLE

Matrix: bulk

Analyte: gangliosides

Sample preparation: Add 50 μL 100 mM pH 5.0 1-ethyl-3-(3-dimethylaminopropyl) carbodiimide in water to <1 μg gangliosides, stir for 1 h, add 50 μL 100 mM sulfanilic acid in water, stir for 1.5 h, inject an aliquot.

CAPILLARY ELECTROPHORESIS

Capillary: 80 cm × 50 μm fused-silica (50 cm to detector) (Polymicro Technologies)

Running buffer: MeCN:10 mM pH 10.0 sodium phosphate buffer 50:50

Voltage/Current: 20 kV

Injection: Hydrodynamic injection at 15 cm for 5 s (1.4 nL)

Detector: F ex 315 em 400 (cutoff filter); UV 247

Model: laboratory constructed

Migration time: 10.5-11 (depending on structure)

REFERENCE

Mechref, Y.; Ostrander, G.K.; El Rassi, Z. Capillary electrophoresis of carboxylated carbohydrates. I. Selective precolumn derivatization of gangliosides with UV absorbing and fluorescent tags, *J.Chromatogr.A*, **1995**, *695*, 83−95.

DIAMINE

9,10-Diaminophenanthrene

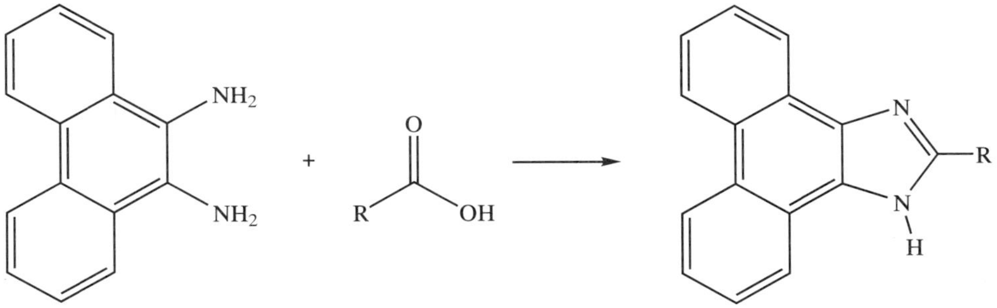

SAMPLE

Matrix: solutions

Analyte: fatty acids

Sample preparation: Mix 10 μL of a toluene or chloroform solution with 500 μL reagent, heat in a sealed ampule at 85° for 3-6 min (optimum heating time should be determined for each batch of polyphosphate), inject an aliquot. (Prepare reagent by dissolving 20 mg 9,10-diaminophenanthrene in 10 mL methylpolyphosphate solution, discard after 4 h. Prepare methylpolyphosphate solution by passing chloroform through aluminum oxide under nitrogen under subdued lighting to remove EtOH and phosgene. Add 37 g phosphoric oxide to 80 mL purified chloroform, then add 47 mL trimethyl phosphate (it is important to do it in this order), allow to cool to room temperature, seal the container, stir for 3 days, use the supernatant as methylpolyphosphate solution.)

HPLC VARIABLES

Column: 250 × 4 ODS-Hypersil

Mobile phase: Gradient. MeOH:10 mM ammonium carbonate from 70:30 to 99:1 over 38 min.

Flow rate: 0.8

Detector: F ex 255 em 382

CHROMATOGRAM

Retention time: 7 (acetic acid), 10 (butyric acid), 14 (caproic acid), 18 (caprylic acid), 21 (capric acid), 23 (lauric acid), 25 (myristic acid), 27 (palmitic acid), 30 (stearic acid), 33 (arachidic acid)

REFERENCE

Lloyd, J.B.F. Phenanthrimidazoles as fluorescent derivatives in the analysis of fatty acids by high-performance liquid chromatography, *J.Chromatogr.*, **1980**, *189*, 359−373.

DIAZO COMPOUND

9-Anthryldiazomethane

RELATED REFERENCES

Aase, B.; Rogstad, A. Optimization of sample cleanup procedure for determination of diarrhoeic shellfish poisoning toxins by use of experimental design. *J.Chromatogr.A* **1997**, *764*, 223-231.

Barker, S.A.; Monti, J.A.; Christian, S.T.; Benington, F.; Morin, R.D. 9-Diazomethylanthracene as a new fluorescence and ultraviolet label for the spectrometric detection of picomole quantities of fatty acids by high-pressure liquid chromatography. *Anal.Biochem.* **1980**, *107*, 116-123.

Betner, I.; Foeldi, P. The FMOC-ADAM approach to amino acid analysis. *LC-GC* **1988**, *6*, 832-840.

Comesana-Losada, M.; Gago-Martínez, A.; Leao-Martins, J.M.; Rodríguez-Vázquez, J.A. High-performance liquid chromatographic methods for determination of marine biotoxins. *Analyst* **1996**, *121*, 1665-1670.

Ghiggeri, G.M.; Candiano, G.; Delfino, G.; Queirolo, C.; Ginevri, F.; Perfamo, F.; Gusmano, R. Separation of the 9-anthryldiazomethane derivates of fatty acids by high-performance liquid chromatography on a Fatty Acid Analysis Column. Application to albumin-bound fatty acid analysis. *J.Chromatogr.* **1986**, *381*, 411-418.

Hatsumi, M.; Kimata, S.; Hirosawa, K. 9-Anthryldiazomethane derivatives of prostaglandins for high-performance liquid chromatographic analysis. *J.Chromatogr.* **1982**, *253*, 271-275.

Hatsumi, M.; Kimata, S.-I.; Hirosawa, K. Microanalysis of free fatty acids in plasma of experimental animals and humans by high-performance liquid chromatography. *J.Chromatogr.* **1986**, *380*, 247-255.

Hayakawa, K.; Oizumi, J. Determination of free biotin in plasma by liquid chromatography with fluorimetric detection. *J.Chromatogr.* **1987**, *413*, 247-250.

Ichinose, N.; Nakamura, K.; Shimizu, C.; Kurokura, H.; Okamoto, K. High-performance liquid chromatography of 5,8,11,14,7-eicosapentaenoic acid in fatty acids (C_{18} and C_{20}) by labelling with 9-anthryldiazomethane as a fluorescent agent. *J.Chromatogr.* **1984**, *295*, 463-469.

Imaoka, S.; Funae, Y.; Sugimoto, T.; Hayahara, N.; Maekawa, M. Specific and rapid assay of urinary oxalic acid using high-performance liquid chromatography. *Anal.Biochem.* **1983**, *128*, 459-464.

James, K.J.; Bishop, A.G.; Gillman, M.; Kelly, S.S.; Roden, C.; Draisci, R.; Lucentini, L.; Giannetti, L.; Boria, P. Liquid chromatography with fluorimetric, mass spectrometric and tandem mass spectrometric detection for the investigation of the seafood-toxin-producing phytoplankton, *Dinophysis acuta*. *J.Chromatogr.A* **1997**, *777*, 213-221.

Kargas, G.; Rudy, T.; Spennetta, T.; Takayama, K.; Querishi, N.; Shrago, E. Separation and quantitation of long-chain free fatty acids in human serum by high-performance liquid chromatography. *J.Chromatogr.* **1990**, *526*, 331-340.

Nakamura, T.; Hatori, Y.; Yamada, K.; Ikeda, M.; Yuzuriha, T. A high-performance liquid chromatographic method for the determination of polyphosphonoinositides in brain. *Anal.Biochem.* **1989**, *179*, 127-130.

Nimura, N.; Kinoshita, T. Fluorescent labeling of fatty acids with 9-anthryldiazomethane (ADAM) for high performance liquid chromatography. *Anal.Lett.* **1980**, *13*, 191-202.

Rodriquez Vazquez, J.A.; Gago Martinez, A.; Paniello, A.I.; Burdaspal Perez, P.; Legarda Gomez, T. High performance liquid chromatography of DSP toxins in bivalves by in situ formation of 9-anthryldiazomethane. *Dev.Mar.Biol.* **1993**, *3*, 571-574.

Shimomura, Y.; Taniguchi, K.; Sugie, T.; Murakami, M.; Sugiyama, S.; Ozawa, T. Analysis of the fatty acid composition of human serum lipids by high performance liquid chromatography. *Clin.Chim.Acta* **1984**, *143*, 361-366.

Shimomura, Y.; Sugiyama, S.; Takamura, T.; Kondo, T.; Ozawa, T. Quantitative determination of the fatty acid composition of human serum lipids by high-performance liquid chromatography. *J.Chromatogr.* **1986**, *383*, 9-17.

Smith, S.H.; Judge, M.D. An HPLC method using FMOC-ADAM for determination of hydroxyproline in muscle. *Meat Sci.* **1991**, *30*, 351-357.

Suzuki, T. High-performance liquid chromatographic resolution of dinophysistoxin-1 and free fatty acids as 9-anthrylmethyl esters. *J.Chromatogr.A* **1994**, *677*, 301-306.

Yoshida, T.; Uetake, A.; Murayama, H.; Nimura, N.; Kinoshita, T. Fluorescent labelling of amino acids with 9-anthryldiazomethane and its applications to high-performance liquid chromatography. *J.Chromatogr.* **1985**, *348*, 425-429.

SAMPLE
Matrix: blood
Analyte: SK&F 106203 (3-(2-carboxyethylthio)-3-[2-(8-phenyloctyl)phenyl]propanoic acid)
Sample preparation: Condition a 1 mL 100 mg C18 SPE cartridge (Analytichem) with 1 mL MeOH and 1 mL water. 250 μL Plasma + 75 μL MeOH:water 50:50 + 50 μL 4 μg/mL IS in MeOH:water 50:50 + 5 mL hexane, shake at low speed for 5 min, centrifuge at 1500 g for 5 min, discard the hexane layer, add 5 mL ethyl acetate:hexane 20:80, add 500 μL 1 M acetic acid, shake at low speed for 30 min, centrifuge at 1500 g for 5 min. Remove the organic layer and evaporate it to dryness under a stream of nitrogen at 40°, reconstitute the residue in 100 μL anhydrous ethyl acetate, add 200 μL reagent solution, mix, let stand at room temperature in the dark for 1 h, evaporate to dryness under a stream of nitrogen, reconstitute with 100 μL MeCN, add to the SPE cartridge, wash with 1 mL 100 mM pH 8.0 phosphate buffer, wash with 1 mL water, wash with 3 mL MeCN:water 85:15, elute with 2 mL MeCN. Evaporate the eluate to dryness under a stream of nitrogen at 40°, reconstitute the residue in 10 μL ethyl acetate, add 90 μL hexane, mix, inject a 10-50 μL aliquot onto column A and elute to waste with mobile phase, after 2.3 min elute the contents of column A onto column B with mobile phase, after 2 min remove column A from the circuit, elute column B with mobile phase, monitor the effluent from column B. Elute column A to waste with mobile phase to remove late eluting components. (9-Anthryldiazomethane is available from Molecular Probes, Eugene OR. Synthesis is as follows. Stir 8.8 g 9-anthraldehyde and 8.5 g 80% hydrazine hydrate in 150 mL EtOH at room temperature for 3 h, filter off the solid 9-anthraldehyde hydrazone and dry under vacuum (mp 124-6°) (Bull. Chem. Soc. Jpn. 1967, 40, 691). Add 1 mL 15.3 mg/mL quinuclidine in anhydrous ethyl acetate and 1 mL 1.84 mg/mL N-chlorosuccinimide in anhydrous ethyl acetate to 1 mL 3.05 mg/mL 9-anthraldehyde hydrazone in anhydrous ethyl acetate, let stand in the dark at room temperature for 30 min, use the resulting solution of 9-anthryldiazomethane immediately. See below for another synthesis.)

HPLC VARIABLES
Column: A 30 × 4.6 5 μm Spheri-5 silica; B 250 × 4.6 5 μm Ultrasphere silica
Mobile phase: Hexane:chloroform:ethyl acetate:glacial acetic acid 93.8:2:4:0.2
Column temperature: 35 (column B only)
Flow rate: A 1; B 1.1
Injection volume: 10-50
Detector: F ex 365 em 415

CHROMATOGRAM
Retention time: 26
Internal standard: 3-(2-carboxyethylthio)-2-methoxy-3-[2-(8-phenyloctyl)phenyl]propanoic acid (SK&F 104736) (36)
Limit of quantitation: 20 ng/mL

KEY WORDS
plasma; SPE; column-switching; heart-cut

REFERENCE
Miller-Stein, C.; Hwang, B.Y.; Rhodes, G.R.; Boppana, V.K. Column-switching high-performance liquid chromatographic method for the determination of SK&F 106203 in human plasma after fluorescence derivatization with 9-anthryldiazomethane, *J.Chromatogr.*, **1993**, *631*, 233−240.

SAMPLE
Matrix: blood, urine
Analyte: quinapril
Sample preparation: Condition a 100 mg 1 mL Bond-Elut C18 SPE cartridge with two 1 mL portions of MeOH, two 1 mL portions of water, and two 1 mL portions of 100 mM HCl. Condition a CBA-Bond-Elut SPE cartridge with MeOH and water. 1 mL Plasma + 1 mL buffer + 250 ng IS (or 50-500 μL urine + 500 ng IS), add to the C18 SPE cartridge, wash with two 1 mL portions of pH 3.4 water, wash with two 1 mL portions of distilled n-hexane, dry under vacuum for 20-30 min, elute with three 1 mL portions of chloroform:MeOH 2:1. Evaporate the eluate to dryness, reconstitute with 50 μL chloroform:MeOH 50:50 and 50 μL 2 mg/mL 9-anthryldiazomethane in MTBE, vortex for a few s, heat at 40° for 90 min, evaporate to dryness, reconstitute with two 100 μL portions of MeCN, add to the CBA SPE cartridge, wash with two 1 mL portions of MeCN, elute with three 1 mL portions of MeCN:triethylamine 99.8:0.2, evaporate the eluate to dryness under reduced pressure, reconstitute with 200 μL MeCN, inject a 20-50 μL aliquot. (9-Anthryldiazomethane is available from Molecular Probes, Eugene OR. Synthesis is as follows. Stir 8.8 g 9-anthraldehyde and 8.5 g 80% hydrazine hydrate in 150 mL EtOH at room temperature for 3 h, filter off the solid 9-anthraldehyde hydrazone and dry under vacuum (mp 124-6°) (Bull. Chem. Soc. Jpn. 1967, 40, 691). Dissolve 220 mg 9-anthraldehyde hydrazone in 100 mL anhydrous ether, add 800 mg activated manganese dioxide, follow the reaction by reverse-phase HPLC using MeCN at 0.4 mL/min and UV 254. At the end of the reaction filter off the manganese and wash it with 20 mL ether, evaporate the filtrate to obtain 9-anthryldiazomethane (mp 64-6°) (Anal.Biochem. 1980, 107, 116 and 1983, 132 456). Prepare activated manganese dioxide as follows. Stir a solution of 20 g potassium permanganate in 250 mL water at room temperature, add 10 g activated carbon (Nuchar C-190 or C-190N), stir for 16 h, filter (Buchner funnel), wash 4 times with 50 mL portions of water, dry in air, dry in an oven at 105-110° for 8-24 h (J.Org.Chem. 1970, 35, 3971). See above for another synthesis.)

HPLC VARIABLES
Column: 125 × 4.6 5 μm Spherisorb ODS II
Mobile phase: MeCN:MeOH:water 45:40:15, containing 0.24% ammonium perchlorate and 0.02% triethylamine
Flow rate: 1.6
Injection volume: 20-50
Detector: F ex 360 em 440

CHROMATOGRAM
Retention time: 3.8 (quinapril), 12 (quinaprilat)
Internal standard: [2S-[1[R*(R*)]],2R*]-1-[2-[[(1-carboxy-3-phenyl)propyl]amino]-1-oxo-propyl]-octahydro-1H-indole-2-carboxylic acid (PD 110021, Warner-Lambert) (16)
Limit of quantitation: 20 ng/mL (plasma), 100 ng/mL (urine)
Limit of detection: 5 ng/mL (plasma)

OTHER SUBSTANCES
Extracted: metabolites
Simultaneously analyzed: enalapril, enalaprilat

KEY WORDS
plasma; pharmacokinetics; SPE

REFERENCE
Hengy, H.; Most, M. Determination of the new ACE-inhibitor quinapril and its active metabolite quinaprilate in plasma and urine by high-performance liquid chromatography and pre-column labelling for fluorescent-detection, *J.Liq.Chromatogr.*, **1988**, *11*, 517−530.

SAMPLE
Matrix: blood, urine
Analyte: imidapril
Sample preparation: Condition a 100 mg Bond Elut C18 SPE cartridge with 3 mL MeOH, with 3 mL water, and with 1 mL 100 mM HCl. Condition a 100 mg Bond Elut SI silica SPE cartridge with 1 mL acetone and 3 mL n-hexane. Plasma. 1 mL Plasma + 200 μL

2 M HCl + 200 μL water + 10 mL acetone, shake for 10 min, centrifuge at 1500 g for 10 min. Remove 10 mL of the supernatant and evaporate it to dryness under reduced pressure at 45°, reconstitute the residue in 1 mL 100 mM HCl, add 10 mL diethyl ether, shake for 5 min, centrifuge at 1500 g for 5 min. Add the aqueous layer to the C18 SPE cartridge, wash with 2 mL 100 mM HCl, wash with 1 mL MeOH:100 mM HCl 20:80, elute with 1 mL MeOH:100 mM HCl 60:40. Evaporate the eluate to dryness, reconstitute in 200 μL MeOH:100 mM HCl 60:40, evaporate to dryness into a vial, reconstitute in 25 μL water and 25 μL acetone, sonicate for 10 min, add 50 μL 0.2% 9-anthryldiazomethane in acetone, heat at 40° for 1 h, evaporate to dryness under reduced pressure at 45°, dissolve in 50 μL acetone and 1 mL hexane, add to the silica SPE cartridge, wash with 1 mL n-hexane:ethyl acetate 70:30, elute with 1 mL n-hexane:ethyl acetate 50:50, evaporate the eluate to dryness, reconstitute in 300 μL 20 ng/mL IS1 in MeCN:water 80:20, inject a 200 μL aliquot. Urine. 1 mL Urine + 200 μL water + 200 μL IS2 + 15 mL 500 mM Na$_2$HPO$_4$ make up to 25 mL with water, remove a 2 mL aliquot and add it to 10 mL diethyl ether, shake for 5 min, centrifuge at 1500 g for 5 min, discard the organic layer, add 10 mL chloroform, shake for 5 min, centrifuge at 1500 g for 5 min. Add 1 mL of the aqueous layer to the C18 SPE cartridge, wash with 2 mL 100 mM HCl, wash with 1 mL MeOH:100 mM HCl 20:80, elute with 1 mL MeOH:100 mM HCl 60:40. Evaporate the eluate to dryness, reconstitute in 200 μL MeOH:100 mM HCl 60:40, evaporate to dryness into a vial, reconstitute in 25 μL water and 25 μL acetone, sonicate for 10 min, add 50 μL 0.2% 9-anthryldiazomethane in acetone, heat at 40° for 1 h, evaporate to dryness under reduced pressure at 45°, dissolve in 50 μL acetone and 1 mL hexane, add to the silica SPE cartridge, wash with 1 mL n-hexane:ethyl acetate 70:30, elute with 1 mL n-hexane:ethyl acetate 50:50, evaporate the eluate to dryness, reconstitute in 250 μL MeCN:water 80:20, inject a 25 μL aliquot. (9-Anthryldiazomethane is available from Molecular Probes, Eugene OR. Synthesis is as described above.)

HPLC VARIABLES
Guard column: 4 × 4 5 μm Lichrosorb Si60
Column: 250 × 4 5 μm Hypersil MOS-5 dimethyloctyl
Mobile phase: MeCN:0.02% triethylamine 80:20, pH adjusted to 3.0 with 10% phosphoric acid
Flow rate: 1
Injection volume: 25-200
Detector: F ex 254 em 412

CHROMATOGRAM
Retention time: 23 (plasma), 20 (urine)
Internal standard: 9-anthrylmethyl myristate (IS1) (17), (S)-3-(N-[(S)-1-ethoxycarbonylundecyl]-L-alanyl)-1-methyl-2-oxoimidazoline-4-carboxylic acid (IS2) (42)
Limit of detection: 0.2 ng/mL (plasma), 10 ng/mL (urine)

OTHER SUBSTANCES
Extracted: metabolites

KEY WORDS
plasma; protect from light during and after derivatization; SPE

REFERENCE
Tagawa, K.; Hayashi, K.; Mizobe, M.; Noda, K. Highly sensitive determination of imidapril, a new angiotensin I-converting enzyme inhibitor, and its active metabolite in human plasma and urine using high-performance liquid chromatography with fluorescent labelling reagent, *J.Chromatogr.*, **1993**, *617*, 95–103.

SAMPLE
Matrix: bulk
Analyte: carboxylic acids
Sample preparation: Dissolve 1 μg in 1 mL pH 7.4 phosphate buffer, add 1 mL MeOH, add 2 mL 5% 9-anthryldiazomethane in ether, mix thoroughly, stir in the dark at 23° for 1 h. Remove the organic layer, extract the aqueous layer with 1 mL ether. Combine the organic layers and wash them with 500 μL water, evaporate to dryness, dissolve in ether, purify by TLC (Merck silica gel G) using ethyl acetate. Remove the appropriate zone and

extract with MeOH:ethyl acetate 50:50, evaporate to dryness, reconstitute, inject an aliquot. (9-Anthryldiazomethane is available from Molecular Probes, Eugene OR. Synthesis is as described above.)

HPLC VARIABLES
Column: 300 × 3.9 Nova-Pak C18
Mobile phase: MeCN:water:triethylamine:acetic acid 75:25:1:1
Detector: UV 256; F ex ≥254 em 400 (cut-off filter)

CHROMATOGRAM
Retention time: 6, 6.5 (trioxilin A_3 (8(S/R),11(R),12(S)-trihydroxyeicosa-5Z,9E,14Z-trienoic acid) (isomers)), 8.5, 9 (trioxilin B_3 (10(S/R),11(R),12(R)-trihydroxyeicosa-5Z,8Z,14Z-trienoic acid) (isomers)), 22, 23 (hepoxilin A_3 (10(S/R)-10-hydroxy-11(S),12(S)-trans-epoxyeicosa-5Z,9E,14Z-trienoic acid) (isomers)), 24.5, 25.5 (hepoxilin B_3 (8(S/R)-10-hydroxy-11(S),12(S)-trans-epoxyeicosa-5Z,9E,14Z-trienoic acid) (isomers)), 39 (12-hydroxyeicosa-5Z,8Z,10E,14Z-tetraenoic acid)
Limit of detection: 50 pg

REFERENCE
Demin, P.; Reynaud, D.; Pace-Asciak, C.R. Extractive derivatization of the 12-lipoxygenase products, hepoxilins, and related compounds into fluorescent anthryl esters for their complete high-performance liquid chromatography profiling in biological systems, *Anal.Biochem.*, **1995**, *226*, 252–255.

SAMPLE
Matrix: cell suspensions
Analyte: prostaglandins
Sample preparation: Centrifuge 1 mL cell suspension at 1600 g for 10 min, add 3 mL EtOH to the supernatant, centrifuge at 0° at 2200 g for 10 min. Dilute the supernatant to an EtOH content of 15%, adjust to pH 3.0 with 1 M HCl, add to a Sep-Pak C18 SPE cartridge, wash with 20 mL water, wash with 20 mL EtOH:water 15:85, wash with 20 mL benzene (Caution! Benzene is a carcinogen!), elute with 20 mL ethyl acetate:MeOH 90:10, evaporate the eluate to dryness under reduced pressure at room temperature, reconstitute with 100 μL MeOH. Remove a 10 μL aliquot and add it to 10 μL 0.2% 9-anthryldiazomethane in ethyl acetate, let stand at room temperature overnight, add to a Sep-Pak silica SPE cartridge (conditioned with chloroform:toluene 50:50), elute with 10 mL MeCN:MeOH 80:20, evaporate to dryness under reduced pressure at room temperature, reconstitute in ethyl acetate:MeOH 50:50, inject an aliquot. (9-Anthryldiazomethane is available from Molecular Probes, Eugene OR. Synthesis is as described above.)

HPLC VARIABLES
Guard column: 30 × 6 Develosil-ODS (Nomura Chemical)
Column: 250 × 4.6 Zorbax ODS
Mobile phase: MeCN:water:phosphoric acid 60:39.9:0.1
Column temperature: 30
Flow rate: 0.7
Detector: F ex 365 em 412

CHROMATOGRAM
Retention time: 17 (6-ketoprostaglandin $F_{1\alpha}$), 20 (6-ketoprostaglandin E_2), 25 (thromboxane B_2), 30 (dinoprostone), 34 (prostaglandin D_2)

KEY WORDS
SPE

REFERENCE
Kiyomiya, K.; Yamaki, K.; Nimura, N.; Kinoshita, T.; Oh-Ishi, S. Phorbol myristate acetate-stimulated release of cyclooxygenase products in rat pleural cells: derivatization of prostaglandins with 9-anthryldiazomethane for fluorometric determination by high performance liquid chromatography, *Prostaglandins*, **1986**, *31*, 71–82.

SAMPLE
Matrix: cell suspensions
Analyte: prostaglandins

Sample preparation: Condition a 1 mL Bond Elut 5 μm ODS SPE cartridge with 2 mL diethyl ether, 2 mL MeOH, and 2 mL buffer. Condition a 1 mL 5 μm Bond Elut silica SPE cartridge with 2 mL MeOH:MeCN 15:85, with 2 mL acetone, and with 2 mL chloroform. Centrifuge cell suspensions at 4° at 12000 g for 10 min, acidify to pH 5.5 with 1% acetic acid, add to the ODS SPE cartridge, wash with 2 mL buffer, wash with 2 mL MeOH:buffer (pH 5.5), suck dry under vacuum for 2 min, wash with 2 mL petroleum ether, elute with 2 mL diethyl ether. Evaporate the eluate to dryness under a stream of nitrogen, reconstitute the residue in 20 μL MeOH, add 20 μL 1 mg/mL 9-anthroyldiazomethane in diethyl ether, heat at 37° for 6 h, evaporate to dryness under a stream of nitrogen, reconstitute with 200 μL chloroform, add to the silica SPE cartridge, wash with 3 mL chloroform, elute with 2 mL MeCN:MeOH 85:15. Evaporate the eluate to dryness under a stream of nitrogen, reconstitute the residue in 50 μL mobile phase, add 50 μL water, inject an aliquot. (Buffer was 0.0001% acetic acid adjusted to pH 5.5 with 1% NaOH. 9-Anthryldiazomethane is available from Molecular Probes, Eugene OR. Synthesis is as described above.)

HPLC VARIABLES
Column: 250 × 4.6 5 μm Nucleosil ODS
Mobile phase: MeCN:water:acetic acid 62:37.9:0.1, pH adjusted to 5.5 with sodium acetate
Flow rate: 1
Detector: F ex 367 em 413

CHROMATOGRAM
Retention time: 12 (6-ketoprostaglandin $F_{1\alpha}$), 16 (thromboxane B_2), 20 (dinoprost, prostaglandin $F_{2\alpha}$), 23 (dinoprostone, prostaglandin E_2), 26 (prostaglandin D_2)

KEY WORDS
SPE

REFERENCE
Wessel, K.; Kaever, V.; Resch, K. Measurement of prostaglandins from biological samples in the subnanogram range by fluorescence labelling and HPLC separation, *J.Liq.Chromatogr.*, **1988**, *11*, 1271–1292.

SAMPLE
Matrix: enzyme incubations
Analyte: fatty acids
Sample preparation: 50 μL Enzyme incubation + 200 μL n-heptane:isopropanol:1 M sulfuric acid 20:80:2, add 120 μL heptane, add 70 μL water, add 5 nmoles margaric acid, vortex for 20 s. Remove a 30-50 μL aliquot of the heptane layer and evaporate it to dryness under reduced pressure, add 50 μL 500 μg/mL 9-anthryldiazomethane in MeOH:ethyl acetate 90:10 (prepare immediately before use), let stand at room temperature for 15 min, inject a 3-10 μL aliquot. (9-Anthryldiazomethane is available from Molecular Probes, Eugene OR or it may be synthesized as described above.)

HPLC VARIABLES
Column: 50 × 4 Superspher RP-18 (Merck)
Mobile phase: MeCN:water 95:5
Column temperature: 20
Flow rate: 1
Injection volume: 3-10
Detector: UV 254

CHROMATOGRAM
Retention time: 4 (linolenic acid), 5 (arachidonic acid), 6 (linoleic acid), 10 (oleic acid), 11 (palmitic acid), 20 (stearic acid)
Internal standard: margaric acid (15)

REFERENCE
Tojo, H.; Ono, T.; Okamoto, M. Reverse-phase high-performance liquid chromatographic assay of phospholipases: application of spectrophotometric detection to rat phospholipase A_2 isozymes, *J.Lipid Res.*, **1993**, *34*, 837–844.

SAMPLE
Matrix: formulations
Analyte: carnitine
Sample preparation: Powder tablets, weigh out amount containing 25 mg carnitine chloride, add 40 mL water, shake for 10 min, make up to 50 mL, centrifuge at 3000 rpm for 5 min, dilute supernatant five times. 10 μL Diluted supernatant + 10 μL 20% aqueous sodium dodecyl sulfate + 10 μL 100 μg/mL N,N-dimethylglycine in water + 370 μL isopropanol + 100 μL 10 mg/mL 9-anthryldiazomethane in ethyl acetate, heat at 50° for 30 min, cool, inject a 55 μL aliquot. (9-Anthryldiazomethane is available from Molecular Probes, Eugene OR. Synthesis is as described above.)

HPLC VARIABLES
Column: 150 × 4 10 μm LiChrosorb Si 100
Mobile phase: MeOH:5% aqueous sodium dodecyl sulfate:phosphoric acid 99:1:0.1
Flow rate: 1
Injection volume: 55
Detector: F ex 365 em 412 or UV 250

CHROMATOGRAM
Retention time: 10
Internal standard: N,N-dimethylglycine (5)
Limit of detection: 1 pg

OTHER SUBSTANCES
Non-interfering: antacids, dicycloverine, ethyl aminobenzoate, methylbenactyzium, nicotinamide, pantothenic acid, papaverine, pyridoxine, riboflavin, thiamine, vitamin C

KEY WORDS
tablets; normal phase

REFERENCE
Yoshida, T.; Aetake, A.; Yamaguchi, H.; Nimura, N.; Kinoshita, T. Determination of carnitine by high-performance liquid chromatography using 9-anthryldiazomethane, *J.Chromatogr.*, **1988**, *445*, 175–182.

SAMPLE
Matrix: solutions
Analyte: arachidonic acid metabolites
Sample preparation: Mix an aliquot of a solution in MeOH with 50 μL purified 9-anthryldiazomethane reagent, after 6 h inject an aliquot. (Purify 9-anthryldiazomethane on a 500 × 7.2 7 μm PG-pak C polystyrene gel column with ethyl acetate at 1 mL/min and UV 350 detection, inject 1 mg, collect the effluent when the purified compound elutes (20-22 min) and use it within 6 h. 9-Anthryldiazomethane is available from Molecular Probes, Eugene OR. Synthesis is as described above.)

HPLC VARIABLES
Column: 250 × 4.6 5 μm PG-Pak B silica gel
Mobile phase: Gradient. Isooctane:ethyl acetate:EtOH:acetic acid 90:10:0:1 for 15 min then 80:15:4:2 for 20 min (step gradient).
Flow rate: 1.2
Detector: F ex 365 em 412

CHROMATOGRAM
Retention time: 10 (hydroxyeicosatetraenoic acid), 15 (HHT), 27 (prostaglandin D_2), 29 (6-ketoprostaglandin $F_{1\alpha}$), 30 (thromboxaneB_2), 31.5 (dinoprostone), 32.5 (alprostadil), 38 (dinoprost), 40 (prostaglandin $F_{1\alpha}$)
Limit of detection: 100 pg

KEY WORDS
normal phase

REFERENCE

Yamauchi, Y.; Tomita, T.; Senda, M.; Hirai, A.; Terano, T.; Tamura, Y.; Yoshida, S. High-performance liquid chromatographic analysis of arachidonic acid metabolites by pre-column derivatization using 9-anthryldiazomethane, *J.Chromatogr.*, **1986**, *357*, 199–205.

SAMPLE

Matrix: solutions
Analyte: fatty acids
Sample preparation: Mix 100 μL of a 0.01-10 μg/mL solution with 100 μL of a 9-anthryl-diazomethane solution, vortex, let stand at room temperature for 1 h, inject a 5 μL aliquot. (Prepare 9-anthryldiazomethane solution as follows. Stir 8.8 g 9-anthraldehyde and 8.5 g 80% hydrazine hydrate in 150 mL EtOH at room temperature for 3 h, filter off the solid 9-anthraldehyde hydrazone and dry under vacuum (mp 124-6°) (Bull. Chem. Soc. Jpn. 1967, 40, 691). Add 500 μL 69 mM quinuclidine in ethyl acetate and 500 μL 6.9 mM N-chlorosuccinimide in ethyl acetate to 500 μL 6.9 mM 9-anthraldehyde hydrazone in ethyl acetate, vortex, let stand for 30 min, use immediately. An alternative synthesis is given above. 9-Anthryldiazomethane is also available from Molecular Probes, Eugene OR.)

HPLC VARIABLES

Column: 150 × 4 5 μm TSK-GEL-120T (Tosoh)
Mobile phase: Gradient. MeCN:water 84:16 for 30 min, to 100:0 over 25 min, maintain at 100:0 for 15 min (A) or MeOH:water 55:45 for 10 min, to 87:13 over 50 min (B) or 55:45 for 10 min, to 100:0 over 65 min (C) or 60:40 (isocratic) (D)
Column temperature: 50
Flow rate: 1
Injection volume: 5
Detector: F ex 255 or 365 em 412

CHROMATOGRAM

Retention time: 20 (linolenic acid (A)), 23 (palmitoleic acid (A)), 24 (myristic acid (A)), 26 (linoleic acid (A)), 36 (palmitic acid (A)), 38 (oleic acid (A)), 53 (stearic acid (A)), 64 (arachidic acid (A)), 18 (pyruvic acid (B)), 22 (levulinic acid (B)), 27 (α-ketobutyric acid (B)), 40 (α-ketovaleric acid (B)), 45 (α-ketocaproic acid (B)), 13 (glycolic acid (C)), 18 (lactic acid (C)), 27 (α-hydroxyisobutyric acid (C)), 38 (α-hydroxyisocaproic acid (C)), 49 (α-hydroxy-caprylic acid (C)), 65 (α-hydroxynaphthoic acid (C)), 68 (α-hydroxymyristic acid (C)), 72 (α-hydroxystearic acid (C)), 11 (lactic acid(D)), 20 (acetic acid (D)), 32 (propionic acid (D))
Limit of quantitation: 125 fmole

REFERENCE

Yoshida, T.; Uetake, A.; Yamaguchi, H.; Nimura, N.; Kinoshita, T. New preparation method for 9-an-thryldiazomethane (ADAM) as a fluorescent labeling reagent for fatty acids and derivatives, *Anal.Biochem.*, **1988**, *173*, 70–74.

SAMPLE

Matrix: solutions
Analyte: carnitine
Sample preparation: 500 μL 5 mg/mL carnitine in water + 100 mL acetone, remove a 1 mL aliquot and add it to 250 μL 1 mg/mL 9-anthryldiazomethane in acetone, let stand at 50° for 20 min, evaporate the solvent under a stream of nitrogen, add 2 mL 0.1% perchloric acid to the residue, wash this solution twice with 6 mL diethyl ether. Remove 1 mL of the aqueous phase and add it to 1 mL MeCN, inject a 20 μL aliquot. (9-Anthryl-diazomethane is available from Molecular Probes, Eugene OR. Synthesis is as described above.)

HPLC VARIABLES

Column: 250 × 4.6 10 μm Chiralcel OD-R (tris(3,5-dimethylphenylcarbamate)
Mobile phase: MeCN:500 mM sodium perchlorate solution 40:60
Flow rate: 0.8
Injection volume: 20
Detector: UV 254; F ex 365 em 412

CHROMATOGRAM

Retention time: 16 (D), 18 (L)

KEY WORDS
also for carnitine esters

REFERENCE
Hirota, T.; Minato, K.; Ishii, K.; Nishimura, N.; Sato, T. High-performance liquid chromatographic determination of the enantiomers of carnitine and acetylcarnitine on a chiral stationary phase, *J.Chromatogr.A*, **1994**, *673*, 37–43.

SAMPLE
Matrix: solutions
Analyte: choline glycerophospholipids
Sample preparation: Dissolve 5 pmole choline glycerophospholipids in 200 μL MeOH, add 200 μL 1% anthryldiazomethane in MeOH, let stand in the dark for 1 h, evaporate to dryness under a stream of nitrogen, purify by TLC (Merck silica gel G, chloroform: MeOH:25% ammonia 65:35:8), reconstitute, inject an aliquot. (9-Anthryldiazomethane is available from Molecular Probes, Eugene OR. Synthesis is as described above. 9-Anthryldiazomethane is also reported to be commercially available from Funakoshi, Tokyo.)

HPLC VARIABLES
Column: LiChrosorb RP-18
Mobile phase: MeCN:MeOH:THF:water 4.9:90.5:10:7 containing 20 mM choline chloride
Flow rate: 1
Detector: F ex 384 em 415

CHROMATOGRAM
Retention time: 10-20 (depending on structure)

REFERENCE
Ou, Z.; Ogamo, A.; Kawai, Y.; Nakagawa, Y. Quantitation of choline glycerophospholipids that contain carboxylate residues by fluorimetric high-performance liquid chromatography, *J.Chromatogr.A*, **1996**, *724*, 131–136.

SAMPLE
Matrix: tissue
Analyte: N-butyl-N-(3-carboxypropyl)nitrosamine
Sample preparation: Homogenize 100 mg tissue with 500 μL normal saline, centrifuge at 10000 g for 20 min. Remove a 1 mL aliquot of the supernatant and add it to 100 μL 2 M HCl and 1 mL ethyl acetate, stir vigorously for 1 min, centrifuge at 3000 rpm for 15 min. Remove a 500 μL aliquot of the organic layer and evaporate it to dryness under a stream of nitrogen, reconstitute the residue in 500 μL 1 mg/mL 9-anthryldiazomethane in ethyl acetate, heat at 40° for 1 h, inject a 15 μL aliquot. (9-Anthryldiazomethane is available from Molecular Probes, Eugene OR. Synthesis is as described above.)

HPLC VARIABLES
Column: ODS LS-120A (Toyo Soda)
Mobile phase: MeOH:water 80:20
Flow rate: 0.6
Injection volume: 15
Detector: F ex 365 em 430

CHROMATOGRAM
Retention time: 29
Limit of detection: <1.5 μg/g

KEY WORDS
rat; thymus; bladder; kidney; liver

REFERENCE
Wada, S.; Funae, Y.; Imaoka, S.; Kawamura, M.; Kinoshita, Y.; Sugimoto, T.; Nishio, S.; Kishimoto, T.; Maekawa, M. Rapid assay of N-butyl-N-(3-carboxypropyl)nitrosamine in rat organs and urine by high-performance liquid chromatography after derivatization, *Jpn.J.Cancer Res.*, **1985**, *76*, 192–196.

SAMPLE
Matrix: tissue
Analyte: polyether antibiotics

Sample preparation: Homogenize (Tissuemizer) 10 g tissue and 25 mL solvent for 1 min, wash blades with 3-4 mL solvent, combine with homogenate, shake for 30 min, centrifuge at 800 g for 15 min, decant. Add 25 mL solvent to residue, mix thoroughly, shake vigorously for 1 min, centrifuge at 800 g for 15 min, add supernatants to a 75 × 20 column of 80-200 mesh alumina (Fisher), rinse container onto column with 25 mL solvent, add 100 mL solvent to the column. Combine all the eluates and add 100 mL 5% NaCl, shake vigorously, let stand 2-3 min, add 30 mL dichloromethane, shake vigorously for 30 s, repeat extraction twice. Combine the dichloromethane layers and evaporate them to dryness under reduced pressure at 48-50°, reconstitute the residue in 1 mL solvent, add to a 75 ×20 column of 25-100 μm Sephadex LH-20, rinse flask with two 3.5 mL portions of solvent and add the rinses to the column, add 10 mL solvent to the column, discard the first 18 mL of eluate, add 10 mL solvent to the column, collect this fraction, evaporate to dryness under a stream of nitrogen at 48-50°, reconstitute the residue in 1 mL dichloromethane, evaporate to dryness under a stream of nitrogen at 48-50°, repeat twice, reconstitute in MeOH, evaporate to dryness under a stream of nitrogen at 48-50°, reconstitute in 100 μL pyridine and 100 μL acetic anhydride, let stand overnight at room temperature, add to 35 mL water, rinse vial with 5 mL water. Combine the aqueous layers and add 40 mL petroleum ether, shake vigorously for 1 min, separate and preserve the organic layer. Add 40 mL anhydrous ethyl ether to the aqueous phase, shake for 1 min. Combine the organic phases and wash with 20 mL 10% HCl, twice with 20 mL water, and with 40 mL saturated NaCl (shake for 1 min each time). Filter the organic layer through a glass fiber filter containing 20-30 g anhydrous sodium sulfate, wash the sodium sulfate with 10-15 mL petroleum ether, evaporate the filtrate to dryness under a stream of nitrogen at 48-50°. Take up the residue in 1 mL solvent and put it in another tube, wash flask into tube with three 2 mL portions of solvent, evaporate almost to dryness under a stream of nitrogen at 48-50°, make up to 1 mL with solvent, mix, add 500 μL 9-anthryldiazomethane solution, let stand in the dark for 30 min, evaporate to dryness under a stream of nitrogen at 48-50°, reconstitute in 1 mL hexane. Condition a Baker-10 silica SPE cartridge with 5-10 mL hexane, do not allow to dry. Add the hexane solution to the SPE cartridge, rinse tube with 9 mL hexane, add rinse to the SPE cartridge. Wash SPE cartridge with 10 mL hexane:dichloromethane 50:50, with 10 mL hexane:dichloromethane 20:80, with 10 mL dichloromethane, and with 1 mL MeOH. Elute with 1 mL MeOH, inject a 20 μL aliquot of the eluate. (Solvent was MeOH:water 80:20. Prepare 9-anthryldiazomethane solution as follows. Add 1100 g manganous sulfate tetrahydrate in 1.5 L water and 1170 mL 40% NaOH over 1 h to a hot stirred solution of 960 g potassium permanganate in 6 L water, stir for 1 h, centrifuge, wash solid with water until washings are colorless, dry solid at 100-120°, grind the activated manganese dioxide to a fine powder. Add 8.5 g 85% hydrazine hydrate (Caution! Hydrazine hydrate is a carcinogen!) to 8.8 g 9-anthraldehyde dissolved in 150 mL EtOH, stir at room temperature for 3 h, filter off solid, dry under vacuum, recrystallize from EtOH to give 9-anthraldehyde hydrazone as red-yellow crystals, mp 124-6°. Dissolve 220 mg 9-anthraldehyde hydrazone in 100 mL anhydrous ethyl ether, add 800 mg activated manganese dioxide, add 600 μL EtOH saturated with KOH, stir vigorously for 30 min, filter (glass fiber), wash solid with 20 mL anhydrous ethyl ether, evaporate to reduce volume, make up to 100 mL with anhydrous ethyl ether, store in a dark flask in the dark in a refrigerator. Discard after 30 days (J.Assoc.Off.Anal.Chem. 1985, 68, 1149). An alternative procedure is given above. 9-Anthryldiazomethane is also available from Molecular Probes, Eugene OR.)

HPLC VARIABLES

Guard column: pellicular C18 (Alltech)
Column: 200 × 4.6 5 μm RP-C8 (Hewlett-Packard)
Mobile phase: Gradient. A was MeCN. B was MeCN:water 10:90. A:B 20:80 for 9 min, 10:90 for 7 min, 20:80 for 1 min.
Column temperature: 40
Flow rate: 1
Injection volume: 20
Detector: F ex 365 em 418 (filter)

CHROMATOGRAM

Retention time: 11.5 (monensin), 12.6 (salinomycin), 13.2 (narasin)

Limit of detection: 0.15 ppm

KEY WORDS
cow; liver; SPE

REFERENCE
Martinez, E.E.; Shimoda, W. Liquid chromatographic determination of multiresidue fluorescent derivatives of ionophore compounds, monensin, salinomycin, narasin, and lasalocid, in beef liver tissue, *J.Assoc.Off.Anal.Chem.*, **1986**, *69*, 637−641.

SAMPLE
Matrix: tissue
Analyte: okadaic acid
Sample preparation: Vortex 1 g hepatopancreas homogenate with 4 mL MeOH:water 80:20, centrifuge at 1200 rpm for 10 min. Remove a 25 mL aliquot and wash it four times with 4 mL portions of n-hexane. Remove a 1 mL aliquot of the MeOH/water phase and add it to 4 mL chloroform, vortex for 20 s, repeat the extraction. Combine the chloroform layers and make up to 10 mL with chloroform. Remove a 500 μL aliquot and evaporate it to dryness under a stream of nitrogen, reconstitute with 100 μL 0.1% 9-anthryldiazomethane in MeOH, let stand in the dark at 25° for 1 h, evaporate to dryness under a stream of nitrogen, reconstitute with three 2 mL portions of chloroform:n-hexane 50:50, add these rinses to a Sep-Pak silica SPE cartridge, wash with 5 mL chloroform, elute with 5 mL chloroform:MeOH 95:5. Evaporate the eluate to dryness under a stream of nitrogen, reconstitute the residue in 100 μL mobile phase, inject a 20 μL aliquot. (9-Anthryldiazomethane is available from Molecular Probes, Eugene OR. Synthesis is as described above.)

HPLC VARIABLES
Column: Nucleosil 7-C18
Mobile phase: MeCN:THF:water 67:3:30
Flow rate: 2
Injection volume: 20
Detector: F ex 365 em 412; UV 254; UV 280

CHROMATOGRAM
Retention time: 17
Limit of detection: 500 ng/g

KEY WORDS
mussels; hepatopancreas; SPE

REFERENCE
Luckas, B.; Meixner, R. Vorkommen und Bestimmung von Okadasäure in Muscheln der deutschen Nordseeküste [Occurrence and determination of okadaic acids in mussels from the German North Sea coast], *Z.Lebensm.Unters.Forsch.*, **1988**, *187*, 421−424.

SAMPLE
Matrix: tissue
Analyte: shellfish toxins
Sample preparation: Condition a 100 mg Bondesil 40 μm silica SPE cartridge (Analytichem) with hexane:dichloromethane 50:50. 1 g Mussel hepatopancreas + 4 mL MeOH:water, homogenize, centrifuge. Remove a 2.5 mL aliquot of the supernatant and wash it twice with 2.5 mL portions of petroleum ether, add 1 mL water, extract with 4 mL chloroform. Dilute the organic layer to 10 mL, remove a 500 μL aliquot and add it to 100 μL 500 ng/mL deoxycholic acid in MeOH, evaporate to dryness, add 100 μL 0.1% 9-anthryldiazomethane in MeOH, sonicate at 37° for 15 min, heat at 37° for 45 min, evaporate to dryness under a stream of nitrogen, reconstitute with three 500 μL portions of hexane:dichloromethane 50:50, add these portions to the SPE cartridge, wash with 1 mL hexane:dichloromethane 50:50, wash with 1 mL dichloromethane:acetone 90:10, elute with 2.5 mL MeCN:dichloromethane 50:50. Evaporate the eluate to dryness under a stream of nitrogen, reconstitute with 100 μL MeOH, inject a 10 μL aliquot. (9-Anthryldiazomethane is available from Molecular Probes, Eugene OR. Synthesis is as described above.)

HPLC VARIABLES
Column: 100 × 4 5 μm Spherisorb ODS-2
Mobile phase: MeCN:MeOH:water 80:10:10
Flow rate: 0.6
Injection volume: 10
Detector: F ex 365 em 412

CHROMATOGRAM
Retention time: 6.5 (okadaic acid), 9 (dinophysistoxin-1)
Internal standard: deoxycholic acid (11.5)
Limit of detection: 40 ng/g

KEY WORDS
mussel; hepatopancreas; SPE

REFERENCE
Stabell, O.B.; Hormazabal, V.; Steffenak, I.; Pedersen, K. Diarrhetic shellfish toxins: improvement of sample clean-up for HPLC determination, *Toxicon*, **1991**, *29*, 21–29.

SAMPLE
Matrix: tissue
Analyte: shellfish toxins
Sample preparation: Condition a 500 mg silica SPE cartridge (Supelco) with 5 mL chloroform and 5 mL hexane:chloroform 50:50. Homogenize 2 g mussel digestive glands with 8 mL MeOH:water 80:20 for 2 min, centrifuge at 4000 rpm for 10 min. Remove a 5 mL aliquot of the supernatant and add it to 5 mL hexane, vortex for 30 s, centrifuge, repeat the hexane wash. Add 1 mL water to the MeOH/water layer, extract twice with 6 mL portions of chloroform. Combine the organic layers and evaporate them to dryness under a stream of nitrogen, reconstitute the residue in 1 mL MeOH. Remove a 10 μL aliquot and add it to 90 μL reagent, heat at 35° for 1 h, evaporate to dryness under a stream of nitrogen, reconstitute with 1 mL hexane:chloroform 50:50, add to the SPE cartridge, wash with 5 mL hexane:chloroform 50:50, wash with 5 mL chloroform, elute with 5 mL chloroform:MeOH 95:5, evaporate the eluate to dryness, reconstitute with 100 μL MeOH, inject an aliquot. (Prepare reagent by mixing 500 μL 35 mM 9-anthraldehyde hydrazone in ethyl acetate with 500 μL 70 mM quinuclidine in ethyl acetate and 500 μL 7 mM N-chlorosuccinimide in ethyl acetate, let stand in the dark at room temperature for 1 h before use. 9-Anthryldiazomethane is prepared in situ. Synthesis of 9-anthraldehyde hydrazone is as described above. 9-Anthryldiazomethane is also available from Molecular Probes, Eugene OR.)

HPLC VARIABLES
Column: 250 × 4 5 μm LiChrospher-100 RP18
Mobile phase: Gradient. MeCN:water 90:10 for 17 min, to 100:0 over 1 min, maintain at 100:0 for 10 min.
Flow rate: 1
Detector: F ex 254 em 412 (280 nm cutoff filter)

CHROMATOGRAM
Retention time: 10 (okadaic acid), 15 (DTX-1)
Limit of detection: 30 ng/g

KEY WORDS
mussels; digestive glands; SPE; comparison with other derivatizing reagents

REFERENCE
Marr, J.C.; McDowell, L.M.; Quilliam, M.A. Investigation of derivatization reagents for the analysis of diarrhetic shellfish poisoning toxins by liquid chromatography with fluorescence detection, *Nat.Toxins.*, **1994**, *2*, 302–311.

SAMPLE
Matrix: tissue
Analyte: shellfish toxins
Sample preparation: Immediately before use prepare an SPE column of 500 mg silica (40 μm Analytichem Bondesil, activate at 130° for 24 h), condition with 6 mL chloroform

(stabilized with EtOH), condition with 3 mL chloroform:hexane 50:50, do not allow to go dry. Homogenize (Polytron Model PT3000) 2 g tissue with 100 μL 100 μg/mL IS in MeOH and 7.9 mL MeOH:water 80:20 at 6000-10000 rpm for 3 min, centrifuge at 4000 g for 10 min. Remove a 5 mL aliquot of the supernatant and add it to 5 mL n-hexane, vortex for 30 s, discard the hexane layer, repeat the wash, add 1 mL water, add 6 mL chloroform, vortex for 30 s, repeat extraction with 6 mL chloroform. Combine the organic layers and evaporate them to dryness under reduced pressure, add 200 μL 35 μg/mL deoxycholic acid in MeOH, add 800 μL MeOH, vortex, add 100 μL 0.2% 9-anthryldiazomethane in MeOH (freshly prepared), sonicate at 37° for 10 min, heat at 37° in the dark for 2 h, evaporate to dryness under reduced pressure, transfer the residue to the SPE column using three 300 μL portions of chloroform:hexane 50:50, wash with 5 mL chloroform:hexane 50:50, wash with 5 mL chloroform:EtOH 98.85:1.15, elute with 5 mL chloroform:MeOH 90:10, evaporate to dryness under reduced pressure, reconstitute with 500 μL MeOH, inject a 10 μL aliquot. (Prepare chloroform:EtOH 98.85:1.15 by passing chloroform through 50 g activated alumina (Woelm basic alumina, activity grade 1, activated at 450° overnight, Alipharm Chemicals, New Orleans), discard first 10 mL, collect the next 50 mL and add it to 575 μL EtOH. This procedure is critical. 9-Anthryldiazomethane is available from Molecular Probes, Eugene OR. Synthesis is as described above.)

HPLC VARIABLES
Column: 250 × 4 5 μm LiChrospher-100 RP-18 octadecylsilica
Mobile phase: MeCN:water 80:20
Column temperature: 40
Flow rate: 1
Injection volume: 10
Detector: F ex 254 em 412 (280 cutoff filter)

CHROMATOGRAM
Retention time: 14 (okadaic acid), 16 (dinophysistoxin-2), 23 (dinophysistoxin-1)
Internal standard: 7-O-acetylokadaic acid (Dissolve 1 mg okadaic acid in 100 μL pyridine and add 50 μL 3.2 mg/mL acetic anhydride in pyridine, let stand at room temperature overnight, evaporate to dryness, reconstitute with MeOH, isolate acetylokadaic acid using 250 × 10 Vydac 201TP column with MeOH:water 60:40 at 4 mL/min and UV 220 detection.) (19), deoxycholic acid (26)
Limit of quantitation: 100 ng/g

KEY WORDS
mussels; protect from light; SPE

REFERENCE
Quilliam, M.A. Analysis of diarrhetic shellfish poisoning toxins in shellfish tissue by liquid chromatography with fluorometric and mass spectrometric detection, *J.AOAC Int.*, **1995**, *78*, 555-570.

SAMPLE
Matrix: tissue
Analyte: okadaic acid
Sample preparation: Extract 1 g mussel hepatopancreas with 4 mL MeOH:water 80:20, centrifuge. Remove a 2.5 mL aliquot of the supernatant and extract it twice with 2.5 mL portions of hexane (A). Add 1 mL water to the MeOH layer, extract twice with 4 mL portions of chloroform. Combine the chloroform layers and make up to 10 mL. Remove a 0.5 mL aliquot and evaporate it to dryness, reconstitute with 200 μL 0.2% 9-anthryldiazomethane in MeOH, shake for 2 min, heat at 40° in the dark for 1 h, evaporate to dryness, reconstitute with 1 mL hexane:chloroform 50:50, add to a Sep-Pak silica SPE cartridge, wash with 5 mL hexane:chloroform 50:50, wash with 5 mL chloroform, elute with 5 mL chloroform:MeOH 95:5. Evaporate the eluate to dryness under a stream of nitrogen, reconstitute the residue in 100 μL MeOH (Agric. Biol. Chem. 1987, 51, 877), inject a 10 μL aliquot. Evaporate the hexane extracts (A) to dryness under a stream of nitrogen, add 500 μL 500 mM NaOH, stir in the dark at room temperature for 1 h to hydrolyze okadaic acid esters. Concentrate, add 2 mL 125 mM HCl, extract three times with 2 mL portions of chloroform. Combine the extracts and dry them over anhydrous sodium sulfate. Remove an aliquot and evaporate it to dryness, reconstitute with 200 μL

0.2% 9-anthryldiazomethane in MeOH, shake for 2 min, heat at 40° in the dark for 1 h, evaporate to dryness, reconstitute with 1 mL hexane:chloroform 50:50, add to a Sep-Pak silica SPE cartridge, wash with 5 mL hexane:chloroform 50:50, wash with 5 mL chloroform, elute with 5 mL chloroform:MeOH 95:5. Evaporate the eluate to dryness under a stream of nitrogen, reconstitute the residue in 100 μL MeOH (Agric. Biol. Chem. 1987, 51, 877), inject a 10 μL aliquot. (9-Anthryldiazomethane is available from Molecular Probes, Eugene OR. Synthesis is as described above.)

HPLC VARIABLES
Column: 250 × 4 4 μm Superspher 100-RP-18 (Merck)
Mobile phase: MeCN:water 80:20
Column temperature: 35
Flow rate: 1.1
Injection volume: 10
Detector: F ex 365 em 412

CHROMATOGRAM
Retention time: 14

KEY WORDS
mussel; hepatopancreas; SPE

REFERENCE
Fernández, M.L.; Míguez, A.; Cacho, E.; Martínez, A. Detection of okadaic acid esters in the hexane extracts of Spanish mussels, *Toxicon*, **1996**, *34*, 381–387.

SAMPLE
Matrix: tissue
Analyte: shellfish toxins
Sample preparation: Homogenize (Ultra-Turrax) 1 g tissue with 4 mL MeOH:water 80:20 for 3 min, centrifuge at 2980 g for 5 min. Remove a 2.5 mL aliquot of the supernatant and extract it twice with 4 mL portions of dichloromethane, combine the organic layers and make up to 10 mL with dichloromethane. Evaporate a 1 mL aliquot of the dichloromethane solution to dryness under a stream of nitrogen, add 200 μL 0.15% 9-anthryldiazomethane in acetone, let stand at 25° in the dark for 1.5 min, inject a 20 μL aliquot on to column A and elute to waste with mobile phase A, divert the fractions eluting from 15-18 min (okadaic acid) and from 29-32 min (dinophysistoxin-1) on to column B. Elute column B with mobile phase B during this process. When both compounds are trapped on column B backflush the contents of column B on to column C with mobile phase B, elute column C with mobile phase B, monitor the effluent from column C. (9-Anthryldiazomethane can be obtained from Molecular Probes, Eugene OR or Serva Feinbiochemica, Germany. It can also be synthesized as described above.)

HPLC VARIABLES
Column: A 250 × 4.6 5 μm Supelcosil LC-8-DB; B 100 × 4 5 μm Nucleosil C18; C 250 × 4.6 5 μm Supelcosil LC-18
Mobile phase: A MeCN:water 65:35; B MeCN:water 90:10
Flow rate: A 1.6; B 1.1
Injection volume: 20
Detector: F ex 365 em 415

CHROMATOGRAM
Retention time: 28 (okadaic acid), 43 (dinophysistoxin-1)
Limit of detection: 0.5 ng

KEY WORDS
shellfish; mussels; column-switching; heart-cut

REFERENCE
Hummert, C.; Shen, J.L.; Luckas, B. Automatic high-performance liquid chromatographic method for the determination of diarrhetic shellfish poison, *J.Chromatogr.A*, **1996**, *729*, 387–392.

SAMPLE
Matrix: water

Analyte: herbicides
Sample preparation: 20 mL Ground water + 100 μL concentrated HCl + 2 mL n-hexane:ethyl acetate 20:80, shake vigorously for 1 min, repeat extraction twice more. Combine the organic layers and evaporate to about 500 μL under reduced pressure at 40°, evaporate to dryness under a stream of nitrogen, add 100 μL 0.025% 9-anthryldiazomethane in acetone, heat at 40° in the dark for 1 h, evaporate to dryness under a stream of nitrogen, reconstitute with 75 μL acetone, inject a 15 μL aliquot. (9-Anthryldiazomethane can be obtained from Molecular Probes, Eugene OR or Serva Feinbiochemica, Germany. It can also be synthesized as described above.)

HPLC VARIABLES
Column: 250 × 4.6 TSK-gel ODS120T (Tosoh)
Mobile phase: MeCN:water:THF 75:25:3
Flow rate: 1
Injection volume: 15
Detector: F ex 365 em 412

CHROMATOGRAM
Retention time: 16 ((2,4-dichlorophenoxy)acetic acid), 19 ((4-chloro-2-methylphenoxy)acetic acid), 26 (2-(4-chloro-2-methylphenoxy)propionic acid), 30 ((4-chloro-2-methylphenoxy)butyric acid)
Limit of detection: 500 pg

KEY WORDS
ground water

REFERENCE
Suzuki, T.; Watanabe, S. Screening method for phenoxy acid herbicides in ground water by high-performance liquid chromatography of 9-anthryldiazomethane derivatives and fluorescence detection, *J.Chromatogr.*, **1991**, *541*, 359–364.

Diazomethane

CH$_2$N$_2$ + R—C(=O)—OH ⟶ CH$_3$—O—C(=O)—R

SAMPLE
Matrix: gastric mucosa
Analyte: prostaglandins
Sample preparation: Condition a Sep-Pak C18 SPE cartridge with 10 mL EtOH and 100 mL water. Homogenize gastric mucosa from three rats with 6 mL EtOH:50 mM pH 3.15 triethylammonium formate buffer 15:85 (cold), centrifuge at 300000 g for 15 min, resuspend the pellet in 3 mL EtOH:50 mM pH 3.15 triethylammonium formate buffer 15:85, centrifuge at 300000 g for 15 min. Combine the supernatants and add them to the SPE cartridge, wash with 50 mM pH 3.15 triethylammonium formate buffer 15:85, wash with 10 mL petroleum ether, elute with 10 mL methyl formate, store at -40°. Evaporate under helium, dissolve the residue in 100 μL diethyl ether with 2-3 drops MeOH, add 500 μL of a fresh solution of diazomethane, let stand in the dark at room temperature for 30 min, store in MeCN at -40°. (Preparation of diazomethane is as follows. Caution! Diazomethane is toxic, explosive, and carcinogenic! A face shield and a safety screen should always be used and the preparation should only be carried out in a properly functioning chemical fume hood. Only smooth glass apparatus with rubber stoppers and plastic tubing should be used. Scratched glassware, ground glass joints, and sharp edges should be avoided (Org. Syn., Coll. Vol. VI; Wiley: New York, 1988, pp. 432-435). Procedures have been reported for the synthesis of diazomethane from N-methyl-N-nitrosourea (Org. Syn., Coll. Vol. II; Wiley: New York, 1943, pp. 165-167), N-nitroso-β-methylaminoisobutyl methyl ketone (Org. Syn., Coll. Vol. III; Wiley: New York, 1955, pp. 244-248), and N,N'-

dimethyl-N,N'-dinitrosoterephthalamide (Org. Syn., Coll. Vol. V; Wiley: New York, 1973, pp. 351-355). Probably the most convenient starting material is N-methyl-N-nitroso-p-toluenesulfonamide (Diazald) (Aldrichimica Acta 1983, 16, 3-10; Org. Syn., Coll. Vol. IV, Wiley: New York, 1963, pp. 250-253). Add 10 mL 95% EtOH to a solution of 5 g KOH in 8 mL water, warm to 65° using a water bath, add a solution of 5 g N-methyl-N-nitroso-p-toluenesulfonamide in 45 mL ether over 20 min at such a rate as to keep the reaction volume constant. Collect the ether and diazomethane that distil in an ice-cooled receiving flask under a dry ice/acetone condenser. When all the N-methyl-N-nitroso-p-toluenesulfonamide has been used up, slowly add 10 mL ether to the reaction flask and continue distillation until the distillate is colorless. A purpose-built distillation apparatus can be purchased from Aldrich (Aldrichimica Acta 1983, 16, 3-10). Excess quantities of diazomethane can be destroyed by adding acetic acid until the yellow color of the diazomethane is discharged. The safe disposal of the nitroso compounds used to generate diazomethane has been discussed (Lunn,G.; Sansone,E.B. Destruction of Hazardous Chemicals in the Laboratory, Second Edition. Wiley: New York, 1994, pp. 277-289).)

HPLC VARIABLES
Column: 150×4 5 µm Spherisorb ODS-2
Mobile phase: Gradient. A was 100 mM ammonium acetate adjusted to pH 3.5 with formic acid. B was MeCN : 200 mM ammonium acetate buffer 2 : 1. A : B from 65 : 35 to 15 : 85 over 27 min.
Injection volume: 20
Detector: MS, Hewlett-Packard 5988A, thermospray, insertion probe 180-185°, SIM m/z 384

CHROMATOGRAM
Retention time: 19.7 (6-keto-prostaglandin $F_{1\alpha}$), 24.8 (dinoprost), 26.7 (dinoprostone), 27.7 (prostaglandin D_2)
Limit of detection: 110-560 pg

KEY WORDS
rat; SPE

REFERENCE
Abián, J.; Bulbena, O.; Gelpí, E. Thermospray liquid chromatography/mass spectrometry of prostaglandin methyl ester derivatives: application to the determination of prostaglandins E_2 and D_2 in rat gastric mucosa, *Biomed.Environ.Mass.Spectrom.*, **1988**, *16*, 215−219.

1-Naphthyldiazomethane

SAMPLE
Matrix: bulk
Analyte: bile acids
Sample preparation: Dissolve 1-5 mg bile acid in 500 µL chloroform with enough MeOH to make a solution, add a solution of 1-naphthyldiazomethane in ether until the reddish-orange color persists, if the color disappears within 1 h add more reagent, add 1 drop acetic acid to decompose excess reagent, make up to 1 mL, inject a 5 µL aliquot. (Preparation of 1-naphthyldiazomethane is as follows. Stir 6.7 g 1-naphthaldehyde and 8.5 g 80% hydrazine hydrate in 150 mL EtOH at room temperature for 3 h (Caution! Hydrazine hydrate is a carcinogen!). Remove the solid by filtration and recrystallize it twice from

EtOH to give 1-naphthaldehyde hydrazone as white crystals (mp 91-92°). Stir 3.1 g 1-naphthaldehyde hydrazone, 5 g anhydrous sodium sulfate, 50 mL ether, 1 mL EtOH saturated with KOH, and 10 g yellow mercuric oxide for 5 h, filter (sintered glass), concentrate the filtrate under reduced pressure to give 1-naphthyldiazomethane as red crystals (mp 40-41°) (Bull. Chem. Soc. Japan 1967, 40, 691).)

HPLC VARIABLES
Column: 300 mm long μPorasil
Mobile phase: Hexane:THF:MeOH 75:30:2
Flow rate: 1
Injection volume: 5
Detector: UV 280

CHROMATOGRAM
Retention time: 7 (lithocholic acid), 10 (deoxycholic acid), 12 (chenodiol), 18 (3,7-dihydroxy-12-ketocholanic acid)
Limit of detection: 20-30 ng

KEY WORDS
normal phase

REFERENCE
Matthees, D.P.; Purdy, W.C. Naphthyldiazomethane as a derivatizing agent for the high-performance liquid chromatography detection of bile acids, *Anal.Chim.Acta*, **1979**, *109*, 161–164.

1-Pyrenyldiazomethane

RELATED REFERENCE
Schneede, J.; Ueland, P.M. Automated assay of methylmalonic acid in serum and urine by derivatization with 1-pyrenyldiazomethane, liquid chromatography, and fluorescence detection. *Clin.Chem.* **1993**, *39*, 392-399.

SAMPLE
Matrix: blood, formulations
Analyte: biotin
Sample preparation: Serum. Condition a Sep-Pak C18 SPE cartridge with 10 mL MeOH, 10 mL water, and 5 mL 1% acetic acid in water. 1 mL Serum + 1 mL 10% trichloroacetic acid, mix, centrifuge at 2000 g for 5 min, add a 1.5 mL aliquot of the supernatant to the SPE cartridge, wash with 10 mL 1% acetic acid in water, wash with 1 mL water, elute with 10 mL MeOH. Evaporate the eluate to dryness, reconstitute with 100 μL MeOH, sonicate, add 100 μL 1 mg/mL 1-pyrenyldiazomethane in ethyl acetate, heat at 40° for 1 h, cool to room temperature, add 300 μL MeOH, inject a 10 μL aliquot. Tablets. Powder tablets, weigh out amount containing 500 μg biotin, add 30 mL MeOH, sonicate for 10 min, shake for 10 min, centrifuge at 3000 rpm for 5 min, remove the supernatant, repeat the extraction three times. Combine the supernatants and make up to 100 mL with MeOH. Remove a 200 μL aliquot and add it to 200 μL 5 mg/mL 1-pyrenyldiazomethane in ethyl acetate, heat at 40° for 1 h, cool to room temperature, add to a Sep-Pak silica

SPE cartridge, wash with 5 mL hexane, elute with 10 mL MeOH. Evaporate the eluate to dryness, reconstitute with 1 mL MeOH, inject a 5 μL aliquot. (1-Pyrenyldiazomethane is available from Molecular Probes, Eugene OR. Synthesis is as follows. Suspend 5 g 1-pyrenecarboxaldehyde in 80 mL EtOH, add 3.4 g hydrazine monohydrate (Caution! Hydrazine monohydrate is a carcinogen!), stir at room temperature for 3 h, filter off the product and wash it with 50 mL cold EtOH, recrystallize from EtOH to obtain 1-pyrenecarboxaldehyde hydrazone as yellow crystals (mp 186-194° d). Add 6.55 g activated manganese dioxide to 2 g 1-pyrenecarboxaldehyde hydrazone in 300 mL diethyl ether, sonicate at room temperature for about 80 min (monitor by HPLC), filter, wash the solid with a little ether, evaporate the filtrate to obtain 1-pyrenyldiazomethane as red crystals (Anal. Chem. 1988, 60, 2067). Prepare activated manganese dioxide as follows. Stir a solution of 20 g potassium permanganate in 250 mL water at room temperature, add 10 g activated carbon (Nuchar C-190 or C-190N), stir for 16 h, filter (Buchner funnel), wash 4 times with 50 mL portions of water, dry in air, dry in an oven at 105-110° for 8-24 h (J.Org.Chem. 1970, 35, 3971).)

HPLC VARIABLES
Column: 150 × 4 5 μm LiChrosorb Si60
Mobile phase: MeCN:water 43:57 (or 57:43 (?))
Flow rate: 1
Injection volume: 5-10
Detector: F ex 340 em 395; UV 240

CHROMATOGRAM
Retention time: 23
Limit of detection: 100 fmole

KEY WORDS
tablets; SPE; serum

REFERENCE
Yoshida, T.; Uetake, A.; Nakai, C.; Nimura, N.; Kinoshita, T. Liquid chromatographic determination of biotin using 1-pyrenyldiazomethane as a pre-column fluorescent labelling reagent, *J.Chromatogr.*, **1988**, *456*, 421–426.

SAMPLE
Matrix: blood, urine
Analyte: S-carboxymethyl-L-cysteine and metabolites
Sample preparation: 25 μL Plasma or urine + 1.25 μg N-acetylalanine, evaporate to dryness under reduced pressure, reconstitute with 150 μL buffer A, add 150 μL 20 μM 1-octanesulfonic acid in isopropanol, add 150 μL 15 mM 1-pyrenyldiazomethane (Molecular Probes, Eugene OR; synthesis also described above) in ethyl acetate, heat at 40° for 30 min, add 150 μL isopropanol, add 400 μL buffer B, add 400 μL 10 mM 9-fluorenylmethyl chloroformate in acetone, let stand at room temperature for 50 s, add 35 μL 500 mM ammonium acetate solution, inject a 10 μL aliquot. (Buffer A was isopropanol:water 50:50 containing 100 mM phosphoric acid, adjusted to pH 4.0 with NaOH. Buffer B was 30.9 g/L boric acid adjusted to pH 7.7 with 30% NaOH. The pyrenyldiazomethane reacts with carboxylic acids and the fluorenylmethyl chloroformate reacts with amines. During the second derivatization the pyrenyl group is removed from amino acids and so amino acids exist only as fluorenyl derivatives.)

HPLC VARIABLES
Column: 150 × 4.6 5 μm Ultracarb ODS (Phenomenex)
Mobile phase: Gradient. MeCN:buffer from 25:75 to 50:50 over 80 min. (Buffer was MeOH:THF:40 mM triethylamine 10:5:85, adjusted to pH 3.2 with phosphoric acid.)
Column temperature: 60
Flow rate: 0.8
Injection volume: 10
Detector: F ex 340 em 380 (pyrenyldiazomethane derivatives); F ex 260 em 305 (fluorenylmethyl chloroformate derivatives)

CHROMATOGRAM
Retention time: 6 (S-carboxymethyl-L-cysteine sulfoxide*), 11 (S-methyl-L-cysteine sulf-oxide*), 17 (N-acetyl-S-carboxymethyl-L-cysteine sulfoxide), 20 (S-carboxymethyl-L-cys-teine*), 22 (thiodiglycolic acid sulfoxide), 25 (N-acetyl-S-methyl-L-cysteine sulfoxide), 28 (thiodiglycolic acid sulfone), 37 (N-acetyl-S-carboxymethyl-L-cysteine), 38 (methylcys-teine*), 58 (thiodiglycolic acid), 64 (N-acetyl-S-methyl-L-cysteine) [* fluorenyl derivatives, others are pyrenyl derivatives]
Internal standard: N-acetylalanine (49)

KEY WORDS
plasma

REFERENCE
Staffeldt, B.; Brockmöller, J.; Roots, I. Determination of S-carboxymethyl-L-cysteine and some of its metabolites in urine and serum by high-performance liquid chromatography using fluorescent pre-column labelling, *J.Chromatogr.*, **1991**, *571*, 133–147.

SAMPLE
Matrix: blood, urine
Analyte: S-carboxymethyl-L-cysteine and metabolites
Sample preparation: 50 µL Plasma or urine + 2.5 µg N-acetylalanine, evaporate to dry-ness under reduced pressure, reconstitute with 300 µL buffer, add 300 µL 20 µM 1-octanesulfonic acid in isopropanol, add 300 µL 15 mM 1-pyrenyldiazomethane (Molecular Probes, Eugene OR; synthesis also described above) in ethyl acetate, shake gently at 40° for 30 min, dilute to 1.5 mL with isopropanol, inject an aliquot. (Buffer was isopropanol: water 50:50 containing 100 mM phosphoric acid, adjusted to pH 4.0 with NaOH.)

HPLC VARIABLES
Column: 150 × 4.6 5 µm Ultracarb ODS (Phenomenex)
Mobile phase: Gradient. MeCN:buffer from 20:80 to 35:65 over 70 min. (Buffer was MeOH:THF:40 mM triethylamine 10:5:85, adjusted to pH 3.2 with phosphoric acid.)
Column temperature: 50
Flow rate: 0.8
Detector: F ex 340 em 380

CHROMATOGRAM
Internal standard: N-acetylalanine

KEY WORDS
plasma

REFERENCE
Staffeldt, B.; Brockmöller, J.; Roots, I. Determination of S-carboxymethyl-L-cysteine and some of its metabolites in urine and serum by high-performance liquid chromatography using fluorescent pre-column labelling, *J.Chromatogr.*, **1991**, *571*, 133–147.

SAMPLE
Matrix: solutions
Analyte: carboxylic acids
Sample preparation: Mix 100 µL of a 0.01-10 µg/mL solution in MeOH with 100 µL 1 mg/mL 1-pyrenyldiazomethane in ethyl acetate, let stand at room temperature for 1.5 h, inject a 5 µL aliquot. (1-Pyrenyldiazomethane is available from Molecular Probes, Eugene OR or may be synthesized as described above.)

HPLC VARIABLES
Column: 150 × 4 5 µm TSK-GEL-120A ODS (TOSOH)
Mobile phase: MeCN:water 50:50 (A) or 75:25 (B) or Gradient. MeCN:water 85:15 for 30 min, to 100:0 over 30 min. (C)
Flow rate: 1
Injection volume: 5
Detector: F ex 340 em 395

CHROMATOGRAM
Retention time: 12 (lactic acid (A)), 15 (formic acid (A)), 30 (propionic acid (A)), 12 (dinoprostone (B)), 13 (dinoprost (B)), 14 (alprostadil (B)), 17 (prostaglandin $F_{1\alpha}$ (B)), 17 (linolenic acid (C)), 18 (palmitoleic acid (C)), 18.5 (myristic acid (C)), 20 (linoleic acid (C)), 25 (palmitic acid (C)), 26 (oleic acid (C)), 40 (stearic acid (C)), 52 (arachidic acid (C))
Limit of detection: 20-30 fmole

REFERENCE
Nimura, N.; Kinoshita, T.; Yoshida, T.; Uetake, A.; Nakai, C. 1-Pyrenyldiazomethane as a fluorescent labeling reagent for liquid chromatographic determination of carboxylic acids, *Anal.Chem.*, **1988**, *60*, 2067–2070.

SAMPLE
Matrix: solutions
Analyte: cortolic and cortolonic acids
Sample preparation: Mix a solution in MeCN with 4 equivalents 1-pyrenyldiazomethane (diazomethylpyrene) dissolved in benzene (10 mg/mL) (Caution! Benzene is a carcinogen!), let stand at room temperature for 2 h, filter (0.2 μm), evaporate the filtrate to dryness under a stream of nitrogen, reconstitute the residue in MeCN, inject an aliquot. (1-Pyrenyldiazomethane is available from Molecular Probes, Eugene OR; a synthesis is also described above.)

HPLC VARIABLES
Column: 30 × 4.6 3 μm Pecosphere CR-C18 end-capped
Mobile phase: MeCN:MeOH:water 34:34:32
Flow rate: 1.5
Detector: UV 242; F ex 340 em 390

CHROMATOGRAM
Retention time: 5.5 (α-cortolic acid), 8 (α-cortolonic acid), 12.5 (β-cortolic acid), 14 (β-cortolonic acid)
Limit of detection: 4-8.4 fmole (F)

REFERENCE
Iohan, F.; Vincze, I.; Monder, C. High-performance liquid chromatographic determination of cortolic and cortolonic acids as pyrenyl ester derivatives, *J.Chromatogr.*, **1991**, *564*, 27–41.

SAMPLE
Matrix: solutions
Analyte: dicarboxylic acids
Sample preparation: Mix 500 μL of a 0.1-1000 μM solution in 10 mM pH 8.0 borate buffer with 500 μL MeCN, 1 mL MeOH, and 500 μL 500 μg/mL 1-pyrenyldiazomethane in ethyl acetate, let stand in the dark at room temperature for 24 h, inject an aliquot. (1-Pyrenyldiazomethane is available from Molecular Probes, Eugene OR or may be synthesized as described above.)

HPLC VARIABLES
Column: 150 × 4.6 3 μm Hypersil
Mobile phase: Gradient. MeCN:MeOH:40 mM pH 2.5 ammonium formate buffer from 40:5:55 to 60:5:35 over 15 min, wash with 0:100:0.
Column temperature: 35
Flow rate: 1.8
Injection volume: 200
Detector: F ex 340 em 376

CHROMATOGRAM
Retention time: 10.1 (malonic acid), 11.2 (succinic acid, α-glutaric acid), 12.2 (methylmalonic acid), 14.5 (ethylmalonic acid, dimethylmalonic acid)

REFERENCE
Schneede, J.; Ueland, P.M. Formation in an aqueous matrix and properties and chromatographic behavior of 1-pyrenyldiazomethane derivatives of methylmalonic acid and other short-chain dicarboxylic acids, *Anal.Chem.*, **1992**, *64*, 315–319.

HYDRAZINE

4-(2-Carbazolylpyrrolidin-1-yl)-7-(N,N-dimethylaminosulfonyl)-2,1,3-benzoxadiazole

SAMPLE
Matrix: blood
Analyte: fatty acids
Sample preparation: 10 μL Serum or plasma + 10 μL 100 μM pentadecanoic acid in DMF + 200 μL 500 mM pH 6.5 phosphate buffer, mix, add 2 mL n-heptane:chloroform 50:50, vortex thoroughly, centrifuge at 1500 g for 10 min. Remove 750 μL of the organic layer and evaporate it to dryness under reduced pressure, reconstitute the residue in 375 μL DMF, add 25 μL 4 M 1-ethyl-3-(3-dimethylaminopropyl)carbodiimide hydrochloride in pyridine:water 40:60, add 100 μL 5 mM 4-(2-carbazolylpyrrolidin-1-yl)-7-(N,N-dimethylaminosulfonyl)-2,1,3-benzoxadiazole in DMF, let stand at room temperature in the dark for 1.5 h, inject an aliquot. (Synthesis of 4-(2-carbazolylpyrrolidin-1-yl)-7-(N,N-dimethylaminosulfonyl)-2,1,3-benzoxadiazole (DBD-ProCZ) is as follows. Dissolve 0.5 g magnesium sulfate heptahydrate and 6 g NaOH in 60 mL water, throughout the reaction keep the flask at about 20° with cold water cooling, add 15 mL 30% hydrogen peroxide, add 75 mL MeOH, add 12.1 g powdered benzoyl peroxide in one go, stir for 10 min, pour into 150 mL 20% sulfuric acid, extract three times with 50 mL portions of chloroform, determine peroxybenzoic acid concentration by iodometric titration (Tetrahedron 1967, 23, 3327). Slowly add 110 mL 1 M peroxybenzoic acid in chloroform to 7 g 2,6-difluoroaniline dissolved in 100 mL chloroform, stir at room temperature, when reaction is complete (iodometric titration) wash with 2% sodium thiosulfate, wash with 5% sodium carbonate, wash with water, dry over anhydrous sodium sulfate, evaporate to dryness under reduced pressure, recrystallize 2,6-difluoronitrosobenzene from EtOH (mp 108.5-109.5). Stir 8.5 g 2,6-difluoronitrosobenzene in 85 mL DMSO at room temperature and add a solution of 3.91 g sodium azide in 85 mL DMSO dropwise, let stand for about 1 h, add to a large volume of water, extract with ether, dry the extracts over anhydrous sodium sulfate, evaporate to dryness under reduced pressure and distil to give 4-fluoro-2,1,3-benzoxadiazole as a colorless oil (bp 83°/12 mm Hg) (J.Chem.Soc.(C) 1970, 1433). Add 11 mL chlorosulfonic acid dropwise to 3 g 4-fluoro-2,1,3-benzoxadiazole in 10 mL chloroform at 0-10° (use a calcium chloride drying tube), stir at room temperature for 1 h, reflux for 2 h, cool, slowly pour into ice water, remove the organic layer, extract the aqueous layer with chloroform, combine the organic layers, wash, dry over anhydrous magnesium sulfate, evaporate under reduced pressure, take up the residue in 5 mL benzene (Caution! Benzene is a carcinogen!), chromatograph on a 150 × 30 column of silica gel (100-200 mesh Kanto Chemical) with n-hexane:benzene 50:50, evaporate the appropriate fractions to give 4-(chlorosulfonyl)-7-fluoro-2,1,3-benzoxadiazole (CBD-F) as pale yellow needles (mp 64-66°)

(Anal. Chem. 1984. 56, 2461). Stir 0.76 g CBD-F in 70 mL MeCN at 0-10° and add 1 g dimethylamine hydrochloride in 10 mL 100 mM pH 10 borax dropwise, adjust pH to 5 with 1 M HCl, concentrate to about 10 mL under reduced pressure, extract three times with 200 mL portions of diethyl ether, wash with water, dry over anhydrous magnesium sulfate, evaporate under reduced pressure, chromatograph on a 500 × 20 column of silica gel with chloroform, isolate the appropriate fraction and re-chromatograph on the same column with ethyl acetate:benzene 1:2 to give 4-(N,N-dimethylaminosulfonyl)-7-fluoro-2,1,3-benzoxadiazole (DBD-F) as white needles (mp 124-125°) (yield 1% !). On a Merck no. 5714 60F$_{254}$ tlc plate eluted with chloroform DBD-F has Rf 0.32 and lies between two other reaction products (Analyst 1989, 114, 413). DBD-F can be purchased from Tokyo Kasei (TCI America, 911 North Harborgate St., Portland, OR 97203; 800-423-8616,503-283-1681, (fax) 503-283-1987; www.tciamerica.com). Add 100 mg DBD-F in 10 mL MeCN to 47 mg proline in 20 mL 250 mM pH 11.5 sodium carbonate solution, stir at room temperature for 30 min, wash with ethyl acetate, adjust the pH of the aqueous layer to 1-2 with 2 M HCl, extract three times with 30 mL ethyl acetate. Combine the extracts and evaporate them under reduced pressure, recrystallize from benzene/ethyl acetate to give 4-(N,N-dimethylaminosulfonyl)-7-(2-carboxypyrrolidin-1-yl)-2,1,3-benzoxadiazole (DBD-Pro) as yellow needles (mp 187-9° d) (Analyst 1989, 114, 1233). Suspend 55 mg (S)-(-)-DBD-Pro in 55 mL anhydrous diethyl ether at 0°, add 110 mg phosphorus pentachloride, stir at 5° for 1 h, filter quickly, evaporate to dryness under reduced pressure, dry under vacuum over phosphorus pentoxide for 12 h to give 4-(N,N-dimethylaminosulfonyl)-7-(2-chloroformylpyrrolidin-1-yl)-2,1,3-benzoxadiazole (DBD-Pro-Cl) as yellow crystals (mp 116-17°) (Analyst 1993, 118, 759). DBD-Pro-Cl is also available from Tokyo Kasei. Add 130 mg DBD-Pro-Cl dissolved in 25 mL anhydrous benzene dropwise to 100 mL MeOH containing 70 mg hydrazine hydrate, stir for 30 min at room temperature, evaporate under reduced pressure, recrystallize from ethyl acetate:MeOH 90:10 to give 4-(2-carbazolylpyrrolidin-1-yl)-7-(N,N-dimethylaminosulfonyl)-2,1,3-benzoxadiazole (DBD-ProCZ) as orange crystals (mp 107-109°) (Anal.Proc. 1994, 31, 265).)

HPLC VARIABLES
Column: 150 × 4.6 5 μm Inertsil ODS-80A
Mobile phase: MeCN:water 55:45 (A) or 70:30 (B)
Column temperature: 40
Flow rate: 1
Detector: F ex 450 em 550

CHROMATOGRAM
Retention time: 25 (palmitic acid (B)), 31 (oleic acid (B)), 57 (stearic acid (B)), 64 (linolenic acid (A)), 70 (myristic acid (A)), 85 (palmitoleic acid (A)), 118 (linoleic acid (A)), 125 (arachidonic acid (A))
Internal standard: heptadecanoic acid (37 (B)), pentadecanoic acid (105 (A))

KEY WORDS
rat; human

REFERENCE
Toyo'oka, T.; Takahashi, M.; Suzuki, A.; Ishii, Y. Determination of free fatty acids in blood, tagged with 4-(2-carbazoylpyrrolidin-1-yl)-7-(N,N-dimethylaminosulfonyl)-2,1,3-benzoxadiazole, by high-pressure liquid chromatography with fluorescence detection, *Biomed.Chromatogr.*, **1995**, *9*, 162–170.

4-(5,6-Dimethoxy-2-benzimidazoyl)benzohydrazide

SAMPLE
Matrix: solutions
Analyte: fatty acids
Sample preparation: Mix 100 μL of a solution of fatty acids in water with 50 μL 2 M 1-(3-dimethylaminopropyl)-3-ethylcarbodiimide in water, 50 μL 10% pyridine in water, and 100 μL 4.9 mM 4-(5,6-dimethoxy-2-benzimidazoyl)benzohydrazide in DMF, heat at 40° for 15 min, inject a 10 μL aliquot. (Preparation of 4-(5,6-dimethoxy-2-benzimida-zoyl)benzohydrazide is as follows. Stir 483 g veratrole in 1.45 L acetic acid at 15°, add 683 g concentrated nitric acid (s.g. 1.05) over 1 h keeping the temperature below 40° (cool if necessary), add 2.127 L fuming nitric acid (s.g. 1.50) over 1 h keeping the temperature below 30°, allow to stand for 2 h, pour into a large volume of cold water, filter, wash the solid until it is free from acid, recrystallize from EtOH to give 4,5-dinitroveratrole (mp 129.5-130.5°) (J. Am. Chem. Soc. 1946, 68, 1536). Reflux 5 g 4,5-dinitroveratrole in 200 mL benzene (Caution! Benzene is a carcinogen!), add 100 g 60 mesh iron powder and 20 mL concentrated HCl in small portions over 1 h, reflux for 4 h, add 10 mL water, reflux for 2 h, cool, make alkaline with 2.5 M NaOH, extract several times with 200 mL portions of benzene. Combine the extracts and evaporate to dryness, add 10 mL concentrated HCl, recrystallize from EtOH to give 1,2-diamino-4,5-dimethoxybenzene hydrochloride as slightly pink needles (mp 240° d) (Anal. Chim. Acta 1982, 134, 39). Another synthesis of 1,2-diamino-4,5-dimethoxybenzene is as follows. Add 9.66 g veratrole in 9.5 mL glacial acetic acid dropwise to 26.6 mL fuming nitric acid over 30 min using an ice bath, stir at room temperature for 30 min, pour into 500 mL ice-cold water, filter to obtain 1,2-dime-thoxy-4,5-dinitrobenzene, recrystallize repeatedly from EtOH. Dissolve 1 g 1,2-dimethoxy-4,5-dinitrobenzene in 50 mL EtOH, add 100 mg 10% palladium on charcoal, stir under hydrogen at atmospheric pressure and room temperature for 4 days, filter through Celite under an atmosphere of nitrogen, saturate the filtrate with hydrogen chloride gas, filter under nitrogen, dry the solid under vacuum to obtain 1,2-diamino-4,5-dimethoxybenzene hydrochloride (mp 240° d (Anal. Chim. Acta 1982, 134, 39)) (Anal. Chim. Acta 1992, 263, 137). 1,2-Diamino-4,5-dimethoxybenzene hydrochloride is also available from Dojindo, Kumamoto, Japan (Dojindo Molecular Technologies, Inc., 3 Bethesda Metro Center, Suite 700, Bethesda MD 20814; (301) 664-8448; www.dojindo.co.jp). Dissolve 5 g 1,2-diamino-4,5-dimethoxybenzene hydrochloride and 4.1 g methyl 4-formylbenzoate (terephthalde-hydic acid methyl ester) in 200 mL MeOH, stir at room temperature for 1 h, add 50 mL of MeOH saturated with HCl, reflux under an inert gas for 2 h, cool, concentrate to about 100 mL under reduced pressure, collect the precipitate, recrystallize from MeOH to obtain 4-(5,6-dimethoxy-2-benzimidazoyl)benzoic acid as colorless needles (yield 13.1%, mp 279-

281°). Dissolve 2 g 4-(5,6-dimethoxy-2-benzimidazoyl)benzoic acid in 100 mL 45% hydrazine hydrate in water (Caution! Hydrazine hydrate is a carcinogen and explodes on distillation in air!), heat at 100° for 1 h, collect the precipitate, recrystallize from EtOH to give 4-(5,6-dimethoxy-2-benzimidazoyl)benzohydrazide (mp 291-293°).)

HPLC VARIABLES

Column: 250 × 4.6 5 μm L-column ODS (Chemicals Inspection and Testing Institute, Tokyo)

Mobile phase: Gradient. MeOH:water from 40:60 to 80:20 over 40 min, 100:0 over 30 min.

Flow rate: 1

Injection volume: 10

Detector: F ex 360 em 460

CHROMATOGRAM

Retention time: 22 (caproic acid), 29 (caprylic acid), 35 (capric acid), 40 (lauric acid), 43 (myristic acid), 46 (palmitic acid), 49 (stearic acid), 52 (arachidic acid), 55 (behenic acid), 59 (lignoceric acid), 62 (hexacosanoic acid)

Limit of detection: 1-3 fmole

OTHER SUBSTANCES

Non-interfering: alcohols, amines, amino acids, α-ketoglutaric acid, ketone, phenols, phenylpyruvic acid, pyruvic acid, sugars, thiols

REFERENCE

Iwata, T.; Nakamura, M.; Yamaguchi, M. 4-(5,6-Dimethoxy-2-benzimidazoyl)benzohydrazide as fluorescence derivatization reagent for carboxylic acids in high-performance liquid chromatography, *Anal.Sci.*, **1992**, *8*, 889–892.

5,6-Dimethoxy-2-(4-hydrazinocarbonylphenyl)benzothiazole

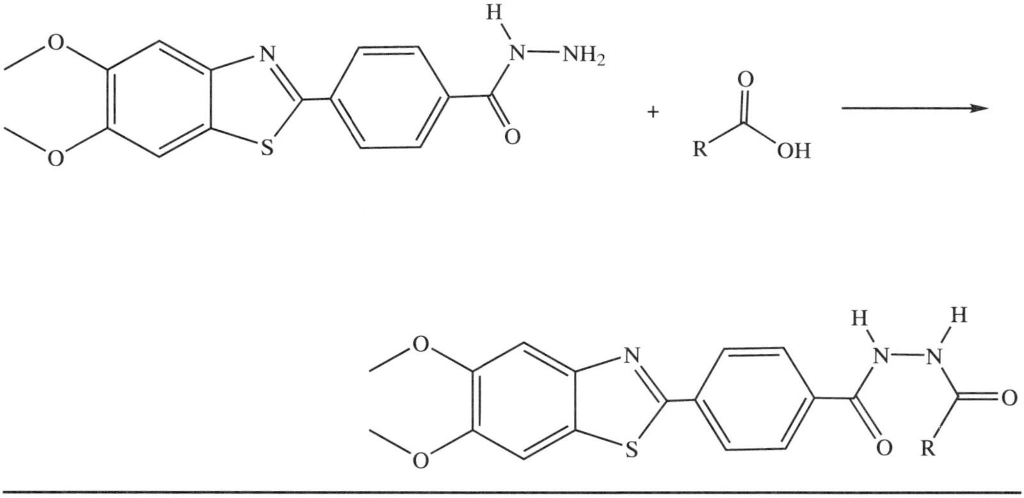

SAMPLE

Matrix: blood

Analyte: fatty acids

Sample preparation: Dilute serum 20 times with water. 100 μL Diluted serum + 100 μL 1 M 1-(3-dimethylaminopropyl)-3-ethylcarbodiimide in MeOH + 100 μL pyridine:MeOH 20:80 + 100 μL 15 mM 5,6-dimethoxy-2-(4-hydrazinocarbonylphenyl)benzothiazole in DMF, heat at 37° for 20 min, inject a 20 μL aliquot. (5,6-Dimethoxy-2-(4-hydrazinocarbonylphenyl)benzothiazole may be purchased from Dojindo Molecular Technologies, Inc., 3 Bethesda Metro Center, Suite 700, Bethesda MD 20814; (301) 664-8448;

www.dojindo.co.jp. Preparation is as follows. Stir 20 g 4-bromoveratrole in 60 mL acetic acid and add 10 mL concentrated nitric acid dropwise, keep at 10-30° with occasional cooling, pour into ice-water, filter. Take up the precipitate in 500mL hot EtOH and add activated charcoal, filter, add 40 mL water to the filtrate, collect 4-bromo-5-nitroveratrole as light yellow crystals (mp 121-2°). Melt 5 g sodium sulfide nonahydrate and 0.7 g sulfur together, add this mixture to 5 g 4-bromo-5-nitroveratrole in 50 mL 95% EtOH, reflux for 30 min, pour into ice-water, filter, recrystallize the product from dichloromethane to give the nitro disulfide as yellow needles (mp 231-2°). Add 8 g tin powder to 2 g of the nitro disulfide in 300 mL EtOH, add 30 mL concentrated HCl dropwise, basify with 4 M NaOH, filter, dilute the filtrate with 200 mL water. Extract the diluted filtrate twice with 100 mL portions of benzene (Caution! Benzene is a carcinogen!). Combine the extracts and evaporate them under reduced pressure, take up the residue in 10 mL benzene and add 2 mL 10% hydrogen peroxide, stir for 30 min, recrystallize the precipitate from EtOH to give 2,2'-dithiobis(1-amino-4,5-dimethoxybenzene) (mp 155-6°) (Anal.Chim.Acta 1994, 291, 189). Add a mixture of 3.7 g 2,2'-dithiobis(1-amino-4,5-dimethoxybenzene) and 1.2 g tri-n-butylphosphine in 40 mL 800 mM disodium hydrogen phosphite in MeOH to 1.2 g 4-carboxybenzaldehyde in 20 mL MeOH, stir at 37° for 2 h, filter the precipitate. Wash the precipitate with MeOH:water 70:30 and dry it under reduced pressure. Dissolve in 50 mL MeOH and treat with ethereal diazomethane, evaporate to dryness under reduced pressure, chromatograph the residue on a 250 × 35 column of 130 g 70-230 mesh silica gel 60 (Merck) with n-hexane:ethyl acetate:chloroform 50:25:25, collect the main fraction and evaporate to obtain the product, 5,6-dimethoxy-2-(4-carbomethoxyphenyl)benzothiazole, as pale yellow needles (mp 223-224°). Add 100 mL aqueous 45% hydrazine hydrate (Caution! Hydrazine hydrate is a carcinogen and may explode on distillation!) to 500 mg of the product in 50 mL EtOH, reflux for 1 h, collect the precipitate and recrystallize it from EtOH to give 5,6-dimethoxy-2-(4-hydrazinocarbonylphenyl)benzothiazole as colorless needles (mp 252-254°). Preparation of diazomethane is as follows. Caution! Diazomethane is toxic, explosive, and carcinogenic! A face shield and a safety screen should always be used and the preparation should only be carried out in a properly functioning chemical fume hood. Only smooth glass apparatus with rubber stoppers and plastic tubing should be used. Scratched glassware, ground glass joints, and sharp edges should be avoided (Org. Syn., Coll. Vol. VI; Wiley: New York, 1988, pp. 432-435). Procedures have been reported for the synthesis of diazomethane from N-methyl-N-nitrosourea (Org. Syn., Coll. Vol. II; Wiley: New York, 1943, pp. 165-167), N-nitroso-β-methylaminoisobutyl methyl ketone (Org. Syn., Coll. Vol. III; Wiley: New York, 1955, pp. 244-248), and N,N'-dimethyl-N,N'-dinitrosoterephthalamide (Org. Syn., Coll. Vol. V; Wiley: New York, 1973, pp. 351-355). Probably the most convenient starting material is N-methyl-N-nitroso-p-toluenesulfonamide (Diazald) (Aldrichimica Acta 1983, 16, 3-10; Org. Syn., Coll. Vol. IV, Wiley: New York, 1963, pp. 250-253). Add 10 mL 95% EtOH to a solution of 5 g KOH in 8 mL water, warm to 65° using a water bath, add a solution of 5 g N-methyl-N-nitroso-p-toluenesulfonamide in 45 mL ether over 20 min at such a rate as to keep the reaction volume constant. Collect the ether and diazomethane that distil in an ice-cooled receiving flask under a dry ice/acetone condenser. When all the N-methyl-N-nitroso-p-toluenesulfonamide has been used up, slowly add 10 mL ether to the reaction flask and continue distillation until the distillate is colorless. A purpose-built distillation apparatus can be purchased from Aldrich (Aldrichimica Acta 1983, 16, 3-10). Diazomethane can also be prepared according to Anal. Chem. 1960, 32, 1412. Excess quantities of diazomethane can be destroyed by adding acetic acid until the yellow color of the diazomethane is discharged. The safe disposal of the nitroso compounds used to generate diazomethane has been discussed (Lunn,G.; Sansone,E.B. Destruction of Hazardous Chemicals in the Laboratory, Second Edition. Wiley: New York, 1994, pp. 277-289).)

HPLC VARIABLES
Column: 250 × 4.6 5 μm TSK gel ODS 120T (Tosoh)
Mobile phase: Gradient. MeCN:water 40:60 for 15 min, to 70:30 over 20 min, maintain at 70:30 for 30 min, to 0:100 over 25 min, maintain at 0:100 for 20 min.
Flow rate: 1
Injection volume: 20
Detector: F ex 365 em 447

CHROMATOGRAM
Retention time: 30 (caprylic acid), 36 (capric acid), 45 (lauric acid), 48 (myristoleic acid), 57 (linolenic acid), 60 (myristic acid), 64 (palmitoleic acid), 67.5 (docosahexaenoic acid), 69 (arachidonic acid), 69 (linoleic acid), 79 (palmitic acid), 82 (oleic acid), 85 (margaric acid), 90 (stearic acid), 98 (arachidic acid)
Limit of detection: 1-2 fmole

KEY WORDS
serum

REFERENCE
Yamaguchi, M.; Hara, S.; Obata, K. 5,6-Dimethoxy-2-(4'-hydrazinocarbonylphenyl)benzothiazole as a highly sensitive and stable fluorescence derivatization reagent for carboxylic acids in high performance liquid chromatography, *J.Liq.Chromatogr.*, **1995**, *18*, 2991–3006.

DMEQ-hydrazide

SAMPLE
Matrix: blood
Analyte: fatty acids
Sample preparation: 5 μL Serum + 5 μL EtOH + 50 μL 4% pyridine in EtOH containing 20 mM HCl, vortex for 10 s, add 25 μL 50 mM reagent in DMF, add 15 μL 2 M 1-ethyl-3-(3-dimethylaminopropyl)carbodiimide in water, heat at 37° for 10 min, centrifuge at 1000 g for 5 min, inject a 10 μL aliquot of the supernatant. (Reagent was DMEQ-hydrazide (6,7-dimethoxy-1-methyl-2(1H)-quinoxalinone-3-propionylcarboxylic acid hydrazide), available from Wako Chemicals, Richmond VA. Synthesis is as follows. Stir 483 g veratrole in 1.45 L acetic acid at 15°, add 683 g concentrated nitric acid (s.g. 1.05) over 1 h keeping the temperature below 40° (cool if necessary), add 2.127 L fuming nitric acid (s.g. 1.50) over 1 h keeping the temperature below 30°, allow to stand for 2 h, pour into a large volume of cold water, filter, wash the solid until it is free from acid, recrystallize from EtOH to give 4,5-dinitroveratrole (mp 129.5-130.5°) (J. Am. Chem. Soc. 1946, 68, 1536). Reflux 5 g 4,5-dinitroveratrole in 200 mL benzene (Caution! Benzene is a carcinogen!), add 100 g 60 mesh iron powder and 20 mL concentrated HCl in small portions over 1 h, reflux for 4 h, add 10 mL water, reflux for 2 h, cool, make alkaline with 2.5 M NaOH, extract several times with 200 mL portions of benzene. Combine the extracts and evaporate to dryness, add 10 mL concentrated HCl, recrystallize from EtOH to give 1,2-diamino-4,5-dimethoxybenzene hydrochloride as slightly pink needles (mp 240° d) (Anal. Chim. Acta 1982, 134, 39). Another synthesis of 1,2-diamino-4,5-dimethoxybenzene is as

follows. Add 9.66 g veratrole in 9.5 mL glacial acetic acid dropwise to 26.6 mL fuming nitric acid over 30 min using an ice bath, stir at room temperature for 30 min, pour into 500 mL ice-cold water, filter to obtain 1,2-dimethoxy-4,5-dinitrobenzene, recrystallize repeatedly from EtOH. Dissolve 1 g 1,2-dimethoxy-4,5-dinitrobenzene in 50 mL EtOH, add 100 mg 10% palladium on charcoal, stir under hydrogen at atmospheric pressure and room temperature for 4 days, filter through Celite under an atmosphere of nitrogen, saturate the filtrate with hydrogen chloride gas, filter under nitrogen, dry the solid under vacuum to obtain 1,2-diamino-4,5-dimethoxybenzene hydrochloride (mp 240° d (Anal. Chim. Acta 1982, 134,39)) (Anal. Chim. Acta 1992, 263, 137). 1,2-Diamino-4,5-dimethoxybenzene hydrochloride is also available from Molecular Probes, Eugene OR or Dojindo Molecular Technologies, Inc., 3 Bethesda Metro Center, Suite 700, Bethesda MD 20814; (301) 664-8448; www.dojindo.co.jp. Dissolve 2.5 mmoles 1,2-diamino-4,5-dimethoxybenzene monohydrochloride and 2.4 mmoles α-ketoglutaric acid in 30 mL 500 mM HCl, heat in a boiling water bath for 2 h, cool in ice, filter, wash the precipitate with water, dry under vacuum, recrystallize from MeOH:water 90:10 to give 6,7-dimethoxy-2(1H)-quinoxalinone-3-propionylcarboxylic acid as yellow needles (mp 240°) (Chem. Pharm. Bull. 1985, 33, 3493). Treat 1.5 g 6,7-dimethoxy-2(1H)-quinoxalinone-3-propionylcarboxylic acid in 100 mL MeOH with ethereal diazomethane, evaporate to dryness under reduced pressure, dissolve the residue in 30 mL chloroform, chromatograph on a 250 × 35 column of 70-230 mesh silica gel 60 (Merck) with hexane:ethyl acetate 50:50 to give methyl 6,7-dimethoxy-1-methyl-2(1H)-quinoxalinone-3-propionylcarboxylate as colorless needles (mp 178-179°). Dissolve 900 mg methyl 6,7-dimethoxy-1-methyl-2(1H)-quinoxalinone-3-propionylcarboxylate in 100 mL 45% hydrazine hydrate in water, heat at 100° for 1 h, recrystallize the precipitate from EtOH to give 6,7-dimethoxy-1-methyl-2(1H)-quinoxalinone-3-propionylcarboxylic acid hydrazide as colorless needles (mp 205-206°) (Analyst 1990, 115, 1363). Preparation of diazomethane is as follows. Caution! Diazomethane is toxic, explosive, and carcinogenic! A face shield and a safety screen should always be used and the preparation should only be carried out in a properly functioning chemical fume hood. Only smooth glass apparatus with rubber stoppers and plastic tubing should be used. Scratched glassware, ground glass joints, and sharp edges should be avoided (Org. Syn., Coll. Vol. VI; Wiley: New York, 1988, pp. 432-435). Procedures have been reported for the synthesis of diazomethane from N-methyl-N-nitrosourea (Org. Syn., Coll. Vol. II; Wiley: New York, 1943, pp. 165-167), N-nitroso-β-methylaminoisobutyl methyl ketone (Org. Syn., Coll. Vol. III; Wiley: New York, 1955, pp. 244-248), and N,N'-dimethyl-N,N'-dinitrosoterephthalamide (Org. Syn., Coll. Vol. V; Wiley: New York, 1973, pp. 351-355). Probably the most convenient starting material is N-methyl-N-nitroso-p-toluenesulfonamide (Diazald) (Aldrichimica Acta 1983, 16, 3-10; Org. Syn., Coll. Vol. IV, Wiley: New York, 1963, pp. 250-253). Add 10 mL 95% EtOH to a solution of 5 g KOH in 8 mL water, warm to 65° using a water bath, add a solution of 5 g N-methyl-N-nitroso-p-toluenesulfonamide in 45 mL ether over 20 min at such a rate as to keep the reaction volume constant. Collect the ether and diazomethane that distil in an ice-cooled receiving flask under a dry ice/acetone condenser. When all the N-methyl-N-nitroso-p-toluenesulfonamide has been used up, slowly add 10 mL ether to the reaction flask and continue distillation until the distillate is colorless. A purpose-built distillation apparatus can be purchased from Aldrich (Aldrichimica Acta 1983, 16, 3-10). Excess quantities of diazomethane can be destroyed by adding acetic acid until the yellow color of the diazomethane is discharged. The safe disposal of the nitroso compounds used to generate diazomethane has been discussed (Lunn,G.; Sansone,E.B. Destruction of Hazardous Chemicals in the Laboratory, Second Edition. Wiley: New York, 1994, pp. 277-289).)

HPLC VARIABLES
Column: 250 × 4.6 10 μm YMC-Pack C8 (Yamamura Chemical Laboratories, Kyoto)
Mobile phase: Gradient. MeCN:water 55:45 for 56 min, to 95:5 over 16 min, maintain at 95:5 for 4 min, re-equilibrate at initial conditions for 4 min.
Column temperature: 30 ± 0.2
Flow rate: 1
Injection volume: 100
Detector: F ex 360 em 435

CHROMATOGRAM
Retention time: 12.5 (lauric acid), 16.5 (myristoleic acid), 24 (myristic acid), 27 (linolenic acid), 29 (eicosapentaenoic acid), 30 (palmitoleic acid), 40 (linoleic acid), 42 (arachidonic acid, docosahexaenoic acid), 49 (palmitic acid), 54 (dihomo-gamma-linolenic acid), 62.5 (oleic acid), 66 (margaric acid), 72 (stearic acid)
Limit of detection: 2-7 fmole

OTHER SUBSTANCES
Non-interfering: adipic acid, alcohols, aldehydes, amines, amino acids, benzoic acid, cinnamic acid, α-keto acids, lactic acid, malic acid, malonic acid, oxalic acid, phenols, salicylic acid, succinic acid, sugars

KEY WORDS
serum

REFERENCE
Iwata, T.; Inoue, K.; Nakamura, M.; Yamaguchi, M. Simple and highly sensitive determination of free fatty acids in human serum by high performance liquid chromatography with fluorescence detection, *Biomed.Chromatogr.*, **1992**, *6*, 120–123.

SAMPLE
Matrix: urine
Analyte: glucuronide conjugates
Sample preparation: 5 µL Urine + 5 µL water + 50 µL 4% pyridine in ethanolic HCl, vortex for 10 s, add 25 µL 50 mM reagent in DMF, add 15 µL 4 M 1-(3-dimethylaminopropyl)-3-ethylcarbodiimide in water, mix, heat at 37° for 20 min, inject a 10 µL aliquot. (Prepare ethanolic HCl by dissolving 1 mL 1.2 M HCl in water in 60 mL EtOH. Reagent was DMEQ-hydrazide (6,7-dimethoxy-1-methyl-2(1H)-quinoxalinone-3-propionylcarboxylic acid hydrazide), available from Wako Chemicals, Richmond VA. Synthesis is as described above.)

HPLC VARIABLES
Guard column: 10 × 4 5 µm TSK gel ODS-120T (Tosoh)
Column: 250 × 4.6 5 µm L-column ODS (Chemicals Inspection and Testing Institute, Tokyo)
Mobile phase: MeCN:MeOH:0.5% triethylamine in water 20:20:40
Column temperature: 40 ± 0.2
Flow rate: 1
Injection volume: 10
Detector: F ex 367 em 445

CHROMATOGRAM
Retention time: 30.3 (etiocholanorone-3-glucuronide), 31.7 (androsterone-3-glucuronide)
Limit of detection: 29.5-33.9 fmole

OTHER SUBSTANCES
Simultaneously analyzed: dehydroisoandrosterone-3-glucuronide, β-estradiol-17-glucuronide, 17β-estradiol-3-glucuronide, 17β-estradiol-3-glucuronide-17-sulfate, estriol-17β-glucuronide, estriol-3-glucuronide, estrone-3-glucuronide, galacturonic acid, glucuronic acid, glycyrrhizin, testosterone-17-glucuronide

KEY WORDS
peaks for other compounds are not resolved from urine peaks

REFERENCE
Iwata, T.; Hirose, T.; Nakamura, M.; Yamaguchi, M. Determination of urinary glucuronide conjugates by high-performanc liquid chromatography with pre-column fluorescence derivatization, *J.Chromatogr.B*, **1994**, *654*, 171–176.

SAMPLE
Matrix: urine
Analyte: estriol 16-glucuronide
Sample preparation: 5 µL Urine + 5 µL water + 50 µL 4% pyridine in EtOH, vortex for 5 min, add 50 µL 50 mM DMEQ-hydrazide in DMF, add 25 µL 4 M 1-(3-dimethylami-

nopropyl)-3-ethylcarbodiimide in water, heat at 37° for 10 min, centrifuge at 1000 g for 5 min, inject a 10 μL aliquot of the supernatant. (DMEQ-hydrazide is 6,7-dimethoxy-1-methyl-2-oxo-1,2-dihydroquinoxalin-3-ylpropionohydrazide. It is available from Wako, Richmond VA; a synthesis is also described above. In this paper a column-switching procedure is described that can also be used to determine estriol 3-glucuronide.)

HPLC VARIABLES
Column: 250 × 4.6 5 μm YMC-Pack C8 (Yamamura)
Mobile phase: Gradient. MeOH:0.05% triethylamine in water 35:65 for 35 min, then 45:55 (step gradient).
Column temperature: 40 ± 0.2
Flow rate: 1
Injection volume: 10
Detector: F ex 367 em 445

CHROMATOGRAM
Retention time: 46
Limit of detection: 17 ng/mL

REFERENCE
Iwata, T.; Hirose, T.; Yamaguchi, M. Direct determination of estriol 3- and 16-glucuronides in pregnancy urine by column-switching high-performance liquid chromatography with fluorescence detection, *J.Chromatogr.B*, **1997**, *695*, 201–207.

Hydrazine

SAMPLE
Matrix: bulk
Analyte: fatty acids
Sample preparation: 10-50 mg Linoleic acid + 4 mL hydrazine hydrate:THF:EtOH:cyclopentene 1:3:3:1 (Caution! Hydrazine hydrate is carcinogenic and may explode when distilled!), heat at 50° under nitrogen in a tightly closed tube for 2 h, add 5 mL acetone, heat at 50° for 30 min, cool. Remove a 100 μL aliquot and evaporate it to dryness under a stream of nitrogen at 50°, reconstitute the residue in 1 mL MeOH, evaporate to dryness under a stream of nitrogen, repeat, reconstitute in 100 μL MeOH, inject a 1-5 μL aliquot.

HPLC VARIABLES
Column: 250 × 4.6 5 μm Hypersil RP-18 ODS
Mobile phase: MeOH:water 85:15 (Wash with 30 mL MeOH after each use.)
Flow rate: 1
Injection volume: 1-5
Detector: UV 229

CHROMATOGRAM
Retention time: 13 (linolenic acid), 14 (palmitoleic acid), 17 (linoleic acid), 22 (palmitic acid), 25 (oleic acid), 27 (elaidic acid), 40 (stearic acid)
Limit of detection: 50 ng

REFERENCE
Agrawal, V.P.; Schulte, E. High-performance liquid chromatography of fatty acid isopropylidene hydrazides and its application in lipid analysis, *Anal.Biochem.*, **1983**, *131*, 356–359.

2-(5-Hydrazinocarbonyl-2-furyl)-5,6-dimethoxybenzothiazole

SAMPLE

Matrix: blood

Analyte: carboxylic acids

Sample preparation: 10 μL Serum + 44 μL MeOH + 1 μL pyridine, sonicate for 5 min, add 25 μL 100 mM reagent in DMF, add 20 μL 400 mM 1-ethyl-3-(3-dimethylaminopropyl)carbodiimide in MeOH, let stand at 25° for 2 h, centrifuge, inject an aliquot. (Reagent was 2-(5-hydrazinocarbonyl-2-furyl)-5,6-dimethoxybenzothiazole which was synthesized as follows. Pass dry hydrogen chloride into a mixture of 12.6 g methyl 2-furoate, 4.5 g paraformaldehyde, and 3.4 g anhydrous zinc chloride in 50 mL dry chloroform for 3 h while holding the reaction temperature at 30°. After cooling pour the contents of the flask into 100 mL cold water, remove the chloroform layer, extract the aqueous layer with chloroform (cf Coll. Czech. Chem. Commun. 1960, 25, 1058). Combine the chloroform layers, neutralize, dry over anhydrous calcium chloride, evaporate, distil to give 5-chloromethyl furyl-2-carboxylic acid methyl ester (bp 108°/4 mm Hg). Reflux 10 g 5-chloromethyl furyl-2-carboxylic acid methyl ester and 25 g silver carbonate in 100 mL THF: water 70:30 for 5 h, filter through Celite, concentrate the filtrate under reduced pressure, chromatograph the product on silica gel with chloroform to give 5-hydroxymethyl furyl-2-carboxylic acid methyl ester as a light yellow oil. Add a solution of 2.9 g 5-hydroxymethyl furyl-2-carboxylic acid methyl ester in 30 mL dichloromethane to 12 g pyridinium chlorochromate in 100 mL, dichloromethane, stir at room temperature for 4 h, evaporate to dryness under reduced pressure, chromatograph on silica with dichloromethane to give 5-formyl furyl-2-carboxylic acid methyl ester as a light yellow powder. Add 10 mL concentrated nitric acid dropwise to 20 g 4-bromoveratrole in 60 mL acetic acid while keeping the temperature at 10-30° with occasional cooling, when the addition is complete pour the reaction mixture into ice-water. Collect the precipitate and dissolve it in 500 mL hot EtOH, add activated charcoal, filter, add 40 mL water to the filtrate to give 4,5-dimethoxy-2-nitrobromobenzene as a light yellow crystalline solid (mp 121-122°). Prepare sodium sulfide by melting together 5 g sodium sulfide nonahydrate and 700 mg sulfur, add this mixture to 5 g 4,5-dimethoxy-2-nitrobromobenzene in 50 mL EtOH:water 95:5, reflux for 30 min, pour into ice-water, collect the solid, recrystallize from dichloromethane to give di(4,5-dimethoxy-2-nitrophenyl)sulfide as yellow needles (mp 231-232°). Add 15 mL concentrated HCl dropwise to 1.5 g di(4,5-dimethoxy-2-nitrophenyl)sulfide and 4.5 g tin powder stirred at 40-50° in 150 mL EtOH, reflux for 1 h, cool to room temperature, filter, add 1.17 g 5-formyl furyl-2-carboxylic acid methyl ester to the filtrate, reflux for 1 h, cool, filter, chromatograph the solid on silica gel with dichloromethane, recrystallize from EtOH to give 5-(5',6'-dimethoxybenzothiazolyl)-N-furan-2-carboxylic acid methyl ester as a yellow powder (mp 192-202°). Add 2 mL hydrazine hydrate (Caution! Hydrazine hydrate is a carcinogen and explodes on distillation in air!) to 800 mg 5-(5',6'-dimethoxybenzothiazolyl)-N-furan-2-carboxylic acid methyl ester in 20 mL EtOH, reflux for 30 min, collect the solid, wash with MeOH, dry under vacuum over phosphorus pentoxide to give 2-(5-hydrazinocarbonyl-2-furyl)-5,6-dimethoxybenzothiazole as a light yellow solid (mp 226-228°).)

HPLC VARIABLES
Column: 250 × 4.6 5 μm Wakosil-II 5C18 HG
Mobile phase: Gradient. MeCN:water from 70:30 to 75:25 over 25 min, to 100:0 over 15 min, maintain at 100:0.
Column temperature: 40
Flow rate: 1
Injection volume: 10
Detector: F ex 363 em 452

CHROMATOGRAM
Retention time: 10 (lauric acid), 12 (myristoleic acid), 17.5 (linolenic acid), 18 (myristic acid), 21 (palmitoleic acid), 24 (arachidonic acid), 25 (linoleic acid), 32 (palmitic acid), 35 (oleic acid), 38 (margaric acid), 40 (prostaglandin $F_{1\alpha}$), 41 (dinoprost), 42 (stearic acid), 50 (dinoprostone), 51 (alprostadil)
Limit of detection: 50 fmole

REFERENCE
Saito, M.; Ushijima, T.; Sasamoto, K.; Ohkura, Y.; Ueno, K. 2-(5-Hydrazinocarbonyl-2-furyl)-5,6-dimethoxybenzothiazole as a precolumn fluorescence derivatization reagent for carboxylic acids in high-performance liquid chromatography and its application to the assay of fatty acids in human serum, *Anal.Sci.*, **1995**, *11*, 103–107.

2-(5-Hydrazinocarbonyl-2-furyl)-5,6-methylenedioxybenzofuran

SAMPLE
Matrix: seminal fluid
Analyte: prostaglandins
Sample preparation: Mix 50 μL seminal fluid with 500 μL dilute HCl (pH 3.0) and 500 μL ethyl acetate, vortex. Remove the organic layer and evaporate it to dryness under reduced pressure, reconstitute the residue in 200 μL water, add to a Toyopak-ODS SPE cartridge, elute with 200 μL MeOH. 100 μL Eluate + 100 μL 100 mM 1-(3-dimethylaminopropyl)-3-ethylcarbodiimide in water + 100 μL 1% aqueous pyridine + 100 μL 15 mM 2-(5-hydrazinocarbonyl-2-furyl)-5,6-methylenedioxybenzofuran in DMF, heat at 37° for 1 h, inject a 10 μL aliquot. (Synthesis of 2-(5-hydrazinocarbonyl-2-furyl)-5,6-methylenedioxybenzofuran is as follows. Slowly add 153 g freshly distilled phosphorus oxychloride to 73 g anhydrous DMF with stirring at room temperature, add 125 g sesamol in portions over 4 h, stir at room temperature overnight, pour into ice water, filter. Dissolve the solid

in ether and wash with water, dry over anhydrous magnesium sulfate, evaporate to dryness, recrystallize from EtOH to give 2-hydroxy-4,5-methylenedioxybenzaldehyde as slightly-yellow crystals (mp 125-126°). Pass HCl gas into 12.6 g methyl 2-furoate, 4.5 g paraformaldehyde, and 3.4 g zinc chloride in 50 mL chloroform with stirring at 30° over 4 h. Pour into ice water and extract with 50 mL chloroform. Wash the chloroform layer 3 times with water, wash twice with aqueous sodium bicarbonate solution, dry over anhydrous sodium sulfate, evaporate to remove the solvent, distil at 108°/4 mm Hg to yield methyl 5-chloromethyl furyl-2-carboxylate as a colorless oil. Heat 3 g 2-hydroxy-4,5-methylenedioxybenzaldehyde, 3.14 g methyl 5-chloromethyl furyl-2-carboxylate, and 2.49 g potassium carbonate in 100 mL anhydrous DMF at 110° for 16 h, filter, evaporate the filtrate to dryness under reduced pressure, chromatograph the residue on silica gel with chloroform, recrystallize from chloroform:hexane 25:75 to give 2-(5-methoxycarbonyl-2-furyl)-5,6-methylenedioxybenzofuran as slightly yellow crystals (mp 195-198°). Heat 1.42 g 2-(5-methoxycarbonyl-2-furyl)-5,6-methylenedioxybenzofuran and 1.2 g hydrazine hydrate in 15 mL DMF at 70° for 1 h (Caution! Hydrazine hydrate is a carcinogen and explodes on distillation in air!), add 10 g hydrazine hydrate, add 20 mL water, filter. Wash the solid with MeOH and dry it under reduced pressure to give 2-(5-hydrazinocarbonyl-2-furyl)-5,6-methylenedioxybenzofuran as a light yellow powder (mp 261-262°).)

HPLC VARIABLES
Column: 250 × 4 5 μm Wakosil ODS-II 5C18 HG
Mobile phase: MeCN:water 34:66
Column temperature: 40
Flow rate: 1
Injection volume: 10
Detector: F ex 362 em 462

CHROMATOGRAM
Retention time: 73 (prostaglandin $F_{1\alpha}$, dinoprost), 89 (dinoprostone), 91 (alprostadil)
Limit of detection: 0.1 pmole

KEY WORDS
SPE

REFERENCE
Saito, M.; Ushijima, T.; Sasamoto, K.; Yakata, K.; Ohkura, Y.; Ueno, K. 2-(5-Hydrazinocarbonyl-2-thienyl)-5,6-methylenedioxybenzofuran and 2-(5-hydrazinocarbonyl-2-furyl)-5,6-methylenedioxybenzofuran as novel fluorescence derivatization reagents for carboxylic acids in liquid chromatography, *Anal.Chim.Acta*, **1995**, *300*, 243–251.

2-(5-Hydrazinocarbonyl-2-oxazolyl)-5,6-dimethoxybenzothiazole

SAMPLE

Matrix: blood

Analyte: carboxylic acids

Sample preparation: 50 μL Serum + 100 μL isopropanol:heptane:0.5 M sulfuric acid 80:20:2, extract. Remove 50 μL of the organic layer and evaporate it to dryness under reduced pressure, reconstitute the residue in 50 μL 15 mM 2-(5-hydrazinocarbonyl-2-oxazolyl)-5,6-dimethoxybenzothiazole in DMSO, 50 μL 100 mM 1-ethyl-3-(3-dimethylam-inopropyl)carbodiimide in MeOH, 50 μL 1% pyridine in MeOH, and 50 μL 100 μM non-adecanoic acid, heat at 37° for 1 h, inject a 10 μL aliquot. (Reagent was 2-(5-hydrazinocarbonyl-2-oxazolyl)-5,6-dimethoxybenzothiazole which was synthesized as follows. Add 10 mL concentrated nitric acid dropwise to 20 g 4-bromoveratrole in 60 mL acetic acid while keeping the temperature at 10-30° with occasional cooling, when the addition is complete pour the reaction mixture into ice-water. Collect the precipitate and dissolve it in 500 mL hot EtOH, add activated charcoal, filter, add 40 mL water to the filtrate to give 4,5-dimethoxy-2-nitrobromobenzene as a light yellow crystalline solid (mp 121-122°). Prepare sodium sulfide by melting together 5 g sodium sulfide nonahydrate and 700 mg sulfur, add this mixture to 5 g 4,5-dimethoxy-2-nitrobromobenzene in 50 mL EtOH:water 95:5, reflux for 30 min, pour into ice-water, collect the solid, recrystallize from dichloromethane to give di(4,5-dimethoxy-2-nitrophenyl)sulfide as yellow needles (mp 231-232°) (Anal. Sci. 1995, 11, 103). Add ethyl oxalyl chloride in ether to a solution of diazomethane in ether at 0° to give ethyl diazopyruvate (Caution! Diazo compounds are explosive and toxic!) (cf. Buehler,C.A.; Pearson,D.E. Survey of Organic Syntheses, Wiley, New York, 1970, p. 179). Heat 100 mg ethyl diazopyruvate, a few mg copper(II) acetylacetonate, and 400 μL chloroacetonitrile in benzene at 60° overnight (Caution! Benzene is a carcinogen!), cool, add to sodium bicarbonate solution, extract with ether, dry the organic layer, evaporate, chromatograph on silica with petroleum ether:ethyl acetate 90:10, distil the product at 90°/12 mm Hg to give ethyl 2-chloromethyl-5-oxazolecarbox-ylate as an oil in 18% yield (US Patent 4 603 209 (July 29, 1986)). Reflux 5.0 g ethyl 2-chloromethyl-5-oxazolecarboxylate and 11.7 g NaI in 80 mL acetone for 1 h, partition the reaction mixture between ethyl acetate and water. Wash the organic layer with water and dry it over anhydrous sodium sulfate, evaporate to give ethyl 2-iodomethyl-5-oxazo-lecarboxylate as a reddish-brown oil. Reflux 7.4 g ethyl 2-iodomethyl-5-oxazolecarboxylate and 21.5 g silver carbonate in 100 mL THF:water 70:30 for 4 h, filter through Celite, evaporate under reduced pressure, chromatograph on silica gel using benzene:ethyl ac-etate 95:5 to give ethyl 2-hydroxymethyl-5-oxazolecarboxylate (mp 60.5-62°). Stir 2.04 g oxalyl chloride in 15 mL dichloromethane at -50° under nitrogen, add 1.54 g DMSO in 3 mL dichloromethane, after 5 min add 1.4 g ethyl 2-hydroxymethyl-5-oxazolecarboxylate in 6 mL dichloromethane, stir for 15 min at -50°, add 5.7 mL triethylamine, allow to warm to room temperature, dilute with dichloromethane, wash with water, dry over anhydrous sodium sulfate, concentrate under reduced pressure, chromatograph on silica gel using benzene:ethyl acetate 95:5 to give ethyl 2-carboxaldehyde-5-oxazolecarboxylate (mp 71.5-73°). Add 11.3 mL concentrated HCl to 750 mg di(4,5-dimethoxy-2-nitrophenyl)sulfide stirred in 100 mL EtOH, add 3.3 g tin powder at 40-45°, stir for 1 h at 40-45°, dilute with 100 mL water, pass hydrogen sulfide gas through this solution (Caution! Hydrogen sulfide is highly toxic!), filter, concentrate the filtrate under reduced pressure to give 4,5-dimethoxy-2-aminothiophenol. Take up this compound in 30 mL EtOH:acetic acid 2:1 and add 750 mg ethyl 2-carboxaldehyde-5-oxazolecarboxylate, reflux for 1 h, collect the precipitate and recrystallize it from EtOH to give 2-(5-ethoxycarbonyl-2-oxazolyl)-5,6-dimethoxybenzothiazole as yellow needles (mp 200-201°). Add 381 mg 2-(5-ethoxycarbonyl-2-oxazolyl)-5,6-dimethoxybenzothiazole to 20 mL EtOH containing 3 mL DMF and 5 mL hydrazine hydrate, reflux for 1 h, collect the precipitate and wash it with EtOH, dry under vacuum to give 2-(5-hydrazinocarbonyl-2-oxazolyl)-5,6-dimethoxybenzothiazole as a yellow powder (mp 255.5-280° (d)). Preparation of diazomethane is as follows. Caution! Diazomethane is toxic, explosive, and carcinogenic! A face shield and a safety screen should always be used and the preparation should only be carried out in a properly functioning chemical fume hood. Only smooth glass apparatus with rubber stoppers and plastic tubing should be used. Scratched glassware, ground glass joints, and sharp edges should be avoided (Org. Syn., Coll. Vol. VI; Wiley: New York, 1988, pp. 432-435). Procedures have been reported for the synthesis of diazomethane from N-methyl-N-nitrosourea (Org. Syn., Coll. Vol. II; Wiley: New York, 1943, pp. 165-167), N-nitroso-β-methylaminoisobutyl methyl ketone (Org. Syn., Coll. Vol. III; Wiley: New York, 1955, pp. 244-248), and N,N'-dimethyl-N,N'-dinitrosoterephthalamide (Org. Syn., Coll. Vol. V; Wiley: New York, 1973,pp. 351-355). Probably the most convenient starting material is N-methyl-N-nitroso-p-toluenesulfonamide (Diazald) (Aldrichimica Acta 1983, 16, 3-10; Org. Syn., Coll. Vol. IV, Wiley: New York, 1963, pp. 250-253). Add 10 mL 95% EtOH to a solution of 5 g KOH in 8 mL water, warm to 65° using a water bath, add a solution of 5 g N-methyl-N-nitroso-p-toluenesulfonamide in 45 mL ether over 20 min at such a rate as to keep the reaction volume constant. Collect the ether and diazomethane that distil in an ice-cooled receiving flask under a dry ice/acetone condenser. When all the N-methyl-N-nitroso-p-toluenesulfonamide has been used up, slowly add 10 mL ether to the reaction flask and continue distillation until the distillate is colorless. A purpose-built distillation apparatus can be purchased from Aldrich (Aldrichimica Acta 1983, 16, 3-10). Excess quantities of diazomethane can be destroyed by adding acetic acid until the yellow color of the diazomethane is discharged. The safe disposal of the nitroso compounds used to generate diazomethane has been discussed (Lunn,G.; Sansone,E.B. Destruction of Hazardous Chemicals in the Laboratory, Second Edition. Wiley: New York, 1994, pp. 277-289).)

HPLC VARIABLES
Column: 250 × 4.6 5 μm Wakosil-II 5C18 HG
Mobile phase: Gradient. MeCN:water from 70:30 to 100:0 over 20 min.
Column temperature: 40
Flow rate: 1
Injection volume: 10
Detector: F ex 369 em 451

CHROMATOGRAM
Retention time: 5 (lauric acid), 5.5 (myristoleic acid), 7.5 (linolenic acid), 8 (myristic acid), 9 (palmitoleic acid), 10 (arachidonic acid, linoleic acid), 12.5 (palmitic acid), 13.5 (oleic acid), 15 (margaric acid), 17 (stearic acid), 25 (prostaglandin $F_{1\alpha}$), 30 (dinoprost), 37 (dinoprostone), 38.5 (alprostadil)
Internal standard: nonadecanoic acid (19)

REFERENCE
Saito, M.; Ushijima, T.; Sasamoto, K.; Ohkura, Y.; Ueno, K. 2-(5-Hydrazinocarbonyl-2-oxazolyl)-5,6-dimethoxybenzothiazole as a precolumn fluorescence derivatization reagent for carboxylic acids in high-performance liquid chromatography and its application to the assay of fatty acids in human serum, *J.Chromatogr.B*, **1995**, *674*, 167–175.

2-(5-Hydrazinocarbonyl-2-oxazolyl)-5,6-methylenedioxybenzofuran

SAMPLE

Matrix: solutions

Analyte: prostaglandins

Sample preparation: Mix 100 μL of a 100 μM solution of the carboxylic acid in water with 100 μL 100 mM 1-(3-methylaminopropyl)-3-ethylcarbodiimide in water, 100 μL 1% pyridine in water, and 100 μL 15 mM 2-(5-hydrazinocarbonyl-2-oxazolyl)-5,6-methylenedioxybenzofuran in DMF, heat at 37° for 1 h, inject a 10 μL aliquot. (Synthesis of 2-(5-hydrazinocarbonyl-2-oxazolyl)-5,6-methylenedioxybenzofuran is as follows. Add ethyl oxalyl chloride in ether to a solution of diazomethane in ether at 0° to give ethyl diazopyruvate (Caution! Diazo compounds are explosive and toxic!) (cf. Buehler, C.A.; Pearson,D.E. Survey of Organic Syntheses, Wiley, New York, 1970, p. 179). Heat 100 mg ethyl diazopyruvate, a few mg copper(II) acetylacetonate, and 400 μL chloroacetonitrile in benzene at 60° overnight (Caution! Benzene is a carcinogen!), cool, add to sodium bicarbonate solution, extract with ether, dry the organic layer, evaporate, chromatograph on silica with petroleum ether:ethyl acetate 90:10, distil the product at 90°/12 mm Hg to give ethyl 2-chloromethyl-5-oxazolecarboxylate as an oil in 18% yield (US Patent 4 603 209 (July 29, 1986)). Add 2 mL phosphorus oxychloride dropwise to a solution of 2 g sesamol in 3 mL DMF at 0°, heat on a steam bath with frequent shaking for 1 h, cool in ice, add 50 mL saturated sodium acetate solution, heat on a steam bath for 30 min, cool, filter, recrystallize the solid from EtOH to give 2-hydroxy-4,5-methylenedioxybenzaldehyde as colorless needles (mp 125-126°) (Bull. Chem. Soc. Jpn. 1962, 35, 1321). Stir 1.4 g ethyl 2-chloromethyl-5-oxazolecarboxylate, 1.5 g 2-hydroxy-4,5-methylenedioxybenzaldehyde, 2 g potassium carbonate, and 50 mL anhydrous DMF at 120° overnight, cool, filter. Evaporate the filtrate to dryness under reduced pressure to give 2-(5-ethoxycarbonyl-2-oxazolyl)-5,6-methylenedioxybenzofuran as a colorless crystalline powder (mp 186°) (yield 39%). Reflux 260 mg 2-(5-ethoxycarbonyl-2-oxazolyl)-5,6-methylenedioxybenzofuran, 100 mg KOH, 20 mL EtOH, and 30 mL water for 2 h, concentrate under reduced pressure, dissolve the residue in 100 mL water, wash with ethyl acetate, treat the aqueous layer with activated carbon, acidify the aqueous layer to pH 2 with 2 M HCl. Filter the precipitate and recrystallize it from EtOH to give 2-(2-oxazole-5-carboxylic acid)-5,6-methylenedioxybenzofuran as a colorless crystalline powder (mp 294-295°). Reflux 150 mg 2-(2-oxazole-5-carboxylic acid)-5,6-methylenedioxybenzofuran and 5 mL thionyl chloride for 2 h, pour the reaction mixture into 300 mL petroleum ether. Filter the precipitate and dry it over KOH to give 2-(5-chlorocarbonyl-2-oxazolyl)-5,6-methylenedioxybenzofuran (mp 290°) (Anal. Sci. 1989, 5, 525). 2-(5-Chlorocarbonyl-2-oxazolyl)-5,6-methylenedioxybenzofuran is also available from Dojindo, Kumamoto, Japan (Dojindo Molecular Technologies, Inc., 3 Bethesda Metro Center, Suite 700, Bethesda MD 20814; (301) 664-8448; www.dojindo.co.jp). Add 2 mL hydrazine hydrate to a stirred solution of 2 g

2-(5-chlorocarbonyl-2-oxazolyl)-5,6-methylenedioxybenzofuran in 20 mL anhydrous DMF (Caution! Hydrazine hydrate is a carcinogen!), stir at room temperature for 4 h, add 20 mL benzene (Caution! Benzene is a carcinogen!). Collect the precipitate and wash it with water and MeCN, recrystallize from DMF:benzene 50:50 to give 2-(5-hydrazinocarbonyl-2-oxazolyl)-5,6-methylenedioxybenzofuran as an off-white crystalline solid (mp >220° d). Preparation of diazomethane is as follows. Caution! Diazomethane is toxic, explosive, and carcinogenic! A face shield and a safety screen should always be used and the preparation should only be carried out in a properly functioning chemical fume hood. Only smooth glass apparatus with rubber stoppers and plastic tubing should be used. Scratched glassware, ground glass joints, and sharp edges should be avoided (Org. Syn., Coll. Vol. VI; Wiley: New York, 1988, pp. 432-435). Procedures have been reported for the synthesis of diazomethane from N-methyl-N-nitrosourea (Org. Syn., Coll. Vol. II; Wiley: New York, 1943, pp. 165-167), N-nitroso-β-methylaminoisobutyl methyl ketone (Org. Syn., Coll. Vol. III; Wiley: New York, 1955, pp. 244-248), and N,N'-dimethyl-N,N'-dinitrosoterephthalamide (Org. Syn., Coll. Vol. V; Wiley: New York, 1973, pp. 351-355). Probably the most convenient starting material is N-methyl-N-nitroso-p-toluenesulfonamide (Diazald) (Aldrichimica Acta 1983, 16, 3-10; Org. Syn., Coll. Vol. IV, Wiley: New York, 1963, pp. 250-253). Add 10 mL 95% EtOH to a solution of 5 g KOH in 8 mL water, warm to 65° using a water bath, add a solution of 5 g N-methyl-N-nitroso-p-toluenesulfonamide in 45 mL ether over 20 min at such a rate as to keep the reaction volume constant. Collect the ether and diazomethane that distil in an ice-cooled receiving flask under a dry ice/acetone condenser. When all the N-methyl-N-nitroso-p-toluenesulfonamide has been used up, slowly add 10 mL ether to the reaction flask and continue distillation until the distillate is colorless. A purpose-built distillation apparatus can be purchased from Aldrich (Aldrichimica Acta 1983, 16, 3-10). Excess quantities of diazomethane can be destroyed by adding acetic acid until the yellow color of the diazomethane is discharged. The safe disposal of the nitroso compounds used to generate diazomethane has been discussed (Lunn,G.; Sansone,E.B. Destruction of Hazardous Chemicals in the Laboratory, Second Edition. Wiley: New York, 1994, pp. 277-289).)

HPLC VARIABLES
Column: 250 × 4.6 5 μm Wakosil ODS-II, WS-II 5C18 HG
Mobile phase: MeCN:water 30:70
Column temperature: 40
Flow rate: 1
Injection volume: 10
Detector: F ex 350 em 450

CHROMATOGRAM
Retention time: 62 (prostaglandin $F_{1\alpha}$), 65 (dinoprost), 76 (alprostadil), 78 (dinoprostone)
Limit of detection: 0.1 pmole

REFERENCE
Saito, M.; Chiyoda, Y.; Ushijima, T.; Sasamoto, K.; Ohkura, Y. 2-(5-Hydrazinocarbonyl-2-oxazolyl)-5,6-methylenedioxybenzofuran as a fluorescence derivatization reagent for carboxylic acids in high-performance liquid chromatography, Anal.Sci., 1994, 10, 679–681.

2-(4-Hydrazinocarbonylphenyl)-4,5-diphenylimidazole

SAMPLE
Matrix: blood
Analyte: fatty acids
Sample preparation: 50 μL Serum + 10 μL 100 μM margaric acid in DMF + 200 μL 500 mM pH 6.5 phosphate buffer, mix, add 2 mL chloroform:n-heptane 50:50, vortex for 1 min, centrifuge at 1500 g for 5 min. Remove a 1.5 mL aliquot of the organic layer and evaporate it to dryness, reconstitute the residue in 200 μL DMF. Remove a 100 μL aliquot and add it to 50 μL 1 M 1-(3-dimethylaminopropyl)-3-ethylcarbodiimide in water, 50 μL 10% pyridine in water, and 100 μL 30 mM 2-(4-hydrazinocarbonylphenyl)-4,5-diphenylimidazole in DMF, mix well, let stand at room temperature for 45 min, inject a 20 μL aliquot. (Synthesis of the reagent, 2-(4-hydrazinocarbonylphenyl)-4,5-diphenylimidazole, is as follows. Add 3.15 g benzil and 2.46 g methyl 4-formylbenzoate (terephthalaldehydic acid methyl ester) to 10 g ammonium acetate in 30 mL acetic acid, stir at 80° for 9 h, cool to room temperature, pour into cold water, filter. Wash the precipitate with water and recrystallize it from EtOH to give 4-(4,5-diphenyl-1H-imidazol-2-yl)benzoic acid methyl ester as pale yellow crystals (mp 245-248°). Reflux 1.47 g 4-(4,5-diphenyl-1H-imidazol-2-yl)benzoic acid methyl ester and 15 mL 80% hydrazine hydrate (Caution! Hydrazine hydrate is a carcinogen and explodes on distillation in air!) in 100 mL EtOH for 4 h, cool, to room temperature, pour into cold water, filter. Wash the precipitate with water and recrystallize it from EtOH:benzene 50:50 (Caution! Benzene is a carcinogen!) to give 2-(4-hydrazinocarbonylphenyl)-4,5-diphenylimidazole as a colorless powder (mp >300°).)

HPLC VARIABLES
Column: 150 × 6 5 μm Shim-pack CLC-ODS
Mobile phase: Gradient. MeOH:water 90:10 for 5 min, to 100:0 over 25 min, maintain at 100:0 for 5 min.
Flow rate: 1
Injection volume: 20
Detector: F ex 335 em 455

CHROMATOGRAM

Retention time: 11.5 (lauric acid), 16.5 (myristic acid), 17.5 (linolenic acid), 18.6 (palmitoleic acid), 20.8 (linoleic acid, arachidonic acid), 22.5 (palmitic acid), 24.5 (oleic acid), 28 (stearic acid)

Internal standard: margaric acid (26)

Limit of detection: 7-57 fmole

KEY WORDS

serum

REFERENCE

Nakashima, K.; Taguchi, Y.; Kuroda, N.; Akiyama, S.; Duan, G. 2-(4-Hydrazinocarbonylphenyl)-4,5-diphenylimidazole as a versatile fluorescent derivatization reagent for the high-performance liquid chromatographic analysis of free fatty acids, *J.Chromatogr.*, **1993**, *619*, 1–8.

SAMPLE

Matrix: blood

Analyte: carnitine and acylcarnitines

Sample preparation: Condition a 100 mg LiChrolut SCX SPE cartridge (Merck) with 2 mL water. Condition a Toyopak IC-SP S SPE cartridge (Tosoh) with 2 mL water. 50 µL Plasma + 10 µL 400 µM IS in water + 1 mL 25 mM pH 1.0 sodium phosphate buffer, mix, add to the LiChrolut SCX SPE cartridge, wash with 5 mL water, elute with 1 mL isopropanol:150 mM pyridine in water 50:50. Evaporate the eluate under reduced pressure, add 20 µL 2.5 mM 2-(4-hydrazinocarbonylphenyl)-4,5-diphenylimidazole in DMF, add 20 µL 100 mM 1-(3-dimethylaminopropyl)-3-ethylcarbodiimide hydrochloride in DMF, add 10 µL pyridine, vortex, let stand at room temperature for 1 h, add 1 mL DMF:water 40:60, add to the Toyopak IC-SP S SPE cartridge, wash with 1 mL DMF:water 40:60, wash with 10 mL water, wash with 100 µL 600 mM KCl in MeOH:water 50:50, elute with 200 µL 600 mM KCl in MeOH:water 50:50, inject an aliquot of the eluate. (Synthesis of 2-(4-hydrazinocarbonylphenyl)-4,5-diphenylimidazole is described above.)

HPLC VARIABLES

Column: 250 × 4.6 5 µm Daisopak SP-120-ODS (Daiso, Osaka)

Mobile phase: Gradient. A was MeCN. B was MeCN:200 mM pH 7.0 Tris-HCl buffer 20:80. A:B 25:75 for 12 min, to 45:55 over 10 min, to 100:0 over 5 min, maintain at 100:0 for 10 min.

Flow rate: 1

Injection volume: 20

Detector: F ex 340 em 475

CHROMATOGRAM

Retention time: 7.3 (carnitine), 8.7 (acetylcarnitine), 10.6 (propionylcarnitine), 22.8 (hexanoylcarnitine), 28.0 (octanoylcarnitine)

Internal standard: cyclohexanoylcarnitine (Synthesis of cyclohexanoylcarnitine is as follows. Dissolve 300 mg carnitine in 500 µL trifluoroacetic acid, add 1 mL cyclohexanecarbonyl chloride, protect from moisture with a calcium chloride tube, mix, heat at 40-45° overnight, cool to room temperature, add 5 mL acetone, cool on ice for a couple of h, centrifuge to remove undissolved material, add diethyl ether to incipient cloudiness, when crystallization starts add 10 mL diethyl ether, cool on ice. Dissolve the crystallization products in 1 mL MeOH, add 4-5 mL acetone, add diethyl ether to incipient cloudiness, when crystallization starts add 5 mL diethyl ether (cf. Biochim. Biophys. Acta 1968, 152, 559) to obtain cyclohexanoylcarnitine hydrochloride (mp 159-160°).) (20)

Limit of detection: 0.24-1.97 µM

KEY WORDS

plasma; SPE

REFERENCE

Kuroda, N.; Ohyama, Y.; Nakashima, K.; Akiyama, S. HPLC determination of carnitine and acylcarnitines in human plasma by means of fluorescence labeling using 2-(4-hydrazinocarbonylphenyl)-4,5-diphenylimidazole, *Chem.Pharm.Bull.*, **1996**, *44*, 1525–1529.

2-(5-Hydrazinocarbonyl-2-thienyl)-5,6-methylenedioxybenzofuran

SAMPLE
Matrix: seminal fluid
Analyte: prostaglandins
Sample preparation: Mix 50 μL seminal fluid with 500 μL dilute HCl (pH 3.0) and 500 μL ethyl acetate, vortex. Remove the organic layer and evaporate it to dryness under reduced pressure, reconstitute the residue in 200 μL water, add to a Toyopak-ODS SPE cartridge, elute with 200 μL MeOH. 100 μL Eluate + 100 μL 100 mM 1-(3-dimethylaminopropyl)-3-ethylcarbodiimide in water + 100 μL 1% aqueous pyridine + 100 μL 15 mM 2-(5-hydrazinocarbonyl-2-thienyl)-5,6-methylenedioxybenzofuran in DMF, heat at 37° for 1 h, inject a 10 μL aliquot. (Synthesis of 2-(5-hydrazinocarbonyl-2-thienyl)-5,6-methylenedioxybenzofuran is as follows. Slowly add 153 g freshly distilled phosphorus oxychloride to 73 g anhydrous DMF with stirring at room temperature, add 125 g sesamol in portions over 4 h, stir at room temperature overnight, pour into ice water, filter. Dissolve the solid in ether and wash with water, dry over anhydrous magnesium sulfate, evaporate to dryness, recrystallize from EtOH to give 2-hydroxy-4,5-methylenedioxybenzaldehyde as slightly-yellow crystals (mp 125-126°). Pass HCl gas into 15.6 g ethyl 2-thiophenecarboxylate, 4.5 g paraformaldehyde, and 3.4 g zinc chloride in 50 mL chloroform with stirring at 30° over 4 h. Pour into ice water and extract with 50 mL chloroform. Wash the chloroform layer 3 times with water, wash twice with aqueous sodium bicarbonate solution, dry over anhydrous sodium sulfate, evaporate to remove the solvent, distil at 86-94°/0.15 mm Hg to yield ethyl 5-chloromethyl thiophene-2-carboxylate as a colorless oil. Heat 3 g 2-hydroxy-4,5-methylenedioxybenzaldehyde, 3.68 g ethyl 5-chloromethyl thiophene-2-carboxylate, and 2.49 g potassium carbonate in 100 mL anhydrous DMF at 110° for 16 h, filter, evaporate the filtrate to dryness under reduced pressure, chromatograph the residue on silica gel with chloroform, recrystallize from chloroform:hexane 25:75 to give 2-(5-ethoxycarbonyl-2-thienyl)-5,6-methylenedioxybenzofuran as yellow crystals (mp 124-126°). Heat 1.5 g 2-(5-ethoxycarbonyl-2-thienyl)-5,6-methylenedioxybenzofuran and 1.2 g hydrazine hydrate in 15 mL DMF at 70° for 1 h (Caution! Hydrazine hydrate is a carcinogen and explodes on distillation in air!), add 10 g hydrazine hydrate, add 20 mL water, filter. Wash the solid with MeOH and dry it under reduced pressure to give 2-(5-hydrazinocarbonyl-2-thienyl)-5,6-methylenedioxybenzofuran as a yellow powder (mp 262-263°).)

HPLC VARIABLES
Column: 250 × 4 5 μm Wakosil ODS-II 5C18 HG
Mobile phase: MeCN:water 34:66
Column temperature: 40
Flow rate: 1
Injection volume: 10
Detector: F ex 373 em 483

CHROMATOGRAM

Retention time: 91 (prostaglandin $F_{1\alpha}$), 95 (dinoprost), 123 (dinoprostone), 125 (alprostadil)
Limit of detection: 0.1 pmole

KEY WORDS

SPE

REFERENCE

Saito, M.; Ushijima, T.; Sasamoto, K.; Yakata, K.; Ohkura, Y.; Ueno, K. 2-(5-Hydrazinocarbonyl-2-thienyl)-5,6-methylenedioxybenzofuran and 2-(5-hydrazinocarbonyl-2-furyl)-5,6-methylenedioxybenzofuran as novel fluorescence derivatization reagents for carboxylic acids in liquid chromatography, *Anal.Chim.Acta*, **1995**, *300*, 243–251.

4-(N-Hydrazinoformylmethyl-N-methyl)amino-7-N,N-dimethylaminosulfonyl-2,1,3-benzoxadiazole

RELATED REFERENCE

Fukushima, T.; Santa, T.; Homma, H.; Al-kindy, S.M.; Imai, K. Enantiomeric separation and detection of 2-arylpropionic acids derivatized with[(N,N-dimethylamino)sulfonyl]benzofurazan reagents on a modified cellulose stationary phase by high-performance liquid chromatography. *Anal.Chem.* **1997**, *69*, 1793-1799

SAMPLE

Matrix: solutions
Analyte: carboxylic acids
Sample preparation: Mix 50 μL of a carboxylic acid solution in DMF with 50 μL 1 M 1-(3-dimethylaminopropyl)-3-ethylcarbodiimide in water, 50 μL pyridine:water 20:80, and 50 μL 20 mM 4-(N-hydrazinoformylmethyl-N-methyl)amino-7-N,N-dimethylaminosulfonyl-2,1,3-benzoxadiazole in DMF, let stand at room temperature for 30 min, inject a 1 μL aliquot. (Synthesis of 4-(N-hydrazinoformylmethyl-N-methyl)amino-7-N,N-dimethylaminosulfonyl-2,1,3-benzoxadiazole is as follows. Dissolve 0.5 g magnesium sulfate heptahydrate and 6 g NaOH in 60 mL water, throughout the reaction keep the flask at about 20° with cold water cooling, add 15 mL 30% hydrogen peroxide, add 75 mL MeOH, add 12.1 g powdered benzoyl peroxide in one go, stir for 10 min, pour into 150 mL 20% sulfuric acid, extract three times with 50 mL portions of chloroform, determine peroxybenzoic acid concentration by iodometric titration (Tetrahedron 1967, 23, 3327). Slowly add 110 mL 1 M peroxybenzoic acid in chloroform to 7 g 2,6-difluoroaniline dissolved in 100 mL chloroform, stir at room temperature, when reaction is complete (iodometric titration) wash with 2% sodium thiosulfate, wash with 5% sodium carbonate, wash with water, dry over anhydrous sodium sulfate, evaporate to dryness under reduced pressure, recrystallize 2,6-difluoronitrosobenzene from EtOH (mp 108.5-109.5). Stir 8.5 g 2,6-difluoronitrosobenzene in 85 mL DMSO at room temperature and add a solution of 3.91 g sodium azide in 85 mL DMSO dropwise, let stand for about 1 h, add to a large volume of water, extract with

ether, dry the extracts over anhydrous sodium sulfate, evaporate to dryness under reduced pressure and distil to give 4-fluoro-2,1,3-benzoxadiazole as a colorless oil (bp 83°/12 mm Hg) (J.Chem.Soc.(C) 1970, 1433). Add 11 mL chlorosulfonic acid dropwise to 3 g 4-fluoro-2,1,3-benzoxadiazole in 10 mL chloroform at 0-10° (use a calcium chloride drying tube), stir at room temperature for 1 h, reflux for 2 h, cool, slowly pour into ice water, remove the organic layer, extract the aqueous layer with chloroform, combine the organic layers, wash, dry over anhydrous magnesium sulfate, evaporate under reduced pressure, take up the residue in 5 mL benzene (Caution! Benzene is a carcinogen!), chromatograph on a 150 × 30 column of silica gel (100-200 mesh Kanto Chemical) with n-hexane:benzene 50:50, evaporate the appropriate fractions to give 4-(chlorosulfonyl)-7-fluoro-2,1,3-benzoxadiazole (CBD-F) as pale yellow needles (mp 64-66°) (Anal. Chem. 1984. 56, 2461). Stir 0.76 g CBD-F in 70 mL MeCN at 0-10° and add 1 g dimethylamine hydrochloride in 10 mL 100 mM pH 10 borax dropwise, adjust pH to 5 with 1 M HCl, concentrate to about 10 mL under reduced pressure, extract three times with 200 mL portions of diethyl ether, wash with water, dry over anhydrous magnesium sulfate, evaporate under reduced pressure, chromatograph on a 500 × 20 column of silica gel with chloroform, isolate the appropriate fraction and re-chromatograph on the same column with ethyl acetate:benzene 1:2 to give 4-(N,N-dimethylaminosulfonyl)-7-fluoro-2,1,3-benzoxadiazole (DBD-F) as white needles (mp 124-125°) (yield 1% !). On a Merck no. 5714 60F$_{254}$ tlc plate eluted with chloroform DBD-F has Rf 0.32 and lies between two other reaction products (Analyst 1989, 114, 413). DBD-F can also be purchased from Tokyo Kasei (TCI America, 911 North Harborgate St., Portland, OR 97203; 800-423-8616, 503-283-1681, (fax) 503-283-1987; www.tciamerica.com). Add 880 mg 4-(N,N-dimethylaminosulfonyl)-7-fluoro-2,1,3-benzoxadiazole in 40 mL MeCN dropwise to N-methylglycine and 2.3 g sodium carbonate in water, stir at room temperature for 1 h, evaporate to remove the MeCN, wash the residue twice with 50 mL portions of ethyl acetate, acidify the aqueous phase with HCl, extract twice with 300 mL portions of ethyl acetate. Combine the organic layers and wash them twice with saturated aqueous NaCl, dry over anhydrous magnesium sulfate, evaporate under reduced pressure, recrystallize from ethyl acetate to give 4-(N-carboxymethyl-N-methyl)amino-7-(N,N-dimethylaminosulfonyl)-2,1,3-benzoxadiazole as orange-yellow crystals (mp 209-210°). Add 3.5 mL oxalyl chloride and 24 μL DMF to 1 g 4-(N-carboxymethyl-N-methyl)amino-7-(N,N-dimethylaminosulfonyl)-2,1,3-benzoxadiazole in anhydrous benzene, stir at room temperature for 30 min, reflux for 1 h, concentrate to dryness, add 20 mL dry benzene to the residue, filter, condense the filtrate to yield 4-(N-chloroformylmethyl-N-methyl)amino-7-(N,N-dimethylaminosulfonyl)-2,1,3-benzoxadiazole as yellow crystals (mp 102°) (Biomed. Chromatogr. 1994, 8, 107). Add 70 mg 4-(N-chloroformylmethyl-N-methyl)amino-7-N,N-dimethylaminosulfonyl-2,1,3-benzoxadiazole in 20 mL MeCN dropwise to 50 mg hydrazine hydrate in 50 mL MeOH (Caution! Hydrazine hydrate is a carcinogen and explodes on distillation in air!), stir at room temperature for 30 min, evaporate to dryness under reduced pressure, dissolve the residue in dichloromethane, chromatograph on silica gel, recrystallize from MeOH to give 4-(N-hydrazinoformylmethyl-N-methyl)amino-7-N,N-dimethylaminosulfonyl-2,1,3-benzoxadiazole as yellow crystals (mp 164-165°).)

HPLC VARIABLES
Column: 150 × 4.6 5 μm TSKgel ODS 80TM (Tosoh)
Mobile phase: Gradient. MeCN:water 50:50 for 10 min, to 100:0 over 20 min, maintain at 100:0 for 15 min.
Flow rate: 1
Injection volume: 1
Detector: F ex 440 em 550

CHROMATOGRAM
Retention time: 15 (capric acid), 22 (lauric acid), 27 (myristic acid), 30 (palmitic acid), 34 (stearic acid), 36 (arachidic acid)
Limit of detection: 3-9 fmole

REFERENCE
Santa, T.; Kimoto, K.; Fukushima, T.; Homma, H.; Imai, K. 4-(*N*-Hydrazinoformylmethyl-*N*-methyl)amino-7-*N,N*-dimethylaminosulphonyl-2,1,3-benzoxadiazole (DBD-CO-Hz) as a precolumn fluorescence derivatization reagent for carboxylic acids in high-performance liquid chromatography, *Biomed.Chromatogr.*, **1996**, *10*, 183–185.

4-(1-Methylphenanthro[9,10-d]imidazol-2-yl)benzohydrazide

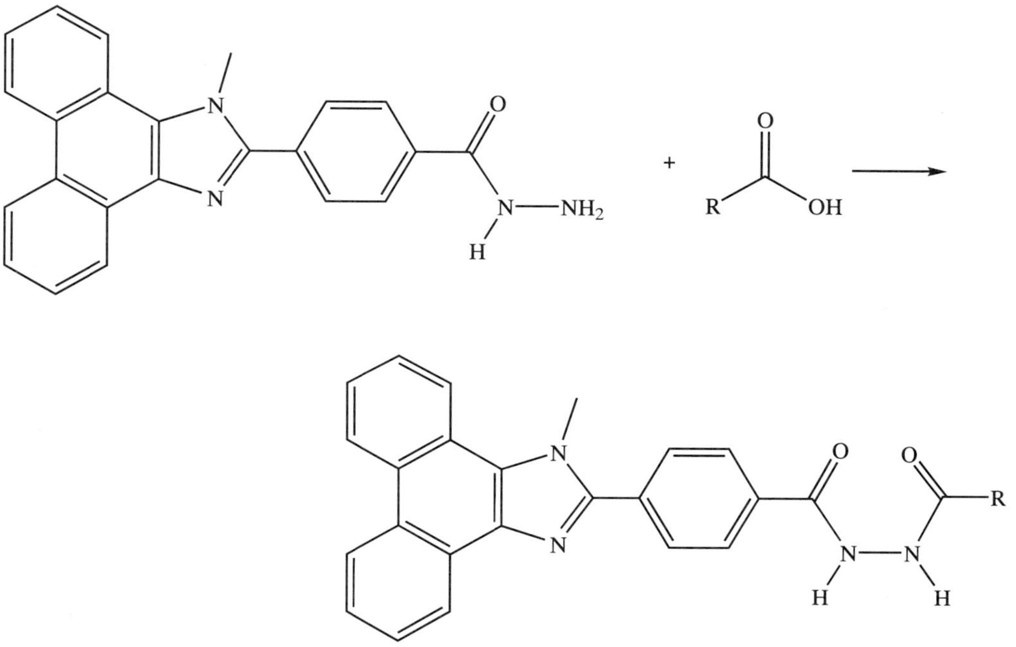

SAMPLE

Matrix: solutions

Analyte: carboxylic acids

Sample preparation: Mix 25 μL of an aqueous solution of fatty acids with 100 μL DMF: pyridine 93:7, add 50 μL 5 mM MPIB-hydrazide in DMF, add 100 μL 4 M (sic) 1-(3-dimethylaminopropyl)-3-ethylcarbodiimide, heat at 40° for 20 min, inject a 10 μL aliquot. (Synthesis of MPIB-hydrazide, 4-(1-methylphenanthro[9,10-d]imidazol-2-yl)benzo-hydrazide, is as follows. Stir 1 g 9,10-diaminophenanthrene and 800 mg methyl 4-for-mylbenzoate (terephthaldehydic acid methyl ester) in 200 mL EtOH at room temperature for 1 h, add 5 mL MeOH saturated with HCl, reflux under an inert gas for 2 h, cool, concentrate to 50 mL under reduced pressure, chromatograph the precipitate on a 200 × 35 column of 70-230 mesh silica gel (ca. 100 g; Merck) with chloroform, recrystallize from MeOH to give methyl 4-(phenanthro[9,10-d]imidazol-2-yl)benzoate as colorless needles (mp 312-315°). Dissolve 500 mg methyl 4-(phenanthro[9,10-d]imidazol-2-yl)benzoate in 100 mL anhydrous MeOH, treat with a solution of diazomethane in ether, evaporate to dryness under reduced pressure, dissolve the residue in 20 mL chloroform, chromatograph on a 200 × 60 column of about 250 g 100 mesh silica gel with chloroform to give methyl 4-(1-methylphenanthro[9,10-d]imidazol-2-yl)benzoate as colorless needles (mp 199-201°). Dissolve 2 g methyl 4-(1-methylphenanthro[9,10-d]imidazol-2-yl)benzoate in 100 mL aqueous hydrazine hydrate (45%) (Caution! Hydrazine hydrate is a carcinogen and ex-plodes on distillation in air!), heat at 100° for 1 h, recrystallize the precipitate from 95% EtOH to give MPIB-hydrazide (4-(1-methylphenanthro[9,10-d]imidazol-2-yl)benzohy-drazide) (mp 291-293°). Preparation of diazomethane is as follows. Caution! Diazomethane is toxic, explosive, and carcinogenic! A face shield and a safety screen should always be used and the preparation should only be carried out in a properly functioning chemical fume hood. Only smooth glass apparatus with rubber stoppers and plastic tubing should be used. Scratched glassware, ground glass joints, and sharp edges should be avoided (Org. Syn., Coll. Vol. VI; Wiley: New York, 1988, pp. 432-435). Procedures have been reported for the synthesis of diazomethane from N-methyl-N-nitrosourea (Org. Syn., Coll. Vol. II; Wiley: New York, 1943, pp. 165-167), N-nitroso-β-methylaminoisobutyl methyl ketone (Org. Syn., Coll. Vol. III; Wiley: New York, 1955, pp. 244-248), and N,N'-dimethyl-N,N'-dinitrosoterephthalamide (Org. Syn., Coll. Vol. V; Wiley: New York, 1973, pp. 351-355). Probably the most convenient starting material is N-methyl-N-nitroso-p-toluene-sulfonamide (Diazald) (Aldrichimica Acta 1983, 16, 3-10; Org. Syn., Coll. Vol. IV, Wiley: New York, 1963, pp. 250-253). Add 10 mL 95% EtOH to a solution of 5 g KOH in 8 mL water, warm to 65° using a water bath, add a solution of 5 g N-methyl-N-nitroso-p-toluenesulfonamide in 45 mL ether over 20 min at such a rate as to keep the reaction volume constant. Collect the ether and diazomethane that distil in an ice-cooled receiving flask under a dry ice/acetone condenser. When all the N-methyl-N-nitroso-p-toluenesulfonamide has been used up, slowly add 10 mL ether to the reaction flask and continue distillation until the distillate is colorless. A purpose-built distillation apparatus can be purchased from Aldrich (Aldrichimica Acta 1983, 16, 3-10). Excess quantities of diazomethane can be destroyed by adding acetic acid until the yellow color of the diazomethane is discharged. The safe disposal of the nitroso compounds used to generate diazomethane has been discussed (Lunn,G.; Sansone,E.B. Destruction of Hazardous Chemicals in the Laboratory, Second Edition. Wiley: New York, 1994, pp. 277-289).)

HPLC VARIABLES

Column: 250 × 4.6 5 μm TSKgel ODS-80Ts (Tosoh)
Mobile phase: MeOH:water 97:3 (A) or 90:10 (B) or 95:5 (C) or 80:20 (D)
Flow rate: 1
Injection volume: 10
Detector: F ex 360 em 460; F ex 325 (10 mW He-Cd laser) em 460

CHROMATOGRAM

Retention time: 8.0 (linoleic acid (C)), 9.6 (docosahexaenoic acid (C)), 9.6 (dehydroisoandrosterone 3-glucuronide (D)), 10.4 (thromboxane A_2 (D)), 10.6 (dinoprostone (D)), 11.6 (testosterone 3-glucuronide (D)), 12.0 (oleic acid (C)), 12.0 (prostaglandin $E_{2\alpha}$ (D)), 12.2 (prostaglandin D_2 (D)), 12.4 (arachidic acid (A)), 14.2 (margaric acid (A)), 16.2 (palmitic acid (A)), 21.6 (eticholanolone 3-glucuronide (D)), 22 (stearic acid (A)), 23.0 (androsterone 3-glucuronide (D)), 25 (arachidonic acid (B))
Limit of detection: 2.2-12.5 fmole (F), 0.4-2.3 fmole (laser F)

REFERENCE

Iwata, T.; Hirose, T.; Nakamura, M.; Yamaguchi, M. 4-(1-Methylphenanthro[9,10-*d*]imidazol-2-yl)benzohydrazide as derivatization reagent for carboxylic acids in high-performance liquid chromatography with conventional and laser-induced fluorescence detection, *Analyst*, **1994**, *119*, 1747-1751.

2-Nitrophenylhydrazine

RELATED REFERENCES

Miwa, H.; Yamamoto, M.; Nishida, T. Assay of free and total fatty acids (as 2-nitrophenylhydrazides) by high performance liquid chromatography. *Clin.Chim.Acta* **1986**, *155*, 95-101.

Miwa, H.; Yamamoto, M. Reversed-phase ion-pair chromatography of straight- and branched-chain dicarboxylic acids in urine as their 2-nitrophenylhydrazides. *Anal.Biochem.* **1988**, *170*, 301-307.

Miwa, H.; Yamamoto, M.; Asano, T. High-performance liquid chromatographic analysis of fatty acid compositions of platelet phospholipids as their 2-nitrophenylhydrazides. *J.Chromatogr.* **1991**, *568*, 25-34.

SAMPLE

Matrix: beverages

Analyte: carboxylic acids

Sample preparation: Measure out 25 μL fruit juice, 50 μL wine or sake, or 100 μL beer, make up to 100 μL with water (if necessary), add 200 μL 2 mM 3-methylglutaric acid in EtOH, add 200 μL 20 mM 2-nitrophenylhydrazine hydrochloride in 100 mM HCl:EtOH 50:50, add 200 μL 250 mM 1-ethyl-3-(3-dimethylaminopropyl)carbodiimide hydrochloride in EtOH:pyridine 97:3, heat at 80° for 5 min, add 200 μL 10% KOH in MeOH:water 50:50, heat at 80° for 5 min, cool, inject a 5-10 μL aliquot.

HPLC VARIABLES

Guard column: 10 × 5 5 μm BBC-5-C8 (Yamamura Chemical Laboratories, Kyoto)

Column: 250 × 4.6 4 μm J'sphere ODS-M 80 (Yamamura Chemical Laboratories, Kyoto)

Mobile phase: MeCN:MeOH:phosphate buffer 10:10:80, pH 7 containing 5 mM tetraethylammonium bromide (Prepare mobile phase by mixing MeCN:MeOH:5 mM KH_2PO_4 10:10:80 and MeCN:MeOH:5 mM Na_2HPO_4 10:10:80 to achieve pH 7 then adding tetraethylammonium bromide to a final concentration of 5 mM.)

Column temperature: 35

Flow rate: 2

Injection volume: 5-10

Detector: UV 400

CHROMATOGRAM

Retention time: 2.8, 3.2 (citric acid isomers), 5 (tartaric acid), 5.3, 5.5 (malic acid isomers), 6 (succinic acid), 7.5 (fumaric acid), 11 (glycolic acid), 14.5 (L-pyroglutamic acid), 15.3 (lactic acid), 16.7 (acetic acid)

Internal standard: 3-methylglutaric acid (9)

Limit of detection: 1-4 pmole

KEY WORDS

wine; fruit juice; beer; sake

REFERENCE

Miwa, H.; Yamamoto, M. Determination of mono-, poly- and hydroxy-carboxylic acid profiles of beverages as their 2-nitrophenylhydrazides by reversed-phase ion-pair chromatography, *J.Chromatogr.A*, **1996**, *721*, 261–268.

SAMPLE

Matrix: blood

Analyte: fatty acids

Sample preparation: 100 μL Serum + 10 μL 100 μM 2-ethylbutyric acid in water + 200 μL 20 mM 2-nitrophenylhydrazine hydrochloride in EtOH:100 mM HCl 50:50 + 400 μL

125 mM 1-(3-dimethylaminopropyl)-3-ethylcarbodiimide hydrochloride in EtOH containing 1.5% pyridine, heat at 60° for 20 min, add 4 mL 33.3 mM pH 6.4 phosphate buffer: 500 mM HCl 3.8:0.4, wash with 5 mL n-hexane. Remove a 3 mL aliquot of the aqueous layer, extract with 4 mL diethyl ether. Wash the ether layer with 33.3 mM pH 6.4 phosphate buffer and evaporate it to dryness under a stream of nitrogen at room temperature, reconstitute the residue in 50 μL MeOH, inject a 5-10 μL aliquot.

HPLC VARIABLES
Column: 250 × 6 5 μm YMC-C8 (Yamamura, Kyoto)
Mobile phase: MeCN:MeOH:water 30:16:54 adjusted to pH 4.5 with MeCN:MeOH:100 mM HCl 30:16:54
Column temperature: 50
Flow rate: 1.2
Injection volume: 5-10
Detector: UV 230

CHROMATOGRAM
Retention time: 5.5 (lactic acid), 6 (acetic acid), 7.5 (propionic acid), 8.5 (crotonic acid), 9 (isobutyric acid), 9.5 (n-butyric acid), 11 (tiglic acid), 12.5 (2-methylbutyric acid), 13 (3-methylcrotonic acid), 13.5 (isovaleric acid), 14 (n-valeric acid), 21 (isocaproic acid), 22 (n-caproic acid)
Internal standard: 2-ethylbutyric acid (17)
Limit of detection: 200-400 fmole

KEY WORDS
serum

REFERENCE
Miwa, H.; Yamamoto, M. High-performance liquid chromatographic analysis of serum short-chain fatty acids by direct derivatization, *J.Chromatogr.*, **1987**, *421*, 33–41.

SAMPLE
Matrix: blood
Analyte: fatty acids
Sample preparation: 25 μL Serum + 25 μL 80 μM margaric acid in EtOH, mix, add 100 μL 20 mM 2-nitrophenylhydrazine hydrochloride in EtOH:40 mM HCl 25:75, add 200 μL 125 mM 1-(3-dimethylaminopropyl)-3-ethylcarbodiimide hydrochloride in EtOH containing 6.5% pyridine, heat at 60° for 20 min, add 2 mL 33.3 mM pH 6.4 phosphate buffer:500 mM HCl 3.8:0.4, add 1.5 mL n-hexane, vortex for 30 s, centrifuge at 1500 g for 5 min. Remove the n-hexane layer and evaporate it to dryness under a stream of nitrogen at room temperature, reconstitute the residue in 50 μL MeOH, inject a 5-10 μL aliquot.

HPLC VARIABLES
Column: 250 × 4.6 5 μm YMC-C8 (Yamamura, Kyoto)
Mobile phase: MeCN:water 85:15 adjusted to pH 4.5 with MeCN:100 mM HCl 85:15
Column temperature: 30
Flow rate: 1.2
Injection volume: 5-10
Detector: UV 400; UV 230

CHROMATOGRAM
Retention time: 4 (capric acid), 4.8 (lauric acid), 5.2 (myristoleic acid), 6.2 (eicosapentaenoic acid), 6.4 (linolenic acid), 6.8 (myristic acid), 7.0 (docosahexaenoic acid), 7.2 (palmitoleic acid), 7.4 (arachidonic acid), 7.6 (linoleic acid), 8.4 (dihomo-gamma-linolenic acid), 9.2 (palmitic acid), 10 (oleic acid), 14 (stearic acid)
Internal standard: margaric acid (11.2)
Limit of detection: 0.4-1 pmole (UV 400), 100-200 fmole (UV 230)

KEY WORDS
serum

REFERENCE

Miwa, H.; Yamamoto, M.; Nishida, T.; Nunoi, K.; Kikuchi, M. High-performance liquid chromatographic analysis of serum long-chain fatty acids by direct derivatization method, *J.Chromatogr.*, **1987**, *416*, 237–245.

SAMPLE

Matrix: fat, oil
Analyte: fatty acids
Sample preparation: Prepare a 0.5-1 mg/mL solution of fat or oil in chloroform containing 0.005% BHT. Remove a 100 μL aliquot and add it to 40 nmole margaric acid, evaporate to dryness under a stream of nitrogen at room temperature, add 100 μL 2.5 M KOH: EtOH 20:80, heat at 90° for 10 min, cool to room temperature, add 400 μL reagent solution, add 200 μL 20 mM 2-nitrophenylhydrazine hydrochloride in 250 mM HCl, heat at 60° for 20 min, add 100 μL 15% KOH in MeOH:water 80:20, heat at 60° for 15 min, cool, inject a 1-5 μL aliquot. (Prepare reagent by mixing equal volumes of 3% pyridine in EtOH and 250 mM 1-(3-dimethylaminopropyl)-3-ethylcarbodiimide hydrochloride in EtOH.)

HPLC VARIABLES

Guard column: 30 × 4.6 ODS
Column: 250 × 4.6 5 μm YMC-C8 (Yamamura Chemical Institute, Kyoto)
Mobile phase: MeCN:water 85:15 adjusted to pH 4.5 with 100 mM HCl in MeCN
Column temperature: 30
Flow rate: 1.2
Injection volume: 1-5
Detector: UV 230

CHROMATOGRAM

Retention time: 4 (capric acid), 4.9 (lauric acid), 5.3 (myristoleic acid), 6 (eicosapentaenoic acid), 6.4 (linolenic acid), 6.7 (myristic acid), 7 (docsahexenoic acid), 7.5 (palmitoleic acid), 7.6 (arachidonic acid), 8 (linoleic acid), 8.7 (eicosatrienoic acid), 9.6 (palmitic acid), 10.7 (oleic acid), 14.4 (stearic acid)
Internal standard: margaric acid (12)
Limit of quantitation: 2.5 pmole

REFERENCE

Miwa, H.; Yamamoto, M. Improved method of determination of biologically important $C_{10:0}$-$C_{22:6}$ fatty acids as their 2-nitrophenylhydrazides by reversed-phase high-performance liquid chromatography, *J.Chromatogr.*, **1986**, *351*, 275–282.

SAMPLE

Matrix: fat, oil
Analyte: carboxylic acids
Sample preparation: Dissolve 1 mg fat or oil in 200 μL 2 mM margaric acid in EtOH, add 100 μL 400 mM KOH:EtOH 50:50, heat at 80° for 20 min, add 200 μL 20 mM 2-nitrophenylhydrazine hydrochloride in 300 mM HCl:EtOH 50:50, add 200 μL 250 mM 1-ethyl-3-(3-dimethylaminopropyl)carbodiimide hydrochloride in EtOH:pyridine 97:3, heat at 80° for 5 min, add 200 μL 10% KOH in MeOH:water 50:50, heat at 80° for 5 min, cool, inject a 5-10 μL aliquot.

HPLC VARIABLES

Guard column: J'sphere ODS-M 80 (Yamamura Chemical Laboratories, Kyoto)
Column: 250 × 4.6 4 μm J'sphere ODS-M 80 (Yamamura Chemical Laboratories, Kyoto)
Mobile phase: MeCN:water 86:14, adjusted to pH 4-5 with 100 mM HCl
Column temperature: 50
Flow rate: 2
Injection volume: 5-10
Detector: UV 400

CHROMATOGRAM

Retention time: 2 (caprylic acid), 2.7 (capric acid), 3.3 (lauric acid), 3.7 (myristoleic acid), 4.5 (octadecatetraenoic acid), 4.65 (eicosapentaenoic acid), 4.8 (α-linolenic acid), 4.95 (gamma-linolenic acid), 5.1 (myristic acid), 5.25 (docosahexaenoic acid), 5.4 (palmitoleic acid), 5.6 (arachidonic acid), 5.8 (linoleic acid), 6.3 (linoelaidic acid), 6.7 (eicosatrienoic acid, dihomo-gamma-linolenic acid), 7.3 (palmitic acid), 7.5 (docosatetraenoic acid), 8 (oleic acid), 8.5 (elaidic acid), 8.8 (eicosadienoic acid), 10 (docosatrienoic acid), 9.7 (stearic acid), 10.3 (eicosenoic (omega-9) acid), 10.8 (eicosenoic (omega-12) acid), 11.7 (docosadienoic acid), 12.2 (eicosenoic(omega-15) acid), 19.7 (arachidic acid), 20.7 (erucic acid)

Internal standard: margaric acid (9.3)

REFERENCE

Miwa, H.; Yamamoto, M. Rapid liquid chromatographic determination of fatty acids as 2-nitrophenyl-hydrazine derivatives, *J.AOAC Int.*, **1996**, *79*, 493–497.

SAMPLE

Matrix: milk products

Analyte: fatty acids

Sample preparation: Free fatty acids. Measure out 100 µL milk, 20 mg butter, 20 mg cheese, 50 mg condensed milk, 50 mg ice cream, or 50 mg yogurt, add 100 µL water, add 200 µL 100 µM 2-ethylbutyric acid in EtOH containing 100 µM margaric acid, add 200 µL 20 mM 2-nitrophenylhydrazine hydrochloride in EtOH:100 mM HCl 50:50, add 200 µL 250 mM 1-ethyl-3-(3-dimethylaminopropyl)carbodiimide hydrochloride in EtOH:pyridine 97:3, heat at 80° for 5 min, add 200 µL 10% KOH in MeOH:water 50:50, heat at 80° for 5 min, cool, add 4 mL 33 mM pH 6.4 phosphate buffer:500 mM HCl 7:1, extract with 5 mL n-hexane. Remove the organic layer and evaporate it to dryness under a stream of nitrogen at room temperature, reconstitute the residue in 200 µL MeOH, filter (0.45 µm), inject a 10-20 µL aliquot of the filtrate to determine long chain fatty acids. Remove a 3 mL aliquot of the aqueous layer and extract it twice with 4 mL portions of diethyl ether. Combine the organic layers and evaporate them to dryness under a stream of nitrogen at room temperature, reconstitute the residue in 200 µL MeOH, filter (0.45 µm), inject a 10-20 µL aliquot of the filtrate to determine short chain fatty acids. Total fatty acids. Measure out 10 µL milk, 1 mg butter, 1 mg cheese, 2 mg condensed milk, 2 mg ice cream, or 10 mg yogurt, add 200 µL 2 mM 2-ethylbutyric acid in EtOH containing 1 mM margaric acid, add 100 µL EtOH:400 mM KOH 50:50, heat at 80° for 20 min, add 200 µL 20 mM 2-nitrophenylhydrazine hydrochloride in EtOH:300 mM HCl 50:50, add 200 µL 250 mM 1-ethyl-3-(3-dimethylaminopropyl)carbodiimide hydrochloride in EtOH:pyridine 97:3, heat at 80° for 5 min, add 200 µL 10% KOH in MeOH:water 50:50, heat at 80° for 5 min, cool, add 4 mL 33 mM pH 6.4 phosphate buffer:500 mM HCl 7:1, extract with 5 mL n-hexane. Remove the organic layer and evaporate to dryness under a stream of nitrogen at room temperature, reconstitute the residue in 200 µL MeOH, filter (0.45 µm), inject a 2-10 µL aliquot of the filtrate to determine long chain fatty acids. Remove a 3 mL aliquot of the aqueous layer and extract it twice with 4 mL portions of diethyl ether. Combine the organic layers and evaporate them to dryness under a stream of nitrogen at room temperature, reconstitute the residue in 200 µL MeOH, filter (0.45 µm), inject a 2-10 µL aliquot of the filtrate to determine short chain fatty acids.

HPLC VARIABLES

Guard column: 10 × 4 5 µm BBC-4-C8 (Yamamura Chemical Labs)

Column: 250 × 6 5 µm YMC-FA C8 (Yamamura Chemical Labs)

Mobile phase: MeCN:MeOH:water 30:20:50, adjusted to pH 4.5 with 100 mM HCl (A; for short chain fatty acids) or MeCN:MeOH:water 75:11:14, adjusted to pH 4.5 with 100 mM HCl (B; for long chain fatty acids)

Column temperature: 35

Flow rate: 1.2

Injection volume: 2-10

Detector: UV 400

CHROMATOGRAM

Retention time: 6.5 (lactic acid (A)), 7 (acetic acid (A)), 8.5 (propionic acid (A)), 10.5 (isobutyric acid (A)), 10.8 (n-butyric acid (A)), 14 (isovaleric acid (A)), 15 (n-valeric acid (A)), 21 (isocaproic acid (A)), 22.5 (n-caproic acid (A)), 5.5 (caprylic acid (B)), 7 (capric acid (B)), 8.5 (lauric acid (B)), 9.5 (myristoleic acid (B)), 10.5 (eicosapentaenoic acid (B)), 11 (linolenic acid (B)), 11.5 (myristic acid (B)), 12 (docosahexaenoic acid (B)), 12.2 (palmitoleic acid (B)), 13 (arachidonic acid (B)), 13.5 (linoleic acid (B)), 14.5 (linoelaidic acid (B)), 15 (eicoatrienoic acid (B)), 16 (palmitic acid(B)), 16.5 (docosatetraenoic acid (B)), 17.5 (oleic acid (B)), 18.5 (elaidic acid (B)), 19 (eicosadienoic acid (B)), 20.5 (docosatrienoic acid (B)), 23.5 (stearic acid (B)), 25 (eicosaenoic acid (B)), 27 (docosadienoic acid (B)), 35.5 (arachidic acid (B)), 37.5 (erucic acid (B))

Internal standard: 2-ethylbutyric acid (17.5 (A)), margaric acid (19.5 (B))
Limit of detection: 0.5-2 pmole

KEY WORDS

milk; butter; cheese; condensed milk; ice cream; yogurt

REFERENCE

Miwa, H.; Yamamoto, M. Liquid chromatographic determination of free and total fatty acids in milk and milk products as their 2-nitrophenylhydrazides, *J.Chromatogr.*, **1990**, *523*, 235–246.

SAMPLE

Matrix: solutions
Analyte: fatty acids
Sample preparation: Mix 100 μL of a solution in EtOH, EtOH/water, or water with 400 μL reagent solution and 200 μL 20 mM 2-nitrophenylhydrazine hydrochloride in water, heat at 60° for 20 min, add 100 μL 15% KOH in MeOH:water 80:20, heat at 60° for 15 min, cool, inject a 1-2 μL aliquot. (Prepare reagent by mixing equal volumes of 3% pyridine in EtOH and 250 mM 1-(3-dimethylaminopropyl)-3-ethylcarbodiimide hydrochloride in EtOH.)

HPLC VARIABLES

Column: 250 × 4.6 5 μm YMC-C8 (Yamamuta Chemical Research, Kyoto)
Mobile phase: MeOH:water 58:42 (A) or 86:14 (B) adjusted to pH 4.5 with 100 mM HCl
Column temperature: 50
Flow rate: 1.2
Injection volume: 1-2
Detector: UV 230; UV 400

CHROMATOGRAM

Retention time: 3 (acetic acid (A)), 3.3 (capric acid (B)), 3.7 (propionic acid (A)), 4.3 (isobutyric acid (A)), 4.7 (butyric acid (A)), 5 (lauric acid (B)), 6 (isovaleric acid (A)), 6.3 (valeric acid (A)), 6.7 (myristic acid (B)), 7 (linolenic acid (B)), 7.5 (palmitoleic acid (B)), 8.5 (linoleic acid (B)), 8.8 (caproic acid (A)), 9.7 (palmitic acid (B)), 10.7 (oleic acid (B)), 14 (heptanoic acid (A)), 14.3 (stearic acid (B)), 23 (caprylic acid (A))

Limit of detection: 2.5-5 pmole (UV 230), 10-15 pmole (UV 400)

REFERENCE

Miwa, H.; Hiyama, C.; Yamamoto, M. High-performance liquid chromatography of short- and long-chain fatty acids as 2-nitrophenylhydrazides, *J.Chromatogr.*, **1985**, *321*, 165–174.

SAMPLE

Matrix: solutions
Analyte: carboxylic acids
Sample preparation: 100 μL Solution + 200 μL 1.25 mM lactic acid in EtOH + 200 μL 20 mM 2-nitrophenylhydrazine hydrochloride in 300 mM HCl:EtOH 50:50 + 200 μL 250 mM 1-ethyl-3-(3-dimethylaminopropyl)carbodiimide hydrochloride in EtOH:pyridine 97:3, heat at 60° for 20 min, add 200 μL 10% KOH in MeOH:water 50:50, heat at 60° for 15 min, cool, add 4 mL 200 mM HCl saturated with NaCl, extract with 4 mL ethyl acetate. Remove the organic layer and dry it over anhydrous sodium sulfate, evaporate to dryness under a stream of nitrogen at room temperature, reconstitute the residue in 100 μL MeOH, inject a 5-20 μL aliquot.

HPLC VARIABLES

Guard column: 10 × 5 5 μm BBC-5-C8 (Yamamura Chemical Laboratories, Kyoto)
Column: 250 × 4.6 5 μm YMC-FA (C8) (Yamamura Chemical Laboratories, Kyoto)
Mobile phase: MeCN:water 25:75, adjusted to pH 4-5 with 100 mM HCl
Column temperature: 45
Flow rate: 2
Injection volume: 5-20

Detector: UV 400

CHROMATOGRAM
Retention time: 6.8 (formic acid), 7.2 (acetic acid)
Internal standard: lactic acid (6.2)
Limit of detection: 0.5-1 pmole

KEY WORDS
acids were degradation products of Tween surfactants

REFERENCE
Miwa, H.; Yamamoto, M. Assay of volatile acids induced by autoxidation of nonionic surfactants by liquid chromatography with direct derivatization, *J.AOAC Int.*, **1996**, *79*, 418–422.

SAMPLE
Matrix: urine
Analyte: hydroxycarboxylic acids
Sample preparation: 200 µL Urine + 200 µL 500 µM 2-hydroxy-2-methylbutyric acid in EtOH + 1 mL EtOH:200 mM HCl 90:10, mix, centrifuge at 450 g for 5 min. Evaporate the supernatant to dryness under a stream of nitrogen at room temperature, reconstitute the residue in 100 µL water, add 200 µL EtOH, add 200 µL 20 mM 2-nitrophenylhydrazine hydrochloride in EtOH:100 mM HCl 50:50, add 200 µL 250 mM 1-ethyl-3-(3-dimethylaminopropyl) carbodiimide hydrochloride in EtOH:pyridine 97:3, mix, heat at 80° for 5 min, add 200 µL 10% KOH in MeOH:water 50:50, heat at 80° for 5 min, cool, add 4 mL 33 mM pH 6.4 phosphate buffer:500 mM HCl 7:1, wash twice with 4 mL portions of n-hexane. Remove the aqueous layer and extract it twice with 4 mL portions of diethyl ether, combine the ether layers and dry them over anhydrous sodium sulfate, evaporate to dryness under a stream of nitrogen at room temperature, reconstitute the residue in 100 µL MeOH, inject a 4-20 µL aliquot.

HPLC VARIABLES
Column: 150 × 6 4 µm J'sphere ODS-M 80 (Yamamura Chemical)
Mobile phase: MeCN:MeOH:water 8:32:60, adjusted to pH 4-5 with 100 mM HCl
Column temperature: 30
Flow rate: 2
Injection volume: 4-20
Detector: UV 400

CHROMATOGRAM
Retention time: 3 (glycolic acid), 3.2 (3-hydroxypropionic acid), 3.7 (lactic acid), 4 (3-hydroxybutyric acid, 3-hydroxyisobutyric acid), 4.5 (2-hydroxyisobutyric acid), 5 (3-hydroxy-2-methylbutyric acid), 5.3 (2-hydroxybutyric acid, 3-hydroxyisovaleric acid), 5.7 (3-hydroxy-2-ethylpropionic acid), 9.2 (2-hydroxyisovaleric acid), 16 (2-hydroxyisocaproic acid), 16.5 (2-hydroxy-3-methylvaleric acid)
Internal standard: 2-hydroxy-2-methylbutyric acid (7)
Limit of detection: 1-2 pmole

REFERENCE
Miwa, H.; Yamamoto, M.; Kan, K.; Futata, T.; Asano, T. High-performance liquid chromatographic measurements of urinary hydroxycarboxylic acids as an index of the metabolic control in non-insulin-dependent diabetic patients, *J.Chromatogr.B*, **1996**, *679*, 1–6.

ISOUREA

N,N'-Diisopropyl-O-(p-nitrobenzyl)isourea

RELATED REFERENCES

Gentile de Illiano, B.; Quintana De Gainzarain, A. HPLC determination of valproic acid in human plasma by derivatization with O-p-nitrobenzyl-N,N'-diisopropylisourea. *J.High Resolut.Chromatogr.* **1989**, *12*, 540-543.

Shaikh, B.; Pontzer, N.J.; Molina, J.E.; Kelsey, M.I. Separation and detection of UV-absorbing derivatives of fecal bile acid metabolites by high-performance liquid chromatography. *Anal.Biochem.* **1978**, *85*, 47-55.

Zou, A.; Xie, M.; Luo, X. Determination of artesunic acid after chemical derivatization with O-p-nitrobenzyl-N,N'-diisopropylisourea by high-performance liquid chromatography and ultraviolet absorption. *J.Chromatogr.* **1987**, *410*, 217-221.

SAMPLE

Matrix: beverages

Analyte: carboxylic acids

Sample preparation: 500 mg Instant coffee powder + 15 mL water + 750 μL 100 mM benzylmalonic acid in water, sonicate at room temperature for 10 min, make up to 25 mL, filter. Filter wine or fruit juice, dilute with water, mix 9 mL with 1 mL 100 μM benzylmalonic acid in water. Remove a 5 mL aliquot and add it to 500 mg 100-200 mesh Dowex 50W-X8, shake gently for a few min. Remove a 50 μL aliquot of the clear supernatant and add it to 500 μL freshly prepared 20 mg/mL N,N'-diisopropyl-O-(4-nitrobenzyl)isourea in dioxane (Caution! Dioxane is a carcinogen!), heat at 80° for 1 h, cool, add 2 mL MeCN, add 500 mg 100-200 mesh Dowex 50W-X8, shake briefly, let stand for at least 15 min, decant, filter (0.2 μm) the supernatant, inject an aliquot of the filtrate. (Distil dioxane from sodium. Recrystallize N,N'-diisopropyl-O-(4-nitrobenzyl)isourea from pentane. Synthesis of N,N'-diisopropyl-O-(p-nitrobenzyl)isourea is as follows. Mix 15.3 g 4-nitrobenzyl alcohol, 12.6 g diisopropylcarbodiimide, 10 mg copper(II) chloride, and 10 mL DMF, let stand at room temperature for 96 h, remove the solvent by distillation (45°/10 mm Hg), dissolve the residue in petroleum ether (bp 40-80°), chromatograph on a 200 × 30 column of aluminum oxide, elute with petroleum ether until the eluate does not turn moistened litmus paper blue. Remove the petroleum ether under reduced pressure to give N,N'-diisopropyl-O-(p-nitrobenzyl)isourea as yellow crystals (mp 42° (softening after 38°)) (Liebigs Ann. Chem. 1965, 685, 161). N,N'-Diisopropyl-O-(p-nitrobenzyl)isourea is also available from Fluka or Regis.)

HPLC VARIABLES

Guard column: 30 × 2.1 Aquapore RP-300 (Brownlee)

Column: 250 × 4 Nucleosil 5 RP-18
Mobile phase: Gradient. MeCN:water from 20:80 to 80:20 over 20 min
Column temperature: 25
Flow rate: 1
Injection volume: 5
Detector: UV 265

CHROMATOGRAM
Retention time: 7.10 (quinic acid), 9.00 (p-nitrobenzoic acid), 9.75 (glycolic acid), 10.15 (pyroglutamic acid), 11.10 (lactic acid), 13.45 (formic acid), 14.40 (acetic acid), 15.25 (tartaric acid), 15.25 (gentisic acid), 15.50 (nicotinic acid),15.75 (tartronic acid), 16.15 (mandelic acid), 16.30 (malic acid), 16.50 (propionic acid), 16.95 (furan-2-carboxylic acid), 18.25 (butyric acid), 18.40 (malonic acid), 18.80 (maleic acid), 18.95 (succinic acid), 19.15 (phosphoric acid), 19.18 (phenylacetic acid), 19.35 (itaconic acid), 19.75 (benzoic acid), 19.80 (glutaric acid), 19.85 (citraconic acid), 19.85 (sorbic acid), 19.95 (citric acid), 20.25 (fumaric acid), 21.05 (mesaconic acid), 21.15 (2-naphthylacetic acid), 21.40 (diphenylacetic acid), 21.75 (cyclohexanecarboxylic acid)
Internal standard: benzylmalonic acid (21.90)

KEY WORDS
coffee; fruit juice; wine

REFERENCE
Badoud, R.; Pratz, G. Improved high-performance liquid chromatographic analysis of some carboxylic acids in food and beverages as their p-nitrobenzyl esters, *J.Chromatogr.*, **1986**, *360*, 119–136.

SAMPLE
Matrix: blood, urine
Analyte: glycolic acid
Sample preparation: Serum. 2 mL Serum + 2 mL water + 1 g NaCl, adjust pH to 2 with 2-3 drops 4 M HCl, add 15 mL methyl ethyl ketone, extract, centrifuge at 2000 g for 10 min. Remove the organic layer and evaporate it to dryness under a stream of air, reconstitute the residue in 3 mL ethyl acetate, add 20 mg N,N'-diisopropyl-O-(p-nitrobenzyl)isourea (Regis; Fluka), heat at 80° for 2 h, cool, inject a 3 μL aliquot. Urine. Dilute 20 mL urine to 100 mL with water, filter (Whatman No. 4 paper), add 25-30 g NaCl, adjust pH to 2 with 8-10 drops concentrated HCl, extract 3 times with equal volumes of NaCl. Combine the organic layers and evaporate them to dryness under a stream of air, reconstitute the residue in 3 mL ethyl acetate, add 20 mg N,N'-diisopropyl-O-(p-nitrobenzyl)isourea, heat at 80° for 2 h, cool, inject a 3 μL aliquot. (Synthesis of N,N'-diisopropyl-O-(p-nitrobenzyl)isourea is also described above.)

HPLC VARIABLES
Column: 250 × 4.6 5 μm Supelcosil LC-Si
Mobile phase: Isooctane:methyl acetate 84:16
Flow rate: 3
Injection volume: 3
Detector: UV 254; UV 280

CHROMATOGRAM
Retention time: 14
Limit of detection: 1-2 ng

KEY WORDS
dog; serum; normal phase; glycolic acid is the major metabolite of ethylene glycol; pharmacokinetics

REFERENCE
Hewlett, T.P.; Ray, A.C.; Reagor, J.C. Diagnosis of ethylene glycol (antifreeze) intoxication in dogs by determination of glycolic acid in serum and urine with high pressure liquid chromatography and gas chromatography-mass spectrometry, *J.Assoc.Off.Anal.Chem.*, **1983**, *66*, 276–283.

SAMPLE
Matrix: bulk
Analyte: stearic acid

Sample preparation: Dissolve 3 μmole compound in 125 μL dichloromethane, add 125 μL 72 mM N,N'-diisopropyl-O-(p-nitrobenzyl)isourea (O-(p-nitrobenzyl)-N,N'-diisopropylisourea) in dichloromethane, heat at 80° for 2 h, cool, inject an aliquot. (Synthesis of N,N'-diisopropyl-O-(p-nitrobenzyl)isourea is described above. It is also available from Fluka or Regis.)

HPLC VARIABLES
Column: 250 mm long 5 μm MicroPak silica
Mobile phase: Hexane:chloroform 80:20
Detector: UV 254

CHROMATOGRAM
Limit of detection: 4 pmole

OTHER SUBSTANCES
Also analyzed: lauric acid, myristic acid, palmitic acid

KEY WORDS
normal phase

REFERENCE
Knapp, D.R.; Krueger, S. Use of O-p-nitrobenzyl-N,N'-diisopropylisourea as a chromogenic reagent for liquid chromatographic analysis of carboxylic acids, *Anal.Lett.*, **1975**, *8*, 603–610.

SAMPLE
Matrix: gastric contents, vomitus
Analyte: sodium fluoroacetate
Sample preparation: Condition a Sep-Pak C18 SPE cartridge with 2 mL MeOH and 5 mL water. Shake 20 g gastric contents and 40-50 mL water on a rotary shaker for 30 min, adjust pH to ≥5 with 4 M NaOH (if necessary), add 10 g Hyflo Super-Cel, mix, filter (Whatman No. 4 paper), wash through with 15-30 mL water, wash filtrate with an equal volume of dichloromethane, adjust the pH of the aqueous layer to 1.6-1.8 with 4 M HCl, extract five times with equal volumes of methyl ethyl ketone. Combine the organic layers and add 20 mL water, add 2 drops 1% phenolphthalein in EtOH, add 4 M NaOH until a permanent pink color is obtained, extract, repeat extraction 4 more times. Combine the aqueous layers, adjust pH to 1.6-1.8 with 4 M HCl, extract 5 times with equal volumes of methyl ethyl ketone. Combine the organic layers, add 2 drops 1% phenolphthalein in EtOH, add 4 M NaOH dropwise until a permanent pink color persists in the small aqueous layer, evaporate to dryness using a stream of air at 80-100°, reconstitute with 1.5 mL 5 mM pH 6.0 sodium phosphate buffer, adjust pH to 6-8 with dilute HCl (if necessary), add to the SPE cartridge, rinse out the container with 500, 500, and 300 μL aliquots of buffer, add the rinses to the SPE cartridge, collect all the effluent from the SPE cartridge. Remove a 500-600 μL aliquot of the eluate and add it to 1 drop 1% phenolphthalein in EtOH, add 1 M NaOH until a permanent pink color is obtained, evaporate to dryness at 80-100° under a stream of air, reconstitute with 1 mL 1 M pH 1.8 sodium phosphate buffer, adjust pH to 1.5-2.5 with 4 M HCl (if necessary), extract three times with 1 mL portions of ethyl acetate. Combine the organic layers and add 25 mg N,N'-diisopropyl-O-(p-nitrobenzyl)isourea, heat at 80 ± 1° for 8-24 h, inject a 3-10 μL aliquot. (N,N'-Diisopropyl-O-(p-nitrobenzyl)isourea is available from Regis or Fluka. Synthesis is also described above.)

HPLC VARIABLES
Column: 300 × 4 10 μm μPorasil silica
Mobile phase: 2,2,4-Trimethylpentane:methyl acetate 95:5
Flow rate: 3
Injection volume: 3-10
Detector: UV 254; UV 280

CHROMATOGRAM
Retention time: 12.9
Limit of detection: 1 ng

OTHER SUBSTANCES
Non-interfering: acetic acid, formic acid, lactic acid, propionic acid, trichloroacetic acid

KEY WORDS
normal phase; dog

REFERENCE
Ray, A.C.; Post, L.O.; Reagor, J.C. High pressure liquid chromatographic determination of sodium fluoroacetate (compound 1080) in canine gastric content, *J.Assoc.Off.Anal.Chem.*, **1981**, *64*, 19–24.

SULFONATE

2-(2-Naphthoxy)ethyl 2-[1-(4-Benzyl)piperazyl]ethanesulfonate

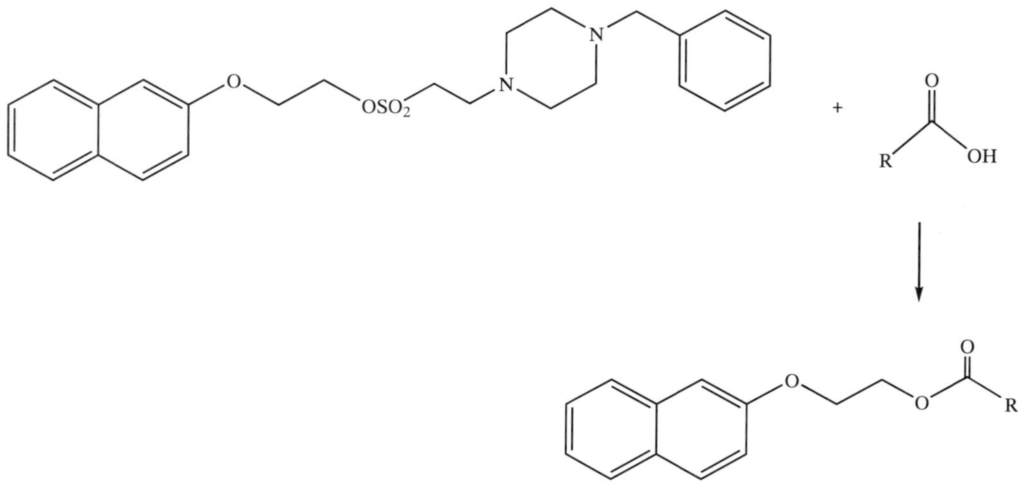

SAMPLE
Matrix: blood
Analyte: caproic acid
Sample preparation: 200 μL Plasma + 50 μL water, mix, add 25 μL 2 M phosphoric acid, add 1 mL dichloromethane, vortex for 2 min, centrifuge at 1800 g for 2 min. Remove a 500 μL aliquot of the dichloromethane layer and add it to 100 μL 9,10-dimethylanthracene solution, evaporate to dryness under a stream of nitrogen, reconstitute with 200 μL toluene, add 300 μL 5 mM reagent in toluene, add 100 μL 100 mM 18-crown-6 in toluene, add 50 mg potassium carbonate, shake at 95° for 1.5 h, cool. Remove a 400 μL aliquot of the reaction mixture and add it to 1 mL 1 M sulfuric acid, vortex for 30 s. Remove a 100 μL aliquot of the toluene layer and evaporate it to dryness under a stream of nitrogen, reconstitute with 100 μL MeCN, inject a 10 μL aliquot. (Synthesis of the reagent, 2-(2-naphthoxy)ethyl 2-[1-(4-benzyl)piperazyl]ethanesulfonate, is as follows. Stir 9.40 g 2-(2-naphthoxy)ethanol (Lancaster Synthesis, Windham NH) in 50 mL dichloromethane at 0°, add 15.78 mL 2-chloro-1-ethanesulfonyl chloride, add 27.72 mL triethylamine, stir at 0° for 2 h, wash with 50 mL 10% sodium carbonate, wash with 50 mL water. Dry the organic layer over 2.5 g anhydrous sodium sulfate, chromatograph the residue with chloroform on about 200 g silica gel 60 in a 270 × 40 column to obtain 2-(2-naphthoxy)ethyl ethanesulfonate as a white powder (mp 82.3-83.2°). Stir 5.56 g 2-(2-naphthoxy)ethyl ethanesulfonate and 4.32 mL 1-benzylpiperazine in 30 mL dichloromethane at 0° for 1 h, evaporate to dryness under reduced pressure, chromatograph with n-hexane:ethyl acetate 75:25 on

about 200 g silica gel in a 270 × 40 column, recrystallize from n-hexane to obtain 2-(2-naphthoxy)ethyl 2-[1-(4-benzyl)piperazyl]ethanesulfonate as plates (mp 74-75°).)

HPLC VARIABLES
Column: 150 × 3.9 4 μm Nova-Pak C18
Mobile phase: MeCN:water 60:40
Flow rate: 1.2
Injection volume: 10
Detector: F ex 305 em 354

CHROMATOGRAM
Retention time: 10
Internal standard: 9,10-dimethylanthracene (14)
Limit of detection: 0.1 pmole

OTHER SUBSTANCES
Simultaneously analyzed: decanoic acid, heptanoic acid, nonanoic acid, octanoic acid, undecanoic acid, valeric acid

KEY WORDS
plasma

REFERENCE
Wu, H.-L.; Shyu, Y.-Y.; Kou, H.-S.; Chen, S.-H.; Wu, S.-M.; Wu, S.-S. Chemically removable derivatization reagent for liquid chromatography: 2-(2-naphthoxy)ethyl 2-[1-(4-benzyl)piperazyl]ethanesulfonate, *J.Chromatogr.A*, **1997**, *769*, 201−207.

TRIAZENE

1-Benzyl-3-p-tolyltriazene

SAMPLE
Matrix: bulk
Analyte: stearic acid
Sample preparation: Dissolve 2 mmoles stearic acid in 10 mL diethyl ether, add 10 mmoles 1-benzyl-3-p-tolyltriazene in 5 mL diethyl ether, stir at 36° for 3 h, cool, wash with two 5 mL portions of 10% HCl, wash with two 5 mL portions of 10% sodium carbonate, dry over anhydrous magnesium sulfate, evaporate to dryness, prepare a solution in mobile phase, inject an aliquot. (1-Benzyl-3-p-tolyltriazene is available from Tokyo Kasei (TCI America, Portland OR). Synthesis is as follows. Stir 50.2 g p-toluidine in an ice/salt bath, add a mixture of 250 g crushed ice and 140 mL concentrated HCl, slowly add a solution of 46.8 g potassium nitrite in 150 mL water over 1-2 h until a positive starch/KI is obtained (stop addition 1-2 min before each test), stir for 1 h, allow to warm to 0°, adjust pH to 6.8-7.2 with cold concentrated sodium carbonate solution to give a

solution of p-toluenediazonium chloride (Org.Syn., Coll.Vol. V, 797). Stir 107 g benzylamine in water in an ice/salt bath and slowly add the diazonium solution, extract with ether, recrystallize 1-benzyl-3-p-tolyltriazene from diethyl ether:n-hexane 50:50 to give yellow crystals (mp 77°) (Berichte 1888, 21, 1016). (Caution! 1-Benzyl-3-p-tolyltriazene explodes when heated to 90-100°!))

HPLC VARIABLES
Column: 1830 mm long Corasil II
Mobile phase: Heptane:chloroform 50:50
Detector: UV 254

OTHER SUBSTANCES
Also analyzed: heptadecanoic acid, palmitic acid

KEY WORDS
normal phase

REFERENCE
Politzer, I.R.; Griffin, G.W.; Dowty, B.J.; Laseter, J.L. Enhancement of ultraviolet detectability of fatty acids for purposes of liquid chromatographic-mass spectrometric analyses, *Anal.Lett.*, **1973**, *6*, 539–546.

TRIFLUOROMETHANESULFONATE

2,3-(Anthracenedicarboximido)ethyl Trifluoromethanesulfonate

SAMPLE
Matrix: solutions
Analyte: fatty acids
Sample preparation: Mix 100 μL of a 0.3 μM solution in MeCN with 50 μL 100 μM tetraethylammonium carbonate in MeCN, add 50 μL freshly-prepared 200 μM 2,3-(anthracenedicarboximido)ethyl trifluoromethanesulfonate in MeCN, vortex for 10 s, let stand at room temperature for 10 min, inject a 1 μL aliquot. (Prepare tetraethylammonium carbonate by adding dry ice to an aqueous solution of tetraethylammonium hydroxide, evaporate to dryness under reduced pressure at 56° over phosphorus pentoxide to give tetraethylammonium carbonate as a white hygroscopic powder (mp 294-288° d)

(Anal. Lett. 1987, 20, 1581). The reagent was 2,3-(anthracenedicarboximido)ethyl trifluoromethanesulfonate, prepared as follows. Add 11.7 g benzoyl chloride dropwise over 30 min to 10 g 1,2,4-trimethylbenzene and 11.7 g aluminum trichloride in 10 mL dichloromethane stirred at 0°, stir at room temperature for 6 h, pour into a mixture of 25 mL concentrated HCl and 50 g ice, remove the organic layer, extract the aqueous layer with dichloromethane. Combine the organic layers and wash them with 5% sodium bicarbonate, dry over anhydrous magnesium sulfate, evaporate to dryness under reduced pressure, distil through a Vigreux column to give 2,4,5-trimethylbenzophenone (bp 130°/0.15 mm Hg). Reflux 12.5 g 2,4,5-trimethylbenzophenone in 75 mL 20% nitric acid for 5 days, cool, decant the aqueous layer, wash the solid with 75 mL water. Dissolve the solid in 125 mL 10% NaOH, reflux this solution while stirring it mechanically, add 35 g potassium permanganate in portions (Caution! Frothing may occur!) over 40 min, reflux for 3 h, allow to cool somewhat, filter. Add the solid that is collected to water, reflux for 6 h, filter while hot. Combine the filtrates and evaporate them to half volume under reduced pressure, cool, acidify slowly with concentrated HCl, filter, dry the solid in air to give benzophenone-2,4,5-tricarboxylic acid as white crystals (mp 281-283°). Stir 2.1 g benzophenone-2,4,5-tricarboxylic acid in 21 g concentrated sulfuric acid at 120° for 3 h, pour onto 30 g of ice, filter, wash the solid with water, dry in air to give anthraquinone-2,3-dicarboxylic acid as a pale yellow solid (mp 342°). Add 1 g anthraquinone-2,3-dicarboxylic acid to 50 mL 20% ammonium hydroxide then add 3.75 g activated zinc dust, reflux, as soon as the blood-red color disappears filter while hot. Add the solid that is collected it to 50 mL 20% ammonium hydroxide, reflux for 2 h, filter while hot. Combine the filtrates and cool them to 0°, acidify to pH 1 with 6 M HCl, let stand at room temperature for 1 day, filter to give 2,3-anthracenedicarboxylic acid as a bright yellow solid (mp 345°) (J. Org. Chem. 1991, 56, 6243). Reflux 2,3-anthracenedicarboxylic acid in acetic anhydride for 2 h to give 2,3-anthracenedicarboxylic anhydride. Vigorously reflux 200 mg 2,3-anthracenedicarboxylic anhydride, 200 mg 2-aminoethanol, 60 mL dry toluene, and 30 mL butanol under a Dean-Stark trap for 1.5 h, evaporate the solvent under reduced pressure until crystallization starts, dissolve these crystals by warming, cool to obtain crystals, recrystallize from toluene/butanol to give N-(hydroxyethyl)-2,3-anthracenedicarboximide as orange crystals (mp 292-294°). Suspend 200 mg N-(hydroxyethyl)-2,3-anthracenedicarboximide in 50 mL dichloromethane and 1 mL pyridine, add this mixture to a solution of 400 mg trifluoromethanesulfonic anhydride in 30 mL dichloromethane at such a rate as to keep the temperature below -5°, stir for 3 h below -5°, add 200 mL cold water. Remove the organic layer and dry it over anhydrous magnesium sulfate, evaporate to dryness under reduced pressure, recrystallize from dichloromethane at -20° to give 2,3-(anthracenedicarboximido)ethyl trifluoromethanesulfonate as pale yellow crystals (mp >300°).)

HPLC VARIABLES
Column: 100 × 4.6 3 μm Develosil ODS-K3
Mobile phase: MeCN:MeOH:water 22.5:67.5:10
Flow rate: 0.8
Injection volume: 1
Detector: F ex 298 em 456

CHROMATOGRAM
Retention time: 4 (caprylic acid), 5.5 (capric acid), 7.5 (lauric acid), 8.5 (myristoleic acid), 10 (cis-5,8,11,14,17-eicosapentaenoic acid), 11 (linolenic acid), 11.5 (myristic acid), 12 (cis-4,7,10,13,16,19-docosahexaenoic acid), 12.5 (palmitoleic acid), 13 (arachidonic acid), 14 (linoleic acid), 16 (cis-8,11,14-eicosatrienoic acid), 17.5 (palmitic acid), 19 (oleic acid), 21 (cis-11,14-eicosadienoic acid), 22 (heptadecanoic acid), 27 (stearic acid), 29 (gondoic acid)
Limit of detection: 1.4-3.8 pmole

REFERENCE
Akasaka, K.; Ohrui, H.; Meguro, H. Determination of carboxylic acids by high-performance liquid chromatography with 2-(2,3-anthracenedicarboximido)ethyl trifluoromethanesulfonate as highly sensitive fluorescent labelling reagent, *Analyst*, **1993**, *118*, 765–768.

SAMPLE
Matrix: tissue
Analyte: shellfish toxins

Sample preparation: Prepare an SPE column containing 100-120 mg LiChrolute Si 60 (Merck), wash with 2 mL dichloromethane. Homogenize 1 g shellfish with 5 mL isopropanol, centrifuge below 5° at 6000 rpm for 5 min, homogenize the precipitate with 5 mL isopropanol, centrifuge. Combine the supernatants and add a 500 μL aliquot to 750 μL n-hexane, 1 mL ethyl acetate, and 1.5 mL 320 mM sodium sulfate solution, vortex for 1 min, centrifuge below 5° at 2000 rpm for 5 min, remove the upper phase, add 500 μL ethyl acetate, extract, centrifuge, combine the two upper phases, dilute to 5 mL with MeOH. Remove a 500 μL aliquot and add it to 75 μL 1.5 mM tetraethylammonium carbonate in MeCN, evaporate under reduced pressure, complete drying under a stream of nitrogen, add 100 μL 2.25 mM reagent in MeCN (freshly prepared), vortex for 30 s, let stand at room temperature for >10 min, evaporate to dryness under reduced pressure, reconstitute with 200 μL dichloromethane, add to the SPE column, wash with 200 μL dichloromethane, wash with 4 mL dichloromethane:acetone 97.5:2.5, elute with 2 mL dichloromethane:acetone:MeOH 95:5:10. Evaporate the eluate to dryness under reduced pressure, reconstitute with 200 μL MeCN, inject a 2 μL aliquot on to column A and elute to waste with mobile phase, after 2.75 min divert the effluent from column A on to column B, after 1.3 min remove column A from the circuit, elute column B with mobile phase, monitor the effluent from column B. (Distil MeCN from phosphorus pentoxide, store over 4 Å molecular sieve. Prepare tetraethylammonium carbonate by adding dry ice to an aqueous solution of tetraethylammonium hydroxide, evaporate to dryness under reduced pressure at 56° over phosphorus pentoxide to give tetraethylammonium carbonate as a white hygroscopic powder (mp 294-288° d) (Anal. Lett. 1987, 20, 1581). The reagent was 2,3-(anthracenedicarboximido)ethyl trifluoromethanesulfonate, prepared as described above.)

HPLC VARIABLES
Column: A 50 × 4.6 5 μm Develosil Ph-5; B 150 × 4.6 5 μm Develosil ODS K-5
Mobile phase: MeOH:water 80:20
Column temperature: 60
Flow rate: 0.8
Injection volume: 2
Detector: F ex 298 em 462

CHROMATOGRAM
Retention time: 16.3 (okadaic acid), 23.3 (dinophysistoxin-1)
Limit of detection: 20 ng/g (okadaic acid), 30 ng/g (dinophysistoxin-1)

KEY WORDS
shellfish; mussels; scallops; SPE; column-switching; heart-cut

REFERENCE
Akasaka, K.; Ohrui, H.; Meguro, H.; Yasumoto, T. Fluorimetric determination of diarrhetic shellfish toxins in scallops and mussels by high-performance liquid chromatography, *J.Chromatogr.A*, **1996**, *729*, 381–386.

4'-Bromophenacyl Trifluoromethanesulfonate

SAMPLE
Matrix: blood, tissue, urine
Analyte: butyrobetaine

Sample preparation: Homogenize liver with 20 volumes MeCN:MeOH 75:25. Mix 100 μL plasma with 1 mL MeCN:MeOH 75:25, mix 25 μL urine with 1 mL MeCN:MeOH 75:25. Centrifuge these mixtures or 500 μL liver homogenate at 13600 g for 2.5 min, evaporate the supernatant to dryness under a stream of air, dissolve the residue in 1 mL water, centrifuge at 13600 g for 2.5 min, add the supernatant to a 1 mL Superclean LC-18 SPE cartridge (Supelco), elute with 2 mL water. Collect all the effluent and evaporate it to dryness under a stream of air, reconstitute with 1 mL MeOH, centrifuge, add the supernatant to a column of 500 mg 230-400 mesh silica gel 60 (Curtin Matheson) in a Pasteur pipette, wash with 2 mL MeOH,wash with 3 mL MeOH:acetic acid 99.5:0.5, elute with 4 mL MeOH:triethylamine:acetic acid 96:2:2. Evaporate the eluate to dryness with a stream of air, reconstitute with 1 mL MeOH, evaporate to dryness under a stream of air, add 50 μL 0.1% N,N-diisopropylethylamine in MeOH, vortex, add 100 μL 45 mM 4'-bromophenacyl trifluoromethanesulfonate in MeCN, vortex for 5 s, let stand for 1 min, add 10 μL 8 mg/mL glycolic acid in MeOH:N,N-diisopropylethylamine 99.9:0.1, inject a 25 μL aliquot. (4'-Bromophenacyl trifluoromethanesulfonate is available from Molecular Probes, Eugene OR or it may be synthesized as described below.)

HPLC VARIABLES
Column: 300 × 3.9 5 μm Resolve-Pak C18 (Waters)
Mobile phase: Gradient. A was MeCN:water 80:20. B was MeCN:water 20:80. C was MeCN:water:triethylamine:phosphoric acid 20:80:0.5:0.4. D was MeCN:water:triethylamine:phosphoric acid 80:20:0.5:0.4. A:B:C:D 100:0:0:0 for 1 min, to 0:100:0:0 (step gradient), maintain at 0:100:0:0 for 4 min, to 0:0:100:0 (step gradient), to 0:0:70:30 over 45 min, to 0:0:0:100(step gradient), maintain at 0:0:0:100 for 10 min, return to initial conditions (step gradient), re-equilibrate for 5 min.
Flow rate: 1
Injection volume: 25
Detector: UV 254

CHROMATOGRAM
Retention time: 33
Limit of detection: 1 pmole

OTHER SUBSTANCES
Extracted: carnitine

KEY WORDS
plasma; liver; rat; human; SPE

REFERENCE
Krahenbuhl, S.; Minkler, P.E.; Hoppel, C.L. Derivatization of isolated endogenous butyrobetaine with 4'-bromophenacyl trifluoromethanesulfonate followed by high-performance liquid chromatography, *J.Chromatogr.*, **1992**, *573*, 3–10.

SAMPLE
Matrix: blood, urine
Analyte: carnitine and acylcarnitines
Sample preparation: Pack a disposable polypropylene chromatography column with 0.5 mL 230-400 mesh Silica gel 60 (Curtin Matheson). 25 μL Urine (or 10 μL urine + 20 μL water or 100 μL plasma) + 25 μL 100 μM IS + 25 μL 1 M KH_2PO_4 + 1 mL MeCN:MeOH 75:25, vortex for 2 s, centrifuge at 13600 g for 5 min, add the supernatant to the silica gel column, wash with 2 mL MeOH, wash with 1 mL 1% acetic acid in MeOH, elute with 4 mL 1% acetic acid in MeOH. Evaporate the eluate to dryness under a stream of nitrogen at 35°, reconstitute the residue in 250 μL MeCN:MeOH 75:25, vortex for 2 s, centrifuge at 13600 g for 5 min. Remove the supernatant and evaporate it to dryness under a stream of nitrogen, reconstitute the residue in 10 μL diisopropylethylamine solution and 20 μL 100 mM 4'-bromophenacyl trifluoromethanesulfonate in MeCN, vortex for 2 s, let stand for 10 min, inject a 6 μL aliquot. (The diisopropylethylamine solution was 25 μL diisopropylethylamine in 10 mL MeOH. 4'-Bromophenacyl trifluoromethanesulfonate is available from Molecular Probes, Eugene OR. Synthesis is as follows. Add 8.8 g p-bromobenzoyl chloride in 40 mL dry ether over 20-30 min to 100 mmoles diazomethane stirred in an ice bath, stir in an ice bath for 8-9 h, let stand at room temperature

for 3 h, evaporate the solvent under reduced pressure, recrystallize 4'-bromo-2-diazoace-
tophenone from ether/hexane (mp 123.5-124° d) (J.Am.Chem.Soc. 1951, 73, 5301). Condense 50 mL anhydrous sulfur dioxide in a flask fitted with a calcium sulfate drying tube,
cool in a dry ice/acetone bath, add 2.25 g 4'-bromo-2-diazoacetophenone, stir for 5 min,
add 900 µL anhydrous trifluoromethanesulfonic acid from a freshly opened bottle in one
portion, stir for 15 min, remove the cooling bath, after 30 min use an ice/water bath to
evaporate the solvent. Dissolve the residue in 100 mL boiling dichloromethane, treat twice
with 5 g portions of decolorizing carbon, filter, evaporate the filtrate, recrystallize the
residue from pentane:dichloromethane 80:20 to give 4'-bromophenacyl trifluoromethanesulfonate as colorless plates (mp 137-8°) (J. Chromatogr. 1984, 299, 365). Preparation
of diazomethane is as follows. Caution! Diazomethane is toxic, explosive, and carcinogenic! A face shield and a safety screen should always be used and the preparation should
only be carried out in a properly functioning chemical fume hood. Only smooth glass
apparatus with rubber stoppers and plastic tubing should be used. Scratched glassware,
ground glass joints, and sharp edges should be avoided (Org. Syn., Coll. Vol. VI; Wiley:
New York, 1988, pp. 432-435). Procedures have been reported for the synthesis of diazomethane from N-methyl-N-nitrosourea (Org. Syn., Coll. Vol. II; Wiley: New York, 1943,
pp. 165-167), N-nitroso-β-methylaminoisobutyl methyl ketone (Org. Syn., Coll. Vol. III;
Wiley: New York, 1955, pp. 244-248), and N,N'-dimethyl-N,N'-dinitrosoterephthalamide
(Org. Syn.,Coll. Vol. V; Wiley: New York, 1973, pp. 351-355). Probably the most convenient
starting material is N-methyl-N-nitroso-p-toluenesulfonamide (Diazald) (Aldrichimica
Acta 1983, 16, 3-10; Org. Syn., Coll. Vol. IV, Wiley: New York, 1963, pp. 250-253). Add
10 mL 95% EtOH to a solution of 5 g KOH in 8 mL water, warm to 65° using a water
bath, add a solution of 5 g N-methyl-N-nitroso-p-toluenesulfonamide in 45 mL ether over
20 min at such a rate as to keep the reaction volume constant. Collect the ether and
diazomethane that distil in an ice-cooled receiving flask under a dry ice/acetone condenser.
When all the N-methyl-N-nitroso-p-toluenesulfonamide has been used up, slowly add 10
mL ether to the reaction flask and continue distillation until the distillate is colorless. A
purpose-built distillation apparatus can be purchased from Aldrich (Aldrichimica Acta
1983, 16, 3-10). Excess quantities of diazomethane can be destroyed by adding acetic acid
until the yellow color of the diazomethane is discharged. The safe disposal of the nitroso
compounds used to generate diazomethane has been discussed (Lunn,G.; Sansone,E.B.
Destruction of Hazardous Chemicals in the Laboratory, Second Edition. Wiley: New York,
1994, pp. 277-289).)

HPLC VARIABLES
Column: 100 × 4.6 3 µm Hypersil (MOS-1) C8
Mobile phase: Gradient. A was MeCN:water 80:20. B was MeCN:water 20:80. C was
  MeCN:water:phosphoric acid:triethylamine 20:80:0.4:0.5. D was MeCN:water:phos-
  phoric acid:triethylamine 90:10:0.2:0.25. A:B:C:D 100:0:0:0 for 0.2 min, 0:100:0:0
  for 0.8 min, then 0:0:100:0 to 0:0:0:100 over 10 min, then 100:0:0:0 for 3 min before
  next run.
Flow rate: 1.75
Injection volume: 6
Detector: UV 260

CHROMATOGRAM
Retention time: 4.25 (carnitine), 4.5 (epsilon-carnitine), 4.8 (betaine, butyrobetaine), 5.1
  (acetylcarnitine), 5.7 (propionylcarnitine), 6.3 (trimethyllysine), 7 (isovalerylcarnitine), 7.8
  (hexanoylcarnitine), 9.2 (octanoylcarnitine)
Internal standard: 4-(N,N-dimethyl-N-ethylammonio)-3-hydroxybutanoate (Prepare by N-
  demethylating carnitine and alkylating the resulting 4-(N,N-dimethylamino)-3-hydroxy-
  butanoic acid with iodoethane. Perform the reaction under nitrogen and protect from
  light. Heat 83 g thiophenol and 20 g NaOH in 100 mL EtOH until they dissolve, add 700
  mL toluene, distil slowly at atmospheric pressure, add 600 mL toluene in 100 mL portions
  to keep the volume in the flask at 500-600 mL, continue distillation for 1 h after the
  distillation head temperature reaches 115°. Sodium thiophenoxide crystallizes out in the
  flask as the distillation progresses (Anal.Chem. 1968, 40, 125). Filter the product under
  nitrogen and wash it with boiling toluene, store in the dark under vacuum. Dissolve 2 g
  l-carnitine chloride in 100 mL DMF with stirring at 80°, cool to room temperature, add

6.6 g sodium thiophenoxide, stir for 20 min, heat at 100° with stirring for 8 h, cool to room temperature, pour into 150 g ice and 4.5 mL concentrated HCl, wash 4 times with 100 mL portions of diethyl ether, evaporate the aqueous phase under reduced pressure at 40°, add more water to the flask until all the DMF is removed. When the volume has been reduced to 5 mL adjust pH to 10 with NaOH, add to a 300 × 20 column of 200-400 mesh Dowex 2X8 (OH⁻) made up in 50 mM NaOH, wash with 200 mL 50 mM NaOH at 1.5 mL/min, wash with water until the pH of the effluent is neutral. Place the resin in a beaker containing ice water, slowly add concentrated HCl with stirring until the supernatant reaches pH 1, pour the resin into the column, wash the column with 200 mL 100 mM HCl. Collect all the effluents and concentrate them almost to dryness under reduced pressure, add to a 500 × 20 column of 200-400 mesh Dowex 50X8 (H⁺ form), elute with 300 mL 500 mM HCl and 300 mL 2.5 M HCl, collect 10 mL fractions, identify fractions containing compound with iodoplatinate spray reagent. Combine fractions containing compound and lyophilize them to give 4-(N,N-dimethylamino)-3-hydroxybutanoic acid as a colorless glass. Dissolve 50 mg 4-(N,N-dimethylamino)-3-hydroxybutanoic acid and 150 mg barium hydroxide octahydrate in 1 mL water, add iodoethane in 5 mL MeOH, stir for 15 h, evaporate to dryness, reconstitute with 2 mL water, add 500 μL 1 M sulfuric acid, centrifuge, wash the solid with 100 mM sulfuric acid. Combine the aqueous layers and adjust the pH to 7 with KOH, add to a 190 × 12 column of 200-400 mesh Dowex 1X8 (OH⁻), elute with water and collect fractions. Adjust the pH of the fractions containing the product to 4 with HCl, evaporate, add the residue to a 190 × 12 column of 200-400 mesh Dowex 50X8 (H⁺) form, elute column with 60 mL 1 M HCl and 200 mL 2 M HCl, collect fractions (J.Labelled Cmpds.Radiopharm. 1982, 9, 535; J.Chromatogr. 1984, 336, 271; Clin.Chim.Acta 1992, 212, 55).) (4.5)

Limit of quantitation: 10 nmole/mL

KEY WORDS
plasma; SPE; plasma details in Anal. Biochem. 1993, 212, 510

REFERENCE
Minkler, P.E.; Hoppel, C.L. Quantification of carnitine and specific acylcarnitines by high-performance liquid chromatography: Application to normal human urine and urine from patients with methylmalonic aciduria, isovaleric acidemia or medium-chain acyl-CoA dehydrogenase deficiency, *J.Chromatogr.*, **1993**, *613*, 203–221.

SAMPLE
Matrix: bulk
Analyte: carboxylic acids
Sample preparation: Neutralize 1-1000 nmoles of the compound with at least 5 equivalents of N,N-diisopropylethylamine in 100 μL MeCN, add 100 μL MeCN containing at least 10 equivalents of 4'-bromophenacyl trifluoromethanesulfonate (freshly prepared), vortex for 10 s, let stand for at least 2 min, add an excess of glycolic acid diisopropylethylammonium salt in MeCN, inject an aliquot. (4'-Bromophenacyl trifluoromethanesulfonate is available from Molecular Probes, Eugene OR or it may be synthesized as described above.)

HPLC VARIABLES
Guard column: 50 × 4 Co:Pell ODS
Column: 100 × 5 10 μm Radial-PAK C18
Mobile phase: MeCN:water 60:40
Flow rate: 1.5
Detector: UV 254

CHROMATOGRAM
Retention time: 3 (propanoic acid), 4 (butanoic acid), 5.5 (pentanoic acid)

REFERENCE
Ingalls, S.T.; Minkler, P.E.; Hoppel, C.L.; Nordlander, J.E. Derivatization of carboxylic acids by reaction with 4'-bromophenacyl trifluoromethanesulfonate prior to determination by high-performance liquid chromatography, *J.Chromatogr.*, **1984**, *299*, 365–376.

SAMPLE
Matrix: solutions

Analyte: betaine
Sample preparation: Mix 600 μL solution with 60 μL IS solution, vortex, add a 550 μL aliquot to a 70 × 5 column of Dowex 1-X8 (OH⁻ form) in a Pasteur pipette, elute with 2 mL water, evaporate the eluate to dryness under a stream of air, reconstitute with 100 μL 2 mM N,N-diisopropylethylamine in MeCN, add 100 μL 5 mM 4'-bromophenacyl trifluoromethanesulfonate in MeCN, vortex for 10 min, add 100 μL 10 mM hydroxyacetic acid N,N-diisopropylethylammonium salt in MeCN, vortex for 2 min, inject a 15 min aliquot. (4'-Bromophenacyl trifluoromethanesulfonate is available from Molecular Probes, Eugene OR or it may be prepared as described above.)

HPLC VARIABLES

Guard column: 50 × 4 Co:Pell ODS
Column: 100 × 8 5 μm Radial-PAK C18
Mobile phase: MeCN:buffer 70:30 (Prepare by dissolving 70 mg sodium dodecyl sulfate, 140 mg $NaH_2PO_4.H_2O$, and 300 μL 3-dimethylamino-1-propanol in 150 mL water, adjusting pH to 6.5 with 85% phosphoric acid, and adding 350 mL MeCN.)
Flow rate: 5
Injection volume: 15
Detector: UV 254

CHROMATOGRAM

Retention time: 7
Internal standard: (4-bromophenyl)carboxymethyl (6-trimethylammonium)hexanoate (Add an aqueous solution of 6-(trimethylammonium)hexanoic acid (Cl⁻ salt, prepared by methylation of 6-aminohexanoic acid) to a column of Dowex 1-X8 (OH⁻ form), elute with 4 column volumes of water. Evaporate the eluate to dryness, dissolve the residue in DMF, add 1.1 equivalents of triethylamine, add 1.1 equivalents of 2,4'-dibromoacetophenone, stir at 40° for 3 h, add ethyl acetate. Collect the precipitate by filtration and recrystallize it from ethanol/acetone to obtain (4-bromophenyl)carboxymethyl (6-trimethylammonium)hexanoate.) (8)
Limit of quantitation: 10 μM

KEY WORDS

SPE

REFERENCE

Minkler, P.E.; Ingalls, S.T.; Kormos, L.S.; Weir, D.E.; Hoppel, C.L. Determination of carnitine, butyrobetaine, and betaine as 4'-bromophenacyl ester derivatives by high-performance liquid chromatography, *J.Chromatogr.*, **1984**, *336*, 271–283.

SAMPLE

Matrix: tissue
Analyte: carnitine and acetylcarnitine
Sample preparation: Pack a disposable polypropylene chromatography column with 0.5 mL 230-400 mesh Silica gel 60 (Curtin Matheson). Homogenize (ground glass homogenizer) 10-20 mg tissue with 1 mL cold water. 50 μL 1 M KOH in MeOH + 25 μL 10% phosphoric acid + 50 μL saturated KH_2PO_4, vortex for 2 s, add 50 μL 25 μM e-carnitine, add 50 μL homogenate, add 1 mL MeCN:MeOH 75:25, vortex for 2 s, centrifuge at 13000 g for 5 min, add the supernatant to the silica gel column, wash with 2 mL MeOH, wash with 1 mL 1% acetic acid in MeOH, elute with 4 mL 1% acetic acid in MeOH. Evaporate the eluate to dryness under a stream of air, reconstitute the residue in 200 μL MeCN:MeOH 75:25, sonicate for 2 min, vortex for 2 s, centrifuge at 13000 g for 5 min. Remove the supernatant and evaporate it to dryness under a stream of air, reconstitute the residue in 10 μL diisopropylethylamine solution and 20 μL 35 mg/mL 4'-bromophenacyl trifluoromethanesulfonate in MeCN, vortex for 2 s, let stand for 10 min, evaporate to dryness under a stream of air, reconstitute with 25 μL MeCN:water 80:10, inject a 2 μL aliquot. (The diisopropylethylamine solution was 10 μL diisopropylethylamine in 10 mL MeOH. 4'-Bromophenacyl trifluoromethanesulfonate is available from Molecular Probes, Eugene OR or it may be synthesized as described above.)

HPLC VARIABLES

Column: 100 × 4.6 3 μm Hypersil (MOS-1) C8

Mobile phase: Gradient. A was MeCN:water 80:20. B was MeCN:water 20:80. C was MeCN:water:phosphoric acid:triethylamine 20:80:0.4:0.5. D was MeCN:water:phosphoric acid:triethylamine 90:10:0.2:0.25. A:B:C:D 100:0:0:0 for 1.2 min, to 0:100:0:0 (step gradient), maintain at 0:100:0:0 for 0.8 min, to 0:0:100:0 (step gradient), to 0:0:65:25 over 5 min, to 0:0:0:100 (step gradient), maintain at 0:0:0:100 for 2 min, re-equilibrate at initial conditions for 2 min before next run.

Flow rate: 1.75

Injection volume: 2

Detector: UV 260

CHROMATOGRAM

Retention time: 5.3 (carnitine), 6.3 (acetylcarnitine)

Internal standard: e-carnitine (4-(N,N-dimethyl-N-ethylammonio)-3-hydroxybutanoate) (Prepare by N-demethylating carnitine and alkylating the resulting 4-(N,N-dimethylamino)-3-hydroxybutanoic acid with iodoethane. Perform the reaction under nitrogen and protect from light. Heat 83 g thiophenol and 20 g NaOH in 100 mL EtOH until they dissolve, add 700 mL toluene, distil slowly at atmospheric pressure, add 600 mL toluene in 100 mL portions to keep the volume in the flask at 500-600 mL, continue distillation for 1 h after the distillation head temperature reaches 115°. Sodium thiophenoxide crystallizes out in the flask as the distillation progresses (Anal. Chem. 1968, 40, 125). Filter the product under nitrogen and wash it with boiling toluene, store in the dark under vacuum. Dissolve 2 g l-carnitine chloride in 100 mL DMF with stirring at 80°, cool to room temperature, add 6.6 g sodium thiophenoxide, stir for 20 min, heat at 100° with stirring for 8 h, cool to room temperature, pour into 150 g ice and 4.5 mL concentrated HCl, wash 4 times with 100 mL portions of diethyl ether, evaporate the aqueous phase under reduced pressure at 40°, add more water to the flask until all the DMF is removed. When the volume has been reduced to 5 mL adjust pH to 10 with NaOH, add to a 300 × 20 column of 200-400 mesh Dowex 2X8 (OH⁻) made up in 50 mM NaOH, wash with 200 mL 50 mM NaOH at 1.5 mL/min, wash with water until the pH of the effluent is neutral. Place the resin in a beaker containing ice water, slowly add concentrated HCl with stirring until the supernatant reaches pH 1, pour the resin into the column, wash the column with 200 mL 100 mM HCl. Collect all the effluents and concentrate them almost to dryness under reduced pressure, add to a 500 × 20 column of 200-400 mesh Dowex 50X8 (H⁺ form), elute with 300 mL 500 mM HCl and 300 mL 2.5 M HCl, collect 10 mL fractions, identify fractions containing compound with iodoplatinate spray reagent. Combine fractions containing compound and lyophilize them to give 4-(N,N-dimethylamino)-3-hydroxybutanoic acid as a colorless glass. Dissolve 50 mg 4-(N,N-dimethylamino)-3-hydroxybutanoic acid and 150 mg barium hydroxide octahydrate in 1 mL water, add iodoethane in 5 mL MeOH, stir for15 h, evaporate to dryness, reconstitute with 2 mL water, add 500 μL 1 M sulfuric acid, centrifuge, wash the solid with 100 mM sulfuric acid. Combine the aqueous layers and adjust the pH to 7 with KOH, add to a 190 × 12 column of 200-400 mesh Dowex 1X8 (OH⁻), elute with water and collect fractions. Adjust the pH of the fractions containing the product to 4 with HCl, evaporate, add the residue to a 190 × 12 column of 200-400 mesh Dowex 50X8 (H⁺) form, elute column with 60 mL 1 M HCl and 200 mL 2 M HCl, collect fractions (J. Labelled Cmpds. Radiopharm. 1982, 9, 535; J. Chromatogr. 1984, 336, 271; Clin. Chim. Acta 1992, 212, 55).) (5.6)

KEY WORDS

heart; muscle; SPE; comparison with other derivatizing reagents (2-(2,3-naphthalimino)ethyl trifluoromethanesulfonate)

REFERENCE

Minkler, P.E.; Brass, E.P.; Hiatt, W.R.; Ingalls, S.T.; Hoppel, C.L. Quantification of carnitine, acetylcarnitine, and total carnitine in tissues by high-performance liquid chromatography: the effect of exercise on carnitine homeostasis in man, *Anal.Biochem.*, **1995**, 231, 315–322.

SAMPLE

Matrix: urine

Analyte: carnitine and butyrobetaine

Sample preparation: 250 μL Urine + 250 μL 200 μM IS + 125 μL 1 M KOH, let stand for 10 min, add to the column, elute with 2.5 mL 500 mM ammonium hydroxide. Evaporate the eluate to dryness, reconstitute the residue in 100 μL 3 mM N,N-diisopropyl-ethylamine in MeCN, vortex for 2 min, add 100 μL 7.5 mM 4'-bromophenacyl trifluoro-methanesulfonate in MeCN, vortex for 2 min, inject a 20 μL aliquot. (The column was 35 × 5 Dowex 50-X8 (NH$_4^+$ form) above 35 × 5 Dowex 1-X8 (OH$^-$ form) in a Pasteur pipette. Convert Dowex 50-X8 (200-400 mesh, H$^+$ form) to the NH$_4^+$ form and 35 × 5 Dowex 1-X8 (200-400 mesh, Cl$^-$ form) to the OH$^-$ form according to instructions from Bio-Rad Labs. In particular conversion of Dowex 1-X8 must continue until tests for Cl$^-$ in the column effluent are negative. 4'-Bromophenacyl trifluoromethanesulfonate is available from Molecular Probes, Eugene OR or it may be synthesized as described above.)

HPLC VARIABLES
Guard column: 50 × 4 Co:Pell ODS
Column: 100 × 5 10 μm Radial-Pak C18
Mobile phase: MeCN:buffer 58:22 (The buffer was 3.56 g sodium dodecyl sulfate, 2.21 g KH$_2$PO$_4$, and 4.75 mL 3-(dimethylamino)-1,2-propanediol in 2.2 L water, adjust pH to 6.5 with concentrated phosphoric acid.)
Flow rate: 3
Injection volume: 20
Detector: UV 254

CHROMATOGRAM
Retention time: 8 (carnitine), 17.5 (butyrobetaine)
Internal standard: 4-(N,N-dimethyl-N-(n-propyl)ammonio)-3-hydroxybutanoate (Prepare by N-demethylating carnitine and alkylating the resulting 4-(N,N-dimethylamino)-3-hydroxybutanoic acid with 1-iodopropane. Perform the reaction under nitrogen and protect from light. Heat 83 g thiophenol and 20 g NaOH in 100 mL EtOH until they dissolve, add 700 mL toluene, distil slowly at atmospheric pressure, add 600 mL toluene in 100 mL portions to keep the volume in the flask at 500-600 mL, continue distillation for 1 h after the distillation head temperature reaches 115°. Sodium thiophenoxide crystallizes out in the flask as the distillation progresses (Anal. Chem. 1968, 40, 125). Filter the product under nitrogen and wash it with boiling toluene, store in the dark under vacuum. Dissolve 2 g l-carnitine chloride in 100 mL DMF with stirring at 80°, cool to room temperature, add 6.6 g sodium thiophenoxide, stir for 20 min, heat at 100° with stirring for 8 h, cool to room temperature, pour into 150 g ice and 4.5 mL concentrated HCl, wash 4 times with 100 mL portions of diethyl ether, evaporate the aqueous phase under reduced pressure at 40°, add more water to the flask until all the DMF is removed. When the volume has been reduced to 5 mL adjust pH to 10 with NaOH, add to a 300 × 20 column of 200-400 mesh Dowex 2X8 (OH$^-$) made up in 50 mM NaOH, wash with 200 mL 50 mM NaOH at 1.5 mL/min, wash with water until the pH of the effluent is neutral. Place the resin in a beaker containing ice water, slowly add concentrated HCl with stirring until the supernatant reaches pH 1, pour the resin into the column, wash the column with 200 mL 100 mM HCl. Collect all the effluents and concentrate them almost to dryness under reduced pressure, add to a 500 × 20 column of 200-400 mesh Dowex 50X8 (H$^+$ form), elute with 300 mL 500 mM HCl and 300 mL 2.5 M HCl, collect 10 mL fractions, identify fractions containing compound with iodoplatinate spray reagent. Combine fractions containing compound and lyophilize them to give 4-(N,N-dimethylamino)-3-hydroxybutanoic acid as a colorless glass. Dissolve 50 mg 4-(N,N-dimethylamino)-3-hydroxybutanoic acid and 150 mg barium hydroxide octahydrate in 1 mL water, add 1-iodopropane in 5 mL MeOH, stir for 15 h, evaporate to dryness, reconstitute with 2 mL water, add 500μL 1 M sulfuric acid, centrifuge, wash the solid with 100 mM sulfuric acid. Combine the aqueous layers and adjust the pH to 7 with KOH, add to a 190 × 12 column of 200-400 mesh Dowex 1X8 (OH$^-$), elute with water and collect fractions. Adjust the pH of the fractions containing the product to 4 with HCl, evaporate, add the residue to a 190 × 12 column of 200-400 mesh Dowex 50X8 (H$^+$) form, elute column with 60 mL 1 M HCl and 200 mL 2 M HCl, collect fractions (J. Labelled Cmpds. Radiopharm. 1982, 9, 535; J. Chromatogr. 1984, 336, 271; Clin. Chim. Acta 1992, 212, 55).) (11.5)

KEY WORDS
SPE

REFERENCE

Minkler, P.E.; Ingalls, S.T.; Hoppel, C.L. Determination of total carnitine in human urine by high-performance liquid chromatography, *J.Chromatogr.*, **1987**, *420*, 385–393.

SAMPLE

Matrix: urine

Analyte: acylcarnitines

Sample preparation: Evaporate 250 μL urine to dryness, reconstitute with 250 μL MeOH, centrifuge at 13600 g for 2.5 min. Add the supernatant to a 500 μL column of Kieselgel 60 in a Pasteur pipette, wash with 2 mL MeOH, elute with 1.25 mL MeOH:water:acetic acid 50:45:5. Collect the last 1 mL of effluent and evaporate it to dryness under a stream of air at room temperature, reconstitute the residue in 250 μL MeOH, centrifuge. Remove the supernatant and evaporate it to dryness under a stream of nitrogen, reconstitute the residue in 250 μL 2 mM N,N-diisopropylethylamine in MeCN, vortex for 2 s, sonicate for 15 min, add 250 μL 5 mM 4'-bromophenacyl trifluoromethanesulfonate in MeCN, vortex for 10 min, add 25 μL glycolic acid solution,vortex for 2 s, inject a 100 μL aliquot. (Prepare glycolic acid solution by dissolving 200 mg glycolic acid and 500 μL N,N-diisopropylethylamine in 25 mL MeCN. 4'-Bromophenacyl trifluoromethanesulfonate is available from Molecular Probes, Eugene OR or it may be synthesized as described above.)

HPLC VARIABLES

Guard column: Guard-PAK Resolve-PAK C18 (Waters)

Column: 100 × 8 5 μm Resolve-Pak C18 (Waters)

Mobile phase: Gradient. A was MeCN:water 70:30. B was MeCN:water:triethylamine:phosphoric acid 10:90:0.5:0.4 containing 7 g KH$_2$PO$_4$. C was THF:water:triethylamine:phosphoric acid 60:40:0.5:0.4. A:B:C 100:0:0 for 0.5 min, to 0:100:0 (step gradient), maintain at 0:100:0 for 4.5 min, to 0:0:100 over 30 min, maintain at 0:0:100 for 25 min, re-equilibrate at initial conditions for 5 min.

Flow rate: 2

Injection volume: 100

Detector: UV 254

CHROMATOGRAM

Retention time: 15 (carnitine), 20 (acetylcarnitine), 23 (propionylcarnitine), 26 (butylcarnitine), 28.5 (valerylcarnitine), 31 (hexanoylcarnitine), 32.5 (heptanoylcarnitine), 34.5 (octanoylcarnitine), 36 (nonanoylcarnitine), 37.5 (decanoylcarnitine), 40 (lauroylcarnitine), 43 (myristoylcarnitine), 48 (palmitoylcarnitine), 53 (stearoylcarnitine)

KEY WORDS

SPE

REFERENCE

Minkler, P.E.; Ingalls, S.T.; Hoppel, C.L. High-performance liquid chromatographic separation of acylcarnitines following derivatization with 4'-bromophenacyl trifluoromethanesulfonate, *Anal.Biochem.*, **1990**, *185*, 29–35.

(S)-(+)-1-Methyl-2-(2,3-naphthalimido)ethyl Trifluoromethanesulfonate

SAMPLE
Matrix: solutions
Analyte: carboxylic acids
Sample preparation: Mix 100 μL of a 1 mM solution of carboxylic acid in MeCN with 100 μL 10 mM 18-crown-6 in MeCN, and about 5 mg anhydrous potassium carbonate, vortex briefly, add 100 μL 10 mM (S)-(+)-1-methyl-2-(2,3-naphthalimido)ethyl trifluoromethanesulfonate in MeCN, vortex for 1 h, dilute 100-fold with MeCN, inject a 5 μL aliquot. (Preparation of (S)-(+)-1-methyl-2-(2,3-naphthalimido)ethyl trifluoromethanesulfonate is as follows. Reflux vigorously 5.2 g 2,3-naphthalenedicarboxylic anhydride (prepare from the diacid by refluxing with acetic anhydride), 2 g (S)-(+)-1-amino-2-propanol, and 200 mL dry toluene under a Dean-Stark trap for 1 h, cool, filter off the solid and wash it 3 times with 50 mL portions of cold water, recrystallize twice from EtOH/water to obtain (S)-(+)-1-(2,3-naphthalimido)-2-propanol as transparent needles (mp 175-176°, $[\alpha]_D^{20} = 52°$ (c = 1, MeOH). Add dropwise a mixture of 1.4 g pyridine and 4.1 g (S)-(+)-1-(2,3-naphthalimido)-2-propanol in 100 mL warm dichloromethane to 5 g trifluoromethanesulfonic anhydride in 100 mL dichloromethane at such a rate as to keep the temperature of the reaction below -5° (about 1 h), stir for 2 h, wash 3 times with cold water, dry over anhydrous magnesium sulfate, evaporate to dryness under reduced pressure, recrystallize the residue twice from dichloromethane/carbon tetrachloride to obtain (S)-(+)-1-methyl-2-(2,3-naphthalimido)ethyl trifluoromethanesulfonate as transparent flakes (mp 139-140° d, $[\alpha]_D^{20} = 20°$ (c = 0.54, dichloromethane) (Caution! Carbon tetrachloride is a carcinogen!).)

HPLC VARIABLES
Column: 150 × 4.6 5 μm ODS-80Tm (Tosoh)
Mobile phase: MeCN:water 50:50
Flow rate: 1
Injection volume: 5
Detector: F ex 259 em 394

CHROMATOGRAM
Retention time: 7 (D-mandelic acid), 10 (L-mandelic acid), 15 (D-α-methoxyphenylacetic acid), 20 (L-α-methoxyphenylacetic acid)

KEY WORDS
chiral

REFERENCE
Yasaka, Y.; Matsumoto, T.; Tanaka, M. S-(+)-1-methyl-2-(2,3-naphthalimido)ethyl trifluoromethanesulfonate as a fluorescence chiral labeling reagent for carboxylic acid enantiomers, *Anal.Sci.*, **1995**, *11*, 295−297.

2-(2,3-Naphthalimino)ethyl Trifluoromethanesulfonate

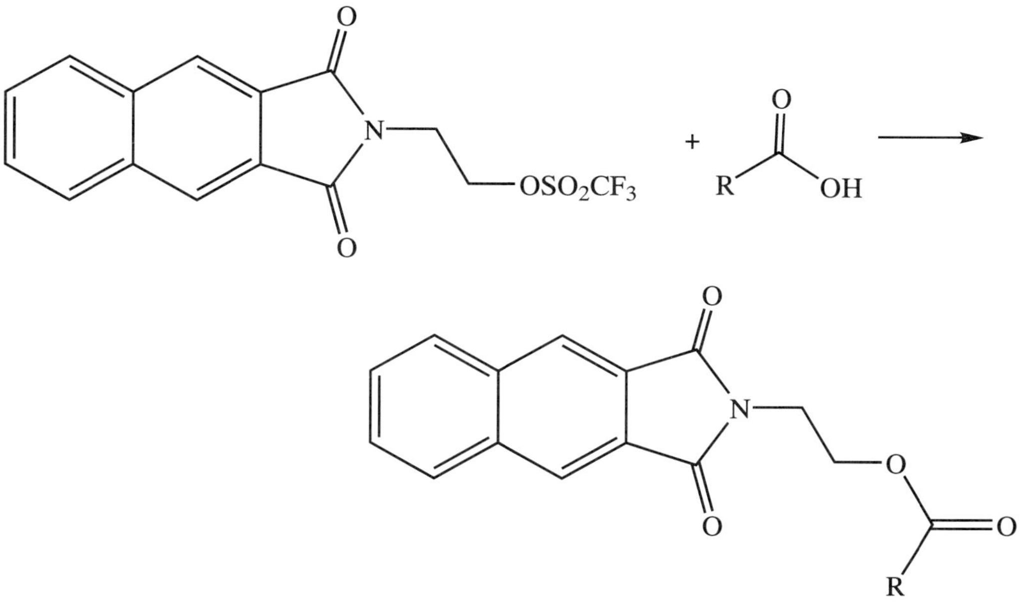

SAMPLE

Matrix: tissue

Analyte: carboxylic acids

Sample preparation: Homogenize mouse cerebrum and margaric acid with 3 mL MeOH, centrifuge at 5000 g for 10 min, remove the supernatant. Homogenize the pellet with MeOH, centrifuge, repeat this process twice more. Combine the supernatants, evaporate to dryness under reduced pressure, reconstitute with 200 μL MeCN, add 40 μL 1 mM 18-crown-6 in MeCN, add 2 mg anhydrous KF, vortex slightly, add 100 μL 1 mM 2-(2,3-naphthalimino)ethyl trifluoromethanesulfonate in MeCN, vortex for 10 min, let stand for 30 s, inject a 10 μL aliquot of the supernatant. (2-(2,3-Naphthalimino)ethyl trifluoromethanesulfonate is available from Molecular Probes, Eugene OR. Synthesis is as follows. 2,3-Naphthalenedicarboxylic anhydride can be prepared from 2,3-naphthalenedicarboxylic acid by heating at 245-250°, by reaction with thionyl chloride, or by heating with acetic anhydride. It is also available from Tokyo Kasei (TCI America, 911 North Harborgate St., Portland, OR 97203; 800-423-8616, 503-283-1681, (fax) 503-283-1987; www.tciamerica.com). Vigorously reflux 9.9 g 2,3-naphthalenedicarboxylic anhydride, 3.1 g ethanolamine, and 300 mL dry toluene for 3 h, cool, filter, wash the solid with three 50 mL portions of cold water, recrystallize from EtOH to give N-(hydroxyethyl)-2,3-naphthalimide as transparent needles (mp 192-195°). Add a suspension of 3.9 g N-(hydroxyethyl)-2,3-naphthalimide and 1.4 g pyridine in 100 mL warm dichloromethane dropwise to a cooled solution of 5 g trifluoromethanesulfonic anhydride in 100 mL dichloromethane at such a rate as to keep the temperature below -5° (about 1 h), stir for 2 h, wash three times with cold water, dry over anhydrous magnesium sulfate, evaporate to dryness under reduced pressure, recrystallize twice from dichloromethane/carbon tetrachloride to give 2-(2,3-naphthalimino)ethyl trifluoromethanesulfonate as transparent flakes (mp 138-140°).)

HPLC VARIABLES

Column: 150 × 4.6 5 μm Chemcosorb 5C8 (Chemco)

Mobile phase: MeOH:water 87:13

Flow rate: 1

Injection volume: 10

Detector: F ex 259 em 394; UV 259

CHROMATOGRAM

Retention time: 8 (docosahexaenoic acid), 9 (arachidonic acid), 12 (palmitic acid), 13 (oleic acid), 20 (stearic acid)

Internal standard: margaric acid (16)

Limit of detection: 4 fmole (F), 100 fmole (UV)

KEY WORDS

mouse; brain

REFERENCE

Yasaka, Y.; Tanaka, M.; Shono, T.; Tetsumi, T.; Katakawa, J. 2-(2,3-Naphthalimino)ethyl trifluoro-methanesulfonate as a highly reactive ultraviolet and fluorescent labelling agent for the liquid chromatographic determination of carboxylic acids, *J.Chromatogr.*, **1990**, *508*, 133–140.

SAMPLE

Matrix: tissue

Analyte: carnitine and acetylcarnitine

Sample preparation: Pack a disposable polypropylene chromatography column with 0.5 mL 230-400 mesh Silica gel 60 (Curtin Matheson). Homogenize (ground glass homogenizer) 10-20 mg tissue with 1 mL cold water. 50 µL 1 M KOH in MeOH + 25 µL 10% phosphoric acid + 50 µL saturated KH_2PO_4, vortex for 2 s, add 50 µL 25 µM e-carnitine, add 50 µL homogenate, add 1 mL MeCN:MeOH 75:25, vortex for 2 s, centrifuge at 13000 g for 5 min, add the supernatant to the silica gel column, wash with 2 mL MeOH, wash with 1 mL 1% acetic acid in MeOH, elute with 4 mL 1% acetic acid in MeOH. Evaporate the eluate to dryness under a stream of air, reconstitute the residue in 200 µL MeCN:MeOH 75:25, sonicate for 2 min, vortex for 2 s, centrifuge at 13000 g for 5 min. Remove the supernatant and evaporate it to dryness under a stream of air, reconstitute the residue in 10 µL diisopropylethylamine solution and 20 µL 13 mg/mL 2-(2,3-naphthalimino)ethyl trifluoromethanesulfonate (Molecular Probes, Eugene OR) in MeCN, vortex for 2 s, let stand for 10 min, evaporate to dryness under a stream of air, reconstitute with 50 µL MeCN:water 50:50, inject a 2 µL aliquot. (The diisopropylethylamine solution was 10 µL diisopropylethylamine in 10 mL MeOH.)

HPLC VARIABLES

Column: 100 × 4.6 3 µm Hypersil (MOS-1) C8

Mobile phase: Gradient. A was MeCN:water 80:20. B was MeCN:water 20:80. C was MeCN:water:phosphoric acid:triethylamine 20:80:0.4:0.5. D was MeCN:water:phosphoric acid:triethylamine 90:10:0.2:0.25. A:B:C:D 100:0:0:0 for 2.2 min, to 0:100:0:0 (step gradient), maintain at 0:100:0:0 for 0.8 min, to 0:0:100:0 (step gradient), to 0:0:80:20 over 12 min, to 0:0:0:100 (step gradient), maintain at 0:0:0:100 for 1 min, re-equilibrate at initial conditions for 2 min before next run.

Flow rate: 1.75

Injection volume: 2

Detector: F ex 259 em 394

CHROMATOGRAM

Retention time: 7.8 (carnitine), 10.3 (acetylcarnitine)

Internal standard: e-carnitine (4-(N,N-dimethyl-N-ethylammonio)-3-hydroxybutanoate) (Prepare by N-demethylating carnitine and alkylating the resulting 4-(N,N-dimethylamino)-3-hydroxybutanoic acid with iodoethane. Perform the reaction under nitrogen and protect from light. Heat 83 g thiophenol and 20 g NaOH in 100 mL EtOH until they dissolve, add 700 mL toluene, distil slowly at atmospheric pressure, add 600 mL toluene in 100 mL portions to keep the volume in the flask at 500-600 mL, continue distillation for 1 h after the distillation head temperature reaches 115°. Sodium thiophenoxide crystallizes out in the flask as the distillation progresses (Anal. Chem. 1968, 40, 125). Filter the product under nitrogen and wash it with boiling toluene, store in the dark under vacuum. Dissolve 2 g l-carnitine chloride in 100 mL DMF with stirring at 80°, cool to room temperature, add 6.6 g sodium thiophenoxide, stir for 20 min, heat at 100° with stirring for 8 h, cool to room temperature, pour into 150 g ice and 4.5 mL concentrated HCl, wash 4 times with 100 mL portions of diethyl ether, evaporate the aqueous phase under reduced pressure at 40°, add more water to the flask until all the DMF is removed.

When the volume has been reduced to 5 mL adjust pH to 10 with NaOH, add to a 300 × 20 column of 200-400 mesh Dowex 2X8 (OH⁻) made up in 50 mM NaOH, wash with 200 mL 50 mM NaOH at 1.5 mL/min, wash with water until the pH of the effluent is neutral. Place the resin in a beaker containing ice water, slowly add concentrated HCl with stirring until the supernatant reaches pH 1, pour the resin into the column, wash the column with 200 mL 100 mM HCl. Collect all the effluents and concentrate them almost to dryness under reduced pressure, add to a 500 × 20 column of 200-400 mesh Dowex 50X8 (H⁺ form), elute with 300 mL 500 mM HCl and 300 mL 2.5 M HCl, collect 10 mL fractions, identify fractions containing compound with iodoplatinate spray reagent. Combine fractions containing compound and lyophilize them to give 4-(N,N-dimethylamino)-3-hydroxybutanoic acid as a colorless glass. Dissolve 50 mg 4-(N,N-dimethylamino)-3-hydroxybutanoic acid and 150 mg barium hydroxide octahydrate in 1 mL water, add iodoethane in 5 mL MeOH, stir for 15 h, evaporate to dryness, reconstitute with 2 mL water, add 500 μL 1 M sulfuric acid, centrifuge, wash the solid with 100 mM sulfuric acid. Combine the aqueous layers and adjust the pH to 7 with KOH, add to a 190 × 12 column of 200-400 mesh Dowex 1X8 (OH⁻), elute with water and collect fractions. Adjust the pH of the fractions containing the product to 4 with HCl, evaporate, add the residue to a 190 × 12 column of 200-400 mesh Dowex 50X8 (H⁺) form, elute column with 60 mL 1 M HCl and 200 mL 2 M HCl, collect fractions (J. Labelled Cmpds. Radiopharm. 1982, 9, 535; J. Chromatogr. 1984, 336, 271; Clin. Chim. Acta 1992, 212, 55).) (8.3)

KEY WORDS
heart; muscle; SPE; comparison with other derivatizing reagents (4'-bromophenacyl trifluoromethanesulfonate)

REFERENCE
Minkler, P.E.; Brass, E.P.; Hiatt, W.R.; Ingalls, S.T.; Hoppel, C.L. Quantification of carnitine, acetylcarnitine, and total carnitine in tissues by high-performance liquid chromatography: the effect of exercise on carnitine homeostasis in man, *Anal.Biochem.*, **1995**, *231*, 315−322.

2-(Phthalimino)ethyl Trifluoromethanesulfonate

SAMPLE
Matrix: tissue
Analyte: fatty acids
Sample preparation: Homogenize mouse brain and margaric acid with 3 mL MeOH, centrifuge at 5000 g for 10 min, resuspend the pellet in MeOH, do the extraction 3 times in total. Combine the supernatants and evaporate them to dryness under reduced pressure, dissolve the residue in 200 μL MeCN, add 100 μL 1 mM 18-crown-6 in MeCN, add 30 mg anhydrous potassium carbonate, vortex slightly, add 100 μL 1 mM 2-(phthalimino)ethyl trifluoromethanesulfonate in MeCN, vortex for 10 min, let stand for 30 s, inject a 10 μL aliquot of the supernatant. (Preparation of 2-(phthalimino)ethyl trifluorometh-

anesulfonate is as follows. Stir 5 g trifluoromethanesulfonic anhydride in 50 mL dichloromethane at -5°, slowly add 3 g N-(2-hydroxyethyl)phthalimide and 1.3 g pyridine in 70 mL dichloromethane, stir at -5° for 2 h, wash with water, dry over anhydrous magnesium sulfate, evaporate to dryness, recrystallize twice from hexane to obtain 2-(phthalimino)ethyl trifluoromethanesulfonate as colorless needles (mp 79°).)

HPLC VARIABLES
Column: 150 × 4.6 5 μm Chemcosorb 5C8 (Chemco, Japan)
Mobile phase: MeCN:water 85:15
Flow rate: 1
Injection volume: 10
Detector: UV 219

CHROMATOGRAM
Retention time: 12 (palmitic acid), 15 (oleic acid), 20 (stearic acid)
Internal standard: margaric acid (15)
Limit of detection: 200 fmole

KEY WORDS
mouse; brain

REFERENCE
Yasaka, Y.; Tanaka, M.; Matsumoto, T.; Katakawa, J.; Tetsumi, T.; Shono, T. 2-(Phthalimino)ethyl trifluoromethanesulfonate as a highly reactive ultraviolet-labeling agent for carboxylic acids in high-performance liquid chromatography, *Anal.Sci.*, **1990**, *6*, 49–52.

CHELATING AGENT

METAL

Copper

SAMPLE
Matrix: blood
Analyte: ethambutol
Sample preparation: 1 mL Plasma + 2 mL 4 M NaOH + 100 μL 1 mg/mL IS + 8 mL chloroform, agitate for 20 min, centrifuge. Remove the organic layer and evaporate it to dryness under a stream of nitrogen, reconstitute the residue in 100 μL MeCN, inject a 50 μL aliquot.

HPLC VARIABLES
Column: 150 × 4.6 5 μm LiChrosorb Si 60
Mobile phase: MeCN:water 50:50 containing 0.04 mM copper sulfate and 2 M ammonia (The mobile phase was saturated with silica using a 250 × 4.6 column packed with 50 μm high-porosity silica (Alltech).)
Flow rate: 1.5
Injection volume: 50
Detector: UV 270

CHROMATOGRAM
Retention time: 5.7
Internal standard: (S,S)-N,N'-bis(hydroxymethyl-1-ethyl)ethylenediamine (7) (Reflux 5 mmoles 1,2-dibromoethane with 50 mmole (S)-2-aminopropan-1-ol at 110° for 25 min, cool to 30°, add 12 mmoles potassium in 5 mL propanol, cool in an ice bath, filter, concentrate the filtrate under reduced pressure, take up the residue in acetone:propanol 1:1, cool in an ice bath, filter, concentrate the filtrate under reduced pressure, add 1 mL EtOH, filter off the product, dry the IS.)
Limit of detection: 150 ng/mL

OTHER SUBSTANCES
Non-interfering: isoniazid, pyrazinamide, rifampin

KEY WORDS
plasma; complexation; copper complexes

REFERENCE
Lacroix, C.; Cerutti, F.; Nouveau, J.; Menager, S.; Lafont, O. Détermination de l'éthambutol plasmatique par chromatographie liquide et détection spectrophotométrique ultraviolette [Determination of ethambutol in plasma by liquid chromatography and ultraviolet spectrophotometric detection], *J.Chromatogr.*, **1987**, *415*, 85–94.

SAMPLE
Matrix: formulations
Analyte: EDTA
Sample preparation: Dilute with water to give an EDTA concentration of 0.01%, mix with an equal volume of 0.02% cupric nitrate, inject a 25 μL aliquot.

HPLC VARIABLES
Guard column: 10 μm Adsorbosphere C18
Column: 150 × 4.5 5 μm Resolve C18 (Waters)
Mobile phase: MeCN:buffer:water 20:20:80 (Buffer was 195 mL water and 6 mL 1 M tetrabutylammonium hydroxide in MeOH adjusted to pH 6.5 ± 0.1 with 1 M phosphoric acid.)

Flow rate: 1.5
Injection volume: 25
Detector: UV 254

CHROMATOGRAM
Retention time: 8.5

OTHER SUBSTANCES
Simultaneously analyzed: sorbic acid
Non-interfering: benzalkonium chloride, hydroxyethyl cellulose, Miranol 2MCA, phenyl-
mercuric nitrate, polyvinyl alcohol, propylene glycol, thimerosal

KEY WORDS
ophthalmic solutions; complexation

REFERENCE
Hall, L.; Takahashi, L. Quantitative determination of disodium edetate in ophthalmic and contact lens
care solutions by reversed-phase high-performance liquid chromatography, *J.Pharm.Sci.*, **1988**, *77*,
247–250.

SAMPLE
Matrix: formulations
Analyte: bisphosphonates
Sample preparation: Dilute injections 100-fold, inject a 20 μL aliquot. Disintegrate a 5
mg tablet in 100 mL water, sonicate for 5 min, centrifuge an aliquot at 3600 g for 4 min,
inject a 20 μL aliquot of the supernatant.

HPLC VARIABLES
Column: 150 × 4.6 10 μm IC-PAK Anion HC (Waters)
Mobile phase: 1.5 mM Nitric acid containing 0.5 mM copper(II) nitrate (Prepare column
by pumping ILC Regenerant A (Waters) and 100 mM nitric acid for 30 min.)
Column temperature: 30
Flow rate: 1
Injection volume: 20
Detector: UV 245

CHROMATOGRAM
Retention time: 1.6 (alendronate), 30.4 (clodronate), 5.8 (etidronate), 1.5 (neridronate), 2
(olpadronate), 2 (pamidronate)

KEY WORDS
complexation; injections; tablets

REFERENCE
Sparidans, R.W.; Den Hartigh, J.; Vermeij, P. High-performance ion-exchange chromatography with in-
line complexation of bisphosphonates and their quality control in pharmaceutical preparations,
J.Pharm.Biomed.Anal., **1995**, *13*, 1545–1550.

SAMPLE
Matrix: solutions
Analyte: ethambutol
Sample preparation: Mix 1 mL of a 16-320 μg/mL solution in pH 4.4 KH_2PO_4 buffer with
100 μL 1.32 mg/mL L-glycine in KH_2PO_4, inject a 20 μL aliquot.

HPLC VARIABLES
Column: 250 × 4.6 5 μm Chiral-Si100 D-ValCu (Serva)
Mobile phase: 50 mM pH 4.4 KH_2PO_4 containing 1 mM copper(II) sulfate
Column temperature: 25
Flow rate: 0.3
Injection volume: 20
Detector: UV 260

CHROMATOGRAM
Retention time: 12.7 (+), 14.2 (meso)
Internal standard: L-glycine

Limit of detection: 1 µg/mL

OTHER SUBSTANCES
Simultaneously analyzed: impurities, 2-amino-1-butanol

KEY WORDS
complexation; chiral; D-enantiomer is not separated from L-enantiomer but they are separated from meso-form

REFERENCE
Ferioli, V.; Gamberini, G.; Rustichelli, C.; Vezzalini, F. Direct determination of non-UV-absorbing compounds by high-performance liquid chromatography, *Farmaco*, **1994**, *49*, 411–413.

Iron

SAMPLE
Matrix: blood, tissue
Analyte: deferoxamine
Sample preparation: Condition a Sep-Pak Plus SPE cartridge with 3 mL MeOH:acetic acid 80:20, 3 mL MeOH, and 9 mL water. Homogenize (motorized Potter-Elvehjem) rat lung with 10 mL buffer. 500 µL Plasma or lung homogenate + 10 µL 250 mM $FeCl_3$ solution, store at -85° for 2 days then in liquid nitrogen until analysis, thaw, add 25 µL 60 µg/mL IS, add 500 µL chloroform, centrifuge at 2000 g for 5 min, add a 400 µL aliquot of the upper aqueous layer to the SPE cartridge, wash with 5 mL water, wash with 500 µL MeOH, elute with 1.5 mL MeOH:acetic acid 80:20. Evaporate the eluate to dryness under reduced pressure, reconstitute with 200 µL 50 mM pH 8.0 NaH_2PO_4, sonicate, centrifuge at 2000 g for 5 min, inject a 100 µL aliquot. (Buffer was 25.6 mM Na_2HPO_4 containing 3.72 mM NaH_2PO_4, 0.74 mM KH_2PO_4, 31 mM NaCl, 1.34 mM KCl, 0.025 mM $CaCl_2$, 0.025 mM $MgCl_2$, and 50 mM dithiothreitol, pH 7.4.)

HPLC VARIABLES
Column: 150 × 3.9 4 µm Novapak C8
Mobile phase: Gradient. MeOH:water:50 mM pH 8.0 NaH_2PO_4 from 0:0:100 to 25:25:50 over 18 min, to 40:60:0 over 4 min, return to initial conditions over 3 min.
Flow rate: 1.2 for 18 min then 1
Injection volume: 100
Detector: UV 427.5

CHROMATOGRAM
Retention time: 12 (as Fe complex, feroxamine)
Internal standard: adrenochrome semicarbazone sulfonate (14)
Limit of quantitation: 330 ng/mL

KEY WORDS
rat; plasma; lung; SPE; complexation

REFERENCE
Merali, S.; Chin, K.; Del Angel, L.; Grady, R.W.; Armstrong, M.; Clarkson, A.B. Jr. Clinically achievable plasma deferoxamine concentrations are therapeutic in a rat model of pneumocystis carinii pneumonia, *Antimicrob.Agents Chemother.*, **1995**, *39*, 2023–2026.

SAMPLE
Matrix: blood, urine
Analyte: deferoxamine
Sample preparation: Plasma. Condition a Sep-Pak C18 SPE cartridge with 5 mL MeOH:acetic acid 80:20, 5 mL MeOH, and 10 mL water. 600 µL Plasma +12 µL 50 mM ferric chloride in water + 600 µL chloroform, vortex for 1 min, centrifuge at 8000 g for 10 min. Remove 500 µL of the upper aqueous layer and add it to the SPE cartridge, wash with three 5 mL portions of water, wash with 400 µL MeOH, elute with 1.5 mL MeOH:acetic acid 80:20. Evaporate the eluate under vacuum, reconstitute in 100 µL mobile phase:

acetic acid 80:20, vortex for 30 s, sonicate for 30 s, centrifuge at 8000 g for 2 min, inject an aliquot of the supernatant. Alternatively, 3 mL Plasma + 50 μL 50 mM ferric chloride, saturate with NaCl, vortex for 30 s, add 600 μL benzyl alcohol, vortex twice for 30 s, centrifuge at 2000 g for 15 min, break up layers with glass rod, centrifuge at 2000 g for 15 min, inject an aliquot of the benzyl alcohol layer. Urine. 5 mL Urine + 100 μL 50 mM ferric chloride, saturate with NaCl, extract with 1 mL benzyl alcohol, inject an aliquot of the benzyl alcohol layer.

HPLC VARIABLES

Column: 150 × 3.9 5 μm Resolve spherical porous silica (Waters)
Mobile phase: MeCN:MeOH:n-butanol:water:5 M sodium perchlorate:perchloric acid 500:300:100:100:1.5:0.05
Injection volume: 20
Detector: UV 435

CHROMATOGRAM

Retention time: 8.70 mL (elution volume; as Fe complex)

OTHER SUBSTANCES

Extracted: metabolites

KEY WORDS

plasma; Use only PTFE and titanium tubing and fittings.; pharmacokinetics; SPE; human; pig; complexation

REFERENCE

Kruck, T.P.A.; Teichert-Kuliszewska, K.; Fisher, E.; Kalow, W.; McLachlan, D.R. High-performance liquid chromatographic analysis of desferrioxamine. Pharmacokinetic and metabolic studies, *J.Chromatogr.*, **1988**, *433*, 207–216.

SAMPLE

Matrix: cell suspensions
Analyte: EDTA
Sample preparation: Centrifuge cell suspension at 14000 rpm for 2 min, dilute if necessary, inject a 30 μL aliquot.

HPLC VARIABLES

Column: 150 × 4.6 LC18 (Supelco)
Mobile phase: MeCN:buffer 0.5:99.5 (Buffer was 1.5 mL/L acetic acid containing 1 g/L ammonium acetate, 0.5 g/L 1,2-diaminopropane-N,N,N',N'-tetraacetic acid, 0.5 mL/L ammonium hydroxide, and 0.1 mL/L triethylamine.) (After 2.5 min wash column with MeCN:water 60:40 containing 0.5 g/L 1,2-diaminopropane-N,N,N',N'-tetraacetic acid, and 0.1 mL/L triethylamine for 5.5 min, re-equilibrate for 4 min.)
Flow rate: 1
Injection volume: 30
Detector: UV 360 (as ferric complex)

KEY WORDS

EDTA measured as ferric complex; complexation

REFERENCE

Lauff, J.J.; Steele, D.B.; Coogan, L.A.; Breitfeller, J.M. Degradation of the ferric chelate of EDTA by a pure culture of an *Agrobacterium sp.*, *Appl.Environ.Microbiol.*, **1990**, *56*, 3346–3353.

SAMPLE

Matrix: formulations
Analyte: EDTA
Sample preparation: Dilute ophthalmic cleanser 1:25 with 100 μM ferric chloride hexahydrate in water, inject a 20 μL aliquot.

HPLC VARIABLES

Column: 150 × 4.6 3 μm Adsorbosphere HS C18
Mobile phase: 50 mM Tetrabutylammonium hydrogen sulfate in water
Flow rate: 1
Injection volume: 20
Detector: UV 254

CHROMATOGRAM
Retention time: 5.2
Limit of detection: 200 ng/mL

OTHER SUBSTANCES
Non-interfering: degradation products

KEY WORDS
complexation; ophthalmic cleanser; stability-indicating

REFERENCE
Tran, G.; Chen, C.; Miller, R.B. HPLC method for the determination of EDTA in an ophthalmic cleanser, *J.Liq.Chromatogr.Rel.Technol.*, **1996**, *19*, 1499–1508.

SAMPLE
Matrix: groundwater
Analyte: EDTA
Sample preparation: Filter (0.45 μm), add 5 mL filtrate to 50 μL reagent, if necessary adjust pH to 3-4, inject a 50 μL aliquot. (Reagent was 4 g ferric chloride and 30 mL glacial acetic acid in 100 mL water.)

HPLC VARIABLES
Guard column: 50 × 4.6 301 SB (Vydac)
Column: 250 × 4.6 Partisil 10 SAX
Mobile phase: 30 g/L NaCl containing 3 mL/L glacial acetic acid, pH 3.1 or 20 g/L sodium nitrate containing 3 mL/L glacial acetic acid
Flow rate: 1.5
Injection volume: 50
Detector: UV 258

CHROMATOGRAM
Retention time: 3
Limit of detection: 200 ng/mL

OTHER SUBSTANCES
Interfering: cobalt
Non-interfering: cadmium, calcium, chromium, copper, lead, magnesium, zinc

KEY WORDS
protect from light; complexation

REFERENCE
Harmsen, J.; Van Den Toorn, A. Determination of EDTA in water by high-performance liquid chromatography, *J.Chromatogr.*, **1982**, *249*, 379–384.

SAMPLE
Matrix: solutions
Analyte: chelating agents
Sample preparation: Dissolve chelating agent in NaOH solution, add a 5% excess of ferric nitrate, adjust pH to 7, let stand overnight, filter (paper), dilute the filtrate with water, filter (0.22 μm), inject a 20 μL aliquot.

HPLC VARIABLES
Column: 150 × 4.6 5 μm LiChrospher RP-18
Mobile phase: MeCN:buffer 30:70 (Prepare mobile phase by adding 20 mL 1.5 M tetrabutylammonium hydroxide solution to 650 mL water, adjust pH to 6.0 with 6 M HCl, add 300 mL MeCN, make up to 1 L with water.)
Flow rate: 1.5
Injection volume: 20
Detector: UV 280

CHROMATOGRAM
Retention time: 1.92 (EDTA), 2.00 (trans-1,2-cyclohexanediaminetetraacetic acid (CDTA) (UV 225)), 3.10 (diethylenetriaminepentaacetic acid (DTPA)), 4.93 (ethylenediaminedi(o-hydroxyphenylacetic) acid (EDDHA) isomer), 6.53 (ethylenediaminedi(o-hydroxyphenyl-acetic) acid (EDDHA) isomer), 7.75 (ethylenediaminedi(o-hydroxy-p-methylphenylacetic) acid (EDDHMA) isomer), 11.78 (ethylenediaminedi(o-hydroxy-p-methylphenylacetic) acid (EDDHMA) isomer), 8.28 (N,N'-bis(2-hydroxybenzyl)ethylenediamine-N,N'-diacetic acid (HBED) isomer), 10.83 (N,N'-bis(2-hydroxybenzyl)ethylenediamine-N,N'-diacetic acid (HBED) isomer), 5.47 (N,N'-bis(2-hydroxybenzyl)ethylenediamine-N,N'-dipropionic acid (HBEP) isomer), 12.26 (N,N'-bis(2-hydroxybenzyl)ethylenediamine-N,N'-dipropionic acid (HBEP) isomer)

KEY WORDS
complexation

REFERENCE
Lucena, J.J.; Barak, P.; Hernández-Apaolaza, L. Isocratic ion-pair high-performance liquid chromatographic method for the determination of various iron(III) chelates, *J.Chromatogr.A*, **1996**, *727*, 253–264.

SAMPLE
Matrix: water
Analyte: EDTA
Sample preparation: Filter (0.2 μm cellulose nitrate) river water, pass the filtrate through a 10 × 5 column filled with SPE 7090 sulfonic acid material (Baker), collect 2-6 mL, evaporate to dryness in an oven at 90°, add 1 mL buffer, add 20 μL iron solution, heat at 90° for 3 h, cool, add 40 μL 50 mM tetrabutylammonium bromide in buffer, inject a 200 μL aliquot. (Buffer was 15 mM formic acid containing 5 mM sodium formate and 1 mM tetrabutylammonium bromide. The iron solution was 1 mM ferric nitrate in 10 mM nitric acid.)

HPLC VARIABLES
Guard column: 4 × 4 Lichrocart (Merck)
Column: 250 × 4 Lichrocart RP-18 (Merck)
Mobile phase: MeCN:buffer 8:92 (Buffer was 15 mM formic acid containing 5 mM sodium formate, pH 3.3.)
Flow rate: 1
Injection volume: 200
Detector: UV 258

CHROMATOGRAM
Retention time: 7.6
Limit of detection: 3 nM

KEY WORDS
protect from light; river water; complexation

REFERENCE
Nowack, B.; Kari, F.G.; Hilger, S.U.; Sigg, L. Determination of dissolved and adsorbed EDTA species in water and sediments by HPLC, *Anal.Chem.*, **1996**, *68*, 561–566.

Lutetium

SAMPLE
Matrix: solution
Analyte: chelating ligands
Sample preparation: Mix a 10 μM solution of the ligand with a 100 μM solution of lutetium chloride, pass through a 70 × 20 column of 70-130 μm AG 50W-X8 cation-exchange resin (sodium form, Bio-Rad), inject an aliquot.

HPLC VARIABLES
Column: 150 × 4.6 5 μm Prodigy 5 ODS-2 (Phenomenex)
Mobile phase: Gradient. A was 1 mM potassium sulfate containing 3 mM tetrapropylammonium bromide. B was MeCN:3 mM potassium sulfate 4:96. A:B 100:0 for 4 min, to 0:100 over 3 min.
Flow rate: 1
Injection volume: 50
Detector: F ex 360 em 500 following post-column reaction. The column effluent mixed with the reagent pumped at 0.3 mL/min and the mixture flowed through a 5.1 m × 0.5 mm ID knitted PTFE coil. The effluent from the coil mixed with 1 M NaOH pumped at 0.3 mL/min and the mixture flowed to the detector. (Reagent was 1 mM trans-1,2-diaminocyclohexane-N,N,N',N'-tetraacetic acid (CDTA) in 1 mM 8-hydroxyquinoline-5-sulfonic acid, adjusted to pH 2.8 with acetic acid.)

CHROMATOGRAM
Retention time: 2.5 (nitrilotriacetic acid (NTA)), 7 (ethylene glycol-bis(β-aminoethylether) N,N,N',N'-tetraacetic acid (EGTA)), 7.5 (EDTA), 14 (diethylenetriaminepentaacetic acid (DTPA)), 17.5 (trans-1,2-diaminocyclohexane-N,N,N',N'-tetraacetic acid (CDTA))
Limit of detection: 25-100 nM

KEY WORDS
complexation; post-column reaction; use a metal-free system

REFERENCE
Ye, L.; Lucy, C.A. Ion chromatographic determination of chelating ligands based on the postcolumn formation of ternary fluorescent complexes, *J.Chromatogr.A*, **1996**, *739*, 307–315.

SAMPLE
Matrix: solutions
Analyte: aminopolycarboxylates
Sample preparation: Mix the test solution with an aqueous solution containing an equimolar amount of lutetium trichloride, inject an aliquot.

CAPILLARY ELECTROPHORESIS
Capillary: 47 cm × 74 μm fused-silica (40 cm to detector) (Polymicro Technologies)
Capillary preparation: Before each run rinse capillary with water for 1 min and with running buffer for 2 min. After each run rinse with 100 mM KOH at high pressure for 2 min. (Rinse new capillaries with 100 mM KOH at high pressure (20 psi) for 10 min, with water at high pressure for 5 min, and with running buffer for 10 min.)
Running buffer: 10 mM pH 11 phosphate buffer containing 1 mM 8-hydroxyquinoline-5-sulfonic acid
Capillary temperature: 50
Voltage/Current: 15 kV
Injection: Pressure injection at 0.5 psi for 3 s
Detector: F ex 325 (He-Cd laser) em 520 ± 20 (filter)
Model: Beckman P/ACE 2100
Migration time: 3.5 (N-(hydroxyethyl)ethylenediaminetriacetic acid (HEDTA)), 4 (trans-1,2-diaminocyclohexane-N,N,N',N'-tetraacetic acid (CDTA)), 4.5 (EDTA)
Limit of detection: 7 ng/mL

KEY WORDS
complexation

REFERENCE
Ye, L.; Wong, J.E.; Lucy, C.A. Determination of aminopolycarboxylate ligands using 8-hydroxyquinoline-5-sulfonic acid-based ternary complexes in capillary zone electrophoresis with laser-induced fluorescence, *Anal.Chem.*, **1997**, *69*, 1837–1843.

Nickel

SAMPLE
Matrix: water
Analyte: EDTA
Sample preparation: Condition a SPEC strong anion-exchange SPE extraction disc (Ansys, Irvine CA) with 200 μL MeOH and 200 μL water adjusted to pH 3 with formic acid. If necessary adjust pH of water to 7-9 with ammonium hydroxide. 5 mL Water + 100 μL 10 mM nickel nitrate, vortex for 1 min, adjust pH to 3 ± 0.2 with about 12 μL 9% formic acid, vortex for 1 min, push through the extraction disc at about 1 mL/min, wash with 1 mL water adjusted to pH 3 with formic acid, wash with 2 mL water, wash with 1 mL MeOH, elute with 300 μL MeOH:water 5:95 containing 50 mM trifluoroacetic acid and 1 mM bromothymol blue. Evaporate the eluate to dryness under a stream of nitrogen at 80°, reconstitute the residue in 100 μL 0.1% ammonium hydroxide, evaporate to dryness, reconstitute with 30 μL water, inject an aliquot. (The bromothymol blue displaces the Ni-EDTA complex from the extraction disc.)

CAPILLARY ELECTROPHORESIS
Capillary: 60 cm × 50 μm CElect-Amine coated capillary (Supelco)
Running buffer: 30 mM Ammonium formate adjusted to pH 3 with formic acid
Voltage/Current: -30 kV/about 32 μA
Injection: Pressure injection of water at 50 mbar for 2 s then electrokinetic injection of the sample at -20 kV for 30 s followed by ramping to -30 kV plus 50 mbar inlet pressure over 1 min.
Detector: MS, PE Sciex API 300 triple quadrupole, makeup liquid MeOH:water 95:5 containing 10 mM ammonium formate at 10 μL/min, electrospray ionization -4 kV, collision gas nitrogen, selective reaction monitoring m/z 347 then m/z 329
Model: Hewlett-Packard 3D HPCE
Migration time: 5
Internal standard: [$^{13}C_4$]EDTA
Limit of quantitation: 300 pg/mL
Limit of detection: 150 pg/mL

KEY WORDS
SPE, coated capillary; complexation

REFERENCE
Sheppard, R.L.; Henion, J. Determination of ethylenediaminetetraacetic acid as the nickel chelate in environmental water by solid-phase extraction and capillary electrophoresis/tandem mass spectrometry, *Electrophoresis*, **1997**, *18*, 287–291.

CHLORAMINE

SULFINIC ACID

Dansylsulfinic Acid

SAMPLE
Matrix: water
Analyte: chloramines
Sample preparation: Dissolve 200 mg sodium bicarbonate in 10 mL sample solution, add 10 mL MeCN (freshly-distilled from anhydrous copper(II) sulfate), add 2 drops 10 M NaOH, add 1 mL reagent solution, mix thoroughly, stopper, store overnight in the dark, inject an aliquot. Alternatively, concentrate on a rotary evaporator to remove MeCN, adjust pH with concentrated phosphoric acid to 3.0, add to an equilibrated Sep-Pak SPE cartridge, rinse the flask with 4 mL water, add the rinse to the SPE cartridge, elute with 2 mL MeCN, inject an aliquot of the eluate. (Prepare the reagent solution by dissolving 1.05 g sodium bicarbonate in 10-15 mL chlorine-demand-free water, add 63.3 mg dansylsulfinic acid, make up to 25 mL. Prepare dansylsulfinic acid as follows. Add 5 g dansyl chloride to a solution of 10.7 g sodium sulfite in 50 mL water vigorously stirred at 70°, stir at 70-80° for 5 h, cool, acidify to pH 4 with sulfuric acid, filter, dissolve the solid in cold dilute aqueous NaOH, filter, precipitate from water or water/EtOH to give dansylsulfinic acid as a white solid (mp 242-244° d (darkens on heating above 150°)). Prepare chlorine-demand-free water by chlorinating HPLC water to 2 mg/L chlorine, stoppering, storing overnight, boiling for 2 h at a low boil, and irradiating for 8 h with a UV light (Ultra-Violet Products Model 11SC-1 pen lamp).)

HPLC VARIABLES
Column: 300 × 3.9 μBondapak C18
Mobile phase: Gradient. MeCN:1% acetic acid 18:82 for 5 min, to 54:46 over 45 min, maintain at 54:46 for 10 min
Detector: F ex 250 em 418 (high bandpass filter)

CHROMATOGRAM
Retention time: 11 (chloramine), 21 (N-chloromethylamine), 34 (N-chlorodimethylamine)

OTHER SUBSTANCES
Non-interfering: α-chloroacetone, chloroform, dibromoacetonitrile, dichloroacetonitrile, hypochlorite, methyl iodide

KEY WORDS
SPE

REFERENCE
Scully, F.E. Jr.; Yang, J.P.; Mazina, K.; Daniel, F.B. Derivatization of organic and inorganic N-chloramines for high-performance liquid chromatographic analysis of chlorinated water, *Environ.Sci.Technol.*, **1984**, *18*, 787–792.

SAMPLE
Matrix: water
Analyte: chloramines
Sample preparation: Dissolve 400 mg sodium bicarbonate in 20 mL sample solution, add 20 mL MeCN (freshly-distilled from anhydrous copper(II) sulfate), add 1 drop 5 M NaOH, add 600 µL reagent solution, mix thoroughly, stopper, store overnight in the dark, inject an aliquot. (Prepare the reagent solution by dissolving 1.05 g sodium bicarbonate in 10-15 mL chlorine-demand-free water, add 63.3 mg dansylsulfinic acid, make up to 25 mL. Prepare dansylsulfinic acid as follows. Add 5 g dansyl chloride to a solution of 10.7 g sodium sulfite in 50 mL water vigorously stirred at 70°, stir at 70-80° for 5 h, cool, acidify to pH 4 with sulfuric acid, filter, dissolve the solid in cold dilute aqueous NaOH, filter, precipitate from water or water/EtOH to give dansylsulfinic acid as a white solid (mp 242-244° d (darkens on heating above 150°)). Prepare chlorine-demand-free water by chlorinating HPLC water to 2 mg/L chlorine, stoppering, storing overnight, boiling for 2 h at a low boil, and irradiating for 8 h with a UV light (Ultra-Violet Products Model 11SC-1 pen lamp).)

HPLC VARIABLES
Column: 250 × 4.6 Zorbax ODS C18
Mobile phase: Gradient. A was 25 mM sodium acetate adjusted to pH 4.5 with acetic acid. B was MeCN. A:B from 90:10 to 0:100 over 108 min.
Flow rate: 1
Detector: F ex 342 em 510

CHROMATOGRAM
Internal standard: dansylnorvaline

KEY WORDS
for chloramine and N-chloroglycine

REFERENCE
Jersey, J.A.; Choshen, E.; Jensen, J.N.; Johnson, J.D.; Scully, F.E. Jr. N-chloramine derivatization mechanism with dansylsulfinic acid; yields and routes of reaction, *Environ.Sci.Technol.*, **1990**, *24*, 1536–1541.

CYANATE

AMINOBENZOIC ACID

2-Aminobenzoic Acid

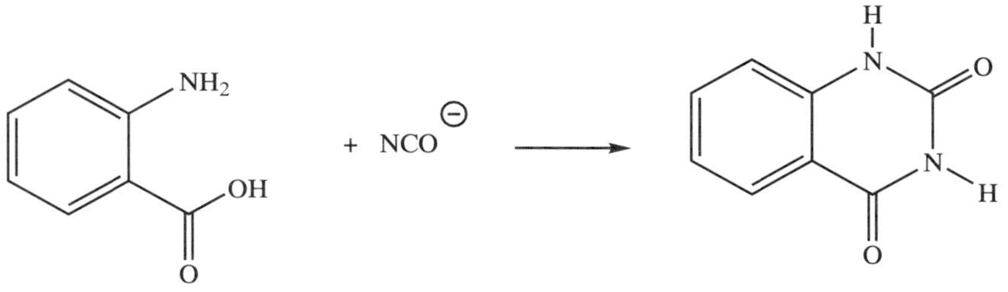

SAMPLE
Matrix: blood
Analyte: cyanate
Sample preparation: Condition a 6 mL 1 g Mega Bond Elut PRS strong cation-exchange cartridge with 5 mL MeOH, 5 mL water, 25 mL 10 mM KOH in 1 M KCl, 10 mL water, 10 mL 1 M HCl, 40 mL water, and 5 mL MeOH then suck air through the cartridge until it is dry. 1 mL Plasma + 1 mL 200 mM pH 7.4 sodium phosphate buffer, adjust pH to 4.7 with 60 μL 4 M acetic acid, add 2.1 mL reagent, mix, heat at 40° for 10 min, cool in ice. Remove a 3 mL aliquot and add it to 3 mL 6 M sulfuric acid, boil in a water bath for 75 s, cool in ice, centrifuge at 2000 g, filter (Minisart) the supernatant. Extract a 4 mL aliquot of the filtrate three times with 3 mL portions of ethyl acetate for 5 min each time. Combine the organic layers and evaporate them to dryness under a stream of nitrogen at 40°, reconstitute the residue in 2 mL water, add 60 μL 5 M HCl, add to the cation-exchange SPE cartridge, elute with two 2 mL portions of water. Add the combined eluate to 500 μL 100 mM pH 7.4 sodium phosphate buffer, adjust the pH to 6-8 with a few drops 3 M NaOH, add to a Sep-Pak C18 SPE cartridge, wash with 3 mL water, elute with 3 mL MeOH. Evaporate the eluate to dryness under a stream of nitrogen at 60°, reconstitute the residue in 500 μL mobile phase, inject an aliquot. (Reagent was 40 mM 2-aminobenzoic acid adjusted to pH 4.7 with sodium acetate. Urea interferes slightly; each 1 mmole urea produces a peak equivalent to 7.5 nmoles cyanate.)

HPLC VARIABLES
Guard column: 10 × 3.2 5 μm Kromsail C-18 (Hicrome)
Column: 250 × 4.6 5 μm Kromasil KR 100-5C-18 (Hicrome)
Mobile phase: MeOH:water 1:2
Flow rate: 0.5
Injection volume: 50
Detector: F ex 313 em 353

CHROMATOGRAM
Retention time: 14
Limit of detection: 8 nM

KEY WORDS
plasma; SPE

REFERENCE
Lundquist, P.; Backman-Gullers, B.; Kågedal, B.; Nilsson, L.; Rosling, H. Fluorometric determination of cyanate in plasma by conversion to 2,4(1*H*,3*H*)-quinazolinedione and separation by high-performance liquid chromatography, *Anal.Biochem.*, **1993**, *211*, 23–27.

CYANIDE

ACYL HALIDE

Benzoyl Chloride/Quinoline

SAMPLE
Matrix: solutions
Analyte: cyanide
Sample preparation: Add 3 mL solution to a tube, add 5 mL 1 mM quinoline in MeCN:
pH 7 phosphate buffer 50:50 and 5 mL 1 mM benzoyl chloride in MeCN:pH 7 phosphate
buffer 50:50 dropwise with continuous stirring and shaking over 15 min, inject a 35 μL
aliquot.

HPLC VARIABLES
Column: 150 × 4.6 5 μm C18
Mobile phase: MeCN:pH 7 phosphate buffer 50:50
Flow rate: 0.75
Injection volume: 35
Detector: UV 225

CHROMATOGRAM
Retention time: 15
Limit of detection: 1 nM

REFERENCE
Madungwe, L.; Zaranyika, M.F.; Gurira, R.C. Reversed-phase liquid chromatographic determination of
cyanide as 1-benzoyl-1,2-dihydroquinaldonitrile, *Anal.Chim.Acta*, **1991**, *251*, 109–114.

DIALDEHYDE

Naphthalene-2,3-dicarboxaldehyde

SAMPLE
Matrix: red blood cells
Analyte: cyanide
Sample preparation: Centrifuge 1 mL blood at 1000 g for 10 min to obtain red blood cells, wash cells twice with 0.9% NaCl, suspend in 1 mL 0.9% NaCl. Remove a 100 μL aliquot and add it to 500 μL water, add 2 mL MeOH, vortex, centrifuge at 1000 g for 10 min. Remove a 500 μL aliquot of the supernatant and add it to 100 μL 5 mM taurine in buffer, add 1 mM 2,3-naphthalenedialdehyde in MeOH:buffer 25:75, let stand at room temperature for 30 min, inject a 20 μL aliquot. (Prepare buffer by mixing 535 mL 50 mM sodium borate and 465 mL 100 mM KH_2PO_4, pH 8.0.)

HPLC VARIABLES
Guard column: LiChroCART RP-18 (Merck)
Column: 125 × 4 5 μm LiChrosorb RP-18
Mobile phase: MeOH:50 mM pH 6.8 potassium phosphate buffer 42:58
Flow rate: 1
Injection volume: 20
Detector: F ex 418 em 460

CHROMATOGRAM
Retention time: 13
Limit of detection: 100 nM

OTHER SUBSTANCES
Non-interfering: thiocyanate

REFERENCE
Sano, A.; Takimoto, N.; Takitani, S. High-performance liquid chromatographic determination of cyanide in human red blood cells by pre-column fluorescence derivatization, *J.Chromatogr.*, **1992**, *582*, 131–135.

Phthalaldehyde

SAMPLE
Matrix: water
Analyte: cyanide
Sample preparation: Inject a 200 μL aliquot.

HPLC VARIABLES
Column: 300 × 8 7 μm Shim-pack SCR-102H polystyrene matrix strong cation-exchanger (Shimadzu)
Mobile phase: 10 mM pH 3.3 Ammonium citrate buffer
Column temperature: 40
Flow rate: 0.8
Injection volume: 200
Detector: F ex 328 em 370 following post-column reaction. The column effluent mixed with the reagent pumped at 0.15 mL/min and the mixture flowed through a 15 m × 0.5 mm ID PTFE coil at 40° to the detector. (Reagent was 54 mM o-phthalaldehyde in 500 mM pH 9.5 borate buffer.)

CHROMATOGRAM
Retention time: 20
Limit of detection: 0.1 ng/mL

OTHER SUBSTANCES
Simultaneously analyzed: 2-mercaptoethanol, 3-mercaptopropionic acid, sulfite

KEY WORDS
post-column reaction; river water

REFERENCE
Sumiyoshi, K.; Yagi, T.; Nakamura, H. Determination of cyanide by high-performance liquid chromatography using postcolumn derivatization with o-phthalaldehyde, *J.Chromatogr.A*, **1995**, *690*, 77–82.

PYRIDINE/BARBITURIC ACID

Isonicotinic Acid/Barbituric Acid

SAMPLE
Matrix: solutions
Analyte: copper(I)-cyanide complex
Sample preparation: Inject a 10 μL aliquot.

HPLC VARIABLES
Column: 150 × 3.9 Nova-Pak C18
Mobile phase: MeCN:5 mM LowUV PIC-A (Waters) 25:75
Flow rate: 1
Injection volume: 10
Detector: UV 515 following post-column reaction. The column effluent mixed with reagent 1 pumped at 0.1 mL/min and the mixture flowed through a stitched 1.4 m × 0.025 mm ID coil at 40°. The effluent from this coil mixed with reagent 2 pumped at 0.2 mL/min and this mixture flowed through three stitched 5 m × 0.025 mm ID coils in series at 40° to the detector. (Reagent 1 was 0.1% N-chlorosuccinimide in 100 mM pH 5.6 succinate buffer containing 2% succinimide. Reagent 2 was 300 mM sodium isonicotinate in water containing 4 mM sodium barbiturate and 10 mM disodium EDTA. Prepare reagent 2 by dissolving sodium isonicotinate and sodium barbiturate in excess NaOH before adding the disodium EDTA, final pH is 7.8.)

CHROMATOGRAM
Retention time: 5

KEY WORDS
post-column reaction

REFERENCE
Fagan, P.A.; Haddad, P.R. Reversed-phase ion-interaction chromatography of Cu(I)-cyanide complexes, *J.Chromatogr.A*, **1997**, *770*, 165–174.

CYANOHYDRIN

MISCELLANEOUS REACTIONS

Hydrolysis

SAMPLE
Matrix: plants
Analyte: cyanogenic glycosides
Sample preparation: Air dry plant material at 40° for 48 h, soak with 20 volumes of boiling MeOH:water 70:30, blend (Waring blender) at 21000 rpm, decant the supernatant, repeat the extraction twice. Combine the extracts and make up to 60 volumes, concentrate to 10 volumes under reduced pressure at 40°, let stand at 5° overnight. Evaporate the supernatant under reduced pressure, reconstitute with 2.5 volumes MeOH:water 15:85, dilute 50-500 times with mobile phase, inject a 20 μL aliquot.

HPLC VARIABLES
Column: 250 × 4.6 Spherisorb S5-C8
Mobile phase: MeOH:pH 5.0 phosphate buffer 15:85
Flow rate: 1
Injection volume: 20
Detector: E, Metrohm Model 656/641, silver working electrode 0 V, glassy carbon auxiliary electrode, Ag/AgCl/KCl reference electrode, following post-column reaction. The column effluent flowed through an immobilized enzyme reactor and then mixed with 2 M NaOH pumped at 0.2 mL/min. This mixture flowed to the detector. The cyanohydrin was hydrolyzed to an aldehyde and cyanide and the cyanide was then detected. (Prepare the immobilized enzyme reactor as follows. Reflux 550 Å controlled-pore glass with 3-aminopropyl triethoxysilane in toluene, dry, pour into a 50 × 3 column, inject five 1 mL portions of glutaraldehyde at 1 mL/min, inject three 100 μL portions of 100-200 mg/mL enzyme at 0.1 mL/min. The mobile phase was either water or 50 mM pH 7.0 phosphate buffer. Prepare enzyme by adding 1 mL Helix pomatia juice (Sigma) to a 400 × 25 Sephadex G-200 column, elute with 50 mM pH 5.0 phosphate buffer, collect 10 mL fractions. Combine fractions 10-24, dialyze overnight, lyophilize.)

CHROMATOGRAM
Retention time: 4 (linamrin), 9 (taxiphyllin), 12 (holocalin), 16 (amygdalin), 21 (prunasin), 23 (sambunigrin)

KEY WORDS
post-column reaction; immobilized enzyme reactor

REFERENCE
Dalgaard, L. Pre- and post-column enzymic hydrolysis and amperometric detection of glycosides. Applications to trimethoprim metabolites and cyanogenic glycosides, *Prog.HPLC*, **1987**, *2*, 219–233.

DIACID

DIAMINE

o-Phenylenediamine

SAMPLE
Matrix: urine
Analyte: oxalic acid
Sample preparation: Adjust pH of urine to 1.0 with 10 M HCl, store frozen. Adjust pH of a 10 mL aliquot to 3.0 with 10 M NaOH. Remove 1 mL aliquots and add them to 10 μL 1 M HCl and (A) 50 μL 1 M pH 4.0 sodium acetate buffer and 50 μL 100 mM pH 4.0 sodium acetate buffer or (B) 50 μL 1 M pH 4.0 sodium acetate buffer and 50 μL 4 U/mL oxalate decarboxylase (Sigma) in 100 mM pH 4.0 sodium acetate buffer. Heat at 37° for 16 h, cool to room temperature, centrifuge, add 1 mL 100 mM o-phenylenediamine in 4 M HCl, mix, heat at 110° for 6 h, cool with tap water, centrifuge briefly, add 4 mL 250 mM pH 6.5 sodium phosphate buffer, add 4 mL 1 M NaOH, mix, let stand overnight at room temperature, filter through a Pasteur pipette plug with glass-fiber filter paper (Whatman GF/A), inject a 100 μL aliquot of the filtrate. (Sample A indicates the total level of oxalic acid and other derivative forming compounds and Sample B indicates the level of other derivative forming compounds, e.g., oxamic acid (after oxalic acid is destroyed by oxalate decarboxylase). The difference between the two results indicates the actual level of oxalic acid.)

HPLC VARIABLES
Column: 150 × 4.6 5 μm Ultrasphere Octyl
Mobile phase: Gradient. MeOH:100 mM pH 6.6 ammonium acetate 5:95 for 10.6 min, to 95:5 (step gradient), maintain at 95:5 for 2 min, re-equilibrate at initial conditions for 2.4 min.
Column temperature: 40
Flow rate: 3
Injection volume: 100
Detector: UV 314

CHROMATOGRAM
Retention time: 7
Limit of detection: 500 ng/mL

REFERENCE
Murray, J.F. Jr.; Nolen, H.W., III; Gordon, G.R.; Peters, J.H. The measurement of urinary oxalic acid by derivatization coupled with liquid chromatography, *Anal.Biochem.*, **1982**, *121*, 301–309.

DIALDEHYDE

GUANIDINE

N-α-Benzoyl-L-arginine Ethyl Ester

SAMPLE
Matrix: blood
Analyte: malondialdehyde
Sample preparation: Add 100 μL serum to 100 μL 400 mM N-α-benzoyl-L-arginine ethyl ester hydrochloride in EtOH, add 2 mL 10 M HCl in EtOH, let stand at 25° overnight, add 2 mL chloroform, neutralize with 7 mL 15% sodium carbonate containing 100 mM Na$_2$HPO$_4$, shake, centrifuge at 1000 g for 5 min. Remove 1 mL of the organic layer and evaporate it to dryness, reconstitute the residue in 200 μL MeCN, inject a 10 μL aliquot.

HPLC VARIABLES
Column: 150 × 4.6 Cosmosil 5C18 (Nakarai tesque)
Mobile phase: MeCN:water 40:60
Flow rate: 1
Injection volume: 10
Detector: UV 309

CHROMATOGRAM
Retention time: 4.5

Limit of quantitation: 1 μM

KEY WORDS
rat; serum

REFERENCE
Kishida, E.; Oribe, M.; Mochizuki, K.; Kojo, S.; Iguchi, H. Determination of malondialdehyde with chemical derivatization into the pyrimidine compound and HPLC, *Biochim.Biophys.Acta*, **1990**, *1045*, 187–188.

SAMPLE
Matrix: rapeseed oil
Analyte: malondialdehyde
Sample preparation: 400 μL Rapeseed oil + 400 μL hexane + 400 μL 100 mM HCl, shake, centrifuge at 2000 g for 5 min. Remove 100 μL of the aqueous layer and add it to 100 μL 400 mM Nα-benzoyl-L-arginine ethyl ester hydrochloride in EtOH, mix, add 2 mL 10 M HCl in EtOH, let stand at 25° overnight, add 2 mL chloroform, add 7 mL 15% sodium carbonate containing 100 mM Na_2HPO_4, shake, centrifuge at 1000 g for 5 min. Remove 1 mL of the chloroform layer and evaporate it to dryness, reconstitute the residue in 200 μL MeCN, inject a 10 μL aliquot.

HPLC VARIABLES
Column: 150 × 4.6 Cosmosil 5C18 (Nakalai Tesque, Kyoto)
Mobile phase: MeCN:water 40:60
Flow rate: 1
Injection volume: 10
Detector: UV 309

REFERENCE
Kishida, E.; Oribe, M.; Kojo, S. Relationship among malondialdehyde, TBA-reactive substances, and tocopherols in the oxidation of rapeseed oil, *J.Nutr.Sci.Vitaminol.(Tokyo)*, **1990**, *36*, 619–623.

HYDRAZINE

Luminarin 3

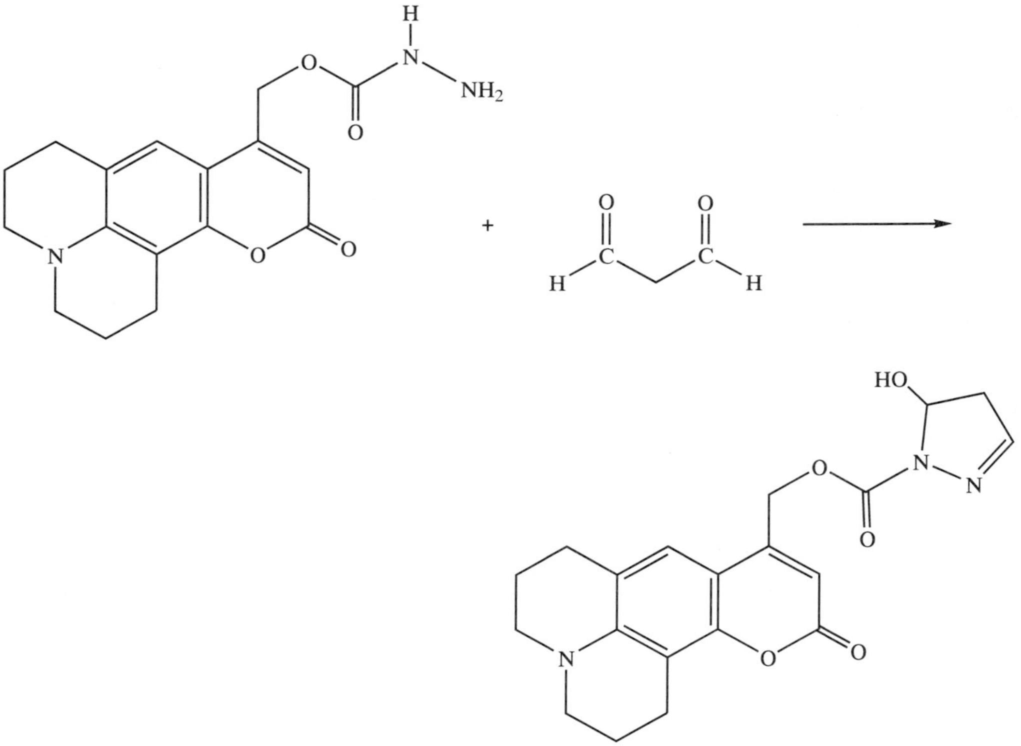

SAMPLE

Matrix: solutions

Analyte: malonaldehyde

Sample preparation: Mix 200 μL of a solution of malonaldehyde in 100 mM ph 4.0 acetate buffer with 10 μL 10 mM luminarin 3 in DMSO and 10 μL 17 μM IS in DMSO, let stand in the dark at room temperature for 30 min, add 10 μL reagent, let stand for 5 min, inject an aliquot. (Reagent was 1 M acetylacetone in 1% sulfuric acid, prepared just before use. Luminarin 3, 2,3,6,7-tetrahydro-11-oxo-1H,5H,11H-[1]benzopyrano[6,7,8-ij]quinolizine-9-acetic acid hydrazide, is available from Eurobio, Les Ulis, France. Synthesis is as follows. Reflux (with protection from moisture and with stirring) 2.12 g 8-hydroxyjulolidine, 2.22 g diethyl 1,3-acetonedicarboxylate (oxo-3-glutaric acid ethyl ester, Fluka), 1.71 g anhydrous zinc chloride, and 6 mL EtOH for 24 h, cool, add to 200 mL water, extract with 200 mL ethyl acetate, extract with 100 mL ethyl acetate. Combine the organic layers and wash them with water, dry over magnesium sulfate, evaporate to dryness, recrystallize from 5 parts ethyl acetate to give ethyl 2,3,6,7-tetrahydro-11-oxo-1H,5H,11H-[1]benzopyrano[6,7,8-ij]quinolizine-9-acetate. Stir 5 g of this compound with 8 mL hydrazine hydrate in 100 mL MeOH for 4 h, filter, wash the solid with 10 mL MeOH, wash with 10 mL dichloromethane to give luminarin 3 (World Pat. 89 12,052; Chem. Abstr. 1990, 113, 23889n).)

HPLC VARIABLES

Column: 150 × 4.6 5 μm Nucleosil C18

Mobile phase: MeCN:buffer 30:70 (Prepare buffer by adjusting the pH of 10 mM imidazole to 7.0 with concentrated nitric acid.)

Flow rate: 1.5
Injection volume: 20
Detector: F ex 395 em 500

CHROMATOGRAM
Retention time: 12.7
Internal standard: luminarin-3-methylmalonaldehyde adduct (Prepare IS by hydrolysing 1 mmole 2-methyl-1,1,3,3-tetraethoxypropane (Eurobio-Seratec, Les Ulis, France) in 100 mL 1% sulfuric acid in the dark at room temperature for 2 h, add 10 mL 100 mM luminarin 3 in DMSO, add 2.5 g sodium bicarbonate, stir gently until the acid is neutralized (pH 7.0), extract with 200 mL dichloromethane. Dry the organic layer over anhydrous magnesium sulfate and evaporate to dryness under reduced pressure, dissolve the residue in 5 mL dichloromethane and chromatograph on 40-63 µm silica gel 60 using a gradient of THF : dichloromethane 0 : 100 to 40 : 60 to give the adduct as yellowish brown needles.) (20.1)
Limit of detection: 0.45 ng/mL

REFERENCE
Traoré, F.; Farinotti, R.; Mahuzier, G. Determination of malonaldehyde by coupled high-performance liquid chromatography-spectrofluorimetry after derivatization with luminarin 3, *J.Chromatogr.*, **1993**, *648*, 111–118.

SAMPLE
Matrix: solutions
Analyte: periodate-oxidized nucleotides and nucleosides
Sample preparation: Mix a 200 µL aliquot of a 25 mM solution of periodate-oxidized nucleotides or nucleosides in 100 mM pH 4.0 (nucleosides) or 3.0 (nucleotides) acetate buffer with 10 µL 10 mM luminarin 3 in DMSO, let stand at room temperature for 2 h, inject a 20 µL aliquot. (Luminarin 3, 2,3,6,7-tetrahydro-11-oxo-1H,5H,11H-[1]benzopyrano[6,7,8-ij]quinolizine-9-acetic acid hydrazide, is available from Eurobio, Les Ulis, France. Synthesis is as described above.)

HPLC VARIABLES
Column: 300 × 4.6 5 µm Spherisorb cyano
Mobile phase: MeCN : buffer 30 : 70 (Buffer was 100 mM K_2HPO_4 containing 10 mM ammonium acetate, adjusted to pH 6.0 with phosphoric acid.)
Flow rate: 1
Injection volume: 20
Detector: F ex 402 em 501

CHROMATOGRAM
Retention time: 6 (periodate-oxidized cytidine triphosphate), 8 (periodate-oxidized cytidine monophosphate), 8 (periodate-oxidized adenosine triphosphate), 10 (periodate-oxidized adenosine monophosphate), 12 (periodate-oxidized cytidine), 21 (periodate-oxidized adenosine)
Limit of detection: 27.7-666.7 fmole

REFERENCE
Traoré, F.; Fente, C.; Prognon, P.; Mahuzier, G. Luminarin 3 as a derivatization reagent for the liquid chromatographic determination of cytidine, adenosine and related nucleotides with fluorimetric detection, *Anal.Chim.Acta*, **1994**, *290*, 94–102.

Luminarin 11

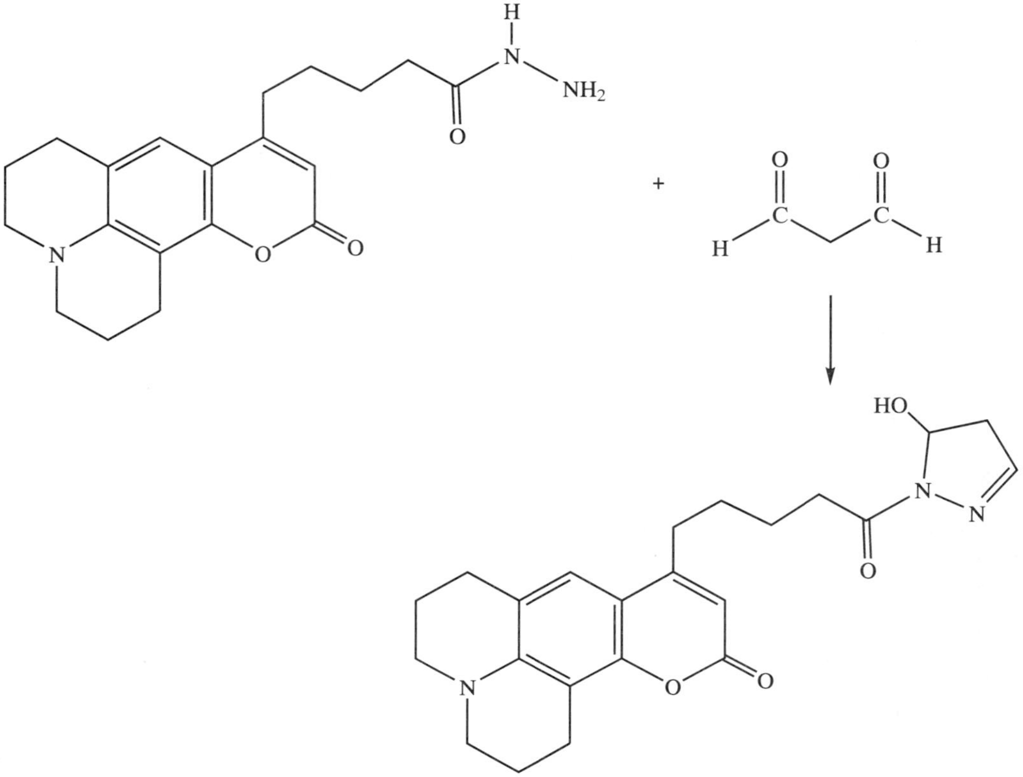

SAMPLE

Matrix: solutions

Analyte: malonaldehyde

Sample preparation: Mix 200 μL of a 100 mM solution in water with 10 μL 50 mM luminarin 11 in DMSO, let stand at room temperature for 2 h, carefully add 100 μL 600 mM sodium bicarbonate to adjust pH to 7.0, agitate gently until gas evolution ceases, add 2 mL dichloromethane, vortex for 2 min, centrifuge. Remove the lower organic layer and evaporate it to dryness under a stream of nitrogen at room temperature, reconstitute the residue in MeCN, inject an aliquot. (Luminarin 11, 2,3,6,7-tetrahydro-11-oxo-1H,5H,11H-[1]benzopyrano[6,7,8-ij]quinolizine-9-pentanoic acid hydrazide, is available from Eurobio, Les Ulis, France.)

HPLC VARIABLES

Column: 150 × 4.6 5 μm Nucleosil ODS

Mobile phase: MeCN:10 mM pH 7.5 imidazole nitrate buffer 60:40

Flow rate: 1-2

Injection volume: 20

Detector: F ex 387 em 450

CHROMATOGRAM

Retention time: 7

Limit of detection: 164 fmole

OTHER SUBSTANCES

Simultaneously analyzed: acetylacetone

KEY WORDS

comparison with other luminarins

REFERENCE

Traoré, F.; Pianetti, G.A.; Dallery, L.; Tod, M.; Chalom, J.; Farinotti, R.; Mahuzier, G. Determination of picomole amounts of carbonyls as luminarin hydrazones by high-performance liquid chromatography with fluorescence detection, *Chromatographia*, **1993**, *36*, 96–104.

Luminarin 12

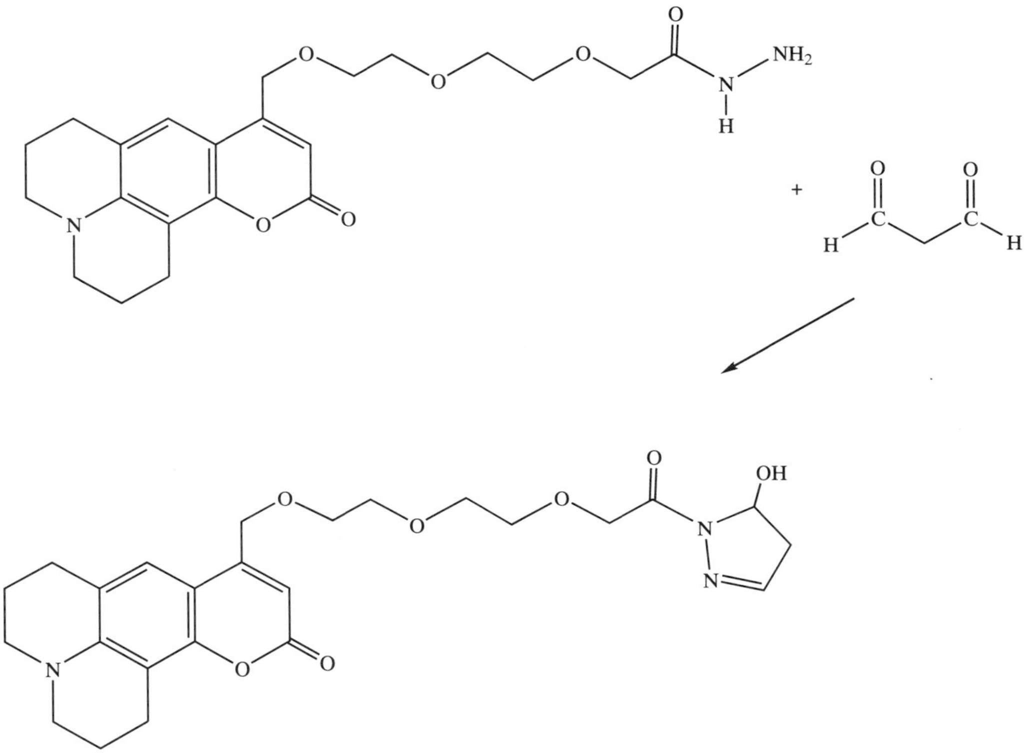

SAMPLE

Matrix: solutions

Analyte: malonaldehyde

Sample preparation: Mix 200 μL of a 100 mM solution in water with 10 μL 50 mM luminarin 12 in DMSO, let stand at room temperature for 30 min, carefully add 100 μL 600 mM sodium bicarbonate to adjust pH to 7.0, agitate gently until gas evolution ceases, add 2 mL dichloromethane, vortex for 2 min, centrifuge. Remove the lower organic layer and evaporate it to dryness under a stream of nitrogen at room temperature, reconstitute the residue in MeCN, inject an aliquot. (Luminarin 12, 2,3,6,7-tetrahydro-11-oxo-1H,5H,11H-[1]benzopyrano[6,7,8-ij]quinolizine-9-methoxyethoxyethoxyacetic acid hydrazide, is available from Eurobio, Les Ulis, France.)

HPLC VARIABLES

Column: 250 × 4.6 5 μm Nucleosil ODS

Mobile phase: MeCN:10 mM pH 7.5 imidazole nitrate buffer 52:48

Flow rate: 1-2

Injection volume: 20

Detector: F ex 399 em 500

CHROMATOGRAM

Retention time: 5

Limit of detection: 425 fmole

OTHER SUBSTANCES
Simultaneously analyzed: acetylacetone

KEY WORDS
comparison with other luminarins

REFERENCE
Traoré, F.; Pianetti, G.A.; Dallery, L.; Tod, M.; Chalom, J.; Farinotti, R.; Mahuzier, G. Determination of picomole amounts of carbonyls as luminarin hydrazones by high-performance liquid chromatography with fluorescence detection, *Chromatographia*, **1993**, *36*, 96–104.

4-Nitrophenylhydrazine

RELATED REFERENCE
Uno, B.; Kawai, K.; Kawasaki, C.; Kawai, S.; Asakura, E.; Tomita, M. High-performance liquid chromatographic determination of urinary malondialdehyde as p-nitrophenylhydrazine derivative. *Anal.Sci.* **1991**, *7*, 963-965

SAMPLE
Matrix: solutions
Analyte: malonaldehyde
Sample preparation: 1 mL Solution + 600 µL 1 M pH 3.76 acetate buffer + 200 µL 500 µg/mL 4-nitrophenylhydrazine hydrochloride in EtOH containing 2-nitroresorcinol, mix, let stand at room temperature for 1 h, inject a 20 µL aliquot. (The product of the derivatization is 1-(p-nitrophenyl)pyrazole.)

HPLC VARIABLES
Column: 250 × 4.6 Develosil C18-5 (Nomura)
Mobile phase: MeCN:isopropanol:10 mM NaH_2PO_4 30:10:60
Flow rate: 1
Injection volume: 20
Detector: UV 315

CHROMATOGRAM
Retention time: 18
Internal standard: 2-nitroresorcinol (7)
Limit of detection: 6 ng/mL

REFERENCE
Kawai, S.; Fuchiwaki, T.; Higashi, T.; Tomita, M. High-performance liquid chromatographic determination of malonaldehyde using p-nitrophenylhydrazine as a derivatizing reagent, *J.Chromatogr.*, **1990**, *514*, 29–35.

DIENE

TRIAZOLINEDIONE

4-(1-Anthryl)-1,2,4-triazoline-3,5-dione

SAMPLE

Matrix: tissue

Analyte: 7-dehydrocholesterol

Sample preparation: Extract 11.34 cm^2 skin with hexane:EtOH 50:50 for 5 min, repeat extraction twice more. Combine the extracts and concentrate them to 500 μL, filter (0.5 μm Millipore FH), evaporate the filtrate to dryness under a stream of nitrogen, reconstitute the residue in 200 μL ethyl acetate, add 65 μg 4-(1-anthryl)-1,2,4-triazoline-3,5-dione in ethyl acetate, mix, let stand at 4° for 10 min, inject an aliquot. (Synthesis of 4-(1-anthryl)-1,2,4-triazoline-3,5-dione is as follows. Dissolve 50 g benzanthrone in 500 mL concentrated sulfuric acid with gentle warming, pour this solution cautiously into 4 L hot water with vigorous stirring. Boil the suspension and slowly add 200 g chromium(VI) oxide (Caution! Chromium oxide is a carcinogen and highly corrosive!), after 6 h cool the mixture, filter, wash the precipitate with hot water. Dissolve the precipitate in dilute ammonia and precipitate with acid, crystallize from boiling concentrated nitric acid to give anthraquinone-1-carboxylic acid (Ber. 1924, 57, 1775). Warm, on a water bath, anthraquinone-1-carboxylic acid in dilute ammonia with twice the amount of zinc dust, when the reaction has ceased (30 min ?) filter the reaction the reaction mixture, add HCl to the filtrate to obtain anthracene-1-carboxylic acid as yellow needles, recrystallize from EtOH (mp 245°) (Ber 1897, 30, 1118). Anthracene-1-carboxylic acid is also available from Tokyo Kasei (TCI America, Portland OR). Treat anthracene-1-carboxylic acid with1 equivalent triethylamine and 1 equivalent diphenylphosphoryl azide in DMF to obtain anthracene-1-carbonyl azide. Reflux the crude azide in toluene for 20 min to obtain 1-isocyanatoanthracene. Add this isocyanate solution to 1 equivalent of ethyl carbazate in benzene cooled in ice (Caution! Benzene is a carcinogen!), let stand at room temperature for 1 h, heat at 100° for 1 h, filter. Wash the precipitate with benzene and dry it under reduced pressure, recrystallize from DMF/water to obtain 1-anthracene ethoxycarbonylsemicarbazide as a pale-yellow amorphous solid (mp 184-185°). Suspend the semicarbazide in 4 M KOH, heat at 90° for 1.5 h, acidify with concentrated HCl, filter, wash the solid with water, dry under reduced pressure, purify by re-precipitation, recrystallize from DMF/water to obtain 4-(1-anthryl)-1,2-dihydro-1,2,4-triazoline-3,5-dione as a yellow amorphous substance (mp 268-269°). Suspend in ethyl acetate, cool in ice, add 1 equivalent tert-butyl hypochlorite (Org. Syn. 1973, Coll. Vol. V, 184) under nitrogen, stir at room temperature for 1 h, centrifuge to obtain an ethyl acetate solution of 4-(1-anthryl)-1,2,4-triazoline-3,5-dione.)

HPLC VARIABLES

Column: 150 × 4.6 5 μm YMC-Gel C8-120-S5

Mobile phase: MeCN:water 72.7:27.3 containing 5 mM methylated β-cyclodextrin (10.5 mol/mol methyl residues (Kao, Tokyo))
Flow rate: 1
Detector: UV 253; F ex 330 em 410

CHROMATOGRAM
Retention time: 21
Limit of detection: 0.06 pmole (UV), 12 pmole (F)

KEY WORDS
skin

REFERENCE
Shimada, K.; Oe, T.; Mizuguchi, T. Cookson-type reagents: highly sensitive derivatization reagents for conjugated dienes in high-performance liquid chromatography, *Analyst*, **1991**, *116*, 1393–1397.

DMEQ-TAD

RELATED REFERENCE
Shimizu, M.; Wang, X; Yamada, S. Fluorimetric assay of 1α,25-dihydroxyvitamin D_3 in human plasma. *J.Chromatogr.B* **1997**, *690*, 15-23.

SAMPLE
Matrix: solutions
Analyte: vitamin D metabolites
Sample preparation: Evaporate solution of calcifediol in EtOH, add 1 mL 7.2 μM DMEQ-TAD in dichloromethane, stir at room temperature for 30 min, add EtOH, evaporate, dissolve residue in MeOH, inject an aliquot. (Dichloromethane should be MeOH free. Wash with concentrated sulfuric acid, water, 5% sodium carbonate, water, dry over calcium chloride, and distil from calcium hydride. DMEQ-TAD was 4-[2-(6,7-dimethoxy-4-methyl-3-oxo-3,4-dihydroquinoxalinyl)ethyl]-1,2,4-triazoline-3,5-dione. Synthesis is as follows. Stir 483 g veratrole in 1.45 L acetic acid at 15° for 1 h, add 683 g concentrated nitric acid (d 1.05) over 1 h (maintain the temperature below 40° by cooling and regulating the rate of addition of the nitric acid). Continue stirring and add 2.127 L fuming nitric acid

(d 1.50) over 1 h while maintaining the temperature below 30°, let stand for 2 h, pour into a large volume of cold water, filter, wash the solid with water until the washings are neutral, recrystallize from EtOH to give 4,5-dinitroveratrole (mp 129.5-130.5°) (J. Am. Chem. Soc. 1946, 68, 1536). Shake a solution of 910 mg 4,5-dinitroveratrole in 80 mL EtOH with 89 mg platinum(IV) oxide under an atmosphere of hydrogen until the theoretical amount of hydrogen (540 mL) is absorbed, filter under nitrogen into a flask containing 580 mg 2-ketoglutaric acid, reflux this mixture for 1.5 h, cool, collect the precipitate, recrystallize from EtOH to obtain 6,7-dimethoxy-3-oxo-3,4-dihydroquinoxaline-2-propionic acid as a crystalline solid (mp 250-252°). Add a solution of 606 mg 6,7-dimethoxy-3-oxo-3,4-dihydroquinoxaline-2-propionic acid in 20 mL DMF under nitrogen to a suspension of 176 mg NaH in 3 mL DMF stirred at 0°, stir for 30 min, add 455 μL methyl iodide, stir at 0° for 1.5 h, pour into ice-water, stir at room temperature for 30 min, acidify with 500 mM HCl, collect the precipitate, recrystallize from chloroform/MeOH to obtain 6,7-dimethoxy-4-methyl-3-oxo-3,4-dihydroquinoxaline-2-propionic acid (mp 239-241°). Add 360 μL triethylamine at room temperature to 500 mg 6,7-dimethoxy-4-methyl-3-oxo-3,4-dihydroquinoxaline-2-propionic acid in 50 mL DMF, add 550 μL diphenylphosphoryl azide, stir at room temperature for 2.5 h, evaporate to dryness under reduced pressure, dissolve the residue in 20 mL benzene (Caution! Benzene is a carcinogen!), reflux for 1 h, cool to room temperature, add 178 mg ethyl carbazate, reflux for 30 min, evaporate, chromatograph on 90 g silica gel, elute with chloroform to remove a by-product then with chloroform:MeOH 96:4 to obtain 1-ethoxycarbonyl-4-[2-(6,7-dimethoxy-4-methyl-3-oxo-3,4-dihydroquinoxalinyl)ethyl]semicarbazide. Reflux a suspension of 272 mg 1-ethoxycarbonyl-4-[2-(6,7-dimethoxy-4-methyl-3-oxo-3,4-dihydroquinoxalinyl)ethyl]semicarbazide and 190 mg potassium carbonate in 20 mL EtOH for 6 h, evaporate the solvent, dissolve the residue in 30 mL water, acidify with 2 M HCl, extract with chloroform:MeOH 90:10, dry over anhydrous sodium sulfate, evaporate, recrystallize from MeOH/chloroform to obtain 4-[2-(6,7-dimethoxy-4-methyl-3-oxo-3,4-dihydroquinoxalinyl)ethyl]-1,2,4-triazolidine-3,5-dione as pale yellow prisms (mp 250-253°). Add 10 mg iodobenzene diacetate to a stirred suspension of 8.6 mg 4-[2-(6,7-dimethoxy-4-methyl-3-oxo-3,4-dihydroquinoxalinyl)ethyl]-1,2,4-triazolidine-3,5-dione in 1.5 mL MeOH-free dichloromethane, stir at room temperature for 3.5 h, filter, store the filtrate at -20° overnight, filter under argon to obtain 4-[2-(6,7-dimethoxy-4-methyl-3-oxo-3,4-dihydroquinoxalinyl)ethyl]-1,2,4-triazoline-3,5-dione as red needles (mp 200-202° d).)

HPLC VARIABLES

Column: 250 × 4 LiChrospher RP-18(e)
Mobile phase: Gradient. MeOH:water from 60:40 to 80:20 over 40 min
Column temperature: 35
Flow rate: 1
Injection volume: 10
Detector: F ex 370 em 440

CHROMATOGRAM

Retention time: 18, 24 (dihydroxyvitamin D_3 C6 epimers), 27, 30 (calcitriol C6 epimers), 33, 36 (calcifediol C6 epimers)

REFERENCE

Shimizu, M.; Kamachi, S.; Nishii, Y.; Yamada, S. Synthesis of a reagent for fluorescence-labeling of vitamin D and its use in assaying vitamin D metabolites, *Anal.Biochem.*, **1991**, *194*, 77−81.

4-[4-(6-Methoxybenzoxazolyl)phenyl]-1,2,4-triazoline-3,5-dione

---

SAMPLE

Matrix: bulk

Analyte: vitamin D_3 glucuronide

Sample preparation: Dissolve vitamin D_3 glucuronide in 100 μL 600 μg/mL MBOTAD in ethyl acetate, let stand at room temperature for 1 h, add 3 drops MeOH, evaporate to dryness under a stream of nitrogen, reconstitute the residue in 1 mL MeOH:water 90:10, add to a 20 × 6 column of PHP-LH-20, wash with 10 mL MeOH:water 90:10, wash with 15 mL 100 mM acetic acid in MeOH:water 90:10, elute with 4 mL 100 ammonium acetate in MeOH:water 90:10, inject an aliquot of the eluate. (Preparation of PHP-LH-20 Sephadex is as follows. Suspend 75.7 g Sephadex LH-20 in 200 mL dichloromethane using a glass stirring rod (not a magnetic stirrer) for 30 min, add 19 mL boron trifluoride ethyl etherate, after 15 min add 50 mL 35% epichlorohydrin in dichloromethane at 1-2 mL/min (Caution! Epichlorohydrin is a carcinogen!), stir for another 30 min, filter, wash with EtOH, dry chlorohydroxypropyl Sephadex LH-20 at 50° (J.Chromatogr. 1971, 59, 45). Stir 27.2 g chlorohydroxypropyl Sephadex LH-20 in 100.5 mL piperidine at room temperature for 30 min, add 5.74 g KOH in 302 mL MeOH, heat at 50-60° for 3 h with occasional shaking, filter, wash with EtOH:water 50:50, wash with 200 mM acetic acid in EtOH:water 70:30, wash with EtOH:water 90:10 until washings become neutral, store in EtOH:water 90:10 (Clin. Chim. Acta 1978 87 141). Synthesize MBOTAD (4-[4-(6-methoxybenzoxazolyl)phenyl]-1,2,4-triazoline-3,5-dione) as follows. Hydrogenate 5-methoxy-2-nitrophenol in EtOH over platinum oxide to give 2-amino-5-methoxyphenol (J. Org. Chem. 1957, 22, 220). Alternatively 6.9 g 5-methoxy-2-nitrophenol can be reduced to 2-amino-5-methoxyphenol with 50 g sodium hydrosulfite in 100 mL boiling water (Anal. Sci. 1994, 10, 697). It should be possible to prepare ethyl 4-methoxycarbonylbenzimidate hydrochloride ($CH_3OCOC_6H_4C(=NH)OC_2H_5.HCl$) by passing dry hydrogen chloride into a mixture of methyl 4-cyanobenzoate and 1.2-1.5 equivalents EtOH in an inert solvent (e.g., benzene, chloroform, dioxane, ether, nitrobenzene (Caution! Benzene, chloroform, and di-oxane are carcinogens!) at 0-5°,the benzimidate should crystallize from the mixture in 7-10 days (J. Chem. Soc. 1942, 103). Add 5.7 g 2-amino-5-methoxyphenol to a stirred sus-pension of 9.6 g ethyl 4-methoxycarbonylbenzimidate hydrochloride in 200 mL EtOH, reflux for 2 h, cool in an ice-water bath, filter, dry the precipitate under reduced pressure. Chromatograph on silica gel with dichloromethane to obtain methyl 4-(6-methoxy-2-ben-zoxazolyl)benzoate. Dissolve the product in 66 mL MeOH:water 90:10, add 3 mL 5 M NaOH, reflux for 3 h, cool in an ice-water bath, acidify with HCl, filter, dry the product under reduced pressure to obtain 4-(6-methoxy-2-benzoxazolyl)benzoic acid (mp 286°) (Anal. Sci. 1994, 10, 697). Treat 4-(6-methoxy-2-benzoxazolyl)benzoic acid with 1 equivalent triethylamine and 1 equivalent diphenylphosphoryl azide in DMF to obtain 4-(6-methoxy-2-benzoxazolyl)benzoyl azide. Reflux the crude azide in benzene or toluene for 20 min to obtain 4-(6-methoxy-2-benzoxazolyl)phenyl isocyanate (Caution! Benzene is a carcinogen!). Add this isocyanate solution to 1 equivalent of ethyl carbazate in benzene cooled in ice, let stand at room temperature for 1 h, heat at 100° for 1 h, filter. Wash the precipitate with benzene and dry it under reduced pressure (Analyst 1991, 116, 393), recrystallize from xylene to obtain 4-(6-methoxy-2-benzoxazolyl)phenyl ethoxycarbonyl-semicarbazide as a colorless amorphous solid (mp 218-220°). Dissolve 10 mg of the semicarbazide in 2 mL 2 M KOH in EtOH:water 50:50, stir at room temperature for 2 h, acidify with 5% HCl, collect the precipitate and wash it with water. Dry the precipitate under reduced pressure to obtain 4-[4-(6-methoxybenzoxazolyl)phenyl]-1,2-dihydro-1,2,4-triazoline-3,5-dione. Suspend 4 mg of this crude material in 1 mL ethyl acetate, cool in ice, add 1.7 μL tert-butyl hypochlorite (Org. Syn. 1973, Coll. Vol. V, 184), after 15 min filter, evaporate the filtrate to dryness, sublime the residue at 140°/0.2 mm Hg to obtain (4-[4-(6-methoxybenzoxazolyl)phenyl]-1,2,4-triazoline-3,5-dione) as purple crystals (mp 148-150°) (J. Chromatogr. 1992, 606, 133). The precursor 4-[4-(6-methoxybenzoxazolyl)phenyl]-1,2-dihydro-1,2,4-triazoline-3,5-dione (4-[4-(6-methoxybenzoxazolyl)phenyl]-1,2,4-triazolidine-3,5-dione) is available from Dojindo, Japan (Dojindo Molecular Technologies, Inc., 3 Bethesda Metro Center, Suite 700, Bethesda MD 20814; (301) 664-8448; www.dojindo.co.jp).)

HPLC VARIABLES
Column: 150 × 4.6 5 μm YMC.GEL C8
Mobile phase: MeCN:0.5% pH 5.0 ammonium acetate 8:5
Flow rate: 1
Detector: F ex 320 em 380

CHROMATOGRAM
Retention time: 9.6

KEY WORDS
SPE

REFERENCE
Shimada, K.; Nakatani, I.; Saito, K.; Mitamura, K. Separation and characterization of monoglucuronides of vitamin D$_3$ and 25-hydroxyvitamin D$_3$ in rat bile by high-performance liquid chromatography, *Biol.Pharm.Bull.*, **1996**, *19*, 491–494.

SAMPLE
Matrix: solutions
Analyte: 7-dehydrocholesterol
Sample preparation: Mix 100 μL 10 μg/mL 7-dehydrocholesterol in ethyl acetate with 100 μL ethyl acetate containing 20 equivalents of MBOTAD at 0°, let stand at 0° for 5 min, inject an aliquot. (MBOTAD is 4-[4-(6-methoxybenzoxazolyl)phenyl]-1,2,4-triazoline-3,5-dione. Synthesis is described above. The precursor 4-[4-(6-methoxybenzoxazolyl)phenyl]-1,2-dihydro-1,2,4-triazoline-3,5-dione (4-[4-(6-methoxybenzoxazolyl)phenyl]-1,2,4-triazolidine-3,5-dione) is available from Dojindo, Japan (Dojindo Molecular Technologies, Inc., 3 Bethesda Metro Center, Suite 700, Bethesda MD 20814; (301) 664-8448; www.dojindo.co.jp). Suspend 4 mg of this material in 1 mL ethyl acetate, cool in ice, add 1.7 μL tert-butyl hypochlorite (Org. Syn. 1973, Coll.Vol. V, 184), after 15 min filter, evaporate the filtrate to dryness, sublime the residue at 140°/0.2 mm Hg to obtain 4-[4-(6-methoxybenzoxazolyl)phenyl]-1,2,4-triazoline-3,5-dione as purple crystals (mp 148-150°).)

HPLC VARIABLES
Column: 150 × 4.6 5 μm YMC-Gel C8-120-S5
Mobile phase: MeCN:water 90:10
Flow rate: 1
Detector: F ex 320 em 380

CHROMATOGRAM
Retention time: 7

Limit of detection: 2 fmole

REFERENCE

Shimada, K.; Mizuguchi, T. Sensitive and stable Cookson-type reagent for derivatization of conjugated dienes for high-performance liquid chromatography with fluorescence detection, *J.Chromatogr.*, **1992**, *606*, 133–135.

SAMPLE

Matrix: solutions
Analyte: vitamin D
Sample preparation: React 52 μmoles vitamin D with 82 μmoles MBOTAD in ethyl acetate at 0°, after 5 min purify by preparative TLC (20 cm × 20 cm × 0.5 mm, silica gel HF254, E.Merck), elute with chloroform:ethyl acetate 2:1, extract the appropriate zone with ethyl acetate. (MBOTAD is 4-[4-(6-methoxybenzoxazolyl)phenyl]-1,2,4-triazoline-3,5-dione. Synthesis is described above. The precursor 4-[4-(6-methoxybenzoxazolyl)phenyl]-1,2-dihydro-1,2,4-triazoline-3,5-dione (4-[4-(6-methoxybenzoxazolyl)phenyl]-1,2,4-triazolidine-3,5-dione) is available from Dojindo, Japan (Dojindo Molecular Technologies, Inc., 3 Bethesda Metro Center, Suite 700, Bethesda MD 20814; (301) 664-8448; www.dojindo.co.jp). Suspend 4 mg of this material in 1 mL ethyl acetate, cool in ice, add 1.7 μL tert-butyl hypochlorite (Org. Syn. 1973, Coll. Vol. V, 184), after 15 min filter, evaporate the filtrate to dryness, sublime the residue at 140°/0.2 mm Hg to obtain 4-[4-(6-methoxybenzoxazolyl)phenyl]-1,2,4-triazoline-3,5-dione as purple crystals (mp 148-150°).)

HPLC VARIABLES

Column: 150 × 4.6 5 μm YMC.GEL C8 (YMC)
Mobile phase: MeOH:water 90:10 containing 7 mM heptakis-(2,6-di-O-methyl)-β-cyclodextrin
Flow rate: 1
Detector: UV 265

CHROMATOGRAM

Retention time: 7.5 (vitamin D3), 8.5 (vitamin D4), 10.0 (vitamin D2), 12 (vitamin D5)

REFERENCE

Shimada, K.; Mitamura, K.; Miura, M.; Miyamoto, A. Retention behavior of vitamin D and related compounds during high-performance liquid chromatography, *J.Liq.Chromatogr.*, **1995**, *18*, 2885–2893.

4-Pentafluorobenzyl-1,2,4-triazoline-3,5-dione

SAMPLE

Matrix: blood
Analyte: Ro 24-2090 ((3β,5Z,7E)-9,10-secocholesta-5,7,10(19),16-tetraen-23-yne-3,25-diol)
Sample preparation: 0.125-1 mL Plasma + IS + 4 mL hexane:dichloromethane 85:15, extract. Remove the organic layer and evaporate it to dryness under a stream of nitrogen at 40°, reconstitute the residue in 150 μL MeCN, add 50 μL 4-pentafluorobenzyl-1,2,4-triazoline-3,5-dione, vortex for 5 s, add 150 μL MeOH, evaporate to dryness under a stream of nitrogen at 40°, reconstitute the residue in 50 μL hexane:ethyl acetate 60:40,

inject a 40 μL aliquot onto column A and elute with mobile phase A, switch the fraction containing the compound onto column B (about 8.5 min), elute column B with mobile phase B. (Preparation of 4-pentafluorobenzyl-1,2,4-triazoline-3,5-dione is as follows. Add 1.8 mL triethylamine and 2.75 mL diphenylphosphoryl azide to a solution of 1.93 g pentafluorophenyl acetic acid in 250 mL DMF, stir at room temperature for 1 h, evaporate to dryness under reduced pressure, dissolve the residue in 100 mL benzene (Caution! Benzene is a carcinogen!), reflux for 1 h, add 900 mg ethyl carbazate, reflux for 1 h, evaporate to dryness under reduced pressure, chromatograph on silica gel with dichloromethane to obtain 1-ethoxycarbonyl-4-pentafluorobenzylsemicarbazide. Add 423 mg potassium carbonate to a suspension of 500 mg 1-ethoxycarbonyl-4-pentafluorobenzylsemicarbazide in 50 mL EtOH, reflux for 3.5 h, evaporate to dryness, dissolve the residue in 40 mL water, acidify to pH 1-2 with 2 M HCl, extract with ethyl acetate. Dry the organic layer over anhydrous sodium sulfate and evaporate it to dryness to obtain 4-pentafluorobenzyl-1,2,4-triazolidine-3,5-dione. Suspend 30 mg 4-pentafluorobenzyl-1,2,4-triazolidine-3,5-dione and 36 mg iodobenzene diacetate in 2 mL dichloromethane (MeOH free, distilled from calcium hydride), shake vigorously for 3.5 h to obtain a clear pink solution of 4-pentafluorobenzyl-1,2,4-triazoline-3,5-dione. Use this solution for derivatization.)

HPLC VARIABLES
Column: A 100 × 3 3 μm Spherisorb silica; B 150 × 3 3 μm Spherisorb amino
Mobile phase: A Hexane:ethyl acetate:THF 60:40:6; B THF
Column temperature: 26
Flow rate: 0.4
Injection volume: 40
Detector: MS, Finnegan MAT SSQ710 single quadrupole, inlet 40°, ion source 250°, negative ionization reagent glass 1% ammonia in methane, SIM m/z 492

CHROMATOGRAM
Retention time: 12.5
Internal standard: hexadeutero Ro 24-2090
Limit of quantitation: 25-200 pg/mL (depending on species)

KEY WORDS
plasma; normal phase; human; dog; hamster; monkey; mouse; pig; rabbit; rat; column-switching

REFERENCE
Wang, K.; Davis, P.P.; Crews, T.; Gabriel, L.; Edom, R.W. An electron-capture dienophile derivatization agent for increasing sensitivity: determination of a vitamin D analog (Ro 24-2090) in plasma samples with liquid chromatography/mass spectrometry, *Anal.Biochem.*, **1996**, *243*, 28–40.

4-Phenyl-1,2,4-triazoline-3,5-dione

SAMPLE
Matrix: solutions
Analyte: 7-dehydrocholesterol
Sample preparation: Mix 100 μL 10 μg/mL 7-dehydrocholesterol in ethyl acetate with 100 μL ethyl acetate containing 20-50 equivalents 4-phenyl-1,2,4-triazoline-3,5-dione at 0°, after 5 min inject an aliquot.

HPLC VARIABLES
Column: 150 × 4.6 5 μm YMC-Gel C8-120-S5
Mobile phase: MeCN:water 90:10
Flow rate: 1
Detector: UV 265

CHROMATOGRAM
Retention time: 8

REFERENCE
Shimada, K.; Oe, T.; Mizuguchi, T. Cookson-type reagents: highly sensitive derivatization reagents for conjugated dienes in high-performance liquid chromatography, *Analyst*, **1991**, *116*, 1393–1397.

4-[2-(1-Pyrenyl)ethyl]-1,2,4-triazoline-3,5-dione

SAMPLE
Matrix: solutions
Analyte: 7-dehydrocholesterol
Sample preparation: Mix 100 μL 10 μg/mL 7-dehydrocholesterol in ethyl acetate with 100 μL ethyl acetate containing 20-50 equivalents 4-[2-(1-pyrenyl)ethyl]-1,2,4-triazoline-3,5-dione at 0°, after 5 min inject an aliquot. (Synthesis of 4-[2-(1-pyrenyl)ethyl]-1,2,4-triazoline-3,5-dione is as follows. Add 400 μL phosphorus trichloride to 2.3 g 1-pyrenemethanol in 20 mL dry benzene (Caution! Benzene is a carcinogen!), stir for 1 h, filter, wash the filtrate with 5% sodium bicarbonate, wash the filtrate with water. Dry the organic layer over anhydrous magnesium sulfate, evaporate to dryness to obtain 3-chloromethylpyrene as cream-colored crystals (mp 144-145°). Reflux 1.5 g finely-powdered sodium with 15.4 mL diethyl malonate in 100 mL benzene for 3 h, cool, add 8 g 3-chloromethylpyrene, warm slowly to boiling with stirring, reflux for 10 h, evaporate to dryness, add 50 mL 40% KOH in water, heat on a steam bath for 1 h, add 100 mL water, warm, filter, acidify, filter to obtain the substituted malonic acid. Heat the substituted malonic acid at 190-200° for 30 min, recrystallize from acetic acid/chlorobenzene to obtain 1-pyrenepropionic acid as pale tan platelets (mp 178-179°) (J. Am. Chem. Soc. 1941, 63, 2494). Treat 1-pyrenepropionic acid with 1 equivalent triethylamine and 1 equivalent diphenylphosphoryl azide in DMF to obtain pyrene-1-propionyl azide. Reflux the crude azide in benzene (Caution! Benzene is a carcinogen!) for 20 min to obtain pyrene-1-ethyl isocyanate. Add this isocyanate solution to 1 equivalent of ethyl carbazate in benzene cooled in ice, let stand at room temperature for 1 h, heat at 100° for 1 h, filter. Wash the precipitate with benzene and dry it under reduced pressure, recrystallize from DMF/water to obtain pyrenylethyl ethoxycarbonylsemicarbazide as a colorless amorphous solid (mp 214-215°). Suspend the semicarbazide in 4 M KOH, heat at 90° for 1.5 h, acidify with concentrated HCl, filter, wash the solid with water, dry under reduced pressure, purify by re-precipitation, recrystallize from water to obtain 4-[2-(1-pyrenyl)ethyl]-1,2-dihydro-1,2,4-triazoline-3,5-dione as a khaki amorphous substance (mp 238-240° (d)). Suspend in ethyl acetate, cool in ice, add 1 equivalent tert-butyl hypochlorite (Org. Syn.1973, Coll. Vol. V, 184) under nitrogen, stir at room temperature for 1 h, centrifuge to obtain an ethyl acetate solution of 4-[2-(1-pyrenyl)ethyl]-1,2,4-triazoline-3,5-dione.)

HPLC VARIABLES
Column: 150 × 4.6 5 μm YMC-Gel C8-120-S5
Mobile phase: MeCN
Flow rate: 1
Detector: UV 245; F ex 342 em 397

CHROMATOGRAM
Retention time: 4.9, 6.1
Limit of detection: 0.4 pmole (UV), 0.025 pmole (F)

REFERENCE
Shimada, K.; Oe, T.; Mizuguchi, T. Cookson-type reagents: highly sensitive derivatization reagents for conjugated dienes in high-performance liquid chromatography, *Analyst*, **1991**, *116*, 1393–1397.

4-(1-Pyrenyl)-1,2,4-triazoline-3,5-dione

RELATED REFERENCE
Shimada, K.; Oe, T. Dienophilic reagent for precolumn derivatization of 7-dehydrocholesterol in high-performance liquid chromatography. *Anal.Sci.* **1990**, *6*, 461-463.

SAMPLE
Matrix: solutions
Analyte: 7-dehydrocholesterol
Sample preparation: Mix 100 μL 10 μg/mL 7-dehydrocholesterol in ethyl acetate with 100 μL ethyl acetate containing 20-50 equivalents 4-(1-pyrenyl)-1,2,4-triazoline-3,5-dione at 0°, after 5 min inject an aliquot. (Synthesis of 4-(1-pyrenyl)-1,2,4-triazoline-3,5-dione is as follows. Treat pyrene-1-carboxylic acid with 1 equivalent triethylamine and 1 equivalent diphenylphosphoryl azide in DMF to obtain pyrene-1-carbonyl azide. Reflux the crude azide in toluene for 20 min to obtain 1-isocyanatopyrene. Add this isocyanate solution to 1 equivalent of ethyl carbazate in benzene cooled in ice (Caution! Benzene is a carcinogen!), let stand at room temperature for 1 h, heat at 100° for 1 h, filter. Wash the precipitate with benzene and dry it under reduced pressure, recrystallize from benzene to obtain

1-pyrene ethoxycarbonylsemicarbazide as a colorless amorphous solid (mp >300°). Suspend the semicarbazide in 4 M KOH, heat at 90° for 1.5 h, acidify with concentrated HCl, filter, wash the solid with water, dry under reduced pressure, purify by re-precipitation, recrystallize from DMF/water to obtain 4-(1-pyrenyl)-1,2-dihydro-1,2,4-triazoline-3,5-dione as a colorless amorphous substance (mp >300°). Suspend in ethyl acetate, cool in ice, add 1 equivalent tert-butyl hypochlorite (Org. Syn. 1973, Coll. Vol. V, 184) under nitrogen, stir at room temperature for 1 h, centrifuge to obtain an ethyl acetate solution of 4-(1-pyrenyl)-1,2,4-triazoline-3,5-dione.)

HPLC VARIABLES
Column: 150 × 4.6 5 μm YMC-Gel C8-120-S5
Mobile phase: MeCN
Flow rate: 1
Detector: UV 240; F ex 270 em 370

CHROMATOGRAM
Retention time: 4.4
Limit of detection: 0.12 pmole (UV), 2.4 pmole (F)

REFERENCE
Shimada, K.; Oe, T.; Mizuguchi, T. Cookson-type reagents: highly sensitive derivatization reagents for conjugated dienes in high-performance liquid chromatography, *Analyst*, **1991**, *116*, 1393–1397.

DIKETONE

BORON FLUORIDE

Boron Trifluoride

SAMPLE
Matrix: solutions
Analyte: β-diketones
Sample preparation: Mix 1 mL of a 100 mM solution in MeOH with 1 mL 4.7 M boron trifluoride in MeOH, make up to 10 mL with MeOH, heat at 65° for 20 min, inject an aliquot.

HPLC VARIABLES
Column: 150 × 4.6 Nucleosil 5C18
Mobile phase: MeCN:water 40:50
Flow rate: 1.5
Detector: UV 300

CHROMATOGRAM
Retention time: 5 (acetylacetone), 6 (3-methyl-2,4-pentanedione), 10 (1-phenyl-1,3-butanedione), 12 (3-phenyl-2,4-pentanedione), 23 (2,2,6,6-tetramethyl-3,5-heptanedione), 26 (1,3-diphenyl-1,3-propanedione)

REFERENCE
Moriyasu, M.; Endo, M.; Ichimaru, M.; Mizutani, T.; Kato, A. High-performance liquid chromatographic behavior of difluoroborane derivatives of β-diketones, *Anal.Sci.*, **1990**, *6*, 45–48.

DIAMINE

1,2-Diamino-4,5-dichlorobenzene

SAMPLE
Matrix: cereal
Analyte: moniloformin
Sample preparation: Condition a 3 mL 500 mg SAX SPE cartridge (International Sorbent Technology) with 3 mL MeOH, 3 mL water, and 3 mL 100 mM phosphoric acid. Stir 5 g

finely ground cereal with 50 mL MeCN:water 90:10 for 30 min, centrifuge a 10 mL aliquot, evaporate the supernatant to dryness under reduced pressure, dissolve the residue twice in 1 mL portions of MeOH and once in MeOH:water 50:50 with vortexing. Combine the extracts and shake them with 7 mL n-hexane, discard the n-hexane layer, add the MeOH/water layer to the SPE cartridge, wash with 3 mL MeOH:water 50:50, wash with 3 mL 100 mM phosphoric acid, wash with 3 mL water, elute with 2 mL 1 M HCl. Mix the eluate with 500 μL 1 mg/mL 1,2-diamino-4,5-dichlorobenzene in 1 M HCl, heat at 60° for 2 h, evaporate to dryness at 40° under a stream of nitrogen, reconstitute with 500 μL water, inject a 20 μL aliquot. (Purify 1,2-diamino-4,5-dichlorobenzene by recrystallization from HCl after treatment with charcoal.)

HPLC VARIABLES
Column: 150 × 4.6 5 μm ODS (Beckman)
Mobile phase: MecN:50 mM ammonium acetate 35:65, pH 7
Column temperature: 40
Flow rate: 0.7
Injection volume: 20
Detector: UV 330; F ex 330 em 440 following post-column reaction. The column effluent mixed with the reagent pumped at 0.5 mL/min and the mixture flowed through a 1 m × 0.25 mm ID coil to the detector. (Reagent was 100 mM glycine containing 100 mM NaCl:100 mM NaOH 39:61, pH 12.)

CHROMATOGRAM
Retention time: 15
Limit of quantitation: 20 ng/g (F)
Limit of detection: 12 ng/g (F), 24 ng/g (UV)

KEY WORDS
post-column reaction; wheat; maize; SPE

REFERENCE
Filek, G.; Lindner, W. Determination of the mycotoxin moniliformin in cereals by high-performance liquid chromatography and fluorescence detection, *J.Chromatogr.A*, **1996**, *732*, 291–298.

2,3-Diaminonaphthalene

SAMPLE
Matrix: butter
Analyte: diacetyl
Sample preparation: 20 g Butter + 50 mL water + 1 drop antifoam, mix, distil at 1 drop/s into an ice-cooled receiver containing 3 mL water, stop when 20 mL distillate is collected. Mix 8 mL distillate with 1 mL 100 μg/mL 2,3-diaminonaphthalene in 100 mM HCl, let stand in the dark for 15 min, inject a 10 μL aliquot.

HPLC VARIABLES
Column: 150 × 4 Micropak MCH-5 C18
Mobile phase: MeCN:water 47:53
Flow rate: 1
Injection volume: 10
Detector: F ex 355 em >420

CHROMATOGRAM
Retention time: 8.96
Limit of quantitation: 0.5 ppm

REFERENCE
Damiani, P.; Burini, G. Determination of diacetyl in butter as 2,3-diaminonaphthalene derivative, using a fluorometric procedure or reverse phase liquid chromatography with fluorescence detection, *J.Assoc.Off.Anal.Chem.*, **1988**, *71*, 462–465.

4,5-Dimethyl-1,2-phenylenediamine

SAMPLE
Matrix: blood, urine
Analyte: ascorbic and dehydroascorbic acid
Sample preparation: Plasma. 1 mL Plasma + 3 mL 10 g/L metaphosphoric acid, mix, centrifuge at 1500 g. Mix a 500 μL aliquot of the supernatant with 50 μL IS solution, inject a 20 μL aliquot. Urine. Dilute 1 mL urine with 3 mL 10 g/L metaphosphoric acid. Mix a 500 μL aliquot with 50 μL IS solution, inject a 20 μL aliquot. (IS solution was 100 mg/L D-isoascorbic acid in 10 g/L metaphosphoric acid containing 2 mM EDTA (nitrogen-saturated).)

HPLC VARIABLES
Guard column: 70 × 2 10 μm PRP-1 (Hamilton)
Column: 75 × 4.6 3 μm Ultrasphere ODS
Mobile phase: 0.15 mM Hexadecyltrimethylammonium bromide containing 0.5 mM sodium acetate and 0.15 mM disodium EDTA, adjusted to pH 4.0 with acetic acid. (Condition system with twenty 20 μL aliquots of 10 g/L metaphosphoric acid.)
Flow rate: 0.5
Injection volume: 20
Detector: F ex 365 em 440 following post-column reaction. The column effluent mixed with the reagent pumped at 1.5 mL/min and flowed through a 20 m × 0.55 mm ID PTFE coil at 65° then a 1.5 m × 0.55 mm ID PTFE coil at 20° to the detector. (Reagent was 2.5 mM cupric sulfate containing 52 mM citric acid and 0.5 mM 4,5-dimethyl-1,2-phenylenediamine dihydrochloride, adjusted to pH 4.1 with saturated NaOH. Prepare fresh each day. Prepare 4,5-dimethyl-1,2-phenylenediamine dihydrochloride by dissolving 4,5-dimethyl-1,2-phenylenediamine in a minimum volume of diethyl ether, pass anhydrous hydrogen chloride through the solution for 20 min, precipitate the salt with diethyl ether. Wash the salt three times with diethyl ether.)

CHROMATOGRAM
Retention time: 4.60 (dehydroisoascorbic acid), 5.61 (dehydroascorbic acid), 15.50 (ascorbic acid)
Internal standard: isoascorbic acid (19.9)
Limit of detection: 4-10 ng

KEY WORDS
post-column reaction; plasma

REFERENCE
Lopez-Anaya, A.; Mayersohn, M. Ascorbic and dehydroascorbic acids simultaneously quantified in biological fluids by liquid chromatography with fluorescence detection, and comparison with a colorimetric assay, *Clin.Chem.*, **1987**, *33*, 1874–1878.

4-Ethoxy-1,2-phenylenediamine

SAMPLE

Matrix: milk, cheese
Analyte: ascorbic and dehydroascorbic acid
Sample preparation: Prepare a C18 SPE cartridge by adding a 1 mL suspension of 25-40 μm Lichroprep C18 (Merck) in EtOH to a filtration tube, allow to drain, wash twice with 1 mL portions of water. Prepare a cation-exchange cartridge by adding 1 mL of a suspension of 30% Aminex 50W-X2 (200-400 mesh, Na$^+$) (Bio-Rad) to a 1 mL filtration tube. Place the C18 cartridge on top of the cation-exchange cartridge. Homogenize 1 mL milk or 1 g cheese with 18 mL 0.5% oxalic acid, add 1 mL 70% perchloric acid, centrifuge at 1600 g. Remove a 500 μL aliquot of the supernatant and add it to 800 μL 1 M KH$_2$PO$_4$ and 500 μL 1 mg/mL 4-ethoxy-1,2-phenylenediamine in pH 2 phosphoric acid buffer, mix, let stand in the dark at room temperature for 45 min, add to the SPE assembly, elute with two 1 mL portions of water. Collect all the eluate, add 50 μL saturated aqueous bromine, remove excess bromine by passing nitrogen through the solution for 10 min, add 500 μL 1 mg/mL 4-methoxy-1,2-phenylenediamine in pH 2 phosphoric acid buffer, mix, let stand in the dark at room temperature for 45 min, pass through the previously used SPE assembly, wash with 1 mL water, elute with 1 mL MeCN:water:trifluoroacetic acid 15:84.5:0.5. Immediately neutralize the eluate with 5 μL 25% ammonia solution, inject a 5 μL aliquot. (Initially dehydroascorbic acid reacts with 4-ethoxy-1,2-phenylenediamine and ascorbic acid is not affected. This derivative remains in the SPE assembly. Ascorbic acid is then oxidized to dehydroascorbic acid with bromine and the dehydroascorbic acid is then derivatized with 4-methoxy-1,2-phenylenediamine. The reaction mixture is added to the SPE assembly and both derivatives are eluted. Thus the ethoxy derivative represents dehydroascorbic acid initially present and the methoxy derivative represents ascorbic acid initially present. Purify 4-methoxy-1,2-phenylenediamine as follows. Dissolve 3 g 4-methoxy-1,2-phenylenediamine dihydrochloride in 100 mL water, decolorize with Florisil, filter (0.45 μm), purify by preparative HPLC (250 × 5 7 μm C18 (Merck), gradient from EtOH:water:trifluoroacetic acid 0:100:0.5 to 60:40:0.5 over 1 h, UV 280), evaporate the appropriate fraction to dryness, dissolve the residue in water, freeze-dry, store under nitrogen. Prepare 4-ethoxy-1,2-phenylenediamine as follows. Dissolve 3 g 4-ethoxy-2-nitroaniline in 200 mL MeOH, hydrogenate over 250 mg palladium phthalocyanine (Merck), filter (0.45 μm), decolorize the filtrate with Florisil, filter (0.45 μm), evaporate to dryness, dissolve the residue in 100 mL water, purify by preparative HPLC (250 × 5 7 μm C18 (Merck), gradient from EtOH:water:trifluoroacetic acid 0:100:0.5 to 80:20:0.5 over 1 h, UV 280), evaporate the appropriate fraction to dryness, dissolve the residue in water, freeze-dry, store under nitrogen. (The appropriate preparative HPLC peak can be identified by adding 1 drop to 5 mL 100 mM disodium α-ketoglutarate in 1 M KH$_2$PO$_4$ and observing blue-green fluorescence under UV light.))

HPLC VARIABLES

Column: 150 × 4.1 5 μm PRP-1 polystyrene-divinylbenzene (Hamilton)
Mobile phase: MeCN:water 16:84 containing 50 mM KH$_2$PO$_4$ and 5 mM 1-propanesulfonic acid, pH adjusted to pH 9 with 150 mM phosphoric acid

Flow rate: 1
Injection volume: 5
Detector: F ex 375 em 475

CHROMATOGRAM
Retention time: 2.03 (ascorbic acid), 3.12 (dehydroascorbic acid)
Limit of detection: 50-70 fmole

KEY WORDS
SPE

REFERENCE
Bilic, N. Assay for both ascorbic and dehydroascorbic acid in dairy foods by high-performance liquid chromatography using precolumn derivatization with methoxy- and ethoxy-1,2-phenylenediamine, *J.Chromatogr.*, **1991**, *543*, 367–374.

4-Methoxy-1,2-phenylenediamine

SAMPLE
Matrix: milk, cheese
Analyte: ascorbic and dehydroascorbic acid
Sample preparation: Prepare a C18 SPE cartridge by adding a 1 mL suspension of 25-40 μm Lichroprep C18 (Merck) in EtOH to a filtration tube, allow to drain, wash twice with 1 mL portions of water. Prepare a cation-exchange cartridge by adding 1 mL of a suspension of 30% Aminex 50W-X2 (200-400 mesh, Na⁺) (Bio-Rad) to a 1 mL filtration tube. Place the C18 cartridge on top of the cation-exchange cartridge. Homogenize 1 mL milk or 1 g cheese with 18 mL 0.5% oxalic acid, add 1 mL 70% perchloric acid, centrifuge at 1600 g. Remove a 500 μL aliquot of the supernatant and add it to 800 μL 1 M KH$_2$PO$_4$ and 500 μL 1 mg/mL 4-ethoxy-1,2-phenylenediamine in pH 2 phosphoric acid buffer, mix, let stand in the dark at room temperature for 45 min, add to the SPE assembly, elute with two 1 mL portions of water. Collect all the eluate, add 50 μL saturated aqueous bromine, remove excess bromine by passing nitrogen through the solution for 10 min, add 500 μL 1 mg/mL 4-methoxy-1,2-phenylenediamine in pH 2 phosphoric acid buffer, mix, let stand in the dark at room temperature for 45 min, pass through the previously used SPE assembly, wash with 1 mL water, elute with 1 mL MeCN:water:trifluoroacetic acid 15:84.5:0.5. Immediately neutralize the eluate with 5 μL 25% ammonia solution, inject a 5 μL aliquot. (Initially dehydroascorbic acid reacts with 4-ethoxy-1,2-phenylenediamine and ascorbic acid is not affected. This derivative remains in the SPE assembly. Ascorbic acid is then oxidized to dehydroascorbic acid with bromine and the dehydroascorbic acid is then derivatized with 4-methoxy-1,2-phenylenediamine. The reaction mixture is added to the SPE assembly and both derivatives are eluted. Thus the ethoxy derivative represents dehydroascorbic acid initially present and the methoxy derivative represents ascorbic acid initially present. Purify 4-methoxy-1,2-phenylenediamine as follows. Dissolve 3 g 4-methoxy-1,2-phenylenediamine dihydrochloride in 100 mL water, decolorize with Florisil, filter (0.45 μm), purify by preparative HPLC (250 × 5 7 μm C18 (Merck), gradient

from EtOH:water:trifluoroacetic acid 0:100:0.5 to 60:40:0.5 over 1 h, UV 280), evaporate the appropriate fraction to dryness, dissolve the residue in water, freeze-dry, store under nitrogen. Prepare 4-ethoxy-1,2-phenylenediamine as follows. Dissolve 3 g 4-ethoxy-2-nitroaniline in 200 mL MeOH, hydrogenate over 250 mg palladium phthalocyanine (Merck), filter (0.45 μm), decolorize the filtrate with Florisil, filter (0.45 μm), evaporate to dryness, dissolve the residue in 100 mL water, purify by preparative HPLC (250 × 5 7 μm C18 (Merck), gradient from EtOH:water:trifluoroacetic acid 0:100:0.5 to 80:20:0.5 over 1 h, UV 280), evaporate the appropriate fraction to dryness, dissolve the residue in water, freeze-dry, store under nitrogen. (The appropriate preparative HPLC peak can be identified by adding 1 drop to 5 mL 100 mM disodium α-ketoglutarate in 1 M KH_2PO_4 and observing blue-green fluorescence under UV light.))

HPLC VARIABLES
Column: 150 × 4.1 5 μm PRP-1 polystyrene-divinylbenzene (Hamilton)
Mobile phase: MeCN:water 16:84 containing 50 mM KH_2PO_4 and 5 mM 1-propanesulfonic acid, pH adjusted to pH 9 with 150 mM phosphoric acid
Flow rate: 1
Injection volume: 5
Detector: F ex 375 em 475

CHROMATOGRAM
Retention time: 2.03 (ascorbic acid), 3.12 (dehydroascorbic acid)
Limit of detection: 50-70 fmole

KEY WORDS
SPE

REFERENCE
Bilic, N. Assay for both ascorbic and dehydroascorbic acid in dairy foods by high-performance liquid chromatography using precolumn derivatization with methoxy- and ethoxy-1,2-phenylenediamine, *J.Chromatogr.*, **1991**, *543*, 367–374.

o-Phenylenediamine

RELATED REFERENCE
Koshiishi, I.; Imanari, T. Measurement of ascorbate and dehydroascorbate contents in biological fluids. *Anal.Chem.* **1997**, *69*, 216-220.

SAMPLE
Matrix: beverages
Analyte: dehydroascorbic acid
Sample preparation: 70 mL Orange juice + 200 mg 1,2-phenylenediamine dihydrochloride, shake, make up to 100 mL with MeOH, centrifuge for 5 min, filter (Whatman 31 paper), let stand for 1 h, filter (0.45 μm) an aliquot, inject a 10 μL aliquot of this filtrate.

HPLC VARIABLES
Column: 300 × 3.9 μBondapak C18
Mobile phase: MeOH:water 60:40 containing 250 μM hexadecyltrimethylammonium bromide
Flow rate: 2
Injection volume: 10
Detector: UV 348

CHROMATOGRAM
Retention time: 5

KEY WORDS
ascorbic acid, which is not derivatized under these conditions, is detected at 2 min with UV 290; protect from light; orange juice

REFERENCE
Keating, R.W.; Haddad, P.R. Simultaneous determination of ascorbic acid and dehydroascorbic acid by reversed-phase ion-pair high-performance liquid chromatography with pre-column derivatization, *J.Chromatogr.*, **1982**, *245*, 249–255.

SAMPLE
Matrix: beverages, vegetables
Analyte: ascorbic acid and dehydroascorbic acid
Sample preparation: 500 µL Juice or homogenized vegetables + 5 mg pyrogallol + 10 mL 100 mM citric acid, vortex under nitrogen for 1 min, add an equal volume of dichloromethane, vortex for 1 min, centrifuge at 4° at 1200 g for 10 min, repeat dichloromethane wash (if necessary to remove excess fat). Filter (0.45 µm) the aqueous layer, pass a 2 mL aliquot through a conditioned Sep-Pak C18 (?) SPE cartridge, inject an aliquot of the eluate.

HPLC VARIABLES
Column: 150 × 5 DA-X8-11 anion-exchange resin (Dionex)
Mobile phase: 100 mM pH 3.8 Citrate buffer containing 10 mM NaCl and 5 mM EDTA
Flow rate: 0.5
Injection volume: 100
Detector: F ex Corning 7-60 filter em Wratten 2-E following post-column reaction. The column effluent mixed with the oxidizer pumped at 0.5 mL/min and this mixture flowed through a 32 cm × 0.25 mm ID stainless-steel coil. The effluent from this coil mixed with the reagent pumped at 0.5 mL/min and this mixture flowed through a 45.7 m × 0.25 mm ID stainless-steel coil at 70° then a 1.5 m × 0.25 mm ID stainless-steel coil at 20° to the detector. (Oxidizer was 2.5 mM mercuric chloride or copper sulfate in mobile phase. Reagent was 3.1 mM o-phenylenediamine in mobile phase.)

CHROMATOGRAM
Retention time: 12 (dehydroascorbic acid), 23 (ascorbic acid)
Limit of detection: 10-20 ng

KEY WORDS
post-column reaction; bean sprouts; beets; broccoli; grape juice; orange juice; potatoes; tomatoes

REFERENCE
Vanderslice, J.T.; Higgs, D.J. HPLC analysis with fluorometric detection of vitamin C in food samples, *J.Chromatogr.Sci.*, **1984**, *22*, 485–489.

SAMPLE
Matrix: beverages, vegetables
Analyte: vitamin C
Sample preparation: Blend 5 g food with 25 mL 300 mM trichloroacetic acid for 1 min, make up to 50 mL with 300 mM trichloroacetic acid. Dilute 5 g beverage to 50 mL with 300 mM trichloroacetic acid. Filter (paper) these solutions, dilute with 300 mM trichloroacetic acid to a vitamin concentration of 1-40 µg/mL. Remove a 3 mL aliquot and add it to 400 µL 4.5 M pH 6.2 sodium acetate buffer, add an ascorbate oxidase spatula (Boehringer Mannheim), heat at 37° for 2 min, mix, heat at 37° for 3 min, remove the spatula (?), add 500 µL 0.1% o-phenylenediamine (freshly prepared), mix, heat at 37° in the dark for 30 min, inject a 30 µL aliquot.

HPLC VARIABLES
Guard column: 20 mm long RP-18 (Bischoff)
Column: 125 × 4.6 3 µm ODS-Hypersil
Mobile phase: MeOH:80 mM KH_2PO_4 20:80, pH 7.8

Flow rate: 1
Injection volume: 30
Detector: F ex 365 (filter) em 418 (filter)

CHROMATOGRAM
Retention time: 8 (vitamin C), 10 (isovitamin C)
Limit of detection: 200 ng/g

KEY WORDS
protect from light; avocado; Brussels sprouts; cabbage; cauliflower; kale; lemon juice; lettuce; orange juice; paprika; parsley; peas

REFERENCE
Speek, A.J.; Schrijver, J.; Schreuers, W.H.P. Fluorometric determination of total vitamin C and total isovitamin C in foodstuffs and beverages by high-performance liquid chromatography with precolumn derivatization, *J.Agric.Food Chem.*, **1984**, *32*, 352–355.

SAMPLE
Matrix: blood
Analyte: dehydroascorbic acid
Sample preparation: 500 μL Plasma + 50 μL 500 μg/mL o-phenylenediamine, mix, let stand for 45 min, inject a 10 μL aliquot.

HPLC VARIABLES
Column: μBondapak C18
Mobile phase: MeOH:buffer 10:90 (Prepare mobile phase by mixing 6.6 g K_2HPO_4, 8.4 g KH_2PO_4, 400 mL MeOH, and 3.6 L water.)
Flow rate: 1.5
Injection volume: 10
Detector: UV 254

CHROMATOGRAM
Retention time: 8.7
Limit of detection: 5 μg/mL

KEY WORDS
plasma

REFERENCE
Baker, J.K.; Kapeghian, J.; Verlangieri, A. Determination of ascorbic acid and dehydroascorbic acid in blood plasma samples, *J.Liq.Chromatogr.*, **1983**, *6*, 1319–1332.

SAMPLE
Matrix: blood
Analyte: vitamin C
Sample preparation: Collect 5 mL whole blood in a tube with 100 μL glutathione solution. Remove a 1 mL aliquot and add it to 4 mL 300 mM trichloroacetic acid, vortex thoroughly for 10 min, let stand in the dark at 4° for 10 min, mix, let stand in the dark at 4° for 10 min, centrifuge at 4° at 2000 g for 10 min. Remove a 1.5 mL aliquot of the supernatant and add it to 200 μL 4.5 M pH 6.2 sodium acetate buffer, add an ascorbate oxidase spatula (Boehringer Mannheim), heat at 37° for 2 min, mix, heat at 37° for 3 min, remove the spatula, add 250 μL 0.1% o-phenylenediamine (freshly prepared), mix, heat at 37° in the dark for 30 min, inject a 20 μL aliquot. (Prepare glutathione solution by dissolving 1.5 g glutathione in 25 mL water, adjust pH to 6.5 with 2M NaOH, add 2.25 g ethyleneglycolbis-(β-aminoethyl ether)-N,N,N',N'-tetraacetic acid (EGTA), adjust pH to 6.5 with 10 M NaOH.)

HPLC VARIABLES
Column: 80 × 4.6 3 μm ODS-Hypersil
Mobile phase: MeOH:80 mM KH_2PO_4 20:80, pH 7.8
Flow rate: 1
Injection volume: 20
Detector: F ex 365 (filter) em 418 (filter)

CHROMATOGRAM
Retention time: 3
Limit of detection: 200 nM

KEY WORDS
protect from light; whole blood

REFERENCE
Speek, A.J.; Schrijver, J.; Schreurs, W.H. Fluorometric determination of total vitamin C in whole blood by high-performance liquid chromatography with pre-column derivatization, *J.Chromatogr.*, **1984**, *305*, 53–60.

HYDRAZINE

Luminarin 11

SAMPLE
Matrix: solutions
Analyte: acetylacetone
Sample preparation: Mix 200 μL of a 100 mM solution in water with 10 μL 50 mM luminarin 11 in DMSO, let stand at room temperature for 2 h, carefully add 100 μL 600 mM sodium bicarbonate to adjust pH to 7.0, agitate gently until gas evolution ceases, add 2 mL dichloromethane, vortex for 2 min, centrifuge. Remove the lower organic layer and evaporate it to dryness under a stream of nitrogen at room temperature, reconstitute the residue in MeCN, inject an aliquot. (Luminarin 11, 2,3,6,7-tetrahydro-11-oxo-1H,5H,11H-

[1]benzopyrano[6,7,8-ij]quinolizine-9-pentanoic acid hydrazide, is available from Eurobio, Les Ulis, France.)

HPLC VARIABLES
Column: 150 × 4.6 5 μm Nucleosil ODS
Mobile phase: MeCN:10 mM pH 7.5 imidazole nitrate buffer 60:40
Flow rate: 1-2
Injection volume: 20
Detector: F ex 387 em 450

CHROMATOGRAM
Retention time: 12.5
Limit of detection: 114 fmole

OTHER SUBSTANCES
Simultaneously analyzed: malonaldehyde

KEY WORDS
comparison with other luminarins

REFERENCE
Traoré, F.; Pianetti, G.A.; Dallery, L.; Tod, M.; Chalom, J.; Farinotti, R.; Mahuzier, G. Determination of picomole amounts of carbonyls as luminarin hydrazones by high-performance liquid chromatography with fluorescence detection, *Chromatographia*, **1993**, *36*, 96–104.

Luminarin 12

SAMPLE
Matrix: solutions
Analyte: acetylacetone

Sample preparation: Mix 200 μL of a 100 mM solution in water with 10 μL 50 mM luminarin 12 in DMSO, let stand at room temperature for 1 h, carefully add 100 μL 600 mM sodium bicarbonate to adjust pH to 7.0, agitate gently until gas evolution ceases, add 2 mL dichloromethane, vortex for 2 min, centrifuge. Remove the lower organic layer and evaporate it to dryness under a stream of nitrogen at room temperature, reconstitute the residue in MeCN, inject an aliquot. (Luminarin 12, 2,3,6,7-tetrahydro-11-oxo-1H,5H,11H-[1]benzopyrano[6,7,8-ij]quinolizine-9-methoxyethoxyethoxyacetic acid hydrazide, is available from Eurobio, Les Ulis, France.)

HPLC VARIABLES
Column: 250 × 4.6 5 μm Nucleosil ODS
Mobile phase: MeCN:10 mM pH 7.5 imidazole nitrate buffer 52:48
Flow rate: 1-2
Injection volume: 20
Detector: F ex 399 em 500

CHROMATOGRAM
Retention time: 21
Limit of detection: 340 fmole

OTHER SUBSTANCES
Simultaneously analyzed: malonaldehyde

KEY WORDS
comparison with other luminarins

REFERENCE
Traoré, F.; Pianetti, G.A.; Dallery, L.; Tod, M.; Chalom, J.; Farinotti, R.; Mahuzier, G. Determination of picomole amounts of carbonyls as luminarin hydrazones by high-performance liquid chromatography with fluorescence detection, *Chromatographia*, **1993**, *36*, 96–104.

DIOL

BORONIC ACID

3-Dansylaminophenylboronic Acid

SAMPLE
Matrix: bulk
Analyte: brassinosteroids
Sample preparation: Dissolve brassinosteroid sample in 100 μL 10 mg/mL 3-dansylaminophenylboronic acid in MeCN:pyridine 99.9:0.1, heat at 70° for 10 min, cool, inject an aliquot. (Synthesis of 3-dansylaminophenylboronic acid is as follows. Stir a solution of 320 mg 3-aminophenylboronic acid in 20 mL acetone, add 10 mL 50 mM potassium carbonate in water, add 20 mL 27 mg/mL dansyl chloride in acetone, stir at room temperature for 20 h, extract with diethyl ether. Dry the organic layer and evaporate it to give an oily residue. Chromatograph on a short column of silica gel with hexane:ethyl acetate 25:75 to give 3-dansylaminophenylboronic acid as an amorphous solid.)

HPLC VARIABLES
Column: 150 × 4.6 Shim-pack CLC-ODS(M) (Shimadzu)
Mobile phase: MeCN:water 80:20
Column temperature: 45
Flow rate: 1
Detector: F ex 345 em 515

CHROMATOGRAM
Retention time: 8 (brassinolide), 8.5 (dolichosterone), 9 (norcastasterone), 10 (homobrassinolide), 11 (castasterone), 14 (homocastasterone)
Limit of detection: 25 pg

REFERENCE

Gamoh, K.; Okamoto, N.; Takatsuto, S.; Tejima, I. Determination of traces of natural brassinosteroids as dansylaminophenylboronates by liquid chromatography with fluorometric detection, *Anal.Chim.Acta*, **1990**, *228*, 101–105.

Ferroceneboronic Acid

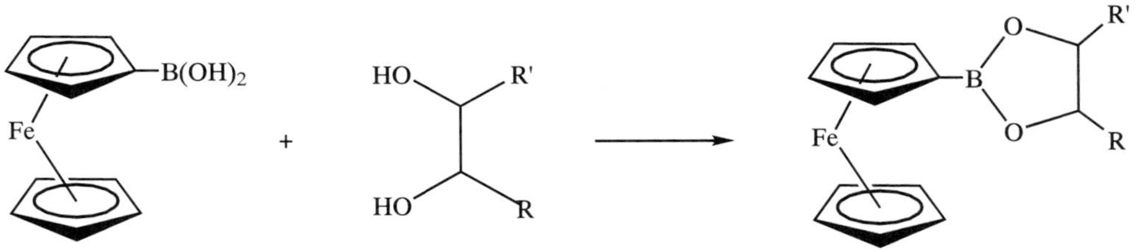

SAMPLE

Matrix: solutions

Analyte: brassinosteroids

Sample preparation: Mix 100 μL of a solution in MeCN with 100 μL 1 mg/mL ferroceneboronic acid in MeCN:pyridine 99:1, heat at 70° for 10 min, cool, inject an aliquot. (Preparation of ferroceneboronic acid is as follows. Add 270 mmoles butyllithium to 16.7 g ferrocene in 220 mL THF:diethyl ether 50:50. Filter the lithium ferrocene so formed through glass wool and add it dropwise over 2 h to 72.5 g tributyl borate in 50 mL diethyl ether at -70°, warm to room temperature, stir for 1.5 h, add 100 mL 10% NaOH, filter. Extract the organic layer 9 times with 45 mL portions of 10% NaOH. Combine all the aqueous layers and acidify them with 10% sulfuric acid at 0°. Collect the precipitate and Soxhlet extract it with diethyl ether for 4 days, evaporate the diethyl ether to obtain ferrocenylboronic acid as a yellow powder (mp 136-140° d) (J. Org. Chem. 1961, 26, 1034).)

HPLC VARIABLES

Column: 250 × 4.6 5 μm Shim-pack CLC-ODS(M)

Mobile phase: MeCN:water 85:15 containing 1 M sodium perchlorate

Column temperature: 45

Flow rate: 1

Detector: E, Shimadzu L-ECD-6A, glassy carbon working electrode +0.6 V, Ag/AgCl reference electrode

CHROMATOGRAM

Retention time: 11 (norbrassinolide), 14 (brassinolide), 15 (dolichosterone), 15.5 (norcastasterone), 18.5 (homobrassinolide), 20 (castasterone), 26 (homocastasterone)

Limit of quantitation: 50 pg

Limit of detection: 25 pg

REFERENCE

Gamoh, K.; Sawamoto, H.; Kakatsuto, S.; Watabe, Y.; Arimoto, H. Ferroceneboronic acid as a derivatization reagent for the determination of brassinosteroids by high-performance liquid chromatography with electrochemical detection, *J.Chromatogr.*, **1990**, *515*, 227–231.

9-Phenanthreneboronic Acid

SAMPLE
Matrix: pollen
Analyte: brassinosteroids
Sample preparation: Extract 25 g broad bean pollen with 200 mL MeOH for 1 week and with 200 mL MeOH:ethyl acetate 50:50 for 1 week. Combine the extracts and concentrate them under reduced pressure below 30°, partition twice between 100 mL portions of ethyl acetate and 100 mL water. Concentrate the organic layer and partition it twice between 100 mL portions of MeOH:water 90:10 and 100 mL n-hexane. Concentrate the MeOH/water layer and partition it between 100 mL ethyl acetate and 100 mL saturated sodium bicarbonate solution. Dry the organic layer over anhydrous magnesium sulfate and concentrate under reduced pressure to give an oil. Chromatograph the oil on a 200 × 10 column of 70-230 mesh Kieselgel 60 (Merck) with 50 mL chloroform, 50 mL chloroform:MeOH 98:2, 50 mL chloroform:MeOH 95:5, 50 mL chloroform:MeOH 90:10, 50 mL chloroform:MeOH 85:15, and 50 mL chloroform:MeOH 80:20, identify the active fraction by bioassay. Evaporate to dryness, take up a portion in 100 μL MeCN, add 100 μL 1 mg/mL 9-phenanthreneboronic acid in MeCN:pyridine 99:1, heat at 70° for 10 min, cool, inject an aliquot. (9-Phenanthreneboronic acid is available from Tokyo Kasei (TCI America, Portland OR). Synthesis is as follows. Perform all reactions under nitrogen. Add a small amount of bromoethane to 100 mmoles magnesium turnings in anhydrous ether, when the reaction starts add 100 mmoles 9-bromophenanthrene in 80 mL anhydrous ether dropwise, reflux until formation of the phenanthrene Grignard reagent is complete (method of entrainment). Stir 100 mL ether in a dry ice/acetone slush bath, add the Grignard reagent and 100 mmoles trimethylborate in 100 mL ether at equal rates over 30 min, stir for 30 min, lower the cooing bath until it is just under the flask, stir overnight, add 50 mL 15% HCl. Remove the ether layer and extract it with NaOH solution. Remove the aqueous layer and add ice, acidify with HCl, extract with ether (J. Chromatogr. 1978, 158, 33), chromatograph on silica gel deactivated with 20% water using hexane:ethyl acetate 50:50 to obtain 9-phenanthreneboronic acid (mp 319°) (J. High Res. Chromatogr. Chromatogr. Commun. 1978, 1, 96).)

HPLC VARIABLES
Column: 150 × 4 5 μm STR ODS-H (Shimadzu)
Mobile phase: MeCN:water 90:10
Column temperature: 45
Flow rate: 0.8
Detector: F ex 305 em 375

CHROMATOGRAM
Retention time: 7.2 (norbrassinolide), 8.6 (brassinolide), 9.1 (dolichosterone), 10.0 (norcastasterone), 10.6 (homobrassinolide), 12 (castasterone), 15.1 (homocastasterone)
Limit of detection: 50 pg

REFERENCE

Gamoh, K.; Omote, K.; Okamoto, N.; Takatsuto, S. High-performance liquid chromatography of brassinosteroids in plants with derivatization using 9-phenanthreneboronic acid, *J.Chromatogr.*, **1989**, *469*, 424–428.

Phenylboric Acid

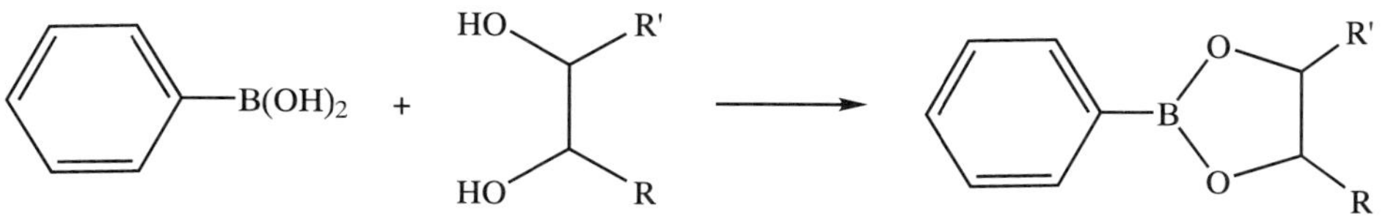

SAMPLE

Matrix: solutions

Analyte: ecdysteroids

Sample preparation: Mix a solution of ecdysteroids in mobile phase with 10 equivalents of phenylboric acid (phenylboronic acid) dissolved in dichloromethane, shake, let stand for 5 min, inject an aliquot. (Only ecdysteroids with diol sidechains react under these conditions.)

HPLC VARIABLES

Column: 250 × 4 7 μm Separon SGX silica gel

Mobile phase: Dichloromethane:MeOH:water 88:11:1

Flow rate: 1.5

Detector: UV (wavelength not specified)

CHROMATOGRAM

Retention time: 2 (20-hydroxyecdysterone-20,22-acetonide), 3 (ponasterone A), 3.5 polypodine B), 5 (ecdysone)

REFERENCE

Pis, J.; Harmatha, J. Phenylboronic acid as a versatile derivatization agent for chromatography of ecdysteroids, *J.Chromatogr.*, **1992**, *596*, 271–275.

DIAMINE

1,2-Bis(4-methoxyphenyl)ethylenediamine

RELATED REFERENCE
Umegae, Y.; Nohta, H.; Ohkura, Y. High-performance liquid chromatographic determination of ribonu-
cleotides by postcolumn derivatization involving oxidation followed by fluorescence reaction, and its
application to human erythrocyte sample. *Anal.Sci.* **1990**, *6*, 519-522

SAMPLE
Matrix: blood, urine
Analyte: catecholamines
Sample preparation: Plasma. Condition a 19-40 μm Toyopak IC-SP S strong cation-exchanger sulfopropyl resin (Na$^+$ form) SPE cartridge with 200 mM pH 5.8 lithium phosphate buffer. 500 μL Plasma + 25 μL 10 nM isoproterenol + 500 μL 200 mM pH 5.8
lithium phosphate buffer, add to the SPE cartridge, wash twice with 5 mL portions of
water, wash with 1 mL MeCN:water 50:50, elute with 500 μL MeCN:600 mM KCl 50:
50 containing 0.6 mM potassium ferricyanide. Add 50 μL 100 mM meso-1,2-bis(4-meth-
oxyphenyl)ethylenediamine to the eluate, heat at 37° for 40 min, inject a 50 μL aliquot.
Urine. 10 μL Urine + 10 μL 500 nM isoproterenol + 500 μL 10 mM meso-1,2-bis(4-
methoxyphenyl)ethylenediamine + 500 μL MeCN:600mM KCl 30:70 + 10 μL 60 mM
potassium ferricyanide, heat at 37° for 45 min, inject a 50 μL aliquot. (Prepare meso-1,2-
bis(4-methoxyphenyl)ethylenediamine as follows. Heat 105 g benzil and 122 g salicaldehyde in 750 mL EtOH at 60° until they dissolve, introduce a weak stream of ammonia
with stirring over 3 h, cool, filter, wash the precipitate with EtOH, dry under vacuum at
80°. Suspend 192 g of this product in 500 mL acetic anhydride, reflux for 14 h using a
constant temperature bath at 148-150°, the product (O,O',N,N'-tetraacetyl-meso-1,2-bis(2-
hydroxyphenyl)ethylenediamine) crystallizes on slow cooling, filter, wash the solid with a
little acetic anhydride. Suspend the product in 250 mL 42% hydrobromic acid:acetic acid
50:50 and reflux for 3 h. Filter off the product that separates on cooling and dissolve
it in 400 mL hot water, neutralize with 20% NaOH. Filter off the product that separates and recrystallize it from MeCN to give meso-1,2-bis(2-hydroxyphenyl)ethylenedia-
mine as a colorless powder (mp 184-186°). Mix 2.44 g meso-1,2-bis(2-hydroxy-
phenyl)ethylenediamine with 20 mmole p-anisaldehyde in 100 mL MeCN, reflux until the
reaction is complete (at least 1 h), reduce the reaction mixture to half its volume by
distillation, cool, filter. Recrystallize the product from toluene to give N,N'-disalicylidene-
meso-1,2-bis(4-methoxyphenyl)ethylenediamine (mp 202-207°) (formed via a diaza-Cope
rearrangement). Suspend 5 mmole N,N'-disalicylidene-meso-1,2-bis(4-methoxyphe-
nyl)ethylenediamine in 50 mL 2 M sulfuric acid and steam distil until no more salicaldehyde comes over, filter the reaction mixture while it is hot, make the filtrate strongly
basic (pH 11) with 20% NaOH. Filter off the crystalline product and recrystallize it from
MeCN to give meso-1,2-bis(4-methoxyphenyl)ethylenediamine (mp 151-152°) (Chem. Ber
1976, 109, 1).)

HPLC VARIABLES
Column: 150 × 4.6 TSK-gel ODS-120T (Tosoh)
Mobile phase: MeCN:MeOH:50 mM pH 7.0 Tris-HCl buffer 50:10:40
Flow rate: 1
Injection volume: 50
Detector: F ex 350 em 460 (norepinephrine); F ex 360 em 470 (epinephrine); F ex 350 em 470 (dopamine)

CHROMATOGRAM
Retention time: 3.1 (norepinephrine), 4.9 (epinephrine), 6.7 (dopamine)
Internal standard: isoproterenol (F ex 365 em 470) (7.5)
Limit of detection: 1-2 nM (urine), 10-20 pM (plasma)

KEY WORDS
plasma; SPE

REFERENCE
Umegae, Y.; Nohta, H.; Lee, M.; Ohkura, Y. 1,2-Diarylethylenediamines as pre-column fluorescence derivatization reagents in high-performance liquid chromatographic determination of catecholamines in urine and plasma, *Chem.Pharm.Bull.*, **1990**, *38*, 2293–2295.

SAMPLE
Matrix: blood, urine
Analyte: pseudouridine
Sample preparation: Serum. 500 μL Serum + 500 μL 50 μM 5-fluorouridine + 500 μL 2 M perchloric acid, centrifuge at 4° at 1000 g for 10 min. Remove a 500 μL aliquot of the supernatant and add it to 65 μL 2 M potassium carbonate, centrifuge briefly, inject a 100 μL aliquot of the supernatant. Urine. 100 μL Urine + 100 μL 500 μM 5-fluorouridine + 800 μL water, mix, inject a 100 μL aliquot.

HPLC VARIABLES
Column: 150 × 4.6 5 μm TSK gel ODS-80TM (Tosoh)
Mobile phase: MeOH:10 mM pH 5.0 sodium phosphate buffer 2:98 (serum) or 4:96 (urine)
Flow rate: 0.5
Injection volume: 100
Detector: F ex 340 em 470 following post-column reaction. The column effluent mixed with 2 mM sodium periodate in water pumped at 0.25 mL/min and this mixture flowed through a 1 m × 0.5 mm ID stainless steel coil. The effluent from this coil mixed with the reagent pumped at 0.25 mL/min and this mixture flowed through a 20 m × 0.5 mm ID stainless steel coil at 140° to the detector. (Reagent was 20 mM 1,2-bis(4-methoxyphenyl) ethylenediamine (p-MOED) in EtOH:water 68:32 containing 140 mM perchloric acid. Prepare meso-1,2-bis(4-methoxyphenyl)ethylenediamine as described above.)

CHROMATOGRAM
Retention time: 10 (serum), 12 (urine)
Internal standard: 5-fluorouridine (21.6 (urine), 23 (serum))
Limit of detection: 8 nM (serum), 40 nM (urine)

KEY WORDS
post-column reaction; serum

REFERENCE
Umegae, Y.; Nohta, H.; Ohkura, Y. Determination of pseudouridine in human urine and serum by high-performance liquid chromatography with post-column fluorescence derivatization, *J.Chromatogr.*, **1990**, *515*, 495–501.

1,2-Diphenylethylenediamine

RELATED REFERENCES
Husek, P.; Malikova, J.; Herzogova, G. Determination of plasma catecholamines via condensation with diphenylethylenediamine: simplification of the procedure. *J.Chromatogr.* **1990**, *533*, 166-170.

Jeon, H.-K.; Nohta, H.; Ohtsubo, K.; Ohkura, Y. Determination of L-threo-3,4-dihydroxyphenylserine and L-α-methyldopa in rat serum by high performance liquid chromatography with fluorescence detection. *Anal.Sci.* **1989**, *5*, 663-666.

Jeon, H.-K.; Nohta, H.; Ohkura, Y. High-performance liquid chromatographic determination of acidic catecholamine metabolites in human urine using postcolumn fluorescence derivatization with *dl*-1,2-diphenylethylenediamine. *Anal.Sci.* **1990**, *6*, 677-682.

Jeon, H.K.; Nohta, H.; Nagaoka, H.; Ohkura, Y. Simultaneous determination of catecholamine-related compounds by high performance liquid chromatography with postcolumn chemical oxidation followed by a fluorescence reaction. *Anal.Sci.* **1991**, *7*, 257-262.

Nohta, H.; Mitsui, A.; Ohkura, Y. High performance liquid chromatography of free catecholamines in urine using 1,2-diphenylethylenediamine as new precolumn fluorescent derivatization reagent. *Bunseki Kagaku* **1984**, *33*, E263-E269.

Nohta, H.; Jeon, H.-K.; Kai, M.; Ohkura, Y. Liquid chromatographic determination of indoleamine, catecholamines and their precursors and metabolites in rat brain tissues with postcolumn fluorescence detection. *Anal.Sci.* **1994**, *10*, 5-9.

Nohta, H.; Mitsui, A.; Umegae, Y.; Ohkura, Y. Determination of free and total catecholamines in human erythrocytes, platelets and plasma by high performance liquid chromatography with fluorescence detection. *Anal.Sci.* **1986**, *2*, 303-308.

Ragab, G.H.; Nohta, H.; Kai, M.; Ohkura, Y. Chemiluminescence determination of catecholamines in human blood plasma and urine using 1,2-diphenylethylenediamine as pre-column derivatization reagent in liquid chromatography. *Anal.Chim.Acta* **1994**, *298*, 431-438.

SAMPLE

Matrix: blood

Analyte: catecholamines

Sample preparation: Plasma. Prepare a SPE column by adding 500 μL of a 20% suspension of 19-40 μm Toyopak SP (strong cation-exchange sulfopropyl resin, Na$^+$ (Toyo Soda)) in water to a 35 × 6 column, wash with two 1 mL portions of 2 M LiOH, wash with two 5 mL portions of water, wash with two 1 mL portions of EtOH:12 M HCl 90:10, wash with two 5 mL portions of water, wash with three 1 mL portions of buffer. 500 μL Plasma + 25 μL 10 nM isoproterenol + 500 μL buffer, mix, add to the SPE column, wash with two 5 mL portions of water, wash with 1 mL MeCN:water 50:50, elute with 300 μL 600 μM potassium ferricyanide in 600 mM KCl:MeCN 50:50, add 50 μL reagent to the eluate, heat at 37° for 40 min, cool in ice-water, inject a 100 μL aliquot. Urine. 10 μL Urine + 1 mL MeCN:500 mM KCl 60:40 + 10 μL 500 nM isoproterenol + 10 μL 75 mM potassium hexacyanoferrate(III) + 100 μL reagent, heat at 37° for 40 min, inject a 100 μL aliquot (J. Chromatogr. 1986, 380, 229). (Prepare buffer by mixing 8 volumes 250 mM LiOH in 200 mM phosphoric acid with 1 volume 200 mM phosphoric acid, pH 5.8. Prepare reagent by dissolving 212 mg 1,2-diphenylethylenediamine in 10 mL 100 mM HCl, pH 6.7. Synthesis of meso-1,2-diphenylethylenediamine is as follows. Reflux 40 g ammonium acetate and 100 mL benzaldehyde for 3 h, cool, collect the precipitate, wash the precipitate with EtOH until it is white, recrystallize from n-butanol to obtain N-benzoyl-N'-benzylidene-meso-1,2-diphenylethylenediamine (mp 259°). Cautiously add 54 mL concentrated sulfuric acid to 100 mL water (Caution! Extremely exothermic!), cautiously add 10 g unpurified N-benzoyl-N'-benzylidene-meso-1,2-diphenylethylenediamine, heat, pass steam through the reaction mixture to remove benzaldehyde and benzoic acid, when the distillate is no longer acid (about 4 h) cool, filter. Cautiously neutralize the filtrate with concentrated ammonium hydroxide with ice-cooling, extract with ether, dry over solid KOH, evaporate to dryness, recrystallize from petroleum ether (bp 60-80°) to obtain meso-1,2-diphenylethylenediamine as white crystals (mp 120°) (J. Inorg. Nucl. Chem. 1965, 27, 270).)

HPLC VARIABLES

Column: 150 × 4.6 5 μm TSK-gel ODS-120T (Toyo Soda)
Mobile phase: MeCN:MeOH:50 mM pH 7.0 Tris-HCl buffer 50:10:40 (Wash with MeCN:
MeOH:water 50:10:40 for 15 min at the end of each day.)
Flow rate: 1
Injection volume: 100
Detector: F ex 345 em 485 (plasma); F ex 350 em 480 (urine)

CHROMATOGRAM

Retention time: 3 (norepinephrine), 5 (epinephrine), 6 (dopamine)
Internal standard: isoproterenol (8)
Limit of detection: 7-10 pM

KEY WORDS

plasma; SPE

REFERENCE

Mitsui, A.; Nohta, H.; Ohkura, Y. High-performance liquid chromatography of plasma catecholamines
using 1,2-diphenylethylenediamine as precolumn fluorescence derivatization reagent, *J.Chromatogr.*,
1985, *344*, 61–70.

SAMPLE

Matrix: blood
Analyte: catecholamines
Sample preparation: 1 mL Plasma + 250 μL 1 ng/mL α-methylnorepinephrine + 1 mL
buffer + 5 mL n-heptane containing 4.6 mM tetraoctylammonium bromide and 10 mL/L
1-octanol, shake for 2 min, centrifuge at 20° at 1000 g for 5 min, freeze in acetone/dry
ice. Remove the organic phase and add it to 2 mL 1-octanol and 200 μL 80 mM acetic
acid, shake, centrifuge at 20° at 1000 g for 5 min, freeze in acetone/dry ice. Discard the
organic phase, thaw the aqueous phase and add it to 1 mL 10 mM HCl, 1 mL buffer, and
5 mL n-heptane containing 4.6 mM tetraoctylammonium bromide and 10 mL/L 1-octanol,
shake for 2 min, centrifuge at 20° at 1000 g for 5 min, freeze in acetone/dry ice. Remove
the organic phase and add it to 2 mL 2 M pH 8.6 ammonia/ammonium chloride buffer
containing 13.4 mM EDTA, shake, freeze in dry ice/acetone. Remove the organic layer
and add it to 2 mL 1-octanol and 150 μL 80 mM acetic acid, shake, centrifuge at 20° at
1000 g for 5 min, freeze in dry ice/acetone, discard the organic layer. Thaw the aqueous
layer and add it to 250 μL MeCN, 50 μL 1.75 M pH 7.05 bicine, and 100 μL 100 mM 1,2-
diphenylethylenediamine in 100 mM HCl, add 20 μL 20 mM potassium ferricyanide in
water, heat at 37° in the dark for 1 h, keep at 20° in the dark, inject a 100 μL aliquot.
(Buffer was 2 M pH 8.6 ammonia/ammonium chloride buffer containing 8.9 mM diphenylborate-ethanolamine complex and 13.4 mM EDTA. Stir buffer with 45 g/L activated
alumina for 2 h before use. Wash 1-octanol with 80 mM acetic acid. Recrystallize 1,2-
diphenylethylenediamine from toluene:light petroleum (bp 60-80°) 10:90, dry overnight
at 60°. Synthesis of meso-1,2-diphenylethylenediamine is as described above.)

HPLC VARIABLES

Column: 100 × 4.6 3 μm Cp MicroSpher C18 (Chrompack)
Mobile phase: MeCN:MeOH:50 mM pH 7.0 sodium acetate buffer 40:8:50
Flow rate: 1
Injection volume: 100
Detector: F ex 350 em 480

CHROMATOGRAM

Retention time: 2 (norepinephrine), 4 (epinephrine), 5 (dopamine), 8 (isoproterenol)
Internal standard: α-methylnorepinephrine (3)
Limit of detection: 2-3 pg/mL

KEY WORDS

plasma; comparison with electrochemical detection

REFERENCE

van der Hoorn, F.A.J.; Boomsma, F.; Man in 't Veld, A.J.; Schalekamp, M.A.D.H. Determination of catecholamines in human plasma by high-performance liquid chromatography: comparison between a

new method with fluorescence detection and an established method with electrochemical detection, *J.Chromatogr.*, **1989**, *487*, 17–28.

SAMPLE
Matrix: blood
Analyte: catecholamines and dobutamine
Sample preparation: Plasma. 1 mL Plasma + 125 µL 4 ng/mL isoproterenol + 1 mL buffer + 5 mL n-heptane containing 4.6 mM tetraoctylammonium bromide and 10 mL/L 1-octanol, shake for 2 min, centrifuge at 1000 g for 5 min, freeze in dry ice/acetone. Remove the organic phase and add it to 2 mL 1-octanol (saturated with 80 mM acetic acid) and 200 µL 80 mM acetic acid, shake, centrifuge at 1000 g for 5 min. Freeze the aqueous layer and remove the organic layer. Add 1 mL 10 mM HCl, 1 mL buffer, and 5 mL n-heptane containing 4.6 mM tetraoctylammonium bromide and 10 mL/L 1-octanol to the aqueous phase. Shake, centrifuge, freeze, remove the organic layer and add it to 2 mL 2 M pH 8.6 ammonia-ammonium chloride buffer containing 13.4 mM EDTA (but no complex). Freeze, remove the organic layer and add it to 2 mL 1-octanol and 150 µL 80 mM acetic acid, shake, centrifuge at 1000 g for 5 min. Freeze, remove the organic layer and add the aqueous layer to 200 µL MeCN, 50 µL 1.75 M pH 6.95 bicine buffer containing 1% EDTA, 100 µL 100 mM 1,2-diphenylethylenediamine in 100 mM HCl, and 20 µL 20 mM potassium ferricyanide in water. Heat at 37° in the dark for 1 h, inject a 50 µL aliquot (keep it in the dark in the autosampler). Urine. 100 µL Urine + 1 mL 10 mM HCl + 125 µL 40 ng/mL isoproterenol + 1 mL buffer + 5 mL n-heptane containing 4.6 mM tetraoctylammonium bromide and 10 mL/L 1-octanol, shake for 2 min, centrifuge at 1000 g for 5 min, freeze in dry ice/acetone. Remove the organic phase and add it to 2 mL 1-octanol (saturated with 80 mM acetic acid) and 200 µL 80 mM acetic acid, shake, centrifuge at 1000 g for 5 min. Freeze the aqueous layer and remove the organic layer. Add 1 mL 10 mM HCl, 1 mL buffer, and 5 mL n-heptane containing 4.6 mM tetraoctylammonium bromide and 10 mL/L 1-octanol to the aqueous phase. Shake, centrifuge, freeze, remove the organic layer and add it to 2 mL 1-octanol and 150 µL 80 mM acetic acid, shake, centrifuge at 1000 g for 5 min. Freeze, remove the organic layer and add the aqueous layer to 200 µL MeCN, 50 µL 1.75 M pH 6.95 bicine buffer containing 1% EDTA, 100 µL 100 mM 1,2-diphenylethylenediamine in 100 mM HCl, and 20 µL 20 mM potassium ferricyanide in water. Heat at 37° in the dark for 1 h, inject a 20 µL aliquot (keep it in the dark in the autosampler). (Buffer was a 2 M pH 8.6 ammonia-ammonium chloride buffer containing 8.9 mM diphenyl borate-ethanolamine complex and 13.4 mM EDTA. Synthesis of meso-1,2-diphenylethylenediamine is as described above.)

HPLC VARIABLES
Column: 100 × 4.6 3 µm Spherisorb ODS2
Mobile phase: Gradient. A was MeCN:MeOH:50 mM pH 7.0 sodium acetate buffer 20:20:60. B was MeCN:MeOH:50 mM pH 7.0 sodium acetate buffer 60:10:30. A:B 52:48 for 6 min, go to 0:100 over 0.1 min, stay at 0:100 for another 10 min. Equilibrate at initial conditions for 4 min before next sample.
Flow rate: 1
Injection volume: 20-50
Detector: F ex 350 em 480

CHROMATOGRAM
Retention time: 3 (norepinephrine), 5.5 (epinephrine), 7.5 (dopamine), 12 (epinine), 16 (dobutamine)
Internal standard: isoproterenol (9)
Limit of detection: 8 pg/mL (plasma), 200 pg/mL (urine)

OTHER SUBSTANCES
Simultaneously analyzed: metabolites
Interfering: α-methyldopa

KEY WORDS
plasma

REFERENCE

Alberts, G.; Boomsma, F.; Man in 't Veld, A.J.; Schalekamp, M.A.D.H. Simultaneous determination of catecholamines and dobutamine in human plasma and urine by high-performance liquid chromatography with fluorimetric detection, *J.Chromatogr.*, **1992**, *583*, 236–240.

SAMPLE

Matrix: blood

Analyte: catecholamines and epinine

Sample preparation: Plasma. 1 mL Plasma + 125 μL 2 ng/mL α-methylnorepinephrine + 1 mL buffer + 5 mL n-heptane containing 4.6 mM tetraoctylammonium bromide and 10 mL/L 1-octanol, shake for 2 min, centrifuge at 1000 g for 5 min, freeze in dry ice/acetone. Remove the organic phase and add it to 2 mL 1-octanol (saturated with 80 mM acetic acid) and 200 μL 80 mM acetic acid, shake, centrifuge at 1000 g for 5 min. Freeze the aqueous layer and remove the organic layer. Add 1 mL 10 mM HCl, 1 mL buffer, and 5 mL n-heptane containing 4.6 mM tetraoctylammonium bromide and 10 mL/L 1-octanol to the aqueous phase. Shake, centrifuge, freeze, remove the organic layer and add it to 2 mL 2 M pH 8.6 ammonia-ammonium chloride buffer containing 13.4 mM EDTA (but no complex). Freeze, remove the organic layer and add it to 2 mL 1-octanol and 150 μL 80 mM acetic acid, shake, centrifuge at 1000 g for 5 min. Freeze, remove the organic layer and add the aqueous layer to 200 μL MeCN, 50 μL 1.75 M pH 6.95 bicine buffer containing 1% EDTA, 100 μL 100 mM 1,2-diphenylethylenediamine in 100 mM HCl, and 20 μL 20 mM potassium ferricyanide in water. Heat at 37° in the dark for 1 h, inject a 75 μL aliquot (keep it in the dark in the autosampler). Urine. 100 μL Urine + 1 mL 10 mM HCl + 125 μL 40 ng/mL α-methylnorepinephrine + 1 mL buffer + 5 mL n-heptane containing 4.6 mM tetraoctylammonium bromide and 10 mL/L 1-octanol, shake for 2 min, centrifuge at 1000 g for 5 min, freeze in dry ice/acetone. Remove the organic phase and add it to 2 mL 1-octanol (saturated with 80 mM acetic acid) and 200 μL 80 mM acetic acid, shake, centrifuge at 1000 g for 5 min. Freeze the aqueous layer and remove the organic layer. Add 1 mL 10 mM HCl, 1 mL buffer, and 5 mL n-heptane containing 4.6 mM tetraoctylammonium bromide and 10 mL/L 1-octanol to the aqueous phase. Shake, centrifuge, freeze, remove the organic layer and add it to 2 mL 1-octanol and 150 μL 80 mM acetic acid, shake, centrifuge at 1000 g for 5 min. Freeze, remove the organic layer and add the aqueous layer to 200 μL MeCN, 50 μL 1.75 M pH 6.95 bicine buffer containing 1% EDTA, 100 μL 100 mM 1,2-diphenylethylenediamine in 100 mM HCl, and 20 μL 20 mM potassium ferricyanide in water. Heat at 37° in the dark for 1 h, inject a 50 μL aliquot (keep it in the dark in the autosampler). (Buffer was a 2 M pH 8.6 ammonia-ammonium chloride buffer containing 8.9 mM diphenyl borate-ethanolamine complex and 13.4 mM EDTA. Synthesis of meso-1,2-diphenylethylenediamine is as described above.)

HPLC VARIABLES

Column: 100 × 4.6 3 μm PhaseSep C18 ODS2

Mobile phase: Gradient. A was MeCN:MeOH:50 mM pH 7.0 sodium acetate buffer 20:4: 76. B was MeCN:MeOH:50 mM pH 7.0 sodium acetate buffer 60:10:30. A:B 40:60 for 3 min, go to 0:100 over 0.5 min, stay at 0:100 for another 4.5 min. (After the last sample flush column with 60 mL MeCN:MeOH:water 70:10:20.)

Flow rate: 1

Injection volume: 50-75

Detector: F ex 350 em 480

CHROMATOGRAM

Retention time: 2 (norepinephrine), 3.5 (epinephrine), 5.5 (dopamine), 10 (epinine)

Internal standard: α-methylnorepinephrine (2.5)

Limit of detection: 0.3-0.6 pg

OTHER SUBSTANCES

Interfering: α-methyldopa

KEY WORDS

plasma

REFERENCE

Boomsma, F.; Alberts, G.; van der Hoorn, F.A.J.; Man in 't Veld, A.J.; Schalekamp, M.A.D.H. Simultaneous determination of free catecholamines and epinine and estimation of total epinine and dopamine in plasma and urine by high-performance liquid chromatography with fluorimetric detection, *J.Chromatogr.*, **1992**, *574*, 109–117.

SAMPLE

Matrix: blood
Analyte: isoproterenol
Sample preparation: Mix plasma and N-methyldopamine, add to a TOYOPAK SP strong cationic exchange SPE cartridge (Tosoh), elute with MeCN:600 mM KCl 50:50 containing 0.6 mM potassium hexacyanoferrate (III), derivatize eluate with 1,2-diphenylethylenediamine, inject an aliquot. (Synthesis of meso-1,2-diphenylethylenediamine is as described above.)

HPLC VARIABLES

Column: 150 × 4.6 Nucleosil 5C18
Mobile phase: MeOH:50 mM Tris-HCl buffer 80:20, adjusted to pH 7.0
Flow rate: 1
Detector: F

KEY WORDS

plasma; guinea pig; SPE; pharmacokinetics

REFERENCE

Ohtani, H.; Yamamoto, K.; Sawada, Y.; Iga, T. Antibronchospasmic, tachycardiac, and hypokalaemic effects of L-isoproterenol in guinea-pigs, *Biopharm.Drug Dispos.*, **1995**, *16*, 745–753.

SAMPLE

Matrix: blood, urine
Analyte: catecholamines
Sample preparation: Condition a 150 μL Toyopak IC-SP S (sulfopropyl resin, H⁺ form) SPE cartridge (Tosoh) with 10 mL water. Plasma. 700 μL Plasma + 50 μL 700 nM 3,4-dihydroxybenzylamine + 350 μL 2 M perchloric acid, mix, centrifuge at 4° at 1000 g for 15 min. Remove a 700 μL aliquot of the supernatant and add it to 30 μL 2 M potassium carbonate, centrifuge at 4° at 1000 g for 5 min, filter (0.2 μM) an aliquot of the supernatant, inject a 50 μL aliquot of the filtrate (A). Add a 500 μL aliquot of the supernatant to the SPE cartridge, wash with 1 mL water, wash with 500 μL EtOH:water 50:50, wash with 5 mL water, elute with 500 μL 2 M sodium perchlorate to give eluate B. Filter (0.2 μm) the eluate and inject a 50 μL aliquot of the filtrate. Urine. Acidify urine collected over 24 h with 10 mL 6 M HCl. 500 μL Urine + 25 μL 10 μM 3,4-dihydroxybenzylamine + 25 μL 40 μM ferulic acid + 500 μL 1 M perchloric acid, mix, centrifuge at 4° at 1000 g for 15 min. Remove a 700 μL aliquot of the supernatant and add it to 30 μL 2 M potassium carbonate, centrifuge at 4° at 1000 g for 5 min, add a 500 μL aliquot of the supernatant to the SPE cartridge, elute with 1.5 mL water, collect all the effluent as eluate A. Wash with 500 μL EtOH:water 50:50, wash with 5 mL water, elute with 500 μL 2 M sodium perchlorate to give eluate B. Filter (0.2 μm) each eluate and inject a 50 μL aliquot of the filtrate.

HPLC VARIABLES

Column: 150 × 4.6 5 μm TSK-gel ODS-80TM (Tosoh)
Mobile phase: Gradient. A was buffer. B was MeCN:MeOH:buffer 8:12:80, pH 3.1. A:B 100:0 for 4 min, to 60:40 over 8 min, to 0:100 over 2 min, maintain at 0:100 for 16 min, return to initial conditions (step gradient), re-equilibrate for 20 min. Buffer was 60 mM pH 3.1 citric acid containing 32 mM Na_2HPO_4, 1.7 mM sodium hexanesulfonate, and 0.1 mM disodium EDTA (J. Chromatogr. 1989, 467, 237).
Flow rate: 1
Injection volume: 50
Detector: F ex 345 em 480 following post-column reaction. The column effluent passed through a Hitachi 655A electrochemical detector with carbon cloth electrodes; working electrode at +0.68 V versus reference electrode (200 mM equimolar mixture of potassium hexacyanoferrate(II) and potassium hexacyanoferrate(III) containing 200 mM potassium

nitrate and 200 mM KOH). The effluent from the electrochemical detector mixed with 20 mM meso-1,2-diphenylethylenediamine in 50 mM HCl pumped at 0.4 mL/min and with 1 M glycine containing 490 mM KOH and 3 mM potassium hexacyanoferrate(III) pumped at 0.4 mL/min. This mixture flowed through a 10 m × 0.47 mm ID coil at 80° to the detector (J. Chromatogr. 1989, 467, 237). (Synthesis of meso-1,2-diphenylethylenediamine is as described above.)

CHROMATOGRAM

Retention time: 6 (norepinephrine (B)), 9.5 (levodopa (B)), 10.5 (epinephrine (B)), 18 (dopamine (B)), 18.5 (metanephrine (B)), 23 (3-methoxytyramine (B)), 4 (3,4-dihydroxymandelic acid (A)), 6 (3,4-dihydroxyphenylethyleneglycol (A)), 7.5 (vanillylmandelic acid (A)), 10 (normetanephrine (A)), 14.5 (4-hydroxy-3-methoxyphenylethanol (A)), 20 (3,4-dihydroxyphenylacetic acid (A)), 27 (homovanillic acid (A)) [A and B refer to fractions A and B in the sample preparation section.]

Internal standard: 3,4-dihydroxybenzylamine (12.5 (B)), ferulic acid (35 (A))

Limit of detection: 0.5-300 nM

KEY WORDS

post-column reaction; plasma; SPE

REFERENCE

Nohta, H.; Yamaguchi, E.; Ohkura, Y.; Watanabe, H. Measurement of catecholamines, their precursor and metabolites in human urine and plasma by solid-phase extraction followed by high-performance liquid chromatography with fluorescence derivatization, *J.Chromatogr.*, **1989**, *493*, 15–26.

SAMPLE

Matrix: blood, urine

Analyte: catecholamine metabolites

Sample preparation: Condition a Toyopak IC-SP S sulfopropyl resin, H$^+$ form, SPE cartridge (Tosoh) with 10 mL water and 2 mL 200 mM pH 5.0 sodium phosphate buffer. Plasma. 700 μL Plasma + 30 μL 700 nM isoproterenol + 50 μL 7 μM 3,4-dihydroxyphenylpropanoic acid + 350 μL 2 M perchloric acid, mix, centrifuge at 4° at 1000 g for 15 min. Remove a 700 μL aliquot of the supernatant and adjust the pH to 1.5-2.0 with about 150 μL 2 M potassium carbonate, centrifuge at 4° at 1000 g for 5 min, add the supernatant to the SPE cartridge, collect all the effluent, filter (cellulose acetate membrane), inject a 100 μL aliquot of the filtrate. Urine. Collect human urine for 24 h in the presence of 10 mL 6 M HCl. 500 μL Urine + 10 μL 15 μM isoproterenol +25 μL 800 μM 3,4-dihydroxyphenylpropanoic acid + 500 μL 1 M perchloric acid, mix, centrifuge at 4° at 1000 g for 15 min. Remove a 700 μL aliquot of the supernatant and adjust the pH to 1.5-2.0 with about 130 μL 2 M potassium carbonate, centrifuge at 4° at 1000 g for 5 min, add the supernatant to the SPE cartridge, elute with 1.5 mL water, collect all the effluent from the SPE cartridge, filter (cellulose acetate membrane), inject a 100 μL aliquot of the filtrate.

HPLC VARIABLES

Column: 250 × 4.6 5 μm TSK-gel ODS-80TM (Tosoh)

Mobile phase: MeOH:buffer 7:93 (Buffer was 30 mM pH 2.5 citrate buffer containing 0.4 mM sodium octanesulfonate.)

Flow rate: 0.8

Injection volume: 100

Detector: F ex 350 em 480 following post-column reaction. The column effluent mixed with reagent A pumped at 0.3 mL/min and the mixture flowed through a 3 m × 0.5 mm ID stainless steel coil at 90°. The effluent from this coil mixed with reagent B pumped at 0.3 mL/min and the mixture flowed through a 10 m × 0.5 mm ID stainless steel coil at 90° and through a 1 m × 0.5 mm ID stainless steel cooling coil to the detector (Anal. Sci. 1991, 7, 257). (Reagent A was 10 mM sodium periodate containing 3 mM potassium ferricyanide. Reagent B was 30 mM meso-1,2-diphenylethylenediamine in EtOH:water 70: 30 containing 130 mM sodium methylate. Synthesis of meso-1,2-diphenylethylenediamine is as described above.)

CHROMATOGRAM

Retention time: 12 (vanillylmandelic acid), 15 ((4-hydroxy-3-methoxyphenyl)ethylene glycol), 26 (3,4-dihydroxyphenylacetic acid), 74 (homovanillic acid)

Internal standard: 3,4-dihydroxyphenylpropanoic acid (52)
Limit of detection: 4-95 nM

KEY WORDS
post-column reaction; plasma; SPE

REFERENCE
Jeon, H.-K.; Nohta, H.; Ohkura, Y. High-performance liquid chromatographic determination of catecholamines and their precursor and metabolites in human urine and plasma by postcolumn derivatization involving chemical oxidation followed by fluorescence reaction, *Anal.Biochem.*, **1992**, *200*, 332–338.

SAMPLE
Matrix: blood, urine
Analyte: catecholamines
Sample preparation: Condition a Toyopak IC-SP S sulfopropyl resin, H$^+$ form, SPE cartridge (Tosoh) with 10 mL water and 2 mL 200 mM pH 5.0 sodium phosphate buffer. Plasma. 700 μL Plasma + 30 μL 700 nM isoproterenol + 50 μL 7 μM 3,4-dihydroxyphenylpropanoic acid + 350 μL 2 M perchloric acid, mix, centrifuge at 4° at 1000 g for 15 min. Remove a 700 μL aliquot of the supernatant and adjust the pH to 1.5-2.0 with about 150 μL 2 M potassium carbonate, centrifuge at 4° at 1000 g for 5 min, add the supernatant to the SPE cartridge, wash with 10 mL water, elute with 300 μL MeOH:2 M sodium perchlorate 7:93, filter (cellulose acetate membrane), inject a 100 μL aliquot of the filtrate. Urine. Collect human urine for 24 h in the presence of 10 mL 6 M HCl. 500 μL Urine + 10 μL 15 μM isoproterenol + 25 μL 800 μM 3,4-dihydroxyphenylpropanoic acid + 500 μL 1 M perchloric acid, mix, centrifuge at 4° at 1000 g for 15 min. Remove a 700 μL aliquot of the supernatant and adjust the pH to 1.5-2.0 with about 130 μL 2 M potassium carbonate, centrifuge at 4° at 1000 g for 5 min, add the supernatant to the SPE cartridge, wash with 1.5 mL water, wash with 500 μL EtOH:water 50:50, wash with 5 mL water, elute with 500 μL 1.5 M KCl in MeOH:100 mM HCl 7:93, filter (cellulose acetate membrane), inject a 100 μL aliquot of the filtrate.

HPLC VARIABLES
Column: 250 × 4.6 5 μm TSK-gel ODS-80TM (Tosoh)
Mobile phase: MeOH:buffer 7:93 (Buffer was 30 mM pH 2.5 citrate buffer containing 0.4 mM sodium octanesulfonate.)
Flow rate: 0.8
Injection volume: 100
Detector: F ex 350 em 480 following post-column reaction. The column effluent mixed with reagent A pumped at 0.3 mL/min and the mixture flowed through a 3 m × 0.5 mm ID stainless steel coil at 90°. The effluent from this coil mixed with reagent B pumped at 0.3 mL/min and the mixture flowed through a 10 m × 0.5 mm ID stainless steel coil at 90° and through a 1 m × 0.5 mm ID stainless steel cooling coil to the detector (Anal. Sci. 1991, 7, 257). (Reagent A was 10 mM sodium periodate containing 3 mM potassium ferricyanide. Reagent B was 30 mM meso-1,2-diphenylethylenediamine in EtOH:water 70:30 containing 130 mM sodium methylate. Synthesis of meso-1,2-diphenylethylenediamine is as described above.)

CHROMATOGRAM
Retention time: 16 (norepinephrine), 20 (epinephrine), 23 (levodopa), 29 (normetanephrine), 37 (dopamine), 41 (metanephrine), 96 (3-methoxytyramine)
Internal standard: isoproterenol (60)
Limit of detection: 0.4-9 nM

KEY WORDS
post-column reaction; plasma; SPE

REFERENCE
Jeon, H.-K.; Nohta, H.; Ohkura, Y. High-performance liquid chromatographic determination of catecholamines and their precursor and metabolites in human urine and plasma by postcolumn derivatization involving chemical oxidation followed by fluorescence reaction, *Anal.Biochem.*, **1992**, *200*, 332–338.

SAMPLE
Matrix: enzyme incubations
Analyte: epinephrine
Sample preparation: Prepare an SPE column by adding 500 μL of a 20% suspension of 19-40 μm Toyopak SP (strong cation-exchange sulfopropyl resin, Na⁺ (Toyo Soda)) in water to a 35 × 6 column, wash with two 2 mL portions of 2 M NaOH, wash with 5 mL water, wash with 2 mL 2 M HCl, wash with 10 mL water. 350 μL Enzyme incubation at pH 8.5 + 50 μL 3 M trichloroacetic acid + 50 μL 2 μM isoproterenol, centrifuge at 4° at 1000 g for 10 min, add a 300 μL aliquot of the supernatant to the SPE column, wash with 10 mL water, wash with 3 mL 200 mM pH 5.5 phosphate buffer, wash with 10 mL water, elute with 2 mL EtOH:1 M NaCl 70:30. Add 100 μL 100 mM pH 6.7 1,2-diphenylethylenediamine in 100 mM HCl and 100 μL 15 mM potassium ferricyanide to the eluate, heat at 37° for 40 min, inject a 100 μL aliquot. (Synthesis of meso-1,2-diphenylethylenediamine is as described above.)

HPLC VARIABLES
Column: 250 × 4.6 TSK-gel ODS-120T (Toyo Soda)
Mobile phase: MeCN:MeOH:50 mM pH 7.0 Tris-HCl buffer 52:3:45 (Wash with MeCN:MeOH:water 52:3:45 for 25 min at the end of each day.)
Flow rate: 1
Injection volume: 100
Detector: F ex 360 em 480

CHROMATOGRAM
Retention time: 8
Internal standard: isoproterenol (13)

KEY WORDS
SPE

REFERENCE
Lee, M.; Nohta, H.; Ohkura, Y.; Yoo, B. Determination of phenylethanolamine N-methyltransferase by high-performance liquid chromatography with fluorescence detection, *J.Chromatogr.*, **1985**, *348*, 407−415.

SAMPLE
Matrix: urine
Analyte: catecholamines
Sample preparation: Add disodium EDTA and sodium metabisulfite to urine. 100 μL Urine + 2 mL water, vortex, add 1 mL reagent 1, add 5 mL reagent 2, shake vigorously for 2 min, centrifuge at 2000 g for 2 min, freeze in dry ice/acetone. Remove the organic layer and add it to 200 μL 80 mM acetic acid and 2 mL n-octanol saturated with acetic acid, shake vigorously for 2 min, centrifuge at 2000 g for 2 min, freeze in dry ice/acetone until the aqueous layer is just solid, remove the organic layer. Thaw out the aqueous layer and add it to 1 mL reagent 1 and 5 mL reagent 2, shake vigorously for 2 min, centrifuge at 2000 g for 2 min, freeze in dry ice/acetone. Remove the organic layer and add it to 200 μL 80 mM acetic acid and 2 mL n-octanol saturated with acetic acid, shake vigorously for 2 min, centrifuge at 2000 g for 2 min, freeze in dry ice/acetone until the aqueous layer is just solid, remove the organic layer. Thaw out the aqueous layer and add 100 μL Bicine buffer, 250 μL MeCN, and 100 μL 100 mM 1,2-diphenylethylenediamine in 100 mM HCl, vortex, add 20 μL 20 mM potassium ferricyanide, vortex, heat at 37° for 40 min, cool to room temperature, inject a 100 μL aliquot. (Prepare reagent 1 by dissolving 214 g ammonium chloride and 10 g disodium EDTA in 2 L water, adjust pH to 8.3-8.5 with concentrated ammonium hydroxide, add 4.0 g diphenylborate-ethanolamine complex, stir for several hours until a clear solution is obtained. Prepare reagent 2 by dissolving 2.5 g tetraoctylammonium bromide and 10 mL n-octanol (saturated with acetic acid) in 1 L n-heptane. Prepare Bicine buffer by dissolving 14.3 g Bicine (N,N-bis(2-hydroxyethyl)glycine) and 359 mg anhydrous sodium acetate in 45 mL water, stir overnight until dissolved, adjust pH to 7.30 with concentrated NaOH, make up to 50 mL with water. Note that concentration of 1,2-diphenylethylenediamine is not given in paper. Other au-

thors have used 100 mM (J.Chromatogr. 1989, 487, 17; 1992, 574, 109; 1992, 583, 236). Synthesis of meso-1,2-diphenylethylenediamine is as described above.)

HPLC VARIABLES
Column: 150 × 4.6 5 μm Ultrasphere ODS
Mobile phase: Gradient. A was MeCN:MeOH:50 mM pH 7.0 sodium acetate buffer 40: 10:50. B was MeCN:MeOH:50 mM pH 7.0 sodium acetate buffer 50:10:40. A:B 75:25 for 1 min, to 10:90 over 7 min, return to initial conditions over 1 min.
Flow rate: 1
Injection volume: 100
Detector: F ex 365 em 418 (cutoff filter)

CHROMATOGRAM
Retention time: 3 (norepinephrine), 5 (epinephrine), 7 (dopamine)
Limit of detection: <0.4 nM

OTHER SUBSTANCES
Simultaneously analyzed: isoproterenol

KEY WORDS
protect from light

REFERENCE
Moleman, P.; van Dijk, J. Determination of urinary norepinephrine and epinephrine by liquid chromatography with fluorescence detection and pre-column derivatization, *Clin.Chem.*, **1990**, *36*, 732−736.

SAMPLE
Matrix: urine
Analyte: catecholamines
Sample preparation: 100 μL Urine + 125 μL 218.6 nM α-methylnorepinephrine in 10 mM HCl + 1 mL 10 mM HCl + 1 mL reagent + 5 mL 4.6 mM tetraoctylammonium bromide in n-heptane:1-octanol 99:1, shake for 2 min, centrifuge at 20° at 1000 g for 5 min, freeze in dry ice/acetone. Remove the organic phase and add it to 2 mL 1-octanol saturated with 80 mM acetic acid and 200 μL 80 mM acetic acid, shake, centrifuge at 20° at 1000 g for 5 min, freeze in dry ice/acetone. Discard the organic layer and add 1 mL 10 mM HCl to the aqueous layer, add 2 mL 1-octanol saturated with 80 mM acetic acid, add 150 μL 80 mM acetic acid, shake, centrifuge at 20° at 1000 g for 5 min, freeze in dry ice/acetone. Discard the organic layer and add 200 μL MeCN and 50 μL buffer to the aqueous layer, add 100 μL 100 mM 1,2-diphenylethylenediamine in 100 mM HCl, add 20 μL 20 mM potassium ferricyanide in water, heat at 37° in the dark for 1 h, inject a 50 μL aliquot. (Reagent was 8.9 mM diphenylborate-ethanolamine complex in 2 M pH 8.6 ammonia/ammonium chloride buffer containing 13.4 mM EDTA. Buffer was 1.75 M pH 7.05 bicine in water containing 1% EDTA. Recrystallize 1,2-diphenylethylenediamine from toluene:light petroleum (bp 60-80°) 10:90, dry overnight at 60°. Synthesis of meso-1,2-diphenylethylenediamine is as described above.)

HPLC VARIABLES
Column: 100 × 4.6 3 μm Cp MicroSpher C18 (Chrompack)
Mobile phase: MeCN:MeOH:50 mM pH 7.0 sodium acetate 40:8:50 (At the end of the day flush column with 60 mL MeCN:MeOH:water 70:10:20.)
Flow rate: 1
Injection volume: 50
Detector: F ex 350 em 480

CHROMATOGRAM
Retention time: 2 (norepinephrine), 4 (epinephrine), 5 (dopamine)
Internal standard: α-methylnorepinephrine (Janssen, Beerse, Belgium) (3)
Limit of quantitation: 1.6-24.8 nM

KEY WORDS
protect from light

REFERENCE

van der Hoorn, F.A.J.; Boomsma, F.; Man in 't Veld, A.J.; Schalekamp, M.A.D.H. Improved measurement of urinary catecholamines by liquid-liquid extraction, derivatization and high-performance liquid chromatography with fluorimetric detection, *J.Chromatogr.*, **1991**, *563*, 348–355.

Ethylenediamine

RELATED REFERENCE

Prados, P.; Higashidate, S.; Imai, K. A fully automated HPLC method for the determination of catecholamines in biological samples utilizing ethylenediamine condensation and peroxyoxalate chemiluminescence detection. *Biomed.Chromatogr.* **1994**, *8*, 1-8.

SAMPLE

Matrix: blood

Analyte: dopamine

Sample preparation: 100 μL Plasma + 5 mg alumina + 10 μL 10 μM IS in 10 mM perchloric acid + 100 μL pH 8.7 Tris-HCl buffer, stir for 10 min, centrifuge at 3000 g for 1 min, discard the supernatant. Wash the alumina with two 500 μL portions of water, add 100 μL 100 mM perchloric acid, mix for 1 min, centrifuge at 3000 g for 1 min, inject a 50 μL aliquot of the supernatant. (Heat 20 g alumina (WA-4, Sigma) with 200 mL 2 M HCl at 100° for 1 h with gentle mixing, decant the supernatant, wash with twenty 200 mL portions of water, filter (Toyo Roshi No. 2 paper), dry at 120° overnight.)

HPLC VARIABLES

Column: 150 × 4.6 catecholpak (JASCO)

Mobile phase: MeCN:50 mM pH 3.20 potassium acetate:50 mM pH 3.20 potassium phosphate buffer 3:92.15:4.85 containing 1 mM sodium hexanesulfonate

Column temperature: 40

Flow rate: 0.5

Injection volume: 50

Detector: Chemiluminescence (Kenko filter Y-46) following post-column reaction. The column effluent mixed with reagent 1 pumped at 0.25 mL/min and the mixture flowed through a 15 m × 0.5 mm i.d. knitted PTFE coil at 80°. The effluent from the coil mixed with reagent 2 pumped at 1.4 mL/min and this mixture flowed to the detector. (Reagent 1 was 105 mM ethylenediamine (semiconductor grade) and 175 mM imidazole in MeCN: EtOH 90:10. Reagent 2 was 0.25 mM bis[4-nitro-2-(3,6,9-trioxadecyloxycarbonyl)phenyl] oxalate (Wako), 150 mM hydrogen peroxide, and 110 mM trifluoroacetic acid in dioxane: ethyl acetate 50:50 (Caution! Dioxane is a carcinogen!).)

CHROMATOGRAM

Retention time: 22 (dopamine), 16 (epinephrine), 14 (norepinephrine)

Internal standard: 3,4-dihydroxybenzylamine (17)

Limit of detection: 1 fmole

KEY WORDS

human; rat; plasma; SPE; post-column reaction

REFERENCE

Higashidate, S.; Imai, K. Determination of femtomole concentrations of catecholamines by high-performance liquid chromatography with peroxyoxalate chemiluminescence detection, *Analyst*, **1992**, *117*, 1863–1868.

o-Phenylenediamine

SAMPLE
Matrix: bulk
Analyte: glucans
Sample preparation: Suspend 2 mg glucan in 200 mM pH 10 carbonate buffer, pass nitrogen through the solution continuously, add 6 mg o-phenylenediamine, and 10 mg sodium sulfite, make up to 2 mL, heat at 100° for 1 h, cool in an ice bath. Remove a 500 μL aliquot, neutralize with 100 mM HCl, make up to 5 mL with water, inject an aliquot. (Derivatization occurs at the reducing-end terminal.)

HPLC VARIABLES
Column: 250 × 4 Irica RP-18
Mobile phase: MeCN:MeOH:water 5:30:65
Flow rate: 0.6
Detector: UV 320

CHROMATOGRAM
Retention time: 8-23 (depending on structure)

REFERENCE
Takagi, M.; Daido, Y.; Kuriyama, M.; Morita, N. Determination of the degree of branching of amylopectin-like glucans evaluated by quinoxaline formation with o-phenylenediamine under alkaline and deoxygenated conditions, *Anal.Sci.*, **1990**, *6*, 529–534.

DICARBOXYLIC ACID

1,8-Naphthalic Dicarboxylic Acid

SAMPLE
Matrix: blood
Analyte: amiprilose
Sample preparation: 250 μL Plasma + 100 μL 120 μg/mL IS in water, vortex for 1 min, add 200 μL 100 mM NaOH, add 6 mL dichloromethane, rotate for 15 min. Remove the lower organic layer and evaporate it to dryness under a stream of nitrogen, add 75 μL 26.67 mg/mL 2-chloro-1-methylpyridinium iodide in MeCN, add 200 μL 2.5 mg/mL 1,8-naphthalic dicarboxylic acid in MeCN containing 6 μL/mL triethylamine, heat at 65° overnight, cool, inject a 20 μL aliquot. (Prepare 1,8-naphthalic dicarboxylic acid by hydrolyzing 1,8-naphthalic anhydride with 5% NaOH (cf. Org. Syn. 1973, Coll. Vol. V, 813).)

HPLC VARIABLES
Column: 250 × 4.6 5 μm Ultrasphere ODS (Altex)
Mobile phase: MeOH:1 M ammonium acetate:N,N-dimethyloctylamine:water 65:2.5: 0.03:32.5
Flow rate: 1
Injection volume: 20
Detector: F ex 280 em 340

CHROMATOGRAM
Retention time: 21 (main peak; a minor peak due to the other conformer is also seen)
Internal standard: 1,2-O-isopropylidene-3-O-[3'-(N,N-diisopropylamino)ethyl]-α-D-gluco-furanose (28)
Limit of quantitation: 185 ng/mL

KEY WORDS
plasma; pharmacokinetics

REFERENCE
Wu, S.T.; Benet, L.Z.; Lin, E.T. Determination of amprilose in human plasma by high-performance liquid chromatography with fluorimetric detection, *J.Chromatogr.B*, **1997**, *692*, 149–156.

MISCELLANEOUS REACTIONS

Alkylation

SAMPLE
Matrix: bulk
Analyte: benz[a]anthracene 5,6-dihydrodiol
Sample preparation: Dissolve in THF (dried over NaH) and add a 500-fold molar excess of iodomethane, add a catalytic amount of sodium hydride, let stand in the dark for 15 min, add MeOH dropwise, inject an aliquot.

HPLC VARIABLES
Column: 80 × 6.2 Golden Series SIL (DuPont)
Mobile phase: Hexane:ethyl acetate:MeOH:THF 94.6:5:0.2:0.2
Flow rate: 2
Detector: UV 254

CHROMATOGRAM
Retention time: 13.3 (5-O-methyl), 14.0 (6-O-methyl)

KEY WORDS
normal phase; many other examples given in paper

REFERENCE
Yang, S.K.; Mushtaq, M.; Bao, Z.; Weems, H.B.; Shou, M.; Lu, X.-L. Improved enantiomeric separation of dihydrodiols of polycyclic aromatic hydrocarbons on chiral stationary phases by derivatization to O-methyl ethers, *J.Chromatogr.*, **1989**, *461*, 377–395.

Oxidation

SAMPLE
Matrix: solutions
Analyte: thymine glycol
Sample preparation: Evaporate to dryness, reconstitute with 1 mL 200 mM sodium periodate, let stand at room temperature for 1 h, evaporate to dryness, extract with MeOH. Evaporate the extract to dryness, reconstitute with 1 mL water, reflux for 3 h, evaporate to dryness under reduced pressure, reconstitute with water, filter (0.2 μm), inject an aliquot.

HPLC VARIABLES
Column: ODS
Mobile phase: water
Detector: radioactivity

CHROMATOGRAM
Retention time: 12

KEY WORDS
^{14}C-labeled; tritium-labeled

REFERENCE
Higgins, S.A.; Frenkel, K.; Cummings, A.; Teebor, G.W. Definitive characterization of human thymine glycol N-glycosylase activity, *Biochem.*, **1987**, *26*, 1683–1688.

SAMPLE
Matrix: urine
Analyte: 3-methoxy-4-hydroxyphenylethylene glycol
Sample preparation: 2 mL Urine + 500 μL buffer + 50 μL 2000 U/mL β-glucuronidase (EC 3.2.1.31, Type H-2S, Helix pomatia, Sigma), heat at 37° for 24 h, add 1 g NaCl, add 10 mL ethyl acetate, shake vigorously for 5 min, centrifuge at 2000 g for 5 min. Remove 5 mL of the organic layer and add it to 2 mL 350 mM acetic acid and 5 mL n-hexane, shake vigorously for 5 min, centrifuge at 2000 g for 5 min, inject a 50 μL aliquot of the aqueous layer. (Prepare buffer by dissolving 96.5 g sodium acetate trihydrate in 900 mL water, adjusting pH to 5.0 with acetic acid, and making up to 1 L with water.)

HPLC VARIABLES
Column: 250 × 4 Nucleosil 5C18
Mobile phase: 10 mM Trichloroacetic acid adjusted to pH 2.0
Flow rate: 1.5
Injection volume: 50
Detector: UV 365 following post-column reaction. The column effluent mixed with the reagent pumped at 0.1 mL/min and the mixture flowed through a 3 m × 0.5 mm ID PTFE coil at 60° to the detector. (Reagent was 20 mM sodium metaperiodate in 1 M NaOH containing 45 mM potassium bicarbonate, stable for 1 month at 4°.)

CHROMATOGRAM
Retention time: 9.7
Limit of detection: 80 ng/mL

OTHER SUBSTANCES
Simultaneously analyzed: levodopa, metanephrine, methyldopa, normetanephrine, vanillin, vanillylmandelic acid

KEY WORDS
post-column reaction

REFERENCE
Abe, K.; Konaka, R. Specific determination of 3-methoxy-4-hydroxyphenylethylene glycol in urine by liquid chromatography with post-column reaction, *Clin.Chem.*, **1988**, *34*, 87–90.

DITHIOCARBAMATE

ALKYL HALIDE

Iodoethane

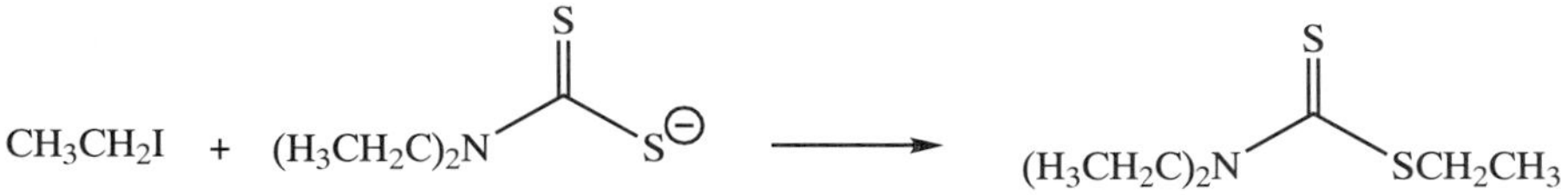

$$CH_3CH_2I \ + \ (H_3CH_2C)_2N-\overset{\overset{\displaystyle S}{\|}}{C}-S^{\ominus} \longrightarrow (H_3CH_2C)_2N-\overset{\overset{\displaystyle S}{\|}}{C}-SCH_2CH_3$$

SAMPLE

Matrix: blood

Analyte: disulfiram

Sample preparation: For disulfiram. Inject a 100 μL aliquot of plasma directly onto column A with mobile phase A and elute to waste, after 4 min backflush the contents of column A onto column B with mobile phase B, after another 4 min remove column A from the circuit, elute column B with mobile phase B and monitor the effluent. For metabolites diethyldithiocarbamate and methyl diethyldithiocarbamate. 1 mL Plasma + 10 mM pH 7.5 phosphate buffer containing 200 mM EDTA + 1 μL 2-mercaptoethanol + 1 μL iodoethane, mix for 30 s, heat at 40° for 30 min, inject a 100 μL aliquot onto column A with mobile phase A and elute to waste, after 4 min backflush the contents of column A onto column B with mobile phase B, after another 4 min remove column A from the circuit, elute column B with mobile phase B and monitor the effluent. (Unchanged disulfiram was not found in plasma but the metabolites were found.)

HPLC VARIABLES

Column: A 50 × 3.9 40 μm Perisorb RP-18 (Merck); B 250 × 3.9 7 μm LiChrosorb RP-18

Mobile phase: A 10 mM pH 7.5 phosphate buffer containing 5 mM EDTA; B MeCN : 10 mM pH 7.5 phosphate buffer 60:40

Flow rate: 1

Injection volume: 100

Detector: UV 254

CHROMATOGRAM

Retention time: 18.3 (disulfiram), 12.5 (diethyldithiocarbamate), 10.5 (methyl diethyldithiocarbamate)

Limit of detection: 3 ng/mL (disulfiram), 3.7 ng/mL (diethyldithiocarbamate), 3.3 ng/mL (methyl diethyldithiocarbamate

OTHER SUBSTANCES

Extracted: metabolites

KEY WORDS

plasma; column-switching

REFERENCE

Johansson, B. Rapid and sensitive on-line precolumn purification and high-performance liquid chromatographic assay for disulfiram and its metabolites, *J.Chromatogr.*, **1986**, *378*, 419–429.

OXONIUM SALT

Triethyloxonium Tetrafluoroborate

$(CH_3CH_2)_3O^{\oplus} \ BF_4^{\ominus}$ + $(H_3CH_2C)_2N$ —C(=S)—$S^{\ominus}$ $\longrightarrow$ $(H_3CH_2C)_2N$ —C(=S)—SCH_2CH_3

SAMPLE
Matrix: blood
Analyte: diethyldithiocarbamate
Sample preparation: Add 10 μL triethyloxonium tetrafluoroborate solution to 500 μL plasma while vortexing, vortex for an additional 5 s, add 500 μL 12 μg/mL IS in chloroform, extract by inverting tube once a second for 3 min, centrifuge at 500 g for 2 min. Remove 100 μL of the organic layer and ad it to 500 μL MeCN, mix, inject an aliquot. (Purify triethyloxonium tetrafluoroborate as follow. Rapidly filter 500 mL of a 1 M solution in dichloromethane through a glass wool plug, evaporate to dryness under reduced pressure, add 100 mL dry dichloromethane, swirl gently for 3 min, decant the colorless solution from the black sludge, store in glass or PTFE container under 10 mL diethyl ether at or below -40°. Triethyloxonium tetrafluoroborate hydrolyzes rapidly so care should be taken to minimize exposure to water.)

HPLC VARIABLES
Column: 250 × 4.6 5 μm Ultrasphere ODS
Mobile phase: MeCN:water 70:30
Flow rate: 1.5
Injection volume: 20
Detector: UV 254

CHROMATOGRAM
Retention time: 4.3
Internal standard: S-isopropyl diethyldithiocarbamate (Dissolve 40 g sodium diethyldithiocarbamate in 100 mL EtOH, rapidly add 250 mmoles isopropyl iodide with stirring, stir at room temperature overnight, add 60 mL water, stir vigorously for 10 min. Remove the lower oily layer and remove the solvent under reduced pressure, distil to give S-isopropyl diethyldithiocarbamate (bp 77°/0.15 mm Hg).) (5.2)
Limit of detection: 5 ng/mL

OTHER SUBSTANCES
Non-interfering: cyclophosphamide, doxorubicin

KEY WORDS
plasma; dog; pharmacokinetics

REFERENCE
Lieder, P.H.; Borch, R.F. Triethyloxonium tetrafluoroborate derivatization and HPLC analysis of diethyldithiocarbamate in plasma, *Anal.Lett.*, **1985**, *18(B1)*, 57–66.

ENAL

AMINE

3-Aminophenol

SAMPLE
Matrix: blood, urine
Analyte: 4-hydroxyifosfamide
Sample preparation: 500 μL Plasma or urine + 500 μL buffer, mix, add 1 mL reagent, heat at 100° for 20 min, cool in ice, filter (0.45 μm), inject a 10 μL aliquot. (Buffer was 0.2% lidocaine hydrochloride in 200 mM citric acid, pH adjusted to 3.5. Reagent was 0.33% 3-aminophenol in 2 M HCl containing 0.4% hydroxylamine (as an antioxidant). 4-Hydroxyifosfamide decomposes to give acrolein which reacts with 3-aminophenol. Endogenous acrolein does not interfere.)

HPLC VARIABLES
Guard column: 10 mm long C18
Column: 150 × 4.6 5 μm Chromspher C18
Mobile phase: MeOH:buffer 10:90 (Buffer was 20 mM tetramethylammonium bromide containing 5 mM heptanesulfonic acid, pH 3.0.)
Flow rate: 1
Injection volume: 10
Detector: F ex 350 em 510

CHROMATOGRAM
Retention time: 8
Internal standard: lidocaine (determined separately by GC)
Limit of detection: 40 nM

OTHER SUBSTANCES
Non-interfering: acrolein (endogenous)

KEY WORDS
plasma

REFERENCE
Kaijser, G.P.; Ter Riet, P.G.J.H.; de Kraker, J.; Bult, A.; Beijnen, J.H.; Underberg, W.J.M. Determination of 4-hydroxyifosfamide in biological matrices by high-performance liquid chromatography, *J.Pharm.Biomed.Anal.*, **1997**, *15*, 773–781.

SAMPLE
Matrix: microsomal incubations
Analyte: acrolein
Sample preparation: 250 μL Microsomal incubation + 250 μL 0.5% 3-aminophenol in 3 M HCl containing 0.6% hydroxylamine hydrochloride, add 10 μL 25 μg/mL 2-butenal, heat at 100° for 10 min, cool, centrifuge at 5000 g for 5 min, inject a 50 μL aliquot of the supernatant.

HPLC VARIABLES
Guard column: 20 × 4 5 μm RP-18 (Merck)
Column: 125 × 3 5 μm LiChrosorb 100 RP-8

Mobile phase: MeCN:0.5% phosphoric acid 4:96
Flow rate: 0.4
Injection volume: 50
Detector: F ex 358 em 505

CHROMATOGRAM
Retention time: 8
Internal standard: 2-butenal (12.3)
Limit of quantitation: 10 ng/mL
Limit of detection: 5 ng/mL

KEY WORDS
human; rat; liver

REFERENCE
Bohnenstengel, F.; Eichelbaum, M.; Golbs, E.; Kroemer, H.K. High-performance liquid chromatographic determination of acrolein as a marker for cyclophosphamide bioactivation in human liver microsomes, *J.Chromatogr.B*, **1997**, *692*, 163–168.

2,4-Diaminotoluene

SAMPLE
Matrix: solutions
Analyte: 2-alkenals
Sample preparation: Mix 500 μL of a solution in EtOH containing 25 μg/mL methacrolein with 1 mL 1% 2,4-diaminotoluene dihydrochloride and 2 mL 5 M HCl, heat in a boiling water bath for 1 h, cool for 10 min, make up to 4 mL with water, inject an aliquot. (Caution! 2,4-Diaminotoluene is a carcinogen!)

HPLC VARIABLES
Column: 150 × 4 STR ODS-H (Shimadzu)
Mobile phase: Gradient. MeCN:50 mM $(NH_4)H_2PO_4$ containing 5 mM sodium 1-octanesulfonate from 20:80 to 60:40 over 1 h.
Flow rate: 0.8
Injection volume: 20
Detector: F ex 396 em 485

CHROMATOGRAM
Retention time: 6 (acrolein), 10 (crotonaldehyde), 16 (2-pentenal), 20 (2-hexenal), 26 (2-heptenal), 34 (2-octenal), 41 (2-nonenal)
Internal standard: methacrolein (13)
Limit of detection: 10-50 pg

REFERENCE
Hirayama, T.; Miura, S.; Mori, Y.; Ueta, M.; Tagami, E.; Yoshizawa, T.; Watanabe, T. High-performance liquid chromatographic determination of 2-alkenals in oxidized lipid as their 7-amino-6-methylquinoline derivatives, *Chem.Pharm.Bull.*, **1991**, *39*, 1253–1257.

EPOXIDE

ALKOXIDE

Sodium Methoxide

SAMPLE
Matrix: bulk
Analyte: benz[a]anthracene 5,6-epoxide
Sample preparation: Dissolve in MeOH saturated with sodium methoxide, heat at 50° for 1 h, store overnight at room temperature, inject an aliquot.

HPLC VARIABLES
Column: 80 × 6.2 Golden Series SIL (DuPont)
Mobile phase: Hexane:ethyl acetate:MeOH:THF 94.6:5:0.2:0.2
Flow rate: 2
Detector: UV 254

CHROMATOGRAM
Retention time: 13.3 (5-O-methyl), 14.0 (6-O-methyl)

KEY WORDS
normal phase; many other examples given in paper

REFERENCE
Yang, S.K.; Mushtaq, M.; Bao, Z.; Weems, H.B.; Shou, M.; Lu, X.-L. Improved enantiomeric separation of dihydrodiols of polycyclic aromatic hydrocarbons on chiral stationary phases by derivatization to O-methyl ethers, *J.Chromatogr.*, **1989**, *461*, 377–395.

AMINE

Isopropylamine

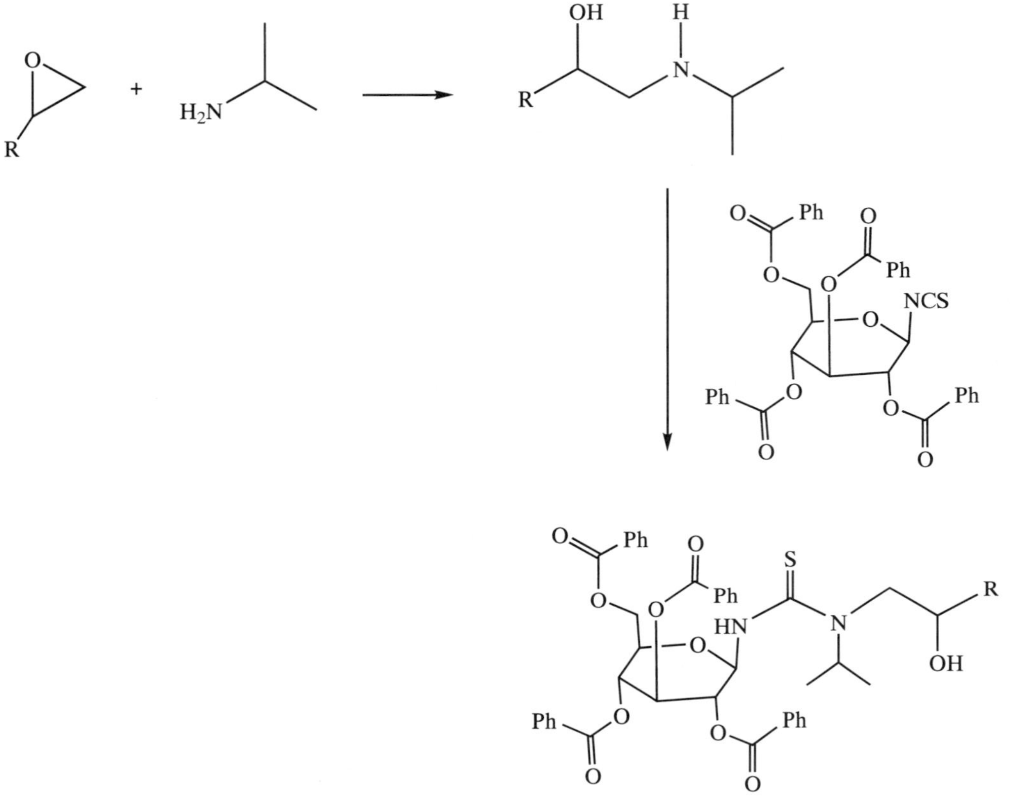

SAMPLE

Matrix: bulk

Analyte: epoxides

Sample preparation: Mix 50 μL epoxide with 200 μL isopropylamine, heat at 100° for 2 h, evaporate excess isopropylamine with a stream of air, add 950 μL MeCN, mix. Remove a 50 μL aliquot and add it to 50 μL 0.66% 2,3,4,6-tetra-O-benzoyl-β-D-glucopyranosyl isothiocyanate (Fluka) in MeCN, mix, let stand at room temperature for 30 min, make up to 1 mL with MeCN, inject a 7 μL aliquot.

HPLC VARIABLES

Column: 25 × 4 (sic) 5 μm LiChrospher 100 RP-18

Mobile phase: MeOH:water 90:10

Flow rate: 0.5

Injection volume: 7

Detector: UV 231

CHROMATOGRAM

Retention time: k' 1.45 (R-propylene oxide), k' 1.65 (S-propylene oxide), k' 9.38 (R-dodecane epoxide), k' 11.35 (S-dodecane epoxide) [All examples of the homologous series between these extremes can be determined.]

KEY WORDS

chiral; comparison with other derivatization reagents

REFERENCE
Lobell, M.; Schneider, M.P. 2,3,4,6-Tetra-O-benzoyl-β-D-glucopyranosyl isothiocyanate: an efficient reagent for the determination of enantiomeric purities of amino acids, β-adrenergic blockers and alkyloxiranes by high-performance liquid chromatography using standard reversed-phase columns, *J.Chromatogr.*, **1993**, *633*, 287–294.

DITHIOCARBAMATE

Diethyldithiocarbamic Acid, Diethylammonium Salt

SAMPLE
Matrix: polymers
Analyte: epoxides
Sample preparation: 500 mg Polymer + 4.5 mL MeOH + 5 mL reagent, shake in a pressure resistant glass vial with a PTFE-lined aluminum snap-cap, heat at 90° for 30 min, cool to room temperature, add 300 μL 85% phosphoric acid, add 10 mL chloroform, shake vigorously for 1 min, let stand for 30 min, inject a 10 μL aliquot of the lower chloroform layer. (Prepare reagent by dissolving 500 mg diethyldithiocarbamic acid, diethylammonium salt (diethylammonium N,N-diethyldithiocarbamate) in 50 mL pH 7 phosphate buffer (Merck), sonicate for 5 min, prepare fresh each day.

HPLC VARIABLES
Column: 250 × 4.6 Zorbax-NH2
Mobile phase: n-Hexane:MeOH:THF 89:7.3:3.7 (Rinse column with 200 mL THF at the end of each week.)
Flow rate: 2
Injection volume: 10
Detector: UV 278

CHROMATOGRAM
Retention time: 4 (propylene oxide), 5 (ethylene oxide)
Limit of detection: 50 ppb

REFERENCE
Van Damme, F.; Oomens, A.C. Determination of residual free epoxide in polyether polyols by derivatization with diethylammonium N,N-diethyldithiocarbamate and liquid chromatography, *J.Chromatogr.A*, **1995**, *696*, 41–47.

Sodium Diethyldithiocarbamate

SAMPLE
Matrix: blood
Analyte: dianhydrogalactitol
Sample preparation: 1 mL Plasma + 500 μL 100 mM pH 7.0 potassium phosphate buffer + 500 μL 5% sodium diethyldithiocarbamate in water (prepared fresh daily) + 2 mL water, mix, let stand at room temperature for 1 h, extract with 10 mL chloroform for 3 min, centrifuge at 1200 g for 5 min. Remove the organic layer and wash it with 2-5 mL 33% NaCl solution. Evaporate an 8 mL aliquot of the organic layer to dryness at 40°, reconstitute the residue in 200 μL heptane:chloroform 70:30, inject a 10 μL aliquot.

HPLC VARIABLES
Column: 300 mm × 6.35 mm o.d. μBondapak CN
Mobile phase: Heptane:chloroform:acetic acid 70:30:1.2
Flow rate: 2.5
Injection volume: 10
Detector: UV 254

CHROMATOGRAM
Retention time: 5
Limit of detection: 50 ng/mL

KEY WORDS
plasma

REFERENCE
Munger, D.; Sternson, L.A.; Repta, A.J.; Higuchi, T. High-performance liquid chromatographic analysis of dianhydrogalactitol in plasma by derivatization with sodium diethyldithiocarbamate, *J.Chromatogr.*, **1977**, *143*, 375–382.

SAMPLE
Matrix: blood, urine
Analyte: teroxirone
Sample preparation: 1 mL Whole blood, plasma, or urine + 500 μL 10 mM pH 7.4 phosphate buffer + 5 mL chloroform, shake mechanically for 15 min, centrifuge at 3000 g. Remove 4 mL of the organic layer and evaporate it to dryness under a stream of nitrogen, reconstitute the residue in 500 μL 10 mM pH 7.4 phosphate buffer, add 500 μL 5% diethyldithiocarbamate in water (freshly prepared), let stand at room temperature for 1 h, add 5 mL chloroform, shake, centrifuge. Remove 4 mL of the organic layer and evaporate it to dryness under a stream of nitrogen, reconstitute the residue in 50-100 μL toluene, inject an aliquot.

HPLC VARIABLES
Column: 10 μm PAC-CN (Whatman)
Mobile phase: Gradient. Heptane:EtOH 60:40 for 10 min, to 10:90 over 6 min.
Flow rate: 2
Detector: UV 254

CHROMATOGRAM
Retention time: 7.7
Limit of detection: 15 ng/mL

KEY WORDS
whole blood; plasma; normal phase; human; rabbit; pharmacokinetics

REFERENCE
Ames, M.M.; Kovach, J.S.; Rubin, J. Pharmacological characterization of teroxirone, a triepoxide anti-tumor agent, in rats, rabbits, and humans, *Cancer Res.*, **1984**, *44*, 4151–4156.

ISOTHIOCYANATE

2,3,4,6-Tetra-O-benzoyl-β-D-glucopyranosyl Isothiocyanate

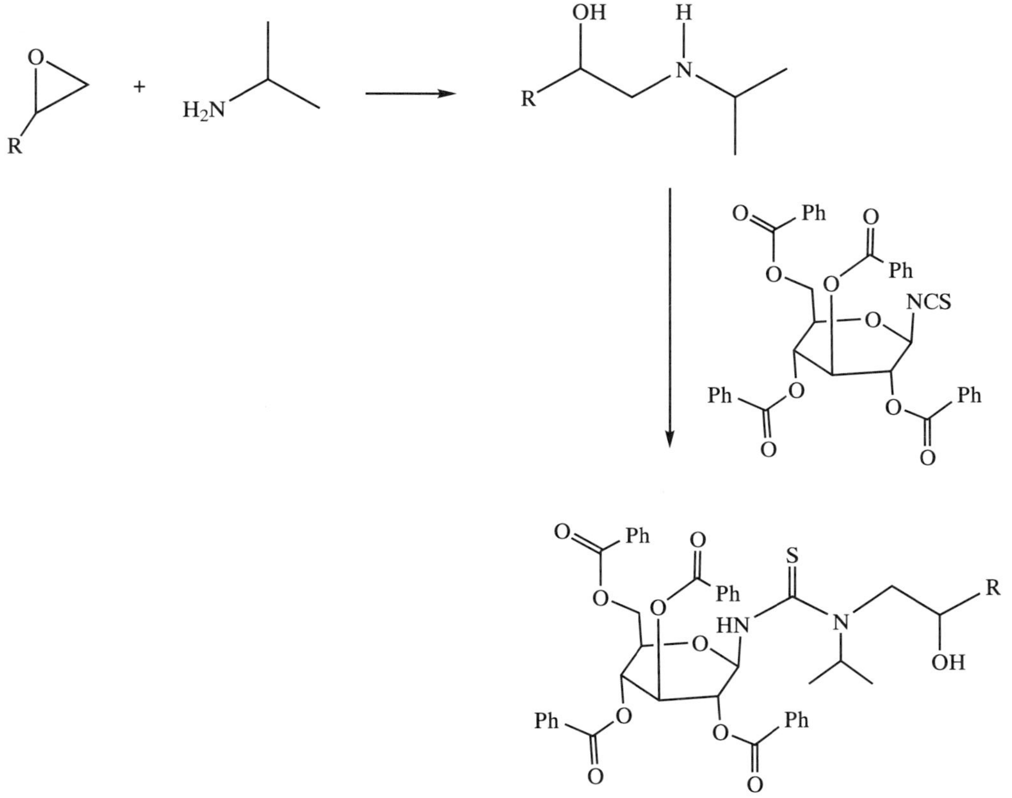

SAMPLE
Matrix: bulk
Analyte: epoxides
Sample preparation: Mix 50 μL epoxide with 200 μL isopropylamine, heat at 100° for 2 h, evaporate excess isopropylamine with a stream of air, add 950 μL MeCN, mix. Remove a 50 μL aliquot and add it to 50 μL 0.66% 2,3,4,6-tetra-O-benzoyl-β-D-glucopyranosyl isothiocyanate (Fluka) in MeCN, mix, let stand at room temperature for 30 min, make up to 1 mL with MeCN, inject a 7 μL aliquot.

HPLC VARIABLES
Column: 25 × 4 (sic) 5 μm LiChrospher 100 RP-18
Mobile phase: MeOH:water 90:10
Flow rate: 0.5
Injection volume: 7
Detector: UV 231

CHROMATOGRAM
Retention time: k' 1.45 (R-propylene oxide), k' 1.65 (S-propylene oxide), k' 9.38 (R-dodecane epoxide), k' 11.35 (S-dodecane epoxide) [All examples of the homologous series between these extremes can be determined.]

KEY WORDS
chiral; comparison with other derivatization reagents

REFERENCE
Lobell, M.; Schneider, M.P. 2,3,4,6-Tetra-O-benzoyl-β-D-glucopyranosyl isothiocyanate: an efficient reagent for the determination of enantiomeric purities of amino acids, β-adrenergic blockers and alkyloxiranes by high-performance liquid chromatography using standard reversed-phase columns, *J.Chromatogr.*, **1993**, *633*, 287–294.

ESTER

MISCELLANEOUS REACTIONS

Hydrolysis

SAMPLE
Matrix: solutions
Analyte: triglycerides
Sample preparation: Inject a 5-40 µL aliquot of a solution in acetone or a <5 µL aliquot of a solution in THF.

HPLC VARIABLES
Column: 250 × 4 3 µm Hitachi gel 3057 ODS
Mobile phase: Gradient. A was MeCN:EtOH 75:25. B was MeCN:EtOH 25:75. A:B from 100:0 to 60:40 over 10 min, to 0:100 over 30 min, maintain at 0:100 for 15 min, return to initial conditions, re-equilibrate for 15 min.
Column temperature: 30
Flow rate: 0.8
Injection volume: 5-40
Detector: UV 410 following post-column reaction. The column effluent mixed with 2.4% KOH in EtOH:water 60:40 pumped at 0.4 mL/min and this mixture flowed through a 20 m × 0.5 mm i.d. PTFE coil at 85° to where it was mixed with 10 mM periodic acid in 4 M ammonium acetate:1.6 M acetic acid 50:50 (pH 5.5) pumped at 0.4 mL/min. This mixture passed through a 1 m × 0.5 mm i.d. coil of PTFE tubing to where it mixed with 200 mM acetylacetone in 4 M ammonium acetate:1.6 M acetic acid 50:50 (pH 5.5) pumped at 0.4 mL/min. This mixture flowed through a 20 m × 0.5 mm i.d. coil of PTFE tubing at 70° to the detector. (The triglyceride is hydrolyzed to glycerol and carboxylic acids and the glycerol is then oxidized to formaldehyde. The formaldehyde reacts with ammonia and acetylacetone in the Hantzsch reaction to give a dihydropyridine which is then detected.)

CHROMATOGRAM
Retention time: 14.5 (tricaprin), 11 (tricaprylin), 22 (trilaurin), 30 (trimyristin), 38 (tri-olein), 42 (tripalmitin), 54 (tristearin)
Limit of quantitation: 0.3 nmole
Limit of detection: 0.1 nmole

KEY WORDS
post-column reaction

REFERENCE
Kondoh, Y.; Takano, S. Determination of triglycerides by high-performance liquid chromatography with postcolumn derivatization, *Anal.Chem.*, **1986**, *58*, 2380–2383.

GUANIDINE

ACYL HALIDE

4-Nitrobenzoyl Chloride

SAMPLE
Matrix: urine
Analyte: metformin
Sample preparation: Mix 2 mL urine with 1 mL 20% NaOH and saturate the mixture with NaCl, add 1 mL MeCN, add 10 mg 4-nitrobenzoyl chloride, let stand at room temperature for 1 h, add 10 mg 4-nitrobenzoyl chloride, let stand for 1 h, inject a 10 μL aliquot of the upper MeCN phase.

HPLC VARIABLES
Column: 914 × 2.2 37-50 μm Bondapak phenyl/Corasil
Mobile phase: MeOH:water 40:60
Flow rate: 1
Injection volume: 10
Detector: UV 280

CHROMATOGRAM
Retention time: 6
Limit of detection: 200 ng/mL (using more urine and less MeCN)

REFERENCE
Ross, M.S.F. Determination of metformin in biological fluids by derivatization followed by high-performance liquid chromatography, *J.Chromatogr.*, **1977**, *133*, 408–411.

DIKETONE

Acetylacetone

RELATED REFERENCE
Bombardt, P.A.; Adams, W.J. Liquid chromatographic determination of guanadrel in laboratory animal diet as the fluorescent acetylacetone derivative. *Anal.Chem.* **1982**, *54*, 1087-1090.

SAMPLE
Matrix: blood, saliva, urine
Analyte: debrisoquine
Sample preparation: Plasma. 1 mL Plasma + IS + 1 mL MeOH + 1 mL 2 M HCl, mix, centrifuge. Remove the supernatant and add it to 1 mL 2 M HCl, cool, wash with 8 mL diethyl ether. Add the aqueous layer to 1 mL 5 M NaOH. Remove a 1 mL aliquot and add it to 500 μL saturated sodium bicarbonate solution, add 500 μL MeOH, add 500 μL acetylacetone, heat at 96° for 2.5 h, cool to room temperature, add 3 mL 5 M NaOH, extract with 8 mL diethyl ether. Remove the organic layer and add it to 500 μL 2 M HCl, shake mechanically for 15 min, centrifuge. Remove the aqueous layer and add it to 500 μL 5 M NaOH, cool, extract with 8 mL diethyl ether. Remove the organic layer and evaporate it to dryness at 45°, reconstitute the residue in 20 μL MeOH, inject an aliquot. Saliva, urine. 1 mL Saliva or urine + 2 (saliva) or 10 (urine) μg IS in MeOH + 500 μL saturated sodium bicarbonate solution + 500 μL MeOH + 500 μL acetylacetone, heat at 96° for 2.5 h, cool to room temperature, add 3 mL 5 M NaOH, extract with 8 mL diethyl ether. Remove the organic layer and add it to 500 μL 2 M HCl, shake mechanically for 15 min, centrifuge. Remove the aqueous layer and add it to 500 μL 5 M NaOH, cool, extract with 8 mL diethyl ether. Remove the organic layer and evaporate it to dryness at 45°, reconstitute the residue in 20 μL MeOH, inject an aliquot.

HPLC VARIABLES
Guard column: C8 (Merck)
Column: 100 × 8 μBondapak C18 in a Z-module
Mobile phase: MeOH:water containing 10 mM sodium 1-pentanesulfonate, adjusted to pH 3.5 with orthophosphoric acid
Flow rate: 1.5
Injection volume: 25-200
Detector: UV 248

CHROMATOGRAM
Retention time: 7.4
Internal standard: guanoxan hemisulfate (Pfizer) (5.6)
Limit of quantitation: 500 ng/mL

OTHER SUBSTANCES
Extracted: metabolites

KEY WORDS
silanize all glassware with hexamethyldisilazane; plasma; comparison with GC; pharmacokinetics

REFERENCE
Chan, K. Comparison of gas chromatographic and high-performance liquid chromatographic assays for the determination of debrisoquine and its 4-hydroxy metabolite in human fluids, *J.Chromatogr.*, **1988**, *425*, 311–321.

SAMPLE
Matrix: bulk
Analyte: guanidine salts
Sample preparation: Prepare a 2.5 mg/mL solution in MeOH. 10 mL Solution + 40 mL MeOH + 50 mL 0.5% acetylacetone in 4 mM NaOH, mix, heat a 10 mL aliquot in a tightly capped 15 mL centrifuge tube at 110° for 4 h (Caution! Pressure may build up in tube!), cool to room temperature, inject a 10 μL aliquot. (For guanidine sulfate heat a 10 mL aliquot of a mixture of 50 mL 0.5 mg/mL guanidine sulfate in MeOH and 50 mL 0.5% acetylacetone in 4 mM NaOH.)

HPLC VARIABLES
Column: μBondapak CN
Mobile phase: MeOH:water:glacial acetic acid 70:30:1
Flow rate: 1
Injection volume: 10
Detector: UV 295

CHROMATOGRAM
Retention time: 9.29

OTHER SUBSTANCES
Non-interfering: urea, thiourea

KEY WORDS
for carbonate, hydrochloride, nitrate, and sulfate salts

REFERENCE
Palaitis, W.; Curran, J.R. HPLC assay for guanidine salts based on pre-column derivatization with acetylacetone, *J.Chromatogr.Sci.*, **1984**, *22*, 99–103.

Phenanthrenequinone

SAMPLE
Matrix: blood
Analyte: famotidine
Sample preparation: 1 mL Plasma + 1 mL saturated potassium carbonate + 5 mL ethyl acetate, mix, centrifuge. Remove the organic layer and add it to 1 mL saturated NaCl solution and 1 mL 1 M HCl, centrifuge. Remove the aqueous phase and add it to 1 mL saturated potassium carbonate solution and 3 mL ethyl acetate, extract. Remove a 2.5 mL aliquot of the organic layer and evaporate it to dryness under reduced pressure, reconstitute the residue with 400 μL 250 μM phenanthrenequinone in MeOH, evaporate to dryness under reduced pressure, reconstitute with 40 μL DMF and 40 μL 2 M NaOH, heat at 60° for 15 min, cool on ice, neutralize with 5 M acetic acid, inject a 20 μL aliquot.

HPLC VARIABLES
Column: reversed phase
Mobile phase: MeCN:10 mM pH 4 citrate buffer 40:50
Flow rate: 1
Injection volume: 20
Detector: F ex 296 em 411

CHROMATOGRAM
Limit of detection: 5 ng/mL

KEY WORDS
plasma; pharmacokinetics

REFERENCE
Echizen, H.; Shoda, R.; Umeda, N.; Ishizaki, T. Plasma famotidine concentration versus intragastric pH in patients with upper gastrointestinal bleeding and in healthy subjects, *Clin.Pharmacol.Ther.*, **1988**, *44*, 690–698.

9,10-Phenanthrenequinone-3-sulfonic Acid

SAMPLE
Matrix: blood
Analyte: guanidino compounds
Sample preparation: 100 μL Serum + 20 μL 20% trichloroacetic acid, vortex for a few s, centrifuge at 10000 g for 2 min. Remove a 60 μL aliquot of the supernatant and add it to 15 μL 400 mM NaOH, adjust pH to 2.5-3.0, inject a 50 μL aliquot of the supernatant.

HPLC VARIABLES
Column: 75 × 4.6 3 μm Nucleosil C8
Mobile phase: MeCN:MeOH:water 2:5:93 containing 5 mM sodium 1-octanesulfonate and 2 mM 9,10-phenanthrenequinone-3-sulfonic acid adjusted to pH 4.0 (A) or 3.5 (B) with acetic acid. (Synthesis of 9,10-phenanthrenequinonesulfonic acid is as follows. Heat 500 g phenanthrene and 600 g concentrated sulfuric acid at 120-130° with frequent shaking for 4.5-5 hours. In order to obtain a good yield this temperature must not be exceeded. Pour the hot reaction mixture into hot water (Caution! Sulfur dioxide is evolved!). Neutralize with barium carbonate until the solution is no longer acid to litmus, make up to 4-8 L, heat strongly for a short time, decant the boiling solution as soon as the barium precipitates, heat the precipitate with 3-4 L water twice more. Filter the decanted solutions, reduce the filtrate somewhat, treat with the amount of sulfuric acid necessary to precipitate the barium. Filter to remove the barium sulfate, extract the barium sulfate with a little boiling water. Add KOH to the filtrate while keeping it clearly acidic, evaporate with heating until potassium 3-phenanthrenesulfonate begins to precipitate, cool slowly to obtain potassium 3-phenanthrenesulfonate as beautiful leaves. Mix 10 g 3-phenanthrenesulfonic acid, 7 g chromium trioxide, and 70 g acetic acid. After a short time a vigorous reaction, encouraged by shaking, starts and the acetic acid boils, warm on a water bath for another 10 min, cool, add EtOH, shake well, filter to recover the crystals,

wash with EtOH until no more green color can be rinsed off. Dissolve the crystals in the minimum amount of hot water, filter while hot, add EtOH to the filtrate (about a fifth the volume of the aqueous solution), allow to cool slowly to obtain potassium phenanthrenequinonesulfonate as fine glittering, orange-yellow needles (Annalen 1902, 321, 248). It is advantageous to use 9,10-phenanthrenequinone-3-sulfonic acid because it is soluble in water whereas 9,10-phenanthrenequinone is almost insoluble in water.)

Flow rate: 1

Injection volume: 50

Detector: F ex 370 em 520 following post-column reaction. The column effluent mixed with 600 mM NaOH pumped at 0.5 mL/min and the mixture flowed through a 10 m $\times$ 0.5 mm ID stainless steel coil at 80° to the detector.

CHROMATOGRAM

Retention time: 2 (guanidinosuccinic acid (A)), 3 (guanidinoacetic acid (A)), 5 (taurocyamine (B)), 6 (guanidinosuccinic acid (B)), 6 (N-acetyl-L-arginine (A)), 7 (β-guanidinopropionic acid (A)), 7 (guanidinoacetic acid (B)), 9.5 (creatinine (A)), 10.5 (guanidine (A)), 13 (arginine (A)), 15 (methylguanidine (a)), 19 (gamma-guanidinobutyric acid (A))

Limit of detection: 6-100 pmole

KEY WORDS

serum

REFERENCE

Kobayashi, Y.; Kubo, H.; Kinoshita, T. Fluorometric determination of guanidino compounds by high-performance liquid chromatography using water-soluble 9,10-phenanthrenequinone-3-sulfonate, *Anal.Sci.*, **1987**, *3*, 363–367.

SAMPLE

Matrix: blood

Analyte: biguanides

Sample preparation: 100 μL Serum + 100 μL 4% perchloric acid, vortex for a few s, centrifuge at 10000 g for 1 min, inject a 100 μL aliquot of the supernatant.

HPLC VARIABLES

Column: 100 $\times$ 8 10 μm Radial-Pak μBondapak C18

Mobile phase: MeCN:water 20:80 (buformin) or 25:75 (phenformin) containing 5 mM sodium hexanesulfonate and 2 mM 9,10-phenanthrenequinonesulfonic acid, adjusted to pH 4.0 with acetic acid (Synthesis of 9,10-phenanthrenequinonesulfonic acid is as described above.)

Flow rate: 2

Injection volume: 100

Detector: F ex 300 em 500 following post-column reaction. The column effluent mixed with 300 mM NaOH pumped at 0.5 mL/min and the mixture flowed through a 15 m $\times$ 0.5 mm ID stainless steel coil at 70° to the detector.

CHROMATOGRAM

Retention time: 6 (buformin), 7 (phenformin)

Limit of detection: 20 ng/mL

KEY WORDS

post-column reaction; serum

REFERENCE

Kobayashi, Y.; Kubo, H.; Kinoshita, T.; Nishikawa, T. Fluorometric determination of biguanides in serum by high-performance liquid chromatography with reagent-containing mobile phase, *J.Chromatogr.*, **1988**, *430*, 65–71.

Potassium 1,2-Naphthoquinone-4-sulfonate

SAMPLE
Matrix: tissue
Analyte: dihydrostreptomycin
Sample preparation: Condition a 6 mL 500 mg Bond Elut Certify II SPE cartridge with 3 mL MeCN, 1 mL water, and three 1 mL aliquots of buffer B, do not allow to dry. Homogenize (Ultra Turrax TP 18/10) 8 g Kidney or meat, 5 mL buffer A, and 1 mL 85% trichloroacetic acid in water for 6 s, centrifuge at 5000 rpm for 3 min, add 2 mL dichloromethane, mix for 6 s, centrifuge at 5000 rpm for 5 min. Remove a 7 mL aliquot of the supernatant and add it to 900 µL 4 M NaOH, blend, centrifuge at 4000 rpm for 5 min. Remove the upper layer and add it to 900 µL 500 mM phosphoric acid, adjust the pH to 5.5-5.8 with 1 M NaOH or 500 mM phosphoric acid, add 2.5 mL buffer B, add to the SPE cartridge at 1 mL/min, rinse out the tube with 1 mL buffer A, add the rinse to the SPE cartridge, wash with two 5 mL portions of buffer A, suck dry for 2 s after each wash, wash with three 5 mL portions of 25% ammonia, suck dry for 2 s after each wash, wash with two 1 mL portions of water, suck dry for 2 s after each wash, wash with 1 mL water, suck dry for 10 s, elute with two 1 mL portions of 20% formic acid in MeOH. Evaporate the eluate to dryness under a stream of nitrogen at 60°, reconstitute the residue in 200 µL MeOH, mix for 3-4 s, evaporate to dryness, reconstitute with 400 µL buffer A, add 200 µL chloroform, mix vigorously for 10 s, centrifuge for 3 min. Filter (Costar Spin X with 0.22 µm nylon membrane) the aqueous layer while centrifuging at 5600 g for 2 min, inject a 25 µL aliquot of the filtrate. (Prepare buffer A by dissolving 4.45 g sodium 1-heptanesulfonate and 1.8 g Na$_2$HPO$_4$ in 750 mL water, adjust pH to 5.9 with 5 M phosphoric acid, adjust pH to 5.5 with 500 mM phosphoric acid, make up to 1 L with water, adjust pH to 5.5 with 500 mM phosphoric acid. Prepare buffer B by dissolving 13.35 g sodium 1-heptanesulfonate and 1.8 g Na$_2$HPO$_4$ in 750 mL water, adjust pH to 5.9 with 5 M phosphoric acid, adjust pH to 5.5 with 500 mM phosphoric acid, make up to 1 L with water, adjust pH to 5.5 with 500 mM phosphoric acid.)

HPLC VARIABLES
Guard column: 20 × 4.6 5 µm Supelcosil LC-ABZ + Plus
Column: 150 × 4.6 5 µm Supelcosil LC-ABZ + Plus
Mobile phase: MeCN:buffer 32:68 (Prepare buffer by dissolving 8.65 g sodium octanesulfonate and 110 mg potassium 1,2-naphthoquinone-4-sulfonate in 750 mL water, make up to 1 L with water, filter (0.45 µm), adjust pH to 3.24 with 1 mL acetic acid.) (Streptomycin can also be determined with MeCN:buffer 30:70.)
Column temperature: 31
Flow rate: 0.6 for 0.5 min, 0.9 for 4 min, 0.6 for 9 min
Injection volume: 25
Detector: F ex 375 em 412 following post-column reaction. The column effluent mixed with 300 mM NaOH pumped at 0.3 mL/min in a 1.2 µL vortex mixer (Kratos PCRS 520) and the mixture flowed through a 15 m × 0.5 mm ID knitted coil at 40° and a room temperature heat exchanger to the detector.

CHROMATOGRAM
Retention time: 12

Limit of quantitation: 40 ppb
Limit of detection: 20 ppb

KEY WORDS
post-column reaction; muscle; kidney; SPE

REFERENCE
Hormazábal, V.; Yndestad, M. High performance liquid chromatographic determination of dihydrostreptomycin sulfate in kidney and meat using post column derivatization, *J.Liq.Chromatogr.Rel.Technol.*,
1997, *20*, 2259–2268.

Sodium 1,2-Naphthoquinone-4-sulfonate

SAMPLE
Matrix: blood
Analyte: guanidino compounds
Sample preparation: 100 μL Serum + 20 μL 20% trichloroacetic acid, vortex for a few s,
centrifuge at 10000 g for 2 min. Remove a 60 μL aliquot of the supernatant, adjust the
pH to 2.5-3.0 with 400 mM NaOH, inject a 50 μL aliquot.

HPLC VARIABLES
Column: 150 × 4.6 5 μm Nucleosil C8
Mobile phase: MeCN:MeOH water 3:5:92 containing 15 mM sodium octanesulfonate and
1 mM sodium 1,2-naphthoquinone-4-sulfonate, adjusted to pH 4.0 with acetic acid
Flow rate: 1
Injection volume: 50
Detector: F ex 355 em 505 following post-column reaction. The column effluent mixed with
1 M NaOH pumped at 0.5 mL/min and the mixture flowed through a 5 m × 0.5 mm ID
stainless steel coil at 65° to the detector.

CHROMATOGRAM
Retention time: 3.5 (taurocyamine), 5 (guanidinosuccinic acid), 6 (guanidinoacetic acid),
14 (N-acetyl-L-arginine), 15.5 (β-guanidinopropionic acid), 17 (creatinine), 18 (guanidine),
20.5 (L-arginine), 23 (methylguanidine), 25 (gamma-guanidinobutyric acid)
Limit of detection: 20 ng

KEY WORDS
post-column reaction; serum

REFERENCE
Kobayashi, Y.; Kubo, H.; Kinoshita, T. Fluorometric determination of guanidino compounds by new
postcolumn derivatization system using reversed-phase ion-pair high-performance liquid chromatography, *Anal.Biochem.*, **1987**, *160*, 392–398.

KETOALCOHOL

Benzoin

RELATED REFERENCE

Hung, Y.L.; Kai, M.; Nohta, H.; Ohkura, Y. High-performance liquid chromatographic analyzer for guanidino compounds using benzoin as a fluorogenic reagent. *J.Chromatogr.* **1984**, *305*, 281-294.

SAMPLE

Matrix: blood

Analyte: angiotensin I

Sample preparation: 50 µL Plasma + 50 µL buffer + 20 µL 28.9 mM 8-hydroxyquinoline in EtOH:water 50:50 containing 13.6 mM dimercaprol + 50 µL 100 mM angiotensinogen fragment 13 (human), heat at 37° for 16 h, cool in ice water, add 50 µL 200 nM IS, add 100 µL water, filter (Tosoh Air Press 30 membrane, cutoff 30000). Remove a 200 µL aliquot of the ultrafiltrate and add it to 100 µL 5 mM benzoin in 2-methoxyethanol, add 100 µL 100 mM 2-mercaptoethanol in water containing 200 mM sodium sulfite, add 200 µL 800 mM KOH, heat in a boiling water bath for 1.5 min, add 200 µL 800 mM HCl in 500 mM pH 8.0 Tris-HCl buffer to adjust the pH to 8.0, inject a 500 µL aliquot onto columns A and B in series and elute to waste with mobile phase A, after 9 min remove column A from the circuit, elute column B to waste with mobile phase A, after 3 min elute the contents of column B onto column C with mobile phase B, after 2 min remove column B from the circuit, elute column C with mobile phase B, monitor the effluent from column C. Wash column A with MeCN:water 50:50 for 5 min then re-equilibrate with mobile phase A. Re-equilibrate column B with mobile phase A. (Prepare buffer by mixing 400 mM citric acid and 800 mM Na$_2$HPO$_4$ to pH 7.4.)

HPLC VARIABLES

Column: A 35 × 4.6 7 µm Nucleosil 7C8; B 10 × 4 5 µm TSKgel ODS-80 TM (Tosoh); C 250 × 4.6 5 µm TSKgel ODS-80 TM (Tosoh)

Mobile phase: A MeCN:200 mM pH 8.1 phosphate buffer:water 25:15:60; B MeCN:200 mM pH 8.1 phosphate buffer:water 35:15:50

Flow rate: 0.6

Injection volume: 500

Detector: F ex 325 em 435

CHROMATOGRAM

Retention time: 25

Internal standard: [Val^5]-angiotensin I (22)

Limit of detection: 820 fmole

KEY WORDS

column-switching; plasma; ultrafiltrate; procedure is an assay for renin activity in plasma

REFERENCE

Miyazaki, T.; Kai, M.; Ohkura, Y. Determination of renin activity in human plasma by column-switching high-performance liquid chromatography with fluorescence detection, *J.Chromatogr.*, **1989**, *490*, 43–51.

SAMPLE
Matrix: blood
Analyte: DMP 728
Sample preparation: Condition a 100 mg Bond Elut C2 SPE cartridge with two 1 mL portions of MeOH and two 1 mL portions of water. Add 500 μL plasma containing IS to the SPE cartridge, wash with two 1 mL portions of water, wash with three 1 mL portions of dichloromethane:MeCN 20:80, dry under vacuum for 5 min, elute with two 1 mL portions of MeOH, evaporate the eluate to dryness under a stream of nitrogen. Reconstitute with 50 μL water, 25 μL 4 mM benzoin in 2-methoxyethanol, 25 μL 100 mM mercaptoethanol in 200 mM aqueous sodium sulfite, and 50 μL 400 mM KOH. Vortex briefly, heat at 45° for 20 min, cool in ice water for 1 min, add 20 μL 500 mM Tris-HCl:800 mM HCl 50:50, vortex for 1 min, inject a 75 μL aliquot.

HPLC VARIABLES
Column: 150 × 4.6 5 μm semi-permeable surface octyl (Regis)
Mobile phase: MeCN:100 mM pH 7.4 potassium phosphate buffer 25:75
Flow rate: 1.5
Injection volume: 75
Detector: F ex 325 em 425

CHROMATOGRAM
Retention time: 10
Internal standard: cyclo(D-2-amino-3-methyl butyryl-L-N-methylarginylglycyl-L-aspartyl-m-aminomethylbenzoic acid) (DMP 757) (12)
Limit of quantitation: 5 ng/mL (rat), 2.5 ng/mL (dog)

KEY WORDS
rat; dog; plasma; SPE; pharmacokinetics

REFERENCE
Wu, S.T.; Stampfli, H.F.; Banks, C.M.; Emm, T.A.; Kapil, R.P.; Padovani, P.K.; Lee, W.M. Jr.; Huang, S.-M. Determination of DMP 728, a IIb/IIIa receptor antagonist, in rat and dog plasma by high-performance liquid chromatography with fluorimetric detection, *J.Chromatogr.B*, **1994**, *657*, 254–260.

SAMPLE
Matrix: blood
Analyte: CG 167 (5-acetylamino-4-guanidino-2,6-anhydro-3,4,5-trideoxy-D-glycero-D-galacto-non-2-enoic acid)
Sample preparation: Condition a Bond Elut SCX SPE cartridge with 2 mL MeOH and 2 mL 10% formic acid. 1 mL Serum + 1 mL 10% formic acid, mix, add to the SPE cartridge, wash with 2 mL 1% trifluoroacetic acid in MeOH, wash with 2 mL water, elute with four 500 μL aliquots of 10% triethylamine in MeOH:water 50:50. Evaporate the eluate to dryness under a stream of nitrogen at 70°, reconstitute the residue in 75 μL 4 mM benzoin in ethylene glycol, add 150 μL 100 mM β-mercaptoethanol in 500 mM KOH containing 200 mM sodium sulfite, vortex, heat at 100° for 3 min, inject a 100 μL aliquot.

HPLC VARIABLES
Column: 100 × 4 ODS Hypersil
Mobile phase: Gradient. MeCN:buffer 20:80 for 12 min, to 80:20 (step gradient), maintain at 80:20 for 5 min, re-equilibrate at initial conditions for 10 min. (Prepare buffer by adding 100 mL 1 M Tris and 29.4 mL 1 M HCl to water and making up to 2 L, pH 8.5.)
Flow rate: 1
Injection volume: 100
Detector: F ex 325 em 442

CHROMATOGRAM
Retention time: 11.5
Limit of quantitation: 10 ng/mL

KEY WORDS
serum; SPE

REFERENCE
Stubbs, R.J.; Harker, A.J. Automated high-performance liquid chromatographic method for the determination of a neuramidase inhibitor (GG167) in human serum by pre-column fluorescence derivatization using benzoin, *J.Chromatogr.B*, **1995**, *670*, 279–285.

SAMPLE
Matrix: blood, tissue
Analyte: leupeptin
Sample preparation: 40 μL Serum + 160 μL water + 200 μL 1.5 M perchloric acid, mix, centrifuge at 800 g for 10 min. Homogenize 100 mg muscle with 1 mL ice-water, centrifuge at 0-2° at 1000 g for 20 min. Remove a 200 μL aliquot of the supernatant and add it to 200 μL 1.5 M perchloric acid, mix, centrifuge at 800 g for 10 min. Remove a 200 μL aliquot of the supernatant and add it to 100 μL 1 M potassium carbonate, mix. Remove a 100 μL aliquot and add it to 50 μL 40 mM sodium borohydride, let stand at room temperature for 5 min, add 50 μL 1.5 M HCl, while cooling in ice add 100 μL 5 mM benzoin in 2-methoxyethanol, add 100 μL mercaptoethanol solution, add 200 μL 1.5 M KOH, heat on a boiling water bath for 1.5 min, add 200 μL 2 M HCl:1 M pH 8.5 Tris-HCl buffer 50:50 (?) to adjust pH to 8.5, inject a 100 μL aliquot. (Prepare mercaptoethanol solution by dissolving 780 mg β-mercaptoethanol and 2.52 g sodium sulfite in 80 mL water, make up to 100 mL with water. The sodium borohydride reduces a leupeptin aldehyde group (that would otherwise interfere with the derivatization) to an alcohol.)

HPLC VARIABLES
Column: 150 × 4 5 μm LiChrosorb RP-18
Mobile phase: MeOH:500 mM pH 8.5 Tris-HCl buffer:40 mM tetra-n-butylammonium chloride 70:10:20
Flow rate: 0.8
Injection volume: 100
Detector: F ex 325 em 425

CHROMATOGRAM
Retention time: 8
Limit of detection: 250 nM (serum), 500 pmole/g (muscle)

KEY WORDS
mouse; serum; muscle; pharmacokinetics

REFERENCE
Kai, M.; Miura, T.; Ishida, J.; Ohkura, Y. High-performance liquid chromatographic method for monitoring leupeptin in mouse serum and muscle by pre-column fluorescence derivatization with benzoin, *J.Chromatogr.*, **1985**, *345*, 259–265.

SAMPLE
Matrix: blood, urine
Analyte: guanidino compounds
Sample preparation: Serum. 300 μL Serum + 300 μL water + 200 μL 2.5 μM IS + 100 μL 900 mM HCl, filter (Amicon UM 05 Diaflo membrane) under nitrogen gas pressure (3 kg/cm^2). Cool in ice while mixing 200 μL ultrafiltrate, 100 μL 4 mM benzoin in 2-methoxyethanol, 100 μL mercaptoethanol solution, and 200 μL 2 M KOH, heat on a boiling water bath for 5 min, cool in ice-water for 2 min, add 200 μL 4 M HCl:buffer 50:50, inject a 100 μL aliquot. Urine. 50 μL Urine + 550 μL water + 200 μL 10 μM IS + 100 μL 900 mM HCl, mix. Cool in ice while mixing 200 μL sample, 100 μL 4 mM benzoin in 2-methoxyethanol, 100 μL mercaptoethanol solution, and 200 μL 2 M KOH, heat on a boiling water bath for 5 min, cool in ice-water for 2 min, add 200 μL 4 M HCl:buffer 50:50, inject a 100 μL aliquot. (Prepare mercaptoethanol solution by dissolving 780 mg β-mercaptoethanol and 2.52 g sodium sulfite in 80 mL water, make up to 100 mL with water. Prepare buffer by dissolving 12.11 g Tris in 80 mL water, adjusting pH to 9.2 with concentrated HCl, and making up to 100 mL with water.)

HPLC VARIABLES
Column: 300 × 3.9 10 μm μBondapak phenyl
Mobile phase: Gradient. MeOH:water:500 mM pH 8.5 Tris-HCl buffer from 50:40:10 to 80:10:10 over 25.

Flow rate: 0.8
Injection volume: 100
Detector: F ex 325 em 425

CHROMATOGRAM
Retention time: 8 (guanidinosuccinic acid), 12 (taurocyamine), 14 (guanidinobutyric acid), 20 (methylguanidine)
Internal standard: phenylguanidine (23)
Limit of detection: 8-16 nM (serum), 30-78 nM (urine)

KEY WORDS
serum; ultrafiltrate

REFERENCE
Kai, M.; Miyazaki, T.; Ohkura, Y. High-performance liquid chromatographic measurement of guanidino compounds of clinical importance in human urine and serum by pre-column fluorescence derivatization using benzoin, *J.Chromatogr.*, **1984**, *311*, 257–266.

SAMPLE
Matrix: enzyme incubations
Analyte: angiotensin II
Sample preparation: 5 μL Serum + 70 μL 200 mM pH 7.5 phosphate buffer containing 30 mM NaCl + 20 μL 800 μM angiotensin I + 10 μL water, heat at 37° for 15 min, add 100 μL 500 mM perchloric acid, centrifuge at 800 g for 5 min. Remove a 100 μL aliquot of the supernatant and cool in ice-water, add 50 μL 5 mM benzoin in 2-methoxyethanol, add 50 μL mercaptoethanol solution, add 100 μL 0.8 M KOH, heat on a boiling water bath for 1.5 min, add 100 μL 1.2 M HCl:1 M pH 8.5 Tris-HCl buffer 50:50 (?) to adjust pH to 8.5, inject a 100 μL aliquot. (Prepare mercaptoethanol solution by dissolving 780 mg β-mercaptoethanol and 2.52 g sodium sulfite in 80 mL water, make up to 100 mL with water. The procedure measures the activity of angiotensin-converting enzyme in serum. The enzyme converts angiotensin I to angiotensin II which is then determined by HPLC.)

HPLC VARIABLES
Column: 250 × 4 5 μm TSK gel ODS-120T (Toyo Soda)
Mobile phase: MeOH:48 mM pH 8.5 phosphate buffer 33:67
Flow rate: 0.8
Injection volume: 100
Detector: F ex 325 em 435

CHROMATOGRAM
Retention time: 10
Limit of detection: 80-300 fmole

OTHER SUBSTANCES
Extracted: angiotensin I

REFERENCE
Sakamoto, Y.; Miyazaki, T.; Kai, M.; Ohkura, Y. Sensitive assay for serum angiotensin-converting enzyme and separation of angiotensin analogues by high-performance liquid chromatography with fluorescence detection, *J.Chromatogr.*, **1986**, *380*, 313–320.

SAMPLE
Matrix: solutions
Analyte: guanidino compounds
Sample preparation: Cool in ice while mixing 200 μL solution, 100 μL 4 mM benzoin in 2-methoxyethanol, 100 μL mercaptoethanol solution, and 200 μL 2 M KOH, heat on a boiling water bath for 5 min, cool in ice-water for 2 min, add 200 μL 4 M HCl:buffer 50:50, inject a 100 μL aliquot. (Prepare mercaptoethanol solution by dissolving 780 mg β-mercaptoethanol and 2.52 g sodium sulfite in 80 mL water, make up to 100 mL with water. Prepare buffer by dissolving 12.11 g Tris in 80 mL water, adjusting pH to 9.2 with concentrated HCl, and making up to 100 mL with water.)

HPLC VARIABLES
Column: 300 × 3.9 10 μm μBondapak phenyl

Mobile phase: Gradient. MeOH:water:500 mM pH 8.5 Tris-HCl buffer 50:35:15 for 2 min, to 80:5:15 over 24 min, maintain at 80:5:15 for 2 min.
Flow rate: 0.8
Injection volume: 100
Detector: F ex 325 em 425

CHROMATOGRAM
Retention time: 8.0 (guanidinosuccinic acid), 11.5 (taurocyamine), 12.0 (guanidinoacetic acid), 12.0 (creatine), 12.0 (creatinine), 13.0 (guanidinopropionic acid), 13.2 (N-α-acetyl-arginine), 13.8 (guanidinobutyric acid), 15.0 (arginine), 4.2, 15.0 (argininosuccinic acid), 16.5 (homoarginine), 17.8 (canavanine), 18.0 (guanidine), 19.0 (methylguanidine), 23.0 (phenylguanidine), 26.0, 31.0 (agmatine), 15.0 (tuftsin), 14.0-26.0 (angiotensin III), 18.0-20.0 (angiotensin II), 14.2, 18.0-23.5 (bradykinin), 19.0-24.0 (angiotensin I), 15.0 (neurotensin), 14.5-18.5 (gonadorelin)
Limit of detection: 20-100 fmole

REFERENCE
Kai, M.; Miyazaki, T.; Yamaguchi, M.; Ohkura, Y. High-performance liquid chromatography of guanidino compounds using benzoin as a pre-column fluorescent derivatization reagent, *J.Chromatogr.*, **1983**, *268*, 417–424.

SAMPLE
Matrix: solutions
Analyte: peptides
Sample preparation: Cool in ice while mixing 100 μL of an aqueous solution, 50 μL 5 mM benzoin in 2-methoxyethanol, 50 μL mercaptoethanol solution, and 100 μL 0.8 M KOH, heat on a boiling water bath for 1.5 min, add 100 μL 1.6 M HCl:1 M pH 8.5 Tris-HCl buffer 50:50, inject a 100 μL aliquot. (Prepare mercaptoethanol solution by dissolving 780 mg β-mercaptoethanol and 2.52 g sodium sulfite in 80 mL water, make up to 100 mL with water.)

HPLC VARIABLES
Column: 15 × 4 (sic) 5 μm LiChrosorb RP-18
Mobile phase: MeCN:50 mM pH 8.5 phosphate buffer 31:69
Flow rate: 0.8
Injection volume: 100
Detector: F ex 325 em 425

CHROMATOGRAM
Retention time: 4 (tuftsin), 5 (angiotensin II), 9 (angiotensin I), 13 (leupeptin acid), 17 (angiotensin III), 22 (gonadorelin), 38.5 (substance P)
Limit of detection: 27-130 fmole

REFERENCE
Kai, M.; Miyazaki, T.; Sakamoto, Y.; Ohkura, Y. Use of benzoin as pre-column fluorescence derivatization reagent for the high-performance liquid chromatography of angiotensins, *J.Chromatogr.*, **1985**, *322*, 473–477.

SAMPLE
Matrix: solutions
Analyte: peptides
Sample preparation: Inject a 10-50 μL aliquot.

HPLC VARIABLES
Column: 200 × 4 5 μm TSKgel ODS-120T (Toyo Soda)
Mobile phase: Gradient. A was MeCN:200 mM pH 2.3 sodium phosphate buffer 5:95. B was MeCN:200 mM pH 2.3 sodium phosphate buffer 40:60. A:B 100:0 for 5 min, to 68.6:31.4 step gradient), to 35.7:64.3 over 45 min.
Flow rate: 1
Injection volume: 10-50
Detector: F ex 325 em 435 following post-column reaction. The column effluent mixed with reagent A pumped at 1 mL/min and the mixture flowed through a 15 m × 0.3 mm ID PTFE coil at 76°. The effluent from this coil mixed with reagent B pumped at 0.4 mL/min

and this mixture passed through a 10 × 4 column packed with 40 mg glass wool. (Prepare reagent A by mixing equal volumes of 6 mM benzoin in 2-methoxyethanol, 4.8 M KOH, and 2.1 M β-mercaptoethanol. Prepare reagent B by mixing equal volumes of 1 M Tris and 4.2 M HCl.)

CHROMATOGRAM
Retention time: 4 (kyotorphin), 16.5 (kallidin), 20 (angiotensin II), 21 (angiotensin III), 28 (angiotensin I), 30 (β-melanocyte stimulating hormone), 35 (substance P)
Limit of detection: 5-15 pmole

OTHER SUBSTANCES
Non-interfering: estriol, estrone-3-sulfate, methionine enkephalin, phenylalanine, phenylpyruvic acid, propionic acid, sorbic acid, tryptophan, tyrosine

KEY WORDS
post-column reaction

REFERENCE
Ohno, M.; Kai, M.; Ohkura, Y. On-line post-column fluorescence derivatization of arginine-containing peptides in high-performance liquid chromatography, *J.Chromatogr.*, **1987**, *392*, 309–316.

SAMPLE
Matrix: solutions
Analyte: antagonist G ([Arg[6],D-Trp[7,9],MePhe[8]]-substance P)
Sample preparation: Mix 80 μL sample solution with 40 μL 4 mM benzoin in methoxyethanol, 40 μL mercaptoethanol solution, and 80 μL 200 mM NaOH with cooling in icewater, heat on a boiling water bath for 10 s, cool in ice for 5 min, add 80 μL buffer, keep at 4° until 10 min before injection, inject a 20 μL aliquot. (Prepare mercaptoethanol solution by dissolving 135.8 μL 2-mercaptoethanol and 630 mg sodium sulfite in 25 mL water. Prepare buffer by mixing equal volumes of 400 mM perchloric acid and 100 mM pH 8.5 phosphate buffer.)

HPLC VARIABLES
Column: 125 × 4 15 μm LiChroCART RP-18 (Merck)
Mobile phase: MeCN:5 mM pH 8.5 phosphate buffer 56:44
Flow rate: 1
Injection volume: 20
Detector: F ex 325 em 425; UV 214

CHROMATOGRAM
Retention time: 5
Limit of detection: 210 nM (F), 840 nM (UV)

OTHER SUBSTANCES
Simultaneously analyzed: degradation products

REFERENCE
Cui, H.; Reubsaet, J.L.E.; Bult, A. Precolumn fluorescence derivatization of the antagonist [Arg[6],D-Trp[7,9],MePhe[8]]-Substance P with benzoin in high-performance liquid chromatography and selective detection of arginine-containing fragments in its degradation products, *J.Chromatogr.A*, **1996**, *736*, 91–96.

SAMPLE
Matrix: tissue
Analyte: peptides
Sample preparation: Condition a Bond Elut C18 SPE cartridge with 3 mL water and 3 mL MeCN (in this order ?). Homogenize 400 mg brain tissue with 2 mL 100 mM HCl, add 20 μL 10 μM IS, add 2 mL acetone, mix, centrifuge at 2450 g for 15 min. Remove the supernatant and add it to 220 μL 1 M sodium bicarbonate and 500 μL 100 mM disodium EDTA, centrifuge at 2450 g for 15 min. Remove the supernatant and evaporate it to remove the acetone, dilute the aqueous residue with 2 mL water, add to the SPE cartridge. wash with 1 mL water, wash with 3 mL 100 mM HCl, wash with two 3 mL portions of dichloromethane, wash with 1 mL water, wash with 2 mL 100 mM pH 8.0 phosphate buffer, wash with 2 mL water, elute with 2 mL MeCN:100 mM pH 2.3 phosphate buffer 70:30. Evaporate the eluate under reduced pressure, make up to 400 μL with water, inject a 100 μL aliquot.

HPLC VARIABLES
Column: 200 × 4 5 μm TSKgel ODS-120T (Tosoh)
Mobile phase: Gradient. A was MeCN:300 mM pH 2.3 sodium phosphate buffer:water 1: 20:79. B was MeCN:300 mM pH 2.3 sodium phosphate buffer:water 60:20:20. A:B from 90:10 to 55:45 over 33 min, maintain at 55:45 for 7 min, to 0:100 (step gradient), maintain at 0:100.
Flow rate: 1
Injection volume: 100
Detector: F ex 325 em 435 following post-column reaction. The column effluent mixed with 2 mM benzoin in 1.6 M KOH containing 700 mM 2-mercaptoethanol and this mixture flowed through a 15 m × 0.33 mm ID PTFE coil at 76 ± 1°. The effluent from this coil mixed with 500 mM Tris containing 2.1 M HCl pumped at 0.4 mL/min and this mixture flowed to the detector.

CHROMATOGRAM
Retention time: 19.1 (vasopressin), 22.0 (α-neoendorphin), 23.6 (kallidin), 24.0 (leucine enkephalin-Arg), 26.2 (bradykinin), 26.8 (β-neoendorphin), 29.2 (gonadorelin), 30.5 (dynorphin 1-8), 32.0 (neurotensin), 33.2 (methionine enkephalin-Arg-Gly-Leu), 35.8 (methionine enkephalin-Arg-Phe), 37.1 (substance P)
Internal standard: [D-Phe^{11}]-neurotensin (40.0)
Limit of detection: 0.5-13.5 pmole

KEY WORDS
post-column reaction; rat; brain

REFERENCE
Ohno, M.; Kai, M.; Ohkura, Y. High-performance liquid chromatographic determination of substance P-like arginine-containing peptide in rat brain by on-line post-column fluorescence derivatization with benzoin, *J.Chromatogr.*, **1989**, *490*, 301–310.

SAMPLE
Matrix: tissue
Analyte: octopine and related compounds
Sample preparation: Homogenize tissue with 3 volumes 1 M perchloric acid, centrifuge at 10000 g for 20 min, neutralize the supernatant with 5 M KOH, centrifuge at 10000 g for 20 min, filter (0.22 μm) the supernatant while centrifuging. 20 μL Filtrate + 10 μL 8 mM benzoin in 2-methoxyethanol + 10 μL 100 mM 2-mercaptoethanol in 200 mM sodium sulfite + 20 μL 4 M KOH, mix, heat at 100° for 5 min, cool in ice-water for 1 min, add 20 μL 4 M HCl in 500 mM pH 9.2 Tris-HCl buffer, mix well, inject a 10 μL aliquot.

HPLC VARIABLES
Column: 250 × 4.6 5 μm Kaseisorb LC ODS-300 (Tokyo Kasei)
Mobile phase: Gradient. A was MeCN:250 mM pH 9.5 Tris-HCl buffer 20:80. B was MeCN:water 80:20. A:B from 100:0 to 50:50 over 20 min, to 0:100 (step gradient), maintain at 100:0 for 5 min.
Column temperature: 40
Flow rate: 1
Injection volume: 10
Detector: F ex 325 em 425

CHROMATOGRAM
Retention time: 7.5 (guanidinosuccinic acid), 11.5 (octopine), 12.5 (guanidinoacetic acid), 12.5 (guanidinopropionic acid), 13.5 (arginine), 17.5 (guanidine), 19 (methylguanidine)
Limit of detection: 5 pmole

KEY WORDS
squid

REFERENCE
Sato, M.; Takeuchi, M.; Kanno, N.; Nagahisa, E.; Sato, Y. Determination of octopine by pre-column fluorescence derivatization using benzoin, *Biochem.Int.*, **1991**, *23*, 1035–1039.

SULFONYL HALIDE

Dansyl Chloride

SAMPLE
Matrix: bulk
Analyte: robenidine
Sample preparation: Prepare a 200 μg/mL solution in dichloromethane:MeOH 90:10, evaporate an aliquot to dryness under a stream of nitrogen at 60°, reconstitute with 1 mL 4 mg/mL 4-dimethylaminopyridine in dichloromethane, add 1 mL 2 mg/mL dansyl chloride in dichloromethane, heat in a capped tube at 80° for 1 h, cool on ice, add to a 3 mL Sep-Pak silica SPE cartridge, rinse the tube with 5 mL dichloromethane, add the rinse to the SPE cartridge, elute with 5 mL ethyl acetate. Evaporate the eluate to dryness under a stream of nitrogen at 60°, reconstitute the residue in 2 mL mobile phase, inject a 20 μL aliquot.

HPLC VARIABLES
Guard column: 150 × 3.2 7 μm silica (Brownlee)
Column: 250 × 4.6 5 μm Zorbax Sil
Mobile phase: Hexane:chloroform:THF:MeOH 50:50:2:1
Flow rate: 2
Injection volume: 20
Detector: F ex 320 em 485

CHROMATOGRAM
Retention time: 4
Limit of detection: 400 ng/mL

KEY WORDS
normal phase; protect from light; SPE

REFERENCE
Cohen, H.; Armstrong, F.; Campbell, H. Sensitive fluorescence detection of robenidine by derivatization with dansyl chloride and high-performance liquid chromatography, *J.Chromatogr.A*, **1995**, *694*, 407–413.

HALIDE

ACTIVATED HALIDE

4-Nitrobenzyl Bromide

SAMPLE
Matrix: solutions
Analyte: iodide
Sample preparation: Protect from light. Mix a 200 μL aliquot of a solution of iodide in water with 1.8 mL 2.5 mM 4-nitrobenzyl bromide in MeCN containing 1.5 mM toluene, stir at 60° for 1 h, filter 0.5 μm, inject a 10 μL aliquot.

HPLC VARIABLES
Column: 150 × 4.6 5 μm YMC A-302 ODS (Yamamura)
Mobile phase: MeOH:water 75:25
Flow rate: 0.5
Injection volume: 10
Detector: UV 254

CHROMATOGRAM
Retention time: 10
Internal standard: toluene (14)
Limit of quantitation: 200 μM

REFERENCE
Funazo, K.; Tanaka, M.; Shono, T. Determination of iodide and sulfide as *p*-nitrobenzyl derivatives by high performance liquid chromatography, *Anal.Sci.*, **1987**, *3*, 41–44.

AROMATIC

Acetanilide

SAMPLE
Matrix: sea water
Analyte: bromide and bromate
Sample preparation: Condition a 500 mg 2.8 mL C18 SPE cartridge (Alltech) with 2 mL MeOH and 2 mL water. Bromide. Dilute with water, filter (0.45 μm). Remove a 5 mL

aliquot of the filtrate and add it to 500 μL reagent, shake for 1 min, add to the SPE cartridge, wash with two 1 mL portions of water, elute with 2 mL MeOH. Evaporate the eluate to dryness under a stream of air, reconstitute the residue in 2 mL mobile phase, inject a 10 μL aliquot. Bromate. Dilute 1 mL reagent to 2 mL with water, stir, add 5 mL filtered (0.45 μm) aqueous sample, add to an SPE cartridge, collect the effluent. Add the effluent to 500 μL 50 mM ascorbic acid, shake for 1 min, add 500 μL 10% hydrogen peroxide, shake for 1 min. Remove a 5 mL aliquot and add it to 500 μL reagent, shake for 1 min, add to a SPE cartridge, wash with two 1 mL portions of water, elute with 2 mL MeOH. Evaporate the eluate to dryness under a stream of air, reconstitute the residue in 2 mL mobile phase, inject a 10 μL aliquot. (Bromide is determined by reaction with acetanilide to produce 4-bromoacetanilide. When bromate is determined the bromide present reacts to produce 4-bromoacetanilide that is left on the first SPE cartridge. The bromate passes through the SPE cartridge and is converted to bromide which is then determined. Prepare the reagent by cautiously adding 1 mL concentrated sulfuric acid to 100 mL MeOH, mix, cool to room temperature, add 150 mg 2-iodosobenzoic acid, add 100 mg acetanilide, dissolve, filter (0.45 μm).)

HPLC VARIABLES
Guard column: 20 × 2 25-40 μm pellicular C18 (Alltech)
Column: 250 × 4.6 5 μm ODS-1 (Anachem)
Mobile phase: MeOH:water 65:35
Flow rate: 1
Injection volume: 10
Detector: UV 240

CHROMATOGRAM
Retention time: 6 (4-bromoacetanilide)
Limit of detection: 2.5 ng/mL

OTHER SUBSTANCES
Interfering: aniline, phenol
Non-interfering: ammonium, bicarbonate, cadmium, calcium, chloride, cobalt, copper, hydroxylamine, iodate, iodide, iron, lead, magnesium, manganese, mercury, nitrate, nitrite, perchlorate, phosphate, sulfate, sulfide, sulfite, thiocyanate, thiosulfate, zinc

KEY WORDS
SPE

REFERENCE
Jain, A.; Chaurasia, A.; Sahasrabuddhey, B.; Verma, K.K. Determination of bromide in complex matrices by pre-column derivatization linked to solid-phase extraction and high-performance liquid chromatography, *J.Chromatogr.A*, **1996**, *746*, 31–41.

SULFONATE

2-(1-Naphthyl)ethyl 2-[1-(4-Benzyl)piperazyl]ethanesulfonate

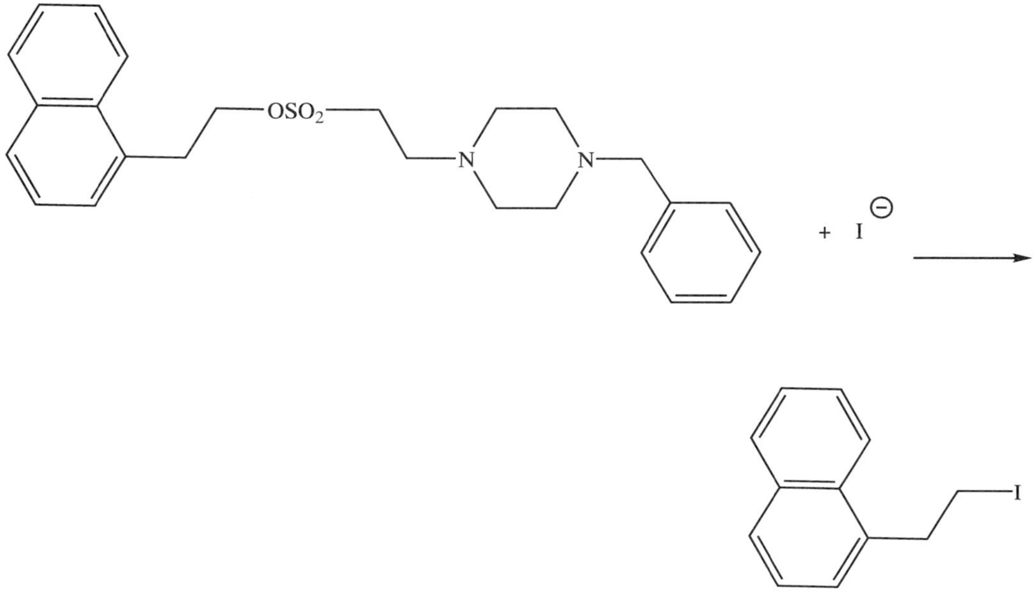

SAMPLE

Matrix: solutions

Analyte: iodide

Sample preparation: Mix 100 µL 250 nM iodide solution in water with 100 µL 200 mM tetrapentylammonium chloride (tetra-n-amylammonium chloride) in water, 500 µL 370 nM 1,2,4,5-tetrachlorobenzene in toluene, 700 µL 28.57 µM 2-(1-naphthyl)ethyl 2-[1-(4-benzyl)piperazyl]ethanesulfonate dihydrochloride in water, and 100 µL 500 mM KOH in water, shake at 95° for 2 h, cool, add 1 mL 1 M sulfuric acid, vortex for 30 s. Remove a 100 µL aliquot of the toluene layer and dilute it 5-fold with MeCN, inject a 25 µL aliquot. (The derivative is 2-(1-naphthyl)ethyl iodide. Excess reagent is protonated by the sulfuric acid and stays in the aqueous layer. The non-basic derivative is in the toluene layer. Preparation of 2-(1-naphthyl)ethyl 2-[1-(4-benzyl)piperazyl]ethanesulfonate dihydrochloride is as follows. Stir 8.61 g 1-naphthaleneethanol (2-(1-naphthyl)ethanol) in 50 mL dichloromethane at 0°, add 15.78 mL 2-chloro-1-ethanesulfonyl chloride, add 31 mL 45% trimethylamine in water (Fluka), stir at 0° for 2 h, wash twice with 50 mL portions of cold 10% sodium carbonate, wash with 50 mL water, dry the organic layer over 2.5 g anhydrous sodium sulfate, concentrate by evaporation under reduced pressure, chromatograph the liquid residue on a 400 × 40 column of 200 g silica gel 60 with ethyl acetate: n-hexane 25:75 to obtain 2-(1-naphthyl)ethyl ethanesulfonate as a colorless oil. Stir 40 mL dichloromethane at 0°, add 7.86 g 2-(1-naphthyl)ethyl ethanesulfonate, add 6.34 g 1-benzylpiperazine, stir at 0° for 30 min, concentrate by evaporation under reduced pressure, chromatograph the liquid residue on a 400 × 40 column of 200 g silica gel 60 with ethyl acetate:n-hexane 1:2 to obtain 2-(1-naphthyl)ethyl 2-[1-(4-benzyl)piperazyl] ethanesulfonate as a colorless oil. Slowly add 12 mL concentrated HCl to a cold solution of 10.95 g 2-(1-naphthyl)ethyl 2-[1-(4-benzyl)piperazyl]ethanesulfonate in 100 mL MeOH, concentrate, recrystallize the residue from EtOH:n-hexane 8:1 to obtain 2-(1-naphthyl)ethyl 2-[1-(4-benzyl)piperazyl]ethanesulfonate dihydrochloride as white needles (mp 183.8-184.5°).)

HPLC VARIABLES
Column: 150 × 3.9 4 μm Nova-Pak C18
Mobile phase: MeCN:water 65:35
Flow rate: 0.8
Injection volume: 25
Detector: UV 282

CHROMATOGRAM
Retention time: 8
Internal standard: 1,2,4,5-tetrachlorobenzene (10)
Limit of detection: 5 pmole

REFERENCE
Kou, H.S.; Wu, H.L.; Chen, S.H.; Wu, S.M. Chemically removable derivatization reagent for chromatography. II. 2-(1-Naphthyl)ethyl 2-[1-(4-benzyl)piperazyl]ethanesulfonate dihydrochloride, *J.Liq.Chromatogr.*, **1995**, *18*, 2323–2335.

HALOALDEHYDE

ADENOSINE-CONTAINING COMPOUND

Adenosine

SAMPLE
Matrix: microsomal incubations
Analyte: 2-bromoacetaldehyde
Sample preparation: Heat rat liver microsomal incubation containing 25 mM adenosine at 37° for 15 min, add zinc sulfate to a final concentration of 30 mM, centrifuge at 4000 g for 10 min, heat the supernatant at 65° for 4 h, inject a 20 μL aliquot.

HPLC VARIABLES
Guard column: 10 × 2 5 μm chromsphere 5C18
Column: two 100 × 3 5 μm chromsphere 5C18 glass columns in series (Chrompack)
Mobile phase: MeCN:50 mM ammonium acetate 5:95, adjusted to pH 5.1 with concentrated HCl
Flow rate: 0.4
Injection volume: 20
Detector: F ex 275 em 410

CHROMATOGRAM
Retention time: 13.1
Limit of quantitation: 50 nM

KEY WORDS
rat; liver

REFERENCE
Wormhoudt, L.W.; Ploemen, J.H.T.M.; Commandeur, J.N.M.; van Ommen, B.; van Bladeren, P.J.; Vermeulen, N.P.E. Cytochrome P450 catalyzed metabolism of 1,2-dibromoethane in liver microsomes of differentially induced rats, *Chem.Biol.Interact.*, **1996**, *99*, 41–53.

THIOUREA

Thiourea

SAMPLE
Matrix: blood
Analyte: chloroacetaldehyde
Sample preparation: Condition a 1 mL Bond Elut SCX SPE cartridge with 1 mL MeOH, 1 mL MeCN, two 1 mL portions of water, and two 1 mL portions of 1% acetic acid, do not allow to dry. Add 100 µL 50 mM formaldehyde to 5 mL heparinized whole blood, centrifuge to obtain plasma. 500 µL Plasma + 60 µL 70% perchloric acid, mix, centrifuge at 3000 g for 2 min. Remove a 200 µL aliquot of the supernatant and add it to 30 µL of a 7.6 g/L thiourea (Caution! Thiourea is a carcinogen!), heat at 90° for 2 h, cool, add to the SPE cartridge, wash four times with 1 mL portions of 1% acetic acid, wash with three 1 mL portions of MeOH:1% acetic acid 50:50, elute with 1 mL MeOH:14 M ammonia 96: 4. Adjust the pH of the eluate to about 6 with 60 µL 3 M HCl, evaporate to dryness under a stream of nitrogen at 35°, reconstitute with 200 µL mobile phase, inject a 20 µL aliquot. (Formaldehyde stabilizes chloroacetaldehyde in plasma.)

HPLC VARIABLES
Column: 150 × 4.6 5 µm ChromSpher C8 (Chrompack)
Mobile phase: MeOH:buffer 5:95 (Buffer was 10 mM pH 7.4 phosphate buffer containing 0.05% triethylamine.)
Flow rate: 1
Injection volume: 20
Detector: UV 254

CHROMATOGRAM
Retention time: 5
Limit of detection: 500 nM

OTHER SUBSTANCES
Non-interfering: acrolein

KEY WORDS
plasma; SPE; pharmacokinetics

REFERENCE
Kaijser, G.P.; Beijnen, J.H.; Jeunink, E.L.; Bult, A.; Keizer, H.J.; de Kraker, J.; Underberg, W.J.M. Determination of chloroacetaldehyde, a metabolite of oxazaphosphorine cytostatic drugs, in plasma, *J.Chromatogr.*, **1993**, *614*, 253–259.

SAMPLE
Matrix: microsomal incubations
Analyte: chloroacetaldehyde
Sample preparation: Condition a 1 mL 100 mg Bond Elut SCX benzenesulfonic acid cation-exchange SPE cartridge with 1 mL MeOH, with 1 mL MeCN, with two 1 mL portions of water, and with two 1 mL portions of 1% acetic acid. 500 µL Microsomal incubation + 200 µL 5% zinc sulfate + 200 µL 2.5% barium hydroxide, mix, centrifuge at 450 g for 10 min. Remove a 600 µL aliquot of the supernatant and add it to 30 µL 100 mM thiourea, heat at 90° for 2 h, cool to room temperature, adjust to pH 3 with 2 M HCl, add to the SPE cartridge, wash with four 1 mL portions of 1% acetic acid, wash with three 1 mL portions of MeOH:1% acetic acid 50:50, elute with 1 mL MeOH:14.5 M ammonium hy-

droxide 96:4, adjust pH of eluate to 6 with about 100 μL 3.2 M HCl, evaporate to dryness under a stream of nitrogen at 35°, add 200 μL mobile phase, vortex, inject an aliquot.

HPLC VARIABLES
Column: 250 × 4.6 10 μm Econosil C18
Mobile phase: MeCN:50 mM pH 7 K_2HPO_4 20:80
Flow rate: 1
Injection volume: 20
Detector: UV 255

CHROMATOGRAM
Retention time: 5.9
Limit of detection: 4 μM

KEY WORDS
rat; liver; SPE

REFERENCE
Ruzicka, J.A.; Ruenitz, P.C. Derivatization-liquid chromatographic assay of chloroacetaldehyde in biological samples, *J.Chromatogr.*, **1990**, *518*, 385–389.

HALOGENATED COMPOUND

DITHIOCARBAMATE

Sodium Diethyldithiocarbamate

$$RX \quad + \quad {}^{\ominus}S \overset{S}{\underset{}{\diagdown}} N(CH_2CH_3)_2 \qquad \longrightarrow \qquad RS \overset{S}{\underset{}{\diagdown}} N(CH_2CH_3)_2$$

SAMPLE
Matrix: blood
Analyte: mitolactol
Sample preparation: 500 µL Plasma + 1 mL ice-cold MeOH, let stand at -20° for 20 min, centrifuge at 1500 g for 5 min. Remove a 1 mL aliquot of the supernatant and add it to 1 mL 5% diethyldithiocarbamate and 2 mL 50 mM pH 7.4 potassium phosphate buffer, heat at 50° for 1 h, extract with 5 mL chloroform. Remove 4 mL of the organic layer and evaporate it to dryness under a stream of nitrogen at 40°, reconstitute, inject an aliquot.

HPLC VARIABLES
Column: 250 × 4.6 5 µm Alltech CN
Mobile phase: Heptane:isopropanol:acetic acid 74.4:21.6:4
Flow rate: 1.3
Detector: UV 254

CHROMATOGRAM
Retention time: 9.9
Limit of detection: 500 nM

KEY WORDS
plasma; human; mouse; pharmacokinetics

REFERENCE
Henner, W.D.; Furlong, E.A.; Kelley, S.L.; Rosowsky, A. Assay for mitolactol and its bifunctional alkylating metabolites in plasma, *J.Pharm.Sci.*, **1985**, *74*, 983–986.

SAMPLE
Matrix: blood
Analyte: nitrogen mustard anticancer drugs
Sample preparation: Condition a 2.4 mL 500 mg Bond Elut phenyl SPE cartridge with 2 mL MeOH and 2 mL water. 1 mL Plasma + 100 µL freshly prepared 100 mg/mL diethyldithiocarbamic acid in 100 mM NaOH, heat at 37° for 30 min (mechlorethamine, galamustine) or 50° for 90 min (melphalan), add to the SPE cartridge, wash with 5 mL water, allow to dry in the dark at room temperature for 1 h, elute with 2 mL MeCN. Evaporate the eluate to dryness under reduced pressure at 40°, reconstitute in 200 µL MeOH, inject a 100 µL aliquot.

HPLC VARIABLES
Column: 300 × 3.8 10 µm µBondapak C18
Mobile phase: Gradient. MeCN:5 mM pH 3.0 orthophosphoric acid 30:70 for 4 min, to 100:0 over 9 min, return to initial conditions over 7 min, re-equilibrate for 5 min.
Column temperature: 40
Flow rate: 1
Injection volume: 100
Detector: UV 276

CHROMATOGRAM
Retention time: 10.7 (galamustine), 13.1 (mechlorethamine), 14.6 (melphalan)

Limit of detection: 1 ng/mL

KEY WORDS
plasma; SPE

REFERENCE
Cummings, J.; MacLellan, A.; Smyth, J.F.; Farmer, P.B. Determination of reactive nitrogen mustard anticancer drugs in plasma by high-performance liquid chromatography using derivatization, *Anal.Chem.*, **1991**, *63*, 1514–1519.

SAMPLE
Matrix: blood
Analyte: ifosforamide mustard
Sample preparation: Condition a Sep-Pak C18 SPE cartridge with three 1 mL portions of MeOH and three 1 mL portions of 25 μM cyclohexylamine in 500 mM pH 8.0 sodium phosphate buffer containing 1 M NaCl. 500 μL Plasma (stabilized with semicarbazide) + 500 μL 500 mM pH 8.0 sodium phosphate buffer containing 1 M NaCl + 100 μL 100 mg/mL sodium diethyldithiocarbamate, heat at 70° for 30 min, cool to 0°, add to the SPE cartridge, wash three times with 25 μM cyclohexylamine in 500 mM pH 8.0 sodium phosphate buffer containing 1 M NaCl, elute with 1 mL MeOH. Evaporate the eluate to dryness under a stream of nitrogen at 30°, reconstitute the residue in 100 μL MeCN:water 27.5:72.5, inject a 10 μL aliquot.

HPLC VARIABLES
Column: 125 × 4 5 μm LichroCART RP8
Mobile phase: MeCN:water 32:68 containing 10 mM sodium phosphate buffer and 20 mM cyclohexylamine, pH 7.0
Injection volume: 10
Detector: UV 280

CHROMATOGRAM
Retention time: 4
Limit of quantitation: 450 nmoles

KEY WORDS
plasma; SPE

REFERENCE
Kaijser, G.P.; Beijnen, J.H.; Rozendom, E.; Bult, A.; Underberg, W.J.M. Analysis of ifosforamide mustard, the active metabolite of ifosfamide, in plasma, *J.Chromatogr.B*, **1996**, *686*, 249–255.

SAMPLE
Matrix: formulations
Analyte: mechlorethamine
Sample preparation: 0.1 g Ointment + 8 mL chloroform + 1.9 mL isopropanol + 100 μL water, shake until homogeneous, add 10 mL 10 mM HCl, invert twice for 2-3 min with a 10 min interval. Remove the aqueous phase and neutralize it with 1 mL 100 mM NaOH, add 1 mL 100 mg/mL diethyldithiocarbamic acid in 100 mM NaOH, heat at 37° for 1 h. Remove a 1 mL aliquot and add it to 1 mL MeOH, inject a 100 μL aliquot.

HPLC VARIABLES
Column: 300 × 3.8 10 μm μBondapak C18
Mobile phase: Gradient. MeCN:5 mM pH 3.0 orthophosphoric acid 30:70 for 4 min, to 100:0 over 9 min, return to initial conditions over 7 min, re-equilibrate for 5 min.
Column temperature: 40
Flow rate: 1
Injection volume: 100
Detector: UV 276

CHROMATOGRAM
Retention time: 13.1

OTHER SUBSTANCES
Extracted: degradation products

KEY WORDS
ointment; chromatographic procedure as Anal. Chem. 1991, 63, 1514

REFERENCE
Cummings, J.; MacLellan, A.; Langdon, S.J.; Smyth, J.F. The long term stability of mechlorethamine hydrochloride (nitrogen mustard) ointment measured by HPLC, *J.Pharm.Pharmacol.*, **1993**, *45*, 6– 9.

SAMPLE
Matrix: tissue
Analyte: prospidin
Sample preparation: Homogenize tissue with an equal volume of water, add 200 μL homogenate to 600 μL MeOH, let stand at room temperature for 1 h, centrifuge. Add 150 μL supernatant to 100 μL 2 mg/mL sodium diethyldithiocarbamate in 50 mM NaOH, heat at 37° for 1.5 h, add an equal volume of MeOH, inject an aliquot.

CAPILLARY ELECTROPHORESIS
Capillary: 50 cm × 70 μm quartz (42 cm to detector)
Capillary preparation: Condition capillary with MeOH:500 mM NaOH 50:50 for 10 min, with water for 10 min, and with running buffer for 10 min.
Running buffer: MeOH:20 mM pH 11.2 borate buffer 50:50
Capillary temperature: 20
Voltage/Current: 14 kV
Injection: Electroosmotic flow injection for 15 s
Detector: UV 254
Migration time: 9.2
Limit of detection: 1 μg/mL (S/N 3)

REFERENCE
Okun, V.M.; Aak, O.V.; Kozlov, V.Y. Determination of the anticancer drug prospidin in human tissue by high-performance capillary electrophoresis using derivatization, *J.Chromatogr.B*, **1996**, *675*, 313– 319.

PHENOL

2,6-Dimethylphenol

SAMPLE
Matrix: blood, solutions
Analyte: bromide, iodate, iodide
Sample preparation: Serum. Centrifuge 2 mL serum at 5000 g, evaporate the clear supernatant to dryness under a stream of nitrogen at 50°, reconstitute with mobile phase. Mix 0.1-1 mL solution with 1 mL buffer, 500 μL 100 μg/mL biphenyl in mobile phase, 1 mL 1 mg/mL 2,6-dimethylphenol in mobile phase, and 1 mL reagent, let stand at 21° for 20 min, make up to 10 mL with mobile phase, inject a 10 μL aliquot. (Iodide is oxidized to iodine and the iodine reacts with 2,6-dimethylphenol to give 2,6-dimethyl-4-iodo-

phenol). Bromide solution. Mix 0.1-1 mL solution with 1 mL buffer, 500 μL 100 μg/mL biphenyl in mobile phase, 1 mL 1 mg/mL 2,6-dimethylphenol in mobile phase, and 1 mL reagent, shake well, let stand for 20 min, add 1 mL sulfuric acid solution, shake for 1 min, make up to 10 mL with mobile phase, inject a 10 μL aliquot. (Bromide is oxidized to bromine and the bromine reacts with 2,6-dimethylphenol to give 2,6-dimethyl-4-bromophenol). Iodate solution. Mix 0.1-1 mL solution with 1 mL 1 mg/mL ascorbic acid in water, swirl for 1-2 min, add 1 mL buffer, 500 μL 100 μg/mL biphenyl in mobile phase, 1 mL 1 mg/mL 2,6-dimethylphenol in mobile phase, and 1 mL reagent, let stand at 21° for 20 min, make up to 10 mL with mobile phase, inject a 10 μL aliquot. (Iodate is reduced to iodide then oxidized to iodine and the iodine reacts with 2,6-dimethylphenol to give 2,6-dimethyl-4-iodophenol). (Prepare buffer by dissolving 10 g KH_2PO_4 and 10 g K_2HPO_4 in 250 mL water, pH 6.4. Prepare reagent by stirring 400 mg 2-iodosobenzoic acid in 7.6 mL 200 mM NaOH, make up to 100 mL with water, filter (0.45 μm). Prepare sulfuric acid solution by cautiously adding 2 mL concentrated sulfuric acid to 80 mL water then making up to 100 mL with water.)

HPLC VARIABLES
Column: 150 × 4.6 5 μm Shim-pack ODS (Shimadzu)
Mobile phase: MeCN:water 60:40 (iodide) or 45:55 (bromide)
Flow rate: 1
Injection volume: 10
Detector: UV 220

CHROMATOGRAM
Retention time: 8 (2,6-dimethyl-4-bromophenol), 4 (2,6-dimethyl-4-iodophenol)
Internal standard: biphenyl (6 (mobile phase 60:40), 15 (mobile phase 45:55))
Limit of detection: 0.5 ng (iodide; 100 μL injection), 0.2 ng (bromide; 100 μL injection)

OTHER SUBSTANCES
Interfering: aniline, phenol (remove by passing through a column of Amberlite XAD-2)
Non-interfering: bicarbonate, cadmium, calcium, chloride, cobalt, iron, magnesium, manganese, nitrate, nitrite, perchlorate, phosphate, sulfate, sulfide, sulfite, thiocyanate, thiosulfate, zinc

KEY WORDS
serum

REFERENCE
Verma, K.K.; Jain, A.; Verma, A. Determination of iodide by high-performance liquid chromatography after precolumn derivatization, *Anal.Chem.*, **1992**, *64*, 1484–1489.

PICRATE

Silver Picrate

SAMPLE
Matrix: gasoline
Analyte: ethylene dibromide
Sample preparation: Dilute gasoline 10-fold with MeCN, add a 200 μL aliquot to the reagent, heat at 160° for 30 min, cool, add 500 μL 100 mM NaCl in MeOH, filter (0.45

μm), wash filter with MeCN:water 50:50, inject a 250 μL aliquot on to column A and elute to waste with water, after 3 min backflush the contents of column A on to column B with mobile phase, elute with mobile phase, monitor the effluent from column B. (Prepare the reagent, silica supported silver picrate, as follows. Reflux silica gel (Waters 37-55 μm for normal phase HPLC) in 2 M HCl for 2 h, wash with water, wash with MeOH, dry under vacuum over phosphorus pentoxide at 110° for 12 h. Add a solution of silver picrate in MeCN (enough to form a 10% loading) to the silica, evaporate to dryness under reduced pressure, dry under vacuum at 61° for 12 h. Prepare silver picrate as follows. Standardize an aqueous solution of picric acid with NaOH, add an equivalent of silver acetate to an aliquot of the picric acid solution, evaporate to dryness, dry under vacuum over phosphorus pentoxide at room temperature (J.Chromatogr. 1985, 333, 349). Caution! Although explosions have not been reported these compounds are potentially explosive and full safety precautions, including a safety shield and eye protection, should be observed!)

HPLC VARIABLES
Column: A Guard-Pak CN (Waters); B 100 × 8 Radial-Pak Resolve CN
Mobile phase: MeOH:200 mM NaCl 65:35
Flow rate: 2
Injection volume: 250
Detector: E, Bioanalytical Systems LC-4A or LC-4B, glassy carbon working electrode -0.8 V, stainless steel counter electrode, Ag/AgCl reference electrode following post-column reaction. The column effluent flowed through a 9.144 m × 0.5 mm i.d. figure eight PTFE coil irradiated with UV light (Photronix Model 816) (maintained at 0-5° with an ice bath) and flowed to the detector.

CHROMATOGRAM
Retention time: 9
Limit of detection: 0.28 ppm

KEY WORDS
column-switching; post-column reaction

REFERENCE
Colgan, S.T.; Krull, I.S.; Dorschel, C.; Bidlingmeyer, B. Derivatization of ethylene dibromide with silica-supported silver picrate for improved high-performance liquid chromatographic detection, *Anal.Chem.*, **1986**, *58*, 2366–2372.

SAMPLE
Matrix: solutions
Analyte: 1-iodopentane
Sample preparation: Heat 200 μL of a solution of 1-iodopentane in MeCN with enough reagent to absorb all the liquid at 90° for 2 h, cool, elute with mobile phase A (?), inject an aliquot of the eluate on to column A and elute to waste with mobile phase A, after 8 min backflush the contents of column A on to column B with mobile phase B, elute with mobile phase B, monitor the effluent from column B. (Prepare the reagent, silica supported silver picrate, as follows. Reflux silica gel (Waters 37-55 μm for normal phase HPLC) in 2 M HCl for 2 h, wash with water, wash with MeOH, dry under vacuum over phosphorus pentoxide at 110° for 12 h. Add a solution of silver picrate in MeCN (enough to form a 10% loading) to the silica, evaporate to dryness under reduced pressure, dry under vacuum at 61° for 12 h. Prepare silver picrate as follows. Standardize an aqueous solution of picric acid with NaOH, add an equivalent of silver acetate to an aliquot of the picric acid solution, evaporate to dryness, dry under vacuum over phosphorus pentoxide at room temperature. Caution! Although explosions have not been reported these compounds are potentially explosive and full safety precautions, including a safety shield and eye protection, should be observed!)

HPLC VARIABLES
Column: A μBondapak CN pre-column; B 300 × 3.9 μBondapak CN
Mobile phase: A water; B MeCN:water 45:55
Flow rate: 2
Detector: UV 220

CHROMATOGRAM
Retention time: 16
Limit of detection: <100 ppb

KEY WORDS
column-switching

REFERENCE
Colgan, S.T.; Krull, I.S.; Neue, U.; Newhart, A.; Dorschel, C.; Stacey, C.; Bidlingmeyer, B. Derivatization of alkyl halides and epoxides with picric acid salts for improved high-performance liquid chromatographic detection, *J.Chromatogr.*, **1985**, *333*, 349–364.

SOLID PHASE REAGENT

Polymer-supported 8-Amino-2-naphthoxide

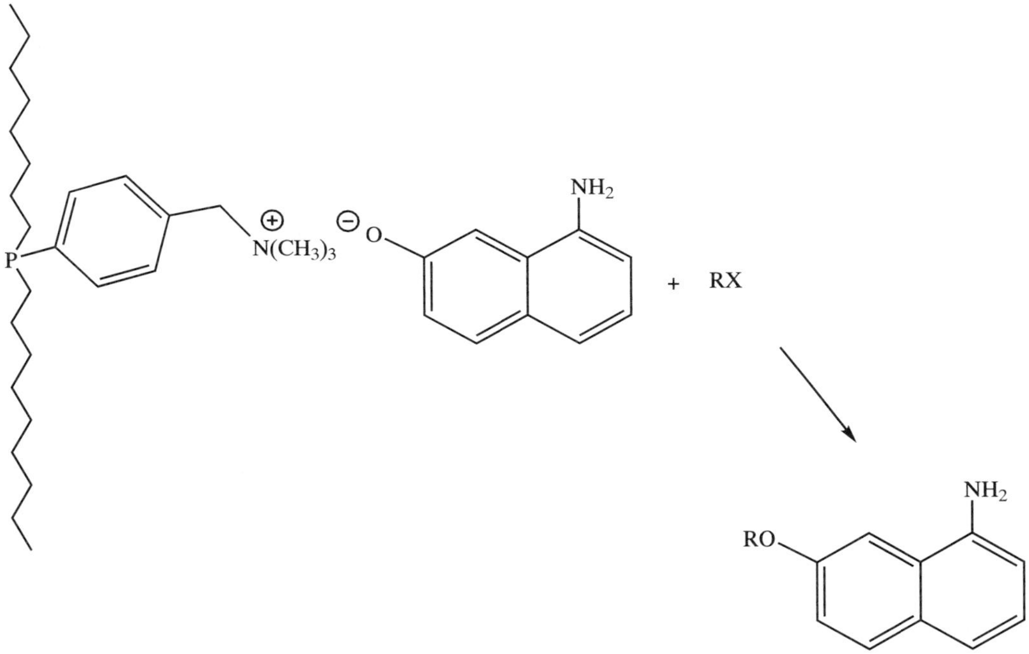

SAMPLE
Matrix: solutions
Analyte: 1-bromopentane
Sample preparation: Mix a 20 μL aliquot of a 153 ppm solution with 40 mg polymer-supported 8-amino-2-naphthoxide, heat at 60° for 30 min, elute with MeCN, inject a 20 μL aliquot of the eluate. (Preparation of polymer-supported 8-amino-2-naphthoxide is as follows. Extract -400 mesh AG 1-X8 anion exchange resin (Cl⁻ form, Bio-Rad) in a Soxhlet extractor with MeCN for 24 h. Pack the resin in a 170 × 12 glass column and wash with 500 mL water, wash with 2.5 L 1 M NaOH. (Check that the effluent is chloride free by acidifying an aliquot with nitric acid and adding silver nitrate.) Wash with 1 L water. Stir 10 g of this resin in basic water with a 0.5 molar excess of 8-amino-2-naphthol for 3 h, add 1 equivalent of NaOH, filter, wash three times with 200 mL portions of water, wash with 200 mL MeCN, dry in air, store in the freezer. 8-Amino-2-naphthol is available from Tokyo Kasei (TCI America, Portland OR). Synthesis is as follows (Caution! Molten

NaOH is extremely corrosive, use appropriate protective equipment!). Melt NaOH with the addition of a small quantity of water, gradually add 8-amino-2-naphthalenesulfonic acid so as to avoid foaming, when addition is complete heat at 265-275° for 1 h, heat at 305° for 5 min, cool to 200°, cautiously dilute with water, neutralize with HCl, boil, filter, add NaOH until no more precipitate of 8-amino-2-naphthol is obtained (J. Am. Chem. Soc. 1929, 51, 1766).)

HPLC VARIABLES
Column: 100 × 8 Radial Pak C18
Mobile phase: MeCN:water 75:25
Column temperature: 60
Flow rate: 2
Injection volume: 20
Detector: F ex 254 em 425; UV 214

CHROMATOGRAM
Retention time: 4.5
Limit of detection: 0.04 ppb (F), 4 ppb (UV)

REFERENCE
Colgan, S.T.; Krull, I.S.; Dorschel, C.; Bidlingmeyer, B.A. Derivatization of alkyl halides, acid chlorides, and other electrophiles with polymer-immobilized 8-amino-2-naphthoxide, *J.Chromatogr.Sci.*, **1988**, 26, 501−512.

THIOL

N-Acetylcysteine

SAMPLE
Matrix: blood
Analyte: melphalan
Sample preparation: 1 mL Plasma + 1 mL 250 mM trichloroacetic acid, mix at 4°, let stand at 4° for 30 min, centrifuge at 1500 g for 10 min. Remove a 1 mL aliquot of the supernatant and add it to 200 μL 1 M N-acetylcysteine, 200 μL 2.65 M NaOH, and 200 μL 500 mM pH 11.0 phosphate buffer, mix, heat at 70° for 15 min, add 200 μL 2 M citric acid, filter (0.2 μm), inject a 100 μL aliquot of the filtrate.

HPLC VARIABLES
Column: 150 × 4 5 μm Nova-Pak C18 Radial-Pak
Mobile phase: EtOH:36 mM pH 4.25 citrate buffer containing 50 μM octanesulfonic acid 4:96 (The mobile phase flowed through a 100 × 4 column of μBondapak C18 before the injector.)
Injection volume: 100
Detector: F ex 260 em 360

CHROMATOGRAM
Retention time: 13
Limit of quantitation: 5 ng/mL

KEY WORDS
plasma; pharmacokinetics

REFERENCE

Ehrsson, H.; Eksborg, S.; Lindfors, A. Quantitative determination of melphalan in plasma by liquid chromatography after derivatization with N-acetylcysteine, *J.Chromatogr.*, **1986**, *380*, 222–228.

HYDRAZINE

ALDEHYDE

p-Anisaldehyde

RELATED REFERENCE
Semple, H.A.; Tam, Y.K.; Croteau, S.M.; Coutts, R.T. Stability problems with hydralazine p-anisaldehyde hydrazone. *J.Pharm.Sci.* **1989**, *78*, 432-434.

SAMPLE
Matrix: blood
Analyte: hydralazine
Sample preparation: 3 mL Whole blood + 20 μL p-anisaldehyde + 8 μL 5 μg/mL 4-methylhydralazine in 10 mM HCl, vortex for 15 s, let stand at room temperature for 10 min, add 10 mL hexane, shake horizontally at 180 strokes/min for 10 min, centrifuge at 1000 g for 10 min. Remove the organic layer and evaporate it to dryness under a stream of nitrogen, reconstitute the residue in 100 μL MeOH, inject the whole amount.

HPLC VARIABLES
Column: 300 × 3.9 μBondapak CN
Mobile phase: MeCN:150 mM pH 3.0 sodium acetate buffer 70:30
Flow rate: 2
Injection volume: 100
Detector: UV 365

CHROMATOGRAM
Retention time: 3.5
Internal standard: 4-methylhydralazine (6)
Limit of quantitation: 1 ng/mL

OTHER SUBSTANCES
Non-interfering: digoxin, furosemide, hydrochlorothiazide, nitroglycerin, propranolol

KEY WORDS
whole blood

REFERENCE
Ludden, T.M.; Ludden, L.K.; Wade, K.E.; Allerheiligen, S.R. Determination of hydralazine in human whole blood, *J.Pharm.Sci.*, **1983**, *72*, 693–695.

Benzaldehyde

SAMPLE
Matrix: formulations
Analyte: hydrazine and phenelzine
Sample preparation: Mix powdered tablet with 5 mL 200 mM pH 6 sodium acetate buffer, rotate at 30 rpm for 30 min, centrifuge. Remove a 1 mL aliquot of the supernatant and add it to 1 mL 15 mg/mL benzaldehyde in MeOH:water 50:50, rotate at 30 rpm for 10 min, add 20 mL 3 μg/mL IS in mobile phase, rotate at 30 rpm for 30 min, inject a 50 μL aliquot of the upper layer.

HPLC VARIABLES
Column: 150 × 4.6 Ultrasphere Si
Mobile phase: n-Hexane:chloroform 95:5
Flow rate: 2
Injection volume: 50
Detector: UV 313

CHROMATOGRAM
Retention time: 7 (hydrazine), 17 (phenelzine)
Internal standard: iminodibenzyl (2)
Limit of detection: 0.6 ng (hydrazine)

KEY WORDS
tablets; normal phase

REFERENCE
Matsui, F.F.; Butterfield, A.G.; Curran, N.M.; Lovering, E.G.; Sears, R.W.; Robertson, D.L. Determination of hydrazine in pharmaceuticals. Part 2. Phenelzine sulfate, *Can.J.Pharm.Sci.*, **1981**, *16*, 20–22.

Cinnamaldehyde

SAMPLE
Matrix: blood
Analyte: isoniazid
Sample preparation: 500 μL Serum + 100 μL 30% trichloroacetic acid, mix, centrifuge at 2000 g for 5 min. Remove a 100 μL aliquot of the supernatant and add it to 20 μL 0.1% trans-cinnamaldehyde in MeOH, let stand for 10 min, add 20 μL 1 M KOH, inject an aliquot.

HPLC VARIABLES
Column: 125 × 3.9 4 μm Nova-pak C18
Mobile phase: MeCN:water:triethylamine:acetic acid 40:60:0.2:0.1, pH 5 ± 1

Flow rate: 1.3
Detector: UV 340

CHROMATOGRAM
Retention time: 1.95
Limit of detection: 20 ng/mL

OTHER SUBSTANCES
Non-interfering: aceprometazine, adrafinil, allopurinol, alprostadil, altretamine, atenolol (tenormine), baclofen, bendroflumethiazide, benserazide, betamethasone, bisoprolol, bromocriptine, caffeine, captopril, chlorpromazine, clomipramine, clonazepam, cortisone, cyamemazine, difebarbamate, dothiepin (dosulepin), ethosuximide, fenspiride, flumazenil, fluoxetine, fluvoxamine, halofantrine, hydrochlorothiazide, hydroxyzine, ibuprofen, imipramine, levamisol, levodopa, maprotiline, medifoxamine, metopimazine, midazolam, nafronyl (naftidrofuryl), naftazone, naproxen, nicergoline, nitrazepam, nordazepam, nortriptyline, penfluridol, phenobarbital, pimozide, pipamperone, pipotizine, primidone, pyrazinamide, pyridoxine, quinine, rifampin, selegiline (deprenyl), streptomycin,tetrazepam, theophylline, thioproperazine, tiapride, triazolam, trihexyphenidyl, trimeprazine (alimemazine), trimipramine, tropatepine, vigabatrin, zopiclone

KEY WORDS
serum

REFERENCE
Sadeg, N.; Pertat, N.; Dutertre, H.; Dumontet, M. Rapid, specific and sensitive method for isoniazid determination in serum, *J.Chromatogr.B*, **1996**, *675*, 113–117.

SAMPLE
Matrix: blood, CSF, urine
Analyte: isoniazid and hydrazine
Sample preparation: Centrifuge 1.5 mL whole blood, CSF, or urine at 3000 g for 3 min, add 500 µL of the supernatant to 500 µL 10% trichloroacetic acid, centrifuge at 10000 g for 1 min. Remove a 200 µL aliquot and add it to 20 µL water, add 40 µL 1% trans-cinnamaldehyde in MeOH, let stand at room temperature for 10 min, inject a 20 µL aliquot.

HPLC VARIABLES
Column: 250 × 4.6 Partisil 5 C8
Mobile phase: Gradient. A was 50 mM KH$_2$PO$_4$. B was MeCN:isopropanol 80:20. A:B 60:40 for 1 min, to 30:70 over 9 min, maintain at 30:70 for 4.5 min, re-equilibrate at initial conditions for 4 min.
Column temperature: 50
Flow rate: 1
Injection volume: 20
Detector: UV 340

CHROMATOGRAM
Retention time: 7 (isoniazid), 13 (hydrazine)
Limit of detection: 500 ng/mL

KEY WORDS
whole blood

REFERENCE
Seifart, H.I.; Gent, W.L.; Parkin, D.P.; van Jaarsveld, P.P.; Donald, P.R. High-performance liquid chromatographic determination of isoniazid, acetylisoniazid and hydrazine in biological fluids, *J.Chromatogr.B*, **1995**, *674*, 269–275.

SAMPLE
Matrix: milk
Analyte: isoniazid
Sample preparation: Condition a 3 mL 500 mg C18 SPE cartridge (J.T. Baker) with 6 mL MeOH and 6 mL water. 80 mL Milk + 20 mL 20% trichloroacetic acid in water, mix, let stand at room temperature for 5 min, centrifuge at 5500 g for 15 min. Filter (0.45 µm)

75 mL of the liquid, add 1 mL 1% cinnamaldehyde in MeOH to the filtrate, mix for a few s, let stand at room temperature for 15 min, add to the SPE cartridge at 5 mL/min, dry for 10 min, wash with 3 mL mobile phase at 0.2 mL/min, wash with 5 mL n-hexane at 0.3 mL/min, dry for 10 min, elute with 6 mL MeOH. Evaporate the eluate under reduced pressure at 35-40°, reconstitute the residue with 200 μL mobile phase, inject an aliquot.

HPLC VARIABLES
Guard column: 4 × 4 5 μm LiChrospher 100-CN
Column: 250 × 4 5 μm LiChrospher 100-CN
Mobile phase: MeOH:water 40:60 containing 0.41 g/L sodium acetate trihydrate and 10 mL/L glacial acetic acid
Flow rate: 1
Injection volume: 100
Detector: UV 330

CHROMATOGRAM
Retention time: 7.4
Limit of detection: 0.1 ng/mL

KEY WORDS
cow; SPE

REFERENCE
Defilippi, A.; Piancone, G.; Costa Laia, R.; Balla, S.; Tibaldi, G.P. High-performance liquid chromatography with UV detection and diode-array UV confirmation of isonicotinic acid hydrazide in cattle milk, *J.Chromatogr.B*, **1994**, *656*, 466−471.

SAMPLE
Matrix: milk
Analyte: isoniazid
Sample preparation: Condition a 3 mL 500 mg 40 μm phenyl SPE cartridge (J.T. Baker) with 6 mL MeOH and 6 mL water. 80 mL Milk + 20 mL 20% trichloroacetic acid in water, let stand at room temperature for 5 min, centrifuge at 5500 g for 15 min, filter (0.45 μm cellulose acetate) a 75 mL aliquot. Add the filtrate to 1 mL 1% cinnamaldehyde in MeOH, mix for a few s, let stand at room temperature for 15 min. Add 70 mL of the sample to the SPE cartridge at 3 mL/min, wash with 3 mL MeOH:water 40:60 at 3 mL/min, dry under vacuum for 1 min, wash with 3 mL MeOH:water 46:54 at 3 mL/min, dry under vacuum for 1 min, wash with 3 mL n-hexane at 3 mL/min, dry under vacuum for 1 min, elute with 6 mL MeOH. Evaporate the eluate under vacuum, reconstitute in 233 μL mobile phase, inject an aliquot.

HPLC VARIABLES
Guard column: 4 × 4 5 μm LiChrosphere 100 RP18
Column: 250 × 4 5 μm LiChrosphere 100 RP18
Mobile phase: MeCN:MeOH:buffer 41:13:46 (Buffer was 1% ammonium acetate adjusted to pH 5.5 with acetic acid.)
Flow rate: 1
Injection volume: 100
Detector: UV 330

CHROMATOGRAM
Retention time: 4.02
Limit of detection: 0.05 ng/mL

KEY WORDS
cow; SPE

REFERENCE
Defilippi, A.; Piancone, G.; Costa Laia, R.; Tibaldi, G.P. An HPLC screening method for the detection of isonicotinic acid hydrazide in cattle milk, *Chromatographia*, **1995**, *40*, 170−174.

2-Hydroxy-1-naphthaldehyde

RELATED REFERENCE
El-Brashy, A.M.; Al-Ghannam, S.M. High-performance liquid chromatographic determination of some amino acids after derivatization with 2-hydroxy-1-naphthaldehyde. *Analyst* **1997**, *122*, 147-150.

SAMPLE
Matrix: blood
Analyte: hydralazine and dihydralazine
Sample preparation: 1 mL Plasma + 1 mL 20 mM disodium EDTA + 1 mL 500 mM HCl + 1 mL 10 mM 2-hydroxy-1-naphthaldehyde, vortex for 15 s, keep at 25° for 90 min, add 50 μL 1 μg/mL methyl red, add 7 mL dichloromethane, vortex for 5 min, centrifuge at 4500 rpm for 15 min. Remove the organic layer and evaporate it to dryness under a stream of nitrogen at 50°, reconstitute the residue in 100 μL MeCN, inject a 20 μL aliquot.

HPLC VARIABLES
Column: 250 × 4 3 μm Spherisorb ODS-2
Mobile phase: MeCN:buffer 80:20, pH 3 (Buffer was 0.75% phosphoric acid and 0.5% triethylamine in water.)
Flow rate: 0.7
Injection volume: 20
Detector: UV 406

CHROMATOGRAM
Retention time: 6.4 (hydralazine), 14.2 (dihydralazine)
Internal standard: methyl red (6)
Limit of detection: 1 ng/mL

KEY WORDS
plasma; pharmacokinetics

REFERENCE
Mañes, J.; Mari, J.; Garcia, R.; Font, G. Liquid chromatographic determination of hydralazine in human plasma with 2-hydroxy-1-naphthaldehyde pre-column derivatization, *J.Pharm.Biomed.Anal.*, **1990**, *8*, 795–798.

2-Nitrobenzaldehyde

RELATED REFERENCES

Hoogenboom, L.A.P.; Berghmans, M.C.J.; Polman, T.H.G.; Parker, R.; Shaw, I.C. Depletion of protein-bound furazolidone metabolites containing the 3-amino-2-oxazolidinone side-chain from liver, kidney and muscle tissues from pigs. *Food Addit.Contam.* **1992**, *9*, 623-630.

McCracken, R.J.; Kennedy, D.G. Determination of the furazolidone metabolite, 3-amino-2-oxazolidinone in porcine tissues using liquid chromatography-thermospray mass spectrometry and the occurrence of residues in pigs produced in Northern Ireland. *J.Chromatogr.B* **1997**, *691*, 87-94

SAMPLE

Matrix: tissue

Analyte: 3-amino-2-oxazolidinone

Sample preparation: Vortex 2 g homogenized liver in 6 mL MeOH:water 2:1, sonicate for 5 min, cool to 4°, centrifuge at 2000 rpm for 10 min, remove the supernatant, vortex the pellet 3 times with 4 mL portions of ice-cold MeOH, vortex the pellet twice with 4 mL portions of EtOH (centrifuge at 4° at 2000 rpm for 10 min after each vortexing), dry the pellet under a stream of nitrogen at room temperature, Vortex the pellet in 6 mL water, sonicate for 5 min, add 500 µL 1 M HCl, add 50 µL 50 mM 2-nitrobenzaldehyde in DMSO, heat at 37° for 16 h, cool, add 500 µL 100 mM pH 7.4 KH$_2$PO$_4$, add 500 µL 1 M NaOH, add 4 mL ethyl acetate, shake on a reciprocal shaker for 10 min, centrifuge at 2000 rpm for 5 min, repeat the extraction twice more. Combine the organic layers and evaporate them to dryness under a stream of nitrogen at 40°, reconstitute the residue in 300 µL water, wash with 2 mL hexane, centrifuge at 4° at 1000 rpm for 5 min, repeat the wash, evaporate traces of hexane under a stream of nitrogen at room temperature for 2 min, add 150 µL MeCN, filter (0.45 µm PVDF), inject an aliquot of the filtrate. (The procedure determines 3-amino-2-oxazolidinone (a furazolidone metabolite) that is bound to protein.)

HPLC VARIABLES

Guard column: µBondapak C18

Column: 250 × 4.6 5 µm Hypersil ODS

Mobile phase: MeCN:10 mM pH 7.4 potassium phosphate buffer 25:75

Column temperature: 18

Flow rate: 0.8

Detector: UV 275

CHROMATOGRAM

Retention time: 16

Limit of detection: 5 ng/g

KEY WORDS

pig; liver

REFERENCE

Horne, E.; Cadogan, A.; O'Keeffe, M.; Hoogenboom, L.A.P. Analysis of protein-bound metabolites of furazolidone and furaltadone in pig liver by high-performance liquid chromatography and liquid chromatography-mass spectrometry, *Analyst*, **1996**, *121*, 1463–1468.

2-Nitrocinnamaldehyde

SAMPLE
Matrix: solutions
Analyte: hydralazine
Sample preparation: Add a 1 mL aliquot of a solution in MeOH:water 70:30 to 1 mL pH 4.5 acetate buffer, add 50 µL acetic acid, add 1 mL 2.2 mM 2-nitrocinnamaldehyde in EtOH, heat at 70° for 50 min, cool to room temperature, inject an aliquot.

HPLC VARIABLES
Column: 250 × 4.5 5 µm Hypersil C18
Mobile phase: MeCN:buffer 65:35 (Prepare buffer by adjusting the pH of 50 mM triethylamine to 3.3 with dilute phosphoric acid.)
Flow rate: 1
Injection volume: 20
Detector: UV 390

CHROMATOGRAM
Retention time: 5.1

OTHER SUBSTANCES
Simultaneously analyzed: hydrazine (UV 350)

REFERENCE
Di Pietra, A.M.; Roveri, P.; Gotti, R.; Cavrini, V. Spectrophotometric and chromatographic (HPLC) analysis of hydralazine, dihydralazine and hydrazine after derivatization with 2-nitrocinnamaldehyde, *Farmaco*, **1993**, *48*, 1555–1567.

SAMPLE
Matrix: solutions
Analyte: hydrazine
Sample preparation: Condition a C18 Bond-Elut SPE cartridge with 10 mL MeOH and 5 mL water. Adjust pH of a 1% aqueous solution to 8.0 with 500 mM NaOH, add 3 mL to the SPE cartridge, elute with three 500 µL portions of water. Collect all the effluent from the SPE cartridge and adjust the pH to 4.5 with dilute phosphoric acid, make up to 5 mL with MeOH, dilute 1+4 with MeOH. Remove a 1 mL aliquot and add it to 1 mL pH 4.5 acetate buffer, add 50 µL acetic acid, add 1 mL 2.2 mM 2-nitrocinnamaldehyde in EtOH, heat at 70° for 50 min, cool to room temperature, inject an aliquot.

HPLC VARIABLES
Column: 250 × 4.5 5 µm Hypersil C18
Mobile phase: MeCN:buffer 65:35 (Prepare buffer by adjusting the pH of 50 mM triethylamine to 3.3 with dilute phosphoric acid.)
Flow rate: 1
Injection volume: 20
Detector: UV 350

CHROMATOGRAM
Retention time: 8
Limit of detection: 800 ng/mL

OTHER SUBSTANCES
Non-interfering: hydralazine

KEY WORDS
SPE

REFERENCE
Di Pietra, A.M.; Roveri, P.; Gotti, R.; Cavrini, V. Spectrophotometric and chromatographic (HPLC) analysis of hydralazine, dihydralazine and hydrazine after derivatization with 2-nitrocinnamaldehyde, *Farmaco*, **1993**, *48*, 1555–1567.

Salicylaldehyde

RELATED REFERENCE
Kester, P.E.; Danielson, N.D. Determination of hydrazine and 1,1-dimethylhydrazine as salicyldehyde derivates by liquid chromatography with electrochemical detection. *Chromatographia* **1984**, *18*, 125-128.

SAMPLE
Matrix: blood
Analyte: hydrazides
Sample preparation: 2 mL Plasma + 50 µL 50 µg/mL phenelzine + 400 µL 10% acetic acid + 7 mL diethyl ether:dichloromethane 2:1, shake, centrifuge at 2059 g for 10 min. Remove the aqueous layer and add it to 600 µL 10% acetic acid, add 300 µL 0.1% salicylaldehyde in EtOH, heat at 60° for 30 min, cool to room temperature, add 1 mL 1 M K_2PO_4 (sic), extract with 7 mL diethyl ether, centrifuge at 2059 g for 10 min, repeat extraction. Combine the organic layers and evaporate them to dryness under a stream of nitrogen at 40°, reconstitute the residue in 200 µL buffer, inject a 20 µL aliquot. (Buffer was 5 mM heptanesulfonic acid in MeCN:water:triethylamine 70:30:0.4.)

HPLC VARIABLES
Guard column: 50 × 4.6 30 µm C8
Column: 250 × 4.6 Spherisorb S5 ODS2 C18
Mobile phase: Gradient. MeCN:buffer:water 0:75:25 for 5 min, 15:85:0 for 12 min (step gradient). (Buffer was 5 mM heptanesulfonic acid in MeCN:water:triethylamine 70:30:0.4.)
Flow rate: 1
Injection volume: 20
Detector: UV 280

CHROMATOGRAM
Retention time: 3.75 (monoacetylhydrazine), 4.68 (isoniazid), 13.77 (hydrazine)
Internal standard: phenelzine (11.09)
Limit of detection: 250 ng/mL

REFERENCE
Walubo, A.; Smith, P.; Folb, P.I. Comprehensive assay for pyrazinamide, rifampicin and isoniazid with its hydrazine metabolites in human plasma by column liquid chromatography, *J.Chromatogr.B*, **1994**, *658*, 391–396.

SAMPLE
Matrix: blood, CSF
Analyte: isoniazid and hydrazine
Sample preparation: 200-500 µL Plasma or CSF + 50 µL 50 µg/mL phenelzine sulfate + 100 µL 10% (?) aqueous acetic acid + 5 mL n-hexane, shake for 30 min, centrifuge at

1870 g for 10 min. Discard the organic layer. Add 300 μL 0.1% salicylaldehyde in EtOH and 400 μL 10% aqueous acetic acid to the aqueous layer, heat at 60° for 30 min, cool, add 1 mL 1 M pH 6.5 K_2HPO_4, shake for 10 s, add 5 mL diethyl ether, shake for 10 min, centrifuge at 1870 g for 10 min. Remove the organic layer and evaporate it to dryness under a stream of nitrogen at 40°, reconstitute the residue in 50 μL mobile phase, inject a 25 μL aliquot.

HPLC VARIABLES
Guard column: 30 × 4.6 30 μm C8 (Waters)
Column: 300 × 3.9 10 μm μBondapak C18
Mobile phase: MeCN:water:triethylamine 70:30:0.4 containing 5 mM heptanesulfonic acid, pH adjusted to 6.0 with acetic acid
Flow rate: 1
Injection volume: 25
Detector: UV 320

CHROMATOGRAM
Retention time: 1.6 (isoniazid), 4 (hydrazine)
Internal standard: phenelzine sulfate (3)
Limit of detection: 200 ng/mL

OTHER SUBSTANCES
Non-interfering: p-aminosalicylic acid, pyrazinamide, rifampin

KEY WORDS
plasma; rabbit

REFERENCE
Walubo, A.; Chan, K.; Wong, C.L. Simultaneous assay for isoniazid and hydrazine metabolite in plasma and cerebrospinal fluid in the rabbit, *J.Chromatogr.*, **1991**, *567*, 261–266.

DIALDEHYDE

Phthalaldehyde/2-Mercaptoethanol

SAMPLE
Matrix: perfusate
Analyte: amines
Sample preparation: 30 μL Perfusate (artificial CSF) + 10 μL 200 mM perchloric acid. Mix a 25 μL aliquot with 12.5 μL reagent, let stand for 2 min, inject an aliquot. (Prepare a stock solution by dissolving 27 mg o-phthalaldehyde in 1 mL MeOH, add 5 μL β-mercaptoethanol, add 9 mL 100 mM pH 9.3 sodium tetraborate containing 10 μM EDTA.

This solution is good for 5 days in a sealed amber bottle at room temperature. Prepare the working reagent by diluting 1 mL of the stock solution with 3 mL 100 mM pH 9.3 sodium tetraborate containing 10 μM EDTA, allow to stand for 24 h before use.)

HPLC VARIABLES
Column: two columns 150 × 4.6 5 μm M.S. Gel C18 (ESA)
Mobile phase: MeOH:buffer 8:92 adjusted to pH 3.0 with phosphoric acid (Buffer was 54 mM NaH_2PO_4 containing 1.24 mM sodium heptanesulfonate.)
Column temperature: 33
Flow rate: 1.2
Detector: E, ESA Coulochem Electrode Array System Model 5500, detector temp 33°, oxidation potential 0-280 mV (depending on drug)

CHROMATOGRAM
Retention time: 2.51 (norepinephrine), 4.00 (morphine-3-glucuronide), 4.50 (apomorphine), 5.00 (methoxamine), 5.57 (dopamine), 6.34 (3,4-dihydroxyphenylacetic acid), 6.60 (5-hydroxytryptophan), 7.60 (phenylephrine), 8.75 (5-hydroxytryptophol), 8.90 (isoproterenol), 10.78 (5-hydroxyindoleacetic acid), 11.10 (morphine), 12.50 (hydralazine)), 14.12 (3-methoxytyramine), 14.13 (5-hydroxytryptamine), 17.58 (homovanillic acid)
Limit of quantitation: 0.125-10.65 ng/mL

KEY WORDS
rat

REFERENCE
Acworth, I.N.; Yu, J.; Ryan, E.; Gariepy, K.C.; Gamache, P.; Hull, K.; Maher, T. Simultaneous measurement of monoamine, amino acid, and drug levels, using high performance liquid chromatography and coulometric array technology: application to in vivo microdialysis perfusate analysis, *J.Liq.Chromatogr.*, **1994**, *17*, 685−705.

MISCELLANEOUS REACTIONS

Cyclization

SAMPLE
Matrix: blood
Analyte: hydralazine
Sample preparation: Add 8-12 mL blood to 125 IU lithium heparin in ice-cold tubes, centrifuge at 8000 g for 30 s, add 1 mL plasma to 75 μL 50% sodium nitrite in a tube kept on ice, add 150 μL 16.7 μM 4-methylhydralazine in 10 mM HCl, add 2 mL 20 mM HCl (perform the preceding procedure as rapidly as possible), vortex briefly, allow to stand at 20 ± 1° for 10.0 min, add 1 mL 1 M NaOH/0.6 M sodium tetraborate buffer (pH 10), add chloroform, shake at 110 rpm for 5 min, centrifuge at 1100 g for 10 min. Remove the organic layer and evaporate it to dryness under a stream of nitrogen at 45°, reconstitute the residue in 500 μL mobile phase, inject a 50 μL aliquot.

HPLC VARIABLES
Column: 10 μm μBondapak phenyl
Mobile phase: MeCN:1.5 mM aqueous phosphoric acid 15:85
Column temperature: 50

Flow rate: 2
Injection volume: 50
Detector: F ex 250 em 360 (cut-off filter)

CHROMATOGRAM
Retention time: 6.7
Internal standard: 4-methylhydralazine (10)
Limit of detection: 1 ng/mL

OTHER SUBSTANCES
Simultaneously analyzed: metabolites, propranolol, quinidine

KEY WORDS
plasma; pharmacokinetics

REFERENCE
Reece, P.A.; Cozamanis, I.; Zacest, R. Selective high-performance liquid chromatographic assays for hydralazine and its metabolites in plasma of man, *J.Chromatogr.*, **1980**, *181*, 427–440.

HYDROXYL RADICAL

SULFOXIDE

Dimethyl Sulfoxide/Fast Yellow GC Salt

SAMPLE
Matrix: reaction mixtures
Analyte: hydroxyl radicals
Sample preparation: Mix 5 mL reaction mixture containing 50 mM DMSO with 1 mL 500 mM pH 4.0 phosphate buffer and 1 mL reagent, shake, let stand for 10 min, add 2 mL ethyl acetate, shake well for 5 min, centrifuge at 1000 g for 5 min. Remove the ethyl acetate layer, filter (5 μm), inject a 20 μL aliquot of the filtrate. (Prepare reagent by dissolving 1 g Fast Yellow GC salt in 100 mL water, filter.)

HPLC VARIABLES
Column: 150 × 4.6 Capcell-Pak NH2 (Shiseido)
Mobile phase: n-Hexane:dilute EtOH 100:3 (Dilute EtOH was ethanol adjusted to specific gravity 0.800 with water.)
Flow rate: 1
Injection volume: 20
Detector: UV 285

CHROMATOGRAM
Retention time: 6
Limit of detection: 8 ng/mL

OTHER SUBSTANCES
Non-interfering: EDTA, iron, phenol, phenylalanine, tryptophan, tyrosine

KEY WORDS
partial interference from large concentration of aniline

REFERENCE
Fukui, S.; Hanasaki, Y.; Ogawa, S. High-performance liquid chromatographic determination of methanesulphinic acid as a method for the determination of hydroxyl radicals, *J.Chromatogr.*, **1993**, *630*, 187–193.

HYDROXYLAMINE

ALKYLATING AGENT

Dimethyl Sulfate

SAMPLE
Matrix: blood
Analyte: ciclopirox
Sample preparation: Condition an Adsorbex CN SPE cartridge (E. Merck) with 1 mL MeCN. 1 mL Plasma + 1 mL 1/15 M pH 5.0 phosphate buffer (KH_2PO_4 and Na_2HPO_4) + 10 μL 40 IU/mL β-glucuronidase, incubate at 37° for 24 h, add 500 μL 200 mM NaOH, add 200 μL dimethyl sulfate (CAUTION! Highly Toxic!), heat at 37° for 20 min, add 200 μL triethylamine, add 10 mL n-hexane, extract. Add 8 mL of the organic phase to the SPE cartridge, wash with 1 mL toluene, aspirate to dryness under reduced pressure for 3 min, elute with 350 μL mobile phase, evaporate eluate at 40° under nitrogen, dissolve residue in 80 μL mobile phase, inject a 50 μL aliquot.

HPLC VARIABLES
Guard column: 5 μm LiChrospher 100 RP18 in a LiChroCART 4-4
Column: 125 × 4 5 μm LiChrospher 100 RP18
Mobile phase: MeCN:water 40:60
Flow rate: 1
Injection volume: 50
Detector: UV 304

CHROMATOGRAM
Retention time: 7.6
Limit of detection: 15 ng/mL

KEY WORDS
plasma; SPE; rabbit; pharmacokinetics

REFERENCE
Coppi, G.; Silingardi, S. HPLC method for pharmacokinetic studies on ciclopirox olamine in rabbits after intravenous and intravaginal administrations, *Farmaco*, **1992**, *47*, 779–786.

Iodomethane

+   CH$_3$I $\longrightarrow$

SAMPLE
Matrix: formulations
Analyte: ciclopirox
Sample preparation: Foam. Weigh out an amount equivalent to about 10 mg ciclopirox, add 2 mL 1 M NaOH, add 30 μL methyl iodide, vortex, keep in an ice bath for 10 min, add 30 μL 25% ammonium hydroxide, dilute to 50 mL with MeCN:water 1:1, inject a 0.2 μL aliquot. Powder. Weigh out an amount equivalent to about 2 mg ciclopirox, add 2 mL MeCN:water 1:1, add 1 mL 1 M NaOH, add 20 μL methyl iodide, vortex, keep in an ice bath for 10 min, add 20 μL 25% ammonium hydroxide, dilute to 50 mL with MeCN:water 1:1, inject a 0.2 μL aliquot.

HPLC VARIABLES
Column: 150 × 0.33 5 μm 300 Å DeltaPak RP-18 fused silica capillary (Fusica, LC Packings)
Mobile phase: MeCN:water 50:50
Column temperature: 20
Flow rate: 0.01
Injection volume: 0.2
Detector: UV 300

CHROMATOGRAM
Retention time: 2
Limit of detection: 10 ng

KEY WORDS
foam; powder; capillary HPLC

REFERENCE
Belliardo, F.; Bertolino, A.; Brandolo, G.; Lucarelli, C. Micro-liquid chromatography method for the determination of ciclopiroxolamine after pre-column derivatization in topical formulations, *J.Chromatogr.*, **1991**, *553*, 41–45.

ISOCYANATE

Methyl Isocyanate

+   CH$_3$NCO $\longrightarrow$

SAMPLE
Matrix: tissue

Analyte: 5-hydroxyaminoindan

Sample preparation: Homogenize (PTFE pestle) with 4 volumes 20 mM pH 7.4 Tris-HCl buffer, centrifuge at 4° at 9000 g for 20 min, dilute the supernatant 10-fold with 20 mM pH 7.4 Tris-HCl buffer. 5 mL Diluted supernatant + 8 mL dichloromethane, extract for 5 min, centrifuge at 800 g for 5 min. Remove a 3 mL aliquot of the organic layer and add 400 μL methyl isocyanate (Caution! Methyl isocyanate is highly toxic!), shake for 1 min, evaporate to dryness under a stream of nitrogen, reconstitute the residue in 250 μL 47.3 μM N,N-dimethylaniline in MeOH:water 50:50, inject a 10 μL aliquot.

HPLC VARIABLES
Column: 300 × 4 μBondapak C18
Mobile phase: MeOH:water 50:50
Flow rate: 2
Injection volume: 10
Detector: UV 254

CHROMATOGRAM
Retention time: 3.9
Internal standard: N,N-dimethylaniline (7.1)
Limit of quantitation: 2 μM

OTHER SUBSTANCES
Non-interfering: 5-aminoindane, 5-azoxyindane, 5-nitroindane, 5,5'-nitrosoindane

KEY WORDS
liver; rat

REFERENCE
Sternson, L.A.; DeWitte, W.J.; Stevens, J.G. High-pressure liquid chromatographic analysis of arylhydroxylamines after derivatization with methyl isocyanate, *J.Chromatogr.*, **1978**, *153*, 481–487.

IMINE

MISCELLANEOUS REACTIONS

Reduction

SAMPLE
Matrix: blood
Analyte: progabide
Sample preparation: Evaporate 20 μL of a 20 μg/mL solution of IS in MeCN into the bottom of a tube using a stream of nitrogen, add 500 μL plasma, add 500 μL 2 M sodium acetate adjusted to pH 4.9 with HCl, add 9 mL hexane:isopropanol 96:4, shake for 5 min, centrifuge. Remove the organic layer and add it to a glass tube (pretreated with MeOH:triethylamine 80:20), add 500 μL 0.4% sodium borohydride in EtOH, vortex, let stand at room temperature for 10 min, add 2 mL 250 mM sodium citrate (adjusted to pH 2 with HCl), shake for 5 min, centrifuge, discard the organic layer, add 9 mL hexane:isopropanol 96:4, shake for 5 min, centrifuge, discard the organic layer. Add 200 μL 5 M NaOH to the aqueous layer, add 500 μL 1 M sodium citrate adjusted to pH 4.8 with HCl, add 9 mL dichloromethane, shake for 5 min, centrifuge. Remove the organic layer and evaporate it to dryness under a stream of nitrogen, reconstitute the residue in 400 μL MeOH, inject a 10 μL aliquot.

HPLC VARIABLES
Guard column: 30 × 4.6 5 μm Spheri-5 RP-18
Column: 100 × 9.4 5 μm RAC Partisil 5 ODS-3
Mobile phase: MeOH:buffer 70:30 (Buffer was 33.3 mM KH_2PO_4 adjusted to pH 5.06 with 33.3 mM K_2HPO_4.)
Flow rate: 2
Injection volume: 10
Detector: E, Bioanalytical Systems BAS LC-4A, TL-5 glassy carbon electrode 1 V

CHROMATOGRAM
Retention time: 4.5
Internal standard: 4-[[(4-chlorophenyl)(5-chloro-2-hydroxyphenyl)methylene]amino]butanamide (SL 78050) (6.5)
Limit of detection: 30 ng/mL

OTHER SUBSTANCES
Extracted: metabolites

KEY WORDS
pharmacokinetics; plasma

REFERENCE
Yonekawa, W.; Kupferberg, H.J.; Lambert, T. Measurement of progabide and its deaminated metabolite in plasma by high-performance liquid chromatography and electrochemical detection, *J.Chromatogr.*, **1983**, *276*, 103–110.

SAMPLE
Matrix: blood, urine
Analyte: progabide

Sample preparation: 1 mL Plasma, whole blood, urine + 50 μL 10 μg/mL IS in MeOH + 500 μL 2 M pH 4.5 acetate buffer + 8 mL toluene, shake on a rotary shaker for 20 min, centrifuge at 4° at 1000 g for 10 min. Remove the organic layer and add it to 500 μL 0.5% sodium borohydride in EtOH, vortex vigorously, let stand at room temperature for 20 min, add 2 mL 250 mM pH 1.8 citrate buffer, extract for 20 min. Remove the aqueous layer and add 200 μL 5 M NaOH, add 500 μL 1 M pH 7.7 citrate buffer, add 7 mL freshly distilled diethyl ether, extract for 20 min. Remove the organic layer and evaporate it to dryness under a stream of nitrogen at 37°, reconstitute the residue in 200 μL MeOH:15 mM pH 7.1 phosphate buffer 40:60, inject a 100 μL aliquot.

HPLC VARIABLES
Column: 150 × 4.6 3 μm Hypersil ODS
Mobile phase: MeCN:MeOH:33 mM pH 5.05 phosphate buffer:1.5 M NaCl 30:30:40:9
Column temperature: 54
Flow rate: 1
Injection volume: 100
Detector: E, Kipp Analytica Model 9205, +850 mV, Ag/AgCl reference electrode

CHROMATOGRAM
Retention time: 4.1
Internal standard: 4-[[(4-chlorophenyl)(5-chloro-2-hydroxyphenyl)methylene]amino]butan-amide (SL 78050) (5.9)
Limit of detection: 10 ng/mL

OTHER SUBSTANCES
Extracted: metabolites
Non-interfering: carbamazepine, carbamazepine epoxide, ethosuximide, phenobarbital, phenytoin, valproic acid

KEY WORDS
plasma; whole blood

REFERENCE
Padovani, P.; Deves, C.; Bianchetti, G.; Thenot, J.P.; Morselli, P.L. Determination of progabide and its main acid metabolite in biological fluids using high-performance liquid chromatography and electrochemical detection. Application to the measurement of blood/plasma partition ratio, *J.Chromatogr.*, **1984**, *308*, 229–239.

ISOCYANATE

ALCOHOL

Ethanol

$$R\!-\!N\!\!=\!\!C\!\!=\!\!O \quad + \quad CH_3CH_2OH \quad \longrightarrow \quad \underset{\underset{R}{|}}{\overset{\overset{O}{\parallel}}{\underset{N}{\overset{C}{\text{H}\diagdown}}}}\!-\!OCH_2CH_3$$

SAMPLE
Matrix: air
Analyte: isocyanates
Sample preparation: Pull air through 10 mL 0.2% KOH in EtOH at 1 L/min for 5 min, add 150 µL 10% phosphoric acid, evaporate to dryness under reduced pressure, reconstitute with 1 mL 100 mM pH 7.5 phosphate buffer. Inject 500 µL water onto column A, inject 0.1-1 mL sample onto column A, inject 500 µL water onto column A, elute the contents of column A onto column B with the mobile phase, monitor the effluent from column B.

HPLC VARIABLES
Column: A RCSS C18 Guard-Pak (Waters); B 250 × 3 5 µm Nucleosil C18
Mobile phase: MeCN:100 mM pH 7.0 sodium phosphate buffer 37:63
Flow rate: 0.5
Injection volume: 100-1000
Detector: E, Bioanalytical Systems, glassy carbon electrode 950 mV, Ag/AgCl reference electrode

CHROMATOGRAM
Retention time: 5.5 (2,6-toluenediisocyanate (mono derivative)), 6.5 (2,4-toluenediisocyanate (mono derivative)), 7.5 (2,4-toluenediisocyanate (mono derivative)), 9.5 (2,6-toluenediisocyanate (bis derivative)), 14.5 (2,4-toluenediisocyanate (bis derivative))
Limit of detection: 0.1 µg/m^3

KEY WORDS
column-switching

REFERENCE
Dalene, M.; Mathiasson, L.; Skarping, G.; Sangö, C.; Sandström, J.F. Trace analysis of airborne aromatic isocyanates and related aminoisocyanates and diamines using high-performance liquid chromatography with ultraviolet and electrochemical detection, *J.Chromatogr.*, **1988**, *435*, 469–481.

AMINE

1-(9-Anthracenylmethyl)piperazine

SAMPLE
Matrix: air
Analyte: 1,6-hexamethylene diisocyanate
Sample preparation: Pull air through 25 mL 62 μg/mL 1-(9-anthracenylmethyl)piperazine in DMSO at 1 L/min for 10-63 min, inject a 20 μL aliquot. (Synthesis of 1-(9-anthracenylmethyl)piperazine is as follows. Add a solution of 931 mg 9-(chloromethyl)anthracene in 150 mL MeCN dropwise over 30 min to a solution of 3.51 g piperazine and 2.08 g triethylamine in 150 mL MeCN, stir overnight, evaporate to dryness under reduced pressure, add 150 mL toluene, add 150 mL water, shake. Remove the organic layer and wash it twice with water. Filter (fritted glass) the organic layer and wash the filter with 30 mL toluene, evaporate the filtrate to 20 mL under reduced pressure, add to a 200 × 20 column of silica gel made up in toluene, wash with 50 mL ethyl acetate, elute with MeOH. Discard the first 100 mL eluate, collect the next 600 mL eluate. Evaporate the eluate to dryness under reduced pressure, sublime at 135°/100 mTorr to yield 1-(9-anthracenylmethyl)piperazine as crystals (mp 146-147°) (Am. Ind. Hyg. Assoc. J. 1996, 57, 905).)

HPLC VARIABLES
Column: 200 × 4.6 10 μm Lichrosorb RP8
Mobile phase: MeCN:40 mM pH 2.85 sodium phosphate buffer 58:42
Flow rate: 1
Injection volume: 20
Detector: UV 257

CHROMATOGRAM
Retention time: 5.7
Limit of detection: 30 ng/mL

REFERENCE

Rudzinski, W.E.; Norman, S.; Dahlquist, B.; Greebon, K.W.; Richardson, A.; Locke, K.; Thomas, T. Evaluation of 1-(9-anthracenylmethyl)piperazine for the analysis of isocyanates in spray-painting operations, *Am.Ind.Hyg.Assoc.J.*, **1996**, *57*, 914–917.

Dibutylamine

RELATED REFERENCES

Tinnerberg, H.; Karlsson, D.; Dalene, M.; Skarping, G. Determination of toluene diisocyanate in air using di-n-butylamine and 9-N-methyl-aminomethyl-anthracene as derivatization reagents. *J.Liq.Chromatogr.Rel.Technol.* **1997**, *20*, 2207-2219.

Tinnerberg, H.; Spanne, M.; Dalene, M.; Skarping, G. Determination of complex mixtures of airborne isocyanates and amines. Part 3. Methylenediphenyl diisocyanate, methylenediphenylamino isocyanate and methylenediphenyldiamine and structural analogues after thermal degradation of polyurethane. *Analyst* **1997**, *122*, 275-278.

$$R{-}N{=}C{=}O + (CH_3CH_2CH_2CH_2)_2NH \longrightarrow$$

(product structure with carbonyl O, N-H, R, and $N(CH_2CH_2CH_2CH_3)_2$)

SAMPLE

Matrix: air

Analyte: isocyanates

Sample preparation: Pull air through 10 mL 10 mM dibutylamine in toluene at 1 L/min, evaporate to dryness, reconstitute with 500 µL mobile phase, inject a 10 µL aliquot.

HPLC VARIABLES

Column: 150 × 2.1 5 µm Supelcosil LC18

Mobile phase: MeCN:water 60:40

Flow rate: 0.5

Injection volume: 10

Detector: UV 240

CHROMATOGRAM

Retention time: 3 (phenyl isocyanate), 5.5 (naphthyl 1,5-diisocyanate), 6 (2,6-toluenediisocyanate), 7 (2,4-toluenediisocyanate), 14 (methylene diphenyl-4,4'-diisocyanate)

Limit of detection: 500-800 ng/cu m (20 µL injection)

OTHER SUBSTANCES

Non-interfering: EtOH, morpholine, phenol, water

REFERENCE

Spanne, M.; Tinnerberg, H.; Dalene, M.; Skarping, G. Determination of complex mixtures of airborne isocyanates and amines. Part 1. Liquid chromatography with ultraviolet detection of monomeric and polymeric isocyanates as their dibutylamine derivatives, *Analyst*, **1996**, *121*, 1095–1099.

SAMPLE

Matrix: air

Analyte: isocyanates

Sample preparation: Pull air through 10 mL 10 mM dibutylamine in toluene at 1 L/min, evaporate to dryness under reduced pressure, reconstitute with 1 mL toluene, add 1 mL 2 M pH 9.5 carbonate buffer, add 50 µL ethyl chloroformate, add 10 µL pyridine, shake for 5 min. Remove the organic layer and evaporate it to dryness, reconstitute the residue in 1 mL MeCN, inject a 10 µL aliquot. (Isocyanates react with dibutylamine to give ureas. Amines that are collected are converted to urethanes by ethyl chloroformate.)

HPLC VARIABLES

Column: 150 × 2.1 5 µm Supelcosil LC18

Mobile phase: Gradient. MeCN:water from 46:54 to 77.5:22.5 over 14 min.
Flow rate: 0.5
Injection volume: 10
Detector: UV 240; MS, VG Organics Quattro quadrupole, atmospheric pressure chemical ionization or electrospray, positive ion monitoring, cone voltage 20 V (APCI) or 40 V (electrospray), ion source 150° (APCI) or 120° (electrospray), probe heater 500° (APCI)

CHROMATOGRAM

Retention time: 2.6 (2,4-toluenediamine), 6 (2-amino-6-isocyanatotoluene), 7 (2-amino-4-isocyanatotoluene), 7.3 (aminoisocyanatotoluene dimer), 9.9 (2,6-toluenediisocyanate), 10.3 (2,4-toluenediisocyanate), 12 (toluenediisocyanate dimer)
Limit of detection: 1-2 nM

REFERENCE

Tinnerberg, H.; Spanne, M.; Dalene, M.; Skarping, G. Determination of complex mixtures of airborne isocyanates and amines. Part 2. Toluene diisocyanate and aminoisocyanate and toluenediamine after thermal degradation of a toluene diisocyanate-polyurethane, *Analyst*, **1996**, *121*, 1101–1106.

1-(2-Methoxyphenyl)piperazine

SAMPLE

Matrix: air
Analyte: 1,6-hexamethylene diisocyanate
Sample preparation: Pull air through 15 mL 43 μg/mL 1-(2-methoxyphenyl)piperazine in DMSO at 1 L/min for 10-63 min, inject a 20 μL aliquot.

HPLC VARIABLES

Column: 200 × 4.6 10 μm Lichrosorb RP18
Mobile phase: MeCN:pH 6.0 methanolic buffer 45:55 (Buffer may be MeOH:water:sodium acetate 50:50:0.1 adjusted to pH 6 with acetic acid (Am. Ind. Hyg. Assoc. J. 1996, 57, 905).)
Flow rate: 1
Injection volume: 20
Detector: UV 245

CHROMATOGRAM

Retention time: 8.6
Limit of detection: 100 ng/mL

REFERENCE

Rudzinski, W.E.; Norman, S.; Dahlquist, B.; Greebon, K.W.; Richardson, A.; Locke, K.; Thomas, T. Evaluation of 1-(9-anthracenylmethyl)piperazine for the analysis of isocyanates in spray-painting operations, *Am.Ind.Hyg.Assoc.J.*, **1996**, *57*, 914–917.

SAMPLE

Matrix: air

Analyte: 1,6-hexamethylene diisocyanate polyisocyanates
Sample preparation: Pull air through 260 μM 1-(2-methoxyphenyl)piperazine in toluene, let stand for 24 h, evaporate to dryness under reduced pressure, reconstitute with 1.5 mL MeCN, sonicate for 5 min, filter (0.45 μm), inject a 30 μL aliquot

HPLC VARIABLES

Column: 300 × 3.9 μBondapak C18
Mobile phase: Gradient. MeCN:buffer 55:45 for 15 min, to 80:20 over 5 min, maintain at 80:20 for 10 min. (Buffer was 2 g/L ammonium acetate adjusted to pH 3.7 with phosphoric acid.)
Flow rate: 1.5
Injection volume: 30
Detector: UV 242

CHROMATOGRAM

Retention time: 13.9 (hexamethylene diisocyanate biuret), 14.1 (hexamethylene diisocyanate isocyanurate)
Limit of detection: <0.25 mg/cu.m.

REFERENCE

Maître, A.; Leplay, A.; Perdrix, A.; Ohl, G.; Boinay, P.; Romazini, S.; Aubrun, J.C. Comparison between solid sampler and impinger for evaluation of occupational exposure to 1,6-hexamethylene diisocyanate polyisocyanates during spray painting, *Am.Ind.Hyg.Assoc.J.*, **1996**, *57*, 153–160.

9-Methylaminomethylanthracene

RELATED REFERENCES

Lind, P.; Dalene, M.; Tinnerberg, H.; Skarping, G. Biomarkers in hydrolysed urine, plasma and erythrocytes among workers exposed to thermal degradation products from toluene diisocyanate foam. *Analyst* **1997**, *122*, 51-56.

Rando, R.J.; Poovey, H.G.; Gibson, R.A. Evaluation of 9-methylamino-methylanthracene as a chemical label for total reactive isocyanate group: application to isocyanate oligomers, polyurethane precursors, and phosgene. *J.Liq.Chromatogr.* **1995**, *18*, 2743-2763.

Tinnerberg, H.; Karlsson, D.; Dalene, M.; Skarping, G. Determination of toluene diisocyanate in air using di-n-butylamine and 9-N-methyl-aminomethyl-anthracene as derivatization reagents. *J.Liq.Chromatogr.Rel.Technol.* **1997**, *20*, 2207-2219.

Tinnerberg, H.; Dalene, M.; Skarping, G. Air and biological monitoring of toluene diisocyanate in a flexible foam plant. *Am.Ind.Hyg.Assoc.J.* **1997**, *58*, 229-235

SAMPLE

Matrix: polymers
Analyte: isocyanates
Sample preparation: Cut 1 g polymer into small pieces, add 15 mL dichloromethane, add 200 μL 1 μg/mL 1-naphthyl isocyanate in dichloromethane, add 1 mL 240 μg/mL 9-(N-methylaminomethyl)anthracene in dichloromethane, shake gently on an orbital shaker in the dark for 12 h. Remove the liquid and evaporate it to dryness under a stream of

nitrogen at 45°, reconstitute the residue in 10 mL DMF:mobile phase 50:50, mix thoroughly, sonicate if necessary, filter (0.45 µm), inject a 20 µL aliquot of the filtrate.

HPLC VARIABLES
Column: 250 × 4.6 5 µm Spherisorb S5ODS1
Mobile phase: MeCN:3% aqueous triethylamine 70:30
Column temperature: 45
Flow rate: 1
Injection volume: 20
Detector: F ex 254 em 412

CHROMATOGRAM
Retention time: 5.5 (phenyl isocyanate), 6.5 (cyclohexyl isocyanate), 8 (2,4-toluene diisocyanate), 9 (2,6-toluene diisocyanate), 9.7 (1,5-naphthalene diisocyanate), 14.5 (diphenylmethane-2,4'-diisocyanate), 15 (isopherone diisocyanate), 17.8 (diphenylmethane-4,4'-diisocyanate), 18.8 (hexamethylene diisocyanate), 20 (2,4-toluene diisocyanate dimer)
Internal standard: 1-naphthyl isocyanate (10.3)
Limit of detection: 30 ng/g

REFERENCE
Damant, A.P.; Jickells, S.M.; Castle, L. Liquid chromatographic determination of residual isocyanate monomers in plastics intended for food contact use, *J.AOAC Int.*, **1995**, *78*, 711–719.

SAMPLE
Matrix: solutions
Analyte: isocyanates
Sample preparation: Add a hexane solution of the isocyanate to a hexane solution of 9-methylaminomethylanthracene, stir for 0.5-2 h, filter, dry, prepare a 50 µM solution in MeCN or 2-methoxyethanol, inject an aliquot.

HPLC VARIABLES
Column: 50 × 4.6 5 µm Supelcosil LC18
Mobile phase: MeCN:buffer 62:38 (Buffer was 0.6% ammonium acetate adjusted to pH 6.5 with acetic acid.)
Flow rate: 1
Injection volume: 10
Detector: F ex 245 em 414; UV 245; UV 370

CHROMATOGRAM
Retention time: 3 (butylisocyanate), 4.5 (p-tolylisocyanate), 21 (4,4'-methylenebis (phenylisocyanate))

OTHER SUBSTANCES
Simultaneously analyzed: benzylisocyanate, 4,4'-dicyclohexylmethanediisocyanate, 4,4'-diphenylmethanediisocyanate, 1,6-hexamethylenediisocyanate, methylisocyanate, phenylisocyanate, 2,4-toluenediisocyanate, 2,6-toluenediisocyanate, o-tolylisocyanate

KEY WORDS
other MeCN:buffer mobile phase ratios may be needed for other isocyanate derivatives

REFERENCE
Rando, R.J.; Poovey, H.G.; Lefante, J.J.; Esmundo, F.R. Evaluation of 9-methylamino-methylanthracene as a chemical label for total reactive isocyanate group; a comparison of mono- and di-isocyanate monomers, *J.Liq.Chromatogr.*, **1993**, *16*, 3977–3996.

4-Nitro-N-propylbenzylamine

SAMPLE
Matrix: air
Analyte: isocyanates
Sample preparation: Draw air through 10 mL reagent at 1-2 L/min, evaporate the reagent to dryness under reduced pressure at 35°, reconstitute with 1 mL dichloromethane, inject a 90 μL aliquot. (Prepare reagent as follows. Dissolve 120 mg 4-nitro-N-propylbenzylamine hydrochloride in 10 mL water, add 13 mL 1 M NaOH, extract with toluene. Dry the extract over anhydrous sodium sulfate, dilute the extract to 2.5 L, store in the dark, discard after 3 weeks.)

HPLC VARIABLES
Column: 610 × 2.4 HC Pellosil (Whatman)
Mobile phase: Gradient. Hexane:EtOH from 99.2:0.8 to 92:8 over 10 min (Waters curve 6).
Flow rate: 2
Injection volume: 90
Detector: UV 254; RI

CHROMATOGRAM
Retention time: 7.8 (2,4-toluene diisocyanate), 8.2 (2,6-toluene diisocyanate), 10 (1,6-hexamethylene diisocyanate), 11 (4,4'-diphenylmethane diisocyanate), 14.5 (Desmodur N-75)
Limit of detection: 0.70 ppb (2,4-toluene diisocyanate)

REFERENCE
Dunlap, K.L.; Sandridge, R.L.; Keller, J. Determination of isocyanates in working atmospheres by high speed liquid chromatography, *Anal.Chem.*, **1976**, *48*, 497-499.

Tryptamine

RELATED REFERENCES
Rudzinski, W.E.; Sutcliffe, R.; Dahlquist, B.; Key-Schwartz, R. Evaluation of tryptamine in an impinger and on XAD-2 for the determination of hexamethylene-based isocyanates in spray-painting operations. *Analyst* **1997**, *122*, 605-608.
Wu, W.S.; Stoyanoff, R.E.; Szklar, R.S.; Gaind, V.S.; Rakanovic, M. Application of tryptamine as a derivatising agent for airborne isocyanate determination. Part 3. Evaluation of total isocyanates analysis by high-performance liquid chromatography with fluorescence and amperometric detection. *Analyst* **1990**, *115*, 801-807.
Wu, W.S.; Stoyanoff, R.E.; Gaind, V.S. Application of tryptamine as a derivatizing agent for airborne isocyanate determination. Part 4. Evaluation of major high-performance liquid chromatographic methods regarding airborne isocyanate determination with specific investigation of the competitive rate of derivatization. *Analyst* **1991**, *116*, 21-25.
Wu, W.S.; Szklar, R.S.; Smith, R. Application of tryptamine as a derivatizing agent for the determination of airborne isocyanates. Part 7. Selection of impinger solvents and the evaluation against dimethyl sulfoxide used in US NIOSH Regulatory Method 5522. *Analyst* **1997**, *122*, 321-323

SAMPLE
Matrix: air
Analyte: methyl isocyanate
Sample preparation: Pass air at 1 L/min through 15 mL 33.3 mg/mL tryptamine in MeCN (or toluene) for 20 min, pass the solution through a 50 × 6 column of silica, inject an aliquot of the eluate. (Pre-wash the silica with MeOH in a Soxhlet extractor.)

HPLC VARIABLES
Column: 220 × 4.6 10 μm LiChrosorb RP-18
Mobile phase: MeCN:0.6% pH 6.5 sodium acetate 60:40
Flow rate: 0.5
Injection volume: 50
Detector: F ex 275 em 320 or E, Bioanalytical Systems Model LC-4A, glassy-carbon working electrode, Ag/AgCl reference electrode, Pt auxiliary electrode, +0.8 V

CHROMATOGRAM
Retention time: 6.1
Limit of detection: <1 ng (F), 1 ng (E)

REFERENCE
Wu, W.S.; Nazar, M.A.; Gaind, V.S.; Calovini, L. Application of tryptamine as a derivatising agent for airborne isocyanates determination. Part 1. Model of derivatisation of methyl isocyanate characterised by fluorescence and amperometric detection in high-performance liquid chromatography, *Analyst*, **1987**, *112*, 863–866.

SAMPLE
Matrix: air
Analyte: poly[methylene(polyphenyl isocyanate)]
Sample preparation: Pass air at 1 L/min through 10 mL 5 μg/mL tryptamine in toluene for 3 h, transfer to a flask with MeCN, evaporate to dryness under reduced pressure with gentle heating, take up in 10 mL MeCN, pass through a 50 × 6 column of silica (40-140 mesh, J.T. Baker), dilute the eluate with MeOH, inject an aliquot.

HPLC VARIABLES
Column: 220 × 4.6 10 μm SI-100 (Brownlee)
Mobile phase: MeOH
Flow rate: 0.5
Injection volume: 50
Detector: F ex 275 em 320

CHROMATOGRAM
Retention time: 6.8

REFERENCE
Wu, W.S.; Szklar, R.S.; Gaind, V.S. Application of tryptamine as a derivatising agent for airborne isocyanate determination. Part 2. Dual function of tryptamine for calibration and derivatisation of poly(methylene(polyphenyl isocyanate)) for quantification by high-performance liquid chromatography, *Analyst*, **1988**, *113*, 1209–1212.

ISOTHIAZOLE

SULFITE

Sodium Metabisulfite

SAMPLE
Matrix: cosmetics
Analyte: preservatives
Sample preparation: Condition a Sep-Pak C18 SPE cartridge with water. Weigh out 1-3 g cosmetics, add 10 mL chloroform, add 5 mL reagent, stir. Remove the upper phase and centrifuge it at 3000 rpm for 10 min, pass through the SPE cartridge (if necessary), inject a 50 µL aliquot. (Prepare reagent by dissolving 400 mg sodium metabisulfite and 1 mg sodium nicotinate in 100 mL water.)

HPLC VARIABLES
Guard column: 50 × 4 5 µm LiChrosorb RP-18
Column: 250 × 4 5 µm LiChrosorb RP-18
Mobile phase: MeOH:water 1:8 containing 20 mM ammonium chloride
Flow rate: 0.7
Injection volume: 50
Detector: UV 260

CHROMATOGRAM
Retention time: 5 (2-methyl-3(2H)-isothiazole), 7.5 (5-chloro-2-methyl-3(2H)-isothiazole)
Limit of detection: <1 ppm

REFERENCE
Matissek, R.; Nagorka, R.; Daase, M. Analytik von Konservierungsmitteln in Cosmetica mittels HPLC-Pre-Column-Derivatisierung von Methylisothiazolonen [Analysis of preservatives in cosmetics using HPLC precolumn derivatization of methylisothiazolones], *Lebensmittelchem.gerichtl.chem.*, **1988**, *42*, 111–112.

KETAL

HYDRAZINE

Dansyl Hydrazine

SAMPLE
Matrix: bulk, feed, premix
Analyte: maduramicin
Sample preparation: Bulk, premix. Dissolve in MeCN to give a maduramicin concentration of 3 µg/mL (sonicate premix). Remove a 1 mL aliquot and mix it with 30 ± 3 mg calcium carbonate, 100 µL 7.5 mg/mL dansyl hydrazine in MeCN, and 100 µL 150 mg/mL trichloroacetic acid in MeCN, shake vigorously by hand for 1 min, centrifuge for 1 min, immediately inject an aliquot. Feed. Add 50 g feed to 250 mL MeCN, shake for 30 min, clarify. Remove a 2 mL aliquot and add it to 600 µL 7.5 mg/mL dansyl hydrazine in MeCN and 600 µL 150 mg/mL trichloroacetic acid in MeCN, shake by hand for 10 s, add to a Florisil Sep-Pak at 1 drop/s, wash with three 5 mL aliquots of MeCN at 1 drop/s, purge with air, elute with MeCN:water 90:10, collect 5 mL eluate, mix eluate, inject an aliquot.

HPLC VARIABLES
Column: 50 × 4.6 5 µm C18 (IBM)
Mobile phase: MeCN:water 75:25 containing 250 mg/L tetrabutylammonium hydrogen sulfate, pH 3.5
Flow rate: 2
Injection volume: 60
Detector: F ex 210 em 320 (cutoff filter)

CHROMATOGRAM
Retention time: 5.8, 7.7 (two possible derivatives are formed)
Limit of quantitation: 3 ppm

KEY WORDS
SPE

REFERENCE

Markantonatos, A. Derivatization and HPLC/fluorescence quantitation of maduramicin ammonium in feed and premixes at levels down to 5 ppm, *J.Liq.Chromatogr.*, **1988**, *11*, 877–890.

KETOACID

DIAMINE

1,2-Diamino-4,5-dimethoxybenzene

RELATED REFERENCES

Dave, K.J.; Riley, C.M.; Vander Velde, D.; Stobaugh, J.F. Improved preparation and structural conformation of the fluorescence labeling reagents 1,2-diamino-4,5-dimethoxybenzene and 3,4-dihydro-6,7-dimethoxy-4-methyl-3-oxoquinoxaline-2-carbonyl chloride. *J.Pharm.Biomed.Anal.* **1990**, *8*, 307-312.

Hara, S.; Takemori, Y.; Iwata, T.; Yamaguchi, M.; Nakamura, M.; Ohkura, Y. Fluorometric determination of α-keto acids with 4,5-dimethoxy-1,2-diaminobenzene and its application to high-performance liquid chromatography. *Anal.Chim.Acta* **1985**, *172*, 167-173.

SAMPLE

Matrix: tissue

Analyte: α-ketoisocaproic acid

Sample preparation: Condition two Bond Elut C18 SPE cartridges with 4 mL MeOH and 10 mL water at 2 mL/min. Homogenize (Polytron) 1.8 g brain tissue with 8 mL 5% perchloric acid, centrifuge, add 7 mL of the supernatant to an SPE cartridge at 2 mL/min, elute with water. Add the fractions containing α-ketoisocaproic acid (about 6 mL) (which have an acid pH) to 3 mL 5% perchloric acid, add to an SPE cartridge at 2 mL/min, wash with 2 mL pH 1.30 HCl, elute with MeOH:30 mM pH 7.4 sodium phosphate buffer 85:15. Add 500 μL freshly prepared 5 mM 1,2-diamino-4,5-dimethoxybenzene in 500 mM HCl containing 210 mM β-mercaptoethanol to the appropriate fraction, heat at 115° for 3.5 h, cool, adjust pH to 6.6-7.4 with 250 μL 300 mM pH 7.4 sodium phosphate buffer, filter (0.45 μm), inject an aliquot of the filtrate. (1,2-Diamino-4,5-dimethoxybenzene is available from Molecular Probes, Eugene OR or Dojindo Molecular Technologies, Inc., 3 Bethesda Metro Center, Suite 700, Bethesda MD 20814; (301) 664-8448; www.dojindo.co.jp. Synthesis is as follows. Stir 483 g veratrole in 1.45 L acetic acid at 15°, add 683 g concentrated nitric acid (s.g. 1.05) over 1 h keeping the temperature below 40° (cool if necessary), add 2.127 L fuming nitric acid (s.g. 1.50) over 1 h keeping the temperature below 30°, allow to stand for 2 h, pour into a large volume of cold water, filter, wash the solid until it is free from acid, recrystallize from EtOH to give 4,5-dinitroveratrole (mp 129.5-130.5°) (J. Am. Chem. Soc. 1946, 68, 1536). Reflux 5 g 4,5-dinitroveratrole in 200 mL benzene (Caution! Benzene is a carcinogen!), add 100 g 60 mesh iron powder and 20 mL concentrated HCl in small portions over 1 h, reflux for 4 h, add 10 mL water, reflux for 2 h, cool, make alkaline with 2.5 M NaOH, extract several times with 200 mL portions of benzene. Combine the extracts and evaporate to dryness, add 10 mL concentrated HCl, recrystallize from EtOH to give 1,2-diamino-4,5-dimethoxybenzene hydrochloride as slightly pink needles (mp 240° d) (Anal. Chim. Acta 1982, 134, 39). Another synthesis of 1,2-diamino-4,5-dimethoxybenzene is as follows. Add 9.66 g veratrole in 9.5 mL glacial acetic acid dropwise to 26.6 mL fuming nitric acid over 30 min using an ice bath,stir at room temperature for 30 min, pour into 500 mL ice-cold water, filter to obtain 1,2-dimethoxy-4,5-dinitrobenzene, recrystallize repeatedly from EtOH. Dissolve 1 g 1,2-dimethoxy-4,5-dinitrobenzene in 50 mL EtOH, add 100 mg 10% palladium on charcoal, stir under hydrogen at atmospheric pressure and room temperature for 4 days, filter through Celite under an atmosphere of nitrogen, saturate the filtrate with hydrogen chloride gas, filter under nitrogen, dry the solid under vacuum to obtain 1,2-diamino-4,5-dimethoxybenzene hydrochloride (mp 240° d (Anal. Chim. Acta 1982, 134, 39)) (Anal. Chim. Acta 1992, 263, 137).)

HPLC VARIABLES
Guard column: 30 × 4.6 5 μm Spheri 5 RP-18
Column: 250 × 4.6 5 μm Ultrasphere ODS
Mobile phase: MeCN:MeOH:THF:40 mM pH 7.4 sodium phosphate buffer 20:45:10:95
Flow rate: 1.2
Detector: F ex 340 em 455

CHROMATOGRAM
Retention time: 12.25
Limit of detection: 300 fmole

KEY WORDS
brain; rat

REFERENCE
Keen, R.E.; Nissenson, C.H.; Barrio, J.R. Analysis of femtomole concentrations of α-ketoisocaproic acid in brain tissue by precolumn fluorescence derivatization with 4,5-dimethoxy-1,2-diaminobenzene, *Anal.Biochem.*, **1993**, *213*, 23–28.

1,2-Diamino-4,5-methylenedioxybenzene

SAMPLE
Matrix: blood, urine
Analyte: α-ketoacids
Sample preparation: Serum. 50 μL Serum + 450 μL 800 mM perchloric acid + 10 μL 30 μM IS, mix, let stand at room temperature for 10 min, centrifuge at 1000 g for 10 min. Remove a 250 μL aliquot of the supernatant and make it up to 500 μL with water, add 500 μL reagent, heat at 100° for 50 min, cool in ice-water, inject a 5 μL aliquot. Urine. Collect human urine for 24 h in the presence of 10 mL 6 M HCl, centrifuge an aliquot at 1000 g for 10 min, dilute the supernatant 10-fold with water. Remove a 100 μL aliquot and add it to 50 μL 10 μM IS, add 900 μL 500 mM HCl. Remove a 500 μL aliquot and add it to 500 μL reagent, heat at 100° for 50 min, cool in ice-water, inject a 5 μL aliquot. (Reagent was 5 mM 1,2-diamino-4,5-methylenedioxybenzene in 400 mM HCl containing 1 M 2-mercaptoethanol and 28 mM sodium hydrosulfite. 1,2-Diamino-4,5-methylenedioxybenzene dihydrochloride is available from Dojindo, Japan (Dojindo Molecular Technologies, Inc., 3 Bethesda Metro Center, Suite 700, Bethesda MD 20814; (301) 664-8448; www.dojindo.co.jp). Synthesis is as follows. Add 50 g piperonal in small portions to 250 mL sulfuric acid (d 1.48), 250 mL nitric acid (d 1.40), and 11.8 g mercury(I) nitrate at 0°, neutralize with KOH (Caution! Exothermic!), add sodium bisulfite solution to remove aldehydes, steam distil to remove volatiles, recrystallize to obtain 1,2-dinitro-4,5-methylenedioxybenzene (J. Am. Chem. Soc. 1928, 50, 2711). Reflux 5 g 1,2-dinitro-4,5-methylenedioxybenzene in 200 mL benzene (Caution! Benzene is a carcinogen!), add 100 g 80 mesh iron powder and 20 mL concentrated HCl in portions over 1 h, reflux for 4 h, add 10 mL water, reflux for 2 h, cool, make alkaline with 2.6 M NaOH, extract 3 times with 200 mL portions of benzene. Combine the extracts and evaporate them to dryness, add 10 mL concentrated HCl, recrystallize from EtOH to give 1,2-diamino-4,5-methylenedioxybenzene dihydrochloride as colorless needles (mp 176-179°) (Chem. Pharm. Bull. 1987,

35, 687). Purify 1,2-diamino-4,5-methylenedioxybenzene dihydrochloride by dissolving 1 g in 28 mL MeOH and adding 500 μL 1 M 2-mercaptoethanol in 400 mM HCl containing 28 mM sodium hydrosulfite, mix slowly with 120 mL MeCN. Wash the resulting needles twice with 2 mL portions of MeOH:MeCN1:5, dry under vacuum overnight.

HPLC VARIABLES
Column: 250 × 4.6 5 μm TSK gel ODS-80TM (Tosoh)
Mobile phase: MeCN:MeOH:40 mM pH 7.0 phosphate buffer 20:32:48
Flow rate: 1
Injection volume: 5
Detector: F ex 365 em 445

CHROMATOGRAM
Retention time: 3 (α-ketoglutaric acid), 4.5 (glyoxylic acid), 5.5 (pyruvic acid), 6 (α-keto-gamma-methylthiobutyric acid), 7.5 (p-hydroxyphenylpyruvic acid), 8 (α-ketobutyric acid), 11.5 (α-ketovaleric acid), 14 (α-ketoisovaleric acid), 16 (α-ketoisocaproic acid), 17 (β-phen-ylpyruvic acid), 20.5 (α-keto-β-methyl-n-valeric acid)
Internal standard: α-ketocaproic acid (19)
Limit of detection: 1.2-180 μM

KEY WORDS
serum

REFERENCE
Wang, Z.J.; Zaitsu, K.; Ohkura, Y. High-performance liquid chromatographic determination of alpha-keto acids in human serum and urine using 1,2-diamino-4,5-methylenedioxybenzene as a precolumn fluorescence derivatization reagent, *J.Chromatogr.*, **1988**, *430*, 223–231.

SAMPLE
Matrix: cell cultures
Analyte: neuraminic acid
Sample preparation: Mix 50 μL cell culture supernatant with 200 μL reagent, heat at 50° for 2.5 h, cool, inject an aliquot. (Prepare reagent by mixing 70 mM 1,2-diamino-4,5-methylenedioxybenzene, 2.8 M acetic acid containing 1.5 M 2-mercaptoethanol and 36 mM sodium hydrosulfite, and water in the ratio 10:50:40. A kit is available from Takara, Tokyo. 1,2-Diamino-4,5-methylenedioxybenzene is available from Dojindo, Japan (Dojindo Molecular Technologies, Inc., 3 Bethesda Metro Center, Suite 700, Bethesda MD 20814; (301) 664-8448; www.dojindo.co.jp). Synthesis is as described above.)

HPLC VARIABLES
Column: 250 × 4.6 5 μm Capcell Pak C18 SG120 (Shiseido)
Mobile phase: MeCN:MeOH:water 9:7:84
Flow rate: 0.9
Detector: F ex 373 em 448

REFERENCE
Arima, H.; Aramaki, Y.; Tsuchiya, S. Contribution of trypsin-sensitive proteins to binding of cationic liposomes to the mouse macrophage-like cell line RAW264.7, *J.Pharm.Sci.*, **1997**, *86*, 786–790.

4,5-Diaminophthalhydrazide

RELATED REFERENCES

Ishida, J.; Sonezaki, S.; Yamaguchi, M. 4,5-Diaminophthalhydrazide as a highly sensitive chemiluminescence derivatization reagent for α-dicarbonyl compounds in high-performance liquid chromatography. *J.Chromatogr.* **1992**, *598*, 203-208.

Yamaguchi, M.; Isokane, M.; Ishida, J. 4,5-Diaminophthalhydrazide as a highly sensitive and selective chemiluminescence derivatization reagent for aldehydes in high-performance liquid chromatography. *Anal.Sci.* **1995**, *11*, 569-573.

SAMPLE

Matrix: blood

Analyte: α-ketoacids

Sample preparation: Condition a 150 μL Toyopak DEAE anion-exchange diethylaminoethyl resin (Tosoh) with two 400 μL portions of 2 M NaOH, with three 600 μL portions of water, with two 400 μL portions of MeCN:concentrated HCl 90:10, with three 600 μL portions of water, and with two 400 μL portions of 10 mM pH 9.0 Tris-HCl buffer. 10 μL Plasma + 20 μL 3.1 μM IS + 400 μL 10 mM pH 9.0 Tris-HCl buffer, mix, add to the SPE cartridge, wash five times with 400 μL aliquots of water, wash five times with 400 μL aliquots of MeCN:water 40:60, elute with 200 μL MeCN:1 M NaCl 40:70. Remove a 100 μL aliquot of the eluate and add it to 100 μL 1.2 mM 4,5-diaminophthalhydrazide dihydrochloride in 600 mM HCl containing 600 mM β-mercaptoethanol (freshly prepared), heat at 100° for 45 min, inject a 20 μL aliquot. (Prepare 4,5-diaminophthalhydrazide dihydrochloride as follows. Reflux 316 g 4-nitrophthalic acid and 50 mL concentrated sulfuric acid in 500 mL MeOH for 10 h, recrystallize the product (dimethyl 4-nitrophthalate) from MeOH (mp 64-65°). Hydrogenate 47.8 g dimethyl 4-nitrophthalate in 300 mL MeOH over 13 g 5% platinum on carbon at an initial hydrogen pressure of 50 psi. When the calculated amount of hydrogen has been absorbed remove the catalyst and evaporate to dryness under reduced pressure, recrystallize the residue from aqueous MeOH to give dimethyl 4-aminophthalate (mp 83-84°). Stir 146.3 g dimethyl 4-aminophthalate in 1.4 L acetic anhydride at 60-70° for 2 h then leave overnight, precipitate product with MeOH. Dry the product and rinse it with sodium carbonate solution, re-dry, recrystallize from benzene/MeOH (Caution! Benzene is a carcinogen!) to give dimethyl 4-acetamidophthalate (mp 138-140°). Add 100.4 g to 600 mL fuming (90%) nitric acid at 0-5° over 30 min, stir at 5-10° for 2.5 h, mix the reaction mixture with 800 mL cold dichloromethane, shake with crushed ice. Remove the organic layer and extract the aqueous layer with 200 mL cold dichloromethane. Combine the organic layers and wash them with ice water, cold sodium bicarbonate solution, and cold water. Dry over anhydrous magnesium sulfate, evaporate to dryness under reduced pressure and, recrystallize repeatedly from MeOH to give dimethyl 4-acetamido-5-nitrophthalate (mp 123-124.5°). Hydrolyze dimethyl 4-acetamido-5-nitrophthalate to dimethyl 4-amino-5-nitrophthalate. Hydrogenate 20.3 g dimethyl 4-amino-5-nitrophthalate in 250 mL MeOH over 1 g 5% platinum on carbon at an initial hydrogen pressure of 50 psi, remove the catalyst, evaporate to dryness under reduced pressure at 25°, recrystallize from chloroform/dichloromethane to give dimethyl 4,5-diaminophthalate (mp 111.5-113°). Add 1.1 g dimethyl 4,5-diaminophthalate to 3 mL hydrazine hydrate (Caution! Hydrazine hydrate is a carcinogen!) and 3 mL triethylamine in 20 mL MeOH, concentrate the resulting solution, triturate with benzene/MeOH, recrystallize from N,N'-dimethylacetamide/acetic acid to give 4,5-diaminophthalhydrazide (6,7-diamino-2,3-dihydrophthalazine-1,4-dione) (mp 407°) (J. Heterocycl. Chem. 1973, 10, 891), mix 4,5-diaminophthalhydrazide with a small amount of concentrated HCl, recrystallize from EtOH to give 4,5-diaminophthalhydrazide dihydrochloride (Anal. Chim. Acta 1990, 231, 1).)

HPLC VARIABLES
Column: 250 × 4.6 5 μm TSKgel ODS-120T (Tosoh)
Mobile phase: MeCN:50 mM pH 7.0 phosphate buffer 12:88
Flow rate: 1
Injection volume: 20
Detector: Chemiluminescence. The column effluent mixed with 50 mM hydrogen peroxide in water pumped at 1 mL/min and with 30 mM potassium ferricyanide in 2 M NaOH pumped at 2 mL/min and the mixture flowed through a 5 cm × 0.5 mm ID tube to the detector.

CHROMATOGRAM
Retention time: 6.5 (α-ketobutyric acid), 8.5 (p-hydroxyphenylpyruvic acid), 12 (α-ketovaleric acid), 13.5 (α-ketoisovaleric acid), 23 (α-ketoisocaproic acid), 29.5 (α-keto-β-methylvaleric acid), 45 (phenylpyruvic acid)
Internal standard: α-ketocaproic acid (30.5)
Limit of detection: 9-92 nM

KEY WORDS
plasma; SPE

REFERENCE
Nakahara, T.; Ishida, J.; Yamaguchi, M.; Nakamura, M. Determination of α-keto acids including phenylpyruvic acid in human plasma by high-performance liquid chromatography with chemiluminescence detection, *Anal.Biochem.*, **1990**, *190*, 309–313.

SAMPLE
Matrix: blood, urine
Analyte: N-acetylneuraminic acid
Sample preparation: Dilute serum 1000 times with water. Dilute urine 200 times with water. Add 10 μL diluted serum or urine to 180 μL 25 mM HCl, heat at 80° for 2 h, cool, add 10 μL 6 μM N-glycolneuraminic acid, add 50 μL 3 mM 4,5-diaminophthalhydrazide dihydrochloride in 400 mM HCl, heat at 100° for 1 h, cool in ice water, inject a 20 μL aliquot. (Prepare 4,5-diaminophthalhydrazide dihydrochloride as described above.)

HPLC VARIABLES
Column: 250 × 4.6 5 μm TSKgel ODS-120T (Tosoh)
Mobile phase: MeCN:50 mM pH 7.0 phosphate buffer 4:96
Flow rate: 1
Injection volume: 20
Detector: Chemiluminescence. The column effluent mixed with 50 mM hydrogen peroxide pumped at 1 mL/min and with 40 mM potassium ferricyanide in 2 M NaOH pumped at 2 mL/min and the mixture flowed immediately to the detector.

CHROMATOGRAM
Retention time: 25.5
Internal standard: N-glycolneuraminic acid (20)
Limit of detection: 9 fmole

OTHER SUBSTANCES
Simultaneously analyzed: α-ketoglutaric acid, pyruvic acid

KEY WORDS
serum

REFERENCE
Ishida, J.; Nakahara, T.; Yamaguchi, M. Measurement of *N*-acetylneuraminic acid in human serum and urine by high performance liquid chromatography with chemiluminescence detection, *Biomed.Chromatogr.*, **1992**, *6*, 135–140.

SAMPLE
Matrix: solutions
Analyte: α-ketoacids
Sample preparation: 100 μL Solution + 100 μL 1.2 mM 4,5-diaminophthalhydrazide dihydrochloride in 600 mM HCl containing 600 mM β-mercaptoethanol (freshly prepared), heat at 100° for 45 min, inject a 20 μL aliquot. (Prepare 4,5-diaminophthalhydrazide dihydrochloride as described above.)

HPLC VARIABLES
Guard column: TSK gel ODS-120T
Column: 250 × 4.6 5 μm TSKgel ODS-120T (Tosoh)
Mobile phase: MeCN:50 mM pH 7.0 phosphate buffer 13:87
Flow rate: 1
Injection volume: 20
Detector: Chemiluminescence. The column effluent mixed with 50 mM hydrogen peroxide in water pumped at 1 mL/min and with 30 mM potassium ferricyanide in 2 M NaOH pumped at 2 mL/min and the mixture flowed through a 5 cm × 0.5 mm ID tube to the detector.

CHROMATOGRAM
Retention time: 6 (α-ketobutyric acid), 8 (p-hydroxyphenylpyruvic acid), 11 (α-ketovaleric acid), 13 (α-ketoisovaleric acid), 22 (α-ketoisocaproic acid), 28 (α-keto-β-methylvaleric acid), 30 (α-ketocaproic acid), 44 (phenylpyruvic acid)
Limit of detection: 4-50 fmole

OTHER SUBSTANCES
Non-interfering: acetone, acetophenone, adenine, aldosterone, amino acids, bilirubin, butyric acid, cholesterol, cortisone, dextrose, epiandrosterone, fructose, glutathione, histamine, inositol, lactic acid, lactose, malic acid, mannose, methoxybenzaldehydes, 4-methylbenzaldehyde, 2-phenylethylamine, ribose, thiamine, tryptamine, tyramine, uracil, urea, uric acid, xylose

REFERENCE
Ishida, J.; Yamaguchi, M.; Nakahara, T.; Nakamura, M. 4,5-Diaminophthalhydrazide as a highly sensitive chemiluminescence reagent for α-keto acids in liquid chromatography, *Anal.Chim.Acta*, **1990**, *231*, 1–6.

o-Phenylenediamine

RELATED REFERENCE
Krishnamurti, C.R.; Janssens, S.M. Quantitation of branched chain α-keto acids in sheep plasma using reversed-phase high-performance liquid chromatography and quinoxalinol derivatization. *J.Liq.Chromatogr.* **1987**, *10*, 2265-2280.

SAMPLE
Matrix: blood, urine
Analyte: α-keto acids

Sample preparation: Serum. 100 µL Serum + 700 µL water + 100 µL 5% sodium tungstate + 100 µL 166 mM sulfuric acid, vortex, let stand for 5 min, centrifuge at 1200 g for 5 min. Remove a 500 µL aliquot of the supernatant and add it to 1.5 mL 0.13% o-phenylenediamine in 3 mL HCl (prepare fresh each day), add 5 µL 2-mercaptoethanol, make up to 3 mL with water, heat in a boiling-water bath for 30 min, cool to room temperature with ice, add 500 mg anhydrous sodium sulfate, extract three times with 3 mL portions of ethyl acetate. Combine the organic layers and dry them over anhydrous sodium sulfate, evaporate to dryness under reduced pressure, reconstitute the residue in 200 µL MeOH, centrifuge, filter (0.45 µm) the supernatant, inject a 10 µL aliquot of the filtrate. Alternatively mix 100 µL serum , 100 µL water, and 1 mL MeOH, centrifuge at 15800 g for 5 min. Remove a 600 µL aliquot of the supernatant and evaporate it to dryness under reduced pressure, reconstitute with 500 µL water, add to 1.5 mL 0.13% o-phenylenediamine in 3 mL HCl (prepare fresh each day), add 5 µL 2-mercaptoethanol, make up to 3 mL with water, heat in a boiling-water bath for 30 min, cool to room temperature with ice, add 500 mg anhydrous sodium sulfate, extract three times with 3 mL portions of ethyl acetate. Combine the organic layers and dry them over anhydrous sodium sulfate, evaporate to dryness under reduced pressure, reconstitute the residue in 200 µL MeOH, centrifuge, filter (0.45 µm) the supernatant, inject a 10µL aliquot of the filtrate. Urine. Dilute with water to a creatinine concentration of ≤1 mg/mL. Remove a 500 µL aliquot and add it to 1.5 mL 0.13% o-phenylenediamine in 3 mL HCl (prepare fresh each day), add 5 µL 2-mercaptoethanol, make up to 3 mL with water, heat in a boiling-water bath for 30 min, cool to room temperature with ice, add 500 mg anhydrous sodium sulfate, extract three times with 3 mL portions of ethyl acetate. Combine the organic layers and dry them over anhydrous sodium sulfate, evaporate to dryness under reduced pressure, reconstitute the residue in 500 µL MeOH, centrifuge, filter (0.45 µm) the supernatant, inject a 10 µL aliquot of the filtrate.

HPLC VARIABLES
Column: 250 × 4.6 Zorbax ODS
Mobile phase: MeOH:water 60:40 (After three analyses of urine samples wash column with 1 column volume THF, 1 column volume MeOH, and 1 column volume MeOH/water.)
Column temperature: 35
Flow rate: 1
Injection volume: 10
Detector: F ex 350 em 410; UV 340

CHROMATOGRAM
Retention time: 4 (α-ketoglutaric acid), 4.5 (glyoxylic acid), 5 (pyruvic acid), 7.5 (α-keto-butyric acid), 14 (α-ketoisovaleric acid), 16 (α-ketoisocaproic acid), 21 (α-keto-β-methyl-valeric acid)
Limit of quantitation: 10 pmole (F)
Limit of detection: 100 pmole (UV)

OTHER SUBSTANCES
Interfering: oxaloacetic acid (remove interference by heating serum with 10 µg citrate synthetase and 100 nmoles acetyl-CoA at 37° for 10 min), phenylpyruvic acid (separate using MeOH:water 50:50 for 15 min then 60:40)

KEY WORDS
serum

REFERENCE
Koike, K.; Koike, M. Fluorescent analysis of α-keto acids in serum and urine by high-performance liquid chromatography, *Anal.Biochem.*, **1984**, *141*, 481–487.

SAMPLE
Matrix: enzyme incubations
Analyte: ketoacids
Sample preparation: Mix enzyme incubation with 3 M HCl, centrifuge. Remove a 250 µL aliquot of the supernatant and add it to 250 µL 10 mg/mL o-phenylenediamine in 3 M HCl, heat in a boiling water bath for 30 min, cool, centrifuge for 30 s, add 500 µL ethyl acetate, vortex vigorously for 30 s. Remove an aliquot of the organic layer and evaporate

it to dryness under a stream of nitrogen, reconstitute the residue in 1 mL mobile phase, inject an aliquot.

HPLC VARIABLES
Guard column: C18
Column: 300 × 3.9 10 μm μBondapak C18
Mobile phase: MeCN:MeOH:40 mM pH 7 phosphate buffer 20:32:48
Flow rate: 1
Detector: UV 254; F ex 340 em 410

CHROMATOGRAM
Retention time: 5.1 (pyruvic acid), 6.6 (2-ketobutyric acid)
Limit of detection: 10 pmole (F)

REFERENCE
Singh, B.K.; Szamosi, I.; Shaner, D. A high-performance liquid chromatography assay for threonine/serine dehydratase, *Anal.Biochem.*, **1993**, *208*, 260−263.

KETOALCOHOL

AMINE

1,2-Diamino-4,5-methylenedioxybenzene

SAMPLE

Matrix: urine

Analyte: hydrocortisone and tetrahydroaldosterone

Sample preparation: Equilibrate a Sephadex G-25M column with 100 mM pH 7.0 phosphate buffer. Condition a Bond-Elut C18 SPE cartridge with 1 mL MeCN, 4 mL acetone: water 20:80, and 4 mL water. 2 mL Urine + 500 μL 500 mM pH 5.0 acetate buffer + 50 μL 1 μg/mL fludrocortisone in MeOH + 160 μL 100000 Fishmann U/mL β-glucuronidase and 800000 Roy U/mL arylsulfatase (from Helix pomatia, Boehringer Mannheim), heat at 37° for 24 h, filter (0.45 μm), add to the Sephadex column, wash with three 2 mL portions of 100 mM pH 7.0 phosphate buffer, elute with four 2 mL portions of 100 mM pH 7.0 phosphate buffer. Add the eluate to the SPE cartridge, wash with 4 mL water, wash with 4 mL acetone:water 20:80, elute with 1 mL MeCN. Evaporate the eluate to dryness under a stream of nitrogen, reconstitute the residue in 100 μL MeOH, add 20 μL cupric acetate solution, let stand at room temperature for 1 h, add 100 μL reagent, heat at 60° for 40 min, cool, centrifuge briefly at 1000 g, inject a 100 μL aliquot of the supernatant. (Cupric acetate solution was 0.7 g cupric acetate in water diluted to 100 mL with MeOH. Reagent was 7 mM 1,2-diamino-4,5-methylenedioxybenzene in water containing 200 mM β-mercaptoethanol and 250 mM sodium hydrosulfite, store in the dark at 4°, stable for at least 2 weeks. 1,2-Diamino-4,5-methylenedioxybenzene is available from Dojindo Molecular Technologies, Inc., 3 Bethesda Metro Center, Suite 700, Bethesda MD 20814; (301) 664-8448; www.dojindo.co.jp. Preparation is as follows. Add 5 g 1,2-(methylenedioxy)-4-nitrobenzene to 37.5 mL concentrated nitric acid and 12.5 mL glacial acetic acid, pour the yellow-colored solution into water, recrystallize the 1,2-dinitro-4,5-methylenedioxybenzene from EtOH (Rec.Trav.Chim.Pays-Bas 1930, 49, 45). Dissolve 5 g 1,2-dinitro-4,5-methylenedioxybenzene in 200 mL benzene (Caution! Benzene is a carcinogen!), add 100 g 80 mesh iron powder, add 20 mL concentrated HCl in small portions over 1 h while heating the mixture under reflux. Reflux for 4 h, add 10 mL water, reflux for 2 h, cool, make alkaline with 2.6 M NaOH, extract three times with 200 mL portions of benzene. Combine the extracts, evaporate to dryness to give 1,2-diamino-4,5-methylenedioxybenzene, mix with 10 mL concentrated HCl, recrystallize from EtOH to give 1,2-diamino-4,5-methylenedioxybenzene dihydrochloride, mp 176-9° (Chem.Pharm.Bull. 1987, 35, 687).)

HPLC VARIABLES

Column: 250 × 4.6 5 μm L-Column ODS (Chemicals Inspection and Testing Institute, Tokyo)

Mobile phase: MeOH:MeCN:500 mM ammonium acetate 50:10:40 (After each injection wash with MeOH:water 80:20 for 20 min for 20 min, re-equilibrate for 20 min.)

Flow rate: 1

Injection volume: 100

Detector: F ex 350 em 390

CHROMATOGRAM

Retention time: 23 (3α,5β-tetrahydroaldosterone), 38.5 (hydrocortisone)

Internal standard: fludrocortisone (35.6)

Limit of detection: 0.45-1.18 ng/mL

OTHER SUBSTANCES

Non-interfering: corticosterone, cortisone, hydroxycorticosteroids

KEY WORDS

SPE

REFERENCE

Yoshitake, T.; Ishida, J.; Sonezaki, S.; Yamaguchi, M. High performance liquid chromatographic determination of 3α,5β-tetrahydroaldosterone and cortisol in human urine with fluorescence detection, *Biomed.Chromatogr.*, **1992**, *6*, 217–221.

KETOALDEHYDE

DIAMINE

1,2-Diamino-4,5-dimethoxybenzene

SAMPLE
Matrix: blood
Analyte: methylglyoxal
Sample preparation: Equilibrate a 2.8 mL 500 mg SPE cartridge with 20 mM pH 2.3 $(NH_4)H_2PO_4$. 1 mL Whole blood + 2 mL 600 mM perchloric acid, vortex, add 20 μL 50 μM IS in 500 mM HCl, let stand on ice for 10 min, centrifuge at 6000 g for 10 min. Remove a 2 mL aliquot of the supernatant and add it to 200 μL 1.1 mM 1,2-diamino-4,5-dimethoxybenzene in 500 mM HCl, mix, let stand in the dark at room temperature for 4 h, adjust the pH to 2.3 with 850 μL 500 mM Na_2HPO_4 add to the SPE cartridge, wash with 6 mL 20 mM pH 2.3 $(NH_4)H_2PO_4$, elute with 3 mL MeOH, evaporate to dryness under reduced pressure at room temperature, reconstitute with 200 μL mobile phase, filter (0.2 μM), inject a 5 (F) or 100 (UV) μL aliquot of the filtrate. (1,2-Diamino-4,5-dimethoxybenzene is available from Molecular Probes, Eugene OR or Dojindo Molecular Technologies, Inc., 3 Bethesda Metro Center, Suite 700, Bethesda MD 20814; (301) 664-8448; www.dojindo.co.jp. Synthesis is as follows. Stir 483 g veratrole in 1.45 L acetic acid at 15°, add 683 g concentrated nitric acid (s.g. 1.05) over 1 h keeping the temperature below 40° (cool if necessary), add 2.127 L fuming nitric acid (s.g. 1.50) over 1 h keeping the temperature below 30°, allow to stand for 2 h, pour into a large volume of cold water, filter, wash the solid until it is free from acid, recrystallize from EtOH to give 4,5-dinitroveratrole (mp 129.5-130.5°) (J. Am. Chem. Soc. 1946, 68, 1536). Reflux 5 g 4,5-dinitroveratrole in 200 mL benzene (Caution! Benzene is a carcinogen!), add 100 g 60 mesh iron powder and 20 mL concentrated HCl in small portions over 1 h, reflux for 4 h, add 10 mL water, reflux for 2 h, cool, make alkaline with 2.5 M NaOH, extract several times with 200 mL portions of benzene. Combine the extracts and evaporate to dryness, add 10 mL concentrated HCl, recrystallize from EtOH to give 1,2-diamino-4,5-dimethoxybenzene hydrochloride as slightly pink needles (mp 240° d) (Anal. Chim. Acta 1982, 134, 39). Another synthesis of 1,2-diamino-4,5-dimethoxybenzene is as follows. Add 9.66 g veratrole in 9.5 mL glacial acetic acid dropwise to 26.6 mL fuming nitric acid over 30 min using an ice bath, stir at room temperature for 30 min, pour into 500 mL ice-cold water, filter to obtain 1,2-dimethoxy-4,5-dinitrobenzene, recrystallize repeatedly from EtOH. Dissolve 1 g 1,2-dimethoxy-4,5-dinitrobenzene in 50 mL EtOH, add 100 mg 10% palladium on charcoal, stir under hydrogen at atmospheric pressure and room temperature for 4 days, filter through Celite under an atmosphere of nitrogen, saturate the filtrate with hydrogen chloride gas, filter under nitrogen, dry the solid under vacuum to obtain 1,2-diamino-4,5-dimethoxybenzene hydrochloride (mp 240° d (Anal. Chim. Acta 1982, 134, 39)) (Anal. Chim. Acta 1992, 263, 137).)

HPLC VARIABLES
Column: 100 × 8 4 μm Nova-Pak ODS radial compression
Mobile phase: MeOH:20 mM pH 3.4 ammonium formate 42:58
Flow rate: 2
Injection volume: 5-100
Detector: UV 352; F ex 352 em 385

CHROMATOGRAM
Retention time: 9
Internal standard: 2,3-dimethyl-6,7-dimethoxyquinoxaline (Dissolve 482 mg 1,2-diamino-4,5-dimethoxybenzene dihydrochloride and 351 μL butan-2,3-dione in 2.5 mL 10.18 M HCL and 15 mL water, stir under nitrogen at room temperature for 6 h, adjust pH to 7.0 with 2 M potassium bicarbonate, collect the product that crystallizes, recrystallize 5 times from MeOH:water 5:95 to obtain 2,3-dimethyl-6,7-dimethoxyquinoxaline (Anal. Chim. Acta. 1992, 263, 137).) (13)
Limit of detection: 10 pmole (F), 45 pmole (UV)

OTHER SUBSTANCES
Non-interfering: dihydroxyacetone phosphate, glucose, glyceraldehyde, glyceraldehyde 3-phosphate, lactate, pyruvate

KEY WORDS
whole blood; SPE

REFERENCE
McLellan, A.C.; Phillips, S.A.; Thornalley, P.J. The assay of methylglyoxal in biological systems by derivatization with 1,2-diamino-4,5-dimethoxybenzene, *Anal.Biochem.*, **1992**, *206*, 17–23.

Ethylenediamine

SAMPLE
Matrix: apple juice
Analyte: saccharides
Sample preparation: Filter apple juice, inject a 20 μL aliquot of the filtrate.

HPLC VARIABLES
Column: 250 × 6 20 μm DA X-4 anion-exchange (Durrum Chemical Co.)
Mobile phase: 700 mM Boric acid containing 7.5 mM ethylenediamine, adjusted to pH 8.6 with 8 M NaOH (Ethylenediamine was Nanochrome II from Breda Scientific, Holland.)
Column temperature: 78
Flow rate: 0.7
Injection volume: 20
Detector: F ex 360 (bandpass filter) em 455 (bandpass filter) following post-column reaction. The column effluent flowed through a 30 m × 0.5 mm ID PTFE coil at 145° to the detector.

CHROMATOGRAM
Retention time: 27 (sucrose), 53 (fructose), 58 (mannose), 71 (galactose), 99 (dextrose)
Limit of detection: 100-400 pmole

KEY WORDS
post-column reaction

REFERENCE
Mopper, K.; Dawson, R.; Liebezelt, G.; Hansen, H.-P. Borate complex ion exchange chromatography with fluorimetric detection for determination of saccharides, *Anal.Chem.*, **1980**, *52*, 2018–2022.

o-Phenylenediamine

SAMPLE
Matrix: cells
Analyte: methylglyoxal
Sample preparation: Condition 2 Sep-Pak tC18 plus SPE cartridges with 6-8 mL MeCN and 6-8 mL 10 mM pH 2.5 KH$_2$PO$_4$. Suspend 450 mg cells in 4.5 mL pH 7.3 phosphate buffered saline, sonicate (Branson Sonifier Cell Disruptor 85) with three 10 W bursts for 5 s, add 450 μL 5 M perchloric acid, mix, let stand at 0° for 10 min, centrifuge at 12000 g for 10 min, pass the supernatant through an SPE cartridge. Add 1.25 nmoles IS and 125 nmoles o-phenylenediamine to the eluate, let stand at 20° for 3.5-4 h, add to an SPE cartridge at 1-2 mL/min, wash with 1-2 mL 10 mM pH 2.5 KH$_2$PO$_4$, elute with 2 mL MeCN, concentrate the eluate to 400 μL under reduced pressure, filter (0.2 μm), inject a 150 μL aliquot of the filtrate.

HPLC VARIABLES
Column: 250 × 4.6 5 μm Adsorbosphere C18
Mobile phase: MeCN:10 mM pH 2.5 KH$_2$PO$_4$ 32:68
Flow rate: 11
Injection volume: 150
Detector: UV 315

CHROMATOGRAM
Retention time: 7.9
Internal standard: 5-methylquinoxaline (12.5)
Limit of detection: 44 pmole

REFERENCE
Chaplen, F.W.R.; Fahl, W.E.; Cameron, D.C. Method for determination of free intracellular and extra-cellular methylglyoxal in animal cells grown in culture, *Anal.Biochem.*, **1996**, *238*, 171−178.

KETONE

AMINE

(R)-(+)-α-Methylbenzylamine

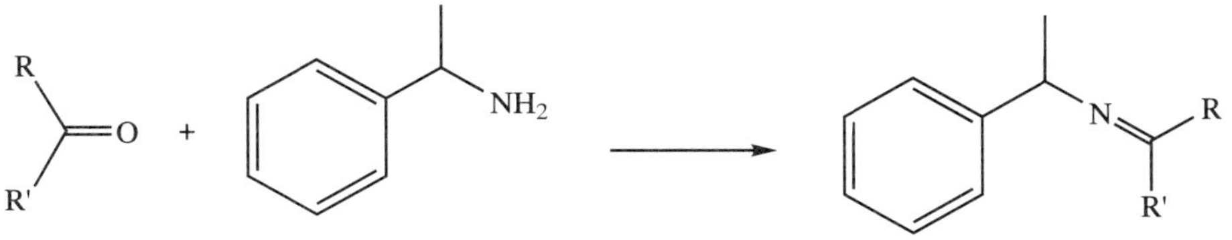

SAMPLE
Matrix: blood

Analyte: 4-[3-(4-acetyl-3-hydroxy-2-propylphenoxy)propylthio]-4-hydroxy-3-methylbenzene-butanoic acid

Sample preparation: Condition two 100 mg Vac-Elut cyano SPE cartridges (Analytichem) with hexane. 1 mL Plasma + 100 μL 100 ng/mL IS + 500 μL 1 M pH 6 sodium acetate + 4 mL MTBE, vortex, centrifuge, freeze at -60°. Remove the organic layer and evaporate it to dryness under a stream of nitrogen, reconstitute the residue in 200 μL 1 mg/mL p-toluenesulfonic acid in MTBE, heat at 80° for 15 min, evaporate to dryness under a stream of nitrogen, reconstitute with 1 mL dichloromethane, add to an SPE cartridge, elute with 250 μL hexane. Collect all the effluent from the cartridge and evaporate it to dryness under a stream of nitrogen, reconstitute in 1 mL hexane, add to an SPE cartridge, wash with 500 μL hexane:isopropanol 90:10, wash with 250 μL hexane, elute with 1 mL dichloromethane. Evaporate the eluate to dryness under a stream of nitrogen, reconstitute the residue in 250 μL DMF, add 50 μL R-(+)-α-methylbenzylamine, heat at 100° for 75 min, evaporate to dryness, reconstitute with 500 μL hexane:isopropanol 90:10, inject a 50 μL aliquot. (Compound is lactonized with p-toluenesulfonic acid and then the Schiff base is formed with methylbenzylamine.)

HPLC VARIABLES
Guard column: 15 × 2.1 40 μm LC-5 pellicular silica (Supelco)

Column: 150 × 4.6 3 μm Supelcosil silica

Mobile phase: Hexane:isopropanol:triethylamine 97.8:2:0.2

Flow rate: 2

Injection volume: 50

Detector: UV 280

CHROMATOGRAM
Retention time: 6 (2S,4R), 8 (2R,4S)

Internal standard: 4-[3-(4-acetyl-3-hydroxy-2-propylphenoxy)propylthio]-4-hydroxybenzene-butanoic acid (9,10 (enantiomers))

Limit of detection: 25 ng/mL

KEY WORDS
normal phase; SPE; pharmacokinetics; chiral; human; rat

REFERENCE
Slobodzian, D.K.; Hsieh, J.Y.-K.; Young, R.N.; Bayne, W.F. Stereoselective determination of (beta S,gamma R and beta R, gamma S)-4-[3-(4-acetyl-3-hydroxy-2-propylphenoxy)propylthio]-gamma-hydroxy-beta-methylbenzenebutanoic acid in human and rat plasma by normal-phase high-performance liquid chromatography, *J.Pharm.Sci.*, **1987**, *76*, 169–173.

HYDRAZINE

4-(Aminosulfonyl)-7-hydrazino-2,1,3-benzoxadiazole

SAMPLE

Matrix: solutions

Analyte: heptan-4-one

Sample preparation: Mix 10 µL of a 2-3 µM solution in MeCN with 20 µL 500 µM 4-(aminosulfonyl)-7-hydrazino-2,1,3-benzoxadiazole in MeCN and 10 µL 1% trifluoroacetic acid in MeCN, let stand at room temperature for 5 h, inject an aliquot. (Synthesis of 4-(aminosulfonyl)-7-hydrazino-2,1,3-benzoxadiazole is as follows. Dissolve 0.5 g magnesium sulfate heptahydrate and 6 g NaOH in 60 mL water, throughout the reaction keep the flask at about 20° with cold water cooling, add 15 mL 30% hydrogen peroxide, add 75 mL MeOH, add 12.1 g powdered benzoyl peroxide in one go, stir for 10 min, pour into 150 mL 20% sulfuric acid, extract three times with 50 mL portions of chloroform, determine peroxybenzoic acid concentration by iodometric titration (Tetrahedron 1967, 23,3327). Slowly add 110 mL 1 M peroxybenzoic acid in chloroform to 7 g 2,6-difluoroaniline dissolved in 100 mL chloroform, stir at room temperature, when reaction is complete (iodometric titration) wash with 2% sodium thiosulfate, wash with 5% sodium carbonate, wash with water, dry over anhydrous sodium sulfate, evaporate to dryness under reduced pressure, recrystallize 2,6-difluoronitrosobenzene from EtOH (mp 108.5-109.5). Stir 8.5 g 2,6-difluoronitrosobenzene in 85 mL DMSO at room temperature and add a solution of 3.91 g sodium azide in 85 mL DMSO dropwise, let stand for about 1 h, add to a large volume of water, extract with ether, dry the extracts over anhydrous sodium sulfate, evaporate to dryness under reduced pressure and distil to give 4-fluoro-2,1,3-benzoxadiazole as a colorless oil (bp 83°/12 mm Hg) (J.Chem.Soc.(C) 1970, 1433). Add 11 mL chlorosulfonic acid dropwise to 3 g 4-fluoro-2,1,3-benzoxadiazole in 10 mL chloroform at 0-10° (use a calcium chloride drying tube), stir at room temperature for 1 h, reflux for 2 h, cool, slowly pour into ice water, remove the organic layer, extract the aqueous layer with chloroform, combine the organic layers, wash, dry over anhydrous magnesium sulfate, evaporate under reduced pressure, take up the residue in 5 mL benzene (Caution! Benzene is a carcinogen!), chromatograph on a 150 × 30 column of silica gel (100-200 mesh Kanto Chemical) with n-hexane:benzene 50:50, evaporate the appropriate fractions to give 4-(chlorosulfonyl)-7-fluoro-2,1,3-benzoxadiazole (CBD-F) as pale yellow needles (mp 64-66°). Add 1 g CBD-F dropwise to 100 mL 6% ammonium hydroxide, neutralize with 10% HCl, evaporate under reduced pressure, add 200 mL MeCN to the residue, filter. Evaporate the filtrate and chromatograph on a 300 × 20 column of 100-200 mesh silica with chloroform, collect the appropriate fractions and evaporate them to give ABD-F (4-(aminosulfonyl)-7-fluoro-2,1,3-benzoxadiazole) as white needles (mp 145-6°) after recrystallization from n-hexane/benzene (Caution! Benzene is a carcinogen!) (Anal. Chem. 1984, 56, 2461). 4-(Aminosulfonyl)-7-fluoro-2,1,3-benzoxadiazole is also available from Wako Chemicals, Richmond VA; Molecular Probes, Eugene OR; or Dojindo Molecular Technologies,

Inc., 3 Bethesda Metro Center, Suite 700, Bethesda MD 20814, (301) 664-8448, www.dojindo.co.jp. Dissolve 24 mg 4-(aminosulfonyl)-7-fluoro-2,1,3-benzoxadiazole in 3 mL MeCN, add 10 μL 98% hydrazine hydrate, heat in the dark at 50-55° for 20 min, evaporate to dryness, recrystallize from MeOH to give 4-(aminosulfonyl)-7-hydrazino-2,1,3-benzoxadiazole as yellow-orange needles (mp 184-185° d).)

HPLC VARIABLES
Column: 150 × 4.6 5 μm TSK-LS 80Tm (Tosoh)
Mobile phase: MeCN:water:trifluoroacetic acid 61:39:0.05
Flow rate: 1
Detector: F ex 450 em 558

CHROMATOGRAM
Retention time: 6
Limit of detection: 18.8 μM

OTHER SUBSTANCES
Also analyzed: acetone, butyraldehyde, 4'-ethylacetophenone, p-hydroxybenzaldehyde, propionaldehyde

REFERENCE
Uzu, S.; Kanda, S.; Imai, K.; Nakashima, K.; Akiyama, S. Fluorogenic reagents: 4-Aminosulphonyl-7-hydrazino-2,1,3-benzoxadiazole, 4-(N,N-dimethylaminosulphonyl)-7-hydrazino-2,1,3-benzoxadiazole and 4-hydrazino-7-nitro-benzoxadiazole hydrazine for aldehydes and ketones, *Analyst*, **1990**, *115*, 1477–1482.

Apmayl Hydrazide

SAMPLE
Matrix: plants
Analyte: abscisic acid
Sample preparation: Homogenize 1 g plant material with 5 mL 100 μg/mL BHT in MeOH, remove MeOH under vacuum, freeze aqueous solution, thaw, centrifuge, adjust the pH of the supernatant to 9.0, wash twice with ether, adjust pH to 2.0, extract with chloroform. Remove the organic layer and evaporate it to dryness. For each 14 g original plant material add 1 mg apmayl hydrazide, for each 1 mmole reagent add 1.1 mmoles HCl in MeOH, purge with argon for 1 min, let stand in the dark at 25° overnight, streak on a TLC plate (Analtech GHR-250 μm), develop with MeOH:chloroform 1:6.7, remove the derivative (between the free acid at the origin and the unreacted hydrazide), extract with

MeOH:chloroform 75:25, evaporate to dryness, reconstitute with 50 μM HCl in MeOH, inject an aliquot. (Prepare 10 mM HCl in MeOH by mixing concentrated aqueous HCl with MeOH. Prepare apmayl hydrazide, N-(4-(6-methyl-2-benzothiazolyl)phenylglycine hydrazide, as follows. Add 960 mg ethyl iodoacetate and 450 mg potassium bicarbonate to 1 g 2-(4-aminophenyl)-6-methylbenzothiazole in 300 mL isopropanol, reflux with stirring for 12 h, cool, evaporate to 50 mL under reduced pressure, dissolve the slurry in chloroform, filter, wash twice with water, dry over anhydrous sodium sulfate, evaporate to dryness under reduced pressure, dry the resultant oil or powder over calcium carbonate to obtain N-(4-(6-methyl-2-benzothiazolyl)phenylglycine ethyl ester. Suspend the N-(4-(6-methyl-2-benzothiazolyl)phenylglycine ethyl ester in 25 mL chloroform:MeOH 50:50 in a 50 mL serum bottle, add a 10-fold molar excess of hydrazine hydrate (Caution! Hydrazine hydrate is a carcinogen!), heat the sealed bottle at 60° for 3 h (Caution! Use a safety screen!), cool, add 200 mL water, add 200 mL chloroform. Remove the organic phase and wash it twice with 60 mL portions of water, evaporate to dryness under reduced pressure, dissolve the residue in 100 mL ethyl acetate, wash twice with 60 mL portions of water, dry over anhydrous sodium sulfate, evaporate to dryness under reduced pressure, purify by TLC (Analtech silica gel G-2 mm) using chloroform:MeOH:petroleum ether 80:10:10 to obtain apmayl hydrazide (N-(4-(6-methyl-2-benzothiazolyl)phenylglycine hydrazide), store in a desiccator at -20°.)

HPLC VARIABLES
Column: 5 μm Radial-Pak C18 radial compression (Waters)
Mobile phase: MeCN:20 mM pH 4.0 phosphate buffer 40:60
Detector: F ex 345 em 413

CHROMATOGRAM
Retention time: 16

KEY WORDS
comparison with other derivatizing reagents

REFERENCE
Anderson, J.M. Fluorescent hydrazides for the high-performance liquid chromatographic determination of biological carbonyls, *Anal.Biochem.*, **1986**, *152*, 146–153.

(R)-(+)-4-(2-Carbazolylpyrrolidin-1-yl)-7-(N,N-dimethylaminosulfonyl)-2,1,3-benzoxadiazole

RELATED REFERENCE
Toyo'oka, T.; Liu, Y-M. Determination of aldehydes by high-performance liquid chromatography with fluorescence detection after labeling with 4-(2- carbazoylpyrrolidin-1-yl)-7-(N,N-dimethylaminosulfonyl)-2,1,3-benzoxadiazole. *J.Chromatogr.A* **1995**, *695*, 11-18.

SAMPLE
Matrix: solutions
Analyte: ketones

Sample preparation: 50 μL 10 mM Reagent in MeCN + 50 μL 0.5% trichloroacetic acid in MeCN + 400 μL 10 μM ketone in MeCN:water 70:30, heat at 65° for 10 min, cool in ice for at least 10 min, inject a 5 μL aliquot. (Synthesis of reagent, (R)-(+)-DBD-ProCZ, is as follows. Dissolve 0.5 g magnesium sulfate heptahydrate and 6 g NaOH in 60 mL water, throughout the reaction keep the flask at about 20° with cold water cooling, add 15 mL 30% hydrogen peroxide, add 75 mL MeOH, add 12.1 g powdered benzoyl peroxide in one go, stir for 10 min, pour into 150 mL 20% sulfuric acid, extract three times with 50 mL portions of chloroform, determine peroxybenzoic acid concentration by iodometric titration (Tetrahedron 1967, 23, 3327). Slowly add 110 mL 1 M peroxybenzoic acid in chloroform to 7 g 2,6-difluoroaniline dissolved in 100 mL chloroform, stir at room temperature, when reaction is complete (iodometric titration) wash with 2% sodium thiosulfate, wash with 5% sodium carbonate, wash with water, dry over anhydrous sodium sulfate, evaporate to dryness under reduced pressure, recrystallize 2,6-difluoronitrosobenzene from EtOH (mp 108.5-109.5). Stir 8.5 g 2,6-difluoronitrosobenzene in 85 mL DMSO at room temperature and add a solution of 3.91 g sodium azide in 85 mL DMSO dropwise, let stand for about 1 h, add to a large volume of water, extract with ether, dry the extracts over anhydrous sodium sulfate, evaporate to dryness under reduced pressure and distil to give 4-fluoro-2,1,3-benzoxadiazole as a colorless oil (bp 83°/12 mm Hg) (J.Chem.Soc.(C) 1970, 1433). Add 11 mL chlorosulfonic acid dropwise to 3 g 4-fluoro-2,1,3-benzoxadiazole in 10 mL chloroform at 0-10° (use a calcium chloride drying tube), stir at room temperature for 1 h, reflux for 2 h, cool, slowly pour into ice water, remove the organic layer, extract the aqueous layer with chloroform, combine the organic layers, wash, dry over anhydrous magnesium sulfate, evaporate under reduced pressure, take up the residue in 5 mL benzene (Caution! Benzene is a carcinogen!), chromatograph on a 150 × 30 column of silica gel (100-200 mesh Kanto Chemical) with n-hexane:benzene 50:50, evaporate the appropriate fractions to give 4-(chlorosulfonyl)-7-fluoro-2,1,3-benzoxadiazole (CBD-F) as pale yellow needles (mp 64-66°) (Anal. Chem. 1984. 56, 2461). Stir 0.76 g CBD-F in 70 mL MeCN at 0-10° and add 1 g dimethylamine hydrochloride in 10 mL 100 mM pH 10 borax dropwise, adjust pH to 5 with 1 M HCl, concentrate to about 10 mL under reduced pressure, extract three times with 200 mL portions of diethyl ether, wash with water, dry over anhydrous magnesium sulfate, evaporate under reduced pressure, chromatograph on a 500 × 20 column of silica gel with chloroform, isolate the appropriate fraction and re-chromatograph on the same column with ethyl acetate:benzene 1:2 to give 4-(N,N-dimethylaminosulfonyl)-7-fluoro-2,1,3-benzoxadiazole (DBD-F) as white needles (mp 124-125°) (yield 1% !). On a Merck no. 5714 60F$_{254}$ tlc plate eluted with chloroform DBD-F has Rf 0.32 and lies between two other reaction products (Analyst 1989, 114, 413). DBD-F can also be purchased from Tokyo Kasei (TCI America, 911 North Harborgate St., Portland, OR 97203; 800-423-8616, 503-283-1681, (fax) 503-283-1987; www.tciamerica.com). Add 100 mg DBD-F in 10 mL MeCN to 47 mg (R)-(+)-proline in 20 mL 250 mM pH 11.5 sodium carbonate solution, stir at room temperature for 30 min, wash with ethyl acetate, adjust the pH of the aqueous layer to 1-2 with 2 M HCl, extract three times with 30 mL ethyl acetate. Combine the extracts and evaporate them under reduced pressure, recrystallize from benzene/ethyl acetate to give (R)-(+)-4-(N,N-dimethylaminosulfonyl)-7-(2-carboxypyrrolidin-1-yl)-2,1,3-benzoxadiazole (DBD-Pro) as yellow needles (mp 187-9° d) (Analyst 1989, 114, 1233). Suspend 55 mg (R)-(+)-DBD-Pro in 55 mL anhydrous diethyl ether at 0°, add 110 mg phosphorus pentachloride, stir at 5° for 1 h, filter quickly, evaporate to dryness under reduced pressure, dry under vacuum over phosphorus pentoxide for 12 h to give (R)-(+)-4-(N,N-dimethylaminosulfonyl)-7-(2-chloroformylpyrrolidin-1-yl)-2,1,3-benzoxadiazole (DBD-Pro-Cl) as yellow crystals (mp 116-17°) (Analyst 1993, 118, 759). DBD-Pro-Cl is also available from Tokyo Kasei. Add 130 mg DBD-Pro-Cl dissolved in 25 mL anhydrous benzene dropwise to 100 mL MeOH containing 70 mg hydrazine hydrate, stir for 30 min at room temperature, evaporate under reduced pressure, recrystallize from ethyl acetate:MeOH 90:10 to give (R)-(+)-4-(2-carbazolylpyrrolidin-1-yl)-7-(N,N-dimethylaminosulfonyl)-2,1,3-benzoxadiazole ((R)-(+)-DBD-ProCZ) as orange crystals (mp 107-109°).)

HPLC VARIABLES

Column: 150 × 4.6 5 μm Inertsil ODS-80A

Mobile phase: MeCN:water 45:55 (2-phenylcyclohexanone, 2-phenylcycloheptanone), 40:60 (2-methylcyclohexanone), or 50:50 (1-decalone)

Column temperature: 40
Flow rate: 1
Injection volume: 5
Detector: F ex 450 em 540

CHROMATOGRAM
Retention time: k' 23.19 (S-2-phenylcyclohexanone), k' 24.44 (R-2-phenylcyclohexanone), k' 33.24 (S-2-phenylcycloheptanone), k' 35.02 (R-2-phenylcycloheptanone), k' 17.58 (S-2-methylcyclohexanone), k' 18.53 (R-2-methylcyclohexanone), k' 20.67 (S-1-decalone), k' 21.64 (R-1-decalone)

KEY WORDS
chiral

REFERENCE
Toyo'oka, T.; Liu, Y.-M. Enantioseparation of carbonyl compounds by high-performance liquid chromatography based on diastereomer formation, *Anal.Proc.*, **1994**, *31*, 265–268.

Dansyl Hydrazine

RELATED REFERENCES
Cartoni, G.; Coccioli, F.; Collalto, A. Determination of 17-hydroxycorticosteroids in urine by HPLC using fluorescence derivatization with dansyl hydrazine. *Ann.Chim.(Rome)* **1990**, *80*, 461-471.

Kawasaki, T.; Maeda, M.; Tsuji, A. Determination of plasma and urinary cortisol by high-performance liquid chromatography using fluorescence derivatization with dansyl hydrazine. *J.Chromatogr.* **1979**, *163*, 143-150.

Mainka, A.; Bächmann, K. UV detection of derivatized carbonyl compounds in rain samples in capillary electrophoresis using sample stacking and a Z-shaped flow cell. *J.Chromatogr.A* **1997**, *767*, 241-247.

Reid, A.D.; Baker, P.R. High-performance liquid chromatography with a reversed-phase radial compression column of serum bile acid dansyl hydrazone derivatives. *Biochem.Soc.Trans.* **1985**, *13*, 1235-1236.

Shimada, K.; Fukuda, N.; Nakagi, T. Studies on neurosteroids V: Separation and characterization of pregnenolone 3-stearate in rat brains using high-performance liquid chromatography. *J.Chromatogr.Sci.* **1997**, *35*, 71-74.

Zheng, G.; Leesman, G.D.; Crane, R.T.; Hagen, S. High performance liquid chromatographic quantitation of 3-O-deacyl-4'-monophosphoryl lipid A obtained from lipopolysaccharides of Salmonella minnesota R595 bacteria (Abstract APQ 1099). *Pharm.Res.* **1996**, *13*, S27.

SAMPLE
Matrix: air
Analyte: ketones
Sample preparation: Slurry Extrelut silica gel with MeCN, let stand for 1 h, filter, dry under vacuum, slurry with dichloromethane, let stand for 1 h, filter, dry under vacuum. Purify dansylhydrazine by injecting 500 μL aliquots of a 20 mg/mL solution in MeCN onto a 500 × 10 10 μm RSI column (Alltech) and eluting with MeCN:water from 20:80 to 50:50 over 5 min, to 58:42 over 5 min, to 80:20 (step gradient), maintain at 80:20 for 15 min, return to initial conditions over 5 min. Elute at 4 mL/min for 10 min, 6 mL/min for 15 min, then 4 mL/min for 5 min. Collect the fraction between 10 and 13.5 min, remove

the MeCN at 30° under vacuum, pour onto silica gel, extract with dichloromethane. Adjust the dansyl hydrazine concentration to 1 mg/mL and add a 1.5 mL aliquot to 700 mg silica gel, dry under vacuum, pack in a 4 mm ID quartz tube. Pull air through a tube at 2 L/min. Add 1.4 mL rain to a tube, let stand for at least 30 min. Extract tubes with 4 mL dichloromethane, evaporate the eluate to dryness under a stream of nitrogen at 35°, reconstitute the residue in 200 µL MeCN, inject a 10 µL aliquot.

HPLC VARIABLES

Column: 250 × 4 5 µm Superspher RP 18 (Merck)
Mobile phase: Gradient. MeCN:water from 20:80 to 50:50 over 5 min, to 65:35 over 15 min, to 80:20 over 13 min, to 75:25 over 7 min, return to initial conditions over 10 min.
Flow rate: 1
Injection volume: 10
Detector: F ex 355 em 525

CHROMATOGRAM

Retention time: 10 (acetaldehyde), 12 (acetone), 13 (propanal), 15 (crotonaldehyde), 16 (n-butanal), 18 (cyclohexanone), 20 (n-pentanal), 20.5 (isopentanal)
Limit of detection: 10-150 pg

KEY WORDS
SPE

REFERENCE
Schmied, W.; Przewosnik, M.; Bachmann, K. Determination of traces of aldehydes and ketones in the troposphere via solid phase derivatization with DNSH, *Fresenius' Z.Anal.Chem.*, **1989**, *335*, 464–468.

SAMPLE

Matrix: blood
Analyte: tetraprenylacetone (6,10,14,18-tetramethylnonandecan-2-one)
Sample preparation: Condition a 200 mg Bond Elut C18 SPE cartridge with 1 column volume of MeOH and 2 column volumes of water. 1 mL Plasma + 1 mL water + 100 µL 10 µg/mL IS in MeOH, vortex, add 1 mL EtOH, add 100 µL 50% KOH, heat in a boiling water bath for 10 min, cool, add 3 mL hexane, shake for 10 min, centrifuge at 1670 g for 5 min, repeat extraction twice more. Combine the organic layers and evaporate to dryness under a stream of nitrogen at 45°, reconstitute the residue in 200 µL 0.5% HCl in isopropanol, add 200 µL 0.2% dansyl hydrazine in isopropanol, mix, heat at 35° for 1 h, add 1 mL water, add to the SPE cartridge, wash twice with 3 mL portions of water, wash with 1 mL MeCN, add with 500 µL MeOH:chloroform 50:50, after 30 s elute by applying a vacuum, repeat this process twice more. Combine the eluates and evaporate them to dryness under a stream of nitrogen at 45°, reconstitute the residue in 200 µL EtOH, sonicate, centrifuge at 1670 g for 5 min, inject a 50 µL aliquot.

HPLC VARIABLES

Column: 150 × 6 5 µm YMC-A312 C18 (Yamamura, Kyoto)
Mobile phase: MeCN:water:triethylamine 92.5:7.5:0.015
Column temperature: 30
Flow rate: 1.8
Injection volume: 50
Detector: F ex 365 em 515

CHROMATOGRAM

Retention time: 12.5 (cis-5), 13 (trans-5)
Internal standard: 2,6,10,14-tetramethyleicosan-18-one (16, 16.5 (syn and anti))
Limit of quantitation: 8 ng/mL (cis-5), 12 ng/mL (trans-5)

KEY WORDS
plasma; SPE; pharmacokinetics

REFERENCE
Seki, T.; Hashida, N.; Kanazawa, T. Determination of tetraprenylacetone in human plasma by high-performance liquid chromatography with fluorescence derivatization using dansylhydrazine, *J.Chromatogr.*, **1988**, *424*, 410–415.

SAMPLE
Matrix: blood
Analyte: tacrolimus
Sample preparation: 1 mL Whole blood + 500 μL 2% saponin in water, vortex for 10 s, after 5 min add 2 mL 180 mM HCl and 6 mL diethyl ether, shake at 60 rpm for 15 rpm, centrifuge at 4000 rpm for 10 min. Remove the organic layer and add it to 2 mL 95 mM NaOH, shake at 60 rpm for 15 min, centrifuge at 4000 rpm for 10 min. Remove the organic layer and evaporate it to dryness under a stream of nitrogen at 37°, reconstitute the residue in 100 μL MeCN, vortex for 1 min, add 100 μL 1% trifluoroacetic acid in MeCN, add 100 μL 0.6 mg/mL dansylhydrazine in MeCN, vortex for 1 min, evaporate to dryness under a stream of nitrogen at 37°, store in the dark at 4°. Just before analysis reconstitute in 100 μL MeCN, inject a 25 μL aliquot onto column A with mobile phase A, elute column A with mobile phase A for 3 min, after 3 min elute contents of column A onto column B with mobile phase B, after 30 s remove column A from circuit, elute column B with mobile phase B for 6.5 min, elute tacrolimus fraction with mobile phase B from column B onto column C for 2 min, elute column C with mobile phase C and monitor the effluent. (Flush column A with MeOH for 5 min then re-equilibrate with mobile phase A for 4 min before next injection.)

HPLC VARIABLES
Column: A 15 × 3.9 25-40 μm C18 (Applied Biosystems); B 150 × 3.9 4 μm Novapack C18; C 100 × 3.2 Hypercarb Ph graphite
Mobile phase: A MeOH:water 50:50; B MeCN:water 80:20; C MeOH:dichloromethane 50:50
Flow rate: A 1; B 1; C 1.5
Injection volume: 25
Detector: F ex 338 em 520 (or 430 nm cut-off filter)

CHROMATOGRAM
Retention time: 3.5 (after the start of the elution of column C with mobile phase C)
Limit of quantitation: 3 ng/mL

KEY WORDS
whole blood; column-switching; heart-cut

REFERENCE
Beysens, A.J.; Beuman, G.H.; van der Heijden, J.J.; Hoogtanders, K.E.J.; Steijger, O.M.; Lingeman, H. Determination of tacrolimus (FK 506) in whole blood using liquid chromatography and fluorescence detection, *Chromatographia*, **1994**, *39*, 490–496.

SAMPLE
Matrix: plants
Analyte: abscisic acid and jasmonic acid
Sample preparation: Homogenize (Polytron PT 20 st) 1 g plant material with 5 mL 100 μg/mL BHT in MeOH, centrifuge at 9200 g for 15 min, remove the supernatant, homogenize the pellet with 5 mL 100 μg/mL BHT in MeOH:water 80:20, centrifuge. Combine the supernatants and remove MeOH under reduced pressure, freeze the aqueous solution in liquid nitrogen, thaw, centrifuge at 26000 g for 20 min, adjust the pH of the supernatant to 9.0 with 10 mM Tris-HCl, wash twice with 0.5 volumes of 100 μg/mL BHT in diethyl ether. Adjust the pH of the aqueous phase to 2.0, extract three times with 0.3 volumes of chloroform, evaporate the extracts to dryness under reduced pressure, reconstitute with MeOH. For each 14 g original plant material add 1 mg dansyl hydrazine, for each 1 mmole reagent add 1.1 mmoles HCl in MeOH, purge with argon, let stand in the dark at 25° overnight, inject an aliquot.

HPLC VARIABLES
Column: 5 μm Radial-Pak C18 radial compression (Waters)
Mobile phase: Gradient. MeCN:20 mM pH 4.0 phosphate buffer from 45:55 for 31 min, to 75:25 over 3 min, maintain at 75:25 for 6 min
Flow rate: 1.5
Detector: F ex 255 em 505 (cut-off filter)

CHROMATOGRAM
Retention time: 13 (abscisic acid), 20 (jasmonic acid)

KEY WORDS
comparison with other derivatizing reagents

REFERENCE
Anderson, J.M. Simultaneous determination of abscisic acid and jasmonic acid in plant extracts using high-performance liquid chromatography, *J.Chromatogr.*, **1985**, *330*, 347–355.

SAMPLE
Matrix: plants
Analyte: abscisic acid
Sample preparation: Homogenize 1 g plant material with 5 mL 100 μg/mL BHT in MeOH, remove MeOH under vacuum, freeze aqueous solution, thaw, centrifuge, adjust the pH of the supernatant to 9.0, wash twice with ether, adjust pH to 2.0, extract with chloroform. Remove the organic layer and evaporate it to dryness. For each 14 g original plant material add 1 mg dansyl hydrazine, for each 1 mmole reagent add 1.1 mmoles HCl in MeOH, purge with argon for 1 min, let stand in the dark at 25° overnight, streak on a TLC plate (Analtech GHR-250 μm), develop with MeOH:chloroform 1:6.7, remove the derivative (between the free acid at the origin and the unreacted hydrazide), extract with MeOH:chloroform 75:25, evaporate to dryness, reconstitute with 50 μM HCl in MeOH, inject an aliquot. (Prepare 10 mM HCl in MeOH by mixing concentrated aqueous HCl with MeOH.)

HPLC VARIABLES
Column: 5 μm Radial-Pak C18 radial compression (Waters)
Mobile phase: MeCN:20 mM pH 4.0 phosphate buffer 45:55
Detector: F ex 340 em 533

CHROMATOGRAM
Retention time: 13.5

KEY WORDS
comparison with other derivatizing reagents

REFERENCE
Anderson, J.M. Fluorescent hydrazides for the high-performance liquid chromatographic determination of biological carbonyls, *Anal.Biochem.*, **1986**, *152*, 146–153.

SAMPLE
Matrix: solutions
Analyte: bile salts
Sample preparation: Condition a Sep-Pak C18 SPE cartridge with water. Treat a bile salt solution with 3α- or 7α-hydroxysteroid dehydrogenase and NAD at 37° for 1 h, add 4 drops concentrated HCl, add to the SPE cartridge, wash with 10 mL water, elute with 5 mL MeOH. Evaporate the eluate to dryness under a stream of nitrogen, reconstitute with 200 μL 650 μL/L concentrated HCl in EtOH and 200 μL 1 mg/mL dansyl hydrazine in EtOH, heat at 60° for 10 min, inject an aliquot. (Hydroxy groups are oxidized to ketones which are then derivatized.)

HPLC VARIABLES
Column: 10 μm Radial-Pak C18 (8 mm ID, Waters)
Mobile phase: Gradient. MeOH:10 mM pH 3.4 KH_2PO_4 buffer from 60:40 to 85:15 over 13 min.
Flow rate: 2
Detector: F ex 360 em 510

CHROMATOGRAM
Retention time: 8 (taurocholic acid), 10 (taurochenodeoxycholic acid, taurodeoxycholic acid), 11.5 (taurolithocholic acid), 12.5 (glycocholic acid), 14 (glycochenodeoxycholic acid, glycodeoxycholic acid), 15.5 (ursodeoxycholic acid, cholic acid), 16 (glycolithocholic acid), 18 (chenodeoxycholic acid, deoxycholic acid), 21.5 (lithocholic acid)
Limit of quantitation: 30 pmole

KEY WORDS
SPE; comparison with 2,4-dinitrophenylhydrazine derivatization

REFERENCE
Reid, A.D.; Baker, P.R. Formation and separation by reversed-phase high-performance liquid chromatography of fluorescent and UV-absorbing bile salt derivatives, *J.Chromatogr.*, **1983**, *260*, 115–121.

SAMPLE
Matrix: solutions
Analyte: budesonide
Sample preparation: Mix 250 μL of a solution of budesonide in MeOH with 625 μL 9% trifluoromethanesulfonic acid in MeOH, immediately add 625 μL 160 μm dansyl hydrazine in MeOH, after 12 h inject a 100 μL aliquot onto column A and elute to waste with mobile phase A, after 100 s backflush the contents of column A onto column B with mobile phase B, elute with mobile phase B, monitor the effluent from column B. (Caution! Trifluoromethanesulfonic acid is highly toxic! Decontaminate surplus solutions with an equal volume of 2 M aqueous ammonia for 1 h!)

HPLC VARIABLES
Column: A 10 × 4.6 5 μm Vydac C18; B 150 × 4.6 5 μm Nucleosil C18
Mobile phase: A MeCN:7.7 mM pH 7 phosphate buffer 6:94; B MeCN:7.7 mM pH 7 phosphate buffer 65:35
Flow rate: 1
Injection volume: 100
Detector: F ex 350 em 520

CHROMATOGRAM
Retention time: 8, 10 (syn and anti isomers of 3-keto derivative)
Limit of detection: 1.5 pmole

KEY WORDS
column-switching; derivatization takes place on the 3-keto group

REFERENCE
Hyytiäinen, M.; Appelblad, P.; Pontén, E.; Stigbrand, M.; Irgum, K.; Jaegfeldt, H. Trifluoromethanesulfonic acid as a catalyst for the formation of dansylhydrazone derivatives, *J.Chromatogr.A*, **1996**, *740*, 279–283.

SAMPLE
Matrix: tissue, urine
Analyte: dinoprostone
Sample preparation: Tissue. Homogenize 100 mg rat tissue with EtOH:100 mM HCl 80 :20, centrifuge, wash the supernatant with petroleum ether, evaporate to dryness under reduced pressure, add 30 μg prednisolone, add 100 μL 2 mg/mL dansyl hydrazine in EtOH, add 100 μL EtOH:HCl 98:2, heat at 40° for 30 min, add 200 μL water, extract three times with 500 μL portions of ether. Combine the organic layers and evaporate them to dryness, reconstitute the residue in 10 μL MeCN, inject a 0.1 μL aliquot. Urine. Activate silica gel (Wako gel C-100) at 120° for 4 h, cool. Suspend 5 g activated silica in toluene:ethyl acetate 90:10, pour into a 200 × 10 glass column. 50 mL Human urine + 30 μg prednisolone, mix, adjust pH to 3 with 1 M HCl, extract twice with 50 mL portions of ethyl acetate, Combine the organic layers and wash them with 100 mL 10 mM HCl, wash with 100 mL water. Evaporate the organic layer to dryness under reduced pressure, reconstitute with a small amount of toluene:ethyl acetate 90:10, add to the column, wash with 10 mL toluene:ethyl acetate 90:10, wash with 10 mL MeOH:ethyl acetate 5:100, elute with 20 mL MeOH:ethyl acetate 50:50, evaporate the eluate to dryness, add 100 μL 2 mg/mL dansyl hydrazine in EtOH, add 100 μL EtOH:HCl 98:2, heat at 40° for 30 min, add 200 μL water, extract three times with 500 μL portions of ether. Combine the organic layers and evaporate them to dryness, reconstitute the residue in 10 μL MeCN, inject a 0.1 μL aliquot.

HPLC VARIABLES
Column: 160 × 1 10 μm μ Fine Pak SIL C18
Mobile phase: MeCN:MeOH:water:acetic acid 60:1:50:0.2

Flow rate: 0.008
Injection volume: 0.1
Detector: F ex 365 em 505

CHROMATOGRAM
Retention time: 11
Internal standard: prednisolone (14)
Limit of quantitation: <10 ng

OTHER SUBSTANCES
Extracted: metabolites

KEY WORDS
microbore; rat; brain; cerebrum; diencephalon; seminal vesicle; prostate; SPE

REFERENCE
Yamada, K.; Onodera, M.; Aizawa, Y. Determination of prostaglandin E_2 and the main prostaglandin E metabolite by micro high-performance liquid chromatography using fluorescence derivatization with dansyl hydrazine, *J.Pharmacol.Methods*, **1983**, *9*, 93–100.

SAMPLE
Matrix: urine
Analyte: 2,5-hexanedione
Sample preparation: Condition a 3 mL Bond-Elut C18 with 3 mL MeOH and 5 mL 1 M HCl. Centrifuge 6.5 mL urine at 1000 g for 10 min, add 50 μL 1.3 mg/mL 1,3-diacetyl-benzene in MeOH, add a 4 mL aliquot to the SPE cartridge, wash with 5 mL water, elute with 2.5 mL MeCN:water:phosphoric acid 50:50:2. Remove a 10 μL aliquot of the eluate and add it to 15 μL reagent, let stand at room temperature for 16 min, inject a 10 μL aliquot. (Hydrolyse 2 mL centrifuged urine with 200 μL 12 M HCl at 100° for 45 min, cool, centrifuge at 1000 g for 10 min, add to the SPE cartridge, proceed as above. Prepare reagent by dissolving 30 mg dansylhydrazine in 100 μL DMF and 250 μL MeCN.)

HPLC VARIABLES
Column: 150 × 4.6 3 μm Supelcosil C18
Mobile phase: Gradient. MeCN:25 mM pH 6.4 phosphate buffer 60:40 for 12 min, to 75:25 over 3 min, maintain at 75:25 for 5 min, return to initial conditions over 3 min, re-equilibrate for 7 min.
Flow rate: 1
Injection volume: 10
Detector: F ex 340 em 525

CHROMATOGRAM
Retention time: 9.5
Internal standard: 1,3-diacetylbenzene (13.7)
Limit of detection: 5 ng/mL (without hydrolysis), 12 ng/mL (with hydrolysis)

KEY WORDS
SPE

REFERENCE
Maestri, L.; Ghittori, S.; Imbriani, M.; Capodaglio, E. Determination of 2,5-hexandione by high-performance liquid chromatography after derivatization with dansylhydrazine, *J.Chromatogr.B*, **1994**, *657*, 111–117.

Darpsyl Hydrazide

SAMPLE
Matrix: plants
Analyte: abscisic acid
Sample preparation: Homogenize 1 g plant material with 5 mL 100 µg/mL BHT in MeOH, remove MeOH under vacuum, freeze aqueous solution, thaw, centrifuge, adjust the pH of the supernatant to 9.0, wash twice with ether, adjust pH to 2.0, extract with chloroform. Remove the organic layer and evaporate it to dryness. For each 14 g original plant material add 1 mg darpsyl hydrazide, for each 1 mmole reagent add 1.1 mmoles HCl in MeOH, purge with argon for 1 min, let stand in the dark at 25° overnight, streak on a TLC plate (Analtech GHR-250 µm), develop with MeOH:chloroform 1:6.7, remove the derivative (between the free acid at the origin and the unreacted hydrazide), extract with MeOH:chloroform 75:25, evaporate to dryness, reconstitute with 50 µM HCl in MeOH, inject an aliquot. (Prepare 10 mM HCl in MeOH by mixing concentrated aqueous HCl with MeOH. Prepare darpsyl hydrazide as follows. Add a solution of 100 mg darpsyl chloride (Polysciences, Inc., Paul Valley Industrial Park, Warrington PA) in 60 mL chloroform dropwise to a solution of 62 mg hydrazine hydrate (Caution! Hydrazine hydrate is a carcinogen!) in 60 mL MeOH at 25°, stir for 30 min, add 200 mL water, add 200 mL chloroform. Remove the organic phase and wash it twice with 60 mL portions of water, evaporate to dryness under reduced pressure, dissolve the residue in 100 mL ethyl acetate, wash twice with 60 mL portions of water, dry over anhydrous sodium sulfate, evaporate to dryness under reduced pressure to obtain darpsyl hydrazide(4-(1-(4,5-dihydro-3-phenylpyrazolyl)benzenesulfonyl hydrazide), store in a desiccator at -20°.)

HPLC VARIABLES
Column: 5 µm Radial-Pak C18 radial compression (Waters)
Mobile phase: MeCN:20 mM pH 4.0 phosphate buffer 45:55
Detector: F ex 330 em 435

CHROMATOGRAM
Retention time: 12

KEY WORDS
comparison with other derivatizing reagents

REFERENCE
Anderson, J.M. Fluorescent hydrazides for the high-performance liquid chromatographic determination of biological carbonyls, *Anal.Biochem.*, **1986**, *152*, 146–153.

Deayl Hydrazide

SAMPLE
Matrix: plants
Analyte: abscisic acid
Sample preparation: Homogenize 1 g plant material with 5 mL 100 μg/mL BHT in MeOH, remove MeOH under vacuum, freeze aqueous solution, thaw, centrifuge, adjust the pH of the supernatant to 9.0, wash twice with ether, adjust pH to 2.0, extract with chloroform. Remove the organic layer and evaporate it to dryness. For each 14 g original plant material add 1 mg deayl hydrazide, for each 1 mmole reagent add 1.1 mmoles HCl in MeOH, purge with argon for 1 min, let stand in the dark at 25° overnight, streak on a TLC plate (Analtech GHR-250 μm), develop with MeOH:chloroform 1:6.7, remove the derivative (between the free acid at the origin and the unreacted hydrazide), extract with MeOH:chloroform 75:25, evaporate to dryness, reconstitute with 50 μM HCl in MeOH, inject an aliquot. (Prepare 10 mM HCl in MeOH by mixing concentrated aqueous HCl with MeOH. Prepare deayl hydrazide, 2-diethylaminobenzoic hydrazide, as follows. Add 2.4 g iodoethane to 2 g ethyl 2-aminobenzoate and 2.4 g potassium bicarbonate in isopropanol, reflux with stirring for 12 h, cool, evaporate to 50 mL under reduced pressure, dissolve the slurry in chloroform, filter, wash twice with water, dry over anhydrous sodium sulfate, evaporate to dryness under reduced pressure, dry the resultant oil or powder over calcium carbonate to obtain ethyl 2-diethylaminobenzoate. Suspend the ethyl 2-diethylaminobenzoate in 25 mL chloroform:MeOH 50:50 in a 50 mL serum bottle, add a 10-fold molar excess of hydrazine hydrate (Caution! Hydrazine hydrate is a carcinogen!), heat the sealed bottle at 60° for 3 h (Caution! Use a safety screen!), cool, add 200 mL water, add 200 mL chloroform. Remove the organic phase and wash it twice with 60 mL portions of water, evaporate to dryness under reduced pressure, dissolve the residue in 100 mL ethyl acetate, wash twice with 60 mL portions of water, dry over anhydrous sodium sulfate, evaporate to dryness under reduced pressure, purify by TLC (silica gel G-2 mm) using chloroform:MeOH 90:10 to obtain deayl hydrazide (2-diethylaminobenzoic hydrazide), store in a desiccator at -20°.)

HPLC VARIABLES
Column: 5 μm Radial-Pak C18 radial compression (Waters)
Mobile phase: MeCN:20 mM pH 4.0 phosphate buffer 40:60
Detector: F ex 355 em 445

CHROMATOGRAM
Retention time: 8.6

KEY WORDS
comparison with other derivatizing reagents

REFERENCE
Anderson, J.M. Fluorescent hydrazides for the high-performance liquid chromatographic determination of biological carbonyls, *Anal.Biochem.*, **1986**, *152*, 146–153.

Deccyl Hydrazide

SAMPLE
Matrix: plants
Analyte: abscisic acid
Sample preparation: Homogenize 1 g plant material with 5 mL 100 μg/mL BHT in MeOH, remove MeOH under vacuum, freeze aqueous solution, thaw, centrifuge, adjust the pH of the supernatant to 9.0, wash twice with ether, adjust pH to 2.0, extract with chloroform. Remove the organic layer and evaporate it to dryness. For each 14 g original plant material add 1 mg deccyl hydrazide (7-diethylaminocoumarin-3-carboxylic hydrazide) (Molecular Probes, Eugene OR), for each 1 mmole reagent add 1.1 mmoles HCl in MeOH, purge with argon for 1 min, let stand in the dark at 25° overnight, streak on a TLC plate (Analtech GHR-250 μm), develop with MeOH:chloroform 1:6.7, remove the derivative (between the free acid at the origin and the unreacted hydrazide),extract with MeOH: chloroform 75:25, evaporate to dryness, reconstitute with 50 μM HCl in MeOH, inject an aliquot. (Prepare 10 mM HCl in MeOH by mixing concentrated aqueous HCl with MeOH.)

HPLC VARIABLES
Column: 5 μm Radial-Pak C18 radial compression (Waters)
Mobile phase: MeCN:20 mM pH 4.0 phosphate buffer 40:60
Detector: F ex 440 em 483

CHROMATOGRAM
Retention time: 22

KEY WORDS
comparison with other derivatizing reagents

REFERENCE
Anderson, J.M. Fluorescent hydrazides for the high-performance liquid chromatographic determination of biological carbonyls, *Anal.Biochem.*, **1986**, *152*, 146–153.

Diayl Hydrazide

SAMPLE
Matrix: plants
Analyte: abscisic acid
Sample preparation: Homogenize 1 g plant material with 5 mL 100 μg/mL BHT in MeOH, remove MeOH under vacuum, freeze aqueous solution, thaw, centrifuge, adjust the pH of the supernatant to 9.0, wash twice with ether, adjust pH to 2.0, extract with chloroform. Remove the organic layer and evaporate it to dryness. For each 14 g original plant material add 1 mg diayl hydrazide, for each 1 mmole reagent add 1.1 mmoles HCl in MeOH, purge with argon for 1 min, let stand in the dark at 25° overnight, streak on a TLC plate (Analtech GHR-250 μm), develop with MeOH:chloroform 1:6.7, remove the derivative (between the free acid at the origin and the unreacted hydrazide), extract with MeOH: chloroform 75:25, evaporate to dryness, reconstitute with 50 μM HCl in MeOH, inject an aliquot. (Prepare 10 mM HCl in MeOH by mixing concentrated aqueous HCl with MeOH. Prepare diayl hydrazide, 2-diisopropylaminobenzoic hydrazide, as follows. Add 4.1 g 2-iodopropane to 2 g ethyl 2-aminobenzoate and 2.4 g potassium bicarbonate in isopropanol, reflux with stirring for 12 h, cool, evaporate to 50 mL under reduced pressure, dissolve the slurry in chloroform, filter, wash twice with water, dry over anhydrous sodium sulfate, evaporate to dryness under reduced pressure, dry the resultant oil or powder over calcium carbonate to obtain ethyl 2-diisopropylaminobenzoate. Suspend the ethyl 2-diisopropyl-aminobenzoate in 25 mL chloroform:MeOH 50:50 in a 50 mL serum bottle, add a 10-fold molar excess of hydrazine hydrate (Caution! Hydrazine hydrate is a carcinogen!), heat the sealed bottle at 60° for 3 h (Caution! Use a safety screen!), cool, add 200 mL water, add 200 mL chloroform. Remove the organic phase and wash it twice with 60 mL portions of water, evaporate to dryness under reduced pressure, dissolve the residue in 100 mL ethyl acetate, wash twice with 60 mL portions of water, dry over anhydrous sodium sulfate, evaporate to dryness under reduced pressure, purify by TLC (silica gel G-2 mm) using chloroform:MeOH 90:10 to obtain diayl hydrazide (2-diisopropylaminobenzoic hydrazide), store in a desiccator at -20°.)

HPLC VARIABLES
Column: 5 μm Radial-Pak C18 radial compression (Waters)
Mobile phase: MeCN:20 mM pH 4.0 phosphate buffer 40:60
Detector: F ex 355 em 445

CHROMATOGRAM
Retention time: 10

KEY WORDS
comparison with other derivatizing reagents

REFERENCE
Anderson, J.M. Fluorescent hydrazides for the high-performance liquid chromatographic determination of biological carbonyls, *Anal.Biochem.*, **1986**, *152*, 146−153.

2-Dimethylamino-6-naphthalenesulfonic Hydrazide

SAMPLE
Matrix: plants
Analyte: abscisic acid
Sample preparation: Homogenize 1 g plant material with 5 mL 100 μg/mL BHT in MeOH, remove MeOH under vacuum, freeze aqueous solution, thaw, centrifuge, adjust the pH of the supernatant to 9.0, wash twice with ether, adjust pH to 2.0, extract with chloroform. Remove the organic layer and evaporate it to dryness. For each 14 g original plant material add 1 mg 2-dimethylamino-6-naphthalenesulfonic hydrazide, for each 1 mmole reagent add 1.1 mmoles HCl in MeOH, purge with argon for 1 min, let stand in the dark at 25° overnight, streak on a TLC plate (Analtech GHR-250 μm), develop with MeOH: chloroform 1:6.7, remove the derivative (between the free acid at the origin and the unreacted hydrazide), extract with MeOH:chloroform 75:25,evaporate to dryness, reconstitute with 50 μM HCl in MeOH, inject an aliquot. (Prepare 10 mM HCl in MeOH by mixing concentrated aqueous HCl with MeOH. Prepare 2-dimethylamino-6-naphthalenesulfonic hydrazide, an isomer of dansyl hydrazine, as follows. Add a solution of 316 μmoles 2-dimethylamino-6-naphthalenesulfonyl chloride (Molecular Probes, Eugene OR) in 60 mL chloroform dropwise to a solution of 62 mg hydrazine hydrate (Caution! Hydrazine hydrate is a carcinogen!) in 60 mL MeOH at 25°, stir for 30 min, add 200 mL water, add 200 mL chloroform. Remove the organic phase and wash it twice with 60 mL portions of water, evaporate to dryness under reduced pressure, dissolve the residue in 100 mL ethyl acetate, wash twice with 60 mL portions of water, dry over anhydrous sodium sulfate, evaporate to dryness under reduced pressure to obtain 2-dimethylamino-6-naphthalenesulfonic hydrazide, store in a desiccator at -20°.)

HPLC VARIABLES
Column: 5 μm Radial-Pak C18 radial compression (Waters)
Mobile phase: MeCN:20 mM pH 4.0 phosphate buffer 45:55
Detector: F ex 330 em 435

CHROMATOGRAM
Retention time: 12

KEY WORDS
comparison with other derivatizing reagents

REFERENCE
Anderson, J.M. Fluorescent hydrazides for the high-performance liquid chromatographic determination of biological carbonyls, *Anal.Biochem.*, **1986**, *152*, 146–153.

4-(N,N-Dimethylaminosulfonyl)-7-hydrazino-2,1,3-benzoxadiazole

RELATED REFERENCES

Shimada, K.; Fukuda, N.; Nakagi, T. Studies on neurosteroids V: Separation and characterization of pregnenolone 3-stearate in rat brains using high-performance liquid chromatography. *J.Chromatogr.Sci.* **1997**, *35*, 71-74.

Uzu, S.; Imai, K.; Nakashima, K.; Akiyama, S. Determination of medroxyprogesterone acetate in serum by HPLC with peroxyoxalate chemiluminescence detection using a fluorogenic reagent, 4-(N,N-di-methylaminosulfonyl)-7-hydrazino-2,1,3- benzoxadiazole. *J.Pharm.Biomed.Anal.* **1992**, *10*, 979-984.

SAMPLE

Matrix: solutions

Analyte: heptan-4-one

Sample preparation: Mix 10 µL of a 2-3 µM solution in MeCN with 20 µL 500 µM 4-(N,N-dimethylaminosulfonyl)-7-hydrazino-2,1,3-benzoxadiazole in MeCN and 10 µL 1% trifluoroacetic acid in MeCN, let stand at room temperature for 5 h, inject an aliquot. (4-(N,N-Dimethylaminosulfonyl)-7-hydrazino-2,1,3-benzoxadiazole (DBD-H) is available from Tokyo Kasei (TCI America, Portland OR). Synthesis is as follows. Dissolve 0.5 g magnesium sulfate heptahydrate and 6 g NaOH in 60 mL water, throughout the reaction keep the flask at about 20° with cold water cooling, add 15 mL 30% hydrogen peroxide, add 75 mL MeOH, add 12.1 g powdered benzoyl peroxide in one go, stir for 10 min, pour into 150 mL 20% sulfuric acid, extract three times with 50 mL portions of chloroform, determine peroxybenzoic acid concentration by iodometric titration (Tetrahedron 1967, 23, 3327). Slowly add 110 mL 1 M peroxybenzoic acid in chloroform to 7 g 2,6-difluoroaniline dissolved in 100 mL chloroform, stir at room temperature, when reaction is complete (iodometric titration) wash with 2% sodium thiosulfate, wash with 5% sodium carbonate, wash with water, dry over anhydrous sodium sulfate, evaporate to dryness under reduced pressure, recrystallize 2,6-difluoronitrosobenzene from EtOH (mp 108.5-109.5). Stir 8.5 g 2,6-difluoronitrosobenzene in 85 mL DMSO at room temperature and add a solution of 3.91 g sodium azide in 85 mL DMSO dropwise, let stand for about 1 h, add to a large volume of water, extract with ether, dry the extracts over anhydrous sodium sulfate, evaporate to dryness under reduced pressure and distil to give 4-fluoro-2,1,3-benzoxadiazole as a colorless oil (bp 83°/12 mm Hg) (J.Chem.Soc.(C) 1970, 1433). Add 11 mL chlorosulfonic acid dropwise to 3 g 4-fluoro-2,1,3-benzoxadiazole in 10 mL chloroform at 0-10° (use a calcium chloride drying tube), stir at room temperature for 1 h, reflux for 2 h, cool, slowly pour into ice water, remove the organic layer, extract the aqueous layer with chloroform, combine the organic layers, wash, dry over anhydrous magnesium sulfate, evaporate under reduced pressure, take up the residue in 5 mL benzene (Caution! Benzene is a carcinogen!), chromatograph on a 150 × 30 column of silica gel (100-200 mesh Kanto Chemical) with n-hexane:benzene 50:50, evaporate the appropriate fractions to give 4-(chlorosulfonyl)-7-fluoro-2,1,3-benzoxadiazole (CBD-F) as pale yellow needles (mp 64-66°) (Anal. Chem. 1984. 56, 2461). Stir 0.76 g CBD-F in 70 mL MeCN at 0-10° and add 1 g dimethylamine hydrochloride in 10 mL 100 mM pH 10 borax dropwise, adjust pH to 5 with 1 M HCl, concentrate to about 10 mL under reduced pressure, extract three times with 200 mL portions of diethyl ether, wash with water, dry over anhydrous magnesium sulfate, evaporate under reduced pressure, chromatograph on a 500 × 20 column of silica gel with chloroform, isolate the appropriate fraction and re-chromatograph on the same column with ethyl acetate:benzene 1:2 to give 4-(N,N-dimethylaminosulfonyl)-7-fluoro-2,1,3-benzoxadiazole (DBD-F) as white needles (mp 124-125°) (yield 1% !) (Analyst 1989, 114, 413). On a Merck no. 5714 60F$_{254}$ tlc plate eluted with chloroform DBD-F has Rf 0.32 and lies between two other reaction products. DBD-F can also be purchased from Tokyo Kasei. Dissolve 80 mg 4-(N,N-dimethylaminosulfonyl)-7-fluoro-2,1,3-benzoxadiazole in 12 mL MeCN, add 40 μL 98% hydrazine hydrate, heat in the dark at 50-55° for 20 min, evaporate to dryness, recrystallize from MeOH to give 4-(N,N-dimethylaminosulfonyl)-7-hydrazino-2,1,3-benzoxadiazole as reddish-brown crystals (mp 138-139° d).)

HPLC VARIABLES
Column: 150 × 4.6 5 μm TSK-LS 80Tm (Tosoh)
Mobile phase: MeCN:water:trifluoroacetic acid 71:29:0.05
Flow rate: 1
Detector: F ex 450 em 550

CHROMATOGRAM
Retention time: 7
Limit of detection: 17.1 μM

OTHER SUBSTANCES
Also analyzed: acetone, butyraldehyde, 4'-ethylacetophenone, p-hydroxybenzaldehyde, propionaldehyde

REFERENCE
Uzu, S.; Kanda, S.; Imai, K.; Nakashima, K.; Akiyama, S. Fluorogenic reagents: 4-Aminosulphonyl-7-hydrazino-2,1,3-benzoxadiazole, 4-(N,N-dimethylaminosulphonyl)-7-hydrazino-2,1,3-benzoxadiazole and 4-hydrazino-7-nitro-benzoxadiazole hydrazine for aldehydes and ketones, *Analyst*, **1990**, *115*, 1477–1482.

2,4-Dinitrophenylhydrazine

RELATED REFERENCES
Beacham, D.B. Analyses of aldehydes and ketones using the DNPH derivatization process. *Proc.,Annu.Meet.- Air Waste Manage.Assoc.* **1994**, *87th*, 17pp 94-TA2CO.11P.

Chiavari, G.; Bergamini, C. High-performance liquid chromatography of carbonyl compounds as 2,4-dinitrophenylhydrazones with electrochemical detection. *J.Chromatogr.* **1985**, *318*, 427-432.

Coutrim, M.X.; Nakamura, L.A.; Collins, C.H. Quantification of 2,4-dinitrophenylhydrazones of low molecular mass aldehydes and ketones using HPLC. *Chromatographia* **1993**, *37*, 185-190.

Ferioli, V.; Vezzalini, F.; Rustichelli, C.; Gamberini, G. High-performance liquid chromatography of dihydroxyacetone as its bis-2,4-dinitrophenylhydrazone derivative. *Chromatographia* **1995**, *41*, 61-65.

Foster, P.; Ferrari, C.; Jacob, V.; Roche, A.; Baussand, P.; Delachaume, J.C. Determination of C1-C5 carbonyls in the atmosphere by 2,4-dinitrophenylhydrazine-coated sorbent sampling and HPLC analysis. The influence of water on ketone analysis. *Analusis* **1996**, *24*, 71-73.

Jacobs, W.A.; Kissinger, P.T. Determination of carbonyl 2,4-dinitrophenylhydrazones by liquid chromatography/electrochemistry. *J.Liq.Chromatogr.* **1982**, *5*, 669-676.

Lange, J.; Eckhoff, S. Determination of carbonyl compounds in exhaust gas by using a modified DNPH-method. *Fresenius' J.Anal.Chem* **1996**, *356*, 385-389.

Liebezeit, G. HPLC gradient elution of dinitrophenylhydrazones of aldehydes and ketones from aqueous samples. *HRC CC,J.High Resolut.Chromatogr.Chromatogr.Commun.* **1982**, *5*, 215-216.

Viras, L.G.; Kotzias, D.; Duane, M. Application of the 2,4-dinitrophenylhydrazine method for measuring carbonyl compounds in a semi-remote and in an urban area. *Fresenius Environ.Bull.* **1992**, *1*, S73-S78,

SAMPLE
Matrix: air
Analyte: ketones
Sample preparation: Pull 5-30 L of air through two 10 mL aliquots of reagent in series at 0.5-1.5 L/min. Combine the samples and extract them twice with 5 mL portions of chloroform. Combine the chloroform layers and wash them with 20 mL 2 M HCl, wash with 20 mL water, evaporate to dryness under reduced pressure with gentle heating, reconstitute with 2 mL MeCN, inject a 4 μL aliquot. (Prepare reagent by dissolving 500 mg 2,4-dinitrophenylhydrazine (recrystallized from EtOH) in 500 mL 2 M HCl and washing twice with 5 mL portions of chloroform.)

HPLC VARIABLES
Column: 200 × 4.6 5 μm LiChrosorb RP-18
Mobile phase: MeCN:water 62:38
Flow rate: 1.5
Injection volume: 4
Detector: UV 254

CHROMATOGRAM
Retention time: 4.03 (formaldehyde), 5.12 (acetaldehyde), 6.69 (acrolein), 6.85 (acetone), 7.42 (propionaldehyde), 9.70 (crotonaldehyde), 10.40 (methyl ethyl ketone), 10.60 (isobutyraldehyde, n-butyraldehyde), 12.41 (benzaldehyde), 14.83 (methyl isopropyl ketone, methyl n-propyl ketone, diethyl ketone), 14.83 (isovaleraldehyde), 15.80 (n-valeraldehyde), 17.53 (o-tolualdehyde), 18.26 (m-tolualdehyde), 18.94 (p-tolualdehyde), 20.55 (methyl isobutyl ketone, methyl sec-butyl ketone), 21.20 (methyl t-butyl ketone), 22.57 (methyl n-butyl ketone), 23.94 (n-caproaldehyde), 34.58 (methyl n-amyl ketone)
Limit of detection: 1.5-2.6 ppb

REFERENCE
Kuwata, K.; Uebori, M.; Yamasaki, Y. Determination of aliphatic and aromatic aldehydes in polluted airs as their 2,4-dinitrophenylhydrazones by high performance liquid chromatography, *J.Chromatogr.Sci.*, **1979**, *17*, 264–268.

SAMPLE
Matrix: air
Analyte: ketones
Sample preparation: Condition a Sep-Pak C18 SPE cartridge with 2 mL MeCN and 2 mL 2 mg/mL 2,4-dinitrophenylhydrazine in MeCN:85% orthophosphoric acid 100:1, dry under nitrogen at 100 mL/min for 30 min, pull air through the cartridge at 0.5-1.5 mL/min for 24 h, elute with 1 mL MeCN, elute with 1 mL water, filter (Zetapor) the eluate, inject a 20 μL aliquot of the filtrate. (Purify nitrogen by passing it over activated charcoal.)

HPLC VARIABLES
Column: two 100 × 4.6 3 μm Microsphere C18 columns in series (Chrompack)
Mobile phase: Gradient. MeCN:MeOH:water 7:60:33 for 6.4 min, to 65:7:28 over 13.6 min, to 100:0:0 over 10 min.
Flow rate: 1
Injection volume: 20
Detector: UV 360

CHROMATOGRAM
Retention time: 4 (formaldehyde), 5.5 (acetaldehyde), 7.5 (acrolein), 7.5 (acetone), 8.5 (propanal), 11 (crotonaldehyde), 12.5 (2-butanone, butanal), 14.5 (benzaldehyde), 15.5 (trans-2-pentenal), 16.9 (cyclohexanone), 17.2 (valeraldehyde), 20.5 (trans-2-hexenal), 21.5 (2-hexanone), 22 (hexanal)

KEY WORDS
SPE

REFERENCE
Dye, C.; Oehme, M. Comments concerning the HPLC separation of acrolein from other C_3 carbonyl compounds as 2,4-dinitrophenylhydrazones; a proposal for improvement, *J.High Res.Chromatogr.*, **1992**, *15*, 5–8.

SAMPLE
Matrix: blood
Analyte: 1-hydroxyacetone (acetol)
Sample preparation: 1 mL Serum + 75 μL 60% perchloric acid, mix, centrifuge at 4° at 18000 g for 20 min. Remove the supernatant and adjust the pH to 5-7 with KOH, centrifuge at 18000 g for 20 min. Remove a 750 μL aliquot of the supernatant and add it to 300 μL reagent, vortex for 5 s, add 1 mL chloroform, shake at 400 cycles/min for 30 min, centrifuge, remove the chloroform layer without removing any water, repeat the extraction. Combine the organic layers and add 10 mg sodium bicarbonate, let stand for 10 min, evaporate to dryness under a stream of nitrogen, reconstitute the residue in 750 μL ethyl acetate, inject a 10 μL aliquot. (Prepare reagent by washing a 0.4% solution of 2,4-dinitrophenylhydrazine in 2 M HCl three times with equal volumes of carbon tetrachloride.)

HPLC VARIABLES
Column: 100 × 8 10 μm C18 radial compression
Mobile phase: Gradient. MeOH:water 55:45 for 20 min, to 100:0 over 5 min, maintain at 100:0 for 5 min, return to initial conditions over 2 min, re-equilibrate for 10 min.
Flow rate: 2 for 20 min, to 4 over 5 min, maintain at 4 for 5 min, to 2 over 2 min
Injection volume: 9.8 (major peak), 15.2 (minor peak)
Detector: UV 365

CHROMATOGRAM
Limit of quantitation: 5 μM

KEY WORDS
serum; human; rat

REFERENCE
Casazza, J.P.; Fu, J.L. Measurement of acetol in serum, *Anal.Biochem.*, **1985**, *148*, 344–348.

SAMPLE
Matrix: blood
Analyte: spectinomycin
Sample preparation: 100 μL Plasma + 400 μL MeCN:trifluoroacetic acid 97:3, vortex, centrifuge at 2000 g for 3 min. Remove a 250 μL aliquot of the supernatant and add it to 200 μL 5 mg/mL 2,4-dinitrophenylhydrazine in MeCN, mix, heat at 70° for 1 h, cool on ice for 2 min, add 30 μL acetone, mix, heat at 70° for 10 min, cool on ice, filter (Ultrafree MC 30000 molecular weight limit) while centrifuging at 2000 g, inject a 20 μL aliquot of the filtrate.

HPLC VARIABLES
Column: 33 × 4.6 3 μm Pecosphere C18 CR
Mobile phase: Gradient. A was MeCN:water 60:40. B was MeCN:MeOH 19:1. A:B 100:0 for 1 min, to 0:100 over 10 min, maintain at 0:100 for 2 min, re-equilibrate at initial conditions for 10 min.
Flow rate: 1.2
Injection volume: 20
Detector: UV 405

CHROMATOGRAM
Retention time: 9
Limit of quantitation: 2 μg/mL

KEY WORDS
plasma; turkey

REFERENCE

Burton, S.D.; Hutchins, J.E.; Fredericksen, T.L.; Ricks, C.; Tyczkowski, J.K. High-performance liquid chromatographic method for the determination of spectinomycin in turkey plasma, *J.Chromatogr.*, **1991**, *571*, 209–216.

SAMPLE

Matrix: blood
Analyte: acetone
Sample preparation: Wet 300 mg Amberlite XAD-2 resin with 300 µL MeCN, add 1 mL of plasma, add 4 mL 4 µM 2,4-dinitrophenylhydrazine in 1-1.5 M HCl, shake for 10 min, filter, wash the resin with four 5 mL portions of water (or until the washings are neutral), dry under vacuum, elute the resin with 5 mL MeCN, inject a 25 µL aliquot.

HPLC VARIABLES

Column: Symmetry octylsilica (Waters)
Mobile phase: Gradient. MeCN:water from 35:65 to 80:20 over 20 min.
Flow rate: 1
Injection volume: 25
Detector: UV (wavelength not given)

CHROMATOGRAM

Retention time: 8
Limit of quantitation: 3 µg/mL

OTHER SUBSTANCES

Simultaneously analyzed: acetaldehyde, formaldehyde

KEY WORDS

plasma

REFERENCE

Breckenridge, S.M.; Yin, X.; Rosenfeld, J.M.; Yu, Y.H. Analytical derivatizations of volatile and hydrophilic carbonyls from aqueous matrix onto a solid phase of a polystyrene-divinylbenzene macroreticular resin, *J.Chromatogr.B*, **1997**, *694*, 289–296.

SAMPLE

Matrix: blood

Analyte: eltanolone (pregnanolone)

Sample preparation: 1 mL Plasma + 2 mL MeCN, mix, centrifuge at 2000 g for 10 min. Remove a 1 mL aliquot of the supernatant and add it to 100 µL reagent, vortex for 10 s, let stand for 1 h, inject a 100-200 µL aliquot. (Prepare the reagent by suspending 2 g 2,4-dinitrophenylhydrazine in 100 mL MeOH and adding 4 mL concentrated sulfuric acid dropwise over 10 min, stir for 10 min, filter.)

HPLC VARIABLES

Column: 150 × 3.9 4 µm Novapak C18
Mobile phase: MeCN:water:trifluoroacetic acid 70:30:0.1
Flow rate: 2
Injection volume: 100-200
Detector: UV 367

CHROMATOGRAM

Retention time: 6.2
Limit of detection: 20 ng/mL (200 µL injection)

KEY WORDS

plasma; pharmacokinetics

REFERENCE

Jones, D.J.; Bjorksten, A.R.; Crankshaw, D.P. Determination of eltanolone in human plasma by high-performance liquid chromatography, *J.Chromatogr.B*, **1997**, *694*, 467–470.

SAMPLE

Matrix: blood, urine
Analyte: camphor

Sample preparation: Plasma. 5 mL Plasma + 10 mL peroxide-free ether, mix. Remove the organic layer and add it to 2 mL isopropanol, evaporate the mixture to 2 mL under a stream of nitrogen at room temperature, add 2 mL of a saturated solution of 2,4-dinitrophenylhydrazine in MeOH, add 1 drop of concentrated HCl, heat at 85-90° for 18 h, cool, evaporate under a stream of nitrogen at 50° to 1 mL, add 2 mL 2 M HCl, evaporate to 2 mL, add 2 mL 2 M HCl, add 4 mL isooctane, rotate mechanically for 30 min. Remove the organic layer and wash it twice with 4 mL water, centrifuge, dry over anhydrous sodium sulfate, add to a florisil column, elute with 10 mL alcohol-free chloroform, evaporate the eluate to dryness under reduced pressure, reconstitute the residue in 2 mL MeCN:water 70:30, inject a 200 µL aliquot. Urine. Adjust 40 mL urine to pH 6.5 with three drops 6 M HCl, saturate with NaCl, extract with 10 mL peroxide-free ether. Remove the organic layer and add it to 2 mL isopropanol, evaporate the mixture to 2 mL under a stream of nitrogen at room temperature, add 2 mL of a saturated solution of 2,4-dinitrophenylhydrazine in MeOH, add 1 drop of concentrated HCl, heat at 85-90° for 18 h, cool, evaporate under a stream of nitrogen at 50° to 1 mL, add 2 mL 2 M HCl, evaporate to 2 mL, add 2 mL 2 M HCl, add 4 mL isooctane, rotate mechanically for 30 min. Remove the organic layer and wash it twice with 4 mL water, centrifuge, dry over anhydrous sodium sulfate, add to a florisil column, elute with 10 mL alcohol-free chloroform, evaporate the eluate to dryness under reduced pressure, reconstitute the residue in 2 mL MeCN:water 70:30, inject a 50 µL aliquot. (The florisil (60-100 mesh; J.T. Baker) was heated at 120° overnight, cooled, and 3% water added. The column was prepared by plugging the end of a Pasteur pipette with glass wool and adding florisil to a height of 6 cm followed by anhydrous sodium sulfate to a height of 1 cm, pre-wash with 3 mL isooctane.)

HPLC VARIABLES
Column: radial compression µBondapak C18
Mobile phase: MeCN:water 82:18
Flow rate: 1
Injection volume: 50-200
Detector: UV 368.5

CHROMATOGRAM
Retention time: 14.5
Limit of detection: 10 ng/mL (urine), 20 ng/mL (plasma)

KEY WORDS
plasma; horse; SPE

REFERENCE
Gallicano, K.D.; Park, H.C.; Young, L.M. A sensitive liquid chromatographic procedure for the analysis of camphor in equine urine and plasma, *J.Anal.Toxicol.*, **1985**, *9*, 24–30.

SAMPLE
Matrix: bulk
Analyte: steroids
Sample preparation: Dissolve in MeOH at a concentration of 0.1-1%, acidify with a few drops of concentrated HCl, add an excess of 2,4-dinitrophenylhydrazine in MeOH, mix thoroughly, heat at 50° for a few min.

HPLC VARIABLES
Column: 1000 × 2.1 Zipax coated with 1% β,β'-oxydipropionitrile
Mobile phase: Heptane
Flow rate: 1
Injection volume: 1
Detector: UV 254

CHROMATOGRAM
Retention time: 8 (dehydroepiandrosterone), 10 (androsterone)
Limit of detection: 1 ng

REFERENCE

Henry, R.A.; Schmit, J.A.; Dieckman, J.F. The analysis of steroids and derivatized steroids by high speed liquid chromatography, *J.Chromatogr.Sci.*, **1971**, *9*, 513–520.

SAMPLE

Matrix: enzyme incubations

Analyte: ketoacids

Sample preparation: 100 μL Enzyme incubation + 20 μL 50% trichloroacetic acid, mix, centrifuge for 2 min. Remove a 50 μL aliquot of the supernatant and add it to 30 μL 0.2% 2,4-dinitrophenylhydrazine in 2 M HCl, vortex thoroughly, let stand at room temperature for 30 min, make up to 1 mL with MeCN:water 60:40, filter (0.45 μm), inject a 20 μL aliquot.

HPLC VARIABLES

Column: 250 × 4.6 5 μm Zorbax ODS

Mobile phase: Gradient. A was MeCN:0.1% triethylamine in water 25:75, adjust pH to 4.5 with glacial acetic acid. B was MeCN. A:B from 80:20 to 50:50 over 20 min.

Flow rate: 1

Injection volume: 20

Detector: UV 254

CHROMATOGRAM

Retention time: 9 (α-ketoisovaleric acid (3-methyl-2-oxobutanoic acid)), 10 (α-keto-β-methylvaleric acid (3-methyl-2-oxopentanoic acid))

Limit of quantitation: 50 pmole

REFERENCE

Smyk-Randall, E.M.; Brown, O.R. A reverse-phase high-performance liquid chromatography assay for dihydroxy-acid dehydratase, *Anal.Biochem.*, **1987**, *164*, 434–438.

SAMPLE

Matrix: feed

Analyte: salinomycin

Sample preparation: Pulverize feed in a grinder, mix with EtOH, sonicate for 5 min, filter (0.45 μm), dilute filtrate with EtOH if necessary. 5 mL Filtrate + 1 mL 600 μg/mL 2,4-dinitrophenylhydrazine in MeOH + 1 drop concentrated HCl, heat at 50° for about 3 min, cool to room temperature, make up to 10 mL with EtOH, inject an aliquot.

HPLC VARIABLES

Column: 150 × 4.6 5 μm Inertsil ODS-2

Mobile phase: MeOH:1.5% aqueous acetic acid 94:6

Flow rate: 1

Injection volume: 20

Detector: UV 380; UV 419

CHROMATOGRAM

Retention time: 7.8

Limit of detection: 2-5 ng

KEY WORDS

maximum sensitivity at 419 nm

REFERENCE

Mathur, A.K. Determination of salinomycin by high-performance liquid chromatography using a pre-column derivatization technique, *J.Chromatogr.A*, **1994**, *664*, 284–288.

SAMPLE

Matrix: perfusate

Analyte: ketones

Sample preparation: 1.5 mL Perfusate + 100 μL 3.1 mg/mL 2,4-dinitrophenylhydrazine in 2 M HCl + 500 μL water, vortex for 15 min, add 10 mL pentane, shake intermittently for 30 min, remove the organic layer and extract the aqueous layer with 20 mL pentane. Combine the pentane layers and evaporate them to dryness under a stream of nitrogen at 30°, reconstitute the residue in 200 μL MeCN, filter (0.2 μm Nylon-66), inject a 25 μL aliquot of the filtrate.

HPLC VARIABLES
Guard column: Bondapak C18 Guard-Pak
Column: 75 × 4.6 3 μm Ultrasphere ODS C18
Mobile phase: MeCN:water:acetic acid 40:60:0.1 (At the start of each day wash column with MeCN:acetic acid 100:0.1.)
Flow rate: 1
Injection volume: 25
Detector: UV 356

CHROMATOGRAM
Retention time: 5.3 (malonaldehyde (UV 307)), 6.6 (formaldehyde), 10.3 (acetaldehyde), 16.5 (acetone), 20.5 (propionaldehyde)

REFERENCE
Cordis, G.A.; Bagchi, D.; Maulik, N.; Das, D.K. High-performance liquid chromatographic method for the simultaneous detection of malonaldehyde, acetaldehyde, formaldehyde, acetone and propional-dehyde to monitor the oxidative stress in heart, *J.Chromatogr.A*, **1994**, *661*, 181–191.

SAMPLE
Matrix: reaction mixture
Analyte: Maillard reaction products
Sample preparation: Reflux 1.35 g glycine, 900 mg glucose monohydrate, 20 drops glacial acetic acid, and 60 mL EtOH for 24 h, filter, evaporate the filtrate to dryness, reconstitute with water. Remove a 1 mL aliquot and add it to 6 mL saturated 2,4-dinitrophenylhydrazine in 2 M HCl, let stand at room temperature for 1 h, filter, wash the solid with water, dissolve the solid in ethyl acetate to a concentration of 5 mg/mL, inject an aliquot.

CAPILLARY ELECTROPHORESIS
Capillary: 60 cm × 75 μm fused-silica (50 cm to detector)
Capillary preparation: Wash with 1 mL MeOH, wash with 1 mL chloroform, wash with 1 mL MeOH, wash with 1 mL 1 M NaOH, wash with 3 M HCl, wash with 2 mL running buffer.
Running buffer: 10 mM Na_2HPO_4 containing 6 mM sodium tetraborate and 50 mM sodium dodecyl sulfate
Voltage/Current: 10 kV
Injection: Electromigration at 10 kV for 3 s
Detector: UV 220
Migration time: 8-21 (depending on nature of compound)

KEY WORDS
Compounds can also be determined by HPLC: 150 × 3.3 7 μm Separon SGX C18 in a glass column (Tessek, Prague); MeOH:water from 30:70 to 65:35 over 15 min, maintain at 65:35 for 15 min, wash with MeOH for 10 min, re-equilibrate at initial conditions for 15 min; column temp 30; UV 360; retention times 6-22 min.

REFERENCE
Deyl, Z.; Miksik, I.; Struzinsky, R. Separation and partial characterization of Maillard reaction products by capillary zone electrophoresis, *J.Chromatogr.*, **1990**, *516*, 287–298.

SAMPLE
Matrix: solutions
Analyte: bile salts
Sample preparation: Condition a Sep-Pak C18 SPE cartridge with water. Treat a bile salt solution with 3α- or 7α-hydroxysteroid dehydrogenase and NAD at 37° for 1 h, add 4 drops concentrated HCl, add to the SPE cartridge, wash with 10 mL water, elute with 5 mL MeOH. Evaporate the eluate to dryness under a stream of nitrogen, reconstitute with 400 μL 1 mg/mL 2,4-dinitrophenylhydrazine in EtOH, add 2 drops concentrated HCl, heat at 60° for 10 min, inject an aliquot. (Hydroxy groups are oxidized to ketones which are then derivatized.)

HPLC VARIABLES
Column: 10 μm Radial-Pak C18 (8 mm ID, Waters)
Mobile phase: Gradient. MeOH:10 mM KH_2PO_4 from 40:60 to 90:10 over 7 min.
Flow rate: 2
Detector: UV 340

CHROMATOGRAM
Retention time: 9 (taurocholic acid), 10 (taurochenodeoxycholic acid, taurodeoxycholic acid), 11.5 (taurolithocholic acid)
Limit of detection: 20 pmole

KEY WORDS
SPE; comparison with dansylhydrazine derivatization

REFERENCE
Reid, A.D.; Baker, P.R. Formation and separation by reversed-phase high-performance liquid chromatography of fluorescent and UV-absorbing bile salt derivatives, *J.Chromatogr.*, **1983**, *260*, 115–121.

SAMPLE
Matrix: solutions
Analyte: bile alcohols
Sample preparation: Condition a Sep-Pak C18 SPE cartridge with water. Dissolve 100 μg bile alcohols in 2 mL 0.06 U/mL 3α-hydroxysteroid dehydrogenase (EC 1.1.1.50, Sigma) in pH 8.9 pyrophosphate buffer containing 19.8 mg/mL NAD, let stand at room temperature for 20 h, add to the C18 SPE cartridge, wash with 10 mL water, elute with 5 mL MeOH. Concentrate the eluate to 1 mL, add 200 μg 2,4-dinitrophenylhydrazine, heat at 60° for 1 h, evaporate to dryness, reconstitute with 2 mL n-hexane:ethyl acetate 2:1, add to a Sep-Pak silica SPE cartridge, wash with 30 mL n-hexane:ethyl acetate 2:1, elute with 20 mL ethyl acetate, inject an aliquot (?). (The 3α-hydroxysteroid dehydrogenase oxidizes the 3α-hydroxy group of the bile alcohol to a ketone. The ketone is subsequently derivatized with 2,4-dinitrophenylhydrazine.)

HPLC VARIABLES
Column: 150 × 3.9 Nova-Pak phenyl
Mobile phase: MeOH:water 83:17
Flow rate: 1
Detector: UV 364

CHROMATOGRAM
Retention time: 7.66 (5α-cyprinol), 7.75 (5α-bufol), 8.77 (5β-cyprinol), 9.12 (5β-bufol)
Limit of detection: 20 ng

OTHER SUBSTANCES
Simultaneously analyzed: cholestanepentols, cholestanetetrols

KEY WORDS
SPE

REFERENCE
Une, M.; Harada, J.-i.; Mikami, T.; Hoshita, T. High-performance liquid chromatographic separation of ultraviolet-absorbing bile alcohol derivatives, *J.Chromatogr.B*, **1996**, *682*, 157–161.

SAMPLE
Matrix: urine
Analyte: methyl ethyl ketone
Sample preparation: 5 mL Urine + 500 μL reagent + 1 mL cyclohexane, shake, let stand in the dark at room temperature for 1 h, shake, centrifuge at 1000 g for 5 min. Remove the organic layer and evaporate it to dryness under a stream of helium at 37°, reconstitute the residue in 500 μL MeCN, inject a 20 μL aliquot. (Prepare reagent as follows. Recrystallize 2,4-dinitrophenylhydrazine twice from MeOH. Dissolve 250 mg 2,4-dinitrophenylhydrazine in 100 mL 4 M HCl, wash twice with 5 mL portions of chloroform, store in the dark under a layer of cyclohexane at 4°, discard after 2 weeks.)

HPLC VARIABLES
Guard column: 30 × 4 30-40 μm Perisorb RP-18
Column: 250 × 4 5 μm LiChrosorb RP-18
Mobile phase: MeCN:water 55:45
Flow rate: 1.5
Injection volume: 20
Detector: UV 360

CHROMATOGRAM
Retention time: 8.8 (syn derivative), 9.7 (anti derivative)
Limit of detection: 100 ng/mL

REFERENCE
Petrarulo, M.; Pellegrino, S.; Testa, E. High-performance liquid chromatographic microassay for methyl ethyl ketone in urine as the 2,4-dinitrophenyl-hydrazone derivative, *J.Chromatogr.*, **1992**, *579*, 324–328.

SAMPLE
Matrix: urine
Analyte: 2,5-hexanedione
Sample preparation: 5 mL Urine + 1 mL concentrated HCl, heat at 100° for 40 min, cool to room temperature, filter (0.2 μm). 1 mL Filtrate + 1 mL 0.2% 2,4-dinitrophenylhydrazine in 2 M HCL (prepare fresh daily) + 1 mL MeCN, heat at 70° for 20 min, cool inject a 50 μL aliquot.

HPLC VARIABLES
Column: 150 × 4.6 5 μm 6-ODS-8-5 SGE
Mobile phase: MeCN:buffer 50:50 (Buffer was 20 mM KH_2PO_4 containing 1 mM sodium octanesulfonate, adjusted to pH 3.3 with phosphoric acid.)
Flow rate: 1
Injection volume: 50
Detector: UV 334

CHROMATOGRAM
Retention time: 18.6
Limit of quantitation: 190 ng/mL

REFERENCE
Gori, G.; Bartolucci, G.B.; Sturaro, A.; Parvoli, G.; Doretti, L.; Troiano, R.; Casetta, B. High-performance liquid chromatographic determination of urinary 2,5-hexanedione as mono-2,4-dinitrophenylhydrazone using ultraviolet detection, *J.Chromatogr.B*, **1995**, *673*, 165–172.

1,1-Diphenylhydrazine

SAMPLE
Matrix: cell cultures
Analyte: dicarbonyl sugars

Sample preparation: Centrifuge cell culture at 500 g. Remove a 50 mL aliquot of the supernatant and add it to 500 mg freshly-distilled 1,1-diphenylhydrazine in 2 mL 40% acetic acid, purify by chromatography on 20 × 20 cm TLC plates (Kavalier, Votice, Czech Republic) with chloroform:MeOH 20:1, recrystallize from EtOH, dissolve in MeOH, inject an aliquot.

CAPILLARY ELECTROPHORESIS
Capillary: 70 cm × 50 μm fused-silica (65 cm to detector) (CElect-UV, Supelco)
Capillary preparation: Between runs wash with 500 mM NaOH for 4 min.
Running buffer: n-Butanol:n-octanol:5 mM pH 8.0 borate buffer:sodium dodecyl sulfate 6.61:0.81:89.27:3.31
Voltage/Current: 20 kV
Injection: Electrokinetic injection at 10 kV for 5 s.
Detector: UV 220
Model: laboratory-constructed
Migration time: 28.8 (D-galactosone), 29.7 (5-keto-D-fructose, bis-derivative), 30.7 (D-glucosone), 43.5 (2-furylglyoxal), 43.8 (D-glucosone, bis-derivative), 5-hydroxy-2,2-dioxopentanal (tris-derivative), 46.0, 46.6 (2-furylglyoxal, bis-derivatve)

REFERENCE
Miksik, I.; Gabriel, J.; Deyl, Z. Microemulsion electrokinetic chromatography of diphenylhydrazones of dicarbonyl sugars, *J.Chromatogr.A*, **1997**, 772, 297–303.

Dmayl Hydrazide

SAMPLE
Matrix: plants
Analyte: abscisic acid and jasmonic acid
Sample preparation: Homogenize (Polytron PT 20 st) 1 g plant material with 5 mL 100 μg/mL BHT in MeOH, centrifuge at 9200 g for 15 min, remove the supernatant, homogenize the pellet with 5 mL 100 μg/mL BHT in MeOH:water 80:20, centrifuge. Combine the supernatants and remove MeOH under reduced pressure, freeze the aqueous solution in liquid nitrogen, thaw, centrifuge at 26000 g for 20 min, adjust the pH of the supernatant to 2 with 20 mM phosphate, extract with chloroform. Concentrate the extract to 5 mL and add it to a Florisil Sep-Pak SPE cartridge, wash with MeOH:chloroform 32:68, elute with MeOH:chloroform 65:45, evaporate to dryness under reduced pressure, reconstitute with MeOH. For each 14 g original plant material add 1 mg dmayl hydrazide (2-dimethylaminobenzoic hydrazide), for each 1 mmole reagent add 1.1 mmoles HCl in MeOH, purge with argon, let stand in the dark at 25° overnight, inject an aliquot. (Prepare 10 mM HCl in MeOH by mixing concentrated aqueous HCl with MeOH. Synthesis of 2-dimethylaminobenzoic hydrazide (dmayl hydrazide) is as follows. Dissolve 340 g anthranilic acid in 500 mL water, add enough KOH to produce a faintly alkaline solution. Add 1.6 L dimethyl sulfate (Caution! Dimethyl sulfate is highly toxic!) with stirring over 2.5 h and at the same time add 40% KOH so as to maintain a slightly alkaline reaction mixture, maintain the temperature at 10° for the first half of the addition then gradually raise it to 65°, extract the resulting product with ether, dry over calcium chloride, distil at 139°/16 mm Hg to obtain methyl 2-dimethylaminobenzoate (J. Am. Chem. Soc. 1935, 57, 1964). It should be possible to react this compound with hydrazine as detailed below. Alternatively prepare 2-dimethylaminobenzoic acid by refluxing methyl 2-dimethylaminobenzoate with excess water for 8 h, recrystallize from ether (mp 68°), prepare the ethyl ester in a conventional fashion. Mix 1 g ethyl o-dimethylaminobenzoate with a 100-fold

excess of anhydrous hydrazine in 10 mL dry MeOH, heat at 60° for 3 h, cool, partition between chloroform and water. Remove the organic layer and evaporate it to dryness under reduced pressure, purify by silica gel TLC using chloroform:MeOH 70:30 to obtain 2-dimethylaminobenzoic hydrazide.)

HPLC VARIABLES
Column: 5 µm Radial-Pak C18 radial compression (Waters)
Mobile phase: Gradient. MeCN:20 mM pH 4.0 phosphate buffer from 30:70 to 80:20 over 30 min.
Flow rate: 3
Detector: F ex 354 em 450

CHROMATOGRAM
Retention time: 14.8 (abscisic acid), 16.4 (jasmonic acid)

KEY WORDS
comparison with other derivatizing reagents; SPE

REFERENCE
Anderson, J.M. Simultaneous determination of abscisic acid and jasmonic acid in plant extracts using high-performance liquid chromatography, *J.Chromatogr.*, **1985**, *330*, 347–355.

SAMPLE
Matrix: plants
Analyte: abscisic acid
Sample preparation: Homogenize 1 g plant material with 5 mL 100 µg/mL BHT in MeOH, remove MeOH under vacuum, freeze aqueous solution, thaw, centrifuge, adjust the pH of the supernatant to 9.0, wash twice with ether, adjust pH to 2.0, extract with chloroform. Remove the organic layer and evaporate it to dryness. For each 14 g original plant material add 1 mg dmayl hydrazide (2-dimethylaminobenzoic hydrazide), for each 1 mmole reagent add 1.1 mmoles HCl in MeOH, purge with argon for 1 min, let stand in the dark at 25° overnight, streak on a TLC plate (Analtech GHR-250 µm), develop with MeOH: chloroform 1:6.7, remove the derivative (between the free acid at the origin and the unreacted hydrazide), extract with MeOH:chloroform 75:25, evaporate to dryness, reconstitute with 50 µM HCl in MeOH, inject an aliquot. (Prepare 10 mM HCl in MeOH by mixing concentrated aqueous HCl with MeOH. Synthesis of 2-dimethylaminobenzoic hydrazide (dmayl hydrazide) is as described above.)

HPLC VARIABLES
Column: 5 µm Radial-Pak C18 radial compression (Waters)
Mobile phase: MeCN:20 mM pH 4.0 phosphate buffer 40:60
Detector: F ex 455 em 450

CHROMATOGRAM
Retention time: 5.6

KEY WORDS
comparison with other derivatizing reagents

REFERENCE
Anderson, J.M. Fluorescent hydrazides for the high-performance liquid chromatographic determination of biological carbonyls, *Anal.Biochem.*, **1986**, *152*, 146–153.

Girard's Reagent T

SAMPLE
Matrix: bulk
Analyte: ketosteroids
Sample preparation: Dissolve compound in 75 mM (carboxymethyl)trimethylammonium chloride hydrazide (Girard's Reagent T) in EtOH:glacial acetic acid 90:10 at a concentration of 15-50 μM, let stand at room temperature for 30 min, dilute 10-fold with running buffer, inject an aliquot.

CAPILLARY ELECTROPHORESIS
Capillary: 64.5 cm × 50 μm fused-silica (56 cm to bubble cell detector) (Hewlett-Packard)
Capillary preparation: Rinse with 100 mM NaOH for 2 min and with running buffer for 5 min. Condition new capillaries with 1 M NaOH for 5 min, with 100 mM NaOH for 30 min, and with running buffer for 30 min.
Running buffer: 20 mM pH 4.8 NaH_2PO_4
Capillary temperature: 15
Voltage/Current: 25 kV
Injection: Pressure injection at 5 kPa for 3 s
Detector: UV 280; UV 250
Model: Hewlett-Packard HP 3DCE
Migration time: 8 (androstenedione (bis derivative) (UV 250)), 10.4 (nandrolone (UV 290)), 10.7 (norethindrone (norethisterone) (UV 280)), 11 (norgestrel (UV 280)), 11 (androstenedione (mono derivative) (UV 250)), 11.5 (dehydroepiandrosterone (UV 250)), 11.8 (ethisterone (UV 250))

REFERENCE
Görög, S.; Gazdag, M.; Kemenes-Bakos, P. Analysis of steroids Part 50. Derivatization of ketosteroids for their separation and determination by capillary electrophoresis, *J.Pharm.Biomed.Anal.*, **1996**, *14*, 1115–1124.

Luminarin 3

SAMPLE
Matrix: solutions
Analyte: ketones
Sample preparation: Mix 100 μL of a 5-500 μM solution in 100 mM sulfuric acid with 100 μL 100 mM sulfuric acid and 10 μL 10 mM luminarin 3 in DMSO, let stand at room temperature for 1 h, add 100 μL 600 mM sodium bicarbonate solution to adjust the pH to 7.0, agitate gently until gas evolution ceases, add 1 mL dichloromethane, vortex for 2 min, centrifuge. Remove the lower organic layer and evaporate it to dryness under a stream of nitrogen at room temperature, reconstitute the residue in 100 μL MeCN, inject an aliquot. (Luminarin 3, 2,3,6,7-tetrahydro-11-oxo-1H,5H,11H-[1]benzopyrano[6,7,8-ij]quinolizine-9-acetic acid hydrazide, is available from Eurobio, Les Ulis, France. Synthesis is as follows. Reflux (with protection from moisture and with stirring) 2.12 g 8-hydroxyjulolidine, 2.22 g diethyl 1,3-acetonedicarboxylate (oxo-3-glutaric acid ethyl ester, Fluka), 1.71 g anhydrous zinc chloride, and 6 mL EtOH for 24 h, cool, add to 200 mL water, extract with 200 mL ethyl acetate, extract with 100 mL ethyl acetate. Combine the organic layers and wash them with water, dry over magnesium sulfate, evaporate to dryness, recrystallize from 5 parts ethyl acetate to give ethyl 2,3,6,7-tetrahydro-11-oxo-1H,5H,11H-[1]benzopyrano[6,7,8-ij]quinolizine-9-acetate. Stir 5 g of this compound with 8 mL hydrazine hydrate in 100 mL MeOH for 4 h, filter, wash the solid with 10 mL MeOH, wash with 10 mL dichloromethane to give luminarin 3 (World Pat. 89 12,052; Chem. Abstr. 1990, 113, 23889n).)

HPLC VARIABLES
Column: 250 × 4.6 5 μm Nucleosil ODS
Mobile phase: MeCN:10 mM pH 7.5 imidazole nitrate buffer 40:60
Flow rate: 1.5
Injection volume: 20
Detector: F ex 399 em 485

CHROMATOGRAM
Retention time: k' 1.53 (acetoin), k' 3.57 (acetone), k' 1.52 (acetoacetic acid), k' 10.51 (acetylacetone), k' 4.33 (diacetyl), k' 6.51 (ethyl acetoacetate)

Limit of detection: 156-1950 fmole

OTHER SUBSTANCES
Also analyzed: aldehydes

REFERENCE
Traoré, F.; Tod, M.; Chalom, J.; Farinotti, R.; Mahuzier, G. 1*H*,5*H*,11*H*-[1]Benzopyrano[6,7,8-*ij*]quinolizine-9-acetic acid 2,3, 6,7-tetrahydro-11-oxohydrazide fluorogenic reagent for liquid chromatographic determination of aldehydes and ketones, *Anal.Chim.Acta*, **1992**, *269*, 211–222.

3-Methyl-2-benzothiazolinone Hydrazone

SAMPLE
Matrix: solutions
Analyte: ketones
Sample preparation: Derivatize aldehydes and ketones following this example. Heat 2 g 4-dimethylaminobenzaldehyde and 2 g 3-methyl-2-benzothiazolinone hydrazone (Fluka) in 50 mL EtOH and 5 mL acetic acid at 100° for 30 min (Annalen 1957, 609, 172), dissolve the product in EtOH, inject an aliquot.

HPLC VARIABLES
Column: 250 × 4.6 Erbasil C18 (Carlo Erba)
Mobile phase: MeOH:20 mM KH_2PO_4 70:30
Flow rate: 1.2
Detector: E, Metrohm 656, glassy carbon electrode +1.05 V, Ag/AgCl reference electrode

CHROMATOGRAM
Retention time: k' 2.22 (formaldehyde), k' 2.67, k' 3.00 (acetaldehyde (Z/E isomers)), k' 3.89, k' 4.89 (propionaldehyde (Z/E isomers)), k' 5.44, k' 6.89 (butyraldehyde (Z/E isomers)), k' 3.30, k' 4.10 (acrolein (Z/E isomers)), k' 4.56, k' 6.22 (crotonaldehyde (Z/E isomers)), k' 7.4, k' 17.8 (benzaldehyde (Z/E isomers)), k' 3.55 (acetone), k' 5.11, k' 6.22 (methyl ethyl ketone (Z/E isomers)), k' 8.22, k' 19.11 (acetophenone (Z/E isomers))
Limit of detection: 0.3-13.5 pmole

REFERENCE
Chiavari, G.; Facchini, M.C.; Fuzzi, S. Behaviour of 3-methyl-2-benzothiazolone azines of carbonyl compounds in high-performance liquid chromatography, *J.Chromatogr.*, **1987**, *387*, 459–466.

1-Methyl-1-(2,4-dinitrophenyl)hydrazine

SAMPLE
Matrix: air
Analyte: ketones

Sample preparation: Pull air through 80 mL reagent at 250 mL/min for 2 min, make up to 100 mL with MeCN, inject a 10 μL aliquot. (Prepare reagent by adding 105 mg 1-methyl-1-(2,4-dinitrophenyl)hydrazine to a solution of 500 μL concentrated sulfuric acid in 99.5 mL MeCN. Synthesis of 1-methyl-1-(2,4-dinitrophenyl)hydrazine is as follows. Add 6 mL methylhydrazine (Caution! Methylhydrazine is a carcinogen!) in 30 mL EtOH to a solution of 10.8 g potassium acetate in 50 mL water, heat to reflux, add a solution of 20.2 g 2,4-dinitrochlorobenzene in 100 mL EtOH dropwise with stirring, reflux for 8 h, cool to 0°, filter. Wash the solid with warm (60°) EtOH, wash with hot water to obtain 1-methyl-1-(2,4-dinitrophenyl)hydrazine (mp 143°) as a yellow-orange solid. The progress of the reaction may be monitored by silica TLC using toluene:MeCN 70:30.)

HPLC VARIABLES
Column: 150 × 4.6 5 μm Deltabond AK
Mobile phase: Gradient. MeCN:water 30:70 for 1 min, to 42:58 over 5.5 min, to 100:0 over 3 min, maintain at 100:0 for 0.5 min, return to initial conditions over 2 min.
Flow rate: 1.5
Injection volume: 10
Detector: UV 368

CHROMATOGRAM
Retention time: 4.2 (acetone), 6 (2-butanone)

OTHER SUBSTANCES
Simultaneously analyzed: acetaldehyde, acrolein, benzaldehyde, butanal, formaldehyde, propanal, p-tolualdehyde

REFERENCE
Büldt, A.; Karst, U. 1-Methyl-1-(2,4-dinitrophenyl)hydrazine as a new reagent for the HPLC determination of aldehydes, *Anal.Chem.*, **1997**, *69*, 3617–3622.

2-Nitrophenylhydrazine

RELATED REFERENCES
Miwa, H.; Yamamoto, M.; Nishida, T. Assay of free and total fatty acids (as 2-nitrophenylhydrazides) by high performance liquid chromatography. *Clin.Chim.Acta* **1986**, *155*, 95-101.
Miwa, H.; Yamamoto, M. Reversed-phase ion-pair chromatography of straight- and branched-chain dicarboxylic acids in urine as their 2-nitrophenylhydrazides. *Anal.Biochem.* **1988**, *170*, 301-307.
Miwa, H.; Yamamoto, M.; Asano, T. High-performance liquid chromatographic analysis of fatty acid compositions of platelet phospholipids as their 2-nitrophenylhydrazides. *J.Chromatogr.* **1991**, *568*, 25-34.

SAMPLE
Matrix: urine
Analyte: ketones
Sample preparation: Condition a 3 mL C18 SPE cartridge (Baker) with 3 mL MeOH and 2 mL cyclohexane. 150 μL Urine + 200 μL reagent + 1 mL cyclohexane, shake in the dark at 60 rpm overnight, centrifuge at 10000 g for 5 min. Add a 900 μL aliquot of the cyclohexane layer to the SPE cartridge, dry under vacuum, elute with two 500 μL aliquots of MeCN, inject a 10 μL aliquot of the eluate. (Prepare reagent by making a 30.9 mM solution of 2-nitrophenylhydrazine in 1 M HCl, wash 5 volumes of reagent with 1 volume chloroform, store in the dark at 4°, prepare fresh each day.)

HPLC VARIABLES
Column: 83 × 4.6 3 μm Pecosphere 3CR-C18
Mobile phase: MeCN:water 55:45
Flow rate: 1.5
Injection volume: 10
Detector: UV 254

CHROMATOGRAM
Retention time: 2.5 (formaldehyde), 3.5 (acetone), 7.3 (methyl ethyl ketone)
Internal standard: cyclohexanone (10)
Limit of detection: 100 ng/mL

KEY WORDS
SPE; pharmacokinetics

REFERENCE
Van Doorn, J.E.; De Cock, J.; Kezic, S.; Monster, A.C. Determination of methyl ethyl ketone in human urine after derivatization with o-nitrophenylhydrazine, using solid-phase extraction and reversed-phase high-performance liquid chromatography and ultraviolet detection, *J.Chromatogr.*, **1989**, *489*, 419–424.

4-Nitrophenylhydrazine

SAMPLE
Matrix: blood
Analyte: 17-ketosteroid sulfates
Sample preparation: 100 μL Serum + 2 mL MeCN, mix, let stand at room temperature for 5 min, centrifuge at 1000 g for 5 min. Remove the supernatant and add it to 100 ng IS, evaporate to dryness below 40°, add 10 μL 50 mg/mL 4-nitrophenylhydrazine in ethyl acetate, add 100 μL 3 mg/mL trichloroacetic acid in benzene (Caution! Benzene is a carcinogen!), heat at 60° for 20 min, evaporate to dryness under a stream of nitrogen, reconstitute with 200 μL MeOH, inject an aliquot. (Purify 4-nitrophenylhydrazine by repeated recrystallization from MeOH, prepare solution just before use.)

HPLC VARIABLES
Column: 305 × 4 5 μm μBondapak C18
Mobile phase: MeOH:0.5% pH 3.0 $(NH_4)H_2PO_4$ 8:3
Flow rate: 1
Detector: E, Yanagimoto Model VMD 101, 0.8 V, Ag/AgCl reference electrode (Divert column effluent from detector for 5 min after injection.)

CHROMATOGRAM
Retention time: 14 (dehydroepiandrosterone sulfate), 17 (epiandrosterone sulfate), 20 (etiocholanolone sulfate), 22 (androsterone sulfate)
Internal standard: 2-hydroxyestrone 3-methyl ether (30)
Limit of detection: 80 ng/mL

KEY WORDS
serum

REFERENCE
Shimada, K.; Tanaka, M.; Nambara, T. Studies on steroids. CC. Determination of 17-ketosteroid sulphates in serum by high-performance liquid chromatography with electrochemical detection using pre-column derivatization, *J.Chromatogr.*, **1984**, *307*, 23–28.

HYDROXYLAMINE

Hydroxylamine

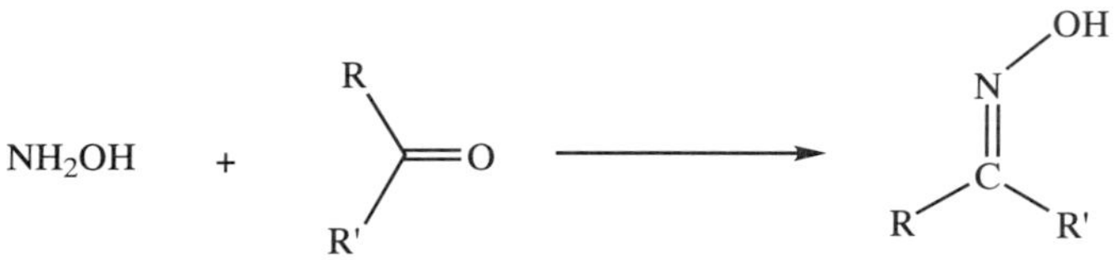

NH₂OH   +   (structure) → (structure)

SAMPLE
Matrix: solutions
Analyte: monosaccharides
Sample preparation: Dissolve 50 mg sugars in 700 μL pyridine, add 700 μL 720 mM hydroxylamine hydrochloride in pyridine, heat at 60° for 10 min, add 250 μL acetic anhydride, heat at 75° for 10 min, evaporate to dryness under reduced pressure, reconstitute with 3 mL chloroform. Wash the organic layer three times with 6 mL portions of water and dry it over anhydrous sodium sulfate, evaporate to dryness under reduced pressure, take up in chloroform, pass through silica gel using chloroform, evaporate the eluate to dryness, reconstitute, inject a 5 μL aliquot.

HPLC VARIABLES
Column: 250 × 4 5 μm μBondapak C18
Mobile phase: Gradient. MeCN:water from 35:75 to 50:50 over 15 min.
Flow rate: 1
Injection volume: 5
Detector: UV 207

CHROMATOGRAM
Retention time: k' 3.1 (xylose), k' 3.3 (lyxose), k' 3.4 (ribose), k' 3.4 (arabinose), k' 4.3 (allose), k' 4.3 (altrose), k' 3.9 (dextrose), k' 4.2 (mannose), k' 4.0 (gulose), k' 3.8 (idose), k' 4.8 (talose), k' 4.2 (galactose), k' 3.0, k' 3.2 (fructose, syn and anti isomers)
Limit of detection: 3 μg

REFERENCE
Velasco, D.; Castells, J.; Lopez-Calahorra, F. High-performance liquid chromatographic separation of monosaccharides as their peracetylated ketoximes and aldononitriles, *J.Chromatogr.*, **1990**, *519*, 228–236.

4-Nitrobenzylhydroxylamine

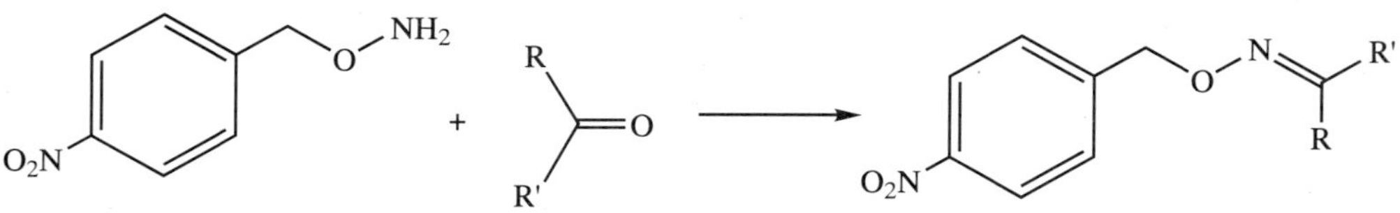

SAMPLE
Matrix: solutions
Analyte: prostaglandins
Sample preparation: Prepare methyl ester by treatment with excess ethereal diazomethane for 5 min, remove excess reagent under a stream of nitrogen. Dissolve 10 μg methyl ester in 200 μL anhydrous pyridine containing a 10-fold molar excess of 4-nitrobenzylhydroxylamine hydrochloride, heat at 40° for 2 h, evaporate to dryness under a stream of nitrogen, reconstitute with MeOH, inject an aliquot. (Preparation of diazomethane is as follows. Caution! Diazomethane is toxic, explosive, and carcinogenic! A face shield and

a safety screen should always be used and the preparation should only be carried out in a properly functioning chemical fume hood. Only smooth glass apparatus with rubber stoppers and plastic tubing should be used. Scratched glassware, ground glass joints, and sharp edges should be avoided (Org. Syn., Coll. Vol. VI; Wiley: New York, 1988, pp. 432-435). Procedures have been reported for the synthesis of diazomethane from N-methyl-N-nitrosourea (Org. Syn., Coll. Vol. II; Wiley: New York, 1943, pp. 165-167), N-nitroso-β-methylaminoisobutyl methyl ketone (Org. Syn., Coll. Vol. III; Wiley: New York, 1955, pp. 244-248), and N,N'-dimethyl-N,N'-dinitrosoterephthalamide (Org. Syn., Coll. Vol. V; Wiley: New York, 1973, pp. 351-355). Probably the most convenient starting material is N-methyl-N-nitroso-p-toluenesulfonamide (Diazald) (Aldrichimica Acta 1983, 16, 3-10; Org. Syn., Coll. Vol. IV, Wiley: New York, 1963, pp. 250-253). Add 10 mL 95% EtOH to a solution of 5 g KOH in 8 mL water, warm to 65° using a water bath, add a solution of 5 g N-methyl-N-nitroso-p-toluenesulfonamide in 45 mL ether over 20 min at such a rate as to keep the reaction volume constant. Collect the ether and diazomethane that distil in an ice-cooled receiving flask under a dry ice/acetone condenser. When all the N-methyl-N-nitroso-p-toluenesulfonamide has been used up, slowly add 10 mL ether to the reaction flask and continue distillation until the distillate is colorless. A purpose-built distillation apparatus can be purchased from Aldrich (Aldrichimica Acta 1983, 16, 3-10). Excess quantities of diazomethane can be destroyed by adding acetic acid until the yellow color of the diazomethane is discharged. The safe disposal of the nitroso compounds used to generate diazomethane has been discussed (Lunn,G.; Sansone,E.B. Destruction of Hazardous Chemicals in the Laboratory, Second Edition. Wiley: New York, 1994, pp. 277-289).)

HPLC VARIABLES
Column: 600 mm long μBondapak C18
Mobile phase: MeCN:water 85:15
Flow rate: 0.75
Detector: UV 254

CHROMATOGRAM
Retention time: 10 (dinoprostone), 11 (alprostadil), 14.5 (prostaglandin A$_2$), 16.5 (prostaglandin A$_1$), 26 (prostaglandin B$_2$), 30 (prostaglandin B$_1$)

REFERENCE
Fitzpatrick, F.A.; Wynalda, M.A.; Kalser, D.G. Oximes for high-performance liquid and electron capture gas chromatography of prostaglandins and thromboxanes, *Anal.Chem.*, **1977**, *49*, 1032–1035.

MISCELLANEOUS REACTIONS

Bromination

SAMPLE
Matrix: enzyme preparations
Analyte: enolpyruvate
Sample preparation: Add 20-100 μL enzyme preparation to an equal volume of 2 M perchloric acid saturated with liquid bromine with rapid vortexing, the bromine color should persist for 30 s, remove the bromine with a stream of nitrogen, centrifuge, neutralize to pH 7 with 2.5 M potassium carbonate, store on ice, add a 3-fold excess of 5-thio-2-nitrobenzoic acid, let stand for 20 min at room temperature, inject an aliquot. (Prepare 5-thio-2-nitrobenzoic acid as follows. Add 5 mL 2-mercaptoethanol to a solution of 1 g 5,5'-

dithiobis(2-nitrobenzoic acid) in 50 mL 500 mM pH 8.0 Tris-HCl buffer, after 5 min acidify to pH 1.5 with 6 M HCl, cool in ice for 24 h, filter, wash the precipitate with dilute HCl, dry over phosphorus pentoxide under vacuum to give 5-thio-2-nitrobenzoic acid as orange crystals (mp 137-8°) (J.Org.Chem. 1971, 36, 2727). 5-Thio-2-nitrobenzoic acid can be further purified by chromatography on Sephadex LH-20 with EtOH at 4° (Anal.Biochem. 1979, 94, 75).)

HPLC VARIABLES
Column: PICO-TAG (Waters)
Mobile phase: Gradient. MeCN : buffer from 5 : 95 to 70 : 30 over 13 min. (Buffer was 10 mM phosphoric acid containing 100 mM NaCl, pH 2.2.)
Flow rate: 1
Detector: UV 342

CHROMATOGRAM
Retention time: 6.2

REFERENCE
Seeholzer, S.H.; Jaworowski, A.; Rose, I.A. Enolpyruvate: chemical determination as a pyruvate kinase intermediate, *Biochem.*, **1991**, *30*, 727–732.

LACTAM

ACTIVATED HALIDE

4-Fluoro-7-nitro-2,1,3-benzoxadiazole

SAMPLE
Matrix: blood
Analyte: penicillins
Sample preparation: Condition a 55 × 5 100-200 mesh AG 50W-X8 (H⁺) column (Bio-Rad) with 10 mL MeCN:water 50:50. 600 μL Serum + 600 μL MeCN, vortex for 1 min, centrifuge at 2000 g for 5 min, add a 1 mL aliquot of the supernatant to the column, discard the first 200 μL effluent, collect the rest of the effluent. Remove a 450 μL aliquot and add it to 50 μL 10% sodium carbonate solution, heat at 60° for 1 h (to hydrolyse the β-lactam ring), cool in an ice bath. Remove a 100 μL aliquot and add it to 15 μL 200 mM pH 6.0 phosphate buffer, add 35 μL 80 mM 4-fluoro-7-nitro-2,1,3-benzoxadiazole in MeCN, heat at 60° for 10 min, cool in an ice bath, add 30 μL 1 M HCl, inject a 5-10 μL aliquot.

HPLC VARIABLES
Column: 150 × 4.6 ODS-80TM (Tosoh)
Mobile phase: MeOH:100 mM pH 3.0 phosphate buffer 40:60 (A) or 55:45 (B)
Flow rate: 1
Injection volume: 5-10
Detector: F ex 470 em 530

CHROMATOGRAM
Retention time: 7 (piperacillin (A)), 10 (penicillin G (A)), 12 (methicillin (A)), 7 (cloxacillin (B)), 9 (dicloxacillin (B))
Limit of detection: 30-85 ng/mL

KEY WORDS
serum; SPE

REFERENCE
Iwaki, K.; Okumura, N.; Yamazaki, M.; Nimura, N.; Kinoshita, T. Precolumn derivatization technique for high-performance liquid chromatographic determination of penicillins with fluorescence detection, *J.Chromatogr.*, **1990**, *504*, 359–367.

ALDEHYDE

Formaldehyde

RELATED REFERENCE
Ang, C.Y.W.; Luo, W. Rapid determination of ampicillin in bovine milk by liquid chromatography with fluorescence detection. *J.AOAC Int.* **1997**, *80*, 25-30.

SAMPLE
Matrix: blood
Analyte: ampicillin
Sample preparation: 500 µL Serum + 1 mL 20% trichloroacetic acid, vortex for 15 s, centrifuge at 1000 g for 20 min. 1 mL Supernatant + 500 µL 7% formaldehyde in 400 mM citric acid, vortex for 15 s, heat at 90° for 2 h, cool to room temperature. Either inject an aliquot of this solution directly or extract it twice with 3 mL portions of diethyl ether. Evaporate the extracts to dryness under reduced pressure, reconstitute with 100 µL mobile phase, inject an aliquot.

HPLC VARIABLES
Guard column: 30 × 4.6 10 µm RP-18 (Pierce)
Column: 100 × 4.6 10 µm RP-18 (Pierce)
Mobile phase: MeCN:100 mM pH 5.6 KH_2PO_4 23:77
Flow rate: 1
Injection volume: 50
Detector: F ex 346 em 422

CHROMATOGRAM
Retention time: 7
Limit of detection: 2 ng/mL (with extraction)

KEY WORDS
serum

REFERENCE
Lal, J.; Paliwal, J.K.; Grover, P.K.; Gupta, R.C. Determination of ampicillin in serum by high-performance liquid chromatography with precolumn derivatization, *J.Chromatogr.B*, **1994**, *655*, 142–146.

SAMPLE
Matrix: tissue
Analyte: amoxicillin

Sample preparation: Condition a 3 mL 500 mg Sep-Pak C18 SPE cartridge with 5 mL MeOH, 2 mL water, and 2 mL 2% trichloroacetic acid. Homogenize (Ultra-Turrax T25) 5 g blended tissue with 20 mL 10 mM pH 4.5 phosphate buffer at 10000 rpm for 1.5 min, centrifuge at 4500 rpm for 10 min, decant supernatant, homogenize residue with another 20 mL buffer, centrifuge. Combine the supernatants and filter them through glass wool, add 1 mL 75% trichloroacetic acid to the filtrate, vortex for 30 s, centrifuge at 4500 rpm for 20 min, filter the supernatant through glass wool. Add the filtrate to the SPE cartridge at 1-2 mL/min, wash with 2 mL 2% trichloroacetic acid, wash with 2 mL water, elute with 1.5 mL MeCN at 0.7 mL/min. Add the eluate to 500 μL water and 3 mL ethyl ether, vortex gently for 30 s, centrifuge at 2000 rpm for 3 min, discard the organic layer. Add 200 μL 20% trichloroacetic acid solution to the aqueous phase, vortex for 15 s, add 200 μL 7% formaldehyde in 400 mM citric acid, vortex for 30 s, heat in boiling water bath for 30 min, cool to room temperature, add 500 mg NaCl, mix briefly, add 3 mL ethyl ether, vortex for 1 min, centrifuge at 2000 rpm for 3 min, repeat extraction twice more. Combine the organic layers and evaporate them to dryness under a stream of nitrogen at 40°, reconstitute the residue in 500 μL mobile phase, vortex thoroughly, inject a 50 μL aliquot.

HPLC VARIABLES
Column: 250 × 4.6 S5 ODS2
Mobile phase: MeCN:buffer 20:80 (Buffer was 50 mM KH_2PO_4 adjusted to pH 5.6 with KOH.)
Flow rate: 1 for 10 min then 2
Injection volume: 50
Detector: F ex 358 em 440

CHROMATOGRAM
Retention time: 6
Limit of quantitation: 1.2 ppb (catfish), 2.0 ppb (salmon)
Limit of detection: 0.5 ppb (catfish), 0.8 ppb (salmon)

KEY WORDS
fish; catfish; salmon; SPE

REFERENCE
Ang, C.Y.W.; Luo, W.; Hansen, E.B. Jr.; Freeman, J.P.; Thompson, H.C., Jr. Determination of amoxicillin in catfish and salmon tissues by liquid chromatography with precolumn formaldehyde derivatization, *J.AOAC Int.*, **1996**, *79*, 389–396.

SAMPLE
Matrix: tissue
Analyte: ampicillin
Sample preparation: Homogenize (Ultra-turrax T25) 5 g muscle with 14 mL 10 mM pH 4.5 sodium phosphate buffer at 10000 rpm for 2 min, add 1 mL 75% trichloroacetic acid in water, shake vigorously for 30 s, centrifuge at 3500 g for 10 min, filter (paper) the supernatant. Remove a 1 mL aliquot of the filtrate, add 200 μL 20% trichloroacetic acid in water, add 200 μL 7% formaldehyde in water, vortex for 20 s, heat at 100° for 30 min, cool to room temperature, make up to 2 mL with MeCN:water 20:80, inject a 100 μL aliquot.

HPLC VARIABLES
Column: 250 × 4.6 5 μm Prodigy ODS-3 (Phenomenex)
Mobile phase: MeCN:20 mM pH 3.5 KH_2PO_4 buffer 25:75
Flow rate: 1
Injection volume: 100
Detector: F ex 346 em 422

CHROMATOGRAM
Retention time: 12.5
Limit of quantitation: 1.5 ng/g
Limit of detection: 0.6 ng/g

KEY WORDS
muscle; cow; pig; chicken; fish; catfish

REFERENCE

Luo, W.; Ang, C.Y.W.; Thompson, H.C. Jr. Rapid method for the determination of ampicillin residues in animal muscle tissues by high-performance liquid chromatography with fluorescence detection, *J.Chromatogr.B*, **1997**, *694*, 401–407.

AZIRIDINE

Dansyl Aziridine

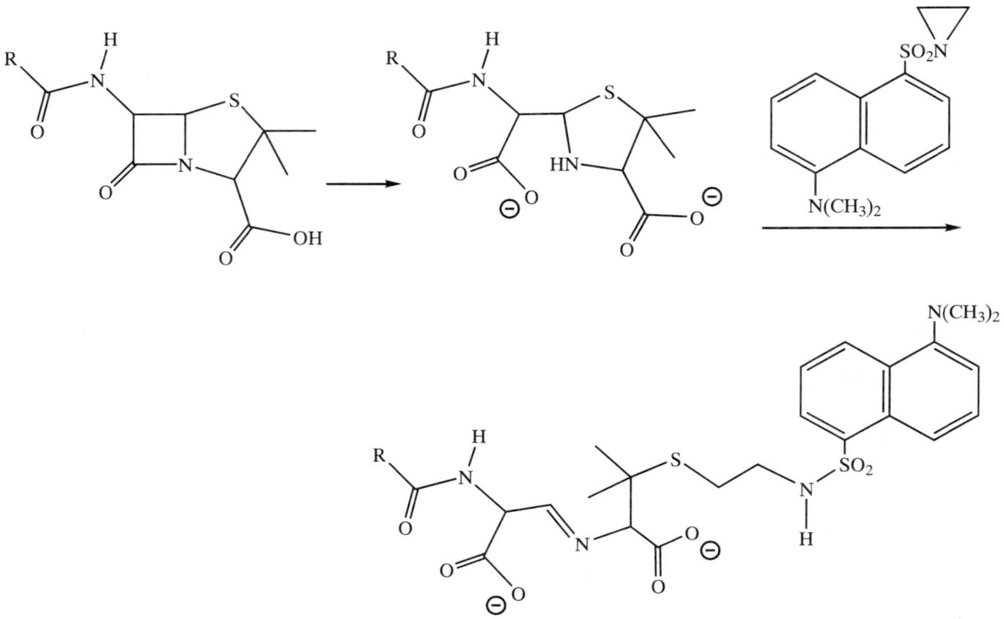

SAMPLE

Matrix: fermentation broth

Analyte: penicillin G

Sample preparation: Centrifuge fermentation broth at 4° at 5000 g for 20 min, filter (0.45 μm) a 1 mL aliquot of the supernatant, add 50 μL 1 M NaOH to the filtrate. Remove a 50 μL aliquot and add it to 50 μL 3 mM N-dansylaziridine (Sigma) in dioxane (Caution! Dioxane is a carcinogen!), heat at 100° for 30 min, cool to room temperature, inject a 20 μL aliquot.

HPLC VARIABLES

Guard column: 50 × 4.6 5 μm Spherisorb C18

Column: 250 × 4.6 5 μm Spherisorb C18

Mobile phase: Gradient. MeCN:20 mM pH 4.4 acetate buffer containing 0.5 mM EDTA from 19:81 to 23:77 over 15 min, to 40:60 over 10 min, to 65:35 over 2.5 min.

Flow rate: 1

Injection volume: 20

Detector: F ex 339 em 540

CHROMATOGRAM

Retention time: 24

Limit of detection: 50 ng/mL

OTHER SUBSTANCES

Extracted: δ-(L-α-aminoadipyl)-L-cysteinyl-D-valine, 6-aminopenicillanic acid, isopenicillin N

REFERENCE

Orford, C.D.; Perry, D.; Adlard, M.W. The determination of naturally produced penicillins and their biosynthetic precursors using pre-column derivatisation with dansylaziridine, *J.Liq.Chromatogr.*, **1991**, *14*, 2665–2684.

AZOLE

Imidazole

RELATED REFERENCE

Eckers, C.; Chalkley, R.; Haskins, N.; Edwards, J.; Griffin, J.; Elson, S. Investigation into the use of derivatization with imidazole for the detection of clavam compounds by liquid chromatography. *Anal.Commun.* **1996**, *33*, 215-218.

SAMPLE

Matrix: blood, urine
Analyte: clavulanic acid
Sample preparation: Serum. Wash Amicon YMB filter membrane by stirring gently in 200 mL 100 mM pH 7.0 sodium phosphate buffer for 30 min, blot dry with filter paper. Dilute serum with an equal volume of 100 mM pH 7.0 sodium phosphate buffer, filter (Amicon YMB) while centrifuging at 5° at 1000 g for 15 min. Add 1 part reagent to 4 parts ultrafiltrate, let stand for 10 min, inject a 25-50 μL aliquot. Urine. Dilute 10-fold with 100 mM pH 7.0 sodium phosphate buffer. Add 1 part reagent to 4 parts diluted urine, let stand for 10 min, inject a 25-50 μL aliquot. (Reagent was 8.25 g imidazole, 24 mL water, and 2 mL 5 M HCl made up to 40 mL with water.)

HPLC VARIABLES

Guard column: CO:PEL ODS C18
Column: 250 × 4.6 μBondapak C18
Mobile phase: MeOH:buffer 6:94 (Use 4:96 for clavulanic acid concentrations of <2 μg/mL in urine.) (Buffer was 100 mM KH_2PO_4 adjusted to pH 3.2 with phosphoric acid.)
Flow rate: 2.5
Injection volume: 25-50
Detector: UV 311

CHROMATOGRAM

Retention time: 4

Limit of detection: 100 ng/mL

OTHER SUBSTANCES
Non-interfering: degradation products, amoxicillin

KEY WORDS
ultrafiltrate; serum; pharmacokinetics

REFERENCE
Foulstone, M.; Reading, C. Assay of amoxicillin and clavulanic acid, the components of Augmentin, in biological fluids with high-performance liquid chromatography, *Antimicrob.Agents Chemother.*, **1982**, *22*, 753–762.

Triazole

SAMPLE
Matrix: blood, urine
Analyte: sulbactam
Sample preparation: Dilute urine ten-fold with water. 50 μL Plasma or 100 μL diluted urine + 150 μL MeCN, vortex for 30 s, incubate at room temperature for 5 min, centrifuge at 1500 g for 10 min. Remove a 100 μL aliquot of the supernatant and add it to 100 μL triazole solution, heat at 50° for 15 min, cool to room temperature, centrifuge at 1500 g for 5 min, inject a 20 μL aliquot of the supernatant. (Triazole solution was 13.81 g 1,2,4-triazole in 60 mL water, pH adjusted to 10.0 ± 0.05 with saturated NaOH solution, made up to 100 mL with water.)

HPLC VARIABLES
Guard column: 30 × 4.6 10 μm Develosil ODS-10
Column: 150 × 4.6 5 μm Develosil ODS-5
Mobile phase: MeCN:5 mM tetrabutylammonium bromide + 1 mM Na_2HPO_4 + 1 mM NaH_2PO_4 25:75
Column temperature: 50
Flow rate: 1
Injection volume: 20
Detector: UV 326

CHROMATOGRAM
Retention time: 7
Limit of detection: 200 ng/mL (plasma), 1000 ng/mL (urine)

OTHER SUBSTANCES
Non-interfering: ampicillin, amoxicillin, cefoperazone, degradation products, penicillin G

KEY WORDS
plasma; human; rat

REFERENCE
Haginaka, J.; Wakai, J.; Yasuda, H.; Uno, T.; Nakagawa, T. High-performance liquid chromatographic assay of sulbactam using pre-column reaction with 1,2,4-triazole, *J.Chromatogr.*, **1985**, *341*, 115–122.

SAMPLE
Matrix: blood, urine
Analyte: clavulanic acid
Sample preparation: Serum. 500 μL Serum + 200 μL 200 mM pH 7.0 phosphate buffer, vortex for 10 s, filter (Amicon YMT membrane) while centrifuging at 5° at 1500 g for 15 min. Mix 250 μL ultrafiltrate with 250 μL reagent, heat at 30° for 5 min, inject a 50 μL aliquot. 500 μL Urine + 4.5 mL water, mix vigorously for 20 s, filter (0.45 μm) an aliquot. Mix 250 μL filtrate with 250 μL reagent, heat at 30° for 5 min, inject a 50 μL aliquot. (Prepare reagent by dissolving 13.81 g 1,2,4-triazole in 70 mL water and adjusting the pH to 9.00 ± 0.05 with 4 M NaOH, make up to 100 mL.)

HPLC VARIABLES
Guard column: 30 × 4.6 μBondapak C18
Column: 300 × 3.9 μBondapak C18
Mobile phase: MeOH:30 mM pH 7.0 phosphate buffer 20:80
Flow rate: 2
Injection volume: 50
Detector: UV 313

CHROMATOGRAM
Retention time: 7
Limit of detection: 100 ng/mL

OTHER SUBSTANCES
Non-interfering: degradation products, penicillins, penicilloic acids

KEY WORDS
serum

REFERENCE
Martín, J.; Méndez, R. High-performance liquid chromatographic determination of clavulanic acid in human serum and urine using a pre-column reaction with 1,2,4-triazole, *J.Liq.Chromatogr.*, **1988**, *11*, 1697–1705.

SAMPLE
Matrix: blood, urine
Analyte: lactamase inhibitors
Sample preparation: Serum. 1 mL Serum + 1 mL MeCN, shake at 0° for 15 min, centrifuge at 8000 g for 10 min. Remove the supernatant and add it to 10 mL dichloromethane, shake at 0° for 15 min, centrifuge at 8000 g for 10 min. Remove a 200 μL aliquot of the aqueous layer and add it to 400 μL reagent, let stand at room temperature for 5 min (clavulanic acid) or 40° for 40 min (sulbactam), inject a 40 μL aliquot. Urine. Filter (0.22 μm) urine, dilute 10-fold, remove a 200 μL aliquot and add it to 400 μL reagent, let stand at room temperature for 5 min (clavulanic acid) or 40° for 40 min (sulbactam), inject a 40 μL aliquot. (Prepare reagent by dissolving 3.45 g 1,2,4-triazole in 15 mL water, adjust pH to 6.0 (sulbactam) or 7.0 (clavulanic acid) with 4 M NaOH, make up to 25 mL.)

HPLC VARIABLES
Guard column: 50 × 4.6 5 μm Spherisorb C18
Column: 250 × 4.6 5 μm Spherisorb C18
Mobile phase: MeCN:20 mM pH 7.0 phosphate buffer 0.2:99.8 (sulbactam) or Gradient. MeCN:20 mM pH 7.0 phosphate buffer from 2:98 to 25:75 over 25 min (clavulanic acid).
Flow rate: 0.5

Injection volume: 40
Detector: UV 325 (sulbactam); UV 315 (clavulanic acid)

CHROMATOGRAM
Retention time: 7 (sulbactam), 20 (clavulanic acid)
Limit of detection: 50 ng/mL (serum), 500 ng/mL (urine)

KEY WORDS
serum

REFERENCE
Shah, A.J.; Adlard, M.W.; Stride, J.D. A sensitive assay for clavulanic acid and sulbactam in biological fluids by high-performance liquid chromatography and precolumn derivatization, *J.Pharm.Biomed.Anal.*, **1990**, *8*, 437–443.

MERCURY SALT

Mercuric Chloride

SAMPLE
Matrix: blood
Analyte: ampicillin
Sample preparation: 500 μL Plasma + 4 mL water + 3 mL 10% trichloroacetic acid, centrifuge at 800-1000 g for 5 min. Remove 3 mL of the supernatant and add it to 500 μL 2 M NaOH, let stand for 5 min, add 500 μL 2 M HCl, add 1 mL 0.1% mercury(II) chloride in buffer, let stand for 5 min, add 2 mL 0.67 M Na_2HPO_4 warmed to 40° to adjust pH to 6.2, heat mixture at 40° for 25 min, add 6 mL ethyl acetate saturated with water, shake vigorously for 5 min, centrifuge. Remove 5 mL of the organic layer and evaporate it to dryness under reduced pressure, reconstitute the residue in 100 μL MeOH containing IS, inject a 20 μL aliquot. (Prepare buffer by dissolving 21 g citric acid in 200 mL 1 M NaOH, make up to 1 L with water, adjust pH to 2.5 with 100 mM HCl.)

HPLC VARIABLES
Column: 250 × 4 5 μm Nucleosil C18
Mobile phase: MeOH:water 60:40
Column temperature: 55
Injection volume: 20
Detector: F ex 345 em 420

CHROMATOGRAM
Retention time: 6 (?)
Internal standard: methyl anthranylate (9 (?))
Limit of detection: 0.5 ng/mL

OTHER SUBSTANCES
Interfering: cephalexin, cephradine

KEY WORDS
plasma

REFERENCE
Miyazaki, K.; Ohtani, K.; Sunada, K.; Arita, T. Determination of ampicillin, amoxicillin, cephalexin, and cephradine in plasma by high-performance liquid chromatography using fluorometric detection, *J.Chromatogr.*, **1983**, *276*, 478–482.

SAMPLE
Matrix: blood, broncho-alveolar lavage fluid
Analyte: ampicillin
Sample preparation: 100 μL Plasma or 500 μL broncho-alveolar lavage fluid + 2 mL water + 1.5 mL 10% trichloroacetic acid, vortex for 30 s, centrifuge at 5000 rpm for 5

min. 1.5 mL Supernatant + 0.2 mL 2 M NaOH + 0.5 mL 0.1% (w/v) $HgCl_2$ in 67 mM pH 4.8 phosphate buffer (Sorensen), mix, after 5 min bring to pH 6.2 with 1 mL 0.67 M Na_2HPO_4, keep at 40° for 25 min. Add 3 mL ethyl acetate, shake horizontally for 5 min, centrifuge at 3015 g for 5 min. Evaporate the organic phase to dryness under nitrogen, reconstitute with 100 μL mobile phase, inject an aliquot.

HPLC VARIABLES
Column: 250 × 4 Spherisorb 5 ODS
Mobile phase: MeOH:buffer 60:40 (Buffer was 5 mM 1-heptanesulfonic acid adjusted to pH 3.7 with acetic acid.)
Flow rate: 1
Injection volume: 20
Detector: F ex 345 em 425

CHROMATOGRAM
Retention time: 3.45
Internal standard: amoxicillin (4.67)
Limit of detection: 50 ng/mL (plasma), 10 ng/mL (broncho-alveolar lavage fluid)

KEY WORDS
plasma

REFERENCE
Rosseel, M.T.; Bogaert, M.G.; Valcke, Y.J. High-performance liquid chromatographic assay of ampicillin in plasma and broncho-alveolar lavage fluid, using fluorescence detection, *Chromatographia*, **1989**, 27, 243–246.

Mercuric Chloride/Dansyl Hydrazine

SAMPLE
Matrix: milk
Analyte: penicillins
Sample preparation: 50 g Milk + 2 drops penicillinase (Difco Laboratories), let stand 1 h at 37°, add 50 mL MeCN, shake vigorously for 1 min, centrifuge at 9000 g for 10 min, decant, add 5 g NaCl, swirl to dissolve, add 100 mL dichloromethane, shake for 1 min, centrifuge at 1000 g for 10 min. Remove top aqueous layer and extract organic layer with 25 mL 10% NaCl by shaking and centrifuging as before. Combine aqueous layers, add 1 mL 0.3% mercuric chloride in water, let stand 30 min, add 1 mL 2 M HCl, extract with three 50 mL portions of dichloromethane by shaking each portion for 1 min and centrifuging at 1000 g for 10 min, filter dichloromethane extracts through 30 g anhydrous sodium sulfate, evaporate to dryness under reduced pressure at 35°, if water remains add 5-10mL MeOH to flask and complete evaporation. Dissolve residue in 1 mL 10% acetic acid, add 0.5 mL 0.08% dansyl hydrazine in 10% acetic acid, let stand 90 min to overnight in the dark, transfer reaction mixture to a separatory funnel with three 25 mL portions of dichloromethane, add 5 mL 2 M HCl, shake for 1 min, wash organic layer with 5 mL 5% $NaHCO_3$ solution, filter through 10-20 g anhydrous sodium sulfate. Extract acid aqueous layer again with 25 mL dichloromethane. Combine dichloromethane layers and evaporate to dryness at 35° under reduced pressure. Dissolve residue in 2 mL IS solution, inject a 20 μL aliquot. (Prepare IS solution by dissolving 10 μL benzaldehyde in 100 mL dichloromethane, evaporate 1 mL to dryness under reduced pressure, dissolve residue in 1 mL 10% acetic acid, add 0.5 mL 0.08% dansyl hydrazine in 10% acetic acid, let stand 90 min to overnight in the dark, transfer reaction mixture to a separatory funnel with three 25 mL portions of dichloromethane, add 5 mL 2 M HCl, shake for 1 min, wash organic layer with 5 mL 5% $NaHCO_3$ solution, filter through 10-20 g anhydrous sodium sulfate. Extract acid aqueous layer again with 25 mL dichloromethane. Combine dichloromethane layers and evaporate to dryness at 35° under reduced pressure. Dissolve residue in 100 mL MeCN then dilute an aliquot 1:4 with MeCN.)

HPLC VARIABLES
Column: 250 × 4 10 μm Lichrosorb RP-18
Mobile phase: MeCN:water 58:42
Flow rate: 1
Injection volume: 20
Detector: F ex 254 em 500 filter

CHROMATOGRAM
Retention time: 5.23 (methicillin), 5.74 (penicillin G), 6.73 (penicillin V), 7.21 (oxacillin), 7.63 (phenethicillin), 8.18 (cloxacillin), 8.61 (nafcillin), 10.73 (dicloxacillin)
Internal standard: benzaldehyde (derivatized) (12.18)
Limit of detection: 5 ng/g

REFERENCE
Munns, R.K.; Shimoda, W.; Roybal, J.E.; Vieira, C. Multiresidue method for determination of eight neutral β-lactam penicillins in milk by fluorescence-liquid chromatography, *J.Assoc.Off.Anal.Chem.*, **1985**, *68*, 968−971.

Mercuric Chloride/1-Hydroxybenzotriazole

SAMPLE
Matrix: fermentation broth
Analyte: penicillins
Sample preparation: Filter (0.22 μm) fermentation broth. Mix 980 μL filtrate with 20 μL 200 mM benzoic anhydride in MeCN at room temperature for 3 min, add 100 μL reagent, mix, heat at 45° for 1 h, cool to room temperature, inject a 40 μL aliquot. (Prepare reagent by dissolving 6.76 g 1-hydroxybenzotriazole hydrate in 10 mL water, add 2.5 mL mercury(II) chloride solution (no concentration given), adjust pH to 9.2 with 4 M NaOH, make up to 25 mL.)

HPLC VARIABLES
Guard column: 10 × 4 5 μm Spherisorb S5ODS2
Column: 259 × 4.9 5 μm Spherisorb S5ODS2
Mobile phase: Gradient. MeCN:20 mM pH 6.5 potassium phosphate buffer:20 mM sodium thiosulfate from 10:45:45 to 25:37.5:37.5 over 25 min, maintain at 25:37.5:37.5 for 10 min (The mobile phase flowed through a 50 × 4 column of 5 μm Spherisorb S5ODS2 before the injector.)
Flow rate: 1 for 25 min then 1.2
Injection volume: 40
Detector: UV 328

CHROMATOGRAM
Retention time: 19 (penicillin K), 20 (penicillin X), 21 (ampicillin), 24 (penicillin G), 25 (6-aminopenicillanic acid), 26 (penicillin V)
Limit of detection: 1-100 ng/mL

REFERENCE
Shah, A.J.; Adlard, M.W.; Holt, G. Determination of natural penicillins in fermentation media by high-performance liquid chromatography using precolumn derivatization with 1-hydroxybenzotriazole, *Analyst*, **1988**, *113*, 1197−1200.

Mercuric Chloride/Imidazole

SAMPLE
Matrix: fermentation broth
Analyte: penicillins
Sample preparation: Adjust pH of fermentation broth to 7, centrifuge at 8000 g for 10 min, add MeCN, centrifuge, add dichloromethane to the supernatant, vortex for 10 s, shake for 15 min, centrifuge at 8000 g for 15 min. Add 1 mL of the aqueous layer to 100 µL reagent, heat at 50° for 50 min, cool in an ice bath, inject a 20 µL aliquot. (Prepare reagent by dissolving 4.125 g imidazole in 2.5 mL water, add 1 mL HCl, add 500 µL 110 mM mercury(II) chloride, add 1.5 mL HCl. Recrystallize imidazole twice from isopropanol.)

HPLC VARIABLES
Guard column: 10 × 4 5 µm Spherisorb C18
Column: 20 × 4.6 5 µm Spherisorb C18 S5ODS2
Mobile phase: Gradient. MeCN:buffer from 16.5:83.5 to 31.5:68.5 over 17 min (Buffer was 10 mM NaH_2PO_4 containing 10 mM EDTA, adjusted to pH 6.5 with 2 M NaOH.)
Flow rate: 2
Injection volume: 20
Detector: UV 325

CHROMATOGRAM
Retention time: 5 (penicillin X), 11 (methicillin), 13 (penicillin G), 14.5 (penicillin V)
Limit of detection: 1 µg/mL

REFERENCE
Rogers, M.E.; Adlard, M.W.; Saunders, G.; Holt, G. High-performance liquid chromatographic determination of penicillins following derivatization to mercury-stabilized penicillenic acids, *J.Liq.Chromatogr.*, **1983**, *6*, 2019–2031.

SAMPLE
Matrix: solutions
Analyte: penicillin G
Sample preparation: Mix a 200 µL aliquot of an aqueous solution with 30 µL reagent, heat at 60° for 20 min, inject an aliquot. (Reagent was imidazole:water:mercury(II) chloride 40:59.9:0.1, adjusted to pH 6.8 with phosphoric acid.)

HPLC VARIABLES
Guard column: 40 × 4 5 µm LiChrosorb RP-18

Column: 150 × 4 5 µm LiChrosorb RP-18
Mobile phase: MeOH:water 45:55 containing 2% imidazole and 50 µM mercury(II) chloride, pH adjusted to 6.6 with phosphoric acid (At the end of each day wash the column with 30 mL MeOH:pH 3.0 phosphate buffer (µ = 0.1) 45:55.)
Column temperature: 50
Detector: UV 325

CHROMATOGRAM
Retention time: 10
Limit of detection: 2 ng

REFERENCE
Wiese, B.; Martin, K. Basic extraction studies of benzylpenicillin and its determination by liquid chromatography with pre-column derivatisation, *J.Pharm.Biomed.Anal.*, **1989**, *7*, 67–78.

Mercuric Chloride/Triazole

RELATED REFERENCE
Rose, M.D.; Tarbin, J.; Farrington, W.H.; Shearer, G. Determination of penicillins in animal tissues at trace residue concentrations: II. Determination of amoxicillin and ampicillin in liver and muscle using cation exchange and porous graphitic carbon solid phase extraction and high-performance liquid chromatography. *Food Addit.Contam.* **1997**, *14*, 127-133

SAMPLE
Matrix: blood
Analyte: penicillin G
Sample preparation: Condition a Baker C18 SPE cartridge with 5 mL water and 5 mL 2% NaCl, do not allow to run dry. 2 mL Plasma + 120 µL 5 µg/mL penicillin V + 30 mL water + 2 mL 170 mM sulfuric acid +2 mL 5% sodium tungstate solution, vortex for 30 s, centrifuge at 2200 g for 10 min, filter supernatant (GF/B glass fiber filter), add 10 mL 20% NaCL, mix, add to SPE cartridge at 3 mL/min, wash with 5 mL 2% NaCl, wash with 5 mL water, draw air through cartridge for 5 min, elute with 500 µL elution solution. Add 500 µL derivatization reagent to the eluate, vortex for 20 s, allow to react at 65° for 30 min, cool to room temperature, vortex, filter (0.45 µm), inject 50-100 µL aliquots. (Prepare derivatization reagent by dissolving 34.45 g 1,2,4-triazole in 150 mL water, add 25 mL 10 mM mercuric chloride solution, mix, adjust the pH to 9.0 ± 0.5 with 5 M NaOH, dilute to 250 mL with water. Prepare elution solution by mixing 60 mL MeCN and 5 mL buffer and making up to 100 mL with water. The buffer was 0.994 g Na_2HPO_4 + 1.794 g $NaH_2PO_4.H_2O$ in 100 mL water, pH 6.5.)

HPLC VARIABLES
Column: 150 × 3.9 4 μm Nova-Pak C18
Mobile phase: MeCN:buffer 25:75 (Buffer contained 4.969 g Na$_2$HPO$_4$ + 8.969 g NaH$_2$PO$_4$ .H$_2$O + 2.482 g anhydrous sodium thiosulfate per liter.)
Flow rate: 1
Injection volume: 50-100
Detector: UV 325

CHROMATOGRAM
Retention time: 4.5
Internal standard: penicillin V (5.8)
Limit of detection: 5 ng/mL

KEY WORDS
plasma; cow; SPE

REFERENCE
Boison, J.O.; Korsrud, G.O.; MacNeil, J.D.; Keng, L.; Papich, M. Determination of penicillin G in bovine plasma by high-performance liquid chromatography after pre-column derivatization, *J.Chromatogr.*, **1992**, *576*, 315–320.

SAMPLE
Matrix: blood
Analyte: amoxicillin
Sample preparation: 100 μL Serum + 100 μL 10 M urea, filter (Amicon MPS-1 with YMT membrane) while centrifuging at 1500 g for 30 min. 80 μL Ultrafiltrate + 80 μL 100 mM pH 9.0 borate buffer + 8 μL 200 mM acetic anhydride in MeCN, let stand for 3 min, add 160 μL reagent, heat at 60° for 10 min, inject a 20 μL aliquot. (Reagent was 13.81 g 1,2,4-triazole in 60 mL water + 10 mL HgCl$_2$ solution (0.27 g HgCl$_2$ in 100 mL water), adjust pH to 9.0 ± 0.05 with 4 M NaOH, dilute to 100 mL (Analyst 1985, 110, 1277).)

HPLC VARIABLES
Guard column: used but not specified
Column: 100 × 3 5 μm Chromospher Spherisorb ODS-2
Mobile phase: MeCN:buffer 15:85 (Buffer was 50 mM pH 4.6 sodium phosphate containing 10 mM thiosulfate.)
Flow rate: 0.8
Injection volume: 20
Detector: UV 328

CHROMATOGRAM
Limit of detection: 1 μg/mL

KEY WORDS
serum; pharmacokinetics; ultrafiltrate

REFERENCE
Huisman-de Boer, J.J.; van den Anker, J.N.; Vogel, M.; Goessens, W.H.F.; Schoemaker, R.C.; de Groot, R. Amoxicillin pharmacokinetics in preterm infants with gestational ages of less than 32 weeks, *Antimicrob.Agents Chemother.*, **1995**, *39*, 431–434.

SAMPLE
Matrix: blood
Analyte: tazobactam
Sample preparation: 1 mL Serum or hemofiltrate + 100 μL 3 M sulfuric acid + 4 mL diethyl ether, vortex for 1 min, centrifuge at 5500 g for 10 min. Remove the organic layer and evaporate it to dryness under a stream of nitrogen at 37°, reconstitute the residue in 100 μL reagent and 10 μL 1% mercuric chloride in water, heat at 37° for 30 min, inject a 20 μL aliquot. (Prepare reagent by dissolving 4 g 1,2,4-triazole in water, adjusting the pH to 9.0 with 10 M NaOH, and making up to 20 mL with water.)

HPLC VARIABLES
Column: 125 × 4 5 μm Lichrospher RP (18)
Mobile phase: MeOH : (NH$_4$)$_2$HPO$_4$ 0.5 : 99.5, pH adjusted to 6.00 with phosphoric acid
Flow rate: 2.2
Injection volume: 20
Detector: UV 325

CHROMATOGRAM
Retention time: 8
Limit of detection: 50 ng/mL

KEY WORDS
serum; hemofiltrate

REFERENCE
Peyrin, E.; Guillaume, Y.; Guinchard, C. High-performance liquid chromatographic determination of tazobactam by precolumn derivatization, *J.Chromatogr.B*, **1995**, *672*, 160–164.

SAMPLE
Matrix: blood, urine
Analyte: antibiotics
Sample preparation: Serum. 200 μL Serum + 100 μL 10 M urea, mix, ultrafilter with Amicon YMT membrane at 1500 g for 10 min. 200 μL Ultrafiltrate + 200 μL 0.1 M pH 9 borate buffer + 20 μL 0.2 M acetic anhydride solution in MeCN, mix, let stand at room temperature for 3 min, add 400 μL reagent, heat at 60° in a water bath for 10 min, cool, inject a 40-80 μL aliquot. Urine. Dilute urine 10 fold with water and filter through 0.45 μm acrylate copolymer membrane. 200 μL Filtrate + 200 μL 0.1 M pH 9 borate buffer + 20 μL 0.2 M acetic anhydride solution in MeCN, mix, let stand at room temperature for 3 min, add 400 μL reagent, heat at 60° in a water bath for 10 min, cool, inject a 20-80 μL aliquot. (Reagent was 13.81 g 1,2,4-triazole in 60 mL water + 10 mL HgCl$_2$ solution (0.27 g HgCl$_2$ in 100 mL water), adjust pH to 9.0 ± 0.05 with 4 M NaOH, dilute to 100 mL.)

HPLC VARIABLES
Column: 150 × 4.6 7 μm Zorbax ODS-7
Mobile phase: MeCN : 20 mM NaH$_2$PO$_4$: 20 mM sodium thiosulfate : MeCN 24 : 38 : 38
Flow rate: 1
Injection volume: 20-80
Detector: UV 328

CHROMATOGRAM
Retention time: 6 (amoxicillin), 6 (ampicillin), 6 (cyclacillin (ciclacillin))
Limit of quantitation: 50 ng/mL

KEY WORDS
serum

REFERENCE
Haginaka, J.; Wakai, J. High-performance liquid chromatographic assay of ampicillin, amoxicillin and ciclacillin in serum and urine using a pre-column reaction with 1,2,4-triazole and mercury(II) chloride, *Analyst*, **1985**, *110*, 1277–1281.

SAMPLE
Matrix: blood, urine
Analyte: carbenicillin
Sample preparation: Serum. 200 μL Serum + 200 μL 10 M urea, mix, filter (Amicon MPS-1 micropartition system with Amicon YMT membranes) while centrifuging at 1500 g for 10 min. Add 200 μL of the ultrafiltrate to 200 μL reagent and heat at 60° for 10 min, cool to room temperature, inject a 30-90 μL aliquot. Urine. Dilute urine 10-fold with water, filter (0.45 μm acrylate copolymer). Add 200 μL of the filtrate to 200 μL reagent and heat at 60° for 10 min, cool to room temperature, inject a 30-60 μL aliquot. (Prepare reagent by dissolving 13.81 g 1,2,4-triazole in 60 mL water, add 10 mL 2.7 mg/mL mercury(II) chloride in water, adjust pH to 9.0 ± 0.05 with saturated NaOH, make up to 100 mL with water.)

HPLC VARIABLES
Column: 150 × 4.6 5 µm Develosil ODS-5 (Nomura Chemicals)
Mobile phase: MeCN:buffer containing 5 mM tetrabutylammonium bromide and 5 mM sodium thiosulfate 1:1.8 (Prepare the buffer by dissolving 14.32 g $Na_2HPO_4.12H_2O$ and 6.240 g $NaH_2PO_4.2H_2O$ in 1 L water then diluting 100-fold.)
Column temperature: 40
Flow rate: 1
Injection volume: 30-90
Detector: UV 328

CHROMATOGRAM
Retention time: 4.5
Limit of detection: 100 ng/mL (plasma), 1 µg/mL (urine)

OTHER SUBSTANCES
Interfering: ticarcillin

KEY WORDS
serum; ultrafiltrate

REFERENCE
Haginaka, J.; Wakai, J. High-performance liquid chromatographic assay of carbenicillin, ticarcillin and sulbenicillin in serum and urine using pre-column reaction with 1,2,4-triazole and mercury(II) chloride, *Analyst*, **1985**, *110*, 1185–1188.

SAMPLE
Matrix: milk
Analyte: ampicillin
Sample preparation: Condition a 3 mL 500 mg Bond Elut C18 SPE cartridge with 20 mL MeOH, 20 mL water, and 10 mL 2% NaCl, do not allow to go dry. 5 mL Milk + 500 µL 30% trichloroacetic acid, vortex thoroughly for 10 s, centrifuge at 0° at 3300 g for 30 min. Remove the liquid supernatant and add it to 100 µL 4 M NaOH (to pH 5.2), vortex for 10 s, add 1 mL 20% NaCl, vortex for 10 s, add to the SPE cartridge at 3 mL/min, wash with 1 mL 2% NaCl, wash with 1 mL water, elute with 1 mL MeCN:100 mM pH 6.5 phosphate buffer 40:60. Add 20 µL 2 M NaOH to the eluate and vortex for 10 s (pH 8), add 10 µL 200 mM acetic anhydride in MeCN, let stand for 3 min, add 500 µL reagent, vortex, heat at 65° for 10 min, cool to room temperature, inject a 200 µL aliquot. (Prepare reagent by dissolving 13.78 g 1,2,4-triazole in 60 mL water, add 10 mL 100 mM mercuric chloride solution, mix, adjust pH to 9.0 ± 0.5 with 4 M NaOH, make up to 100 mL with water.)

HPLC VARIABLES
Column: 150 × 3.9 4 µm Nova-Pak C18
Mobile phase: MeCN:MeOH:buffer 18:12:70 (Prepare buffer by dissolving 4.969 g NaH_2PO_4, 10.139 g $Na_2HPO_4.2H_2O$, 3.894 g sodium thiosulfate pentahydrate, and 6.791 g tetrabutylammonium hydrogen sulfate in 800 mL water, make up to 1 L with water.)
Flow rate: 0.8
Injection volume: 200
Detector: UV 325

CHROMATOGRAM
Retention time: 16
Limit of quantitation: 10 ng/mL
Limit of detection: 3 ng/mL

OTHER SUBSTANCES
Simultaneously analyzed: amoxicillin
Non-interfering: cloxacillin, oxacillin, penicillin G

KEY WORDS
SPE; cow

REFERENCE

Zukowski, J.; Pawlowska, M.; Armstrong, D.W. Efficient enantioselective separation and determination of trace impurities in secondary amino acids (i.e. imino acids), *J.Chromatogr.*, **1992**, *623*, 33–41.

SAMPLE

Matrix: milk

Analyte: penicillin G

Sample preparation: Condition a 6 mL 500 mg Bond Elut C18 SPE cartridge with 20 mL MeOH, 20 mL water, and 10 mL 2% NaCl, do not allow to go dry. 5 mL Milk (or 5 g yogurt or cottage cheese + 4 mL 1 M pH 6 phosphate buffer) + 20 μL 20 μg/mL penicillin V in water + 25 mL water + 4 mL 170 mM sulfuric acid + 40 mL 5% sodium tungstate, vortex for 30 s, centrifuge at 1500 g for 10 min, remove the supernatant, add 10 mL 20% NaCl to the residue, vortex for 10 s, centrifuge. Combine the supernatants and add them to the SPE cartridge, wash with 10 mL 2% NaCl, wash with 10 mL water, elute with 1 mL MeCN:200 mM pH 6.5 sodium phosphate buffer:water 60:5:35. Add 1 mL reagent to the eluate, vortex for 10 s, heat at 65° for 30 min, cool to room temperature, vortex, filter (Acro 0.45 μm), inject a 50-100 μL aliquot of the filtrate. (Prepare reagent by dissolving 34.45 g 1,2,4-triazole in 150 mL water, add 25 mL 10 mM mercuric chloride solution, mix, adjust pH to 9.0 ± 0.5 with 5 M NaOH, make up to 250 mL with water.)

HPLC VARIABLES

Column: 150 × 3.9 4 μm Nova-Pak C18

Mobile phase: MeCN:buffer 25:75 (Buffer was 4.696 g Na_2HPO_4, 8.969 g $NaH_2PO_4.H_2O$, and 2.482 g anhydrous sodium thiosulfate in 1 L water.)

Flow rate: 0.8

Injection volume: 50-100

Detector: UV 325

CHROMATOGRAM

Retention time: 5.5

Internal standard: penicillin V (7)

Limit of detection: 3 ng/mL

KEY WORDS

cow; SPE

REFERENCE

Boison, J.O.K.; Keng, L.J.-Y.; MacNeil, J.D. Analysis of penicillin G in milk by liquid chromatography, *J.AOAC Int.*, **1994**, *77*, 565–570.

SAMPLE

Matrix: milk

Analyte: ampicillin

Sample preparation: Condition a 3 mL 500 mg Bond Elut C18 SPE cartridge with 20 mL MeOH, 20 mL water, and 10 mL 2% NaCl, do not allow to go dry. 5 mL Milk + 500 μL 30% trichloroacetic acid, vortex thoroughly for 10 s, centrifuge at 0° at 3300 g for 30 min. Remove the liquid supernatant and add it to 100 μL 4 M NaOH (to pH 5.2), vortex for 10 s, add 1 mL 20% NaCl, vortex for 10 s, add to the SPE cartridge at 3 mL/min, wash with 1 mL 2% NaCl, wash with 1 mL water, elute with 1 mL MeCN:100 mM pH 6.5 phosphate buffer 40:60. Add 20 μL 2 M NaOH to the eluate and vortex for 10 s (pH 8), add 10 μL 200 mM acetic anhydride in MeCN, let stand for 3 min, add 500 μL reagent, vortex, heat at 65° for 10 min, cool to room temperature, inject a 200 μL aliquot. (Prepare reagent by dissolving 13.78 g 1,2,4-triazole in 60 mL water, add 10 mL 100 mM mercuric chloride solution, mix, adjust pH to 9.0 ± 0.5 with 4 M NaOH, make up to 100 mL with water.)

HPLC VARIABLES

Column: 150 × 3.9 4 μm Nova-Pak C18

Mobile phase: MeCN:MeOH:buffer 18:12:70 (Prepare buffer by dissolving 4.969 g NaH_2PO_4, 10.139 g $Na_2HPO_4.2H_2O$, 3.894 g sodium thiosulfate pentahydrate, and 6.791 g tetrabutylammonium hydrogen sulfate in 800 mL water, make up to 1 L with water.)

Flow rate: 0.8

Injection volume: 200
Detector: UV 325

CHROMATOGRAM
Retention time: 16
Limit of quantitation: 10 ng/mL
Limit of detection: 3 ng/mL

OTHER SUBSTANCES
Simultaneously analyzed: amoxicillin
Non-interfering: cloxacillin, oxacillin, penicillin G

KEY WORDS
SPE; cow

REFERENCE
Verdon, E.; Couedor, P. Determination of ampicillin residues in milk by ion-pair reversed phase high performance liquid chromatography after precolumn derivatization, *J.Pharm.Biomed.Anal.*, **1996**, *14*, 1201–1207.

SAMPLE
Matrix: milk
Analyte: penicillins
Sample preparation: Condition a 500 mg tC18 SPE cartridge (Waters) with 20 mL MeOH, 20 mL water, and 10 mL 2% NaCl. Centrifuge 30 mL milk at 1500 g for 10 min. Dilute a 10 mL portion of the defatted milk with 20 mL water, add 200 μL 2 μg/mL penicillin V in pH 9.0 buffer, add 6 mL 170 mM sulfuric acid, add 5.6 mL 5% sodium tungstate, shake vigorously for 1 min, allow to stand for 5 min, check that the pH is in the range 4.6-4.8 (if it is outside this range start again using a different volume of sodium tungstate solution), centrifuge at 1500 g for 10 min, adjust the pH of the supernatant to 8.1-8.2 with 5 M and 0.1 M NaOH, filter (glass fiber) the clear liquid phase. Pass the filtrate through the SPE cartridge at 2 mL/min, wash with 2 mL water, dry by pulling air through the cartridge for 1 min, elute with 2 mL MeCN. Add 150 μL pH 9.0 buffer to the eluate and evaporate to about 100 μL under a stream of nitrogen at 45-50°, add 400 μL pH 9.0 buffer, add 75 μL reagent I, vortex for 30 s, let stand at room temperature for 10 min, use 500 μL water to transfer the mixture to a separatory funnel, add 20 mL dichloromethane, add 5 mL pH 2.45 buffer, shake for 1 min, let stand for no more than 5 min. Remove the organic layer and evaporate it to dryness under reduced pressure at 35-40°, dissolve the residue in 500 μL pH 9.0 buffer, add 75 μL reagent I, vortex for 30 s, let stand at room temperature for 10 min, add 450 μL reagent II, vortex for 1 min, heat at 55 ± 1° for 30 min, cool, filter (0.45 μm), inject a 150 μL aliquot. (Prepare pH 9.0 buffer by dissolving 0.34 g KH_2PO_4 in water, adjusting the pH to 9.0 with NaOH, and making up to 100 mL with water. Prepare pH 2.45 buffer by dissolving 2.72 g KH_2PO_4 in water, adjusting the pH to 2.45 with phosphoric acid, and making up to 100 mL with water. Prepare reagent 1 by dissolving 1.13 g benzoic anhydride in MeCN, make up to 25 mL with MeCN. Prepare reagent II by dissolving 6.905 g 1,2,4-triazole in 30 mL water and adding 5 mL 26 mM mercuric chloride in water, adjust pH to 9.0 ± 0.05 with 5 M NaOH, make up to 50 mL. Prepare reagents I and II 1-4 h before use. Silanize glassware with dichlorodimethylsilane.)

HPLC VARIABLES
Column: 150 × 3.9 4 μm Nova-Pak C18
Mobile phase: Gradient. A as MeCN:buffer 10:90. B was MeCN:buffer 30:70. A:B from 100:0 to 0:100 over 30 min, maintain at 0:100 for 13 min, return to initial conditions over 2 min, re-equilibrate at initial conditions for 5 min. (Prepare buffer by dissolving 9.938 g Na_2HPO_4, 17.938 g $NaH_2PO_4.H_2O$, and 4.964 g sodium thiosulfate in water, make up to 2 L with water, pH 6.5.)
Column temperature: 30
Flow rate: 1
Injection volume: 150
Detector: UV 323

CHROMATOGRAM
Retention time: 25 (amoxicillin), 27 (penicillin G), 32.5 (ampicillin), 34 (oxacillin), 35.5 (cloxacillin), 40 (dicloxacillin)
Internal standard: penicillin V (28.5)
Limit of quantitation: 1.9-3.7 ng/mL
Limit of detection: 1.3-2.7 ng/mL

KEY WORDS
cow; SPE

REFERENCE
Sorensen, L.K.; Rasmussen, B.M.; Boison, J.O.; Keng, L. Simultaneous determination of six penicillins in cows' raw milk by a multiresidue high-performance liquid chromatographic method, *J.Chromatogr.B*, **1997**, *694*, 383–391.

SAMPLE
Matrix: tissue
Analyte: antibiotics
Sample preparation: Condition a 6 mL 500 mg Bond Elut C18 SPE cartridge with 20 mL MeOH, 20 mL water, and 20 mL 2% NaCl. Shake 10 g tissue and 20 mL MeCN on a mechanical shaker for 30 min, centrifuge, remove the supernatant, repeat the extraction with 20 and 10 mL portions of MeCN. Combine the extracts and add them to 30 mL 4% NaCl, remove the MeCN under reduced pressure at 40°, filter (Whatman GF/C and Gelman 0.45 μm membrane) the remaining aqueous mixture, add the filtrate to the SPE cartridge at <2 mL/min, wash with 15 mL 2% NaCl, elute with 5 mL MeCN. Add 100 μL 20 μg/mL penicillin V in MeCN to the eluate, evaporate to dryness under a stream of nitrogen at 37°, reconstitute the residue in 1 mL water, vortex, add 1 mL 2 M pH 9 1,2,4-triazole containing 1 mM mercuric chloride, vortex, heat at 65° for 30 min, cool, filter (0.45 μm), inject a 50 μL aliquot.

HPLC VARIABLES
Column: 150 × 3.9 4 μm Nova-Pak C18
Mobile phase: MeCN:buffer 22.5:77.5 (Prepare buffer by dissolving 4.96 g Na_2HPO_4, 10.14 g $NaH_2PO_4.2H_2O$, and 3.90 g sodium thiosulfate in 1 L water.)
Flow rate: 1.2
Injection volume: 50
Detector: UV 325

CHROMATOGRAM
Retention time: 5 (penicillin G), 23 (cloxacillin)
Internal standard: penicillin V (6.5)
Limit of detection: 5 ng/g

KEY WORDS
cow; sheep; kidney; liver; muscle; SPE

REFERENCE
Gee, H.-E.; Ho, K.-B.; Toothill, J. Liquid chromatographic determination of benzylpenicillin and cloxacillin in animal tissues and its application to a study of the stability at -20°C of spiked and incurred residues of benzylpenicillin in ovine liver, *J.AOAC Int.*, **1996**, *79*, 640–644.

SAMPLE
Matrix: tissue
Analyte: penicillin G
Sample preparation: Condition a 6 mL 500 mg Bond Elut C18 SPE cartridge with 20 mL MeOH, 20 mL water, and 10 mL 2% NaCl, do not allow to go dry. 5 g Tissue + 75 μL 20 μg/mL penicillin V in water + 20 mL water, homogenize (Polytron, 20 mm probe), rinse probe with water so that total volume is 35 mL, shake mechanically for 5 min, add 5 mL 170 mM sulfuric acid, add 5 mL 5% sodium tungstate, vortex for 20 s, centrifuge at 2200 g for 10 min, remove the supernatant, add 15 mL water to the residue, shake for 5 min, centrifuge at 2200 g for 10 min. Combine the supernatants and filter (Whatman GF/B) them, add 10 mL 20% NaCl to the filtrate, mix thoroughly, add to the SPE cartridge at 3 mL/min, wash with 10 mL 2% NaCl, wash with 10 mL water, draw air through the

cartridge for 5 min, immediately elute with 1 mL MeCN:200 mM pH 6.5 sodium phosphate buffer:water 60:5:35. Add 1 mL reagent to the eluate, vortex, heat at 65° for 30 min, cool rapidly to room temperature, vortex, filter (Acro 0.45 μm), inject a 80-100 μL aliquot of the filtrate. (Prepare reagent by dissolving 34.45 g 1,2,4-triazole in 150 mL water, add 25 mL 10 mM mercuric chloride solution, mix, adjust pH to 9.0 ± 0.5 with 5 M NaOH, make up to 250 mL with water.)

HPLC VARIABLES
Column: 150 × 3.9 4 μm Nova-Pak C18
Mobile phase: MeCN:buffer 25:75 (Buffer was 4.969 g Na_2HPO_4, 8.969 g $NaH_2PO_4.H_2O$, and 2.482 g anhydrous sodium thiosulfate in 1 L water.)
Flow rate: 0.8
Injection volume: 80-100
Detector: UV 325

CHROMATOGRAM
Retention time: 5.8
Internal standard: penicillin V (7.6)
Limit of detection: 5 ng/g

OTHER SUBSTANCES
Simultaneously analyzed: ampicillin, chloramphenicol

KEY WORDS
muscle; liver; kidney; cow; SPE

REFERENCE
Boison, J.O.; Salisbury, C.D.C.; Chan, W.; MacNeil, J.D. Determination of penicillin G residues in edible animal tissues by liquid chromatography, *J.Assoc.Off.Anal.Chem.*, **1991**, *74*, 497–501.

MISCELLANEOUS REACTIONS

Hydrolysis

SAMPLE
Matrix: blood, urine
Analyte: clavulanic acid
Sample preparation: Plasma. Vortex plasma with 2 volumes of MeCN for 30 s, centrifuge at 3600 rpm for 5 min, inject a 50 μL aliquot of the supernatant. Urine. Filter (0.45 μm) urine, inject a 25 μL aliquot of the filtrate.

HPLC VARIABLES
Guard column: 50 × 4.6 LiChrosorb RP-2
Column: 250 × 4.6 Develosil ODS-10 (Nomura Chemicals)
Mobile phase: MeOH:buffer A 25:75 (plasma) or MeOH:buffer B 20:100 (urine) (Buffer A was 0.1 mM Na_2HPO_4 containing 0.1 mM NaH_2PO_4 and 5 mM tetrabutylammonium bro-

mide. Buffer B was 1 mM Na_2HPO_4 containing 1 mM NaH_2PO_4 and 5 mM tetrabutylammonium bromide.)

Flow rate: 1.2

Injection volume: 25-50

Detector: UV 270 following post-column reaction. The column effluent mixed with 500 mM NaOH pumped at 0.6 mL/min and the mixture flowed through a 2 m × 0.25 mm ID coil to the detector.)

CHROMATOGRAM

Retention time: 13 (plasma), 20 (urine)

Limit of detection: 100 ng/mL

KEY WORDS

post-column reaction; plasma

REFERENCE

Haginaka, J.; Yasuda, H.; Uno, T.; Nakagawa, T. Alkaline degradation of clavulanic acid and high performance liquid chromatographic determination by post-column alkaline degradation, *Chem.Pharm.Bull.(Tokyo)*, **1983**, *31*, 4436–4447.

LACTONE

MISCELLANEOUS REACTIONS

Degradation

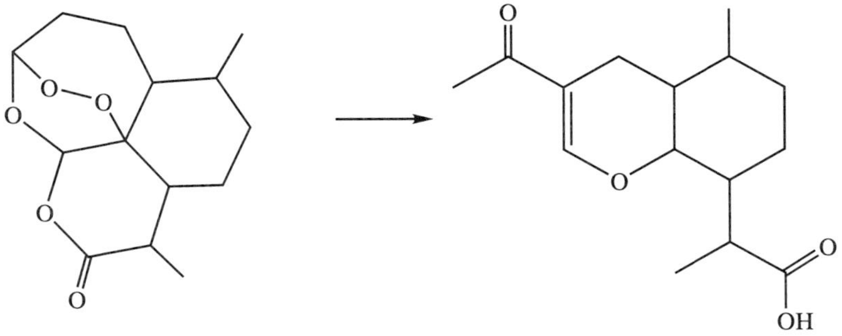

SAMPLE
Matrix: blood
Analyte: artemisinin
Sample preparation: 1 mL Serum + 4 mL n-chlorobutane, vortex for 30 s, centrifuge at 5000 rpm for 15 min, freeze in acetone/dry ice. Remove the organic layer and evaporate it to dryness under a stream of nitrogen at 40°, reconstitute the residue in 100 μL MeOH, add 400 μL 0.2% NaOH, mix, heat at 45° for 30 min, cool, add 50 μL 100 mM acetic acid in MeOH, inject a 200 μL aliquot.

HPLC VARIABLES
Column: 100 mm long 5 μm LiChrosorb C18
Mobile phase: MeOH:10 mM pH 4.5 phosphate buffer 45:55
Flow rate: 0.8
Injection volume: 200
Detector: UV (wavelength not specified)

CHROMATOGRAM
Retention time: 10
Limit of detection: 2.5 ng/mL

KEY WORDS
serum; pharmacokinetics

REFERENCE
Titulaer, H.A.C.; Zuidema, J.; Kager, P.A.; Wetsteyn, J.C.F.M.; Lugt, C.B.; Merkus, F.W.H.M. The pharmacokinetics of artemisinin after oral, intramuscular and rectal administration to volunteers, *J.Pharm.Pharmacol.*, **1990**, *42*, 810−813.

SAMPLE
Matrix: blood, saliva
Analyte: artemisinin
Sample preparation: 1.5 mL Plasma or saliva + 500 μL 0.9% NaCl + 2.5 mL ethyl acetate, vortex, centrifuge at 2800 rpm for 3 min, repeat extraction twice more. Combine the organic layers and evaporate them to dryness under a stream of air at room temperature, reconstitute the residue in 100 μL EtOH, add 400 μL 0.2% NaOH, heat at 50° for 30 min, cool rapidly in water, wash twice with 500 μL aliquots of ethyl acetate, centrifuge, evaporate traces of ethyl acetate with a stream of air at room temperature, add 40 μL 2.5 M acetic acid in EtOH, make up to 500 μL with MeOH:water 20:80, inject a 200 μL aliquot.

HPLC VARIABLES
Column: 250 × 4 10 μm LiChrosorb RP-18
Mobile phase: MeOH:water 40:60 containing 10 mM Na_2HPO_4-NaH_2PO_4
Column temperature: 35 ± 1
Flow rate: 1.3
Injection volume: 200
Detector: UV 260

CHROMATOGRAM
Retention time: 12
Limit of detection: 2.5 ng/mL

KEY WORDS
plasma; pharmacokinetics

REFERENCE
Zhao, S. High-performance liquid chromatographic determination of artemisinine (Qinghaosu) in human plasma and saliva, *Analyst*, **1987**, *112*, 661–664.

METAL

COMPLEXATION

2-Acetylpyridine-4-phenyl-3-thiosemicarbazone

SAMPLE
Matrix: solutions
Analyte: metals
Sample preparation: Mix a 5-50 mL aliquot of an aqueous solution with 3-5 mg ascorbic acid, 2 mL 1 M pH 6 sodium acetate buffer, and 2 mL 0.2% 2-acetylpyridine-4-phenyl-3-thiosemicarbazone in MeOH, add 2 mL chloroform, mix well, repeat the extraction, combine the organic layers. Remove a 2 mL aliquot and evaporate it to dryness, reconstitute the residue in 1 mL MeOH, inject a 5 μL aliquot. (Preparation of 2-acetylpyridine-4-phenyl-3-thiosemicarbazone is as follows. Add 1.45 g 2-acetylpyridine in 25 mL EtOH to 2 g 4-phenyl-3-thiosemicarbazide in 75 mL hot EtOH, reflux for 2 h, cool to room temperature. Filter off the crystals and recrystallize them twice from EtOH to obtain 2-acetylpyridine-4-phenyl-3-thiosemicarbazone as yellow crystals (mp 187-189°) (Anal. Chim. Acta 1977, 90, 335).)

HPLC VARIABLES
Column: 150 × 4.6 5 μm Microsorb C18
Mobile phase: MeCN:water:1 mM sodium acetate:1 mM tetrabutylammonium bromide 86:12:1:1
Flow rate: 1
Injection volume: 5
Detector: UV 254

CHROMATOGRAM
Retention time: 3 (Co(II)), 4 (Fe(II)), 5 Cu(II))
Limit of detection: 120-250 ng

KEY WORDS
complexation

REFERENCE
Khuhawar, M.Y.; Memon, Z.P.; Lanjwani, S.N. HPLC determination of copper(II), cobalt(II) and iron(II) in pharmaceutical preparations using 2-acetylpyridine-4-phenyl-3-thiosemicarbazone derivatizing agent, *Chromatographia*, **1995**, *41*, 236–237.

Ammonium Bis(2-hydroxyethyl)dithiocarbamate

$$H_4N^{\oplus} \quad S^{\ominus} \overset{\overset{S}{\|}}{C} N(CH_2CH_2OH)_2 \longrightarrow complex$$

SAMPLE
Matrix: solutions
Analyte: metals
Sample preparation: Add 250 μL 100 mM ammonium bis(2-hydroxyethyl) dithiocarbamate to 25 mL of a solution of metal ions in mobile phase buffer, inject an aliquot. (Synthesis of ammonium bis(2-hydroxyethyl)dithiocarbamate is as follows. Dissolve 10.5 g diethanolamine in 50 mL MeOH, add 150 mL THF, add 10 mL concentrated ammonium hydroxide, cool in an ice bath to below 10°, add 5 mL carbon disulfide dropwise with stirring, let stand in the ice bath for several h, filter, wash the solid with THF, dry under reduced pressure at room temperature to obtain ammonium bis(2-hydroxyethyl)dithiocarbamate (mp 106°).)

HPLC VARIABLES
Guard column: 20 × 4.6 5 μm Supelcosil C18
Column: 250 × 4.6 5 μm Supelcosil C18
Mobile phase: MeOH:buffer 40:60 (Prepare buffer by dissolving 25.3 g triethylamine in 300 mL water and adjusting the pH to 6.5 with acetic acid, pass through a 1 cm bed of 74-105 μm Amberlite XAD-4 to remove impurities, make up to 500 mL with water.)
Flow rate: 1
Injection volume: 10
Detector: UV 255; UV 405 (Co, Ni, Cu only)

CHROMATOGRAM
Retention time: 4 (Co(II)), 6 (Ni(II)), 8 (Cu(II)), 10 (Hg(II))
Limit of quantitation: 5 ng/mL

KEY WORDS
complexation

REFERENCE
King, J.N.; Fritz, J.S. Determination of cobalt, copper, mercury, and nickel as bis(2-hydroxyethyl)dithiocarbamate complexes by high-performance liquid chromatography, *Anal.Chem.*, **1987**, *59*, 703–708.

4-Benzoyl-3-methyl-1-phenyl-2-pyrazolin-5-one

RELATED REFERENCE
Akama, Y.; Tong, A. Determination of aluminum and indium by high-performance liquid chromatography after precolumn complexation with 1-phenyl-3-methyl-4-benzoyl-5-pyrazolone. *Anal.Sci.* **1991**, 7, 745-747

SAMPLE
Matrix: salt
Analyte: metals
Sample preparation: Dissolve 40 g salt in 150 mL water containing 5 mL concentrated nitric acid at 100°, dilute to 200 mL with water. Remove a 20 mL aliquot and adjust the pH to 3 with 1 M sodium acetate, add 10 mL 20 mM 4-benzoyl-3-methyl-1-phenyl-2-pyrazolin-5-one in MeOH, heat at 90° with stirring for 1 h. Remove the solid by filtration and wash it with water. Dissolve the solid in 25 mL dioxane (Caution! Dioxane is a carcinogen!), inject a 10 μL aliquot.

HPLC VARIABLES
Column: 250 × 4.6 Chemcosorb 5-ODS-UH
Mobile phase: MeCN
Flow rate: 1
Injection volume: 10
Detector: UV 245

CHROMATOGRAM
Retention time: 14.5 (Al(III)), 15.5 (Fe(III))
Limit of detection: 13-28 ng/mL

OTHER SUBSTANCES
Non-interfering: Co, Mn, Ni, Pb, Zn

KEY WORDS
complexation

REFERENCE
Akama, Y.; Tong, A. High-performance liquid chromatographic determination of aluminum and iron(III) in solar salt in the form of their 1-phenyl-3-methyl-4-benzoyl-5-pyrazolone chelates, *J.Chromatogr.*, **1993**, *633*, 129–133.

N-Benzoyl-N-phenylhydroxylamine

SAMPLE
Matrix: solutions
Analyte: vanadium and molybdenum
Sample preparation: Add an aliquot of an aqueous solution to 20 mL water and 36.4 mL concentrated HCl, make up to 100 mL with water, add 5 mL 0.1% N-benzoyl-N-phenylhydroxylamine in chloroform, shake for 3 min. Remove the organic phase and dry it over anhydrous sodium sulfate, filter (0.2 μm), inject a 20 μL aliquot.

HPLC VARIABLES
Guard column: 10 × 4.6 Spherisorb
Column: 250 × 4.6 5 μm Spherisorb S5 nitrile
Mobile phase: Chloroform containing 590 μM N-benzoyl-N-phenylhydroxylamine
Column temperature: 10
Flow rate: 0.6
Injection volume: 20

Detector: UV 360

CHROMATOGRAM
Retention time: 11 (V(V)), 11.5 (Mo(VI))
Limit of detection: 2.1 ng/mL (V), 3.3 ng/mL (Mo)

OTHER SUBSTANCES
Interfering: W(VI), iodide
Non-interfering: other metals

KEY WORDS
complexation

REFERENCE
Gracia Bagur, M.; Gázquez Evangelista, D.; Sanchez-Viñas, M. Simultaneous determination of vanadium and molybdenum as N-benzoyl-N-phenylhydroxamate complexes by combining solvent extraction and liquid chromatography, *J.Chromatogr.A*, **1996**, *730*, 241–246.

Bis(salicaldehyde)tetramethylethylenediamine

SAMPLE
Matrix: blood, formulations
Analyte: cisplatin
Sample preparation: Whole blood. 5 mL Whole blood + 2 mL 10% trichloroacetic acid in water, mix thoroughly, let stand for 15 min, centrifuge for 20 min. Remove the supernatant and add it to 2 mL concentrated HCl, add 8 mL 10% trichloroacetic acid in water, mix, centrifuge for 15 min. Remove the supernatant and add it to 2 mL concentrated HCl, evaporate to near dryness with heat, dissolve the residue in 3 mL water, adjust the pH to 6, add 2 mL pH 8 sodium bicarbonate buffer, add 2 mL 1% bis(salicaldehyde)tetramethylethylenediamine in EtOH, warm for 15 min, cool, add 4 mL chloroform, mix well. Remove the chloroform layer and evaporate it to dryness, reconstitute with 100 µL MeOH, inject a 5 µL aliquot. Injections. Add 1 g cisplatin injection to 30 mL concentrated HCl, evaporate to dryness with heat, reconstitute with 10 mL water. Remove a 5 mL aliquot, add 2 mL pH 8 sodium bicarbonate buffer, add 2 mL 1% bis(salicaldehyde)tetramethylethylenediamine in EtOH, warm for 15 min, cool, add 4 mL chloroform, mix well. Remove a 2 mL aliquot of the chloroform layer and evaporate it to dryness, reconstitute with 1 mL EtOH, inject a 5 µL aliquot. (Preparation of bis(salicaldehyde)tetramethylethylenediamine is as follows. Stir 44.5 g 2-nitropropane and 84 mL 6 M NaOH with cooling, add 40 g bromine dropwise, add 165 mL EtOH, reflux gently for 3 h, pour into 500 mL ice-water, filter to obtain 2,3-dimethyl-2,3-dinitrobutane. Vigorously stir 17.6 2,3-dimethyl-2,3-dinitrobutane with 150 mL concentrated HCl at 50-60°, slowly add 75 g 20-mesh granular tin, reflux for 15 min, cool, make strongly alkaline with NaOH (Caution! Exothermic!), steam distil, collect 350 mL distillate. Add 100 g solid NaOH to the distillate to obtain 2,3-diamino-2,3-dimethylbutane as a separate layer (mp of oxalate 323-324°) (J. Am. Chem. Soc. 1955, 77, 6689). Stir 20 mmoles salicylaldehyde in 10 mL MeOH, add 10 mmoles 2,3-diamino-2,3-dimethylbutane in 6 mL MeOH dropwise, let stand for several h, collect the precipitate, recrystallize twice from MeOH to obtain bis (salicaldehyde)tetramethylethylenediamine as yellow needles (mp 117°) (Inorg. Chem. 1978, 7, 3389).)

HPLC VARIABLES
Column: 150 × 4.6 3 μm Hypersil ODS
Mobile phase: MeCN:MeOH:water 30:50:20
Flow rate: 0.4
Injection volume: 5
Detector: UV 254

CHROMATOGRAM
Retention time: 7
Limit of detection: 1 μg/mL

OTHER SUBSTANCES
Simultaneously analyzed: copper(II), iron(II), nickel(II), palladium(II), uranyl

KEY WORDS
complexation; injections; whole blood

REFERENCE
Khuhawar, M.Y.; Lanjwani, S.N.; Memon, S.A. High-performance liquid chromatographic determination of cisplatin as platinum(II) in a pharmaceutical preparation and blood samples of cancer patients, *J.Chromatogr.B*, **1997**, *693*, 175–179.

2-(5-Bromo-2-pyridylazo)-5-(diethylamino)phenol

SAMPLE
Matrix: enzymes
Analyte: metals
Sample preparation: Heat 4 mL Enzyme solution (obtained from dialysis) with 4 mL concentrated nitric acid, evaporate nearly to dryness, add 1 drop concentrated nitric acid, make up to 5 mL with water. Remove a 1 mL aliquot and add it to 1 mL reagent, add 2 mL buffer, make up to 10 mL with water, heat on a boiling water bath for 30 min, cool, inject a 100 μL aliquot. (Prepare reagent by dissolving 75 mg 2-(5-bromo-2-pyridylazo)-5-(diethylamino)phenol in 25 g poly(oxyethylene) 4-nonylphenyl ether with 10 oxyethylene units (PONPE-10, Tokyo Chemical Industry or from Tokyo Kasei (TCI America, Portland OR)) with warming and stirring, dilute with about 150 mL water with warming and gentle stirring, make up to 250 mL with water. Prepare buffer by adjusting the pH of 2 M acetic acid to 4.5 with NaOH.)

HPLC VARIABLES
Column: 250 × 4.6 Cosmosil 5TMS (Nacalai Tesque, Kyoto)
Mobile phase: MeCN:water 46:54 containing 2 mM tetrabutylammonium bromide, 5 mM sodium acetate, and 100 μM EDTA
Flow rate: 1
Injection volume: 100
Detector: UV 595

CHROMATOGRAM
Retention time: 5 (V), 7 (Co), 9 (Ni), 12 (Fe)
Limit of detection: 0.72-98.8 pg

OTHER SUBSTANCES
Non-interfering: aluminum, cadmium, copper, manganese, zinc

KEY WORDS
complexation

REFERENCE
Miura, J.; Itoh, N. Determination of vanadium, cobalt, nickel, and iron in bromoperoxidases from *Pseudomonas putida* and *Corallina pilulifera* by high performance liquid chromatography with spectrophotometric detection, *J.Liq.Chromatogr.Rel.Technol.*, **1997**, *20*, 2367−2376.

SAMPLE
Matrix: solutions

Analyte: cobalt

Sample preparation: Mix 1 mL 1 mM cobalt chloride in water and 2 mL 1 mM reagent with buffer, make up to 100 mL with buffer, inject an aliquot. (Buffer was running buffer diluted 10-fold with water. Prepare reagent by dissolving 34.9 mg 2-(5'-bromo-2-pyridylazo)-5-(diethylamino)phenol in 50 mL EtOH, make up to 100 mL with water.)

CAPILLARY ELECTROPHORESIS
Capillary: 70 cm × 75 μm fused-silica (63 cm to detector) (Composite Metal Services, Hallow, UK)

Capillary preparation: Wash with running buffer for 5 min before each injection. Condition new capillaries by purging with 100 mM NaOH at 60° for 5 min and with water at 60° for 5 min. Equilibrate with run buffer for 5 min then apply voltage for 5 min a number of times.

Running buffer: 50 mM pH 8 Sodium acetate containing 2 mM cetyltrimethylammonium bromide and 100 nM 2-(5'-bromo-2-pyridylazo)-5-(diethylamino)phenol

Voltage/Current: -25 kV

Injection: Make three hydrodynamic (vacuum) injection for 30 s each then apply + 25 kV (to remove solvent). When the current reaches 95% of its pre-injection level switch polarity to begin the electrophoresis.

Detector: UV 587

Model: Spectra Phoresis 1000

Migration time: 4.6

Limit of detection: 50 nM (10 injections)

KEY WORDS
complexation

REFERENCE
Smyth, W.F.; Harland, G.B.; McClean, S.; McGrath, G.; Oxspring, D. Effect of on-capillary large volume sample stacking on limits of detection in the capillary zone electrophoretic determination of selected drugs, dyes and metal chelates, *J.Chromatogr.A*, **1997**, *772*, 161−169.

2-(5-Bromo-2-pyridylazo)-5-(N-propyl-N-sulfopropylamino)phenol

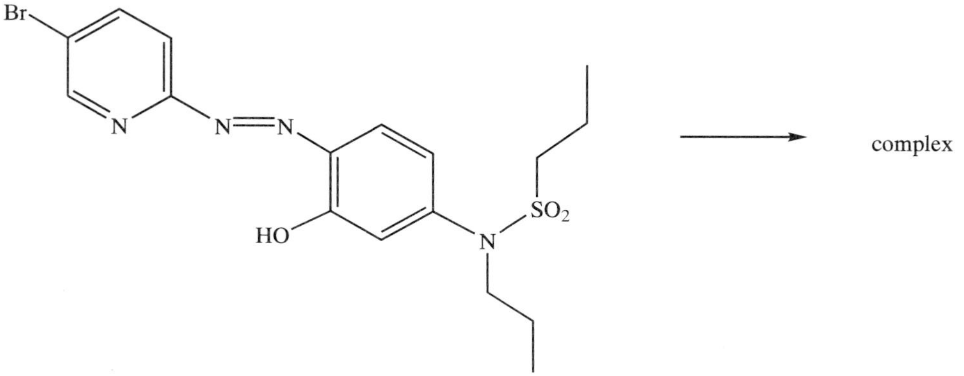

SAMPLE

Matrix: sea water

Analyte: metals

Sample preparation: Filter (0.45 μm) sea water, add HCl to a final concentration of 100 mM. Mix a 50 mL aliquot with 500 μL 5 mM 2-(5-bromo-2-pyridylazo)-5-(N-propyl-N-sulfopropylamino)phenol in water and 2.5 mL 2 M sodium acetate solution, adjust pH to 4.5 with 4 M ammonia, add 5 mL 20 mM methyltrioctylammonium chloride (Capriquat, Aliquat 336) in xylene, shake for 5 min, centrifuge at 2000 rpm for 2 min. remove a 4 mL aliquot of the organic layer and add it to 500 μL MeOH:water 50:50 containing 2 M sodium perchlorate and 10 mM pH 4.0 sodium acetate, extract for 10 min, centrifuge at 2000 rpm for 2 min, remove a 20 μL aliquot of the aqueous phase. (Reagent was 2-(5-bromo-2-pyridylazo)-5-(N-propyl-N-sulfopropylamino)phenol available from Dojindo, Kumamoto, Japan (Dojindo Molecular Technologies, 3 Bethesda Metro Center, Suite 700, Bethesda MD 20814; (301) 664 8448; www.dojindo.co.jp). Prepare as follows. React 3-aminophenol with 1-bromopropane to give 3-(N-n-propylamino)phenol (bp 129-135°/1 mm Hg). Treat 3-(N-n-propylamino)phenol with an equimolar amount of 1,3-propane sultone (Caution! 1,3-Propane sultone is a carcinogen!) to give 3-(N-n-propyl-N-3-sulfopropylamino)phenol and couple this compound with the diazotate from 2-amino-5-bromopyridine in cooled MeOH, recrystallize the reagent from water (mp >250°) (Clin.Chim.Acta 1982, 120, 127).)

HPLC VARIABLES

Column: 250 × 4 7 μm LiChrosorb RP-18

Mobile phase: MeOH:water 65:35 containing 50 μM 2-(5-bromo-2-pyridylazo)-5-(N-propyl-N-sulfopropylamino)phenol, 100 mM LiCl, and 10 mM pH 3.5 sodium acetate buffer

Flow rate: 0.6

Injection volume: 20

Detector: UV 575

CHROMATOGRAM

Retention time: 5 (V), 8 (Cu), 27 (Ni)

Limit of detection: 26-90 ng/L

KEY WORDS

complexation

REFERENCE

Shijo, Y.; Sato, H.; Uehara, N.; Aratake, S. Simultaneous determination of trace amounts of copper, nickel and vanadium in sea-water by high-performance liquid chromatography after extraction and back-extraction, *Analyst*, **1996**, *121*, 325−328.

Chlorophosphonazo III

SAMPLE

Matrix: solutions

Analyte: metals

Sample preparation: Inject a 50 μL aliquot of a 5 μg/mL solution.

HPLC VARIABLES

Column: 150 × 4.6 Nitrilotriacetate chelating resin (Prepare chelating resin as follows. Add 5 g GMA gel (cross-linked glycidyl methacrylate gel, Hitachi) to 100 mL 25 mM lysine-N^α,N^α-diacetic acid in water, adjust pH to 11, heat at 50° for 24 h, filter wash the beads with 2 M HCl, wash with water, dry. Before use wash with 50 mM nitric acid, wash with water, wash with MeOH, wash with water, slurry pack in a 150 × 4.6 column. Prepare lysine-N^α,N^α-diacetic acid as follows. Reflux 695 g lysine hydrochloride in 7 L 2 M anhydrous HCl in MeOH for 3 h, cool to -30°, filter to obtain methyl lysinate dihydrochloride, wash the crystals with 1 L MeOH. Reflux 2 moles methyl lysinate dihydrochloride with 4.35 moles sodium methoxide in 8 L MeOH for 4 h, add 23.4 g ammonium chloride, allow to stand overnight, filter, evaporate to remove the solvent, extract the thick residue with 500 mL boiling dimethoxyethane. Filter this extract and remove the solvent by evaporation, take up the residue in EtOH, acidify with HCl in EtOH, cool in ice for 20 min, filter, wash the solid with EtOH, dry under vacuum at 60° to give α-aminocaprolactam, recrystallize from toluene (mp 300-305° d (L enantiomer); mp 75-79.5° (DL)) (J. Org. Chem. 1979, 44, 4841). α-Aminocaprolactam is also available from Tokyo Kasei (TCI America, Portland OR). React α-aminocaprolactam with sodium chloroacetate in alkaline aqueous solution, adjust pH to 1.8, let stand at room temperature, filter to obtain α-aminocaprolactam-N^α,N^α-diacetic acid. Reflux 61 g α-aminocaprolactam-N^α,N^α-diacetic acid with 300 mL 6 M KOH, cool, acidify to pH 1.8 with 70% perchloric acid, filter to remove potassium perchlorate, concentrate the filtrate to give lysine-N^α,N^α-diacetic acid (Chem. Lett. 1990, 693).)

Mobile phase: Gradient. A was 20 mM nitric acid. B was 80 mM nitric acid. A:B from 100:0 to 0:100 over 25 min

Column temperature: 40

Flow rate: 1

Injection volume: 50

Detector: UV 660 following post column reaction. The column effluent mixed with 50 μg/mL chlorophosphonazo III in 20 mM nitric acid and the mixture flowed to the detector. (Chlorophosphonazo III (3,6-bis[(4-chloro-2-phosphonophenyl)azo]-4,5-dihydroxy-2,7-naphthalenedisulfonic acid) is commercially available from Dojindo Chemical Laboratories, Kumamoto, Japan (Dojindo Molecular Technologies, 3 Bethesda Metro Center, Suite 700, Bethesda MD 20814; (301) 664 8448; www.dojindo.co.jp). It can also be prepared as follows. Diazotize 5-chloro-2-nitroaniline with sodium nitrite at 0° in the presence of tetrafluoroboric acid to give 5-chloro-2-nitrobenzenediazonium tetrafluoroborate (Buehler, C.A.; Pearson, D.E. Survey of Organic Syntheses, Wiley:New York, 1970, pp. 345-347). Add 25 g cuprous chloride to 635 mL ethyl acetate and 155 mL phosphorus trichloride, add 118.5 g 5-chloro-2-nitrobenzenediazonium tetrafluoroborate in portions, at the same time add 25 g cuprous chloride in portions, keep below 60° for 3 h, add 10 mL water at 15-35°, heat at 50-65° for 1 h, add 590 mL water at 50°, adjust pH to 4.5 with sodium carbonate, heat to 45°, filter to remove a copper complex, treat the filtrate to recover more complex. Treat the complex with hydrogen sulfide in 1:2 HCl (Caution! Hydrogen sulfide is highly toxic!), filter, evaporate, adjust pH to 3 with sodium carbonate, filter to obtain 2-amino-5-chlorobenzenephosphonic acid as a colorless amorphous solid (Yield = 11 g) (Chem. Abstr. 1962, 56, 1474f). Add 2.1 g sodium nitrite in 10 mL water to 6.2 g 2-amino-5-chlorobenzenephosphonic acid, 5 mL concentrated HCl, and 20 mL water stirred at -5°, add dropwise a solution of 3.65 g 4,5-dihydroxynaphthalene-2,7-disulfonic acid disodium salt dihydrate (chromotropic acid) in 20 mL water and 5 mL pyridine, add dropwise a suspension of 4 g finely powdered calcium oxide in 20 mL water and 5 mL pyridine, stir at -3 to -5° for 30-60 min, acidify with HCl below 0°, let stand at 0° overnight, filter to obtain chlorophosphonazo III (Fr. Pat. 1 474 471 (1967); Chem. Abstr. 1967, 67, 73688y).

CHROMATOGRAM
Retention time: 3 (La), 4.2 (Ce), 5.7 (Pr), 7 (Nd), 10 (Sm), 10.7 (Eu, Gd), 12 (Tb), 12.9 (Dy), 13.4 (Ho), 14.4 (Er), 16 (Tm), 18 (Yb), 19 (Lu)

KEY WORDS
post-column reaction; complexation

REFERENCE
Inoue, Y.; Kumagai, H.; Shimomura, Y.; Yokoyama, T.; Suzuki, T.M. Ion chromatographic separation of rare-earth elements using a nitrilotriacetate-type chelating resin as the stationary phase, *Anal.Chem.*, **1996**, *68*, 1517–1520.

DBF-Arsenazo

SAMPLE
Matrix: solutions
Analyte: metals
Sample preparation: Mix an aliquot of a solution with 2 mL pH 4.7 acetate buffer and 2 mL 0.1% DBF-Arsenazo (2-(2-arsenophenylazo)-1,8-dihydroxy-7-(2,6-dibromo-4-fluorophenylazo)naphthalene-3,6-disulfonic acid) in water, make up to 10 mL, heat on a boiling water bath for 3 min, cool to room temperature, inject a 20 μL aliquot.

HPLC VARIABLES
Column: 150 × 4.6 5 μm Shimadzu RP-18
Mobile phase: MeOH:water 35:65 containing 2 mM tetrabutylammonium bromide and 64 mM acetate buffer, pH 4.7
Column temperature: 35
Flow rate: 1
Injection volume: 20
Detector: UV 630

CHROMATOGRAM
Retention time: 6 (rare earths), 10 (thorium), 12 (chromium)
Limit of detection: 2-10 ng/mL

KEY WORDS
complexation

REFERENCE
Zhang, X.; Wang, M.; Cheng, J. Determination of total rare earth elements, thorium and chromium with 2-(2-arsenophenylazo)-1,8-dihydroxy-7-(2,6-dibromo-4-fluorophenylazo)naphthalene-3,6-disulfonic acid by reversed-phase ion-pair liquid chromatography, *Anal.Chim.Acta*, **1990**, *237*, 311–315.

2,6-Diacetylpyridine Bis(benzoylhydrazone)

RELATED REFERENCE
Martin-Daguet, V.; Gasnier, P.; Caude, M. Packed-column supercritical fluid chromatography: Quantitative determination of uranium without liquid waste generation. *Anal.Chem.* **1997**, *69*, 536–541.

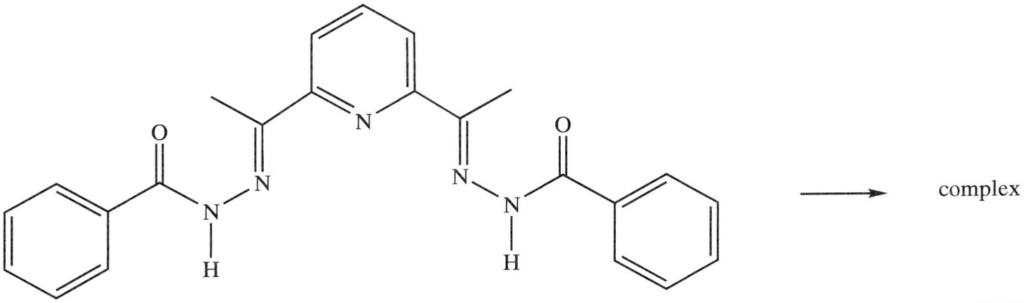

SAMPLE
Matrix: solutions
Analyte: uranyl salts
Sample preparation: Mix equimolar amounts of the uranyl salts in MeOH solution and 2,6-diacetylpyridine bis(benzoylhydrazone) in EtOH solution, collect the precipitate, dissolve the precipitate in dichloromethane, inject a 5-10 µL aliquot. (Synthesis of 2,6-diacetylpyridine bis(benzoylhydrazone) is as follows. Reflux benzoic hydrazide and 2,6-diacetylpyridine in a 2:1 molar ratio in EtOH for 3 h, cool, recrystallize from MeOH to obtain 2,6-diacetylpyridine bis(benzoylhydrazone) as white-yellow microcrystals (mp 212°) (J. Chem. Soc. Dalton 1983, 721).)

HPLC VARIABLES
Column: 250 × 4 10 µm LiChrosorb RP2
Mobile phase: MeOH:water 60:40
Flow rate: 2
Injection volume: 5-10
Detector: UV 265

CHROMATOGRAM
Retention time: 4
Limit of quantitation: 5 ng

OTHER SUBSTANCES
Non-interfering: copper

KEY WORDS
complexation

REFERENCE
Casoli, A.; Mangia, A.; Predieri, G. Determination of uranium by reversed-phase high-performance liquid chromatography, *Anal.Chem.*, **1985**, *57*, 561–563.

2,6-Diacetylpyridine Bis(N-methylenepyridiniohydrazone) Dichloride

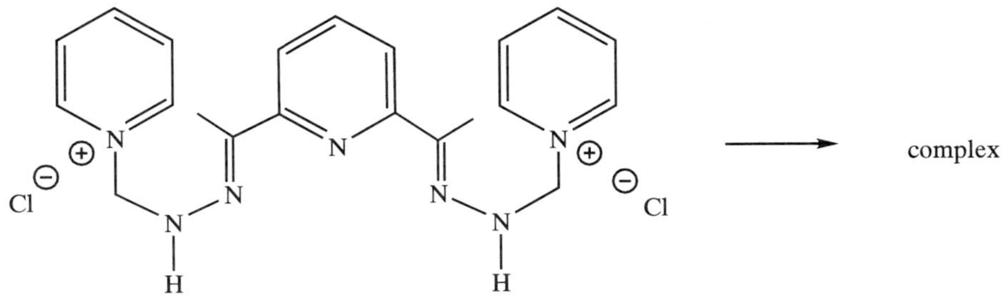

SAMPLE

Matrix: solutions

Analyte: metals

Sample preparation: Adjust pH to 7.0 with ammonium acetate and NaOH, add enough 2,6-diacetylpyridine bis(N-methylenepyridiniohydrazone) dichloride to form a 1-5 mM solution, let stand at pH 7.0 for 10 min, inject an aliquot. (If Ti(IV) or U(VI) are present let stand at pH 2-3 for 30 min before increasing the pH to 2-3. Synthesis of 2,6-diacetylpyridine bis(N-methylenepyridiniohydrazone) dichloride is as follows. Reflux 12 mmoles Girard's Reagent P and 6 mmoles 2,6-diacetylpyridine in 250 mL EtOH for 3 h, recrystallize the resulting white precipitate from EtOH:water 85:15 to obtain 2,6-diacetylpyridine bis(N-methylenepyridiniohydrazone) dichloride (Anal. Chem. 1989, 61, 1272).)

HPLC VARIABLES

Column: 150 × 4.6 5 μm PLRP-S polystyrene-divinylbenzene (Polymer Laboratories)

Mobile phase: Gradient. A was MeCN:water 10:90 containing 20 mM ammonium acetate and 50 mM sodium perchlorate. B was MeCN:water 40:60 containing 20 mM ammonium acetate and 50 mM sodium perchlorate. A:B from 95:5 to 30:70 over 18 min (X) or from 82:18 to 70:30 over 10 min, to 0:100 (step gradient), maintain at 0:100 for 4 min (Y). (Mobile phase components were prepared from 1 M pH 7.0 ammonium acetate containing 1 M sodium perchlorate.)

Flow rate: 1

Injection volume: 10

Detector: UV 340

CHROMATOGRAM

Retention time: 3 (Sn (X)), 5 (Ti (X)), 8 (Mn (X)), 9 (Co (X)), 10 (Sb (X)), 12 (Cu (X)), 13 (Ni (X)), 19 (U (X)), 7 (Sm (Y)), 8 (Nd (Y)), 8.5 (Pr (Y)), 9 (Ce (Y)), 10 (La (Y))

Limit of detection: 0.2-5 μM

KEY WORDS

complexation

REFERENCE

Main, M.V.; Fritz, J.S. Chromatographic determination of metal chelates of 2,6-diacetylpyridine bis(N-methylenepyridiniohydrazone), *Anal.Chim.Acta*, **1990**, *229*, 101–106.

4,5-Dihydroxybenzene-1,3-disulfonic Acid

(structure: benzene ring with SO₃H at top, HO at left, OH at bottom, SO₃H at lower right) $\longrightarrow$ complex

SAMPLE

Matrix: solutions
Analyte: metals
Sample preparation: Mix 1 mL 15 mM 4,5-dihydroxybenzene-1,3-disulfonic acid disodium salt solution with 3 mL 100 mM pH 4.0 sodium acetate buffer, add 50 μL of a solution containing metals, make up to 10 mL with water, let stand for 10 min, inject a 100 μL aliquot.

HPLC VARIABLES

Guard column: Waters no. 88141
Column: 300 × 3.9 μBondapak C18
Mobile phase: MeOH:buffer 63:37 (Prepare mobile phase by mixing 315 mL MeOH, 50 mL 15 mM 4,5-dihydroxybenzene-1,3-disulfonic acid disodium salt solution, 4.9805 g tetrabutylammonium bromide, and 7.5 mL 100 mM pH 4.0 sodium acetate buffer, make up to 500 mL with water.)
Flow rate: 0.7
Injection volume: 100
Detector: UV 315

CHROMATOGRAM

Retention time: 9.4 (Mo(VI)), 12.8 (Zr(IV))
Limit of detection: 3.6-9 ppb

OTHER SUBSTANCES

Non-interfering: calcium, chromium, cobalt, copper, iron, magnesium, nickel, niobium, selenium, tantalum, titanium, zinc

KEY WORDS

complexation

REFERENCE

Tsai, S.J.J.; Yan, H.T. Determination of zirconium and molybdenum with 4,5-dihydroxybenzene-1,3-disulfonic acid disodium salt by ion-pair reversed-phase high-performance liquid chromatography, *Analyst*, **1993**, *118*, 521–527.

1-(2,4-Dihydroxy-1-phenylazo)-8-hydroxy-3,6-naphthalenedisulfonate

SAMPLE
Matrix: solutions
Analyte: beryllium(II) ion
Sample preparation: Add 5 mL 10 mM pH 6 disodium EDTA to <40 mL of a slightly acidic solution, heat in a boiling water bath for 25 min, add 1 mL 1 mM reagent, add 5 mL 100 mM pH 8.5 Tris-HCl buffer, heat in a boiling water bath for 25 min, cool, make up to 50 mL, inject a 100 μL aliquot. (Reagent was 1-(2,4-dihydroxy-1-phenylazo)-8-hydroxy-3,6-naphthalenedisulfonate (H-resorcinol) (Tokyo Kasei (TCI America, 911 North Harborgate St., Portland, OR 97203; 800-423-8616, 503-283-1681, (fax) 503-283-1987; www.tciamerica.com)).)

HPLC VARIABLES
Column: 150 × 4.6 5 μm Asahipak ODP-50 C18 bonded poly(vinyl alcohol) copolymer
Mobile phase: MeOH:water 35:65 (w/w) containing 25 mM Tris-HCl, 13 mM tetrabutyl-ammonium bromide, and 100 μM disodium EDTA, pH 8.8
Flow rate: 0.5
Injection volume: 100
Detector: UV 500

CHROMATOGRAM
Retention time: 8
Limit of quantitation: 23 pg/mL
Limit of detection: 7.2 pg/mL

OTHER SUBSTANCES
Simultaneously analyzed: aluminum, iron

KEY WORDS
complexation

REFERENCE
Hoshino, H.; Nomura, T.; Nakano, K.; Yotsuyanagi, T. Complexation of Be(II) ion with 1-(2,4-dihydroxy-1-phenylazo)-8-hydroxy-3,6-naphthalenedisulfonate as the chemical basis of the selective detection of utratrace Be by reversed-phase high-performance liquid chromatography, *Anal.Chem.*, **1996**, *68*, 1960–1965.

EDTA

RELATED REFERENCE
Conradi, S.; Vogt, C.; Wittrisch, H.; Knobloch, G.; Werner, G. Capillary electrophoretic separation of metal ions using complex forming equilibria of different stabilities. *J.Chromatogr.A* **1996**, *745*, 103–109.

M^{+2} + EDTA ⟶ complex

SAMPLE
Matrix: solutions
Analyte: metals

CAPILLARY ELECTROPHORESIS
Capillary: 55 cm × 75 μm fused-silica (Scientific Glass Engině)
Capillary preparation: Rinse capillary with running buffer for 2 min before each run. At the beginning of each day rinse with 100 mM NaOH for 5 min, with water for 5 min, and with running buffer for 5 min.
Running buffer: 10 mM pH 4.8 Sodium acetate containing 2 mM tartaric acid and 0.2 mM tetradecyltrimethylammonium bromide
Voltage/Current: 30 kV
Injection: Hydrostatic injection of 1 mM EDTA at 10 cm for 30 s at the cathode and hydrostatic injection of the sample at 10 cm for 30 s at the anode
Detector: UV 242
Model: Laboratory constructed
Migration time: 4.5 (Cu^{++}), 5 (Pb^{++}), 5.5 (Ni^{++}), 8 (Fe^{++})
Limit of detection: 0.5-3 ppm (S/N 3)

KEY WORDS
complexation; cobalt can be determined under different conditions; derivatization occurs in the capillary

REFERENCE
Haumann, I.; Bächmann, K. On-column chelation of metal ions in capillary zone electrophoresis, *J.Chromatogr.A*, **1995**, *717*, 385–391.

N,N'-Ethylenebis(salicylaldimine)

SAMPLE
Matrix: ore
Analyte: metals
Sample preparation: Gently heat 1 g ore, 60 mL concentrated HCl and 40 mL 65% nitric acid to near dryness, add 30 mL 65% nitric acid, heat to near dryness, dissolve the residue in 100 mM nitric acid, filter, adjust the volume to 50 mL, adjust the pH to 6. Remove a 5 mL aliquot and add it to 2 mL 500 mM pH 6 sodium acetate buffer, add 2 mL 1% N,N'-ethylenebis(salicylaldimine) in MeOH, heat on a water bath for 15 min, cool to room temperature, add 2 mL chloroform, mix for 2-3 min. Remove a 1 mL aliquot of the organic layer and evaporate it to dryness, reconstitute with 1 mL MeOH, inject a 5 μL aliquot. (Prepare N,N'-ethylenebis(salicylaldimine by condensing 2 equivalents of salicylaldehyde with 1 equivalent of ethylenediamine in EtOH.)

HPLC VARIABLES
Column: 150 × 4.6 3 μm Hypersil ODS
Mobile phase: MeCN:MeOH:water 60:20:20

Flow rate: 0.6
Injection volume: 5
Detector: UV 260

CHROMATOGRAM
Retention time: 7 (uranyl), 8 (Pt(II)), 9 (Fe(II)), 11 (Pd(II)), 12 (Ni(II)), 13 (Cu(II))
Limit of detection: 2.5 µg/mL

KEY WORDS
complexation

REFERENCE
Khuhawar, M.Y.; Lanjwani, S.N. Simultaneous solvent extraction and high-performance liquid chromatographic determination of uranium, iron, nickel and copper in mineral ore samples and phosphate rock residues using N,N'-ethylenebis(salicylaldimine) as complexing reagent, *J.Chromatogr.A*, **1996**, *740*, 296–301.

2-Hydroxyisobutyric Acid

RELATED REFERENCES
Conradi, S.; Vogt, C.; Wittrisch, H.; Knobloch, G.; Werner, G. Capillary electrophoretic separation of metal ions using complex forming equilibria of different stabilities. *J.Chromatogr.A* **1996**, *745*, 103-109.

Hao, F.; Paull, B.; Haddad, P.R. Determination of trace levels of thorium(IV) and uranyl by reversed-phase chromatography with on-line preconcentration and ligand exchange. *J.Chromatogr.A* **1996**, *749*, 103-113.

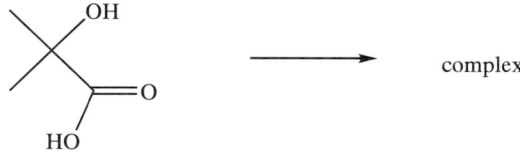 complex

SAMPLE
Matrix: solutions
Analyte: dysprosium
Sample preparation: Inject an aliquot of a 16 µg/mL solution of dysprosium chloride in mobile phase or a 30 mg/mL solution of Dy(HP-DO3A) in mobile phase.

HPLC VARIABLES
Column: 150 × 3.9 5 µm Delta-Pak HPI C18
Mobile phase: 10 mM pH 5.5 Sodium 1-octanesulfonate containing 60 mM 2-hydroxyisobutyric acid (PEEK tubing was used throughout.)
Flow rate: 1
Injection volume: 20 (titanium loop)
Detector: UV 658 following post-column reaction. The column effluent mixed with reagent pumped at 0.6 mL/min and the mixture flowed through a Dionex Reagent Delivery Module with an IonPac membrane reactor (48 cm reactor) to the detector. (Reagent was 31 mM acetic acid containing 5 mM urea and 65 µM Arsenazo III.)

CHROMATOGRAM
Retention time: 3.7

KEY WORDS
post-column reaction; complexation

REFERENCE
Troskosky, J.A.; Katona, T.; Zodda, J.P.; Eakins, M.N. Determination of trace levels of Dy^{3+} in Dy(HP-DO3A) by ion-pair liquid chromatography with post-column reaction, *J.Pharm.Biomed.Anal.*, **1995**, *13*, 1421–1426.

Lumogallion

SAMPLE
Matrix: blood
Analyte: aluminum
Sample preparation: 500 µL Serum + 500 µL 1.2 M trichloroacetic acid in water, vortex for 1 min, centrifuge at 1500 g for 10 min. Remove the supernatant, add 500 µL 300 mM trichloroacetic acid in water to the residue, vortex vigorously for 2 min, centrifuge at 1500 g for 10 min. Combine the supernatants, add 1 mL 200 mM potassium hydrogen phthalate in water, adjust the pH to 4.7 with 1 M NaOH, add 250 µL 200 mM sodium thiosulfate in water, mix, let stand for 15 min, add 250 µL 3 mM lumogallion in water, make up to 5 mL with water, vortex vigorously for 2 min, let stand at room temperature for 15 min, inject a 100 µL aliquot. (Lumogallion is 4-chloro-3-(2,4-dihydroxyphenylazo)-2-hydroxybenzene-1-sulfonic acid, obtainable from Tokyo Kasei (TCI America, 911 North Harborgate St., Portland, OR 97203; 800-423-8616, 503-283-1681, (fax) 503-283-1987; www.tciamerica.com).)

HPLC VARIABLES
Column: 250 × 4.6 10 µm LiChrosorb RP18
Mobile phase: MeCN:buffer 30:70 (Prepare buffer by adding 2.04 g potassium hydrogen phthalate and 1.7 mg lumogallion to 500 mL water, adjust pH to 4.2 with 1 M NaOH.)
Flow rate: 1
Injection volume: 100
Detector: F ex 505 em 574

CHROMATOGRAM
Retention time: 5.5
Limit of detection: 2.2 ng/mL

KEY WORDS
complexation; serum

REFERENCE
Wu, J.; Zhou, C.Y.; Chi, H.; Wong, M.K.; Lee, H.K.; Ong, H.Y.; Ong, C.N. Determination of serum aluminium using an ion-pair reversed-phase high-performance liquid chromatographic-fluorimetric system with lumogallion, *J.Chromatogr.B*, **1995**, *663*, 247–253.

SAMPLE
Matrix: blood, urine
Analyte: aluminum
Sample preparation: Serum. 80 µL Serum + 80 µL 830 mM perchloric acid, vortex for 30 s, sonicate for 15 min, centrifuge at 15800 g for 4 min. Remove a 75 µL aliquot of the supernatant and add it to 30 µL freshly-prepared 300 µM lumogallion in 200 mM potassium hydrogen phthalate containing 1 M NaOH, mix, let stand for >1 h, inject a 5 µL aliquot. Urine. 50 µL Urine + 50 µL 200 mM potassium hydrogen phthalate containing 2 M NaOH, mix, let stand for 10 min, add 1 mL MeOH, mix, let stand at -20° for 1 h, centrifuge at 15800 g for 4 min, discard the liquid phase. Dissolve the precipitate in 100 µL 830 mM perchloric acid, add 200 µL freshly-prepared 150 µM lumogallion in 100 mM potassium hydrogen phthalate containing 0.5 M NaOH, mix, let stand for >1 h, inject a 5 µL aliquot. (Lumogallion is 4-chloro-3-(2,4-dihydroxyphenylazo)-2-hydroxybenzene-1-sulfonic acid, obtainable from Tokyo Kasei (TCI America, 911 North Harborgate St., Portland, OR 97203; 800-423-8616, 503-283-1681, (fax) 503-283-1987; www.tciamerica.com).)

HPLC VARIABLES
Guard column: C18
Column: 150 × 3.9 4 μm Nova-Pak C18
Mobile phase: MeCN:200 mM potassium hydrogen phthalate:water 22:10:68, pH 4.8
Flow rate: 1
Injection volume: 5
Detector: F ex 500 em 575

CHROMATOGRAM
Retention time: 3.5
Limit of detection: 0.5 ng/mL (serum), 2 ng/mL (urine)

KEY WORDS
complexation; serum

REFERENCE
Lee, B.-L.; Chua, L.-H.; Ong, H.-Y.; Yang, H.G.; Wu, J.; Ong, C.-N. Determination of serum and urinary aluminum by HPLC with fluorometric detection of Al-lumogallion complex, *Clin.Chem.*, **1996**, *42*, 1405–1411.

SAMPLE
Matrix: water
Analyte: aluminum
Sample preparation: 300 mL Filtered water + 50 μL HCl (super-pure), evaporate to 5 mL, filter, dilute the filtrate to 10 mL with water. Remove a 200 μL aliquot and add it to 250 μL 0.1% lumogallion in water, make up to 5 mL with running buffer, heat at 100° for 15 min, cool, inject an aliquot. (Lumogallion is 4-chloro-3-(2,4-dihydroxyphenylazo)-2-hydroxybenzene-1-sulfonic acid, obtainable from Tokyo Kasei (TCI America, 911 North Harborgate St., Portland, OR 97203; 800-423-8616, 503-283-1681, (fax) 503-283-1987; www.tciamerica.com).)

CAPILLARY ELECTROPHORESIS
Capillary: 50 cm × 50 μm fused-silica (50 cm to detector) (Polymicro Technologies)
Capillary preparation: Rinse with running buffer at 2 bar for 15 min.
Running buffer: 40 mM Acetic acid containing 10 mM ammonium acetate, pH 4.0
Voltage/Current: 15 kV
Injection: Gravity injection at 15 cm for 10 s.
Detector: F ex 491 em 576
Migration time: 4
Limit of detection: 19 ppb

KEY WORDS
comparison with ICP-MS; complexation

REFERENCE
He, H.-B.; Lee, H.-K.; Li, S.F.Y.; Hsieh, A.-K.; Chi, H.; Siow, K.-S. Determination of trace levels of aluminum by capillary electrophoresis with lumogallion fluorometric detection, *J.Chromatogr.Sci.*, **1997**, *35*, 333–326.

5,5'-Methylenedisalicylohydroxamic Acid

SAMPLE
Matrix: solutions

Analyte: titanium(IV)

Sample preparation: Homogenize 200 μL of a solution of titanium in 5 mM sulfuric acid with 4.8 mL 10 mM 5,5'-methylenedisalicylohydroxamic acid in MeOH, filter (0.2 μm), inject a 20 μL aliquot of the filtrate. (Preparation of 5,5'-methylenedisalicylohydroxamic acid is as follows. Heat 50 g salicylic acid, 18.3 g 34% formaldehyde, and 281.5 g 25% sulfuric acid at 90-95° for 40 min, wash with benzene (Caution! Benzene is a carcinogen!), purify to obtain 5,5'-methylenedisalicylic acid (mp 243°) (Ann. chim. farm. May 1940, 35 (Chem. Abs. 1940, 34, 7883[7])). Slowly add 30 mL concentrated sulfuric acid to 90 g 5,5'-methylenedisalicylic acid in 400 mL EtOH, reflux for 3-4 h, add 100 mL benzene. Remove water by distilling 125 mL of azeotropic EtOH/water/benzene, repeat this process 5 times, neutralize with sodium bicarbonate to obtain crude ethyl 5,5'-methylenedisalicylate, extract with benzene, recrystallize from EtOH to obtain ethyl 5,5'-methylenedisalicylate (mp 100-103°). Slowly add 5 g ethyl 5,5'-methylenedisalicylate to a stirred solution of 3.5 g hydroxylamine hydrochloride in 50 mL 12% NaOH, let stand overnight, add 2 M HCl to precipitate product, recrystallize from water and petroleum ether/ethyl acetate to obtain 5,5'-methylenedisalicylohydroxamic acid as a colorless powder (mp 169-171°) (Proc. Indian Acad. Sci. (Chem. Sci.) 1982, 91, 399).)

HPLC VARIABLES

Guard column: 20 × 4.6 Polyspher guard column
Column: 150 × 4.6 Polyspher RP18 (Merck)
Mobile phase: MeOH:5 mM sulfuric acid 96:4 containing 1.5 mM 5,5'-methylenedisalicylohydroxamic acid
Flow rate: 0.3
Injection volume: 20
Detector: UV 360

CHROMATOGRAM

Retention time: 3.8
Limit of detection: 12-18 ng/mL

OTHER SUBSTANCES

Non-interfering: calcium, chlorate, chloride, copper, iron, lithium, magnesium, molybdenum, nitrate, phosphate, potassium, sodium, vanadium

KEY WORDS

complexation

REFERENCE

Bagur, G.; Sánchez-Viñas, M.; Gázquez, D. Determination of titanium(IV) as an additive in organic matrices by reversed-phase high-performance liquid chromatography with 5, 5'-methylenedisalicylohydroxamic acid, *J.Chromatogr.Sci.*, **1997**, *35*, 131–134.

Nitrilotriacetic Acid

SAMPLE

Matrix: blood
Analyte: iron
Sample preparation: 900 μL Serum + 100 μL 800 mM pH 7 nitrilotriacetic acid, let stand at 25° for 15 min, filter (Amicon Centriflo CF 25), inject a 20 μL aliquot of the ultrafiltrate. (The nitrilotriacetic acid complexes free iron but not iron bound to transferrin, ferritin, or deferoxamine. Bound iron is left behind on the filter and the free iron becomes complexed with the 3-hydroxy-1-propyl-2-methylpyridin-4-one in the mobile phase.)

HPLC VARIABLES
Column: 100 × 3 5 μm Chrom-Spher glass column (Chrompack)
Mobile phase: MeCN:5 mM pH 7 MOPS containing 3 mM 3-hydroxy-2-methyl-1-propyl-pyridin-4-one (Synthesis of 3-hydroxy-2-methyl-1-propylpyridin-4-one is as follows. Add 22.2 g 3-hydroxy-2-methyl-4-pyrone in 225 mL MeOH to 7.5 g NaOH in 25 mL water, add 25.5 g benzyl chloride, reflux for 6 h, allow to cool overnight, evaporate under reduced pressure to remove MeOH, add 50 mL water to the residue, extract three times with 25 mL portions of dichloromethane. Combine the organic layers and wash them twice with 25 mL portions of 5% NaOH, wash twice with 25 mL portions of water, dry over anhydrous magnesium sulfate, evaporate, distil under nitrogen under reduced pressure to obtain 3-benzyloxy-2-methyl-4-pyrone. Dissolve 4.8 g 3-benzyloxy-2-methyl-4-pyrone and 1.36 g propylamine in 200 mL water, add 2 g NaOH dissolved in 100 mL EtOH, stir at room temperature for 6 days, acidify to pH 2 with concentrated HCl, evaporate to dryness, wash the resulting solid with water, extract twice with 50 mL portions of chloroform. Combine the extracts and dry them over anhydrous magnesium sulfate, evaporate to obtain 3-benzyloxy-2-methyl-1-propylpyridin-4-one. Add 2 g 3-benzyloxy-2-methyl-1-pro-pylpyridin-4-one to 10 mL concentrated hydrobromic acid, heat on a steam bath for 30 min, recrystallize from water to obtain 3-hydroxy-2-methyl-1-propylpyridin-4-one (mp 182-183°) (US Pat. 4 550 101).)
Flow rate: 0.8
Injection volume: 20
Detector: UV 450

CHROMATOGRAM
Retention time: 3.1
Limit of quantitation: 2 μM

KEY WORDS
complexation; serum; ultrafiltrate; pharmacokinetics

REFERENCE
Singh, S.; Hider, R.C.; Porter, J.B. A direct method for quantification of non-transferrin-bound iron, *Anal.Biochem.*, **1990**, *186*, 320–323.

1,10-Phenanthroline

RELATED REFERENCE
Xu, J.; Che, P.; Ma, Y. More sensitive way to determine iron using an iron(II)-1,10-phenanthroline complex and capillary electrophoresis. *J.Chromatogr.A* **1996**, *749*, 287-294.

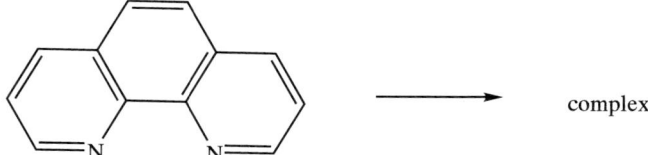

complex

SAMPLE
Matrix: solutions
Analyte: metals
Sample preparation: Add a small excess of 1,10-phenanthroline to an aqueous solution of the metal salt, dilute if necessary, inject an aliquot.

HPLC VARIABLES
Column: 250 × 4.6 10 μm μPartisil-SCX
Mobile phase: MeCN:water 48:52 containing 100 μM 1,10-phenanthroline and 15.8 mM lithium perchlorate
Flow rate: 0.5-8
Detector: UV 265

CHROMATOGRAM
Retention time: 16.9 (cobalt), 17.7 (nickel), 18.1 (zinc), 19.1 (ruthenium), 21.0 (iron), 25.6 (copper), 26.0 (cadmium) (Values are retention volumes in mL.)

KEY WORDS
complexation

REFERENCE
O'Laughlin, J.W. Separation of cationic metal chelates of 1,10-phenanthroline by liquid chromatography, *Anal.Chem.*, **1982**, *54*, 178–181.

SAMPLE
Matrix: solutions
Analyte: iron
Sample preparation: Prepare a 1 mM solution in the reagent, inject an aliquot. (Prepare the reagent by mixing equal volumes of 20 mM EDTA in water and 20 mM o-phenanthroline in EtOH.)

CAPILLARY ELECTROPHORESIS
Capillary: 38.5 cm × 50 μm fused-silica (30 cm to detector)
Running buffer: 100 mM pH 9.2 sodium tetraborate containing 200 μM EDTA and 200 μM o-phenanthroline
Capillary temperature: 25
Voltage/Current: 30 kV
Injection: Pressure injection at 50 mbar for 2-5 s (4.5-11.3 nL)
Detector: UV 200
Model: Hewlett-Packard HP3D
Migration time: 0.6 (Fe(II)), 2 (Fe(III))
Limit of detection: 0.3 ppm (Fe(II)), 0.6 ppm (Fe(III))

KEY WORDS
complexation

REFERENCE
Schäffer, S.; Gareil, P.; Dezael, C.; Richard, D. Direct determination of iron(II), iron(III) and total iron as UV-absorbing complexes by capillary electrophoresis, *J.Chromatogr.A*, **1996**, *740*, 151–157.

N,N'-o-Phenylenebis(3,5-dichlorosalicylaldimine)

SAMPLE
Matrix: solutions
Analyte: metals

Sample preparation: 1 mL Sample + 1 mL 10 mM reagent in DMF + 2 mL 1 M pH 5.0 sodium acetate buffer, make up to 50 mL with water and DMF so that the DMF concentration is 50%, heat at 60° for 30 min, cool, inject a 100 μL aliquot. (Prepare the reagent N,N'-o-phenylenebis(3,5-dichlorosalicylaldimine), by mixing 20 mL 1 M o-phenylenediamine in EtOH and 40 mL 1 M 3,5-dichlorosalicylaldehyde in EtOH, reflux for 1 h, cool, collect the precipitate by filtration and recrystallize it from hot EtOH to give the reagent.)

HPLC VARIABLES
Column: 50 × 4.6 3 μm Shim-pack FLC-ODS (Shimadzu)
Mobile phase: MeCN:water 65:35 containing 5 mM sodium acetate and 100 μM disodium EDTA
Flow rate: 0.8
Injection volume: 100
Detector: UV 376

CHROMATOGRAM
Retention time: 5 (Cu^{2+}), 11 (Ni^{2+})
Limit of quantitation: 100 nM (Cu), 40 nM (Ni)

KEY WORDS
complexation; comparison with other complexing agents

REFERENCE
Kanbayashi, M.; Hoshino, H.; Yotsuyanagi, T. Highly selective determination of trace amounts of copper(II), nickel(II) and vanadium(V) ions with tetradentate Schiff-base ligands by reversed-phase high-performance liquid chromatography and spectrophotometric determination, *J.Chromatogr.*, **1987**, *386*, 191–197.

N,N'-o-Phenylenebis(4-diethylaminosalicylaldimine)

SAMPLE
Matrix: solutions
Analyte: metals
Sample preparation: 1 mL Sample + 1 mL 10 mM reagent in DMF + 2 mL 1 M pH 5.0 sodium acetate buffer, make up to 50 mL with water and DMF so that the DMF concentration is 50%, heat at 60° for 30 min, cool, inject a 100 μL aliquot. (Prepare the reagent N,N'-o-phenylenebis(4-diethylaminosalicylaldimine), by mixing 20 mL 1 M o-phenylenediamine in EtOH and 40 mL 1 M 4-diethylaminosalicylaldehyde in EtOH, reflux for 1 h, cool, collect the precipitate by filtration and recrystallize it from hot EtOH to give the reagent.)

HPLC VARIABLES
Column: 50 × 4.6 3 μm Shim-pack FLC-ODS (Shimadzu)
Mobile phase: MeCN:water 65:35 containing 5 mM sodium acetate, 100 μM disodium EDTA, and 6.2 mM tetra-n-butylammonium bromide
Flow rate: 0.6
Injection volume: 100
Detector: UV 448

CHROMATOGRAM
Retention time: 7 (V^{5+}), 11 (Cu^{2+}), 12 (Ni^{2+})
Limit of quantitation: 10 nM (V), 40 nM (Cu), 40 nM (Ni)
Limit of detection: 0.3 ppb (V)

OTHER SUBSTANCES
Non-interfering: Al, Cd, Cr, Fe, Mn, Pb

KEY WORDS
complexation; comparison with other complexing agents

REFERENCE
Kanbayashi, M.; Hoshino, H.; Yotsuyanagi, T. Highly selective determination of trace amounts of copper(II), nickel(II) and vanadium(V) ions with tetradentate Schiff-base ligands by reversed-phase high-performance liquid chromatography and spectrophotometric determination, *J.Chromatogr.*, **1987**, *386*, 191–197.

N,N'-o-Phenylenebis(salicylaldimine)

SAMPLE
Matrix: solutions
Analyte: metals
Sample preparation: 1 mL Sample + 1 mL 10 mM reagent in DMF + 2 mL 1 M pH 5.0 sodium acetate buffer, make up to 50 mL with water and DMF so that the DMF concentration is 50%, heat at 60° for 30 min, cool, inject a 100 μL aliquot. (Prepare the reagent N,N'-o-phenylenebis(salicylaldimine), by mixing 20 mL 1 M o-phenylenediamine in EtOH and 40 mL 1 M salicylaldehyde in EtOH, reflux for 1 h, cool, collect the precipitate by filtration and recrystallize it from hot EtOH to give the reagent.)

HPLC VARIABLES
Column: 50 × 4.6 3 μm Shim-pack FLC-ODS (Shimadzu)
Mobile phase: MeCN:water 58:42 containing 5 mM sodium acetate and 100 μM disodium EDTA

Flow rate: 0.5
Injection volume: 100
Detector: UV 380

CHROMATOGRAM
Retention time: 4 (V^{5+}), 7 (Co^{2+}), 11 (Cu^{2+}), 16 (Ni^{2+})
Limit of detection: 0.6 ppb (Cu)

KEY WORDS
complexation; comparison with other complexing agents

REFERENCE
Kanbayashi, M.; Hoshino, H.; Yotsuyanagi, T. Highly selective determination of trace amounts of copper(II), nickel(II) and vanadium(V) ions with tetradentate Schiff-base ligands by reversed-phase high-performance liquid chromatography and spectrophotometric determination, *J.Chromatogr.*, **1987**, *386*, 191–197.

o-Phenylenediamine

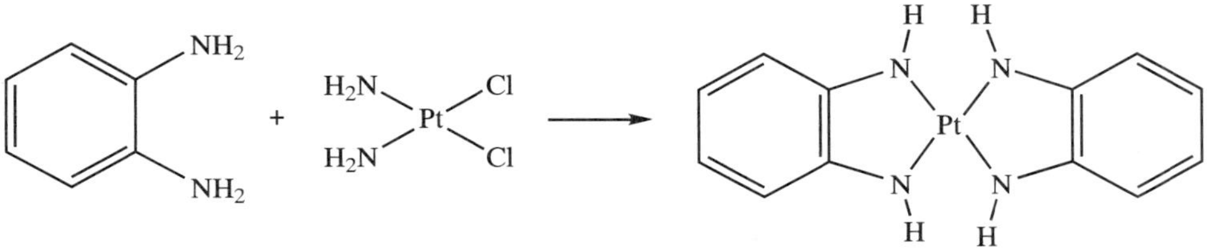

SAMPLE
Matrix: solutions
Analyte: cisplatin
Sample preparation: Mix 2 mL of a 750 µg/mL solution of cisplatin in water with 1 mL 1.8 mg/mL o-phenylenediamine in water, heat at 100° for 20 min, filter (paper) the precipitate and wash it with water. Dissolve the precipitate in DMF and inject a 100 µL aliquot.

HPLC VARIABLES
Column: 300 × 4.6 10 µm µBondapak C18
Mobile phase: Chloroform
Flow rate: 1
Injection volume: 100
Detector: UV 703

CHROMATOGRAM
Retention time: 3
Limit of detection: 400 ng/mL Pt

KEY WORDS
complexation

REFERENCE
Hasson, H.; Warshawsky, A. High performance liquid chromatographic determination of *cis*-diamminedichloroplatinum(II) (cisplatin) as the *o*-phenylenediamine complex, *J.Chromatogr.*, **1990**, *530*, 219–221.

Picolinaldehyde 4-Phenyl-3-thiosemicarbazone

complex

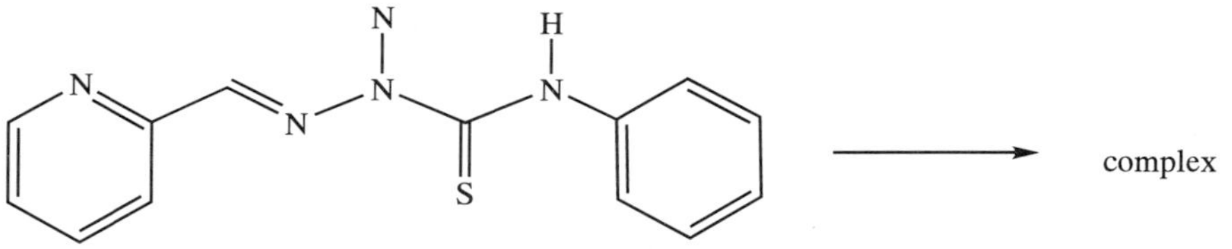

SAMPLE
Matrix: water
Analyte: metals
Sample preparation: 5 mL Water + 500 μL 5 mM picolinaldehyde 4-phenyl-3-thiosemi-
carbazone in MeOH + 500 μL 2 M pH 8.0 Tris buffer, make up to 10 mL with MeOH,
inject a 20-100 μL aliquot. (Preparation of picolinaldehyde 4-phenyl-3-thiosemicarbazone
is as follows. Dissolve 4.6 g 4-phenyl-3-thiosemicarbazide in 150 mL hot EtOH, add 3 g
2-pyridinecarboxaldehyde (picolinaldehyde) in 35 mL EtOH, reflux for 2 h, allow to cool
to room temperature. Filter off the yellow crystals and recrystallize from EtOH to obtain
picolinaldehyde 4-phenyl-3-thiosemicarbazone (mp 204-205°) (Talanta 1976, 23, 460).)

HPLC VARIABLES
Column: 125 × 4 5 μm LiChrospher RP-18
Mobile phase: MeCN:water:MeOH:pyridine 50:46:1.5:2.5 containing 75 μM picolinal-
dehyde 4-phenyl-3-thiosemicarbazone and 20 μM acetate (Adjust the pH to 6.0 before
adding the MeCN.)
Flow rate: 1
Injection volume: 20-100
Detector: UV 390

CHROMATOGRAM
Retention time: 3.5 (Cu), 5.5 (Co), 6 (Fe), 7.5 (Ni), 8 (Bi), 15 (Cd)
Limit of detection: 0.6-3 ng/mL

KEY WORDS
complexation; river water

REFERENCE
Uehara, N.; Morimoto, K.; Shijo, Y. Separation and determination of metal ions as picolinaldehyde 4-
phenyl-3-thiosemicarbazone chelates by reversed-phase high-performance liquid chromatography,
Analyst, **1992**, *117*, 977–979.

2-(2-Pyridylazo)-1-naphthol-4-sulfonate

complex

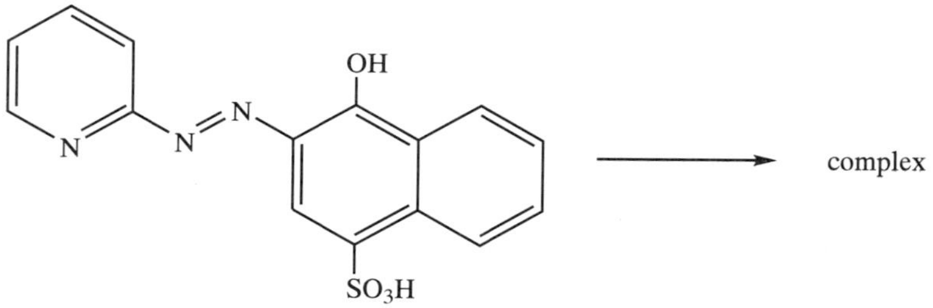

SAMPLE
Matrix: solutions
Analyte: metals
Sample preparation: Mix 1 mL of a 100 μM solution with 4 mL 250 μM 2-(2-pyridylazo)-
1-naphthol-4-sulfonate in MeCN/water, add 1 mL 100 mM pH 5 acetate buffer, inject an

8 µL aliquot. (Preparation of 2-(2-pyridylazo)-1-naphthol-4-sulfonate is as follows. Add a solution of 1.1 g 2-hydrazinopyridine in 20 mL water to a solution of 2.6 g sodium 1,2-naphthoquinone-4-sulfonate and 12 mL 72% perchloric acid in 100 mL water, dissolve the resulting orange precipitate in NaOH solution, precipitate with HCl to obtain 2-(2-pyridylazo)-1-naphthol-4-sulfonate as an orange-red powder (Analyst 1968, 93, 13).)

HPLC VARIABLES
Column: 250 × 4.6 5 µm Inertsil ODS-2
Mobile phase: MeCN:water 40:60 containing 10 mM tetrabutylammonium bromide and 1 mM pH 5.0 aqueous acetate buffer
Column temperature: 40
Flow rate: 0.8
Injection volume: 8
Detector: UV

CHROMATOGRAM
Retention time: 7 (cobalt), 21 (iron), 23 (nickel)

KEY WORDS
complexation

REFERENCE
Niwa, H.; Yasui, T.; Yuchi, A.; Yamada, H.; Wada, H. Reversed-phase ion-pair high-performance liquid chromatography of metal chelates with sulfonated 2-(2-pyridylazo)-1-naphthols. Positional effect of sulfonate group, *Anal.Sci.*, **1997**, *13*, 137–140.

2-(2-Pyridylazo)-1-naphthol-7-sulfonate

SAMPLE
Matrix: solutions
Analyte: metals
Sample preparation: Mix 1 mL of a 100 µM solution with 4 mL 250 µM 2-(2-pyridylazo)-1-naphthol-7-sulfonate in MeCN/water, add 1 mL 100 mM pH 5 acetate buffer, inject an 8 µL aliquot. (Preparation of 2-(2-pyridylazo)-1-naphthol-7-sulfonate is as follows. Couple benzenediazonium chloride with sodium 2-naphthol-7-sulfonate in weakly alkaline conditions, dissolve the product in NaOH solution, cleave with sodium dithionite, remove the aniline, neutralize with HCl to obtain 1-amino-2-naphthol-7-sulfonic acid. Dissolve 0.98 g potassium dichromate in 2 M HCl, add 2.4 g 1-amino-2-naphthol-7-sulfonic acid (to form 1,2-naphthoquinone-7-sulfonic acid), add a solution of 1.1 g 2-hydrazinopyridine in 20 mL water, dissolve the resulting orange precipitate in NaOH solution, precipitate with HCl to obtain 2-(2-pyridylazo)-1-naphthol-7-sulfonate as an orange-red powder (Analyst 1968, 93, 13).)

HPLC VARIABLES
Column: 250 × 4.6 5 µm Inertsil ODS-2
Mobile phase: MeCN:water 40:60 containing 10 mM tetrabutylammonium bromide and 1 mM pH 5.0 aqueous acetate buffer
Column temperature: 40
Flow rate: 0.8
Injection volume: 8
Detector: UV

CHROMATOGRAM
Retention time: 5 (cobalt), 34 (iron), 41 (nickel)

KEY WORDS
complexation

REFERENCE
Niwa, H.; Yasui, T.; Yuchi, A.; Yamada, H.; Wada, H. Reversed-phase ion-pair high-performance liquid chromatography of metal chelates with sulfonated 2-(2-pyridylazo)-1-naphthols. Positional effect of sulfonate group, *Anal.Sci.*, **1997**, *13*, 137–140.

4-(2-Pyridylazo)resorcinol

RELATED REFERENCES
Co, A.C.; Ko, A.N.; Ye, L.; Lucy, C.A. Modification of 4-(2-pyridylazo)-resorcinol postcolumn reagent selectivity through competitive equilibria with chelating ligands. *J.Chromatogr.A* **1997**, *770*, 69-74.

Mazzucotelli, A.; Frache, R.; Viarengo, A.; Martino, G. The speciation of trace amounts of organometallics in marine organisms by gel-permeation high-pressure liquid chromatography with PAR derivatization. *Talanta* **1988**, *35*, 693-696.

Ming, X.; Wu, Y.; Schwedt, G. HPLC analysis for vanadium, cobalt, iron, and nickel by 4-(2-pyridylazo)resorcinol (PAR) and hydrogen peroxide and studies on complex properties influencing retention. *Fresenius' J.Anal.Chem.* **1992**, *342*, 556-559.

Nagaosa, Y.; Tanizaki, M. Simultaneous determination of zinc(II) and iron(III) in human serum by liquid chromatography using post-column derivatization with 4-(2-pyridylazo)resorcinol. *J.Liq.Chromatogr.Rel.Technol.* **1997**, *20*, 2357-2366.

Saraswati, R.; Rao, T.H. Reversed-phase high-performance liquid chromatographic separation of some trace impurities in oxygen-free electronic copper by post-column chelation with 4-(2-pyridylazo)rescorcinol and Arsenazo-III. *J.Chromatogr.* **1992**, *605*, 63-68.

Simo Alfonso, E.F.; Ramis Ramos, G.; Estela Ripoll, J.M.; Hernandez Guerra, L.; Cerda Martin, V. Determination of metals with pyridylazoresorcinol by thermal lensing spectrometry in static and flow conditions. *Quim.Anal.(Barcelona)* **1990**, *9*, 245-254.

SAMPLE
Matrix: air
Analyte: metals
Sample preparation: Pull air through an 18×22.5 cm glass-fiber filter at 70 m^3/h for 24 h, cut a 4×17.5 cm portion of the filter into small pieces, add to 26 mL HF:aqua regia:boric acid 5:15:6 in a PTFE container, microwave (630 W) at 50% power for 8 min, 80% power for 4 min, and 50% power for 7 min, make up to 100 mL with 10% nitric acid, inject a 25 µL aliquot.

HPLC VARIABLES
Guard column: IonPac CG2 (Dionex)
Column: IonPac CS2 polystyrene-divinylbenzene with 2% sulfonyl functional groups (Dionex)
Mobile phase: 20 mM Oxalic acid containing 20 mM citric acid, adjusted to pH 3.6 with LiOH
Flow rate: 1
Injection volume: 25
Detector: UV 520 following post-column reaction. The column effluent mixed with the reagent pumped at 0.7 mL/min and the mixture flowed to the detector. (Reagent was 200 µM 4-(2-pyridylazo)resorcinol in 1.5 M ammonia containing 1 M acetic acid, pH 9.0.)

CHROMATOGRAM
Retention time: 2 (Cu^{++}), 3.5 (Ni^{++}), 5 (Zn^{++}), 7.5 (Co^{++}), 10 (Pb^{++}), 20 (Fe^{++})
Limit of detection: 0.9-6.6 ng/mL

KEY WORDS
post-column reaction; complexation

REFERENCE
Rahmalan, A.; Abdullah, M.Z.; Sanagi, M.M.; Rashid, M. Determination of heavy metals in air particulate matter by ion chromatography, *J.Chromatogr.A*, **1996**, *739*, 233–239.

SAMPLE
Matrix: blood, tissue
Analyte: metals
Sample preparation: Homogenize (glass-PTFE potter) tissue with 250 mM sucrose. Heat 500 µL plasma or tissue homogenate at 100° for 2 h, cool to room temperature, add 5.5 mL concentrated nitric acid:70% perchloric acid 10:1, heat at 80° for 12 h, at 140° for 2 h, at 180° for 2 h, at 190° for 1 h, cool to room temperature, add 500 µL 10 mM nitric acid, rotate at 32 rpm for 1 h, inject a 100 µL aliquot.

HPLC VARIABLES
Column: 50 × 4.5 5 µm TSKgel IC-Cation SW (Tosoh)
Mobile phase: 350 mM pH 3.0 Lactic acid containing 350 mM sodium lactate
Column temperature: 30
Flow rate: 0.7
Injection volume: 100
Detector: UV 520 following post-column reaction. The column effluent mixed with the reagent pumped at 0.7 mL/min and the mixture flowed through a 1 m × 0.25 mm ID PTFE coil to the detector. (Reagent was 100 µg/mL 4-(2-pyridylazo)resorcinol in 40 mg/mL sodium carbonate.)

CHROMATOGRAM
Retention time: 3.1 (Cu^{++}), 6.6 (Zn^{++}), 8.0 (Ni^{++}), 11.6 (Co^{++}), 16.2 (Cd^{++}), 20.1 (Mn^{++})
Limit of quantitation: 1-10 ppb
Limit of detection: 0.3-7 ppb

OTHER SUBSTANCES
Non-interfering: bismuth, iron, hafnium, mercury, lanthanum, lead, vanadium, zirconium

KEY WORDS
post-column reaction; rat; plasma; liver; pancreas; kidney; complexation

REFERENCE
Sudo, J.; Hayashi, T.; Suzue, T.; Terui, J. Application of biological samples treated by wet-digested mineralization to cation-exchange high-performance liquid chromatography with post-column reaction by 4-(2-pyridylazo)-resorcinol, *J.Toxicol.Sci.*, **1991**, *16*, 155–166.

SAMPLE
Matrix: semiconductors
Analyte: metals
Sample preparation: Photolyze 15-70 mg finely ground semiconductor with 50 µL concentrated nitric acid and 500 µL 30% nitric acid with a 500 W high-pressure mercury lamp in a quartz tube using cooling to keep the temperature between 80 and 90° for 1-2 h, cool, add 500 µL 2 M ammonium acetate (for mobile phase A only), make up to 10 mL with water, inject a 150 µL aliquot.

HPLC VARIABLES
Guard column: IonPac CG5 (Dionex)
Column: IonPac CS5 (Dionex)
Mobile phase: 50 mM pH 4.6 Acetic acid containing 50 mM sodium acetate and 6 mM pyridine-2,6-dicarboxylic acid (A) or 50 mM pH 4.8 oxalic acid containing 95 mM LiOH (B)
Flow rate: 1
Injection volume: 150

Detector: UV 520 following post-column reaction. The column effluent mixed with the reagent pumped at 0.5 mL/min and the mixture flowed to the detector. (Reagent was 300 μM 4-(2-pyridylazo)resorcinol in 1 M 2-dimethylaminoethanol containing 500 mM ammonium hydroxide and 500 mM sodium bicarbonate.)

CHROMATOGRAM
Retention time: 3 (Pb(II) (B)), 4 (Cu(II) (B)), 4.5 (Cd(II) (B)), 6 (Fe(III) (A)), 8.5 (Co(II) (B)), 9 (Cu(II) (A)), 10 (Ni(II) (A)), 11 (Zn(II) (A)), 12.5 (Zn(II) (B)), 13 (Cd(II) (A)), 15 (Ni(II) (B))
Limit of detection: 5 ng/mL

KEY WORDS
post-column reaction; use a metal free system; complexation

REFERENCE
Buldini, P.L.; Cavalli, S.; Mevoli, A.; Milella, E. Application of ion chromatography to the analysis of high-purity CdTe, *J.Chromatogr.A*, **1996**, *739*, 131–137.

SAMPLE
Matrix: solutions
Analyte: metals
Sample preparation: Prepare a 1 ng/mL solution acidified to pH 3 with HCl. Inject a 25 mL aliquot onto column A (pre-equilibrated with mobile phase) at 4.4 mL/min, elute the contents of column A onto column B with mobile phase, monitor the effluent from column B.

HPLC VARIABLES
Column: A MetPac-CC1 (Dionex); B CG5 (Dionex) + CS5 (Dionex)
Mobile phase: 50 mM pH 4.5 Acetic acid containing 50 mM sodium acetate and 6 mM pyridine-2,6-dicarboxylic acid (Purify the mobile phase by passing it through an MFC-1 column (Dionex) placed before the injector.)
Flow rate: 1
Injection volume: 25000
Detector: UV 520 following post-column reaction. The column effluent mixed with the reagent pumped at 0.4 mL/min and flowed to the detector. (Reagent was 400 μM pyridylazo-resorcinol in 3 M ammonia containing 1 M acetic acid, pH 9.7.)

CHROMATOGRAM
Retention time: 6 (Fe(III)), 8.8 (Cu), 9.6 (Ni), 10.6 (Zn), 12 (Co), 14.2 (Mn)
Limit of detection: 22-64 pg/mL

KEY WORDS
post-column reaction; complexation; column-switching

REFERENCE
Motellier, S.; Pitsch, H. Simultaneous analysis of some transition metals at ultra-trace level by ion-exchange chromatography with on-line preconcentration, *J.Chromatogr.A*, **1996**, *739*, 119–130.

SAMPLE
Matrix: solutions
Analyte: metals
Sample preparation: Mix metal solution with 1 mM 4-(2-pyridylazo)resorcinol in slightly alkaline medium, heat at 60° for 15 min, inject an aliquot.

CAPILLARY ELECTROPHORESIS
Capillary: 60 cm × 50 μm fused-silica (46 cm to detector) (GL Science, Japan)
Capillary preparation: Wash capillary with running buffer for 1 min between runs. Purge new capillaries with water for 5 min, with 100 mM NaOH for 10 min, with water for 10 min, and with running buffer for 30 min.
Running buffer: 4 mM pH 8.0 Sodium phosphate buffer containing 1 mM EDTA and 20 mM sodium sulfate
Voltage/Current: -20 kV
Injection: Hydrostatic injection at 13 cm for 15 s.
Detector: UV 490

Model: laboratory constructed
Migration time: 4.7 (Ni), 5 (Co), 5.2 (Fe)

KEY WORDS
complexation

REFERENCE
Krokhin, O.V.; Hoshino, H.; Shpigun, O.A.; Yotsuyanagi, T. Use of cationic polymers for the simultaneous determination of inorganic anions and metal-4-(2-pyridylazo)resorcinolato chelates in kinetic differentiation-mode capillary electrophoresis, *J.Chromatogr.A*, **1997**, *776*, 329–336.

Sodium Diethyldithiocarbamate

SAMPLE
Matrix: blood
Analyte: platinum
Sample preparation: Filter (Amicon Centriflo CF-50A, 50000 MW cut-off) 0.1-2 mL plasma while centrifuging at 21° at 1000 g for 30 min. Remove a 0.1-1 mL aliquot of the ultrafiltrate, make up to 1 mL with saline, add 100 μL 10 μg/mL palladium chloride in saline, add 100 μL 10% diethyldithiocarbamate in 100 mM NaOH, add 200 μL saturated aqueous sodium nitrate, mix, add 3 mL chloroform, shake at 1500 rpm for 5 min, centrifuge. Remove the organic layer and add it to 200 mg anhydrous sodium sulfate, evaporate to dryness under a stream of nitrogen at 40°, reconstitute the residue in 25 μL chloroform, inject a 10 μL aliquot.

HPLC VARIABLES
Column: 250 × 4.6 5 μm Cyano Spheri-5
Mobile phase: Heptane:isopropanol 90:10
Column temperature: 40
Flow rate: 1.5
Injection volume: 10
Detector: UV 254

CHROMATOGRAM
Retention time: 8.1
Internal standard: palladium chloride (5.3)
Limit of detection: 2.5 ng/mL

OTHER SUBSTANCES
Simultaneously analyzed: antimony, cadmium, cobalt, gold, iron(II), iron(III), mercury, nickel, silver
Non-interfering: beryllium, germanium, molybdenum, selenium, strontium, tungsten

KEY WORDS
plasma; ultrafiltrate; pharmacokinetics; complexation

REFERENCE
Reece, P.A.; McCall, J.T.; Powis, G.; Richardson, R.L. Sensitive high-performance liquid chromatographic assay for platinum in plasma ultrafiltrate, *J.Chromatogr.*, **1984**, *306*, 417–423.

SAMPLE
Matrix: blood
Analyte: platinum
Sample preparation: Filter (Amicon CF 25 A filter cone) while centrifuging at 4° at 991 g for 30 min. Remove a 500 μL aliquot of the ultrafiltrate and add it to 5 μL 841 μM nickel

chloride, add 50 μL 10% diethyldithiocarbamate in 100 mM NaOH, heat at 37° for 30 min, chill on ice, extract with 200 μL chloroform, inject a 20 μL aliquot of the chloroform layer.

HPLC VARIABLES
Column: 100 × 8 10 μm Radial-Pak C18
Mobile phase: MeOH:water 80:20
Flow rate: 1.5
Injection volume: 20
Detector: UV 254

CHROMATOGRAM
Internal standard: nickel

KEY WORDS
plasma; pharmacokinetics; ultrafiltrate; method determines cisplatin and other platinum-containing species; complexation

REFERENCE
Goel, R.; Andrews, P.A.; Pfeifle, C.E.; Abramson, I.S.; Kirmani, S.; Howell, S.B. Comparison of the pharmacokinetics of ultrafilterable cisplatin species detectable by derivatization with diethyldithiocarbamate or atomic absorption spectroscopy, *Eur.J.Cancer*, **1990**, *26*, 21–27.

SAMPLE
Matrix: plants, yeast
Analyte: metals
Sample preparation: Reflux 150 mg leaves or 500 mg yeast with 30 mL nitric acid:perchloric acid 75:15 for 4 h, evaporate to a small volume, add 5 mL 15 M nitric acid, reflux briefly, cool, add 25 mL water, add 10 mL 10% ammonium citrate, neutralize, adjust pH to 9.5 with 20 mL ammonia/ammonium chloride buffer, add 3 mL 35 mM sodium diethyldithiocarbamate, heat at 60° for 15 min, cool, extract 4 times with 5 mL portions of chloroform. Combine the extracts and wash them with an equal volume of water, evaporate to dryness at room temperature, reconstitute with 2 mL MeOH, inject a 10 μL aliquot. (The ammonia/ammonium chloride buffer contained a small quantity of sodium diethyldithiocarbamate (250 μM ?) and was washed with chloroform before use.)

HPLC VARIABLES
Column: 250 × 4.6 10 μm μBondapak C18
Mobile phase: MeOH:water 66:34
Flow rate: 2
Injection volume: 10
Detector: UV 254

CHROMATOGRAM
Retention time: 11 (nickel), 15 (cobalt), 16 (chromium), 19 (copper)
Limit of detection: 0.2-8 ng

OTHER SUBSTANCES
Also analyzed: arsenic, cadmium, iron, lead, mercury, selenium, zinc

KEY WORDS
complexation; leaves

REFERENCE
Dilli, S.; Haddad, P.R.; Htoon, A.K. Further studies of diethyldithiocarbamate complexes by high-performance liquid chromatography, *J.Chromatogr.*, **1990**, *500*, 313–328.

SAMPLE
Matrix: urine
Analyte: platinum species
Sample preparation: 9 mL Urine + 1 mL 10% sodium diethylthiocarbamate in 100 mM NaOH + 2 mL saturated sodium nitrate in water, mix, let stand at room temperature for 1 h, extract with 1 mL water-saturated chloroform for 3 min, centrifuge at 1200 g for 5 min, vortex briefly, centrifuge for 10 min, inject a 30 μL aliquot of the chloroform layer.

(Purify sodium diethylthiocarbamate by washing with chloroform, filtering under nitrogen, and drying under reduced pressure. Store at -10°. Derivatization of cisplatin or its metabolites gives the same complex which is then chromatographed.)

HPLC VARIABLES
Column: two 300 mm long μBondapak columns in series
Mobile phase: Heptane:isopropanol 82:18
Flow rate: 2
Injection volume: 30
Detector: UV 254

CHROMATOGRAM
Retention time: 25
Limit of detection: 25 ng/mL

KEY WORDS
complexation, normal phase

REFERENCE
Bannister, S.J.; Sternson, L.A.; Repta, A.J. Urine analysis of platinum species derived from *cis*-dichlorodiammineplatinum(II) by high-performance liquid chromatography following derivatization with sodium diethyldithiocarbamate, *J.Chromatogr.*, **1979**, *173*, 333–342.

Sodium Pyrrolidinedithiocarbamate

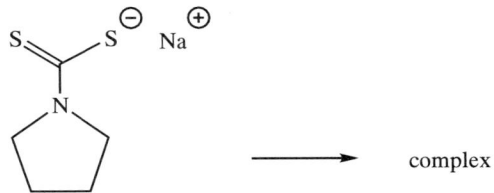

SAMPLE
Matrix: water
Analyte: mercury species
Sample preparation: Condition a Sep-Pak C18 SPE cartridge with MeOH and water. Add 100 μmoles sodium pyrrolidinedithiocarbamate (Fluka) and 1 mM potassium nitrate to 100 mL water, mix, let stand in the dark for 30 min, add to the SPE cartridge, wash with 5 mL water, elute with 2 mL MeOH. Evaporate the eluate to dryness, reconstitute the residue in 100 μL mobile phase, inject a 20 μL aliquot. (Caution! Organomercury compounds are highly toxic! Deaths have occurred amongst laboratory workers!)

HPLC VARIABLES
Column: 150 × 4 5 μm Spherisorb C18 polyether ketone column
Mobile phase: MeOH:water 75:25 containing 50 mM potassium nitrate
Injection volume: 20
Detector: E, Metrohm Model 461-VA, Model 656 amperometric flow cell, carbon paste electrode + 1.15 V, Ag/AgCl reference electrode

CHROMATOGRAM
Retention time: 4 (MeHg$^+$), 6 (EtHg$^+$), 9 (PhHg$^+$), 11 (Hg^{++})
Limit of detection: 30-100 ng/mL

KEY WORDS
SPE; complexation

REFERENCE
Pilar da Silva, M.; Procopio, J.R.; Hernández, L. Reversed-phase high-performance liquid chromatography of pyrrolidinethiocarbamate complexes of mercuric species using amperometric and coulometric detection, *J.Chromatogr.A*, **1997**, *761*, 139–146.

Tartaric Acid

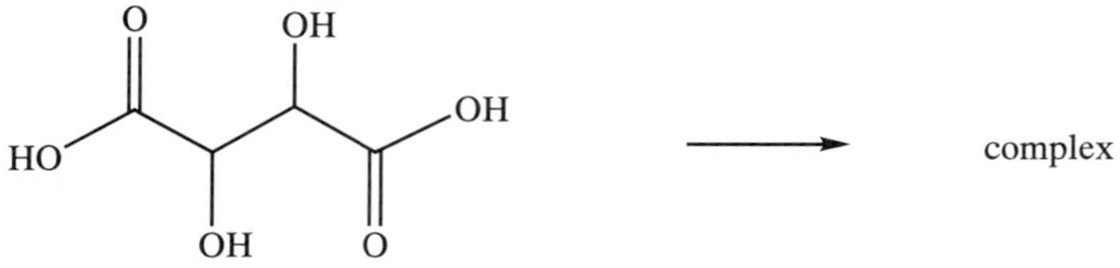

→ complex

SAMPLE
Matrix: water
Analyte: metals
Sample preparation: Add 1 mL 1 M tartaric acid to 50 mL water, filter (0.22μm nylon), pass the filtrate through a C18 SPE cartridge (HPLC Technology), inject a 500 μL aliquot of the eluate.

HPLC VARIABLES
Column: 300 × 3.9 10 μm Bondclone C18 (Phenomenex)
Mobile phase: MeCN:50 mM tartaric acid containing 2 mM octanesulfonic acid 5:95, pH adjusted to 3.4 with NaOH
Column temperature: 30
Flow rate: 1
Injection volume: 500
Detector: UV 510 following post-column reaction. The column effluent mixed with the reagent pumped at 1 mL/min (using a Waters delivery system module) and flowed to the detector. (Reagent was 200 μM 4-(2-pyridylazo)resorcinol in 2 M ammonia containing 1 M ammonium acetate.)

CHROMATOGRAM
Retention time: 4 (Fe(III)), 7.5 (Cu), 10 (Pb), 17.5 (Zn), 20 (Ni), 28.5 (Co), 37 (Cd), 49 (Mn)
Limit of detection: 2-100 ng/mL

KEY WORDS
post-column reaction; tap water; rain water; fog water; complexation

REFERENCE
Zappoli, S.; Morselli, L.; Osti, F. Application of ion interaction chromatography to the determination of metal ions in natural water samples, *J.Chromatogr.A*, **1996**, *721*, 269–277.

Zinc n-Butyl-2-naphthylmethyldithiocarbamate Salt

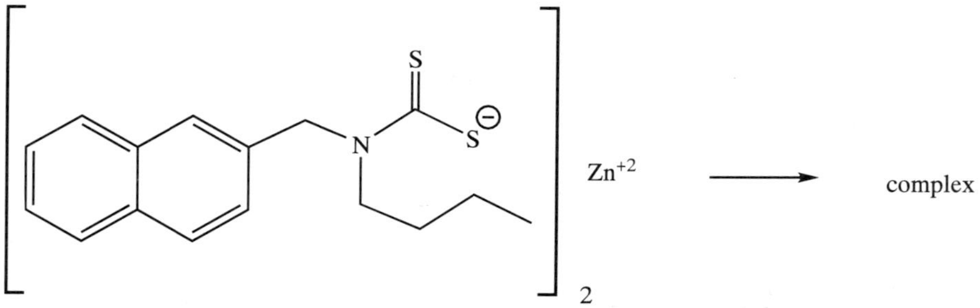

→ complex

SAMPLE
Matrix: solutions
Analyte: metals
Sample preparation: Mix 1 L of a 100 nM solution of metal in 1 mM pH 7.0 Tris buffer with 1.5 mL 1 mM reagent in MeOH, stir at room temperature for 1 h, add 1 g 20-60 mesh XAD resin, stir for 3 h, filter, elute the resin with 10 mL MeOH, inject an aliquot

of the eluate. (Prepare the reagent, zinc n-butyl-2-naphthylmethyldithiocarbamate salt, as follows. Reflux 100 mmoles 2-(bromomethyl)naphthalene and 200 mmoles n-butylamine in 100 mL dry benzene with stirring for 6 h (Caution! Benzene is a carcinogen!), filter, evaporate the filtrate to dryness under reduced pressure, take up the residue in 100 mL ether, pass hydrogen bromide gas through the solution until precipitation is complete, recrystallize from boiling water containing a little HBr to give n-butyl-2-naphthylmethylamine hydrobromide (cf. J. Chem. Soc. (A) 1966, 1573). Add 8.6 g n-butyl-2-naphthylmethylamine hydrobromide to 60 mL 1 M NaOH, add the minimum quantity of isopropanol (about 5 mL) required to dissolve the amine, slowly add 1.75 mL carbon disulfide over 10 min, stir at room temperature for 2 h, add 100 mL 100 mM pH 5.5 sodium acetate buffer, add 40 mL 400 mM zinc nitrate with occasional stirring, digest overnight, filter. Dissolve the solid in chloroform, filter, evaporate the filtrate to 5 mL with gentle heating, add MeOH, filter to obtain zinc n-butyl-2-naphthylmethyldithiocarbamate salt (mp 115-118°), dry in air.)

HPLC VARIABLES

Column: 100 × 5 10 μm RCM-100 C18 (Waters)

Mobile phase: MeOH:water 95:5 containing 1 mM Tris, pH adjusted to 8.25 with phosphoric acid (The mobile phase was pre-saturated with silica using a 250 × 4.6 column packed with silica. The mobile phase also passed through a 100 × 4.6 diaminosilane-derivatized silica column to remove metal ions. These columns were placed before the injector.)

Flow rate: 2

Detector: UV 221 [Post-column photochemical derivatization using a 10 m × 0.25 mm ID PTFE coil irradiated by a 175 W medium-pressure mercury lamp gives detection limits of 5-50 nM (Anal. Chim. Acta 1984, 159, 211).]

CHROMATOGRAM

Retention time: 4 (Fe), 6 (Ni), 8 (Cu), 9 (Hg), 11 (Co)

Limit of detection: 1-10 ng

KEY WORDS

SPE; complexation

REFERENCE

Shih, Y.-T.; Carr, P.W. Determination of metals at trace levels via precolumn derivatization and nonpolar stationary-phase high-performance liquid chromatography with n-butyl-2-naphthylmethyldithiocarbamate complexes, *Anal.Chim.Acta*, **1982**, *142*, 55–62.

MISCELLANEOUS REACTIONS

COMPLEXATION

Aluminum

SAMPLE

Matrix: tissue

Analyte: tetracyclines

Sample preparation: Condition a 500 mg Bond Elut cyclohexyl (CH) SPE cartridge with 10 mL MeOH and 10 mL water. Powder (domestic food blender) frozen kidney or muscle. Homogenize (Silverson Machines) 5 g powdered tissue and 45 mL 100 mM glycine in 1 M HCl for 1 min, add 5 g ammonium sulfate, shake for 30 s, let stand for 10 min, centrifuge at 2000 rpm for 15 min, filter (glass wool) the supernatant, repeat the extraction with 50 mL 100 mM glycine in 1 M HCl. Combine the filtrates and centrifuge an aliquot at 2200 rpm for 10 min, add a 20 mL aliquot of the supernatant to the SPE cartridge, wash with 10 mL water, elute with 7 mL MeOH. Evaporate the eluate to dryness under a stream of nitrogen at 65°, reconstitute the residue in 500 μL MeCN:20 mM oxalic acid 20:80, inject a 50 μL aliquot.

HPLC VARIABLES

Guard column: Chromspher C8 (Chrompack)

Column: 200 × 3 5 μm Chromspher C8 glass column (Chrompack)

Mobile phase: Gradient. A was MeCN. B was MeCN:20 mM oxalic acid 10:90. A:B from 10:90 to 20:80 over 2 min, maintain at 20:80 for 8 min, to 25:75 over 1 min, maintain at 25:75 for 9 min, return to initial conditions over 5 min, re-equilibrate for 10 min.

Flow rate: 0.4

Injection volume: 50

Detector: F ex 390 em 490 following post-column reaction. The column effluent mixed with 750 mM aluminum chloride (degas by sonication, store in a brown bottle) pumped at 0.6 mL/min and flowed through a 13.7 m × 0.3 mm ID PTFE column at 60° to the detector.

CHROMATOGRAM

Retention time: 10.7 (oxytetracycline), 13.7 (tetracycline), 20.4 (chlortetracycline)

Limit of detection: 20 ng/g (muscle), 230 ng/g (kidney)

KEY WORDS

pig; cow; poultry; kidney; muscle; SPE; post-column reaction; complexation

REFERENCE

McCracken, R.J.; Blanchflower, W.J.; Haggan, S.A.; Kennedy, D.G. Simultaneous determination of oxytetracycline, tetracycline and chlortetracycline in animal tissues using liquid chromatography, post-column derivatization with aluminium, and fluorescence detection, *Analyst*, **1995**, *120*, 1763–1766.

2-Anilinonaphthalene-6-sulfonic Acid

RELATED REFERENCE
Ishida, J.; Abe, K.; Nakamura, M.; Yamaguchi, M. High-performance liquid chromatographic determination of human serum albumin in plasma and urine by post-column fluorescence enhancement detection using 8-anilino-1-naphthalenesulfonic acid. *Biol.Pharm.Bull.* **1996**, *19*, 1391-1395.

Complex

SAMPLE
Matrix: solutions
Analyte: cyclodextrins

CAPILLARY ELECTROPHORESIS
Capillary: 100 cm × 50 μm fused-silica (Polymicro Technologies)
Capillary preparation: At the start of each day rinse capillary with 1 M NaOH, with water, and with running buffer.
Running buffer: 40 mM pH 11.76 Sodium phosphate buffer containing 1 mM 2-anilinonaphthalene-6-sulfonic acid (Molecular Probes, Eugene OR) (Better resolution of the components of 2,6-di-O-methyl-β-cyclodextrin can be obtained using 30 mM sodium benzoate containing 40 μM tetrabutylammonium bromide and 1 mM 2-anilinonaphthalene-6-sulfonic acid, adjusted to pH 4.0 with 5 M HCl. Cover inlet electrode with a PTFE sleeve so that only the tip of the electrode is exposed to minimize electrolysis of reagent.)
Voltage/Current: 300 V/cm
Injection: Electrokinetic injection at 5 kV for 2 s
Detector: F ex 363.8 (14 mW argon ion laser) em 424 (bandpass filter)
Model: laboratory constructed
Migration time: 6.8 (α), 7.05 (gamma), 7.4 (β), 17.3 (2,6-di-O-methyl-β-cyclodextrin)
Limit of detection: 2.4-62 μM

KEY WORDS
complexation

REFERENCE
Penn, S.G.; Chiu, R.W.; Monnig, C.A. Separation and analysis of cyclodextrins by capillary electrophoresis with dynamic fluorescence labelling and detection, *J.Chromatogr.A*, **1994**, *680*, 233–241.

Avidin Labeled with Fluorescein

SAMPLE
Matrix: formulations, feed
Analyte: biotin and biocytin
Sample preparation: Formulations. Dilute 50 μL liquid vitamin to 2 mL with mobile phase, filter, inject a 20 μL aliquot. Feed. Extract 100 mg horse feed with 10 mL 1 M NaOH. Remove a 6 mL aliquot and adjust the pH to 6-7 with 1 M HCl, dilute to 100 mL with 100 mM pH 6.0 phosphate buffer, filter, inject a 20 μL aliquot.

HPLC VARIABLES
Guard column: 15 × 4.6 5 μm Microsorb C18
Column: 250 × 4.6 5 μm Microsorb C18
Mobile phase: A was 100 mM pH 6.0 phosphate buffer. B was MeOH:200 mM pH 6.0 phosphate buffer 50:50. A:B was 54:46.

Flow rate: 0.4
Injection volume: 20
Detector: F ex 490 em 520 (photon-counting) following post-column reaction. The column effluent mixed with a 2 μg/mL solution of avidin-FITC (Sigma) in 100 mM pH 7.0 phosphate buffer pumped at 1 mL/min and the mixture flowed through a 10 m × 0.5 mm ID knitted open tubular PTFE tubing to the detector.

CHROMATOGRAM
Retention time: 15 (biotin), 16.6 (biocytin)
Limit of detection: 4.45 ng/mL

OTHER SUBSTANCES
Non-interfering: acetone, DMF, methyl ethyl ketone, niacin, pantothenic acid, riboflavin, thiamine, vitamin A, vitamin B6, vitamin B12, vitamin C, vitamin D, vitamin E

KEY WORDS
liquid vitamin; horse feed; complexation

REFERENCE
Przyjazny, A.; Hentz, N.G.; Bachas, L.G. Sensitive and selective liquid chromatographic postcolumn reaction detection system for biotin and biocytin using a homogeneous fluorophore-linked assay, *J.Chromatogr.A*, **1993**, *654*, 79–86.

Copper

SAMPLE
Matrix: blood
Analyte: TP9201
Sample preparation: 3 mL Serum + 6 mL ice-cold MeCN, vortex, invert vigorously, let stand on ice for 10 min, centrifuge at 4° at 2050 g for 20 min. Remove the supernatant and centrifuge it at 4° at 3050 g for 2 h. Remove 6 mL of the supernatant and evaporate it to dryness under a stream of argon at 50°, reconstitute the residue in 1 mL water, filter (0.2 μm) while centrifuging at 2050 g, inject a 100 μL aliquot of the filtrate.

HPLC VARIABLES
Guard column: Bondapak C18
Column: 150 × 2 5 μm 300 Å Delta-Pak C18 (Waters)
Mobile phase: Gradient. A was n-propanol:0.1% trifluoroacetic acid 3:97. B was MeCN:n-propanol:trifluoroacetic acid:water 60:3:0.1:36.9. A:B from 100:0 to 50:50 over 30 min.
Column temperature: 50
Flow rate: 0.25
Injection volume: 100
Detector: E, BAS LC-4C, BAS CC-5 thin-cell detector with dual glassy carbon electrodes, 16 μm PTFE gasket, upstream electrode 410 mV, downstream electrode 85 mV, Ag/AgCl reference electrode, following post-column reaction. The column effluent mixed with reagent pumped 0.1 mL/min and the mixture flowed through a 0.25 mm i.d. coil of knitted PTFE tubing (residence time 1.1 min) at 50° to the detector. (Reagent was 1.2 M pH 9.9 carbonate buffer containing 0.5 mM copper sulfate and 3 mM disodium tartrate.)

CHROMATOGRAM
Retention time: 15.8
Limit of detection: 20 nM

KEY WORDS
serum; cow; post-column reaction; complexation

REFERENCE
Woltman, S.J.; Chen, J.-G.; Weber, S.G.; Tolley, J.O. Determination of the pharmaceutical peptide TP9201 by post-column reaction with copper(II) followed by electrochemical detection, *J.Pharm.Biomed.Anal.*, **1996**, *14*, 155–164.

SAMPLE
Matrix: fermentation medium
Analyte: amino acids
Sample preparation: Centrifuge 1 mL fermentation medium at 15000 g for 3 min, dilute the supernatant 10-100-fold with water, inject an aliquot.

HPLC VARIABLES
Column: 150 × 3.3 5 μm Separon C18 glass column (Tessek, Prague)
Mobile phase: MeOH:water 25:75 containing 1 mM copper sulfate
Column temperature: 45
Flow rate: 0.5
Injection volume: 1
Detector: UV 245

CHROMATOGRAM
Retention time: 0.5 (alanine), 0.6 (α-aminobutyric acid), 0.9 (valine), 1.65 (leucine), 3.5 (homoleucine)

KEY WORDS
complexation

REFERENCE
Polanuer, B.M.; Ivanov, S.V. High-performance liquid chromatography of amino acids in copper(II) complex form: application to valine fermentation samples, *J.Chromatogr.A*, **1996**, *722*, 311–315.

SAMPLE
Matrix: milk
Analyte: tetracyclines
Sample preparation: Prepare a column by adding 1.5 mL of thoroughly mixed Chelating Sepharose Fast-Flow suspension in EtOH:water 20:80 (Pharmacia) to a 150 × 10 glass column, allow to drain, wash with three 2 mL portions of water, add 2 mL 10 mM copper(II) sulfate in water, wash with two 2 mL portions of water. Centrifuge 10 mL milk at 2100 g for 5 min, decant the skimmed milk, rinse the tube with two 1 mL portions of water. Add 10 mL pH 4.0 buffer to the milk and rinses, sonicate for 3 min, filter (Whatman 541 paper) the supernatant. Add the filtrate to the column, wash with 2 mL pH 4.0 buffer, wash with 2 mL water, wash with 2 mL MeOH, wash with 2 mL water, add 700 μL EDTA buffer to the column, elute with 3 mL EDTA buffer, add 20 μL 25 μg/mL demeclocycline hydrochloride in MeOH to the eluate, inject a 100 μL aliquot. (Prepare pH 4.0 buffer by adjusting 100 mM succinic acid to pH 4.0 with 10 M NaOH. Prepare EDTA buffer by dissolving 12.9 g citric acid monohydrate, 10.9 g Na_2HPO_4, 29.2 g NaCl, and 100 mmoles EDTA in 1 L water.)

HPLC VARIABLES
Guard column: 5 × 3 PLRP-S (Polymer Laboratories)
Column: 250 × 4.6 5 μm 100 Å PLRP-S (Polymer Laboratories)
Mobile phase: MeCN:MeOH:buffer 15:10:60 (Buffer was 10 mM oxalic acid adjusted to pH 2.0 with 4 M HCl.)
Flow rate: 1
Injection volume: 100
Detector: F ex 406 em 515 following post-column reaction. The column effluent mixed with reagent pumped at 1 mL/min and the mixture flowed through a 600 μL reaction coil to the detector. (Reagent was 5% zirconyl chloride octahydrate in water.)

CHROMATOGRAM
Retention time: 4.9 (oxytetracycline), 5.9 (tetracycline), 11.7 (chlortetracycline)
Internal standard: demeclocycline (8.3)
Limit of detection: 1-4 ng/mL

KEY WORDS
protect from light; cow; post-column reaction; SPE; complexation

REFERENCE
Croubels, S.; Van Peteghem, C.; Baeyens, W. Sensitive spectrofluorimetric determination of tetracycline residues in bovine milk, *Analyst*, **1994**, *119*, 2713–2716.

Iron

SAMPLE

Matrix: food

Analyte: phytic acid

Sample preparation: Shake 2 g food with 20 mL 800 mM HCl at room temperature for 2 h, centrifuge at 1800 g for 10 min, filter (paper) the supernatant. Remove a 1 mL aliquot of the filtrate and add it to 250 μL 2.8 M NaOH, add 750 μL 167 mg/mL sodium acetate solution, add 1 mL EDTA solution, let stand at room temperature for 15 min, make up to 10 mL with water, dilute with 15 g/L pH 6 sodium acetate solution (if necessary), filter (0.45 μm), inject a 50 μL aliquot. (Prepare EDTA solution by adjusting pH of a 40 g/L EDTA solution to 6.0 with 2.8 M NaOH.)

HPLC VARIABLES

Guard column: AG-3 (Dionex)

Column: 250 × 5 HPIC AS-3 (Dionex)

Mobile phase: Gradient. A was 2.5 g sodium nitrate in 1 L water containing 500 μL 40 mg/mL EDTA solution and 200 μL 5 mg/mL pentachlorophenol in EtOH, pH adjusted to 6.0 with 2.8 M NaOH. B was 1 L 100 mM sodium nitrate containing 200 μL 5 mg/mL pentachlorophenol in EtOH, pH adjusted to 3.5 with dilute nitric acid (9 mL/L). C was 9 mL concentrated nitric acid in 1 L water. A:B:C 100:0:0 for 5 min, to 0:100:0 (step gradient), maintain at 0:100:0 for 5 min, to 0:0:100 (step gradient), maintain at 0:0:100 for 13 min, re-equilibrate at initial conditions for 7 min.

Flow rate: 1

Injection volume: 50

Detector: UV 300 following post-column reaction. The column effluent mixed with the reagent pumped at 0.5 mL/min and the mixture flowed through a 3 m × 0.3 mm ID PTFE reaction coil at 50° to the detector. (Prepare the reagent by dissolving 2.2 g iron(III) perchlorate nonahydrate and 12.8 mL 70% perchloric acid in 1 L water, store in a brown flask.)

CHROMATOGRAM

Retention time: 18

Limit of quantitation: 250 ng

KEY WORDS

post-column reaction; soybean meal; maize; wheat; barley; chyme; wheat bran; coffee; complexation

REFERENCE

Bos, K.D.; Verbeek, C.; Van Eeden, C.H.P.; Slump, P.; Wolters, M.G.E. Improved determination of phytate by ion-exchange chromatography, *J.Agric.Food Chem.*, **1991**, *39*, 1770–1772.

SAMPLE

Matrix: water

Analyte: phosphonates

Sample preparation: Rapidly mix 500 mL 200 mM calcium chloride solution with 500 mL 200 mM sodium carbonate solution, let stand for 10 min, centrifuge at 2000 rpm for 3 min, suspend the calcium carbonate precipitate in 100 mL water. Add a 100 μL aliquot of the suspension to a 10-100 mL aliquot of the test solution, filter (0.45 μm) after 1 h. The phosphonates are adsorbed on the calcium carbonate which is then caught on the filter. Dissolve the calcium carbonate on the filter in 6 M HCl, make up to 10 mL with water, pass through a column of cation-exchange resin (Dionex OnGuard H, H⁺ form). Remove a 5 mL portion of the effluent and adjust the pH to 3 with 10 M KOH. Remove a 3 mL aliquot and add it to 100 μL iron solution, mix, let stand at room temperature for 2 h, add 100 μL 1 mM trisodium nitrilotriacetic acid in water, mix, let stand for 30 min, add 300 μL buffer, mix, inject an aliquot. (Prepare iron solution by dissolving 1 mM $Fe(NO_3)_3.9H_2O$ and 10 mmoles nitric acid in 1 L water. Buffer was 600 mM sodium bicarbonate solution containing 10 mM tetrabutylammonium bromide.)

HPLC VARIABLES

Column: 150 × 4 5 μm PLRP-S (polystyrene-divinylbenzene copolymer) (Polymer Labs)

Mobile phase: MeCN:buffer 14:86 (Buffer was 20 mM sodium bicarbonate containing 1 mM tetrabutylammonium bromide, pH 8.3.)
Flow rate: 1
Injection volume: 200
Detector: UV 260

CHROMATOGRAM
Retention time: 8 (1-hydroxyethylene-(1,1-diphosphonic acid)), 10 (aminotris(methyl-enephosphonic acid)), 12 (ethylenediaminetetra(methylenephosphonic acid)), 17 (diethylenetriaminepenta(methylenephosphonic acid))
Limit of detection: 10 nM

KEY WORDS
complexation; SPE; wastewater

REFERENCE
Nowack, B. Determination of phosphonates in natural waters by ion-pair high-performance liquid chromatography, *J.Chromatogr.A*, **1997**, *773*, 139–146.

Nickel

SAMPLE
Matrix: solutions
Analyte: amino drugs
Sample preparation: Mix 1 mL of an aqueous solution with 1 mL 100 mM nickel sulfate in water, 1 mL 20% aqueous ammonia, and 5 mL chloroform:carbon disulfide 98:2, shake vigorously for 1 min, wash the organic layer with three 2 mL portions of water, filter (phase-separation paper). Evaporate the filtrate to dryness under a stream of nitrogen, reconstitute with 1 mL mobile phase, inject a 10 µL aliquot. (Copper may also be used with electrochemical detection or UV detection at 270 nm.)

HPLC VARIABLES
Guard column: 30 × 4 40 µm LiChrosorb RP-18
Column: 250 × 4 7 µm LiChrosorb RP-18
Mobile phase: EtOH:dichloromethane:20 mM pH 5.8 sodium acetate buffer 90:5:5 containing 5 mM lithium perchlorate (perhexiline) or MeOH:20 mM pH 5.8 sodium acetate buffer 80:20 containing 5 mM lithium perchlorate (others)
Flow rate: 1 (perhexiline), 1.5 (others)
Injection volume: 10
Detector: UV 325; E, Merck-Clevenot E 230, Model LCC 231 thin-layer electrolytic cell with a glassy carbon electrode at +0.7 V, standard calomel reference electrode

CHROMATOGRAM
Retention time: k' 2.48 (acebutolol), k' 21.96 (alprenolol), k' 3.07 (ephedrine), k' 2.93 (flecainide), k' 9.67 (methamphetamine), k' 21.96 (propranolol), k' 3.28 (perhexiline)
Limit of detection: 1 fmole (E), 1 nmole (UV)

KEY WORDS
complexation

REFERENCE
Leroy, P.; Nicolas, A. Determination of secondary amino drugs as their metal dithiocarbamate complexes by reversed-phase high-performance liquid chromatography with electrochemical detection, *J.Chromatogr.*, **1984**, *317*, 513–521.

SAMPLE
Matrix: solutions
Analyte: cyanide
Sample preparation: Pass the solution through a 50 × 10 column of resin, inject an aliquot of the eluate. (Prepare resin by cleaning Bio-Rex 70 cation-exchange resin with sulfonate functionality (Bio-Rad) with several volumes of water, adding 10 mM nickel sulfate

solution, and rinsing with water. Remove sulfide interference by adding ferric ions to the solution.)

HPLC VARIABLES
Column: 250 × 4.6 Transition Metal U7 (C18) (Alltech)
Mobile phase: MeCN:water 35:65 containing 5 mM tetrabutylammonium perchlorate
Flow rate: 0.7
Detector: UV 268

CHROMATOGRAM
Retention time: 12
Limit of detection: 80 ng/mL (of NaCN)

OTHER SUBSTANCES
Non-interfering: ammonium, cadmium, iron, nitrate, sulfate, sulfide, zinc

KEY WORDS
complexation

REFERENCE
Cox, J.A.; Novak, H.L.; Montgomery, R.M. High-performance liquid chromatographic determination of cyanide using a pre-column conversion to a transition metal complex, *J.Chromatogr.A*, **1996**, *739*, 239–232.

Palladium Chloride

SAMPLE
Matrix: solutions
Analyte: malathion
Sample preparation: Thoroughly mix 2 mL of a solution in MeOH with 2 mL 6 M NaOH for 5-7 min, add 4 mL 7 M HCl, add 1 mL 1 mg/mL palladium(II) chloride, add 2 mL chloroform, mix well. Remove a 1 mL aliquot of the chloroform layer and evaporate it to dryness, reconstitute the residue in 1 mL MeOH, inject a 2 μL aliquot. (Malathion is hydrolyzed to dimethyldithiophosphate which then forms a complex with palladium.)

HPLC VARIABLES
Column: 150 × 4.6 5 μm YMC Pack ODS-AQ
Mobile phase: MeCN:MeOH:water 10:20:70
Flow rate: 0.6
Injection volume: 2
Detector: UV 295

CHROMATOGRAM
Retention time: 4
Limit of detection: 2.4 μg/mL

KEY WORDS
complexation

REFERENCE
Khuhawar, M.Y.; Channar, A.H.; Lanjwani, S.N. Indirect liquid chromatographic determination of malathion in formulations, based on the formation of palladium(II)-dimethyldithiophosphate complex, *J.Chromatogr.A*, **1997**, *758*, 159–162.

Streptavidin Labeled with Fluorescein

SAMPLE
Matrix: formula
Analyte: biotin and related compounds
Sample preparation: 20 mL Infant formula + 150 μL concentrated HCl, filter (Whatman No. 1 paper), rinse precipitate with 1 mL water, adjust the pH of the filtrate to 7.0 with 6 M NaOH, extract four times with 8 mL n-hexane, dilute the aqueous layer to 25 mL with water, inject an aliquot.

HPLC VARIABLES
Guard column: 15 × 4.6 5 μm Microsorb C18
Column: 250 × 4.6 5 μm Microsorb C18
Mobile phase: MeOH:100 mM pH 7.0 phosphate buffer 20:80
Flow rate: 0.4
Injection volume: 20
Detector: F ex 495 em 518 following post-column reaction. The effluent from the column mixed with reagent pumped at 0.1 mL/min and flowed through a 10 m × 0.5 mm ID PTFE tube in a knitted open-tubular reactor to the detector. (Reagent was 2 μg/mL streptavidin-FITC (streptavidin labeled with fluorescein isothiocyanate 1:3.6, Vector Laboratories) in 100 mM pH 9.5 phosphate buffer, prepared fresh daily.)

CHROMATOGRAM
Retention time: 13 (ethylenediamine), 14 (biotin), 16 (biocytin), 27.5 (6-(biotinoylamino)caproic acid), 34.5 (6-(biotinoylamino)caproic acid hydrazide)
Limit of detection: 97 pg

OTHER SUBSTANCES
Non-interfering: niacinamide

KEY WORDS
post-column reaction; complexation

REFERENCE
Hentz, N.G.; Bachas, L.G. Class-selective detection system for liquid chromatography based on the streptavidin-biotin interaction, *Anal.Chem.*, **1995**, *67*, 1014–1018.

Yttrium

SAMPLE
Matrix: cell suspensions
Analyte: inositol polyphosphates
Sample preparation: Centrifuge 10 mL cell suspension at 450 g for 2 min, discard the supernatant, add 1 mL ice-cold 10% perchloric acid, vortex, freeze-thaw twice with liquid nitrogen, let stand at 0° for 30 min, centrifuge at 8800 g for 10 min. Remove the supernatant and adjust the pH to 4-5 with KOH solutions of varying concentration (7-0.875 M), let stand at 0° for 30 min, centrifuge at 8800 g for 10 min, freeze-dry the supernatant, dissolve in 1 mL water, adjust pH to <5, add 50 μL Norit suspension, vortex 5 times over 15 min, centrifuge for 3 min, treat the supernatant again in the same fashion, centrifuge, extract the pellets with 1 mL 100 mM NaCl (Biochem. J. 1988, 254, 585). Combine all the supernatants, filter (0.45 μm), inject an aliquot of the filtrate. (Prepare Norit suspension by boiling Norit A in 3 M HCl for 2 h, wash with water until the washings are neutral, dry at 120°, prepare a 20% suspension in 100 mM pH 4.0 NaCl containing 50 mM sodium acetate (Biochem. J. 1988, 254, 585).)

HPLC VARIABLES
Column: two 50 × 5 MonoQ (Pharmacia) columns in series

Mobile phase: Gradient. A was 20 μM HCl containing 13.5 μM yttrium chloride. B was 400 mM HCl containing 21 μM yttrium chloride. A:B 0:100 for 5 min, to 95:5 over 4 min, to 90:10 over 11 min, to 82:18 over 11 min, to 34:66 over 13 min, to 40:60 over 12 min, to 20:80 over 8 min, to 10:90 over 8 min (Biochem. J. 1988, 254, 585).

Flow rate: 1.2

Detector: UV 520 following post column reaction. The column effluent mixed with the reagent pumped at 0.6 mL/min in a 800 μL mixing chamber and this mixture flowed to the detector. (Reagent was 1.6 M pH 9.1 triethanolamine containing 350 μM 2-(4-pyridylazo)resorcinol.)

CHROMATOGRAM
Retention time: 20-75 (depending on structure)

KEY WORDS
post-column reaction; indirect UV detection; complexation. The yttrium ions complex with the pyridylazoresorcinol to form a colored complex. Inositol polyphosphates interfere with this complexation so the UV signal is reduced.

REFERENCE
Guse, A.H.; Emmrich, F. Determination of inositol polyphosphates from human T-lymphocyte cell lines by anion-exchange high-performance liquid chromatography and post-column derivatization, *J.Chromatogr.*, **1992**, *593*, 157–163.

POST-COLUMN EXTRACTION

1-Cyano-[2-(2-trimethylammonio)ethyl]benz(f)isoindole

SAMPLE
Matrix: solutions
Analyte: anionic surfactants

HPLC VARIABLES
Column: 40 × 4 Spherisorb S5-C1

Mobile phase: Gradient. A was 10 mM trisodium citrate containing 1 mL/L 5 M HCl. B was 10 mM trisodium citrate in MeCN:water 50:50 containing 1 mL/L 5 M HCl. A:B from 80:20 to 0:100 over 20 min, maintain at 0:100 for 2 min, return to initial conditions over 1 min, re-equilibrate at initial conditions for 2 min.

Flow rate: 0.8

Injection volume: 50

Detector: F ex 285 em 485 following post-column extraction. The column effluent mixed with the reagent pumped at 0.8 mL/min and the mixture flowed through a 50 cm × 0.5 mm ID PTFE coil. The effluent from this coil mixed with chloroform pumped at 0.8 mL/min and the mixture flowed through a 1 m × 0.5 mm ID Tefzel coil to a phase separator. The aqueous phase flowed to waste and the chloroform phase flowed to the detector. (At the end of each run flush the post-column extraction system with MeOH at 0.8 mL/min for 2 min. The reagent was 14 μM 1-cyano-[2-(2-trimethylammonio)ethyl]benz(f)isoindole in 11 mM trisodium citrate dihydrate. Synthesis of 1-cyano-[2-(2-trimethylammonio)ethyl]benz(f)isoindole is as follows. Dissolve 9.5 mg naphthalene-2,3-dicarboxaldehyde in 100 mL MeOH with sonication for 10 min. Dissolve 9 mg 2-aminoethyltrimethylammonium chloride hydrochloride in 100 mL 10 mM pH 9.5 sodium borate buffer, add 5.5 mL 10 mM NaCN, add the solution of naphthalene-2,3-dicarboxaldehyde in MeOH, shake for 20 s, let stand at room temperature for 2 h, wash twice with 200 mL portions of chloroform, store this 250 μM stock solution of 1-cyano-[2-(2-trimethylammonio)ethyl]benz(f)isoindole at room temperature in the dark, discard after 1 week (Int. J. Environ. Anal. Chem. 1991, 43, 79). Dilute with the sodium citrate solution before use.)

CHROMATOGRAM

Retention time: 6.5 (sodium 1-decanesulfonate), 10.5 (sodium 1-dodecanesulfonate), 12 (sodium 1-dodecanesulfate), 13 (sodium 1-tetradecanesulfonate), 14.5 (sodium 1-tetradecanesulfate), 16.5 (sodium 1-hexadecanesulfate)

Limit of quantitation: 12-66 ng

Limit of detection: 4-27 ng

KEY WORDS

post-column reaction; post-column extraction

REFERENCE

Schoester, M.; Kloster, G. HPLC separation and quantification of anionic surfactants using an automated on-line ion pair extraction system, *Fresenius J.Anal.Chem.*, **1993**, *345*, 767–772.

SAMPLE

Matrix: water

Analyte: herbicides

Sample preparation: Pump 20 mL aqueous sample through column A at 2.5 mL/min, backflush the contents of column A onto column B with mobile phase.

HPLC VARIABLES

Column: A 10 × 2 15-25 μm PLRP-S divinylbenzene-styrene copolymer (Polymer Laboratories); B 150 × 3.1 3 μm Hypersil MOS C8

Mobile phase: MeOH:buffer 55:45 (Buffer was water adjusted to pH 2.5 with phosphoric acid.)

Flow rate: 0.6

Injection volume: 20000

Detector: F ex 260 em 470 (cut-off filter) following post-column extraction. The column effluent mixed with the reagent pumped at 0.6 mL/min and the mixture flowed through a 25 cm × 0.5 mm ID stainless steel coil. The effluent from this coil mixed with chloroform:1-butanol 80:20 pumped at 1 mL/min and the mixture flowed through a 1.5 m × 0.5 mm stainless steel coil to a phase separator (Anal. Chim. Acta 1987, 192, 267). The aqueous phase flowed to waste and the chloroform phase flowed (at about 0.3 mL/min) to the detector. (The reagent was 12.5 μM 1-cyano-[2-(2-trimethylammonio)ethyl]benz(f)isoindole in 10 mM pH 8.0 sodium phosphate buffer. Synthesis of 1-cyano-[2-(2-trimethylammonio)ethyl]benz(f)isoindole is as follows. Dissolve 9.5 mg naphthalene-2,3-dicarboxaldehyde in 100 mL MeOH with sonication for 10 min. Dissolve 9 mg 2-aminoethyltrimethylammonium chloride hydrochloride in 100 mL 10 mM pH 9.5 sodium borate buffer, add 5.5 mL 10 mM NaCN, add the solution of naphthalene-2,3-dicarboxaldehyde in MeOH, shake for 20 s, let stand at room temperature for 2 h, wash twice with 200 mL portions of chloroform, store this 250 μM stock solution of 1-cyano-[2-(2-trimethylammonio)ethyl]benz(f)isoindole at room temperature in the dark, discard after 1 week. Dilute the stock solution with 19 volumes of 10 mM pH 8.0 sodium phosphate buffer before use.)

CHROMATOGRAM

Retention time: 2 (2,3,6-trichlorobenzoic acid), 5 (2,4-D), 6 (4-chloro-2-methylphenoxyacetic acid, MCPA), 8 (2,4,5-T), 9 (2-(4-chloro-2-methylphenoxy)propionic acid, MCPP), 11 (4-(2,4-dichlorophenoxy)butyric acid), 12 (4-(4-chloro-2-methylphenoxy)butyric acid, MCPB)

Limit of detection: 50-100 ppt

KEY WORDS

post-column reaction; post-column extraction; column-switching

REFERENCE

de Ruiter, C.; Minnaard, W.A.; Lingeman, H.; Kirk, E.M.; Brinkman, U.A.T.; Otten, R.R. Sensitive and selective fluorescence detection of phenoxyacid herbicides by liquid chromatography with on-line ion-pair extraction, *Int.J.Environ.Anal.Chem.*, **1991**, *43*, 79–90.

α-(3,4-Dimethoxyphenyl)-4'-trimethylammoniummethylcinnamonitrile Methosulfate

SAMPLE

Matrix: formulations

Analyte: drugs

Sample preparation: Add capsule contents to 80 mL MeOH, sonicate for 10 min, make up to 100 mL with MeOH, filter, allow to settle for 30 min. Dilute a 4 mL aliquot to 100 mL with mobile phase. Mix a 20 mL aliquot of the diluted solution with 10 mL 50 µg/mL ibuprofen in mobile phase, make up to 100 mL with mobile phase, inject a 50 µL aliquot.

HPLC VARIABLES

Column: 100 × 4.6 5 µm Spheri-5 RP-8

Mobile phase: MeOH:buffer 30:70 (Prepare buffer by mixing 4 mM Na_2HPO_4 and 7 mM KH_2PO_4 to achieve pH 7.)

Flow rate: 1

Injection volume: 50

Detector: F ex 355 em 460 (408 nm cutoff filter) following post-column extraction. The column effluent mixed with 50 µg/mL reagent in mobile phase pumped at 0.5 mL/min and then with chloroform pumped at 1 mL/min and the mixture flowed through a 1.8 m × 0.3 mm ID knitted PTFE coil to a 50 µL membrane phase separator using a polyethylene-backed 0.5 µm Fluoropore membrane filter (design in paper). The organic phase flowed to the detector. (Synthesize the reagent, α-(3,4-dimethoxyphenyl)-4'-trimethylammoniummethylcinnamonitrile methosulfate, as follows. Stir 20 mmoles 3,4-dimethoxyphenylacetonitrile and 20 mmoles p-toluamide in 50 mL EtOH at 50°, add 5 mL 50% aqueous KOH slowly, stir at 50° for 5 min, cool to room temperature, filter, dry the precipitate of α-(3,4-dimethoxyphenyl)-4'-methylcinnamonitrile. Dissolve 20 mmoles α-(3,4-dimethoxyphenyl)-4'-methylcinnamonitrile, 20 mmoles N-bromosuccinimide, and 20 mg benzoyl peroxide in 100 mL carbon tetrachloride (Caution! Carbon tetrachloride is a carcinogen!), reflux with stirring for 1.5 h , cool, filter, evaporate to dryness under reduced pressure, recrystallize from MeOH to give α-(3,4-dimethoxyphenyl)-4'-bromomethylcinnamonitrile. Vigorously stir 30 mmoles anhydrous dimethylamine in 100 mL dry benzene (Caution! Benzene is a carcinogen!) at 0°, very slowly add 10 mmoles α-(3,4-dimethoxyphenyl)-4'-bromomethylcinnamonitrile while stirring at 0°, stir at room temperature overnight, add 150 mL water, remove the organic phase, extract the aqueous phase twice with 100 mL portions of diethyl ether, wash the organic layers with saturated NaCl solution, dry over anhydrous magnesium sulfate, evaporate under reduced pressure to give α-(3,4-dimethoxyphenyl)-4'-dimethylaminomethylcinnamonitrile (J.Chem.Eng.Data 1987, 32, 387). Reflux 10 mmoles α-(3,4-dimethoxyphenyl)-4'-dimethylaminomethylcinnamonitrile, 20 mmoles dimethyl sulfate (Caution! Dimethyl sulfate is a carcinogen and acutely toxic!), and 5 g potassium carbonate in 50 mL acetone for 1 h, cool to room temperature, filter, dry the precipitate under vacuum at room temperature overnight, recrystallize from chloroform containing 2-3 drops of 95% EtOH to give α-(3,4-dimethoxyphenyl)-4'-trimethylammoniummethylcinnamonitrile methosulfate (mp 212-215°). Protect solutions from light.)

CHROMATOGRAM
Retention time: k' 0.2791 (salicylic acid), k' 0.3256 (valproic acid), k' 0.9690 (naproxen), k' 1.2791 (probenecid), k' 1.5504 (ketoprofen), k' 7.5039 (mefenamic acid)
Internal standard: ibuprofen (k' 4.124)
Limit of detection: 0.01-2 μg/mL

KEY WORDS
capsules; post-column extraction; post-column reaction

REFERENCE
Kim, M.; Stewart, J.T. HPLC post-column ion-pair extraction of acidic drugs using a substituted α-phenylcinnamonitrile quaternary ammonium salt as a new fluorescent ion-pair reagent, *J.Liq.Chromatogr.*, **1990**, *13*, 213–237.

Methylene Blue

SAMPLE
Matrix: dry beverage base
Analyte: sulfonic acids and sodium dioctylsulfosuccinate
Sample preparation: Dissolve dry beverage base in mobile phase, filter (0.45 μm), inject a 25 μL aliquot of the filtrate.

HPLC VARIABLES
Column: 100 × 4.6 10 μm C18 (CSC, Montreal)
Mobile phase: Acetone:20 mM KH_2PO_4 55:45
Flow rate: 1.5
Injection volume: 25
Detector: UV 546 following post-column reaction. The column effluent mixed with 10 μm methylene blue in 500 mM phosphoric acid pumped at 0.1 mL/min and this mixture then mixed with chloroform pumped at 1 mL/min. This mixture flowed through a 50 cm × 1 mm ID coil to a phase separator. The aqueous phase flowed upwards to waste and the organic phase flowed downwards through a wad of hydrophobic 1 μm Millipore FA filter material to the detector. (Greater sensitivity can be achieved at 657 nm. The acid/methylene blue ion pair is extracted into the chloroform.)

CHROMATOGRAM
Retention time: 3 (octanesulfonic acid), 4 (dodecanesulfonic acid), 8 (sodium dioctylsulfosuccinate), 10 (hexadecanesulfonic acid)
Limit of detection: 100 ng/g

KEY WORDS
post-column reaction; post-column extraction

REFERENCE

Lawrence, J.F. Determination of sodium dioctylsulfosuccinate in dry beverage bases by liquid chromatography with post-column ion-pair extraction and absorbance detection, *J.Assoc.Off.Anal.Chem.*, **1987**, *70*, 15–17.

SAMPLE

Matrix: food
Analyte: fatty acids
Sample preparation: Orange juice. 10 mL Orange juice + 70 mL water + 8 g NaCl, mix until homogeneous, add 50 mL dichloromethane, shake for 1 min, centrifuge at 400 g for 5 min, repeat extraction twice more. Combine the organic layers and evaporate them to dryness under reduced pressure at 40°, dissolve the residue in five 3 mL portions of MeCN. Combine the MeCN extracts and evaporate them to 1 mL under a stream of nitrogen at 40°, filter (0.45 μm), inject an aliquot. Butter, margarine. Warm 1 g butter of margarine at 40° until it melts, add 10 mL MeCN, vortex for 5 min, filter (0.45 μm), inject an aliquot.

HPLC VARIABLES

Column: 150 × 4.6 5 μm Spherisorb ODS-2
Mobile phase: Gradient. MeCN:buffer from 79:21 to 87:13 over 10 min, to 99:1 over 5 min, maintain at 99:1 for 5 min, return to initial conditions over 2 min, re-equilibrate for 6 min. (Buffer was water adjusted to pH 4.0 with 5% orthophosphoric acid.)
Flow rate: 0.8
Injection volume: 10
Detector: UV 651 following post-column extraction. The column effluent mixed with 2 μg/mL methylene blue in 20 mM Na_2HPO_4 pumped at 2 mL/min and the mixture flowed through a 45 cm × 0.5 mm ID stainless steel coil. The effluent from this coil mixed with chloroform pumped at 1 mL/min and the mixture flowed through a 50 cm × 1 mm ID stainless steel coil to a CTFE polymer phase separator fitted with a polyethylene filter disc from a SPE cartridge (construction details in paper). The organic phase flowed downwards to the detector at 0.5 mL/min and the aqueous phase flowed to waste. Every 2-3 injections rinse detector cell with 2-3 mL MeCN. (The acid/methylene blue ion pair is extracted into the chloroform.)

CHROMATOGRAM

Retention time: 3.0 (caprylic acid), 4.0 (capric acid), 6.0 (lauric acid), 7.3 (linolenic acid), 9.1 (myristic acid, palmitoleic acid), 9.7 (linoleic acid), 10.8 (linoledaidic acid), 13.8 (cis-oleic acid), 14.5 (palmitic acid, trans-elaidic acid), 20.5 (stearic acid)
Limit of detection: 26-83 ng

KEY WORDS

post-column extraction; post-column reaction; butter; margarine; orange juice

REFERENCE

Lawrence, J.F.; Charbonneau, C.F. Direct, sensitive and selective detection of free fatty acids by high-performance liquid chromatography with post-column ion-pair extraction and absorbance detection, *J.Chromatogr.*, **1988**, *445*, 189–197.

Sodium 9,10-Dimethoxyanthracene-2-sulfonate

SAMPLE
Matrix: blood
Analyte: rocuronium
Sample preparation: Condition a 1 mL Bond-Elut C18 SPE cartridge with 1 mL MeOH:
MeCN 2:1 and 1 mL water. Acidify 1 mL plasma with 200 μL 1 M NaH_2PO_4. Add 20-200
ng IS to 1 mL acidified plasma, add to the SPE cartridge, wash with 1 mL water, wash
with 1 mL 100 mM pH 3 NaH_2PO_4, elute with 400 μL mobile phase, discard first 100 μL
eluate, inject a 200 μL aliquot of the remaining eluate (from J. Chromatogr. 1987, 421,
327; modifications may be necessary).

HPLC VARIABLES
Column: 150 × 3.9 4 μm Nova-Pak C18
Mobile phase: Dioxane:water 20:80 containing 100 mM NaH_2PO_4 and 0.44 mM 9,10-di-
methoxyanthracene-2-sulfonate, pH adjusted to 3 with phosphoric acid. (Caution! Dioxane
is a carcinogen!) (After each series of analyses flush column with 200 mL MeOH then re-
equilibrate with 120 mL mobile phase.)
Flow rate: 1
Injection volume: 200
Detector: F ex 380 em 452 following post-column extraction. The column effluent mixed
with dichloroethane pumped at 1.6 mL/min and the mixture flowed through a 1 m × 0.25
mm ID stainless steel coil to a phase separator (Anal. Chim. Acta 1987, 192, 267) then
the organic phase flowed through the detector.

CHROMATOGRAM
Internal standard: 3,17-didesacetyl vecuronium (Org 7402)
Limit of quantitation: 10 ng/mL

OTHER SUBSTANCES
Extracted: metabolites

KEY WORDS
SPE; plasma; pharmacokinetics; post-column extraction; post-column reaction

REFERENCE
Cooper, R.A.; Maddineni, V.R.; Mirakhur, R.K.; Wierda, J.M.K.H.; Brady, M.; Fitzpatrick, K.T.J. Time
course of neuromuscular effects and pharmacokinetics of rocuronium bromide (Org 9426) during
isoflurane anaesthesia in patients with and without renal failure, *Br.J.Anaesth.*, **1993**, *71*, 222–226.

SAMPLE
Matrix: blood
Analyte: dopaminergic prodrugs
Sample preparation: 500 μL Plasma + 500 μL pH 7.4 phosphate buffer, mix, add 9 mL
dioxane (Caution! Dioxane is a carcinogen!), mix, centrifuge at high speed for 30 s, inject
a 20 μL aliquot of the supernatant.

HPLC VARIABLES
Guard column: μBondapak C18 Guard-Pak
Column: 150 × 3.9 Nova-Pak C18

Mobile phase: Dioxane:buffer 35:65 (Caution! Dioxane is a carcinogen!) (Buffer was 100 mM NaH_2PO_4 containing 0.11 mM sodium 9,10-dimethoxyanthracene-2-sulfonate and 0.11 mM 1-heptanesulfonic acid, pH adjusted to 3 with phosphoric acid. Purify sodium 9,10-dimethoxyanthracene-2-sulfonate (Fluka) by Soxhlet extraction with dichloroethane before use.)

Flow rate: 0.5

Injection volume: 20

Detector: F ex 380 em 452 following post-column extraction. The column effluent mixed with dichloroethane pumped at 1 mL/min and the mixture flowed through a 1 m × 0.25 mm i.d. stainless steel coil to a phase separator (Anal. Chim. Acta 1987, 192, 267) then the organic phase flowed through the detector.

CHROMATOGRAM

Retention time: k' 3.3 (ibopamine), k' 3.4 (dibudop), k' 10.9 (dipivaloyl dopamine), k' 11.7 (dibenzoyl dopamine)

Limit of detection: 10 ng

KEY WORDS

plasma; post-column extraction; post-column reaction

REFERENCE

Haas, M.; Moolenaar, F.; Kluppel, A.C.A.; Dijkstra, D.; Meijer, K.F.; De Zeeuw, D. Determination of dopaminergic prodrugs by high-performance liquid chromatography followed by post-column ion-pair extraction, *J.Chromatogr.B*, **1997**, *693*, 484–488.

SAMPLE

Matrix: blood, bile, perfusate, tissue, urine

Analyte: vecuronium

Sample preparation: Condition a 1 mL Bond-Elut C18 SPE cartridge with 1 mL MeOH: MeCN 2:1 and 1 mL water. Acidify 1 mL plasma with 150 μL 1 M NaH_2PO_4. Homogenize liver, centrifuge. Dilute 1 μL urine or bile to 1 mL with 100 mM pH 3 NaH_2PO_4. Add 20-200 ng IS to 1 mL acidified plasma, liver homogenate supernatant, diluted urine, diluted bile, or liver perfusate, add to the SPE cartridge, wash with 1 mL water, wash with 1 mL 100 mM pH 3 NaH_2PO_4, elute with 400 μL mobile phase, discard first 100 μL eluate, inject a 200 μL aliquot of the remaining eluate.

HPLC VARIABLES

Column: 150 × 3.9 4 μm Nova-Pak C18

Mobile phase: Dioxane:water 20:80 containing 100 mM NaH_2PO_4 and 0.44 mM 9,10-dimethoxyanthracene-2-sulfonate, pH adjusted to 3 with phosphoric acid. (Caution! Dioxane is a carcinogen!) (After each series of analyses flush column with 200 mL MeOH then re-equilibrate with 120 mL mobile phase.)

Flow rate: 1

Injection volume: 200

Detector: F ex 380 em 452 following post-column extraction. The column effluent mixed with dichloroethane pumped at 1.6 mL/min and the mixture flowed through a 1 m × 0.25 mm ID stainless steel coil to a phase separator (Anal. Chim. Acta 1987, 192, 267) then the organic phase flowed through the detector.

CHROMATOGRAM

Retention time: 13

Internal standard: 1-(3α,17β-diacetoxy-2β-piperidino-5α-androstan-16β,5α-yl)piperidine (16)

Limit of detection: 5 ng/mL

KEY WORDS

SPE; pharmacokinetics; rat; dog; human; cat; plasma; liver; post-column extraction; post-column reaction

REFERENCE

Paanakker, J.E.; Thio, J.M.; Van den Wildenberg, H.M.; Kaspersen, F.M. Assay of vecuronium in plasma using solid-phase extraction, high-performance liquid chromatography and post-column ion-pair extraction with fluorimetric detection, *J.Chromatogr.*, **1987**, *421*, 327–335.

SAMPLE
Matrix: blood, urine
Analyte: erythromycin A
Sample preparation: Urine. Centrifuge urine at 2500 g for 5 min, inject a 100 μL aliquot. Whole blood. 200 μL Whole blood + 100 μL 10% EDTA + 50 μL 20 μg/mL josamycin in MeOH:water 50:50, centrifuge to separate plasma. 200 μL Plasma + 20 μL saturated potassium carbonate + 1 mL MTBE, mix, centrifuge. Remove 800 μL of the MTBE layer and evaporate it to dryness, reconstitute the residue in 200 μL MeOH:water 50:50, inject a 100 μL aliquot.

HPLC VARIABLES
Column: 250 × 4.6 8 μm PLRP-S 1000 Å (Polymer Labs)
Mobile phase: MeCN:t-butanol:200 mM pH 9.0 phosphate buffer:water 3:19:5:73
Column temperature: 70
Flow rate: 1.5
Injection volume: 100
Detector: F ex 365 em 450 following post-column extraction. The column effluent mixed with reagent pumped at 0.7 mL/min and this mixture flowed through a 1.5 m × 0.5 mm ID stainless steel coil. The effluent from the coil mixed with chloroform pumped at 1.5 mL/min and this mixture flowed through a 1.5 m × 0.5 mm ID stainless steel coil to a sandwich-type phase separator with a 40 μL groove volume (Vrije Universiteit, Amsterdam). Part of the organic layer was separated and flowed through the detector at 0.5 mL/min. (Reagent was 5 μM sodium 9,10-dimethoxyanthracene-2-sulfonate in 100 mM citric acid.)

CHROMATOGRAM
Retention time: 11
Internal standard: josamycin (28)
Limit of detection: 12.5 ng/mL (plasma), 50 ng/mL (urine)

OTHER SUBSTANCES
Extracted: metabolites
Simultaneously analyzed: midecamycin, troleandomycin

KEY WORDS
plasma; whole blood; post-column extraction; post-column reaction

REFERENCE
Khan, K.; Paesen, J.; Roets, E.; Hoogmartens, J. Analysis of erythromycin A and its metabolites in biological samples by liquid chromatography with post-column ion-pair extraction, *J.Liq.Chromatogr.*, **1994**, *17*, 4195–4213.

SAMPLE
Matrix: sediment, sludge
Analyte: ditallowdimethylammonium chloride
Sample preparation: Extract 500 mg dry homogenized sludge or 5 g dry sediment with four 50 mL portions of hot 1 M HCl in MeOH. Evaporate the extracts to dryness, add IS, extract residue with three 50 mL portions of chloroform. Evaporate the extracts to dryness, reconstitute with 120 mL MeOH, pass through 100 g 50-100 mesh Dowex 1x2 Cl⁻ form, evaporate the eluate to dryness, reconstitute, inject an aliquot. Alternatively, extract 0.1-1 g sludge or sediment together with 30 mg acid-cleaned Cu powder with supercritical carbon dioxide:MeOH 70:30 at 380 atm at 100° at 0.5-0.8 mL/min for at least 15 min using a 10 cm × 30 μm (sludge) or 12 cm × 40 μm (sediment) fused-silica capillary, collect in 3-4 mL chloroform or ethyl acetate, evaporate to dryness under a stream of nitrogen, reconstitute, inject an aliquot.

HPLC VARIABLES
Column: 125 × 4 5 μm Nucleosil NH2
Mobile phase: Gradient. A was chloroform:MeOH:acetic acid 99:1:0.25. B was chloroform:MeOH:acetic acid 75:25:0.25. A:B 95:5 for 2 min, to 75:25 over 13 min, to 60:40 over 2 min, maintain at 60:40 for 2 min, return to initial conditions over 3 min, re-equilibrate for 8 min.
Flow rate: 1.2

Detector: UV 420 following post-column extraction. The column effluent mixed with 100 μg/mL methyl orange in water pumped at 0.4 mL/min and the mixture flowed through a 1 m × 0.7 mm PEEK (1.5 cm dia) or stainless steel (3 cm dia) coil to a phase separator (Int.J.Environ.Anal.Chem. 1987, 31, 325). The organic phase flowed to the detector. After each run clean post-column extraction system with 5% ammonia in MeOH at 1 mL/min for 5 min.; F ex 360 em 450 following post-column extraction. The column effluent mixed with 30 μg/mL 9,10-dimethoxyanthracene-2-sulfonic acid in water pumped at 0.4 mL/min and the mixture flowed through a 1 m × 0.7 mm PEEK (1.5 cm dia) or stainless steel (3 cm dia) coil to a phase separator (Int.J.Environ.Anal.Chem. 1987, 31, 325). The organic phase flowed to the detector. After each run clean post-column extraction system with 5% ammonia in MeOH at 1 mL/min for 5 min.

CHROMATOGRAM
Retention time: 6
Internal standard: dimethyldioctylammonium iodide (9)
Limit of quantitation: 80 μg/g (UV)
Limit of detection: 16 μg/g (F)

KEY WORDS
condition glassware overnight with 5 μg/mL ditallowdimethylammonium chloride in water, rinse with hot water, distilled water, MeOH, dry; SFE; normal phase; post-column extraction; post-column reaction

REFERENCE
Fernández, P.; Alder, A.C.; Suter, M.J.-F.; Giger, W. Determination of the quaternary ammonium surfactant ditallowdimethylammonium in digested sludges and marine sediments by supercritical fluid extraction and liquid chromatography with postcolumn ion-pair formation, *Anal.Chem.*, **1996**, *68*, 921–929.

SAMPLE
Matrix: sewage
Analyte: benzalkonium chloride
Sample preparation: Condition a C18ec SPE cartridge (Macherey-Nagel) with three bed volumes of MeOH and three bed volumes of water. Allow sewage to settle, add an 8 mL aliquot of the supernatant to the SPE cartridge, wash with 3 bed volumes of water, wash with 2 bed volumes of ethyl acetate, elute with 2 bed volumes of 1% calcium chloride in MeOH:ethyl acetate 50:50, inject a 15 μL aliquot of the eluate. (For all steps involving the SPE cartridge the flow rate should be 3 mL/min.)

HPLC VARIABLES
Column: 250 × 4 5 μm Partisil PAC CN-NH2
Mobile phase: Chloroform:MeOH 80:20
Column temperature: 15
Flow rate: 1
Injection volume: 15
Detector: F ex 383 em 459 following post-column extraction. The column effluent mixed with 37 μg/mL sodium 9,10-dimethoxyanthracene-2-sulfonate in water pumped at 0.3 mL/min and the mixture flowed to a phase separator. The organic phase flowed to the detector and the aqueous phase was discarded. (9,10-Dimethoxyanthracene-2-sulfonate forms a hydrophobic fluorescent ion-pair with benzalkonium chloride.)

CHROMATOGRAM
Retention time: 2.49
Limit of detection: 50 ng/mL

OTHER SUBSTANCES
Simultaneously analyzed: dimethyldidecylammonium chloride (DDMAC)

KEY WORDS
post-column reaction; post-column extraction; SPE

REFERENCE

Kümmerer, K.; Eitel, A.; Braun, U.; Hubner, P.; Daschner, F.; Mascart, G.; Milandri, M.; Reinthaler, F.; Verhoef, J. Analysis of benzalkonium chloride in the effluent from European hospitals by solid-phase extraction and high-performance liquid chromatography with post-column ion-pairing and fluorescence detection, *J.Chromatogr.A*, **1997**, *774*, 281–286.

Sodium α-(3,4-Dimethoxyphenyl)cinnamonitrile-2'-sulfonate

SAMPLE

Matrix: blood

Analyte: physostigmine

Sample preparation: Condition a 3 mL 200 mg octadecyl SPE cartridge with 2 column volumes of MeOH and 2 column volumes of 100 mM pH 4 phosphate buffer. 1 mL Serum + 5 μL 1 mg/mL neostigmine bromide in 100 mM pH 4 NaH_2PO_4, vortex for 15 s, add 1 mL 1 mM reagent in 100 mM pH 4 phosphate buffer, mix for 30 s, add to the SPE cartridge, wash with 4 mL water, elute with 200 μL MeOH:water 95:5. Evaporate the eluate to dryness under a stream of nitrogen at 30°, reconstitute the residue in 100 μL mobile phase, inject a 50 μL aliquot. (Synthesize reagent, sodium α-(3,4-dimethoxyphenyl) cinnamonitrile-2'-sulfonate, as follows. Add 5 mL 10% KOH in water to a stirred solution of 20 mmoles 3,4-(dimethoxyphenyl)acetonitrile and 20 mmoles 2-formylbenzenesulfonic acid, sodium salt hydrate (sodium benzaldehyde-2-sulfonate) in 50 mL EtOH at 50°, stir at 50° for 5 min, cool (evaporate to near dryness, if necessary), filter to obtain sodium α-(3,4-dimethoxyphenyl) cinnamonitrile-2'-sulfonate (mp of p-toluidine salt is 218-223°) (J. Chem. Eng. Data 1975, 20, 215).)

HPLC VARIABLES

Column: 250 × 4.6 5 μm diol (ES Industries, Marlton NJ)

Mobile phase: MeOH:50 mM pH 4 NaH_2PO_4 20:80 containing 500 μM sodium α-(3,4-dimethoxyphenyl)cinnamonitrile-2'-sulfonate

Flow rate: 1

Injection volume: 50

Detector: F ex 243 em 418 (cutoff filter) following post-column extraction. The column effluent mixed with dichloromethane pumped at 1 mL/min and the mixture flowed through a 90 cm × 0.3 mm ID knitted PTFE coil to a 50 μL membrane phase separator using a polyethylene-backed 0.5 μm Fluoropore membrane filter (design in paper). The organic phase flowed to the detector.

CHROMATOGRAM

Retention time: 6.42

Internal standard: neostigmine (11.16)

Limit of detection: 2 ng/mL

OTHER SUBSTANCES

Extracted: metabolites, eseroline

Also analyzed: amantadine, amphetamine, atropine, chlorpheniramine, clidinium bromide, N,N-dimethyl-N-benzyltetradecylammonium chloride, guanethidine, hydralazine, imipramine, malachite green, promazine, propantheline bromide

Non-interfering: chlordiazepoxide

KEY WORDS
post-column extraction; SPE; serum; silanize glassware; post-column reaction

REFERENCE
Quinn, K.D.; Stewart, J.T. A high performance liquid chromatographic post-column fluorescent ion pair extraction system: application to physostigmine and its metabolite eseroline in human serum, *Biomed.Chromatogr.*, **1991**, *5*, 8–13.

POST-COLUMN PHOTOCHEMICAL DERIVATIZATION

Chemiluminescence Detection

SAMPLE
Matrix: water
Analyte: phenols
Sample preparation: Condition a Baker amino SPE cartridge with dichloromethane, dry with nitrogen. Filter (2 μm) water, acidify to pH 3.0 with 1 M nitric acid, add a 3 mL aliquot to a Baker C18 SPE cartridge, elute with 3 mL dichloromethane, evaporate the eluate to dryness under a stream of nitrogen. Reconstitute with 100 μL 30 mg/mL pH 12 tetrabutylammonium bromide in water, add 500 μL pH 12 water, add 600 μL 100 μg/mL dansyl chloride in dichloromethane, vortex vigorously for 2 min, add a 500 μL aliquot of the dichloromethane layer to the amino SPE cartridge, let stand for 10 min, elute with 3 mL dichloromethane. Evaporate the eluate to dryness, reconstitute the residue in 500 μL MeOH:water 50:50, inject a 100 μL aliquot. (Initial SPE extraction is only validated for 2,4-dimethylphenol, 4-chloro-3-methylphenol, 2,4-dichlorophenol, 2,4,6-trichlorophenol, and pentachlorophenol. Excess dansyl chloride reacts with the amino groups in the SPE cartridge.)

HPLC VARIABLES
Column: 200 × 3.1 3 μm LiChrosorb RP-18
Mobile phase: Gradient. A was MeOH:100 mM pH 7.0 imidazole buffer 97.5:2.5. B was MeOH:2.5 mM pH 7.0 imidazole buffer 2.5:97.5. A:B 75:25 for 9.5 min, to 85:15 over 0.5 min, maintain at 85:15 for 4.5 min, to 95:5 over 0.5 min, maintain at 95:5 for 4.5 min, to 100:0 over 0.5 min, maintain at 100:0 for 20 min, return to initial conditions over 1 min, re-equilibrate for 20 min.
Flow rate: 0.5
Injection volume: 100
Detector: Chemiluminescence (470 nm cut-off filter) following post-column reaction. The column effluent flowed through a 130 × 0.3 mm ID PTFE coil irradiated with a fan-cooled 90 w Philips Model 93110E mercury lamp. The effluent from this coil mixed with 50 mM hydrogen peroxide in MeCN containing 5 mM bis(2-nitrophenyl) oxalate pumped at 0.3 mL/min and this mixture flowed to the detector.

CHROMATOGRAM
Retention time: 9.5 (2-nitrophenol), 14 (phenol), 16.5 (4-nitrophenol), 19 (2-chlorophenol), 23 (2,4-dimethylphenol), 24.5 (4-chloro-3-methylphenol), 25.5 (2,4-dichlorophenol), 28 (2,4,6-trichlorophenol), 38 (pentachlorophenol)
Limit of detection: 0.1 ng/mL

KEY WORDS
post-column reaction; SPE; post-column photochemical derivatization

REFERENCE
Kwakman, P.J.M.; Kamminga, D.A.; Brinkman, U.A.T.; de Jong, G.J. Sensitive liquid chromatographic determination of alkyl-, nitro- and chlorophenols by precolumn derivatization with dansyl chloride, postcolumn photolysis and peroxyoxalate chemiluminescence detection, *J.Chromatogr.*, **1991**, *553*, 345–356.

Electrochemical Detection

RELATED REFERENCES
Bachman, W.J.; Stewart, J.T. HPLC-photolysis-electrochemical detection in pharmaceutical analysis: application to the determination of spironolactone and hydrochlorothiazide in tablets. *J.Chromatogr.Sci.* **1990**, *28*, 123-128.

Childress, W.L.; Krull, I.S.; Selavka, C.M. Determination of deoxynivalenol (DON, vomitoxin) in wheat by high-performance liquid chromatography with photolysis and electrochemical detection (HPLC-hv-EC). *J.Chromatogr.Sci.* **1990**, *28*, 76-82.

Dou, L.; Krull, I.S. Identification of photochemical products of amino acids, peptides, and proteins in on-line, postcolumn photolytic derivatization detection by HPLC-electrochemistry. *Electroanalysis* **1992**, *4*, 381-391.

Krull, I.S.; Selavka, C.M.; Nelson, R.J.; Bratin, K.; Lurie, I. Trace analysis for drugs and biologically active materials using LC-photolysis-electrochemistry. *Curr.Sep.* **1985**, *7*, 11-14.

Krull, I.S.; Dou, L. Determination of peptides and proteins using on-line, post-column photolytic derivatization in liquid chromatography/electrochemistry (LC-hv-EC). *Curr.Sep.* **1992**, *11*, 7-11.

Lookabaugh, M.; Krull, I.S. Determination of nitrite and nitrate by reversed-phase high-performance liquid chromatography using on-line post-column photolysis with ultraviolet absorbance and electrochemical detection. *J.Chromatogr.* **1988**, *452*, 295-308.

Selavka, C.M.; Krull, I.S.; Bratin, K. Analysis for penicillins and cefoperazone by HPLC-photolysis-electrochemical detection (HPLC-hv-EC). *J.Pharm.Biomed.Anal.* **1986**, *4*, 83-93.

SAMPLE
Matrix: beverages, formulations
Analyte: aspartame
Sample preparation: Dilute beverages and formulations with water, inject an aliquot.

HPLC VARIABLES
Column: 150 × 4.6 5 µm Spherisorb Hexyl
Mobile phase: MeOH:0.1% perchloric acid 15:85, pH 2.8
Flow rate: 1
Detector: E, ESA Coulochem Model 5100A, first electrode (screen mode) +0.10 V, second electrode (measuring electrode) +0.80 V, following post-column reaction. The column effluent flowed through a 20 m × 0.3 mm ID PTFE coil irradiated with a UV 254 lamp to the detector.

CHROMATOGRAM
Retention time: 13
Limit of quantitation: 1 µg/mL
Limit of detection: 500 ng/mL

OTHER SUBSTANCES
Simultaneously analyzed: caffeine

KEY WORDS
post-column reaction; soft drinks; post-column photochemical derivatization

REFERENCE
Galletti, G.C.; Bocchini, P. High-performance liquid chromatography with electrochemical detection of aspartame with a post-column photochemical reactor, *J.Chromatogr.A*, **1996**, *729*, 393–398.

SAMPLE
Matrix: blood
Analyte: clofibric acid
Sample preparation: Condition a 1 mL 100 mg C2 SPE cartridge (Analytichem) with 2 mL MeOH and 1 mL water. 100 µL Plasma + 20 µL 200 µg/mL ketoprofen in MeOH + 500 µL 1 M HCl, vortex for 15 s, add to the SPE cartridge, rinse out tube with 1 mL water, add rinse to the SPE cartridge, elute with 1 mL mobile phase, vortex the eluate, inject a 10 µL aliquot.

HPLC VARIABLES
Guard column: 30-40 µm pellicular Vydac Reversed-Phase
Column: 75 × 3.9 4 µm Nova-Pak phenyl

Mobile phase: MeOH:buffer 42:58 (Buffer was 100 mM NaH_2PO_4 adjusted to pH 7.0 with 50% aqueous NaOH.)

Flow rate: 1

Injection volume: 10

Detector: E, Bioanalytical Systems LC-4B, LC-17 thin-layer glassy carbon working electrode +1.10 V, Ag/AgCl reference electrode following post-column reaction. The column effluent flowed through an air-cooled 7.9 m × 0.3 mm ID PTFE coil irradiated by an SC3-9 ultraviolet lamp (UVP, Inc.) to the detector.

CHROMATOGRAM

Retention time: 2

Internal standard: ketoprofen (2.7)

Limit of detection: 200 ng/mL

KEY WORDS

post-column reaction; plasma; SPE; post-column photochemical derivatization

REFERENCE

Bachman, W.J.; Stewart, J.T. HPLC-photolysis-electrochemical detection in pharmaceutical analysis: Application to the determination of clofibric acid in human plasma, *J.Liq.Chromatogr.*, **1989**, *12*, 2947–2959.

SAMPLE

Matrix: blood

Analyte: Ro 42-5892 ((S)-α-[(S)-α-[(tert-butylsulfonyl)methyl]hydrocinnamido]-N-[(1S,2R, 3S)-1-(cyclohexylmethyl)-3-cyclopropyl-2,3-dihydroxylpropyl]imidazole-4-propionamide methanesulfonate)

Sample preparation: 1 mL Plasma + 85 μL 1 M NaOH + 800 μL 8.75 ng/mL IS in MeCN, mix, let stand for 10 min, centrifuge at 1500 g for 10 min. Remove the liquid phase and add it to 2 mL dichloromethane, rotate head-over-head at 20 rpm for 10 min, centrifuge at 1700 g for 20 min. Remove the lower organic layer and evaporate it to dryness under a stream of helium at 70°, reconstitute the residue in 70 μL MeCN:water 50:50, centrifuge at 4600 g for 10 min. Remove a 65 μL aliquot and cool it to 10°, add 15 μL 200 mM pH 8.0 sodium borate buffer at 10°, add 15 μL 133 mM 2,4-dinitrofluorobenzene in MeCN at 10°, heat at 80° for 25 min, inject a 50 μL aliquot while still hot.

HPLC VARIABLES

Guard column: 4 × 4 5 μm LiChrospher 100 RP-18

Column: 150 × 3.9 4 μm Novapak C18

Mobile phase: MeCN:buffer 85:100 (Buffer was 92.5 mg/L disodium EDTA in 100 mM acetic acid, adjusted to pH 7.0 with 1 M NaOH.)

Flow rate: 1.2

Injection volume: 25

Detector: E, ESA Coulochem II, Model 5011 measuring cell, first electrode 300 mV second electrode 550 mV, Model 5020 guard cell 800 mV (before injector) following post-column reaction. The column effluent flowed through a 1 m × 0.3 mm ID PTFE coil irradiated at 254 nm to the detector.

CHROMATOGRAM

Retention time: 10.5

Internal standard: (S)-α-[(S)-α-[(tert-butylsulfonyl)methyl]hydrocinnamido]-N-[(1S,2R,3S)-1-(cyclohexylmethyl)-3-isopropyl-2,3-dihydroxylpropyl]imidazole-4-propionamide (Ro 42-4661/000) (16.5)

Limit of quantitation: 0.3 ng/mL

KEY WORDS

post-column reaction; plasma; post-column photochemical derivatization

REFERENCE

Leube, J.; Fischer, G. Determination of the renin inhibitor Ro 42-5892 in human plasma by automated pre-column derivatization, reversed-phase high-performance liquid chromatographic separation and electrochemical detection after post-column irradiation, *J.Chromatogr.B*, **1995**, *665*, 373–381.

SAMPLE

Matrix: broncho-alveolar lavage fluid

Analyte: aspoxicillin
Sample preparation: 1 mL Broncho-alveolar lavage fluid + 100 μL 5 μg/mL amoxicillin, vortex for 10 s, filter (Tosoh Ultracent-30 with a molecular mass cut-off at 30000) while centrifuging at 1500 g at 5° for 30 min, inject a 100 μL aliquot of the ultrafiltrate.

HPLC VARIABLES
Column: 150 × 4.6 5 μm Shodex C18 5A (Showa Denko)
Mobile phase: MeCN:50 mM pH 3.0 potassium hydrogen phosphate containing 20 mM sodium 1-heptanesulfonate and 5 mg/L EDTA 10:100
Column temperature: 40
Flow rate: 1.2
Injection volume: 100
Detector: E, Irica Σ875, glassy carbon electrode 800 mV, Ag/AgCl reference electrode, following post-column reaction. The column effluent flowed through a 10 m × 0.3 mm PTFE coil irradiated by a GL-10 10 W mercury lamp to the detector.

CHROMATOGRAM
Retention time: 24
Internal standard: amoxicillin (17)
Limit of detection: 1 ng/mL

KEY WORDS
ultrafiltrate; post-column reaction; post-column photochemical derivatization

REFERENCE
Yamazaki, T.; Ishikawa, T.; Nakai, H.; Miyai, M.; Tsubota, T.; Asano, K. Determination of aspoxicillin in broncho-alveolar lavage fluid by high-performance liquid chromatography with photolysis and electrochemical detection, *J.Chromatogr.*, **1993**, *615*, 180–185.

SAMPLE
Matrix: bulk
Analyte: cocaine
Sample preparation: Inject a 200 μL aliquot.

HPLC VARIABLES
Column: 100 × 5 10 μm Radial-Pak C18 radial compression (Waters)
Mobile phase: MeOH:100 mM NaCl:butylamine 50:50:0.7, adjusted to pH 3 with sulfuric acid
Flow rate: 1
Injection volume: 200
Detector: E, Bioanalytical systems LC4B, MF 1000 dual glassy carbon working electrode + 1.1 V and +0.75 V, MF 1018 stainless steel auxiliary electrode, MF 2020 Ag/AgCl reference electrode following post-column reaction. The column flowed through a knitted 9.1 m × 0.5 mm ID PTFE coil irradiated at 254 nm by a Photronix medium-pressure mercury lamp (cooled with ice-water) to the detector.

CHROMATOGRAM
Retention time: k' 0.53
Limit of detection: 500 ppb

OTHER SUBSTANCES
Simultaneously analyzed: amylocaine, ascorbic acid, caffeine, chloroprocaine, lidocaine, niacinamide, procaine
Non-interfering: inositol, lactose, mannitol

KEY WORDS
post-column reaction; post-column photochemical derivatization

REFERENCE
Selavka, C.M.; Krull, I.S.; Lurie, I.S. An improved method for the rapid screening of illicit cocaine preparations using high performance liquid chromatography with electrochemical detection, *Forensic Sci.Int.*, **1986**, *31*, 103–117.

SAMPLE
Matrix: formulations
Analyte: insulin

HPLC VARIABLES
Column: 50 × 4.6 C4 wide pore (300 Å) (Supelco)
Mobile phase: Gradient. A was MeOH:isopropanol:water 4:1:95 containing 2.8 g/L NaCl. B was MeOH:isopropanol:water 60:10:30 containing 4.2 g/L NaCl. A:B from 45:55 to 30:70 over 5 min.
Flow rate: 1
Injection volume: 20
Detector: E, Bioanalytical systems Model LC-4B, dual glassy-carbon working electrode used in parallel mode, +0.65 V and +0.80 V (monitored), stainless steel auxiliary electrode, Ag/AgCl reference electrode following post-column reaction. The column effluent flowed at 0-5° through a 2 mL knitted coil of 0.5 mm i.d. PTFE tubing irradiated with a low pressure mercury lamp (Photronix Model 816) to the detector.

CHROMATOGRAM
Retention time: 8

OTHER SUBSTANCES
Also analyzed: b-lactoglobulin A, lysozyme, phenylalanine, ribonuclease A, tryptophan, tyramine

KEY WORDS
post-column reaction; post-column photochemical derivatization

REFERENCE
Dou, L.; Krull, I.S. Determination of aromatic and sulfur-containing amino acids, peptides, and proteins using high-performance liquid chromatography with photolytic electrochemical detection, *Anal.Chem.*, **1990**, *62*, 2599–2606.

SAMPLE
Matrix: solutions
Analyte: L-658,758

HPLC VARIABLES
Column: 100 × 4.6 3 μm Hypersil ODS
Mobile phase: MeOH:100 mM pH 3.0 lithium perchlorate 20:80
Column temperature: 50
Flow rate: 1.5
Detector: E, Bioanalytical Systems LC-4A, single glassy carbon working electrode, stainless steel auxiliary electrode, Ag/AgCl reference electrode following post-column reaction. The column effluent flowed through a knotted coil of 40 cm × 0.025 mm ID (sic, 0.25 mm ID?) PTFE tubing irradiated with an uncooled shortwave UV lamp (Mineralight UVS-11, Ultraviolet Products, Inc.) to the detector.

CHROMATOGRAM
Limit of detection: 1 ng

KEY WORDS
post-column reaction; post-column photochemical derivatization

REFERENCE
McClintock, S.A.; Cotton, M.L. Analysis of a modified neutral cephalosporin by HPLC-photolysis-electrochemical detection, *J.Liq.Chromatogr.*, **1989**, *12*, 2961–2970.

SAMPLE
Matrix: solutions
Analyte: chromate and dichromate
Sample preparation: Inject a 50 μL aliquot.

HPLC VARIABLES
Column: 250 × 4 10 μm LiChrospher C18
Mobile phase: MeOH:buffer 30:70, pH 6.8 containing 3 mM tetrabutylammonium hydrogen sulfate (Buffer was 7 mM Na_2HPO_4 and 7.2 mM NaH_2PO_4.)
Flow rate: 0.8
Injection volume: 50
Detector: E, Bioanalytical Systems Model LC-4B, dual glassy carbon working electrode at +1.15 V (monitored) and +1.00 V, stainless steel auxiliary electrode, Ag/AgCl reference

electrode following post-column reaction. The column effluent flowed through a 1.8 mL quartz tube irradiated by a UV 214 zinc lamp (BHK, Monrovia CA) to the detector.

CHROMATOGRAM
Retention time: 9
Limit of detection: 500 ppb

KEY WORDS
post-column reaction; no chromatographic separation of chromate and dichromate; post-column photochemical derivatization

REFERENCE
Dou, L.; Krull, I.S. Determination of inorganic anions by flow injection analysis and high-performance liquid chromatography combined with photolytic-electrochemical detection, *J.Chromatogr.*, **1990**, *499*, 685–697.

SAMPLE
Matrix: solutions
Analyte: perchlorate, periodate, and thiocyanate
Sample preparation: Inject a 20 µL aliquot.

HPLC VARIABLES
Column: 100 × 4 10 µm LiChrospher C18
Mobile phase: MeOH:buffer 15:85, pH 6.7 containing 1 mM tetrabutylammonium hydrogen sulfate (Buffer was 14 mM Na_2HPO_4 and 14.4 mM NaH_2PO_4.)
Flow rate: 0.8
Injection volume: 20
Detector: E, Bioanalytical Systems Model LC-4B, dual glassy carbon working electrode at +1.10 V (monitored) and +1.00 V, stainless steel auxiliary electrode, Ag/AgCl reference electrode following post-column reaction. The column effluent flowed through a 1.8 mL quartz tube irradiated by a UV 214 zinc lamp (BHK, Monrovia CA) to the detector.

CHROMATOGRAM
Retention time: 4 (periodate), 9 (perchlorate), 10 (thiocyanate)
Limit of detection: 70-300 ppb

KEY WORDS
post-column reaction; post-column photochemical derivatization

REFERENCE
Dou, L.; Krull, I.S. Determination of inorganic anions by flow injection analysis and high-performance liquid chromatography combined with photolytic-electrochemical detection, *J.Chromatogr.*, **1990**, *499*, 685–697.

SAMPLE
Matrix: solutions
Analyte: bendroflumethiazide
Sample preparation: Prepare a solution in MeOH:water 80:20, inject a 6 µL aliquot.

HPLC VARIABLES
Guard column: 5 × 4 10 µm LiChrosorb RP-8
Column: 100 × 4.6 5 µm Spheri RP-18 (Brownlee)
Mobile phase: MeOH:water 80:20 containing 2 g/L lithium perchlorate
Flow rate: 0.5
Injection volume: 6
Detector: E, ESA Model 5100A Coulochem, model 5020 guard cell +950 mV, Model 5010 analytical cell +400 mV, palladium reference electrode, following post-column reaction. The effluent from the column flowed through a 20 m × 0.3 mm coil of PTFE tubing irradiated at 254 nm with a Sylvania GTE 8 W low-pressure lamp to the detector.

CHROMATOGRAM
Retention time: 5
Limit of detection: 267 ng/mL

OTHER SUBSTANCES
Also analyzed: butizide, chlorthalidone, ethacrynic acid, furosemide, hydrochlorothiazide

KEY WORDS

post-column reaction; post-column photochemical derivatization

REFERENCE

Macher, M.; Wintersteiger, R. Improved electrochemical detection of diuretics in high-performance liquid chromatographic analysis by postcolumn on-line photolysis, *J.Chromatogr.A*, **1995**, *709*, 257–264.

SAMPLE

Matrix: tissue
Analyte: penicillins
Sample preparation: Homogenize (Ultra-Turrax) 25 g tissue with 25 mL MeCN for 1 min, add 5 mL 500 mM pH 2.2 phosphate buffer while the homogenizer is still running, add 65 mL MeCN, homogenize for 1 min, centrifuge at 4000 g for 10 min. Remove the supernatant and add it to 7 g NaCl and 50 mL dichloromethane, shake for 2 min, allow to stand for 30 min. Remove the upper organic layer and add it to 5 g anhydrous sodium sulfate, shake for 30 s, filter through a cotton-wool plug, evaporate to about 4 mL under reduced pressure at 30°, add 3 mL dichloromethane, evaporate to about 4 mL, add 3 mL light petroleum, evaporate to about 0.5 mL, Suspend this residue with sonication in three 3 mL portions of light petroleum and place these fractions in a separate tube, rinse the original tube with 2 mL pH 7 phosphate buffer. Add the phosphate buffer rinse to the light petroleum extracts, vortex for 30 s, centrifuge, remove the aqueous layer. Extract the light petroleum layer with 2 mL pH 7 phosphate buffer and with two 1.5 mL portions of pH 7 phosphate buffer, combine all the aqueous phase, centrifuge, inject a 200 μL aliquot on to column A and elute to waste with mobile phase B, after 15 min elute to waste with mobile phase C at 2 mL/min, after 10 min elute the contents of column A on to column B with mobile phase D, after 2 min remove column A from the circuit, elute column B with mobile phase D, monitor the effluent from column B. (Wash column A with mobile phase A at 2 mL/min for 7 min, with mobile phase A at 1 mL/min for 5 min, with mobile phase B at 2 mL/min for 8 min, and with mobile phase B at 1 mL/min for 6 min.)

HPLC VARIABLES

Column: A 4 × 4 5 μm LiChrospher 100 RP-18e; B 250 × 4 5 μm LiChrospher 100 RP-18e
Mobile phase: A MeCN:water 50:50; B 20 mM pH 7 phosphate buffer; C MeCN:20 mM pH 3 phosphate buffer 10:90; D MeCN:200 mM pH 3.0 phosphate buffer 35:65 containing 2 mM disodium EDTA
Column temperature: 35
Flow rate: 1 (except where indicated)
Injection volume: 200
Detector: E, Merck Model L3500, glassy carbon working electrode +0.65 V, stainless-steel auxiliary electrode, Ag/AgCl reference electrode following post-column reaction. The column effluent flowed through a 10 m × 0.3 mm ID woven PTFE coil illuminated by a UV 254 low-pressure mercury lamp to the detector.

CHROMATOGRAM

Retention time: 5.2 (penicillin G), 6.1 (penicillin V), 7.1 (oxacillin), 8.9 (cloxacillin), 13.0 (dicloxacillin)
Limit of detection: 1.2-4.6 ng

KEY WORDS

post-column reaction; cow; muscle; column-switching; post-column photochemical derivatization

REFERENCE

Lihl, S.; Rehorek, A.; Petz, M. High-performance liquid chromatographic determination of penicillins by means of automated solid-phase extraction and photochemical degradation with electrochemical detection, *J.Chromatogr.A*, **1996**, *729*, 229–235.

SAMPLE

Matrix: urine
Analyte: drugs
Sample preparation: 2 mL Urine + 3 mL 5 M NaOH, vortex 30 s, add 12 mL diethyl ether, rotate for 5 min, centrifuge at 2500 rpm for 5 min. Remove the ether layer and

evaporate it to dryness at 40° under a stream of nitrogen, reconstitute in 2 mL mobile phase, inject a 200 µL aliquot.

HPLC VARIABLES
Column: 250 × 4.6 10 µm Alltech C18
Mobile phase: MeOH:water 50:50 containing 7 mL/L butylamine, adjusted to pH 3.2 with sulfuric acid
Flow rate: 1.8
Injection volume: 200
Detector: E, Bioanalytical Systems Model LC4B, dual glassy carbon working electrode cell half operated in the parallel mode + 1.0 V and +0.9 V, stainless steel auxiliary electrode cell half, Ag/AgCl reference electrode. The detector was preceded by a Photronix Model 816 UV irradiator which irradiated the mobile phase in a 9.144 m length of 0.5 mm ID × 1.6 mm o.d. Teflon tubing in a three-dimensional figure eight configuration. The irradiation apparatus was maintained at 0-5° using an ice bath.

CHROMATOGRAM
Retention time: 4 (methylphenidate), 4.5 (phenobarbital), 5.5 (chlordiazepoxide), 6 (cocaine), 8 (nitrazepam)
Limit of detection: 2-750 ppb

KEY WORDS
post-column reaction; post-column photochemical derivatization

REFERENCE
Selavka, C.M.; Krull, I.S.; Lurie, I.S. Photolytic derivatization for improved LCEC determinations of pharmaceuticals in biological fluids, *J.Chromatogr.Sci.*, **1985**, *23*, 499–508.

Fluorescence Detection

RELATED REFERENCES
Borner, K.; Borner, E.; Hartwig, H.; Lode, H. Fluorimetric detection of non-fluorescent quinolones after post-column photoreaction. *Methodol.Surv.Bioanal.Drugs* **1996**, *24*, 147-150.

de Ruiter, C.; Bohle, J.F.; de Jong, G.J.; Brinkman, U.A.T.; Frei, R.W. Enhanced fluorescence detection of dansyl derivatives of phenolic compounds using a postcolumn photochemical reactor and application to chlorophenols in river water. *Anal.Chem.* **1988**, *60*, 666-670.

Jansen, H.; Vreuls, J.J.; Van der Heide, T.A.J.; de Jong, G.J.; Brinkman, U.A.T.; Frei, R.W. In-column and post-column photochemical fluorescence enhancement in packed capillary liquid chromatography. *J.Liq.Chromatogr.* **1988**, *11*, 1855-1863.

Joshua, H. Analysis of aflatoxins in naturally contaminated corn by HPLC, post-column photochemical derivatization, and fluorescence detection. *Am.Lab.* **1995**, *27*, 36J-36M.

Lores, M.; Garcia, C.M.; Cela, R. High-performance liquid chromatography of phenolic aldehydes with highly selective fluorimetric detection by means of postcolumn photochemical derivatization. *J.Chromatogr.A* **1994**, *683*, 31-44.

Mawatari, K.; Iinuma, F.; Watanabe, M. Fluorometric determination of kynurenine in human serum by high-performance liquid chromatography coupled with post-column photochemical reaction with hydrogen peroxide. *J.Chromatogr.* **1989**, *488*, 349-355.

Mawatari, K.; Iinuma, F.; Watanabe, M. Determination of isoniazid, acetylisoniazid and isonicotinic acid in human urine by high-performance liquid chromatography coupled with postcolumn photochemical reaction and fluorescence detection. *Anal.Sci.* **1990**, *6*, 515-518.

Mawatari, K.; Iinuma, F.; Watanabe, M. Determination of nicotinic acid and nicotinamide in human serum by high-performance liquid chromatography with postcolumn ultraviolet-irradiation and fluorescence detection. *Anal.Sci.* **1991**, *7*, 733-736.

Mawatari, K.; Iinuma, F.; Watanabe, M. Fluorometric determination of nalidixic acid in human urine using high-performance liquid chromatography with postcolumn photochemical reaction. *Rinsho Kagaku (Nippon Rinsho Kagakkai)* **1992**, *21*, 13-17.

Miles, C.J.; Moye, H.A. High performance liquid chromatography with post-column photolysis of pesticides for generation of fluorophores. *Chromatographia* **1987**, *24*, 628-632.

Miles, C.J.; Moye, H.A. Postcolumn photolysis of pesticides for fluorometric determination by high-performance liquid chromatography. *Anal.Chem.* **1988**, *60*, 220-226.

Poulsen, J.R.; Birks, J.W. Photoreduction fluorescence detection of quinones in high-performance liquid chromatography. *Anal.Chem.* **1989**, *61*, 2267-2276.

Salamoun, J.; Macka, M.; Nechvatal, M.; Matousek, M.; Knesel, L. Identification of products formed during UV irradiation of tamoxifen and their use for fluorescence detection in high-performance liquid chromatography. *J.Chromatogr.* **1990**, *514*, 179-187.

Takeda, N.; Tsuji, M.; Akiyama, Y. HPLC analysis of pesticides in brown rice using UV absorption and postcolumn photolysis-fluorescence detection. *Shokuhin Eiseigaku Zasshi* **1995**, *36*, 601-606.

Toreson, H.; Eriksson, B.-M. Liquid chromatographic determination of fluvastatin and its enantiomers in blood plasma by automated solid-phase extraction. *Chromatographia* **1997**, *45*, 29-34.

Urmos, I.; Benko, S.M.; Klebovich, I. Simple and rapid determination of clomiphene cis and trans isomers in human plasma by high-performance liquid chromatography using on-line post-column photochemical derivatization and fluorescence detection. *J.Chromatogr.* **1993**, *617*, 168-172.

Yu, Z.; Westerlund, D.; Boos, K.-S. Determination of methotrexate and its metabolite 7-hydroxymethotrexate by direct injection of human plasma into a column-switching liquid chromatographic system using post-column photochemical reaction with fluorimetric detection. *J.Chromatogr.B* **1997**, *689*, 379-386.

SAMPLE

Matrix: bile, blood, feces, urine

Analyte: ciprofloxacin and metabolites

Sample preparation: Plasma, serum, urine, bile. Filter (0.45 μm), homogenize, dilute with 33 mM phosphoric acid, inject a 10 μm aliquot. Feces. Homogenize (Ultra-Turrax) feces with 2-4 volumes of water. Mix a 100 mL aliquot with 500 mL dichloromethane: isopropanol:300 mM phosphoric acid 10:50:40, shake mechanically for 1.5 h, centrifuge a 4 mL aliquot, dilute the supernatant with 33 mM phosphoric acid, inject a 10 μL aliquot. Tissue. Homogenize (Ultra-Turrax) 200 mg tissue with 5 mL extraction solution, centrifuge. Filter the supernatant, inject a 10 μL aliquot of the filtrate. (Extraction solution was 12.5 mL concentrated perchloric acid and 2 g concentrated phosphoric acid made up to 1 L with MeOH:water 50:50.)

HPLC VARIABLES

Column: 250 × 4 5 μm Nucleosil 120-5 C18

Mobile phase: MeCN:40 g/L tetrabutylammonium bisulfate 7:93 (A) or 25:75 (B)

Column temperature: 40

Flow rate: 2

Injection volume: 10

Detector: F ex 278 em 445 following post-column reaction. The column effluent flowed through a 250 mm × 0.2 mm stainless steel coil heated to just below the boiling point of the eluent, through a 200 mm × 0.5 mm ID PTFE coil irradiated by a water-cooled high-pressure mercury lamp (Philips HPK 125 WL), and through a cooling coil to the detector. (Fluorescence of metabolites increases greatly with post-column reaction but fluorescence of ciprofloxacin increases only a little.)

CHROMATOGRAM

Retention time: 2.5 (1-cyclopropyl-6-fluoro-1,4-dihydro-4-oxo-7-(1-(3-oxopiperazinyl))-3-quinolinecarboxylic acid, M3, BAY Q 3542 (B)), 3.5 (1-cyclopropyl-6-fluoro-1,4-dihydro-4-oxo-7-(1-(piperazine-4-carboxaldehyde))-3-quinolinecarboxylic acid, M4, BAY P 9357 (B)), 3.5 (1-cyclopropyl-6-fluoro-1,4-dihydro-4-oxo-7-(N-(2-aminoethyl)amine)-3-quinolinecarboxylic acid, M1, BAY R 3964 (A)), 4.5 (ciprofloxacin (A)), 6 (1-cyclopropyl-6-fluoro-1,4-dihydro-4-oxo-7-(1-(piperazine-4-sulfonic acid))-3-quinolinecarboxylic acid, M2, BAY S 9435 (B))

Limit of detection: 0.23-1.1 ng/mL

KEY WORDS

post-column reaction; plasma; serum; post-column photochemical derivatization

REFERENCE

Scholl, H.; Schmidt, K.; Weber, B. Sensitive and selective determination of picogram amounts of ciprofloxacin and its metabolites in biological samples using high-performance liquid chromatography and photothermal post-column derivatization, *J.Chromatogr.*, **1987**, *416*, 321–330.

SAMPLE

Matrix: blood

Analyte: clomiphene

Sample preparation: 3 mL Plasma + 1 mL pH 9 borate buffer, vortex, add 9 mL redistilled ether, vortex for 2 min, centrifuge at 1600 g for 15 min. Remove the organic layer and dry it over 1 g anhydrous sodium sulfate, centrifuge at 1600 g for 10 min. Remove the organic layer and evaporate it to dryness under a stream of air, reconstitute the residue in 75 µL mobile phase, vortex for 2 min, inject a 10 µL aliquot.

HPLC VARIABLES
Column: 250 × 4.6 6 µm Zorbax Sil
Mobile phase: Chloroform:MeOH 80:20
Flow rate: 0.8
Injection volume: 10
Detector: F following post-column reaction. The column effluent flowed through a 3 m × 0.3 mm ID PTFE coil irradiated by a medium-pressure mercury lamp (Hanovia) with water cooling to the detector.

CHROMATOGRAM
Retention time: 7 (cis-clomiphene), 7.5 (trans-clomiphene)
Limit of detection: 350 pg/mL

OTHER SUBSTANCES
Extracted: metabolites

KEY WORDS
post-column reaction; plasma; normal phase; post-column photochemical derivatization

REFERENCE
Harman, P.J.; Blackman, G.L.; Phillipou, G. High-performance liquid chromatographic determination of clomiphene using post-column on-line photolysis and fluorescence detection, *J.Chromatogr.*, **1981**, *225*, 131–138.

SAMPLE
Matrix: blood
Analyte: demoxepam
Sample preparation: 1 mL Serum + 1 mL MeOH, shake, centrifuge at 7000 g for 7 min, inject a 20 µL aliquot.

HPLC VARIABLES
Column: 150 × 4.6 5 µm Supelcosil LC-18
Mobile phase: MeOH:100 mM pH 8 phosphate buffer 60:40
Flow rate: 1
Injection volume: 20
Detector: F ex 340 em 410 following post-column photolysis. The column effluent was mixed with bubbles (at 0.7 mL/min) and flowed through a 3.8 m × 1.1 mm ID PTFE coil irradiated by a mercury-xenon lamp to a debubbler, the liquid phase then flowed to the detector. (The air bubbles suppress band broadening in the PTFE coil.)

CHROMATOGRAM
Retention time: 8
Limit of detection: 100 pg

KEY WORDS
serum; post-column reaction; post-column photochemical derivatization

REFERENCE
Brinkman, U.A.T.; Welling, P.L.M.; De Vries, G.; Scholten, A.H.M.T.; Frei, R.W. Liquid chromatography of demoxepam and phenothiazines using a post-column photochemical reactor and fluorescence detection, *J.Chromatogr.*, **1981**, *217*, 463–471.

SAMPLE
Matrix: blood
Analyte: tamoxifen
Sample preparation: Vortex serum with 10 volumes hexane:butanol 98:2 for 15 s, centrifuge for 10 min. Remove an aliquot of the organic layer and evaporate it to dryness under a stream of nitrogen at 55°, reconstitute the residue in 50-100 µL mobile phase, inject an aliquot.

HPLC VARIABLES

Column: 250 × 4.6 10 μm ODS-2 (Whatman)
Mobile phase: MeOH containing 0.04% diethylamine acetate
Flow rate: 2
Detector: F ex 220 or 254 em 360 following post-column reaction. The column effluent flowed through a 70 cm × 0.2 mm ID quartz coil irradiated with two Mineralite short-wave UV lamps to the detector.

CHROMATOGRAM

Retention time: 5.4
Limit of detection: 0.2 ng

OTHER SUBSTANCES

Extracted: metabolites

KEY WORDS

post-column reaction; serum; protect from light; post-column photochemical derivatization

REFERENCE

Brown, R.R.; Bain, R.; Jordan, V.C. Determination of tamoxifen and metabolites in human serum by high-performance liquid chromatography with post-column fluorescence activation, *J.Chromatogr.*, **1983**, *272*, 351–358.

SAMPLE

Matrix: blood
Analyte: tamoxifen
Sample preparation: 500 μL Plasma + 2 μg clomiphene + 2 mL diethyl ether, extract, centrifuge at 2000 rpm at 4° for 10 min, repeat extraction. Combine the organic phases and evaporate them to dryness under a stream of nitrogen at 37°. Reconstitute the residue in 250 μL MeOH, centrifuge at 2000 rpm at 4° for 10 min, inject a 10-100 μL aliquot.

HPLC VARIABLES

Column: 250 × 4.6 Zorbax CN
Mobile phase: MeCN:10 mM KH_2PO_4:300 mM phosphoric acid :water 42:20:10:28
Flow rate: 2.8
Injection volume: 10-100
Detector: F ex 258 em 378 following post-column reaction. The column effluent flowed through a 6.5 m × 0.35 mm × 1.5 mm o.d. PTFE tube irradiated with a Philips HPK 125 watt high pressure mercury lamp to the detector.

CHROMATOGRAM

Retention time: 9
Internal standard: clomiphene (12)
Limit of detection: 2 ng/mL

OTHER SUBSTANCES

Simultaneously analyzed: metabolites

KEY WORDS

plasma; post-column reaction; post-column photochemical derivatization

REFERENCE

Milano, G.; Etienne, M.C.; Frenay, M.; Khater, R.; Formento, J.L.; Renee, N.; Moll, J.L.; Francoual, M.; Berto, M.; Namer, M. Optimised analysis of tamoxifen and its main metabolites in the plasma and cytosol of mammary tumours, *Br.J.Cancer*, **1987**, *55*, 509–512.

SAMPLE

Matrix: blood
Analyte: tamoxifen
Sample preparation: 800 μL Plasma + 200 μL 100 mM HCl, centrifuge at 12000 g for 2 min. Inject the following solutions onto column A; 500 μL MeOH, 700 μL water, 300 μL 100 mM HCl, 500 μL supernatant, and 1 mL water. Flush the contents of column A onto column B with mobile phase, elute with mobile phase, monitor the effluent from column B.

HPLC VARIABLES
Column: A 35 × 2 40 μm Sepralyte CN-propyl modified silica (Analytichem); B 110 × 4.6 Partisil Si
Mobile phase: MeOH:5 mM ammonium acetate 90:10
Flow rate: 1.4
Injection volume: 500
Detector: F ex 256 em 380 following post-column reaction. The column effluent flowed through a 15 m × 0.25 mm ID crocheted PTFE tube irradiated with a Sylvania G8 UV lamp at 254 nm to the detector.

CHROMATOGRAM
Retention time: 3
Limit of detection: 100 pg/mL

OTHER SUBSTANCES
Extracted: metabolites

KEY WORDS
post-column reaction; plasma; column-switching; post-column photochemical derivatization

REFERENCE
Kikuta, C.; Schmid, R. Specific high-performance liquid chromatographic analysis of tamoxifen and its major metabolites by "on-line" extraction and post-column photochemical reaction, *J.Pharm.Biomed.Anal.*, **1989**, 7, 329–337.

SAMPLE
Matrix: blood
Analyte: diclofenac
Sample preparation: 50 μL Plasma + 50 μL MeCN, mix, centrifuge at 12000 g for 2 min, inject a 50 μL aliquot of the supernatant.

HPLC VARIABLES
Guard column: 15 × 3.2 5 μm NewGuard ODS
Column: 150 × 4 5 μm Nucleosil C18
Mobile phase: MeCN:buffer 32:68 (Buffer was 40 mL 1 M NaH_2PO_4 and 40 mL 500 mM Na_2HPO_4 made up to 680 mL with water, pH 6.6.)
Flow rate: 0.7
Injection volume: 50
Detector: F ex 288 em 360 following post-column reaction. The column effluent flowed through a 1.3 m × 1 mm ID PTFE tube irradiated by a UV 254 lamp (Philips TUV 6W, TYP 103314) to the detector.

CHROMATOGRAM
Retention time: 10
Limit of detection: 6 ng/mL

KEY WORDS
post-column reaction; plasma; human; rat; post-column photochemical derivatization

REFERENCE
Wiese, B.; Hermansson, J. Bioanalysis of diclofenac as its fluorescent carbazole acetic acid derivative by a post-column photoderivatization high performance liquid chromatographic method, *J.Chromatogr.*, **1991**, 567, 175–183.

SAMPLE
Matrix: blood
Analyte: CL 275,838 (4,5-dihydro-4-[4-(phenylmethyl)-1-piperazinyl)acetyl]-7-[3-(trifluoro-methyl)phenyl]-pyrazolo[1,5-a]pyrimidine-3-carbonitrile)
Sample preparation: Condition a Sep-Pak C18 SPE cartridge with 5 mL MeCN, 5 mL MeCN:water 50:50, and 5 mL water. 2 mL Plasma + 2.5 ng IS, add to the SPE cartridge, wash with 5 mL water, wash with 5 mL MeCN:water 20:80, wash with 350 μL MeCN, elute with 2 mL MeCN. Evaporate the eluate to dryness under reduced pressure, reconstitute the residue in 200 μL mobile phase, centrifuge at 12000 g for 5 min, inject a 190 μL aliquot of the supernatant.

HPLC VARIABLES
Guard column: Newguard RP18
Column: 150 × 4.6 LC 18 DB (Supelco)
Mobile phase: MeCN:MeOH:100 mM NaH$_2$PO$_4$:100 mM phosphoric acid 30:22:24:24
Flow rate: 1
Injection volume: 190
Detector: F ex 335 em 416-436 (filter) following post-column reaction. The column effluent flowed through a 10 m × 0.3 mm ID coil irradiated by a UV 254 lamp to the detector.

CHROMATOGRAM
Retention time: 6
Internal standard: 5-methyl-7-[3-(trifluoromethyl)phenyl]-pyrazolo[1,5-a]pyrimidine-3-carbonitrile (American Cyanamid) (16)
Limit of quantitation: 1.25 ng/mL

KEY WORDS
post-column reaction; SPE; plasma; pharmacokinetics; post-column photochemical derivatization

REFERENCE
Guiso, G.; Confalonieri, S.; Gherardi, S.; Guido, M.; Caccia, S. Liquid chromatographic determination of the potential memory enhancing agent CL 275,838 using a post-column photolysis and fluorimetric detection, *J.Liq.Chromatogr.*, **1992**, *15*, 1463–1472.

SAMPLE
Matrix: blood
Analyte: BAY y 3118
Sample preparation: 500 µL Plasma + 500 µL 50 mM phosphoric acid, mix, centrifuge at 1500 g for 10 min, inject a 20 µL aliquot of the supernatant. Saliva. Inject 20 µL saliva directly. Urine. Dilute urine 50-1000-fold with 100 mM pH 7.5 phosphate buffer, inject a 20 µL aliquot.

HPLC VARIABLES
Column: 250 × 4.6 5 µm Nucleosil 100 C18
Mobile phase: MeCN:buffer 19:81 (Prepare buffer by mixing 500 mL 100 mM tetrabutyl-ammonium bromide with 500 mL 50 mM phosphoric acid, adjust pH to 2.0.)
Column temperature: 50
Flow rate: 1
Injection volume: 20
Detector: F ex 277 em 418 post-column reaction. The column effluent passed through a 20 m long PTFE coil irradiated with a UV lamp at 254 nm (Beam Boost) to the detector. Alternatively, for greater sensitivity, the column effluent passed through a 1.9 m long stainless steel coil at 275°, a 10 cm × 0.3 mm ID PTFE coil irradiated with an HPK 125 W UV lamp (Philips), and a cooler to the detector.

CHROMATOGRAM
Retention time: 5.3
Limit of quantitation: 10 ng/mL

KEY WORDS
post-column reaction; plasma; post-column photochemical derivatization

REFERENCE
Schaefer, H.G. Determination of BAY y 3118, a novel 4-quinolone, in biological fluids using high-performance liquid chromatography and photothermal post-column derivatization, *J.Chromatogr.*, **1993**, *616*, 87–93.

SAMPLE
Matrix: blood
Analyte: clomiphene
Sample preparation: Condition a 1 mL 40 µm Bond Elut C18 SPE cartridge with 1 mL MeOH and 1 mL water. 1 mL Plasma + 500 ng IS + 1 mL water + 500 µL 3 M NaCl, mix, add to the SPE cartridge, wash with 3 mL water, elute with 3 mL MeOH. Evaporate the eluate to dryness under a stream of nitrogen at 50°, reconstitute the residue in 1 mL mobile phase, inject a 20 µL aliquot.

HPLC VARIABLES
Column: 250 × 4 LiChrospher 100 RP-18
Mobile phase: MeCN:MeOH:water:1% ammonium chloride:1% potassium carbonate 95:3:2:0.4:0.8
Column temperature: 30
Flow rate: 1
Injection volume: 20
Detector: F ex 247 em 378 following post-column reaction. The column effluent flowed through a knitted 15 m × 0.3 mm ID PTFE reaction coil irradiated by a mercury lamp at 254 nm (Beam Boost) to the detector.

CHROMATOGRAM
Retention time: 6.8 (E), 7.2 (Z)
Internal standard: 1-(4-diethylaminoethoxy)phenyl-1,2-diphenylethanol (EGIS, Budapest) (4.7)
Limit of quantitation: 0.75 ng/mL (Z), 1.25 ng/mL (E)
Limit of detection: 0.4 ng/mL

KEY WORDS
post-column reaction; plasma; SPE; post-column photochemical derivatization

REFERENCE
Ürmös, I.; Benkö, S.M.; Klebovich, I. Simple and rapid determination of clomiphene *cis* and *trans* isomers in human plasma by high-performance liquid chromatography using on-line post-column photochemical derivatization and fluorescence detection, *J.Chromatogr.*, **1993**, *617*, 168–172.

SAMPLE
Matrix: blood
Analyte: BRL 46470 (endo-N-(8-methyl-8-azabicyclo[3.2.1]oct-3-yl)-2,3-dihydro-3,3-dimethylindole-1-carboxamide)
Sample preparation: 1 mL Plasma + 100 µL water + 100 µL 5 µg/mL IS in water + 500 µL 100 mM NaOH, vortex briefly, add 3.5 mL dichloromethane, shake mechanically for 30 min, centrifuge at 15° at 3000 g for 6 min. Remove the organic layer and evaporate it to dryness under a stream of nitrogen at 55°, reconstitute the residue in 100 µL mobile phase, vortex for 1.5 min, inject an 80 µL aliquot.

HPLC VARIABLES
Guard column: Novapak C18 Guard-Pak
Column: 100 × 5 Novapak CN radial compression
Mobile phase: MeCN:50 mM pH 7 ammonium acetate 80:20
Flow rate: 0.75
Injection volume: 80
Detector: F ex 363 em >400 (filter) following post-column reaction. The column effluent flowed through a 5 m reaction coil irradiated with a 254 nm UV lamp to the detector.

CHROMATOGRAM
Retention time: 10.75
Internal standard: BRL 43704 (endo-N-(8-methyl-8-azabicyclo[3.2.1]oct-3-yl)-1-methylindazole-3-carboxylate) (9.0)
Limit of quantitation: 0.1 ng/mL

KEY WORDS
silanize all glassware; plasma; post-column reaction; post-column photochemical derivatization

REFERENCE
Deeks, N.J.; Abbott, R.W.; Allen, G.D.; Hollis, F.J.; Rhodes, G. Determination of BRL 46470 in human plasma by high performance liquid chromatography with ultraviolet absorbance detection followed by post-column photochemical reaction and fluorescence detection, *Analyst*, **1994**, *119*, 2043–2050.

SAMPLE
Matrix: blood
Analyte: maprotiline
Sample preparation: 1 mL Serum + 1 mL 2 M sodium bicarbonate + 6 mL hexane, extract, centrifuge at 2000 g for 10 min. Remove 5 mL of the hexane layer and add it to

150 µL 100 mM phosphoric acid, mix, centrifuge at 2000 g for 10 min, inject a 100 µL aliquot of the aqueous layer.

HPLC VARIABLES
Column: 250 × 4 4 µm Supersphere Select B (Merck)
Mobile phase: MeCN:pH 5.8 phosphate buffer 25:75 (Buffer was 2 mL 85% phosphoric acid and 4 mL triethylamine in 1 L water.)
Flow rate: 0.3-1
Injection volume: 100
Detector: F ex 275 em 315 following post-column reaction. The effluent from the column flowed through a Beam Boost photochemical reactor equipped with a 20 m coil and then to the detector.

CHROMATOGRAM
Retention time: 28.1
Limit of detection: 0.1 ng/mL

OTHER SUBSTANCES
Extracted: metabolites

KEY WORDS
post-column reaction; serum; post-column photochemical derivatization; pharmacokinetics

REFERENCE
Kuss, H.-J.; Sirch, S.; Zhao, D.Y. Assay for maprotiline in human serum with improved sensitivity and selectivity, *J.Chromatogr.B*, **1994**, *656*, 245–249.

SAMPLE
Matrix: blood
Analyte: panomifene
Sample preparation: Condition a 1 mL phenyl SPE cartridge with five 1 mL portions of MeCN, 1 mL water, and 1 mL 2.5 mL/L triethylamine in 50 mM pH 3.0 phosphate buffer. 980 µL Plasma + 20 µL MeOH:water 50:50 + 10 µL 2 µg/mL tamoxifen in MeOH:water 50:50 + 1 mL MeCN, vortex for 1 min, centrifuge at -10° at 2500 g for 1 h. Remove a 1.6 mL aliquot of the supernatant and add it to 400 µL 2% heptanesulfonic acid in 50 mM pH 3.0 KH_2PO_4/phosphoric acid buffer, add to the SPE cartridge, wash with two 100 µL portions of MeCN:buffer 80:20, wash with 50 µL 25 mM sulfuric acid, elute with five 100 µL portions of MeCN:buffer 80:20. Evaporate the eluate to dryness under a stream of nitrogen at room temperature, reconstitute the residue in 100 µL MeCN:buffer 70:30, inject a 10-30 µL aliquot. (Buffer was 5 mM heptanesulfonic acid in 50 mM pH 3.0 phosphate buffer.)

HPLC VARIABLES
Guard column: 20 × 4.6 10 µm Si-100-S Phenyl (BST, Budapest)
Column: 250 × 4.6 10 µm Si-100-S Phenyl (BST, Budapest)
Mobile phase: MeCN:buffer 75:25 (Buffer was 50 mM pH 3.0 KH_2PO_4/phosphoric acid buffer containing 5 mM heptanesulfonic acid and 300 µL/L triethylamine. Temperature of MeCN was 60° and temperature of buffer was 80°.)
Flow rate: 1.2
Injection volume: 10-30
Detector: F ex 257 em 378 following post-column reaction. The column effluent flowed through a 10 m × 0.3 mm ID knitted PTFE coil irradiated by a mercury lamp at 254 nm to the detector.

CHROMATOGRAM
Retention time: 5.73
Internal standard: tamoxifen (7.03)
Limit of detection: 1 ng/mL

KEY WORDS
post-column reaction; plasma; pharmacokinetics; SPE; post-column photochemical derivatization

REFERENCE
Erdélyi-Tóth, V.; Pap, E.; Kralovánszky, J.; Bojti, E.; Klebovich, I. Determination of panomifene in human plasma by high-performance liquid chromatography, *J.Chromatogr.A*, **1994**, *668*, 419–425.

SAMPLE
Matrix: blood
Analyte: tamoxifen
Sample preparation: 150 μL Plasma + 150 μL MeCN, vortex for 2 min, centrifuge at 13000 g for 5 min, inject 50 μL supernatant onto column A with mobile phase A, elute column A to waste for 2 min with mobile phase A, elute column A to waste for another 2 min with mobile phase B, elute column A onto column B with mobile phase B, analyze effluent from column B. (Single pump used. Switch from mobile phase A to mobile phase B by switching solvent reservoirs.)

HPLC VARIABLES
Column: A SPS CN guard column (Regis); B Regis C18 guard column + 250 × 4.6 5 μm Regis Rexchrom CN
Mobile phase: A water; B MeCN: 20 mM pH 3.1 K_2HPO_4 35:65
Flow rate: A 1; B 1
Injection volume: 50
Detector: F ex 250 em 370 (cut-off filter) following post-column reaction. The column effluent flowed through a photochemical reactor (ICT Beam Boost, 254 nm UV lamp) with a 5 m reaction coil to the detector.

CHROMATOGRAM
Retention time: 47
Limit of detection: 8 ng/mL

OTHER SUBSTANCES
Simultaneously analyzed: metabolites, 4-hydroxytamoxifen, N-desdimethyltamoxifen, N-desmethyltamoxifen, tamoxifen-ol

KEY WORDS
plasma; rugged; post-column photochemical derivatization; post-column reaction

REFERENCE
Fried, K.M.; Wainer, I.W. Direct determination of tamoxifen and its four major metabolites in plasma using coupled column high-performance liquid chromatography, *J.Chromatogr.B*, **1994**, *655*, 261–268.

SAMPLE
Matrix: blood
Analyte: BAY X 7195 ([(S)-4-(4-carboxyphenylthio)-7-[4-(4-phenoxybutoxy)phenyl]-5-(Z)-heptenoic acid)
Sample preparation: Condition a 500 mg Sep-Pak RP-18 SPE cartridge with three 2.5 mL portions of MeOH and two 2.5 mL portions of 0.025% phosphoric acid. 1 mL Plasma + 1 mL 5% phosphoric acid + 4 mL 460 pg/mL IS in MeCN:water 60:40, shake, add to the SPE cartridge, wash with eight 2.5 mL portions of MeOH:0.025% phosphoric acid 70:30, air dry for 5 min, elute with 1.5 mL MeCN (under gravity). Evaporate the eluate to dryness at 40°, reconstitute the residue in 250 μL MeCN:water 60:40, sonicate, inject a 50 μL aliquot.

HPLC VARIABLES
Guard column: 4 × 4 LichroCART (Merck)
Column: 250 × 4 5 μm Lichrosphere RP-18
Mobile phase: Gradient. MeCN:0.025% phosphoric acid from 55:45 to 65:35 over 4.5 min, maintain at 65:35 for 8.5 min, to 100:0 over 0.1 min, maintain at 100:0 for 6.9 min, return to initial conditions over 0.1 min.
Column temperature: 45
Flow rate: 1
Injection volume: 50

Detector: F ex 230 em 310 following post-column reaction. The column effluent flowed through a 5 m × 0.3 mm ID PTFE coil irradiated by a 254 nm lamp (Beam Boost) to the detector.

CHROMATOGRAM
Retention time: 11.5
Internal standard: [(R,S)-5-(4-carboxyphenylthio)-7-[4-(4-phenoxybutoxy)phenyl]-6-(Z)-octenoic acid (BAY X 1308) (12.5)
Limit of quantitation: 0.6 ng/mL
Limit of detection: 0.2 ng/mL

KEY WORDS
post-column reaction; post-column photochemical derivatization; protect from light; SPE; plasma; pharmacokinetics

REFERENCE
Heinig, R. Determination of BAY X 7195, a novel leukotriene D4 antagonist, in human plasma by high-performance liquid chromatography with post-column photo derivatization and fluorescence detection, *J.Chromatogr.B*, **1995**, *667*, 137–147.

SAMPLE
Matrix: blood
Analyte: droloxifene
Sample preparation: Mix serum with an equal volume of MeCN, mix, centrifuge, inject a 50 μL aliquot of the supernatant.

HPLC VARIABLES
Column: 150 × 4.6 3 μm ODS Hypersil
Mobile phase: MeCN:water 66:34 containing 3 mM acetic acid and 2 mM diethylamine
Flow rate: 1
Injection volume: 50
Detector: F ex 260 em 360 following post-column reaction. The column effluent flowed through a 10 m × 0.3 mm ID knitted PTFE coil irradiated with UV light at 254 nm to the detector.

CHROMATOGRAM
Retention time: 18.8
Limit of detection: 5 ng/mL

OTHER SUBSTANCES
Extracted: metabolites
Non-interfering: acetaminophen, atenolol, captopril, clodronate, dexamethasone, dextropropoxyphene, diazepam, doxycycline, econazole, enoxaparin, felodipine, flunitrazepam, furosemide, glibenclamide, indomethacin, insulin, isosorbide mononitrate, megestrol acetate, metoclopramide, mianserin, morphine, nitroglycerin, oxazepam, perphenazine, phenytoin, pivmecillinam, prochlorperazine, promethazine, ranitidine, tamoxifen

KEY WORDS
post-column reaction; serum; pharmacokinetics; post-column photochemical derivatization

REFERENCE
Lien, E.A.; Anker, G.; Lonning, P.E.; Ueland, P.M. Determination of droloxifene and two metabolites in serum by high-performance liquid chromatography, *Ther.Drug Monit.*, **1995**, *17*, 259–265.

SAMPLE
Matrix: blood
Analyte: droloxifene
Sample preparation: Condition a 100 mg Bond Elut benzenesulfonic acid (SCX) SPE cartridge with 1 mL MeOH:30% ammonium hydroxide 96.5:3.5, 1 mL MeOH, and 1 mL 1% acetic acid at 1 mL/min. Add 500 μL 1% acetic acid, 200 μL 2 ng/mL IS in mobile phase, and 200 μL plasma or serum sequentially to the SPE cartridge, wash with 1 mL water, wash with 1 mL MeOH, elute with 1 mL MeOH:30% ammonium hydroxide 96.5:3.5. Evaporate the eluate to dryness under a stream of nitrogen at 50° or reduced pressure at 75°, reconstitute with 100 μL mobile phase, vortex, centrifuge at 1000 g for 30 s, inject an 80 μL aliquot.

HPLC VARIABLES
Column: 100 × 4.6 3 μm C18 (Rainin)
Mobile phase: MeCN:50 mM sodium phosphate buffer 45:55, adjusted to pH 3.5 with phosphoric acid
Column temperature: 40
Flow rate: 2
Injection volume: 80
Detector: F ex 260 em 375 following post-column reaction. The column effluent passed through a 3.1 m × 0.25 mm ID PTFE coil irradiated by a Beam Boost 254 nm photochemical reaction lamp to the detector.

CHROMATOGRAM
Retention time: 3.1
Internal standard: (E)-α-[p-[2-(diethylamino)ethoxy]phenyl]-α'-ethyl-3-stilbenol (K 21.089 E, Klinge Pharma, Munich) (4.8)
Limit of quantitation: 25 pg/mL
Limit of detection: 10 pg/mL

OTHER SUBSTANCES
Extracted: metabolites

KEY WORDS
rat; monkey; human; plasma; serum; post-column reaction; post-column photochemical derivatization; pharmacokinetics; SPE

REFERENCE
Tess, D.A.; Cole, R.O.; Toler, S.M. Sensitive method for the quantitation of droloxifene in plasma and serum by high-performance liquid chromatography employing fluorimetric detection, *J.Chromatogr.B*, **1995**, *674*, 253–260.

SAMPLE
Matrix: blood
Analyte: methotrexate
Sample preparation: Condition a 100 mg Bond-Elut SPE cartridge with 1 mL MeOH and 1.5 mL 50 mM pH 2.7 phosphate buffer. 1 mL Plasma + 1 mL 50 mM pH 6.5 phosphate buffer, mix, add to the SPE cartridge at 2 mL/min, wash with 2 mL 50 mM pH 2.7 phosphate buffer, wash with 1 mL 100 mM NaOH, wash with 1 mL 50 mM pH 2.7 phosphate buffer, elute with 1.5 mL MeOH. Evaporate the eluate to dryness under a stream of nitrogen, reconstitute the residue in 200 μL mobile phase, inject a 100 μL aliquot.

HPLC VARIABLES
Guard column: 30 × 3.2 30 μm Spherisorb C18
Column: 150 × 3.9 5 μm Novapak C18
Mobile phase: DMF:MeCN:3% hydrogen peroxide:14 mM pH 6.5 phosphate buffer 4:3.3 :0.5:92.2
Flow rate: 1
Injection volume: 100
Detector: F ex 350 em 465 following post-column reaction. The column effluent flowed through a 3 m long × 0.38 mm ID length of polyethylene (PE-20) tubing irradiated by a Spectroline pencil UV lamp (UV 254 nm) to the detector.

CHROMATOGRAM
Retention time: 4
Limit of detection: 0.05 ng/mL

OTHER SUBSTANCES
Extracted: metabolites

KEY WORDS
plasma; SPE; human; dog; pharmacokinetics; post-column photochemical derivatization; post-column reaction

REFERENCE
Lu, G.; Jun, H.W. Determination of trace methotrexate and 7-OH-methotrexate in plasma by high-performance liquid chromatography with fluorimetric detection, *J.Liq.Chromatogr.*, **1995**, *18*, 155–171.

SAMPLE
Matrix: blood
Analyte: sulindac
Sample preparation: Condition a Bond-Elut C2 SPE cartridge with 1 mL MeOH and 1 mL mobile phase. 1 mL Serum + 75 µL 100 µg/mL indomethacin in MeOH + 1 drop saturated ammonium sulfate solution + 1 drop 1 M HCl, vortex for 3 min, add to the SPE cartridge, wash with six 500 µL portions of wash solvent, elute with four 250 µL aliquots of mobile phase, combine the eluates, vortex, inject a 100 µL aliquot. (Wash solvent was MeCN:water adjusted to pH 3.0 with phosphoric acid 20:80.)

HPLC VARIABLES
Column: 150 × 4.6 5 µm Ultrasphere C8
Mobile phase: MeCN:68 mM pH 2.5 phosphate buffer 55:45
Flow rate: 0.5
Injection volume: 100
Detector: F ex 232 em 335 (filter) following post-column reaction. The effluent from the column flowed through a 7.9 m × 0.3 mm ID coil of PTFE irradiated by an SC3-9 UV lamp (UVP) (cooled with air) to the detector.

CHROMATOGRAM
Retention time: 7
Internal standard: indomethacin (12)
Limit of quantitation: 50 ng/mL
Limit of detection: 10 ng/mL

OTHER SUBSTANCES
Extracted: metabolites

KEY WORDS
serum; post-column reaction; post-column photochemical derivatization; SPE

REFERENCE
Siluveru, M.; Stewart, J.T. Determination of sulindac and its metabolites in human serum by reversed-phase high-performance liquid chromatography using on-line post-column ultraviolet irradiation and fluorescence detection, *J.Chromatogr.B*, **1995**, *673*, 91–96.

SAMPLE
Matrix: blood
Analyte: fenbufen
Sample preparation: Condition a 1 mL Bond-Elut C18 SPE cartridge with 1 mL MeOH. 1 mL Serum + 6 µL 1 mg/mL ketoprofen in MeOH + 1 mL water + 20 µL saturated ammonium sulfate solution + 60 µL concentrated HCl, vortex for 3 min, add to the SPE cartridge, wash with three 1 mL portions of water, allow to dry for 3 min, elute with five 500 µL portions of MeOH. Evaporate the eluate to dryness under a stream of nitrogen, reconstitute the residue in 1 mL mobile phase, inject a 100 µL aliquot.

HPLC VARIABLES
Column: 100 × 4.6 5 µm Spheri-5 cyano
Mobile phase: MeCN:MeOH:water:phosphoric acid 21:22:56.5:0.5
Flow rate: 0.5
Injection volume: 100
Detector: F ex 248 em 335 (filter) following post-column reaction. The column effluent flowed through a knitted 7.9 m × 0.3 mm ID PTFE coil irradiated with an SC3-9 UV lamp (UVP, San Gabriel CA) and cooled with a fan to the detector.

CHROMATOGRAM
Retention time: 8.5
Internal standard: ketoprofen (6.5)

Limit of detection: 10 ng/mL

OTHER SUBSTANCES
Extracted: metabolites

KEY WORDS
post-column reaction; serum; SPE; post-column photochemical derivatization

REFERENCE
Siluveru, M.; Stewart, J.T. Determination of fenbufen and its metabolites in serum by reversed-phase high-performance liquid chromatography using solid-phase extraction and on-line post-column ultra-violet irradiation and fluorescence detection, *J.Chromatogr.B*, **1996**, *682*, 89–94.

SAMPLE
Matrix: blood

Analyte: methotrexate

Sample preparation: Condition an Isolute HAX 200 mg SPE cartridge (International Sorbent Technology) with 3 mL MeOH and 3 mL 100 mM phosphoric acid. Dilute 0.01-1 mL plasma with 2 mL 100 mM phosphoric acid, add to the SPE cartridge at 2.5 mL/min, wash with 2 mL MeOH:100 mM phosphoric acid 5:95, wash with 3 mL pH 8.6 Na_2HPO_4 wash with MeOH:water 5:95, air dry for 2 min, elute with 2 mL 20 g/L trifluoroacetic acid in MeOH. Evaporate the eluate to dryness under a stream of nitrogen at 40°, reconstitute the residue in 100 μL 100 mM phosphoric acid, inject a 20 μL aliquot.

HPLC VARIABLES
Column: 80 × 4.6 3 μm C18 (Perkin-Elmer)

Mobile phase: MeCN:100 mM pH 6.5 phosphate buffer:30% hydrogen peroxide 6.5:93.3:0.2

Column temperature: 22

Flow rate: 1

Injection volume: 20

Detector: F ex 350 em 435 following post-column reaction. The column effluent flowed through a 10 m × 0.3 mm coil irradiated with a 254 nm UV lamp at 37° and flowed to the detector.

CHROMATOGRAM
Limit of quantitation: 0.1 nM

KEY WORDS
post-column reaction; post-column photochemical derivatization; SPE; plasma; comparison with immunoassays

REFERENCE
Albertioni, F.; Rask, C.; Eksborg, E.; Poulsen, J.H.; Pettersson, B.; Beck, O.; Schroeder, H.; Peterson, C. Evaluation of clinical assays for measuring high-dose methotrexate in plasma, *Clin.Chem.*, **1996**, *42*, 39–44.

SAMPLE
Matrix: blood

Analyte: nicorandil

Sample preparation: 200 μL Plasma + 100 μL 1.5 M perchloric acid, vortex, centrifuge at 9600 g for 1 min, add to 100 μL 3 M potassium carbonate, vortex, centrifuge for 1 min, inject a 100-200 μL aliquot of the supernatant.

HPLC VARIABLES
Column: 150 × 4.6 5 μm Capcell Pak C8 AG-120 endcapped (Shiseido)

Mobile phase: MeOH:buffer 20:80 (Prepare by mixing 60 mL 500 mM acetic acid and 940 mL 500 mM pH 6.0 sodium acetate containing 20% MeOH and 20 mM hydrogen peroxide.)

Flow rate: 0.8

Injection volume: 100-200

Detector: F ex 323 em 386 following post-column reaction. The column effluent flowed through a 3 m × 0.25 mm ID Tefzel coil irradiated by two 4 W germicidal lamps (Nippon Denki, Tokyo) at about 40°, a 0.4 m × 0.1 mm ID stainless steel coil, and a 2 m × 0.25 mm ID PTFE coil to the detector.

CHROMATOGRAM
Retention time: 10
Limit of detection: 7 ng/mL

OTHER SUBSTANCES
Non-interfering: histidine, kynurenic acid, nicotinic acid, phenylalanine, pyridoxine, thiamine, tyrosine

KEY WORDS
post-column reaction; plasma; pharmacokinetics; post-column photochemical derivatization

REFERENCE
Mawatari, K.-i.; Nakamura, Y.; Shimizu, R.; Sate, S.; Iinuma, F.; Watanabe, M. Fluorimetric determination of nicorandil in human plasma by a high-performance liquid chromatographic-postcolumn ultraviolet detection system equipped with on-line back-pressure tubing, *J.Chromatogr.B*, **1996**, *679*, 155–159.

SAMPLE
Matrix: blood
Analyte: tamoxifen
Sample preparation: 1 mL Plasma + 1 mL 50% urea + 5 mL diethyl ether, extract. Remove the organic layer and evaporate it to dryness under a stream of air at 40°, reconstitute the residue in 200 μL mobile phase, inject a 50 μL aliquot.

HPLC VARIABLES
Column: 150 × 4.6 5 μm Inertsil ODS-2
Mobile phase: MeCN:67 mM pH 2.2 phosphate buffer 50:50
Flow rate: 1
Injection volume: 50
Detector: F ex 260 em 375 following post-column reaction. The column effluent flowed though a 7 m × 0.3 mm ID knitted PTFE coil irradiated by a Sylvana G8T6 UV lamp to the detector.

CHROMATOGRAM
Limit of quantitation: 0.51 ng/mL

KEY WORDS
post-column reaction; pharmacokinetics; plasma; post-column photochemical derivatization

REFERENCE
Fuchs, W.S.; Leary, W.P.; Van der Meer, M.J.; Gay, S.; Witschital, K.; von Nieciecki, A. Pharmacokinetics and bioavailability of tamoxifen in postmenopausal healthy women, *Arzneimittelforschung*, **1996**, *46*, 418–422.

SAMPLE
Matrix: blood, saliva, urine
Analyte: methotrexate
Sample preparation: Condition a 3 mL Isolute C8 SPE cartridge (International Sorbent Technology) with 3 mL MeOH and 3 mL 100 mM phosphoric acid at about 3 mL/min, do not allow to dry. 1 mL Plasma, 10 μL urine, or 100 μL saliva + 2 mL 100 mM phosphoric acid, wash with 2 mL MeOH:water 5:95, dry with air, elute with 2 mL MeOH:trifluoroacetic acid 98:2. Evaporate the eluate to dryness under a stream of nitrogen at 40°, reconstitute the residue in 100 μL water, inject a 10 μL aliquot.

HPLC VARIABLES
Column: 80 × 4.6 3 μm C18 (Perkin-Elmer)
Mobile phase: MeCN:10 mM pH 6.5 phosphate buffer:30% hydrogen peroxide 6:93.8:0.2
Flow rate: 1
Injection volume: 10
Detector: F ex 350 em 435 following post-column reaction. The column effluent flowed through a 10 m × 0.3 mm ID coil, illuminated with a UV 254 lamp at 37°, to the detector.

CHROMATOGRAM
Retention time: 10
Limit of detection: 0.1 nM (plasma, saliva), 10 nM (urine)

OTHER SUBSTANCES
Extracted: metabolites
Non-interfering: acetaminophen, allopurinol, amoxicillin, ara-C, aspirin, azathioprine, chloroquine, ciprofloxacin, F-ara-A, ibuprofen, 6-mercaptopurine, naproxen, sulfamethoxazole, trimethoprim

KEY WORDS
plasma; pharmacokinetics; SPE; post-column photochemical derivatization; post-column reaction

REFERENCE
Albertioni, F.; Pettersson, B.; Beck, O.; Rask, C.; Seideman, P.; Peterson, C. Simultaneous quantitation of methotrexate and its two main metabolites in biological fluids by a novel solid-phase extraction procedure using high-performance liquid chromatography, *J.Chromatogr.B*, **1995**, *665*, 163–170.

SAMPLE
Matrix: blood, urine
Analyte: methotrexate
Sample preparation: Centrifuge heparinized blood at 16000 g for 15 min, inject a 10-80 μL aliquot. Inject a 10 μL aliquot of urine.

HPLC VARIABLES
Guard column: 40 μm LC-18 (Supelco)
Column: 250 × 4.6 10 μm LiChrosorb RP-18
Mobile phase: MeCN:DMF:50 mM pH 6.2 phosphate buffer:30% hydrogen peroxide 7: 5.6:100:0.15 (Prepare and sonicate 1 day before use. Degas with helium during use.)
Column temperature: 45
Flow rate: 1
Injection volume: 10-80
Detector: F ex 370 em 417 (cut-off filter) following post-column reaction. The column effluent flowed through a 1.59 m × 0.25 mm ID PTFE coil irradiated by a Sylvania G8T5 germicidal lamp at 254 nm to the detector.

CHROMATOGRAM
Retention time: 4
Limit of detection: 0.4 ng

OTHER SUBSTANCES
Extracted: metabolites

KEY WORDS
post-column reaction; whole blood; post-column photochemical derivatization

REFERENCE
Salamoun, J.; Smrz, M.; Kiss, F.; Salamounova, A. Column liquid chromatography of methotrexate and its metabolites using a post column photochemical reactor and fluorescence detection, *J.Chromatogr.*, **1987**, *419*, 213–223.

SAMPLE
Matrix: blood, urine
Analyte: DU-6859a   (7-[(7S)-7-amino-5-azaspiro[2,4]heptan-5-yl]-8-chloro-6-fluoro-1-[(1R, 2S)-2-fluoro-1-cyclopropyl]-1,4-dihydro-4-oxo-3-quinolinecarboxylic acid sesquihydrate)
Sample preparation: Condition a 200 mg Bond Elut C8 LRC SPE cartridge with 4 mL MeOH, 6 mL water, and 6 mL 50 mM KH_2PO_4. 200 μL Serum + 400 μL 50 mM KH_2PO_4 + 200 μL 0.1 (serum) or 1 (urine) μg/mL IS in water, mix, add to the SPE cartridge, wash with 6 mL 50 mM KH_2PO_4, wash with 2 mL THF:water 20:80, elute with 3 mL THF: 0.15% phosphoric acid 50:50. Evaporate the eluate to dryness under reduced pressure, reconstitute the residue in 300 μL THF:water 20:80, inject a 50 μL aliquot.

HPLC VARIABLES
Column: 150 × 4.6 5 μm Inertsil ODS-2
Mobile phase: THF:buffer:1 M ammonium acetate 19:81:1 (Prepare buffer by adjusting the pH of 50 mM NaH_2PO_4 to 2 with orthophosphoric acid.)
Flow rate: 1

Injection volume: 50
Detector: F ex 280 em 430 following post-column reaction. The column flowed through a 20 m × 0.3 mm ID PTFE coil irradiated with a UV lamp at 254 nm to the detector.

CHROMATOGRAM
Retention time: 9.6
Internal standard: 7-[5-amino-3-azaspiro[4,5]decan-3-yl]-6-fluoro-1-cyclopropyl-1,4-dihydro-4-oxo-3-quinolinecarboxylic acid (Daichi Pharmaceutical DX-9484) (14.4)
Limit of detection: 3.43 ng/mL (serum), 4.39 ng/mL (urine)

KEY WORDS
post-column reaction; serum; SPE; pharmacokinetics; post-column photochemical derivatization

REFERENCE
Aoki, H.; Ohshima, Y.; Tanaka, M.; Okazaki, O.; Hakusui, H. High-performance liquid chromatographic determination of the new quinolone antibacterial agent DU-6859a in human serum and urine using solid-phase extraction with photolysis-fluorescence detection, *J.Chromatogr.B*, **1994**, *660*, 365–374.

SAMPLE
Matrix: blood, urine
Analyte: zuclopenthixol
Sample preparation: Condition a 1 mL Bond Elut CN with 2 mL MeCN and 2 mL water. Add 2 mL plasma or urine to the SPE cartridge, wash with 2 mL water, elute with MeCN:n-butylamine 90:10. Evaporate the eluate to dryness under a stream of nitrogen at 40°, reconstitute the residue in 200 μL mobile phase, inject a 100 μL aliquot.

HPLC VARIABLES
Column: 120 × 4.6 Spherisorb S5 CN
Mobile phase: MeCN:200 mM pH 6.5 potassium phosphate buffer:water 36:5:59 containing 6 mM dodecyl-N,N,N-trimethylammonium bromide
Column temperature: 40
Flow rate: 1
Injection volume: 100
Detector: F ex 260 em 435 following post-column reaction. The column effluent flowed through a 5 m × 0.5 mm IDcoil of PTFE tubing irradiated by a low-pressure 8 W mercury UV light to the detector.

CHROMATOGRAM
Retention time: 9
Limit of detection: 0.05 ng/mL

OTHER SUBSTANCES
Extracted: metabolites

KEY WORDS
protect from light; plasma; SPE; post-column reaction; post-column photochemical derivatization

REFERENCE
Hansen, B.B.; Hansen, S.H. Determination of zuclopenthixol and its main N-dealkylated metabolite in biological fluids using high-performance liquid chromatography with post-column photochemical derivatization and fluorescence detection, *J.Chromatogr.B*, **1994**, *658*, 319–325.

SAMPLE
Matrix: blood, urine
Analyte: ciprofloxacin
Sample preparation: Blood. 500 μL Serum or plasma + 100 μL 20 μg/mL IS in 100 mM phosphoric acid + 300 μL MeCN:5 M trichloroacetic acid 50:50, vortex, add 100 μL MeCN, add 300 μL water, vortex, centrifuge at 1500 g for 15 min, inject a 10 μL aliquot of the supernatant. Urine. Dilute urine 1:20 (or more) with 50 mM pH 3.0 KH_2PO_4, remove a 500 μL aliquot and add it to 100 μL 20 μg/mL IS in 100 mM phosphoric acid, add 700 μL 100 mM trichloroacetic acid, vortex, inject a 10 μL aliquot.

HPLC VARIABLES
Guard column: 5 × 3 PLRP-S (Polymer Laboratories)
Column: 150 × 4.6 PLRP-S (Polymer Laboratories)
Mobile phase: MeCN:MeOH:20 mM pH 3.0 trichloroacetic acid 22:4:74
Column temperature: 30
Flow rate: 0.7
Injection volume: 10
Detector: F ex 277 em 418 following post-column photolysis. The column effluent flowed
through a 10 m × 0.25 mm knitted PTFE coil irradiated with a UV 254 low pressure
lamp and flowed to the detector.

CHROMATOGRAM
Retention time: 8
Internal standard: 1-isopropyl-6-fluoro-1,4-dihydro-4-oxo-7-(1-piperazinyl)-3-quinolinecar-
boxylic acid (13)
Limit of quantitation: 50 ng/mL

OTHER SUBSTANCES
Extracted: metabolites

KEY WORDS
serum; plasma; post-column reaction; post-column photochemical derivatization

REFERENCE
Krol, G.J.; Beck, G.W.; Benham, T. HPLC analysis of ciprofloxacin and ciprofloxacin metabolites in body
fluids, *J.Pharm.Biomed.Anal.*, **1996**, *14*, 181–190.

SAMPLE
Matrix: eggs, milk, tissue
Analyte: oxytetracycline
Sample preparation: Muscle. Homogenize (Ultra-Turrax) 5 g muscle, 200 µL 125 µg/mL
tetracycline in 10 mM HCl, 50 mL 50 mM HCl, and 10 mL hexane for 2 min, sonicate
for 3 min, centrifuge at 1920 g for 5 min. Dialyze (Cuprophan cellulose dialysis membrane,
15 kD cutoff) a 530 µL aliquot of the aqueous layer against seven 875 µL aliquots of
recipient solution. Pump the recipient solution into the recipient channel at 1.7 mL/min,
allow to remain stationary for 36 s, then pump onto column A. Finally, backflush the
contents of column A onto the analytical column with mobile phase. After 2 min remove
column A from the circuit and flush it with 2 mL recipient solution. Flush the recipient
channel with 2 mL recipient solution, flush the donor channel with 2 mL 0.01% Triton X-
100. (Recipient solution was 20 mM pH 5 phosphate buffer containing 5 mM sodium
heptanesulfonate.) Liver. Homogenize (Ultra-Turrax) 2 g liver, 200 µL 25 µg/mL tetracycline in 10 mM HCl, 20 mL 10 mM HCl, and 10 mL hexane:dichloromethane 25:75 for
1 min, sonicate for 3 min, centrifuge at 1920 g for 5 min. Dialyze (Cuprophan cellulose
dialysis membrane, 15 kD cutoff) a 530 µL aliquot of the aqueous layer against seven
875 µL aliquots of recipient solution. Pump the recipient solution into the recipient channel at 1.7 mL/min, allow to remain stationary for 36 s, then pump onto column A. Finally,
backflush the contents of column A onto the analytical column with mobile phase. After
2 min remove column A from the circuit and flush it with 2 mL recipient solution. Flush
the recipient channel with 2 mL recipient solution, flush the donor channel with 2 mL
0.01% Triton X-100. (Recipient solution was 20 mM pH 5 phosphate buffer containing 5
mM sodium heptanesulfonate.) Eggs. 2 g Homogenized whole egg + 100 µL 50 µg/mL
tetracycline in 10 mM HCl + 2 mL 0.9% NaCl + 600 µL 10% sodium azide (Caution!
Sodium azide is highly toxic and may be carcinogenic!), shake manually for 10 s. Dialyze
(Cuprophan cellulose dialysis membrane, 15 kD cutoff) a 530 µL aliquot against seven
875 µL aliquots of recipient solution. Pump the recipient solution into the recipient channel at 1.7 mL/min, allow to remain stationary for 36 s, then pump onto column A. Finally,
backflush the contents of column A onto the analytical column with mobile phase. After
2 min remove column A from the circuit and flush it with 2 mL recipient solution. Flush
the recipient channel with 2 mL recipient solution, flush the donor channel with 2 mL
0.01% Triton X-100 containing 18 g/L NaCl. (Recipient solution was 20 mM pH 5 phosphate buffer containing 5 mM sodium heptanesulfonate and 16.5 g/L NaCl.) Milk. 5 mL

Milk + 100 μL 50 μg/mL tetracycline in 10 mM HCl + 1 mL 100 mM EDTA, shake manually for 10 s, centrifuge for 10 min, freeze at -20° for 10 min, remove the decreamed milk solution. Dialyze (Cuprophan cellulose dialysis membrane, 15 kD cutoff) a 530 μL aliquot of the decreamed milk solution against seven 875 μL aliquots of recipient solution. Pump the recipient solution into the recipient channel at 1.7 mL/min, allow to remain stationary for 36 s, then pump onto column A. Finally, backflush the contents of column A onto the analytical column with mobile phase. After 2 min remove column A from the circuit and flush it with 2 mL recipient solution. Flush the recipient channel with 2 mL recipient solution, flush the donor channel with 2 mL 0.01% Triton X-100. (Recipient solution was 20 mM pH 5 phosphate buffer containing 10 mM sodium heptanesulfonate.)

HPLC VARIABLES
Column: A 10 × 2 36 μm Dynospheres polystyrene beads (Dyno Particles, Lillestrom, Norway); B 150 × 4.6 5 μm PLRP-S (Polymer Labs)
Mobile phase: MeCN:buffer 23:77 (Buffer was 20 mM orthophosphoric acid containing 5 (muscle, liver, eggs) or 10 (milk) mM sodium heptanesulfonate.)
Flow rate: 0.7
Detector: F ex 358 em 460 following post-column reaction. The column effluent mixed with 2 M NaOH pumped at 0.15 mL/min and the mixture flowed through a knitted 10 m × 0.3 mm ID reaction coil irradiated at 366 nm to the detector.

CHROMATOGRAM
Retention time: 8
Internal standard: tetracycline (11)
Limit of detection: 4 ng/g (chicken muscle), 3 ng/g (cow muscle), 8 ng/g (salmon liver), 1 ng/g (eggs), 1 ng/mL (milk)

KEY WORDS
post-column reaction; post-column photochemical derivatization; column-switching; dialysis; salmon; cow; chicken; muscle; liver

REFERENCE
Agasoster, T. Automated determination of oxytetracycline residues in muscle, liver, milk and egg by on-line dialysis and post-column reaction detection HPLC, *Food Addit.Contam.*, **1992**, *9*, 615–622.

SAMPLE
Matrix: milk
Analyte: dihydrostreptomycin
Sample preparation: Condition a Sep-Pak tC18 Vac (trifunctional) SPE cartridge with two 5 mL portions of MeOH, two 5 mL portions of water, three 5 mL portions of MeOH, three 5 mL portions of water, and 2 mL buffer. 6 mL Whole milk + 500 μL buffer + 1.5 mL 85% trichloroacetic acid in water, shake vigorously for 10 s, centrifuge at 4000 rpm for 3 min, add 2 mL dichloromethane, mix for 5 s, centrifuge at 4000 rpm for 5 min. Remove 6 mL of the supernatant and add it to 2.5 mL 4 M NaOH, mix for 2 s, centrifuge at 4000 rpm for 10 min. Remove the upper layer and add it to 2.5 mL 500 mM phosphoric acid, adjust pH to 6 ± 0.03 with 1 M NaOH or 500 mM phosphoric acid, add 1.5 mL buffer, mix, add to the SPE cartridge, wash with 1 mL buffer, wash with 1 mL MeOH:water 3:97, wash with 16 mL MeOH:water 30:70 (gravity only), elute with 3.5 mL MeOH:formic acid 80:20 (under vacuum). Evaporate the eluate to dryness under a stream of nitrogen at 60°, reconstitute the residue with 300 μL buffer, add 200 μL chloroform, mix vigorously for 10 s, centrifuge for 3 min, inject a 75 μL aliquot of the aqueous layer. (Prepare buffer by dissolving 4.45 g sodium 1-heptanesulfonate and 1.8 g Na$_2$HPO$_4$.2H$_2$O in 750 mL water, adjust pH to 6.3 with 5 M phosphoric acid, make up to 1 L with water, adjust pH to 6.0 with 1 M phosphoric acid.)

HPLC VARIABLES
Guard column: 20 × 4.6 5 μm Supelcosil LC-ABZ
Column: 150 × 4.6 5 μm Supelcosil LC-ABZ
Mobile phase: MeCN:MeOH:triethylamine:buffer 18.943:18:0.057:63 (Prepare buffer by dissolving 8.65 g sodium octanesulfonate and 4.68 g disodium 1,2-ethanedisulfonate in 750 mL water, adjust pH to 3.2 with acetic acid, add 0.891 g ninhydrin, make up to 1 L with water, adjust pH to 3.2.)
Flow rate: 1.4

Injection volume: 75
Detector: F ex 305 em 500 following post-column reaction. The column effluent mixed with
300 mM NaOH pumped at 0.5 mL/min in a 1.2 μL vortex mixer, the mixture was illuminated with UV light as it flowed through a 10 m × 0.3 mm ID PTFE coil at 80°, the
effluent from the photochemical reactor was cooled to room temperature and passed to
the detector.

CHROMATOGRAM
Retention time: 9
Limit of quantitation: 25 ng/mL
Limit of detection: 15 ng/mL

KEY WORDS
post-column reaction; post-column photochemical derivatization; SPE

REFERENCE
Hormazábal, V.; Yndestad, M. Determination of dihydrostreptomycin sulfate in milk by HPLC using
ion-pair and postcolumn derivatization, *J.Liq.Chromatogr.*, **1995**, *18*, 2695–2702.

SAMPLE
Matrix: sediment
Analyte: whitening agents
Sample preparation: Shake 200 mg dry sediment briefly with 9 mL 30 mM tetrabutylammonium hydrogen sulfate in MeOH, sonicate for 30 min, centrifuge at 2500 rpm for 5
min, repeat the extraction twice. Combine the extracts and evaporate them to dryness
under reduced pressure at 50°, reconstitute with 3 mL DMF:water 50:50, add 30 μL 1
mg/mL IS in water:DMF 50:50, centrifuge at 2500 rpm for 5 min, inject a 10 μL aliquot.

HPLC VARIABLES
Guard column: present but not specified
Column: 100 × 2.1 3 μm Hypersil ODS
Mobile phase: Gradient. A was MeCN:MeOH 60:40. B was 100 mM pH 6.5 ammonium
acetate buffer. A:B from 30:70 to 60:40 over 22 min, to 90:10 over 2 min, maintain at
90:10 for 8 min, return to initial conditions over 3 min, re-equilibrate at initial conditions
for 5 min.
Flow rate: 0.4
Injection volume: 10
Detector: F ex 350 em 430 following post-column reaction. The column effluent flowed
through a 0.5 m × 0.3 mm ID PTFE coil irradiated at 254 nm by a UV lamp (Beam Boost)
to the detector. (UV irradiation converts non-fluorescent Z-isomers to fluorescent E-
isomers.)

CHROMATOGRAM
Retention time: 13 ((E,Z)-4,4'-bis(2-sulfostyryl)biphenyl), 15 ((E,E)-4,4'-bis(2-sulfosty-
ryl)biphenyl), 16 ((E)-4,4'-[4-anilino-6-morpholino-1,3,5-triazin-2-yl)amino]stilbene-2,2'-
disulfonate), 16.5 (4,4'-bis(4-chloro-3-sulfostyryl)biphenyl), ((Z)-4,4'-[4-anilino-6-morpho-
lino-1,3,5-triazin-2-yl)amino]stilbene-2,2'-disulfonate)
Internal standard: 4,4'-bis(5-ethyl-3-sulfobenzofur-2-yl)biphenyl (Ciba-Geigy) (24)
Limit of quantitation: 10.5 ng/g

KEY WORDS
post-column reaction; microbore; post-column photochemical derivatization

REFERENCE
Stoll, J.-M.A.; Giger, W. Determination of detergent-derived fluorescent whitening agent isomers in lake
sediments and surface waters by liquid chromatography, *Anal.Chem.*, **1997**, *69*, 2594–2599.

SAMPLE
Matrix: solutions
Analyte: clobazam

HPLC VARIABLES
Column: 100 × 4.6 10 μm LiChrosorb RP8
Mobile phase: MeOH:100 mM pH 7.0 tetraethylammonium phosphate buffer 39:48
Flow rate: 2

Injection volume: 75
Detector: UV 230 following post-column reaction. The column effluent flowed through a 20 m × 0.5 mm ID knitted PTFE coil irradiated by a 15 w low-pressure mercury lamp (Original Hanau TNN 15/32) to the detector.; F ex 364 em 400 following post-column reaction. The column effluent flowed through a 20 m × 0.5 mm ID knitted PTFE coil irradiated by a 15 w low-pressure mercury lamp (Original Hanau TNN 15/32) to the detector.

CHROMATOGRAM
Retention time: 5
Limit of detection: 1 ng

OTHER SUBSTANCES
Extracted: metabolites

KEY WORDS
post-column reaction; post-column photochemical derivatization

REFERENCE
Uihlein, M.; Schwab, E. A novel reactor for photochemical post-column derivatization in HPLC, *Chromatographia*, **1982**, *15*, 140–146.

SAMPLE
Matrix: solutions
Analyte: fenbendazole

HPLC VARIABLES
Column: 120 × 4.6 7 μm LiChrosorb RP8
Mobile phase: MeOH:33 mM phosphoric acid 50:50
Flow rate: 2
Injection volume: 75
Detector: UV 245 following post-column reaction. The column effluent flowed through a 20 m × 0.5 mm ID knitted PTFE coil irradiated by a 15 w low-pressure mercury lamp (Original Hanau TNN 15/32) to the detector.; F ex 300 em 342 following post-column reaction. The column effluent flowed through a 20 m × 0.5 mm ID knitted PTFE coil irradiated by a 15 w low-pressure mercury lamp (Original Hanau TNN 15/32) to the detector.

CHROMATOGRAM
Retention time: 10
Limit of detection: 5 ng

OTHER SUBSTANCES
Extracted: metabolites

KEY WORDS
post-column reaction; post-column photochemical derivatization

REFERENCE
Uihlein, M.; Schwab, E. A novel reactor for photochemical post-column derivatization in HPLC, *Chromatographia*, **1982**, *15*, 140–146.

SAMPLE
Matrix: solutions
Analyte: digoxin

HPLC VARIABLES
Column: 250 × 4.6 10 μm C18 (Alltech)
Mobile phase: MeCN:water 60:40 containing 510 μM 2-tert-butylanthraquinone
Flow rate: 1
Injection volume: 20
Detector: F ex 375 em 470 following post-column reaction. The column effluent flowed through a 0.8 m × 0.5 mm ID PTFE coil irradiated with a UV pen-ray lamp (Ultra-Violet Products) to the detector. (The detector was continually flushed with nitrogen. Full construction details are given in the paper.)

CHROMATOGRAM
Retention time: 3.5
Limit of detection: 2 ng

KEY WORDS
post-column reaction; post-column photochemical derivatization

REFERENCE
Gandelman, M.S.; Birks, J.W. Liquid chromatographic detection of cardiac glycosides, saccharides and hydrocortisone based on the photoreduction of 2-tert-butylanthraquinone, *Anal.Chim.Acta*, **1983**, *155*, 159–171.

SAMPLE
Matrix: solutions
Analyte: digoxin

HPLC VARIABLES
Column: 250 × 4.6 10 μm C18 (Alltech)
Mobile phase: MeCN:water 40:60 containing 200 μM anthraquinone-2,6-disulfonate, disodium salt
Flow rate: 1
Injection volume: 20
Detector: F ex 275 em >475 following post-column photochemical derivatization. The column effluent flowed through a 12.5 m × 0.33 mm ID knitted PTFE coil irradiated by an 8 W fluorescent lamp (Sylvania Model E 8TS 1BLB) to the detector. (The anthraquinone-2,6-disulfonate absorbs UV and reacts with the analyte to produce a fluorescent product (dihydroxyanthracene-2,6-disulfonate).)

CHROMATOGRAM
Retention time: 6

OTHER SUBSTANCES
Simultaneously analyzed: lactose

KEY WORDS
post-column reaction; post-column photochemical derivatization

REFERENCE
Gandelman, M.S.; Birks, J.W.; Brinkman, U.A.T.; Frei, R.W. Liquid chromatographic detection of cardiac glycosides and saccharides based on the photoreduction of anthraquinone-2,6-disulfonate, *J.Chromatogr.*, **1983**, *282*, 193–209.

SAMPLE
Matrix: solutions
Analyte: aflatoxins
Sample preparation: Inject a 20 μL aliquot of a solution in MeOH.

HPLC VARIABLES
Column: 250 × 4.6 Microsorb MV C18
Mobile phase: MeCN:MeOH:water 22:15:63
Column temperature: 40
Flow rate: 1
Injection volume: 20
Detector: F ex 365 em >415 following post-column reaction. The column effluent flowed through a knitted 15 m × 0.25 mm ID coil irradiated by a low-pressure mercury lamp to the detector. (The peak heights of aflatoxins G_1 and B_1 are greatly increased by photolysis, the other aflatoxins are not affected.)

CHROMATOGRAM
Retention time: 9.54 (G_2), 11.40 (G_1), 12.37 (B_2), 15.04 (B_1)
Limit of detection: 0.25-1 ppb

KEY WORDS
post-column reaction; post-column photochemical derivatization

REFERENCE

Joshua, H. Determination of aflatoxins by reversed-phase high-performance liquid chromatography with post-column in-line photochemical derivatization and fluorescence detection, *J.Chromatogr.A*, **1993**, *654*, 247–254.

SAMPLE

Matrix: solutions
Analyte: phenolic aldehydes
Sample preparation: Inject a 10 μL aliquot of a solution in MeOH:water 50:50.

HPLC VARIABLES

Column: 150 × 3.9 4 μm Novapack C18
Mobile phase: Gradient. A was 0.5% formic acid in water. B was 0.5% formic acid in MeOH. A:B 95:5 for 5 min, to 70:30 over 15 min (Waters curve 9), to 60:40 over 10 min (Waters curve 9), to 55:45 over 10 min (Waters curve 9), to 50:50 over 10 min (Waters curve 9) (J.Chromatogr.A, 1994, 683, 31).
Flow rate: 1
Injection volume: 10
Detector: F following post-column reaction. The column effluent flowed through a 10 m × 0.33 mm ID coil of PTFE tubing irradiated with five 8 w low-pressure mercury lamps to the detector. The design of the photoreactor is detailed in the paper.

CHROMATOGRAM

Retention time: 4.5 (protocatechualdehyde), 6.5 (2,5-dihydroxybenzaldehyde), 7.5 (4-hydroxybenzaldehyde), 8.5 (3-hydroxybenzaldehyde), 12.5 (salicaldehyde), 15 (vanillin), 17 (isovanillin), 24 (o-vanillin), 31.5 (m-anisaldehyde), 40 (veratraldehyde), 42.5 (2,4-dimethoxybenzaldehyde), 43.5 (3,5-dimethoxybenzaldehyde)

KEY WORDS

post-column reaction; post-column photochemical derivatization

REFERENCE

Lores, M.; Garcia, C.M.; Cela, R. Selectable-power photoreactor for flow-injection analysis systems and high-performance liquid chromatography post-column photochemical derivatization, *J.Chromatogr.A*, **1996**, *724*, 55–65.

SAMPLE

Matrix: tissue
Analyte: oxytetracycline
Sample preparation: Homogenize (Ultra-Turrax T25) 5 g tissue, 200 μL 125 μg/mL tetracycline in 10 mM HCl, 50 mL 50 mM HCl, and 10 mL hexane for 2 min, sonicate for 3 min, centrifuge at 1920 g for 5 min. Dialyze (Cuprophan cellulose dialysis membrane, 15 kD cutoff) a 530 μL aliquot of the aqueous layer against seven 875 μL aliquots of recipient solution. Pump the recipient solution into the recipient channel at 1.7 mL/min, allow to remain stationary for 36 s, then pump onto column A. Finally, backflush the contents of column A onto the analytical column with mobile phase. After 2 min remove column A from the circuit and flush it with 2 mL recipient solution. Flush the recipient channel with 2 mL recipient solution, flush the donor channel with 2 mL 0.01% Triton X-100. (Recipient solution was 20 mM pH 5 phosphate buffer containing 5 mM sodium heptanesulfonate.)

HPLC VARIABLES

Column: A 10 × 2 36 μm Dynospheres polystyrene beads (Dyno Particles, Lillestrom, Norway); B 150 × 4.6 5 μm PLRP-S (Polymer Labs)
Mobile phase: MeCN:buffer 23:77 (Buffer was 20 mM orthophosphoric acid containing 5 mM sodium heptanesulfonate.)
Flow rate: 0.7
Detector: F ex 358 em 460 following post-column reaction. The column effluent mixed with 2 M NaOH pumped at 0.15 mL/min and the mixture flowed through a knitted 10 m × 0.3 mm ID reaction coil irradiated at 366 nm to the detector.

CHROMATOGRAM

Retention time: 8
Internal standard: tetracycline (10)

Limit of detection: 5 ng/g

KEY WORDS
post-column reaction; muscle; fish; salmon; column-switching; dialysis; post-column photochemical derivatization

REFERENCE
Agasoster, T.; Rasmussen, K.E. On-line dialysis, liquid chromatography and post-column reaction detection of oxytetracycline in salmon muscle extracts, *J.Pharm.Biomed.Anal.*, **1992**, *10*, 349–354.

SAMPLE
Matrix: urine
Analyte: cromolyn
Sample preparation: Dilute urine 10-fold with water, filter (Chromatodisc 25A, Kurabou, Osaka), inject a 50-200 μL aliquot of the filtrate.

HPLC VARIABLES
Column: 250 × 4.6 5 μm Capcell Pak C18 SG-120 (Shiseido)
Mobile phase: MeOH:35 mM pH 8 phosphate buffer 30:70, containing 75 mM hydrogen peroxide and 20 mM 18-crown-6
Flow rate: 0.6
Injection volume: 50-200
Detector: F ex 325 em 448 following post-column reaction. The column effluent flowed through a 3 m × 0.25 mm ID Tefzel coil irradiated by two 4 W germicidal lamps. The effluent from this coil flowed through a 50 cm × 0.13 mm ID PEEK coil and a 2 m × 0.25 mm ID PTFE coil to the detector.

CHROMATOGRAM
Retention time: 18
Limit of detection: 400 ng/mL

KEY WORDS
post-column reaction; post-column photochemical derivatization

REFERENCE
Mawatari, K.-i.; Mashiko, S.; Sate, Y.; Usui, Y.; Iinuma, F.; Watanabe, M. Determination of disodium cromoglycate in human urine by high-performance liquid chromatography with post-column photoirradiation-fluorescence detection, *Analyst*, **1997**, *122*, 715–717.

MS Detection

SAMPLE
Matrix: beverages
Analyte: nitrosamines
Sample preparation: 100 mL Beer + 1 mL 200 ng/mL N-nitrosodimethylamine + 20 mL dichloromethane, shake for 1 min, centrifuge at 500 g for 10 min, repeat extraction. Filter extracts through anhydrous sodium sulfate, concentrate (Kuderna-Danish with 3 ball Snyder condenser) to 4 mL at 65°, evaporate to 200 μL under a stream of nitrogen, inject a 10-20 μL aliquot.

HPLC VARIABLES
Column: 300 × 3.9 5 μm μBondapak C18
Mobile phase: Gradient. MeCN:water 5:95 for 5 min, to 95:5 over 10 min, maintain at 95:5 for 10 min.
Flow rate: 1
Injection volume: 10-20
Detector: MS, Finnigan MAT Model TSQ 7000, capillary temp 240°, electrospray 4.5 kV, sheath gas 400 kPa, following post-column reaction. The column effluent flowed through a 15 m × 0.25 mm i.d. knitted PTFE coil irradiated by an 8 W 366 nm UV light and was then split so that 180 μL/min entered the detector.

CHROMATOGRAM

Retention time: 7 (N-nitrosomorpholine), 9.2 (N-nitrospyrrolidine), 13.5 (N-nitrosodiethyl-amine), 14.5 (N-nitrosopiperidine), 17 (N-nitrosobutylethylamine)
Internal standard: N-nitrosodimethylamine (5.6)
Limit of detection: 2-6 ng

KEY WORDS

beer; post-column reaction; post-column photochemical derivatization

REFERENCE

Volmer, D.A.; Lay, J.O. Jr.; Billedeau, S.M.; Vollmer, D.L. Detection and confirmation of *N*-nitrosodi-alkylamines using liquid chromatography-electrospray ionization coupled on-line with a photolysis reactor, *Anal.Chem.*, **1996**, *68*, 546−552.

UV Detection

RELATED REFERENCES

Andrisano, V.; Gotti, R.; Roveri, P.; Cavrini, V. Analysis of semipermanent hair dyes by HPLC with on-line post-column photochemical derivatization. *Chromatographia* **1997**, *44*, 431-437.

Cela, R.; Lores, M.; Garcia, C.M. Applicability of a postcolumn photochemical reactor in the high-performance liquid chromatography of 34 polyphenolic compounds with UV detection. *J.Chromatogr.* **1992**, *626*, 117-126.

Lookabaugh, M.; Krull, I.S. Determination of nitrite and nitrate by reversed-phase high-performance liquid chromatography using on-line post-column photolysis with ultraviolet absorbance and electrochemical detection. *J.Chromatogr.* **1988**, *452*, 295-308.

SAMPLE

Matrix: blood
Analyte: barbiturates
Sample preparation: Vigorously shake equal volumes of plasma and MeCN, centrifuge at 10000 g for 3 min, inject a 20 μL aliquot of the supernatant.

HPLC VARIABLES

Column: 110 × 4.6 PartiSphere C8 (Whatman)
Mobile phase: MeCN:120 mM pH 6.2 phosphate buffer 50:50
Flow rate: 1
Injection volume: 20
Detector: UV 270 following post-column reaction. The column effluent flowed through a 6 m × 0.25 mm ID crocheted PTFE coil irradiated with a Sylvania G8-T5 lamp at 254 nm to the detector.

CHROMATOGRAM

Retention time: 3.28 (pentobarbital), 4.25 (thiopental)
Limit of detection: 200 ng/mL

KEY WORDS

post-column reaction; plasma; post-column photochemical derivatization

REFERENCE

Schmid, R.W.; Wolf, C. Simultaneous determination of thiopental and its metabolite, pentobarbital, in blood by high-performance liquid chromatography and post-column photochemical reaction, *J.Pharm.Biomed.Anal.*, **1989**, *7*, 1749−1755.

SAMPLE

Matrix: blood
Analyte: barbiturates
Sample preparation: Mix plasma with an equal volume of MeCN, centrifuge at 10000 g, dilute supernatant with an equal volume of water, inject a 50 μL aliquot.

HPLC VARIABLES

Column: 110 × 4.7 5 μm PartiSphere C18 (Whatman)
Mobile phase: MeCN:15 mM pH 7.0 phosphate buffer 30:70

Flow rate: 0.8
Injection volume: 50
Detector: UV 270 following post-column reaction. The column effluent flowed through a 6 m × 0.25 mm ID crocheted coil of PTFE tubing irradiated by an 8 W low-pressure mercury lamp to the detector.

CHROMATOGRAM
Retention time: 4.4 (aprobarbital), 5.3 (butethal), 7 (pentobarbital), 8.7 (secobarbital)

KEY WORDS
plasma; post-column reaction; post-column photochemical derivatization

REFERENCE
Wolf, C.; Schmid, R.W. Enhanced UV-detection of barbiturates in HPLC analysis by on-line photochemical reaction, *J.Liq.Chromatogr.*, **1990**, *13*, 2207–2216.

SAMPLE
Matrix: blood
Analyte: busulfan
Sample preparation: Condition a 1 mL Bond Elut C18 SPE cartridge with 1 mL MeOH and 3 mL water. Add 1 mL plasma to the cartridge, wash with 2 mL water, dry, elute with 1 mL MeOH by centrifuging. Remove the eluate and add it to 1 mL 4 M NaI in water and 400 μL n-heptane, heat the mixture in a closed vial with stirring at 70° for 40 min, cool to room temperature. Remove 350 μL of the upper layer and centrifuge it at 12000 g. Remove 250 μL of the upper organic layer and add it to 100 μL 2-methoxyethanol, evaporate the n-heptane without heating under vacuum for 8 min, inject a 20 μL aliquot of the residue.

HPLC VARIABLES
Guard column: 50 × 4.6 5 μm LiChrosorb CN
Column: 250 × 4.6 5 μm LiChrosorb CN
Mobile phase: MeOH:water 20:80
Flow rate: 1
Injection volume: 20
Detector: UV 226 following post-column reaction. The column effluent flowed through a 0.8 mm ID PTFE knitted tube reactor, internal volume 2.4 mL (or a 25 m × 0.3 mm ID tube in a commercial Beam Booster reactor) irradiated by a GTE G8T5 germicidal lamp to the detector.

CHROMATOGRAM
Retention time: 16.5
Limit of detection: 20 ng/mL

KEY WORDS
plasma; post-column reaction; post-column photochemical derivatization; SPE; pharmacokinetics

REFERENCE
Blanz, J.; Rosenfeld, C.; Proksch, B.; Ehninger, G.; Zeller, K.-P. Quantitation of busulfan in plasma by high-performance liquid chromatography using postcolumn photolysis, *J.Chromatogr.*, **1990**, *532*, 429–437.

SAMPLE
Matrix: formulations
Analyte: imidazole antimycotic drugs
Sample preparation: Tablets. Powder tablets, weigh out amount equivalent to about 30 mg ketoconazole, add 100 mL MeOH, sonicate for 5 min, filter. Add a 2 mL aliquot of filtrate to 5 mL of 200 μg/mL clotrimazole in MeOH, make up to 25 mL with MeOH, inject 20 μL aliquot. Cream. Condition a 500 mg Bond-Elut diol cartridge with 6 mL dichloromethane. Weigh out cream equivalent to about 5 mg of drug, add 30 mL dichloromethane, sonicate for 3 min, make up to 100 mL with dichloromethane, filter. Add a 2 mL aliquot to the cartridge, wash with 2 mL dichloromethane:methanol 4:1, wash with 1 mL MeOH, elute with 3 mL MeOH:buffer 85:15. Add eluate to 1 mL 200 μg/mL clo-

trimazole in MeOH, make up to 5 mL with MeOH, inject 20 µL aliquot. (Buffer was 50 mM triethylamine adjusted to pH 7.0 with phosphoric acid.)

HPLC VARIABLES
Column: 250 × 4.6 5 µm Spherisorb CN
Mobile phase: THF:buffer 30:70 (Buffer was 50 mM triethylamine adjusted to pH 3.0 with phosphoric acid.)
Flow rate: 1
Injection volume: 20
Detector: UV 230; UV 270 following post-column reaction. The column effluent flowed through a Beam Boost model C6808 with 10 m × 0.3 mm reaction coil to the detector.

CHROMATOGRAM
Retention time: 7 (ketoconazole), 11 (bifonazole), 12 (tioconazole), 15 (econazole), 16.5 (isoconazole), 19 (miconazole), 31 (fenticonazole)
Internal standard: clotrimazole (9.5)

KEY WORDS
tablets; creams; post-column reaction; post-column photochemical derivatization; enhanced sensitivity with post-column reaction

REFERENCE
Di Pietra, A.M.; Cavrini, V.; Andrisano, V.; Gatti, R. HPLC analysis of imidazole antimycotic drugs in pharmaceutical formulations, *J.Pharm.Biomed.Anal.*, **1992**, *10*, 873–879.

SAMPLE
Matrix: formulations
Analyte: aspirin
Sample preparation: Weigh out powdered sample containing 68 mg aspirin, add 80 mL MeOH, sonicate for 10 min, dilute to 100 mL with MeOH, centrifuge. Remove a 5 mL aliquot of the supernatant and add it to 1 mL 2 mg/mL resorcinol, add 2 mL MeOH, make up to 20 mL with 50 mM pH 3.0 triethylamine phosphate, inject an aliquot.

HPLC VARIABLES
Column: 150 × 3.2 5 µm Hypersil ODS
Mobile phase: THF:50 mM pH 3.0 triethylamine phosphate 12:88
Flow rate: 0.6
Injection volume: 20
Detector: UV 275 following post-column reaction. The column effluent flowed through a 10 m × 0.3 mm ID crocheted PTFE coil irradiated with an 8 W low-pressure mercury lamp at 254 nm to the detector.

CHROMATOGRAM
Retention time: 15
Internal standard: resorcinol (9)

OTHER SUBSTANCES
Simultaneously analyzed: acetaminophen (post-column irradiation gives little increase in peak height), caffeine (post-column irradiation gives little increase in peak height), propyphenazone (post-column irradiation gives a decrease in peak height)

KEY WORDS
post-column reaction; post-column photochemical derivatization

REFERENCE
Di Pietra, A.M.; Gatti, R.; Andrisano, V.; Cavrini, V. Application of high-performance liquid chromatography with diode-array detection and on-line post-column photochemical derivatization to the determination of analgesics, *J.Chromatogr.A*, **1996**, *729*, 355–361.

SAMPLE
Matrix: formulations
Analyte: chlorpheniramine
Sample preparation: Condition a 500 mg Bond Elut SCX strong cation-exchange SPE cartridge with 6 mL MeOH and 3 mL 10 mM pH 4.5 phosphate buffer. Weigh out pow-

dered tablet containing 0.45 mg chlorpheniramine, add 25 mL MeCN:10 mM pH 4.5 phosphate buffer 25:75, sonicate for 10 min, dilute to 50 mL with 10 mM pH 4.5 phosphate buffer, centrifuge, add a 10 mL aliquot of the supernatant to the SPE cartridge, wash with four 3 mL portions of 10 mM pH 4.5 phosphate buffer, elute with 4 mL MeCN:100 mM pH 8.0 triethylamine phosphate 40:60, inject an aliquot.

HPLC VARIABLES
Column: 150 × 4.6 5 μm Spherisorb CN
Mobile phase: MeCN:100 mM pH 3.0 triethylamine phosphate 5:95
Flow rate: 1
Injection volume: 20
Detector: UV 275 or UV 330 following post-column reaction. The column effluent flowed through a 10 m × 0.3 mm ID crocheted PTFE coil irradiated with an 8 W low-pressure mercury lamp at 254 nm to the detector.

CHROMATOGRAM
Retention time: 4.7

OTHER SUBSTANCES
Non-interfering: acetaminophen, caffeine

KEY WORDS
post-column reaction; SPE; tablets; post-column photochemical derivatization

REFERENCE
Di Pietra, A.M.; Gatti, R.; Andrisano, V.; Cavrini, V. Application of high-performance liquid chromatography with diode-array detection and on-line post-column photochemical derivatization to the determination of analgesics, *J.Chromatogr.A*, **1996**, *729*, 355–361.

SAMPLE
Matrix: formulations
Analyte: resorcinol
Sample preparation: Inject an aliquot of a solution in MeOH:50 mM pH 3.0 triethylamine phosphate 40:60.

HPLC VARIABLES
Column: 150 × 3.2 5 μm Hypersil ODS
Mobile phase: THF:50 mM pH 3.0 triethylamine phosphate 12:88
Flow rate: 0.6
Injection volume: 20
Detector: UV 275 following post-column reaction. The column effluent flowed through a 10 m × 0.3 mm ID crocheted PTFE coil irradiated with an 8 W low-pressure mercury lamp at 254 nm to the detector.

CHROMATOGRAM
Retention time: 9

OTHER SUBSTANCES
Simultaneously analyzed: acetaminophen (post-column irradiation gives little increase in peak height), aspirin, caffeine (post-column irradiation gives little increase in peak height), propyphenazone (post-column irradiation gives a decrease in peak height)

KEY WORDS
post-column reaction; post-column photochemical derivatization

REFERENCE
Di Pietra, A.M.; Gatti, R.; Andrisano, V.; Cavrini, V. Application of high-performance liquid chromatography with diode-array detection and on-line post-column photochemical derivatization to the determination of analgesics, *J.Chromatogr.A*, **1996**, *729*, 355–361.

POST-COLUMN REACTION

Acetic Acid

CH_3COOH $\longrightarrow$ post-column reaction

SAMPLE
Matrix: blood
Analyte: pimobendan
Sample preparation: Condition a 1 mL Bond Elut PH SPE cartridge with 3 mL MeOH and 3 mL water. 100 μL Plasma + 1 mL 100 mM pH 9.5 phosphate buffer, add to the SPE cartridge, wash with 3 mL water, dry by pulling air through the cartridge for 15 min, elute with 1 mL MeOH. Evaporate the eluate to dryness under a stream of nitrogen at 60°, reconstitute the residue in 200 μL EtOH:n-hexane 50:50, inject a 100 μL aliquot.

HPLC VARIABLES
Column: 250 × 4.6 5 μm Sumchiral OA-4400 (Sumika)
Mobile phase: n-Hexane:EtOH:acetic acid 300:120:1
Column temperature: 40
Flow rate: 0.8
Injection volume: 100
Detector: F ex 330 em 415 following post-column reaction. The column effluent mixed with EtOH:acetic acid pumped at 0.3 mL/min and the mixture flowed to the detector.

CHROMATOGRAM
Retention time: 14.4 (-), 15.1 (+)
Limit of detection: 1.25 ng/mL

OTHER SUBSTANCES
Extracted: metabolites

KEY WORDS
chiral; post-column reaction; plasma; rat; SPE; pharmacokinetics

REFERENCE
Asakura, M.; Nagakura, A.; Tarui, S.; Matsumura, R. Simultaneous determination of the enantiomers of pimobendan and its main metabolite in rat plasma by high-performance liquid chromatography, *J.Chromatogr.*, **1993**, *614*, 135–141.

SAMPLE
Matrix: urine
Analyte: oxazepam
Sample preparation: Condition a 100 mg Bond-Elut C2 SPE cartridge with MeOH and 10 mM pH 6.0 phosphate buffer. 5 mL Urine + 1250 U β-glucuronidase, adjust pH to 5.0 with HCl, heat at 37° for 24 h. Add 5 μg nordiazepam, buffer to pH 6.0 with 500 μL 100 mM pH 6.0 phosphate buffer, add to the SPE cartridge, wash with 3 volumes of water, wash with 1 mL MeOH:water 25:75, wash with 1 mL water, elute with 1 mL MeOH:water 60:40. Evaporate the eluate to dryness, reconstitute in 200 μL mobile phase, inject an aliquot.

HPLC VARIABLES
Column: 35 × 4.6 5 μm Ultrabase C18 (Scharlau)
Mobile phase: MeOH:water 60:40
Flow rate: 0.5
Injection volume: 20
Detector: F ex 364 em 469 following post-column reaction. The effluent from the column mixed with acetic acid pumped at 1.1 mL/min and the mixture flowed through a 15 m × 0.5 mm ID coil of PTFE tubing at 100° to the detector.

CHROMATOGRAM
Retention time: 6
Internal standard: nordiazepam (8.5)
Limit of quantitation: 10 ng/mL

Limit of detection: 4 ng/mL

KEY WORDS
post-column reaction; SPE

REFERENCE
Berrueta, L.A.; Gallo, B.; Vicente, F. Analysis of oxazepam in urine using solid-phase extraction and high-performance liquid chromatography with fluorescence detection by post-column derivatization, *J.Chromatogr.*, **1993**, *616*, 344–348.

Aluminum Nitrate

$Al(NO_3)_3$ ⟶ post-column reaction

SAMPLE
Matrix: solutions
Analyte: flavonols
Sample preparation: Inject a 20 µL aliquot.

HPLC VARIABLES
Guard column: 15 × 3.2 7 µm Newguard RP-18
Column: 150 × 4.6 5 µm Inertsil ODS-2
Mobile phase: MeCN:25 mM pH 2.4 phosphate buffer 31:69
Column temperature: 30
Flow rate: 1
Injection volume: 20
Detector: F ex 422 em 485 following post-column reaction. The column effluent mixed with 1.5 M aluminum nitrate in MeOH:acetic acid 92.8:7.2 pumped at 0.4 mL/min and the mixture flowed through a 15 m × 0.25 mm ID coil of PTFE tubing at 30° to the detector.

CHROMATOGRAM
Limit of detection: 50-450 pg/mL

KEY WORDS
post-column reaction; for isorhamnetin, kaempferol, myricetin, quercetin

REFERENCE
Hollman, P.C.H.; van Trijp, J.M.P.; Buysman, M.N.C.P. Fluorescence detection of flavonols in HPLC by postcolumn chelation with aluminum, *Anal.Chem.*, **1996**, *68*, 3511–3515.

Aluminum Trichloride

$AlCl_3$ ⟶ post-column reaction

SAMPLE
Matrix: food
Analyte: sterigmatocystin
Sample preparation: Bread, maize. Shake 25 g bread or maize with 100 mL MeCN:4% KCl in water 90:10 on a wrist-action shaker for 30 min, filter (paper). Remove a 50 mL aliquot of the filtrate and wash it twice with 25 mL portions of n-hexane. Add 12.5 mL water to the MeCN/water layer, extract with 25 mL chloroform, extract twice with 12.5 mL portions of chloroform. Pass the chloroform extracts through anhydrous sodium sulfate and evaporate them to dryness under reduced pressure at 35°, make up the residue to 5 mL with chloroform. Remove a 1 mL aliquot and evaporate it to dryness under a stream of nitrogen, reconstitute with 1 mL MeCN:water 60:40, inject an aliquot. Cheese. Homogenize (Ultra-Turrax) 20 g grated cheese with 200 mL MeOH:4% KCl in water 90:10 at medium speed for 1.5 min, filter (paper). Remove a 100 mL aliquot of the filtrate and wash it twice with 50 mL portions of n-hexane. Add 25 mL water to the MeOH/water layer, extract with 100 mL chloroform, extract twice with 25 mL portions of chloroform. Pass the chloroform extracts through anhydrous sodium sulfate and evaporate them to

dryness under reduced pressure at 40°, dissolve the residue in chloroform/MeOH, evaporate, make up the residue to 5 mL with chloroform. Remove a 1.25 mL aliquot and evaporate it to dryness under a stream of nitrogen, reconstitute with 1 mL MeCN:water 60:40, inject an aliquot.

HPLC VARIABLES
Column: 250 × 4.6 5 μm Hypersil ODS-1
Mobile phase: MeCN:MeOH:water 15:50:35 (bread, maize) or 13.5:47.5:39 (cheese)
Column temperature: 35
Flow rate: 1
Detector: F ex 355 em 525 following post-column reaction. The column effluent mixed with 100 mM aluminum chloride hexahydrate in MeOH:water 50:50 pumped at 0.1 mL/min and the mixture flowed through a 5 m × 0.3 mm ID PTFE coil at 50° to the detector.

CHROMATOGRAM
Retention time: 10 (bread, maize), 14 (cheese)
Limit of detection: 30-40 ng/g (bread), 10 ng/g (maize), 20 ng/g (cheese)

KEY WORDS
post-column reaction; bread; maize; cheese; comparison with LC-MS

REFERENCE
Scudamore, K.A.; Hetmanski, M.T.; Clarke, P.A.; Barnes, K.A.; Startin, J.R. Analytical methods for the determination of sterigmatocystin in cheese, bread and corn products using HPLC with atmospheric pressure ionization mass spectrometric detection, *Food Addit.Contam*, **1996**, *13*, 343–358.

SAMPLE
Matrix: solutions
Analyte: zearalenone

HPLC VARIABLES
Column: 250 × 4.6 5 μm Spherisorb ODS-1
Mobile phase: MeOH:water 80:20
Flow rate: 1
Detector: F ex 285 em 440 following post-column reaction. The column effluent mixed with 250 mM aluminum chloride hexahydrate in MeOH:water 75:25 pumped at 0.5 mL/min and the mixture flowed through a 5 m × 0.3 mm ID PTFE coil at 50° to the detector. (At the end of each day flush system with MeOH:water 75:25.)

CHROMATOGRAM
Retention time: 6.5

KEY WORDS
post-column reaction

REFERENCE
Hetmanski, M.T.; Scudamore, K.A. Detection of zearalenone in cereal extracts using high-performance liquid chromatography with post-column derivatization, *J.Chromatogr.*, **1991**, *588*, 47–52.

Ammonium Hydroxide

NH$_4$OH $\longrightarrow$ post-column reaction

SAMPLE
Matrix: blood, milk
Analyte: ochratoxin A
Sample preparation: Condition an Easy-Extract immunoaffinity SPE cartridge (Biocode, Heslington, York, UK) with 20 mL PBS. Blood. 0.5-2 g Serum or whole blood + 10 mL buffer, vortex for 1 min, add 5 mL chloroform, mix intensively for 2-3 min, centrifuge at 2500 g for 15 min, repeat the extraction. Combine the lower organic layers and evaporate them to dryness under reduced pressure at 30-40°, reconstitute with 5 mL MeOH:PBS 15:85, add to the SPE cartridge at 1-2 mL/min, rinse the flask three times with 5 mL

portions of MeOH:PBS 15:85, add the rinses to the SPE cartridge, wash with 10 mL water, force air through the SPE cartridge, elute with 3 mL MeOH at 0.5 mL/min, force out the last of the eluate with air. Evaporate the eluate to dryness under a stream of nitrogen at 40-45°, reconstitute the residue in 50 μL mobile phase, inject a 20 μL aliquot. For confirmation prepare the methyl ester by evaporating a 30 μL aliquot to dryness under a stream of nitrogen, reconstitute with 2.5 mL MeOH, add 100 μL concentrated HCl, let stand at room temperature overnight, evaporate to dryness, reconstitute with 30 μL mobile phase, inject a 20 μL aliquot. Milk. 5 g Milk + 10 mL buffer, mix intensively for 1 min, add 5 mL chloroform, mix intensively for 2-3 min, centrifuge at 2500 g for 15 min, repeat the extraction three times. Combine the lower organic layers and evaporate them to dryness under reduced pressure at 30-40°, reconstitute with 5 mL chloroform, extract twice with 5 mL aliquots of sodium bicarbonate solution. Combine the aqueous layers and add them to 500 μL concentrated formic acid and 1 mL chloroform, extract the aqueous phase four times with 1 mL portions of chloroform, combine the organic layers and evaporate them to dryness under reduced pressure at 30-40°, reconstitute with 5 mL MeOH:PBS 15:85, add to the SPE cartridge at 1-2 mL/min, rinse the flask three times with 5 mL portions of MeOH:PBS 15:85, add the rinses to the SPE cartridge, wash with 10 mL water, force air through the SPE cartridge, elute with 3 mL MeOH at 0.5 mL/min, force out the last of the eluate with air. Evaporate the eluate to dryness under a stream of nitrogen at 40-45°, reconstitute the residue in 50 μL mobile phase, inject a 20 μL aliquot. For confirmation prepare the methyl ester by evaporating a 30 μL aliquot to dryness under a stream of nitrogen, reconstitute with 2.5 mL MeOH, add 100 μL concentrated HCl, let stand at room temperature overnight, evaporate to dryness, reconstitute with 30 μL mobile phase, inject a 20 μL aliquot. (PBS was 120 mM pH 7.4 NaCl containing 2.7 mM KCl, 10 mM phosphate, and 0.5 g/L sodium azide (Caution! Sodium azide is highly toxic!). Buffer was 33.7 mL 85% orthophosphoric acid and 118 g NaCl in 1 L water, pH 1.6. The sodium bicarbonate solution was 10 mg/mL. 200 mL Aliquots were washed three times with 10 mL aliquots of chloroform before use.)

HPLC VARIABLES
Guard column: 30 × 4.6 5 μm Spherisorb ODS-2
Column: 250 × 4.6 5 μm Spherisorb ODS-1
Mobile phase: MeOH:9% acetic acid in water 72:28
Column temperature: 50
Flow rate: 1
Injection volume: 20
Detector: F ex 390 em 440 following post-column reaction. The column effluent mixed with 25% ammonium hydroxide solution pumped at 0.1 mL/min and the mixture flowed through a 20 cm × 0.18 mm ID stainless steel coil to the detector.

CHROMATOGRAM
Retention time: 5 (ochratoxin), 6.5 (ochratoxin methyl ester)
Limit of quantitation: 5-10 pg/g (ochratoxin)
Limit of detection: 2 pg (ochratoxin), 5 pg (ochratoxin methyl ester)

KEY WORDS
post-column reaction; protect from light; serum; whole blood; SPE

REFERENCE
Zimmerli, B.; Dick, R. Determination of ochratoxin A at the ppt level in human blood, serum, milk and some foodstuffs by high-performance liquid chromatography with enhanced fluorescence detection and immunoaffinity column cleanup: methodology and Swiss data, *J.Chromatogr.B*, **1995**, *666*, 85–99.

SAMPLE
Matrix: saliva
Analyte: amobarbital
Sample preparation: Saliva sample obtained by chewing on 100 sq. cm Parafilm, centrifuge at 3000 g for 5 min. Remove a 1 mL aliquot of the supernatant and add it to 100 μL 5 μg/mL hexobarbital solution, add the mixture to a 3 mL octadecyl Baker-10 SPE cartridge, wash with three 3 mL portions of water, elute with 800 μL MeOH, inject a 50 μL aliquot of the eluate.

HPLC VARIABLES
Guard column: 30 × 4 5 μm Develosil ODS-5 (Nomura)
Column: 150 × 4 5 μm Develosil ODS-5 (Nomura)
Mobile phase: MeOH:water 50:50
Flow rate: 0.8
Injection volume: 50
Detector: UV 240 following post-column reaction. The column effluent flowed through a 150 × 3 sulfonated hollow-fiber membrane reactor (Dionex AFS-2) which was surrounded by 50 mM pH 10.2 ammonium hydroxide solution to the detector.

CHROMATOGRAM
Retention time: 14
Internal standard: hexobarbital (10)
Limit of detection: 0.5-2.5 ng

OTHER SUBSTANCES
Simultaneously analyzed: barbital, phenobarbital

KEY WORDS
SPE; post-column reaction

REFERENCE
Haginaka, J.; Wakai, J. Liquid chromatographic determination of barbiturates using a hollow-fibre membrane for postcolumn pH modification, *J.Chromatogr.*, **1987**, *390*, 421–428.

Ammonium Molybdate/Tin Chloride/Hydrazine

$(NH_4)_2MoO_4$ + $SnCl_2$ + H_2NNH_2 $\longrightarrow$ post-column reaction

SAMPLE
Matrix: sewage
Analyte: polyphosphates
Sample preparation: Filter (Whatman GF/C), filter (0.22 μm), inject a 500 μL aliquot of the filtrate.

HPLC VARIABLES
Column: 250 × 4.1 Wescan Anion/R poly(styrene-divinylbenzene)trimethylammonium anion-exchange (Alltech)
Mobile phase: 115 mM KCl containing 1.32 mM tetrasodium EDTA (Prepare fresh each day.)
Flow rate: 1.2
Injection volume: 500
Detector: UV 690 following post-column reaction. The column effluent mixed with reagent A pumped at 0.25 mL/min and the mixture flowed through a 10 m × 0.5 mm ID PTFE coil at 120°. The effluent from this coil passed through a cooled (Peltier) 75 cm × 0.5 mm ID PTFE coil to a 0.5 μm 210 sq mm PTFE membrane debubbler with a slight vacuum on the other side). The effluent from the debubbler mixed with reagent B pumped at 0.25 mL/min and this mixture flowed through a 300 mm × 0.5 mm ID coil to the detector. (Reagent A was 8.90 mM ammonium molybdate in 1.4 M sulfuric acid, store refrigerated, discard after 1 month. Reagent B was 1.05 mM tin(II) chloride in 520 mM sulfuric acid containing 15.4 mM hydrazine sulfate, store refrigerated, discard after 1 month.)

CHROMATOGRAM
Retention time: 5.76 (orthophosphate), 8.51 (pyrophosphate), 12.24 (triphosphate)
Limit of detection: 10-20 ng/mL (phosphorus)

KEY WORDS
post-column reaction

REFERENCE

Halliwell, D.J.; McKelvie, I.D.; Hart, B.T.; Dunhill, R.H. Separation and detection of condensed phosphates in waste waters by ion chromatography coupled with flow injection, *Analyst*, **1996**, *121*, 1089–1093.

Arsenazo III

RELATED REFERENCES

Hayakawa, T.; Moriyasu, M.; Hashimoto, Y. High-performance liquid chromatography of rare earth metals and post-column color reaction with Arsenazo III. *Bunseki Kagaku* **1983**, *32*, 136-138.

Kuban, V.; Gladilovich, D.B. Determination of rare-earth elements by ion-pair HPLC with post-column derivatization using Arsenazo III. *Collect.Czech.Chem.Commun.* **1988**, *53*, 1664-1677.

Saraswati, R.; Rao, T.H. Reversed-phase high-performance liquid chromatographic separation of some trace impurities in oxygen-free electronic copper by post-column chelation with 4-(2-pyridylazo)rescorcinol and Arsenazo-III. *J.Chromatogr.* **1992**, *605*, 63-68.

SAMPLE

Matrix: blood, urine

Analyte: gadolinium chelates

Sample preparation: Serum. Filter (Millipore Ultrafree-MC, type PTGC, 10 000 NMWL Filter unit) serum while centrifuging at 4° at 5000 g for 1 h, inject a 10 μL aliquot of the ultrafiltrate. Urine. Centrifuge urine at 4° at 15000 g for 10 min, dilute a 100 μL aliquot of the supernatant with 400 μL water, inject a 10 μL aliquot.

HPLC VARIABLES

Guard column: 20 × 2.1 5 μm Supelguard LC-19-DB (Supelco)

Column: 250 × 2.1 5 μm Supelcosil LC-19-DB

Mobile phase: 10 mM Triethylammonium acetate containing 2 mM EDTA, pH adjusted to 6.5-7.0 with 1 M acetic acid or 1 M NaOH

Column temperature: 30

Flow rate: 0.3

Injection volume: 10

Detector: UV 658 following post-column reaction with the reagent pumped at 0.3 mL/min. (Reagent was 100 mM nitric acid containing 0.15 mM Arsenazo III and 10 mM urea, filter (0.45 μm), sonicate, discard after 2 days.)

CHROMATOGRAM

Retention time: 5 (gadodiamide), 11 (gadopentetate dimeglumine)

KEY WORDS

serum; ultrafiltrate; post-column reaction

REFERENCE

Hvattum, E.; Normann, P.T.; Jamieson, G.C.; Lai, J.-J.; Skotland, T. Detection and quantitation of gadolinium chelates in human serum and urine by high-performance liquid chromatography and postcolumn derivatization of gadolinium with Arsenazo III, *J.Pharm.Biomed.Anal.*, **1995**, *13*, 927–932.

SAMPLE
Matrix: seawater
Analyte: thorium and uranyl ions
Sample preparation: Inject an aliquot onto column A and elute to waste with mobile phase A, after 3 min flush the contents of column A onto column B with mobile phase B, after 10 min flush the contents of column B onto column C with mobile phase C, elute with mobile phase C, monitor the effluent from column C.

HPLC VARIABLES
Column: A Ion-Exclusion Guard-Pak insert, sulfonic acid functionalized, 200 mg 5 mequiv/g resin (Waters); B 50 × 3.9 Nova-Pak C18; C 150 × 3.9 Nova-Pak C18
Mobile phase: A 80 mM nitric acid; B MeOH:water 1:99 containing 400 mM mandelic acid, pH 4.0; C MeOH:water 5:95 containing 200 mM α-hydroxyisobutyric acid, adjusted to pH 4.0 with NaOH
Flow rate: A 0.5, B 0.5, C 2
Injection volume: 100
Detector: UV 658 following post-column reaction. The column effluent mixed with the reagent and the mixture flowed to the detector. (Reagent was 130 μM Arsenazo III in 62 mM acetic acid containing 1 mM urea, prepare fresh daily.)

CHROMATOGRAM
Retention time: 5 (thorium (IV)), 15 (uranyl)

OTHER SUBSTANCES
Non-interfering: phosphate, lanthanides

KEY WORDS
post-column reaction; column-switching

REFERENCE
Hao, F.; Paull, B.; Haddad, P.R. Determination of thorium and uranyl in nitrophosphate solution by on-line matrix-elimination reversed-phase chromatography, *Chromatographia*, **1996**, *42*, 690–696.

SAMPLE
Matrix: solutions
Analyte: uranium
Sample preparation: Prepare a solution in mobile phase A, inject a 2 mL aliquot onto column A and elute to waste with mobile phase A, after 5 min backflush the contents of column A onto column B with mobile phase B, monitor the effluent from column B.

HPLC VARIABLES
Column: A 30 × 2.1 OD-032 (Brownlee); B 50 mm long LC18DB (Supelco)
Mobile phase: A 110 mM α-hydroxyisobutyric acid adjusted to pH 5.5 with 5% NaOH; B 110 mM pH 3.5 α-hydroxyisobutyric acid containing 25 mM pentanesulfonic acid
Flow rate: 1
Injection volume: 2000
Detector: UV 658 following post-column reaction. The column effluent mixed with the reagent pumped at 1 mL/min and the mixture flowed through a 4 m × 0.8 mm ID PTFE coil to the detector. (Reagent was 75 μM Arsenazo III in 100 mM nitric acid.)

CHROMATOGRAM
Retention time: 2
Limit of quantitation: 1-2 ng/mL

OTHER SUBSTANCES
Non-interfering: bicarbonate, calcium, carbonate, chloride, lanthanides, magnesium, nitrate, potassium, silicate, sodium, strontium, sulfate

KEY WORDS
post-column reaction; column-switching

REFERENCE
Kerr, A.; Kupferschmidt, W.; Attas, M. Determination of uranium in natural groundwaters using high-performance liquid chromatography, *Anal.Chem.*, **1988**, *60*, 2729–2733.

SAMPLE
Matrix: solutions
Analyte: thorium and uranyl ions
Sample preparation: Prepare a solution in 20 mM pH 4.0 α-hydroxyisobutyric acid and inject a 1 mL aliquot onto column A, elute to waste with mobile phase A, backflush the contents of column A onto column B with mobile phase B, monitor the effluent from column B.

HPLC VARIABLES
Column: A Guard-Pak C18; B 300 × 3.9 μBondapak C18
Mobile phase: A 40 mM pH 4.0 acetic acid; B MeOH:water 10:90 containing 400 mM α-hydroxyisobutyric acid, adjusted to pH 4.0 with NaOH
Flow rate: 1
Injection volume: 1000
Detector: UV 658 following post-column reaction. The column effluent mixed with the reagent and the mixture flowed to the detector. (Reagent was 130 μM Arsenazo III in 62 mM acetic acid containing 10 mM urea.)

CHROMATOGRAM
Retention time: 7.5 (thorium (IV)), 13 (uranyl)
Limit of detection: 2.2 ng/mL (thorium(IV)), 5.8 ng/mL (uranyl)

KEY WORDS
post-column reaction; column-switching

REFERENCE
Fuping, H.; Haddad, P.R.; Jackson, P.E.; Carnevale, J. Studies on the retention behaviour of α-hydroxy-isobutyric acid complexes of thorium(IV) and uranyl ion in reversed-phase high-performance liquid chromatography, *J.Chromatogr.*, **1993**, *640*, 187–194.

Benzamidine

SAMPLE
Matrix: beverages, plants
Analyte: sugars
Sample preparation: Beverages. Dilute 50-fold, filter (0.22 μm), inject an aliquot of the filtrate. Plants. Heat 1 g barley leaves and 10 mL EtOH:water 80:20 at 100° in a sealed tube for 15-30 min. Evaporate the liquid phase to dryness, reconstitute with water, pass through Analytichem trimethylaminopropyl and cyclohexyl SPE cartridges, inject an aliquot.

HPLC VARIABLES
Column: 300 × 6.5 Sugar-Pak I (Waters)
Mobile phase: water
Column temperature: 70
Flow rate: 0.4
Injection volume: 10
Detector: F ex 360 em 470 following post-column reaction. The effluent from the column passed through a 75 × 3.8 reactor containing Dowex 50 W × 2 sulfonic-acid type styrene divinylbenzene copolymer at 100° and mixed with 30 mM benzamidine in 1 M KOH pumped at 1 mL/min. This mixture flowed through a 530 μL reaction coil (Varian PCR-1) at 100° to the detector.

CHROMATOGRAM
Retention time: 15.30 (sucrose), 18.27 (dextrose), 22.77 (fructose)
Limit of detection: 60 pmole

KEY WORDS
post-column reaction; barley

REFERENCE
Coquet, A.; Haerdi, W.; Degli Agosti, R.; Veuthey, J.-L. Determination of sugars by liquid chromatography with post-column catalytic derivatization and fluorescence detection, *Chromatographia*, **1994**, *38*, 12–16.

SAMPLE
Matrix: plants
Analyte: reducing saccharides
Sample preparation: Condition trimethylaminopropylsilica SAX and cyclohexylsilica SPE cartridges (Analytichem) with 4 mL MeOH and 4 mL water. Heat 1 g plant material with 10 mL EtOH:water 80:20 in a sealed tube at 100° for 15-30 min, evaporate the extract to dryness, reconstitute with water, pass through the SPE cartridges, inject a 50 μL aliquot of the eluate.

HPLC VARIABLES
Column: 300 × 6.5 Sugar Pak-1 microparticulate gel, calcium form (Waters)
Mobile phase: 100 μM Calcium EDTA
Column temperature: 70
Flow rate: 0.4
Injection volume: 50
Detector: F ex 360 em 470 following post-column reaction. The column effluent mixed with 30 mM benzamidine in 1 M KOH pumped at 1 mL/min and the mixture flowed through a 530 μL reaction coil (Varian PCR1) at 100° to the detector.

CHROMATOGRAM
Retention time: 8.94 (lactose), 10.92 (dextrose), 12.16 (rhamnose, xylose), 12.25 (galactose), 12.53 (mannose), 13.48 (fructose), 13.69 (fucose), 13.99 (arabinose)
Limit of detection: 15.8-62.5 pmole

KEY WORDS
post-column reaction; SPE

REFERENCE
Coquet, A.; Veuthey, J.-L.; Haerdi, W. Selective post-column fluorigenic reaction with benzamidine for trace level detection of reducing saccharides in liquid chromatography, *J.Chromatogr.*, **1991**, *553*, 255–263.

SAMPLE
Matrix: solutions
Analyte: disaccharides
Sample preparation: Inject a 20 μL aliquot of an aqueous solution.

HPLC VARIABLES
Column: 100 × 4.7 Hypercarb No. 0498 (Shandon)
Mobile phase: MeCN:water 1:99
Column temperature: 95
Flow rate: 1
Injection volume: 20
Detector: F ex 360 em 470 following postcolumn reaction. The column effluent mixed with the reagent pumped at 0.2 mL/min and the mixture flowed through a 10 m × 0.3 mm ID knitted PTFE tube at 100° then a 10 m × 0.4 mm ID coil of stainless steel tubing immersed in ice-cold water to the detector. (The reagent was 60 mM benzamidine in 500 mM KOH.)

CHROMATOGRAM
Retention time: 1.5 (dextrose), 4.5 (melibiose), 5.5 (maltose), 7.5 (lactose), 10 (gentibiose), 16 (cellobiose)

Limit of quantitation: 2 μM
Limit of detection: 0.5-1 μM

KEY WORDS
post-column reaction

REFERENCE
Koimur, M.; Lu, B.; Westerlund, D. High performance liquid chromatography of disaccharides on a porous graphitic column applying post-column derivatization with benzamidine, *Chromatographia*, **1996**, *43*, 254–260.

SAMPLE
Matrix: urine
Analyte: hydrocortisone
Sample preparation: Adjust pH of 2 mL urine to 6.5, add 100 μL 500 Fishman U/mL β-glucuronidase (from E. coli), add 200 μL 200 mM pH 6.5 phosphate buffer, add 1 drop chloroform, mix well, heat at 37° for 24 h, add 20 μL 100 μg/mL betamethasone in MeOH, add 4 mL dichloromethane, shake for 3 min. Remove the organic layer and wash it with 500 μL 100 mM NaOH, wash with 500 μL water. Remove the organic layer and evaporate it to dryness at 80°, reconstitute the residue in 100 μL mobile phase, inject a 10 μL aliquot.

HPLC VARIABLES
Column: 250 × 4.6 10 μm Finepak C18
Mobile phase: MeOH:water 50:50
Column temperature: 40
Flow rate: 0.8
Injection volume: 10
Detector: F ex 370 em 480 following post-column reaction. The column effluent mixed with 400 mM NaOH and reagent pumped at 0.5 mL/min and the mixture flowed through a 30 m × 0.5 mm ID PTFE coil at 95° and another coil immersed in water to the detector. (Reagent was 0.5% benzamidine hydrochloride in isopropanol:water 50:50.)

CHROMATOGRAM
Retention time: 18
Internal standard: betamethasone (20)

OTHER SUBSTANCES
Simultaneously analyzed: tetrahydrocortisol, tetrahydrocortisone, tetrahydro-11-deoxycortisol
Non-interfering: aldosterone, androsterone, corticosterone, dehydroepiandrosterone, 11-deoxycorticosterone, 16-hydroxydehydroepiandrosterone, progesterone

KEY WORDS
post-column reaction

REFERENCE
Seki, T.; Yamaguchi, Y. New fluorimetric determination of 17-hydroxycorticosteroids after high-performance liquid chromatography using post-column derivatization with benzamidine, *J.Chromatogr.*, **1984**, *305*, 188–193.

SAMPLE
Matrix: urine
Analyte: ascorbic acid
Sample preparation: Prepare a column by suspending 200-400 mesh Dowex 50W-X8 resin in water and pouring it into a 100 × 7 column, allow to settle, wash with 10 mL 2 M HCl, wash with water until the washings are neutral. Mix urine with an equal volume of 5% metaphosphoric acid containing 0.5% β-thiodiglycol. Add a 1 mL aliquot to the column, wash with 3.95 mL 2 mM tartaric acid containing 0.05% β-thiodiglycol, collect all the effluent from the column in a tube containing 50 μL 5% disodium EDTA cooled in ice, filter (0.45 μm) the eluate, inject a 50-250 μL aliquot.

HPLC VARIABLES
Column: two 50 × 7.6 Asahipak GS-320 hydrophilic gel columns in series

Mobile phase: 2.25 g/L Tartaric acid containing 0.75 g/L disodium EDTA and 0.5 g/L β-thiodiglycol, adjusted to pH 3.00-3.03 with 4 M NaOH
Column temperature: 30
Flow rate: 1
Injection volume: 50-250
Detector: F ex 325 em 400 following post-column reaction. The column effluent mixed with 20 mM benzamidine hydrochloride pumped at 0.36 mL/min and with 750 mM pH 10 potassium borate buffer containing 200 mM potassium sulfite pumped at 0.36 mL/min and the mixture flowed through a 50 m × 0.5 mm ID PTFE tube at 90° to the detector.

CHROMATOGRAM
Retention time: 48

OTHER SUBSTANCES
Extracted: isoascorbic acid
Simultaneously analyzed: dehydroascorbic acid, diketogluconic acid, diketogulonic acid

KEY WORDS
post-column reaction; SPE

REFERENCE
Seki, T.; Yamaguchi, Y.; Noguchi, K.; Yanagihara, Y. Determination of ascorbic acid in human urine by high-performance liquid chromatography coupled with fluorimetry after post-column derivatization with benzamidine, *J.Chromatogr.*, **1987**, *385*, 287–291.

1,2-Bis(4-methoxyphenyl)ethylenediamine

RELATED REFERENCE
Umegae, Y.; Nohta, H.; Ohkura, Y. Determination of reducing sugars by high-performance liquid chromatography with postcolumn fluorescence derivatization using 1,2-bis(4-methoxyphenyl) ethylenediamine. *Anal.Sci.* **1989**, *5*, 675-680.

SAMPLE
Matrix: blood
Analyte: glucose and deoxyglucose
Sample preparation: 100 μL Serum + 10 μL 1 μM L-fucose + 400 μL 500 mM trichloroacetic acid, centrifuge at 4° at 1500 g for 5 min. Remove a 100 μL aliquot of the supernatant and add it to 100 μL 1 M pH 11.0 borate buffer, mix, inject a 100 μL aliquot.

HPLC VARIABLES
Column: 150 × 4.6 TSK gel Sugar AXG trimethylammonium-bonded styrene-divinylbenzene copolymer resin strong anion-exchange (Tosoh)
Mobile phase: 1 M pH 9.0 Borate buffer
Column temperature: 60
Flow rate: 0.4

Injection volume: 100

Detector: F ex 330 em 460 following post-column reaction. 600 mM NaOH pumped at 0.15 mL/min mixed with 15 mM meso-1,2-bis(4-methoxyphenyl)ethylenediamine in EtOH: water 30:70 pumped at 0.15 mL/min and the mixture flowed through a 1 m × 0.5 mm ID stainless steel coil. This mixture mixed with the column effluent and the mixture flowed through a 20 m × 0.5 mm ID stainless steel coil at 140° and a 1 m × 0.5 mm ID air-cooled stainless steel coil to the detector. (Prepare meso-1,2-bis(4-methoxyphenyl)ethylenediamine as follows. Heat 105 g benzil and 122 g salicaldehyde in 750 mL EtOH at 60° until they dissolve, introduce a weak stream of ammonia with stirring over 3 h, cool, filter, wash the precipitate with EtOH, dry under vacuum at 80°. Suspend 192 g of this product in 500 mL acetic anhydride, reflux for 14 h using a constant temperature bath at 148-150°, the product (O,O',N,N'-tetraacetyl-meso-1,2-bis(2-hydroxyphenyl)ethylenediamine) crystallizes on slow cooling, filter, wash the solid with a little acetic anhydride. Suspend the product in 250 mL 42% hydrobromic acid:acetic acid 50: 50 and reflux for 3 h. Filter off the product that separates on cooling and dissolve it in 400 mL hot water, neutralize with 20% NaOH. Filter off the product that separates and recrystallize it from MeCN to give meso-1,2-bis(2-hydroxyphenyl)ethylenediamine as a colorless powder (mp 184-186°). Mix 2.44 g meso-1,2-bis(2-hydroxyphenyl)ethylenediamine with 20 mmole p-anisaldehyde in 100 mL MeCN, reflux until the reaction is complete (at least 1 h), reduce the reaction mixture to half its volume by distillation, cool, filter. Recrystallize the product from toluene to give N,N'-disalicylidene-meso-1,2-bis(4-methoxyphenyl)ethylenediamine (mp 202-207°) (formed via a diaza-Cope rearrangement). Suspend 5 mmole N,N'-disalicylidene-meso-1,2-bis(4-methoxyphenyl)ethylenediamine in 50 mL 2 M sulfuric acid and steam distil until no more salicaldehyde comes over, filter the reaction mixture while it is hot, make the filtrate strongly basic (pH 11) with 20% NaOH. Filter off the crystalline product and recrystallize it from MeCN to give meso-1,2-bis(4-methoxyphenyl)ethylenediamine (mp 151-152°) (Chem. Ber. 1976, 109, 1).)

CHROMATOGRAM

Retention time: 11.2 (2-deoxy-D-ribose), 12.4 (2-deoxy-D-glucose), 15.2 (L-rhamnose), 28.0 (D-glucose)

Internal standard: L-fucose (22.2)

Limit of detection: 520-560 nM

KEY WORDS

post-column reaction; serum; rat; pharmacokinetics

REFERENCE

Umegae, Y.; Nohta, H.; Ohkura, Y. Simultaneous determination of 2-deoxy-D-glucose and D-glucose in rat serum by high-performance liquid chromatography with post-column fluorescence derivatization, *Chem.Pharm.Bull.*, **1990**, *38*, 963–965.

Blue Tetrazolium

SAMPLE

Matrix: urine

Analyte: monosaccharides
Sample preparation: Mix acetone with urine so as to make a 63:47 acetone:urine mixture, centrifuge a 6 mL aliquot. Evaporate the acetone from the supernatant under a stream of helium at 35°, add 30 mg Dowex 50W-X8, add 30 mg Dowex 1-X8, agitate, centrifuge, inject a 1-10 μL aliquot of the supernatant.

HPLC VARIABLES
Column: 150 × 4.3 Hitachi 3013 N anion-exchange resin, phosphate form
Mobile phase: MeCN:water 83:17
Column temperature: 60
Flow rate: 1
Injection volume: 1-10
Detector: UV 530 following post-column reaction. The column effluent mixed with the reagent pumped at 1.5 mL/min and the mixture flowed through a 3 m × 0.5 mm ID coil of PTFE tubing at 85° and a 1 m × 0.5 mm ID coil of PTFE tubing at room temperature to the detector. (Reagent was 2 g/L blue tetrazolium in EtOH:water 50:50 containing 180 mM NaOH. The sugar is reduced and the blue tetrazolium is oxidized to colored products.)

CHROMATOGRAM
Retention time: 7 (ribose, fucose), 10 (xylose), 11 (arabinose), 13 (fructose), 20 (galactose), 63 (lactose), 23 (dextrose)
Limit of detection: 10 ng

KEY WORDS
post-column reaction

REFERENCE
D'Amboise, M.; Hanai, T.; Noël, D. Liquid-chromatographic measurement of urinary monosaccharides, *Clin.Chem.*, **1980**, *26*, 1348–1350.

Borate Buffer

RELATED REFERENCE
Paibir, S.G.; Soine, W.H. High-performance liquid chromatographic analysis of phenobarbital and phenobarbital metabolites in human urine. *J.Chromatogr.B* **1997**, *691*, 111-117.

SAMPLE
Matrix: urine
Analyte: catecholamines
Sample preparation: Acidify urine by adding 1% 6 M HCl. 5 mL Acidified urine + 1 mL 7.5% disodium EDTA, adjust pH to 8.5 with 1 M NaOH, add 250 mg alumina (previously treated with 2 M HCl), shake for 5 min, decant the supernatant, wash the alumina three times with 5 mL portions of water. Place the alumina in a 4 mm ID glass column, elute with 250 mM acetic acid in water, collect 2.5 mL eluate, inject a 100 μL aliquot.

HPLC VARIABLES
Column: 1000 × 2.1 Zipax SCX (DuPont)
Mobile phase: MeCN:50 mM NaH_2PO_4 5:95
Column temperature: 40
Flow rate: 0.8
Injection volume: 100
Detector: F ex 400 em 490 following post-column reaction. The column effluent mixed with the reagent pumped at 0.4 mL/min and the mixture flowed through a 10 m × 0.5 mm PTFE coil at 75 ± 0.1° to the detector. (Reagent was 500 mM borate buffer adjusted to pH 9.7 with NaOH.)

CHROMATOGRAM
Retention time: 5.5 (norepinephrine), 9 (epinephrine)
Limit of quantitation: 0.25 ng

KEY WORDS
post-column reaction; SPE

REFERENCE
Nimura, N.; Ishida, K.; Kinoshita, T. Novel post-column derivatization method for the fluorimetric determination of norepinephrine and epinephrine, *J.Chromatogr.*, **1980**, *221*, 249–255.

Bromine

RELATED REFERENCES
Kok, W.T.; Halvax, J.J.; Voogt, W.H.; Brinkman, U.A.T.; Frei, R.W. Detection of thioethers of pharmaceutical importance by liquid chromatography with on-line generated bromine. *Anal.Chem.* **1985**, *57*, 2580-2583.

Palmisano, F.; Sibilia, A.; Visconti, A. Determination of altenuene and isoaltenuene by liquid chromatography - electrochemical detection with on-line generated bromine. *Chromatographia* **1990**, *29*, 333-337.

SAMPLE
Matrix: feed
Analyte: aflatoxin B_1
Sample preparation: 25 g Ground feed + 125 mL chloroform + 12.5 mL water + 12.5 g Celite, extract, filter (paper), add a 50 mL aliquot of the filtrate to a Sep-Pak Florisil SPE cartridge, wash with 5 mL chloroform, wash with 20 mL MeOH, elute with 15 mL acetone:water 90:10. Evaporate the eluate to dryness under reduced pressure, take up the residue in 20 mL water, add to a Sep-Pak C18 SPE cartridge, elute with 50 mL acetone:water 15:85, inject a 250 μL aliquot of the filtrate.

HPLC VARIABLES
Column: two 100 × 3 5 μm LiChrosorb RP-18 columns in series
Mobile phase: MeCN:MeOH:water 16:28:52 containing 1 mM KBr and 1 mM nitric acid
Flow rate: 0.5
Injection volume: 250
Detector: F ex 360 em >420 following post-column reaction. The column effluent passed through an electrochemical cell (construction details in Anal.Chim.Acta 1984, 162, 19) and the bromide was oxidized to bromine. The mixture flowed through a 4 s reaction coil to the detector.

CHROMATOGRAM
Limit of detection: 10 ng/g

KEY WORDS
post-column reaction; SPE

REFERENCE
Traag, W.A.; van Trijp, J.M.P.; Tuinstra, L.G.M.T.; Kok, W.T. Sample clean-up and post-column derivatization for the determination of aflatoxin B1 in feedstuffs by liquid chromatography, *J.Chromatogr.*, **1987**, *396*, 389–394.

SAMPLE
Matrix: solutions
Analyte: opiates

HPLC VARIABLES
Column: Supelco LC-8
Mobile phase: MeOH:water:acetic acid 40:59:1 containing 100 mM potassium nitrate, 10 mM tetramethylammonium bromide, and 2.5 mM heptanesulfonic acid
Flow rate: 1
Detector: E, Metrohm 1096/2, platinum working electrode +0.4 V, Ag/AgCl reference electrode following post-column reaction. The column effluent passed through an electrochemical cell (construction details in paper) and the bromide was oxidized to bromine (mor-

phine 1.5 µA; codeine 3 µA; noscapine, papaverine 100 µA). The mixture flowed through a 20 s reaction coil (3.9 m (?) × 0.33 mm ID) to the detector.

CHROMATOGRAM
Retention time: 4 (morphine), 5 (codeine), 10.5 (noscapine), 12 (papaverine)
Limit of detection: 0.4-300 ng

KEY WORDS
post-column reaction

REFERENCE
Kok, W.T.; Brinkman, U.A.T.; Frei, R.W. On-line electrochemical reagent production for detection in liquid chromatography and continuous flow systems, *J.Chromatogr.*, **1984**, *162*, 19–32.

SAMPLE
Matrix: urine
Analyte: aflatoxins
Sample preparation: Condition a 500 mg Bond Elut C2 SPE cartridge with 5 mL MeCN and 10 mL water. 10 mL Urine + 10 mL 100 mM pH 5.0 sodium acetate buffer, add to the column, wash with 10 mL MeCN:water 10:90, dry with nitrogen at 100 kPa for 5 min, wash with 3 mL hexane:THF 90:10, dry with nitrogen at 100 kPa for 1 min, elute with 3 mL dichloromethane, evaporate the eluate to dryness under a stream of nitrogen at 100 mL/min at room temperature over 45 min, reconstitute the residue in 200 µL MeCN:water 50:50, vortex for 15 s, inject a 20 µL aliquot.

HPLC VARIABLES
Column: 250 × 4.6 5 µm Supelcosil LC-18
Mobile phase: MeCN:MeOH:water 20:20:60 containing 1 mM KBr and 1 mM nitric acid
Flow rate: 1
Injection volume: 20
Detector: F ex 365 em 440 following post-column reaction. The column effluent passed through a KOBRA cell (Lamers & Pleuger, Den Bosch, Netherlands; cf Anal. Chim. Acta 1984, 162, 19) operating at 20 µA which oxidized the bromide to bromine and the reaction mixture then flowed through a 0.5 m × 0.55 mm ID PTFE coil to the detector.

CHROMATOGRAM
Retention time: 8 (G_2), 10 (G_1), 11 (B_2), 14 (B_1)
Limit of detection: 20-29 pg/mL

KEY WORDS
post-column reaction; SPE; silanize glassware with chlorotrimethylsilane

REFERENCE
Kussak, A.; Andersson, B.; Andersson, K. Automated sample clean-up with solid-phase extraction for the determination of aflatoxins in urine by liquid chromatography, *J.Chromatogr.*, **1993**, *616*, 235–241.

SAMPLE
Matrix: urine
Analyte: aflatoxin Q_1
Sample preparation: Condition an Aflaprep M immunoaffinity column (Rhône-Poulenc) with 10 mL water at 20 mL/min. 10 mL Urine + 10 mL 100 mM pH 5.0 sodium acetate buffer, add to the column at 2 mL/min, wash with 10 mL water at 2 mL/min, dry with nitrogen for 5 min, wash with 3 mL hexane:THF 75:25 at 2 mL/min, dry with nitrogen for 1 min, elute with 0.5 mL MeCN at 2 mL/min, pause for 30 s, elute with 1 mL MeCN at 2 mL/min. Combine the eluates and evaporate them to dryness under a stream of nitrogen at 80 mL/min, reconstitute the residue in 200 µL MeCN:water 50:50, vortex for 15 s, inject a 30 µL aliquot.

HPLC VARIABLES
Column: 100 × 5 4 µm Nova Pak C18
Mobile phase: MeCN:water 25:75 containing 1 mM KBr and 1 mM nitric acid
Flow rate: 1
Injection volume: 30

Detector: F ex 365 em 440 following post-column reaction. The column effluent passed through a KOBRA cell (Lamers & Pleuger, Den Bosch, Netherlands; cf Anal. Chim. Acta 1984, 162, 19) operating at 100 μA which oxidized the bromide to bromine and the reaction mixture then flowed through a 0.5 m × 0.55 mm ID PTFE coil to the detector.

CHROMATOGRAM
Retention time: 5
Limit of detection: 49.5 pg/mL

KEY WORDS
post-column reaction; SPE; immunoaffinity; silanize glassware with chlorotrimethylsilane

REFERENCE
Kussak, A.; Andersson, B.; Andersson, K. Determination of aflatoxin Q_1 in urine by automated immunoaffinity column clean-up and liquid chromatography, *J.Chromatogr.B*, **1994**, *656*, 329–334.

SAMPLE
Matrix: urine
Analyte: aflatoxins
Sample preparation: Condition an Aflaprep immunoaffinity column (Rhône-Poulenc) with 10 mL water. 10 mL Urine + 10 mL 100 mM pH 5.0 sodium acetate buffer, add to the column, wash with 10 mL water, dry with nitrogen at 1.2 L/min for 1 min, elute with 1 mL MeCN, pause for 30 s, elute with 500 μL MeCN, pause for 30 s, elute with 500 μL MeCN. Combine the eluates and evaporate them to dryness under a stream of nitrogen at 120 mL/min, reconstitute the residue in 200 μL MeCN:water 30:70, vortex, inject a 150 μL aliquot.

HPLC VARIABLES
Column: 100 × 8 4 μm Nova Pak phenyl
Mobile phase: MeCN:water 30:70 containing 1 mM KBr and 1 mM nitric acid
Flow rate: 1.5
Injection volume: 150
Detector: F ex 365 em 440 following post-column reaction. The column effluent passed through a KOBRA cell (Lamers & Pleuger, Den Bosch, Netherlands; cf Anal. Chim. Acta 1984, 162, 19) operating at 20 μA which oxidized the bromide to bromine and the reaction mixture then flowed through a 0.5 m × 0.55 mm ID PTFE coil to the detector.

CHROMATOGRAM
Retention time: 6 (M_1), 6.7 (Q_1), 8 (Q_2), 10 (B_2), 11.3 (G_1), 14 (B_1)
Limit of detection: 6.8-18 pg/mL

KEY WORDS
post-column reaction; SPE; immunoaffinity; silanize glassware with chlorotrimethylsilane

REFERENCE
Kussak, A.; Andersson, B.; Anderrson, K. Immunoaffinity column clean-up for the high-performance liquid chromatographic determination of aflatoxins B_1, B_2, G_1, G_2, M_1 and Q_1 in urine, *J.Chromatogr.B*, **1995**, *672*, 253–259.

Butylamine

 post-column reaction

SAMPLE
Matrix: blood
Analyte: warfarin
Sample preparation: Plasma. 1 mL Plasma + 100 μL 8.5-68 μg/mL IS in water + 200 μL NaOH + 4 mL peroxide-free ether, shake mechanically for 5 min, centrifuge at 3000 rpm for 5 min, discard the organic phase. Acidify the aqueous phase with 500 μL 3 M HCl, add 8 mL peroxide-free ether, shake for 2 min, centrifuge at 3000 rpm for 10 min. Remove the organic layer and add it to 5 mL pH 4 phosphate buffer, shake for 5 min.

Dry the organic layer over 1 g calcium chloride, centrifuge at 3000 rpm for 3 min, evaporate to dryness under a stream of nitrogen at 45°, reconstitute the residue in 10 μL 200 mg/mL carbobenzyloxy-L-proline in MeCN, 10 μL 1 mg/mL imidazole in MeCN, and 10 μL 200 mg/mL dicyclohexylcarbodiimide in MeCN, vortex for 10 s, let stand at room temperature for 2 h, add 10 μL ethyl acetate, mix, inject an aliquot. Urine. 500 μL Urine + 100 μL 8-70 μg/mL IS in water + 500 μL 3 M HCl + 8 mL chloroform, rotate gently for 45 min, centrifuge at 3000 rpm for 10 min, repeat the extraction. Combine the organic layers and add them to 1.7 mL 100 mM NaOH, mix. Remove the aqueous layer and acidify it with 500 μL 3 M HCl, add 8 mL chloroform, shake for 45 min, centrifuge for 10 min. Remove the organic layer and add it to 1 g calcium chloride, vortex for 10 s, centrifuge at 3000 rpm for 5 min, evaporate to dryness under a stream of nitrogen at 65°, reconstitute the residue in 10 μL 200 mg/mL carbobenzyloxy-L-proline in MeCN, 10 μL 1 mg/mL imidazole in MeCN, and 10 μL 200 mg/mL dicyclohexylcarbodiimide in MeCN, vortex for 10 s, let stand at room temperature for 2 h, add 10 μL ethyl acetate, mix, inject an aliquot.

HPLC VARIABLES

Column: 250 × 5 5 μm Spherisorb Si
Mobile phase: Hexane:ethyl acetate:MeOH:acetic acid 74.75:25:0.25:0.3
Flow rate: 0.8
Detector: F ex 313 em (No. 370 filter) following post-column reaction. The column effluent mixed with MeOH:n-butylamine 50:50 pumped at 0.4 mL/min and the mixture flowed through a 250 × 3 column packed with 40 μm glass beads to the detector.

CHROMATOGRAM

Retention time: 15.5 (S), 17 (R)
Internal standard: 4'-fluorowarfarin (16, 19 (enantiomers))
Limit of detection: 100 ng

OTHER SUBSTANCES

Extracted: metabolites, hydroxywarfarin
Interfering: salicylic acid (with hydroxywarfarin only)
Non-interfering: cimetidine

KEY WORDS

post-column reaction; plasma; chiral; normal phase

REFERENCE

Banfield, C.; Rowland, M. Stereospecific fluorescence high-performance liquid chromatographic analysis of warfarin and its metabolites in plasma and urine, *J.Pharm.Sci.*, **1984**, *73*, 1392–1396.

Calcium Hypochlorite

$Ca(OCl)_2$ ⟶ post-column reaction

SAMPLE

Matrix: solutions
Analyte: folic acid

HPLC VARIABLES

Guard column: 30 × 4.6 10 μm octadecylsilica (Brownlee)
Column: 300 × 3.9 10 μm μBondapak phenyl
Mobile phase: Gradient. MeCN:33 mM pH 2.3 sodium phosphate buffer from 7.2:92.8 to 11.3:88.7 over 15 min. (At the end of each day flush system with water then 50-75 mL MeOH. Place a column of 37-53 μm silica (Whatman) between pump and injection valve.)
Flow rate: 1
Injection volume: 100
Detector: F ex 365 em >415 (filter) following post-column reaction. The column effluent mixed with the reagent pumped at 0.23 mL/min and the mixture flowed through a 5 m × 0.8 mm ID coil of PTFE tubing at 60° to the detector. (Reagent was 0.005% calcium

hypochlorite (HTH dry chlorine, Olin, Overland KS) in 100 mM K_2HPO_4 containing 200 mM NaCl.)

CHROMATOGRAM
Retention time: 22

KEY WORDS
post-column reaction

REFERENCE
Gregory, J.F. III; Sartain, D.B.; Day, B.P.F. Fluorometric determination of folacin in biological materials using high performance liquid chromatography, *J.Nutr.*, **1984**, *114*, 341–353.

Chemiluminescence

SAMPLE
Matrix: blood
Analyte: metoprolol
Sample preparation: Condition a Bond-Elut C18 SPE cartridge with 2 mL MeCN and 1 mL MeCN:water 40:60. Add 20 µL serum to 500 µL MeCN:water 40:60, vortex, add to SPE cartridge, wash with 2 mL MeCN:water 40:60, elute with 1 mL MeOH:water 1:1 containing 0.05% trifluoroacetic acid. Evaporate eluate to dryness under reduced pressure at 40°, reconstitute with 50 µL 100 mM pH 9.0 borate buffer containing 2 mM EDTA, add 50 µL 50 mM DBD-F in MeCN, heat at 45° for 8 h, add 100 µL 100 mM acetic acid in MeCN:water 50:50, inject a 20 µL aliquot. (DBD-F can be purchased from Tokyo Kasei (TCI America, 911 North Harborgate St., Portland, OR 97203; 800-423-8616, 503-283-1681, (fax) 503-283-1987; www.tciamerica.com). Synthesis of DBD-F is as follows. Dissolve 0.5 g magnesium sulfate heptahydrate and 6 g NaOH in 60 mL water, throughout the reaction keep the flask at about 20° with cold water cooling, add 15 mL 30% hydrogen peroxide, add 75 mL MeOH, add 12.1 g powdered benzoyl peroxide in one go, stir for 10 min, pour into 150 mL 20% sulfuric acid, extract three times with 50 mL portions of chloroform, determine peroxybenzoic acid concentration by iodometric titration (Tetrahedron 1967, 23, 3327). Slowly add 110 mL 1 M peroxybenzoic acid in chloroform to 7 g 2,6-difluoroaniline dissolved in 100 mL chloroform, stir at room temperature, when reaction is complete (iodometric titration) wash with 2% sodium thiosulfate, wash with 5% sodium carbonate, wash with water, dry over anhydrous sodium sulfate, evaporate to dryness under reduced pressure, recrystallize 2,6-difluoronitrosobenzene from EtOH (mp 108.5-109.5). Stir 8.5 g 2,6-difluoronitrosobenzene in 85 mL DMSO at room temperature and add a solution of 3.91 g sodium azide in 85 mL DMSO dropwise, let stand for about 1 h, add to a large volume of water, extract with ether, dry the extracts over anhydrous sodium sulfate, evaporate to dryness under reduced pressure and distil to give 4-fluoro-2,1,3-benzoxadiazole as a colorless oil (bp 83°/12 mm Hg) (J.Chem.Soc.(C) 1970, 1433). Add 11 mL chlorosulfonic acid dropwise to 3 g 4-fluoro-2,1,3-benzoxadiazole in 10 mL chloroform at 0-10° (use a calcium chloride drying tube), stir at room temperature for 1 h, reflux for 2 h, cool, slowly pour into ice water, remove the organic layer, extract the aqueous layer with chloroform, combine the organic layers, wash, dry over anhydrous magnesium sulfate, evaporate under reduced pressure, take up the residue in 5 mL benzene (Caution! Benzene is a carcinogen!), chromatograph on a 150 × 30 column of silica gel (100-200 mesh Kanto Chemical) with n-hexane:benzene 50:50, evaporate the appropriate fractions to give 4-(chlorosulfonyl)-7-fluoro-2,1,3-benzoxadiazole (CBD-F) as pale yellow needles (mp 64-66°) (Anal. Chem. 1984, 56, 2461). Stir 0.76 g CBD-F in 70 mL MeCN at 0-10° and add 1 g dimethylamine hydrochloride in 10 mL 100 mM pH 10 borax dropwise, adjust pH to 5 with 1 M HCl, concentrate to about 10 mL under reduced pressure, extract three times with 200 mL portions of diethyl ether, wash with water, dry over anhydrous magnesium sulfate, evaporate under reduced pressure, chromatograph on a 500 × 20 column of silica gel with chloroform, isolate the appropriate fraction and re-chromatograph on the same column with ethyl acetate:benzene 1:2 to give 4-(N,N-dimethylaminosulfonyl)-7-fluoro-2,1,3-benzoxadiazole (DBD-F) as white needles (mp 124-

125°) (yield 1% !). On a Merck no. 5714 60F$_{254}$ tlc plate eluted with chloroform DBD-F has Rf 0.32 and lies between two other reaction products (Analyst 1989, 114, 413).)

HPLC VARIABLES
Column: 250 × 4.6 5 μm TSK gel ODS 80Tm (Tosoh)
Mobile phase: MeCN:THF:50 mM pH 6.0 imidazole nitrate buffer 28:20:52
Column temperature: 40
Flow rate: 0.8
Injection volume: 20
Detector: Chemiluminescence following post-column reaction. The column effluent mixed with the reagent pumped at 1.4 mL/min and the mixture flowed to the detector. (Reagent was 0.25 mM bis[4-nitro-2-(3,6,9-trioxadecyloxycarbonyl)phenyl] oxalate (Wako) and 37.5 mM hydrogen peroxide in MeCN:ethyl acetate 50:50.)

CHROMATOGRAM
Retention time: 9
Limit of detection: 0.8 ng/mL

KEY WORDS
serum; SPE; post-column reaction

REFERENCE
Uzu, S.; Imai, K.; Nakashima, K.; Akiyama, S. Use of 4-(*N,N*-dimethylaminosulphonyl)-7-fluoro-2,1,3-benzoxadiazole as a labelling reagent for peroxyoxalate chemiluminescence detection and its application to the determination of the β-blocker metoprolol in serum by high-performance liquid chromatography, *Analyst*, **1991**, *116*, 1353–1357.

SAMPLE
Matrix: blood
Analyte: artemisinin
Sample preparation: Condition a 4 mm diameter Empore C18 SPE membrane with 0.5 mL MeOH and 0.5 mL water, do not allow to dry. Centrifuge 1 mL serum, add to SPE membrane, wash with 300 μL water, elute with 100 μL MeCN:water 65:35, inject a 50 μL aliquot of the eluate.

HPLC VARIABLES
Column: 150 × 2 5 μm Ultrasphere ODS
Mobile phase: MeCN:water 50:50
Flow rate: 0.3
Injection volume: 50
Detector: Chemiluminescence in a fluorescence detector with no light source, emission wavelength 425 nm. The effluent from the column mixed with reagent pumped at 0.5 mL/min and flowed through a convoluted mixing coil (1.1 mL dead volume) to the detector. (Reagent was 15 μg/mL luminol and 30 μg/mL hematin in 100 mM NaOH. Let stand for 30 min before use. Protect from light. Prepare daily.)

CHROMATOGRAM
Retention time: 7
Internal standard: dihydroartemisinin (5)
Limit of detection: 10 ng/mL

OTHER SUBSTANCES
Non-interfering: arteether, artemether

KEY WORDS
serum; post-column reaction; SPE

REFERENCE
Green, M.D.; Mount, D.L.; Todd, G.D.; Capomacchia, A.C. Chemiluminescent detection of artemisinin. Novel endoperoxide analysis using luminol without hydrogen peroxide, *J.Chromatogr.A*, **1995**, *695*, 237–242.

SAMPLE
Matrix: formulations, urine
Analyte: antihistamines

Sample preparation: Tablets. Crush tablets, add 100 mL water and 30-40 mL MeCN, dissolve, add N,N-dimethylbenzylamine, make up to 250 or 500 mL with water, centrifuge an aliquot, inject a 20 μL aliquot of the supernatant. Urine. Inject a 100 μL aliquot of urine directly.

HPLC VARIABLES
Column: 150 × 4.6 Asahipak ODP-50 C18
Mobile phase: MeCN:200 mM pH 7.0 phosphate buffer 27:73
Flow rate: 0.8
Injection volume: 20-100
Detector: Chemiluminescence following post-column reaction. Oxidize a 1 mM tris(2,2'-bipyridine) ruthenium(II) hexachloride solution in 50 mM pH 5.5 acetate buffer to Ru(III) using a Princeton Applied Research polarographic analyzer with a platinum gauze working electrode, platinum wire auxiliary electrode, and a silver wire reference electrode, +950 mV. Pump the reagent solution at 0.28 mL/min and mix with the column effluent, allow to flow through detector. The chemiluminescence detector was a fluorescence detector with the light source removed.

CHROMATOGRAM
Retention time: 4 (pheniramine), 7 (chlorpheniramine), 8 (brompheniramine), 9 (pyrilamine), 13 (diphenhydramine)
Internal standard: N,N-dimethylbenzylamine
Limit of detection: 140 ng/mL

KEY WORDS
tablets; post-column reaction

REFERENCE
Holeman, J.A.; Danielson, N.D. Liquid chromatography of antihistamines using post-column tris(2, 2'-bipyridine) ruthenium(III) chemiluminescence detection, *J.Chromatogr.A*, **1994**, *679*, 277−284.

SAMPLE
Matrix: solutions
Analyte: polycyclic aromatic hydrocarbons
Sample preparation: Inject an aliquot of a solution in mobile phase.

HPLC VARIABLES
Column: 150 × 3.9 Nova-pack C18
Mobile phase: MeCN:water 75:25
Flow rate: 0.5
Injection volume: 20
Detector: Chemiluminescence at 438 nm. The column effluent was mixed with 1.57 mg/mL bis(2,4,6-trichlorophenyl) oxalate in ethyl acetate pumped at 0.5 mL/min and oxidant pumped at 0.2 mL/min with magnetic stirring in a 10 mm quartz cuvette (reaction volume 350 μL and the chemiluminescence detected at 438 nm. (Prepare the oxidant solution by adding 40 mL concentrated hydrogen peroxide to 1 mL 150 mM pH 9.5 Tris-HCl buffer and diluting to 100 mL with isopropanol.)

CHROMATOGRAM
Retention time: 8 (anthracene), 12 (pyrene), 21 (perylene), 24 (benzo[a]pyrene), 46 (9,10-diphenylanthracene)

KEY WORDS
post-column reaction

REFERENCE
Cepas, J.; Silva, M.; Pérez-Bendito, D. Zero-dead-volume peroxyoxalate chemiluminescence detection in liquid chromatography, *Anal.Chem.*, **1995**, *67*, 4376−4379.

SAMPLE
Matrix: solutions
Analyte: amino acids

CAPILLARY ELECTROPHORESIS
Capillary: 50 cm × 50 μm (Polymicro Technologies)

Running buffer: 15 mM pH 10.0 Carbonate buffer containing 5 mM luminol and 25 mM hydrogen peroxide
Voltage/Current: 21 kV
Injection: Electrokinetic at 21 kV for 2 s
Detector: Chemiluminescence following post-column reaction. The capillary effluent mixed in a PEEK tee connector with 30 μM copper sulfate in 15 mM pH 10.0 carbonate buffer containing 30 μM tartaric acid delivered by gravity (at 40 cm above the buffer reservoir) through a 100 cm × 180 μm ID capillary and this mixture flowed past the detector and through a 30 cm × 180 μm ID capillary to a grounded reservoir. (The copper catalyzes the chemiluminescent reaction between luminol and hydrogen peroxide. The amino acids complex the copper and thus reduce the chemiluminescent light output.)
Migration time: 2 (arginine), 2.6 (proline), 3 (valine), 3.2 (leucine), 3.5 (glutamine), 3.8 (asparagine), 4 (serine), 5.3 (cysteine), 6 (glutamic acid), 6.5 (aspartic acid)
Limit of detection: 120-440 fmole (S/N 3)

KEY WORDS
post-column reaction; indirect chemiluminescence detection

REFERENCE
Liao, S.-Y.; Whang, C.-W. Indirect chemiluminescence detection of amino acids separated by capillary electrophoresis, *J.Chromatogr.A*, **1996**, *736*, 247–254.

Chloramine T

SO$_2$N with Cl and Na substituents → post-column reaction

SAMPLE
Matrix: tissue
Analyte: vitamin B3
Sample preparation: Homogenize rat intestinal tissue in physiological saline. Removal a 3 mL aliquot and add it to 18 mL acetone, centrifuge at 10000 g for 10 min. Add the supernatant to 14 mL chloroform, mix intensively, centrifuge at 1000 g for 5 min. Filter (0.45 μm) a 500 μL aliquot of the upper aqueous phase, inject a 100 μL aliquot.

HPLC VARIABLES
Guard column: 10 μm ODS Hypersil in an Upchurch C-135 B pre-column kit
Column: 250 × 4.6 5 μm ODS Hypersil
Mobile phase: Gradient. MeOH:5 mM pH 7.0 tetrabutylammonium phosphate from 10:90 to 70:30 over 10 min.
Flow rate: 1.2
Injection volume: 100
Detector: UV 410 following post-column derivatization. The column effluent mixed with 2% chloramine-T pumped at 0.5 mL/min and flowed through a 2 m × 0.5 mm ID coil of PTFE tubing kept at 60°. This mixture was mixed with 0.25% KCN containing 25 mM Tris and 40 mM HCl pumped at 0.5 mL/min. The mixture flowed through an 8 m × 0.5 mm ID coil of PTFE tubing kept at 60° to the detector.

CHROMATOGRAM
Retention time: 4.95 (niacinamide), 9.25 (niacin)
Limit of detection: 21 pmole (niacinamide), 12 pmole (niacin)

KEY WORDS
post-column reaction; rat; intestine

REFERENCE
Stein, J.; Hahn, A.; Rehner, G. High-performance liquid chromatographic determination of nicotinic acid and nicotinamide in biological samples applying post-column derivatization resulting in bathmochrome absorption shifts, *J.Chromatogr.B*, **1995**, *665*, 71–78.

Cobalt Acetate/Hydroxylamine

$$Co(CH_3COO)_2 \quad + \quad H_2NOH \quad \longrightarrow \quad \text{post-column reaction}$$

SAMPLE
Matrix: solutions
Analyte: enkephalins
Sample preparation: Inject a 100 μL aliquot of a pH 7-8 solution.

HPLC VARIABLES
Column: 150 × 6 5 μm Asahipak ODP-50
Mobile phase: Gradient. A was MeCN:50 mM pH 7.0 sodium borate buffer:water 5:20: 75. B was MeCN:50 mM pH 7.0 sodium borate buffer:water 60:20:20. A:B from 93:7 to 55:45 over 40 min. (Wash column with MeCN:water 50:50 for at least 30 min after each run.)
Flow rate: 1
Injection volume: 100
Detector: F ex 330 em 430 following post-column reaction. The column effluent mixed with 300 mM pH 9.0 borate buffer pumped at 0.4 mL/min and with 8 mM hydroxylamine oxalate containing 0.2 mM cobalt(II) acetate pumped at 0.4 mL/min and this mixture flowed through a 10 m × 0.5 mm ID stainless steel coil at 100° to the detector.

CHROMATOGRAM
Retention time: 15 (methionine enkephalin), 17 (leucine enkephalin)

OTHER SUBSTANCES
Simultaneously analyzed: numerous other peptides

KEY WORDS
post-column reaction

REFERENCE
Zhang, G.-Q.; Kai, M.; Ohkura, Y. Assay for enkephalin-generating enzyme in rat brain tissues by high-performance liquid chromatography with postcolumn fluorescence derivatization, *Anal.Sci.*, **1991**, 7, 561−565.

Copper Acetate/Sulfuric Acid

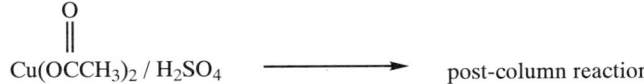

$$Cu(O\overset{O}{\overset{\|}{C}}CH_3)_2 \; / \; H_2SO_4 \quad \longrightarrow \quad \text{post-column reaction}$$

SAMPLE
Matrix: blood
Analyte: cardenolides
Sample preparation: Prepare an SPE column by packing 200 mg LCA-SiO6 short-chain alkyl silica material (Sykam) into a 4 mL column, wash with 5 mL MeCN, wash with 4 mL water. 1 mL Plasma or serum + 100 μL (1 mL) water, add to the SPE cartridge, wash with 2 (1) mL MeOH:water 20:80, wash with 3 mL water, wash with 2 (1) mL acetone water 15:85, suck dry, wash with 1.5 (1) mL hexane:ethyl acetate 90:10, (wash with 2 mL hexane:acetone 90:10), dry with a stream of nitrogen, elute with 1.5 (0.5) mL hexane:acetone 60:40. Evaporate the eluate to dryness, reconstitute the residue in 300 μL MeOH:water 20:80, inject the whole amount. (Values in parentheses represent those given for the schematic procedure which differs from the written procedure.)

HPLC VARIABLES
Guard column: 25 × 4 40-63 μm LiChrolut RP18 (Merck)
Column: 100 × 2 4 μm Superspher RP18
Mobile phase: Gradient. MeCN:water 10:90 for 0.1 min, to 18:82 over 3.9 min, to 26:74 over 15 min, to 75:25 over 7 min, return to initial conditions over 3 min, re-equilibrate for 15 min.

Flow rate: 0.2
Injection volume: 300
Detector: F ex 379 em 423 following post-column reaction. The column effluent mixed with 0.08% copper(II) acetate in 95% sulfuric acid pumped at 0.35 mL/min and the mixture flowed through a knitted 5 m × 0.3 mm ID PTFE coil at 75° to the detector.

CHROMATOGRAM

Retention time: 12.6 (digoxigenin), 14.5 (digoxigenin monodigitoxoside), 18.11 (digoxigenin bisdigitoxoside), 21.1 (digoxin)
Limit of quantitation: 280-510 pg
Limit of detection: 70-140 pg

KEY WORDS

post-column reaction; plasma; serum; SPE

REFERENCE

Belsner, K.; Büchele, B. Fluorescence detection of cardenolides in reversed-phase high-performance liquid chromatography after post-column derivatization, *J.Chromatogr.B*, **1996**, *682*, 95–107.

Copper Sulfate/Ammonium Hydroxide

SAMPLE

Matrix: perfusate
Analyte: sugars
Sample preparation: 200 μL Perfusate + 200 μL MeCN, centrifuge at 2000 rpm, inject a 20 μL aliquot of the supernatant.

HPLC VARIABLES

Column: 110 × 4.6 5 μm irregular silica (HSCP, Bourne End, UK)
Mobile phase: MeCN:water:1,4-diaminobutane 67:33:0.03
Flow rate: 0.8
Injection volume: 20
Detector: UV 285 following post-column reaction. The column effluent mixed with the reagent pumped at 1 mL/min and the mixture flowed to the detector. (Prepare stock reagent solution by dissolving 25 g cupric sulfate in 500 mL water, add 500 mL 28% ammonium hydroxide. Prepare working reagent by mixing 100 mL stock solution, 250 mL 28% ammonium hydroxide, and 650 mL water, filter (Whatman GF/C). Place the waste solvent bottle in a chemical fume hood.)

CHROMATOGRAM

Retention time: 5.5 (fructose), 6.5 (glucose), 8.5 (sucrose)
Limit of detection: 2.5 nmole

KEY WORDS

post-column reaction

REFERENCE

Grimble, G.K.; Barker, H.M.; Taylor, R.H. Chromatographic analysis of sugars in physiological fluids by postcolumn reaction with cuprammonium: a new and highly sensitive method, *Anal.Biochem.*, **1983**, *128*, 422–428.

Copper Sulfate/Sodium Bicinchoninate

+ $CuSO_4$ ⟶ post-column reaction

SAMPLE
Matrix: solutions
Analyte: sugars

HPLC VARIABLES
Column: 160 × 4 8 μm DAX8 anion-exchange resin, sulfate form (Durrum Chemical Co.) (Regenerate resin outside the column by washing 10 g resin with 400 mL water, 400 mL 500 mM NaCl at 50°, water, 0.5 N sodium sulfate at 50° (until a negative chloride test is obtained), water, and EtOH:water 95:5. Slurry pack below 70° with mobile phase at 1.14 mL/min.)
Mobile phase: EtOH:water 87.6:12.4
Column temperature: 88
Flow rate: 0.57
Injection volume: 5-100
Detector: UV 562 following post-column reaction. The column effluent mixed with the reagent pumped at 0.3 mL/min and the mixture flowed for 5 min through a coil of 0.3 mm ID PTFE at 100° to the detector. (Prepare reagent by mixing equal volumes of solution A and solution B, the mixture is stable for at least 1 month. Solution A is 1 g of copper sulfate pentahydrate and 3.7 g aspartic acid in 1 L water. Solution B is 38 g sodium carbonate decahydrate and 2 g sodium bicinchoninate (Pierce Chemical Co.; Fluka) in 1 L water.)

CHROMATOGRAM
Retention time: 11 (digitose), 19 (2-d-ribose), 28 (2-d-galactose), 31 (3-O-methylglucose), 35 (rhamnose), 40 (fucose), 47 (ribose), 53 (6-d-glucose), 60 (lyxose), 68 (arabinose), 83 (xylose), 90 (fructose), 93 (tagatose), 97 (sorbose), 104 (mannose), 120 (gulose), 142 (galactose), 168 (dextrose)
Limit of detection: <500 pmole

KEY WORDS
post-column reaction

REFERENCE
Mopper, K. Improved chromatographic separations on anion-exchange resins. I. Partition chromatography of sugars in ethanol, *Anal.Biochem.*, **1978**, *85*, 528–532.

Cyanoacetamide

RELATED REFERENCES

Honda, S.; Konishi, T.; Suzuki, S.; Takahashi, M.; Kakehi, K.; Ganno, S. Automated analysis of hexosamines by high-performance liquid chromatography with photometric and fluorometric postcolumn labeling using 2-cyanoacetamide. *Anal.Biochem.* **1983**, *134*, 483-488.

Honda, S.; Suzuki, S.; Takahashi, M.; Kakehi, K.; Ganno, S. Automated analysis of uronic acids by high-performance liquid chromatography with photometric and fluorometric postcolumn labeling using 2-cyanoacetamide. *Anal.Biochem.* **1983**, *134*, 34-39.

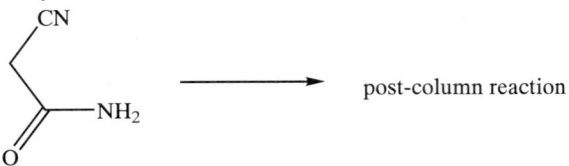

post-column reaction

SAMPLE

Matrix: blood, urine

Analyte: catecholamines

Sample preparation: Serum. Add 1 mL serum to 100 mg activated aluminum oxide suspended in 1 mL pH 8.7 Tris-HCl buffer, stir, let stand for 10 min. Discard the supernatant and wash the solid three times with 5 mL portions of water, wash the solid with 3 mL MeOH, dry under reduced pressure, elute with 3 mL 4 M acetic acid. Evaporate the eluate to dryness under reduced pressure, reconstitute the residue with 90 μL water, inject a 10 μL aliquot. Urine. 5 mL Urine + 5.3 mL 2 M HCl, heat at 100° for 20 min, cool to room temperature, add 1 mL 50 mM disodium EDTA, adjust the pH to 8.5 with dilute ammonia, add 500 mg 200 mesh aluminum oxide (Wako), shake for 10 min, filter, wash the solid with 10 mL water, elute with 5 mL 300 mM acetic acid, inject an aliquot of the eluate.

HPLC VARIABLES

Column: 250 × 3.6 10-25 μm Hitachi 3011 C resin

Mobile phase: 50 mM K_2HPO_4 containing 0.05% phosphoric acid

Column temperature: 45

Flow rate: 0.6

Injection volume: 10

Detector: F ex 383 em 486 following post-column reaction. The column effluent mixed with 1% 2-cyanoacetamide in water pumped at 0.5 mL/min and with buffer pumped at 1 mL/min and the mixture flowed through a 5 m × 0.5 mm ID PTFE coil at 100 ± 1° to the detector. (Buffer was 600 mM boric acid containing 750 mM KOH.)

CHROMATOGRAM

Retention time: 6 (norepinephrine), 7 (epinephrine), 10 (dopamine)

Limit of detection: 0.098-0.28 pmole

KEY WORDS

post-column reaction; serum; SPE

REFERENCE

Honda, S.; Takahashi, M.; Araki, Y.; Kakehi, K. Postcolumn derivatization of catecholamines with 2-cyanoacetamide for fluorimetric monitoring in high-performance liquid chromatography, *J.Chromatogr.*, **1983**, *274*, 45–52.

Dehydroascorbic Acid

SAMPLE
Matrix: blood
Analyte: digoxin
Sample preparation: 3 mL Plasma + 3 mL acetone, vortex, centrifuge at 1000 g for 5 min, remove the supernatant and add it to 2 mL isooctane, vortex, centrifuge at 1000 g for 5 min. Remove the acetone/water layer and evaporate it to 3 mL under a stream of nitrogen at 37°, add 10 mL dichloromethane:n-propanol 98:2, rotate for 10 min, centrifuge at 1000 g for 5 min, filter the organic layer and evaporate it to dryness under a stream of nitrogen at 37°, reconstitute the residue in 100 μL MeOH:water 50:50, inject the whole amount.

HPLC VARIABLES
Guard column: 37 μm ODS
Column: 150 × 4.6 3 μm Spherisorb ODS II
Mobile phase: MeOH:EtOH:isopropanol:water 52:3:1:45
Flow rate: 0.3
Injection volume: 100
Detector: F ex 360 (filter) em 425 (filter) following post-column reaction. The column effluent mixed with the reagent and flowed through a 10 m × 0.3 mm i.d. knitted PTFE coil at 79 ± 1° to the detector. (The reagent was generated by mixing 1.1 mM hydrogen peroxide in 0.1% ascorbic acid pumped at 0.038 mL/min and concentrated HCl pumped at 0.192 mL/min and allowing this mixture to flow through a 2 m × 0.8 mm i.d. PTFE coil to the point where it mixed with the column effluent (J.Chromatogr. 1986, 377, 233). Hydrogen peroxide and ascorbic acid form dehydroascorbic acid which enhances the fluorescence reaction between HCl and the steroid portion of the cardiac glycoside.)

CHROMATOGRAM
Retention time: 33
Internal standard: digitoxigenin (42)
Limit of detection: 0.5 ng/mL

OTHER SUBSTANCES
Simultaneously analyzed: digoxigenin, digoxigenin bisdigitoxoside, digoxigenin monodigitoxoside, dihydrodigoxigenin, dihydrodigoxin, furosemide, spironolactone
Non-interfering: captopril, dipyridamole, disopyramide, procainamide, propafenone, quinidine, sulfamethoxazole, trimethoprim, verapamil

KEY WORDS
plasma; post-column reaction; comparison with RIA

REFERENCE
Kwong, E.; McErlane, K.M. Analysis of digoxin at therapeutic concentrations using high-performance liquid chromatography with post-column derivatization, *J.Chromatogr.*, **1986**, *381*, 357–363.

SAMPLE
Matrix: solutions
Analyte: digoxin and metabolites

HPLC VARIABLES
Guard column: 37 μm ODS
Column: 150 × 4.6 3 μm Spherisorb ODS II

Mobile phase: MeOH:EtOH:isopropanol:water 52:3:1:45
Flow rate: 0.3
Detector: F ex 360 (filter) em 425 (filter) following post-column reaction. The column effluent mixed with the reagent and flowed through a 10 m × 0.3 mm i.d. knitted PTFE coil
at 79 ± 1° to the detector. (The reagent was generated by mixing 1.1 mM hydrogen
peroxide in 0.1% ascorbic acid pumped at 0.038 mL/min and concentrated HCl pumped
at 0.192 mL/min and allowing this mixture to flow through a 2 m × 0.8 mm i.d. PTFE
coil to the point where it mixed with the column effluent. Hydrogen peroxide and ascorbic
acid form dehydroascorbic acid which enhances the fluorescence reaction between HCl
and the steroid portion of the cardiac glycoside.)

CHROMATOGRAM
Retention time: 8 (furosemide), 11 (dihydrodigoxigenin, digoxigenin), 14 (digoxigenin monodigitoxoside), 23 (digoxigenin bisdigitoxoside), 31 (dihydrodigoxin), 33 (digoxin), 43 (digitoxigenin), 57 (spironolactone)

KEY WORDS
post-column reaction

REFERENCE
Kwong, E.; McErlane, K.M. Development of a high-performance liquid chromatographic assay for digoxin using post-column fluorogenic derivatization, *J.Chromatogr.*, **1986**, *377*, 233–242.

o-Dianisidine

SAMPLE
Matrix: water
Analyte: bromate
Sample preparation: Degas sparkling water, filter, inject an aliquot. (High levels of chlo-
rite may interfere with the determination of bromate.)

HPLC VARIABLES
Guard column: Zorbax SB C-18
Column: 250 × 4.6 Zorbax SB C-18
Mobile phase: MeOH:buffer 20:80 (Prepare mobile phase by adding 3.1 g glacial acetic
acid and 29.8 g tetrabutylammonium hydroxide to 700 mL water, adjust pH to 6.3-6.5
with 40% tetrabutylammonium hydroxide solution or acetic acid, add 200 mL MeOH, cool
to room temperature, make up to 1 L with water.)
Flow rate: 1
Injection volume: 250
Detector: UV 500 following post-column reaction. The column effluent mixed with the reagent pumped at 0.5 mL/min and the mixture flowed through a 9 m × 0.53 mm ID
crocheted PTFE coil at 60° to the detector. (Prepare reagent by adding 90 g nitric acid to
600 mL water, add 10 g KBr, add 200 mL 2.5 mg/mL o-dianisidine in MeOH, mix, make
up to 1 L with water, stable indefinitely. Caution! o-Dianisidine is a carcinogen! Disposal
procedures can be found in Lunn, G.; Sansone, E.B. Destruction of Hazardous Chemicals
in the Laboratory, 2nd Edition, Wiley, New York: 1994.)

CHROMATOGRAM
Retention time: 8.5
Limit of quantitation: 3 ng/mL
Limit of detection: 1 ng/mL

OTHER SUBSTANCES
Simultaneously analyzed: chlorite (high levels may interfere), iodate, nitrite
Non-interfering: bromide, chloride

KEY WORDS
post-column reaction; bottled water

REFERENCE
Warner, C.R.; Daniels, D.H.; Joe, F.L. Jr.; Diachenko, G.W. Measurement of bromate in bottled water by high-performance liquid chromatography with post-column flow reactor detection, *Food Addit.Contam.*, **1996**, *13*, 633–638.

Diazo Coupling

SAMPLE
Matrix: tissue
Analyte: clenbuterol and cimaterol
Sample preparation: Homogenize (Ultra Turrax T 25) for 3 min 5 g minced tissue, 100 µL 300 ng/mL IS, and 10 mL 500 mM HCl saturated with ethyl acetate, centrifuge at 4000 g for 20 min, remove the supernatant, add 10 mL 500 mM HCl saturated with ethyl acetate to the residue, homogenize for 1 min, centrifuge at 4000 g for 20 min. Combine the supernatants and adjust the pH to 12 with 32% NaOH, add to a Chem Elut 1020 SPE cartridge (Analytichem), let stand for 10 min, rinse the flask with 20 mL toluene: dichloromethane 75:25, add the rinse to the SPE cartridge, after 5 min elute with two 20 mL portions of toluene:dichloromethane 75:25, add 160 µL 100 mM HCl to the eluate, shake vigorously for 1 min, sonicate for 5 min, centrifuge at 3000 g for 10 min, inject a 100 µL aliquot of the aqueous phase.

HPLC VARIABLES
Column: 150 × 4.6 4 µm Nova-Pak C18
Mobile phase: MeCN:buffer 47:53 (Buffer was 25 mM sodium n-dodecyl sulfate containing 20 mM glacial acetic acid, adjusted to pH 3.5 with 1 M NaOH.)
Flow rate: 1.3
Injection volume: 100
Detector: UV 537 (cimaterol) or 493 (clenbuterol) following post-column reaction. The column effluent mixed with reagent A pumped at 0.2 mL/min and the mixture flowed through a 1 m × 0.33 mm ID PTFE coil. The effluent from this coil mixed with reagent B pumped at 0.2 mL/min and the mixture flowed through a 3 m × 0.33 mm ID PTFE coil. The effluent from this coil mixed with reagent C pumped at 0.2 mL/min and the mixture flowed through a 1 m × 0.33 mm ID PTFE coil. (Reagent A was 0.3 g sodium nitrite, 125 mL water, and 25 mL 65% nitric acid. Reagent B was 2.25 g ammonium amidosulfonate in 150 mL water. Reagent C was 150 mg N-(1-naphthyl)ethylenediamine dihydrochloride in 150 mL water.)

CHROMATOGRAM
Retention time: 7.08 (clenbuterol)
Internal standard: 2-cyclohexylamino-1-(4-amino-3,5-dichlorophenyl)ethanol (11.54)
Limit of detection: 0.2 ng/g

KEY WORDS
post-column reaction; SPE

REFERENCE
Courtheyn, D.; Desaever, C.; Verhe, R. High-performance liquid chromatographic determination of clenbuterol and cimaterol using post-column derivatization, *J.Chromatogr.*, **1991**, *564*, 537–549.

2,6-Dibromo-N-chloro-p-benzoquinoneimine

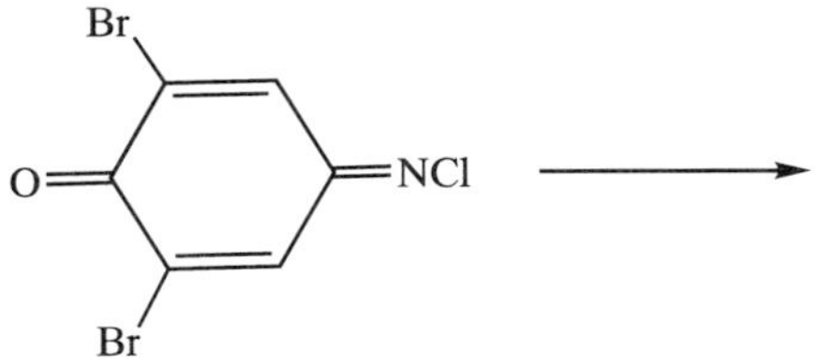

post-column reaction

SAMPLE
Matrix: solutions
Analyte: vitamin B6
Sample preparation: Inject a 10 μL aliquot of a 1 mg/mL solution in water.

HPLC VARIABLES
Column: 250 × 4.6 5 μm Zorbax C8
Mobile phase: 5 mM Sodium perchlorate containing 10 mM sodium 1-hexanesulfonate, pH adjusted to 2.5 with perchloric acid
Column temperature: 45
Flow rate: 0.8
Injection volume: 10
Detector: UV 650 following post-column reaction. The column effluent mixed with 0.5 mg/mL 2,6-dibromo-N-chloro-p-benzoquinoneimine (2,6-dibromoquinone-4-chlorimide) (Eastman) at 1.4 mL/min and flowed through a 2 m × 0.5 mm ID stainless steel coil. The effluent from this coil mixed with 2.5% ammonia solution pumped at 1 mL/min and this mixture flowed through a 2 m × 0.5 mm ID stainless steel coil to the detector.

CHROMATOGRAM
Retention time: 4 (4-pyridoxic acid), 6 (pyridoxal), 8 (pyridoxine), 12 (pyridoxamine)
Limit of detection: 10 ng

KEY WORDS
post-column reaction

REFERENCE
Kawamoto, T.; Okada, E.; Fujita, T. Post-column derivatization of vitamin B_6 using 2,6-dibromoquinone-4-chlorimide, *J.Chromatogr.*, **1983**, *267*, 414–419.

4-Dimethylaminocinnamaldehyde

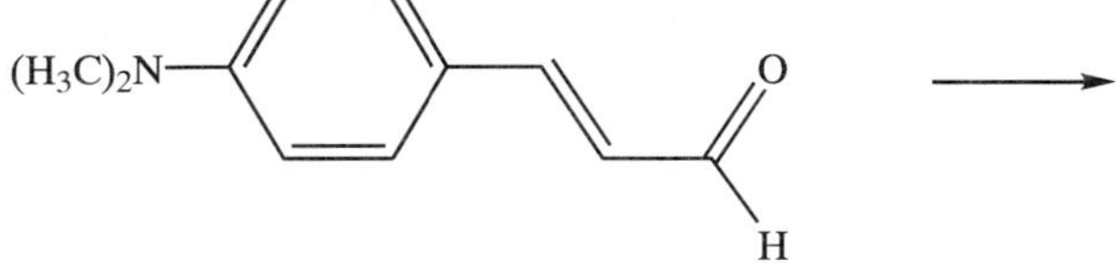

post-column reaction

SAMPLE
Matrix: solutions
Analyte: catechins and proanthocyanidins

HPLC VARIABLES
Column: 250 × 4 3 μm Hypersil ODS
Mobile phase: Gradient. MeOH:5% acetic acid 5:95 for 5 min, to 10:90 over 5 min, maintain at 10:90 for 5 min, to 15:85 over 10 min, maintain at 15:85 for 10 min, to 20:80 over 2 min, maintain at 20:80 for 8 min, to 30:70 over 10 min, to 45:55 over 15 min, to 90:10 over 20 min.

Detector: UV 640 following post-column reaction. The column effluent mixed with 1% 4-dimethylaminocinnamaldehyde in 1.5 M sulfuric acid in MeOH and the mixture flowed through a 9 m × 0.5 mm ID knitted PTFE coil to the detector. (The combined eluent and reagent flow rates were 1.2 mL/min.)

CHROMATOGRAM
Limit of detection: 2.5 ng (epicatechin)

KEY WORDS
post-column reaction

REFERENCE
Treutter, D. Chemical reaction detection of catechins and proanthocyanidins with 4-dimethylaminocinnamaldehyde, *J.Chromatogr.*, **1989**, *467*, 185–193.

Diphenylcarbazide

post-column reaction

SAMPLE
Matrix: solutions
Analyte: chromium
Sample preparation: Add 1 mL 100 mM 1,2-diaminocyclohexane-N,N,N',N'-tetracetic acid disodium salt to 50 mL sample solution, heat at 80° for 15 min, cool to room temperature, make up to 100 mL, inject a 200 μL aliquot.

HPLC VARIABLES
Column: 100 × 4.1 PRP-X100 anion trimethylammonium copolymer (Hamilton)
Mobile phase: 2 mM pH 3.5 potassium hydrogen phthalate containing 10 mM sodium sulfate
Flow rate: 2
Injection volume: 200
Detector: UV 540 following post-column reaction. The column effluent mixed with 0.5 g/L cerium(IV) sulfate in 800 mM sulfuric acid pumped at 0.5 mL/min and the mixture flowed through a 6 m coil at 50°. 800 mM Sulfuric acid pumped at 0.5 mL/min mixed with 0.05% diphenylcarbazide pumped at 0.5 mL/min and this mixture flowed through a 1 m coil. The effluents from these two coils mixed and flowed through a 1 m coil to the detector.

CHROMATOGRAM
Retention time: 2.5 (Cr(VI)), 6.7 (Cr(III))
Limit of detection: 1.5 ng/mL (Cr(VI)), 4.5 ng/mL (Cr(III))

KEY WORDS
post-column reaction; complexation

REFERENCE
Pobozy, E.; Wojasinska, E.; Trojanowicz, M. Ion chromatographic speciation of chromium with diphenylcarbazide-based spectrophotometric detection, *J.Chromatogr.A*, **1996**, *736*, 141–150.

Diphenyl-1-pyrenyl Phosphine

RELATED REFERENCES

Meguro, H.; Akasaka, K.; Ohrui, H. Determination of hydroperoxides with fluorometric reagent diphenyl-1-pyrenylphosphine. *Methods Enzymol.* **1990**, *186*, 157-161.

Ohshima, T.; Hopia, A.; German, J.B.; Frankel, E.N. Determination of hydroperoxides and structures by high-performance liquid chromatography with post-column detection with diphenyl-1-pyrenylphosphine. *Lipids* **1996**, *31*, 1091-1096.

SAMPLE

Matrix: blood

Analyte: hydroperoxides

Sample preparation: 750 μL Plasma + 50 μL 10 mg/mL BHT in dichloromethane + 50 μL 150 μg/mL IS in dichloromethane + 750 μL MeOH + 2 mL dichloromethane, shake vigorously for 1 min, centrifuge at 1000 g for 5 min, repeat the extraction with 2 mL dichloromethane. Combine the organic layers and evaporate them to dryness under reduced pressure at 20°, bleed nitrogen into the system to restore the pressure, reconstitute the residue in 50 μL MeOH:chloroform 50:50, inject the whole amount onto column A and elute onto column B with mobile phase, after 0.4 min divert the effluent from column A onto column C, after 1.05 min divert the effluent from column A onto column B. Elute both column B and column C with mobile phase, for 9.5 min monitor the effluent from column B, monitor the effluent from column C for 16.5 min, monitor the effluent from column B for 6 min.

HPLC VARIABLES

Column: A 30 × 4.6 3 μm Develosil 60-3; B 50 × 4.6 3 μm Develosil 60-3; C 100 × 4.6 5 μm Develosil ODS HG-5

Mobile phase: n-Hexane:1-butanol:MeOH:water 40:100:500:35

Column temperature: 30 (column C only)

Flow rate: 0.8

Injection volume: 50

Detector: F ex 352 em 380 following post-column reaction. The column effluent mixed with the reagent pumped at 0.3 mL/min and the mixture flowed through a 30 m × 0.5 mm ID stainless-steel coil at 80° and a 50 cm × 0.5 mm ID stainless-steel coil in a water jacket to the detector. (Reagent was 7.5 mg/mL diphenyl-1-pyrenyl phosphine in MeOH:1-butanol 50:50, sonicate under reduced pressure, store at -10° for at least 5 h before use, keep at 0° when in use. Diphenyl-1-pyrenyl phosphine is available from Molecular Probes, Eugene OR. Synthesis is as follows. Stir 5.09 g triphenylphosphine and 410 mg lithium in 200 mL THF under nitrogen for 3 h, add 1.8 g tert-butyl chloride in 20 mL THF, reflux for 10 min, cool, filter through glass wool. Add the filtrate to 5.47 g 1-bromopyrene and reflux for 3.5 h, pour into water, extract with chloroform, chromatograph on silica gel with hexane:benzene 70:30 (Caution! Benzene is a carcinogen!), recrystallize from chloroform to give diphenyl-1-pyrenyl phosphine as pale-yellow crystals (mp 171-174°; yield 13.6%) (Anal.Lett. 1987, 20, 731). The peroxides convert the non-fluorescent diphenyl-1-pyrenyl phosphine to the fluorescent diphenyl-1-pyrenyl phosphine oxide.)

CHROMATOGRAM
Retention time: 11.3 (phosphatidylcholine hydroperoxide), 20 (trilinolein hydroperoxide), 20.4 (cholesteryl linolate hydroperoxide), 22.5 (cholesteryl olate hydroperoxide), 29 (triolein hydroperoxide)
Internal standard: cholesteryl cinnamate (monitored at UV 268 without post-column reaction) (14.2)
Limit of detection: 2-4 pmole

KEY WORDS
post-column reaction; plasma; normal phase; column-switching

REFERENCE
Akasaka, K.; Ohrui, H.; Meguro, H. Simultaneous determination of hydroperoxides of phosphatidylcholine, cholesterol esters and triacylglycerols by column-switching high-performance liquid chromatography with a post-column detection system, *J.Chromatogr.*, **1993**, *622*, 153–159.

SAMPLE
Matrix: blood
Analyte: hydroperoxides
Sample preparation: 250 µL Plasma + 50 µL 131 µM N-stearylcinnamide in hexane + 50 µL 20 mg/mL BHT in hexane (freshly prepared) + 1 mL MeOH + 2 mL n-hexane, shake vigorously for 1 min, centrifuge at 4° at 1000 g for 5 min. Remove the upper layer and evaporate it to dryness under reduced pressure, bleed nitrogen into the system to restore the pressure, reconstitute the residue in 100 µL n-hexane, inject a 70-100 µL aliquot.

HPLC VARIABLES
Column: 100 × 4.6 3 µm Develosil 60-3
Mobile phase: Gradient. A was n-hexane. B was n-hexane:1-butanol 100:5. A:B 92:8 for 2 min, to 33:67 over 18 min, to 10:90 over 1 min, maintain at 10:90 for 1 min, return to initial conditions over 1 min, re-equilibrate for 2 min.
Flow rate: 0.6
Injection volume: 70-100
Detector: F ex 352 em 380 following post-column reaction. The column effluent mixed with the reagent pumped at 0.3 mL/min and the mixture flowed through a 20 m × 0.5 mm ID stainless-steel coil at 80° and a 50 cm × 0.5 mm ID stainless-steel coil in a water jacket to the detector. (Reagent was 7.5 mg/mL diphenyl-1-pyrenyl phosphine in MeOH:1-butanol 50:50 containing 0.5 mg/mL BHT. Diphenyl-1-pyrenyl phosphine is available from Molecular Probes, Eugene OR. Synthesis is as described above. The peroxides convert the non-fluorescent diphenyl-1-pyrenyl phosphine to the fluorescent diphenyl-1-pyrenyl phosphine oxide.)

CHROMATOGRAM
Retention time: 8-13 (cholesterol ester hydroperoxides), 13.8-14.5 (triacylglycerol hydroperoxides)
Internal standard: N-stearylcinnamide (Prepare N-stearylcinnamide by refluxing 1 g octadecylamine (stearylamine) and 600 mg cinnamoyl chloride in 30 mL pyridine, pour into 300 mL hexane:ethyl acetate 50:50, wash twice with aqueous phosphoric acid, wash twice with water, evaporate the organic layer to dryness, recrystallize from hexane:ethyl acetate 50:50 to give N-stearylcinnamide (mp 89.8°).) (19.8)
Limit of detection: 1 pmole

KEY WORDS
post-column reaction; plasma; normal phase

REFERENCE
Akasaka, K.; Ohrui, H.; Meguro, H.; Tamura, M. Determination of triacylglycerol and cholesterol ester hydroperoxides in human plasma by high-performance liquid chromatography with fluorometric post-column detection, *J.Chromatogr.*, **1993**, *617*, 205–211.

SAMPLE
Matrix: food
Analyte: hydroperoxides

Sample preparation: Dilute 200-400 mg edible oil to 20 mL with chloroform, inject an aliquot. Add 500 µL water to 0.3-1 g mayonnaise, butter, or margarine, extract twice with 1.5 mL portions of chloroform:MeOH 2:1, dilute the extracts to 5 mL with chloroform, inject a 10 µL aliquot.

HPLC VARIABLES
Column: 150 × 4.6 5 µm TSK-gel ODS 80Tm (Tosoh) (A) or 150 × 4.6 5 µm Develosil Ph-5 (Nomura Chemical) (B)
Mobile phase: MeOH:1-butanol 90:10 (A) or MeOH:water 95:5 (B)
Flow rate: 0.6
Injection volume: 10
Detector: F ex 352 em 380 following post-column reaction. The column effluent mixed with the reagent pumped at 0.3 mL/min and the mixture flowed through a 20 m × 0.5 mm ID stainless-steel coil at 80° and a 50 cm × 0.5 mm ID stainless-steel coil in a water jacket at 20° to the detector. (Reagent was 7.5 mg/mL diphenyl-1-pyrenyl phosphine in MeOH:acetone 75:25, kept in the dark at 0°. Diphenyl-1-pyrenyl phosphine is available from Molecular Probes, Eugene OR. Synthesis is as described above. The peroxides convert the non-fluorescent diphenyl-1-pyrenyl phosphine to the fluorescent diphenyl-1-pyrenyl phosphine oxide.)

CHROMATOGRAM
Retention time: 15-25 (triacylglycerol monohydroperoxides (A)), 18 (trilinolein monohydroperoxides (B))
Limit of quantitation: 2 pmole

KEY WORDS
post-column reaction; oil; mayonnaise; butter; margarine

REFERENCE
Akasaka, K.; Ijichi, S.; Watanabe, K.; Ohrui, H.; Meguro, H. High-performance liquid chromatography and post-column derivatization with diphenyl-1-pyrenylphosphine for fluorimetric determination of triacylglycerol hydroperoxides, *J.Chromatogr.*, **1992**, *596*, 197–202.

Electrochemical Oxidation

SAMPLE
Matrix: tissue
Analyte: malachite green and leucomalachite green
Sample preparation: Condition a 500 mg Bakerbond C18 SPE cartridge with 10 mL MeCN:water 20:80. Homogenize 9 g tissue with 2 mL dichloromethane, 16 mL MeCN, and 1 mL 400 mM perchloric acid in MeCN at high speed for 1 min, stir the mixture at 60 rpm in the dark for 3 h, centrifuge at 3200 rpm for 5 min. Remove 15 mL of the supernatant and evaporate it to 5 mL under a stream of nitrogen at 60°, dilute with 20 mL water, add to the SPE cartridge, dry the cartridge with a stream of air for 10 min, elute with 2 mL mobile phase, inject a 25 µL aliquot.

HPLC VARIABLES
Column: 250 × 4.6 5 µm Econosphere C18 (Alltech)
Mobile phase: MeCN:50 mM aqueous phosphoric acid 94:6, containing 10 mM pentanesulfonic acid
Injection volume: 25
Detector: UV 610 following post-column reaction. The column effluent flowed through an ESA 5010 analytical cell (ESA Coulochem model 5100A) operating at 450 mV to the detector. (Leucomalachite green (colorless) is oxidized to malachite green (colored) by the electrochemical cell.)

CHROMATOGRAM
Retention time: 11 (malachite green), 17 (leucomalachite green)
Limit of quantitation: 6-12 ng/g

Limit of detection: 3-6 ng/g

KEY WORDS
post-column reaction; trout; fish; muscle; SPE

REFERENCE
Swarbrick, A.; Murby, E.J.; Hume, P. Post-column electrochemical oxidation of leuco malachite green for the analysis of rainbow trout flesh using HPLC with absorbance detection, *J.Liq.Chromatogr.Rel.Technol.*, **1997**, *20*, 2269–2280.

Electrochemical Reduction

SAMPLE
Matrix: blood
Analyte: vitamin K_1
Sample preparation: 1 mL Plasma + 50 ng IS + 1 mL 0.9% NaCl + 3 mL isopropanol + 10 mL n-hexane, mix by inversion at 20 rpm for 1 h, centrifuge at 1000 g for 10 min. Remove the upper layer and evaporate it to dryness under reduced pressure at room temperature, reconstitute the residue in 1 mL MeOH, inject a 50 µL aliquot.

HPLC VARIABLES
Column: 100×3 5 µm Hypersil MOS
Mobile phase: MeOH:water 92.5:7.5 containing 30 mM sodium chlorate (Continuously bubble oxygen-free nitrogen (presaturated with mobile phase) through the mobile phase reservoir.)
Flow rate: 0.9
Injection volume: 50
Detector: F ex 320 em 420 following post-column reaction. The mobile phase flowed through an electrochemical cell with coulometric reduction at -400 mV to the detector.

CHROMATOGRAM
Retention time: 10
Internal standard: vitamin $K_{2(30)}$ (17)
Limit of quantitation: 1 ng/mL

KEY WORDS
post-column reaction; plasma; protect from light

REFERENCE
Langenberg, J.P.; Tjaden, U.R. Determination of (endogenous) vitamin K_1 in human plasma by reversed-phase high-performance liquid chromatography using fluorometric detection after post-column electrochemical reduction. Comparison with ultraviolet, single and dual electrochemical detection, *J.Chromatogr.*, **1984**, *305*, 61–72.

Enzyme

SAMPLE
Matrix: tissue
Analyte: acetylcholine and choline
Sample preparation: Homogenize brain tissue with 3 mL 400 mM perchloric acid containing 2 nmoles ethylhomocholine, centrifuge at 35000 g for 20 min, adjust pH of supernatant to about 4.2 with about 200 µL 7.5 M potassium acetate, centrifuge at 35000 g for 20 min. Add the supernatant to 100 µL 5 mM tetramethylammonium chloride, add 3 mL 2% ice-cold reineckate solution (Reinecke salt), let stand on ice for 1 h, centrifuge at 1000 g at 0° for 10 min. Remove the supernatant and dry the precipitate under vacuum overnight, add about 1 mL 5 mM silver tosylate (silver p-toluenesulfonate) in MeCN until the pink color disappears, centrifuge at 1000 g at 0° for 10 min. Remove the supernatant

and evaporate it to dryness under a stream of nitrogen, reconstitute the residue in 200 μL 20 mM pH 3.5 citrate-phosphate buffer, inject an aliquot.

HPLC VARIABLES
Guard column: ODS-5 (Bio-Rad)
Column: 150 mm long Bio-Sil ODS-5S (Bio-Rad)
Mobile phase: Buffer (Buffer was 10 mM sodium acetate buffered to pH 5 with 20 mM citric acid, containing 4.5 mg/L sodium octyl sulfate and 1.2 mM tetramethylammonium chloride.)
Flow rate: 0.8
Injection volume: 20
Detector: E, Bio Analytical Systems LC-4A, Pt electrode +0.5 V, Ag/AgCl reference electrode following post-column reaction. The column effluent mixed with 1 U/mL choline oxidase and 2 U/mL acetylcholinesterase in 200 mM pH 8.5 phosphate buffer pumped at 0.5 mL/min and the mixture flowed through a 30 m × 0.3 mm ID PTFE tube (2.5 min) to the detector.

CHROMATOGRAM
Retention time: 3.6 (choline), 4.2 (homocholine, dimethyl-3-amino-1-propanol), 5.5 (ethylcholine), 10 (acetylcholine, diethylcholine), 13 (diethylhomocholine)
Internal standard: ethylhomocholine (7.2)
Limit of detection: 1-2 pmole

KEY WORDS
post-column reaction; rat; brain

REFERENCE
Potter, P.E.; Meek, J.L.; Neff, N.H. Acetylcholine and choline in neuronal tissue measured by HPLC with electrochemical detection, *J.Neurochem.*, **1983**, *41*, 188–194.

Ethanolamine

RELATED REFERENCE
Del Nozal, M.J.; Bernal, J.L.; Gomez, F.J.; Antolin, A.; Toribio, L. Post-column derivatization of carbohydrates with ethanolamine-boric acid prior to their detection by high-performance liquid chromatography. *J.Chromatogr.* **1992**, *607*, 191-198.

SAMPLE
Matrix: blood
Analyte: carbohydrates
Sample preparation: 100 μL Serum + 500 μL MeOH, shake for 1 min, centrifuge at 10000 rpm for 1 min, inject a 10 μL aliquot of the supernatant.

HPLC VARIABLES
Column: 300 × 4 Aminex A-27
Mobile phase: 500 mM Boric acid adjusted to pH 8.7 with KOH
Flow rate: 2
Injection volume: 10
Detector: F ex 357 (low-pressure mercury lamp) em 436 (420 nm cutoff filter) following post-column reaction. The column effluent mixed with the reagent pumped at 0.5 mL/min and the mixture flowed through a 10 m × 0.8 mm ID stainless steel coil at 150° to the detector. (Reagent was 20 g boric acid and 20 g ethanolamine in 1 L water.)

CHROMATOGRAM
Retention time: 6 (maltose), 9 (ribose), 12.5 (fructose), 16 (galactose), 20 (xylose), 27 (dextrose)
Limit of quantitation: 5 nmoles

KEY WORDS
post-column reaction; serum

REFERENCE
Kato, T.; Kinoshita, T. Fluorometric detection and determination of carbohydrates by high-performance liquid chromatography using ethanolamine, *Anal.Biochem.*, **1980**, *106*, 238−243.

SAMPLE
Matrix: solutions
Analyte: amines
Sample preparation: Mix a 100 μL aliquot of an aqueous solution with 100 μL 50 mM pH 9.6 sodium borate buffer, add 100 μL 10 mM 2-methoxy-2,4-diphenyl-3(2H)-furanone (Fluka) in MeCN while vortexing vigorously, vortex for 10 s, let stand at 20° for 30 min, inject an aliquot.

HPLC VARIABLES
Column: 300 × 4 5 μm TSK LS-410 K (Toyo Soda)
Mobile phase: MeOH:50 mM pH 7.0 phosphate buffer 70:30
Flow rate: 0.45
Injection volume: 10
Detector: F ex 390 em 480 following post-column reaction. The column effluent mixed with reagent pumped at 0.25 mL/min and the mixture flowed through a 3 m × 0.5 mm ID stainless steel coil at 60° and a 50 cm × 0.5 mm ID stainless steel coil at 0° to the detector. (The reagent was ethanolamine adjusted to pH 10.5 with concentrated HCl. If only primary amines are to be determined the post-column reaction system is not necessary.)

CHROMATOGRAM
Retention time: 14 (methylamine), 17 (diethylamine), 25 (di-n-propylamine), 30 (pentylamine), 42 (di-n-butylamine)
Limit of detection: 0.5-50 pmole

KEY WORDS
post-column reaction

REFERENCE
Nakamura, H.; Takagi, K.; Tamura, Z.; Yoda, R.; Yamamoto, Y. Stepwise fluorometric determination of primary and secondary amines by liquid chromatography after derivatization with 2-methoxy-2,4-diphenyl-3(2H)-furanone, *Anal.Chem.*, **1984**, *56*, 919−922.

Ethylenediamine

post-column reaction

SAMPLE
Matrix: solutions
Analyte: reducing carbohydrates

HPLC VARIABLES
Column: 80 × 3 Hitachi 2633
Mobile phase: 700 mM pH 8.5 Borate buffer containing 0.01% EDTA
Column temperature: 60
Flow rate: 0.7
Injection volume: 20
Detector: E, Irika E-502, glassy carbon working electrode 350 mV, Ag/AgCl reference electrode, following post-column reaction. The column effluent mixed with 100 mM ethylenediamine sulfate pumped at 0.25 mL/min and 700 mM pH 9.0 borate buffer pumped at 0.25 mL/min and the mixture flowed through a 30 m × 0.5 mm ID PTFE coil at 140° and a 10 m × 0.2 mm ID cooling coil to the detector.

CHROMATOGRAM
Retention time: 19 (rhamnose), 22 (mannose), 27 (galactose), 30 (xylose), 36 (dextrose)

Limit of detection: 1 pmole

KEY WORDS
post-column reaction

REFERENCE
Honda, S.; Enami, K.; Konishi, T.; Suzuki, S.; Kakehi, K. Use of ethylenediamine sulphate for post-column derivatization of reducing carbohydrates to electrochemically oxidizable compounds in high-performance liquid chromatography, *J.Chromatogr.*, **1986**, *361*, 321–329.

Fast Blue B Salt

post-column reaction

SAMPLE
Matrix: blood, tissue
Analyte: dronabinol
Sample preparation: Plasma. 1-2 mL Plasma + 10-20 μL (?) 55 μg/mL cannabinol + 5 mL MeOH, mix vigorously at 0°, centrifuge at 0° at 1000 g for 5 min, suspend the white gel-like precipitate in 3 mL ice-cold MeOH, centrifuge. Combine the supernatants and evaporate them to 2 mL under a stream of air at 65°, cool in ice-water for 10-20 min, centrifuge at 0° at 1000 g for 5 min. Combine the supernatants and add 2 mL dichloromethane, vortex, add 5 mL hexanes, vortex, centrifuge at 500 g for 5 min, remove the upper organic layer, extract the lower MeOH/water layer with 2 mL dichloromethane and 5 mL hexanes. Combine the upper organic layers and evaporate them to dryness under a stream of air at 65°, reconstitute the residue in 1 mL hexanes, evaporate to dryness under reduced pressure, reconstitute the residue in 200 μL hexanes, evaporate to dryness under reduced pressure, reconstitute with 20 μL MeOH (warm slightly if necessary), add 100 μL acetone, mix vigorously, cool in ice-water for 10 min, centrifuge at 0° at 1000 g for 5 min. Remove the supernatant and add it to 75 μL EtOH, dry under reduced pressure, reconstitute with 100 μL MeOH, inject a 75 μL aliquot. Tissue. Homogenize (tissue grinder mortar) 500 mg mouse brain with 3 mL ice-cold MeOH and 10-20 μL 55 μg/mL cannabinol at 0°, centrifuge at 0° at 1000 g for 5 min, suspend the pellet in 3 mL ice cold MeOH, centrifuge at 0° at 1000 g for 5 min. Combine the supernatants and evaporate them to 1.5 mL under a stream of air at 65°, cool in ice-water for 10-20 min, centrifuge at 0° at 1000 g for 5 min, suspend the pellet in 500 μL ice-cold MeOH:water 80:20, centrifuge. Combine the supernatants and add 2 mL dichloromethane, vortex, add 1 mL water, vortex, add 5 mL hexanes, vortex, centrifuge at 500 g for 5 min, remove the upper organic layer, extract the lower MeOH/water layer with 2 mL dichloromethane and 5 mL hexanes. Combine the upper organic layers and evaporate them to dryness under a stream of air at 65°, reconstitute the residue in 1 mL hexanes, evaporate to dryness under reduced pressure, reconstitute the residue in 200 μL hexanes, evaporate to dryness under reduced pressure, reconstitute with 20 μL MeOH (warm slightly if necessary), add 100 μL acetone, mix vigorously, cool in ice-water for 10 min, centrifuge at 0° at 1000 g for 5 min. Remove the supernatant and add it to 75 μL EtOH, dry under reduced pressure, reconstitute with 100 μL MeOH, inject a 75 μL aliquot.

HPLC VARIABLES
Guard column: 50 × 3.9 37-50 μm μBondapak C18/Corasil
Column: 300 × 3.9 μBondapak C18
Mobile phase: MeOH:water 85:15
Flow rate: 1.5

Injection volume: 75

Detector: UV 490 following post-column reaction. The column effluent mixed with the reagent pumped at 0.5 mL/min and the mixture flowed through a 4 m × 0.8 mm PTFE coil to the detector. (Prepare the reagent by dissolving 1 g Fast Blue B Salt in 160 mL water, adding 200 mL MeOH, and adding 40 mL 10% sodium nitrite in water. Stir for 30 min, filter (Whatman GF/C), filter (0.45 μm Gelman Metricel GA-6), discard after 5 h. Protect from light. At the end of each day flush the post-column reaction system with MeOH: water 50:50, acetone, and MeOH:water 50:50.)

CHROMATOGRAM
Retention time: 7
Internal standard: cannabinol (6)
Limit of detection: 50 ng

OTHER SUBSTANCES
Extracted: 11-hydroxytetrahydrocannabinol

KEY WORDS
post-column reaction; treat glassware with Dri-Film SC-87 (Pierce); mouse; plasma; brain

REFERENCE
Borys, H.K.; Karler, R. Post-column derivatization procedure for the colorimetric analysis of tissue cannabinoids separated by high-performance liquid chromatography, *J.Chromatogr.*, **1981**, *205*, 303–323.

Ferric Ion

RELATED REFERENCE
Witts, D.; Wilson, I.D. Post-column reaction with ferric chloride for the detection of D-penicillamine. *Methodol.Surv.Biochem.Anal.* **1984**, *14*, 185-188.

$$Fe^{+++} \longrightarrow \text{post-column reaction}$$

SAMPLE
Matrix: fertilizer
Analyte: EDTA
Sample preparation: Prepare an aqueous solution, filter (paper), filter (0.2 μm), inject an aliquot.

HPLC VARIABLES
Guard column: Ion Pac AG7 (Dionex)
Column: 10 μm Ion Pac AS7 (Dionex)
Mobile phase: 70 mM nitric acid
Flow rate: 0.5
Injection volume: 50
Detector: UV 330 following post-column reaction with 1 g/L $Fe(NO_3)_3.9H_2O$ in 2% perchloric acid

CHROMATOGRAM
Retention time: 5

KEY WORDS
post-column reaction

REFERENCE
Vande Gucht, I. Determination of chelating agents in fertilizers by ion chromatography, *J.Chromatogr.A*, **1994**, *671*, 359–365.

SAMPLE
Matrix: plants
Analyte: inositol phosphates

Sample preparation: 1 g Finely-ground dry plant material + 10 mL 500 mM HCl, shake mechanically at 150 rpm at room temperature for 2 h, centrifuge at 48400 g for 20 min. Filter (0.2 µm) the supernatant and inject an aliquot of the filtrate.

HPLC VARIABLES

Guard column: 5 × 3 1000 Å PL-SAX polystyrene-based strong anion-exchange column (Polymer Labs)

Column: 50 × 4.6 1000 Å PL-SAX polystyrene-based strong anion-exchange column (Polymer Labs)

Mobile phase: Gradient. A was 10 mM pH 4.0 1-methylpiperazine. B was 500 mM sodium nitrate in 10 mM pH 4.0 1-methylpiperazine. A:B from 100:0 to 0:100 over 30 min.

Flow rate: 1

Detector: UV 500 following post-column reaction. The column effluent mixed with the reagent pumped at 1 mL/min and the mixture flowed through a 2.9 m × 0.76 mm ID coil of PEEK tubing to the detector. (Reagent was 0.015% ferric chloride hexahydrate in 0.015% sulfosalicylic acid. Complexation of the eluting phosphates with the iron leads to a decrease in absorbance.)

CHROMATOGRAM

Retention time: 2-26 (depending on degree of substitution)

KEY WORDS

post-column reaction

REFERENCE

Rounds, M.A.; Nielsen, S.S. Anion-exchange high-performance liquid chromatography with post-column detection for the analysis of phytic acid and other inositol phosphates, *J.Chromatogr.A*, **1993**, *653*, 148–152.

SAMPLE

Matrix: plants, tissue

Analyte: inositol phosphates

Sample preparation: Seeds. Pulverize 10 g soybean seeds, add 200 mL 1.2% HCl, stir for 30 min, filter (Whatman No. 1 paper). Dilute the filtrate to 500 mL with water, add to a 4 g column of 200-400 mesh AG 1-X8 (Bio-Rad), elute with 50 mL 1 M HCl. Lyophilize the eluate, reconstitute with 2 mL water, inject a 50 µL aliquot. Tissue. Homogenize (Waring blender) 50 g calf brains and 500 µg IS with 60 mL 4% HCl for 2 min, centrifuge at 39000 g for 20 min. Remove a 50 mL aliquot of the aqueous layer and dilute 5-fold with water, add to a 2 g column of 200-400 mesh AG 1-X8 (Bio-Rad), elute with 50 mL 1 M HCl. Lyophilize the eluate, reconstitute with 500 µL water, inject a 50 µL aliquot.

HPLC VARIABLES

Guard column: AG3

Column: AS3 (Dionex)

Mobile phase: Gradient. A was 180 µg/mL pyrocatechol. B was 155 mM nitric acid. A:B from 100:0 to 0:100 over 20 min, maintain at 0:100 for 10 min.

Flow rate: 1

Injection volume: 50

Detector: UV 290 following post-column reaction. The column effluent mixed with 0.1% ferric nitrate nonahydrate in 2% perchloric acid pumped at 0.5 mL/min and the mixture flowed to the detector.

CHROMATOGRAM

Retention time: 7-28 (depending on structure)

Internal standard: L-I(1,2,3,4)P$_4$ (17)

Limit of detection: 1-2 nmole

KEY WORDS

post-column reaction; use only PTFE tubing; cow; brain; soybean; seeds

REFERENCE

Phillippy, B.Q.; Bland, J.M. Gradient ion chromatography of inositol phosphates, *Anal.Biochem.*, **1988**, *175*, 162–166.

Guanidine

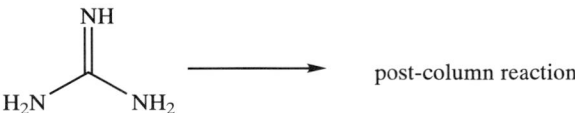

$\longrightarrow$ post-column reaction

SAMPLE
Matrix: blood
Analyte: dermatan sulfate
Sample preparation: 100 µL Plasma + 100 µL 1% actinase (Kaken Pharmaceutical, To-kyo) in 50 mM pH 8.0 Tris-HCl buffer, mix, heat at 45° for 3 h, add 1.5 mL 15 mM acetic acid containing 10% NaCl, heat in a boiling water bath for 5 min, cool in an ice bath for 1 min, centrifuge at 2300 g for 15 min. Remove the supernatant and add it to 5 mL cooled EtOH saturated with sodium acetate, mix, let stand at 0° for 1 h, centrifuge at 4° at 2300 g for 15 min. Lyophilize the precipitate and dissolve it in 50 µL water, inject a 10-20 µL aliquot.

HPLC VARIABLES
Column: 250 × 4.6 Asahipak NH2P-50
Mobile phase: Gradient. A was 100 mM pH 10.0 phosphate buffer containing 100 mM NaCl. B was 100 mM pH 10.0 phosphate buffer containing 500 mM NaCl. A:B from 100:0 to 0:100 over 10 min, maintain at 0:100 for 20 min, re-equilibrate at initial con-ditions for 15 min.
Flow rate: 0.5
Injection volume: 10-20
Detector: F ex 320 em 425 following post-column reaction. The column effluent mixed with 0.5 M NaOH pumped at 0.125 mL/min and with 50 mM guanidine pumped at 0.125 mL/min and the mixture flowed through a 10 m × 0.5 mm ID PTFE tube at 110° and a 2 m × 0.25 mm ID PTFE tube at room temperature to the detector.

CHROMATOGRAM
Retention time: 15 (dermatan sulfate), 18 (oversulfated dermatan sulfate)
Limit of detection: 10-20 ng

KEY WORDS
post-column reaction; plasma; rat; pharmacokinetics

REFERENCE
Huang, Y.; Washio, Y.; Hara, M.; Toyoda, H.; Koshiishi, I.; Toida, T.; Imanari, T. Simultaneous deter-mination of dermatan sulfate and oversulfated dermatan sulfate in plasma by high-performance liquid chromatography with postcolumn fluorescence derivatization, *Anal.Biochem.*, **1996**, *240*, 227–234.

Heptakis(2,6-di-O-methyl)-β-cyclodextrin

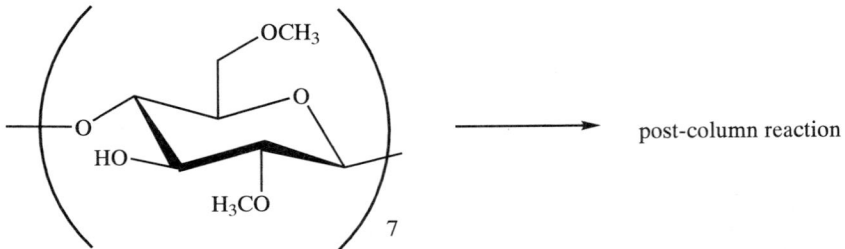

SAMPLE
Matrix: solutions
Analyte: aflatoxins
Sample preparation: Filter (0.45 μm) a solution of aflatoxins in mobile phase, inject an aliquot of the filtrate.

HPLC VARIABLES
Column: 220 × 4.6 5 μm Spheri-5 C18
Mobile phase: MeOH:water 40:60
Flow rate: 1
Injection volume: 20
Detector: F ex 365 em 418 following post-column reaction. The column effluent mixed with a 10 mM solution of heptakis(2,6-di-O-methyl)-β-cyclodextrin in water pumped at 1 mL/min and the mixture flowed through a 300 mm × 0.016 mm i.d. (sic) steel tube to the detector.

CHROMATOGRAM
Retention time: 7.2 (aflatoxin G_2), 8.8 (aflatoxin G_1), 10 (aflatoxin B_2), 13.8 (aflatoxin B_1)
Limit of detection: 4-6 ng/mL

KEY WORDS
post-column reaction

REFERENCE
Cepeda, A.; Franco, C.M.; Fente, C.A.; Vázquez, B.I.; Rodriguez, J.L.; Prognon, P.; Mahuzier, G. Postcolumn excitation of aflatoxins using cyclodextrins in liquid chromatography for food analysis, *J.Chromatogr.A*, **1996**, *721*, 69–74.

Hydrochloric Acid

HCl ⟶ post-column reaction

SAMPLE
Matrix: blood
Analyte: digoxin
Sample preparation: 3 mL Serum + 20 μL 8 μg/mL IS in EtOH + 3 mL acetone, vortex for 20 s, centrifuge at 1000 g for 5 min, remove the supernatant and add it to 2 mL isooctane:dichloromethane 80:20, vortex for 1 min, centrifuge at 1000 g for 5 min. Remove the acetone/water layer and evaporate it to 3 mL under a stream of nitrogen at 37°, add 10 mL dichloromethane:n-propanol 98:2, rotate for 10 min, centrifuge at 1000 g for 5 min, repeat extraction, filter the organic layers and evaporate them to dryness under a stream of nitrogen at 37°, reconstitute the residue in 100 μL MeOH:water 50:50, inject the whole amount.

HPLC VARIABLES
Guard column: 15 × 3.2 ODS (Brownlee)

Column: 150 × 4.6 3 μm Spherisorb ODS II
Mobile phase: MeOH:EtOH:isopropanol:buffer 52:3:1:45 (Prepare buffer by mixing 12.5 mL 0.15% hydrogen peroxide in water with 500 mL 500 μg/mL L-ascorbic acid in water, stir for 2 h. Prepare fresh each week.)
Flow rate: 0.4
Injection volume: 100
Detector: F ex 360 (filter) em 425 (filter) following post-column reaction. The column effluent mixed with concentrated HCl pumped at 0.5 mL/min and flowed through a 20 m × 0.3 mm i.d. PTFE coil at 79 ± 1° to the detector. (The flow of concentrated HCl was generated by displacing concentrated HCl from a pressure vessel with hexane. The hexane was pumped into the pressure vessel by an HPLC pump.)

CHROMATOGRAM
Retention time: 18.5
Internal standard: digitoxigenin (25.5)
Limit of quantitation: 0.5 ng/mL

OTHER SUBSTANCES
Simultaneously analyzed: digoxigenin, digoxigenin bisdigitoxoside, digoxigenin mono-digitoxoside, dihydrodigoxigenin, dihydrodigoxin, furosemide, spironolactone
Non-interfering: mexiletine, captopril, dipyridamole, disopyramide, procainamide, propafenone, quinidine, sulfamethoxazole, trimethoprim, verapamil

KEY WORDS
serum; post-column reaction

REFERENCE
Embree, L.; McErlane, K.M. Development of a high-performance liquid chromatographic-post-column fluorogenic assay for digoxin in serum, *J.Chromatogr.*, **1989**, *496*, 321–334.

SAMPLE
Matrix: solutions
Analyte: citrinin

HPLC VARIABLES
Column: 250 × 4.6 5 μm Nucleosil C18
Mobile phase: MeOH containing 570 μM tetrabutylammonium hydroxide, pH adjusted to 5.5 with 1 M HCl in water
Column temperature: 21 ± 0.1
Flow rate: 0.8
Injection volume: 20
Detector: F ex 331 em 500 following post-column reaction. The column effluent mixed with 1 M HCl in water pumped at 0.2 mL/min and the mixture flowed immediately to the detector.

CHROMATOGRAM
Retention time: 11.5
Limit of quantitation: 200 nM
Limit of detection: 90 nM

KEY WORDS
post-column reaction

REFERENCE
Franco, C.M.; Fente, C.A.; Vazquez, B.; Cepeda, A.; Lallaoui, L.; Prognon, P.; Mahuzier, G. Simple and sensitive high-performance liquid chromatography-fluorescence method for the determination of citrinin. Application to the analysis of fungal cultures and cheese extracts, *J.Chromatogr.A*, **1996**, *723*, 69–75.

Hydrogen Peroxide

RELATED REFERENCES

Iinuma, F.; Tabara, M.; Mawatari, K.; Suzuki, M.; Watanabe, M. Fluorometric determination of kynurenine derivatives by high performance liquid chromatography with hydrogen peroxide and sodium carbonate. *Bunseki Kagaku* **1984**, *33*, E315-E322.

Kubo, H.; Umiguchi, Y.; Fukumoto, M.; Kinoshita, T. Fluorometric determination of methotrexate in serum by high-performance liquid chromatography using in-line oxidation with hydrogen peroxide. *Anal.Sci.* **1992**, *8*, 789-792.

Mawatari, K.; Iinuma, F.; Watanabe, M. Fluorometric determination of kynurenine in human serum by high-performance liquid chromatography coupled with post-column photochemical reaction with hydrogen peroxide. *J.Chromatogr.* **1989**, *488*, 349-355.

Yu, Z.; Westerlund, D.; Boos, K.-S. Determination of methotrexate and its metabolite 7-hydroxymethotrexate by direct injection of human plasma into a column-switching liquid chromatographic system using post-column photochemical reaction with fluorimetric detection. *J.Chromatogr.B* **1997**, *689*, 379-386.

H_2O_2 ———————→ post-column reaction

SAMPLE

Matrix: blood

Analyte: indomethacin

Sample preparation: 20 µL serum + 20 µL MeCN, vortex for a few s, centrifuge at 10000 g for 2 min, inject a 20 µL aliquot of the supernatant.

HPLC VARIABLES

Column: 150 × 4.6 5 µm Inertsil ODS-2

Mobile phase: MeCN:25 mM phosphate buffer 35:65 containing 30 mM hydrogen peroxide, adjusted to pH 7.0 with 1 M NaOH

Flow rate: 1

Injection volume: 20

Detector: F ex 358 em 462 following post-column reaction. The column effluent flowed through a 15 m × 0.5 mm ID stainless steel coil at 180° then a 3 m × 0.5 mm ID stainless steel coil at 15° to the detector.

CHROMATOGRAM

Retention time: 10

Limit of detection: 500 ng/mL

KEY WORDS

post-column reaction; serum; pharmacokinetics

REFERENCE

Kubo, H.; Umiguchi, Y.; Kinoshita, T. Fluorometric determination of indomethacin in serum by high performance liquid chromatography using in-line oxidation with hydrogen peroxide, *J.Liq.Chromatogr.*, **1993**, *16*, 465–474.

SAMPLE

Matrix: solutions

Analyte: carboxylic acids

Sample preparation: Mix a 100 µL aliquot of a 5-1000 nM solution of a carboxylic acid in MeCN with 100 µL 18-crown-6 solution, add 100 µL 100 µM dansyl-BAP in MeCN, mix, let stand at room temperature (aliphatic acids) or 55° (aromatic acids) for 30 min, add 100 µL 3 mM thymine in MeCN, add 5 mg potassium bicarbonate, vortex for 30 s, let stand for 30 min, evaporate to dryness under a stream of nitrogen. Reconstitute with dichloromethane, add to a Bond-Elut silica SPE cartridge, elute with 1.5 mL MeCN: dichloromethane 50:50. Evaporate the eluate to dryness and reconstitute the residue with 500 µL mobile phase, inject a 25 µL aliquot. (Prepare 18-crown-6 solution by sonicating a 1 mg/mL solution of 18-crown-6 in MeCN containing 1 mg/mL potassium bicarbonate for 20 min. Prepare dansyl-BAP (N-(bromoacetyl)-N'-[5-(dimethylamino)naphthalene-1-sulfonyl]piperazine) as follows. Slowly add a solution of 135 mg dansyl chloride in 30 mL acetone to a 10-fold molar excess of piperazine in acetone:water 75:25, stir at 50° for 1 h, evaporate the acetone. Acidify the remaining aqueous layer with concentrated nitric acid, wash 3 times with 15 mL portions of dichloromethane, adjust the pH of the aqueous layer to 11 with concentrated NaOH, extract three times with 15 mL portions of dichloromethane. Combine the extracts and dry them over anhydrous calcium chloride, concentrate to about 5 mL, chromatograph on a 400 × 25 column of 60-200 μm silica gel Si-60, wash with about 20 mL dichloromethane:MeOH 99:1 to remove a small fluorescent band, elute with about 30 mL dichloromethane:MeOH 95:5-94:6 to obtain a solution of dansylpiperazine, determine the concentration by UV absorption at 340 nm (extinction coefficient = 4300 in MeOH). Stir a solution of 700 mg bromoacetic acid and 1.1 g dicyclohexylcarbodiimide in 100 mL MeCN at room temperature for 1 min, slowly add 470 μmoles dansylpiperazine in MeCN, stir for 1 h, evaporate to dryness, reconstitute the residue with 10 mL dichloromethane, filter. Chromatograph the filtrate on a 400 × 25 column of 40-60 μm Si-60 silica gel with dichloromethane, when the first-eluting, strongly-fluorescent yellow band reaches the outlet change the eluent to dichloromethane:MeOH 99:1, collect about 50 mL eluate to obtain dansyl-BAP.)

HPLC VARIABLES

Column: 150 × 3.1 5 μm LiChrosorb RP-18
Mobile phase: MeCN:water 60:40, containing 2.5 mM pH 7.0 imidazole buffer
Flow rate: 0.5
Injection volume: 25
Detector: F ex 246 em 490 (cut-off filter) following post-column reaction. The column effluent mixed with 50 mM hydrogen peroxide in MeCN containing 5 mM bis(2-nitrophenyl)oxalate pumped at 0.3 mL/min and the mixture flowed immediately to the detector. (Prepare bis(2-nitrophenyl)oxalate by dissolving 13.9 g 2-nitrophenol in 250 mL benzene (Caution! Benzene is a carcinogen!), remove 50 mL benzene by azeotropic distillation, cool to 10°, add 10.1 g freshly distilled triethylamine, add 7 g oxalyl chloride dropwise, allow to warm to room temperature, let stand overnight, evaporate to dryness under reduced pressure, recrystallize to give bis(2-nitrophenyl)oxalate (J. Chem. Educ. 1974, 51, 529).)

CHROMATOGRAM

Retention time: 6.5 (benzoic acid), 7 (2-methoxybenzoic acid), 9 (2,4-dichlorobenzoic acid), 10 (lipoic acid), 12 (naproxen), 16 (octanoic acid), 24 (ibuprofen)
Limit of detection: 0.8-1 pmole

KEY WORDS

post-column reaction; SPE

REFERENCE

Kwakman, P.J.M.; Van Schaik, H.P.; Brinkman, U.A.T.; de Jong, G.J. *N*-(Bromoacetyl)-*N'*-[5-(dimethylamino)naphthalene-1-sulfonyl]piperazine as a sensitive labeling reagent for the determination of carboxylic acids by liquid chromatography with peroxyoxalate chemiluminescence and fluorescence detection, *Analyst*, **1991**, *116*, 1385–1391.

8-Hydroxyquinoline-5-sulfonic Acid

RELATED REFERENCES

Conradi, S.; Vogt, C.; Wittrisch, H.; Knobloch, G.; Werner, G. Capillary electrophoretic separation of metal ions using complex forming equilibria of different stabilities. *J.Chromatogr.A* **1996**, *745*, 103-109.

Shijo, Y.; Saitoh, A.; Suzuki, K. Separation of aluminum(III), copper(II), gallium(III), and iron(III) as 8-hydroxyquinoline-5-sulfonic acid complexes by high-performance liquid chromatography. *Chem.Lett.* **1989**, 181-184.

post-column reaction

SAMPLE

Matrix: solution

Analyte: chelating ligands

Sample preparation: Mix a 10 μM solution of the ligand with a 100 μM solution of lutetium chloride, pass through a 70 × 20 column of 70-130 μm AG 50W-X8 cation-exchange resin (sodium form, Bio-Rad), inject an aliquot.

HPLC VARIABLES

Column: 150 × 4.6 5 μm Prodigy 5 ODS-2 (Phenomenex)

Mobile phase: Gradient. A was 1 mM potassium sulfate containing 3 mM tetrapropylammonium bromide. B was MeCN:3 mM potassium sulfate 4:96. A:B 100:0 for 4 min, to 0:100 over 3 min.

Flow rate: 1

Injection volume: 50

Detector: F ex 360 em 500 following post-column reaction. The column effluent mixed with the reagent pumped at 0.3 mL/min and the mixture flowed through a 5.1 m × 0.5 mm ID knitted PTFE coil. The effluent from the coil mixed with 1 M NaOH pumped at 0.3 mL/min and the mixture flowed to the detector. (Reagent was 1 mM trans-1,2-diaminocyclohexane-N,N,N',N'-tetraacetic acid (CDTA) in 1 mM 8-hydroxyquinoline-5-sulfonic acid, adjusted to pH 2.8 with acetic acid.)

CHROMATOGRAM

Retention time: 2.5 (nitrilotriacetic acid (NTA)), 7 (ethylene glycol-bis(β-aminoethylether) N,N,N',N'-tetraacetic acid (EGTA)), 7.5 (EDTA), 14 (diethylenetriaminepentaacetic acid (DTPA)), 17.5 (trans-1,2-diaminocyclohexane-N,N,N',N'-tetraacetic acid (CDTA))

Limit of detection: 25-100 nM

KEY WORDS

complexation; post-column reaction; use a metal-free system

REFERENCE

Ye, L.; Lucy, C.A. Ion chromatographic determination of chelating ligands based on the postcolumn formation of ternary fluorescent complexes, *J.Chromatogr.A*, **1996**, *739*, 307–315.

SAMPLE

Matrix: solutions

Analyte: gallium and indium

Sample preparation: Filter (0.45 μm nylon) the acidic sample solution, inject a 100 μL aliquot of the filtrate.

HPLC VARIABLES

Guard column: 10 μm Partisil 10 SCX

Column: 250 × 4.6 10 μm Partisil 10 SCX

Mobile phase: 100 mM Lactic acid containing 1.2 mM ethylenediamine, adjusted to pH 3.4-3.5 with NaOH
Flow rate: 1.5
Injection volume: 100
Detector: F ex 389 em 529 following post-column reaction. The column effluent mixed with the reagent pumped at 1.5 mL/min and the mixture flowed through a 30 cm × 0.5 mm ID reaction coil to the detector. (Reagent was 1 mM 8-hydroxyquinoline-5-sulfonic acid (quinolin-8-ol-5-sulfonic acid) containing 1.2 mM cetyltrimethylammonium bromide.)

CHROMATOGRAM
Retention time: 3 (gallium), 4.5 (indium)
Limit of detection: 1-1.5 ng/mL

OTHER SUBSTANCES
Simultaneously analyzed: aluminum
Non-interfering: arsenic, beryllium, cadmium, copper, iron, magnesium, nickel, zinc

KEY WORDS
post-column reaction

REFERENCE
Prat, M.D.; Compañó, R.; Granados, M.; Miralles, E. Liquid chromatographic determination of gallium and indium with fluorimetric detection, *J.Chromatogr.A*, **1996**, *746*, 239−245.

Immobilized Enzyme Reactor

SAMPLE
Matrix: CSF, dialysate, tissue
Analyte: acetylcholine and choline
Sample preparation: Dialysate. Inject dialysate directly. Tissue. Homogenize brain tissue with 10 volumes 100 mM perchloric acid (Potter-Elvehjem), let stand on ice for 15 min, centrifuge at 4000 g for 15 min, inject a 0.5 μL aliquot. CSF. Deproteinize by passing through a 0.02 μm Anatop 10 syringe filter (Alltech), inject a 0.5 μL aliquot of the filtrate.

HPLC VARIABLES
Column: 100 × 1 PEEK column packed with Aminex A-9 (Bio-Rad)
Mobile phase: 200 mM pH 8.0 K/Na 3/1 phosphate buffer containing 5 mM NaCl and 0.1% Kathon CG
Column temperature: 25
Flow rate: 0.06 (obtained with a flow splitter)
Injection volume: 0.5
Detector: E, AMOR (Spark Holland), platinum working electrode + 250 mV, carbon composite auxiliary electrode, Ag/AgCl reference electrode, following post-column reaction. The column effluent flowed through a reactor which had 4 U acetylcholine esterase (EC 3.1.1.7 type VI-S from electric eel, 260 IU/mg) and 4 U choline oxidase (EC 1.1.3.17 from Alcaligenes sp., 12.7 IU/mg) enclosed between two 0.01 μm cellulose nitrate filters (Sartorius) (construction details given) to the detector. (Acetylcholine and choline are converted to hydrogen peroxide which is then detected.)

CHROMATOGRAM
Retention time: 8 (choline), 12 (acetylcholine)
Limit of detection: 10 fmole

KEY WORDS
use metal-free tubing and connectors; solvent reservoir, column, reactor, and detector maintained at 25°; rat; brain; human; post-column reaction; immobilized enzyme reactor

REFERENCE
Flentge, F.; Venema, K.; Koch, T.; Korf, J. An enzyme-reactor for electrochemical monitoring of choline and acetylcholine: applications in high-performance liquid chromatography, brain tissue, microdialysis and cerebrospinal fluid, *Anal.Biochem.*, **1992**, *204*, 305−310.

SAMPLE
Matrix: blood
Analyte: bile acids
Sample preparation: Deproteinize 20 μL serum with a pretreatment column (Autoserumout, Sekisui), inject an aliquot.

HPLC VARIABLES
Column: 150 × 4.6 Medipola Bile column (Sekisui)
Mobile phase: Gradient. A was MeCN:MeOH:30 mM ammonium acetate 20:20:60. B was MeCN:MeOH:30 mM ammonium acetate 30:30:40. A:B from 100:0 to 80:20 over 10 min, to 0:100 over 27 min, maintain at 0:100 for 30 min.
Flow rate: 1
Detector: F ex 340 em 460 following post-column reaction. The effluent from the column mixed with reagent pumped at 1 mL/min and the mixture flowed through a 20 × 4 3α-HSD column (Sekisui) containing bound 3α-hydroxysteroid dehydrogenase to the detector. (The reagent was 1.36 g/L KH_2PO_4, 372 mg/L disodium EDTA, 140 mg/L βNAD, and 450 μL/L 2-mercaptoethanol in water adjusted to pH 7.8 with 5 M KOH.)

CHROMATOGRAM
Retention time: 15 (glycoursodeoxycholic acid), 16 (taurineursodeoxycholic acid), 18 (ursodiol), 20 (glycocholic acid), 21 (taurocholic acid), 22 (cholic acid), 28 (glycochenodeoxycholic acid), 29 (taurochenodeoxycholic acid), 30 (glycodeoxycholic acid), 31 (taurodeoxycholic acid), 35 (chenodiol (chenodeoxycholic acid)), 37 (deoxycholic acid), 42 (glycolithocholic acid), 43 (taurolithocholic acid), 54 (lithocholic acid)

KEY WORDS
serum; post-column reaction; immobilized enzyme reactor

REFERENCE
Adachi, Y.; Nanno, T.; Itoh, T.; Kurumi, Y.; Yamazaki, K.; Sawada, Y.; Yamamoto, T. Determination of individual serum bile acids in chronic liver diseases: fasting levels and results of oral chenodeoxycholic acid tolerance test, *Gastroenterol.Jpn.*, **1988**, *23*, 401–407.

SAMPLE
Matrix: blood
Analyte: choline
Sample preparation: 100 μL Plasma + 2 mL 300 ng/mL ethylhomocholine in MeCN, vortex for 1 min, centrifuge at 2000 g for 15 min. Remove 1.5 mL of the supernatant and evaporate it to dryness under vacuum, reconstitute the residue in 500 μL mobile phase, inject a 50 μL aliquot.

HPLC VARIABLES
Column: 100 × 4.6 5 μm Cyano Spheri-5
Mobile phase: 20 mM Sodium hydrogen phosphate and 10 mM tetramethylammonium chloride adjusted to pH 7.1 with 65% phosphoric acid
Flow rate: 0.7
Injection volume: 50
Detector: E, ESA Coulochem II, Model 5040 Pt analytical cell + 300 mV, following an enzyme reactor. (The reactor was a 30 × 2.1 7 μm Aquapore AX-300 anion-exchange cartridge (Brownlee), inject slowly 50 μL 100 U/mL choline oxidase (EC 1.1.3.17, Alcaligenes) and catalase (EC 1.11.1.6) (Sigma), wash with mobile phase for several minutes before use, reload after 100 samples.)

CHROMATOGRAM
Retention time: 5
Internal standard: ethylhomocholine (8)
Limit of quantitation: 3.58 μM

KEY WORDS
plasma; pharmacokinetics; post-column reaction; immobilized enzyme reactor

REFERENCE

Fossati, T.; Colombo, M.; Castiglioni, C.; Abbiati, G. Determination of plasma choline by high-performance liquid chromatography with a postcolumn enzyme reactor and electrochemical detection, *J.Chromatogr.B*, **1994**, *656*, 59–64.

SAMPLE
Matrix: blood
Analyte: amino acids
Sample preparation: 5 μL Plasma + 200 μL 20 mM pH 8.0 phosphate buffer, filter (Advantec Q0100, molecular mass cutoff), inject a 20 μL aliquot of the ultrafiltrate.

HPLC VARIABLES
Column: 150 × 4 5 μm Capcell C18 SG120
Mobile phase: 20 mM pH 7.5 phosphate buffer
Column temperature: 50
Flow rate: 0.4
Injection volume: 20
Detector: Chemiluminescence following post-column reaction. The column effluent mixed with 5 mM nicotinamide adenine dinucleotide in 20 mM pH 7.0 phosphate buffer pumped at 0.15 mL/min and 5 mM luminol in carbonate buffer pumped at 0.15 mL/min and this mixture flowed through an immobilized-enzyme reactor at 50°. The effluent from the reactor mixed with 20 mM potassium hexacyanoferrate(III) in water pumped at 0.4 mL/min and this mixture flowed through a 90 cm × 0.5 mm ID tube to the detector. (The carbonate buffer was 400 mM sodium carbonate containing 400 mM sodium bicarbonate, pH 10.5. Prepare the immobilized enzyme reactor as follows. Wash 1 g 13 μm poly(vinyl alcohol) beads (GS-520, Showa Denko, Tokyo) with 50 mL dry acetone, suspend in 20 mL dry acetone:pyridine 50:50 with vigorous stirring, add 1 mL 2,2,2-trifluoroethanesulfonyl chloride dropwise over 2 min, stir for 10 min, wash beads with 10 mL acetone, wash beads with 20 mL 1 mM HCl, slurry pack in a 50 × 4 column, circulate enzyme solution through the column at 0.2 mL/min for 4 h, monitor the immobilization process at 380 nm. The enzyme solution consisted of 5 mg (325 U) LeuDH (from Bacillus stearothermophilus, Unitika, Osaka) and 5 mg (280 U) NAOD (from Bacillus megaterium, Unitika, Osaka) in 10 mL 100 mM pH 7.0 phosphate buffer. When not in use store the reactor in 5 mM nicotinamide adenine dinucleotide in 20 mM pH 7.0 phosphate buffer at 5°.)

CHROMATOGRAM
Retention time: 8 (valine), 10.8 (isoleucine), 11.6 (leucine)
Limit of detection: 100 nM

KEY WORDS
post-column reaction; plasma; ultrafiltrate; immobilized enzyme reactor

REFERENCE

Kiba, N.; Oyama, Y.; Kato, A.; Furusawa, M. Postcolumn co-immobilized leucine dehydrogenase-NADH oxidase reactor for the determination of branched-chain amino acids by high-performance liquid chromatography with chemiluminescence detection, *J.Chromatogr.A*, **1996**, *724*, 354–357.

SAMPLE
Matrix: blood
Analyte: glucose, 1,5-anhydroglucitol, and 3-hydroxybutyrate
Sample preparation: 10 μL Plasma + 190 μL 50 mM NaOH, filter (Advantec Q0100, 10000 molecular mass cut-off), inject a 50 μL aliquot of the ultrafiltrate.

HPLC VARIABLES
Column: 40 × 4 5 μm TSKgel SAX (Cl type) (Showa Denko, Tokyo)
Mobile phase: 50 mM NaOH containing 30 mM sodium butyrate (Protect mobile phase from carbon dioxide absorption with a soda lime guard column. At the start of the day wash column with 500 mM NaOH at 0.5 mL/min for 20 min and with mobile phase at 0.7 mL/min for 10 min.)
Column temperature: 40
Flow rate: 0.7
Injection volume: 50

Detector: Chemiluminescence following post-column reaction. The mobile phase flowed through an immobilized pyranose oxidase reactor and a co-immobilized 3-hydroxybutyrate dehydrogenase/NADH oxidase reactor at 40°. The effluent from the reactors mixed with 2 mM luminol in 400 mM pH 10 carbonate buffer pumped at 0.5 mL/min and 10 mM potassium ferricyanide in water pumped at 0.5 mL/min and the mixture flowed to the detector. (Store luminol solution in a refrigerator for 1 day before use. Prepare immobilized pyranose oxidase reactor as follows. Circulate 10 mL 1 mg/mL pyranose oxidase (EC 1.1.3.10 from Polyporus obtusus, 10 U/mg, Takara Shuzo, Kyoto) in 50 mM pH 7.0 phosphate buffer through a 40 × 4 column of tresylated poly(vinyl alcohol) beads at 0.2mL/min for 4 h. Prepare a co-immobilized 3-hydroxybutyrate dehydrogenase/NADH oxidase reactor as follows. Circulate 10 mL 0.5 mg/mL 3-hydroxybutyrate dehydrogenase (EC 1.1.1.30, Pseudomonas sp., grade III, 120 U/mg) in 50 mM pH 7.0 phosphate buffer containing 0.5 mg/mL NADH oxidase (from Bacillus megaterium, 50 U/mg, Toyobo, Osaka) through a 40 × 4 column of tresylated poly(vinyl alcohol) beads at 0.2 mL/min for 4 h. Prepare the columns of poly(vinyl alcohol) beads as follows. Wash 1 g 13 μm poly(vinyl alcohol) beads (GS-520, Showa Denko, Tokyo) with 50 mL dry acetone, suspend in 20 mL dry acetone:pyridine 50:50 with vigorous stirring, add 1 mL 2,2,2-trifluoroethanesulfonyl chloride dropwise over 2 min, stir for 10 min, wash beads with 10 mL acetone, wash beads with 20 mL 1 mM HCl, slurry pack in a 40 × 4 column (J. Chromatogr.A 1996, 724, 354).)

CHROMATOGRAM
Retention time: 1.5 (1,5-anhydroglucitol), 3 (glucose), 9 (3-hydroxybutyrate)
Limit of detection: 2 μg/mL

KEY WORDS
post-column reaction; plasma; ultrafiltrate; immobilized enzyme reactor

REFERENCE
Kiba, N.; Saegusa, K.; Furusawa, M. Post-column enzyme reactors for chemiluminometric detection of glucose, 1,5-anhydroglucitol and 3-hydroxybutyrate in an anion-exchange chromatographic system, *J.Chromatogr.B*, **1997**, *689*, 393–398.

SAMPLE
Matrix: dialysate
Analyte: acetylcholine
Sample preparation: Inject a 10 μL aliquot of dialysate onto a 55 × 1 reactor containing immobilized choline oxidase and catalase (BAS). (The enzymes destroy choline but not acetylcholine. Acetylcholine is eluted from the reactor onto the analytical column.)

HPLC VARIABLES
Column: 530 × 1 10 μm ACh (BAS)
Mobile phase: 28 mM pH 8.5 Na_2HPO_4 containing 0.5% of 1% Kathon antimicrobial solution (BAS)
Flow rate: 0.1
Injection volume: 10
Detector: E, Pt electrode +0.5 V, Ag/AgCl reference electrode following post-column reaction. The column effluent flowed through a 50 × 1 reactor packed with immobilized acetylcholine oxidase and choline oxidase (BAS) to the detector. (Acetylcholine was converted to hydrogen peroxide which was then detected.)

CHROMATOGRAM
Retention time: 13
Limit of quantitation: 10 nM
Limit of detection: 5 nM

KEY WORDS
post-column reaction; rat; immobilized enzyme reactor

REFERENCE
Tsai, T.-R.; Cham, T.-M.; Chen, K.-C.; Chen, C.-F.; Tsai, T.-H. Determination of acetylcholine by on-line microdialysis coupled with pre- and post-microbore column enzyme reactors with electrochemical detection, *J.Chromatogr.B*, **1996**, *678*, 151–155.

SAMPLE
Matrix: solutions
Analyte: acetylcholine
Sample preparation: solutions

HPLC VARIABLES
Column: 30 × 2.1 Aquapore AX300 (Brownlee) (Prepare the column by injecting ten 20 μL aliquots of 10 mg/mL choline oxidase (from Alcaligenes sp, 33 U/mg) and 100 μL cholinesterase (Type III, from electric eel, 970 U/mg, 0.65 mg/mL). Replenish enzymes every 1-2 weeks. Acetylcholine is converted to choline and choline is converted to hydrogen peroxide on the column.)
Mobile phase: 20 mM pH 7 Tris-acetate buffer containing 1 mM tetramethylammonium chloride and 200 μM octanesulfonate
Flow rate: 2
Detector: E, BAS LC4B, Pt working electrode +0.5 V, Ag/AgCl reference electrode

KEY WORDS
post-column reaction; immobilized enzyme reactor

REFERENCE
Meek, J.L.; Eva, C. Enzymes adsorbed on an ion exchanger as a post-column reactor: application to acetylcholine measurement, *J.Chromatogr.*, **1984**, *317*, 343–347.

SAMPLE
Matrix: tissue
Analyte: acetylcholine and choline
Sample preparation: Sonicate 250 mg rat brain tissue with 6 mL 1 M formic acid containing 10 nmoles IS for 5 min, centrifuge at 4° at 10000 g for 20 min, add the supernatant to an equal volume of diethyl ether, add 5 mL water, shake, centrifuge at 2000 g for 5 min, discard the organic layer. Lyophilize the aqueous layer, dissolve the residue in 400 μL water, filter (0.45 μm). Add 30 μL reagent to the filtrate, mix, centrifuge at 10000 g for 5 min. Dissolve the precipitate in 300 μL water, add 50 mg Dowex 1x8, shake, centrifuge at 10000 g for 5 min, inject a 10 μL aliquot of the supernatant. (Reagent contained 20% KI and 18% iodine in water.)

HPLC VARIABLES
Column: 150 × 4.6 Nucleosil C18
Mobile phase: Buffer (Prepare buffer by dissolving 1.36 g sodium acetate. 3.72 disodium EDTA, 25 mg sodium octyl sulfate, and 1.2 mmoles tetramethylammonium chloride in 900 mL water, adjust pH to 5.0 with 200 mM citric acid, make up to 1 L.)
Column temperature: 37
Flow rate: 0.8
Injection volume: 10
Detector: E, Bioanalytical Systems LC-4B/17, TL-10A platinum electrode +500 mV, Ag/AgCl reference electrode following post-column reaction. The effluent from the column mixed with buffer pumped at 0.5 mL/min and the mixture flowed through an immobilized enzyme reactor to the detector. (Prepare buffer by dissolving 71.6 g $Na_2HPO_4.12H_2O$ and 372 mg disodium EDTA in 900 mL water, adjust pH to 8.5 with NaH_2PO_4, make up to 1 L. Prepare reactor by heating 200-400 mesh porous glass beads (pore size 400 Å, Electronucleonics, Fairfield NJ) in 5% nitric acid, wash with water, dry, add to 10% 3-aminopropyltriethoxysilane in toluene, reflux overnight. Suspend the beads in 2% glutaraldehyde in water at room temperature for 2 h. Dissolve 0.45 mg acetylcholinesterase (type III, EC.3.1.1.7, Sigma) and 16.6 mg choline oxidase (EC.1.1.3.17, Sigma) in 200 μL 50 mM pH 7.0 phosphate buffer containing 10 mM disodium EDTA, add 500 mg activated beads, pack in a 10 × 4 tube. Acetylcholine and choline are converted to hydrogen peroxide which is then detected.)

CHROMATOGRAM
Retention time: 6 (choline), 10 (acetylcholine)
Internal standard: ethylhomocholine (Prepare ethylhomocholine by adding 3-dimethylamino-1-propanol to EtOH, add bromoethane. When reaction is complete add ether, filter off the precipitate and wash it with ether.) (8)

Limit of detection: 100 fmoles

KEY WORDS
post-column reaction; rat; brain; immobilized enzyme reactor

REFERENCE
Asano, M.; Miyauchi, T.; Kato, T.; Fujimori, K.; Yamamoto, K. Determination of acetylcholine and choline in rat brain tissue by liquid chromatography/electrochemistry using an immobilized enzyme post column reactor, *J.Liq.Chromatogr.*, **1986**, *9*, 199–215.

SAMPLE
Matrix: tissue
Analyte: acetylcholine and choline
Sample preparation: Sonicate rat brain with ten volumes of 1 M formic acid:acetone 15:85 containing IS, centrifuge at 4° at 20000 g. Remove a 500 μL aliquot of the supernatant and add it to 2 mL heptane:chloroform 80:10, vortex. Remove the aqueous layer and add it to 250 μL 3 mg/mL sodium tetraphenylboron in 3-heptanone, vortex. Remove a 200 μL aliquot of the upper organic layer and add it to 50 μL 1 M HCl, vortex. Remove the aqueous layer and evaporate it to dryness under reduced pressure, reconstitute with mobile phase, inject a 30 μL aliquot.

HPLC VARIABLES
Guard column: C18 (Waters)
Column: 250 × 4.6 5 μm Hypersil ODS
Mobile phase: 100 mM pH 7 KH_2PO_4 containing 10 μg/mL sodium octane sulfate and 600 μg/mL tetramethylammonium chloride
Flow rate: 1
Injection volume: 30
Detector: E, Chromatofield, Pt electrode +0.5 V following post-column reaction. The column effluent flowed through an immobilized enzyme reactor to the detector. (Prepare reactor by heating 200-400 mesh porous glass beads (pore size 350 Å, Sigma) in 5% nitric acid at 100° for 1 min, wash with water, dry, add to 10% 3-aminopropyltriethoxysilane in toluene, heat at 115° for 12 h. Suspend the beads in 2% glutaraldehyde in water at room temperature for 2 h. Dissolve 100 U acetylcholinesterase (type III, electric eel, Sigma) and 100 U choline oxidase (Alcaligenes, Sigma) in 1 mL 50 mM pH 7 phosphate buffer, add 120 mg activated beads, shake periodically, pack in a 20 × 2 tube. Acetylcholine and choline are converted to hydrogen peroxide which is then detected.)

CHROMATOGRAM
Retention time: 4.5 (choline), 9.5 (acetylcholine)
Internal standard: ethylhomocholine bromide (Prepare ethylhomocholine by adding 3-dimethylamino-1-propanol to EtOH, add bromoethane, when reaction is complete add ether, filter off the precipitate and wash it with ether.) (7)

KEY WORDS
rat; brain; post-column reaction; immobilized enzyme reactor

REFERENCE
Beley, A.; Zekhnini, A.; Lartillot, S.; Fage, D.; Bralet, J. Improved method for determination of acetylcholine, choline, and other biogenic amines in a single brain tissue sample using high performance liquid chromatography and electrochemical detection, *J.Liq.Chromatogr.*, **1987**, *10*, 2977–2992.

SAMPLE
Matrix: tissue
Analyte: acetylcholine and choline
Sample preparation: Homogenize (Nissei US-300T ultrasonic cell disrupter at 300 W and 20 kHz for 1 min) rat brain striatal tissue with 1 mL 1 μM ethylhomocholine in 50 mM perchloric acid, centrifuge at 4° at 20000 g for 15 min, filter (0.45 μm) the supernatant, inject a 10 μL aliquot.

HPLC VARIABLES
Guard column: 10 × 4 74-149 μm plastic formed carbon (details in paper)
Column: 60 × 4 3 μm Acetylcholine Separation polymeric styrene column (Bioanalytical Systems)

Mobile phase: 50 mM pH 8.40 Phosphate buffer containing 1 mM disodium EDTA and 0.40 mM sodium 1-octanesulfonate
Column temperature: 35 ± 1
Flow rate: 0.7
Injection volume: 10
Detector: E, Bioanalytical Systems LC-4A, dual Pt working electrode +500 mV, Ag/AgCl reference electrode following post-column reaction. The effluent from the column flowed through a 5 × 4 immobilized enzyme reactor containing acetylcholinesterase and choline oxidase (Bioanalytical Systems) to the detector.

CHROMATOGRAM
Retention time: 3 (choline), 12.5 (acetylcholine)
Internal standard: ethylhomocholine (6.5)

OTHER SUBSTANCES
Non-interfering: 3,4-dihydroxybenzylamine, dopamine, norepinephrine

KEY WORDS
rat; brain; guard-column removes interferences from catecholamines; post-column reaction; immobilized enzyme reactor

REFERENCE
Ikarashi, Y.; Blank, C.L.; Suda, Y.; Kawakubo, T.; Maruyama, Y. Application of a novel, plastic formed carbon as a precolumn packing material for the liquid chromatographic determination of acetylcholine and choline in biological samples, *J.Chromatogr.A*, **1995**, *718*, 267–272.

SAMPLE
Matrix: tissue (fish muscle)
Analyte: 17α-methyltestosterone
Sample preparation: Homogenize 1-5 g tissue in 10-50 mL chloroform:MeOH 2:1, add 0.2 volume water, centrifuge at 1800 g for 10 min, extract aqueous phase again with 2 volumes of chloroform, evaporate organic phases under vacuum, dissolve residues in MeCN, pass through a Lipidex 5000 column and a Sep Pak silica cartridge, inject an aliquot.

HPLC VARIABLES
Guard column: 25 × 3.8 5 μm ODS 2
Column: 150 × 4.6 5 μm Hypersil C8 plus 150 × 4.6 5 μm Nucleosil C18
Mobile phase: MeOH:water 75:25
Flow rate: 0.8
Injection volume: 50
Detector: F ex 340 em 470 following post-column reaction. The eluate from the column mixed with freshly prepared 0.6 mM nicotinamide-adenine dinucleotide (NAD) in 20 mM pH 9 pyrophosphate buffer pumped at 1.2 mL/min in a 950 μL static mixing chamber, the mixture then flowed through an immobilized enzyme reactor to the detector. The Chrom Sep immobilized enzyme reactor was prepared by pumping 10 mM pH 7.8 KH_2PO_4 through it at 0.8 mL/min. Ten 100 μL samples of a solution of 25 U/mL 3α-hydroxysteroid dehydrogenase in 10 mM pH 7.8 KH_2PO_4 were injected onto the reactor which was then equilibrated with 10 mM pH 7.8 KH_2PO_4 for 20 min. The reactor was then attached to the chromatographic system.

CHROMATOGRAM
Retention time: 21.2 (5β-17-methyltestosterone), 22.6 (5α-17-methyltestosterone)
Limit of detection: 1 ng/g

KEY WORDS
immobilized enzyme reactor; fish; muscle; SPE; post-column reaction

REFERENCE
Cravedi, J.P.; Delous, G. Use of an immobilized enzyme reactor for the analysis of residues of 17α-methyltestosterone in trout by high-performance liquid chromatography, *J.Chromatogr.*, **1991**, *564*, 461–467.

SAMPLE
Matrix: tissue, urine

Analyte: polyamines
Sample preparation: Urine. 1 mL Urine + 1 mL 6 M HCl, reflux with stirring for 3 h, filter. Neutralize the filtrate with 2 M NaOH, dilute to 20 mL with water, add 50 mg sodium octanesulfonate, add to a Sep Pak C18 SPE cartridge, wash with 50 mL water, elute with 500 µL 0.5 N KH$_2$PO$_4$, inject a 20 µL aliquot of the eluate. Tissue. Homogenize rat brain with 200 mM perchloric acid containing 0.002% EDTA, centrifuge at 10000 g for 20 min, filter (0.45 µm) the supernatant, inject an aliquot.

HPLC VARIABLES

Column: 150 × 6 Biophase-III polymer-based C18 (BAS)
Mobile phase: 200 mM pH 7.7 phosphate buffer containing 0.05% sodium octanesulfonate
Flow rate: 0.4
Injection volume: 20
Detector: E, Bioanalytical Systems LC-4B, Pt working electrode following post-column reaction. The column effluent flowed through a 50 × 2 immobilized enzyme reactor to the detector and the amines were converted to hydrogen peroxide which was then detected. (Prepare reactor as follows. Dissolve 25 mg polyamine oxidase (from soybean seedlings, 30000 U/g (Nagoya J. Med. Sci. 1984, 46, 87) in 10 mL 100 mM pH 6.8 phosphate buffer, add 500 mg 200-400 mesh 400 Å controlled-pore glass beads (Electro Nucleonics, Fairfield NJ) activated with glutaraldehyde, stir for 24 h, filter, wash, store in the refrigerator.)

CHROMATOGRAM

Retention time: 12 (putrescine), 18 (spermidine), 33 (spermine)
Limit of detection: 0.3-4 pmole

KEY WORDS

post-column reaction; brain; rat; human; SPE; immobilized enzyme reactor

REFERENCE

Watanabe, N.; Asano, M.; Yamamoto, K.; Nagatsu, T.; Matsumoto, T.; Fujita, K. High performance liquid chromatography of biological polyamines using immobilized enzyme as post-column reactor followed by electrochemical detection, *Biomed.Chromatogr.*, **1989**, *3*, 187–191.

SAMPLE

Matrix: urine
Analyte: trimethoprim metabolites
Sample preparation: Dilute urine 10 times with water, inject a 10 µL aliquot.

HPLC VARIABLES

Column: 250 × 4.6 Spherisorb S5-C8
Mobile phase: MeCN:buffer 15:85 containing 700 µM tetrabutylammonium hydrogen sulfate (Buffer was 100 mM KH$_2$PO$_4$ adjusted to pH 7.5.)
Flow rate: 1
Injection volume: 10
Detector: E, Metrohm Model 656/641, glassy carbon working electrode 700 mV, glassy carbon auxiliary electrode, Ag/AgCl/KCl reference electrode, following post-column reaction. The column effluent flowed through an immobilized enzyme reactor to the detector. (Prepare the immobilized enzyme reactor as follows. Reflux 550 Å controlled-pore glass with 3-aminopropyl triethoxysilane in toluene, dry, pour into a 50 × 3 column, inject five 1 mL portions of glutaraldehyde at 1 mL/min, inject three 100 µL portions of 100-200 mg/mL E. coli glucuronidase (Boehringer) at 0.1 mL/min. The mobile phase was either water or 50 mM pH 7.0 phosphate buffer.)

CHROMATOGRAM

Retention time: 11.75 (trimethoprim M1 glucuronide), 14.55 (trimethoprim M4 glucuronide)
Internal standard: sulfamethoxazole

KEY WORDS

post-column reaction; immobilized enzyme reactor

REFERENCE

Dalgaard, L. Pre- and post-column enzymic hydrolysis and amperometric detection of glycosides. Applications to trimethoprim metabolites and cyanogenic glycosides, *Prog.HPLC*, **1987**, *2*, 219–233.

Iodine

RELATED REFERENCES

Dorner, J.W.; Cole, R.J. Rapid determination of aflatoxins in raw peanuts by liquid chromatography with postcolumn iodination and modified minicolumn cleanup. *J.Assoc.Off.Anal.Chem.* **1988**, *71*, 43-47.

Garner, R.C.; Whattam, M.M.; Taylor, P.J.L.; Stow, M.W. Analysis of United Kingdom purchased spices for aflatoxins using an immunoaffinity column clean-up procedure followed by high-performance liquid chromatographic analysis and post-column derivatization with pyridinium bromide perbromide. *J.Chromatogr.* **1993**, *648*, 485-490.

Jansen, H.; Jansen, R.; Brinkman, U.A.T.; Frei, R.W. Fluorescence enhancement for aflatoxins in HPLC by post-column split-flow iodine addition from a solid-phase iodine reservoir. *Chromatographia* **1987**, *24*, 555-559.

I_2 ⟶ post-column reaction

SAMPLE

Matrix: feed

Analyte: aflatoxin B_1

Sample preparation: 25 g Feed + 12.5 g Celite + 12.5 mL water + 125 mL chloroform, shake for 30 min, filter (paper). Evaporate the filtrate to dryness under a stream of nitrogen at 50°, reconstitute the residue in 200 μL chloroform, vortex, inject a 20 μL aliquot. (Treat Celite 545 overnight with HCl:water 1:1, filter (paper), wash with water until the washings are neutral, dry at 110°.)

HPLC VARIABLES

Guard column: present but not specified

Column: 100 × 4.6 Cp Microsphere C18 (Chrompack)

Mobile phase: MeCN:water 30:70

Column temperature: 60

Flow rate: 0.5

Injection volume: 20

Detector: F ex 360 (filter) em >420 (filter) following post-column reaction. The column effluent mixed with the reagent pumped at 0.4 mL/min and the mixture flowed through a 3 m × 0.5 mm ID PTFE reaction coil at 60° to the detector. (Prepare the reagent by stirring 1 g iodine in 200 mL water for 15 min then filtering (paper).)

CHROMATOGRAM

Retention time: 7.5

Limit of detection: 1 ng/g

KEY WORDS

post-column reaction; protect from light

REFERENCE

Tuinstra, L.G.M.T.; Haasnoot, W. Rapid determination of aflatoxin B_1 in Dutch feeding stuffs by high-performance liquid chromatography and post-column derivatization, *J.Chromatogr.*, **1983**, *282*, 457–462.

SAMPLE

Matrix: feed

Analyte: aflatoxins

Sample preparation: Condition a Florisil Sep-Pak SPE cartridge with 10 mL chloroform. Condition a C18 Sep-Pak SPE cartridge with 10 mL acetone and 10 mL water. 50 g Ground feed + 25 g acid-washed Celite 545 + 250 mL chloroform + 25 mL water, shake on a wrist-action shaker for 30 min, filter (paper), add 2 mL chloroform to the top of the Florisil cartridge, add the filtrate on top of this, pass the mixture through the cartridge, wash with 5 mL chloroform, wash with 20 mL MeOH, elute with 15 mL acetone:water 90:10. Remove the acetone under reduced pressure at 40-50°, add 20 mL water to the concentrated eluate. Add 2 mL water to the top of the SPE cartridge, add the diluted eluate on top of this, pass the mixture through the cartridge, rinse flask with two 2.5 mL portions of acetone:water 10:90, add the rinses to the cartridge, elute with 50 mL acetone:water 15:85, make up the eluate to 50 mL, inject a 250 μL aliquot.

HPLC VARIABLES
Column: 150 × 4.6 5 μm Lichrosorb RP-18
Mobile phase: MeCN:MeOH:water 20:35:65
Flow rate: 0.5
Injection volume: 250
Detector: F ex 365 em 400 following post-column reaction. The column effluent mixed with the reagent pumped at 0.4 mL/min and the mixture flowed through a 3 m × 0.5 mm ID PTFE coil at 60° to the detector. (Prepare reagent by mixing 2 g iodine in 400 mL water for 30 min, filter (0.45 μm).)

CHROMATOGRAM
Retention time: 13 (aflatoxin G_2), 15 (aflatoxin G_1), 17 (aflatoxin B_2), 20 (aflatoxin B_1)
Limit of detection: <1 ng/g

KEY WORDS
post-column reaction; Soak glassware in dilute acid for several h, wash with water until the washings are acid free.; protect from light; SPE

REFERENCE
Paulsch, W.E.; Sizoo, E.A.; van Egmond, H.P. Liquid chromatographic determination of aflatoxins in feedstuffs containing citrus pulp, *J.Assoc.Off.Anal.Chem.*, **1988**, 71, 957–961.

SAMPLE
Matrix: feed
Analyte: aflatoxins
Sample preparation: Shake 50 g finely ground feed with 250 mL MeCN:4% KCl in water:20% phosphoric acid 89:10:1 at high speed on a wrist-action shaker for 1 h, filter (paper). Remove a 10 mL aliquot of the filtrate and add it to 36 mL water, mix well, filter (0.45 μm), inject a 300 μL aliquot on to column A and elute to waste with mobile phase A, after 2 min elute the contents of column A on to column B with mobile phase B and start the gradient, after 1 min remove column A from the circuit, elute column B with mobile phase B and monitor the effluent from column B. Backflush column A with MeCN:MeOH:water 5:5:90 for 27 min, re-equilibrate column A with mobile phase A for 9 min.

HPLC VARIABLES
Column: A 50 × 4.6 10 μm Adsorbosphere C18; B 150 × 4.6 5 μm Econosphere C18
Mobile phase: A MeCN:MeOH:buffer 5:5:90; B Gradient. X was MeCN:MeOH:buffer 25:25:50. Y was MeCN:MeOH 50:50. X:Y 95:5 for 7 min, to 5:95 over 7 min (Waters curve no. 10), to 0:100 over 2 min (Waters curve no. 6), maintain at 0:100 for 14 min, return to initial conditions over 2 min (Waters curve no. 6), maintain at initial conditions for 8 min. (Buffer was 5 mM Na_2HPO_4 adjusted to pH 3.7 ± 0.1 with 20% phosphoric acid. At the end of the day flush with 100 mL water and store in MeOH:water 80:20.)
Flow rate: A 1.5; B 0.5 for 32 min, to 1 over 1 min (Waters curve no. 6), maintain at 1 for 6 min, return to 0.5 over 1 min (Waters curve no. 6)
Injection volume: 300
Detector: F ex 360 em 425 following post-column reaction. The column effluent mixed the reagent pumped at 0.5 mL/min and the mixture flowed through a 25 cm × 0.5 mm ID knitted PTFE coil at 90° to the detector. (Prepare the reagent, a saturated solution of iodine in water, by stirring 200 mg iodine in 500 mL water for 10 min and then filtering (0.45 μm).)

CHROMATOGRAM
Retention time: 11.5 (G_2), 12.5 (G_{2a}), 13 (B_2), 14.5 (B_{2a})
Limit of quantitation: 5 ppb

OTHER SUBSTANCES
Simultaneously analyzed: ochratoxin (response is greater without post-column reaction)
Non-interfering: zearalenone

KEY WORDS
post-column reaction

REFERENCE
Chamkasem, N.; Cobb, W.Y.; Latimer, G.W.; Salinas, C.; Clement, B.A. Liquid chromatographic determination of aflatoxins, ochratoxin A, and zearalenone in grains, oilseeds, and animal feeds by post-
column derivatization and on-line sample cleanup, *J.Assoc.Off.Anal.Chem.*, **1989**, *72*, 336–341.

SAMPLE
Matrix: feed
Analyte: aflatoxins
Sample preparation: Blend 50 g feed, 5 g NaCl, and 140 mL MeOH:water 70:30 at high
speed for 2 min, filter (Whatman No. 4 paper). Dilute a 12 mL aliquot of the filtrate with
30 mL water, add to an Aflatest P monoclonal affinity SPE column (Vicam, Somerville
MA), wash with water, elute with 1 mL MeOH. Evaporate the eluate to 0.5 mL with a
stream of nitrogen, make up to 1 mL with water, inject a 50 μL aliquot.

HPLC VARIABLES
Guard column: present but not specified
Column: 250 × 4.6 5 μm C18 (Supelco)
Mobile phase: MeCN:MeOH:water 10:40:50
Flow rate: 0.8
Injection volume: 50
Detector: F ex 365 em 440 following post-column reaction. The column effluent mixed with
the reagent pumped at 0.5 mL/min and the mixture flowed through a 5 m × 0.3 mm ID
stainless steel coil at 68° to the detector. (Prepare reagent by shaking 1 g iodine in 500
mL water in the dark for 2 h, filter (0.45 μm).)

CHROMATOGRAM
Retention time: 9 (aflatoxin G_2), 10 (aflatoxin G_1), 12 (aflatoxin B_2), 14 (aflatoxin B_1)
Limit of detection: 0.25 ppb

KEY WORDS
post-column reaction; SPE; post-column reaction only enhances peaks for B_1; and G_1

REFERENCE
Holcomb, M.; Thompson, H.C. Jr. Analysis of aflatoxins (B_1, B_2, G_1, and G_2) in rodent feed by HPLC
using postcolumn derivatization and fluorescence detection, *J.Agric.Food Chem.*, **1991**, *39*, 137–140.

Lead(IV) Oxide

PbO_2 $\longrightarrow$ post-column reaction

SAMPLE
Matrix: tissue
Analyte: malachite green and gentian violet
Sample preparation: Condition a 6 mL 1 g neutral alumina SPE cartridge (J.T. Baker)
with 5 mL MeCN. Condition a 2.8 mL 0.5 g Bond Elut PRS SPE cartridge with 5 mL
MeCN. 20 g Muscle tissue + 3 mL 250 mg/mL hydroxylamine hydrochloride in water +
5 mL 50 mM p-toluenesulfonic acid in water + 20 mL 100 mM pH 4.5 ammonium acetate
buffer, homogenize (Tekmar Ultra-Turrax T25 tissuemizer) at 20000 rpm for 1 min, add
90 mL MeCN, homogenize for 10 s, shake vigorously by hand for 1 min, add 20 g basic
alumina (Brockman activity I, Fisher Scientific), shake vigorously for 1 min, centrifuge,
decant, add 30 mL MeCN to the residue, extract, centrifuge, decant. Combine the supernatants and add 100 mL water, add 50 mL dichloromethane, add 2 mL diethylene glycol,
shake vigorously by hand for 1 min, let stand for 45 min, remove the lower organic layer,
add 50 mL dichloromethane, shake for 1 min, let stand for 5 min, remove the lower
organic layer, repeat the extraction with 50 mL dichloromethane. Combine the organic
layers and evaporate them to 2-5 mL under reduced pressure at 65°, add 2 mL dichloromethane, add 5 mL MeCN, add the mixture to the alumina SPE cartridge on top of the
PRS SPE cartridge, rinse the flask with two 5 mL portions of MeCN, add the rinses to
the SPE cartridges, wash the SPE cartridges with 5 mL MeCN. Discard the alumina
cartridge, wash the PRS cartridge with 2 mL water, wash with 1 mL solvent, elute with

2 mL solvent, add 500 µL 2.5 mg/mL hydroxylamine hydrochloride in water to the eluate, inject a 100 µL aliquot. (Solvent was MeCN:100 mM ammonium acetate 50:50, adjusted to pH 4.5 with glacial acetic acid.)

HPLC VARIABLES

Guard column: 20 × 2 pellicular C18
Column: 150 × 4.6 5 µm SynChropak SCD-100 (SynChrom)
Mobile phase: MeCN:buffer 55:45 (Prepare buffer by dissolving 400 mg ammonium acetate and 1 mL triethylamine in 400 mL water, adjust to pH 3.6 with glacial acetic acid, make up to 450 mL with water.)
Flow rate: 2
Injection volume: 100
Detector: UV 588 following post-column oxidation. The column effluent passed through a 20 × 2 column packed with lead(IV) oxide (Mallinckrodt) to the detector.

CHROMATOGRAM

Retention time: 4.0 (leucomalachite green), 4.3 (leucogentian violet), 5.3 (malachite green), 6.1 (methyl violet), 8.2 (gentian violet)
Limit of quantitation: 0.5-3 ng/g
Limit of detection: 0.3-1.8 ng/g

KEY WORDS

fish; muscle; catfish; trout; SPE; post-column reaction

REFERENCE

Rushing, L.G.; Thompson, H.C. Jr. Simultaneous determination of malachite green, gentian violet and their leuco metabolites in catfish and trout tissue by high-performance liquid chromatography with visible detection, *J.Chromatogr.B*, **1997**, *688*, 325–330.

Manganese Dioxide

$MnO_2 \longrightarrow$ post-column reaction

SAMPLE

Matrix: tissue
Analyte: shellfish toxins
Sample preparation: Boil 10 g shellfish tissue with 10 mL 100 mM HCl for 5 min, cool, centrifuge. Pass a 1 mL aliquot of the supernatant through a C18 SPE cartridge. wash with 2 mL water. Collect all the effluent from the SPE cartridge, dilute 7-10 times, inject an aliquot.

HPLC VARIABLES

Column: 150 × 4.1 10 µm Hamilton PRP-1
Mobile phase: Gradient A was MeCN:water 1:99 containing 1.5 mM ammonium phosphate and 3 mM heptanesulfonic acid, pH 7. B was MeCN:water 25:75 containing 6.25 mM ammonium phosphate and 3 mM heptanesulfonic acid, pH 7. A:B from 97:3 to 77:23 over 7.5 min, to 0:100 over 8 min.
Column temperature: 45
Flow rate: 0.8
Injection volume: 50
Detector: F ex 330 em 400 following post-column reaction. The column effluent mixed with buffer pumped at 0.3 mL/min and the mixture flowed through a 75 × 4.6 column packed with 45-75 µm manganese dioxide heated to 85°. The effluent from this column mixed with 80 mM nitric acid pumped at 0.3 mL/min and flowed to the detector. (Buffer was 100 mM Na_2HPO_4 adjusted to pH 11.5 with KOH.)

CHROMATOGRAM

Retention time: 6 (gonyautoxin-3), 9 (gonyautoxin-2), 14 (neosaxitoxin), 17 (saxitoxin)
Limit of detection: 0.1-2 ng

KEY WORDS
post-column reaction; shellfish

REFERENCE
Lawrence, J.F.; Wong, B. Development of a manganese dioxide solid-phase reactor for oxidation of toxins associated with paralytic shellfish poisoning, *J.Chromatogr.A*, **1996**, *755*, 227–233.

2-Methyl-1,4-naphthoquinone

post-column reaction

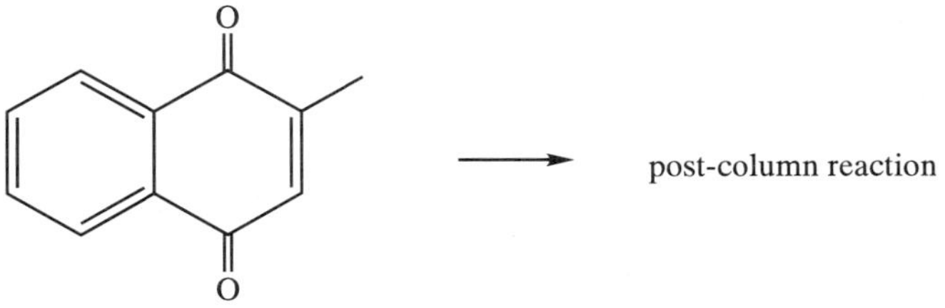

SAMPLE
Matrix: oil
Analyte: fatty acids
Sample preparation: Mix oil with mobile phase, centrifuge, inject a 20 μL aliquot of the supernatant.

HPLC VARIABLES
Column: 250 × 4 5 μm LiChrospher 100 RP-18
Mobile phase: MeCN:EtOH 90:10
Flow rate: 1.1
Injection volume: 20
Detector: E, Jasco Model EC-840, glassy carbon working electrode -415 mV, saturated calomel reference electrode, stainless steel auxiliary electrode following post-column reaction. The column effluent mixed with 76 mM lithium perchlorate in MeCN:EtOH 90:10 containing 6 mM 2-methyl-1,4-naphthoquinone (menadione, vitamin K_3) pumped at 1.1 mL/min and the mixture flowed through a 50 cm × 0.5 mm ID coil to the detector.

CHROMATOGRAM
Retention time: 5.5 (linoleic acid), 7.5 (oleic acid), 8 (palmitic acid), 11.5 (stearic acid)
Limit of detection: 20 pmole

KEY WORDS
post-column reaction; camellia oil; corn oil; rapeseed oil; olive oil; soy bean oil

REFERENCE
Fuse, T.; Kusu, F.; Takamura, K. Determination of higher fatty acids in oils by high-performance liquid chromatography with electrochemical detection, *J.Chromatogr.A*, **1997**, *764*, 177–182.

2-Naphthoyltrifluoroacetone

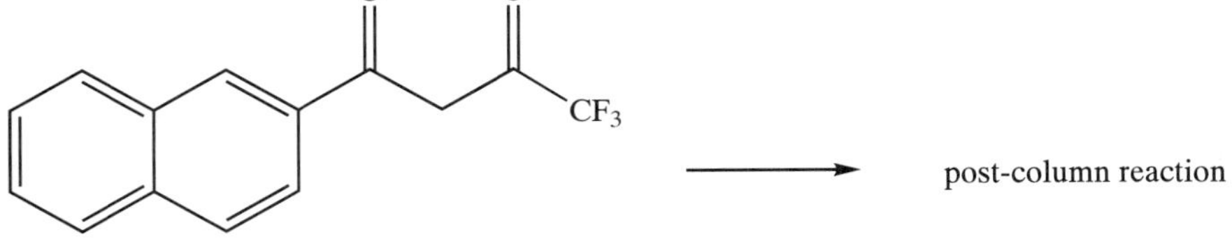

post-column reaction

SAMPLE
Matrix: solutions

Analyte: amines
Sample preparation: Mix a 100 μL aliquot of a 0.1-1 μM solution of amines with 50 μL 100-500 μM reagent and 50 μL 100 mM pH 9.8 carbonate buffer, heat at 40° for 8 h, inject a 20 μL aliquot. (The reagent was the Eu^{3+} chelate of N^1-(p-isothiocyanatobenzyl)diethylenetriaminetetraacetic acid; synthesis is as follows. Dissolve 100 g diethylenetriamine in toluene, add 48 g 4-nitrobenzyl bromide, stir at room temperature for 5 h, filter, extract the filtrate with water. Make the aqueous base alkaline and saturate it with NaCl, extract with chloroform, evaporate to dryness to give a mixture of N-(p-nitrobenzyl)diethylenetriamines. Dissolve this mixture in toluene, stir at 0°, slowly add 55 g salicylaldehyde (2-hydroxybenzaldehyde), reflux under a Dean and Stark trap to remove water, evaporate to remove the toluene, dissolve the residue in hot EtOH, allow to cool slowly to room temperature with vigorous stirring, filter. Dissolve the solid in warm EtOH, add HCl, add water, wash with diethyl ether. Make the aqueous layer alkaline and saturate it with NaCl, extract with chloroform, evaporate to dryness to obtain N^1-(p-nitrobenzyl)diethylenetriamine. Dissolve 53 g N^1-(p-nitrobenzyl)diethylenetriamine in water, adjust pH to 9-11, slowly add a solution of 160 g bromoacetic acid in water while adding 7 M KOH to maintain the pH at 9-11, stir at pH 9-11 for 24 h, acidify, concentrate by evaporation, add acetone to precipitate most of the salts, evaporate the precipitate to dryness to obtain crude N^1-(p-nitrobenzyl)diethylenetriaminetetraacetic acid. Dissolve N^1-(p-nitrobenzyl)diethylenetriaminetetraacetic acid in water, adjust pH to 5-6, add an equimolar amount of Eu^{3+} in water, maintain pH at 5-6, stir at room temperature for 30 min, raise the pH to 9, filter, concentrate the filtrate by evaporation, purify by preparative HPLC using water as eluent to obtain the Eu^{3+} chelate of N^1-(p-nitrobenzyl)diethylenetriaminetetraacetic acid. Dissolve 2 g of the chelate in water, add 200 mg 5% palladium on activated charcoal, hydrogenate at 0-5° under 1.3 MPa hydrogen, monitor the reaction by UV spectroscopy. When the reaction is complete filter and evaporate the filtrate to dryness to obtain the Eu^{3+} chelate of N^1-(p-aminobenzyl)diethylenetriaminetetraacetic acid. Add a solution of 2.3 g of the Eu^{3+} chelate of N^1-(p-aminobenzyl)diethylenetriaminetetraacetic acid in water to a mixture of 1.7 g thiophosgene, 1 g sodium bicarbonate, and 15 mL chloroform, stir vigorously for 30 min. Wash the aqueous phase twice with chloroform and evaporate it to dryness, wash the product with EtOH to obtain the Eu^{3+} chelate of N^1-(p-isothiocyanatobenzyl)diethylenetriaminetetraacetic acid (Anal. Biochem. 1989, 176, 319).)

HPLC VARIABLES

Column: 250 × 4.6 5 μm Capcellpak SG-120 C18 (Shiseido)
Mobile phase: MeCN:20 mM pH 7.0 phosphate buffer 23:77
Flow rate: 1
Injection volume: 20
Detector: F ex 344 em 617 following post-column reaction. The column effluent mixed with the reagent pumped at 0.2 mL/min and the mixture flowed through a 2 m coil to the detector. (Prepare reagent by dissolving 15 g Triton X-100 in 500 mL water, add 58 mL acetic acid, add 20 mL 50 mM tri-n-octylphosphine oxide in MeCN, add 1 mL 100 mM 2-naphthoyltrifluoroacetone in MeCN, make up to 1 L with water. 2-Naphthoyltrifluoroacetone (4,4,4-trifluoro-1-(2-naphthyl)-1,3-butanedione) is available from Tokyo Kasei (TCI America, Portland OR). Prepare by adding 1 mole ethyl trifluoroacetate dropwise with stirring to 1.05 moles sodium methoxide in 100 mL anhydrous ether, add 1 mole 2'-acetonaphthone, let stand overnight, evaporate to dryness under reduced pressure, heat at 90° under reduced pressure to remove EtOH, add 1 mole 10% sulfuric acid, recrystallize from EtOH/water to obtain 2-naphthoyltrifluoroacetone (mp 70.1-71.1°) (J. Am. Chem. Soc. 1950, 72, 2948).); UV 254

CHROMATOGRAM

Retention time: 6 (phenethylamine), 10 (phenylpropylamine), 16 (phenylbutylamine)

KEY WORDS

post-column reaction

REFERENCE

Okabayashi, Y.; Kitagawa, T. High-performance liquid chromatography for amino compounds and thiol compounds derivatized with europium chelate, *Anal.Chem.*, **1994**, *66*, 1448–1453.

SAMPLE
Matrix: solutions
Analyte: thiols
Sample preparation: Mix a 100 μL aliquot of a 10-100 nM solution of thiols with 50 μL 10-100 μM reagent and 50 μL 100 mM pH 9 borate buffer, let stand at room temperature for 2 min, add 50 μL 1 M Bis-Tris, inject a 20 μL aliquot. (The reagent was the Eu^{3+} chelate of 1-[p-(((5-maleimidopentyl)carbonyl)amino)benzyl]ethylenediaminetetraacetic acid (Dojindo, Kumamoto, Japan (Dojindo Molecular Technologies, Inc., 3 Bethesda Metro Center, Suite 700, Bethesda MD 20814; (301) 664-8448; www.dojindo.co.jp).)

HPLC VARIABLES
Column: 250 × 4.6 5 μm Capcellpak SG-120 C18 (Shiseido)
Mobile phase: MeCN:20 mM pH 7.0 phosphate buffer 8:92
Flow rate: 1
Injection volume: 20
Detector: F ex 344 em 617 following post-column reaction. The column effluent mixed with the reagent pumped at 0.2 mL/min and the mixture flowed through a 2 m coil to the detector. (Prepare reagent by dissolving 15 g Triton X-100 in 500 mL water, add 58 mL acetic acid, add 20 mL 50 mM tri-n-octylphosphine oxide in MeCN, add 1 mL 100 mM 3-naphthoyltrifluoroacetone in MeCN, make up to 1 L with water. 2-Naphthoyltrifluoroacetone (4,4,4-trifluoro-1-(2-naphthyl)-1,3-butanedione) is available from Tokyo Kasei (TCI America, Portland OR). Prepare by adding 1 mole ethyl trifluoroacetate dropwise with stirring to 1.05 moles sodium methoxide in 100 mL anhydrous ether, add 1 mole 2'-acetonaphthone, let stand overnight, evaporate to dryness under reduced pressure, heat at 90° under reduced pressure to remove EtOH, add 1 mole 10% sulfuric acid, recrystallize from EtOH/water to obtain 2-naphthoyltrifluoroacetone (mp 70.1-71.1°) (J. Am. Chem. Soc. 1950, 72, 2948).)

CHROMATOGRAM
Retention time: 12 (cysteine), 15 (homocysteine), 19 (coenzyme A)
Limit of detection: 13.2-18.8 fmole

KEY WORDS
post-column reaction

REFERENCE
Okabayashi, Y.; Kitagawa, T. High-performance liquid chromatography for amino compounds and thiol compounds derivatized with europium chelate, *Anal.Chem.*, **1994**, *66*, 1448–1453.

Ninhydrin

RELATED REFERENCES
Candito, M.; Bedoucha, P.; Mahagne, M.H.; Scavini, G.; Chatel, M. Total plasma homocysteine by liquid chromatography before and after methionine loading. Results in cerebrovascular disease. *J.Chromatogr.B* **1997**, *692*, 213-216.

Cunico, R.L.; Schlabach, T. Comparison of ninhydrin and o-phthalaldehyde post-column detection techniques for high-performance liquid chromatography of free amino acids. *J.Chromatogr.* **1983**, *266*, 461-470.

Dhoot, J.S.; Appel, B.R.; Del Rosario, A.R. Confirmation of domoic acid in seafood using reverse phase liquid chromatography with ninhydrin post-column derivatization. *Int.J.Environ.Anal.Chem.* **1993**, *53*, 269-279.

Mendoza, C.B.; Dixon, S.A.; Lods, M.M.; Ma, M.G.; Nguyen, K.T.; Nutt, R.F.; Tran, H.S.; Nolan, T.G. Quantitation of an orally available thrombin inhibitor in rat, monkey and human plasma and in human urine by high-performance liquid chromatography and fluorescent post-column derivatization of arginine. *J.Chromatogr.A* **1997**, *762*, 299-310.

Okayama, A.; Kitada, Y.; Aoki, Y.; Umesako, S.; Ono, H.; Nishii, Y.; Kubo, H. Fluorescence HPLC determination of streptomycin in meat using ninhydrin as a postcolumn labeling agent. *Bunseki Kagaku* **1988**, *37*, 221-224.

Seracu, D.I. The study of UV and visible absorption spectra of the complexes of amino acids with ninhydrin. *Anal.Lett.* **1987**, *20*, 1417-1428.

Sundaram, K.M.S.; Curry, J. A comparison of UV and fluorescence detectors in the liquid chromatographic analysis of glyphosate deposits after post-column derivatization. *J.Liq.Chromatogr.Rel.Technol.* **1997**, *20*, 511-524.

post-column reaction

SAMPLE
Matrix: blood
Analyte: guanidino compounds
Sample preparation: 100 μL Serum + 20 μL 20% trichloroacetic acid, vortex for a few s, centrifuge at 10000 g for 2 min. Remove a 60 μL aliquot of the supernatant and add it to 15 μL 400 mM NaOH to adjust pH to 2.5-3.0, inject a 50 μL aliquot of the supernatant.

HPLC VARIABLES
Column: 150 × 4.6 5 μm Nucleosil C8
Mobile phase: MeCN:MeOH:water containing 3:5:92 containing 15 mM sodium octane-sulfonate and 5 mM ninhydrin, adjusted to pH 4.0 with acetic acid
Flow rate: 1
Injection volume: 50
Detector: F ex 395 em 500 following post-column reaction. The column effluent mixed with 500 mM NaOH pumped at 0.5 mL/min and the mixture flowed through a 5 m × 0.5 mm ID stainless-steel coil at 75° to the detector.

CHROMATOGRAM
Retention time: 3.8 (guanidinosuccinic acid), 4.6 (guanidinoacetic acid), 11 (N-acetyl-L-arginine), 14 (β-guanidinopropionic acid), 15.5 (creatinine), 17.5 (guanidine), 20 (arginine), 24.5 (methylguanidine), 27 (gamma-guanidinobutyric acid), 3 (taurocyamine (with mobile phase adjusted to pH 3.5))
Limit of detection: 4-500 pmole

KEY WORDS
post-column reaction; serum

REFERENCE
Kobayashi, Y.; Kubo, H.; Kinoshita, T. Post-column derivatization system for the fluorimetric determination of guanidino compounds with ninhydrin by reversed-phase ion-pair high-performance liquid chromatography, *J.Chromatogr.*, **1987**, *400*, 113–121.

SAMPLE
Matrix: blood
Analyte: guanidino compounds
Sample preparation: 200 μL Plasma + 100 μL 30% trichloroacetic acid, mix, centrifuge at 1000 g for 10 min. Adjust the pH of the supernatant to 2.0 with 400 mM NaOH, inject a 150 μL aliquot.

HPLC VARIABLES
Column: 38 × 4.2 5 μm ISC-05/S0504 strong cation-exchange (Shimadzu)
Mobile phase: Gradient. A was 14.7 g/L sodium citrate dihydrate containing 10.5 mL/L 60% perchloric acid, adjusted to pH 3.5 with NaOH or perchloric acid. B was 34.3 g/L sodium citrate dihydrate containing 11.0 mL/L 60% perchloric acid, adjusted to pH 5.0 with NaOH or perchloric acid. C was 34.3 g/L sodium citrate dihydrate containing 3.0 mL/L 60% perchloric acid, adjusted to pH 6.0 with NaOH or perchloric acid. D was 34.3 g/L sodium citrate dihydrate containing 6.2 g/L boric acid and 4.0 g/L NaOH, adjusted to pH 11.4 with NaOH or perchloric acid. E was 200 mM NaOH. F was water. A for 5 min, B for 8 min, C for 2 min, D for 14 min, E for 2 min, F for 3 min (all step gradients).
Column temperature: 50
Flow rate: 0.7

Injection volume: 150

Detector: F ex 395 em 500 following post-column reaction. The column effluent mixed with 750 mM NaOH pumped at 0.6 mL/min and with 0.6% ninhydrin in water pumped at 0.4 mL/min and the mixture flowed through a 10 m × 0.5 mm ID PTFE coil at 50 ± 0.05° to the detector.

CHROMATOGRAM

Retention time: 2.5 (taurocyamine), 5 (guanidinosuccinic acid), 7 (creatine), 10 (guanidinoacetic acid), 12 (guanidinopropionic acid), 14 (creatinine), 17 (guanidinobutyric acid), 20 (arginine), 27 (guanidine), 29 (methylguanidine)

Limit of detection: 1-1000 pmole

KEY WORDS

post-column reaction; plasma

REFERENCE

Hiraga, Y.; Konoshita, T. Post-column derivatization of guanidino compounds in high-performance liquid chromatography using ninhydrin, *J.Chromatogr.*, **1981**, *226*, 43–51.

SAMPLE

Matrix: blood

Analyte: guanidino compounds

Sample preparation: 100 μL Serum + 20 μL 60% trichloroacetic acid, stir, centrifuge at 3000 g for 5 min. Remove a 60 μL aliquot of the supernatant and add it to 25 μL 800 mM NaOH, inject an aliquot.

HPLC VARIABLES

Column: 38 × 4.6 ISC-05 ion-exchange (Shimadzu)

Mobile phase: Gradient. A was 50 mM trisodium citrate adjusted to pH 3.5 with 60% perchloric acid. B was 120 mM trisodium citrate adjusted to pH 5.0 with 60% perchloric acid. C was 120 mM trisodium citrate adjusted to pH 6.0 with 60% perchloric acid. D was 120 mM trisodium citrate containing 500 mM NaCl and 100 mM boric acid, adjusted to pH 11.4 with 1 M NaOH. E was 200 mM NaOH. F was water. A for 7 min, B for 7 min (step gradient), C for 3 min (step gradient), D for 15 min (step gradient), E for 2 min (step gradient), F for 3 min (step gradient).

Flow rate: 0.7

Detector: F ex 395 em 500 following post-column reaction. The column effluent mixed with the 0.6% ninhydrin in water pumped at 0.4 mL/min and 1.5 M NaOH pumped at 0.4 mL/min and this mixture flowed through a 7 m × 0.5 mm ID PTFE coil at 50° to the detector.

KEY WORDS

post-column reaction; serum; For taurocyamine, guanidinosuccinic acid, creatine, guanidinoacetic acid, guanidinopropionic acid, creatinine, guanidinobutyric acid, arginine, guanidine, methylguanidine

REFERENCE

Hiraga, Y.; Kinoshita, T. High-performance liquid chromatographic analysis of guanidino compounds using ninhydrin reagent. II. Guanidino compounds in blood of patients on haemodialysis therapy, *J.Chromatogr.*, **1985**, *342*, 269–275.

SAMPLE

Matrix: blood

Analyte: peptides

Sample preparation: Condition a 1 mL Analytichem weak cation-exchange (carboxymethylhydrogen form, CBA) SPE cartridge with 1 mL 1% trifluoroacetic acid in MeOH, 1 mL MeOH, and 2 mL water. Add 1 mL plasma to the SPE cartridge, rinse the tube with 1 mL water, add the rinse to the SPE cartridge, wash with 1 mL 1% trifluoroacetic acid in water, wash with 2 mL water, wash with 2 mL MeOH, elute with 2 mL 1% trifluoroacetic acid in MeOH. Evaporate the eluate to dryness under a stream of nitrogen, reconstitute the residue in 100 μL MeOH:buffer 50:50, inject a 5-75 μL aliquot. (Buffer was 5.7 g monochloroacetic acid, 2.0 g NaOH, and 0.2 g disodium EDTA in 1 L water, pH 3.2.) [Procedure was not necessarily validated for all compounds.]

HPLC VARIABLES

Column: 250 × 2 5 μm Ultrasphere octyl

Mobile phase: Gradient. A was MeOH containing 10 mM sodium octanesulfonate. B was buffer containing 10 mM sodium octanesulfonate. A:B from 45:55 to 70:30 over 30 min, maintain at 70:30 for 1 h. Alternatively isocratic at A:B 55:45 (Buffer was 5.7 g monochloroacetic acid, 2.0 g NaOH, and 0.2 g disodium EDTA in 1 L water, pH 3.2.)

Column temperature: 60

Flow rate: 0.3

Injection volume: 5-75

Detector: F ex 390 em 470 following post-column reaction. The column effluent mixed with 400 mM NaOH pumped at 0.15 mL/min and 0.05% ninhydrin pumped at 0.05 mL/min and the mixture flowed through a 12 m × 0.33 mm ID reaction coil at 70° to the detector.

CHROMATOGRAM

Retention time: 6.6 (angiotensin II (isocratic)), 8.5 (bradykinin (isocratic)), 11 (vasopressin (gradient)), 16 (gonadorelin (LHRH (gradient)), 21 (angiotensin II (gradient)), 26 (bombesin (gradient)), 27 (bradykinin (gradient)), 30 (angiotensin III (gradient)), 33 (angiotensin I (gradient)), 39 (atrial natriuretic peptide (gradient)), 44 (adrenocorticotropin (gradient)), 54 (somatoliberin (gradient))

Limit of detection: 100 fmole

KEY WORDS

plasma; SPE; post-column reaction

REFERENCE

Rhodes, G.R.; Boppana, V.K. High-performance liquid chromatographic analysis of arginine-containing peptides in biological fluids by means of a selective post-column reaction with fluorescence detection, *J.Chromatogr.*, **1988**, *444*, 123–131.

SAMPLE

Matrix: blood

Analyte: SK&F 105494 (octapeptide antagonist of vasopressin)

Sample preparation: Condition a weak cation-exchange (carboxymethylhydrogen form, CBA) (Analytichem) SPE cartridge with 1 mL 1% trifluoroacetic acid in MeOH, 1 mL MeOH, and 2 mL water. 1 mL Plasma + 50 μL 1 μg/mL IS in MeOH:water 50:50, add to the SPE cartridge, rinse the tube with 1 mL water, add the rinse to the SPE cartridge, wash with 1 mL 1% trifluoroacetic acid in water, wash with 2 mL water, wash with 2 mL MeOH, elute with 2 mL 1% trifluoroacetic acid in MeOH. Evaporate the eluate to dryness under a stream of nitrogen at 40°, reconstitute with 100 μL MeOH:buffer 50:50, inject a 5-50 μL aliquot. (Prepare buffer by dissolving 5.7 g monochloroacetic acid, 2.0 g NaOH, and 200 mg EDTA in 1 L water, pH 3.2.)

HPLC VARIABLES

Column: 250 × 2.1 5 μm Ultrasphere octyl

Mobile phase: Gradient. MeOH:buffer from 55:45 to 70:30 over 10 min, maintain at 70:30 for 5 min, return to initial conditions over 2 min, re-equilibrate for 13 min. (Prepare buffer by dissolving 5.7 g monochloroacetic acid, 2.0 g NaOH, and 200 mg EDTA in 1 L water, pH 3.2.)

Column temperature: 60

Flow rate: 0.3

Injection volume: 5-50

Detector: F ex 390 em 470 (cutoff filter) following post-column reaction. The column effluent mixed with 400 mM NaOH pumped at 0.15 mL/min and 0.05% ninhydrin pumped at 0.05 mL/min and the mixture flowed through a 12 m × 0.33 mm i.d. reaction coil at 70° to the detector (J.Chromatogr. 1988, 123, 444).

CHROMATOGRAM

Retention time: 18.6

Internal standard: SK&F 104146 (14.0)

Limit of quantitation: 0.5 ng/mL

KEY WORDS

post-column reaction; plasma; SPE

REFERENCE

Boppana, V.K.; Rhodes, G.R. High-performance liquid chromatographic determination of an arginine-containing octapeptide antagonist of vasopressin in human plasma by means of a selective post-column reaction with fluorescence detection, *J.Chromatogr.*, **1990**, *507*, 79–84.

SAMPLE

Matrix: blood

Analyte: leucine

Sample preparation: Vortex plasma with an equal volume of cold 6% sulfosalicylic acid, centrifuge at 4° at 2000 g for 10 min, inject a 750 µL aliquot of the supernatant.

HPLC VARIABLES

Column: 320 × 9 glass column containing 7-20 µm 7.25% cross-linked sulfonated polystyrene cation-exchange resin BP-AN6 (Benson, Reno NV) (Slurry pack at 2 mL/min with resin:200 mM NaOH 50:50 to a resin bed height of 13 ± 0.5 cm, condition with 200 mM NaOH for 3 min and with mobile phase for 15 min.)

Mobile phase: 85 mM Sodium citrate containing 5 mL/L 30% Brij-35 solution, adjusted to pH 3.30 with concentrated HCl (After each run regenerate column with 200 mM NaOH containing 1% EDTA for 3 min then re-equilibrate with mobile phase for 15 min.)

Column temperature: 53.5

Flow rate: 2

Injection volume: 750

Detector: UV 570 following post-column reaction. A portion (20%) of the column effluent mixed with reagent A, reagent B, and nitrogen gas and the segmented mixture flowed through a mixing coil (Technicon 178-G196-02) and a coil heated to 95° to the detector (total reaction time 11 min). (Reagent A was 1% ninhydrin in DMSO:4 M pH 5.2 lithium acetate buffer 75:25. Reagent B was 2 mM hydrazine sulfate in water containing 3 drops concentrated sulfuric acid. The other 80% of the column effluent was collected by a fraction collector.)

CHROMATOGRAM

Retention time: 26

KEY WORDS

post-column reaction; plasma

REFERENCE

Donahue, E.P.; Brown, L.L.; Flakoll, P.J.; Abumrad, N.N. Rapid measurement of leucine-specific activity in biological fluids by ion-exchange chromatography and post-column ninhydrin detection, *J.Chromatogr.*, **1991**, *571*, 29–36.

SAMPLE

Matrix: blood

Analyte: guanethidine

Sample preparation: Inject 200 µL serum onto column A and elute to waste with mobile phase A, after 2 min elute the contents of column A onto column B with mobile phase B, after another 3 min remove column A from the circuit, elute column B with mobile phase B, monitor the effluent from column B. Flush column A with mobile phase C for 10 min, re-equilibrate with mobile phase A for 7 min.

HPLC VARIABLES

Column: A 35 × 4.6 10 µm TSK precolumn BSA-ODS (Tosoh); B 150 × 4.6 5 µm TSKgel ODS-80TM (Tosoh)

Mobile phase: A 50 mM NaH_2PO_4 adjusted to pH 3.0 with 50 mM phosphoric acid; B MeCN:buffer 30:70, containing 7 g/L sodium 1-octanesulfonate (Buffer was 50 mM NaH_2PO_4 adjusted to pH 3.0 with 50 mM phosphoric acid.); C MeCN:water 50:50.

Flow rate: 1

Injection volume: 200

Detector: F ex 392 em 500 following post-column reaction. The effluent from column B mixed with 1 M NaOH pumped at 0.3 mL/min and with 6 g/L ninhydrin in water pumped at 0.3 mL/min and the mixture flowed through a 10 m × 0.5 mm ID PTFE coil at 56° to the detector.

CHROMATOGRAM
Retention time: 16
Limit of quantitation: 3.1 ng/mL
Limit of detection: 1 ng/mL

KEY WORDS
serum; post-column reaction; column-switching; rat; human; pharmacokinetics

REFERENCE
Inamoto, Y.; Inamoto, S.; Hanai, T.; Takahashi, Y.; Kadowaki, K.; Kinoshita, T. Development of automated highly sensitive analytical system for guanethidine sulfate in serum, *J.Liq.Chromatogr.Rel.Technol.*, **1997**, *20*, 2099–2108.

SAMPLE
Matrix: blood, urine
Analyte: methylguanidine
Sample preparation: Plasma. Condition a 1 mL weak cation-exchange (carboxymethyl-hydrogen form, CBA) SPE cartridge (Analytichem) with 1 mL 1% trifluoroacetic acid in MeOH, with 1 mL MeOH, and with 2 mL water. 1 mL Plasma + 20 μL 10 μg/mL ethylguanidine, mix, adjust pH to 11 with 26 μL 1 M NaOH, add to the SPE cartridge, wash with 3 mL water, elute with 2 mL 1% trifluoroacetic acid in MeOH. Evaporate the eluate to dryness under a stream of nitrogen at 40°, reconstitute the residue in 200 μL mobile phase, inject a 100 μL aliquot. Urine. Dilute 1:10 with water, inject a 20 μL aliquot.

HPLC VARIABLES
Column: 250 × 4.6 10 μm Ultrasil cation-exchange (Beckman)
Mobile phase: MeOH:buffer 15:85 (Prepare buffer by dissolving 11.32 g monochloroacetic acid, 4.86 g NaOH, and 400 mg disodium EDTA in 2 L water.)
Column temperature: 60
Flow rate: 0.75
Injection volume: 20-100
Detector: F ex 390 em 470 (cut-off filter) following post-column reaction. The column effluent mixed with 400 mM NaOH pumped at 0.25 mL/min and 0.2% ninhydrin pumped at 0.05 mL/min and the mixture flowed through a 24 m × 0.33 mm ID coil at 70° to the detector.

CHROMATOGRAM
Retention time: 13.2
Internal standard: ethylguanidine (14.8)
Limit of detection: 1 ng/mL (plasma), 100 ng/mL (urine)

KEY WORDS
post-column reaction; plasma; SPE

REFERENCE
Boppana, V.K.; Rhodes, G.R.; Brooks, D.P. Determination of methylguanidine in plasma and urine by high-performance liquid chromatography with fluorescence detection following postcolumn derivatization, *Anal.Biochem.*, **1990**, *184*, 213–218.

SAMPLE
Matrix: food
Analyte: amines
Sample preparation: Cheese, chocolate. Homogenize (stomacher) 5 g ground cheese or chocolate with 45 mL 70 mM trisodium citrate at 45° for 5 min. Remove a 3 mL aliquot and add it to 3 mL 600 mM trichloroacetic acid, mix, centrifuge at 4° at 10000 g for 10 min, suspend the pellet in 3 mL 300 mM trichloroacetic acid, centrifuge. Combine the supernatants, filter (0.45 μm), make up the filtrate to 10 mL with water, inject an aliquot. Wine. 3 mL Wine + 3 mL 600 mM trichloroacetic acid, centrifuge, filter the supernatant, inject an aliquot of the filtrate. Fish, sauerkraut. Blend 200 g fish or sauerkraut with 200 mL water for 3 min. Remove a 3 mL aliquot and add it to 3 mL 600 mM trichloroacetic acid, centrifuge, filter the supernatant, inject an aliquot of the filtrate.

HPLC VARIABLES
Guard column: 30 × 3 Corasil C18

Column: 100 × 8 10 µm Nucleosil C18 radial-compression

Mobile phase: DMSO:buffer 47:53 (Prepare mobile phase by dissolving 16 g ninhydrin and 1.2 g hydrindantin in 322 mL DMSO with sonication for 10 min, add 350 mL 1.8 M pH 5.00 sodium acetate buffer, add 2 g sodium dodecyl sulfate dissolved in a mixture of 618 mL DMSO and 710 mL water, flush constantly with nitrogen. Flush column with DMSO:water 50:50 at the end of each day.)

Column temperature: 29

Flow rate: 1

Detector: UV 546 following post-column reaction. The column effluent flowed through a 10 m × 0.25 mm PTFE tube coiled in a figure 8 at 145° to the detector.

CHROMATOGRAM

Retention time: 13 (tyramine), 16 (histamine), 23 (putrescine), 26 (cadaverine), 54 (tryptamine), 65 (phenylethylamine)

Limit of detection: 800 ng/g (sauerkraut), 300 ng/g (wine)

KEY WORDS

post-column reaction; cheese; chocolate; wine; fish; sauerkraut; tuna

REFERENCE

Joosten, H.M.L.J.; Olieman, C. Determination of biogenic amines in cheese and some other food products by high-performance liquid chromatography in combination with thermo-sensitized reaction detection, *J.Chromatogr.*, **1986**, *356*, 311–319.

SAMPLE

Matrix: soil

Analyte: glyphosate

Sample preparation: Shake 5 g soil with 150 mL 500 mM ammonium hydroxide at 180 oscillations/min for 15 min, centrifuge at 2000 rpm for 15 min, filter (glass wool) the supernatant. repeat the extraction twice more. Combine the filtrates and add them to 20 g resin, stir vigorously for 20 min, allow to settle, pass the supernatant through a 1 cm bed of resin. Combine the resin portions and elute them with two 50 mL portions of ammonium bicarbonate, evaporate to dryness under reduced pressure at 60°, add water, evaporate to dryness under reduced pressure at 60°, repeat 2-3 times, take up residue in 5 mL water, add to a column filled with 10 mL 200-400 mesh Dowex 50-W-X8 (H$^+$ form) cation-exchange resin, rinse flask with three 5 mL portions of water, add rinses to column, elute with 55 mL water. Collect all the eluate and evaporate to dryness under reduced pressure at 60°, reconstitute with 5 mM KH$_2$PO$_4$, filter (0.45 µm), inject a 100 µL aliquot. After 3-3.5 min remove the guard column from the circuit and flush it with mobile phase. (Resin was 200-400 mesh AG-IX-8 anion-exchange resin (chloride form; Bio-Rad). Prepare as follows. Slurry 1 kg resin with 2 L 1 M ammonium bicarbonate solution, stir for 1 h, discard the liquid, repeat the process five more times, finally slurry with water and refrigerate.)

HPLC VARIABLES

Guard column: Aminex A-9 cation exchange K$^+$ guard column (Bio-Rad)

Column: 100 × 4.6 Aminex A-9 cation exchange (Bio-Rad)

Mobile phase: MeOH:water 4:96 containing 680.3 mg/L KH$_2$PO$_4$, adjust pH to 1.9 with concentrated phosphoric acid

Column temperature: 50 (analytical column)

Flow rate: 0.5

Injection volume: 100

Detector: UV 570 following post-column reaction. The column effluent mixed with the reagent pumped at 0.5 mL/min and the mixture flowed through a 2.84 m × 0.2 mm ID coil at 100° to the detector. (Prepare reagent by mixing 416.66 mL DMSO, 416.66 mL water, and 83.33 mL 4 M sodium acetate, stir while bubbling nitrogen through the mixture, add 13.33 g ninhydrin in 4 portions 10 min apart over 30 min, add a solution of 416 mg hydrindantin in 83.33 mL DMSO, stir while passing nitrogen through the solution for 45 min. Store at 4° under nitrogen for up to 5 days.)

CHROMATOGRAM

Retention time: 8 (glyphosate), 15 (aminomethylphosphonic acid (metabolite))

Limit of quantitation: 100 ng/g
Limit of detection: 50 ng/g

KEY WORDS
post-column reaction; column-switching

REFERENCE
Thompson, D.G.; Cowell, J.E.; Daniels, R.J.; Staznik, B.; MacDonald, L.M. Liquid chromatographic method for quantitation of glyphosate and metabolite residues in organic and mineral soils, stream sediments, and hardwood foliage, *J.Assoc.Off.Anal.Chem.*, **1989**, *72*, 355–360.

Nitric Acid

HNO_3 $\longrightarrow$ post-column reaction

SAMPLE
Matrix: solutions
Analyte: alkylamines
Sample preparation: Reflux a 10 mL aliquot of a 50-100 nM solution of amine in MeOH containing 170 equivalents of 5-(4-pyridyl)-2-thiophenecarbaldehyde for 5 h, dilute to 50 mL, inject a 10 μL aliquot. (Preparation of 5-(4-pyridyl)-2-thiophenecarbaldehyde is as follows. Dissolve 2 g 4-aminopyridine in 15 mL acetic acid, 1.5 mL concentrated sulfuric acid, and 2 mL water, stir at 0°, diazotize with 1 equivalent (?) sodium nitrite, add an excess of potassium iodide, keep at room temperature for 1 day, heat at 50° for 30 min, make alkaline, extract with ether. Evaporate the extract to dryness and recrystallize from EtOH:water to give 4-iodopyridine (yield 15%; mp 100° d) (J. Chem. Soc. 1953, 3226). Heat 22.2 g iodine, 6.66 g periodic acid dihydrate, and 20 g 2-methylthiophene in 200 mL 80% acetic acid at 65° for 8 h, cool, remove the oil, basify the solution, extract with chloroform. Combine the oil and the residue from the extract, add to a NaOH solution containing sodium thiosulfate, steam distil. Distil the crude product to give 2-iodo-5-methylthiophene as a yellowish oil (bp 90°/4.1 kPa). Reflux 460 mg palladium amalgam, 14.7 g NaOH, 7.5 g 4-iodopyridine, 16.4 g 2-iodo-5-methylthiophene, and 2.34 g hydrazine hydrate in 47 mL water with stirring for 6 h (Caution! Hydrazine hydrate is a carcinogen and explodes on distillation in air!), filter, wash the precipitate with hot chloroform and hot water. Remove the organic layer from the filtrate and reduce its volume to 50 mL, extract with 15 mL 25% HCl. Remove the aqueous layer and basify it, extract with chloroform. Evaporate the chloroform layer to dryness, distil the residue at 150°/0.4 kPa, chromatograph the distillate on silica gel with acetone:chloroform 50:50 to give 4-(5-methyl-2-thienyl)pyridine as a white solid (mp 136.0-136.6°) in the first fraction. Reflux 1 g 4-(5-methyl-2-thienyl)pyridine and 957 mg selenium(IV) oxide in 35 mL m-xylene with stirring for 8 h, evaporate to dryness. Distil the residue at 200°/0.8 kPa and chromatograph the distillate on silica gel with ethyl acetate to give 5-(4-pyridyl)-2-thiophenecarbaldehyde as white plate crystals (mp 135-136°).)

HPLC VARIABLES
Column: 150 × 6 Shim-pack CLC-ODS (Shimadzu)
Mobile phase: MeOH:water 100:0 (A) or 90:10 (B)
Column temperature: 40
Flow rate: 1
Injection volume: 10
Detector: F ex 340 em 395 following post-column reaction. The column effluent mixed with 1 M nitric acid pumped at 0.2 mL/min and the mixture flowed through a 300 × 0.3 mm ID coil to the detector. (The original fluorescent 5-(4-pyridyl)-2-thiophenecarbaldehyde is regenerated by hydrolysis.)

CHROMATOGRAM
Retention time: 3.5 (1,3-propanediamine (A)), 3.7 (2-phenylethylamine (A)), 4.0 (benzylamine (B)), 4.0 (p-chloroaniline (A)), 4.3 (1,8-octanediamine (A)), 4.5 (1-phenylethylamine (B)), 4.6 (butylamine (B)), 4.7 (octylamine (A)), 5.0 (cyclohexylamine (B)), 5.1 (p-toluidine

(B)), 5.5 (decylamine (A)), 6.1 (hexylamine (B)), 6.7 (dodecylamine (A)), 8.3 (tetradecylamine (A)), 10.8 (hexadecylamine (A)), 21.3 (octadecylamine (A))
Limit of detection: 0.1-0.15 pmole

KEY WORDS
post-column reaction

REFERENCE
Nakajima, R.; Yamamoto, A.; Hara, T. Thienylpyridines as a new fluorescent reagent. I. Determination of primary alkylamines with 5-(4-pyridyl)-2-thiophenecarbaldehyde using HPLC, *Bull.Chem.Soc.Jpn.*, **1990**, *63*, 1968–1972.

Orcinol

$\longrightarrow$ post-column reaction

SAMPLE
Matrix: solutions
Analyte: oligosaccharides
Sample preparation: Inject a 20 μL aliquot into the mobile phase pre-heated to 90°.

HPLC VARIABLES
Column: 300 × 7.8 Aminex HPX-42A (Bio-Rad)
Mobile phase: Water
Column temperature: 85
Flow rate: 0.5-0.6
Injection volume: 20
Detector: UV 420 following post-column reaction. The column effluent mixed with the reagent pumped at 1.5-1.6 mL/min and was air-segmented. The mixture flowed through a mixing coil, a 3 min reaction coil at 125°, and a cooling coil to a debubbler. The liquid phase flowed to the detector. (Reagent was 3.75 g orcinol in 3.7 L 75% sulfuric acid. All coils were 2 mm ID glass.)

KEY WORDS
post-column reaction

REFERENCE
Schmid, G.; Wandrey, C. An automated method for the quasi-continuous analysis of degradation and transfer products during the enzymatic hydrolysis of oligosaccharides, *Anal.Biochem.*, **1986**, *153*, 144–150.

Orthophosphoric Acid

H_3PO_4 $\longrightarrow$ post-column reaction

SAMPLE
Matrix: blood
Analyte: pimobendan
Sample preparation: Extract from plasma using SPE.

HPLC VARIABLES
Column: 5 μm ODS-Hypersil
Mobile phase: MeOH:water 59:46 containing 2.5 g/L ammonium acetate

Detector: F ex 332 em 405 following post-column reaction. The column effluent mixed with MeOH:85% orthophosphoric acid:water 60:20:20 pumped at 0.2 mL/min and the mixture flowed to the detector.

CHROMATOGRAM
Limit of detection: 1 ng/mL

OTHER SUBSTANCES
Extracted: metabolites

KEY WORDS
pig; plasma; SPE; post-column reaction

REFERENCE
Verdouw, P.D.; Hartog, J.M.; Duncker, D.J.; Roth, W.; Saxena, P.R. Cardiovascular profile of pimobendan, a benzimidazole-pyridazinone derivative with vasodilating and inotropic properties, *Eur.J.Pharmacol.*, **1986**, *126*, 21–30.

SAMPLE
Matrix: cell incubations
Analyte: pimobendan
Sample preparation: Inject 200 µL cell incubation on to column A with mobile phase A then switch to mobile phase B (start the gradient) and elute to waste, after 4 min direct the effluent from column A on to column B, after 30 min remove column A from the circuit.

HPLC VARIABLES
Column: A 40 × 4.6 37-75 µm Porasil B; B 30 mm long 5 µm Hypersil ODS + 125 × 4.6 5 µm Hypersil ODS
Mobile phase: A 1% pH 6.8 ammonium acetate buffer; B Gradient. X was water:25% ammonia 100:0.02. Y was MeOH. X:Y 100:0 for 6.9 min, to 95:5 over 0.1 min, to 85:15 over 11 min, to 0:100 over 9 min, maintain at 0:100 for 3 min, re-equilibrate at initial conditions for 2 min.
Column temperature: 28
Flow rate: 1
Injection volume: 200
Detector: F ex 332 em 405 following post-column reaction. The column effluent mixed with MeOH:water:85% orthophosphoric acid 60:20:20 pumped at 0.2 mL/min and flowed to the detector.

CHROMATOGRAM
Retention time: 28

OTHER SUBSTANCES
Extracted: metabolites

KEY WORDS
human; liver; hepatocytes; post-column reaction; column switching

REFERENCE
Pahernik, S.A.; Schmid, J.; Sauter, T.; Schildberg, F.W.; Koebe, H.-G. Metabolism of pimobendan in long-term human hepatocyte culture: *in vivo-in vitro* comparison, *Xenobiotica*, **1995**, *25*, 811–823.

Periodic Acid

SAMPLE
Matrix: algae
Analyte: PSP toxins
Sample preparation: Homogenize (Ultra-Turrax) 1 g algae with 3 mL 30 mM acetic acid, centrifuge at 2980 g for 10 min, filter (0.45 µm nylon), inject a 10 µL aliquot of the filtrate.

HPLC VARIABLES
Column: 250 × 4.6 5 µm Supelcosil-C18 DB

Mobile phase: Gradient. A was THF:buffer 1.5:98.5. (Prepare buffer by dissolving 40 mL 1 M phosphoric acid in 900 mL water and adding 2.379 g sodium octanesulfonate, adjust pH to 6.6 with 25% ammonia solution, make up to 1 L with water.) B was MeCN:THF: buffer 15:2.5:82.5. (Prepare buffer by dissolving 25 mL 1 M phosphoric acid in 450 mL water and adding 1.406 g sodium octanesulfonate, adjust pH to 6.6 with 25% ammonia solution, make up to 500 mL with water.) A:B 100:0 for 13.5 min, to 0:100 over 1 min, maintain at 0:100 for 22.5 min, return to initial conditions over 1 min, re-equilibrate for 10 min.

Flow rate: 1.1 for 13.5 min, to 1.2 over 1 min, maintain at 1.2 for 22.5 min, to 1.1 over 1 min, maintain at 1.1

Injection volume: 10

Detector: F ex 333 em 390 following post-column reaction. The column effluent mixed with reagent 1 and flowed through a 1 mL reaction coil (CRX 390, Pickering Laboratories, Mountain View CA) at 50°. The effluent from this coil mixed with reagent 2 and the mixture flowed to the detector. (Prepare reagent 1 by dissolving 40 mL 25% ammonia solution and 1.14 g periodic acid in 500 mL water. Prepare reagent 2 by dissolving 30 mL glacial acetic acid in 500 mL water.)

CHROMATOGRAM

Retention time: 12.5 (gonyautoxin-2), 15 (gonyautoxin-1), 32.5 (decarbamoylsaxitoxin), 34 (saxitoxin)

Limit of detection: 0.01 ng

KEY WORDS

post-column reaction

REFERENCE

Hummert, C.; Ritscher, M.; Reinhardt, K.; Luckas, B. Analysis of the characteristic PSP profiles of *Pyrodinium bahamense* and several strains of *Alexandrium* by HPLC based on ion-pair chromatographic separation, post-column oxidation, and fluorescence detection, *Chromatographia*, **1997**, *45*, 312–316.

Phenolphthalein

post-column reaction

SAMPLE

Matrix: blood, tissue, urine

Analyte: cyclodextrins

Sample preparation: Plasma. 1 mL Plasma + 250 µL 20% trichloroacetic acid in water, vortex for 30 s, centrifuge at 3000 g for 5 min. Remove the upper layer and add it to 1 drop 4 M NaOH, shake for a few s, inject a 250 µL aliquot. Alternatively, shake 100 µL plasma with 100 µL water for a few s, add 50 µL 20% trichloroacetic acid, mix, centrifuge. Remove the upper layer and add it to 1 drop 1 M NaOH, shake for a few s, inject the whole amount (180 µL) (LOD 10 µg/mL). Tissue. Homogenize (ultra turrax) 2.5 g tissue in 5 mL water, centrifuge at 3000 g for 30 min. Remove a 1 mL aliquot of the supernatant, add 250 µL 20% trichloroacetic acid in water, vortex for 30 s, centrifuge at 3000 g for 5 min. Remove the upper layer and add it to 1 drop 4M NaOH, shake for a few s, inject a 250 μL aliquot. Urine. Dilute, if necessary, inject an aliquot.

HPLC VARIABLES
Column: 300 × 3.9 10 μm μBondapak phenyl
Mobile phase: MeOH:water 10:90
Flow rate: 2
Injection volume: 180-250
Detector: UV 546 following post-column reaction. The column effluent mixed with the reagent pumped at 2 mL/min and the mixture flowed through a 1.5 m × 1 mm ID PTFE coil to the detector. (Prepare reagent by mixing 10 mL 6 mM phenolphthalein in EtOH: water 96:4 with 990 mL 8 mM sodium carbonate in water, adjust to pH 10.5 with 1 M NaOH. Heat reagent to 60° before delivery. Phenolphthalein-cyclodextrin complexes absorb less than phenolphthalein and so procedure is based on indirect UV detection.)

CHROMATOGRAM
Retention time: 4.0 (gamma-cyclodextrin), 5.1 (β-cyclodextrin)
Limit of detection: 1 μg/mL

KEY WORDS
post-column reaction; rat; plasma; indirect UV direction; pharmacokinetics

REFERENCE
Frijlink, H.W.; Visser, J.; Drenth, B.F.H. Determination of cyclodextrins in biological fluids by high-performance liquid chromatography with negative colorimetric detection using post-column complexation with phenolphthalein, *J.Chromatogr.*, **1987**, *415*, 325−333.

Phenylglyoxal

post-column reaction

SAMPLE
Matrix: blood
Analyte: guanine nucleosides and nucleotides
Sample preparation: Separate erythrocytes from 5 mL blood, wash three times with 5 mL portions of ice-cold saline, centrifuge at 4° at 1400 g for 10 min. Remove a 100 μL aliquot of the erythrocyte layer and add it to 400 μL water, add 100 μL 3 M perchloric acid, add 20 μL 200 μM 9-ethylguanine in water, homogenize, centrifuge at 1400 g for 10 min. Adjust the pH of the supernatant to 6.0 with about 190 μL 1 M potassium bicarbonate, pass through a Toyopak ODS-M SPE cartridge, collect the initial eluate, inject a 50 μL aliquot.

HPLC VARIABLES
Column: 150 × 4.6 5 μm TSKgel ODS-120T (Tosoh)
Mobile phase: Gradient. A was 10 mM pH 6.0 tetra-n-propylammonium phosphate:50 mM pH 6.0 sodium phosphate buffer:water 17:20:63. B was MeOH:10 mM pH 6.0 tetra-n-propylammonium phosphate:50 mM pH 6.0 sodium phosphate buffer:water 20:17:20: 43. A:B 90:10 for 5 min, to 50:50 over 35 min.
Flow rate: 1
Injection volume: 50
Detector: F ex 365 em 515 following post-column reaction. The column effluent mixed with 60 nM phenylgloxal monohydrate in 2-methoxyethanol:water 2:1 pumped at 0.5 mL/min and the mixture flowed through a 3 m × 0.5 mm ID stainless steel coil at 80° to the detector.

CHROMATOGRAM

Retention time: 7 (guanine), 12 (guanosine monophosphate), 15 (guanosine), 18 (deoxyguanosine), 19 (guanosine diphosphate), 22 (deoxyguanosine monophosphate), 28.5 (guanosine triphosphate), 30 (deoxyguanosine diphosphate), 33.5 (cyclic guanosine monophosphate), 35.5 (deoxyguanosine triphosphate)

Internal standard: 9-ethylguanine (26.5)

Limit of detection: 3.2-10 pmole

OTHER SUBSTANCES

Non-interfering: N-acetylgalactosamine, amino acids, cholesterol, dextrose, estrone, galactosamine, galactose, histamine, niacin, niacinamide, spermidine, spermine, vitamin C

KEY WORDS

post-column reaction; erythrocytes; SPE

REFERENCE

Yonekura, S.; Iwasaki, M.; Kai, M.; Ohkura, Y. Determination of guanine and its nucleosides and nucleotides in human erythrocytes by high-performance liquid chromatography with postcolumn fluorescence derivatization using phenylglyoxal reagent, *J.Chromatogr.B*, **1994**, *654*, 19-24.

Phthalaldehyde

post-column reaction

SAMPLE

Matrix: blood, urine

Analyte: citrulline and homocitrulline

Sample preparation: Plasma (for citrulline). 100 μL Plasma + 200 μL 10% trichloroacetic acid, mix, let stand for 10 min, centrifuge at 10000 g for 3 min, inject an 80 μL aliquot of the supernatant. Urine (for homocitrulline). Adjust pH of urine to 2 with 8 M HCl, add a 4 mL aliquot to a 20 × 5 column of Amberlite CG-120 (H$^+$ form), wash with 2 mL water, elute with 100 mM pH 7.6 Tris-HCl buffer containing 1 M NaCl, discard the first 1 mL eluate, collect the next 1 mL. Add 10 μL 8 M HCl to this sample, inject an 80 μL aliquot.

HPLC VARIABLES

Column: 150 × 4 TSK gel SCX

Mobile phase: 50 mM pH 3.5 Citrate buffer containing 300 mM NaCl

Flow rate: 0.4

Injection volume: 80

Detector: UV 520 following post-column reaction. The column effluent mixed with reagent A pumped at 0.2 mL/min and reagent B pumped at 0.2 mL/min and the mixture flowed through a 10 m × 0.5 mm ID coil at 50° to the detector. (Reagent A was 1.5 g/L o-phthalaldehyde in 1 M sulfuric acid containing 0.1% Brij 35. Reagent B was 0.8 g/L N-(1-naphthyl)ethylenediamine in 1 M sulfuric acid containing 0.1% Brij 35 and 5 g/L boric acid.)

CHROMATOGRAM

Retention time: 11 (citrulline), 20 (homocitrulline)

Limit of quantitation: 2.6-4.3 μM

KEY WORDS

post-column reaction; plasma; SPE

REFERENCE
Koshiishi, I.; Kobori, Y.; Imanari, T. Determination of citrulline and homocitrulline by high-performance liquid chromatography with post-column derivatization, *J.Chromatogr.*, **1990**, *532*, 37–43.

Polyclonal Antibody

SAMPLE
Matrix: blood
Analyte: digoxin
Sample preparation: Add 5% MeCN to serum, centrifuge at 5200 g for 10 min, inject a 1 mL aliquot of the supernatant on to column A and elute to waste with mobile phase A, after 18 min elute the contents of column A on to column B with mobile phase B, monitor the effluent from column B.

HPLC VARIABLES
Column: A C18 alkyl-diol restricted access column (Ger. Pat. DE 41 30 475 A1 (1991)); B 125 × 4.6 5 μm LiChroCART C18 (Merck)
Mobile phase: A 200 mM pH 7.0 sodium acetate; B MeCN:50 mM pH 7.0 sodium acetate 30:70
Flow rate: A 1; B 0.5
Injection volume: 1000
Detector: F ex 480 em 514 following post-column reaction. The column effluent mixed with the reagent pumped at 1 mL/min and the mixture flowed through a 1.6 mL coil of 0.5 mm ID PTFE tubing at 20° to a 10 × 4 column of Carbolink-hydrazide (Pierce) which removed free antibodies. The effluent from this column flowed to the detector. (Reagent was 1.3 nM Fab-DIG (affinity-purified fluorescein-labelled Fab fragments of polyclonal anti-digoxigenin (Boehringer Mannheim)) in phosphate-buffered saline containing 0.5% Tween 20.)

CHROMATOGRAM
Retention time: 9.5
Limit of detection: 200 pM

OTHER SUBSTANCES
Extracted: metabolites
Interfering: dihydrodigoxin (although response is much lower)
Non-interfering: dehydroepiandrosterone-3-sulfate, progesterone, spironolactone

KEY WORDS
post-column reaction; serum; column-switching; pharmacokinetics

REFERENCE
Oosterkamp, A.J.; Irth, H.; Beth, M; Unger, K.K.; Tjaden, U.R.; van de Greef, J. Bioanalysis of digoxin and its metabolites using direct serum injection combined with liquid chromatography and on-line immunochemical detection, *J.Chromatogr.B*, **1994**, *653*, 55–61.

Potassium Chromate

$$K_2CrO_4 \longrightarrow \text{post-column reaction}$$

SAMPLE
Matrix: blood
Analyte: 6-mercaptopurine
Sample preparation: 1 mL Plasma + 25 μL 4 μg/mL 6-thioguanine in water + 100 μL 400 mM NaOH + 1 mL 0.3% phenylmercuric acetate in ethyl acetate + 3 mL diethyl ether, shake on a tumble mixer for 10 min, centrifuge for 5 min. Remove the organic layer and add it to 500 μL 100 mM HCl, whirlmix for 2 min, centrifuge for 5 min, discard the

organic layer, evaporate traces of organic solvent under a stream of nitrogen at room temperature for 15 min, add 10 μL 3 mg/mL dithioerythritol in water, inject an aliquot.

HPLC VARIABLES
Column: 250 × 4.6 LiChrosorb 10 RP-18

Mobile phase: Isopropanol:water 3:97 containing 13.80 g/L $NaH_2PO_4.H_2O$, 200 μL/L 85% phosphoric acid, 60 mg/L dithioerythritol, and 500 mg/L sodium octanesulfonate, pH 3.6-3.7

Flow rate: 1.5

Detector: F ex 295 em 380 following post-column reaction. The column effluent mixed with 8 mM potassium chromate in 500 mM HCl pumped at 0.16 mL/min and with air flowing at 0.32 mL/min and the mixture flowed through a single mixing coil. The effluent from this coil mixed with 1.6% sodium metabisulfite pumped at 0.16 mL/min and this mixture flowed through a single mixing coil. The effluent from this coil mixed with 4 M ammonium hydroxide pumped at 0.23 mL/min and this mixture flowed through a double mixing coil to a debubbler. The liquid effluent from the debubbler flowed to the detector.

CHROMATOGRAM
Retention time: 6

Internal standard: 6-thioguanine (8)

Limit of detection: <2 ng/mL

KEY WORDS
post-column reaction; plasma; pharmacokinetics

REFERENCE
Jonkers, R.E.; Oosterhuis, B.; ten Berge, R.J.M.; van Boxtel, C.J. Analysis of 6-mercaptopurine in human plasma with a high-performance liquid chromatographic method including post-column derivatization and fluorimetric detection, *J.Chromatogr.*, **1982**, *233*, 249–255.

SAMPLE
Matrix: blood

Analyte: cisplatin

Sample preparation: 400 μL Plasma + 400 μL MeCN, mix, centrifuge at 4° at 3500 g for 5 min. Remove a 200 μL aliquot of the supernatant and add it to 100 μL 10 mM citric acid containing 100 μM cetyltrimethylammonium bromide and 700 μL dichloromethane, rotate for 10 min, centrifuge at 4° at 3500 g for 5 min, inject a 40 μL aliquot of the aqueous layer.

HPLC VARIABLES
Guard column: 15 × 3.2 7 μm Polymer RP (Brownlee)

Column: 100 × 4.6 BDS-Hypersil C18 (Before analyses pump 120 mL 50 mM cetyltrimethylammonium bromide in isopropanol:water 5:95 through the column at 30° to coat the column with cetyltrimethylammonium bromide.)

Mobile phase: 10 mM Citric acid containing 100 μM cetyltrimethylammonium bromide, adjusted to pH 5.0 with 5 M NaOH

Column temperature: 25

Flow rate: 0.7

Injection volume: 40

Detector: UV 290 following post-column reaction. The column effluent mixed with 117 μM potassium dichromate pumped at 0.2 mL/min and the mixture flowed through a 200 μL knitted coil of PTFE tubing at 30° and then mixed with 28.16 mM sodium bisulfite pumped at 0.2 mL/min. This mixture flowed through a 1 mL knitted coil of PTFE tubing at 30° to the detector. (The reaction coils were contained in a PCX 5000 Post Column Reaction Module (Pickering Laboratories, Mountain View CA).)

CHROMATOGRAM
Retention time: 9

Limit of quantitation: 60 ng/mL

KEY WORDS
post-column reaction; plasma; dog; pharmacokinetics

REFERENCE

Farrish, H.H.; Hsyu, P.-H.; Pritchard, J.F.; Brouwer, K.R.; Jarrett, J. Validation of a liquid chromatography post-column derivatization assay for the determination of cisplatin in plasma, *J.Pharm.Biomed.Anal.*, **1994**, *12*, 265–271.

SAMPLE

Matrix: blood, urine

Analyte: cisplatin

Sample preparation: Plasma. Filter using a UFC 3GC membrane with a 10000 molecular weight cut-off (Japan Millipore) at 4000 g at 4° for 30 min, inject a 100 μL aliquot of the ultrafiltrate. Urine. Centrifuge at 1000 g for 1 min, dilute a 50 μL aliquot of the upper layer 1:10 with distilled water, inject a 100 μL aliquot.

HPLC VARIABLES

Guard column: Cyano Guard-Pak (Waters)

Column: 150 × 4.6 5 μm anionic exchange resin (Hitachi No. 3013-N, Chromato Research)

Mobile phase: MeCN:10 mM NaCl 85:15

Column temperature: 40

Flow rate: 0.7

Injection volume: 100

Detector: UV 290 following post-column reaction. The column effluent mixed with 0.026 mM potassium dichromate pumped at 0.14 mL/min and 6.6 mM sodium hydrogen sulfite pumped at 0.07 mL/min and flowed through a 7 m × 0.5 mm or 30 m × 0.25 mm PTFE tube reactor to the detector.

CHROMATOGRAM

Retention time: 11

Limit of detection: 80 ng/mL

KEY WORDS

plasma; post-column reaction

REFERENCE

Kinoshita, M.; Yoshimura, N.; Ogata, H.; Tsujino, D.; Takahashi, T.; Takahashi, S.; Wada, Y.; Someya, K.; Ohno, T.; Masuhara, K.; Tanaka, Y. High-performance liquid chromatographic analysis of unchanged cis-diamminedichloroplatinum (cisplatin) in plasma and urine with post-column derivatization, *J.Chromatogr.*, **1990**, *529*, 462–467.

SAMPLE

Matrix: tissue

Analyte: cisplatin

Sample preparation: Homogenize with four volumes saline at 4°, centrifuge at 100000 g at 4° for 1 h. Remove a 450 μL aliquot of the supernatant, filter (Millipore UFC 3GC 10000 molecular mass cut-off) with centrifuging at 4000 g at 4° for 30 min, inject a 20 μL aliquot of the supernatant

HPLC VARIABLES

Guard column: 50 × 4.6 5 μm Hitachi No. 3013-N

Column: 150 × 4.6 5 μm Hitachi No. 3013-N

Mobile phase: MeCN:10 mM NaCl 15:85

Column temperature: 40

Flow rate: 0.9

Injection volume: 100

Detector: UV 290 following post-column reaction. The column effluent mixed with 0.026 mM potassium dichromate pumped at 0.6 mL/min and with 6.6 mM sodium hydrogen sulfite pumped at 0.3 mL/min and the mixture flowed through a 7 m × 0.5 mm or 30 m × 0.25 mm PTFE coil to the detector.

CHROMATOGRAM

Retention time: 9

Limit of detection: 100 ng/g

KEY WORDS

rat; liver; kidney; ultrafiltrate; pharmacokinetics; post-column reaction

REFERENCE
Hanada, K.; Nagai, N.; Ogata, H. Quantitative determination of unchanged cisplatin in rat kidney and liver by high-performance liquid chromatography, *J.Chromatogr.B*, **1995**, *663*, 181–186.

Potassium Ferricyanide

RELATED REFERENCE
Ohura, H.; Imato, T.; Yamasaki, S.; Ishibashi, N. Potentiometric determination of reducing sugars as borate complexes using hexacyanoferrate(III)-hexacyanoferrate(II) potential buffer and its application to liquid chromatography. *Anal.Sci.* **1990**, *6*, 777-779

$K_3Fe(CN)_6$ $\longrightarrow$ post-column reaction

SAMPLE
Matrix: blood
Analyte: thiamine
Sample preparation: 200 µL Plasma, whole blood, or erythrocytes + 200 µL 100 mg/mL trichloroacetic acid, vortex vigorously, centrifuge at 35000 g for 5 min, inject a 100 µL aliquot of the supernatant.

HPLC VARIABLES
Column: 250 × 4 µBondapak C18
Mobile phase: MeCN:buffer 3.8:96.2 (Mobile phase was 200 mM NaH_2PO_4 in 3 g/L MeCN in water.)
Flow rate: 1
Injection volume: 100
Detector: F ex 375 em 450 following post-column reaction. The column effluent mixed with the reagent pumped at 0.5 mL/min and flowed to the detector. (Reagent was 100 µg/mL potassium ferricyanide in 150 g/L NaOH.)

CHROMATOGRAM
Retention time: 8.0 (thiamine), 3.1 (thiamine triphosphate), 3.8 (thiamine pyrophosphate), 5.0 (thiamine monophosphate)
Limit of detection: 30 fmole

KEY WORDS
post-column reaction; pharmacokinetics; plasma; whole blood; erythrocytes

REFERENCE
Kimura, M.; Itokawa, Y. Determination of thiamin and thiamin phosphate esters in blood by liquid chromatography with post-column derivatization, *Clin.Chem.*, **1983**, *29*, 2073–2075.

SAMPLE
Matrix: blood
Analyte: thiamine
Sample preparation: Hemolyze whole blood by freezing at -20° for 20 min, thaw, homogenize. Add a 200 µL aliquot of hemolyzed blood or serum to 200 µL chilled 10% perchloric acid, let stand below 4° for 15 min, centrifuge at 10000 g for 1 min. Remove a 200 µL aliquot of the supernatant and add it to 1.8 M sodium acetate containing 600 mM NaOH, mix, filter (Costar filter unit) while centrifuging for 30 s, inject a 20 µL aliquot of the filtrate.

HPLC VARIABLES
Guard column: present but not specified
Column: 110 × 4.7 Partisphere 5 C18 (Whatman)
Mobile phase: Gradient. A was 15 mM citric acid adjusted to pH 4.2 with 50% ammonium hydroxide, prepare fresh each day. B was 100 mM formic acid containing 4% diethylamine, pH 3.2. A:B from 90:10 to 50:50 over 2.5 min, to 5:95 over 0.5 min, maintain at 5:95 over 3.5 min, to 95:5 over 0.5 min, maintain at 95:5 for 3 min.
Flow rate: 1
Injection volume: 20

Detector: F ex 365 em 435 following post-column reaction. The column effluent mixed with the reagent pumped at 0.2 mL/min and the mixture flowed through a 90 cm × 0.5 mm ID PTFE coil to the detector. (Prepare reagent by dissolving 100 mg potassium ferricyanide in 120 mL 3 M NaOH.)

CHROMATOGRAM
Retention time: 3 (thiamine triphosphate), 5 (thiamine pyrophosphate), 5.8 (thiamine monophosphate), 7.6 (thiamine)
Limit of quantitation: 2 nM

KEY WORDS
post-column reaction; whole blood; serum

REFERENCE
Lee, B.L.; Ong, H.Y.; Ong, C.N. Determination of thiamine and its phosphate esters by gradient-elution high-performance liquid chromatography, *J.Chromatogr.*, **1991**, *567*, 71–80.

SAMPLE
Matrix: blood
Analyte: vitamin B$_1$ (thiamine)
Sample preparation: 1 mL Plasma + 150 μL 3 M perchloric acid, vortex, centrifuge at >1500 g. Remove 500 μL of the supernatant and add it to 300 μL 1 M pH 4.6 acetate buffer, add 100 μL 10 mg/mL acidic phosphatase (2 U/mg, grade II, Boehringer Mannheim) in water, heat at 40° for 16 h, add 150 μL 3 M perchloric acid, vortex, centrifuge at >1500 g, inject a 20 μL aliquot of the supernatant.

HPLC VARIABLES
Column: 125 × 4 Nucleosil 120 5 C18
Mobile phase: MeCN:buffer 25:75 (Buffer was 10 mM perchloric acid containing 10 mM octanesulfonic acid.)
Flow rate: 2
Injection volume: 20
Detector: F ex 365 em 435 following post-column reaction. The column effluent mixed with 0.8 g/L potassium ferricyanide in 3 M NaOH pumped at 1 mL/min and the mixture flowed through a 10 m × 0.3 mm ID PTFE coil at 30° to the detector.

CHROMATOGRAM
Retention time: 2.3
Limit of detection: 2 ng/mL

KEY WORDS
post-column reaction; plasma; pharmacokinetics

REFERENCE
Mascher, H.; Kikuta, C. High-performance liquid chromatographic determination of total thiamine in human plasma for oral bioavailability studies, *J.Pharm.Sci.*, **1993**, *82*, 56–59.

SAMPLE
Matrix: blood
Analyte: amprolium
Sample preparation: 200 μL Plasma + 100 μL 100 ng/mL IS in 330 mM perchloric acid + 500 μL 330 mM perchloric acid, vortex for 30 s, centrifuge at 2150 g for 10 min. Remove the supernatant and allow it to stand for 3 h, inject a 30 μL aliquot of the supernatant.

HPLC VARIABLES
Guard column: Sumipax PG-ODS-filter (Sumika, Osaka)
Column: 250 × 4.6 5 μm Capcell pack C18 UG-120 (Shiseido, Tokyo)
Mobile phase: MeCN:200 mM KH$_2$PO$_4$ 10:90 containing 5 mM sodium 1-hexanesulfonate
Column temperature: 40
Flow rate: 0.6
Injection volume: 30
Detector: F ex 400 em 460 following post-column reaction. The column effluent mixed with the reagent pumped at 0.6 mL/min and the mixture flowed through a 10 m × 0.25 mm

ID stainless steel coil at 40° to the detector. (Prepare reagent by dissolving 50 g NaOH and 800 mg potassium ferricyanide in 1 L water, store in the dark, discard after 24 h.)

CHROMATOGRAM
Retention time: 13
Internal standard: beclotiamine (3-[(4-amino-2-methyl-5-pyrimidinyl)methyl-5-(2-chloro-ethyl)-4-methylthiazolium chloride; Sankyo, Tokyo) (12)
Limit of quantitation: 5 ng/mL
Limit of detection: 2 ng/mL

OTHER SUBSTANCES
Extracted: metabolites, thiamine
Non-interfering: ethopabate

KEY WORDS
post-column reaction; chicken; plasma; pharmacokinetics

REFERENCE
Hamamoto, K.; Koike, R.; Shirakura, A.; Sasaki, N.; Machida, Y. Rapid and sensitive determination of amprolium in chicken plasma by high-performance liquid chromatography with post-column reaction, *J.Chromatogr.B*, **1997**, *693*, 489–492.

SAMPLE
Matrix: blood, tissue
Analyte: amprolium
Sample preparation: Slurry 6 g alumina (alumina B Akt. I, ICN Biomedicals) in MeCN: MeOH 60:40, add to a 300 × 15 column, wash with 30 mL MeCN:MeOH 60:40. Homogenize (Niti-on Bio-mixer BM-2) 5 g chopped tissue or plasma with 25 mL MeCN for 2 min, wash twice with 20 mL portions of MeCN, filter (cotton plug), wash filter with 30 mL n-hexane saturated with MeCN, add 30 g anhydrous sodium sulfate to the filtrate, let stand at room temperature for 30 min, filter (cotton plug), add 30 mL isopropanol to the filtrate. Evaporate the filtrate to dryness at 35°, reconstitute with 5 mL MeCN:MeOH 60:40, sonicate, add to the column, elute with 35 mL MeCN:MeOH 60:40. Add 10 mL isopropanol to the eluate and evaporate it to dryness at 40°, reconstitute with 1 µg/mL chloramphenicol in mobile phase, filter (Gelman Ekikurodisk 13 CR), inject a 20 µL aliquot of the filtrate.

HPLC VARIABLES
Column: 250 × 4.6 L-column ODS (Chemicals Inspection and Testing Institute, Tokyo)
Mobile phase: MeCN:200 mM KH_2PO_4 15:85 containing 5 mM sodium 1-hexane-sulfonate
Column temperature: 40
Flow rate: 0.7
Injection volume: 20
Detector: F ex 367 em 470 following post-column reaction. The column effluent mixed with the reagent pumped at 0.7 mL/min and the mixture flowed through a 10 m × 0.25 mm ID stainless steel coil at 40° to the detector. (Prepare reagent by dissolving 50 g NaOH in water, adding 800 mg potassium ferricyanide, and making up to 1 L with water.)

CHROMATOGRAM
Retention time: 8
Limit of detection: 2-4 ng/g

KEY WORDS
post-column reaction; chicken; muscle; liver; kidney; skin; plasma; SPE

REFERENCE
Takahashi, Y.; Sekiya, T.; Nishikawa, M.; Endoh, Y.S. Simultaneous high-performance liquid chromatographic determination of amprolium, ethopabate, sulfaquinoxaline, and N4-acetylsulfaquinoxaline in chicken tissues, *J.Liq.Chromatogr.*, **1994**, *17*, 4489–4512.

SAMPLE
Matrix: eggs, tissue
Analyte: amprolium

Sample preparation: Mix egg yolk with an equal amount of water, homogenize (Ultra-turrax) for 30 s. Blend (Lameris Lab Blender 400 stomacher) 20-30 g tissue with twice the amount of water for 5 min, centrifuge at 460 g for 10 min. Dialyze (24″ dialyzer with membrane Type C (Technicon, Tarrytown NY)) diluted egg yolk or tissue supernatant against water (both pumped at 0.6 mL/min), inject a 2 mL aliquot of the dialysate onto column A at 1 mL/min, wash with water at 1 mL/min for 6 min, backflush the contents of column A onto column B with mobile phase, after 2 min remove column A from the circuit, elute column B with mobile phase, monitor the effluent from column B.

HPLC VARIABLES
Column: A 50 × 4.6 37-50 μm Corasil C18; B 20 mm long LC8 (Supelco) + 150 × 4.6 5 μm Supelcosil LC8-DB
Mobile phase: MeOH:water:acetic acid:triethylamine 25:75:1:0.5 containing 5 mM heptanesulfonate
Column temperature: 40
Flow rate: 1
Injection volume: 2000
Detector: F ex 365 em 470 following post-column reaction. The column effluent mixed with the reagent pumped at 0.4 mL/min and the mixture flowed through a 3 m × 0.5 mm ID knitted PTFE coil to the detector. (Prepare reagent by dissolving 25 g NaOH and 160 mg potassium ferricyanide in 100 mL water.)

CHROMATOGRAM
Retention time: 10
Limit of detection: 3 ng/g

KEY WORDS
post-column reaction; column-switching; dialysis; yolk; muscle; chicken

REFERENCE
van Leeuwen, W.; Wilhelmus van Gend, H. Determination of amprolium in egg yolk and muscle tissue (chicken) by HPLC with post-column reaction and fluorometric detection, using on-line sample clean-up and pre-concentration steps, *Z.Lebensm.Unters.Forsch.*, **1988**, *186*, 500–504.

SAMPLE
Matrix: feed, premix
Analyte: amprolium
Sample preparation: Feed. 1 g Feed + 2 mL 5 μg/mL thiamine monophosphate + 10 mL 5% sulfosalicylic acid + 10 mL hexane, vortex for 1 min, centrifuge at 2400 g for 10 min. Remove the aqueous layer and filter it (0.45 μm, Gelman Acro LC 13). Inject a 360 μL aliquot onto a 300 × 6 glass column packed with 200-400 mesh Dowex AG 2-X8 anion exchange resin, elute with 100 mM HCl at 1.2 mL/min, after about 2 min collect a 6-10 mL fraction, neutralize with 1 M NaOH (pH 7.0 ± 0.5), inject a 500 μL aliquot of this fraction (J. Agric. Food Chem. 1980, 28, 1145). After 15-20 samples clean the sulfosalicylic acid from the column by backflushing with 700 mM NaCl containing 100 mM HCl and 2% ferric chloride at 1.2 mL/min, flush for 20-30 min after the last of the iron chelate has gone (about 3 h). Re-equilibrate for 30 min. Premix. 0.8 mg Premix + 0.1 mg pyrithiamine + 10 mL water, grind (Omni-Mixer), add 10 mL hexane, grind for 5 min, centrifuge at 2400 g at 4° for 10 min. Filter the aqueous layer (0.45 μm, Millipore), dilute 20 times, inject an aliquot.

HPLC VARIABLES
Guard column: 30 × 4.6 5 μm Rainin RP-18 guard column
Column: 30 × 3 3 μm Perkin-Elmer C18
Mobile phase: Feed. Gradient. 100 mM pH 5.5 sodium phosphate buffer for 6 min then 100 mM pH 2.6 sodium phosphate buffer for 19 min, re-equilibrate with original buffer for 15 min. Premix. Isocratic. 100 mM pH 2.6 Sodium phosphate buffer.
Flow rate: 1
Injection volume: 10
Detector: F ex 339 em 432 following post-column reaction. The column effluent mixed with 0.01% potassium ferricyanide in 15% NaOH pumped at 1 mL/min and flowed through a 7 m × 0.4 mm ID PTFE tube at 32° to the detector.

CHROMATOGRAM
Retention time: 13.98 (feed), 4.03 (premix)
Internal standard: thiamine monophosphate (5.26), pyrithiamine (1.62)

OTHER SUBSTANCES
Simultaneously analyzed: thiamine, thiamine diphosphate

KEY WORDS
post-column reaction

REFERENCE
Vanderslice, J.T.; Huang, M.-H.A. Liquid chromatographic determination of amprolium in poultry feed and premixes using postcolumn chemistry with fluorometric detection, *J.Assoc.Off.Anal.Chem.*, **1987**, *70*, 920–922.

SAMPLE
Matrix: rice
Analyte: thiamine
Sample preparation: Heat ground rice with at least 10 volumes of 100 mM HCl at 95-100° for 30 min with frequent mixing, cool, dilute to a thiamine concentration of 200 ng/mL with 100 mM HCl, adjust the pH of a 65 mL aliquot to 4.0-4.5 with about 5 mL 2 M sodium acetate, add 5 mL 10% takadiastase (Pfaltz and Bauer, Stamford CT) in water, heat at 45-50° for 3 h, adjust pH to 3.5, make up to 100 mL with water, filter (paper) (AOAC Official Methods of Analysis, 1990, 1049), centrifuge at 3500 rpm for 20 min, inject a 25 µL aliquot of the supernatant.

HPLC VARIABLES
Column: 150 × 4 Nucleosil 5 C18
Mobile phase: 10 mM NaH_2PO_4 containing 150 mM sodium perchlorate, adjusted to pH 2.2 with perchloric acid
Column temperature: 55
Flow rate: 0.6
Injection volume: 25
Detector: F ex 375 em 435 following post-column reaction. The column effluent mixed with 0.1% potassium ferricyanide (potassium hexacyanoferrate(III)) in 12% NaOH pumped at 0.6 mL/min and this mixture flowed through a 30 cm × 0.8 mm ID stainless steel coil at 55° to the detector. (Thiamine is converted to thiochrome.)

CHROMATOGRAM
Retention time: 8.5

KEY WORDS
post-column reaction

REFERENCE
Ohta, H.; Baba, T.; Suzuki, Y.; Okada, E. High-performance liquid chromatographic analysis of thiamine in rice flour with fluorimetric post-column derivatization, *J.Chromatogr.*, **1984**, *284*, 281–284.

SAMPLE
Matrix: tissue
Analyte: amprolium
Sample preparation: Prepare a cleanup column by plugging a 10 mm ID column with glass wool and adding 5 g activity I alumina, prewash with 70 mL MeCN:MeOH 90:10. Homogenize 10 g minced chicken muscle with 50 mL MeOH at maximum speed (Ultra-Turrax T-18), filter through cotton, repeat extraction with another 50 mL MeOH. Combine filtrates, add 20 mL 1-propanol, concentrate to 3-4 mL under vacuum at 45°. Add the residue to 20 mL MeCN and 50 mL n-hexane, shake vigorously by hand for 5 s, discard n-hexane layer, add another 50 mL n-hexane, shake for 5 min on a mechanical shaker, discard n-hexane layer. Evaporate the lower phase to dryness under vacuum at 45°, dissolve the residue in 1 mL MeOH, add to cleanup column, wash with 30 mL MeCN:MeOH 90:10,elute with 30 mL MeCN:water 95:5. Evaporate the eluate to dryness under vacuum at 45°, reconstitute the residue in 1 mL water, inject a 10 µL aliquot.

HPLC VARIABLES
Column: 250 × 4 7 µm LiChrosorb RP-8
Mobile phase: MeCN : 200 mM KH_2PO_4 20:80 containing 5 mM sodium 1-hexanesulfonate
Column temperature: 30
Flow rate: 0.7
Injection volume: 10
Detector: F ex 367 em 470 following post-column reaction. The column effluent mixed with 0.8 g/L potassium ferricyanide in 50 g/L NaOH pumped at 0.7 mL/min and flowed through a 3 m × 0.3 mm ID stainless steel reaction coil to the detector.

CHROMATOGRAM
Retention time: 10
Limit of detection: 0.01 ppm

OTHER SUBSTANCES
Non-interfering: ethopabate, sulfonamides

KEY WORDS
chicken; muscle; SPE; post-column reaction

REFERENCE
Nagata, T.; Saeki, M. Liquid chromatographic determination of amprolium in chicken tissues, using post-column reaction and fluorometric detection, *J.Assoc.Off.Anal.Chem.*, **1986**, *69*, 941–943.

SAMPLE
Matrix: urine
Analyte: corticosteroids
Sample preparation: Adjust pH of 2 mL urine to 6.5, add 100 µL β-glucuronidase (500 Fishmann U/mL, from E. coli, EC 3.2.1.31), add 200 µL 200 mM pH 6.5 phosphate buffer, add 1 drop chloroform, mix well, heat at 37° for 24 h, add 4 mL dichloromethane, shake for 3 min. Remove the organic layer and wash it with 500 µL 100 mM NaOH and 500 µL water, centrifuge. Remove 2 mL of the organic layer and evaporate it to dryness at 80°, reconstitute the residue in 100 µL MeOH : water 50:50, inject a 10 µL aliquot.

HPLC VARIABLES
Column: 250 × 4.6 10 µm Finepak SIL C18
Mobile phase: MeOH : water 50:50 (?)
Column temperature: 40
Flow rate: 0.8
Injection volume: 10
Detector: F ex 325 em 385 following post-column reaction. The column effluent mixed with the reagent pumped at 1 mL/min and the mixture flowed through a 30 m × 0.5 mm ID PTFE coil at 90° then a cooling coil to the detector. (The reagent was 100 µg/mL potassium ferricyanide in 300 mM pH 9.8 potassium borate buffer containing 1.67 mg/mL glycinamide.)

CHROMATOGRAM
Retention time: 22 (tetrahydrocortisone), 26 (tetrahydrocortisol)
Limit of detection: 5 ng

KEY WORDS
post-column reaction

REFERENCE
Seki, T.; Yamaguchi, Y. New fluorimetric detection method of corticosteroids after high-performance liquid chromatography using post-column derivatization with glycinamide, *J.Liq.Chromatogr.*, **1983**, *6*, 1131–1138.

Potassium Hydroxide

KOH $\longrightarrow$ post-column reaction

SAMPLE
Matrix: blood
Analyte: artesunic acid and dihydroartemisinin
Sample preparation: Condition a 1 mL 100 mg Bond-Elut phenyl SPE cartridge with 1 mL MeOH and 1 mL 1 M acetic acid. 1 mL Plasma + 20 µL 12.6 µg/mL artemisinin in MeOH, mix, add to the SPE cartridge, wash with two 1 mL portions of 1 M acetic acid, wash with 1 mL MeOH:1 M acetic acid 20:80, elute with two 1 mL portions of ethyl acetate:butyl chloride 20:80. Remove any portion of aqueous phase from the eluate and evaporate the organic portion to dryness under a stream of nitrogen at 40°, reconstitute the residue in 200 µL mobile phase, inject a 50-75 µL aliquot.

HPLC VARIABLES
Guard column: Symmetry C8 (Waters)
Column: 150 × 3.9 5 µm Symmetry C8 (Waters)
Mobile phase: MeCN:buffer 50:50 (Buffer was 40 mL 1 M acetic acid and 60 mL 1 M sodium acetate per liter, pH 4.8.)
Flow rate: 0.7
Injection volume: 50-75
Detector: UV 290 following post-column reaction. The column effluent mixed with 1.2 M KOH MeOH:water 90:10 pumped at 0.3 mL/min and the mixture flowed through a 1 mL reaction coil (Waters) at 69° to the detector.

CHROMATOGRAM
Retention time: 6.3 (artesunic acid), 7.5 (α-dihydroartemisinin), 10 (β-dihydroartemisinin)
Internal standard: artemisinin (13)
Limit of detection: 20-30 ng/mL

KEY WORDS
post-column reaction; plasma; SPE; pharmacokinetics

REFERENCE
Batty, K.T.; Davis, T.M.E.; Thu, L.T.A.; Binh, T.Q.; Anh, T.K.; Ilett, K.F. Selective high-performance liquid chromatographic determination of artesunate and α- and β-dihydroartemisinin in patients with falciparum malaria, *J.Chromatogr.B*, **1996**, *677*, 345–350.

SAMPLE
Matrix: formulations
Analyte: misoprostol
Sample preparation: Extract polymeric controlled-release formulations under supercritical fluid conditions using carbon dioxide:formic acid 95:5 at 330 atmospheres at 75° in a 500 µL cell, restrictor temperature 100°, collection solvent 15 mL hexane:EtOH 2:1, collection solvent temperature 0°, extraction time 1 h. Evaporate the collection solvent to dryness, reconstitute with 1 mL mobile phase, inject a 10 µL aliquot.

HPLC VARIABLES
Column: 250 × 4.6 Supelco ODS
Mobile phase: MeCN:MeOH:water 45:20:35
Flow rate: 1.5
Injection volume: 10
Detector: UV 280 following post-column reaction. The column effluent mixed with 4 M KOH pumped at 0.5 mL/min and the mixture flowed at 80° to the detector. (The secondary alcohol on the ring undergoes dehydration and the double bond so formed moves into conjugation with the ketone and pre-existing double bond.)

CHROMATOGRAM
Retention time: 7.5
Limit of detection: <0.1 ppt

OTHER SUBSTANCES
Simultaneously analyzed: degradation products

KEY WORDS
polymeric controlled-release formulation; SFE; post-column reaction; normal phase chromatography also described

REFERENCE
Roston, D.A.; Sun, J.J.; Collins, P.W.; Perkins, W.E.; Tremont, S.J. Supercritical fluid extraction-liquid chromatography method development for a polymeric controlled-release drug formulation, *J.Pharm.Biomed.Anal.*, **1995**, *13*, 1513–1520.

SAMPLE
Matrix: leaves
Analyte: artemisinin
Sample preparation: Dry Artemisia leaves at 40° for 24 h, crush, reflux with 100 mL hexane for 15 min, filter, evaporate hexane to dryness under vacuum at 40°. Add 25 mL MeCN to the residue, sonicate, filter (0.45 μm). Add 100 μL filtrate to 100 μL 1 mg/mL acetophenone in MeCN, inject a 50 μL aliquot.

HPLC VARIABLES
Guard column: 15 × 3.2 Brownlee C18
Column: 300 × 3.9 10 μm μBondapak C18
Mobile phase: MeCN:buffer 55:45 (Buffer was 8.3 g sodium acetate and 4 mL glacial acetic acid in 1 mL water, pH 5.1.)
Flow rate: 0.45
Injection volume: 50
Detector: UV 289 following post-column reaction. The column effluent mixed with 1 M KOH in MeOH:water 9:1 pumped at 0.2 mL/min and the mixture flowed through a 4.4 × 0.5 mm (sic) knitted PTFE capillary at 70° to the detector.

CHROMATOGRAM
Retention time: 16.5
Internal standard: acetophenone (11.5)
Limit of detection: 25 ng

KEY WORDS
post-column reaction

REFERENCE
ElSohly, H.N.; Croom, E.M.; ElSohly, M.A. Analysis of the antimalarial sesquiterpene artemisinin in *Artemisia annua* by high-performance liquid chromatography (HPLC) with postcolumn derivatization and ultraviolet detection, *Pharm.Res.*, **1987**, *4*, 258–260.

Pyridinium Bromide Perbromide

RELATED REFERENCE
Patel, S.; Hazel, C.M.; Winterton, A.G.; Mortby, E. Survey of ethnic foods for mycotoxins. *Food Addit.Contam.* **1996**, *13*, 833-841

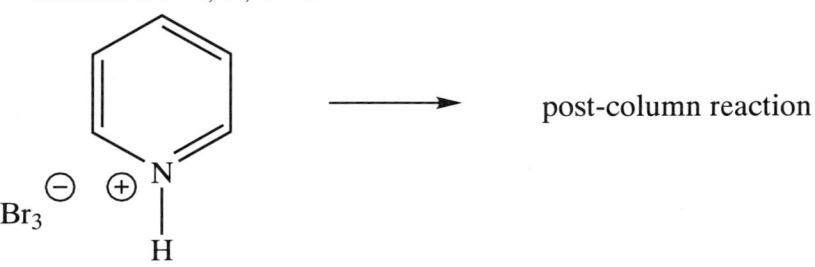

post-column reaction

SAMPLE
Matrix: spice
Analyte: aflatoxins
Sample preparation: Homogenize (Ultra Turrax) 10 g powdered spice with 40 mL
MeOH:water 80:20, centrifuge at 1600 g for 15 min. Dilute the supernatant to a MeOH
concentration of 10% with PBS, centrifuge at 1660 g for 15 min, add the supernatant to
an EASY-EXTRACT aflatoxin SPE cartridge (Biocode, York UK) at 5 mL/min, wash with
20 mL water, elute slowly with 1.5 mL MeOH, dilute the eluate with 2 mL water, inject
a 500 µL aliquot.

HPLC VARIABLES
Column: 250 × 4.6 5 µm Spherisorb ODS2
Mobile phase: MeCN:MeOH:water 22.2:17.8:60
Flow rate: 0.75
Injection volume: 500
Detector: F ex 362 em 418 following post-column reaction. The column effluent mixed with
50 µg/mL pyridinium bromide perbromide in water pumped at 0.3 mL/min and the mix-
ture flowed through a 30 cm × 0.25 mm ID coil to the detector.

CHROMATOGRAM
Retention time: 15 (aflatoxin G_2), 18 (aflatoxin G_1), 20 (aflatoxin B_2), 24 (aflatoxin B_1)
Limit of quantitation: 0.14 ppb
Limit of detection: 0.06 ppb

KEY WORDS
post-column reaction; SPE; comparison with iodine as a post-column reagent

REFERENCE
Garner, R.C.; Whattam, M.M.; Taylor, P.J.L.; Stow, M.W. Analysis of United Kingdom purchased spices
for aflatoxins using an immunoaffinity column clean-up procedure followed by high-performance liquid chromatographic analysis and post-column derivatization with pyridinium bromide perbromide,
J.Chromatogr., **1993**, *648*, 485–490.

Sodium Borohydride

RELATED REFERENCES
Le, X.C.; Ma, M. Speciation of arsenic compounds by using ion-pair chromatography with atomic spec-
troscopy and mass spectrometry detection. *J.Chromatogr.A* **1997**, *764*, 55-64.
Le, X.C.; Ma, M.; Wong, N.A. Speciation of arsenic compounds using high-performance liquid chroma-
tography at elevated temperature and selective hydride generation atomic fluorescence detection.
Anal.Chem. **1996**, *68*, 4501-4506.
López-Gonzálvez, M.; Gómez, M.M.; Palacios, M.A.; Cámara, C. Urine clean-up method for determina-
tion of six arsenic species by LC-AAS involving microwave assisted oxidation and hydride generation.
Chromatographia **1996**, *43*, 507-512.

$NaBH_4$ ⟶ post-column reaction

SAMPLE
Matrix: blood
Analyte: arsenic-containing compounds
Sample preparation: Filter (Filtron Microsep 10 kDa cutoff membrane) while centrifuging
at 3000 g for 30-50 min, mix the ultrafiltrate with concentrated mobile phase to obtain a
similar pH and ionic concentration to that of the mobile phase, filter (0.45 µm), inject a
100 µL aliquot.

HPLC VARIABLES
Guard column: Ionpac CS 10 polystyrene/divinylbenzene cation-exchange guard column
(Dionex)
Column: 250 × 4 Ionpac CS 10 polystyrene/divinylbenzene cation-exchange (Dionex)
Mobile phase: 50 mM NaH_2PO_4 containing 100 mM HCl
Flow rate: 1

Injection volume: 100
Detector: Atomic absorption, Perkin-Elmer 3030, 8 W As electrodeless discharge lamp, wavelength 193.7 nm, slit width 0.7 nm, quartz tube atomizer at 900° following post-column reaction. The column effluent mixed with argon gas flowing at 0.5 mL/min and with 15 g/L sodium persulfate in 1 M NaOH flowing at 0.5 mL/min and the mixture flowed through a 5 m × 0.8 mm ID PTFE coil illuminated by a 6 W 12 × 3 cm low-pressure UV light (Philips). The effluent from this coil mixed with five-fold diluted concentrated HCl pumped at 0.5 mL/min, with 20 g/L sodium borohydride in 0.5 g/L NaOH pumped at 0.5 mL/min, and with argon and this mixture flowed through a coil to a gas/liquid separator. The arsine gas that is generated flows to the detector.

CHROMATOGRAM
Retention time: 3.5 (monomethylarsonic acid), 5.5 (dimethylarsinic acid), 7 (arsenobetaine), 8 (arsenocholine)
Limit of detection: 1-1.5 ng/mL

KEY WORDS
post-column reaction; serum; use a metal-free pump and metal-free injection valve.

REFERENCE
Zhang, X.; Cornelis, R.; De Kimpe, J.; Mees, L.; Vanderbiesen, V.; De Cubber, A.; Vanholder, R. Accumulation of arsenic species in serum of patients with chronic renal disease, *Clin.Chem.*, **1996**, *42*, 1231−1237.

SAMPLE
Matrix: blood
Analyte: arsenic-containing compounds
Sample preparation: Filter (Filtron Microsep 10 kDa cutoff membrane) while centrifuging at 3000 g for 30-50 min, mix the ultrafiltrate with concentrated mobile phase to obtain a similar pH and ionic concentration to that of the mobile phase, filter (0.45 μm), inject a 100 μL aliquot.

HPLC VARIABLES
Guard column: C18 guard column (Bio-Rad)
Column: 250 × 4.6 C18 (Bio-Rad)
Mobile phase: 20 mM pH 6.0 Phosphate buffer containing 10 mM tetrabutylammonium
Injection volume: 100
Detector: Atomic absorption, Perkin-Elmer 3030, 8 W As electrodeless discharge lamp, wavelength 193.7 nm, slit width 0.7 nm, quartz tube atomizer at 900° following post-column reaction. The column effluent mixed with argon gas flowing at 0.5 mL/min and with 15 g/L sodium persulfate in 1 M NaOH flowing at 0.5 mL/min and the mixture flowed through a 5 m × 0.8 mm ID PTFE coil illuminated by a 6 W 12 × 3 cm low-pressure UV light (Philips). The effluent from this coil mixed with five-fold diluted concentrated HCl pumped at 0.5 mL/min, with 20 g/L sodium borohydride in 0.5 g/L NaOH pumped at 0.5 mL/min, and with argon and this mixture flowed through a coil to a gas/liquid separator. The arsine gas that is generated flows to the detector.

CHROMATOGRAM
Retention time: 3 (As(III)), 4 (dimethylarsinic acid), 6 (monomethylarsonic acid), 7 (As(V))
Limit of detection: 1-1.5 ng/mL

KEY WORDS
post-column reaction; serum; use a metal-free pump and metal-free injection valve.

REFERENCE
Zhang, X.; Cornelis, R.; De Kimpe, J.; Mees, L.; Vanderbiesen, V.; De Cubber, A.; Vanholder, R. Accumulation of arsenic species in serum of patients with chronic renal disease, *Clin.Chem.*, **1996**, *42*, 1231−1237.

Sodium Diethyldithiocarbamate

$$(H_3CH_2C)_2N \overset{\overset{\displaystyle S}{\|}}{} SNa \longrightarrow \quad \text{post-column reaction}$$

SAMPLE
Matrix: blood
Analyte: cisplatin
Sample preparation: Filter through a 10000 molecular mass cut-off filter (Filtron) by centrifuging at 4000 g at 4° for 30 min, inject an aliquot of the ultrafiltrate.

HPLC VARIABLES
Column: 150 × 4.6 5 μm Nucleosil SB strong anion exchanger
Mobile phase: MeOH:125 mM succinic acid adjusted to pH 5.2 with NaOH 60:40
Flow rate: 0.5
Detector: UV 344 following post-column derivatization with 20 mM sodium diethyldithiocarbamate in MeOH:water 60:40 pumped at 0.17 mL/min. The mixture flowed through a 500 × 3.5 mm packed-bed reactor packed with 75 μm silanized glass beads at 115° then to the detector.

CHROMATOGRAM
Retention time: 13
Limit of detection: about 200 ng/mL

KEY WORDS
plasma; human; guinea pig; post-column reaction

REFERENCE
Andersson, A.; Ehrsson, H. Determination of cisplatin and cis-diammineaquachloroplatinum(II) ion by liquid chromatography using post-column derivatization with diethyldithiocarbamate, *J.Chromatogr.B*, **1994**, *652*, 203–210.

Sodium Hydroxide

RELATED REFERENCE
Adams, E.; Dalle, J.; De Bie, E.; De Smelt, I.; Roets, E.; Hoogmartens, J. Analysis of kanamycin sulfate by liquid chromatography with pulsed electrochemical detection. *J.Chromatogr.A* **1997**, *766*, 133-139

NaOH $\longrightarrow$ post-column reaction

SAMPLE
Matrix: beverages, fruit juice, milk
Analyte: sugars
Sample preparation: Orange juice. Dilute orange juice 100-fold with water, filter (Millipore HV, 0.45 μm), dilute filtrate 10-fold, inject an aliquot. Beverages. Dilute soft drinks 1000-fold with water, inject an aliquot. Milk. Dilute 5 mL milk to 100 mL with mobile phase, filter (Millipore HV, 0.45 μm), dilute filtrate 50-fold, inject an aliquot.

HPLC VARIABLES
Guard column: 30 × 4.6 Cation H (Bio-Rad)
Column: 300 × 3.8 9 μm HPX 87-H Aminex (Bio-Rad)
Mobile phase: 10 mM Sulfuric acid
Column temperature: 50
Flow rate: 0.5
Injection volume: 40
Detector: E following post-column reaction, Hewlett-Packard 1049A programmable electrochemical detector, Metrohm detector cell, cuprous oxide working electrode +550 mV, glassy carbon auxiliary electrode, Ag/AgCl (3 M KCl) reference electrode. The column

effluent mixed with 200 mM NaOH pumped at 0.4 mL/min, the mixture flowed through a 220 × 0.8 single-bead string reactor packed with 0.6 mm glass beads to the detector. (Prepare cuprous oxide electrode as follows. Stir 300 mg conductive carbon cement (Gerhard Neubauer, Münster), 60 mg cuprous oxide (Fluka), and 300 µL acetone until a thick paste forms as the acetone evaporates. Pack conductive carbon cement into the base of a 3 mm diameter cavity carbon paste electrode base (Metrohm), allow to dry, polish with dry emery paper (grade 2/0, Oakey), remove surface layer with an acetone-soaked tissue, pack the paste into the cavity, allow to dry overnight, polish with dry emery paper (grade 2/0), 3 µm imperial micro finishing film sheet (3M), 0.3 µm imperial micro finishing film sheet (3M), and 0.05 µm alumina particles on a Buehler pad, sonicate for 2 min in water (Anal. Chim. Acta 1995, 300, 5).)

CHROMATOGRAM
Retention time: 8.03 (raffinose), 8.63 (cellobiose), 8.85 (sucrose), 8.85 (maltose), 9.03 (lactose), 9.50 (lactulose), 10.28 (galacturonic acid), 10.47 (dextrose), 10.63 (sorbose), 10.98 (myo-inositol), 11.23 (mannose), 11.25 (xylose), 11.35 (galactose), 11.55 (fructose), 11.65 (mannitol), 11.77 (rhamnose), 11.85 (lyxose), 11.98 (galactitol), 12.50 (arabinose), 12.97 (ribose), 12.98 (fucose)
Limit of detection: 1 µM

KEY WORDS
orange juice; soft drinks; post-column reaction

REFERENCE
Huang, X.; Pot, J.J.; Kok, W.T. Determination of sugars by liquid chromatography and amperometric detection with a cuprous oxide modified electrode, *Chromatographia*, **1995**, *40*, 684–689.

SAMPLE
Matrix: blood
Analyte: clavulanic acid
Sample preparation: Filter (Amicon MPS-1 with YMT membrane) while centrifuging at 1500 g for 10 min, inject a 20 µL aliquot of the ultrafiltrate.

HPLC VARIABLES
Guard column: 30 × 4.6 10 µm Develosil ODS-10
Column: 7 µm Zorbax ODS-7
Mobile phase: MeOH:buffer 1:2.7 (1:3.5 for concentrations <100 ng/mL) (Prepare buffer by dissolving 1.791 g $Na_2HPO_4.12H_2O$ and 0.780 g $NaH_2PO_4.2H_2O$ in 1 L water, add tetrabutylammonium bromide to a final concentration of 5 mM.)
Flow rate: 0.8
Injection volume: 20
Detector: UV 272 following post-column reaction. The column effluent mixed with MeOH: 500 mM NaOH 1:2.7 (1:3.5 for concentrations <100 ng/mL) pumped at 0.2 mL/min and this mixture flowed through a 1 m × 0.5 mm ID coil to the detector.

CHROMATOGRAM
Retention time: 5
Limit of quantitation: 50 ng/mL
Limit of detection: 25 ng/mL

OTHER SUBSTANCES
Non-interfering: ampicillin, cefoperazone, ticarcillin

KEY WORDS
plasma; post-column reaction; ultrafiltrate

REFERENCE
Haginaka, J.; Wakai, J.; Yasuda, H.; Uno, T.; Nakagawa, T. Improved high-performance liquid chromatographic assay of clavulanic acid and sulbactam by postcolumn alkaline degradation, *J.Liq.Chromatogr.*, **1985**, *8*, 2521–2534.

SAMPLE
Matrix: blood
Analyte: sulbactam

Sample preparation: Filter (Amicon MPS-1 with YMT membrane) while centrifuging at 1500 g for 10 min, inject a 20 μL aliquot of the ultrafiltrate.

HPLC VARIABLES
Guard column: 30 × 4.6 10 μm Develosil ODS-10
Column: 150 × 4.6 5 μm Develosil ODS-5 (Nomura Chemicals)
Mobile phase: MeOH:buffer 1:1.5 (Prepare buffer by dissolving 1.791 g $Na_2HPO_4.12H_2O$ and 0.780 g $NaH_2PO_4.2H_2O$ in 10 L water, add tetrabutylammonium bromide to a final concentration of 5 mM.)
Flow rate: 0.8
Injection volume: 20
Detector: UV 278 following post-column reaction. The column effluent mixed with MeOH: 750 mM NaOH 1:1.5 pumped at 0.2 mL/min and this mixture flowed through a 1 m × 0.5 mm ID coil to the detector.

CHROMATOGRAM
Retention time: 6
Limit of quantitation: 50 ng/mL
Limit of detection: 25 ng/mL

OTHER SUBSTANCES
Non-interfering: ampicillin, cefoperazone

KEY WORDS
ultrafiltrate; plasma; post-column reaction

REFERENCE
Haginaka, J.; Wakai, J.; Yasuda, H.; Uno, T.; Nakagawa, T. Improved high-performance liquid chromatographic assay of clavulanic acid and sulbactam by postcolumn alkaline degradation, *J.Liq.Chromatogr.*, **1985**, 8, 2521–2534.

SAMPLE
Matrix: blood
Analyte: guanidino compounds
Sample preparation: 100 μL Serum + 20 μL 20% trichloroacetic acid, vortex for a few s, centrifuge at 10000 g for 2 min. Remove a 60 μL aliquot of the supernatant and add it to 15 μL 400 mM NaOH, adjust pH to 2.5-3.0, inject a 50 μL aliquot of the supernatant.

HPLC VARIABLES
Column: 75 × 4.6 3 μm Nucleosil C8
Mobile phase: MeCN:MeOH:water 2:5:93 containing 5 mM sodium 1-octanesulfonate and 2 mM 9,10-phenanthrenequinone-3-sulfonic acid adjusted to pH 4.0 (A) or 3.5 (B) with acetic acid. (Synthesis of 9,10-phenanthrenequinonesulfonic acid is as follows. Heat 500 g phenanthrene and 600 g concentrated sulfuric acid at 120-130° with frequent shaking for 4.5-5 hours. In order to obtain a good yield this temperature must not be exceeded. Pour the hot reaction mixture into hot water (Caution! Sulfur dioxide is evolved!). Neutralize with barium carbonate until the solution is no longer acid to litmus, make up to 4-8 L, heat strongly for a short time, decant the boiling solution as soon as the barium precipitates, heat the precipitate with 3-4 L water twice more. Filter the decanted solutions, reduce the filtrate somewhat, treat with the amount of sulfuric acid necessary to precipitate the barium. Filter to remove the barium sulfate, extract the barium sulfate with a little boiling water. Add KOH to the filtrate while keeping it clearly acidic, evaporate with heating until potassium 3-phenanthrenesulfonate begins to precipitate, cool slowly to obtain potassium 3-phenanthrenesulfonate as beautiful leaves. Mix 10 g 3-phenanthrenesulfonic acid, 7 g chromium trioxide, and 70 g acetic acid. After a short time a vigorous reaction, encouraged by shaking, starts and the acetic acid boils, warm on a water bath for another 10 min, cool, add EtOH, shake well, filter to recover the crystals, wash with EtOH until no more green color can be rinsed off. Dissolve the crystals in the minimum amount of hot water, filter while hot, add EtOH to the filtrate (about a fifth the volume of the aqueous solution), allow to cool slowly to obtain potassium phanthrenequinonesulfonate as fine glittering, orange-yellow needles (Annalen 1902, 321, 248). It is advantageous to use 9,10-phenanthrenequinone-3-sulfonic acid because it is soluble in water whereas 9,10-phenanthrenequinone is almost insoluble in water.)

Flow rate: 1
Injection volume: 50
Detector: F ex 370 em 520 following post-column reaction. The column effluent mixed with 600 mM NaOH pumped at 0.5 mL/min and the mixture flowed through a 10 m × 0.5 mm ID stainless steel coil at 80° to the detector.

CHROMATOGRAM
Retention time: 2 (guanidinosuccinic acid (A)), 3 (guanidinoacetic acid (A)), 5 (taurocyamine (B)), 6 (guanidinosuccinic acid (B)), 6 (N-acetyl-L-arginine (A)), 7 (β-guanidinopropionic acid (A)), 7 (guanidinoacetic acid (B)), 9.5 (creatinine (A)), 10.5 (guanidine (A)), 13 (arginine (A)), 15 (methylguanidine (a)), 19 (gamma-guanidinobutyric acid (A))
Limit of detection: 6-100 pmole

KEY WORDS
post-column reaction; serum

REFERENCE
Kobayashi, Y.; Kubo, H.; Kinoshita, T. Fluorometric determination of guanidino compounds by high-performance liquid chromatography using water-soluble 9,10-phenanthrenequinone-3-sulfonate, *Anal.Sci.*, **1987**, *3*, 363–367.

SAMPLE
Matrix: blood
Analyte: quinacrine
Sample preparation: 1 mL Plasma + 300 μL 200 mM pH 8 Na$_2$HPO$_4$ containing 0.2 mM decylamine + 4 mL dichloromethane, extract (rotamixer), centrifuge at 1200 g. Remove 3 mL of the organic layer and evaporate it to dryness under a stream of air, reconstitute the residue in 200 μL mobile phase, inject a 40 μL aliquot.

HPLC VARIABLES
Column: 250 × 4 7 μm LiChroCart RP-18 (Merck)
Mobile phase: MeOH:10 mM phosphate buffer 65:35 containing 0.1 mM decylamine and 8 mM ethanolamine, adjust pH to 3.0 with HCl
Flow rate: 1
Injection volume: 40
Detector: F ex 270 em 495 following post-column reaction. The column effluent mixed with MeOH:500 mM NaOH 65:35 pumped at 0.3 mL/min and flowed to the detector. (Without post-column reaction use F 285 em 500, sensitivity 3 times less.)

CHROMATOGRAM
Retention time: 4
Limit of detection: 0.2 ng/mL

KEY WORDS
plasma; protect from light; post-column reaction; rinse glassware with mobile phase; decylamine prevents glass adsorption

REFERENCE
Bjorkman, S.; Elisson, L.O. Determination of quinacrine (mepacrine) in plasma by high-performance liquid chromatography with fluorimetric detection, *J.Chromatogr.*, **1987**, *420*, 341–348.

SAMPLE
Matrix: blood
Analyte: streptomycin
Sample preparation: 100 μL Serum + 50 μL 3.5% perchloric acid, vortex for a few s, centrifuge at 10000 g for 1 min, inject a 50 μL aliquot of the supernatant.

HPLC VARIABLES
Column: 100 × 8 10 μm Radial-PAK C18 radial compression (Waters)
Mobile phase: MeCN:water 20:80 containing 20 mM disodium 1,2-ethanedisulfonate, 5 mM sodium octanesulfonate, and 5 mM ninhydrin, adjusted to pH 3.3 with acetic acid
Flow rate: 1.5
Injection volume: 50

Detector: F ex 302 em 420 (cut-off filter) following post-column reaction. The column effluent mixed with 300 mM NaOH pumped at 0.5 mL/min and flowed through a 10 m × 0.5
mm ID stainless steel coil at 80° to the detector.

CHROMATOGRAM
Retention time: 12
Limit of detection: 1 μg/mL

KEY WORDS
serum; post-column reaction

REFERENCE
Kubo, H.; Li, H.; Kobayashi, Y.; Kinoshita, T. Fluorometric determination of streptomycin in serum by
high-performance liquid chromatography using mobile phase containing fluorogenic reagent,
Anal.Biochem., **1987**, *162*, 219–223.

SAMPLE
Matrix: blood
Analyte: biguanides
Sample preparation: 100 μL Serum + 100 μL 4% perchloric acid, vortex for a few s,
centrifuge at 10000 g for 1 min, inject a 100 μL aliquot of the supernatant.

HPLC VARIABLES
Column: 100 × 8 10 μm Radial-Pak μBondapak C18
Mobile phase: MeCN:water 20:80 (buformin) or 25:75 (phenformin) containing 5 mM sodium hexanesulfonate and 2 mM 9,10-phenanthrenequinonesulfonic acid, adjusted to pH
4.0 with acetic acid (Synthesis of 9,10-phenanthrenequinonesulfonic acid is as follows.
Heat 500 g phenanthrene and 600 g concentrated sulfuric acid at 120-130° with frequent
shaking for 4.5-5 hours. In order to obtain a good yield this temperature must not be
exceeded. Pour the hot reaction mixture into hot water (Caution! Sulfur dioxide is
evolved!). Neutralize with barium carbonate until the solution is no longer acid to litmus,
make up to 4-8 L, heat strongly for a short time, decant the boiling solution as soon as
the barium precipitates, heat the precipitate with 3-4 L water twice more. Filter the
decanted solutions, reduce the filtrate somewhat, treat with the amount of sulfuric acid
necessary to precipitate the barium. Filter to remove the barium sulfate, extract the
barium sulfate with a little boiling water. Add KOH to the filtrate while keeping it clearly
acidic, evaporate with heating until potassium 3-phenanthrenesulfonate begins to precipitate, cool slowly to obtain potassium 3-phenanthrenesulfonate as beautiful leaves. Mix
10 g 3-phenanthrenesulfonic acid, 7 g chromium trioxide, and 70 g acetic acid. After a
short time a vigorous reaction, encouraged by shaking, starts and the acetic acid boils,
warm on a water bath for another 10 min, cool, add EtOH, shake well, filter to recover
the crystals, wash with EtOH until no more green color can be rinsed off. Dissolve the
crystals in the minimum amount of hot water, filter while hot, add EtOH to the filtrate
(about a fifth the volume of the aqueous solution), allow to cool slowly to obtain potassium
phanthrenequinonesulfonate as fine glittering, orange-yellow needles (Annalen 1902, 321,
248). It is advantageous to use 9,10-phenanthrenequinone-3-sulfonic acid because it is
soluble in water whereas 9,10-phenanthrenequinone is almost insoluble in water.)
Flow rate: 2
Injection volume: 100
Detector: F ex 300 em 500 following post-column reaction. The column effluent mixed with
300 mM NaOH pumped at 0.5 mL/min and the mixture flowed through a 15 m × 0.5 mm
ID stainless steel coil at 70° to the detector.

CHROMATOGRAM
Retention time: 6 (buformin), 7 (phenformin)
Limit of detection: 20 ng/mL

KEY WORDS
post-column reaction; serum

REFERENCE
Kobayashi, Y.; Kubo, H.; Kinoshita, T.; Nishikawa, T. Fluorometric determination of biguanides in serum
by high-performance liquid chromatography with reagent-containing mobile phase, *J.Chromatogr.*,
1988, *430*, 65–71.

SAMPLE

Matrix: blood, eggs, tissue

Analyte: carbadox

Sample preparation: Blend 10 g muscle, liver, kidney, plasma, or eggs with 40 mL MeCN:MeOH 50:50 for 3 min (stomacher blender), centrifuge at 2000 g for 5 min, pass the supernatant through a 400 × 10 glass column containing 8 g alumina (Woelm neutral, activity 1, lower layer) and 2 g florisil (75-150 μm, upper layer), collect the eluate. Evaporate 10 mL of the eluate to 0.9-1.1 mL under a stream of nitrogen at 40-50°, dilute to 4 mL with water, mix, wash with 2 mL isooctane, centrifuge at 2000 g for 5 min, inject 1 mL of the aqueous phase onto column A with mobile phase A, after 20 min backflush the contents of column A onto column B with mobile phase B, after 5 min remove column A from the circuit, monitor the effluent from column B.

HPLC VARIABLES

Column: A 10 × 2.1 column was slurry-packed with 55-105 μm material from a C18 Sep-Pak, 20 μm screens were used; B 10 × 2.1 37-50 μm Bondapak C18/Corasil + 100 × 3 5 μm ChromSpher C18 (Chrompack)

Mobile phase: A water; B MeCN:10 mM pH 6 acetate buffer 14:86

Flow rate: A 0.3; B 0.5

Injection volume: 1000

Detector: UV 390 following post-column reaction. The column effluent mixed with 0.5 M NaOH pumped at 0.23 mL/min. The mixture flowed through a 2 m × 0.5 mm ID knitted PTFE reaction coil to the detector.

CHROMATOGRAM

Retention time: 9

Limit of detection: 1 ng/g

OTHER SUBSTANCES

Extracted: metabolites

Non-interfering: chloramphenicol, chlortetracycline, clopidol, dapsone, decoquinate, dimetridazole, dinitolmide, doxycycline, ethopabate, fenbendazole, furaltadone, furazolidone, furnicozone, halofuginone, ipronidazole, methylbenzoquate, nicarbazin, nifursol, nitrofurantoin, nitrofurazone, nitrovin, olaquindox, oxytetracycline, pyrantel, robenidine, ronidazole, sulfadiazine, sulfanilamide, sulfadimethoxine, sulfadoxine, sulfamerazine, sulfamethazine, sulfamethoxazole, sulfaquinoxaline, tetracycline, thiophanate, trimethoprim

KEY WORDS

plasma; column-switching; post-column reaction; use yellow light and amber glassware; muscle; liver; kidney; pig

REFERENCE

Binnendijk, G.M.; Aerts, M.M.L.; Keukens, H.J.; Brinkman, U.A.T. Optimization and ruggedness testing of the determination of residues of carbadox and metabolites in products of animal origin. Stability studies in animal tissues, *J.Chromatogr.*, **1991**, *541*, 401−410.

SAMPLE

Matrix: blood, tissue

Analyte: amoxicillin

Sample preparation: 9 mL Whole blood + 1 mL 1 M pH 2.6 phosphate buffer, centrifuge at 4000 g for 20 min, remove a 1.33 mL aliquot and add it to 70 μL 400 μg/mL cefadroxil in 10 mM pH 4.7 phosphate buffer. Homogenize 25 g sliced muscle with 25 mL 100 mM pH 4.7 phosphate buffer, remove a 10 g aliquot of the homogenate and add it to 50 μg cefadroxil, mix, centrifuge at 4000 g for 20 min. Condition column A with 5 mL MeOH, 5 mL water, and 5 mL 100 mM pH 8.5 phosphate buffer containing 0.5 mM hexadecyltrimethylammonium chloride. Dialyze 750 μL sample in three 250 μL portions for 6.67 min per portion against 100 mM pH 8.5 phosphate buffer pumped at 0.2 mL/min using a Cuprophan membrane (15 kDa cut-off), pump the buffer through column A to waste, wash column A with 2 mL 100 mM pH 8.5 phosphate buffer containing 0.5 mM hexadecyltrimethylammonium chloride, backflush the contents of column A onto column B with mobile phase, elute with mobile phase, monitor the effluent from column B. Flush the donor channel with 100 μL 0.01% Triton X-10 and 5 mL water. Flush the acceptor channel with 5 mL 100 mM pH 8.5 buffer.

HPLC VARIABLES
Column: A 20 × 4.6 10 μm C18 (J.T. Baker); B 150 × 4.6 Type ABZ (Supelco)
Mobile phase: MeCN:10 mM pH 7.0 phosphate buffer 13:87 containing 0.5 mM hexadecyltrimethylammonium chloride
Flow rate: 1
Detector: UV 260 following post-column reaction. The column effluent mixed with 100 mM NaOH pumped at 0.1 mL/min and the mixture flowed to the detector.

CHROMATOGRAM
Retention time: 16
Internal standard: cefadroxil (10)
Limit of detection: 50 ng/mL (blood), 200 ng/g (muscle)

KEY WORDS
post-column reaction; cow; whole blood; muscle; column-switching

REFERENCE
Snippe, N.; Van de Merbel, N.C.; Ruiter, F.P.M.; Steijer, O.M.; Lingeman, H.; Brinkman, U.A.T. Automated column liquid chromatographic determination of amoxicillin and cefadroxil in bovine serum and muscle tissue using on-line dialysis for sample preparation, *J.Chromatogr.B*, **1994**, *662*, 61–70.

SAMPLE
Matrix: bulk
Analyte: trigalacturonic acid
Sample preparation: Dissolve 10 mg trigalacturonic acid in 100 μL water, add 350 μL reagent, heat at 80° for 1 h, cool to room temperature, add glacial acetic acid until the evolution of hydrogen ceases, evaporate to dryness under a stream of air at 25°, reconstitute with 500 μL water, add to a 500 × 10 column of Sephadex G-25, elute with water at 0.4 mL/min, collect 1.2 mL fractions, assay for derivatized product at UV 274. Concentrate the appropriate fractions to 1 mL under reduced pressure, inject a 100 μL aliquot. (Prepare reagent by dissolving 137 mg tyramine and 35 mg sodium cyanoborohydride in 400 μL acetic acid:MeOH 10:90 at 100°.)

HPLC VARIABLES
Column: CarboPac PA-1 (Dionex)
Mobile phase: Gradient. 200 mM pH 8 sodium acetate:500 mM pH 8 sodium acetate from 100:0 to 0:100 over 30 min.
Flow rate: 1
Injection volume: 100
Detector: UV 274; E, Dionex, gold working electrode, pulsed amperometric mode, E1 150 mV, E2 700 mV, E3 -300 mV, T1 480 ms, T2 120 ms, T3 360 ms, 3 μA sensitivity following post-column reaction. The column effluent mixed with 400 mM NaOH pumped at 0.5 mL/min and the mixture flowed to the detector.

CHROMATOGRAM
Retention time: 20 (lactone form), 27 (free acid form)

KEY WORDS
post-column reaction

REFERENCE
Spiro, M.D.; Ridley, B.L.; Glushka, J.; Darvill, A.G.; Albersheim, P. Synthesis and characterization of tyramine-derivatized (1–>4)-linked α-D-oligogalacturonides, *Carbohydr.Res.*, **1996**, *290*, 147–157.

SAMPLE
Matrix: bulk, formulations
Analyte: gentamicin
Sample preparation: Prepare a 100-200 μg/mL solution in water, inject a 20 μL aliquot.

HPLC VARIABLES
Guard column: 50 × 4 Carbopac PA-1 anion-exchange (Dionex)

Column: 250 × 4 Carbopac PA-1 anion-exchange (Dionex)
Mobile phase: Gradient. Water: 10 mM NaOH from 70:30 to 50:50 over 15 min, re-equilibrate for 5 min.
Flow rate: 1
Injection volume: 20
Detector: E, Dionex PED-1 pulsed amperometric detector, gold working electrode, amperometry mode, E1 0.10 V, t1 300 ms, E2 0.60 V t2 120 ms, E3 -0.80 V t3 300 ms following post-column reaction. The effluent from the column mixed with 500 mM NaOH pumped at 0.5 mL/min and flowed through a mixing coil (Dionex RDM) to the detector.

CHROMATOGRAM
Retention time: 5.91 (C_{1a}), 6.66 (C_2), 8.04 (C_{2a}), 10.05 (C_1)
Limit of detection: 20 ng

OTHER SUBSTANCES
Non-interfering: cefazolin, clindamycin, cloxacillin, kanamycin, neomycin, penicillin G, tobramycin

KEY WORDS
post-column reaction; injections

REFERENCE
Kaine, L.A.; Wolnik, K.A. Forensic investigation of gentamicin sulfates by anion-exchange ion chromatography with pulsed electrochemical detection, *J.Chromatogr.A*, **1994**, *674*, 255–261.

SAMPLE
Matrix: cereal
Analyte: trichothecene mycotoxins
Sample preparation: Activate 60-100 mesh Florisil-PR at 130° for 2 h, store in a desiccator. Add 10 g anhydrous sodium sulfate to a 300 × 22 glass column, add 40 mL n-hexane, add a slurry of 20 g Florisil in n-hexane, add 15 g anhydrous sodium sulfate. Shake 20 g finely-ground cereal with 200 mL MeCN:water 75:25 for 30 min, filter (paper), wash a 125 mL aliquot of the filtrate with 100 mL n-hexane for 10 s. Evaporate the MeCN/water layer to dryness under reduced pressure at 45°, reconstitute with 5 mL MeOH with sonication, add a 4 mL aliquot to the Florisil column, wash with 200 mL n-hexane, elute with 250 mL chloroform:MeOH 90:10 at 10 mL/min (Food Addit. Contam. 1985, 2, 125). Evaporate the eluate to near dryness under reduced pressure at 45°, reconstitute with 1 mL MeOH:water 20:80, add to a Sep-Pak CN SPE cartridge, elute with 4 mL MeOH:water 20:80. Evaporate the eluate to dryness and reconstitute the residue in 200 μL MeCN:water 15:85, inject a 10 μL aliquot.

HPLC VARIABLES
Guard column: LiChroCART RP-18 (Merck)
Column: 250 × 4 10 μm LiChrosorb RP-18
Mobile phase: MeCN:water 15:85
Flow rate: 1
Injection volume: 10
Detector: F ex 370 em 460 following post-column reaction. The effluent from the column mixed with 150 mM NaOH pumped at 0.5 mL/min and this mixture flowed through an 8 m × 0.5 mm ID PTFE coil at 115°. The effluent from this coil mixed with 2 M ammonium acetate containing 30 mM methyl acetoacetate (prepared fresh each day) and this mixture flowed through a 6 m × 0.5 mm ID PTFE coil at 115° and a 1 m × 0.5 mm ID PTFE cooling coil in a water bath to the detector. (The tricothecene was degraded by NaOH to formaldehyde. The formaldehyde then reacted with ammonia and methyl acetoacetate to form a pyridine compound via the Hantzch reaction.)

CHROMATOGRAM
Retention time: 4.7 (nivalenol), 6.0 (deoxynivalenol), 8.4 (fusarenon-X)
Limit of detection: 5-10 ng

KEY WORDS
post-column reaction; barley; corn; wheat; SPE

REFERENCE

Sano, A.; Matsutani, S.; Suzuki, M.; Takitani, S. High-performance liquid chromatographic method for determining trichothecene mycotoxins by post-column fluorescence derivatization, *J.Chromatogr.*, **1987**, *410*, 427–436.

SAMPLE

Matrix: food

Analyte: pesticides

Sample preparation: Prepare a column by adding 500 mg silanized Celite 545 to a 22 mm ID column, add 5 g Nuchar S-N:silanized Celite 545 20:80, wash with 50 mL MeCN: toluene 75:25, do not allow to go dry. Homogenize (Polytron) 150 g sample with 300 mL MeOH at half speed for 30 s and full speed for 1 min, filter, remove a portion equivalent to 100 g sample, add water so that there is 100 mL water in the flask, concentrate to 75 mL under reduced pressure at 35°, add 15 g NaCl, add 75 mL MeCN, shake for 30 s, extract the aqueous layer with 50 mL MeCN for 20 s. Combine the organic layers and add them to 25 mL 20% NaCl, shake, discard the aqueous layer, wash the MeCN layer with 100 mL petroleum ether, extract the petroleum ether layer with 10 mL MeCN. Combine the MeCN layers and add 50 mL 2% NaCl, extract with 100 mL dichloromethane, extract twice with 25 mL portions of dichloromethane, filter the dichloromethane layers through 5 g anhydrous sodium sulfate, evaporate the dichloromethane extracts to dryness under reduced pressure, immediately reconstitute with 10 mL dichloromethane, add to the column, elute at 5 mL/min, rinse flask with 10 mL dichloromethane, add the rinse to the column, elute with 125 mL MeCN:toluene 75:25, collect all the effluent from the column and evaporate it just to dryness under reduced pressure, reconstitute with 5 mL MeOH, filter (5 μm), inject an aliquot. (Silanize Celite 545 by boiling 150 g Celite 545 in 1 L HCl:water 50:50 for 10 min, cool, filter, wash with water until the filtrate is neutral, wash with 500 mL MeOH, wash with 500 mL dichloromethane, air dry, heat to 120°, cool in desiccator, add 3 mL dichlorodimethylsilane, mix well, let stand at room temperature for 4 h, add 500 mL MeOH, mix, let stand for 15 min, filter, wash with isopropanol until neutral, air dry, dry at 105° for 2 h, cool in desiccator. Test for total silanization by placing 1 g in 20 mL toluene saturated with methyl red. Silanized Celite should appear yellow. Silanized Celite should also float on water. If silanization is not complete, repeat process. Boil 100 g Nuchar S-N with 700 mL HCl for 1 h, add 700 mL water, boil for 30 min, cool, filter, wash with water until the filtrate is neutral,wash with 500 mL MeOH, wash with 500 mL dichloromethane, air dry, dry at 120° for 4 h, cool in desiccator.) (J. Assoc. Off. Anal. Chem. 1985, 68, 726).

HPLC VARIABLES

Guard column: 20 × 2 30-40 μm Perisorb RP-8

Column: 250 × 4.6 6 μm Zorbax C8

Mobile phase: Gradient. MeCN:water from 20:80 to 70:30 over 25 min, re-equilibrate at initial conditions for 10 min.

Column temperature: 35

Flow rate: 1.5

Injection volume: 20

Detector: E, ESA Model 5100A, Model 5010 dual analytical cell, detector 1 + 0.20 V, detector 2 +0.60 V (monitored), Model 5020 guard cell in NaOH stream at +0.70 V, following post-column reaction. The column effluent mixed with 100 mM NaOH pumped at 0.5 ± 0.02 mL/min and the mixture flowed through a 3 m × 0.48 mm ID stainless-steel coil at 100° to the detector.

CHROMATOGRAM

Retention time: 7 (3-hydroxycarbofuran), 14 carbofuran), 15 (carbaryl), 16.7 (isoprocarb), 18.8 (methiocarb), 22.4 (bufencarb)

Limit of quantitation: 10 ppb

Limit of detection: 0.25-0.65 ng

KEY WORDS

post-column reaction; apples; cabbages; grapes; tomatoes

REFERENCE

Krause, R.T. High-performance liquid chromatographic determination of aryl N-methylcarbamate residues using post-column hydrolysis electrochemical detection, *J.Chromatogr.*, **1988**, *442*, 333–343.

SAMPLE

Matrix: fruit, grain, vegetables

Analyte: carbamate insecticides

Sample preparation: Homogenize (Polytron) 150 g high moisture sample and 300 mL MeOH at half speed for 30 s and full speed for 1 min, filter (paper) under vacuum, remove portion of filtrate equal to 100 g sample and make up to 100 mL with water. Homogenize (Polytron) 75 g low moisture sample and 300 mL MeOH at half speed for 30 s and full speed for 1 min, filter (paper) under vacuum, remove portion of filtrate equal to 50 g sample and make up to 100 mL with water. Concentrate samples to 75 mL under reduced pressure at 35°, add 15 g NaCl, add 75 mL MeCN, shake for 30 s, let stand for 5 min. Remove the aqueous phase and add it to 50 mL MeCN, shake for 20 s, let layers separate, discard the aqueous layer. Combine the MeCN layers, wash with 25 mL 20% NaCl, wash with 100 mL petroleum ether, extract petroleum ether layer with 10 mL MeCN. Combine the MeCN layers and add them to 50 mL 2% NaCl, extract with 100 mL dichloromethane, extract twice with 25 mL portions of dichloromethane. Combine the dichloromethane layers and pass them through a 22 mm ID column containing 5 g anhydrous sodium sulfate. Evaporate the eluate to dryness under reduced pressure at 35°, reconstitute in 10 mL dichloromethane, add to the charcoal column, rinse flask with 10 mL dichloromethane, rinse flask with 25 mL MeCN:toluene 75:25. Evaporate the eluate to dryness under reduced pressure at 35°, reconstitute with 5 mL MeOH, filter (5 μm), inject a 10 μL aliquot (J. Assoc. Off. Anal. Chem. 1980, 63, 1114). (Charcoal column was 5 g silanized Celite 545:Nuchar S-N 4:1 on top of 0.5 g silanized Celite 545 in a 300 × 22 glass column, wash with 50 mL MeCN:toluene 75:25, do not allow to go dry. Prepare silanized Celite 545 as follows. Boil 150 g Celite 545 in 1 L 6 M HCl with stirring for 10 min, cool, filter, wash with water until filtrate is neutral, wash with 500 mL MeOH, wash with 500 mL dichloromethane, air dry in hood, heat to 120° in a flask, cool in a desiccator, add 3 mL dichlorodimethylsilane, mix well, let stand at room temperature for 4 h, add 500 mL MeOH, mix, let stand for 15 min, filter, wash with isopropanol until neutral, air dry in hood, dry at 105° for 2 h, cool in desiccator, store in stoppered container. Totally silanized Celite should float on water and appear yellow (not pink) in toluene saturated with methyl red (J. Assoc. Off. Anal. Chem. 1980, 63, 1114).)

HPLC VARIABLES

Guard column: 70 × 2.1 25-37 μm Co-Pell ODS

Column: 250 × 4.6 6 μm Zorbax C8

Mobile phase: Gradient. MeCN:water from 12:88 to 70:30 over 30 min, 100:0 for 5 min.

Column temperature: 35

Flow rate: 1.5

Injection volume: 10

Detector: F ex 288 em 330 following post-column reaction. The column effluent mixed with 200 mM NaOH pumped at 0.5 mL/min and flowed through a 3 m × 0.48 mm stainless steel column to the detector.

CHROMATOGRAM

Retention time: 20 (carbofuran), 21 (carbaryl), 26 (azinphos methyl), 27 (napropamide), 29 (azinphos ethyl), 34 (phosalone), 35 (piperonyl butoxide)

Limit of quantitation: 20 ppb

KEY WORDS

post-column reaction; pears; green beans

REFERENCE

Krause, R.T.; August, E.M. Applicability of a carbamate insecticide multiresidue method for determining additional types of pesticides in fruits and vegetables, *J.Assoc.Off.Anal.Chem.*, **1983**, *66*, 234–240.

SAMPLE

Matrix: milk, tissue

Analyte: streptomycin and dihydrostreptomycin

Sample preparation: Tissue. Condition a 5 mL 500 mg 40 μm Bakerbond aromatic sulfonic acid SPE cartridge with 5 mL water. Homogenize (Polytron Model PT 10/35) 5 g tissue with 10 mL 3.6% perchloric acid for 10-15 s, shake horizontally for 5 min, centrifuge at 2000 g for 5 min, add the supernatant to the SPE cartridge at 2 mL/min, wash with 3 mL water, allow to dry, elute with 9 mL buffer. Add 500 μL 200 mM 1-hexanesulfonic acid in water and 150 μL 70% perchloric acid to the eluate, make up to 10 mL with water, filter (0.45 μm), inject a 2 mL aliquot of the filtrate on to column A and elute to waste with mobile phase A, after 5 min elute the contents of column A on to column B with mobile phase B, after 5 min remove column A from the circuit, elute column B with mobile phase B, monitor the effluent from column B. Milk. 10 g Milk + 3 mL 3.6% perchloric acid, shake horizontally for 30 s, centrifuge at 2000 g for 5 min. Add the supernatant to 70 μL 5 M NaOH and 500 μL 200 mM 1-hexanesulfonic acid, make up to 10 mL with water, filter (0.45 μm), inject a 2 mL aliquot of the filtrate on to column A and elute to waste with mobile phase A, after 5 min elute the contents of column A on to column B with mobile phase B, after 5 min remove column A from the circuit, elute column B with mobile phase B, monitor the effluent from column B (J. AOAC Int. 1994, 77, 765). (Prepare the buffer by dissolving 33.46 g K_2HPO_4 and 1.05 g KH_2PO_4 in 1 L water, pH 8.0. At the end of each day wash with MeCN:water 50:50 at 1.5 mL/min for 10 min.)

HPLC VARIABLES
Column: A 40 × 4.6 5 μm Inertsil C8; B 250 × 4.6 5 μm LC8-DB (Supelco)
Mobile phase: A 10 mM 1-Hexanesulfonic acid adjusted to pH 3.3 with acetic acid; B MeCN:water 17:83 containing 10 mM 1-hexanesulfonic acid and 400 μM 1,2-naphthoquinone-4-sulfonic acid, adjust pH to 3.3 with acetic acid
Flow rate: A 1; B 1.5
Injection volume: 2000
Detector: F ex 347 em 418 (tissue) or ex 365 em 418 (milk) following post-column reaction. The column effluent mixed with 500 mM NaOH pumped at 0.5 mL/min and the mixture flowed through a 2 mL reaction coil (Pickering, Mountain View CA) at 50° to the detector. (At the end of each day wash the post-column reaction system with 1% acetic acid for 10 min and with water for 10 min.)

CHROMATOGRAM
Retention time: 22 (streptomycin), 23 (dihydrostreptomycin)
Limit of detection: 10-20 ppb

KEY WORDS
post-column reaction; column-switching; SPE; pig; cow; muscle; kidney

REFERENCE
Gerhardt, G.C.; Salisbury, C.D.C.; MacNeil, J.D. Determination of streptomycin and dihydrostreptomycin in animal tissue by on-line sample enrichment liquid chromatography, *J.AOAC Int.*, **1994**, 77, 334–337.

SAMPLE
Matrix: perfusate
Analyte: indomethacin
Sample preparation: Mix 100 nL perfusate with 600 nL perfusion fluid containing IS, inject an aliquot. (Perfusion fluid contained 104 mM NaCl, 25 mM sodium bicarbonate, 2.3 mM sodium biphosphate, 10 mM sodium acetate, 1.2 mM calcium chloride, 1 mM magnesium sulfate, 5 mM KCl, 5 mM dextrose, and 5 mM alanine.)

HPLC VARIABLES
Column: 300 × 2 10 μm μBondapak C18
Mobile phase: MeOH:water 52:48
Flow rate: 0.13
Injection volume: 0.2
Detector: F ex 295 em 376 following post-column reaction. The column effluent mixed with 4 M NaOH pumped at 0.0013 mL/min and the mixture flowed through a 130 μL PTFE coil at 64° to the detector.

CHROMATOGRAM
Internal standard: phenylbutazone
Limit of detection: 25 ng/mL

KEY WORDS
post-column reaction; microbore

REFERENCE
De Zeeuw, D.; Leinfelder, J.L.; Brater, D.C. Highly sensitive measurement of indomethacin using a high performance liquid chromatographic technique combined with post column in-line hydrolysis, *J.Chromatogr.*, **1986**, *380*, 157−162.

SAMPLE
Matrix: solutions
Analyte: monosaccharides and inositols

HPLC VARIABLES
Column: 300 × 8.7 Aminex HPX-87C Ca^{++} (Bio-rad)
Mobile phase: Water
Column temperature: 50
Flow rate: 0.6
Detector: E, pulsed amperometric detector (Dionex ?), E1 0.1 V, T1 300 ms, E2 0.6 V, T2 120 ms, E3 -0.8 V, T3 300 ms, following post-column reaction. The column effluent mixed with 100 mM NaOH pumped at 0.2 mL/min and the mixture flowed to the detector.

CHROMATOGRAM
Retention time: 10.02 (β-glucose), 11.16 (scyllo-inositol), 11.22 (α-glucose), 11.55 (β-galactose), 12.24 (mannose), 13.20 (α-galactose), 13.37 (fucose), 13.64 (myo-inositol), 14.16 (chiro-inositol), 14.73 (2-deoxyribositol), 18.37 (2-deoxygalactitol), 19.58 (neo-inositol), 19.82 (mannitol), 24.62 (galactitol), 25.10 (perseitol), 25.50 (fucitol), 26.16 (sorbitol)

KEY WORDS
post-column reaction

REFERENCE
Wang, W.T.; Safar, J.; Zopf, D. Analysis of inositol by high-performance liquid chromatography, *Anal.Biochem.*, **1990**, *188*, 432−435.

SAMPLE
Matrix: solutions
Analyte: N-butyldeoxynojirimycin
Sample preparation: Inject a 15 μL aliquot of a 1 mg/mL solution in mobile phase.

HPLC VARIABLES
Column: 250 × 4.6 Zorbax Rx C8
Mobile phase: MeCN:buffer 10 :90 (Buffer was 5 mM sodium heptanesulfonate adjusted to pH 3.0 with glacial acetic acid.)
Column temperature: 35
Flow rate: 1
Injection volume: 15
Detector: E, Dionex Pulsed Amperometric Detector, PAD-II cell, E1 +0.05 V, t1 300 ms, E2 +0.06 V, t2 120 ms, E3 -0.8 V, t3 300 ms, gold working electrode, silver reference electrode, stainless steel counter electrode following post-column reaction. The column effluent mixed with 300 mM NaOH pumped at 0.5 mL/min and the mixture flowed to the detector.

CHROMATOGRAM
Retention time: 18
Limit of quantitation: 500 ng/mL

KEY WORDS
post-column reaction

REFERENCE
Roston, D.A.; Rhinebarger, R.R. Evaluation of HPLC with pulsed-amperometric detection for analysis of an aminosugar drug substance, *J.Liq.Chromatogr.*, **1991**, *14*, 539−556.

SAMPLE
Matrix: solutions
Analyte: spectinomycin and impurities
Sample preparation: Inject 50 μL of a solution in 150 mM sodium acetate.

HPLC VARIABLES
Guard column: 25 × 4 IonPac CG3 (Dionex)
Column: 250 × 4 IonPac CS3 (Dionex)
Mobile phase: 150 mM pH 6 sodium acetate buffer
Column temperature: 21
Flow rate: 1
Injection volume: 50
Detector: E following post-column reaction, Dionex pulsed-amperometric detector, gold working electrode, stainless steel auxiliary electrode, Ag/AgCl reference electrode, rise-time filter 3.0 s, output range 1 μA. Pulse sequence was +0.05 V sampling potential with 100 ms delay-time and 380 ms measuring time followed by +0.60 V oxidation potential to clean the electrode for 120 ms and -0.70 V reduction potential to activate the electrode for 60 ms. The column effluent mixed with 100 mM NaOH pumped at 0.5 mL/min and the mixture flowed to the detector.

CHROMATOGRAM
Retention time: 2.5 (actinospecinoic acid), 10.7 (spectinomycin), 12.5 (actinamine)
Limit of quantitation: 60 ng
Limit of detection: 20 ng

KEY WORDS
post-column reaction

REFERENCE
Phillips, J.G.; Simmonds, C. Determination of spectinomycin using cation-exchange chromatography with pulsed amperometric detection, *J.Chromatogr.A*, **1994**, *675*, 123–128.

SAMPLE
Matrix: solutions
Analyte: digoxin and metabolites
Sample preparation: Inject a 10 μL aliquot of a solution in mobile phase.

HPLC VARIABLES
Column: 250 × 4.6 Deltabond C18 (Keystone)
Mobile phase: Gradient. MeCN:water from 10:90 to 45:55 over 8 min.
Flow rate: 1.3
Injection volume: 10
Detector: E, Dionex pulsed electrochemical detector, integrated amperometry mode, 1.4 mm gold working electrode with 0.005 inch gasket, E1 +0.07 V, t1 400 ms, E2 +0.70 V, t2 120 ms, E3 1.00 V, t3 300 ms, stainless steel counter electrode, Ag/AgCl reference electrode, following post-column reaction. The column effluent mixed with 1 M NaOH pumped at 0.5 mL/min and the mixture flowed through a 500 μL reaction coil (Dionex) to the detector.

CHROMATOGRAM
Retention time: 7 (digoxigenin), 8 (digoxigenin monodigitoxoside), 9 (digoxigenin bisdigitoxoside), 9.5 (digoxin), 10.5 (digitoxigenin), 11 (digitoxigenin monodigitoxoside), 11.5 (digitoxigenin bisdigitoxoside), 12 (digitoxin)
Limit of detection: 160-400 ng/mL

KEY WORDS
post-column reaction

REFERENCE
Kelly, K.L.; Kimball, B.A.; Johnston, J.J. Quantitation of digitoxin, digoxin, and their metabolites by high-performance liquid chromatography using pulsed amperometric detection, *J.Chromatogr.A*, **1995**, *711*, 289–295.

SAMPLE
Matrix: solutions
Analyte: aminoglycoside antibiotics
Sample preparation: Inject an aliquot of an aqueous solution.

HPLC VARIABLES
Column: 250 × 4.6 8 μm PLRP-S 1000 Å poly(styrene-divinylbenzene) (Polymer Laboratories)
Mobile phase: Water containing 70 g/L sodium sulfate, 1.4 g/L sodium 1-octanesulfonate, and 50 mL/L 200 mM pH 3.0 phosphate buffer
Column temperature: 35
Flow rate: 1
Injection volume: 20
Detector: E, Dionex PED-1 pulsed electrochemical detector at 35°, 3 mm dia. gold working electrode, E_1 +0.05 V, E_2 +0.75 V, E_3 -0.15 V, t_1 0-0.40 s, t_2 0.41-0.60 s, t_3 0.61-1.00 s, measure signal between 0.2 and 0.4 s, stainless steel counter electrode, Ag/AgCl reference electrode, following post-column reaction. The column effluent mixed with 500 mM NaOH pumped at 0.3 mL/min and the mixture flowed through a 1.2 m long 500 μL coil to the detector. (Prepare 500 mM NaOH solution by diluting 50% NaOH with helium-degassed water. Clean gold electrode after each 60 analyses.)

CHROMATOGRAM
Retention time: 4 (paromamine), 5 (neomycin LP-A), 6 (neamine), 9 (paromomycin II), 11 (paromomycin I), 15 (neomycin LP-B), 17 (neomycin C), 23 (neomycin B)
Limit of quantitation: 15 ng
Limit of detection: 5 ng

KEY WORDS
post-column reaction

REFERENCE
Adams, E.; Schepers, R.; Roets, E.; Hoogmartens, J. Determination of neomycin sulfate by liquid chromatography with pulsed electrochemical detection, *J.Chromatogr.A*, **1996**, *741*, 233–240.

SAMPLE
Matrix: solutions
Analyte: disaccharides
Sample preparation: Inject a 10-20 μL aliquot.

HPLC VARIABLES
Column: 250 × 4.6 TSK gel SAX
Mobile phase: 100 mM Acetic acid containing 6 mM KCl
Column temperature: 80
Flow rate: 0.5
Injection volume: 10-20
Detector: F ex 335 em 390 following post-column reaction. The column effluent mixed with 2 M NaOH pumped at 0.2 mL/min and with 1% 2-cyanoacetamide pumped at 0.2 mL/min and the mixture flowed through a 20 m × 0.5 mm ID coil at 110° and a 2 m × 0.25 mm ID cooling coil to the detector.

CHROMATOGRAM
Retention time: 27 (N-acetyldermosine), 39 (N-acetylchondrosine), 41 (N-acetylhyalobiuronic acid)

KEY WORDS
post-column reaction

REFERENCE
Qiu, G.; Toyoda, H.; Toida, T.; Koshiishi, I.; Imanari, T. Compositional analysis of hyaluronan, chondroitin sulfate and dermatan sulfate: HPLC of disaccharides produced from the glycosaminoglycans by solvolysis, *Chem.Pharm.Bull.*, **1996**, *44*, 1017–1020.

SAMPLE
Matrix: tissue

Analyte: carbadox
Sample preparation: Blend (stomacher) 10 g tissue with 40 mL MeCN:MeOH for 3 min, centrifuge at 2000 g for 5 min, add the supernatant to a 400 × 10 column having 2 g 75-150 μm Florisil on top of 8 g alumina (Woelm neutral, activity 1). Collect the first 10 mL of eluate and evaporate it to 1-1.5 mL under a stream of nitrogen at 40-50°, make up to 4 mL with water, mix, add 2 mL isooctane, extract, centrifuge at 2000 g for 5 min, inject a 2 mL aliquot of the aqueous phase onto column A and elute to waste with mobile phase A, after 20 min backflush the contents of column A onto column B, after 5 min remove column A from the circuit, elute column B with mobile phase B, monitor the effluent from column B.

HPLC VARIABLES
Column: A 60 × 4.6 37-50 μm Bondapak C18/Corasil; B 10 × 2.1 37-50 μm Bondapak C18/Corasil + 200 × 3 5 μm ChromSpher C18 (Chrompack)
Mobile phase: A Water; B MeCN:10 mM pH 6 sodium acetate buffer 15:85
Flow rate: A 0.5; B 0.6
Injection volume: 2000
Detector: UV 420 following post-column reaction. The column effluent mixed with 500 mM NaOH pumped at 0.23 mL/min and the mixture flowed through a 2 m × 0.5 mm ID knitted PTFE coil to the detector.

CHROMATOGRAM
Retention time: 6
Limit of detection: 0.5-1 ng/g

OTHER SUBSTANCES
Extracted: metabolites
Simultaneously analyzed: furaltadone, furazolidone, nitrofurantoin, nitrofurazone
Non-interfering: chloramphenicol, chlortetracycline, clopidol, dapsone, decoquinate, dimetridazole, dinitolmide, doxycycline, ethopabate, fenbendazole, furnicozone, halofuginone, ipronidazole, methylbenzoquate, nicarbazin, nifursol, nitrovin, olaquindox, oxytetracycline, pyrantel, robenidine, ronidazole, sulfadiazine, sulfadimethoxine, sulfadoxine, sulfamerazine, sulfamethazine, sulfamethoxazole, sulfanilamide, sulfaquinoxaline, tetracycline, thiophanate, trimethoprim

KEY WORDS
post-column reaction; column-switching; protect from light; muscle; liver; kidney; pig; SPE

REFERENCE
Aerts, M.M.L.; Beek, W.M.J.; Keukens, H.J.; Brinkman, U.A.T. Determination of residues of carbadox and some of its metabolites in swine tissues by high-performance liquid chromatography using on-line pre-column enrichment and post-column derivatization with UV-VIS detection, *J.Chromatogr.*, **1988**, *456*, 105–119.

SAMPLE
Matrix: urine
Analyte: 2,5-hexanedione
Sample preparation: 2.5 mL Urine + 150 μL concentrated HCl, heat at 95° for 45 min, cool, add 750 mg NaCl, extract with 1 mL dichloromethane, add to a silica SPE cartridge, elute with 1 mL mobile phase, inject a 20 μL aliquot of the eluate.

HPLC VARIABLES
Column: 250 × 4.6 Nucleosil 120-5C18
Mobile phase: EtOH:buffer 4.6:95.4 (w/v) (Mobile phase was 5 mM pH 7.2 phosphate buffer containing 1 M EtOH. At the end of each run purge column with 6 M EtOH.)
Flow rate: 1
Injection volume: 20
Detector: UV 235 following post-column reaction. (The column effluent mixed with 6.25 M NaOH pumped at 0.2 mL/min and the mixture flowed through a 4.2 m × 0.5 mm PEEK coil at 165° to the detector.)

CHROMATOGRAM
Limit of detection: <700 ng/mL

KEY WORDS

post-column reaction; SPE

REFERENCE

Dietz, W.; Vujtovic-Ockenga, N.; Gans, G. Determination of urinary 2,5-hexanedione by high-performance liquid chromatography HPLC with post-column derivatisation to 3-methyl-2-cyclopentenone, *Naunyn-Schmiedeberg's Arch.Pharmacol.*, **1992**, *345*, R31.

Sodium Hypochlorite

RELATED REFERENCES

Haginaka, J.; Wakai, J. Liquid chromatographic determination of amoxicillin and its metabolites in human urine by postcolumn degradation with sodium hypochlorite. *J.Chromatogr.* **1987**, *413*, 219-226.

Haginaka, J.; Wakai, J.; Nishimura, Y.; Yasuda, H. Liquid chromatographic determination of penicillins by postcolumn degradation with sodium hypochlorite using a hollow-fiber membrane reactor. *J.Chromatogr.* **1988**, *447*, 365-372.

NaOCl $\longrightarrow$ post-column reaction

SAMPLE

Matrix: bile, blood, urine

Analyte: ampicillin

Sample preparation: Plasma. 100 μL Plasma + 100 μL 10 M urea, mix, filter (Amicon MPS-1) while centrifuging at 1500 g for 10 min, inject a 20-50 μL aliquot of the ultrafiltrate. Bile. Dilute bile 5-10-fold with mobile phase, centrifuge at 8000 g for 5 min, inject a 20-50 μL aliquot of the supernatant. Urine. Dilute urine 5-10-fold with water, filter (0.45 μm), inject a 20-50 μL aliquot of the filtrate.

HPLC VARIABLES

Guard column: 30 × 4.6 5 μm Nucleosil 5C18

Column: 150 × 4.6 5 μm Nucleosil 5C18

Mobile phase: MeOH:10 mM NaH_2PO_4:10 mM Na_2HPO_4 1:1:1 (plasma) or MeOH:15 mM sodium heptanesulfonate:3 mM NaH_2PO_4:27 mM phosphoric acid 30:16:16:16 (bile) or MeOH:10 mM sodium heptanesulfonate:100 mM phosphoric acid 10:8:8 (urine)

Flow rate: 0.8

Injection volume: 20-50

Detector: UV 270 following post-column reaction. The column effluent mixed with 0.02% sodium hypochlorite in 1.5 M NaOH pumped at 0.2 mL/min and the mixture flowed through a 1 m × 0.5 mm ID PTFE coil to the detector.

CHROMATOGRAM

Retention time: 7.5 (plasma), 30 (bile, urine)

Limit of detection: 25 ng

OTHER SUBSTANCES

Extracted: metabolites

KEY WORDS

post-column reaction; rat; plasma; ultrafiltrate; pharmacokinetics

REFERENCE

Haginaka, J.; Wakai, J.; Yasuda, H.; Uno, T.; Takahashi, K.; Katagi, T. High-performance liquid chromatographic determination of ampicillin and its metabolites in rat plasma, bile and urine by post-column degradation with sodium hypochlorite, *J.Chromatogr.*, **1987**, *400*, 101–111.

SAMPLE

Matrix: blood

Analyte: CL 275,838 (4,5-dihydro-4-[[(4-phenylmethyl)-1-piperazinyl]acetyl]-7-[3-(trifluoro-methyl)phenyl]pyrazolo[1,5-a]pyrimidine-3-carbonitrile dihydrochloride)

Sample preparation: Condition a C18 Sep-Pak SPE cartridge with 2 mL MeCN:water 50:50 and 5 mL water. Add 2 mL plasma containing IS to the SPE cartridge, wash with

5 mL water, wash with 5 mL MeCN:water 20:80, wash with 250 μL MeCN, elute with 2 mL MeCN. Evaporate the eluate to dryness under reduced pressure, reconstitute with 200 μL mobile phase, centrifuge at 12000 g for 5 min, inject an aliquot.

HPLC VARIABLES

Guard column: Newguard RP 18
Column: 150 × 4.6 LC 18 DB (Supelco)
Mobile phase: MeCN:MeOH:100 mM NaH_2PO_4:100 mM phosphoric acid 30:33:24:24
Flow rate: 1
Detector: F ex 335 em 416-436 (filter) following post-column reaction. The column effluent mixed with the reagent pumped at 0.5 mL/min and the mixture flowed through a 3 m × 0.5 mm ID PTFE coil at 70° and a heat exchanger (to cool it to room temperature) to the detector. (Prepare reagent by dissolving 38 g $Na_3PO_4.12H_2O$ in 400 mL water, adding 37.5 mL 8% sodium hypochlorite, and making up to 500 mL with water.)

CHROMATOGRAM

Retention time: 6.3
Internal standard: CL 259,858 (5-methyl-7-[3-(trifluoromethyl)phenyl]pyrazolo[1,5-a]pyrimidine-3-carbonitrile (15.6)
Limit of detection: 1.25 ng/ml

KEY WORDS
plasma; SPE; post-column reaction

REFERENCE
Guiso, G.; Confalonieri, S.; Caccia, S.; Kelly, R.G. Post-column derivatization and fluorimetric detection for the liquid chromatographic analysis of the potential memory-enhancing agent CL 275,838 in human plasma, *Farmaco*, **1992**, *47*, 1215–1223.

SAMPLE
Matrix: blood, urine
Analyte: ampicillin and sulbactam
Sample preparation: Serum. Filter using Molcut II (Millipore), inject a 50 μL aliquot of the ultrafiltrate. Urine. Dilute ten-fold with water, filter (Gelman acrylate copolymer 0.45 μm), inject a 20 μL aliquot of the filtrate.

HPLC VARIABLES
Guard column: 4 × 4 5 μm LiChrospher RP-18(e)
Column: 250 × 4 5 μm LiChrospher RP-18(e)
Mobile phase: MeOH:20 mM tetrabutylammonium bromide + 5 mM Na_2HPO_4 + 5 mM NaH_2PO_4 1:1.75
Flow rate: 0.8
Injection volume: 20-50
Detector: UV 270 following post-column reaction. The column effluent mixed with 0.05% sodium hypochlorite in 2 M NaOH pumped at 0.1 mL/min in a 400 × 0.5 mm hollow fiber membrane reactor at 40° and this mixture flowed through a 1.4 m × 0.3 mm knitted open tubular reactor at 50° to the detector.

CHROMATOGRAM
Retention time: 8 (sulbactam), 30 (ampicillin)
Limit of detection: 5-20 ng

KEY WORDS
serum; post-column reaction

REFERENCE
Haginaka, J.; Nishimura, Y. Simultaneous determination of ampicillin and sulbactam by liquid chromatography: post-column reaction with sodium hydroxide and sodium hypochlorite using an active hollow-fiber membrane reactor, *J.Chromatogr.*, **1990**, *532*, 87–94.

SAMPLE
Matrix: blood, urine
Analyte: ranitidine

Sample preparation: Plasma. 500 μL Plasma + 100 μL 100 mM NaOH, mix, add 3 mL dichloromethane, shake for 10 min, centrifuge at 2000 g for 10 min, repeat the extraction. Combine the organic layers and evaporate them to dryness under a stream of argon, reconstitute with 500 μL mobile phase, inject a 100 μL aliquot. Urine. Dilute 1 mL urine with 25 mL water, filter (0.2 μm), inject a 100 μL aliquot.

HPLC VARIABLES
Guard column: 5 μm Spherisorb ODS-2
Column: 150 × 4 5 μm Spherisorb ODS-2
Mobile phase: Gradient. MeCN:7.5 mM pH 6 phosphate buffer 7:93 for 8 min, to 25:75 over 1 min, maintain at 25:75 for 6 min, return to initial conditions over 1 min, re-equilibrate for 10 min. (At the end of each day wash column with MeCN then water then re-equilibrate with mobile phase.)
Flow rate: 1
Injection volume: 100
Detector: F ex 350 em 450 following post-column reaction. The column effluent mixed with 5 mM sodium hypochlorite in 50 mM pH 4.5 sodium acetate buffer pumped at 1 mL/min and this mixture flowed through a 0.8 m × 0.5 mm ID PTFE coil at 25°. The effluent from this coil mixed with 20 mM o-phthalaldehyde in EtOH:500 mM pH 10.5 borate buffer 2:98 pumped at 1 mL/min and 100 mM 2-mercaptoethanol in 500 mM pH 10.5 borate buffer pumped at 1 mL/min and this mixture flowed through a 2.5 m × 0.5 mm ID PTFE coil at 25° to the detector. (The hypochlorite oxidizes the secondary to a primary amine which then reacts with the o-phthalaldehyde and 2-mercaptoethanol.)

CHROMATOGRAM
Retention time: 13
Limit of quantitation: 106 ng/mL
Limit of detection: 32 ng/mL

OTHER SUBSTANCES
Extracted: metabolites

KEY WORDS
post-column reaction; plasma

REFERENCE
Viñas, P.; Campillo, N.; López-Erroz, C.; Hernández-Córdoba, M. Use of post-column fluorescence derivatization to develop a liquid chromatographic assay for ranitidine and its metabolites in biological fluids, *J.Chromatogr.B*, **1997**, *693*, 443–449.

Sodium Metabisulfite

NaHSO₃ ⟶ post-column reaction

SAMPLE
Matrix: blood
Analyte: cisplatin
Sample preparation: Inject an aliquot of plasma ultrafiltrate.

HPLC VARIABLES
Column: 100 × 4.6 5 μm Hypersil ODS coated with hexadecyltrimethylammonium bromide (Coat the column by passing 75 mL of a 27 mM aqueous solution of hexadecyltrimethylammonium bromide through the column at 30° at 1 mL/min (J.Chromatogr. 1981, 217, 405).)
Mobile phase: 10 mM pH 4.5 Citrate buffer containing 100 μM hexadecyltrimethylammonium bromide
Flow rate: 1
Injection volume: 20
Detector: UV 290 following post-column reaction. The column effluent mixed with 26 μM potassium dichromate pumped at 0.1 mL/min and this mixture flowed through a 3.2 m

× 0.3 mm i.d. knitted coil of PTFE tubing. The effluent from the coil mixed with 3.3 mM sodium bisulfite solution pumped at 0.1 mL/min and this mixture flowed through a 44.6 m × 0.3 mm i.d. knitted coil of 0.3 mm i.d. PTFE tubing to the detector.

CHROMATOGRAM
Retention time: 9
Limit of detection: 40 ng/mL

KEY WORDS
plasma; ultrafiltrate; post-column reaction

REFERENCE
Marsh, K.C.; Sternson, L.A.; Repta, A.J. Post-column reaction detector for platinum(II) antineoplastic agents, *Anal.Chem.*, **1984**, *56*, 491–497.

SAMPLE
Matrix: blood, CSF, tissue
Analyte: vitamin B6
Sample preparation: Dilute rat plasma 1:10. Dilute human plasma 1:25. Homogenize 10 mg liver, 20 mg brain, or 250 μL CSF or diluted plasma with 250 μL 5 or 10% metaphosphoric acid by sonication at 300 W for 30 s, centrifuge at 0-4° at 10000 g for 20 min. Remove the supernatant and add it to 250 μL dichloromethane, vortex, centrifuge at 0-4° at for 15 min, filter (0.22 μm) the aqueous layer, inject a 25 μL aliquot.

HPLC VARIABLES
Guard column: 30 × 4.6 3 μm Ultramex C18
Column: 150 × 4.6 3 μm Ultramex C18
Mobile phase: Gradient. A was 33 mM phosphoric acid containing 10 mM 1-octanesulfonic acid adjusted to pH 2.2 with 6 M KOH. B was isopropanol:330 mM phosphoric acid adjusted to pH 2.2 with 6 M KOH. A:B from 100:0 to 0:100 over 10 min, maintain at 0:100 for 15 min, return to initial conditions over 4.5 min, re-equilibrate for 5.5 min. (Every 30 samples flush with water at 0.5 mL/min for 1 h and with isopropanol at 0.2 mL/min for 1 h. Every morning flush with water at 0.5 mL/min for 1 h.)
Flow rate: 1.2
Injection volume: 25
Detector: F ex 328 em 393 following post-column reaction with the reagent. (Reagent was 1 mg/mL sodium bisulfite in 100 mM potassium phosphate buffer adjusted to pH 7.4 with 6 M KOH.)

CHROMATOGRAM
Retention time: 2 (pyridoxal phosphate), 6 (4-pyridoxic acid), 11 (pyridoxamine phosphate), 15 (pyridoxal), 18.5 (pyridoxine), 22 (4-deoxypyridoxine), 26 (pyridoxamine)
Limit of quantitation: 7.5 pmole
Limit of detection: 2.6 pmole

KEY WORDS
protect from light; rat; human; liver; brain; plasma; post-column reaction

REFERENCE
Sharma, S.K.; Dakshinamurti, K. Determination of vitamin B$_6$ vitamers and pyridoxic acid in biological samples, *J.Chromatogr.*, **1992**, *578*, 45–51.

SAMPLE
Matrix: blood, urine
Analyte: cisplatin
Sample preparation: Plasma. Filter (Amicon MPS-1 with a YMT membrane) while centrifuging at 3000 g at 4° for 15 min, inject an aliquot of the ultrafiltrate. Urine. Inject an aliquot directly.

HPLC VARIABLES
Column: 80 × 4.6 MCI gel CDR10
Mobile phase: MeCN:buffer 30:70 (Buffer was 100 mM sodium sulfate and 10 mM acetate, pH 5.5.)
Column temperature: 40

Flow rate: 1
Injection volume: 100
Detector: UV 290 following post-column derivatization. The column effluent mixed with the reagent pumped at 0.3 mL/min, the mixture flowed through a 10 m × 0.5 mm ID coil of PTFE tubing held at 60° to the detector. The reagent was 40 mM sodium bisulfite and 10 mM acetate buffer, pH 5.5.

CHROMATOGRAM
Retention time: 10
Limit of detection: 20 nM

KEY WORDS
plasma; ultrafiltrate; rabbit; human; post-column reaction

REFERENCE
Kizu, R.; Yamamoto, T.; Yokoyama, T.; Tanaka, M.; Miyazaki, M. A sensitive postcolumn derivatization/UV detection system for HPLC determination of antitumor divalent and quadrivalent platinum complexes, *Chem.Pharm.Bull.*, **1995**, *43*, 108–114.

Sodium Metaperiodate

SAMPLE
Matrix: urine
Analyte: alditols
Sample preparation: Pass 1 mL urine through a column containing 500 μL Amberlite CG-120 (hydrogen form) and 500 μL Amberlite CG-400 (acetate form), elute with 2 mL water, collect all the effluent from the column, inject a 300 μL aliquot.

HPLC VARIABLES
Column: 80 × 8 11 μm Hitachi 2633 spherical anion-exchange quaternary ammonium resin
Mobile phase: Gradient. A was 500 mM boric acid adjusted to pH 7.1 with KOH. B was 300 mM boric acid adjusted to pH 8.0 with KOH. C was 500 mM boric acid adjusted to pH 10.5 with KOH. A:B:C 100:0:0 for 35 min, to 0:100:0 (step gradient), maintain at 0:100:0 for 30 min, to 0:0:100 (step gradient), maintain at 0:0:100 for 25 min, re-equilibrate at initial conditions for 20 min.
Column temperature: 65
Flow rate: 1
Injection volume: 300
Detector: UV 412 following post-column reaction. The column effluent mixed with 50 mM sodium metaperiodate solution pumped at 0.5 mL/min and the mixture flowed through a 10 m × 0.5 mm PTFE coil at room temperature. The effluent from this coil mixed with reagent pumped at 0.5 mL/min and this mixture flowed through a 10 m × 0.5 mm PTFE coil at 100 ± 0.2° to the detector. (Reagent was 15% ammonium acetate containing 2% acetylacetone and 200 mM sodium thiosulfate.); F ex 410 em 503 following post-column reaction. The column effluent mixed with 50 mM sodium metaperiodate solution pumped at 0.5 mL/min and the mixture flowed through a 10 m × 0.5 mm PTFE coil at room temperature. The effluent from this coil mixed with reagent pumped at 0.5 mL/min and this mixture flowed through a 10 m × 0.5 mm PTFE coil at 100 ± 0.2° to the detector. (Reagent was 15% ammonium acetate containing 2% acetylacetone and 200 mM sodium thiosulfate. The alditol is probably oxidized to formaldehyde that then reacts with ammonia and acetylacetone in the Hantzch reaction.)

CHROMATOGRAM
Retention time: 35 (xylitol), 37 (ribitol), 43 (arabinitol), 46 (rhamnitol), 60 (glucitol), 77 (fucitol), 80 (mannitol), 83 (galactitol)
Limit of detection: 2 nmole (UV), 0.5 nmole (F)

OTHER SUBSTANCES
Non-interfering: aldoses

KEY WORDS
post-column reaction

REFERENCE
Honda, S.; Takahashi, M.; Shimada, S.; Kakehi, K.; Ganno, S. Automated analysis by anion-exchange chromatography with photometric and fluorimetric postcolumn derivatization, *Anal.Biochem.*, **1983**, *128*, 429–437.

Sodium Nitrite

$NaNO_2$ $\longrightarrow$ post-column reaction

SAMPLE
Matrix: tissue
Analyte: pterins
Sample preparation: Homogenize tissue in 1-3 mL 500 mM perchloric acid containing 0.1 mM disodium EDTA and 0.1 mM sodium thiosulfate, centrifuge at 1700 g for 15 min. Filter (0.45 μm) the supernatant and inject a 10-100 μL aliquot of the filtrate

HPLC VARIABLES
Guard column: 50 × 4.6 Cosmosil 5C18
Column: 250 × 4.6 Cosmosil 5C18
Mobile phase: MeOH:water 5:95 containing 100 mM sodium phosphate, 3 mM sodium octylsulfate, 0.1 mM disodium EDTA, and 0.1 mM ascorbic acid, pH 3.0
Column temperature: 40
Flow rate: 1
Injection volume: 10-100
Detector: F ex 350 em 440 following post-column reaction. The column effluent mixed with 5 mM sodium nitrite pumped at 1 mL/min and the mixture flowed through a coil at 80° and a cooling coil to the detector.

CHROMATOGRAM
Retention time: 8 (neopterin), 15 (biopterin), 16 (pterin), 20.5 (dihydrobiopterin), 22.5 (6R-tetrahydrobiopterin), 25.5 (6S-tetrahydrobiopterin)
Limit of detection: 10-20 pg

KEY WORDS
post-column reaction; rat; liver; brain; kidney

REFERENCE
Tani, Y.; Ohno, T. Analysis of 6*R*- and 6*S*-tetrahydrobiopterin and other pterins by reversed-phase ion-pair liquid-chromatography with fluorimetric detection by post-column sodium nitrite oxidation, *J.Chromatogr.*, **1993**, *617*, 249–255.

Sodium Sulfite

Na_2SO_3 $\longrightarrow$ post-column reaction

SAMPLE
Matrix: blood
Analyte: vitamin B_6 vitamers

Sample preparation: 2 mL Serum + 100 μL 1 μM IS + 100 (200 ?) μL 4 M perchloric acid, mix, centrifuge at 1500 g for 5 min. Filter (0.45 μm) the supernatant, inject an aliquot of the filtrate.

HPLC VARIABLES

Column: 250 × 4.6 5 μm Spherisorb ODS2 (12% C, endcapped)

Mobile phase: 67 mM KH_2PO_4 containing 125 μM sodium hexanesulfonate, adjusted to pH 2.5 with concentrated orthophosphoric acid (As column ages it may be necessary to increase concentration of sodium hexanesulfonate to 250 μM to maintain separation.)

Flow rate: 1

Injection volume: 200

Detector: F ex 325 em 400 following post-column reaction. The column effluent mixed with the reagent pumped at 0.5 mL/min and flowed through a 300 mm tube to the detector. (The reagent was 67 mM KH_2PO_4 containing 1 g/L sodium sulfite adjusted to pH 7.5 with Na_2HPO_4.)

CHROMATOGRAM

Retention time: 5.5 (pyridoxamine), 8.1 (pyridoxal-5'-phosphate), 13.4 (pyridoxal), 18.1 (pyridoxine)

Internal standard: pyridoxamine-5'-phosphate (4.6)

Limit of detection: 1.5-12.5 nM

KEY WORDS

post-column reaction; serum

REFERENCE

Reynolds, T.M.; Brain, A. A simple internally-standardised isocratic HPLC assay for vitamin B_6 in human serum, *J.Liq.Chromatogr.*, **1992**, *15*, 897–914.

Solid Phase Reactor

SAMPLE

Matrix: blood

Analyte: vitamin K1

Sample preparation: 0.5-1 mL Plasma + 20 μL 87.5 ng/mL IS in EtOH + 2 mL EtOH + 6 mL hexane, mix, centrifuge at 3500 g for 5 min. Remove the upper hexane layer and evaporate it to dryness at 30° in a vortex-type evaporator, reconstitute the residue in hexane, add to a Sep-Pak silica SPE cartridge, wash with 8 mL hexane, elute with 8 mL hexane:diethyl ether 97:3, evaporate the eluate to dryness in a vortex-type evaporator, reconstitute with 1 mL hexane, add 4 mL reagent, add 5-10 mg zinc metal, vortex for 2 min, centrifuge, discard the hexane layer. Remove the lower (MeCN) layer and evaporate it to dryness, reconstitute the residue in 6 mL hexane and 2 mL water, vortex, centrifuge. Remove the upper hexane layer and evaporate it to dryness under a stream of air at 60°, reconstitute the residue in 250 μL mobile phase, inject a 100 μL aliquot. (Reagent was 70 mM zinc chloride in MeCN:acetic 97:3.)

HPLC VARIABLES

Column: 250 × 4.6 5 μm Hypersil-ODS

Mobile phase: MeOH:dichloromethane:buffer 80:20:0.5 (Buffer was 2 M zinc chloride containing 1 M sodium acetate and 1 M acetic acid.)

Flow rate: 1

Injection volume: 100

Detector: F ex 248 em 418 (longpass cut-off filter) following post-column reaction. The column effluent flowed through a 20 × 3.9 column packed with 200 mesh zinc particles (to remove oxygen and reduce the vitamin K) to the detector.

CHROMATOGRAM

Retention time: 9.5

Internal standard: dihydro-vitamin K_1 (10.5)

Limit of detection: 50 pg/mL

OTHER SUBSTANCES
Extracted: vitamin K_1 epoxide

KEY WORDS
plasma; SPE; pharmacokinetics; post-column reaction

REFERENCE
Haroon, Y.; Bacon, D.S.; Sadowski, J.A. Liquid-chromatographic determination of vitamin K_1 in plasma, with fluorimetric detection, *Clin.Chem.*, **1986**, *32*, 1925–1929.

SAMPLE
Matrix: blood
Analyte: malachite green and leucomalachite green
Sample preparation: Condition a 1 mL 100 mg Bakerbond sulfonic acid SPE cartridge with 1 mL MeOH and 1 mL reagent, do not allow to go dry. 100 µL Plasma + 2.5 mL reagent, vortex for 10 s, shake mechanically at 500 rpm for 10 min, centrifuge at 4400 g for 10 min, add the supernatant to the SPE cartridge, rinse the tube with 1 mL reagent, add the rinse to the SPE cartridge, wash with 1 mL water, wash with 500 µL MeOH, allow to run dry, dry with a stream of nitrogen for 30 min, pass a stream of anhydrous ammonia gas through the cartridge, elute with 500 µL MeOH. Evaporate the eluate to dryness under a stream of nitrogen at 37°, reconstitute the residue in 250 µL mobile phase:1 mg/mL methanolic L-(+)-ascorbic acid 99:1, vortex for 10 s, inject a 25 µL aliquot. (Reagent was 1 mg/mL methanolic L-(+)-ascorbic acid:MeOH:buffer 1:75:25. Prepare buffer by dissolving 5.26 g citric acid monohydrate in 225 mL water, adjusting the pH to 3.0 with 1 M NaOH, and making up to 250 mL with water.)

HPLC VARIABLES
Guard column: 10 × 2.1 40 µm pellicular reversed-phase
Column: 100 × 3 5 µm Chromspher B glass column
Mobile phase: MeCN:buffer 60:40 (Prepare buffer by dissolving 4.36 g sodium 1-pentanesulfonate, 3.40 g sodium acetate trihydrate, and 7.02 g sodium perchlorate monohydrate in 950 mL water, adjusting the pH to 4.0 with acetic acid, and making up to 1 L with water. Before the injector the mobile phase flowed through a 50 × 4.6 column packed with a lead oxide/Celite mixture prepared by vortexing 3 g lead(IV) oxide with 3 g high-purity analytical-grade Celite for 1 min.)
Flow rate: 0.6
Injection volume: 25
Detector: UV 610 following post-column reaction. The column effluent flowed through a 10 × 2 PEEK reactor to the detector. (Pack the reactor with a lead oxide/Celite mixture prepared by vortexing 1 g lead(IV) oxide with 3 g high-purity analytical-grade Celite for 1 min, flush with MeOH for 10 min before use.)

CHROMATOGRAM
Retention time: 5.5 (leucomalachite green), 9.5 (malachite green)
Limit of quantitation: 0.9-5 ng/mL
Limit of detection: 0.6-3.5 ng/mL

KEY WORDS
post-column reaction; plasma; eel; SPE

REFERENCE
Hajee, C.A.; Haagsma, N. Simultaneous determination of malachite green and its metabolite leucomalachite green in eel plasma using post-column oxidation, *J.Chromatogr.B*, **1995**, *669*, 219–227.

SAMPLE
Matrix: blood
Analyte: vitamin K_1 (phylloquinone)
Sample preparation: Condition a 500 mg Enviroprep Inert silica SPE cartridge (Baxter) with 5 mL hexane. 500 µL Plasma + 1 mL 1 ng/mL IS in isopropanol, vortex for 5 s, add 2 mL hexane, vortex for 30 s, centrifuge at 2000 g for 5 min, remove the hexane layer, repeat extraction twice more. Combine the organic layers and evaporate them to dryness

under a stream of nitrogen (make sure all alcohol is removed), reconstitute the residue in 500 μL hexane, vortex thoroughly, add to the SPE cartridge, wash with 10 mL hexane, elute with 5 mL hexane:ether 97:3, discard the first 1 mL, evaporate the next 4 mL eluate to dryness under a stream of nitrogen, reconstitute with 100-500 μL EtOH, inject a 50 μL aliquot.

HPLC VARIABLES
Column: 250 × 4.6 5 μm 201TP 54 (Vydac)
Mobile phase: EtOH:MeOH 40:60
Flow rate: 1
Injection volume: 50
Detector: F ex 242 em 430 (or 280 cut-off filter) following post-column reaction. The column effluent flowed through a 50 × 4 column packed with 10% Pt on alumina catalyst (Alfa) to the detector. (Caution! Catalyst/flammable solvent mixtures may ignite!)

CHROMATOGRAM
Retention time: 11
Internal standard: vitamin K$_2$ (menaquinone-4) (8)
Limit of detection: 20 pg/mL

KEY WORDS
serum; post-column reaction

REFERENCE
MacCrehan, W.A.; Schönberger, E. Determination of vitamin K1 in serum using catalytic-reduction liquid chromatography with fluorescence detection, *J.Chromatogr.B*, **1995**, *670*, 209–217.

SAMPLE
Matrix: blood, tissue
Analyte: idebenone
Sample preparation: Serum. 100 μL serum + 100 μL 100 ng/mL vitamin K3 in EtOH, shake, add 4 mL cyclohexane:benzene 20:80 (Caution! Benzene is a carcinogen!), vortex for 2 min, centrifuge at 1500 g for 10 min. Remove the organic layer and evaporate it to dryness under a stream of nitrogen, reconstitute the residue in 100 μL EtOH, filter (0.45 μm), inject a 20 μL aliquot. Tissue. 50-100 mg Brain tissue + 500 μL 100 mM perchloric acid + 20 μL 100 ng/mL vitamin K3 in EtOH, homogenize in a glass homogenizer at 4°, add 6 mL cyclohexane:benzene 80:20 (Caution! Benzene is a carcinogen!), vortex for 2 min, centrifuge at 1500 g for 10 min. Remove the organic layer and evaporate it to dryness under a stream of nitrogen, reconstitute the residue in 100 μL EtOH, filter (0.45 μm), inject a 20 μL aliquot.

HPLC VARIABLES
Column: 150 × 6 5 μm Shim-pack CLC-ODS (Shimadzu)
Mobile phase: MeOH:water 85:15 containing 50 mM sodium perchlorate
Flow rate: 1
Injection volume: 20
Detector: E, EICOM ECD 100, glassy carbon working electrode +0.7 V, Ag/AgCl reference electrode following post-column reaction. The column effluent flowed through a 10 × 4.6 column packed with 10 μm 5% platinum on alumina catalyst (Toa Electronics) to the detector. (Purge catalyst column with water at 10 mL/min for 5 min before use.)

CHROMATOGRAM
Retention time: 7.8
Internal standard: vitamin K3 (4.9)
Limit of detection: 5 pg

KEY WORDS
rat; serum; brain; post-column reaction

REFERENCE
Wakabayashi, H.; Nakajima, M.; Yamato, S.; Shimada, K. Determination of idebenone in rat serum and brain by high-performance liquid chromatography using platinum catalyst reduction and electrochemical detection, *J.Chromatogr.*, **1992**, *573*, 154–157.

SAMPLE

Matrix: blood, tissue

Analyte: malachite green

Sample preparation: Plasma. 200 µL Plasma + 1 mL MeCN, vortex for 20 s, centrifuge for 5 min. Add the supernatant to 2 mL MeOH, evaporate to dryness, reconstitute with 500 µL diluting solvent with sonication, centrifuge for 5 min, filter (0.2 µm), inject a 50 µL aliquot of the filtrate. Muscle. Condition a 6 mL Bakerbond neutral alumina SPE cartridge with 1 column volume MeCN and a 3 mL Bond Elut LRC propylsulfonic acid (PRS) SPE cartridge with 1 column volume MeCN, place the alumina cartridge on top of the PRS cartridge. 5 g Muscle + 1.5 mL 250 mg/mL hydroxylamine hydrochloride in water + 2.5 mL 1 M p-toluenesulfonic acid in water + 5 mL buffer, mix, homogenize (Tekmar SDT Tissuemizer) at medium speed for 30 s, remove contents of container, add 45 mL MeCN to container and shake vigorously for 30 s, add 10 g alumina (activated 80-200 mesh, Alcoa type F-20) to the container and shake vigorously for 30 s. Combine all these mixtures and centrifuge at 4° at 2700 g for 10 min, remove the supernatant, add 45 mL MeCN to the pellet, shake vigorously for 30 s. Combine the supernatants and add them to 100 mL water and 2 mL diethylene glycol, add 50 mL dichloromethane, shake vigorously for 30 s, allow to separate for 15 min, repeat extraction. Combine the organic layers and evaporate to about 10 mL under reduced pressure at 40°, add 15 mL dichloromethane and evaporate to about 10 mL, add 10 mL MeCN and evaporate to about 1 mL, add 2 mL dichloromethane, mix, add 5 mL MeCN, mix, add to the SPE cartridges, rinse container 3 times with MeCN and add the rinses to the SPE cartridges, discard the alumina cartridge. Elute the PRS cartridge with two 1 mL portions of mobile phase then 1 mL 2.5 mg/mL hydroxylamine hydrochloride in MeOH, make up the eluate to 3 mL with mobile phase (if necessary), mix, inject a 50 µL aliquot. (Diluting solvent was MeCN:buffer 60:40 containing 700 µL/L 250 mg/mL hydroxylamine hydrochloride in water. Prepare buffer by dissolving 8.2 g sodium acetate trihydrate and 0.95 g p-toluenesulfonic acid monohydrate in 900 mL water, adjust pH to 4.5 with glacial acetic acid, make up to 1 L with water.)

HPLC VARIABLES

Column: 250 × 4.6 5 µm Ultremex 5 CN

Mobile phase: MeCN:buffer 50:50 (Prepare buffer by dissolving 8.2 g sodium acetate trihydrate and 0.95 g p-toluenesulfonic acid monohydrate in 900 mL water, adjust pH to 4.5 with glacial acetic acid, make up to 1 L with water.)

Flow rate: 1

Injection volume: 50

Detector: UV 618 following post-column reaction. The column effluent passed through a column (Alltech direct-connect rfefillable guard column) packed with lead(IV) oxide:Celite 545-AW (25:75) to the detector.

CHROMATOGRAM

Retention time: 16.2 (malachite green), 6.3 (leucomalachite green)

Limit of detection: 10 ppb (plasma), 2 ppb (muscle)

OTHER SUBSTANCES

Extracted: metabolites

KEY WORDS

plasma; muscle; fish; catfish; SPE; post-column reaction

REFERENCE

Plakas, S.M.; El Said, K.R.; Stehly, G.R.; Roybal, J.E. Optimization of a liquid chromatographic method for determination of malachite green and its metabolites in fish tissues, *J.AOAC Int.*, **1995**, *78*, 1388–1393.

SAMPLE

Matrix: feed

Analyte: menadione

Sample preparation: 1 g Ground feed + 10 mL MeOH:water 40:60, shake for 30 min on a mechanical shaker, centrifuge at 1000 g for 10 min. Remove 5 mL of the supernatant and add it to 10 mL 5% sodium carbonate and 10 mL n-pentane, shake for 1 min, centrifuge at 1000 g for 1 min, repeat extraction twice. Combine the n-pentane layers and evaporate to dryness under reduced pressure at room temperature, reconstitute the residue in 10 mL MeOH, inject a 10 μL aliquot.

HPLC VARIABLES
Column: 250 × 4.6 5 μm Supelcosil LC-18
Mobile phase: MeOH:water 75:25
Flow rate: 0.9
Injection volume: 10
Detector: F ex 325 em 425 following post-column reaction. The effluent from the column flowed through a 20 × 2 column packed with powdered zinc (≤ 38 μm) and a 0.5 μm filter and then to the detector.

CHROMATOGRAM
Retention time: 6
Limit of quantitation: 1 ng

OTHER SUBSTANCES
Also analyzed: menadione sodium bisulfite

KEY WORDS
post-column reaction; post-column reduction

REFERENCE
Billedeau, S.M. Fluorimetric determination of vitamin K3 (menadione sodium bisulfite) in synthetic animal feed by high-performance liquid chromatography using a post-column zinc reducer, *J.Chromatogr.*, **1989**, *472*, 371–379.

SAMPLE
Matrix: solutions
Analyte: catecholamines
Sample preparation: Inject a 250 μL aliquot.

HPLC VARIABLES
Column: 250 × 4.6 Nucleosil 5-C18
Mobile phase: 50 mM Potassium perchlorate containing 250 μL/L 3% copper acetate in water and 10 g/L sodium acetate, pH adjusted to 4.45 with acetic acid
Flow rate: 1
Injection volume: 250
Detector: F ex 400 em 500 following post-column reaction. The column effluent flowed through the reactor then through a 3.5 m × 0.8 mm ID coil of PTFE tubing at 30°. The effluent from the coil mixed with reducing solution pumped at 1 (?) mL/min and this mixture flowed through a 2 m × 0.8 mm ID PTFE coil at 30° to the detector. (Prepare the reactor as follows. Dissolve 75.3 g manganese nitrate in 500 mL water, add 50 g 18-35 mesh silica gel (Macherey-Nagel), stir vigorously, slowly add 31.6 g potassium permanganate in 500 mL water, stir for 30 min, filter (500 μm sieve), wash until no permanganate color is left, dry in a desiccator, pack in a 50 × 2.1 stainless steel tube. The reducing solution was 266 g NaOH, 13.4 g anhydrous sodium sulfite, and 9 mL 2-mercaptoethanol in 1 L water. Note that some Nucleosil 5-C18 column packings do not give separation at pH 4.45. In this case it is necessary to use 50 mM perchloric acid as mobile phase and mix the column effluent with pH 4.4 sodium acetate buffer before it enters the reactor.)

CHROMATOGRAM
Retention time: 7 (norepinephrine), 10 (epinephrine)
Limit of detection: 0.4-1 ng/mL

KEY WORDS
post-column reaction

REFERENCE
Rüter, J.; Kurz, U.P.; Neidhart, B. Solid phase reactors as an analytical tool in the determination of urinary noradrenaline and adrenaline, *J.Liq.Chromatogr.*, **1985**, *8*, 2475–2496.

SAMPLE
Matrix: tissue
Analyte: gentian violet and leucogentian violet
Sample preparation: Condition at 6 mL 1 g neutral alumina SPE cartridge (J.T. Baker) with 5 mL MeCN. Condition a 2.8 mL 0.5 g Bond Elut PRS SPE cartridge with 5 mL MeCN. 10 g Muscle tissue + 3 mL 250 mg/mL hydroxylamine hydrochloride in water + 5 mL 50 mM p-toluenesulfonic acid + 10 mL 100 mM pH 4.5 ammonium acetate buffer, homogenize (Tekmar Ultra-Turrax T25 tissuemizer) at 20000 rpm for 1 min, add 90 mL MeCN, homogenize for 10 s, shake vigorously by hand for 1 min, add 20 g basic alumina (Brockman activity I, Fisher Scientific), shake vigorously for 1 min, centrifuge, decant, add 30 mL MeCN to the residue, extract, centrifuge, decant. Combine the supernatants and add 100 mL water, add 50 mL dichloromethane, add 2 mL diethylene glycol, shake vigorously by hand for 1 min, let stand for 45 min, remove the lower organic layer, add 50 mL dichloromethane, shake for 1 min, let stand for 5 min, remove the lower organic layers. Combine the organic layers and evaporate them to 2-5 mL under reduced pressure at 65°, add 2 mL dichloromethane, add 5 mL MeCN, add the mixture to the alumina SPE cartridge on top of the PRS SPE cartridge, rinse the flask with two 5 mL portions of MeCN, add the rinses to the SPE cartridges, wash the SPE cartridges with 5 mL MeCN, wash with 1 mL solvent, elute with 1.5 mL solvent, add 500 µL water to the eluate, inject a 100 µL aliquot. (Solvent was MeCN:100 mM ammonium acetate 50:50, adjusted to pH 4.5 with glacial acetic acid.)

HPLC VARIABLES
Guard column: 20 × 2 pellicular CN
Column: 250 × 4.6 5 µm LC-CN (Supelco)
Mobile phase: MeCN:buffer 60:40 (Prepare buffer by dissolving 3.85 g ammonium acetate in 380 mL water, adjust to pH 4.5 with glacial acetic acid, make up to 400 mL with water.)
Flow rate: 1
Detector: UV 588 following post-column reaction. The column effluent passed through a 20 × 2 column packed with lead(IV) oxide (Mallinckrodt) to the detector.

CHROMATOGRAM
Retention time: 6.2 (leucogentian violet), 12.6 (gentian violet)
Limit of quantitation: 1 ng/g

OTHER SUBSTANCES
Simultaneously analyzed: methyl violet
Interfering: 100

KEY WORDS
fish; muscle; catfish; SPE; post-column reaction

REFERENCE
Rushing, L.G.; Webb, S.F.; Thompson, H.C. Jr. Determination of leucogentian violet and gentian violet in catfish tissue by high-performance liquid chromatography with visible detection, *J.Chromatogr.B*, **1995**, *674*, 125–131.

SAMPLE
Matrix: urine
Analyte: disulfiram
Sample preparation: Adjust pH of 1-5 mL urine to 6.8 using 10 mM phosphate buffer, add 500 µL 100 mM lead (II) acetate solution, pump mixture onto column A at 1 mL/min and elute to waste, elute the contents of column A onto column B with mobile phase B, elute column B with mobile phase B and monitor the effluent from column B. (Before the next injection flush column A with 5 mL MeCN then 5 mL of a buffer containing 10 mM potassium citrate and 10 mM EDTA.)

HPLC VARIABLES
Column: A 4 × 2.1 5 µm Hypersil ODS; B 200 × 2.1 5 µm Hypersil ODS
Mobile phase: MeCN:10 mM pH 5.3 acetate buffer 65:35
Flow rate: A 1; B 0.5
Injection volume: 1000-5000

Detector: UV 435 following post-column reaction. The column effluent flowed through a 2 × 4.6 reactor filled with metallic copper and a 4 × 4.6 reactor filled with copper(II) phosphate to the detector. (Metallic copper was prepared by adding 1 g Cu(I)Cl to 2 g sodium borohydride in 20 mL water, wash precipitate twice with water and MeOH, suspend in MeOH, sonicate, dry on tissue paper, pack firmly into column, use immediately (J.Chromatogr. 1986, 370, 439) (reactor lifetime 2 days). Copper(II) phosphate was prepared by mixing equal volumes of 100 mM potassium hydrogenphosphate (sic) and 100 mM copper(II) sulfate, wash precipitate twice with water and MeOH, suspend precipitate in MeOH, sonicate for 20 min, dry, pack in column (reactor lifetime 2 weeks).)

CHROMATOGRAM
Retention time: 5
Limit of detection: <1 ppm

OTHER SUBSTANCES
Extracted: metabolites, diethyldithiocarbamate (as copper complex)

KEY WORDS
post-column reaction; column-switching

REFERENCE
Irth, H.; de Jong, G.J.; Brinkman, U.A.T.; Frei, R.W. Determination of disulfiram and two of its metabolites in urine by reversed-phase liquid chromatography and spectrophotometric detection after post-column complexation, *J.Chromatogr.*, **1988**, *424*, 95–102.

Sulfuric Acid

RELATED REFERENCE
Sudo, A. Analysis of corticosterone in rat urine by high-performance liquid chromatography and fluorimetry using post-column reaction with sulfuric acid. *J.Chromatogr.* **1990**, *528*, 453-458.

H_2SO_4 $\longrightarrow$ post-column reaction

SAMPLE
Matrix: saliva
Analyte: hydrocortisone
Sample preparation: 0.5 mL Saliva + 0.5 mL water + 1 mL mobile phase A, filter, inject a 400 μL aliquot on to column A and elute to waste with mobile phase A, after 7 min elute the contents of column A on to column B with mobile phase A, after 8 min remove column A from the circuit and elute column B with mobile phase B, start the gradient, monitor the effluent from column B.

HPLC VARIABLES
Column: A 100 × 4.6 Capcell pak MF [PCMF, silicone polymer-coated silica with diol and phenyl groups] (Shiseido); B Capcell pak CN (Shiseido)
Mobile phase: A MeCN:water 10:90 containing 2 mM trisodium citrate, adjusted to pH 6.5 with HCl; B Gradient. X was MeCN:water 10:90. Y was MeCN. X:Y from 100:0 to 72.3:27.7 over 5 min, maintain at 72.3:27.7 for 16 min, return to initial conditions over 1 min.
Column temperature: 40
Flow rate: A 0.5; B 0.5
Injection volume: 400
Detector: F ex 488 (10 mW Ar$^+$ laser) em 537 following post-column reaction. The column effluent mixed with concentrated sulfuric acid pumped at 0.75 mL/min and flowed through a 2.5 m × 0.25 mm ID Dyflon reaction coil at 105° to the detector.

CHROMATOGRAM
Retention time: 29.4
Limit of quantitation: 0.5 nM

KEY WORDS
column-switching; heart-cut; post-column reaction

REFERENCE
Okumura, T.; Nakajima, Y.; Takamatsu, T.; Matsuoka, M. Column-switching high-performance liquid chromatographic system with a laser-induced fluorimetric detector for direct, automated assay of salivary cortisol, *J.Chromatogr.B*, **1995**, *670*, 11–20.

Terbium Chloride

$TbCl_3$ ⟶ post-column reaction

SAMPLE
Matrix: solutions
Analyte: mycotoxins

HPLC VARIABLES
Column: 250 × 4.6 5 μm Inertsyl OSD-2 (Interchrom, France)
Mobile phase: MeOH:water 70:30 containing 1 mM tetrabutylammonium hydroxide, pH adjusted to 5.5 with HCl
Column temperature: 21 ± 0.1
Flow rate: 0.8
Injection volume: 20
Detector: Luminescence ex 331 em 545, delay time 0.03 ms, gating time 1 ms, following post-column reaction. The column effluent mixed with the reagent pumped at 0.2 mL/min and flowed to the detector. (Reagent was 5 mM terbium(III) chloride, 0.5 mM trioctylphosphine oxide, and 25 mM triethylamine in butanol.)

CHROMATOGRAM
Retention time: 10 (citrinin), 12 (ochratoxin A)
Limit of detection: 2-3 μM

KEY WORDS
post-column reaction

REFERENCE
Vazquez, B.I.; Fente, C.; Franco, C.; Cepeda, A.; Prognon, P.; Mahuzier, G. Simultaneous high-performance liquid chromatographic determination of ochratoxin A and citrinin in cheese by time-resolved luminescence using terbium, *J.Chromatogr.A*, **1996**, *727*, 185–193.

Tetramethylammonium Octahydridotriborate

$(CH_3)_4NB_3H_8$ ⟶ post-column reaction

SAMPLE
Matrix: milk
Analyte: vitamin K1 (phylloquinone)
Sample preparation: 500 μL Milk + 50 μL 54 ng/mL IS in EtOH + 10 μL 200 mg/mL albumin + 200 μL water containing 50 mM sodium taurocholate, 100 mM calcium chloride, and 150 mM NaCl, sonicate (titanium probe MSE Scientific Instruments) for 2 min (30 s on, 30 s wait) in an ice bath, add 1.2 mL 3.4 mg/mL crude lipase (porcine pancreas Type II (EC 3.1.1.3), Sigma), in 200 mM pH 7.7 Tris buffer, shake at 37° at 100 strokes/min for 45 min, add 4 mL EtOH, add 2 mL water, add 200 μL 50 g/L ammonium hydroxide, add 7.5 mL n-hexane, vortex for 2 min, inject an aliquot of the hexane layer on to a 200 × 4.6 5 μm RSIL column (RSL, Eke, Belgium) and elute with n-hexane:diisopropyl ether 98.5:1.5 at 0.85 mL/min (Caution! Diisopropyl ether readily forms explosive peroxides!), collect the fraction containing vitamin K and the IS, evaporate to dryness, reconstitute with 75 μL MeOH:ethyl acetate 96:4, inject a 50 μL aliquot.

HPLC VARIABLES

Column: 150 × 3.2 5 μm RoSIL HL (RSL, Eke, Belgium)
Mobile phase: MeOH:ethyl acetate 96:4 containing 2.25 g/L tetramethylammonium octahydridotriborate (Alfa)
Flow rate: 0.7
Injection volume: 50
Detector: F ex 325 em 430 following post-column reaction. The column effluent flowed through a 5 m × 0.5 mm i.d. knitted PTFE coil at 80° to the detector (the phylloquinone is reduced).

CHROMATOGRAM

Retention time: 7
Internal standard: a structural analog of phylloquinone with one more isoprene unit (Hoffman-La Roche) (11)
Limit of quantitation: 0.08 ng/mL
Limit of detection: 0.035 ng/mL

KEY WORDS

protect from light; post-column reaction

REFERENCE

Lambert, W.E.; Vanneste, L.; De Leenheer, A.P. Enzymatic sample hydrolysis and HPLC in a study of phylloquinone concentration in human milk, *Clin.Chem.*, **1992**, *38*, 1743–1748.

1,1,3,3-Tetramethyl-2-thiourea

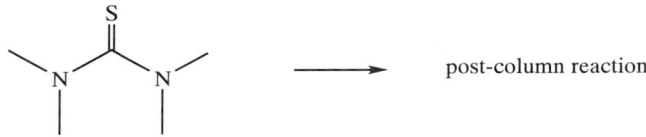

post-column reaction

SAMPLE

Matrix: solutions
Analyte: selenium and tellurium
Sample preparation: Inject a 100 μL aliquot of a solution in mobile phase.

HPLC VARIABLES

Column: 15 μm IonPac AS5 (Dionex)
Mobile phase: 30 mM HCl
Flow rate: 1
Injection volume: 100
Detector: UV 325 following post-column reaction. The column effluent mixed with the reagent pumped at 1 mL/min and the mixture flowed through a 500 μL packed-bed reaction coil to the detector. (Prepare the reagent by dissolving 5 g 1,1,3,3-tetramethyl-2-thiourea in 500 mL 800 mM HCl. Do not use until at least 3 h after preparation.)

CHROMATOGRAM

Retention time: 1.5 (Se), 4.5 (Te)
Limit of detection: 60-80 ng

KEY WORDS

post-column reaction; complexation

REFERENCE

Bruzzoniti, M.C.; Mentasti, E.; Sarzanini, C.; Cavalli, S. Behaviour of selenium and tellurium species and their determination by ion chromatography, *Chromatographia*, **1997**, *46*, 49–56.

Thienoyltrifluoroacetone

SAMPLE
Matrix: solutions
Analyte: 3-phenyl-1-propylamine
Sample preparation: Mix 10 μL of a 1-1000 μM solution with 45 μL reagent and 45 μL 100 mM pH 9.8 sodium carbonate buffer, heat at 40° for 8 h, inject an aliquot. (Prepare the reagent by adding 1 mg isothiocyanobenzyl-EDTA (Dojindo Molecular Technologies, Inc., 3 Bethesda Metro Center, Suite 700, Bethesda MD 20814; (301) 664-8448; www.dojindo.co.jp) to 5 mL 500 μM europium acetate solution.)

HPLC VARIABLES
Column: 50 × 0.15 Crest C18T-5
Mobile phase: MeCN:20 mM pH 7.0 sodium phosphate buffer 20:80
Flow rate: 0.1
Injection volume: 10
Detector: F ex 290 em 615 following post-column reaction. The column effluent mixed with reagent pumped at 0.02 mL/min and the mixture flowed through a 4 m × 0.25 mm ID tefzel coil at 40° to the detector. (Prepare reagent as follows. Mix 200 mL 5% Triton X-100, 100 mL 2 M pH 6.5 sodium acetate buffer, 10 mL 50 mM tri-n-octylphosphine oxide, and 32 mL 2.5 mM thienoyltrifluoroacetone (Dojindo Molecular Technologies, Inc., 3 Bethesda Metro Center, Suite 700, Bethesda MD 20814; (301) 664-8448; www.dojindo.co.jp), make up to 1 L with water. The construction of the detector is discussed.)

CHROMATOGRAM
Retention time: 8
Limit of quantitation: 100 nM

KEY WORDS
post-column reaction; microbore

REFERENCE
Iwata, T.; Senda, M.; Kurosu, Y.; Tsuji, A.; Maeda, M. Construction of time-resolved fluorescence detector for amino compounds after high-performance liquid chromatography using europium chelate, *Anal.Chem.*, **1997**, *69*, 1861–1865.

Thorium Nitrate

Th(NO$_3$)$_3$ ⟶ post-column reaction

SAMPLE
Matrix: urine
Analyte: clodronate
Sample preparation: Inject directly.

HPLC VARIABLES
Guard column: 50 × 4 10 μm HPLC AG7 anion-exchange (Dionex)
Column: 250 × 4 10 μm HPIC AS7 anion-exchange (Dionex)
Mobile phase: 30 mM nitric acid adjusted to pH 2.4 with dilute NaOH
Flow rate: 1
Injection volume: 100
Detector: UV 550 following post-column reaction. The column effluent mixed with reagent pumped at 1 mL/min and flowed through a 1.2 m × 0.25 mm ID coil of PEEK tubing to

the detector. (Reagent was 4.5 mL 0.2 mM thorium solution + 100 mL 0.4 mM disodium EDTA in water, mix, add 60 mL buffer, add 1.9 mL 500 µg/mL xylenol orange in water, shake vigorously, adjust pH so that the combined pH of the reagent and the column effluent will be 6.0-6.3, dilute to 500 mL with water, allow to stand for at least 1 h before use. Thorium solution was 1.104 g $Th(NO_3)_3.4H_2O$ in 20 mL 2 M nitric acid, dilute to 500 mL with water to give a 4 mM solution, dilute an aliquot further to give a 0.2 mM solution. Buffer was 10.1 mL ethylenediamine in ice-cold water, add 15 mL concentrated HCl slowly with stirring so as to keep the temperature below 20°, adjust pH to 7.3 with dilute HCl, dilute to 250 mL with water.)

CHROMATOGRAM
Retention time: 12
Limit of detection: 1.1 µg/mL

OTHER SUBSTANCES
Extracted: clodronate esters

KEY WORDS
post-column reaction

REFERENCE
Virtanen, V.; Lajunen, L.H. High-performance liquid chromatographic method for simultaneous determination of clodronate and some clodronate esters, *J.Chromatogr.*, **1993**, *617*, 291–298.

Thymol

post-column reaction

SAMPLE
Matrix: solutions
Analyte: sugars

HPLC VARIABLES
Column: HPIC-AS6 anion-exchange (Dionex)
Mobile phase: 150 mM NaOH
Column temperature: 36
Flow rate: 0.5
Injection volume: 20
Detector: UV 500 following post-column reaction. The column effluent mixed with the reagent pumped at 0.2 mL/min (?) and the mixture flowed through a knitted 10 m × 0.3 mm ID PTFE coil at 90° and then a short knitted PTFE coil at 22° to the detector. (Reagent was 2 mg/mL thymol in concentrated sulfuric acid, let stand for 30 min after preparation, discard after 48 h. (The reagent was displaced from a pressure vessel into the post-column reaction system by pumping n-heptane into the vessel.)

CHROMATOGRAM
Retention time: 6.5 (desoxyribose), 8.1 (arabinose), 8.7 (mannose), 9.0 (galactose, dextrose), 9.3 (xylose), 10.8 (fructose), 12.4 (ribose), 13.6 (lactose), 14.9 (saccharose), 22.4 (raffinose), 28.5 (maltose)
Limit of detection: 100 ng

OTHER SUBSTANCES
Non-interfering: methyl arabinose, methyl glycose, rhamnose, rutinose, trehalose

KEY WORDS
post-column reaction

REFERENCE

Engelhardt, H.; Ohs, P. Trace analysis of sugars by HPLC and post-column derivatization, *Chromatographia*, **1987**, *23*, 657–662.

Trifluoroacetic Acid

CF_3COOH $\longrightarrow$ post-column reaction

SAMPLE

Matrix: blood
Analyte: 9-aminocamptothecin
Sample preparation: Mix 50 μL Plasma with 150 μL 2 μg/mL camptothecin in MeOH at -70°, vortex for 10 s, centrifuge for 2 min. Remove a 150 μL aliquot of the supernatant and add it to 125 μL ice-cold 100 mM pH 5.5 ammonium acetate buffer, inject a 200 μL aliquot immediately.

HPLC VARIABLES

Guard column: 15 mm long RP-18 (Brownlee)
Column: 250 × 4.6 5 μm Ultrasphere ODS
Mobile phase: MeCN:MeOH:100 mM pH 5.5 ammonium acetate buffer 23:10:67
Flow rate: 1
Injection volume: 200
Detector: F ex 352 em 418 (cutoff filter) following post-column reaction. The column effluent mixed with 300 mM trifluoroacetic acid in water pumped at 0.3 mL/min and the mixture flowed to the detector.

CHROMATOGRAM

Retention time: 7
Internal standard: camptothecin (12)
Limit of quantitation: 5 ng/mL

KEY WORDS

post-column reaction; plasma

REFERENCE

Supko, J.G.; Malspeis, L. Liquid chromatographic analysis of 9-aminocamptothecin in plasma monitored by fluorescence induced upon postcolumn acidification, *J.Liq.Chromatogr.*, **1992**, *15*, 3261–3283.

Vanillin

RELATED REFERENCES

Akhtar, A.H.; abou el-Sooud, K.; Shehata, M.A. Concentrations of salinomycin in eggs and tissues of laying chickens fed medicated feed for 14 days followed by withdrawal for 3 days. *Food Addit.Contam.* **1996**, *13*, 897-907.

Bridges, D.A.; Roth, D.M.; Cleveland, C.M.; Moran, J.W.; Coleman, M.R. Determination of monensin in high-moisture cattle rations by liquid chromatography with postcolumn derivatization. *J.AOAC Int.* **1996**, *79*, 1255-1259.

Coleman, M.R.; Moran, J.W.; Mowrey, D.H. Liquid chromatographic determination of monensin in premix and animal feeds: Collaborative study. *J.AOAC Int* **1997**, *80*, 693-702.

Ferdous, A.J.; Singh, M. Development of a modified HPLC method for monensin analysis and its comparison with spectrophotometric and radioactive methods (Abstract APQ 1098). *Pharm.Res.* **1996**, *13*, S27.

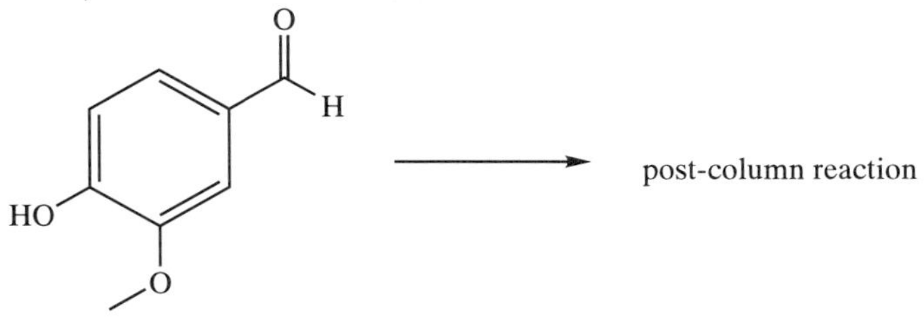

post-column reaction

SAMPLE

Matrix: bulk, feed, premix

Analyte: narasin

Sample preparation: 100 mg Bulk or 5 g premix or feed + 200 mL MeOH:water 90:10, shake on a gyratory shaker at 200 rpm for 1 h, allow to settle. Dilute an aliquot to 20 μg/mL with MeOH:water 90:10, filter (0.45 μm), inject a 200 μL aliquot.

HPLC VARIABLES

Guard column: C18

Column: 250 × 4.6 Partisil 5 ODS-3 25 LC

Mobile phase: MeOH:water:acetic acid 94:6:0.1

Flow rate: 0.7

Injection volume: 200

Detector: UV 520 following post-column reaction. The column effluent mixed with the reagent pumped at 0.7 mL/min and the mixture flowed through a 6.1 m × 0.5 mm ID stainless steel coil at 98° to the detector. (Prepare reagent by cautiously adding 20 mL concentrated sulfuric acid to 950 mL MeOH, allow to cool to room temperature, allow 30 g vanillin while stirring.)

CHROMATOGRAM

Retention time: 14 (narasin), 11 (monensin)

Limit of quantitation: 2.5 ppm

OTHER SUBSTANCES

Non-interfering: bacitracin, bambermycin, lincomycin, nicarbazin, tylosin

KEY WORDS

post-column reaction

REFERENCE

Rodewald, J.M.; Moran, J.W.; Donoho, A.L.; Coleman, M.R. Determination of narasin in raw material, premix, and animal feeds by liquid chromatography and correlation to microbiological assay, *J.AOAC Int.*, **1994**, *77*, 821–828.

SAMPLE

Matrix: feed

Analyte: polyether antibiotics

Sample preparation: 20 g Ground feed + 50 mL MeOH, shake on a reciprocating shaker for 1 h, allow to settle, centrifuge an aliquot of the supernatant at 2500 rpm for 5 min, inject a 50 μL aliquot of the supernatant.

HPLC VARIABLES

Column: 250 × 4.6 Partisil 5ODS-3

Mobile phase: MeOH:water:glacial acetic acid 94:5.9:0.1

Flow rate: 0.7

Injection volume: 50

Detector: UV 520 following post-column reaction. The column effluent mixed with the reagent pumped at 0.7 mL/min and the mixture flowed through a 7.6 m × 0.05 mm ID coil of stainless steel tubing at 70° to the detector. (Prepare reagent by stirring 900 mL MeOH and slowly and cautiously adding 20 mL concentrated sulfuric acid (Caution! Corrosive! Exothermic reaction!), cool in an ice bath, add 100 g vanillin, mix to dissolve, make up to 1 L with MeOH. Degas under vacuum with sonication for 2 min.)

CHROMATOGRAM

Retention time: 11.5 (monensin), 13.5 (salinomycin), 15.5 (narasin)

Limit of detection: 250 ng/g

KEY WORDS

post-column reaction

REFERENCE

Blanchflower, W.J.; Rice, D.A.; Hamilton, J.T.G. Simultaneous high-performance liquid chromatographic determination of monensin, narasin and salinomycin in feeds using post-column derivatisation, *Analyst*, **1985**, *110*, 1283–1287.

SAMPLE
Matrix: feed
Analyte: polyether antibiotics
Sample preparation: 20 g Ground Feed + 200 mL hexane:ethyl acetate 90:10, stir at high speed for 2 h, let stand. Remove an aliquot equivalent to 1 g feed and evaporate it to dryness under reduced pressure at 40°, reconstitute with 2 mL MeOH, filter (0.45 μm), inject an aliquot of the filtrate.

HPLC VARIABLES
Column: 60 × 4.6 3 μm C18 (Hewlett-Packard)
Mobile phase: MeOH:5% acetic acid 90:10
Flow rate: 0.5
Injection volume: 20-25
Detector: UV 520 following post-column reaction. The column effluent mixed with the reagent pumped at 1 mL/min and the mixture flowed through a 1.5 mL reaction coil (Kratos Model 510) at 95° to the detector. (Reagent was 40 g/L vanillin in MeOH:sulfuric acid 100:2. Keep in an ice bath and prepare fresh daily.)

CHROMATOGRAM
Retention time: 5.5 (monensin), 6.7 (salinomycin), 8.2 (narasin)
Limit of detection: 0.5 ppm

KEY WORDS
post-column reaction

REFERENCE
Lapointe, M.R.; Cohen, H. High-speed liquid chromatographic determination of monensin, narasin, and salinomycin in feeds, using post-column derivatization, *J.Assoc.Off.Anal.Chem.*, **1988**, *71*, 480–484.

SAMPLE
Matrix: fermentation broth
Analyte: narasin
Sample preparation: Homogenize 5 mL fermentation broth with 25 mL MeOH, filter (0.45 μm) the supernatant, inject a 20 μL aliquot of the filtrate.

HPLC VARIABLES
Column: 50 × 4.6 3 μm Little Champ ODS (Regis)
Mobile phase: MeOH:100 mM pH 4 KH_2PO_4 buffer 93:7
Flow rate: 2
Injection volume: 20
Detector: UV 520 following post-column reaction. The column effluent mixed with 10% sulfuric acid in MeOH pumped at 1 mL/min and with 6% vanillin in MeOH pumped at 1 mL/min and the mixture flowed through a 1.5 m × 0.25 mm ID stainless steel coil at 120° to the detector.

CHROMATOGRAM
Retention time: 2

KEY WORDS
post-column reaction

REFERENCE
Neely, F.L. A rapid HPLC determination of narasin in fermentation broth, *Chromatographia*, **1991**, *31*, 277–280.

SAMPLE
Matrix: fermentation broth
Analyte: monensin
Sample preparation: Homogenize (mechanical grinder) 5 mL fermentation broth with 45 mL MeOH, allow to settle at 5° for 1 h, filter (0.45 μm) the supernatant, dilute the filtrate with MeOH (if necessary), inject a 10 μL aliquot.

HPLC VARIABLES
Column: Little Champ C18 (Regis)

Mobile phase: MeOH:buffer 85:15 (Buffer was 1 g/L (?) $(NH_4)H_2PO_4$ adjusted to pH 3.0 with phosphoric acid.)

Flow rate: 1.5

Injection volume: 10

Detector: UV 520 following post-column reaction. The column effluent mixed with the reagent pumped at 1.5 mL/min and the mixture flowed through a 3 m × 0.25 mm ID stainless steel coil at 120° to the detector. (Prepare reagent by making a 3% solution of vanillin in MeOH, chill solution, slowly add concentrated sulfuric acid to a final concentration of 3%. Store at 0°, discard after 2 weeks.)

CHROMATOGRAM

Retention time: 2 (monensin B), 2.7 (monensin A)

KEY WORDS

post-column reaction

REFERENCE

Neely, F.W. Determination of monensin in fermentation broth by HPLC with post-column derivatization, *J.Liq.Chromatogr.*, **1992**, *15*, 1513–1522.

SAMPLE

Matrix: formulations

Analyte: monensin

Sample preparation: 100 μL Liposome + 50 μL Tween 20, vortex for 5 min, add 600 μL MeOH, centrifuge at 10000 rpm for 30 min, dilute the supernatant 10-fold with MeOH, inject an aliquot.

HPLC VARIABLES

Column: 70 × 4.6 3 μm Ultrasphere XL-ODS (Beckman)

Mobile phase: MeCN:MeOH:dichloromethane:water:acetic acid 20:45:25:9.5:0.5

Flow rate: 0.33

Detector: UV 520 following post-column reaction. The column effluent mixed with the reagent pumped at 0.67 mL/min and the mixture flowed through a Beckman 231 reactor at 70° to the detector. (Prepare the reagent by dissolving 8 g vanillin in 100 mL cooled MeOH, add 4 g concentrated sulfuric acid, mix, keep in an ice bath.)

CHROMATOGRAM

Retention time: 6.05

Limit of quantitation: 100 ng/mL

KEY WORDS

post-column reaction; liposomes

REFERENCE

Ferdous, A.J.; Bennefield, S.D.; Singh, M. A modified HPLC method for monensin analysis in liposomes and nanocapsules and its comparison with spectrophotometric and radioactive methods, *J.Pharm.Biomed.Anal.*, **1997**, *15*, 1775–1780.

SAMPLE

Matrix: milk, tissue

Analyte: monensin

Sample preparation: Tissue. Condition a Sep-Pak silica gel SPE cartridge (No. 51900) with 3 mL dichloromethane, do not allow it to go dry. Sonicate or blend 10 g ground or minced tissue and 75 mL MeOH:water 85:15, centrifuge at 2000 rpm for 10 min. Remove the supernatant and add it to 50 (muscle, liver), 40 (kidney) or 60 (fat) mL 100 mg/mL NaCl, extract twice with 35 mL portions of dichloromethane. Combine the organic layers and evaporate them to dryness under vacuum at 45°, reconstitute the residue in 7 mL dichloromethane, add it to the SPE cartridge at no more than 5 mL/min, rinse out the flask with 3 mL dichloromethane, add the rinse to the SPE cartridge, wash with 10 mL dichloromethane, elute with 5 mL dichloromethane:MeOH 95:5. Evaporate the eluate to dryness under a stream of nitrogen or air at 45°, reconstitute the residue in 1 mL MeOH:water 90:10, filter (Gelman Acrodisc CR PTFE 0.45 μm), inject a 100 μL aliquot. Milk. Condition a Sep-Pak silica gel SPE cartridge (No. 51900) with 3 mL dichloromethane, do not allow it to go dry. Sonicate (ultrasonic cell disrupter) or blend 40 mL milk and 160 mL MeOH for 30 s, let stand for 10-15 min, centrifuge at 2000 rpm for 10 min. Decant the supernatant and add it to 60 mL 100 mg/mL NaCl, extract twice with 70 mL portions of dichloromethane. Combine the organic layers and evaporate them to dryness under vacuum at 45°, reconstitute the residue in 7 mL dichloromethane, add it to the SPE cartridge at no more than 5 mL/min, rinse out the flask with 3 mL dichloromethane, add the rinse to the SPE cartridge, wash with 10 mL dichloromethane, elute with 5 mL dichloromethane:MeOH 95:5. Evaporate the eluate to dryness under a stream of nitrogen or air at 45°, reconstitute the residue in 1 mL MeOH:water 90:10, filter (Gelman Acrodisc CR PTFE 0.45 μm), inject a 100 μL aliquot.

HPLC VARIABLES
Column: 250 × 4.6 Partisil 5 ODS-3 25 LC
Mobile phase: MeOH:water:acetic acid 94:6:0.1
Flow rate: 0.7
Injection volume: 100
Detector: UV 520 following post-column reaction. The column effluent mixed with the reagent pumped at 0.7 mL/min and flowed through a 6 m × 0.5 mm ID stainless steel tube at 98° to the detector. (The reagent was prepared by slowly adding 20 mL concentrated sulfuric acid to 950 mL MeOH (Caution! Exothermic!), allow to cool to room temperature, add 30 g vanillin while stirring, degas under vacuum or with helium for 5-10 min.)

CHROMATOGRAM
Retention time: 8
Limit of quantitation: 5 ppb (milk), 25 ppb (tissue)

OTHER SUBSTANCES
Simultaneously analyzed: narasin, salinomycin
Non-interfering: bacitracin, bambermycin, lasalocid, lincomycin, nicarbazin, tylosin

KEY WORDS
cow; muscle; liver; fat; kidney; post-column reaction

REFERENCE
Moran, J.W.; Turner, J.M.; Coleman, M.R. Determination of monensin in edible bovine tissues and milk by liquid chromatography, *J.AOAC Int.*, **1995**, *78*, 668–673.

SAMPLE
Matrix: tissue
Analyte: monensin
Sample preparation: Blend 20 g ground or minced tissue with 150 mL MeOH:water 85:15 for 30 s or until homogeneous, centrifuge at 2500 rpm for 15 min. Remove the supernatant and mix it with 40 mL 10% NaCl in water, extract three times with 25 mL portions of carbon tetrachloride (Caution! Carbon tetrachloride is a carcinogen!), evaporate the extracts to dryness under reduced pressure at ≤40°, reconstitute with 5 mL chloroform, add to a SepPak silica SPE cartridge at 2 mL/min, rinse flask with 3 mL chloroform, add the rinse to the SPE cartridge, wash with 10 mL hexane:chloroform 10:90, elute with 10 mL chloroform:MeOH 95:5, evaporate the eluate to dryness under reduced pressure or a stream of air at ≤40°, reconstitute with 2 mL MeOH:water 90:10, filter (0.45 μm), inject a 200 μL aliquot.

HPLC VARIABLES
Column: 250 × 4.6 Partisil 5 ODS-3 25 LC
Mobile phase: MeOH:water:acetic acid 94:6:0.1
Flow rate: 0.7
Injection volume: 200
Detector: UV 520 following post-column reaction. The column effluent mixed with the reagent pumped at 0.7 mL/min and the mixture flowed through a 6.1 m × 0.5 mm ID stainless steel coil at 98° to the detector. (Prepare reagent by cautiously adding 20 mL concentrated sulfuric acid to 950 mL MeOH, allow to cool to room temperature, allow 30 g vanillin while stirring.)

CHROMATOGRAM
Retention time: 11

Limit of quantitation: 25 ng/g
Limit of detection: 5 ng/g

OTHER SUBSTANCES
Simultaneously analyzed: narasin, salinomycin
Non-interfering: bacitracin, bambermycin, lasalocid, lincomycin, nicarbazin, tylosin

KEY WORDS
post-column reaction; SPE; chicken; muscle; liver; skin

REFERENCE
Moran, J.W.; Rodewald, J.M.; Donoho, A.L.; Coleman, M.R. Determination of monensin in chicken tissues by liquid chromatography with postcolumn derivatization, *J.AOAC Int.*, **1994**, *77*, 885–890.

SAMPLE
Matrix: tissue
Analyte: semduramicin
Sample preparation: Condition a 200 mg BondElut LRC C8 SPE cartridge with 5 mL MeCN, 5 mL MeOH, add 5 mL water, do not allow to dry. Condition a 500 mg BondElut LRC silica SPE cartridge with 5 mL chloroform and 5 mL isooctane:dichloromethane 50:50, do not allow to dry. Vortex 1.25 g homogenized liver with 7.5 mL MeOH:water: ammonium hydroxide 80:20:1 for 3 min, heat at 55° for 1 h, centrifuge at 4000 rpm for 5 min, decant the supernatant, rinse the tube with 1-2 mL MeOH. Combine the rinse and the supernatant and evaporate them to 2-3 mL under a stream of nitrogen at 55°, add 5 mL water, vortex, sonicate for 5 min, add to the C8 SPE cartridge, rinse the tube with 1-2 mL water, add the rinse to the C8 SPE cartridge, wash with 3 mL water, wash with 1 mL MeOH:water 25:75, elute with 5 mL ethyl acetate, evaporate to dryness under a stream of nitrogen at 55°, reconstitute with 6 mL isooctane:dichloromethane 50:50, vortex, sonicate for 5 min, add to the silica SPE cartridge, rinse the tube with 1.5 mL isooctane:dichloromethane 50:50, add the rinse to the SPE cartridge, wash with 2.5 mL isooctane:dichloromethane 50:50, wash with 1 mL ethyl acetate, elute with 5 mL dichloromethane:MeOH 90:10, evaporate to dryness under a stream of nitrogen at 55°, reconstitute with 150 µL isooctane:ethyl acetate 60:40, vortex, sonicate for 2 min, inject a 75 µL aliquot.

HPLC VARIABLES
Guard column: 20 mm long 40 µm LC-Si (Supelco)
Column: 250 × 4.6 Zorbax silica
Mobile phase: Isooctane:ethyl acetate:glacial acetic acid:triethylamine:MeOH 35:65: 0.4:0.2:0.1 (At the end of each day flush column with isooctane:ethyl acetate 40:60 for at least 3 h.)
Flow rate: 0.6
Injection volume: 75
Detector: UV 522 following post-column reaction. The column effluent mixed with the reagent pumped at 0.3 mL/min and the mixture flowed through a 15 m × 0.25 mm ID stainless steel coil at 95 ± 1° to the detector. At the end of each day flush the pump with MeOH for at least 3 h. (Prepare reagent by cautiously adding 20 mL concentrated sulfuric acid to 500 mL EtOH, cool to room temperature, add 30 g vanillin dissolved in 500 mL EtOH, mix, protect from light, prepare every other day.)

CHROMATOGRAM
Retention time: 11.3
Limit of detection: 25 ng/g

OTHER SUBSTANCES
Simultaneously analyzed: maduramicin, monensin, narasin, salinomycin

KEY WORDS
post-column reaction; normal phase; SPE; chicken; liver

REFERENCE
Ericson, J.F.; Calcagni, A.; Lynch, M.J. Determination of senduramicin sodium in poultry liver by liquid chromatography with vanillin postcolumn derivatization, *J.AOAC Int.*, **1994**, *77*, 577–582.

SAMPLE

Matrix: tissue

Analyte: coccidiostats

Sample preparation: Condition a 500 mg Bond Elut silica SPE cartridge with 5 mL isooctane:ethyl acetate 90:10. Homogenize (Polytron) 5 g tissue with 10 mL iso-octane:ethyl acetate 90:10 for 30 s, shake mechanically for 5 min, centrifuge at 1500 g for 5 min, remove the supernatant, extract twice more with 10 mL portions of isooctane:ethyl acetate 90:10. Combine the supernatants and dry them over 1 g anhydrous sodium sulfate, add to the SPE cartridge at 3 mL/min, wash with 10 mL dichloromethane, dry under vacuum for 30 s, elute with two 3 mL portions of dichloromethane:MeOH 90:10. Evaporate the eluate to dryness under a stream of nitrogen at 40°, reconstitute the residue in 500 μL MeOH:water 90:10, vortex for 30 s, sonicate for 5 min, inject a 100 μL aliquot.

HPLC VARIABLES

Column: 250 × 3.2 5 μm Inertsil ODS-2

Mobile phase: MeOH:10 mM pH 4.0 ammonium acetate 6:94

Flow rate: 0.5

Injection volume: 100

Detector: UV 520 following post-column reaction. The column effluent mixed with the reagent pumped at 0.3 mL/min and the mixture flowed through a 150 × 4.6 column packed with 75 μm glass beads at 100° to the detector. (Prepare reagent by adding 2.5 mL sulfuric acid to 125 mL cold MeOH, add 9.5 g vanillin, prepare fresh each day.)

CHROMATOGRAM

Retention time: 10.0 (monensin), 12.5 (salinomycin), 14.5 (narasin)

Limit of quantitation: 3-10 ppb

Limit of detection: 2-5 ppb

OTHER SUBSTANCES

Non-interfering: bacitracin, chlortetracycline, furazolidone, lasalocid, lincomycin, melengestrol acetate, oxytetracycline, penicillin G, roxarsone, tetracycline, tilmicosin, tylosin, virginiamycin

KEY WORDS

post-column reaction; cow; chicken; muscle; fat; kidney; liver; SPE

REFERENCE

Gerhardt, G.C.; Salisbury, C.D.C.; Campbell, H.M. Determination of ionophores in the tissues of food animals by liquid chromatography, *Food Addit.Contam.*, **1995**, *12*, 731−737.

Xylenol Orange

post-column reaction

SAMPLE
Matrix: solutions
Analyte: lanthanides

HPLC VARIABLES
Column: 150 × 4.6 Vydac 201 TP C18
Mobile phase: Gradient. A was 50 mM 2-hydroxyisobutyric acid in water containing 10 mM 1-octanesulfonic acid, pH adjusted to 3.8 with concentrated ammonium hydroxide. B was 240 mM 2-hydroxyisobutyric acid in water containing 10 mM 1-octanesulfonic acid, pH adjusted to 3.8 with concentrated ammonium hydroxide. A:B from 100:0 to 0:100 over 10 min.
Column temperature: 30
Flow rate: 1
Injection volume: 100
Detector: UV 618 following post-column reaction. The column effluent mixed with 5 mM xylenol orange in MeOH:water 40:60 containing 4 mM cetyltrimethylammonium bromide pumped at 1.5 mL/min and the mixture flowed through a 30 cm × 0.25 mm ID stainless steel tube to the detector.

CHROMATOGRAM
Retention time: 3.9 (lutetium), 4.2 (ytterbium), 4.4 (thulium), 4.8 (erbium), 5 (holmium), 5.3 (dysprosium), 5.8 (terbium), 6.6 (gadolinium), 7 (europium), 7.8 (samarium), 9.6 (neodymium), 10.5 (praseodymium), 12.1 (cerium), 15.7 (lanthanum)
Limit of quantitation: 50 ng/mL
Limit of detection: 3 ng

OTHER SUBSTANCES
Interfering: cadmium (interferes with praseodymium and cerium), yttrium (interferes with dysprosium)
Non-interfering: aluminum, barium, calcium, copper, iron, magnesium, mercury, strontium, uranium

KEY WORDS
post-column reaction

REFERENCE
Gautier, E.A.; Gettar, R.T.; Servant, R.E.; Batistoni, D.A. Surfactant-sensitized post-column reaction with xylenol orange for the determination of lanthanides by ion chromatography, *J.Chromatogr.A*, **1997**, *770*, 75–83.

Zinc

Zn $\longrightarrow$ post-column reaction

SAMPLE
Matrix: milk
Analyte: vitamin K
Sample preparation: Dissolve 1 g powdered milk or infant formula in 15 mL warm (<40°) water. Dilute 10 g liquid milk to 15 mL. Add 5 mL buffer and 1 g lipase (Type VII from Candida cylindracea, L1754, Sigma) to each solution, vortex and shake for 7 min, heat at 37° for 2 h with brief shaking at 20 min intervals, cool to room temperature, add 10 mL EtOH:MeOH 95:5 with inversion, add 1 g potassium carbonate, mix, add 30 mL hexane, shake for 7 min, centrifuge at 180 g for 10 min. Remove a 0.5-5 mL aliquot of the hexane layer and evaporate it to dryness under a stream of nitrogen, reconstitute with 1 mL MeOH, filter (0.45 μm), inject a 20 μL aliquot. (Protect from light. Prepare buffer by dissolving 54 g KH_2PO_4 in water, adjusting pH to 7.9-8.0 with NaOH, and making up to 500 mL with water.)

HPLC VARIABLES
Guard column: 4 μm Nova-Pak C18
Column: 100 × 8 5 μm Resolve C18 (Waters)
Mobile phase: Dichloromethane:MeOH:10 mM zinc chloride in MeOH containing 5 mM anhydrous sodium acetate and 5 mM glacial acetic acid 10:90:0.5
Flow rate: 1.5
Injection volume: 20
Detector: F ex 243 em 430 following post-column reaction. The column effluent flowed through a 20 × 4 column dry-packed with zinc powder to the detector.

CHROMATOGRAM
Retention time: 7 (menaquinone MK4), 10 (menaquinone MK5), 11 (vitamin K$_1$), 14 (2',3'-dihydrophylloquinone), 15 (menaquinone MK6)
Limit of detection: 1.5 ng/mL

KEY WORDS
post-column reaction; formula; cow; pig; mouse; cat; dog; antelope

REFERENCE
Indyk, H.E.; Woollard, D.C. Vitamin K in milk and infant formulas: Determination and distribution of phylloquinone and menaquinone-4, *Analyst*, **1997**, *122*, 465–469.

PRE-COLUMN PHOTOCHEMICAL DERIVATIZATION

Fluorescence Detection

SAMPLE
Matrix: blood
Analyte: tamoxifen
Sample preparation: 30 μL Plasma + 12 μL 15 μg/mL quinine bisulfate in MeOH, mix, make up to 90 μL with 600 mM orthophosphoric acid in MeCN, illuminate in a shortwave UV transilluminator (UVP, San Gabriel CA) with 0.25 J/min for 2 min, inject a 50 μL aliquot.

HPLC VARIABLES
Guard column: CN Resolve (Waters)
Column: 100 × 8 10 μm Radial Pak CN radial compression (Waters)
Mobile phase: MeOH:buffer 70:30 (Buffer was 100 mM sodium acetate containing 1 mM tetrabutylammonium phosphate, adjusted to pH 6 with orthophosphoric acid.)

Flow rate: 4
Injection volume: 50
Detector: F ex 258 em 378

CHROMATOGRAM
Retention time: 10.13
Internal standard: quinine bisulfate (4.36)
Limit of quantitation: 10 ng/mL

OTHER SUBSTANCES
Simultaneously analyzed: metabolites, aspirin, azathioprine, carmustine, chlorambucil, cytarabine, dacarbazine, N-desmethyltamoxifen, etoposide, 4-hydroxytamoxifen, indomethacin, lomustine, methotrexate, procarbazine, salicylic acid, tenoposide, thioguanine
Non-interfering: cyclophosphamide, doxorubicin, ifosfamide, mitomycin C, prednisone, taxol, vincristine

KEY WORDS
plasma

REFERENCE
el-Yazigi, A.; Legayada, E. Direct liquid chromatographic micro-measurement of tamoxifen in plasma of cancer patients, *J.Chromatogr.B*, **1997**, *691*, 457–462.

REDUCTION

Nitro Group

$RNO_2 \longrightarrow RNH_2$

SAMPLE
Matrix: meat
Analyte: nitro-polycyclic aromatic hydrocarbons
Sample preparation: Homogenize 10-20 g meat, Soxhlet extract with 150 mL MeCN at 98° for 4 h. Reduce the extract to 2 mL by evaporation under reduced pressure, evaporate to 500 µL using a stream of nitrogen at room temperature, take up in 3 mL cyclohexane:ethyl acetate 50:50, add to a 190 × 30 column of biobeads SX-3 (size exclusion chromatography), elute with cyclohexane:ethyl acetate 50:50 at 6 mL/min, collect the fraction from 110 mL to 210 mL. Concentrate the collected fraction under reduced pressure and evaporate it to dryness under a stream of nitrogen, reconstitute with 500 µL MeOH, reduce flow rate to 0.1 mL/min. Inject an aliquot into the system, after 30 s increase the flow rate to 0.9 mL/min. The nitro-compounds are reduced to the corresponding amines in the guard column and the amines are then separated on the main column.

HPLC VARIABLES
Guard column: 50 × 4 5 µm catalyst (The guard column was heated at 80°. Prepare the catalyst as follows. Treat 1 g 5 µm alumina (Macherey Nagel) with tensid (?) Triton X-100 to increase surface activity, dry at 300° for 2 h, mix with 400 µL 50% aqueous ammonia containing 23 mg platinum(II) chloride and 14 mg rhodium(II) acetate, add 7 mL 2 g/L sodium borohydride in 5 g/L NaOH. The reaction is complete when the color changes from pink to grey. Wash with water several times, dry at 120° for 6 h. Dry pack the column and pass MeOH:water 35:65 through at 5 mL/min.)
Column: 250 × 4 5 µm Lichrospher 60 RP-select B
Mobile phase: MeOH:water 65:35
Flow rate: 0.9
Detector: F ex 243 em 429 (1-nitronaphthalene), ex 285 em 370 (2-nitrofluorene), ex 360 em 430 (1-nitropyrene)

CHROMATOGRAM
Retention time: 4.3 (1-nitronaphthalene), 6.1 (2-nitrofluorene), 7.8 (1-nitropyrene)

Limit of detection: 0.5-4 pg

REFERENCE

Schlemitz, S.; Pfannhauser, W. Analysis of nitro-PAHs in food matrices by on-line reduction and high performance liquid chromatography, *Food Addit.Contam.*, **1996**, *13*, 969–977.

UNDEFINED REACTIONS

Antibody

SAMPLE
Matrix: blood
Analyte: theophylline
Sample preparation: Mix 25 μL serum, 100 μL solution T, and 100 μL solution S, vortex for 10 s, let stand at room temperature for 10 min, inject an aliquot. (Solution T was theophylline fluorescein tracer solution and solution S was monoclonal theophylline antiserum solution from Abbott TDxFLx theophylline FPIA assay kit.)

CAPILLARY ELECTROPHORESIS
Capillary: 67 cm × 50 μm fused silica (60 cm to detector) (Polymicro Technologies)
Capillary preparation: Between runs wash capillary with 100 mM NaOH at 5 psi for 3 min, with water at 5 psi for 3 min, and with running buffer at 5 psi for 3 min.
Running buffer: 6 mM Sodium tetraborate containing 10 mM Na_2HPO_4 and 75 mM sodium dodecyl sulfate, pH 9.2
Capillary temperature: 20
Voltage/Current: 20 kV/about 29 μA
Injection: Pressure injection at 0.5 psi for 1 s
Detector: F ex 488 (4 mW Ar ion laser) 520 (emission filter)
Model: Beckman P/ACE System 5510
Migration time: 13.4
Limit of detection: 1 μg/mL

KEY WORDS
serum; peak obtained even with no theophylline present; non-linear calibration curve

REFERENCE
Steinmann, L.; Caslavska, J.; Thormann, W. Feasibility study of a drug immunoassay based on micellar electrokinetic capillary chromatography with laser induced fluorescence detection: Determination of theophylline in serum, *Electrophoresis*, **1995**, *16*, 1912–1916.

2,6-Dichloroquinone-4-chloroimide

SAMPLE
Matrix: blood
Analyte: methimazole

Sample preparation: 1 mL Plasma + 4 mL buffer + 200 μL 100 mg/mL 2,6-dichloroqui-none-4-chloroimide in MeOH, vortex for 1 min, let stand at room temperature for 5 min, add 2 mL chloroform, vortex for 5 min, centrifuge at 1000 g for 10 min, filter (Whatman 1PS phase-separating paper) the lower organic layer, evaporate the filtrate to dryness, reconstitute with 50 μL chloroform, inject a 20 μL aliquot. (Prepare buffer by dissolving 3.1 g boric acid, 3.75 g KCl, and 100 g NaCl in water, add 40 mL 100 mM NaOH, make up to 1 L with water, adjust pH to 8 with 6 M NaOH.)

HPLC VARIABLES
Column: 300 × 3.9 μBondapak-NH2
Mobile phase: chloroform
Flow rate: 1.5
Injection volume: 20
Detector: UV 405

CHROMATOGRAM
Retention time: 5
Limit of detection: 5 ng/mL

KEY WORDS
plasma; pharmacokinetics

REFERENCE
Meulemans, A.; Manuel, C.; Ferriere, C.; Vulpillat, M. Determination of methimazole in plasma by high performance liquid chromatography, *J.Liq.Chromatogr.*, **1980**, *3*, 287–298.

Sulfuric Acid

H_2SO_4

SAMPLE
Matrix: blood
Analyte: hydrocortisone
Sample preparation: 250 μL Serum + 250 μL EtOH, vortex, centrifuge at 12100 g for 10 min. Remove a 250 μL aliquot of the supernatant and add it to 480 μL sulfuric acid, vortex, let stand at room temperature in the dark for 10 min, let stand at 0° in the dark for 15 min, add 700 μL mobile phase A, vortex, let stand at 0° in the dark for 15 min, store at 4-5°, within 15 h inject a 1 mL aliquot onto column A and elute to waste with mobile phase A, after 5 min elute the contents of column A onto column B with mobile phase B, monitor the effluent from column B.

HPLC VARIABLES
Column: A 10 × 4.6 5 μm ODS-80Tm (TSK); 150 × 4.6 5 μm ODS-80Tm (TSK)
Mobile phase: A 10 mM potassium biphthalate adjusted to pH 1.85 with trifluoroacetic acid; B MeCN:THF:19 mM potassium biphthalate 40:6:54 adjusted to pH 1.85 with trifluoroacetic acid
Flow rate: 1
Injection volume: 1000
Detector: F ex 365 em 520

CHROMATOGRAM
Retention time: 12.5
Limit of detection: 3 ng/mL

OTHER SUBSTANCES
Simultaneously analyzed: corticosterone, estradiol, testosterone, tetrahydrocortisol
Non-interfering: androstenediol, androstenedione, betamethasone, cortol, cortolone, dehydroepiandrosterone, dexamethasone, estriol, estrone, prednisolone, prednisone, progesterone, tetrahydrocortisone

KEY WORDS
column-switching; serum

REFERENCE

Nozaki, O.; Ohata, T.; Ohba, Y.; Moriyama, H.; Kato, Y. Determination of serum cortisol by reversed-phase liquid chromatography using precolumn sulphuric acid-ethanol fluorescence derivatization and column switching, *J.Chromatogr.*, **1991**, *570*, 1–11.

SAMPLE

Matrix: urine

Analyte: hydrocortisone

Sample preparation: Condition a 1 mL 20 μm bovine serum albumin coated ODS SPE cartridge (TOSOH TSK gel BSA-ODS) with 4 mL MeOH and 4 mL water. Add 2 mL urine to the SPE cartridge, wash with 2 mL EtOH:57 mM pH 8.0 potassium phosphate buffer 15:85, wash with two 2 mL portions of water, elute with 2 mL EtOH:water 90:10. Remove a 320 μL aliquot and add it to 480 μL sulfuric acid, vortex for 1 min, let stand in the dark at room temperature for 10 min, let stand at 0° in the dark for 15 min, add 700 μL buffer, let stand in the dark at 0° for 15 min, inject an aliquot. Inject a 1 mL aliquot on to column A and elute to waste with mobile phase A, after 5 min elute the contents of column A on to column B with mobile phase C, after 4 min remove column A from the circuit, elute column B with mobile phase C, monitor the effluent from column B. Wash column A with mobile phase B for 4 min then re-equilibrate with mobile phase A for 12 min before the next injection. (Buffer was 20 mM KH_2PO_4 adjusted to pH 1.85 with trifluoroacetic acid.)

HPLC VARIABLES

Column: A 10 × 4.6 5 μm ODS-80 Tm; B 150 × 4.6 5 μm ODS-80 Tm

Mobile phase: A was MeCN:23.5 mM KH_2PO_4 15:85, pH adjusted to 1.85 with trifluoroacetic acid. B was MeCN:water 50:50. C was MeCN:36.4 mM KH_2PO_4 45:55, pH adjusted to 1.85 with trifluoroacetic acid.

Flow rate: A 1; B 2; C 1

Injection volume: 1000

Detector: F ex 365 em 520

CHROMATOGRAM

Retention time: 15.3

Limit of detection: 2.6 ng/mL

OTHER SUBSTANCES

Simultaneously analyzed: corticosterone, testosterone

Non-interfering: allotetrahydrocortisol, androstenediol, androstenedione, betamethasone, cortisone, cortol, cortolone, dehydroepiandrosterone, dexamethasone, estradiol, estriol, estrone, prednisolone, prednisone, progesterone, tetrahydrocortisone

KEY WORDS

SPE; column-switching

REFERENCE

Nozaki, O.; Ohata, T.; Ohba, Y.; Moriyama, H.; Kato, Y. Determination of urinary free cortisol by high performance liquid chromatography with sulphuric acid-ethanol derivatization and column switching, *Biomed.Chromatogr.*, **1992**, *6*, 109–114.

NITRATE

MISCELLANEOUS REACTIONS

Naphthylethylenediamine/Sulfanilamide

SAMPLE

Matrix: blood, cell cultures, urine

Analyte: nitrate

Sample preparation: Plasma. Dilute plasma 6-fold with water, filter (Amicon 3 KDa cutoff membrane, inject a 400 µL aliquot of the ultrafiltrate. Cell cultures, urine. Condition a 200 mg Bond Elut C18 SPE cartridge with 3 mL MeOH and 6 mL water. Add this SPE cartridge to the top of a 300 mg Alltech IC-AG$^+$ SPE cartridge, wash the assembly with 6 mL water. Dilute cell culture supernatant and urine 6-fold with water, add 3 mL to the SPE cartridge assembly, discard the first 1 mL, inject a 400 µL aliquot of the next 1 mL of eluate (J. Liq. Chromatogr. 1992, 15, 3315).

HPLC VARIABLES

Column: 250 × 4.6 5 µm Spherisorb SAX quaternary ammonium strong anion exchange

Mobile phase: 60 mM pH 2.8 Ammonium chloride

Flow rate: 0.7

Injection volume: 400

Detector: UV 540 following post-column reaction. The column effluent mixed with reagent 1 pumped at 0.7 mL/min and this mixture flowed through the solid-phase reactor (to reduce nitrate to nitrite). The effluent from the reactor mixed with reagent 2 pumped at 0.7 mL/min and this mixture flowed through a 30 cm × 0.5 mm ID coil of PTFE tubing filled with acid-washed 250 µm (?) beads (Supelco) and a 3 m × 0.5 mm ID coil to the detector. (Reagent 1 was 100 g ammonium chloride, 20 g sodium tetraborate decahydrate, 1 g disodium EDTA, and 200 mg copper(II) sulfate pentahydrate in 1 L water. Reagent 2 was prepared by mixing equal volumes of 0.1% naphthylethylenediamine dihydrochloride in 5% phosphoric acid with 1% sulfanilamide in 5% phosphoric acid immediately before use. Reagent 3 was 100 g ammonium chloride, 20 g sodium tetraborate decahydrate, and 1 g disodium EDTA in 1 L water. Prepare the solid phase reactor by suspending 15 g 300-340 mesh cadmium powder in 100 mL water, add 600 mL 0.5% copper(II) sulfate pentahydrate with continuous stirring, filter (Whatman No. 3 paper), wash the precipitate three times with water, wash with 1 M HCl, wash with water (after each wash decant the supernatant). Suspend the precipitate in 50 mL reagent 3 and pack into a 100 × 4.6 column, condition the reactor with reagent 1 at 0.7 mL/min until the pressure is stable.)

CHROMATOGRAM

Retention time: 9.8 (nitrite), 13.2 (nitrate)

Limit of detection: 75 nM

KEY WORDS

post-column reaction; rat; plasma; ultrafiltrate; SPE

REFERENCE

Muscará, M.N.; de Nucci, G. Simultaneous determination of nitrite and nitrate anions in plasma, urine and cell culture supernatants by high-performance liquid chromatography with post-column reaction, *J.Chromatogr.B*, **1996**, *686*, 157–164.

NITRITE

AMINE

7-Amino-4-methylcoumarin

SAMPLE
Matrix: solutions
Analyte: nitrite
Sample preparation: Mix 50 μL 200 μM 7-amino-4-methylcoumarin (Coumarin 120) in MeOH with 950 μL 500 mM sulfuric acid, add 500 μL aqueous nitrite solution, mix, let stand at room temperature for 20 min, heat at 100° for 70 min, cool to room temperature, extract twice with 2 mL portions of ethyl acetate. Combine the organic layers and evaporate them to dryness under a stream of nitrogen, reconstitute the residue in 400 μL EtOH, inject a 20 μL aliquot. (Nitrite diazotizes 7-amino-4-methylcoumarin. On heating the diazo compound is converted to the corresponding hydroxy compound (7-hydroxy-4-methylcoumarin, 4-methylumbelliferone) and this is detected.)

HPLC VARIABLES
Column: 100×4.6 5 μm Spheri-5 cyano
Mobile phase: Hexane:isopropanol 95:5
Flow rate: 1.8
Injection volume: 20
Detector: F ex 325 em 380

CHROMATOGRAM
Retention time: 2
Limit of quantitation: 1 ng/mL
Limit of detection: 0.5 ng/mL

REFERENCE
Zhou, J.Y.; Prognon, P.; Dauphin, C.; Hamon, M. HPLC fluorescence determination of nitrites in water using precolumn derivatization with 4-methyl-7-aminocoumarin, *Chromatographia*, **1993**, *36*, 57–60.

MISCELLANEOUS REACTIONS

Naphthylethylenediamine/Sulfanilamide

RELATED REFERENCE
Alonso, A.; Etxaniz, B.; Martinez, M.D. The determination of nitrate in cured meat products. A comparison of the HPLC UV/VIS and Cd/spectrophotometric methods. *Food Addit.Contam.* **1992**, *9*, 111-117.

SAMPLE
Matrix: blood, cell cultures, urine
Analyte: nitrite
Sample preparation: Plasma. Dilute plasma 6-fold with water, filter (Amicon 3 KDa cutoff membrane, inject a 400 μL aliquot of the ultrafiltrate. Cell cultures, urine. Condition a 200 mg Bond Elut C18 SPE cartridge with 3 mL MeOH and 6 mL water. Add this SPE cartridge to the top of a 300 mg Alltech IC-AG$^+$ SPE cartridge, wash the assembly with 6 mL water. Dilute cell culture supernatant and urine 6-fold with water, add 3 mL to the SPE cartridge assembly, discard the first 1 mL, inject a 400 μL aliquot of the next 1 mL of eluate (J. Liq. Chromatogr. 1992, 15, 3315).

HPLC VARIABLES
Column: 250 × 4.6 5 μm Spherisorb SAX quaternary ammonium strong anion exchange
Mobile phase: 60 mM pH 2.8 Ammonium chloride
Flow rate: 0.7
Injection volume: 400
Detector: UV 540 following post-column reaction. The column effluent mixed with reagent 1 pumped at 0.7 mL/min and this mixture flowed through the solid-phase reactor (to reduce nitrate to nitrite). The effluent from the reactor mixed with reagent 2 pumped at 0.7 mL/min and this mixture flowed through a 30 cm × 0.5 mm ID coil of PTFE tubing filled with acid-washed 250 μm (?) beads (Supelco) and a 3 m × 0.5 mm ID coil to the detector. (Reagent 1 was 100 g ammonium chloride, 20 g sodium tetraborate decahydrate, 1 g disodium EDTA, and 200 mg copper(II) sulfate pentahydrate in 1 L water. Reagent 2 was prepared by mixing equal volumes of 0.1% naphthylethylenediamine dihydrochloride in 5% phosphoric acid with 1% sulfanilamide in 5% phosphoric acid immediately before use. Reagent 3 was 100 g ammonium chloride, 20 g sodium tetraborate decahydrate, and 1 g disodium EDTA in 1 L water. Prepare the solid phase reactor by suspending 15 g 300-340 mesh cadmium powder in 100 mL water, add 600 mL 0.5% copper(II) sulfate pentahydrate with continuous stirring, filter (Whatman No. 3 paper), wash the precipitate three times with water, wash with 1 M HCl, wash with water (after each wash decant the supernatant). Suspend the precipitate in 50 mL reagent 3 and pack into a 100 × 4.6 column, condition the reactor with reagent 1 at 0.7 mL/min until the pressure is stable.)

CHROMATOGRAM
Retention time: 9.8 (nitrite), 13.2 (nitrate)
Limit of detection: 75 nM

KEY WORDS
post-column reaction; rat; plasma; ultrafiltrate; SPE

REFERENCE
Muscará, M.N.; de Nucci, G. Simultaneous determination of nitrite and nitrate anions in plasma, urine and cell culture supernatants by high-performance liquid chromatography with post-column reaction, *J.Chromatogr.B*, **1996**, *686*, 157–164.

NITRO GROUP

MISCELLANEOUS REACTIONS

Reduction

$RNO_2 \longrightarrow RNH_2$

SAMPLE
Matrix: air
Analyte: nitropolycyclic aromatic hydrocarbons
Sample preparation: Collect air particulates by pulling air through a filter, cut up the filter (10 cm^2), add 32 mL benzene:EtOH 75:25 (Caution! Benzene is a carcinogen!), sonicate, filter, wash the filtrate with 30 mL 5% NaOH, evaporate the organic layer to dryness, reconstitute the residue with 400 μL MeCN, dilute 8-fold, inject an aliquot.

HPLC VARIABLES
Guard column: 10 × 4 ODS
Column: 250 × 4.6 STR ODS-II (Shimadzu)
Mobile phase: MeCN:10 mM pH 6.8 imidazole-perchlorate buffer 75:25
Column temperature: 40
Flow rate: 1
Injection volume: 100
Detector: Chemiluminescence (Shimadzu CLD-10A) following post-column reaction. The mobile phase flowed through a reaction column then mixed with 200 μM bis(2,4,6-trichlorophenyl) oxalate in MeCN containing 15 mM hydrogen peroxide pumped at 1 mL/min. The mixture flowed through a 300 mm × 0.5 mm ID PTFE coil to the detector. (The reaction column was a 10 × 4 column packed with 100-200 mesh Zn powder:200-240 mesh 75 Å CPG-10 controlled-pore glass beads (CPG Inc., Fairfield NJ) 30:70. It was repacked each week. The nitro compounds were reduced to the corresponding amines which were then detected in the chemiluminescence detector.)

CHROMATOGRAM
Retention time: 8 (1-nitropyrene), 15 (1-nitrosopyrene), 20 (1-aminopyrene)

KEY WORDS
post-column reaction

REFERENCE
Hayakawa, K.; Terai, N.; Dinning, P.G.; Akutsu, K.; Iwamoto, Y.; Etoh, R.; Murahashi, T. An on-line reduction HPLC/chemiluminescence detection system for nitropolycyclic aromatic hydrocarbons and metabolites, *Biomed.Chromatogr.*, **1996**, *10*, 346–350.

SAMPLE
Matrix: solutions
Analyte: nitro-polycyclic aromatic hydrocarbons
Sample preparation: Inject a 0.5 μL aliquot.

HPLC VARIABLES
Column: 250 × 1 3 μm Nucleosil C18
Mobile phase: MeOH:THF:water 49:21:30
Flow rate: 0.03
Injection volume: 0.5
Detector: F ex 289 em 370 (cut-off filter) following post-column reaction. The column effluent flowed through a 60 mm tube packed with 5% platinum on alumina (Merck) at 60° to the detector. (The catalytic column reduced nitro-polycyclic aromatic hydrocarbons to fluorescent amino-polycyclic aromatic hydrocarbons.)

CHROMATOGRAM
Retention time: 7 (1,5-dinitronaphthalene), 10 (2-nitronaphthalene), 13 (9-nitroanthracene), 15 (1-nitropyrene), 17 (3-nitrofluoranthrene), 23 (6-nitrochrysene)

KEY WORDS
post-column reaction; microbore

REFERENCE
Götze, H.-J.; Schneider, J.; Herzog, H.-G. Determination of polycyclic aromatic hydrocarbons in Diesel soot by high-performance liquid chromatography, *Fresenius' J.Anal.Chem.*, **1991**, *340*, 27–30.

SAMPLE
Matrix: water
Analyte: nitroaromatics
Sample preparation: Adjust the pH of 100 mL water to 5.5 ± 0.1 with 1 mL 500 mM pH 5.5 citrate buffer, pass the solution through the reducing column and the SPE cartridge at about 10 mL/min. Remove the SPE cartridge and wash it with 1 mL 2% formic acid, elute with 1 mL THF:water:formic acid 60:40:2, inject a 20 μL aliquot of the eluate. (Prepare the reducing column by packing a 60 × 10 polypropylene tube with 500 mg iron powder (150 μm, Merck) and washing with 5 mL water. Condition a 100 mg octadecylsilane SPE cartridge (J.T. Baker) with two 1 mL portions of MeOH and two 1 mL portions of water. Mix 5 mL 2% formic acid and 500 μL 0.05% fluorescamine in acetone, add to the SPE cartridge. The nitro-compounds are reduced on the iron column to the corresponding amino-compounds which are then derivatized on the SPE cartridge.)

HPLC VARIABLES
Column: 250 × 4.6 5 μm Spherisorb octyl C8
Mobile phase: THF:water:formic acid 50:50:2
Flow rate: 0.7
Injection volume: 20
Detector: F ex 395 em 495

CHROMATOGRAM
Retention time: 8.5 (3-nitrophenol), 10 (nitrobenzene), 11 (3-nitrotoluene, 4-nitrotoluene), 12.5 (2-nitrotoluene)
Limit of detection: 12-280 pg/mL

KEY WORDS
SPE

REFERENCE
Djozan, D.; Faraz-Zadeh, M.A. Liquid chromatographic determination of nitroaromatics in water following solid phase extraction, pre-column reduction and derivatization, *Chromatographia*, **1996**, *43*, 25–30.

NITROPRUSSIDE

MISCELLANEOUS REACTIONS

Pyridine/Barbituric Acid

$$[Fe(CN)_5NO]^{\ominus} \longrightarrow CN^{\ominus} \xrightarrow{\text{chloramine T}} CNCl \longrightarrow$$

SAMPLE
Matrix: blood
Analyte: nitroprusside
Sample preparation: 100 µL Whole blood or plasma + 100 µL 3 M perchloric acid, mix, let stand at 0° for 5 min, centrifuge at 8000 g for 5 min. Remove a 150 µL aliquot of the supernatant and add it to 250 µL 1 M K_2HPO_4, mix vigorously, centrifuge for 8000 g for 5 min, inject a 20 µL aliquot of the supernatant.

HPLC VARIABLES
Column: 150 × 4 Asahipak BEST 502Q
Mobile phase: 100 mM pH 6.0 Acetate buffer containing 300 mM sodium perchlorate
Flow rate: 0.5
Injection volume: 20
Detector: F ex 583 em 607 following post-column reaction. The column effluent mixed with 5 mM dithiothreitol in 50 mM pH 9.0 Tris-HCl buffer containing 5 mM EDTA pumped at 0.1 mL/min and this mixture flowed through a 4.5 m × 0.5 mm ID coil at 90°. The effluent from this coil mixed with 0.5% Chloramine T in water pumped at 0.1 mL/min and flowed through a mixing coil. The effluent from this coil mixed with reagent pumped at 0.1 mL/min and this mixture flowed through a mixing coil to the detector. (Reagent was a mixture of 1.5 g barbituric acid, 15 mL pyridine, 3 mL concentrated HCl, and 82 mL water. Nitroprusside is reduced to cyanide with dithiothreitol. The cyanide is converted to cyanogen chloride with chloramine T, the cyanogen chloride reacts with the pyridine

to form pent-2-en-1,5-dial, and this compound reacts with barbituric acid to form the fluorescent 5,5'-(1,3-pentadiene-1-yl-5-ylidene)dibarbituric acid.)

CHROMATOGRAM
Retention time: 17
Limit of detection: 200 fmole

OTHER SUBSTANCES
Non-interfering: cyanide, thiocyanate

KEY WORDS
post-column reaction; whole blood; plasma; rat; pharmacokinetics

REFERENCE
Watanabe, T.; Nagamine, Y.; Toida, T.; Koshiishi, I.; Imanari, T. Sensitive determination of nitroprusside in blood by high performance liquid chromatography, *Anal.Sci.*, **1988**, *4*, 203–206.

NITROSAMIDE

SULFIDE

Sodium Sulfide

SAMPLE
Matrix: blood
Analyte: N-methyl-N-nitrosourea
Sample preparation: 400 μL Whole blood + 100 μL 100 mM pH 3.5 potassium citrate buffer + 2 mL 625 nM N-n-butyl-N-nitrosourea in MeCN, add 1 g anhydrous calcium chloride, vortex vigorously, centrifuge at 1000 g for 10 min. Remove a 200 μL aliquot of the upper layer and add it to 20 μL 30 mM sodium sulfide in borate-citrate buffer, let stand at room temperature for 10 min, cool to 0°, add 400 μL 300 μM taurine in borate-phosphate buffer, add 400 μL 300 μM o-phthalaldehyde in borate-phosphate buffer, let stand at room temperature for 5-10 min, inject a 20 μL aliquot. (Caution! N-Alkyl-N-nitrosoureas are carcinogens! Perform initial procedure at 0° until the MeCN is added. Prepare borate-citrate buffer by mixing 250 mL 50 mM sodium borate solution with 350 mL 100 mM potassium dihydrogen citrate, pH 5.5. Prepare borate-phosphate buffer by mixing 535 mL 50 mM sodium borate solution with 465 mL 100 mM KH_2PO_4, pH 8.0. Alkylnitrosoureas react with sulfide to give the corresponding alkylthiols which are then derivatized with o-phthalaldehyde and taurine.)

HPLC VARIABLES
Guard column: 10 × 4.6 Cosmosil 5C18-AR (Nacalai Tesque)
Column: 150 × 4.6 5 μm Cosmosil 5C18-AR (Nacalai Tesque)
Mobile phase: MeOH:THF:100 mM pH 6.8 potassium phosphate buffer 53:5:42
Flow rate: 1
Injection volume: 20
Detector: F ex 340 em 445

CHROMATOGRAM
Retention time: 4
Internal standard: N-n-butyl-N-nitrosourea (10)
Limit of detection: 2.5 ppb

OTHER SUBSTANCES
Also analyzed: N-ethyl-N-nitrosourea

KEY WORDS
whole blood; pharmacokinetics; rabbit

REFERENCE
Sano, A.; Ohashi, M.; Kodama, M.; Takitani, S. High-performance liquid chromatographic determination of N-nitroso-N-alkylureas by pre-column fluorescence derivatization and application to blood analysis, *J.Chromatogr.*, **1993**, *621*, 157–163.

NITROSAMINE

ACYL HALIDE

4-(2-Phthalimidyl)benzoyl Chloride

SAMPLE
Matrix: solutions
Analyte: nitrosamines
Sample preparation: Mix 100 μL of a 0.64-200 μM nitrosamine solution with 70 μL HBr reagent, heat at 40° for 5 min, evaporate to dryness under a stream of nitrogen, reconstitute with 100 μL 400 mM sodium bicarbonate solution, add 100 μL 5 mM 4-(2-phthalimidyl)benzoyl chloride in MeCN, mix, let stand at room temperature for 1 min, inject a 10 μL aliquot. (Prepare the HBr reagent by diluting 5 mL 47% HBr in water to 26 mL with acetic anhydride. The nitrosamine is denitrosated by the HBr to give the corresponding amine which is then acylated with 4-(2-phthalimidyl)benzoyl chloride. Preparation of 4-(2-phthalimidyl)benzoyl chloride is as follows. Mix 1.34 g o-phthalaldehyde in 50 mL diethyl ether with 1.37 g 4-aminobenzoic acid in 150 mL diethyl ether, stir at room temperature overnight. Filter and dry the solid to obtain 4-(2-phthalimidyl)benzoic acid. Suspend the 4-(2-phthalimidyl)benzoic acid in 100 mL chloroform, add 6 mL thionyl chloride, reflux for 1 h, evaporate to dryness, wash the residue twice with 10 mL portions of chloroform, recrystallize from benzene (Caution! Benzene is a carcinogen!) to obtain 4-(2-phthalimidyl)benzoyl chloride as fine colorless needles (mp >230°) (Anal. Chim. Acta 1987, 192, 309).)

HPLC VARIABLES
Column: 125 × 4.6 5 μm Nucleosil C18
Mobile phase: MeCN:water 48:52
Flow rate: 0.8
Injection volume: 10
Detector: F ex 299 em 426

CHROMATOGRAM
Retention time: 3.5 (N-nitrosodimethylamine), 4 (N-nitrosopyrrolidine), 4.5 (N-nitrosodiethylamine), 5 (N-nitrosopiperidine), 8 (N-nitrosodipropylamine), 17 (N-nitrosodibutylamine)

Limit of detection: 0.4-1.6 pmole

OTHER SUBSTANCES
Interfering: amines

REFERENCE
Zheng, M.; Fu, C.; Xu, H. High-performance liquid chromatographic detection of trace N-nitrosoamines by precolumn derivatization with 4-(2-phthalimidyl)benzoyl chloride, *Analyst*, **1993**, *118*, 269–271.

AMINE

7-Amino-4-methylcoumarin

R$_2$NNO   +

SAMPLE
Matrix: solutions
Analyte: N-nitrosodiethanolamine
Sample preparation: Mix 20 mL aqueous solution with 200 μL freshly prepared 10 mM ascorbic acid solution, adjust pH to 1.0 with 1 M HCl, vortex, extract three times with 50 mL portions of ethyl acetate. Combine the organic layers and dry them by passing them through a 1 cm layer of anhydrous sodium sulfate, evaporate almost to dryness under reduced pressure at room temperature, evaporate to dryness under a stream of nitrogen, reconstitute with 200 μL buffer, vortex, let stand in the dark for 15 min, add 1 mL reagent, heat at 100° for 70 min, cool to room temperature, extract twice with 2 mL portions of ethyl acetate. Combine the organic layers and evaporate them to dryness under a stream of nitrogen, reconstitute with 400 μL EtOH, inject a 20 μL aliquot. (Note that a small amount of the derivative, 4-methylumbelliferone is formed spontaneously so that it is always necessary to prepare a blank. Prepare pH 12.4 buffer by dissolving 3.8 g Na$_3$PO$_4$.12H$_2$O in 100 mL previously boiled water adjusted to pH 12.4 with 100 mM HCl. Prepare reagent by diluting 50 μL 200 μM 7-amino-4-methylcoumarin in MeOH with 950 μL 500 mM sulfuric acid.)

HPLC VARIABLES
Column: 100 × 4.6 Spheri-5 cyano-bonded
Mobile phase: Hexane:isopropanol 95:5
Column temperature: 20 ± 0.5
Flow rate: 1.8
Injection volume: 20
Detector: F ex 325 em 380

CHROMATOGRAM
Retention time: 3
Limit of quantitation: 1 ng/mL
Limit of detection: 0.8 ng/mL

REFERENCE
Diallo, S.; Zhou, J.Y.; Dauphin, C.; Prognon, P.; Hamon, M. Determination of N-nitrosodiethanolamine as nitrite in ethanolamine derivative raw materials by high-performance liquid chromatography with fluorescence detection after alkaline denitrosation, *J.Chromatogr.A*, **1996**, *721*, 75–81.

Sulfanilamide

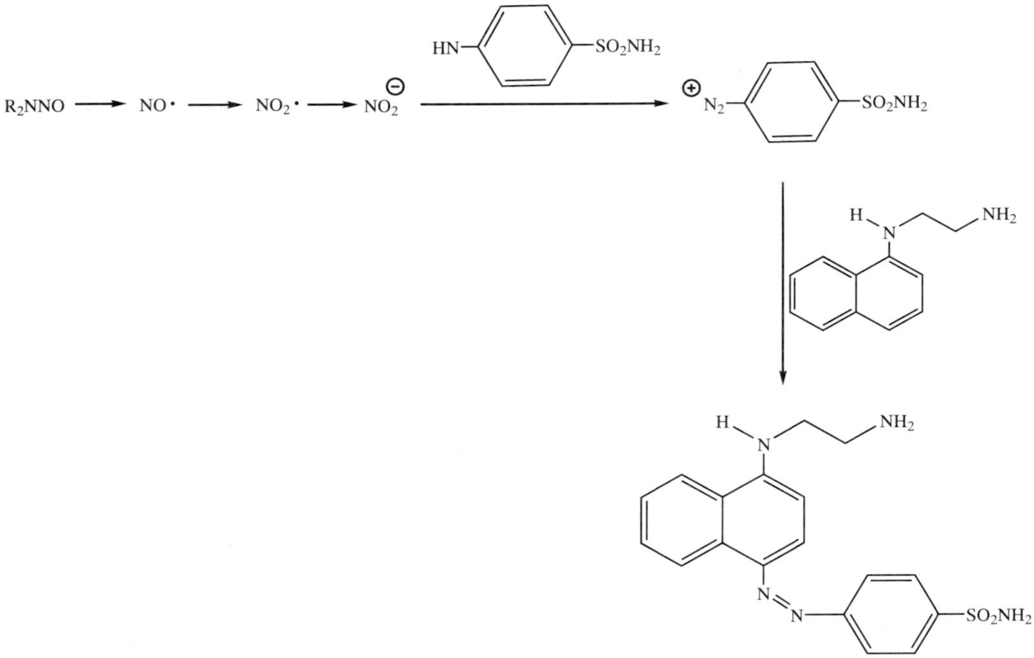

SAMPLE

Matrix: beverages, gastric juice

Analyte: nitrosamines

Sample preparation: Gastric juice. Extract 5 mL gastric juice containing N-nitrosodipro-pylamine three times with 10 mL portions of dichloromethane for 30 min. Combine the extracts and reduce the volume to 0.5 mL in a Kuderna-Danish evaporator at 55° with stirring, evaporate this sample to dryness under a slow stream of nitrogen, reconstitute with mobile phase, inject an aliquot. Beer. Add 50 mL beer and 100 ng N-nitrosomethyl-propylamine to a Chem Elut SPE cartridge, elute with five 30 mL portions of dichloro-methane, reduce the volume to 0.5 mL in a Kuderna-Danish evaporator at 55° with stir-ring, evaporate this sample to dryness under a slow stream of nitrogen, reconstitute with mobile phase, inject an aliquot.

HPLC VARIABLES

Column: 250 × 4.6 5 μm Nucleosil C18

Mobile phase: Gradient. A was MeCN:water:glacial acetic acid 5:94:1. B was MeCN:water:glacial acetic acid 94:5:1. A:B 80:20 for 10 min, to 40:60 over 10 min, to 10:90 over 20 min, maintain at 10:90 for 20 min.

Flow rate: 0.8

Detector: UV 546 following post-column reaction. The column effluent flowed through a 6 m × 0.3 mm ID knitted PTFE coil irradiated with a 1.2 W low pressure UV lamp and then mixed with reagent pumped at 0.8 mL/min. This mixture flowed through a 1 m × 0.3 mm ID PTFE coil at 60° then a 0.5 m × 0.3 mm ID PTFE coil at room temperature to the detector. (The reagent was a Griess reagent consisting of equal volumes of 0.1% naphthylethylenediamine dihydrochloride in water and 1% sulfanilamide in 5% phos-phoric acid, mixed fresh each day. The nitrosamines were photohydrolyzed to nitrite which then reacted with the Griess reagent to give a colored product.)

CHROMATOGRAM

Retention time: 6.5 (N-nitrosodimethylamine), 7.3 (N-nitrosomorpholine), 7.8 (N-nitroso-ethylmethylamine), 8.1 (N-nitrosopyrrolidine), 11.4 (N-nitrosodiethylamine), 12.8 (N-ni-trosopiperidine), 33.3 (N-nitrosodibutylamine), 34.5 (N-nitrosodiphenylamine)

Internal standard: N-nitrosodipropylamine (26.5)
Limit of detection: 1 ng

KEY WORDS
post-column reaction; beer

REFERENCE
Bellec, G.; Cauvin, J.M.; Salaun, M.C.; Le Calvé, K.; Dréano, Y.; Gouérou, H.; Ménez, J.F.; Berthou, F. Analysis of N-nitrosamines by high-performance liquid chromatography with post-column photohydrolysis and colorimetric detection, *J.Chromatogr.A*, **1996**, *727*, 83–92.

PERACID

SULFIDE

Methyl p-Tolyl Sulfide

SAMPLE
Matrix: solutions
Analyte: peroxyacetic acid
Sample preparation: Add 1 mL solution (containing not more than 2 g/L peroxyacetic acid) to 2 mL 10 mM methyl p-tolyl sulfide in MeOH, make up to 10 mL with water, mix, after 1 min add 100 mg manganese dioxide, centrifuge at 3000 rpm for 15 min, inject an aliquot of the supernatant. (Peroxyacetic acid oxidizes methyl p-tolyl sulfide to methyl p-tolyl sulfoxide which is then determined by HPLC.)

HPLC VARIABLES
Column: 50×4.6 3 μm Nucleosil C18
Mobile phase: MeCN:water 40:60
Flow rate: 2
Injection volume: 5
Detector: UV 230

CHROMATOGRAM
Retention time: 0.35 (for methyl p-tolyl sulfoxide)
Limit of detection: 30 ppb

OTHER SUBSTANCES
Non-interfering: hydrogen peroxide

REFERENCE
Pinkernell, U.; Effemann, S.; Nitzsche, F.; Karst, U. Rapid high-performance liquid chromatographic method for the determination of peroxyacetic acid, *J.Chromatogr.A*, **1996**, *730*, 203–208.

SAMPLE
Matrix: solutions
Analyte: peroxyacetic acid
Sample preparation: 100 μL Sample + 100 μL 20 mM methyl p-tolyl sulfide in MeCN + 300 μL water, mix, let stand for 10 min, add 400 μL MeCN, add 100 μL 10 mM triphenylphosphine in MeCN, mix, let stand in the dark for 30 min, inject a 10 μL aliquot. (Caution! Peroxyacetic acid and hydrogen peroxide are strong oxidizing agents. Do not mix concentrated solutions with reducing agents, organic compounds, or organic solvents! These procedures have only been tested with peroxide concentrations up to 2 g/L. Peroxyacetic acid oxidizes methyl p-tolyl sulfide to methyl p-tolyl sulfoxide which elutes at 1 min. Hydrogen peroxide oxidizes triphenylphosphine to triphenylphosphine oxide which elutes at 2.6 min. Blanks should always be run.)

HPLC VARIABLES
Column: 70×3 5 μm Nucleosil C8
Mobile phase: Gradient. MeCN:water 40:60 for 3 min, to 100:0 over 10 s, maintain at 100:0 for 50 s, return to initial conditions over 10 s, re-equilibrate for 1 min 50 s.
Flow rate: 1
Injection volume: 10

Detector: UV 225

CHROMATOGRAM
Retention time: 1 (peroxyacetic acid), 2.6 (hydrogen peroxide)
Limit of detection: 1 mM

REFERENCE
Pinkernell, U.; Effkemann, S.; Karst, U. Simultaneous HPLC determination of peroxyacetic acid and hydrogen peroxide, *Anal.Chem.*, **1997**, *69*, 3623–3627.

PEROXIDE

ALCOHOL

Methanol

$$H_2O_2 + CH_3OH \longrightarrow HCHO \longrightarrow$$

SAMPLE
Matrix: beverages
Analyte: hydrogen peroxide
Sample preparation: Mix 3.45 mL 500 mM pH 5.0 potassium phosphate buffer, 250 μL MeOH, and 100 μL antifoaming reagent, pass nitrogen gas through the mixture for a few min, add 1 mL beverage, add 100 μL 1000 U/mL catalase (Boehringer Mannheim) in water (purge with nitrogen before use), add 100 μL 100 mg/mL 4-amino-3-penten-2-one (Fluoral-P, Eastman) in MeCN, bubble nitrogen at 200 mL/min through the mixture, heat at 30° for 10 min, add to a Sep-Pak C18 SPE cartridge, wash with a little water, elute with 5 mL MeCN:water 50:50, inject a 20 μL aliquot of the eluate. (Prepare antifoaming reagent by diluting concentrated silicon polymer (Sigma) with water to give a 0.1% suspension. The hydrogen peroxide reacts with methanol to form formaldehyde. The formaldehyde reacts with 4-amino-3-penten-2-one to give a dihydropyridine in a Hantzch reaction.)

HPLC VARIABLES
Column: 250 × 4.6 5 μm Zorbax ODS
Mobile phase: MeCN:water 50:50
Flow rate: 1
Injection volume: 20
Detector: F ex 410 em 510

CHROMATOGRAM
Retention time: 3.5
Limit of detection: 50 ppb

OTHER SUBSTANCES
Non-interfering: ascorbic acid

REFERENCE
Hamano, T.; Mitsuhashi, Y.; Yamamoto, S. Determination of hydrogen peroxide in beverages by high-performance liquid chromatography with fluorescence detection, *J.Chromatogr.*, **1987**, *411*, 423–429.

PHOSPHINE

Triphenylphosphine

SAMPLE

Matrix: solutions
Analyte: hydrogen peroxide
Sample preparation: 100 µL Sample + 100 µL 20 mM methyl p-tolyl sulfide in MeCN + 300 µL water, mix, let stand for 10 min, add 400 µL MeCN, add 100 µL 10 mM triphenylphosphine in MeCN, mix, let stand in the dark for 30 min, inject a 10 µL aliquot. (Caution! Peroxyacetic acid and hydrogen peroxide are strong oxidizing agents. Do not mix concentrated solutions with reducing agents, organic compounds, or organic solvents! These procedures have only been tested with peroxide concentrations up to 2 g/L. Peroxyacetic acid oxidizes methyl p-tolyl sulfide to methyl p-tolyl sulfoxide which elutes at 1 min. Hydrogen peroxide oxidizes triphenylphosphine to triphenylphosphine oxide which elutes at 2.6 min. Blanks should always be run.)

HPLC VARIABLES

Column: 70 × 3 5 µm Nucleosil C8
Mobile phase: Gradient. MeCN:water 40:60 for 3 min, to 100:0 over 10 s, maintain at 100:0 for 50 s, return to initial conditions over 10 s, re-equilibrate for 1 min 50 s.
Flow rate: 1
Injection volume: 10
Detector: UV 225

CHROMATOGRAM

Retention time: 1 (peroxyacetic acid), 2.6 (hydrogen peroxide)
Limit of detection: 3 mM

REFERENCE

Pinkernell, U.; Effkemann, S.; Karst, U. Simultaneous HPLC determination of peroxyacetic acid and hydrogen peroxide, *Anal.Chem.*, **1997**, *69*, 3623–3627.

PHOSPHATE

SULFITE

Sodium Metabisulfite

SAMPLE
Matrix: blood
Analyte: pyridoxal
Sample preparation: 500 µL Plasma + 500 µL 800 mM perchloric acid, vortex vigorously, centrifuge at 35000 g for 5 min, inject a 50-500 µL aliquot of the supernatant.

HPLC VARIABLES
Column: 150 × 4.6 AQ-302 ODS (YMC)
Mobile phase: 100 mM pH 3.0 KH_2PO_4 containing 100 mM sodium perchlorate and 0.5 g/L sodium bisulfite
Flow rate: 1
Injection volume: 50-500
Detector: F ex 300 em 400

CHROMATOGRAM
Retention time: 3 (pyridoxal phosphate), 5 (pyridoxal), 11 (4-pyridoxic acid)
Limit of detection: 1 pmole (pyridoxal), 0.5 pmole (pyridoxal phosphate)

OTHER SUBSTANCES
Simultaneously analyzed: pyridoxamine, pyridoxamine phosphate, pyridoxine

---

KEY WORDS
plasma; rat; human; protect from light

REFERENCE
Kimura, M.; Kanehira, K.; Yokoi, K. Highly sensitive and simple liquid chromatographic determination in plasma of B_6 vitamers, especially pyridoxal 5'-phosphate, *J.Chromatogr.A*, **1996**, *722*, 295–301.

THIOL

1,2-Ethanedithiol

SAMPLE
Matrix: phosphopeptides
Analyte: phosphoserine residues on peptides
Sample preparation: Evaporate 40 μL 236 μg/mL phosphopeptide to dryness under reduced pressure, add 50 μL derivatization solution, heat at 55° under nitrogen for 1 h, add 1 mL 1.75 M acetic acid, add to a Sep-Pak C18 SPE cartridge, elute with MeCN:water 50:50, concentrate to 40 μL, inject an aliquot. (Derivatization solution was 200 μL water, 65 μL 5 M NaOH, 6 μL 250 mM EDTA, 245 μL DMSO, 90 μL EtOH, and 15 μL 1,2-ethanedithiol, thoroughly gassed with nitrogen.)

CAPILLARY ELECTROPHORESIS
Capillary: 47 cm × 50 μm fused-silica (Beckman)
Running buffer: 20 mM pH 9.15 Phosphate/borate buffer
Capillary temperature: 22
Voltage/Current: 30 kV
Injection: 6 nL
Detector: UV 200
Model: Beckman P/ACE System 2100
Migration time: 5.63 (Ala-Thr-SerP-Asn-Val-Phe)

KEY WORDS
SPE

REFERENCE

Fadden, P.; Haystead, T.A. Quantitative and selective fluorophore labeling of phosphoserine on peptides and proteins: characterization at the attomole level by capillary electrophoresis and laser-induced fluorescence, *Anal.Biochem.*, **1995**, *225*, 81–88.

SAMPLE

Matrix: phosphopeptides

Analyte: phosphoserine residues on peptides

Sample preparation: Evaporate 40 µL 236 µg/mL phosphopeptide to dryness under reduced pressure, add 50 µL derivatization solution, heat at 55° under nitrogen for 1 h, add 1 mL 1.75 M acetic acid, add to a Sep-Pak C18 SPE cartridge, elute with MeCN:water 50:50, concentrate to 40 µL, add a 2- to 4-fold excess of 6-iodoacetamidofluorescein (Molecular Probes, Eugene OR), inject an aliquot. (Derivatization solution was 200 µL water, 65 µL 5 M NaOH, 6 µL 250 mM EDTA, 245 µL DMSO, 90 µL EtOH, and 15 µL 1,2-ethanedithiol, thoroughly gassed with nitrogen.)

CAPILLARY ELECTROPHORESIS

Capillary: 47 cm × 50 µm fused-silica (Beckman)

Running buffer: 20 mM pH 9.15 Phosphate/borate buffer

Capillary temperature: 22

Voltage/Current: 30 kV

Injection: 6 nL

Detector: F ex 493 (laser) em 522

Model: Beckman P/ACE System 2100

Migration time: 4.71 (Ala-Thr-SerP-Asn-Val-Phe)

Limit of quantitation: <75 amole

KEY WORDS

SPE

REFERENCE

Fadden, P.; Haystead, T.A. Quantitative and selective fluorophore labeling of phosphoserine on peptides and proteins: characterization at the attomole level by capillary electrophoresis and laser-induced fluorescence, *Anal.Biochem.*, **1995**, *225*, 81–88.

PHOSPHATIDYLCHOLINE

DIAZO COMPOUND

Diazomethane

SAMPLE
Matrix: bulk
Analyte: phosphatidylcholine
Sample preparation: Emulsify 150 μg phosphatidylcholine in 8 mL 100 mM pH 5.5 acetate buffer and 1.5 mL 1 M calcium chloride, add 100 μg cabbage phospholipase D, add 4 mL diethyl ether, stir at room temperature for 20 h, add 3 mL 500 mM EDTA, extract with chloroform/MeOH. Dissolve the extracted phosphatidate in 2 mL chloroform: MeOH:water 63.5:31.5:5, add 500 μL 100 mM HCl, mix rapidly. Remove the lower phase and evaporate it to dryness, reconstitute the residue in 500 μL diethyl ether, add 1 mL diazomethane in diethyl ether, evaporate to dryness under a stream of nitrogen, dissolve the residue in 20 μL MeOH, inject an aliquot. (Preparation of diazomethane is as follows. Caution! Diazomethane is toxic, explosive, and carcinogenic! A face shield and a safety screen should always be used and the preparation should only be carried out in a properly functioning chemical fume hood. Only smooth glass apparatus with rubber stoppers and plastic tubing should be used. Scratched glassware, ground glass joints, and sharp edges should be avoided (Org. Syn., Coll. Vol. VI; Wiley: New York, 1988, pp. 432-435). Procedures have been reported for the synthesis of diazomethane from N-methyl-N-nitrosourea (Org. Syn., Coll. Vol. II; Wiley: New York, 1943, pp. 165-167), N-nitroso-β-methylamino-isobutyl methyl ketone (Org. Syn., Coll. Vol. III; Wiley: New York, 1955, pp. 244-248), and N,N'-dimethyl-N,N'-dinitrosoterephthalamide (Org. Syn., Coll. Vol. V; Wiley:New York, 1973, pp. 351-355). Probably the most convenient starting material is N-methyl-N-nitroso-p-toluenesulfonamide (Diazald) (Aldrichimica Acta 1983, 16, 3-10; Org. Syn., Coll. Vol. IV, Wiley: New York, 1963, pp. 250-253). Add 10 mL 95% EtOH to a solution of 5 g KOH in 8 mL water, warm to 65° using a water bath, add a solution of 5 g N-methyl-N-nitroso-p-toluenesulfonamide in 45 mL ether over 20 min at such a rate as to keep the reaction volume constant. Collect the ether and diazomethane that distil in an ice-cooled receiving flask under a dry ice/acetone condenser. When all the N-methyl-N-nitroso-p-toluenesulfonamide has been used up, slowly add 10 mL ether to the reaction flask and continue distillation until the distillate is colorless. A purpose-built distillation apparatus can be purchased from Aldrich (Aldrichimica Acta 1983, 16, 3-10). Excess quantities of diazomethane can be destroyed by adding acetic acid until the yellow color of the diazomethane is discharged. The safe disposal of the nitroso compounds used to generate diazomethane has been discussed (Lunn,G.; Sansone,E.B. Destruction of Hazardous Chemicals in the Laboratory, Second Edition. Wiley: New York, 1994, pp. 277-289).)

HPLC VARIABLES
Column: 250 × 4 5 μm LiChrosorb RP-18
Mobile phase: MeCN:MeOH:isopropanol:water 50:18:27:5
Flow rate: 1.5
Detector: UV 205

CHROMATOGRAM
Retention time: 9.4-36.0 (depending on structure)

REFERENCE

Nakagawa, Y.; Waku, K. Improved procedure for the separation of the molecular species of dimethylphosphatidate by high-performance liquid chromatography, *J.Chromatogr.*, **1986**, *381*, 225–231.

MISCELLANEOUS REACTIONS

Micellization

SAMPLE

Matrix: solutions

Analyte: pumactant

Sample preparation: Centrifuge 10 mL rabbit lung or Eustachian tube washing saline at 1000 g for 5 min, lyophilize the supernatant, reconstitute with 5 mL chloroform, filter (0.45 μm). Evaporate the filtrate to dryness under a stream of nitrogen, reconstitute the residue in 500 μL MeOH, inject a 100 μL aliquot.

HPLC VARIABLES

Column: 100 × 4.6 5 μm ODS Hypersil

Mobile phase: MeOH:THF 97:3 containing 40 mM choline chloride

Flow rate: 1

Injection volume: 100

Detector: F 340 ex 460 following post-column derivatization. The column effluent mixed with the reagent pumped at 1.2 mL/min and the mixture flowed through a 300 × 5 PTFE column filled with 250 μm glass beads and a 3 m × 0.5 mm ID coil of PTFE tubing at 50° to the detector. (The reagent was 150 μL 3 μM 1,6-diphenyl-1,3,5-hexatriene in THF made up to 1 L with water containing 0.001% Tween 20. The phospholipids form micelles and the diphenylhexatriene fluoresces in the micelles.)

CHROMATOGRAM

Retention time: 5 (dipalmitoylphosphatidylcholine)

Limit of detection: 1.5 μg/mL

KEY WORDS

post-column reaction; rabbit

REFERENCE

Kitsos, M.; Gandini, C.; Massolini, G.; De Lorenzi, E.; Caccialanza, G. High-performance liquid chromatography post-column derivatization with fluorescence detection to study the influence of ambroxol on dipalmitoylphosphatidylcholine levels in rabbit eustachian tube washings, *J.Chromatogr.*, **1991**, *553*, 1–6.

SAMPLE

Matrix: tissue

Analyte: phosphatidylcholine

Sample preparation: Homogenize (Potter-Elvehjem) liver or lung with 20 volumes chloroform:MeOH 2:1, filter (paper), wash with a volume of 50 mM NaCl equal to one-fifth the volume of extract, centrifuge (J.Biol.Chem. 1957, 226, 497). Remove the lower organic layer and evaporate it to dryness under a stream of nitrogen, reconstitute the residue in chloroform, add 100 μL 1 mM IS in chloroform. Purify by normal phase HPLC using hexane:isopropanol:water 42:56:6.3 at 1 mL/min on a 300 × 4.9 10 μm μPorasil column using UV 200 detection, collect fraction eluting between 26 and 32 min, evaporate, dissolve in the minimum amount of trifluoroethanol, inject an aliquot.

HPLC VARIABLES

Column: 250 × 4.6 5 μm Apex ODS 2

Mobile phase: MeOH:water 92.5:7.5 containing 40 mM choline chloride

Column temperature: 50

Flow rate: 1

Detector: F ex 340 em 460 following post-column reaction. The column effluent mixed with reagent pumped at 3 mL/min in a 15 μL mixing chamber and the mixture flowed through a 3 m × 0.5 mm ID PTFE coil at 50° to the detector. (Reagent was water containing 150 μL/L 3 mM 1,6-diphenyl-1,3,5-hexatriene in THF:Tween 20 99.999:0.001. The phospholipids form micelles and the diphenylhexatriene fluoresces in the micelles.)

CHROMATOGRAM
Retention time: 35-57
Internal standard: phosphatidylcholine 14:0/14:0 (20)

KEY WORDS
rat; liver; lung; post-column reaction

REFERENCE
Postle, A.D. Method for the sensitive analysis of individual molecular species of phosphatidylcholine by high-performance liquid chromatography using post-column fluorescence detection, *J.Chromatogr.*, **1987**, *415*, 241−251.

SAMPLE
Matrix: tissue
Analyte: pumactant
Sample preparation: Homogenize (Potter-Elvehjem) liver or lung with 20 volumes chloroform:MeOH 2:1, filter (paper), wash with a volume of 50 mM NaCl equal to one-fifth the volume of extract, centrifuge (J.Biol.Chem. 1957, 226, 497). Remove the lower organic layer and evaporate it to dryness under a stream of nitrogen, reconstitute with chloroform:MeOH 25:75, inject an aliquot.

HPLC VARIABLES
Guard column: 10 × 4.6 5 μm Nucleosil 5 NH2
Column: 50 × 4.6 5 μm Nucleosil 5 NH2 + 175 × 4.6 5 μm Nucleosil 5 NH2
Mobile phase: MeCN:MeOH:water:50% methylphosphonic acid in water 73:25:1.5:0.03, adjusted to pH 3 with 25% ammonium hydroxide in water
Flow rate: 1
Detector: F ex 340 em 460 following post-column reaction. The column effluent mixed with the reagent pumped at 4.5 mL/min and the mixture flowed through a 2 m × 0.8 mm ID coil of PTFE tubing at 50° to the detector. (Reagent was water containing 150 μL/L 3 mM 1,6-diphenyl-1,3,5-hexatriene in THF:Tween 20 99.999:0.001. The phospholipids form micelles and the diphenylhexatriene fluoresces in the micelles.)

CHROMATOGRAM
Retention time: 22 (phosphatidylglycerol) (dipalmitoylphosphatidylcholine is separated but retention time is not given)

KEY WORDS
pig; lung; liver; gastric mucosa; post-column reaction

REFERENCE
Bernhard, W.; Linck, M.; Creutzburg, H.; Postle, A.D.; Arning, A.; Martin-Carrera, I.; Sewing, K.-F. High-performance liquid chromatographic analysis of phospholipids from different sources with combined fluorescence and ultraviolet detection, *Anal.Biochem.*, **1994**, *220*, 172−180.

PHOSPHONIC ACID

ACTIVATED HALIDE

Panacyl Bromide

SAMPLE
Matrix: solutions
Analyte: alkylphosphonic acids
Sample preparation: Aqueous solutions. Adjust the pH of an aqueous solution containing 10 μmoles to 7.1 with tetra-n-butylammonium hydroxide, evaporate to dryness under reduced pressure, add toluene, evaporate to dryness under reduced pressure. Add 2 mL 25 mM panacyl bromide (Molecular Probes, Eugene OR) in DMF, stir at 40° for 2 h, dilute with MeCN, inject an aliquot. DMF solutions. Mix 10 μmoles dissolved in DMF with 21 mg panacyl bromide and 4 mg diisopropylethylamine, make up to 2 mL with DMF, stir at 80° for 30 min, dilute with MeCN, inject an aliquot.

HPLC VARIABLES
Column: 250 × 1 Adsorbosphere C18
Mobile phase: Gradient. MeCN:water from 60:40 to 100:0 over 30 min.
Flow rate: 0.05
Detector: F ex 325 (10 mW He-Cd laser)

CHROMATOGRAM
Retention time: 21 (methylphosphonic acid), 22 (ethylphosphonic acid), 24 (isopropylphosphonic acid), 27 (n-hexylphosphonic acid)
Limit of detection: 20 fmole

KEY WORDS
laser

REFERENCE
Roach, M.C.; Ungar, L.W.; Zare, R.N.; Reimer, L.M.; Pompliano, D.L.; Frost, J.W. Fluorescence detection of alkylphosphonic acids using p-(9-anthroyloxy)phenacyl bromide, *Anal.Chem.*, **1987**, *59*, 1056–1059.

PHOSPHORIC ACID

AMINE

Dansylethylenediamine

SAMPLE
Matrix: solutions
Analyte: 2'-deoxynucleoside 5'-monophosphates
Sample preparation: 10 μL Solution + 200 μL buffer + 10 μL 100 mM 1-(3-dimethylam-inopropyl)-3-ethylcarbodiimide in buffer + 40 μL 50 mM dansylethylenediamine in DMSO, mix, let stand in the dark at room temperature for 18 h, inject a 10 μL aliquot. (Buffer was 100 mM pH 7.0 1-methylimidazole buffer. Synthesis of dansylethylenediamine is as follows. Add a solution of dansyl chloride in DMF to a 10% molar excess of ethylenediamine in DMF, evaporate to dryness, dissolve the residue in pH 10 carbonate buffer, purify by TLC, extract with MeOH.)

HPLC VARIABLES
Column: 250 × 4.6 Finepak ODP-50 octadecyl-bonded polyvinyl alcohol gel (Asahikasei, Tokyo)
Mobile phase: Gradient. A was MeCN:10 mM pH 10.3 phosphate buffer 15:85. B was MeCN:10 mM pH 10.3 phosphate buffer 30:70. A:B 100:0 for 10 min, to 0:100 over 30 min.
Column temperature: 40
Flow rate: 0.6
Injection volume: 10
Detector: F ex 270 em 546

CHROMATOGRAM

Retention time: 15 (2'-deoxyguanosine 5'-monophosphate), 20 (2'-deoxycytidine 5'-mono-phosphate), 25 (2'-deoxythymidine 5'-monophosphate), 28 (2'-deoxyadenosine 5'-monophosphate)

Limit of detection: 6.4-6.7 pmole

REFERENCE

Sonoki, S.; Kadoike, Y.; Kiyokawa, M.; Hisamatsu, S. High-performance liquid chromatographic analysis of 2'-deoxynucleoside 5'-monophosphate using N-(dansyl)ethylenediamine as a fluorescent derivatizing reagent, *J.Liq.Chromatogr.*, **1993**, *16*, 2731–2739.

SAMPLE

Matrix: solutions

Analyte: nucleotides

Sample preparation: 10 μL Solution + 200 μL buffer + 10 μL 100 mM 1-(3-dimethylam-inopropyl)-3-ethylcarbodiimide in buffer + 40 μL 50 mM dansylethylenediamine in DMSO, mix, let stand in the dark at 27° for 18 h, inject a 10 μL aliquot. (Buffer was 100 mM pH 7.5 1-methylimidazole buffer. Synthesis of dansylethylenediamine is as follows. Add a solution of dansyl chloride in DMF to a 10% molar excess of ethylenediamine in DMF, evaporate to dryness, dissolve the residue in pH 10 carbonate buffer, purify by TLC, extract with MeOH (J. Liq. Chromatogr. 1993, 16, 2731).)

HPLC VARIABLES

Column: 250 × 4.6 Finepak ODP-50 octadecyl-bonded polyvinyl alcohol gel (Asahikasei, Tokyo)

Mobile phase: Gradient. A was MeCN:10 mM pH 10.3 phosphate buffer 12:88. B was MeCN:10 mM pH 10.3 phosphate buffer 22:78. C was MeCN:10 mM pH 10.3 phosphate buffer 40:60. A:B:C 100:0:0 for 10 min, to 0:100:0 over 18 min, to 0:0:100 (step gradient), maintain at 0:0:100 for 10 min.

Column temperature: 40

Flow rate: 0.6

Injection volume: 10

Detector: F ex 270 em 546

CHROMATOGRAM

Retention time: 10 (guanosine triphosphate), 11 (adenosine triphosphate), 15 (guanosine diphosphate), 21 (adenosine diphosphate), 29 (guanosine monophosphate), 30 (uridine monophosphate), 31 (cytidine monophosphate), 35 (adenosine monophosphate)

Limit of detection: 4.7-20.3 pmole

REFERENCE

Sonoki, S.; Sanda, A.; Hisamatsu, S. Simultaneous determination of mono-, di-, and trinucleotides by high-performance liquid chromatography using *N*-(dansyl)ethylenediamine as a fluorescent derivatizing reagent, *J.Liq.Chromatogr.*, **1994**, *17*, 1057–1064.

PYRIDINE RING

BARBITURIC ACID

Barbituric Acid

SAMPLE
Matrix: urine
Analyte: nicotine
Sample preparation: 200 μL Urine + 100 μL 4 M pH 4.7 sodium acetate buffer + 40 μL 1.5 M KCN in water + 40 μL 400 mM chloramine T in water + 200 μL 78 mM barbituric acid in acetone:water 50:50, mix for 10 s, let stand at room temperature for 15 min, add 40 μL 1 M sodium metabisulfite in water (Clin. Chim. Acta 1987, 165, 45), inject a 50 μL aliquot.

HPLC VARIABLES
Guard column: 10 × 4.6 10 μm reversed-phase (Whatman)
Column: 250 × 4.6 5 μm Spherisorb S 5ODS2
Mobile phase: MeCN:50 mM pH 5.2 sodium acetate buffer 22:78
Flow rate: 1.5
Injection volume: 50
Detector: UV 490

CHROMATOGRAM
Retention time: 4.1
Limit of detection: 10 nM

OTHER SUBSTANCES
Extracted: cotinine

REFERENCE
Ubbink, J.B.; Lagendijk, J.; Vermaak, W.H.H. Simple high-performance liquid chromatographic method to verify the direct barbituric acid assay for urinary cotinine, *J.Chromatogr.*, **1993**, *620*, 254–259.

1,3-Diethyl-2-thiobarbituric Acid

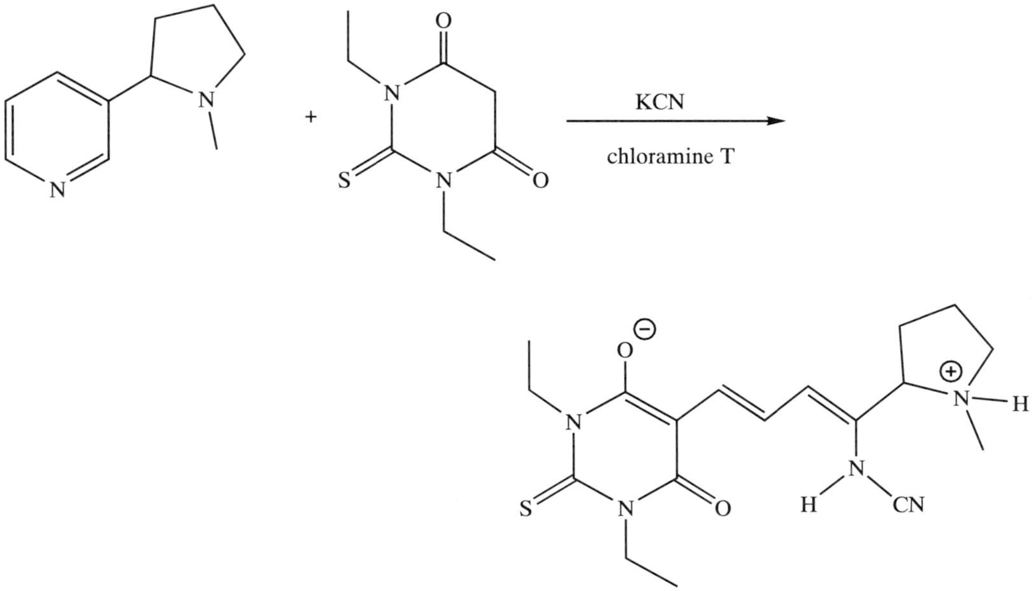

SAMPLE

Matrix: urine

Analyte: nicotine

Sample preparation: Centrifuge urine at 10000 g for 5 min, filter (Millipore Ultrafree-MC, 30 000 NMWL). Dilute a 10-150 μL aliquot to 150 μL with water, add 60 μL 5 μg/mL IS in buffer, add 30 μL 1.5 M KCN, add 15 μL 1 M chloramine T, add 75 μL 75 mM in 1,3-diethyl-2-thiobarbituric acid in water:acetone 50:50, mix vigorously for 30 s, centrifuge for 2 min, let stand for 11-12 min, inject a 50-250 μL aliquot. (Buffer was 6 M sodium acetate adjusted to pH 4.7 with 6 M trichloroacetic acid. The chloramine T and 1,3-diethyl-2-thiobarbituric acid solutions should be stored at 40°.)

HPLC VARIABLES

Guard column: 25 × 4 5 μm Nucleosil 100 C18

Column: 150 × 3.9 4 μm Nova Pak RP18

Mobile phase: Gradient. A was 60 mM sodium 1-pentanesulfonate adjusted to pH 3.8 with 1% orthophosphoric acid. B was MeOH:THF 95:5. C was MeCN:THF 97:3. A:B:C from 80:10:10 to 74:15:11 over 5 min, to 65:20:15 over 5 min, to 65:15:20 over 5 min, to 65:10:25 over 5 min, to 60:0:40 over 10 min, to 35:0:65 over 3 min.

Flow rate: From 1.55 to 1.45 over 15 min, maintain at 1.45.

Injection volume: 50-250

Detector: UV 529

CHROMATOGRAM

Retention time: 26

Internal standard: N'-ethylnorcotinine (21)

Limit of quantitation: 1.5 μM

OTHER SUBSTANCES

Extracted: metabolites

KEY WORDS

human; hamster; rat

REFERENCE

Rustemeier, K.; Demetriou, D.; Schepers, G.; Voncken, P. High-performance liquid chromatographic determination of nicotine and its urinary metabolites via their 1,3-diethyl-2-thiobarbituric acid derivatives, *J.Chromatogr.*, **1993**, *613*, 95–103.

CYANIDE

Potassium Cyanide

SAMPLE
Matrix: urine
Analyte: nicotine and cotinine
Sample preparation: 500 μL Urine + 200 μL 4 M pH 4.7 acetate buffer + 100 μL 1.5 M KCN in water + 100 μL 400 mM chloramine T in water + 500 μL 50 mM 1,3-diethyl-2-thiobarbituric acid in acetone:water 50:50, mix, let stand at room temperature for 15 min, inject a 5 μL aliquot immediately.

HPLC VARIABLES
Guard column: 10 × 4.6 10 μm reversed-phase (Whatman)
Column: 300 × 3.9 μBondapak C18
Mobile phase: MeOH:water 2:1 containing 20 mM pentanesulfonic acid
Flow rate: 2
Injection volume: 5
Detector: UV 546

CHROMATOGRAM
Retention time: 4 (cotinine), 6 (nicotine)
Limit of detection: 10 ng/mL (20 μL injection)

OTHER SUBSTANCES
Extracted: metabolites

REFERENCE

Barlow, R.D.; Thompson, P.A.; Stone, R.B. Simultaneous determination of nicotine, cotinine and five additional nicotine metabolites in the urine of smokers using pre-column derivatization and high performance liquid chromatography, *J.Chromatogr.*, **1987**, *419*, 375–380.

QUINONE

MISCELLANEOUS REACTIONS

Reduction

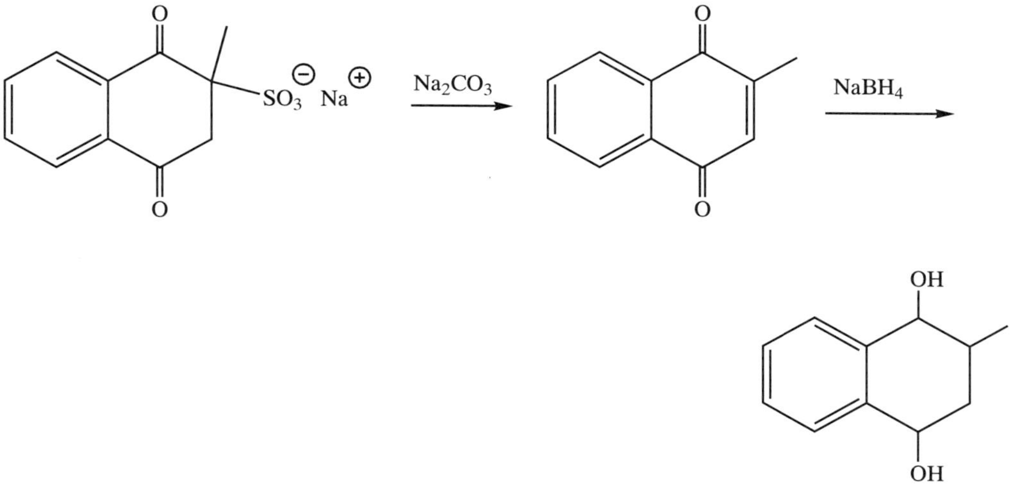

SAMPLE
Matrix: feed, premix
Analyte: menadione
Sample preparation: 1-10 g Ground feed or premix + 96 mL EtOH:water 40:60, shake for 10 min, add 4 mL 10% tannin solution, shake for 1 min, centrifuge at 2000 g for 5 min, filter through a fine-pore glass filter. Add a 40 mL portion of the filtrate to 50 mL n-hexane and 20 mL 10% sodium carbonate, shake for 1 min, discard the lower layer. Wash the n-hexane layer twice with 100 mL portions of water and dry it with strips of blue-ribbon filter paper. Evaporate an aliquot to dryness under nitrogen with a rotary evaporator, reconstitute in mobile phase, sonicate for 1 min, filter (0.2 μm) (if necessary), inject a 100 μL aliquot. (Use a brown glass separatory funnel.)

HPLC VARIABLES
Column: 250 × 4.6 5 μm ODS-Hypersil
Mobile phase: EtOH:water 60:40
Flow rate: 0.6
Injection volume: 100
Detector: F ex 325 em 425 following post-column reaction. The column effluent mixed with the reagent pumped at 0.27 mL/min and the mixture flowed through a 1.4 m long stainless steel reaction coil to the detector. (Reagent was 0.8 g/L sodium borohydride in EtOH, sonicate for 5 min, stable for 2 h. The reagent was passed through a debubbler before mixing with the column effluent.)

CHROMATOGRAM
Retention time: 12
Limit of quantitation: 20 ng/g

KEY WORDS
protect from light; post-column reaction

REFERENCE

Speek, A.J.; Schrijver, J.; Schreurs, W.H.P. Fluorimetric determination of menadione sodium bisulphite (vitamin K3) in animal feed and premixes by high-performance liquid chromatography with post-column derivatization, *J.Chromatogr.*, **1984**, *301*, 441–447.

SELENIUM

ACTIVATED HALIDE

2,4-Dinitrofluorobenzene

SAMPLE
Matrix: solutions
Analyte: selenium compounds
Sample preparation: Mix 20-10000 nmole selenious acid or dimethyl diselenide with 2 mL water, 350 mg zinc dust, and 2 drops n-octanol, pass nitrogen over the reaction mixture at 200 mL/min for 10 min, pass the exhaust through a mixture of 100 µmoles 2,4-dinitrofluorobenzene, 166 µmoles sodium bicarbonate, 1.4 mL DMF, and 600 µL water. At the end of this time add 3 mL 12 M HCl to the sample vial and flush with nitrogen for 10 min. Extract the trap three times with 4 mL portions of benzene (Caution! Benzene is a carcinogen!), combine the organic layers, evaporate to dryness under reduced pressure, reconstitute, inject an aliquot. (Selenic acid does not react.)

HPLC VARIABLES
Column: 200 × 4 5 µm Nucleosil NH2
Mobile phase: n-Heptane:chloroform 80:20
Flow rate: 1.25
Detector: UV 254

CHROMATOGRAM
Retention time: 4.0 (methyl 2,4-dinitrophenyl selenide (from dimethyl diselenide)), 16.0 (di(2,4-dinitrophenyl)selenide (from selenious acid))

REFERENCE
Ganther, H.E.; Kraus, R.J. Identification of hydrogen selenide and other volatile selenols by derivatization with 1-fluoro-2,4-dinitrobenzene, *Anal.Biochem.*, **1984**, *138*, 396–403.

N-Iodoacetyl-N'-(5-sulfo-1-naphthyl)ethylenediamine

SAMPLE
Matrix: blood
Analyte: selenocysteine
Sample preparation: Dissolve 1 g urea in 1 mL plasma or serum. Remove a 500 μL aliquot and add it to 5 μL 2-octanol and 50 μL 1.65 M potassium borohydride in 8 M urea, let stand at room temperature for 30 min, add 125 μL 1 M pH 3.0 NaH_2PO_4 containing 2 mM disodium EDTA, vortex vigorously, again add 125 μL 1 M pH 3.0 NaH_2PO_4 containing 2 mM disodium EDTA, vortex vigorously, immediately add 50 μL 3.4 mM N-iodoacetyl-N'-(5-sulfo-1-naphthyl)ethylenediamine in DMF (freshly prepared), let stand at room temperature for 15-60 min, add 1.15 mL 1.06 M trichloroacetic acid containing 600 mM HCl, let stand at 0° for 20 min, centrifuge at 5° at 3000 g for 15 min, filter (Acro LC-13) the supernatant, inject a 5-50 μL aliquot of the filtrate.

HPLC VARIABLES
Guard column: 30 × 4.6 5 μm RP-18 (Brownlee)
Column: 250 × 4.6 5 μm Ultrasphere ODS
Mobile phase: Gradient. A was buffer. B was isopropanol:buffer 3:97. A:B 80:20 for 20 min, to 10:90 over 3 min, maintain at 10:90 for 10 min, return to initial conditions over 3 min, re-equilibrate for 10 min. (Buffer was 50 mM formic acid containing 60 mM acetic acid, adjusted to pH 5.1 with NaOH.)
Flow rate: 1.5
Injection volume: 5-50
Detector: F ex 315 em 418

CHROMATOGRAM
Retention time: 19
Limit of detection: 400 nM

KEY WORDS
plasma; serum; protect from light

REFERENCE
Hawkes, W.C.; Kutnink, M.A. High-performance liquid chromatographic determination of selenocysteine with the fluorescent reagent, N-(iodoacetylaminoethyl)-5-naphthylamine-1-sulfonic acid, *J.Chromatogr.*, **1992**, *576*, 263–270.

SULFATE

MISCELLANEOUS REACTIONS

Hydrolysis

SAMPLE
Matrix: wastewater
Analyte: alkyl sulfates
Sample preparation: Cautiously add 50 mL sulfuric acid to 1 L wastewater, heat at 90° for 1 h, cool to room temperature, add 25 g NaCl, add 100 µg IS, extract twice with 75 mL portions of dichloromethane. Combine the organic layers and evaporate them to dryness, reconstitute the residue in 1 mL ethylene chloride, add 10 µL phenyl isocyanate. Cap the vial loosely and heat it at 55 ± 2° in a vacuum oven at 70-100 kPa below atmospheric pressure for 45 min (J. Am. Oil Chem. Soc. 1990, 67, 103), reconstitute with 250 µL ethylene dichloride, inject an aliquot.

HPLC VARIABLES
Column: 250 × 4.6 5 µm Hypersil ODS
Mobile phase: Gradient. MeOH:water 80:20 for 30 min, to 100:0 over 15 min.
Flow rate: 2
Detector: UV 240

CHROMATOGRAM
Retention time: 12-15 (depending on structure)
Internal standard: octaethyleneglycol monohexadecyl ether (49)

REFERENCE
Nitschke, I.; Huber, L. Determination of alkyl sulfates in the influent and effluent of a laboratory activated sludge plant by HPLC, *Fresenius' J.Anal.Chem.*, **1993**, *346*, 453–454.

SULFIDE

ACTIVATED HALIDE

Iodoacetamide

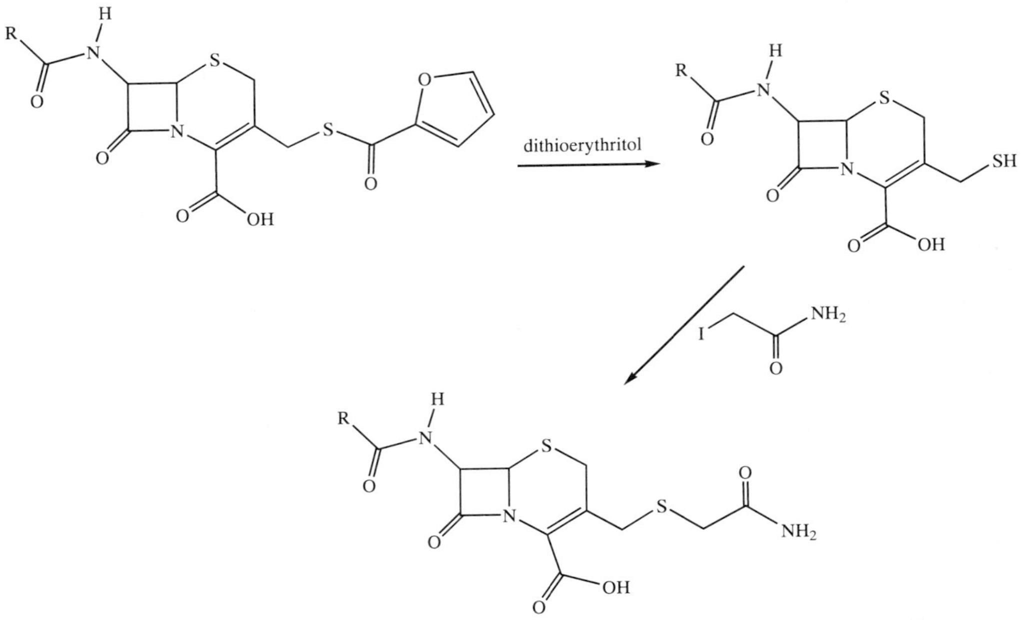

SAMPLE

Matrix: tissue

Analyte: ceftiofur

Sample preparation: Condition a 6 mL 1 g Mega Bond Elut C18 SPE cartridge with 4 mL MeOH and 5 mL phosphate buffer. Condition a 10 mL 500 mg Bond Elut LRC SAX SPE cartridge with 2 mL MeOH, 2 mL MeOH:100 mM NaCl 25:75, and two 1 mL portions of water. Condition a 10 mL 100 mg Bond Elut LRC SCX SPE cartridge with 1 mL MeOH, 2 mL MeOH:100 mM calcium chloride 25:75, and two 1 mL portions of water. Fat. Homogenize (Waring blender) 10 g fat, 20 mL 0.4% dithioerythritol in borate buffer, and 20 mL hexane at medium speed for 5 min, centrifuge at 3000 g for 10 min. Remove a 2 mL aliquot of the aqueous (bottom) layer and add it to 13 mL 0.4% dithioerythritol in borate buffer, shake at 50° for 15 min, add 3 mL 14% iodoacetamide in phosphate buffer, mix well, let stand at room temperature, adjust pH to 2.5-2.6 with 5% phosphoric acid, centrifuge at 4° at 48000 g for 20 min, add the supernatant to the C18 SPE cartridge, wash with 5 mL phosphate buffer, wash with 3 mL 10 mM NaOH, elute with 3 mL MeCN:water 15:85. Add the eluate to 15 mL water, add this mixture to the SAX SPE cartridge, wash with 1 mL water, elute with 2.5 mL MeCN:5% acetic acid 5:95, inject a 500 µL aliquot of the eluate. Muscle, liver, kidney. Homogenize (Waring blender) 10 g tissue and 140 mL 0.4% dithioerythritol in borate buffer at medium speed for 5 min, centrifuge at 3000 g for 10 min. Remove a 15 mL aliquot of the homogenate and shake it at 50° for 15 min, add 3 mL 14% iodoacetamide in phosphate buffer, mix well, let stand at room temperature, adjust pH to 2.5-2.6 with 5% phosphoric acid, centrifuge at 4° at 48000 g for 20 min, add the supernatant to the C18 SPE cartridge, wash with 5 mL phosphate buffer, wash with 3 mL 10 mM NaOH, elute with 3 mL MeCN:water 15:85.

Add the eluate to 15 mL water, add this mixture to the SAX SPE cartridge, wash with 1 mL water, elute with 2.5 mL MeCN:5% acetic acid 5:95. Add the eluate to 10 mL water, mix well, add the mixture to the SCX SPE cartridge, wash with 1 mL water, elute with 2.5 mL MeCN:100 mM NaCl 5:95 (muscle, kidney) or 2 mL MeCN:100 mM NaCl 10:90 (liver), inject a 500 μL aliquot of the eluate. (Borate buffer was 19 g sodium borate and 3.7 g KCl in 1 L water, pH 9. Phosphate buffer was 3.4 g KH_2PO_4 in 700 mL water, pH adjusted to 7 with KOH, made up to 1 L with water.)

HPLC VARIABLES
Guard column: BDS Hypersil C18
Column: 250 × 4.6 BDS Hypersil C18
Mobile phase: Gradient. A was 0.1% trifluoroacetic acid in water. B was 0.1% trifluoroacetic acid in MeCN. A:B from 100:0 to 65:35 over 35 min, wash with 50:50 at 1.5 mL/min for 15 min, re-equilibrate for 20 min (muscle, kidney) or A:B 85:15 for 5 min, to 75:25 over 5 min, wash with 50:50 at 1.5 mL/min for 15 min, re-equilibrate for 20 min (liver, fat).
Flow rate: 1
Injection volume: 500
Detector: UV 266

CHROMATOGRAM
Retention time: 7 (liver, fat), 26 (muscle, kidney)
Limit of quantitation: 100 ng/g
Limit of detection: 10-30 ng/g

OTHER SUBSTANCES
Extracted: cephapirin
Non-interfering: cefoperazone, cefquinome, cephacetril, dihydrostreptomycin, neomycin, penicillin G, spectinomycin, tetracycline

KEY WORDS
pig; muscle; kidney; liver; fat; SPE; rugged

REFERENCE
Beconi-Barker, M.G.; Roof, R.D.; Millerioux, L.; Kausche, F.M.; Vidmar, T.J.; Smith, E.B.; Callahan, J.K.; Hubbard, V.L.; Smith, G.A.; Gilbertson, T.J. Determination of ceftiofur and its desfuroylceftiofur-related metabolites in swine tissues by high-performance liquid chromatography, *J.Chromatogr.B*, **1995**, *673*, 231−244.

2-Iodo-1-methylpyridinium Chloride

SAMPLE
Matrix: solutions
Analyte: sulfide
Sample preparation: Mix an aliquot of the solution with 1 mL 10 mM 2-iodo-1-methyl-pyridinium chloride in water and 3 mL 100 mM pH 8 buffer, mix by inversion, let stand at room temperature for 30 min, make up to 10 mL with water, inject a 20 μL aliquot. (Prepare 2-iodo-1-methylpyridinium chloride by heating 15 g 2-chloropyridine with 20 g iodomethane at 75° overnight, cool, filter, wash the solid with diethyl ether, recrystallize three times from EtOH to give 2-iodo-1-methylpyridinium chloride as white needles (mp 208-209°).)

HPLC VARIABLES
Column: 125 × 4 5 μm ODS-2

Mobile phase: MeOH:5 mM pH 4.7 KH$_2$PO$_4$ 20:80
Flow rate: 0.7
Injection volume: 20
Detector: UV 340

CHROMATOGRAM
Retention time: 4.2
Limit of detection: 2 ng/mL

OTHER SUBSTANCES
Non-interfering: acetate, carbonate, chloride, glutathione, iodide, nitrate, phosphate, sulfite, thiocyanate, thiols, thiosulfate

REFERENCE
Bald, E.; Sypniewski, S. Reversed-phase high-performance liquid chromatographic determination of sulphide in an aqueous matrix using 2-iodo-1-methylpyridinium chloride as a precolumn ultraviolet derivatization reagent, *J.Chromatogr.*, **1993**, *641*, 184–188.

Monobromobimane

SAMPLE
Matrix: blood
Analyte: sulfide, sulfite, thiosulfate
Sample preparation: Sulfide, sulfite. 500 μL Serum + 500 μL 20 mM dithiothreitol in 50 mM pH 9 borate buffer containing 5 mM disodium EDTA, mix, heat at 37° for 10 min. Inject a 500 μL aliquot into 20% phosphoric acid pumped at 0.5 mL/min through a 160 μL gas dialysis cell with a PTFE membrane (design in paper), pump absorbing solution at 0.5 mL/min through the other side of the cell. Starting 75 s after injection collect absorbing solution for 20 s. Remove a 50 μL aliquot of this sample and adjust the pH to 8.0 with 50 μL 100 mM pH 8.0 borate buffer:1 M HCl 90:10, add 100 μL 5 mM monobromobimane (Calbiochem; Molecular Probes) in MeCN, let stand at room temperature in the dark for 1 h, add 100 μL 50 mM pH 2 KCl/HCl buffer, inject a 20 μL aliquot. Thiosulfate. 250 μL Serum + 250 μL 50 mM pH 8.0 borate buffer, mix, filter (Amicon MPS-1) while centrifuging at 1000 g for 15 min. Remove a 100 μL aliquot of the ultrafiltrate and add it to 100 μL 5 mM monobromobimane in MeCN, let stand at room temperature in the dark for 1 h, add 100 μL 50 mM pH 2 KCl/HCl buffer, inject a 20 μL aliquot. (Absorbing solution was 100 mM NaOH containing 10 mM glycerol and 5 mM disodium EDTA. The 20% phosphoric acid converts sulfide to hydrogen sulfide and sulfite to sulfur dioxide. These gases diffuse through the membrane to the absorbing solution. Purify monobromobimane by column chromatography using silica and eluting with dichloromethane:petroleum ether from 70:30 to 100:0 over 10-30 h. The dibromo impurity elutes first followed by monobromobimane (J. Am. Chem. Soc. 1980, 102, 4983).)

HPLC VARIABLES
Guard column: 50 × 4 Nucleosil 100-5 C18
Column: 250 × 4 Nucleosil 5 dimethylamino-bonded silica
Mobile phase: MeCN:pH 3 acetic acid solution 18:78 containing 25 mM sodium perchlorate
Flow rate: 1

Injection volume: 20
Detector: F ex 396 em 476

CHROMATOGRAM
Retention time: 13 (sulfide), 16 (sulfite), 19 (thiosulfate)
Limit of detection: 20-40 nM

KEY WORDS
serum; ultrafiltrate

REFERENCE
Togawa, T.; Ogawa, M.; Nawata, M.; Ogasawara, Y.; Kawanabe, K.; Tanabe, S. High performance liquid chromatographic determination of bound sulfide and sulfite and thiosulfate at their low levels in human serum by pre-column fluorescence derivatization with monobromobimane, *Chem.Pharm.Bull.*, **1992**, *40*, 3000–3004.

SAMPLE
Matrix: sea water
Analyte: sulfide
Sample preparation: Mix 50 μL sea water with 210 μL reagent in the dark, let stand in the dark for 30 min, add 100 μL 65 mM methanesulfonic acid, inject an aliquot. (Prepare the reagent by mixing 50 μL 50 mM pH 8.0 HEPES buffer containing 5 mM EDTA, 50 μL MeCN, and 10 μL 48 mM monobromobimane (Calbiochem; Molecular Probes) in MeCN.)

HPLC VARIABLES
Column: 125 × 4 5 μm LiChrospher 60 RP select B
Mobile phase: Gradient. MeOH:0.25% pH 4 acetic acid 12:88 for 7 min, to 30:70 over 8 min, maintain at 30:70 for 4 min, to 50:50 over 4 min, to 100:0 over 7 min, maintain at 100:0 for 3 min, return to initial conditions over 0.1 min, re-equilibrate for 2.9 min.
Column temperature: 35
Flow rate: 1
Detector: F ex 380 em 480

CHROMATOGRAM
Retention time: 24.55

OTHER SUBSTANCES
Simultaneously analyzed: sulfite, thiosulfate

REFERENCE
Rethmeier, J.; Rabenstein, A.; Langer, M.; Fischer, U. Detection of traces of oxidized and reduced sulfur compounds in small samples by combination of different high-performance liquid chromatography methods, *J.Chromatogr.A*, **1997**, *760*, 295–302.

4-Nitrobenzyl Bromide

SAMPLE
Matrix: solutions
Analyte: sulfide
Sample preparation: Protect from light. Mix a 200 μL aliquot of a solution of sulfide in 50 mM NaOH with 1.8 mL 2.5 mM 4-nitrobenzyl bromide in MeCN containing 100 μM naphthalene, stir at room temperature for 15 min, filter (0.5 μm), inject a 10 μL aliquot.

HPLC VARIABLES
Column: 150×4.6 5 µm YMC A-302 ODS (Yamamura)
Mobile phase: MeOH:water 85:15
Flow rate: 0.6
Injection volume: 10
Detector: UV 254

CHROMATOGRAM
Retention time: 6
Internal standard: naphthalene (7)
Limit of quantitation: 60 µM

REFERENCE
Funazo, K.; Tanaka, M.; Shono, T. Determination of iodide and sulfide as *p*-nitrobenzyl derivatives by high performance liquid chromatography, *Anal.Sci.*, **1987**, *3*, 41–44.

AMINE

N⁴,N⁴-Diethyl-2-methyl-1,4-phenylenediamine

SAMPLE
Matrix: blood
Analyte: sulfide
Sample preparation: 100 µL Serum + 4 mL 10% zinc acetate in water + 400 µL 20 mg/mL N⁴,N⁴-diethyl-2-methyl-1,4-phenylenediamine monohydrochloride (2-amino-5-N,N-diethylaminotoluene hydrochloride) in 350 mM sulfuric acid + 200 µL 100 µM iron(III) chloride in 50 mM sulfuric acid, mix, make up to 5 mL with water, add 500 µL 21.6 mg/mL sodium 1-octanesulfonate in water, add 1 mL 2-octanol, mix for 10 min, centrifuge at 1820 g, inject a 10 µL aliquot of the organic layer.

HPLC VARIABLES
Column: 250×4.6 Inertsil ODS-2
Mobile phase: MeCN:water 90:10 containing 2.16 mg/mL sodium 1-octanesulfonate
Flow rate: 0.5
Injection volume: 10
Detector: F ex 640 em 675

CHROMATOGRAM
Retention time: 13
Limit of detection: 0.44 nM

OTHER SUBSTANCES
Non-interfering: calcium, carbonate, chloride, cysteine, cystine, magnesium, methionine, nitrate, phosphate, potassium, silicate, sulfate

KEY WORDS
serum

REFERENCE
Nagashima, K.; Fukushima, K.; Kamaya, M. Determination of trace amounts of sulfide in human serum by high-performance liquid chromatography with fluorometric detection after derivatization with 2-amino-5-N,N-diethylaminotoluene and iron (III), *J.Liq.Chromatogr.*, **1995**, *18*, 515–526.

N,N-Dimethyl-1,4-phenylenediamine

SAMPLE
Matrix: solutions
Analyte: sulfide
Sample preparation: Add 7 mL sample, 500 μL 4 mg/mL N,N-dimethyl-1,4-phenylenedi-
amine in 1:1 sulfuric acid, and 3 drops 2.5 g/mL ferric nitrate nonahydrate in water to
a 10 mL volumetric flask, invert once only, make up to 10 mL with water, let stand for 1
min, inject a 100 μL aliquot.

HPLC VARIABLES
Column: 150 × 3.9 μBondapak C18
Mobile phase: MeCN:buffer 40:60 (Prepare buffer by mixing 400 mL MeCN, 300 mL wa-
ter, 5 mL glacial acetic acid, and a vial of low-UV PIC B5 (pentanesulfonic acid, Waters),
make up to 1 L with water.)
Flow rate: 1
Injection volume: 100
Detector: UV 664

CHROMATOGRAM
Retention time: 4.6
Limit of detection: 0.8 ppb

OTHER SUBSTANCES
Interfering: ferrocyanide, iodide, thiosulfate
Non-interfering: bromide, carbonate, chloride, cyanide, ferricyanide, nitrate, nitrite, phos-
phate, sulfate, sulfite, thiocyanate

KEY WORDS
product of derivatization reaction is methylene blue

REFERENCE
Haddad, P.R.; Heckenberg, A.L. Trace determination of sulfide by reversed-phase ion-interaction chromatography using pre-column derivatization, *J.Chromatogr.*, **1988**, *447*, 415–420.

p-Phenylenediamine

SAMPLE
Matrix: blood
Analyte: sulfide
Sample preparation: Mix 1 mL red blood cells with 1 mL concentrated sulfuric acid:water
1:9, allow hydrogen sulfide to diffuse (Conway microdiffusion cell) into 1 mL 100 mM
NaOH containing 5 mM EDTA and 2 mM glycerol. Remove a 500 μL aliquot of the NaOH
solution and add it to 400 μL 12.5 mM p-phenylenediamine in water and 100 μL iron

solution, invert tube twice carefully (vigorous agitation causes loss of hydrogen sulfide), let stand for 10 min, inject a 100 μL aliquot. (Prepare iron solution by dissolving 540 mg iron(III) chloride hexahydrate in 50 mL concentrated HCl:water 50:50.)

HPLC VARIABLES
Guard column: 5 μm LiChrospher RP-8 (end-capped)
Column: 250 × 4 5 μm LiChrospher RP-8 (end-capped)
Mobile phase: MeCN:50 mM pH 4.0 sodium phosphate buffer 50:50 containing 40 mM sodium dodecyl sulfate
Flow rate: 1
Injection volume: 100
Detector: F ex 600 em 623; UV 600

CHROMATOGRAM
Retention time: 6.9
Limit of detection: 1.5 nM (F)

KEY WORDS
red blood cells

REFERENCE
Ogasawara, Y.; Ishii, K.; Togawa, T.; Tanabe, S. Determination of trace amounts of sulphide in human red blood cells by high-performance liquid chromatography with fluorimetric detection after derivatization with p-phenylenediamine and iron(III), *Analyst*, **1991**, *116*, 1359–1363.

DISULFIDE

5,5'-Dithiobis(2-nitrobenzoic acid)

SAMPLE
Matrix: urine
Analyte: auranofin
Sample preparation: Rabbit urine. Centrifuge at 1000 g at 3° for 1 min, inject an aliquot. Human urine. Inject directly.

HPLC VARIABLES
Column: 150 × 4.6 YMC AM-302 octadecylsilyl (YMC)
Mobile phase: MeOH:water 65:35
Flow rate: 1
Injection volume: 100
Detector: UV 412 following post-column reaction. The column effluent mixed with the reagent pumped at 0.5 mL/min and the mixture flowed through a 5 m × 0.5 mm ID PTFE

coil at 60° to the detector. (Reagent was 50 μM 5,5'-dithiobis(2-nitrobenzoic acid) in 50 mM pH 7.4 phosphate buffer containing 300 mM KI.)

CHROMATOGRAM
Retention time: 6

KEY WORDS
human; rabbit; post-column reaction

REFERENCE
Kizu, R.; Kaneda, M.; Yamauchi, Y.; Miyazaki, M. Determination of auranofin, a chrysotherapy agent, in urine by HPLC with a postcolumn reaction and visible detection, *Chem.Pharm.Bull.*, **1993**, *41*, 1261–1265.

MISCELLANEOUS REACTIONS

Desulfurization

SAMPLE
Matrix: solutions
Analyte: phenothiazine
Sample preparation: Mix 100 μL of a 0.05-1 μg/mL solution in MeOH with 20 μL 5% sodium carbonate and about 35 mg of a suspension of W-2 Raney nickel in EtOH (80 μL), stir at 70° for 15 min, cool in ice-water, add 50 μL 400 ng/mL 1-naphthol in MeOH, mix well, filter (0.45 μm), inject a 5 μL aliquot of the filtrate. (Raney nickel can be purchased from Aldrich or prepared from aluminum-nickel alloy (Org. Syn. 1955, Coll. Vol. 3, 181). Phenothiazine is desulfurized to diphenylamine.)

HPLC VARIABLES
Column: 150 × 4.6 5 μm Shim-pack CLC-ODS (Shimadzu)
Mobile phase: MeOH:water 75:25 containing 50 mM sodium perchlorate and 5 mL/L glacial acetic acid
Column temperature: 35
Flow rate: 0.6
Injection volume: 5
Detector: E, Showa Denko Shodex EC-1, glassy carbon working electrode 0.8 V, Ag/AgCl reference electrode

CHROMATOGRAM
Retention time: 12.5
Internal standard: 1-naphthol (8.5)
Limit of detection: 10 pg

KEY WORDS
desulfurization

REFERENCE
Shimada, K.; Mino, T.; Nakajima, M.; Wakabayashi, H.; Yamato, S. Application of the desulfurization of phenothiazines for a sensitive detection method by high-performance liquid chromatography, *J.Chromatogr.B*, **1994**, *661*, 85–91.

SULFITE

ACTIVATED HALIDE

Monobromobimane

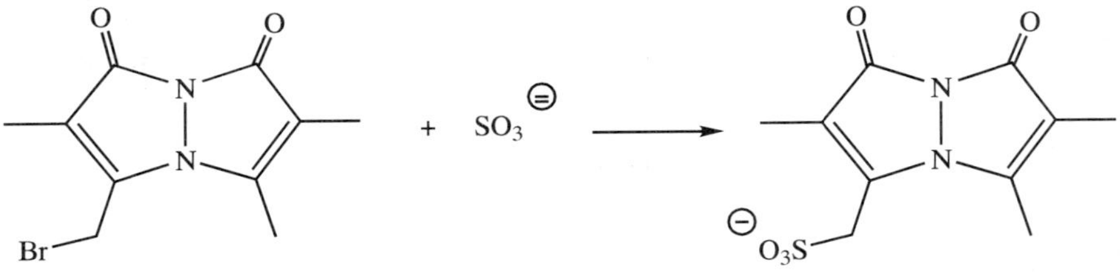

SAMPLE

Matrix: blood

Analyte: sulfide, sulfite, thiosulfate

Sample preparation: Sulfide, sulfite. 500 μL Serum + 500 μL 20 mM dithiothreitol in 50 mM pH 9 borate buffer containing 5 mM disodium EDTA, mix, heat at 37° for 10 min. Inject a 500 μL aliquot into 20% phosphoric acid pumped at 0.5 mL/min through a 160 μL gas dialysis cell with a PTFE membrane (design in paper), pump absorbing solution at 0.5 mL/min through the other side of the cell. Starting 75 s after injection collect absorbing solution for 20 s. Remove a 50 μL aliquot of this sample and adjust the pH to 8.0 with 50 μL 100 mM pH 8.0 borate buffer:1 M HCl 90:10, add 100 μL 5 mM monobromobimane (Calbiochem; Molecular Probes) in MeCN, let stand at room temperature in the dark for 1 h, add 100 μL 50 mM pH 2 KCl/HCl buffer, inject a 20 μL aliquot. Thiosulfate. 250 μL Serum + 250 μL 50 mM pH 8.0 borate buffer, mix, filter (Amicon MPS-1) while centrifuging at 1000 g for 15 min. Remove a 100 μL aliquot of the ultrafiltrate and add it to 100 μL 5 mM monobromobimane in MeCN, let stand at room temperature in the dark for 1 h, add 100 μL 50 mM pH 2 KCl/HCl buffer, inject a 20 μL aliquot. (Absorbing solution was 100 mM NaOH containing 10 mM glycerol and 5 mM disodium EDTA. The 20% phosphoric acid converts sulfide to hydrogen sulfide and sulfite to sulfur dioxide. These gases diffuse through the membrane to the absorbing solution. Purify monobromobimane by column chromatography using silica and eluting with dichloromethane:petroleum ether from 70:30 to 100:0 over 10-30 h. The dibromo impurity elutes first followed by monobromobimane (J. Am. Chem. Soc. 1980, 102, 4983).)

HPLC VARIABLES

Guard column: 50 × 4 Nucleosil 100-5 C18

Column: 250 × 4 Nucleosil 5 dimethylamino-bonded silica

Mobile phase: MeCN:pH 3 acetic acid solution 18:78 containing 25 mM sodium perchlorate

Flow rate: 1

Injection volume: 20

Detector: F ex 396 em 476

CHROMATOGRAM

Retention time: 13 (sulfide), 16 (sulfite), 19 (thiosulfate)

Limit of detection: 20-40 nM

KEY WORDS

serum; ultrafiltrate

REFERENCE
Togawa, T.; Ogawa, M.; Nawata, M.; Ogasawara, Y.; Kawanabe, K.; Tanabe, S. High performance liquid chromatographic determination of bound sulfide and sulfite and thiosulfate at their low levels in human serum by pre-column fluorescence derivatization with monobromobimane, *Chem.Pharm.Bull.*, **1992**, *40*, 3000–3004.

SAMPLE
Matrix: blood
Analyte: sulfite
Sample preparation: 100 μL Serum + 70 μL 212 mM sodium borohydride in 50 mM pH 8.5 Tris buffer + 10 μL 70 mM monobromobimane (Calbiochem) in MeCN, heat at 42° for 7 min, add 50 μL 1.5 M perchloric acid with vortexing, centrifuge at 12400 g for 10 min. Remove the supernatant and gently mix it with 20 μL 2 M Tris, centrifuge at 12400 g for 8 min, inject a 20 μL aliquot of the supernatant

HPLC VARIABLES
Guard column: 15 × 3.2 7 μm NewGuard RP-8
Column: 250 × 4.6 5 μm C8 (Beckman)
Mobile phase: Gradient. MeOH:water (0.25% acetic acid ?) 3:97 for 5 min, to 13:87 over 4 min, to 15:85 over 2 min, to 100:0 over 2 min, maintain at 100:0 for 2 min, return to 3:97 over 2 min, maintain at 3:97 for 3 min. Equilibrate with MeOH:acetic acid:water 5:0.25:94.75
Flow rate: 1.5
Injection volume: 20
Detector: F ex 390 em >418 (cutoff filter)

CHROMATOGRAM
Retention time: 6.70 (cysteine), 7.39 (cysteinylglycine), 7.82 (sulfite), 9.67 (homocysteine), 10.70 (glutathione)
Limit of detection: 440 nM (for sulfite)

KEY WORDS
serum; pharmacokinetics

REFERENCE
Ji, A.J.; Savon, S.R.; Jacobsen, D.W. Determination of total serum sulfite by HPLC with fluorescence detection, *Clin.Chem.*, **1995**, *41*, 897–903.

SAMPLE
Matrix: sea water
Analyte: sulfite
Sample preparation: Mix 50 μL sea water with 210 μL reagent in the dark, let stand in the dark for 30 min, add 100 μL 65 mM methanesulfonic acid, inject an aliquot. (Prepare the reagent by mixing 50 μL 50 mM pH 8.0 HEPES buffer containing 5 mM EDTA, 50 μL MeCN, and 10 μL 48 mM monobromobimane (Calbiochem; Molecular Probes) in MeCN.)

HPLC VARIABLES
Column: 125 × 4 5 μm LiChrospher 60 RP select B
Mobile phase: Gradient. MeOH:0.25% pH 4 acetic acid 12:88 for 7 min, to 30:70 over 8 min, maintain at 30:70 for 4 min, to 50:50 over 4 min, to 100:0 over 7 min, maintain at 100:0 for 3 min, return to initial conditions over 0.1 min, re-equilibrate for 2.9 min.
Column temperature: 35
Flow rate: 1
Detector: F ex 380 em 480

CHROMATOGRAM
Retention time: 2.38

OTHER SUBSTANCES
Simultaneously analyzed: sulfide, thiosulfate

REFERENCE

Rethmeier, J.; Rabenstein, A.; Langer, M.; Fischer, U. Detection of traces of oxidized and reduced sulfur compounds in small samples by combination of different high-performance liquid chromatography methods, *J.Chromatogr.A*, **1997**, *760*, 295–302.

DISULFIDE

5,5'-Dithiobis(2-nitrobenzoic acid)

SAMPLE

Matrix: food

Analyte: sulfite

Sample preparation: Condition a Supelclean C-18 SPE cartridge with three 3 mL portions of MeOH and 3 mL reagent:water 25:75, force out solvents with air pressure. Grind (Sorvall mixer) a portion of food in 25 mL reagent, make up to 100 mL with water. Centrifuge a 15 mL aliquot, filter (Versapor 800) the supernatant. Remove a 2 mL aliquot of the filtrate and add it to the SPE cartridge, force through with air pressure, discard the first 1 mL, inject a 100 µL aliquot of the second mL. (Prepare reagent by dissolving 1.02 g potassium hydrogen phthalate in 700 mL water and adding 54 mL 37% formaldehyde, adjust pH to 5.1 with concentrated KOH, make up to 1 L with water.)

HPLC VARIABLES

Column: 250 × 4.6 Spherisorb 10 ODS

Mobile phase: Buffer (Prepare buffer by dissolving 34 g 40% tetrabutylammonium hydroxide in water and 3 g acetic acid in 700 mL water, adjust pH to 5.8 with tetrabutylammonium hydroxide or acetic acid, make up to 1 L with water.)

Column temperature: 50

Flow rate: 1

Injection volume: 100

Detector: UV 412 following post-column reaction. The column effluent mixed with 64 mM KOH pumped at 1 mL/min and this mixture flowed through a 3 m × 0.1 mm ID coil of PTFE tubing. The effluent from this coil mixed with reagent and this mixture flowed through a 1 mL reaction coil at 50° to the detector. (Prepare reagent by dissolving 360 mg 5,5'-dithiobis(2-nitrobenzoic acid) in 15 mL EtOH:water 95:5 and adding it slowly to 700 mL 280 mM KH_2PO_4 with rapid stirring, make up to 1 L with water, filter (Versapor 800). Bisulfite ion reacts with formaldehyde to form hydroxymethylsulfonate and this compound is chromatographed. Hydroxymethylsulfonate is decomposed to formaldehyde and bisulfite ion by KOH and the bisulfite reacts with the reagent to form strongly absorbing 3-carboxy-4-nitrothiophenolate.)

CHROMATOGRAM

Retention time: 6

Limit of quantitation: 5 ppm

OTHER SUBSTANCES

Non-interfering: chloride, nitrite, sulfamate, sulfite, thiosulfate

KEY WORDS

post-column reaction; SPE

REFERENCE

Warner, C.R.; Daniels, D.H.; Pratt, D.E.; Joe, F.L. Jr.; Fazio, T.; Diachenko, G.W. Sulphite stabilization and high-performance liquid chromatographic determination: a reference method for free and reversibly bound sulphite in food, *Food Addit.Contam.*, **1987**, *4*, 437–445.

2,2'-Dithiobis(5-nitropyridine)

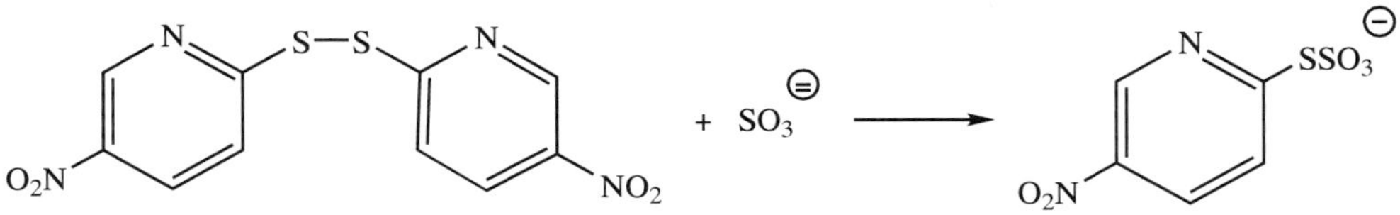

SAMPLE

Matrix: seawater

Analyte: sulfite

Sample preparation: Condition a Sep-Pak C18 SPE cartridge with 5 mL MeOH, 5 mL water, and 5 mL 20 mM pH 6 sodium acetate buffer containing 10 mM tetrabutylammonium sulfate. Filter seawater, sparge with nitrogen. 20 mL Sample + 1 mL 2 mM 2,2'-dithiobis(5-nitropyridine) in MeCN + 1 mL 200 mM pH 6 acetate buffer, let stand at room temperature for 5 min, add to the SPE cartridge at 3-4 mL/min, wash with 3-5 mL water, blow nitrogen through the SPE cartridge, elute with 1 mL MeOH. Dilute the eluate with an equal volume of water, inject a 20 μL aliquot. Alternatively, evaporate the eluate to dryness, reconstitute with 200 μL MeOH:water 20:80, inject a 20 μL aliquot.

HPLC VARIABLES

Guard column: 5 μm Spherisorb ODS-2 (microbore)

Column: 150 × 2 5 μm Hypersil ODS

Mobile phase: Gradient. A was 50 mM sodium acetate containing 7.5 mM tetrabutylammonium hydrogen sulfate, adjusted to pH 3.5 ± 0.05 with 6 M HCl. B was MeCN. A:B 90:10 for 1 min, to 66:34 over 8 min, to 45:55 over 14 min, to 0:100 over 5 min, maintain at 0:100 for 2 min, return to initial conditions over 2 min.

Flow rate: 0.1-0.2

Injection volume: 20

Detector: UV 320

CHROMATOGRAM

Retention time: 19

Limit of detection: 6-8 nM

OTHER SUBSTANCES

Extracted: thiosulfate

Simultaneously analyzed: cysteine, glutathione, mercaptoacetic acid, 2-mercaptoethanesulfonic acid, 2-mercaptoethanol, 2-mercaptopropionic acid, 3-mercaptopropionic acid, monothioglycerol, thiomalic acid

KEY WORDS

microbore; SPE

REFERENCE

Valravamurthy, A.; Mopper, K. Determination of sulfite and thiosulfate in aqueous samples including anoxic seawater by liquid chromatography after derivatization with 2,2'-dithiobis(5-nitropyridine), *Environ.Sci.Technol.*, **1990**, *24*, 333–337.

SULFONATE

DITHIOCARBAMATE

Sodium Diethyldithiocarbamate

RELATED REFERENCE

Rifai, N.; Sakamoto, M.; Lafi, M.; Guinan, E. Measurement of plasma busulfan concentration by high-performance liquid chromatography with ultraviolet detection. *Ther.Drug Monit.* **1997**, *19*, 169-174

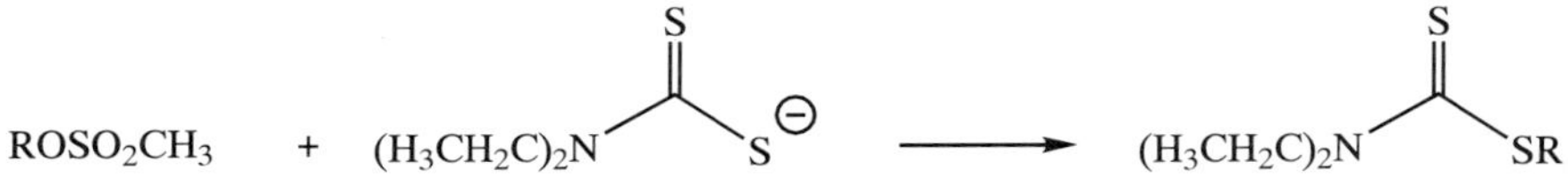

$$ROSO_2CH_3 \quad + \quad (H_3CH_2C)_2N \overset{S}{\underset{S^{\ominus}}{\diagup\!\!\diagdown}} \quad \longrightarrow \quad (H_3CH_2C)_2N \overset{S}{\underset{SR}{\diagup\!\!\diagdown}}$$

SAMPLE

Matrix: blood

Analyte: busulfan

Sample preparation: 300 μL Plasma + 600 μL MeOH, mix, let stand at -20° for 20 min, centrifuge at 1500 g for 10 min. Add 600 μL supernatant to 150 μL 5% diethyldithiocarbamate and 600 μL 100 mM pH 5.5 ammonium acetate, mix, extract with 1.5 mL ethyl acetate, centrifuge for 1.5 min. Lyophilize a 1 mL aliquot of the extract, reconstitute with 200 μL MeOH, inject an aliquot.

HPLC VARIABLES

Column: 300 × 3.9 10 μm μBondapak C18

Mobile phase: MeOH:water 80:20

Flow rate: 1

Detector: UV 251

CHROMATOGRAM

Retention time: 12.0

Limit of detection: 200 nM

KEY WORDS

plasma; pharmacokinetics

REFERENCE

Henner, W.D.; Furlong, E.A.; Flaherty, M.D.; Shea, T.C.; Peters, W.P. Measurement of busulfan in plasma by high-performance liquid chromatography, *J.Chromatogr.*, **1987**, *416*, 426−432.

SAMPLE

Matrix: blood

Analyte: busulfan

Sample preparation: Condition a 1 mL Bakerbond C18 SPE cartridge with three 1 mL portions of MeOH and two 1 mL portions of water, do not allow to dry. 300 μL Plasma + 30 μL 25 μg/mL IS in MeOH + 150 μL reagent solution, vortex for 10 s, add 2 mL ethyl acetate, vortex briefly, rock for 10 min on a blood mixer, centrifuge at 1500 g for 10 min. Remove 1 mL of the organic layer and evaporate it to dryness under a stream of air at 70°, reconstitute the residue in 500 μL MeOH, add 500 μL water, add to SPE cartridge, wash with two 1 mL portions of MeOH:water 50:50, elute with two 250 μL portions of MeOH, inject a 20 μL aliquot. (The reagent solution was 8.2 g sodium diethyldithiocarbamate in 100 mL water.)

HPLC VARIABLES

Column: 150 × 4 3 μm MicroPak-SP-C18

Mobile phase: MeCN:water:THF 55:25:20

Flow rate: 0.8

Injection volume: 20
Detector: UV 278

CHROMATOGRAM
Retention time: 4.5
Internal standard: N-(2,6-difluorobenzoyl)-N'-[3,5-dichloro-4-(3-chloro-5-trifluoromethyl-pyridin-2-yloxy)phenyl]urea (CGA-112913) (5.5)
Limit of detection: 0.4 ng/mL

KEY WORDS
plasma; SPE

REFERENCE
MacKichan, J.J.; Bechtel, T.P. Quantitation of busulfan in plasma by high-performance liquid chromatography, *J.Chromatogr.*, **1990**, *532*, 424–428.

SAMPLE
Matrix: blood
Analyte: busulfan
Sample preparation: 500 µL serum + 100 µL water + 5 mL diethyl ether:dichloromethane 70:30, shake for 5 min, centrifuge at 1500 g for 5 min, freeze in dry ice/acetone. Remove the organic layer and evaporate it to dryness under a stream of nitrogen at 40°, reconstitute the residue in 250 µL water, inject a 100 µL aliquot onto column A and elute to waste with mobile phase A, after 5 min backflush the contents of Column A onto column B with mobile phase B and elute to waste, after 3 min remove column A from the circuit, after 2 min direct the effluent from column B onto column C, after 0.6 min divert the effluent from column B to waste, elute column C with mobile phase C, monitor the effluent from column C. (Derivatization of busulfan with sodium diethyldithiocarbamate occurs on column A.)

HPLC VARIABLES
Column: A 10 × 4 5 µm Inertsil ODS-80A; B 150 × 4.6 5 µm Inertsil ODS-2; C 150 × 6 5 µm Capcell Pak C18 (Shiseido)
Mobile phase: A 1% sodium diethyldithiocarbamate in water; B MeCN:20 mM pH 4.6 KH_2PO_4 60:40; C MeCN:20 mM pH 4.6 KH_2PO_4 65:35 (Pass mobile phase A through a 10 × 4 5 µm Inertsil ODS-80A column before it enters the system.)
Flow rate: A 0.5; B 1; C 1
Injection volume: 100
Detector: UV 278

CHROMATOGRAM
Retention time: 16.1
Limit of quantitation: 10 ng/mL

KEY WORDS
serum; column-switching; heart-cut

REFERENCE
Funakoshi, K.-i.; Yamashita, K.; Chao, W.-f.; Yamaguchi, M.; Yashiki, T. High-performance liquid chromatographic determination of busulfan in human serum with on-line derivatization, column switching and ultraviolet absorbance detection, *J.Chromatogr.B*, **1994**, *660*, 200–204.

SULFONIUM COMPOUND

HYDRIDE

Sodium Borohydride

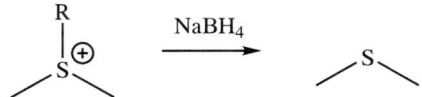

SAMPLE
Matrix: plants
Analyte: sulfonium compounds
Sample preparation: Extract 10 g finely-ground plant material with 10 mL 2% HCl, centrifuge at 2800 g for 10 min, filter (Whatman GF/C) the supernatant. Immediately before analysis buffer the filtrate to pH 5.66, dilute, inject an aliquot.

HPLC VARIABLES
Column: 250×4.6 5 μm Techsphere SCX (HPLC Technology Co., Macclesfield UK)
Mobile phase: 50 mM pH 5.66 Phosphate buffer (Adjust pH of 50 mM KH_2PO_4 to 5.66 with NaOH.)
Flow rate: 1
Injection volume: 200
Detector: Flame photometric detection following post-column reaction. The column effluent mixed with 2 M sodium borohydride solution pumped at 0.75 mL/min and with 100 mM HCl pumped at 0.75 mL/min and the mixture flowed through a 2.1 m $\times$ 0.74 mm ID PTFE coil at 85°. The effluent from this coil mixed with nitrogen flowing at 40 mL/min to a gas-liquid separator. The liquid flowed to waste and the gas stream flowed through an empty 2 mL chamber and through a 2 mL chamber filled with granular anhydrous magnesium perchlorate to the detector. (Details of detector construction are given.)

CHROMATOGRAM
Retention time: 5 ((dimethylsulfonio)acetate), 6 (dimethylsulfoniopropionate), 7 ((dimethylsulfonio)-2-methylpropionate), 9 ((dimethylsulfonio)butanoate), 11 ((dimethylsulfonio)pentanoate), 17 (S-methylmethionine)

KEY WORDS
post-column reaction

REFERENCE
Howard, A.G.; Russell, D.W. Borohydride-coupled HPLC-FPD instrumentation and its use in the determination of dimethylsulfonium compounds, *Anal.Chem.*, **1997**, *69*, 2882–2997.

SULFUR DIOXIDE

MISCELLANEOUS REACTIONS

Oxidation

RELATED REFERENCE
Pizzoferrato, L.; Quattrucci, E.; Di Lullo, G. Evaluation of an HPLC method for the determination of sulphiting agents in foods. *Food Addit.Contam.* **1990**, *7*, 189-195

$$SO_2 \longrightarrow SO_4^{\ominus}$$

SAMPLE
Matrix: air
Analyte: sulfur dioxide
Sample preparation: Pull air through a 5 μm PTFE membrane filter then 50 mL reagent at 185 mL/min for 24 h, dilute to 50 mL with water, filter (0.45 μm nylon), inject a 100 μL aliquot of the filtrate. (Prepare reagent by mixing 20 mL 30% hydrogen peroxide with 100 μL 600 mM HCl and making up to 1 L with sulfate-free water. Airborne sulfur dioxide is oxidized to sulfate.)

HPLC VARIABLES
Guard column: IonPac AG4A (Dionex)
Column: IonPac AS4A (Dionex)
Mobile phase: 1.8 mM Sodium carbonate containing 1.7 mM sodium bicarbonate (Regenerant was 12.5 mM sulfuric acid at 3.3 mL/min.)
Flow rate: 1.3
Injection volume: 100
Detector: Conductivity (with Dionex AMMS-II suppressor)

CHROMATOGRAM
Retention time: 7
Limit of detection: 44 ng/mL (sulfate), 5.5 μg/m^3 (sulfur dioxide)

REFERENCE
Velásquez, H.; Ramírez, H.; Díaz, J.; González de Nava, M.; Sosa de Borrego, B.; Morales, J. Determination of atmospheric sulfur dioxide by ion chromatography in the city of Cabimas, Venezuela, *J.Chromatogr.A*, **1996**, *739*, 295–299.

TETRAHYDROPYRIMIDINE

MISCELLANEOUS REACTIONS

Hydrolysis

SAMPLE
Matrix: milk
Analyte: morantel
Sample preparation: 20 mL Milk + 1.5 mL HCl + 1 mL 37.4 ng/mL pyrantel in water, heat at 95° with stirring for 2 h, cool in an ice bath, basify with 2.5 mL 12 M KOH, add 25 mL toluene, shake, centrifuge, repeat extraction. Combine the toluene layers and extract them vigorously twice with 4 mL portions of 100 mM HCl. Combine the aqueous layers and basify them with 2.5 mL 12 M KOH, heat at 95° for 5 h, acidify with 3 mL HCl, cool in an ice bath, extract twice with 5 mL portions of chloroform. Combine the organic layers, add 500 μL 100 mM NaOH, mix, centrifuge. Remove the aqueous layer and place it on a vortex evaporator for 5 min to remove residual chloroform, inject an aliquot. (Prepare 12 M KOH by cautiously dissolving 790 g KOH pellets in enough water to make 1 L in a fume hood, use an ice bath.)

HPLC VARIABLES
Guard column: 30 mm long μC18 Corasil Bondapak
Column: 100 × 5 Radial-Pak A (Waters)
Mobile phase: MeOH:MeCN:water:acetic acid 40:10:49:1
Flow rate: 1
Injection volume: 100
Detector: UV 313

CHROMATOGRAM
Retention time: 6.6 (as 3-(3-methyl-2-thienyl)acrylic acid, the hydrolysis product)
Internal standard: pyrantel (as 3-(2-thienyl)acrylic acid, the hydrolysis product) (4.2)
Limit of quantitation: 1 ppb

KEY WORDS
protect from light

REFERENCE
Lynch, M.J.; Mosher, F.R.; Brunner, L.A.; Bartolucci, S.R. Liquid chromatographic determination and identification of morantel-related residues as precursors of 3-(3-methyl-2-thienyl) acrylic acid (CP-20,107) in bovine milk, *J.Assoc.Off.Anal.Chem.*, **1986**, *69*, 931–935.

THIOCYANATE

ACTIVATED HALIDE

3-Bromomethyl-7-methoxy-1,4-benzoxazin-2-one

SAMPLE

Matrix: blood, saliva

Analyte: thiocyanate anion

Sample preparation: Saliva. Dilute 200-fold with water, add a 100 μL aliquot to 300 μL 2 mM 3-bromomethyl-7-methoxy-1,4-benzoxazin-2-one in MeCN and 600 μL 11 mM β-naphthol in MeCN, shake mechanically at 70° for 1 h, inject a 5 μL aliquot. Plasma. Condition a 0.8 mL 360 mg 37-55 μm Sep-pak Accell plus QMA anion-exchange SPE cartridge with 10 mL 16.13 mM potassium nitrate in MeCN:water 80:20 and 10 mL water. Add 500 μL plasma to the SPE cartridge, wash with 20 mL 1.61 mM potassium nitrate in MeCN:water 80:20, elute with 6.45 mM potassium nitrate in MeCN:water 80:20. Evaporate a 1 mL aliquot of the eluate to dryness under a stream of nitrogen at 50°, reconstitute with 100 μL water, add 300 μL 2 mM 3-bromomethyl-7-methoxy-1,4-benzoxazin-2-one in MeCN and 600 μL 11 mM β-naphthol in MeCN, shake mechanically at 70° for 1 h, inject a 5 μL aliquot. (3-Bromomethyl-7-methoxy-1,4-benzoxazin-2-one is available from Tokyo Kasei (TCI America, Portland OR). Synthesis is as follows. Add 36 g sodium hydrosulfite to 5.07 g 5-methoxy-2-nitrophenol in 60 mL water, reflux under an inert gas for 30 min, cool in ice, filter to obtain 2-amino-5-methoxyphenol as amber prisms. Add 3.48 g ethyl pyruvate to a solution of 4.17 g 2-amino-5-methoxyphenol in 45 mL EtOH and 11 mL acetic acid, stir at room temperature in the dark overnight, filter, recrystallize the solid from EtOH to obtain 3-methyl-7-methoxy-1,4-benzoxazin-2-one as orange-red crystals (mp 126°). Add 3.76 g phenyltrimethylammonium tribromide to a solution of 1.09 g 3-methyl-7-methoxy-1,4-benzoxazin-2-one in 20 mL THF stirred at 0°, stir gently overnight in a refrigerator, dilute with 50 mL diethyl ether, wash with 25 mL 100 mM sodium bisulfite, wash with 25 mL 100 mM sodium bicarbonate, wash with water. Dry the organic layer over anhydrous sodium sulfate, evaporate to dryness under reduced pressure, chromatograph the residue on 50 g silica gel with n-hexane:dichloromethane 50:50, recrystallize the product from n-hexane:dichloromethane 50:50 to obtain 3-bromomethyl-7-methoxy-1,4-benzoxazin-2-one as faint yellow prisms (mp 146°) (J.Chromatogr. 1992, 591, 159).)

HPLC VARIABLES

Guard column: 10 μm Resolve C18 (Waters)

Column: 150 × 3.9 4 μm Nova-Pak C18

Mobile phase: MeCN:acetone:water 30:3:67

Flow rate: 1

Injection volume: 5

Detector: F ex 355 em 430

CHROMATOGRAM
Retention time: 9
Internal standard: β-naphthol (7)
Limit of detection: 3.3 fmole (20 μL injection)

KEY WORDS
plasma; SPE

REFERENCE
Chen, S.-H.; Yang, Z.-Y.; Wu, H.-L.; Kou, H.-S.; Lin, S.-J. Determination of thiocyanate anion by high-performance liquid chromatography with fluorimetric detection, *J.Anal.Toxicol.*, **1996**, *20*, 38−42.

THIOL

ACTIVATED HALIDE

4-(Aminosulfonyl)-7-fluoro-2,1,3-benzoxadiazole

RELATED REFERENCES

Okabayashi, Y.; Kitagawa, T. High-performance liquid chromatography for amino compounds and thiol compounds derivatized with europium chelate. *Anal.Chem.* **1994**, *66*, 1448-1453.

Reddy, M.N.; Behnke, C. A rapid and simple assay to determine total homocysteine and other thiols in pediatric samples by high pressure liquid chromatography and fluorescence detection. *J.Liq.Chromatogr.Rel.Technol.* **1997**, *20*, 1391-1408.

Toyooka, T.; Miyano, H.; Imai, K. Amino acid composition analysis of minute amounts of cysteine-containing proteins using 4-(aminosulfonyl)-7-fluoro-2,1,3-benzoxadiazole and 4-fluoro-7-nitro-2,1,3-benzoxadiazole in combination with HPLC. *Biomed.Chromatogr.* **1986**, *1*, 15-20.

Toyooka, T.; Furukawa, F.; Suzuki, T.; Saito, Y.; Takahashi, M.; Hayashi, Y.; Uzu, S.; Imai, K. Determination of thiols and disulfides in normal rat tissues and hamster pancreas treated with N-nitrosobis(2-oxopropyl)amine using 4-(aminosulfonyl)-7-fluoro-2,1,3-benzoxadiazole and ammonium 7-fluoro-2,1,3-benzoxadiazole-4-sulfonate. *Biomed.Chromatogr.* **1989**, *3*, 166-172.

SAMPLE

Matrix: blood

Analyte: thiols

Sample preparation: 500 μL Plasma + 50 μL 10% tri-n-butylphosphine in DMF, mix, let stand at 4° for 30 min to reduce disulfides and decouple them from proteins, add 500 μL 10% trichloroacetic acid containing 1 mM disodium EDTA, vortex, centrifuge at 4° at 12500 rpm for 5 min. Remove a 200 μL aliquot of the supernatant and add it to 200 μL 100 mM pH 8.0 sodium borate containing 2 mM disodium EDTA and 400 μL 1 mM 4-(aminosulfonyl)-7-fluoro-2,1,3-benzoxadiazole in 100 mM pH 8.0 sodium borate buffer, mix vigorously, heat at 50° for 5 min, cool in ice water, filter (0.45 μm), inject an aliquot. (4-(Aminosulfonyl)-7-fluoro-2,1,3-benzoxadiazole (ABD-F) is available from Molecular Probes, Eugene OR; Wako Chemicals, Richmond VA; or Dojindo Molecular Technologies, Inc., 3 Bethesda Metro Center, Suite 700, Bethesda MD 20814, (301) 664-8448, www.dojindo.co.jp. Synthesis is as follows. Dissolve 0.5 g magnesium sulfate heptahydrate and 6 g NaOH in 60 mL water, throughout the reaction keep the flask at about 20° with cold water cooling, add 15 mL 30% hydrogen peroxide, add 75 mL MeOH, add 12.1 g powdered benzoyl peroxide in one go, stir for 10 min, pour into 150 mL 20% sulfuric acid, extract three times with 50 mL portions of chloroform, determine peroxybenzoic acid concentration by iodometric titration (Tetrahedron 1967, 23, 3327). Slowly add 110 mL 1 M peroxybenzoic acid in chloroform to 7 g 2,6-difluoroaniline dissolved in 100 mL chloroform, stir at room temperature, when reaction is complete (iodometric titration) wash with 2% sodium thiosulfate, wash with 5% sodium carbonate, wash with water, dry over anhydrous sodium sulfate, evaporate to dryness under reduced pressure, recrystallize 2,6-difluoronitrosobenzene from EtOH (mp 108.5-109.5). Stir 8.5 g 2,6-difluoronitrosobenzene in 85 mL DMSO at room temperature and add a solution of 3.91 g sodium azide in 85 mL DMSO dropwise, let stand for about 1 h, add to a large volume of water, extract with ether, dry the extracts over anhydrous sodium sulfate, evaporate to dryness under reduced pressure and distil to give 4-fluoro-2,1,3-benzoxadiazole as a colorless oil (bp 83°/12 mm Hg) (J.Chem.Soc.(C) 1970, 1433). Add 11 mL chlorosulfonic acid dropwise to 3 g 4-fluoro-2,1,3-benzoxadiazole in 10 mL chloroform at 0-10° (use a calcium chloride drying tube), stir at room temperature for 1 h, reflux for 2 h, cool, slowly pour into ice water, remove the organic layer, extract the aqueous layer with chloroform, combine the organic layers, wash, dry over anhydrous magnesium sulfate, evaporate under reduced pressure, take up the residue in 5 mL benzene (Caution! Benzene is a carcinogen!), chromatograph on a 150 × 30 column of silica gel (100-200 mesh Kanto Chemical) with n-hexane:benzene 50:50, evaporate the appropriate fractions to give 4-(chlorosulfonyl)-7-fluoro-2,1,3-benzoxadiazole (CBD-F) as pale yellow needles (mp 64-66°). Add 1 g CBD-F dropwise to 100 mL 6% ammonium hydroxide, neutralize with 10% HCl, evaporate under reduced pressure, add 200 mL MeCN to the residue, filter. Evaporate the filtrate and chromatograph on a 300 × 20 column of 100-200 mesh silica with chloroform, collect the appropriate fractions and evaporate them to give ABD-F (4-(aminosulfonyl)-7-fluoro-2,1,3-benzoxadiazole) as white needles (mp 145-6°) after recrystallization from n-hexane/benzene (Caution! Benzene is a carcinogen!).)

CAPILLARY ELECTROPHORESIS
Capillary: 27 cm × 50 μm fused-silica (20 cm to detector) (Polymicro Technologies)
Capillary preparation: Before each run rinse with 100 mM NaOH, water, and running buffer for 10 min.
Running buffer: 50 mM pH 2.1 Sodium phosphate buffer
Capillary temperature: 25
Voltage/Current: 15 kV
Injection: Hydrodynamic injection at 0.5 psi for 5 s.
Detector: UV 220
Model: Beckman P/ACE 5000
Migration time: 5 (homocysteine), 6 (glutathione), 6.5 (cysteine)
Limit of detection: 0.5-2 μM

KEY WORDS
plasma; comparison with ammonium 7-fluoro-2,1,3-benzoxadiazole-4-sulfonate

REFERENCE
Kang, S.H.; Kim, J.-W.; Chung, D.S. Determination of homocysteine and other thiols in human plasma by capillary electrophoresis, *J.Pharm.Biomed.Anal.*, **1997**, *15*, 1435–1441.

SAMPLE
Matrix: solutions
Analyte: thiols
Sample preparation: Mix 1 mL of a 2 μM solution of thiols in 100 mM pH 8.0 sodium borate buffer containing 2 mM disodium EDTA with 1 mL 1 mM ABD-F in 100 mM pH 8.0 sodium borate buffer, vortex, heat at 50° for 5 min, cool in ice, add 600 μL 100 mM HCl, inject a 10 μL aliquot. (ABD-F is available from Molecular Probes, Eugene OR; Wako Chemicals, Richmond VA; or Dojindo Molecular Technologies, Inc., 3 Bethesda Metro Center, Suite 700, Bethesda MD 20814, (301) 664-8448, www.dojindo.co.jp. Synthesis is as described above.)

HPLC VARIABLES
Guard column: 20 × 3.9 37-50 μm Bondapak C18/Corasil
Column: 300 × 3.9 8-10 μm μBondapak C18
Mobile phase: MeCN:50 mM pH 4.0 potassium biphthalate buffer 8:92
Flow rate: 1
Injection volume: 10
Detector: F ex 380 em 510

CHROMATOGRAM
Retention time: 10 (N-acetylcysteine), 32 (cysteamine), 7.5 (cysteine), 9 (glutathione), 12 (homocysteine)
Limit of detection: 1.9 pmole

REFERENCE
Toyo'oka, T.; Imai, K. New fluorigenic reagent having halogenobenzofurazan structure for thiols: 4-(Aminosulfonyl)-7-fluoro-2,1,3-benzoxadiazole, *Anal.Chem.*, **1984**, *56*, 2461–2464.

1-Benzyl-2-chloropyridinium bromide

SAMPLE
Matrix: blood
Analyte: captopril
Sample preparation: Condition a 100 mg Bakerbond C18 SPE cartridge with two 200 µL portions of MeOH and two 200 µL portions of water. 1 mL Plasma + 100 µL 200 mM disodium EDTA + 100 µL 200 mM ascorbic acid + 200 µL 1 µg/mL IS + 400 µL 3 M perchloric acid, mix, centrifuge at 4000 g for 15 min, wash the precipitate three times with 500 µL portions of water. Combine the supernatant and the washings and neutralize (indicator paper) them with 1 M NaOH, add 3 mL 1 M pH 8.2 Tris buffer, add 100 µL 20 µg/mL 1-benzyl-2-chloropyridinium bromide, let stand for 15 min, adjust pH to 2.5–3.0 (indicator paper) with 4 M phosphoric acid, centrifuge, add to the SPE cartridge, wash with 1 mL water, dry under vacuum suction for 10 min, wash with three 100 µL portions of MeCN, dry under vacuum for 5 min, elute with 200 µL MeOH:acetic acid 80:20, elute with two 200 µL portions of MeOH:water 80:20. Combine the eluates and evaporate them to dryness at 60°, reconstitute with 50 µL mobile phase, inject a 20 µL aliquot. To measure total captopril proceed as follows. 1 mL Plasma + 100 µL 200 mM disodium EDTA + 100 µL 200 mM ascorbic acid + 200 µL 1 µg/mL IS + 2 mL 100 mM perchloric acid + 100 µL 40 mg/mL triphenylphosphine in MeCN, mix, heat at 50° for 40 min, cool, add 400 µL 3 M perchloric acid, mix, centrifuge at 4000 g for 15 min, wash the precipitate three times with 500 µL portions of water. Combine the supernatant and the washings and neutralize (indicator paper) them with 1 M NaOH, add 3 mL 1 M pH 8.2 Tris buffer, add 100 µL 20 µg/mL 1-benzyl-2-chloropyridinium bromide, let stand for 15 min, adjust pH to 2.5–3.0 (indicator paper) with 4 M phosphoric acid, centrifuge, add. to the SPE cartridge, wash with 1 mL water, dry under vacuum suction for 10 min, wash with three 100 µL portions of MeCN, dry under vacuum for 5 min, elute with 200 µL MeOH:acetic acid 80:20, elute with two 200 µL portions of MeOH:water 80:20. Combine the eluates and evaporate them to dryness at 60°, reconstitute with 50 µL mobile phase, inject a 20 µL aliquot. To measure protein-conjugated captopril proceed as follows. 1 mL Plasma + 100 µL 200 mM disodium EDTA +100 µL 200 mM ascorbic acid + 200 µL 1 µg/mL IS + 400 µL 3 M perchloric acid, mix, centrifuge at 4000 g for 15 min, wash the precipitate three times with 500 µL portions of water. Suspend the precipitate in 2 mL 100 mM perchloric acid, add 200 µL 1 µg/mL IS, add 100 µL 40 mg/mL triphenylphosphine in MeCN, heat at 50° with occasional shaking for 40 min, cool, add 400 µL 3 M perchloric acid, mix, centrifuge at 4000 g for 15 min, wash the precipitate three times with 500 µL portions of water. Combine the supernatant and the washings and neutralize (indicator paper) them with 1 M NaOH, add 3 mL 1 M pH 8.2 Tris buffer, add 100 µL 20 µg/mL 1-benzyl-2-chloropyridinium bromide, let stand for 15 min, adjust pH to 2.5–3.0 (indicator paper) with 4 M phosphoric acid, centrifuge, add to the SPE cartridge, wash with 1 mL water, dry under vacuum suction for 10 min, wash with three 100 µL portions of MeCN,

dry under vacuum for 5 min, elute with 200 μL MeOH:acetic acid 80:20, elute with two 200 μL portions of MeOH:water 80:20. Combine the eluates and evaporate them to dryness at 60°, reconstitute with 50 μL mobile phase, inject a 20 μL aliquot. (Prepare 1-benzyl-2-chloropyridinium bromide as follows. Add 8.5 g benzyl bromide to 4.5 g 2-chloropyridine with stirring, stir at 60° overnight, cool, filter, wash with acetone, dry under vacuum to give 1-benzyl-2-chloropyridinium bromide (mp 187-191°).)

HPLC VARIABLES
Guard column: 20 × 2.1 5 μm Hypersil
Column: 100 × 2.1 5 μm ODS Hypersil
Mobile phase: Acetone:buffer 25:75 (Buffer was 200 mM pH 2.5 citrate buffer containing 10 mM sodium octanesulfonate and 15 mM NaCl.)
Column temperature: 50
Flow rate: 0.2
Injection volume: 20
Detector: UV 314

CHROMATOGRAM
Retention time: 10
Internal standard: 1-benzyl-2-chloro-4-methylpyridinium bromide-captopril adduct (Preparation is as follows. Add 8.5 g benzyl bromide to 5.1 g 2-chloro-4-methylpyridine (Loba-Chemie, Vienna) with stirring, stir at 60° overnight, cool, filter, wash with acetone, dry under vacuum to give 1-benzyl-2-chloro-4-methylpyridinium bromide. Condition a 1 g C18 SPE cartridge with two 1 mL portions of MeOH and with 1 mL water. Stir 30 mg captopril and 80 mg 1-benzyl-2-chloro-4-methylpyridinium bromide in 5 mL 500 mM pH 8.2 Tris buffer for 1 h, add 1.6 mL 100 mM sodium sulfide, let stand for 10 min, add to the SPE cartridge, wash with two 1 mL portions of water, dry under vacuum for 10 min, wash with three 1 mL portions of MeCN, dry under vacuum for 5 min, elute with 5 mL MeOH:water 80:20. Dilute the eluate to 10 mL with water, dilute with water to an adduct concentration of 3 μg/mL. 2-Chloro-4-methylpyridine can also be prepared as follows. Stir 3.6 g 2-amino-4-picoline (2-amino-4-methylpyridine) in 75 g concentrated HCl at between -15° and -20°, saturate with hydrogen chloride gas, add 3.5 g pulverized sodium nitrite in small portions, let stand overnight, neutralize, steam distil to give 2-chloro-4-methylpyridine as a colorless liquid (bp 194-195°) (Ber. 1924, 57, 791).) (12)
Limit of quantitation: 10 ng/mL

KEY WORDS
plasma; SPE

REFERENCE
Bald, E.; Sypniewski, S.; Drzewoski, J.; Stepien, M. Application of 2-halopyridinium salts as ultraviolet derivatization reagents and solid-phase extraction for determination of captopril in human plasma by high-performance liquid chromatography, *J.Chromatogr.B*, **1996**, *681*, 283–289.

SAMPLE
Matrix: blood, urine
Analyte: captopril
Sample preparation: Condition a 100 mg Bakerbond C18 SPE cartridge with two 200 μL portions of MeOH and two 200 μL portions of water. Whole blood. 3 mL Whole blood + 100 μL 100 mM EDTA + 100 μL 200 mM ascorbic acid + 2 mL 1 M pH 8.2 Tris buffer + 200 μL 3 μg/mL IS + 100 μL 20 μg/mL 1-benzyl-2-chloropyridinium bromide, vortex for 15 min, centrifuge at 3000 g for 10 min. Remove a 1 mL aliquot of the supernatant and add it to 400 μL 3 M perchloric acid, centrifuge for 15 min, rinse the precipitate with 500 μL portions of water. Combine the supernatant and the rinses and adjust the pH to 2.5-3.0 (indicator paper) with 100 mM NaOH, add to the SPE cartridge, wash with 1 mL water, dry under vacuum suction for 10 min, elute with 200 μL MeOH:acetic acid 80:20, elute with two 200 μL portions of MeOH:water 80:20. Combine the eluates and evaporate them to dryness at 60°, reconstitute with 50 μL water, inject a 20 μL aliquot. Urine. 500 μL Urine + 100 μL 200 mM EDTA + 100 μL 200 mM ascorbic acid + 3 mL 1 M pH 8.2 Tris buffer + 200 μL 3 μg/mL IS + 100 μL 20 μg/mL 1-benzyl-2-chloropyridinium bromide, vortex for 15 min, adjust the pH to 2.5-3.0 with 4 M phosphoric acid, add to the SPE cartridge, wash with 1 mL water, dry under vacuum suction for 10 min, elute with

200 μL MeOH:acetic acid 80:20, elute with two 200 μL portions of MeOH:water 80:20. Combine the eluates and evaporate them to dryness at 60°, reconstitute with 50 μL water, inject a 20 μL aliquot. (Prepare 1-benzyl-2-chloropyridinium bromide as follows. Add 8.5 g benzyl bromide to 4.5 g 2-chloropyridine with stirring, stir at 60° overnight, cool, filter, wash with acetone, dry under vacuum to give 1-benzyl-2-chloropyridinium bromide (mp 187-191°) (J. Chromatogr. B 1996, 681, 283).)

HPLC VARIABLES
Guard column: 20 × 2.1 5 μm Hypersil
Column: 150 × 3.3 7 μm Separon SGX (Struzeni, Prague)
Mobile phase: Gradient. A was MeCN:100 mM pH 2.5 citric acid buffer containing 20 mM sodium octanesulfonate 25:75. B was MeCN:MeOH 50:50. A:B 100:0 for 10 min, to 80:20 over 10 min, maintain at 80:20 for 5 min, to 60:40 over 5 min, return to initial conditions over 7 min.
Column temperature: 50
Flow rate: 0.5
Injection volume: 20
Detector: UV 314

CHROMATOGRAM
Retention time: 23
Internal standard: 1-benzyl-2-chloro-4-methylpyridinium bromide-captopril adduct (Preparation is as described above.) (27)
Limit of quantitation: 10 ng/mL
Limit of detection: 0.3 ng/mL

KEY WORDS
whole blood; SPE

REFERENCE
Sypniewski, S.; Bald, E. Determination of captopril and its disulphides in whole human blood and urine by high-performance liquid chromatography with ultraviolet detection and precolumn derivatization, *J.Chromatogr.A*, **1996**, *729*, 335–340.

1,1-Bis(phenylsulfonyl)ethylene

SAMPLE
Matrix: formulations
Analyte: thiol drugs
Sample preparation: Dissolve solutions, powders, or powdered tablets in water to give a 200-325 μM solution, filter if necessary. Mix 500 μL solution with 500 μL 170 mM pH 7.5 borate buffer and 500 μL 2 mM 1,1-bis(phenylsulfonyl)ethylene (1,1'-ethenylidene-bis(sulfonyl)bis-benzene; Fluka, Merck) in MeOH, let stand at room temperature for 2 min, add 500 μL water, add 300 μL chloroform, vortex for 1 min, centrifuge for 2 min. Remove a 1 mL aliquot of the aqueous layer and add it to 500 μL 300 mM orthophosphoric acid, add 100-200 μL 30 μM IS in MeCN, mix, inject a 50 μL aliquot.

HPLC VARIABLES
Column: 150 × 4.6 5 μm Hypersil C18
Mobile phase: MeOH:buffer 42:58 (A) or MeOH:buffer 36:64 (B) or MeCN:buffer 51:49 (C) (Buffer was 50 mM pH 4.0 triethylamine-phosphate buffer.)
Flow rate: 1
Injection volume: 50
Detector: UV 254

CHROMATOGRAM
Retention time: 8 (N-2-mercaptopropionylglycine (thiola) (A)), 12 (N-acetylcysteine (B)), 3.5 (captopril (C))
Internal standard: methyl p-hydroxybenzoate (6 (A), 10 (B)), 1(2H)-acenaphthylenone (Synthesis is as follows. Prepare dichloroacenaphthylenone by refluxing acenaphthene-quinone with phosphorus pentachloride in anhydrous toluene. Mix 200 g glacial acetic acid, 20 g crude dichloroacenaphthylenone, and 30 g iron powder, heat gently until re-fluxing, heat for about 3 h, dilute with 500 mL water, steam distil, filter the distillate to recover the product, recrystallize from benzene (Caution! Benzene is a carcinogen!) to obtain 1(2H)-acenaphthylenone (mp 119-121°) (Gazz. Chim. Italia 1938, 68, 184).) (5.5 (C))
Limit of detection: 100 pmole

KEY WORDS
tablets; powders

REFERENCE
Cavrini, V.; Gotti, R.; Andrisano, V.; Gatti, R. 1,1'-[Ethylidenebis(sulfonyl)]bis-benzene: A useful pre-chromatographic derivatization reagent for HPLC analyses of thiol drugs, *Chromatographia*, **1996**, *42*, 515–520.

7-Chloro-2,1,3-benzoxadiazole-4-sulfonic acid, Ammonium Salt

SAMPLE
Matrix: blood
Analyte: SK&F 101 926 (1-pentamethylene propionic acid-2-(D)Tyr(Et)-4-Val-9-des-Gly)-vasopressin)
Sample preparation: Condition two 1 mL carboxymethyl-hydrogen form weak cation-exchange SPE cartridges (Analytichem) with 1 mL MeOH and 1 mL water. 1 mL Plasma + 1 mL water + 50 μL 500 ng/mL IS, mix, let stand at room temperature for 30 min, add to the SPE cartridge, wash with 1 mL 1% trifluoroacetic acid in MeOH, wash with 1 mL water, wash with 1 mL 100 mM borate buffer containing 1 mM EDTA, wash with 1 mL water, wash with 500 μL MeOH, elute with 400 μL 1% trifluoroacetic acid in MeOH. Evaporate the eluate to dryness under a stream of nitrogen at 40°, reconstitute the residue in 90 μL borate buffer, mix, let stand for 15 min, add 5 μL 200 mM dithiothreitol, heat at 40° for 30 min, add 20 μL 200 mM 7-chloro-2,1,3-benzoxadiazole-4-sulfonic acid, ammonium salt (ammonium 4-chloro-7-sulfobenzofuran), heat at 60° for 2 h, add to an-other SPE cartridge, wash with 1 mL water, wash with 1 mL borate buffer, wash with 1 mL water, wash with 200 μL hexane, dry, elute with two 150 μL aliquots of 500 mM ammonium acetate in MeOH. Evaporate the eluate to dryness under a stream of nitrogen

at 40°, reconstitute the residue in 100 μL water, mix, let stand for 30 min, inject a 20 μL aliquot. (The disulfide bond is reduced with dithiothreitol and the thiol groups are then derivatized.)

HPLC VARIABLES
Guard column: 30 × 2.1 10 μm RP-8
Column: 220 × 2.1 5 μm RP-8 (Brownlee)
Mobile phase: Gradient. A was MeOH:20 mM tetrabutylammonium phosphate (PIC A) 50:50, adjusted to pH 5.2 with phosphoric acid. B was MeOH. A:B from 90:10 to 45:55 over 22 min.
Column temperature: 50
Flow rate: 0.3
Injection volume: 20
Detector: F ex 380 em 470

CHROMATOGRAM
Retention time: 13.8
Internal standard: SK&F 103 784 (12.8)
Limit of quantitation: 0.5 ng/mL

KEY WORDS
plasma; SPE; silylate glassware with dichlorodimethylsilane; pharmacokinetics; narrow-bore

REFERENCE
Rhodes, G.R.; Rubenfield, M.J.; Garvie, C.T. Determination of a disulfide-containing octapeptide antagonist of vasopressin in human plasma utilizing high-performance liquid chromatography with fluorescence detection, *J.Chromatogr.*, **1989**, *488*, 456−462.

1-Chloro-2,4-dinitrobenzene

O_2N—Cl + RSH ⟶ O_2N—SR
NO$_2$ ... NO$_2$

SAMPLE
Matrix: tissue
Analyte: thiols
Sample preparation: Homogenize rat liver in 10 mM pH 7.4 sodium phosphate buffer containing 250 mM sucrose, centrifuge at 100000 g for 1 h. Precipitate the homogenate with 5% trichloroacetic acid, mix with dithioerythritol, adjust pH to 6.5, add 1-chloro-2,4-dinitrobenzene, heat at 37° with gentle shaking for 1 h, add 2 mL MeOH, evaporate to dryness, reconstitute with 1 mL water, inject a 250 μL aliquot.

HPLC VARIABLES
Column: 250 × 4.6 5 μm C18 (Vydac)
Mobile phase: Gradient. A was 0.1% trifluoroacetic acid. B was 0.1% trifluoroacetic acid in MeCN:water 50:50. A:B 70:30 for 1 min, to 50:50 over 15 min.
Flow rate: 1
Injection volume: 250
Detector: UV 340

CHROMATOGRAM
Retention time: 10.2 (cysteine), 11.5 (cysteinylglycine), 13.4 (glutathione), 15.0 (gamma-glutamylcysteine)

KEY WORDS
rat; liver

REFERENCE
Sugimoto, M.; Kuhlenkamp, J.; Ookhtens, M.; Aw, T.Y.; Reeve, J. Jr.; Kaplowitz, N. Gamma-glutamylcysteine: a substrate for glutathione S-transferases, *Biochem.Pharmacol.*, **1985**, *34*, 3643–3647.

4-Chloro-7-nitro-2,1,3-benzoxadiazole

SAMPLE
Matrix: solutions
Analyte: sodium 2-pyridinethiol-1-oxide
Sample preparation: Mix 250 µL of an aqueous solution of sodium 2-pyridinethiol-1-oxide with 250 µL 10 mg/mL 7-chloro-4-nitro-2,1,3-benzoxadiazole (NBD-Cl) in MeCN, let stand at room temperature for 5 min, inject a 10 µL aliquot.

HPLC VARIABLES
Guard column: Upchurch
Column: 250 × 4.6 LC-8 (Altex)
Mobile phase: MeCN:water 40:60
Column temperature: 60
Flow rate: 1
Injection volume: 10
Detector: UV 425

CHROMATOGRAM
Retention time: 4.5
Limit of detection: 1 ppm

REFERENCE
Valdez, D.; Reier, J.C. HPLC analysis of sodium 2-pyridinethiol-1-oxide in aqueous metalworking fluids via derivatization with NBD-Cl, *J.Liq.Chromatogr.*, **1987**, *10*, 2133–2144.

SAMPLE
Matrix: urine
Analyte: acetylcysteine
Sample preparation: 250 µL Urine + 10 µL 14.5 mg/mL dithioerythritol in water + 100 µL phosphate buffer. After 30 min remove a 250 µL aliquot and add it to 250 µL citrate buffer and 50 µL 1 mg/mL 4-chloro-7-nitrobenzo-2-oxa-1,3-diazole in MeOH, after 20 min inject a 20 µL aliquot. (Phosphate buffer was 17.8 g $Na_2HPO_4.2H_2O$ in 200 mL water, pH adjusted to 8.3 with NaOH. Citrate buffer was 29.4 g sodium citrate and 744 mg EDTA in 200 mL water, pH adjusted to 8.3 with NaOH.)

HPLC VARIABLES
Column: 300 × 4.6 7 µm Nucleosil C18
Mobile phase: MeCN:0.5% Na_2HPO_4 30:70
Column temperature: 30
Flow rate: 1.6
Injection volume: 20
Detector: UV 470

CHROMATOGRAM
Retention time: 8

Limit of detection: 5 µM

REFERENCE

Frank, H.; Thiel, D.; Langer, K. Determination of N-acetyl-L-cysteine in biological fluids, *J.Chromatogr.*, **1984**, *309*, 261–267.

CY5.4a-IA

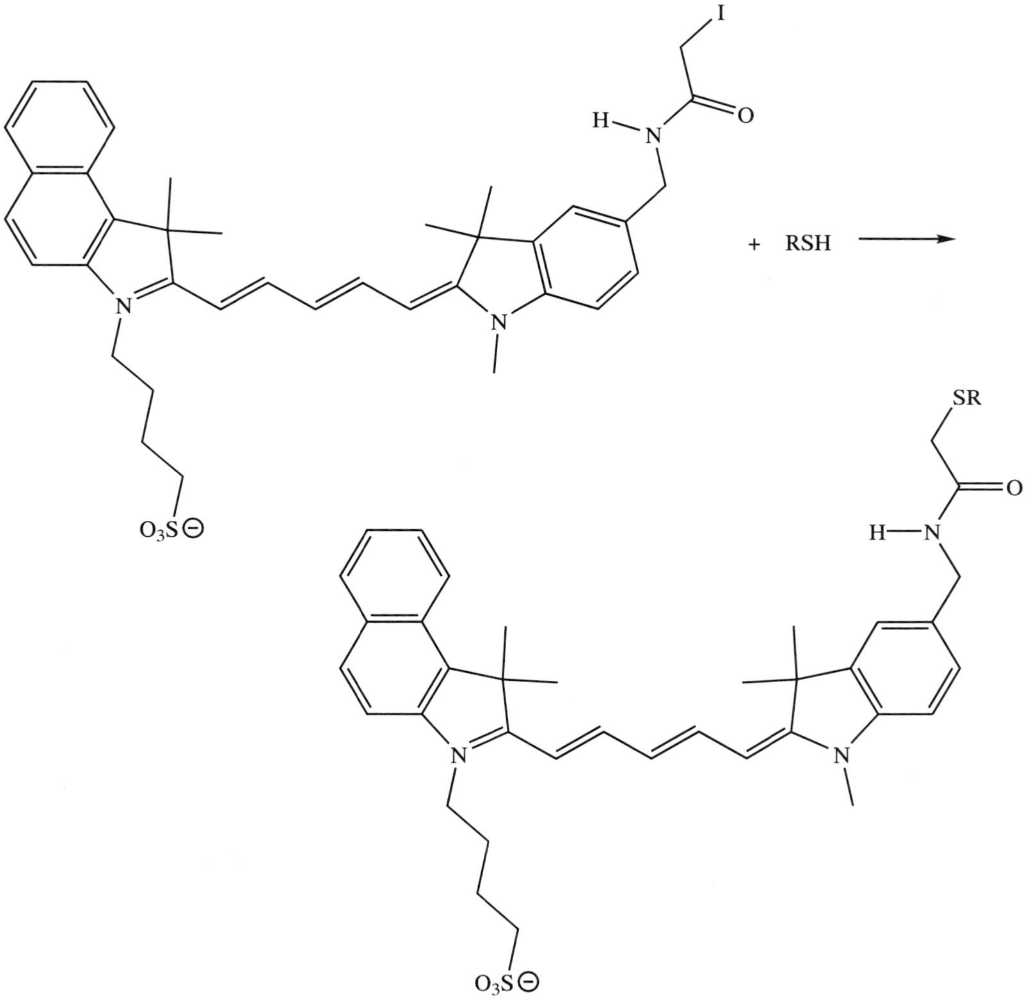

SAMPLE

Matrix: blood, urine

Analyte: WR1065 (2-(3-aminopropylamino)ethanethiol)

Sample preparation: Plasma. Filter (0.2 µm) urine. Mix plasma with 4 volumes of MeOH, centrifuge at 2500 rpm for 15 min. Remove a 500 µL aliquot of the supernatant and add it to 250 µL 200 mM pH 8.0 ammonium carbonate buffer and 250 µL reagent, pass dry nitrogen through the mixture for 1 min, heat at 65° for 1 h, cool, dilute with mobile phase, inject a 25 µL aliquot on to column A and elute to waste with mobile phase A, after 3 min backflush the contents of column A on to column B with mobile phase B, monitor the effluent from column B. Urine. Filter (0.2 µm) urine, remove a 500 µL aliquot of the filtrate and add it to 500 µL reagent, pass dry nitrogen through the mixture for 1 min, heat at 65° for 1 h, cool, dilute with mobile phase, inject a 25 µL aliquot on to column A and elute to waste with mobile phase A, after 3 min backflush the contents of column A on to column B with mobile phase B, monitor the effluent from column B. (Reagent was 200 µm CY5.4a-IA in MeOH:200 mM pH 8.0 ammonium carbonate buffer 75:25, store in the dark at room temperature, stable for at least a week. Incomplete details for the synthesis of CY5.4a-IA are given in Anal.Chem. 1993, 65, 2197.)

HPLC VARIABLES
Column: A 10 × 3.1 20 µm Alox T20 (Merck) (change after 10 analyses); B 100 × 3.1 10 µm Alox T10
Mobile phase: A MeOH:10 mM pH 5.5 phosphate buffer 50:50; B MeOH:250 mM pH 5.5 acetate buffer 80:20
Column temperature: 50 (column B only)
Flow rate: A 1; B 0.75
Injection volume: 25
Detector: F ex 670 (9.5 mW LAS 200-670-10 diode laser, Lasermax, Rochester NJ)

CHROMATOGRAM
Retention time: 14
Limit of detection: 5 nM

KEY WORDS
column-switching; plasma

REFERENCE
Mank, A.J.G.; Molenaar, E.J.; Gooijer, C.; Lingeman, H.; Velthorst, N.H.; Brinkman, U.A.T. Determination of the anticancer drug metabolite WR1065 using pre-column derivatization and diode laser induced fluorescence detection, *J.Pharm.Biomed.Anal.*, **1995**, *13*, 255–263.

SAMPLE
Matrix: solutions
Analyte: mercaptobenzothiazole
Sample preparation: Mix 500 µL of a solution in MeOH:water 90:10 with 500 µL 50 µM CY5.4a-IA in MeOH:water 70:30, pass dry nitrogen through the mixture for 1 min, heat at 65° for 1.5 h, inject a 25 µL aliquot. (Some details for the synthesis of CY5.4a-IA are given in the paper.)

HPLC VARIABLES
Column: 150 × 3.1 5 µm LiChrosorb RP-8
Mobile phase: MeOH:10 mM pH 6.8 phosphate buffer 65:35 containing 1 mM triethylamine
Flow rate: 0.75
Injection volume: 25
Detector: F ex 670 (9.5 mW Lasermax LAS200-670-10 diode laser)

CHROMATOGRAM
Retention time: 7
Limit of detection: 1 nM

REFERENCE
Mank, A.J.G.; Molenaar, E.J.; Lingeman, H.; Gooijer, C.; Brinkman, U.A.T.; Velthorst, N.H. Visible diode laser induced fluorescence detection in liquid chromatography after precolumn derivatization of thiols, *Anal.Chem.*, **1993**, *65*, 2197–2203.

2,4'-Dibromoacetophenone

Br—⟨ ⟩—C(=O)—CH₂—Br + RSH → Br—⟨ ⟩—C(=O)—CH₂—SR

SAMPLE
Matrix: blood
Analyte: captopril
Sample preparation: 3 mL Blood + 1.5 mL 0.5% 2,4'-dibromoacetophenone (p-bromo-phenacyl bromide) in acetone, vortex for 30 s, let stand for 5 min, acidify with 300 mM HCl, store below -15°, extract with 16 mL benzene (Caution! Benzene is a carcinogen!), extract with 8 mL benzene. Combine the organic layers and evaporate them to dryness under reduced pressure, reconstitute the residue in 4 mL 50 mM pH 7 phosphate buffer, add 20 mL hexane, sonicate, wash with 6 mL hexane, discard the hexane layer, acidify the aqueous layer with 100 μL 2 M HCl, extract with 6 mL benzene, extract with 2 mL benzene. Combine the organic layers and add 500 ng IS, evaporate to dryness under reduced pressure, reconstitute the residue in 200 μL MeCN, inject a 5-25 μL aliquot. (2,4'-Dibromoacetophenone reacted with the thiol group but not the carboxylic acid.)

HPLC VARIABLES
Column: μBondapak C18
Mobile phase: MeCN:water:acetic acid 48:51.5:0.5
Injection volume: 5-25
Detector: UV 254

CHROMATOGRAM
Internal standard: thiosalicylic acid-p-bromophenacyl bromide adduct (Prepare by dis-solving 2.4 mmoles thiosalicylic acid and 2.4 mmoles p-bromophenacyl bromide in 40 mL MeOH, adjust to pH 7 by the dropwise addition of 1 M NaOH, allow to stand at room temperature for 10 min, evaporate to dryness under reduced pressure, reconstitute with 40 mL 50 mM pH 7.0 phosphate buffer, wash twice with 20 mL portions of hexane, adjust pH to 2 with dilute HCl, extract with 40 mL ethyl acetate, evaporate to dryness under reduced pressure, recrystallize the residue from benzene to give the adduct as pale yellow plates.)
Limit of quantitation: 5 ng/mL

KEY WORDS
whole blood

REFERENCE
Kawahara, Y.; Hisaoka, M.; Yamazaki, Y.; Inage, A.; Morioka, T. Determination of captopril in blood and urine by high-performance liquid chromatography, *Chem.Pharm.Bull.*, **1981**, *29*, 150–157.

SAMPLE
Matrix: blood
Analyte: captopril
Sample preparation: 500 μL Plasma + 30 μL 1 mg/mL 2,4'-dibromoacetophenone (p-bromophenacyl bromide) in MeCN + 50 μL 100 mM NaOH, shake for 15 min, add 75 μL 1 M HCl, add 500 ng nitrazepam for each mL, add 150 μL 200 mM pH 4.0 acetate buffer. Extract with 4 mL benzene (Caution! Benzene is a carcinogen!), centrifuge. Remove the organic layer and evaporate it to dryness under a stream of nitrogen, reconstitute the residue in 200 μL mobile phase, inject a 20 μL aliquot.

HPLC VARIABLES
Column: 250 × 4.6 Kontron Analytical S5 ODS2
Mobile phase: MeCN:1% acetic acid 60:40
Flow rate: 1.3
Injection volume: 20
Detector: UV 260

CHROMATOGRAM
Retention time: 4
Internal standard: nitrazepam (4.5)
Limit of quantitation: 30 ng/mL
Limit of detection: 15 ng/mL

KEY WORDS
plasma; pharmacokinetics

REFERENCE
Jankowski, A.; Skorek, A.; Krzysko, K.; Zarzyxki, P.K.; Ochocka, R.J.; Lamparczyk, H. Captopril: determination in blood and pharmacokinetics after single oral dose, *J.Pharm.Biomed.Anal.*, **1995**, *13*, 655–660.

SAMPLE
Matrix: blood
Analyte: captopril
Sample preparation: 1 mL Plasma + 150 μL 1 mg/mL 2,4'-dibromoacetophenone (p-bromophenacyl bromide) in MeOH + 20 μL 50 μg/mL 4-chloro-2-nitroaniline in MeOH, vortex for 30 s, let stand at room temperature for 15 min, add 200 μL 2 M HCl, vortex for 10 s, add 5 mL benzene:ethyl acetate 50:50 (Caution! Benzene is a carcinogen!), vortex for 2 min. Remove a 4.5 mL aliquot of the organic layer and evaporate it to dryness under a stream of nitrogen at 50°, reconstitute the residue in 100 μL MeCN, vortex for 20 s, inject a 25 μL aliquot.

HPLC VARIABLES
Column: 250 × 4.6 10 μm Spherisorb C18
Mobile phase: MeCN:water:acetic acid 44:55:0.2
Flow rate: 1.4
Injection volume: 25
Detector: UV 258

CHROMATOGRAM
Retention time: 3.6
Internal standard: 4-chloro-2-nitroaniline (4.7)
Limit of quantitation: 5 ng/mL
Limit of detection: 2 ng/mL

KEY WORDS
plasma; pharmacokinetics

REFERENCE
Li, K.; Zhou, J.-h. HPLC determination of captopril in human plasma and its pharmacokinetic study, *Biomed.Chromatogr.*, **1996**, *10*, 237–239.

SAMPLE
Matrix: blood
Analyte: captopril
Sample preparation: Condition a 1 mL silica SPE cartridge with 1 mL benzene (Caution! Benzene is a carcinogen!). Collect 10 mL blood in a tube containing 30 mg 2,4'-dibromoacetophenone (p-bromophenacyl bromide), vortex for 30 s, let stand at room temperature for at least 15 min. Remove the serum and freeze it at -20°. Thaw serum and add 1 mL to 200 μL 1 M HCl, mix, extract twice with 2 mL portions of benzene. Add the organic layers to the SPE cartridge, wash with 4 mL benzene, elute with 500 μL MeCN, add 50 μL 10 μg/mL IS in acetone to the eluate, make up to 1 mL with mobile phase, inject a 100 μL aliquot. (2,4'-Dibromoacetophenone reacted with the thiol group but not the carboxylic acid (Chem. Pharm. Bull. 1981, 29, 150).)

HPLC VARIABLES
Guard column: Spheri-5 ODS-GU
Column: 100 × 4.6 5 μm Spheri-5 ODS
Mobile phase: MeCN:water:acetic acid 45:54:1
Flow rate: 1

Injection volume: 100
Detector: UV 263

CHROMATOGRAM
Retention time: 4
Internal standard: thiosalicylic acid-p-bromophenacyl bromide adduct (Prepare by dissolving 2.4 mmoles thiosalicylic acid and 2.4 mmoles p-bromophenacyl bromide in 40 mL MeOH, adjust to pH 7 by the dropwise addition of 1 M NaOH, allow to stand at room temperature for 10 min, evaporate to dryness under reduced pressure, reconstitute with 40 mL 50 mM pH 7.0 phosphate buffer, wash twice with 20 mL portions of hexane, adjust pH to 2 with dilute HCl, extract with 40 mL ethyl acetate, evaporate to dryness under reduced pressure, recrystallize the residue from benzene to give the adduct as pale yellow plates (Chem. Pharm. Bull. 1981, 29, 150).) (8)
Limit of detection: 15 ng/mL

KEY WORDS
SPE; whole blood; pharmacokinetics

REFERENCE
Bahmaei, M.; Khosravi, A.; Zamiri, C.; Massoumi, A.; Mahmoudian, M. Determination of captopril in human serum by high performance liquid chromatography using solid-phase extraction, *J.Pharm.Biomed.Anal.*, **1997**, *15*, 1181–1186.

SAMPLE
Matrix: blood, CSF
Analyte: captopril
Sample preparation: 1 mL blood + 50 μL of a solution containing 100 mM EDTA and 100 mM ascorbic acid, centrifuge at 13000 g for 2 min. Proceed immediately. 500 μL Plasma or CSF + 50 μL 1 mg/mL 2,4'-dibromoacetophenone (p-bromophenacyl bromide) in MeCN, vortex for 15 s, leave at room temperature for 30 min, add 100 μL 1 M HCl, add 50 μL 0.02% phenylacetic acid in methanol, add 5 mL ethyl acetate : benzene (Caution! Benzene is a carcinogen!), vortex for 3 min, shake gently for 15 min. Saturate the aqueous phase with NaCl, centrifuge. Remove the organic layer and evaporate it to dryness under a stream of nitrogen at 50°, reconstitute in 250 μL MeCN, inject a 20 μL aliquot.

HPLC VARIABLES
Column: 300 × 4 5 μm Lichrosorb RP-18
Mobile phase: MeCN : water : acetic acid 220 : 180 : 2.5
Flow rate: 1 mL/min increasing to 3 mL/min after elution of captopril adduct to get rid of excess derivatizing agent
Injection volume: 20
Detector: UV 260

CHROMATOGRAM
Retention time: 4.7
Internal standard: phenylacetic acid (3.2)
Limit of detection: 5 ng/mL

KEY WORDS
plasma; with modification can be used to determine captopril disulfides.

REFERENCE
Colin, P.; Scherer, E. Simple high-performance liquid chromatography determination of captopril in human plasma and cerebrospinal fluid, *J.Liq.Chromatogr.*, **1989**, *12*, 629–643.

SAMPLE
Matrix: blood, urine
Analyte: captopril
Sample preparation: Plasma. 1.5 mL Blood + 50 μL 100 mM ascorbic acid and 100 mM disodium ethylenediaminetetraacetate, centrifuge. Remove 0.5 mL plasma and immediately add it to 50 μL 1 mg/mL 2,4'-dibromoacetophenone (p-bromophenacyl bromide), vortex 30 s, let stand at room temperature for 20 min, add 300 μL 6% perchloric acid, vortex for 30 s, centrifuge at 10000 g for 10 min, inject a 500 μL aliquot onto column A with mobile phase A and elute to waste, after 3 min elute the contents of column A onto

column B with mobile phase B, after 2 min remove column A, elute column B with mobile phase B, monitor the effluent from column B. Urine. 50 μL Urine + 50 μL 1 mg/mL p-bromophenacyl bromide + 600 μL water, vortex 30 s, let stand at room temperature for 20 min, inject a 500 μL aliquot onto column A with mobile phase A and elute to waste, after 3 min elute the contents of column A onto column B with mobile phase B, after 2 min remove column A, elute column B with mobile phase B, monitor the effluent from column B. (Column A should be washed with MeOH for 2 min then re-equilibrated with mobile phase A for 2 min.)

HPLC VARIABLES
Column: A 50 × 5 37-50 μm μBondapak C18; B 150 × 5 5 μm Tianjing silica gel YWG-C18
Mobile phase: A 0.2% Acetic acid in water; B MeCN:water:acetic acid 35:65:0.4
Flow rate: A 3; B 2
Injection volume: 500
Detector: UV 260

CHROMATOGRAM
Retention time: 9
Limit of detection: 10 ng/mL

KEY WORDS
plasma; column-switching; pharmacokinetics

REFERENCE
Gao, S.; Tian, W.; Wang, S. Simple high-performance liquid chromatographic method for the determination of captopril in biological fluids, *J.Chromatogr.*, **1992**, *582*, 258–262.

SAMPLE
Matrix: urine
Analyte: captopril
Sample preparation: 5 mL Urine + 2 mL 500 mM pH 7.0 phosphate buffer + 0.5 mL 20 mg/mL 2,4'-dibromoacetophenone (p-bromophenacyl bromide) in MeOH, shake vigorously. Remove a 1 mL aliquot and add it to 1.5 mL water and 6 mL hexane, mix, discard the hexane layer. Remove a 2 mL aliquot of the aqueous layer and add it to 100 μL 2% tributylphosphine in MeOH, heat at 50° for 30 min, wash with 6 mL hexane, add 200 μL 0.2% N-(4-dimethylamino-3,5-dinitrophenyl)maleimide in acetone to the aqueous layer, mix, let stand at room temperature for 5 min, wash with 6 mL hexane, discard the hexane layer, acidify the aqueous layer with about 200 μL 2 M HCl, extract twice with 6 mL portions of benzene (Caution! Benzene is a carcinogen!). Combine the organic layers and add 10 μg IS, evaporate to dryness under reduced pressure, reconstitute the residue in 200 μL MeOH, inject a 5-20 μL aliquot. (Free captopril is derivatized as its 2,4'-dibromoacetophenone (p-bromophenacylbromide adduct) then oxidized captopril is reduced and derivatized as its N-(4-dimethylamino-3,5-dinitrophenyl)maleimide adduct.)

HPLC VARIABLES
Column: μBondapak C18
Mobile phase: MeCN:water:acetic acid 46.5:53:0.5
Injection volume: 5-20
Detector: UV 254

CHROMATOGRAM
Retention time: 7 (free captopril (as 2,4'-dibromoacetophenone (p-bromophenacyl bromide) adduct)), 8 (oxidized captopril (as N-(4-dimethylamino-3,5-dinitrophenyl)maleimide adduct))
Internal standard: thiosalicylic acid-p-bromophenacyl bromide adduct (Prepare by dissolving 2.4 mmoles thiosalicylic acid and 2.4 mmoles 2,4'-dibromoacetophenone (p-bromophenacyl bromide) in 40 mL MeOH, adjust to pH 7 by the dropwise addition of 1 M NaOH, allow to stand at room temperature for 10 min, evaporate to dryness under reduced pressure, reconstitute with 40 mL 50 mM pH 7.0 phosphate buffer, wash twice with 20 mL portions of hexane, adjust pH to 2 with dilute HCl, extract with 40 mL ethyl

acetate, evaporate to dryness under reduced pressure, recrystallize the residue from benzene to give the adduct as pale yellow plates.) (12.5)
Limit of quantitation: 100 ng/mL

REFERENCE

Kawahara, Y.; Hisaoka, M.; Yamazaki, Y.; Inage, A.; Morioka, T. Determination of captopril in blood and urine by high-performance liquid chromatography, *Chem.Pharm.Bull.*, **1981**, *29*, 150–157.

4-(N,N-Dimethylaminosulfonyl)-7-fluoro-2,1,3-benzoxadiazole

SAMPLE

Matrix: tissue
Analyte: thiols
Sample preparation: Homogenize tissue with 4 volumes of 5% trichloroacetic acid containing 5 mM disodium EDTA at 0°, centrifuge at 3500 rpm for 10 min. Remove a 10 μL aliquot of the supernatant and add it to 150 μL 27 mM DBD-F in MeCN, add 440 μL 100 mM pH 8.0 borax buffer containing 1 mM disodium EDTA, heat at 50° for 10 min, cool in ice, inject a 20 μL aliquot. (DBD-F can be purchased from Tokyo Kasei (TCI America, 911 North Harborgate St., Portland, OR 97203; 800-423-8616, 503-283-1681, (fax) 503-283-1987; www.tciamerica.com). Synthesis of DBD-F is as follows. Dissolve 0.5 g magnesium sulfate heptahydrate and 6 g NaOH in 60 mL water, throughout the reaction keep the flask at about 20° with cold water cooling, add 15 mL 30% hydrogen peroxide, add 75 mL MeOH, add 12.1 g powdered benzoyl peroxide in one go, stir for 10 min, pour into 150 mL 20% sulfuric acid, extract three times with 50 mL portions of chloroform, determine peroxybenzoic acid concentration by iodometric titration (Tetrahedron 1967, 23, 3327). Slowly add 110 mL 1 M peroxybenzoic acid in chloroform to 7 g 2,6-difluoroaniline dissolved in 100 mL chloroform, stir at room temperature, when reaction is complete (iodometric titration) wash with 2% sodium thiosulfate, wash with 5% sodium carbonate, wash with water, dry over anhydrous sodium sulfate, evaporate to dryness under reduced pressure, recrystallize 2,6-difluoronitrosobenzene from EtOH (mp 108.5-109.5). Stir 8.5 g 2,6-difluoronitrosobenzene in 85 mL DMSO at room temperature and add a solution of 3.91 g sodium azide in 85 mL DMSO dropwise, let stand for about 1 h, add to a large volume of water, extract with ether, dry the extracts over anhydrous sodium sulfate, evaporate to dryness under reduced pressure and distil to give 4-fluoro-2,1,3-benzoxadiazole as a colorless oil (bp 83°/12 mm Hg) (J.Chem.Soc.(C) 1970, 1433). Add 11 mL chlorosulfonic acid dropwise to 3 g 4-fluoro-2,1,3-benzoxadiazole in 10 mL chloroform at 0-10° (use a calcium chloride drying tube), stir at room temperature for 1 h, reflux for 2 h, cool, slowly pour into ice water, remove the organic layer, extract the aqueous layer with chloroform, combine the organic layers, wash, dry over anhydrous magnesium sulfate, evaporate under reduced pressure, take up the residue in 5 mL benzene (Caution! Benzene is a carcinogen!), chromatograph on a 150 × 30 column of silica gel (100-200 mesh Kanto Chemical) with n-hexane:benzene 50:50, evaporate the appropriate fractions to give 4-(chlorosulfonyl)-7-fluoro-2,1,3-benzoxadiazole (CBD-F) as pale yellow needles (mp 64-66°) (Anal. Chem. 1984. 56, 2461). Stir 0.76 g CBD-F in 70 mL MeCN at 0-10° and add 1 g dimethylamine hydrochloride in 10 mL 100 mM pH 10 borax dropwise, adjust pH to 5 with 1 M HCl, concentrate to about 10 mL under reduced pressure, extract three times with 200 mL portions of diethyl ether, wash with water, dry over anhydrous magnesium sulfate, evaporate under reduced pressure, chromatograph on a 500 × 20 column of silica

gel with chloroform, isolate the appropriate fraction and re-chromatograph on the same column with ethyl acetate:benzene 1:2 to give 4-(N,N-dimethylaminosulfonyl)-7-fluoro-2,1,3-benzoxadiazole (DBD-F) as white needles (mp 124-125°) (yield 1% !). On a Merck no. 5714 $60F_{254}$ tlc plate eluted with chloroform DBD-F has Rf 0.32 and lies between two other reaction products.)

HPLC VARIABLES
Column: 150 × 4.6 5 μm Inertsil ODS
Mobile phase: Gradient. MeCN:150 mM phosphoric acid from 20:80 to 50:50 over 30 min.
Column temperature: 40
Flow rate: 1
Injection volume: 20
Detector: F ex 390 em 520

CHROMATOGRAM
Retention time: 5 (cysteine), 6 (glutathione), 7 (homocysteine), 12.4 (N-acetylcysteine), 13.8 (α-mercaptopropionylglycine)
Limit of detection: 0.13-0.92 pmole

KEY WORDS
rat; liver; kidney; spleen; testes; lung; pancreas

REFERENCE
Toyo'oka, T.; Suzuki, T.; Saito, Y.; Uzu, S.; Imai, K. Fluorigenic reagent for thiols: 4-(*N,N*-Dimethylaminosulphonyl)-7-fluoro-2,1,3-benzoxadiazole, *Analyst*, **1989**, *114*, 413–419.

2,4-Dinitrofluorobenzene

SAMPLE
Matrix: blood, urine
Analyte: acetylcysteine
Sample preparation: Plasma. Condition CF50A Centriflo ultrafiltration cones (Amicon) by immersing in water for 1 h and centrifuging at 1000 g for 5 min. 3 mL Plasma + 1 mL water + 500 μL 5 mg/mL dithiothreitol, vortex for 10 s, incubate at 37° for 30 min, add 1 mL 2% sodium bicarbonate, add 350 μL 5% 2,4-dinitro-1-fluorobenzene in EtOH, vortex for 10 s, incubate at 60° for 30 min, place in ultrafiltration cone, centrifuge at 1000 g at 20° for 20 min. Remove 1 mL of the ultrafiltrate and add it to 5 mL water and 10 mL ether, shake mechanically in the dark at 250 cycles/min for 5 min, centrifuge at 1000 g at 10° for 5 min. Remove the organic layer and evaporate it to dryness under a stream of nitrogen, reconstitute the residue in 75 μL mobile phase. Urine. 100 μL urine + 1 mL water + 500 μL 5 mg/mL dithiothreitol, vortex for 10 s, incubate at 37° for 30 min, add 1 mL 2% sodium bicarbonate, add 350 μL 5% 2,4-dinitro-1-fluorobenzene in EtOH, vortex for 10 s, incubate at 60° for 30 min. Remove 1 mL and add it to 5 mL water and 10 mL ether, shake mechanically in the dark at 250 cycles/min for 5 min, centrifuge at 1000 g at 10° for 5 min. Remove the organic layer and evaporate it to dryness under a stream of nitrogen, reconstitute the residue in 1 mL mobile phase.

HPLC VARIABLES
Column: 250 × 4.6 5 μm Hypersil ODS
Mobile phase: MeOH:50 mM trisodium citrate and 1 mM EDTA adjusted to pH 7.0 with citric acid solution 30:70 (plasma) or 35:65 (urine)
Flow rate: 1
Injection volume: 50

Detector: UV 360

CHROMATOGRAM
Retention time: 13
Limit of detection: 50 ng/mL

KEY WORDS
plasma; ultrafiltrate

REFERENCE
Lewis, P.A.; Woodward, A.J.; Maddock, J. High-performance liquid chromatographic assay for N-acetyl-cysteine in plasma and urine, *J.Pharm.Sci.*, **1984**, *73*, 996–998.

7-Fluoro-2,1,3-benzoxadiazole-4-sulfonic Acid, Ammonium Salt

RELATED REFERENCES
Garg, U.C.; Zheng, Z.-J.; Folsom, A.R.; Moyer, Y.S.; Tsai, M.Y.; McGovern, P.; Eckfeldt, J.H. Short-term and long-term variability of plasma homocysteine measurement. *Clin.Chem.* **1997**, *43*, 141-145.
Gilfix, B.M.; Blank, D.W.; Rosenblatt, D.S. Novel reductant for determination of total plasma homocysteine. *Clin.Chem.* **1997**, *43*, 687-688.
Toyo'oka, T.; Imai, K. High-performance liquid chromatography and fluorometric detection of biologically important thiols, derivatized with ammonium 7-fluorobenzo-2-oxa-1,3-diazole-4-sulfonate (SBD-F). *J.Chromatogr.* **1983**, *282*, 495-500.
Toyo'oka, T.; Imai, K.; Kawahara, Y. Determination of total captopril in dog plasma by HPLC after prelabeling with ammonium 7-fluorobenzo-2-oxa-1,3-diazole-4-sulfonate (SBD-F). *J.Pharm.Biomed.Anal.* **1984**, *2*, 473-479.
Toyooka, T.; Furukawa, F.; Suzuki, T.; Saito, Y.; Takahashi, M.; Hayashi, Y.; Uzu, S.; Ima, K. Determination of thiols and disulfides in normal rat tissues and hamster pancreas treated with N-nitroso-bis(2-oxopropyl)amine using 4-(aminosulfonyl)-7-fluoro-2,1,3-benzoxadiazole and ammonium 7-fluoro-2,1,3-benzoxadiazole-4-sulfonate. *Biomed.Chromatogr.* **1989**, *3*, 166-172.
Vilaseca, M.A.; Moyano, D.; Ferrer, I.; Artuch, R. Total homocysteine in pediatric patients. *Clin.Chem.* **1997**, *43*, 690-692

SAMPLE
Matrix: blood
Analyte: thiols
Sample preparation: 100 μL Plasma + 50 μL 12.5 μM IS + 15 μL freshly-prepared 10% tri-n-butylphosphine, mix, let stand at +4° for 30 min, add 150 μL trichloroacetic acid, mix, let stand at room temperature for 10 min, centrifuge at 4000 g for 10 min. Remove a 50 μL aliquot of the supernatant and add it to a mixture of 125 μL 125 mM pH 9.5 borate buffer, 15 μL 1.25 mM NaOH, and 50 μL freshly-prepared 1 mg/mL 7-fluoro-2,1,3-benzoxadiazole-4-sulfonic acid, ammonium salt (7-fluorobenzo-2-oxa-1,3-diazole-4-sulfonic acid, ammonium salt; SBD-F; Fluka), mix, heat at 60° for 1 h, cool to +4°, inject a 20 μL aliquot.

HPLC VARIABLES
Guard column: 5 μm Lichrospher C18
Column: 150 × 4.6 3 μm Supelco LC-18 base deactivated
Mobile phase: MeCN:100 mM KH_2PO_4, adjusted to pH 2.15 with orthophosphoric acid
Flow rate: 0.5

Injection volume: 20
Detector: F ex 385 em 515

CHROMATOGRAM
Retention time: 4.8 (cysteine), 8.0 (Cys-Gly), 9.0 (homocysteine)
Internal standard: 2-mercaptoethylamine (2-aminoethanethiol) (6.0)

KEY WORDS
plasma

REFERENCE
Daskalakis, I.; Lucock, M.D.; Anderson, A.; Wild, J.; Schorah, C.J.; Levene, M.I. Determination of plasma total homocysteine and cysteine using HPLC with fluorescence detection and an ammonium 7-fluoro-2,1,3-benzoxadiazole-4-sulphonate (SBD-F) derivatization protocol optimized for antioxidant concentration, derivatization reagent concentration, temperature and matrix pH, *Biomed.Chromatogr.*, **1996**, *10*, 205–212.

SAMPLE
Matrix: blood
Analyte: homocysteine
Sample preparation: 120 µL Plasma + 30 µL 50 µM cysteamine + 15 µL 100 mL/L tri-n-butylphosphine in DMF, mix, let stand at 4° for 30 min, add 150 µL 100 g/L trichloroacetic acid, mix, centrifuge. Remove a 50 µL aliquot of the supernatant and add it to 10 µL 1.55 M NaOH, add 125 µL 4 mM EDTA in 125 mM pH 9.5 borate buffer, add 50 µL 1 g/L 7-fluoro-2,1,3-benzoxadiazole-4-sulfonic acid, ammonium salt (Fluka), heat at 60° for 1 h, inject a 20 µL aliquot of the supernatant.

HPLC VARIABLES
Column: 4.6 mm ID 5 µm Spherisorb ODS
Mobile phase: MeCN:100 mM pH 2.0 KH_2PO_4 4:96
Flow rate: 0.8
Injection volume: 20
Detector: F (wavelengths not given)

CHROMATOGRAM
Retention time: 9
Internal standard: cysteamine (5)
Limit of quantitation: 2 µM

OTHER SUBSTANCES
Extracted: cysteine, cysteinyl glycine, glutathione

KEY WORDS
plasma

REFERENCE
Kuo, K.; Still, R.; Cale, S.; McDowell, I. Standardization (external and internal) of HPLC assay for plasma homocysteine, *Clin.Chem.*, **1997**, *43*, 1653–1655.

SAMPLE
Matrix: solutions
Analyte: thiols
Sample preparation: Mix 1 mL of a 150 µM solution in 100 mM pH 9.5 sodium borate buffer containing 2 mM disodium EDTA with 1 mL 1 mM 7-fluoro-2,1,3-benzoxadiazole-4-sulfonic acid, ammonium salt (ammonium 7-fluoro-2,1,3-benzoxadiazole-4-sulfonate) in 100 mM H 9.5 sodium borate buffer, vortex, heat at 60° for 1 h, cool, add 100 µL 2 M HCl, inject an aliquot.

CAPILLARY ELECTROPHORESIS
Capillary: 57 cm × 50 µm fused-silica (Polymicro Technologies)
Capillary preparation: Before each run rinse with 100 mM NaOH, water, and running buffer for 10 min.
Running buffer: 20 mM pH 9.8 Sodium borate buffer
Capillary temperature: 25
Voltage/Current: 25 kV

Injection: Hydrodynamic injection at 0.5 psi for 5 s.
Detector: UV 220
Model: Beckman P/ACE 5000
Migration time: 5 (mercaptoethanol), 6.5 (dithioerythritol), 8 (homocysteine), 8.5 (glutathione), 9 (dithiothreitol), 9.5 (cysteine)

KEY WORDS
comparison with 4-(aminosulfonyl)-7-fluoro-2,1,3-benzoxadiazole

REFERENCE
Kang, S.H.; Kim, J.-W.; Chung, D.S. Determination of homocysteine and other thiols in human plasma by capillary electrophoresis, *J.Pharm.Biomed.Anal.*, **1997**, *15*, 1435–1441.

4-Fluoro-7-nitro-2,1,3-benzoxadiazole

SAMPLE
Matrix: solutions
Analyte: thiols

HPLC VARIABLES
Column: 150 × 4.6 7 μm Kyowa Gel 62210F 10% cross-linked sulfonated polystyrene resin strong cation exchange (Kyowa Seimitsu)
Mobile phase: 200 mM pH 4.25 sodium citrate buffer
Column temperature: 50
Flow rate: 0.3
Injection volume: 15
Detector: F ex 450 em 520 following post-column reaction. The column effluent mixed with 0.05% 4-fluoro-7-nitro-2,1,3-benzoxadiazole in EtOH pumped at 0.13 mL/min and this mixture flowed through a 2 m × 0.25 mm ID PTFE coil at 60° and a 1 m × 0.25 mm ID PTFE coil at 5-10°. The effluent from this coil mixed with 1.5 M HCl in MeOH:water 50:50 pumped at 0.8 mL/min and this mixture flowed to the detector.

CHROMATOGRAM
Retention time: 6.1 (glutathione), 11.3 (cysteine), 13.9 (homocysteine)
Limit of detection: 10-150 pmole

KEY WORDS
post-column reaction

REFERENCE
Watanabe, Y.; Imai, K. Liquid chromatographic determination of amino and imino acids and thiols by postcolumn derivatization with 4-fluoro-7-nitrobenzo-2,1,3-oxadiazole, *Anal.Chem.*, **1983**, *55*, 1786–1791.

Iodoacetic Acid

SAMPLE
Matrix: blood
Analyte: glutathione
Sample preparation: Mix 3 mL blood with 30 mg powdered 1,10-phenanthroline, centrifuge a 500 μL aliquot at 0-5° at 7000 g for 10 min, discard the supernatant. Add 400 μL 20 mM 1,10-phenanthroline to the remaining erythrocytes, add 300 μL 10% metaphosphoric acid, vortex, centrifuge at 15000 g for 15 min. Remove a 400 μL aliquot of the supernatant and add it to 100 μL 400 mM iodoacetic acid in water (freshly prepared), neutralize with 20 mg sodium bicarbonate, shake in the dark at room temperature for 1 h, add 500 μL 5% 2,4-dinitrofluorobenzene in EtOH, let stand in the dark at room temperature for 20 h, centrifuge at 15000 g for 15 min, inject a 20 μL aliquot.

HPLC VARIABLES
Guard column: 4 × 4 5 μm LiChrospher 100 NH2
Column: 250 × 4 5 μm LiChrospher 100 NH2
Mobile phase: Gradient. A was MeOH:water 80:20. B was MeOH:2 M pH 4.6 sodium acetate 64:36. A:B from 75:25 to 5:95 over 30 min, maintain at 5:95 for 15 min, re-equilibrate at initial conditions for 15 min.
Column temperature: 40
Flow rate: 1.2
Injection volume: 20
Detector: UV 365

CHROMATOGRAM
Retention time: 31.7 (glutathione), 40.6 (oxidized glutathione)

KEY WORDS
erythrocytes

REFERENCE
Yoshida, T. Determination of reduced and oxidized glutathione in erythrocytes by high-performance liquid chromatography with ultraviolet absorbance detection, *J.Chromatogr.B*, **1996**, *678*, 157–164.

SAMPLE
Matrix: cells
Analyte: thiols
Sample preparation: Suspend hepatocytes in 1 mL 0.9% saline, add 50 μL 70% perchloric acid, centrifuge. Remove a 500 μL aliquot of the supernatant and add it to 50 μL 80 mM iodoacetic acid, neutralize with solid sodium bicarbonate, let stand in the dark at room temperature for 1 h, add 500 μL 1.5% 1-fluoro-2,4-dinitrobenzene in EtOH, let stand in the dark for 4 h, inject an aliquot.

HPLC VARIABLES
Column: 300 × 4 Aminopropyl Spherisorb (Prepare packing as follows. Dry 10 g Spherisorb S5-W under vacuum at 125° for 3 h, suspend in 150 mL toluene containing 80 mmole 3-aminopropyltriethoxysilane, reflux for 24 h, cool, wash the solid with five 200 mL portions of toluene, with five 200 mL portions of diethyl ether, and five 200 mL portions of MeOH, dry at under vacuum at 125° for 1.5 h, store in a desiccator.)
Mobile phase: Gradient. A was MeOH:water 80:20. B was MeOH:water:buffer 64:16:20. (Buffer was 272 g sodium acetate trihydrate, 122 mL water, and 378 mL glacial acetic acid.) A:B 75:25 for 10 min, to 5:95 over 20 min.
Detector: UV 365

CHROMATOGRAM
Retention time: 2.21 (cysteinesulfinic acid, N-derivative), 2.28 (cysteine, N,S-bis-derivative), 4.43 (glutathione, S-derivative), 6.68 (homocysteine, N,N-bis-derivative), 7.51

(glutamic acid, N-derivative), 7.79 (cystinyl-bis-glycine, N,N-bis-derivative), 7.84 (cysta-thionine, N,N-bis-derivative), 8.20 (glutathione, N,S-bis-derivative), 8.31 (cystine, N,N-bis-derivative), 9.33 (glutathione, N-derivative), 12.13 (homocysteic acid, N-derivative), 13.18 (homocysteine, N-derivative), 13.21 (cysteinylglycine, N-derivative), 13.28 (carbox-ymethylpenicillamine, N-derivative), 13.49 (glutamyl-S-carboxymethyl-cysteine, N-deriv-ative), 14.74 (aspartic acid, N-derivative), 15.57 (cysteic acid, N-derivative), 17.13 (S-car-boxymethyl-L-cysteine,N-derivative),   21.46   (homocysteine-glutathione   disulfide, N,N-bis-derivative), 22.95 (cysteine-glutathione disulfide, N,N-bis-derivative), 25.95 (S-carboxymethylglutathione, N-derivative), 30.15 (glutathione disulfide, N,N-bis-derivative)

KEY WORDS
rat; hepatocytes

REFERENCE
Reed, D.J.; Babson, J.R.; Beatty, P.W.; Brodie, A.E.; Ellis, W.W.; Potter, D.W. High-performance liquid chromatography analysis of nonomole levels of glutathione, glutathione disulfide, and related thiols and disulfides, *Anal.Biochem.*, **1980**, *106*, 55–62.

SAMPLE
Matrix: tissue
Analyte: glutathione
Sample preparation: Homogenize 1 g rat liver with 10 mL 10% perchloric acid containing 1 mM bathophenanthrolinedisulfonic acid, centrifuge at 4° at 15000 g for 3 min. Remove a 500 µL aliquot of the supernatant and add it to 50 µL 15 mM gamma-glutamylgluta-mate in 0.3% perchloric acid, add 50 µL 100 mM iodoacetic acid in 200 µM m-cresol purple solution, adjust pH to 8.0 with 2 M KOH containing 2.4 M potassium bicarbonate, let stand in the dark at room temperature for 10 min, add 1 mL 1% 2,4-dinitrofluorobenzene in EtOH, mix, inject a 100 µL aliquot.

HPLC VARIABLES
Guard column: 20 × 4.6 Spherisorb NH2
Column: 250 × 4.6 5 µm Spherisorb NH2
Mobile phase: Gradient. A was MeOH:water 80:20. B was MeOH:water 64:36 containing 500 mM sodium acetate. A:B 80:20 for 5 min, to 1:99 over 10 min, maintain at 1:99.
Flow rate: 1
Injection volume: 100
Detector: UV 365

CHROMATOGRAM
Retention time: 16
Internal standard: gamma-glutamylglutamate (15)

KEY WORDS
rat; liver

REFERENCE
Santori, G.; Domenicotti, C.; Bellocchio, A.; Pronzato, M.A.; Marinari, U.M.; Cottalasso, D. Different efficacy of iodoacetic acid and N-ethylmaleimide in high-performance liquid chromatographic mea-surement of liver glutathione, *J.Chromatogr.B*, **1997**, *695*, 427–433.

Iodomethane

$$CH_3I + RSH \longrightarrow CH_3SR$$

SAMPLE
Matrix: peptides
Analyte: cysteine
Sample preparation: Heat 0.5-5 mg peptide with 1 mL 1% indolylpropionic acid in 6 M HCl under vacuum in a sealed tube at 115° for 4 h, evaporate to dryness, reconstitute with 10-100 mL water, remove a 1 mL aliquot, add 100 µL 20 mM dithioerythritol, add 100 µL 400 mM iodomethane in MeOH:water 50:50, add 200 µL 3 M NaOH, let stand

for 10 min, add 200 μL 3 M HCl, mix briefly, add 400 μL 400 mM pH 10 sodium borate buffer, mix, check that pH is about 10. Remove a 100 μL aliquot, add 400 μL reagent, mix, let stand for 10 min, inject a 25 μL aliquot. (Iodomethane is used to protect cysteine as its S-methyl derivative. Prepare reagent by dissolving 30 mg o-phthaldialdehyde in 1 mL EtOH, add 22 mL 400 mM pH 10 sodium borate buffer, add 30 mg Boc-L-cysteine. Prepare Boc-L-cysteine by adding Boc-S-benzyl-L-cysteine to liquid ammonia, add metallic sodium until a blue color persists for 15 min, add ammonium chloride until all the excess sodium is destroyed, allow the ammonia to evaporate, add ice and water, extract with ethyl acetate.)

HPLC VARIABLES
Column: 120 × 4 3 μm Hypersil ODS
Mobile phase: Gradient. A was 50 mM pH 7.0 phosphate buffer. B was MeOH:THF:50 mM pH 7.0 phosphate buffer 65:1:35. A:B from 70:30 to 0:100 over 48 min.
Flow rate: 1.4
Injection volume: 25
Detector: F ex 344 em 443

CHROMATOGRAM
Retention time: 20.5 (L-Thr), 22.5 (D-Thr), 28.5 (S-methyl-L-Cys), 31 (L-threoninol), 31.5 (S-methyl-D-Cys), 32 (D-threoninol), 34 (L-Trp), 36.5 (L-Phe), 37 (D-Trp), 38 (D-Phe), 47 (L-Lys), 48 (D-Lys)

KEY WORDS
chiral; comparison with the results obtained with other thiols in the derivatization reagent.

REFERENCE
Buck, R.H.; Krummen, K. High-performance liquid chromatographic determination of enantiomeric amino acids and amino alcohols after derivatization with o-phthaldialdehyde and various chiral mercaptans. Application to peptide hydrolysates, *J.Chromatogr.*, **1987**, *387*, 255–265.

Monobromobimane

RELATED REFERENCES

Baeyens, W.; Van Der Weken, G.; Moerloose, P.D. Effects of reducing agents on the determination of thiolic compounds in the presence of their disulfides using bimane precolumn derivatization. *Chromatographia* **1987**, *23*, 717-721.

Baeyens, W.; Van Der Weken, G.; Ling, B.L.; Moerloose, P.D. HPLC determination of N-acetylcysteine in pharmaceutical preparations after precolumn derivatization with Thiolyte MB using fluorescence detection. *Anal.Lett.* **1988**, *21*, 741-757.

Baeyens, W.R.G.; Van Der Weken, G.; De Moerloose, P. Reversed-phase high-performance liquid chromatography of thiol-bimane derivatives. *Anal.Chim.Acta* **1988**, *205*, 43-51.

Fahey, R.C.; Newton, G.L. Determination of low-molecular-weight thiols using monobromobimane fluorescent labeling and high-performance liquid chromatography. *Methods Enzymol.* **1987**, *143*, 85-96.

Kok, R.J.; Visser, J.; Moolenaar, F.; De Zeeuw, D.; Meijer, D.K.F. Bioanalysis of captopril: two sensitive high-performance liquid chromatographic methods with pre- or postcolumn fluorescent labeling. *J.Chromatogr.B* **1997**, *693*, 181-189.

Livesey, J.F.; Donnelly, J.G.; Ooi, D.S. HPLC screening method for cystinuria. *Clin.Chem.* **1996**, *42*, 1714-1715.

O'Keefe, D.O. Process clearance of dithiothreitol monitored by reversed-phase high-performance liquid chromatography with fluorescence detection. Assay development and validation. *J.Chromatogr.A* **1997**, *775*, 151-156

SAMPLE
Matrix: DNA
Analyte: 2'-deoxy-6-thioguanosine
Sample preparation: Heat 5-25 µg DNA in 50 µL 10 mM pH 8.0 Tris-HCl containing 1 mM EDTA in boiling water for 5 min, cool in ice, add 162 µL 20 mM pH 5.2 sodium acetate containing 100 µM zinc chloride, add 1 µL 1 mg/mL P1 nuclease (Sigma), heat at 52° for 1 h, cool to room temperature, add 25 µL 1 M pH 8.0 Tris-HCl, add 1 µL 20 mM monobromobimane (Molecular Probes, Eugene OR) in MeCN, mix, let stand at room temperature for 10 min, add 1 µL 1000 U/mL calf intestinal alkaline phosphatase (Sigma), mix, heat at 37° for 30 min, add 10 µL 2 M acetic acid, inject a 50 µL aliquot.

HPLC VARIABLES
Guard column: 20 mm long Supelguard
Column: 150 × 4.6 3 µm Supelcosil LC-18
Mobile phase: Acetone:buffer 11:89 (Between runs rinse column with acetone:buffer 41:59 for 3 min. Prepare buffer by adjusting the pH of 200 mM formic acid to 4.0 with 10 mM NaOH.)
Column temperature: 40
Flow rate: 1
Injection volume: 50
Detector: F ex 377 em 478

CHROMATOGRAM
Retention time: 16
Limit of quantitation: 560 nmoles/g DNA

REFERENCE
Warren, D.J.; Slordal, L. A high-performance liquid chromatographic method for the determination of 6-thioguanine residues in DNA using precolumn derivatization and fluorescence detection, *Anal.Biochem.*, **1993**, *215*, 278–283.

SAMPLE
Matrix: blood
Analyte: thiols
Sample preparation: Vortex 750 µL red blood cells with 3.2 mL 100 mM HCl, centrifuge at 4° at 15000 rpm for 5 min. Remove the supernatant and mix it with an equal volume of cold 4 M sodium methanesulfonate, freeze in dry ice/isopropanol, thaw, centrifuge at 20000 rpm for 10 min. Determine the thiol content of the supernatant by titrating with 5,5'-dithiobis(2-nitrobenzoic acid). Add N-ethylmorpholine to a final concentration of 10 mM, adjust pH to 8.0 with 1 M NaOH, add a 1 molar equivalent of dithiothreitol, mix, let stand at room temperature for 5 min, add 6 equivalents of monobromobimane (Calbiochem; Molecular Probes), mix, let stand for 15 min, add 12 equivalents of thiol agarose, let stand for 20 min, add acetic acid to a final concentration of 3% (Anal. Biochem.1981, 111, 357), dilute with 200 mM pH 2.2 sodium citrate, inject an aliquot. (Prepare 4 M sodium methanesulfonate by adjusting the pH of methanesulfonic acid to 1.5 with 50% NaOH then diluting to 4 M.)

HPLC VARIABLES
Column: 150 × 4.6 5 µm Ultrasphere-ODS C18
Mobile phase: Gradient. A was MeOH:water:acetic acid 10:89.75:0.25, adjusted to pH 3.9 with 50% NaOH. B was MeOH:water:acetic acid 90:9.75:0.25, adjusted to pH 3.9 with 50% NaOH. A:B 92:8 for 10 min, to 60:40 over 10 min, maintain at 60:40 for 5 min, to 10:90 over 5 min, to 0:100 over 2 min.
Flow rate: 1.5
Detector: F (o-phthalaldehyde filters)

CHROMATOGRAM
Retention time: 3.7 (cysteine), 4.3 (thiosulfate), 4.5 (gamma-glutamylcysteine), 5.2 (cysteinylglycine), 6.1 (glutathione), 6.7 (coenzyme M), 7.2 (homocysteine), 9.0 (ergothioneine), 11.8 (N-acetylcysteine), 12.4 (cysteamine), 17.7 (4'-phosphopantetheine), 19.5 (2-mercaptoethanol), 19.8 (2-thiouracil), 20.9 (coenzyme A), 22.2 (pantetheine), 24.3 (methanethiol), 24.6 (hydrogen sulfide), 25.1 (mercaptopyrimidine), 29.2 (ethanethiol)
Limit of detection: 2-20 pmole

KEY WORDS
red blood cells

REFERENCE
Newton, G.L.; Dorian, R.; Fahey, R.C. Analysis of biological thiols: derivatization with monobromobimane and separation by reverse-phase high-performance liquid chromatography, *Anal.Biochem.*, **1981**, *114*, 383–387.

SAMPLE
Matrix: blood
Analyte: acetylcysteine
Sample preparation: 100 μL Plasma + 5 μL 100 mM dithiothreitol in 10% Triton X-100, vortex, allow to stand for 30 min, add 100 μL 30 mM monobromobimane (Calbiochem; Molecular Probes) in 50 mM pH 8.0 N-ethylmorpholine, store in the dark for 5 min, add 10 μL 100% (w/v) trichloroacetic acid, centrifuge at 3000 g for 3 min, inject a 25 μL aliquot of the supernatant. (Dissolve monobromobimane in the minimum amount of MeCN before making up aqueous solutions (J. Biochem. Biophys. Methods 1986, 13, 231).)

HPLC VARIABLES
Column: 75 × 4.5 3 μm Supelco octadecylsilica
Mobile phase: Gradient. A was MeCN:acetic acid:perchloric acid:water 9:0.25:0.25:90.75, pH 3.7. B was MeCN:water:perchloric acid 75:25:0.25. A:B 100:0 for 7 min then 0:100 for 4 min then re-equilibrate at 100:0 for 7 min.
Flow rate: 1
Injection volume: 25
Detector: F ex 394 em 480

CHROMATOGRAM
Retention time: 7.4
Limit of detection: 0.5 nM

KEY WORDS
plasma

REFERENCE
Cotgreave, I.A.; Moldéus, P. Methodologies for the analysis of reduced and oxidized N-acetylcysteine in biological systems, *Biopharm.Drug Dispos.*, **1987**, *8*, 365–375.

SAMPLE
Matrix: blood
Analyte: thiols
Sample preparation: 1 Volume plasma + 0.4 volume 10 mM monobromobimane (Calbiochem; Molecular Probes) + 0.01 volume 1 M acetic acid + 2.5 volume MeCN, mix, centrifuge at 4° at 1000 g for 10 min, filter (5 μm) the supernatant, inject a 100 μL aliquot.

HPLC VARIABLES
Column: μBondapak C18 radial compression
Mobile phase: MeOH:water:glacial acetic acid 22:77.75:0.25, pH adjusted to 3.9 with NaOH. (After each injection wash column with MeOH at 4 mL/min for 10 min.)
Flow rate: 1.5
Injection volume: 100
Detector: F ex 365 em 418

CHROMATOGRAM
Retention time: 4.60 (cysteine), 5.50 (glutathione), 6.80 (homocysteine)
Internal standard: d-penicillamine (13.60)
Limit of detection: 10 nM

KEY WORDS
plasma

REFERENCE
Velury, S.; Howell, S.B. Measurement of plasma thiols after derivatization with monobromobimane, *J.Chromatogr.*, **1988**, *424*, 141–146.

SAMPLE
Matrix: blood
Analyte: glutathione
Sample preparation: 30 μL Plasma + 30 μL 2 M sodium borohydride + 30 μL 50 μM dithioerythritol in DMSO containing 5% sulfosalicylic acid + 130 μL physiological salt solution containing 140 mM HBr and 44% DMSO + 50 μL 1 M pH 9.0 N-ethylmorpholine + 10 μL 100 mM monobromobimane (Calbiochem; Molecular Probes) in MeCN, mix, let stand in the dark at room temperature for 20 min, add 20 μL 8.15 M perchloric acid, mix, let stand at 4° for 2 h, centrifuge at 10000 g for 2 min, inject a 25 μL aliquot. (This procedure determines total plasma glutathione; alternative procedures for determining free oxidized glutathione, total free glutathione, protein-bound glutathione, and reduced glutathione are described in the paper.)

HPLC VARIABLES
Column: 150 × 4.6 3 μm MOS-Hypersil C8
Mobile phase: Gradient. A was 0.25% pH 3.9 acetic acid in water. B was MeOH:water 20:80. C was MeOH:water 90:10. A:B:C 60:40:0 for 8 min, to 20:80:0 over 4 min, to 0:0:100 (step gradient), maintain at 0:0:100 for 5 min.
Flow rate: 1.5
Injection volume: 25
Detector: F ex 400 em 475

CHROMATOGRAM
Retention time: 12
Limit of detection: 2 pmole

KEY WORDS
plasma

REFERENCE
Svardal, A.M.; Mansoor, M.A.; Ueland, P.M. Determination of reduced, oxidized, and protein-bound glutathione in human plasma with precolumn derivatization with monobromobimane and liquid chromatography, *Anal.Biochem.*, **1990**, *184*, 338–346.

SAMPLE
Matrix: blood
Analyte: mesna
Sample preparation: 50 μL Plasma (within 3 min of collection) + penicillamine + 10 μL 25 mM monobromobimane (Calbiochem; Molecular Probes) in MeCN, let stand for 5 min at room temperature, add 20 μL 20% perchloric acid. (For total mesna add 100 μL 20 mM dithiothreitol in 200 mM pH 8.5 Tris/HCl buffer to 50 μL plasma, let stand for 40 min at room temperature, add 50 μL 15% sulfosalicylic acid, centrifuge. Remove 200 μL of the supernatant and wash it three times with ethyl acetate. Remove 60 μL of the aqueous phase and add it to 300 μL 200 mM pH 8.5 Tris/HCl buffer and 10 μL 15 mM monobromobimane in MeCN, let stand at room temperature for 5 min, add 20 μL 20% perchloric acid.)

HPLC VARIABLES
Column: 150 × 4.6 7 μm Nucleosil RP-18
Mobile phase: Gradient. A was MeCN. B was 1% aqueous acetic acid containing 1 g/L octanesulfonic acid. A:B from 5:95 to 8:92 over 2 min, to 10:90 over 13 min (Waters convex), to 30:70 over 20 min (Waters convex), maintain at 30:70 for 6 min, re-equilibrate at initial conditions for 9 min.
Flow rate: 1.4
Detector: F (wavelengths not given)

CHROMATOGRAM
Internal standard: penicillamine
Limit of detection: 10 μM

KEY WORDS
plasma; pharmacokinetics

REFERENCE
Stofer-Vogel, B.; Cerny, T.; Borner, M.; Lauterburg, B.H. Oral bioavailability of mesna tablets, *Cancer .Chemother.Pharmacol.*, **1993**, *32*, 78–81.

SAMPLE
Matrix: blood
Analyte: glutathione
Sample preparation: Mix plasma with 2 volumes of 2 mM monobromobimane (Calbiochem; Molecular Probes) in MeOH, store at 4° for 20-30 min, store at room temperature for 20 min, centrifuge twice at 13500 g for 20 min, inject an aliquot.

HPLC VARIABLES
Column: 150 × 4.6 5 μm Econosphere C18
Mobile phase: Gradient. A was MeOH:water 25:75 containing 30 mM tetrabutylammonium phosphate, pH 3.1-3.4. B was MeOH containing 30 mM tetrabutylammonium phosphate, pH 3.1-3.4. A:B 100:0 for 10 min, 90:10 for 5 min (step gradient). Re-equilibrate at initial conditions for 5 min. (A stock solution of tetrabutylammonium phosphate was prepared by adjusting the pH of 40% tetrabutylammonium hydroxide in water to pH 6.0 with 85% phosphoric acid.)
Flow rate: 1.5
Detector: F ex 260 em 474

CHROMATOGRAM
Retention time: 5.7
Limit of detection: 10 nM

KEY WORDS
plasma

REFERENCE
Yang, C.-S.; Chou, S.-T.; Liu, L.; Tsai, P.-J.; Kuo, J.-S. Effect of ageing on human plasma glutathione concentrations as determined by high-performance liquid chromatography with fluorimetric detection, *J.Chromatogr.B*, **1995**, *674*, 23–30.

SAMPLE
Matrix: blood, cell suspensions, tissue
Analyte: thiols
Sample preparation: Homogenize tissue with 6.5% trichloroacetic acid, centrifuge at 4° at 3000 g for 5 min, neutralize supernatant with sodium bicarbonate. Add a 100 μL aliquot of tissue homogenate, plasma, or cell suspension to 100 μL 8-30 mM monobromobimane (Calbiochem; Molecular Probes) in 50 mM pH 8.0 N-ethylmorpholine buffer, let stand in the dark at room temperature for 5 min, add 10 μL 100% trichloroacetic acid, centrifuge at 3000 g for 3 min, inject a 25 μL aliquot of the supernatant.

HPLC VARIABLES
Column: 75 × 4.5 3 μm octadecylsilica (Supelco)
Mobile phase: Gradient. A was MeCN:water:acetic acid 9:90.75:0.25, pH 3.7. B was MeCN:water 75:25. A:B 100:0 for 7 min, to 0:100 (step gradient), maintain at 0:100 for 4 min, return to initial conditions (step gradient), re-equilibrate at initial conditions for 7 min.
Flow rate: 1
Injection volume: 25
Detector: F ex 394 em 480

CHROMATOGRAM
Retention time: 2.3 (cysteine), 3.5 (glutathione)
Limit of detection: 250 nM

KEY WORDS
rat; hamster; rabbit; human; lung; liver; kidney

REFERENCE
Cotgreave, I.A.; Moldéus, P. Methodologies for the application of monobromobimane to the simultaneous analysis of soluble and protein thiol components of biological systems, *J.Biochem.Biophys.Methods*, **1986**, *13*, 231–249.

SAMPLE
Matrix: blood, tissue
Analyte: thiols
Sample preparation: Plasma. 400 μL Plasma + 20 μL 69% trichloroacetic acid, cool on ice, centrifuge at 0° at 10000 g for 15 min. Remove a 100 μL aliquot and add it to 6 μL 0.5 mM dithiothreitol in 1 mM EDTA, add 190 μL buffer, add 10 μL 10 mM monobromobimane (Behring Diagnostics, LaJolla CA; Calbiochem; Molecular Probes) in MeCN, let stand in the dark at room temperature for 30 min, add 50 μL glacial acetic acid, inject a 20-80 μL aliquot. Liver. Homogenize (Ten Broeck) 0.5 g liver with 4.5 mL 10% trichloroacetic acid at 0°, centrifuge at 0° at 10000 g for 15 min, dilute with 2 volumes of water. Remove a 100 μL aliquot and add it to 6 μL 0.5 mM dithiothreitol in 1 mM EDTA, add 190 μL buffer, add 10 μL 10 mM monobromobimane (Behring Diagnostics, LaJolla CA; Molecular Probes) in MeCN, let stand in the dark at room temperature for 30 min, add 50 μL glacial acetic acid, inject a 20-80 μL aliquot. (Buffer was 400 mM potassium borate adjusted to pH 10.5 with NaOH.)

HPLC VARIABLES
Column: 150 × 3.9 4 μm Novapack C18
Mobile phase: Gradient. A was MeOH:0.25% acetic acid 10:90, adjusted to pH 4 with 50% NaOH. B was MeOH:0.25% acetic acid 90:10, adjusted to pH 4 with 1 M NaOH. A:B 92:8 for 10 min, to 0:100 (step gradient), maintain at 0:100 for 10 min, re-equilibrate at initial conditions for 10 min.
Column temperature: 28
Flow rate: 0.65
Injection volume: 20-80
Detector: F ex 395 em 455

CHROMATOGRAM
Retention time: 3.79 (cysteine), 5.32 (glutathione), 6.21 (homocysteine)

KEY WORDS
plasma; rat; liver; protect from light

REFERENCE
Hum, S.; Robitaille, L.; Hoffer, L.J. Plasma glutathione turnover in the rat: effect of fasting and buthionine sulfoximine, *Can.J.Physiol.Pharmacol.*, **1991**, *69*, 581–587.

SAMPLE
Matrix: blood, urine
Analyte: thiols
Sample preparation: 30 μL Plasma, serum, or urine + 30 μL 4 M sodium borohydride in 66 mM NaOH containing 333 mL/L DMSO + 10 μL 2 mM EDTA in 1.65 mM dithioerythritol + 10 μL 1-octanol + 20 μL 1.8 M HCl, mix, let stand for 3 min, add 100 μL 1.5 M N-ethylmorpholine buffer, add 400 μL water, add 20 μL 25 mM monobromobimane (Molecular Probes, Eugene OR), mix, let stand for 3 min, add 40 μL glacial acetic acid, mix, inject a 20 μL aliquot.

HPLC VARIABLES
Guard column: 25 × 4.6 Pelliguard LC 18
Column: 150 × 4.6 3 μm Hypersil ODS
Mobile phase: Gradient. MeCN:buffer from 0:100 to 10.5:89.5 over 11 min. (Buffer was 40 mM ammonium formate containing 30 mM ammonium nitrate, pH 3.65 (plasma) or pH 3.50 (urine).)
Flow rate: 2
Injection volume: 20
Detector: F ex 365 em 475

CHROMATOGRAM
Retention time: 8 (cysteine), 9.5 (cysteinylglycine), 10.5 (homocysteine)
Limit of detection: 0.05 pmole

KEY WORDS
plasma; serum

REFERENCE

Fiskerstrand, T.; Refsum, H.; Kvalheim, G.; Ueland, P.M. Homocysteine and other thiols in plasma and urine: automated determination and sample stability, *Clin.Chem.*, **1993**, *39*, 263–271.

SAMPLE

Matrix: blood, urine

Analyte: sodium mercaptoundecahydrodecaborate

Sample preparation: Plasma. 150 μL Plasma + 600 μL MeCN, vortex, centrifuge at 1650 g for 5 min. Remove a 600 μL aliquot of the supernatant and evaporate it to dryness under a stream of nitrogen, reconstitute the residue in 100 μL buffer, add 20 μL 5 mg/mL monobromobimane (Molecular Probes, Eugene OR) in MeCN, vortex for 30 s, let stand in the dark for 4 h, inject a 50 μL aliquot. Urine. Dilute urine with 2 volumes of buffer, filter (0.22 μm). Remove a 500 μL aliquot of the filtrate and add it to 50 μL 5 mg/mL monobromobimane (Molecular Probes, Eugene OR) in MeCN, vortex, let stand in the dark for 4 h, inject a 50 μL aliquot. (Prepare buffer by dissolving 12.1 g Tris in 50 mL water, adjusting the pH to 8.8 with 5 M HCl, and making up to 100mL with water.)

HPLC VARIABLES

Guard column: 10 × 4.6 5 μm Hypersil ODS

Column: 150 × 4.6 5 μm Hypersil ODS

Mobile phase: MeOH:20 mM pH 3.0 potassium phosphate buffer containing 50 mM tetramethylammonium chloride 20:80 (Run time 45 min to allow for late-eluting components.)

Flow rate: 2

Injection volume: 50

Detector: UV 373; F ex 315 em 370-390

CHROMATOGRAM

Retention time: 8.5

Limit of quantitation: 1 μg/mL (urine; UV), 200 ng/mL (plasma; UV)

KEY WORDS

rat; plasma

REFERENCE

Abu-Izza, K.; Lu, D.R. Liquid chromatographic determination of sodium mercaptoundecahydrododecaborate in rat urine and plasma after precolumn derivatization, *J.Chromatogr.B*, **1994**, *660*, 347–352.

SAMPLE

Matrix: cytosol

Analyte: glutathione, gamma-glutamylcysteine

Sample preparation: 200 μL Hepatic cytosol + 100 μL 5% sulfosalicylic acid, mix, centrifuge for 3 min. Remove a 100 μL aliquot of the supernatant and add it to 20 μL 50 mM pH 8.5 N-ethylmorpholine and 10 μL 50 mM monobromobimane in MeCN, let stand in the dark at room temperature for 15 min, add 10 M HCl, dilute with water, inject a 20 μL aliquot.

HPLC VARIABLES

Guard column: 5 × 4.6 Supelguard pellicular reversed phase (Supelco)

Column: 250 × 4.6 5 μm Ultrasphere ODS

Mobile phase: Gradient. A was 0.25% acetic acid in water, pH 3.9. B was MeOH:0.25% acetic acid in water 30:70, pH 3.9. A:B 65:35 for 1 min, to 20:80 over 7 min, to 0:100 (step gradient, maintain at 0:100 for 5 min, re-equilibrate at initial conditions for 5 min

Flow rate: 1.5

Injection volume: 20

Detector: F ex 303-395 em 420-470

CHROMATOGRAM

Retention time: 9 (gamma-glutamylcysteine), 10 (glutathione)

Limit of detection: 6.25 pmole

KEY WORDS

comparison with o-phthalaldehyde derivatization

REFERENCE

Yan, C.C.; Huxtable, R.J. Fluorimetric determination of monobromobimane and o-phthalaldehyde adducts of gamma-glutamylcysteine and glutathione: application to assay of gamma-glutamylcysteinyl synthetase activity and glutathione concentration in liver, *J.Chromatogr.B*, **1995**, *672*, 217–224.

SAMPLE

Matrix: solutions
Analyte: thiols
Sample preparation: 100-200 μL Solution + 1.75-1.85 mL buffer + 50 μL reagent (final volume 2 mL), purge head space with nitrogen, shake at room temperature in the dark for 5 min, add 2 mL dichloromethane, shake for 10 s, centrifuge for 2 min. Remove the aqueous phase and adjust the pH to 7 with 15 μL 6 M HCl, inject a 20 μL aliquot. (Buffer was 100 mM pH 8.0 ammonium bicarbonate purged with nitrogen for at least 1 h before use. Reagent was 40 mM bromobimane (Calbiochem; Molecular Probes) in MeCN.)

HPLC VARIABLES

Guard column: 45 × 4.6 10 μm Ultrasphere IP C18
Column: 250 × 4.6 5 μm Ultrasphere IP C18
Mobile phase: Gradient. A was 20 mM tetrabutylammonium bromide in MeOH. B was 20 mM tetrabutylammonium bromide in water. A:B 55:45 for 11 min, to 75:25 in 1 min, maintain at 75:25 for 7 min, return to initial conditions over 1 min, re-equilibrate for 15 min.
Flow rate: 1
Injection volume: 20
Detector: F ex 356 em 450)

CHROMATOGRAM

Retention time: k' 0.58 (mercaptosuccinic acid), k' 0.78 (sodium sulfite), k' 1.19 (2-mercaptoethanesulfonic acid), k' 2.55 (succimer), k' 3.05 (2,3-dimercaptopropane-1-sulfonic acid), k' 4.93 (N-(2,3-dimercaptopropyl)phthalamidic acid), k' 5.49 (4-sec-butyl-5-ethyl-2-thiobarbituric acid)

KEY WORDS

protect from light

REFERENCE

Maiorino, R.M.; Weber, G.L.; Aposhian, H.V. Fluorometric determination of 2,3-dimercaptopropane-1-sulfonic acid and other dithiols by precolumn derivatization with bromobimane and column liquid chromatography, *J.Chromatogr.*, **1986**, *374*, 297–310.

SAMPLE

Matrix: solutions
Analyte: glutathione and esters
Sample preparation: Mix 2 μL of a 100 mM solution with 194 μL 10 mM pH 7.6 ammonium bicarbonate containing 1 mM EDTA, add 4 μL 100 mM monobromobimane (Calbiochem) in MeCN, mix, let stand for 1 min, inject a 10-20 μL aliquot.

HPLC VARIABLES

Column: 250 × 4.6 5 μm ODS2 fully capped C18 (Phase Separations)
Mobile phase: Gradient. A was MeOH:water:acetic acid 10:89.75:0.25, adjusted to pH 3.9 with 50% NaOH. B was MeOH:water:acetic acid 90:9.75:0.25, adjusted to pH 3.9 with 50% NaOH. A:B 92:8 for 10 min, to 60:40 over 10 min, maintain at 60:40 for 5 min, to 10:90 over 5 min, to 0:100 over 2 min.
Flow rate: 1.5
Injection volume: 10-20
Detector: F (o-phthalaldehyde filters)

CHROMATOGRAM

Retention time: 8 (glutathione), 21 (glutathione monoethyl ester (glycyl carboxylate esterified)), 23 (glutathione monoethyl ester (glutamyl α-carboxylate esterified)), 34 (glutathione diethyl ester)

REFERENCE

Campbell, E.B.; Griffith, O.W. Glutathione monoethyl ester: high-performance liquid chromatographic analysis and direct preparation of the free base form, *Anal.Biochem.*, **1989**, *183*, 21–25.

SAMPLE
Matrix: tissue
Analyte: sodium mercaptoundecahydrodecaborate
Sample preparation: Liver, kidney. Homogenize with an equal volume of phosphate-buffered saline, add 4 volumes of MeCN, centrifuge at 3172 g for 10 min. Remove a 1 mL aliquot of the supernatant, filter (0.22 μm), evaporate to dryness under a stream of nitrogen, reconstitute the residue in 200 μL buffer, add 25 μL 5 mg/mL monobromobimane (Molecular Probes, Eugene OR) in MeCN, vortex, let stand in the dark for 4 h, inject a 50 μL aliquot. Brain. Homogenize with an equal volume of phosphate-buffered saline, add 4 volumes of MeCN, centrifuge at 3172 g for 10 min. Remove a 2 mL aliquot of the supernatant, wash with 8 mL hexane, centrifuge at 3172 g for 5 min. Remove a 1 mL aliquot of the aqueous phase, filter (0.22 μm), evaporate to dryness under a stream of nitrogen, reconstitute the residue in 200 μL buffer, add 25 μL 5 mg/mL monobromobimane (Molecular Probes, Eugene OR) in MeCN, vortex, let stand in the dark for 4 h, inject a 50 μL aliquot. (Prepare buffer by dissolving 12.1 g Tris in 50 mL water, adjusting the pH to 8.8 with 5 M HCl, and making up to 100 mL with water.)

HPLC VARIABLES
Guard column: 10 × 4.6 Hypersil ODS
Column: 150 × 4.6 5 μm Econosphere ODS
Mobile phase: MeOH:20 mM pH 7.0 potassium phosphate buffer 43:57 containing 10 mM tetrabutylammonium dihydrogen phosphate (Run time is 15 min.)
Flow rate: 2
Injection volume: 50
Detector: UV 373

CHROMATOGRAM
Retention time: 6.5
Limit of detection: 500 ng/mL

KEY WORDS
rat; liver; kidney; brain

REFERENCE
Saini, P.; Abu-Izza, K.; Lu, D.R. High-performance liquid chromatographic assay for sodium mercaptoundecahydrododecaborate in rat tissues, *J.Chromatogr.B*, **1995**, *665*, 155–161.

SAMPLE
Matrix: urine
Analyte: succimer
Sample preparation: 50 μL Urine + 900 μL buffer + 50 μL reagent, stir under nitrogen for 45 min, add 50 μL 80 mM monobromobimane (Calbiochem; Molecular Probes) in MeCN, let stand under nitrogen in the dark for 10 min, add 2 mL dichloromethane, extract for 15 s, centrifuge for 2 min, repeat extraction. Remove the aqueous phase and add 5 μL 6 M HCl, inject a 20 μL aliquot. (Buffer was 100 mM ammonium bicarbonate purged with nitrogen for 1 h. Reagent was 50 mM dithiothreitol in nitrogen-purged water.) [Procedures for electrolytic reduction and reduction with sodium borohydride are also described.]

HPLC VARIABLES
Column: 150 × 4.6 5 μm Spherisorb ODS RP-18
Mobile phase: Gradient. A was 20 mM tetrabutylammonium bromide in MeOH. B was 20 mM tetrabutylammonium bromide in 10 mM pH 4.1 acetate buffer. A:B 47.5:52.5 for 12 min, to 90:10 over 5 min, maintain at 90:10 for 7 min, re-equilibrate at initial conditions for 11 min.
Flow rate: 1
Injection volume: 20
Detector: F ex 356 em 450

CHROMATOGRAM
Retention time: 10

KEY WORDS
pharmacokinetics

REFERENCE

Maiorino, R.M.; Barry, T.J.; Aposhian, H.V. Determination and metabolism of dithiol-chelating agents: electrolytic and chemical reduction of oxidized dithiols in urine, *Anal.Biochem.*, **1987**, *160*, 217–226.

SAMPLE

Matrix: urine

Analyte: S-phenylmercapturic acid

Sample preparation: Condition a 500 mg Sep-Pak C18 with 5 mL MeOH and 10 mL 1% acetic acid. 2 mL Urine + 50 μL 15 μM IS in MeOH, adjust to pH 1.0 with 25% HCl, add to the SPE cartridge, wash with 2 mL 1% acetic acid, elute with 2 mL MeOH:100 mM pH 7.0 ammonium acetate buffer 80:20. Add 100 μL 10 M NaOH to the eluate and concentrate to about 100 μL under nitrogen at 78°, heat at 95° for 30 min, add 2 mL 400 mM pH 9.0 glycine/NaOH buffer, add 350 μL phosphoric acid:water 50:50, add 500 μL 1 mM monobromobimane (Calbiochem; Molecular Probes) in MeCN:water 10:90, inject an aliquot onto column A and elute to waste with mobile phase, after 22 min direct the effluent from column A onto column B, monitor the effluent from column B.

HPLC VARIABLES

Column: A 11 × 2 5 μm Nucleosil 120 C18 + 50 × 2 5 μm Nucleosil 120 C18; B 125 × 2 3 μm Nucleosil 120 C18

Mobile phase: Gradient. MeOH:0.1% acetic acid 10:90 for 2 min, to 50:50 over 3 min, maintain at 50:50 for 17 min, to 60:40 over 1 min. (Re-equilibrate column A with 10:90 for 10 min at 0.2 mL/min; re-equilibrate column B with 50:50 for 10 min at 0.1 mL/min.)

Flow rate: 0.2 for 22 min, to 0.1 over 1 min, maintain at 0.1

Detector: F ex 395 em 470

CHROMATOGRAM

Retention time: 48

Internal standard: S-acetyl-4-methylthiophenol (Dissolve 500 mg 4-methylthiophenol in 20 mL MeCN, add 1 mL acetic anhydride, add 1 mL 500 mM N,N-dimethyl-4-aminopyridine in MeCN, reflux for 1 h, cool to room temperature, pour into 500 mL ice-water, extract three times with 100 mL portions of dichloromethane. Combine the extracts and evaporate them to dryness to obtain S-acetyl-4-methylthiophenol.) (65)

Limit of quantitation: 1 ng/mL

KEY WORDS

SPE; column-switching

REFERENCE

Einig, T.; Dehnen, W. Sensitive determination of the benzene metabolite S-phenylmercapturic acid in urine by high-performance liquid chromatography with fluorescence detection, *J.Chromatogr.A*, **1995**, *697*, 371–375.

Monobromotrimethylammoniobimane

SAMPLE

Matrix: blood

Analyte: thiols

Sample preparation: Vortex 750 μL red blood cells with 3.2 mL 100 mM HCl, centrifuge at 4° at 15000 rpm for 5 min. Remove the supernatant and mix it with an equal volume of cold 4 M sodium methanesulfonate, freeze in dry ice/isopropanol, thaw, centrifuge at 20000 rpm for 10 min. Determine the thiol content of the supernatant by titrating with

5,5'-dithiobis(2-nitrobenzoic acid). Add N-ethylmorpholine to a final concentration of 10 mM, adjust pH to 8.0 with 1 M NaOH, add a 1 molar equivalent of dithiothreitol, mix, let stand at room temperature for 5 min, add 6 equivalents of monobromotrimethylammoniobimane (Calbiochem), mix, let stand for 15 min, add 12 equivalents of thiol agarose, let stand for 20 min, add acetic acid to a final concentration of 3%, inject an aliquot. (Prepare 4 M sodium methanesulfonate by adjusting the pH of methanesulfonic acid to 1.5 with 50% NaOH then diluting to 4 M.)

HPLC VARIABLES
Column: 150 × 4 AA-10 resin (Beckman)
Mobile phase: Gradient. A was 2-methoxyethanol:200 mM pH 3.20 buffer 10:90 at 45°. B was 2-methoxyethanol:200 mM pH 4.40 buffer 10:90 at 45°. C was 2-methoxyethanol:200 mM pH 4.75 buffer 10:90 at 45°. D was 2-methoxyethanol:0.2 N pH 6.40 trisodium citrate containing 800 mM NaCl 10:90 at 55°. E was 2-methoxyethanol:100 mM NaOH containing 100 mM NaCl 10:90 at 55°. A for 10 min, B for 20 min, C for 5 min, D for 90 min, E for 10 min, re-equilibrate at initial conditions for 20 min.
Flow rate: 0.2
Detector: F (o-phthalaldehyde filters)

CHROMATOGRAM
Retention time: 5.2 (coenzyme A), 7.8 (4'-phosphopantetheine), 10.2 (thiosulfate), 10.5 (coenzyme M), 27.5 (N-acetylcysteine), 31.5 (gamma-glutamylcysteine), 40 (glutathione), 51.5 (cysteinylglycine), 52 (ergothioneine), 55 (pantetheine), 57 (thiouracil), 61.5 (cysteine), 70 (homocysteine), 86 (2-mercaptoethanol), 110 (methanethiol), 110 (dithiothreitol)
Limit of quantitation: 10 pmole
Limit of detection: 1 pmole

KEY WORDS
red blood cells

REFERENCE
Fahey, R.C.; Newton, G.L.; Dorian, R.; Kosower, E.M. Analysis of biological thiols: Quantitative determination of thiols at the picomole level based upon derivatization with monobromobimanes and separation by cation-exchange chromatography, *Anal.Biochem.*, **1981**, *111*, 357–365.

SAOX-Cl

SAMPLE
Matrix: solutions
Analyte: thiols
Sample preparation: Mix 250 μL of a solution in 50 mM pH 9.3 borate buffer containing 1 mM disodium EDTA with 250 μL 1 mM SAOX-Cl in MeCN, let stand in the dark at room temperature for 1 h, add 500 μL MeCN:1 M HCl 50:50, inject an aliquot. (Prepare

SAOX-Cl as follows. Gently reflux 21 g benzoin and 45 g urethane (Caution! Urethane is a carcinogen!) in 300 mL DMF for 6 h, cool, pour into water, filter, recrystallize to give 4,5-diphenyl-2-oxazolone (mp 211°) (Ber. 1956, 89, 1749). Carefully add 60 mL dimethyl-sulfamoyl chloride (?) to 7.3 g 4,5-diphenyl-2-oxazolone at 0°, heat at 55-60° for 4 h, cool, add dropwise to 500 g ice-water, filter, wash the solid with 4 L water. Add 100 mL dry benzene (Caution! Benzene is a carcinogen!) to 2 g of the crude material (4,5-bis(p-N,N-dimethylaminosulfonylphenyl)-2-oxazolone) and evaporate to dryness to remove traces of moisture, suspend the residue in 30 mL phosphorus oxychloride, stir at 0°, add 610 µL triethylamine dropwise, heat at 100° for 7 h, remove excess phosphorus oxychloride using a rotary evaporator. Dissolve the residue in dichloromethane and wash with cold satu-rated sodium bicarbonate, dry the organic layer over anhydrous magnesium sulfate, evap-orate to dryness, chromatograph on 100 g silica gel with dichloromethane:ethyl acetate 90:10 to give SAOX-Cl (2-chloro-4,5-bis(p-N,N-dimethylaminosulfonylphenyl)oxazole) as a white solid (mp 222-224°) (Analyst 1993, 118, 257).)

HPLC VARIABLES
Column: 250 × 4.6 5 µm LC-8 (Supelco)
Mobile phase: MeCN:100 mM phosphoric acid 35:65
Flow rate: 1
Detector: F ex 330 em 425

CHROMATOGRAM
Retention time: 12.5 (glutathione), 15 (cysteine), 20 (homocysteine), 22.5 (N-acetylcys-teine), 30 (2-mercaptopropionylglycine), 54 (captopril)
Limit of detection: 1.2-7.9 fmole

REFERENCE
Toyo'oka, T.; Chokshi, H.P.; Givens, R.S.; Carlson, R.G.; Lunte, S.M.; Kuwana, T. Fluorescence and chemiluminescence detection of oxazole-labelled amines and thiols, *Biomed.Chromatogr.*, **1993**, 7, 208–216.

ACYL HALIDE

4-(N-Chloroformylmethyl-N-methyl)amino-7-N,N-dimethylaminosulfonyl-2,1,3-benzoxadiazole

SAMPLE
Matrix: solutions
Analyte: α-mercapto-N,2-naphthylacetamide
Sample preparation: Mix 10 µL 0.5 mM compound in anhydrous benzene containing 0.5 mM quinuclidine with 10 µL 25 mM DBD-COCl in anhydrous benzene (Caution! Benzene is a carcinogen!), stir at room temperature for 3 h, add 980 µL MeCN:water:acetic acid

50:50:1, inject a 2 µL aliquot. (Purify quinuclidine by sublimation. DBD-COCl is 4-(N-chloroformylmethyl-N-methyl)amino-7-N,N-dimethylaminosulfonyl-2,1,3-benzoxadiazole. DBD-COCl is available from Tokyo Kasei (TCI America, Portland OR). Synthesis is as follows. Dissolve 0.5 g magnesium sulfate heptahydrate and 6 g NaOH in 60 mL water, throughout the reaction keep the flask at about 20° with cold water cooling, add 15 mL 30% hydrogen peroxide, add 75 mL MeOH, add 12.1 g powdered benzoyl peroxide in one go, stir for 10 min, pour into 150 mL 20% sulfuric acid, extract three times with 50 mL portions of chloroform, determine peroxybenzoic acid concentration by iodometric titration (Tetrahedron 1967, 23, 3327). Slowly add 110 mL 1 M peroxybenzoic acid in chloroform to 7 g 2,6-difluoroaniline dissolved in 100 mL chloroform, stir at room temperature, when reaction is complete (iodometric titration) wash with 2% sodium thiosulfate, wash with 5% sodium carbonate, wash with water, dry over anhydrous sodium sulfate, evaporate to dryness under reduced pressure, recrystallize 2,6-difluoronitrosobenzene from EtOH (mp 108.5-109.5). Stir 8.5 g 2,6-difluoronitrosobenzene in 85 mL DMSO at room temperature and add a solution of 3.91 g sodium azide in 85 mL DMSO dropwise, let stand for about 1 h, add to a large volume of water, extract with ether, dry the extracts over anhydrous sodium sulfate, evaporate to dryness under reduced pressure and distil to give 4-fluoro-2,1,3-benzoxadiazole as a colorless oil (bp 83°/12 mm Hg) (J.Chem.Soc.(C) 1970, 1433). Add 11 mL chlorosulfonic acid dropwise to 3 g 4-fluoro-2,1,3-benzoxadiazole in 10 mL chloroform at 0-10° (use a calcium chloride drying tube), stir at room temperature for 1 h, reflux for 2 h, cool, slowly pour into ice water, remove the organic layer, extract the aqueous layer with chloroform, combine the organic layers, wash, dry over anhydrous magnesium sulfate, evaporate under reduced pressure, take up the residue in 5 mL benzene (Caution! Benzene is a carcinogen!), chromatograph on a 150 × 30 column of silica gel (100-200 mesh Kanto Chemical) with n-hexane:benzene 50:50, evaporate the appropriate fractions to give 4-(chlorosulfonyl)-7-fluoro-2,1,3-benzoxadiazole (CBD-F) as pale yellow needles (mp 64-66°) (Anal. Chem. 1984, 56, 2461). Stir 0.76 g CBD-F in 70 mL MeCN at 0-10° and add 1 g dimethylamine hydrochloride in 10 mL 100 mM pH 10 borax dropwise, adjust pH to 5 with 1 M HCl, concentrate to about 10 mL under reduced pressure, extract three times with 200 mL portions of diethyl ether, wash with water, dry over anhydrous magnesium sulfate, evaporate under reduced pressure, chromatograph on a 500 × 20 column of silica gel with chloroform, isolate the appropriate fraction and re-chromatograph on the same column with ethyl acetate:benzene 1:2 to give 4-(N,N-dimethylaminosulfonyl)-7-fluoro-2,1,3-benzoxadiazole (DBD-F) as white needles (mp 124-125°) (yield = 1% !) (Analyst 1989, 114, 413). On a Merck no. 5714 60F$_{254}$ tlc plate eluted with chloroform DBD-F has Rf 0.32 and lies between two other reaction products. DBD-F can also be purchased from Tokyo Kasei. Stir N-methylglycine and 2.3 g sodium carbonate in water at room temperature, add 880 mg DBD-F in 40 mL MeCN dropwise, stir for 1 h, evaporate to remove the MeCN, wash twice with 50 mL portions of ethyl acetate. Acidify the aqueous phase with HCl and extract it twice with 300 mL portions of ethyl acetate. Wash the organic layer twice with 100 mL portions of saturated aqueous NaCl, dry over anhydrous magnesium sulfate, evaporate to dryness under reduced pressure, recrystallize from ethyl acetate to give 4-(N-carboxymethyl-N-methyl)amino-7-dimethyl-aminosulfonyl-2,1,3-benzoxadiazole (DBD-COOH) as orange-yellow crystals (mp 209-210°). Add 3.5 mL oxalyl chloride and 24 µL DMF to 1 g DBD-COOH in anhydrous benzene, stir at room temperature for 30 min, reflux for 1 h, evaporate to dryness, add 20 mL dry benzene to the residue, filter, evaporate the filtrate to give 4-(N-chloroformyl-methyl-N-methyl)amino-7-N,N-dimethylaminosulfonyl-2,1,3-benzoxadiazole (DBD-COCl) as yellow crystals (mp 102°).)

HPLC VARIABLES
Column: 150 × 4.6 5 µm Cosmosil 5C18
Mobile phase: Gradient. MeCN:water from 30:70 to 100:0 over 1 h.
Flow rate: 1
Injection volume: 2
Detector: F ex 437 em 544

CHROMATOGRAM
Retention time: 27
Limit of detection: 103 fmole

REFERENCE
Imai, K.; Fukushima, T.; Yokosu, H. A novel electrophilic reagent, 4-(*N*-chloroformylmethyl-*N*-methyl)amino-7-*N*,*N*-dimethylaminosulphonyl-2,1,3-benzoxadiazole (DBD-COCl) for fluorometric detection of alcohols, phenols, amines and thiols, *Biomed.Chromatogr.*, **1994**, *8*, 107–113.

AZIRIDINE

Dansyl Aziridine

SAMPLE
Matrix: fermentation broth
Analyte: δ-(L-α-aminoadipyl)-L-cysteine-D-valine
Sample preparation: Centrifuge fermentation broth at 4° at 5600 g for 20 min, treat the supernatant with 2 volumes of acetone, let stand for 20 min, centrifuge, filter (0.45 μm) the supernatant. For each 1 mL add 3.5 mg sodium borohydride, heat at 60° for 30 min, add glacial acetic acid until the pH reaches 4 (usually 20 μL/mL), adjust the pH to 8.8 with 2 M KOH. Remove a 50 μL aliquot and add it to 50 μL 3 mM N-dansylaziridine (Sigma) in MeOH, heat at 60° for 30 min, cool rapidly to room temperature, inject a 20 μL aliquot.

HPLC VARIABLES
Guard column: 50 × 4.6 5 μm C18 (Hichrom)
Column: 250 × 4.6 5 μm C18 (Hichrom)
Mobile phase: Gradient. MeCN:20 mM pH 4.0 acetate buffer containing 0.5 mM EDTA from 16:74 to 24:76 over 15 min, to 100:0 over 5 min.
Flow rate: 1
Injection volume: 20
Detector: F ex 339 em 540

CHROMATOGRAM
Retention time: 12
Limit of quantitation: 5.6 ng/mL
Limit of detection: 120 pg/mL

REFERENCE
Orford, C.D.; Perry, D.; Adlard, M.W. High-performance liquid chromatographic determination of δ-(L-α-aminoadipyl)-L-cysteine-D-valine in complex media by precolumn derivatization with dansylaziridine, *J.Chromatogr.*, **1989**, *481*, 245–254.

CHLORAMINE

N-Chlorodansylamide

non-fluorescent   fluorescent

SAMPLE
Matrix: blood
Analyte: glutathione
Sample preparation: 100 μL Whole blood + 700 μL 0.2% EDTA, shake gently for 1 min, add 200 μL 100 mg/mL metaphosphoric acid, mix for 10 min, centrifuge at 2500 rpm for 10 min, filter (0.5 μm) an aliquot of the supernatant, inject a 20 μL aliquot of the filtrate.

HPLC VARIABLES
Column: 300 × 4 μBondapak C18
Mobile phase: MeOH:water 25:75 containing 10 mM tetra-n-butylammonium bromide, pH adjusted to 7.0 with 1 M NaOH
Flow rate: 1
Injection volume: 20
Detector: F ex 360 em 510 following post-column reaction. The column effluent mixed with reagent pumped at 0.54 mL/min and the mixture flowed through a 15 m × 0.5 mm ID PTFE coil at 42° to the detector. (Prepare reagent (N-chlorodansylamide) by adding aqueous sodium hypochlorite solution (3% available chlorine) dropwise to 100 mL 200 μg/mL dansylamide in MeOH under a UV light, stop the addition when the fluorescence has almost disappeared.)

CHROMATOGRAM
Retention time: 12
Limit of detection: 0.1 nmoles

KEY WORDS
post-column reaction; whole blood

REFERENCE
Murayama, K.; Kinoshita, T. Determination of glutathione by high performance liquid chromatography using N-chlorodansylamide (NCDA), *Anal.Lett.*, **1981**, *14*, 1221–1232.

DIALDEHYDE

Phthalaldehyde

RELATED REFERENCE
Kok, R.J.; Visser, J.; Moolenaar, F.; De Zeeuw, D.; Meijer, D.K.F. Bioanalysis of captopril: two sensitive high-performance liquid chromatographic methods with pre- or postcolumn fluorescent labeling. *J.Chromatogr.B* **1997**, *693*, 181-189

SAMPLE
Matrix: solutions
Analyte: acetylcysteine
Sample preparation: 5 mL 10 mM N-acetylcysteine in water + 5 mL 10.1 mM o-phthalaldehyde in 123 mM pH 10.4 borate buffer, mix, add 5 mL 10.1 mM L-valine in 10 mM HCl, shake, inject an aliquot.

CAPILLARY ELECTROPHORESIS
Capillary: 37 cm × 50 μm fused-silica (30 cm to detector)
Capillary preparation: Condition a new capillary with 100 mM NaOH at 50° for 30 min then equilibrate with running buffer under running conditions for 40 min.
Running buffer: 155 mM pH 8.98 Borate buffer containing 5% polyethylene glycol 20000 (Prepare by dissolving 153 mg boric acid, 1.242 mg sodium tetraborate, and 5 g polyethylene glycol 20000 in water, make up to 100 mL with water.)
Capillary temperature: 25
Injection: Inject 100 mM NaOH for 2 s, inject sample for 4 s, inject 100 mM HCl for 2 s (at 0.5 psi)
Detector: UV 214
Model: Beckman P/ACE 2050 or 2100
Migration time: 3.38 (D), 3.55 (L)
Limit of detection: 0.4% (of major enantiomer)

OTHER SUBSTANCES
Simultaneously analyzed: N,N-diacetylcystine

KEY WORDS
chiral

REFERENCE
Dette, C.; Watzig, H. Separation of enantiomers of *N*-acetylcysteine by capillary electrophoresis after derivatization by *o*-phthaldialdehyde, *Electrophoresis*, **1994**, *15*, 763–768.

DISULFIDE

N-[6-(7-Amino-4-methylcoumarin-3-acetamido)hexyl]-3'-(2'-pyridyldithio)propionamide

SAMPLE
Matrix: blood
Analyte: 6-mercaptopurine
Sample preparation: Treat whole blood with EDTA, freeze a 2 mL aliquot at -20°, thaw, add 2 mL PBS, centrifuge at 3000 g for 15 min, suspend the pellet in 200 μL PBS, treat with proteinase K at 70° for 10 min, add to a disposable spin column (Diagen QIAamp Blood Kit, Hilden, Germany), centrifuge for 1 min, wash twice with buffer for 1 min, elute with 200 μL 10 mM pH 9.0 Tris-HCl buffer containing 0.1 mM EDTA, heat at 100° for 5 min, cool in ice. Remove a 100 μl aliquot and add it to 10 μL buffer, add 20 μL 25 μg/mL P$_1$ nuclease (Boehringer Mannheim) and 12.5 U/mL acid phosphatase (Sigma) in buffer:water 10:90, heat at 42° for 1 h, add 10 μL 400 mM formic acid, add 60 μL MeOH, add 1 μL 5 mM N-[6-(7-amino-4-methylcoumarin-3-acetamido)hexyl]-3'-(2'-pyridyldithio)propionamide (AMCA-HPDP, Pierce) in DMF, inject a 25 μL aliquot. (Buffer was 500 mM pH 4.5 sodium acetate buffer containing 10 mM magnesium chloride.)

HPLC VARIABLES
Guard column: 20 mm long Supelguard (Supelco)
Column: 150 × 4.6 3 μm Supelcosil LC-8
Mobile phase: MeOH:buffer 37:63 (Between analyses wash column with MeOH:buffer 80:20 for 3 min. Buffer was 200 mM formic acid adjusted to pH 4.0 with 10 M NaOH.)
Column temperature: 45
Flow rate: 1
Injection volume: 25
Detector: F ex 345 em 450

CHROMATOGRAM
Retention time: 11.5 (as 2'-deoxy-6-thioguanosine metabolite)
Limit of detection: 60 pmole/g DNA

KEY WORDS
whole blood

REFERENCE
Warren, D.J.; Andersen, A.; Slordal, L. Quantitation of 6-thioguanine residues in peripheral blood leukocyte DNA obtained from patients receiving 6-mercaptopurine-based maintenance therapy, *Cancer Res.*, **1995**, *55*, 1670–1674.

5,5'-Dithiobis(2-nitrobenzoic acid)

RELATED REFERENCES

Kuwata, K.; Uebori, M.; Yamada, K.; Yamazaki, Y. Liquid chromatographic determination of alkylthiols via derivatization with 5,5'-dithiobis(2-nitrobenzoic acid). *Anal.Chem.* **1982**, *54*, 1082-1087.

Nozal, M.J.; Bernal, J.L.; Toribio, L.; Marinero, P.; Moral, O.; Manzanas, L.; Rodriguez, E. Determination of glutathione, cysteine and N-acetylcysteine in rabbit eye tissues using high-performance liquid chromatography and post-column derivatization with 5,5'-dithiobis(2-nitrobenzoic acid). *J.Chromatogr.A* **1997**, *778*, 347-353

SAMPLE

Matrix: enzyme incubations
Analyte: thiols
Sample preparation: 1 mL Enzyme incubation + 25 µL 200 mM dithioerythritol + 100 µL 200 mM 5,5'-dithiobis(2-nitrobenzoic acid) (adjusted to pH 8 with NaOH), mix, inject an aliquot.

HPLC VARIABLES

Column: 250 × 4.6 10 µm LiChrosorb RP-18
Mobile phase: Gradient. A was 23 mM pH 5.0 ammonium formate. B was MeOH. A:B from 100:0 to 90:10 over 5 min, maintain at 90:10 for 5 min, to 70:30 (step gradient), maintain at 70:30 for 8 min, re-equilibrate at initial conditions for 5 min.
Column temperature: 20
Flow rate: 1.5
Detector: UV 280

CHROMATOGRAM

Retention time: 7.32 (cysteine), 8.01 (gamma-L-glutamyl-L-cysteine), 9.12 (glutathione)
Limit of detection: 0.2-0.5 nmole

REFERENCE

Dennda, G.; Kula, M.-R. Assay of the glutathione-synthesizing enzymes by high-performance liquid chromatography, *Biotechnol.Appl.Biochem.*, **1986**, *8*, 459–464.

SAMPLE

Matrix: formulations
Analyte: glutathione
Sample preparation: Dissolve formulation in pH 8.00 phosphate buffer containing 50 µM ascorbic acid to give a glutathione concentration of 100 µM. Remove a 2 mL aliquot and add it to 200 µL 10 mM 5,5'-dithiobis(2-nitrobenzoic acid) in pH 7.00 phosphate buffer, inject an aliquot within 2 min.

HPLC VARIABLES

Column: 250 × 4.6 Hypersil 5 ODS
Mobile phase: MeOH:23 mM pH 5 ammonium formate 10:90
Flow rate: 1
Injection volume: 20
Detector: UV 280

CHROMATOGRAM

Retention time: 5.6
Internal standard: ascorbic acid (2.8)

KEY WORDS
comparison with capillary electrophoresis

REFERENCE
Raggi, M.A.; Mandrioli, R.; Bugamelli, F.; Sabbioni, C. Comparison of analytical methods for quality control of pharmaceutical formulations containing glutathione, *Chromatographia*, **1997**, *46*, 17–22.

SAMPLE
Matrix: formulations
Analyte: glutathione
Sample preparation: Dissolve formulation in pH 8.00 phosphate buffer to give a glutathione concentration of 10 μM. Remove a 2 mL aliquot and add it to 200 μL 10 mM 5,5'-dithiobis(2-nitrobenzoic acid) in pH 7.00 phosphate buffer, inject an aliquot within 2 min.

CAPILLARY ELECTROPHORESIS
Capillary: 27 cm × 75 μm fused-silica (20 cm to detector)
Capillary preparation: Rinse with 1 M NaOH for 20 min then equilibrate with running buffer.
Running buffer: 100 mM pH 7.00 Sodium phosphate buffer
Capillary temperature: 25
Voltage/Current: 10 kV
Injection: Pressure injection at 0.5 psi for 5 s.
Detector: UV 200
Model: Beckman P/ACE 5000
Migration time: 4

KEY WORDS
comparison with HPLC

REFERENCE
Raggi, M.A.; Mandrioli, R.; Bugamelli, F.; Sabbioni, C. Comparison of analytical methods for quality control of pharmaceutical formulations containing glutathione, *Chromatographia*, **1997**, *46*, 17–22.

SAMPLE
Matrix: formulations
Analyte: thiols
Sample preparation: Dissolve contents of vial in 0.3% phosphoric acid, inject a 10 μL aliquot.

HPLC VARIABLES
Column: 100 × 4.65 μm Hypersil ODS C18
Mobile phase: MeCN:buffer 10:90 (Buffer was 0.3% phosphoric acid containing 0.1% sodium octyl sulfate.)
Flow rate: 1
Injection volume: 10
Detector: UV 412 following post-column reaction. The column effluent mixed with reagent pumped at 0.5 mL/min and the mixture flowed through a 1 mL coil to the detector. (Reagent was 500 mM pH 7 sodium citrate/sodium phosphate buffer containing 0.03% 5,5'-dithiobis(2-nitrobenzoic acid) and 0.025% EDTA.)

CHROMATOGRAM
Retention time: 4 (cysteine), 2 (sodium bisulfite), 1.5 (sulfocysteine)
Limit of quantitation: <1 ppm

OTHER SUBSTANCES
Non-interfering: cysteic acid, cystine

KEY WORDS
post-column reaction; injections

REFERENCE
Jenke, D.R.; Brown, D.S. Determination of cysteine in pharmaceuticals via liquid chromatography with postcolumn derivatization, *Anal.Chem.*, **1987**, *59*, 1509–1512.

SAMPLE
Matrix: solutions
Analyte: thiols
Sample preparation: Mix a 200 μM solution in buffer with three volumes of a 400 μM solution of 5,5'-dithio-(bis-2-nitrobenzoic acid) in buffer, let stand at room temperature for 30 min, inject a 75 μL aliquot. (Buffer was 125 mM NaH_2PO_4 containing 154 mM NaCl, pH adjusted to 7.4 with NaOH.)

HPLC VARIABLES
Column: 250 × 4.6 Hypersil ODS1
Mobile phase: Gradient. MeCN:buffer 0:100 for 20 min, to 17.5:82.5 over 40 min. (Buffer was 125 mM NaH_2PO_4 containing 154 mM NaCl, pH adjusted to 7.4 with NaOH.)
Flow rate: 0.25 for 20 min, to 1 over 40 min
Injection volume: 75
Detector: UV 357

CHROMATOGRAM
Retention time: 21 (thiomalic acid), 30 (cysteine), 36 (glutathione), 41 (N-acetylcysteine), 45 (penicillamine), 56 (N-acetylpenicillamine), 61 (captopril)
Limit of detection: 10 μM

REFERENCE
Russell, J.; McKeown, J.A.; Hensman, C.; Smith, W.E.; Reglinski, J. HPLC determination of biologically active thiols using pre-column derivatization with 5,5'-dithio-(bis-2-nitrobenzoic acid), *J.Pharm.Biomed.Anal.*, **1997**, *15*, 1757–1763.

SAMPLE
Matrix: urine
Analyte: thiols
Sample preparation: For each 1 mL urine add 1-2 mg EDTA and 5 μg homocysteine, inject a 20 μL aliquot.

HPLC VARIABLES
Guard column: 45 × 4.6 5 μm ODS Hypersil
Column: 100 × 4.6 5 μm ODS Hypersil
Mobile phase: 1 g/L pH 4 Heptanesulfonic acid in water containing 150 mg/L sodium EDTA
Flow rate: 1
Injection volume: 20
Detector: UV 412 following post-column reaction. The column effluent mixed with the reagent pumped at 0.5 mL/min and the mixture flowed through a 150 × 2 column filled with 40 μm glass beads to the detector. (Prepare reagent by dissolving 200 mg 5,5'-dithiobis(2-nitrobenzoic acid) and 10 g tripotassium citrate in 100 mL 250 mM pH 7.4 phosphate buffer, dilute 10-fold with water immediately before use.)

CHROMATOGRAM
Retention time: 0.4 (cysteine), 1.3 (penicillamine)
Internal standard: homocysteine (0.8)
Limit of quantitation: 10 ng

KEY WORDS
post-column reaction

REFERENCE
Beales, D.; Finch, R.; McLean, A.E.M.; Smith, M.; Wilson, I.D. Determination of penicillamine and other thiols by combined high-performance liquid chromatography and post-column reaction with Ellman's reagent: application to human urine, *J.Chromatogr.*, **1981**, *226*, 498–503.

SAMPLE
Matrix: urine
Analyte: mesna
Sample preparation: Dilute urine 1:3 to 1:39, inject a 50 μL aliquot.

HPLC VARIABLES
Column: 250 × 4 5 μm Hypersil ODS

Mobile phase: MeOH:250 mM pH 7.4 phosphate buffer 5:95 containing 5 mM tetrabutyl-ammonium phosphate
Flow rate: 1
Injection volume: 50
Detector: UV 412 following post-column reaction. The column effluent mixed with the reagent pumped at 0.5 mL/min and the mixture flowed through a 200 × 4 column packed with 100-120 mesh glass beads (dichlorodimethylsilane treated) to the detector. (Prepare reagent by diluting 0.2% 5,5'-dithiobis(2-nitrobenzoic acid) in 250 mM pH 7.4 phosphate buffer containing 10% tripotassium citrate 1:10 with water.)

CHROMATOGRAM
Retention time: 7.5
Limit of detection: 75 ng

KEY WORDS
post-column reaction; comparison with electrochemical detection without derivatization

REFERENCE
Sidau, B.; Shaw, I.C. Determination of sodium 2-mercaptoethanesulphonate by high-performance liquid chromatography using post-column reaction colorimetry or electrochemical detection, *J.Chromatogr.*, **1984**, *311*, 234–238.

SAMPLE
Matrix: urine
Analyte: disulfides
Sample preparation: Acidify urine to pH 1 with concentrated HCl, filter (0.45 μm), dilute 10-fold with 0.1% EDTA, add N-acetyl cysteine, inject a 100 μL aliquot.

HPLC VARIABLES
Column: 250 × 4.5 5 μm Spherisorb ODS II
Mobile phase: MeCN:33 mM pH 4.0 KH_2PO_4 containing 2 mM tetrabutylammonium hydroxide and 0.2 mM EDTA 13:87
Flow rate: 0.8
Injection volume: 100
Detector: UV 412 following post-column reaction. The column effluent mixed with the reagent pumped at 1 mL/min and the mixture flowed through a 2 m × 0.6 mm ID stainless steel coil at 60° to the detector. (Prepare the reagent by dissolving 100 mg 5,5'-dithiobis(2-nitrobenzoic acid) in 10 mL 1 M sodium sulfite at 38°, adjust pH to 7.5, pass oxygen through the solution for 45 min to form 2-nitro-5-thiosulfobenzoate, store frozen. Before use dilute 100-fold with 200 mM pH 9.5 Tris containing 100 mM sodium sulfite and 3 mM EDTA.)

CHROMATOGRAM
Retention time: 3 (cysteamine, cystamine), 4 (cysteine, cystine, homocysteine), 5 (glutathione), 6 (glutathione disulfide), 10 (Thiola)
Internal standard: N-acetylcysteine (8)

OTHER SUBSTANCES
Non-interfering: penicillamine

KEY WORDS
post-column reaction

REFERENCE
Crawhall, J.C.; Kalant, D. Measurement of biological disulfides by postcolumn sulfitolysis following separation by HPLC, *Anal.Biochem.*, **1988**, *172*, 479–483.

6,6'-Dithiodinicotinic Acid

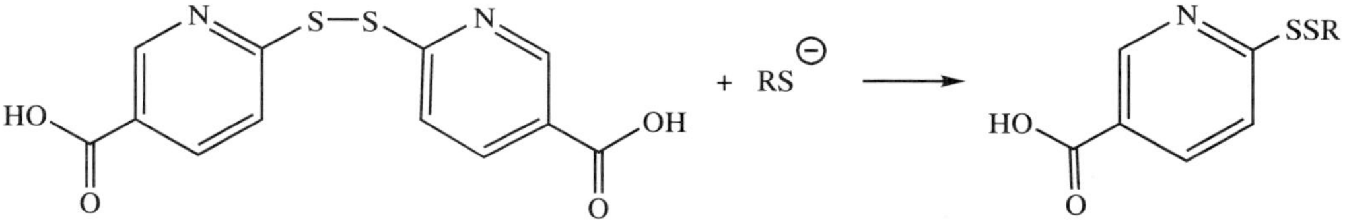

SAMPLE

Matrix: blood, cells

Analyte: thiols

Sample preparation: Blood. Mix 9 mL whole blood with 1 mL 3.8% sodium citrate, centrifuge at 4° at 150 g for 15 min, wash the erythrocytes three times with isotonic saline. Suspend 100 μL erythrocytes in 700 μL 6 mM EDTA, mix gently for 1 min, add 200 μL 25% metaphosphoric acid, mix for 10 min, centrifuge at 5000 g for 15 min, filter (0.45 μm) the supernatant, inject a 10 μL aliquot of the filtrate. Cells. Wash 500 mg (wet weight) E. coli cells with water, add 2 mL 5% metaphosphoric acid, sonicate, centrifuge at 4° at 15000 g for 15 min, filter (0.45 μm) the supernatant, inject an aliquot of the filtrate.

HPLC VARIABLES

Column: 250 × 4.6 Fine Sil C18-10 (Japan Spectroscopic)

Mobile phase: 33 mM KH_2PO_4 adjusted to pH 2.2 with phosphoric acid

Flow rate: 1

Injection volume: 10

Detector: UV 344 following post-column reaction. The column effluent mixed with the 1.5 mM 6,6'-dithiodinicotinic acid in 200 mM pH 7.0 sodium phosphate buffer pumped at 1 mL/min and the mixture flowed through a 60 cm × 0.5 mm ID stainless steel coil to the detector.

CHROMATOGRAM

Retention time: 3.7 (cysteine), 4 (cysteamine), 4.8 (homocysteine), 7 (glutathione), 8.2 (penicillamine)

Limit of detection: 0.1 nmole

KEY WORDS

post-column reaction; whole blood; erythrocytes

REFERENCE

Nishiyama, J.; Kuninori, T. Assay of biological thiols by a combination of high-performance liquid chromatography and postcolumn reaction with 6,6'-dithiodinicotinic acid, *Anal.Biochem.*, **1984**, *138*, 95–98.

4,4'-Dithiodipyridine

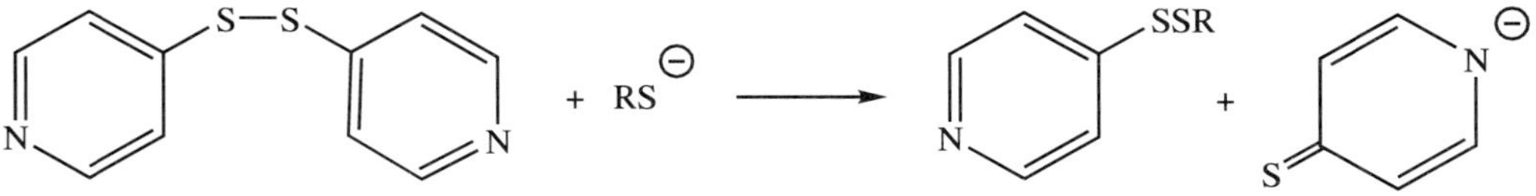

SAMPLE

Matrix: blood, urine

Analyte: thiols

Sample preparation: Plasma. 400 μL Plasma + 100 μL 100 mM dithiothreitol, heat at 37° for 15 min, add 100 μL 150 mg/mL sulfosalicylic acid, mix, let stand at room temperature for 30 min, centrifuge at 4000 g for 15 min, inject a 30 μL aliquot. Urine. 400 μL Urine + 30 μL 100 mM dithiothreitol + 70 μL 300 mM pH 8.5 Tris buffer, mix, heat at 37° for 15 min, add 100 μL 100 mg/mL sulfosalicylic acid, mix, let stand at room temperature for 30 min, centrifuge at 4000 g for 15 min, inject a 30 μL aliquot.

HPLC VARIABLES
Guard column: 10 × 3.2 7 μm G18-013 C18 (Brownlee)
Column: 100 × 3.2 3.2 Velosep RP-18 (Brownlee, Applied Biosystems)
Mobile phase: MeOH:buffer 4.3:100 (Buffer was 16 mM NaH_2PO_4, 19 mM phosphoric acid, and 8 mM octyl sulfate, pH 2.36 ± 0.02. Use a 30 × 3.2 5 μm Brownlee SS-GU silica column before the injector.)
Flow rate: 0.8
Injection volume: 30
Detector: UV 324 following post-column reaction. The effluent from the column mixed with the reagent pumped at 0.4 mL/min and the mixture flowed through a 2 m × 0.5 mm stainless steel coil to the detector. (Prepare reagent by mixing (at 4°) 3 mL 10 mM 4,4'-dithiodipyridine in MeOH:10 mM HCl 3:97 with 300 mL 300 mM Tris base containing 1 mM EDTA (adjusted to pH 8.5 with phosphoric acid). Sparge with helium before use, keep in an ice bath during use.)

CHROMATOGRAM
Retention time: 5.4 (cysteine), 7.2 (glutathione), 10.4 (gamma-glutamylcysteine), 13.2 (cysteinylglycine), 15.9 (homocysteine)
Limit of detection: 50 nM

KEY WORDS
plasma; post-column reaction

REFERENCE
Andersson, A.; Isaksson, A.; Brattström, L.; Hultberg, B. Homocysteine and other thiols determined in plasma by HPLC and thiol-specific postcolumn derivatization, *Clin.Chem.*, **1993**, *39*, 1590–1597.

ENONE

Ethacrynic Acid

RELATED REFERENCES
Cavrini, V.; Gatti, R.; DiPietra, A.M.; Raggi, M.A. HPLC determination of thio drugs in pharmaceutical formulations using ethacrynic acid as a precolumn ultraviolet derivatization reagent. *Chromatographia* **1987**, *23*, 680-683.
Cavrini, V.; Andrisano, V.; Gatti, R.; Scapini, G. HPLC determination of thioglycolic acid and other aliphatic thiols in cosmetic formulations using ethacrynic acid as precolumn derivatization reagent. *Int.J.Cosmet.Sci.* **1990**, *12*, 141-150.

SAMPLE
Matrix: formulations
Analyte: glutathione and cysteine
Sample preparation: Glutathione. Stir powdered tablets with EtOH:water 50:50 for 15 min, filter, dilute with water to a final glutathione concentration of 60 μg/mL. Dilute solutions or lyophilized products with water to a glutathione concentration of 75 μg/mL. 1 mL Solution + 1 mL 1.2 mg/mL ethacrynic acid in pH 7.4 buffer, mix, let stand at room temperature for 25 min, add 300 μL 8.5% phosphoric acid, add 1.5 mL dichloromethane, vortex for 1 min, centrifuge for 2 min. Remove a 1 mL aliquot of the aqueous layer and add it to 500 μL 80 μg/mL propyl p-hydroxybenzoate in MeOH:pH 3.0 buffer 60:40, mix, inject an aliquot. L-Cysteine. Extract powder with EtOH:water 50:50 for 20 min, filter, dilute with 1 mg/mL sodium EDTA solution to give a cysteine concentration of 40 μg/mL. 1 mL Solution + 1 mL 1.2 mg/mL ethacrynic acid in pH 7.4 buffer, mix, let stand at room temperature for 25 min, add 300 μL 8.5% phosphoric acid, add 1.5 mL dichloromethane, vortex for 1 min, centrifuge for 2 min. Remove a 1 mL aliquot of the aqueous layer and add it to 500 μL 80 μg/mL propyl p-hydroxybenzoate in MeOH:pH 3.0 buffer 60:40, mix, inject an aliquot. Oxidized glutathione. 1 mL Solution + 200 μL pH 7.4 buffer + 100 μL 1 mM dithiothreitol in water, mix, let stand at room temperature in the dark for 30 min, add 200 μL 1.2 mg/mL ethacrynic acid in pH 7.4 buffer, mix, let stand at room temperature for 30 min, add 1 mL water, add 300 μL 8.5% phosphoric acid, add 1 mL dichloromethane, vortex for 1 min, centrifuge for 2 min. Remove a 1 mL aliquot of the aqueous layer and add it to 500 μL 80 μg/mL propyl p-hydroxybenzoate in MeOH:pH 3.0 buffer 60:40, mix, inject an aliquot. (Prepare pH 7.4 buffer by adjusting pH of 300 mM KH_2PO_4 to 7.4 with 2 M NaOH. Prepare pH 3.0 buffer by adjusting pH of 50 mM triethylamine to 3.0 with phosphoric acid.)

HPLC VARIABLES
Column: 150 or 250 × 4.6 5 μm Hypersil C18
Mobile phase: MeOH:buffer 58:42 (A) or MeCN:buffer 47:53 (B) (Prepare buffer by adjusting pH of 50 mM triethylamine to 3.0 with phosphoric acid.)
Flow rate: 1
Injection volume: 20
Detector: UV 270

CHROMATOGRAM
Retention time: 2.7 (glutathione and cysteine (A)), 2.4 (glutathione (B)), 2.7 (cysteine (B))
Internal standard: propyl p-hydroxybenzoate (5.2 (A))
Limit of detection: 500 ng/mL

KEY WORDS
powder; tablets

REFERENCE
Di Pietra, A.M.; Gotti, R.; Bonazzi, D.; Andrisano, V.; Cavrini, V. HPLC determination of glutathione and L-cysteine in pharmaceuticals after derivatization with ethacrynic acid, *J.Pharm.Biomed.Anal.*, **1994**, *12*, 91–98.

2-Hexenal

SAMPLE
Matrix: blood
Analyte: glutathione
Sample preparation: Suspend red blood cells in 19 volumes 50 mM pH 7.4 phosphate-buffered saline, add (E)-2-hexenal to a concentration of 50 μM, heat at 37° for 1 h. Remove

a 500 μL aliquot and add it to 1.5 mL MeOH, centrifuge at 8000 g for 15 min, filter (0.45 μm), inject an aliquot of the filtrate.

HPLC VARIABLES
Column: 250 × 4.6 5 μm Capcellpak C18 (Shiseido)
Mobile phase: Gradient. MeOH:25 mM pH 7.0 phosphate buffer 0:100 for 5 min, to 70:30 over 20 min, maintain at 70:30 for 5 min.
Column temperature: 40
Flow rate: 1
Detector: F ex 340 em 450 following post-column reaction. The column effluent mixed with the reagent pumped at 0.3 mL/min and the mixture flowed through a 1 m × 0.8 mm i.d. coil at 40° to the detector. (Reagent contained 0.7% o-phthalaldehyde and 2% 2-mercaptoethanol (Wako Pure Chemical Industries).)

CHROMATOGRAM
Retention time: 17.5
Limit of detection: 11.9 pmole

KEY WORDS
post-column reaction; rat; red blood cells

REFERENCE
Fujita, M.; Sano, M.; Takeda, K.; Tomita, I. Fluorescence detection of glutathione S conjugate with aldehyde by high-performance liquid chromatography with post-column derivatization, *Analyst*, **1993**, *118*, 1289–1292.

Methyl Acrylate

SAMPLE
Matrix: blood
Analyte: BMS186716
Sample preparation: Immediately add 1 mL blood to a tube containing tripotassium EDTA and 10 μL methyl acrylate, invert gently, let stand in ice for 10-15 min, centrifuge, immediately remove 500 μL plasma and add it to 15 μL 21.6 μg/mL IS in MeCN:10 mM Na_2HPO_4 75:25, add 500 μL 100 mM HCl, mix, add 3 mL MTBE, shake at high speed for 10 min, centrifuge at 2000 g. Remove the organic layer and evaporate it to dryness under a stream of nitrogen at 40° for 10 min (Turbovap), reconstitute with 60 μL mobile phase, vortex for 1 min, centrifuge at 3000 g for 5 min, inject a 15 μL aliquot.

HPLC VARIABLES
Guard column: Hypersil C18
Column: 100 × 2 5 μm Hypersil C18
Mobile phase: MeOH:buffer 35:65 (Prepare mobile phase as follows. A is 770 mg ammonium acetate in 250 mL MeOH and 750 mL water, pH adjusted to 5.5 with glacial acetic acid. B is 770 mg ammonium acetate in 1 L MeOH. A:B is 65:35.)
Flow rate: 0.4
Injection volume: 15
Detector: MS, Sciex API I, articulated ion-spray interface, 14% of column flow entered MS, nebulizing gas nitrogen at 60 psi, curtain gas ultra-high purity nitrogen at 1.2 L/min, interface 60°, sprayer +3500 V, orifice +45 V, SIM, m/z 512

CHROMATOGRAM
Retention time: 2.7
Internal standard: BMS188035 (m/z 498) (1.7)
Limit of quantitation: 2.5 ng/mL

KEY WORDS
dog; whole blood

REFERENCE
Jemal, M.; Hawthorne, D.J. Quantitative determination of BMS186716, a thiol compound, in dog plasma by high-performance liquid chromatography-positive ion electrospray mass spectrometry after formation of the methyl acrylate adduct, *J.Chromatogr.B*, **1997**, *693*, 109–116.

Methyl 4-(6-Methoxynaphthalen-2-yl)-4-oxo-2-butenoate

SAMPLE
Matrix: blood, tissue
Analyte: thiols
Sample preparation: Plasma. Dilute rat plasma with 4 volumes 20 mM disodium EDTA, remove a 500 µL aliquot and add it to 200 µL water and 300 µL reagent, let stand at room temperature for 10 min, add 500 µL 300 mM phosphoric acid, add 50 µL 4 µg/mL IS solution, inject a 50 µL aliquot. Tissue. Homogenize tissue with 20 (spleen) or 100 (liver) volumes of disodium EDTA solution at 0°. Mix 1 mL homogenate with 300 µL reagent, let stand at room temperature for 10 min, add 500 µL 300 mM phosphoric acid, add 3 mL 4 µg/mL IS solution, make up to 10 mL with water, centrifuge at 3000 rpm for 5 min, filter (0.45 µm), inject a 50 µL aliquot of the filtrate. (Prepare reagent by dissolving 3.5 mg methyl 4-(6-methoxynaphthalen-2-yl)-4-oxo-2-butenoate in 10 mL THF and diluting to 25 mL with 200 mM pH 7.5 borate buffer. Prepare methyl 4-(6-methoxynaphthalen-2-yl)-4-oxo-2-butenoate as follows. Dissolve 5 g 6'-methoxy-2'-acetonaphthone in warm glacial acetic acid and add 2.5 g glyoxylic acid, reflux for 24 h, evaporate to dryness under reduced pressure. Take up the residue in chloroform and extract it three times with 5% sodium carbonate solution. Combine the aqueous layers and acidify them with concentrated HCl, collect the product by filtration, recrystallize from MeOH/water or acetic acid to give 4-(6-methoxynaphthalen-2-yl)-4-oxo-2-butenoic acid (mp 167-9°) (Farmaco, Ed. Sci. 1982, 37, 171). Reflux 0.5 g 4-(6-methoxynaphthalen-2-yl)-4-oxo-2-butenoic acid, 2.5 mL MeOH, and 2-3 drops sulfuric acid in 25 mL anhydrous benzene (Caution! Benzene is a carcinogen!) for 1 h, add 20 mL water, wash the organic layer with 10 mL 5% sodium bicarbonate solution, wash the organic layer with 20 mL water. Dry the organic layer over anhydrous sodium sulfate, evaporate to dryness under reduced pressure, purify by flash chromatography on silica gel using ethyl acetate:light petroleum (bp 40-70°) 40:60 to give methyl 4-(6-methoxynaphthalen-2-yl)-4-oxo-2-butenoate as a pale yellow compound (mp 116-120°) (J. Chromatogr. 1990, 507, 451).)

HPLC VARIABLES
Column: 150 × 4.6 C-18 Hypersil
Mobile phase: MeCN:50 mM pH 4 triethylammonium phosphate 32:68

Flow rate: 1
Injection volume: 50
Detector: F ex 310 em 450

CHROMATOGRAM
Retention time: 3 (glutathione), 4 (cysteine), 5 (homocysteine)
Internal standard: 4-(6-methoxynaphthalen-2-yl)-4-oxobutanoic acid (10)
Limit of detection: 0.5 pmole

KEY WORDS
rat; plasma; liver; spleen

REFERENCE
Gotti, R.; Andrisano, V.; Gatti, R.; Cavrini, V.; Candeletti, S. Determination of glutathione in biological samples by high performance liquid chromatography with fluorescence detection, *Biomed.Chromatogr.*, **1994**, *8*, 306–308.

SAMPLE
Matrix: formulations
Analyte: thiols
Sample preparation: Powders. Weigh out powder equivalent to about 200 µg cysteine, dissolve in 200 mL 1 mg/mL disodium EDTA solution. Remove a 1 mL aliquot and add it to 300 µL reagent solution, let stand at room temperature for 20 min, add 500 µL 300 mM phosphoric acid solution, add 3 mL 240 µg/mL IS solution, make up to 10 mL with water, inject a 50 µL aliquot. Tablets. Weigh out powdered tablet equivalent to about 7.7 mg cysteine, add 250 mL 1 mg/mL disodium EDTA solution, sonicate for 5 min, filter (paper). Dilute a 3 mL aliquot of the filtrate to 100 mL with 1 mg/mL disodium EDTA solution. Remove a 1 mL aliquot and add it to 300 µL reagent solution, let stand at room temperature for 20 min, add 500 µL 300 mM phosphoric acid solution, add 3 mL 240 µg/mL IS solution, make up to 10 mL with water, inject a 50 µL aliquot. (Prepare the reagent solution by dissolving 3.5 mg methyl 4-(6-methoxynaphthalen-2-yl)-4-oxo-2-butenoate in 10 mL THF, make up to 25 mL with pH 7.5 borate buffer. Prepare methyl 4-(6-methoxynaphthalen-2-yl)-4-oxo-2-butenoate as described above.)

HPLC VARIABLES
Column: 150 × 4 5 µm Spherisorb RP-8
Mobile phase: MeCN:50 mM pH 3.0 triethylammonium phosphate 32:68
Flow rate: 1
Injection volume: 50
Detector: F ex 310 em 450

CHROMATOGRAM
Retention time: 3 (glutathione), 3.5 (cysteine), 4 (homocysteine), 5.5 (cysteamine), 6 (mesna), 9 (acetylcysteine)
Internal standard: 4-(naphthalen-2-yl)-4-oxobutanoic acid

OTHER SUBSTANCES
Non-interfering: bacitracin, biotin, calcium pantothenate, cystine, glycine, magnesium oxide, neomycin, starch, threonine, vitamin E, vitamin B6, vitamin B2 phosphate

KEY WORDS
powders; tablets

REFERENCE
Gatti, R.; Cavrini, V.; Roveri, P.; Pinzauti, S. High-performance liquid chromatographic determination of aliphatic thiols with aroylacrylic acids as fluorogenic precolumn derivatization reagents, *J.Chromatogr.*, **1990**, *507*, 451–458.

4-(6-Methylnaphthalen-2-yl)-4-oxo-2-butenoic Acid

RELATED REFERENCE

Cavrini, V.; Gatti, R.; Roveri, P.; Di Pietra, A. Analysis of thiols by HPLC after fluorescent prelabelling with 4-(6-methylnaphthalen-2-yl)-4-oxobuten-2-oic acid. *Chromatographia* **1989**, *27*, 185-190.

SAMPLE

Matrix: formulations

Analyte: glutathione

Sample preparation: Solutions, tablets. Dilute solutions with water, extract tablets with water. 500 µL Solution + 1 mL reagent, sonicate for 1 min, let stand at room temperature for 20 min, add 500 µL 3 M phosphoric acid, add 1 mL IS solution, make up to 10 mL with water, filter (0.45 µm), inject a 50 µL aliquot. Cream, gel. 1 g Cream or gel + 50 mL 5 M NaCl (cream) or water (gel), shake vigorously, centrifuge at 4000 rpm for 20 min. 500 µL Supernatant + 1 mL reagent, sonicate for 1 min, let stand at room temperature for 20 min, add 500 µL 3 M phosphoric acid, add 1 mL IS solution, make up to 10 mL with water, filter (0.45 µm), inject a 50 µL aliquot. (Reagent was 125 µg/mL 4-(6-methylnaphthalen-2-yl)-4-oxo-2-butenoic acid in pH 7.5 borate buffer. Prepare of 4-(6-methylnaphthalen-2-yl)-4-oxo-2-butenoic acid as follows. Slowly add 60 g finely-powdered aluminum chloride to a stirred solution of 24.5 g maleic anhydride in 250 mL 1,2-dichloroethane, stir for 30 min at room temperature. Gradually decant the solution away from undissolved aluminum trichloride into a solution of 35.5 g 2-methylnaphthalene in 100 mL 1,2-dichloroethane, after 30 min pour this mixture on to ice and dilute HCl. Remove the organic layer and remove the solvent by distillation, wash the residue (which solidifies overnight) with water, dry under reduced pressure, crystallize from acetic acid as golden-yellow needles (mp 169-170°) (J.Chem.Soc. 1958, 2437).)

HPLC VARIABLES

Column: 150 × 4 5 µm Hypersil C-8

Mobile phase: MeCN:buffer 32:68 (Buffer was 50 mM triethylamine adjusted to pH 4.0 with dilute phosphoric acid.)

Flow rate: 1

Injection volume: 50

Detector: F ex 300 em 460

CHROMATOGRAM

Retention time: 3

Internal standard: adduct of reagent with homocysteine (4.5)

OTHER SUBSTANCES

Non-interfering: acetyl methionine, BHA, glucuronolactone, glycine, inositol, oxidized glutathione, parabens, starch saccharose, uridine 5'-diphosphate, vitamin C, vitamin B12, vitamin A palmitate

KEY WORDS

solutions; tablets; cream; gels

REFERENCE
Gotti, R.; Andrisano, V.; Cavrini, V.; Bongini, A. Determination of glutathione in pharmaceuticals and cosmetics by HPLC with UV and fluorescence detection, *Chromatographia*, **1994**, *39*, 23–28.

ISOTHIOCYANATE

R-(-)-4-(3-Isothiocyanatopyrrolidin-1-yl)-7-nitro-2,1,3-benzoxadiazole

SAMPLE

Matrix: solutions

Analyte: thiols

Sample preparation: Mix 20 μL of a 30 μM solution in 2 mM disodium EDTA containing 1.5% pyridine (tiopronin) or 3% triethylamine (penicillamine) with 10 μL 12 mM R-(-)-4-(3-isothiocyanatopyrrolidin-1-yl)-7-nitro-2,1,3-benzoxadiazole in MeCN, let stand for 40 min, inject a 10 μL aliquot. (Synthesis of (R)-(-)-4-(3-isothiocyanatopyrrolidin-1-yl)-7-nitro-2,1,3-benzoxadiazole, (R)-(-)-NBD-PyNCS, is as follows. Cool a solution of 16.4 g (S)-(-)-1-benzyl-3-pyrrolidinol in 164 mL pyridine to +5°, add 19.35 g p-toluenesulfonyl chloride, stir at +10° for 48 h, evaporate to dryness, chromatograph using dichloromethane:acetone 95:5 to obtain (3S)-3-[(4-tolylsulfonyl)oxy]-1-(phenylmethyl)pyrrolidine (mp 68°). Heat a solution of (3S)-3-[(4-tolylsulfonyl)oxy]-1-(phenylmethyl)pyrrolidine in 200 mL anhydrous DMF to 65°, add 33.5 g sodium azide (Caution! Sodium azide is highly toxic!), stir at 60° for 7 h, filter, evaporate the filtrate to dryness under reduced pressure, dissolve the residue in ethyl acetate, wash twice with water, dry over anhydrous magnesium sulfate, evaporate to obtain (3R)-3-azido-1-(phenylmethyl)pyrrolidine as an oil. Add 3.5 g 10% palladium on carbon under nitrogen to a solution of 7.05 g (3R)-3-azido-1-(phenylmethyl)pyrrolidine in 34.8 mL 1 M HCl in water and 245 mL EtOH, hydrogenate at atmospheric pressure for 30 min, add 3.5 g catalyst, hydrogenate for 2 h, filter, add 34.8 mL 1 M HCl to the filtrate, evaporate to dryness under reduced pressure, take up the residue in 70 mL EtOH, filter, evaporate the filtrate to dryness under reduced pressure, repeat this operation twice, crystallize with the minimum amount of EtOH to obtain (3R)-3-aminopyrrolidine dihydrochloride (J. Med. Chem. 1992, 35, 4205). 3R-(+)-Aminopyrrolidine is also available from Tokyo Kasei (TCI America, 911 North Harborgate St., Portland, OR 97203; 800-423-8616, 503-283-1681, (fax) 503-283-1987; www.tciamerica.com). Add 100 mg 4-fluoro-7-nitro-2,1,3-benzoxadiazole in 20 mL MeCN dropwise to a stirred solution of 200 mg 3R-(+)-aminopyrrolidine in 20 mL MeCN at 0-10°, stir at room temperature for 30 min, remove the MeCN by evaporation under reduced pressure, dissolve the residue in 50 mL water, extract 4 times with 80 mL portions of ethyl acetate. Combine the organic layers and wash them with 20 mL water, dry over anhydrous sodium sulfate, evaporate to dryness under reduced pressure, recrystallize from hexane to obtain (R)-(-)-4-(3-aminopyrrolidin-1-yl)-7-nitro-2,1,3-benzoxadiazole ((R)-(-)-NBD-APy) as dark red crystals (mp 178-181°) (Analyst 1992, 117, 727). (R)-(-)-NBD-APy is also available from Tokyo Kasei. Add 100 μL thiophosgene in 10 mL benzene (Caution! Benzene is a carcinogen!) to 100 mg (R)-(-)-4-(3-aminopyrrolidin-1-yl)-7-nitro-2,1,3-benzoxadiazole in 100 mL acetone, reflux for 1 h, remove the solvent by evaporation under reduced pressure, suspend the residue in 100 mL water, extract 4 times with 25 mL portions of benzene. Combine the extracts and wash them with 20 mL water, dry over anhydrous sodium sulfate, evaporate to dryness under reduced pressure, recrystallize from hexane:benzene 1:2 to obtain (R)-(-)-4-(3-isothiocyanatopyrrolidin-1-yl)-7-nitro-2,1,3-benzoxadiazole as red crystals (mp 165-170°) (Analyst 1995, 120, 385).)

HPLC VARIABLES
Column: 150 × 4.6 5 μm Ultron VX-ODS (Shinwa, Kyoto)
Mobile phase: MeCN:water:trifluoroacetic acid 30:70:0.1 (tiopronin) or 35:65:0.1 (penicillamine)
Column temperature: 40
Flow rate: 1
Injection volume: 10
Detector: F ex 455 em 568

CHROMATOGRAM
Retention time: 21.5 (R-tiopronin), 23 (S-tiopronin), 20, 23 (penicillamine enantiomers)
Limit of detection: 0.5 pmole (tiopronin)

KEY WORDS
chiral

REFERENCE
Jin, D.; Takehana, K.; Toyo'oka, T. Chiral separation of racemic thiols based on diastereomer formation with a fluorescent chiral tagging reagent by reversed-phase liquid chromatography, *Anal.Sci.*, **1997**, *13*, 113–115.

2,3,4,6-Tetra-O-acetyl-β-D-glucopyranosyl Isothiocyanate

SAMPLE
Matrix: solutions
Analyte: thiols
Sample preparation: Add 1.05-3 equivalents 2,3,4,6-tetra-O-acetyl-β-D-glucopyranosyl isothiocyanate to 10 mL of a 100 μM solution of the thiol in MeCN:water 50:50 containing 1-3 equivalents triethylamine, vortex briefly, let stand at room temperature for 30 min, dilute with mobile phase, inject a 20 μL aliquot.

HPLC VARIABLES
Column: 150 × 4.6 5 μm TSKgel ODS-80TM (Tosoh)
Mobile phase: MeCN:10 mM pH 2.8 potassium phosphate buffer 50:50 (A) or 60:40 (B) or 53:47 (C)
Column temperature: 40
Flow rate: 1
Injection volume: 20

Detector: UV 250

CHROMATOGRAM
Retention time: k' 1.41 (L-alanine (A)), k' 1.88 (D-alanine (A)), k' 2.12 (S-tiopronin (A)), k' 2.63 (R-tiopronin (A)), k' 3.02 (L-cysteine (B)), k' 3.25, 3.81 (homocysteine (enantiomers) (B)), k' 3.50 (S-bucillamine (C)), k' 3.99 (L-penicillamine (B)), k' 4.07 (R-bucillamine (C)), k' 4.23, 5.14 (2-mercaptopropionic acid (enantiomers) (A)), k' 4.54 (D-cysteine (B)), k' 5.24 (D-penicillamine (B))

KEY WORDS
chiral

REFERENCE
Ito, S.; Ota, A.; Yamamoto, K.; Kawashima, Y. Resolution of the enantiomers of thiol compounds by reversed-phase liquid chromatography using chiral derivatization with 2,3,4,6-tetra-O-acetyl-β-D-glucopyranosyl isothiocyanate, *J.Chromatogr.*, **1992**, *626*, 187−196.

MALEIMIDE

N-(4-Anilinophenyl)maleimide

SAMPLE
Matrix: blood
Analyte: (2R,4R)-2-(2-hydroxyphenyl)-3-(3-mercaptopropionyl)-4-thiazolidinecarboxylic acid (SA 446)
Sample preparation: 500 µL Whole blood + 20 µL 10 mg/mL N-(4-anilinophenyl)maleimide in acetone + 300 µL 33 mM pH 6.85 sodium/potassium phosphate buffer + 100 ng IS, vortex, let stand at 0° for 30 min, extract twice with 2 mL portions of ether:hexane 50:50, add 500 µg glutathione, let stand at 0° for 20 min, add 3 mL acetone, centrifuge at 1500 g for 5 min, wash the precipitate with 3 mL acetone. Combine the supernatant and the wash and evaporate them to about 1 mL under reduced pressure, add 6 mL water, mix, add to a Sep-Pak C18 SPE cartridge, wash with 2 mL water, elute with 8 mL MeCN. Evaporate the eluate to dryness below 40° under reduced pressure, reconstitute the residue in 200 µL MeOH, inject an aliquot. (Prepare N-(4-anilinophenyl)maleimide as follows. Add dropwise 1.1 g maleic anhydride in 10 mL chloroform to 1 g N-phenylphenylenediamine (4-aminodiphenylamine) stirred in 10 mL chloroform at 0°, filter, dry to give N-(4-anilinophenyl)maleamic acid. Heat 100 mg N-(4-anilinophenyl)maleamic acid and 25 mg sodium acetate in 400 µL acetic anhydride on a water bath for 2 h, cool, pour into ice-water, filter, recrystallize from ethyl acetate/hexane to give N-(4-anilinophenyl)maleimide as yellow needles (mp 135-6°) (Anal. Chim. Acta 1983, 147, 375).)

HPLC VARIABLES
Column: 305 × 4 µBondapak C18
Mobile phase: MeCN:buffer 50:50 (Buffer was 0.8% (NH₄)H₂PO₄ adjusted to pH 3.0 with phosphoric acid.)
Flow rate: 1
Detector: E, Yanangimoto Model VMD 101, +1.0 V, Ag/AgCl reference electrode

CHROMATOGRAM
Retention time: 8
Internal standard: (2R,4R)-2-(2-methoxyphenyl)-3-(3-mercaptopropionyl)-4-thiazolidinecarboxylic acid (SA 427) (13)
Limit of quantitation: 2 ng/mL

KEY WORDS
whole blood; pharmacokinetics

REFERENCE
Shimada, K.; Tanaka, M.; Nambara, T.; Imai, Y. Determination of a new antihypertensive agent (2R,4R)-2-(2-hydroxyphenyl)-3-(3-mercaptopropionyl)-4-thiazolidinecarboxylic acid in blood by high performance liquid chromatography with electrochemical detection, *J.Pharm.Sci.*, **1984**, *73*, 119–121.

SAMPLE
Matrix: solutions
Analyte: thiols
Sample preparation: Add 1 mL 50 µg/mL N-(4-anilinophenyl)maleimide in 33 mM pH 6.85 phosphate buffer to 0.1-4 µg thiol, let stand at 0° for 90 min, wash twice with 2 mL portions of ether, heat the aqueous phase to 50° for 20 min, inject an aliquot. (Prepare N-(4-anilinophenyl)maleimide as described above.)

HPLC VARIABLES
Column: 305 × 6.3 µBondapak C18
Mobile phase: MeCN:0.5% pH 3.0 (NH₄)H₂PO₄ 4:7
Flow rate: 1
Injection volume: 10
Detector: E, Yanagimoto model VMD-101, glassy carbon electrode +1.0 V, Ag/AgCl reference electrode

CHROMATOGRAM
Retention time: 4 (glutathione), 8 (L-cysteine), 10 (N-acetyl-L-cysteine), 12 (D-penicillamine)

REFERENCE
Shimada, K.; Tanaka, M.; Nambara, T. Sensitive derivatization reagents for thiol compounds in high-performance liquid chromatography with electrochemical detection, *Anal.Chim.Acta*, **1983**, *147*, 375–380.

N-[p-(2-Benzoxazolyl)phenyl]maleimide

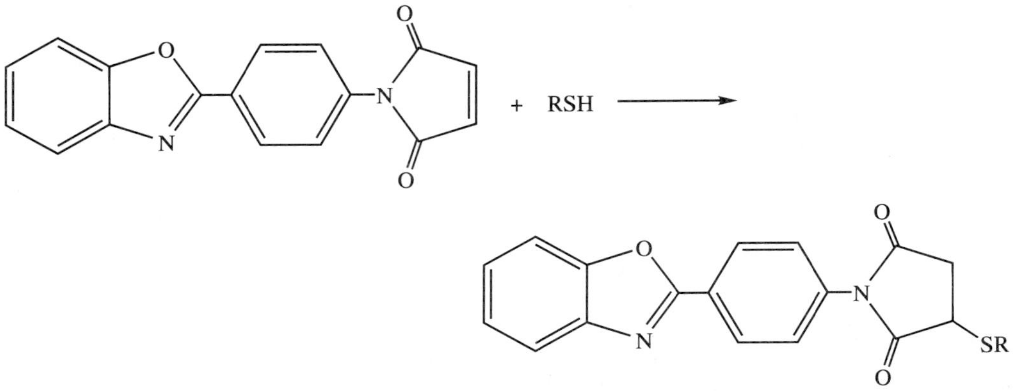

SAMPLE
Matrix: blood
Analyte: penicillamine
Sample preparation: 1 mL Plasma + 150 μL 25% trichloroacetic acid, vortex, cool on ice for 10 min, centrifuge at 6500 g for 2 min. Remove a 500 μL aliquot of the supernatant and add it to 200 μL 1% NaOH in water, add 250 μL buffer, add 1 mL 1 mM N-[p-(2-benzoxazolyl)phenyl]maleimide (Eastman) in EtOH, heat at 37° overnight, inject a 50 μL aliquot. (Buffer was 500 mM sodium citrate adjusted to pH 5.0 with perchloric acid.)

HPLC VARIABLES
Column: 300 × 3.9 10 μm μBondapak C18
Mobile phase: MeOH:100 μM sodium acetate 48:52
Flow rate: 2
Injection volume: 50
Detector: F ex 319 em 360 (cutoff filter)

CHROMATOGRAM
Retention time: 5.5
Limit of quantitation: 1 μM
Limit of detection: 250 nM

KEY WORDS
plasma

REFERENCE
Miners, J.O.; Fearnley, I.; Smith, K.J.; Birkett, D.J.; Brooks, P.M.; Whitehouse, M.W. Analysis of D-penicillamine in plasma by fluorescence derivatisation with N-[p-(2-benzoxazolyl)-phenyl] maleimide and high-performance liquid chromatography, *J.Chromatogr.*, **1983**, *275*, 89–96.

N-(4-Benzoylphenyl)maleimide

SAMPLE
Matrix: blood, urine
Analyte: captopril
Sample preparation: Plasma. 1 mL Plasma + 2 mL 100 mM pH 6.0 phosphate buffer + 500 μL 0.5% N-(4-benzoylphenyl)maleimide in acetone, vortex for 15 s, let stand at room temperature for 10 min, add 2 mL 500 mM pH 7.0 phosphate buffer, add 100 μL 40 μg/mL IS1 in acetone, wash twice with 4 mL portions of ether, acidify the aqueous phase with 500 μL 6 M HCl, extract with 7 mL chloroform. Remove the organic layer and evaporate it to dryness, reconstitute the residue in 100 μL MeOH, inject a 20 μL aliquot. Urine. 200 μL Urine + 200 μL 0.5% N-(4-benzoylphenyl)maleimide in acetone + 200 μL 100 mM pH 6.5 phosphate buffer, mix, let stand at room temperature for 15 min, add 2.5 mL 500 mM pH 7.0 phosphate buffer, wash with 4 mL diethyl ether, add 100 μL 10 μg/mL IS2 in acetone to the aqueous phase, acidify with 500 μL 6 M HCl, extract with 6 mL chloroform. Remove the organic layer and evaporate it to dryness, reconstitute the residue in 200 μL MeCN, inject a 20 μL aliquot. (Prepare N-(4-benzoylphenyl)maleimide by adding 5.3 g maleic anhydride to 9.6 g 4-aminobenzophenone in dioxane (Caution! Dioxane is a carcinogen!), stir at room temperature (Japan Pat. 59,204,171 (19 Nov. 1984); Chem. Abstr. 1985, 102, 113288t).)

HPLC VARIABLES
Column: 300 × 4 10 μm μBondapak C18
Mobile phase: MeCN:MeOH:1% acetic acid 45:11:75 (plasma) or 42.5:8.2:47.3 (urine)
Flow rate: 1
Injection volume: 20
Detector: UV 254

CHROMATOGRAM
Retention time: 21 (plasma), 10 (urine
Internal standard: adduct of N-(4-benzoylphenyl)maleimide with (4R)-2-(2-hydroxy-phenyl)-3-(3-mercaptopropionyl)-4-thiazolidinecarboxylic acid (IS1) (30 min), adduct of N-(4-benzoylphenyl)maleimide with thiosalicylic acid (IS2) (14 min) (Prepare adducts as follows. Add 150 mg N-(4-benzoylphenyl)maleimide in 2 mL acetone to 500 μmoles compound in 2 mL water, add 1 drop triethylamine, let stand at room temperature for 15 min, evaporate to dryness under reduced pressure.)
Limit of detection: 50 ng/mL

KEY WORDS
plasma; pharmacokinetics

REFERENCE
Hayashi, K.; Miyamoto, M.; Sekine, Y. Determination of captopril and its mixed disulfides in plasma and urine by high-performance liquid chromatography, *J.Chromatogr.*, **1985**, *338*, 161–169.

7-Diethylamino-3-(4'-maleimidylphenyl)-4-methyl Coumarin

SAMPLE
Matrix: cell suspensions
Analyte: thiols
Sample preparation: Suspend cells in 50 μL 40 mM pH 7.8 sodium Hepes buffer, add 50 μL 1.25 mM 7-diethylamino-3-(4'-maleimidylphenyl)-4-methylcoumarin (Molecular Probes, Eugene OR) in acetone, let stand at room temperature for 5 min, centrifuge, inject an aliquot.

HPLC VARIABLES
Column: 250 × 4.6 phenyl (Vydac)
Mobile phase: Gradient. A was 50 mM pH 5.5 ammonium acetate. B was MeCN:water 60:40. C was MeCN. A:B:C 50:50:0 for 5 min, to 30:70:0 over 30 min, to 10:0:90 over 15 min, maintain at 10:0:90 for 5 min.
Flow rate: 0.7
Detector: F ex 387 em 465

CHROMATOGRAM
Retention time: 14 (CoA), 15.5 (glutathione), 26 (CoM), 28 (homocysteine), 40 (pantetheine), 42 (trypanothione), 51 (dithiothreitol), 52 (lipoic acid)

REFERENCE
Steenkamp, D.J. Simple methods for the detection and quantification of thiols from *Crithidia fasciculata* and for the isolation of trypanothione, *Biochem.J.*, **1993**, *292*, 295–301.

SAMPLE
Matrix: tissue
Analyte: thiols
Sample preparation: Homogenize tissue with 4 volumes of ice-cold 5% 5-sulfosalicylic acid, centrifuge at 10000 g for 10 min. Adjust the pH of the supernatant to 7.5-8.5 with 10 M KOH, add 300 mM dithiothreitol so that the final concentration of dithiothreitol is 3 mM, let stand at room temperature for 1 h, dilute 10-fold. Remove an aliquot and mix it with an equal volume of reagent, heat at 37° for 1 h, inject a 50 μL aliquot. (Prepare reagent by mixing 100 μL 5 mg/mL 7-diethylamino-3-(4'-maleimidylphenyl)-4-methyl coumarin (Molecular Probes, Eugene OR) in MeCN with 900 μL 50 mM pH 7.5 sodium acetate in MeCN:water 40:60. Prepare just before use.)

HPLC VARIABLES
Guard column: μBondapak C18
Column: 150 × 4.6 4 μm Nova-Pak C18

Mobile phase: Gradient. A was 50 mM pH 6.0 sodium acetate. B was MeCN:water 80:20. A:B from 63:37 to 55:45 over 15 min, to 51:49 over 5 min, to 49:51 over 20 min, to 0: 100 over 5 min
Flow rate: 0.9
Injection volume: 50
Detector: F ex 387 em 465

CHROMATOGRAM
Retention time: 7 (glutathione), 11 (cysteine), 29 (cysteamine)
Limit of detection: 0.7 nmole/g

KEY WORDS
rat; pig; sheep; cow; liver; heart; kidney

REFERENCE
Garcia, R.A.G.; Hirschberger, L.L.; Stipanuk, M.H. Measurement of cyst(e)amine in physiological samples by high performance liquid chromatography, *Anal.Biochem.*, **1988**, *170*, 432–440.

N-[4-(6-Dimethylamino-2-benzofuranyl)phenyl]maleimide

SAMPLE
Matrix: formulations
Analyte: penicillamine
Sample preparation: Dissolve about 1 mg of the contents of a capsule in 10 mL water and dilute with MeCN to a penicillamine concentration of 8 mM. 50 µL Solution + 200 µL buffer + 150 µL reagent, heat at 60° for 30 min, cool, inject a 10 µL aliquot. (Buffer was prepared by adjusting the pH of a solution containing 100 mM boric acid and 100 mM KCl to 8.5 with 100 mM sodium carbonate. Reagent was 200 µM N-[4-(6-dimethyl-amino-2-benzofuranyl)phenyl]maleimide (DBPM) in MeCN. DBPM is available from Molecular Probes, Eugene OR or Dojindo Molecular Technologies, Inc. (3 Bethesda Metro Center, Suite 700, Bethesda MD 20814; (301) 664-8448; www.dojindo.co.jp). Synthesis of DBPM is as described below.)

HPLC VARIABLES
Column: 250 × 4.6 5 µm Sumichiral OA-2500S Pirkle-type (Sumika Chemical Analysis Service)
Mobile phase: MeOH:water 75:25 containing 150 mM ammonium acetate and 50 mM tetra-n-butylammonium bromide
Flow rate: 1
Injection volume: 10
Detector: F ex 360 em 455

CHROMATOGRAM
Retention time: 27 (D), 31 (L) (Each enantiomer gives 2 peaks, the later peaks are used for quantitation.)
Limit of detection: 290 fmole (D), 350 fmole (L)

KEY WORDS
capsules; chiral

REFERENCE
Nakashima, K.; Ishimaru, T.; Kuroda, N.; Akiyama, S. High-performance liquid chromatographic separation of penicillamine enantiomers labelled with N-[4-(6-dimethylamino-2-benzofuranyl)phenyl]maleimide on a chiral stationary phase, *Biomed.Chromatogr.*, **1995**, *9*, 90–93.

SAMPLE
Matrix: blood, tissue
Analyte: thiols
Sample preparation: Tissue. Homogenize (Potter-Elvehjem PTFE-glass homogenizer) tissue in 20 mM EDTA, adjust to 1% (w/v) (kidney) or 2.5% (w/v) (spleen). Remove 100 μL of this solution and add it to 400 μL 100 mM pH 8.5 borate buffer, add 300 μL 0.24 mM N-[4-(6-dimethylamino-2-benzofuranyl)phenyl]maleimide in MeCN, add 200 μL IS, heat at 60° for 30 min, cool for 5 min, centrifuge at 4° at 2000 g for 15 min, filter (Millipore 2 μm) the supernatant, inject a 20 μL aliquot of the supernatant. Plasma. Dilute rat plasma to 20% (v/v) with 20 mM EDTA. Remove 100 μL of this solution and add it to 400 μL 100 mM pH 8.5 borate buffer, add 300 μL 0.24 mM N-[4-(6-dimethylamino-2-benzofuranyl)phenyl]maleimide in MeCN, add 200 μL IS, heat at 60° for 30 min, cool for 5 min, centrifuge at 4° at 2000 g for 15 min, filter (Millipore 2 μm) the supernatant, inject a 20 μL aliquot of the supernatant. Serum. Dilute human serum to 10% (v/v) with 20 mM EDTA. 1 mL Diluted serum + 200 μL 30% metaphosphoric acid, centrifuge at 2000 g at 4° for 20 min. Remove 500 μL of the supernatant and add it to 240 μL 2 M KOH. Remove 100 μL of this solution and add it to 400 μL 100 mM pH 8.5 borate buffer, add 300 μL 0.24 mM N-[4-(6-dimethylamino-2-benzofuranyl)phenyl]maleimide in MeCN, add 200 μL IS, heat at 60° for 30 min, cool for 5 min, centrifuge at 4° at 2000 g for 15 min, filter (Millipore 2 μm) the supernatant, inject a 20 μL aliquot of the supernatant. (N-[4-(6-Dimethylamino-2-benzofuranyl)phenyl]maleimide is available from Molecular Probes, Eugene OR or Dojindo Molecular Technologies, Inc., 3 Bethesda Metro Center, Suite 700, Bethesda MD 20814; (301) 664-8448; www.dojindo.co.jp. Synthesis is as follows. Add 8.8 g aluminum trichloride to 12.50 g 3-dimethylaminophenol in 185 mL chloroform and 84 g triethyl orthoformate, mix at room temperature for 10 min, when the exothermic reaction ceases add 50 mL 10% HCl, stir to hydrolyze the acetal, neutralize with 10% NaOH, filter through a short column of Celite, wash through with chloroform, wash the filtrate with saturated aqueous NaCl, dry over magnesium sulfate, concentrate under reduced pressure, recrystallize from chloroform to give 4-(dimethylamino)salicylaldehyde (mp 78-79°). Add 400 mg KOH in 3 mL EtOH to a solution of 1 g 4-(dimethylamino)salicylaldehyde and 1.3 g (?) 4-nitrobenzyl bromide in 12 mL EtOH, reflux for 7 h, cool, filter to recover the crystals, wash with water, dry under vacuum, recrystallize from EtOH to give 4-dimethylamino-2-(4-nitrobenzyloxy)benzaldehyde (mp 179-180°). Add a solution of 900 mg 4-dimethylamino-2-(4-nitrobenzyloxy)benzaldehyde in 6 mL DMF to a sodium methoxide solution (prepared from 69 mg sodium in 1 mL MeOH), reflux for 20 min, add 1 mL MeOH, filter the crystals, recrystallize from EtOH to give 6-dimethylamino-2-(4-nitrophenyl)benzofuran as red needles (mp 209.5-210.5°). Reflux 1 g 6-dimethylamino-2-(4-nitrophenyl)benzofuran in 20 mL benzene (Caution!Benzene is a carcinogen!) and 18 mL MeOH containing 80 mg active carbon and a catalytic amount of ferric chloride hexahydrate for 10 min, add 2.30 g 98% hydrazine hydrate (Caution! Hydrazine hydrate is a carcinogen!) dropwise, reflux for 7 h, filter, concentrate the filtrate, recrystallize from cyclohexane to give 6-dimethylamino-2-(4-aminophenyl)benzofuran as orange needles (mp 198.5-200°). Stir 605 mg 6-dimethylamino-2-(4-aminophenyl)benzofuran and 230 mg maleic anhydride in 5 mL chloroform at room temperature for 3 h, filter the crystals, wash with a small amount of chloroform, recrystallize from EtOH to obtain N-[4-(6-dimethylamino-2-benzofuranyl)phenyl]maleamic acid (mp 219.5-221°). Reflux a mixture of 1.17 g N-[4-(6-dimethylamino-2-benzofuranyl)phenyl]maleamic acid and 30 mg sodium acetate

in 18 mL acetic anhydride, cool in an ice bath, collect the crystals of product, wash with water. Neutralize the filtrate with 20% NaOH, extract twice with 30 mL portions of chloroform, wash the organic layers with saturated aqueous NaCl, dry over anhydrous magnesium sulfate, evaporate to give more product. Combine the products and recrystallize them from acetone to give N-[4-(6-dimethylamino-2-benzofuranyl)phenyl]maleimide as reddish purple crystals (mp 203-204°) (Bull.Chem.Soc.Jpn. 1985, 58, 2192).)

HPLC VARIABLES
Column: 150 × 4.6 5 μm Toyo Soda ODS-80
Mobile phase: MeCN:10 mM pH 7.7 phosphate buffer 50:50 containing 30 mM tetrabutylammonium bromide
Flow rate: 0.8
Injection volume: 20
Detector: F ex 355 em 457

CHROMATOGRAM
Retention time: 4.5 (homocysteine), 5 (reduced glutathione (GSH)), 7 (N-acetylcysteine), 9 (cysteine), 10 (cysteamine), 11 (coenzyme A)
Internal standard: disodium 6-amino-1,3-naphthalene disulfonate (3.5)
Limit of detection: 50 fmole

KEY WORDS
plasma; serum; rat; human; liver; kidney; spleen

REFERENCE
Nakashima, K.; Umekawa, C.; Yoshida, H.; Nakatsuji, S.; Akiyama, S. High-performance liquid chromatography-fluorometry for the determination of thiols in biological samples using N-[4-(6-dimethylamino-2-benzofuranyl)phenyl]-maleimide, *J.Chromatogr.*, **1987**, *414*, 11–17.

N-(4-Dimethylamino-3,5-dinitrophenyl)maleimide

RELATED REFERENCES
Nakashima, K.; Muraoka, M.; Nakatsuji, S.; Akiyama, S. Determination of D-penicillamine in serum by high-performance liquid chromatography using N-[4-(6-dimethylamino-2-benzofuranyl)phenyl] maleimide. *Rinsho Kagaku (Nippon Rinsho Kagakkai)* **1988**, *17*, 51-54.
Nakashima, K.; Umekawa, C.; Nakatsuji, S.; Akiyama, S.; Givens, R.S. High-performance liquid chromatography/chemiluminescence determination of biological thiols with N-[4-(6-dimethylamino-2-benzofuranyl)phenyl]maleimide. *Biomed.Chromatogr.* **1989**, *3*, 39-42.

SAMPLE
Matrix: urine
Analyte: captopril
Sample preparation: 5 mL Urine + 2 mL 500 mM pH 7.0 phosphate buffer + 0.5 mL 20 mg/mL 2,4'-dibromoacetophenone (p-bromophenacyl bromide) in MeOH, shake vigorously. Remove a 1 mL aliquot and add it to 1.5 mL water and 6 mL hexane, mix, discard the hexane layer. Remove a 2 mL aliquot of the aqueous layer and add it to 100 μL 2% tributylphosphine in MeOH, heat at 50° for 30 min, wash with 6 mL hexane, add 200 μL 0.2% N-(4-dimethylamino-3,5-dinitrophenyl)maleimide in acetone to the aqueous layer, mix, let stand at room temperature for 5 min, wash with 6 mL hexane, discard the hexane layer, acidify the aqueous layer with about 200 μL 2 M HCl, extract twice with 6 mL portions of benzene (Caution! Benzene is a carcinogen!). Combine the organic layers and

add 10 μg IS, evaporate to dryness under reduced pressure, reconstitute the residue in 200 μL MeOH, inject a 5-20 μL aliquot. (Free captopril is derivatized as its 2,4'-dibromoacetophenone (p-bromophenacyl bromide) adduct then oxidized captopril is reduced and derivatized as its N-(4-dimethylamino-3,5-dinitrophenyl)maleimide adduct.)

HPLC VARIABLES
Column: μBondapak C18
Mobile phase: MeCN:water:acetic acid 46.5:53:0.5
Injection volume: 5-20
Detector: UV 254

CHROMATOGRAM
Retention time: 7 (free captopril (as 2,4'-dibromoacetophenone (p-bromophenacyl bromide adduct))), 8 (oxidized captopril (as N-(4-dimethylamino-3,5-dinitrophenyl)maleimide adduct))
Internal standard: thiosalicylic acid-p-bromophenacyl bromide adduct (Prepare by dissolving 2.4 mmoles thiosalicylic acid and 2.4 mmoles 2,4'-dibromoacetophenone (p-bromophenacyl bromide) in 40 mL MeOH, adjust to pH 7 by the dropwise addition of 1 M NaOH, allow to stand at room temperature for 10 min, evaporate to dryness under reduced pressure, reconstitute with 40 mL 50 mM pH 7.0 phosphate buffer, wash twice with 20 mL portions of hexane, adjust pH to 2 with dilute HCl, extract with 40 mL ethyl acetate, evaporate to dryness under reduced pressure, recrystallize the residue from benzene to give the adduct as pale yellow plates.) (12.5)
Limit of quantitation: 100 ng/mL

REFERENCE
Kawahara, Y.; Hisaoka, M.; Yamazaki, Y.; Inage, A.; Morioka, T. Determination of captopril in blood and urine by high-performance liquid chromatography, *Chem.Pharm.Bull.*, **1981**, *29*, 150–157.

N-(7-Dimethylamino-4-methyl-3-coumarinyl)maleimide

SAMPLE
Matrix: urine
Analyte: acetylcysteine and mercaptoacetate
Sample preparation: 5 mL Urine + EDTA, adjust pH to 9.8-10.0, react with thiopropyl-Sepharose, acidify with acetic acid, centrifuge. Remove a portion of the supernatant containing no more than 2 μmoles of thiols and add it to p-acetoxymercurianiline-Sepharose 4-B (Biochim. Biophys. Acta 1970, 200, 593), wash, elute with cysteine. Pass eluate through a small cation-exchange column (AG 50 W) (Bio-Rad) to remove cysteine, elute this column with 4 mL 10 mM HCl. (See Clin. Chim. Acta 1979, 95, 189.) Neutralize 250 μL eluate with 25 μL 100 mM NaOH, add 5 mL 50 mM pH 9.0 carbonate buffer containing 10 mM disodium EDTA, add 500 μL 0.5 mM N-(7-dimethylamino-4-methyl-3-

coumarinyl)maleimide in acetone, heat at 37° for 20 h, dilute 1:5 with mobile phase, inject a 100 μL aliquot.

HPLC VARIABLES

Column: 250 × 4 5 μm LiChrosorb RP-8
Mobile phase: MeOH:2 mM sodium phosphate 15:85, containing 10 mM tetramethylammonium hydroxide, pH adjusted to 7.4 with 6 M HCl
Flow rate: 0.75
Injection volume: 100
Detector: F ex 400 em 480

CHROMATOGRAM

Retention time: 15 (mercaptoacetate), 20 (acetylcysteine)
Limit of detection: 53 fmole

REFERENCE

Kågedal, B.; Källberg, M. Reversed-phase ion-pair high-performance liquid chromatography of mercaptoacetate and N-acetylcysteine after derivatization with N-(1-pyrene)maleimide and N-(7-dimethylamino-4-methyl-3-coumarinyl)maleimide, *J.Chromatogr.*, **1982**, *229*, 409–415.

SAMPLE

Matrix: urine
Analyte: thiols
Sample preparation: Mix 5 mL urine with 200 μL 130 mM disodium EDTA, adjust pH to 9.8-10.0 with 5 M ammonia, make up to 6 mL with water, add 1 mL thiopropyl-Sepharose 6-B suspension, shake mechanically for 30 min, acidify to pH 3.5-4.0 with 1 mL 4 M acetic acid, centrifuge, add a 3-5 mL aliquot of the supernatant (containing up to 2 μmoles thiol) to a 13 × 7 p-acetoxymercurianiline Sepharose 4-B column, wash with 2 mL water, elute with 3 mL 10 mM cysteine hydrochloride. Add the eluate to a 25 × 5 column of 100-200 mesh AG 50 W-X8 (hydrogen form, Bio-Rad), elute with 1 mL 10 mM HCl, collect all the effluent (Clin. Chim. Acta 1979, 95, 189), add 200 μL 180 mM disodium EDTA. Remove a 250 μL aliquot and neutralize it with 25 μL 100 mM NaOH, add 5 mL 50 mM pH 9.0 carbonate buffer containing 10 mM disodium EDTA, add 500 μL 500 μM N-(7-dimethylamino-4-methyl-3-coumarinyl)maleimide in acetone, heat at 37° for 20 h, dilute 1:5 with mobile phase, inject a 100 μL aliquot. (Thiopropyl-Sepharose 6-B suspension contains 20 μmoles thiol/mL. Before use convert to free thiol form with dithiothreitol according to the manufacturer's instructions. Prepare p-acetoxymercurianiline-Sepharose 4B as follows. Mix 100 g Sepharose 4B with an equal volume of water, for each 1 mL Sepharose add 100 mg cyanogen bromide in an equal volume of water, adjust pH to 11 with 4 M NaOH, maintain at pH 11 with 4 M NaOH (Proc. Natl. Acad. Sci. USA 1968, 61, 636), at the end of the reaction (about 8 min), wash with 1.5 L 100 mM pH 9.0 sodium bicarbonate. Suspend in 100 mL DMSO:water 10:90 at 0°, slowly add 1.3 g 4-aminophenylmercuric acetate in 20 mL DMSO, stir slowly at 0° for 20 h, warm to 30°, filter, resuspend in 130 mL DMSO:water 20:80 at 35° for 5 min, filter, repeat this procedure 4 times, pack in a column, slowly wash with DMSO:water 20:80 until no mercury appears in the effluent (about 500 mL), store as a slurry in DMSO:water 20:80 (Biochim. Biophys. Acta 1970, 200, 593).)

HPLC VARIABLES

Column: 250 × 4 5 μm LiChrosorb RP-8
Mobile phase: MeOH:buffer 15:85 (Buffer was 2 mM sodium phosphate buffer containing 11.8 mM tetramethylammonium hydroxide, pH adjusted to 7.4 with 6 M HCl.)
Flow rate: 0.75
Injection volume: 100
Detector: F ex 400 em 480

CHROMATOGRAM

Retention time: 15 (mercaptoacetate), 20 (N-acetylcysteine)
Limit of detection: 38-53 fmole

KEY WORDS

SPE

REFERENCE

Kågedal, B.; Källberg, M. Reversed-phase ion-pair high-performance liquid chromatography of mercaptoacetate and N-acetylcysteine after derivatization with N-(1-pyrene)maleimide and N-(7-dimethylamino-4-methyl-3-coumarinyl)maleimide, *J.Chromatogr.*, **1982**, *229*, 409–415.

SAMPLE

Matrix: urine

Analyte: 2-aminothiazoline-4-carboxylic acid

Sample preparation: 2.5 mL Urine + 100 µL 180 mM disodium EDTA, mix, adjust pH to 2-3 with 500 µL 8.7 M acetic acid, add to a 1 mL 26 × 7 column of 100-200 mesh AG 50W-X8 strong cation exchanger (Bio-Rad) equilibrated with 10 mM HCl, wash with 2 mL water, wash with 1 mL 2 M ammonium hydroxide, elute with 3 mL 2 M ammonium hydroxide. Adjust the eluate to pH 9.8-10.0 with 175 µL 8.7 M acetic acid, add 500 µL thiopropyl-Sepharose 6B, mix at room temperature for 30 min, adjust pH to 4.5 with 700 µL glacial acetic acid, centrifuge at 1000 g for 10 min, add the supernatant to a 500 µL 13 × 7 column of p-acetoxymercurianiline-Sepharose 4B, elute with 1 mL water. Collect all the effluent from the column, add 125 µL 20 g/L human serum albumin (ORHA 20/21, Behringwerke, Marburg, Germany), mix. Remove a 3 mL aliquot and add it to 3 mL 10 M NaOH, heat at 100° for 12.5 min, cool in ice. Remove a 100 µL aliquot and add it to 800 µL water and 1.5 mL 300 mM pH 9.7 carbonate/bicarbonate buffer containing 10 mM disodium EDTA, add 80 µL 5 M HCl, add 250 µL 20 µM N-(7-dimethylamino-4-methyl-3-coumarinyl)maleimide in acetone, mix vigorously, heat at 37° for 1.5 h, let stand at 4-8° overnight, dilute with mobile phase, inject an aliquot. (Convert thiopropyl-Sepharose 6B to the free thiol form by treating with dithiothreitol according to the manufacturer's directions. Prepare p-acetoxymercurianiline-Sepharose 4B as follows. Mix 100 g Sepharose 4B with an equal volume of water, for each 1 mL Sepharose add 100 mg cyanogen bromide in an equal volume of water, adjust pH to 11 with 4 M NaOH, maintain at pH 11 with 4 M NaOH (Proc. Natl. Acad. Sci. USA 1968, 61, 636), at the end of the reaction (about 8 min), wash with 1.5 L 100 mM pH 9.0 sodium bicarbonate. Suspend in 100 mL DMSO:water 10:90 at 0°, slowly add 1.3 g 4-aminophenylmercuric acetate in 20 mL DMSO, stir slowly at 0° for 20 h, warm to 30°, filter, resuspend in 130 mL DMSO:water 20:80 at 35° for 5 min, filter, repeat this procedure 4 times, pack in a column, slowly wash with DMSO:water 20:80 until no mercury appears in the effluent (about 500 mL), store as a slurry in DMSO:water 20:80 (Biochim. Biophys. Acta 1970, 200, 593).)

HPLC VARIABLES

Guard column: 10 × 3.2 5 µm Kromasil C18 (Hicrome)

Column: 250 × 4.6 5 µm Kromasil KR 100-5 C18 (Hicrome)

Mobile phase: MeOH:50 mM pH 2.3 orthophosphoric acid 10:13

Flow rate: 0.5

Injection volume: 50

Detector: F ex 385 em 476

CHROMATOGRAM

Retention time: 15

Limit of detection: 300 nM

OTHER SUBSTANCES

Simultaneously analyzed: cysteine

KEY WORDS

rat; human; SPE

REFERENCE

Lundquist, P.; Kågedal, B.; Nilsson, L.; Rosling, H. Analysis of the cyanide metabolite 2-aminothiazoline-4-carboxylic acid in urine by high-performance liquid chromatography, *Anal.Biochem.*, **1995**, *228*, 27–34.

N-(4-Dimethylaminophenyl)maleimide

SAMPLE
Matrix: blood
Analyte: captopril
Sample preparation: 500 µL Whole blood + 20 µL 1% N-(4-dimethylamino-phenyl)maleimide in acetone + 300 µL 33.3 mM pH 6.85 phosphate buffer, add 100 ng IS, vortex, let stand at 0° for 30 min, freeze in dry ice/acetone, thaw, wash twice with 2 mL portions of ether. Add 500 µg glutathione to the aqueous layer, let stand at 0° for 20 min, add 3 mL acetone, centrifuge at 1580 g for 5 min. Remove the supernatant and wash the precipitate with 3 mL acetone. Combine the supernatant and the wash and evaporate them to about 1 mL under reduced pressure at room temperature, dilute the residue with 6 mL water, add to a Sep-Pak C18 SPE cartridge, wash with 2 mL water, elute with 8 mL MeCN. Evaporate the eluate to dryness under reduced pressure below 40°, reconstitute the residue in 200 µL MeOH, inject an aliquot. (Prepare N-(4-dimethylamino-phenyl)maleimide as follows. Mix equimolar amounts of maleic anhydride and N,N-dimethyl-1,4-phenylenediamine in ether or THF with cooling in ice and stirring, allow to stand overnight, remove the maleamic acid by filtration, wash with THF. Heat 1 mmole of the maleamic acid with 3 mmole acetic anhydride and 0.3 mmole sodium acetate at 100° for 5-10 min (until the solution goes clear), cool, add ice water, neutralize with sodium bicarbonate, extract with ethyl acetate. Wash the organic layer with saturated sodium chloride and dry over anhydrous sodium sulfate (Chem. Pharm. Bull. 1976, 24, 3045; 1977, 25, 2739). Evaporate to dryness and recrystallize from acetone to give N-(4-dimethylaminophenyl)maleimide as reddish-brown crystals (mp 153-154° (J. Org. Chem. 1963, 28, 2018)).)

HPLC VARIABLES
Column: 300 × 3.9 8-10 µm µBondapak C18
Mobile phase: MeCN:0.8% pH 3.0 $(NH_4)H_2PO_4$ 1:2
Flow rate: 1
Detector: E, Yanagimoto VMD 101, +0.9 V, Ag/AgCl reference electrode

CHROMATOGRAM
Retention time: 10
Internal standard: (4R)-2-(2-hydroxyphenyl)-3-(3-mercaptopropionyl)-4-thiazolidinecarboxylic acid (Sankyo SA 446) (19)
Limit of detection: 10 ng/mL

KEY WORDS
whole blood; SPE; pharmacokinetics

REFERENCE
Shimada, K.; Tanaka, M.; Nambara, T.; Imai, Y.; Abe, K.; Yoshinaga, K. Determination of captopril in human blood by high-performance liquid chromatography with electrochemical detection, *J.Chromatogr.*, **1982**, *227*, 445–451.

N-Ethylmaleimide

SAMPLE
Matrix: blood
Analyte: 6-mercaptopurine
Sample preparation: 200 µL Serum + 5 µL 50 µg/mL 2-ethyl-4-oxoquinazoline in EtOH + 100 µL reagent, let stand at room temperature for 1 h, add 1.8 mL ethyl acetate, mix, centrifuge at 1800 g for 5 min, remove 1.5 mL of the supernatant, repeat the extraction. Combine the organic layers and evaporate them to dryness under reduced pressure below 30°, reconstitute the residue in 100 µL initial mobile phase, inject a 90 µL aliquot. (Reagent was 30 mg N-ethylmaleimide in 2 mL 50 mM pH 7.0 phosphate buffer, prepare fresh daily.)

HPLC VARIABLES
Column: 10 µm µBondapak C18
Mobile phase: Gradient. MeCN:10 mM KH_2PO_4 9:91 for 26 min, then 50:50 for 1 min (step gradient).
Flow rate: 1.5
Injection volume: 90
Detector: UV 280

CHROMATOGRAM
Retention time: 18.6
Internal standard: 2-ethyl-4-oxoquinazoline (28)
Limit of quantitation: 10 ng/mL

OTHER SUBSTANCES
Extracted: azathioprine (not derivatized)

KEY WORDS
serum

REFERENCE
Tsutsumi, K.; Otsuki, Y.; Kinoshita, T. Simultaneous determination of azathioprine and 6-mercaptopurine in serum by reversed-phase high-performance liquid chromatography, *J.Chromatogr.*, **1982**, *231*, 393–399.

Eu³⁺ Chelate of 1-[p-(((5-Maleimidopentyl)carbonyl)amino)benzyl]ethylenediaminetetraacetic Acid

RSH +

SAMPLE
Matrix: solutions
Analyte: thiols
Sample preparation: Mix a 100 μL aliquot of a 10-100 nM solution of thiols with 50 μL 10-100 μM reagent and 50 μL 100 mM pH 9 borate buffer, let stand at room temperature for 2 min, add 50 μL 1 M Bis-Tris, inject a 20 μL aliquot. (The reagent was the Eu³⁺ chelate of 1-[p-(((5-maleimidopentyl)carbonyl)amino)benzyl]ethylenediaminetetraacetic acid (Dojindo, Kumamoto, Japan (Dojindo Molecular Technologies, Inc., 3 Bethesda Metro Center, Suite 700, Bethesda MD 20814; (301) 664-8448; www.dojindo.co.jp).)

HPLC VARIABLES
Column: 250 × 4.6 5 μm Capcellpak SG-120 C18 (Shiseido)
Mobile phase: MeCN:20 mM pH 7.0 phosphate buffer 8:92
Flow rate: 1
Injection volume: 20
Detector: F ex 344 em 617 following post-column reaction. The column effluent mixed with the reagent pumped at 0.2 mL/min and the mixture flowed through a 2 m coil to the detector. (Prepare reagent by dissolving 15 g Triton X-100 in 500 mL water, add 58 mL acetic acid, add 20 mL 50 mM tri-n-octylphosphine oxide in MeCN, add 1 mL 100 mM 3-naphthoyltrifluoroacetone in MeCN, make up to 1 L with water. 2-Naphthoyltrifluoroacetone (4,4,4-trifluoro-1-(2-naphthyl)-1,3-butanedione) is available from Tokyo Kasei (TCI America, Portland OR). Prepare by adding 1 mole ethyl trifluoroacetate dropwise with stirring to 1.05 moles sodium methoxide in 100 mL anhydrous ether, add 1 mole 2'-acetonaphthone, let stand overnight, evaporate to dryness under reduced pressure, heat at 90° under reduced pressure to remove EtOH, add 1 mole 10% sulfuric acid, recrystallize from EtOH/water to obtain 2-naphthoyltrifluoroacetone (mp 70.1-71.1°) (J. Am. Chem. Soc. 1950, 72, 2948).)

CHROMATOGRAM
Retention time: 12 (cysteine), 15 (homocysteine), 19 (coenzyme A)
Limit of detection: 13.2-18.8 fmole

KEY WORDS
post-column reaction

REFERENCE
Okabayashi, Y.; Kitagawa, T. High-performance liquid chromatography for amino compounds and thiol compounds derivatized with europium chelate, *Anal.Chem.*, **1994**, *66*, 1448–1453.

N-(Ferrocenyl)maleimide

SAMPLE
Matrix: blood, tissue
Analyte: thiols
Sample preparation: Homogenize liver with buffer. Dilute whole blood 20-fold with ice-cold buffer. Remove a 100 μL aliquot of diluted blood or liver homogenate and add it to 50 μL 0.5% N-(ferrocenyl)maleimide in acetone and 440 μL 3.4 μg/mL N-acetyl-L-cysteine in buffer, vortex, let stand at 0° for 30 min, wash with three 2 mL portions of diethyl ether:hexane 50:50, add 2 mL acetone to the aqueous layer, centrifuge at 1100 g for 5 min, inject an aliquot of the supernatant. (Buffer was 67 mM pH 6.8 phosphate buffer containing 1 mM EDTA. Prepare N-(ferrocenyl)maleimide as follows. Stir 13 g ferrocene in 200 mL anhydrous THF at -30° under nitrogen, add 160 mL 1.3 M butyllithium in ether dropwise over 25 min, stir at 0° for 2 h, stir at room temperature for 4 h. Stir at -20° and add 10.3 g methoxylamine in 75 mL anhydrous ether dropwise over 30 min, allow to warm gradually to room temperature, stir for 4 h, slowly add 10% HCl with stirring until the pH of the aqueous layer is 2, discard the organic layer. Make the aqueous layer strongly basic with KOH, extract with ether, extract the ether layer with 2 M HCl. Make the aqueous layer basic with KOH, extract with ether, dry over anhydrous magnesium sulfate, evaporate to dryness, recrystallize from ether/petroleum ether to obtain ferrocenylamine (8%, mp 140-145°) (J. Org. Chem. 1959, 24, 1487). Dissolve 300 mg ferrocenylamine in the minimum amount of chloroform, add 150 mg maleic anhydride in a little chloroform. Collect the compound that crystallizes and heat 300 mg with 1.5 g acetic anhydride and 60 mg anhydrous sodium acetate at 50-60° for 3 h with the exclusion of moisture (Chem. Zvesti 1963, 17, 21), recrystallize the product from diethyl ether to give N-(ferrocenyl)maleimide as deep purple prisms (mp 151-152°).

HPLC VARIABLES
Column: 150 × 4.6 5 μm YMC-GEL C8 (Yamamura, Kyoto)
Mobile phase: MeCN:buffer 20:50 (Buffer was 0.32% Na_2HPO_4 adjusted to pH 5.0 with phosphoric acid.)
Flow rate: 1
Detector: E, Environmental Sciences Associates 5100A, 5011 porous graphite dual electrode analytical cell, upstream electrode +150 mV, downstream electrode -100 mV, 5020 guard cell +200 mV, palladium reference electrode

CHROMATOGRAM
Retention time: 4 (glutathione), 5 (L-cysteine)
Internal standard: N-acetyl-L-cysteine (6)
Limit of detection: 0.06 pmole

KEY WORDS
whole blood; liver; human; mouse

REFERENCE
Shimada, K.; Oe, T.; Nambara, T. Sensitive ferrocene reagents for derivatization of thiol compounds in high-performance liquid chromatography with dual-electrode coulometric detection, *J.Chromatogr.*, **1987**, *419*, 17–25.

2-(4-N-Maleimidephenyl)-6-methylbenzothiazole

SAMPLE
Matrix: blood, tissue
Analyte: thiols
Sample preparation: Plasma. 1 mL Plasma + 1 mL ice-cold 2 M perchloric acid containing 4 mM EDTA, vortex, centrifuge at 4° at 4000 g for 10 min. Neutralize an aliquot of the supernatant with cold 2 M LiOH solution, adjust to 10% with pH 8.4 borate buffer. Remove a 100 μL aliquot and add it to 100 μL 50 mM N-acetylcysteine solution, 100 μL reagent solution, and 700 μL pH 8.4 borate buffer, vortex for a few s, let stand at room temperature for 1 h, add an equal volume of the mobile phase, inject a 10 μL aliquot. Tissue. Powder tissue at low temperature. Add 4 volumes ice-cold 2 M perchloric acid containing 25 mM EDTA to 0.1-0.5 g powdered tissue, vortex quickly, homogenize (Brinkman). Neutralize an aliquot with cold 2 M LiOH solution, adjust to 1% (brain, lung, liver, kidney, testes, small intestine) or 2.5% (aorta, heart, spleen) with pH 8.4 borate buffer. Remove a 100 μL aliquot and add it to 100 μL 50 mM N-acetylcysteine solution, 100 μL reagent solution, and 700 μL pH 8.4 borate buffer, vortex for a few s, let stand at room temperature for 1 h, add an equal volume of the mobile phase, inject a 10 μL aliquot. (Prepare reagent, 2-(4-N-maleimidephenyl)-6-methylbenzothiazole, as follows. Recrystallize 2-(4-aminophenyl)-6-methylbenzothiazole from chloroform before use. Add 500 mg maleic anhydride in 2 mL chloroform dropwise to 1.2 g 2-(4-aminophenyl)-6-methylbenzothiazole in 10 mL DMF, stir at room temperature for 2 h, filter, wash with 30 mL chloroform, recrystallize from DMF to give 2-(4-N-phenylmaleamic acid)-6-methylbenzothiazole as yellow crystals (mp 242°). Reflux 2 g 2-(4-N-phenylmaleamic acid)-6-methylbenzothiazole, 100 mg anhydrous sodium acetate, and 25 mL acetic anhydride for 2 h, cool on ice, filter, wash the solid with water. Neutralize the filtrate with cold 10% NaOH, extract with chloroform. Dry the organic layer over anhydrous magnesium sulfate and evaporate it to dryness under reduced pressure. Combine this product with the solid obtained earlier and recrystallize from isopropanol to give 2-(4-N-maleimidephenyl)-6-

methylbenzothiazole as yellow needles (mp 254-6°). Prepare the reagent solution by dissolving 50 μmoles of this compound in 10 mL DMF and diluting 25-fold with MeCN.)

HPLC VARIABLES

Column: 250 × 4.6 5 μm Ultrasphere ODS
Mobile phase: MeCN:buffer 35:65, pH 4.5 (Buffer was 10 mM KH_2PO_4 containing 0.1% sodium hexanesulfonate.)
Flow rate: 1.5
Injection volume: 10
Detector: F ex 320 em 405

CHROMATOGRAM

Retention time: 2 (coenzyme A), 3 (homocysteine), 5 (glutathione), 10 (cysteine), 12 (N-acetylpenicillamine), 17 (penicillamine)
Internal standard: N-acetylcysteine (7)
Limit of detection: 2-20 fmole

KEY WORDS

plasma; rat; aorta; heart; lung; liver; kidney; testes; spleen; brain; small intestine

REFERENCE

Haj-Yehia, A.I.; Benet, L.Z. Determination of aliphatic thiols by fluorometric high-performance liquid chromatography after precolumn derivatization with 2-(4-N-maleimidophenyl)-6-methylbenzothiazole, *Pharm.Res.*, **1995**, *12*, 155–160.

N-(1-Pyrenyl)maleimide

RELATED REFERENCES

Arroyo, C.; López-Calull, C.; García-Capdevila, L.; Gich, I.; Barbanoj, M.; Bonal, J. Determination of captopril in plasma by high-performance liquid chromatography for pharmacokinetic studies. *J.Chromatogr.B* **1997**, *688*, 339-344.

Ercal, N.; Oztezcan, S.; Hammond, T.C.; Matthews, R.H.; Spitz, D.R. High-performance liquid chromatography assay for N-acetylcysteine in biological samples following derivatization with N-(1-pyrenyl)maleimide. *J.Chromatogr.B* **1996**, *685*, 329-334.

SAMPLE

Matrix: blood
Analyte: captopril
Sample preparation: 1 mL Blood + 50 μL solution containing 100 mM EDTA and 100 mM ascorbic acid, centrifuge at 13000 g for 2 min. 500 μL Supernatant + 2 mL 100 mM pH 7 phosphate buffer + 200 ng IS + 200 μL 1.5 mg/mL N-(1-pyrenyl)maleimide in MeCN, shake for 15 min, acidify with 100 μL 11 M HCl, add 6 mL ethyl acetate, vortex for 20 min, centrifuge at 2500 g for 5 min. Remove the organic layer, dry it under nitrogen, dissolve in 50 or 200 μL MeCN, inject a 5-15 μL aliquot.

HPLC VARIABLES
Column: 100 × 4.6 5 μm Partisil ODS-3 C18
Mobile phase: MeCN:1% acetic acid 37:63
Flow rate: 1.5
Injection volume: 5-15
Detector: F ex 340 em 389

CHROMATOGRAM
Retention time: 8
Internal standard: (4R)-2-(2-hydroxyphenyl)-3-(3-mercaptopropionyl)-4-thiazolidinecar-boxylic acid (SA 446) (14)
Limit of detection: 10 ng/mL

OTHER SUBSTANCES
Non-interfering: chloral hydrate, chlorpromazine, digoxin, furosemide, promethazine

KEY WORDS
with modifications can be used to determine captopril disulfide; pharmacokinetics

REFERENCE
Pereira, C.M.; Tam, Y.K.; Collins-Nakai, R.L.; Ng, P. Simplified determination of captopril in plasma by high-performance liquid chromatography, *J.Chromatogr.*, **1988**, *425*, 208–213.

SAMPLE
Matrix: blood
Analyte: tiopronin
Sample preparation: 1 mL Plasma + 200 μL 200 mM pH 8.0 phosphate buffer + 200 μL chloroform:tributylphosphine 90:10, vortex for 15 s, heat at 50° for 30 min, cool in an ice bath, add 2 mL EtOH, vortex for 15 s, centrifuge at 5° at 1800 g for 10 min, inject a 50 μL aliquot of the supernatant within 15 min.

HPLC VARIABLES
Guard column: 4 × 4 5 μm LiChrospher 100 RP-18e
Column: 125 × 4 5 μm LiChrospher 100 RP-18e
Mobile phase: MeCN:10 mM pH 7.0 phosphate buffer 25:75 containing 5 mM cetrimonium bromide
Column temperature: 35
Flow rate: 1
Injection volume: 50
Detector: F ex 260 em 370 (long-pass filter) following post-column reaction. The column effluent mixed with MeCN:water 30:70 containing 2% triethylamine and 1% Brij-35 pumped at 0.35 mL/min and 50 μM pyrenemaleimide in MeCN pumped at 0.35 mL/min and flowed through a 3 m × 0.5 mm ID PTFE knitted open tubular reactor to the detector.

CHROMATOGRAM
Retention time: 6.5
Limit of detection: 50 ng/mL

OTHER SUBSTANCES
Extracted: metabolites

KEY WORDS
plasma; pharmacokinetics; post-column reaction

REFERENCE
Leroy, P.; Nicolas, A.; Gavriloff, C.; Matt, M.; Netter, P.; Bannwarth, B.; Hercelin, B.; Mazza, M. Determination of 2-mercaptopropionylglycine and its metabolite, 2-mercaptopropionic acid, in plasma by ion-pair reversed-phase high-performance liquid chromatography with post-column derivatization, *J.Chromatogr.*, **1991**, *564*, 258–265.

SAMPLE
Matrix: cell suspensions
Analyte: thiols
Sample preparation: Centrifuge at 4° at 400 g for 5 min, store the pellet at -20° overnight, lyse in 100 mM pH 8.0 Tris-HCl buffer containing 10 mM borate and 5 mM L-serine

(Proc. Natl. Acad. Sci. 1978, 75, 4806) with four 5 s sonicator (Biosonik III, Bronwill Scientific) bursts at 30% output at 0°, make up to 250 μL with water, add 250 μL MeCN, add 500 μL 1.5 mM N-(1-pyrenyl)maleimide in MeCN, mix, let stand at room temperature for 5 min, add 10 μL 50% acetic acid, filter (0.2 μm), store the filtrate at 4°, inject a 20 μL aliquot. (To measure glutathione disulfide make up to 98 μL with water, add 2 μL 2-vinylpyridine, let stand at room temperature for 1 h, add 95 μL 2 mg/mL NADPH in 100 mM pH 7.5 sodium phosphate buffer containing 6.3mM EDTA, add 5 μL 5 mg/mL glutathione reductase suspension (Boehringer-Mannheim), mix, remove a 100 μL aliquot, proceed as above.)

HPLC VARIABLES
Column: 100 × 4.6 3 μm C18 (Astec)
Mobile phase: MeCN:water:acetic acid:phosphoric acid 65:35:0.1:0.1
Flow rate: 0.5
Injection volume: 20
Detector: F ex 330 em 380

CHROMATOGRAM
Retention time: 8.5 (glutathione), 10 (gamma-glutamylcysteine), 11.5 (cysteine), 14.5 (cysteinylglycine), 15 (homocysteine)
Limit of quantitation: 10 nM
Limit of detection: 50 fmole

KEY WORDS
comparison with other derivatizing reagents

REFERENCE
Winters, R.A.; Zukowski, J.; Ercal, N.; Matthews, R.H.; Spitz, D.R. Analysis of glutathione, glutathione disulfide, cysteine, homocysteine, and other biological thiols by high-performance liquid chromatography following derivatization by N-(1-pyrenyl)maleimide, *Anal.Biochem.*, **1995**, *227*, 14–21.

SAMPLE
Matrix: perfusate
Analyte: captopril
Sample preparation: Centrifuge perfusate, add 100 μL supernatant to 50 μL 2 mM N-(1-pyrenyl)maleimide in acetone, add mixture to 2 mL pH 7 phosphate buffer, stir for 15 min at room temperature, inject an aliquot.

HPLC VARIABLES
Column: 250 × 4.6 Nucleosil 5C18
Mobile phase: MeCN:0.1% phosphoric acid 47:53
Detector: F ex 340 em 390

REFERENCE
Kobayashi, D.; Matsuzawa, T.; Sugibayashi, K.; Morimoto, Y.; Kobayashi, M.; Kimura, M. Feasibility of use of several cardiovascular agents in transdermal therapeutic systems with *l*-menthol-ethanol system on hairless rat and human skin, *Biol.Pharm.Bull.*, **1993**, *16*, 254–258.

SAMPLE
Matrix: urine
Analyte: thiols
Sample preparation: Mix 5 mL urine with 200 μL 130 mM disodium EDTA, adjust pH to 9.8-10.0 with 5 M ammonia, make up to 6 mL with water, add 1 mL thiopropyl-Sepharose 6-B suspension, shake mechanically for 30 min, acidify to pH 3.5-4.0 with 1 mL 4 M acetic acid, centrifuge, add a 3-5 mL aliquot of the supernatant (containing up to 2 μmoles thiol) to a 13 × 7 p-acetoxymercurianiline Sepharose 4-B column, wash with 2 mL water, elute with 3 mL 10 mM cysteine hydrochloride. Add the eluate to a 25 × 5 column of 100-200 mesh AG 50 W-X8 (hydrogen form, Bio-Rad), elute with 1 mL 10 mM HCl, collect all the effluent (Clin. Chim. Acta 1979, 95, 189), add 200 μL 180 mM disodium EDTA. Remove a 250 μL aliquot and neutralize it with 25 μL 100 mM NaOH, add 5 mL 50 mM pH 9.0 carbonate buffer containing 10 mM disodium EDTA, add 500 μL 500 μM N-(1-pyrenyl)maleimide in EtOH:acetone 50:50, heat at 37° for 20 h, dilute 1:5 with mobile phase, inject a 100 μL aliquot. (Thiopropyl-Sepharose 6-B suspension contains 20 μmoles

thiol/mL. Before use convert to free thiol form with dithiothreitol according to the manufacturer's instructions. Prepare p-acetoxymercurianiline-Sepharose 4B as follows. Mix 100 g Sepharose 4B with an equal volume of water, for each 1 mL Sepharose add 100 mg cyanogen bromide in an equal volume of water, adjust pH to 11 with 4 M NaOH, maintain at pH 11 with 4 M NaOH (Proc. Natl. Acad. Sci. USA 1968, 61, 636), at the end of the reaction (about 8 min), wash with 1.5 L 100 mM pH 9.0 sodium bicarbonate. Suspend in 100 mL DMSO:water 10:90 at 0°, slowly add 1.3 g 4-aminophenylmercuric acetate in 20 mL DMSO, stir slowly at 0° for 20 h, warm to 30°, filter, resuspend in 130 mL DMSO:water 20:80 at 35° for 5 min, filter, repeat this procedure 4 times, pack in a column, slowly wash with DMSO:water 20:80 until no mercury appears in the effluent (about 500 mL), store as a slurry in DMSO:water 20:80 (Biochim. Biophys. Acta 1970, 200, 593).)

HPLC VARIABLES
Column: 250 × 4 5 μm LiChrosorb RP-8
Mobile phase: MeOH:buffer 30:70 (Buffer was 2 mM sodium phosphate buffer containing 14.3 mM tetramethylammonium hydroxide, pH adjusted to 7.4 with 6 M HCl.)
Flow rate: 0.75
Injection volume: 100
Detector: F ex 342 em 396

CHROMATOGRAM
Retention time: 20 (mercaptoacetate), 22 (N-acetylcysteine)
Limit of detection: 320-380 fmole

KEY WORDS
SPE

REFERENCE
Kågedal, B.; Källberg, M. Reversed-phase ion-pair high-performance liquid chromatography of mercaptoacetate and N-acetylcysteine after derivatization with N-(1-pyrene)maleimide and N-(7-dimethylamino-4-methyl-3-coumarinyl)maleimide, *J.Chromatogr.*, **1982**, *229*, 409–415.

SAMPLE
Matrix: urine
Analyte: cysteinylglycine
Sample preparation: Collect 24 h urine with 5 mL 700 mM thymol. 5 mL Urine + 200 μL 130 mM disodium EDTA + 500 μL 1 M dithiothreitol, adjust pH to 6.0 with 1 M NaOH or 1 M HCl, make up to 6 mL with water, mix, let stand at room temperature for 1.5 h, adjust pH to 1.5 ± 0.1 with 6 M HCl, make up to 6.2 mL with water. Remove a 1 mL aliquot and add it to 5 mL ethyl acetate saturated with water, shake horizontally for 5 min, centrifuge, discard the organic phase, repeat the ethyl acetate wash three times. Evaporate the aqueous phase to dryness at 40°, add 10 mL 50 mM pH 9.0 carbonate buffer containing 10 mM disodium EDTA, adjust pH to 9.0 ± 0.1 with 1 M NaOH, make up to 10.2 mL with water. Remove a 5 mL aliquot and add it to 500 μL 1 mM N-(1-pyrenyl)maleimide in EtOH:acetone 50:50, heat at 37° for 20 h, dilute 5-fold with mobile phase, inject an aliquot. (Purify 5 mg N-(1-pyrenyl)maleimide in 3 mL toluene:acetone 90:10 on a Lobar LiChroprep Si-60 column (size B, Merck) by eluting with toluene:acetone 90:10 at 1 mL/min with detection at 339 nM. Collect 5 mL fractions and evaporate them to dryness, prepare 1 mM solutions in EtOH:acetone 50:50.)

HPLC VARIABLES
Guard column: 37-53 μm Solvecon Silica (Pierce)
Column: 250 × 4.6 5 μm Supelcosil LC-8
Mobile phase: MeOH:50 mM phosphoric acid 45:55
Column temperature: 25
Flow rate: 1
Injection volume: 100
Detector: F ex 342 em 396

CHROMATOGRAM
Retention time: 40

REFERENCE
Kågedal, B.; Källberg, M. Cysteinylglycine in urine determined by high-performance liquid chromatography, *J.Chromatogr.*, **1984**, *308*, 75–82.

QUINONE

3,5-Di-tert-butyl-1,2-benzoquinone

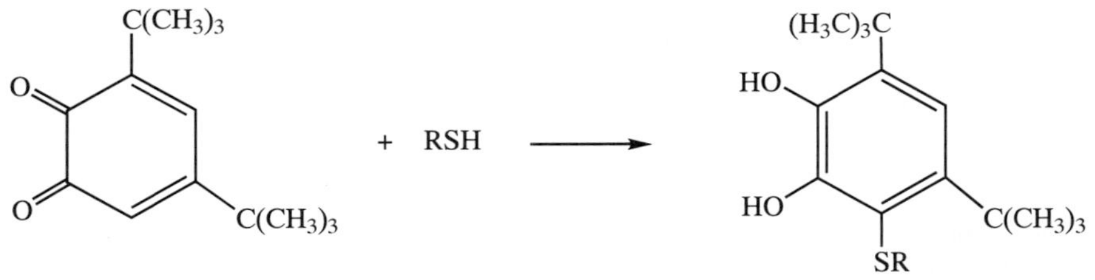

SAMPLE
Matrix: tissue
Analyte: thiols
Sample preparation: Homogenize (Ten-Brocke glass homogenizer) tissue with 9 volumes of cold 400 mM perchloric acid, centrifuge at 4° at 12000 g for 20 min, dilute the clear supernatant 10-fold with 400 mM perchloric acid. Remove a 100 µL aliquot and add it to 100 µL 1 mM 3,5-di-tert-butyl-1,2-benzoquinone in EtOH, shake for 30 min, add 1 mL hexane, shake for 2 min, centrifuge at 10000 g for 30 s, discard the hexane layer, repeat the hexane wash twice more, inject a 10 µL aliquot of the aqueous layer.

HPLC VARIABLES
Column: 250 × 4.6 5 µm Develosil C8 (Nomura Chemical)
Mobile phase: MeOH:water:1 M perchloric acid 70:30:1.5
Column temperature: 60
Flow rate: 0.8
Injection volume: 10
Detector: E, Yanagimoto Yanaco Model VMD-101A, glassy carbon electrode +1200 mV, Ag/AgCl reference electrode

CHROMATOGRAM
Retention time: 6.3 (ergothioneine), 8.4 (glutathione), 9.7 (cysteine), 10.4 (homocysteine)
Limit of quantitation: 100 nM

OTHER SUBSTANCES
Non-interfering: arginine, ascorbic acid, cystine, lysine, methionine, tryptophan, tyrosine, uric acid

KEY WORDS
rat; liver; lung; kidney

REFERENCE
Imai, Y.; Ito, S.; Fujita, K. Determination of natural thiols by liquid chromatography after derivatization with 3,5-di-tert.-butyl-1,2-benzoquinone, *J.Chromatogr.*, **1987**, *420*, 404–410.

THIOSULFATE

ACTIVATED HALIDE

Monobromobimane

SAMPLE
Matrix: blood
Analyte: thiosulfate
Sample preparation: 250 μL Serum + 250 μL 50 mM pH 8.0 borate buffer, mix, filter (Amicon MPS-1) while centrifuging at 1000 g for 15 min. Remove a 100 μL aliquot of the ultrafiltrate and add it to 100 μL 5 mM monobromobimane (Calbiochem; Molecular Probes) in MeCN, let stand at room temperature in the dark for 1 h, add 100 μL 50 mM pH 2 KCl/HCl buffer, inject a 20 μL aliquot. (Purify monobromobimane by column chromatography using silica and eluting with dichloromethane:petroleum ether from 70:30 to 100:0 over 10-30 h. The dibromo impurity elutes first followed by monobromobimane (J. Am. Chem. Soc. 1980, 102, 4983).)

HPLC VARIABLES
Guard column: 50 × 4 Nucleosil 100-5 C18
Column: 250 × 4 Nucleosil 5 dimethylamino-bonded silica
Mobile phase: MeCN:pH 3 acetic acid solution 18:78 containing 25 mM sodium perchlorate
Flow rate: 1
Injection volume: 20
Detector: F ex 396 em 476

CHROMATOGRAM
Retention time: 19
Limit of detection: 20-40 nM

KEY WORDS
serum; ultrafiltrate

REFERENCE
Togawa, T.; Ogawa, M.; Navata, M.; Ogasawara, Y.; Kawanabe, K.; Tanabe, S. High performance liquid chromatographic determination of bound sulfide and sulfite and thiosulfate at their low levels in human serum by pre-column fluorescence derivatization with monobromobimane, *Chem.Pharm.Bull.*, **1992**, *40*, 3000–3004.

SAMPLE
Matrix: blood, urine
Analyte: thiosulfate
Sample preparation: Plasma. Filter (Amicon CF-25) while centrifuging. Remove a 100 μL aliquot of the ultrafiltrate, add 40 μL 10 mM monobromobimane (Calbiochem; Molecular Probes) in MeCN, mix, let stand in the dark for at least 10 min, add 10 μL 1% HCl, dry under a stream of air at 50°, reconstitute with 100 μL mobile phase A, inject a 50 μL aliquot. Urine. Dilute urine 100-fold with 10 mM pH 8.0 ammonium bicarbonate contain-

ing 1 mM EDTA. Remove a 100 μL aliquot, add 40 μL 10 mM monobromobimane in MeCN, mix, let stand in the dark for at least 10 min, dry under a stream of air at 50°, reconstitute with 100 μL mobile phase A, inject a 50 μL aliquot.

HPLC VARIABLES
Guard column: C18/Corasil
Column: 300 × 3.9 μBondapak C18
Mobile phase: Gradient. A was MeOH:water 10:90 containing 10 mM tetrabutylammonium hydroxide, adjusted to pH 3.5 with phosphoric acid. B was MeOH:water 90:10 containing 10 mM tetrabutylammonium hydroxide, pH 5.5. A:B from 80:20 to 50:50 over 10 min, maintain at 50:50.
Flow rate: 1.5
Injection volume: 50
Detector: F ex 365 em 420 (cut-off filter)

CHROMATOGRAM
Retention time: 15
Limit of detection: 3.2 μM

KEY WORDS
plasma; ultrafiltrate

REFERENCE
Shea, M.; Howell, S. High-performance liquid chromatographic measurement of exogenous thiosulfate in urine and plasma, *Anal.Biochem.*, **1984**, *140*, 589–594.

SAMPLE
Matrix: sea water
Analyte: thiosulfate
Sample preparation: Mix 50 μL sea water with 210 μL reagent in the dark, let stand in the dark for 30 min, add 100 μL 65 mM methanesulfonic acid, inject an aliquot. (Prepare the reagent by mixing 50 μL 50 mM pH 8.0 HEPES buffer containing 5 mM EDTA, 50 μL MeCN, and 10 μL 48 mM monobromobimane (Calbiochem; Molecular Probes) in MeCN.)

HPLC VARIABLES
Column: 125 × 4 5 μm LiChrospher 60 RP select B
Mobile phase: Gradient. MeOH:0.25% pH 4 acetic acid 12:88 for 7 min, to 30:70 over 8 min, maintain at 30:70 for 4 min, to 50:50 over 4 min, to 100:0 over 7 min, maintain at 100:0 for 3 min, return to initial conditions over 0.1 min, re-equilibrate for 2.9 min.
Column temperature: 35
Flow rate: 1
Detector: F ex 380 em 480

CHROMATOGRAM
Retention time: 5.08

OTHER SUBSTANCES
Simultaneously analyzed: sulfide, sulfite

REFERENCE
Rethmeier, J.; Rabenstein, A.; Langer, M.; Fischer, U. Detection of traces of oxidized and reduced sulfur compounds in small samples by combination of different high-performance liquid chromatography methods, *J.Chromatogr.A*, **1997**, *760*, 295–302.

Monobromotrimethylammoniobimane

SAMPLE
Matrix: blood
Analyte: thiosulfate
Sample preparation: Vortex 750 µL red blood cells with 3.2 mL 100 mM HCl, centrifuge at 4° at 15000 rpm for 5 min. Remove the supernatant and mix it with an equal volume of cold 4 M sodium methanesulfonate, freeze in dry ice/isopropanol, thaw, centrifuge at 20000 rpm for 10 min. Determine the thiol content of the supernatant by titrating with 5,5'-dithiobis(2-nitrobenzoic acid). Add N-ethylmorpholine to a final concentration of 10 mM, adjust pH to 8.0 with 1 M NaOH, add a 1 molar equivalent of dithiothreitol, mix, let stand at room temperature for 5 min, add 6 equivalents of monobromotrimethylammoniobimane (Calbiochem), mix, let stand for 15 min, add 12 equivalents of thiol agarose, let stand for 20 min, add acetic acid to a final concentration of 3%, inject an aliquot. (Prepare 4 M sodium methanesulfonate by adjusting the pH of methanesulfonic acid to 1.5 with 50% NaOH then diluting to 4 M.)

HPLC VARIABLES
Column: 150 × 4 AA-10 resin (Beckman)
Mobile phase: Gradient. A was 2-methoxyethanol:200 mM pH 3.20 buffer 10:90 at 45°. B was 2-methoxyethanol:200 mM pH 4.40 buffer 10:90 at 45°. C was 2-methoxyethanol: 200 mM pH 4.75 buffer 10:90 at 45°. D was 2-methoxyethanol:0.2 N pH 6.40 trisodium citrate containing 800 mM NaCl 10:90 at 55°. E was 2-methoxyethanol:100 mM NaOH containing 100 mM NaCl 10:90 at 55°. A for 10 min, B for 20 min, C for 5 min, D for 90 min, E for 10 min, re-equilibrate at initial conditions for 20 min.
Flow rate: 0.2
Detector: F (o-phthalaldehyde filters)

CHROMATOGRAM
Retention time: 5.2 (coenzyme A), 7.8 (4'-phosphopantetheine), 10.2 (thiosulfate), 10.5 (coenzyme M), 27.5 (N-acetylcysteine), 31.5 (gamma-glutamylcysteine), 40 (glutathione), 51.5 (cysteinylglycine), 52 (ergothioneine), 55 (pantetheine), 57 (thiouracil), 61.5 (cysteine), 70 (homocysteine), 86 (2-mercaptoethanol), 110 (methanethiol), 110 (dithiothreitol)
Limit of quantitation: 10 pmole
Limit of detection: 1 pmole

KEY WORDS
red blood cells

REFERENCE
Fahey, R.C.; Newton, G.L.; Dorian, R.; Kosower, E.M. Analysis of biological thiols: Quantitative determination of thiols at the picomole level based upon derivatization with monobromobimanes and separation by cation-exchange chromatography, *Anal.Biochem.*, **1981**, *111*, 357–365.

DISULFIDE

2,2'-Dithiobis(5-nitropyridine)

SAMPLE
Matrix: seawater
Analyte: thiosulfate
Sample preparation: Condition a Sep-Pak C18 SPE cartridge with 5 mL MeOH, 5 mL water, and 5 mL 20 mM pH 6 sodium acetate buffer containing 10 mM tetrabutylammonium sulfate. Filter seawater, sparge with nitrogen. 20 mL Sample + 1 mL 2 mM 2,2'-dithiobis(5-nitropyridine) in MeCN + 1 mL 200 mM pH 6 acetate buffer, let stand at room temperature for 5 min, add to the SPE cartridge at 3-4 mL/min, wash with 3-5 mL water, blow nitrogen through the SPE cartridge, elute with 1 mL MeOH. Dilute the eluate with an equal volume of water, inject a 20 μL aliquot. Alternatively, evaporate the eluate to dryness, reconstitute with 200 μL MeOH:water 20:80, inject a 20 μL aliquot.

HPLC VARIABLES
Guard column: 5 μm Spherisorb ODS-2 (microbore)
Column: 150 × 2 5 μm Hypersil ODS
Mobile phase: Gradient. A was 50 mM sodium acetate containing 7.5 mM tetrabutylammonium hydrogen sulfate, adjusted to pH 3.5 ± 0.05 with 6 M HCl. B was MeCN. A:B 90:10 for 1 min, to 66:34 over 8 min, to 45:55 over 14 min, to 0:100 over 5 min, maintain at 0:100 for 2 min, return to initial conditions over 2 min.
Flow rate: 0.1-0.2
Injection volume: 20
Detector: UV 320

CHROMATOGRAM
Retention time: 21
Limit of detection: 6-8 nM

OTHER SUBSTANCES
Extracted: sulfite
Simultaneously analyzed: cysteine, glutathione, mercaptoacetic acid, 2-mercaptoethanesulfonic acid, 2-mercaptoethanol, 2-mercaptopropionic acid, 3-mercaptopropionic acid, monothioglycerol, thiomalic acid

KEY WORDS
microbore; SPE

REFERENCE
Valravamurthy, A.; Mopper, K. Determination of sulfite and thiosulfate in aqueous samples including anoxic seawater by liquid chromatography after derivatization with 2,2'-dithiobis(5-nitropyridine), *Environ.Sci.Technol.*, **1990**, *24*, 333–337.

APPENDIX I:
BIBLIOGRAPHY OF
RELATED REFERENCES

Abamectin

Chamkasem, N.; Papathakis, M.L.; Lee, S.M. Liquid chromatographic determination of abamectin in fruits and vegetables. *J.AOAC Int.* **1993**, *76*, 691–694.
Prabhu, S.V.; Varsolona, R.J.; Wehner, T.A.; Egan, R.S.; Tway, P.C. Rapid and sensitive high-performance liquid chromatographic method for the quantitation of abamectin and its delta 8,9-isomer. *J.Agric.Food Chem.* **1992**, *40*, 622–625.

Abamectin-8,9-oxide (using post-column reaction detection)

Demchak, R.J.; MacConnell, J.G. Normal-phase chromatography and post-column colorimetric detection of abamectin-8,9-oxide. *J.Chromatogr.* **1990**, *511*, 353–358.

Acenocoumarin

Thijssen, H.H.W.; Baars, L.G.; Reijnders, M.J. Analysis of acenocoumarin and its amino and acetamido metabolites in body fluids by high-performance liquid chromatography. *J.Chromatogr.* **1983**, *274*, 231–238.

Acephylline

Zuidema, J.; Merkus, F.W.H.M. Rapid method for the high-performance liquid-chromatographic determination of acephylline in human serum. *J.Chromatogr.* **1978**, *145*, 489–491.

Acetaldehyde

Wen, Y.H.; Lin, S.J.; Wu, S.S.; Wu, H.L. Trace analysis of acetaldehyde by derivatization and high performance liquid chromatography. *Zhonghua Yaoxue Zazhi* **1993**, *45*, 473–481.

Acetaminophen

Palmer, J.L. Novel method of sample preparation for the determination of paracetamol in plasma by high-performance liquid chromatography with electrochemical detection. *J.Chromatogr.* **1986**, *382*, 338–342.

Acetaminophen (using post-column reaction detection)

Manno, B.R.; Manno, J.E.; Dempsey, C.A.; Wood, M.A. A high-pressure liquid chromatographic method for the determination of N-acetyl-p-aminophenol (acetaminophen) in serum or plasma using a direct injection technique. *J.Anal.Toxicol.* **1981**, *5*, 24–28.

Acetic Anhydride

Wu, W.S.; Gaind, V.S. High-performance liquid chromatography with a specific dual detection system for the determination of acetic anhydride: an effective approach to analytical confirmation. *J.High Resolut.Chromatogr.* **1992**, *15*, 479–481.

Acetone

Brega, A.; Villa, P.; Quadrini, G.; Quadri, A.; Lucarelli, C. High-performance liquid chromatographic determination of acetone in blood and urine in the clinical diagnostic laboratory. *J.Chromatogr.* **1991**, *553*, 249–254.

N-Acetylaspartylglutamate

Brovia, V.; Ricciardi, A.; Barbeito, L. N-Acetylaspartylglutamate (NAAG) in human cerebrospinal fluid: determination by high performance liquid chromatography, and influence of biological variables. *Amino Acids* **1995**, *9*, 175–184.

Acetylcarbethoxycysteine

Toyo'oka, T.; Okudaira, K.; Kurihara, M.; Miyata, N.; Takahashi, A.; Suzuki, T.; Saito, Y. Determination of N-acetyl-S-carbethoxycysteine in rat and mouse urine by liquid chromatography with fluorescence detection. *Anal.Chim.Acta* **1989**, *226*, 109–119.

Acetylcholine and Choline

Kendrick, K.M. Simultaneous measurement of acetylcholine and amino acid and monoamine transmitters in microdialysis experiments using rapid sampling. *Curr.Sep.* **1993**, *12*, 3–7.

Tyrefors, N.; Gillberg, P.G. Determination of acetylcholine and choline in microdialysates from spinal cord of rat using liquid chromatography with electrochemical detection. *J.Chromatogr.* **1987**, *423*, 85–91.

Acetylcholine and Choline (using post-column reaction detection)

Barnes, N.M.; Costall, B.; Fell, A.F.; Naylor, R.J. An HPLC assay procedure of sensitivity and stability for measurement of acetylcholine and choline in neuronal tissue. *J.Pharm.Pharmacol.* **1987**, *39*, 727–731.

Bertrand, N.; Bralet, J.; Beley, A. Turnover rate of brain acetylcholine using HPLC separation of the transmitter. *J.Neurochem.* **1990**, *55*, 27–30.

Bymaster, F.P.; Perry, K.W.; Wong, D.T. Measurement of acetylcholine and choline in brain by HPLC with electrochemical detection. *Life Sci.* **1985**, *37*, 1775–1781.

Damsma, G.; Westerink, B.H.C.; de Vries, J.B.; Van den Berg, C.J.; Horn, A.S. Measurement of acetylcholine release in freely moving rats by means of automated intracerebral dialysis. *J.Neurochem.* **1987**, *48*, 1523–1528.

Damsma, G.; Westerink, B.H.C.; Horn, A.S. A simple, sensitive, and economic assay for choline and acetylcholine using HPLC, an enzyme reactor, and an electrochemical detector. *J.Neurochem.* **1985**, *45*, 1649–1652.

Eva, C.; Meek, J.L.; Costa, E. Vasoactive intestinal peptide which coexists with acetylcholine decreases acetylcholine turnover in mouse salivary glands. *J.Pharmacol.Exp.Ther.* **1985**, *232*, 670–674.

Eva, C.; Hadjiconstantinou, M.; Neff, N.H.; Meek, J.L. Acetylcholine measurement by high-performance liquid chromatography using an enzyme-loaded postcolumn reactor. *Anal.Biochem.* **1984**, *143*, 320–324.

Gunaratna, P.C.; Wilson, G.S. Optimization of multienzyme flow reactors for determination of acetylcholine. *Anal.Chem.* **1990**, *62*, 402–407.

Lund, D.D.; Oda, R.P.; Pardini, B.J.; Schmid, P.G. Vagus nerve stimulation alters regional acetylcholine turnover in rat heart. *Circ.Res.* **1986**, *58*, 372–377.

Maysinger, D.; Herrera-Marschitz, M.; Carlsson, A.; Garofalo, L.; Cuello, A.C.; Ungerstedt, U. Striatal and cortical acetylcholine release in vivo in rats with unilateral decortication: effects of treatment with monosialoganglioside GM1. *Brain Res.* **1988**, *461*, 355–360.

Nilsson, O.G.; Kalen, P.; Rosengren, E.; Bjoerklund, A. Acetylcholine release in the rat hippocampus as studied by microdialysis is dependent on axonal impulse flow and increases during behavioral activation. *Neuroscience (Oxford)* **1990**, *36*, 325–338.

Niwa, O.; Horiuchi, T.; Morita, M.; Huang, T.; Kissinger, P.T. Determination of acetylcholine and choline with platinum-black ultramicroarray electrodes using liquid chromatography with a post-column enzyme reactor. *Anal.Chim.Acta* **1996**, *318*, 167–173.

Potter, P.E.; Hadjiconstantinou, M.; Meek, J.L.; Neff, N.H. Measurement of acetylcholine turnover rate in brain: an adjunct to a simple HPLC method for choline and acetylcholine. *J.Neurochem.* **1984**, *43*, 288–290.

Ricny, J.; Tucek, S.; Vins, I. Sensitive method for HPLC determination of acetylcholine, choline and their analogues using fluorometric detection. *J.Neurosci.Methods* **1992**, *41*, 11–17.

Teelken, A.W.; Schuring, H.F.; Trieling, W.B.; Damsma, G. Measurement of acetylcholine and choline in cerebrospinal fluid by high-performance liquid chromatography: failure to detect acetylcholine in normal human cerebrospinal fluid. *J.Chromatogr.* **1990**, *529*, 408–416.

Webb, L.E.; Johnson, R.C. Choline in plasma measured by liquid-chromatography with electrochemical detection. *Clin.Biochem.(Ottawa)* **1986**, *19*, 212–215.

Acetyl-coenzyme A (using post-column reaction detection)

Yamato, S.; Nakajima, M.; Wakabayashi, H.; Shimada, K. Specific detection of acetyl-coenzyme A by reversed-phase ion-pair high-performance liquid chromatography with an immobilized enzyme reactor. *J.Chromatogr.* **1992**, *590*, 241–245.

Acetylcysteine

Dehnen, W. Thioethers excretion in urine: detection of N-acetylcysteine and thiophenol after alkaline hydrolysis. *Zentralbl.Hyg.Umweltmed.* **1990**, *189*, 441–451.

Gabard, B.; Mascher, H. Endogenous plasma N-acetylcysteine and single dose oral bioavailability from two different formulations as determined by a new analytical method. *Biopharm.Drug Dispos.* **1991**, *12*, 343–353.

Holdiness, M.R.; Morgan, L.R., Jr.; Gillen, L.E.; Harrison, E.F. High-performance liquid chromatographic determination of N-acetylcysteine in human serum following acetaminophen overdosage. *J.Chromatogr.* **1986**, *382*, 99–106.

Kaagedal, B.; Kaellberg, M.; Maartensson, J. Determination of non-protein-bound N-acetylcysteine in plasma by high-performance liquid chromatography. *J.Chromatogr.* **1984**, *311*, 170–175.

Acetylcysteine (using post-column reaction detection)

Johansson, M.; Lenngren, S. Determination of cysteine, glutathione and N-acetylcysteine in plasma by ion-pair reversed-phase liquid chromatography and post-column derivatization. *J.Chromatogr.* **1988**, *432*, 65–74.

Johansson, M.; Westerlund, D. Determination of N-acetylcysteine, intact and oxidized, in plasma by column liquid chromatography and post-column derivatization. *J.Chromatogr.* **1987**, *385*, 343–356.

Acetylglutamate

Rubio, V.; Alonso, E.; Portoles, M.; Grisolia, S. Determination of a key activator (acetylglutamate) and of a metabolite (argininosuccinate) of the urea cycle exemplifies impact of HPLC techniques. *Adv.Clin.Enzymol.* **1986**, *4*, 169–184.

Acetylglutamine

Hellmann, U.; Luederwald, I.; Neuhaeuser, M. Determination of N-acetyl-L-glutamine in urine by HPLC. *Fresenius' Z.Anal.Chem.* **1986**, *325*, 290–292.

Acetylhomotaurine

Chabenat, C.; Ladure, P.; Blanc-Continsouza, D.; Boismare, F.; Boucly, P. Determination of calcium acetylhomotaurinate by liquid chromatography with fluorometric and electrochemical detection. *J.Chromatogr.* **1987**, *414*, 417–422.

Acetylmorphine

Kerks, H.J.G.M.; van Twillert, K.; Pereboom-de Fauw, D.P.K.H.; Zomer, G.; Loeber, J.G. Determination of the heroin metabolite 6-acetylmorphine by high-performance liquid chromatography using automated pre-column derivatization and fluorescence detection. *J.Chromatogr.* **1986**, *370*, 173–178.

Acrolein

Al-Rawithi, S.; el-Yazigi, A.; Nicholls, P.J. Rapid liquid chromatographic analysis of acrolein in plasma following precolumn derivatization. *Pharm.Sci.* **1995**, *1*, 201–203.

Acyl Halides

Bissinger, J.M.; Rullo, K.T.; Stoklosa, J.T.; Shearer, C.M.; DeAngelis, N.J. High-performance liquid chromatographic analysis of acid chlorides by prederivatization. *J.Chromatogr.* **1983**, *268*, 102–106.

Adenosylmethionine

Shugart, L. Identification of fluorescent derivatives of adenosylmethionine and related analogs with high-pressure liquid chromatography. *J.Chromatogr.* **1979**, *174*, 250–253.

Wagner, J.; Danzin, C.; Mamont, P. Reversed-phase ion-pair liquid chromatographic procedure for the simultaneous analysis of S-adenosylmethionine, its metabolites and the natural polyamines. *J.Chromatogr.* **1982**, *227*, 349–368.

Wagner, J.; Danzin, C.; Huot-Olivier, S.; Claverie, N.; Palfreyman, M.G. High-performance liquid chromatographic analysis of S-adenosylmethionine and its metabolites in rat tissues: interrelationship with changes in biogenic catechol levels following treatment with L-dopa. *J.Chromatogr.* **1984**, *290*, 247–262.

Wagner, J.; Hirth, Y.; Claverie, N.; Danzin, C. HPLC with fluorescence detection for the analysis of decarboxylated S-adenosylmethionine and its analogs. *Symp.Biol.Hung.* **1986**, *34*, 85–102.

Adipic Acid

Yoshioka, K.; Shimojo, N.; Nakanishi, T.; Naka, K.; Okuda, K. Measurements of urinary adipic acid and suberic acid using high-performance liquid chromatography. *J.Chromatogr.B* **1994**, *655*, 189–193.

Aflatoxins

Akiyama, H.; Chen, D.; Miyahara, M.; Toyoda, M.; Saito, Y. Simple HPLC determination of aflatoxins B1, B2, G1, and G2 in nuts and corn. *Shokuhin Eiseigaku Zasshi* **1996**, *37*, 195–201.

Carisano, A.; Della Torre, G. Sensitive reversed-phase high-performance liquid chromatographic determination of aflatoxin M1 in dry milk. *J.Chromatogr.* **1986**, *355*, 340–344.

de Sylos, C.M.; Amaya, D.R. Comparative study of methods for aflatoxin M1 determination. *Rev.Inst.Adolfo Lutz* **1996**, *56*, 87–97.

Fremy, J-M.; Boursier, B. Rapid determination of aflatoxin M1 in dairy products by reversed-phase high-performance liquid chromatography. *J.Chromatogr.* **1981**, *219*, 156–161.

Hutchins, J.E.; Hagler, W.M., Jr. Rapid liquid chromatographic determination of aflatoxins in heavily contaminated corn. *J.Assoc.Off.Anal.Chem.* **1983**, *66*, 1458–1465.

Liu, Z.; Tu, W.; Li, D.; Li, Y.; Xie, C.; Yang, Y.; Qin, B. A new method for the quantitation of aflatoxin M1 in urine by high performance liquid chromatography and its application to the etiologic study of hepatoma. *Biomed.Chromatogr.* **1990**, *4*, 83–86.

Park, D.L.; Nesheim, S.; Trucksess, M.W.; Stack, M.E.; Newell, R.F. Liquid chromatographic method for determination of aflatoxins B1, B2, G1, and G2 in corn and peanut products: collaborative study. *J.Assoc.Off.Anal.Chem.* **1990**, *73*, 260–266.

Patey, A.L.; Sharman, M.; Gilbert, J. Liquid chromatographic determination of aflatoxin levels in peanut butters using an immunoaffinity column cleanup method: international collaborative trial. *J.Assoc.Off.Anal.Chem.* **1991**, *74*, 76–81.

Roch, O.G.; Blunden, G.; Haig, D.J.; Coker, R.D.; Gay, C. Determination of aflatoxins in groundnut meal by high-performance liquid chromatography: a comparison of two methods of derivatisation of aflatoxin B1. *Br.J.Biomed.Sci.* **1995**, *52*, 312–316.

Simonella, A.; Torreti, L.; Filipponi, C.; Falgiani, A.; Ambrosii, L. Simultaneous determination of aflatoxins G1, B1, G2, B2, in animal feed by HPTLC and RP-HPLC. *HRC CC, J.High Resolut.Chromatogr. Chromatogr.Commun.* **1988**, *10*, 626–628.

Tarter, E.J.; Hanchay, J.P.; Scott, P.M. Improved liquid chromatographic method for determination of aflatoxins in peanut butter and other commodities. *J.Assoc.Off.Anal.Chem.* **1984**, *67*, 597–600.

Trucksess, M.W.; Stack, M.E.; Nesheim, S.; Albert, R.H.; Romer, T.R. Multifunctional column coupled with liquid chromatography for determination of aflatoxins B1, B2, G1, and G2 in corn, almonds, Brazil nuts, peanuts, and pistachio nuts: collaborative study. *J.AOAC Int.* **1994**, *77*, 1512–1521.

Vesely, Z.; Bohacenko, I. Immunoaffinity extraction of aflatoxins and their HPLC determination in raw materials, foods and feeds. *Potravin.Vedy* **1995**, *13*, 451–460.

Yen, I.C.; Bidasee, K.R. Liquid chromatographic determination of aflatoxins in animal feeds and feed components. *J.AOAC Int.* **1993**, *76*, 366–370.

Aflatoxins (using post-column reaction detection)

Beaver, R.W.; Wilson, D.M.; Trucksess, M.W. Comparison of postcolumn derivatization-liquid chromatography with thin-layer chromatography for determination of aflatoxins in naturally contaminated corn. *J.Assoc.Off.Anal.Chem.* **1990**, *73*, 579–581.

Buchholz, H. Aflatoxins in feed studies. *Kraftfutter* **1991**, 395–399.

Holcomb, M.; Thompson, H.C., Jr.; Cooper, W.M.; Hopper, M.L. SFE extraction of aflatoxins (B1, B2, G1, and G2) from Corn and analysis by HPLC. *J.Supercrit.Fluids* **1996**, *9*, 118–121.

Hurst, W.J.; Snyder, K.P.; Martin, R.A., Jr. Determination of aflatoxins in peanut products using disposable bonded-phase columns and post-column reaction detection. *J.Chromatogr.* **1987**, *409*, 413–418.

Kussak, A.; Andersson, B.; Andersson, K. Determination of aflatoxins in airborne dust from feed factories by automated immunoaffinity column clean-up and liquid chromatography. *J.Chromatogr.A* **1995**, *708*, 55–60.

Lazaro, F.; Luque de Castro, M.D.; Valcarcel, M. Fluorimetric determination of aflatoxins in foodstuffs by high-performance liquid chromatography with flow injection analysis. *J.Chromatogr.* **1988**, *448*, 173–181.

Niedwetzki, G.; Lach, G.; Geschwill, K. Determination of aflatoxins in food by use of an automatic work station. *J.Chromatogr.A* **1994**, *661*, 175–180.

Pieta, L. Robotic automation in the analysis of aflatoxins. *Adv.Lab.Autom.Rob.* **1991**, *7*, 303–313.

Reif, K.; Metzger, W. Determination of aflatoxins in medicinal herbs and plant extracts. *J.Chromatogr.A* **1995**, *692*, 131–136.

Roch, O.G.; Blunden, G.; Coker, R.D.; Nawaz, S. The validation of a solid phase clean-up procedure for the analysis of aflatoxins in groundnut cake using HPLC. *Food Chem.* **1995**, *52*, 93–98.

Rudat, B.; Thiry, E. Determination of aflatoxins. HPLC with post column derivatization and fluorometric detection. *LaborPraxis* **1991**, *15*, 380–383.

Sharman, M.; Gilbert, J. Automated aflatoxin analysis of foods and animal feeds using immunoaffinity column clean-up and high-performance liquid chromatographic determination. *J.Chromatogr.* **1991**, *543*, 220–225.

Shepherd, M.J.; Gilbert, J. An investigation of HPLC post-column iodination conditions for the enhancement of aflatoxin B1 fluorescence. *Food Addit.Contam.* **1984**, *1*, 325–335.

Thiel, P.G. HPLC determination of aflatoxins and mammalian aflatoxin metabolites. *Bioact.Mol.* **1986**, *1*, 329–340.

Thiel, P.G.; Stockenstroem, S.; Gathercole, P.S. Aflatoxin analysis by reverse phase HPLC using post-column derivatization for enhancement of fluorescence. *J.Liq.Chromatogr.* **1986**, *9*, 103–112.

Trucksess, M.W.; Stack, M.E.; Nesheim, S.; Page, S.W.; Albert, R.H.; Hansen, T.J.; Donahue, K.F. Immunoaffinity column coupled with solution fluorometry or liquid chromatography postcolumn derivatization for determination of aflatoxins in corn, peanuts, and peanut butter: collaborative study. *J.Assoc.Off.Anal.Chem.* **1991**, *74*, 81–88.

Werner, G. Determination of feed aflatoxins following selective clean-up on an immunoaffinity column. *Agribiol.Res.* **1991**, *44*, 289–298.

Worner, F.M.; Patey, A.L.; Wood, R. Determination of the levels of aflatoxin in peanut butter using the Aflaprep immunoaffinity column clean-up procedure: collaborative trial. *J.Assoc.Public Anal.* **1993**, *28*, 1–10.

Agmatine

Patchett, M.L.; Monk, C.R.; Daniel, R.M.; Morgan, H.W. Determination of agmatine, arginine, citrulline and ornithine by reversed-phase liquid chromatography using automated pre-column derivatization with o-phthalaldehyde. *J.Chromatogr.* **1988**, *425*, 269–276.

Alafosfalin

Huber, J.E..I.; Calabrese, K.L. Derivatization of glyphosate and alafosfalin for reversed-phase HPLC analysis. *LC Mag.* **1985**, *3*, 888–890.

Aliphatic Alcohols Polyethylenglycol Ethers

Marcomini, A.; Zanette, M. Derivatization procedures and HPLC separations for environmental analysis of aliphatic alcohols polyethylenglycol ethers (AE). *Riv.Ital.Sostanze Grasse* **1994**, *71*, 203–208.

Alcohols

Boppana, V.K.; Simpson, R.C.; Anderson, K.; Miller-Stein, C.; Blake, T.J.A.; Hwang, B.Y.H.; Rhodes, G.R. High-performance liquid chromatographic determination of monohydroxy compounds by a combination of precolumn derivatization and post-column reaction detection. *J.Chromatogr.* **1992**, *593*, 29–36.

Czichocki, G.; Mueller, P.; Vollhardt, D.; Krueger, M. Determination of trace amounts of alcohols in sodium alkyl sulfate mixtures using high-performance liquid chromatography and surface tension measurements. *J.Chromatogr.* **1992**, *604*, 213–218.

Fujino, H.; Hidaka, M.; Goya, S. [Synthesis and reactivity of anthracene-1-carbonyl azide as a fluorescent derivatization reagent for alcohols]. *Yakugaku.Zasshi.* **1989**, *109*, 606–610.

Goss, J. Reverse phase high-performance liquid chromatographic determination of primary and secondary aliphatic alcohols as phthalate monoesters by UV detection. *Chromatographia* **1994**, *38*, 417–420.

Goto, J.; Komatsu, S.; Goto, N.; Nambara, T. A new sensitive derivatization reagent for liquid chromatographic separation of hydroxyl compounds. *Chem.Pharm.Bull.* **1981**, *29*, 899–901.

Shimada, K.; Orii, S.; Tanaka, M.; Nambara, T. New ferrocene reagents for derivatization of alcohols in high-performance liquid chromatography with electrochemical detection. *J.Chromatogr.* **1986**, *352*, 329–335.

Stong, J.D.; Witchey-Lakshmanan, L.C. Determination and characterization of unreacted hydroxyl groups in a cross-linked poly(ortho ester). *Anal.Chem.* **1994**, *66*, 3505–3511.

Takagi, T.; Aoyanagi, N.; Nishimura, K.; Ando, Y.; Ota, T. Enantiomer separations of secondary alkanols with little asymmetry by high-performance liquid chromatography on chiral columns. *J.Chromatogr.* **1993**, *629*, 385–388.

Tsuruta, Y.; Kohashi, K. Sensitive derivatization reagents for hydroxyl and amino compounds for thin-layer or high-performance liquid chromatography with fluorescence detection. *Anal.Chim.Acta* **1987**, *192*, 309–313.

Xie, K.H.; Santasania, C.T.; Krull, I.S.; Neue, U.; Bidlingmeyer, B.; Newhart, A. Solid phase derivatizations in HPLC: polymeric permanganate oxidations of alcohols and aldehydes in HPLC-SPR. *J.Liq.Chromatogr.* **1983**, *6*, 2109–2127.

Yoshida, T.; Moriyama, Y.; Taniguchi, H. 6-Methoxy-2-methylsulfonylquinoline-4-carbonyl chloride as a fluorescence derivatization reagent for alcohols in high-performance liquid chromatography. *Anal.Sci.* **1992**, *8*, 355–359.

Alcohols (using post-column reaction detection)

Tagliaro, F.; Schiavon, G.; Dorizzi, R.; Marigo, M. Development of post-column enzymic reactors with immobilized alcohol oxidase for use in the high-performance liquid chromatographic assay of alcohols with electrochemical detection. *J.Chromatogr.* **1991**, *563*, 11–21.

Aldehydes

Bagchi, D.; Dickson, P.H.; Stohs, S.J. The identification and quantitation of malondialdehyde, formaldehyde, acetaldehyde, and acetone in serum of rats. *Toxicol.Methods* **1992**, *2*, 270–279.

Beckman, J.K.; Morley, S.A., Jr.; Greene, H.L. Analysis of aldehydic lipid peroxidation products by TLC/densitometry. *Lipids* **1991**, *26*, 155–161.

Chiavari, G.; Torsi, G.; Asmundsdottir, A.M. Different methods for HPLC analysis for aldehydes in aqueous solutions. *Ann.Chim.(Rome)* **1992**, *82*, 349–356.

Edlund, P.O. On-column catalysis of hydration equilibria during liquid chromatographic analysis for acetaldehyde and formaldehyde. *Chromatographia* **1987**, *23*, 709–712.

Goetze, H.J.; Harke, S. Determination of aldehydes and ketones in natural gas combustion in the ppb range by high-performance liquid chromatography. *Fresenius' Z.Anal.Chem.* **1989**, *335*, 286–288.

Hirayama, T.; Kamata, K.; Kasai, T.; Watanabe, T. Determination of saturated aliphatic aldehydes in edible oil by acetylacetone method high-performance liquid chromatography. *Japan.J.Toxicol.Env. Health* **1994**, *40*, 574–581.

Keuken, M.P.; Schoonebeek, C.A.M. Simultaneous determination of C1-C2 aldehydes and organic acids in aqueous atmospheric samples by high-performance liquid chromatography. *Int.J.Environ.Anal. Chem.* **1989**, *35*, 227–239.

Kirschmer, P. Aldehydes in air. *Analusis* **1992**, *20, M48-M50,*

Kuntz, R.; Lonneman, W.; Namie, G.; Hull, L.A. Rapid determination of aldehydes in air analyses. *Anal.Lett.* **1980**, *13*, 1409–1415.

Mann, B.; Grayeski, M.L. New chemiluminescent derivatizing agent for the analysis of aldehydes and ketones by high-performance liquid chromatography with peroxyoxalate chemiluminescence. *J.Chromatogr.* **1987**, *386*, 149–158.

Miller, B.E.; Danielson, N.D. Derivatization of vinyl aldehydes with anthrone prior to high-performance liquid chromatography with fluorometric detection. *Anal.Chem.* **1988**, *60*, 622–626.

Nohta, H.; Sakai, F.; Kai, M.; Ohkura, Y.; Hara, S.; Yamaguchi, M. 2-Aminothiophenols as fluorogenic reagents for aromatic aldehydes. *Anal.Chim.Acta* **1993**, *282*, 625–631.

Nuijens, M.J.; Zomer, M.; Mank, A.J.G.; Gooijer, C.; Velthorst, N.H.; Hofstraat, J.W. A cyanine fluorophore with a hydrazide functionality as labeling reagent for aldehydes in liquid chromatography. *Anal.Chim.Acta* **1995**, *311*, 47–55.

Reichert, D.; Schutz, S.; Benedetti, A.; Pompella, A.; Fulceri, R.; Romani, A.; Comporti, M. Mercapturic acid formation is an activation and intermediary step in the metabolism of hexachlorobutadiene 4-hydroxynonenal and other aldehydes produced in the liver in vivo after bromobenzene intoxication. *Biochem.Pharmacol. Toxicol.Pathol.* **1986**, *14*, 457–461.

Reindl, B.; Grossklaus, R.; Stan, H.J. Determination of aliphatic aldehydes in biological samples at the ppb level using reversed-phase high-performance liquid chromatography. *Anal.Chem.Symp.Ser.* **1983**, *13*, 185–191.

Risner, C.H.; Martin, P. Quantitation of formaldehyde, acetaldehyde, and acetone in sidestream cigarette smoke by high-performance liquid chromatography. *J.Chromatogr.Sci.* **1994**, *32*, 76–82.

Sesana, G.; Nano, G.; Baj, A.; Balestreri, S. New sampling tool for airborne volatile aldehydes. *Fresenius' J.Anal.Chem.* **1991**, *339*, 485–487.

Van Hoof, F.; Wittocx, A.; Van Buggenhout, E.; Janssens, J. Determination of aliphatic aldehydes in waters by high-performance liquid chromatography. *Anal.Chim.Acta* **1985**, *169*, 419–424.

Werner, A.; Grune, T.; Siems, W.; Schneider, W.; Shimasaki, H.; Esterbauer, H.; Gerber, G. Nucleotide and aldehyde analysis by HPLC for determination of radical induced damage. *Chromatographia* **1989**, *28*, 65–68.

Wu, R.; White, L.B. Automated procedure for determination of trace amounts of aldehydes in drinking water. *J.Chromatogr.A* **1995**, *692*, 1–9.

Yoshino, K.; Matsuura, T.; Sano, M.; Saito, S.; Tomita, I. Fluorometric liquid chromatographic determination of aliphatic aldehydes arising from lipid peroxides. *Chem.Pharm.Bull.* **1986**, *34*, 1694–1700.

Aldehydes (using post-column reaction detection)

Gachanja, A.N.; Lewis, S.W.; Worsfold, P.J. Determination of aldehydes in used engine oils by liquid chromatography with chemiluminescence detection. *J.Chromatogr.A* **1995**, *704*, 329–337.

Igawa, M.; Munger, J.W.; Hoffmann, M.R. Analysis of aldehydes in cloud- and fogwater samples by HPLC with a postcolumn reaction detector. *Environ.Sci.Technol.* **1989**, *23*, 556–561.

Kaneda, H.; Takashio, M.; Osawa, T.; Kawakishi, S.; Koshino, S.; Tamaki, T. Analysis of aldehyde-bisulfites in beer by HPLC-fluorescence detection with post-column derivatization. *J.Food Sci.* **1996**, *61*, 105–108.

Koizumi, H.; Suzuki, Y. High-performance liquid chromatography of aliphatic aldehydes by means of post-column extraction with fluorometric detection. *J.Chromatogr.* **1988**, *457*, 299–307.

Suzaki, H.; Igawa, M. Analysis for aldehydes in fogwater by HPLC with postcolumn reaction detector. *Anal.Sci.* **1991**, *7*, 133–134.

Takeuchi, T.; Ishii, D.; Nakanishi, A. Determination of aldehydes by micro high-performance liquid chromatography with post-column derivatization on enzyme-immobilized glass beads. *Chromatographia* **1988**, *25*, 507–510.

Alditols (using post-column reaction detection)

Dona, A-M.; Verchere, J-F. High-performance liquid chromatography of alditols with indirect photometric detection. *J.Chromatogr.A* **1995**, *689*, 13–21.

Alentamol

Schwende, F.J.; Rykert, U.M. Determination of alentamol hydrobromide, a novel antipsychotic agent, in human blood plasma and urine by high-performance liquid chromatography with fluorescence detection and solid-phase extraction. *J.Chromatogr.* **1991**, *565*, 488–496.

Alkanolamines

Serbin, L.; Birkholz, D. A sensitive analytical procedure for the determination of primary and secondary alkanolamines in air. *Am.Ind.Hyg.Assoc.J.* **1995**, *56*, 66–69.

Alkanolamines (using post-column reaction detection)

Nakae, A.; Mansho, K.; Tsuji, K. Determination of alkanolamines by high-performance liquid chromatography with post-column derivatization. *Bunseki Kagaku* **1981**, *30*, 353–357.

Alkenes

D'Avila, L.A.; Colin, H.M.; Guiochon, G. In-column derivatization for the separation of heavy aza-arenes and olefins by liquid chromatography. *J.Liq.Chromatogr.* **1987**, *10*, 71–82.

Alkenylcysteine

Muetsch-Eckner, M.; Sticher, O.; Meier, B. Reversed-phase high-performance liquid chromatography of S-alk(en)yl-L-cysteine derivatives in Allium sativum including the determination of (+)-S-allyl-L-cysteine sulfoxide, gamma-L-glutamyl-S-allyl-L-cysteine and gamma-L-glutamyl-S-(trans-1-propenyl)-L-cysteine. *J.Chromatogr.* **1992**, *625*, 183–190.

Alkylbenzenesulfonates

Taylor, P.W.; Nickless, G. Paired-ion high-performance liquid chromatography of partially biodegraded linear alkylbenzenesulfonate. *J.Chromatogr.* **1979**, *178*, 259–269.

Alkyl Methylphosphonic Acids

Bossle, P.C.; Martin, J.J.; Sarver, E.W.; Sommer, H.Z. High-performance liquid chromatographic determination of alkyl methylphosphonic acids by derivatization. *J.Chromatogr.* **1983**, *267*, 209–212.

Allantoin

Nakao, K.; Honda, K.; Yoneya, T. High-pressure liquid chromatographic determination of allantoin in cosmetic creams and lotions. *J.Assoc.Off.Anal.Chem.* **1982**, *65*, 1362–1365.

Allantoin (using post-column reaction detection)

Yokoyama, T.; Kinoshita, T. Simultaneous determination of allantoin and taurine in eye lotion by HPLC with postcolumn reaction. *Bunseki Kagaku* **1991**, *40*, 349–353.

Alliin

Velisek, J.; de Vos, R.H.; Schouten, A. HPLC determination of alliin and its transformation products in garlic and garlic-containing phytomedical preparations. *Potravin.Vedy* **1993**, *11*, 445–453.
Ziegler, S.J.; Sticher, O. HPLC of S-alk(en)yl-L-cysteine derivatives in garlic including quantitative determination of (+)-S-allyl-L-cysteine sulfoxide (alliin). *Planta Med.* **1989**, *55*, 372–378.

AL-toxin

Kodama, M.; Otani, H.; Kohmoto, K. A rapid and sensitive procedure for the quantitative detection of AL-toxin by fluorescence derivatization and separation by high performance liquid chromatography. *Nippon Shokubutsu Byori Gakkaiho* **1995**, *61*, 477–480.

Aluminum (using post-column reaction detection)

Jackson, P.E.; Carnevale, J. Analysis of aluminum in pharmaceutical products by ion chromatography using post-column derivatization and fluorescence detection. *Am.Lab.* **1995**, *27*, 24.

Amadori Compounds (using post-column reaction detection)

Reutter, M.; Eichner, K. Separation and determination of Amadori Compounds by high pressure liquid chromatography and post-column reaction. *Z.Lebensm.-Unters.Forsch.* **1989**, *188*, 28–35.

6-Amidino-2-naphthyl [4-(4,5-dihydro-1H-imidazol-2-yl)amino]benzoate dimethanesulfonate (using post-column reaction detection)

Marunaka, T.; Maniwa, M.; Matsushima, E.; Yoshida, K.; Azuma, R.; Kurotori, M.; Komatsu, S. High-performance liquid chromatographic determination of 6-amidino-2-naphthyl [4-(4,5-dihydro-1H-imidazol-2-yl)amino]benzoate dimethanesulfonate and its metabolites in biological fluids. *J.Chromatogr.* **1988**, *433*, 177–186.

Amikacin

Barends, D.M.; Blauw, J.S.; Smits, M.H.; Hulshoff, A. Determination of amikacin in serum by high-performance liquid chromatography with ultraviolet detection. *J.Chromatogr.* **1983**, *276*, 385–394.

Maitra, S.K.; Yoshikawa, T.T.; Steyn, C.M.; Guze, L.B.; Schotz, M.C. Determination of amikacin isomers by high-pressure liquid chromatography. *J.Liq.Chromatogr.* **1979**, *2*, 823–836.

Maitra, S.K.; Yoshikawa, T.T.; Steyn, C.M.; Guze, L.B.; Schotz, M.C. Amikacin assay in serum by high-performance liquid chromatography. *Antimicrob.Agents Chemother.* **1978**, *14*, 880–885.

Amikacin (using post-column reaction detection)

Sar, F.; Leroy, P.; Nicolas, A.; Archimbault, P.; Ambroggi, G. Determination of amikacin in dog plasma by reversed-phase ion-pairing liquid chromatography with post-column derivatization. *Anal.Lett.* **1992**, *25*, 1235–1250.

Wichert, B.; Schreier, H.; Derendorf, H. Sensitive liquid chromatography assay for the determination of amikacin in human plasma. *J.Pharm.Biomed.Anal.* **1991**, *9*, 251–254.

Amines

Achilli, G.; Cellerino, G.P.; Melzi d'Eril, G. Determination of amines in wines by high-performance liquid chromatography with electrochemical coulometric detection after precolumn derivatization. *J.Chromatogr.A* **1994**, *661*, 201–205.

Adachi, K.; Ichinose, N. Fluorescent high-performance liquid chromatographic determination of primary aromatic amines by formation of Schiff base. *Fresenius' J.Anal.Chem.* **1990**, *338*, 265–268.

Allison, L.A.; Mayer, G.S.; Shoup, R.E. The o-phthalaldehyde derivatives of amines for high-speed liquid chromatography/electrochemistry. *Anal.Chem.* **1984**, *56*, 1089–1096.

Beale, S.C.; Savage, J.C.; Wiesler, D.; Wietstock, S.M.; Novotny, M. Fluorescence reagents for high-sensitivity chromatographic measurements of primary amines. *Anal.Chem.* **1988**, *60*, 1765–1769.

Bellatti, M.; Parolari, G. Aliphatic secondary amines in meat and fish products and their analysis by high pressure liquid chromatography. *Meat Sci.* **1982**, *7*, 59–65.

Bernstein, S.C.; Abrams, S.K.; Leckrone, K.J.; Paul, L.A. Chloroisothiocyanatoquinolines as fluorogenic derivatizing agents for primary and secondary amines. *J.Pharm.Biomed.Anal.* **1993**, *11*, 61–69.

Bjorkqvist, B. Separation and determination of aliphatic and aromatic amines by high-performance liquid chromatography with ultraviolet detection. *J.Chromatogr.* **1981**, *204*, 109–114.

Bostick, J.M.; Strojek, J.W.; Metcalf, T.; Kuwana, T. Development of a helium-cadmium laser-based fluorescence detector for analysis of derivatized biogenic amines. *Appl.Spectrosc.* **1992**, *46*, 1532–1539.

Busto, O.; Valero, Y.; Guasch, J.; Borrull, F. Solid phase extraction applied to the determination of biogenic amines in wines by HPLC. *Chromatographia* **1994**, *38*, 571–578.

Choi, S.H.; Kim, H.G.; Park, H.I.; Chun, B.G. A simple, sensitive, and specific HPLC analysis of tissue polyamines using FNBT derivatization: its application on the study of polyamine metabolism in regenerating rat liver. *Korean J.Pharmacol.* **1988**, *24*, 233–240.

Claas, K.E.; Hohaus, E.; Monien, H. Boron chelates and boron metal chelates. Part XVII. Liquid column chromatographic separation and determination of long-chain primary n-alkylamines (n-C10H21NH2 to n-C22H45NH2) after their fluorescence derivatization with salicylaldehyde diphenylboron chelate. *Fresenius' Z.Anal.Chem.* **1986**, *325*, 15–19.

Clark, C.R.; Wells, M.M. Precolumn derivatization of amines for enhanced detectability in liquid chromatography. *J.Chromatogr.Sci.* **1978**, *16*, 332–339.

Corbin, J.L.; Marsh, B.H.; Peters, G.A. An improved method for analysis of polyamines in plant tissue by precolumn derivatization with o-phthalaldehyde and separation by high performance liquid chromatography. *Plant Physiol.* **1989**, *90*, 434–439.

Davis, T.P.; Gehrke, C.W.; Gehrke, C.W., Jr.; Cunningham, T.D.; Kuo, K.C.; Gerhardt, K.O.; Johnson, H.D.; Williams, C.H. High-performance liquid-chromatographic separation and fluorescence measurement of biogenic amines in plasma, urine, and tissue. *Clin.Chem.* **1978**, *24*, 1317–1324.

Davis, T.P.; Gehrke, C.W.; Gehrke, C.W., Jr.; Cunningham, T.D.; Kuo, K.C.; Gerhardt, K.O.; Johnson, H.D.; Williams, C.H. High-performance liquid chromatographic analysis of biogenic amines in biological materials as o-phthalaldehyde derivatives. *J.Chromatogr.* **1979**, *162*, 293–310.

Duchateau, A.L.L.; Guns, J.J.; Kubben, R.G.R.; van Tilburg, A.F.P. high-performance liquid chromatography of diamine enantiomers as Schiff bases on a chiral stationary phase. *J.Chromatogr.A* **1994**, *664*, 169–176.

Eerola, S.; Hinkkanen, R.; Lindfors, E.; Hirvi, T. Liquid chromatographic determination of biogenic amines in dry sausages. *J.AOAC Int.* **1993**, *76*, 575–577.

Feth, F.; Wagner, K.G. Determination of ornithine, putrescine, N-methylputrescine and N-methylpyrroline pools in tobacco tissue by high-performance liquid chromatography. *Physiol.Plant.* **1989**, *75*, 71–74.

Gallo, A.A.; Walters, F.H. Indirect fluorescent visualization of aliphatic amines. *Anal.Lett.* **1986**, *19*, 979–985.

Garthwaite, I.; Stead, A.D.; Rider, C.C. Assay of the polyamine spermine by a monoclonal antibody-based ELISA. *J.Immunol.Methods* **1993**, *162*, 175–178.

Gennaro, M.C.; Marengo, E. Separation of aliphatic and aromatic amines by the ion interaction reagent RP-HPLC technique. *Chromatographia* **1988**, *25*, 603–608.

Goldfinger, M.D. Detection by HPLC-EC of primary amines recovered in aqueous push-pull perfusates from cat cuneate nucleus. *Life Sci.* **1985**, *37*, 1765–1774.

Gonzalez Lomba, J.R.; Gil, E.P.; Moreno, P.C.; Carra, R.M.G-M.; Misiego, A.S. Anodic oxidation of thioureido derivatives of biogenic amines at a glassy carbon electrode in an aqueous medium. *J.Electroanal.Chem.* **1996**, *410*, 87–92.

Gross, G.A. Simple methods for quantifying mutagenic heterocyclic aromatic amines in food products. *Carcinogenesis* **1990**, *11*, 1597–1603.

Hernandez-Jover, T.; Izquierdo-Pulido, M.; Veciana-Nogues, M.T.; Vidal-Carou, M.C. Ion-pair high-performance liquid chromatographic determination of biogenic amines in meat and meat products. *J.Agric.Food Chem.* **1996**, *44*, 2217–2222.

Hohaus, E. Boron chelates and boron metal chelates, part XV. Liquid column-chromatographic separation and fluorometric determination of primary n-alkylamines after derivatization with 2,2-diphenyl-1-oxa-3-oxonia-2-boratanaphthalene (DOOB). *Fresenius' Z.Anal.Chem.* **1984**, *319*, 533–539.

Hornero-Mendez, D.; Garrido-Fernandez, A. Biogenic amines in table olives. Analysis by high-performance liquid chromatography. *Analyst* **1994**, *119*, 2037–2041.

Hui, J.Y.; Taylor, S.L. High pressure liquid chromatographic determination of putrefactive amines in foods. *J.Assoc.Off.Anal.Chem.* **1983**, *66*, 853–857.

Hurst, W.J.; Toomey, P.B. High-performance liquid chromatographic determination of four biogenic amines in chocolate. *Analyst* **1981**, *106*, 394–402.

Ibe, A.; Saito, K.; Nakazato, M.; Kikuchi, Y.; Fujinuma, K.; Nishima, T. Quantitative determination of amines in wine by liquid chromatography. *J.Assoc.Off.Anal.Chem.* **1991**, *74*, 695–698.

Iniguez Crespo, M.; Vazquez Lasa, B. Determination of biogenic amines and other amines in wine by an optimized HPLC method with polarity gradient elution. *Am.J.Enol.Vitic.* **1994**, *45*, 460–463.

Iwaki, K.; Yoshida, S.; Nimura, N.; Kinoshita, T.; Takeda, K.; Ogura, H. Activated carbamate reagent as chiral derivatizing agent for liquid chromatographic optical resolution of enantiomeric amino compounds. *Chromatographia* **1987**, *23*, 899–902.

Jandera, P.; Ventura, K.; Hladonikova, R.; Churacek, J. Comparison of various sorbents for the enrichment of samples of aliphatic amines using solid-phase extraction prior to the determination by HPLC with fluorometric detection. *J.Liq.Chromatogr.* **1994**, *17*, 69–95.

Jandera, P.; Pechova, H.; Tocksteinova, D.; Churacek, J.; Kralovsky, J. Separation and identification of aliphatic amine derivatives with a new fluorescent reagent using high-performance liquid chromatography and thin-layer chromatography. *Chromatographia* **1982**, *16*, 275–281.

Jedrzejczak, K.; Gaind, V.S. Polymers with reactive functions as sampling and derivatizing agents. Part 1. Effective sampling and simultaneous derivatization of an airborne amine. *Analyst* **1990**, *115*, 1359–1362.

Kawasaki, T.; Wong, O.; Wang, C.; Kuwana, T. Trace biogenic amine analysis with precolumn derivatization and with fluorescent and chemiluminescent detection in HPLC. *J.Res.Natl.Bur.Stand.(U.S.)* **1988**, *93*, 504–506.

Khuhawar, M.Y.; Rind, M.B.; Qureshi, G.A. High performance liquid chromatographic determination of putrescine in tomato using 2-hydroxynaphthaldehydes as derivatizing reagents. *J.Chem.Soc.Pak.* **1996**, *18*, 119–122.

Koga, M.; Akiyama, T. Determination of trace heterocyclic amines in water as their 1,2-naphthoquinone derivatives by high performance liquid chromatography. *Anal.Sci.* **1985**, *1*, 285–288.

Koizumi, H.; Suzuki, Y. Micro high-performance liquid chromatography of aliphatic amines by means of resonance Raman detection. *HRC CC, J.High Resolut.Chromatogr.Chromatogr.Commun.* **1987**, *10*, 173–176.

Krizek, M. Determination of biogenic amines in silage. *Arch.Anim.Nutr.* **1991**, *41*, 97–104.

Kvannes, J.; Flatmark, T. Rapid and sensitive assay of ornithine decarboxylase activity by high-performance liquid chromatography of the o-phthalaldehyde derivative of putrescine. *J.Chromatogr.* **1987**, *419*, 291–295.

Lauren, D.R.; Parker, C.H.; Agnew, M.P.; Smith, G.S. The analysis of putrescine in plant samples by automated HPLC. *J.Liq.Chromatogr.* **1981**, *4*, 1269–1280.

Lehtonen, P.; Saarinen, M.; Vesanto, M.; Riekkola, M.L. Determination of wine amines by HPLC using automated precolumn derivatization with o-phthalaldehyde and fluorescence detection. *Z.Lebensm.-Unters.Forsch.* **1992**, *194*, 434–437.

Maibaum, J. Indirect high-performance liquid chromatographic resolution of racemic tertiary amines as their diastereomeric urea derivatives after N-dealkylation. *J.Chromatogr.* **1988**, *436*, 269–278.

Mank, A.J.G.; van der Laan, H.T.C.; Lingeman, H.; Gooijer, C.; Brinkman, U.A.T.; Velthorst, N.H. Visible diode laser-induced fluorescence detection in liquid chromatography after precolumn derivatization of amines. *Anal.Chem.* **1995**, *67*, 1742–1748.

Matuszewski, B.K.; Givens, R.S.; Srinivasachar, K.; Carlson, R.G.; Higuchi, T. N-substituted 1-cyano-benz[f]isoindole: evaluation of fluorescence efficiencies of a new fluorogenic label for primary amines and amino acids. *Anal.Chem.* **1987**, *59*, 1102–1105.

McCrossen, S.D.; Giles, R.G.; Oxley, P.W.; McArdle, J.V. High-performance liquid chromatography procedure for the determination of purity of di-N-n-propylamine. *J.Chromatogr.* **1992**, *623*, 229–235.

Mentasti, E.; Sarzanini, C.; Abollino, O.; Porta, V. Chromatographic behavior of homologous amines: comparison of derivatizing agents, columns and mobile phases. *Chromatographia* **1991**, *31*, 41–49.

Morier-Teissier, E.; Drieu, K.; Rips, R. Determination of polyamines by pre-column derivatization and electrochemical detection. *J.Liq.Chromatogr.* **1988**, *11*, 1627–1650.

Moriyasu, M.; Hashimoto, Y.; Endo, M. High-performance liquid chromatographic determination of organic substances by metal chelate derivatization. I. Dithiocarbamate chelates of aliphatic amines. *Bull.Chem.Soc.Jpn.* **1981**, *54*, 3369–3373.

Pawlowska, M.; Zukowski, J.; Armstrong, D.W. Sensitive enantiomeric separation of aliphatic and aromatic amines using aromatic anhydrides as nonchiral derivatizing agents. *J.Chromatogr.A* **1994**, *666*, 485–491.

Pietsch, J.; Hampel, S.; Schmidt, W.; Brauch, H.-J.; Worch, E. Determination of aliphatic and alicyclic amines in water by gas and liquid chromatography after derivatization by chloroformates. *Fresenius' J.Anal.Chem.* **1996**, *355*, 164–173.

Plessi, M.; Ferioli, V.; Gamberini, G.; Monzani, A. High-performance liquid chromatographic assay of bioactive amines in Parmigiano-Reggiano cheese. *Atti Soc.Nat.Mat.Modena* **1990**, *121*, 1–9.

Possanzini, M.; Di Palo, V. Improved HPLC determination of aliphatic amines in air by diffusion and derivatization techniques. *Chromatographia* **1990**, *29*, 151–154.

Rissler, K. Separation of acetylated polypropylene glycol di- and triamines by gradient reversed-phase high-performance liquid chromatography and evaporative light scattering detection. *J.Chromatogr.A* **1994**, *667*, 167–174.

Saito, K.; Horie, M.; Nose, N.; Nakagomi, K.; Nakazawa, H. Determination of polyamines in foods by liquid chromatography with on-column fluorescence derivatization. *Anal.Sci.* **1992**, *8*, 675–680.

Sanchez-Rodas, D.; Hohaus, E.; Wenclawiak, B. High-performance liquid chromatographic determination of primary amines in aqueous solutions after extraction and derivatization with 2,2-diphenyl-1-oxa-3-oxonia-2-boratanaphthalene (DOOB). *Fresenius' J.Anal.Chem.* **1996**, *355*, 187–189.

Seiler, N.; Knoedgen, B. High-performance liquid chromatographic procedure for the simultaneous determination of the natural polyamines and their monoacetyl derivatives. *J.Chromatogr.* **1980**, *221*, 227–235.

Shimada, K.; Tanaka, M.; Nambara, T. New derivatization of amines for electrochemical detection in liquid chromatography. *Chem.Pharm.Bull.* **1979**, *27*, 2259–2260.

Siegler, R.; Sternson, L.A.; Stobaugh, J.F. Suitability of DTAF as a fluorescent labeling reagent for direct analysis for primary and secondary amines - spectral and chemical reactivity considerations. *J.Pharm.Biomed.Anal.* **1989**, *7*, 45–55.

Simon, P.; Lemacon, C. Determination of aliphatic primary and secondary amines and polyamines in air by high-performance liquid chromatography. *Anal.Chem.* **1987**, *59*, 480–484.

Simpson, R.C.; Spriggle, J.E.; Veening, H. Off-line liquid chromatographic-mass spectrometric studies of o-phthalaldehyde-primary amine derivatives. *J.Chromatogr.* **1983**, *261*, 407–414.

Simpson, R.C.; Mohammed, H.Y.; Veening, H. Reversed-phase ion-pairing liquid-chromatographic separation and fluorometric detection of polyamines. *J.Liq.Chromatogr.* **1982**, *5*, 245–264.

Slocum, R.D.; Flores, H.E.; Galston, A.W.; Weinstein, L.H. Improved method for HPLC analysis of polyamines, agmatine and aromatic monoamines in plant tissue. *Plant Physiol.* **1989**, *89*, 512–517.

Smith, J.R.L.; Smart, A.U.; Hancock, F.E.; Twigg, M.V. High-performance liquid chromatographic determination of low levels of primary and secondary amines in aqueous solutions including 2-amino-2-methylpropanol by pre-column derivatization to sulfonamides. *J.Chromatogr.* **1991**, *547*, 447–451.

Smith, R.M.; Ghani, A.A. Electrochemical detection of aliphatic amines by HPLC after derivatization with 3-(4-hydroxyphenyl)propionic acid ester. *Electroanalysis (N.Y.)* **1990**, *2*, 167–169.

Stobaugh, J.F.; Repta, A.J.; Sternson, L.A.; Garren, K.W. Factors affecting the stability of fluorescent isoindoles derived from reaction of o-phthalaldehyde and hydroxyalkylthiols with primary amines. *Anal.Biochem.* **1983**, *135*, 495–504.

Steinert, J.; Khalaf, H.; Keese, W.; Rimpler, M. Sensitive liquid chromatograhic determination of hydrophobic primary and secondary amines by derivatization to form highly fluorescent thiazoles. *Anal.Chim.Acta* **1996**, *327*, 153–159.

Szulc, M.E.; Krull, I.S. Mixed-bed polymeric o-nitrobenzophenone reagents for the on-line derivatization of amines in high performance liquid chromatography. *Biomed.Chromatogr.* **1992**, *6*, 269–277.

Tanaka, M.; Shimada, K.; Nambara, T. Novel ferrocene reagent for pre-column labeling of amines in high-performance liquid chromatography with electrochemical detection. *J.Chromatogr.* **1984**, *292*, 410–411.

Uzunov, D.; Stoev, G. Some aspects of the enantiorecognition of derivatized primary amines on a Pirkle-type chiral stationary phase utilizing tocainide and mexiletine as model compounds. *J.Chromatogr.* **1993**, *645*, 233–239.

Vandemark, F.L.; Schmidt, G.J.; Slavin, W. Determination of polyamines by liquid chromatography and precolumn labeling for fluorescence detection. *J.Chromatogr.Sci.* **1978**, *16*, 465–469.

Warman, P.R.; Bishop, C. The use of reverse-phase HPLC for soil amino-nitrogen analysis. *J.Liq.Chromatogr.* **1985**, *8*, 2595–2606.

Wehr, J.B. Purification of plant polyamines with anion-exchange column clean-up prior to high-performance liquid chromatographic analysis. *J.Chromatogr.A* **1995**, *709*, 241–247.

Westerholm, R.; Li, H.; Almen, J. Estimation of aliphatic amine emissions in automobile exhausts. *Chemosphere* **1993**, *27*, 1381–1384.

Wu, W.S.; Gaind, V.S. Quantification of solid sorbent-sampled airborne aliphatic polyamines on HPLC using a common calibration standard - application of the concept of isolation of a selected -system of a derivative for specific detection. *J.Liq.Chromatogr.* **1992**, *15*, 267–282.

Yen, G.C.; Hsieh, C.L. Simultaneous analysis of biogenic amines in canned fish by HPLC. *J.Food Sci.* **1991**, *56*, 158–160.

Yonekura, T.; Kamata, S.; Wasa, M.; Okada, A.; Kawata, S.; Tarui, S. Simultaneous analysis of plasma phenethylamine, phenylethanolamine, tyramine and octopamine in patients with hepatic encephalopathy. *Clin.Chim.Acta* **1991**, *199*, 91–97.

Yun, Z.; Zhang, R. HPLC determination of polyamines in the femtomole range. *Biomed.Chromatogr.* **1987**, *2*, 173–176.

Amines (using post-column reaction detection)

Dzido, T.; Wawrzynowicz, T. Selection of adsorption systems for the analysis of aromatic amines by thin-layer and column chromatography on alumina with detection based on post-column derivatization. *Chem.Anal.(Warsaw)* **1979**, *24*, 953–964.

Himuro, A.; Nakamura, H.; Tamura, Z. Fluorometric determination of secondary amines by high-performance liquid chromatography with post-column derivatization. *J.Chromatogr.* **1983**, *264*, 423–433.

Izquierdo-Pulido, M.L.; Vidal-Carou, M.C.; Marine-Font, A. Determination of biogenic amines in beers and their raw materials by ion-pair liquid chromatography with postcolumn derivatization. *J.AOAC Int.* **1993**, *76*, 1027–1032.

Katayama, M.; Takeuchi, H.; Taniguchi, H. Determination of polyamines by liquid chromatography with aryl oxalate-sulforhodamine 101 chemiluminescence detection. *Anal.Chim.Acta* **1994**, *287*, 83–88.

Kim, M.; Stewart, J.T. HPLC post column derivatization of aromatic amines using N-methyl-9-chloroacridinium triflate. *Mikrochim.Acta* **1990**, *3*, 221–232.

Kudoh, M.; Matoh, I.; Fudano, S. Determination of tertiary aliphatic amines by high-performance liquid chromatography. *J.Chromatogr.* **1983**, *261*, 293–297.

Kwakman, P.J.M.; Brinkman, U.A.T.; Frei, R.W.; de Jong, G.J.; Spruit, F.J.; Lammers, N.G.F.M.; Van den Berg, J.H.M. A post-column extraction system for the determination of tertiary amines by liquid chromatography with chemiluminescence detection. *Chromatographia* **1987**, *24*, 395–399.

Luten, J.B.; Bouquet, W.; Seuren, L.A.J.; Burggraaf, M.M.; Riekwel-Booy, G.; Durand, P.; Etienne, M.; Gouyou, J.P.; Landrein, A.; et al. Biogenic amines in fishery products: standardization methods within EC. *Dev.Food Sci.* **1992**, *30*, 427–439.

Matsunaga, A.; Yamamoto, A.; Saito, Y.; Makino, M. Determination of biologically active amines in fermented foods and fish products by high performance liquid chromatography. *Hokuriku Koshu Eisei Gakkaishi* **1985**, *12*, 19–26.

Matsunaga, A.; Yamamoto, A.; Sekiguchi, H.; Shimizu, R. Enzymic decarboxylation products in fermented foods: simultaneous analysis of amines by HPLC [high-performance liquid chromatography]. *Toyama-ken Eisei Kenkyusho Nenpo* **1983**, 62–69.

Mellbin, G.; Smith, B.E.F. Trace determination of aliphatic amines using high-performance liquid chromatography with chemiluminescence excitation and photon counting. *J.Chromatogr.* **1984**, *312*, 203–210.

Noffsinger, J.B.; Danielson, N.D. Liquid chromatography of aliphatic trialkylamines with post-column chemiluminescent detection using tris(2,2'-bipyridine)ruthenium(III). *J.Chromatogr.* **1987**, *387*, 520–524.

Ohta, H.; Takeda, Y.; Yoza, K-I.; Nogata, Y. High-performance liquid chromatographic determination of polyamines in selected vegetables with postcolumn fluorimetric derivatization. *J.Chromatogr.* **1993**, *628*, 199–204.

Schreuder, R.H.; Martijn, A.; Poppe, H.; Kraak, J.C. Determination of the composition of ethoxylated alkylamines in pesticide formulations by high-performance liquid chromatography using ion-pair extraction detection. *J.Chromatogr.* **1986**, *368*, 339–350.

Shaw, G.G.; Al-Deen, I.H.S.; Elworthy, P.M. The construction and performance of a low cost automated HPLC system for polyamine assay. *J.Chromatogr.Sci.* **1980**, *18*, 166–170.

Shinozaki, O.; Kunezaki, N.; Yamazaki, S.; Tanimura, T. Chemiluminescent method for aliphatic tertiary amines using Ru(bipy)$_3$$^{2+}$-S$_2O_8$ reagent. *Kuromatogurafi* **1992**, *13*, 275–276.

Straub, B.; Schollenberger, M.; Kicherer, M.; Luckas, B.; Hammes, W.P. Extraction and determination of biogenic amines in fermented sausages and other meat products with reversed-phase HPLC. *Z.Lebensm.-Unters.Forsch.* **1993**, *197*, 230–232.

Takeda, Y.; Yoza, K.; Nogata, Y.; Ohta, H. Effects of storage temperatures on polyamine content of some leafy vegetables. *Engei Gakkai Zasshi* **1993**, *62*, 425–430.

Tomoki, S.; Shigeo, Y.; Takenori, T. Detection of aliphatic tertiary amines with chemiluminescence method using [Ru(bipy)$_3$]$^{2+}$-S$_2$O$_8$$^{2-}$ reagent. *Kuromatogurafi* **1993**, *14*, 60–61.

Toyooka, T.; Chokshi, H.P.; Givens, R.S.; Carlson, R.G.; Lunte, S.M.; Kuwana, T. Fluorescence and chemiluminescence detection of oxazole-labeled amines and thiols. *Biomed.Chromatogr.* **1993**, *7*, 208–216.

Uchikura, K.; Kirisawa, M.; Sugii, A. Electrochemiluminescence detection of primary amines using tris(bipyridine)ruthenium(III) after derivatization with divinylsulfone. *Anal.Sci.* **1993**, *9*, 121–123.

Watanabe, N.; Asano, M.; Yamamoto, K.; Nagatsu, T.; Matsumoto, T.; Fujita, K. Electrochemical detection of polyamines using immobilized enzyme as post-column reactor in high performance liquid chromatography. *Chem.Lett.* **1988**, 1169–1170.

Watanabe, S.; Sato, S.; Nagase, S.; Tomita, M.; Saito, T.; Ishizu, H. Automated quantitation of polyamines by improved cation-exchange high-performance liquid chromatography using a pump equipped with a plunger washing system. *J.Liq.Chromatogr.* **1993**, *16*, 619–632.

Whiteside, I.R.C.; Worsfold, P.J.; McKerrell, E.H. Determination of alkylamines by high-performance liquid chromatography with postcolumn fluorescence derivatization. *Anal.Chim.Acta* **1988**, *212*, 155–163.

Yamazaki, S.; Chiba, R.; Uchikura, K.; Tanimura, T. Simultaneous determination of primary, secondary and tertiary amines by postcolumn reaction. *Kuromatogurafi* **1995**, *16*, 322–323.

Amino Acids

Aminuddin, M.; Miller, J.N. A fast liquid chromatographic method for the determination of amino acids. *J.Chem.Soc.Pak.* **1994**, *16*, 17–20.

Ang, S.G.; Low, S.H. Studies on the stereochemical characterization of N-methylated amino acids. *Aust.J.Chem.* **1991**, *44*, 1591–1601.

Astephen, N.; Wheat, T. An amino acid analysis method for assessing nutritional quality of infant formulas. *Am.Lab.(Shelton, Conn.)* **1993**, *25*, 24–27.

Auger, J.; Mellouki, F.; Vannereau, A.; Boscher, J.; Cosson, L.; Mandon, N. Analysis of Allium for sulfur amino acids by HPLC after derivatization. *Chromatographia* **1993**, *36*, 347–350.

Baffi, F. HPLC determination of α-amino acids in phytoplanktonic marine cells. *Int.J.Environ. Anal.Chem.* **1990**, *41*, 173–176.

Berne, F.D.; Mwelet, N.; Cauchi, B.; Legube, B. Evolution of amino acids and dissolved organic matter in a drinking water treatment plant: Correlations with biodegradable dissolved organic carbon and long-term chlorine demand. *Revue Sciences de l'eau* **1996**, *9*, 115–133.

Berne, F.D.; Panais, B.; Merlet, N.; Cauchi, B.; Legube, B. Total dissolved amino acid analysis in natural and drinking waters. *Environ.Technol.* **1994**, *15*, 901–916.

Bhown, A.S.; Cornelius, T.W.; Bennett, J.C. A rapid separation method for precolumn derivatized amino acids using reversed-phase HPLC. *LC, Liq.Chromatogr.HPLC Mag.* **1983**, *1*, 50–52.

Blundell, G.; Brydon, W.G. High performance liquid chromatography of plasma amino acids using orthophthalaldehyde derivatization. *Clin.Chim.Acta* **1987**, *170*, 79–84.

Brueckner, H.; Keller-Hoehl, C. HPLC separation of DL-amino acids derivatized with N2-(5-fluoro-2,4-dinitrophenyl)-L-amino acid amides. *Chromatographia* **1990**, *30*, 621–629.

Brueckner, H.; Luepke, M. Derivatization of DL-amino acids by urethane-protected α-amino acid N-carboxyanhydrides for separation by LC. *Chromatographia* **1995**, *40*, 601–606.

Brueckner, H.; Strecker, B. Chiral monochloro-s-triazines as derivatizing reagents for resolving DL-amino acids by HPLC. *Chromatographia* **1992**, *33*, 586–587.

Brueckner, H.; Strecker, B. Use of chiral monohalo-s-triazine reagents for the liquid chromatographic resolution of DL-amino acids. *J.Chromatogr.* **1992**, *627*, 97–105.

Brueckner, H.; Zivny, S. High-performance liquid chromatographic resolution of (R,S)-α-alkyl-α-amino acids as diastereomeric derivatives. *Amino Acids* **1993**, *4*, 157–167.

Brueckner, H.; Jaek, P.; Langer, M.; Godel, H. Liquid chromatographic determination of D-amino acids in cheese and cow milk. Implication of starter cultures, amino acid racemases, and rumen microorganisms on formation, and nutritional considerations. *Amino Acids* **1992**, *2*, 271–284.

Brueckner, H.; Gah, C. High-performance liquid chromatographic separation of DL-amino acids derivatized with chiral variants of Sanger's reagent. *J.Chromatogr.* **1991**, *555*, 81–95.

Buck, R.H.; Krummen, K. High-performance liquid chromatography with automated pre-column derivatization for amino acids. *J.Chromatogr.* **1984**, *303*, 238–243.

Burgoyne, R.F. Determining amino acid composition. *Bio/Technology* **1993**, *11*, 1302, 1304.

Canepa, A.; Perfumo, F.; Carrea, A.; Sanguineti, A.; Piccardo, M.T.; Gusmano, R. Measurement of free amino acids in polymorphonuclear leukocytes by high-performance liquid chromatography. *J.Chromatogr.* **1989**, *491*, 200–208.

Canevari, L.; Vieira, R.; Aldegunde, M.; Dagani, F. High performance liquid chromatographic separation with electrochemical detection of amino acids focusing on neurochemical application. *Pharmacol.Res.* **1992**, *25*, 115–116.

Cann-Moisan, C.; Caroff, J.; Girin, E. Rapid high-performance liquid chromatographic determination of gamma-aminobutyric acid and some other amino acids: application to rat brain. *J.Chromatogr.* **1990**, *532*, 438–441.

Chang, J.Y.; Knecht, R.; Braun, D.G. Amino acid analysis in the picomole range by precolumn derivatization and high-performance liquid chromatography. *Methods Enzymol.* **1983**, *91*, 41–48.

Chang, J.Y.; Martin, P.; Bernasconi, R.; Braun, D.G. High-sensitivity amino acid analysis: measurement of amino acid neurotransmitter in mouse brain. *FEBS Lett.* **1981**, *132*, 117–120.

Chang, J.Y.; Aebersold, R.; Gruetter, T.; Rosenfelder, G.; Braun, D.G. High sensitivity analysis of amino acids, peptides and proteins by color labeling techniques. *Protides Biol.Fluids* **1984**, *32*, 955–960.

Chang, J.Y. A complete quantitative N-terminal analysis method. *Anal.Biochem.* **1988**, *170*, 542–556.

Chang, S.C.; Wang, L.R.; Armstrong, D.W. Facile resolution of N-tert-butoxycarbonyl amino acids: The importance of enantiomeric purity in peptide synthesis. *J.Liq.Chromatogr.* **1992**, *15*, 1411–1429.

Chen, X.; Sato, M. HPLC determination of primary amino acids using electrochemiluminescence of ruthenium complex. *J.Flow Injection Anal.* **1995**, *12*, 216–222.

Chow, J.; Orenberg, J.B.; Nugent, K.D. Comparison of automated pre-column and post-column analysis of amino acid oligomers. *J.Chromatogr.* **1987**, *386*, 243–249.

Cloete, C. Automated optimised high performance liquid chromatographic analysis of pre-column o-phthaldialdehyde-amino acid derivatives. *J.Liq.Chromatogr.* **1984**, *7*, 1979–1990.

Confer, D.R.; Logan, B.E.; Aiken, B.S.; Kirchman, D.L. Measurement of dissolved free and combined amino acids in unconcentrated wastewaters using high performance liquid chromatography. *Water Environ.Res.* **1995**, *67*, 118–125.

Cooper, J.D.H.; Lewis, M.T.; Turnell, D.C. Pre-column o-phthalaldehyde derivatization of amino acids and their separation using reversed-phase high-performance liquid chromatography. II. Simultaneous determination of amino and imino acids in protein hydrolyzates. *J.Chromatogr.* **1984**, *285*, 490–494.

Corona, G.; Mattioli, S.; Di Martino, A.; Callegaro, L. A simple method to detect contaminating peptides and amino acids in large-scale ganglioside preparations. *Biologicals* **1991**, *19*, 311–316.

Cunico, R.; Mayer, A.G.; Wehr, C.T.; Sheehan, T.L. High sensitivity amino acid analysis using a novel automated precolumn derivatization system. *BioChromatography* **1986**, *1*, 608–614.

Dawson, R.; Kalbfleisch, J.; Liebezeit, G.; Llewellyn, C.A.; Mantoura, R.F.C.; Moreau, F.; Poulet, S.A. HPLC analyses of dissolved free amino acids, pigments and vitamins in plankton and particles. *Oceanis* **1985**, *11*, 521–531.

DeSilva, K.; Strojek, J.; Kuwana, T. Dual wavelength excitation method for the off-line liquid chromatographic analysis of derivatized amino acids. *Anal.Sci.* **1994**, *10*, 573–578.

Do, K.Q.; Tappaz, M.L. Specificity of cysteine sulfinate decarboxylase (CSD) for sulfur-containing amino-acids. *Neurochem.Int.* **1996**, *28*, 363–371.

Dong, M.W.; DiCesare, J.L. Amino acid analysis by liquid chromatography. *LC, Liq.Chromatogr.HPLC Mag.* **1983**, *1*, 222–228.

Dong, M.W. Analytical derivatization using robotics: amino acid analysis by precolumn o-phthalaldehyde. *LC-GC* **1987**, *5*, 255–260.

Duchateau, A.L.L.; Heemels, G.M.P.; Maesen, L.W.; De Vries, N.K. Separation of benz[f]isoindole derivatives of amino acid and amino acid amide enantiomers on a β-cyclodextrin-bonded phase. *J.Chromatogr.* **1992**, *603*, 151–156.

Dyremark, A.; Ericsson, M. An LCEC method for the analysis of synthetic amino acids in fruit juices. *Chromatographia* **1990**, *29*, 51–53.

Einarsson, S.; Folestad, S.; Josefsson, B.; Lagerkvist, S. High-resolution reversed-phase liquid chromatography system for the analysis of complex solutions of primary and secondary amino acids. *Anal.Chem.* **1986**, *58*, 1638–1643.

Elkin, R.G. Quantitative amino acid analysis of feedstuff hydrolyzates by reverse phase liquid chromatography and conventional ion-exchange chromatography. *J.Assoc.Off.Anal.Chem.* **1984**, *67*, 1024–1026.

Endo, T.; Nishida, A.; Kano, S. Role of neurotransmitter in forensic medicine (report III). A new analysis of inhibitory neurotransmitter amino acids in tissues and biological fluids by high performance liquid chromatography with electrochemical detection. *Nippon Hoigaku Zasshi* **1987**, *41*, 308–315.

Eslami, M.; Hashemi, P.; Sarbolouki, M.N. Separation and indirect detection of amino acids by reversed-phase ion-pair chromatography. *J.Chromatogr.Sci.* **1993**, *31*, 480–485.

Euerby, M.R.; Partridge, L.Z.; Nunn, P.B. Resolution of neuroactive non-protein amino acid enantiomers by high-performance liquid chromatography utilizing pre-column derivatisation with o-phthaldialdehyde-chiral thiols. Application to 2-amino-omega-phosphonoalkanoic acid homologs and α-amino-β-N-methylaminopropanoic acid (β-methylaminoalanine). *J.Chromatogr.* **1989**, *469*, 412–419.

Euerby, M.R.; Partridge, L.Z.; Gibbons, W.A. Study of the chromatographic behavior and resolution of α-amino acid enantiomers by high-performance liquid chromatography utilizing pre-column derivatization with o-phthaldialdehyde and new chiral thiols. *J.Chromatogr.* **1989**, *483*, 239–252.

Gorog, S.; Herenyi, B.; Low, M. Derivatization of N-protected amino acids for chiral separation by high-performance liquid chromatography. *J.Chromatogr.* **1986**, *353*, 417–424.

Griffin, M.; Price, S.J.; Palmer, T. A rapid and sensitive procedure for the quantitative determination of plasma amino acids. *Clin.Chim.Acta* **1982**, *125*, 89–95.

Haginaka, J.; Wakai, J. Automated precolumn derivatization of amino acids with ortho-phthalaldehyde using a hollow-fiber membrane reactor. *J.Chromatogr.* **1990**, *502*, 317–324.

Halfpenny, A.P.; Brown, P.R. Evaluation of conditions for the HPLC determination of plasma amino acids using precolumn derivatization. *HRC CC, J.High Resolut.Chromatogr.Chromatogr.Commun.* **1985**, *8*, 243–247.

Halpine, S.M. Amino acid analysis of proteinaceous media from Cosimo Tura's 'The Annunciation with Saint Francis and Saint Louis of Toulouse'. *Stud.Conserv.* **1992**, *37*, 22–38.

Hancock, W.S.; Bishop, C.A.; Hearn, M.T.W. The analysis of nanogram levels of free amino acids by reverse-phase high-pressure liquid chromatography. *Anal.Biochem.* **1979**, *92*, 170–173.

Hasegawa, Y.; Jette, D.C.; Miyamoto, A.; Kawasaki, H.; Yuki, H. Chemiluminescent determination of amino-terminal residues of peptides after derivatization of anilinothiazolinone from Edman degradation with N-(4-aminohexyl)-N-ethylisoluminol. *Anal.Sci.* **1991**, *7*, 945–948.

Hayakawa, K.; Schilpp, T.; Imai, K.; Higuchi, T.; Wong, O.S. Determination of aspartic acid, phenylalanine, and aspartylphenylalanine in aspartame-containing samples using a precolumn derivatization HPLC method. *J.Agric.Food Chem.* **1990**, *38*, 1256–1260.

Hikal, A.H.; Lipe, G.W.; Slikker, W.J.; Scallet, A.C.; Ali, S.F.; Newport, G.D. Determination of amino acids in different regions of the rat brain. Application to the acute effects of tetrahydrocannabinol (THC) and trimethyltin (TMT). *Life Sci.* **1988**, *42*, 2029–2035.

Hill, D.W.; Walters, F.H.; Wilson, T.D.; Stuart, J.D. High performance liquid chromatographic determination of amino acids in the picomole range. *Anal.Chem.* **1979**, *51*, 1338–1341.

Hodgin, J.C. The separation of pre-column o-phthalaldehyde derivatized amino acids by high performance liquid chromatography. *J.Liq.Chromatogr.* **1979**, *2*, 1047–1059.

Hogan, D.L.; Kraemer, K.L.; Isenberg, J.I. The use of high-performance liquid chromatography for quantitation of plasma amino acids in man. *Anal.Biochem.* **1982**, *127*, 17–24.

Ho, J.W. Microassay for amino acids in human liquid blood and dried blood. *Clin.Chim.Acta* **1989**, *185*, 197–201.

Hoskins, J.A.; Holliday, S.B.; Davies, F.F. Problems in the analysis of low levels of amino acids in physiological fluids and tissues using o-phthalaldehyde derivatization and reversed-phase high-performance liquid chromatography with electrochemical detection. *J.Chromatogr.* **1986**, *375*, 129–133.

Hsieh, Y.Z.; Beale, S.C.; Wiesler, D.; Novotny, M. 3-Benzoyl-2-naphthaldehyde, a new fluorogenic reagent for microcolumn liquid chromatography-laser-induced fluorescence detection of amino acids. *J.Microcolumn Sep.* **1989**, *1*, 96–100.

Huang, Z.; Ough, C.S. Determination of amino acid hydantoins by HPLC with diode array detection. *J.Agric.Food Chem.* **1991**, *39*, 2218–2222.

Hurst, W.J. The use of automated derivative formation in the determination of amino acids in urine. *Lab.Rob.Autom.* **1990**, *2*, 235–238.

Hurst, W.J.; Martin, R.A., Jr. Use of o-phthalaldehyde derivatives and high-pressure liquid chromatography in determining the free amino acids in cocoa beans. *J.Agric.Food Chem.* **1980**, *28*, 1039.

Irvine, G.B. Amino acid analysis. *Methods Mol.Biol.(Totowa, N.J.)* **1994**, *32*, 257–265.

Iwaki, K.; Yoshida, S.; Nimura, N.; Kinoshita, T.; Takeda, K.; Ogura, H. Optical resolution of enantiomeric amino acid derivatives on a naphthylethylurea multiple-bonded chiral stationary phase prepared via an activated carbamate intermediate. *J.Chromatogr.* **1987**, *404*, 117–122.

Iwase, H.; Ono, I. Precolumn deproteinization method of human plasma using a hydroxyapatite cartridge for high-performance liquid chromatographic amino acid analysis. *Anal.Sci.* **1995**, *11*, 73–77.

Jacobs, W. o-Phthalaldehyde derivatization of amino acids and peptides for LCEC. *Curr.Sep.* **1986**, *7*, 39–42.

Jadaud, P.; Thelohan, S.; Schonbaum, G.R.; Wainer, I.W. The stereochemical resolution of enantiomeric free and derivatized amino acids using an HPLC chiral stationary phase based on immobilized alpha-chymotrypsin: chiral separation due to solute structure or enzyme activity. *Chirality* **1989**, *1*, 38–44.

Jones, B.N.; Paabo, S.; Stein, S. Amino acid analysis and enzymic sequence determination of peptides by an improved o-phthaldialdehyde precolumn labeling procedure. *J.Liq.Chromatogr.* **1981**, *4*, 565–586.

Jones, B.N.; Gilligan, J.P. Amino acid analysis by o-phthaldialdehyde precolumn derivatization and reversed-phase HPLC. *Am.Biotechnol.Lab.* **1983**, 46–51.

Kabus, P.; Koch, G. Quantitative determination of amino acids in tissue culture cells by high performance liquid chromatography. *Biochem.Biophys.Res.Commun.* **1982**, *108*, 783–790.

Kam, J.A.; Mohabbat, T.; Greenhagen, R.D.; Decedue, C.J.; Lunte, S.M. Amino acid analysis using naphthalenedialdehyde and the BAS 200A with fluorescence detection. *Curr.Sep.* **1992**, *11*, 57–60.

Kan, T.A.; Shipe, W.F. Modification and evaluation of a reverse phase high performance liquid chromatographic method for amino acid analysis. *J.Food Sci.* **1981**, *47*, 338–341.

Kasimura, H.; Kishikawa, K.; Kohmoto, S.; Yamamoto, M.; Yamada, K. Diastereomeric separation of α-amino acid derivatives using a chiral carbodiimide. *Anal.Chim.Acta* **1990**, *239*, 297–299.

Kawakami, Y.; Ohmori, S. Microidentification of N-terminal-blocked amino acid residues of proteins and peptides. *Anal.Biochem.* **1994**, *220*, 66–72.

Kawakami, Y.; Ohga, T.; Shimamoto, C.; Satoh, N.; Ohmori, S. Determination of free N-acetylamino acids in biological samples and N-terminal acetylamino acids of proteins. *J.Chromatogr.* **1992**, *576*, 63–70.

Khim-Heang, S.; Haerdi, W. Application of HPLC in the determination of amino acids in natural waters. *Int.J.Environ.Anal.Chem.* **1983**, *15*, 309–318.

Korte, E. Amino acid determination by HPLC. *GIT-Suppl.* **1987**, 44–46.

Lammens, J.; Verzele, M. Rapid and easy HPLC analysis of amino acids. *Chromatographia* **1978**, *11*, 376–378.

Lam, S. Resolution of D- and L-amino acids after precolumn derivatization with o-phthalaldehyde by mixed chelation with copper(II)-L-proline. *J.Chromatogr.* **1986**, *355*, 157–164.

Larsen, B.R.; West, F.G. A method for quantitative amino acid analysis using precolumn o-phthalaldehyde derivatization and high performance liquid chromatography. *J.Chromatogr.Sci.* **1981**, *19*, 259–265.

Lau, O-W.; Mok, C-S. Indirect conductometric detection of amino acids after liquid chromatographic separation. Part II. Determination of monosodium glutamate in foods. *Anal.Chim.Acta* **1995**, *302*, 45–52.

Lee, S.H.; Berthod, A.; Armstrong, D.W. Systematic study on the resolution of derivatized amino acids enantiomers on different cyclodextrin-bonded stationary phases. *J.Chromatogr.* **1992**, *603*, 83–93.

Levin, S.; Grushka, E. Factors controlling the separation of amino acids in isocratic reversed-phase liquid chromatography. *J.Chromatogr.* **1987**, *384*, 249–258.

Liebezeit, G.; Dawson, R. Isoindole derivatives of amino acids for HPLC separations - effect of reaction pH and time on fluorescence yield. *HRC CC, J.High Resolut.Chromatogr.Chromatogr.Commun.* **1981**, *4*, 354–356.

Lindroth, P.; Mopper, K. High performance liquid chromatographic determination of subpicomole amounts of amino acids by precolumn fluorescence derivatization with o-phthaldialdehyde. *Anal.Chem.* **1979**, *51*, 1667–1674.

Liu, C-S.; Lin, C-C.; Chen, J-M.; Chang, C-H.; Lo, T-B. A convenient amino acid analysis of proteins electroblotted onto polyvinylidene difluoride [PVDF] membrane from sodium dodecyl sulfate-polyacrylamide gel - a direct in situ derivatization of amino acids after gas-phase hydrolysis. *J.Chin.Biochem.Soc.* **1993**, *22*, 69–76.

Lookhart, G.L.; Jones, B.L. High performance liquid chromatography analysis of amino acids at the picomole level. *Cereal Chem.* **1985**, *62*, 97–102.

Lottspeich, F. Amino acids. *Fresenius' Z.Anal.Chem.* **1987**, *327*, 23–24.

Mancheva, I.N.; Simeonova, R.A. On the regeneration of amino acids from their diphenylindonyl-substituted thiohydantoin derivatives. *Dokl.Bolg.Akad.Nauk* **1982**, *35*, 1685–1688.

Mank, A.J.G.; Yeung, E.S. Diode laser-induced fluorescence detection in capillary electrophoresis after pre-column derivatization of amino acids and small peptides. *J.Chromatogr.A* **1995**, *708*, 309–321.

Martinez-Force, E.; Benitez, T. Separation of o-phthalaldehyde derivatives of amino acids of the internal pool of yeast by reverse-phase liquid chromatography. *Biotechnol.Tech.* **1991**, *5*, 209–214.

Matsoukas, J.; Moharir, Y.E.; Findlay, J.A. A convenient method for the isolation of free amino acids from marine invertebrates. *J.Nat.Prod.* **1983**, *46*, 582–585.

McCourt, D.W.; Leykam, J.F.; Schwartz, B.D. Analysis of sulfate and phosphate esters of amino acids by ion-exchange chromatography on polymeric DEAE. *J.Chromatogr.* **1985**, *327*, 9–15.

McGowan, C.; Wiley, V.A.; Bates, R.P. Application of methodology for RP-HPLC amino acid analysis to the measurement of hypoglycin A. *BioChromatography* **1989**, *4*, 161–164.

Mehra, A.; Narang, C.K.; Mathur, N.K. Analysis of amino acid mixtures after pre-column derivatization by N-methylisatoic anhydride using HPLC. *Indian J.Biochem.Biophys.* **1988**, *25*, 360–363.

Merino Merino, I.; Blanco Gonzalez, E.; Sanz-Medel, A. Liquid chromatographic enantiomeric resolution of amino acids with β-cyclodextrin bonded phases and derivatization with o-phthalaldehyde. *Anal.Chim.Acta* **1990**, *234*, 127–131.

Montanarella, L. Relative merits of pre- and post-column derivatization in amino acid analysis by HPLC. *Chim.Oggi* **1987**, 27–32.

Mopper, K.; Dawson, R. Determination of amino acids in sea water-recent chromatographic developments and future directions. *Sci.Total Environ.* **1986**, *49*, 115–131.

Murai, S.; Saito, H.; Abe, E.; Masuda, Y.; Itoh, T. A rapid assay for neurotransmitter amino acids, aspartate, glutamate, glycine, taurine and gamma-aminobutyric acid in the brain by high-performance liquid chromatography with electrochemical detection. *J.Neural Transm.: Gen.Sect.* **1992**, *87*, 145–153.

Nagata, Y.; Yamamoto, K.; Shimojo, T. Determination of D- and L-amino acids in mouse kidney by high-performance liquid chromatography. *J.Chromatogr.* **1992**, *575*, 147–152.

Neitz, A.W.; Yunker, C.E. Amino acid and protein depletion in medium of cell cultures infected with Cowdria ruminantium. *Ann.N.Y.Acad.Sci.* **1996**, *791*, 24–34.

Nishi, H.; Ishii, K.; Taku, K.; Shimizu, R.; Tsumagari, N. New chiral derivatization reagent for the resolution of amino acids as diastereomers by TLC and HPLC. *Chromatographia* **1989**, *27*, 301–305.

Norma, C.R.; Naccha Torres, L.; Torres de Navarro, E.; Salazar, C.; de la Luz, M. Determination of free amino acids in orange juice using high-resolution liquid chromatography. *Alimentaria (Madrid)* **1996**, *272*, 47–52.

Oates, M.D.; Cooper, B.R.; Jorgenson, J.W. Quantitative amino acid analysis of individual snail neurons by open tubular liquid chromatography. *Anal.Chem.* **1990**, *62*, 1573–1577.

Oates, M.D.; Jorgenson, J.W. Determination of naphthalene-2,3-dicarboxaldehyde-labeled amino acids by open tubular liquid chromatography with electrochemical detection. *Anal.Chem.* **1989**, *61*, 432–435.

Oates, M.D.; Jorgenson, J.W. Quantitative amino acid analysis of subnanogram levels of protein by open tubular liquid chromatography. *Anal.Chem.* **1990**, *62*, 1577–1580.

Olson, D.C.; Schmidt, G.J.; Slavin, W. The determination of amino acids in physiological fluids using liquid chromatography and fluorescence detection. *Chromatogr.Newsl.* **1979**, *7*, 22–25.

Palmero, S.; De Marchis, M.; Prati, M.; Fugassa, E. HPLC analysis of free amino acids and amino acids of total proteins in cultured cells: an application to the study of rat Sertoli cell protein metabolism. *Anal.Biochem.* **1992**, *202*, 152–158.

Pecavar, A.; Prosek, M.; Fercej-Temeljotov, D.; Marsel, J. Quantitative evaluation of amino acids using microwave accelerated hydrolysis. *Chromatographia* **1990**, *30*, 159–162.

Peter, A.; Toth, G.; Olajos, E.; Fueloep, F.; Tourwe, D. Monitoring of optical isomers of some conformationally constrained amino acids with tetrahydroisoquinoline or tetraline ring structures. Part II. *J.Chromatogr.A* **1995**, *705*, 257–265.

Peter, A.; Toth, G.; Toeroek, G.; Tourwe, D. Separation of enantiomeric β-methyl amino acids and of β-methyl amino acid containing peptides. *J.Chromatogr.A* **1996**, *728*, 455–465.

Peter, A.; Toth, G.; Tourwe, D. Monitoring of optical isomers of some conformationally constrained amino acids with tetrahydroisoquinoline or tetraline ring structures. *J.Chromatogr.A* **1994**, *668*, 331–335.

Pinter, G.; Kovacs, A.L. Amino acid analysis within 25 minutes using C1 and C18 reversed-phase high-performance liquid chromatography columns with o-phthalaldehyde precolumn derivatization; application for biological samples. *Symp.Biol.Hung.* **1986**, *31*, 475–488.

Price, S.J.; Palmer, T.; Griffin, M. High-speed assay of amino acids using reversed-phase liquid chromatography. *Chromatographia* **1984**, *18*, 62–64.

Prieto, J.A.; Collar, C.; Benedito de Barber, C. Reversed-phase high-performance liquid chromatographic determination of biochemical changes in free amino acids during wheat flour mixing and bread baking. *J.Chromatogr.Sci.* **1990**, *28*, 572–577.

Qureshi, G.A.; Soedersten, P. Determination of free amino acids in the cerebrospinal fluid of male rats by high-performance liquid chromatography with fluorescence detection. Alterations in amino acid concentrations during sexual activity. *J.Chromatogr.* **1987**, *400*, 247–251.

Qureshi, G.A.; Gokmen, S. Determination of amino acids by liquid chromatography based on pre- and post-column derivatization. *Chim.Acta Turc.* **1987**, *15*, 227–241.

Qureshi, G.A. High-performance liquid chromatographic methods with fluorescence detection for the determination of branched-chain amino acids and their alpha-keto analogs in plasma samples of healthy subjects and uremic patients. *J.Chromatogr.* **1987**, *400*, 91–99.

Qureshi, G.A.; Fohlin, L.; Bergström, J. Application of high-performance liquid chromatography to the determination of free amino acids in physiological fluids. *J.Chromatogr.* **1984**, *297*, 91–100.

Razal, R.A.; Lewis, N.G.; Towers, G.H.N. Pico-Tag analysis of arogenic acid and related free amino acids from plant and fungal extracts. *Phytochem.Anal.* **1994**, *5*, 98–104.

Reid, S.; Randerson, D.H.; Greenfield, P.F. Amino acid determination in mammalian cell culture supernatants. *Aust.J.Biotechnol.* **1987**, *1*, 69–72.

Renlund, S.; Klintrot, I.M.; Nunn, M.; Schrimsher, J.L.; Wernstedt, C.; Hellman, U. Peptide mapping of HIV polypeptides expressed in Escherichia coli. Quality control of different batches and identification of tryptic fragments containing residues of aromatic amino acids or cysteine. *J.Chromatogr.* **1990**, *512*, 325–335.

Roach, M.C.; Harmony, M.D. Determination of amino acids at subfemtomole levels by high-performance liquid chromatography with laser-induced fluorescence detection. *Anal.Chem.* **1987**, *59*, 411–415.

Sandberg, M.; Hagberg, H.; Jacobson, I.; Karlsson, B.; Lehmann, A.; Hamberger, A. Analysis of amino acids: neurochemical application. *Life Sci.* **1987**, *41*, 829–832.

Sanders, E.M.; Ough, C.S. Determination of free amino acids in wine by HPLC. *Am.J.Enol.Vitic.* **1985**, *36*, 43–46.

Schmidt, G.J.; Olson, D.C.; Slavin, W. Amino acid profiling of protein hydrolyzates using liquid chromatography and fluorescence detection. *J.Liq.Chromatogr.* **1979**, *2*, 1031–1045.

Schreiber, J.; Lohmann, W.; Berthold, F.; Lampert, F. A new, fast method for the determination of 26 free amino acids in blood plasma from patients with different types of leukemia within 60 min. *Fresenius' Z.Anal.Chem.* **1986**, *325*, 476–477.

Shah, A.J.; Younie, D.A.; Adlard, M.W.; Evans, C.S. High performance liquid chromatography of nonprotein amino acids extracted from Acacia seeds. *Phytochem.Anal.* **1992**, *3*, 20–25.

Shenoy, N.R.; Shively, J.E.; Bailey, J.M. Studies in C-terminal sequencing: New reagents for the synthesis of peptidylthiohydantoins. *J.Protein Chem.* **1993**, *12*, 195–205.

Sherwood, R.A. Amino acid measurement by high-performance liquid chromatography using electrochemical detection. *J.Neurosci.Methods* **1990**, *34*, 17–22.

Shoji, S.; Ichikawa, M.; Yamaoka, T.; Funakoshi, T.; Kubota, Y. High-sensitivity amino acid analysis of stained peptides and proteins from a sodium dodecyl sulfate-polyacrylamide slab gel. *J.Chromatogr.* **1986**, *354*, 463–470.

Singh, B.K.; Tecle, B.; Shaner, D.L. Determination of 2-keto acids and amino acids in plant extracts. *J.Liq.Chromatogr.* **1994**, *17*, 4469–4477.

Sista, H.S. Sensitive amino acid analysis by reversed-phase high-performance liquid chromatography. Optimization of the o-phthalaldehyde method for composition of picomole amounts of acid hydrolyzates. *J.Chromatogr.* **1986**, *359*, 231–240.

Smith, R.J.; Panico, K.A. Automated analysis of o-phthalaldehyde derivatives of amino acids in physiological fluids by reverse phase high performance liquid chromatography. *J.Liq.Chromatogr.* **1985**, *8*, 1783–1795.

Smolders, I.; Sarre, S.; Michotte, Y.; Ebinger, G. The analysis of excitatory, inhibitory and other amino acids in rat brain microdialysates using microbore liquid chromatography. *J.Neurosci.Methods* **1995**, *57*, 47–53.

Soper, S.A.; Lunte, S.M.; Kuwana, T. High sensitivity fluorescence detection of naphthalenedialdehyde-derivatized amino acids with a low power helium-cadmium laser for high-performance liquid chromatographic analysis. *Anal.Sci.* **1989**, *5*, 23–29.

Spurlin, S.R.; Cooper, M.M. A chemiluminescent precolumn labeling reagent for high-performance liquid chromatography of amino acids. *Anal.Lett.* **1986**, *19*, 2277–2283.

Szokan, G.; Mezo, G.; Hudecz, F.; Majer, Z.; Schon, I.; Nyeki, O.; Szirtes, T.; Doelling, R. Racemization analyses of peptides and amino acid derivatives by chromatography with pre-column derivatization. *J.Liq.Chromatogr.* **1989**, *12*, 2855–2875.

Szokan, G.; Hadfi, S.; Krizsan, K.; Liembeck, A.; Krecz, I.; Almas, M.; Somlai, C. HPLC determination of enantiomeric purity of protected amino acid derivatives used in peptide synthesis. *J.Liq.Chromatogr.* **1994**, *17*, 2759–2775.

Takaya, T.; Sakakibara, S. Determination of the optical purity of amino acids by reversed phase high-performance liquid chromatography. *Pept.Chem.* **1979**, *17th*, 139–144.

Tatar, E.; Khalifa, M.; Zaray, G.; Molnar-Perl, I. Comparison of the recovery of amino acids in vapor-phase hydrolyzates of proteins performed in a Pico Tag work station and in a microwave hydrolysis system. *J.Chromatogr.A* **1994**, *672*, 109–115.

Thelohan, S.; Jadaud, P.; Wainer, I.W. Immobilized enzymes as chromatographic phases for HPLC: the chromatography of free and derivatized amino acids on immobilized trypsin. *Chromatographia* **1989**, *28*, 551–555.

Thomas, D.J.; Parkin, K.L. Quantification of alk(en)yl-L-cysteine sulfoxides and related amino acids in alliums by high-performance liquid chromatography. *J.Agric.Food Chem.* **1994**, *42*, 1632–1638.

Tivesten, A.; Oernskov, E.; Folestad, S. On-column chiral derivatization of D- and L-amino acids in micellar electrokinetic chromatography by zone mixing. *J.High Resolut.Chromatogr.* **1996**, *19*, 229–233.

Tsugita, A.; Kamo, M.; Jone, C.S.; Shikama, N. Sensitization of Edman amino acid derivatives using the fluorescent reagent, 4-aminofluorescein. *J.Biochem.* **1989**, *106*, 60–65.

Turnell, D.C.; Cooper, J.D.H. Automated pre-column derivatization and its application to amino acid analysis using high-performance liquid chromatography. *J.Autom.Chem.* **1983**, *5*, 36–39.

Turnell, D.C.; Cooper, J.D.H. Removal of ammonia from urine by tetraphenylboron before amino acid analysis. *Clin.Chem.* **1984**, *30*, 588–589.

Turnell, D.C.; Cooper, J.D.H. Automated preparation of biological samples prior to high pressure liquid chromatography: Part I - The use of dialysis for deproteinizing serum for amino-acid analysis. *J.Autom.Chem.* **1985**, *7*, 177–180.

Umagat, H.; Kucera, P.; Wen, L.F. Total amino acid analysis using precolumn fluorescence derivatization. *J.Chromatogr.* **1982**, *239*, 463–474.

Van der Meer, J.M. Amino acid analysis of feeds in The Netherlands: four-year proficiency study. *J.Assoc.Off.Anal.Chem.* **1990**, *73*, 394–398.

van Eijk, H.M.H.; Rooyakkers, D.R.; Deutz, N.E.P. Rapid routine determination of amino acids in plasma by high-performance liquid chromatography with a 2–3-μm Spherisorb ODS II column. *J.Chromatogr.* **1993**, *620*, 143–148.

Van der Boon, J.; Van den Thillart, G.E.E.J.M.; Addink, A.D.F. Reversed-phase liquid chromatographic analysis of o-phthaldialdehyde-derivatized free amino acids in two types of goldfish muscles. *J.Pharm.Biomed.Anal.* **1989**, *7*, 471–481.

Vance, N.C.; Zaerr, J.B. Analysis by high-performance liquid chromatography of free amino acids extracted from needles of drought-stressed and shaded Pinus ponderosa seedlings. *Physiol.Plant.* **1990**, *79*, 23–30.

Veuthey, J.L.; Caude, M.; Rosset, R. Separation of some amino acids by supercritical fluid chromatography after a prederivatization step with classical reagents. *Chromatographia* **1989**, *27*, 105–108.

Vollmer, D.W.; Jinks, D.C.; Guthrie, R. Isocratic reverse-phase liquid chromatography assay for amino acid metabolic disorders using eluates of dried blood spots. *Anal.Biochem.* **1990**, *189*, 115–121.

Wassner, S.J.; Li, J.B. High-performance liquid chromatographic separation of six essential amino acids and its use as an aid in the diagnosis of branched-chain ketoaciduria. *J.Chromatogr.* **1982**, *227*, 497–502.

Watanabe, Y.; Imai, K. Sensitive detection of amino acids in human serum and dried blood disk of 3 mm diameter for diagnosis of inborn errors of metabolism. *J.Chromatogr.* **1984**, *309*, 279–286.

Weinstein, S.; Engel, M.H.; Hare, P.E. The enantiomeric analysis of a mixture of all common protein amino acids by high-performance liquid chromatography using a new chiral mobile phase. *Anal.Biochem.* **1982**, *121*, 370–377.

Wenck, A.; Mintrop, L.; Duinker, J.C. Automated determination of amino acids in seawater. *Mar.Chem.* **1991**, *33*, 1–7.

White, J.A.; Hart, R.J. Derivatization methods for liquid chromatographic separation of amino acids. *Food Sci.Technol.* **1992**, *52*, 53–74.

White, J.A.; Hart, R.J.; Fry, J.C. An evaluation of the Waters Pico-Tag system for the amino acid analysis of food materials. *J.Autom.Chem.* **1986**, *8*, 170–177.

Willis, D.E. Automated precolumn derivatization of amino acids with o-phthalaldehyde by a reagent sandwiching technique. *J.Chromatogr.* **1987**, *408*, 217–225.

Winspear, M.J.; Oaks, A. Automated precolumn amino acid analyses by reversed-phase high-performance liquid chromatography. *J.Chromatogr.* **1983**, *270*, 378–382.

Woodward, C.N.; Sur, P.; Capizzi, R.L.; Modest, E.J. Serum amino acid levels in leukemic mice after L-asparaginase treatment. *Biochem.Med.Metab.Biol.* **1988**, *39*, 199–207.

Xia, L.Y.; Lu, X.L.; Zhang, S.L. Determination of compound amino acid injections by precolumn o-phthaldehyde derivatization and HPLC. *Chin.J.Pharm.Anal.(Yaowu.Fenxi.Zazhi).* **1990**, *10*, 68–72.

Xu, Y.; Xu, X.; Wang, Y. High performance liquid chromatographic determination of amino acids by automatic precolumn. *Hebei Yixueyuan Xuebao* **1996**, *17*, 132–134.

Yamada, T.; Nonomura, S.; Fujiwara, H.; Miyazawa, T.; Kuwata, S. Separation of peptide diastereomers by reversed-phase high-performance liquid chromatography and its applications. IV. New derivatization reagent for the enantiomeric analysis of α- and β-amino acids. *J.Chromatogr.* **1990**, *515*, 475–482.

Yamada, T.; Dejima, K.; Shimamura, M.; Miyazawa, T.; Kuwata, S. Gly-L-Phe-OMe, a useful derivatization reagent for chiral separation of N-(benzyloxycarbonyl) amino acids by reversed-phase high-performance liquid chromatography. *Chem.Express* **1989**, *4*, 725–728.

Yamada, T.; Shimamura, M.; Miyazawa, T.; Kuwata, S. Separation of diastereomers of protected glycine-middled tripeptides. (III). Application to the determination of the absolute configuration of constituent amino acids of peptides. *Pept.Chem.* **1983**, *21st.*, 31–36.

Yang, S.S.; Smetena, I. Determination of free amino acids in tobacco by HPLC with fluorescence detection and precolumn derivatization. *Chromatographia* **1993**, *37*, 593–598.

Ye, J.; Baldwin, R.P. Determination of amino acids and peptides by capillary electrophoresis and electrochemical detection at a copper electrode. *Anal.Chem.* **1994**, *66*, 2669–2674.

Ye, W.L. [High performance liquid chromatographic determination of amino acids in brain by precolumn derivatization with o-phthaldialdehyde]. *Sheng.Li.Hsueh.Pao.* **1988**, *40*, 308–313.

Young, P.R.; Grynspan, F. Analysis of methylated amino acids by high-performance liquid chromatography: methylation of myelin basic protein. *J.Chromatogr.* **1987**, *421*, 130–135.

Zecca, L.; Ferrario, P. Determination of neurotransmitter amino acids by high-performance liquid chromatography with fluorescence detection. *J.Chromatogr.* **1985**, *337*, 391–396.

Zhang, X.S.; Chu, Z.Y.; Huang, L.; Lin, M.; Qu, J. Analysis of compound amino acid injection by precolumn derivatization with o-phthalidialdehyde and RP-HPLC. *Chin.J.Pharm.Anal.(Yaowu Fenxi Zazhi)* **1996**, *16*, 6–9.

Zhao, M.; Bada, J.L. Determination of α-dialkylamino acids and their enantiomers in geological samples by high-performance liquid chromatography after derivatization with a chiral adduct of o-phthaldialdehyde. *J.Chromatogr.A* **1995**, *690*, 55–63.

Amino Acids (using post-column reaction detection)

Al-Najafi, S.; Wellington, C.A.; Wade, A.P.; Sly, T.J.; Betteridge, D. Computer-assisted optimization of HPLC with post-column reaction for the determination of amino acids. *Talanta* **1987**, *34*, 749–756.

Ashworth, R.B. Ion-exchange separation of amino acids with postcolumn orthophthalaldehyde detection. *J.Assoc.Off.Anal.Chem.* **1987**, *70*, 248–252.

Baeyens, W.; Bruggeman, J.; Lin, B. Enhanced chemiluminescence detection of some dansyl amino acids in liquid chromatography. *Chromatographia* **1989**, *27*, 191–193.

Baeyens, W.; Bruggeman, J.; Dewaele, C.; Lin, B.; Imai, K. Optimization of an HPLC peroxyoxalate chemiluminescence detection system for some dansyl amino acids. *J.Biolumin.Chemilumin.* **1990**, *5*, 13–23.

Baffi, F.; Ianni, M.C.; Magi, E.; Ravera, M. Comparison of the performance of RP C18 on silica and polymeric supports for the liquid chromatographic separation of metal-amino acid complexes with postcolumn derivatization. *Anal.Chim.Acta* **1993**, *278*, 83–89.

Belliardo, J.J.; Faure, U.; Ooghe, W. Determination of free amino acids in orange juice - a European collaborative study. *Fluess.Obst* **1991**, *58*, 368.

Bonomi, A.; Anghinetti, A.; Lucchelli, L.; Darecchio, D. Amino acid determination in feeds by high-performance liquid chromatography (HPLC). *Riv.Soc.Ital.Sci.Aliment.* **1989**, *18*, 155–162.

Brune, S.N.; Bobbitt, D.R. Role of electron-donating/withdrawing character, pH, and stoichiometry on the chemiluminescent reaction of tris(2,2'-bipyridyl)ruthenium(III) with amino acids. *Anal.Chem.* **1992**, *64*, 166–170.

Ci, Y.; Tie, J.; Wang, Q.; Chang, W. Flow injection and liquid chromatographic postcolumn detection of amino acids by mimetic peroxidase-catalyzed chemiluminescence reaction. *Anal.Chim.Acta* **1992**, *269*, 109–114.

Dong, M.W.; Gant, J.R.; Benson, J.R. Characterization and performance of a high speed amino acid analysis column. *Am.Biotechnol.Lab.* **1985**, *3*, 34–43.

Dong, M.W.; Gant, J.R. High-speed liquid chromatographic analysis of amino acids by post-column sodium hypochlorite-o-phthalaldehyde reaction. *J.Chromatogr.* **1985**, *327*, 17–25.

Dong, M.W.; DiCesare, J.I.; Steinwand, M. HPLC analysis of amino acids-comparison of six traditional methods. *Angew.Chromatogr.* **1985**, *42*, 27.

Dossena, A.; Galaverna, G.; Corradini, R.; Marchelli, R. Two-dimensional high-performance liquid chromatographic system for the determination of enantiomeric excess in complex amino acid mixtures. Single amino acid analysis. *J.Chromatogr.* **1993**, *653*, 229–234.

Duchateau, A.L.L.; Crombach, M.G. Determination of α-aminonitriles, α-amino acid amides, and α-amino acids by means of HPLC, post-column reaction and fluorescence detection. *Chromatographia* **1987**, *24*, 339–343.

Elkin, R.G. Measurement of free amino acids in avian blood serum by reversed-phase high-performance liquid chromatography as compared to ion-exchange chromatography. *J.Agric.Food Chem.* **1984**, *32*, 53–57.

Fernstrom, M.H.; Fernstrom, J.D. Rapid measurement of free amino acids in serum and CSF using high-performance liquid chromatography. *Life Sci.* **1981**, *29*, 2119–2130.

Franzen, K.H. HPLC amino acid analysis. *GIT Fachz.Lab.* **1983**, *27*, 610–612.

Frey, J.; Chamson, A.; Raby, N. Separation of amino acids using ion-paired reversed-phase high-performance liquid chromatography with special reference to collagen hydrolysate. *Amino Acids* **1993**, *4*, 45–51.

Fujinari, E.M.; Ribble, E.; Piserchio, M.V. Nitrogen-specific liquid chromatography detection of amino acids, peptides, and proteins by HPLC-CLND. *Dev.Food Sci.* **1993**, *32*, 75–89.

Grunau, J.A.; Swiader, J.M. Chromatographic quantitation of free amino acids: S-methylmethionine, methionine and lysine in corn. *J.Plant Nutr.* **1991**, *14*, 653–662.

Grunau, J.A.; Swiader, J.M. Chromatographic quantitation of free amino acids: S-methylmethionine, methionine and lysine in corn. *Commun.Soil Sci.Plant Anal.* **1991**, *22*, 1873–1882.

Henshall, A.; Pickering, M.J.; Soto, D. Analysis of amino acids by HPLC with a ternary gradient system. *Spectra 2000 [Deux Mille]* **1984**, *12*, 29–32.

Kakehi, K.; Konishi, T.; Sugimoto, I.; Honda, S. Post-column reaction of amino acids with the pentane-2,4-dione-formaldehyde system for their automated analysis. *J.Chromatogr.* **1985**, *318*, 367–372.

Kiba, N.; Oyama, Y.; Kato, A.; Furusawa, M. Postcolumn co-immobilized leucine dehydrogenase-NADH oxidase reactor for the determination of branched-chain amino acids by high-performance liquid chromatography with chemiluminescence detection. *J.Chromatogr.A* **1996**, *724*, 355–357.

Kiba, N.; Kaneko, M. Use of immobilized amino acid oxidase as post-column reactor in the high-performance liquid chromatography of amino acids. *J.Chromatogr.* **1984**, *303*, 396–403.

Kiba, N.; Oyama, Y.; Furusawa, M. Determination of aliphatic amino acids in serum by HPLC with fluorometric detection using coimmobilized enzyme reactor. *Talanta* **1993**, *40*, 657–660.

Kiba, N.; Hori, S.; Furusawa, M. A post-column immobilized leucine dehydrogenase reactor for determination of branched chain amino acids by high-performance liquid chromatography with fluorescence detection. *J.Chromatogr.* **1989**, *463*, 177–182.

Kim, H. Electrochemical activity of o-phthalaldehyde derivatives of large neutral amino acids in plasma. *Korean J.Biochem.* **1987**, *19*, 83–88.

Koerner, P.J., Jr.; Nieman, T.A. Luminol chemiluminescence HPLC reaction detector for amino acids and other ligands. *Mikrochim.Acta* **1987**, *2*, 79–90.

Konomi, H.; Hayashi, T.; Takehana, Y.; Arima, M. Rapid and sensitive amino acid analysis of human collagens using high-performance liquid chromatography. *J.Chromatogr.* **1984**, *311*, 375–379.

Lee, W-Y.; Nieman, T.A. Determination of dansyl amino acids using tris(2,2'-bipyridyl)ruthenium(II) chemiluminescence for post-column reaction detection in high-performance liquid chromatography. *J.Chromatogr.A* **1994**, *659*, 111–118.

LePage, J.N.; Rocha, E.M. Liquid chromatography with postcolumn reaction for detection of amines and amino acids. *Anal.Chem.* **1983**, *55*, 1360–1364.

Luo, X.; Yu, Z.; Wang, X. Quantitative amino acid analysis at picomole level. *Shengwu Huaxue Yu Shengwu Wuli Jinzhan* **1986**, 58–60.

Macek, J.; Miterova, L.; Adam, M. Separation of collagen hydrolysate amino acids by ion-pair reversed-phase high-performance liquid chromatography. *J.Chromatogr.* **1986**, *364*, 253–257.

Milakofsky, L.; Hare, T.A.; Miller, J.M.; Vogel, W.H. Comparison of amino acid levels in rat blood obtained by catheterization and decapitation. *Life Sci.* **1984**, *34*, 1333–1340.

Miyaguchi, K.; Honda, K.; Toyo'oka, T.; Imai, K. Application of a microbore high-performance liquid chromatography-chemiluminescence detection system to the N-terminal amino acid analysis of bradykinin. *J.Chromatogr.* **1986**, *352*, 255–260.

Miyaguchi, K.; Honda, K.; Imai, K. Microbore high-performance liquid chromatography and chemiluminescence detection of Dns-amino acids. *J.Chromatogr.* **1984**, *316*, 501–505.

Mohan, A.G.; Rosenkrans, A.M.; Shahied, S.I. HPLC analysis for phenylalanine and tyrosine in serum for PKU diagnosis and treatment. *Am.Lab.* **1989**, *21*, 29–32.

Molina-Holgado, E.; Dewar, K.M.; Grondin, L.; van Gelder, N.M.; Reader, T.A. Changes of amino acid and monoamine levels after neonatal 6-hydroxydopamine denervation in rat basal ganglia, substantia nigra, and Raphe nuclei. *J.Neurosci.Res.* **1993**, *35*, 409–418.

Nimura, N.; Suzuki, T.; Kasahara, Y.; Kinoshita, T. Reversed-phase liquid chromatographic resolution of amino acid enantiomers by mixed chelate complexation. *Anal.Chem.* **1981**, *53*, 1380–1383.

Radjai, M.K.; Hatch, R.T. Fast determination of free amino acids by ion-pair high-performance liquid chromatography using on-line post-column derivatization. *J.Chromatogr.* **1980**, *196*, 319–322.

Schlabach, T.; Cunico, B. Post column derivatization in LC: determination of amino acids and VMA. *Analytika (Johannesburg)* **1982**, 7–9.

Schlabach, T.; Cunico, B. Post-column derivatization in LC: determination of amino acids and VMA. *VIA, Varian Instrum.Appl.* **1981**, *15*, 6–8.

Sirai, Y. Determination of L-lysine hydrochloride in premix by high performance liquid chromatography. *Chikusan no Kenkyu* **1987**, *41*, 1363–1367.

Tous, G.I.; Fausnaugh, J.L.; Akinyosoye, O.; Lackland, H.; Winter-Cash, P.; Vitorica, F.J.; Stein, S. Amino acid analysis on polyvinylidene difluoride membranes. *Anal.Biochem.* **1989**, *179*, 50–55.

Uchikura, K.; Kirisawa, M. Ru(bpy)$_3^{2+}$ chemiluminescence detection HPLC for D,L-amino acid. *Kuromatogurafi* **1995**, *15*, 232–233.

Wajcman, H. The role of chromatographic techniques in amino acid analysis: post-column or pre-column derivatizations? *Bio-Sciences* **1989**, *8*, 46–51.

Walton, D.J.; McPherson, J.D. Non-enzymic glycation of proteins: analysis of N-(1-deoxyhexitol-1-yl)amino acids by high-performance liquid chromatography. *Carbohydr.Res.* **1986**, *153*, 285–293.

Wang, W.; Liu, H. HPLC separation of amino acids in hydrolyzates of plants with postcolumn derivatization. *Fenxi Huaxue* **1986**, *14*, 700–703.

Yao, T.; Sato, M.; Wasa, T. Optically specific detection of L- and D-amino acids by a high-performance liquid chromatographic system with immobilized amino acid oxidase reactors. *Chem.Express* **1988**, *3*, 559–562.

Yokoyama, Y.; Ozaki, O.; Sato, H. Simultaneous determination of amino acids and creatinine by dual-mode gradient ion-pair chromatography. *Kuromatogurafi* **1995**, *16*, 138–141.

Zhou, Y. Determination of compound amino acid injections by post-column derivatization of HPLC. *Yaowu Fenxi Zazhi* **1987**, *7*, 267–270.

Ziegler, F.; Le Boucher, J.; Coudray-Lucas, C.; Cynober, L. Plasma amino acid determinations by reversed-phase HPLC: improvement of the ortho-phthalaldehyde method and comparison with ion-exchange chromatography. *J.Autom.Chem.* **1992**, *14*, 145–149.

Amino Alcohols

Claas, K.E.; Hohaus, E. Boron chelates and boron metal chelates. Part XVI. Precolumn fluorescent derivatization for high-performance liquid chromatography with 2,2-diphenyl-1-oxa-3-oxonia-2-boratanaphthalene (DOOB): determination of aminoalkanols. *Fresenius' Z.Anal.Chem.* **1985**, *322*, 343–347.

Engelhardt, H.; Goetzinger, W.; Kraemer, M.; Wintringer, R. Trace analysis of quaternary amino alcohols as degradation products from softeners in wastewater. *Acta Hydrochim.Hydrobiol.* **1995**, *23*, 173–179.

Gal, J. Determination of the enantiomeric composition of chiral amino alcohols using chiral derivatization and reversed-phase liquid chromatography. *J.Liq.Chromatogr.* **1986**, *9*, 673–681.

Gelber, L.R.; Karger, B.L.; Neumeyer, J.L.; Feibush, B. Ligand exchange chromatography of amino alcohols. Use of Schiff bases in enantiomer resolution. *J.Am.Chem.Soc.* **1984**, *106*, 7729–7734.

Guebitz, G.; Pierer, B.; Wendelin, W. Resolution of the enantiomers of drugs containing amino alcohol structure after derivatization with bromoacetic acid. *Chirality* **1992**, *4*, 333–337.

Jegorov, A.; Trnka, T.; Stuchlik, J. High-performance liquid chromatographic detection of enantiomeric amino alcohols after derivatization with o-phthaldialdehyde and various thiosugars. *J.Chromatogr.* **1991**, *558*, 311–317.

Amino Alcohols (using post-column reaction detection)

Yamazaki, O.; Ozaki, K.; Tanimura, T. Enantiomeric separation of β-aminoalcohol using (R)-mandelato-Cu(II) complex. *Kuromatogurafi* **1995**, *15*, 56–57.

p-Aminobenzoic Acid

Esteve-Romero, J.S.; Simo-Alfonso, E.F.; Garcia-Alverez-Coque, M.C.; Ramis-Ramos, G. Conventional and thermal-lens spectrophotometric determination of p-aminobenzoic acid and arylamine diuretics after previous azo dye formation in a micellar medium. *Talanta* **1993**, *40*, 1711–1718.

Aminobisphosphonates (using post-column reaction detection)

Kwong, E.; Chiu, A.M.Y.; McClintock, S.A.; Cotton, M.L. HPLC analysis of an amino bisphosphonate in pharmaceutical formulations using postcolumn derivatization and fluorescence detection. *J.Chromatogr.Sci.* **1990**, *28*, 563–566.

gamma-Aminobutyric Acid

Murai, S.; Saito, H.; Nagahama, H.; Miyate, H.; Masuda, Y.; Itoh, T. Ultra-rapid assay of brain gamma-aminobutyric acid by liquid chromatography with electrochemical detection. *J.Chromatogr.* **1989**, *497*, 363–366.

Murai, S.; Nagahama, H.; Saito, H.; Miyate, H.; Masuda, Y.; Itoh, T. Very rapid assay of gamma-aminobutyric acid in mouse brain regions within 3 minutes by high-performance liquid chromatography with electrochemical detection. *J.Pharmacol.Methods* **1989**, *21*, 115–121.

Herranz, A.S.; Lerma, J.; Del Rio, R.M. Determination of gamma-aminobutyric acid in physiological samples by a simple, rapid high-performance liquid chromatographic method. *J.Chromatogr.* **1984**, *309*, 139–144.

Lasley, S.M.; Greenland, R.D.; Michaelson, I.A. Determination of gamma-aminobutyric and glutamic acids in rat brain by liquid chromatography with electrochemical detection. *Life Sci.* **1984**, *35*, 1921–1930.

Westerink, B.H.C.; de Vries, J.B. On the origin of extracellular GABA collected by brain microdialysis and assayed by a simplified on-line method. *Naunyn-Schmiedeberg's Arch.Pharmacol.* **1989**, *339*, 603–607.

2-Aminocyclopentane-1-carboxylic Acid

Peter, A.; Fueloep, F. High-performance liquid chromatographic method for the separation of isomers of cis- and trans-2-amino-cyclopentane-1-carboxylic acid. *J.Chromatogr.A* **1996**, *715*, 219–226.

1-Aminocyclopropanecarboxylic Acid

Miller, R.; La Grone, J.; Skolnick, P.; Boje, K.M. High-performance liquid chromatographic assay for 1-aminocyclopropanecarboxylic acid from plasma and brain. *J.Chromatogr.* **1992**, *578*, 103–108.

7-Aminoflunitrazepam

Sumirtapura, Y.C.; Aubert, C.; Coassolo, P.; Cano, J.P. Determination of 7-aminoflunitrazepam (Ro 20–1815) and 7-aminodesmethylflunitrazepam (Ro 5–4650) in plasma by high-performance liquid chromatography and fluorescence detection. *J.Chromatogr.* **1982**, *232*, 111–118.

Aminoglycosides

Caturla, M.C.; Cusido, E.; Westerlund, D. High-performance liquid chromatography method for the determination of aminoglycosides based on automated pre-column derivatization with o-phthalaldehyde. *J.Chromatogr.* **1992**, *593*, 69–72.

Aminoglycosides (using post-column reaction detection)

Fabre, H.; Sekkat, M.; Blanchin, M.D.; Mandrou, B. Determination of aminoglycosides in pharmaceutical formulations—II. High-performance liquid chromatography. *J.Pharm.Biomed.Anal.* **1989**, *7*, 1711–1718.

δ-Aminolevulinic Acid

Endo, Y.; Okayama, A.; Endo, G.; Horiguchi, S.; Nakazono, N. [Improvement of urinary delta-aminolevulinic acid determination by HPLC-fluorometry using pre-column derivatization]. *Sangyo.Igaku.* **1993**, *35*, 126–127.

Ho, J.; Guthrie, R.; Tieckelmann, H. Detection of δ-aminolevulinic acid, porphobilinogen and porphyrins related to heme biosynthesis by high-performance liquid chromatography. *J.Chromatogr.* **1986**, *375*, 57–63.

Meisch, H.U.; Wannemacher, B. Fluorometric determination of 5-aminolevulinic acid after derivatization with o-phthaldialdehyde and separation by reversed-phase high-performance liquid chromatography. *Anal.Chem.* **1986**, *58*, 1372–1375.

Minder, E.I. Measurement of 5-aminolevulinic acid by reversed phase HPLC and fluorescence detection. *Clin.Chim.Acta* **1986**, *161*, 11–18.

Morita, Y.; Sakai, T.; Araki, S.; Araki, T.; Masuyama, Y. [Usefulness of delta-aminolevulinic acid in blood as an indicator of lead exposure]. *Sangyo.Igaku.* **1993**, *35*, 112–118.

Tomokuni, K. [Delta-Aminolevulinic acid in urine (screening method)]. *Nippon.Rinsho.* **1995**, *53*, 1383–1388.

Aminophospholipids

Kaneko, T.; Ohta, Y.; Machida, Y. Sensitive analysis of aminophospholipids with precolumn fluorescent derivatization by high-performance liquid chromatography. *Agric.Biol.Chem.* **1987**, *51*, 2023–2024.

Resmini, P.; Pellegrino, L.; Hogenboom, J.A.; Sadini, V.; Rampilli, M. Detection of buttermilk solids in skim milk powder by HPLC quantification of aminophospholipids. *Sci.Tec.Latt.-Casearia* **1988**, *39*, 395–412.

Aminophosphonic Acids

Huber, J.W..I.; Calabrese, K.L. Derivatization of aminophosphonic acids for HPLC analysis. *J.Liq.Chromatogr.* **1985**, *8*, 1989–2001.

Aminotriazole

Dugay, J.; Hennion, M.-C. Evaluation of the performance of analytical procedures for the trace-level determination of aminotriazole in drinking waters. *Trends Anal.Chem.* **1995**, *14*, 407–414.

Amitrole

Ternes, W.; Ruessel-Sinn, H.A. Determination of 3-amino-1,2,4-triazole (amitrole) in tissues with electrochemical detection by ion-pair high-performance liquid chromatography. *Fresenius' Z.Anal.Chem.* **1987**, *326*, 757–759.

Ammonia

Goyal, S.S.; Rains, D.W.; Huffaker, R.C. Determination of ammonium ion by fluorometry or spectrophotometry after on-line derivatization with o-phthalaldehyde. *Anal.Chem.* **1988**, *60*, 175–179.

Possanzini, M.; Di Palo, V. Determination of atmospheric ammonia as m-toluamide by denuder sampling and HPLC-UV detection. *Chromatographia* **1989**, *28*, 27–30.

Amoxicillin (using post-column reaction detection)

Carlqvist, J.; Westerlund, D. Automated determination of amoxycillin in biological fluids by column switching in ion-pair reversed-phase liquid chromatographic systems with post-column derivatization. *J.Chromatogr.* **1985**, *344*, 285–296.

Carlqvist, J.; Westerlund, D. Determination of amoxicillin in body fluids by reversed-phase liquid chromatography coupled with a post-column derivatization procedure. *J.Chromatogr.* **1979**, *164*, 373–381.

Haginaka, J.; Wakai, J. Liquid chromatographic determination of amoxicillin and its metabolites in human urine by postcolumn degradation with sodium hypochlorite. *J.Chromatogr.* **1987**, *413*, 219–226.

Lee, T.L.; D'Arconte, L.; Brooks, M.A. High-pressure liquid chromatographic determination of amoxicillin in urine. *J.Pharm.Sci.* **1979**, *68*, 454–458.

Leroy, P.; Gavriloff, C.; Nicolas, A.; Archimbault, P.; Ambroggi, G. Comparative assay of amoxicillin by high-performance liquid chromatography and microbiological methods for pharmacokinetic studies in calves. *Int.J.Pharm.* **1992**, *82*, 157–164.

Amphetamine and Methamphetamine

Herraez-Hernandez, R.; Campins-Falco, P.; Sevillano-Cabeza, A. On-line derivatization into precolumns for the determination of drugs by liquid chromatography and column switching: determination of amphetamines in urine. *Anal.Chem.* **1996**, *68*, 734–739.

Kram, T.C.; Lurie, I.S. The determination of enantiomeric composition of methamphetamine by proton-NMR spectroscopy. *Forensic Sci.Int.* **1992**, *55*, 131–137.

Lurie, I.S. Micellar electrokinetic capillary chromatography of the enantiomers of amphetamine, methamphetamine and their hydroxyphenethylamine precursors. *J.Chromatogr.* **1992**, *605*, 269–275.

Moriyasu, M.; Endo, M.; Hashimoto, Y.; Koeda, T. High-performance liquid chromatographic determination of organic substances by metal chelate derivatization. II. Microdetermination of methamphetamine and amphetamine. *Chem.Pharm.Bull.* **1984**, *32*, 600–608.

Noggle, F.T., Jr.; DeRuiter, J.; Clark, C.R. Methods for the analysis and characterization of forensic samples containing amphetamines and related amines. *J.Chromatogr.Sci.* **1990**, *28*, 529–536.

Pfordt, J. Separation of enantiomers of amphetamine and related amines by HPLC. *Fresenius'* *Z.Anal.Chem.* **1986**, *325*, 625–626.

Zhou, F.X.; Krull, I.S. Direct enantiomeric analysis of amphetamine in plasma by simultaneous solid phase extraction and chiral derivatization. *Chromatographia* **1993**, *35*, 153–159.

Amphetamine and Methamphetamine (using post-column reaction detection)

Hayakawa, K.; Miyoshi, Y.; Kurimoto, H.; Matsushima, Y.; Takayama, N.; Tanaka, S.; Miyazaki, M. Simultaneous determination of methamphetamine and its metabolites in the urine samples of abusers by high performance liquid chromatography with chemiluminescence detection. *Biol.Pharm.Bull.* **1993**, *16*, 817–821.

Takayama, N.; Tanaka, S.; Hayakawa, K.; Miyazaki, M. Determination of methamphetamine metabolites in suspected human urine by high-performance liquid chromatography with chemiluminescence detection. *Hochudoku* **1992**, *10*, 138–139.

Takayama; Tanaka, S.; Hayakawa, K.; Miyazaki, M. Comparison of extraction methods of methamphetamine and its metabolites in human hair for HPLC using chemiluminescence detection. *Hochudoku* **1995**, *13*, 134–135.

Ampicillin

Nelis, H.J.; Vandenbranden, J.; Verhaeghe, B.; De Kruif, A.; Mattheeuws, D.; De Leenheer, A.P. Liquid chromatographic determination of ampicillin in bovine and dog plasma by using a tandem solid-phase extraction method. *Antimicrob.Agents Chemother.* **1992**, *36*, 1606–1610.

Ampicillin (using post-column reaction detection)

Haginaka, J.; Wakai, J. Liquid chromatographic determination of ampicillin and its metabolites in human urine by postcolumn alkaline degradation. *J.Pharm.Pharmacol.* **1987**, *39*, 5–8.

Nakagawa, H.; Nishiyama, K.; Higashitani, T.; Ishikawa, S.; Fukui, Y. Application of high performance liquid chromatography with fluorescence detection to determination of ampicillin in plasma deproteinized by phenol method. *Yakugaku Zasshi* **1985**, *105*, 1096–1099.

Amprolium (using post-column reaction detection)

van Leeuwen, W.; van Gend, H.W. Determination of amprolium in egg yolk and muscle tissue (chicken) by HPLC with postcolumn reaction and fluorometric detection, using on-line sample clean-up and preconcentration steps. *Z.Lebensm.-Unters.Forsch.* **1988**, *186*, 500–504.

Aniline

Geerdink, R.B. Determination of aniline derivatives by high-performance liquid chromatography with fluorescence detection. *J.Chromatogr.* **1988**, *445*, 273–281.

Radzik, D.M.; Kissinger, P.T. Determination of aniline and metabolites produced in vitro by liquid chromatography/electrochemistry. *Anal.Biochem.* **1984**, *140*, 74–83.

Anions

Padarauskas, A.; Schwedt, G. Simultaneous ion-pair chromatography of inorganic anions and cations using on-column derivatization with chelating agents and UV detection. *Fresenius' J.Anal.Chem.* **1995**, *351*, 708–713.

Anions (using post-column reaction detection)

Jupille, T.; Burge, D.; Togami, D. High-speed analysis for acid-rain anions by single-column ion chromatography (SCIC). *Chromatographia* **1982**, *16*, 312–316.

Anthraquinones (using post-column reaction detection)

Kiba, N.; Takamatsu, M.; Furusawa, M. Determination of anthraquinones in pulping materials by high-performance liquid chromatography using on-line post-column derivatization. *J.Chromatogr.* **1985**, *328*, 309–315.

Anticholinergics (using post-column reaction detection)

Holeman, J.A.; Danielson, N.D. Microbore liquid chromatography of tertiary amine anticholinergic pharmaceuticals with tris(2,2'-bipyridine)ruthenium(III) chemiluminescence detection. *J.Chromatogr.Sci.* **1995**, *33*, 297–302.

Antioxidants

Anderson, J.; Van Niekerk, P.J. High-performance liquid chromatographic determination of antioxidants in fats and oils. *J.Chromatogr.* **1987**, *394*, 400–402.

Rose, R.C.; Bode, A.M. Analysis of water-soluble antioxidants by high-pressure liquid chromatography. *Biochem.J.* **1995**, *306*, 101–105.

Apomorphine

Smith, R.V.; Humphrey, D.W.; Szeinbach, S.; Glade, J.C. High performance liquid chromatographic determination of apomorphine in blood serum. *Anal.Lett.* **1979**, *12*, 371–379.

Arbaprostil

Cox, J.W.; Pullen, R.H.; Royer, M.E. Isolation of plasma components by double antibody precipitation and filtration: application to the chromatographic determination of arbaprostil [(15R)-15-methylprostaglandin E2]. *Anal.Chem.* **1985**, *57*, 2365–2369.

Arginine

Saito, K.; Itaya, T.; Horie, M.; Nakazawa, H.; Imanari, T. Estimation of arginine metabolism in putrefactive bacteria using liquid chromatography. *Jpn.J.Toxicol.Environ.Health* **1994**, *40*, 140–146.

Arginine Aldehyde

Tomori, E. Pre-column fluorescence derivatization of peptides containing arginine aldehyde moiety in high-performance liquid chromatography. *Chromatographia* **1993**, *36*, 105–109.

Arginine Derivatives (using post-column reaction detection)

Gruebler, G.; Gutjahr, F.; Echner, H.F.; Voelter, W.; Bauer, H. High-performance liquid chromatographic investigations on the cleavage kinetics of side-chain-protected arginine derivatives with a sophisticated post-column reaction detector. *J.Chromatogr.* **1990**, *512*, 249–254.

Arsenic

Fish, R.H.; Brinckman, F.E. Organometallic geochemistry. Isolation and identification of organoarsenic and inorganic arsenic compounds from Green River Formation oil shale. *Prepr.Pap.-Am.Chem.Soc., Div.Fuel Chem.* **1983**, *28*, 177–180.

Arsenic (using post-column reaction detection)

Bushee, D.S.; Krull, I.S.; Demko, P.R.; Smith, S.B., Jr. Trace analysis and speciation for arsenic anions by HPLC-hydride generation-inductively coupled plasma emission spectroscopy. *J.Liq.Chromatogr.* **1984**, *7*, 861–876.

Colon, L.A.; Barry, E.F. The determination of arsenic and selenium by HPLC combined with alternating current plasma detection and post-column hydride formation. *J.High Resolut.Chromatogr.* **1991**, *14*, 608–612.

Stummeyer, J.; Harazim, B.; Wippermann, . Speciation of arsenic in water samples by high-performance liquid chromatography-hydride generation-atomic absorption spectrometry at trace levels using a post-column reaction system. *Fresenius' J.Anal.Chem.* **1996**, *354*, 344–351.

Artemether

Muhia, D.K.; Mberu, E.K.; Watkins, W.M. Differential extraction of artemether and its metabolite dihydroartemisinin from plasma and determination by high-performance liquid chromatography. *J.Chromatogr.B* **1994**, *660*, 196–199.

Artemisinin

Edwards, G. Measurement of artemisinin and its derivatives in biological fluids. *Trans.R.Soc. Trop.Med.Hyg.* **1994**, *88*, 37–39.

Ferreir, J.F.S.; Janick, J. Immunoquantitative analysis of artemisinin from Artemisia annua using polyclonal antibodies. *Phytochemistry* **1996**, *41*, 97–104.

Artesunate (using post-column reaction detection)

Edlund, P.O.; Westerlund, D.; Carlqvist, J.; Wu, B.; Jin, Y. Determination of artesunate and dihydroartemisinine in plasma by liquid chromatography with post-column derivatization and UV-detection. *Acta Pharm.Suec.* **1984**, *21*, 223–234.

2-Arylpropionic Acids

Van Overbeke, A.; Baeyens, W.; Van de Voorde, I.; Van Der Weken, G.; Dewaele, C. Enantiomeric separation of 2-arylpropionic acids after precolumn derivatization on a cellulose-based chiral stationary phase. *Biomed.Chromatogr.* **1995**, *9*, 292–294.

Ascorbic Acid

Tausz, M.; Kranner, I.; Grill, D. Simultaneous determination of ascorbic acid and dehydroascorbic acid in plant materials by high performance liquid chromatography. *Phytochem.Anal.* **1996**, *7*, 69–72.

Benlloch, R.; Farré, R.; Frigola, A. A quantitative estimate of ascorbic and isoascorbic acid by high performance liquid chromatgography: Application to citric juices. *J.Liq.Chromatogr.* **1993**, *16*, 3113–3122.

Huang, T.; Kissinger, P.T. Simultaneous LC determination of ascorbic acid and dehydroascorbic acid in foods and biological fluids. *Curr.Sep.* **1989**, *9*, 19–23.

Vander Jagt, D.J.; Garry, P.J.; Hunt, W.C. Ascorbate in plasma as measured by liquid chromatography and by dichlorophenolindophenol colorimetry. *Clin.Chem.* **1986**, *32*, 1004–1006.

Ascorbic Acid (using post-column reaction detection)

Capellmann, M.; Bolt, H.M. Simultaneous determination of ascorbic acid and dehydroascorbic acid by HPLC with postcolumn derivatization and fluorometric detection. *Fresenius' J.Anal.Chem.* **1992**, *342*, 462–466.

Kacem, B.; Marshall, M.R.; Matthews, R.F.; Gregory, J.F. Simultaneous analysis of ascorbic and dehydroascorbic acid by high-performance liquid chromatography with post column derivatization and UV absorbance. *J.Agric.Food Chem.* **1986**, *34*, 271–274.

Karp, S.; Ciambra, C.M.; Miklean, S. High-performance liquid chromatographic post-column reaction system for the electrochemical detection of ascorbic acid and dehydroascorbic acid. *J.Chromatogr.* **1990**, *504*, 434–440.

Karp, S.; Helt, C.S.; Soujari, N.H. Solid state postcolumn reactor for the electrochemical detection of ascorbic and dehydroascorbic acids in high performance liquid chromatography. *Microchem.J.* **1993**, *47*, 157–162.

Yasui, Y.; Hayashi, M. Simultaneous determination of ascorbic acid and dehydroascorbic acid by high performance liquid chromatography. *Anal.Sci.* **1991**, *7*, 125–128.

Ziegler, S.J.; Meier, B.; Sticher, O. Rapid and sensitive determination of dehydroascorbic acid in addition to ascorbic acid by reversed-phase high-performance liquid chromatography using a post-column reduction system. *J.Chromatogr.* **1987**, *391*, 419–426.

Aspartic Acid

Fu, S-J.; Fan, C-C.; Song, H-W.; Wei, F-Q. Age estimation using a modified HPLC determination of ratio of aspartic acid in dentin. *Forensic Sci.Int.* **1995**, *73*, 35–40.

Gillard, R.D.; Pollard, A.M.; Sutton, P.A.; Whittaker, D.K. An improved method for age at death determination from the measurement of D-aspartic acid in dental collagen. *Archaeometry* **1990**, *32*, 61–70.

Aspartylglucosamine

Kaartinen, V.; Mononen, I. Analysis of aspartylglucosamine at the picomole level by high-performance liquid chromatography. *J.Chromatogr.* **1989**, *490*, 293–299.

Aspirin

Kmetec, V. Simultaneous determination of acetylsalicylic, salicylic, ascorbic and dehydroascorbic acid by HPLC. *J.Pharm.Biomed.Anal.* **1992**, *10*, 1073–1076.

Aspirin (using post-column reaction detection)

Siebert, D.M.; Bochner, F. Determination of plasma aspirin and salicylic acid concentrations after low aspirin doses by high-performance liquid chromatography with postcolumn hydrolysis and fluorescence detection. *J.Chromatogr.* **1987**, *420*, 425–431.

Venema, D.P.; Hollman, P.C.H.; Janssen, K.P.L.T.M.; Katan, M.B. Determination of acetylsalicylic and salicylic acids in foods, using hplc with fluorescence detection. *J.Agric.Food Chem.* **1996**, *44*, 1762–1767.

Astromicin

Kobayashi, S.; Takai, K.; Masuda, S.; Inoue, A. [High performance liquid chromatographic determination of astromicin and piperacillin used in combination in blood samples]. *Jpn.J.Antibiot.* **1986**, *39*, 3156–3163.

Kobayashi, S.; Takai, K.; Inoue, A. [High performance liquid chromatographic determination of astromicin and cefsulodin used in combination in blood samples]. *Jpn.J.Antibiot.* **1986**, *39*, 3140–3147.

Atenolol

Chin, S.K.; Hui, A.C.; Giacomini, K.M. High-performance liquid chromatographic determination of the enantiomers of beta-adrenoceptor blocking agents in biological fluids. II. Studies with atenolol. *J.Chromatogr.* **1989**, *489*, 438–445.

Azo Dye Metabolites

Radzik, D.M.; Brodbelt, J.S.; Kissinger, P.T. Determination of toxic azo dye metabolites in vitro by liquid chromatography/electrochemistry with a dual-electrode detector. *Anal.Chem.* **1984**, *56*, 2927–2931.

B24/76

Zschiesche, M.; Baumann, A. Determination of the β-adrenoceptor blocking drug B 24/76 in serum by high-performance liquid chromatography with fluorimetric detection after pre-column dansylation. *J.Chromatogr.* **1989**, *488*, 482–486.

Baclofen

Chen, C.M.; Chen, Y.P. Determination of baclofen enantiomers in plasma by high-performance liquid chromatography. *Chung-hua Yao Hsueh Tsa Chih* **1989**, *41*, 267–277.

Wuis, E.W.; Dirks, R.J.M.; Vree, T.B.; van der Kleyn, E. High-performance liquid chromatographic analysis of baclofen in plasma and urine of man after precolumn extraction and derivatization with o-phthaldialdehyde. *J.Chromatogr.* **1985**, *337*, 341–350.

Barbiturates (using post-column reaction detection)

Jansen, H.; Vermunt, C.J.M.; Brinkman, U.A.T.; Frei, R.W. Parallel column ion exchange for post-separation pH modification in liquid chromatography. Application to barbiturates and miniaturization. *J.Chromatogr.* **1986**, *366*, 135–144.

Minder, E.I.; Schaubhut, R.; Vonderschmitt, D.J. Screening for drugs in clinical toxicology by high-performance liquid chromatography: identification of barbiturates by post-column ionization and detection by a multiplace photodiode array spectrophotometer. *J.Chromatogr.* **1988**, *428*, 369–376.

Bendazac

Scalia, S.; Massaccesi, M. Simultaneous determination of bendazac and underivatized lysine in ophthalmic preparations by reversed-phase ion-pair high-performance liquid chromatography. *Int.J.Pharm.* **1992**, *82*, 179–183.

Benzylideneacetone (using post-column reaction detection)

Yates, R.L.; Wenninger, J.A. Fluorometric determination of benzylideneacetone in fragrance products by liquid chromatography with post-column derivatization. *J.Assoc.Off.Anal.Chem.* **1988**, *71*, 965–967.

β-Blockers

Hyotylainen, T.; Pilvio, R.; Riekkola, M.L. Screening of four beta-blockers and codeine in urine by online coupled RPLC-GC with online derivatization. *J.High Resolut.Chromatogr.* **1996**, *19*, 439–443.

Betaines

Gorham, J. Separation of plant betaines and their sulfur analogs by cation-exchange high-performance liquid chromatography. *J.Chromatogr.* **1984**, *287*, 345–351.

Konosu, S.; Shinagawa, A.; Yamaguchi, K. Determination of omega-betaines in aquatic animals by high performance liquid chromatography. *Nippon Suisan Gakkaishi* **1986**, *52*, 869–873.

Mar, M-H.H.; Ridky, T.W.; Garner, S.C.; Zeisel, S.H. A method for the determination of betaine in tissues using high performance liquid chromatography. *J.Nutr.Biochem.* **1995**, *6*, 392–398.

Betamethasone

Wu, S-M.; Wu, H-L.; Chen, S-H. Determination of betamethasone and dexamethasone in plasma by fluorogenic derivatization and liquid chromatography. *Anal.Chim.Acta* **1995**, *307*, 103–107.

Wu, S-M.; Wu, H-L.; Chen, S-H.; Kou, H-S. Determination of betamethasone and dexamethasone by fluorogenic derivatization and high-performance liquid chromatography. *Chin.Pharm.J.(Taipei)* **1994**, *46*, 393–400.

Wu, S.M.; Chen, S.H.; Wu, H.L. Determination of betamethasone and dexamethasone by derivatization and liquid chromatography. *Anal.Chim.Acta* **1992**, *268*, 255–260.

Bile Acids

Baker, P.R.; Siow, Y.; Reid, A.D. Application of bonded silica sorbent extraction and selective derivatization to the high-performance liquid chromatographic analysis of hydroxy and keto bile acids in human serum. *Biochem.Soc.Trans.* **1986**, *14*, 971–972.

Campbell, G.R.; Odling-Smee, W.; Irvine, G.B. Comparison of methods for detection of bile acids and their conjugates in serum after analysis by high-performance liquid chromatography. *Biochem.Soc.Trans.* **1989**, *17*, 374–375.

Goldsmith, R.F.; Gruca, M.; O'Halloran, M.T.; Earl, J.W.; Gaskin, K. Determination of conjugated and unconjugated serum 3 alpha-OH bile acids by high-performance liquid chromatography. *Ann.Clin.Biochem.* **1994**, *31*, 479–484.

Goto, J. [Chromatographic determination of bile acids in biological fluids with sensitive and selective detection]. *Yakugaku.Zasshi.* **1990**, *110*, 807–821.

Goto, J.; Saisho, Y.; Nambara, T. Studies on steroids. CCLII. Separation and characterization of 3-oxo bile acids in serum by high-performance liquid chromatography with fluorescence detection. *J.Chromatogr.* **1991**, *567*, 343–349.

Goto, J.; Chikai, T.; Nambara, T. Separation of bile acid 7- and 12-sulfates by high-performance liquid chromatography with precolumn fluorescence labeling. *Anal.Sci.* **1986**, *2*, 175–178.

Goto, J.; Chikai, T.; Nambara, T. Studies on steroids. CCXXVII. Separation and determination of bile acid 7- and 12-sulfates in urine by high-performance liquid chromatography with fluorescence labeling. *J.Chromatogr.* **1987**, *415*, 45–52.

Hayashi, M.; Imai, Y.; Minami, Y.; Kawata, S.; Matsuzawa, Y.; Tarui, S.; Uchida, K. Determination of bile acids in rat bile by high-performance liquid chromatography. *J.Chromatogr.* **1985**, *338*, 195–200.

Iida, T.; Chang, F.C.; Goto, J.; Nambara, T. High-performance liquid chromatographic behavior of 2-,4- and 6-hydroxylated bile acid stereoisomers. *J.Liq.Chromatogr.* **1991**, *14*, 2527–2539.

Iida, T.; Ohnuki, Y.; Chang, F.C.; Goto, J.; Nambara, T. High performance liquid chromatographic separation of stereoisomeric bile acids as their UV-sensitive esters. *Lipids* **1985**, *20*, 187–194.

Ikawa, S.; Miyake, M.; Mura, T.; Ikeguchi, M. High-performance liquid chromatographic-fluorescence determination of human fecal bile acids. *J.Chromatogr.* **1987**, *400*, 149–161.

Niwa, T.; Fujita, K.; Goto, J.; Nambara, T. Separation of bile acid N-acetylglucosaminides by high-performance liquid chromatography with precolumn fluorescence labeling. *Anal.Sci.* **1992**, *8*, 659–662.

Niwa, T.; Fujita, K.; Goto, J.; Nambara, T. Separation and characterization of ursodeoxycholate 7-N-acetylglucosaminides in human urine by high-performance liquid chromatography with fluorescence detection. *J.Liq.Chromatogr.* **1993**, *16*, 2531–2544.

Paciotti, M.; Perinati, L.; Gori, F.; Rampazzo, P. High-performance liquid chromatographic determination of bile acids involved in the synthesis of ursodeoxycholic acid. *J.Chromatogr.* **1983**, *270*, 402–406.

Shimada, K.; Komine, Y.; Mitamura, K. High-performance liquid chromatographic separation of bile acid pyrenacyl esters with cyclodextrin-containing mobile phase. *J.Chromatogr.* **1991**, *565*, 111–118.

Takeuchi, T.; Ishii, D. Analysis of bile acids in a serum by micro-HPLC. *HRC CC, J.High Resolut.Chromatogr.Chromatogr.Commun.* **1983**, *6*, 571–572.

Wildgrube, H.J.; Stockhausen, H.; Petri, J.; Fuessel, U.; Lauer, H. Naturally occurring conjugated bile acids, measured by high-performance liquid chromatography, in human, dog, and rabbit bile. *J.Chromatogr.* **1986**, *353*, 207–213.

Bile Acids (using post-column reaction detection)

Arisue, K.; Ogawa, Z.; Koda, K.; Hayashi, C.; Ishida, Y. HPLC analysis of bile acids using a new technique of post column reaction. *Rinsho Kagaku* **1980**, *9*, 104–110.

Ishii, D.; Murata, S.; Takeuchi, T. Analysis of bile acids by micro high-performance liquid chromatography with postcolumn enzyme reaction and fluorometric detection. *J.Chromatogr.* **1984**, *282*, 569–577.

Li, B.; Rui, L.; Chen, Y.; Sheng, H. High performance liquid chromatographic quantitation of conjugated bile acids in human body fluids with immobilized enzyme detection system. *Sepu* **1989**, *7*, 165–167.

Osredkar, J. The quantitative determination of conjugated bile acids in serum. *Farm.Vestn.(Ljubljana)* **1992**, *43*, 147–150.

Sakakura, H.; Suzuki, M.; Kimura, N.; Takeda, H.; Nagata, S.; Maeda, M. Simultaneous determination of bile acids in rat bile and serum by high-performance liquid chromatography. *J.Chromatogr.* **1993**, *621*, 123–131.

Swobodnik, W.; Zhang, Y.Y.; Klueppelberg, U.; Janowitz, P.; Wechsler, J.G.; Fuerst, P.; Ditschuneit, H. Analysis of conjugated bile acids in human serum by high-performance liquid chromatography using post-column enzyme reaction and off-line fluorometric determination. *J.Chromatogr.* **1987**, *423*, 75–84.

Takeuchi, T.; Saito, S.; Ishii, D. Micro high-performance liquid chromatographic separations of bile acids. *J.Chromatogr.* **1983**, *258*, 125–134.

Yamade, N. Determination of esterified and nonesterified 3β-hydroxylated bile acids and the clinical significance. *Wakayama Igaku* **1995**, *46*, 317–333.

Bilirubin

Little, G.H. Separation of bilirubin azopigments from bile by high-performance liquid chromatography. *J.Chromatogr.* **1979**, *163*, 81–85.

Rothuizen, J.; Heirwegh, K.P.M.; Van Kouwen, A.M. Novel method for high-performance liquid chromatography of azo derivatives of conjugated and unconjugated bilirubin. *J.Chromatogr.* **1988**, *427*, 19–28.

Biliverdin-IX

Hirota, K. Urinary excretion of α- and β-isomers of biliverdin-IX in humans. *Biol.Pharm.Bull.* **1995**, *18*, 481–484.

Biopolymers

Freiser, H.H.; Gooding, K.M. Silica-based supports for high-performance liquid chromatography of biopolymers using non-denaturing conditions. *J.Chromatogr.* **1991**, *544*, 125–135.

Biotin (using post-column reaction detection)

Yokoyama, T.; Kinoshita, T. High-performance liquid chromatographic determination of biotin in pharmaceutical preparations by post-column fluorescence reaction with thiamin reagent. *J.Chromatogr.* **1991**, *542*, 365–372.

1,19-Bis(ethylamino)-5,10,15-triazanonadecane

Eiseman, J.L.; Yuan, Z-M.; Eddington, N.D.; Sentz, D.L.; Callery, P.S.; Egorin, M.J. Plasma pharmacokinetics and urinary excretion of the polyamine analog 1,19-bis(ethylamino)-5,10,15-triazanonadecane in CD2F1 mice. *Cancer Chemother.Pharmacol.* **1996**, *38*, 13–20.

Boric Acid (using post-column reaction detection)

Zou, J.; Motomizu, S.; Oshima, M.; Fukutomi, H. Sensitive determination of boric acid by on-line complexation/anion-exchange chromatography with H-resorcinol. *Anal.Sci.* **1992**, *8*, 719–722.

Boronates

Gamoh, K.; Brooks, C.J.W. Stability and reversed-phase liquid chromatographic studies of cyclic boronates. *Anal.Sci.* **1993**, *9*, 549–545.

Bradykinin (using post-column reaction detection)

Omori, H.; Watanabe, N.; Nakashizuka, T.; Yamazaki, S. Fluorometric determination of saliva bradykinin by HPLC with a postcolumn reaction using o-phthaldehyde. *HRC CC, J.High Resolut.Chromatogr.Chromatogr.Commun.* **1986**, *9*, 306–307.

Brassinolide

Motegi, C.; Takatsuto, S.; Gamoh, K. Identification of brassinolide and castasterone in the pollen of orange (Citrus sinensis Osbeck) by high-performance liquid chromatography. *J.Chromatogr.A* **1994**, *658*, 27–30.

Brassinosteroids

Gamoh, K.; Takatsuto, S.; Ikekawa, N. Effective separation of C-24 epimeric brassinosteroids by liquid chromatography. *Anal.Chim.Acta* **1992**, *256*, 319–322.

Gamoh, K.; Takatsuto, S. A boronic acid derivative as a highly sensitive fluorescence derivatization reagent for brassinosteroids in liquid chromatography. *Anal.Chim.Acta* **1989**, *222*, 201–204.

Bredinin

Takada, K.; Nakae, H.; Asada, S.; Ichikawa, Y.; Fukunishi, T.; Sonoda, T. Rapid method for the high-performance liquid chromatographic determination of Bredinin in human serum. *J.Chromatogr.* **1981**, *222*, 156–159.

Brobactam (using post-column reaction detection)

Kissmeyer-Nielsen, A.M. Determination of 6β-bromopenicillanic acid (brobactam) in human serum and urine by high-performance liquid chromatography. *J.Chromatogr.* **1988**, *426*, 425–430.

Brofaromine

Horne, C.; Mutschler, E. Quantitative determination of brofaromine in human plasma after high performance liquid chromatographic separation. *Arch.Pharm.(Weinheim, Ger.)* **1988**, *321*, 359–361.

Bromide

Verma, K.K.; Sanghi, S.K.; Jain, A.; Gupta, D. Liquid chromatographic determination of bromide after pre-column derivatization to 4-bromoacetanilide. *J.Chromatogr.* **1988**, *457*, 345–353.

Bromural

Okamoto, M.; Yamada, F.; Ishiguro, M.; Yasue, T. High-performance liquid chromatographic determination of bromural in serum upon hemoperfusion. *J.Chromatogr.* **1981**, *223*, 473–478.

Busulfan

Kazemifard, A.G.; Morgan, D.J. Determination of busulfan in plasma by high-performance liquid chromatography. *J.Chromatogr.* **1990**, *528*, 274–276.

Butabarbital (using post-column reaction detection)

Scott, E.P. Application of postcolumn ionization in the high-performance liquid chromatographic analysis of butabarbital sodium elixir. *J.Pharm.Sci.* **1983**, *72*, 1089–1091.

Butylated Hydroxyanisole

Ansari, G.A.S. High-performance liquid chromatographic separation of the isomers of butylated hydroxyanisole. *J.Chromatogr.* **1983**, *262*, 393–396.

N-(n-Butyl)thiophosphoric triamide (using post-column reaction detection)

Creason, G.L.; Schmitt, M.R.; Douglass, E.A.; Hendrickson, L.L. Urease inhibitory activity associated with N-(n-butyl)thiophosphoric triamide is due to formation of its oxon analog. *Soil Biol.Biochem.* **1990**, *22*, 209–211.

Cadralazine

Schuetz, H.; Faigle, J.W. Multiple inverse isotope dilution assay for cadralazine and four metabolites in biological fluids. *J.Chromatogr.* **1983**, *281*, 273–280.

Caerulein (using post-column reaction detection)

Feurle, G.E.; Ohnheiser, G.; Loser, C. Dissimilar trophic effects of caerulein and xenopsin in the rat pancreas. *Int.J.Pancreatol.* **1990**, *6*, 129–137.

Calcitonin

Windisch, V.; Karpenko, C.; Daruwala, A. LC assay for salmon calcitonin in aerosol formulations using fluorescence derivatization and size exclusion chromatography. *J.Pharm.Biomed.Anal.* **1992**, *10*, 71–76.

Calcitonin (using post-column reaction detection)

Fukuda, T.; Yun, A.; Imai, K. A rapid and sensitive determination of salmon calcitonin in solutions containing bovine serum albumin or gelatin. *Biomed.Chromatogr.* **1993**, *7*, 229–230.

Lee, K.C.; Yoon, J.Y.; Woo, B.H.; Kim, C-K.; DeLuca, P.P. Post-column fluorescence HPLC for salmon calcitonin formulations. *Int.J.Pharm.* **1995**, *114*, 215–220.

Cannabinoids (using post-column reaction detection)

Twitchett, P.J.; Williams, P.L.; Moffat, A.C. Photochemical detection in high-performance liquid chromatography and its application to cannabinoid analysis. *J.Chromatogr.* **1978**, *149*, 683–691.

Cantharidin

Ray, A.C.; Tamulinas, S.H.; Reagor, J.C. High pressure liquid chromatographic determination of cantharidin, using a derivatization method in specimens from animals acutely poisoned by ingestion of blister beetles, Epicauta lemniscata. *Am.J.Vet.Res.* **1979**, *40*, 498–504.

Captopril

Jarrott, B.; Anderson, A.; Hooper, R.; Louis, W.J. High-performance liquid chromatographic analysis of captopril in plasma. *J.Pharm.Sci.* **1981**, *70*, 665–667.

Carbohydrates

Kang, E.Y.J.; Coleman, R.D.; Pownall, H.J.; Gotto, A.M., Jr.; Yang, C.Y. Analysis of the carbohydrate composition of glycoproteins by high-performance liquid chromatography. *J.Protein Chem.* **1990**, *9*, 31–35.

Okada, T.; Kuwamoto, T. High-performance liquid chromatographic determination of electrically neutral carbohydrates with conductivity detection. *Anal.Chem.* **1986**, *58*, 1375–1379.

White, C.A.; Kennedy, J.F.; Golding, B.T. Analysis of derivatives of carbohydrates by high-pressure liquid chromatography. *Carbohydr.Res.* **1979**, *76*, 1–10.

Carbohydrates (using post-column reaction detection)

Haginaka, J.; Nomura, T. Liquid chromatographic determination of carbohydrates with pulsed amperometric detection and a membrane reactor. *J.Chromatogr.* **1988**, *447*, 268–271.

Kraemer, M.; Engelhardt, H. Analysis of carbohydrates by HPLC with post-column derivatization. *J.High Resolut.Chromatogr.* **1992**, *15*, 24–29.

Soga, T.; Inoue, Y.; Yamaguchi, K. Determination of carbohydrates by hydrophilic interaction chromatography with pulsed amperometric detection using postcolumn pH adjustment. *J.Chromatogr.* **1992**, *625*, 151–155.

Tabata, S.; Ide, T. Electrochemical detection of reducing carbohydrates produced by the transferase action of yeast debranching enzyme on maltosaccharides. *Carbohydr.Res.* **1988**, *176*, 245–251.

Tougas, T.P.; DeBenedetto, M.J.; DeMott, J.M., Jr. Postchromatographic electrochemical detection of carbohydrates at a silver oxide electrode. *Electroanalysis* **1993**, *5*, 669–675.

Van den Berg, J.H.M.; Horsels, H.W.M.; Deelder, R.S. Column liquid chromatography of reducing carbohydrates by fluorometric reaction detection with a pressurized reactor outlet. *J.Liq.Chromatogr.* **1984**, *7*, 2351–2365.

Watanabe, N. Amperometric detection of reducing carbohydrates in high-performance liquid chromatography using an amino-bonded column and acetonitrile-water as the eluent. *J.Chromatogr.* **1985**, *330*, 333–338.

Yamauchi, S.; Nakai, C.; Nimura, N.; Kinoshita, T.; Hanai, T. Highly sensitive detection of non-reducing carbohydrates by liquid chromatography. *Analyst* **1993**, *118*, 769–771.

Yamauchi, S.; Nakai, C.; Nimura, N.; Kinoshita, T.; Hanai, T. Development of a highly sensitive fluorescence reaction detection system for liquid chromatographic analysis of reducing carbohydrates. *Analyst* **1993**, *118*, 773–776.

Carbonyl Compounds

Baechmann, K.; Haag, I.; Han, K.Y.; Schmitzer, R.Q. Determination of carbonyl compounds in the low ppb-range by capillary electrophoresis. *Fresenius' J.Anal.Chem.* **1993**, *346*, 786–788.

Goto, J.; Saisho, Y.; Nambara, T. Studies on steroids. Part CCXLVII. Sensitive fluorescence labeling reagents for high-performance liquid chromatography of carbonyl compounds. *Anal.Sci.* **1989**, *5*, 399–402.

Kieber, R.J.; Mopper, K. Determination of picomolar concentrations of carbonyl compounds in natural waters, including seawater, by liquid chromatography. *Environ.Sci.Technol.* **1990**, *24*, 1477–1481.

Krull, I.S.; Xie, K.H.; Colgan, S.; Neue, U.; Izod, T.; King, R.; Bidlingmeyer, B. Solid phase derivatization reactions in HPLC polymeric reductions for carbonyl compounds. *J.Liq.Chromatogr.* **1983**, *6*, 605–626.

Krull, I.S.; Colgan, S.; Xie, K.H.; Neue, U.; King, R.; Bidlingmeyer, B. Solid phase derivatizations in HPLC: borohydride/silica reductions for carbonyl compounds. *J.Liq.Chromatogr.* **1983**, *6*, 1015–1036.

Raymer, J.H.; Novotny, M.V. Recent developments of the determination of carbonyl compounds in biological fluids and tissues. *Trace Anal.* **1984**, *3*, 3–29.

Raymer, J.; Holland, M.L.; Wiesler, D.P.; Novotny, M. Preconcentration and multicomponent chromatographic determination of biological carbonyl compounds. *Anal.Chem.* **1984**, *56*, 962–966.

Ueno, K.; Umeda, T. Electrochemical and chromatographic properties of selected hydrazine and hydrazide derivatives of carbonyl compounds. *J.Chromatogr.* **1991**, *585*, 225–231.

Zhou, X.; Mopper, K. Measurement of sub-parts-per-billion levels of carbonyl compounds in marine air by a simple cartridge trapping procedure followed by liquid chromatography. *Environ.Sci.Technol.* **1990**, *24*, 1482–1485.

Carbonyl Compounds (using post-column reaction detection)

Nondek, L.; Milofsky, R.E.; Birks, J.W. Determination of carbonyl compounds in air by HPLC using on-line analyzed microcartridges, fluorescence and chemiluminescence detection. *Chromatographia* **1991**, *32*, 33–39.

Carboprost

Brown, L.W.; Carpenter, B.E. Comparison of two high-pressure liquid chromatographic assays for carboprost, a synthetic prostaglandin. *J.Pharm.Sci.* **1980**, *69*, 1396–1399.

gamma-Carboxyglutamic Acid

Muramoto, K.; Kamiya, H. Analysis for gamma-carboxyglutamic acid in marine invertebrate shell matrixes. *Anal.Sci.* **1991**, *7*, 939–940.

Peter, A.; Somlai, C.; Penke, B. High-performance liquid chromatographic method for the separation of the optical isomers of gamma,gamma'-di-tert.-butyl-D,L-gamma-carboxyglutamic acid and D,L-gamma-carboxyglutamic acid. *J.Chromatogr.A* **1995**, *710*, 297–302.

Carboxylic Acids

Allenmark, S.; Chelminska-Bertilsson, M. Determination of carboxylic acids by liquid chromatography after phase-transfer-catalyzed fluorogenic labeling. *J.Chromatogr.* **1988**, *456*, 410–416.

Baiocchi, C.; Campi, E.; Gennaro, M.C.; Mentasti, E.; Sarzanini, C. Preconcentration and HPLC separation of carboxylic acids as phenacyl esters. *Ann.Chim.(Rome)* **1983**, *73*, 659–673.

Coenen, A.J.J.M.; Kerkhoff, M.J.G.; Heringa, R.M.; van der Wal, S. Comparison of several methods for the determination of trace amounts of polar aliphatic monocarboxylic acids by high-performance liquid chromatography. *J.Chromatogr.* **1991**, *593*, 243–252.

Funazo, K.; Tanaka, M.; Yasaka, Y.; Takigawa, H.; Shono, T. New ultraviolet labeling agents for high-performance liquid chromatographic determination of monocarboxylic acids. *J.Chromatogr.* **1989**, *481*, 211–219.

Fujimoto, Y.; Ishi, K.; Nishi, H.; Tsumagari, N.; Kakimoto, T.; Shimizu, R. New derivatization reagents for the resolution of carboxylic acid enantiomers by high-performance liquid chromatography. *J.Chromatogr.* **1987**, *402*, 344–348.

Grayeski, M.L.; DeVasto, J.K. Coumarin derivatizing agents for carboxylic acid detection using peroxyoxalate chemiluminescence with liquid chromatography. *Anal.Chem.* **1987**, *59*, 1203–1206.

Hurlbut, J.A.; Dahm, C.E. HPLC separation and UV detection of carboxylic acids using 2-bromo-1-(9-phenanthrenyl)ethanone as a derivatizing agent. *Chromatogram* **1990**, *11*, 7–8.

Iwata, T.; Nakamura, M.; Yamaguchi, M. 4-(5,6-Dimethoxy-2-benzimidazoyl)benzohydrazide as fluorescence derivatization reagent for carboxylic acids in high-performance liquid chromatography. *Anal.Sci.* **1992**, *8*, 889–892.

Kondo, J.; Suzuki, N.; Imaoka, T.; Kawasaki, T.; Nakanishi, A.; Kawahara, Y. 6-Methoxy-2-(4-substituted phenyl)benzoxazoles as fluorescent chiral derivatization reagents for carboxylic acid enantiomers. *Anal.Sci.* **1994**, *10*, 17–23.

Levai, F.; Liu, C-M.; Tse, M.M.; Lin, E.T. Pre-column fluorescence derivatization using leucine-coumarinylamide for HPLC determination of mono- and dicarboxylic acids in plasma. *Acta Physiol.Hung.* **1995**, *83*, 39–46.

Lewis, S.W.; Worsfold, P.J.; McKerrell, E.H. Monitoring carboxylic acid formation in engine oils by liquid chromatography with fluorescence detection. *J.Chromatogr.A* **1994**, *667*, 91–98.

Ladanyi, L.; Sztruhar, I.; Slegel, P.; Vereczekey-Donath, G. Determination of the enantiomeric composition of chiral carboxylic acids using chiral derivatization and HPLC. *Chromatographia* **1987**, *24*, 477–481.

Marce, R.M.; Calull, M.; Olucha, J.C.; Borrull, F.; Rius, F.X. Optimization of the derivatization method for the liquid chromatographic determination of carboxylic acids in wines. *Anal.Chim.Acta* **1991**, *242*, 25–30.

Miller, J.M.; Brindle, I.D.; Cater, S.R.; So, K-H.; Clark, J.H. Potassium fluoride assisted derivatization of carboxylic acids to phenacyl esters for determination by high-performance liquid chromatography. *Anal.Chem.* **1980**, *52*, 2430–2432.

Nishida, Y.; Itoh, E.; Abe, M.; Ohrui, H.; Meguro, H. Syntheses of a series of fluorescent carboxylic acids with a 1,3-benzodioxole skeleton and their evaluation as chiral derivatizing reagents. *Anal.Sci.* **1995**, *11*, 213–220.

Narita, S.; Kitagawa, T. S-(-)-2-[4-(1-Aminoethyl)naphthyl]-6-methoxy-N-methyl-2H-benzotriazolyl-5-amine dihydrochloride as a fluorescence chiral derivatization reagent for carboxylic acids in high-performance liquid chromatography. *Anal.Sci.* **1989**, *5*, 361–362.

Narita, S.; Kitagawa, T. Application of fluorescent triazoles to analytical chemistry. II. Fluorescent derivatization of carboxylic acids. *Chem.Pharm.Bull.* **1989**, *37*, 831–833.

Saito, M.; Chiyoda, Y.; Ushijima, T.; Sasamoto, K.; Ohkura, Y. 2-(5-Hydrazinocarbonyl-2-oxazolyl)-5,6-methylenedioxybenzofuran as a fluorescence derivatization reagent for carboxylic acids in high-performance liquid chromatography. *Anal.Sci.* **1994**, *10*, 679–681.

Samo, A.R.; Arbani, S.A.; Chippa, M.A.; Khahwer, M.Y.; Qureshi, G.A. Micellar phase-transfer catalysis: a suitable procedure to derivatize carboxylic acids in biological matrixes. *J.Chem.Soc.Pak.* **1994**, *16*, 108–111.

Sasamoto, K.; Ushijima, T.; Saito, M.; Ohkura, Y. Precolumn fluorescence derivatization of carboxylic acids using 4-aminomethyl-6,7-dimethoxycoumarin in a two-phase medium. *Anal.Sci.* **1996**, *12*, 189–193.

Shimada, K.; Haniuda, E.; Oe, T.; Nambara, T. Ferrocene derivatization reagents for optical resolution of carboxylic acids by high-performance liquid chromatography with electrochemical detection. *J.Liq.Chromatogr.* **1987**, *10*, 3161–3172.

Takadate, A.; Masuda, T.; Tajima, C.; Murata, C.; Irikura, M.; Goya, S. 3-Bromoacetyl-7-methoxycoumarin as a new fluorescent derivatization reagent for carboxylic acids in high-performance liquid chromatography. *Anal.Sci.* **1992**, *8*, 663–668.

Tanaka, M.; Muramatsu, H.; Yasaka, Y. 2-(6,7-Dimethoxy-2,3-naphthalimido)ethyl trifluoromethanesulfonate as a labeling reagent for carboxylic acids in liquid chromatography. *Microchem.J.* **1994**, *49*, 159–164.

Ushijima, T.; Saito, M.; Sasamoto, K.; Ohkura, Y.; Ueno, K. Sensitive and mild fluorogenic reagents for biogenic carboxylic acids in HPLC. *Anal.Sci.Technol.* **1995**, *8*, 545–551.

van der Horst, F.A.L. Micellar catalysis for the derivatization of carboxylic acids in physiological matrixes. *TrAC, Trends Anal.Chem.(Pers.Ed.)* **1989**, *8*, 268–273.

Van den Beld, C.M.B.; Lingeman, H.; Van Ringen, G.J.; Tjaden, U.R.; van der Greef, J. Laser-induced fluorescence detection in liquid chromatography after preliminary derivatization of carboxylic acid and primary amino groups. *Anal.Chim.Acta* **1988**, *205*, 15–27.

Wahlund, K.G.; Edlen, B. Tributyl phosphate as stationary phase in reversed phase liquid chromatographic separations of hydrophilic carboxylic acids, amino acids and dipeptides. *J.Liq.Chromatogr.* **1981**, *4*, 309–323.

Carboxylic Acids (using post-column reaction detection)

Tsai, C.P.; Sahil, A.; McGuire, J.M.; Karger, B.L.; Vouros, P. High-performance liquid chromatographic mass spectrometric determination of volatile carboxylic acids using ion-pair extraction and thermally introduced alkyl alkylation. *Anal.Chem.* **1986**, *58*, 2–6.

S-Carboxymethylcysteine

Brockmoeller, J.; Simane, Z.J.; Roots, I. HPLC-analysis of S-carboxymethylcysteine and its sulfoxide metabolites. *Drug Metab.Drug Interact.* **1988**, *6*, 447–456.

De Schutter, J.A.; Van Der Weken, G.; Van den Bossche, W.; De Moerloose, P. Determination of S-carboxymethylcysteine in serum by reversed-phase ion-pair liquid chromatography with column switching following pre-column derivatization with o-phthalaldehyde. *J.Chromatogr.* **1988**, *428*, 301–310.

Guinebault, P.; Dubruc, C.; Haddouche, A.; Colafranceschi, C.; Bianchetti, G. An improved method for the determination of S-carboxymethyl-L-cysteine in biological fluids with precolumn derivatization and on-line clean-up. *Chromatographia* **1988**, *26*, 377–382.

Woolfson, A.D.; Millership, J.S.; Karim, E.F.I.A. Determination of the sulfoxide metabolites of S-carboxymethyl-L-cysteine by high-performance liquid chromatography with electrochemical detection. *Analyst* **1987**, *112*, 1421–1425.

Carboxymethyllysine

Hartkopf, J.; Pahlke, C.; Luedemann, G.; Erbersdobler, H.F. Determination of N-epsilon-carboxymethyllysine by a reversed-phase high-performance liquid chromatography method. *J.Chromatogr.A* **1994**, *672*, 242–246.

3-Carboxy-4-methyl-5-propyl-2-furanpropanoic acid

Mabuchi, H.; Nakahashi, H. Determination of 3-carboxy-4-methyl-5-propyl-2-furanpropanoic acid, a major endogenous ligand substance in uremic serum, by high-performance liquid chromatography with ultraviolet detection. *J.Chromatogr.* **1987**, *415*, 110–117.

Cardenolides (using post-column reaction detection)

Maekawa, B.; Morimoto, K. Development of analytical method for selective detection of cardenolides by high-performance liquid chromatography. *Biosci., Biotechnol., Biochem.* **1992**, *56*, 967.

Cardiolipin

Schlame, M.; Otten, D. Analysis of cardiolipin molecular species by high-performance liquid chromatography of its derivative 1,3-bisphosphatidyl-2-benzoyl-sn-glycerol dimethyl ester. *Anal.Biochem.* **1991**, *195*, 290–295.

Carnitine

Bhuiyan, A.K.M.J.; Causey, A.G.; Bartlett, K. Studies of [U-14C]ketoisoleucine metabolism by rat liver mitochondria: identification of labeled acyl-carnitines by reverse-phase high-performance thin-layer chromatography. *Biochem.Soc.Trans.* **1986**, *14*, 1072–1073.

Kamimori, H.; Hamashima, Y.; Konishi, M. Determination of carnitine and saturated-acyl group carnitines in human urine by high-performance liquid chromatography with fluorescence detection. *Anal.Biochem.* **1994**, *218*, 417–424.

Konishi, M.; Hashimoto, H. Determination of pivaloylcarnitine in human plasma and urine by high-performance liquid chromatography with fluorescence detection. *J.Pharm.Sci.* **1992**, *81*, 1038–1041.

Minkler, P.E.; Hoppel, C.L. Quantification of free carnitine, individual short- and medium-chain acylcarnitines, and total carnitine in plasma by high-performance liquid chromatography. *Anal.Biochem.* **1993**, *212*, 510–518.

Pourfarzam, M.; Bartlett, K. Synthesis, characterisation and high-performance liquid chromatography of C6-C16 dicarboxylyl-mono-coenzyme A and -mono-carnitine esters. *J.Chromatogr.* **1991**, *570*, 253–276.

Van Kempen, T.A.T.G.; Odle, J. Quantification of carnitine esters by high-performance liquid chromatography. Effect of feeding medium-chain triglycerides on plasma carnitine ester profile. *J.Chromatogr.* **1992**, *584*, 157–165.

β-Carotene (using post-column reaction detection)

Meydani, M.; Martin, A.; Ribaya-Mercado, J.D.; Gong, J.; Blumberg, J.B.; Russell, R.M. β-Carotene supplementation increases antioxidant capacity of plasma in older women. *J.Nutr.* **1994**, *124*, 2397–2403.

Catechin

Cheynier, V.; Rigaud, J.; Moutounet, M. High-performance liquid chromatographic determination of the free o-quinones of trans caffeoyltartaric acid, 2-S-glutathionylcaffeoyltartaric acid and catechin in grape must. *J.Chromatogr.* **1989**, *472*, 428–432.

Catecholamines

Allgire, J.F.; Juenge Eric C ;, ; Damo, C.P.; Sullivan, G.M.; Kirchhoefer, R.D. High-performance liquid chromatographic determination of d-/l-epinephrine enantiomer ratio in lidocaine-epinephrine local anesthetics. *J.Chromatogr.* **1985**, *325*, 249–254.

Froehlich, P.M.; Cunningham, T.D. An H.P.L.C.-fluorimetric analysis for L-dopa, noradrenalin and dopamine. *Anal.Chim.Acta* **1978**, *97*, 357–363.

Fyllingen, G.; Langvik, T.A.; Hasselgard, P.; Roksvaag, P.O. Racemization and oxidation in adrenaline injections. *Acta Pharm.Nord.* **1990**, *2*, 355–362.

Gurkan, T.; Ozdemir, O. Simultaneous determination of TNP-derivatized dopamine, noradrenaline and 5-hydroxytryptamine in rat brain tissue by high-performance liquid chromatography. *Sci.Pharm.* **1988**, *56*, 97–103.

Imai, K.; Tamura, Z. Liquid chromatographic determination of urinary dopamine and norepinephrine as fluorescamine derivatives. *Clin.Chim.Acta* **1978**, *85*, 1–6.

Jackman, G.P.; Carson, V.J.; Bobik, A.; Skews, H. Simple and sensitive procedure for the assay of serotonin and catecholamines in brain by high-performance liquid chromatography using fluorescence detection. *J.Chromatogr.* **1980**, *182*, 277–284.

Kamahori, M.; Taki, M.; Watanabe, Y.; Miura, J. Analysis of plasma catecholamines by high-performance liquid chromatography with fluorescence detection: simple sample preparation for pre-column fluorescence derivatization. *J.Chromatogr.* **1991**, *567*, 351–358.

Kehr, J. Determination of catecholamines by automated pre-column derivatization and reversed-phase column liquid chromatography with fluorescence detection. *J.Chromatogr.A* **1994**, *661*, 137–142.

Kirchhoefer, R.D.; Sullivan, G.M.; Allgire, J.F. Analysis of USP epinephrine injections for potency, impurities, degradation products, and d-enantiomer by liquid chromatography, using ultraviolet and electrochemical detectors. *J.Assoc.Off.Anal.Chem.* **1985**, *68*, 163–165.

Kubo, H.; Umiguchi, Y.; Kinoshita, T. In-line derivatization method for fluorometric determination of catecholamines by high performance liquid chromatography. *Chromatographia* **1994**, *38*, 591–594.

Mell, L.D., Jr.; Dasler, A.R.; Gustafson, A.B. Pre-column fluorescent derivatization for high pressure liquid chromatography with o-phthalaldehyde. Separation of urinary catecholamines. *J.Liq. Chromatogr.* **1978**, *1*, 261–277.

Nozaki, O.; Ohba, Y. [Recent development of determination methods for adrenocortical hormones and catecholamines]. *Rinsho.Byori.* **1990**, *38*, 1126–1133.

Ragab, G.H.; Nohta, H.; Kai, M.; Ohkura, Y.; Zaitsu, K. 1,2-Diarylethylenediamines as sensitive pre-column derivatizing reagents for chemiluminescence detection of catecholamines in HPLC. *J.Pharm.Biomed.Anal.* **1995**, *13*, 645–650.

Sasa, S.; Blank, C.LR. Simultaneous determination of norepinephrine, dopamine, and serotonin in brain tissue by high-pressure liquid chromatography with electrochemical detection. *Anal.Chim.Acta* **1979**, *104*, 29–45.

Ukai, T.; Ogawa, K.; Satake, T.; Kaneda, N.; Nagatsu, T. A new HPLC-fluorescence assay of adrenaline in brain: changes of brain adrenaline in spontaneously hypertensive rats. *Biog.Amines* **1989**, *6*, 87–90.

Umegae, Y.; Nohta, H.; Lee, M.; Ohkura, Y. 1,2-Diarylethylenediamines as pre-column fluorescence derivatization reagents in high-performance liquid chromatographic determination of catecholamines in urine and plasma. *Chem.Pharm.Bull.* **1990**, *38*, 2293–2295.

Yang, Z.; Xu, R. Investigation on the enantiomeric impurity of epinephrine hydrochloride injections. *Chirality* **1989**, *1*, 92–93.

Catecholamines (using post-column reaction detection)

Arakawa, Y.; Imai, K.; Tamura, Z. Determination of catecholamine sulfoconjugate isomers in normal human urine by use of high-performance liquid chromatography with a photochemical detector. *Anal.Biochem.* **1983**, *132*, 389–399.

Boos, K.S.; Wilmers, B.; Sauerbrey, R.; Schlimme, E. Development and performance of an automated HPLC analyzer for catecholamines. *Chromatographia* **1987**, *24*, 363–370.

Boos, K.S.; Wilmers, B.; Sauerbrey, R.; Schlimme, E. Catecholamine: on-line sample handling and post-column derivatization. *Chromatographia* **1988**, *25*, 199–204.

Edlund, P.O.; Westerlund, D. Direct injection of plasma and urine in automated analysis of catecholamines by coupled-column liquid chromatography with post-column derivatization. *J.Pharm. Biomed.Anal.* **1984**, *2*, 315–333.

Higashidate, S.; Imai, K.; Prados, P.; Adachi-Akahane, S.; Nagao, T. Relations between blood pressure and plasma norepinephrine concentrations after administration of diltiazem to rats: HPLC-peroxy-oxalate chemiluminescence determination on an individual basis. *Biomed.Chromatogr.* **1994**, *8*, 19–21.

Iwaeda, T.; Kuroki, M.; Ohta, K.; Ishimura, S.; Takahashi, H.; Watanabe, H. Development of fully automated catecholamine analyzer, HLC-8030. *Toso Kenkyu Hokoku* **1988**, *32*, 59–64.

Kawasaki, T.; Imai, K.; Higuchi, T.; Wong, O.S. Determination of fluorescent cyanobenz[f]isoindole derivatives of dopamine and norepinephrine using high performance liquid chromatography with chemiluminescence detection. *Biomed.Chromatogr.* **1990**, *4*, 113–118.

Kamidate, T.; Yoshida, K.; Segawa, T.; Watanabe, H. Determination of catecholamines with lucigenin chemiluminescence. *Anal.Sci.* **1989**, *5*, 359–360.

Kakizaki, T.; Hasebe, K.; Yoshida, H. Analysis of catecholamines by high-performance liquid chromatography with polarographic detection by use of catalytic maximum wave. *Nippon Kagaku Kaishi* **1987**, 1009–1015.

Kondo, N.; Hibi, K. Fully automatic urinary catecholamine analyzer. *Kuromatogurafi* **1992**, *13*, 253–254.

Nohta, H.; Jeon, H.K.; Kai, M.; Ohkura, Y. Liquid chromatographic determination of indoleamine, catecholamines and their precursors and metabolites in rat brain tissues with postcolumn fluorescence derivatization. *Anal.Sci.* **1994**, *10*, 5–9.

Nohta, H.; Jeon, H.K.; Kai, M.; Ohkura, Y. High-performance liquid chromatographic determination of total catecholamines and metabolites in human urine by application of the fluorescence derivatization method for their free form. *Anal.Sci.* **1993**, *9*, 537–540.

Okamoto, K.; Ishida, Y.; Asai, K. Separation and detection of small amounts of catecholamines by high-performance liquid chromatography. *J.Chromatogr.* **1978**, *167*, 205–217.

Piemonte, G.; Perina, L. Determination of catecholamines in biological samples by high performance liquid chromatography. *G.Ital.Chim.Clin.* **1983**, *8*, 57–73.

Ramos, B.L.; Mike, J.H. An electrochemical reactor for postcolumn fluorescence detection of catecholamines using HPLC. *Microchem.J.* **1993**, *47*, 33–40.

Yamamoto, T.; Yamatodani, A.; Nishimura, M.; Wada, H. Determination of dopamine-3- and -4-O-sulfate in human plasma and urine by anion-exchange high-performance liquid chromatography with fluorimetric detection. *J.Chromatogr.* **1985**, *342*, 261–267.

Yui, Y.; Kawai, C. Comparison of the sensitivity of various post-column methods for catecholamine analysis by high-performance liquid chromatography. *J.Chromatogr.* **1981**, *206*, 586–588.

Cefatrizine (using post-column reaction detection)

Crombez, E.; Van den Bossche, W.; De Moerloose, P. Quantitative high performance liquid chromatography determination of cefatrizine in serum and urine by UV detection and fluorescence detection after post column derivatization. *Anal.Chem.Symp.Ser.* **1980**, *3*, 261–266.

Crombez, E.; Van Der Weken, G.; Van den Bossche, W.; De Moerloose, P. Quantitative liquid chromatographic determination of cefatrizine in serum and urine by fluorescence detection after post-column derivatization. *J.Chromatogr.* **1979**, *177*, 323–332.

Ceforanide

Antibiotic drugs; ceforanide for injection. *Fed.Regist.* **1984**, *49*, 25845–9, 25 Jun 1.

Ceftiofur

Beconi-Barker, M.G.; Roof, R.D.; Vidmar, T.J.; Hornish, R.E.; Smith, E.B.; Gatchell, C.L.; Gibertson, T.J. Ceftiofur sodium: absorption, distribution, metabolism, and excretion in target animals and its determination by high-performance liquid chromatography. *ACS Symp.Ser.* **1996**, *636*, 70–84.

Jaglan, P.S.; Cox, B.L.; Arnold, T.S.; Kubicek, M.F.; Stuart, D.J.; Gilbertson, T.J. Liquid chromatographic determination of desfuroylceftiofur metabolite of ceftiofur as residue in cattle plasma. *J.Assoc.Off.Anal.Chem.* **1990**, *73*, 26–30.

Robb, E.J.; Kausche, F.M.; Belschner, A.P. Methods for detection of residues in milk after parenteral treatment with ceftiofur sodium - an evaluation using common milk residue screening tests and a quantitative HPLC method following either a single injection (Study I) or multiple injections (Study II) of 2.2 mg of ceftiofur free acid equivalents per kg of body weight. *Int.Dairy Fed.Spec.Issue* **1995**, *9505*, 195–200.

Cellulose Ethers

Sachse, K.; Metzner, K.; Welsch, T. Substitution in cellulose ethers. Part II. Determination of the distribution of alkoxyl substituents on the glucose units using high-performance liquid chromatography. *Analyst* **1983**, *108*, 597–602.

Chemiluminescence—General References

Calokerinos, A.C.; Defteros, N.T.; Baeyens, W.R. Chemiluminescence in drug assay. *J.Pharm.Biomed.Anal.* **1995**, *13*, 1063–1071.

DeAngelis, J.J.; Barkley, R.M.; Sievers, R.E. Coupling liquid chromatography with redox chemiluminescence detection. *J.Chromatogr.* **1988**, *441*, 125–134.

García Campaña, A.M.; Baeyens, W.R.G.; Zhao, Y. Chemiluminescence detection in capillary electrophoresis. *Anal.Chem.* **1997**, *69*, 83A-88A.

Hanaoka, N. Post-column adjustment of conditions for peroxyoxalate chemiluminescence detection for high-performance liquid chromatography. *J.Chromatogr.* **1990**, *503*, 155–165.

Hayakawa, K.; Imaizumi, N.; Miyazaki, M. Effect of a dual-head short-stroke pump on postcolumn peroxyoxalate chemiluminescence detection. *Biomed.Chromatogr.* **1991**, *5*, 148–152.

Hayakawa, K.; Minogawa, E.; Yokoyama, T.; Miyazaki, M.; Imai, K. A universal peroxyoxalate-chemiluminescence detection system for mobile phases of differing pH. *Biomed.Chromatogr.* **1992**, *6*, 84–87.

Honda, K.; Miyaguchi, K.; Imai, K. Evaluation of fluorescent compounds for peroxyoxalate chemiluminescence detection. *Anal.Chim.Acta* **1985**, *177*, 111–120.

Honda, K.; Miyaguchi, K.; Imai, K. Evaluation of aryl oxalates for chemiluminescence detection in high-performance liquid chromatography. *Anal.Chim.Acta* **1985**, *177*, 103–110.

Huang, B.; Li, J.-j.; Zhang, L.; Cheng, J.-k. On-line chemiluminescence detection for capillary ion analysis. *Anal.Chem.* **1996**, *68*, 2366–2369.

Imai, K.; Matsunaga, Y.; Tsukamoto, Y.; Nishitani, A. Application of bis[4-nitro-2-(3,6,9-trioxadecyl-oxycarbonyl)phenyl] oxalate to post-column chemiluminescence detection in high-performance liquid chromatography. *J.Chromatogr.* **1987**, *400*, 169–176.

Kwakman, P.J.M.; de Jong, G.J. Peroxyoxalate chemiluminescence detection in HPLC. *Methodol. Surv.Biochem.Anal.* **1992**, *22*, 261–268.

Kwakman, P.J.M.; Brinkman, U.A.T. Peroxyoxalate chemiluminescence detection in liquid chromatography. *Anal.Chim.Acta* **1992**, *266*, 175–192.

MacFarlane, J.D. Chemiluminescence and the challenge to attomole detection. *Adv.Exp.Med.Biol.* **1994**, *366*, 403.

Maeda, M.; Shimada, S.; Tsuji, A. High-performance liquid chromatography with a 3α-hydroxysteroid dehydrogenase postcolumn reactor and isoluminol-microperoxidase chemiluminescence detection. *J.Chromatogr.* **1990**, *515*, 329–335.

Nakashima, K.; Nagata, M.; Takahashi, M.; Akiyama, S. Peroxyoxalate chemiluminescence detection of condensates of malondialdehyde with thiobarbituric acids using a flow system. *Biomed.Chromatogr.* **1992**, *6*, 55–58.

Nakashima, K.; Maki, K.; Akiyama, S.; Wang, W.H.; Tsukamoto, Y.; Imai, K. Synthesis and evaluation of aryl oxalates as peroxyoxalate chemiluminescence reagents. *Analyst* **1989**, *114*, 1413–1416.

Poulsen, J.R.; Birks, J.W.; Guebitz, G.; van Zoonen, P.; Gooijer, C.; Velthorst, N.H.; Frei, R.W. Solid-state peroxyoxalate chemiluminescence detection of hydrogen peroxide generated in a post-column reaction. *J.Chromatogr.* **1986**, *360*, 371–383.

Sandmann, B.W.; Grayeski, M.L. Peroxyoxalate chemiluminescence detection with packed column supercritical fluid chromatography. *Chromatographia* **1994**, *38*, 163–167.

Sievers, R.E.; Nyarady, S.A.; Shearer, R.L.; DeAngelis, J.J.; Barkley, R.M.; Hutte, R.S. Selectivity of the redox chemiluminescence detector for complex sample analysis. *J.Chromatogr.* **1986**, *349*, 395–403.

Yan, B.; Lewis, S.W.; Worsfold, P.J.; Lancaster, J.S.; Gachanja, A. Procedures for the enhancement of selectivity in liquid phase chemiluminescence detection. *Anal.Chim.Acta* **1991**, *250*, 145–155.

Chloramines

Lukasewycz, M.T.; Bieringer, C.M.; Liukkonen, R.J.; Fitzsimmons, M.E.; Corcoran, H.F.; Lin, S.; Carlson, R.M. Analysis of inorganic and organic chloramines: derivatization with 2-mercaptobenzothiazole. *Environ.Sci.Technol.* **1989**, *23*, 196–199.

Scully, F.E., Jr.; Yang, J.P.; Bempong, M.A.; Daniel, F.B. Analysis of organic N-chloramines. *Water Chlorination: Environ.Impact Health Eff.* **1983**, *4*, 555–561.

Scully, F.E., Jr.; Burns, K.; Speed, M.A.; Arber, R.P. Analysis of organic N-chloramines in a municipal drinking water supply. *Proc.-AWWA Water Qual.Technol.Conf.* **1984**, *11th*, 197–205.

Chloramines (using post-column reaction detection)

Jersey, J.A.; Johnson, J.D. Analysis of N-chloramines in chlorinated wastewater and surface water by on-line enrichment HPLC with post-column reaction electrochemical detection. *Proc.-Water Qual.Technol.Conf.* **1991**, 1071–1081.

Yoon, J.; Jensen, J.N. Analysis of organic and inorganic monochloramines by HPLC. *Proc.-Water Qual.Technol.Conf.* **1992**, 475–488.

Chloramphenicol

LeBelle, M.J.; Young, D.C.; Graham, K.C.; Wilson, W.L. High-performance liquid chromatographic determination of chloramphenicol and 2-amino-1-(p-nitrophenyl)-1,3-propanediol in pharmaceutical formulations. *J.Chromatogr.* **1979**, *170*, 282–287.

Long, A.R.; Hsieh, L.C.; Bello, A.C.; Malbrough, M.S.; Short, C.R.; Barker, S.A. Method for the isolation and liquid chromatographic determination of chloramphenicol in milk. *J.Agric.Food Chem.* **1990**, *38*, 427–429.

Chloride

Nguyen Thanh T , Liquid chromatographic determination of flavor enhancers and chloride in food. *J.Assoc.Off.Anal.Chem.* **1984**, *67*, 747–751.

Chlorine

Jain, A.; Verma, K.K. HPLC determination of chlorine in air and water samples following precolumn derivatization to 4-bromoacetanilide. *Chromatographia* **1993**, *37*, 492–496.

2-Chloroethyl Ethyl Sulfide

Bossle, P.C.; Martin, J.J.; Sarver, E.W.; Sommer, H.Z. High-performance liquid chromatography analysis of 2-chloroethyl ethyl sulfide and its decomposition by-products by derivatization. *J.Chromatogr.* **1984**, *283*, 412–416.

Chlorohydroxyfuranones

Franzen, R. Synthesis and analyses of mutagenic chlorohydroxyfuranones and related compounds: By-products of chlorine disinfection of drinking water and of chlorine bleaching of pulp. *Acta Academiae Aboensis Ser B Mathematica et Physica Matematik Naturvetenskaper Teknik* **1995**, *55*, 1–52.

Chlorophenols

Kwakman, P.J.M.; Mol, J.G.J.; Kamminga, D.A.; Frei, R.W.; Brinkman, U.A.T.; de Jong, G.J. Rhodamine labeling reagent for the determination of chlorophenols by liquid chromatography with peroxyoxalate chemiluminescence detection. *J.Chromatogr.* **1988**, *459*, 139–149.

Chlorophenols (using post-column reaction detection)

Novak, T.J.; Grayeski, M.L. Acridinium-based chemiluminescence for high-performance liquid chromatographic detection of chlorophenols. *Microchem.J.* **1994**, *50*, 151–160.

Werkhoven-Goewie, C.E.; Boon, W.M.; Praat, A.J.J.; Frei, R.W.; Brinkman, U.A.T.; Little, C.J. Preconcentration and LC analysis of chlorophenols, using a styrene-divinylbenzene copolymeric sorbent and photochemical reaction detection. *Chromatographia* **1982**, *16*, 53–59.

Chlorothiazide (using post-column reaction detection)

Hessey, G.A.; Constanzer, M.L.; Bayne, W.F. Determination of chlorothiazide in urine using reversed-phase high-performance liquid chromatography with ultraviolet detection. *J.Chromatogr.* **1986**, *380*, 450–454.

Chloroxuron

Maris, F.A.; Geerdink, R.B.; Van Delft, R.; Brinkman, U.A.T. Determination of chloroxuron in strawberries using liquid chromatography with electron-capture detection. *Bull.Environ.Contam.Toxicol.* **1985**, *35*, 711–715.

Chlorpheniramine

Miyamoto, Y. Highly sensitive determination of chlorpheniramine as fluorescence derivative by high-performance liquid chromatography. *J.Chromatogr.* **1987**, *420*, 63–72.

Chlorpromazine

Midha, K.K.; Hubbard, J.W.; Cooper, J.K.; Gurnsey, T.; Hawes, E.M.; McKay, G.; Chakraborty, B.S.; Yeung, P.K. Therapeutic monitoring of chlorpromazine. IV: Comparison of a new high-performance liquid chromatographic method with radioimmunoassays for parent drug and some of its major metabolites. *Ther.Drug Monit.* **1987**, *9*, 358–365.

Chlortetracycline (using post-column reaction detection)

Bryan, P.D.; Hawkins, K.R.; Stewart, J.T.; Capomacchia, A.C. Analysis of chlortetracycline by high performance liquid chromatography with postcolumn alkaline-induced fluorescence detection. *Biomed.Chromatogr.* **1992**, *6*, 305–310.

Cholesterol

Chen, W.X.; Li, P.Y.; Wang, S.; Dong, J.; Li, J.Z. Serum cholesterol determined by liquid chromatography with 6-chlorostigmasterol as internal standard. *Clin.Chem.* **1993**, *39*, 1602–1607.

Matsuoka, C.; Nohta, H.; Kuroda, N.; Ohkura, Y. Simultaneous determination of cholestanol and cholesterol in human serum by high-performance liquid chromatography with fluorescence detection. *J.Chromatogr.* **1985**, *341*, 432–436.

Sugino, K.; Terao, J.; Murakami, H.; Matsushita, S. High-performance liquid chromatographic method for the quantification of cholesterol epoxides in spray-dried egg. *J.Agric.Food Chem.* **1985**, *34*, 36–39.

Chondroitin Sulfates (using post-column reaction detection)

Toyoda, H.; Yamanashi, S.; Hakamada, Y.; Shinomiya, K.; Imanari, T. Profile analysis of chondroitin sulfates in human urine and serum. *Chem.Pharm.Bull.* **1989**, *37*, 1627–1628.

Ciclopirox

Lehr, K.H.; Damm, P. Quantification of ciclopirox by high-performance liquid chromatography after pre-column derivatization. An example of efficient clean-up using silica-bonded cyano phases. *J.Chromatogr.* **1985**, *339*, 451–456.

Cilastatin (using post-column reaction detection)

Kamei, K.; Okazaki, A.; Okada, N.; Hamajima, K. Assay methods for cilastatin sodium in body fluids and tissues. *Chemotherapy (Tokyo)* **1985**, *33*, 282–289.

Cinchonine

Diaz, A.N.; Sanchez, F.G.; Gallardo, A.A.; Pareja, A.G. HPLC enantiomeric resolution of (+)-cinchonine and (-)-cinchonidine with diode-laser polarimetric detection. *Instrum.Sci.Technol.* **1996**, *24*, 47–56.

Ciprofloxacin

Katagiri, Y.; Naora, K.; Ichikawa, N.; Hayashibara, M.; Iwamoto, K. High-performance liquid chromatographic determination of ciprofloxacin in rat brain and cerebrospinal fluid. *Chem.Pharm.Bull.* **1990**, *38*, 2884–2886.

Ciprofloxacin (using post-column reaction detection)

Krol, G.J.; Beck, G.W.; Benham, T. HPLC analysis of ciprofloxacin and ciprofloxacin metabolites in body fluids. *J.Pharm.Biomed.Anal.* **1995**, *14*, 181–190.

Cispentacin

Zanol, M.; Hermann, R.; Bernareggi, A.; Borgonovi, M.; Taglietti, E.; Zerilli, L.F. HPLC method for the quantitation of cispentacin enantiomers in rat urine. *Boll.Chim.Farm.* **1995**, *134*, 390–393.

Cisplatin

Andrews, P.A.; Wung, W.E.; Howell, S.B. A high-performance liquid chromatographic assay with improved selectivity for cisplatin and active platinum (II) complexes in plasma ultrafiltrate. *Anal.Biochem.* **1984**, *143*, 46–56.

Krull, I.S.; Ding, X.D.; Braverman, S.; Selavka, C.; Hochberg, F.; Sternson, L.A. Trace analysis for cis-platinum anti-cancer drugs via LCEC. *J.Chromatogr.Sci.* **1983**, *21*, 166–173.

Citalopram

Rochat, B.; Amey, M.; Van Gelderen, H.; Testa, B.; Baumann, P. Determination of the enantiomers of citalopram, its demethylated and propionic acid metabolites in human plasma by chiral HPLC. *Chirality* **1995**, *7*, 389–395.

Citrulline

Carlberg, M. Assay of neuronal nitric oxide synthase by HPLC determination of citrulline. *J.Neurosci.Methods* **1994**, *52*, 165–167.

Clavulanic Acid

Low, A.S.; Taylor, R.B.; Gould, I.M. Determination of clavulanic acid by a sensitive HPLC method. *J.Antimicrob.Chemother.* **1989**, *24 Suppl B*, 83–86.

Clavulanic Acid (using post-column reaction detection)

Haginaka, J.; Wakai, J.; Yasuda, H. Liquid chromatographic assay of clavulanic acid using a hollow-fiber postcolumn reactor. *Chem.Pharm.Bull.* **1986**, *34*, 1850–1852.

Jehl, F.; Monteil, H.; Brogard, J.M. Direct determination of clavulanic acid in human biological fluids by HPLC. *Pathol.Biol.* **1987**, *35*, 702–706.

Nakagawa, T.; Shibukawa, A.; Tsuchiya, Y.; Tanaka, H.; Haginaka, J. High-performance liquid chromatographic assay and pharmacokinetic study of BRL 28500 (clavulanic acid-ticarcillin) in man. *Chemotherapy (Tokyo)* **1986**, *34*, 341–347.

Clenbuterol

Biondi, P.A.; Guidotti, L.; Montana, M.; Manca, F.; Brambilla, G.; Lucarelli, C. A derivatization procedure suitable for HPLC analysis for clenbuterol. *J.Chromatogr.Sci.* **1991**, *29*, 190–193.

Brambilla, G.F.; Agrimi, U.; Pierdominici, E. Clenbuterol residues in vitreous humor and urine of calves. *Ital.J.Food Sci.* **1991**, *3*, 303–306.

Clinofibrate

Nakazawa, H.; Kanamaru, Y.; Murano, A. Determination of optical isomers of clinofibrate by high-performance liquid chromatography. *Chem.Pharm.Bull.* **1979**, *27*, 1694–1696.

Clodronate (using post-column reaction detection)

Kosonen, J.P. Determination of disodium clodronate in bulk material and pharmaceuticals by ion chromatography with post-column derivatization. *J.Pharm.Biomed.Anal.* **1992**, *10*, 881–887.

Virtanen, V.; Lajunen, L.H.J. Determination of clodronate in aqueous solutions by HPLC using postcolumn derivatization. *Talanta* **1993**, *40*, 661–667.

Clomiphene (using post-column reaction detection)

Baustian, C.L.; Mikkelson, T.J. Analysis of clomiphene isomers in human plasma and detection of metabolites using reversed-phase chromatography and fluorescence detection. *J.Pharm.Biomed.Anal.* **1986**, *4*, 237–246.

Clorsulon

Schenck, F.J.; Barker, S.A.; Long, A.R. Matrix solid phase dispersion (MSPD) isolation and liquid chromatographic determination of clorsulon in milk. *J.Liq.Chromatogr.* **1991**, *14*, 2827–2834.

Cocaine (using post-column reaction detection)

Roy, I.M.; Jefferies, T.M.; Threadgill, M.D.; Dewar, G.H. Analysis of cocaine, benzoylecgonine, ecgonine methyl ester, ethylcocaine and norcocaine in human urine using HPLC with post-column ion-pair extraction and fluorescence detection. *J.Pharm.Biomed.Anal.* **1992**, *10*, 943–948.

Coenzyme A

Corkey, B.E.; Brandt, M.; Williams, R.J.; Williamson, J.R. Assay of short-chain acyl coenzyme A intermediates in tissue extracts by high-pressure liquid chromatography. *Anal.Biochem.* **1981**, *118*, 30–41.

Demoz, A.; Netteland, B.; Svardal, A.; Mansoor, M.A.; Berge, R.K. Separation and detection of tissue CoASH and long-chain acyl-CoA by reversed-phase high-performance liquid chromatography after precolumn derivatization with monobromobimane. *J.Chromatogr.* **1993**, *635*, 251–256.

Colchicine

Klein, A.E.; Davis, P.J. Determination of colchicine and colchiceine in microbial cultures by high-performance liquid chromatography. *Anal.Chem.* **1980**, *52*, 2432–2435.

Collagen

Accinni, R.; Belingheri, L.; Giglioni, A.; Micelli, G.; Quaranta, M.; Wei, J.; Lucarelli, C. Quantitation of collagen/total protein ratio in biological samples by isocratic high performance liquid chromatography 4-hydroxyproline assay. *G.Ital.Chim.Clin.* **1992**, *17*, 27–33.

Monbiosse, V.; Monboisse, J.C.; Borel, J.P.; Randoux, A. Nonisotopic evaluation of collagen in fibroblasts cultures. *Anal.Biochem.* **1989**, *176*, 395–399.

Reiser, K.M.; Last, J.A. Analysis of collagen composition and biosynthesis by HPLC. *LC, Liq. Chromatogr.HPLC Mag.* **1983**, *1*, 498–502.

Takahashi, S.; Zhao, M.; Eng, C. Isolation and characterization of insoluble collagen of dog hearts. *Protein Expression Purif.* **1991**, *2*, 304–312.

Complexes

Da, S.; Xu, W.; Han, H.; Wang, Z. Analysis of dicarboxylic acid-copper complexes by high performance liquid chromatography with UV detection. *Fenxi Huaxue* **1996**, *24*, 777–781.

Liu, Q.; Zhao, T.; Liu, J.; Cheng, J. Determination of platinum metals complexes with 2-(6-methyl-2-benzothiazolylazo)-5-diethylaminophenol by reversed-phase high performance liquid chromatography. *Mikrochim.Acta* **1996**, *122*, 27–33.

Corticosteroids (using post-column reaction detection)

Takeda, M.; Maeda, M.; Tsuji, A. Chemiluminescence high performance liquid chromatography of corticosteroids using lucigenin as post-column reagent. *Biomed.Chromatogr.* **1990**, *4*, 119–122.

Creatine (using post-column reaction detection)

Webb, L.E. Determination of creatine in serum by ion-pair high-performance liquid chromatography with fluorometric detection. *J.Chromatogr.* **1986**, *381*, 406–410.

Creatinine

Gennaro, M.C.; Abrigo, C. Simultaneous determination of creatinine, uric, L(+)-ascorbic and orotic acids in milk by reversed-phase ion-interaction HPLC chromatography. *Fresenius' J.Anal.Chem.* **1991**, *340*, 422–425.

CS-670

Takasaki, W.; Asami, M.; Muramatsu, S.; Hayashi, R.; Tanaka, Y.; Kawabata, K.; Hoshiyama, K. Stereoselective determination of the active metabolites of a new anti-inflammatory agent (CS-670) in human and rat plasma using antibody-mediated extraction and high-performance liquid chromatography. *J.Chromatogr.* **1993**, *613*, 67–77.

Cucurbitin

Duez, P.; Chamart, S.; Hanocq, M.; Sawadogo, M. Determination of cucurbitin in Cucurbita spp seeds by HPTLC-densitometry and HPLC. *J.Planar Chromatogr.—Mod.TLC* **1988**, *1*, 313–316.

Cyanide

Gasparrini, F.; Misiti, D.; Pierini, M.; Meloni, D. Microdetermination of cyanides by reverse phase liquid chromatography. *Chim.Oggi* **1986**, 70–71.

Cyanide (using post-column reaction detection)

Gamoh, K.; Imamichi, S. Postcolumn liquid chromatographic method for the determination of cyanide with fluorometric detection. *Anal.Chim.Acta* **1991**, *251*, 255–259.

Cyanuric Acid

Jessee, J.A.; Valerias, C.; Benoit, R.E.; Hendricks, A.C.; McNair, H.M. Determination of cyanuric acid by high-performance liquid chromatography. *J.Chromatogr.* **1981**, *207*, 454–456.

Cyclodextrins (using post-column reaction detection)

Frijlink, H.W.; Visser, J.; Hefting, N.R.; Oosting, R.; Meijer, D.K.; Lerk, C.F. The pharmacokinetics of beta-cyclodextrin and hydroxypropyl-beta-cyclodextrin in the rat. *Pharm.Res.* **1990**, *7*, 1248–1252.
Kubota, Y.; Fukuda, M.; Ohtsuji, K.; Koizumi, K. Microanalysis of β-cyclodextrin and glucosyl-β-cyclodextrin in human plasma by high-performance liquid chromatography with pulsed amperometric detection. *Anal.Biochem.* **1992**, *201*, 99–102.

Cyclopentenones

Miller, L.; Weyker, C. Effects of compound structure and temperature on the resolution of enantiomers of cyclopentenones by liquid chromatography on derivatized cellulose chiral stationary phases. *J.Chromatogr.* **1993**, *653*, 219–228.

Cyclophosphamide

Reid, J.M.; Stobaugh, J.F.; Sternson, L.A. Liquid chromatographic determination of cyclophosphamide enantiomers in plasma by precolumn chiral derivatization. *Anal.Chem.* **1989**, *61*, 441–446.

Cyclosporine

Fu, I.; Bowers, L.D. Micro-quantification of cyclosporine and its metabolites and determination of their spectral absorptivities. *Clin.Chem.* **1991**, *37*, 1185–1190.

Cypermethrin

Chen, A.W.; Fink, J.M.; Letinski, D.J. Analytical methods to determine residual cypermethrin and its major acid metabolites in bovine milk and tissues. *J.Agric.Food Chem.* **1996**, *44*, 3534–3539.

Cysteine

Cooper, J.D.H.; Turnell, D.C. Cysteine and cystine: high-performance liquid chromatography of o-phthalaldehyde derivatives. *Methods Enzymol.* **1987**, *143*, 141–143.

Mansoor, M.A.; Svardal, A.M.; Ueland, P.M. Determination of the in vivo redox status of cysteine, cysteinylglycine, homocysteine, and glutathione in human plasma. *Anal.Biochem.* **1992**, *200*, 218–229.

Matsui, S.; Kitabatake, K.; Takahashi, H.; Meguro, H. Fluorometric determination of cysteine in beer by high-performance liquid chromatography with precolumn derivatization. *J.Inst.Brew.* **1984**, *90*, 20–23.

Podhradsky, D.; Mihalovova, H.; Toth, G. Determination of cysteine, glutathione, coenzyme A and homocysteine by high-performance liquid chromatography. *Biologia (Bratislava)* **1983**, *38*, 787–793.

Saetre, R.; Rabenstein, D.L. Determination of cysteine in plasma and urine and homocysteine in plasma by high-pressure liquid chromatography. *Anal.Biochem.* **1978**, *90*, 684–692.

Sypniewski, S.; Bald, E. Ion-pair high-performance liquid chromatography of cysteine and metabolically related compounds in the form of their S-pyridinium derivatives. *J.Chromatogr.A* **1994**, *676*, 321–330.

Ziegler, S.J.; Sticher, O. Electrochemical, fluorescence, and UV detection for HPLC analysis of various cysteine derivatives. *HRC CC, J.High Resolut.Chromatogr.Chromatogr.Commun.* **1988**, *11*, 639–646.

Cystine

Cooper, J.D.H.; Turnell, D.C. Fluorescence detection of cystine by o-phthalaldehyde derivatization and its separation using high-performance liquid chromatography. *J.Chromatogr.* **1982**, *227*, 158–161.

Cytokinins

Sonoki, S.; Sono, Y.; Hisamatsu, S.; Sugiyama, T. High-performance liquid chromatographic analysis of cytokinins using isatoic anhydride as a fluorescent derivatizing reagent. *J.Liq.Chromatogr.* **1993**, *16*, 343–352.

Danthron (using post-column reaction detection)

Miller, B.E.; Danielson, N.D. Fluorometric determination of danthron in pharmaceutical tablets and in urine. *Anal.Chim.Acta* **1987**, *192*, 293–299.

Debrisoquine

Gerova, Z.; Vlahov, V. Debrisoquine hydroxylation phenotype: liquid chromatographic determination of debrisoquine and its 4-hydroxy metabolite in human urine. *Anal.Lab.* **1994**, *3*, 162–167.

Goubier, C.; Girard, I.; Ferry, S. High-performance liquid chromatographic assay of debrisoquine and its 4-hydroxy metabolite in human urine. *Chromatographia* **1991**, *32*, 523–526.

Lee, E.J.D.; Ang, S.B. Measurement of debrisoquine and 4-hydroxydebrisoquine in urine by liquid chromatography. *J.Pharm.Biomed.Anal.* **1987**, *5*, 435–439.

Dehydrocholesterol

Oe, T.; Mizuguchi, T.; Shimada, K. Cookson-type reagents: application to the determination of 7-dehydrocholesterol in human skin surface. *Anal.Sci.* **1991**, *7*, 171–172.

Iwata, T.; Hanazono, H.; Yamaguchi, M.; Nakamura, M.; Ohkura, Y. Determination of 7-dehydrocholesterol in human skin surface lipids by high-performance liquid chromatography with fluorescence detection. *Anal.Sci.* **1989**, *5*, 671–673.

Iwata, T.; Hanazono, H.; Yamaguchi, M.; Nakamura, M.; Ohkura, Y. Ultramicro determination of 7-dehydrocholesterol in rat skin by high-performance liquid chromatography with fluorescence detection. *J.Chromatogr.* **1989**, *491*, 404–409.

Delmopinol

Egginger, G.; Blaschke, E.; Lindner, W.; Olsson, A.-M. Stereoselective high-performance liquid chromatographic assay of (±)-delmopinol in plasma using solid-phase extraction, a chiral derivatizing agent and electrochemical detection. *J.Chromatogr.A* **1994**, *666*, 275–282.

Denopamine

Nishi, H.; Fujimura, N.; Yamaguchi, H.; Jyomori, W.; Fukuyama, T. Reversed-phase HPLC separation of the enantiomers of denopamine after derivatization with GITC chiral reagent. *Chromatographia* **1990**, *30*, 186–190.

3-Deoxyglucusone

Fujii, E.; Iwase, H.; Ishii-Karakasa, I.; Yajima, Y.; Hotta, K. Quantitation of the glycation intermediate 3-deoxyglucosone by oxidation with rabbit liver oxoaldehyde dehydrogenase to 2-keto-3-deoxygluconic acid followed by high-performance liquid chromatography. *J.Chromatogr.B* **1994**, *660*, 265–270.

Dermatan Sulfate (using post-column reaction detection)

Akiyama, H.; Shidawara, S.; Mada, A.; Toyoda, H.; Toida, T.; Imanari, T. Chemiluminescence high-performance liquid chromatography for the determination of hyaluronic acid, chondroitin sulfate and dermatan sulfate. *J.Chromatogr.* **1992**, *579*, 203–207.

Qiu, G.; Tanikawa, M.; Akiyama, H.; Toida, T.; Koshiishi, I.; Imanari, T. Separation and characterization of dermatan sulfate in normal human urine. *Biol.Pharm.Bull.* **1993**, *16*, 340–342.

Dextrins (using post-column reaction detection)

Ortega, F. Selective determination of dextrins by liquid chromatography with postcolumn enzymic reaction, using co-immobilized enzymes. *Anal.Chim.Acta* **1992**, *257*, 79–87.

Dibekacin (using post-column reaction detection)

Baillarie, D.; Chavez, J.; Arancibia, A.; Bravo, M. Method for dibekacin determination in plasma and urine using high performance liquid chromathography. *An.R.Acad.Farm.* **1987**, *53*, 596–601.

Diethylstilbestrol

Kenyhercz, T.M.; Kissinger, P.T. Determination of diethylstilbestrol (DES) in animal tissue via liquid chromatography with electrochemical detection. *J.Anal.Toxicol.* **1978**, *2*, 1–2.

Lea, A.R.; Kayaba, W.J.; Hailey, D.M. Analysis of diethylstilbestrol and its impurities in tablets using reversed-phase high-performance liquid chromatography. *J.Chromatogr.* **1979**, *177*, 61–68.

Desmosine

Charpiot, P.; Calaf, R.; Chareyre, C.; Rolland, P.H.; Garcon, D. Rapid determination of desmosine and isodesmosine in tissue hydrolyzates by isocratic high performance liquid chromatography and precolumn derivatization. *Amino Acids* **1994**, *6*, 57–63.

2, 4-Diaminobutyric Acid

Foster, J.G. High performance liquid chromatographic analysis of 2,4-diaminobutyric acid in flatpea extracts. *J.Liq.Chromatogr.* **1989**, *12*, 3033–3050.

Foster, J.G. Optimal extraction procedure for high-performance liquid chromatographic determination of α,gamma-diaminobutyric acid in flatpea, Lathyrus sylvestris L. *J.Agric.Food Chem.* **1990**, *38*, 748–752.

Diaminopimelic Acid

Dugan, M.E.R.; Sauer, W.C.; Lien, K.A.; Fenton, T.W. Ion-pair high-performance liquid chromatography of diaminopimelic acid in hydrolyzates of physiological samples. *J.Chromatogr.* **1992**, *582*, 242–245.

Webster, P.M.; Hoover, W.H.; Miller, T.K. Determination of 2,6-diaminopimelic acid in biological materials using high-performance liquid chromatography. *Anim.Feed Sci.Technol.* **1990**, *30*, 11–20.

Dianhydrodisuccinylgalactitol

Szokan, G.; Elekes, I.; Taborhegyi, E.; Csanadi, G.; Bencze, J. RP-HPLC assay for 1,2–5,6-dianhydro-3,4-disuccinylgalactitol and its metabolites in blood plasma and liver. *Chromatographia* **1987**, *24*, 839–841.

Dideoxyinosine

Nagaoka, H.; Nohta, H.; Saito, M.; Ohkura, Y. Determination of 2',3'-dideoxyinosine and 2',3'-dideoxyadenosine in rat plasma by high-performance liquid chromatography with precolumn fluorescence derivatization. *Chem.Pharm.Bull.* **1992**, *40*, 2202–2204.

Difluoromethylarginine

Hunter, K.J.; Fairlamb, A.H. Separation and quantitation of the polyamine biosynthesis inhibitor D,L-α-difluoromethylarginine and other guanidine-containing compounds by high-performance liquid chromatography. *Anal.Biochem.* **1990**, *190*, 281–285.

Difluoromethylornithine

Reavis, F.J.; Contario, J.J. High performance liquid chromatographic assay for alpha-(difluoromethyl)ornithine in rodent feed. *J.Liq.Chromatogr.* **1984**, *7*, 2179–2191.

Digitalis Glycosides (using post-column reaction detection)

Gfeller, J.C.; Frey, G.; Frei, R.W. Post-column derivatization in high-performance liquid chromatography using the air segmentation principle: application to digitalis glycosides. *J.Chromatogr.* **1977**, *142*, 271–281.

Digoxin (using post-column reaction detection)

Desta, B. Separation of digoxin from dihydrodigoxin and the other metabolites by high-performance liquid chromatography with post-column derivatization. *J.Chromatogr.* **1987**, *421*, 381–386.

Embree, L.; McErlane, K.M. Comparison of digoxin analysis by high-performance liquid chromatography/postcolumn derivatization and fluorescence polarization immunoassay. *Xenobiotica* **1990**, *20*, 635–643.

Dihydrostreptomycin (using post-column reaction detection)

Hormazabal, V.; Yndestad, M. Determination of dihydrostreptomycin sulfate in milk by HPLC using ion-pair and postcolumn derivatization. *J.Liq.Chromatogr.* **1995**, *18*, 2695–2702.

5,6-Dihydroxy-2-methylaminotetralin

Rondelli, I.; Corsaletti, R.; Redenti, E.; Acerbi, D.; Delcanale, M.; Amari, G.; Ventura, P. New method for the resolution of the enantiomers of 5,6-dihydroxy-2-methyl-aminotetralin by selective derivatization and HPLC analysis: application to biological fluids. *Chirality* **1996**, *8*, 381–389.

Dihydroxyproline

Lindblad, W.J.; Diegelmann, R.F. Chromatographic separation of preparative quantities of the stereoisomers of trans-2,3-cis-3,4-dihydroxy-L-proline. *J.Chromatogr.* **1984**, *315*, 447–450.

β-Diketones (using post-column reaction detection)

Moriyasu, M.; Kato, A.; Hashimoto, Y. Kinetic studies of fast equilibrium by means of high-performance liquid chromatography. XVI. Separation of tautomers of β-diketones and β-diketoamides. *J.Chromatogr.* **1987**, *411*, 466–471.

Dimer Acid

Veazey, R.L. Rapid analysis of dimer acid by HPLC/FID. *JAOCS, J.Am.Oil Chem.Soc.* **1986**, *63*, 1043–1046.

3,5-Dimethyl-4,6-diphenyltetrahydro-2H-1,3,5-thiadiazine-2-thione (using post-column reaction detection)

Nakajima, A.; Ushijima, T.; Migita, J. Fluorimetric determination of 3,5-dimethyl-4,6-diphenyltetrahydro-2H-1,3,5-thiadiazine-2-thione in rat serum by high performance liquid chromatography with post column derivatization. *Bunseki Kagaku* **1986**, *35*, 384–388.

5,5-Dimethyl-2,4-oxazolidinone

Gazdzik, W.R.; Filipek, M.; Kmiotek, W. Rapid and simple HPLC determination of 5,5-dimethyl-2,4-oxazolidinedione in rabbit serum. *Biomed.Chromatogr.* **1987**, *2*, 137–138.

β-(Dimethylsulfonio)propionate (using post-column reaction detection)

Howard, A.G.; Russell, D.W. HPLC-FPD Instrumentation for the measurement of the atmospheric dimethyl sulfide precursor β-(dimethylsulfonio)propionate. *Anal.Chem.* **1995**, *67*, 1293–1295.

Diols

Boos, K.S.; Wilmers, B.; Schlimme, E.; Sauerbrey, R. On-line sample processing and analysis of diol compounds in biological fluids. *J.Chromatogr.* **1988**, *456*, 93–104.

Dipentaerythritol

Schirra, R. Determination of dipentaerythritol in commercial pentaerythritol by means of HPLC. A closer look at the stoichiometry of. *Int.Annu.Conf.ICT* **1996**, *27th*, 117.1–117.12.

2,3-Diphosphoglycerate (using post-column reaction detection)

Takebayashi, Y.; Mitsuma, R.; Imanari, T. Determination of 2,3-diphosphoglycerate in red blood cells by high-performance liquid chromatography. *Anal.Sci.* **1987**, *3*, 569–572.

Disulfides

Meredith, M.J. Analysis of protein-glutathione mixed disulfides by high performance liquid chromatography. *Anal.Biochem.* **1983**, *131*, 504–509.

Srivastava, S.K.; Piper, J.T.; Singhal, S.S.; Awasthi, S. A new method for specific determination of glutathione-protein mixed disulfides in ocular lens. *Biochem.Arch.* **1995**, *11*, 229–236.

Ditallowdimethylammonium Chloride (using post-column reaction detection)

Gort, S.M.; Hogendoorn, E.A.; Baumann, R.A.; van Zoonen, P. Rapid screening method for ditallowdimethylammonium chloride at the low ppb level in surface water using solid phase extraction and normal-phase liquid chromatography with online post-column ion-pair extraction and fluorescence detection. *Int.J.Environ.Anal.Chem.* **1993**, *53*, 289–296.

Dithianon

Baker, P.G.; Clarke, P.G. Determination of residues of dithianon in apples by high-performance liquid chromatography. *Analyst* **1984**, *109*, 81–83.

Dithiocarbamates

Schwedt, G.; Schneider, P. Pre- and post-chromatographic derivatization by water soluble dithiocarbamates for HPLC determination of metal ions. *Fresenius' Z.Anal.Chem.* **1986**, *325*, 116–120.

DNA

Sonoki, S.; Hisamatsu, S.; Kiuchi, A. High-performance liquid chromatographic determination of DNA base composition with fluorescence detection. *Nucleic Acids Res.* **1993**, *21*, 2776.

Dolichol

Yasugi, E.; Oshima, M. Sequential microanalyses of free dolichol, dolichyl fatty acid ester and dolichyl phosphate levels in human serum. *Biochim.Biophys.Acta* **1994**, *1211*, 107–113.

Dolichyl Phosphates

Yamada, K.; Abe, S.; Suzuki, T.; Katayama, K.; Sato, T. A high-performance liquid chromatographic method for the determination of dolichyl phosphates in tissues. *Anal.Biochem.* **1986**, *156*, 380–385.

Domoic Acid

Lawrence, J.F.; Charbonneau, C.F.; Page, B.D.; Lacroix, G.M.A. Confirmation of domoic acid in molluskan shellfish by chemical derivatization and reversed-phase liquid chromatography. *J.Chromatogr.* **1989**, *462*, 419–425.

Doxorubicin

Configliacchi, E.; Razzano, G.; Rizzo, V.; Vigevani, A. HPLC methods for the determination of bound and free doxorubicin, and of bound and free galactosamine, in methacrylamide polymer-drug conjugates. *J.Pharm.Biomed.Anal.* **1996**, *15*, 123–129.

Drugs

Adams, R.F.; Schmidt, G.J.; Vandemark, F.L. A micro liquid column chromatography procedure for twelve anticonvulsants and some of their metabolites. *J.Chromatogr.* **1978**, *145*, 275–284.

Asukabe, H.; Murata, H.; Harada, K.; Suzuki, M.; Oka, H.; Ikai, Y. Improvement of chemical analysis of antibiotics. 21. Simultaneous determination of three polyether antibiotics in feeds using high-performance liquid chromatography with fluorescence detection. *J.Agric.Food Chem.* **1994**, *42*, 112–117.

Baeck, S.E.; Nilsson-Ehle, I.; Nilsson-Ehle, P. Chemical assay, involving liquid chromatography, for aminoglycoside antibiotics in serum. *Clin.Chem.* **1979**, *25*, 1222–1225.

Bopp, R.J.; Kennedy, J.H. Practical considerations for chiral separations of pharmaceutical compounds. *LC-GC* **1988**, *6*, 514–516.

Carson, M.C.; Chu, P-S.; Von Bredow, J. Toward a regulatory method: comparison and validation of multiresidue procedures for determination of β-lactam antibiotics in milk. *ACS Symp.Ser.* **1996**, *636*, 108–120.

Decolin, D.; Gavriloff, C.; Archimbault, P.; Leroy, P.; Nicolas, A. Analysis of antibiotic residues on edible tissues by HPLC-fluorimetry. *Ann.Falsif.Expert.Chim.Toxicol.* **1996**, *89*, 11–23.

Endo, M.; Imamichi, H.; Moriyasu, M.; Hashimoto, Y. Microdetermination of stimulant drugs in urine by high-performance liquid chromatography. *J.Chromatogr.* **1980**, *196*, 334–336.

Falk, A.J. Thyroid compounds and x-ray contrast agents. *Chromatogr.Sci.* **1979**, *9*, 1129–1148.

Fukushima, T.; Santa, T.; Homma, H.; Imai, K. Enantiomeric separation of drugs on chiral stationary phases in HPLC. Effect of mobile phase composition on the enantiomeric separation and the detection. *Kuromatogurafi* **1996**, *17*, 152–153.

Gal, J.; Desai, D.M.; Meyer-Lehnert, S. Reversed-phase LC resolutions of chiral antiarrhythmic agents via derivatization with homochiral isothiocyanates. *Chirality* **1990**, *2*, 43–51.

Gal, J.; Brown, T.R. Liquid chromatographic separation of enantiomers of adrenergic agonists. *J.Pharmacol.Methods* **1986**, *16*, 261–269.

Gambardella, P.; Punziano, R.; Gionti, M.; Guadalupi, C.; Mancini, G.; Mangia, A. Quantitative determination and separation of analogues of aminoglycoside antibiotics by high-performance liquid chromatography. *J.Chromatogr.* **1985**, *348*, 229–240.

Gfeller, J.C.; Haas, R.; Troendl, J.M.; Erni, F. Practical aspects of speed in high-performance liquid chromatography for the analysis of pharmaceutical preparations. *J.Chromatogr.* **1984**, *294*, 247–259.

Gullo, V.P.; Goegelman, R.T.; Putter, I.; Lam, Y-K. High-performance liquid chromatographic analysis of derivatized hypocholesteremic agents from fermentation broths. *J.Chromatogr.* **1981**, *212*, 234–238.

Leroy, P.; Bellucci, L.; Nicolas, A. Chiral derivatization for separation of racemic amino and thiol drugs by liquid chromatography and capillary electrophoresis. *Chirality* **1995**, *7*, 235–242.

Leroy, P.; Nicolas, A.; Moreau, A. Electrochemical detection of sympathomimetic drugs, following precolumn o-phthalaldehyde derivatization and reversed-phase high-performance liquid chromatography. *J.Chromatogr.* **1983**, *282*, 561–568.

Lillsunde, P.; Michelson, L.; Forsstrom, T.; Korte, T.; Schultz, E.; Ariniemi, K.; Portman, M.; Sihvonen, M.L.; Seppala, T. Comprehensive drug screening in blood for detecting abused drugs or drugs potentially hazardous for traffic safety. *Forensic Sci.Int.* **1996**, *77*, 191–210.

Long, A.R.; Hsieh, L.C.; Malbrough, M.S.; Short, C.R.; Barker, S.A. Multiresidue method for isolation and liquid chromatographic determination of seven benzimidazole anthelmintics in milk. *J.Assoc.Off.Anal.Chem.* **1989**, *72*, 739–741.

Long, A.R.; Hsieh, L.C.; Malbrough, M.S.; Short, C.R.; Barker, S.A. Matrix solid phase dispersion (MSPD) extraction and liquid chromatographic determination of five benzimidazole anthelmintics in pork muscle tissue. *J.Food Compos.Anal.* **1990**, *3*, 20–26.

Matsunaga, H.; Fujimoto, T.; Tawa, R.; Hirose, S. An on-line clean-up procedure for large sample volume analysis of serum aminoglycoside antibiotics by reversed-phase high-performance liquid chromatography. *Chem.Pharm.Bull.* **1988**, *36*, 1565–1570.

Merriken, R.A. Tissue analysis for thiazide diuretics. *Aviat., Space Environ.Med.* **1980**, *51*, 996–998.

Olsen, L.; Bronnum-Hansen, K.; Helboe, P.; Jorgensen, G.H.; Kryger, S. Chiral separations of β-blocking drug substances using derivatization with chiral reagents and normal-phase high-performance liquid chromatography. *J.Chromatogr.* **1993**, *636*, 231–241.

Rogers, M.E.; Adlard, M.W.; Saunders, G.; Holt, G. High-performance liquid-chromatographic determination of β-lactam antibiotics using precolumn and postcolumn derivatization techniques. *Biochem.Soc.Trans.* **1984**, *12*, 640–641.

Rogers, M.E.; Adlard, M.W.; Saunders, G.; Holt, G. Derivatization techniques for high-performance liquid chromatographic analysis of β-lactams. *J.Chromatogr.* **1984**, *297*, 385–391.

Schmidt, G.J.; Slavin, W. The evaluation of coupled-column liquid chromatography for determining aminoglycoside antibiotics. *Chromatogr.Newsl.* **1981**, 9 21–4,

Tang, Y. Significance of mobile phase composition in enantioseparation of chiral drugs by HPLC on a cellulose-based chiral stationary phase. *Chirality* **1996**, *8*, 136–142.

Thibonnier, M.; Chehade, N.; Hinko, A. A V1-vascular vasopressin antagonist suitable for radioiodination and photoaffinity labeling. *Am.J.Hypertens.* **1990**, *3*, 471–475.

Venn, R.F.; Barnard, G.; Macrae, P.V.; Saunders, K.C. Analytical strategies for a new peptide drug. *Methodol.Surv.Bioanal.Drugs* **1996**, *24*, 20–34.

Ward, K.D.; Manes, L.V. Separation of the four optical isomers of a dihydropyridine calcium channel antagonist. *J.Chromatogr.* **1989**, *478*, 169–179.

Drugs (using post-column reaction detection)

Fejglova, Z.; Dolezal, J.; Hrdlicka, A.; Frgalova, K. Microbore HPLC determination of polyether antibiotics using postcolumn derivatization with benzaldehyde reagents. *J.Liq.Chromatogr.* **1994**, *17*, 359–372.

Haginaka, J.; Wakai, J.; Yasuda, H. Liquid chromatographic assay of beta-lactamase inhibitors in human serum and urine using a hollow-fiber postcolumn reactor. *Anal.Chem.* **1987**, *59*, 324–327.

Johannsen, F.H. Analysis of feed additives. Part 2. Determination of ionophor antibiotics in feeds and foods by HPLC with post-column derivatization. *Agribiol.Res.* **1991**, *44*, 79–89.

Kim, M.; Stewart, J.T. HPLC post-column ion-pair extraction of acidic drugs using a new fluorescent ion-pair reagent. *Anal.Sci.Technol.* **1991**, *4*, 273–284.

Nicolas, A.; Leroy, P.; Decolin, D.; Siest, G. Improvements in the HPLC measurement of drug and metabolite levels in biological fluids. *Methodol.Surv.Biochem.Anal.* **1990**, *20*, 271–278.

Ortega, F.; Cuevas, J.L.; Centenera, J.I.; Dominguez, E. Liquid chromatographic separation of phenolic drugs using catalytic detection: Comparison of an enzyme reactor and enzyme electrode. *J.Pharm.Biomed.Anal.* **1992**, *10*, 789–796.

Rogers, M.E.; Adlard, M.W.; Saunders, G.; Holt, G. High-performance liquid chromatographic determination of β-lactam antibiotics, using fluorescence detection following post-column derivatization. *J.Chromatogr.* **1983**, *257*, 91–100.

Shah, S.; Taylor, L.T. Application of on-line reversed-phase HPLC with Fourier-transform infrared detection for analysis of analgesics. *LC-GC* **1989**, *7*, 340–346.

Dyes (using post-column reaction detection)

Adams, C.D.; Fusco, W.; Kanzelmeyer, T. Ozone, hydrogen peroxide/ozone and UV/ozone treatment of chromium- and copper-complex dyes: decolorization and metal release. *Ozone: Sci.Eng.* **1995**, *17*, 149–162.

Andrisano, V.; Gotti, R.; DiPietra, A.M.; Cavrini, V. Analysis of basic hair dyes by HPLC with online post-column photochemical derivatization. *Chromatographia* **1994**, *39*, 138–145.

Kimoto, K.; Gohda, R.; Murayama, K.; Santa, F.; Fukushima, T.; Homma, H.; Imai, K. Sensitive detection of near-infrared fluorescent dyes using high-performance liquid chromatography with peroxyoxalate chemiluminescence detection system. *Biomed.Chromatogr.* **1996**, *10*, 189–190.

Eicosanoids

Do, U.H.; Ahern, D.G.; Iles, J.; Maniscalco, M.; Tutunjian, M. Specific radioactivity determination of labeled eicosanoids. *J.Chromatogr.* **1989**, *489*, 359–363.

Yamaki, K.; Ohishi, S. Comparison of eicosanoids production between rat polymorphonuclear leukocytes and macrophages: detection by high-performance liquid chromatography with precolumn fluorescence labeling. *Jpn.J.Pharmacol.* **1992**, *58*, 299–307.

Eicosapentaenoic Acid

Kuroda, N.; Taguchi, Y.; Nakashima, K.; Akiyama, S. Fluorometric determination of eicosapentaenoic and docosahexaenoic acids by high-performance liquid chromatography. *Anal.Sci.* **1995**, *11*, 989–993.

Emulsifiers

Schindlbauer, H.; Scheuer, F. Determination of basic adhesive additives and cationic emulsifiers used in road building. *Fresenius' Z.Anal.Chem.* **1985**, *321*, 163–165.

Endralazine

Reece, P.A.; Cozamanis, I.; Zacest, R. Sensitive high-performance liquid chromatographic assay for endralazine and two of its metabolites in human plasma. *J.Chromatogr.* **1981**, *225*, 151–160.

Enkephalins

Hendrickson, T.L.; Wilson, G.S. Improved clean-up method for the enkephalins in plasma using immunoaffinity chromatography. *J.Chromatogr.B* **1994**, *653*, 147–154.

Enzyme Assays

Allenmark, S.; Ali Qureshi, G. Studies on dopamine-converting enzymes in human plasma. *J.Chromatogr.* **1981**, *223*, 188–192.

Bailey, S.W.; Ayling, J.E. An assay for picomole levels of tyrosine and related phenols and its application to the measurement of phenylalanine hydroxylase activity. *Anal.Biochem.* **1980**, *107*, 156–164.

Campa, J.S.; Cambrey, A.D.; McAnulty, R.J.; Laurent, G.J. Measurement of fibroblast collagen synthesis and degradation by reverse-phase high-pressure liquid chromatography. *Biochem.Soc.Trans.* **1989**, *17*, 1127–1128.

Cook, N.D.; Peters, T.J. A sensitive high-performance liquid chromatography assay for gamma-glutamyl hydrolase. *Biochem.Soc.Trans.* **1985**, *13*, 1226–1227.

Davis, A.T. Assay for lysine-ketoglutarate reductase by reversed-phase high-performance liquid chromatography. *J.Chromatogr.* **1989**, *497*, 263–267.

Fink, R.M.; Elstner, E.F. Comparison of different methods for the determination of phenylalanine hydroxylase activity in rat liver and Euglena gracilis. *Z.Naturforsch., C: Biosci.* **1984**, *39C*, 728–733.

Heintze, A.; Schultz, G. High-performance liquid chromatographic determination of 3-hydroxy-3-methylglutaryl coenzyme A as diphenacyl ester of 3-hydroxy-3-methylglutarate. *J.Chromatogr.B* **1994**, *655*, 117–120.

Holdiness, M.R. Determination of glutamic acid decarboxylase activity in subregions of rat brain by high-pressure liquid chromatography. *J.Liq.Chromatogr.* **1982**, *5*, 479–487.

Kaartinen, V.; Mononen, I. Assay of aspartylglycosylaminase by high-performance liquid chromatography. *Anal.Biochem.* **1990**, *190*, 98–101.

Kunitoh, S.; Tanaka, T.; Imaoka, S.; Funae, Y.; Monna, Y. Contribution of cytochrome P450s to MEOS (microsomal ethanol-oxidizing system): A specific and sensitive assay of MEOS activity by HPLC with fluorescence labeling. *Alcohol Alcohol.* **1993**, *28*, 63–68.

Lattmann, R.; Ghisalba, O.; Gygax, D.; Schaer, H.P.; Schmidt, E. Screening and application of microbial esterases for the enantioselective synthesis of chiral glycerol derivatives. *Biocatalysis* **1990**, *3*, 137–144.

Luque de Castro, M.D.; Fernandez-Romero, J.M. Total and individual determination of creatine kinase isoenzyme activities by flow injection and liquid chromatography. *Anal.Chim.Acta* **1992**, *263*, 43–52.

Lutz, M.P.; Pinon, D.I.; Miller, L.J. A nonradioactive fluorescent gel-shift assay for the analysis of protein phosphatase and kinase activities toward protein-specific peptide substrates. *Anal.Biochem.* **1994**, *220*, 268–274.

Martin, F.; Suzuki, A.; Hirel, B. A new high-performance liquid chromatography assay for glutamine synthetase and glutamate synthase in plant tissues. *Anal.Biochem.* **1982**, *125*, 24–29.

Mita, H.; Yasueda, H.; Hayakawa, T.; Shida, T. Quantitation of platelet-activating factor by high-performance liquid chromatography with fluorescent detection. *Anal.Biochem.* **1989**, *180*, 131–135.

Nakano, M.; Inaba, T. A radiometric assay for debrisoquine 4-hydroxylase in human liver microsomes. *Can.J.Physiol.Pharmacol.* **1984**, *62*, 84–88.

Neels, H.M.; Van Sande, M.E.; Scharpe, S.L. Sensitive colorimetric assay for angiotensin converting enzyme in serum. *Clin.Chem.* **1983**, *29*, 1399–1403.

O'Neill, F.H.; Christov, L.P.; Botes, P.J.; Prior, B.A. Rapid and simple assay for feruloyl and p-coumaroyl esterases. *World J.Microbiol.Biotechnol.* **1996**, *12*, 239–242.

O'Neill, R.A.; Darvill, A.; Albersheim, P. A fluorescence assay for enzymes that cleave glycosidic linkages to produce reducing sugars. *Anal.Biochem.* **1989**, *177*, 11–15.

Palmerini, C.A.; Datti, A.; Alunni, S.; VanderElst, I.E.; Orlacchio, A. A fluorescent assay for the determination of UDP-GlcNAc: Galβ1,3GalNAc-R (GlcNAc to GalNAc) β1,6 N-acetylglucosaminyltransferase activity. *Anal.Biochem.* **1995**, *225*, 315–320.

Pfeiffer, H.F.; Waldhoff, H.; Worsfold, P.J.; Whiteside, I.R.C. Automated flow-injection procedures for the determination of hydrolytic enzymes in bioreactor preparations. *Chromatographia* **1992**, *33*, 49–52.

Safarik, I. Detection of proteolytic enzymes in fractions after liquid chromatography. *J.Chromatogr.* **1989**, *463*, 212–215.

Saris, N.E.L.; Somerharju, P. Fluorimetric assay of phospholipase A acting on biomembrane phospholipids. *Acta Chem.Scand.* **1989**, *43*, 82–85.

Tsuruta, Y.; Ishida, S.; Kohashi, K.; Ohkura, Y. Fluorimetric assay of histamine N-methyltransferase by high-performance liquid chromatography. *Chem.Pharm.Bull.* **1981**, *29*, 3398–3400.

van Dijk, J.; Boomsma, F.; Alberts, G.; Man in 't Veld, A.J.; Schalekamp, M.A. Determination of semicarbazide-sensitive amine oxidase activity in human plasma by high-performance liquid chromatography with fluorimetric detection. *J.Chromatogr.B* **1995**, *663*, 43–50.

Vanderlaan, M.; Lotti, R.; Siek, G.; King, D.; Goldstein, M. Perfusion immunoassay for acetylcholinesterase: analyte detection based on intrinsic activity. *J.Chromatogr.A* **1995**, *711*, 23–31.

Wang, Y.; Rotundo, R.F.; Nimec, Z.; Ryan, T.J.; Galivan, J. Two novel HPLC methods which rapidly detect the substrates and cleavage products of gamma-glutamyl hydrolase. *Adv.Exp.Med.Biol.* **1993**, *338*, 655–658.

Yamamoto, K.; Kadowaki, S.; Takegawa, K.; Fan, J-Q.; Ashida, H.; Kumagai, H.; Tochikura, T. Production and application of microbial endoglycosidases acting on oligosaccharides of glycoconjugates. *Microb.Util.Renewable Resour.* **1995**, *9*, 420–429.

Yokohama, H.; Ohtsuka, I.; Shiojiri, H.; Katayama, K.; Ishikawa, S. A high-performance liquid chromatographic method for the assay of cholesterol 7α-hydroxylase activity. *Anal.Biochem.* **1986**, *157*, 186–190.

Yu, P.H.; Bailey, B.A. New sensitive high-performance liquid chromatographic method for p-tyrosine aminotransferase assay. *J.Chromatogr.* **1986**, *362*, 55–59.

Enzyme Assays (using post-column reaction detection)

Brotherton, J.E.; Widholm, J.M. High-performance liquid chromatography of anthranilate synthase using gel filtration and a post-column reactor. *J.Chromatogr.* **1985**, *350*, 332–335.

Edani, M.; Imai, H. Automated enzyme analysis by high-performance liquid chromatography postcolumn substrate reaction: application to trypsin and chymotrypsin. *Anal.Sci.* **1993**, *9*, 15–17.

Fulton, J.A.; Schlabach, T.D.; Kerl, J.E.; Toren, E.C., Jr.; Miller, A.R. Dual-detector-post-column reactor system for the detection of isoenzymes separated by high-performance liquid chromatography. I. Description and theory. *J.Chromatogr.* **1979**, *175*, 269–281.

Gonchoroff, D.G.; Branum, E.L.; Cedel, S.L.; Riggs, B.L.; O'Brien, J.F. Clinical evaluation of high-performance affinity chromatography for the separation of bone and liver alkaline phosphatase isoenzymes. *Clin.Chim.Acta* **1991**, *199*, 43–50.

Hunt, A.N.; Kinkaid, A.R.; Wilton, D.C.; Postle, A.D. Phospholipase A2 specificities determined in mixed substrate vesicles using a combination of continuous fluorescence displacement and quantitative HPLC analyses. *Biochem.Soc.Trans.* **1992**, *20*, 298S.

Kaneda, N.; Noro, Y.; Nagatsu, T. Highly sensitive assay for acetylcholinesterase activity by high-performance liquid chromatography with electrochemical detection. *J.Chromatogr.* **1985**, *344*, 93–100.

Kaneda, N.; Nagatsu, T. Highly sensitive assay for choline acetyltransferase activity by high-performance liquid chromatography with electrochemical detection. *J.Chromatogr.* **1985**, *341*, 23–30.

Kato, S.; Negishi, K.; Honma, K.; Sakai, K.; Shimada, Y. A high performance liquid chromatography assay for glutamine synthetase. *Neurochem.Int.* **1989**, *14*, 491–496.

Kennerly, D.A. Quantitative analysis of water-soluble products of cell-associated phospholipase C- and phospholipase D-catalyzed hydrolysis of phosphatidylcholine. *Methods Enzymol.* **1991**, *197*, 191–197.

Kudoh, S.; Nakamura, H. Simultaneous separation of β-lactamase isoenzymes by high performance hydroxyapatite chromatography with selective postcolumn enzymic reaction. *Anal.Sci.* **1991**, 7, 267–272.

Kudoh, S.; Nakamura, H. Facile measurement of β-lactamase activity by high-performance liquid chromatography with postcolumn enzymatic reaction. *Anal.Sci.* **1988**, *4*, 111–113.

Magnusson, P.; Loefman, O.; Larsson, L. Methodological aspects on separation and reaction conditions of bone and liver alkaline phosphatase isoform analysis by high-performance liquid chromatography. *Anal.Biochem.* **1993**, *211*, 156–163.

Magnusson, P.; Loefman, O.; Larsson, L. Determination of alkaline phosphatase isoenzymes in serum by high-performance liquid chromatography with post-column reaction detection. *J.Chromatogr.* **1992**, *576*, 79–86.

Matsukura, H.; Suzuki, Y.; Okada, T.; Naiki, S.; Sakuragawa, N. Automated separation and measurement of urinary isoenzymes and protein by ion-exchange liquid chromatography. *J.Chromatogr.* **1987**, *414*, 47–54.

Matsukura, H.; Suzuki, Y.; Okada, T.; Naiki, S.; Sakuragawa, N. Multiple urinary isoenzyme assay by high-performance liquid chromatography. *HRC CC, J.High Resolut.Chromatogr.Chromatogr.Commun.* **1986**, *9*, 479–480.

Minakuchi, K.; Kujime, T.; Umeda, T.; Fushitani, S.; Yoshisaka, T.; Takasugi, M. Quality testing for kallidinogenase preparations by HPLC-post-column analysis. *Byoin Yakugaku* **1990**, *16*, 86–93.

Morimoto, S.; Akiyama, H.; Takaoka, M.; Okamura, H.; Imaoka, S.; Funae, Y. High-performance liquid chromatography with a continuous-flow enzyme detector for the demonstration of the conversion of rat urinary inactive kallikrein into its active form. *J.Chromatogr.* **1984**, *295*, 226–230.

Naoi, M.; Kondoh, M.; Mutoh, T.; Takahashi, T.; Kojima, T.; Hirooka, T.; Nagatsu, T. Microassay for GM1 ganglioside β-galactosidase activity using high-performance liquid chromatography. *J.Chromatogr.* **1988**, *426*, 75–82.

Nicot, G.; Lachatre, G.; Gonnet, C.; Dupuy, J.L.; Valette, J.P. Rapid assay of N-acetyl-β-D-glucosaminidase isoenzymes in urine by ion-exchange chromatography. *Clin.Chem.* **1987**, *33*, 1796–1800.

Ohno, M.; Kai, M.; Ohkura, Y. Assay for enkephalin-degrading peptidases in rat brain tissues by high-performance liquid chromatography with on-line post-column fluorescence detection. *J.Chromatogr.* **1988**, *430*, 291–298.

Omichi, K.; Ikenaka, T. Separation of human salivary α-amylase isozymes by high-performance liquid chromatography with a continuous monitor system of the activity. *Anal.Biochem.* **1988**, *168*, 332–336.

Sanny, C.G.; Rymas, K. In vivo effects of disulfiram and cyanamide on canine liver aldehyde dehydrogenase isoenzymes as detected by high-performance (pressure) liquid chromatography. *Alcohol.: Clin.Exp.Res.* **1993**, *17*, 982–987.

Stein, T.A.; Cohen, J.R.; Mandell, C.; Wise, L. Measurement of elastin hydrolysis by reverse-phase HPLC with on-line post-column derivatization. *Chromatographia* **1989**, *27*, 225–227.

Sun, F.Z.; Prentice, D.A.; Steinberger, A. The effect of castration and androgen replacement on rat pituitary follicle-stimulating hormone investigated by reversed-phase high-performance liquid chromatography. *Andrologia* **1986**, *18*, 259–267.

Ephedra Bases

Moriyasu, M.; Endo, M.; Kanazawa, R.; Hashimoto, Y.; Kato, A.; Mizuno, M. High-performance liquid chromatographic determination of organic substances by metal chelate derivatization. III. Analysis of Ephedra bases. *Chem.Pharm.Bull.* **1984**, *32*, 744–747.

Ephedrine

Gal, J. Resolution of the enantiomers of ephedrine, norephedrine and pseudoephedrine by high-performance liquid chromatography. *J.Chromatogr.* **1984**, *307*, 220–223.

Moriyasu, M.; Hashimoto, Y.; Endo, M. High-performance liquid chromatographic determination of optically active and racemic ephedrines by derivatization and metal chelate formation. *Chem.Lett.* **1980**, 761–764.

Shao, G.; Wu, F.; Wang, D.S.; Zhu, R.; Luo, X. Quantitative analysis of (l)-ephedrine and (d)-pseudoephedrine in plasma by high-performance liquid chromatography with fluorescence detection. *Yaoxue Xuebao* **1995**, *30*, 384–389.

Wainer, I.W.; Doyle, T.D.; Fry, F.S., Jr.; Hamidzadeh, Z. Chiral recognition model for the resolution of ephedrine and related α,β-aminoalcohols as enantiomeric oxazolidine derivatives. *J.Chromatogr.* **1986**, *355*, 149–156.

Epoxides

Duchateau, A.L.L.; Jacquemin, N.M.J.; Straatman, H.; Noorduin, A.J. Liquid chromatographic determination of chiral epoxides by derivatization with sodium sulfide, o-phthalaldehyde and an amino acid. *J.Chromatogr.* **1993**, *637*, 29–34.

Gal, J. Determination of the enantiomeric composition of chiral epoxides using chiral derivatization and liquid chromatography. *J.Chromatogr.* **1985**, *331*, 349–357.

Epoxy Resins

Hinton, I.G. Chromatographic studies of epoxy resins. *Br.Polym.J.* **1983**, *15*, 47–49.

Erythromycin (using post-column reaction detection)

Tsuji, K. Fluorimetric determination of erythromycin and erythromycin ethylsuccinate in serum by a high-performance liquid chromatographic post-column, on-stream derivatization and extraction method. *J.Chromatogr.* **1978**, *158*, 337–343.

Ethambutol

Suzuki, M.; Ono, T.; Takitani, S. Determination of ethambutol in plasma by high-performance liquid chromatography with fluorescence detection. *Anal.Sci.* **1991**, *7*, 949–950.

Ethanol (using post-column reaction detection)

Tagliaro, F.; Dorizzi, R.; Ghielmi, S.; Marigo, M. Direct injection high-performance liquid chromatographic method with electrochemical detection for the determination of ethanol and methanol in plasma using an alcohol oxidase reactor. *J.Chromatogr.* **1991**, *566*, 333–339.

Tagliaro, F.; De Leo, D.; Marigo, M.; Dorizzi, R.; Schiavon, G. Use of enzymic reactors in the high performance liquid chromatographic determination of ethanol and methanol with electrochemical detection. *Biomed.Chromatogr.* **1990**, *4*, 224–228.

Ethanolamine

Levin, J.O.; Andersson, K.; Hallgren, C. Determination of monoethanolamine and diethanolamine in air. *Ann.Occup.Hyg.* **1989**, *33*, 175–180.

McMaster, C.R.; Choy, P.C. The determination of tissue ethanolamine levels by reverse-phase high-performance liquid chromatography. *Lipids* **1992**, *27*, 560–563.

Ethosuximide

Wu, J-K.; Chen, S-H.; Kou, H-S.; Wu, S-M.; Wu, H-L. Derivatization and high performance liquid chromatographic determination of ethosuximide. *Chin.Pharm.J.(Taipei)* **1994**, *46*, 413–421.

Ethylenethiourea

Smith, R.M.; Madahar, K.C.; Salt, W.G.; Smart, N.A. Determination of trace levels of ethylenethiourea by HPLC following derivatization with phenacyl halides. *Chromatographia* **1984**, *19*, 411–414.

Ethylenimine

Morales, R.; Stampfer, J.F.; Hermes, R.E. Air sampling and liquid chromatographic determination of ethylenimine. *Anal.Chem.* **1982**, *54*, 1340–1344.

Etoposide

van Opstal, M.A.J.; Krabbenborg, P.; Holthuis, J.J.M.; Van Bennekom, W.P.; Bult, A. Comparison of flow-injection analysis with high-performance liquid chromatography for the determination of etoposide in plasma. *J.Chromatogr.* **1988**, *432*, 395–400.

Explosives (using post-column reaction detection)

Engelhardt, H.; Meister, J.; Kolla, P. Optimization of post-column reaction detector for HPLC of explosives. *Chromatographia* **1993**, *35*, 5–12.

Selavka, C.M.; Krull, I.S. Liquid chromatography with photolysis - electrochemical detection for nitro-based high explosives and water gel formulation sensitizers. *J.Energ.Mater.* **1986**, *4*, 273–303.

Fatty Acid Esters

Nikolova-Damyanova, B.; Christie, W.W.; Hersloef, B. Silver ion high-performance liquid chromatography of esters of isomeric octadecenoic fatty acids with short-chain monounsaturated alcohols. *J.Chromatogr.A* **1995**, *693*, 235–239.

Weaner, L.E.; Hoerr, D.C. Separation of fatty acid ester and amide enantiomers by high-performance liquid chromatography on chiral stationary phases. *J.Chromatogr.* **1988**, *437*, 109–119.

Fatty Acids

Aserin, A.; Garti, N.; Frenkel, M. HPLC analysis of nonionic surfactants - Part V; ethoxylated fatty acids. *J.Liq.Chromatogr.* **1984**, *7*, 1545–1557.

Bandi, Z.L.; Ansari, G.A.S. High-performance liquid chromatographic analysis of saturated monohydroxy fatty acid mixtures containing positional isomers of various chain-lengths. *J.Chromatogr.* **1986**, *363*, 402–406.

Callahan, F.E.; Norman, H.A.; Srinath, T.; St John, J.B.; Dhar, R.; Mattoo, A.K. Identification of covalently bound fatty acids on acylated proteins immobilized on nitrocellulose paper. *Anal.Biochem.* **1989**, *183*, 220–224.

Calull, M.; Borrull, F.; marce, R.M.; Zamora, F. HPLC analysis of fatty acids in wine. *Am.J.Enol.Vitic.* **1991**, *42*, 268–273.

D'Amboise, M.; Gendreau, M. Isocratic separation of human blood plasma long chain free fatty acid derivatives by reversed phase liquid chromatography. *Anal.Lett.* **1979**, *12*, 381–395.

Engelmann, G.J.; Esmans, E.L.; Alderweireldt, F.C.; Rillaerts, E. Rapid method for the analysis of red blood cell fatty acids by reversed-phase high-performance liquid chromatography. *J.Chromatogr.* **1988**, *432*, 29–36.

Hordijk, K.A.; Cappenberg, T.E. Quantitative high-pressure liquid chromatography-fluorescence determination of some important lower fatty acids in lake sediments. *Appl.Environ.Microbiol.* **1983**, *46*, 361–369.

Ichinose, N.; Abe, S.; Akishige, Y.; Yoshimura, H.; Adachi, K. Fluorescence high-performance liquid chromatography of arachidonic, eicosapentaenoic, and some other higher fatty acids in ovary of several fishes. *Fresenius' Z.Anal.Chem.* **1987**, *329*, 47–49.

Ikeda, M.; Kusaka, T. An application of liquid chromatography/mass spectrometry to determination of fatty acids with a wide range of carbon chain lengths (C12-C54) in Mycobacteria. *Kawasaki Med.J.* **1996**, *22*, 1–8.

Ikenoya, S.; Hiroshima, O.; Ohmae, M.; Kawabe, K. Electrochemical detector for high-performance liquid chromatography. IV. Analysis of fatty acids, bile acids and prostaglandins by derivatization to an electrochemically active form. *Chem.Pharm.Bull.* **1980**, *28*, 2941–2947.

Juengling, E.J.; Kammermeier, H. A one-vial method for routine extraction and quantification of free fatty acids in blood and tissue by HPLC. *Anal.Biochem.* **1988**, *171*, 150–157.

Kobayashi, T.; Katayama, M.; Suzuki, S.; Tomoda, H.; Goto, I.; Kuroiwa, Y. Adrenoleukodystrophy: detection of increased very-long-chain fatty acids by high-performance liquid chromatography. *J.Neurol.* **1983**, *230*, 209–215.

Kuehn, H.; Wiesner, R. Separation of hydroxylated polyenoic fatty acid enantiomers on Pirkle-type chiral phase high-performance liquid chromatographic columns. *J.Chromatogr.* **1990**, *520*, 391–401.

Marini, D.; Balestrieri, F. Liquid chromatographic determination of free fatty acids and triglycerides in butter. *Riv.Soc.Ital.Sci.Aliment.* **1988**, *17*, 477–480.

Masuda, T.; Isobe, A.; Murata, C.; Takadate, A.; Komai, M.; Kimura, S. Anion-exchange resin catalyzed derivatization of fatty acids with 3-bromoacetyl-6,7-methylenedioxycoumarin. *Tohoku J.Agric.Res.* **1993**, *43*, 111–117.

Mueller-Harvey, I.; Parkes, R.J. Measurement of volatile fatty acids in pore water from marine sediments by HPLC. *Estuarine, Coastal Shelf Sci.* **1987**, *25*, 567–579.

Ohshima, T.; Miyamoto, K.; Sakai, R. Simultaneous separation and sensitive measurement of free fatty acids in ancient pottery by high performance liquid chromatography. *J.Liq.Chromatogr.* **1993**, *16*, 3217–3227.

Ryu, J-H.; Park, M-K. Derivatization of fatty acids with 2-bromoacetyltriphenylene for high-performance liquid chromatography. *Anal.Sci.Technol.* **1993**, *6*, 411–415.

Shimada, K.; Sakayori, C.; Nambara, T. Determination of fatty acids by high-performance liquid chromatography with electrochemical detection using a ferrocene derivatization reagent. *J.Liq.Chromatogr.* **1987**, *10*, 2177–2187.

Tanaka, M.; Yasaka, Y.; Funazo, K.; Shono, T. Poly[2-(1-naphthyl)ethyl p-styrenesulfonate] as polymeric derivatizing reagent for liquid chromatography of fatty acids. *Fresenius' Z.Anal.Chem.* **1989**, *335*, 311–312.

Tweenten, T.N.; Wetzel, D.L. Selective derivatization and high-performance liquid chromatographic analysis of free fatty acids in lipid extracts. *Cereal Chem.* **1983**, *60*, 411–413.

Yamaguchi, M.; Matsunaga, R.; Fukuda, K.; Nakamura, M.; Ohkura, Y. Highly sensitive determination of free polyunsaturated, long-chain fatty acids in human serum by high-performance liquid chromatography with fluorescence detection. *Anal.Biochem.* **1986**, *155*, 256–261.

Yasaka, Y.; Tanaka, M.; Matsumoto, T.; Funazo, K.; Shono, T. Poly(styrenesulfonates) as polymeric derivatizing reagents for fatty acids in liquid chromatography. *Anal.Sci.* **1989**, *5*, 611–612.

Zamir, I.; Grushka, E.; Cividalli, G. High-performance liquid chromatographic analysis of free palmitic and stearic acids in cerebrospinal fluid. *J.Chromatogr.* **1991**, *565*, 424–429.

Zamir, I.; Grushka, E.; Chemke, J. Separation and determination of saturated very-long-chain free fatty acids in plasma of patients with adrenoleukodystrophy using solid-phase extraction and high-performance liquid chromatographic analysis. *J.Chromatogr.* **1991**, *567*, 319–330.

Fatty Acids (using post-column reaction detection)

Bigley, F.P.; Grob, R.L. Detection and quantification of the straight chain fatty acids from propionic to decanoic using post-column reaction detection high performance liquid chromatography. *J.Environ.Sci.Health* **1986**, *A21*, 289–303.

Fenoldopam (using post-column reaction detection)

Boppana, V.K.; Fong, K-L.L.; Ziemniak, J.A.; Lynn, R.K. Use of a post-column immobilized β-glucuronidase enzyme reactor for the determination of diastereomeric glucuronides of fenoldopam in plasma and urine by high-performance liquid chromatography with electrochemical detection. *J.Chromatogr.* **1986**, *353*, 231–247.

Fenoprofen

Sallustio, B.C.; Abas, A.; Hayball, P.J.; Purdie, Y.J.; Meffin, P.J. Enantiospecific high-performance liquid chromatographic analysis of 2-phenylpropionic acid, ketoprofen and fenoprofen. *J.Chromatogr.* **1986**, *374*, 329–337.

Fentanyl

Kumar, K.; Morgan, D.J.; Crankshaw, D.P. Determination of fentanyl and alfentanil in plasma by high-performance liquid chromatography with ultraviolet detection. *J.Chromatogr.* **1987**, *419*, 464–468.

FK565 (using post-column reaction detection)

Tashiro, Y.; Suzuki, A.; Noda, K.; Noguchi, H. Determination of immunostimulative acylpeptide (FK565) in plasma by high performance liquid chromatography with fluorogenic post-column derivatization. *Bunseki Kagaku* **1987**, *36*, 787–791.

Flavonoids (using post-column reaction detection)

Galensa, R. Application possibilities of post column reactors in the high-performance liquid chromatographic analysis of foods, in particular, flavonoids. *Lebensmittelchem.gerichtl.chem.* **1982**, *36*, 49–51.

Sabatier, S.; Amiot, M.J.; Tacchini, M.; Aubert, S. Identification of flavonoids in sunflower honey. *J.Food Sci.* **1992**, *57*, 773–777.

Flavonols (using post-column reaction detection)

Treutter, D.; Santos-Buelga, C.; Gutmann, M.; Kolodziej, H. Identification of flavan-3-ols and procyanidins by high-performance liquid chromatography and chemical reaction detection. *J.Chromatogr.A* **1994**, *667*, 290–297.

Flecainide

Lie-A-Huen, L.; Stuurman, R.M.; IJdenberg, F.N.; Kingma, J.H.; Meijer, D.K. High-performance liquid chromatographic assay of flecainide and its enantiomers in serum. *Ther.Drug Monit.* **1989**, *11*, 708–711.

Turgeon, J.; Kroemer, H.K.; Prakash, C.; Blair, I.A.; Roden, D.M. Stereoselective determination of flecainide in human plasma by high-performance liquid chromatography with fluorescence detection. *J.Pharm.Sci.* **1990**, *79*, 91–95.

Fludarabine

Kemena, A.; Fernandez, M.; Bauman, J.; Keating, M.; Plunkett, W. A sensitive fluorescence assay for quantitation of fludarabine and metabolites in biological fluids. *Clin.Chim.Acta* **1991**, *200*, 95–106.

Fluoride (using post-column reaction detection)

Matsui, M.; Nishikawa, M.; Morita, M. Determination of fluoride ion in water samples by HPLC using lanthanum alizarin complexon method. *Kankyo Kagaku* **1994**, *4*, 665–670.

Matsui, M.; Nishikawa, M.; Morita, M. Determination of fluoride ion in environmental samples by HPLC using lanthanum alizarin complexone reaction. *Kankyo Kagaku* **1993**, *3*, 464–465.

Fluorodeoxyuridine

Michaelis, H.C.; Foth, H.; Kahl, G.F. Determination of 5-fluoro-2'-deoxyuridine in human plasma by high-performance liquid chromatography with pre-column fluorimetric derivatization. *J.Chromatogr.* **1987**, *416*, 176–182.

Fluoropyrimidines

Yoshida, S.; Urakami, K.; Kito, M.; Takeshima, S.; Hirose, S. Precolumn derivatization for the determination of fluoropyrimidines by liquid chromatography with chemiluminescence detection. *Anal.Chim.Acta* **1990**, *239*, 181–187.

Yoshida, S.; Urakami, K.; Kito, M.; Takeshima, S.; Hirose, S. High-performance liquid chromatography with chemiluminescence detection of serum levels of pre-column derivatized fluoropyrimidine compounds. *J.Chromatogr.* **1990**, *530*, 57–64.

Fluorotyrosine

Kehry, M.R.; Wilson, M.L.; Dahlquist, F.W. A simple quantitative method for the determination of 3-fluorotyrosine substitution in proteins. *Anal.Biochem.* **1983**, *131*, 236–241.

Fluoxetine

Potts, B.D.; Parli, C.J. Analysis of the enantiomers of fluoxetine and norfluoxetine in plasma and tissue using chiral derivatization and normal-phase liquid chromatography. *J.Liq.Chromatogr.* **1991**, *15*, 665–681.

Fluphenazine (using post-column reaction detection)

Mann, B.; Grayeski, M.L. Evaluation of peroxyoxalate chemiluminescence postcolumn detection of fluphenazine in urine and blood plasma using high performance liquid chromatography. *Biomed.Chromatogr.* **1991**, *5*, 47–52.

Fluvastatin

Smith, H.T.; Kalafsky, G. Analysis of fluvastatin and its enantiomers in plasma by HPLC. *Methodol.Surv.Bioanal.Drugs* **1994**, *23*, 203–209.

Folacins

Day, B.P.; Gregory, J.F..I. Determination of folacin derivatives in selected foods by high-performance liquid chromatography. *J.Agric.Food Chem.* **1981**, *29*, 374–377.

Folacins (using post-column reaction detection)

Hahn, A.; Stein, J.; Rump, U.; Rehner, G. Optimized high-performance liquid chromatographic procedure for the separation and quantification of the main folacins and some derivatives. I. Chromatographic system. *J.Chromatogr.* **1991**, *540*, 207–215.

Folates

McNulty, H.; McPartlin, J.; Weir, D.; Scott, J. Reversed-phase high-performance liquid chromatographic method for the quantitation of endogenous folate catabolites in rat urine. *J.Chromatogr.* **1993**, *614*, 59–66.

Folates (using post-column reaction detection)

Holt, D.L.; Wehling, R.L.; Zeece, M.G. Determination of native folates in milk and other dairy products by high-performance liquid chromatography. *J.Chromatogr.* **1988**, *449*, 271–279.

Formaldehyde

Benassi, C.A.; Semenzato, A.; Zaccaria, F.; Bettero, A. High-performance liquid chromatographic determination of free formaldehyde in cosmetics preserved with Dowicil 200. *J.Chromatogr.* **1990**, *502*, 193–200.

Bicking, M.K.L.; Cooke, W.M.; Kawahara, F.K.; Longbottom, J.E. Method development for the determination of formaldehyde in samples of environmental origin. *ASTM Spec.Tech.Publ.* **1988**, *976*, 159–175.

Cohen, H.P.; Tway, P.C. High performance liquid chromatographic detection of residual formaldehyde in a hepatitis-A vaccine by use of hydralazine. *J.Liq.Chromatogr.* **1993**, *16*, 1667–1684.

Cotsaris, E.; Nicholson, B.C. Low level determination of formaldehyde in water by high-performance liquid chromatography. *Analyst* **1993**, *118*, 265–268.

Kaminski, J.; Atwal, A.S.; Mahadevan, S. Determination of formaldehyde in fresh and retail milk by liquid column chromatography. *J.AOAC Int.* **1993**, *76*, 1010–1013.

Kaminski, J.; Atwal, A.S.; Mahadevan, S. High performance liquid chromatographic determination of formaldehyde in milk. *J.Liq.Chromatogr.* **1993**, *16*, 521–526.

Kleindienst, T.E.; Shepson, P.B.; Nero, C.M.; Arnts, R.R.; Tejada, S.B.; Mackay, G.I.; Mayne, L.K.; Schiff, H.I.; Lind, J.A.; et al., An intercomparison of formaldehyde measurement techniques at ambient concentration. *Atmos.Environ.* **1988**, *22*, 1931–1939.

Lam, S.K.; Margiasso, V.A. Determination of formaldehyde in the polymerized ragweed antigen by HPLC. *J.Liq.Chromatogr.* **1984**, *7*, 2643–2651.

Renon, P.; Biondi, P.A.; Malandra, R. HPLC determination of formaldehyde in frozen crustaceans. *Ital.J.Food Sci.* **1991**, *3*, 59–62.

Tomkins, B.A.; McMahon, J.M.; Caldwell, W.M.; Wilson, D.L. Liquid chromatographic determination of total formaldehyde in drinking water. *J.Assoc.Off.Anal.Chem.* **1989**, *72*, 835–839.

Formaldehyde (using post-column reaction detection)

Engelhardt, H.; Klinkner, R. Determination of free formaldehyde in nail varnish by HPLC. *Chromatographia* **1985**, *20*, 729–731.

Engelhardt, H.; Klinkner, R. Determination of free formaldehyde in the presence of donators in cosmetics by HPLC and post-column derivation. *Chromatographia* **1985**, *20*, 559–565.

Kijima, K.; Takeda, M.; Okaya, Y.; Takamatsu, T.; Murase, M.; Sawamura, K.; Nomura, T.; Koba, T. A study on release of formaldehyde from its donor-type preservatives. *Anal.Sci.* **1991**, *7*, 913–916.

Meister, J.; Engelhardt, H. Analysis of free formaldehyde in formaldehyde-donator solution. *LaborPraxis* **1995**, *19*, 28–31.

Ftorafur

Uematsu, T.; Miyazawa, N.; Wada, K. 1-(Tetrahydro-2-furanyl)-5-fluorouracil (ftorafur) determined in rat hair as an index of drug exposure. *J.Pharm.Sci.* **1993**, *82*, 1272–1274.

Fucose

Fleming, S.C.; Gill, M.; Laker, M.F. A rapid HPLC method for the determination of free fucose in urine. A marker of malignancy in children. *J.Liq.Chromatogr.* **1995**, *18*, 173–180.

Fumonisins

Ackermann, T. Fast thin-layer chromatography systems for fumonisin isolation and identification. *J.Appl.Toxicol.* **1991**, *11*, 451.

Alberts, J.F.; Gelderblom, W.C.A.; Marasas, W.F.O. Evaluation of the extraction and purification procedures of the maleyl derivatization HPLC technique for the quantification of the fumonisin B mycotoxins in corn cultures. *Mycotoxin Res.* **1993**, *9*, 2–12.

Bennett, G.A.; Richard, J.L. Liquid chromatographic method for analysis of the naphthalene dicarboxaldehyde derivative of fumonisins. *J.AOAC Int.* **1994**, *77*, 501–506.

Fukuda, S.; Nagahara, A.; Kikuchi, M.; Kumagai, S. Comparison of HPLC and ELISA methods in the detection of fumonisins. *Maikotokishin (Tokyo)* **1994**, *39*, 19–22.

Maragos, C.M.; Richard, J.L. Quantitation and stability of fumonisins B1 and B2 in milk. *J.AOAC Int.* **1994**, *77*, 1162–1167.

Maragos, C.M.; Bennett, G.A.; Richard, J.L. Analysis of fumonisin B1 in corn by capillary electrophoresis. *Adv.Exp.Med.Biol.* **1996**, *392*, 105–112.

Scott, P.M.; Lawrence, G.A. Analysis of beer for fumonisins. *J.Food Prot.* **1995**, *58*, 1379–1382.

Shephard, G.S.; Sydenham, E.W.; Thiel, P.G.; Gelderblom, W.C.A. Quantitative determination of fumonisins B1 and B2 by high-performance liquid chromatography with fluorescence detection. *J.Liq.Chromatogr.* **1990**, *13*, 2077–2087.

Thiel, P.G.; Sydenham, E.W.; Shephard, G.S.; Van Schalkwyk, D.J. Study of the reproducibility characteristics of a liquid chromatographic method for the determination of fumonisins B1 and B2 in Corn: IUPAC collaborative study. *J.AOAC Int.* **1993**, *76*, 361–366.

Thiel, P.G.; Sydenham, E.W.; Shephard, G.S. The reliability and significance of analytical data on the natural occurrence of fumonisins in food. *Adv.Exp.Med.Biol.* **1996**, *392*, 145–151.

Trucksess, M.W.; Stack, M.E.; Allen, S.; Barrion, N. Immunoaffinity column coupled with liquid chromatography for determination of fumonisin B1 in canned and frozen sweet corn. *J.AOAC Int.* **1995**, *78*, 705–710.

Fura-2 (using post-column reaction detection)

Tran, N.N.P.; Leroy, P.; Bellucci, L.; Robert, A.; Nicolas, A.; Atkinson, J.; Capdeville-Atkinson, C. Intracellular concentrations of Fura-2 and Fura-2/AM in vascular smooth muscle cells following perfusion loading of Fura-2/AM in arterial segments. *Cell Calcium* **1995**, *18*, 420–428.

Furaldehyde

Lo Coco, F.; Valentini, C.; Novelli, V.; Ceccon, L. Liquid chromatographic determination of 2-furaldehyde and 5-hydroxymethyl-2-furaldehyde in beer. *Anal.Chim.Acta* **1995**, *306*, 57–64.

Furazolidone

Long, A.R.; Hsieh, L.C.; Malbrough, M.S.; Short, C.R.; Barker, S.A. Matrix solid phase dispersion (MSPD) isolation and liquid chromatographic determination of furazolidone in pork muscle tissue. *J.Assoc.Off.Anal.Chem.* **1991**, *74*, 292–294.

Furosemide

Lin, E.T.; Smith, D.E.; Benet, L.Z.; Hoener, B.A. High-performance liquid chromatographic assays for furosemide in plasma and urine. *J.Chromatogr.* **1979**, *163*, 315–321.

Furprofen

Carlucci, G.; Mazzeo, P.; Palumbo, G. Indirect stereoselective determination of enantiomers of furprofen in human plasma by high-performance liquid chromatography. *Chromatographia* **1992**, *34*, 618–620.

Galacturonic Acid

Maness, N.O.; Mort, A.J. Separation and quantitation of galacturonic acid oligomers from 3 to over 25 residues in length by anion-exchange high-performance liquid chromatography. *Anal.Biochem.* **1989**, *178*, 248–254.

Gangliosides

Wagener, R.; Kobbe, B.; Stoffel, W. Quantification of gangliosides by microbore high-performance liquid chromatography. *J.Lipid Res.* **1996**, *37*, 1823–1829.

Yamaguchi, M.; Hara, S.; Takemori, Y.; Nakamura, M. High-performance liquid chromatography of monosialogangliosides in human plasma with fluorescence detection. *Anal.Sci.* **1989**, *5*, 35–38.

Gelatin (using post-column reaction detection)

Riedl, J.; Moll, F. HPLC analysis of gelatins in pharmaceutical formulations. *Arch.Pharm.* **1986**, *319*, 89–91.

Gentamicin

Anhalt, J.P.; Sancilio, F.D.; McCorkle, T. Gentamicin C-component ratio determination by high-pressure liquid chromatography. *J.Chromatogr.* **1978**, *153*, 489–493.

Barends, D.M.; Van der Sandt, J.S.F.; Hulshoff, A. Micro determination of gentamicin in serum by high-performance liquid chromatography with ultraviolet detection. *J.Chromatogr.* **1980**, *182*, 201–210.

Busson, R.; Claes, P.J.; Vanderhaeghe, H. Determination of the composition of gentamicin sulfates by proton and carbon-13 nuclear magnetic spectroscopy. *J.Assoc.Off.Anal.Chem.* **1986**, *69*, 601–608.

Claes, P.J.; Busson, R.; Vanderhaeghe, H. Determination of the component ratio of commercial gentamicins by high-performance liquid chromatography using pre-column derivatization. *J.Chromatogr.* **1984**, *298*, 445–457.

D'Souza, J.; Ogilvie, R.I. Determination of gentamicin components C1a, C2 and C1 in plasma and urine by high-performance liquid chromatography. *J.Chromatogr.* **1982**, *232*, 212–218.

Essers, L. Comparison of a simplified liquid chromatographic assay of gentamicin in serum with enzyme immunoassay and bioassay. *Eur.J.Clin.Microbiol.* **1982**, *1*, 367–370.

Fennell, M.A.; Uboh, C.E.; Sweeney, R.W.; Soma, L.R. Gentamicin in tissue and whole milk: an improved method for extraction and cleanup of samples for quantitation on HPLC. *J.Agric.Food Chem.* **1995**, *43*, 1849–1852.

Kraisintu, K.; Parfitt, R.T.; Rowman, M.G. A high-performance liquid chromatographic method for the determination and control of the composition of gentamicin sulfate. *Int.J.Pharm.* **1982**, *10*, 67–75.

Sar, F.; Leroy, P.; Nicolas, A.; Archimbault, P. Development and optimization of a liquid chromatographic method for the determination of gentamicin in calf tissues. *Anal.Chim.Acta* **1993**, *275*, 285–293.

Tay, L-F.; Khor, E.; Lee, K-B. High performance liquid chromatography quantitation of gentamicin: Optimization for routine analysis. *Bull.Singapore Natl.Inst.Chem.* **1995**, *23*, 47–53.

Weigand, R.; Coombes, R.J. Gentamicin determination by high-performance liquid chromatography. *J.Chromatogr.* **1983**, *281*, 381–385.

Gentamicin (using post-column reaction detection)

Guggisberg, D.; Koch, H. Method for the quantitative determination of gentamicin in meat, liver, and kidney by HPLC and post-column derivatization. *Mitt.geb.Lebensmittelunters.Hyg.* **1995**, *86*, 14–28.

Patel, C.P. A comparison of Ames TDA, Syva Emit, and HPLC methods in the determination of serum gentamicin. *LC Mag.* **1985**, *3*, 148.

Seidl, G.; Nerad, H.P. Gentamicin C: separation of C1, C1a, C2, C2a, and C2b components by HPLC using isocratic ion-exchange chromatography and post-column derivatization. *Chromatographia* **1988**, *25*, 169–171.

Tamai, G.; Imai, H.; Yoshida, H. On-line deproteinization of serum sample for HPLC analysis of hydrophilic compounds and its application to gentamicin. *Chromatographia* **1986**, *21*, 519–522.

Gibberellins

Barendse, G.W.M.; Van de Werken, P.H.; Takahashi, N. High-performance liquid chromatography of gibberellins. *J.Chromatogr.* **1980**, *198*, 449–455.

Turnbull, C.G.N.; Crozier, A.; Schneider, G. HPLC-based methods for the identification of gibberellin conjugates: metabolism of [3H]gibberellin A4 in seedlings of Phaseolus coccineus. *Phytochemistry* **1986**, *25*, 1823–1828.

Ginseng Saponin

Besso, H.; Saruwatari, Y.; Futamura, K.; Kunihiro, K.; Fuwa, T.; Tanaka, O. High-performance liquid chromatographic determination of ginseng saponin by ultraviolet derivatization. *Planta Med.* **1979**, *37*, 226–233.

Ginsenosides

Soldati, F.; Sticher, O. HPLC separation and quantitative determination of ginsenosides from Panax ginseng, Panax quinquefolium and from ginseng drug preparations. 2nd Communication. *Planta Med.* **1980**, *39*, 348–357.

Gizzerosine

Ito, Y.; Noguchi, T.; Naito, H. Fluorometric determination of gizzerosine, a histamine H2-receptor agonist discovered in feedstuffs, employing high-performance liquid chromatography. *Anal.Biochem.* **1985**, *151*, 28–31.

Wagener, W.W.D.; Koch, K.R.; Wessels, J.P.H.; Post, B.J. An investigation into the occurrence of the toxic amino acid gizzerosine in fish meal. *J.Sci.Food Agric.* **1991**, *54*, 147–152.

Gizzerosine (using post-column reaction detection)

Murakita, H.; Gotoh, T. o-Phthalaldehyde post-column derivatization for the determination of gizzerosine in fish meal by high-performance liquid chromatography. *J.Chromatogr.* **1990**, *515*, 527–530.

Ohta, Y.; Ohashi, H.; Enomoto, S.; Machida, Y. Simple measurement of gizzerosine in fish meals by high-performance liquid chromatography. *Agric.Biol.Chem.* **1988**, *52*, 2817–2821.

Glibenclamide

Besenfelder, E. Glibenclamide in serum: HPLC determination with pre-column derivatization. *HRC CC, J.High Resolut.Chromatogr.Chromatogr.Commun.* **1981**, *4*, 237–239.

Glucosamine

Jahnel, J.B.; Frimmel, F.H. Detection of glucosamine in the acid hydrolysis solution of humic substances. *Fresenius' J.Anal.Chem.* **1996**, *354*, 886–888.

Glucosides (using post-column reaction detection)

Koerner, P.J., Jr.; Nieman, T.A. High-performance liquid chromatographic determination of glucosides (glucose conjugates) with post-column reaction detection combining immobilized enzyme reactors and luminol chemiluminescence. *J.Chromatogr.* **1988**, *449*, 217–228.

Glucosinolates

Heaney, R.K. Chromatographic analysis of glucosinolates - a potential aid to the plant breeder. *Anal.Proc.* **1984**, *21*, 482–484.

Glucuronides

Lingeman, H.; Meussen, G.W.M.; Van der Zouwen, C.; Underberg, W.J.M.; Hulshoff, A. Pre-column (HPLC) fluorescence labeling of glucuronides. *Methodol.Surv.Biochem.Anal.* **1986**, *16*, 343–353.

Glutamic Acid

Bergeron, E.; Jolivet, P. Quantitative determination of glutamate in a Rhodophyceae (Chondrus crispus) and four Phaeophyceae (Fucus vesiculosus, Fucus serratus, Cystoseira elegans, Cystoseira barbata). *J.Appl.Phycol.* **1991**, *3*, 115–120.

Kuwada, M.; Katayama, K. A high-performance liquid chromatographic method for the simultaneous determination of gamma-carboxyglutamic acid and glutamic acid in proteins, bone, and urine. *Anal.Biochem.* **1981**, *117*, 259–265.

Glutamine

Darmaun, D.; D'Amore, D.; Haymond, M.W. Determination of glutamine and alpha-ketoglutarate concentration and specific activity in plasma using high-performance liquid chromatography. *J.Chromatogr.* **1993**, *620*, 33–38.

Kuhn, K.S.; Stehle, P.; Fuerst, P. Quantitative analyses of glutamine in peptides and proteins. *J.Agric.Food Chem.* **1996**, *44*, 1808–1811.

Teerlink, T.; Hennekes, M.W.T.; Van Leeuwen, P.A.M.; Houdijk, A. Rapid determination of glutamine in biological samples by high-performance liquid chromatography. *Clin.Chim.Acta* **1993**, *218*, 159–168.

Wen, A.; Jiang, Y.; Fan, Y.; Geng, X.; Guo, Z. Rapid determination of glutamine in human plasma and muscle tissue by high performance liquid chromatography with fluorescence detection. *Sepu* **1995**, *13*, 406–407.

Glutaraldehyde

Barnes, A.R.; Bradley, C.R.; Nash, S. Comparison of methods for the determination of glutaraldehyde in solution. *Pharm.Acta Helv.* **1994**, *69*, 21–24.

Glutathione

Buchberger, W.; Winsauer, K. Determination of glutathione in biological material by high-performance liquid chromatography with electrochemical detection. *Anal.Chim.Acta* **1987**, *196*, 251–254.

Burton, N.K.; Aherne, G.W. Sensitive measurement of glutathione using isocratic high-performance liquid chromatography with fluorescence detection. *J.Chromatogr.* **1986**, *382*, 253–257.

Chaney, W.G.; Spector, A. HPLC analysis of lens GSH and GSSG. *Curr.Eye Res.* **1984**, *3*, 345–350.

Cheynier, V.; Souquet, J.M.; Moutounet, M. Glutathione content and glutathione to hydroxycinnamic acid ratio in Vitis vinifera grapes and musts. *Am.J.Enol.Vitic.* **1989**, *40*, 320–324.

Dizdar, N.; Kaagedal, B.; Smeds, S.; Aastrand, K. A high-sensitivity fluorometric high-performance liquid chromatographic method for determination of glutathione and other thiols in cultured melanoma cells, microdialysis samples from melanoma tissue, and blood plasma. *Melanoma Res.* **1991**, *1*, 33–42.

Kranner, I.; Grill, D. Determination of glutathione and glutathione disulfide in lichens: a comparison of frequently used methods. *Phytochem.Anal.* **1996**, *7*, 24–28.

Luo, J-L.; Hammarqvist, F.; Cotgreave, I.A.; Lind, C.; Andersson, K.; Wernerman, J. Determination of intracellular glutathione in human skeletal muscle by reversed-phase high-performance liquid chromatography. *J.Chromatogr.B* **1995**, *670*, 29–36.

Maartensson, J. Method for determination of free and total glutathione and gamma-glutamylcysteine concentrations in human leukocytes and plasma. *J.Chromatogr.* **1987**, *420*, 152–157.

Morier-Teissier, E.; Mestdagh, N.; Bernier, J.L.; Henichart, J.P. Reduced and oxidized glutathione ratio in tumor cells: comparison of two measurement methods using HPLC and electrochemical detection. *J.Liq.Chromatogr.* **1993**, *16*, 573–596.

Noguchi, K.; Higuchi, S. Determination of glutathione isopropyl ester in rat, dog and human blood by high-performance liquid chromatography with fluorescence detection. *J.Pharm.Biomed.Anal.* **1992**, *10*, 515–520.

Siller-Cepeda, J.H.; Chen, T.H.H.; Fuchigami, L.H. High performance liquid chromatography analysis of reduced and oxidized glutathione in woody plant tissues. *Plant Cell Physiol.* **1991**, *32*, 1179–1185.

Slordal, L.; Andersen, A.; Dajani, L.; Warren, D.J. A simple HPLC method for the determination of cellular glutathione. *Pharmacol.Toxicol.* **1993**, *73*, 124–126.

Tracy, J.W.; O'Leary, K.A. Analysis of glutathione S-transferase-catalyzed S-alkylglutathione formation by high-performance liquid chromatography. *Anal.Biochem.* **1991**, *193*, 1–5.

Yang, C-S.; Chou, S-T.; Lin, N-N.; Liu, L.; Tsai, P-J.; Kuo, J-S.; Lai, J-S. Determination of extracellular glutathione in rat brain by microdialysis and high-performance liquid chromatography with fluorescence detection. *J.Chromatogr.B* **1994**, *661*, 231–235.

Glutathione (using post-column reaction detection)

Alpert, A.J.; Gilbert, H.F. Detection of oxidized and reduced glutathione with a recycling postcolumn reaction. *Anal.Biochem.* **1985**, *144*, 553–562.

Keller, D.A.; Menzel, D.B. Picomole analysis of glutathione, glutathione disulfide, glutathione S-sulfonate, and cysteine S-sulfonate by high-performance liquid chromatography. *Anal.Biochem.* **1985**, *151*, 418–423.

Leroy, P.; Nicolas, A.; Wellmann, M.; Michelet, F.; Oster, T.; Siest, G. Evaluation of o-phthalaldehyde as bifunctional fluorogenic post-column reagent for glutathione in LC. *Chromatographia* **1993**, *36*, 130–134.

Glyceric Acid

Petrarulo, M.; Marangella, M.; Cosseddu, D.; Linari, F. High-performance liquid chromatographic assay for L-glyceric acid in body fluids. Application in primary hyperoxaluria type 2. *Clin.Chim.Acta* **1992**, *211*, 143–153.

Glycerol

Hamano, T.; Mitsuhashi, Y.; Aoki, N.; Yamamoto, S.; Tsuji, S.; Ito, Y.; Oji, Y. Determination of glycerol in foods by high-performance liquid chromatography with fluorescence detection. *J.Chromatogr.* **1991**, *541*, 265–272.

L-α-Glycerophosphorylcholine (using post-column reaction detection)

Abbiati, G.; Fossati, T.; Arrigoni, M.; Rolle, P.; Dognini, G.L.; Castiglioni, C. High-performance liquid chromatographic assay of L-α-glycerophosphorylcholine using a two-step enzymic conversion. *J.Chromatogr.* **1991**, *566*, 445–451.

Glycoforms

Treuheit, M.; Halsall, H.B. HPLC of individual site glycoforms in serum orosomucoid. *Chromatographia* **1991**, *31*, 478–480.

Glycolic Acid

Petrarulo, M.; Marangella, M.; Linari, F. High-performance liquid chromatographic determination of plasma glycolic acid in healthy subjects and in cases of hyperoxaluria syndromes. *Clin.Chim.Acta* **1991**, *196*, 17–26.

Glycolipids

McCluer, R.H.; Ullman, M.D. Preparative and analytical high performance liquid chromatography of glycolipids. *ACS Symp.Ser.* **1980**, *128*, 1–13.

Pick, J.; Vajda, J.; Leisztner, L. Isolation of glycolipids from blood elements. *J.Liq.Chromatogr.* **1983**, *6*, 2647–2660.

Snada, S.; Uchida, Y.; Anraku, Y.; Izawa, A.; Iwamori, M.; Nagai, Y. Analysis of ceramide and monohexaosyl glycolipid derivatives by high-performance liquid chromatography and its application to the determination of the molecular species in tissues. *J.Chromatogr.* **1987**, *400*, 223–231.

Glycopeptides

Joos, B.; Luethy, R. Identification of fluorescent glycopeptide derivatives by two consecutive high pressure liquid chromatographic procedures. *J.Antibiot.* **1988**, *41*, 302–307.

Rohrer, J.S.; Cooper, G.A.; Townsend, R.R. Identification, quantification, and characterization of glycopeptides in reversed-phase HPLC separations of glycoprotein proteolytic digests. *Anal.Biochem.* **1993**, *212*, 7–16.

Glycoproteins

Hase, S.; Ikenaka, K.; Mikoshiba, K.; Kenaka, T. Analysis of tissue glycoprotein sugar chains by two-dimensional high-performance liquid chromatographic mapping. *J.Chromatogr.* **1988**, *434*, 51–60.

Morehead, H.W.; Talmadge, K.W.; O'Shannessy, D.J.; Siebert, C.J. Optimization of oxidation of glyco-proteins: an assay for predicting coupling to hydrazide chromatographic supports. *J.Chromatogr.* **1991**, *587*, 171–176.

Glycosaminoglycans (using post-column reaction detection)

Huang, Y.; Toyoda, H.; Koshiishi, I.; Toida, T.; Imanari, T. Determination of a depolymerized holothurian glycosaminoglycan in plasma after intravenous administration by postcolumn HPLC. *Chem.Pharm.Bull.* **1995**, *43*, 2182–2186.

Glycosides

Moro, L.; Modricky, C.; Stagni, N.; Vittur, F.; De Bernard, B. High-performance liquid chromatographic analysis of urinary hydroxylysyl glycosides as indicators of collagen turnover. *Analyst* **1984**, *109*, 1621–1622.

Murui, T.; Wanaka, K. Measurement of sterylglycosides by high performance liquid chromatography with 1-anthroylnitrile derivatives. *Biosci., Biotechnol., Biochem.* **1993**, *57*, 614–617.

Napoli, R.M.; Middleditch, B.S.; Cintron, N.M.; Chen, Y.M. Isolation and quantitative analysis of hy-droxylysine glycosides. *Chromatographia* **1989**, *28*, 497–501.

Tenni, R.; Rimoldi, D.; Zanaboni, G.; Cetta, G.; Castellani, A.A. Hydroxylysine glycosides: preparation and analysis by reverse phase high performance liquid chromatography. *Ital.J.Biochem.* **1984**, *33*, 117–127.

Tor, E.R.; Holstege, D.M.; Galey, F.D. Determination of oleander glycosides in biological matrixes by high-performance liquid chromatography. *J.Agric.Food Chem.* **1996**, *44*, 2223–2226.

Yoshihara, K.; Mochidome, N.; Hara, T.; Osada, S.; Takayama, A.; Nagata, M. Urinary excretion levels of hydroxylysine glycosides in osteoporotic patients. *Biol.Pharm.Bull.* **1994**, *17*, 836–839.

Glycosides (using post-column reaction detection)

Dalgaard, L.; Brimer, L. Electrochemical detection of cyanogenic glycosides after enzymic post-column cleavage. *J.Chromatogr.* **1984**, *303*, 67–76.

Schaufelberger, D.; Hostettmann, K. Analytical and preparative reversed-phase liquid chromatography of secoiridoid glycosides. *J.Chromatogr.* **1985**, *346*, 396–400.

Schaufelberger, D.; Hostettmann, K. High-performance liquid chromatographic analysis of secoiridoid and flavone glycosides in closely related Gentiana species. *J.Chromatogr.* **1987**, *389*, 450–455.

Glycosphingolipids

Tomono, Y.; Abe, K.; Watanabe, K. High-performance liquid affinity chromatography and in situ fluo-rescent labeling on thin-layer chromatography of glycosphingolipids. *Anal.Biochem.* **1990**, *184*, 360–368.

Kadowaki, H.; Rys-Sikora, K.E.; Koff, R.S. Separation of derivatized glycosphingolipids into individual molecular species by high performance liquid chromatography. *J.Lipid Res.* **1989**, *30*, 616–627.

Watanabe, K.; Tomono, Y. One-step fractionation of neutral and acidic glycosphingolipids by high-performance liquid chromatography. *Anal.Biochem.* **1984**, *139*, 367–372.

Glyoxylic Acid

Mentasti, E.; Savigliano, M.; Marangella, M.; Petrarulo, M.; Linari, F. High-performance liquid chro-matographic determination of glyoxylic acid and other carbonyl compounds in urine. *J.Chromatogr.* **1987**, *417*, 253–260.

Glyphosate

Glass, R.L. Liquid chromatographic determination of glyphosate in fortified soil and water samples. *J.Agric.Food Chem.* **1983**, *31*, 280–282.

Lundgren, L.N. A new method for the determination of glyphosate and (aminomethyl)phosphonic acid residues in soils. *J.Agric.Food Chem.* **1986**, *34*, 535–538.

Oppenhuizen, M.E.; Cowell, J.E. Liquid chromatographic determination of glyphosate and aminomethylphosphonic acid (AMPA) in environmental water: collaborative study. *J.Assoc.Off.Anal.Chem.* **1991**, *74*, 317–323.

Seiber, J.N.; McChesney, M.M.; Kon, R.; Leavitt, R.A. Analysis of glyphosate residues in kiwi fruit and asparagus using high-performance liquid chromatography of derivatized glyphosate as a cleanup step. *J.Agric.Food Chem.* **1984**, *32*, 678–681.

Glyphosate (using post-column reaction detection)

Cessna, A.J.; Cain, N.P. Residues of glyphosphate and its metabolite AMPA in strawberry fruit following spot and wiper applications. *Can.J.Plant Sci.* **1992**, *72*, 1359–1365.

Cowell, J.E.; Kunstman, J.L.; Nord, P.J.; Steinmetz, J.R.; Wilson, G.R. Validation of an analytical residue method for analysis of glyphosphate and metabolite: an interlaboratory study. *J.Agric.Food Chem.* **1986**, *34*, 955–960.

Sen, N.P.; Baddoo, P.A. Determination of glyphosate as N-nitroso derivative by high performance liquid chromatography with chemiluminescence detection. *Int.J.Environ.Anal.Chem.* **1996**, *63*, 107–117.

Tuinstra, L.G.M.T.; Kienhuis, P.G.M. Automated two-dimensional HPLC residue procedure for glyphosate on cereals and vegetables with postcolumn fluoregenic labeling. *Chromatographia* **1987**, *24*, 696–700.

Wigfield, Y.Y.; Lanouette, M. Residue analysis of glyphosate and its principal metabolite in certain cereals, oilseeds, and pulses by liquid chromatography and postcolumn fluorescence detection. *J.Assoc.Off.Anal.Chem.* **1991**, *74*, 842–847.

Gossypol

Kim, H.L.; Calhoun, M.C.; Stipanovic, R.D. Accumulation of gossypol enantiomers in ovine tissues. *Comp.Biochem.Physiol., B: Biochem.Mol.Biol.* **1996**, *113B*, 417–420.

Wu, D.F.; Reidenberg, M.M.; Drayer, D.E. Determination of gossypol enantiomers in plasma after administration of racemate using high-performance liquid chromatography with precolumn chemical derivatization. *J.Chromatogr.* **1988**, *433*, 141–148.

Wu, D.F.; Yu, Y.W.; Zheng, D.K. [Determination of (+)- and (-)-gossypol in human plasma using high performance liquid chromatography with pre-column chemical derivatization]. *Yao.Hsueh.Hsueh.Pao.* **1988**, *23*, 927–932.

Growth Regulators

Shiao, M.S.; Shih, M. Determination of plant growth regulators and their changes in relation to the biosynthesis of storage proteins in developing rice seeds. *Proc.Natl.Sci.Counc., Repub.China, Part B: Life Sci.* **1987**, *11*, 1–9.

Guaifenesin

Demian, I. High-performance liquid chromatography (HPLC) chiral separations of guaifenesin, methocarbamol, and racemorphan. *Chirality* **1993**, *5*, 238–240.

Guanethidine

Patel, M.B.S.; Mason, J.P. The development of a sensitive high-performance liquid-chromatographic method for the analysis of guanethidine. *Pharm.Sci.* **1995**, *1*, 95–98.

Guanidino Compounds

Baker, M.D.; Mohammed, H.Y.; Veening, H. Reversed-phase ion-pairing liquid chromatographic separation and fluorometric detection of guanidino compounds. *Anal.Chem.* **1981**, *53*, 1658–1662.

Boppana, V.K.; Rhodes, G.R. High-performance liquid chromatographic determination of guanidino compounds by automated pre-column fluorescence derivatization. *J.Chromatogr.* **1990**, *506*, 279–288.

Guanidino Compounds (using post-column reaction detection)

Jansen, H.; van der Velde, E.G.; Brinkman, U.A.T.; Frei, R.W.; Veening, H. Liquid chromatographic determination of guanidines with an anion exchange column used simultaneously as separator and postcolumn reagent generator. *Anal.Chem.* **1986**, *58*, 1380–1383.

Halocarbons

Street, K.W.; Hocson, V. β-Naphthol as a chromophore and fluorophore forming reagent in the HPLC determination of alkyl halides. *Anal.Lett.* **1983**, *16*, 1403–1425.

Wintersteiger, R.; Macher, M. Improvement of selectivity of electrochemical detection for the determination of compounds with chloro aromatic rings. *Chromatographia* **1995**, *40*, 247–252.

Halocarbons (using post-column reaction detection)

Guenther, F.R.; Chesler, S.N. A post column solvent trapping technique for the analysis of very volatile halocarbons. *HRC CC, J.High Resolut.Chromatogr.Chromatogr.Commun.* **1983**, *6*, 684–685.

Selavka, C.M.; Jiao, K.S.; Krull, I.S.; Sheih, P.; Yu, W.; Wolf, M. Liquid chromatography-photolysis-electrochemical detection for organobromides and organochlorides. *Anal.Chem.* **1988**, *60*, 250–254.

Haloperidol (using post-column reaction detection)

Tsuneyoshi, T. Fluorometric determination of haloperidol by high performance liquid chromatography. *HRC CC, J.High Resolut.Chromatogr.Chromatogr.Commun.* **1986**, *9*, 252–254.

Hepatitis Antigen

O'Keefe, D.O.; Paiva, A.M. Assay for recombinant hepatitis B surface antigen using reversed-phase high-performance liquid chromatography. *Anal.Biochem.* **1995**, *230*, 48–54.

Heptaminol

Brodie, R.R.; Chasseaud, L.F.; Rooney, L.; Darragh, A.; Lambe, R.F. Determination of heptaminol in human plasma and urine by high-performance liquid chromatography. *J.Chromatogr.* **1983**, *274*, 179–186.

Nicolas, A.; Leroy, P.; Moreau, A.; Mirjolet, M. Determination of heptaminol in pharmaceutical preparations by high-performance liquid chromatography. *J.Chromatogr.* **1982**, *244*, 148–152.

Herbicides

Lauren, D.R.; Taylor, H.J.; Rahman, A. Analysis of the herbicides dicamba, clopyralid and bromacil in asparagus by high-perfoprmance liquid chromatography. *J.Chromatogr.* **1988**, *439*, 470–475.

Luchtefeld, R.G. Multiresidue method for determining substituted urea herbicides in foods by liquid chromatography. *J.Assoc.Off.Anal.Chem.* **1987**, *70*, 740–745.

Suzuki, T.; Watanabe, S. Liquid chromatographic screening method for fluorescent derivatives of chlorophenoxy acid herbicides in water. *J.AOAC Int.* **1992**, *75*, 720–724.

Herbicides (using post-column reaction detection)

Luchtefeld, R.G. An HPLC detection system for phenylurea herbicides using post-column photolysis and chemical derivatization. *J.Chromatogr.Sci.* **1985**, *23*, 516–520.

Simon, V.A.; Taylor, A. High-sensitivity high-performance liquid chromatographic analysis of diquat and paraquat with confirmation. *J.Chromatogr.* **1989**, *479*, 153–158.

Simon, V.A. HPLC assay with confirmation of low levels of paraquat and diquat in well water. *LC-GC* **1987**, *5*, 899–903.

Hexamethylenetetramine

Levin, J.O.; Fangmark, I. High-performance liquid chromatographic determination of hexamethylenetetramine in air. *Analyst* **1988**, *113*, 511–513.

Histamine

Angi, M.R.; Bettero, A.; Benassi, C.A. Histamine in tears: developments in collection and HPLC-fluorimetric detection. *Agents Actions* **1985**, *16*, 84–86.

Bettero, A.; Angi, M.R.; Galiano, F.; Benassi, C.A. Histamine determination in tears as an index of safety in use of cosmetic products. *Int.J.Cosmet.Sci.* **1985**, *7*, 1–8.

Buteau, C.; Duitschaever, C.L.; Ashton, G.C. Stability of the o-phthalaldehyde-histamine complex. *J.Chromatogr.* **1981**, *212*, 23–27.

Czerwonka, R.; Tsikas, D.; Brunner, G. High-performance liquid chromatographic determination of plasma histamine after precolumn derivatization with o-phthaldialdehyde. *Chromatographia* **1988**, *25*, 219–222.

Gouygou, J.P.; Sinquin, C.; Durand, P. High pressure liquid chromatography determination of histamine in fish. *J.Food Sci.* **1987**, *52*, 925–927.

Harsing, L.G., Jr.; Nagashima, H.; Duncalf, D.; Vizi, E.S.; Goldiner, P.L. Determination of histamine concentrations in plasma by liquid chromatography/electrochemistry. *Clin.Chem.* **1986**, *32*, 1823–1827.

Mahendradatta, M.; Schwedt, G. Determination of histamine in foods by capillary electrophoresis and HPLC in comparison. *Dtsch.Lebensm.-Rundsch.* **1996**, *92*, 218–221.

Mell, L.D., Jr.; Hawkins, R.N.; Thompson, R.S. Fluorometric determination of histamine in biological fluids and tissue by high-performance liquid chromatography. *J.Liq.Chromatogr.* **1979**, *2*, 1393–1406.

Mett, C.L.; Sturgeon, R.J. Cation-exchange chromatography of histamine in the presence of ethylammonium chloride. *J.Chromatogr.* **1982**, *235*, 536–538.

Ohkura, Y.; Kohashi, K.; Tsuruta, Y.; Ishida, S. Ultramicrodetermination of histamines by high-performance liquid chromatography. *Adv.Biosci.* **1982**, *33*, 243–253.

Saito, K.; Horie, M.; Nakazawa, H. Determination of urinary excretion of histamine and 1-methylhistamine by liquid chromatography. *J.Chromatogr.B* **1994**, *654*, 270–275.

Saito, K.; Yamada, F.; Horie, M.; Nakazawa, H. Determination of histamine and 1-methylhistamine by liquid chromatography using on-column derivatization and a column-switching technique. *Anal.Sci.* **1993**, *9*, 803–806.

Serrar, D.; Brebant, R.; Bruneau, S.; Denoyel, G.A. The development of a monoclonal antibody-based ELISA for the determination of histamine in food: application to fishery products and comparison with the HPLC assay. *Food Chem.* **1995**, *54*, 85–91.

Skofitsch, G.; Saria, A.; Holzer, P.; Lembeck, F. Histamine in tissue: determination by high-performance liquid chromatography after condensation with o-phthaldialdehyde. *J.Chromatogr.* **1981**, *226*, 53–59.

Takagi, K.; Watanabe, T.; Yamaki, K.; Satake, T.; Suzuki, R.; Hasegawa, T.; Yamatodani, A. [Application of HPLC to measurement of plasma histamine in bronchial asthma]. *Arerugi.* **1989**, *38*, 1070–1076.

Tsuruta, Y.; Kohashi, K.; Ohkura, Y. Simultaneous determination of histamine and Ntau-methylhistamine in human urine and rat brain by high-performance liquid chromatography with fluorescence detection. *J.Chromatogr.* **1981**, *224*, 105–110.

Histamine (using post-column reaction detection)

Adachi, N.; Itoh, Y.; Oishi, R.; Saeki, K. Direct evidence for increased continuous histamine release in the striatum of conscious freely moving rats produced by middle cerebral artery occlusion. *J.Cereb.Blood Flow Metab.* **1992**, *12*, 477–483.

Alam, M.K.; Sasaki, M.; Watanabe, T.; Maeyama, K. Simultaneous determinations of histamine and N tau-methylhistamine by high-performance liquid chromatography-chemiluminescence coupled with immobilized diamine oxidase. *Anal.Biochem.* **1995**, *229*, 26–34.

Arakawa, Y.; Tachibana, S. A direct and sensitive determination of histamine in acid-deproteinized biological samples by high-performance liquid chromatography. *Anal.Biochem.* **1986**, *158*, 20–27.

Kuruma, K.; Hirai, E.; Uchida, K.; Kikuchi, J.; Terui, Y. Quantification of histamine by postcolumn fluorescence detection high-performance liquid chromatography using orthophthalaldehyde in tetrahydrofuran and reaction mechanism. *Anal.Sci.* **1994**, *10*, 259–265.

Histidine

Ali Qureshi, G.; Gutierrez, A.; Bergstroem, J. Determination of histidine, 1-methylhistidine and 3-methylhistidine in biological samples by high-performance liquid chromatography. Clinical application of urinary 3-methylhistidine in evaluating the muscle protein breakdown in uremic patients. *J.Chromatogr.* **1986**, *374*, 363–369.

Qureshi, G.A.; Gutierrez, A.; Bergstroem, J. Determination of histidine, 1-methylhistidine and 3-methylhistidine in biological samples by HPLC. Clinical application of urinary 3-methylhistidine in evaluating the muscle protein breakdown in uremic patients. *Symp.Biol.Hung.* **1986**, *34*, 103–115.

Raghavan, M.; Smith, C.K.; Schutt, C.E. Analytical determination of methylated histidine in proteins: actin methylation. *Anal.Biochem.* **1989**, *178*, 194–197.

Homocysteine

Araki, A.; Sako, Y. Determination of free and total homocysteine in human plasma by high-performance liquid chromatography with fluorescence detection. *J.Chromatogr.* **1987**, *422*, 43–52.

Fiskerstrand, T.; Refsum, H.; Kvalheim, G.; Ueland, P.M. Homocysteine and other thiols in plasma and urine: automated determination and sample stability. *Clin.Chem.* **1992**, *39*, 263–271.

Hagan, R.L. Determination of plasma homocysteine by HPLC with fluorescence detection: a survey of current methods. *J.Liq.Chromatogr.* **1993**, *16*, 2701–2714.

Ubbink, J.B.; Vermaak, W.J.H.; Bissbort, S. Rapid high-performance liquid chromatographic assay for total homocysteine levels in human serum. *J.Chromatogr.* **1991**, *565*, 441–446.

Vester, B.; Rasmussen, K. High performance liquid chromatography method for rapid and accurate determination of homocysteine in plasma and serum. *Eur.J.Clin.Chem.Clin.Biochem.* **1991**, *29*, 549–554.

Young, P.B.; Molloy, A.M.; Scott, J.M.; Kennedy, D.G. A rapid high performance liquid chromatographic method for determination of homocysteine in porcine tissue. *J.Liq.Chromatogr.* **1994**, *17*, 3553–3561.

Homomevalonate

Bergot, B.J.; Baker, F.C.; Lee, E.; Schooley, D.A. Absolute configuration of homomevalonate and 3-hydroxy-3-ethylglutaryl- and 3-hydroxy-3-methylglutaryl CoA, produced by cell-free extracts of insect corpora allata; cautionary note on prediction of absolute stereochemistry based on liquid chromatographic elution order of diastereomeric derivatives. *J.Am.Chem.Soc.* **1979**, *101*, 7432–7434.

Homoserine

Maier, K.; Costabel, U.; Lenz, A.G.; Leuschel, L. Simultaneous determination of L-homoserine and L homoserine lactone by reversed-phase liquid chromatoraphy in acid hydrolyzates of proteins after cyanogen bromide treatment. *J.Chromatogr.* **1989**, *493*, 380–387.

Hyaluronic Acid (using post-column reaction detection)

Mitsuma, R.; Yamanashi, S.; Toyoda, H.; Imanari, T. Determination of hyaluronic acid in rabbit plasma by HPLC with fluorometric postcolumn detection. *Bunseki Kagaku* **1989**, *38*, 92–93.

Hydralazine

Ludden, T.M.; Ludden, L.K.; McNay, J.L.; Skrdlant, H.B.; Swaggerty, P.J.; Shepherd, A.M.M. Improved assays for hydralazine and hydralazine pyruvic acid hydrazone in human plasma. *Anal.Chim.Acta* **1980**, *120*, 297–304.

Ravichandran, K.; Baldwin, R.P. Determination of hydralazine and metabolites in urine by liquid chromatography with electrochemical detection. *J.Chromatogr.* **1985**, *343*, 99–108.

Wong, J.K.; Joyce, T.H..I.; Morrow, D.H. Determination of hydralazine in human plasma by high-performance liquid chromatography with electrochemical detection. *J.Chromatogr.* **1987**, *385*, 261–266.

Hydrazine

Fiala, E.S.; Kulakis, C. Separation of hydrazine, monomethylhydrazine, 1, 1-dimethylhydrazine and 1, 2-dimethylhydrazine by high-performance liquid chromatography with electrochemical detection. *J.Chromatogr.* **1981**, *214*, 229–233.

George, G.D.; Stewart, J.T. HPLC determination of trace hydrazine levels in phenelzine sulfate drug substance. *Anal.Lett.* **1990**, *23*, 1417–1429.

Jackson, P.E.; Kahler, B. The determination of hydrazine in sulfuric acid by precolumn derivatization liquid chromatography. *J.High Resolut.Chromatogr.* **1992**, *15*, 620–621.

Kirchherr, H. Determination of hydrazine in human plasma by high-performance liquid chromatography. *J.Chromatogr.* **1993**, *617*, 157–162.

Matsui, F.; Sears, R.W.; Lovering, E.G. Liquid chromatographic determination of hydrazine in polyvinylpyrrolidone. *J.Assoc.Off.Anal.Chem.* **1986**, *69*, 521–523.

Hydrocortisone

Goehl, T.J.; Sundaresan, G.M.; Prasad, V.K. Fluorometric high-pressure liquid chromatographic determination of hydrocortisone in human plasma. *J.Pharm.Sci.* **1979**, *68*, 1374–1376.

Goto, J.; Shamsa, F.; Goto, N.; Nambara, T. The simultaneous determination of serum cortisol and cortisone by high-performance liquid chromatography with fluorimetric detection. *J.Pharm. Biomed.Anal.* **1983**, *1*, 83–88.

Passingham, B.J.; Barton, R.N. Application of high-performance liquid chromatography to the measurement of cortisol secretion rate. *J.Chromatogr.* **1987**, *416*, 25–35.

Hydrocortisone (using post-column reaction detection)

Van Ingen, H.E.; Endert, E. Improved method for the determination of the cortisol production rate using high-performance liquid chromatography and liquid scintillation counting. *J.Chromatogr.* **1988**, *430*, 233–239.

Hydrogen Azide

Van Wambeke, E.; Decoster, P.; Van Assche, C.; Vanachter, A. HPLC analysis of the potential disinfestant hydrogen azide in soil samples. *Meded.Fac.Landbouwwet., Rijksuniv.Gent* **1991**, *56*, 961–967.

Hydroperoxides and Peroxides (using post-column reaction detection)

Akasaka, K.; Ohrui, H.; Meguro, H. An aromatic phosphine reagent for the HPLC-fluorescence determination of hydroperoxides - determination of phosphatidylcholine hydroperoxides in human plasma. *Anal.Lett.* **1988**, *21*, 965–975.

Akasaka, K.; Ohrui, H.; Meguro, H. Measurement of cholesterol ester hydroperoxides of high and combined low and very low density lipoprotein in human plasma. *Biosci., Biotechnol., Biochem.* **1994**, *58*, 396–399.

Akasaka, K.; Ohata, A.; Ohrui, H.; Meguro, H. Automatic determination of hydroperoxides of phosphatidylcholine and phosphatidylethanolamine in human plasma. *J.Chromatogr.B* **1995**, *665*, 37–43.

Akasaka, K.; Morimune, K.; Ohrui, H.; Meguro, H.; Ohta, M. Determination of phosphatidylcholine hydroperoxide in plasma and serum by HPLC/postcolumn method. *Bunseki Kagaku* **1993**, *42*, 27–31.

Chan, Y.; Schultz, G.R.; Shulsky, M.L. A sampling and analytical method for methyl ethyl ketone peroxide. *Appl.Occup.Environ.Hyg.* **1991**, *6*, 309–314.

Frei, B.; Stocker, R.; Ames, B.N. Antioxidant defenses and lipid peroxidation in human blood plasma. *Proc.Natl.Acad.Sci.U.S.A.* **1988**, *85*, 9748–9752.

Holley, A.E.; Slater, T.F. Measurement of lipid hydroperoxides in normal human blood plasma using HPLC-chemiluminescence linked to a diode array detector for measuring conjugated dienes. *Free Radical Res.Commun.* **1991**, *15*, 51–63.

Kohno, Y.; Sakamoto, O.; Nakamura, T.; Miyazawa, T. Determination of human skin surface lipid peroxides by chemiluminescence-HPLC. II. Detection of squalene hydroperoxide. *Yukagaku* **1993**, *42*, 204–209.

Miyazawa, T. Determination of phospholipid hydroperoxides in human blood plasma by a chemiluminescence-HPLC assay. *Free Radical Biol.Med.* **1989**, *7*, 209–217.

Miyazawa, T.; Suzuki, T.; Yasuda, K.; Fujimoto, K..; Meguro, K.; Sasaki, H. Accumulation of phospholipid hydroperoxides in red blood cell membranes of senile dementia. *Int.Congr.Ser.-Excerpta Med.* **1992**, *998*, 327–330.

Miyazawa, T.; Suzuki, T.; Fujimoto, K.; Yasuda, K. Chemiluminescent simultaneous determination of phosphatidylcholine hydroperoxide and phosphatidylethanolamine hydroperoxide in the liver and brain of the rat. *J.Lipid Res.* **1992**, *33*, 1051–1059.

Miyazawa, T.; Fujimoto, K.; Oikawa, S. Determination of lipid hydroperoxides in low density lipoprotein from human plasma using high performance liquid chromatography with chemiluminescence detection. *Biomed.Chromatogr.* **1990**, *4*, 131–134.

Patel, C.P.; Lilly, S. Postcolumn high performance liquid chromatographic assessment of organic hydro-peroxide. *LC-GC* **1988**, *6*, 424–428.

Saeki, A.; Tsuchida, T.; Yamada, R.; Miyazawa, T. Measurement of lipid hydroperoxides in human blood by using chemiluminescence. *Optronics* **1995**, *162*, 142–146.

Yamamoto, Y.; Ames, B.N. Detection of lipid hydroperoxides and hydrogen peroxide at picomole levels by an HPLC and isoluminol chemiluminescence assay. *Free Radical Biol.Med.* **1987**, *3*, 359–361.

Yamamoto, Y.; Frei, B.; Ames, B.N. Assay of lipid hydroperoxides using high-performance liquid chro-matography with isoluminal chemiluminescence detection. *Methods Enzymol.* **1990**, *186*, 371–380.

Hydroxy Acids

Lu, D.S.; Feng, W.Y.; Ling, D.H.; Hua, W.Z. Determination of hydroxy acids as their copper(II) complexes by reversed-phase liquid chromatography with UV detection. *J.Chromatogr.* **1992**, *623*, 55–62.

Hydroxy Acids (using post-column reaction detection)

Katoh, H.; Ishida, T.; Kuwata, S.; Kiniwa, H. Optical resolution of 2-hydroxy acids by high-performance ligand exchange chromatography. *Chromatographia* **1990**, *28*, 481–486.

Hydroxybenzoates

Burini, G. Determination of alkyl esters of p-hydroxybenzoic acid in mayonnaise by high-performance liquid chromatography and fluorescence labeling. *J.Chromatogr.A* **1994**, *664*, 213–219.

Hydroxychloroquine

Brocks, D.R.; Pasutto, F.M.; Jamali, F. Analytical and semi-preparative high-performance liquid chro-matographic separation and assay of hydroxychloroquine enantiomers. *J.Chromatogr.* **1992**, *581*, 83–92.

4-Hydroxycyclophosphamide (using post-column reaction detection)

Wright, J.E.; Tretyakov, O.; Ayash, L.J.; Elias, A.; Rosowsky, A.; Frei, E.I. Analysis of 4-hydroxycyclo-phosphamide in human blood. *Anal.Biochem.* **1995**, *224*, 154–158.

Hydroxydecenoic Acid

Bloodworth, B.C.; Harn, C.S.; Hock, C.T.; Boon, Y.O. Liquid chromatographic determination of trans-10-hydroxy-2-decenoic acid content of commercial products containing royal jelly. *J.AOAC Int.* **1995**, *78*, 1019–1023.

Hydroxyeicosatetraenoic Acids

Williamson, P.K.; Zurier, R.B. Derivatization, separation and direct quantification of monohydroxy-eicosatetraenoic acids using reversed phase high performance liquid chromatography. *J.Liq.Chromatogr.* **1984**, *7*, 2193–2201.

Hydroxy Fatty Acids

Azerad, R.; Boucher, J.L.; Dansette, P.; Delaforge, M. High-performance liquid chromatographic separa-tion of 11-hydroxylauric acid enantiomers. Application to the determination of the stereochemistry of microsomal lauric acid (omega-1) hydroxylation. *J.Chromatogr.* **1990**, *498*, 293–302.

Takagi, T.; Itabashi, Y.; Tsuda, T. High-performance liquid chromatographic separation of 2-hydroxy fatty acid enantiomers on a chiral slurry-packed capillary column. *J.Chromatogr.Sci.* **1989**, *27*, 574–577.

Hydroxyindoles

Viell, B.; Vestweber, K.H.; Krause, B. Analysis of 5-OH-indoles in human gut biopsy tissues by reversed-phase high-performance liquid chromatography with fluorometric detection. *J.Pharm.Biomed.Anal.* **1988**, *6*, 939–944.

5-Hydroxyindole Derivatives (using post-column reaction detection)

Iinuma, F.; Mawatari, K.; Tabara, M.; Watanabe, M. Fluorometric determination of 5-hydroxyindole derivatives by high performance liquid chromatography with cobalt(II) chloride, sodium carbonate, and sodium hydroxide. *Bunseki Kagaku* **1984**, *33*, E323-E330.

Hydroxylamines

Musson, D.G.; Sternson, L.A. Conversion of arylhydroxylamines to electrochemically-active derivatives suitable for high-performance liquid chromatographic analysis with amperometric detection. *J.Chromatogr.* **1980**, *188*, 159−167.

Hydroxymethanesulfonate

Ang, C.C.; Lipari, F.; Swarin, S.J. Determination of hydroxymethanesulfonate in wet deposition samples. *Environ.Sci.Technol.* **1987**, *21*, 102−105.

Hydroxyproline and Proline

Calabrese, M.; Stancher, B.; Riccobon, P. High-performance liquid chromatography determination of proline isomers in Italian wines. *J.Sci.Food Agric.* **1995**, *69*, 361−366.

Castelain, S.; Kamel, S.; Picard, C.; Desmet, G.; Sebert, J.L.; Brazier, M. A simple and automated HPLC method for determination of total hydroxyproline in urine. Comparison with excretion of pyridinolines. *Clin.Chim.Acta* **1995**, *235*, 81−90.

Cooper, J.D.H.; Lewis, M.T.; Turnell, D.C. Pre-column o-phthalaldehyde derivatization of amino acids and their separation using reversed-phase high-performance liquid chromatography. I. Detection of the imino acids hyroxyproline and proline. *J.Chromatogr.* **1984**, *285*, 484−489.

Dawson, C.D.; Jewell, S.; Driskell, W.J. Liquid-chromatographic determination of total hydroxyproline in urine. *Clin.Chem.* **1988**, *34*, 1572−1574.

Green, G.D.; Reagan, K. Determination of hydroxyproline by high pressure liquid chromatography. *Anal.Biochem.* **1992**, *201*, 265−269.

Hughes, H.; Hagen, L.; Sutton, R.A.L. Liquid-chromatographic determination of 4-hydroxyproline in urine. *Clin.Chem.* **1986**, *32*, 1002−1004.

Lange, M.; Malyusz, M. Improved determination of small amounts of free hydroxyproline in biological fluids. *Clin.Chem.* **1994**, *40*, 1735−1738.

Morleo, M.A.; Musi, E. Evaluation of new product for hydroxyproline determination in HPLC. *Boll.Chim.Farm.* **1993**, *132*, 281−284.

Palmerini, C.A.; Fini, C.; Floridi, A.; Morelli, A.; Vedovelli, A. High-performance liquid chromatographic analysis of free hydroxyproline and proline in blood plasma and of free and peptide-bound hydroxyproline in urine. *J.Chromatogr.* **1985**, *339*, 285−292.

Palmerini, C.A.; Vedovelli, A.; Morelli, A.; Fini, C.; Floridi, A. Analysis of acid-soluble hydroxyproline, free proline and collagen-bound hydroxyproline in rat liver by high performance liquid chromatography with pre-column derivatization. *J.Liq.Chromatogr.* **1985**, *8*, 1853−1868.

Reed, P.; Holbrook, I.B.; Gardner, M.L.G.; McMurray, J.R. Simple, optimized liquid-chromatographic method for measuring total hydroxyproline in urine evaluated. *Clin.Chem.* **1991**, *37*, 285−290.

Schilb, L.A.; Fiegel, V.D.; Knighton, D.R. Hydroxyproline measurement by high performance liquid chromatography: an improved method of derivatization. *J.Liq.Chromatogr.* **1990**, *13*, 557−567.

Schlabach, T.D. Dual-detector methods for selective identification of prolyl residues and amide-blocked N-terminal groups in chromatographically separated peptides. *J.Chromatogr.* **1983**, *266*, 427−437.

Svanberg, G.K. Hydroxyproline determination in serum and gingival crevicular fluid. *J.Periodontal Res.* **1987**, *22*, 133−138.

Svanberg, G.K. Biological variations in serum total hydroxyproline concentration in the beagle dog. *Lab.Anim.* **1988**, *22*, 157−161.

Turpeinen, U.; Pomoell, U.M. Liquid-chromatographic determination of total hydroxyproline in urine. *Clin.Chem.* **1985**, *31*, 828−830.

Wu, G. Determination of proline by reversed-phase high-performance liquid chromatography with automated pre-column o-phthaldialdehyde derivatization. *J.Chromatogr.* **1993**, *641*, 168−175.

Hydroxyproline and Proline (using post-column reaction detection)

Macek, J.; Adam, M. Method for rapid determination of hydroxyproline by high-performance liquid chromatography and its exploitation for the study of collagen formation. *J.Chromatogr.* **1986**, *374*, 125–128.

Ozaki, A.; Shibasaki, T.; Mori, H. Specific proline and hydroxyproline detection method by post-column derivatization for high-performance liquid chromatography. *Biosci., Biotechnol., Biochem.* **1995**, *59*, 1764–1765.

8-Hydroxyquinoline (using post-column reaction detection)

Miura, K.; Nakamura, H.; Tanaka, H.; Tamura, Z. Fluorescence detection of 8-hydroxyquinoline and some of its halogenated derivatives using post-column derivatization in high-performance liquid chromatography. *J.Chromatogr.* **1981**, *210*, 536–539.

Hydroxytriazine

Ramsteiner, K.A.; Hoermann, W.D. High-pressure liquid chromatographic determination of hydroxy-s-triazine residues in plant material. *J.Agric.Food Chem.* **1979**, *27*, 934–938.

5-Hydroxytryptophol (using post-column reaction detection)

Helander, A.; Loewenmo, C.; Beck, O. Determination of 5-hydroxytryptophol in urine by high-performance liquid chromatography: application of a new post-column derivatization method with fluorometric detection. *J.Pharm.Biomed.Anal.* **1995**, *13*, 651–654.

Hypoglycin

Sarwar, G.; Botting, H.G. Reversed-phase liquid chromatographic determination of hypoglycin A (HG-A) in canned ackee fruit samples. *J.AOAC Int.* **1994**, *77*, 1175–1179.

Hypoxanthine (using post-column reaction detection)

Tawa, R.; Kitoh, M.; Hirose, S.; Adachi, K. Determination of kinetic parameters for immobilized xanthine oxidase and its application to a post-column enzyme reactor in high-performance liquid chromatographic analysis of hypoxanthine, xanthine and uric acid. *Chem.Pharm.Bull.* **1982**, *30*, 615–621.

Hypusine

Beninati, S.; Abbruzzese, A.; Folk, J.E. High-performance liquid chromatographic method for determination of hypusine and deoxyhypusine. *Anal.Biochem.* **1989**, *184*, 16–20.

IAA

Hall, J.F.; Brown, S.J.; Gartland, K.M.A. IAA analysis in transgenic plants. *Methods Mol.Biol.(Totowa, N.J.)* **1995**, *44*, 237–244.

Ibuprofen

Rudy, A.C.; Anliker, K.S.; Hall, S.D. High-performance liquid chromatographic determination of the stereoisomeric metabolites of ibuprofen. *J.Chromatogr.* **1990**, *528*, 395–405.

Sen, A.K.; Bandyopadhyay, A.; Podder, G.; Chowdhury, B. Reversed-phase high performance liquid chromatographic determination of ibuprofen and ethambutol in pharmaceutical dosage form. *J.Indian Chem.Soc.* **1990**, *67*, 443–444.

Idapril

Lippi, A.; Criscuoli, M.; Sardelli, G.; Subissi, A. High-performance liquid chromatographic method with electrochemical detection for the determination of idapril, a novel angiotensin-converting enzyme inhibitor, in biological matrices. *J.Chromatogr.B* **1994**, *660*, 127–134.

Imidazoline

Newsome, W.H.; Panopio, L.G. A method for the determination of 2-imidazoline residues in food crops. *J.Agric.Food Chem.* **1978**, *26*, 638–640.

Imipenem (using post-column reaction detection)

Musson, D.G.; Hajdu, R.; Bayne, W.F.; Rogers, J.D. Quantification of imipenem's primary metabolite in plasma by postcolumn chemical rearrangement and UV detection. *Pharm.Res.* **1991**, *8*, 33–39.

Indoleacetic Acid

Blakesley, D.; Hall, J.F.; Weston, G.D.; Elliott, M.C. Simultaneous analysis of indole-3-acetic acid and detection of 4-chloroindole-3-acetic acid and 5-hydroxyindole-3-acetic acid in plant tissues by high-performance liquid chromatography of their 2-methylindolo-α-pyrone derivatives. *J.Chromatogr.* **1983**, *258*, 155–164.

Lebuhn, M.; Hartmann, A. Method for the determination of indole-3-acetic acid and related compounds of L-tryptophan catabolism in soils. *J.Chromatogr.* **1993**, *629*, 255–266.

Indomethacin (using post-column reaction detection)

Kubo, H.; Umiguchi, Y.; Kinoshita, T. Fluorometric determination of indomethacin in serum by high performance liquid chromatography with in-line alkaline hydrolysis. *Chromatographia* **1992**, *33*, 321–324.

Pratzel, H.; Dittrich, P.; Kukovetz, W. Spontaneous and forced cutaneous absorption of indomethacin in pigs and humans. *J.Rheumatol.* **1986**, *13*, 1122–1125.

Stubbs, R.J.; Schwartz, M.S.; Chiou, R.; Entwistle, L.A.; Bayne, W.F. Improved method for the determination of indomethacin in plasma and urine by reversed-phase high-performance liquid chromatography. *J.Chromatogr.* **1986**, *383*, 432–437.

Inorganic Anions

De Kleijn, J.P. Simple, sensitive, and simultaneous determination of some selected inorganic anions by high-performance liquid chromatography. *Analyst* **1982**, *107*, 223–225.

Leuenberger, U.; Gauch, R.; Rieder, K.; Baumgartner, E. Determination of nitrate and bromide in food-stuffs by high-performance liquid chromatography. *J.Chromatogr.* **1980**, *202*, 461–468.

Inositol

Indyk, H.E.; Woollard, D.C. Determination of free myo-inositol in milk and infant formula by high-performance liquid chromatography. *Analyst* **1994**, *119*, 397–402.

Kargacin, M.E.; Bassell, G.; Ryan, P.J.; Honeyman, T.W. Separation and analysis of fluorescent derivatives of myo-inositol and myo-inositol 2-phosphate by high-performance liquid chromatography. *J.Chromatogr.* **1987**, *393*, 454–458.

Inositol Phosphates (using post-column reaction detection)

Clarkin, C.M.; Minear, R.A.; Kim, S.; Elwood, J.W. An HPLC postcolumn reaction system for phosphorus-specific detection in the complete separation of inositol phosphate congeners in aqueous samples. *Environ.Sci.Technol.* **1992**, *26*, 199–204.

Inamoto, Y.; Hiraga, Y.; Hanai, T.; Kinosita, T. The development of a sensitive myo-inositol analyzer using a liquid chromatograph with a post-label fluorescence detector. *Biomed.Chromatogr.* **1995**, *9*, 146–149.

Mayr, G.W. A novel metal-dye detection system permits picomolar-range h.p.l.c. analysis of inositol polyphosphates from non-radioactively labelled cell or tissue specimens. *Biochem.J.* **1988**, *254*, 585–591.

Franz, H.; Maier, H.G. Inositol phosphates in coffee and coffee products. Part 1. Identification and method of determination. *Dtsch.Lebensm.-Rundsch.* **1993**, *89*, 276–282.

Guse, A.H.; Emmrich, F. T-cell receptor-mediated metabolism of inositol polyphosphates in Jurkat T-lymphocytes. Identification of a D-myo-inositol 1,2,3,4,6-pentakisphosphate-2-phosphomonoesterase activity, a D-myo-inositol 1,3,4,5,6-pentakisphosphate-1/3-phosphatase activity and a D/L-myo-inositol 1,2,4,5,6-pentakisphosphate-1/3-kinase activity. *J.Biol.Chem.* **1991**, *266*, 24498–24502.

Guse, A.H.; Roth, E.; Broeker, B.M.; Emmrich, F. Complex inositol polyphosphate response induced by co-cross-linking of CD4 and Fcγ receptors in the human monocytoid cell line U937. *J.Immunol.* **1992**, *149*, 2452–2458.

Guse, A.H.; Goldwich, A.; Weber, K.; Mayr, G.W. Non-radioactive, isomer-specific inositol phosphate mass determinations: high-performance liquid chromatography-micro-metal-dye detection strongly improves speed and sensitivity of analyses from cells and micro-enzyme assays. *J.Chromatogr.B* **1995**, *672*, 189–198.

Guse, A.H.; Greiner, E.; Emmrich, F.; Brand, K. Mass changes of inositol 1,3,4,5,6-pentakisphosphate and inositol hexakisphosphate during cell cycle progression in rat thymocytes. *J.Biol.Chem.* **1993**, *268*, 7129–7133.

Radenberg, T.; Scholz, P.; Bergmann, G.; Mayr, G.W. The quantitative spectrum of inositol phosphate metabolites in avian erythrocytes, analyzed by proton NMR and HPLC with direct isomer detection. *Biochem.J.* **1989**, *264*, 323–333.

Ye, W.; Cheng, K.; Zhu, P.; Li, S.; Yuan, Z. Assay of inositol triphosphate by HPLC using an immobilized alkaline phosphatase-loaded post-column reactor. *Sepu* **1991**, *9*, 111–113.

Iodine (using post-column reaction detection)

Migliuolo, G.; Ruggeri, P. Quantitative determination of organic iodine levels in Feijoa sellowiana fruit. *Riv.Merceol.* **1995**, *33*, 29–36.

Yamada, H.; Kajiyama, S.; Yonebayashi, K. Determination of trace amounts of iodine in soils by HPLC with fluorescence detection. *Bunseki Kagaku* **1995**, *44*, 1027–1032.

Isocyanates

Beasley, R.K.; Warner, J.M. Determination of polymethylenepolyphenylene isocyanate in air by size exclusion chromatography. *Anal.Chem.* **1984**, *56*, 1604–1608.

Brenner, K.S.; Bosscher, F. TDI- and MDI-analysis during fire simulation tests - filter- and high-volume impinger-sampling. *Fresenius' J.Anal.Chem.* **1995**, *351*, 216–220.

Chang, S.N.; Burg, W.R. Determination of airborne 2,4-toluenediisocyanate vapors. *J.Chromatogr.* **1982**, *246*, 113–120.

Graham, J.D. Simplified sample handling procedure for monitoring industrial isocyanates in air. *J.Chromatogr.Sci.* **1980**, *18*, 384–387.

Hanus, F.; Merz, W.; Oldeweme, J.; Randt, C. Method for determining airborne diisocyanates. *Mikrochim.Acta* **1988**, *3*, 197–206.

Hardy, H.L.; Walker, R.F. Novel reagent for the determination of atmospheric isocyanate monomer concentrations. *Analyst* **1979**, *104*, 890–891.

Lesage, J.; Goyer, N.; Desjardins, F.; Vincent, J.Y.; Perrault, G. Workers' exposure to isocyanates. *Am.Ind.Hyg.Assoc.J.* **1992**, *53*, 146–153.

Meyer, S.D.; Tallman, D.E. The determination of toluene diisocyanate in air by high-performance liquid chromatography with electrochemical detection. *Anal.Chim.Acta* **1983**, *146*, 227–236.

Raghuveeran, C.D.; Kaushik, M.P. Reversed-phase high-performance liquid chromatography of methyl isocyanate. *J.Chromatogr.* **1985**, *346*, 446–449.

Rando, R.J.; Poovey, H.G. Dichotomous sampling of vapor and aerosol of methylene-bis-(phenylisocyanate) with an annular diffusional denuder. *Am.Ind.Hyg.Assoc.J.* **1994**, *55*, 716–721.

Robert, A.; Simon, P. A solvent-free sampling method for airborne toluene diisocyanate. *Chromatographia* **1987**, *23*, 507–511.

Rosenberg, C.; Tuomi, T. Airborne isocyanates in polyurethane spray painting: determination and respirator efficiency. *Am.Ind.Hyg.Assoc.J.* **1984**, *45*, 117–121.

Schmidtke, F.; Seifert, B. A highly sensitive high-performance liquid chromatographic procedure for the determination of isocyanates in air. *Fresenius' J.Anal.Chem.* **1990**, *336*, 647–654.

Simon, P.; Moulut, O. Separation of the urea piperazine derivatives of polyisocyanate monomers and prepolymers by normal phase chromatography. *J.Liq.Chromatogr.* **1988**, *11*, 2071–2089.

Streicher, R.P.; Arnold, J.E.; Cooper, C.V.; Fischbach, T.J. Investigation of the ability of MDHS Method 25 to determine urethane-bound isocyanate groups. *Am.Ind.Hyg.Assoc.J.* **1995**, *56*, 437–442.

Wu, W.S.; Huang, L.K.; Gaind, V.S. High performance liquid chromatographic analysis of airborne isophorone diisocyanate and the authentication of analytical standards. *Am.Ind.Hyg.Assoc.J.* **1986**, *47*, 482–487.

Isoniazid

Kohno, H.; Kubo, H.; Furukawa, K.; Yoshino, N.; Nishikawa, T. Fluorometric determination of isoniazid and its metabolites in urine by high-performance liquid chromatography using in-line derivatization. *Ther.Drug Monit.* **1991**, *13*, 428–432.

Svensson, J.O.; Muchtar, A.; Ericsson, O. Ion-pair high-performance liquid chromatographic determination of isoniazid and acetylisoniazid in plasma and urine. Application for acetylator phenotyping. *J.Chromatogr.* **1985**, *341*, 193–197.

von Sassen, W.; Castro-Parra, M.; Musch, E.; Eichelbaum, M. Determination of isoniazid, acetylisoniazid, acetylhydrazine and diacetylhydrazine in biological fluids by high-performance liquid chromatography. *J.Chromatogr.* **1985**, *338*, 113–122.

Isoniazid (using post-column reaction detection)

Kubo, H.; Kinoshita, T.; Matsumoto, K.; Nishikawa, T. Fluorometric determination of isoniazid and its metabolites in urine by high-performance liquid chromatography. *Chromatographia* **1990**, *30*, 69–72.

N-(*trans*-4-Isopropylcyclohexylcarbonyl)-D-phenylalanine

Shinkai, H.; Nashikawa, M.; Sato, Y. Separation of a new antidiabetic agent, N-(trans-4-isopropylcyclohexylcarbonyl)-D-phenylalanine, and its isomers by chiral high-performance liquid chromatography. *J.Liq.Chromatogr.* **1989**, *12*, 457–464.

Ivermectin

Chiou, R.; Stubbs, R.J.; Bayne, W.F. Determination of ivermectin in human plasma and milk by high-performance liquid chromatography with fluorescence detection. *J.Chromatogr.* **1987**, *416*, 196–202.

Iosifidou, E.; Shearan, P.; O'Keeffe, M. Application of the matrix solid phase dispersion technique for the determination of ivermectin residues in fish muscle tissue. *Analyst* **1994**, *119*, 2227.

Schenck, F.J.; Barker, S.A.; Long, A.R. Matrix solid-phase dispersion extraction and liquid chromatographic determination of ivermectin in bovine liver tissue. *J.AOAC Int.* **1992**, *75*, 655–658.

Josamycin

Tod, M.; Biarez, O.; Nicolas, P.; Petitjean, O. Sensitive determination of josamycin and rokitamycin in plasma by high-performance liquid chromatography with fluorescence detection. *J.Chromatogr.* **1991**, *575*, 171–176.

Kanamycin (using post-column reaction detection)

Kubo, H.; Kobayashi, Y.; Nishikawa, T. Rapid method for determination of kanamycin and dibekacin in serum by use of high-pressure liquid chromatography. *Antimicrob.Agents Chemother.* **1985**, *28*, 521–523.

Ketamine

Aboul-Enein, H.Y.; Islam, M.R. Enantiomeric separation of ketamine hydrochloride in enantiomeric formulation and human serum by chiral pharmaceutical liquid chromatography. *J.Liq.Chromatogr.* **1992**, *15*, 3285–3293.

Keto Acids

Buchanan, D.N.; Thoene, J.G. Analysis of α-ketocarboxylic acids by ion exchange HPLC with UV and amperometric detection. *J.Liq.Chromatogr.* **1981**, *4*, 1219–1224.

Garibotto, G.; Ancarani, P.; Russo, R.; Sala, M.R.; Fiorini, F.; Paoletti, E. Reversed-phase high-performance liquid chromatographic analysis of branched-chain keto acid hydrazone derivatives: optimization of techniques and application to branched-chain keto acid balance studies across the forearm. *J.Chromatogr.* **1991**, *572*, 11–23.

Gasking, A.L.; Edwards, W.T.E.; Hobson-Frohock, A.; Elia, M.; Livesey, G. Quantitative high-performance liquid chromatographic analysis of branched-chain 2-keto acids in biological samples. *Methods Enzymol.* **1988**, *166*, 20–27.

Hara, S.; Takemori, Y.; Yamaguchi, M.; Nakamura, M.; Ohkura, Y. Determination of α-keto acids in serum and urine by high-performance liquid chromatography with fluorescence detection. *J.Chromatogr.* **1985**, *344*, 33–39.

Hayashi, T.; Tsuchiya, H.; Todoriki, H.; Naruse, H. High-performance liquid chromatographic determination of α-keto acids in human urine and plasma. *Anal.Biochem.* **1982**, *122*, 173–179.

Hayashi, T.; Todoriki, H.; Naruse, H. High-performance liquid chromatographic determination of α-keto acids. *J.Chromatogr.* **1981**, *224*, 197–204.

Hayashi, T.; Tsuchiya, H.; Naruse, H. High-performance liquid chromatographic determination of α-keto acids in plasma with fluorometric detection. *J.Chromatogr.* **1983**, *273*, 245–252.

Kieber, D.J.; Mopper, K. Reversed-phase high-performance liquid chromatographic analysis of α-keto acid quinoxalinol derivatives. Optimization of technique and application to natural samples. *J.Chromatogr.* **1983**, *281*, 135–149.

Kieber, D.J.; Mopper, K. Trace determination of α-keto acids in natural waters. *Anal.Chim.Acta* **1986**, *183*, 129–140.

Livesey, G.; Edwards, W.T.E. Quantification of branched-chain α-keto acids as quinoxalinols: importance of excluding oxygen during derivatization. *J.Chromatogr.* **1985**, *337*, 98–102.

Keto Acids (using post-column reaction detection)

Kiba, N.; Muto, M.; Furusawa, M. High-performance liquid chromatographic determination of branched-chain α-keto acids in serum using immobilized leucine dehydrogenase as post-column reactor. *J.Chromatogr.* **1989**, *497*, 236–242.

Tanabe, S.; Toida, T.; Kawanishi, T.; Togawa, T.; Imanari, T. Determination of α-keto and hydroxy acids by high performance liquid chromatography using ferric perchlorate as a detection reagent. *Anal.Sci.* **1986**, *1*, 281–284.

Ketoconazole (using post-column reaction detection)

Hoffman, D.W.; Jones-King, K.L.; Ravaris, C.L.; Edkins, R.D. Electrochemical detection for high-performance liquid chromatography of ketoconazole in plasma and saliva. *Anal.Biochem.* **1988**, *172*, 495–498.

Kynurenic Acid (using post-column reaction detection)

Heyes, M.P.; Quearry, B.J. Quantification of kynurenic acid in cerebrospinal fluid: effects of systemic and central L-kynurenine administration. *J.Chromatogr.* **1990**, *530*, 108–115.

Shibata, K. Fluorimetric microdetermination of kynurenic acid, an endogenous blocker of neurotoxicity, by high-performance liquid chromatography. *J.Chromatogr.* **1988**, *430*, 376–380.

Swartz, K.J.; Matson, W.R.; MacGarvey, U.; Ryan, E.A.; Beal, M.F. Measurement of kynurenic acid in mammalian brain extracts and cerebrospinal fluid by high-performance liquid chromatography with fluorometric and coulometric electrode array detection. *Anal.Biochem.* **1990**, *185*, 363–376.

Labetalol (using post-column reaction detection)

Luke, D.R.; Matzke, G.R.; Clarkson, J.T.; Awni, W.M. Improved liquid-chromatographic assay of labetalol in plasma. *Clin.Chem.* **1987**, *33*, 1450–1452.

Lactic Acid

Simonides, W.S.; Zaremba, R.; Van Hardeveld, C.; Van der Laarse, W.J. A nonenzymic method for the determination of picomole amounts of lactate using HPLC: its application to single muscle fibers. *Anal.Biochem.* **1988**, *169*, 268–273.

LC-MS (using post-column reaction detection)

Kohler, M.; Leary, J.A. LC/MS/MS of carbohydrates with postcolumn addition of metal chlorides using triaxial electrospray probe. *Anal.Chem.* **1995**, *67*, 3501–3508.

Kuhlmann, F.E.; Apffel, A.; Fischer, S.M.; Goldberg, G.; Goodley, P.C. Signal enhancement for gradient reverse-phase high-performance liquid chromatography-electrospray ionization mass spectrometry analysis with trifluoroacetic and other strong acid modifiers by postcolumn addition of propionic acid and isopropanol. *J.Am.Soc.Mass Spectrom.* **1995**, *6*, 1221–1225.

Leucine

Balagopal, P.; Ford, G.C.; Ebenstein, D.B.; Nadeau, D.A.; Nair, K.S. Mass spectrometric methods for determination of [13C]leucine enrichment in human muscle protein. *Anal.Biochem.* **1996**, *239*, 77–85.

Brown, L.L.; Williams, P.E.; Becker, T.A.; Ensley, R.J.; May, M.E.; Abumrad, N.N. Rapid high-performance liquid chromatographic method to measure plasma leucine: importance in the study of leucine kinetics in vivo. *J.Chromatogr.* **1988**, *426*, 370–375.

Jegorov, A.; Matha, V.; Trnka, T.; Cerny, M. Enantiomeric separation of leucines. *J.High Resolut.Chromatogr.* **1990**, *13*, 718–720.

Leukotrienes

Dobrowsky, R.T.; O'Sullivan, G.; Ballas, L.M.; Fleisher, L.N.; Olson, N.C. Formation of isoindole derivatives of sulfidopeptide leukotrienes by reaction with o-phthalaldehyde and separation by reverse phase high performance liquid chromatography. *J.Liq.Chromatogr.* **1987**, *10*, 137–160.

Tsikas, D.; Brunner, G. Ion-exchange high-performance liquid chromatographic determination of leukotrienes after precolumn derivatization with O-phthaldialdehyde. *Fresenius' Z.Anal.Chem.* **1988**, *332*, 369–370.

Willamson, P.K.; Zurier, R.B.; Godfrey, R.; Bomalski, J.S.; Clark, M.A. Quantification of leukotriene B4 using high-performance liquid chromatography. *J.Liq.Chromatogr.* **1987**, *10*, 2205–2212.

Yamaguchi, M.; Takehiro, O.; Ishida, J.; Nakamura, M. High-performance liquid chromatography of arachidonic acid metabolites and its application to the determination of leukotriene B4 in stimulated leukocytes. *Chem.Pharm.Bull.* **1989**, *37*, 2846–2848.

Leukotrienes (using post-column reaction detection)

Nicoll-Griffith, D.; Zamboni, R.; Rasmussen, J.B.; Ethier, D.; Charleson, S.; Tagari, P. BIO-fully automated sample treatment high-performance liquid chromatography and radioimmunoassay for leukotriene E4 in human urine from asthmatics. *J.Chromatogr.* **1990**, *526*, 341–354.

Leucovorin (using post-column reaction detection)

Mandl, A.; Lindner, W. Improved detection of leucovorin in mixed folates and antifolates by reversed-phase liquid chromatography and on-line post-column UV irradiation. *Chromatographia* **1996**, *43*, 327–330.

Levoprotiline

Horne, C.; Spahn, H.; Mutschler, E. Fluorometric determination of levoprotiline in human plasma after thin-layer chromatographic or high performance liquid chromatographic separation. *Arzneim.-Forsch.* **1987**, *37*, 1179–1181.

Lidocaine

Tam, Y.K.; Tawfik, S.R.; Ke, J.; Coutts, R.T.; Gray, M.R.; Wyse, D.G. High-performance liquid chromatography of lidocaine and nine of its metabolites in human plasma and urine. *J.Chromatogr.* **1987**, *423*, 199–206.

Lipids

Blank, M.L.; Cress, E.A.; Lee, P.; Stephens, N.; Piantadosi, C.; Snyder, F. Quantitative analysis of ether-linked lipids as alkyl- and alk-1-enyl-glycerol benzoates by high-performance liquid chromatography. *Anal.Biochem.* **1983**, *133*, 430–436.

Eaton, S.; Shmueli, E.; al-Mardini, H.; Bartlett, K.; Record, C.O. An HPLC assay for sn-1,2-diacylglycerol. *Clin.Chim.Acta* **1995**, *234*, 71–78.

Gutnikov, G.; Streng, J.R. Rapid high-performance liquid chromatographic determination of fatty acid profiles of lipids by conversion to their hydroxamic acids. *J.Chromatogr.* **1991**, *587*, 292–296.

Hamilton, R.J.; Mitchell, S.F.; Sewell, P.A. Techniques for the detection of lipids in high-performance liquid chromatography. *J.Chromatogr.* **1987**, *395*, 33–46.

Laakso, P.; Christie, W.W. Chromatographic resolution of chiral diacylglycerol derivatives: potential in the stereospecific analysis of triacyl-sn-glycerols. *Lipids* **1990**, *25*, 349–353.

Saeed, T.; Ali, S.G.; Abdul Rahman, H.A.; Sawaya, W.N. Detection of pork and lard as adulterants in processed meat: liquid chromatographic analysis of derivatized triglycerides. *J.Assoc.Off.Anal.Chem.* **1989**, *72*, 921–925.

Sempore, B.G.; Bezard, J.A. Separation of monoacylglycerol enantiomers as urethane derivatives by chiral-phase high performance liquid chromatography. *J.Liq.Chromatogr.* **1994**, *17*, 1679–1694.

Sempore, B.G.; Bezard, J.A. Enantiomer separation by chiral-phase liquid chromatography of urethane derivatives of natural diacylglycerols previously fractionated by reversed-phase liquid chromatography. *J.Chromatogr.* **1991**, *557*, 227–240.

Takagi, T.; Nishimura, K. HPLC (high-performance liquid chromatographic) separation of acyl reversed isomers of 1,2-diacylglycerol as dinitrobenzoates on a silver ion column. *Yukagaku* **1991**, *40*, 678–679.

Warne, T.R.; Robinson, M. A method for the simultaneous determination of alkylacylglycerol, diacylglycerol, monoalkylglycerol, monoacylglycerol, and cholesterol by high-performance liquid chromatography. *Anal.Biochem.* **1991**, *198*, 302–307.

Lipids (using post-column reaction detection)

Christensen, T.C.; Hoelmer, G. Lipid oxidation determination in butter and dairy spreads by HPLC. *J.Food Sci.* **1996**, *61*, 486–489.

Kiba, N.; Goto, K.; Furusawa, M. Determination of glycerol, 1,2-propanediol, and triglycerides by high-performance liquid chromatography and a post-column reactor containing immobilized glycerol dehydrogenase. *Anal.Chim.Acta* **1986**, *185*, 287–294.

Kondoh, Y.; Takano, S. Analysis of acylglycerols by high-performance liquid chromatography with post-column derivatization. IV. Simultaneous determination of mono-, di- and triacylglycerols. *J.Chromatogr.* **1987**, *393*, 427–432.

Maekinen, M.; Piironen, V.; Hopia, A. Postcolumn chemiluminescence, ultraviolet and evaporative light-scattering detectors in high-performance liquid chromatographic determination of triacylglycerol oxidation products. *J.Chromatogr.A* **1996**, *734*, 221–229.

Miyazawa, T.; Kunika, H.; Fujimoto, K.; Endo, Y.; Kaneda, T. Chemiluminescence detection of mono-, bis-, and tris-hydroperoxy triacylglycerols present in vegetable oils. *Lipids* **1995**, *30*, 1001–1006.

Takano, S.; Kondo, Y. Triglyceride analysis by combined argentation/nonaqueous reversed phase high performance liquid chromatography. *JAOCS, J.Am.Oil Chem.Soc.* **1987**, *64*, 380–383.

Takano, S.; Kondoh, Y. Monoglyceride analysis with reversed phase HPLC. *JAOCS, J.Am.Oil Chem.Soc.* **1987**, *64*, 1001–1003.

Yasaei, P.M.; Yang, G.C.; Warner, C.R.; Daniels, D.H.; Ku, Y. Singlet oxygen oxidation of lipids resulting from photochemical sensitizers in the presence of antioxidants. *J.Am.Oil Chem.Soc.* **1996**, *73*, 1177–1181.

Lipoproteins (using post-column reaction detection)

Hazell, L.J.; Stocker, R. Oxidation of low-density lipoprotein with hypochlorite causes transformation of the lipoprotein into a high-uptake form for macrophages. *Biochem.J.* **1993**, *290*, 165–172.

Sattler, W.; Mohr, D.; Stocker, R. Rapid isolation of lipoproteins and assessment of their peroxidation by high-performance liquid chromatography postcolumn chemiluminescence. *Methods Enzymol.* **1994**, *233*, 469–489.

LSD

Nakahara, Y.; Kikura, R.; Takahashi, K.; Foltz, R.L.; Mieczkowski, T. Detection of LSD and metabolite in rat hair and human hair. *J.Anal.Toxicol.* **1996**, *20*, 323–329.

Lysine

Cottingham, L.S.; Smallidge, R.T. Reverse-phase liquid chromatographic determination of lysine in complete feeds and premixes, using manual precolumn derivatization. *J.Assoc.Off.Anal.Chem.* **1988**, *71*, 1012–1016.

Muhammad, N.; Bodnar, J.A. Stability-indicating high-pressure liquid chromatographic assay for L-lysine. *J.Liq.Chromatogr.* **1980**, *3*, 529–536.

Soper, S.A.; Chamberlin, S.; Johnson, C.K.; Kuwana, T. The intramolecular loss of fluorescence by lysine derivatized with naphthalenedialdehyde. *Appl.Spectrosc.* **1990**, *44*, 858–863.

Thio, A.P.; Tompkins, D.H. Regulatory approach to determination of lysine in feedstuffs by liquid chromatography with fluorescence detection via precolumn dansylation. *J.Assoc.Off.Anal.Chem.* **1989**, *72*, 609–613.

Lysinoalanine

Moret, S.; Cherubin, S.; Rodriguez-Estrada, M.T.; Lercker, G. Determination of lysinoalanine by high performance liquid chromatography. *J.High Resolut.Chromatogr.* **1994**, *17*, 827–830.

Lysocellin

Williams, W.H.; Ash, G.D.; Heady, M.A. High-performance liquid chromatographic technique for the determination of the polyether antibiotic lysocellin sodium. *Analyst* **1989**, *114*, 887–889.

Lyso-platelet-activating Factor

Salari, H.; Eigendorf, G.K. Detection of lyso-platelet-activating factor by high-performance liquid chromatography after derivatization with fluorescent fatty acids. *J.Chromatogr.* **1990**, *527*, 303–314.

Maduramicin

Gliddon, M.J.; Wright, D.; Markantonatos, A.; Groth, W. Determination of maduramicin ammonium in poultry feed by high-performance liquid chromatography. *Analyst* **1988**, *113*, 813–816.

Malic Acid (using post-column reaction detection)

Doner, L.W.; Cavender, P.J. Chiral liquid chromatography for resolving malic acid enantiomers in adulterated apple juice. *J.Food Sci.* **1988**, *53*, 1898–1899.

Malonaldehyde

Draper, H.H.; Squires, E.J.; Mahmoodi, H.; Wu, J.; Agarwal, S.; Hadley, M. A comparative evaluation of thiobarbituric acid methods for the determination of malondialdehyde in biological materials. *Free Radical Biol.Med.* **1993**, *15*, 353–363.

Fukunaga, K.; Takama, K.; Suzuki, T. High-performance liquid chromatographic determination of plasma malondialdehyde level without a solvent extraction procedure. *Anal.Biochem.* **1995**, *230*, 20–23.

Guichardant, M.; Valette-Talbi, L.; Cavadini, C.; Crozier, G.; Berger, M. Malondialdehyde measurement in urine. *J.Chromatogr.B* **1994**, *655*, 112–116.

Kawai, S.; Kasashima, K.; Tomita, M. High-performance liquid chromatographic determination of malonaldehyde in serum. *J.Chromatogr.* **1989**, *495*, 235–238.

Osawa, T.; Shibamoto, T. Analysis of free malonaldehyde formed in lipid peroxidation systems via a pyrimidine derivative. *J.Am.Oil Chem.Soc.* **1992**, *69*, 466–468.

Mannitol

Davis, G.E.; Garwood, V.W.; Barfuss, D.L.; Husaini, S.A.; Blanc, M.B.; Viani, R. Chromatographic profile of carbohydrates in commercial coffees. 2. Identification of mannitol. *J.Agric.Food Chem.* **1990**, *38*, 1347–1350.

Samarco, E.C.; Parente, E.S. Automated high pressure liquid chromatographic system for determination of mannitol, sorbitol, and xylitol in chewing gums and confections. *J.Assoc.Off.Anal.Chem.* **1982**, *65*, 76–78.

Melphalan

Sweeney, D.J.; Greig, N.H.; Rapoport, S.I. High-performance liquid chromatographic analysis of melphalan in plasma, brain and peripheral tissue by o-phthalaldehyde derivatization and fluorescence detection. *J.Chromatogr.* **1985**, *339*, 434–439.

6-Mercaptopurine

Warren, D.J.; Slordal, L. A sensitive high-performance liquid chromatographic method for the determination of 6-mercaptopurine in plasma using precolumn derivatization and fluorescence detection. *Ther.Drug Monit.* **1993**, *15*, 25–30.

Whalen, C.E.; Tamary, H.; Greenberg, M.; Zipursky, A.; Soldin, S.J. Analysis of 6-mercaptopurine in serum or plasma using high performance liquid chromatography. *Ther.Drug Monit.* **1985**, *7*, 315–320.

6-Mercaptopurine (using post-column reaction detection)

Jonkers, R.E.; Oosterhuis, B.; Berge, R.J.M.T.; van Boxtel, C.J. Analysis of 6-mercaptopurine in human plasma with a high-performance liquid chromatographic method including post-column derivatization and fluorimetric detection. *J.Chromatogr.* **1982**, *233*, 249–255.

Metal Complexes (using post-column reaction detection)

Ehrling, C.; Schmidt, U.; Liebscher, H. Analysis of chromium(III)-fluoride complexes by ion chromatography. *Fresenius' J.Anal.Chem.* **1996**, *354*, 870–873.

Metals

Bettmer, J.; Cammann, K.; Robecke, M. Determination of organic ionic lead and mercury species with high-performance liquid chromatography using sulfur reagents. *J.Chromatogr.* **1993**, *654*, 177–182.

Borch, R.F.; Markovitz, J.H.; Pleasants, M.E. A new method for the HPLC analysis of platinum(II) in urine. *Anal.Lett.* **1979**, *12*, 917–926.

Cammann, K.; Robecke, M.; Bettner, J. Simultaneous determination of organic ionic lead and mercury species using HPLC. *Fresenius' J.Anal.Chem.* **1994**, *350*, 30–33.

Fabbri, D.; Trombini, C. A novel derivatization procedure for inorganic mercury(II) for HPLC analysis. *Chromatographia* **1994**, *39*, 246–248.

Hoffmann, B.W.; Schwedt, G. Application of HPLC to inorganic analysis. Part VII. Comparison between pre-column- and on-column derivatization for separation of different metal oxinates; quantitative determination of manganese(II) besides manganese(III) ions. *HRC CC, J.High Resolut.Chromatogr. Chromatogr.Commun.* **1982**, *5*, 439–440.

Hoshino, H.; Nakano, K.; Yotsuyanagi, T. Formazan derivatives as the precolumn derivatization reagents in a coupled high-performance liquid chromatographic-spectrophotometric system for trace metal determination. *J.Chromatogr.* **1990**, *515*, 603–610.

Iki, N.; Hoshino, H.; Yotsuyanagi, T. 2-Pyridylaldehyde benzoylhydrazone derivatives as highly selective precolumn chelating reagents for nickel(II) ion in kinetic differentiation mode high-performance liquid chromatography. *Mikrochim.Acta* **1994**, *113*, 137–152.

Inoue, H.; Ito, K. Determination of trace amounts of iron(II, III) in natural water by reversed-phase high-performance liquid chromatography. *Microchem.J.* **1994**, *49*, 249–255.

Irth, H.; de Jong, G.J.; Brinkman, U.A.T.; Frei, R.W. Trace enrichment and separation of metal ions as dithiocarbamate complexes by liquid chromatography. *Anal.Chem.* **1987**, *59*, 98–101.

Jancarova, I.; Krizova, H.; Kuban, V. Determination of uranium in technological waters by ion-pair liquid chromatography. *Talanta* **1991**, *38*, 1093–1097.

Janos, P.; Stulik, K.; Pacakova, V. An ion-exchange separation of copper(2+), cadmium(2+), lead(2+), and thallium(1+) on silica gel with polarographic detection. *Talanta* **1991**, *38*, 1445–1452.

Karcher, B.D.; Krull, I.S. The use of complexing eluents for the high performance liquid chromatographic determination of metal species. *J.Chromatogr.Libr.* **1991**, *47*, 123–166.

Kivimaki, P.R.; Lajunen, L.H.J. The simultaneous determination of chromium and cobalt in steel samples by HPLC. *Finn.Chem.Lett.* **1988**, *15*, 81–89.

Li, L.Y.; Gui, M-D.; Zhao, Y-Q. Reversed-phase HPLC determination of Co(II), Ni(II) and Fe(III) as their 2-(2-thiazolylazo)-5-dimethylaminophenol chelates. *Talanta* **1995**, *42*, 89–92.

Lin, Y.; Smart, N.G.; Wai, C.M. Supercritical fluid extraction and chromatography of metal chelates and organometallic compounds. *Trends Anal.Chem.* **1995**, *14*, 123–132.

Liu, Q.; Zhang, H.; Cheng, J. Separation and determination of cobalt, rhodium, nickel and iridium with 2-(6-methyl-2-benzothiazolylazo)-5-diethylaminophenol by HPLC. *Chem.Res.Chin.Univ.* **1993**, *9*, 18–23.

Liu, Q.; Zhang, H.; Cheng, J. Reversed-phase HPLC determination of some noble metals as their thiazolylazoresorcinol chelates. *Talanta* **1991**, *38*, 669–672.

Liu, Q.; Liu, J.; Tong, Y.; Cheng, J. Separation and determination of platinum(II), rhodium(III), palladium(II), osmium(IV), nickel(II) and cobalt(II) complexes by reversed-phase liquid chromatography. *Anal.Chim.Acta* **1992**, *269*, 223–228.

Liu, Q.; Zhang, H.; Cheng, J. Separation of cobalt(II), nickel(II), and palladium(II) with 2-(5-nitro-2-pyridylazo)-5-dimethylaminobenzoic acid by high-performance liquid chromatography. *Fresenius' J.Anal.Chem.* **1992**, *344*, 356–357.

Liu, Q.; Wang, Y.; Liu, J.; Cheng, J. Separation and determination of platinum metals and some transition metals by reversed-phase high-performance liquid chromatography. *Anal.Sci.* **1993**, *9*, 523–528.

Nagaosa, Y.; Suenaga, T.; Bond, A.M. Extraction-liquid chromatography with electrochemical and spectrophotometric detection for the determination of copper and iron in biological and river water samples. *Anal.Chim.Acta* **1990**, *235*, 279–285.

Nagaosa, Y.; Mizuyuki, T. Determination of cobalt(II) by reversed-phase liquid chromatography with electrochemical and spectrophotometric detection. *J.Liq.Chromatogr.* **1995**, *18*, 3139–3146.

Nagaosa, Y.; Kobayashi, T. Comparison of on-column and precolumn derivatization for liquid chromatographic determination of molybdenum in seawater and bovine liver. *J.AOAC Int.* **1995**, *78*, 1307–1311.

Palmisano, F.; Zambonin, P.G.; Cardellicchio, N. Speciation and simultaneous determination of mercury species in dolphin liver by liquid chromatography with on-line cold vapor atomic absorption spectrometry. *Fresenius' J.Anal.Chem.* **1993**, *346*, 648–652.

Qiping, L.; Yuanchao, W.; Jinchun, L.; Jieke, C. Separation mechanism exploration on metal-MBTAE-salicylic acid mixed ligand complexes by reversed-phase high performance liquid chromatography. *Talanta* **1995**, *42*, 901–907.

Ramesh, A. Simultaneous determination of ruthenium, osmium and palladium by reversed-phase HPLC using 4-(2'-thiazolylazo)resacetophenone oxime as chelating reagent. *Talanta* **1994**, *41*, 355–358.

Saitoh, K.; Suzuki, N. High-performance liquid chromatographic determination of nickel, copper, and zinc as their tetraphenylporphine chelates. *Anal.Chim.Acta* **1985**, *178*, 169–177.

Saraswati, R.; Desikan, N.R.; Rao, T.H. Determination of transition and rare earth elements in low-alloy steels as chelates with 4-(2-thiazolylazo)resorcinol by reversed-phase high-performance liquid chromatography. *Mikrochim.Acta* **1992**, *109*, 253–260.

Schuster, G.; Hampel, W.A. Determination of trace levels of transition metal ions in fermentation broth by ion-pair chromatography. *Anal.Chim.Acta* **1992**, *258*, 275–280.

Siren, H. Effect of ion-pairing modifiers in the separation of cobalt, copper, iron, and palladium by precolumn derivatization and high-performance liquid chromatography. *Chromatographia* **1990**, *29*, 144–150.

Smith, R.M.; Butt, A.M.; Thakur, A. Determination of lead, mercury, and cadmium by liquid chromatography using on-column derivatization with dithiocarbamates. *Analyst* **1985**, *110*, 35–37.

Steenkamp, P.A.; Coetzee, P.P. Simultaneous determination of toxic heavy metals in Metformin hydrochloride using reversed-phase high-performance liquid chromatography. *Fresenius' J.Anal.Chem.* **1993**, *346*, 1017–1021.

Vrchlabsky, M.; Pollakova, N.; Hrdlicka, A. A study of the separation of molybdenum(VI) and tungsten(VI) by reversed phase HPLC. *Collect.Czech.Chem.Commun.* **1989**, *54*, 2133–2140.

Wang, E.; Liu, A. Determination of iron, titanium, osmium, and aluminum with tiron by reverse-phase high performance liquid chromatography/electrochemistry. *Microchem.J.* **1991**, *43*, 191–197.

Wang, H.; Miao, Y-.; Zhang, H-.; Cheng, J-. Determination of V(V), Nb(V) and Ta(V) as their 2-(5-bromo-2-pyridylazo)-5-diethylaminophenol chelates by reversed-phase HPLC. *Talanta* **1994**, *41*, 685–689.

Wang, H.; Miao, Y-X.; Mou, W-.; Zhang, H-S.; Cheng, J-K. Determination of Nb(V), V(V), Co(II), Fe(III), Ni(II), Ru(III) and Pd(II) as their 4-(5-nitro-2-pyridylazo) resorcinol chelates by reversed-phase high performance liquid chromatography. *Mikrochim.Acta* **1994**, *117*, 65–74.

Wenclawiak, B.; Bickman, F. Liquid chromatographic separation of some platinum-group metal 8-hydroxyquinolinates. *Bunseki Kagaku* **1984**, *33*, E67-E72,

Wu, Y.; Schwedt, G. HPLC determination of chromium, vanadium, and molybdenum using precolumn in combination with on-column derivatization by oxine, bipyridine, and hydrogen peroxide. *Fresenius' Z.Anal.Chem.* **1987**, *329*, 39–42.

Yao, X.; Liu, J.; Cheng, J.; Zeng, Y. Speciation study of trace elements by reversed phase-high performance liquid chromatography. I. Simultaneous determination of chromium(III) and chromium(VI). *Chem.Res.Chin.Univ.* **1991**, *7*, 37–41.

Yao, X.; Liu, J.; Cheng, J.; Zeng, Y. Speciation of trace mercury in natural water by reversed-phase high-performance liquid chromatography. Part II. *Anal.Sci.* **1992**, *8*, 255–258.

Metals (using post-column reaction detection)

Adachi, M.; Oguma, K.; Kuroda, R. Reversed-phase chromatographic separation of the rare earth elements. *Chromatographia* **1990**, *29*, 579–582.

Barkley, D.J.; Blanchette, M.; Cassidy, R.M.; Elchuk, S. Dynamic chromatographic systems for the determination of rare earths and thorium in samples from uranium ore refining processes. *Anal.Chem.* **1986**, *58*, 2222–2226.

Bauer, H.; Ottenlinger, D.; Yan, D. Trace analysis of heavy metal ions by ion chromatography. *Chromatographia* **1989**, *28*, 315–316.

Beckett, J.R.; Nelson, D.A. Trace metal determinations by liquid chromatography and fluorescence detection. *Anal.Chem.* **1981**, *53*, 909–911.

Bowles, C.J.; Bader, L.W.; Jackson, K.W. Ion chromatography of metals with post-column ion displacement. *Talanta* **1990**, *37*, 835–840.

Bruzzoniti, M.C.; Mentasti, E.; Sarzanini, C.; Braglia, M.; Cocito, G.; Kraus, J. Determination of rare earth elements by ion chromatography. Separation procedure optimization. *Anal.Chim.Acta* **1996**, *322*, 49–54.

Cassidy, R.M.; Elchuk, S. Trace enrichment methods for the determination of metal ions by high performance liquid chromatography. *J.Chromatogr.Sci.* **1980**, *18*, 217–223.

Cassidy, R.M. Determination of rare-earth elements in rocks by liquid chromatography. *Chem.Geol.* **1988**, *67*, 185–195.

Cassidy, R.M.; Elchuk, S.; Dasgupta, P.K. Performance of annular membrane and screen-tee reactors for postcolumn-reaction detection of metal ions separated by liquid chromatography. *Anal.Chem.* **1987**, *59*, 85–90.

Chambaz, D.; Edder, P.; Haerdi, W. Preconcentration of divalent trace metals on chelating silicas followed by on-line ion chromatography. *J.Chromatogr.* **1991**, *541*, 443–452.

Chiba, M.; Shinohara, A.; Saiki, M.; Inaba, Y. Comparative study of methods for determining lanthanide elements in biological materials by using NAA, HPLC postcolumn reaction, and ICP-MS. *Biol.Trace Elem.Res.* **1994**, *43–45*, 561–569.

Duggan, J.X. Phosphorimetric detection in HPLC via trivalent lanthanides: high sensitivity time-resolved luminescence detection of tetracyclines using europium in a micellar post column reagent. *J.Liq.Chromatogr.* **1991**, *14*, 2499–2525.

Elchuk, S.; Cassidy, R.M. Separation of the lanthanides on high-efficiency bonded phases and conventional ion-exchange resins. *Anal.Chem.* **1979**, *51*, 1434–1438.

Falter, R.; Schoeler, H.F. A new pyrrolidinedithiocarbamate screening method for the determination of methylmercury and inorganic mercury relation in hair samples by HPLC-UV-PCO-CVAAS. *Fresenius' J.Anal.Chem.* **1996**, *354*, 492–493.

Falter, R.; Schoeler, H.F. Determination of mercury species in natural waters at picogram level with online RP C18 preconcentration and HPLC-UV-PCO-CVAAS. *Fresenius' J.Anal.Chem.* **1995**, *353*, 34–38.

Ferreira, J.I.; Ferronato, C.R.; Shihomatsu, H.M.; de Moraes, N.M.P. Study of the separation of rare earth elements by reversed phase liquid chromatography. *Publ.ACIESP* **1994**, *89*, 209–219.

Fu, C.; Zuo, B. Determination of copper(II), zinc, nickel(II), cobalt(II) and iron(II) by post-column reaction high-performance ion-exchange chromatography. *Fenxi Huaxue* **1981**, *9*, 635–639.

Gan, W.; Hu, W.; Dong, C.; Dong, M. Determination of rare earths by high-performance liquid chromatography with post-column derivatization. *Fenxi Huaxue* **1985**, *13*, 569–573.

Hirose, A.; Iwasaki, Y.; Iwata, I.; Ueda, K.; Ishii, D. Post-column colorimetric detection with Xylenol Orange in micro-HPLC of rare earth metals. *HRC CC, J.High Resolut.Chromatogr. Chromatogr. Commun.* **1981**, *4*, 530–531.

Hrdlicka, A.; Havel, J.; Valiente, M. Detection of rare earth elements by post-column reaction with Xylenol Orange and cetylpyridinium bromide. *J.High Resolut.Chromatogr.* **1992**, *15*, 423–427.

Hwang, J.M.; Chang, F.C.; Yeh, Y.C. Determination of metal ions by high performance liquid chromatography. *J.Chin.Chem.Soc.(Taipei)* **1983**, *30*, 167–172.

Karcher, B.D.; Krull, I.S.; Schleicher, R.G.; Smith, S.B., Jr. On-line extraction of metal ions using liquid-liquid segmentation and a membrane type phase separator for fluorescence detection. *Chromatographia* **1987**, *24*, 705–712.

Karcher, B.D.; Krull, I.S. Fluorescence detection of metal ions separated on a silica-based HPLC reversed-phase support. *J.Chromatogr.Sci.* **1987**, *25*, 472–478.

Kawase, J.; Nakae, A.; Tsuji, K. Determination of zeolite-A in detergent powders as acid-soluble aluminum by high-performance liquid chromatography with post-column derivatization. *Anal.Chim.Acta* **1981**, *131*, 213–222.

Kuroda, R.; Wada, T.; Kokubo, Y.; Oguma, K. Ion-interaction chromatography of nitrilotriacetato complexes of the rare earth elements with post-column reaction detection. *Talanta* **1993**, *40*, 237–241.

Lucy, C.A.; Dinh, H.N. Kinetics and equilibria of the Zn-EDTA-PAR postcolumn reaction detection system for the determination of alkaline earth metals. *Anal.Chem.* **1994**, *66*, 793–797.

Moraes, N.M.P.; Shihomatsu, H.M. Dynamic ion-exchange chromatography for the determination of lanthanides in rock standards. *J.Chromatogr.A* **1994**, *679*, 387–391.

Na, C.; Nakano, T.; Tazawa, K.; Sakagawa, M.; Ito, T. A systematic and practical method of liquid chromatography for the determination of Sr and Nd isotopic ratios and REE concentrations in geological samples. *Chem.Geol.* **1995**, *123*, 225–237.

Orlov, V.I.; Aratskova, A.A.; Timberbaev, A.R.; Petrukhin, O.M. Ion chromatography of transition and heavy metals. Determination of metal ions by cation-exchange high-performance liquid chromatography. *Zh.Anal.Khim.* **1992**, *47*, 686–692.

Rehkaemper, M. A highly sensitive HPLC method for the determination of Th and U concentrations in geological samples. *Chem.Geol.* **1995**, *119*, 1–12.

Saito, M.; Tanzawa, H.; Yamane, T. High-performance liquid chromatography of cobalt, nickel, and copper with postcolumn reaction system using 2-(5-bromo-2-pyridylazo)-5-(N-propyl-N-sulfopropylamino)phenol. *Mem.Fac.Lib.Arts Educ., Part 2 (Yamanashi Univ.)* **1991**, *42*, 32–38.

Schmidt, G.J.; Scott, R.P.W. Simple and sensitive ion chromatograph for trace metal determination. *Analyst (London)* **1984**, *109*, 997–1002.

Schreurs, M.; Somsen, G.W.; Gooijer, C.; Velthorst, N.H.; Frei, R.W. Lanthanide luminescence quenching as a detection method in ion chromatography. Chromate in surface and drinking water. *J.Chromatogr.* **1989**, *482*, 351–359.

Shinohara, A.; Chiba, M.; Inaba, Y. Effects of administration of rare earth elements on concentrations of essential elements in organs of mice. *Biomed.Res.Trace Elem.* **1993**, *4*, 115–116.

Smirnov, I.P.; Nesterenko, P.N. Use of high-performance liquid chromatography for determination of metals in wastewater. 2. HPLC of metals on silica modified by thiazolylazo compounds. *Khim.Volokna* **1992**, 56–57.

Tielrooy, J.A.; Kraak, J.C.; Maessen, F.J.M.J. High-performance liquid chromatography with post-column reaction detection for the determination of rare-earth elements in phosphoric acids produced for the manufacture of phosphate fertilizers. *Anal.Chim.Acta* **1985**, *176*, 161–174.

Trojanowicz, M.; Pobozy, E.; Worsfold, P.J. Speciation of chromium by ion-pair chromatography with postcolumn spectrophotometric detection. *Anal.Lett.* **1992**, *25*, 1373–1387.

Vacha, P.; Strouhal, R. Optimization of reaction conditions of postcolumn detection in the chromatographic determination of rare earth elements. *Sb.Vys.Sk.Chem.-Technol.Praze, H: Anal.Chem., H24* **1991**, 73–84.

Vortmueller, T.; Wuensch, G. Fluorescence detection system for the multielement determination of cations after separation by HPLC. *J.Prakt.Chem./Chem.-Ztg.* **1994**, *336*, 11–15.

Wang, E.; Liu, A. In situ electrochemical complex formation for selected metal ions based on controlled release of the pyrrolidine dithiocarbamate ligand from polypyrrole polymer. *Microchem.J.* **1991**, *44*, 327–334.

Wang, Q.; Gao, F.; Yang, L.; Shi, T.; Xie, Q. Study of the thermodynamic behavior of rare earths in the system of R-SO$_3$H-α-HIBA on HPLC and its application to analytical chemistry. *Microchem.J.* **1995**, *52*, 236–245.

Williams, T.; Barnett, N.W. 8-Quinolinol-5-sulfonic acid as a nonselective post-column reagent for fluorometric detection of trace metals in ion chromatography. *Anal.Chim.Acta* **1992**, *264*, 297–301.

Yamazaki, S.; Omori, H.; Oh, C.E. High-performance liquid chromatography of alkaline earth metal ions using reversed-phase column coated with N-n-dodecyliminodiacetic acid. *HRC CC, J.High Resolut.Chromatogr.Chromatogr.Commun.* **1986**, *9*, 765–767.

Yan, D.; Schwedt, G. Optimization of ion chromatographic trace analysis for heavy and alkaline earth metals by post-chromatographic derivatization. *Fresenius' Z.Anal.Chem.* **1987**, *327*, 503–508.

Yuan, Y.; Liu, H.; Wang, Y. High-performance liquid chromatography of micellar solubilization complex. VIII. Ion chromatography/post-column derivatization spectrophotometry of seven heavy metals. *Fenxi Huaxue* **1991**, *19*, 460–463.

Zhang, X.-X.; Cheng, J.-K.; Zeng, Y.-E. Determination of trace amount of metal ions by reversed-phase ion-pair liquid chromatography with post-column derivatization and chemiluminescence detection. *Gaodeng Xuexiao Huaxue Xuebao* **1993**, *14*, 1653–1657.

Metformin

Ohta, M.; Iwasaki, M.; Kai, M.; Ohkura, Y. Determination of a biguanide, metformin, by high-performance liquid chromatography with precolumn fluorescence derivatization. *Anal.Sci.* **1993**, *9*, 217–230.

Methadone

Schmidt, N.; Brune, K.; Geisslinger, G. Stereoselective determination of the enantiomers of methadone in plasma using high-performance liquid chromatography. *J.Chromatogr.* **1992**, *583*, 195–200.

Methionine Sulfoxide

Scislowski, P.W.D.; Harris, I.; Pickard, K.; Brown, D.S.; Buchan, V. Determination of L-methionine-dl-sulfoxide in tissue extracts. *J.Chromatogr.* **1993**, *619*, 299–305.

Puchala, R.; Pior, H.; von Keyserlingk, M.; Shelford, J.A.; Barej, W. Determination of methionine sulfoxide in biological materials using HPLC and its degradability in the rumen of cattle. *Anim.Feed Sci.Technol.* **1994**, *48*, 121–130.

Methotrexate (using post-column reaction detection)

Salamoun, J.; Frantisek, J. Determination of methotrexate and its metabolites 7-hydroxymethotrexate and 2,4-diamino-N10-methylpteroic acid in biological fluids by liquid chromatography with fluorimetric detection. *J.Chromatogr.* **1986**, *378*, 173–181.

Methoxuron

Lantos, J.; Brinkman, U.A.T.; Frei, R.W. Residue analysis of methoxuron and its breakdown product (3-chloro-4-methoxyaniline) by thin-layer and high-performance liquid chromatography with fluorescence detection. *J.Chromatogr.* **1984**, *292*, 117–127.

Methoxy(polyethylene glycol)

Leister, W.H.; Weaner, L.E.; Walker, D.G. Analysis and purification of modified methoxy(polyethylene glycol) compounds of similar molecular mass by high-performance liquid chromatography. *J.Chromatogr.A* **1995**, *704*, 369–376.

3-Methoxy-4-hydroxyphenylethyleneglycol

Hirata, K.; Inagaki, H.; Minami, M. A new solid-phase extraction method for human urinary 3-methoxy-4-hydroxyphenylethyleneglycol. *Yakugaku Zasshi* **1996**, *116*, 813–821.

α-(Methoxyimino)-1-azabicyclo[2.2.2]octane-3-acetonitrile

Ennis, J.N.; Buxton, P.C. Differential pulse polarographic determination of [R-(Z)]-α-(methoxyimino)-1-azabicyclo[2.2.2]octane-3-acetonitrile monohydrochloride in tablets following derivatization with dimethyldioxirane. *Anal.Commun.* **1996**, *33*, 261–263.

Methylarginines

Ueno, S.; Sano, A.; Kotani, K.; Kondoh, K.; Kakimoto, Y. Distribution of free methylarginines in rat tissues and in the bovine brain. *J.Neurochem.* **1992**, *59*, 2012–2016.

Methylaziridine

Gaind, V.S.; Jedrzejczak, K.; Huang, L.; Vohra, K. Determination of 2-methylaziridine in workplace atmospheres. *Analyst* **1990**, *115*, 925–928.

Methylcysteine Sulfoxide

Gustine, D.L. Determination of S-methyl cysteine sulfoxide in Brassica extracts by high-performance liquid chromatography. *J.Chromatogr.* **1985**, *319*, 450–453.

4,4'-Methylenedianiline

Tiljander, A.; Skarping, G.; Dalene, M. Determination of 4,4'-methylenedianiline in hydrolyzed human urine as chloroformate derivative using column switching and liquid chromatography with UV detection. *J.Liq.Chromatogr.* **1990**, *13*, 803–820.

Brunmark, P.; Persson, P.; Skarping, G. Determination of 4,4'-methylenedianiline in hydrolyzed human urine by micro liquid chromatography with ultraviolet detection. *J.Chromatogr.* **1992**, *579*, 350–354.

Methylglyoxal

Ohmori, S.; Mori, M.; Kawase, M.; Tsuboi, S. Determination of methylglyoxal as 2-methylquinoxaline by high-performance liquid chromatography and its application to biological samples. *J.Chromatogr.* **1987**, *414*, 149–155.

Yamaguchi, M.; Hara, S.; Nakamura, M. Determination of methylglyoxal in mouse blood by liquid chromatography with fluorescence detection. *Anal.Chim.Acta* **1989**, *221*, 163–166.

Methylguanidine (using post-column reaction detection)

Brooks, D.P.; Rhodes, G.R.; Woodward, P.; Boppana, V.K.; Mallon, F.M.; Griffin, H.E.; Kinter, L.B. Production of methylguanidine in dogs with acute and chronic renal failure. *Clin.Sci.* **1989**, *77*, 637–641.

Methylhistidine

Fermo, I.; De Vecchi, E.; Arcelloni, C.; Brambilla, P.; Pastoris, A.; Paroni, R. Improved high performance liquid chromatographic method for the quantification of 3-methylhistidine in serum and urine. *J.Liq.Chromatogr.* **1991**, *14*, 1715–1728.

Shen, M.; Yuan, Y.; Yin, L. Determination of urinary 3-methylhistidine by high-performance liquid chromatography with o-phthaldialdehyde precolumn derivatization. *J.Chromatogr.* **1992**, *581*, 272–276.

Wassner, S.J.; Schlitzer, J.L.; Li, J.B. A rapid, sensitive method for the determination of 3-methylhistidine levels in urine and plasma using high-pressure liquid chromatography. *Anal.Biochem.* **1980**, *104*, 284–289.

Yuan, Y.; Shen, M.; Li, J. Determination of 3-methylhistidine in urine by high performance liquid chromatography-fluorescence detection. *Sepu* **1996**, *14*, 274–276.

3-Methylhistidine (using post-column reaction detection)

Friedman, Z.; Smith, H.W.; Hancock, W.S. Rapid high-performance liquid chromatographic method for quantitation of 3-methylhistidine. *J.Chromatogr.* **1980**, *182*, 414–418.

Hancock, W.S.; Harding, D.R.K.; Friedman, Z. Rapid analysis of 3-methylhistidine in urine, plasma, muscle and amniotic fluid with a single high-performance liquid chromatographic system but with different ion-pairing reagents. *J.Chromatogr.* **1982**, *228*, 273–278.

Rifai, Z.; Kingston, W.J.; McCraith, B.; Moxley, R.T..I. Forearm 3-methylhistidine efflux in myotonic dystrophy. *Ann.Neurol.* **1994**, *34*, 682–686.

Methylisothiazolones

Matissek, R.; Nagorka, R.; Daase, M.; Wengatz, I. Determination of methylisothiazolones by ion-pair high-performance liquid chromatography after precolumn derivatization. Part II. Determination of 2-methyl-3(2H)-isothiazolone and 5-chloro-2-methyl-3(2H)-isothiazolone in cosmetics. *Fresenius' Z.Anal.Chem.* **1989**, *333*, 806–809.

Matissek, R.; Nagorka, R.; Wengatz, I.; Rohde, J. Analysis of methylisothiazolones by ion-pair high-performance liquid chromatography after precolumn derivatization. Part I. Derivatization of 2-methyl-3(2H)-isothiazolone and 5-chloro-2-methyl-3(2H)-isothiazolone. *Fresenius' Z.Anal.Chem.* **1988**, *332*, 813–816.

Methylmalonic Acid

Babidge, P.J.; Babidge, W.J. Determination of methylmalonic acid by high-performance liquid chromatography. *Anal.Biochem.* **1994**, *216*, 424–426.

Morgan, D.K.; Danielson, N.D. Determination of methylmalonic acid after diazonium derivatization by high-performance liquid chromatography. *Anal.Chim.Acta* **1985**, *170*, 301–310.

Methylmethioninesulfonium

Sumitani, H.; Suekane, S.; Sakai, Y.; Tatsuka, K. Precolumn o-phthalaldehyde derivatization and reversed-phase liquid chromatography of S-methylmethioninesulfonium in Satsuma mandarin juice. *J.AOAC Int.* **1992**, *75*, 77–79.

Methylthiolincosaminide

Yurek, D.A.; Kuo, M.S.; Li, G.P. Assay of methylthiolincosaminide in fermentations by high-performance liquid chromatography with fluorescence detection. *J.Chromatogr.* **1990**, *502*, 184–188.

Metolachlor

Chang, C.D.; Armstrong, D.W.; Fleischmann, T.J. Fractionation of [14C]metolachlor metabolites by centrifugal partition chromatography. *J.Liq.Chromatogr.* **1994**, *17*, 19–32.

Metoprolol

Ahnoff, M.; Chen, S.; Green, A.; Grundevik, I. Chiral chloroformates as transparent reagents for the resolution of metoprolol enantiomers by reversed-phase liquid chromatography. *J.Chromatogr.* **1990**, *506*, 593–599.

Mexiletine

Farid, N.A.; White, S.M. Determination of mexiletine and its metabolites in serum by liquid chromatography with fluorescence detection. *J.Chromatogr.* **1983**, *275*, 458–462.
Grech-Belanger, O.; Turgeon, J.; Gilbert, M. High-performance liquid chromatographic assay for mexiletine enantiomers in human plasma. *J.Chromatogr.* **1985**, *337*, 172–177.

Microcystins (using post-column reaction detection)

Kaya, K. HPLC with post-column derivatization for determination of microcystins in blue-green algae. *Kankyo Kagaku* **1994**, *4*, 532–533.
Murata, H.; Shoji, H.; Oshikata, M.; Harada, K-I.; Suzuki, M.; Kondo, F.; Goto, H. High-performance liquid chromatography with chemiluminescence detection of derivatized microcystins. *J.Chromatogr.A* **1995**, *693*, 263–270.

Micronomicin (using post-column reaction detection)

Mi, X.; Cheng, H.; Fan, J.; Zhou, C. Separation of micronomicin and related components by high performance liquid chromatography. *Zhongguo Kangshengsu Zazhi* **1990**, *15*, 260–263.

Midazolam

Drewe, J.; Kuesters, E. High-performance liquid chromatographic method for the enantiomeric separation of the chiral metabolites of midazolam. *J.Chromatogr.* **1992**, *609*, 395–398.

Milacemide

Semba, J.; Ratnaraj, N.; Patsalos, P.N. Simple and rapid microanalytical procedures for the estimation of milacemide and its metabolite glycinamide in rat plasma and cerebrospinal fluid by high-performance liquid chromatography. *J.Chromatogr.* **1991**, *565*, 357–362.

Miscellaneous Papers

Agbekodo, K.; Legube, B.; Cote, P.; Bourbigot, M.-M. Nanofiltration performance for the removal of natural organic matter; the Mery/Oise case. *Revue des Sciences de l'Eau* **1994**, *7*, 183–200.
Baenziger, J.U. Preparation of glycopeptide and oligosaccharide probes for receptor studies. *Methods Enzymol.* **1983**, *98*, 154–159.
Barisci, J.N.; Wallace, G.G. Advances in electrochemical detection for HPLC. *Chem.Aust.* **1993**, *60*, 538–546.

Barisci, J.N.; Wallace, G.G. Development of an improved online chromatographic monitor with new methods for environmental and process control. *Anal.Chim.Acta* **1995**, *310*, 79–92.

Benincasa, M.; Cartoni, G.; Coccioli, F. Preparation and applications of microbore columns in HPLC. *Ann.Chim.(Rome)* **1987**, *77*, 801–811.

Bowers, L.D.; Johnson, P.R. Immobilized β-glucuronidase as an on-line precolumn modification reagent for high-performance liquid chromatography. *Anal.Biochem.* **1981**, *116*, 111–115.

Bryant, R.A.R.; Hansen, D.E. Direct measurement of the uncatalyzed rate of hydrolysis of a peptide bond. *J.Am.Chem.Soc.* **1996**, *118*, 5498–5499.

Burger, K.; Koehler, J.; Jork, H. Application of AMD to the determination of crop protection agents in drinking water; Part 1. Fundamentals and method. *J.Planar Chromatogr.Mod.TLC* **1990**, *3*, 504–510.

Chang, M.; Chen, L.; Ding, X.; Selavka, C.M.; Krull, I.S.; Bratin, K. Precolumn derivatization for improved detection in liquid chromatography-photolysis-electrochemistry. *J.Chromatogr.Sci.* **1987**, *25*, 460–467.

Colgan, S.T.; Chou, T.Y.; Krull, I.S. Solid phase derivatizations for liquid chromatography/electrochemistry. *Curr.Sep.* **1986**, *7*, 80–83.

Dardoize, F.; Goasdoue, N.; Goasdoue, C.; Couffignal, R. High-performance liquid chromatography of chemical hybridizing agent in wheat. *J.Liq.Chromatogr.* **1993**, *16*, 1517–1528.

Einarsson, O.; Hansen, L. A PC-controlled module system for automatic sample preparation and analysis. *J.Autom.Chem.* **1995**, *17*, 21–24.

Findlay, J.B.C. Uses of high-performance liquid chromatographic methods in micro sequencing. *Biochem.Soc.Trans.* **1985**, *13*, 1071–1073.

Frei, R.W. Pre-concentration and chemical derivatization techniques in HPLC. *Methodol.Surv.Biochem.* **1978**, *7*, 243–255.

Frei, R.W.; Velthorst, N.H.; Gooijer, C. Detection in liquid chromatography based on different luminescence principles. *Pure Appl.Chem.* **1985**, *57*, 483–494.

Frei, R.W.; Brinkman, U.A.T. Trace enrichment in the HPLC analysis of aqueous samples. *Methodol.Surv.* **1981**, *10*, 86–95.

Fujino, H.; Koya, S. [Synthesis of ethyl 1-(Difluoro-1,3,5,-triazinyl)-2-methylindolizine-3-carboxylate as a fluorescent derivatization reagent and its reactivity]. *Yakugaku.Zasshi.* **1994**, *114*, 794–798.

Gasparrini, F.; Misiti, D.; Villani, C. Chromatographic optical resolution on trans-1,2-diaminocyclohexane derivatives: Theory and applications. *Chirality* **1992**, *4*, 447–458.

Gfeller, J.C.; Huen, J.M.; Thevenin, J.P. Automated pre-column derivatization HPLC. *Chromatographia* **1979**, *12*, 368–370.

Gfeller, J.C.; Huen, J.M.; Thevenin, J.P. Automation of pre-column derivatization in high-performance liquid chromatography. Application to ion-pair partition chromatography. *J.Chromatogr.* **1978**, *166*, 133–140.

Gluckman, J.; Shelly, D.; Novotny, M. Laser fluorimetry for capillary column liquid chromatography: high-sensitivity detection of derivatized biological compounds. *J.Chromatogr.* **1984**, *317*, 443–453.

Harpold, D.J.; Wasilauskas, B.L. Rapid identification of obligately anaerobic gram-positive cocci using high-performance liquid chromatography. *J.Clin.Microbiol.* **1987**, *25*, 996–1001.

Hill, E.; Humphreys, E.; Malcolme-Lawes, D.J. Electrochemiluminescence as a detection technique for reversed-phase high-performance liquid chromatography. IV. Detection of fluorescent derivatives. *J.Chromatogr.* **1988**, *441*, 394–399.

Honda, S.; Kuwada, H. Precolumn labeling device for liquid chromatography. *Anal.Chem.* **1983**, *55*, 2466–2468.

Houben, R.J.H.; Gielen, H.; van der Wal, S. Automated preseparation derivatization on a capillary electrophoresis instrument. *J.Chromatogr.* **1993**, *634*, 317–322.

Huff, P.B.; Tromberg, B.J.; Sepaniak, M.J. Sequentially excited fluorescence detection in liquid chromatography. *Anal.Chem.* **1982**, *54*, 946–950.

Jandera, P.; Prokes, B. Possibilities of enhancing the sensitivity of the determination of UV-absorbing compounds in high-performance liquid chromatography. *J.Chromatogr.* **1991**, *550*, 495–506.

Jellum, E.; Thorsrud, A.K.; Time, E. Capillary electrophoresis for diagnosis and studies of human disease, particularly metabolic disorders. *J.Chromatogr.* **1991**, *559*, 455–465.

Kaliszan, R. High performance liquid chromatographic methods and procedures of hydrophobicity determination. *Quant.Struct.-Act.Relat.* **1990**, *9*, 83–87.

Karnes, H.T.; Rahavendran, S.V.; Gui, M. Long-wavelength derivatization reagents for use in diode laser induced fluorescence detection. *Proc.SPIE-Int.Soc.Opt.Eng.* **1995**, *2388*, 21–31.

Katrukha, S.P.; Kukes, V.G. Derivatization of endogenous and exogenous compounds in plasma for high-performance liquid chromatographic analysis. *J.Chromatogr.* **1986**, *365*, 105–110.

Kempe, M.; Glad, M.; Mosbach, K. An approach towards surface imprinting using the enzyme ribonuclease A. *J.Mol.Recognit.* **1995**, *8*, 35–39.

Lam, S.; Karmen, A. Resolution of optical isomers as the mixed chelate copper(II) complexes by reversed phase chromatography. *J.Liq.Chromatogr.* **1986**, *9*, 291–311.

Lee, S.H.; Kim, B.H.; Lee, Y.C. Chiral recognition models of enantiomeric separation on cyclodextrin chiral stationary phases. *Bull.Korean Chem.Soc.* **1995**, *16*, 305–309.

Lijun, X.; Maohua, B.; Jinhua, W.; Haifu, W.; Zhaohan, W. Determination of metabolites of fat emulsion-Intralipid by high-performance liquid chromatography. *Fenxi Huaxue* **1994**, *22*, 759–762.

Marko-Varga, G.A. Column liquid chromatography in combination with immobilized enzymes and electrochemical detection and its applications in some industrial processes. *Electroanalysis (N.Y.)* **1992**, *4*, 403–427.

Marko-Varga, G.; Dominguez, E.; Hahn-Haegerdal, B.; Gorton, L. Bioselective detection in liquid chromatography by the use of immobilized enzymes. *J.Pharm.Biomed.Anal.* **1990**, *8*, 817–823.

Martz, R.F.; Sebacher, D.I.; White, D.C. Biomass measurement of methane forming bacteria in environmental samples. *J.Microbiol.Methods* **1983**, *1*, 53–61.

Maskarinec, M.P.; Sepaniak, M.J.; Balchunas, A.T.; Vargo, J.D. Liquid chromatography in open tubes. *Clin.Chem.* **1984**, *30*, 1473–1476.

McGuffin, V.L.; Evans, C.E.; Chen, S.H. Direct examination of separation processes in liquid chromatography: effect of temperature and pressure on solute retention. *J.Microcolumn Sep.* **1993**, *5*, 3–10.

Meulendijk, J.A.P.; Underberg, W.J.M. Ultraviolet-visible derivatization. *Chromatogr.Sci.* **1990**, *48*, 247–281.

Nakajima, R.; Nakamoto, E.; Kamiuchi, H.; Ueda, Y.; Ogura, S.; Tamura, T.; Hara, T. The utility of 4-(2-thienyl)pyridines as a derivatization reagent for HPLC and CE. *Anal.Sci.* **1991**, *7*, 177–180.

Nelson, D.A.; Bush, J.H.; Beckett, J.R.; Lenz, D.M.; Rowe, D.W. Use of fluorescent 1, 3-disubstituted 2(1H)-pyridones for environmental analysis. *ACS Symp.Ser.* **1989**, *383*, 206–227.

Nicolas, A.; Leroy, P. New chromatographic methods for in vitro glucuronidation studies. *Inst.Natl.Sante Rech.Med., [Colloq.]* **1988**, *173*, 201–209.

Novotny, M.; Karlsson, K.E.; Konishi, M.; Alasandro, M. New biochemical separations using precolumn derivatization and microcolumn liquid chromatography. *J.Chromatogr.* **1984**, *292*, 159–167.

Onuska, F.I. Techniques and procedures for preparation of aquatic samples for chromatographic analyses. *J.High Resolut.Chromatogr.* **1990**, *12*, 4–11.

Palmieri, G.; Cassani, G.; Fassina, G. Peptide immobilization on calcium alginate beads: applications to antibody purification and assay. *J.Chromatogr.B* **1995**, *664*, 127–135.

Ramos, L.S. Characterization of mycobacteria species by HPLC and pattern recognition. *J.Chromatogr.Sci.* **1994**, *32*, 219–227.

Reid, T.S.; Gisch, D.J. Applications for group specific ligands in fast, high performance affinity chromatography. *BioChromatography* **1988**, *3*, 201–204.

Saito, K.; Horie, M.; Nakazawa, H. Kinetic study of the stability of the o-phthalaldehyde-spermine fluorophore formed by on-column derivatization. *Anal.Chem.* **1994**, *66*, 134–138.

Sakamoto, H.; Kumazawa, Y.; Kawajiri, H.; Motoki, M. epsilon-(gamma-Glutamyl)lysine crosslink distribution in foods as determined by improved method. *J.Food Sci.* **1995**, *60*, 416–419.

Schieffer, G.W. Precolumn coulometric cell for high-performance liquid chromatography. *Anal.Chem.* **1981**, *53*, 126–127.

Schreurs, M.; Gooijer, C.; Velthorst, N.H. Application of long-lived luminescence for detection in liquid chromatography. *Fresenius' J.Anal.Chem.* **1991**, *339*, 499–503.

Sepaniak, M.J.; Vargo, J.D.; Kettler, C.N.; Maskarinec, M.P. Open tubular liquid chromatography with thermal lens detection. *Anal.Chem.* **1984**, *56*, 1252–1257.

Shea, P.A.; Jacobs, W.A. Improved gradient method for LCEC of putative transmitter amino acids. *Curr.Sep.* **1989**, *9*, 53–55.

Simon, V.A.; Hale, Y.; Taylor, A. Identification of Mycobacterium tuberculosis and related organisms by HPLC with fluorescence detection. *LC-GC* **1993**, *11*, 444–446.

Smyth, W.F.; Smyth, M.R. Electrochemical analysis of organic pollutants. *Pure Appl.Chem.* **1987**, *59*, 245–256.

Strojek, J.; Soper, S.A.; Ratzlaff, K.L.; Kuwana, T. "Thin-layer chromatography diskette" storage of high-performance liquid chromatographic effluents with off-line laser-induced fluorescence detection. *Anal.Sci.* **1990**, *6*, 121–129.

Takeuchi, T.; Niwa, T.; Ishii, D. Stepwise gradient elution using switching valves in micro high-performance liquid chromatography. *J.Chromatogr.* **1987**, *405*, 117–124.

Tanner, R.L.; Daum, P.H.; Kelly, T.J. New instrumentation for airborne acid rain research. *Int.J.Environ.Anal.Chem.* **1983**, *13*, 323–335.

Toyooka, T.; Uchiyama, S.; Saito, Y. Effect of gamma-irradiation on thiol compounds in grapefruit. *J.Agric.Food Chem.* **1989**, *37*, 769–775.

Tsunoda, K.; Nomura, A.; Yamada, J.; Nishi, S. Pulsed laser-induced fluorescence detector for liquid chromatography with a fiber optic-based flow cell. *Anal.Chim.Acta* **1990**, *229*, 3–7.

Van de Merbel, N.C.; Lingeman, H.; Brinkman, U.A.T.; Kolhorn, A.; de Rijke, L.C. Automated monitoring of biotechnological processes using online ultrafiltration and column liquid chromatography. *Anal.Chim.Acta* **1993**, *279*, 39–50.

Wedekind, F.; Baer-Pontzen, K.; Bala-Mohan, S.; Choli, D.; Zahn, H.; Brandenburg, D. Hormone binding site of the insulin receptor: analysis using photoaffinity-mediated avidin complexing. *Biol.Chem.Hoppe-Seyler* **1989**, *370*, 251–258.

White, J.G.; Jorgenson, J.W. Improvements in scanning voltammetric detection for open-tubular liquid chromatography. *Anal.Chem.* **1986**, *58*, 2992–2995.

Wu, W.S.; Stoyanoff, R.E.; Gaind, V.S. Isolation of a selected -system of a derivative for specific detection: an unconventional approach to HPLC quantification. *J.High Resolut.Chromatogr.* **1991**, *14*, 294–296.

Zuehlke, J.; Knopp, D.; Niessner, R. Sol-gel glass as a new support matrix in immunoaffinity chromatography. *Fresenius' J.Anal.Chem.* **1995**, *352*, 654–659.

Miscellaneous Papers (using post-column reaction detection)

Apffel, J.A.; Brinkman, U.A.T.; Frei, R.W. Use of nonsegmented flow, post-columns reaction detection with miniaturized HPLC systems. *Chromatographia* **1983**, *17*, 125–131.

Assenza, S.P. Detection systems. *Chromatogr.Sci.* **1984**, *28*, 139–160.

Barkley, D.J.; Bennett, L.A.; Charbonneau, J.R.; Pokrajac, L.A. Applications of high-performance ion chromatography in the mineral processing industry. *J.Chromatogr.* **1992**, *606*, 195–201.

Batley, G.E. Use of Teflon components in photochemical reactors. *Anal.Chem.* **1984**, *56*, 2261–2262.

Beresford, A.P.; Caswell, K.; Chambers, R.; Kirk, I.P. Advantages of achiral HPLC as a preparative step for chiral analysis in biological samples and its use in toxicokinetic studies. *Xenobiotica* **1992**, *22*, 789–798.

Birks, J.W.; Poulsen, J.R.; Shellum, C.L. Photochemical reaction detection in HPLC. *ASTM Spec.Tech.Publ.* **1988**, *1009*, 26–40.

Blair, P.C.; Popp, J.A.; Bryant-Varela, B.J.; Thompson, M.B. Promotion of hepatocellular foci in female rats by chenodeoxycholic acid. *Carcinogenesis (London)* **1991**, *12*, 59–63.

Brinkman, U.A.T.; Frei, R.W. Selective detection in liquid chromatography through postcolumn derivatization and fluorescence monitoring. *Anal.Proc.(London)* **1983**, *20*, 354–356.

Brinkman, U.A.T.; Frei, R.W. Post-column reaction detectors for trace analysis in HPLC. *Methodol.Surv.Biochem.Anal.* **1984**, *14*, 183–184.

Chen, D.; Zhong, Y.; Liang, S. Continuous monitoring of elution process in extraction chromatography with post-column spectrometry. *Fenxi Ceshi Tongbao* **1991**, *10*, 15–22.

Curtis, M.A.; Shahwan, G.J. A high-efficiency open-tubular postcolumn HPLC reactor using PTFE tubing in a modified serpentine geometry. *LC-GC* **1988**, *6*, 158.

Davis, J.C.; Peterson, D.P. Hollow fiber post-column reactor for liquid chromatography. *Anal.Chem.* **1985**, *57*, 768–771.

Deelder, R.S.; Kroll, M.G.F.; Beeren, A.J.B.; Van den Berg, J.H.M. Post-column reactor systems in liquid chromatography. *J.Chromatogr.* **1978**, *149*, 669–682.

Deelder, R.S.; Kuijpers, A.T.J.M.; Van den Berg, J.H.M. Evaluation and comparison of reaction detectors. *J.Chromatogr.* **1983**, *255*, 545–561.

Duchateau, A.L.L.; Hillemans, M.G.; Schepers, C.H.M.; Ketelaar, P.E.F.; Hermes, H.F.M.; Kamphuis, J. Determination of biocatalyst consumption in an aminopeptidase process using automated sample preparation and high-performance liquid chromatography. *J.Chromatogr.* **1991**, *566*, 493–498.

Engelhardt, H.; Lillig, B. Practical applications of open tubes in liquid chromatography. *HRC CC, J.High Resolut.Chromatogr.Chromatogr.Commun.* **1985**, *8*, 531–534.

Engelhardt, H.; Neue, U.D. Reaction detector with three dimensional coiled open tubes in HPLC. *Chromatographia* **1982**, *15*, 403–408.

Folestad, S.; Johnson, L.; Josefsson, B.; Galle, B. Laser-induced fluorescence detection for conventional and microcolumn liquid chromatography. *Anal.Chem.* **1982**, *54*, 925–929.

Freeman, R.D.; Hammaker, R.M.; Meloan, C.E.; Fateley, W.G. A detector for liquid chromatography and flow injection analysis using surface-enhanced Raman spectroscopy. *Appl.Spectrosc.* **1988**, *42*, 456–460.

Frei, R.W.; de Jong, G.J.; Brinkman, U.A.T. Detection of trace analytes with liquid chromatography. *Chem.Mag.(Rijswijk, Neth.)* **1986**, 695–699.

Frei, R.W.; Michel, L.; Santi, W. New aspects of post-column derivatization in high-performance liquid chromatography. *J.Chromatogr.* **1977**, *142*, 261–270.

Gjerde, D.T.; Benson, J.V. Suspension postcolumn reaction detection method for liquid chromatography. *Anal.Chem.* **1990**, *62*, 612–615.

Hippe, H. HPLC post-column derivatization with electrochemical detection. *Bioengineering* **1993**, *9*, 47.

Hsu, S.H.; Raglione, T.; Tomellini, S.A.; Floyd, T.R.; Sagliano, N., Jr.; Hartwick, R.A. Zone compression effects in high-performance liquid chromatography. *J.Chromatogr.* **1986**, *367*, 293–300.

Irth, H.; Oosterkamp, A.J.; van der Welle, W.; Tjaden, U.R.; van der Greef, J. Online immunochemical detection in liquid chromatography using fluorescein-labeled antibodies. *J.Chromatogr.* **1993**, *633*, 65–72.

Ishida, J.; Abe, K.; Nakamura, M.; Yamaguchi, M. Evaluation of bovine serum albumin as a fluorescence enhancement reagent in liquid chromatography. *Anal.Sci.* **1995**, *11*, 743–747.

Jansen, H.; Brinkman, U.A.T.; Frei, R.W. Miniaturization of solid-phase reactors for on-line post-column derivatization in narrow-bore liquid chromatography. *Chromatographia* **1985**, *20*, 453–460.

Jones, P.; Barron, K.; Ebdon, L. Recent advances in trace analysis by high-performance liquid chromatography. *Anal.Proc.(London)* **1985**, *22*, 373–375.

Kobayashi, S.; Imai, K. Rotating flow mixing device for post column reaction in high performance liquid chromatography. *Anal.Chem.* **1980**, *52*, 1548–1549.

Kucera, P.; Umagat, H. Design of a post-column fluorescence derivatization system for use with microbore columns. *J.Chromatogr.* **1983**, *255*, 563–579.

Lin, Y.P.; Riley, P.J.; Wallace, G.G. Controlled release of the dithiocarbamate ligand from a polypyrrole polymer. A basis for on-line electrochemically controlled derivatization. *Anal.Lett.* **1989**, *22*, 669–681.

Little, C.J.; Whatley, J.A.; Dale, A.D. Detection involving post-chromatographic addition of reagents. *J.Chromatogr.* **1979**, *171*, 63–72.

Lucy, C.A.; Ye, L. Displacement post-column detection reagents based on the fluorescent magnesium 8-hydroxyquinoline-5-sulfonic acid complex. *J.Chromatogr.A* **1994**, *671*, 121–129.

Momose, A.; Uchikura, K.; Kabasawa, Y. New postcolumn detection system using electrochemical reaction for high performance liquid chromatography. *Bunseki Kagaku* **1983**, *32*, 142–143.

Moor, E.; de Boer, P.; Beldhuis, H.J.A.; Westerink, B.H.C. A novel approach for studying septo-hippocampal cholinergic neurons in freely moving rats: a microdialysis study with dual-probe design. *Brain Res.* **1994**, *648*, 32–38.

Nakamura, H.; Matsumoto, K.; Ishikawa, O.; Yamamoto, M.; Sato, T. Post-column derivatization type high speed liquid chromatograph HLC-805. *Toyo Soda Kenkyu Hokoku* **1979**, *23*, 119–130.

Nakamura, K.; Satomura, S.; Matsuura, S. Analysis of antigen-antibody reaction in buffer solution by high-pressure liquid chromatography. *Anal.Sci.* **1991**, *7*, 905–908.

Nakamura, K.; Satomura, S.; Matsuura, S. Liquid-phase binding assay of human chorionic gonadotropin using high-performance liquid chromatography. *Anal.Chem.* **1993**, *65*, 613–616.

Oelrich, E.; Preusch, H.; Wilhelm, E.; Theuerkauf, D. HPLC cassettes - a new modular separation system with high flexibility. *J.Chromatogr.Sci.* **1979**, *17*, 289–296.

Oelrich, E.; Theuerkauf, D. Post-column derivatization using HPLC cassettes. *HRC CC, J.High Resolut.Chromatogr.Chromatogr.Commun.* **1979**, *2*, 256–258.

Omichi, K.; Ikenaka, T. Use of a post-column enzymic reaction for detection of macroamylasemia by high-performance liquid chromatography. *J.Chromatogr.* **1988**, *428*, 415–418.

Pickering, M.V. Assembling an HPLC postcolumn system: practical considerations. *LC-GC* **1988**, *6*, 994–997.

Popovich, D.J.; Dixon, J.B.; Ehrlich, B.J. The photo[chemical]conductivity detector - a new selective detector for HPLC. *J.Chromatogr.Sci.* **1979**, *17*, 643–650.

Poulsen, J.R.; Birks, K.S.; Gandelman, M.S.; Birks, J.W. Crocheted PTFE reactors for post-column photochemistry in HPLC. *Chromatographia* **1986**, *22*, 231–234.

Przyjazny, A.; Bachas, L.G. Competitive-binding approach to liquid chromatographic postcolumn reactions with fluorimetric detection. *Anal.Chim.Acta* **1991**, *246*, 103–112.

Przyjazny, A.; Kjellstrom, T.L.; Bachas, L.G. High-performance liquid chromatographic postcolumn reaction detection based on a competitive binding system. *Anal.Chem.* **1990**, *62*, 2536–2540.

Reim, R.E. Postcolumn deoxygenator for liquid chromatography with reductive electrochemical detection. *Anal.Chem.* **1983**, *55*, 1188–1191.

Risby, T.H.; Long, J. Physiologically relevant pseudophase high-performance liquid-liquid chromatography. *Anal.Chem.* **1987**, *59*, 200–202.

Rouseff, R.L. Post-column derivatization in liquid chromatographic analysis of citrus products. *Food Technol.(Chicago)* **1985**, *39*, 87–90.

Ruchti, B.; Schramm, C.; Kubitschko, S.; Neidhart, B. A new screening procedure for the estimation of oxidizable organic compounds in water samples. *Fresenius' J.Anal.Chem.* **1992**, *342*, 822–826.

Schleich, W.; Engelhardt, H. Post-column derivatization. *GIT Fachz.Lab.* **1988**, *32*, 401–405.

Scholten, A.H.M.T.; Welling, P.L.M.; Brinkman, U.A.T.; Frei, R.W. Use of PTFE coils in post-column photochemical reactors for liquid chromatography - application to pharmaceuticals. *J.Chromatogr.* **1980**, *199*, 239–248.

Schwedt, G.; Reh, E. Construction and optimization of fluorimetric reaction detectors with air-segmented liquid streams for high-performance liquid chromatography. Part 2. Reactor materials and effects of variable dynamic dimensions. *Chromatographia* **1981**, *14*, 123–126.

Schwedt, G.; Reh, E. Fluorometric reaction detectors for high-performance liquid chromatography. Part 1. Selection of chemical reactions. *Chromatographia* **1981**, *14*, 249–253.

Selavka, C.M.; Jiao, K.S.; Krull, I.S. Construction and comparison of open tubular reactors for post-column reaction detection in liquid chromatography. *Anal.Chem.* **1987**, *59*, 2221–2224.

Selavka, C.M.; Krull, I.S. Filling a void in the applicability of oxidative LCEC using post column photolysis. *Curr.Sep.* **1986**, *7*, 85–87.

Shellum, C.L.; Birks, J.W. Photochemical amplifier for liquid chromatography based on singlet oxygen sensitization. *Anal.Chem.* **1987**, *59*, 1834–1841.

Shih, Y.T.; Carr, P.W. Flow-rate dependence of post-column reaction chromatographic detectors and optimization of reactor length for slow chemical reactions. *Anal.Chim.Acta* **1985**, *167*, 137–144.

Stocker, R.; Bowry, V.W.; Frei, B. Ubiquinol-10 protects human low density lipoprotein more efficiently against lipid peroxidation than does α-tocopherol. *Proc.Natl.Acad.Sci.U.S.A.* **1991**, *88*, 1646–1650.

Tijssen, R. Axial dispersion and flow phenomena in helically coiled tubular reactors for flow analysis and chromatography. *Anal.Chim.Acta* **1980**, *114*, 71–89.

Van Vliet, H.P.M.; Bruin, G.J.M.; Kraak, J.C.; Poppe, H. Post-column reaction detection for open-tubular liquid chromatography using laser-induced fluorescence. *J.Chromatogr.* **1986**, *363*, 187–198.

Voelter, W.; Gruebler, G.; Gutjahr, F.; Echner, H. An HPLC system for use in the development of new arginine protecting groups. *Chromatographia* **1990**, *30*, 719–723.

Weinberger, R.; Yarmchuk, P.; Love, L.J.C. Liquid chromatographic phosphorescence detection with micellar chromatography and postcolumn reaction modes. *Anal.Chem.* **1982**, *54*, 1552–1558.

Xia, F.; Cassidy, R.M. Application of micelles in postcolumn reaction systems. *Anal.Chem.* **1991**, *63*, 2883–2887.

Yoden, K.; Iio, T. Determination of thiobarbituric acid-reactive substances in oxidized lipids by high-performance liquid chromatography with a postcolumn reaction system. *Anal.Biochem.* **1989**, *182*, 116–120.

Monensin

Rodewald, J.M.; Moran, J.W.; Donoho, A.L.; Coleman, M.R. Determination of monensin in raw material, premix, and animal feeds by liquid chromatography with correlation to microbiological assay. *J.AOAC Int.* **1992**, *75*, 272–279.

Monolignols

Lewis, N.G.; Inciong, M.E.J.; Dhara, K.P.; Yamamoto, E. High-performance liquid chromatographic separation of E- and Z-monolignols and their glucosides. *J.Chromatogr.* **1989**, *479*, 345–352.

Morphine

Combie, J.; Blake, J.W.; Ramey, B.E.; Tobin, T. Pharmacology of narcotic analgesics in the horse: quantitative detection of morphine in equine blood and urine and logit-log transformations of this data. *Am.J.Vet.Res.* **1981**, *42*, 1523–1530.

Garrett, E.R.; Gurkan, T. Pharmacokinetics of morphine and its surrogates. I. Comparisons of sensitive assays of morphine in biological fluids and application to morphine pharmacokinetics in the dog. *J.Pharm.Sci.* **1978**, *67*, 1512–1517.

Morphine (using post-column reaction detection)

Garrett, E.R.; Derendorf, H.; Mattha, A.G. Pharmacokinetics of morphine and its surrogates. VII: High-performance liquid chromatographic analyses and pharmacokinetics of methadone and its derived metabolites in dogs. *J.Pharm.Sci.* **1985**, *74*, 1203–1214.

Nelson, P.E. High-performance liquid chromatography detection of morphine by fluorescence after post-column derivatization. II. The effect of micelle formation. *J.Chromatogr.* **1984**, *298*, 59–65.

Nelson, P.E.; Nolan, S.L.; Bedford, K.R. High-performance liquid chromatography detection of morphine by fluorescence after postcolumn derivatization. *J.Chromatogr.* **1982**, *234*, 407–414.

Morpholine

Joseph, M.; Kagdiyal, V.; Tuli, D.K.; Rai, M.M.; Jain, S.K.; Srivastava, S.P.; Bhatnagar, A.K. Determination of trace amounts of morpholine and its thermal degradation products in boiler water by HPLC. *Chromatographia* **1993**, *35*, 173–176.

Lamarre, C.; Gilbert, R.; Gendron, A. Liquid chromatographic determination of morpholine and its thermal breakdown products in steam-water cycles at nuclear power plants. *J.Chromatogr.* **1989**, *467*, 249–258.

Muramic Acid

Ronkko, R.; Pennanen, T.; Smolander, A.; Kitunen, V.; Kortemaa, H.; Haahtela, K. Quantification of Frankia strains and other root-associated bacteria in pure cultures and in the rhizosphere of axenic seedlings by high-performance liquid chromatography-based muramic acid assay. *Appl.Environ. Microbiol.* **1994**, *60*, 3672–3678.

Muramic Acid (using post-column reaction detection)

Watanabe, N. Amperometric detection of muramic acid in high-performance liquid chromatography with a post-column reaction. *J.Chromatogr.* **1984**, *316*, 495–500.

Mycolic Acids

Butler, W.R.; Kilburn, J.O. High-performance liquid chromatography patterns of mycolic acids as criteria for identification of Mycobacterium chelonae, Mycobacterium fortuitum, and Mycobacterium smegmatis. *J.Clin.Microbiol.* **1990**, *28*, 2094–2098.

Hagen, S.R.; Thompson, J.D. Analysis of mycolic acids by high-performance liquid chromatography and fluorimetric detection. Implications for the identification of mycobacteria in clinical samples. *J.Chromatogr.A* **1995**, *692*, 167–172.

Mycotoxins

Jordan, L.; Hansen, T.J.; Zabe, N.A. Automated mycotoxin analysis. *Am.Lab.* **1994**, *26*, 18–24.

N-0437

Gerding, T.K.; Drenth, B.F.H.; de Zeeuw, R.A. Separation of N-0437 enantiomers by RP-HPLC after precolumn derivatization with D(+)-glucuronic acid. *HRC CC, J.High Resolut.Chromatogr. Chromatogr.Commun.* **1987**, *10*, 523–525.

NADH (using post-column reaction detection)

Eisenberg, E.J.; Cundy, K.C. Amperometric high-performance liquid chromatographic detection of NADH at a base-activated glassy carbon electrode. *Anal.Chem.* **1991**, *63*, 845–847.

Nadolol

Lee, C.R.; Porziemsky, J.P.; Aubert, M.C.; Krstulovic, A.M. Liquid and high-pressure carbon dioxide chromatography of β-blockers: resolution of the enantiomers of nadolol. *J.Chromatogr.* **1991**, *539*, 55–69.

Narciclasine (using post-column reaction detection)

Svagrova, I.; Stulik, K.; Pacakova, V.; Caliceti, P.; Veronese, F.M. Determination of narciclasine in serum by reversed-phase high-performance liquid chromatography: comparison of amperometric, ultraviolet photometric and fluorescence detection. *J.Chromatogr.* **1991**, *563*, 95–102.

Neomycin

Funk, W.; Kuepper, T.; Wirtz, A.; Netz, S. Quantitative TLC/HPTLC determination of neomycins A, B, and C. Part 1: Chromatographic separation and postchromatographic derivatization. *J.Planar Chromatogr.—Mod.TLC* **1994**, *7*, 10–13.
Shaikh, B.; Jackson, J. Improved liquid chromatographic determination of neomycin B in bovine kidney. *J.AOAC Int.* **1993**, *76*, 543–548.

Neomycin (using post-column reaction detection)

Apffel, J.A.; Van der Louw, J.; Lammers, K.R.; Kok, W.T.; Brinkman, U.A.T.; Frei, R.W.; Burgess, C. Analysis of neomycins A, B and C by high-performance liquid chromatography with post-column reaction detection. *J.Pharm.Biomed.Anal.* **1985**, *3*, 259–267.
Guggisberg, D.; Koch, H. Method for the fluorimetric determination of neomycin in kidney and liver by HPLC and post-column derivatization. *Mitt.geb.Lebensmittelunters.Hyg.* **1995**, *86*, 449–457.

Neuraminic Acids

Hara, S.; Yamaguchi, M.; Takemori, Y.; Nakamura, M.; Ohkura, Y. Highly sensitive determination of N-acetyl- and N-glycolylneuraminic acids in human serum and urine and rat serum by reversed-phase liquid chromatography with fluorescence detection. *J.Chromatogr.* **1986**, *377*, 111–119.
Hara, S.; Takemori, Y.; Yamaguchi, M.; Nakamura, M.; Ohkura, Y. Fluorometric high-performance liquid chromatography of N-acetyl-and N-glycolylneuraminic acids and its application to their microdetermination in human and animal sera, glycoproteins, and glycolipids. *Anal.Biochem.* **1987**, *164*, 138–145.
Karamanos, N.K.; Wikstroem, B.; Antonopoulos, C.A.; Hjerpe, A. Determination of N-acetyl- and N-glycolylneuraminic acids in glycoconjugates by reversed-phase high-performance liquid chromatography with ultraviolet detection. *J.Chromatogr.* **1990**, *503*, 421–429.
Kobayashi, K.; Akiyama, Y.; Kawaguchi, K.; Tanabe, S.; Imanari, T. Fluorometric determination of N-acetyl and N-glycolyl neuraminic acids by high-performance liquid chromatography as their 4'-hydrazino-2-stilbazole derivatives. *Anal.Sci.* **1985**, *1*, 81–84.
Lagana, A.; Marino, A.; Fago, G.; Martinez, B.P. A hydrolysis method using microwaves: determination of N-acetyl-and N-glycolylneuraminic acids in biological systems by fluorometric high-performance liquid chromatography. *Anal.Biochem.* **1993**, *215*, 266–272.
McNicholas, P.A.; Batley, M.; Redmond, J.W. Reversed-phase high-performance liquid chromatography of hydrazones of 3-deoxy-D-manno-oct-2-ulosonic acid and neuraminic acids. *J.Chromatogr.* **1984**, *315*, 451–456.

Niacinamide

Hirayama, T.; Nakata, H.; Watanabe, T. The effect of nicotinamide on the mutagenicity of various alkylating agents. *Jpn.J.Toxicol.Environ.Health* **1992**, *38*, 431–436.

Nicarbazin

Schenck, F.J.; Barker, S.A.; Long, A.R. Matrix solid-phase dispersion extraction and liquid chromatographic determination of nicarbazin in chicken tissue. *J.AOAC Int.* **1992**, *75*, 659–662.

Nicotine

O'Doherty, S.; Cooke, M.; Roberts, D.J. Enhancing the LC analysis of nicotine and its metabolites in urine using Meldrum's acid as a complexing agent. *J.High Resolut.Chromatogr.* **1990**, *13*, 74–77.

Nifedipine

Eiden, F.; Braatz Greeske, K. Contribution to the analysis of calcium antagonists. Part 2. Nifedipine. *Dtsch.Apoth.Ztg.* **1983**, *120*, 2003–2009.

Nitrilotriacetic Acid

Momoki, K.; Yokoyama, Y. Liquid chromatographic determination of microquantities of nitrilotriacetic acid in water by phenacyl derivatization. *Bull.Fac.Eng., Yokohama Natl.Univ.* **1982**, *31*, 195–203.

Nitrite

Kieber, R.J.; Seaton, P.J. Determination of subnanomolar concentrations of nitrite in natural waters. *Anal.Chem.* **1995**, *67*, 3261–3264.

Tsikas, D.; Gutzki, F.M.; Froelich, J.C. Analysis for nitrite and nitrate as 1-nitro-2,4,6-trimethoxybenzene derivatives by reversed-phase HPLC with UV-detection. *Fresenius' J.Anal.Chem.* **1992**, *342*, 95–97.

Tsikas, D.; Froelich, J.C. Determination of nitrite as pentafluorobenzyl derivative by RP-HPLC and UV detection with and without ion-pair extraction. *Fresenius' J.Anal.Chem.* **1992**, *344*, 256–260.

Verma, K.K.; Verma, A. Determination of nitrite by precolumn derivatization and high-performance liquid chromatography. *Anal.Lett.* **1992**, *25*, 2083–2093.

Nitrite (using post-column reaction detection)

Lee, S.H.; Field, L.R. Postcolumn fluorescence detection of nitrite, nitrate, thiosulfate, and iodide anions in high-performance liquid chromatography. *Anal.Chem.* **1984**, *56*, 2647–2653.

Matsui, M.; Ishibashi, T.; Mizuno, T.; Nishikawa, M. Study on analysis of N-nitroso compounds in the environment. (II). Determination of nitrite in environmental samples by HPLC using diazotization reaction. *Kankyo Kagaku* **1992**, *2*, 779–785.

Sen, N.P.; Baddoo, P.A.; Seaman, S.W. Rapid and sensitive determination of nitrite in foods and biological materials by flow injection or high-performance liquid chromatography with chemiluminescence detection. *J.Chromatogr.A* **1994**, *673*, 77–84.

Nitroarginine (using post-column reaction detection)

Whiting, M.J.; Rutten, A.J.; Williams, P.; Bersten, A.D. Determination of NG-nitro-L-arginine and NG-nitro-L-arginine methyl ester in plasma by high-performance liquid chromatography. *J.Chromatogr.B* **1994**, *660*, 170–175.

Nitrofurantoin

Hoener, B.A.; Wolff, J.L. High-performance liquid chromatographic assay for the metabolites of nitrofurantoin in plasma and urine. *J.Chromatogr.* **1980**, *182*, 246–251.

3-Nitro-4-hydroxyphenylarsonic acid (using post-column reaction detection)

Croteau, L.G.; Akhtar, M.H.; Belanger, J.M.R.; Pare, J.R.J. High performance liquid chromatography determination following microwave assisted extraction of 3-nitro-4-hydroxyphenylarsonic acid from swine liver, kidney, and muscle. *J.Liq.Chromatogr.* **1994**, *17*, 2971–2981.

Nitrophenols

Mussmann, P.; Eisert, R.; Levsen, K.; Wuensch, G. Determination of nitrophenols, diaminotoluenes, and chloroaromatics in ammunition wastewater. *Acta Hydrochim.Hydrobiol.* **1995**, *23*, 13–19.

Nitro Polycyclic Aromatic Hydrocarbons (using post-column reaction detection)

Imaizumi, N.; Hayakawa, K.; Suzuki, Y.; Miyazaki, M. Determination of nitrated pyrenes and their derivatives by high performance liquid chromatography with chemiluminescence detection after online electrochemical reduction. *Biomed.Chromatogr.* **1990**, *4*, 108–112.

Sigvardson, K.W.; Birks, J.W. Detection of nitro-polycyclic aromatic hydrocarbons in liquid chromatography by zinc reduction and peroxyoxalate chemiluminescence. *J.Chromatogr.* **1984**, *316*, 507–518.

Veigl, E.; Posch, W.; Lindner, W.; Tritthart, P. Selective and sensitive analysis of 1-nitropyrene in diesel exhaust particulate extract by multidimensional HPLC. *Chromatographia* **1994**, *38*, 199–206.

Nitrosamines

Massey, R.C.; Key, P.E.; McWeeny, D.J.; Knowles, M.E. N-Nitrosamine analysis in foods: N-nitrosoamino acids by high-performance liquid chromatography/thermal energy analysis and total N-nitroso compounds by chemical denitrosation/thermal energy analysis. *IARC Sci.Publ.* **1984**, *57*, 131–136.

Verhagen, L.C.; Strating, J. Quantitative determination of N-nitrosodimethylamine in malt and beer by high-performance liquid chromatography using fluorescence detection. *J.Inst.Brew.* **1981**, *87*, 57–60.

Nitrosamines (using post-column reaction detection)

Favaro, G.; Sacchetto, G.A.; Pastore, P.; Fiorani, M. Liquid chromatographic determination of non-volatile nitrosamines by post-column redox reactions and voltammetric detection at solid electrodes. Study of a flow reactor system based on cerium(IV) reagent. *Anal.Chim.Acta* **1993**, *273*, 457–467.

Fu, C-G.; Xu, H-D. High-performance liquid chromatography with post-column chemiluminescence detection for simultaneous determination of trace N-nitrosamines and corresponding secondary amines in groundwater. *Analyst* **1995**, *120*, 1147–1151.

Havery, D.C. Determination of N-nitroso compounds by high-performance liquid chromatography with postcolumn reaction and a thermal energy analyzer. *J.Anal.Toxicol.* **1990**, *14*, 181–185.

Lee, S.H.; Field, L.R. Fluorescence detection of some nitrosoamines in high-performance liquid chromatography after post-column reaction. *J.Chromatogr.* **1987**, *386*, 137–148.

MacMillan, W.D. Separation and direct chemical determination of nitrosamines by high performance liquid chromatography (HPLC). *Anal.Lett.* **1983**, *16*, 957–968.

Sacchetto, G.A.; Pastore, P.; Favaro, G.; Fiorani, M. Liquid chromatographic determination of nonvolatile nitrosamines by post-column redox reactions and voltammetric detection at solid electrodes. Behavior of the cerium(IV)-cerium(III) couple at gold, platinum and glassy carbon electrodes and suitability of the cerium(IV) reagent. *Anal.Chim.Acta* **1992**, *258*, 99–108.

Sacchetto, G.A.; Favaro, G.; Pastore, P.; Fiorani, M. Optimization of the amperometric detection of nitrite by reaction with iodide in a post-column reactor for liquid chromatography of non-volatile nitrosamines. *Anal.Chim.Acta* **1994**, *294*, 251–260.

Shuker, D.E.G.; Tannenbaum, S.R. Determination of nonvolatile N-nitroso compounds in biological fluids by liquid chromatography with postcolumn photohydrolysis detection. *Anal.Chem.* **1983**, *55*, 2152–2155.

Nonenal

Verhagen, L.C.; Strating, J.; Tjaden, U.R. Development of an HPLC-method for the determination of E-2-nonenal in beer using column switching techniques. *Int.J.Environ.Anal.Chem.* **1986**, *25*, 67–79.

Verhagen, L.C.; Strating, J.; Tjaden, U.R. Analysis of E-2-nonenal in beer at the ultra trace level by high-performance liquid chromatography using precolumn derivatization and column switching techniques. *J.Chromatogr.* **1987**, *393*, 85–96.

Nucleosides and Nucleotides

Alders, D.; Hoefnagel, R.M.; Rhemrev-Boom, M.M. Determination of nucleosides and their adducts with RP-HPLC and (laser-)fluorescence detection. *Nucleosides Nucleotides* **1990**, *9*, 395–405.

Davis, W.M.; White, D.C. Fluorometric determination of adenosine nucleotide derivatives as measures of the microfouling, detrital, and sedimentary microbial biomass and physiological status. *Appl.Environ.Microbiol.* **1980**, *40*, 539–548.

Fujimori, H.; Sasaki, T.; Hibi, K.; Senda, M.; Yoshioka, M. Measurement of adenine nucleotide levels with an adenine analyzer as an index of freshness of porgy. *J.Chromatogr.* **1990**, *528*, 305–314.

Henderson, R.J., Jr.; Griffin, C.A. Electrochemical detection of adenosine and other purine metabolites during high-performance liquid chromatographic analysis. *J.Chromatogr.* **1984**, *298*, 231–242.

Iwamoto, M.; Yoshida, S.; Hirose, S. Fluorescence labelling of pyrimidine nucleobases and their related compounds for high-performance liquid chromatography. *Nucleic Acids Symp.Ser.* **1984**, 21–24.

Kuttesch, J.F.; Schmalstieg, F.C.; Nelson, J.A. Analysis of adenosine and other adenine compounds in patients with immunodeficiency diseases. *J.Liq.Chromatogr.* **1978**, *1*, 97–109.

Linden, J.; Taylor, H.E.; Feldman, M.D.; Woodward, E.B.; Ayers, C.R.; Ripley, M.L.; Iflah, S.; Patel, A. The precise radioimmunoassay of adenosine: minimization of sample collection artifacts and immunocrossreactivity. *Anal.Biochem.* **1992**, *201*, 246–254.

Martens, R. A comparison of soil adenine nucleotide measurements by HPLC and enzymic analysis. *Soil Biol.Biochem.* **1992**, *24*, 639–645.

Munholland, J.M.; Bright, K.A.; Nazar, R.N. Use of a volatile buffer system in ion-exchange high-performance liquid chromatography of oligonucleotides. *Anal.Biochem.* **1989**, *178*, 320–323.

Nagaoka, H.; Nohta, H.; Saito, M.; Ohkura, Y. Precolumn fluorescence derivatization of nucleosides for mono-and oligonucleotides. *Anal.Sci.* **1992**, *8*, 565–566.

Nagaoka, H.; Nohta, H.; Saito, M.; Ohkura, Y. Precolumn derivatization of nucleotides based on fluorescent carbamate formation on the sugar moieties in high-performance liquid chromatography. *Chem.Pharm.Bull.* **1992**, *40*, 2559–2561.

Ohba, Y.; Kai, M.; Nohta, H.; Ohkura, Y. Alkoxyphenylglyoxals as fluorogenic reagents selective for guanine and its nucleosides and nucleotides in liquid chromatography. *Anal.Chim.Acta* **1994**, *287*, 215–221.

Ramos-Salazar, A.; Baines, A.D. Fluorometric determination of adenine nucleotides and adenosine by ion-paired, reverse-phase, high-performance liquid chromatography. *Anal.Biochem.* **1985**, *145*, 9–13.

Walker, G.S.; Coveney, M.F.; Klug, M.J.; Wetzel, R.G. Isocratic HPLC analysis of adenine nucleotides in environmental samples. *J.Microbiol.Methods* **1986**, *5*, 255–264.

Yonekura, S.; Iwasaki, M.; Kai, M.; Ohkura, Y. High-performance liquid chromatographic determination of guanine and its nucleosides and nucleotides in biospecimens with precolumn fluorescence derivation using phenylglyoxal. *Anal.Sci.* **1994**, *10*, 247–251.

Yoshioka, M.; Yamada, K.; Abu-Zeid, M.M.; Fujimori, H.; Fuke, A.; Hirai, K.; Goto, A.; Ishii, M.; Sugimoto, T.; Parvez, H. Fluorescent-HPLC for adenine nucleosides and nucleotides in life science. *Prog.HPLC* **1989**, *4*, 181–209.

Yoshioka, M.; Yamada, K.; Abu-Zeid, M.M.; Fujimori, H.; Fuke, A.; Hirai, K.; Goto, A.; Ishii, M.; Sugimoto, T.; Parvez, H. Analyses of adenosine and adenine nucleotides in biological materials by fluorescence reaction-high-performance liquid chromatography. *J.Chromatogr.* **1987**, *400*, 133–141.

Yoshioka, M.; Nakamura, A.; Iizuka, H.; Nishidate, K.; Tamura, Z.; Miyazaki, T. Sensitive fluorometry of adenine, its nucleosides and nucleotides. *Nucleic Acids Symp.Ser.* **1980**, *8*, s61-s63,

Nucleosides and Nucleotides (using post-column reaction detection)

Kito, M.; Tawa, R.; Takeshima, S.; Hirose, S. Determination of purine nucleosides and their bases by high-performance liquid chromatography using co-immobilized enzyme reactors. *J.Chromatogr.* **1990**, *528*, 91–99.

Yao, T.; Matsumoto, Y.; Wasa, T. Development of a FIA system with immobilized enzymes for specific post-column detection of purine bases and their nucleosides separated by HPLC column. *J.Biotechnol.* **1990**, *14*, 89–97.

Yoshioka, M.; Tamura, Z.; Senda, M.; Miyazaki, T. Analyzer for adenine nucleotides. *J.Chromatogr.* **1985**, *344*, 345–350.

Ochratoxin A

Castegnaro, M.; Maru, V.; Maru, G.; Ruiz-Lopez, M.D. High-performance liquid chromatographic determination of ochratoxin A and its 4R-4-hydroxy metabolite in human urine. *Analyst* **1990**, *115*, 129–131.

Nakajima, M.; Terada, H.; Hisada, K.; Tsubouchi, H.; Yamamoto, K.; Uda, T.; Itoh, Y.; Kawamura, O.; Ueno, Y. Determination of ochratoxin A in coffee beans and coffee products by monoclonal antibody affinity chromatography. *Food Agric.Immunol.* **1990**, *2*, 189–196.

Nakajima, M.; Terada, H.; Hisada, K.; Tsubouchi, H.; Yamamoto, K.; Uda, T.; Itoh, Y.; Kawamura, O.; Ueno, Y. Determination of ochratoxin A in coffee beans and coffee products by monoclonal antibody affinity chromatography. *Food Agric.Immunol* **1990**, *2*, 189–195.

Nesheim, S.; Stack, M.E.; Trucksess, M.W.; Eppley, R.M.; Krogh, P. Rapid solvent-efficient method for liquid chromatographic determination of ochratoxin A in corn, barley, and kidney: collaborative study. *J.AOAC Int.* **1992**, *75*, 481–487.

Ochratoxin A (using post-column reaction detection)

Hunt, D.C.; Philp, L.A.; Crosby, N.T. Determination of ochratoxin A in pig's kidney using enzymic digestion, dialysis, and high-performance liquid chromatography with postcolumn derivatization. *Analyst* **1979**, *104*, 1171–1175.

Langseth, W.; Nymoen, U.; Bergsjo, B. Ochratoxin A in plasma of Norwegian swine determined by an HPLC column-switching method. *Nat.Toxins* **1993**, *1*, 216–221.

Octopamine

Mell, L.D., Jr.; Carpenter, D.O. Fluorometric determination of octopamine in tissue homogenates by high-performance liquid chromatography. *Neurochem.Res.* **1980**, *5*, 1089–1096.

Ofloxacin

Lehr, K.H.; Damm, P. Chiral stationary phases versus chiral derivatization for the quantitation of ofloxacin enantiomers. *Methodol.Surv.Biochem.Anal.* **1990**, *20*, 205–206.

Ondansetron

Kirkham, J.C.; Rutherford, E.T. Compatibility of ondansetron hydrochloride with concomitant amino acid analysis in a standard parenteral nutrition solution. *ASHP Midyear Clinical Meeting Abstracts* **1993**, *28*, 454(R).

OPC-18790

Kitani, M.; Miyamoto, G.; Odomi, M. Stereoselective high-performance liquid chromatographic assay for the determination of OPC-18790 enantiomers in human plasma and urine. *J.Chromatogr.* **1993**, *620*, 97–104.

Organic Acids

Blanco Gomis, D. HPLC analysis of organic acids. *Food Sci.Technol.* **1992**, *52*, 371–385.

Lagoutte, D.; Lombard, G.; Nisseron, S.; Papet, M.P.; Saint-Jalm, Y. Determination of organic acids in cigaret smoke by high-performance liquid chromatography and capillary electrophoresis. *J.Chromatogr.A* **1994**, *684*, 251–257.

Stanek, R.; Gain, R.E.; Glover, D.D.; Larsen, B. High performance ion exclusion chromatographic characterization of the vaginal organic acids in women with bacterial vaginosis. *Biomed.Chromatogr.* **1992**, *6*, 231–235.

Organic Acids (using post-column reaction detection)

Hashimoto, A.; Yamasaki, K.; Kokusenya, Y.; Miyamoto, T.; Nakai, H.; Sato, T. Study on "signal" constituents for the evaluation of animal crude drugs. II. Organic acids. *Chem.Pharm.Bull.* **1995**, *43*, 2195–2199.

Hayashi, M. Determination of organic acids in foods by HPLC with postcolumn pH buffered electric conductivity detection. *GIT Spez.Chromatogr.* **1995**, *15*, 64–67.

Hayashi, M. Determination of organic acids in foods by HPLC with post-column pH buffered electroconductivity detection. *Shimadzu Hyoron* **1992**, *49*, 59–64.

Moegele, R.; Pabel, B.; Galensa, R. Determination of organic acids, amino acids and saccharides by high-performance liquid chromatography and a postcolumn enzyme reactor with amperometric detection. *J.Chromatogr.* **1992**, *591*, 165–173.

Wada, A.; Bonoshita, M.; Tanaka, Y.; Hibi, K. A study of a reaction system for organic acid analysis using a pH indicator as post-column reagent. *J.Chromatogr.* **1984**, *291*, 111–118.

Organic Phosphates (using post-column reaction detection)

Ikeguchi, Y.; Nakamura, H.; Nakajima, T. Organic phosphates analyzer using high-performance anion-exchange chromatography and a postcolumn phosphomolybdic acid reaction. *Anal.Sci.* **1993**, *9*, 653–655.

Organometallics

Attar, K.M. Analytical methods for speciation of organotins in the environment. *Appl. Organometal.Chem.* **1996**, *10*, 317–337.

Blais, J.S.; Marshall, W.D. Determination of ionic alkyllead compounds in water, soil and sediment by high-performance liquid chromatography-quartz tube atomic absorption spectrometry. *J.Anal.At.Spectrom.* **1989**, *4*, 271–277.

Blais, J.S.; Momplaisir, G.M.; Marshall, W.D. Determination of arsenobetaine, arsenocholine, and tetra-methylarsonium cations by HPLC thermochemical hydride generation-atomic absorption spectrome-try. *Anal.Chem.* **1990**, *62*, 1161–1166.

Nakashima, H.; Hori, S.; Nakazawa, H. Determination of dibutyltin and dioctyltin compounds in PVC food containers, wrappage and clothes by reversed phase HPLC with column switching. *Eisei.Kagaku.* **1990**, *36*, 15–20.

Parkin, J.E. Interference by disodium edetate with the reaction of phenylmercury ions with dithiocar-bamate derivatising agent. *J.Liq.Chromatogr.* **1992**, *15*, 441–449.

Praet, A.; Dewaele, C.; Verdonck, L.; Van der Kelen, G.P. Liquid chromatography of organotin compounds on cyanopropyl silica gel. *J.Chromatogr.* **1990**, *507*, 427–437.

Szpunar-Lobinska, J.; Witte, C.; Lobinski, R.; Adams, F.C. Separation techniques in speciation analysis for organometallic species. *Fresenius' J.Anal.Chem.* **1995**, *351*, 351–377.

Schickling, C.; Broekaert, J.A.C. Determination of mercury species in gas condensates by online coupled high-performance liquid chromatography and cold-vapor atomic absorption spectrometry. *Appl.Organomet.Chem.* **1995**, *9*, 29–36.

Yu, T.H.; Arakawa, Y. High-performance liquid chromatographic determination of dialkyltin homologs using fluorescence detection. *J.Chromatogr.* **1983**, *258*, 189–197.

Organometallics (using post-column reaction detection)

Blais, J.S.; Marshall, W.D. Post-column ethylation for the determination of ionic alkyllead compounds by high-performance liquid chromatography-atomic absorption spectrometry. *J.Anal.At.Spectrom.* **1989**, *4*, 641–645.

Bushee, D.S.; Krull, I.S.; Smith, S.B., Jr.; Schleicher, R.G. Determination of organolead compounds by liquid chromatography with on-line extraction and ultraviolet detection. *Anal.Chim.Acta* **1987**, *194*, 235–245.

Epstein, R.L.; Phillippo, E.T.; Harr, R.; Koscinski, W.; Vasco, G. Organotin residue determination in poultry and turkey sample survey in the United States. *J.Agric.Food Chem.* **1991**, *39*, 917–921.

Gast, C.H.; Kraak, J.C. Phase systems and post-column dithizone reaction detection for the analysis of organomercurials by HPLC. *Int.J.Environ.Anal.Chem.* **1979**, *6*, 297–312.

Kleiboehmer, W.; Cammann, K. Separation and determination of organotin compounds by HPLC and fluorometric detection in micellar systems. *Fresenius' Z.Anal.Chem.* **1989**, *335*, 780–784.

Nakashima, H.; Hori, S.; Nakazawa, H. Determination of dibutyltin and dioctyltin compounds in PVC food containers, packagings and clothes by reversed-phase HPLC with column switching. *Eisei Kagaku* **1990**, *36*, 15–20.

Nakashima, H.; Hori, S.; Iwagami, S.; Nakazawa, H.; Fujita, M. Determination of dialkyltin compounds in textiles by reversed phase HPLC based on post-column fluorescence derivatization. *Bunseki Kagaku* **1987**, *36*, 867–871.

Parks, E.J.; Brinckman, F.E.; Jewett, K.L.; Blair, W.R.; Weiss, C.S. Trace speciation by HPLC-graphite furnace atomic absorption spectroscopy for tin- and lead-bearing organometallic compounds, with sig-nal increases induced by transition-metal ions. *Appl.Organomet.Chem.* **1988**, *2*, 441–450.

Pfeffer, M.; Gelbe, B.; Hampe, P.; Steinberg, B.; Walenciak-Reddel, E.; Woicke, B.; Wykhoff, B. Sensitive fluorescence labeling for analysis of organotin compounds with morin. *Fresenius' J.Anal.Chem.* **1992**, *342*, 839–845.

Pfeffer, M.; Gelbe, B.; Woicke, B.; Wykhoff, B. High-performance liquid chromatography of a mixture of two dodecyltins by sensitive fluorescence tagging with morin. *J.Chromatogr.A* **1994**, *662*, 407–413.

Staeb, J.A.; Rozing, M.J.M.; van Hattum, B.; Cofino, W.P.; Brinkman, U.A.T. Normal-phase high-performance liquid chromatography with UV irradiation, morin complexation and fluorescence detec-tion for the determination of organotin pesticides. *J.Chromatogr.* **1992**, *609*, 195–203.

Organophosphorus Compounds (using post-column reaction detection)

Priebe, S.R.; Howell, J.A. Post-column reaction detection system for the determination of organophos-phorus compounds by liquid chromatography. *J.Chromatogr.* **1985**, *324*, 53–63.

Priebe, S.R.; Howell, J.A. Preliminary study of a potential post-column photodegradation reaction scheme for the detection of organophosphorus compounds. *Anal.Lett.* **1983**, *16*, 1219–1233.

Schieffer, G.W. Preliminary examination of a new post-column photolysis-molybdate reaction detection system for the determination of organophosphorus compounds by high performance liquid chromatography. *Instrum.Sci.Technol.* **1995**, *23*, 255–263.

Orthophosphate (using post-column reaction detection)

Hirai, Y.; Yoza, N.; Ohashi, S. Flow injection system as a post-column reaction detector for high-performance liquid chromatography of phosphinate, phosphonate and orthophosphate. *J.Chromatogr.* **1981**, *206*, 501–509.

Oxalic Acid

Martz, F.A.; Weiss, M.F.; Belyea, R.L. Determination of oxalate in forage by reverse-phase high pressure liquid chromatography. *J.Dairy Sci.* **1990**, *73*, 474–479.

Brega, A.; Quadri, A.; Villa, P.; Prandini, P.; Wei, J.Q.; Lucarelli, C. Improved HPLC determination of plasma and urine oxalate in the clinical diagnostic laboratory. *J.Liq.Chromatogr.* **1992**, *15*, 501–511.

McWhinney, B.C.; Cowley, D.W.; Chalmers, A.H. Simplified column liquid chromatographic method for measuring urinary oxalate. *J.Chromatogr.* **1986**, *383*, 137–141.

Oxalic Acid (using post-column reaction detection)

Millan, A.; Grases, J.M.; Grases, F. Rapid determination of urinary oxalate by high-performance liquid chromatography. *J.Chromatogr.* **1990**, *529*, 402–407.

Skotty, D.R.; Nieman, T.A. Determination of oxalate in urine and plasma using reversed-phase ion-pair high-performance liquid chromatography with tris(2,2'-bipyridyl)ruthenium(II)-electrogenerated chemiluminescence detection. *J.Chromatogr.B* **1995**, *665*, 27–36.

Wessels, D.; Von der Brelie, A.; Gromus, J.; Glaensa, R. Determination of sulfurous acid and oxalic acid in beverages by HPLC-biosensor coupling. *Monatsschr.Brauwiss.* **1995**, *48*, 96–101.

Yamato, S.; Wakabayashi, H.; Nakajima, M.; Shimada, K. Amperometric determination of oxalate in plasma and urine by liquid chromatography with immobilized oxalate oxidase. *J.Chromatogr.B* **1994**, *656*, 29–35.

Oxalyl Thiolesters

Skorczynski, S.S.; Yang, C.S.; Hamilton, G.A. Determination of oxalyl thiolesters, N-oxalylcysteine and N-oxalylcysteamine in biological materials. *Anal.Biochem.* **1991**, *192*, 403–410.

Oxamniquine

Fell, A.F.; Noctor, T.A.G.; Kaye, B. *In vitro* metabolism studies on oxamniquine and related compounds by chiral liquid chromatography. *J.Pharm.Biomed.Anal.* **1989**, *7*, 1743–1748.

Oxiracetam

Visconti, M.; Spalluto, R.; Crolla, T.; Pifferi, G.; Pinza, M. Determination of oxiracetam in human serum and urine by high-performance liquid chromatography. *J.Chromatogr.* **1987**, *416*, 433–438.

2-Oxo Acids

Lange, M.; Malyusz, M. Fast method for the simultaneous determination of 2-oxo acids in biological fluids by high-performance liquid chromatography. *J.Chromatogr.B* **1994**, *662*, 97–102.

2-Oxoglutarate

Kaysinger, K.K.; Pierce, W.M., Jr.; Nerland, D.E. Quantitative analysis of 2-oxoglutarate in biological samples using liquid chromatography with electrochemical detection. *Anal.Biochem.* **1994**, *222*, 81–85.

Oxprenolol

Tsuei, S.E.; Thomas, J.; Moore, R.G. Quantification of oxprenolol in biological fluids using high-performance liquid chromatography. *J.Chromatogr.* **1980**, *181*, 135–140.

Oxyphenonium (using post-column reaction detection)

Drenth, B.F.H.; Bosman, J.; Feitsma, K.G.; Van Nijhuis, A. Direct determination of the enantiomeric purity of oxyphenonium using chiral HPLC with post-column extraction detection. *Chromatographia* **1989**, *26*, 281–284.

Oxytocin

Nachtmann, F. High-performance liquid chromatography of intermediates in the oxytocin synthesis. *J.Chromatogr.* **1979**, *176*, 391–397.

Palatinit

Klingebiel, L.; Krueger, D.; Grossklaus, R. Analysis of Palatinit (isomalt) and its monomers in rat intestinal samples by high-performance liquid chromatography with ultraviolet detection. *J.Chromatogr.* **1990**, *527*, 238–243.

Pamidronate

Schnaare, R.L.; Franckowiak, J.A.; Sugita, E.T. Compatibility and stability of pamidronate disodium in IV fluids and in the presence of selected chemotherapeutic agents. *ASHP Annual Meeting* **1993**, *50*, P-136E.

Parathyroid Hormone (using post-column reaction detection)

Hage, D.S.; Taylor, B.; Kao, P.C. Intact parathyroid hormone: performance and clinical utility of an automated assay based on high-performance immunoaffinity chromatography and chemiluminescence detection. *Clin.Chem.* **1992**, *38*, 1494–1500.

Pareptide

Hui, K-S.; Hui, M.; Cheng, K-P.; Lajtha, A.; Boksay, I.; Fencik, M. Determination of the melanotropin-inhibiting factor analog pareptide in urine by high-performance liquid chromatography. *J.Chromatogr.* **1981**, *222*, 512–517.

Parthenolide

Dolman, D.M.; Knight, D.W.; Salan, U.; Toplis, D. A quantitative method for the estimation of parthenolide and other sesquiterpene lactones containing α-methylenebutyrolactone functions present in feverfew, Tanacetum parthenium. *Phytochem.Anal.* **1992**, *3*, 26–31.

PCBs

Haglund, P. Isolation and characterization of polychlorinated biphenyl (PCB) atropisomers. *Chemosphere* **1996**, *32*, 2133–2140.

Haglund, P. Isolation of PCB atropisomers for toxicological testing using chiral high-performance liquid chromatography. *Organohalogen Compd.* **1995**, *23*, 35–40.

PEG600

Kinahan, I.M.; Smyth, M.R. High-performance liquid chromatographic determination of PEG 600 in human urine. *J.Chromatogr.* **1991**, *565*, 297–307.

Penicillamine

Lankmayr, E.P.; Budna, K.W.; Mueller, K.; Nachtmann, F.; Rainer, F. Determination of D-penicillamine in serum by fluorescence derivatization and liquid column chromatography. *J.Chromatogr.* **1981**, *222*, 249–255.

Nachtmann, F. Determination of the L-isomer in D-penicillamine by derivatization liquid chromatography. *Int.J.Pharm.* **1980**, *4*, 337–345.

Penicillin

Christensen, L.H.; Mandrup, G.; Nielsen, J.; Villadsen, J. A robust liquid chromatographic method for measurement of medium components during penicillin fermentations. *Anal.Chim.Acta* **1994**, *296*, 51–62.

Penicillin G

Rumble, R.H.; Roberts, M.S. Determination of benzylpenicillin in plasma and urine by high-performance liquid chromatography. *J.Chromatogr.* **1985**, *342*, 436–441.

Tarbin, J.A.; Farrington, W.H.H.; Shearer, G. Determination of penicillins in animal tissues at trace residue concentrations. Part I. Determination of benzylpenicillin in milk by reversed-phase liquid chromatography with solid phase extraction and liquid chromatographic fractionation clean-up. *Anal.Chim.Acta* **1995**, *318*, 95–101.

Wiese, B.; Martin, K. Determination of benzylpenicillin in plasma and lymph at the ng ml^{-1} level by reversed-phase liquid chromatography in combination with digital subtraction chromatography technique. *J.Pharm.Biomed.Anal.* **1989**, *7*, 107–118.

Penicillins (using post-column reaction detection)

Bygrave, J.; Rose, M.; Tarbin, J. A comparison of screening and quantitative test methods for the determination of penicillin in milk. *Int.Dairy Fed.Spec.Issue* **1995**, *9505*, 266–268.

Haginaka, J.; Wakai, J. Liquid chromatographic determination of penicillins by postcolumn alkaline degradation using a hollow-fiber membrane reactor. *Anal.Biochem.* **1988**, *168*, 132–140.

Haginaka, J.; Wakai, J. Liquid chromatographic determination of penicillins by postcolumn degradation with sodium hypochlorite. *Anal.Chem.* **1986**, *58*, 1896–1898.

Haginaka, J.; Wakai, J. Liquid chromatographic determination of penicillins by postcolumn alkaline degradation. *Anal.Chem.* **1985**, *57*, 1568–1571.

Nakashima, K.; Kawaguchi, S.; Akiyama, S.; Schulman, S.G. High performance liquid chromatography of penicillins with penicillin-enhanced luminol chemiluminescence detection. *Biomed.Chromatogr.* **1993**, *7*, 217–219.

Westerlund, D.; Carlqvist, J.; Theodorsen, A. Analysis of penicillins in biological material by reversed phase liquid chromatography and post-column derivatization. *Acta Pharm.Suec.* **1979**, *16*, 187–214.

Pentobarbital (using post-column reaction detection)

Soine, W.H.; Soine, P.J.; England, T.M.; Graham, R.M.; Capps, G. Identification of the diastereomers of pentobarbital N-glucosides excreted in human urine. *Pharm.Res.* **1994**, *11*, 1535–1539.

Peptides and Proteins

Alpert, A.J. Cation-exchange high-performance liquid chromatography of proteins on poly(aspartic acid)-silica. *J.Chromatogr.* **1983**, *266*, 23–37.

Bennett, H.P.J.; Solomon, S. Use of Pico-Tag methodology in the chemical analysis of peptides with carboxyl-terminal amides. *J.Chromatogr.* **1986**, *359*, 221–230.

Bernhard, W.R.; Kagi, J.H. Purification and characterization of atypical cadmium-binding polypeptides from Zea mays. *EXS.* **1987**, *52*, 309–315.

Burnouf, T.; Bietz, J.A. Reversed-phase high-performance liquid chromatography of reduced glutenin, a disulfide-bonded protein of wheat endosperm. *J.Chromatogr.* **1984**, *299*, 185–199.

Ceccarelli, E.A.; Chan, R.L.; Arana, J.L.; Carrillo, N. A fast and sensitive micromethod for the manual sequencing of peptides using o-phthalaldehyde as derivatizing reagent. *J.Biochem.Biophys.Methods* **1984**, *10*, 49–54.

Chang, J.Y.; Knecht, R.; Ball, R.; Alkan, S.S.; Braun, D.G. A sensitive peptide mapping method. Identification of three amino acid substitutions within two anti-azobenzenearsonate monoclonal antibody light chains. *Eur.J.Biochem.* **1982**, *127*, 625–629.

Chaufer, B.; Rollin, M.; Sebille, B. High-performance liquid chromatography and ultrafiltration of whey proteins with inorganic porous materials coated with polyvinylimidazole derivatives. *J.Chromatogr.* **1991**, *548*, 215–228.

Eckert, H.; Koller, M. Derivatizing reagents based on ferrocene for HPLC-ECD determination of peptides and proteins. *J.Liq.Chromatogr.* **1990**, *13*, 3399–3414.

Eriksson, K.O. The influence of mobile phase pH on the retention and selectivity of peptides and proteins on a C-18 derivatized reversed phase matrix based on nonporous glycidol-filled agarose beads. *J.Liq.Chromatogr.* **1990**, *13*, 71–79.

Feldman, J.A.; Cohn, M.L.; Blair, D. Neuroendocrine peptides - analysis by reversed phase high performance liquid chromatography. *J.Liq.Chromatogr.* **1978**, *1*, 833–848.

Fullmer, C.S. Identification of cysteine-containing peptides in protein digests by high-performance liquid chromatography. *Anal.Biochem.* **1984**, *142*, 336–339.

Hill, J.C.; Flannery, G.M.; Fraser, B.A. Identification of α-carboxamidated and carboxy-terminal glycine forms of peptides in bovine hypothalamus, bovine pituitary and porcine heart extracts. *Neuropeptides (Edinburgh)* **1993**, *25*, 255–264.

Kai, M.; Nakano, M.; Zhang, G.Q.; Ohkura, Y. Determination of leucine- and methionine-enkephalins in rat brains by high-performance liquid chromatography with precolumn fluorescence derivatization. *Anal.Sci.* **1989**, *5*, 289–293.

Kasziba, E.; Flancbaum, L.; Fitzpatrick, J.C.; Schneiderman, J.; Fisher, H. Simultaneous determination of histidine-containing dipeptides, histamine, methylhistamine and histidine by high-performance liquid chromatography. *J.Chromatogr.* **1988**, *432*, 315–320.

Kehl, M.; Lottspeich, F.; Henschen, A. High-performance liquid chromatography of proteins as applied to fibrinogen chains. *Hoppe-Seyler's Z.Physiol.Chem.* **1982**, *363*, 1501–1505.

Ko, M.K.; Cole, K.D.; Pellegrino, J. Determination of total protein adsorbed on solid (membrane) surface by a hydrolysis technique: single protein adsorption. *J.Membr.Sci.* **1994**, *93*, 21–30.

Ko, M.K.; Cole, K.; Pellegrino, J.J. Determination of total protein adsorbed on solid (membrane) surface: Single protein adsorption. *Adv.Filtr.Sep.Technol.* **1993**, *7*, 440–445.

Kuliopulos, A.; Walsh, C.T. Production, purification, and cleavage of tandem repeats of recombinant peptides. *J.Am.Chem.Soc.* **1994**, *116*, 4599–4607.

Kumari, K.; Khanna, P.; Ansari, N.H.; Srivastava, S.K. High-performance liquid chromatography method for the determination of protein-glutathione mixed disulfide. *Anal.Biochem.* **1994**, *220*, 374–376.

Le Marechal, P.; Decottignies, P.; Jacquot, J.P.; Miginiac-Maslow, M. Separation by high-performance liquid chromatography of the ferredoxin-thioredoxin system proteins. *J.Chromatogr.* **1989**, *477*, 305–314.

Liu, D.; Mc Adoo, D.J. High-performance liquid chromatography of arginine-containing neuropeptides by precolumn derivatization and fluorimetric detection. *J.Liq.Chromatogr.* **1990**, *13*, 2049–2057.

Llorens-Cortes, C.; Schwartz, J.C.; Gros, C. Detection of the tripeptide Tyr-Gly-Gly, a putative enkephalin metabolite in brain, using a sensitive radioimmunoassay. *FEBS Lett.* **1985**, *189*, 325–328.

Lu, H.S.; Lai, P.H. Use of narrow-bore high-performance liquid chromatography for microanalysis of protein structure. *J.Chromatogr.* **1986**, *368*, 215–231.

Meek, J.L. Derivatizing reagents for high-performance liquid chromatography detection of peptides at the picomole level. *J.Chromatogr.* **1983**, *266*, 401–408.

Mendez, E.; Matas, R.; Soriano, F. Complete automatization of peptide maps by reversed-phase liquid chromatography using o-phthalaldehyde pre-column derivatization. *J.Chromatogr.* **1985**, *323*, 373–382.

Mousa, S.; Mullet, D.; Couri, D. Sensitive and specific high-performance liquid chromatographic method for methionine and leucine enkephalins. *Life Sci.* **1981**, *29*, 61–68.

Ng, D.S.; Yip, C.C. Peptide mapping of the insulin-binding site of the 130-kDa subunit of the insulin receptor by means of a novel cleavable radioactive photoprobe. *Biochem.Biophys.Res.Commun.* **1985**, *133*, 154–160.

Peter, A.; Toth, G.; Olajos, E.; Tourwe, D. Chromatographic behaviour of opioid peptides containing beta-methylphenylalanine isomers. *J.Chromatogr.A* **1995**, *705*, 267–273.

Polo, M.C.; Gonzalez de Llano, D.; Ramos, M. Derivatization and liquid chromatographic separation of peptides. *Food Sci.Technol.* **1992**, *52*, 117–140.

Rivier, J.; McClintock, R.; Galyean, R.; Anderson, H. Reversed-phase high-performance liquid chromatography: preparative purification of synthetic peptides. *J.Chromatogr.* **1984**, *288*, 303–328.

Ronca, F. Protein determination in polychromed stone sculptures, stuccoes and gesso grounds. *Stud.Conserv.* **1994**, *39*, 107–120.

Sandberg, M.; Li, X.; Folestad, S.; Weber, S.G.; Orwar, O. Liquid chromatographic determination of acidic β-aspartyl and gamma-glutamyl peptides in extracts of rat brain. *Anal.Biochem.* **1994**, *217*, 48–61.

Schepky, A.G.; Schulz-Knappe, P.; Forssmann, W-G. High-performance liquid chromatographic determination of sulfated peptides in human hemofiltrate using a radioactivity monitor. *J.Chromatogr.A* **1995**, *691*, 255–261.

Shively, J.E.; Hawke, D.; Jones, B.N. Microsequence analysis of peptides and proteins. III. Artifacts and the effects of impurities on analysis. *Anal.Biochem.* **1982**, *120*, 312.

Soskel, N.T.; Sandberg, L.B. Detection of urinary valylproline as an indicator of elastin degradation. *LC, Liq.Chromatogr.HPLC Mag.* **1983**, *1*, 434–436.

Tsai, H.; Weber, S.G. Electrochemical detection of dipeptides and dipeptide amides. *J.Chromatogr.* **1990**, *515*, 451–457.

Wu, K.M.; Sloan, J.W.; Martin, W.R. Development of a combined high-performance liquid chromatographic-fluorometric quantitative assay for enkephalins. *J.Chromatogr.* **1980**, *202*, 500–503.

Xu, Q.Y.; Shively, J.E. Microsequence analysis of peptides and proteins. VIII. Improved electroblotting of proteins onto membranes and derivatized glass-fiber sheets. *Anal.Biochem.* **1988**, *170*, 19–30.

Yang, X.; Huang, L.; Wang, Q.; Ci, Y. Study of derivatization of protein with a europium chelate and application to high-performance liquid chromatography. *Chin.Chem.Lett.* **1993**, *4*, 233–234.Yuan, D.; Pietrzyk, D.J. Separation and indirect detection of small-chain peptides using chromophoric mobile phase additives. *J.Chromatogr.* **1990**, *509*, 357–368.

Zhang, G.Q.; Kai, M.; Ohkura, Y. High-performance liquid chromatographic determination of peptides released by tryptic degradation from opioid peptide precursors in rat brain. *Chem.Pharm.Bull.* **1991**, *39*, 2369–2372.

Zhao, R.; Liu, G.; Feng, F.; Zhang, R. Determination of enkephalins in fetus brain tissues by means of HPLC with pre-column fluorescence derivatization. *Anal.Sci.* **1991**, *7*, 951–954.

Peptides and Proteins (using post-column reaction detection)

Castledine, J.B.; Fell, A.F.; Sellberg, B.; Modin, R.; Luque de Castro, M.D.; Valcarcel, M. Background noise reduction in post-column continuous-flow analysis combined with RPLC and computer-aided detection for the characterization of peptides. *J.Pharm.Biomed.Anal.* **1990**, *8*, 1079–1082.

Ding, W.; Xu, S.; Zhang, R. Fluorometric assay for proteins by o-phthaldialdehyde. *Fenxi Huaxue* **1987**, *15*, 17–21.

Fell, A.F.; Castledine, J.B.; Sellberg, B.; Modin, R.; Weinberger, R. Post-column continuous-flow analysis combined with reversed-phase liquid chromatography and computer-aided detection for the characterization of peptides. *J.Chromatogr.* **1990**, *535*, 33–39.

Fujinari, E.M.; Manes, J.D. Nitrogen-specific detection of peptides in liquid chromatography with a chemiluminescent nitrogen detector. *J.Chromatogr.A* **1994**, *676*, 113–120.

Herraiz, T.; Casal, V.; Polo, M.C. Reversed-phase HPLC analysis of peptides in standard and dairy samples using online absorbance and post-column OPA-fluorescence detection. *Z.Lebensm.-Unters.Forsch.* **1995**, *199*, 265–269.

Newcomb, R. High-sensitivity detection of peptides by liquid chromatography using postcolumn derivatization with fluorescamine. *LC-GC* **1992**, *10*, 34–39.

Noteborn, H.P.J.M.; Weusten, J.J.A.M.; Bartsch, H.; Bartsch, C.; Flehmig, B.; Ebels, I.; Salemink, C.A. Partial purification of a polypeptide extract derived from ovine pineal that suppresses the growth of human melanoma cells in vitro. *J.Pineal Res.* **1989**, *6*, 385–396.

Noteborn, H.P.J.M.; Reinharz, A.C.; Pevet, P.; Ebels, I.; Salemink, C.A. Neurohypophyseal hormone-like peptides in the ovine pineal gland using reverse-phase liquid chromatography and radioimmunoassay. *Peptides* **1988**, *9*, 455–462.

Noteborn, H.P.J.M.; Bartsch, H.; Bartsch, C.; Mans, D.R.A.; Weusten, J.J.A.M.; Flehmig, B.; Ebels, I.; Salemink, C.A. Partial purification of (a) low-molecular-weight ovine pineal compound(s) with an inhibiting effect on the growth of human melanoma cells in vitro. *J.Neural Transm.* **1988**, *73*, 135–155.

Shioya, Y.; Yoshida, H.; Nakajima, T. Estimation of molecular weights of peptides by determination of height equivalent to a theoretical plate in size-exclusion chromatography. *J.Chromatogr.* **1982**, *240*, 341–348.

Stegehuis, D.S.; Tjaden, U.R.; Van den Beld, C.M.B.; van der Greef, J. Bioanalysis of the neuropeptide des-enkephalin-gamma-endorphin by high-performance liquid chromatography with on-line sample pretreatment using gel permeation and solid-phase isolation. *J.Chromatogr.* **1991**, *549*, 185–193.

Tsai, H.; Weber, S.G. Influence of tyrosine on the dual electrode electrochemical detection of copper(II)-peptide complexes. *Anal.Chem.* **1992**, *64*, 2897–2903.

van Riel, J.A.; Olieman, C. Selective detection in RP-HPLC of Tyr-, Trp-, and sulfur-containing peptides by pulsed amperometry at platinum. *Anal.Chem.* **1995**, *67*, 3911–3915.

Xia, Q.; Wu, G.; Wang, H.; Xu, G. A modified solvent system for fluram postcolumn derivatization and RP-HPLC of proteins and peptides. *Shengwu Huaxue Yu Shengwu Wuli Xuebao* **1988**, *20*, 424–427.

Zhang, G.Q.; Kai, M.; Ohkura, Y. Determination of opioid peptides in rat brain by post-column fluorescence derivatization HPLC. *Kyushu Daigaku Chuo Bunseki Senta Hokoku* **1989**, 23–28.

Peptidoleukotrienes

Westcott, J.Y.; Johnston, K.; Batt, R.A.; Wenzel, S.E.; Voelkel, N.F. Measurement of peptidoleukotrienes in biological fluids. *J.Appl.Physiol.* **1990**, *68*, 2640–2648.

Perhexiline

Amoah, A.G.B.; Gould, B.J.; Parke, D.V. Single-dose pharmacokinetics of perhexiline administered orally to humans. *J.Chromatogr.* **1984**, *305*, 401–409.

Pesticides

Bushway, R.J. High-performance liquid chromatographic determination of carbaryl and 1-naphthol at residue levels in various water sources by direct injection and trace enrichment. *J.Chromatogr.* **1981**, *211*, 135–143.

Cabras, P.; Meloni, M.; Plumitallo, A.; Gennari, M. High-performance liquid chromatographic determination of ethiofencarb and its metabolic products. *J.Chromatogr.* **1989**, *462*, 430–434.

Engelhardt, H.; Lillig, B. Optimization of a reaction detector for analysis at the femtomol level of carbamate insecticides by HPLC. *Chromatographia* **1986**, *21*, 136–142.

Lawrence, J.F.; Leduc, R. High pressure liquid chromatography with ultraviolet absorbance or fluorescence detection of carbaryl in potato and corn. *J.Assoc.Off.Anal.Chem.* **1978**, *61*, 872–876.

Marco, M-P.; Chiron, S.; Gascon, J.; Hammock, B.D.; Barcelo, D. Validation of two immunoassay methods for environmental monitoring of carbaryl and 1-naphthol in ground water samples. *Anal.Chim.Acta* **1995**, *311*, 319–329.

Mount, M.E.; Oehme, F.W. Microprocedure for determination of carbaryl in blood and tissues. *J.Anal.Toxicol.* **1980**, *4*, 286–292.

Moye, H.A.; St John, P.A. A critical comparison of pre-column and post-column fluorogenic labeling for the HPLC analysis of pesticide residues. *ACS Symp.Ser.* **1980**, *136*, 89–102.

Tena, M.T.; Luque Castro, M.D.; Valcarcel, M. Sensitivity enhancement by using an HPLC flow-through sensor for determination of pesticide mixtures. *J.Liq.Chromatogr.* **1992**, *15*, 2373–2383.

Tsumura, Y.; Ujita, K.; Nakamura, Y.; Tonogai, Y.; Ito, Y. Simultaneous determination of aldicarb, ethiofencarb, methiocarb and their oxidized metabolites in grains, fruits and vegetables by high-performance liquid chromatography. *J.Food Prot.* **1994**, *57*, 1001–1007.

Pesticides (using post-column reaction detection)

Aharonson, N.; Muszkat, L.; Klein, M. Residue analysis of butocarboxim and aldicarb applied in a drip-irrigated peach grove by the HPLC post-column fluorogenic labeling technique. *Phytoparasitica* **1985**, *13*, 129–138.

Blass, W. Determination of methyl carbamate residues using on-line coupling of HPLC with a post column fluorimetric labeling technique. *Fresenius' J.Anal.Chem.* **1991**, *339*, 340–343.

Blass, W.; Philipowski, C. Determination of N-methyl carbamate residues using HPLC and on-line coupling of a post-column reactor in food of plant origin and soil. *Pflanzenschutz-Nachr.Bayer (Ger.Ed.)* **1992**, *45*, 277–318.

Chaput, D. On-line trace enrichment for determination of aldicarb species in water, using liquid chromatography with post-column derivatization. *J.Assoc.Off.Anal.Chem.* **1986**, *69*, 985–989.

de Kok, A.; Hiemstra, M.; Vreeker, C.P. Optimization of the postcolumn hydrolysis reaction on solid phases for the routine high-performance liquid chromatographic determination of N-methylcarbamate pesticides in food products. *J.Chromatogr.* **1990**, *507*, 459–472.

de Kok, A.; Hiemstra, M.; Vreeker, C.P. Improved cleanup method for the multiresidue analysis of N-methylcarbamates in grains, fruits, and vegetables by means of HPLC with postcolumn reaction and fluorescence detection. *Chromatographia* **1987**, *24*, 469–476.

Fucci, G.; Ciaravolo, S.; Mazza, G. Optimization of an analysis method for N-methylcarbamate pesticide residue determinations. *Ind.Aliment.(Pinerolo, Italy)* **1995**, *34*, 1160–1163.

Fukuda, Y.; Sasaki, T.; Murano, S.; Nakashima, M.; Seo, Y.; Funakoshi, A.; Yano, Y.; Kubota, A.; Nakano, I.; Mito, M. Studies on the analytical method of newly regulated pesticide residues by high-performance liquid chromatography. *Hiroshima-shi Eisei Kenkyusho Nenpo* **1994**, *14*, 33–45.

Goewie, C.E.; Hogendoorn, E.A. Pre-column clean-up and liquid chromatographic determination of residues of N-methylcarbamate pesticides in extracts of total diet. *J.Chromatogr.* **1987**, *404*, 352–358.

Hill, K.M.; Hollowell, R.H.; Dal Cortivo, L.A. Determination of N-methylcarbamate pesticides in well water by liquid chromatography with post-column fluorescence derivatization. *Anal.Chem.* **1984**, *56*, 2465–2468.

Jiang, X.; Cai, D.; Hua, X. Determination of carbamate pesticides by high performance liquid chromatography with post-column derivatization. *Sepu* **1994**, *12*, 32–34.

Koeduka, K.; Kenmotsu, K.; Ogino, Y.; Matsunaga, K.; Mori, T. Determination of methomyl and oxamyl in the environmental and biological samples by HPLC with post column fluorescence derivation. *Kankyo Kagaku* **1993**, *3*, 771–782.

Krause, R.T. Resolution, sensitivity and selectivity of a high-performance liquid chromatographic post-column fluorometric labeling technique for determination of carbamate insecticides. *J.Chromatogr.* **1979**, *185*, 615–624.

Krause, R.T. Multiresidue method for determining N-methylcarbamate insecticides in crops, using high performance liquid chromatography. *J.Assoc.Off.Anal.Chem.* **1980**, *63*, 1114–1124.

Krause, R.T. Further characterization and refinement of an HPLC post-column fluorometric labeling technique for the determination of carbamate insecticides. *J.Chromatogr.Sci.* **1978**, *16*, 281–288.

McDonald, P.D.; Leveille, W.P.; Sims, A.E.; Wildman, W.J.; Zener, V.R.; Scarchilli, A.D. Optimization of a method for the analysis of carbamate pesticides and their metabolites in drinking water using HPLC with post-column derivatization. *Proc.-Water Qual.Technol.Conf.* **1989**, *17*, 631–649.

McGarvey, B.D. Liquid chromatograph determination of N-methylcarbamate pesticides using a single-stage post-column derivatization reaction and fluorescence detection. *J.Chromatogr.* **1989**, *481*, 445–451.

Miyanaga, A.; Uchida, K. HPLC analysis of N-methylcarbamate pesticides using post-column electrochemical and fluorescence. *Kankyo Kagaku* **1995**, *5*, 336–337.

Muth, J.; Giles, J. Post column derivatization in the analysis of N-methyl carbamate insecticides. *Altex Chromatogram* **1980**, *3*, 5–6.

Nagayama, T.; Kobayashi, M.; Shioda, H.; Morino, M.; Ito, M.; Tamura, Y. Simultaneous determination of N-methylcarbamate pesticides in agricultural products by liquid chromatography. *Shokuhin Eisei-gaku Zasshi* **1994**, *35*, 470–478.

Nondek, L.; Frei, R.W.; Brinkman, U.A.T. Heterogeneous catalytic post-column reaction detectors for high-performance liquid chromatography application to N-methylcarbamates. *J.Chromatogr.* **1983**, *282*, 141–150.

Okamoto, H.S.; Wijekoon, D.; Esperanza, C.E.; Cheng, J.C.; Park, S.L.; Garcha, J.S.; Gill, S.S.; Perera, K.S. Analysis for N-methylcarbamate pesticides by high performance liquid chromatography in environmental samples. *ASTM Spec.Tech.Publ., STP 1075(Waste Test.Qual.Assur.: Third Vol.)* **1992**, 115–28.

Stafford, S.C.; Lin, W. Determination of oxamyl and methomyl by high-performance liquid chromatography using a single-stage postcolumn derivatization reaction and fluorescence detection. *J.Agric.Food Chem.* **1992**, *40*, 1026–1029.

Sundaram, K.M.S.; Curry, J. High-performance liquid chromatographic method for the analysis of aminocarb, mexacarbate and some of their N-methylcarbamate metabolites by post-column derivatization with fluorescence detection. *J.Chromatogr.A* **1994**, *672*, 117–124.

Tena, M.T.; Linares, P.; Luque de Castro, M.D.; Valcarcel, M. Total and individual determination of carbamate pesticides by use of an integrated flow-injection/HPLC system. *Chromatographia* **1992**, *33*, 449–453.

Tena, M.T.; De Castro, L.; Valcarcel, M. HPLC-postcolumn derivatizing-integrated retention-detection system for the determination of carbaryl and its hydrolysis product. *J.Chromatogr.Sci.* **1992**, *30*, 276–279.

Ting, K.C.; Kho, P.K.; Musselman, A.S.; Root, G.A.; Tichelaar, G.R. High performance liquid chromatographic method for determination of six N-methylcarbamates in vegetables and fruits. *Bull.Environ.Contam.Toxicol.* **1984**, *33*, 538–547.

Walters, S.M.; Westerby, B.C.; Gilvydis, D.M. Determination of phenylurea pesticides by high-performance liquid chromatography with UV and photoconductivity detectors in series. *J.Chromatogr.* **1984**, *317*, 533–544.

Phenols

Charpentier, B.A.; Cowles, J.R. Rapid method of analyzing phenolic compounds in Pinus elliotti using high-performance liquid chromatography. *J.Chromatogr.* **1981**, *208*, 132–136.

Di Nunzio, C.; Parisi, G.; Santoro, P.; Ricci, P.A. Determination of phenols and cresols in aluminum-backed paper by high-performance liquid chromatography. *J.Chromatogr.* **1987**, *392*, 454–459.

Erdmann, K.; Mohan, T.; Verkade, J.G. HPLC and 31P NMR analysis of phenols in coal liquefaction oils. *Energy Fuels* **1996**, *10*, 378–385.

Koch, J.; Voelker, P. Solid phase extraction of phenols from water by highly porous cross-linked polystyrene. *Acta Hydrochim.Hydrobiol.* **1995**, *23*, 66–71.

Kuwata, K.; Tanaka, S. Liquid chromatographic determination of traces of phenols in air. *J.Chromatogr.* **1988**, *442*, 407–411.

Marko-Varga, G.A. Liquid chromatographic determination of phenols and substituted derivatives in water samples. *Tech.Instrum.Anal.Chem.* **1993**, *13*, 225–271.

Piangerelli, V.; Nerini, F.; Cavalli, S. Phenols determination in water samples by SPE using keto-derivatized poly(styrene-divinylbenzene) copolymer and HPLC with amperometric detection. *Ann.Chim.(Rome)* **1993**, *83*, 331–343.

Wen, Y.H.; Wu, S.S.; Lin, S.J.; Wu, H.L. Trace analysis for phenol by derivatization and high performance liquid chromatography. *Zhonghua Yaoxue Zazhi* **1992**, *44*, 221–227.

Phenols (using post-column reaction detection)

Hostettmann, K.; Domon, B.; Schaufelberger, D.; Hostettmann, M. On-line high-performance liquid chromatography. Ultraviolet-visible spectroscopy of phenolic compounds in plant extracts using postcolumn derivatization. *J.Chromatogr.* **1984**, *283*, 137–147.

Noack, J.; Mattusch, J.; Werner, G. HPLC-EC-dual detection of phenols by photochemical post column derivatization. *Z.Chem.* **1990**, *30*, 448–450.

Ratanathanawongs, S.K.; Crouch, S.R. Development of a selective post-column detector for phenols separated by high-performance liquid chromatography. *Anal.Chim.Acta* **1987**, *192*, 277–287.

Phenothiazines

Kok, W.T.; Voogt, W.H.; Brinkman, U.A.T.; Frei, R.W. On-line electrochemical reagent production for fluorescence detection of phenothiazines in liquid chromatography. *J.Chromatogr.* **1986**, *354*, 249–257.

Phenylacetic Acid

Iwata, T.; Ishimaru, T.; Nakamura, M.; Yamaguchi, M. Direct determination of free phenylacetic acid in human plasma and urine by column-switching high performance liquid chromatography with fluorescence detection. *Biomed.Chromatogr.* **1994**, *8*, 283–287.

Phenylalanine

Rennie, P.J. Determination of phenylalanine in river water by high-performance liquid chromatography. *J.Chromatogr.* **1989**, *461*, 277–280.

Schmidt, G.J.; Olson, D.C.; Slavin, W. Determination of phenylalanine in serum using reversed-phase liquid chromatography and fluorescence detection. *J.Chromatogr.* **1979**, *164*, 355–362.

Schmidt, G.J.; Olson, D.C.; Slavin, W. A liquid chromatography procedure for determining phenylalanine using pre-column fluorescence derivatization. *Chromatogr.Newsl.* **1979**, *7*, 10–11.

Phenylenediamine

Burg, W.R.; Winner, P.C.; Saltzman, B.E.; Elia, V.J. The development of an air sampling and analytic method for o-phenylenediamine in an industrial environment. *Am.Ind.Hyg.Assoc.J.* **1980**, *41*, 557–562.

Phenylethylamine

McAllister, T.A.; Samuels, S.E.; Sedgwick, G.W.; Fenton, T.W.; Thompson, J.R. High-performance liquid chromatographic analysis of beta-phenylethylamine for the estimation of in vivo protein synthesis. *J.Chromatogr.B* **1995**, *666*, 336–341.

Huebert, N.D.; Schuurmans Schwach, V.; Richter, G.; Zreika, M.; Hinze, C.; Haegele, K.D. The measurement of beta-phenylethylamine in human plasma and rat brain. *Anal.Biochem.* **1994**, *221*, 42–47.

Gusovsky, F.; Jacobson, K.A.; Kirk, K.L.; Marshall, T.; Linnoila, M. New high-performance liquid chromatographic procedure for the detection and quantification of β-phenylethylamine. *J.Chromatogr.* **1987**, *415*, 124–128.

Phenylpropanolamine

Stockley, C.S.; Wing, L.M.H.; Miners, J.O. Stereospecific high-performance liquid chromatographic assay for the enantiomers of phenylpropanolamine in human plasma. *Ther.Drug Monit.* **1991**, *13*, 332–338.

Mason, W.D.; Mason, J.S. Improved high-pressure liquid chromatographic method for phenylpropanolamine in human plasma. *Anal.Lett.* **1983**, *16*, 693–699.

Phenylpropanolamine (using post-column reaction detection)

Mike, J.H.; Ramos, B.L.; Zupp, T.A. Electrochemical enhancement of high-performance liquid chromatography-UV detection for determination of phenylpropanolamine. *J.Chromatogr.* **1990**, *518*, 167–177.

Phosphatidic Acid

Yamada, K.; Abe, S.; Katayama, K.; Sato, T. Sensitive high-performance liquid chromatographic method for the determination of phosphatidic acid. *J.Chromatogr.* **1988**, *424*, 367–372.

Phosphatidylbutanol (using post-column reaction detection)

Heung, Y.M.M.; Postle, A.D. Synthesis of phosphatidyl[3H]butanol molecular species by phospholipase D in HL60 granulocytes. *Biochem.Soc.Trans.* **1995**, *23*, 275S.

Phosphatidylcholine

Rabe, H.; Kruger, J.; Reichmann, G.; Rustov, B. Separation and determination of phosphatidylcholine species as diacylglycerol naphthylurethanes by high-performance liquid chromatography. *Symp.Biol.Hung.* **1986**, *31*, 443–448.

Phosphatidylcholine (using post-column reaction detection)

Hunt, A.N.; Kelly, F.J.; Postle, A.D. Developmental changes in individual molecular species of phosphatidylcholine from fetal lungs of rat, guinea pig and man. *Biochem.Soc.Trans.* **1989**, *17*, 729–730.

Phosphatidylinositol

Nakagawa, Y. Application of paired-ion high-performance liquid chromatography to the separation of molecular species of phosphatidylinositol. *Lipids* **1993**, *28*, 1033–1035.

Phosphinothricin

Fiedler, H.P.; Plaga, A.; Schuster, R. Automated on-column derivatization of the antibiotics phosphinothricin and phosphinothricyl-alanyl-alanine with o-phthalaldehyde and microbore column high-performance liquid chromatography for quantitative determination in biological cultures. *J.Chromatogr.* **1986**, *353*, 201–206.

Phosphoamino Acids

Carlomagno, L.; Huebner, V.D.; Matthews, H.R. Rapid separation of phosphoamino acids including the phosphohistidines by isocratic high-performance liquid chromatography of the orthophthalaldehyde derivatives. *Anal.Biochem.* **1985**, *149*, 344–348.

Phospholipids

Abidi, S.L.; Mounts, T.L.; Rennick, K.A. Reversed-phase high-performance liquid chromatography of phospholipids with fluorescence detection. *J.Chromatogr.* **1993**, *639*, 175–184.

Batley, M.; Packer, N.H.; Redmond, J.W. High-performance liquid chromatography of diglyceride p-nitrobenzoates. An approach to molecular analysis of phospholipids. *J.Chromatogr.* **1980**, *198*, 520–525.

Dugan, L.L.; Demediuk, P.; Pendley, C.E.; Horrocks, L.A. Separation of phospholipids by high-performance liquid chromatography: all major classes, including ethanolamine and choline plasmalogens, and most minor classes, including lysophosphatidylethanolamine. *J.Chromatogr.* **1986**, *378*, 317–327.

Gasser, H.; Hallstroem, S.; Redl, H.; Schlag, G. Oxidatively modified plasma phospholipids containing reactive carbonyl functions measured by HPLC: evidence for phosphatidylcholine-bound aldehydes in plasma of burn patients. *Free Radical Res.* **1995**, *22*, 327–336.

Kito, M.; Takamura, H.; Narita, H.; Urade, R. A sensitive method for quantitative analysis of phospholipid molecular species by high-performance liquid chromatography. *J.Biochem.* **1985**, *98*, 327–331.

Matthees, D.P. Precolumn derivatization of amino phospholipids for liquid chromatography. *Proc.S.D.Acad.Sci.* **1980**, *59*, 62–64.

Ramesha, C.S.; Pickett, W.C.; Murthy, D.V.K. Sensitive method for the analysis of phospholipid subclasses and molecular species as 1-anthroyl derivatives of their diglycerides. *J.Chromatogr.* **1989**, *491*, 37–48.

Sax, S.M.; Moore, J.J.; Oley, A.; Amenta, J.S.; Silverman, J.A. Liquid-chromatographic estimation of saturated phospholipid palmitate in amniotic fluid compared with a thin-layer chromatographic method for acetone-precipitated lecithin. *Clin.Chem.* **1982**, *28*, 2264–2268.

Takamura, H.; Kito, M. A highly sensitive method for quantitative analysis of phospholipid molecular species by high-performance liquid chromatography. *J.Biochem.* **1991**, *109*, 436–439.

Takamura, H.; Narita, H.; Urade, R.; Kito, M. Quantitative analysis of polyenoic phospholipid molecular species by high performance liquid chromatography. *Lipids* **1986**, *21*, 356–361.

Phospholipids (using post-column reaction detection)

Burdge, G.; Creaney, A.; Postle, A.; Wilton, D.C. Mammalian secreted and cytosolic phospholipase A2 show different specificities for phospholipid molecular species. *Int.J.Biochem.Cell Biol.* **1995**, *27*, 1027–1032.

Kondoh, Y.; Yamada, A.; Takano, S. Analysis for phospholipids by HPLC with postcolumn derivatization. *Bunseki Kagaku* **1991**, *40*, 57–63.

Ouhazza, M.; Siouffi, A.M. Liquid chromatography analysis of some phospholipids with fluorescence detection. *Analusis* **1992**, *20*, 185–188.

Yamamoto, M. A new method for analysis of phospholipids by high-performance liquid chromatography and biliary phospholipids in patients with hepatolithiasis. *Wakayama Igaku (1995)* **1996**, *46*, 483–492.

Phosphorus (using post-column reaction detection)

Nanny, M.A.; Kim, S.; Minear, R.A. Aquatic soluble unreactive phosphorus: HPLC studies on concentrated water samples. *Water Res.* **1995**, *29*, 2138–2148.

Phosphorus Oxo Acids (using post-column reaction detection)

Meek, S.E.; Pietrzyk, D.J. Liquid chromatographic separation of phosphorus oxo acids and other anions with post-column indirect fluorescence detection by aluminum-morin. *Anal.Chem.* **1988**, *60*, 1397–1400.

Phosphotyrosine

Malencik, D.A.; Zhao, Z.; Anderson, S.R. Determination of dityrosine, phosphotyrosine, phosphothreonine, and phosphoserine by high-performance liquid chromatography. *Anal.Biochem.* **1990**, *184*, 353–359.

Ringer, D.P. Separation of phosphotyrosine, phosphoserine, and phosphothreonine by high-performance liquid chromatography. *Methods Enzymol.* **1991**, *201*, 3–10.

Phthalates

Neville, G.A.; Benning, B.J.; Black, D.B.; Ethier, J.-C.; Chaloner-Larsson, G.; Zamecnik, J. Analysis of commercial albumin (ALB) preparations for mono-2-ethylhexyl phthalate (MEHP) and di-2-ethylhexyl phthalate (DEHP). *Can.J.Appl.Spectrosc.* **1996**, *41*, 66–70.

Phytic Acid (using post-column reaction detection)

Cilliers, J.J.L.; Van Niekerk, P.J. HPLC determination of phytic acid in food by postcolumn colorimetric detection. *J.Agric.Food Chem.* **1986**, *34*, 680–683.

Phytochelatins (using post-column reaction detection)

Kubota, H.; Sato, K.; Yamada, T.; Maitani, T. Phytochelatins (class III metallothioneins) and their des-glycyl peptides induced by cadmium in normal root cultures of Rubia tinctorum L. *Plant Sci.* **1995**, *106*, 157–166.

Pindolol

Hasegawa, R.; Murai-Kushiya, M.; Komuro, T.; Kimura, T. Stereoselective determination of plasma pindolol in endotoxin-pretreated rats by high-performance liquid chromatography. *J.Chromatogr.* **1989**, *494*, 381–388.

Pipecolic Acid

Nishio, H.; Segawa, T. Determination of pipecolic acid in human blood plasma by high-performance liquid chromatography. *Clin.Chim.Acta* **1984**, *143*, 57–63.

Pipemidic Acid

Ito, H.; Inoue, M.; Morikawa, M.; Tsuboi, M.; Oka, K. Determination of pipemidic acid in plasma by normal-phase high-pressure liquid chromatography. *Antimicrob.Agents Chemother.* **1985**, *28*, 192–194.

Piperine

Verzele, M.; Van Damme, F.; Schuddinck, G.; Vyncke, P. Quantitative microscale liquid chromatography of piperine in pepper and pepper extracts. *J.Chromatogr.* **1989**, *471*, 335–346.

Pirprofen

Sioufi, A.; Colussi, D.; Marfil, F.; Dubois, J.P. Determination of the (+)- and (-)-enantiomers of pirprofen in human plasma by high-performance liquid chromatography. *J.Chromatogr.* **1987**, *414*, 131–137.

Polycyclic Aromatic Heterocycles

Bodzek, D.; Janoszka, B.; Warzecha, L. The analysis of PAHs nitrogen derivatives in the sewage sludges of Upper Silesia, Poland. *Water Air Soil Pollut.* **1996**, *89*, 417–427.

Polycyclic Aromatic Hydrocarbons

Ariese, F.; Verkaik, M.; Hoornweg, G.P.; van de Nesse, R.J.; Jukema-Leenstra, S.R.; Hofstraat, J.W.; Gooijer, C.; Velthorst, N.H. Trace analysis of 3-hydroxy benzo[a]pyrene in urine for the biomonitorng of human exposure to polycyclic aromatic hydrocarbons. *J.Anal.Toxicol.* **1994**, *18*, 195–204.

Fetzer, J.C.; Biggs, W.R. Liquid chromatographic retention behavior of large, fused polycyclic aromatics. Normal bonded phases. *J.Chromatogr.* **1985**, *346*, 81–92.

Mazzeo, J.R.; Krull, I.S.; Kissinger, P.T. Use of cerium(IV) oxidizing agent for the derivatization of polycyclic aromatic hydrocarbons for liquid chromatography-electrochemical detection. *J.Chromatogr.* **1991**, *550*, 585–594.

Polycyclic Aromatic Hydrocarbons (using post-column reaction detection)

Li, Q.; Fu, C. Determination of perylene in petroleum-source rocks by HPLC-post-column chemilumi-nescence. *Fenxi Ceshi Tongbao* **1990**, *9*, 33–37.

Murahashi, T.; Hayakawa, K.; Iwamoto, Y.; Miyazaki, M. Simultaneous determination of polycyclic aromatic hydrocarbons and their nitrated derivatives in airborne particulates by HPLC with fluorescence/chemiluminescence detection. *Bunseki Kagaku* **1994**, *43*, 1017–1020.

Schulman, S.G.; Rutledge, J.M. Enhancement of the fluorimetric properties of some carboxyl substituted anthracenes by complexation with bovine serum albumin. *Anal.Lett.* **1986**, *19*, 2141–2145.

Polyethylene Glycol Esters (using post-column reaction detection)

Shah, B.; Watson, E. Determination of N-hydroxysuccinimidyl-activated polyethylene glycol esters by gel permeation chromatography with post-column alkaline hydrolysis. *J.Chromatogr.* **1993**, *629*, 398–400.

Polylysine

Hishiyama, T.; Nakamura, M.; Ujiie, T. Determination of epsilon-polylysine in foods by high performance liquid chromatography with precolumn method of dansyl derivatization. *Nippon Shokuhin Kagaku Kogaku Kaishi* **1996**, *43*, 1105–1109.

Polymethylenephosphonic Acids (using post-column reaction detection)

Tschaebunin, G.; Fischer, P.; Schwedt, G. Analysis for polymethylenephosphonic acids. II. A systematic survey of post-column derivatization in ion chromatography. *Fresenius' Z.Anal.Chem.* **1989**, *333*, 117–122.

Polyols (using post-column reaction detection)

Niwa, T.; Kajita, M.; Wada, Y.; Maeda, K.; Tohyama, K.; Kato, Y. LC/MS analysis of polyols in uremic serum. *Nippon Iyo Masu Supekutoru Gakkai Koenshu* **1992**, *17*, 175–178.

Poly(oxy-1,2-ethanediyl) Oligomers (using post-column reaction detection)

Warner, C.R.; Selim, S.; Daniels, D.H. Post-column complexation technique for the spectrophotometric detection of poly(oxy-1,2-ethanediyl) oligomers in steric exclusion chromatography. *J.Chromatogr.* **1979**, *173*, 357–363.

Polyphosphates (using post-column reaction detection)

Linares, P.; Luque de Castro, M.D.; Valcarcel, M. Determination of polyphosphates in intermediate materials for detergent manufacture by ion high-performance liquid chromatography with post-column derivatization. *J.Chromatogr.* **1991**, *585*, 267–271.

Matsunaga, A.; Ooizumi, T.; Yamamoto, A.; Kawasaki, K.; Mizukami, E. Degradation of polyphosphates during manufacturing process of surimi-based products. *Nippon Suisan Gakkaishi* **1990**, *56*, 2077–2082.

Matsunaga, A.; Yamamoto, A.; Mizukami, E.; Kawasaki, K.; Ooizumi, T. Determination of polyphosphates in food by high performance liquid chromatography. *Nippon Shokuhin Kogyo Gakkaishi* **1990**, *37*, 20–25.

Muessig-Zufika, M.; Kornmueller, A.; Merkelbach, B.; Jekel, M. Isolation and analysis of intact polyphosphate chains from activated sludges associated with biological phosphate removal. *Water Res.* **1994**, *28*, 1725–1733.

Post-Column Extraction (using post-column reaction detection)

Apffel, J.A.; Brinkman, U.A.T.; Frei, R.W. Design and application of a post-column extraction system compatible with miniaturized liquid chromatography. *Chromatographia* **1984**, *18*, 5–10.

Bigley, F.; Grob, R.L.; Brenner, G. Post-column extraction system for use with high-performance liquid chromatography. *J.Chromatogr.* **1984**, *288*, 293–302.

Brinkman, U.A.T.; Lawrence, J.F.; Van Buuren, C.; Frei, R.W. Coupling of post column ion-pair extraction and fluorescence detection for the analysis of basic organic compounds. *Anal.Chem.Symp.Ser.* **1980**, *3*, 247–259.

de Ruiter, C.; Wolf, J.H.; Brinkman, U.A.T.; Frei, R.W. Design and evaluation of a sandwich phase separator for on-line liquid/liquid extraction. *Anal.Chim.Acta* **1987**, *192*, 267–275.

Lawrence, J.F. The use of post-column dynamic ion-pair extraction for the HPLC detection of anionic additives in foods. *Int.J.Environ.Anal.Chem.* **1990**, *38*, 115–126.

Maris, F.A.; Nijenhuis, M.; Frei, R.W.; de Jong, G.J.; Brinkman, U.A.T. On-line post-column extraction in column liquid chromatography with electron-capture detection. *J.Chromatogr.* **1988**, *435*, 297–306.

Maris, F.A.; Nijenhuis, M.; Frei, R.W.; de Jong, G.J.; Brinkman, U.A.T. On-line post-column extraction in column liquid chromatography with electron-capture detection. Part I: Design and analytical characteristics. *Chromatographia* **1986**, *22*, 235–240.

Roy, I.M.; Jefferies, T.M. Performance evaluation of an aqueous-organic phase separator for post-column reactions in high-performance liquid chromatography, and its application to the enhanced detection of some basic drugs of abuse. *J.Pharm.Biomed.Anal.* **1990**, *8*, 831–835.

Veltkamp, A.C.; Das, H.A.; Frei, R.W.; Brinkman, U.A.T. Flow-through radioactivity detection of tritium- and carbon-14-labeled compounds in reversed-phase liquid chromatography by a liquid scintillation counting technique based on post-column extraction. *Anal.Chim.Acta* **1990**, *233*, 181–189.

Pravastatin

Dumousseaux, C.; Muramatsu, S.; Takasaki, W.; Takahagi, H. Highly sensitive and specific determination of pravastatin sodium in plasma by high-performance liquid chromatography with laser-induced fluorescence detection after immobilized antibody extraction. *J.Pharm.Sci.* **1994**, *83*, 1630–1636.

Praziquanamine

Li, H.Z.; Zhao, H.; Yu, Y.; Qiu, Z.Y. [Determination of the optical purity of praziquanamine by reversed phase HPLC after derivatization with GITC]. *Yao.Hsueh.Hsueh.Pao.* **1993**, *28*, 450–454.

Prizidilol

Pearce, J.C.; Murkitt, G.S.; Taylor, D.C.; Cresswell, P.R. An assay for prizidilol in plasma and urine by high-performance liquid chromatography. *J.Pharm.Biomed.Anal.* **1986**, *4*, 115–121.

Procyanidins

Rigaud, J.; Escribano-Bailon, M.T.; Prieur, C.; Souquet, J.-M.; Cheynier, V. Normal-phase high-performance liquid chromatographic separation of procyanidins from cacao beans and grape seeds. *J.Chromatogr.* **1993**, *654*, 255–260.

Propafenone

Brode, E.; Kripp, U.; Hollmann, M. Simultaneous determination of propafenone and 5-hydroxypropafenone in plasma by means of high pressure liquid chromatography. *Arzneim.-Forsch.* **1984**, *34*, 1455–1460.

Propranolol

Gupta, M.B.; Hubbard, J.W.; Midha, K.K. Separation of enantiomers of derivatized or underivatized propranolol by means of high-performance liquid chromatography. *J.Chromatogr.* **1988**, *424*, 189–194.

Miller, R.B. High-performance liquid chromatographic assay for the derivatized enantiomers of propranolol and 4-hydroxypropranolol in human plasma. *J.Pharm.Biomed.Anal.* **1991**, *9*, 953–958.

Pham-Huy, C.; Sahui-Gnassi, A.; Massicot, F.; Galons, H.; Dutertre-Catella, H.; Radenen, B.; Warnet, J.-M., Jr.; Duc, H.T.; Claude, J.R., Jr. Analysis of (S)- and (R)-propranolol in human plasma and urine by a specific immunoenzymic assay versus HPLC. *Arch.Toxicol., Suppl.* **1995**, *17*, 522–527.

Silber, B.; Riegelman, S. Stereospecific assay for (-)- and (+)-propranolol in human and dog plasma. *J.Pharmacol.Exp.Ther.* **1980**, *215*, 643–648.

N-Propylnorapomorphine

Smith, R.V.; Klein, A.E.; Thompson, D.O. High-performance liquid chromatographic determination of N-n-propylnorapomorphine and its glucuronide metabolite in blood plasma. *Mikrochim.Acta* **1980**, *1*, 151–157.

Prostaglandins

Beck, G.M.; Roston, D.A.; Jaselskis, B. Derivatization procedures for detection of in biological matrices by liquid chromatography/electrochemistry. *Talanta* **1989**, *36*, 373–377.

Cox, J.W.; Pullen, R.H. Determination of prostaglandin E1 in plasma with picogram-per-milliliter sensitivity by double antibody extraction and column-switching high-performance liquid chromatography. *J.Pharm.Biomed.Anal.* **1986**, *4*, 653–662.

Goto, H.; Sugiyama, S.; Kawabe, Y.; Kuroiwa, M.; Ohara, A.; Tsukamoto, Y.; Nakazawa, S.; Ozawa, T. Quantitative determination of prostaglandin levels in human gastric mucosa — analysis by microcolumn high performance liquid chromatography with laser induced fluorescence detection. *Biochem.Int.* **1990**, *20*, 1119–1125.

Kubo, I.; Komatsu, S. Simultaneous measurement of and ecdysteroids in insects by high-performance liquid chromatography with fluorescence labeling. *Agric.Biol.Chem.* **1987**, *51*, 1305–1309.

Kubo, I.; Komatsu, S. Micro analysis of and ecdysteroids in insects by high-performance liquid chromatography and fluorescence labeling. *J.Chromatogr.* **1986**, *362*, 61–70.

Omori, H.; Okabe, K.; Nakashizuka, T.; Yamazaki, S. Determination of in human saliva by high performance liquid chromatography using a column switching technique. *HRC CC, J.High Resolut.Chromatogr.Chromatogr.Commun.* **1986**, *9*, 477–479.

Plaisted, S.M.; Zwier, T.A.; Snider, B.G. High-performance liquid chromatographic separation of 15-methyl PGF2α methyl ester isomers. *J.Chromatogr.* **1983**, *281*, 151–157.

Sakae, S.; Harata, A.; Kitamori, T.; Okubo, A.; Shimizu, T.; Watanabe, H.; Toda, S.; Sawada, T. Highly sensitive simultaneous determination of by using HPLC/laser induced fluorescence (LIF). *Anal.Sci.* **1991**, *7*, 957–958.

Sakae, S.; Harata, A.; Kitamori, T.; Sawada, T.; Okubo, A.; Toda, S.; Shimizu, T. Simultaneous determination of ultratrace in biological samples with high-performance liquid chromatography/laser-induced fluorometry. *Microchem.J.* **1994**, *49*, 355–361.

Wang, Z.S.; Zhu, Y.H.; Zhang, S.L.; Zhu, J.Y.; Wang, M.L. [Determination of 15-methyl prostaglandin F2a by derivatization high pressure liquid chromatography]. *Yao.Hsueh.Hsueh.Pao.* **1982**, *17*, 603–608.

Proxyphylline

Ruud-Christensen, M.; Aasen, A.J.; Rasmussen, K.E.; Salvesen, B. High-performance liquid chromatographic determination of (R)-and (S)-proxyphylline in human plasma. *J.Chromatogr.* **1989**, *491*, 355–366.

Pseudoephedrine

Duddu, S.P.; Mehvar, R.; Grant, D.J.W. Liquid chromatographic analysis of the enantiomeric impurities in various (+)-pseudoephedrine samples. *Pharm.Res.* **1991**, *8*, 1430–1433.

Findlay, J.W.A.; Warren, J.T.; Hill, J.A.; Welch, R.M. Stereospecific radioimmunoassays for d-pseudoephedrine in human plasma and their application to bioequivalency studies. *J.Pharm.Sci.* **1981**, *70*, 624–631.

Pseudouridine

Yoshida, S.; Adachi, T.; Hirose, S. Determination of pseudouridine and 5-fluoropyrimidines in human serum by high-performance liquid chromatography with precolumn fluorometric derivatization. *Microchem.J.* **1989**, *39*, 351–360.

Psychosine

Nozawa, M.; Iwamoto, T.; Tokoro, T.; Eto, Y. Novel procedure for measuring psychosine derivatives by a HPLC method. *J.Neurochem.* **1992**, *59*, 607–609.

Pterins (using post-column reaction detection)

Howells, D.W.; Smith, I.; Hyland, K. Estimation of tetrahydrobiopterin and other pterins in cerebrospinal fluid using reversed-phase high-performance liquid chromatography with electrochemical and fluorescence detection. *J.Chromatogr.* **1986**, *381*, 285–294.

Purines and Pyrimidines (using post-column reaction detection)

Assenza, S.P.; Brown, P.R. Ultraviolet and fluorescence characterization of purines and pyrimidines by post-column pH manipulation. *J.Chromatogr.* **1984**, *289*, 355–365.

Pyridoxal-5'-phosphate

Ubbink, J.B.; Serfontein, W.J.; de Villiers, L.S. Analytical recovery of protein-bound pyridoxal-5'-phosphate in plasma analysis. *J.Chromatogr.* **1986**, *375*, 399–404.

Pyroligneous

Menard, H.; Gaboury, A.; Belanger, D.; Roy, C. High-performance liquid chromatographic analysis of carboxylic acids in pyroligneous liquors. *J.Anal.Appl.Pyrolysis* **1984**, *6*, 45–57.

Pyrrolidinone

Dell, D.; Wendt, G.; Bucheli, F.; Trautmann, K.H. Determination of 2-pyrrolidinone in plasma. *Methodol.Surv.Biochem.Anal.* **1986**, *16*, 113–116.

Pyrophosphate (using post-column reaction detection)

Yoza, N.; Akazaki, I.; Nakazato, T.; Ueda, N.; Kodama, H.; Tateda, A. High-performance liquid chromatographic determination of pyrophosphate in the presence of a 20,000-fold excess of orthophosphate. *Anal.Biochem.* **1991**, *199*, 279–285.

Pyruvic Acid

Buckberry, L.D.; Gallagher, S.D.; Shaw, P.N. The development of a high performance liquid chromatographic assay method to determine pyruvate generated during the C-S lysis of cysteine conjugated xenobiotics. *J.Liq.Chromatogr.* **1993**, *16*, 3073–3081.

Koike, K.; Urata, Y.; Hiraoka, N. Microquantitative simultaneous analysis of lactate and pyruvate in blood plasma and urine by high-performance liquid chromatography. *J.Clin.Biochem.Nutr.* **1990**, *9*, 151–161.

Nakamura, H.; Tamura, Z. Fluorometric determination of pyruvic acid and α-ketoglutaric acid by high performance liquid chromatography. *Anal.Chem.* **1979**, *51*, 1679–1683.

Quinones (using post-column reaction detection)

Poulsen, J.R.; Birks, J.W. Photocatalytic chemiluminescence detection of quinones in high-performance liquid chromatography. *Anal.Chem.* **1990**, *62*, 1242–1251.

Radicals

Kieber, D.J.; Blough, N.V. Determination of carbon-centered radicals in aqueous solution by liquid chromatography with fluorescence detection. *Anal.Chem.* **1990**, *62*, 2275–2283.

Ramipril

Aboul-Enein, H.Y.; Bakr, S.A. High-performance liquid chromatographic identification of ramipril, and its precursor enantiomers using a Chiralpak OT(+) column. *Drug Dev.Ind.Pharm.* **1992**, *18*, 1013–1022.

Reserpine (using post-column reaction detection)

Lang, J.R.; Honigberg, I.L.; Stewart, J.T. Post-column air-segmentation reactor for the fluorometric detection of reserpine by liquid chromatography. *J.Chromatogr.* **1982**, *252*, 288–292.

Resin Acids (using post-column reaction detection)

Richardson, D.E.; Bremner, J.B.; O'Grady, B.V. Quantitative analysis of total resin acids by high-performance liquid chromatography of their coumarin ester derivatives. *J.Chromatogr.* **1992**, *595*, 155–162.

Retinoids Acid

Biesalski, H.K.; Weiser, H. Microdetermination of retinyl esters in guinea pig tissues under different vitamin-A-status conditions. *J.Micronutr.Anal.* **1990**, *7*, 97–116.

El Mansouri, S.; Tod, M.; Leclercq, M.; Porthault, M.; Chalom, J. Precolumn derivatization of retinoic acid for liquid chromatography with fluorescence and coulometric detection. *Anal.Chim.Acta* **1994**, *293*, 245–250.

Retinoids (using post-column reaction detection)

Bryan, P.D.; Capomacchia, A.C. Use of stop-flow oxalate ester chemiluminescence as a means to determine conditions for high-performance liquid chromatography chemiluminescence detection of retinoids using normal-phase chromatography. *J.Pharm.Biomed.Anal.* **1991**, *9*, 855–860.

Rhamnolipids

Schenk, T.; Schuphan, I.; Schmidt, B. High-performance liquid chromatographic determination of the rhamnolipids produced by Pseudomonas aeruginosa. *J.Chromatogr.A* **1995**, *693*, 7–13.

Rivastatin (using post-column reaction detection)

Krol, G.J.; Beck, G.W.; Ritter, W.; Lettieri, J.T. LC separation and induced fluorometric detection of rivastatin in blood plasma. *J.Pharm.Biomed.Anal.* **1993**, *11*, 1269–1275.

Krol, G.J.; Beck, G.W.; Ritter, W.; Lettieri, J.T.; Ness, G.C.; Lopez, D.; Heller, A.H. Determination of rivastatin levels in plasma samples by HPLC and enzyme inhibition assays. *Methodol.Surv. Bioanal.Drugs* **1994**, *23*, 147–156.

RNA

McFarland, G.D.; Borer, P.N. Separation of oligo-RNA by reverse-phase HPLC. *Nucleic Acids Res.* **1979**, *7*, 1067–1080.

Zhang, S.B.; Bronskii, P.M.; Wang, Q.S.; Wong, J.T.F. Separation of tRNA by high-performance liquid chromatography at ambient temperature. *J.Chromatogr.* **1986**, *360*, 282–287.

Rodenticide (using post-column reaction detection)

Hunter, K. High-performance liquid chromatographic strategies for the determination and confirmation of anticoagulant rodenticide residues in animal tissues. *J.Chromatogr.* **1985**, *321*, 255–272.

Hunter, K. Determination of coumarin anticoagulant rodenticide residues in animal tissue by high-performance liquid chromatography. I. Fluorescence detection using post-column techniques. *J.Chromatogr.* **1983**, *270*, 267–276.

RTX-III

Makhnyr', V.M.; Kozlovskaia, E.P. [Modification of the neurotoxin RTX-III from the sea anemone Radianthus macrodactylus]. *Bioorg.Khim.* **1990**, *16*, 643–648.

Rutin

Romero, M.L.; Escobar, L.I.; Lozoya, X.; Enriquez, R.G. High-performance liquid chromatographic study of Casimiroa edulis. I. Determination of imidazole derivatives and rutin in aqueous and organic extracts. *J.Chromatogr.* **1983**, *281*, 245–251.

SA446 (using post-column reaction detection)

Horiuchi, M.; Takashina, H.; Fujimura, K.; Iso, T. High-performance liquid chromatography of organosulfur compounds by the postcolumn ligand exchange reaction with iodoplatinate. Application to the simultaneous determination of (2R,4R)-2-(o-hydroxyphenyl)-3-(3-mercaptopropionyl)-4-thiazolidinecarboxylic acid (SA446), an angiotensin-converting enzyme inhibitor, and its urinary metabolites. *Yakugaku Zasshi* **1986**, *106*, 1028–1033.

Suzuki, T.; Akasaka, K.; Meguro, H. Determination of antihypertensive agent, (2R,4R)-2-(2-hydroxy-phenyl)-3-(3-mercaptopropionyl)-4-thiazolidinecarboxylic acid by high performance liquid chromatography with post-column derivatization with N-(9-acridinyl)maleimide. *Bunseki Kagaku* **1984**, *33*, E207-E210.

Saccharides

Anumula, K.R. Quantitative determination of monosaccharides in glycoproteins by high-performance liquid chromatography with highly sensitive fluorescence detection. *Anal.Biochem.* **1994**, *220*, 275–283.

Chang, M.; Meyers, H.V.; Nakanishi, K.; Ojika, M.; Park, J.H.; Park, M.H.; Takeda, R.; Vazquez, J.T.; Wiesler, W.T. Microscale structure determination of oligosaccharides by the exciton chirality method. *Pure Appl.Chem.* **1989**, *61*, 1193–1200.

Corti, F.; Luzzani, F.; Ventura, P. The determination of hexoses in rat gastric mucus by high-performance liquid chromatography. *J.Pharm.Biomed.Anal.* **1988**, *6*, 1049–1054.

Daniel, P.F.; De Feudis, D.F.; Lott, I.T.; McCluer, R.H. Quantitative microanalysis of oligosaccharides by high-performance liquid chromatography. *Carbohydr.Res.* **1981**, *97*, 161–180.

Finden, D.A.S.; Fysh, R.R.; White, P.C. Quantitative method for the detection of glucose in body fluids by high-performance liquid chromatography with fluorescence detection. *J.Chromatogr.* **1985**, *342*, 179–185.

Formato, M.; Senes, A.; Soccolini, F.; Coinu, R.; Cherchi, G.M. A reversed phase HPLC method for the simultaneous determination of all monosaccharides contained in galactosaminoglycan isomers from human aorta proteoglycans. *Carbohydr.Res.* **1994**, *255*, 27–39.

Gaberc-Porekar, V.; Socic, H. Automated qualitative and quantitative analysis of individual sugars in fermentation media. *Vestn.Slov.Kem.Drus.* **1984**, *31*, 369–379.

Ghosh, D.; Mathur, N.K.; Narang, C.K. Separation of anthranilate derivatives of sugars by reverse-phase chromatography. *Chromatographia* **1993**, *37*, 543–545.

Hong, S-P.; Nakamura, H.; Nakajima, T. 8-Amino-2-naphthalenesulfonic acid as a precolumn fluorescence derivatization reagent for reversed-phase high-performance liquid chromatography of reducing sugars. *Anal.Sci.* **1994**, *10*, 647–648.

Jinno, K.; Takayama, K. Separation of saccharides on cross-linked chitosan beads with microcolumn liquid chromatography. *J.Microcolumn Sep.* **1989**, *1*, 195–199.

Kallin, E.; Loenn, H.; Norberg, T. New derivatization and separation procedures for reducing oligosaccharides. *Glycoconjugate J.* **1986**, *3*, 311–319.

Lawson, M.A.; Russell, G.F. Trace level analysis of reducing sugars by high-performance liquid chromatography. *J.Food Sci.* **1980**, *45*, 1256–1258.

Linhardt, R.J.; Gu, K.N.; Loganathan, D.; Carter, S.R. Analysis of glycosaminoglycan-derived oligosaccharides using reversed-phase ion-pairing and ion-exchange chromatography with suppressed conductivity detection. *Anal.Biochem.* **1989**, *181*, 288–296.

Lloyd, P.; Crabbe, M.J.C. High-performance liquid chromatographic analysis of glucose, fructose, sorbitol and low levels of the aldose reductase inhibitor sorbinil in human lens and plasma. *J.Chromatogr.* **1985**, *343*, 402–406.

Mentasti, E.; Gennaro, M.C.; Sarzanini, C.; Porta, V. Derivatization, separation and determination of oligosaccharides by reversed-phase HPLC and fluorometric detection. *Ann.Chim.(Rome)* **1987**, *77*, 579–590.

Muramoto, K.; Goto, R.; Kamiya, H. Analysis of reducing sugars as their chromophoric hydrazones by high-performance liquid chromatography. *Anal.Biochem.* **1987**, *162*, 435–442.

Nakagawa, H.; Kawamura, Y.; Kato, K.; Shimada, I.; Arata, Y.; Takahashi, N. Identification of neutral and sialyl N-linked oligosaccharide structures from human serum glycoproteins using three kinds of high-performance liquid chromatography. *Anal.Biochem.* **1995**, *226*, 130–138.

Suzuki, J.; Kondo, A.; Kato, I.; Hase, S.; Ikenaka, T. Analysis by high-performance anion-exchange chromatography of component sugars as their fluorescent pyridylamino derivatives. *Agric.Biol.Chem.* **1991**, *55*, 283–284.

Vazquez, J.T.; Wiesler, W.T.; Nakanishi, K. Circular dichroism spectra of bichromophorically derivatized methyl-D-galactopyranosides, calculable by pairwise additivity, provide a basis for novel microanalysis of oligosaccharides. *Carbohydr.Res.* **1988**, *176*, 175–194.

Yadav, M.P.; BeMiller, J.N.; Embuscado, M.E. Compositional analysis of polysaccharides via solvolysis with liquid hydrogen fluoride. *Carbohydr.Polym.* **1994**, *25*, 315–318.

Saccharides (using post-column reaction detection)

Chen, F.; Liu, Y.; Lu, J.; Hwang, K.J.; Lee, V.H.L. A sensitive fluorometric assay for reducing sugars. *Life Sci.* **1992**, *50*, 651–659.

Clarke, A.J.; Sarabia, V.; Keenleyside, W.; MacLachlan, P.R.; Whitfield, C. The compositional analysis of bacterial extracellular polysaccharides by high-performance anion-exchange chromatography. *Anal.Biochem.* **1991**, *199*, 68–74.

Coquet, A.; Veuthey, J.L.; Haerdi, W. Comparison of post-column fluorescence derivatization and evaporative light-scattering detection to analyze saccharides selectively by LC. *Chromatographia* **1992**, *34*, 651–654.

Escott, R.E.A.; Taylor, A.F. The determination of sugars by high performance liquid chromatography using a novel bonded phase column and selective post column colorimetric detection. *HRC CC, J.High Resolut.Chromatogr.Chromatogr.Commun.* **1985**, *8*, 290–292.

Franta, B.D.; Mattick, L.R.; Sherbon, J.W. The analysis of pentoses in dry wine by high performance liquid chromatography with post-column derivatization. *Am.J.Enol.Vitic.* **1986**, *37*, 269–274.

Ikuta, K.; Ohtani, Y.; Isobe, Y.; Harada, H. Quantitative determination of oligosaccharides in commercial soybean lecithin products by high-performance liquid chromatography. *Nippon Shokuhin Kogyo Gakkaishi* **1994**, *41*, 515–518.

Ishikawa, M.; Hamano, Y.; Ganno, S.; Takaki, M. HPLC analysis of non-reducing sugars by post column derivatization without concentrated inorganic acid. *Kuromatogurafi* **1995**, *15*, 238–239.

Joergensen, N.O.G. Assimilation of free monosaccharides and amino acids relative to bacterial production in eutrophic lake water. *Ergeb.Limnol.* **1988**, *34*, 99–110.

Kai, M.; Tamura, K.; Watanabe, H.; Ohkura, Y. HPLC of reducing sugars with postcolumn fluorescence derivatization using methoxybenzamidine. *Bunseki Kagaku* **1989**, *38*, 568–572.

Kiba, N.; Shitara, K.; Furusawa, M.; Takata, Y. Post-column oligosaccharide dehydrogenase reactor for coulometric detection of malto-oligosaccharides in a liquid chromatographic system. *J.Chromatogr.* **1991**, *537*, 443–448.

Kiba, N.; Shitara, K.; Furusawa, M. A post-column coimmobilized galactose oxidase/peroxidase reactor for fluorometric detection of saccharides in a liquid chromatographic system. *J.Chromatogr.* **1989**, *463*, 183–187.

Kikuchi, J.; Nakamura, K.; Nakata, O.; Morikawa, Y. Determination of monosaccharides constituting the glycosides in saponins by high-performance liquid chromatography. *J.Chromatogr.* **1987**, *403*, 319–323.

Kinoshita, T.; Kamitani, Y.; Yoshida, J.; Urano, T.; Nimura, N.; Hanai, T. Ultramicro analysis of reducing and non-reducing sugars by liquid chromatography. *J.Liq.Chromatogr.* **1991**, *14*, 1929–1938.

Kobayashi, H.; Sugimoto, K.; Ohno, T.; Mizusawa, S. Electrochemical detection of reducing sugars in photographic gelatins. *J.Photogr.Sci.* **1990**, *37*, 49–51.

Mada, A.; Toyoda, H.; Imanari, T. Utility of a carbon column for high-performance liquid chromatographic separation of unsaturated disaccharides produced from glycosaminoglycans. *Anal.Sci.* **1992**, *8*, 793–797.

Maes, P.C.; Nagels, L.J.; Dewaele, C.; Alderweireldt, F.C. Characterization of cellulase-based enzyme reactors for the high-performance liquid chromatographic determination of β-D-glucan oligosaccharides. *J.Chromatogr.* **1991**, *558*, 343–355.

Maes, P.C.; Nagels, L.J.; Spanoghe, B.R. High-performance liquid chromatographic determination of native and reduced α-glucan oligosaccharides on reversed phase columns using post-column immobilized enzyme reactors. *Chromatographia* **1993**, *37*, 511–516.

McKay, D.B.; Tanner, G.P.; Maclean, D.J.; Scott, K.J. Detection of polyols and sugars by cuprammonium ion in the presence of strong base. *Anal.Biochem.* **1987**, *165*, 392–398.

Mikami, H.; Ishida, Y. Post-column fluorometric detection of reducing sugars in high performance liquid chromatography using arginine. *Bunseki Kagaku* **1983**, *32*, E207-E210, (using post-column reaction detection)

Nordling, M.; Elmgren, M.; Staahlberg, J.; Pettersson, G.; Lindquist, S.E. A combined cellobiose oxidase/glucose oxidase biosensor for HPLC determination on-line of glucose and soluble cellodextrins. *Anal.Biochem.* **1993**, *214*, 389–396.

Ohtsuki, K.; Kawabata, M.; Taguchi, K. Analysis of sugars in kiwifruit. *Kyoto-furitsu Daigaku Gakujutsu Hokoku, Rigaku, Seikatsu Kagaku* **1984**, 21–25.

Peelen, G.O.H.; de Jong, J.G.N.; Wevers, R.A. HPLC analysis of oligosaccharides in urine from oligosaccharidosis patients. *Clin.Chem.* **1994**, *40*, 914–921.

Peelen, G.O.H.; de Jong, J.G.N.; Wevers, R.A. High-performance liquid chromatography of monosaccharides and oligosaccharides in a complex biological matrix. *Anal.Biochem.* **1991**, *198*, 334–341.

Saska, M.; Wang, J. Minor component interference in HPLC analysis of molasses. *Int.Sugar J.* **1995**, *96*, 469–471.

Sorensen, S.H.; Proud, F.J.; Adam, A.; Rutgers, H.C.; Batt, R.M. A novel HPLC method for the simultaneous quantification of monosaccharides and disaccharides used in tests of intestinal function and permeability. *Clin.Chim.Acta* **1993**, *221*, 115–125.

Suon, K.N.; Vialle, J.; Rocca, J.L. Post-column derivatization of sugars by liquid chromatography. *Analusis* **1979**, *7*, 381–385.

Tanaka, H.; Hamada, R.; Kondoh, A.; Sakagami, K. Determination of component sugars in soil organic matter by HPLC. *Zentralbl.Mikrobiol.* **1990**, *145*, 621–628.

Thomas, J.; Mort, A.J. Continuous postcolumn detection of underivatized polysaccharides in high-performance liquid chromatography by reaction with permanganate. *Anal.Biochem.* **1994**, *223*, 99–104.

Toyoda, H.; Shinomiya, K.; Yamanashi, S.; Koshiishi, I.; Imanari, T. Microdetermination of unsaturated disaccharides produced from chondroitin sulfates in rabbit plasma by high-performance liquid chromatography with fluorometric detection. *Anal.Sci.* **1988**, *4*, 381–384.

Umegae, Y.; Nohta, H.; Ohkura, Y. Simultaneous determination of 2-deoxy-D-glucose and D-glucose in rat serum by high-performance liquid chromatography with post-column fluorescence derivatization. *Chem.Pharm.Bull.* **1990**, *38*, 963–965.

Vratny, P.; Brinkman, U.A.T.; Frei, R.W. Comparative study of post-column reactions for the detection of saccharides in liquid chromatography. *Anal.Chem.* **1985**, *57*, 224–229.

Wnukowski, M. Cane sugar invert analysis by HPLC utilizing a post column derivatization reaction. *Publ.Tech.Pap.Proc.Annu.Meet.Sugar Ind.Technol.* **1984**, *42*, 143–163.

Wnukowski, M. Cane sugar invert analysis by HPLC utilizing a post-column derivatization reaction. *Int.Sugar J.* **1984**, *86*, 170–175.

Zhang, G.Q.; Kai, M.; Nohta, H.; Umegae, Y.; Ohkura, Y. Simultaneous determination of glycated albumin and D-glucose in human serum by high-performance liquid chromatography with postcolumn fluorescence derivatization. *Anal.Sci.* **1993**, *9*, 9–14.

Zou, G.; Wen, H. Determination of glucose in serum and whole blood by micro-HPLC-immobilized enzyme post column reactor-electrochemical detection. *Fenxi Ceshi Tongbao* **1992**, *11*, 30–34.

Salinomycin

Dimenna, G.P.; Creegan, J.A.; Turnbull, L.B.; Wright, G.J. Determination of sodium salinomycin in chicken skin/fat by high-performance liquid chromatography utilizing column switching and UV detection. *J.Agric.Food Chem.* **1986**, *34*, 805–810.

Salinomycin (using post-column reaction detection)

Akhtar Humayoun, M.; Croteau, L.G. Microwave extraction of incurred salinomycin from chicken tissues. *J.Environ.Sci.Health, Part B, B31(1)* **1996**, *117*–33.

Akhtar, M.H.; Croteau, L.G. Extraction of salinomycin from finished layers ration by microwave solvent extraction followed by liquid chromatography. *Analyst* **1996**, *121*, 803–806.

Salsolinol

Stammel, W.; Thomas, H. A simple and rapid method for the separation of the (R)- and (S)-enantiomers of the tetrahydroisoquinoline alkaloid salsolinol by high-performance liquid chromatography. *Anal.Lett.* **1993**, *26*, 2513–2524.

Sapogenins

Tai, B.; Goldberg, I. High-performance liquid chromatographic separation of steroidal sapogenins. *J.Nat.Prod.* **1981**, *44*, 750–751.

Saponins

Oleszek, W.; Jurzysta, M.; Price, K.R.; Fenwick, G.R. High-performance liquid chromatography of alfalfa root saponins. *J.Chromatogr.* **1990**, *519*, 109–116.

Saxitoxin (using post-column reaction detection)

Janiszewski, J.; Boyer, G.L. The electrochemical oxidation of saxitoxin and derivatives: its application to the HPLC analysis of PSP toxins. *Dev.Mar.Biol.* **1993**, *3*, 889–894.

Luckas, B. Determination of saxitoxin in canned mussels by HPLC with postcolumn derivatization and fluorescence detection. *Dtsch.Lebensm.-Rundsch.* **1987**, *83*, 379–381.

Selenium

Khuhawar, M.Y.; Bozdar, R.B.; Babar, M.A. High-performance liquid chromatographic determination of selenium in coal after derivatization to 2,1,3-benzoselenadiazoles. *Analyst* **1992**, *117*, 1725–1727.

Nakagawa, T.; Aoyama, E.; Hasegawa, N.; Kobayashi, N.; Tanaka, H. High-performance liquid chromatography-fluorometric determination of selenium based on selenotrisulfide formation reaction. *Anal.Chem.* **1989**, *61*, 233–236.

Tanaka, H.; Nakagawa, T.; Aoyama, E.; Chikuma, M.; Nakayama, M.; Tanaka, T.; Ito, K. Determination of selenium in water samples by HPLC with fluorescent detection. *Stud.Environ.Sci.* **1988**, *34*, 347–351.

Selenium (using post-column reaction detection)

Shibata, Y.; Morita, M.; Fuwa, K. Determination of ultratrace levels of selenite and selenate in water using high-performance liquid chromatography with automated fluorometric detection and an online reduction system. *Analyst (London)* **1985**, *110*, 1269–1270.

Selenomethionine

Hansen, S.H.; Poulsen, M.N. Assay of L-selenomethionine in raw materials and tablets by high-performance liquid chromatography. *Acta Pharm.Nord.* **1991**, *3*, 95–97.

Semduramicin (using post-column reaction detection)

Lynch, M.J.; Frame, G.M.; Ericson, J.F.; Illyes, E.F.; Nowakowski, M.A. Semduramicin in the chicken. Tissue residue depletion studies. *ACS Symp.Ser.* **1992**, *503*, 49–69.

Serotonin

Linnoila, M.; Jacobson, K.A.; Marshall, T.H.; Miller, T.L.; Kirk, K.L. Liquid chromatographic assay for cerebrospinal fluid serotonin. *Life Sci.* **1986**, *38*, 687–694.

Shellfish Toxins

Dickey, R.W.; Granade, H.R.; Bencsath, F.A. Improved analytical methodology for the derivatization and HPLC-fluorometric determination of okadaic acid in phytoplankton and shellfish. *Dev.Mar.Biol.* **1993**, *3*, 495–499.

Flynn, K.; Flynn, K.J. An automated HPLC method for the rapid analysis of paralytic shellfish toxins from dinoflagellates and bacteria using precolumn oxidation at low temperature. *J.Exp.Mar.Biol.Ecol.* **1996**, *197*, 145–157.

Marr, J.C.; McDowell, L.M.; Quilliam, M.A. Investigation of derivatization reagents for the analysis of diarrhetic shellfish poisoning toxins by liquid chromatography with fluorescence detection. *Nat.Toxins* **1994**, *2*, 302–311.

Sciacchitano, C.J.; Mopper, B. Analysis of paralytic shellfish toxin (saxitoxin) in mollusks by capillary zone electrophoresis. *J.Liq.Chromatogr.* **1993**, *16*, 2081–2088.

Shellfish Toxins (using post-column reaction detection)

Franco, J.M.; Fernandez-Vila, P. Separation of paralytic shellfish toxins by reversed-phase high-performance liquid chromatography, with postcolumn reaction and fluorimetric detection. *Chromatographia* **1993**, *35*, 613–620.

Lawrence, J.F.; Wong, B. Evaluation of a postcolumn electrochemical reactor for oxidation of paralytic shellfish poison toxins. *J.AOAC Int.* **1995**, *78*, 698–704.

Oshima, Y. Postcolumn derivatization liquid chromatographic method for paralytic shellfish toxins. *J.AOAC Int.* **1995**, *78*, 528–532.

Sialic Acids

Anumula, K.R. Rapid quantitative determination of sialic acids in glycoproteins by high-performance liquid chromatography with a sensitive fluorescence detection. *Anal.Biochem.* **1995**, *230*, 24–30.

Stanton, P.G.; Shen, Z.; Kecorius, E.A.; Burgon, P.G.; Robertson, D.M.; Hearn, M.T. Application of a sensitive HPLC-based fluorometric assay to determine the sialic acid content of human gonadotropin isoforms. *J.Biochem.Biophys.Methods* **1995**, *30*, 37–48.

Unland, F.; Muething, J. An improved method for preparation of perbenzoylated ganglioside-derived sialic acids and nanogram detection of N-acetyl- and N-glycolylneuraminic acid by high performance liquid chromatography. *Biomed.Chromatogr.* **1992**, *6*, 155–159.

Sialic Acids (using post-column reaction detection)

Honda, S.; Iwase, S.; Suzuki, S.; Kakehi, K. Fluorometric determination of sialic acids using malono-nitrile in weakly alkaline media and its application to postcolumn labeling in high-performance liquid chromatography. *Anal.Biochem.* **1987**, *160*, 455–461.

Manzi, A.E.; Diaz, S.; Varki, A. High-pressure liquid chromatography of sialic acids on a pellicular resin anion-exchange column with pulsed amperometric detection; a comparison with six other systems. *Anal.Biochem.* **1990**, *188*, 20–32.

Sisomicin

Kawamoto, T.; Mashimo, I.; Yamauchi, S.; Watanabe, M. Determination of sisomicin, netilmicin, astro-micin and micronomicin in serum by high-performance liquid chromatography. *J.Chromatogr.* **1984**, *305*, 373–379.

Tawa, R.; Hirose, S.; Fujimoto, T. Determination of the aminoglycoside antibiotics sisomicin and netil-micin in dried blood spots on filter discs, by high-performance liquid chromatography with pre-column derivatization and fluorimetric detection. *J.Chromatogr.* **1989**, *490*, 125–132.

Sitosterol

Byrne, K.J.; Reazin, G.H.; Andreasen, A.A. High-pressure liquid chromatographic determination of β-sitosterol and β-sitosterol-D-glucoside in whiskey. *J.Assoc.Off.Anal.Chem.* **1981**, *64*, 181–185.

Sodium Fluoroacetate

Collins, D.M.; Fawcett, J.P.; Rammell, C.G. Determination of sodium fluoroacetate (compound 1080) in poison baits by HPLC. *Bull.Environ.Contam.Toxicol.* **1981**, *26*, 669–673.

Solasodine

Drewes, F.E.; Van Staden, J. Aspects of the extraction and purification of solasodine from Solanum aculeastrum tissues. *Phytochem.Anal.* **1995**, *6*, 203–206.

Drewes, F.E.; Van Staden, J.; Drewes, S.E. Benzoylation and high performance liquid chromatographic detection of solasodine. *Phytochem.Anal.* **1992**, *3*, 85–87.

Sorbic Acid

Burini, G.; Damiani, P. Determination of sorbic acid in margarine and butter by high-performance liquid chromatography with fluorescence detection. *J.Chromatogr.* **1991**, *543*, 69–80.

Sorbitan Fatty Acid Esters

Garti, N.; Wellner, E.; Aserin, A.; Sarig, S. Analysis of sorbitan fatty acid esters by HPLC. *JAOCS, J.Am.Oil Chem.Soc.* **1983**, *60*, 1151–1154.

Sorbitol

Dethy, J.M.; Callaert-Deveen, B.; Janssens, M.; Lenaers, A. Determination of sorbitol and galactitol at the nanogram level in biological samples by high-performance liquid chromatography. *Anal.Biochem.* **1984**, *143*, 119–124.

Sotalol

Carr, R.A.; Foster, R.T.; Bhanji, N.H. Stereospecific high-performance liquid chromatographic assay of sotalol in plasma. *Pharm.Res.* **1991**, *8*, 1195–1198.

Sallustio, B.C.; Morris, R.G.; Horowitz, J.D. High-performance liquid chromatographic determination of sotalol in plasma. I. Application to the disposition of sotalol enantiomers in humans. *J.Chromatogr.* **1992**, *576*, 321–327.

Soyasaponins

Ireland, P.A.; Dziedzic, S.Z. High-performance liquid chromatography of soyasaponins on silica phase with evaporative light-scattering detection. *J.Chromatogr.* **1986**, *361*, 410–416.

Sphingosine

Lagana, A.; Marino, A.; Fago, G.; Miccheli, A. Determination of free sphingosine in biological system by high performance liquid chromatography. *Ann.Chim.(Rome)* **1991**, *81*, 721–734.

Spinacine

Anastasia, M.; Colombo, D.; Fiecchi, A. Quantitative high-performance liquid chromatographic determination of the amino acid spinacine in blood and chow of rats. *J.Chromatogr.* **1987**, *410*, 504–508.

Resmini, P.; Pellegrino, L.; Saracchi, S.; Chiodi, J. A sensitive method to determine spinacine (6-carboxy-1,2,3,4-tetrahydroimidazopyridine) in ripened cheese by HPLC. *Ital.J.Food Sci.* **1989**, *1*, 35–44.

Starch (using post-column reaction detection)

Autio, K.; Suortti, T.; Hamunen, A.; Poutanen, K. Heat-induced structural changes of acid hydrolyzed and hypochlorite-oxidized barley starches. *Carbohydr.Polym.* **1996**, *29*, 155–161.

Autio, K.; Poutanen, K.; Suortti, T.; Pessa, E. Heat-induced structural changes in acid-modified barley starch dispersions. *Food Struct.* **1992**, *11*, 315–322.

Autio, K.; Pessa, E.; Suortti, T.; Poutanen, K. Characterization of acid-hydrolyzed starches for the confectionery industry. *Food Hydrocolloids* **1992**, *6*, 371–377.

Sterigmatocystin (using post-column reaction detection)

Neely, F.L.; Emerson, C.S. Determination of sterigmatocystin in fermentation broths by reversed-phase high-performance liquid chromatography using post-column fluorescence enhancement. *J.Chromatogr.* **1990**, *523*, 305–311.

Steroids

Baumeister, M.; Stalla, G.K.; Mueller, O.A. Problems in quantitative analysis of five steroid hormones by HPLC. *Fresenius' Z.Anal.Chem.* **1984**, *317*, 683–684.

Cartoni, G.P.; Coccioli, F. High-performance liquid chromatographic determination of estrogens in human urine. *J.Chromatogr.* **1983**, *278*, 144–150.

Fujino, H.; Goya, S. [Synthesis of 1-(dichloro-1,3,5-triazinyl)-2-methylisoindole as a new fluorescent derivatization reagent and its reactivity for estrogens. II. Determination of estriol in pregnancy urine]. *Yakugaku.Zasshi.* **1988**, *108*, 801–803.

Goto, J.; Shao, G.; Miura, H.; Nambara, T. Studies on steroids. CCXLII. Separation of C-25 epimers of 5β-cholestanoic acids by high-performance liquid chromatography with precolumn fluorescence labeling. *Anal.Sci.* **1989**, *5*, 19–22.

Goto, J.; Goto, N.; Shamsa, F.; Saito, M.; Komatsu, S.; Suzaki, K.; Nambara, T. New sensitive derivatization of hydroxy steroids for high-performance liquid chromatography with fluorescence detection. *Anal.Chim.Acta* **1983**, *147*, 397–400.

Ishida, J.; Kai, M.; Ohkura, Y. Determination of estrogens in pregnancy urine by high-performance liquid chromatography with fluorescence detection. *J.Chromatogr.* **1988**, *431*, 249–257.

Kurosawa, S.; Yoshimura, T.; Kurosawa, T.; Chiba, H.; Kobayashi, K.; Koike, T.; Tohma, M. Simultaneous determination of 18-oxygenated corticosteroids by high-performance liquid chromatography with fluorescence detection. *J.Liq.Chromatogr.* **1995**, *18*, 2383–2396.

Schmidt, G.J.; Vandemark, F.L.; Slavin, W. Estrogen determination using liquid chromatography with precolumn fluorescence labeling. *Anal.Biochem.* **1978**, *91*, 636–645.

Shimada, K.; Satoh, Y.; Nishimura, S. Studies on neurosteroids. I. Retention behavior of derivatized 17-oxosteroids using high-performance liquid chromatography. *J.Liq.Chromatogr.* **1995**, *18*, 713–723.

Shimada, K.; Oe, T.; Suzuki, M. Effect of derivatization of steroids on their retention behavior in inclusion chromatography using cyclodextrin as a mobile phase additive. *J.Chromatogr.* **1991**, *558*, 306–310.

Shimada, K.; Nishimura, S. Studies on neurosteroids. II. Retention behavior of derivatized 20-oxosteroids and their sulfates using high-performance liquid chromatography. *J.Liq.Chromatogr.* **1995**, *18*, 1691–1701.

Shimada, K.; Nakagi, T.; Yoshizawas, M.; Ueda, T. Studies on neurosteroids. III. Separation and characterization of dehydroepiandrosterone and pregnenolone in rat brains by high-performance liquid chromatography. *Anal.Sci.* **1995**, *11*, 445–447.

Shimada, K.; Tanaka, M.; Nambara, T. Derivatization of ketosteroids for high-performance liquid chromatography with electrochemical detection. *Anal.Lett.* **1980**, *13*, 1129–1136.

Ueno, K.; Morimoto, A.; Umeda, T. Determination of two antiandrogenic anthrasteroids in human serum by column switching high-performance liquid chromatography with electrochemical detection. *Anal.Sci.* **1992**, *8*, 13–17.

Yoshitake, T.; Hara, S.; Yamaguchi, M.; Nakamura, M.; Ohkura, Y.; Gorog S, Measurement of 21-hydroxycorticosteroids in human and rat sera by high-performance liquid chromatography with fluorimetric detection. *J.Chromatogr.* **1989**, *489*, 364–370.

Steroids (using post-column reaction detection)

de Ruiter, C.; Tsoi, J.N.L.T.T.; Brinkman, U.A.T.; Frei, R.W. On-line procedures for the phase-transfer-catalyzed dansylation of phenolic steroids - application to biological samples. *Chromatographia* **1989**, *26*, 267–273.

Kawasaki, T.; Maeda, M.; Tsuji, A. Fluorescence high performance liquid chromatographic determination of 3α-hydroxysteroids in urine of 21-hydroxylase deficiency. *Biomed.Chromatogr.* **1986**, *1*, 1–6.

Lam, S.; Malikin, G. Strategy combining chiral chromatography with β-cyclodextrin and stereospecific enzyme reaction detection with hydroxysteroid dehydrogenases for resolving steroid epimers. *Chirality* **1992**, *4*, 395–399.

Lam, S.; Malikin, G.; Karmen, A. High-performance liquid chromatography of hydroxy steroids detected with post-column immobilized enzyme reactors. *J.Chromatogr.* **1988**, *441*, 81–87.

Maeda, M.; Tsuji, A. Chemiluminescence with lucigenin as post-column reagent in high-performance liquid chromatography of corticosteroids and p-nitrophenacyl esters. *J.Chromatogr.* **1986**, *352*, 213–220.

Yamaguchi, Y.; Iyama, S.; Hata, N.; Kohda, K.; Amino, N.; Miyai, K. Fluorimetric detection method of steroids carring ketol group on D ring. *Seibutsu Shiryo Bunseki* **1989**, *12*, 90–92.

Sterols

Kasama, T.; Byun, D.S.; Seyama, Y. Quantitative analysis of sterols in serum by high-performance liquid chromatography. Application to the biochemical diagnosis of cerebrotendinous xanthomatosis. *J.Chromatogr.* **1987**, *400*, 241–246.

Streptomycin

Kubo, H.; Kobayashi, Y.; Kinoshita, T. Fluorescence determination of streptomycin in serum by reversed-phase ion-pairing liquid chromatography. *Anal.Chem.* **1986**, *58*, 2653–2655.

Streptomycin (using post-column reaction detection)

Kocher, U. Detection of streptomycin residues in honey by Charm II-test and verification of the results by HPLC with post-column derivatization and fluorescence detection. *Lebensmittelchemie* **1996**, *50*, 115–117.

Succinylacetone

Jakobs, C.; Dorland, L.; Wikkerink, B.; Kok, R.M.; De Jong, A.P.J.M.; Wadman, S.K. Stable isotope dilution analysis of succinylacetone using electron capture negative ion mass fragmentography: an accurate approach to the pre- and neonatal diagnosis of hereditary tyrosinemia type I. *Clin.Chim.Acta* **1988**, *171*, 223–231.

Sulbactam (using post-column reaction detection)

Haginaka, J.; Yasuda, H.; Uno, T.; Nakagawa, T. Sulbactam: alkaline degradation and determination by high-performance liquid chromatography. *Chem.Pharm.Bull.* **1984**, *32*, 2752–2758.

Sulbutiamine (using post-column reaction detection)

Gele, P.; Boursier-Neyret, C.; Lesourd, M.; Sauveur, C. Determination of sulbutiamine and its disulfide derivatives in human plasma by HPLC using on-line post-column reactors and fluorimetric detection. *Chromatographia* **1993**, *36*, 167–173.

Sulfadiazine

Gehring, T.A.; Rushing, L.G.; Churchwell, M.I.; Doerge, D.R.; McErlane, K.M.; Thompson, H.C.J. HPLC determination of sulfadiazine residues in Coho salmon (Oncorhynchus kisutch) with confirmation by liquid chromatography with atmospheric pressure chemical ionization mass spectrometry. *J.Agric.Food Chem.* **1996**, *44*, 3164–3169.

Sulfamethazine

Vilim, A.B.; Larocque, L.; MacIntosh, A.I. A HPLC screening procedure for sulfamethazine residues in pork tissues. *J.Liq.Chromatogr.* **1980**, *3*, 1725–1736.

Sulfide

Savage, J.C.; Gould, D.H. Determination of sulfide in brain tissue and rumen fluid by ion-interaction reversed-phase high-performance liquid chromatography. *J.Chromatogr.* **1990**, *526*, 540–545.

Sulfocysteine

Johnson, J.L.; Rajagopalan, K.V. An HPLC assay for detection of elevated urinary S-sulfocysteine, a metabolic marker of sulfite oxidase deficiency. *J.Inherited Metab.Dis.* **1995**, *18*, 40–47.
Kaagedal, B.; Kaellberg, M.; Soerbo, B. Determination of S-sulfocysteine in urine by high-performance liquid chromatography. *J.Chromatogr.* **1983**, *276*, 418–422.

Sulfonamides

Barbieri, G.; Bergamini, C.; Ori, E.; Resca, P. Determination of sulfonamides in meat and meat products. *Ind.Aliment.(Pinerolo, Italy)* **1995**, *34*, 1273–1276.
Long, A.R.; Hsieh, L.C.; Malbrough, M.S.; Short, C.R.; Barker, S.A. Multiresidue method for the determination of sulfonamides in pork tissue. *J.Agric.Food Chem.* **1990**, *38*, 423–426.
Tena, M.T.; Luque de Castro, M.D.; Valcarcel, M. Improved supercritical fluid extraction of sulfonamides. *Chromatographia* **1995**, *40*, 197–203.
Van Poucke, L.S.G.; Depourcq, G.C.I.; Van Peteghem, C.H. A quantitative method for the detection of sulfonamide residues in meat and milk samples with a high-performance thin-layer chromatographic method. *J.Chromatogr.Sci.* **1991**, *29*, 423–427.

Sulfonamides (using post-column reaction detection)

Aerts, M.M.L.; Beek, W.M.J.; Kan, C.A.; Nouws, J.F.M. Detection of sulfaguanidine residues in eggs with a fully automated liquid-chromatographic method using post-column derivatization. Drug depletion study in eggs after a single oral dose to laying hens. *Arch.Lebensmittelhyg.* **1986**, *37*, 142–145.

Guggisberg, D.; Mooser, A.E.; Koch, H. Screening method for the quantitative determination of twelve sulfonamides in meat, liver, and kidney by HPLC with online post-column derivatization. *Mitt.geb.Lebensmittelunters.Hyg.* **1993**, *84*, 263–273.

Pacciarelli, B.; Reber, S.; Douglas, C.; Dietrich, S.; Etter, R. Determination of 12 sulfonamides in meat and kidney by HPLC with post-column derivatization. *Mitt.geb.Lebensmittelunters.Hyg.* **1991**, *82*, 45–55.

Sulforidazine

Ganes, D.A.; Midha, K.K. Sensitive and specific high-performance liquid chromatographic assay for the quantification of sulforidazine and two diastereomeric sulforidazine-5-sulfoxide metabolites in plasma. *J.Chromatogr.* **1987**, *423*, 227–237.

Sulfur Compounds

Hwang, Y.; Matsuo, T.; Hanaki, K.; Suzuki, N. Identification and quantification of sulfur and nitrogen containing odorous compounds in wastewater. *Water Res.* **1995**, *29*, 711–718.

LaCourse, W.R.; Owens, G.S. Pulsed electrochemical detection of thiocompounds following microchromatographic separations. *Anal.Chim.Acta* **1995**, *307*, 301–319.

Poulson, R.E.; Borg, H.M. Separation and detection of sulfur-containing anions using single-column ion chromatography. *J.Chromatogr.Sci.* **1987**, *25*, 409–414.

Sinkkonen, S. Liquid chromatographic determination of planar aromatic sulfur compounds in crude oil. *J.Chromatogr.* **1989**, *475*, 421–425.

Surfactants

Cavalli, L.; Gellera, A.; Lazzarin, A.; Nucci, G.C.; Romano, P.; Ranzani, M.; Lorenzi, E. Linear alkylbenzene sulfonate removal and biodegradation in a metropolitan plant from water treatment. *Riv.Ital.Sostanze.Grasse.* **1991**, *68*, 75–81.

Dubey, S.T.; Kravetz, L.; Salanitro, J.P. Analysis of nonionic surfactants in bench-scale biotreater samples. *J.Am.Oil Chem.Soc.* **1995**, *72*, 23–30.

Field, J.A.; Field, T.M.; Poiger, T.; Siegrist, H.; Giger, W. Fate of secondary alkane sulfonate surfactants during municipal wastewater treatment. *Water Res.* **1995**, *29*, 1301–1307.

Fujita, I.; Nishiyama, K.; Nagano, Y.; Harada, K.; Nakayama, M.; Sugii, A. Determination of nonionic surfactants in river water using a chemically modified styrene-divinylbenzene resin. *Int.J.Environ.Anal.Chem.* **1994**, *56*, 57–62.

Kloster, G.; Schoester, M.; Schwuger, M.J. HPLC analysis of aliphatic ionic surfactants at trace levels. *Comun.Jorn.Com.Esp.Deterg.* **1993**, *24*, 25–33.

Moreno, A.; Bravo, J.; Ferrer, J.; Bengoechea, C. Soap determination in sewage sludge by high-performance liquid chromatography. *J.Am.Oil Chem.Soc.* **1993**, *70*, 667–671.

Nitschke, L.; Huber, L. Analysis for ethoxylated alcohol surfactants in water by HPLC. *Fresenius' J.Anal.Chem.* **1993**, *345*, 585–588.

Surfactants (using post-column reaction detection)

de Ruiter, C.; Hefkens, J.C.H.F.; Brinkman, U.A.T.; Frei, R.W.; Evers, M.; Matthijs, E.; Meijer, J.A. Liquid chromatographic determination of cationic surfactants in environmental samples using a continuous post-column ion-pair extraction detector with a sandwich phase separator. *Int.J.Environ.Anal.Chem.* **1987**, *31*, 325–339.

Fernandez, P.; Alder, A.C.; Giger, W. Quantitative determination of cationic surfactants in sewage sludges and sediments by supercritical fluid extraction and HPLC applying post-column ion-pair extraction. *Natl.Meet.-Am.Chem.Soc., Div.Environ.Chem.* **1993**, *33*, 303–306.

Kanesato, M.; Nakamura, K.; Nakata, O.; Morikawa, Y. Analysis of ionic surfactants by HPLC with ion-pair extraction detector. *JAOCS, J.Am.Oil Chem.Soc.* **1987**, *64*, 434–438.

Kondoh, Y.; Yamada, A.; Takano, S. Determination of nonionic surfactants with ester groups by high-performance liquid chromatography with post-column derivatization. *J.Chromatogr.* **1991**, *541*, 431–441.

Schoester, M.; Koster, G. HPLC separation and quantification of anionic surfactants using an automated on-line ion pair extraction system. *Fresenius' J.Anal.Chem.* **1993**, *345*, 767–772.

Tamoxifen (using post-column reaction detection)

Jordan, V.C.; Bain, R.R.; Lyman, S.D.; Brown, R.R. Analysis of tamoxifen and its metabolites. *Methodol.Surv.Biochem.Anal.* **1984**, *14*, 219–225.

Nieder, M.; Jaeger, H. Quantification of tamoxifen and N-desmethyltamoxifen in human plasma by high-performance liquid chromatography, photochemical reaction and fluorescence detection, and its application to biopharmaceutic investigations. *J.Chromatogr.* **1987**, *413*, 207–217.

Taurine

Anderson, G.M.; Durkin, T.A.; Chakraborty, M.; Cohen, D.J. Liquid chromatographic determination of taurine in whole blood, plasma and platelets. *J.Chromatogr.* **1988**, *431*, 400–405.

Guo, L.J.; Athineos, P. Effects of hemodynamic changes on taurine release from posterior hypothalamus of freely moving rats. *Chung.Kuo.Yao.Li.Hsueh.Pao.* **1995**, *16*, 405–408.

Hirai, T.; Ohyama, H.; Kido, R. A direct determination of taurine in perchloric acid-deproteinized biological samples. *Anal.Biochem.* **1987**, *163*, 339–342.

Hischenhuber, C. High-performance liquid-chromatographic and thin-layer-chromatographic determination of taurine in infant formulas. *Dtsch.Lebensm.-Rundsch.* **1988**, *84*, 117–120.

Hopkins, P.C.; Kay, I.S.; Davies, W.E. A rapid method for the determination of taurine in biological tissue. *Neurochem.Int.* **1989**, *15*, 429–432.

Murai, S.; Saito, H.; Masuda, Y.; Itoh, T. Very rapid and simple assay of taurine in the brain within two minutes by high-performance liquid chromatography with electrochemical detection. *J.Pharmacol.Methods* **1990**, *23*, 195–202.

Porter, D.W.; Banks, M.A.; Castranova, V.; Martin, W.G. Reversed-phase high-performance liquid chromatography technique for taurine quantitation. *J.Chromatogr.* **1988**, *454*, 311–316.

Rao, G.N.S. Liquid chromatographic determination of taurine in vitamin premix formulations. *J.Assoc.Off.Anal.Chem.* **1987**, *70*, 799–801.

Schmidt, G.J.; Olson, D.C.; Atwood, J.G. The determination of nanogram amounts of taurine using liquid chromatography and fluorescence detection. *Chromatogr.Newsl.* **1980**, *8*, 13–14.

Trautwein, E.A.; Hayes, K.C. Evaluating taurine status: determination of plasma and whole blood taurine concentration. *J.Nutr.Biochem.* **1991**, *2*, 571–576.

Wheler, G.H.T.; Russell, J.T. Separation and quantitation of o-phthalaldehyde derivatives of taurine and related compounds in a high performance liquid chromatography (HPLC) system. *J.Liq.Chromatogr.* **1981**, *4*, 1281–1291.

Taurolin

Woolfson, A.D.; Gorman, S.P.; McCafferty, D.F.; Jones, D.S. Assay procedures for Taurolin solutions using pre-column derivatization and high-performance liquid chromatography with fluorescence detection. *Int.J.Pharm.* **1989**, *49*, 135–140.

Terpenes

Burnouf-Radosevich, M.; Delfel, N.E. High-performance liquid chromatography of oleanane-type triterpenes. *J.Chromatogr.* **1984**, *292*, 403–409.

Lin, J-T.; Nes, W.D.; Heftmann, E. High-performance liquid chromatography of triterpenoids. *J.Chromatogr.* **1981**, *207*, 457–463.

Tetracyclines (using post-column reaction detection)

Croubels, S.; Baeyens, W.; Van Peteghem, C. Post-column zirconium chelation and fluorescence detection for the liquid chromatographic determination of tetracyclines. *Anal.Chim.Acta* **1995**, *303*, 11–16.

Fu, C.; Xu, H. Fluorescence detection of tetracyclines by reversed-phase high performance liquid chromatography/post-column pH adjustment and fluorescence enhancement with micelle. *Sepu* **1995**, *13*, 365–367.

Tetraethylene

Baudrand, V.; Mouloungui, Z.; Gaset, A. Separation and quantification of tetraethylene glycol monoheptanoate and diheptanoate by HPLC. *J.Chromatogr.A* **1995**, *704*, 524–529.

Tetrahydrocarbolines

Bosin, T.R.; Jarvis, C.A. Derivatization in aqueous solution, isolation and separation of tetrahydro-β-carbolines and their precursors by liquid chromatography. *J.Chromatogr.* **1985**, *341*, 287–293.

Herraiz, T.; Ough, C.S. Separation and characterization of 1,2,3,4-tetrahydro-beta-carboline-3-carboxylic acids by HPLC and GC-MS. Identification in wine samples. *Am.J.Enology Viticulture* **1994**, *45*, 92–101.

Tetrahydropyrimidines

Kunte, H.J.; Galinski, E.A.; Trueper, H.G. A modified FMOC-method for the detection of amino acid-type osmolytes and tetrahydropyrimidines (ectoines). *J.Microbiol.Methods* **1993**, *17*, 129–136.

Tetrodotoxin (using post-column reaction detection)

Yotsu, M.; Endo, A.; Yasumoto, T. An improved tetrodotoxin analyzer. *Agric.Biol.Chem.* **1989**, *53*, 893–895.

Thiamine

Bettendorff, L.; Peeters, M.; Jouan, C.; Wins, P.; Schoffeniels, E. Determination of thiamin and its phosphate esters in cultured neurons and astrocytes using an ion-pair reversed-phase high-performance liquid chromatographic method. *Anal.Biochem.* **1991**, *198*, 52–59.

Bettendorff, L.; Grandfils, C.; De Rycker, C.; Schoffeniels, E. Determination of thiamine and its phosphate esters in human blood serum at femtomole levels. *J.Chromatogr.* **1986**, *382*, 297–302.

Bontemps, J.; Bettendorff, L.; Lombet, J.; Dandrifosse, G.; Schoffeniels, E.; Nevejans, F.; Yang, Y.; Verzele, M. Determination of thiamin and thiamin phosphates as thiochrome derivatives by reversed-phase chromatography on polystyrene packing materials. *Chromatographia* **1984**, *18*, 424–426.

Bontemps, J.; Bettendorff, L.; Dandrifosse, G.; Schoffeniels, E.; Nevejans, F. Sensitization of thiamin analysis by the peak compression technique. *HRC CC, J.High Resolut.Chromatogr.Chromatogr. Commun.* **1984**, *7*, 490–491.

Brunnekreeft, J.W.I.; Eidhof, H.; Gerrits, J. Optimized determination of thiochrome derivatives of thiamine and thiamine phosphates in whole blood by reversed-phase liquid chromatography with pre-column derivatization. *J.Chromatogr.* **1989**, *491*, 89–96.

Hasselmann, C.; Franck, D.; Grimm, P.; Diop, P.A.; Soules, C. High-performance liquid chromatographic analysis of thiamin and riboflavin in dietetic foods. *J.Micronutr.Anal.* **1989**, *5*, 269–279.

Herve, C.; Beyne, P.; Delacoux, E. Determination of thiamine and its phosphate esters in human erythrocytes by high-performance liquid chromatography with isocratic elution. *J.Chromatogr.B* **1994**, *653*, 217–220.

Iwata, H.; Matsuda, T.; Tonomura, H. Improved high-performance liquid chromatographic determination of thiamine and its phosphate esters in animal tissues. *J.Chromatogr.* **1988**, *450*, 317–323.

Nicolas, E.C.; Pfender, K.A. Fast and simple liquid chromatographic determination of nonphosphorylated thiamine in infant formula, milk, and other foods. *J.Assoc.Off.Anal.Chem.* **1990**, *73*, 792–798.

Sander, S.; Hahn, A.; Stein, J.; Rehner, G. Comparative studies on the high-performance liquid chromatographic determination of thiamin and its phosphate esters with chloroethylthiamin as an internal standard using pre- and post-column derivatization procedures. *J.Chromatogr.* **1991**, *558*, 115–124.

Sims, A.; Shoemaker, D. Simultaneous liquid chromatographic determination of thiamine and riboflavin in selected foods. *J.AOAC Int.* **1993**, *76*, 1156–1160.

Thiamine (using post-column reaction detection)

Kimura, M. Methods for determination of total thiamin, thiamin and its phosphate esters in blood by high-performance liquid chromatography with post-column derivatization. *Bitamin* **1989**, *63*, 15–24.

Kimura, M.; Itokawa, Y. Determination of thiamine and its phosphate esters in human and rat blood by high-performance liquid chromatography with post-column derivatization. *J.Chromatogr.* **1985**, *332*, 181–188.

Ujiie, T.; Tsutake, Y.; Morita, K.; Matsuno, M.; Kodaka, K. Distribution and stability of 2-(1-hydroxyethyl) thiamin and thiamin in foods. *Bitamin* **1991**, *65*, 249–256.

Ujiie, T.; Tsutake, Y.; Morita, K.; Tamura, M.; Kodaka, K. Simultaneous determination of 2-(1-hydrox-yethyl)thiamin and thiamin in foods by high performance liquid chromatography with post-column derivatization. *Bitamin* **1990**, *64*, 379–385.

Yamaguchi, T.; Uchimura, K.; Takenaka, H.; Fukumoto, K.; Kurokawa, Y.; Hamada, M.; Sugimoto, S.; Matsukura, S.; Inoue, K.; Shibuya, N. Quantitative determination of thiamin and its phosphate esters in human whole blood by high-performance liquid chromatography and clinical application. *J.Clin.Biochem.Nutr.* **1987**, *2*, 203–216.

Thienothiopyransulfonamide

Matuszewski, B.K.; Constanzer, M.L.; Hessey, G.A..I.; Bayne, W.F. Indirect stereoselective determination of the enantiomers of a thieno[2,3-b]thiopyran-2-sulfonamide in biological fluids. *Anal.Chem.* **1990**, *62*, 1308–1315.

Thiocyanate

Yang, Z-Y.; Chen, S-H.; Kou, H-S.; Wu, S-M.; Wu, H-L. Determination of thiocyanate anion by high-performance liquid chromatography with ultraviolet detection. *Chin.Pharm.J.(Taipei)* **1994**, *46*, 505–515.

Thioglycolic Acid

Rooselaar, J.; Liem, D.H. High pressure liquid chromatography determination of thioglycolic acid in cold wave fluids and depilating creams. *Int.J.Cosmet.Sci.* **1981**, *3*, 37–47.

Thiols

Aw, T.Y. Assay of thiols and disulfides in intestinal lymph. *Methods Enzymol.* **1995**, *251*, 221–228.

Dupuy, D.; Szabo, S. Measurement of tissue sulfhydryls and disulfides in tissue protein and nonprotein fractions by high performance liquid chromatography using electrochemical detection. *J.Liq. Chromatogr.* **1987**, *10*, 107–119.

Euerby, M.R. Effect of differing thiols on the reversed-phase high-performance liquid chromatographic behavior of o-phthaldialdehyde-thiol-amino acids. *J.Chromatogr.* **1988**, *454*, 398–405.

Fariss, M.W.; Reed, D.J. High-performance liquid chromatography of thiols and disulfides: dinitrophenol derivatives. *Methods Enzymol.* **1987**, *143*, 101–109.

Jacobsen, D.W.; Gatautis, V.J.; Green, R.; Robinson, K.; Savon, S.R.; Secic, M.; Ji, J.; Otto, J.M.; Taylor, L.M., Jr. Rapid HPLC determination of total homocysteine and other thiols in serum and plasma: sex differences and correlation with cobalamin and folate concentrations in healthy subjects. *Clin.Chem.* **1994**, *40*, 873–881.

Ling, B.L.; Baeyens, W.R.G.; Dewaele, C. Capillary zone electrophoresis with ultraviolet and fluorescence detection for the analysis of thiol. Application to mixtures and blood. *Anal.Chim.Acta* **1991**, *255*, 283–288.

Ling, B.L.; Dewaele, C.; Baeyens, W.R.G. Micro liquid chromatography with fluorescence detection of thiols and disulfides. *J.Chromatogr.* **1991**, *553*, 433–439.

Ling, B.L.; Baeyens, W.R.G.; Dewaele, C. Comparison of micro-LC and capillary zone electrophoresis for the analysis of thiols. *J.High Resolut.Chromatogr.* **1991**, *14*, 169–173.

Ling, B.L.; Dewaele, C.; Baeyens, W.R.G. Application of micro-scale liquid chromatography with fluorescence detection to the determination of thiols. *J.Chromatogr.* **1990**, *514*, 189–198.

Millot, M.C.; Sebille, B.; Mahieu, J.P. Solid-phase derivatization of thiols. Application to high-performance liquid chromatography. *J.Chromatogr.* **1986**, *354*, 155–167.

Mopper, K.; Delmas, D. Trace determination of biological thiols by liquid chromatography and precolumn fluorometric labeling with o-phthalaldehyde. *Anal.Chem.* **1984**, *56*, 2557–2560.

Nishikawa, Y.; Kuwata, K. Liquid chromatographic determination of low molecular weight alkylthiols in air via derivatization with 7-chloro-4-nitro-2,1,3-benzoxadiazole. *Anal.Chem.* **1985**, *57*, 1864–1868.

Pheifer, J.H.; Briggs, D.E. The estimation of thiols and disulfides in barley. *J.Inst.Brew.* **1995**, *101*, 5–10.

Sano, A.; Takitani, S.; Nakamura, H. Optical resolution of thiol enantiomers by high-performance liquid chromatography using precolumn fluorescence derivatization with o-phthalaldehyde and α-amino acids. *Anal.Sci.* **1995**, *11*, 299–301.

Toyooka, T.; Uchiyama, S.; Saito, Y.; Imai, K. Simultaneous determination of thiols and disulfides by high-performance liquid chromatography with fluorescence detection. *Anal.Chim.Acta* **1988**, *205*, 29–41.

Tsuruta, Y.; Tomida, H.; Kohashi, K. N-[4-(2-Phthalimidyl)phenyl]maleimide as fluorescence derivatization reagent for thiols. *Anal.Sci.* **1988**, *4*, 531–532.

Vairavamurthy, A.; Mopper, K. Field method for determination of traces of thiols in natural waters. *Anal.Chim.Acta* **1990**, *236*, 363–370.

Thiols (using post-column reaction detection)

Nakamura, H.; Tamura, Z. Simultaneous determination of biogenic thiols and disulfides by HPLC with fluorescence detection. *Bunseki Kagaku* **1988**, *37*, 35–40.

Nakamura, H.; Tamura, Z. Fluorometric determination of thiols by liquid chromatography with postcolumn derivatization. *Anal.Chem.* **1981**, *53*, 2190–2193.

Takahashi, H.; Yoshida, T.; Meguro, H. Fluorometric analysis of thiols by high performance liquid chromatography with post column derivatization. *Bunseki Kagaku* **1981**, *30*, 339–341.

Thioureas

Kobayashi, H.; Matano, O.; Goto, S. Simultaneous quantitation of thioureas in rat plasma by high-performance liquid chromatography. *J.Chromatogr.* **1981**, *207*, 281–285.

THIP

Madsen, S.M. Quantitative determination of the gamma-aminobutyric acid agonist, 4,5,6,7-tetrahydroisoxazolo[5,4-c]pyridin-3-ol, in serum by high-performance liquid chromatography. *J.Chromatogr.* **1983**, *274*, 209–218.

Schultz, B.; Hansen, S.H. Determination of the gamma-aminobutyric acid agonist 4,5,6,7-tetrahydroisoxazolo[5,4-c]pyridine-3-ol in urine by high-performance liquid chromatography. *J.Chromatogr.* **1982**, *228*, 279–284.

Thromboxane

Ben Gueddour, R.; Matt, M.; Nicolas, A.; Donner, M.; Stoltz, J.F. Determination of thromboxane B2 by normal-phase high-performance liquid chromatography after pre-column derivatization with fluorescent reagents. *Anal.Lett.* **1993**, *26*, 429–443.

Thymosin

Hannappel, E.; Leibold, W. Determination of thymosin β4 in cultured cell lines. *Protides Biol.Fluids* **1984**, *32*, 1093–1096.

Thyronines

Lankmayr, E.P.; Maichin, B.; Knapp, G.; Nachtmann, F. Catalytic detection principle for high-performance liquid chromatography: determination of enantiomeric iodinated thyronines in blood serum. *J.Chromatogr.* **1981**, *224*, 239–248.

Thyroxines

Leb, G.; Lankmayr, E.P.; Goebel, R.; Pristautz, H.; Nachtmann, F.; Knapp, G. Stereospecific determination of D-thyroxin and L-thyroxin in serum. *Klin.Wochenschr.* **1981**, *59*, 861–863.

Lankmayr, E.; Maichin, B.; Knapp, G. High-pressure liquid chromatography for the determination of enantiomeric thyroxines in serum using catalytic detection. *Fresenius' Z.Anal.Chem.* **1980**, *301*, 187.

Tiopronin

Kaagedal, B.; Carlsson, M.; Denneberg, T. Determination of 2-mercaptopropionylglycine in plasma and urine by high-performance liquid chromatography. *J.Chromatogr.* **1986**, *380*, 301–311.

Tiopronin (using post-column reaction detection)

Hercelin, B.; Leroy, P.; Nicolas, A.; Gavriloff, C.; Chassard, D.; Thebault, J.J.; Reveillaud, M.T.; Salles, M.E.; Netter, P. The pharmacokinetics of tiopronin and its principal metabolite (2-mercaptopropionic acid) after oral administration to healthy volunteers. *Eur.J.Clin.Pharmacol.* **1992**, *43*, 93–95.

TNP-470

Figg, W.D.; Yeh, H.J.C.; Thibault, A.; Pluda, J.M.; Itoh, F.; Yarchoan, R.; Cooper, M.R. Assay of the antiangiogenic compound TNP-470, and one of its metabolites, AGM-1883, by reversed-phase high-performance liquid chromatography in plasma. *J.Chromatogr.B* **1994**, *652*, 187–194.

Tobramycin

Barends, D.M.; Zwaan, C.L.; Hulshoff, A. Micro-determination of tobramycin in serum by high-performance liquid chromatography with ultraviolet detection. *J.Chromatogr.* **1981**, *225*, 417–426.
Kubo, H.; Kobayashi, Y.; Kinoshita, T.; Nishikawa, T. Micro determination of tobramycin in serum by high-performance liquid chromatography. *Bunseki Kagaku* **1982**, *31*, E263-E268,
Maitra, S.K.; Yoshikawa, T.T.; Hansen, J.L.; Schotz, M.C.; Guze, L.B. Quantitation of serum tobramycin concentration using high-pressure liquid chromatography. *Am.J.Clin.Pathol.* **1979**, *71*, 428–432.

Tobramycin (using post-column reaction detection)

Krugers Dagneaux, P.G.L.C.; Klein Elhorst, J.T. HPLC-estimation of tobramycin in serum. *Pharm.Weekbl., Sci.Ed.* **1981**, *3*, 66–70.

TP9201 (using post-column reaction detection)

Woltman, S.J.; Chen, J-G.; Weber, S.G.; Tolley, J.O. Determination of the pharmaceutical peptide TP9201 by post-column reaction with copper(II) followed by electrochemical detection. *J.Pharm.Biomed.Anal.* **1995**, *14*, 155–164.

Triazines (using post-column reaction detection)

Mattusch, J.; Baran, H.; Schwedt, G. HPLC of triazines after on-line photoderivatization. *Fresenius' J.Anal.Chem.* **1991**, *340*, 782–784.

Trichlorfon

Slahck, S.C. Determination of trichlorfon in technical and powder formulations. *J.Assoc.Off.Anal.Chem.* **1988**, *71*, 317–320.

Trientine

Zai, E.S. Approaches to the assay of trientine dihydrochloride. *Anal.Proc.* **1982**, *19*, 125–126.

Triethylenetetramine (using post-column reaction detection)

Meguro, Y.; Kodama, H.; Abe, T. Detection of triethylenetetramine dihydrochloride in serum and urine samples. *Biomed.Res.Trace Elem.* **1991**, *2*, 197–198.

Trimethyllysine

Davis, A.T.; Ingalls, S.T.; Hoppel, C.L. Determination of free trimethyllysine in plasma and tissue specimens by high-performance liquid chromatography. *J.Chromatogr.* **1984**, *306*, 79–87.
Davis, A.T. Assay for trimethyllysine hydroxylase by high-performance liquid chromatography. *J.Chromatogr.* **1987**, *422*, 253–256.
Hoppel, C.L.; Weir, D.E.; Gibbons, A.P.; Ingalls, S.T.; Brittain, A.T.; Brown, F.M. Determination of 6-N-trimethyllysine in urine by high-performance liquid chromatography. *J.Chromatogr.* **1983**, *272*, 43–50.
Lehman, L.J.; Olson, A.L.; Rebouche, C.J. Measurement of epsilon-N-trimethyllysine in human blood plasma and urine. *Anal.Biochem.* **1987**, *162*, 137–142.

Trimethyllysine (using post-column reaction detection)

Kohse, K.P.; Graser, T.A.; Godel, H.G.; Roessle, C.; Franz, H.E.; Fuerst, P. High-performance liquid chromatographic determination of plasma free trimethyllysine in humans. *J.Chromatogr.* **1985**, *344*, 319–324.

Minkler, P.E.; Erdos, E.A.; Ingalls, S.T.; Griffin, R.L.; Hoppel, C.L. Improved high-performance liquid chromatographic method for the determination of 6-N,N,N-trimethyllysine in plasma and urine: biomedical application of chromatographic figures of merit and amine mobile phase modifiers. *J.Chromatogr.* **1986**, *380*, 285–299.

Trinitrotoluene

Preslan, J.E.; Hatrel, B.B.; Emerson, M.; White, LA.; George, W.J. An improved method for analysis of 2,4,6-trinitrotoluene and its metabolites from compost and contaminated soils. *J.Hazard.Mater.* **1993**, *33*, 329–337.

Trospectomycin (using post-column reaction detection)

Wood, S.A.; Simmonds, R.J. Development of a sensitive and robust assay for the aminocyclitol antibiotic, trospectomycin. *Methodol.Surv.Biochem.Anal.* **1990**, *20*, 103–108.

Trypanothione

Krauth-Siegel, R.L.; Jacoby, E.M.; Schirmer, R.H. Trypanothione and N1-glutathionylspermidine: Isolation and determination. *Methods Enzymol.* **1995**, *251*, 287–294.

Tryptic Digests

Hogan, B.L.; Yeung, E.S. Indirect fluorometric detection of tryptic digests separated by capillary zone electrophoresis. *J.Chromatogr.Sci.* **1990**, *28*, 15–18.

Tryptophan

Alegria, A.; Barbera, R.; Farre, R.; Ferreres, M.; Lagarda, M.J.; Lopez, J.C. Isocratic high-performance liquid chromatographic determination of tryptophan in infant formulas. *J.Chromatogr.A* **1996**, *721*, 83–88.

Kim, H.S.; Angyal, G. Comparison of liquid chromatographic method to AOAC microbiological method for determination of L-tryptophan in tablets and capsules. *J.AOAC Int.* **1993**, *76*, 414–417.

Molnar-Perl, I.; Pinter-Szakacs, M.; Khalifa, M. High-performance liquid chromatography of tryptophan and other amino acids in hydrochloric acid hydrolyzates. *J.Chromatogr.* **1993**, *632*, 57–61.

Richards, D.A. Electrochemical detection of tryptophan metabolites following high-performance liquid chromatography. *J.Chromatogr.* **1979**, *175*, 293–299.

Tyramine (using post-column reaction detection)

Reuvers, T.B.A.; Martin de Pozuelo, M.; Ramos, M.; Jimenez, R. A rapid ion-pair HPLC procedure for the determination of tyramine in dairy products. *J.Food Sci.* **1986**, *51*, 84–86.

Tyrosine

Abecassis, J.; David-Eteve, C.; Soun, A. The separation of 24 OPA-AA of natural origin and quantitative analysis of tyrosine by means of HPLC. *J.Liq.Chromatogr.* **1985**, *8*, 135–153.

Crow, J.P.; Beckman, J.S. Quantitation of protein tyrosine, 3-nitrotyrosine, and 3-aminotyrosine utilizing HPLC and intrinsic ultraviolet absorbance. *Methods (San Diego)* **1995**, *7*, 116–120.

Tyrosine Derivatives (using post-column reaction detection)

Carne, T.J.; Huber, R.E.; Davitt, P.; Edwards, L.A. Analysis of fluorine and iodine derivatives of tyrosine. *J.Chromatogr.* **1986**, *367*, 393–403.

Tyrosine-O-sulfate (using post-column reaction detection)

Suiko, M.; Fernando, P.H.P.; Nakamura, T.; Ohshima, T.; Liu, M.C.; Nakatsu, S. Quantitation of tyrosine O-sulfate in human urine by ion-pair reversed-phase high-performance liquid chromatography. *Kenkyu Hokoku - Miyazaki Daigaku Nogakubu* **1992**, *39*, 141–146.

Suiko, M.; Fernando, P.H.P.; Arino, Y.; Terada, M.; Nakatsu, S.; Liu, M.C. Quantitation of tyrosine-O-sulfate in human urine by ion-pair reverse-phase high-performance liquid chromatography. *Clin.Chim.Acta* **1990**, *193*, 193–197.

Urapidil (using post-column reaction detection)

Veltkamp, A.C.; Das, H.A.; Frei, R.W.; Brinkman, U.A.T. On-line radiometric determination of [14C]-urapidil and its main metabolites in rat plasma, using post-column ion-pair extraction and solvent segmentation techniques. *J.Pharm.Biomed.Anal.* **1987**, *6*, 609–622.

Urea

Almy, J.; Ough, C.S. Urea analysis for wines. *J.Agric.Food Chem.* **1989**, *37*, 968–970.

Urea (using post-column reaction detection)

Abshahi, A.; Goyal, S.S.; Mikkelsen, D.S. Simultaneous determination of urea and ammonia nitrogen in soil extracts and water by high performance liquid chromatography. *Soil Sci.Soc.Am.J.* **1988**, *52*, 969–973.

Jansen, H.; Frei, R.W.; Brinkman, U.A.; Deelder, R.S.; Snellings, R.P.J. Determination of urea and ammonia using ion-pair liquid chromatography with on-line post-column derivatization in an enzymic solid-phase reactor. *J.Chromatogr.* **1985**, *325*, 255–263.

Kawase, J.; Ueno, H.; Nakae, A.; Tsuji, K. High-performance liquid chromatography of urea and related compounds with post-column derivatization. *J.Chromatogr.* **1982**, *252*, 209–216.

Matsudo, T.; Sasaki, M. Determination of urea and citrulline in fermented foods and beverages. *Biosci., Biotechnol., Biochem.* **1995**, *59*, 827–830.

Valine

Bongiovanni, R.; Dutton, W. High performance liquid chromatographic analysis for free valine in plasma. *J.Liq.Chromatogr.* **1978**, *1*, 617–630.

Valproic Acid

Bousquet, E.; Tirendi, S.; Santagati, N.A.; Salvo, M.; Castello, P.M.P. Clean-up and determination of sodium valproate in serum by high-performance liquid chromatography with fluorimetric detection. *Pharmazie* **1992**, *46*, 257–258.

Gupta, R.N.; Keane, P.M.; Gupta, M.L. Valproic acid in plasma, as determined by liquid chromatography. *Clin.Chem.* **1979**, *25*, 1984–1985.

Lucarelli, C.; Villa, P.; Lombaradi, E.; Prandini, P.; Brega, A. HPLC method for the simultaneous analysis for valproic acid and other common anticonvulsant drugs in human plasma or serum. *Chromatographia* **1992**, *33*, 37–40.

Schmidt, G.J.; Slavin, W. Determination of dipropylacetic acid (sodium valproate) by liquid chromatography with pre-column labeling and ultraviolet detection. *Chromatogr.Newsl.* **1978**, *6*, 22–24.

Wolf, J.H.; Veenma-van der Duin, L.; Korf, J. The extracellular concentration of the anti-epileptic drug valproate in the rat brain as determined with microdialysis and an automated HPLC procedure. *J.Pharm.Pharmacol.* **1991**, *43*, 101–106.

Vancomycin

McClain, J.B.L.; Bongiovanni, R.; Brown, S. Vancomycin quantitation by high-performance liquid chromatography in human serum. *J.Chromatogr.* **1982**, *231*, 463–466.

Vanilmandelic Acid (using post-column reaction detection)

Buergi, W.; Kaufmann, H. Vanilmandelic acid assay by liquid chromatography using a LC pump system for post column periodate oxidation. *Fresenius' Z.Anal.Chem.* **1984**, *317*, 681–682.

Vitamin B1 (using post-column reaction detection)

Motoe, K.; Nabeshima, H. Determination of purified vitamin B1 solution by high-performance liquid chromatography (determination of vitamin B1 in foods by high-performance liquid chromatography Part I). *Toyama-ken Shokuhin Kenkyusho Kenkyu Hokoku* **1993**, *1*, 25–30.

Vitamin B6

Edwards, P.; Liu, P.K.S.; Rose, G.A. A simple liquid-chromatographic method for measuring vitamin B6 compounds in plasma. *Clin.Chem.* **1989**, *35*, 241–245.

Gregory, J.F.-I.; Manley, D.B.; Kirk, J.R. Determination of vitamin B-6 in animal tissues by reverse-phase high-performance liquid chromatography. *J.Agric.Food Chem.* **1981**, *29*, 921–927.

Kurioka, S.; Ishioka, N.; Sato, J.; Nakamura, J.; Ohkubo, T.; Matsuda, M. Assay of vitamin B6 in human plasma with graphitic carbon column. *Biomed.Chromatogr.* **1993**, *7*, 162–165.

Reitzer-Bergaentzle, M.; Marchioni, E.; Hasselmann, C. HPLC determination of vitamin B6 in foods after pre-column derivatization of free and phosphorylated vitamers into pyridoxol. *Food Chem.* **1993**, *48*, 321–324.

Vitamin B6 (using post-column reaction detection)

Gregory, J.F., III; Feldstein, D. Determination of vitamin B-6 in foods and other biological materials by paired-ion high-performance liquid chromatography. *J.Agric.Food Chem.* **1985**, *33*, 359–363.

Hamaker, B.; Kirksey, A.; Ekanayake, A.; Borschel, M. Analysis of B-6 vitamers in human milk by reverse-phase liquid chromatography. *Am.J.Clin.Nutr.* **1985**, *42*, 650–655.

Shephard, G.S.; Van der Westhuizen, L.; Labadarios, D. Analysis of vitamin B6 vitamers in human tissue by cation-exchange high-performance liquid chromatography. *J.Chromatogr.* **1989**, *491*, 226–234.

Vitamin D3

Koskinen, T.; Valtonen, P. Comparison of HPLC separation of vitamin D3 metabolites and their iso-tachysterol3 derivatives. *J.Liq.Chromatogr.* **1985**, *8*, 463–472.

Vitamin D3 (using post-column reaction detection)

Vreeken, R.J.; Honing, M.; van Baar, B.L.M.; Ghijsen, R.T.; de Jong, G.J.; Brinkman, U.A.T. Online post-column Diels-Alder derivatization for the determination of vitamin D3 and its metabolites by liquid chromatography/thermospray mass spectrometry. *Biol.Mass Spectrom.* **1993**, *22*, 621–632.

Vitamin K

Kusube, K.; Abe, K.; Hiroshima, O.; Ishiguro, Y.; Ishikawa, S.; Hoshida, H. Determination of vitamin K analogs by high performance liquid chromatography with electrochemical derivatization. *Chem.Pharm.Bull.* **1984**, *32*, 179–184.

Vitamin K (using post-column reaction detection)

Booth, S.L.; Davidson, K.W.; Sadowski, J.A. Evaluation of an HPLC method for the determination of phylloquinone (vitamin K1) in various food matrixes. *J.Agric.Food Chem.* **1994**, *42*, 295–300.

Cham, B.E.; Roeser, H.P.; Kamst, T.W. Simultaneous liquid-chromatographic determination of vitamin K1 and vitamin E in serum. *Clin.Chem.* **1989**, *35*, 2285–2289.

Gao, Z.H.; Ackman, R.G. Determination of vitamin K1 in canola oils by high performance liquid chromatography with menaquinone-4 as an internal standard. *Food Res.Int.* **1995**, *28*, 61–69.

Guillaumont, M.; Leclercq, M.; Gosselet, H.; Makala, K.; Vignal, B. HPLC determination of serum vitamin K1 by fluorometric detection after post-column electrochemical reduction. *J.Micronutr.Anal.* **1988**, *4*, 285–294.

Haroon, Y.; Bacon, D.S.; Sadowski, J.A. Chemical reduction system for the detection of phylloquinone (vitamin K1) and menaquinones (vitamin K2). *J.Chromatogr.* **1987**, *384*, 383–389.

Jakob, E.; Elmadfa, I. Rapid HPLC assay for the assessment of vitamin K1, A, E and beta-carotene status in children (7–19 years). *Int.J.Vitam.Nutr.Res.* **1995**, *65*, 31–35.

Jakob, E.; Elmadfa, I. Application of a simplified HPLC assay for the determination of phylloquinone (vitamin K1) in animal and plant food items. *Food Chem.* **1996**, *56*, 87–91.

Kitahashi, T. Studies on the measurement of blood vitamin K. *Acta Med.Kinki Univ.* **1987**, *12*, 101–119.

Lambert, W.E.; De Leenheer, A.P. Simplified postcolumn reduction and fluorescence detection for the high-performance liquid chromatographic determination of vitamin K1(20). *Anal.Chim.Acta* **1987**, *196*, 247–250.

Lambert, W.E.; De Leenheer, A.P.; Lefevere, M.F. Determination of vitamin K in serum using HPLC with post-column reaction and fluorescence detection. *J.Chromatogr.Sci.* **1986**, *24*, 76–79.

Lambert, W.E.; De Leenheer, A.P.; Baert, E.J. Wet-chemical postcolumn reaction and fluorescence detection analysis of the reference interval of endogenous serum vitamin K1(20). *Anal.Biochem.* **1986**, *158*, 257–261.

Langenberg, J.P.; Tjaden, U.R.; De Vogel, E.M.; Langerak, D.I. Determination of phylloquinone (vitamin K1) in raw and processed vegetables using reversed phase HPLC with electrofluorometric detection. *Acta Aliment.* **1986**, *15*, 187–198.

Shino, M. Determination of endogenous Vitamin K (phylloquinone and menaquinone-n) in plasma by high-performance liquid chromatography using platinum oxide catalyst reduction and fluorescence detection. *Analyst* **1988**, *113*, 393–397.

Soedirman, J.R.; de Bruijn, E.A.; Maes, R.A.A.; Hanck, A.; Gruter, J. Pharmacokinetics and tolerance of intravenous and intramuscular phylloquinone (vitamin K1) mixed micelles formulation. *Br.J.Clin.Pharmacol.* **1996**, *41*, 517–523.

Van Haard, P.M.M.; Engel, R.; Pietersma-De Bruyn, A.L.J.M. Quantitation of trans-vitamin K1 in small serum samples by off-line multidimensional liquid chromatography. *Clin.Chim.Acta* **1986**, *157*, 221–230.

Vitamins

van der Horst, A.; Martens, H.J.M.; De Goede, P.N.F.C. Analysis of water-soluble vitamins in total parenteral nutrition solution by high-pressure liquid chromatography. *Pharm.Weekbl., Sci.Ed.* **1989**, *11*, 169–174.

Warfarin (using post-column reaction detection)

Lee, S.H.; Field, L.R.; Howald, W.N.; Trager, W.F. High-performance liquid chromatographic separation and fluorescence detection of warfarin and its metabolites by postcolumn acid/base manipulation. *Anal.Chem.* **1981**, *53*, 467–471.

Wong, Y.W.J.; Davis, P.J. Analysis of warfarin and its metabolites by reversed-phase ion-pair liquid chromatography with fluorescence detection. *J.Chromatogr.* **1989**, *469*, 281–291.

Whitening Agents (using post-column reaction detection)

Poiger, T.; Field, J.A.; Field, T.M.; Giger, W. Occurrence of fluorescent whitening agents in sewage and river water determined by solid-phase extraction and high-performance liquid chromatography. *Environ.Sci.Technol.* **1996**, *30*, 2220–2226.

Poiger, T.; Field, J.A.; Field, T.M.; Giger, W. Determination of detergent-derived fluorescent whitening agents in sewage sludges by liquid chromatography. *Anal.Methods Instrum.* **1993**, *1*, 104–113.

Xanthomegnin

Kuan, S.S.; Carman, A.S.; Francis, O.J.; Ware, G.M. Derivatization of xanthomegnin for fluorometric determination. *Anal.Lett.* **1985**, *18*, 837–846.

Xylenediamine

Paseiro Losada, P.; Paz Abuin, S.; Vazquez Oderiz, L.; Simal Lozano, J.; Simal Gandara, J. Quality control of cured epoxy resins. Determination of residual free monomers (m-xylylenediamine and bisphenol A diglycidyl ether) in the finished product. *J.Chromatogr.* **1991**, *585*, 75–81.

YM529 (using post-column reaction detection)

Usui, T.; Kawakami, R.; Watanabe, T.; Higuchi, S. Sensitive determination of a novel bisphosphonate, YM529, in plasma, urine and bone by high-performance liquid chromatography with fluorescence detection. *J.Chromatogr.B* **1994**, *652*, 67–72.

Zanhic Acid Glycoside

Nowacka, J.; Oleszek, W. High performance liquid chromatography of zanhic acid glycoside in alfalfa (Medicago sativa). *Phytochem.Anal.* **1992**, *3*, 227–230.

APPENDIX II:
BIBLIOGRAPHY
OF REVIEWS OF
DERIVATIZATION REACTIONS

Abidi, S.L.; Mounts, T.L. High-performance liquid chromatography of phosphatidic acid. *J.Chromatogr.B* **1995**, *671*, 281–297.

Beaver, R.W. Determination of aflatoxins in corn and peanuts using high performance liquid chromatography. *Arch.Environ.Contam.Toxicol.* **1989**, *18*, 315–318.

Brinkman, U.A.T.; Frei, R.W.; Lingeman, H. Post-column reactors for sensitive and selective detection in high-performance liquid chromatography: Categorization and applications. *J.Chromatogr.* **1989**, *492*, 251–298.

Campíns-Falcó, P.; Sevillano-Cabeza, A.; Molins-Legua, C. Amphetamine and methamphetamine determinations in biological samples by high performance liquid chromatography. A review. *J.Liq.Chromatogr.* **1994**, *17*, 731–747.

Cohen, S.A.; Strydom, D.J. Amino acid analysis utilizing phenylisothiocyanate derivatives. *Anal.Biochem.* **1988**, *174*, 1–16.

Danielson, N.D.; Targove, M.A.; Miller, B.E. Pre- and postcolumn derivatization chemistry in conjunction with HPLC for pharmaceutical analysis. *J.Chromatogr.Sci.* **1988**, *26*, 362–371.

Davies, N.M. Methods of analysis of chiral non-steroidal anti-inflammatory drugs. *J.Chromatogr.B* **1997**, *691*, 229–261.

El Rassi, Z.; Mechref, Y. Recent advances in capillary electrophoresis of carbohydrates. *Electrophoresis* **1996**, *17*, 275–301.

Fekkes, D. State-of-the-art of high-performance liquid chromatographic analysis of amino acids in physiological samples. *J.Chromatogr.B* **1996**, *682*, 3–22.

Fürst, P.; Stehle, P.; Graser, T.A. [Progress in the analysis of amino acids with special reference to the determination of the intracellular amino acid pattern]. *Infusionsther.Klin.Ernahr.* **1987**, *14*, 137–146.

Fürst, P.; Pollack, L.; Graser, T.A.; Godel, H.; Stehle, P. HPLC analysis of free amino acids in biological material - An appraisal of four pre-column derivatization methods. *J.Liq.Chromatogr.* **1989**, *12*, 2733–2760.

Fürst, P.; Pollack, L.; Graser, T.A.; Godel, H.; Stehle, P. Appraisal of four pre-column derivatization methods for the high-performance liquid chromatographic determination of free amino acids in biological materials. *J.Chromatogr.* **1990**, *499*, 557–569.

Gamoh, K.; Takatsuto, S. Liquid chromatographic assay of brassinosteroids in plants. *J.Chromatogr.A* **1994**, *658*, 17–25.

Görög, S.; Gazdag, M. Enantiomeric derivatization for biomedical chromatography. *J.Chromatogr.B* **1994**, *659*, 51–84.

Hase, S. Precolumn derivatization for chromatographic and electrophoretic analyses of carbohydrates. *J.Chromatogr.A* **1996**, *720*, 173–182.

Hietpas, P.B.; Ewing, A.G. On-column and post-column derivatization for capillary electrophoresis with laser-induced fluorescence for the analysis of single cells. *J.Liq.Chromatogr.* **1995**, *18*, 3557–3576.

Hjemdahl, P. Catecholamine measurements by high-performance liquid chromatography. *Am.J.Physiol.* **1984**, *247*, E13-E20.

Honda, S. Postcolumn derivatization for chromatographic analysis of carbohydrates. *J.Chromatogr.A* **1996**, *720*, 183–199.

Imai, K. Derivatization in liquid chromatography. *Adv.Chromatogr.* **1987**, *27*, 215–245.

Imai, K.; Fukushima, T.; Santa, T.; Homma, H.; Hamase, K.; Sakai, K.; Kato, M. Analytical chemistry and biochemistry of D-amino acids. *Biomed.Chromatogr.* **1996**, *10*, 303–312.

Kai, M.; Ohkura, Y. Fluorescence derivatization of bioactive peptides for high performance liquid chromatography. *Trends Anal.Chem.* **1987**, *6*, 116–120.

Kakehi, K.; Honda, S. Analysis of glycoproteins, glycopeptides and glycoprotein-derived oligosaccharides by high-performance capillary electrophoresis. *J.Chromatogr.A* **1996**, *720*, 377–393.

Karnes, H.T.; Sarkar, M.A. Enantiomeric resolution of drug compounds by liquid chromatography. *Pharm.Res.* **1987**, *4*, 285–292.

Kok, W.T. Derivatization reactions for the determination of aflatoxins by liquid chromatography with fluorescence detection. *J.Chromatogr.B* **1994**, *659*, 127–137.

Krull, I.S.; Deyl, Z.; Lingeman, H. General strategies and selection of derivatization reactions for liquid chromatography and capillary electrophoresis. *J.Chromatogr.B* **1994a**, *659*, 1–17.

Krull, I.S.; Szulc, M.E.; Bourque, A.J.; Zhou, F.X.; Yu, J.; Strong, R. Solid-phase derivatization reactions for biomedical liquid chromatography. *J.Chromatogr.B* **1994b**, *659*, 19–50.

Lehtonen, P. Determination of amines and amino acids in wine - a review. *Am.J.Enol.Vitic.* **1996**, *47*, 127–133.

Lingeman, H.; Underberg, W.J.M.; Takadate, A.; Hulshoff, A. Fluorescence detection in high performance liquid chromatography. *J.Liq.Chromatogr.* **1985**, *8*, 789–874.

Luckas, B. Phycotoxins in seafood—toxicological and chromatographic aspects. *J.Chromatogr.* **1992**, *624*, 439–456.

Marcomini, A.; Zanette, M. Chromatographic determination of non-ionic aliphatic surfactants of the alcohol polyethoxylate type in the environment. *J.Chromatogr.A* **1996**, *733*, 193–206.

McGarvey, B.D. Derivatization reactions applicable to pesticide determination by high-performance liquid chromatography. *J.Chromatogr.B* **1994**, *659*, 243–257.

Miners, J.O. The analysis of penicillins in biological fluids and pharmaceutical preparations by high-performance liquid chromatography: A review. *J.Liq.Chromatogr.* **1985**, *8*, 2827–2843.

Molnár-Perl, I. Advances in the high-performance liquid chromatographic determination of phenylthiocarbamyl amino acids. *J.Chromatogr.A* **1994**, *661*, 43–50.

Mori, K. Liquid chromatography/luminescence techniques. *Life Sci.* **1987**, *41*, 901–904.

Mount, M.E.; Oehme, F.W. Carbaryl; a literature review. *Residue Rev.* **1981**, *77*, 1–64.

Mukherjee, P.S.; Karnes, H.T. Ultraviolet and fluorescence derivatization reagents for carboxylic acids suitable for high performance liquid chromatography: a review. *Biomed.Chromatogr.* **1996**, *10*, 193–204.

Myher, J.J.; Kuksis, A. General strategies in chromatographic analysis of lipids. *J.Chromatogr.B* **1995**, *671*, 3–33.

Nagatsu, T. Application of high-performance liquid chromatography to the study of biogenic amine-related enzymes. *J.Chromatogr.* **1991**, *566*, 287–307.

Nishikawa, H.; Sakai, T. Derivatization and chromatographic determination of aldehydes in gaseous and air samples. *J.Chromatogr.A* **1995**, *710*, 159–165.

Ohkura, Y.; Nohta, H. Fluorescence derivatization in high-performance liquid chromatography. *Adv.Chromatogr.* **1989**, *29*, 221–258.

Otson, R.; Fellin, P. A review of techniques for measurement of airborne aldehydes. *Sci.Total Environ.* **1988**, *77*, 95–131.

Paulus, A.; Klockow, A. Detection of carbohydrates in capillary electrophoresis. *J.Chromatogr.A* **1996**, *720*, 353–376.

Rao, M.S.; Hidajat, K.; Ching, C.B. Reversed-phase HPLC: the separation method for the characterization and purification of long chain polyunsaturated fatty acids - a review. *J.Chromatogr.Sci.* **1995**, *33*, 9–21.

Rice, K.G.; Corradi Da Silva, M.L. Preparative purification of tyrosinamide N-linked oligosaccharides. *J.Chromatogr.A* **1996**, *720*, 235–249.

Scalia, S. Bile acid separation. *J.Chromatogr.B* **1995**, *671*, 299–317.

Seiler, N. Liquid chromatographic methods for assaying polyamines using prechromatographic derivatization. *Methods Enzymol.* **1983**, *94*, 10–25.

Shimada, K.; Mitamura, K. Derivatization of thiol-containing compounds. *J.Chromatogr.B* **1994**, *659*, 227–241.

Srinivas, N.R.; Igwemezie, L.N. Chiral separation by high performance liquid chromatography. I. Review on indirect separation of enantiomers as diastereomeric derivatives using ultraviolet, fluorescence and electrochemical detection. *Biomed.Chromatogr.* **1992**, *6*, 163–167.

Szulc, M.E.; Krull, I.S. Improved detection and derivatization in capillary electrophoresis. *J.Chromatogr.A* **1994**, *659*, 231–245.

Timerbaev, A.R. Analysis of inorganic pollutants by capillary electrophoresis. *Electrophoresis* **1997**, *18*, 185–195.

Toyo'oka, T. Use of derivatization to improve the chromatographic properties and detection selectivity of physiologically important carboxylic acids. *J.Chromatogr.B* **1995**, *671*, 91–112.

Toyo'oka, T. Recent progress in liquid chromatographic enantioseparation based upon diastereomer formation with fluorescent chiral derivatization reagents. *Biomed.Chromatogr.* **1996**, *10*, 265–277.

Verzele, M.; Simoens, G.; Van Damme, F. A critical review of some liquid chromatography systems for the separation of sugars. *Chromatographia* **1987**, *23*, 292–300.

Vetticaden, S.J.; Chandrasekaran, A. Chromatography of cardiac glycosides. *J.Chromatogr.* **1990**, *531*, 215–234.

Wolf, J.H.; Korf, J. 4-Bromomethyl-7-methoxycoumarin and analogues as derivatization agents for high-performance liquid chromatography determinations: a review. *J.Pharm.Biomed.Anal.* **1992**, *10*, 99–107.

Wyss, R. Chromatography of retinoids. *J.Chromatogr.* **1990**, *531*, 481–508.

Yasaka, Y.; Tanaka, M. Labeling of free carboxyl groups. *J.Chromatogr.B* **1994**, *659*, 139–155.

Zhou, Y.; Luan, P.; Liu, L.; Sun, Z.P. Chiral derivatizing reagents for drug enantiomers bearing hydroxyl groups. *J.Chromatogr.B* **1994**, *659*, 109–126.

REAGENT MOLECULAR FORMULA INDEX

Molecular Formula	Compound	Page
Al	aluminum	1304
AlCl$_3$	aluminum trichloride	1357
AlN$_3$O$_9$	aluminum nitrate	1357
BF$_3$	boron trifluoride	1101
Br$_2$	bromine	1369
CCl$_2$O	phosgene	822
CKN	potassium cyanide	1514
CH$_2$N$_2$	diazomethane	1012, 1505
CH$_2$O	formaldehyde	1250
CH$_3$I	iodomethane	1185, 1559
CH$_3$NaO	sodium methoxide	1135
CH$_4$N$_2$S	thiourea	1162
CH$_4$O	methanol	914, 1500
CH$_5$N$_3$	guanidine	1395
CH$_5$N$_3$O	semicarbazide	234
C$_2$HF$_3$O$_2$	trifluoroacetic acid	1473
C$_2$H$_2$O$_2$	glyoxal	502
C$_2$H$_3$BrO	bromoacetaldehyde	431
C$_2$H$_3$ClO	acetyl chloride	10
C$_2$H$_3$ClO	chloroacetaldehyde	432
C$_2$H$_3$IO$_2$	iodoacetic acid	1558
C$_2$H$_3$NO	methyl isocyanate	1185
C$_2$H$_3$N$_3$	mercuric chloride/triazole	1260
C$_2$H$_3$N$_3$	triazole	1254
C$_2$H$_4$INO	iodoacetamide	1520
C$_2$H$_4$O$_2$	acetic acid	1356
C$_2$H$_4$O$_2$S	phthalaldehyde/mercaptoacetic acid	402
C$_2$H$_6$N$_6$S	purpald	233
C$_2$H$_6$O	ethanol	1189
C$_2$H$_6$OS	dimethyl sulfoxide/Fast Yellow GC salt	1183
C$_2$H$_6$OS	phthalaldehyde/2-mercaptoethanol	403, 703, 1180
C$_2$H$_6$O$_4$S	dimethyl sulfate	1184
C$_2$H$_6$S	phthalaldehyde/ethanethiol	399, 699
C$_2$H$_6$S$_2$	1,2-ethanedithiol	1503

Molecular Formula	Compound	Page
C_2H_7NO	ethanolamine	1390
C_2H_7NS	naphthalene-2,3-dicarboxaldehyde/ 2-mercaptoethanol	390
$C_2H_8N_2$	ethylenediamine	1127, 1211, 1391
$C_3H_2Cl_4O_2$	trichloroethyl chloroformate	353
$C_3H_4N_2$	imidazole	1253
$C_3H_4N_2$	mercuric chloride/imidazole	1259
$C_3H_4N_2O$	2-cyanoacetamide	154, 1380
$C_3H_5ClO_2$	ethyl chloroformate	307
$C_3H_6SO_2$	phthalaldehyde/3-mercaptopropionic acid	426, 713
C_3H_9N	isopropylamine	1136
$C_4F_6O_3$	trifluoroacetic anhydride	85, 369
$C_4H_2O_3$	maleic anhydride	366
$C_4H_4N_2O_2S$	thiobarbituric acid	160
$C_4H_4N_2O_3$	barbituric acid	1512
$C_4H_4N_2O_3$	isonicotinic acid/barbituric acid	1081
$C_4H_4N_2O_3$	pyridine/barbituric acid	1491
$C_4H_5N_3$	aminopyrazine	167
$C_4H_6CoO_4$	cobalt acetate/hydroxylamine	1377
$C_4H_6CuO_4$	copper acetate/sulfuric acid	1377
$C_4H_6O_2S_2$	thioacetylthioglycolic acid	820
$C_4H_6O_3$	acetic anhydride	72, 361
$C_4H_6O_6$	tartaric acid	1302
$C_4H_7ClO_2$	propyl chloroformate	350
C_4H_7NO	propyl isocyanate	118
$C_4H_7O_2$	methyl acrylate	1584
$C_4H_8O_3$	2-hydroxyisobutyric acid	1285
$C_4H_{10}N_2Si$	trimethylsilyldiazomethane	103
$C_4H_{10}O_2S_2$	naphthalene-2,3-dicarboxaldehyde/ dithiothreitol	389
$C_4H_{10}S$	phthalaldehyde/t-butanethiol	396
$C_4H_{11}N$	butylamine	1371
$C_4H_{11}NS$	phthalaldehyde/N,N-dimethyl-2- mercaptoethylamine	397, 698
$C_4H_{20}B_3N$	tetramethylammonium octahydrido- triborate	1469
C_5H_5N	pyridine/barbituric acid	1491
$C_5H_6Br_3N$	pyridinium bromide perbromide	1438
$C_5H_6N_2$	2-aminopyridine	169
$C_5H_8NNaS_2$	sodium pyrrolidinedithiocarbamate	1301
$C_5H_8O_2$	acetylacetone	189, 823, 1143
$C_5H_9NO_3S$	N-acetylcysteine	1170
$C_5H_9NO_3S$	phthalaldehyde/acetylcysteine	393, 691
C_5H_9NS	butyl isothiocyanate	728
$C_5H_{10}NNaS_2$	sodium diethyldithiocarbamate	1138, 1164, 1299, 1441, 1532
$C_5H_{12}N_2S$	1,1,3,3-tetramethyl-2-thiourea	1470
$C_5H_{14}ClN_3O$	Girard's Reagent T	1241
$C_5H_{14}NO_2S_2$	ammonium bis(2-hydroxyethyl)- dithiocarbamate	1272
$C_6F_{10}O_3$	pentafluoropropionic anhydride	367
$C_6FeK_3N_6$	potassium ferricyanide	1431

MOLECULAR FORMULA	COMPOUND	PAGE
$C_8H_6ClNO_4$	2-nitrobenzyloxycarbonyl chloride	673
$C_8H_6ClNO_4$	4-nitrobenzyloxycarbonyl chloride	674
$C_8H_6O_2$	phenylglyoxal	503, 1426
$C_8H_6O_2$	phthalaldehyde	391, 690, 1080, 1427, 1575
$C_8H_6O_2$	phthalaldehyde/acetylcysteine	393, 691
$C_8H_6O_2$	phthalaldehyde/acetylpenicillamine	696
$C_8H_6O_2$	phthalaldehyde/Boc-L-cysteine	697
$C_8H_6O_2$	phthalaldehyde/t-butanethiol	396
$C_8H_6O_2$	phthalaldehyde/N,N-dimethyl-2-mercaptoethylamine	397, 698
$C_8H_6O_2$	phthalaldehyde/ethanethiol	399, 699
$C_8H_6O_2$	phthalaldehyde/N-isobutyryl-L-cysteine	700
$C_8H_6O_2$	phthalaldehyde/mercaptoacetic acid	402
$C_8H_6O_2$	phthalaldehyde/2-mercaptoethanol	403, 703, 1180
$C_8H_6O_2$	phthalaldehyde/3-mercaptopropionic acid	426, 713
$C_8H_6O_2$	phthalaldehyde/sodium sulfite	429, 718
$C_8H_6O_2$	phthalaldehyde/1-thio-β-D-galactopyranose	719
$C_8H_6O_2$	phthalaldehyde/1-thio-β-D-glucose	720
$C_8H_6O_2$	phthalaldehyde/thioglucose tetraacetate	430, 721
C_8H_7BrO	2-bromoacetophenone	253, 842
$C_8H_7BrO_3$	1-(2,5-dihydroxyphenyl)-2-bromoethanone	903
C_8H_7ClO	m-toluoyl chloride	353
$C_8H_7ClO_2$	p-anisoyl chloride	11
$C_8H_7ClO_2$	benzyl chloroformate	649
C_8H_7NS	benzyl isothiocyanate	464, 727
$C_8H_8FN_3O_3S$	4-(N,N-dimethylaminosulfonyl)-7-fluoro-2,1,3-benzoxadiazole	264, 629, 1553
$C_8H_8N_4O_2$	4,5-diaminophthalhydrazide	1203
$C_8H_8O_2$	p-anisaldehyde	1172
$C_8H_8O_3$	vanillin	1473
$C_8H_8O_7$	diacetyl-L-tartaric anhydride	81
C_8H_9NO	acetanilide	1157
$C_8H_9NO_3$	2-amino-4,5-ethylenedioxyphenol	184
$C_8H_9N_3S$	3-methyl-2-benzothiazolinone hydrazone	104, 228, 1243
$C_8H_9N_5O_3$	4-(2-aminoethylamino)-7-nitro-2,1,3-benzoxadiazole	920
$C_8H_{10}N_2O_2$	1,2-diamino-4,5-ethylenedioxybenzene	186
$C_8H_{10}N_2O_2$	(R)-(+)-α-methyl-4-nitrobenzylamine	985
$C_8H_{10}N_2O_2$	(S)-(-)-α-methyl-4-nitrobenzylamine	986
$C_8H_{10}O$	2,6-dimethylphenol	1166
$C_8H_{10}S$	methyl p-tolyl sulfide	1498
$C_8H_{11}ClSi$	chlorodimethylphenylsilane	139
$C_8H_{11}N$	3,5-dimethylaniline	963
$C_8H_{11}N$	(R)-(+)-α-methylbenzylamine	983, 1213
$C_8H_{11}N$	(S)-(-)-α-methylbenzylamine	978
$C_8H_{11}NO$	tyramine	183
$C_8H_{11}NO_2$	2,4-dimethoxyaniline	953
$C_8H_{11}N_5O_3S$	4-(aminosulfonyl)-7-(2-aminoethylamino)-2,1,3-benzoxadiazole	939
$C_8H_{11}N_5O_3S$	4-(N,N-dimethylaminosulfonyl)-7-hydrazino-2,1,3-benzoxadiazole	208, 1229

REAGENT NAME INDEX